Handbook of the
Birds of Europe the Middle East and North Africa
The Birds of the Western Palearctic
Volume IV

Handbook of the Birds of Europe the Middle East and North Africa

The Birds of the Western Palearctic

Volume IV · Terns to Woodpeckers

Stanley Cramp *Chief Editor*
Duncan J Brooks   Euan Dunn   Robert Gillmor
P A D Hollom   Robert Hudson   E M Nicholson
M A Ogilvie   P J S Olney   C S Roselaar
K E L Simmons   K H Voous   D I M Wallace
Jan Wattel   M G Wilson

OXFORD   NEW YORK
OXFORD UNIVERSITY PRESS

DEDICATED TO THE MEMORY OF

H F WITHERBY

(1873–1943)

EDITOR OF *THE HANDBOOK OF BRITISH BIRDS*

(1938–41)

Oxford University Press, Walton Street, Oxford OX2 6DP
London New York Toronto
Delhi Bombay Calcutta Madras Karachi
Petaling Jaya Singapore Hong Kong Tokyo
Nairobi Dar es Salaam Cape Town
Melbourne Auckland

and associated companies in
Berlin Ibadan

Published in the United States
by Oxford University Press, New York

© *Oxford University Press 1985*

First published 1985
Reprinted (with corrections) 1989

All rights reserved. No part of this publication may be reproduced,
stored in a retrieval system, or transmitted, in any form or by any means,
electronic, mechanical, photocopying, recording, or otherwise, without
the prior permission of Oxford University Press

British Library Cataloguing in Publication Data
Handbook of the birds of Europe, the Middle East,
    and North Africa.
    Vol. 4: Terns to woodpeckers
    1. Birds—Europe
    I. Cramp, Stanley
    598.294   QL690.A1
    ISBN 0-19-857507-6

Library of Congress Cataloging in Publication Data
(Revised for volume IV)
Main entry under title:

Handbook of the birds of Europe, the Middle East and
    North Africa.

    includes bibliographical references and indexes.
    Contents: —v. 2. Hawks to bustards.—v. 3.
Waders to gulls.—v. 4. Terns to woodpeckers.
    1. Birds—Europe—Collected works. 2. Birds—
Mediterranean Region—Collected works. I. Cramp, Stanley.
QL690.A1H25   598.29'182'2   79-42914
ISBN 0-19-857505-X (v. 2)

Printed in Hong Kong

CONTENTS

	Page
INTRODUCTION	1
ACKNOWLEDGEMENTS	2

CHARADRIIFORMES (*continued*)

STERNIDAE terns		5
	Gelochelidon nilotica **Gull-billed Tern**	6
	Sterna caspia **Caspian Tern**	17
	Sterna maxima **Royal Tern**	27
	Sterna bergii **Swift Tern**	36
	Sterna bengalensis **Lesser Crested Tern**	42
	Sterna sandvicensis **Sandwich Tern**	48
	Sterna dougallii **Roseate Tern**	62
	Sterna hirundo **Common Tern**	71
	Sterna paradisaea **Arctic Tern**	87
	Sterna aleutica **Aleutian Tern**	100
	Sterna forsteri **Forster's Tern**	102
	Sterna repressa **White-cheeked Tern**	105
	Sterna anaethetus **Bridled Tern**	109
	Sterna fuscata **Sooty Tern**	116
	Sterna albifrons **Little Tern**	120
	Sterna saundersi **Saunders' Little Tern**	133
	Chlidonias hybridus **Whiskered Tern**	133
	Chlidonias niger **Black Tern**	143
	Chlidonias leucopterus **White-winged Black Tern**	155
	Anous stolidus **Brown Noddy**	163
RYNCHOPIDAE skimmers		166
	Rynchops flavirostris **African Skimmer**	166
ALCIDAE auks		168
	Uria aalge **Guillemot**	170
	Uria lomvia **Brünnich's Guillemot**	184
	Alca torda **Razorbill**	195
	Pinguinus impennis **Great Auk**	207
	Cepphus grylle **Black Guillemot**	208
	Alle alle **Little Auk**	219
	Aethia cristatella **Crested Auklet**	229
	Cyclorrhynchus psittacula **Parakeet Auklet**	230
	Fratercula arctica **Puffin**	231

Contents

PTEROCLIDIFORMES 244

PTEROCLIDIDAE sandgrouse 244
- *Pterocles lichtensteinii* Lichtenstein's Sandgrouse — 245
- *Pterocles coronatus* Crowned Sandgrouse — 249
- *Pterocles senegallus* Spotted Sandgrouse — 253
- *Pterocles exustus* Chestnut-bellied Sandgrouse — 259
- *Pterocles orientalis* Black-bellied Sandgrouse — 263
- *Pterocles alchata* Pin-tailed Sandgrouse — 269
- *Syrrhaptes paradoxus* Pallas's Sandgrouse — 277

COLUMBIFORMES 283

COLUMBIDAE pigeons 283
- *Columba livia* Rock Dove — 285
- *Columba oenas* Stock Dove — 298
- *Columba eversmanni* Yellow-eyed Stock Dove — 309
- *Columba palumbus* Woodpigeon — 311
- *Columba trocaz* Long-toed Pigeon — 329
- *Columba bollii* Bolle's Laurel Pigeon — 331
- *Columba junoniae* Laurel Pigeon — 334
- *Streptopelia roseogrisea* African Collared Dove (Pink-headed Turtle Dove) — 336
- *Streptopelia decaocto* Collared Dove — 340
- *Streptopelia turtur* Turtle Dove — 353
- *Streptopelia orientalis* Rufous Turtle Dove — 363
- *Streptopelia senegalensis* Laughing Dove — 366
- *Oena capensis* Namaqua Dove — 374
- *Ectopistes migratorius* Passenger Pigeon — 378

PSITTACIFORMES 378

PSITTACIDAE parrots 378
- *Psittacula krameri* Ring-necked Parakeet (Rose-ringed Parakeet) — 379

CUCULIFORMES 388

CUCULIDAE cuckoos 388

CUCULINAE parasitic cuckoos 388
- *Clamator jacobinus* Jacobin Cuckoo — 388
- *Clamator glandarius* Great Spotted Cuckoo — 391
- *Chrysococcyx caprius* Didric Cuckoo — 400
- *Cuculus canorus* Cuckoo — 402
- *Cuculus saturatus* Oriental Cuckoo — 417

PHAENICOPHAEINAE non-parasitic cuckoos 423
- *Coccyzus erythrophthalmus* Black-billed Cuckoo — 423
- *Coccyzus americanus* Yellow-billed Cuckoo — 425

CENTROPODINAE coucals 427
- *Centropus senegalensis* Senegal Coucal — 427

Contents

STRIGIFORMES — 432

TYTONIDAE barn owls and allies — 432
    Tyto alba **Barn Owl** — 432

STRIGIDAE typical owls — 449
    BUBONINAE eagle owls and allies — 449
        Otus brucei **Striated Scops Owl** — 450
        Otus scops **Scops Owl** — 454
        Bubo bubo **Eagle Owl** — 466
        Ketupa zeylonensis **Brown Fish Owl** — 481
        Nyctea scandiaca **Snowy Owl** — 485
        Surnia ulula **Hawk Owl** — 496
        Glaucidium passerinum **Pygmy Owl** — 505
        Athene noctua **Little Owl** — 514
    STRIGINAE wood owls and allies — 525
        Strix aluco **Tawny Owl** — 526
        Strix butleri **Hume's Tawny Owl** — 547
        Strix uralensis **Ural Owl** — 550
        Strix nebulosa **Great Grey Owl** — 561
        Asio otus **Long-eared Owl** — 572
        Asio flammeus **Short-eared Owl** — 588
        Asio capensis **Marsh Owl** — 601
        Aegolius funereus **Tengmalm's Owl** — 606

CAPRIMULGIFORMES — 616

CAPRIMULGIDAE nightjars, nighthawks — 616
    CAPRIMULGINAE nightjars — 617
        Caprimulgus nubicus **Nubian Nightjar** — 617
        Caprimulgus europaeus **Nightjar** — 620
        Caprimulgus ruficollis **Red-necked Nightjar** — 636
        Caprimulgus eximius **Golden Nightjar** — 641
        Caprimulgus aegyptius **Egyptian Nightjar** — 641
    CHORDEILINAE nighthawks — 646
        Chordeiles minor **Common Nighthawk** — 646

APODIFORMES — 649

APODIDAE swifts — 649
    CHAETURINAE spine-tailed swifts — 649
        Hirundapus caudacutus **Needle-tailed Swift** — 649
    APODINAE typical swifts — 652
        Apus alexandri **Cape Verde Swift** — 652
        Apus unicolor **Plain Swift** — 653
        Apus apus **Swift** — 657
        Apus pallidus **Pallid Swift** — 670

Contents

	Apus pacificus **Pacific Swift (Fork-tailed Swift)**	676
	Apus melba **Alpine Swift**	678
	Apus caffer **White-rumped Swift**	687
	Apus affinis **Little Swift**	692
	Cypsiurus parvus **Palm Swift**	698

CORACIIFORMES — 699

ALCEDINIDAE kingfishers — 700

    DACELONINAE forest kingfishers and allies — 700

        Halcyon smyrnensis **White-breasted Kingfisher** — 701

        Halcyon leucocephala **Grey-headed Kingfisher** — 705

    ALCEDININAE small kingfishers — 710

        Alcedo atthis **Kingfisher** — 711

    CERYLINAE pied kingfishers and allies — 723

        Ceryle rudis **Pied Kingfisher** — 723

        Ceryle alcyon **Belted Kingfisher** — 731

MEROPIDAE bee-eaters — 733

    Merops orientalis **Little Green Bee-eater** — 734

    Merops superciliosus **Blue-cheeked Bee-eater** — 740

    Merops apiaster **Bee-eater** — 748

CORACIIDAE rollers — 763

    Coracias garrulus **Roller** — 764

    Coracias abyssinicus **Abyssinian Roller** — 776

    Coracias benghalensis **Indian Roller** — 778

    Eurystomus glaucurus **Broad-billed Roller** — 783

UPUPIDAE hoopoes — 786

    Upupa epops **Hoopoe** — 786

PICIFORMES — 799

PICIDAE wrynecks, woodpeckers, and allies — 799

    JYNGINAE wrynecks — 800

        Jynx torquilla **Wryneck** — 800

    PICINAE woodpeckers — 812

        Colaptes auratus **Northern Flicker** — 813

        Picus canus **Grey-headed Woodpecker** — 813

        Picus viridis **Green Woodpecker** — 824

        Picus vaillantii **Levaillant's Green Woodpecker** — 837

        Dryocopus martius **Black Woodpecker** — 840

        Sphyrapicus varius **Yellow-bellied Sapsucker** — 853

        Dendrocopos major **Great Spotted Woodpecker** — 856

        Dendrocopos syriacus **Syrian Woodpecker** — 874

        Dendrocopos medius **Middle Spotted Woodpecker** — 882

Contents

Dendrocopos leucotos	**White-backed Woodpecker**	891
Dendrocopos minor	**Lesser Spotted Woodpecker**	901
Picoides tridactylus	**Three-toed Woodpecker**	913

REFERENCES		924
CORRECTIONS		954
INDEXES	Scientific names	955
	English names	958
	Noms français	959
	Deutsche Namen	960

INTRODUCTION

This volume completes the treatment of the non-passerines. The first four volumes have dealt with 439 species (plus 37 species occurring only before 1900 or otherwise meriting lesser treatment). A number of species in this volume are relatively unknown. Some desert-dwellers such as sandgrouse, nightjars, and owls have a distribution which has not been fully recorded. The Passeriformes, to date some 323 species deserving full treatment, will be covered in the last three volumes. This number is certain to increase as new accidental species continue to be discovered annually—and even one entirely new breeding species, Kabylian Nuthatch *Sitta ledanti*, has been described from our area recently. Meanwhile we are well past the halfway mark in our arduous task and continue to be encouraged by the generally warm welcome from readers. In the past, serious errors or omissions have been relatively few but we welcome all corrections supplied.

The editors responsible for the main sections of the species accounts in this volume are:

Field Characters   D I M Wallace
Habitat   E M Nicholson
Distribution and Population   S Cramp
Movements   R Hudson
Food   D J Brooks and P J S Olney
Social Pattern and Behaviour   Dr E K Dunn and M G Wilson
Voice   Dr E K Dunn and M G Wilson
Breeding   Dr M A Ogilvie
Plumages, Bare Parts, Moults, Measurements, Weights, Structure, and Geographical Variation   C S Roselaar and Dr J Wattel

D J Brooks has again been responsible for the bulk of the editing of all texts.

In some cases, major assistance has been given by various experts for certain species and the sections concerned then bear their initials at the end; full details are given under Acknowledgements. Once again, the Voice treatments owe much to Mrs J Hall-Craggs, who prepared the sonagrams, and to P J Sellar, who searched vigorously for suitable material and compiled tapes; both also provided further invaluable help with expert comments and advice. Their joint efforts have been of major importance.

The paintings in this volume are the work of Norman Arlott, Dr C J F Coombs, Dr N W Cusa, Håkan Delin, Robert Gillmor, Miss C E Talbot Kelly, and D I M Wallace; their initials appear at the end of the caption for each plate. Norman Arlott, Ms Hilary Burn, Dr C J F Coombs, and Robert Gillmor have prepared the line drawings in the Social Pattern and Behaviour sections, J Zaagman the annual cycle diagrams and a drawing in Plumages, and D J Brooks a figure in Food. R J Connor and A C Parker have once again most generously provided the photographs of the eggs.

The Introduction to Volume I gave a full account of the scope and treatment of the various sections, including, in particular, the glossaries, guidance on voice illustrations, and diagrams showing the names used in plumage descriptions. Some modifications and amplifications were included in the Introductions to Volumes II and III. The following notes contain some additional material but should be read in conjunction with the Introductions to the earlier volumes.

DISTRIBUTION AND POPULATION

The basic information on the distribution and populations of species in the west Palearctic has again been supplied by the correspondents for each country, whose names are listed in the Acknowledgements and their initials given in the relevant texts after any data supplied by them. It must again be stressed that the maps represent the state of our current knowledge, which is still limited for some areas even in the west Palearctic; in this volume, information is particularly limited for those species living in desert areas, especially Pteroclididae and Caprimulgidae.

On the maps, RED is used to indicate breeding distribution and GREY, for most species, areas where the species is regularly present in normal winters. For terns (Sternidae)—except the *Chlidonias* species where most passage is overland—and all auks (Alcidae), however, GREY indicates their marine distribution at one or more seasons of the year, though over land areas it marks only regular winter occurrence. In all cases, the maps should be used in conjunction with the relevant sections on Movements.

We are particularly indebted to G G Buzzard for much information on the distribution of species in some east European countries, and, for world distributions, to the late C W Benson for data on species in Madagascar and the neighbouring islands, to S C Madge and Dr D A Scott for details from Afghanistan and Iran respectively, and to G Bundy, M D Gallagher, M C Jennings, and Mrs F E Warr for data for various parts of Arabia.

Finally, it should perhaps be stressed again that the Accidental paragraph under Distribution covers only accidental occurrences within the west Palearctic region.

ACKNOWLEDGEMENTS

We are once again greatly indebted to those who have provided essential assistance to this project, either financial aid or expert help and advice. For this volume, further grants or loans have been most generously given by the Commission of the European Economic Communities, the Royal Society for the Protection of Birds, the World Wildlife Fund, the Late Lord Rootes Charity Trust, the Ernest Kleinwort Charitable Trust, the Kleinwort Benson Charitable Trust, and the Inchcape Charitable Trust.

The members of the Editorial Board have once again made major contributions, often under difficult circumstances, and we are most grateful to the British Trust for Ornithology, the Edward Grey Institute of Field Ornithology at the University of Oxford, the University of Amsterdam, and the Wildfowl Trust for facilitating their labours. We should also like to thank the British Museum (Natural History), Tring, and the Sub-department of Animal Behaviour, Cambridge, for much valuable assistance.

The editors have received much help from amateur and professional ornithologists throughout the world, who are listed in the general Acknowledgements below. In several sections, major assistance was given by various experts whose initials are given at the end of the relevant accounts. Thus, in Movements, some accounts were compiled by Dr E K Dunn, and in Food by Dr E K Dunn, J R Parrott, Dr A S Richford, H A Robertson, B D S Smith, and M G Wilson. In Social Pattern and Behaviour and Voice some accounts were compiled by D J Brooks and Dr K E L Simmons, and in Plumages, etc., by Dr G P Hekstra and R Sluys.

The British Library of Wildlife Sounds, the Transvaal Museum, and recordists in many countries most generously provided tape-recordings; details are given in the captions to the sonagrams used. Where the recordings are available as gramophone records or cassettes, references are given as follows:

Alauda (1979) *Supplément sonore* disc 11.
BBC (1971) *Sea and island birds*. BBC Records, Wildlife Series 12.
J Kirby (1973) *Wild tracks* 2. Aysgarth, England.
J-C Roché (1966) *Guide sonore des oiseaux d'Europe* 2; (1970) 3. Institut Echo, Aubenas-les-Alpes, France.
J-C Roché (1967) *Guide sonore des oiseaux nicheurs du Maghreb*. Institut Echo, Aubenas-les-Alpes, France.
Sveriges Ornitologiska Förening (1982) *Skogsfåglar* cassette SWPAB-8.
Sveriges Radio (1972) *A field guide to the bird songs of Britain and Europe* by S Palmér and J Boswall 1–12 (1972), 13 (1973) (discs).
Swedish Radio Company (1981) *A field guide to the bird songs of Britain and Europe* by S Palmér and J Boswall 1–16 (cassettes).

For recordings which have not been published commercially, assistance in contacting the original recordists may often be obtained from the Curator, The British Library of Wildlife Sounds, 29 Exhibition Road, London SW7 2AS.

A vital part was once again undertaken by the correspondents in the following countries or areas who, as stated earlier, provided much of the basic data on status, distribution, and population of the various species:

ALBANIA Dr E Nowak
ALGERIA E D H Johnson
AUSTRIA Dr H Schifter, P Prokop
AZORES G Le Grand
BELGIUM AND LUXEMBOURG Dr P Devillers
BRITAIN Dr J T R Sharrock, Dr P Lack
BULGARIA T Michev, Dr M Boev, S Simeonov, P Simeonov, Z Spiridonov, J L Roberts
CANARY ISLANDS K W Emmerson
CHAD Prof J Vielliard
CYPRUS P Flint, P F Stewart
CZECHOSLOVAKIA Dr K Hudec
DENMARK T Dybbro
EGYPT P L Meininger, W C Mullié, S M Goodman
FINLAND Dr O Hildén
FRANCE R Cruon, P Nicolau-Gillaumet
GERMANY (EAST) DDR the late Dr W Makatsch
GERMANY (WEST) BRD Dr G Rheinwald
GREECE W Bauer, Dr H J Böhr, G Müller
HUNGARY Dr L Horváth
ICELAND Dr A Petersen
IRELAND C D Hutchinson
ISRAEL Prof H Mendelssohn, the late Y Paran
ITALY Prof S Frugis, P Brichetti, B Massa
JORDAN P A D Hollom, D I M Wallace
KUWAIT P R Haynes
LEBANON Dr H Kumerloeve, Lt-Col A M Macfarlane
LIBYA G Bundy
MALI Dr J M Thiollay
MALTA J Sultana, C Gauci
MAURITANIA Abbé R de Naurois, J Trotignon, R A Williams
MOROCCO J D R Vernon
NETHERLANDS C S Roselaar
NIGER Dr J M Thiollay
NORWAY Prof S Haftorn, the late G Lid, V Ree
POLAND Dr A Dyrcz, Dr L Tomiałojć
PORTUGAL R Rufino, N G Oliveira
RUMANIA Dr V Ciochia, Prof W Klemm
SPAIN J A G Morales, A Noval
SVALBARD Dr H Holgersen, I Byrkjedal
SWEDEN L Risberg
SWITZERLAND R Winkler

SYRIA Dr H Kumerloeve, Lt-Col A M Macfarlane
TUNISIA M Smart
TURKEY M A S Beaman, R F Porter
USSR Prof V E Flint, Dr V M Galushin, H Veroman
YUGOSLAVIA V F Vasić

We also wish to thank all those who made available photographs, sketches, and published material on which the drawings illustrating the Social Pattern and Behaviour sections were based; their names are given at the end of the relevant sections.

We are grateful to Ms J Barreiro, Mrs H Harkness, P Hersteinsson, Dr H Källander, Dr R E Kenward, Mrs A Kerkel, Ms H Moss, Ms I Silva, and Mrs P Stephensson for valuable help with translations.

Finally, we are greatly indebted to the following, who assisted us in many ways too diverse to specify in detail, though credits are given in the text where appropriate:

I Alexander, Dr S Ali, L Arias-de-Reyna, Dr J S Ash, N K Atkinson, P J Bacon, Dr S Baillie, F Bárcena, Dr R T Barrett, K Beloleru, Prof G Bergman, Dr H-H Bergmann, A Bermejo, R Berry, Dr C J Bibby, A Biber, Dr T R Birkhead, Dr B F Blake, H Blokpoel, D Blume, A Boldo, A J Borisovna, Dr W R P Bourne, C G R Bowden, Dr M S W Bradstreet, M H Brinkworth, P L Britton, R Broad, Dr M de L Brooke, R K Brooke, C C Brown, Dr R G B Brown, the late L H Brown, S Bude, G Bundy, G G Buzzard, Dr C J Cadbury, C J Camphuysen, J Cayford, Dr C Chappuis, R Chestney, D Choussy, N Cleere, T Cleeves, Dr N J Collar, Dr J Cooper, Dr L Cornwallis, Dr F Creutz, I K Dawson, Dr W R J Dean, H Delin, J Desselberger, Dr W J A Dick, G van Dijk, A J Dijksen, S Dijksen, Dr R J Douthwaite, Dr D C Duffy, G van Duin, A R Dupuy, T G Easterbrook, Prof C S Elton, K W Emmerson, R E Emmett, P Etienne, Dr P G H Evans, Dr K-M Exo, T Fagerström, Dr C J Feare, G D Field, J A van Franeker, Dr C H Fry, I C J Galbraith, Lt-Col M D Gallagher, Dr K Gärtner, Dr A J Gaston, C Gauci, A Gebauer, P Gloe, D E Glue, Dr S Goodman, D Goodwin, A F Gosler, P J Grant, Dr P W Greig-Smith, A Gretton, F C Gribble, A I Grieve, C I Griffiths, F Hansen, S Hansen, Dr M P Harris, B Hawkes, K Hazevoet, S Hedgren, Dr G P Hekstra, Dr H-W Helb, A Hill, K Hinrichs, F Hiraldo, Dr O Hogstad, Dr T Holmberg, Dr J A Horsfall, K Hulsman, R E Hutchison, M C Jennings, P E Jónsson, P H Jones, Dr H E Källander, R Jervis, M Kilpi, Dr B King, B King, A B Kitson, Dr A Knox, Dr C König, F J Koning, J de Korte, Dr J Krebs, Dr R Kuhk, A Kuznetsov, Dr N P E Langham, Dr J-D Lebreton, M J Leloup, R Lévêque, Y Leshem, J C Lidgate, M D Linsley, M G Lobb, Dr H Löhrl, Dr S Lovari, Dr J P Ludwig, Dr A Lundberg, J A McGeoch, Dr H M C McLannahan, M Máñez Rodríguez, S Marchant, A Martin, Dr B W Massey, C J Mead, Dr G F Mees, P L Meininger, R Möckel, A P Møller, W C Mullié, Abbé R de Naurois, Dr V A Nechaev, Dr D N Nettleship, Dr I A Neufeldt, D B Nicolai, G Nikolaus, Dr I C T Nisbet, the late M E W North, W E Oddie, K M Olsen, V Olsson, R Overall, L Palma, J L F Parslow, E Pascal, R Perry, Dr A Petersen, Dr B Petterson, A L Pieters, A J Prater, T G Prins, R J Prytherch, A M Rackham, Dr J V Ramsen, Dr A-U Reyer, J F Reynolds, Dr G Rheinwald, N Riddiford, G Rinnhofer, D Robel, T Roberts, H A Robertson, Dr P Robin, M Robinson, M D Rogers, M J Rogers, T D Rogers, H Sandee, P Saurola, J G Scharringa, Dr E R Scherner, H Schlegel, R K Schmidt, W Schubert, Dr D A Scott, Sir P Scott, Dr H F Sears, Dr D C Seel, L L Semago, Dr G W Shugart, N Sills, A J M Smith, G A Smith, the late K D Smith, Dr D W Snow, Prof M Soikelli, H N Southern, Z Spiridonov, R Steer, O Stefansson, Dr J E Strauch, S Su-Aretz, J Sultana, E Sutton, R Sutton, L Svensson, J Swaab, C Swennen, M Tasker, Dr G K Taylor, Dr I R Taylor, Dr M Thain, Dr J-C Thibault, Dr G J Thomas, P Thompson, W Thönen, W Tilgner, Dr P Tomkovich, O Tostain, G A Tyler, Dr B Ullrich, A Vale, A Vittery, R Vroman, Dr F E Vuilleumier, H E Wall, C S Waller, J C Walmsley, Dr V Wendland, J S Weske, B Westman, B Wettin, W J R de Wijs, C G Wiklund, Dr D Willard, A J Williams, Dr I Wyllie, D W Yalden, Prof A Zahavi, P A Zino.

CITATION

The editors recommend that for references to this volume in scientific publications the following citation should be used: Cramp, S (ed.) (1985) *The Birds of the Western Palearctic*, Vol. IV.

Order CHARADRIIFORMES (continued)

Family STERNIDAE terns

Small to moderately large charadriiform seabirds (Lari) with mostly 'light' plumages (in sense of Simmons 1972), though some 'dark'. Similar to gulls (Laridae) but more aerial in habits. About 43 species in 7 genera, though Moynihan (1959) recognized only 3 genera and Wolters (1975) as many as 12: (1) *Gelochelidon*, single species—Gull-billed Tern *G. nilotica*; (2) *Sterna* (sea terns), some 33 species (see further, below); (3) *Chlidonias* (marsh terns), 3 species; (4) *Larosterna*, single species—extralimital Inca Tern *L. inca* (seas off Peru and northern Chile); (5) *Procelsterna* (pale noddies), 2 extralimital species; (6) *Anous* (dark noddies), 2 species—one (Black Noddy *A. tenuirostris*) extralimital, other (Brown Noddy *A. stolidus*) accidental in west Palearctic; (7) *Gygis*, single species—extralimital Fairy Tern *G. alba* (pan-tropical). 18 species represented in west Palearctic, 15 breeding. Cosmopolitan in distribution, but majority of species (including most pelagic seabird of whole order, Sooty Tern *S. fuscata*) in warm and tropical regions.

Large genus *Sterna* contains species differing greatly in size and other features, comprising following sub-groups (some often recognized as distinct genera): (1) large, short-crested, and heavy-billed Caspian Tern *S. caspia* ('*Hydroprogne*'); (2) large to medium-large, crested terns—Royal Tern *S. maxima*, Swift Tern *S. bergii*, Lesser Crested Tern *S. bengalensis*, Sandwich Tern *S. sandvicensis*, and 2 other species ('*Thalasseus*'); (3) medium-large, riverine Large-billed Tern *S. simplex* of South America ('*Phaetusa*'); (4) medium-sized to fairly small typical sea terns—Roseate Tern *S. dougallii*, Common Tern *S. hirundo*, Arctic Tern *S. paradisaea*, and numerous other species (core *Sterna*); (4) medium-sized, dusky-backed oceanic terns—*S. fuscata*, Bridled Tern *S. anaethetus*, (extralimital) Spectacled Tern *S. lunata*, and Aleutian Tern *S. aleutica*; (5) small, white-fronted, partially riverine species—Little Tern *S. albifrons* and 3–4 other species ('*Sternula*').

Bodies gull-like but, in all except largest species, more delicately built (slimmer and more elongated). Sexes similar in size. Necks short. Wings long and pointed, typically narrower than in Laridae. Flight more buoyant, with faster and usually deeper wing-beats; rarely if ever glide except in display and, apart from *S. fuscata* (which remains entirely in flight when not at breeding station), never soar. About 20–24 secondaries. Tails with 12 feathers: often deeply forked, sometimes with outermost feathers elongated as streamers; short and with shallow fork in *Chlidonias*; wedge-shaped in *Anous*. Bills straight, pointed, and slender in most species, but heavy in larger ones (e.g. *S. caspia*) and short and thick in *Gelochelidon*; not hooked. No cere. Rhamphotheca simple. Tarsi short or very short, weak; scutellated. Toes weak; 3 front ones partially or fully webbed; hind toe raised, vestigial. Legs used much less for running or swimming than in Laridae. Oil-gland less well developed; vestigial in *S. fuscata* (which lacks waterproof plumage and never settles voluntarily on surface of sea). Caeca poorly developed. Down on both pterylae and apteria. Supra-orbital salt-glands well developed. Sternidae, though sharing most anatomical and structural characters with Laridae, differ in shorter tarsi, flight musculature (lacking *expansor secondarium*), shorter ulnar part of wing, and moult schedule of flight- and tail-feathers; see also Charadriiformes (Volume III).

Plumages usually pale grey above and white or pale grey below, often with contrasting black cap and dusky wing-tips (*Gelochelidon*, most *Sterna*)—but dusky above in some (*S. fuscata* and allies), or largely dark grey, black, or brown above and below (*Anous*, *Procelsterna*, *Larosterna*, *Chlidonias*), or all-white (*Gygis*); contrasting white or pale grey cap in *Anous*. Seasonal changes slight in black-capped species, most of which have cap reduced or flecked white in non-breeding plumage, develop dark carpal bars, and show shallower fork in tail; winter plumage of *Chlidonias* mainly white. Sexes alike. Feathers long and dense on underparts. Irises normally dark brown. No coloured bare skin round eye as in Laridae. Bills red, yellow, orange, or black—sometimes with contrasting tip; colour often changes seasonally. Fleshy wattles around gape in *Larosterna*. Legs and feet red, pink, yellow, orange, brown, or black. Post-breeding moult of head, body, and wing-coverts complete; pre-breeding partial. Moult of flight-feathers peculiar (Stresemann and Stresemann 1966); for details, see species accounts. Young precocial and, if undisturbed, semi-nidifugous (most species) or nidicolous (*Anous*, *Gygis*); food-dependent on parents for long period, until well after fledging in many species (even in far-distant winter quarters), food being given in bill (most species) or by regurgitation (*S. fuscata*, *S. anaethetus*, *Anous*). Down ramose and woolly in most species; terminal barbs matted in *S. sandvicensis*, *S. dougallii*, and several tropical species, giving spiny look, similar traces sometimes being found on nape in *S. hirundo*, *S. paradisaea*, and *S. albifrons* (down of *S. albifrons* very short); down long, straight, silky, and extremely soft in *Chlidonias* (Fjeldså 1977). Colours varied, from whitish to grey, yellowish, and brown (most species); rich cinnamon-orange in *Chlidonias*. Dark markings present in many species; varied and not fully studied. Faint or bold, dense or sparse on body; particularly bold in *Chlidonias*. Usually 3 radiating lines on crown, with additional arched line above each eye. Back pattern variable—complex but vague; resembles that of certain Laridae (Sabine's Gull *Larus sabini*, Ross's Gull *Rhodostethia rosea*) and pratincoles *Glareola* (see Glareolinae, Volume III). Individual variation in both colour and markings extreme, even within same brood; continuous within most species, but of distinct types in some—frequency of which varies geographically (Fjeldså 1977). Downy chick of *A. tenuirostris* resembles adults (even to presence of white patch on forehead and crown) but polymorphic chick of *A. stolidus* dark all over, or mostly white, or intermediate (Ashmole 1962; Dorward and Ashmole 1963). Juvenile plumages like adult but usually more barred with buff, especially on back and wings; only *S. fuscata* has mainly brown juvenile plumage, but that of *Anous* much like adult. Adult non-breeding plumage usually acquired at 5–10 months, adult breeding at 21 months (rarely 9, frequently 33).

Gelochelidon nilotica Gull-billed Tern

PLATES 1 and 2
[between pages 134 and 135]

Du. Lachstern   Fr. Sterne hansel   Ge. Lachseeschwalbe
Ru. Чайконосая крачка   Sp. Pagaza piconegra   Sw. Sandtärna

Sterna nilotica Gmelin, 1789

Polytypic. Nominate *nilotica* (Gmelin, 1789), Europe, north-west Africa, and Middle East, eastwards to Kazakhstan (USSR), Manchuria, Pakistan, and perhaps Ceylon. Extralimital: *affinis* (Horsfield, 1820), eastern Asia; *aranea* (Wilson, 1814), eastern USA, Bahamas, and Greater Antilles; 3 further races in western North America, South Ameica, and Australia.

Field characters. 35–38 cm (bill 3·4–4·1, legs 3·3–3·7, tail 9–12 cm); wing-span 100–115 cm. Close in size to Sandwich Tern *Sterna sandvicensis*, looking shorter in length due to less slender outline and 20% shorter tail and bill, but with 10% longer and broader wings. Medium-sized tern, with, compared to *Sterna* terns of same size, proportionately thicker and blunter bill, rather heavy head and body, rather broad wings, shorter, less forked tail, and longer legs. Adult in summer shows wholly black bill and crown, ash-grey upperparts with rump and tail paler, and white underparts. Immature shows paler head than *Sterna* terns. Undersurface of primaries noticeably dusky. Calls in flight deeper-toned, less harsh than those of *S. sandvicensis*. Sexes similar; some seasonal variation. Juvenile and 1st-year immature separable.

ADULT BREEDING. Crown and nape black, forming deeper cap than that of *S. sandvicensis*. Lower face, lower neck, and whole of underparts white. Back, rump, and tail ash-grey, somewhat paler on last two tracts; tail fringed off-white. Uppersurface of primaries appears uniform dusky, lacking pale fringes and tips usually obvious on *S. sandvicensis* and contrasting slightly with more pearly-grey wing-coverts and secondaries. Most of underwing white but dusky inner webs of outer 5–6 primaries combine to form distinctly dark wedge on end of wing, again more obvious than on *S. sandvicensis*. Bill and legs black. ADULT NON-BREEDING. Whole of crown becomes white or pale-grey (only sparsely streaked black at rear and on nape), but retains obvious dark panel through eye. Back slightly paler than in summer, so that pearly tone becomes uniform across upperparts. On both surfaces, outer primaries become darker through wear. JUVENILE. Top of head buff-white, with pattern of streaks on crown and nape and dark panel through eye similar to winter adult (but eye-panel less distinct). Back dusky-grey, all feathers with sandy tips, dark brown streaks, and bolder subterminal patches (wedge-shaped and most obvious as bars on scapulars): rump and tail grey, with buff tips and faint streaks less obvious than back but brown spots on ends of tail-feathers distinctive at close range. Wings darker and duller than adult, with tertials and larger inner wing-coverts marked as scapulars, and primaries pale-fringed, looking paler than those of adults in same season. Legs dark red-brown. FIRST WINTER. From September onwards, becoming more like winter adult but retaining paler head, much duskier primaries (almost black above when fully worn), and less uniform and paler appearance over scapulars and wings. FIRST SUMMER. Closely resembles winter adult but subject to erratic moult, often retaining some old flight-feathers (notably dark outer primaries). No record of exaggerated '*portlandica*'-type plumage pattern.

Generic differences from *Sterna* obvious in prolonged observation, with stubby bill, rather stouter body, shorter tail, broader wings, and characteristic aerial feeding from usually leisurely flight. Liable, however, to high risk of confusion with adult and immature *S. sandvicensis* in short view or at longer ranges, particularly on coastal passage during which all medium-sized terns may show similar flight actions. First indications at distance of *G. nilotica* likely to be shallow wing-beats and heavy appearance, but separation from *S. sandvicensis* safely based only on (1) shorter, deeper bill, often appearing as dark blob on front of white forehead, (2) darker undersurface to ends of broader wings, (3) grey of back extending over rump and tail. Other helpful characters are generally less white appearance than *S. sandvicensis* and shorter tail, but latter difficult to ascertain at most angles of view. In distinguishing *G. nilotica* from *S. sandvicensis* beware particularly short, often apparently black bill of juvenile *S. sandvicensis*. Normal flight actions much less active than those of *Sterna*, suggesting large leisurely marsh tern *Chlidonias*, as does habit of hawking for insects and swooping down to take prey from ground or water surface; in comparison with *S. sandvicensis*, wing-beats slower and shallower (with less flexing of carpal joints and more changes in body plane); nevertheless capable of sudden acceleration after prey. Gait free, with rather long legs allowing both easy walk and run (and hence feeding on ground or in shallow water). Carriage less horizontal than *Sterna*, with long leg length responsible for higher-chested appearance. Usually shuns contact with deep water, only rarely plunging after fish.

Less marine than *Sterna* terns. Voice distinctive, with absence of high-pitched notes; commonest calls written 'ack-ack' or 'kaahk', 'k-k-k', and 'charock'; though metallic and sometimes harsh, tone generally lower and less sharp and ringing than *S. sandvicensis*.

Habitat. In west Palearctic breeds in lower middle and middle latitudes, in temperate, steppe, Mediterranean, and subtropical zones. Inhabits lowland coasts, estuaries, deltas, and lagoons, and also inland lakes, rivers, and marshes, ascending to mountain lakes in Armeniya at *c*. 2000 m (Dementiev and Gladkov 1951c) and to similar altitudes in Turkey and Spain (Glutz and Bauer 1982). Generally near water, but less aquatic and less marine than most Sternidae. Normally avoids exposed ocean coasts, but switched to them in North America when driven by persecution from preferred inner dunes, salt-marshes, and grassy islands in lagoons (Bent 1921). Requires access to extensive and biologically productive landward foraging areas, choosing generally flat and dry or slightly moist terrain, sandy rather than muddy, free of rocky, broken, or steep ground and of trees or unbroken stands of dense tall vegetation. For spectrum of breeding habitats used in Camargue (France), see Hoffmann (1958), illustrated in Ferguson-Lees (1952). In USSR, on sand-banks or pebbly spits, usually on islands by shallow brackish, salt, or fresh water, sometimes in the open but also among shrubs (Dementiev and Gladkov 1951c). Also on rather dry littoral meadows cropped earlier by wild geese (Anserinae), on mud in open spaces on lagoon islet covered with *Suaeda* and *Salicornia* (Glegg 1925), and on sand-banks in rivers, etc. (Ali and Ripley 1969b). Formerly on bare shingly islets in torrents in Bayern (West Germany); also near plants, logs, and rocks (Niethammer 1942). Found nesting in Turkey among salt-pans (Seebohm 1885) and in Essex (England) on temporary islet exposed by falling water-level in recently created freshwater reservoir (Pyman and Wainwright 1952). Sometimes rests with other Sternidae above high-tide level, awaiting the ebb, but infrequently fishes either in sea or in sheltered waters. Flies low over grasslands and crops such as potatoes, standing cereals, or groundnuts *Arachis*, or stubbles, to pick up food from surface or to catch flying insects. Will follow the plough.

Outside breeding season met with far inland in Africa on large tropical rivers, lakes, ricefields, lagoons, and estuaries (Serle *et al.* 1977). As a transient in Venezuela, chiefly inland on fresh water (Schauensee and Phelps 1978). Also in India along sheltered coasts and in harbours, where recorded as attracted to sewage outflow (Ali and Ripley 1969b). In Argentina, recorded diving for small fish in tidal channels and feeding in shallow bays or resting on muddy points (Wetmore 1926). Not normally seen far out at sea. In Africa, often gathers at grass fires to take insects (Archer and Godman 1937; McLachlan and Liversidge 1970).

Flies strongly, normally in lower airspace but much higher on passage. Walks better than most Sternidae. Vulnerability to human impacts in natural habitat offset to a limited extent by adaptability to irrigation and other artificial changes.

Distribution. Colonies often changed, with erratic breeding outside main areas.

BRITAIN. Bred 1950 (1 pair) and possibly 1949 (Pyman and Wainwright 1952). FRANCE. Also bred Brittany 1946. NETHERLANDS. 1–3 pairs bred 1931, 1944–5, 1949–56, and 1958 (CSR). WEST GERMANY. Bred 19th century and early 20th century in Bayern, and at various sites on north coast in 2nd half of 19th century (Niethammer 1942; Møller 1975a). EAST GERMANY. Bred Rügen early 19th century (Klafs and Stubs 1977). AUSTRIA. Bred near Vienna to 1901 and Seewinkel to 1942 (HS). CZECHOSLOVAKIA. Perhaps nested 19th century (Hudec and Černý 1977). HUNGARY. Has bred rarely (Keve 1960). PORTUGAL. Formerly bred (Tait 1924); according to H W Coverley last bred Aviero 1932 (MDE). No recent proof (NGO). SPAIN. Small colonies also occur elsewhere, especially in wet seasons (AN). ALBANIA. Possibly nesting Lake Durazzo (Ticehurst and Whistler 1932); said to breed (Lamani and Puzanov 1962). SYRIA. Bred Lake Djabboul 1919 (Kumerloeve 1968b). IRAQ. May also breed inland (PVGK). ALGERIA. Bred 19th century, apparently in considerable numbers (Heim de Balsac and Mayaud 1962); breeds Lake Boughzoul (Jacob and Jacob 1980; EDHJ) and probably nests in some years at Macta (Metzmacher 1979). MOROCCO. Bred Iriki 1966 and 1968 (Robin 1968), possibly elsewhere 1963 and 1975 (JDRV).

Accidental. Azores, Ireland, Britain, Belgium, Luxembourg, Norway, Sweden, East Germany, Poland, Czechoslovakia, Hungary, Switzerland, Yugoslavia, Syria, Malta.

Population. Marked fluctuations, but decreased many areas; ascribed to habitat changes (Møller 1975a).

FRANCE. Camargue: fluctuating numbers, e.g. 250 pairs 1956, 262 pairs 1962, 73 pairs 1970, and 200 pairs 1976 and 1979 (A Johnson). WEST GERMANY. Over 100 pairs

8 Sternidae

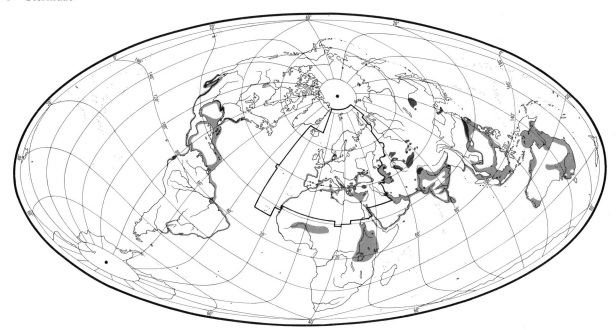

1917, then marked decline (none breeding some years); increased but fluctuating in 1960s, only 3 pairs 1972 (Møller 1975a); 50 pairs 1979 Schleswig-Holstein, and 1–3 pairs 1976 Niedersachsen (Bauer and Thielcke 1982). DENMARK. Considerable fluctuations but recent marked decline, e.g. c. 650 pairs 1865–1900, 292 pairs 1901–40, 398 pairs 1941–50, 229 pairs 1951–60, 106 pairs 1961–70, 37 pairs 1971–4 (Møller 1975a); c. 30 pairs 1979 (TD). ITALY. About 25 pairs 1972 (Møller 1975a); c. 160 pairs 1981 (Brichetti and Isenmann 1981). SPAIN. About 900 pairs 1972 (Møller 1975a). Malaga: 3500 birds at Fuenta Pidra colony (*Ardeola* 1978, **24**, 227–31). Ebro delta: 20 pairs 1979 (Y Bourgaut), 54 pairs 1980 (A Martinez). GREECE. About 400 pairs 1972 (Møller 1975a). Much reduced: e.g. in Nestos delta 300–400 pairs 1965, c. 10 pairs 1981 (WB, HJB, GM). BULGARIA. First proved breeding 1956; regular since at this site in fluctuating numbers, e.g. 1 nest 1976, 36 nests 1977 (JLR). RUMANIA. Marked decrease (Møller 1975a; Van Impe 1977; MT). USSR. Marked decrease, estimated 2000 pairs at most (Møller 1975a). At Tendra Bay (Black Sea), 773 pairs 1980, 573 pairs 1981 (A T Borisovna). TURKEY. Decreased, c. 400 pairs (Møller 1975a); fairly common (OSME). TUNISIA. Breeds in small numbers, e.g. 25 pairs 1972 (MS), 22 pairs 1976, 25–30 pairs 1977 (C A Czajkowski). MAURITANIA. Banc d'Arguin: 1750–1950 pairs 1959–65 (Naurois 1969a); 1600 pairs 1974 (Trotignon 1976).

Survival. Europe: mortality in 1st year of life 52·3%, in later years 22·8% (Møller 1975c). Oldest ringed bird 15 years 9 months (Rydzewski 1978).

Movements. Migratory. Wintering range of north and west European and probably Tunisian birds extends from Mauritania east to Nigeria and Chad (Guichard 1947; Heim de Balsac and Mayaud 1962; Malzy 1962; Elgood *et al.* 1966; Moreau 1972). Populations from Balkan peninsula and Ukraine (USSR) probably winter from Sudan south to Botswana; those from Aral–Caspian region of USSR probably account for wintering birds in Persian Gulf, Iran, Saudi Arabia, Pakistan (where also breeds), and India (Meinertzhagen 1930; Dementiev and Gladkov 1951c; Møller 1975c), though no recoveries. Completely deserts colonies on Banc d'Arguin (Mauritania) in July, many apparently wintering in Sénégal with European birds (Morel and Roux 1966); observation at Banc d'Arguin of 630 (maximum) in September, 946 in October, and 35 in November suggests another exodus in late October, probably including passage birds (Dick 1975). Nearctic populations winter from west coast of Mexico to north-west Peru, on east coast through Panama and West Indies to Surinam (American Ornithologists' Union 1957; Blake 1977). East Asian populations winter south to Greater Sunda Islands. Australian population nomadic, occurring north to New Guinea and Moluccas.

DANISH AND WEST GERMAN POPULATIONS. Mainly from Møller (1975c). Post-breeding dispersal of family groups begins at end of June, c. 1–3 weeks after their occupation of feeding grounds near colony (Møller 1981c); groups move initially in all directions, e.g. many reach northern parts of Jutland and occasionally Norway; also numerous June records from southern England and a few from eastern Scotland and western Ireland (Gloe and Møller 1978). During July, predominantly south to southwesterly movement asserts itself, passage reaching peak in Denmark in 2nd week of July. Movement may be partly nocturnal (Gloe and Møller 1978). On leaving Denmark, most birds follow west coast of Schleswig-Holstein (West Germany) but sightings from Helgoland suggest a few cut

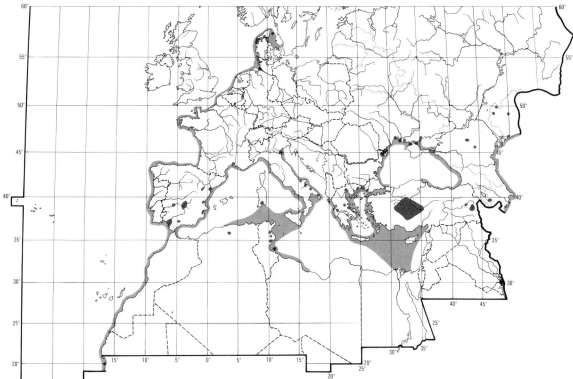

south-west across German Bight. A small proportion, having crossed the Kattegat, travel south along west coast of Sweden, then cross eastern end of Baltic Sea to Poland. Initial migration may thus be largely coastal. Movement through Schleswig-Holstein peaks at beginning of August, Netherlands and Belgium in mid-August, Britain (where rare) at end of August (latest, beginning of November), France late August to mid-September. Most appear to follow French coast and there are several recoveries along English Channel and Bay of Biscay; inland recoveries generally coincide with river systems, e.g. 3rd-year bird in September at Villefranche on River Cher. Recoveries indicate that *c.* 75% of Danish population migrates through Iberia. Recoveries from beginning of September to mid-October indicate broad-front inland movement down the peninsula: records include Estarrea (Portugal), Coto Doñana, Cadiz, Valencia (Spain); one juvenile found near Madrid. Juvenile and 1st-winter birds are occasional vagrants far to west of normal range, reaching Barbados at least 3 times. According to recoveries, about 25% of Danish population takes more easterly route, crossing central Europe on broad front; eventually funnels through Italy and to lesser extent neighbouring countries. These presumably include those that initially migrate to Poland via Sweden. Inland recoveries mostly associated with major river systems, notably Vistula, Elbe, Rhine, and Danube (Gloe and Møller 1978). Bulk of birds passing through Iberia probably proceed to West African winter quarters along Atlantic coast. There are 1st-winter recoveries of Danish-bred birds from Nouadhibou (Mauritania) in February, and Cassamance river (Sénégal) in November. Despite these coastal recoveries (which probably reflect distribution of human population) winter quarters are most likely inland. Some birds may go inland via Sénégal river, cross to Niger at Bamako, and thus reach upper Niger inundation zone (Guichard 1947). Birds following route through Italy cross Mediterranean to meet North African coast around Tunisia where passage conspicuous August–September (Heim de Balsac and Mayaud 1962). Thereafter, probably cross Sahara to Chad where species occurs from August (earliest 9 August), most arriving after 1 September (Malbrant 1952; Elgood *et al.* 1966; Etchécopar and Hüe 1967; Newby 1979). However, no recoveries in Chad.

Spring movement across Sahara indicated by records in south-east Morocco, first half April (Smith 1968*b*). From middle of April (exceptionally end of March) spring migration occurs throughout Europe, but precise routes poorly known (Gloe and Møller 1978). Recoveries and sightings suggest more rapid movement than in autumn, and reversal of proportions following major routes—*c.* 75% travelling north across Sahara and on to Italy, and *c.* 25% going via Iberia. In north-west Algeria, 2 peaks of spring passage: end of March to beginning of April, and (more important) in second half of May (Metzmacher 1979). Passage in Tunisia occurs April–May (Heim de Balsac and Mayaud 1962). Migration reaches peak in Italy at end of April. Occurs Britain (rarely) March–May, Netherlands and Schleswig-Holstein April–May. Average date of arrival of adults in Denmark 28 April. One bird ringed Italy in April recovered 4 months later in Denmark. Danish immatures generally remain south of natal area, distributed widely throughout Europe; summer recoveries of birds 1–3 years old in France, Balearic Islands, Italy, Yugoslavia, and Poland. Earliest known (from ringing) age of return to breeding colony (Schleswig-Holstein) is 3 years (see Møller 1978).

OTHER POPULATIONS. Though some may follow Iberian peninsula, recoveries of birds ringed Camargue (France) concentrated in Italy and Greece, coincident with eastern migration route of Danish and West German birds. 2 chicks ringed Camargue in same nest recovered together 3 months later in Greece. No African recoveries of French or Italian birds, but Møller (1975c) suggested trans-Saharan migration to Chad region. However, also possible that migration route and winter quarters lie further east, coincident with Balkan and USSR populations. The little available evidence suggests birds from Spain probably follow western route along Atlantic seaboard to winter in West Africa; one juvenile recovered Cadiz, and one 1st-winter in Guinea-Bissau, January (Møller 1975c; Gloe and Møller 1978). Populations breeding Greece, Rumania, and Ukraine (USSR) migrate south in autumn, travelling via Nile valley to winter from Sudan south to Botswana. Access to Ethiopia gained via Blue Nile, and to Great Lakes (where many birds winter) via White Nile (Archer and Godman 1937; Jackson 1938; Benson 1956; Smith 1957; Smithers 1964; Ardamatskaya 1977). According to Meinertzhagen (1930), common winter visitor to Egypt, November–March, but now only in very small numbers (P L Meininger, W C Mullié). Turkish birds presumably winter in East Africa; northward movement in Turkey occurs April (Wadley 1951c). For timing of movements in USSR, see Dementiev and Gladkov (1951c) and Borodulina (1960). Spring passage in Iraq 23 March–21 April, autumn passage September–November (Moore and Boswell 1956a). EKD

Food. Wide range of vertebrates and invertebrates, with considerable geographical and seasonal variation. Feeds mostly inland over dry and wet areas, and in relatively few places also on coast. Usually feeds singly, or in small parties, rarely of more than 2–3 (Ticehurst 1924a; Henry 1955; Erwin 1978); for larger flocks, see Møller (1982). 5 feeding methods. (1) Most commonly, forages by flying slowly at c. 3–6 m into wind with head down, dipping-to-surface at intervals to snatch prey from water, ground, or vegetation (Ticehurst 1924a; Witherby et al. 1941; Ali and Ripley 1969b; Tait 1970; Flint 1975; Britton 1977)—either spreads tail and drops down, or spreads tail and wings, drifts up and back with wind, and swoops, stretching neck down to grab prey before regaining height (Jensen 1946; Sears 1976). (2) Aerial-pursuit of insects (Witherby et al. 1941; Bannerman 1953; Sears 1976). (3) Hovers briefly, lands, and captures and eats prey on ground (Sears 1976). (4) Picks up food whilst walking on ground or in shallow water (Witherby et al. 1941). (5) Very occasionally plunge-dives (Witherby et al. 1941; Lévêque 1955; Sears 1976). Foraging distances from colonies vary: up to c. 10 km (Sears 1976); on average c. 15–20 km, but up to 30 km (Lévêque 1955); 2–31 km, mainly c. 9·6 km (Møller 1978c); see also Møller (1982).

Diverse diet recorded in west Palearctic: insects (adult and larvae) include: grasshoppers, crickets, and locusts (*Gryllotalpa*, *Gryllus*, *Calliptamus*, *Stenobothrus*, *Tettigonia*, *Oedipoda*, *Chorthippus*, *Dociostaurus*, *Locusta*), dragonflies *Aeshna*, beetles (*Carabus*, *Geotrupes*, *Anomala*, *Scarabaeus*, *Pentodon*, *Polyphilla*, *Aphodius*, *Melolontha*, *Necrophorus*, *Hydrophilus*, *Hydrous*, *Cybister*, *Dytiscus*, *Rhantus*, *Macrodytes*, *Prosodes*, *Blaps*, *Pimelia*, *Hylobius*), ant-lions (Myrmeleonidae), bugs (*Naucoris*, *Corixa*, *Cicada*, *Eurygaster*), butterflies and moths (*Lasiocampa*, *Saturnia*, *Mamestra*, *Loxostege*, *Sphinx*), flies *Selidopogan*, ants *Myrmica*. Other invertebrates include spiders *Trochosa*, crustaceans (*Triops*, *Carcinus*, *Cancer*, *Uca tangeri*, shrimps), molluscs (*Cardium*, *Macoma*, *Abra*), and worms (*Lumbricus*, *Nereis*, *Arenicola*). Vertebrates include mice (*Mus musculus*, *Apodemus sylvaticus*, *Micromys minutus*), shrews (*Sorex araneus*, *S. minutus*, *Neomys fodiens*), field vole *Microtus agrestis*, eggs and young (including full-grown) of birds (e.g. Avocet *Recurvirostra avosetta*, Common Tern *Sterna hirundo*, Skylark *Alauda arvensis*), lizards (*Lacerta agilis*, *L. vivipara*, *L. viridis*, *Eremias arguta*, *Phrynocephalus*), amphibians (*Rana temporaria*, *R. arvalis*, *R. ridibunda*, *Bufo calamita*, *B. viridis*, *Pelobates cultripes*, *Triturus*), fish (stickleback *Gasterosteus aculeatus*, rudd *Scardinius erythrophthalmus*, pike *Esox lucius*, gobies *Gobius*, sand-eel *Ammodytes*, Pleuronectidae, mudskippers *Periophthalmus* and *Scartelaos*). (Salvin 1859; Alléon 1886; Jäckel 1891; Whitaker 1905; Glegg 1925; Van der Meer 1930; Dircksen 1932b; Emeis 1932; Witherby et al. 1941; Andersen 1945; Jensen 1946; IJzendoorn 1947; Brandolini 1950; Dementiev and Gladkov 1951c; Løppenthin 1951c; Lévêque 1955; Kistyakivski 1957; Borodulina 1960; Dragesco 1961b; König 1961b; Bannerman 1962; Duhart and Descamps 1963; Schlenker 1966; Gloe 1974, 1976; Flint 1975; Møller 1977; Zubakin and Kostin 1977; Vargas et al. 1978; Altenburg et al. 1982.)

Analysis of pellets from colonies in northern Jutland (Denmark) and Camargue (France), and review of literature, showed considerable variations in diet. In northern Jutland takes greater diversity of prey than further south, with main foods field vole (36% by weight, 15% by number), frogs (22%, 9%), young birds (14%, 7%), lizards (12%, 8%), and insects, chiefly beetles (6%, 49%); lizards may be main prey in dry weather, and more frogs taken in wet weather. In Camargue, has more restricted diet, and average weight of prey items only 4·4 g compared with 8·3 g in Jutland; main foods frogs (76%, 17%), insects, chiefly orthopterans (17%, 66%), with very few lizards and mice, and no fish, but some crustaceans, chiefly

Triops (Lévêque 1955; Møller 1975b, 1977, 1982). In western Jutland similar diet (based on regurgitations) to northern Jutland: mainly frogs, mice, and shrews, and some lizards, birds, and insects (Andersen 1945). In Spain, pellets from colony contained almost entirely crickets (91% frequency) and fewer bugs (7%), beetles (2%), and toads (0·2%) (Vargas *et al.* 1978). Considerable variation also in USSR: in Chernomorski, where fed mainly over land, 20 stomachs and 56 regurgitations contained mainly insects, especially beetles, dragonflies, and crickets (63% by weight, 48% frequency), lizards (34%, 26%), shrimps (15%, 11%), and mice (13%, 10%). In Sivash, 12 stomachs contained chiefly lizards *L. agilis* and *Eremias arguta*. 25 stomachs and regurgitations from near Lebyazhie islands contained mainly insects (especially cicadas and beetles *Anisoplia* and *Pentodon*) and fewer lizards. At fish farms, Akhtarsk, fed over marshy lowlands, and stomach analyses showed: late July and August, chiefly frogs (95% frequency, 65% weight) and some fish (21% and 20%) but only species not bred in fisheries, i.e. pike, rudd, and gobies. At end of August rarely seen over water and took mostly terrestrial prey, mainly locusts (Borodulina 1960). In Ukraine, mainly lizards (*L. agilis* and *E. arguta*) and insects, especially beetles and orthopterans (Kistyakivski 1957). During June–August, Black Sea, chiefly orthopterans, and rarely lizards (Flint 1975). At Chongarskiye islands, Black Sea, 91 meals for chicks comprised 41·8% insects (mostly *Gryllotalpa* and *Tettigonia*), 25·3% fish (mostly Syngnathidae), 24·2% amphibians (adults and larvae), 6·6% lizards *Lacerta agilis*, and 1·1% each of mice and conspecific chicks. Adults ate large numbers of conspecific chicks (see also Zubakin 1975), as well as chicks of waders (Charadrii), Little Tern *S. albifrons*, and *S. hirundo* (Zubakin and Kostin 1977). In Persian Gulf, April, often fed on mudskippers (*Periophthalmus*, *Scartelaos*) and small crustaceans in tidal zone; also took insects over land (Løppenthin 1951). On Banc d'Arguin (Mauritania), mainly fiddler crabs *Uca* (Altenburg *et al.* 1982).
EKD

Social pattern and behaviour. Compiled partly from material supplied by A P Møller and H F Sears.

1. Mostly gregarious throughout the year, but less so than most Sternidae. In Denmark, mean flock size increases from 1·73 (sample 22) in April to 2·65 (sample 91) in June, and to 3·24 (sample 66) in August, declining to 1·67 (sample 13) in September; in Netherlands, mean group size for April–June 1·4, increasing to 4·57 (sample 35) by August, and falling to 2·19 (sample 26) in September (Gloe and Møller 1978). Generally more gregarious in winter quarters: e.g. flock of 30 in November, Iraq (Moore and Boswell 1956a). In breeding season, usually feeds alone, rarely more than 2–3 birds together. Spring and autumn migrants travel in small groups (often 2 birds) or singly (Audubon 1840; Moore and Boswell 1956a; Feeny *et al.* 1968; Gloe and Møller 1978). BONDS. Essentially monogamous, pair-bond persisting from year to year (A P Møller) though not necessarily outside breeding season. Regular occurrence of 2-bird groups on spring migration suggests some pairing before arrival at breeding grounds (H F Sears). In Danish population, immature non-breeders remain widely scattered through Europe south of breeding grounds (Møller 1975c; Gloe and Møller 1978), and may form pairs before starting to breed (Sears 1981; A P Møller). At 4 years old, rarely earlier, birds may take up nesting territories, but do not breed, so age of first breeding usually at least 5 years (Møller 1975b; A P Møller). Both sexes care for young. Family parties, sometimes including only 1 parent, stay in close contact for at least 2–3 months after fledging, during which parents feed young; no observations of parents accompanying or feeding young in winter quarters (Møller 1975b, c; Gloe and Møller 1978; A P Møller). BREEDING DISPERSION. Generally colonial, though usually lacks dense nesting of some Sternidae. Fidelity to colony-site rather weak; shifts common between years, especially following disturbance by quadruped predators (Møller 1978b, 1982). Typical colony contains c. 10–20 nests (Bent 1921; Jensen 1946; Stewart and Robbins 1947; Lind 1963a), but up to 1000 recorded (Vargas *et al.* 1978). Large colonies commonly subdivided into sub-colonies, e.g. of 10–20 nests (Borodulina 1960). 260 colonies, Denmark, of 1–200 pairs, mean 36 (Møller 1978b); 57 colonies West Germany, of 1–55 pairs, mean 5 (P Gloe). Solitary birds more frequent at fringe of range in northern Europe (A P Møller). For colony sizes, USA, see Erwin (1978) and Sears (1978). Distance between nests variable: in Denmark, 2–92 m; mean $22 \cdot 4 \pm 14 \cdot 3$ m, sample 37 (A P Møller); mean 1·5 m recorded at one colony (Lind 1963a). At Chernomorski (USSR), nests generally 1·5–2 m apart, mean 1·73 m (Borodulina 1960; in North Carolina, 2–114 m, mean $21 \pm 19 \cdot 3$ m, sample 50 (H F Sears); in California (USA), mean 6·1 m (Pemberton 1927); in one colony, North Carolina, minimum c. 12 m (Sears 1978). Nests occasionally very close: 0·8 m (Borodulina 1960; Lind 1963a). Over 18 years, mean pre-laying period spent in colony area 19 days (Møller 1975b). ♂ establishes nest-area territory, later defended by pair. Minimum diameter, Camargue, c. 40–50 cm (Lévêque 1955); usually 3–8 m² in area, Denmark (A P Møller), minimum 3–4 m² (Lind 1963b). Serves for nesting, and resting of off-duty bird, while courtship and copulation occur largely outside it (Borodulina 1960; Lind 1963a; H F Sears). Non-breeding immatures also known to defend territories in breeding season (A P Møller). Common nesting associates are other Sternidae, gulls (Laridae), and waders (Charadrii), though conspecifics preferred as nearest neighbours (Jensen 1946; Lévêque 1955; Bourke *et al.* 1973; Sears 1978; A P Møller). If colony area restricted (e.g. small island), strong communal and individual territorial defence mounted against other species competing for nesting space (Lévêque 1955; Vargas *et al.* 1978). After young disperse from nest, adopt temporary refuge sites which are defended by parents against intruders (see below). ROOSTING. Especially outside breeding season, readily assembles in roosts, typically in areas of low vegetation, e.g. *Spartina* grass, near coast or other large waterbodies (Gloe 1977a; A P Møller); flocks of up to 800 recorded (Borodulina 1960) though generally smaller. Fledged young often loaf in regular spot where parents can locate them with food; after breeding, adults and young roost together on colony site (A P Møller). During autumn migration in northwest Europe, temporary roosts of adults and juveniles assemble early August to mid-September where good feeding encountered (König 1956a; P Gloe, A P Møller). Birds generally arrive about sunset and depart at sunrise, with preening and bathing occurring around arrival and departure. Individuals stand a few metres apart, favouring conspecifics as nearest neighbours in mixed roosts with other Sternidae and Laridae (A P Møller).

2. FLOCK BEHAVIOUR. In foraging flocks, birds keep several metres apart (A P Møller). In colony, first bird on ground to detect danger assumes Aggressive-upright (see Antagonistic

Behaviour, below) giving Ack- or Chip-call (see 1–2 in Voice); elicits upflight ('alarm') among neighbours. Birds silent on take-off, but give Ack- or Chip-calls once in flock over colony; up to 15 upflights recorded in 15 min (Lévêque 1955; H F Sears). 'Dreads' occur, mostly in response to like behaviour of other Sternidae or Laridae nesting nearby, but sometimes performed independently. Small colonies perhaps less prone to spontaneous dreads (Glegg 1925; Lind 1963a, b). ANTAGONISTIC BEHAVIOUR. (1) Ground encounters. Defends nest-territory vigorously. Initial reaction to conspecific intruder usually to adopt Aggressive-upright (Lind 1963a; Sears 1981), in which standing bird stretches neck upwards and a little forward towards adversary, with fore-body held relatively high and wings slightly out but not drooped (Fig A). Posture varies with intensity of situation, a higher neck indicating greater alarm and escape tendency; if aggressor incubating or about to retreat, adopts high Aggressive-upright (neck vertical), but if running towards, or about to fly at, intruder, adopts low Aggressive-upright (neck horizontal) (Sears 1981). Bill is pointed at intruder and often opened—silently or to utter Ack-call or, in more hostile encounters, Rattle-call (Sears 1981: see 3 in Voice). Occasionally, body horizontal and head withdrawn close to shoulders, or bird performs Head-nodding—head quickly and repeatedly rotated upwards, often a few times in rapid succession—by which bird may draw attention to itself while also being mildly intimidating (Sears 1981). Conspecifics challenged with Aggressive-upright often withdraw (especially at start of nest-scraping period), sometimes presaged by Anxiety-upright: neck stretched upwards, with head and bill held horizontal, wings held tightly against body, and plumage sleeked (Lind 1963a; Sears 1981). Appeasement probably also denoted by single down-and-up movement of head so that bird seems to be glancing at feet (Looking-down); this occasionally performed by disturbed birds (Goethe 1957a; Sears 1981). Sitting bird defends nest vigorously, bill-snapping at intruders, including strange chicks, which suffer blows to the head (Lévêque 1955) and are sometimes lifted into the air, flown, and deposited further away (Dragesco 1961b). Fights common (Borodulina 1960) but birds with hatching eggs reluctant to leave nest (Sears 1978); (2) Aerial encounters. Bird in Aggressive-upright may make short walking approach towards intruder, sometimes (H F Sears) or often (Lind 1963a) proceeding to flying attack; if intruder airborne, it is chased, commonly with Rattle-call; bird also makes dive-attacks, sometimes from great height, variously uttering Ack-, Chip-, and Rattle-calls, aiming at head or back of target; often strikes with bill or feet, sometimes defecating simultaneously (Lind 1963a; Gloe 1978; Sears 1981; A P Møller, J S Weske). Occasionally, fighting birds may fly up, hovering for a while, but sustained Upward-flutter, as in *S. hirundo*, not observed (Lind 1963a). HETEROSEXUAL BEHAVIOUR. (1) General. Pair-formation involves aerial and ground behaviour at colony. Engages in much more ground display early in season than *S. hirundo*, suggesting that, in keeping with more terrestrial existence, pairing behaviour may have shifted from aerial to ground arena (Sears 1981). (2) Aerial courtship. Following

account compiled largely from H F Sears. Though High-flight and Low-flight recognized in *G. nilotica* by Lind (1963a), not discernible as predictable behaviour sequences, some components being rarely performed. Early in season, ♂ typically invites one or more birds to join him in aerial display by Head-down posture (Fig B): holds head stiffly down for 1–2 s immediately after take-off (Lind 1963a; Sears 1981). Interpreted as specific 'invitation to flying up' by Lind (1963a), by analogy with Sandwich Tern *S. sandvicensis*; since it often occurs at nest, even during chick-rearing, where mate fails to respond, may indicate that exit is not in response to danger (H F Sears). Food-carrying by advertising ♂ considered frequent by H F Sears, uncommon by A P Møller. In 104 out of 128 observations (H F Sears), sequence as follows. Sometimes just 1, but commonly up to 6–8 birds attracted to follow advertising ♂, and group then flies in and around colony, giving Ack-, Chip-, and Advertising-calls (see 5 in Voice), most calls apparently given by group leader. Pass-ceremony, in which leader overtaken by following bird, is frequent. Having been passed above, below, or to side, former leader almost always initiates 180° turn, so regaining front position along other side of hairpin course. By repeating sequence after straight flight of 200–300 m, group may thus fly back and forth, though frequently breaks off to land and display, this ground behaviour being confined much more closely to prospective colony area than aerial behaviour. Other ritualized aerial displays common in some other Sternidae—but rare in *G. nilotica* (Sears 1981)—as follows. (a) Aerial version of Bent-posture ('the Aerial Bent'): in paired flight, one bird flies with wings in raised V, and with head bent down. (b) Aerial version of Erect ('the Straight'): one bird glides, usually with body and head in line, wings up to 30° below horizontal. (c) Swaying: one bird flies back and forth in front of other's relatively smooth course, though follower may also sway (A P Møller). (d) Complex Pass-ceremony: leader adopts aerial Bent-posture as bird following overtakes in aerial Erect-posture; as participants separate, resume flapping flight and usually turn. (e) Arc-soar: both birds, usually close together and high up, glide in aerial Erect-posture, describing wide, level arc. (f) V-flying: contains elements of Low-flight (see below); lone bird flies low and slowly, wings moving from horizontal to *c*. 45° above; this truncated flapping alternates with V-gliding (wings in stationary V). Low-flight (Lind 1963a) occurs when ♂, paired or not, arrives in colony with food, usually alights near mate or other bird, and is followed into the air in chasing fashion, sometimes by other birds also; Ack-, Chip-, and Advertising-calls heard, especially from food-carrying bird. As no great height achieved, flight lacks glide and associated behaviour. If ♂ initially transfers food on ground, no paired flight ensues. (3) Ground courtship. Rate at which birds come to restrict aerial and ground activities to prospective nesting areas varies with locality and season. Colony area may be occupied rapidly, but in one instance, North Carolina, ground activity initially widely scattered in colony environs, concentrated in morning hours only, and each bout of display lasted only 5–10 min before group took off again. Gradually attention confined more to selected areas and longer periods spent there

C

(Sears 1976). Nest-territories established only 1–2 days before 1st egg laid (Sears 1978). Main ground courtship as follows; information largely from Sears (1981) and H F Sears, though Lind (1963a) has described most postures. Variations of Erect-posture commonest in ground repertoire. In Erect-posture, neck upstretched, fore-body somewhat raised on straight legs, and wings held free of body, slightly forwards and sometimes downwards, tips crossing over tail. Bird alighting near another prior to courtship may, during final part of descent, and briefly after alighting, extend neck and point bill downwards (landing version of Bent-posture), almost always followed by either Forward-erect, with bill horizontal (Fig C, right), or Erect-posture, with bill almost vertically upwards (Fig D, left). Mate usually faces landing bird and likewise adopts Forward-erect, or may not posture at all (Lind 1963a). Forward-erect relatively static; may be sustained by both birds, or one may so continue while other parades round it in Bent-posture, with body about horizontal, wings slightly out, tail horizontal to vertically upwards, and bill as much as vertically downwards, often accompanied by Cooing-call (see 7 in Voice) and Head-nodding. Full ground courtship a dynamic interaction involving other, briefer variations of Erect-posture. Bird lands, approaches mate, and then adopts Erect-posture, often (85% of observations) with Head-turning, in which head rotated rapidly from side to side (Fig D, right). Each bird walks around the other, moves a few steps away, and returns, tending to perform Down-erect, with bill pointing downwards, sometimes vertically (Fig C, left), on approach. When pair are side by side in Down-erect, one bird may make single tilt of head away from partner, who may reciprocate; this, and Head-turning, also seen in Forward-erect, which often ends courtship sequence. Both members of pair participate in scrape-making which occurs several times a day before 1st egg laid. When both birds present on territory, one starts walking, commonly in Bent-posture, then drops forward, neck extended, on to breast, hind-body and tail elevated, and extended legs kicking alternately backwards; this, with turning of body, creates shallow depression. Mate may reciprocate with Erect-posture, then Bent-posture with (sometimes) Advertising-call (Lind 1963a) and proceed to sit alongside, or relieve scraping mate, and both may start Cooing-call. May persist at same spot for some time, or birds may select another (Lind 1963a; Sears 1978; H F Sears). (4) Mating. Starts in 3 ways. (a) ♀ adopts Hunched-posture, as in other Sternidae—either silent or uttering soft Begging-calls (see 6 in Voice), while tossing head upwards with open bill as if grabbing for food (Head-tossing); ♂ approaches and mounts. (b) Rarely, ♀ continues begging after courtship-feeding (see below) and ♂ mounts. (c) ♂ approaches ♀ with intention to mount and ♀ starts begging. In all cases, ♂ approaches in Down-erect, usually mounting if ♀ stands still. If ♀ moves away, ♂ assumes Erect-posture, or, keeping his breast against her side, moves with her (Herding). ♂ lowers bill when he moves or when ♀ begs, and raises it again when he pauses or when ♀ passive. Herding often followed by mounting (Sears 1981). Just before and after mounting, ♂ may wag tail a few times. During copulation, ♂ holds head down while ♀ may perform Head-tossing. ♂ may attempt copulation several times with pauses up to several minutes between each. After mating, pair sometimes perform Aggressive-upright or Forward-erect, tilting heads from each other. Copulation rare after 1st egg laid (Lind 1963a; H F Sears). (5) Courtship-feeding. ♂ plies ♀ with food during brief pre-laying period and during egg-laying, but rarely afterwards. As ♂ appears with food in bill, giving Advertising-call, ♀ begs, adopting Hunched-posture and uttering Begging-call. Alighting ♂ adopts Erect- or Bent-posture. After food transferred, both adopt Erect-posture, and ♂ often Herds ♀, though copulation rarely follows, ♀ rebuffing ♂ after later feedings especially (Lind 1963a; H F Sears). (6) Nest-relief. During laying, begging or courtship-feeding may anticipate nest-relief (Jensen 1946; Lind 1963a). Following sequence from Lind (1963a) and Sears (1981). Relieving bird may alight in landing version of Bent-posture, then switch to Forward-erect or Erect-posture, and usually approaches nest in Bent-posture, giving Advertising-call, which sitting bird may reciprocate; may also indulge in side-throwing, in which sticks, often shells, and other substrate material tossed sideways and backwards from nearby towards nest; this different from nest-building, where performer stands nearer nest and places material on rim. Once close, Cooing-call signals intention to sit. Sitting bird may answer softly, stretching flat on nest at approach of mate, or may fly off suddenly (Waard 1952); often starts nest-building, rises, and, if it then adopts Down-erect, usually performs side-throwing. Relieved bird commonly defecates, never however in or very near nest (Sears 1978). Bird often scrapes before settling on eggs (Lind 1963a; Sears 1978). RELATIONS WITHIN FAMILY GROUP. Chicks beg by pecking at parent's bill, and giving Food-call. Parent arriving with food often performs same sequence of postures as used for nest-relief (see above), summoning young from beneath brooding bird (either sex) or from refuge (usually surrounding vegetation) with Advertising- or Cooing-call and Bent-posture (bill pointing at chick), which also stimulates young to accept meal (Jensen 1946; Lind 1963a; H F Sears). While chick swallows food, adult waits in Down-erect, sometimes uttering mild Ack-call. When disturbed, adult may deliver food in flight, without alighting (Dragesco 1961b). Adult usually retrieves and re-presents food dropped on ground. Young often refuse oversize items, which parent may then swallow (Jensen 1946; Lind 1963a; A P Møller). ♂ often delivers food to ♀ (for young) but converse not observed. Before all eggs hatched, parent may coax young back to nest by scraping (Lind 1963a) but if coaxing unsuccessful, may brood them where they are (Borodulina 1960). After a few days, parents lead young away from nest by walking in desired direction, repeating Advertising-call, and occasionally pausing to scrape. Young follow, usually calling. Sometimes chick spontaneously leaves nest on parent's arrival (Lind 1963a; Gloe 1977b; Sears 1978). 3 factors apparently control timing, distance, and direction of movement from nest. (a) Age of young: interval from hatching of 1st egg to departure similar (2·5 days) irrespective of clutch size, so

D

youngest chick to leave often only 1 day old; however, broods with 1-day-old chicks not led as far (mean 26 m) within 1st hour of leaving nest as those with older chicks (80 m). (b) Disturbance: often precipitates departure. (c) Vegetation: tend to settle in thick vegetation, staying there longer than in bare areas (Sears 1978). After moving from a few hundred metres to 1 km (Sears 1978), or several km (Møller 1975b), young 2 weeks and older led to water's edge, beaches, mudflats, etc., spending most time in one spot to which both parents return with food, though often wander between feedings (Lévêque 1955); young often run, calling, with upraised wings, to meet alighting parent, which it then crouches beside, begging in manner similar to adult Hunched-posture. Young later accompany parents in flight, and flock with adults c. 2 weeks after fledging (Bourke et al. 1973; Vargas et al. 1978; A P Møller). ANTI-PREDATOR RESPONSES OF YOUNG. In response to Ack- or Chip-calls by parents, or, in more imminent danger, Rattle-call, small young prostrate themselves, neck outstretched, seeking cover of vegetation where available (Cain 1933). Larger chicks, not yet fledged, may flee to water, swimming across quite deep stretches if pressed (Møller 1975b). PARENTAL ANTI-PREDATOR STRATEGIES. (1) Passive measures. At approach of danger, adults give Ack-, Chip-, and, when intrusion especially hazardous, Rattle-calls (see above for response of young). (2) Active measures: against birds. Approaching aerial predator elicits Aggressive-upright and Ack-, Chip-, or Rattle-calls. One or more birds then fly up and chase intruder, calling continuously (Møller 1978a; Sears 1978, 1979). At feeding grounds, Denmark, birds chased Kestrel *Falco tinnunculus* in zigzag flight, cuffing it with wings and uttering Ack-calls (Møller 1978b). Communal defence marked in dense colonies (Vargas et al. 1978), and all birds (c. 600) at colony, Camargue, flew up to attack Marsh Harrier *Circus aeruginosus* and Purple Heron *Ardea purpurea* (Lévêque 1955). Intruder on ground elicits upflight; birds hover over intruder and, one by one, dive towards it. Ack-, Chip-, and sometimes Attack-calls uttered during dive, especially at lowest point, when bird aims at head or back of target, and may also defecate on it, or strike it with bill or feet. Communal attack usually repels intruder (Dragesco 1961b; Sears 1978). For reactions to other predators, North America, see Sears (1978). (3) Active measures: against other predators. Dogs and domestic livestock encroaching on colony subjected to dive-attacks, as above (Gloe 1978, 1979; Sears 1978).

(Figs A–D adapted from Lind 1963a.) EKD

Voice. Highly vocal, especially at breeding grounds where both sexes have extensive repertoire. Many vocalizations have features of calls 1, 2, and 5, these often given together (Sears 1981). Most data from colony in North Carolina, USA (Sears 1981), but comparison of sonagrams for Europe and USA suggests that vocalizations closely similar across this part of range (E K Dunn).

CALLS OF ADULTS. (1) Ack-call. A loud, metallic 'ack-ack-ack' (Sears 1981); 'gaa-gaa-gaa' (Lind 1963a); 'kaahka-kaahka-kaahka-kaahk' (Archer and Godman 1973); 'kak-kak' (J Hall-Craggs: Fig I, first 2 and last 2 notes). Comprises 1–4 or more rapidly repeated (8·7–10·2 notes per s) vibrating sounds: expresses mild threat, alarm, excitement in courtship, disputes, etc.; the greater the disturbance, the longer the series (Sears 1981). Mostly at colony, but also at feeding grounds, in or out of season (Christensen 1912; Møller 1978a; E K Dunn). (2) Chip-call. A rippling or tittering 'uk-uk-uk' or 'k-k-k' (J Hall-Craggs: Fig I, middle 4 notes); notes of lower pitch, slightly shorter duration (0·3 s compared with 0·7 s), and more rapid delivery (11·5 per s) than those of Ack-call (see 1). Given in similar situations to Ack-call, with which often combined in single performance (Sears 1981). (3) Rattle-call. A harsh, churring 'arrr' (Waard 1952); variable number (up to 30) of quickly repeated notes of similar form, each of shorter duration, and lower intensity than in Ack-call (see 1). Rate of delivery varies from less than 10, to more than 20 notes per s (Sears 1981: Fig II). (4) Attack-call. Emphatic, harsh 'graack' or 'groock' (Sears 1981); 'rrr-a-a' or 'a-a-a' (Lind 1963a); variable duration (mean 0–15 s) and structure. Short 'graack' frequently followed immediately by series of Ack-call notes (see 1). Used almost exclusively during aggressive swoops on large

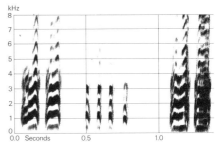

I J-C Roché (1966) Spain May 1965

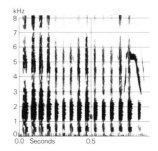

II H F Sears USA May 1973

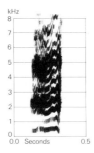

III H F Sears USA May 1973

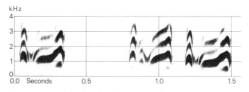

IV J-C Roché (1966) Spain May 1965

ground intruders at colony (Sears 1981: Fig III). (5) Advertising-call. 'Charock' or 'chareck' (Lind 1963a); 'chirup' (Sears 1981); 'kurruk' (Coward 1920); 'gorrok' (Jensen 1946); 'ka-huk', 'tirrUCK', or 'kay-vek' (Witherby et al. 1941: Fig IV); see also Moore and Boswell (1956a). Usually uttered singly (Lind 1963a; Sears 1981); 2nd syllable longer and more emphatic than 1st (H F Sears). Often given in combination with call 1. Strongly associated with nest, and almost invariably given when bird, of either sex, lands there for whatever purpose; also for luring brood to or from nest (H F Sears; see also Lind 1963a). Heard in migrant flocks (König 1956a). (6) Begging-call of ♀. A single, soft, relatively long (c. 0·3 s), nasal 'bäää-bäää' or 'bäät' (Lind 1963a), described as whine (Sears 1981), closely associated with Hunched-posture and Head-tossing. (7) Cooing-call. A series of short, subdued notes, described as a murmur; resembles Rattle-call (see 3) but lacks harsh quality; sometimes persists for several minutes at average delivery of 13·4 notes per s, and uttered with bill barely open (Sears 1981). Associated with scrape-making, nest-relief, and summoning young for brooding (Jensen 1946; Lind 1963a; Sears 1981).

CALLS OF YOUNG. Food-call of downy chicks a clear, high-pitched cheeping—'iihkch' (Gloe 1982). Sample frequencies of calls from 3 chicks 1–3 days old were 5·58–5·82 kHz, duration 0·074–0·549 s (H F Sears). At 7 days, disyllabic 'iih-ihkch' (Gloe 1982). After fledging, shrill, quavering whistle (Bourke et al. 1973)—rendered 'DIL-Lip' (see Witherby et al. 1941) or 'psillip'—heard at feeding grounds from young waiting to be fed by approaching parents, or when following parents in flight (König 1956; A P Møller). EKD

Breeding. SEASON. Denmark: see diagram. Mediterranean and Baltic regions: laying begins end of April, extending to June (Lévêque 1955; Borodulina 1960). SITE. On ground in the open, though usually close to tuft of vegetation, stick, or other object; less often in scattered vegetation. Colonial. Nest: shallow depression excavated in soil, often with rim of soil or sand, to which small pieces of available vegetation and debris added; size of nest depends on substrate and weather. Range of measurements, Crimea (USSR), no sample size given: external diameter 19–29 cm; internal diameter 12–16 cm; depth of cup 2·5–4 cm; height of nest 3·5–8 cm (Borodulina 1960). Weight of 6 nests average 3·3 g (1–6·5) (Gloe 1978a; see also for wide variety of material used). Building: by both sexes, using scraping, tossing, and building movements (Sears 1978). EGGS. See Plate 87. Sub-elliptical, smooth and very slightly or not glossy; very pale yellow-buff or cream, with small spots, speckles, and blotches of blackish- or dark brown. 49–35 mm (45–66 × 32–40), sample 200; calculated weight 32 g (Schönwetter 1967). Clutch: 1–4, rarely 5–6, but then probably by 2 ♀♀. Of 339 clutches, Denmark: 1 egg, 17%; 2, 34%; 3, 47%; 4, 2%; mean 2·35; clutch size increased to peak c. 1 month after laying commenced, then declined (Møller 1975b); of 303 clutches, France: 1 egg, 4%; 2, 10%; 3, 75%; 4, 7%; 5, 3%; 6, 1%; mean 3·0; clutches of 1 probably had eggs lost from them (Lévêque 1955). Eggs laid daily. One brood. Replacements laid after egg loss. INCUBATION. 22–23 days. By both parents, though ♀ may do more; in spells of 1 to over 194 min (mean 42, sample 36) (Sears 1978); shifts of 15–60 min, Denmark (Jensen 1946). Changeovers occur 3–7 times per day (Borodulina 1960); interval between incubation spells ranges from a few seconds to a few minutes (Sears 1978). Begins with 1st egg, rarely 2nd (Lévêque 1955); ♂ shows complete incubation from 1st egg, ♀ not until 2nd (A P Møller, H F Sears). Hatching asynchronous, though over period shorter than laying (Lévêque 1955). YOUNG. Precocial and semi-nidifugous. Cared for and fed by both parents; each chick fed 9–19 times a day (Borodulina 1960). Leave nest 1–4 days after hatching and taken to shelter of nearest thick vegetation, if available (Sears 1978). FLEDGING TO MATURITY. Fledging period 28–35 days. Become independent within 3 months, though may be more as families migrate together (Møller 1975b, c; A P Møller). BREEDING SUCCESS. 94 pairs, Denmark, raised mean 1·48 young per pair; 62·9% of eggs laid produced flying young (Møller 1975b).

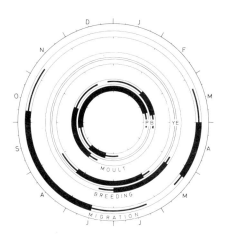

Plumages (nominate *nilotica*). ADULT BREEDING. Forehead, upper half of lores, crown down to slightly below middle of eye, and hindneck black. No white on lower hindneck or upper mantle. Mantle, scapulars, tertials, back, and all upper wing-coverts pale grey, similar in colour to Sandwich Tern *Sterna sandvicensis*; longer feathers slightly paler at tips when fresh. Rump, upper tail-coverts, and tail usually similar in colour to rest of upperparts, but sometimes slightly paler grey, especially lateral tail-feathers; never contrasting white as in *S. sandvicensis*. Sides of head and neck below black cap and all underparts white. Outer webs of primaries, primary coverts, and all secondaries grey like upperparts, but slightly paler and with silvery tinge when fresh; inner primaries and all secondaries narrowly edged white. Inner web of primaries light grey with long and broad white wedge to base. Exposed parts of primaries wear to medium-grey by midsummer, especially p4–p7; by early autumn, primaries mostly dull black except for white wedge to inner web, but innermost new and pale grey. ADULT NON-BREEDING. Forehead and crown down to eye white, nape pale grey or off-white; crown and nape frequently with fine dull black shaft-streaks of variable width, especially on sides of crown and on nape. Lores white, sometimes narrowly streaked black. Distinct dark grey to dull black patch from just in front of and below eye to over ear-coverts. Lower hindneck and upper mantle white. Remainder of plumage as in adult breeding; colour and contrast between inner and outer flight-feathers depends on wear and stage of moult. DOWNY YOUNG. Down rather long, soft; tips hair-like, not spiny as in *S. sandvicensis*. Head and upperparts pale grey-buff to deep ochre with off-white patches at gape and round and behind eye; underparts off-white. Head and upperparts dotted brown to variable extent: either uniform grey-buff or pale ochre except for a few specks on back, or heavily dotted from crown to tail, marks joining to form 3 lines on crown and 2 on back; occasionally, head and upperparts buff-brown dotted with dull black, throat brown-black. Much variation between these types. (Fjeldså 1977.) JUVENILE. Forehead off-white, crown and nape off-white or pale grey, feather-tips tinged buff when fresh; often with narrow grey-brown shaft-streaks, widening towards nape. Hindneck dull grey, white of feather-bases shining through. Lores white, often partly speckled dull black, patch in front of eye dull black. Narrow eye-ring white. Streak below eye and patch behind dull grey, often less conspicuous than in adult non-breeding. Upperparts and tail rather variable: feathers grey as in adult, but slightly darker and tinged brown towards base, feather-tips washed white and with variable amount of buff freckling; in some birds, many feathers of mantle, scapulars, tertials, tail, and inner upper wing-coverts show dull black to sepia-brown subterminal V-mark or dot; in others, dark marks restricted to some tertials, longer scapulars, or central tail-feathers. Underparts white. Fresh flight-feathers darker grey than fresh adult ones, less silvery; greater number of primaries edged white at tips and inner webs, these edges occasionally tinged buff-brown subterminally and often wider and less distinctly defined from grey centres than in adult. In worn plumage, closely similar to adult non-breeding, especially when dark subterminal marks restricted to tertials or tail; frayed feather-tips of upperparts and upper wing-coverts extensively whitish, and fringes to secondaries, most primaries, and primary coverts wide and ill-defined. FIRST IMMATURE NON-BREEDING AND SUBSEQUENT PLUMAGES. First immature non-breeding like adult non-breeding, but many worn juvenile outer primaries and primary coverts, some secondaries, and often some tail-feathers or tertials retained up to May–June of 2nd calendar year. No adult breeding acquired in summer of 2nd calendar year, but some show broad black drops to feather-centres of crown and nape. Head, body, and wing in fresh adult non-breeding by late August or September of 2nd autumn (c. 2 months ahead of adult). Subsequent plumages like adult, though birds in breeding plumage showing narrow white fringes to black feathers of cap are probably in 3rd calendar year.

Bare parts. ADULT. Iris dark brown. Bill black, gape orange. Foot black, often with red-brown tinge; soles pink-brown, orange-yellow, orange-red, or black. DOWNY YOUNG. Iris dark brown. Bill pink or apricot-orange, frequently with faint dusky tip. Foot orange-yellow, rufous, or vinaceous-brown. JUVENILE. Iris dark grey-brown to dark brown. Bill pale orange with black tip at fledging, gradually darkening to dark horn-brown and black in autumn, but base of lower mandible often tinged orange or olive-brown up to September. Foot dull flesh, dull pink-brown, or dark reddish-grey, soles yellow or red-brown; gradually darkening to adult colour in 1st autumn and winter. (Heinroth and Heinroth 1931–3; Van der Meer 1931; Fjeldså 1977; BMNH, RMNH, ZMA.)

Moults. ADULT POST-BREEDING. Complete; primaries descendant or serially descendant. Starts with innermost primary late June to mid-August when in or near breeding area; moult suspended during migration with inner 2–4 primaries new. Moult on head and body starts with scattered feathers on crown late July to mid-August (often not with forehead, unlike many *Sterna*), followed by feathering of mantle, scapulars, and underparts. Moult of head completed late September to mid-October, body and wing-coverts October–December. Tail starts with t1 early July to mid-August, soon followed by t2, t3, and t6; moult suspended during autumn migration, remainder usually replaced by November. Moult of flight-feathers usually continued directly after arrival in winter quarters, and completed with p10 mid-December to late January; a few resume later, completing February–March, when inner primaries also often moulted, thus showing serially descendant moult. ADULT PRE-BREEDING. Partial. Starts with inner primaries between late December and late February, usually after completion of post-breeding moult of outer primaries. Primary moult arrested from January to March, when inner 2–3 (1–5) primaries new in nominate *nilotica*, inner 3–5(–6) in *aranea* and *affinis*; about ⅓ of European nominate *nilotica* apparently do not change inner primaries, however. Arrested moult not continued at end of breeding season. Tail replaced late January to April, usually complete but frequently excluding t4–t5. Head occasionally starts late December, more often from mid-February, completing March; underparts and some feathers of mantle and scapulars at same time; remainder of body apparently not moulted. POST-JUVENILE. Complete. Starts with head late August to late September, followed by underparts, part of mantle and scapulars, and some lesser and median upper wing-coverts late September to November; all head and body and most tertials and wing-coverts in 1st non-breeding by December–February. Tail moulted October–February, t4–t5 last. Flight-feathers start with p1 between early December and late March; moult often suspended in early summer and completed during 1st immature post-breeding July to early September, when also 1st immature non-breeding on head and body replaced by adult non-breeding, and inner primaries and tail start with new moult series. Subsequent moults probably as in adult.

Measurements. Nominate *nilotica*. Central and southern Europe and Persian Gulf, summer, and Africa, winter; skins (BMNH, RMNH, ZMA). Tail to tip of t6; adult tail includes fresh spring birds only.

WING	AD ♂	326	(7·36; 35)	309–341	♀ 319	(7·41; 24)	307–333
	JUV	303	(9·15; 11)	289–317	302	(12·00; 8)	287–315
TAIL	AD	132	(5·27; 22)	123–143	127	(5·32; 14)	118–136
	JUV	107	(6·90; 8)	100–120	103	(4·92; 5)	95–107
BILL	AD	39·8	(1·44; 36)	38–42	37·7	(1·37; 28)	35–40
TARSUS		34·8	(1·31; 34)	33–38	33·4	(1·14; 27)	31–35
TOE		31·1	(1·24; 28)	29–33	30·2	(1·41; 25)	28–32

Sex differences significant, except juveniles. Growth of juvenile wing halts at age of c. 53 days, or c. 18 days after fledging (Heinroth and Heinroth 1931–3). Juvenile bill shorter and more slender than adult, reaching full size about November–February. Fresh adult non-breeding tail similar to adult breeding; tail of immature in 2nd calendar year on average 13 mm shorter than adult.

G.n. affinis. China and Java; skins (BMNH, RMNH, ZMA).

WING AD	♂	312	(7·12; 10)	302–321	♀ 295 (12·50; 8)	280–308
TAIL AD		124	(8·29; 8)	116–140	113 (5·74; 7)	106–120
BILL AD		38·4	(1·12; 12)	37–40	35·5 (1·42; 10)	33–37
TARSUS		32·0	(0·85; 13)	31–33	30·7 (0·97; 10)	30–32

Weights. Nominate *nilotica*. Netherlands: adult ♂♂ 204, 220, 229, 230, 239, 245 (August); adult ♀♀ 214 (June), 189, 224, 245, 264, 292 (August); juvenile ♂♂ 220, 229, 243 (August) (Van der Meer 1931; RMNH, ZMA). Central Europe: lean adult 155; fat adult, mid-September, 250; pullus on 1st day 21, at 2 weeks 136, at 5 weeks 175, at 9 weeks 250, at 13 weeks 220 (Heinroth and Heinroth 1931–3). Turkey, July: ♂ 198, ♀ 221 (Rokitansky and Schifter 1971). Mongolia, late May to June: ♂♂ 170, 185, 190, 190, 210; ♀♀ 180, 184 (Piechocki 1968c).
G.n. affinis. Hopeh (China): ♂ 223 (10) 178–320, ♀ 212 (11) 185–231 (Shaw 1936).
G.n. aranea. Middle and northern South America, September–April: ♂ 165 (4) 152–174; ♀ 165 (10) 140–181 (Russell 1964; RMNH, ZMA).

Structure. Wing long and narrow, pointed. 11 primaries: p10 longest; p9 16–24 shorter in adult, 10–18 in juvenile; p8 34–50 shorter, p7 54–76, p6 79–100, p5 105–128, p1 179–202; p11 minute, concealed by primary coverts. Longest tertials reach to about tip of p5 in folded wing. Nape not crested as in some larger *Sterna*. Tail forked, but less so than in *Sterna*; 12 feathers, length of t1 85·3 (34) 79–93 in adult, 81·8 (10) 78–88 in juvenile; depth of fork 45 (17) 38–55 in adult ♂, 41 (11) 34–52 in adult ♀, 24 (9) 19–33 in juvenile, 31 (5) 28–33 in 2nd calendar year. Fresh t6 gradually attenuated towards rounded tip, more pointed when worn. Bill stout, deep at base, about as long as head; lower mandible relatively heavy and with distinct gonys angle. Legs relatively longer than in *Sterna*; about ¼ of tibia bare, tarsus relatively stout. Front toes webbed; outer toe c. 78% of middle, inner c. 67%, hind c. 37%.

Geographical variation. Involves colour of upperparts and wings, and size. Grey of upperparts and wings of east Asiatic and North American races similar to nominate *nilotica* from Europe, North Africa, and central Asia; back to tail rather pale grey in some, but this also not uncommon in nominate *nilotica*. Races from South America and Australia distinctly paler grey on upperparts and upper wing-coverts; back, rump, and tail white or nearly so, outer webs of primaries silvery-white. Size of east Asiatic *affinis* on average smaller than nominate *nilotica* (see Measurements), but bill nearly similar and as heavy at base as latter: depth at basal corner of nostril in both 12·0 (11·4–12·8) in adult ♂, 11·2 (10·6–11·8) in adult ♀. *G.n. aranea* of eastern North America and Caribbean only slightly smaller than *affinis* in wing, tail, and tarsus, but length of bill similar and depth of bill at base distinctly less: 11·1 (10·5–11·5) in adult ♂, 10·7 (10·4–11·0) in adult ♀. Races of western North America and of South America intermediate in size between *aranea* and nominate *nilotica*, but bill relatively long; Australian race distinctly larger than others, foot and bill especially heavy. CSR

Sterna caspia Caspian Tern

PLATES 1 and 3
[between pages 134 and 135]

Du. Reuzenstern   Fr. Sterne caspienne   Ge. Raubseeschwalbe
Ru. Чеграва   Sp. Pagaza piquirroja   Sw. Skräntärna

Sterna caspia Pallas, 1770. Synonym: *Hydroprogne tschegrava*.

Monotypic

Field characters. 47–54 cm (bill 6·4–7·2, legs 4·5–4·7, tail 11–13 cm); wing-span 130–145 cm. 20% larger than Royal Tern *S. maxima*, with broader, proportionately shorter primaries, heavier head, and deeper, blunter-tipped bill; 25% larger than any other *Sterna* and looking twice as large as common medium-sized terns of that genus. Huge, gull-like tern, with deep, dagger-shaped bill on heavy head, noticeably round body, large, rather blunt wings (with proportionately long 'arm'), and relatively short forked tail. Flight recalls that of large *Larus* gull, with slow and heavy wing-beats. Adult shows prominent blood-red bill, thick black crown, grey upperparts, and white underparts; wing-tip shows large dusky area. Immature has speckled white crown, as winter adult, and bold, almost black marks across base of neck and scapulars. Sexes similar; some seasonal variation. Juvenile and immature separable.

ADULT BREEDING. Whole of crown, including line under eye, and shaggy nape black. Hindneck, face, and all underparts white. Back and inner wings silvery-grey; rump and tail paler grey, latter sometimes almost white. In flight, upper surface of 3–4 outer primaries appears darker grey than inner ones and secondaries but contrast not striking when feathers fresh. Underwing white with dark grey central areas of 5–6 outer primaries forming dusky area. Bill red; legs rather long, black. ADULT NON-BREEDING. Head markings change to give essentially grey cap (including almost whole of ear coverts and all lores), with black speckles or streaks increasing in density over crown to nape and with black patch around and behind eye. Bill-tip dusky, making bill appear blunter. With wear, primaries darken on both surfaces. JUVENILE. Basic plumage pattern as winter adult, but head markings less heavy, not extending so much below line of eye and less black (more grey, flecked buff); black panel through eye more isolated. Dark brown shaft-streaks and V-shaped patches scattered over back (most obvious on scapulars), forming less regular pattern than in other terns. Except for dark grey lesser coverts, most of upperwing appears unmarked, though fringes paler. Primaries duller and darker on both surfaces

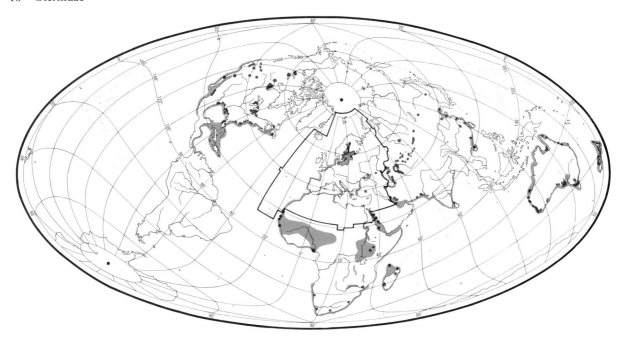

than those of adult. Tail shows dusky ends to feathers. Bill tipped dusky; legs dark red-brown. FIRST WINTER. Closely resembles winter adult but cap usually blacker, mantle duller, and worn wings show much darker lesser coverts, rather dark secondaries, almost black primaries, and dark tips on primary coverts and tail-feathers. FIRST SUMMER. Most retain traces of immature plumage, noticeably in speckled cap, less silvery upperparts, darker primaries, and less white tail. With wear, dark centres of lesser coverts and secondaries may appear as bars across front and rear of inner wing, and, when associated with faded central coverts and dark primaries, may show exaggerated '*portlandica*'-type plumage pattern. Bill sometimes dark-tipped.

Unmistakable. No other tern shows such pronounced similarities of size and structure to large *Larus* gulls or has as large a cap or bill. Next largest species, *S. maxima*, generally much paler, shaped like large Sandwich Tern *S. sandvicensis*, and has yellow or orange bill. Flight actions more gull-like with powerful wing-beats of (usually) rather deep stroke; soars, wheels, and makes dramatic plunges after fish; actions and attitudes lack grace of other terns. Gait also less free, with rather waddling walk and lurching run. Carriage usually noticeably horizontal, with stocky outline showing heavy bill and head, but markedly upright when alert or in display. Settles freely on water, often feeding like gull and riding buoyantly.

Strays widely on passage, appearing suddenly off capes and coasts and (rarely) inland. Call given in flight a deep and raucous 'kra-kra-kraa-uh'; in alarm gives barking, single elements of same (reminiscent of Jay *Garrulus glandarius*). Other calls include gull-like 'kuk-kuk-kuk'.

Habitat. Breeds in west Palearctic highly selectively on sheltered continental coasts and near substantial standing or slow-flowing inland waters, saline or fresh, from boreal through temperate to Mediterranean, steppe, and semi-desert zones. Prefers clear and fairly shallow undisturbed water with flat or gently sloping margins. Nests on islets or skerries, reefs or flat rock surfaces, sand, including sand-dunes, shingle, or occasionally on peninsulas of low rocky coasts (Niethammer 1942); also on coastal lagoons, estuaries, and floodlands. May be seen at some distance from water (Prozesky 1970). In North America, nests among driftwood and debris on low flat sandy islands, and on floating masses of dead marsh vegetation at lakesides (Bent 1921; see also Dementiev and Gladkov 1951*c*, Ferguson-Lees 1954, Merikallio 1958, Bannerman 1962). Forages largely in lagoons and other sheltered or inshore waters, and on lakes near the coast, in Sweden up to several tens of km from breeding sites. Fishes along freshwater canals and over marshes in winter in Louisiana (USA), in contrast to strictly coastal Royal Tern *S. maxima* (Lowery 1955). On the Nile, forages some distance from river banks, but in Somalia, as elsewhere, use made of tidal sand-spits as base for fishing over shoal water (Archer and Godman 1937). Exceptionally strong on the wing, accustomed to flying in upper as well as lower airspace, and able also to swim and dive even in rough water. Infrequently found at sea beyond inshore zone, nor does it venture into icy waters.

Distribution. Marked decrease in range; now extinct or occasional in areas where formerly apparently bred more

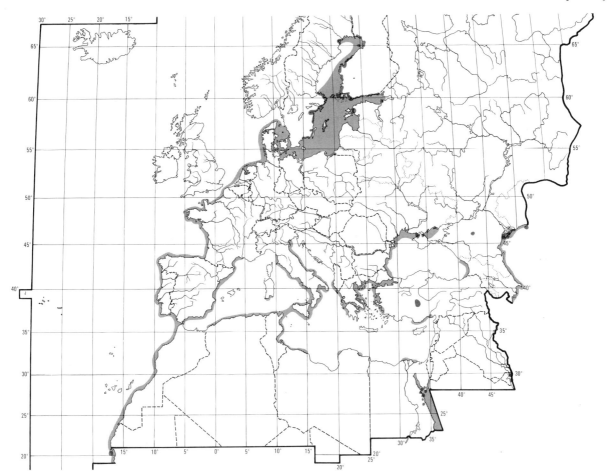

or less regularly—Denmark, West Germany, East Germany, Rumania, and Tunisia.

FRANCE. Reported breeding Corsica in 19th century (Yeatman 1976). WEST GERMANY. Bred 19th century, 1900–14, 1918, and 1928 (Schmidt and Brehm 1974). DENMARK. Bred several areas 19th century, when a few hundred pairs; decreased, last 1918; possibly bred 1931 and attempted 1944 (TD). NORWAY. Bred 1969 (SH). EAST GERMANY. Formerly regular breeder on some Baltic islands; sporadic nesting or attempted breeding one site 1956–72 (WM). POLAND. Bred 1969 (Tomiałojć 1976a). ITALY. Bred Sardinia 1837 (Schenk 1976) and Po valley 1978 (SF). YUGOSLAVIA. Probably bred 1965–9 (Matvejev and Vasić 1973). RUMANIA. Formerly bred on coast; marked decrease 20th century and no recent proof of nesting (Vasiliu 1968; MT). USSR. Latvia: bred 1976 (HV). SYRIA. Bred 1919 and 1955 (Kumerloeve 1968b). TUNISIA. Breeding in south, declining (Heim de Balsac and Mayaud 1962); no recent proof, but occurs in summer (MS).

Accidental. Faeroes, Britain, Ireland, Belgium, Norway, Yugoslavia, Syria, Cyprus, Malta, Algeria.

Population. Increased recently Sweden, Finland, and Estonia.

SWEDEN. Increasing; 850–950 pairs 1971 (Staav et al. 1972), c. 1050 pairs late 1970s (R Staav). FINLAND. Increased from c. 12 pairs late 1920s, 60 pairs 1930s, 200 pairs 1949, 500 pairs early 1950s (Hildén 1966) to c. 700 pairs (Merikallio 1958); now stable, c. 1000 pairs 1970s (OH). USSR. Estonia: increasing, 200 pairs (Onno 1966), 356 pairs 1971 (Staav et al. 1972), 500 pairs (Randla 1978). Crimea: 122 Pairs on Lebazhie islands, 1979; decreased (Kostin and Tarina 1981). TURKEY. Under 50 pairs (OSME). KUWAIT. 300–350 pairs 1958 (Sales 1965). EGYPT. Sinai: small colonies in Gulfs of Suez and Aqaba, with 25 pairs Tiran island 1979 (YP). MAURITANIA. Banc d'Arguin: 1200–1800 pairs 1959–65 (Naurois 1969a); over 1320 pairs 1974 (Trotignon 1976).

Survival. Sweden: mortality in 1st year 38·4%, in first 3 years 55·9% (R Staav). Finland: annual mortality after $1\frac{1}{2}$ years old $12 \pm 3\%$ (Soikkeli 1970). USSR: mortality in 1st year 36% (Joséfik 1969). North America (Great Lakes): 62% died prior to first breeding at 3 years old, thereafter annual mortality almost constant to 15 years old at 11·3% (Ludwig 1965). Oldest ringed bird 26 years 2 months (Clapp et al. 1982).

Movements. Migratory throughout most of west Palearctic. Movements well documented for populations of

Baltic and USSR, but not for those of Gulf of Suez and Banc d'Arguin (Mauritania); in latter area, however, an exodus in October has been recorded (Dick 1975; Trotignon et al. 1980). Movements of Eurasian populations have been reviewed by Mayaud (1956, 1958), Staav (1977), Tamantseva (1955), and Shevareva (1962). Finnish, Swedish, Estonian, and Black Sea (USSR) populations, though differing in migration routes, appear to have similar winter quarters, notably upper Niger inundation zone and Gulf of Guinea (especially Ghana coast), with a few in Mediterranean and upper Nile, Sudan (M Soikkeli, R Staav). Birds from Caspian, Aral, and Kazakhstan regions winter in Persian Gulf and on coasts of Iran, Pakistan, India, and Bangladesh (Dementiev and Gladkov 1951c; Tamantseva 1955; Shevareva 1962). Numerous at Masirah island (Oman) late August to mid-May. Nearctic populations winter south through Baja California, along shores of Gulf of Mexico and Caribbean (American Ornithologists' Union 1957), south to Lake Maracaibo, Venezuela (Schauensee and Phelps 1978). Recovery in Yorkshire (England) of bird ringed in Michigan (USA) demonstrates transatlantic vagrancy.

Dispersal from Baltic breeding grounds begins soon after fledging, from mid-July to August, and continues through Europe until October, occasionally November. Main arrival in Baltic late April to May (M Soikkeli, R Staav). Knowledge of movements derived from both ringing and sightings. Recoveries of Swedish birds soon after fledging indicate initial movement in various directions—usually in family parties comprising single parents and 1–2 young—to traditional feeding grounds (eutrophic lakes and inlets) up to $c.$ 100 km from colony. Dispersal persists for $c.$ 1 month before onset of southward migration (Staav 1980). By September, prevailing movement is south along broad spreading front from Portugal to Black Sea. Minority follow North Sea and Atlantic coast, and may stray: e.g. to Britain, July–November, where more often found inland than on coast. However, majority cross central Europe and USSR rapidly via major river systems and inland waterbodies, with most recoveries east of line from Genoa (Italy) to Denmark (Nordström 1963; Isenmann 1973; Staav 1980), mostly in specially productive deltas and lagoons (Staav 1977).

In August, most recoveries of Finnish and Swedish birds are at 50–60°N, but some reach 40–50°N, exceptionally south coast of Mediterranean (e.g. 1 Swedish bird found Egypt). By September, prevailing latitude still 40–50°N, and recoveries on south coast of Mediterranean most frequent (Staav 1977; M Soikkeli). Most Swedish birds follow a southerly route, $c.$ 10–20°E, and are thus often recovered in Italy, Sicily, Yugoslavia, and Greece (Moltoni 1954a; Baglieri and Fagotto 1977; Staav 1977); others fly south-east along river valleys to Danube delta—several Rumanian recoveries (Paşcovschi 1974; Staav 1977); yet others take south-westerly route along Mediterranean coast of Spain (Martinez and Muntaner 1979) and several recovered in Coto Doñana (Staav 1977). Finnish birds fly mostly farther east (mainly 20–30°E), following rivers running south to south-east from Gulf of Finland and south-east coast of Baltic. Many thus arrive (e.g. via Hungary) at Balkan peninsula or western Black Sea (Tamantseva 1955; M Soikkeli). Since very few recoveries between (a) Hungary and Yugoslavia, and (b) eastern Rumania and south-west USSR, this easterly passage apparently follows 2 distinct routes (Paşcovschi 1974), though shooting and resultant recoveries may tend to exaggerate sectors of migration path (M Soikkeli). Recoveries in Rumania confirm easterly bias of Finnish, and perhaps Estonian, birds: almost all are of Baltic origin, mostly Finnish, followed by Swedish and Estonian birds (Paşcovschi 1974).

Timing and strength of European passage well documented: for Sweden, see Rudebeck (1974), Staav (1980); for Denmark, Kjaer and Rosendahl (1975); for Germany, Bezzel and Reichholf (1965), Schwarz and Krägenow (1968), Gloe (1980); for Hungary (notably Lake Balaton), Keve (1948–51, 1959, 1964–5), Stein-Spiess (1956), Beretzk and Keve (1971); for Switzerland (especially Lakes Léman and Neuchâtel, with some evidence of flight over the Alps), and for summary of movements through Netherlands, Britain, France, Spain, Czechoslovakia, and Yugoslavia, see Mayaud (1956).

In Poland, inland migration appears to have increased with expansion of north-east Baltic population; for detailed study see Jósefik (1969), also Grenquist (1965) and Dobrowolski (1970). Before about 10 August, there is a fairly rapid movement due south, birds covering $c.$ 40–50 km per day. Thereafter, more leisurely 2nd wave of migration begins, characterized initially by dispersive movements, both on the coast and inland along north-south river valleys, notably the Vistula and eastern tributaries. One route skirts Carpathians to the east, reaching Dniester valley from River San. Others use tributaries of Vistula to cross central Carpathians and a few penetrate south via Moravian Gate. Dispersive movements continue until September on the coast, by which time inland movements strictly southerly. Significant percentage of late migrants use coastal route.

Small proportion of Baltic and Black Sea birds winter in Mediterranean, especially Tunisia, Nile delta, and Greece (Tamantseva 1955; Mayaud 1956; Nordström 1963; Staav 1977; G I Handrinos). Occasionally reaches south along Nile to eastern Sudan (2 Swedish birds recovered Blue Nile, including juvenile in October). Most birds from Baltic and Black Sea (Shevareva 1962) winter in tropical West Africa. Of 83 Swedish birds recovered up to 1979, 52 in Mali, 16 Ghana, 3 Sénégal, 2 in each of Chad, Sudan, and Niger, and 1 in each of Ivory Coast, Upper Volta, Nigeria, Liberia, Sierra Leone, and Zaïre (R Staav). Recoveries of Finnish birds in West Africa begin on 3 October and are relatively numerous by second half of month. In November, few recoveries from 40–

50°N, most from 30–40°N and West Africa (M Soikkeli). Most, having first spent several weeks in Mediterranean (Staav 1977), thought to cross Sahara by heading south to south-west from Gulf of Gabés (Tunisia). Indirect evidence for this from: (a) concentration of migrants and recoveries on this stretch of Tunisian coast, (b) absence of recoveries between Morocco and Sénégal, indicating little passage towards Atlantic (Moreau 1967; R Staav); (c) after build-up in Tunisia, rapid appearance (sightings and recoveries) inland in Niger inundation zone (Mali) and, to much lesser extent, Chad, where 3 recoveries of Baltic birds (Vielliard 1972; M Soikkeli, R Staav). A few, however, follow north-west coast of Africa to Gulf of Guinea. Most birds wintering Mauritania, Sénégal, and Gambia are probably local breeders (Morel and Roux 1966; Kjaer and Rosendahl 1975; R Staav). Occurrence at Rosso on Sénégal river, and recovery of Swedish bird *c.* 750 km upstream suggest that upper Niger may be reached via Sénégal river (Guichard 1947; R Staav). Recoveries suggest that reverse route followed through Africa in spring (R Staav). Bulk of Finnish and Swedish birds in winter are recovered from 2 regions: (a) inland Niger inundation zone, especially Bamako to Gao, and (b) Gulf of Guinea, notably coastal lagoons from Ghana to Ivory Coast; recoveries of Swedish birds in Ghana are confined to Keta lagoon, with only 1 recovery from Ivory Coast (R Staav). 1 Finnish bird recovered Central African Republic.

A few Finnish and Swedish 1-year-olds remain in winter quarters, e.g. 15-month-old Finnish bird at Lake Chad (Nordström 1963). However, many others have clearly moved north by late summer of 2nd calendar year. Most do not venture beyond north shore of Mediterranean, though there are several August recoveries at 40–50°N. A few reach Baltic (4 out of 62 recoveries of Finnish and Swedish birds in March–November of 2nd calendar year) but appear not to visit colonies (of 630 controls in Swedish colonies, none was 1 year old). For Swedish birds, no difference in May–July recovery regions between 1- and 2-year-old birds, but if recoveries of Finnish and Swedish birds combined, fewer in March–November of 1st summer than in same period of 2nd. Rarity of recoveries of 2-year-olds south of Sahara, March–November, indicates widespread exodus towards breeding areas. Most move north, majority probably reaching Baltic (a few visiting colonies), though later in summer than older birds. Adults reach 40–50°N, exceptionally 50–60°N by March, though main period of arrival in Baltic is late April. Most visit colonies first at 3 years old, some not until 4 (M Soikkeli, R Staav).

Spring migration (March–May) through Europe and USSR apparently follows reverse of autumn routes, with some significant differences. Passage through most countries, including those in which breeding occurs, generally lighter and less protracted than in autumn, partly because adults travel faster and roost less without accompanying offspring in (e.g.) Spain, Denmark, Switzerland, West Germany, Poland, Estonia (Mayaud 1956; Niklus 1957; Bezzel and Reichholf 1965; Jósefik 1969; Kjaer and Rosendahl 1975; Martinez and Muntaner 1979). No spring passage in Camargue (Isenmann 1973). Occurs Israel in spring (Hovel 1970). In Rumania, 64% of recoveries are in autumn, 24% in spring, 12% in summer. In southern USSR, passage heavy in autumn, but virtually nil in spring. In Hungary, however, spring passage heavier than autumn. Finnish and Swedish birds recovered Czechoslovakia and Hungary in April (Tamantseva 1955; R Staav). In Sweden, first arrivals usually 10–15 April, a few days later in Finland (R Staav). Thus in spring may take shorter, more westerly route which takes advantage of prevailing weather (Paşcovschi 1974).

Interchange of birds, especially first-time breeders, commonly occurs between colonies in Sweden, Finland, and Estonia; one bird bred *c.* 800 km from natal colony (Niklus 1957; Staav 1979); similar movements occur between Great Lakes colonies, North America (Ludwig 1979). EKD

Food. Mainly fish, occasionally invertebrates. Usually feeds singly or in pairs, though when many fish available numbers will collect, often with other Sternidae and gulls (Laridae). Feeding flight, back and forth when prey plentiful, at heights recorded as *c.* 3–10 m (Ferguson-Lees 1971), *c.* 5–15 m (G Bergman), *c.* 5–20 m (Whitfield and Blaber 1978), *c.* 3–15 m, adults averaging 6–9 m and juveniles in same area 4·5–6 m (E K Dunn). Flies with bill vertically down, hovers, flexes wings, and plunge-dives, frequently checking dive at last moment; usually, but not always, submerges completely. Fish swallowed head first, in flight, though recorded taking them (probably large ones) to river bank to kill before eating (Borodulina 1960; Whitfield 1977; Whitfield and Blaber 1978; G Bergman, E K Dunn). Less often, takes prey in shallow water by dipping-to-surface (Bird 1937; Suchantke 1960; Allan 1978). Also recorded aerial-skimming for prey like skimmer *Rynchops*, but more likely to be form of bill-washing (Tomkins 1963; Buckley and Hailman 1970; G Bergman). Practices food-piracy on Laridae and other Sternidae (Meinertzhagen 1930; Henry 1955; Ferguson-Lees 1971; Allan 1978), and scavenges dead fish from nets (Whitfield and Blaber 1978). Foraging distances from colonies recorded as *c.* 10–12 km and up to 30 km (Borodulina 1960), up to 30 km (Whitfield and Blaber 1978), up to 62 km (Gill 1976), and 30–100 km (Bergman 1953*a*; Soikkeli 1973*a*, *b*). In South Africa, fishes throughout day with peak activity early morning (Whitfield and Blaber 1978). In Texas (USA), feeds only until mid-morning if fishing success high (D Willard).

In Palearctic, fish include Cyprinidae (including roach *Rutilus rutilus*, dace *Leuciscus leuciscus*, ide *L. idus*, bleak *Alburnus alburnus*, rudd *Scardinius erythrophthalmus*), perch *Perca fluviatilis*, white bream *Blicca bjoerkna*, salmon and trout *Salmo*, ruffe *Gymnocephalus cernua*, pike *Esox lucius*, smelt *Osmerus eperlanus*, Clupeidae (including herring *Clupea harengus*), eel *Anguilla*, plaice *Pleuronectes*, and mackerel *Scomber scombrus*. May occasionally take eggs and young of birds (e.g. half-grown Lapwing *Vanellus vanellus*), though few definite records. Insects, especially locusts (Orthoptera), recorded, but infrequently (Witherby *et al.* 1941; Bergman 1953a; Borodulina 1960; Bannerman 1962; Soikkeli 1973b; Koli and Soikkeli 1974.) Average length of fish 15–20 cm, occasionally 30 cm (Borodulina 1960); 9–25·5 cm (Koli and Soikkeli 1974); up to 25 cm (Bergman 1953a).

On Åland archipelago (Finland), analysis of pellets showed both adults and young fed only on fish; at least 12 species, but mainly roach (9–25·5 cm, 57% of items in pellets), perch (10·5–21·5 cm, 24%), and clupeids, especially herring (11%); also white bream, salmon, and pike (4%). Roach dominant early May, both roach and perch in June and early July; spawning clupeids most frequent in diet late May and early June. Thus an opportunist, feeding on available fish of suitable size. No statistically significant difference between fish sizes during incubation and rearing periods (Koli and Soikkeli 1974). Similar species (roach, perch, ide, *Salmo*, and unidentified Cyprinidae) taken in Uppland and Småland, Sweden (see Koli and Soikkeli 1974). In Finnish and Swedish Baltic, also takes salmon and trout smolts—based on tags found in nest debris (Soikkeli 1973b). In Crimea (USSR), stomachs contained only fish; in lower Syr-Darya river area seen catching locusts (Borodulina 1960). Dependence on few species of suitable size noted in Finland also apparent in North American and South African studies: in San Diego bay, 76% of diet consisted of shiner perch *Cymatogaster aggregatus* and 12% topsmelt *Atherinops affinis* (Martini 1964), and in another Californian area 84% shiner perch and 16% northern anchovy *Engraulis mordax* (Baltz 1979); in Great Lakes, 74% alewife *Alosa pseudoharengus* and 15% American smelt *Osmerus mordax* (Ludwig 1965); in St Lucia Bay (South Africa), chiefly mini-kob *Johnius belengerii*—probably eaten as carrion (Whitfield and Blaber 1978).

Chicks fed fish brought whole by parents. In Finland, fish species same as those eaten by adults, with no apparent size difference (Koli and Soikkeli 1974). Adults occasionally bring fish too large for chicks to swallow, e.g. 25 cm (Bergman 1953a; Dragesco 1961b). EKD

Social pattern and behaviour.
1. Compared with most other *Sterna*, not highly gregarious outside breeding season, except when roosting. Commonly occurs singly or in groups of 2–5 (Bent 1921; Milon 1950). Separate pairs often seen as breeding season approaches, also on spring migration, though single migrants and groups of 2–4 (Jósefik 1969)—7 at most (Mayaud 1956)—occur. On spring migration, Poland, flocks generally smaller in south (average 2) than in north (average 5) (Jósefik 1969). Coastal migrant flocks generally larger (average 4·5) than inland ones (average 2) (Dobrowolski 1970). Late-summer and autumn flocks typically of 2–3 (62% of observations) comprising family parties of 1 adult with 1–2 off-spring, though both adults sometimes present (25% of observations) (Jósefik 1969; G Bergman). BONDS. Monogamous pair-bond, persisting from year to year. 1- and 2-year-old birds generally remain south of breeding range in summer (Mayaud 1956, 1958; R Staav and K Forssgren), some returning to breed at 3 years, most not until 4 (Ludwig 1979) or 5 (R Staav). Not known if pairs remain in close contact in winter quarters but majority are already paired during spring migration (Jósefik 1969). Both sexes care for young, feeding them for 6–8 months after fledging, 1 adult usually accompanying 1 juvenile. In Texas (USA), juveniles largely dependent until November, at which time foraging success still relatively poor. During winter parents provide less and less; juveniles usually independent by March (D Willard). One juvenile *c.* 9 months old seen accompanying parent and occasionally obtaining food from it (E K Dunn). BREEDING DISPERSION. Generally colonial, but solitary pairs and groups of 2–3 pairs also occur (Ferguson-Lees 1971); *c.* 5% of Baltic population solitary (Bergman 1980). In Baltic, colonies typically of 10–200 pairs (Bergman 1953a; Merikallio 1958; Aumees 1967; Kumar 1967; Staav *et al.* 1972; Väisänen 1973). Of 26 colonies, Finland, 1971, mean 63 pairs (5–132); solitary pairs 6·5% of total (Väisänen 1973). In areas where colonies established, frequency of single pairs decreases (Hildén *et al.* 1978; G Bergman). On Banc d'Arguin (Mauritania), colonies of *c.* 75–400 pairs (Naurois 1959). Elsewhere, e.g. Australia (Cooper 1964), assemblages similar. In North America, maximum colony size greater than in Europe: in 1923, colony of 1000–1500 pairs, Lake Michigan (Wood 1951); in 1961–4, Great Lakes colonies of 35–500 pairs, average 154 (Ludwig 1965); one colony, Oregon, contained *c.* 500 nests (Bent 1921). Colonies often divided into smaller groups: e.g. 4 groups, each 20–50 m apart (Milon 1950). If terrain open and even, typical distance between nests in Baltic 1·2–2 m (G Bergman), California 2–3 m (Miller 1943), Madagascar at least 1·8 m (Milon 1950), south-west Cape (South Africa) 0·65–2·90 m (Hockey and Hockey 1980). Minimum distances 0·53 m (Ludwig 1965), 0·6 m (Miller 1943; Milon 1950). In Baltic may be 0·6 m on rocky terrain where nests dispersed along suitable cracks, often with several metres of unoccupied rock between adjacent lines (G Bergman). On Banc d'Arguin, loosely colonial, nests 2–15 m apart (Naurois 1959). First-time breeders often choose colony other than natal one, many birds raised in Finland breeding in Sweden and Estonia (R Staav and K Forssgren); once established, shows strong fidelity to nest-site between years (G Bergman). However, entire colonies may desert and seek new nesting grounds (in Finland at least 80 km distant), especially—though not always—if disturbed during egg-laying (Väisänen 1973; G Bergman). Defends small, roughly circular nest-area territory; in 6 colonies, Great Lakes, area 1–1·4 m² (Ludwig 1965). Solitary pairs, Baltic, more aggressive than colonial nesters, defending whole islet (Bergman 1980). Territory used for courtship, mating, nesting, and location of young up to 30 days after hatching (see Relations within Family Group, below). Solitary pairs often nest inside colonies of associates—e.g. other Sternidae, gulls (Laridae), skuas (Stercorariidae), skimmers (Rynchopidae), pelicans (Pelecanidae), cormorants (Phalacrocoracidae)—whereas colonies of *S. caspia* usually only adjoin them, without mixing (Pettingill 1958; Naurois 1959; Tordoff and Southern 1959; Dragesco 1961b; Woolfenden and Meyerriecks 1963; Cooper 1964; Chaniot 1970; Schreiber and Dinsmore 1972; G Bergman). In establishing new colony, Finland, often settles in colony of other Sternidae or gulls, especially Lesser Black-backed Gull *L. fuscus*. Colony may develop from single nesting *S. caspia* attracting others (Grenquist 1965). ROOSTING. Forms flocks for overnight roosting and day-time loafing, at all times of year. Favours rocks, beaches, flats, and sand-

bars in open areas, not necessarily remote from man: e.g. frequents mud-banks in South Arabian harbour (Browne 1950), and oyster and shrimp canneries in USA (Bent 1921). At breeding colonies in spring, early arrivals roost together, especially on reefs well offshore but close to preferred fishing grounds, up to 50 km from intended colony site. Courtship behaviour, as well as sleeping and preening, takes place. At colony, pre-breeding roost occurs only for a few days before arrival of main body of birds, largely on same ground near colony used for loafing by family groups prior to dispersal from colony. During breeding season, off-duty birds spend $c.\ \frac{3}{4}$ of the day and all night beside incubating or brooding mate. Roosting partner commonly leaves well before sunrise (G Bergman). Often loafs near feeding grounds on rising or high tide (G Bergman, D Willard). On Texas coast, when fishing success high, adults feed until mid-morning and spend rest of daylight hours loafing; less efficient immatures (up to 2 years old) often hunt for longer and may make sporadic sorties in afternoon. 2nd-year birds spend longer loafing than 1st-years, but less than older birds, and in spring this limits opportunities for immatures to join in pairing activities at roosts (D Willard). After quitting colony, one or more family parties roost together on migration, coastally or inland. Roosting migrants occur up to 100 km from coast beside lakes and on river flats (G Bergman). Regular loafing sites allow parents to locate offspring easily for feeding (E K Dunn). Dried-out brackish lagoons in Camargue (France) attract roosts (up to 42 birds) of family parties on autumn migration (Isenmann 1973). Autumn migrants, Poland, commonly loaf around midday but most have embarked on feeding flights by 15.00 hrs, often moving south in evening. Not all roost at dusk, as nocturnal migration occurs. Less roosting on faster spring migration (Jósefik 1969). In winter quarters, roosts of 1–140 birds, with groups of 1–3 common (Bent 1921; Marchant 1941; Browne 1950; Mayaud 1956). At least early in wintering period, small roosts comprise parents and offspring (E K Dunn). Especially on migration and in winter quarters, readily roosts with other Sternidae; also with Laridae and waders (Charadrii). Favours conspecifics as nearest neighbours, birds standing $c.\ 1$ m apart (Bent 1921; Suchantke 1958; Jósefik 1969; Isenmann 1973; G Bergman). Other Sternidae sometimes shun their company, roosts departing at their approach (Bird 1973).

2. FLOCK BEHAVIOUR. Based mostly on Bergman (1953a) and information supplied by G Bergman. First bird in colony to detect danger gives Alarm-call (see 1 in Voice), and, if on ground, takes off; thus alerted, most sitting birds likewise take flight ('upflight' or 'alarm'). 'Dreads' or 'panics' from pre-breeding roost adjacent to colony site (but not from colony itself) occasionally occur in spring; departing birds initially silent as they fly up *en masse*, but start calling 5–8 s after take-off (see 1–2 in Voice). Unlike some other *Sterna*, only slight tendency to fly low over water during dread. Not known if such flights spontaneous (G Bergman). ANTAGONISTIC BEHAVIOUR. Highly aggressive, more so than in most Sternidae, whether in colony or in pre-breeding roost nearby. Territorial aggression, however, not marked at initial occupation of colony (Dragesco 1961b). Initial hostile reaction to conspecific intruding on nest-territory usually a repeated bowing between Aggressive-upright (Fig A, right) and Forward-posture (Fig A, left), sequence of Aggressive-upright/Forward-posture/Aggressive-upright taking from 2 to 5 s. Forward-posture invariably accompanied by Gakkering-call (G W Shugart: see 2 in Voice). In Aggressive-upright, stands upright, stretching neck upwards and slightly forwards towards adversary, with fore-body relatively high and plumage somewhat ruffled; carpal joints may be held only slightly away from body or well out (indicating strong escape

A

tendency). In open aggression, front of black cap remains flat while crest raised. In more intense encounters, aggressor adopts exaggerated Forward-posture with body almost prone and wings half open or, in extreme cases, raised outspread over back. Body plumage noticeably ruffled, especially just before, or in lapses during, fight. Forward-posture may be sustained up to 1 min. Sitting birds threaten intruders with raised crest and neck craned in direction of threat (Woolfenden and Meyerriecks 1963). Participants in disputes vociferous, but bill may also be opened without calling. Most disputes on ground, but combatants occasionally fly a few metres vertically into the air, sometimes with bills locked (Schüz 1940; G Bergman); dive-attacks on conspecifics rare. Participates vigorously in disputes between neighbours (G Bergman) and birds also seen attacking own mate (Schüz 1940). Egg breakage (probably accidental) not uncommon in disputes (Milon 1950; G Bergman). Parents may attack violently or kill strange chicks (conspecific or otherwise) encroaching on territory (Bent 1921; Milon 1950; Chaniot 1970). Adult conspecifics challenged with threat postures often precede withdrawal by appeasing Erect-posture (Fig B)—called 'Straight' by Bergman (1953a): neck upstretched, head and bill tilted almost vertically upwards, and wings held out sideways to extent dependent on tendency to flee—often half or more open with wing-tips almost touching ground. Raised forehead feathers (rest of cap sleeked) also denotes appeasement and alarm (G Bergman). Occasionally, disturbed bird performs Looking-down in which, with down-up movement of head, bird seems to be glancing at feet or ground just in front; probably denotes appeasement (H F Sears) or may be displacement activity (Bergman 1956; Goethe 1957a). HETEROSEXUAL BEHAVIOUR. (1) General. Pair-formation involves aerial and ground behaviour at colony. Though many birds arrive at colony apparently already paired, much courtship at colony and, initially, at pre-breeding roost adjoining it. (2) Aerial courtship. Earliest display is High-flight which occasionally starts many km from colony, at temporary shore roosts on spring migration (E K Dunn). More commonly initiated from pre-breeding roost at colony; later from colony itself once territories established. However, High-flight much less common than in (e.g.) Common Tern *S. hirundo*. Sequence of behaviour immediately prior to and possibly

B

stimulating High-flight variable: sometimes follows—and continuous with—Low-flight (see below); often results from some disturbance such as lone bird arriving at roost, or some aggressive interaction within it. No fish ever carried. 2 birds (never more) suddenly take off and ascend rapidly with fast wing-beats, one pursuing other at 2–20 m, uttering intermittent Alarm-calls (see 1 in Voice). Ascent to *c.* 200 m (maximum) alternates with several (5–12) shorter, gently descending glides—with wings usually half to three-quarters extended—in which relatively little height lost. During glide, both birds adopt aerial version of Erect-posture: stretch neck forwards, bill in line with body or slightly upwards; neck extension more marked in pursuer; Pass-ceremony (see *S. hirundo*) does not occur. Sex of leader and pursuer not generally known, but ♂ sometimes leads. Final glide downwards often brings birds low over sea, where they separate in pronounced Erect-posture; rarely land close together. High-flight generally lasts *c.* 5 min. Since ascent is straight (not spiral as in some Sternidae) with little change of direction, pair may terminate flight up to 5 km from starting point. Once ♂ has attracted ♀ and fed her a few times, they may perform Low-flight; represents advanced pairing behaviour, preliminary to nest-site occupation. ♂ flies low over colony, usually—but not always—carrying fish, inducing ♀ to follow. Both birds perform V-flying in which wings describe higher, shallower arc than in normal flight, rarely dropping below horizontal; more pronounced in ♂. Pair may overtake one another (Pass-ceremony) 2–5 times. Eventually, ♂ alights in territory; ♀ alights nearby, and takes fish suddenly and without ceremony; ♀'s cap sometimes sleeked, denoting mild alarm and appeasement. Near end of breeding season, some adults which have lost young show some resurgence of Low- and High-flights. (3) Ground courtship. ♂ approaches pre-breeding roost with fish, giving Advertising-call (see 3 in Voice), lands, and adopts Forward-erect, not unlike Aggressive-upright but body plumage less ruffled, and also, except in a bird's earliest presentations, with crest less raised. Fish offered to birds that do not react aggressively. ♀ willing to accept fish appeases ♂ by raising forehead feathers. ♂ then struts to and fro in front of her, maintaining Forward-erect, before passing fish. Unlike some Sternidae, no special posture adopted by ♀ after accepting fish. Often ♀ actively solicits fish by begging in Hunched-posture: stoops low and horizontally, plumage sleeked, with neck retracted, Head-tossing from horizontal upwards; weak Begging-calls (see 4 in Voice) given throughout. Both members of pair take part in scrape-making before egg-laying. Bird drops forwards on to breast, hindbody and wing-tips elevated, extended legs kicking alternately backwards; rotates body, creating shallow depression. The little nest-material accumulated is obtained almost entirely by incubating bird reaching out and collecting it from around scrape. (4) Copulation. May follow Courtship-feeding (see below) at pre-breeding roost or nest-territory, or be invited at other times by either sex. When ♂ takes initiative, approaches in Forward-erect, making intention movements of aggressive bowing (see Antagonistic Behaviour, above), and with whole cap raised to give rounded profile (denoting appeasement). Consenting ♀ responds by begging (see above). Alternatively, ♀ may take initiative by begging from outset, and ♂ responds with Forward-erect and mild bowing before mounting. After dismounting, ♂ (but not ♀) often adopts Erect-posture. (5) Courtship-feeding. ♂ regularly provides ♀ with fish in territory during pre-laying period and during laying; progressively less as incubation proceeds. ♂ flying in with fish in bill starts giving Advertising-call *c.* 10–100 m before reaching mate. With each call, bird simultaneously performs aerial version of Fish-bowing (Fig C)—a sudden downwards (*c.* 40° below horizontal) and then upwards (*c.* 30° above horizontal) rotation of head. ♀, recognizing ♂'s call,

C

often begs. Food commonly transferred in mid-flight with low, rapid swoop. (6) Nest-relief. Early in incubation, relieving bird of either sex commonly arrives with fish for mate (G Bergman, G W Shugart); ♀ not known to feed ♂ in other Sternidae. Bird arriving without fish approaches nest in mild Erect-posture. Both usually raise whole of black cap while maintaining its rounded profile (denoting appeasement). RELATIONS WITHIN FAMILY GROUP. Parents apparently recognize own young within 3 days of hatching (Shugart 1977). Not known when young recognize parents, though even before hatching they respond with weak food-calls to Advertisement-calls of adults (not necessarily parent); later distinguish both parents. Chicks 1–7 days old remain in or close to nest (Bergman 1956; Shugart 1977). Until young *c.* 7 days old, parents alight near nest to feed young which beg by pecking at bill, making forward bowing movements, and uttering food-call. Hungry young run towards alighting parent, bowing frantically and snatching at fish. By scraping, parent summons young for brooding (G Bergman), or sheltering against sun in south of range (Dragesco 1961b). In colony undisturbed by man, Lake Michigan (USA), families remained around nest-site until at least 1 week after fledging (G W Shugart), but in disturbed colonies territories abandoned 2–30 days after hatching—usually 10–20 (Bergman 1953a; Shugart 1977; Zubakin and Kostin 1977)—and birds gather at shore roost, where parents defend area of at least 1 m radius aroung young (Bergman 1953). Chicks initially coaxed from nest and—once free of it—summoned from danger (see Parental Anti-Predator Strategies, below) by parent simultaneously delivering Advertising-call and performing Fish-bowing, usually without fish (Fig D): bows forward so that bill lies along ground facing chick, then raises forebody, throws head back (still calling), and resumes normal stance; crest slightly raised, indicating aggression towards source of danger (e.g. strange adult near offspring), but no other aggression shown (G Bergman); if trying to summon slow or reluctant chick, prone posture sometimes maintained (silently) for several minutes before standing upright (Cooper 1964). At *c.* 15 days, young begin to beg in crouched posture similar to Hunched-posture of adult, which they reinforce with gaping bill (erectile white feathers edging it accentuate gape) and loud juvenile food-call. Older young often fed from the air, parent making low pass. If fish too big to swallow and rejected, parent often flies off to wash it before offering it again; some parents offer fish several times (Woolfenden and Meyerriecks 1963) but others swallow it themselves if rejected only twice (G

D

Bergman). Once moved to shore, older young sometimes wash fish themselves (Bergman 1956). Young develop aggressive behaviour at early age, showing Forward-posture, calling loudly and (after fledging) raising crest. Fights between chicks common and older ones also join in disputes between adults. Captive-reared young first showed Aggressive-upright at c. 37 days old; fishing skills regularly exercised from 30 days old, juveniles repeatedly picking up objects from ground and shaking them (Bergman 1956). 1 week after fledging, young begin to follow parents on distant fishing trips (Soikkeli 1973c) and are often fed on sea surface (E K Dunn), but also return to shore to be fed. Head-tossing (as in adult ♀) now accompanies juvenile food-call. At c. 45 days, captive-reared young caught fish by dipping to sea surface; by 60 days, diving skills developing fast and by 65 days self-sufficiency achieved (Bergman 1956). Most, however, remain partly dependent on parents for much longer (see Bonds, above). ANTI-PREDATOR RESPONSES OF YOUNG. On approach of danger, downy young crouch in response to Alarm-call of parent (Bergman 1955). If closely approached, however, chicks often run away and may swim if water nearby (Chaniot 1970; G Bergman). From c. 10 days, disturbed chicks run to seek regular refuge in patch of vegetation or other suitable spot, but this habit ceases after 25 days (Bergman 1956). PARENTAL ANTI-PREDATOR STRATEGIES. (1) Passive measures. At approach of danger, adults give Alarm-call (see above for response of young). (2) Active measures: against birds. Sitting birds meet mild threat, e.g. gull flying too close, with raised crest and Gakkering-call. Also pursues—singly or collectively—aerial predators such as gulls and skuas, successfully driving them off (Audubon 1840; Chaniot 1970). Other species, such as eagles (Accipitridae) not thus repulsed, and Hooded Crow *Corvus corone* robbed eggs from Baltic colony despite mass aerial pursuit starting when predator over 200 m from colony. (3) Active measures: against man and other ground predators. Most sitting birds take flight on approach of intruder, doing so when human appears 100–200 m away (De Groot 1931; Soikkeli 1973c). Some—never more than 2–3 at once (G Bergman)—leave colony to confront intruder, swooping to within a few cm of head, and giving loud Gakkering-call (see 2 in Voice). In Baltic colonies, rarely strikes or defecates on humans (G Bergman), but strikes intruders on Banc d'Arguin (Dragesco 1961b).

(Figs A–D after photographs in Bergman 1953a.) EKD

Voice. Highly vocal, especially at breeding grounds where sexes share repertoire except for Begging-call of ♀ (see 4).
CALLS OF ADULTS. (1) Alarm-call. A low, loud, barking 'ra' (G Bergman); 'kraa' (Bent 1921); 'rāb', 'gráab', 'grāap', 'krrāpp', 'kráah' (Deitrich 1921); also disyllabic 'kraa-uh' (E K Dunn: Fig I). Commonly heard on approach of intruder towards colony; also during High-flight. (2) Gakkering-call. A vehement, rasping 'ra ra ra-ra-ra-rau' (G Bergman); 'rra rra rra rra RRRÄÄ (Schüz 1940); disyllabic final note observed in 'ca ca ca crau-au' (Miller 1943); comprises elements of call 1 run together to produce sequence of increasing frequency, ending in note louder and more extended than call 1 (Fig II) at closest point to intruder during air-to-ground, or aerial swoop. (3) Advertising-call; named 'Fish-call' by Bergman (1953a). A 4–5 note 'ra-ra-ra-ratschrau', probably providing individual recognition (mate–mate, parent–offspring). Delivered mostly in flight, especially by ♂ arriving with fish for mate, when given usually 2–5 times at

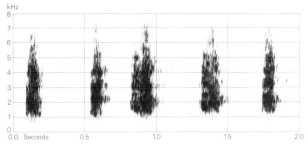

I L B McPherson New Zealand January 1973

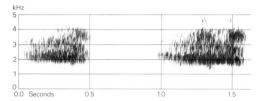

II J-C Roché (1970) Finland June 1963

c. 5-s intervals, or by both sexes when they have young; also on ground to summon young from hiding or danger (G Bergman). Sometimes preceded by Alarm-call (see 1). (4) Begging-call of ♀. A high, weak, reedy 'reee' or 'räää'; used to solicit courtship-feeding and copulation from ♂ (G Bergman). (5) Contact-call. Commonest call heard in undisturbed colony a loud, relatively protracted 'rau' or 'rrau' (G Bergman). Heard from mates at nest or from bird with brood.

CALLS OF YOUNG. Beg with vibrating, high-pitched whistling note 'i-i-i-i-' (as in 'see') of varying length; grows stronger towards fledging. Hunger without begging in older chicks expressed by 'vui' or 'vi', changing to 'ivi' when almost fledged. After fledging, food-call a more protracted and penetrating 'uivi' or 'uiviii' (G Bergman); 'weeeee' (E K Dunn), heard when offspring accompany parents in flight or on ground (G Bergman); 'wiri-wiri', 'swie', or 'swirje' probably the same call (Frieling 1932).

EKD

Breeding. SEASON. Baltic: see diagram. Black, Caspian, and Aral Seas: laying begins early to mid-May (Dementiev and Gladkov 1951c; Zubakin and Kostin 1977). Iraq: eggs found early April (Ticehurst et al. 1922). Banc d'Arguin (Mauritania): nests found April–May (Naurois 1969a); nests with eggs found November–December, 1978 (Trotignon 1979; Trotignon et al. 1980). SITE. On ground in the open, on sand, gravel, stony beaches, or flat rocks. Colonial or solitary. Nest: shallow depression; usually unlined, sometimes with rim of available pieces of vegetation or debris. Range of dimensions, Crimea, sample size not given: external diameter 16–27 cm; internal diameter 19–24 cm; depth of cup 4–7 cm (Borodulina 1960). Building: by both sexes, turning and scraping. EGGS. See Plate 88. Sub-elliptical, smooth and slightly or not glossy; cream

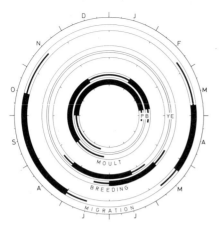

or pale buff, with small dark brown or black spots and speckles. 64 × 44 mm (55–72 × 41–47), sample 180; calculated weight 65 g (Schönwetter 1967). Clutch: 1–3. Of 365 clutches, Finland: 1 egg, 11%, 2, 40%; 3, 48%; 4, 1%; 5, 1%; mean 2·39; clutch size declined during season from 2·7 in first 2 weeks to 1·6 towards end. One brood. Replacements (smaller clutch) laid after egg loss (Soikkeli 1973c). Laying interval c. 1 day. INCUBATION. 20–22 days. By both parents, though ♀ does more (Dietrich 1921). Begins with 1st egg (G Bergman). Hatching occurs over 1–2 days (Zubakin and Kostin 1977). YOUNG. Precocial and semi-nidifugous. Cared for and fed by both parents. Usually stay in or near nest for 1–7 days, then move to shore where they have regular refuge sites (Bergman 1953a); if completely undisturbed may stay at nest until ready to fledge (G W Shugart). FLEDGING TO MATURITY. Fledging period 30–35 days. Age of independence up to 8 months, as still partly dependent on parental feeding until this age (D Willard). Age of first breeding 3 years or older (Ludwig 1965). BREEDING SUCCESS. 380 pairs, Finland, hatched c. 1·83 eggs per pair and fledged c. 1·57 young per pair; 13–20% of eggs failed to hatch, and 13–14% of young died before fledging (Soikkeli 1973c). For comparable data, Great Lakes (Canada and USA), see Ludwig (1965) and Shugart et al. (1978).

Plumages. ADULT BREEDING. Forehead, upper half of lores, crown to just below eye, and slightly elongated feathers of nape black, latter forming short rough crest. Hindneck white. Mantle, scapulars, tertials, secondaries, and all upper wing-coverts pale grey; back pale grey, shading to white towards rump and upper tail-coverts. Tail white, often with faint pale grey tinge. Outer webs of outer primaries and both webs of innermost pale silvery-grey; tips and inner webs of outer 5–6 blackish-grey, inner webs with ill-defined pale grey wedges towards bases; undersurface of outer feathers mostly dark grey. Underparts white, occasionally with slight grey tinge from lower throat to belly. Axillaries and under wing-coverts white. As in other *Sterna*, silvery-grey primary-tips change to blackish by wear, especially on exposed parts; centres and outer webs of worn secondaries and tertials and outer webs and tips of tail-feathers become dark grey or brown-grey. ADULT NON-BREEDING. Forehead, upper half of lores, and crown heavily streaked black and white: streaks on crown formed by rows of black tear-drops, on forehead and lores by smaller black speckles. Small patch in front of and below eye, ear-coverts, and slightly elongated feathers of nape mostly black, feathers faintly fringed white only. Remainder of head, body, tail, and wing as in adult breeding; no grey tinge to underparts; colour of flight-feathers and contrast between inner and outer feathers variable, depending on moult and wear. DOWNY YOUNG. Much variation (see Chaniot 1970). Head and upperparts off-white, glaucous-grey, pale cream, or buff-brown; underparts paler, chin and belly white; lores and lower throat often grey; variable amount of dark freckling on back, nape, and wing (Fjeldså 1977). JUVENILE. Forehead, lores, crown, and nape closely streaked pale buff or off-white and dull black; spot before and below eye to ear-coverts dark grey to dull black. Upper mantle, scapulars, and tertials pale grey, feathers usually shading into broad white tips (latter partly tinged buff when fresh); variable number of feathers with dull black or sepia subterminal V-mark—in some, all upperparts marked, but marks frequently absent (especially on mantle and upper scapulars) or restricted to tertials. Back and rump like adult breeding, but feather-tips often slightly tinged buff when fresh. Hindneck, underparts, axillaries, and under wing-coverts white. Tail-feathers pale grey to nearly white, wide margins along tips and inner webs white; central 1–3 pairs often with dull black subterminal dots, others occasionally shaded dull grey subterminally. Flight-feathers like adult, but grey less silvery; inner edges and tips of inner 5–7 primaries and of primary coverts bordered white, grey of feather-centres slightly darker subterminally; wedges to inner webs of primaries duller grey, indistinct. Upper wing-coverts like adult, but longer lesser and most median coverts tinged off-white or buff at tips when not too worn, shorter lesser slightly darker grey subterminally, forming indistinct carpal bar; some inner occasionally with dull black subterminal mark. In worn plumage, forehead and crown paler and less heavily streaked black than in adult non-breeding. FIRST IMMATURE NON-BREEDING AND SUBSEQUENT PLUMAGES. 1st immature non-breeding like adult non-breeding; often indistinguishable when juvenile feathers with characteristic subterminal marks lost, but usually still with indistinct carpal bar and juvenile outer primaries sepia and heavily worn, strongly contrasting with fresh inner feathers. Primary moult cycle in 1st immature non-breeding and following immature plumages usually different from rather rigid adult moult cycle (see Moults). During spring and summer of 2nd calendar year, no adult breeding obtained and 1st immature non-breeding directly replaced by 2nd. Spring and summer birds with forehead and crown black except for small white spot or short streak at side of each feather-tip are probably in 3rd, perhaps 4th, calendar year; primary moult scores of these often still different from adults (e.g. still actively moulting April–May, or suspending with inner 6 new in June).

Bare parts. ADULT. Iris red-brown, sepia, or dark brown. Bill bright coral-red to carmine during breeding, extreme tip yellow, sometimes shaded brown subterminally; bill paler orange-red in non-breeding (about August–March), 1–2 cm of tip tinged horn-black. Foot black, soles occasionally orange- or yellow-flesh. DOWNY YOUNG. Iris dark brown. Bill apricot-orange, usually with black tip. Foot fuscous or yellow-brown. JUVENILE. Iris dark brown. Bill dull orange-yellow to orange-red, tip extensively shaded horn-black. Foot yellow with blackish tinge, gradually darkening to black. (Saunders 1896; Hartert 1912–21; Fjeldså 1977; BMNH.)

Moults. ADULT POST-BREEDING. Complete, primaries descendant. Starts with forehead and crown between late July and late

September; inner primaries either start at same time (suspending during migration) or after arrival in winter quarters. Body, tail, and wing completed January–February. ADULT PRE-BREEDING. Partial. Head and part of body, often a few inner primaries, and part or all of tail. Head and body February–March; tail and primaries start at termination of previous post-breeding primary moult, arresting from about March with inner (1–)2–4 primaries new, but birds which terminate post-breeding late do not moult any flight-feathers. Arrested primary moult not continued after breeding season. POST-JUVENILE AND SUBSEQUENT MOULTS. Post-juvenile complete, in winter quarters. Starts (October–)November–December; completed by March–April, but outer primaries often not until June. Next series of primaries starts with innermost from about March–April; completed with p10 about September–October, during which 1st immature non-breeding on head, body, tail, and wing-coverts replaced by 2nd immature non-breeding. Subsequent immature moults much like adult, but post-breeding may start earlier or last longer, and pre-breeding may involve greater number of inner primaries; primary moult score may be different from adult at same time of year, or, when arrested, part of primaries often relatively more worn than adult.

Measurements. Europe, Middle East, and Africa, all year; skins (BMNH, RMNH, ZFMK).

WING AD	♂ 421	(9·14; 18)	404–441	♀ 412	(12·1; 22)	387–429
TAIL AD	141	(7·63; 13)	130–155	135	(6·79; 13)	125–147
BILL AD	72·4	(2·73; 18)	69–79	67·8	(3·17; 22)	62–73
TARSUS	46·2	(2·13; 19)	43–50	44·4	(2·59; 23)	40–49
TOE	42·4	(1·80; 15)	40–45	40·4	(1·77; 15)	38–43

Sex differences significant, except tail. Juvenile wing and tail both $c.$ 20 shorter than adult; juvenile bill not full-grown until $c.$ 1 year old.

Weights. Kazakhstan (USSR): ♂♂ 700 (April), 600, 700 (June), 700, 700 (August), 650, 650 (September); ♀♀ 500, 625 (June), 640 (August) (Gavrin et al. 1962). Mongolia: ♂♂ 630, 670, 680, 725, 760 (late May to early June), 780 (late July) (Piechocki 1968c). Hopeh (China): ♀ 560 (Shaw 1936). Captive juveniles at age of nearly 1 week 85, 90, 96, 106; 2 reached maxima of 520 and 610 just before fledging at $c.$ 1 month; when wings full-grown at $c.$ 2 months, 475, 485, and 520 (Heinroth and Heinroth 1931–3). Australian adults $c.$ 680 (Serventy et al. 1971).

Structure. Wing long and rather narrow; pointed. 11 primaries: p10 longest, p9 12–25 shorter, p8 36–54, p7 70–86, p6 100–115, p1 240–256; p11 minute, concealed by primary coverts. Tail forked, but less so than in other west Palearctic *Sterna*: length of t1 102 (96–110), average depth of fork in adult breeding $c.$ 36, in juvenile $c.$ 16 only. Bill straight and heavy, culmen decurved; stouter than in any other *Sterna*; $c.$ 20–23 mm deep at base. Feathers of nape elongated, but shorter and tips less pointed than in (e.g.) Royal Tern *S. maxima* and Sandwich Tern *S. sandvicensis*. Foot rather heavy; tarsus relatively longer than in other *Sterna*, exceeding length of middle toe with claw. Toes fully webbed; outer toe $c.$ 84% of middle, inner $c.$ 70%, hind $c.$ 30%.

Geographical variation. Slight. Several races formerly recognized (see, e.g., Hartert 1912–21, Witherby et al. 1941), mainly based on apparent variation in size in small samples. Larger series show that birds from North America, Red Sea, and Persian Gulf slightly smaller than main west Palearctic population, and those from South Africa and Australia a little larger, but overlap nearly complete (Ridgway 1919; Serventy et al. 1971; BMNH), stated difference in colour of some populations, e.g. South Africa (Clancey 1971), non-existent and recognition of subspecies not warranted. CSR

Sterna maxima Royal Tern

PLATES 1 and 3
[between pages 134 and 135]

Du. Koningsstern   Fr. Sterne royale   Ge. Rotschnabelseeschwalbe
Ru. Королевская крачка   Sp. Charrán real   Sw. Kungstärna

Sterna maxima Boddaert, 1783. Synonym: *Thalasseus maximus*.

Polytypic. *S. m. albididorsalis* Hartert, 1921, West Africa; nominate *maxima* Boddaert, 1783, North, Central, northern South America, and (perhaps this race) coastal Uruguay and Argentina, accidental in west Palearctic.

Field characters. 45–50 cm (bill 5·3–6·7, legs 2·6–3·4, tail 10–15 cm); wing-span 125–135 cm. Length of head, body, tail, and wings just overlap with Caspian Tern *S. caspia*, but size generally 20% less, with narrower bill and wings, and longer, more forked tail; close in size and structure to Swift Tern *S. bergii* but with heavier bill and head and less slender body (than all races except *S. bergii velox*); slightly larger than Common Gull *Larus canus*. Very large tern, with narrow, dagger-like bill, rather long, oval head and body, long, pointed wings, and less deeply forked tail than smaller terns. Adult in summer shows orange bill, long black crown, grey-white upperparts, and white underparts except for restricted dusky wing-tip (far less obvious than in *S. caspia*). Immature shows bolder, more variegated inner wing pattern than *S. caspia*. Sexes similar; little seasonal variation. Juvenile and immature separable.

ADULT BREEDING. When fresh, lores, forehead, and crown jet-black, extending into loose nape of long black feathers, but with moult (from late May) forehead and front part of lores quickly become white. Back and most of upperwing pale grey, looking in even light slightly paler than that of *S. caspia*; rump and tail grey-white. In flight, upperwing largely uniform with silvery tone on flight-feathers except for darkening outer primaries which show dusky lines at close range; underwing white but with dusky webs on outer primaries forming diffuse wedge under wing-point, duller and more restricted than in *S. caspia*. Rest of underparts white. Bill almost as long as head, with depth obscured by markedly even taper of both mandibles to much finer tip than in *S. caspia*; usually deep orange. Legs black. ADULT NON-BREEDING. Head mainly white, with black panel through eye broadening into speckled rear crown and shaggy black nape. Upperparts rather paler, with noticeably white appearance in strong light; with wear, primaries darker and dark wedge on outer wing becomes more prominent above and below. Bill often paler, yellow or pale orange. JUVENILE. Not noticeably smaller than adult but less bulky. Head as winter adult but forehead initially spotted and streaked. Back cream-grey, with dusky spots on mantle and bolder wedge-shaped marks or bars on scapulars and tertials. Dark grey lesser coverts show as bold line across folded wing. Tail with bold dusky spots and wedges on tips, forming obviously dark points to shallow fork. In flight, upperwing shows strong, variegated pattern, with dark dusky-brown band across secondaries and similarly coloured outer primaries, dusky tips to greater coverts and primary coverts, and dark grey lesser wing-coverts all obvious; underwing has noticeably dark wedge over most primaries, dull dusky band across secondaries, and narrow white trailing edge from base of wing to innermost primaries. Inner wing pattern as well marked as that of juvenile gull (e.g. Common Gull *Larus canus*) and much bolder than that of *S. caspia* or Sandwich Tern *S. sandvicensis*. Bare-part colours variable, but bill and legs usually dull yellow. FIRST WINTER. Head usually extremely white, with black panel through eye narrow and widening only on lower nape. When replacement of body plumage and wing-coverts complete, inner wing pattern less variegated, with bar on lesser wing-coverts narrow and indistinct, other inner coverts more uniform grey, and dark grey band across secondaries narrower and broken on outer feathers; conversely, outerwing even darker with both primary coverts and primaries almost black, relieved only by paler tips to coverts and innermost primaries. Underwing pattern little changed from juvenile. Tail whiter, with dusky terminal tips to feathers more contrasting. Bill yellow; legs usually darker than juvenile, often mottled yellow and black, sometimes black. FIRST SUMMER. Some retain pattern of 1st-winter plumage, others become more like adult with blacker crown and paler wings. Bill more orange, legs usually black. Judging by strength of juvenile wing pattern, some birds should show exaggerated '*portlandica*'-type plumage pattern, but no certain record of this on 2nd-year bird.

In west Palearctic, large size and uniformly pale bill preclude any prolonged confusion with medium-sized or small congeners. Important to recognize that *S. maxima* is 25% larger than *S. sandvicensis* and Lesser Crested Tern *S. bengalensis*, with much heavier bill and shallower tail-fork. Only serious risk of confusion is with *S. caspia* and *S. bergii*. Compared to *S. caspia*, *S. maxima* has typically tern-like, not gull-like configuration, much paler, whiter upperparts (particularly head in winter and 1st and 2nd years), less deep and less blunt bill (never red or dusky-tipped), less dusky wing-tips in adult (but not in immature), shorter legs, and higher-pitched, less gruff calls. For distinctions from *S. bergii*, see that species. See also Moon (1983). Flight powerful but graceful, suggesting large *S. sandvicensis* (and not *S. caspia*); wing-beats noticeably deep when on passage or into wind but shallower over shorter distance or when searching for food; dives for fish, with less impact than *S. caspia*. Gait free, but rather pattering (due to proportionately short legs). Carriage markedly horizontal, with head usually carried low and folded primaries extending well beyond tail.

Vagrants in west Palearctic have all occurred on coasts. Voice strongly recalls that of *S. sandvicensis* with common calls similarly phrased and sounding like bass versions of that species. Commonest call in flight 'krryuk', deeper and more rolling than *S. sandvicensis*.

Habitat. Breeds in tropical and subtropical lower latitudes, on coasts and inshore or offshore islands almost at sea-level, where surrounding warm waters yield ample fish supplies. Most often on sandy islets or bars which may be flooded by exceptionally high tides or tropical storms; sometimes on bare crumbling rock. In Venezuela, occurs on sand beaches, estuaries, lagoons, and mangrove coasts (Schauensee and Phelps 1978). Detailed study of 5 colonies in USA revealed 4 apparent prerequisites for establishment: (1) absence of quadruped predators, (2) general inaccessibility and excellent visibility of surroundings, (3) extensive areas of adjacent shallows for feeding. (4) location at or near inlet between bay and ocean. All colonies but one were on sandy substrate but sand did not seem to be an essential requirement. Only 2 rose to 1 m or more above mean high water, the rest being liable to flooding. Many neighbouring locations having some but not all of the above 4 prerequisites were never used. 4 occupied sites were spoil banks, the other a shoal sand-bar, but some breeding occurred on sand-spits attached to a spoil bank. All were near water, on white or grey sand substrates, in one case overlaid with shells of oysters. Vegetation varied from almost none to heavy. (Buckley and Buckley 1972*a*.)

Young may withdraw to pools among marsh vegetation. While most foraging done by plunging down into water, often with complete submersion, will also feed with waders at shallow pools or inlets (Bent 1921).

Outside breeding season, will sometimes travel short distances up larger rivers and backwaters (Lowery 1955). In West Africa, frequents shore, lagoons, and harbours (Bannerman 1953; Serle *et al.* 1977). Flies strongly in lower and upper airspace and dives readily but swims infrequently. Insulated from most human impacts except marine pollution by remoteness and inaccessibility of main breeding stations, at least in west Palearctic.

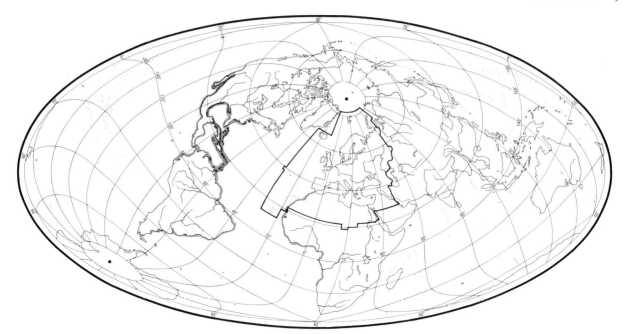

Distribution. In west Palearctic, breeds only in Mauritania (Banc d'Arguin).

Accidental. Spain: April 1970 (2), April 1971 (2), July 1977, September 1977. Ireland: March 1954. Britain: July 1965, September 1971, September 1974, November 1979 (immature nominate *maxima*) (Moon 1983). Norway: June 1976.

Population. MAURITANIA. Banc d'Arguin: fluctuating numbers, estimated 3000–4500 pairs (Naurois 1969a; Westernhagen 1970), and 4600–5300 pairs 1974 (Trotignon 1976).

Oldest ringed bird 17 years (Clapp *et al.* 1982).

Movements. West African race *albididorsalis*, to which the following refers, is mainly migratory. Reported to be absent from Banc d'Arguin (Mauritania) in winter (Heim de Balsac and Mayaud 1962), but recent observers have proved that some winter locally: 2160 in December 1978 to January 1979 (Trotignon *et al.* 1980), and 3318 in January–March 1980 (M Engelmoer), though latter figure may reflect some early return movement. However, majority do leave in autumn (over 85% in 1978–9), and migrate south along West African coast to winter from Sénégal (small numbers: Morel and Roux 1973) to Angola. Largest numbers occur around Gulf of Guinea, where Wallace (1973) found them in all months due to summering immatures; also noted purposeful westerly passage of adults on Nigerian coast in March. 17 distant recoveries for birds ringed as chicks on Banc d'Arguin (*CRMMO Bull.* 1960–1, 14–15): 8 found as juveniles in September and early October in Sénégal, the others in 1st winter or summer in Sierra Leone (1), Liberia (2), Ivory Coast (5), and Ghana (1).

Immatures obtained at Nouadhibou (northern Mauritania) in March, and birds can occur there in hundreds during May–August (Heim de Balsac and Mayaud 1962; Duhautois *et al.* 1974); reduced numbers into December (Pététin and Trotignon 1972). Also limited northward dispersal, with small numbers occurring along Atlantic coast of Morocco north to Tangier. These occur mainly August–October, but also June–July (Heim de Balsac and Mayaud 1962; Pineau and Giraud-Audine 1974, 1975); Smith (1965) noted 1–2 in January 1963–4. Doubtless from here that birds reach southern Spain occasionally. However, cannot be assumed that this applies also to more northerly records (e.g. from Britain and Ireland), since these might equally be transatlantic vagrants.

Nearctic race, nominate *maxima*, largely migratory, wintering south to Argentina and Ecuador (Lévêque 1964).

EKD

Food. Mainly fish and crustaceans, especially crabs. Feeds mostly by plunge-diving; sometimes by aerial-skimming like skimmer *Rynchops* (Buckley and Buckley 1972a). On east coast of USA, forages over shallow inshore waters, usually either singly or in groups of 2–3 (Buckley and Buckley 1972a; Erwin 1978), but sometimes in flocks of up to 150 (Bent 1921; Erwin 1978); will forage up to 65 km from breeding colonies (Buckley and Buckley 1972a). Feeding-flight up to 100 m out from shore, usually covering several 100-m stretches parallel to beach several times, at height of *c*. 5–10 m (Buckley and Buckley 1972a). Successful dives for fish or, more especially, for crabs, often followed by aerial-skimming (43% of 76 dives), as if to soothe or clean bill after being pecked or irritated by crabs' claws; aerial-skimming used also for drinking and, on occasion, for feeding, with fish seen to be caught (Buckley and Hailman 1970; Buckley and Buckley 1972a). In Bonaire (Lesser Antilles), juveniles (which tended to associate together during day-time) took twice as long as adults to forage over given length of beach; diving frequency of adults *c*. 1·7 times greater. Adults' success (ratio of number of dives resulting in prey capture to number of completed dives) similar to that of juveniles, but capture rate 1·6 times greater. Juveniles dropped and recovered prey 13·5 times more than adults, although number of times prey dropped and lost and number of aborted dives similar in both age groups. Adults hovered precisely, plunged neatly and rapidly, and left water immediately; juveniles circled repeatedly above prey, made many intention movements before diving, and often entered water obliquely, sometimes flopping on to surface. Significantly more juveniles than adults trapped on baited hooks of fishermen. (Buckley and Buckley 1974.) Flies low over water to catch flying fish (Exocoetidae) (R van Halewijn). In Sierra Leone, winter, often dips-to-surface for offal from fishing vessels (Dunn 1972a). 20–25 birds may give chase in intraspecific food-piracy (Dragesco 1961b), usually without success (F G Buckley). In nominate *maxima*, most fishing activity at mid-tide and in early morning and late afternoon; nocturnal fishing occurs during breeding season (F G Buckley).

5 stomachs, Sierra Leone and Ghana, winter, contained 15 fish 3·3–18·5 cm long, mean 6·7 cm: sardines (Clupeidae), mullet (Mugilidae), Carangidae, Pomadasyidae, Ephippidae (Dunn 1972a). In USA, recorded taking fish up to 10 cm long, and crabs (Bent 1921). In Virginia and North Carolina (USA), diet includes locally abundant soft-shelled blue crabs (especially small instars); also squid *Loligo*, shrimps (probably *Crangon*), and fish—silversides *Menidia*, killifish *Fundulus*, anchovy *Anchoviella*, menhaden *Brevoortia*, toadfish *Ossanus*, pipefish *Syngnathus*, jacks *Caranx*, flounders (Pleuronectidae), and eels *Anguilla*, mostly *c*. 5–10 cm long (Buckley and Buckley 1972a).

Chicks fed on large numbers of regurgitated crabs (mainly blue crabs 2–4 cm across carapace), and on fish—always swallowed head first (Buckley and Buckley 1972a). Adults each collect 0·17±0·11 prey (average 15·4 g) per min; as adults eat some, young each receive 0·36 prey per hr (Erwin 1977).   EKD

Social pattern and behaviour. Compiled largely from studies by F G and P A Buckley on nominate *maxima*; in USA at Fisherman's Island (Virginia), outer Banks (North Carolina), and Texas, and in Puerto Rico and Netherlands Antilles.

1. Mostly gregarious throughout the year, occurring in flocks of up to several thousand, but occasionally solitary outside breeding season (Naurois 1959; Dragesco 1961a,b; Buckley and Buckley 1972a, 1974, 1976, 1977). Winter flocks vary in size, usually 50 to several hundred (Lévêque 1964; Serle 1965; Ashmole and Tovar 1968; E K Dunn); once, 900 (Trotignon *et al*. 1980). Foraging flocks may comprise adults and juveniles in varying proportions (Buckley and Buckley 1974, 1976). Associates readily with other Sternidae or gulls (Laridae) (Nelson 1962; Buckley and Buckley 1972a, 1980; E K Dunn). BONDS. Monogamous, but not known if pair-bond persists for more than one season (F G and P A Buckley). First breeding usually at 4 years, less commonly 3, sometimes 2; birds first visit colony area at 2 years (J S Weske). Parents share incubation, and brooding and feeding of young. From early age to fledging, crèche formation appears obligatory in young, not facultative as in (e.g.) Sandwich Tern *S. sandvicensis* (F G and P A Buckley). Parents feed only their own chick in crèche, and continue to accompany and feed it during migration and often in winter quarters, generally for at least 5–8 months after hatching in nominate *maxima* (Ashmole and Tovar 1968; Buckley and Buckley 1974, 1976). BREEDING DISPERSION. Densely colonial—from a few to 10000 or more pairs; colonies often divided into several adjacent sub-colonies (Buckley and Buckley 1972a; Dupuy 1976; Boswall and Barrett 1978; Blus *et al*. 1979). Fidelity to colony site very weak, entire colony often deserting if disturbed—especially by predators early in season—and moving to nearby site or leaving area completely (Buckley and Buckley 1972a, 1976; Blus *et al*. 1979; Loftin and Sutton 1979). Fidelity to nest-site also weak, probably due to habitat changes and/or human disturbance (Buckley and Buckley 1972a, 1976, 1980; Blus *et al*. 1979). Nest density high: 6–9 per m^2 at Banc d'Arguin, Mauritania (Naurois 1959); mean 7·4±0·3 and 6·8±0·3 per m^2 in consecutive years at colony, Virginia (Buckley and Buckley 1972a). Distance between neighbouring nests of *albididorsalis* 0·25–0·4 m (Naurois 1959), of nominate *maxima* in Virginia and North Carolina 0·08–0·7 m, mean 0·37 m; some colonies contained areas where sitting neighbours could touch one another at will, creating hexagonal dispersion whereby nests shared common boundary with 4–8 others; minimum distance between nests dictated by distance bird can stretch neck and lunge at neighbour (Buckley and Buckley 1972a, 1977). Small nest-area territory, limited virtually to nest-site, thus defended, serving for nesting and raising of young up to a few days old. At Banc d'Arguin, colonies often adjacent to nesting Caspian Terns *S. caspia* (Dragesco 1961b). In USA, colonies frequently mixed with *S. sandvicensis* and, in the south, alongside *S. caspia* (see Buckley and Buckley 1972a, Blus *et al*. 1979). ROOSTING. Assembles in compact flocks on open areas such as sand-bars, mudflats, reefs, undisturbed beaches near high-water mark, and saltpans. In winter, juveniles often loaf on fishing piers, buoys, etc., in harbours (Young 1946; Serle 1965; Buckley and Buckley 1974; F G and P A Buckley, E K Dunn). Sometimes rests briefly on water (R van Halewijn). Loafing flocks of *albididorsalis*, Sierra Leone, up to 260 (E K Dunn). Often associated with other Sternidae and Laridae, notably *S. sandvicensis*, Common Tern *S. hirundo*, and Grey-headed Gull *Larus cirrocephalus*. For roosting associates of nominate *maxima*, and interactions with them, see Nelson (1962). By day, loafs between foraging bouts, especially when state of tide makes fishing less successful; nominate *maxima* thus less active (a) at high and low tide, and (b) between foraging peaks in early morning and late afternoon; since nocturnal fishing occurs in breeding season, may not roost throughout night (F G Buckley). At onset of breeding season, pre-breeding roost—adjacent to colony site or up to 100 m from it, generally on shoreline—becomes focus of courtship and copulatory behaviour, though courtship also occurs in winter and spring roosts, far from breeding grounds (Kale *et al*. 1965; Buckley and Buckley 1972a; E K Dunn). Off-duty birds often fly to communal loafing area to bathe and preen. Chicks

in crèche generally loaf and roost near shoreline, or sit in shade of vegetation, adults standing nearby (F G and P A Buckley). Winter roosts often comprise family groups (Ashmole and Tovar 1968) though juveniles tend to associate more closely with one another than with adults (Buckley and Buckley 1974).

2. FLOCK BEHAVIOUR. Birds in colony respond collectively to Alarm-call (see 1 in Voice) of 1 or more birds, either overhead or on ground by adopting mild Aggressive-upright (see Antagonistic Behaviour, below) in which standing bird scans around with neck craned upwards, bill slightly elevated, crest feathers raised, and neck plumage slightly ruffled. Depending on imminence of danger, may then fly up one by one, in successive groups, or hurriedly *en masse*, and circle overhead uttering Alarm-calls ('upflight' or so-called 'alarm'). Once alarm ceases, settling birds give 'kleer' call (F G Buckley: see 2 in Voice). Silent 'dreads' and 'panics' occur, as in *S. sandvicensis* and *S. hirundo* at roost and colony. Nesting birds in dread circle silently over colony for up to 20 min before resuming calling and settling on nest. In nominate *maxima*, birds flew up from nests much sooner when disturbed, and took longer to return than *S. sandvicensis* nesting among them, some resorting to loafing sites (Buckley and Buckley 1972a, 1976). Nocturnal disturbance at colony, North Carolina, caused adults to perform dread to nearby island; temporarily-deserted chicks in nest or crèche remained silent (F G Buckley). ANTAGONISTIC BEHAVIOUR. Similar to other 'crested' *Sterna*, but conspicuously non-aggressive; most aggressive towards conspecifics at nest-site. Antagonistic repertoire varied, combining degrees of crest erection, differing body postures, vocalizations, and neck and bill movements. Aggressive-upright (Fig A) adopted by adult protecting offspring, negotiating route—with or without food—to nest-site, or to chick in crèche, and sometimes when approached by conspecifics during courtship activities: crest feathers fully ruffled, bill elevated above horizontal, neck held vertically or angled slightly backwards, neck plumage ruffled, back feathers raised (distinctly ruffled at extreme intensities: Fig B, right), and carpal joints and wings held away from body (F G and P A Buckley). Aggressive-upright often reciprocated by combatant, and 2 birds will then often circle one another until one sidles off in Anxiety-upright (feathers sleeked, wings and carpal joints held close to body: Fig B, left), continuing so to retreat, or flying off. Erect-posture, which similarly indicates appeasement and fear in other 'crested' *Sterna* not recorded in nominate *maxima* (F G and P A Buckley). Occasionally, 2 birds may lunge at each other with necks outstretched, crest feathers ruffled, grabbing for each other's bill. May lead to ground or aerial chase, but not with frequency or intensity of 'non-crested' *Sterna* (F G and P A Buckley). Sitting birds often lunge (as above), with Gakkering-call (see 3 in Voice) at wing and tail of passing bird; neighbours frequently bicker in like manner, duelling with bills, Gakkering, or silently gaping at one another; one bird may thus bicker with several others simultaneously. Birds

B

often crane neck upwards at 45–60° to horizontal, raise crest and neck feathers, and gape at birds flying low over colony. Aggression towards strange chicks is less pronounced than in 'non-crested' *Sterna*. A chick wandering too close to, or trying to grab fish from strange adult may, however, be poked or pecked on head (F G and P A Buckley). At Banc d'Arguin, straying chicks occasionally grabbed by bill, wing, or foot, and dragged along ground; if parents intervene, aerial chase involving 3–6 individuals may ensue (Dragesco 1961b). Aggression between chicks uncommon, though nominate *maxima* c. 20 days old recorded killing 7-day-old chick (Buckley and Buckley 1969). HETEROSEXUAL BEHAVIOUR. (1) General. Pair-formation, courtship, and copulation can occur away from, or *en route* to, colony-site (Kale *et al.* 1965; Buckley and Buckley 1972a), and on wintering grounds (F G and P A Buckley, E K Dunn), e.g. ground and aerial courtship on 12 March at roost in Freetown Peninsula (Sierra Leone), 900 km south of nearest known breeding grounds (E K Dunn). Most pair-formation, however, occurs at colony-site (Buckley and Buckley 1972a). Both aerial and ground courtship described here for nominate *maxima* (F G and P A Buckley). (2) Aerial courtship. Similar to *S. sandvicensis*. Dreads occur early in season at pre-breeding roost. High-flights begin in various ways. (a) After ground courtship. (b) ♂, carrying fish, flies slowly and repeatedly over pre-breeding roost, often giving Advertising-call (see 4 in Voice); receptive bird may leave flock and follow fish-carrier to nearby spot, where ground courtship follows, or into High-flight. (c) From a dread. In High-flight, 1 bird, presumably ♂, spirals upwards giving Advertising-call, pursued by 1 or more birds, commonly 2. During ascent, leader also gives Aack-calls (see 5 in Voice), as do pursuers. Where more than 1 pursuer, these may pair off and continue their own High-flight. Fish sometimes passed between 2 flying birds that have paired off, one dropping it a short distance to other below. A 3rd bird may join, and replace one of the 2 original participants, or cause the 2 birds to break off their aerial courtship. At peak of ascent (to over 100 m), leader initiates downward glide, closely followed by pursuer. The 2 birds describe a spiral descent, during which Pass-ceremony may occur, as in *S. sandvicensis*. High-flight may last up to 25 min (F G Buckley), and lead to ground courtship, but birds more often retire to fish, or to loafing flock to rest, preen, or bathe. Aerial courtship diminishes at hatching time, but occurs throughout breeding season. (3) Ground courtship. May precede, follow, or occur independently of aerial courtship. Often, though not always, involves presentation of food. Typically, ♂ alights in pre-breeding roost with food item (fish, crab, or shrimp in nominate *maxima*) in bill held at, or slightly below, horizontal. ♂ walks slowly amongst other birds, giving Advertising-call in Forward-erect (similar to Aggressive-upright but neck feathers sleeked and bill pointed downwards). On encountering interested ♀, ♂, Nodding head, draws attention

A

to food if present, and continues to give Advertising-call. Receptive ♀ first adopts Forward-erect facing ♂, and the 2 birds may proceed to strut head to tail, Circling tightly, or else ♂ may circle ♀ who turns on the spot, keeping side on to him. Sometimes, displaying pair may Parade through colony, side by side, or in line. Up to 15 birds may stand in circle in Forward-erect, Nodding and Circling one another. Most frequently, such groups comprise 3 birds, resulting in 2 moving away pursued by 3rd, or else 2 chasing each other, leaving 3rd behind. Food item may finally be offered to ♀ who retracts her neck and carpal joints, and sleeks her plumage (while keeping crest raised), thus assuming more relaxed, submissive posture. If ♀ accepts food and swallows it, both birds then fly off or preen. Sometimes food dropped, or stolen by another bird, causing confrontation in which food may be pulled to pieces. If, on initial arrival, ♂ has no food, ground courtship may still proceed as above. Both members of pair select nest-site. As they inspect chosen area of colony in Forward-erect or normal walking posture, there is much pointing at feet and ground by both birds. After circling particular spot several times, pointing to ground assumes extreme form with bill directed between legs while body plumage relaxed, and wings folded or held slightly out. Both birds spend much time sitting at one spot, or one sits while other stands nearby. Several scrapes may be made (as in other Sternidae), before final choice. (4) Mating. Description by F G Buckley; see also Kilham (1981). Occurs at or away from colony-site, especially at pre-breeding roost, and at nest during incubation. Usually initiated after ground courtship or Courtship-feeding (see below). Successful mating generally dependent on ♀'s adoption of Hunched-posture: ♀ crouches, crest feathers slightly raised, body and neck plumage ruffled, wings held in or slightly out, and gives Whinny-call (see 6 in Voice). Adopting mild Forward-erect (wings out slightly), ♂ approaches ♀ from side or rear and mounts. ♂ often stands or treads on ♀'s back or neck for up to 10 min without copulating. During copulation, ♀'s crest and body plumage sleeked, tail raised, and Whinny-call continues. ♂'s plumage sleeked (except for raised crest), tail and bill directed downwards; ♂ may give undescribed call. Birds usually remain mounted for 2–5 min, and conclude with ♀ shrugging off or pecking ♂; preening usually follows. (5) Courtship-feeding. ♂ feeds ♀ during courtship and incubation. Especially in early stages of courtship, food may be withheld by ♂, or rejected by ♀; ♂ may then fly off with it, or swallow it. May be passed back and forth several times between courting birds before one finally eats it. (6) Nest-relief. No elaborate ceremony. Incoming bird, with or without fish, and usually giving version of Advertising-call (see 4 in Voice), alights by nest or nearby and walks up to it. Sitting bird generally looks up on hearing mate, and answers with version of Aack-call (see 5 in Voice). Sitting bird may leave nest just before mate arrives, or continue sitting briefly, or longer, causing newly arrived bird either to depart again, or to peck at sitting bird. Usually, pair stand over, or near nest for several minutes before one leaves. Especially on hot days, sitting bird will often quit nest to drink or forage before mate has returned. Bird hesitant about sitting on egg gives version of Aack-call (see 5 in Voice). RELATIONS WITHIN FAMILY GROUP. Parents recognize own young by voice and appearance within 1–2 days of hatching, while offspring recognize parents by voice in same period (Buckley and Buckley 1970a, 1972b, 1976). Downy chick solicits food by pecking at prey item or bill. If chick rejects food, brooding parent may eat it. At Banc d'Arguin, 1 parent remains with small young, brooding it under wing in heat of day, while other parent forages (Dragesco 1961b). When small chicks of nominate *maxima* are fed, one parent generally attends and defends it with Aggressive-upright, while other offers fish in appeasing posture (bill low, plumage sleeked) typical of other Sternidae. Once food successfully transferred, both parents adopt Aggressive-upright (Fig A). Older chicks beg with a hunched posture and call resembling Whinny-call of adult ♀ (Buckley and Buckley 1976), this display continuing into winter quarters. Parents coax young to leave nest with undescribed call (F G and P A Buckley). At Banc d'Arguin, young of *albididorsalis* did not disperse from nest until 1 week old, and did not crèche until 15 days (Dragesco 1961b); in nominate *maxima*, led to crèche from 2–3 days old where they remain until fledging (Buckley and Buckley 1972a). Age at which young recorded leaving nest may reflect differing levels of disturbance in these studies (E K Dunn). At fledging, parents abandon colony and young accompany them to fishing grounds where they are fed on sea surface or shore. During winter, young increasingly fish for themselves (Buckley and Buckley 1974, 1976). ANTI-PREDATOR RESPONSES OF YOUNG. Chicks still in the egg, or up to 4 days after hatching, fall silent in response to parental Alarm-call. Young in nest freeze there, or run and hide under vegetation, if available. Chicks occasionally crouch, or make scrape for themselves (Buckley and Buckley 1972a, 1976; Blus et al. 1979). Young variously lunge, gape, peck, regurgitate, or defecate if handled. During an alarm, crèche coalesces and moves rapidly away from disturbance, often taking to water where young swim in formation behind noticeable leaders, arcing first away from, then back to shore. If closely approached on land or water, crèche scatters and re-forms further off (Buckley and Buckley 1972a, 1976). PARENTAL ANTI-PREDATOR STRATEGIES. Behaviour intensified on hatching, and peaks in nominate *maxima* when young leave nest to join crèche (Buckley and Buckley 1972a, 1976). (1) Passive measures. If danger approaches, adults give Alarm-calls; see above for response of young (Buckley and Buckley 1972a, b, 1976, 1980). (2) Active measures: against birds. Sitting birds confront intruders passing too close to nest as they do conspecifics (see Antagonistic Behaviour, above), relying on density to ward off attack. Only directly threatened birds respond to intrusion; unguarded nests not defended by neighbours (F G Buckley). When *L. cirrocephalus* flew over colony of *albididorsalis*, birds gave Alarm-calls and assumed Aggressive-upright; *S. caspia* passing too close were chased (Dragesco 1961b). Aerial chasing unusual in nominate *maxima* (F G Buckley), though one bird chased and struck Herring Gull *L. argentatus* which stole egg (Loftin and Sutton 1979). Usually out-flies kleptoparasitic species, including conspecifics (Dragesco 1961a, b; F G and P A Buckley). (3) Active measures: against man and other mammals. Mammalian intruders elicit no ground defence, only a fleeing response. Birds may flush from colony when intruder quite far off (Loftin and Sutton 1979). As human intruder approaches, birds in upflight intensify Alarm-calls and perform dive-attacks, uttering version of Gakkering-call; may defecate on intruder, but rarely deliver blow (P A Buckley), more often veering off c. 5–10 m short (J S Weske). Similar behaviour sometimes shown by nominate *maxima* towards dogs (F G and P A Buckley).

(Figs A–B from photographs by F G and P A Buckley.)

FGB, PAB, EKD

Voice. Highly vocal, especially at breeding grounds. Similar to Sandwich Tern *S. sandvicensis*, but lower pitched and louder.

CALLS OF ADULTS. (1) Alarm-call. Loud, hard 'keet keet' (F G and P A Buckley), or 'krit krit' (Fig I). Given in flight or on ground when threatened (F G and P A Buckley). (2) Kleer-call. 'Kleer-kleer', given when danger has passed

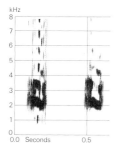

I R C Stein and R B Augstadt/Sveriges Radio (1973) USA April 1961

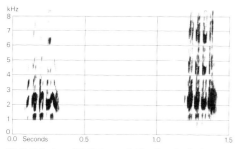

II R C Stein and R B Augstadt/Sveriges Radio (1973) USA April 1961

as birds return to land on nest-sites (F G and P A Buckley). (3) Gakkering-call. A series of rapidly repeated, hoarse 'ack-ack-ack-ack' notes, expressing threat or alarm; given in aggressive encounters by sitting or standing birds, together with lunging, gaping, and duelling; also directed at aerial intruders; similar but shorter series given by 2 birds encountering each other when fishing or flying over colony (F G and P A Buckley). (4) Advertising-call. A throaty 'kireet' or 'keereet' (F G Buckley), 'krrryuk' (P J Sellar; Fig II), 'krerink' (E K Dunn). Given in ground and aerial courtship, especially by advertising ♂, and by both sexes arriving at colony with fish and/or for nest-relief. Probably identifies mates to each other and to offspring. (5) Aack-call. Deep, guttural 'aack-aack'. Given by both members of pair during aerial courtship; apparently a recognition call answering other's Advertising-call. Variable in intensity; more clipped version given by incubating bird in response to Advertising-call of approaching mate. Soft 'ack-ack' often given in ambivalent situation, e.g. when bird seems hesitant about sitting on egg. A long 'aack' may grade into a soft 'ack' (F G and P A Buckley). (6) Whinny-call. A shrill, whinnying 'whee-whee-whee-whee', given by ♀ in Hunched-posture before copulation (F G and P A Buckley). (7) Undescribed call given by ♂ during mounting and copulation (F G and P A Buckley). (8) Undescribed call used to coax young from nest (F G and P A Buckley).

CALLS OF YOUNG. Downy young beg with clear, sharp, repeated 'peep'; length of each note increases with age. As young approach fledging, they adopt juvenile food-call similar to adult ♀'s Whinny-call (see 6, above) which is used for several months after fledging (Buckley and Buckley 1976).

FGB, PAB, EKD

Breeding. SEASON. Banc d'Arguin (Mauritania): main egg-laying period last 10 days of April (Naurois 1959). SITE. On ground in the open, sometimes on tideline of seaweed, etc. Colonial. Nest: shallow depression, unlined or with a few pieces of debris, fishbones, etc. Building: probably by both parents with scraping movements. EGGS. See Plate 88. Sub-elliptical, smooth and slightly glossy; variable ground-colour, from white to medium brown, spotted and blotched red-brown, sometimes gathered at broad end. Nominate *maxima*: 64–45 mm (58–72 × 41–48), sample 100 (Schönwetter 1967); weight 64 g (58–70), sample 25 (Buckley and Buckley 1972a). *S.m. albididorsalis*: 55 g (53–56), sample 6 (A R Dupuy). Clutch: 1, very rarely 2. Of 1000 clutches. Mauritania, only 2–3 of 2 eggs (Naurois 1959); of 911 clutches, USA (nominate *maxima*), 1·4% of 2 (Buckley and Buckley 1972a). One brood. Replacements laid after egg loss. INCUBATION. 30–31 days (28–35) (nominate *maxima*). By both parents for varying spells, sometimes leaving eggs untended for hours (Buckley and Buckley 1972a). Begins immediately on laying of egg. YOUNG. Precocial and semi-nidifugous. Cared for and fed by both parents. Chicks of nominate *maxima* able to leave nest within 24 hrs of hatching; move into crèche with other young by 2–3 days (Buckley and Buckley 1972a). Reported that *albididorsalis* chicks remain in or near nest until *c*. 1 week and do not crèche until 15 days (Dragesco 1961b). FLEDGING TO MATURITY. Fledging period not precisely recorded but lies between 28 and 35 days, with 30–31 probably average (Buckley and Buckley 1972a). Age of independence probably at least 5–8 months (Ashmole and Tovar 1968; Buckley and Buckley 1974, 1976). Age of first breeding 4 years, rarely 2 (Buckley and Buckley 1970; J S Weske). BREEDING SUCCESS. Rather variable, e.g. tidal flooding may reduce colony's success to nil. In nominate *maxima*, South Carolina, mean 0·36 and 0·44 young fledged per nest in 2 years (Blus *et al.* 1979); figures perhaps abnormally low due to flooding and disturbance (F G Buckley). Most mortality in egg stage; almost 100% of hatched young fledge in good years (Buckley and Buckley 1972a). In Sénégambia, sharks inflict heavy losses on young gathered on strandline (Dupuy 1975).

Plumages (*S. m. albididorsalis*). ADULT BREEDING. Forehead, upper half of lores, crown down to eye, and rough crest to nape glossy black. Sides of head and hindneck white. Mantle, scapulars, upper wing-coverts, and secondaries pale grey, slightly paler than Sandwich Tern *S. sandvicensis*, occasionally almost white. Inner webs and tips of tertials and secondaries broadly fringed white, outer webs fringed more narrowly. Back grey like mantle, but rump, upper tail-coverts, and tail gradually paler; coverts and tail white with faint grey tinge only. Underparts, under wing-coverts, and axillaries white. Primaries pale grey with silvery tinge to outer webs; innermost feathers fringed white on inner webs and tips; fringes become gradually narrower towards outer primaries, faint or absent on outer 2–3. Inner webs of primaries less silvery than outer webs, grey with large white wedges to bases, reaching to c. 4–7 cm from tips; undersurface of primaries appears mostly white. Grey of flight-feathers, tertials, and tail strongly subject to wear, especially on those parts not protected by adjacent feathers: tips and proximal edges of primaries change to black (dark outer primaries then contrasting with 4–6 fresh inner ones), centres and outer webs of secondaries and tertials to dull grey or sepia-grey, and tips and outer webs of tail-feathers to dull grey or brown-grey. ADULT NON-BREEDING. Forehead and lores white. Crown white, often with small black spots or shaft-streaks, larger towards nape. Spot in front of eye, line below eye, and upper ear-coverts black, slightly mottled white; short crest on nape black, feathers edged white. Body, wing, and tail as in adult breeding (though outer tail shorter); variable number of flight- and tail-feathers blackish, depending on progression of moult. DOWNY YOUNG. Nominate *maxima* strongly variable, and *albididorsalis* probably also: ground-colour of upperparts, nape, wings, and throat white or various shades of buff-brown, different parts varying independently; upperparts variably marked with dark spots of varying size and number, some all uniform or dots restricted to back, others heavily marked on all upperparts and throat; underparts white or pale buff (Buckley and Buckley 1972a). Some rather similar to Caspian Tern *S. caspia*, but bill more slender and middle toe with claw shorter than tarsus instead of longer. JUVENILE. Forehead white, crown white with variable number of dull black shaft-streaks. Feathers of nape and ear-coverts dull black, bases and sides dull grey, narrow fringes white. Lores white with black dot in front of eye. Neck white. Mantle white, feathers grey at centres and sometimes with sepia-black subterminal dot, fringes at tips tinged buff. Scapulars and tertials grey, laterally fringed white and tinged buff to tips, often with small subterminal sepia spots. Back, rump, and upper tail-coverts mouse-grey, feathers fringed white; much white to sides of rump and tail-base. Tail-feathers grey, rather broad fringes to tips white with slight buff tinge, inner webs edged white, bases of outer webs of t1–t5 and all outer web of t6 pale grey. Flight-feathers pale grey with broad white tips and narrow fringes to secondaries and c. 4 inner primaries; p5–p8 narrowly tipped white; outer primaries with broad white wedge on grey-brown inner webs; outer webs of primaries frosted silvery-grey. Lesser upper wing-coverts (at border of white leading edge of wing) dark grey, slightly frosted silvery, longer lesser and all median coverts contrastingly whitish (not as grey as in adult), greater coverts ash-grey with slight brown clouding towards middle portion of outer webs. Primary coverts dark grey, tips fringed white. Underparts, axillaries, and under wing-coverts white. Plumage strongly subject to wear, grey changing to sepia-brown or black, especially on primaries, tail, and on centres of secondaries and tertials. IMMATURE SECOND AND THIRD CALENDAR YEAR. 1st non-breeding (early 2nd calendar year) mostly similar to adult non-breeding, but part of worn juvenile feathering retained; new lesser upper wing-coverts with indistinct dark grey blotches, forming faint carpal bar across forewing. During summer of 2nd calendar year, 1st non-breeding replaced by fresh 2nd non-breeding similar to adult non-breeding and indistinguishable from latter except for differences in timing of moult; occasionally, faint carpal bar still visible until spring of 3rd calendar year. Immature breeding acquired in spring of 3rd calendar year intermediate between adult breeding and adult non-breeding: feathers of forehead and crown white with tips and centres black to variable extent; some appear black like adult but with some white mottling, others still mainly as in adult non-breeding. Ageing best done by state of wear and moult. (1) Winter (November–February). 1st winter: single moult series in primaries, score 0–25; outer primaries and secondaries worn, all or partly blackish; dark grey carpal bar across lesser upper wing-coverts; tarsus usually yellow with some black dots. 2nd winter: 2 moult series, or 2nd about to start, scores 20–46 and 0–15 respectively; outer primaries rather worn and dark; occasionally, some faint dark dots on lesser coverts or some yellow on tarsus. Adult: 2 moult series or 2nd about to start, scores 20–46 and 0–15; outer primaries rather worn, still partly grey; no dark bar on lesser coverts; tarsus black. (2) Spring and early summer (March–June). 2nd calendar year: single primary moult series or 2nd just started, scores 10–40 and 0–15 respectively; outer primaries juvenile, heavily worn, sepia-black; wing-coverts and tarsus as in 1st winter; head in non-breeding, occasionally with black mottling on crown. 3rd calendar year: 2 moult series, scores 38–49 and 15–25 respectively, occasionally with moult suspended with scores (e.g.) 20 and 45, 1–2 old outer primaries often still present; lesser coverts and tarsus as in 2nd winter; forehead usually white, crown white with black dots or black mottled white. Adult: 2 moult series, but outer series just completed (primaries fresh and pale), inner series scores 20–30, but arrested April–June; no dark grey on lesser coverts; tarsus black; forehead and crown black (former at onset of breeding only). (3) Autumn (July–October). 1st autumn: plumage juvenile; all primaries equally worn, no sign of moult or suspension; indistinct dark grey bar across lesser upper wing-coverts; tarsus usually mainly yellow; forehead and crown off-white and grey, mottled and streaked dusky. 2nd autumn: 2 moult series, scores 5–25 and 35–50 respectively; occasionally, some heavily worn juvenile outer primaries or inner secondaries still present; lesser coverts and tarsus as in 2nd winter; forehead and crown all or partly in worn non-breeding. 3rd autumn: 2 moult series active or one active and one just finished, scores 15–30 and 35–50 respectively; otherwise like adult and often indistinguishable. Adult: usually only one moulting series active, starting with inner primaries (score 1–28); moult of arrested pre-breeding moult series (score 20–30) usually not continued; outer primaries rather old and worn; crown still with some black of breeding or all in fresh non-breeding; no dark grey bar across lesser coverts; tarsus black.

Bare parts. ADULT. Iris dark brown. Bill orange-red; paler red or yellow-orange from July, orange-yellow (perhaps occasionally yellow) in winter. Foot black; undersurface of toes often ochre-yellow, pink-yellow, or orange-yellow. DOWNY YOUNG. Iris dark brown. Bill yellow-orange or pink, occasionally olive-green, usually with black tip. Foot varies from uniform pink, yellow-orange, greenish, or black to bicoloured with any of these colours in spots of varying extent; variation in bill and foot colour mostly unrelated to strongly variable colour of body (Buckley and Buckley 1970a). JUVENILE. Iris dark brown. Bill pinkish-yellow, dull orange-yellow, or greenish-yellow; brighter yellow or orange-yellow from 2nd autumn or winter. Foot dull yellow, sometimes with pink or olive-grey tinge, occasionally black, except for soles; brighter yellow and with round black spots developing from early in 2nd calendar year, first on tibia, front of tarsus, joints, and upper sides of toes; foot black from May–July of 2nd calendar year, except for yellow soles; occasionally, some yellow on rear of tarsus until September. (BMNH, RMNH, ZMA.)

Moults. Based on Atlantic populations (east Pacific ones deviate slightly). ADULT POST-BREEDING. Complete, primaries descendant or serially descendant. Starts with some feathers on forehead late May or June; forehead and crown mainly in non-breeding late June or early July. Remainder of head, part of mantle and scapulars, some tertials, and median upper wing-coverts follow July–August, remainder of body and most wing-coverts replaced August–November, some coverts not until February. Tail starts about late August, completed November–December; sequence of replacement approximately t1-t2-t3-t5-t6-t4, but rather variable. Primaries start with p1 July or early August, completed with p10 late January to early March; an earlier series, started during previous pre-breeding, completed August–October, but moult of this series usually arrested and outer primaries replaced once a year only. ADULT PRE-BREEDING. Partial. Head and some feathering of body replaced early February to early April; tail-feathers start with t1 from about November, followed by t2 (about January), t3 and t6 (January–February), and sometimes t4–t5 (February–March). Primaries start with p1 late November to early January; moult arrested about April, when inner 5–6 primaries (*albididorsalis*) or 4–5 (west Atlantic nominate *maxima*) new; this moult only occasionally resumed in post-breeding. POST-JUVENILE. Complete. Starts with head and some feathers of mantle and scapulars October–November; all head, body, and wing-coverts in 1st non-breeding by February, except for most primary coverts and frequently some greater or lesser upper wing-coverts or tertials. Tail starts about same time; sequence centrifugal, but sometimes irregular; completed April–May. Flight-feathers start with p1 from late November to January; moult regularly descendant, usually without suspending, finishing August–October of 2nd calendar year. FIRST IMMATURE POST-BREEDING. Complete. Starts April–May of 2nd calendar year; head, central tail-feathers, and inner primaries first. Moult rather slow, occasionally temporarily suspended, and 1st non-breeding replaced directly by 2nd. Tail complete by December–January, outer primaries February–March of 3rd calendar year, but wing moult occasionally suspended early spring and 1–2 old outer primaries retained during summer. Following pre-breeding as in adult pre-breeding, but tail and wing start December–January, often suspended in summer, completed about September.

Measurements. *S. m. albididorsalis*. West Africa (Mauritania to Ghana), February–May; Morocco, October, and Gulf of Guinea (Sierra Leone to Angola), November–January; skins (BMNH, RMNH). Tail is length of t6 of adult breeding.

WING	AD	♂ 360	(4.19; 7)	354–366	♀ 352	(4.08; 10)	346–358
	JUV	349	(5.50; 5)	343–355	340	(— ; 3)	334–348
TAIL	AD	166	(5.89; 6)	158–173	165	(8.60; 7)	151–174
BILL	AD	66.3	(2.00; 9)	64–69	63.0	(1.14; 11)	61–65
TARSUS		33.5	(1.31; 13)	32–35	32.2	(1.15; 12)	31–34
TOE		34.3	(1.53; 12)	33–36	33.9	(1.32; 10)	32–36

Sex differences significant for adult wing, bill, and tarsus. Adult breeding tails cited above rather worn; when fresh, perhaps $c.$ 10 more; full-grown adult non-breeding tail 144 (6) 138–158, juvenile tail 127 (4) 120–130. Bill not full-grown before age of 1–2 years. Banc d'Arguin (Mauritania), adult bill 65.7 (2.10; 20) 62–69, immature 64.8 (2.72; 28) 61–70, juvenile bill 61.2 (3.67; 13) 56–67 (W J A Dick).

Nominate *maxima*. Lesser Antilles and northern South America, all year; skins (RMNH, ZMA).

WING	AD	♂ 362	(8.84; 7)	351–375	♀ 360	(3.29; 8)	357–367
TAIL	AD	182	(7.16; 5)	174–193	180	(— ; 3)	175–188
BILL	AD	64.2	(2.50; 7)	61–71	58.8	(1.96; 11)	56–61
TARSUS		33.4	(2.11; 7)	31–36	31.6	(1.11; 11)	30–33

Sex differences significant for tarsus and bill. Tail of adult non-breeding 153 (6) 144–164.

Weights. *S. m. albididorsalis*. Banc d'Arguin (Mauritania), October: adult 367 (14.6; 22) 350–395, immature 353 (23.2; 28) 310–410, juvenile 341 (13.6; 13) 320–360 (W J A Dick). Adult, at sea off West Africa, late October: ♂ 400; ♀♀ 420, 440 (RMNH). Ghana, immature, October: 324 (15.4; 4) 309–345 (E K Dunn).

Nominate *maxima*. Northern South America: adult ♂♂ 390 (April), 475 (January); adult ♀♀ 300 (November), 346 (December), 415 (February), 449 (August); 2nd calendar year ♀♀ 368, 385 (May), 390 (July); 3rd calendar year ♂♂ 500 (January), 430 (May) (RMNH, ZMA). At hatching, Virginia (USA), 52.9 (7) 50–58 (Buckley and Buckley 1972a).

Structure. Wing long and narrow, pointed. 11 primaries: p10 longest; in adult, p9 12–24 shorter, p8 37–45, p7 60–72, p6 86–102, p1 214–240; in juvenile, p9 2–16 shorter, p8 24–38, p7 50–64, p6 79–96, p1 200–216; p11 minute, concealed by primary coverts. Tail deeply forked; length of t1 91 (88–94) in *albididorsalis*, 88 (80–99) in nominate *maxima*; fork (tip of t1 to tip of t6) on average 75 in adult breeding *albididorsalis*, 53 in adult non-breeding, 36 in juvenile; 93 in adult breeding nominate *maxima*, 65 in adult non-breeding, 50 in juvenile. Bill long and rather heavy, slightly decurved; stouter than any other Palearctic *Sterna* except Caspian Tern *S. caspia*; bill lengths of *S. maxima* and *S. caspia* nearly similar, but depth at base $c.$ 15–17 and $c.$ 20–23 respectively. Bill of Swift Tern *S. bergii* closely similar to *S. maxima* in length and depth. Feathers of nape form distinct rough crest. Tarsus and toes rather heavy, but short; front toes connected by webs. Outer toe $c.$ 84% of middle, inner $c.$ 66%, hind $c.$ 30%.

Geographical variation. Slight, mainly involving size; recognition of separate races perhaps unwarranted. West African *albididorsalis* described as being paler on upperparts and having bill somewhat longer and more slender (Hartert 1912–21), but comparison of upperparts of a series of *albididorsalis* with western Atlantic nominate *maxima* shows most to be identical in colour, with only a few old West African skins paler and less blue-grey; bill depth at base similar in both, and *albididorsalis* upheld as a race only because of slightly different wing/bill ratio. Pacific populations of nominate *maxima* from western USA and western Mexico similar to Atlantic ones, except for slightly longer wing: Mexican ♂ 387 (5.18;5) 379–390 (RMNH); average of 5 ♀♀ 383 (Ridgway 1919). Birds of coastal Argentina, Uruguay, and southern Brazil large, bill about as long as in *albididorsalis*, wing close to birds of western Mexico (Escalante 1968; BMNH, RMNH).

Forms superspecies with *S. bergii*, being closely similar in size and structural characters; latter mainly differs by darker upperparts, white forehead in breeding plumage, and duller bill colour, but differences largely bridged by *S. b. bergii* from South Africa and Moçambique, which is rather pale above and shows only limited white on forehead (Clancey 1975; C S Roselaar).CSR

Sterna bergii Swift Tern

PLATES 1 and 4
[between pages 134 and 135]

Du. Grote Kuifstern   Fr. Sterne huppée   Ge. Gelbschnabelseeschwalbe/Eilseeschwalbe
Ru. Болщая хохлатая крачка   Sp. Charrán de Berg   Sw. Iltärna

Sterna Bergii Lichtenstein, 1823. Synonym: *Thalasseus bergii*.

Polytypic. *S. b. velox* Cretzschmar, 1826, Red Sea, Persian Gulf, and Indian Ocean south to Somalia, Maldive Islands, northern Sumatra, and western Malay peninsula. Extralimital: nominate *bergii* Lichtenstein, 1823, Namibia and South Africa; *thalassina* Stresemann, 1914, islands off East Africa from Moçambique to Tanzania, Madagascar, Mascarenes, Chagos, Amirantes Islands, Aldabra, and Comoros archipelago; *cristata* Stephens, 1826, eastern south-east Asia, Indonesia, and Australia, north to Ryukyu and Marcus Islands, east to Tuamotu Islands, south to Tasmania.

Field characters. 46–49 cm (bill 5·4–6·5, legs 2·8–3·3, tail of adult 14–15, of juvenile 11 cm); wing-span 125–130 cm. Close in size to Royal Tern *S. maxima*; much smaller than Caspian Tern *S. caspia*, with much narrower wings and more sharply pointed bill; 25% longer than Lesser Crested Tern *S. bengalensis*, with longer bill, longer, heavier head, and bulkier body. Large sea-tern, with long, dagger-like bill, prominent white forehead (at all seasons), dark dusky-grey mantle and inner wings, and (when adult) pale silvery primaries; rear crown noticeably shaggy. Immature shows bold, variegated wing pattern, recalling *S. maxima*. Bill shows distinct uplift from gonys; usually pale green-yellow. Flight strong and swift, with wings often sharply angled at carpal joints. Sexes similar though ♂ slightly larger; little seasonal variation. Juvenile and immature separable.

ADULT BREEDING. Wide band on forehead (from bill to eye) white; angled forecrown, crown, and nape jet-black, with rear feathers loose and flowing. Face, neck, and underparts white. Back, most of inner wing, and primary coverts dark dusky-grey, fading into white trailing edge to secondaries and silvery-white uppersurface of primaries (only faintly lined darker on tips of outermost feathers). Pattern of upperwing thus rather the reverse of that normal in genus; that of underwing less striking, with no obvious dusky border to primaries. Rump and tail grey, paler than back and with white on sides (and on tips of tail-feathers). Bill yellow, with green tone at base. Legs dull black. ADULT NON-BREEDING. Forecrown and centre of crown white, but black on rear of head usually more extensive than on *S. maxima*. Upperparts less dusky, except along leading edge of inner wing; when worn, primaries less silvery (and contrast with inner wing less marked). Bill-tip dusky on some. JUVENILE. Forehead and crown brown-white, streaked black particularly at rear. Back buff-grey, strongly mottled and barred brown-black. Throat and neck sometimes streaked brown. Otherwise as 1st-winter. FIRST WINTER. Plumage pattern close to that of *S. maxima*, but with all wing-coverts darker grey, inner primaries paler, and undersurface of outer primaries less dusky, showing marked dark edge only round wing-point. FIRST SUMMER. Wear of 1st-winter dress often produces exaggerated '*portlandica*'-type plumage pattern, with dark band across secondaries outlined white on both edges, almost white inner primaries creating wide white wedge inside almost black outer primaries, and border to undersurface of wing-point darker, appearing black. Central wing-coverts dusky-grey, much darker than in *S. maxima*. Following moult, apparently resembles winter adult but field study of this plumage and final progression to adult inadequate.

Unlikely to be confused with *S. caspia* or, when adult, *S. maxima*. However, faded *S. bergii* in immature plumage occasionally resemble *S. maxima* closely, though none known to show such generally white plumage; distinction best based on more pointed and paler, yellower bill, greyer mantle and wing-coverts, less extensive markings on undersurface of primaries, and different voice. Flight powerful and rapid, with easy, deep, and usually flexed wing-beats; much addicted to shearing across wind and track thus less direct than that of *S. maxima* or *S. sandvicensis*. Gait free and quite high-stepping, in both walk and run. Carriage markedly horizontal and rather hunched, with head appearing large and angular, particularly when held up.

Advertising-call loud, raucous, even recalling crow *Corvus*—'kerrak'; anxiety or excitement note at nest 'korrkorrkorr', in the air a hard 'wep wep'.

Habitat. Tropical and warm temperate, coastal and marine; in parts of range, also in lagoons and estuaries, but hardly ever on tidal creeks or inland waters. Breeds mainly on low-lying islands, sandy, rocky, or coral; sometimes among dwarf shrubs or other stunted vegetation, but often without shelter of any kind. Rests on rocks, sand-banks or sand-spits, beaches, or artefacts such as buoys or masts of moored ships.

When fishing, flies a few metres above sea but apparently does not normally swim, and prefers to rest standing, perching, or sitting. Flight strong and wide-ranging, mostly far out to sea, in lower airspace. Locally vulnerable to excessive harvesting of eggs, but not much exposed to human impacts. (Archer and Godman 1937; Ali and Ripley 1969*b*.)

Distribution. Breeding areas inadequately known.

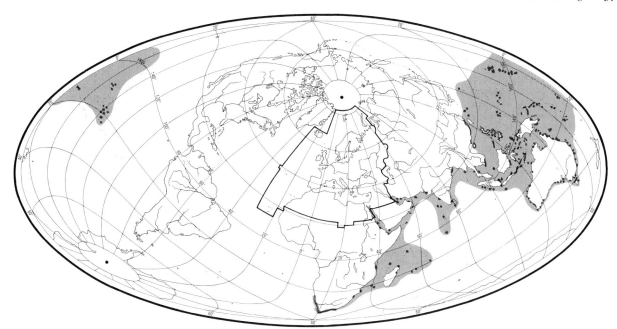

EGYPT. Occurs coasts of Red Sea from Suez south, but no nests found north of tropics (Meinertzhagen 1930, 1954); fairly common on coast near Hurghada, but no signs of breeding (Al-Hussaini 1938); possibly breeding Ras Gemsa (Marchant 1941) and Ghanim island 1983 (M C Jennings). IRAQ. Breeds Boonah island, near Al Faw (Ticehurst et al. 1926); no recent information. KUWAIT. Present in summer; no confirmed breeding (P R Haynes).

Population. No information.
Oldest ringed bird 20 years 4 months (Rydzewski 1979).

Movements. Northern race *velox* (Red Sea to Bay of Bengal), to which the following refers, is dispersive within breeding range, though a few penetrate south of it.

In west (Red Sea and Gulf of Aden), colonies known only off Eritrea, Aden, and Somalia. From these, small numbers disperse northward as far as Gulf of Suez, with regularity that has led to speculation about breeding there (Etchécopar and Hüe 1967). Apparently occurs off Egyptian coasts mainly in summer (April–August), though also a record of c. 30 in Jemsah Bay in January–February 1940 (Al-Hussaini 1938, 1939; Marchant 1941); can occur all year (P L Meininger). Once obtained, 19 March, at Port Said on Mediterranean coast (Meinertzhagen 1930). In limited dispersal south of Gulf of Aden, small numbers of *velox* occur regularly on Kenyan coast south to Mombasa; present in all months, though flocks of 20 or more seen only January–June (Britton and Brown 1974; Britton and Osborne 1976); by comparison, maximum numbers occur in Aden between June and October (Browne 1950). Peak counts reach 180 birds at Sabaki estuary (Kenya), March–May, though only single birds recorded south of Malindi (Britton and Britton 1976). Birds occurring Tanzanian coast (more commonly than in Kenya) believed to be of Indian Ocean race *thalassina* (Britton and Osborne 1976; Britton 1977). In Persian Gulf, Arabian Sea, and Bay of Bengal, *velox* present all year, occurring along entire seaboard of Arabia, southern Iran, Indian subcontinent, and Burma to western Malaya (Vaurie 1965; Ali and Ripley 1969b), but movements in relation to natal colony unknown in absence of ringing.

Food. Primarily fish, caught mainly by plunge-diving, sometimes rather obliquely; usually taken by gripping just behind head (Ali and Ripley 1969b; Serventy et al. 1971), and usually swallowed in mid-air (Archer and Godman 1937). Will also dip-to-surface to feed (Domm and Recher 1973). Usually singly, in pairs, or loose parties (Clapham 1964; Ali and Ripley 1969b), also in mixed flocks of Sternidae (Ali and Ripley 1969b), especially when hunting small fish driven to surface by shoals of predatory fish, e.g. bonito *Sarda sarda*, Spanish mackerel *Scomberomorus commerson* (Domm and Recher 1973). On One Tree Island (Great Barrier Reef, Australia), intraspecific kleptoparasitism occurs though with negligible success. Silver Gulls *Larus novaehollandiae* and Reef Herons *Egretta sacra* stole fish from adults returning to feed chicks: gulls stole 11·8% in December–January, decreasing to 3·5% in February when chicks could fly (Domm and Recher 1973). Juveniles tend to feed mostly by incomplete submersion rather than plunge-diving (Feare 1975). Will carry fish at least 3·2 km to breeding colony, Australia (Serventy et al. 1971).

No information on diet in west Palearctic. In Indian coastal waters, takes fish and prawns (Crustacea) (Ali and Ripley 1969b). Off Somalia, feeds mainly on sprats

(Clupeidae) (Archer and Godman 1937). In Australia, diet includes various Clupeidae up to 5 cm in length, pilchards *Sardina pilchardus* especially from coastal shoals, and young mackerel *Scomber scombrus*. Will take fish, usually tommy ruff *Arripis georgianus* up to 11·5 cm long, from fishing nets; occasionally takes eggs and young of turtles (Chelonia). Recorded feeding on kitchen waste from ships, Pacific Ocean (Lacan and Mougin 1974).

Chicks fed on whole fish (Diamond 1971a; C J Feare); on Aldabra (Indian Ocean), chicks fed mainly on Labridae (up to 12 cm long) and Acanthuridae (Diamond 1971b). On One Tree Island, regurgitated stomach contents of chicks contained principally Engraulidae 2–10 cm long (chiefly *Engraulis australis*), Scombridae over 6 cm (*Euthynnus affinis*), and Balistidae of 6–10 cm (mainly *Arotrolepis filicauda*, less *Catherines fronticinctus*—in both species, large lockable dorsal fin, which would make swallowing difficult, removed by adults before being fed to chicks); other species included *Parexocoetus brachypteris, Stethojulis axillaris, S. strigiventer, Caranx sexfasciatus, Carangoides, Pomacentrus, Scarus, Eleotriodes strigatus, E. longipinnus, Arothron hispidus, Lagocephalus,* and, in small numbers, prawns and squid (cephalopod mollusc) (Domm and Recher 1973). Mean of 248 intervals between feeds, One Tree Island, 62·5 ($\pm 46·4$) min (K Hulsman).

Social pattern and behaviour. Based largely on observations by K Hulsman at One Tree Island (Great Barrier Reef, Australia).

1. Gregarious at all seasons (Ticehurst 1924c; Archer and Godman 1937; Mackworth-Praed and Grant 1952; MacDonald 1973). Winter flocks—whether fishing or roosting—vary in size: 'hundreds' in Gulf of Aden (Archer and Godman 1937) and up to 250 off southern Australia (Stirling *et al.* 1970), though most records are of smaller numbers—from single birds up to 50 (Milon 1950; Erard and Etchécopar 1970; McLachlan and Liversidge 1970; Britton 1977; J Cooper). When few in number, associates readily with other Sternidae or gulls (Laridae) (Archer and Godman 1937; McLachlan and Liversidge 1970; J Cooper). BONDS. Monogamous pair-bond (J Cooper), persisting from year to year. First breeding usually at 2 years, though recorded at 1 year (Purchase 1973). Young may form crèches until fledging (Domm 1977). On Christmas Island (Pacific Ocean), crèches contained up to 40 chicks less than 2 weeks old (R Perry); these chaperoned by several adults, though chicks fed only by own parents (Vincent 1934, 1946a; Keast 1942; Domm and Recher 1973; Domm 1977). Parents continue to feed offspring for variable period after fledging, probably 2–4 months, or more (Meiklejohn 1951; Holyoak 1973; Purchase 1973; Feare 1975; see Hulsman 1977a). BREEDING DISPERSION. Nests in dense colonies, with frequent changes of location between years (Domm 1977). Colony sizes: only 6 pairs, in sub-colonies of 2 and 4, Amirantes, Indian Ocean (Feare 1979a; C J Feare); total of up to 300 pairs in 2–3 sub-colonies, Christmas Island (R Perry); 270 pairs on Masthead Island, Great Barrier Reef (Jahnke 1977); 181–342 (in different years) at One Tree Island (Hulsman 1977c); 1000 or more nests, South Africa (J Cooper); 6000 pairs, South Neptune Islands, Australia (Stirling *et al.* 1970); *c.* 8000 pairs at North Solitary Island, New South Wales (Lane 1979). Even where associated with other nesting Sternidae and Laridae, usually forms one or more discrete and compact groupings (Phillips 1923; Uys 1978; C J Feare, R Perry): e.g. in Gulf of Aden, small sub-colonies of 10–40 pairs, separate from Lesser Crested Tern *S. bengalensis* (Archer and Godman 1937). Nest density varies between localities and years. In some years at One Tree Island, neighbouring birds can touch bills, in others density only 0·97–1·04 per m^2 (Hulsman 1977c); on Christmas Island, 3·9 nests per m^2 (R Perry). Typically *c.* 30 cm between nests (Milon 1950); in Gulf of Oman, 30·5 cm (Cott 1954), though C J Feare recorded 1·2$\pm$0·4 m, minimum 20 cm; at Aldabra (Indian Ocean) 50 cm (A W Diamond); at One Tree Island, distance (centre to centre) between nests 33·3$\pm$10·0 cm, minimum 20 cm (Hulsman 1977c; K Hulsman); at Masthead Island, 38–46 cm (Cooper 1948); at colony of 12 nests, Tuamotu archipelago (Pacific Ocean), mean 90 cm (50–175 cm) (Lacan and Mougin 1974); mean 1·3 m, Western Australia (Keast 1942); at South Neptune Islands, nests 32·5$\pm$5·0 cm apart at colony edge, 31·4$\pm$4·9 cm at centre (Stirling *et al.* 1970). Distance between nests partly determined by size of defended area; if neighbours face each other while scrape-making, distance will be at least 30 cm, but closer if oriented at right angles. Small, roughly circular nest-area territory thus established with nest at centre; generally 0·03–0·13 m^2 (K Hulsman). ROOSTING. Roosts in tight flocks, regularly in company with other Sternidae (especially *S. bengalensis*) and Laridae, on beaches, rocky headlands, lagoon islets, etc. In flocks of over 500 on Masirah (C J Feare) but often in relatively small numbers: up to 60 (Gallagher 1960), less than 10 (J Cooper), and even singly on rocks, buoys, ships' masts, or piles (Ticehurst 1924c; Archer and Godman 1937; Mackworth-Praed and Grant 1952; Meinertzhagen 1954; Gallagher 1960; Simpson 1972). By day, loafs on shore between foraging bouts, especially at times less suitable for fishing (Hulsman 1977b), and at low or high tide at different sites on Aldabra atoll (Diamond 1971a). Up to *c.* 1300 roosted at night on saltpans in Aden (Browne 1950), arrival typically continuing well after dark (Paige 1960). Outside breeding season, little activity at roost except preening (J Cooper). At onset of breeding season, pre-breeding roost near colony-site becomes centre of interactive behaviour, involving display by whole group and between prospective mates, including initiation of aerial courtship, ground courtship, and copulation (Paige 1960; C J Feare, K Hulsman). Off-duty birds usually loaf outside, though near, their territories, especially if territory small (distance to nearest neighbour less than 30 cm) (K Hulsman). After fledging, juveniles usually fed by parents in shore roosts. Adults and juveniles roost together (Schreiber and Ashmole 1970; Feare 1975).

2. FLOCK BEHAVIOUR. First bird in colony to detect danger gives Alarm-call (see 1 in Voice)—if on ground, accompanied by Aggressive-upright (see Antagonistic Behaviour, below)—and flies up. Alerted neighbours then fly up, and wheel overhead (so-called 'alarm'). 'Dreads' and 'panics' occur, as in *S. hirundo*, but only from pre-breeding roost (K Hulsman). ANTAGONISTIC BEHAVIOUR. Well-developed ground-based behaviour but, as in related 'crested' Sternidae (see Sandwich Tern *S. sandvicensis*), aerial attack poorly developed compared with other *Sterna* (see Cullen 1960b). Aggressive bird adopts intimidating Aggressive-upright posture (Tinbergen and Broekhuysen 1954)—in which plumage of stretched neck ruffled, bill pointed down, and carpal joints and crest raised (Fig A); maintaining posture, may charge opponent. Much bickering occurs in dense colonies, especially between neighbours with minimum distance between nests. Incubating birds duel with bills, giving Gakkering-call (Hulsman 1977c: see 2 in Voice). Standing birds may fight in Aggressive-upright (as in *S. sandvicensis*) with Gakkering-call. Head-bobbing, in which head and neck bobbed up and down, indicates heightened threat (K Hulsman). While incoming birds hover

A

overhead looking for place to land, birds below point bills up at them and raise crests in threat (Warham 1956). Adults aggressive towards strange chicks, vigorously pecking any straying too close to own offspring, thus often scattering other chicks in crèche before feeding their own. Chick which does not immediately escape often grabbed by nape (K Hulsman). Adult challenged or attacked on ground may appease by sidling away in Erect-posture ('pole stance': Van den Assem 1954), neck stretched, bill directed vertically upwards, crest and body plumage sleeked, wings held out, and wing-tips sometimes held below tail touching ground (Fig B); this indicates high escape tendency. HETEROSEXUAL BEHAVIOUR. (1) General. Pair-formation, courtship, and copulation seen up to 120 km from nearest known breeding grounds (C J Feare) but concentrated at traditional pre-breeding roost usually only 80 m to 8 km from prospective colony-site (Gibson 1956; J Cooper, K Hulsman). Pairs do not frequent nesting area until clutch due to be laid, and birds may enter colony, establish nest-site, and lay, all within 1 day (K Hulsman, A J Williams). Both aerial and ground courtship occur. (2) Aerial courtship. Begins at least 1 month before laying (R Perry); usually elicited by unpaired ♂ arriving at pre-breeding roost, uttering Advertising-call (see 3 in Voice) with or without fish in bill. Bird often uses Butterfly-flight—slow and emphatic, with unusually deep wing-beats (K Hulsman; see Van den Assem 1954). Arrival of bird often precipitates a dread. Thereafter, fish-carrier may attract no further attention, or be chased, sometimes by several birds which it seems intent on evading by twisting, turning, and swooping. When pursuer overtakes fish-carrier (Pass-ceremony), former extends neck (aerial version of Erect-posture) and sways from side to side in front of fish-carrier which may give Karr-call (see 4 in Voice). As pre-breeding period progresses, dreads increasingly lead to ground courtship (see below), and to High-flights; latter may originate directly from dread, without intervening ground courtship, or from ground courtship. In High-flight, one bird (usually ♂) generally carries fish. If initiated from ground, ♂—without noticeable inviting posture (K Hulsman)—takes off, followed by ♀ (or sometimes 2 birds); participants spiral upwards c. 20 m apart, accelerating, until, at

c. 100–130 m their wing-beats are 3 times normal rate; call frequently (see 3 in Voice) during ascent. Calling subsides as gap between them closes, often at great height (K Hulsman). Then, one slightly ahead of the other, apparently almost touching, the 2 birds bank and, with wings held half open, glide swiftly downwards, weaving from side to side, sometimes rising till they almost halt, then falling again, every slight change in flight attitude and direction of one matched by the other (Gibson 1956). During glide, leader and pursuer may swap positions (Pass-ceremony) one or more times; rapid Kek-calls (see 5 in Voice) heard. Aerial version of Bent-posture—occurring in *S. sandvicensis* during Pass-ceremony—not recorded (K Hulsman). Pair eventually return to flock, go off fishing, or engage—with Butterfly-flight (this generally starting directly after glide) and aerial version of Erect-posture—in paired flight, in which Pass-ceremony occurs a few times before landing; as leader overtaken, occasionally gives Karr-call. As pre-laying period progresses, High-flights increasingly initiated without prior collective upflight. Both upflights and High-flights diminish rapidly when most birds have started nesting. However, major resurgence of courtship—mostly advertising ♂♂ chased by other birds, with Pass-ceremony, and less commonly High-flights—occurs when first chicks fledge; participants mostly unpaired adults, immatures, and failed breeders (K Hulsman). (3) Ground courtship. Usually precedes or follows High-flight. In former case, ♂ carrying fish flies into social roost; walks amongst others, uttering Advertising-call and adopting Forward-erect—wings and plumage as in Erect-posture, body and neck more or less upright, and bill slightly below horizontal (Fig C); bill raised with each call. On encountering co-operative ♀, both adopt Forward-erect, and Strut around each other head-to-tail in tight circle, with short mincing steps. On returning to flock after High-flight, pair stand quietly and display to one another, initially adopting Erect-posture, turning upraised head quite jerkily from side to side (Head-turning). Then both relax to Forward-erect (K Hulsman). Both birds of pair participate in nest-site selection. (4) Mating. Often stimulated by ♀ begging: adopts crouching Hunched-posture with crest moderately raised, uttering Begging-calls, while pointing at ♂ and trying to keep in front of him; ♂ adopts Forward-erect, or Down-erect (lowered bill) and tries to approach from side or rear and mount; many attempts fail because ♀'s main object often to beg food. Most copulation occurs in pre-breeding roost (C J Feare, K Hulsman). (5) Courtship-feeding. From pre-breeding roost to early incubation, ♂ feeds mate or prospective mate, ♀ sometimes begging food (as above). ♀ occasionally pecks at bill or snatches at fish (Warham 1956; Hulsman 1976, 1977a; J Cooper). Fish offered may eventually be eaten by ♂ if he fails to find recipient; if transferred to ♀, she usually eats it immediately or else Struts about with it in bill, giving Advertising-call; rarely, ♀ returns fish to ♂ who eats it or returns it again—may thus change custody several times

B

C

before one bird, usually ♀, eats it (Hulsman 1977a). (6) Nest-relief. Especially during first few days of incubation, incoming bird sometimes brings fish (see above) and gives Advertising-call on approach; lands near nest, adopts Forward-erect (Warham 1956), and gives disyllabic version of Advertising-call, sometimes changing to Kuk-call (see 7 in Voice) when very close to nest. Sitting bird recognizes mate's call, stands and adopts Forward-erect, and usually leaves nest just before, or as, mate arrives (Hulsman 1977a; A W Diamond); if reluctant to leave, incoming bird may move around it, touching bills (A W Diamond). RELATIONS WITHIN FAMILY GROUP. Parents do not recognize own eggs or newly hatched young, but mutual recognition established by the time young ready to leave nest at 2 days old (Davies and Carrick 1962). Small young often brooded under wing for protection against sun (R Perry). Induced by parental calls (see 7 in Voice), chicks leave nest (Stirling et al. 1970) and scatter, or else form crèches (see Bonds, above). Parents seek own chicks in crèche, forcibly rejecting others. Initially chicks remain passive when pecked by strange adults, but begin to retaliate from c. 1 week old (K Hulsman). Both chicks and fledged juveniles beg by calling, and may later be fed by parent hovering above. In 1st week after fledging, juveniles awaiting return of parents with food often practise fishing, dipping-to-surface to pick up floating objects (Feare 1975; Hulsman 1977a). From 1 week after fledging, juveniles accompany one parent to feeding grounds; 2 juveniles may follow 1 parent (Holyoak 1973). Adults and young leave colony c. 13–19 days after fledging (Hulsman 1977a). As parents feed young less and less, young may try to steal fish from parent: in Tahiti, 1st-winter bird seen vociferously chasing fish-carrying adult which zigzagged before swallowing fish itself (E K Dunn). ANTI-PREDATOR RESPONSES OF YOUNG. In response to Alarm-call of parent, or other adult, small chicks crouch prone; after a few days, however, they scramble for cover, and may also gather regularly under vegetation for shade and safety (Hulsman 1977a; R Perry). Alarmed older young run in the open or, assembling near shoreline, take to water (Keast 1942; Gallagher 1960; Stirling et al. 1970; Hulsman 1976, 1977a; C J Feare). PARENTAL ANTI-PREDATOR STRATEGIES. (1) Passive measures. At approach of danger, adults give Alarm-calls. (2) Active measures: against birds. Co-operates only weakly to mob predators in the air, and individuals rarely fly far from nest to challenge them. Birds on One Tree Island took flight and buzzed approaching White-breasted Fish-eagle Haliaeetus leucogaster, but not vigorously enough to prevent it taking a chick. Sitting birds lunged and pecked (giving Gakkering-call) at Silver Gulls Larus novaehollandiae which passed within 0·75 m; thus, if nests close enough, birds can effectively ward off even groups of gulls (Hulsman 1977a). Excited birds may occasionally fly up and dive on nearby gull but this generally less effective as defence since it leaves clutch exposed (Hulsman 1974, 1977c). Adults drive gulls from their own chicks, but not from chicks of others unless their own are close by. Bird may pass fish quickly to young and then shield it with own body while threatening aggressor hovering overhead (Hulsman 1976). (3) Active measures: against man. At One Tree Island, birds raised wings and some gave Alarm-calls when human approached within 30 m of colony; when c. 20 m away, flew overhead and swooped (not closer than c. 1 m to person's head), sometimes defecating (Hulsman 1977a). Incubating birds approached to within 3 m jump up and down with Gakkering- and intense Alarm-calls, crests raised (Bark Jones 1946). Adults attending crèche likewise raise crests and give Gakkering-call, and may remain on ground or take flight (Vincent 1946). (4) Active measures: against other predators. Parent swooped on sand shark (probably Eulamia) as it approached (and took) chick on water (Gallagher 1960).

(Figs A and B after Tinbergen and Broekhuysen 1954 and photographs by R Perry; Fig C after photograph by T Kendall in Wade 1975.) EKD

Voice. Highly vocal, especially at breeding grounds.
CALLS OF ADULTS. (1) Alarm-call. A hard 'wep-wep' (Warham 1956); given (e.g.) in flight at approach of danger. (2) Gakkering-call. Throaty 'korrkorrkorr' (K Hulsman); 'ka-kurr ka-ka-ka-ka-ka-kurr' (J Hall-Craggs: Fig I), expressing threat or alarm; heard among birds densely packed on nests, duelling with neighbours or in standing fight with wings partly spread, or directed at aerial intruders. A harsh 'krow-krow' is a related call (Warham 1956; K Hulsman); probably also the grating 'kik-kik-kik' given continuously by birds in fishing flocks (Erard and Etchécopar 1970). (3) Advertising-call. Raucous 'kirrak', 'kirrik', or 'kirrah' (K Hulsman); also rendered as 'krrik' (Fig II, left), 'kerwah' (Fig II, right) (E K Dunn, P J Sellar); 'kree kree' (Slater 1971). Deeper than Advertising-call of S. sandvicensis. Given by both sexes (but especially ♂), often announcing arrival at colony with fish and/or for nest-relief; also by ♂ in flight advertising for ♀ to follow him, and by each of 2 birds (calls often alternating) in ascent phase of High-flight. (4) Karr-call. Often given after some interaction has finished: at end of billing fight between sitting birds; after ground courtship parade; from both birds on ground after ♂ has swallowed fish instead of presenting it to ♀; sometimes from both birds during nest-relief just before relieved bird departs; rarely during Pass-ceremony (K Hulsman). Probably homologous with 'wah'-call of S. sandvicensis; interpreted as expressing

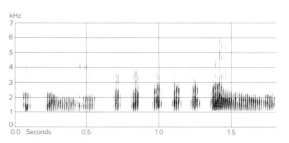

I P A D Hollom Iran April 1972

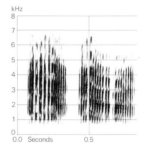

II P A D Hollom Iran April 1972

strong escape tendencies (Van Iersel and Bol 1958; Smith 1975*a*). (5) Kek-call. Rapid 'kekekekekerr', last syllable dropping in pitch (K Hulsman); 'krit-krit-krit' (P A D Hollom); given during gliding phase of High-flight (K Hulsman). (6) Kuk-call. A soft 'kuk', uttered once, or slowly repeated, invites close approach; given (e.g.) by arriving bird to sitting bird at nest-relief, or when summoning young to be brooded or to follow. ♂ has apparently related call, a deep 'kré-kré', given during copulation (K Hulsman). (7) Begging-call of ♀. A persistent, penetrating 'quee-quee-quee' to beg fish from ♂ or solicit copulation.

CALLS OF YOUNG. Downy young beg with a repeated 'peep' which becomes clearer and sharper over first 3 days (see Davies and Carrick 1962 for sonagrams). Food-call of juvenile a petulant squeaky or whistling call (Holyoak 1973); probably the shrill, carrying 'tssriii' of Erard and Etchécopar (1970). EKD

Breeding. Very little information from west Palearctic. SEASON. Sudan: eggs and young in June (Moore and Balzarotti 1983). Jebel Teir (Gulf of Aden): well incubated eggs and 1 newly hatched young found mid-August (Bark Jones 1946). Iran: eggs found Sheedvar Island late June (L Cornwallis). Somalia: breeding starts August (Archer and Godman 1937). In Indian and Pacific Oceans, may breed at 6-month intervals or irregularly. SITE. On bare ground or among scattered bushes. Colonial. Nest: shallow scrape, unlined. Building: no information on role of sexes, though both participate in selection of site (K Hulsman). EGGS. See Plate 88. Sub-elliptical, smooth, not glossy; variable, from olive-yellow to pink, buff, or cream, variably blotched, spotted, and scrawled black, brown, chestnut, and paler grey. 62×43 mm ($54-69 \times 40-46$), sample 250; calculated weight 60 g (Schönwetter 1967). 5 eggs, Bird Island (Seychelles), 50–57 g (Feare 1979*b*). Clutch: 1–2 (–3); clutches of 3 probably by 2 ♀♀ (E K Dunn); on Jebel Teir, most clutches of 1; a few clutches of 2, but only 1 certainly by 1 ♀ (Bark Jones 1946). One brood. No information on replacements, but probable after egg loss. INCUBATION. 25–30 days. By both parents in very variable spells, mean 184 min (sample 73). (K Hulsman.) YOUNG. Precocial and semi-nidifugous. Cared for and fed by both parents. Young leave nest at *c*. 2 days, either scattering with one parent looking after each chick, or forming crèche (Domm and Recher 1973; Domm 1977; K Hulsman). FLEDGING TO MATURITY. Fledging period 38–40 days (K Hulsman). Dependent on parents for food for up to 4 months (Feare 1975; Hulsman 1977*a*). Age of first breeding usually 2 years (Purchase 1973). BREEDING SUCCESS. In Australia, much predation by gulls *Larus* (Hulsman 1977*c*).

Plumages (*S. b. velox*). ADULT BREEDING. 3–8 mm of central forehead (varying individually) and band 7–10 mm wide between base of upper mandible and sides of crown white. Crown to just below eye and elongated feathers of hindcrown black. Hindneck and upper mantle white. Lower mantle, scapulars, tertials, and upper wing-coverts slate-grey, darker than any other west Palearctic *Sterna*; small coverts at wing-bend and part of inner webs of tertials white. Back, rump, upper tail-coverts, and tail slate-grey like mantle and scapulars, but sometimes slightly paler and more silvery, especially on tail; outer web of t6 sometimes pale grey; basal inner webs of tail-feathers white. Sides of head below black cap and all underparts white. Secondaries dark blackish-grey (darker than coverts) with white tips and inner webs; 4–5 fresh inner primaries silvery-grey, inner webs bordered white, outer primaries with dark silvery-grey outer webs and blackish-grey inner webs, latter with narrow white edges and off-white wedges to base. Underwing and axillaries white, with grey band *c*. 6 cm wide towards end of primaries. In worn plumage, slate-grey of upperparts with pale brown-grey shading, secondaries, tertials, and longer tail-feathers shaded grey-brown; inner primaries dull silvery-grey, contrasting with sepia-black outer primaries; latter with only some silvery-grey remaining on those parts protected by neighbouring feathers, outer webs and borders of tips of inner webs especially blackish, white wedges to inner webs more contrasting. ADULT NON-BREEDING. Like adult breeding, but feathers of crown black with white fringes, variable in width: in some, crown black with indistinct white mottling only, others white with heavy black spotting; patch in front of eye and streak from behind eye across nape black. Often some grey dots or short streaks from below eye across ear-coverts to sides of nape, occasionally extending to sides of throat. Forehead, lores, and hindneck white as in adult breeding. Remainder of body as in adult breeding, but upperparts and upperwing more variable, dependent on wear and moult; often mixture of old grey-brown and fresh slate-grey feathers, while moult may reveal part of white feather-bases. Depending on moult, primaries either with inner new and pale silvery-grey, outer blackish (shortly after breeding season), or all about equally new and silvery-grey (about midwinter and early spring). DOWNY YOUNG. Variable; silvery-grey to olive-green, mottled to varying extent (Harrison 1975). JUVENILE. Forehead, lores, and sides of head from lower mandible to below ear white, densely spotted black. Crown heavily streaked black and white, ear-coverts and nape nearly uniform black. Mantle and scapulars white with large subterminal black spot, U-shaped on lower scapulars. Tertials dull black, broadly fringed white. Back to upper tail-coverts white, spotted black; spots largest on lower rump, forming subterminal streak on upper tail-coverts. Underparts white, often with grey freckles below ear-coverts, occasionally extending to sides of chest and forming broken chest-band. Tail black with faint grey tinge, contrasting with white upper tail-coverts; feathers narrowly fringed and tipped white. Primaries black with faint grey tinge; bases of inner webs with white wedges, shading to grey towards tip, reaching to *c*. 6 cm from tip; narrow edges of inner 4–6 primaries white. Secondaries dull black, tips narrowly and inner webs broadly bordered white. Primary coverts and greater upper wing-coverts dull black with ill-defined white tips and borders, especially innermost; median and longer lesser coverts contrastingly white (median often with faint dull black smudges to centre), shorter lesser contrastingly black again, forming dark carpal bar. Leading edge of wing, axillaries, and under wing-coverts white. Some variation in size and extent of black marks on upperparts: spots and other marks sometimes small and feather-centres greyish, but generally more heavily marked than other large *Sterna*. FIRST IMMATURE NON-BREEDING. Head like adult non-breeding, but crown often blackish with narrow white feather-fringes, hardly contrasting with black patch from front of eye over ear to nape; grey or dull black spots or short streaks below ear sometimes extend down sides of hind-cheeks to form half-collar. Upperparts and upperwing with variable mixture of worn grey-brown and fresh grey feathers, depending on wear; carpal bar across lesser wing-coverts dull black, dark grey, or dark sepia. By midwinter, outer tail-feathers and most flight-feathers still juvenile, sepia-black and heavily worn; by spring, only abraded juvenile outer primaries remain; during summer, 1st non-breeding directly replaced by 2nd non-breeding. SUBSEQUENT IMMATURE PLUMAGES. Recognition difficult, except when dark carpal bar still present. Ageing as in Royal Tern *S. maxima*.

Bare parts. ADULT. Iris dark brown. Bill deep chrome-yellow with paler tip in breeding, duller lemon-yellow or green-yellow with dark olive or horn-brown tinge at base in non-breeding. Foot black with all or partly yellow soles. DOWNY YOUNG. No information. JUVENILE AND IMMATURE. Like adult non-breeding, but foot either black with yellow soles or dull yellow with black gradually spreading with age, starting on front of tarsus and uppersurface of toes. (Clancey 1975; BMNH, RMNH, ZMA, ZMO.)

Moults. *S. b. velox* similar to *S. maxima*. Moult of populations breeding near equator and in southern hemisphere more variable due to different or more prolonged breeding seasons (C S Roselaar.)

Measurements. Sexes combined. (1) *S. b. velox*, Red Sea and northern Indian Ocean; (2) *S. b. cristata*, Greater Sunda Islands; skins (BMNH, RMNH, ZMA). Depth of fork for adult breeding; t6 is t1 plus fork.

WING AD	(1) 366	(9·87; 12)	354–381	(2) 344	(9·91; 11)	333–365
TAIL (t1)	86·9	(6·99; 8)	81– 98	81·1	(3·89; 17)	76– 88
FORK	89·7	(— ; 3)	83–100	72·2	(10·8; 6)	66– 94
BILL AD	63·9	(2·58; 14)	59– 68	60·3	(3·64; 20)	54– 65
TARSUS	32·2	(0·79; 11)	31– 33	28·8	(0·92; 20)	28– 30
TOE	33·7	(1·76; 11)	31– 36	31·0	(1·00; 10)	29– 33

Depth of fork of adult non-breeding and immatures on average *c.* 12 below adult breeding. Juvenile wing on average *c.* 10 below adult; depth of juvenile fork 44·5(4)37–51; juvenile bill not full-grown until *c.* 1 year old. Average bill length of adult ♂ *c.* 4 mm greater than adult ♀.

Weights. *S. b. velox*. Laccadive Islands, February, adult ♀ 397 (BMNH). Kenya, late April: adult 360, non-breeding ♀ 340 (Britton and Osborne 1976). Burma, February, adult ♀ 340; off Bangladesh, January, 2nd-calendar-year ♀ 335 (ZMO). Smaller race *thalassina* from Amirante Islands, September: ♂♂ 325, 345, 350; ♀ 350 (Britton and Osborne 1976). Australian *cristata* 325–383 (Serventy *et al.* 1971), but one beached adult ♀, Western Australia, May, 274 (still some fat); adult ♂, May, 480 including single fish of 87 g in stomach (RMNH).

Structure. Closely similar to *S. maxima*.

Geographical variation. Rather complex; involves size and colour of upperparts. *S. b. velox* of Red Sea, Persian Gulf, and northern Indian Ocean largest (average wing 366, average bill 64); abruptly replaced by distinctly smaller *thalassina* (wing 340, bill 56) in western Indian Ocean from Kenya and Chagos Islands to Madagascar and Mascarenes. Further south-west, size gradually increases: nominate *bergii* of Namibia and South Africa rather large (wing 362, bill 62). Further east, *cristata* intermediate or rather small, somewhat variable, though with much overlap. Typical *cristata* from Philippines, Ryukyu Islands, Taiwan, and China small (wing 331, bill 58); birds from eastern Malay peninsula and Greater Sunda Islands larger (wing 344, bill 60), similar to populations from Moluccas, New Guinea, and eastern Australia (wing 344, bill 59), Micronesia (wing 343, bill 60), and Polynesia (wing 344, bill 58); birds of western and north-west Australia have slightly larger wing (354), but similar bill (58). *S. b. velox* not only the largest, but also the darkest of all races: upperparts slate-grey (similar to those of British race of Lesser Black-backed Gull *Larus fuscus graellsi*). Nominate *bergii* of southern Africa and *cristata* eastward from China, Indonesia, and eastern Australia both paler, upperparts pale mouse-grey. *S. b. thalassina* from western Indian Ocean and *cristata* from Western Australia both slightly paler still, like Common Tern *S. hirundo*, but upperparts slightly less bluish, more chalky, and upper tail-coverts and tail greyish-white. Amount of white on forehead rather variable in all races, perhaps least in South African nominate *bergii*. (Baker 1951; Clancey 1975; BMNH, RMNH.) CSR

Sterna bengalensis Lesser Crested Tern

PLATES 1 and 4
[between pages 134 and 135]

Du. Bengaalse Stern Fr. Sterne voyageuse Ge. Rüppellseeschwalbe
Ru. Малая хохлатая крачка Sp. Charrán bengales Sw. Mindre iltärna

Sterna bengalensis Lesson, 1831. Synonym: *Thalasseus bengalensis*.

Polytypic. Nominate *bengalensis* Lesson, 1831, Red Sea, East Africa, Madagascar, Maldives, southern and eastern India, and Ceylon; *torresii* (Gould, 1843), Mediterranean, Persian Gulf, Pakistan, southern New Guinea, and Australia.

Field characters. 35–37 cm (bill 5·0–5·7, legs 2·4–2·6, tail of adult 13–14, of juvenile 10 cm); wing-span 92–105 cm. Size overlaps with Sandwich Tern *S. sandvicensis*, but usually smaller and slighter; distinctly smaller and shorter winged than both Royal Tern *S. maxima* and Swift Tern *S. bergii*, but with proportionately longer tail. Elegant sea tern, with structure and appearance between that of *S. sandvicensis* and Common Tern *S. hirundo*. Most individual characters are (1) slightly drooping, pale orange bill, rather long and slender; (2) pale blue-grey mantle, rump, and tail contrasting with pale silvery-white primaries (in fresh breeding plumage); (3) fast, graceful flight, recalling that of *S. hirundo*. Sexes similar; little seasonal variation. Juvenile and immature separable.
ADULT BREEDING. Lores white, forming narrow band between bill and crown, unlike broad panel of *S. bergii*. Most of forehead, crown, and long nape jet-black, appearing less crested than *S. bergii* and extending further down

neck than in most terns. Lower hindneck white. Back, most of inner wing, primary coverts, rump, and tail blue-grey, with white tertial-fringes, white trailing edge to secondaries, and silvery-white primaries contrasting strongly in most lights. Underparts and underwing white, underwing showing narrow black line on tips of outer primaries at close range. Bill usually orange, paler than in *S. maxima*, but always lacking green-yellow tone of *S. bergii*. Legs black. ADULT NON-BREEDING. Forehead becomes white; mid-crown also speckled white. Rest of plumage little changed but upper mantle and tail whiter than in summer. Bill yellower, appearing waxy. JUVENILE. Most likely to recall similarly aged *S. sandvicensis* but differs (in fresh plumage) in (1) less heavily spotted back and less heavily marked tertials and inner wing-coverts, (2) greyer tail, (3) duskier flight-feathers, and (4) duskier band across upper wing-coverts. Bill pale olive-yellow to orange-yellow; legs yellow. FIRST WINTER. Head and body as winter adult, but wings retain juvenile pattern, with dusky band across secondaries darker and more sharply outlined white, inner primaries almost white (forming pale wedge), and outer primaries increasingly dusky towards leading edge; part of tail and sometimes back to rump remain dark until February. Underwing noticeably pale, with outer primaries edged dark (but lacking dusky wedge of *S. maxima*). FIRST SUMMER. Some resemble winter adult but others retain most of 1st-winter wing pattern, except for dark lesser coverts. Even when feathers worn, no '*portlandica*'-type plumage pattern assumed.

Has been confused with *S. maxima* (in circumstances where unusual light made comparison of size and plumage tones difficult) but usually unmistakable in adult plumage. Immature appears much slighter and more graceful than those of *S. maxima* and *S. bergii* even at long range, with plumage pattern never as variegated. Flight fast and direct, with actions recalling those of both *S. sandvicensis* and *S. hirundo*. Gait and carriage recall *S. hirundo*, with stance noticeably lower than *S. sandvicensis* but with folded primaries extending well past tail. Readily settles on water to rest.

Calls recall those of *S. sandvicensis* but are more scratchy and less ringing in tone.

Habitat. Breeds in lower middle and low latitudes from Mediterranean through subtropical and tropical warm seas. Associates commonly with Swift Tern *S. bergii*, sharing nest-sites on flat sandy upper beaches, especially on low-lying islands, among dwarf or stunted and sparse vegetation, and on bare sand-spits, flat rocks, or coral reefs. Forages in surf, but ranges well offshore, much as *S. bergii* from which no clear distinctions recorded. Swims and dives freely, flying mainly in lowest airspace. (Archer and Godman 1937.)

Distribution. Single birds paired with Sandwich Tern *S. sandvicensis* in Banc d'Arguin (France) 1974–6 (Yeatman 1976; Géroudet and Landenbergue 1977) and Ebro delta (Spain) 1979 (Y Bourgaut).

IRAQ. Breeds Boonah island, off Al Faw (Ticehurst *et al.* 1926); no recent information. KUWAIT. Present in summer; no confirmed breeding records (P R Haynes). EGYPT. Breeds on islands off Hurghada, Red Sea (Meinertzhagen 1954); abundant Abu Mingar Island (Al-Hussaini 1939). LEBANON. Reported bred Nakl Island end 19th century (Vere Benson 1970). TUNISIA. Occurs on coast May–August; no proof of breeding (MS).

Accidental. France, Austria, Switzerland, Italy, Israel.

Population. EGYPT. About 50 pairs in southern Sinai, sites often changed from year to year (HM). Bred Tiran island 1975 and 1979 (PLM, WCM). LIBYA. Estimated 2000 birds breeding on islets in Gulf of Sirte 1937; no recent counts (Bundy 1976).

Movements. Migratory in Mediterranean, partially so elsewhere. Extent of movements little known in absence of ringing; only those affecting west Palearctic considered here.

MEDITERRANEAN POPULATION. Despite rare January and February specimens from northern Morocco (Heim de Balsac and Mayaud 1962), now established as being summer migrant to Mediterranean. Regular seasonal movements occur along North African coasts (east in spring, west in autumn), but only a straggler to European side, despite traversing Straits of Gibraltar. Passage occurs along Atlantic coast of Morocco (Smith 1965; Bundy 1970; Pineau and Giraud-Audine 1974, 1975), and now known to winter in West Africa: limits not ascertained but regular in coastal Gambia where discovered 1974 (Batten 1975; Nielsen 1975); 2–3 summering immatures at Bargny (Sénégal) in June 1974 (Erard 1975). Evidently a late spring migrant. Passes Tangier in late May (Pineau and Giraud-Audine 1974, 1975); on central Algerian coast, seen 3 times 4 May–18 June 1978 (Jacob 1983); 250–300 seen at Thyna-Chaffar (Tunisia) on 15–16 June though only 7 on 17 July (Gaugris 1968), and eastward movement in Libya also in June (Bundy and Morgan 1969). Some present Libya into November, but westerly passage on Algerian coast noted 19 September–27 October (Schmitt 1963; see also Jacob 1983); on Atlantic coast of Morocco, where some non-breeders summer, protracted autumn passage recorded 17 August–27 October, largest numbers in October (Smith 1965; Pineau and Giraud-Audine 1974, 1975; Dubois 1979). 1 specimen considered Mediterranean *torresii* collected Gulf of Suez in May (C S Roselaar); significance unknown.

INDIAN OCEAN POPULATIONS. Breeds Red Sea, Gulf of Aden, and Persian Gulf, where present all year though numbers reduced in winter. Probably wholly migratory as far north as Gulf of Suez, where absent Hurghada (Egypt) until 17 May (Marchant 1941). Abundant off Aden in summer though scarce there in winter (Bailey 1966a); big

44 Sternidae

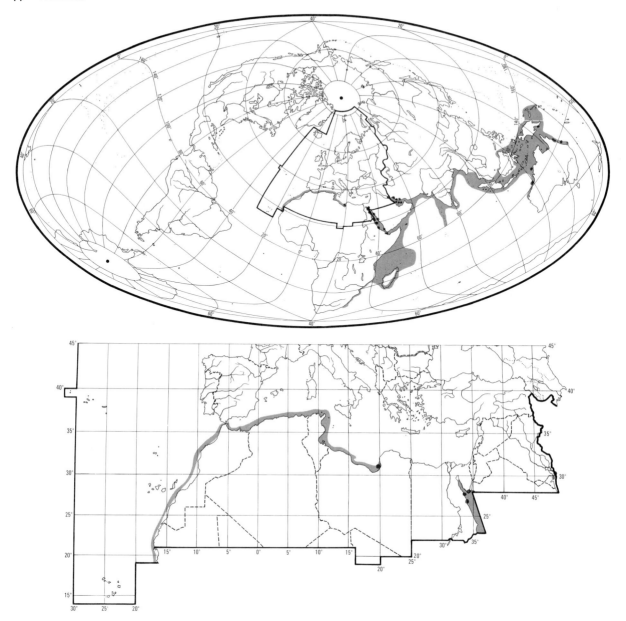

flocks on Masirah Island (Oman) in March and September–October with few at other times (Griffiths and Rogers 1975) probably reflect entry to and exodus from Persian Gulf, where few present in winter (Hallam 1980). Spreads south in winter (September–April) along East African coast from Kenya to Madagascar and Natal, rarely to Cape Province (also 5 inland records from Rift Valley lakes in August–September), and along coasts from southern Iran to India and Ceylon. Biometric evidence indicates Red Sea population occurs East Africa in winter, and that Persian Gulf birds reach western India, but not known whether wintering ranges discrete (C S Roselaar). Non-breeders frequently summer in southern wintering areas. See Ali and Ripley (1969b), Britton and Brown (1974), Britton and Britton (1976), and Britton (1977).

Food. Primarily small fish; also crustaceans. Feeds in shallow and deep water, mainly by plunge-diving or by taking items from water surface (Archer and Godman 1937; Mackworth-Praed and Grant 1952; Hitchcock 1976). Flies c. 4·5–9 m above surface; dives vertically, usually submerging completely; emerges after a few seconds and swallows fish in flight (Henry 1955). Often feeds in mixed flocks of up to 2000 Sternidae (Mackworth-Praed and Grant 1952; Clapham 1964). At One Tree Island (Great Barrier Reef, Australia), subject to food-piracy by Silver Gull *Larus novaehollandiae* and 23·4% of fish caught were thus lost (Hulsman 1976).

In harbours, Ceylon, caught small herring (Clupeidae) (Henry 1955). In Sind harbour (Pakistan), took prawns

(Crustacea) and what appeared to be fry of clupeid fish (Ticehurst 1924a). In Australia, took fish of up to c. 10 cm (Serventy et al. 1971). On One Tree Island, preyed on shoals of antherinid fish *Pranesus capricornensis* (Hulsman 1976).

On islands in Red Sea, chicks fed on whole small fish (Clapham 1964). Of 19 feeds, One Tree Island, mean interval 37 min (K Hulsman).   EKD

Social pattern and behaviour. Based largely on material supplied by K Hulsman from One Tree Island (Great Barrier Reef, Australia).

1. Gregarious at all seasons in small or large flocks up to 200 (Ticehurst 1924a; Archer and Godman 1937; Mackworth-Praed and Grant 1952; McLachlan and Liversidge 1970; MacDonald 1973; Britton 1977); roosting flocks (see below) from 10 (K Hulsman) to 400 (Milon 1950). Often occurs in company with other *Sterna*, especially Swift Tern *S. bergii* (McLachlan and Liversidge 1970; Britton and Brown 1974; Hulsman 1977a). BONDS. Apparently essentially monogamous pair-bond; not known if pairs persist for more than one breeding season. 3 birds sometimes share incubation duties (K Hulsman). From c. 7 days old to 1 week before fledging, young usually reared in crèches watched over by several adults on ground, although fed only by own parents, at least one of which continues to feed them up to 5 months after fledging (Clapham 1964; Domm and Recher 1973; Hulsman 1974; K Hulsman). For pairing with Sandwich Tern *S. sandvicensis*, see that species. BREEDING DISPERSION. Colonial, in groups of less than 10 to 2000 pairs (Moltoni 1938; Hulsman 1979). Fidelity to colony-site weak, so colony size may vary markedly between seasons: e.g. 3, 48, and 120 pairs in consecutive years, One Tree Island, where periodic alternation with another island site likely (Hulsman 1977c, 1979; Jahnke 1977). In colony of 120 pairs, mean nest density 2·76 per m² (Hulsman 1977c); much greater densities recorded, however, adjacent nests sometimes almost touching (Bark Jones 1946; Clapham 1964), e.g. 37 nests in c. 9 m² (Ticehurst et al. 1926). Small nest-area territory defended, but courtship and copulation occur outside colony, birds rarely visiting nest area until clutch about to be laid. Young led from nest after a few days old and fed outside it (K Hulsman). Often nests in company with other *Sterna*, especially *S. bergii* (Archer and Godman 1937; Bark Jones 1946; Domm and Recher 1973; MacDonald 1973; Domm 1977); also White-cheeked Tern *S. repressa* on Sheedvar Island, Iran (D A Scott), and, in Australia, in sub-colonies amongst Roseate Tern *S. dougallii* and Black-naped Tern *S. sumatrana* (Hulsman 1977a, c). Typically forms discrete sub-colonies near *S. bergii*, etc., but sometimes mixed—more likely if greatly outnumbered by other species (Domm 1977; Hulsman 1977c; L Cornwallis). ROOSTING. Assembles in flocks at all times of year—often with *S. bergii* or other *Sterna* (see *S. bergii*), also with gulls *Larus*—on open, undisturbed areas such as beaches, mudflats, coral reefs, saltpans, and harbours; also loafs singly on posts and driftwood (Ticehurst 1924a; Browne 1950; Milon 1950; Mackworth-Praed and Grant 1952; Paige 1960; Schmitt 1963; K Hulsman). Because site chosen is close to water's edge, actual location varies with height of tide (Hulsman 1977b): e.g. on Aldabra (Indian Ocean), roosts on mangroves at high tide, and on sand or reef flats at low tide (Diamond 1971a). Also settles readily on water surface to rest (Serventy et al. 1971). In mixed roosts, prefers conspecifics as neighbours. Single bird in roost aggressively maintained individual-distance from Black-headed Gulls *Larus ridibundus* (Géroudet and Landenbergue 1977). During day, loafs communally between spells of feeding; at One Tree Island, during rising or high tide (Hulsman 1977b). In Gulf of Aden, generally gathers to loaf at hottest time of day, squatting or standing on sand (Archer and Godman 1937). Birds assemble near prospective colony-site in pre-breeding roost which serves as focus for courtship and mating. On Sheedvar Island, c. 1000 birds arrived each evening to roost on sand-dunes, indulging in much courtship display and showing aggression towards human intruders (D A Scott). Flights into roost continue until well after sunset (Paige 1960). After laying, off-duty birds loaf by day and roost by night near colony (not in nest-territories). Once crèche of young established on shore, this used as roosting-site by young and adults alike (see Archer and Godman 1937; K Hulsman).

2. FLOCK BEHAVIOUR. As in *S. bergii*. ANTAGONISTIC BEHAVIOUR. See *S. bergii* for description of postures named below. Outside breeding season, sometimes aggressive towards gulls (Laridae) on feeding grounds, making flying attacks if gull attempts to steal food (Schmitt 1963). At colony, basic threat posture on ground is Aggressive-upright. Where distance between nests small, sitting neighbours frequently involved in disputes in which they stretch towards one another, plumage ruffled and crest raised, duelling with bills, silent gaping alternating with Gakkering-call (see 2 in Voice of *S. bergii*). Sitting birds increase threat towards intruding conspecifics by Head-bobbing; higher frequency of Head-bobbing indicates greater intensity of threat (K Hulsman). Threatened bird appeases aggressor with Erect-posture. HETEROSEXUAL BEHAVIOUR. (1) General. Very similar to *S. bergii* and *S. sandvicensis* (Isenmann 1972c; Petit 1976; Géroudet and Landenbergue 1977; K Hulsman). See *S. bergii* for full descriptions of displays and vocalizations. Following account mostly from K Hulsman. At One Tree Island, pair-formation, courtship, and copulation concentrated at pre-breeding roost up to 3 km from prospective colony-site. (2) Aerial courtship. Early courtship characterized by 'dreads', apparently stimulated by arrival at roost of ♂, with or without fish, uttering Advertising-call. ♂ may attract one or more birds, presumably mostly ♀, to pursue him. Pursuer may overtake leader (Pass-ceremony), adopting aerial version of Erect-posture, and swaying from side to side after passing. As leader is overtaken, utters Karr-call (see 4 in Voice of *S. bergii*). Ground courtship (see 3, below) may follow. After dreads have progressed for a few days, High-flights begin, initially originating directly from dreads without intervening ground courtship. High-flight similar to that of *S. bergii* but much less synchrony in twists and turns during downward glide. After the glide, both birds may adopt aerial version of Erect-posture and perform paired display, characterized by 'butterfly flight', Pass-ceremony, and Karr-call (Paige 1960; K Hulsman). Frequency of aerial courtship decreases after onset of laying, but, when chicks c. 4–6 weeks old, marked resurgence of aerial display by failed and non-breeders, with crèche as focal point (K Hulsman). (3) Ground courtship. As pre-breeding period continues, collective upflights and associated aerial display increasingly followed by ground display, from which High-flights often originate. In ground courtship, ♂ lands in roost, and Struts about (Parades) in Forward-erect, offering fish to individuals. If a ♀ is receptive, both Parade in Forward-erect, and fish may or may not be transferred. Early in pairing, ♂ does not readily surrender fish to ♀ who may attempt to seize it. After Parade, ♂ may initiate High-flight, after which birds go off fishing or return to roost, adopting Erect-posture on landing, then relaxing to Forward-erect. If ♂ advertising with fish at roost attracts several birds to pursue, and then land beside him, these may display one to another, Parading and calling. In years when no breeding colony at One Tree Island, *S. bengalensis* present participated in this behaviour with *S. bergii* (Hulsman 1977a). Both members of pair participate in nest-site selection

(K Hulsman). (4) Mating. As in *S. bergii*. Begging by ♀ (see Courtship-feeding, below) will sometimes stimulate ♂ to approach in Forward-erect and mount instead of feeding her. First copulation attempts often fail, ♀ walking off and forcing ♂ to dismount. (5) Courtship-feeding. Continues from start of pairing, up to, and including, incubation (Clancey 1971a; K Hulsman), ♀ often soliciting food in Hunched-posture and uttering Begging-call (see 8 in Voice of *S. bergii*). (6) Nest-relief. Incoming bird gives Advertising-call in flight, alights near nest, and adopts Forward-erect; sitting bird rises and adopts same posture, and both posture for *c*. 5–15 s before relieved bird departs. Alternatively, before incoming bird alights, sitting bird may rise and adopt Forward-erect first; departing bird occasionally leaves before mate lands (Hulsman 1977a). RELATIONS WITHIN FAMILY GROUP. Parents appear not to recognize newly hatched young, but have strong nest-site recognition. After dispersal from nest (see Bonds, above), young recognized by voice, and visually at close quarters (Hulsman 1974). Both parents brood downy young. If crèche not formed, chicks scatter, and one parent remains in attendance while other forages. Crèche comprises young of various ages; at One Tree Island, one contained 85 young of 2–4 weeks old. As chicks age, crèches assemble on shoreline and range more widely, shifting with tide; if stranded on reef by incoming tide, young readily swim in compact formation, usually with a few adults flying overhead, calling. Chicks beg in crouched posture, similar to Hunched-posture of adult, and utter food-call (as in *S. bergii*); up to 3–4 weeks old, beg from any adult carrying fish, but by 5–6 weeks, only from own parents (Hulsman 1974). Fledged young follow at least one parent to feeding grounds where, at least initially, they spend much time standing on exposed reefs near foraging parent. Young mostly fed on ground, but if harassed by gulls, alight on sea to receive food from parent hovering overhead. During loafing, fledged young periodically practise feeding skills: drop and pick up inanimate objects floating on surface, and make zigzag flights with steep dives, climbs, and other evasive movements (Hulsman 1977a). ANTI-PREDATOR RESPONSES OF YOUNG. In response to Alarm-call of parent (see 1 in Voice of *S. bergii*), small chicks seek refuge in vegetation; older chicks in crèche close ranks, run, and may take to water. Closely approached crèche may scatter and re-group further off (Moltoni 1938; Hulsman 1977a). PARENTAL ANTI-PREDATOR STRATEGIES. (1) Passive measures. No information. (2) Active measures: against birds. Sitting bird threatens intruder much as for conspecific birds, lunging and pecking at gulls that approach too closely, but not leaving nest to attack (Hulsman 1977c); thus, gulls effectively warded off if nests close enough. Co-operative aerial mobbing less a feature of defence and attack than in some other *Sterna*. From airborne flock, attack by individuals on White-breasted Fish Eagle *Haliaeetus leucogaster* flying overhead, One Tree Island, ineffective in preventing it from taking chick of *S. bergii*. However, Lesser Frigatebird *Fregata ariel* attempting to rob bird of fish over crèche was mobbed and driven off by 10–15 others. Parents carrying fish to young, One Tree Island, klepto-parasitized by Silver Gull *Larus novaehollandiae*, either at sea immediately after fish capture, or at crèches as they attempt to deliver fish. Birds carrying fish seldom attack gulls, but sometimes do if dispossessed or after delivering food to chick; such attacks sometimes drive gulls off, but only briefly (Hulsman 1976). (3) Active measures: against man. Birds flush early from nest on approach of intruder and wheel overhead (Bark Jones 1946); swoop low over intruder, feinting as if to strike (Archer and Godman 1937). If crèche approached, guarding adults fly overhead, giving Alarm-calls, some diving at intruder, others circling above fleeing chicks (Hulsman 1977a).   EKD

Voice. Calls of adults and young apparently very similar to Swift Tern *S. bergii* though higher pitched (K Hulsman); see that species. Advertising-call (3) rendered 'kir-rit' (Clapham 1964), 'kriik' while fishing (Géroudet and Landenbergue 1977).   EKD

Breeding. SEASON. Libya: eggs, and young up to 3 weeks old on 21 August (Moltoni 1938). Red Sea: well-grown young in nests at end of August (Clapham 1964). Somali coast: breeding August to mid-September (Archer and Godman 1937). Persian Gulf: breeds May–June (Ali and Ripley 1969b); on Sheedvar Island (Iran), eggs found late June and late July (L Cornwallis). SITE. On ground in the open. Colonial. Nest: shallow scrape; unlined. Building: no information, but both sexes share in nest-site selection (Hulsman 1977). EGGS. See Plate 87. Sub-elliptical, smooth, not glossy; white, cream, or buff, variably spotted or speckled black or dark brown. 52×36 mm ($48–55 \times 34–37$), sample 40; calculated weight 35 g (Schönwetter 1967). Clutch: 1–2 (–3); clutches of 3 eggs presumably by 2 ♀♀ (E K Dunn); only 1 clutch of 2 seen for every 200–300 clutches of 1, Jebel Teir, Gulf of Aden (Bark Jones 1946). At One Tree Island (Great Barrier Reef, Australia), clutch 1 (K Hulsman). One brood. No information on replacements. INCUBATION. Generally 21–26 days (K Hulsman); up to 30 (Anon 1976). By both parents, Australia; mean of 17 spells $284 \pm 58 \cdot 1$ min (Hulsman 1977a; K Hulsman). YOUNG. Precocial and semi-nidifugous. Cared for and fed by both parents. Leave nest after a few days and sometimes scatter, one parent looking after each young, or may form crèche (Archer and Godman 1937; Moltoni 1938; Clapham 1964; Domm and Recher 1973; K Hulsman). FLEDGING TO MATURITY. Fledging period 32–35 days (K Hulsman). BREEDING SUCCESS. No information.

Plumages. ADULT BREEDING. Forehead, upper third of lores, crown down to middle of eye, and elongated feathers of nape black. Lower hindneck and upper mantle white, sometimes suffused grey. Lower mantle, scapulars, tertials, back, and all upper wing-coverts light grey, closely similar to colour of Common Tern *S. hirundo*, but sometimes slightly less bluish and slightly duller grey when worn; darker grey than Sandwich Tern *S. sandvicensis*; distinctly paler than Swift Tern *S. bergii*, except for *S. bergii thalassina*. Rump and tail similar to mantle and back or sometimes slightly paler grey, but not contrasting with back and never whitish; outer web of t6 and basal inner webs of all tail-feathers sometimes nearly white. Sides of head, underparts, axillaries, and under wing-coverts white. Outer webs of primaries pale silvery-grey with dark grey streak *c*. 5 mm wide along shafts of inner webs, remainder of inner webs white. Narrow white borders to tips of outer primaries, widening towards inner primaries. Secondaries light grey like upper wing-coverts and mantle, inner webs and tips broadly bordered white. In fresh plumage, inner 5–6 primaries slightly paler than older outer ones; contrast more marked when plumage worn. ADULT NON-BREEDING. Forehead and lores white, except for small black spot in front of eye; crown white with small black dots, varying in size and extent. Upper ear-coverts and nape black, slightly mottled white just behind and below eye. Remainder of body as in adult breeding, but upperparts often slightly duller grey; colour of flight-feathers and presence of contrast between inner and outer primaries or secondaries depends on wear and stage of moult. DOWNY YOUNG. Pale grey, pale buff, or off-white on head and upperparts, white below; variable amount of black spots on crown, nape, wings, and back, tending to form streaks on back (Hartert and Steinbacher 1932–8; Harrison 1975; BMNH). JUVENILE. Like adult non-breeding, but dark dots on crown less contrasting and white of forehead and crown often tinged buff when not too worn; grey feather-centres on upperparts and upper wing-coverts merge into off-white of feather-tips, latter sometimes tinged or freckled buff; variable number of feathers of lower mantle, scapulars, tertials, or inner upper wing-coverts with sepia or brown subterminal mark, but only exceptionally as heavily marked as juvenile *S. bergii* and Palearctic *S. sandvicensis*. Tail-feathers dark grey, inner webs and tips broadly bordered white; grey usually slightly darker subterminally, fading to pale sepia-brown when worn. Flight-feathers darker grey than in adult, tips and inner webs of secondaries and inner primaries more broadly edged white; outer webs of primaries less silvery. Indistinct carpal bar across forewing formed by darker grey centres to shorter lesser upper wing-coverts. FIRST IMMATURE NON-BREEDING AND SUBSEQUENT PLUMAGES. 1st immature non-breeding like adult non-breeding, but most primaries and wing-coverts, all secondaries, part of tail, and sometimes tertials and back to rump still juvenile up to February of 1st winter; these rather worn and tinged with brown, primaries mostly dark sepia, but a few inner usually contrastingly new. By May–June of 2nd calendar year, only some heavily worn juvenile outer primaries remain—plumage a mixture of old 1st immature non-breeding and 2nd immature non-breeding (both resembling adult non-breeding), no adult breeding acquired. Second series of primaries starts to moult from May–July (when moult of adults arrested); outer primaries new September–December, when adults have these old; indistinct dark grey dots on lesser upper wing-coverts usually present until 2nd winter, but sometimes lost by June of 2nd calendar year. Full breeding usually develops in spring of 3rd calendar year, but often much white on bases of black feathers of forehead, lores, and forecrown, especially visible when plumage worn.

Bare parts. ADULT. Iris dark brown. Bill orange-yellow when breeding, paler yellow for rest of year. Foot black, soles usually yellow. DOWNY YOUNG. Iris dark brown. Bill olive-yellow. Foot orange-yellow (Harrison 1975). JUVENILE. Iris dark brown. Bill pale olive-yellow to pale orange-yellow. Foot yellow, black gradually spreading across tarsus and over uppersurface of toes; like adult from age of 1 year, except for occasional yellow spots on joints or rear of tarsus.

Moults (northern-hemisphere populations). ADULT POST-BREEDING. Complete; primaries descendant. Starts with inner primaries, central tail-feathers, and scattered feathers on forehead, lores, crown, or scapulars from late July or August; forehead and crown do not usually start especially early (unlike *S. sandvicensis*). Head and much of body and tail in fresh non-breeding by October–November; moult completed with p10 late February to early March. ADULT PRE-BREEDING. Involves head, underparts, part of mantle and scapulars, tail, and inner primaries. Starts with p1 between late November and late January; head and body from late February; completed mid-March to early April, when primary moult arrested with inner (4–)5–6 primaries new. POST-JUVENILE. Complete. Starts November–January with head, body, central tail-feathers, and inner primaries. Mostly in 1st immature non-breeding by March–April, but outer primaries not until July–August. FIRST IMMATURE POST-BREEDING. Complete; as no 1st immature breeding attained, 1st immature non-breeding replaced directly by 2nd immature non-breeding. Inner primaries start May–July (before previous post-juvenile moult in outer primaries completed); all head, body, and tail new by October, outer primaries and inner secondaries by January–April. In contrast to adults, primary moult only occasionally suspended in early summer. Following pre-breeding and subsequent moults like adult.

Measurements. ADULT. Skins (BMNH, RMNH). Depth is depth of bill at basal corner of nostril; tail is to tip of t6; fork is tip of t6 to tip of t1 in adult breeding. Sexes combined (on average, bill length of ♀ 3·0 below ♂, bill depth at base 0·9 below, tarsus 0·9 below).

S. b. torresii. Mediterranean; May–August.

WING	314	(7·63; 6) 305–325	BILL	56·3 (3·95; 6) 50·7–61·2
TAIL	147	(7·81; 5) 132–158	DEPTH	12·0 (0·52; 6) 11·4–12·9
FORK	66	(4·15; 5) 60–70	TARSUS	26·9 (0·50; 6) 26·1–27·4

S. b. torresii. Persian Gulf and Pakistan; March–June.

WING	310	(7·07; 11) 303–321	BILL	55·7 (3·76; 10) 51·3–60·8
TAIL	146	(5·43; 9) 137–157	DEPTH	11·7 (0·49; 7) 11·2–12·7
FORK	69	(5·17; 9) 61–80	TARSUS	26·4 (0·61; 11) 25·4–27·3

S. b. torresii. Indonesia; August–May (wintering).

WING	309	(5·81; 26) 301–318	BILL	55·1 (2·21; 26) 51·3–59·3
TAIL	149	(7·05; 8) 139–159	DEPTH	11·6 (0·52; 27) 10·6–12·8
FORK	72	(6·94; 9) 61–83	TARSUS	26·0 (0·80; 37) 24·9–27·5

Nominate *bengalensis*. Red Sea; May–July (one ♂ Suez, early May, wing 316, bill 55·1, and depth 13·2 excluded—probably migrant or straggler from Mediterranean).

WING	296	(4·08; 12) 291–305	BILL	52·2 (2·39; 13) 47·5–55·3
TAIL	150	(5·89; 9) 138–160	DEPTH	10·9 (0·57; 12) 9·9–11·9
FORK	74	(6·71; 9) 65–86	TARSUS	25·3 (0·88; 12) 24·1–26·6

Nominate *bengalensis*. Southern and eastern India, Ceylon, Madagascar, and East Africa; July–March.

WING	297	(6·48; 12) 287–306	BILL	50·5 (1·65; 12) 47·5–55·3
TAIL	136	(7·39; 6) 125–145	DEPTH	10·7 (0·59; 9) 9·9–11·9
FORK	63	(4·73; 6) 56–69	TARSUS	25·6 (0·91; 12) 24·1–26·6

Toe of all *torresii* combined, 28·2 (1·68; 12) 27–31; of all nominate *bengalensis* combined, 26·8 (1·26; 16) 25–29. Tail and fork in adult non-breeding average 16 mm less.

JUVENILE AND IMMATURE. Average juvenile wing 20 mm less than adult, wing in late 2nd and early 3rd calendar year 8 mm less. Juvenile tail and fork 33 mm less than adult. Bill probably not full-grown until age of 2 years: average of 14 birds late 2nd calendar year still 3 mm below adult.

Weights. Kenya, April: ♂♂ 185, 190; ♀♀ 205, 235 (Britton 1970). Average weight, Australia, c. 240 (Serventy et al. 1971).

Structure. Closely similar to *S. sandvicensis* in all respects, but bill of *torresii* slightly heavier at base: depth at rear corner of nostril in ♂ 11·8 (0·37; 20) 11·2–12·8, in ♀ 10·9 (0·22; 7) 10·6–11·2; depth at base in nominate *bengalensis* similar to *S. sandvicensis*.

Geographical variation. Slight, mainly involving size. Mediterranean birds (named *emigrata* Neumann, 1934) distinctly larger and slightly paler on upperparts than nominate *bengalensis* from eastern and southern India, Ceylon, and westward to East Africa. As Australian and Indonesian birds (for which name *torresii* Gould, 1843 available) inseparable by size and plumage from Mediterranean breeders, latter have to be included in *torresii*. Colour of upperparts of breeders Red Sea and of Persian Gulf east to Karachi intermediate between typical nominate *bengalensis* and typical *torresii*; as colour differences too slight to warrant recognition of separate races, small-sized Red Sea birds included in nominate *bengalensis*, larger Persian Gulf breeders in *torresii*.

Forms superspecies with *S. sandvicensis*; probably also with slender-billed Elegant Tern *S. elegans* from western Mexico and heavy-billed Chinese Crested Tern *S. bernsteini* (synonym *S. zimmermanni*) from eastern China, as these both comparable in size and plumage with former group and complementary in distribution.
CSR

Sterna sandvicensis Sandwich Tern

PLATES 1 and 2
[between pages 134 and 135]

Du. Grote Stern   Fr. Sterne caugek   Ge. Brandseeschwalbe
Ru. Пестроносая крачка   Sp. Charrán patinegro   Sw. Kentsk tärna

Sterna Sandvicensis Latham, 1787. Synonym: *Thalasseus sandvicensis*.

Polytypic. Nominate *sandvicensis* Latham, 1787, coasts of western Europe, north-west Mediterranean, Black Sea, and Caspian Sea; *acuflavida* Cabot, 1847, coastal eastern USA and Caribbean, south to Bahamas, Cuba, and Yucatan, accidental in west Palearctic. Extralimital: *eurygnatha* Saunders, 1876, islands off Venezuela and coasts of northern and eastern South America south to Patagonia.

Field characters. 36–41 cm (bill 5·2–5·8, legs 2·7–3, tail of adult 11–14, of juvenile 9 cm); wing-span 95–105 cm. About 20% smaller than Royal Tern *S. maxima* and Swift Tern *S. bergii* but at least 10% larger and longer winged than Common Tern *S. hirundo*; closely matched in size only by Gull-billed Tern *Gelochelidon nilotica* and Lesser Crested Tern *S. bengalensis*. Long-billed sea tern, with slender and angular silhouette, looking noticeably bulkier than other common sea terns only on ground or at close range. Length of bill and head exaggerate short-tailed appearance; wings long and narrow. At all ages, noticeably whiter in appearance than all other terns except faded *S. maxima* and Roseate Tern *S. dougallii*. Adult shows long black, yellow-tipped bill and long black crown. Immature less boldly patterned than other medium-sized terns, again appearing mainly white; smaller and shorter-billed than adult, inviting confusion with *G. nilotica*. Markings on fresh primaries indistinct, but become dark and bold with wear. Contact call loud and far-carrying, grating in tone; that of immature distinctly higher pitched. Sexes similar; little seasonal variation. Juvenile and 1st-winter separable.

ADULT BREEDING. Whole of crown and nape, upper lores and line under eye jet-black, with feathers of rear crown and nape long and pointed (often blown up by wind). Lower face and whole of underparts, rump, and tail white, but rump and tail clouded grey when fresh. Back and most of upperwing pale ash-grey, latter fading to white on both edges; flight-feathers silver-grey, fringed white and only rarely showing obvious dark lines on outer primaries (formed by grey-black lines by shaft on inner webs) when fully spread. Underwing pure white except for restricted pale dusky panel on outer primaries (again formed by grey-black lines on inner webs); pattern lacks obvious dark tips to primaries noticeable on *S. hirundo* (and is closely matched only by that of *S. dougallii*). Outer tail-feathers pure white. On some, pink flush on lower underparts. Bill long (exceeding distance between base and eye), narrow, tapering, and lacking dagger shape of most terns; black with small yellow tip. Legs rather long, up to 50% more so than those of *S. hirundo*; black. ADULT NON-BREEDING. Forehead white, mid-crown mainly white speckled black, and rest of crown and nape black streaked white; long-headed appearance diminished but bill length even more obvious. Rest of upperparts usually fade to grey-white; all but innermost primaries become darker (due to loss of grey bloom and white fringes), showing broad dusky or almost black lines (above and below) and contrasting strongly with inner wing. Outer tail-feathers show grey streak at close range. Due to variable moult and bleaching, pattern of outer wing not constant. JUVENILE. Noticeably smaller than adult, with more compact silhouette and shorter, less narrow bill. At first, strongly patterned, with short blackish bill (yellow tip almost absent), dusky cap with a few white flecks, white collar on lower neck, sandy-buff back with dark brown-black flecks and bars particularly obvious on scapulars and tertials, and grey-white upperwing, rump, and tail (with almost black spots most obvious on longer coverts, and ends of flight-feathers and tail). Upperwing lacks dark leading edge obvious on *S. hirundo*. By mid-August, less strongly patterned, more closely resembling winter adult, though still with rather short, almost black bill, darker mid- and rear crown, many black spots and flecks on upperparts, and darker primaries. FIRST WINTER. Variegation of even faded juvenile plumage lost by December, as bird becomes fully grown and resembles winter adult except for less black on nape, and worn wings which often show narrow dusky leading edge, line of dusky marks across secondaries, and almost black primary coverts and primaries (contrasting with new pale grey central wing-coverts). Tail-feathers tipped black or dusky. FIRST SUMMER. Also resembles winter adult, though usually with more complete black crown; when completely moulted, distinction not safe except by grey outer tail-feathers on some. During often erratic moult, retention of dark outer primaries, narrow dark line on lesser coverts, and rather patchy appearance of otherwise grey upperparts obvious. Bare parts as adult. Few birds in such plumage appear in west Palearctic; no exaggerated '*portlandica*'-type plumage pattern recorded, but some have secondaries and inner primaries forming shallow white triangle, inside black primaries and behind grey coverts.

Adult unmistakable except at long range but juvenile and 1st-winter birds subject to much confusion with Gull-billed Tern *Gelochelidon nilotica* and even *S. dougallii*; see those species. In most conditions, wing-beats noticeably stiffer and less elastic than those of *S. hirundo*, and, particularly at longer ranges, seeming to come from further to rear of body (function of different silhouette, as noted above); action into strong winds more powerful. Flight turns less agile but display glides more dramatic (with more momentum) than in smaller terns. Gait free, with rather long legs allowing relatively high-stepping (not pattering) walk or run. Carriage variable; noticeably horizontal at rest but markedly upright when alert or in display. Feeding behaviour bold and dashing; dives usually from greater height and with less pronounced preceding hover, and submerges longer than smaller terns. Juvenile at first constantly begs food from adult on autumn migration, where loud duet of adult and juvenile is often first signal of presence off coasts.

In western Europe, migrants stray inland infrequently, unlike *S. hirundo*. Commonest call of flying adult loud, grating or rasping 'kireet', 'kirrik', or 'kirrink'; of accompanying juvenile a thinner, more plaintive 'chee-chee-chee'. Alarmed bird utters repeated, insistent 'krit' or 'krik'.

Habitat. Breeds in west Palearctic in 2 widely dispersed groups: (1) lower middle latitude Mediterranean steppe and desert zones on low-lying coasts of land-locked seas and other mainly saline waters; (2) upper middle latitudes of north-east Atlantic seas, infrequently on more exposed ocean coasts, and very locally (in Ireland) on islands in freshwater lakes. Support of colonies requires immediate access to clear water, rich in surface-level fish and usually fairly shallow with sandy bottom. Nest-sites either on inshore sandy islands or rocky usually calcareous islets, sand-spits, sand-dunes, shingle beaches, extensive deltas, or sandy eminences with dry bare patches among herbage, dense or tall vegetation being avoided (Voous 1960). Such sites often ecologically unstable, and liable to become unsuitable through vegetation growth, inundation, erosion, blown sand, or other natural causes, as well as through human disturbance. These factors believed to account for sudden conspicuous fluctuations in numbers and site changes, especially in north-west Europe (Marples and Marples 1934; Smith 1975a). Protection against predators apparently also sought through proximity of more aggressive colonial breeding *Sterna* and gulls *Larus*. In USA, colonies almost invariably associated with those of Royal Tern *S. maxima* (Buckley and Buckley 1972a; Blus *et al.* 1979). One of these, Grand Cochère islet (Louisiana) is 183 km offshore. Inland breeders on Irish loughs bring back food from marine fishing grounds 20 km or more distant; in Scotland evidence of foraging to *c.* 70 km from colony (Cramp *et al.* 1974).

Outside breeding season, prefers warm waters, travelling sometimes well offshore to winter close inshore in marine or estuarine waters along coastlines, including mudflats fringed by mangroves, sandy beaches, and rocky shores (Dunn 1972b). American birds roam about outer islands and sand-bars in company with other terns and gulls, following schools of fish or resting in flocks on sand (Bent 1921). In some circumstances, as in display, rises into upper airspace, and normally flies somewhat higher over water than most seabirds. Adults swim rarely. Sensitive to disturbance at breeding places and benefits immediately from effective conservation measures, including provision of artificial breeding sites (Axell 1977).

Distribution. Sites frequently changed because of disturbance and habitat changes. Some signs of recent spread, especially in Baltic.

BRITAIN AND IRELAND. Colonies often unstable due to disturbance or habitat changes; breeds occasionally Shetland, Isles of Scilly, and Channel Isles (Cramp *et al.* 1974; R Long). FRANCE. Some sites deserted but major new colony Bassin d'Arcachon since 1966 (Yeatman 1976; Campredon 1978). SPAIN. Bred from 1961; also nested coast of Asturias 1971 (AN). WEST GERMANY. Breeding from at least early 19th century; sites often changed; see Schmidt and Brehm (1974). NORWAY. 1–5 pairs nested 1974–6 (SH); 1 pair 1981 (Toft 1983). SWEDEN. First bred 1911 and spread (SM). POLAND. Bred near Gdańsk 1929–36 and 1977–82 (Tomiałojć 1976a; AD). EAST GERMANY. First bred 1957 and spread (Klafs and Stübs 1977). ITALY. Possibly bred Sardinia in 19th century (Schenk 1976). Bred Comacchio 1979 (Brichetti 1979). Bred Evros delta 1979 and Lovras delta 1980 (HJB, WB, GM). GREECE. Occurs regularly in breeding season Evros delta but no proof of breeding (HJB, WB, GM). USSR. Sites frequently varied with habitat changes; for Volga delta and Caspian, see Poslavski and Krivonosov (1976). KUWAIT. Present all year; no proof of breeding (PRH). ALGERIA. Said to have bred mid-19th century (Heim de Balsac and

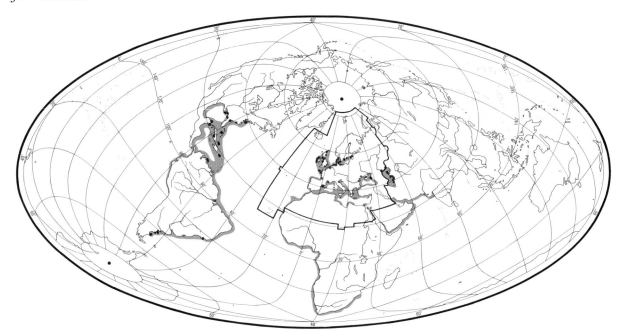

Mayaud 1962). TUNISIA. Said to have bred late 19th century, and probably bred 1959 (Heim de Balsac and Mayaud 1962); no recent evidence (MS). MADEIRA. For discussion of reported breeding see Bannerman and Bannerman (1965).

Accidental. Jan Mayen, Finland, Czechoslovakia, Hungary, Lebanon, Cyprus, Azores, Madeira.

Population. Fluctuating numbers and frequent change of sites make trends often difficult to ascertain; after probable declines in 19th century, has recently increased in several areas of northern and western Europe, helped by protection, though large population in Netherlands still below previous peaks after pesticide deaths in 1960s.

BRITAIN AND IRELAND. Considerable fluctuations, but decreased 19th century and increased markedly 20th century. Perhaps nearly 6000 pairs 1962, when probably highest in 20th century (Parslow 1967); 11 860 pairs 1960–70 (Cramp et al. 1974); over 14 300 pairs 1971, 12 267 pairs 1974 (Lloyd et al. 1975), 13 484 pairs 1977, 14 416 pairs 1978, and 15 464 pairs 1979 (Thomas 1982). FRANCE. Under 4000 pairs (Yeatman 1976); nearly 5500 pairs 1978 (J-YM); 5300 pairs 1979 (Thomas 1982). Marked fluctuations: e.g. in Camargue increased from 3 pairs 1956 to 1050 pairs 1976 and 700 pairs 1979 (Isenmann 1972a; A Johnson); in Bassin d'Arcachon c. 1200 pairs 1966, 2810 pairs 1976 (Campredon 1978); in Brittany declined by 58% over 1968–78 (J-YM). SPAIN. Bred, at first irregularly, Ebro delta from 1961; c. 150 pairs 1979 (AN) and 23 pairs 1980 (A Martinez). NETHERLANDS. Increased early 20th century, fluctuated between 25 000 and 40 000 pairs 1940–57 (Rooth and Mörzer Bruijns 1959), declined to 12 000 pairs 1961–2, 2500 pairs 1964, and 650 pairs 1965 due to poisoning by chlorinated hydrocarbons (Koeman 1971), then slow increase to 1500 pairs 1968, 2900 pairs 1971, and 6100 pairs 1978 (Rooth and Jonkers 1972; Teixeira 1979; D A Jonkers, J Rooth). WEST GERMANY. Fluctuating from 2000 to 6000 pairs 1954–78 (Smit and Wolff 1981); c. 7000 pairs 1979 (Thomas 1982). DENMARK. Fluctuating, some decrease since 1940s; c. 4000 pairs (Mardal 1974), now varying c. 2000–4000 pairs (TD). SWEDEN. Fluctuating, but increased since colonization; c. 400 pairs (Rooth and Jonkers 1972), c. 500 pairs (Ulfstrand and Högstedt 1976), c. 1100 pairs (SM). EAST GERMANY. Increased; 27 pairs 1957, 290 pairs 1961, 711 pairs 1971, then little change to 1975 (Klafs and Stübs 1977); c. 950 pairs 1977 (Thomas 1982). POLAND. About 300 pairs 1982 (AD). ITALY. 53 pairs 1981 (Brichetti and Isenmann 1981); 82 pairs 1982 (PB). RUMANIA. Declined since 19th century (Vasiliu 1968). USSR. Estonia: first bred 1962; c. 500 pairs 1968 (HV); c. 100 pairs 1973–7 (Thomas 1982). Volga delta: for fluctuations 1963–71, see Poslavski and Krivonosov (1976). About 26 000 pairs Black Sea and under 1400 pairs Sea of Azov (A N Golovkin).

Oldest ringed bird 23 years 7 months (Rydzewski 1978).

Movements. Migratory throughout west Palearctic. Movements reviewed by Dircksen (1932a), Thomson (1943), Müller (1959), Langham (1971), Ardamatskaya (1977), and Møller (1981a, b); information also supplied by C J Mead, largely from ringing studies. West European, East German, and probably Estonian birds share similar winter quarters, mainly on west coast of Africa from Mauritania south to Cape of Good Hope, South Africa (Møller 1981a, b; C J Mead). Birds from Black Sea winter principally in eastern Black Sea and

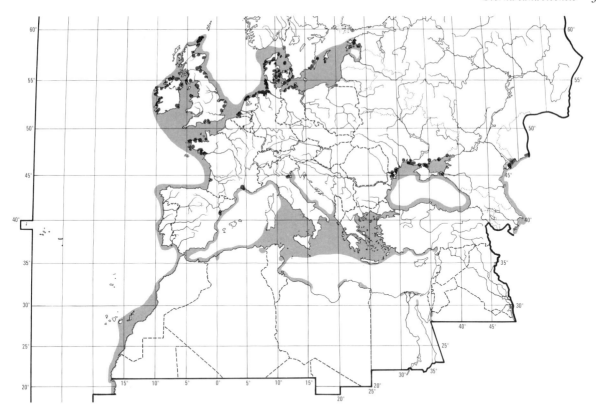

central and south-east Mediterranean; also along coasts of Spain and Portugal, occasionally reaching West Africa (Müller 1959; Ardamatskaya 1977). Caspian Sea population has separate winter quarters, mainly in Persian Gulf and Arabian Sea (Müller 1959; Erard and Etchécopar 1970); birds recovered in Iran (April, October), Persian Gulf (January, May), and Ceylon (December). Passage negligible, Red Sea. On Masirah island (Oman), common throughout the year; numbers at peak August (Griffiths and Rogers 1975; F E Warr). In Dhofar (Oman), 1978, autumn passage started 8 September, increasing to 1000 or more birds on 4 October, and declining to 200 on 14 November (Walker 1981b). Winter flocks at Masirah often up to 1000, exceptionally 2000 (Griffiths and Rogers 1975). Spring passage, northern Oman, January to end of June; in 1976, peak c. 2500 on 23 April (Walker 1981a). Presumably some movement along south Arabian coast, since regular on passage in Aden (Bailey 1966a); this the likely origin of birds reaching Kenya and Tanzania irregularly or in very small numbers (Harvey 1972; Britton 1977b). Birds entering Gulf of Aden may be those which subsequently occur on spring passage in Gulf of Aqaba (Pihl 1978) and Azraq, Jordan (Nelson 1973), taking alternative return route to Caspian. North American *acuflavida* said to winter south to Argentina and Ecuador (Schauensee 1970; Blake 1977), though wintering from Brazil south doubted by Voous (1977b). Juvenile *acuflavida* ringed North Carolina recovered Netherlands, December (Scharringa 1979). Northern populations of South American *eurygnatha* partly winter to south; birds from Patagonian colonies probably move north (Schauensee 1970; Blake 1977; C S Roselaar).

EUROPEAN POPULATIONS. Differences in distribution due to age exceed any due to breeding area (Müller 1959). Seasonal movements within Gulf of Guinea described by Grimes (1977), recoveries in South Africa by Elliott (1971). Spring migration probably follows autumn routes in reverse, though sightings in south-east Morocco, early April, suggest some inland Saharan movement on return passage (Smith 1968b). (1) Fledging to 2nd winter. Post-fledging dispersal of family groups to traditional feeding grounds begins late June and typically precedes southerly migration. Dispersal from breeding grounds may be rapid, e.g. juveniles found up to 65 km from colony 3 days after fledging (Smith 1975a). Colonies furthest north tend to have highest dispersal rates, and commence migration first (Møller 1981a). In July–August, birds from Britain and Ireland are recovered almost equally north and south of natal colonies (Langham 1971); some disperse to Netherlands, and vice versa (Møller 1981a). Northward dispersal also occurs in other north European countries (Müller 1959; Speek 1969) but no preference for this direction (A P Møller). Baltic birds, of all ages, generally go round north tip of Denmark (Müller 1959). French birds from Gironde disperse principally north (Campredon 1977, 1978); from Camargue, partly east as far as Adriatic (Isenmann 1972a).

In September, British and Irish juveniles remain equally distributed north and south within 480 km of natal colonies, but beyond this limit, southward migration predominates. Movement follows coasts of Netherlands, France, and Iberia. Irish juvenile in Ghana as early as 2 August (Spencer and Hudson 1980). By early September, some have moved 1600 km to Iberia, 10% in Sénégal, and a few reach Ghana (Langham 1971; Møller 1981a). Sight records indicate that a proportion of birds from Europe stop at Banc d'Arguin (Mauritania) and winter there from September onwards; over 29 000 December 1978 to January 1979 (Pététin and Trotignon 1972; Gandrille and Trotignon 1973; Dick 1975; Trotignon et al. 1980). In October, recoveries of British and Irish birds extend from France to Angola—mostly 0–20°N, with 30% of recoveries in Sénégal (Langham 1971; A P Møller). By November most juveniles are in tropical West Africa, few remaining in Europe; median distances of British and Irish birds from colony are 500 km in September, 4600 km in October, 5300 km in November (A P Møller). In December, rare within 500 km of colony; most are at 0–10°N, notably in Sierra Leone, Liberia, Ivory Coast, Ghana, and Togo. In December–May, median distance of British birds from colony is 5300–5400 km (A P Møller). Significantly more birds from Denmark, western France, and West Germany (compared with Britain, Ireland, and Netherlands) winter in West Africa north of Gulf of Guinea. Winter distribution of Swedish and East German birds similar to distribution of Danish (etc.) populations (Tåning 1944; Schloss 1966; Rosendahl and Skovgaard 1971; Møller 1981a). Recoveries partly reflect density of human population and popularity of trapping, 63% of British recoveries from West Africa coming from vicinity of large towns (Allison 1959; Langham 1971). In particular, extent of wintering in Angola and Zaïre may be over-represented (C J Mead). Notable paucity of recoveries from Nigeria, Cameroun, and Gabon reflects genuine scarcity of wintering birds (of all ages), probably associated with absence of upwellings and associated fish stocks (A P Møller). Gambia, Guinea, and Guinea-Bissau also apparently avoided (C J Mead). Upwellings off Sénégal and Ghana used especially in 1st winter, less so in 2nd and 3rd, whereas Benguela upwelling (south-west Africa) used in 2nd and 3rd winters and to even greater extent subsequently; some birds may spend 1st year in West African waters and move to southern Africa in 2nd year (Møller 1981b). Of birds ringed Britain, Ireland, Netherlands, West Germany, and Denmark, 10–15% of 1st-winter recoveries are from southern Africa (most Danish, least Netherlands); proportion of birds spending 1st winter in southern Africa appears to be de-creasing (Møller 1981b). 1st-winter British bird recovered St Lucia Estuary on east coast of South Africa (Langham 1971). A few recovered in European, notably Iberian, waters throughout 1st winter (see below). Great majority of 1-year-olds remain in winter quarters during summer; small proportion move north, mainly to Mediterranean or Iberia, and a few further north (Müller 1959; Langham 1971; Isenmann 1972a. In May, only 4·5% of recoveries of 1-year-olds from Britain, Netherlands, West Germany, and Denmark are within 500 km of natal colony. Several records on English south coast, July–August (A J M Smith), and 1-year-olds occasionally seen at colonies, e.g. Grampian, Scotland (A J M Smith), and Gironde (Campredon 1977). (2) 2 years old and older. Some 2-year-olds remain in tropics throughout summer, but majority migrate to European waters, many reaching colonies about June when, exceptionally, they breed (Veen 1977; A J M Smith). In autumn, most return to tropics. Before moving south, these and other immatures exhibit dispersal movements resembling those of juveniles. At 3 years old, minority still summer in tropics, but majority migrate to breeding grounds, arriving at colonies from late April (Denmark) to May (Britain and Ireland) (Langham 1971; Rosendahl and Skovgaard 1971). Variable proportion breeds, depending on colony (e.g. Nehls 1969). Bulk of 2- and 3-year-olds from Denmark and West Germany present at colonies earlier (May–September) than those for Britain, Ireland, and Netherlands (July–October). Depending on country of origin, 31·8–45·2% of 2- and 3-year-olds were within 500 km of natal colony in June (Møller 1981b). British 3-year-olds mostly at 40–60°N, June–September, thereafter returning to winter quarters (Langham 1971). Most birds of 4 years and older visit colonies; British birds begin to leave wintering grounds in February, with recoveries mainly north of 40°N June–October (Langham 1971); 80–100% of recoveries of European birds of this age were within 500 km of natal colony in June, though birds of up to 7 years old occasionally occur in winter quarters then (Møller 1981b). Adults occupy much the same winter range as younger birds, though some evidence suggests they stay further north (Langham 1971). British adult recovered Moçambique, January, is most northerly recovery of any age class on east coast of Africa. Exceptionally, some birds winter in Europe, apparently more so in some recent years, e.g. Netherlands (Ouweneel 1975, 1979) and Scotland (Prato et al. 1981); of 62 recoveries of British, Netherlands, Danish, and West German birds in Europe, December–February, 48% were 1st-winter, 18% 2nd–3rd, 34% 4th and older (Møller 1981b). About 50 birds, mainly adults of unknown origin, regularly winter at Gironde (Campredon 1977), while c. 700 remained off Portugal in January 1975 and over 300 in January 1976 (A J Prater and A Grieve). Minimum of 3000 birds estimated to winter in Mediterranean (Vilagrasa et al. 1982).

First-time breeders regularly settle far from natal colony. Of 93 birds caught during 1965–8 at new colony, Langenwerder (East Germany), 41 were from countries throughout north European range, though mostly of Baltic origin; about 50% were first-time breeders, but recovery of birds of up to 16 years old indicates that experienced breeders also change colony-site (Nehls 1969). British birds, probably first-time breeders, recovered at colonies in Camargue (Isenmann 1972a) and Rumania (Spencer and Hudson 1978b); also 1 Danish bird in Ukraine (Møller 1981a).

BLACK SEA POPULATION. Based on Müller (1959) and Ardamatskaya (1977). Post-fledging dispersal (late July to late October) from north coast of Black Sea takes most juveniles (with parents) south, often via Rumanian coast, through Sea of Marmara to Mediterranean. Large numbers also move directly east from breeding grounds, producing recoveries in eastern Sea of Azov in September, and occasionally inland. Considerable numbers winter on eastern shore of Black Sea, and recoveries there continue into following summer. On reaching Mediterranean, majority of juveniles disperse west and south (some crossing Greece *en route*), less commonly east. 1 juvenile reached Egypt on 3 August, travelling over 3000 km in 14–18 days. 80% of all recoveries throughout the year (mostly 1st-winter) are in western Mediterranean from Gibraltar to Italy and Tunisia, especially Algeria; less commonly (14%) east of this, especially on south coast of Mediterranean (Ardamatskaya 1977; Jacob 1979). 6% of recoveries (all ages) are west of Gibraltar, notably in Portugal and Morocco, occasionally further south. Distribution along north Mediterranean coast well documented (Besson 1969a; MacIvor and Navarro Medina 1972; Isenmann 1976c, 1980; Isenmann and Czajkowski 1978). By January–February, some evidence of easterly shift in Mediterranean by adults and immatures. Major route during spring migration, which begins March, is across Greece and Balkans (Ardamatskaya 1977). Up to 2nd winter, birds mostly frequent coasts of Iberia, France, Morocco (Mediterranean), Algeria, and Tunisia (Tait 1960, 1961; Ardamatskaya 1977). Birds are recovered in winter quarters during July–August up to 3 years old. Exceptionally, adults winter in West Africa, reaching Ivory Coast (Ardamatskaya 1977). 1 bird ringed Odessa (Ukraine) recovered in Danish colony 3 years later (A P Møller). EKD

Food. Chiefly surface-dwelling marine fish. Hunts by plunge-diving, usually from 5–10 m; mean of 346 dives, Coquet Island (Northumberland, England), 5·6 m (Dunn 1972a). Dives mostly vertical, sometimes angled; often preceded by hovering, though this less frequent than in smaller *Sterna* (Taylor 1975; Campredon 1977; E K Dunn). In especially high dives, orientation and impact on entry may be adjusted by a few wing-beats during descent. Immersion usually just complete, but only partial when prey or visibility of prey restricted to surface; probably not deeper than 1·5 m (Dunn 1972a) or 2 m (Borodulina 1960), deeper than most Palearctic *Sterna*. Immersion time directly proportional to dive height and length of prey captured; mean time 1·3 s (sample 241), rarely exceeding 2 s (Dunn 1972a); maximum 3 s (Campredon 1977). No evidence that prey pursued underwater. Prey always brought to surface held crosswise in bill unless very small; immediately swallowed head first unless intended for mate or young. Long pendulous fish, especially sand-eels *Ammodytes*, generally carried just behind head with rest of fish's body orientated away from direction of side wind (R Chestney). Usually only one fish caught but successive and simultaneous captures of more than one recorded, and multiple loads of up to 4 sand-eels observed in 2·2% of fish-carrying birds at Coquet Island (Hays *et al.* 1973; Taylor 1975). Success of dive affected by size, abundance, depth, and visibility of food, and age and experience of bird. In winter quarters (Sierra Leone), mean fishing success (proportion of completed dives yielding prey) of 1st-winter birds (then 7–9 months) 13·4%, significantly less than older birds (16·6%); no difference in diving rate (1·7 completed dives per min) or size of fish captured. 1st-winter birds took average 10 fish per hr of sustained foraging, older birds 14 per hr. Dive height and fishing success increased between fledging and end of 1st winter, but not quite to level of older birds (Dunn 1972b). In breeding season, Northumberland (Dunn 1973a), fishing success and overall capture rates positively correlated with wind speed up to *c*. 26 km per hr such that 1 km per hr increase associated with 0·16 more fish caught per min; can fish at wind speeds up to 80 km per hr, though with reduced capture rate (Campredon 1977). Sea surface conditions apparently had independent influence, moderately choppy surface yielding higher success than either flat calm or very rough water (Dunn 1973a, 1975). At Ythan estuary (Scotland), however (Taylor 1975, 1983), success (including aborted dives) not altered by increasing wind speed up to 56 km per hr, though diving rate decreased, causing 0·4 prey per min reduction for each 16 km per hr increase in wind speed; in high winds, birds congregated in sheltered areas to fish. Unlike open sea, therefore, windless conditions apparently optimum. See Common Tern *S. hirundo* for similar results. Along seashore or over offshore reefs and sand-banks, fishing success and diving rate 2–3 times higher (and more birds fish) at low tide—when water shallower and fish more accessible—than at high tide (Dunn 1972a). At Ythan estuary, including aborted dives, fishing success (54%) and diving rate highest on rising spring tides (0·63 prey per min) when prey penetrate furthest up estuary; at neap tides, 0·14 prey per min. Numbers feeding peaked at high spring tides (Taylor 1975). Tidal and diurnal patterns interact; generally feeds most actively early morning and late afternoon or evening (Pearson 1968; Dunn 1972a; Campredon 1977). Shoals of fish may attract large flocks and lead to higher diving rate and fishing success at sea. For complete dives, mean fishing success of flock fishing at sea was 45·6% (480 dives); 27·1% (427 dives) by solitary individuals on same 4 days (Dunn 1972a). At Ythan estuary, aerial interference in high flock densities increased rate of aborted dives and reduced fishing success (Taylor 1975).

Less versatile in fishing techniques than smaller *Sterna*. Hawking for insects and robbing fish from other *Sterna* not recorded in west Palearctic, though former not uncommon in North America (Blus *et al.* 1979). Taylor (1975) found increased use of 'partial submersion plunges' as wind speed increased; less apt than *S. hirundo* to switch to dipping-to-surface, though does so to catch crustaceans and polychaete worms. Dipping-to-surface and shallow dives (from 1·5–3 m) common and highly successful (66–90%) in Sierra Leone where birds took fish escaping from nets hauled ashore, or driven to surface by predators, especially tuna *Thunnus* (Dunn 1972b). Dipping thus for fish left on beach or cast out on line predisposes them to snaring (Bourne and Smith 1974; Mead 1978). For juveniles, dipping-to-surface is first step in learning plunge-diving (Dunn 1972b). In Bay of Biscay (France), catches fish driven to surface by bass *Morone labrax* (Campredon 1977), and, in Britain, by porpoises *Phocoena phocoena* (E K Dunn). Recorded scavenging at fishing boats in Irish Sea (Watson 1981). In breeding season, most feed within a few km of colony, but may travel much further to obtain food for young (e.g. Pearson 1968, Andrews 1971, Campredon 1977); *c.* 67 km recorded (see Cramp *et al.* 1974).

Predominantly small marine fish, including, in North Sea and North Atlantic, sand-eels (*Ammodytes marinus*, *A. tobianus*, *A. lanceolatus*, *A. americanus*), herring *Clupea harengus*, sprat *C. sprattus*, whiting *Gadus merlangus*, cod *G. morhua*, Norway pout *G. esmarkii*, rockling *Onos cimbrius*, sticklebacks (*Gasterosteus aculeatus*, *Spinachia vulgaris*), butterfish *Pholis gunnellus*; occasionally, Salmonidae and small Pleuronectidae. Polychaete worms (Annelida) locally important. Shrimps *Crangon* and squid *Loligo* occasionally taken; also insects (unidentified) in North America. (Niethammer 1942; Dementiev and Gladkov 1951c; Koeman *et al.* 1967; Langham 1968; Pearson 1968; Chestney 1970; Dunn 1972a; Cramp *et al.* 1974; Fuchs 1977a; Veen 1977; Blus *et al.* 1979). On east coast of North America, almost wholly small fish such as mullets (Mugilidae), sand-eels, young garfish (Belonidae), and occasionally shrimps or squid (Bent 1921). In Sea of Azov (Black Sea), adults fed almost exclusively on sand-smelt *Atherina*. At Chernomorski (Black Sea), mainly on gobies (Gobiidae); occasionally, isopod *Idotea baltica*, prawn *Leander*, bush-cricket *Tettigonia caudata*, and beetle *Anisoplia austriaca* (Borodulina 1960). In Camargue (France), chicks were fed consistently on sardines *Sardina pilchardus* (Lévêque 1957; Isenmann 1972a). Other Mediterranean species from stomach contents include *Cepola*, sand-smelt, pandora *Pagellus erythrinus*, and squid (Terry 1952; Isenmann 1972a). At Banc d'Arguin (France), 1966, main prey fed to young was sand-eel *A. tobianus*; also anchovy *Engraulis encrasicholus* and scad *Trachurus trachurus* (Davant 1967). At Banc d' Arguin, 1975, 39 regurgitated fish comprised 77% anchovies, also sand smelt *Atherina presbiter*; in 1976, of 318 fish, 74% anchovies, 21% sand-smelt, *c.* 2% sand-eels (Campredon 1977).

Most quantitative data for North Sea populations. Stomachs from 9 adults (7♂, 2♀), Norfolk (England), contained (by number) 35% sand-eels, 31% other 'food' fishes, 33% Annelida, 1% marine molluscs (Collinge 1924–7). From over 900 sight records and regurgitations, prey fed to chicks at Farne Islands comprised (by number) 74% Ammodytidae, 15% Clupeidae (all *C. harengus*), 6% Gadidae, 2% Gasterosteidae, and 1% each of *Pholis gunnellus*, crustaceans, and cephalopod molluscs. Relative proportions of Ammodytidae and Clupeidae vary through season, and between years and localities. At Coquet Island (50 km further south), Langham (1968) found 88% Clupeidae and 11% Ammodytidae in 1965, but 54% and 46% respectively in 1966. Dunn (1972a) found about equal proportions at Coquet Island, as did Taylor (1975) at Ythan estuary. At latter colony, 5% Gadidae recorded (Fuchs 1977a) and numerous tags from salmon smolt and silvering parr (Salmonidae) found (see Cramp *et al.* 1974). At all these east-coast colonies, Ammodytidae usually predominate April–May, Clupeidae in July, and Ammodytidae sometimes again in August; the switch to Clupeidae in July, and growth of these fish through summer, yields estimated 4-fold increase in wet weight of prey caught per unit time from late May to early August (Taylor 1975). Thus, sand-eels more commonly used in courtship and supplied to incubating mates than later on to small chicks (Fuchs 1977a), though selectivity may override seasonal availability: e.g. Taylor (1975) found that of Ammodytidae and Clupeidae caught, Clupeidae preferentially brought back for incubating mates; Langham (1968) found young chicks fed on small, easily swallowed sand-eels, but given more Clupeidae later on. Weather also affects diet, e.g. sand-eels predominate at Scolt Head Island (Norfolk) in bad weather, whitebait *Clupea harengus* or *C. sprattus* at other times (R Chestney). Of 33 stomachs, 1904–23, South Carolina (USA), anchovies *Anchoviella* occurred in 27·3%, silversides *Menidia* 15·2%, rough silverside *Membras vagrans* 9·1%, menhaden *Brevoortia* 9·1%, other fish 12·1%, shrimps *Peneus* 36·4%, squid *Loligo* 3·0%, insects 18·2% (Blus *et al.* 1979). In winter quarters, Sierra Leone, 3 stomachs contained sardines *Sardinella eba*, *Brachydeuterus auritus* (Pomadasyidae), and a mullet; anchovy *Engraulis guineensis* also taken (Dunn 1972b). In Ghana, sardines, mainly *Sardinella aurita* are important, June–November (Grimes 1977).

Size of prey more variable, and significantly larger fish caught and fed to chicks, compared with sympatric *Sterna* in summer quarters; at Coquet Island, mean length 12 cm (sample 515) in 1965 (Langham 1968), 6·7 cm in 1969 (Dunn 1972a); at Ythan estuary, Ammodytidae up to 10·5 cm, Clupeidae maximum monthly mean 12·8 cm (Taylor 1975). Of 130 fish seen caught, Coquet Island, those swallowed immediately often smaller (mean 5·3 cm) than those taken back for chicks (mean 7·6 cm) (Dunn 1972a); similarly for incubating mates, Ythan estuary (Taylor 1975); very young chicks, however, fed smaller fish (Langham 1968; Fuchs 1977a). Stomachs of adults at Sea of Azov contained only 15-cm *Atherina* (Borodulina 1960). At Farne Islands, average prey weight (calculated from regression equation) 2·8 g (Pearson 1968). In winter quarters, Sierra Leone, 3 stomachs contained 4 fish of 3–10 cm (mean 7·1); in 134 sightings, those caught were generally shorter—5 cm estimated for both 1st-winter and older birds (Dunn 1972a, b).

At Chernomorski (no brood size specified), chicks fed 8–12 times per day, average 0·5 times per hr, maximum 2 per hr. In Camargue, each chick fed on average 15 times per day on prey 5–8 cm long, each weighing 7–11 g, a daily intake of c. 125 g (Isenmann 1975a). At Banc d'Arguin (France), each chick fed 7–8 times per day, on prey 4–15 cm long, each weighing 8–9 g, total daily intake c. 65 g (Campredon 1977). At Farne Islands, broods of 1 fed 14 times per day on items of c. 3 g each—daily intake 42 g (Pearson 1968). Not clear how much of this variation between studies due to experimental error.   EKD

Social pattern and behaviour. Compiled largely from Smith (1975a), Campredon (1977), and Veen (1977).

1. Gregarious throughout the year. When feeding, mostly solitary, though forms flocks, occasionally of hundreds, when prey abundant and concentrated (E K Dunn). During post-breeding dispersal, large assemblages of adults and juveniles roost near good feeding grounds: e.g. over 5000, Black Sea (Ardamatskaya 1977); c. 12 000 at Scolt Head (England), end of July to beginning of August (Chestney 1970). Passage flocks in autumn at Bosporus (Turkey) of 3 to c. 20 birds (Kumerloeve 1980), often comprising family parties. Large flocks (e.g. over 1500) recorded on migration, Banc d'Arguin (Mauritania), at end of November (Dick 1975). On spring migration at Beachy Head (England), 5620 flocks over 7 years averaged 3·1 birds, 91% in flocks of 12 or less (M J Rogers). Many groups of 2 are probably established pairs (E K Dunn). Associates readily with other Sternidae and gulls (Laridae), especially Black-headed Gull *L. ridibundus*. BONDS. Monogamous pair-bond, persisting from year to year. One pair nested in 4 successive seasons (Smith 1975a). Not known if pairs maintain close contact in winter quarters but, from ringing, occurrence of birds from same colony in same winter locality suggests this possible (E K Dunn). Colonies sometimes visited by 1-year-olds, commonly by older immatures, usually from mid-June; these court, make scrapes, and regularly attempt to feed young of established breeders (Smith 1975a). First breeding occasionally at 2 years, mostly 3–4, sometimes not until 5 (Nehls 1969; Campredon 1977; Veen 1977; A J M Smith). In some years, no 2-year-olds bred at colony, Grampian, Scotland (A J M Smith). Birds tend to pair with other of same age (Veen 1977). Birds which have bred before generally return to colony earlier, and breed earlier than first-time breeders (Nehls 1969; Veen 1977). Most arrive at colony already paired; in colony at Griend (Netherlands), 70% in pairs by 20 April (Langham 1974; Veen 1977). Birds in main influx are almost always paired, but many early arrivals are unpaired or have pair-bond poorly developed (Smith 1975a). During incubation, 3rd bird, of either sex, may attach itself to established pair, and incubate, though attachment does not persist until hatching. One case of 2 ♂♂ and 2 ♀♀ thus associated (Smith 1975a). Not uncommon, especially among 3-year-old first-time breeders, for 2 ♀♀ to lay in same nest. Both sexes incubate (♀ taking greater share early on) and care for young (Veen 1977). If brood of 2 becomes separated after leaving nest, parents may divide brood care, each feeding 1 chick (Smith 1975a). Young may form crèche, sometimes exceeding 1000 birds, from c. 14 days (Campredon 1977; Veen 1977); crèche chaperoned by several adults (usually c. 1 adult to 10 young), though young fed exclusively by own parents (Steinbacher 1931; Smith 1975a). Dependence on parental feeding continues at least until October in winter quarters, possibly longer. Young wintering in Scotland fed by adult 31 January and 14 March (Prato *et al.* 1981); begging seen late January, West Africa (Dunn 1972b). Recorded pairing with Lesser Crested Tern *S. bengalensis* over 3 successive seasons; young seen being brooded (Géroudet and Landenbergue 1977). BREEDING DISPERSION. Densely colonial. If colony-site is disturbed, especially by mammalian predators early in season, birds readily quit area and settle elsewhere. However, fickleness sometimes overstressed—withstands considerable disturbance once incubation begun. Colony sizes in Britain and Ireland, 1975–9, ranged from 2 to c. 4000 pairs, mean 853 in 1979 (G J Thomas); similar range over most of north-west Europe and North America, though maximum usually less (Blus *et al.* 1979). Prior to 1960, Netherlands, one colony of up to 35 000 pairs in some years (C S Roselaar). Colony at Smalenyi Island (Black Sea) 1258 pairs (Borodulina 1960), at Osushnoi Island (Krasnovodsk Gulf, USSR) 1841 nests (Poslavski and Krivonosov 1976). Typically forms sub-colonies, laying highly synchronized within each; large sub-colonies contain spatially distinct sub-groups distinguished by synchronized laying dates (Langham 1974). At Griend, sub-colonies of 15–26 pairs (Veen 1977); at Osushnoi Island, 9 sub-colonies of 11–478 pairs (Poslavski and Krivonosov 1976). At Coquet Island (England), mean number of pairs in sub-colonies, 1965–7, was 29·4, 56·9, and 94·8 respectively (Langham 1974). Bad weather (notably high winds) at start of colony occupation tends to result in more sub-colonies (Chestney 1970; Campredon 1977). Nest density varies between and within sub-colonies. Commonest distance between nests, Griend, 31–40 cm, minimum 21 (Veen 1977); at Osushnoi Island, 18–25 cm (Poslavski and Krivonosov 1976); range of 39 measurements, Norderoog (West Germany), 24–45, mean 34·9 (Dircksen 1932). Nest density at centre of sub-colony 10 per m^2, at edge 2 per m^2, mean 5–7 per m^2 (Campredon 1977). Density lower in vegetation than on open ground (Dircksen 1932), and among nests of first-time breeders (Veen 1977) which are more likely to occupy edge (Smith 1975a). However, sub-colonies often well-mixed assemblages of birds 3–11 years old or more (A J M Smith). Small, roughly circular, nest-area territory established and defended by both sexes from 2–4 days before laying (Langham 1974), minimum radius from nest-centre to bill-tip of sitting bird (Steinbacher 1931), though usually greater. Mean radius physically defended 57 cm (50–60), though sitting bird responds aggressively to intrusion into area of radius 150 cm, on ground or in the air (Veen 1977). Distance between nests greater if birds face one another while scrape-making, less if at right angles (Veen 1977). Territory serves for nesting, and raising of young—only until a few days old in disturbed colonies, but until fledging if undisturbed (R Chestney). Faithful to previous nest-site, sometimes to precise scrape position, but extensive shifts also occur within colony area between years (Smith 1975a). May also defend temporary feeding territory, especially along shore, or in narrow estuary (Campredon 1977; E K Dunn): e.g. bird (not known if always same one) frequently defended prime fishing position below weir (E K Dunn). Nests in close association with Common Tern *S. hirundo*, Arctic Tern *S. paradisaea*, and *L. ridibundus* in north-west Europe (Veen 1977); in Caspian, also with Slender-billed Gull *L. genei*, Little Tern *S. albifrons*, and Caspian Tern *S. caspia* (Dementiev and Gladkov 1951c;

Poslavski and Krivonosov 1976). Seeks association with *L. ridibundus* (nesting within its colonies), and probably also other Sternidae, gaining protection from greater aggressiveness (Van den Assem 1954; Rooth 1958; Cullen 1960b; Lind 1963b; Croze 1970; Fuchs 1977b; Veen 1977). Of 25 colonies, Netherlands, 22 associated with *L. ridibundus*, 20 with *S. hirundo*, 19 with both (Veen 1977). At Griend, *c.* 90% nested within 4 m of *S. hirundo* or *L. ridibundus* (Veen 1977). If *L. ridibundus* moves to new site, *S. sandvicensis* often follows (Nehls 1969; Campredon 1977). ROOSTING. Roosts in dense flocks, regularly with other Sternidae, less so with Laridae, on beaches, mudflats, saltpans, and occasionally grassy fields; also on piers, buoys, and driftwood; rarely on water. Several thousand may occur, more commonly tens or hundreds (E K Dunn). By day, loafs, bathes, and preens on shore. At feeding grounds, loafing group largest when most birds foraging nearby (Dunn 1972a). Most forage early morning and late evening, resting during middle of day. In night roost at colony, much calling and flying (Dunn 1972a; Campredon 1977). On arrival at breeding grounds, pre-breeding roost forms near (but always outside) colony-site, and becomes centre of interactive behaviour, involving display by whole group and between prospective mates, including initiation of aerial courtship and copulation (E K Dunn). At Gironde (France), pre-breeding roost largest, and display most intense, 3 hrs before and after low tide (Campredon 1977). Unpaired ♀♀ tend to form groups within roost (Veen 1977); within flock, particular displaying groups (both sexes) later form discrete sub-colonies (Smith 1975a). At colony, Grampian (Scotland), birds visited pre-breeding roost on foreshore of colony area mostly around midday and in large numbers at dusk. Overnight roosting in colony area began *c.* 2 weeks before laying (Smith 1975a) and at Langenwerder (East Germany) 21 days before (Nehls 1969), but period may be much shorter: 2–4 days (Langham 1974), 3–4 days (Isenmann 1972a). Presence through most of day signals full occupation of colony area (Veen 1977). From incubation onwards, off-duty bird roosts outside but near colony. Fledged young assemble on flats near colony, often in same area as pre-breeding roost. One or both parents may roost at night with offspring (Steinbacher 1931).

2. FLOCK BEHAVIOUR. First birds in colony to detect danger give Alarm-call (see 1 in Voice)—if on ground accompanied by Aggressive-upright (see Antagonistic Behaviour, below)—and fly up. Alerted neighbours fly up in staggered fashion (unlike 'dread', below), and wheel overhead, giving Alarm-calls from outset (so-called 'alarm'). 'Dreads' or 'panics' ('collective upflight': Lind 1963b) occur—as in *S. hirundo* but much more frequently—in pre-breeding roost and from early in colony occupation until egg-laying, after which uncommon. Dreads observed every 7–8 min at colony (Campredon 1977). At Griend, commonest at sunset when most birds present. Dread often appears spontaneous, with no obvious stimulus, but also precipitated by variety of mild disturbances, e.g. sudden wash of tide upshore, distant approach of intruder (Smith 1975a). Birds (often with other Sternidae and Laridae) make simultaneous, silent exodus, and exhibit synchronized, swerving flight away from roost or colony area, usually low towards, or over, water. After a few seconds or minutes, birds wheel and make noisy return (mostly involving Gakkering-calls: see 2 in Voice) to former positions. Dread may be confined to single sub-colony, or spread to others (Langham 1974; Veen 1977). ANTAGONISTIC BEHAVIOUR. Ground-based behaviour well-developed, but aerial attack weak compared with most 'non-crested' *Sterna*. On ground, standing bird adopts intimidating Aggressive-upright posture (see Fig A in Swift Tern *S. bergii*), in which plumage (notably of neck) ruffled, crest raised, bill pointed at source of threat, and carpal joints exposed or held away from body (Smith 1975a; Veen 1977). In high-intensity threat, rhythmically raises and lowers head (Head-bobbing). If threatening conspecific (or other) bird overhead, raises head to track it, and may gape silently or utter Gakkering-call. ♂ soliciting ♀ in courtship often repulsed with Aggressive-upright. Physical contact uncommon, but standing birds sometimes fight, grappling with bills, pulling, and sometimes raising wings (Veen 1977). Much bickering in colony between neighbours, especially by nesting birds towards those establishing new nest or arriving to relieve mate or feed young; may occasionally cause desertion by one or more pairs (Dircksen 1932; Smith 1975a). Sitting bird stretches neck and bill towards intruder, ruffles plumage, raises crest, and sweeps bill from side to side (delineating perimeter of territory), close neighbours frequently pecking and duelling with bills. Bird also gapes silently, or utters Gakkering-call; at high intensity, with Head-bobbing (Van Iersel and Bol 1958; Smith 1975a; Veen 1977). At end of a bout of Gakkering, bill is usually lowered to point at ground, and head may be shaken (Van Iersel and Bol 1958). Virtually all birds initially aggressive towards neighbours when colony settles after disturbance (Campredon 1977). Likelihood of aggressive response increases strongly when intruder within 1·5 m of nest (Veen 1977). Normal posture when sitting seems to inhibit aggression by 'house-hunting' birds. Adult challenged or attacked on ground may appease by sidling off in Erect-posture ('pole stance', a variation of 'stretch' posture: Van den Assem 1954)—fore-body tilted up to 60° above horizontal, bill upwards in line with body, tail variously uptilted, and wings closed or held out (Smith 1975a; Veen 1977: see Fig B, right, for posture in different context). Chicks wandering in colony are struck on head by nesting birds; occasionally killed, especially in very dense colonies (see Panov *et al.* 1980). May be carried off and dropped from up to *c.* 12 m above ground (Campredon 1977). Adults defend temporary feeding territories by flying hard at intruder, giving variant of Gakkering-call. HETEROSEXUAL BEHAVIOUR. (1) General. Pair-formation and courtship begin in winter quarters; aerial courtship seen, Ghana, from mid-February (Smith 1975a). Most birds, except first-time breeders, arrive already paired at breeding grounds. Pairs refrain from entering nesting grounds and selecting nest-site until shortly before egg-laying, all preliminaries taking place at pre-breeding roost where display involves paired as well as unpaired birds (Veen 1977). Both aerial and ground courtship occur. (2) Aerial courtship. Predominates at start of season, giving way to ground courtship later. Consists of High-flight which is precipitated in various ways (compiled largely from Van den Assem 1954, Cullen 1960a, and Smith 1975a): (a) birds break away from dread, mainly at pre-breeding roost, to perform High-flight; (b) 2 or more birds fly in from sea to roost or colony area, and perform High-flight; (c) advertising ♂ attracts 1 or more ♀♀ to follow into High-flight; (d) birds proceed to High-flight from ground

A

courtship within pre-breeding flock. In (c), ♂ flies around over flock, usually—though not always—carrying fish, and giving Advertising-call (see 3 in Voice) to attract ♀♀. On ground, several birds, paired and unpaired, usually respond with Advertising-calls in Forward-erect posture (as Erect-posture, but bill about horizontal). Posture varies greatly in intensity, notably in extension of neck and wings. General excitement pervades group, and several High-flights may start simultaneously (Veen 1977). If advertising ♂ initiates flight from ground, takes off in Head-down posture, with head and bill pointed down, and back slightly arched—possibly homologous with aerial version of Bent-posture during gliding phase (see below). Both ♂ and pursuing ♀ may sometimes carry fish (A J M Smith). Up to 5 ♀♀ may follow ♂, possibly explaining later attachment of extra birds to mated pair at nest (Smith 1975a; A J M Smith). Birds make rapid, broadly circling ascent to considerable height (Smith 1975a). Especially when no fish carried, role of leader and pursuer may be exchanged. Birds give intermittent Advertising-calls, one apparently answering other. Before reaching peak of ascent, some (occasionally all) followers may drop out, but typically, ♂, pursued by 1 ♀, initiates fast downward glide with wings half bent, often after several false starts. During glide, birds may change main direction of descent a few times, and zigzag ('swaying': Cullen 1960a) within this broad pattern; also perform several Pass-ceremonies in which one bird overtakes other, sometimes touching as it does so (Smith 1975a). Overtaken bird adopts aerial version of Bent-posture, as in *S. hirundo*, while overtaking bird retains normal gliding attitude (Cullen 1960a), but where several birds in glide, the 2 in Pass-ceremony may both adopt aerial Bent-posture. Overtaken bird, especially if carrying fish, may raise wings over back, sometimes beating them slightly (V-flying). During glide, both birds utter Kek-call (see 5 in Voice), especially bird in aerial Bent-posture (Smith 1975a). Glide may be interrupted before reaching ground, birds resuming flapping flight and ascending for further High-flight (Campredon 1977). Towards end of glide, birds resume flapping or perform rudimentary Low-flight in which wings thrust emphatically through air, body rising and falling conspicuously with each stroke. Pass-ceremony occurs, and, especially early in pairing, leading bird, with head lifted slightly, may utter Wah-call (see 4 in Voice) just before being overtaken. Depending on development of bond, birds may then break off (with Advertising-calls) to fish, or land, usually apart from flock, either to preen or initiate ground courtship, in which ♂ usually offers fish. High-flight more prevalent in sunny than overcast weather (Smith 1975a; Veen 1977), but often commonest in evening when roost largest; sometimes occurs at night, especially in moonlight, up to 02.30 hrs (Campredon 1977; Veen 1977). Resurgence of High-flights at end of season mainly attributable to late-arriving immatures (Campredon 1977). Juvenile at colony, and 1st-winter bird in Ghana, November, recorded pursuing adult, presumably parent,

C

in typical High-flight ascent, to *c.* 100 m, but no glide (A J M Smith). (2) Ground courtship. Occasionally begins in winter quarters, e.g. seen 20 January, Sierra Leone, among birds still in winter plumage (Smith 1975a). At pre-breeding roost near colony, often initiated by fish-bearing ♂ landing in or near flock and adopting Forward-erect; gives Advertising-call while flicking bill upwards to draw attention to fish. Thus approaches several ♀♀ in succession. ♂'s posture may be rather low and hunched, initially with crest sleeked, in appeasing frontal advance on ♀ (Fig A). ♀ likewise adopts Forward-erect, often stretching higher than ♂ (Fig B, left), with bill rarely angled below his (Smith 1975a). If ♀ co-operative, birds parade around each other head-to-tail and with characteristic mincing steps, in small circle. If fish transferred, both adopt (often moderate) Erect-posture, bill slightly above horizontal or higher (Fig C, ♂ on left). Both Head-turn from side to side. ♂ may continue Advertising-call and ♀ may reciprocate; further Circling may occur. Aerial courtship or departure to feeding grounds may follow, depending on stage of pairing (Veen 1977). In early courtship, ♂ commonly withholds fish and initiates High-flight. After High-flight, landing birds typically adopt High Erect-posture, often with Head-turning, wings sometimes widely spread and tips touching ground (Fig D). ♀ may then begin to solicit fish by lowering to moderate Erect-posture and Circling stationary ♂ (Smith 1975a). As ground courtship progresses, fish transfer more rapid, postures briefer, Circling less prominent, and sequence increasingly leads, via begging, to copulation or, in colony, scrape-making. ♂ may lead ♀ to prospective site in low Forward-erect with Advertising-call, raising bill at each delivery. ♀ follows in moderate Forward-erect. Both birds examine potential site with bill down and carpal joints usually raised (see Veen 1977 for subdivision of postures); both scrape, though ♂ takes initiative (Smith 1975a). ♀ may then take over scraping while ♂ defends (Campredon 1977). Immatures also make scrapes (Smith 1975a). (4) Copulation. Typically occurs in or near pre-breeding roost, rarely in colony. Fish transfer may or may not precede it. Soliciting ♀ lowers body to Hunched-posture (Fig E, right), as in other Sternidae, and utters Begging-call (see 7 in Voice), and confronts ♂ who adopts Down-erect, bill below horizontal (Fig E, left), uttering repeated Kuk-call (see 6 in Voice). ♂'s attempts to take

B

D

E

up position at rear rather hindered by ♀'s attempts to keep facing him. On cloacal contact, ♂ flaps wings for balance. Up to 8 or more cloacal contacts occur (Van den Assem 1954; Smith 1975a). If ♀ dislodges ♂ before copulating, ♂ may press cloaca on ground, waggle tail, and ejaculate. Erect-posture and/or preening may follow successful copulation (Smith 1975a). (5) Courtship-feeding. From pre-breeding roost to incubation, ♂ feeds ♀, sometimes on demand. ♀ may beg, in semi-Hunched-posture, to fish-bearing ♂ flying overhead and calling. ♀ may peck at ♂'s bill or snatch at fish. On receiving it, may strut with it, giving Advertising-call. Fish may change custody several times. If thwarted in finding recipient, ♂ often eats fish. Transition from pairing to laying relatively short, so courtship-feeding mainly after copulation and most in first few days of incubation when feeding frequency often c. 1 fish per ½ hr. Frequency wanes as incubation proceeds (E K Dunn). (6) Nest-relief. In early incubation, ♀ relieved by ♂ arriving with food, giving Advertising-call on aerial approach, and landing in Forward-erect, sometimes uttering soft Kuk-call. ♀ relieving ♂ similar, but without fish. As relieved bird leaves nest, may give Wah-call (Van den Assem 1954), and ground courtship may follow, though less as incubation proceeds, exchange sometimes occurring without ceremony before half-way stage in incubation. Relieved bird often picks up material (even grains of sand) on leaving nest, and drops or tosses it sideways and to rear (side-throwing). As this subsequently billed into nest by sitting bird, scrape may become substantially lined during incubation (Smith 1975a). RELATIONS WITHIN FAMILY GROUP. Parents do not recognize own eggs; newly hatched young not recognized until 4–5 days old (Tinbergen 1953). Age at which young leave nest varies from 2nd day in disturbed colonies to fledging age in completely undisturbed ones (Desselberger 1929; Smith 1975a; Veen 1977; R Chestney); commonly c. 3–5 days. Where exodus rapid, 2nd egg or chick may be left behind in nest (Veen 1977). One or both parents coax young to leave nest with behaviour similar to ♂ leading ♀ to scrape. Parents guide young between neighbouring nests to inhibit neighbours' aggression (Campredon 1977). Usually disperse widely into surrounding vegetation, seeking regular refuge sites up to 1·5 km from nest (Poslavski and Krivonosov 1976). Chicks periodically brooded and guarded constantly by 1 or both parents for 1st week, after which left increasingly on their own. Defence includes repulsing immatures and perhaps failed breeders which attempt (unsuccessfully) to feed young (Dircksen 1932; Smith 1975a). In some colonies, young remain scattered until fledging; in others, form mobile crèche from c. 15 days which occupies more open ground. Crèching appears facultative: in some colonies, scarcely ever occurs, habitual in others (Smith 1975a). Parents feed only their own young in crèche (Steinbacher 1931; Smith 1975a); offspring recognize Advertising-call of approaching parent and generally break away from crèche to be fed (Hutchison et al. 1968). Adult may wet fish before presenting to young (Campredon 1977). Near fledging, young assemble in flock on shore where they continue to be fed; practise mandibulation by picking up objects. Once fledged, do same by dipping-to-surface from flight at sea. Juveniles abandon shore flock 1–2 weeks after fledging, and then follow parents (often only 1, probably ♀: Smith 1975a) to feeding grounds; fed there on land or water. Juvenile begs in Hunched-posture (often with wings partly open), giving food-call. Parents may alight to pass food, or make transfer in flight, especially if young on water, or on land if potential food-pirates nearby (Smith 1975a). ANTI-PREDATOR RESPONSES OF YOUNG. In response to Alarm-call of parent or other adult, small chicks crouch prone in nest, on sand, or amongst vegetation. On leaving nest, scattered young seek regular refuge, often tunnelling into dense vegetation; site indicated, however, by accumulation of droppings at entrance. Siblings usually stay together, but may become widely separated. Often remain faithful to one refuge unless disturbed, in which case may move considerable distance, under parental guidance, to another; may thus occupy numerous refuges (Smith 1975a; Veen 1977; E K Dunn). Alternatively may join crèche which relies on density, mobility, and proximity to water to avoid danger. Members of crèche usually loosely dispersed until alarmed; young then bunch into one or more groups, each moving as unit away from disturbance. When danger imminent, birds in crèche run and may swim (Dircksen 1932; Smith 1975a; Veen 1977). PARENTAL ANTI-PREDATOR STRATEGIES. (1) Passive measures. At approach of danger, adults give Alarm-calls. (2) Active measures: against birds. From Veen (1977). Like other 'crested' Sterna, not highly aggressive towards intruders. Avoids predation of nest contents by sitting tight, and never seen to leave nest to attack L. ridibundus; gull coming too close on foot or in flight threatened (in same manner as conspecific bird) with Gakkering. Clutch temporarily left gains protection from neighbours within 50 cm, also from S. hirundo and L. ridibundus within 4 m. When large Larus flies close overhead, sitting bird sleeks plumage, raises head, points bill at intruder, and gives Wah-call. If gull swoops, many birds stand up, raise wings, and all or part of colony may fly up in alarm. Incubating birds usually alight quickly (slower in small colonies), never separating from flock to pursue or attack flying gull. If gull tries to alight, or alights, may be subjected to weak dive-attacks by a few birds; rarely approach within 1 m and never strike or defecate. Birds with young may fly up when gull over 50 m away, and pursue it for 10–100 m, making fast, horizontal approach from behind, veering off at minimum of 50 cm (Fuchs 1977b; Veen 1977). Birds with young occasionally strike grounded gull. Birds of prey and herons (Ardeidae) induce mostly fleeing response, especially if only adults threatened. (3) Active measures: against man. When intruder 100 m distant, birds flew up and alighted again; flew up en masse when intruder 20–25 m away, and concentrated in noisy flock above colony, from which a few made shallow dive-attacks, never coming closer than 1–2 m, sometimes giving Gakkering-call; does not defecate on intruder (Veen 1977). Aggressive birds fly out from colony with Butterfly-flight to meet intruder, giving Wah-call before swooping with Gakkering-call (Smith 1975a). Adults (on foot and in flight) lead crèche away from intruder. (4) Active measures: against other predators. Virtually defenceless against mammalian predators (Veen 1977). Dense flock gathered over stoat Mustela erminea, and a few birds made weak dive-attacks; as stoat moved through colony, successive sub-colonies performed upflights and then alighted again (Fuchs 1977b). If newly settled colony visited by dog or other quadruped, colony area may be abandoned (Veen 1977; R Chestney).

(Figs A–E after photographs by J van de Kam in Smith 1975a.)

EKD

Sterna sandvicensis 59

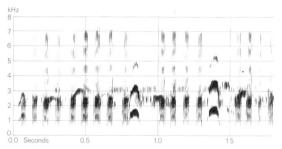

I J-C Roché (1970) France June 1965

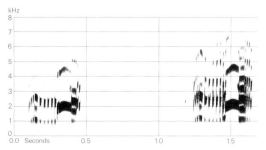

II P J Sellar France May 1977

Voice. Highly vocal, especially at breeding grounds.
CALLS OF ADULTS. (1) Alarm-call. Clipped 'krit krit' or 'krik krik' (E K Dunn, P J Sellar), 'wiet wiet' (Van den Assem 1954), mostly in flight, like attenuated call 3. (2) Gakkering-call. Hoarse, rhythmic 'ga-ga-ga' or 'gègègè' (Van Iersel and Bol 1958), 'gakgakgak' (Veen 1977); given mostly by sitting birds towards intruders, conspecific or otherwise (Fig I). Related to 'kekekek' given in dive-attack (Smith 1975a); shrill 'kree' also given in dive-attack (Campredon 1977). Version of Gakkering-call described as 'fat quacking' (Smith 1975a) probably maintains individual distance when birds fishing in fog (Smith 1975a); also given by bird defending feeding territory (Campredon 1977). (3) Advertising-call. Strident, metallic 'kireet' or 'kirrik'; permits recognition between individuals (Hutchison *et al.* 1968); given by both sexes, approaching in flight or on ground, to mate or young (Smith 1975a). ♀ may have higher-pitched call (R Chestney). More hollow 'koreet' (Smith 1975a) or 'keeryuk' (P J Sellar: Fig II) when fish carried (given about once per second by bird approaching colony edge: Hutchison *et al.* 1968), or by ♂ leading ♀ to scrape, or young from it (Smith 1975a). Degenerates to 'djeet' in winter quarters (Smith 1975a). (4) Wah-call. Single, loud, low-pitched 'wah' (Van den Assem 1954; Smith 1975a; Veen 1977), or 'whaa' (Campredon 1977) expressing alarm and strong escape tendencies (Van Iersel and Bol 1958); given by bird on nest to gull *Larus* overhead (Campredon 1977; Veen 1977), in horizontal flight towards human intruder (Smith 1975a; Campredon 1977), and sometimes in low Pass-ceremony or by relieved bird at nest-relief (Van Iersel and Bol 1958). Always given with head slightly uptilted and bill wide open. (5) Kek-call. During Pass-ceremony in gliding phase of High-flight, a rapid 'kekekekekekek' (Van den Assem 1954; Smith 1975a) or 'krékrékré' (Campredon 1977); possibly related to call 2. (6) Kuk-call. Soft slow 'kuk', used singly or repeated ad lib, e.g. 'kwēkkwēkkrōkkrōkkrōkrōkrō' (Marples and Marples 1934); given during incubation when arranging eggs, when brooding, or to small young outside nest. Probably invites close approach of young (Marples and Marples 1934). ♂'s deep-throated 'kuk-kuk' (Smith 1975a), 'ké-ké' or 'kré-kré' (Campredon 1977) to ♀ just prior to copulation (Smith 1975a) may serve similar purpose. (7) Begging-call of ♀. Persistent, penetrating 'quee-quee-quee' used in begging fish from ♂ or inviting copulation; delivered in Hunched-posture (Smith 1975a).

CALLS OF YOUNG. Downy young beg with high-pitched peeping. Food-call of juvenile and 1st-winter bird a shrill, petulant 'chee-chee-chee'. 1st-winter birds also call 'kjeet'—probably incipient Advertising-call of adult, since similar (Smith 1975a).   EKD

Breeding. SEASON. North-west Europe: see diagram. Mediterranean France: laying begins end of April to early May (Isenmann 1972a). Crimea: main laying period mid-May (Borodulina 1960). SITE. On ground in the open. Colonial. Nest: shallow scrape, unlined or with a few pieces of available material. Range of dimensions, Crimea, sample size not given: external diameter 20–28·5 cm; internal, 11–20 cm; depth of cup 2–4·5 cm (Borodulina 1960). Building: by both sexes, with turning and scraping movements. EGGS. See Plate 87. Sub-elliptical, smooth and slightly glossy; creamy-white to very pale yellow, variably streaked, spotted, and blotched black, dark brown, and grey. 51 × 36 mm (44–59 × 33–40), sample 250; calculated weight 35 g (Schönwetter 1967). Clutch: 1–2, rarely 3. Of 3831 clutches, Baltic: 1 egg, 55%; 2, 45%; 3, 0·25%; mean 1·43 (Dircksen 1932). Of 1258 clutches, Crimea: 1 egg, 52%; 2, 44%; 3, 4%; average 1·5

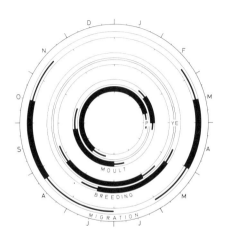

(Borodulina 1960). Range of average clutch sizes, Northumberland (England), 1965–70, 1·05–1·5, from samples of 206–1750 (Dunn 1972a; Langham 1974). Of 4995 clutches, France: 1 egg, 13·6%; 2, 86·0%; 3, 0·4%; average 1·87 (Campredon 1978). Of 1523 clutches, Sweden: 1 egg, 34%; 2, 65·5%; 3, 0·5%; average 1·66. In 7 large colonies, averaging 175 pairs, 2-egg clutches comprised 69%, average 2.07 eggs per clutch, while in 7 small colonies, averaging 42 pairs, 2-egg clutches comprised 50%, average 1·60 eggs per clutch. As number of pairs breeding in Sweden, and average colony size, declined during 1947–77, so average clutch declined from 1·72 to 1·47 (Mathiasson 1980.) Clutch size increases with age of bird (Veen 1977). One brood. Replacements laid after egg loss. Laying interval 2–5 days (Langham 1974; Veen 1977). INCUBATION. 21–29 days, average 25 days (Langham 1974; Smith 1975a). By both parents, with ♀ taking larger share at beginning (Campredon 1977; Veen 1977). Begins with 1st egg. 2 eggs hatch over 1–5 days, mean 2·45 (Veen 1977). YOUNG. Precocial and semi-nidifugous. Cared for and fed by both parents. Fed 8–12 times per day per chick (Borodulina 1960). Nest exodus behaviour varies with colony: if colony totally undisturbed, young may remain at nest-site until fledging (Chestney 1970); in some colonies, young may crèche, in others young disperse throughout colony area, seeking refuges where parents locate them for feeding; latter may include separation of siblings (Smith 1975a). FLEDGING TO MATURITY. Fledging period 28–30 days (Smith 1975a), though proper flying may not occur for a further day or two (Campredon 1977). Age of independence probably c. 4 months, perhaps longer. Age of first breeding usually 3–4 years (see Social Pattern and Behaviour). BREEDING SUCCESS. Hatching success, Northumberland, varied between years: 56·6% of 1102 eggs hatched in 1965, but 95·7% of 1982 eggs in 1967; hatching success also increased with colony size. Fledging success less variable: 88·1% of 235 young in 1965, 85·9% of 608 young in 1966, and 95·2% of 1897 young in 1967; 2nd chick of 2-chick broods least successful, and main mortality probably starvation of 2nd chick in 1st week after hatching (Langham 1974). In same colony, 1968–70, hatching success was 39·1% from 564 eggs in 1968, 21·7% from 206 eggs in 1969, and 65·4% from 670 eggs in 1970; fledging success 89·9% from 247 young in 1968, 87·8% from 49 young in 1969, and 83·7% from 657 young in 1970 (Dunn 1972a). In Netherlands, better fledging success from 1-egg clutches than 2-egg: in 1-egg clutches, 81·8% of 121 eggs hatched and 82·8% of 99 young fledged; in 2-egg clutches, 76·0% of 604 eggs hatched and 54·8% of 459 young fledged; over 3 years, fledging success of 2nd chick 11·8–20·0% compared with 49·3–83·8% for 1st-hatched. 1st-hatched chicks from 2-egg clutches survived slightly better than chicks from 1-egg clutches; in general, older, early breeders more successful at fledging young than younger, later birds; birds over 5 years old had best breeding success (Veen 1977).

For details on nest success related to laying synchrony, position within colony, distances between nests, colony size, etc., see Langham (1974) and Veen (1977).

Plumages (nominate *sandvicensis*). ADULT BREEDING. Forehead, upper half of lores, crown down to lower edge of eye, and elongated feathers of nape deep black. Hindneck and upper mantle white. Lower mantle, scapulars, tertials, back, and all upper wing-coverts pale grey, distinctly paler than in Common Tern *S. hirundo*, Arctic Tern *S. paradisaea*, and Lesser Crested Tern *S. bengalensis*, but close to Royal Tern *S. maxima*, Roseate Tern *S. dougallii*, and nominate race of Gull-billed Tern *Gelochelidon n. nilotica*; tips and inner webs of tertials and some longer scapulars as well as small coverts at leading edge of wing white. Rump, upper tail-coverts, sides of head below black cap, and all underparts (including axillaries and under wing-coverts) white. Chest and belly rarely flushed pink at start of nesting season. Tail white, outer webs of outer feathers often faintly shaded pale grey. Primaries pale silvery-grey, fresh 5–6 innermost palest, outer web of p10 and streak along shafts of inner webs of outer 4–5 primaries slightly darker grey; margins of inner webs of all primaries broadly white, narrowing into border c. 2–4 mm wide along tip. Secondaries pale grey like upper wing-coverts, broad tip and nearly all of inner webs white. In abraded plumage, silvery-grey of primaries wears to dull grey-brown, especially at tips and proximal inner webs; white edges to tips of primaries lost. ADULT NON-BREEDING. Forehead and lores white; crown white, each feather usually with black droplet at centre, largest towards hindcrown. Distinct spot in front of eye and band from behind eye across nape black, latter streaked to variable extent with some white, especially on nape and upper ear-coverts, nape sometimes appearing mainly white. Remainder of head, body, tail, and wings as adult breeding; outer primaries and inner secondaries mostly dull black or dark grey September–October, contrasting with fresh others, all new and silvery-grey December–May. DOWNY YOUNG. Down spiny on head, back, and wing, a character shared only with *S. dougallii*. Ground-colour of head and upperparts variable, from grey-brown to pale buff or cream; of chin and underparts pale cream to off-white, throat often darker towards bases of down. Crown, sides of head, and upperparts usually rather heavily dotted sepia-black, dots sometimes coalescing into streaks or bars. JUVENILE. Forehead, lores, crown down to just below eye, and nape black, feathers fringed buff when fresh, paler and mottled or streaked white when worn, especially forehead and lores. Hindneck and upper mantle white, some feathers with dull grey subterminal mark, faintly fringed buff when fresh. Lower mantle, scapulars, tertials, and back pale grey, feathers shading to off-white at tips (tinged buff when fresh), each with black-brown subterminal arc; in some, upperparts mainly grey with rather narrow and broken arcs, others with heavy black-brown arcs and pale grey and off-white rather restricted, tertials mostly black with white notches at sides. Rump and upper tail-coverts white, variably (usually slightly) barred or dotted dull black. All underparts white. Bases and inner webs of tail-feathers white, tips of outer webs grading from pale grey (t1) to dull black (t6); bold subterminal marks sepia-black. Flight-feathers like adult, but grey darker and less silvery; white borders along tips of outer primaries narrower, grey on tips of inner webs more extensive, projecting into a point along border of inner web. Lesser upper wing-coverts along wing-bend dark grey with pale grey fringes, sometimes subterminally bordered black; greater coverts and primary coverts medium grey, shading to white at tips. Median and longer lesser wing-coverts variable: in some, boldly though variably marked dull black like mantle and scapulars; in others, pale grey to off-white, contrasting with darker wing-bend and greater coverts. Unlike most Sternidae, post-juvenile moult starts early, often when wings hardly full-grown; much of head, body, and wing-coverts replaced by immature non-breeding before autumn migration. FIRST IMMATURE NON-BREEDING AND SUBSEQUENT PLUMAGES. 1st immature non-breeding like adult non-breeding and hard to distinguish when last juvenile tail-feathers and upper wing-coverts lost by about December–January. Flight-feathers still juvenile in December, rather worn and dark (when adults have these mostly new and pale), or, when in moult about January–June of 2nd calendar year, primaries show contrast between pale inner and dark outer feathers; juvenile outer primary coverts (present up to about May) dull grey with ill-defined off-white borders instead of uniform pale silvery-grey. New shorter lesser upper wing-coverts pale grey with darker grey centres, forming poorly defined and indistinct bar across forewing, usually hardly visible. During summer, 1st immature non-breeding directly replaced by 2nd immature non-breeding; no breeding attained except sometimes for some dark blotches on crown. In 2nd calendar year, moult of primaries usually not arrested in spring as in adult, but continued, while a 2nd series starts about June–July, thus showing serial moult with 2 active centres June–September. In spring of 3rd calendar year, partial breeding plumage attained, usually mixed with some white non-breeding feathers on lores, forehead, and crown; primary moult June–September serially descendant as in 2nd calendar year, or, when arrested, this sometimes with different primary moult score from adult.

Bare parts. ADULT. Iris dark brown. Bill black, 8–12 mm of tip yellow. Foot black, soles often yellow; rarely some yellow spots on tarsus or joints. DOWNY YOUNG. Iris dark brown. Bill light blue-grey, sometimes tinged yellow or pink, frequently with black subterminal spot. Foot dusky grey, sometimes with pink or yellow-brown tinge. (Fjeldså 1977; RMNH, ZMA.) JUVENILE. Like adult, but relatively short bill often almost lacks yellow tip; cutting edges grey, pink, or yellowish at fledging.

Moults. ADULT POST-BREEDING. Complete; primaries descendant. Starts with scattered feathers on lores, forehead, or central crown between mid-June and mid-July; on head, completed late August to late September; those birds with head in fresh non-breeding by late July probably non-breeders. Tail moult starts mostly second half July, usually completed by early October; approximate sequence of replacement t2–t1–t3–t6–t4–t5. Body, tertials, and wing-coverts start with some feathers of mantle and scapulars from late July, moulting mainly late August and September, completed late September to late October. P1 shed mid-July to late August; all primaries new by late October–December. ADULT PRE-BREEDING. Partial: head, inner primaries, and probably all body and tail. Head mainly mid-February to late March, but sometimes not entirely completed on arrival at breeding grounds. Inner primaries from about November–December (only a few examined), arrested late January or February with up to p4 (in 13% of 32 west European birds examined), p5 (59%), or p6 (28%) new; feathers of body and tail probably replaced in same period. POST-JUVENILE. Complete. Starts rather soon after fledging, mainly late August to mid-September (some from late July). By late September or early October, head, body, most median upper wing-coverts, and part of tail and tertials usually in 1st non-breeding; by midwinter, no juvenile left, except for flight-feathers, primary coverts, and an occasional tail-feather (usually t4 or t5). Flight-feathers start with p1 from about December–January, completed with p10 May–July of 2nd calendar year, when also a 2nd series starts with p1 from about June–July. During summer of 2nd calendar year, 1st non-breeding directly replaced by 2nd non-breeding. SUBSEQUENT MOULTS. Like adult, but in 3rd calendar year not always synchronous with adult. See also Plumages.

Measurements. Netherlands, April–October; skins (RMNH, ZMA). Tail is length of central pair (t1); fork is tip of t1 to tip of t6.

WING AD	♂ 309	(4.61; 21)	302–317	♀ 304	(6.03; 16)	294–320
JUV	302	(3.96; 5)	297–307	296	(4.43; 11)	290–307
TAIL	73.9	(2.47; 26)	70–79	73.2	(2.63; 30)	67–78
FORK AD	73.1	(6.43; 12)	65–84	68.9	(6.17; 10)	58–76
JUV	39.0	(4.20; 6)	34–45	39.5	(3.82; 13)	34–45
BILL AD	55.5	(1.78; 21)	53–58	53.1	(2.19; 16)	49–56
TARSUS	26.9	(0.70; 21)	26–28	25.9	(0.86; 16)	24–27
TOE	27.8	(1.53; 12)	26–30	26.5	(1.23; 13)	25–29

Sex differences significant for bill, tarsus, and toe. Juvenile wing not full-grown until age of at least 3 months, when p10 exceeds p9 by at least 9 mm; juvenile tail at c. 2 months; juvenile bill reaches adult length after c. 1 year.

Weights. Nominate *sandvicensis*. ADULT. Coquet Islands, Northumberland (Britain), May–July: 229 (12.3; 20) (Langham 1968). Netherlands: April, 280, 285; August, 237 (14.4; 30) 215–275 (Smit and Wolff 1981). Britain, late September and early October, ♂♂, 262, 284, 291 (BMNH). Caspian Sea, July: ♂♂ 248, 260, 260, 275, ♀♀ 225, 248 (Gavrin et al. 1962). Iran: ♂♂ 198, 203 (March) Schüz 1959).

JUVENILE. Netherlands: August, 219 (13.6; 17) 200–245; September, 234, 240, 250 (Smit and Wolff 1981).

S. s. acuflavida. Panama: ♀ 202 (May) (Strauch 1977). Lesser Antilles: ♀ 177 (February 2nd calendar year) (ZMA). Surinam: ♀ 176 (July 2nd calendar year) (RMNH).

S. s. eurygnatha. Aruba, Bonaire, and Curaçao (off northern Venezuela): ♂♂ 170, 172 (March), 182, 190, 202, 210 (April), 206 (August); ♀♀ 180, 201 (March), 185 (April), 182, 190 (August) (ZMA). Surinam: adult ♂ 186 (February); 2nd-calendar-year ♂♂ 185, 195 (July); ♀♀ 187 (May), 195 (July) (RMNH).

Structure. Wing long and narrow, pointed. 11 primaries: p10 longest, p9 16–18 shorter in adult, 8–14 in juvenile; p8 36–43 shorter, p7 59–65, p6 80–89, p1 192–203; p11 minute, concealed by primary coverts. Tertials short, reaching to about p4 in closed wing. Tail deeply forked, but less so than (e.g.) *S. hirundo*; depth of fork on average 71 in adult breeding, 61 in adult and immature non-breeding, 39 in juvenile. Bill slightly longer than head, compressed laterally, rather heavy at base, especially in ♂: depth at basal corner of nostril in ♂ 11.2 (0.46; 21) 10.6–12.0, in ♀ 10.4 (0.47; 16) 9.5–11.0. Gonys with distinct angle about half way along lower mandible. Juvenile bill distinctly shorter than adult, tip especially, not reaching full adult length until c. 1 year old; depth at base like adult from age of c. 3 months. Foot short and rather slender. Outer toe c. 83% of middle, inner c. 66%, hind c. 29%.

Geographical variation. Rather slight, except for bill colour; grey of upperparts similar in all races. North and Central American *acuflavida* differs from nominate *sandvicensis* by slightly smaller size (averages for adults: wing ♂ 298, ♀ 295; bill ♂ 53.3, ♀ 51.4) and by different pattern on outer 3–4 primaries in adult, which is apparent only when feathers are quite new: tips of inner webs grey with narrow white border only (width at most c. 1 mm), and this border does not reach shaft at tip, while nominate *sandvicensis* has white border along tips of inner webs 2–

4 mm wide and extending to tip of outer web; hence, outer tips of outer primaries white in latter race, grey in former. S. s. *eurygnatha* of South America (sometimes considered separate species) differs markedly by bill being all yellow instead of mostly black, but populations of islands off Venezuela show variably intermediate pattern on bill: depending on locality, year, and month, 38–84% of breeding birds have bill uniform pale straw-yellow or greenish-yellow, 11–36% yellow with variable amount of black at base, and 5–26% black with yellow tip; pattern of latter birds similar to *acuflavida* and nominate *sandvicensis*, though yellow often slightly more extensive; besides these, a few (mostly below 1%) have bill uniform orange-red or red (Ansingh et al. 1960; Voous 1963). Even as far south as Argentina a few *eurygnatha* show black marks at base of yellow bill. Foot of *eurygnatha* usually black with black or yellow soles, as in both other races, but occasionally all yellow or mixed yellow and black. Size of Caribbean *eurygnatha* rather close to *acuflavida*: average wing of adult ♂ 294, of ♀ 288, bill ♂ 55·0, ♀ 52·1; slightly larger towards Argentina. White edges to tips of outer primaries as in *acuflavida*, but outer tips sometimes faintly bordered white. Juveniles and downy young of *acuflavida* and *eurygnatha*, though strongly variable, usually less heavily marked black than nominate *sandvicensis*, wing-coverts of juveniles often unmarked (except for dark carpal bar), and forehead often white. CSR

Sterna dougallii Roseate Tern

PLATES 5, 6, and 7
[between pages 134 and 135, and facing page 158]

Du. Dougalls Stern Fr. Sterne de Dougall Ge. Rosenseeschwalbe
Ru. Розовая крачка Sp. Charrán rosado Sw. Rosentärna

Sterna Dougallii Montagu, 1813

Polytypic. Nominate *dougallii* Montagu, 1813, Atlantic Ocean and Caribbean from Britain and USA to South Africa; *bangsi* Mathews, 1912, Arabian Sea, western Indian Ocean, and Pacific region from Ryukyu Islands, China, and Greater Sunda Islands eastward. Extralimital: *korustes* (Hume, 1874), Bay of Bengal from India and Ceylon to Burma; *gracilis* Gould, 1845, Australia and New Caledonia.

Field characters. 33–38 cm (bill 3·7–4, legs c. 2, tail of adult up to 17, of juvenile 9 cm); wing-span 72–80 cm. Similar in size to Common Tern *S. hirundo*, but wings at least 10% shorter and form more slender, culminating (in adult) in long tail-streamers. Medium-sized, slender, sea tern, with rather long bill and head, relatively short, rather narrow, and slightly blunt wings, narrow body, and marked tail-fork with noticeably whipping streamers. Appearance of adult likely to recall Sandwich Tern *S. sandvicensis* before *S. hirundo*—noticeably white in general tone and bill appears black. When fresh, outer wing paler than other similarly sized terns, but when worn, primaries show dark central lines. Juvenile also more like *S. sandvicensis* than *S. hirundo*, with noticeable dark spotting on upperparts retained into 1st winter. Calls distinctive, including one reminiscent of typical call of Spotted Redshank *Tringa erythropus*. Sexes similar; little seasonal variation. Juvenile and 1st-year immature separable.

ADULT BREEDING. Long, rather shallow cap on head jet-black; long, narrow black bill with red base and fine, drooping tip. Lower face, underparts, rump, and tail white, with faint grey wash on centre of last two tracts and strong rosy wash on underparts, particularly on breast and belly. Back and wings pearl- or pale blue-grey, noticeably paler than in *S. hirundo*. Primaries much paler silvery- or pearl-grey than in *S. hirundo* or *S. sandvicensis*, with white on inner webs extending to tip on all feathers of wing-point, narrow dusky lines visible on only 2 outermost primaries, and no dusky tips or panel on undersurface. Against strong light, broad trailing edge on all flight-feathers translucent (but effect less marked than in Arctic Tern *S. paradisaea* due to generally paler wing). Outer tail-feathers pure white. Basal half of bill becomes scarlet in summer. Legs red. ADULT NON-BREEDING. Forehead white, but with more black speckles over eye than on top of lesser coverts and loss of rosy wash on underparts (though still visible on 80% of birds, at least on centre of belly). On some, loss of bloom and wear (from July) darkens outermost primaries so that 4–5 stronger lines and darker tips appear on uppersurface of wing-point. Grey of upperparts fades so that many appear noticeably white at distance or in flight. Bill darkens. Legs orange or vinaceous-red. JUVENILE. Slightly smaller than adult, with even blunter, more rounded wing-point and slightly blunter bill. Cap mainly black and noticeably more complete than on *S. hirundo* and *S. paradisaea*; when fresh, forehead brown (only slightly paler than crown and little streaked). Back, scapulars, and tertials buff-grey, with brown-black barring (noticeably on lower part next to rump) as obvious as on *S. sandvicensis*. Wings pale grey on inner half, with narrow dusky leading edge (formed by lesser coverts) and obvious white trailing edge on secondaries, but usually duskier grey on outer half, with dark lines obvious on primaries and white rim to same feathers extending from white secondaries. Complete white trailing edge and tips to all flight-feathers particularly obvious from below, and translucent against strong light. Upper tail-coverts and tail grey, but latter with outermost feathers pure white (and both tracts fading to almost white). Change to 1st-winter plumage known to be well advanced by late September when white-speckled forehead obvious on most, centre of back largely unmarked, and wing-coverts broadly speckled (due to wear and moult). Tail-feathers longer and fork more obvious than in other similarly sized terns. Bill and legs wholly black. FIRST WINTER. Loses pattern of juvenile marks on head and body but retains similar wing and tail pattern, though markings on lesser coverts reduced (forming even narrower bar on leading edge) and dusky areas on primary coverts and primaries never as dark nor as uniform as in *S. hirundo*. Legs black becoming orange-red. FIRST SUMMER. Resembles winter adult, but fore-crown white, mid-crown hoary rather than streaked, lesser coverts and secondaries slightly darker, and all but outer tail-feathers greyer. Birds of this appearance not certainly present in west Palearctic during summer; no record of '*portlandica*'-type plumage pattern, but, on much faded birds, dark outer panel on upperwing (formed by 4–5 lines on outer primaries) striking.

Long subject to confusion with paler adult *S. hirundo* and, more recently, with noticeably white-winged immature *S. paradisaea*, but this largely due to inadequate study. *S. dougallii* actually of most individual appearance, which stems from (1) configuration of body, wings, and tail-base being mid-way between that of *S. sandvicensis* and *S. hirundo*, (2) paler, whiter appearance than *S. hirundo*, and (3) flight action being (usually) much less fluid and elastic. Long and narrow tail-streamers often difficult to see and do not extend body line noticeably backwards; so important to recognize that (1) most of tail of *S. dougallii* actually shorter and less full than that of *S. hirundo* and *S. paradisaea*, and (2) bill and head protrude more in silhouette, like *S. sandvicensis*. Thus, balance of *S. dougalli* in flight (and also at times on ground) distinctly different from *S. hirundo* and *S. paradisaea*, with wings appearing to be set further back on body as in *S. sandvicensis*. Flight extremely light and buoyant, with arc of wing-beats usually less deep than in *S. hirundo* and *S. paradisaea* and stroke stiffer, recalling both *S. sandvicensis* and even Little Tern *S. albifrons*; when hovering, wings fluttered rapidly, in action again recalling *S. albifrons*. Gait, normal carriage, and feeding actions hardly distinguishable from *S. hirundo*, but when diving for fish tends to create splash like *S. albifrons*. Infrequently observed on passage along coasts; exceptional inland.

Commonest call throughout the year 'chivy', 'chiv-ik', or 'chew-ik', recalling typical call of *Tringa erythropus*. Calls given in alarm or excitement always loud; most distinctive is 'aakh' or 'kraak', more guttural than commonest note of *S. hirundo* and *S. paradisaea* (though not as grating as disyllabic note of *S. sandvicensis*); one note of quarrelling birds, written 'kroi-IK', suggests *S. sandvicensis*.

Habitat. Breeds from tropics and subtropics through Mediterranean zone and in North Atlantic into temperate higher middle latitudes, exclusively on maritime coasts and especially islands, avoiding land-locked seas and stormy or cold waters. Preferred nest-sites close to clear shallow sandy fishing grounds; may consist of coral reefs, low rocky coasts and islands, sand-dunes and sand-spits, or shingle beaches with dead and growing herbage and marine debris, sometimes in tidal bays and often in protected inshore waters. In temperate regions, occasionally nests in open areas commanding wide views but more often in enclosed small spaces made by tunnelling into marram grass *Ammophila* or other rank littoral vegetation, in quite deep crevices between and under rocks, or even partially underground at or near entrance to burrow of rabbit *Oryctolagus cuniculus* or Puffin *Fratercula arctica*.

After breeding, avoidance of inland areas and of fresh waters continues, as movement to warm tropical coasts progresses without delay. Habitat remains similar; in Australia, where also breeds, typically frequents rocky coastal areas (Slater 1970). In West African winter quarters, many, especially young birds, caught from beaches with bait or fishing tackle. This slaughter has added to problems of maintaining small north-west European and Nearctic stocks attempting to survive near margin of range in face of increasing encroachment of expanding gull *Larus* populations on the few suitable breeding habitats. In North America, thus forced to nest on inshore islands or mainland where subject to attentions of ground predators especially (Nisbet 1980).

Distribution. Colony-sites often occupied erratically; map shows most recent known distribution in west Palearctic. World breeding and marine distribution inadequately known in some areas, especially in Pacific (see Nisbet 1980).

BRITAIN AND IRELAND. Local, with frequent changes in colony-sites; became almost extinct Britain and extinct Ireland late 19th century, recolonized from 1906 (Parslow 1967). BELGIUM. Mixed pair (with Common Tern *S. hirundo*) bred Knokke since 1976 (PD). FRANCE. Formerly also bred occasionally in Camargue (Mayaud 1953). SPAIN. Bred Ebro delta 1961 (AN). WEST GERMANY. Reported breeding 1819, 1875, and 1877; possibly nested 1965 (Schmidt and Brehm 1974). EGYPT. Said to breed Red Sea coast (Etchécopar and Hüe 1967); no recent records (PLM, WCM). TUNISIA. Bred Djerba end 19th century (Heim de Balsac and Mayaud 1962). MADEIRA. Almost certainly breeds, at least irregularly, but no proof (Bannerman and Bannerman 1965). 2 small colonies on Selvagens 1980 and 1982 (F Roux, P A Zino).

Accidental. Netherlands, West Germany, Denmark, Sweden, Austria, Switzerland, Italy, Malta.

Population. Decreased during 19th century in north-west Europe; marked increase 20th century due to protection. Decline in recent years (precise cause obscure, but considerable human predation in winter in West Africa).

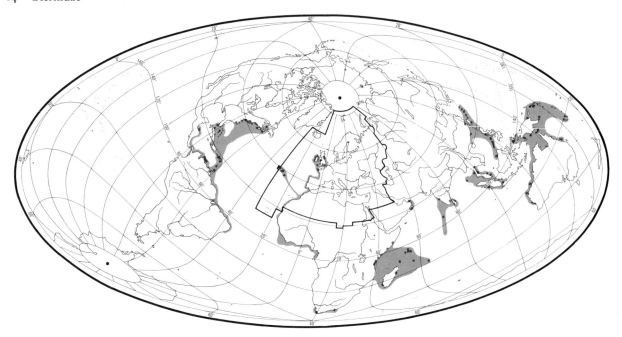

Similar trends in north-east North America—declined 19th century due to plumage trade, reached peak in 1930s after protection, declined since, due, at least in part, to human activity in wintering areas (Nisbet 1980).

BRITAIN AND IRELAND. Great decrease first half 19th century, marked increase 20th century (Alexander and Lack 1944). Irregular counts, frequently changing colonies, and suppression of data for security reasons prevent any exact summary of changes, but by 1962 marked recovery from near extinction c. 1900 to perhaps 3500 pairs (Parslow 1967). Since then declined, despite protection, to c. 2500 pairs 1969–70, over 1786 pairs 1971, 1414 pairs 1974 (Cramp et al. 1974; Lloyd et al. 1975), 1123 pairs 1975, c. 976 pairs 1979 (Thomas 1982), and c. 800 pairs 1980 (G J Thomas). FRANCE. Had become rare by 1936 (Yeatman 1976); fluctuating, but marked decline from c. 500 pairs 1973 to 78 pairs 1976, c. 50 pairs 1977, 88–92 pairs 1978, and c. 120 pairs 1979 (J-YM). AZORES. Marked decrease; estimated over 300 pairs 1980 (G Le G).

Oldest ringed bird 12 years 11 months (BTO).

Movements. Migratory. At all seasons the most thoroughly marine of European Sternidae; only a vagrant inland anywhere. European population winters exclusively in West Africa (Langham 1971; C J Mead). Movements of birds from North America and West Indies reviewed by Nisbet (1980); most winter along north coast of South America from Pacific coast of Colombia to eastern Brazil, and a few in Trinidad (West Indies). Elsewhere, most or all populations migratory though movements poorly understood (Nisbet 1980).

EAST ATLANTIC POPULATION. No information available on movements of French birds. Movements of British and Irish birds analysed from ringing recoveries by Langham (1971) and C J Mead. There is a brief post-fledging dispersal of juveniles and adults in August, including northward movement in Britain, but by September all recoveries are from 290–8000 km south of breeding grounds, indicating rapid movement along Atlantic seaboard towards winter quarters. By November, all recoveries are from coast of West Africa at 0–10°N, and until following May this appears to be common wintering zone for birds of all ages. Recoveries indicate that coast of Ghana is by far the most important wintering region, though this may be biased by density of human population and popularity of trapping (Mead 1978; E K Dunn). From September 1967 to August 1973, 82% of 136 African recoveries (all ages) were from Ghana; rest from Mauritania (1), Sénégal (4), Sierra Leone (4), Liberia (3), Ivory Coast (5), Togo (7), and Nigeria (1). Over 1955–72, relative frequencies of 152 recoveries of birds up to 1 year old in West Africa were: Mauritania and Sénégal 4·6%, Guinea and Sierra Leone 4·6%, Liberia and Ivory Coast 8·6%, Ghana 75·7%, Inner Gulf of Guinea 6·6%. Said to occur rarely in upper Niger inundation zone (Mali), November–December (Lamarche 1980).

1st-summer birds remain in tropics, though there is evidence of slight movement northwards, 5 of 15 recoveries occurring north of 10°N; most northerly was on Virginia Island (22°12′N), in August; most southerly in Ghana. At 2 years old, at least some return to Europe, and may visit breeding grounds, but rarely breed; most breed first at 3 years old (Langham 1971). No evidence that adults ever spend summer south of Sahara (C J

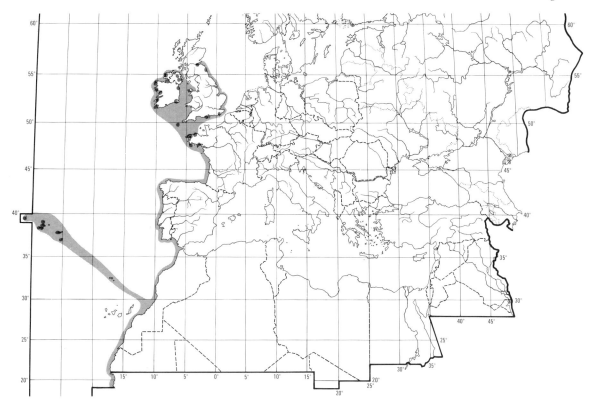

Mead). Arrival on breeding grounds, mostly in mid-May, indicates spring migration in April. Passage through Mauritania recorded early May (Bird 1937).

WEST ATLANTIC POPULATION. Age-related movements of North American birds similar (Nisbet 1980). From mid-July until mid-September, recoveries indicate juveniles and adults disperse up to 500 km throughout breeding area and beyond. Most migrate across western part of North Atlantic, juveniles reaching West Indies by mid-August. Most have left summer quarters by mid-September, latest mid-October. Recoveries indicate wintering range well-occupied by October, though no records in Brazil until November. Little evidence of northward migration in 2nd calendar year, majority remaining in winter quarters. Some migrate north in summer of 3rd calendar year, but not all reach colonies, and only a few attempt to breed. Many, if not most, migrate north to breed in following summer at 3 years old. In subsequent summers, almost all recoveries derive from breeding area, though some older birds may occasionally remain in South America. Spring migration takes place rapidly in late April to early May, and birds begin to reach colonies in 1st week of May.

EKD

Food. Chiefly marine fish. Less versatile in feeding methods than Common Tern *S. hirundo*. Food caught mostly offshore by plunge-diving from the air after flying upwind; may hover, though tends to do so less than other Palearctic *Sterna* (Dunn 1972a). In Massachusetts (USA), usually hunts by flying in wide circles and sweeps, covering area up to 1 km across between dives (I C T Nisbet). In Northumberland (England), mean dive height 4·2 m (*c*. 1–6, sample 231), significantly higher than *S. hirundo* and Arctic Tern *S. paradisaea* in same place (Dunn 1972a). In Massachusetts, dives from *c*. 2–12 m—higher than *S. hirundo* (I C T Nisbet); likewise at Great Gull Island (New York, USA), where mean height 4·4 m (sample 31) (D C Duffy). In South Africa, often dives forcibly into shoals from 2 m (Randall and Randall 1978). Immersion time directly proportional to dive height; mean time 1·2 s at both Coquet Island (Northumberland) and Great Gull Island (samples 100 and 202) (Dunn 1972a; D C Duffy). Immersion usually complete—probably up to 0·75 m (Dunn 1972a) or 0·5 m (D C Duffy)—but only partial when prey near surface. Prey always brought to surface, usually crosswise in bill, and either swallowed forthwith in air or carried to mate or young. More than one fish occasionally carried, resulting presumably from either simultaneous or successive captures, both witnessed in other *Sterna* (Hays *et al.* 1973; Taylor 1975). Of 265 birds carrying fish, Coquet Island, 1·5% had more than one; as many as 9 fish in a load, New York (USA), but 2–4 more common (Hays *et al.* 1973). Occasionally catches small prey like crustaceans, or any prey on surface, by dipping-to-surface, but less so than *S. hirundo* in breeding season. Dipping similarly for fish (especially sardines

Sardinella aurita and *S. eba*) left on beach or cast out on line predisposes them to snaring in winter quarters (Bourne and Smith 1974; Mead 1978; E K Dunn). Off East Africa, dips-to-surface for small fish and invertebrates driven to surface by shoals of bonito *Sarda* and tuna *Thunnus*, etc., joining in large mixed flocks of other *Sterna* (Britton and Brown 1974; L H Brown); likewise, North America, over schools of predatory bluefish *Pomatomus saltatrix* (Bent 1921; Erwin 1978). Skimming behaviour, Australia, may function to wet fish for swallowing (Hulsman 1975)—but see also Breeding. Only once recorded in aerial-pursuit of insects; took cicadas *Tibicina* during heavy outbreak (Forbush 1924). Practises food-piracy on other *Sterna*. At Farne Islands and Coquet Island (Northumberland) habitually robs Sandwich Tern *S. sandvicensis*, *S. hirundo*, *S. paradisaea*, and sometimes even other *S. dougallii* returning with fish, mostly by swooping on them in flight from vertically above and wresting fish from bill (Watt 1951; Langham 1968; Dunn 1973*b*). Food-pirates hunt alone, depending on surprise for success. Success low (7·5% of attempts), and behaviour probably confined to a few specialists (Dunn 1973*b*). In Massachusetts, occasionally attempts to rob *S. hirundo* and other *S. dougallii* by chasing them in the air; birds alighting at nest with fish continually harassed by neighbouring adults attempting to steal fish (I C T Nisbet). At One Tree Island (Great Barrier Reef, Australia), regularly dives on and chases other *S. dougallii* and Black-naped Terns *S. sumatrana* in colony carrying fish, but with little success (2% or less). Most pursuits of *S. sumatrana* last less than 10 s, but some up to 80 s. Also chases Swift Tern *S. bergii*, both in colony and in foraging flocks at sea edge, with more success. One *S. bergii* carrying 3 fish was chased by 3 birds which retrieved fish from sea surface when dropped (Hulsman 1976). Recorded seizing—before they reached the water—small fish thrown into the air by fishermen (see Marples and Marples 1934). Recorded scavenging at fishing boats in Irish Sea, September (Watson 1981). Feeds mainly offshore in open water, but sometimes inshore (Langham 1968; Dunn 1972*a*; I C T Nisbet). Daily weight changes of chicks vary, indicating environmental influence on foraging success of parents. Chick growth rate declines linearly with increasing wind speed; 19 km per hr wind depresses weight increase by 55% (Langham 1968) to 67% (Dunn 1975) over 24-hr period. May be poorly adapted for foraging in windy conditions of temperate latitudes (Dunn 1975). Shoals of fish attract feeding flocks and yield higher capture success than more dispersed prey. Capture success of 46 dives over one shoal was 47·8% (Dunn 1972*a*). In USA, capture rate lower in dense flocks than in dispersed birds, unlike *S. hirundo* (D C Duffy). Generally feeds in small parties—smaller and more diffuse than those of *S. hirundo* (Duffy 1975; see Erwin 1978). Tidal rhythm influences capture success inshore where water depth affects visibility and accessibility of prey. Fish capture rate increased from zero at high tide to 0·5 per min 2 hrs after low tide. Increase in capture rate a combination of better capture success and faster diving rate. Improved success at low tide only weakly reflected in rate of fish input to colony since most birds feed offshore beyond tidal influence. Day-time feeder only, with peaks of foraging activity just after dawn and early evening (Dunn 1972*a*).

Fish caught in Europe mainly sand-eels *Ammodytes marinus*, *A. tobianus*, herring *Clupea harengus*, sprat *C. sprattus* (Langham 1968; Dunn 1972*a*). In North America, diet comprises 50–95% (depending on colony) *Ammodytes americanus*, fewer (as available) Clupeidae (*Clupea harengus*, *Alosa aestivalis*, *Etrumeus teres*, menhaden *Brevoortia tyrannus*, mackerel *Scomber scombrus*), rarely silversides *Menidia menidia*, cunner *Tautogolabrus adspersus*, and invertebrates (Hays *et al.* 1973; I C T Nisbet). In South Africa, common prey species are ratfish *Gonorhynchus gonorhynchus*, sardine *Sardinella*, and Cheilodactylidae (Randall and Randall 1978). Rarely takes invertebrates: in Europe, shrimps *Crangon*; in North America, pteropods (pelagic molluscs) and, once, cicadas *Tibicina septemdecem* (Audubon 1840; Forbush 1924; I C T Nisbet). Fish generally 5–7·5 cm, though sand-eels up to 10 cm robbed from other *Sterna* on Coquet Island (Dunn 1973*b*); mean length of fish seen caught on fishing grounds 4·5 cm (34 fish over 3 days) (Dunn 1972*a*). In Massachusetts, fish taken up to 11 cm, median *c*. 6 cm (I C T Nisbet). *Brevoortia tyrannus* up to 2·5 cm brought in to Great Gull Island (New York, USA) over 2-day period (Hays *et al.* 1973). At One Tree Island, most kleptoparasitic attempts on *S. sumatrana* (74%) involved birds carrying fish 2–6 cm long, but not infrequently (13%) birds with fish of 8–10 cm; *S. dougallii* otherwise usually fed on open water for fish 2–6 cm long (Hulsman 1976). At St Croix Island (South Africa), from fish lying in colony or regurgitated, mean length of 20 of commonest prey *Gonorhynchus gonorhynchus* was 8·35 cm; remaining 23 items 4–6·5 cm (Randall and Randall 1978).

Clupeidae chief food of chicks, Coquet Island, outnumbering sand-eels 9:1 (Langham 1968), though proportions almost equal in other years (Dunn 1972*a*). At Massachusetts colony, diet in one year almost exclusively sand-eels (50%) and 2 species of Clupeidae (I C T Nisbet). Mean length of 100 fish fed to chicks, Coquet Island, 7·2 cm; smaller and more easily manipulated fish (e.g. sand-eels and small Clupeidae) fed to smaller young whereas older chicks got mostly larger Clupeidae (Langham 1968). Mean of 6 prey taxa fed to young per year, Great Gull Island (D C Duffy). EKD

Social pattern and behaviour.
1. Usually gregarious throughout the year, especially when breeding and roosting, less so when foraging; flocks tend to be smaller and more diffuse than in Common Tern *S. hirundo* (Duffy 1975), and solitary birds often seen feeding. Off East Africa, joins mixed feeding flocks with Little Tern *S. albifrons*, Bridled Tern *S. anaethetus*, and White-cheeked Tern *S. repressa*,

totalling several hundred birds (L H Brown). Off South Africa, seen feeding in compact flocks of c. 80 often with Sandwich Tern *S. sandvicensis* and *S. hirundo* (Randall and Randall 1978). Gregarious when bathing, forming compact groups of up to 60 in shallow water (I C T Nisbet). BONDS. Not known when pairing first occurs; on east coast of USA, some 2-year-old birds and most 3-year-olds attempt to breed (Donaldson 1971; Harlow 1971; Nisbet 1980). Both sexes incubate and care for young. Juveniles fed by parents for several weeks, probably months; single parents with young seen, Ghana, October–November (E K Dunn). For occasional pairing with *S. hirundo*, see that species. BREEDING DISPERSION. Forms small to medium-sized colonies—in west Palearctic, with other Sternidae. In Britain and Ireland, 1974, colonies of less than 10 pairs to over 600, mean 129 (Lloyd *et al.* 1975); formerly (1962) 2000 pairs on Tern Island (Ireland) (see Cramp *et al.* 1974). In Massachusetts (USA), 1972, 8 colonies of 6–1100 pairs, mean 285 (Nisbet 1973b). Characteristically forms sub-colonies within colony; sub-colonies usually discrete from, though close to, other Sternidae. At Coquet Island (Northumberland, England), mean 16 nests per sub-colony; nest density 0·4 per m² in one group of 20 pairs (Langham 1974). At Bird Island (Massachusetts), 50 nests in 28 m² (Nisbet and Drury 1972). Marples and Marples (1934) estimated 0·5–0·6 m between nests; on Jazirat Shaghaf (Oman), mean 0·8±0·05 m (0·3–3·2, sample 102) (C J Feare); on Madagascar, as close as 0·3 m (Milon 1950); on Aride (Seychelles), 0·31 nests per m² on woodland floor, 0·97 in grassland (Warman 1979). Small, roughly circular nest-area territory established and defended by both members of pair, from 1–3 days before egg-laying (I C T Nisbet); used for nesting, and concealment and location of young; mean size, Massachusetts, 0·6 m² (Nisbet and Drury 1972; I C T Nisbet). Probably capable of strong fidelity to previous nest-site, but can also be collectively fickle, with large numbers or whole colonies shifting colony-site in mid-season (Floyd 1932; Sharrock 1976). In North America, more prone to shift colony-site than *S. hirundo* or Arctic Tern *S. paradisaea*, but major shifts observed only after heavy predation, especially by mammals (I C T Nisbet). In Europe and North America, always associated with *S. hirundo* (I C T Nisbet), also variously with *S. paradisaea* and *S. sandvicensis*. In tropics, however, may nest in isolation from other species. Colonies usually separate from associates, but in Oman (C J Feare) and less so in Kenya (Britton and Brown 1974), mixes with *S. repressa*; on Amirantes (Indian Ocean), nests border sub-colonies of Swift Tern *S. bergii* (C J Feare). ROOSTING. Both during and outside breeding season, forms communal roosts on flat, open areas, such as beaches, mudflats, reefs, and saltpans; also undisturbed quays, ship's moorings, etc. At start of breeding season, assembles to roost near colony area (see Flock Behaviour, below); also loafs, by day, on shore between feeding bouts with own, or associate nesting species (Dunn 1972a). Birds occupy colony gradually, initially spending time at sea or in areas outside, but near, colony where they gather to court and roost overnight (I C T Nisbet). From incubation onwards, off-duty birds roost, preen, etc., on areas of common ground at edge of colony, usually within sight of own nest. Fledged juveniles loaf in flocks on shore, Aride (Warman 1979). Post-breeding and, initially, winter roosts comprise family parties, usually single parents with dependent offspring (E K Dunn). Black Tern *Chlidonias niger* and other west Palearctic Sternidae are common associates in winter roosts (e.g. Bird 1937).

2. FLOCK BEHAVIOUR. At Aride, birds formed flock (so called 'social upflight') at dawn, each morning till laying began (*c.* 20 days); flew over island in close formation, calling continuously. Less coherent flight also at dusk (Warman 1979). Exhibits upflights (of alarm), 'dreads', and 'panics' as in *S. hirundo*, though dreads said by Marples and Marples (1934) to be less frequent than in *S. hirundo*, not always involving whole colony. First birds to detect danger give Alarm-call (see 1 in Voice), and collective upflight follows. In dread, Australia, each sub-colony may depart as discrete flock, not necessarily co-ordinated with rest of colony (Serventy and White 1951). Birds on nests hidden under rocks, North America, emerged to join dread (Palmer 1941a). ANTAGONISTIC BEHAVIOUR. Vigorous in defence of nest-area territory (Bent 1921; Randall and Randall 1978; Warman 1979), spending more time in disputes than *S. hirundo*, although intensity of aggression declines sharply with distance from territory. Aggressive towards many non-predatory bird species (I C T Nisbet). (1) Ground encounters. Threat behaviour homologous with 'gakkering' of 'crested' terns, not with Bent-posture of *S. hirundo*. Threatening bird crouches facing intruder, with crest and back feathers raised, wings almost closed, and tail cocked high; head raised and lowered rapidly, while Gakkering-call (see 2 in Voice) given (I C T Nisbet; also Cullen 1962). In Australia, disputes often involve birds pecking one another (Serventy and White 1951). Defeated bird signals appeasement with Erect-posture, as in *S. hirundo*, but with neck and tail held higher. (2) Aerial encounters. In dive-attacks, accompanied by Anger-calls (see 2–3 in Voice), rarely strikes victim (unlike *S. hirundo*). Upward-flutter of *S. hirundo* does not occur, though birds occasionally hover together and grab each other's bills (E K Dunn). Recorded swooping on and severely mauling juveniles (conspecific or otherwise) wandering too near own young (Bent 1921). Picks up small chicks of same or other species and carries them away to drop them outside nest-territory (I C T Nisbet). HETEROSEXUAL BEHAVIOUR. (1) General. Pair-formation includes both ground and aerial displays. Birds first arrived *en masse* at breeding grounds, Aride, many already paired (Warman 1979). A few briefly explore potential nest area 10–15 days (I C T Nisbet) or up to 21 days (Langham 1974) before laying. (2) Aerial courtship. High-flight resembles that of *S. hirundo* in consisting of ascent and glide, but more lively with quicker tempo (Cullen 1960a), and more birds (usually 3–8, rarely 2) involved (I C T Nisbet). Make spiral, jerk-flying ascent to 30–150 m, wings angled back and beating faster than normal; may be silent, or call (see 4–6 in Voice, also 4 for possible origin of display). Display appears to be flying contest, and only first 2 birds to reach highest point in ascent take part in subsequent downward glide (I C T Nisbet); during glide, pursuer overtakes leader (Pass-ceremony) adopting aerial Erect-posture as it does so, while overtaken bird adopts aerial Bent-posture, both postures apparently like those of *S. hirundo* and *S. paradisaea*. Several Pass-ceremonies may occur, but not known if participants switch roles. After passing, bird in Erect-posture sways from side to side, as in *S. hirundo* and *S. paradisaea*, but much more pronounced and regular; course of bird in aerial Bent-posture usually relatively straight (Cullen 1960a). In Low-flight, single bird, apparently unpaired, flies in wide circles low over colony, often with fish,

A

uttering Advertising-call 'chiVIK' (see 6 in Voice); advertising bird often pursued by excited group of up to 6 others which compete for position nearest it (I C T Nisbet). Low-flight analogous to that of *S. hirundo* and *S. paradisaea*, but V-flying by advertising bird never seen, and whole display seems less elaborate, as in 'crested' terns (Cullen 1960a; I C T Nisbet). (3) Ground courtship. Most occurs in pre-breeding roost outside colony, and most birds thus pair before establishing territories (I C T Nisbet). Sequence of display as in *S. hirundo*. Ground Bent-posture more accentuated than in *S. hirundo*, with neck stretched much higher, and tail nearer vertical; head arched down and half-folded wings held well clear of body (Fig A). As in aerial display, ground display initially often involves more than 2 birds, especially when 1 has fish. Calls equivalent, but different-sounding, to those of *S. hirundo* are given. Scrape-making important part of courtship, restricted to 2 birds; pair move around territory together, often side by side, even touching, scraping alternately at various places until nest-site chosen. (4) Mating. Most occurs outside colony area from 6 days before 1st egg laid to laying of 2nd. Behaviour similar to that in *S. hirundo*; ♂ raises back feathers during copulation and gives Quacking-call (see 9 in Voice). Frequently interferes with other copulating pairs (I C T Nisbet). (5) Courtship-feeding. Occurs much less often than in *S. hirundo*, mostly outside colony area and, unlike *S. hirundo*, little at nest-site though continuing sporadically throughout incubation (I C T Nisbet) (6) Nest-relief. Side-throwing occurs, as in *S. hirundo*, but not obligatory as in that species (I C T Nisbet). RELATIONS WITHIN FAMILY GROUP. Parents dipped in flight to wet breast feathers, Australia, for cooling brood (Serventy and White 1951). Up to 2 days old, chicks stimulate feeding response by pecking at parents' bills. Chicks in refuge sites (see Anti-predator Responses of Young, below) respond to parents' Advertising-calls with high-pitched 'ki-VIK' call. Unlike *S. hirundo*, little competition for food between siblings (I C T Nisbet). Fledged young assemble in flocks on shore to be fed, Aride (Warman 1979). Newly fledged young closely attended by parents and, unlike *S. hirundo*, do not return to territory to be fed. When 2 chicks raised, 1 parent accompanies 1st chick after fledging, while other parent stays behind up to 4 days to attend 2nd chick, remaining with it after fledging. Juveniles sometimes fly out to sea on first flight and may accompany parent to feeding grounds within 2–3 days (I C T Nisbet). Juveniles seen alighting on water to be fed up till mid-August in North America (Jones 1903; I C T Nisbet), and accompanied, if not always fed, into winter quarters (see Bonds, above). ANTI-PREDATOR RESPONSES OF YOUNG. In nests under rocks, chicks often remain crouched in hiding until 15–20 days old, after which often led by parents to edge of cover (E K Dunn, I C T Nisbet). If nest cover poor, parents usually move young to safety of vegetation when 2–7 days old (usually 3–5), and up to 100 m from nest-site by fledging time (Jones 1903; LeCroy and Collins 1972; I C T Nisbet). Chicks tunnel into vegetation, emerging only to be fed or if molested. Chick seeks cover on hearing Alarm-call of parent or other conspecific bird. PARENTAL ANTI-PREDATOR STRATEGIES. Noticeably less pugnacious towards ground predators than *S. hirundo*, less still than *S. paradisaea*, but more so than *S. sandvicensis*, the 4 species representing descending sequence of aggression, correlated with nest density and decreasing use of nest camouflage. By nesting in association with *S. hirundo*, benefits from its greater pugnacity (I C T Nisbet). However, highly aggressive on—or in close vicinity of—own territory (see Antagonistic Behaviour, above). May respond to heavy predation by shifting to new colony-site (see Breeding Dispersion, above). (1) Passive measures. At approach of danger, adults give Alarm-call (see above for response of young). Though deserts colonies in North America at night (returning in morning) in response to predation by Great Horned Owl *Bubo virginianus*, more prone to remain than *S. hirundo* and thus suffers proportionately more predation (I C T Nisbet). (2) Active measures: against birds. Aggressive primarily to aerial predators, relying on concealment to thwart ground predators. Makes flying attacks on gulls, crows (Corvidae) and, in North America, Night Herons *Nycticorax nycticorax*. Owls (Strigiformes) can be important predators (Warman 1979; I C T Nisbet). (2) Active measures: against man and other predators. Anxiety-call (see 5 in Voice) given in flight above intruder near, but not at, nest. Nearer nest, Alarm-call given and territory-owner makes half-hearted swoops at intruder's head, sometimes uttering Anger-calls; only occasionally strikes, unlike *S. hirundo* and *S. paradisaea*. More aggressive after chicks fledge. (E K Dunn, I C T Nisbet.)

(Fig A from Bannerman 1962 and photograph by E K Dunn.)

EKD

Voice. Highly vocal, especially at breeding grounds where both sexes have large vocabulary. Calls generally quieter but sharper than those of Common Tern *S. hirundo*.

CALLS OF ADULTS. (1) Alarm-call. Harsh, drawn-out 'kra' or 'kraak' of uniform pitch, and thus distinct from *S. hirundo* and Arctic Tern *S. paradisaea* (Marples and Marples 1934: Fig I). Also described as a grating or rasping 'kraak' (Bent 1921), or 'aakh' (Watson 1966); in less intense form, a quieter 'kaaa' (Marples and Marples 1934); likened to 'forcibly tearing a strong piece of cotton cloth' (see Bent 1921). (2) Gakkering-call. A hoarse, cackling 'ge-KEkekekeKEkekeke' (accent every 4th syllable) given in aggressive confrontation between birds on ground (I C T Nisbet). Related call 'kekekekeke-KAAAAK' in aerial swoop at intruder, usually predator (Marples and Marples 1934). (3) Anger-call. A slowly repeated 'choi-IK' or 'kro-IK', with creaky quality reminiscent of Sandwich Tern *S. sandvicensis*; sometimes given during aerial swoop at conspecific on ground (Marples and Marples 1934). (4) Chik-call. A clipped 'chik' uttered by leading bird in jerk-flying ascent of High-flight, interpreted as mild alarm-call (reflecting origins of High-flight in hostile pursuit: Cullen 1960a) or variation of call 6 (I C T Nisbet). Also given with strong nasal quality, from parent to young in winter quarters, Ghana, when flying close together (E K Dunn). (5) Kliu-call. A musical 'kliu', given in flight above intruder near, but not at nest, and interpreted as anxiety-call (I C T Nisbet). Similar call said to be given, occasion-

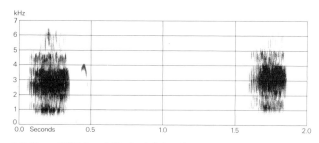

I E Simms/BBC (1971) England July 1960

ally, by pursuer in High-flight (Cullen 1960a). (6) Advertising-call. Described as 'chivy' (Cullen 1960a), 'chewy' (E K Dunn). Given by bird returning to colony with fish for mate or young. Many variations in pitch and intensity depending on context (I C T Nisbet): 'chiVIK-chiVIK-chiVIK-chiVIK' given by advertising ♂ in Low-flight; excited 'kileek-kileek-kileek-kileek' at start of High-flight. Other renderings express liquid quality, e.g. 'hew-it', reminiscent of Spotted Redshank *Tringa erythropus* (E K Dunn) or a musical 'kulick' (Bent 1921); in Ghana, October, parent regularly called 'chew-ik' when offspring flying some distance away (E K Dunn). (7) Begging-call 'ki-ki-ki-ki', similar to *S. hirundo* and *S. paradisaea* but thinner and higher pitched (I C T Nisbet); most often heard on nest-territory from ♀ begging fish from ♂. (8) Kruk-call. A soft, conversational 'kruk-k-k' by one or both members of pair in ground courtship, invariably during scrape-making (I C T Nisbet); probably the chattering, gurgling notes of Bent (1921). Similar call given when settling on newly hatched young (Palmer 1941a); probably also summons young for brooding. (9) Quacking-call. Duck-like 'gwa-gwa-gwa. . . .' given by ♂ during copulation (I C T Nisbet).

CALLS OF YOUNG. Cheeping or peeping food-calls of downy young slower, weaker (Palmer 1941a) but more rasping (E K Dunn) than *S. hirundo* or *S. paradisaea*. At *c.* 8–9 days, cheeping gives way to shrill, rapidly repeated 'ki-ki-ki', similar to adult Begging-call (see 7). At colony, older young in hiding reply to parent's Advertising-call (see 6) with high-pitched 'ki-VIK', or squeaky 'k-leek k-leek', which helps parent locate young. Immediately after fledging, and in winter quarters, juveniles use same call in flight, answering parent's more liquid 'chew-ik' (E K Dunn, I C T Nisbet).   EKD

Breeding. SEASON. See diagram for British Isles. SITE. Under cover of vegetation or in shelter of hollow or rock; occasionally on open sand or among sparse dead grass; also in burrow of (e.g.) rabbit *Oryctolagus cuniculus* or Puffin *Fratercula arctica*; of 85 sites in Northumberland (England): 33 in burrows, 23 under vegetation, 15 in hollows, 14 among rocks (Langham 1974). Colonial. Nest: shallow scrape, unlined or with a few pieces of available debris. Building: by both sexes; debris added during incubation. EGGS. See Plate 89. Markedly sub-elliptical, smooth and not glossy; pale cream, sometimes tinted buff to olive, variably spotted, blotched, and scrawled black-brown, sometimes gathered at broad end. 43 × 30 mm (38–48 × 27–32), sample 180 (Schönwetter 1967). In 50 single-egg clutches 44.7 × 29.7; in 53 2-egg clutches, 1st egg 44.3 × 29.8, 2nd (significantly smaller) 43.2 × 29.3 (Dunn 1972a). Weight, North America, 21 g (19–24), sample 20 (Collins and LeCroy 1972); mean weight of 76 1st eggs, North America, 20.4 g, and of 63 2nd eggs 19.5 g (Nisbet and Cohen 1975). Clutch: 1–2(–4). Mean clutch size, Northumberland, 1965–70, 1.38–1.59 with samples 74–228

(Dunn 1972a; Langham 1974); mean 1.76 (sample 2743), including 1.1% 3-egg clutches, Massachusetts, USA (I C T Nisbet). One brood. One replacement clutch laid. Laying interval *c.* 2–4 days, tending to lengthen during season (Nisbet and Cohen 1975). INCUBATION. 23 (21–26) days; increases to 24–31 when predators cause night desertion (I C T Nisbet). By both parents with ♀ taking larger share. Begins with 1st egg; hatching asynchronous, with mean interval 2–5 days, tending to lengthen during season (Nisbet and Cohen 1975). YOUNG. Precocial and semi-nidifugous. Cared for and fed by both parents. If nest cover good, chicks remain by nest until 15–20 days old; move within a few days if cover poor. FLEDGING TO MATURITY. Fledging period 22–30 days, usually 27–30 (Nisbet and Drury 1972). Age of independence not known but dependence continues for at least 8 weeks, probably longer (Bent 1921; I C T Nisbet). Age of first breeding 2–3 years (Donaldson 1971; Harlow 1971), most commonly 3 (Nisbet 1980). BREEDING SUCCESS. In Northumberland, 1965–70, 82.5–97.0% of eggs hatched and 79.8–94.4% of chicks fledged, range of samples 102–285 (Dunn 1972a; Langham 1974); production of young better from 1-egg clutches than 2-egg, with 87–90% of young fledging from 1-egg clutches in 1965, and 77–82% from 2-egg clutches in same year; 0.87 and 0.90 young fledged per pair from 1-egg clutches, and 1.54 and 1.63 from 2 egg clutches (Langham 1974). For breeding success in North American colonies, see LeCroy and Collins (1972) and Nisbet and Drury (1972); greater fledging success achieved from larger eggs (Nisbet 1978b).

Plumages (nominate *dougallii*). ADULT BREEDING. Forehead, upper half of lores, crown down to lower eyelid, and nape black. Upper mantle white; lower mantle, scapulars, tertials, back, and all upper wing-coverts pale grey, similar to upperparts of (e.g.) Sandwich Tern *S. sandvicensis* and distinctly paler than Common Tern *S. hirundo* and Little Tern *S. albifrons*. Rump and upper tail-coverts gradually paler grey, latter greyish-white; not contrasting with back. All underparts, axillaries, under wing-coverts, and lesser wing-coverts along wing-bend white, underparts with variable amount of pink tinge at start of breeding, gradually fading during summer, but sometimes still discernible in autumn.

Tail-feathers greyish-white, often indistinctly fringed pure white, outer web and tip of t6 largely white. Primaries pale silvery-grey; outer webs narrowly margined white (except p10), inner webs broadly, white margins of inner webs reach tip of feathers; hence primaries much paler than in *S. hirundo* or Arctic Tern *S. paradisaea*, which have webs grey up to feather-tip with white on inner webs of outer primaries restricted to broad basal wedge. 3 (2–4) outer primaries older than neighbouring inners (5–6 in *S. hirundo*), grey duller and less silvery, nearly greyish-black when worn, but usually still with silvery-grey bloom and not as blackish as worn *S. hirundo*. Secondaries pale grey to greyish-white; inner webs and tips broadly bordered white, outer webs narrowly. ADULT NON-BREEDING. Like adult breeding, but forehead white (sometimes dotted black) and crown white marked with black drops and bars, gradually merging into black nape patch, extending from eye backwards. Upper mantle pale grey. Underparts white, pink tinge usually faint or absent. Contrast between darker outer and paler inner primaries, if present, depends on moult and wear; outer primaries often distinctly blackish July–October, all primaries fresh and pale November–April, slightly darker May–June. DOWNY YOUNG. Down markedly spiny on upperparts, throat, and flanks, as in *S. sandvicensis*. Ground-colour either cinnamon-buff to pale buff, or pale grey to off-white, heavily (though individually variable in density) marked with diffuse brown-black specks—without detectable pattern or forming indistinct streaks. Forehead sometimes nearly uniform black-brown, throat and sides of head grey-brown to nearly black. Breast and belly uniform white or pale buff. (Witherby *et al.* 1941; Fjeldså 1977.) JUVENILE. Forehead and lores heavily mottled buff, white, and brown, crown streaked black, brown, and white. Patch from just in front of eye over ear to across nape black, feathers faintly fringed white or pale buff. Some black specks below eye. Upper mantle white, freckled brown. Lower mantle, scapulars, back, and tertials pale grey, fading to white at tip, latter variably freckled buff; each feather with one or more distinct dull black subterminal crescents (sometimes partly lacking). Rump and upper tail-coverts pale grey, mottled brown or buff to variable extent. Underparts, under wing-coverts, and axillaries white. T1 pale grey with dark subterminal crescent, like scapulars; t2–t5 grey with much white on inner web, white fringe to tip, and black-brown subterminal crescent, dot, or streak; t6 largely white. Flight-feathers as in adult, but grey duller, less silvery; white border of inner web of primaries extends to tip. Greater and median upper wing-coverts pale grey, grading to white at tip, mottled brown or buff subterminally to variable extent; lesser coverts dark grey with white fringes. When worn, forehead paler, buff-brown, but upperparts darker with more contrastingly sepia-black marks; grey of primaries and tail darker, especially at feather-tips. Differs from juvenile *S. hirundo* and *S. paradisaea* by dark forehead and heavily streaked crown, cap appearing nearly uniform dark, or, when worn, forehead buff-brown rather than white (though occasionally all-white); mantle, scapulars, tertials, and back with distinct blackish crescents, approaching juvenile *S. sandvicensis* in heaviness of marks; rump and upper tail-coverts pale grey to nearly white; t6 largely white; white of basal inner borders of primaries reaches up to tip; dark grey carpal bar across lesser upper wing-coverts slightly narrower and paler grey. FIRST IMMATURE NON-BREEDING AND SUBSEQUENT PLUMAGES. 1st immature non-breeding like adult non-breeding, but part of juvenile tail and tertials retained up to at least December and part of juvenile flight-feathers and primary coverts up to April–May. Only immatures of non-Atlantic tropical populations examined; these point to further plumage sequence similar to *S. hirundo*.

Bare parts (nominate *dougallii*). ADULT. Iris dark brown. Bill black on arrival at breeding grounds, dark red appearing at base during incubation from late May or June; red spreads and brightens after eggs hatch, reaching maximum extent with basal half of bill vermilion or coral-red by early or mid-July when young about to fledge; bill gradually duller again and with black spreading from base late July and August; all-black September–May. Well-fed summer adults show more red on bill than birds in poor condition. In tropical and subtropical breeding localities, bill often more extensively red than in temperate North Atlantic, and red appears earlier in breeding cycle. Foot vermilion or coral-red during nesting, orange-red or dull red in winter. (Witherby *et al.* 1941; Donaldson Cormons 1976; BMNH.) DOWNY YOUNG. Iris dark brown. Bill blue-grey to flesh-grey, tip dark red-brown to horn black. Foot vinaceous-flesh to leaden-grey, blackish in larger young. (Witherby *et al.* 1941; Fjeldså 1977.) JUVENILE. Iris dark brown. Bill pink or grey with black tip at fledging, soon darkening to black. Foot black.

Moults. Not fully known, as only a few winter specimens available; apparently rather similar to *S. hirundo*. ADULT POST-BREEDING. Usually starts with inner primaries when feeding young June–July; head, body (forehead first), and tail start from mid-July, followed by remaining feathers. Before 1st series of primaries completed, another series starts with p1 from early winter, while occasionally a 3rd series starts late winter, birds thus sometimes showing 3 series on arrival at breeding grounds: 1st series completed with p10; 2nd arrested up to about p6–p8 (p6–p7 in North Atlantic, p7–p8 in Caribbean); 3rd, when present, arrested with up to p1–p4 new. Non-breeding on body and tail complete by midwinter. ADULT PRE-BREEDING. Partial: head, underparts, tail, probably part of upperparts, and inner primaries (see above); completed on arrival at breeding grounds. POST-JUVENILE. In winter quarters. No information for Atlantic populations, but see Plumages.

Measurements. Nominate *dougallii*. ADULT. Wing (1) north-east USA, Britain, France, and Tunisia, (2) Caribbean; other measurements combined, as basically similar. All May–August; skins (BMNH, RMNH, ZFMK, ZMA). Tail is to fresh t6.

WING (1)	♂ 236	(3·69; 12) 230–242	♀ 233	(4·04; 9) 228–242	
(2)	229	(3·41; 7) 225–235	229	(1·89; 4) 228–232	
TAIL	180	(12·7 ; 13) 165–205	179	(14·8 ; 12) 158–201	
BILL	38·8	(1·19; 20) 37–40	37·1	(1·75; 13) 35–40	
TARSUS	20·2	(0·90; 16) 19–21	19·6	(0·54; 10) 19–21	
TOE	23·4	(1·25; 16) 21–25	22·7	(0·83; 9) 21–24	

Sex differences significant for wing(1) and bill.
JUVENILE. Wing on average *c.* 10 shorter than adult, tail *c.* 55.

Weights (nominate *dougallii*). ADULT. Coquet Island (Northumberland, England), May–July: 123·5 (6·9; 11) (Langham 1968). New York State (USA), mid-June to early July, 110 (8·08; 345) 92–133 (Collins and LeCroy 1972; Donaldson Cormons 1976).
JUVENILE. At hatching, 14·8 (9) 13·1–16·6; adult weight reached from *c.* 3 weeks (when wings hardly half-grown); at 22–27 days, 105 (9) 94·5–116; slightly lighter at fledging (LeCroy and Collins 1972).

Structure. Wing relatively short and narrow, pointed. 11 primaries: p10 longest, p9 9–16 shorter, p8 27–33, p7 44–52, p6 59–63, p1 140–149; p11 minute, concealed by primary coverts. Tail deeply forked, t6 much elongated and acutely pointed; length of t1 in adult breeding 66 (62–74); fork in adult breeding 95–135, in juvenile 42–54. Structure of bill and foot closely similar to *S. hirundo*, but bill relatively slightly longer, especially tip, hence angle of gonys relatively closer to base, and depth of bill slightly less, 7·4 (10) 7·1–7·6 at basal corner of nostril.

Geographical variation. Slight, mainly involving size of wing and bill, and colour of bill during breeding. Both size and bill colour dependent on latitude; birds in tropical areas smaller and with more red on bill than those of temperate populations; no pink tinge on underparts, or present only at start of breeding. Number of recognizable races and their boundaries difficult to establish, and variation in need of more thorough revision. Within Atlantic region, where all populations usually considered to comprise single subspecies (nominate *dougallii*), variation nearly as large as in whole species, with average size (expressed in wing length) varying from 234 (north-east USA) to 226 (some Lesser Antilles) and bill from largely black (north-east USA and Europe) to mainly red (some islands in Caribbean). *S. d. korustes* from India, Ceylon, Andaman Islands, and Burma perhaps the only subspecies for which recognition warranted, as size not only small (wing mainly 212–224, bill 31–37), but upperparts also slightly darker grey on average. *S. d. gracilis* of Australia and New Caledonia separated here on combination of short wing (208–228) and long bill (33–42). All remaining populations tentatively combined in *bangsi*: wing of Indonesian birds (Barussan Islands to New Guinea) 226 (26) 218–234, bill 36·8 (28) 33–41, depth of bill at basal corner of nostril 7·15 (27) 6·4–8·0 (similar to *gracilis* and *korustes*, less deep than nominate *dougallii*), colour of bill during breeding half-red to fully red. Other populations of *bangsi* apparently similar, hence hardly different from Caribbean populations of nominate *dougallii*, nor from *gracilis*. (Hartert 1912–21; Serventy *et al.* 1971; C S Roselaar.) CSR

Sterna hirundo Common Tern

PLATES 5, 6, 7, and 8
[between pages 134 and 135, and facing page 158]

Du. Visdiefje   Fr. Sterne pierregarin   Ge. Flussseeschwalbe
Ru. Обыкновенная крачка   Sp. Charrán común   Sw. Fisktärna

Sterna Hirundo Linnaeus, 1758

Polytypic. Nominate *hirundo* Linnaeus, 1758, eastern North America, Caribbean, Europe, North and West Africa, and Middle East, east to plains of Kazakhstan (USSR) and western Siberia. Extralimital: *longipennis* Nordmann, 1835, eastern Siberia south to northern Kuril Islands, Sakhalin, and north-east China, grading into nominate *hirundo* over wide area in central Siberia between about Ob river and *c.* 110°E; *tibetana* Saunders, 1876, eastern Kashmir, Pamir, Tien Shan, and Dzungaria through Tibet east to western Mongolia and Kansu (China).

Field characters. 31–35 cm (bill 3·2–4·0, legs *c.* 2, tail of adult 7–12, of juvenile 9 cm); wing-span 77–98 cm. Only a little shorter than Black-headed Gull *Larus ridibundus* but with much slighter and more attenuated form; distinctly smaller than Sandwich Tern *S. sandvicensis*, with shorter, proportionately less narrow bill, shorter, more compact head, and *c.* 15% shorter wings, but proportionately longer tail. Size overlaps that of Arctic Tern *S. paradisaea*, but bill and head more prominent, body less slender, and (in adult) tail-streamers shorter. Elegant sea tern, with pointed bill, evenly curved line to upper head, long wings, slim, rather oval body, and obviously forked tail. Adult in summer shows black-tipped red bill, neat black crown, grey upperparts, and mostly white underparts; in winter, forehead white. Immature has upperparts less uniform, with brown saddle in juvenile and broad dusky leading edge to upperwing and dusky outer tail-feathers obvious in juvenile and 1st-winter. At all ages, shows quite broad dusky or black tips to undersurface of outer primaries which form diffuse but marked dark border to rear edge of wing-point. For most of year, adult also shows dusky uppersurface to outer primaries, contrasting with wholly paler grey inner ones. Sexes similar; little seasonal variation. Juvenile and 1st-year immature separable.

ADULT BREEDING. Forehead, crown, and nape, down to level of eye (but not below), jet-black. Hindneck grey-white. Back, scapulars, and most of wings grey (with blue tinge), wings with narrow white leading and trailing edges. Uppersurface of primaries blue-grey, with outermost appearing pale but it and next 4–5 showing obvious dusky lines (formed by almost black outer and part of inner webs) which usually form distinctly darker wedge on end of wing. Underwing white, with long black tips (not webs) to primaries clearly visible and forming marked but diffuse trailing band from wing-tip (terminating abruptly on innermost primaries); against strong light, only innermost primaries translucent, with rest of flight-feathers noticeably opaque. Rump white; tail grey-white, with outermost feathers showing pale grey outer webs. Lower face and most of upper neck white, but rest of neck, chest, and underbody grey-white or pale grey, sometimes distinctly tinged mauve; vent paler, almost white. Grey tone on underparts much affected by quality of light, but only occasionally looking as dark as in *S. paradisaea*. Bill quite long, with upper mandible slightly decurved towards tip and lower showing slight upwards angle from gonys (total outline like narrow dagger); scarlet to orange-red with black tip usually obvious but sometimes missing. Legs bright red. ADULT NON-BREEDING. Breeding plumage lost earlier than in *S. paradisaea*. Forehead and lores white, front of crown over eye mottled grey-brown; rest of crown and nape dull black. Black spot in front of eye. Hindneck paler, almost white. Upperparts slightly paler grey, with less contrast between back, rump, and tail, but with dusky bar obvious between shoulder and carpal joint. Before wing moult, outer primaries show blacker tones above and more ragged band along trailing edge below; afterwards (December–February at least), all primaries evenly coloured above and almost as translucent as in *S. paradisaea*. Bill black, with variable red at base. Legs duller red or red-brown. JUVENILE. Slightly smaller than adult, with blunter wing-points and shorter tail. Initially, forehead white, often with buff or even gingery wash. Crown and large spot before eye dull black. Hindneck white, mottled buff-grey. Back and scapulars basically buff-grey, with bright gingery tone and darker brown bars showing as dull mottling. Upperwing strongly marked, with (1) inner half showing broadly black-edged, buff-brown panel on lesser and median coverts (forming obvious leading edge to inner wing), paler, greyer greater coverts (forming pale central panel), and dark grey secondaries (forming subterminal dark line), and (2) outer half showing dusky-grey primary coverts and primaries, latter with diffuse black tips on all but innermost whose narrow white tips join those on secondaries to form thin pale trailing edge to whole wing. Underwing duller than in adult, with dusky trailing band on outer primaries obvious (and grey line across secondaries visible at close range). Rump and tail as winter adult, but outer webs of latter darker. Face white, with buff tinge on chin and throat; rest of underparts white. Later, plumage pattern cleaner and paler, with marked loss of buff and ginger tones on forehead and mantle, and wing pattern even more obvious. Bill-base flesh to yellow-brown, with distal half black. Legs orange. FIRST WINTER. Head, back, and body as winter adult, but sometimes grey plumage of back duller and rump whiter. Wings become worn and faded, with dark leading edge to inner half, tips of primary coverts and primaries even duskier, central panel paler (even almost grey-white), and bases of secondaries dusky; pattern thus more variegated than in juvenile. Tail-feathers wear greyer, with outermost dusky. In some birds, juvenile back feathers retained until December (by when mantle noticeably patchy). FIRST SUMMER. Resembles adult non-breeding, but if moult delayed, plumage often subject to exceptional wear and bleaching, creating '*portlandica*'-type plumage pattern. By 2nd year, all sub-adults show dark brown-black or black rear crown, leading edge to inner wing, primary coverts, primaries, band across secondaries, and tips to outer tail-feathers; all such features contrast boldly with rest of dirty grey-white or dull cream upperparts and dusky-white underparts. Underwing shows broader dusky band along tips of outer primaries and across secondaries. Most birds with this markedly variegated appearance remain south of Europe but a few reach vicinity of most southern breeding haunts.

Long subject to confusion with *S. paradisaea* but differentiation much studied in recent years, so that distinction in all plumages now regularly achieved by practised observers. Characters particularly associated with *S. hirundo* are (1) longer bill and less rounded head, protruding further ahead of wings, (2) rather fuller body, (3) longer legs, (4) paler underparts in summer, (5) darker outer primaries, contrasting with inner on uppersurface (except December–February, after post-breeding moult), (6) diffuse dusky trailing edge on lower surface of primaries (not tapering inwards), (7) less contrasting rump, (8) darker secondaries, preventing marked translucency or obvious white trailing edge of wing as in *S. paradisaea*, and restricting translucency to discrete patch on inner primaries, (9) darker and wider carpal bar (when present), (10) paler bill colour, (11) usually more fluid and exaggerated wing-beats, (12) less rapid hover, and (13) less splashing dive. Flight actions varied, with that preceding fishing at first slow with head lowered, then quick in establishing best position for preceding hover or final dive; flight slower and steadier on passage, with less exaggerated wing-strokes. Hovering ability marked, but action less rapid than in *S. paradisaea*. Carriage usually horizontal, with marked extension of wing and tail-streamers; more upright when excited or in display. Swims well, but settles regularly on water only in winter quarters.

Vocabulary extensive, with many calls multisyllabic and varying from growls to harsh swearing and squeaks; close in phrasing and form to *S. paradisaea* but most notes pitched lower, with most distinct a grating 'kee-yah' (in alarm, with emphasis on 1st syllable), angry 'kek-kek' and 'karrr' (in interspecific conflicts). In contact (between birds in fishing or migrant parties), 'kip' or 'kik', as *S. paradisaea*.

Habitat. Breeds over wider spectrum of habitats than other *Sterna*: from arctic fringe through boreal, temperate, steppe, Mediterranean, and semi-desert zones to tropics, both along coasts and on inland fresh waters, mainly in lowland but up to 300 m or more in Scotland (Baxter and Rintoul 1953), to 2000 m in Armeniya (Dementiev and Gladkov 1951c), and 4400–4800 m extralimitally in Asia (Voous 1960). Avoids icy waters, sites exposed to strong winds and heavy rainfall, stands of dense or tall vegetation, and precipitous or broken terrain. Along maritime coasts, favours flat rock surfaces on inshore islands or islets, or open shingle and sand on upper beaches or in dunes. Also uses undisturbed mainland peninsulas, sand or shingle spits, or salt-marshes, even below level of highest spring tides. In Shetland (Scotland), nests typically on shingle or rocky coasts; in Orkney, nests more evenly divided between shingle, rough pasture, and heath (Bullock and Gomersall 1981). Sometimes nests between windrows of seaweed or drifted plant remains and driftwood, or on floating masses of dead plants in shallow water (Pough 1951). Usually chooses bare ground or short grass, but occasionally persists on areas where grass or shrubs have grown up, often stimulated by fertilizing effect of earlier occupancy. In parts of range, inland breeding is regular, especially on rocky or stony islands in freshwater lakes, on shingle banks in rivers (Sharrock 1976), or on marshes, ponds, or grassy areas (Niethammer 1942); also on islands in coastal lagoons, and on artificial sites such as patches of dumped soil from dredging, gravel-pits, and rafts or islets; even factory roof (Hakala and Jokinen 1971; Axell 1977). Where necessary, will fly some distance to feed (see Food) and, after breeding season, up to 50 km between foraging and roosting areas (Bent 1921). Winters mainly along coasts south of main breeding range.

Relies mainly on powers of sustained, easy flight, commonly in lower airspace but rising high as occasion demands. Perches readily, especially on convenient artefacts and other positions free of clutter. Vulnerable at breeding places to inclement weather, flooding, predators, and human disturbance; in winter quarters, to snaring by man.

Distribution. Little marked change, though decreased inland in some areas.

FAEROES. First bred 1968, then 1972–4 (BO). PORTUGAL. Bred 1937 (Coverley 1939). EAST GERMANY. Mecklenburg: breeds irregularly elsewhere, especially inland, where formerly more frequent (Klafs and Stübs 1977). POLAND. Now largely extinct Lower Silesia (Tomiałojć 1976a). AUSTRIA. Widely distributed on Danube and larger tributaries in 19th century; now only 2 regular sites, but nests irregularly in Waldviertel and on lower Inn river (HS). GREECE. Formerly bred Corfu, Paxos, and Siros (Reiser 1905). SYRIA. Possibly breeds in north but no proof; formerly nested on coast (Kumerloeve 1968b; Macfarlane 1978; HK). LEBANON. Formerly bred on coast (HK). KUWAIT. Bred 1959 (F E Warr). CYPRUS. Bred 1905 (PRF, PFS). LIBYA. Bred 1937 (Bundy 1976). ALGERIA. Said to breed mid-19th century (Heim de Balsac and Mayaud 1962); no recent records (EDHJ). MOROCCO. Bred Puerto Cansado 1967 (P Robin). CANARY ISLANDS. Has nested Gran Canaria, Tenerife, and Fuertaventura (Bannerman 1963a).

Accidental. Iceland, Malta, Cape Verde Islands.

Population. Limited information for most of range but probably declined many areas in 19th century, followed by some increase under protection. Recent declines in France, Netherlands, West Germany, Denmark, Czechoslovakia, Greece, and possibly elsewhere.

BRITAIN AND IRELAND. No complete census. Over 14 700 pairs at coastal colonies 1969–70—numbers nesting inland unknown, but increasing in Scotland and England, decreasing Ireland; partial counts to 1979 show marked fluctuations since, but little overall change (Cramp et al. 1974; Lloyd et al. 1975; Thomas 1982). Probably declined in 19th century and increased under protection in 20th century reaching peak in 1930s (Parslow 1967). FRANCE. Declined in 20th century to c. 4500 pairs (Yeatman 1976); c. 3000–3500 pairs 1978 (J-YM). In Brittany, decreased by 46% 1970–8 (J-YM); in Camargue, declined from 2500–2800 pairs 1956 to 1100 pairs 1979 (A Johnson). SPAIN. Decreasing (AN); c. 11000 pairs 1976 (Thomas 1982). BELGIUM. Some increase; c. 105 pairs 1969 (Lippens and Wille 1972); c. 220 pairs 1979 (PD). NETHERLANDS. Before 1908 probably over 50 000 pairs, then brought to near extinction by plumage trade, recovering to former level from 1920 to 1950 (CSR). Marked decline since: 42 000–48 000 pairs 1954, 21 500–26 000 pairs 1957 (Braaksma 1958), 7600–8300 pairs 1971 (E R Osieck), with slight recovery to over 10 000 pairs 1978 (Teixeira 1979). WEST GERMANY. Declined inland, but coastal colonies well over 10 000 pairs in 1930s (Niethammer 1942); since declined on coast also (Schmidt and Brehm 1974), but increased Bayern (Bauer and Thielcke 1982). Perhaps c. 6000 pairs (Thomas 1982). DENMARK. Decreased; 800–1000 pairs (Mardal 1974), 600–800 pairs 1970–6 (Rasmussen 1979), now c. 600 pairs (TD). NORWAY. Estimated 13 000 pairs 1970 (E Brun). SWEDEN. About 40 000 pairs (Ulfstrand and Högstedt 1976). FINLAND. About 6000 pairs (Merikallio 1958); numbers stable in some areas, slow decline in others (Hildén 1966a; OH). EAST GERMANY. Mecklenburg: 1400–1500 pairs, fluctuating; decreased inland (Klafs and Stübs 1977). POLAND. Increased in some areas (AD). About 2000 pairs 1979 (Thomas 1982). CZECHOSLOVAKIA. Decreasing (KH). AUSTRIA. Marked fluctuations, depending on water levels; Rheindelta varying 30–170 pairs with 115 pairs 1980 but Seewinkel decreased from c. 200 pairs mid-1960s to 77 nests 1979 (HS). HUNGARY. Over 200 pairs in 1970s (Thomas 1982). SWITZERLAND. Marked decrease since early 20th century to c. 70 pairs early 1950s; now c. 310 pairs, aided by provision of artificial nest-sites (RW). ITALY. 2500–3000 pairs 1982 (G Bogliani, PB). Sardinia: 60–80 pairs, declined (Schenk 1976). GREECE. Over 300 pairs 1978 (Thomas 1982); 1100 pairs (Muselet 1982). Decreasing (HJB, WB, GM). RUMANIA. Dobrogea: some decline (Van Impe 1977). USSR. Estonia: c. 3000 pairs (Onno 1966); 5000 pairs 1973–7 (Thomas 1982). Latvia: 1000–2000 pairs (Viksne 1978). About 14000 pairs Black Sea and c. 4000 pairs Sea of Azov 1981 (A N Golovkin). TURKEY. 1000–10 000 pairs (OSME). ISRAEL. Fluctuating, 300–475 pairs 1967–76 (AP). TUNISIA. On Kneiss islands over 200 pairs 1971, 150–200 pairs 1977, and Djerba 100 pairs 1976 (M A Czajkowski, MS). MADEIRA. Breeding in small numbers (Bannerman and Bannerman 1965); no recent information. MAURITANIA. 500–900 pairs 1959–65 (Naurois 1969a), 500 pairs (Westernhagen 1970), 185 pairs 1974 (Trotignon 1976). AZORES. Abundant (Bannerman and Bannerman 1966); probably decreasing due to disturbance, etc. (G Le G).

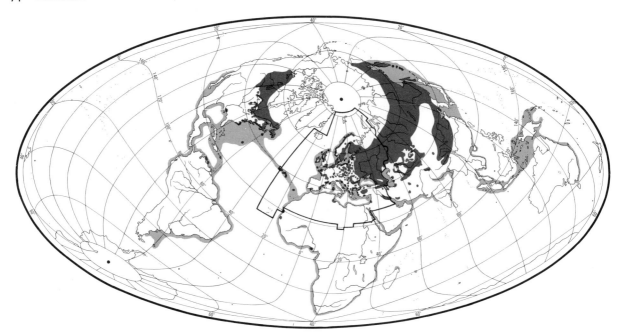

Survival. Population model for declining population, Massachusetts (USA), suggested 7–13% of fledged young survived to first breeding at 4 years old, with thereafter annual mortality 7·5–11% 1940–56, increasing to 13–21% 1970–5 (Nisbet 1978a). At colony New York (USA), over 14·3% survived to first breeding; annual adult mortality c. 8% (DiCostanzo 1980). Oldest ringed bird 25 years (BTO).

Movements. For review, see Muselet (1982). Migratory throughout most of west Palearctic. Majority winter on western seaboard of Africa, principally West and southern Africa; minority appear to winter off Portugal and southern Spain (Muselet 1982). Movements, if any, of population on Banc d'Arguin (Mauritania) not known. Common on passage, Egypt (P L Meininger, W C Mullié). Small spring and autumn passage occurs along coast of Eritrea (Moreau 1972). Heavy spring and autumn passage (thousands together) along coast, also inland, Somalia, where immatures also summer (J S Ash; see below). Regular on coasts of Kenya and Tanzania (Backhurst et al. 1973; Britton 1980); birds ringed in West Germany and Austria recovered in Tanzania, though reached perhaps via Cape of Good Hope (see below). Also recorded, October–February, at Lake Tanganyika (Gaugris 1979). Notable airstrike 2 hrs after sunset on 22 October with flock at 600 m over south-east Gezira, Sudan (14°13′N 33°22′E) further suggests passage along Nile/Rift valleys (E K Dunn). Origin of these birds not known, but possibly Asia Minor; 1-year-old and 2-year-old, ringed Odessa (Black Sea), recovered May and August in Somalia (J S Ash); also one exceptional recovery of Norwegian 2-year-old in upper Nile, Sudan (11°58′N 32°49′E). Common on passage at Elat (Israel) late spring (Safriel 1968). Persian Gulf possibly wintering grounds for birds from central USSR; single birds ringed Krasnovodsk (Turkmeniya) and Novosibirsk recovered on south Caspian coast in January and October respectively, indicating possible wintering there or passage towards Persian Gulf. Peak passages in Gulf of Oman occur April and October (Walker 1981a). Central Asian populations also winter on coasts of western Pakistan and India, Ceylon, and possibly Malay peninsula. South-east Asian race *tibetana* probably winters from east coast of India to Burma and Malay peninsula; east Siberian race *longipennis* from Malay peninsula east through Indonesia and New Guinea, and to lesser extent Australia (Hitchcock 1965), probably straggling to New Zealand (Latham 1979). Birds recovered on passage in China and Japan (Kullenberg 1946; Dementiev and Gladkov 1951c). Nearctic populations winter from southern limits of breeding range through Caribbean and Central America to Peru, southern Argentina, and Falkland Islands (Austin 1953; American Ornithologists' Union 1957; Meyer and Schauensee 1966; Haymes and Blokpoel 1978). Bird ringed as pullus, Massachusetts (USA), recovered in October on Azores (J V Dennis); another, from Great Lakes, recovered in mid-Atlantic off Azores (Haymes and Blokpoel 1978); another from New York, off Ivory Coast in December (Raynor 1970).

European birds show partial differentiation of winter quarters according to country of origin: the more southerly and westerly populations tend to winter north of equator, more northerly and easterly ones south of it (Salomonsen 1955; Isenmann 1972b; Muselet 1982). Birds breeding Britain, Netherlands, Belgium, France, Spain, Switzerland, Austria, and West Germany winter principally along

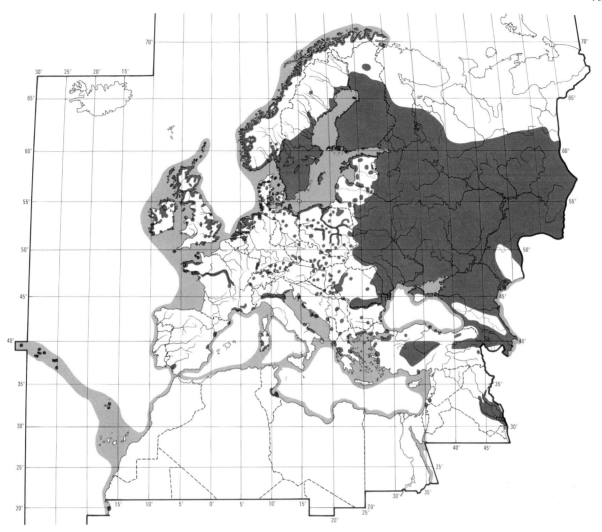

West African coast at c. 10°S–20°N, mostly Mauritania to Nigeria. A few penetrate south to Angola, but rather few recoveries in South Africa (e.g. Rowan 1962, Elliott 1971). Birds from Fenno-Scandia, East Germany, Hungary (probably), and European USSR winter further south in Angola, South Africa, and, to lesser extent, Moçambique (Dementiev and Gladkov 1951c; Milenz 1962; Pátkai 1966; Elliott 1971; Isenmann 1972b; Muselet 1982).

Dispersal begins soon after fledging, as early as July, and continues into October; southward movement most marked August–October, juveniles accompanying adults or travelling alone. Spring migration late March to late May, at peak late April.

SOUTH AND WEST EUROPEAN POPULATIONS. Knowledge of movements derived mainly from ringing. Recoveries of British-ringed birds first analyzed by Radford (1961); following account based largely on study by Langham (1971), with additional material from C J Mead. In Britain (as in eastern North America: see Austin 1953, Nisbet 1976), juveniles and parents initially show north-south dispersal; all recoveries in July were within 160 km of natal colony, suggesting more leisurely dispersal than Arctic Tern *S. paradisaea* or Sandwich Tern *S. sandvicensis*. By September, few recoveries to north, indicating strong southward movement along coast of south-west Europe. By October, mean latitude of recoveries 45°N; in November, rapid movement down West African coast so that recoveries then rarely north of 20°N (Mauritania); in December, most recoveries at 0–10°N (Langham 1971). Ghana apparently the major wintering area, although trapping for food and recreation is particularly popular and consequent high recovery rate may overestimate its importance. Sénégal, Sierra Leone, Ivory Coast, and Liberia next in importance; Mauritania, Gambia, Guinea-Bissau, and Guinea less so (see Table A, not corrected for length of coastline available for recovery). Similar distribution for birds from Netherlands (Speek 1969; C J Mead), France and Switzerland (Isenmann 1972b; Muselet 1982), West Ger-

Table A Relative frequency of recoveries of British-bred Common Terns *S. hirundo* in Afrotropical region up to 1981 (R Hudson).

	%
Mauritania to Guinea	31·6
Sierra Leone to Ivory Coast	27·1
Ghana	34·7
Togo, Dahomey, Nigeria	3·5
Gabon, Angola, Namibia, South Africa	3·1

many (Schloss 1962), and Spain; sparse evidence suggests Belgian population follows suit (Verheyen 1970). British and Dutch birds spend 1st summer and 2nd winter predominantly in northern tropical belt, 10°S–20°N (Langham 1971). Little evidence that 2-year-olds remain south of Sahara in summer (C J Mead); some reach European waters, but not until after mid-June, so that mean latitude of 55°N not attained until August. Many breed at 3 years old and some not until 4 when mean latitude of 55°N maintained April–September. Birds occasionally winter in European waters (Hudson 1973). Birds from south-central Europe (e.g. Switzerland, Austria) generally move south to south-west overland in autumn, some across Swiss Alps, though recoveries in upper valleys of Rhône and upper Loire suggest others, perhaps most, follow lakes and major river systems. Occurs on Swiss lakes up to mid-November. Autumn recoveries of Hungarian-ringed birds in Greece and Yugoslavia imply movement initially south towards Mediterranean, then presumably west to Atlantic (as Black Sea population); one spring recovery in Bulgaria.

NORTH AND EAST EUROPEAN POPULATIONS. Juveniles from Finland, Sweden, and Latvia and Estonia (USSR) disperse south-west towards southern Sweden, Denmark, and southern Baltic; post-fledging dispersal (as in British population) not evident (Tamantseva 1955; Lebedeva 1962; Lemmetyinen 1968). Thereafter, route follows other populations along west European seaboard. A few birds from Black Sea region winter in Turkey; most juveniles and adults travel west through Mediterranean via Balkan peninsula and Sicily to Straits of Gibraltar, then down west coast of Africa as far as Cape Province (Lebedeva 1962; Ardamatskaya 1977). One bird hatched in Black Sea region recovered there in 2nd winter (Tamantseva 1955). Populations from northern and eastern Europe overlap less in winter quarters than those of southern and western Europe, having somewhat different but partly overlapping ranges along western seaboard of Africa. Though Norwegian population reaches South Africa, majority winter in West Africa, notably Ghana and Ivory Coast (Holgersen 1975); Ghana accounted for 52·6% of Norwegian recoveries in Africa over 1956–65 (calculated from Elliott 1971). Danish birds may show similar distribution. Origins of birds recovered in South Africa discussed by Nordström (1961), Elliott (1971), and Isenmann (1972b). ⅔ of recoveries are on west coast of South Africa, ⅓ on east coast (opposite pattern to *S. paradisaea*). Most frequently represented country of origin (taking ringing totals into account) is Finland; over 1956–65, 63% of South African recoveries were of Finnish birds, and 65% of African recoveries of Finnish birds were from South Africa (after Elliott 1971; see also Nordström 1961). After Finland, areas of origin represented in South African recoveries were (in order of importance) West Germany, Norway, USSR (Baltic and Black Seas), and Sweden. West German and Norwegian birds comprised respectively 15·6% and 8·9% of South African recoveries over 1956–65; of African recoveries of West German and Norwegian birds only 8·5% and 15·8% respectively were in South Africa, indicative of wider distribution of these 2 populations to the north. More Swedish birds are recovered in South Africa than elsewhere in Africa, though they, like populations from other Fenno-Scandian countries, also winter in West Africa. Proportion of Russian birds wintering in South Africa not known, but birds from Latvia and Ukraine are regular (Broekhuysen 1965; Donnelly 1966; Elliott 1971). Origins of birds wintering in South Africa confirmed by periodic 'wrecks', e.g. November 1976; such recoveries suggest high incidence of birds from Finland, West Germany, and Estonia (USSR), with smaller numbers from Sweden and Denmark (Rowan 1962; Clancey 1977a). One Swedish bird ringed as pullus was recovered 6 months later near Fremantle, Western Australia (Dunnet 1956), consistent with westerly airstream prevalent in latitude of South Africa. Irish bird also recovered in Victoria, Australia (Rogers 1969).

Interchange of birds, especially first-time breeders, between neighbouring colonies may be frequent, but adults highly faithful to breeding colony (Austin 1949; Haymes and Blokpoel 1978; Di Costanzo 1980). EKD

Food. Chiefly marine fish, though crustaceans predominate at some colonies, Massachusetts, USA (Nisbet 1973a). Opportunist feeder, switching rapidly between prey types and feeding methods as circumstances change. Considerable variation in diet between colonies, years, and even from hour to hour within colonies (I C T Nisbet). Fish caught largely by plunge-diving from the air; often preceded by hovering. Dive height 1–6 m, mean 2·6 (D C Duffy), 3·1 m (Dunn 1972a). Occasionally—regularly with some individuals—perch-plunges from (e.g.) bridge, pier, masthead, moored boat (Walters 1966; Gloe 1980b; B Campbell, I C T Nisbet). Immersion either complete—depth typically 0·2–0·3 m (Boecker 1967), probably not exceeding 0·3 m (D C Duffy), 0·5 m (Dunn 1972a)—or only partial when prey restricted to surface; mean submersion time 1·1 s (164 dives) (Dunn 1972a) or 0·8 s (sample 213), increasing with dive height, maximum 1·6 (D C Duffy). Prey always brought to surface (held crosswise in bill unless very small) and swallowed immediately unless intended for mate or young. Occasionally, more than one caught and carried. Both successive (Hays *et al.* 1973) and simultaneous (Taylor 1975) captures of more than one fish recorded. Multiple loads of up to 5 sand-eels *Ammodytes* observed in 1·7% of fish-carrying birds at colony (Hays et al. 1973). High prey density induces more changes of flight direction, more 180° turns, stronger tendency to search into wind, and slower searching speed than for low prey density (Taylor 1975). Success of plunge-dives affected by size, abundance, depth, and visibility of prey, and probably age and experience of bird (Dunn 1972a). At Ythan estuary (Scotland), capture rate (including aborted dives) declined by 0·2 prey items per min for every 16 km per hr increase in wind speed. In high winds, sought sheltered water (Taylor 1975, 1983). Strong gales reduce rate of feeding young (Boecker 1967). In open sea, however, intermediate wind speeds and surface conditions can be more favourable than either calm or rough conditions. In moderate sea conditions, fishing success (complete dives) 39%, diving rate 1·5 per min, and capture rate 0·5 prey items per min; in calm seas, corresponding figures 22%, 1·0, and 0·23 (Dunn 1973a). In Virginia (USA), overall capture rate 0·34 prey items per min (Erwin 1977). When feeding along seashore over submerged reefs and sand-banks, capture rates higher at low tide than high tide (Dunn 1975) and more food brought into colony at low tide (Boecker 1967; Dunn 1972a; Mes and Schuckard 1976), or during falling tide (I C T Nisbet). Situation differs in estuaries where marine fish advance upriver on incoming tide. At Ythan estuary, including aborted dives, fishing success (54·7%) and diving rate highest on rising spring tides (0·66 prey per min); at neap tides, 0·12 prey per min (Taylor 1975). Shoals of fish attract flocks and induce faster diving rates and yield higher fishing success than more dispersed prey. Flocks had 40–57% fishing success compared with 27–44% for solitary birds on same days (Dunn 1972a; D C Duffy). 3 other main feeding methods. (1) Dipping-to-surface for crustaceans, insects, and other small aquatic invertebrates. At Ythan estuary, increasing wind speed and wave amplitude favour switch to dipping-to-surface and partial submersion, especially when taking sand-eels *Ammodytes*; more successful than complete submersion (Taylor 1975, 1983). In northern England, dipping-to-surface for shrimps *Crangon* resorted to when fish scarce (Dunn 1972a). Used in West Africa to catch prey escaping from seine nets or driven to surface by predators, especially tuna *Thunnus* (Dunn 1972b; Nisbet 1973b; Grimes 1977). For same reason, occasionally associates at sea with (e.g.) auks (Alcidae) which chase prey underwater (Cantelo and Gregory 1975). (2) Aerial-pursuit of insects. (3) Food-piracy of other *Sterna*. Usually occurs sporadically, and most commonly involves one or a few individuals chasing fish-carrying conspecifics, although at Petit Manan Island (Maine, USA) Arctic Terns *S. paradisaea* regularly, though infrequently, robbed of large fish after long chases (Hopkins and Wiley 1972) and Roseate Terns *S. dougallii* frequently robbed at sea, Great Gull Island (New York, USA) (D C Duffy). Robbing can be commonest if food shortage coincides with peak food demand in colony (Chestney 1970; Dunn 1972a). Often occurs as food passed to young (Hays 1970). 2 birds at colony, Massachusetts, were regular food-pirates of adults and chicks, and had unusually good breeding success. 1 of these, and other birds, also picked up fish freshly dropped on ground (I C T Nisbet). In West Africa, takes fish escaping from nets hauled ashore, behaviour which makes them susceptible to capture by snare traps and lines baited with dead fish (Dunn 1972a; Bourne and Smith 1974; Mead 1978). Once recorded scavenging on foot on upper beach, Sierra Leone, winter (Dunn 1972a); in breeding areas, sometimes hunts on foot for beetles (Coleoptera) (Bauer 1965). In times of scarcity in breeding season, and regularly in winter quarters, may eat fish offal thrown into water by fishermen (Forbush 1925; E K Dunn; see also Watson 1981). Recorded following boats and catching minnows tossed to them in the air (Bretsch 1926). Day-time feeder, though most active early morning and evening (Langham 1968; Pearson 1968; Dunn 1972a; Taylor 1975; Mes and Schuckard 1976). Peak foraging times at Wangerooge (West Germany) dictated more by tidal than diurnal cycle (Boecker 1967). Feeds mostly within 3–10 km of nest-site (Borodulina 1960; Boecker 1967), up to 22 km (Pearson 1968), 37 km (Andrews 1971), or even 'some scores' of km (Borodulina 1960). ♂ may forage further than ♀, especially in first few days after hatching. (Nisbet 1973a).

Marine fish in diet include herring *Clupea harengus*, sprat *C. sprattus*, sand-eels *Ammodytes marinus*, *A. tobianus*, pollock *Pollachius virens*, haddock *Melanogrammus aeglefinus*, whiting *Gadus merlangus*, cod *G. morhua*, Norway pout *Trisopterus esmarkii*, rockling *Onos cimbrius*, mackerel *Scomber scombrus*, sticklebacks (*Gasterosteus aculeatus*, *Spinachia vulgaris*); less commonly, pipefish *Siphonostoma typhle*, weever *Trachinus vipera*, lumpsucker *Cyclopterus lumpus*, gurnard *Trigla*, squirrel hake *Urophysis chuss*, gobies (Gobiidae), butterfish (Stromateidae), flatfish (Pleuronectiformes). North American species include herrings *Etrumeus*, *Alosa*, menhaden *Brevoortia*, sand-eels *Ammodytes americanus*, silversides *Menidia menidia*, cunner *Tautogolabrus adspersus*, chubb mackerel *Scomber colias*, anchovies (*Anchoa*, *Stolephorus*), killifish *Fundulus*, bluefish *Pomatomus saltatrix*, pipefish *Siphonostoma fuscum*, lamprey *Petromyzon marinus*, winter flounder *Pseudopleuronectes americanus*, yellow perch *Perca flavescens*. In winter quarters, Sierra Leone, anchovy *Engraulis guineensis* and mullet (Mugilidae) commonly taken (Dunn 1972a). In Ghana, sardines, notably *Sardinella aurita* and *S. eba*, are important (Grimes 1977). Fish of fresh and brackish water include roach *Rutilus rutilus*, perch (*Perca fluviatilis*, *P. flavescens*), ruffe *Gymnocephalus cernua*, pike-perch *Stizostedion lucioperca*, bleak *Alburnus alburnus*, bream *Abramis brama*, rudd *Scardinius erythrophthalmus*, carp *Cyprinus carpio*, freshwater minnows (Cyprinidae), goldfish *Carassius carassius*, chub *Squalius cephalus*, smelt *Osmerus eperlanus*, viviparous blenny *Zoarces viviparus*, goby *Gobius minutus*, salmon *Salmo salar*, trout *S. trutta*, whitefish *Coregonus albula*, eel *Anguilla anguilla*, and flatfish (Pleuronectiformes). Crustaceans include shrimp *Crangon vulgaris*, prawns (*Leander serratus*, *Palaemonetes varians*, *Penaeus*, in USA), shore crab *Carcinus maenas*, mole crab *Emerita talpoida* (in USA), and isopod *Idotea baltica*. Insects include beetles (Coleoptera), especially cockchafers (Melolonthinae) and water-beetle larvae, moths and occasionally butterflies (Lepidoptera), flies (Diptera), caddisflies (Trichoptera), ants and even bees (Hymenoptera), grasshoppers and crickets (Orthoptera), mayflies (Ephemeroptera), dragonflies (Odonata), cicadas and other bugs (Hemiptera). Small squid (Cephalopoda), polychaete worms (*Arenicola*, *Nereis*), and leeches (Hirudinea) also occasionally taken. Berries and vegetation fragments occasionally recorded. (Bent 1921; Forbush 1924; Collinge 1924–7; Marples and Marples 1934; Mendall 1935; Palmer 1941a; Dementiev and Gladkov 1951c; Baxter and Rintoul 1953; Moore and Boswell 1956a; Mills 1957; Borodulina 1960; Bauer 1965; Boecker 1967; Langham 1968; Pearson 1968; Bartlet 1971; Dunn 1972a; Hays and Risebrough 1972; Hopkins and Wiley 1972; Hays *et al.* 1973; Lemmetyinen 1973b; Nisbet 1973a; Unhola 1973; Greenhalgh and Greenwood 1975; Taylor 1975; Ladhams 1976; Mueller 1976; Lehtonen 1981; I C T Nisbet.) Fish prey 2·5–8 cm (mean 7·5) in length (Langham 1968); mean 5·5 cm (Dunn 1972a). Shrimps caught were larger than average of those available (Taylor 1975).

48 adults, Norfolk (England), contained 25·5% (by number) whiting, haddock, herring, or sprats, 14·8% sand-eels, 14·2% crustaceans, 15·4% annelids, 10·2% molluscs, 14·7% insects, mostly cockchafers (Collinge 1924–7). From over 500 sight records, prey fed to chicks at Farne Islands (Northumberland, England) comprised 44% (by number) Ammodytidae, 38% Clupeidae, 11% Gadidae, 2% Gasterosteidae, 3% Trachinidae, and 1% each of crustaceans and cephalopods (Pearson 1968). Clupeidae outnumbered Ammodytidae 7:3 in diet at Coquet Island, Northumberland (Langham 1968); there, at Farne Islands, and at Ythan estuary, Clupeidae relatively more important as breeding season progressed. At Ythan estuary, sand-eels and Clupeidae main prey at high tide, shrimps and blennies at low tide (Taylor 1975). At Ribble estuary (England), sand-eels main prey at high tide, sprats at low tide (Greenhalgh and Greenwood 1975). At Wangerooge, 75% of items fed to chicks were fish (85% by weight, half of fish Clupeidae), 25% invertebrates; crustaceans 18% of total (mostly shrimps and shore crabs), also polychaete worms, cephalopods, and adult caddisflies *Stenophylas permistus* (Boecker 1967). At Petit Manan Island, young fed exclusively on herring (Hopkins and Wiley 1972). At the Sugarloaves (Maine, USA), diet of adults and young comprised largely herring and sand-eels (Palmer 1941a). 116 stomachs from North American colonies contained 95·5% fish (mostly freshwater minnows and sand-eels), 3·5% insects (mostly moths), and 1% all other invertebrates (molluscs, crustaceans, and worms) (McAtee and Beal 1912). 155 stomachs, Maine, contained mainly herring, mackerel, and shrimps; 3 contained berries and vegetation (Mandall 1935). 17 stomachs, River Elbe (West Germany), contained mostly smelt, sand-eel, and a few mayfish *Alosa vulgaris* and grundling *Gobio fluviatilis* (Peters 1933). In Volga delta (USSR), predominantly fish, proportions of bream, roach, rudd, bleak, sticklebacks, and carp varying seasonally (from 529 stomach regurgitations). Elsewhere in Black Sea region (USSR), smelts, *Clupeonella*, pike-perch, sticklebacks, and shrimps important (see Borodulina 1960). For freshwater diet at Austrian colony, from pellet analysis, see Bauer (1965). In Kustavi archipelago (Finland), mainly fish: mostly sticklebacks, but also Cyprinidae, especially bleak and perch (Lemmetyinen 1973b). Among 12 pairs at Bird Island (Massachusetts, USA), 14% of items ♂♂ fed to mates during courtship were fish (52% by weight), and 86% shrimps (48% by weight) (Nisbet 1973a). At Ythan estuary, ratio of fish to shrimps presented to mate was higher than ratio caught. Moreover, significantly larger fish (of both Ammodytidae and Clupeidae) retained for mates and chicks (Taylor 1975). At Bird Island, ♂♂ that performed well at both courtship and early chick-rearing periods were those that switched most successfully from shrimps and full-grown silversides at time of laying, to silverside fry and cunners at hatching (Nisbet 1973a). Rates of provisioning ♀ during egg-laying and 'honeymoon' phases of courtship were 1·9 g per hr (mean) and 6–7 g per hr (estimate) respectively (Nisbet 1977).

Chicks less than 1 week old receive smaller and probably more easily handled prey than that fed to older chicks (Boecker 1967). Older chicks, Kustavi archipelago, received proportionally more, and larger, sticklebacks than did younger chicks (Lemmetyinen 1973b). In Northumberland, sand-eels and small Clupeidae fed to young chicks, but larger Clupeidae more prevalent as chicks grew (Langham 1968); as chicks aged, proportionally more fish caught close to colony fed to them rather than being eaten by parents (Taylor 1975).

Broods of 1, 2, and 3 at Farne Islands received respectively 15, 23, and 38 feeds per day, 0·8, 1·8, and 2·1 feeds per hr (Pearson 1968); at Wangerooge, 0·6, 1·8, and 2·7 feeds per hr (Boecker 1967). In Virginia (USA), on fish diet, broods of 3 received estimated 4 feeds per hr (Erwin 1977); in Volga delta, 0·8 feeds per hr, equivalent to 18 g per day, whereas at Chernomorski (Black Sea), where 34% of diet was shrimps, broods of 3 received 1–1·6 feeds per hr (Borodulina 1960). Smaller, especially invertebrate, prey produce higher feeding rates (Boecker 1967; Langham 1968). In Finland, young ate average 798g up to fledging (Lehtonen 1981; see also for daily food requirements).

EKD

Social pattern and behaviour.

1. Mostly gregarious throughout the year. Fish shoals attract dense feeding flocks but feeding otherwise often solitary, or in small loose hunting parties. 63% fed in dense flocks, 23% in medium flocks, and 14% dispersed, New York, USA (D C Duffy). Feeding flocks of 200 recorded; most commonly 2–10 (Erwin 1978). In winter quarters, 1st-winter and older birds associate freely, commonly in company with other Palearctic Sternidae, notably Sandwich Tern *S. sandvicensis*, Roseate Tern *S. dougallii*, and Black Tern *Chlidonias niger*. After breeding season, usually encountered in family parties or small flocks of adults and juveniles; all-juvenile flocks occur on migration (see Bonds, below). BONDS. Monogamous pair-bond, persisting from year to year, with records of pairings maintained for 4 years of study (Albertson 1934; Austin 1947). Of 122 pairs studied over 4 seasons, Massachusetts (USA), 79% of pairings persisted for at least 2 seasons (Austin 1947). Less than 0·1% return to colony at 1 year (I C T Nisbet; see also Haymes and Blokpoel 1978), many at 2 years, some probably not until 3. Some probably pair, or attempt to do so in colony in season prior to first breeding (I C T Nisbet). Pairing and breeding by 1-year-olds highly exceptional and some apparent records may concern individuals attaching themselves to established breeding pairs, helping to incubate eggs and feed young, but probably incapable of egg-laying themselves (Austin 1932; Palmer 1941b). From thousands of ringing recoveries in colonies in Massachusetts, c. 16% bred at 2 years old, most at 3–4 (Austin 1945; Langham 1971). In Great Lakes (North America), breeding at 2 years more common (Haymes and Blokpoel 1978). Mates do not associate closely in winter quarters, but majority appear to arrive at nest-site already paired, having joined up in the few days beforehand in and around colony, or even earlier when on passage (Marples and Marples 1934; Palmer 1941a; Austin 1947). Both sexes incubate (♀ taking greater share) and care for young. For 4 days after hatching, ♂ does most fishing while ♀ stays to brood (Nisbet and Drury 1972; Nisbet 1973a). After 5–7 days, both forage, though one parent (♂ in recorded cases) may raise part or whole of brood to fledging if mate dies (Nisbet et al. 1978). Juveniles generally spend most of first 3–4 days after fledging in parents' territory, thereafter resorting increasingly to shore to be fed, though feeding on territory seen up to 23 days after fledging; after c. 5 days, juveniles make first short flights to sea and begin to learn fishing; after 9–10 days, start accompanying parents to feeding grounds up to 2–6 km from colony but return at nightfall. Most remain attached to colony for 10–15 days after fledging, dispersing thereafter (Nisbet 1976); fed by parents (usually both) for at least 6 weeks (Nisbet 1976), possibly months, after fledging (see also Roosting, below); some, perhaps late-hatched, offspring at least partly dependent for food in winter quarters (E K Dunn), though family bonds may be broken before or during migration—since some migratory flocks, USA, apparently composed entirely of juveniles (Palmer 1941a). Birds, especially ♂♂, occasionally pair with Roseate Tern *S. dougallii* (Robbins 1974; Burggraeve 1977); such pairs possibly stable over several seasons (e.g. Berg 1979). At colony in New York (USA), 2 cases of hybrids pairing and fledging young (Hays 1975). BREEDING DISPERSION. Sometimes solitary (Bergman 1980) but usually colonial, typically now in groups of a few to 1800 pairs: mean 263 pairs in 1974, Britain and Ireland (Lloyd et al. 1975); mean 95 pairs (31 colonies) on coast of Virginia, USA (Erwin 1978); 1–554 pairs (34 colonies) in New Jersey, USA (Burger and Lesser 1977). Islet of Griend (Netherlands) held up to 25 000 pairs until c. 1955 (C S Roselaar). Inland colonies on rivers generally smaller and less dense than those on coasts (Cramp et al. 1974). Breeds mostly in homogeneous groups, but sub-colonies occur in salt-marsh habitats (Greenhalgh 1974). Mean distance between nests $160·4 \pm 49$ cm in Virginia colony of 55 nests (Erwin 1977); mean 350 cm, minimum 50 cm at Chernomorski, USSR (Borodulina 1960); may be as close as 43 cm—thought to be minimum tolerable (Palmer 1941a). Average density of nests on Coquet Island (Northumberland, England) 0·06–0·13 per m^2 in different years (Langham 1974). Defends small, roughly circular nest-area territory; mean area, Massachusetts, 2 m^2 (Nisbet and Drury 1972) or c. 3 m^2 (Austin 1929), but may be much smaller—and not necessarily circular, especially on uneven terrain. Boundaries not absolutely rigid, varying with antagonism between neighbours (Palmer 1941a), especially early in breeding season. Territory used for ground phase of courtship and pair-formation, copulation, nesting, and concealment and location of young up to 23 days after fledging (Nisbet 1976). Fidelity to previous nest-site strong (Austin 1940; Mänd 1982); one pair occupied same site for 17 years, curtailed by death of one bird (Marples and Marples 1934). Older, experienced birds both arrive at colony and establish territories earlier than younger birds (Hays 1978); in one study, over 90% returned to territory of previous year. Younger birds tend to seek site in natal area, and more likely than older birds to change site and even colony between years (Austin 1940); in Great Lakes, however, birds breeding for first time tended to settle in colony other than natal one (Haymes and Blokpoel 1978); in unpaired birds, ♂ establishes territory. Occasionally, early display territories slightly separate from nest-territories, especially in dense vegetation (as in Arctic Tern *S. paradisaea*). At Bird Island (Massachusetts), at all stages of tide, individual ♂♂ also defended fixed feeding territories along 200–300 m of shore, serving 'honeymoon' (courtship-feeding) phase of cycle; ♀♀ did not participate in defence (Nisbet 1977). At low tide, Ythan estuary (Grampian, Scotland), feeding territories at water's edge defended daily from conspecific intruders by individuals or pairs, birds feeding solitarily (at least 50 m apart) on shrimps *Crangon* (Taylor 1975). Commonly nests close to other Sternidae, notably *S. paradisaea*, *S. dougallii*, and *S. sandvicensis*; in USSR, sometimes close to Little Tern *S. albifrons* and Little Ringed Plover *Charadrius dubius* (Borodulina 1960; Bergman 1980; Blinov and Blinova 1980). ROOSTING. Both during and outside breeding season, readily forms communal roosts and loafing groups (see also Flock Behaviour, below), especially when conditions unsuitable for fishing—e.g. at high tide or at night (Conrad 1979a)—or after bouts of fishing; usually on flat open land, like beaches, mudflats, rocks, or saltpans—occasionally on open sea (Dunn 1972a). In winter quarters, February–April, feeds at upwellings up to 600 km off Guinea coast from dawn to dusk; may thus roost on sea or else remain airborne all night (see Grimes 1977). Bathing, followed by preening, commonly occurs at shore roosts. Individuals may regularly occupy and defend fixed loafing sites such as piles and buoys (Palmer 1941a); fledged juveniles habitually occupy similar sites where parents locate them for feeding. Where summer feeding territories established (see Breeding Dispersion, above), one member of pair (♀ during courtship-feeding phase) often loafs inside territory on beach, rock, boat, or other prominent spot, while other fishes in it (Taylor 1975; Nisbet 1977).

2. Following account based largely on Palmer (1941a), Cullen (1960a), and information from I C T Nisbet; for earlier studies, see Tinbergen (1931, 1938), Marples and Marples (1934), and Southern (1938). FLOCK BEHAVIOUR. On first arrival at colony, birds roost together at night, often on shore near (but not in) colony, favouring areas just above tideline; arrive after northerly migration in small groups, seldom exceeding a dozen birds, some already paired. Members of pair stand closer to each other than to other birds in flock, and may show courtship behaviour. Later,

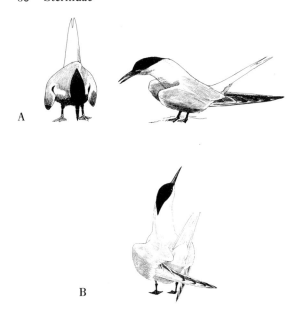

during incubation, off-duty birds form shore roost near fishing grounds, also joining failed breeders and non-breeders in club areas at edge of colony, usually below high-water mark. Just prior to and during migration, roosting flocks contain family parties, and mixed-age roosts persist into winter quarters. In colony, first bird on ground to detect danger gives 'kyàr' call (see 1 in Voice); may raise wings, and take off with rapid downstrokes. Alerted neighbouring birds fall silent and, if they take off ('upflight', so-called 'alarm'), remain silent till in flock over colony. Chick leaving cover may attract adult flock hovering overhead from which individuals periodically swoop as if attacking or driving it back to cover. Flock may also hover over injured or freshly dead adult — usually silently if it is merely enfeebled (or dead) but noisily (with Alarm-calls: see 1 in Voice) if it is evidently injured and attempting to move about; in latter case, flock typically tracks ground movements of bird, and may collectively attack and even kill it. 'Dreads' involve some or all birds suddenly quitting colony to fly low, swiftly, and silently seawards in dense flock; after a few seconds or minutes, birds begin to call again and return individually to former positions in colony. Dreads involving birds mostly already in flight called 'panics' (Marples and Marples 1934). Behaviour usually spontaneous, with no obvious stimulus, and function therefore obscure, but probably associated with protection against aerial predators (Cullen 1960a). Dreads also show diurnal and seasonal patterns of frequency, associated with incidence of courtship display, and are most intense in densest nesting groups (as in *S. paradisaea*). For flock reaction to predators, see Parental Anti-predator Strategies (below). ANTAGONISTIC BEHAVIOUR. Competition for nest-sites normally involves threat and fighting; this pronounced at start of season and when nests close together. (1) Ground encounters. ♂ on territory invariably challenges intruder regardless of sex, but ♀ generally more tolerant, especially if intruder has fish (I C T Nisbet). ♀ may help mate to repulse lone intruder but usually only briefly (Palmer 1941a). Territorial ♂ or mate first warns off potential ground-trespasser with Alarm- or repeated Advertising-calls (see 1 and 4 in Voice); given more rapidly, with wing-raising, if advance persists. From outset, rivals face each other directly in ground version of intimidating Bent-posture ('the Ground Bent'), with head down, neck plumage ruffled, breast low, carpal joints drooped, and tail up (Fig A). As Bent-posture becomes more intense (Fig A, left), bill pointed almost vertically downwards so that tip nears ground, wings more drooped and outspread, and tail more raised (I C T Nisbet). If warning fails, defender may suddenly become silent and dart at intruder; they grapple with bills interlocked or else try to seize each other's head or neck, sometimes accompanied by Anger-calls (see 2 in Voice). Each tries to pull other towards it. Loser, usually intruder, signals submission by adopting ground version of appeasing Erect-posture ('the Erect') with bill skywards, wings drooped, and tail lifted (Fig B). Defence of territory by pair against other pairs strongly ritualized: involves repeated bowing, synchronized in the defending pair, between intense Bent- and Erect-postures (I C T Nisbet), and often accompanied by Advertising-calls. Favourite loafing sites on feeding territory (see Breeding Dispersion, above) and elsewhere also defended from ground intruders using Bent-posture. ♂ standing with fish, even in another's territory, may elicit mixed response from neighbouring incubating bird of either sex (I C T Nisbet). Latter may first adopt submissive and appeasing Hunched-posture: crouches horizontally, head retracted (Fig C, right), and then rushes aggressively at other, attempting to wrest fish from bill. Older chicks trespassing outside territory vigorously pecked by neighbouring adults; react by crouching submissively and thus inviting brooding. (2) Aerial encounters. Owner returning to find stranger on territory gains height and swoops bill-first on intruder, giving Anger-calls. Adjacent neighbours in flight may fight in air, or perform highly characteristic Upward-flutter (see Marples and Marples 1934) in which participants rise vertically from ground (sometimes till nearly out of sight), ascending gradually in half-hover with rapid wing-beats and revolving spirally, mostly one above the other; then separate and fly away leisurely, though still with rapid wing-beats. Such encounters apparently hostile, with birds intermittently flying bill-first at each other, accompanied by Anger- and Kip-calls (see 2 and 6 in Voice). A 2nd ascent may follow if 1st brief, or if one bird swoops aggressively at other on return to ground (Palmer 1941a; Hawksley 1950; Cullen 1960a). Conspecific birds and *S. sandvicensis* intruding on feeding territories (see Breeding Dispersion, above) evicted in the air by owner adopting variant of aerial Bent-posture (see Heterosexual Behaviour, below), either with arched back and stiff wing-beats and uttering Advertising- and Anger-calls (Taylor 1975), or, commonly, by flying down towards intruder and accelerating when low over water, often causing it to retreat when still 200 m from feeding territory (I C T Nisbet). Upward-flutter also occurs between owners of neighbouring feeding territories (Taylor 1975). HETEROSEXUAL BEHAVIOUR. (1) General. Pair-formation involves both aerial and ground behaviour at colony. See also Bonds and Flock Behaviour, above. (2) Aerial courtship. Consists of 2 performances — High-flight and Low-flight — both of which culminate in Pass-ceremony ('the Pass': Fig D); descriptions based largely on Cullen (1960a).

D

High-flight characterizes early stages of pair-formation; often starts after a dread, but also occasionally follows Upward-flutter between territorial ♂♂. Starts with circling ascent and rapid wing-beats (Jerk-flying) up to 200 m (Bergman 1953a) by 2 birds 20–50 m apart, usually diametrically opposite each other. ♂ (never ♀: I C T Nisbet) usually carries fish and leads, while ♀ pursues. During ascent, leader sometimes gives Kip-call. At peak of ascent, pursuer closes gap and flies either low over or to side of leader (Pass-ceremony); during the pass, pursuer adopts aerial version of appeasing Erect-posture ('the Straight': Fig D, right), with bill forwards, head tilted away from partner, and wings arched downwards. Pursuer in Pass-ceremony may give Begging-call (see 7 in Voice), in this case without any marked posture. Leader being overtaken adopts aerial version of Bent-posture ('the Aerial Bent') (Fig D, left), head often tilted away from bird in aerial Erect-posture, wings raised in V, and may give Advertising-call before Pass-ceremony, Kor-call during it, and sometimes—after pass—the Kruk-call (see 4, 5, and 8 in Voice) (Palmer 1941a). Once in front, bird in aerial Erect-posture initiates fast, downward glide, swaying from side to side in front of bird in aerial Bent-posture. Several passes may occur during glide which is steep at first, flattening out towards ground. At each pass, respective postures are accentuated, but there is no switching of postures, and fish-bearer (♂) almost always adopts aerial Bent-posture (Cullen 1960a; I C T Nisbet). After one or several High-flights, birds usually separate or retire together to fish. High-flight occasionally performed by several birds (more than one pursuer) or even solitary one. Low-flight nearly always initiated by ♂—often as advertising display by unpaired bird, but also when pair-bond already established; commonly called 'Fish-flight' (Tinbergen 1938) but food often not carried (Cullen 1960a). ♂ starts V-flying with wings held high, beating with slow, shallow strokes, and giving Advertising-call. Another bird (♂ or ♀ but always ♀ if advertising ♂ has no fish) may be attracted, and advertising ♂ then adopts aerial Bent-posture, inviting a pass, and other adopts aerial Erect posture; 1–2 passes, with calls as above, may follow and may develop into High-flight, especially if ♂ has fish. If partner leaves, advertising ♂ may go into High-flight alone or return to territory; alternatively, other bird may follow and ground courtship ensues. (3) Ground courtship. Early in season, behaviour on ground involves peculiar shuffling gait and rudimentary posturing. After Low-flights have progressed for several days, birds come to ground more frequently as ♂♂ succeed in enticing ♀♀ into territories. Ground courtship typically occurs within, but also away from territory; complex and variable. Involves birds of varying degrees of relationship—from strangers, even of same sex, to well-adjusted mates (I C T Nisbet). ♂, with or without fish, usually adopts inhibited ground Bent-posture—directed to one side of ♀, often with head tilted away from her (at most intense, ♂ leans body so that carpal joint furthest from ♀ touches ground)—and walks ('Parades': Palmer 1941a) in circle around her with short, mincing steps, neck craned forwards, bill slightly downwards. Both birds may give Kruk-call. ♀ stands in centre, often with head slightly bowed and tilted away from ♂. If unresponsive to advertising ♂, ♀ rotates on spot so as always to face him. Otherwise, ♀ retreats, or walks in Erect-posture round ♂, to which ♂ may respond by halting and also assuming Erect-posture, but with bill less elevated; if ♀ lowers head again, ♂ resumes circling. Sequence and patterns of ground courtship change and become ritualized as pair-bond develops. Later, bird on territory may only dip and raise bill (mild Erect-posture) at return of mate. If ♂ in ground courtship has fish, ♀'s approach is towards fish which ♂ is then careful to avoid having taken too quickly (♂♂ sometimes play ♀ role till they get fish, then switch back to ♂ role and court ♀♀ with same fish). If advertising ♂ has no fish, ground courtship usually followed immediately by scrape-making, initiated by ♂, advertising possession of territory: bird bends forwards (breast touching ground), pivoting and kicking dirt backwards. If ♀ is interested in territory, she also scrapes, pair moving around territory, scraping alternately in same places (so-called 'House-hunting'), giving Gurr-call (see 9 in Voice). One scrape so made may later be used for laying. (4) Mating. Copulation may begin well in advance of egg-laying and is often 'abortive' at outset. Pre-copulatory displays more consistent than ground courtship (I C T Nisbet): commonly, ♂ walks in circles or 'horseshoes' in front of ♀, his neck stretched up, head craned forward, breast forward, and wings folded close to body. ♀ may respond aggressively, retreat, or adopt submissive Hunched-posture (crouching horizontally with head lowered and tucked into body) and give Begging-call (see 7 in Voice) with bill open between calls. Intensity of posture and calls indicates readiness to copulate. ♂ usually mounts only when ♀ halts and crouches. If undisturbed, ♂ mounts for c. 1–3 min, copulating several times while still mounted. ♀ raises head up to ♂ at moment of copulation. When ♂ steps down after successful copulation, both birds adopt extreme Erect-posture, then preen; if copulation incomplete, further pre-copulatory display commonly follows. (5) Courtship-feeding. Copulation frequently follows Courtship-feeding and some newly paired ♀♀ will copulate with strange ♂♂ if they have accepted fish from them (I C T Nisbet). ♀ solicits food from ♂ by begging (see above). Early in courtship, ♂ may present fish, then give own Begging-call to elicit its return. 3 stages of Courtship-feeding first recognized by Cullen (1962), and described in detail by Nisbet (1973a, 1977): significance of fish in display-flights probably mostly symbolic but, once paired, ♂ feeds ♀ regularly, making major nutritional contribution prior to egg-laying. After preliminary display stages, pair retire to fishing grounds for 5–10 days ('honeymoon period') where ♂ feeds mate frequently; both visit nest-territory occasionally and return to colony in evenings. Finally, 6 hrs to 6 days before 1st egg laid, ♀ remains on territory and is fed there by ♂ until clutch complete, doing little or no fishing for herself, and starting incubation after 1st egg laid. Rate of Courtship-feeding declines sharply from 2 days after 1st egg laid when ♂ begins to share incubation. (6) Nest-relief. Precipitated by fish-bearing ♂ approaching with Advertising-call, alighting, and walking towards nest with head held low, and sometimes giving Kruk-call (see 8 in Voice); sitting ♀ may give Begging-call, whether or not ♂ has food, and rise to crouch near him. Later, relief may occur without ceremony, though relieving bird sometimes raises wings briefly on alighting near nest (Palmer 1941a). Relieved bird, before or on leaving nest, almost invariably performs strongly ritualized side-throwing (reaching out to pick up loose material, passing it backward over back and wings, and dropping it) which brings material towards nest. Relieved bird nearly always gives Kip-call on take-off. Incubating bird may bill previously tossed material into nest structure. Off-duty bird frequently allopreens tail and wing-tips of incubating mate

(I C T Nisbet). RELATIONS WITHIN FAMILY GROUP. Based mostly on Palmer (1941a) and information from I C T Nisbet. Chicks stimulate parents to feed them by pecking at bill, supplemented by simultaneous food-calls though these only weak at hatching and mainly effective from 2 days. Parents often wash fish dropped by chicks. ♂♂ may forage further than ♀♀ (Borodulina 1960), especially in first few days after hatching (Nisbet 1973a). Pecking of adult's bill—silently or with subdued food-call—stimulates brooding. Under heat stress, parents sometimes dip belly and feet in water and return to wet young, pair alternating rapidly at nest to do so (I C T Nisbet). Parents recognize own young by 2nd day. From 4th day, young distinguish parents' Advertising-calls (Stevenson et al. 1970). At this stage, young seek refuge separately in vegetation within nest-territory, emerging when they hear parent approaching. Older young sometimes gain significant proportion of diet by stealing fish from neighbouring (usually younger) chicks (Nisbet et al. 1978). Parents may peck their own dead or dying chicks out of nest, especially undersized 3rd chicks or runts. Older young and juveniles often pick up objects such as pebbles and sticks, sometimes repeatedly dropping or tossing them aside and retrieving them. Juveniles sometimes make scrapes to lie in for cooling themselves. ANTI-PREDATOR RESPONSES OF YOUNG. On approach of danger, downy young crouch in response to Alarm- or Kip-call of parents. Broods up to 5 days old tend to crouch together outside nest, but thereafter seek separate refuges in vegetation, among stones, driftwood, etc. From c. 12 days old, young exhibit escape reactions, taking to water if nearby, and also showing slight tendency to flock with others of same age if harassed by humans. Juvenile occasionally performs erratic flight; resembles predator avoidance of gulls (Laridae) (Moynihan 1955, 1959). PARENTAL ANTI-PREDATOR STRATEGIES. (1) Passive measures. At approach of danger, adults give Alarm- and Kip-calls (see above for response of young). In response to predation by Great Horned Owl Bubo virginianus, readily deserts colony at night, North America, returning in morning (Nisbet 1975). (2) Active measures. Avian predators on ground, birds that resemble them (e.g. pigeons Columba), and man subjected to concerted aerial attacks in manner used against conspecific intruders on nest-territory (see Antagonistic Behaviour, above). From flock hovering and wheeling above predator, and uttering Alarm-calls, constant succession of birds makes swoops, many striking with bill (sometimes drawing blood, even killing) and occasionally with hind-body or feet, accompanied (at lowest point of swoop) by defecation and Anger- and Growl-calls (Moffat 1941; Rankin et al. 1942; Manville 1949: see 2–3 in Voice). Airborne predators pursued at speed and buzzed (Veen 1977). For aerial predators, Massachusetts, see Nisbet (1975). Attacks other mammalian predators much more vigorously than man, forming dense, vociferous, mobbing flock above intruder, swooping at it. Attacks stoat Mustela erminea, uttering Anger-call, not Alarm-call (see Simmons 1952).

(Fig A from photographs in Southern 1938; Figs B–C from Palmer 1941a; Fig D from Cullen 1960a.) EKD

Voice. Most vocal at breeding grounds; except where indicated, calls attributable to both sexes. Calls generally harsher, more rasping, and lower pitched than those of Arctic Tern S. paradisaea.

CALLS OF ADULTS. (1) Alarm-call. Usual version a harsh, shrill, often extended 'keee-arrr' or 'ke-arr' (Palmer 1941a); given on ground or in the air. Emphasis variable—commonly on 1st, but also occurs on 2nd syllable, alternatively, may be heard as single 'keeeerr' (Fig I). Generally starts at lower pitch than in S. paradisaea (J Hall-Craggs). Given slowly and at intervals at sight of conspecific bird starting to trespass on roost or feeding territory (Taylor 1975), more rapidly if encroachment continues, here containing element of threat. Uttered in lazy, drawn-out fashion when partly habituated to object of alarm (Marples and Marples 1934). Related 'kyàr'—shriller and more clipped than usual version—denotes fear; given just before taking flight, immediately silencing and mobilizing surrounding birds (I C T Nisbet). (2) Anger-call. A staccato 'kek-kek-kek-k-k' in confrontations and fights, especially among ♂♂ in territorial disputes, notably in dive-attacks on conspecific birds and predators. More subdued, shorter run of calls—'kek-k'—may accompany physical contact in Upward-flutter (Palmer 1941a). (3) Growl-call. A 'kaaarrr' of extreme anger and aggression, heard in aerial and ground confrontations. (4) Advertising-call. Basic version described as 'keeur' (I C T Nisbet) or 'kierr' (Palmer 1941a); commonly a more extended 'keeuri', 'keeri' (I C T Nisbet), or 'kee-eri' (Marples and Marples 1934). Given by bird flying towards mate or young with food, often beginning far from nest when bird first makes catch; probably identifies bird to mate and offspring (Stevenson et al. 1970). Also used in defence of nest or feeding territory; heard outside breeding season (Tinbergen 1931). A mild, subdued 'kearii-kearii', widespread in newly arrived flocks and in fishing flocks containing juveniles; perhaps to maintain contact in both contexts (Palmer 1941a). More elaborate 'keeur keeuri keeri keeri keeri' (first 2 notes descending, successive ones increasingly ascending) given by advertising ♂ in High-flight or Low-flight especially before Pass-ceremony; also when flying around looking for a mate, or standing on territory awaiting her return (I C T Nisbet: see Fig II for 'keeur', III for 'keeuri'). (5) Kor-call. A rapid, low 'kor kor korkrrr' or 'kerkerker', given by bird in Bent-posture (normally ♂) during aerial Pass-ceremony in High- and Low-flights (Palmer 1941a; Cullen 1960a); often joined to initial 'kierr' (call 4) (Fig IV). Also given by ♂ in reply to Begging-call (see 7) (of either sex but usually mate) on nest-territory (Palmer 1941a). (6) Kip-call. Short, sharp, high-pitched note given in variety of contexts, generally denoting excitement. Sometimes heard in Upward-flutter, and almost invariably on take-off from territory (I C T Nisbet). Also heard in fishing flocks, especially where capture rate high or if bird makes determined effort at difficult catch—then often uttered during prolonged hovering, or while sweeping round to make new attempt (E K Dunn); almost inaudible squeaking sounds also produced in same context. 'Kip', 'ki', or 'chi' variously stimulates young to crouch or hide, summons them from hiding, or answers juvenile's food-calls at fishing grounds (Marples and Marples 1934). (7) Begging-call. A shrill 'ki-ki-ki' (Palmer 1941a) or 'kye-kye-kye', hoarser than S. paradisaea (Cullen 1960a); given typically by ♀ soliciting fish from ♂, but also (shorter and

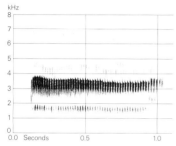

I E Simms/BBC (1971) England May 1959

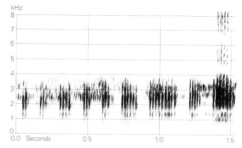

IV S Palmér and R Edberg/Sveriges Radio (1972) Sweden May 1960

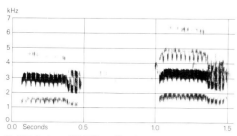

II S Palmér and R Edberg/Sveriges Radio (1972) Sweden May 1960

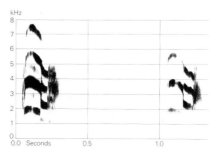

V P J Sellar Scotland June 1974

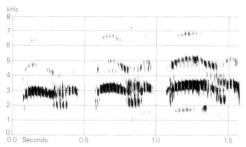

III S Palmér and R Edberg/Sveriges Radio (1972) Sweden May 1960

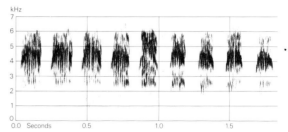

VI P A D Hollom England September 1978

of lower intensity) from ♂ soliciting fish from ♀ or other ♂ (Palmer 1941a; I C T Nisbet). Also used during Pass-ceremony by bird in aerial Erect-posture (usually ♀ in Low- and High-flight). May accompany Hunched-posture preceding mating. (8) Kruk-call. Soft, chuckling 'kruk-k-k' notes; given by one or both birds during ground courtship, though not in Erect-posture. Also given by bird with wings held aloft in V on downward glide during High-flight (Palmer 1941a). Similar, soft, low 'ker-uk-k-k' invites close approach; given by arriving bird at nest-relief, summoning young for brooding, and during delivery of fish to mate or young (Marples and Marples 1934). (9) Gurr-call. An intermittent, growling 'kruu-krurr-krurr' (Palmer 1941a) or 'kerkerker' (Marples and Marples 1934) given by both birds during scrape-making. (10) Kleea-call. A musical 'kleea' (Fig V), lacking harsh quality of most other calls, heard from bird in solo High-flight (P J Sellar).

CALLS OF YOUNG. Chicks highly vocal. Food-calls given from hatching up to 8–9 days; described as feeble, hoarse chirps, chatter, or squeaky cheeping at first, increasing in strength and volume in first 2–3 days after hatching (Marples and Marples 1934; Palmer 1941a). After 8–9 days, when juvenile feathers begin to show, cheeping gives way to shrill, rasping, rapidly repeated 'ki' notes similar to adult Begging-call (Palmer 1941a) but coarser and higher pitched (I C T Nisbet). Most vociferous from c. 35–40 days, when juvenile starts following parent on fishing trips (Fig VI); call continues intermittently on migration up to independence, alternating rapidly with parent's answering Kip-call.

EKD

Breeding. SEASON. See diagram for north-west Europe. Little variation throughout range. Older birds lay earlier (Hays 1978; Nisbet 1978b). SITE. On ground in the open; usually on bare substrate, occasionally near vegetation or in it, or on floating mat of vegetation. Of 83 nests in Poland: 45 on bare substrate or in very low vegetation; 30 on floating vegetation; 3 by tuft of grass; 2 surrounded

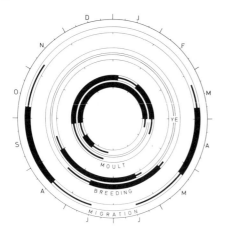

by high vegetation; 3 against stones or tree stumps (Bocheński 1966). Colonial; sometimes solitary. Nest: shallow depression, unlined or with lining and/or rim of available pieces of debris and vegetation. Of 53 nests, Poland: 2 unlined; 15 with a little material forming rim; 31 with thin lining and rim; 5 with thick lining and rim (Bocheński 1966). Mean external diameter 15 cm (11·5–24), sample 44; mean internal diameter 10 cm (8–13), sample 50; depth up to 4 cm (Bocheński 1966). Building: by both sexes with initial scraping, then building during incubation. EGGS. See Plate 89. Sub-elliptical, smooth, not glossy; cream to buff, variably tinted yellow, green, blue, or olive, very variably marked with spots, streaks, blotches, and fine lines of dark brown to black, and some grey. 41 × 31 mm (35–48 × 27–33), sample 400 (Schönwetter 1967). Weight 21 g (19–24), sample 20 (Collins and LeCroy 1972). Mean fresh weights of 76 1st eggs, 21·5 g; of 75 2nd eggs, 21.3 g; of 64 3rd eggs, 20·8 g (Nisbet and Cohen 1975). In clutches of 2, 2nd egg smaller; in clutches of 3, 3rd egg smallest (Dunn 1972a). Clutch: 1–3, varying between colonies (Nisbet and Davy 1972). Of 420 clutches, England; 1 egg, 4%; 2, 37%; 3, 59%; mean 2·65 (Langham 1972). Mean of 143 clutches, Orkney and Shetland, 2·55 (Bullock and Gomersall 1981). Mean of 151 clutches, West Germany, 2·84; of 67 clutches, April, 2·94; of 69 clutches, May, 2·77; of 16 clutches, June, 2·75 (Witt 1970). Of 1588 clutches, Crimea (USSR): 1 egg, 5%; 2, 16%; 3, 77%; 4, 2%; 5, 0·1%; mean 2·88 (Borodulina 1960). Some clutches of 4 by 1♀ (I C T Nisbet), others and clutches of 5 possibly by 2. Older birds lay bigger clutches (Hays 1978; Nisbet 1978b). One brood. Replacements (usually of smaller clutches and eggs) laid after egg loss (Nisbet and Cohen 1975). Of 236 re-layings, Crimea, only 48% had 3 eggs (Borodulina 1960). Laying interval between eggs in clutch of 2 less than 12 hrs; in clutch of 3, c. 1 day between 1st and 2nd, then c. 2 days before 3rd (Nisbet and Cohen 1975). INCUBATION. 21–22 days, increasing to 25–31 when predators caused night desertion (Nisbet and Cohen 1975). By both parents though ♀ takes larger share (75%), especially early on. Relief takes place about 7 times a day (Borodulina 1960). Incubation only partial until last egg laid; hatching interval shorter than laying interval (Nisbet and Cohen 1975). YOUNG. Precocial and semi-nidifugous. Cared for and fed by both parents. Normally brooded by one parent (usually ♀) during 1st week of life while other hunts (Nisbet 1973a). Subsequently both parents collect food. 14–41 feeds per day for 2 nestlings (Borodulina 1960). In 2 cases where ♀ died with young 7–11 days old, ♂ successfully reared at least 1 young (Nisbet et al. 1978). Chicks seek separate refuges inside territory, but outside nest, from 3–4 days old; if disturbed, may wander more when older but return to territory to be fed, even up to 23 days after fledging (Nisbet 1976). FLEDGING TO MATURITY. Fledging period 22–28 days, usually 25–26, occasionally up to 33 (Nisbet and Drury 1972). Age of independence not certainly known but probably often 2–3 months. Age of first breeding usually 3–4 years, some at 2. BREEDING SUCCESS. Of 420 broods, England, mean fledging success 69%. Success varied according to size of brood and order of hatching: for broods of 1, fledging success 62% (sample 39); for broods of 2, success of 1st chick 84% (sample 39); for broods of 2, success of 1st chick 84% (sample 204), of 2nd chick 57% (sample 167); for broods of 3, success of 1st chick 89% (sample 181), of 2nd chick 77% (sample 172), of 3rd chick 22% (sample 125); of 242 chicks dying before fledging, 80% died in first 5 days, with 2nd and 3rd chicks dying younger than 1st; starvation main cause of death (Langham 1972). Of 468 eggs laid, West Germany, 77% hatched and c. 57% fledged, or c. 1·6 young reared per pair (Witt 1970). Of 152 eggs laid, Finland, 80% hatched, with 59% of those hatching surviving to 14 days, and 1·46 young fledging per pair (Lemmetyinen 1973c). Where no adverse factors, hatching success typically over 90%. In colony New York (USA), 6-year-old birds fledged 0·83 young per pair and 7-year-old birds 1·03 young per pair (Hays 1978). Over 11 years at 9 colonies, Massachusetts (USA), birds fledged 0–2·2 young per pair; average 0·9 per pair, 1970–5 (Nisbet and Drury 1972; Nisbet 1978b; I C T Nisbet).

Plumages (nominate *hirundo*). ADULT BREEDING. Forehead, upper half of lores, crown down to lower eyelid, and nape black. Lower half of lores, sides of head below eye, and sides of neck bordering black nape white; hindneck pale grey, nearly white when worn. Mantle, scapulars, tertials, back, and upper wing-coverts light grey; some lower scapulars and outer tertials edged white; leading edge of wing white. Rump and upper tail-coverts white, contrasting with grey of back. Chest and belly pale grey, slightly paler than upperparts, gradually paler towards white of chin and upper sides of neck and towards white under tail-coverts. Occasionally, slight pink or mauve tinge to grey of chest and belly. Inner webs of most tail-feathers white and outer webs grey, but outer web of t6 dark grey and inner web tinged pale grey towards tip; t1(–t3) sometimes pale grey on both webs. Outer webs of primaries light silvery-grey; inner webs with extensive white wedge, grey restricted to streak c. 3–5 mm wide along shafts and at tip; inner edges of inner primaries bordered white. By late spring, outer 5–6 primaries often slightly worn, outer web of p10 and tips and inner edges of others hardly silvery, sometimes dusky grey; inner 4–5 primaries fresh and pale grey; in summer and early autumn, outer primaries mostly tinged blackish on those parts not protected by adjacent feathers, more distinctly contrasting with light grey inner primaries. Secondaries light grey, tips and wide borders to inner webs white. Axillaries and under wing-coverts white. In worn plumage, some white of feather-bases on forehead and lores sometimes visible; grey of underparts much faded, lower throat and chest nearly white; grey of secondaries and on centres of some larger upper wing-coverts darker; outer 5–6 primaries blackish and contrasting with pale grey inner ones; outer web of t6 blackish. ADULT NON-BREEDING. Forehead and lores white, feathers of crown white with tips suffused dark grey or dull black to variable extent. Patch from front of eye over ear-coverts and across hindcrown and nape black. Upper mantle, sides of head and neck below black patch, and all underparts white. Lesser upper wing-coverts dark grey with indistinct off-white fringes, forming distinct carpal bar. Remainder of upperparts, wing, and tail as in adult breeding, but rump and upper tail-coverts sometimes slightly suffused grey. Colour of flight-feathers dependent on moult and wear: 5–6 outer primaries and most secondaries (especially inner) blackish in late autumn; all new and silvery by late winter; slightly worn and darker grey with fresh inner 4–5 primaries and a few outer secondaries silvery-grey in early spring, when much adult breeding already present. DOWNY YOUNG. Upperparts, wing, and flanks pink-buff or cinnamon-buff, spotted or streaked to variable extent with black (except for face and forehead), forming indistinct lines over crown and back. Throat, lores, and occasionally forehead pale brown, dark sepia-brown, or brown-black. Breast, belly, and usually chin and spot at gape white. (Fjeldså 1977; ZMA.) JUVENILE. Head and upperparts strongly variable. Forehead and lores white, often tinged pale buff or buff-brown to variable extent; crown dotted or streaked black and white, pale buff, grey-buff, or warm buff. Patch from front of eye, over eye, slightly below eye, over ear-coverts, and across nape black. Hindneck and sides of head below black patch and buff lores white. Head of bird lacking buff tinge and head in worn plumage strongly similar to adult non-breeding. Mantle, scapulars, tertials, back, and most upper wing-coverts light grey, feathers tipped off-white or with various shades of buff or rufous-brown; grey of centres often separated from tip by buff-brown, sepia, or dull black crescent or mottling, especially on scapulars and tertials; grey of upper wing-coverts more usually gradually merges into white or buff of tips and outer edges. Rump and upper tail-coverts pale grey, but feather-tips often extensively white, especially on tail-coverts, feathers appearing whitish; some feathers bordered buff when fresh. Underparts, under wing-coverts, and axillaries white. Tail pale grey, outer webs of outer 2–3 feathers blackish, broad borders to inner webs and narrow tips to all white; tips often variable suffused buff and tinged dark grey subterminally. Flight-feathers grey, darker than in adult, especially inner secondaries; broad tips and inner edges white; in contrast to adult, up to p7–p8 bordered white at tip. Distinct black carpal bar $c.$ 1 cm wide across lesser upper wing-coverts from shoulder to median primary coverts; black lesser secondary coverts narrowly bordered white, black lesser primary coverts more broadly and diffusely. In worn plumage, buff tinges on head and upperparts faded or lost, but dark subterminal marks on scapulars and tertials usually still present and dark carpal bar prominent. FIRST IMMATURE NON-BREEDING AND SUBSEQUENT PLUMAGES. Immatures closely similar to adult non-breeding, differing mainly in moult cycles not being synchronized with that of adult. Some juvenile feathers of upperparts and tertials with characteristic dark crescents, brown tinge, or indistinctly defined white fringes usually present up to January of 2nd calendar year. Juvenile flight-feathers all rather worn by September–November (when adults actively moulting or showing suspended moult), all heavily worn December–January of 1st winter (when adults all new or nearly so). From about January to July, juvenile flight-feathers gradually replaced, with outer primaries sepia-black and heavily abraded February–June, contrasting with fresh inner primaries; no breeding plumage obtained in spring, in contrast to adults which besides this also show arrested primary moult with outer 5–6 fairly fresh and inners very fresh. Immatures in summer of 2nd calendar year thus characterized by non-breeding plumage (white forehead and part of crown, white underparts, dark carpal bar, short tail, black bill, and dull legs), fresh outer primaries (duller grey than adult, outermost 1–2 occasionally still juvenile and heavily abraded), and gradually darker grey older inner primaries (1–3 inners often new, pale grey). During August–December, body in fresh non-breeding without traces of old breeding (body moult earlier than adult, slightly worn by December, when adults either still show some breeding or are in fresh non-breeding). In 3rd calendar year, moult cycle similar to adult, but breeding plumage (if any) obtained later in spring and primary moult either not arrested or, sometimes, arrested with score other than those found in adult. Immatures in summer of 3rd calendar year strongly variable: some similar to 2nd calendar year, showing mainly non-breeding plumage (though often mixed with some breeding on head), bill often red, no juvenile outer primaries present, and 3–6 inner primaries new, often still moulting May–July; others closely similar to adult breeding, but usually retain part of non-breeding—forehead and underparts mixed with white and traces of carpal bar present (often showing as scattered dark feathers only); some similar to adult, but grey of underparts usually paler. In summer of 4th calendar year, usually similar to adult breeding, but some have a few all or partly white feathers on lores and forehead (showing white mottling especially when worn), or dark feathers of carpal bar partly present; as a few older adults show these features, ageing impossible.

Bare parts. ADULT. Iris dark brown. Bill scarlet or coral-red, with $c.$ 12–18 mm of tip black, but this occasionally absent; in winter, bill black with crimson-red restricted to patch below nostril and to base of lower mandible, sometimes all-black; black may appear from July and some largely black by late July and August, others still red by early October; on arrival at breeding grounds in spring, bill usually already red with black tip. Foot vermilion-red; in winter, foot pale orange-red, dull red, or red-brown. DOWNY YOUNG. Iris dark brown. Bill pink to orange or scarlet with distinctly defined black tip. Foot pink to apricot-orange. JUVENILE. Iris dark brown. Basal half of bill pale flesh or pale yellow-orange at fledging, tip black; during late summer and autumn, bill gradually darkens to horn-black with pink-flesh or orange-red confined to basal cutting edge of upper mandible and to base of lower; by September, often largely black. Foot pink-orange to pale yellow-orange. IMMATURE. In 2nd calendar year, bill usually completely black and foot like adult winter; in spring and summer of 3rd calendar year, bill either all black (sometimes with variable amount of pink or red at base) or similar to adult, though sometimes duller red or with partly black culmen ridge. (Witherby *et al.* 1941; Grant and Scott 1969; Fjeldså 1977; RMNH, ZMA.)

Moults. ADULT POST-BREEDING. Complete. Starts when in or near breeding area, suspended during autumn migration. Wing starts with p1 from early July to late August, rarely late May and June or early September, moult suspended during migration late August and September with 1–4 inner primaries new. Remaining primaries replaced in winter quarters, completed late January to early March. Tail starts about same time as wing; t1 first, soon followed by t2 (t2 sometimes before t1), occasionally followed by t6 and t3 before autumn migration, but t3–t6 usually replaced in winter quarters. Amount of moult on head, body, and wing-coverts before autumn migration limited; about $\frac{1}{3}$ of all birds examined showed limited non-breeding on head by August–September, and a few some new feathers on mantle or scapulars, these presumably mainly failed breeders; main body moult October–January. Some birds delay start of post-breeding until arrival in winter quarters, starting with p1, tail, and head from October or early November, completing February–March. ADULT PRE-BREEDING. Partial: head, body, tail, upper wing-coverts, outer secondaries, and inner primaries. Starts with p1 and tail December–February; halted March–April, when moult on head, body, and tail completed and flight-feathers arrested with up to p3 (10% of 130 west European spring adults examined), p4 (40%), p5 (39%), p6 (5%), or p7 (6%) new; a few birds with up to p5–p7 new had also started another series of inner primaries in late winter, arrested with p1–p2 new. POST-JUVENILE. Complete. Exceptionally, moult starts (with some feathers of head and body) before autumn migration: usually starts in winter quarters, November–December(–February). Head, body, tail, and most wing-coverts new by February–March. Flight-feathers start with p1 in (December–)January–February, completing June–August; a few birds visiting breeding grounds suspend with up to p7–p9 new. A 2nd series of primaries starts May–July of 2nd calendar year; moult of this series slow and frequently temporarily suspended (e.g. with up to p1–p3 new in birds visiting breeding grounds), completed March–June of 3rd calendar year. During June–August of 2nd calendar year, 1st non-breeding on head, body, and tail replaced directly by 2nd non-breeding. 3rd series of primaries starts December–February of 2nd winter, arrested May–June in birds visiting breeding colonies, but only occasionally suspended in those staying in winter quarters; completed August–October; partial breeding obtained February–June, involving 40–90% of feathers of head and underparts, but often less of upperparts and tail and a variable number (nil to all) of upper wing-coverts. Subsequent moults like adult.

Measurements. Netherlands, May–September; skins (RMNH, ZMA). Tail is length of central pair (t1), fork is tip of t1 to tip of t6.

WING AD	♂ 272	(7·01; 73)	257–287	♀ 270	(6·52; 39)	259–290
JUV	255	(8·84; 19)	248–270	256	(8·37; 21)	244–268
TAIL (t1)	70·5	(3·06; 45)	66–76	69·7	(2·92; 33)	66–75
FORK AD	76·9	(7·42; 46)	64–94	78·4	(7·20; 27)	66–92
JUV	43·0	(5·47; 20)	36–54	42·5	(3·68; 21)	37–48
BILL AD	37·1	(1·40; 66)	35–40	35·2	(1·24; 36)	32–37
TARSUS	20·2	(0·77; 45)	19–22	19·8	(0·52; 36)	19–21
TOE	22·8	(0·98; 25)	21–24	22·1	(1·07; 13)	21–24

Sex differences significant for bill and tarsus. Adult non-breeding fork 19 shorter than adult breeding cited above. Juvenile tarsus and toe full-grown at c. 2 weeks old; wing at $2\frac{1}{2}$–3 months, when p10 exceeds p9 by at least 5 mm. Wing in 3rd calendar year on average still c. 7 shorter than adult, average fork c. 10, average bill c. 1.

Weights. Nominate *hirundo*. Coquet Island (Northumberland, England), adults, May–July: 126·2 (10·0; 30) (Langham 1968). Netherlands, adults, May and early June: ♂ 124 (8·96; 5) 112–137, ♀ 126 (11·0; 6) 110–141 (ZMA); July, 123 (6·7; 7) 114–135 (Smit and Wolff 1981). Germany, 140 (20) 101–175 (Niethammer 1942). Poland: average of summer adults 128, of juveniles at fledging 126 (Szulc-Olechowa 1964).

Sexes combined, autumn: (1) Netherlands, August (Smit and Wolff 1981; A L Pieters, ZMA); (2) Netherlands, September (Smit and Wolff 1981; ZMA); (3) Shetland, September (BTO).

	AD			JUV		
(1)	133	(12·8; 67)	89–165	121	(18·8; 33)	80–160
(2)	125	(13·7; 4)	114–145	122	(22·7; 7)	86–156
(3)	110	(17·0; 8)	93–150	116	(—; 2)	101–130

USA: adults, spring and summer, 120 (265) 103–145 (LeCroy and LeCroy 1974); late June and early July 116 (10·9; 56) 103–129 (Collins and LeCroy 1972); juveniles gain weight quickly up to 15 days (in years of abundant food) or 20 days (years of scarcity), reaching peak of 128 (68) 108–148 at c. 3 weeks, decreasing slightly later on, and fledging after c. 4 weeks at 112 (68) 95–128 (LeCroy and LeCroy 1974); for growth rates see Langham (1972) and LeCroy and Collins (1972). Surinam (RMNH) and Panama (Strauch 1977): wintering birds, December–May (partly immature), 103 (10·1; 8) 87–122; summering immatures, June–July, 120 (13·0; 5) 98–131. Exhausted birds, Netherlands, mainly 72–80 (ZMA).

Mongolia, early June to early August: adult ♂ 119 (12) 110–138, ♀ 129 (10) 105–150 (Piechocki 1968c).

Structure. Wing long and narrow, pointed. 11 primaries: p10 longest, p9 14–20 shorter in adult, 5–13 in juvenile; p8 30–42 shorter, p7 52–65, p6 74–90, p5 95–113, p1 160–182; p11 minute, concealed by primary coverts. Longest tertials reach to about tip of p4 in folded wing. Tail deeply forked, especially in adult breeding; for variation in depth of fork with season and age, see Measurements. Bill straight, about equal to head length, compressed laterally, though wide at gape; culmen slightly and gradually sloping towards sharply pointed bill-tip, rather distinct gonys angle about half-way along lower mandible. Depth of bill at basal corner of nostril 8·30 (0·38; 24) 7·8–9·1 in west European ♂♂, 7·62 (0·29; 11) 7·3–8·0 in ♀♀. Leg short and slender, 6–10 mm of lower tibia bare. Outer toe c. 90% of middle, inner c. 66%, hind c. 36%.

Geographical variation. Mainly involves colour of bill and foot; to lesser extent also wing and bill length and colour of body. Wing length of other European breeders similar to those of Netherlands cited in Measurements; wing of birds breeding Canary Islands and North Africa 273 (5) 265–279 (BMNH, ZFMK). Wing of nominate *hirundo* breeding North and Central America slightly shorter than European populations, e.g. USA birds average 6·5 mm shorter (Ridgway 1919; Vaurie 1965; RMNH, ZMA). *S.h. longipennis* from eastern Siberia differs in breeding plumage by uniform black bill, dark brown-red or blackish foot, slightly darker grey body, and more white on central tail-feathers; wing slightly longer—278 (16) 268–286 for adults of both sexes; bill shorter—35·4 (1·54; 10) 33–37 in adult ♂, 33·9 (0·97; 6) 32–35 in adult ♀. Nominate *hirundo* and *longipennis* grade into each other over wide area in central Siberia, a few birds with *longipennis* characters breeding as far west as Ob basin. These variable central Siberian populations have bill either red (usually with more black on tip than in nominate *hirundo*) or all-black, proportion of birds with black bills increasing eastwards; foot dull red or brownish-red; colour of body and length of bill and wing closer to *longipennis*. Sometimes separated as *minussensis* Sushkin, 1925, but as individuals are either similar to *longipennis*, to nominate *hirundo*, or to *tibetana*, recognition not warranted. *S. h. tibetana* from central Asiatic mountains and highlands as dark as *longipennis* and length of wing and bill similar, but bill and leg red in summer, as in nominate *hirundo*.

In non-breeding and juvenile plumages, when all races have blackish bill and dull leg, identification difficult; see (e.g.) Clancey (1976) for differences between *tibetana* and nominate *hirundo*.
CSR

Sterna paradisaea Arctic Tern

PLATES 5, 6, and 7
[between pages 134 and 135, and facing page 158]

Du. Noordse Stern   Fr. Sterne arctique   Ge. Küstenseeschwalbe
Ru. Полярная крачка   Sp. Charrán ártico   Sw. Silvertärna

Sterna paradisaea Pontoppidan, 1763

Monotypic

Field characters. 33–35 cm (bill 2·9–3·5, legs 1·5–1·7, tail of adult 10–18, of juvenile 8–10 cm); wing-span 75–85 cm. Slightly smaller than Common Tern *S. hirundo*, with proportionately shorter bill and head, more slender body, much shorter legs, and narrower wings, but (in breeding adult) longer tail-streamers; body size close to White-cheeked Tern *S. repressa*, but with proportionately longer wings and tail and distinctly shorter bill. Delicate sea tern, with plumage intermediate between *S. hirundo* and *S. repressa* but with individual structure most obvious in smaller, shorter bill, more rounded head (set closer to wings in flight), relatively shorter-armed but longer-handed wings, and less angular wing attitudes. At all ages, shows markedly translucent flight-feathers and sharp narrow black trailing edge to outer primaries (most visible from below); in immature, also broad panel of white on rear half of wing. Contrast between grey back and pure white rump and mostly white tail always marked. Wing-beats in active flight or hover faster than in *S. hirundo*. Calls often more drawn-out with sharper inflexion than in *S. hirundo*. Sexes similar; marked seasonal variation. Juvenile and immature separable.

ADULT BREEDING. In full fresh plumage, differs from *S. hirundo* in rather shorter, more rounded jet-black crown, greyer nape, slightly bluer upperparts, bolder white ends to longest scapulars, whiter tail (particularly in centre), and colder, more intensely grey underparts, last tending to make white lower face appear as contrasting oval panel below black crown. When both closed and open, primaries appear paler or more uniform than in *S. hirundo*, with uppersurface of outermost showing no dusky lines nor any mark contrasting with other flight-feathers, and undersurface showing only narrow black line along trailing edge on tips of 6–7 outer primaries. White trailing edge to secondaries and inner primaries obvious (unlike *S. hirundo*) and whole undersurface of flight-feathers markedly translucent against strong light (and this partly responsible for apparent narrowness of wings). Bill shaped as *S. hirundo* but, compared with head length, distinctly shorter, thus appearing to be stubbier and recalling that of White-winged Black Tern *Chlidonias leucopterus*; dark, blood-red, never with orange tone of some *S. hirundo* and only rarely showing small dusky tip. Legs coral-red. Moults later than *S. hirundo*, so that above plumage retained well into autumn by which time many *S. hirundo* worn and patchy. ADULT NON-BREEDING. Differs less from *S. hirundo*, becoming similarly white below. Most constant characters are: (1) deeper black patch on rear crown (adding to roundness of head) contrasting with obvious collar formed by white hindneck; (3) slightly greyer upperparts (still with bolder white tips to scapulars); (3) fainter, dusky bar between shoulder and carpal joint; (4) purer white rump and tail-centre. All such marks difficult to prove in harsh light, so that separation still best based on wing pattern—except December–February when similarly aged *S. hirundo* have all flight-feathers pale and translucent. With wear, a few outer primaries may show dusky lines (but neither these nor dark tips on undersurface as obvious as in *S. hirundo* and white-winged appearance remains pronounced). JUVENILE. Often distinctly smaller than adult, at longer range inviting confusion with *Chlidonias* terns. Differs distinctly from *S. hirundo* in much colder grey and white appearance (with buff or brown tones of fledgling's upperparts worn off quickly and rarely visible on autumn migrant) and more slender, yet often more compact silhouette. Most obvious plumage characters are: (1) bold white forehead sharply contrasting with compact, black patch on crown and rear cheeks; (2) wider white neck-collar; (3) less barring on back; (4) dusky (not black) leading edge to inner wing; (5) wide silvery-grey area over centre of wing fading into, (6) wide and brightly white trailing panel over secondaries and inner primaries; (7) bold white rump and tail-centre. In flight, narrow black tips to most primaries form obvious tapering line, clearly visible from above and very striking from below (where it emphasizes wide translucent area on rear half of under-wing). Bill appears small and wholly black at most ranges. Legs orange but darkening quickly. FIRST WINTER.

Change from juvenile plumage less marked than in *S. hirundo*, but white forehead usually increased, mantle completely without faint bars, and contrasts of wing pattern increased with wear and bleaching. Bill and legs usually black. FIRST SUMMER. Plumage changes similar to those of *S. hirundo*. Most resemble winter adult but some show exaggerated '*portlandica*'-type plumage pattern. In all non-breeding plumages differs from *S. hirundo* in showing less black lesser coverts and primaries, only patchy dusky marks across secondaries (not usually showing on undersurface), and greyer back. Bill black, with variable red base; legs black or dusky red.

Distinction from *S. hirundo* fully analysed above; for differences from *S. repressa*, see that species. Immature also subject to confusion with Roseate Tern *S. dougallii* of similar age, but latter longer billed, shorter winged, much paler (never intensely grey above), and always lacking narrow black line along tips of outer primaries. Rare individuals of *S. paradisaea*, however, lack last mark (probably through excessive wear) and then suggest *S. dougallii*, pale winter adult Black Tern *C. niger*, and immature and winter adult *C. leucopterus* and Whiskered Tern *C. hybridus*. Flight as *S. hirundo*, but in most actions wing-beats lack exaggerated elastic stroke of that species, appearing stiffer and creating, when translucency of flight-feathers evident, characteristic flickering appearance; when hovering, depresses tail more fully than *S. hirundo*. Gait less free than *S. hirundo*, with markedly short legs. Carriage essentially horizontal, but looks smaller headed and rounder chested than *S. hirundo* in most attitudes.

Voice similar in phrasing and form to *S. hirundo* but all notes pitched higher, with most distinct a harsh 'kee-arr' (in alarm, with emphasis on 2nd syllable), a scolding 'kit-it-it-kaar' (with emphasis on last syllable), and whistled or squeaked notes such as 'kee' and 'peet'.

Habitat. Breeds up to higher latitudes than any other tern, from temperate through boreal to high Arctic, mainly on coasts and inshore islands, but also in some regions inland, even up rivers to at least 700 m (Murie 1963). In USSR, frequently on grassy flats, and also on shingle (Dementiev and Gladkov 1951c). In Norway, also nests fully 100 km up rivers, on islets or banks or adjoining peat mosses, and even at a pool surrounded by well-grown pines, perching on topmost sprays of these with ease (Bannerman 1962). In Greenland and Spitsbergen, nests on low, rocky or grass-covered skerries and islands off the coast; occasionally on low forelands or on beaches of lagoons and less often on islands in fjords; always less than 20–25 m above sea-level. Often fishes near large icebergs, in ice-filled bays, and in offshore zone, by night as well as by day (Salomonsen 1950–1). In Canada, nests near and forages over salt or fresh water on sand-spits, dunes, deltas, sand and gravel beaches, rocky shores and islands, and marshy tundra (Godfrey 1966). In Orkney and Shetland (Scotland), nests mostly on shingle and sand; storm beaches and low rocky shores also important in Shetland. Most sites less than 10 m above sea level, highest 90 m; most within 100 m of water's edge, maximum 7 km. However, most extensive colonies are inland on heath, rough pasture, sedge grassland, or islets in lochs. Large colonies often close to deeper water where tide rips attract fish (Bullock and Gomersall 1981). In Alaska, occupies moraines or islets below glaciers (Gabrielson and Lincoln 1959), and also sandy or rocky islands, salt-marshes, bars, islets in small pools, and even tundra 400 m from nearest lake (Bent 1921). In mixed colonies, Common Tern *S. hirundo* appears to nest in somewhat longer vegetation and Roseate Tern *S. dougallii* even more so, difference being associated with their longer legs (Fisher and Lockley 1954). Where, however, there is no competition for sites, *S. paradisaea* may nest in lush vegetation, as well as on grazed islets and on open tundra (Cramp *et al.* 1974). Adverse weather, floods, high tides, ecological and topographic changes in successive seasons, pressures of ground and aerial predators, human disturbance and persecution, and other factors may lead to marked instability in occupancy of sites or fluctuations in numbers. In Greenland and elsewhere, stretches of apparently suitable coastline recorded as left vacant over long periods. Commensalism with eiders *Somateria* has also been credited with affecting habitat choice (Burton and Thurston 1959).

After breeding season, becomes almost entirely marine, and on migration coastal (and sometimes inland) or pelagic, resting on floating objects at sea, extending to antarctic pack-ice in winter as well as arctic pack-ice in summer quarters (Fisher and Lockley 1954), incidentally encountering fewer annual hours of darkness and covering greater distances than any other bird. Normally flies in lower airspace.

Distribution. Retreating slightly in recent years in south of range.

JAN MAYEN. Usually present in summer but recorded breeding only in 1938, when few nests though many birds (Seligman and Willcox 1940). BRITAIN AND IRELAND. Some evidence of recent range contraction in southern Britain (e.g. Isles of Scilly now deserted) and Ireland (where many inland sites deserted), perhaps due to climatic amelioration (Sharrock 1976). FRANCE. Declined; only 2–3 pairs nested 1970–4, and none since (Yeatman 1976; J-YM). SPAIN. Reports of breeding Ebro delta incorrect (Maluquer Maluquer 1971). BELGIUM. Bred 1976 and perhaps other years (PD). NORWAY. Range extended in south-east in 1970s (GL, VR). POLAND. Bred 1929, 1936, and 1972 (Tomiałojć 1976a; AD).

Accidental. Poland, Czechoslovakia, Austria, Switzerland, Italy, Turkey, Cyprus, Algeria.

Population. No evidence of marked changes over most of range, but some recent declines in Ireland, Britain, France, West Germany, and Finland; for recent declines in south of eastern North American range see Nisbet (1973b).

BEAR ISLAND. 250 pairs 1932 (Bertram and Lack 1933), 160–180 pairs (Lütken 1969), c. 40 pairs 1980 (R Luttik and J A van Franeker). ICELAND. Over 100 000 pairs (A Gardarsson). No apparent recent change (AP). BRITAIN. Census difficulties, especially in Orkney and Shetland, make assessment of recent trends uncertain. In 1969–70, conservatively estimated at 29 500 pairs (Cramp et al. 1974), but on reassessment considered to be probably over 50% higher (Lloyd et al. 1975). In 1980, when full counts made in Orkney and Shetland, British population estimated at 75 000 pairs (Thomas 1982). IRELAND. Population small and declining (Ruttledge 1966; Cramp et al. 1974). Counts incomplete, but probably over 1000 pairs (Lloyd et al. 1975), and 3000 pairs 1979 (Thomas 1982). NETHERLANDS. 2500 pairs 1965, 1200–1500 pairs 1978 (Teixeira 1979). WEST GERMANY. Fluctuating, declined. In 1930s over 4000 pairs (Niethammer 1942); total over 1400 pairs of which c. 380 pairs on Baltic coasts (Schmidt and Brehm 1974): estimated 370–520 pairs (Szijj 1977); 2500 pairs 1978–9 (Thomas 1982). DENMARK. Estimated 5500–6000 pairs 1973 (Mardal 1974); no marked changes (TD). NORWAY. Estimated 21 000 pairs 1970 (E Brun). Numbers fluctuate, especially in north (Haftorn 1971). SWEDEN. Estimated 10 000 pairs (Ulfstrand and Högstedt 1976). FINLAND. About 6000 pairs (Merikallio 1958); increased c. 1950–70, some recent decline (OH); c. 10 000 pairs 1978 (Thomas 1982). EAST GERMANY. Fluctuating numbers (90–227 at 2 main colonies 1957–75), but no marked trends (Klafs and Stübs 1977). About 2000 pairs 1979 (Thomas 1982). USSR. Fairly rare Franz Josef Land and Novaya Zemlya; numbers fluctuate (Dementiev and Gladkov 1951c). At least 25 000 pairs White Sea and c. 10 000 pairs on coast of Murmansk 1963 (Bianki 1967). Estonia: estimated 5700 pairs (Onno 1966); 12 500 pairs 1973–5 (Thomas 1982). Latvia: breeds occasionally (Viksne 1978).

Survival. Britain: estimated survival rate in 1st year of life estimated 80·5%; adult survival rate 87% and 88% by 2 independent methods (Coulson and Horobin 1976) and in earlier study 86% (Cullen 1957). Oldest ringed bird 34 years (Clapp et al. 1982).

Movements. Migratory, performing most extensive movements of any bird; reviewed by Kullenberg (1946), Stoor (1958), Salomonsen (1967b), and Eckhardt (1969) mostly from sightings, reinforced by recoveries—numerous for Eurasian population as far as South Africa, sparse otherwise. Main wintering zone lies in Antarctic pack-ice between c. 55°E and c. 150°E, though western border of range extends to 30°W (Falla 1937; Salomonsen 1967b). Movements described by Salomonsen (1967b) in Southern Ocean largely speculative.

ADULT MOVEMENTS. Based on Salomonsen (1967b). Post-breeding migration, from late July in south of breeding range to early October in north, occurs along 2 principal routes. Birds from Bering Sea, Bering Straits, eastern Siberia, and western Alaska thought to move southwards along eastern Pacific coast of North and South America, crossing Drake Passage to Antarctic zone. Portenko (1959) suggested that birds from north-east Siberia fly north towards Arctic Ocean, east along arctic coast of North America, and thence to Atlantic route (see below), but this highly speculative. Those breeding in Canada, north-east USA, Greenland, Spitsbergen, Faeroes, Iceland, northern Europe, and coast of northern Siberia (excluding easternmost parts) converge on route along west coast of Europe and Africa. Movement through Britain, Ireland, and north-west Europe mostly coastal or offshore, but regularly occurs inland in small numbers, though less so than in spring. To join North Sea/Atlantic route, Siberian birds initially travel west along coasts of Arctic Ocean, then south along seaboard of northern Europe. Those from Nearctic perform wind-aided transatlantic crossing at c. 50–60°N; route south of east, producing recoveries in Scotland, England, and western France in September–October; since fewer recoveries north of Biscay (France) than south of it, this suggests much south-east passage across open ocean, bypassing north-west Europe (Spencer and Hudson 1980). Small proportion believed to move south-west offshore from West Africa, crossing Atlantic and reaching South America in region of Argentina, where they proceed southwards along east coast. Majority, however, keep to West African coast, probably further offshore than Common Tern *S. hirundo*; recoveries from most coastal countries (often from trapping activities) from Morocco to South Africa. Some reach southern Africa by September, most October–November. Hereafter, some adults apparently continue due south to pack-ice, but majority move ESE–SE under influence of cyclonic westerlies, bringing them to edge of pack-ice at 50–110°E in October–December. A few cross southern Indian Ocean (see Mörzer Bruyns and Voous 1964) due east towards Australasian waters where there are several recoveries (Stoor 1958; Gwynn 1968), notably of birds ringed in Sweden, White Sea (USSR), and Anglesey (Wales). In combination, those routes indicate initial distribution in pack-ice of c. 30–150°E, with greatest concentration at c. 50–110°E. One Danish-bred adult recovered February in Southern Ocean at 65°08′S 111°15′E. Hundreds seen Weddell Sea, February (Parmelee 1977). Those leaving South America said to reach Antarctica in region of western Weddell Sea, south to c. 74°S, or move east to join birds that migrated via South Africa. Birds regularly seen November–December at subantarctic Heard Island (53°07′S 73°20′E) on northern boundary of migration route. Spring migration begins early March; most thought to move west along edge of pack-ice which, now receded, generally lies in zone of easterly winds. In region of Weddell Sea, turn north to north-east towards South Africa. Most of those which followed coast of Chile in autumn

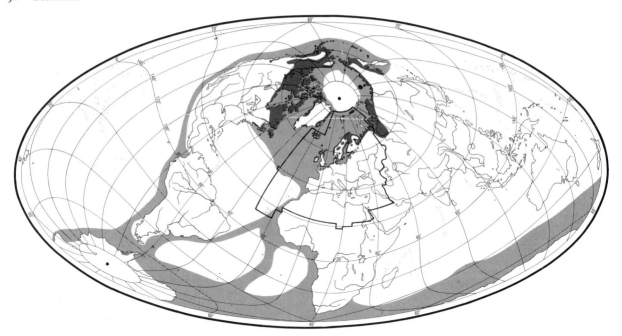

and wintered in Weddell Sea are also thought to head towards South Africa in spring. At least part of population wintering further west, from Amundsen Sea to Ross Sea, probably goes north-east to follow coast of Chile north. This possibly the origin of bird raised western Greenland and recovered in June on Pacific coast of Colombia (6°25′N 72°30′W). Sight records almost absent thereafter in spring, but Eckhardt (1969) suggested routes on basis of minimum time paths, given prevailing winds. From South Africa, birds of Eurasian origin thought to travel north up eastern Atlantic, roughly retracing autumn route. However, recent evidence of small April–May passage of adults off Somalia (where some birds also stay, May–July) suggests some return by Indian Ocean and then perhaps fly overland to northern breeding grounds (J S Ash); see also next 2 paragraphs. Eckhardt (1969) agreed with Salomonsen (1967b) in believing that many birds from eastern North America and Greenland initially head north-east towards southern Africa; after following eastern Atlantic route northwards until south-west of West Africa, most efficient route is to turn west towards northern Brazil, and thence north along pelagic Atlantic route. Breeding grounds reoccupied early May in south of range, sometimes not until near end of June in north (e.g. Dementiev and Gladkov 1951c). Regular inland on spring passage, Britain, sometimes in abundance (Gibb 1948).

POST-FLEDGING DISPERSAL. Ringing shows that dispersal of juveniles from breeding colonies in Britain involves slight northward, as well as southward movement (Lemmetyinen 1968); recoveries encompass West Germany, Sweden, and Netherlands. Juveniles ringed in Europe undergo comparable movements, and birds from southern Sweden have been recovered on Baltic coasts to the east. Populations from Gulf of Bothnia and south-west archipelago of Finland show clear tendency to migrate across Scandinavia before turning south (Lemmetyinen 1968); 3 juveniles ringed in Estonia (USSR) recovered in August in eastern Britain. Birds from Kandalaksha Bay (White Sea, USSR) disperse in various directions. Those from colonies in north of archipelago known to travel north through lake system towards Murmansk, then north-west round northern end of Norway before heading south; birds from Barents Sea also follow coast around Norway. Some birds from southern archipelago head south-west, overland, to Gulf of Bothnia, or perhaps due west to Norwegian Sea before heading south. Minority of juveniles, and rarely older birds, move due south from Kandalaksha, penetrating deep into interior of Russia; notable recoveries are August in Chelyabinsk region (c. 55°12′N 61°25′E) and in Khmelnitsky region of western Ukraine (c. 49°48′N 23°39′E). Occasional occurrences on Black Sea assumed to arise from overland migration, probably via Dnieper valley (Bianki 1967); 3 seen on the Bosporus, early September (Ballance and Lee 1961).

SUBSEQUENT IMMATURE MOVEMENTS. After August, recoveries of birds ringed in Britain are predominantly south of breeding colonies (Radford 1961; Langham 1971). Some appear to migrate earlier and faster than *S. hirundo*, as shown by wide latitudinal spread of recoveries early in autumn, and notably distant recoveries south of breeding area in August: e.g. 2 in Liberia, 1 in Cameroun (C J Mead). A few 1st-winter birds appear to remain in equatorial region. Many more winter in South Africa, particularly east coast as far north as Madagascar. Recoveries show that this population comprises birds from Europe, north-west Russia, Faeroes, Greenland, and Canada

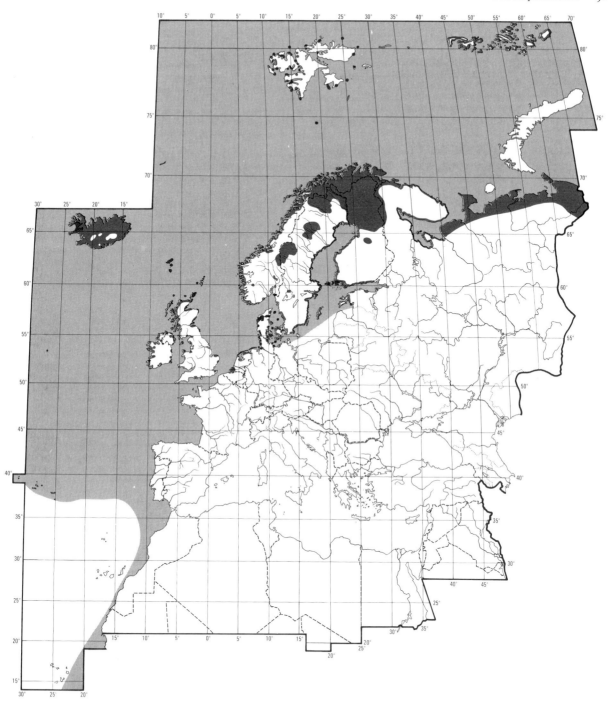

(Liversidge 1959; Salomonsen 1967b; Elliott 1971); a Greenland bird had travelled 18 000 km in less than 3 months (Salomonsen 1956). Humboldt Current off Chile provides equivalent wintering grounds for birds that have crossed to South America from eastern Atlantic, or followed eastern Pacific coast south (Salomonsen 1967b). Other juveniles penetrate antarctic zone with adults; English-bred bird recovered in December of 1st winter at 56°20′S 39°30′E. May also drift from main wintering area eastwards into Amundsen and Bellinghausen Seas (80–120°W). Those that then return ENE to South Africa to join spring migration of adults thus circumnavigate Antarctica in their 1st year. Others, perhaps majority, make shorter journey from Amundsen and Bellinghausen Seas north-east to coast of Chile where they spend summer (Salomonsen 1967b). Ringing data suggest northward movement in 1st summer; thus, recoveries near equator in August–October of birds ringed Britain. 1-year-olds have occurred on breeding grounds: e.g. in Iceland (Gudmundsson 1956), Spitsbergen, and Kandalaksha Bay (Bianki 1967) late in season (usually late June onwards), though none recorded on Farne Islands (England) despite extensive ringing (Coulson and Horobin 1976); 1-year-olds commonly recorded at colonies, Massachusetts, USA (I C T Nisbet). Recoveries of birds ringed in Britain indicate movement south again in following autumn, though wintering area not well documented. At this time, some of those that spent 1st summer off Chile are thought to move south-east into Weddell Sea, and the following spring north-east to South Africa, having thus circumnavigated Antarctica in almost 2 years (Salomonsen 1967b). In 2nd spring, generally penetrates further north than in previous year, and a few visit colonies without breeding (Langham 1971; Coulson and Horobin 1976). 2 unusual recoveries of English-bred 2-year-olds: in June at Bashkir, USSR (55°50′N 55°25′E), and in July at Perm, USSR (59°36′N, 54°23′E), showing that movement inland, in these cases to west of Urals (perhaps via Mediterranean and Black Seas, then Volga) occasionally occurs. Wintering area of 3rd-year birds not well known but presumably approximates that of adults. Although 3rd-years move north to Farne Islands in spring and a proportion breed, some appear not to visit colony at all; most breed at 4 years old. At Farne Islands, progressive advance with age in date of spring return from 2-year-olds to birds over 8 years. Latter arrive on average *c.* 17 days earlier than youngest breeding group (3-year-olds); 2-year-olds a month later than oldest group (Coulson and Horobin 1976).

Inter-colony exchanges (especially of first-time breeders nesting in colony other than natal one) are common and may involve movements of a few to several hundred km. Only a third of birds from Kandalaksha Bay, breeding for first time, settled in natal colony; colonies there included birds raised in Gulf of Bothnia, Barents Sea, and Onega Bay, White Sea (Bianki 1977). One bird hatched Kandalaksha was recovered July, 2 years later in Greenland (68°40′N 53°15′W) (Salomonsen 1971). EKD

Food. Marine fish, crustaceans, and insects, proportions varying regionally. Fish caught mostly by plunge-diving after searching flight upwind; often preceded by hovering. Mean dive height 3 m (1–6). Dives to just below water surface, probably not deeper than 0·5 m; mean immersion time 1·1 s, varying with dive height to suit depth at which prey located (Dunn 1972a). Immersion normally just complete, but only partial when prey or visibility of prey restricted to surface. Always emerges with prey held crosswise in bill before either swallowing it immediately or taking it to mate or young. Occasionally more than one fish caught and carried; probably capable of both simultaneous and successive captures (Hays *et al.* 1973; Taylor 1975). 3 fish recorded being carried by 1·7% of birds with food at colony (Hays *et al.* 1973). Small prey items (e.g. crustaceans, insects), caught by dipping-to-surface or by oblique plunge-dive (with partial immersion). Also takes insects in aerial-pursuit, exploiting swarms (Bertram and Lack 1933, 1938; Baxter *et al.* 1949; Lemmetyinen 1973b). Occasionally hunts over fields in Iceland, taking caterpillars off grass stems (Bannerman 1962): more commonly, alights periodically to pick up earthworms (Annelida), notably in Iceland (Gudmundsson 1956) and Aberdeenshire, Scotland (Feare 1969). On Faeroes, regularly takes food from Puffins *Fratercula arctica* on ground by swooping and forcing them to drop fish (Nørrevang 1960). In Iceland and Faeroes, recorded aerially defending areas of ground against conspecific birds in order to take fish dropped in interactions between *F. arctica* and skuas *Stercorarius* or gulls (Laridae) (Williamson 1948; G K Taylor). In Iceland, takes food from surfacing Slavonian Grebes *Podiceps auritus* (Bengtson 1966) and Black Guillemots *Cepphus grylle* (Bardarson 1975). Scavenges at fishing boats in Irish Sea (Watson 1981). Daily weight changes of chicks vary, suggesting fishing success of parents influenced by environmental factors. At Lake Mývatn (Iceland), proportion of dives yielding prey and dive-rate highest (69·2%, 4·6 dives per min) in dry, calm weather, declining with increasing wind and rain (lowest 16·3%, 2·6) (Bengtson 1966). Hawksley (1950, 1957) recorded weight gains on clear days and losses on foggy days. Growth rates also retarded with overcast skies and rain but effect of wind varies. Often no influence detected (Anon 1968a; Langham 1968; Lemmetyinen 1972a). However, foraging birds may prefer waters sheltered from wind (Burton and Thurston 1959), as strong gales reduce fishing success (Bianki 1977) and rate of feeding young (Boecker 1967). Newly fledged young much less successful at catching fish than adults (Withers 1973). Shoals of fish attract dense feeding flocks and yield higher fishing success than more dispersed prey; 48·2% fishing success recorded at one dense shoal of sand-eels (Ammodytidae) (Dunn 1972a). Otherwise hunts alone or in small loose parties. No tidal component in rate of food input to colony at Coquet Island (Northumberland, England) where most feeding in deep water offshore (Dunn 1972a). At Wangerooge (West Germany), numbers feeding and frequency of provisioning young greater around low tide, when water shallowest over submerged reef (Boecker 1967). In Waddenzee (Netherlands), maximum activity 1 hr before low tide and least at high tide (Mes and Schuckard 1976). Day-time feeder only, with peaks of feeding flights in early morning and before sunset (Dunn 1972a; Mes and Schuckard 1976), though feeding frequency at Wangerooge influenced more by state of tide during daylight hours (Boecker 1967). In continuous daylight of Spitsbergen summer, marked diurnal rhythm of activity observed: peak visiting rates to colony around 09.30 and 18.30 hrs, with lull around midnight and periods of reduced activity about 4 hrs either side (Burton and Thurston 1959). Most feeding done within 3 km of colony; maximum 10 km (Boecker 1967), 20 km (Pearson 1968).

Fish include sand-eels (*Ammodytes marinus, A. tobianus, A. americanus, A. hexapterus, A. lanceolatus*), herring *Clupea harengus*, sprat *C. sprattus*, capelin *Mallotus villosus*, sticklebacks (*Gasterosteus aculeatus, Spinachia vulgaris, Pungitius pungitius*), pipefish *Siphonostoma typhle*, flounder *Pleuronectes flesus*, sole *Solea vulgaris*, hake *Urophysis*, lumpsucker *Cyclopterus lumpus*, dollarfish *Poronotus triacanthus*, rockling *Onos cimbrius*, arctic char *Salvelinus alpinus*, haddock *Melanogrammus aeglefinus*, coalfish *Pollachius virens*, butterfish *Pholis gunnellus*, eelpout *Lycodes jugoricus*, rosefish *Sebastes marinus*, barrelfish *Palinurichthys pereiformis*, bleak *Alburnus alburnus*, perch *Perca fluviatilis*, salmon *Salmo salar*; also other Gadidae, Scorpaenidae, Trachinidae, and Blennidae. Crustaceans include isopod *Idotea baltica*, amphipods (*Gammarus setosus, G. locusta, Gammaracanthus loricatus, Atylus carinatus, Anonyx nugax*), euphausiids (*Thyanoessa inermis, Meganyctiphanes norvegica*), mysid *Mysis oculata*, shore crab *Carcinus maenas*, shrimps (*Crangon vulgaris, Hippolyte polaris*), tadpole shrimp *Lepidurus arcticus*, and other branchiopods and copepods. Molluscs include cephalopods (*Allotheutis subulata, Loligo pealei*), gastropods *Hydrobia ulvae*, and pteropods (*Clio, Limacina*). Also polychaete worms *Nereis*, especially *N. pelagica*. Insects include flying ants (Hymenoptera), chironomids and craneflies (Diptera), moths (Lepidoptera), mayflies (Ephemeroptera), aphids and cicadas (Hemiptera), dragonflies (Odonata), and beetles (Coleoptera). (Fielden 1877; Wright 1909; Bent 1921; Dircksen 1932; Roberts 1934; Hartley and Fisher 1936; Bird and Bird 1941; Hawksley 1950; Dementiev and Gladkov 1951c; Belopol'ski 1957; Burton and Thurston 1959; Parmelee and MacDonald 1960; Løvenskiold 1964; Boecker 1967; Langham 1968; Pearson 1968; Dunn 1972a; Korte 1972; Lemmetyinen 1972a, 1973.) Recorded swooping on scraps of fish thrown into Godthaab harbour, Greenland (Nicholson 1930). Small numbers of berries taken by ♂♂, Barents Sea, USSR (Belopol'ski 1957); plant material also taken at Kandalaksha Bay, USSR (Bianki 1967). Recorded taking pieces of biscuit thrown overboard from ship (Anon 1936). Tends to take smaller fish than Common Tern *S. hirundo* (Boecker 1967; Langham 1968; Lemmetyinen 1973b). For chicks up to 1 week old, parents may select smaller fish and thereafter larger fish, than they eat themselves (Hawksley 1950; Quine and Cullen 1964; Boecker 1967; Lemmetyinen 1973b). At Farne Islands (Northumberland), chicks less than 1 week old fed fish 5·6 cm long, against 6·4 cm for chicks older than 2 weeks; mean (all ages) 6·1 cm (Horobin 1971). Prey of 4–5·5 cm fed to smaller young at Wangerooge (Boecker 1967). At Coquet Island, chicks received sand-eels of mean length 4·5 cm and clupeids of 6·8 cm (Langham 1968). At same colony, Dunn (1972a) recorded mean length of 4·8 cm for all fish prey fed to young.

Fish or crustaceans/insects predominate in diet according to location (especially latitude) and availability of freshwater habitat. At Petit Manan Island (Maine, USA), young fed almost exclusively on herring (Hopkins and Wiley 1972). At Farne Islands, diet of young 95% fish (by number), 2% crustaceans, 2% cephalopods, and 1% insects; most important fish were Ammodytidae (65% of total food) and Clupeidae (22%) (Pearson 1968; see also Quine and Cullen 1964). At Coquet Island, 60·8% of fish Ammodytidae, 39·2% Clupeidae (Langham 1968). Clupeidae became relatively more important in late July as they moved inshore, and Ammodytidae offshore. At Wangerooge, about half diet fish, mostly Clupeidae, and half crustaceans, mainly shore crabs (by weight) and shrimps; also a few cephalopods and gastropods (Boecker 1967). On Murmansk coast of Barents Sea, 67 adult stomachs contained 50·6% (by number) fish (mostly sand-eels and herring), 25·3% crustaceans, 21·5% insects (obtained mostly at sea), and the rest polychaetes and plant material (berries); 18 stomachs of chicks revealed proportionately more fish (88·2%), fewer insects (11·8%), and no crustaceans though these provided sometimes. 45 and 22 stomachs respectively showed adult ♂♂ ate relatively more fish (56·4%) than ♀♀ (37·5%), and fewer insects (16·4% and 33·3%). In ♀♀ from Barents Sea, sharp reduction in sea food and increase in terrestrial insects in July towards end of incubation and during chick rearing; ♂♂ less attached to nest and reduction in sea food less marked. In general, insects and sand-eels become relatively more important in late summer, crustaceans and herring less so (Belopol'ski 1957). In Kandalaksha Bay, fish likewise important; of 63 stomachs of adults, 59% contained fish, 15% crustaceans (especially amphipods), 5% *Nereis pelagica*, and 19·5% insects; also traces of plant material. Unlike diet from Barents Sea, however, fish predominantly three-spined sticklebacks *Gasterosteus aculeatus*; before these migrate inshore to spawn, adult diet included only 19% fish and 63% crustaceans, but afterwards 66% and 4% respectively. Chicks' and juveniles' diet reflected dependence on sticklebacks; diet of 93 juveniles 62% fish, 8% crustaceans, 20% *Nereis*, and 2% plant material (Bianki 1977). Pellet analysis, inner archipelago of Kustavi (Finland), showed sticklebacks comprised 70% of adult diet. In outer islands, however, fish, isopods, and insects evenly represented in laying period, though fish predominated later (Lemmetyinen 1973b). In pack-ice off Alaska (USA), Arctic cod *Boreogadus saida* and amphipod *Apheruse glacialis* important prey (Boeckelheide 1978). In Kongsfjord area (Spitsbergen), *Gammarus setosus* main food of chicks (Lemmetyinen 1972a). Crustaceans, especially *Thyanoessa inermis* and *Gammarus locusta* dominate diet of young in Spitsbergen (Hartley and Fisher 1936; Burton and Thurston 1959; Korte 1972), and pteropods also taken by adults (Løvenskiold 1964). In Iceland, antler moth *Cerapteryx graminis* can be important prey; about 60% of birds seen feeding on them; 1 stomach contained remains of 56 moths, while 15 stomachs contained 1 each of spider (Araneae), weevil (Curculionidae), fly (Diptera), and parasitic wasp (Hymenoptera); maggots infesting fish heads also regularly taken there (Roberts 1934).

Broods of 1 and 2, Farne Islands, received respectively 1·4 and 2·6 feeds of fish per hr (Pearson 1968); 2·3 and 3·0 at Wangerooge (Boecker 1967). In Kustavi archipelago, chicks 5–9 days old received 1·9 feeds per hr, while those 12–16 days old got 1·2 (Lemmetyinen 1973). At Kodiak Island (Alaska), chicks apparently fed on average only 3·5 times per day (Baird 1978). Feeding rates almost twice as fast as for *S. hirundo*, reflecting higher frequency of invertebrates and other small prey in diet of *S. paradisaea*. However, proportionately more fish than crustaceans (lower calorific value) fed to younger chicks (Belopol'ski 1957; Boecker 1967). Horobin (1971) calculated fish requirement of 26·6 g per day to maintain mean growth rate of 6·8 g per day, Farne Islands. In south-west Finland, daily food intake of chicks less than 10 days old 27 g; 36 g for older chicks. Total intake to fledging (23 days) was 747 g (Lemmetyinen 1973b); generally, however, 778 g obtained from feeding experiments (Unhola 1973). EKD

Social pattern and behaviour.
1. Mostly gregarious throughout the year when breeding, roosting, and often when foraging (Bent 1921; Hawksley 1950, 1957; Salomonsen 1967a; Hopkins and Wiley 1972). In breeding season, bathing flocks of 10–30, and feeding flocks of up to several hundred recorded (Hawksley 1950; E K Dunn). Especially large flocks in winter quarters (Bierman and Voous 1950; Salomonsen 1967a). Also occurs solitarily or in small parties, especially on migration, when flocks may be of up to 20 (Wynne-Edwards 1935); of 30–100 when departing from breeding grounds (Bianki 1967). BONDS. Monogamous pair-bond, tending to persist from year to year. Divorce more common in younger birds. After death of mate, survivor pairs sometimes with bird previously known to it; also, survivor is later in breeding the following season, or misses a season (Busse 1983; see also for homosexual pairs). Age at which birds return and start breeding varies between colonies: 1-year-olds commonly seen at Massachusetts colonies (I C T Nisbet), in Iceland (Gudmundsson 1956), and elsewhere (Bianki 1967), and some may select mates at end of season (I C T Nisbet) (see Heterosexual Behaviour, below). Occasionally breeds—or attempts to do so—at 2 years (Cullen 1957; Belopol'ski 1957; Hawksley 1957; Bianki 1967), but usually not until 3 or older (Drost 1953; Gudmundsson 1956; Grosskopf 1957; Bianki 1967; Langham 1971). 3-year-olds may show incomplete breeding behaviour, occasionally attaching themselves to established pair, helping to incubate and rarely to feed young (Cullen 1957), though no evidence of polygamy. On Farne Islands (Northumberland, England), breeding apparently deferred longer than in some other colonies; no 1-year-olds returned to colony, only a few 2-year-olds (none bred), and though most 3-year-olds returned, c. 70% deferred breeding till following year. Most 4-year-olds bred, but c. 5% not until 5 years, even some 5-year-old ♂♂ failing to pair. Ages of pair members showed high positive correlation. Older birds first to return to colony, those over 8 years old arriving c. 17 days ahead of 3-year-olds, and 2-year-olds c. 1 month later still (Cullen 1957; Horobin 1971; Coulson and Horobin 1976). No evidence that mates associate in winter quarters, and little evidence of courtship occurring on spring migration. Members of many pairs thought to re-establish contact in social flocks that form near colony prior to occupation (Hawksley 1950; Bianki 1967; Horobin 1971). For pattern of colony reoccupation, see Roosting, below. Pre-laying period of 10 days spent in nest vicinity, Coquet Island (Northumberland, England) (Langham 1974), 5 days shorter than in Common Tern *S. hirundo*. In high Arctic, birds arrive relatively late at breeding grounds and interval between arrival and nesting much shorter; Manniche (1910) and Belopol'ski (1957) recorded birds resorting to prospective nesting sites immediately on arrival. Interval between first arrival and first egg 13 days (shortest recorded) at Machias Seal Island (New Brunswick, Canada), 20 days at Churchill (Manitoba, Canada) (Hawksley 1950), and 14 days at Southampton 480 km further north (Sutton 1932). Both sexes incubate and care for young; parent-young bonds maintained for several weeks, probably covering at least part of migration south. BREEDING DISPERSION. Solitary to densely colonial (Bent 1921; Løvenskiold 1964; Bergman 1980). No distinct sub-colony structure apparent, though some tendency to nest in groups (Hawksley 1950). Flocks of up to 100 1-year-olds, discrete from older birds but joining them in alarm upflights (see Flock Behaviour, below), form at edge of colonies, Iceland (Gudmundsson 1956). Scattered dispersion characteristic of arctic regions, especially inland (Dementiev and Gladkov 1951c): no colonies recorded in Alaska, pairs (often 1 per islet) seldom nesting closer than 92 m; in scattered pairs and small colonies along both shores of Bering Sea (Bent 1921); numerous small colonies—sometimes 2–3 pairs (Løvenskiold 1964)—with a few larger ones (over 100 pairs) in Spitsbergen (Burton et al. 1960); colonies usually of 30–40 pairs, Finland (Lemmetyinen 1971), 100–400 pairs exceptional (Haartman et al. 1963–72); usually not more than 100 pairs, maximum 3000, Kandalaksha Bay, USSR (Bianki 1967); up to 10000 pairs, Iceland (Gudmundsson 1956); c. 10% of Baltic breeding records comprised single pairs (Bergman 1980). Further south, more gregarious and colonies generally larger: in Britain and Ireland, from less than 10 to 17500 pairs (on Papa Westray, Orkney, 1969), mean 1483 pairs in 1974 (Lloyd et al. 1975). On east coast of North America, 1968–73, 2–5000 pairs, mean 668 (11 colonies) (Erwin 1978), though Massachusetts colonies generally less than 30 pairs (Nisbet 1973b). Nesting density in colonies similar to that of *S. hirundo* but generally more dispersed in keeping with more aggressive nature (Cullen 1960b; Langham 1971). In Spitsbergen, 9·2–21·5 m between nests, mean 12·3 m (Summerhayes and Elton 1928); one colony, Bear Island, had nests only 2 m apart (Bertram and Lack 1933); while 2 on islands had some not more than c. 30–60 cm apart (Løvenskiold 1964); usually 5–10 m, minimum 75 cm, Kharlova Island (USSR) (see Bianki 1977); 1–10 m or more, Kandalaksha Bay (Bianki 1967); 7–15 m, Barents Sea (Dementiev and Gladkov 1951c). Further south, colonies generally denser in keeping with larger size. At Machias Seal Island, mean distance between 27 nests 1·35 m (0·38–4·88) (Hawksley 1950). On Farne Islands, nests c. 3 m (1–5) apart (Bullough 1942); 40 pairs at density of c. 0·3 nests per m² (Cullen 1957); c. 2000 pairs in colony 1966–8. Density at egg-laying, relative to age, measured by number of birds in 4 m² around nests of known-age birds: range 6 nests per 4 m² around 3-year-olds, 10·3 around birds of over 15 years. Young birds (less than 4 years) tended to nest in more open, exposed places, often nearer tideline (Coulson and Horobin 1976). At nearby Coquet Island, mean density 0·02 nests per m² (175 nests) but much higher where ground littered with debris (Langham 1974). Territorial, defending small, roughly circular nest-area territory—in dense colonies at least; serves for ground phases of courtship and pair-formation, copulation, nesting, and concealment and location of young. In 25 out of 35 cases in dense vegetation, nest-territory

ultimately adopted was displaced from initial display-territory; in 19 out of 23 cases on open ground, pre-laying and nest-territories identical (Horobin 1971). As in *S. hirundo*, strong year-to-year fidelity to territory (Mänd 1982), though at Kandalaksha Bay at least 16·2% of adults change colony-site each year and first breeding not uncommonly at site distant from natal one (Bianki 1967). Various nesting associates—often other *Sterna* in south, but also gulls (Laridae), ducks (Anatidae), grebes (Podicipedidae), and waders (Charadrii) (Koskimies 1957)—benefit from pugnacity of *S. paradisaea* (Bailey 1925; Marples and Marples 1934; Burton et al. 1960; Løvenskiold 1964; Dunn 1972a). Sabine's Gull *Larus sabini* always breeds with *S. paradisaea* (Larson 1960b). Shares mutual attraction with *S. hirundo* (Bergman 1980). On Baltic coast of Sweden, if *S. paradisaea* shifts breeding grounds, Little Terns *S. albifrons* and Turnstones *Arenaria interpres* follow suit (Durango 1945a). Toleration of associates may vary: e.g. Aleutian Tern *S. aleutica*, Alaska, sometimes nests within 0·6 m of *S. paradisaea*, but on Kodiak Island, landing *S. aleutica* may be delayed up to 30 min and forced to drop fish loads by harassment from kleptoparasitic *S. paradisaea* (Baird 1978). ROOSTING. Readily congregates in roosts both during and outside breeding season, preferring open areas near low tide mark, sometimes with other *Sterna* (Dunn 1972a). Occasionally roosts on trees (Bianki 1967); on migration, often rests on floating debris, or on sea surface (Wynne-Edwards 1935; Garrison 1942); in winter quarters, often rests on ice floes between bouts of feeding, and even on powdered brash ice during moult (Routh 1949; Salomonsen 1967a). Especially reluctant to fly during wing moult (late December to January in adults), mostly roosting in flocks of less than a dozen up to several thousand. Flocks particularly large when gales compress large areas of ice floes (Bierman and Voous 1950; Salomonsen 1967a). Birds often assemble to preen, rest, or sleep after bathing or fishing. Usually arrives at breeding grounds in small flocks, up to 30–40 birds, though, if held up, may arrive *en masse*. In temperate latitudes, less so in Arctic, birds occupy breeding grounds gradually, initially visiting colony only at dawn, spending rest of day at sea, and roosting away from colony at night (Dircksen 1932; Drury 1960). Overnight roosting in colony signals full occupation (Horobin 1971). At edge of colony, patches of open, often bare ground, e.g. flat rock, become regular loafing areas for failed and non-breeders, also attracting off-duty nesters and late arrivals which court there (Hawksley 1950; Palmer 1941a). In colony, off-duty bird may loaf inside territory, by nest. Apparently more noisy at night than *S. hirundo* (Hawksley 1950), perhaps reflecting more vigorous territorial defence.

2. FLOCK BEHAVIOUR. In colony, upflights, 'dreads', and 'panics' as in *S. hirundo*. On Farne Islands, normal clamour of colony increases directly preceding dread, followed by sudden silence signalling departure; flight over sea and returning to colony usually lasts 5–25 s. Almost whole colony participates, though dread usually starts locally, often from densest nesting group. Diurnal and waning seasonal patterns occur (Emmerson 1969); dreads at intervals of 5–10 min (Bianki 1967), presumably at very start of season; mean rate of 2 per hr (Thorley 1963). For flock reactions to predators, see Parental Anti-predator Strategies, below. ANTAGONISTIC BEHAVIOUR. Apparently very similar to *S. hirundo*, and various studies failed to demonstrate major differences in postures and ceremonies associated with advertisement, establishment, and defence of territory (Hawksley 1950; Cullen 1960a; Busse 1975); calls are equivalent, though different-sounding. (1) Ground encounters. As in *S. hirundo*. (2) Aerial encounters. As in *S. hirundo*, but generally more aggressive (see also Parental Anti-predator Strategies, below). Tendency for birds nesting in given area to drive strange individuals out of neighbours' territories; common (loafing) ground also sometimes defended by several birds with adjacent territories against trespass by those nesting further away. Adults with young chicks particularly pugnacious, occasionally attacking neighbour on nest or straying (strange) chicks (Hawksley 1950). HETEROSEXUAL BEHAVIOUR. (1) General. Almost identical to *S. hirundo*. See also Bonds, above. (2) Aerial courtship. Habitually spends much more time on the wing than *S. hirundo*, so aerial courtship relatively more important: e.g. ♀ *S. paradisaea* tends to fly when approached by ♂, whereas ♀ *S. hirundo* walks away. Resurgence of aerial courtship in late July, Massachusetts, just prior to departure from breeding grounds, may be part of mate selection for following year. Some 1-year-olds court at this time, often joining courting pairs and displaying with them (see Bonds, above) (I C T Nisbet). High-flight apparently as in *S. hirundo* except that pair usually switch roles (aerial Bent-posture and aerial Erect-posture) after one or more Pass-ceremonies, with no further switching thereafter (Cullen 1960a). In Low-flight, advertising ♂ performs V-flying—wings held high, beating with slow, shallow strokes—much more regularly than does *S. hirundo* in which V-flying rare (I C T Nisbet). (3) Ground courtship. As in *S. hirundo*; often raises and spreads wings, especially in early pair-formation (Drury 1960). Scrape-making usually precedes copulation (Hawksley 1950). (4) Mating. Copulation, and associated display, as in *S. hirundo*. (5) Courtship-feeding. No detailed study, but apparently consists of same 3 stages as in *S. hirundo* (see Cullen 1962; I C T Nisbet). (6) Nest-relief. Often follows a dread. Unlike *S. hirundo*, relieving bird rarely approaches on foot, but hovers 2–3 m above nest, giving Advertising-call (see 4 in Voice), and alights directly on to it (Bianki 1967). Similar to *S. hirundo* in other respects, including side-throwing. RELATIONS WITHIN FAMILY GROUP. Chicks up to a few days old beg from parents by giving food-call and pecking at bill; may do so from a few hours old before parents have offered any food. Parent becomes agitated in response, and as soon as relieved by mate, departs to forage. Fish nearly always offered, and swallowed, head first. If dropped, parents may retrieve and offer fish again, this time to whichever chick grabs it most greedily (Quine and Cullen 1964). As in *S. hirundo*, one parent, always ♀ (I C T Nisbet), stays to brood for first few days (minimum 3) after hatching, while other forages (Hawksley 1950). Thereafter, both forage and chicks hide in vegetation within nest-territory, emerging only when they hear and recognize Advertising-call of returning parent; call recognition by chicks begins at 2–3 days (Busse 1975). Parent once observed attempting (unsuccessfully) to fly back to shore carrying newly hatched chick which had swum out to sea (Løvenskiold 1964), though, unlike *S. hirundo*, no evidence that parents recognize their own chicks even after several days (Hawksley 1950). In several Spitsbergen colonies with only 1–2 surviving downy young, several adults apparently tended them indiscriminately (Longstaff 1924; Løvenskiold 1964). This prevented in older chicks by their voice recognition of parents; thus 10-day-old chick accepted by foster parents in lieu of own chick but foster chick rejected parents, successfully homing by next day through several hostile territories to own nest 13 m away (Hawksley 1950). Parents occasionally kill own runt chick by pecking, removing from nest, and dropping on rocks (E K Dunn). Similar behaviour recorded for unpaired adult killing healthy chick in parents' absence (Green 1977). Where birds nest in open (unvegetated) areas, Massachusetts, chicks usually led away from nest before 3 days old (I C T Nisbet). Otherwise, begin to venture outside own nest-territory from 12–15 days (Hawksley 1950), taking up and defending territories nearby, but initially returning to nest-territory to be fed (Withers 1973). Start accompanying parents on short feeding trips 2–3 days after first

flight (Bianki 1967). Some adults teach fledged young to feed by dropping fish into water for them to retrieve (Løvenskiold 1964). Adults and offspring make early exodus from colonies, north-east USA, leaving 5–10 days after young fledge (I C T Nisbet). At Vikingavatn (Iceland), parents abandoned fledged young for increasingly long spells over 6-day period after fledging; young partly fed themselves, partly fed by parents; families left colony, without returning, after c. 6 days (Withers 1973). ANTI-PREDATOR RESPONSES OF YOUNG. On approach of danger, downy young crouch motionless, hide in vegetation or under rock, driftwood, etc., or swim (from 2 days old), in response to Alarm-call of parents. Juvenile occasionally performs erratic flight, as in *S. hirundo*. At colony, Massachusetts, 1-year-olds aggressive to predators such as humans and dogs, and to wide variety of non-predatory animals and inanimate objects (I C T Nisbet). PARENTAL ANTI-PREDATOR STRATEGIES. Similar to *S. hirundo*, but generally more aggressive (Suomalainen 1939; Rankin *et al.* 1942; Cullen 1960*b*; Tenovuo 1963; Boecker 1967; Haartman *et al.* 1963–72; Langham 1968), especially in northern parts of range where dispersed nesting common (see above) and pressure from predators high (e.g. Rankin *et al.* 1942; Lemmetyinen 1972*b*). In Arctic, birds aggressive only near own nest (Drury 1960). (1) Passive measures. At approach of danger, adults give Alarm-call (see above for response of young). Variant of Kip-call (see 6 in Voice) summons young from hiding when danger passed (Hawksley 1950). (2) Active measures: against birds. As in *S. hirundo*. Lemmetyinen (1971) showed no difference in frequency or violence of attacks by *S. paradisaea* and *S. hirundo* on dummy crows (Corvidae) and gulls (Laridae) at any stage in nesting cycle. Flying crows, gulls, and skuas (Stercorariidae) actively chased from nest, sometimes with Kliucalls (see 10 in Voice). Collective mobbing generally unsuccessful in repelling Ravens *Corvus corax*, which took unfledged young, Vikingavatn (Withers 1973). Peregrine *Falco peregrinus* is serious predator, Kandalaksha Bay, and White-tailed Eagle *Haliaeetus albicilla* is mobbed there most aggressively of all (Bianki 1967). May swoop at Eider *Somateria mollissima* crossing colony, Coquet Island, especially if taking ducklings to water (E K Dunn). (3) Active measures: against man. Probably attacks more boldly than *S. hirundo* (Lemmetyinen 1971). From small flock hovering overhead (smaller than for quadruped predators: Bianki 1967, see below), individuals swoop at intruder, uttering Anger-calls 'kek-kek-kek' (see 2 in Voice), often striking with bill and drawing blood; may also defecate at low point of swoop (Løvenskiold 1964), but rarely strike with feet (Rankin *et al.* 1942). Solitary nesters mount fiercer, more frequent attacks. Also, those nesting in Arctic near human settlements attacked man more often and more violently than in less disturbed areas (Larson 1960*a*; Løvenskiold 1964; Lemmetyinen 1972*b*). (4) Active measures: against other predators. Attacks quadruped mammalian predators much more vigorously than man, but otherwise in like fashion. Stoat *Mustela erminea* tends to attract densest, most vociferous, mobbing flock. Arctic fox *Alopex lagopus* or dog attracts smaller flock (Bianki 1967). Co-operative attack usually sufficient to confine fox to outskirts of larger colonies (Larson 1960*b*; Løvenskiold 1964). EKD

Voice. Both sexes highly vocal, especially at breeding grounds. Similar repertoire to Common Tern *S. hirundo* but calls generally slighter, softer, and higher pitched (I C T Nisbet). Order of listing reflects homology with *S. hirundo*; for further descriptions of contexts, see that species.

CALLS OF ADULTS. (1) Alarm-call. A shrill, extended 'kee-arr' (Hawksley 1950); emphasis often on 1st syllable (Fig I). Less harsh and higher-pitched, especially at beginning of call, than *S. hirundo* (Marples and Marples 1934; J Hall-Craggs, I C T Nisbet). Denotes imminent danger, especially approach of ground intruder. (2) Anger-call. A staccato 'kek-kek-kek' of mounting anger and aggression, in dive-attacks on humans and mammals; followed by 'tuck-keer' or 'tuck-keé-yah' as bird strikes victim, 2nd syllable rasping, extended, and uttered at lowest point of swoop (Hawksley 1950). Accelerating 'jik-jik-jik', given by bird swooping on another in hostile aerial chase, often after ground fight, Upward-flutter (see *S. hirundo*), or robbery of fish; probably same as 'kek'. Isolated 'jik' sometimes uttered by leading bird in ascent phase of High-flight; interpreted as alarm-call by Cullen (1960*a*). (3) Growl-call. A 'kurr' of extreme anger as birds close in Upward-flutter (Cullen 1960*a*). In fights, a guttural 'wrraa' or 'grraa' of about uniform pitch, expressing anger (Hawksley 1950: Fig II). (4) Advertising-call. Highly variable. Basic version a shrill 'kree-ah' (Hawksley 1950); 'keeri' or 'keeari' (Marples and Marples 1934) given by bird returning to colony with food, probably identifying bird to mate and offspring. Also delivered from territory by advertising ♂. A more subdued 'kaaar' given by bird returning to territory and calling for mate not in sight (Hawksley 1950). A loud, repeated (16 times in one observation: E K Dunn) 'kitikéeyer' given by bird in aerial Bent-posture before and during Pass-ceremony—also by V-flying ♂ in Low-flight—is elaboration of basic call (Cullen 1960*a*). Bird dropping to nest with food may begin with 'kreeah kreeah', or 'kiterr-kiterr-kiterr', followed immediately by 'kititikeeri-kititikeeri-kititikeeri' or repeated 'kititikeeari' (E K Dunn, P J Sellar: Fig III); pitch drops slightly on 'a' of 'keeari'. The 'kititi-' (also rendered 'K-K-K' or 'T-T-T') preceding '-keeari' equivalent to rapidly repeated call 6, and probably serves to help young to locate approaching parent (J Hall-Craggs). (5) Kor-call. A gruff 'kiyor-yor-yor' given by advertising ♂ in aerial Bent-posture as other bird passes it in High-flight or Low-flight, following on from call 4. (6) Kip-call. A high-pitched, clipped 'kip' (Hawksley 1950), or 'teuk' (P J Sellar) given singly or quickly repeated; see Fig IV for various 'kip' and 'teuk' (higher-pitched) notes; see also short notes in Fig II. Heard in variety of contexts, but usually denotes excitement or anxiety, e.g. a rapid 'tyeek-tyeek-tyeek' of alarm on approach of Great Black-backed Gull *Larus marinus* (P J Sellar); 'tchip' given by birds fishing in flock over shoal (Drury 1960). Also used as location signal to mate; at nest-relief, arriving bird may give subdued version, 'chib-chib-chib' (Marples and Marples 1934: Fig II, lower notes). Parent summons chicks from hiding with 'kip-kip' or modified 'kup-kup' (Hawksley 1950), often closely combined with call 4 (Fig III, short notes). (7) Begging-call. A shrill 'kee-kee-kee', more piercing than in *S. hirundo*, given typically by ♀ soliciting fish from ♂;

Sterna paradisaea

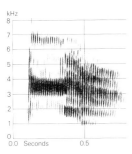

I J Gordon Finland July 1976

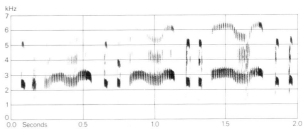

III P J Sellar Iceland August 1979

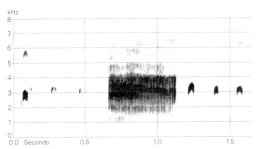

II S Palmér and R Edberg/Sveriges Radio (1972) Norway June 1960

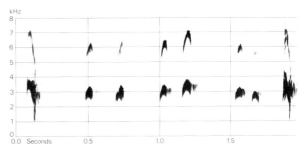

IV J Gordon Finland July 1976

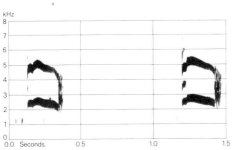

V P J Sellar Iceland August 1979

sometimes prelude to mating. Occasionally uttered by overtaking bird (in aerial Erect-posture) during Pass-ceremony (Cullen 1960a). (8) Kruk-call. A soft, murmuring 'kruk-kruk-kruk' given during ground courtship, and especially by ♂, often eliciting Begging-call from mate (Hawksley 1950) invites close approach, e.g. given by parent summoning wandering chick for brooding (Marples and Marples 1934). (9) Gurr-call. A guttural, growling 'kukuk-kurr-kukuk-kurr' heard during nest-scraping (Marples and Marples 1934). (10) Kliu-call. A musical (i.e. not harsh) 'kliu' (Cullen 1960a) or 'peea' (P J Sellar) sometimes given by pursuer in High-flight; otherwise accompanies attacks on intruders such as gulls (Cullen 1960a: Fig V).

CALLS OF YOUNG. No significant differences noted from. S. hirundo. Alarm-call of 1-year-old bird higher pitched than that of adult (I C T Nisbet).

Breeding. SEASON. North Sea area: see diagram. Arctic regions: laying begins mid- to end of June. SITE. On ground in the open. Colonial. Nest: shallow scrape, unlined or with a few pieces of available debris. Building: by both sexes, often alternately (Hawksley 1950); with scraping movements. EGGS. See Plate 89. Sub-elliptical, smooth and not glossy; pale buff to olive, rarely brown, also variably blotched, spotted, and scrawled black and dark brown. 41 × 30 mm (36–46 × 26–33), sample 300. Calculated weight 19 g (Schönwetter 1967). 143 eggs (corrected to fresh laying weight) weighed 16·7–22·1 g (I C T Nisbet). 1st egg usually bigger than 2nd (E K Dunn). Clutch: 1–3. Mean 2·66, Barents Sea (Belopol'ski et al. 1977). Of 511 clutches, south-west Finland: 1 egg, 10%; 2, 76%; 3, 13%; mean 2·03; decreases through summer, and some variation depending on age (Coulson and Horobin 1976) and food supply, diet of fish inducing bigger clutches than one of crustaceans (Lemmetyinen 1973a). Of 152 clutches, Spitsbergen: 1 egg, 30%; 2, 69%; 3, 1%; mean 1·7 (Bengtson 1971; see also Korte 1972). Of 453 clutches, Orkney and Shetland, mean 1·95 (Bullock and Gomersall 1981). Of 182 clutches over 3 years, Coquet Island (England), mean 1·82 (Langham 1974). One brood. Replacements laid after loss of clutch at lower latitudes, though not after 10 days from start of incubation (Bianki 1967); unlikely in high Arctic, though no information. Laying interval usually 1–2 days, up to 5, mean 2·8 (Norderhaug 1964). INCUBATION. 20–24 days. Mean 22·2 days (Bianki 1967); increases to 25–33 days when predators cause night desertion (I C T Nisbet). By both parents; sharing fairly evenly according to Dircksen (1932), though Horobin (1971) found ♂ incubating on

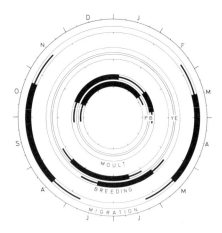

46% of occasions, ♀ on 54%, and Hawksley (1950) noted ♀ took greater share; mainly by ♀ according to Dementiev and Gladkov (1951c). Begins with 1st or 2nd egg; hatching near synchronous; hatching over 1·19 days in Finland, 1·78 days in Spitsbergen (Lemmetyinen 1972a); in clutches of 3, first 2 chicks appear on same day, 3rd chick on following day (Bianki 1967). YOUNG. Precocial and semi-nidifugous. Cared for and fed by both parents. Brooded while small. Stay in or near nest for 1–3 days, then leave for shelter of vegetation or stones. FLEDGING TO MATURITY. Fledging period 21–24 days. Age of independence uncertain but not for at least a further 1–2 months. Age of first breeding 2–5 years, mainly 4. BREEDING SUCCESS. Of 90 eggs laid in 50 nests, Spitsbergen, 66% hatched; 29% alive after 2 weeks, or 0·52 young per nest (Bengtson 1971). Of 832 eggs laid, south-west Finland, 68% hatched; 57% of those hatching survived to 2 weeks, or 0·91 young per pair; survival best in clutches of 2; compared with siblings, 3rd chicks in broods of 3 survived poorly; main losses (73%) to predators, especially crows *Corvus* and gulls *Larus* (Lemmetyinen 1973b). In Arctic, also predation by fox *Alopex*. Hatching and fledging success, Farne Islands (England), related to age of parents: 31% of eggs hatch from 3-year-olds, rising to 53% from birds over 8 years though this low due to predation by Starlings *Sturnus vulgaris*; fledging success 54% with 3-year-olds, rising to 70% with parents of 8 years; mean number of young reared per pair varied from 0·24 for 3-year-olds, to 0·58 for birds of 6–8 years, and 0·59 for those over 8 years (Coulson and Horobin 1976). At Coquet Island, where predation less, mean hatching success in 2 years 69·1% (217 clutches) and 87·7% (163 clutches) (Langham 1974).

Plumages. ADULT BREEDING. Closely similar to Common Tern *S. hirundo*, but black cap slightly more extensive, often leaving narrower white strip between upper lores and upper mandible and reaching slightly further down on sides of neck. Underparts slightly more bluish-grey, less ashy, concolorous with upperparts, lacking pink or mauve tinge of some *S. hirundo*; grey of underparts reaches up to chin and lower cheeks, separated from black cap by more obvious white area from lower lores to sides of nape, rather than gradually merging into white of cheeks and upper throat; white under tail-coverts more sharply demarcated from grey vent. Tail white, except for pale grey outer web of (t4–)t5 and darker one of t6; occasionally, all outer webs grey and t6 nearly black, as in *S. hirundo*. Flight-feathers close to those of *S. hirundo*, but without showing contrast between fresh inner primaries and older outer: all equally fresh and pale silvery-grey in spring, all darker grey with blackish tips in autumn; white wedges to inner webs of primaries more extensive, leaving grey streak 1½–2½ mm wide next to shaft, against 3–5 mm in *S. hirundo* (in both, measured at 10 cm from feather-tip); grey on tip of each feather forms narrow dark fringe beneath trailing edge of primaries, narrower than in *S. hirundo* and terminating less abruptly on innermost primaries. Secondaries more broadly tipped white than in *S. hirundo*, white on inner webs more extensive. In worn plumage, grey on underparts fades and throat and chest especially appear whitish. ADULT NON-BREEDING. Lores and forehead white; crown white mottled and fringed with some black, merging into black patch extending round and above eye over ear across hindcrown and nape. Upper mantle, sides of head below black patch, and all underparts white. Upperparts, upper wing-coverts, and tail as in adult breeding, but lesser upper wing-coverts and lesser primary coverts slate-grey forming dark carpal bar, pale grey on outer webs of outer tail-feathers often extending to t1–t4, and upper tail-coverts sometimes tinged grey, leaving white rump only. Sometimes indistinguishable from *S. hirundo* on plumage alone; *S. hirundo* often has more extensive grey tinge to rump, upper tail-coverts, and tail, white wedge on inner webs of primaries less extensive, and white on tips and inner webs of secondaries more restricted, but characters overlap. Moulting birds, December–January, have inner primaries new and pale and outer ones old and blackish, showing same contrast as *S. hirundo* has March–November. DOWNY YOUNG. Upperparts either pearl-grey or various shades of buff, densely speckled or streaked black; lower forehead, lores, and throat suffused sepia-brown to black-brown, chin occasionally white; underparts white, tinged or freckled grey or brown on flanks and belly. Closely similar to *S. hirundo* and difficult to distinguish by plumage alone; speckling on upperparts of *S. paradisaea* usually heavier and less clear-cut, upper chin and forehead often dark sepia rather than white or buff; forecrown usually spotted, throat less often pale brown; ground-colour frequently grey instead of buff, especially where breeding range overlaps; underparts often suffused or speckled grey or brown. (Fjeldså 1977, which see also for other differences.) Greyish individuals more frequent in areas dominated by bare grey rocks than where rocks reddish and/or covered with brown sandy soil, often with dense vegetation (Lemmetyinen *et al.* 1974). JUVENILE. Head and body closely similar to *S. hirundo* and equally variable, though usually less tinged with gingery-buff above and often without distinct dark scaling: forehead and lores off-white or buffish-grey (white when worn); crown and nape deeper black; mantle, scapulars, and tertials light grey with off-white tips and variable amount of dusky grey to black subterminal marks or freckling; feather-tips only occasionally tinged gingery-buff; sides of breast and flanks sometimes suffused grey-buff. Rump and upper tail-coverts white, unlike most *S. hirundo*. Outer webs of tail-feathers grey, tips and inner webs white, inner web of t1 sometimes grey; occasionally, dark subterminal spot or buff tinge to some feather-tips. Wing rather different from *S. hirundo*: carpal bar over lesser wing-coverts slate-grey (less blackish), usually narrower, and less sharply defined from median coverts; secondaries paler grey than greater upper wing-coverts instead of darker, and more broadly tipped white, sometimes almost completely white; inner webs of primaries with more extensive white wedge, reaching closer to shaft (as in adult), dark streak along distal edge of inner web often narrower and shorter, forming narrow and clear-cut trailing-edge to primaries when seen from below. FIRST IMMATURE NON-BREEDING. Like adult non-breeding and indistinguishable when last juvenile feathers lost January—only juvenile outer primaries and inner secondaries then remain, and these as worn as those of adult. However, adults attain breeding February–March, while immatures retain non-breeding throughout 2nd calendar year, showing white forehead, white crown grizzled with black, white underparts, dark slate carpal bar, more extensive grey on tail, and largely black bill. Some such birds visit northern hemisphere. Differs from 2nd calendar year *S. hirundo* in same way as adult non-breeding; besides this, primaries either all fresh, or actively moulting in one centre (January–April and October–December only), while *S. hirundo* shows arrested moult with strong contrast between inner and outer primaries, or primaries actively moulting in 2 centres (May–August, December); during January–April, 1st-winter *S. hirundo* moults primaries in single centre only and is thus similar to *S. paradisaea* of same age in this respect. SUBSEQUENT PLUMAGES. No certain 2-year-olds examined, but of *c.* 90 summer birds from northern hemisphere, 11% retained slightly worn non-breeding upper wing-coverts with distinct or spotted slate-grey carpal bar, and these presumably in 3rd calendar year. Besides these non-breeding wing-coverts, some retained scattered white non-breeding feathers on forehead, lores, or belly also, as well as a few old scapulars or tail-feathers; belly sometimes paler grey; one bird had started 2nd series of primaries, arresting with p1–p2 new.

Bare parts. ADULT. Iris dark brown. Bill deep carmine or blood-red during breeding, tip of upper mandible rarely black; bill black in non-breeding, often with some vermilion or scarlet to basal cutting edges of upper mandible and to base of lower; red usually fully attained March when still in winter quarters, but black occasionally partly retained in summer birds in poor condition; black reappears September–October. Foot coral-red in summer, dark brown-red or blackish in winter, often with vermilion sole. DOWNY YOUNG. Iris dark brown. Bill light coral-red or orange-red with diffuse dark tip. Foot pink to orange-red with slight grey tinge. JUVENILE. Iris dark brown. Bill apricot-pink or orange, distal $\frac{1}{3}$–$\frac{2}{3}$ black at fledging, darkening to adult non-breeding colour within *c.* 1 month; bill often largely black by mid-August; usually all-black by September. Foot orange-red, changing through dull red, pinkish-grey, and greyish-red to dark brown-red or blackish October–December. IMMATURES. In 1st and following winters, like adult non-breeding. In summer, bill of 1-year-old like adult non-breeding, but red at base sometimes slightly more extensive; foot like adult non-breeding, but tinged light vermilion to variable extent, some largely black with a few red dots, others red with at least joints dark red-brown or dark grey. 2-year-old apparently like adult breeding, but perhaps occasionally as 1-year-old. (Witherby *et al.* 1941; Bierman and Voous 1950; Bundy 1974; Fjeldså 1977; RMNH, ZMA.)

Moults. ADULT POST-BREEDING. Complete, primaries descendant. Usually no moult on breeding grounds, nor during migration; a few birds showing limited moult on body in late summer perhaps failed or non-breeders. Exact date of start of moult not certain, as no early winter specimens from southern oceans seen; apparently starts with p1, tail, head, and underparts late September to early November; late October migrants still in worn breeding. Main moult in pack-ice during southern summer; head, body, wing-coverts, and tail in fresh non-breeding January, primaries completed early February to early March. Flight-feather moult thus more rapid than in *S. hirundo*: single cycle completed in *c.* 5 months, against $6\frac{1}{2}$–7 in *S. hirundo*. ADULT PRE-BREEDING. Partial: late February and March, mostly completed before spring migration. Involves head, body, tail (occasionally inners and t6 only), and upper wing-coverts; unlike *S. hirundo*, no inner primaries and outer secondaries. POST-JUVENILE. Complete. Starts on arrival in winter quarters, but head, underparts and part of mantle and scapulars sometimes moulted during migration, starting from late October. By February, usually in full non-breeding, including wing-coverts and tail; moult of flight-feathers starts approximately December–January, completed about May, but only a few in moult seen and these appear to be somewhat retarded; healthy birds perhaps moult flight-feathers earlier, at about same time as adult, as the few birds reaching northern hemisphere in 2nd calendar year have flight-feathers all fresh in June, with moult probably completed before migration started about April. SUBSEQUENT MOULTS. Like adult; restricted to short period of southern hemisphere summer, October–March. In 2nd winter, post-breeding and flight-feather moult as in adult, but pre-breeding starts later and is more restricted, as part of non-breeding upperparts, apparently all non-breeding wing-coverts, and sometimes scattered feathers on forehead and belly retained. Exceptionally, 1–2 inner primaries replaced in pre-breeding.

Measurements. Netherlands and Scandinavia; breeding adults May–July, juveniles September–October; skins (RMNH, ZMA). Tail is length of t1 in adult; fork is tip of t1 to tip of t6—in adult, fresh breeding only.

WING AD	♂ 279	(5·44; 20)	270–290	♀ 274	(10·1; 16)	261–288
JUV	246	(2·68; 5)	244–250	244	(4·42; 15)	238–253
TAIL (t1)	72·2	(3·57; 11)	67–78	72·0	(2·91; 12)	68–76
FORK AD	111	(10·6; 19)	96–130	97·8	(13·1; 18)	72–118
JUV	46·8	(4·66; 5)	41–52	50·3	(4·96; 14)	44–59
BILL AD	33·0	(1·40; 19)	31–35	30·8	(1·16; 20)	29–33
TARSUS	15·9	(0·49; 20)	15·0–17·0	15·6	(0·45; 19)	14·8–16·5
TOE	21·7	(1·11; 10)	20–23	22·0	(0·82; 10)	21–23

Sex differences significant for adult fork and bill. Adult non-breeding fork 69 (4) 62–75. Juvenile t1 averages 7·5 shorter than adult, bill in 1st autumn 4·9 shorter; tarsus and toe similar to adult from *c.* 2 weeks old. Wing in 3rd calendar year 266 (11) 255–273, fork 85 (5) 69–103, bill 30·2 (7) 27–33.

Slight geographical variation occurs in wing, bill, and tarsus, as shown by adults of following Atlantic populations (no data for Pacific birds available):

Greenland and Iceland, summer (RMNH, ZMA).

WING AD	♂ 274	(6·30; 26)	262–286	♀ 272	(2·71; 12)	266–276
BILL AD	32·8	(1·37; 19)	31–35	30·9	(1·39; 9)	29–33
TARSUS	16·0	(0·90; 17)	14·8–17·2	16·1	(0·40; 9)	15·7–16·8

Spitsbergen and Bear Island, summer (RMNH, ZMA).

WING AD	♂ 279	(5·60; 11)	268–286	♀ 272	(4·22; 7)	267–278
BILL AD	31·2	(1·30; 10)	29–32	30·6	(0·84; 7)	29–32
TARSUS	15·3	(0·59; 14)	14·6–16·1	14·8	(0·64; 10)	13·9–15·5

Weights. ADULT. North-east Greenland (J de Korte), Iceland (Timmermann 1938–49; ZMA), Svalbard (J de Korte), northern Scotland (BTO), Netherlands (ZMA), and Alaska (Bee 1958); combined.

MAY   ♂ 112   (4·97; 5) 106–119   ♀ 107   (8·32; 5) 98–118
JUL   102   (9·77; 11) 87–118   109   (7·02; 8) 99–119
AUG   103   (6·95; 7) 96–115   111   (— ; 2) 105–117

Summer: north-east Greenland 99·1 (5·37; 56) (Asbirk and Franzmann 1978); Yakutia (USSR) 91·2 (9·66; 6) 80–105 (Uspenski *et al.* 1962). In poor condition, May, 66 (RMNH). Winter: in pack-ice, South Atlantic, late February and early March, ♂♂ 140, 145; ♀ 125 (ZMA).

JUVENILE. Western and northern Europe: September 76, 92, 105, 110 (BTO, ZMA); October 98 (ZMA). South Atlantic, ♀♀: 69 (November, retarded), 110 (January), 100 (March) (ZMA).

For growth rate of chicks, see Lemmetyinen (1972a).

Structure. Closely similar to *S. hirundo*, but p9 in adult 14–19 shorter than p10, p8 34–45 shorter, p7 56–69, p6 77–97, p5 98–118, p1 163–188; depth of tail-fork greater in all plumages; bill relatively shorter, culmen straighter, and gonys less distinctly angled; depth of closed bill at basal corner of nostril 8·41 (0·33; 10) 7·9–8·9 in ♂, 7·56 (0·46; 10) 7·2–8·2 in ♀; tarsus distinctly and toes slightly shorter.

Geographical variation. Negligible; see Measurements.   CSR

Sterna aleutica Aleutian Tern

PLATE 15
[facing page 206]

DU. Aleoetenstern   FR. Sterne aléoute   GE. Aleuten-Seeschwalbe
RU. Алеутская крачка   SP. Charrán de las Aleutianas   SW. Aleutisk tärna

Sterna aleutica Baird, 1869

Monotypic

Field characters. 32–34 cm (bill 3·3–4·3, legs 1·7–2 cm); wing-span 75–80 cm. Close in size to Bridled Tern *S. anaethetus*, but with shorter bill and legs; slightly bulkier, and shorter tailed than Arctic Tern *S. paradisaea*. Medium-sized tern, with character and plumage pattern recalling *S. anaethetus* (and '*portlandica*' types of Common Tern and *S. paradisaea*). Adult shares large white forehead of *S. anaethetus* but shows wholly white rump and tail and striking dark bar on undersurface of secondaries; rest of wing also strikingly patterned, unlike other small or medium-sized terns. Immature unstudied but appears to lack obvious diagnostic marks. Flight action distinctive—noticeably deep wing-beats with occasional but characteristic double flick on upstroke. Whistling call, easily distinguished from grating chorus of congeners. Sexes similar; little seasonal variation. Juvenile and immature separable.

ADULT BREEDING. Differs most noticeably from *S. anaethetus* in (1) lack of pale grey-white collar at base of nape, (2) paler though still slate-toned upperparts, (3) wholly white rump, vent, and tail, and (4) dark tip to underwing. Wing pattern obvious on both surfaces and recalls particularly that of '*portlandica*'-type *S. hirundo*: above shows bold white leading edge, broad dusky lines on at least 5–6 outer primaries (when spread), silvery leading edge on same feathers when closed, paler central flight-feathers, and dusky band over inner secondaries; below shows broad, almost black tips to 5–6 outer primaries, translucent inner primaries, and white coverts contrasting sharply with almost black band across all secondaries and grey body. Underpart pattern usually also marked, with intense grey wash on foreneck, breast, flanks, and belly isolating white panel below crown and white vent and rump. White forehead (with extension to just behind eye) sharply delineated, appearing as triangular V head-on and contrasting boldly with quite heavy black crown and loral stripe. Rather stubby bill and legs black. Tibia feathered. ADULT NON-BREEDING. Crown speckled white, mantle also when fresh. Underparts white. JUVENILE. More closely resembles similarly aged *S. anaethetus* but rump and tail not uniform with back, being pale grey with white margins (and thus recalling pattern of adult *S. anaethetus*). Bill dusky yellow; legs red-yellow. Other plumages not studied, but likely to show strongly variegated wing pattern including dark panel along lesser coverts.

Oceanic tern, incompletely studied in non-breeding plumages. Solitary vagrant to England announced presence with distinctive call and could be easily followed by unusual flight action and dark plumage. Distinction from *S. anaethetus* covered above; differentiation of adult from exceptionally dark '*portlandica*' types of *S. hirundo* and *S. paradisaea* not difficult but risk of confusion between these and immature *S. aleutica* unexplored. Flight action amongst most flowing of all terns; wing-beats noticeably slow, deep, and powerful with continuous emphasis on downstroke and occasional interjection of double flick on upstroke, last action accompanied by call in English vagrant. Gait and stance as similarly sized congeners.

Migrations and behaviour away from breeding colonies virtually unknown; English vagrant associated with 4 congeners on island colony (and was much chivvied). Commonest call has soft, whistling (not grating) tone: short repeated phrase of 4–5 syllables recalling wader (Charadrii) before congener.

Habitat. Breeds on subarctic and boreal island coasts, sometimes on raised low plateaux overlooking sea at *c.* 10 m, on marshy patches of soft low grey moss scattered among matted dry grass of previous season's growth. Also on sparse vegetation on dry sandy ground and once on rotten wood from decayed driftwood (Bent 1921; Dementiev and Gladkov 1951c; Buckley and Buckley 1979). Often associated with Arctic Tern *S. paradisaea* and Kittiwake *Rissa tridactyla* (Pough 1957). Colonies vulnerable and often impermanent. Forages in tidal waters and over 50 km from shore during breeding season (Kessel and Gibson 1978), but probably pelagic during rest of year (see Movements).

Distribution. Breeds coasts and islands of Bering Sea and extreme North Pacific: in eastern Siberia (Sakhalin, Kamchatka, Sea of Okhotsk), Aleutians, and in west, south-west, and southern Alaska (locally from Kotzebue Sound to Yakutat). Winter distribution unknown (see Movements).

Accidental. Britain: adult photographed Farne Islands (Northumberland), 28–29 May 1979 (Dixey *et al.* 1981).

Movements. Evidently pelagic outside breeding season, but extent of movements unknown (Kessel and Gibson 1978). No records at all from Canada or USA south of Gulf of Alaska, though stragglers reported Commander Islands (1) and Japan (6), dated records spanning May–September (Dementiev and Gladkov 1951c; Austin and Kuroda 1953; Ornithological Society of Japan 1974). Said by various authors (e.g. American Ornithologists' Union 1957, Vaurie 1965, Tuck and Heinzel 1978) to winter in North Pacific within or close to breeding latitudes. This apparently based on statement by Hartert (1912–21) that birds had been collected Sakhalin and northern Japan in winter; but this assertion suspect since 'winter' not defined, and sole Japanese record then was an undated specimen (though in non-breeding plumage). No confirmed midwinter records anywhere, and may yet be found to winter far to the south like sympatric Arctic Tern *S. paradisaea*.

Fragmentary information on timing of movements indicates arrival Alaskan coasts in second half of May, and departure by mid-September (Bent 1921; Gabrielson and Lincoln 1959; Gibson 1981).

Voice. See Field Characters.

Plumages. ADULT BREEDING. Forehead and streak over eye to above rear of eye white, contrasting with black streak from lores through eye over ear-coverts and with black of crown down to ear-coverts and central nape. Mantle, scapulars, back, tertials, and upper wing-coverts pale slate-grey, contrasting with white rump, upper tail-coverts, and tail; central upper tail-coverts and t1 sometimes distinctly tinged grey. Tips of tertials white. Chin and cheeks below black lores and ear-coverts white, grading into grey of underparts on throat. Belly grey, under tail-coverts contrastingly white. Inner primaries and outer webs of secondaries medium grey, latter with white edge to tip, inner webs of secondaries dark grey. Outer primaries dark grey with slight silvery tinge on outer webs, dark grey with broad contrastingly white wedge to bases on inner webs; wedge extends to *c*. 5 cm from tip on outer primaries; dark grey streak *c*. 2 mm wide along distal half of inner edge of outer primaries, readily visible from below. Leading edge of wing, under wing-coverts, and axillaries white, latter sometimes slightly suffused grey. At start of breeding, underparts sometimes slightly tinged pink. ADULT NON-BREEDING. Like adult breeding, but lores finely speckled white, forecrown streaked white, hindcrown finely spotted white. Fresh feathers of mantle and scapulars narrowly tipped white, underparts white. Not known whether central pairs of tail-feathers are more extensively tinged grey, as in related *Sterna*. JUVENILE. Forehead and crown down to lores and eye dull grey-buff, streaked black on hindcrown; black bar on nape extending down to sides of neck; dark spot in front of eye. Mantle, scapulars, tertials, and lesser and median upper wing-coverts dull black, feathers fringed buff on tips (*c*. 1 mm wide on mantle, *c*. 3 mm on tertials). Back dark grey with faint buff feather-fringes; rump and upper tail-coverts medium grey with broad but ill-defined white feather-tips, latter partly suffused buff on edge. Underparts white, except for vinous or lavender-grey spot on side of breast, forming broken breast-band, contiguous with similar coloured band round neck behind black nape-band. Tail-feathers medium or pale grey, tip broadly buff with black subterminal V-mark, inner webs of t2–t5 mainly white; t6 almost fully white, except for grey wash on tip of inner web, and buff-tinged tip with black subterminal spot on outer web. Greater upper wing-coverts medium grey with ill-defined white tips. Primaries dark grey (darkest on tips) with slightly silvery bloom, tips narrowly bordered pale buff (wider on inner primaries); secondaries dull black with broad white border to inner webs and tips. Upper primary coverts dark grey, narrowly tipped buff. Leading edge of wing, under wing-coverts, and axillaries white. In worn plumage, white of feather-bases visible on forehead and crown; buff feather-fringes of upperparts abraded and bleached to pale buff or off-white. SUBSEQUENT PLUMAGES. None examined. Undoubtedly as in other *Sterna*, showing adult non-breeding plumage during 2nd calendar year and with partial or complete adult breeding developing in spring of 3rd. (Hartert 1912–21; Ridgway 1919; C S Roselaar.)

Bare parts. ADULT. Iris dark brown. Bill and foot black. JUVENILE. Iris dark brown. Bill black with much red to base of lower mandible. Foot light reddish. (Ridgway 1919; BMNH.)

Moults. ADULT. None of 19 examined May to late August showed active moult. No winter specimens seen, but undoubtedly a complete post-breeding moult in winter quarters as in related *Sterna*, followed by a partial pre-breeding in late winter and spring; latter involves at least head, underparts, mantle, tail-feathers, and inner primaries (up to p5 in 17 birds, to p4 in 2). JUVENILE. None of 5 from July–August in breeding area showed moult. No further details known; see also Plumages.

Measurements. ADULT. Western Alaska, summer; skins (BMNH). Fork is tip of t1 to tip of t6.

WING	♂ 272	(7.08; 10)	262–282	♀ 274	(6.21; 9)	260–281
TAIL (t1)	65.6	(3.13; 10)	61–72	66.1	(4.54; 9)	58–73
FORK	91.1	(11.6; 8)	76–104	98.0	(9.64; 7)	87–113
BILL	33.6	(0.94; 10)	32.0–34.6	33.4	(0.86; 9)	32.3–34.8
TARSUS	18.8	(0.76; 10)	18.1–20.3	19.3	(0.86; 9)	18.3–20.7
TOE	27.0	(0.87; 10)	25.2–28.6	27.0	(0.87; 9)	25.6–28.1

Sex differences not significant.

Weights. Eastern Siberia, juveniles about to fledge: 104, 114 (V A Nechaev).

Structure. Wing long and narrow, pointed. 11 primaries: p10 longest, p9 14–16 shorter in adult, less in juvenile; p8 32–36 shorter, p7 55–59, p6 71–83, p1 161–177; p11 minute, concealed by primary coverts. Tail deeply forked in adult, less in juvenile; 12 feathers, t6 142–181 in fresh adult breeding; depth of fork 76–113 in adult breeding, 40–60 in juvenile, probably intermediate between these in adult non-breeding and immature. Bill length about similar to Arctic Tern *S. paradisaea*, but base even more slender, depth 7·17 (6·7–7·6) at basal corner of nostril. Tarsus rather short and slender. Toes rather long; outer toe *c.* 82% of middle, inner *c.* 62%, hind *c.* 31%.

Geographical variation. None.

Probably forms superspecies with Spectacled Tern *S. lunata* of central Pacific and Bridled Tern *S. anaethetus* of Atlantic, Indian, and western Pacific Oceans (see also Hartert 1912–21).

In particular, *S. lunata* closely similar, showing same head pattern and similar colour of mantle, scapulars, and wing in both adult breeding and non-breeding, and similar foot and bill structure; in breeding plumage, *S. aleutica* differs from *S. lunata* by white back to tail and grey underparts (these mainly grey and fully white, respectively, in *S. lunata*), but in non-breeding season (when *S. aleutica* probably winters in breeding range of *S. lunata*) both probably closely similar, except for longer bill of *S. lunata* (38–46 mm). *S. anaethetus* slightly darker grey on upperparts and tail than *S. lunata*, less extensively white on inner webs of primaries, wings relatively shorter and broader, foot and bill heavier. Sooty Tern *S. fuscata* darker still than *S. anaethetus*, foot and bill even heavier, but as it widely overlaps in breeding range with both *S. anaethetus* and *S. lunata*, not included in the same superspecies. CSR

Sterna forsteri Forster's Tern

PLATES 5, 7, and 8
[between pages 134 and 135, and facing page 158]

Du. Forsters Stern   Fr. Sterne de Forster   Ge. Sumpfseeschwalbe
Ru. Американская речная крачка   Sp. Charrán de Forster   Sw. Forsters tärna

Sterna forsteri Nuttall, 1834

Monotypic

Field characters. 33–36 cm (bill 3·6–4·2 cm, legs 2–2·5 cm, tail of adult 13–20, of juvenile 8 cm); wing-span 73–82 cm. Slightly larger than Common Tern *S. hirundo*, with longer bill, legs, and tail, but 5% shorter wings. Graceful, grey and white tern of rather similar structure to *S. hirundo*, but differing in characteristic pale silvery primaries and grey inside border of tail-fork. Juvenile, immature, and winter adult show diagnostic white head, strongly marked by black panel through eye (not extending on to buff rear crown or nape) making bill more obvious; also lack dark lesser wing-coverts (obvious on *S. hirundo*). Flight action faster, with shallower wing-beats than *S. hirundo*. Flight call distinctly nasal and relatively low-pitched, quite unlike *S. hirundo*. Bill of adult more orange than red. Sexes similar; little seasonal variation. Juvenile and immature separable.

ADULT BREEDING. On ground, differs most obviously from *S. hirundo* in white underparts. In flight, all primaries appear silvery and form distinctly paler area which contrasts with inner wing and back; outer feathers show dusky lines and tips (almost black on outermost) at close range, but do not form broad dusky wedge above nor obvious dusky trailing edge below. Tail pattern also distinctive, with centre and base almost white contrasting with pale grey rump but inside of fork faintly rimmed dusky (due to uniquely grey inner webs of longest feathers). Bill shaped as *S. hirundo* but with slightly longer and narrower point; orange-red with black tip (and looking paler than in *S. hirundo*). Legs orange. ADULT NON-BREEDING. Crown and nape grey-white, only faintly streaked black at rear; black panel through eye, forming conspicuous mark like that of immature Gull-billed Tern *Gelochelidon nilotica*. Wing pattern unchanged but silvery tone on primaries reduced by wear; rump whiter. Bill mostly black, length and weight accentuated by black eye-panel; legs browner. JUVENILE. Differs from *S. hirundo* in generally paler plumage, with diagnostic lack of both black crown and bold dusky leading edge to inner wing. Rest of wing lacks obvious silvery area over primaries and shows dusky rim to primaries and darker secondaries (but neither of latter marks as striking as in juvenile *S. hirundo*). Tail shows darker rim to inside of fork than adult. Bill as winter adult; legs red-brown. FIRST WINTER. Resembles winter adult but shows conspicuous pale mid-wing panel, with white bases to primaries and outer secondaries contrasting with dark (worn) outer webs and tips of at least outer primaries. Lack of dark leading edge to inner wing remains further obvious distinction from *S. hirundo*. 1st-summer plumage not studied in the field.

Subject to confusion with *S. hirundo* in Nearctic but separation not difficult; differences in adult wing pattern and in head pattern in all non-breeding plumages diagnostic. Flight typical of genus, but actions show adaptation to aerial feeding on insects, with frequent turns, swoops, and generally faster and shallower wing-beats. Gait and carriage as *S. hirundo*, but one vagrant immature looked noticeably long-legged.

As likely to occur over marshes as along coasts. Calls

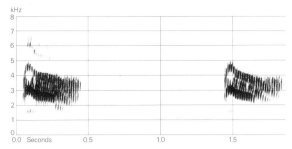

I W W H Gunn/Sveriges Radio (1973) Canada June 1960

subject to wide variation in transcription, but all described as nasal, wooden, or toneless, and relatively lower pitched than those of *S. hirundo*; alarm call a harsh descending 'peorr' or 'keorr' (Fig I), fuller and deeper than *S. hirundo* or Arctic Tern *S. paradisaea*, and often linked to series of 'ker' notes.

Habitat. Breeds mainly inland but also on coasts, in warm middle and lower middle Nearctic latitudes. Mainly in lowland areas, but in Colorado and elsewhere in Rocky Mountain region up to 1600 m or higher. Nests in marshes bordering reservoirs and lakes, on houses of muskrats *Ondatra*, drift, matted vegetation among tules, in bulrush or club-rush *Scirpus*, upon mud hummocks (Bent 1921; Niedrach and Rockwell 1959), or on floating mats of rotting reeds. Also on floating logs and on edge of water some 1·5 m deep. Along coast of Virginia, on inward side of long wide islands, bordering small pools, nests on tide-drifted masses of dry dead sedges, or on extensive salt-marshes intersected by creeks. Sometimes in Gulf of Mexico, on small grassy islands in bays and lagoons (Lowery 1955); in Texas, on wet mud some way from water on low nearly submerged island in riverside salt-marshes, or in sand on islands in a bay (Bent 1921). Outside breeding season, still avoids marine waters except those close inshore. Hunts over water, sometimes making diving plunge but often picking up food from surface; also catches insects on the wing, sometimes hawking at up to 100–200 m (Bent 1921).

Distribution. Breeds in North America from southern Alberta, Saskatchewan, and Manitoba south to south-central California, central Idaho, south-east Wyoming, eastern Colorado, South Dakota, central Iowa, and south-east Wisconsin; also from south-east Maryland and eastern Virginia to southern Louisiana and Texas. Winters from central California to Guatemala, and from Virginia to northern Florida and along Gulf of Mexico from Florida to Mexico.

Accidental. Iceland: ♂, probably adult, taken Vestmannaeyjar, 22 October 1959 (AP). Britain: 1st-winter, Cornwall, February–March 1980 (Cave 1982); 1st-winter, Cornwall, October 1982 (Madge and Madge 1983).

Movements. Migratory in north, but no more than dispersive in southern breeding area (Gulf of Mexico). 2 discrete breeding populations: (1) inland in prairies of Canada and USA, migrating extensively overland, mainly towards Pacific and Gulf of Mexico; (2) along east and south USA seaboard (Maryland to Texas), with migratory elements following coasts. Combined winter range: central California to Guatemala, Virginia to Florida and thence west through Gulf of Mexico to eastern Mexico (American Ornithologists' Union 1957). Small numbers may penetrate a little further south; several seen Honduras in April 1953 (Munroe 1968), and though stated as rare in West Indies (Bond 1971), small parties seen Puerto Rico, January 1969, indicate more than casual there (Buckley and Buckley 1970b). Hence winters relatively far north for a tern (mainly 20–37°N), not reaching South America, or southern Central America. 2 juveniles ringed Maryland in June recovered Florida in December and North Carolina in January (Stewart and Robbins 1958).

Autumn passage begins mid-July as far apart as Minnesota and Maryland. Mostly gone from inland prairies by late September, though in Chesapeake Bay exodus continues in strength to *c.* 20 October and in dwindling numbers to mid-November (Stewart and Robbins 1958; Green and Janssen 1975). Fairly common on Great Lakes September–October; in Massachusetts occurs from mid-September in erratic but increasing numbers which peak in mid-October (seldom more than 12 together up to 1970, but groups of up to 200 in 1979 and 1980), with stragglers remaining into November or even December (R A Forster, I C T Nisbet). As last 2 regions are north of Atlantic breeding range, eastern movement by part of inland (prairies) breeding population is presumed. Only irregular visitor to Canadian maritime provinces, including several late-autumn hurricane-related records from Nova Scotia (Mills 1969).

Return movement begins early April. In Minnesota arrives mainly second half April, passage continuing into late May, though spring migration in Maryland completed by 10 May (Stewart and Robbins 1958; Green and Janssen 1975). Rare in north-east USA in spring.

Voice. See Field Characters.

Plumages. ADULT BREEDING. Forehead, upper half of lores, crown down to just below eye, and nape black, reaching down to upper mantle and to upper sides of neck. Feathers of nape not elongated; no white across upper mantle. Mantle, scapulars, tertials, back, and most upper wing-coverts light grey, slightly darker and less bluish than in Common Tern *S. hirundo*; sides of mantle grey rather than white as in latter. Rump white; upper tail-coverts and tail pale grey, outer web and extreme tip of t6 white. Lower half of lores, small spot below eye, cheeks from below ear-coverts, sides of neck, leading edge of wing, and all underparts, including axillaries and under wing-coverts white. Flight-feathers silvery-white, in fresh plumage contrasting with darker grey of wing-coverts and upperparts; rather broad tips and inner edges of secondaries white, inner primaries with similar but narrower edges. Silvery-white on inner webs of outer primaries restricted to streak along shaft, remainder of inner webs dull grey with indistinct pale grey wedge to bases. Greater primary coverts light grey, shading to silvery-white at tips. Colours strongly influenced by bleaching and wear: mantle and scapulars become duller grey, tertials grade to brown at centres, indistinct grey half-collar at sides of neck develops. Pale grey and silvery tips and inner edges of primaries and tail-feathers become dull black on exposed parts: inner web of t6, outer web of p10, and tips of c. p5–p10 change to dark grey, deep sepia, or black; contrast between silvery flight-feathers and darker grey wing-coverts less pronounced, but contrast between white outer web and dark inner of t6 distinct. ADULT NON-BREEDING. Like adult breeding, but forehead white, crown white with fine black streaks or small black dots, nape streaked and mottled off-white, grey, and black; large patch round eye to ear-coverts uniform black. Head pattern similar to Gull-billed Tern *Gelochelidon nilotica*, lacking black nape of *S. hirundo*. Flight-feathers and tail variable, depending on moult: in early autumn, mostly worn and blackish, except for contrastingly pale new inner primaries and central tail-feathers; by midwinter, mostly fresh and pale, except for some inner secondaries or tail-feathers. Breeding plumage starts to appear on lores when outer primaries in moult. JUVENILE. Like adult non-breeding, but tips and sides of feathers of forehead, crown, mantle, scapulars, tertials, and shorter inner wing-coverts white or pale grey, freckled or tinged with buff or pale cinnamon; flight-feathers and tail pale grey, less silvery, tail-feathers fringed with white and mottled buff at tips; faint dusky grey carpal bar across lesser upper wing-coverts. By late autumn, plumage usually strongly abraded and bleached; buff tinges on crown, upperparts, and upper wing-coverts lost, these appearing heavily mottled dusky-grey and off-white; centres of secondaries and tertials sepia; primaries and tail-feathers sepia-black, except for an occasional new pale inner feather. FIRST IMMATURE NON-BREEDING. Like adult non-breeding, but forehead and crown more heavily mottled dusky; indistinct darker grey carpal bar across lesser upper wing-coverts. Usually only distinguishable from adult non-breeding by difference in primary moult schedule and by relatively more heavily worn plumage on part of body and upperwing. FIRST IMMATURE BREEDING. Like 1st immature non-breeding; some show black to forehead, crown, and nape, but this usually restricted to feather-tips or -centres and much white on bases of feathers visible. Tail shorter than in adult breeding. SECOND IMMATURE NON-BREEDING AND BREEDING. Like adult non-breeding and breeding, but as primaries in 2nd calendar year moulted earlier than those of adults, outers relatively more worn and darker in spring and summer of 3rd calendar year. Breeding as in adult, but more white towards bases of feathers of forehead, shining through when worn.

Bare parts. ADULT. Iris dark brown. Bill orange-yellow in breeding, tip black from angle of gonys; in non-breeding, dull black with some dull orange at base. Foot orange-red, duller in non-breeding. JUVENILE. Iris dark brown. Bill blackish-horn with dull yellow base. Foot yellow-brown. IMMATURE. Like adult, but foot and base of bill paler orange-yellow. (Saunders 1896; Ridgway 1919; BMNH, RMNH.)

Moults. ADULT POST-BREEDING. Complete, primaries descendant. Starts with inner primaries and scattered feathers on forehead and crown from about mid-July to mid-August, soon followed by remainder of head, body, and tail. Outer primaries, secondaries, and about t4–t5 last; moult completed November–December. ADULT PRE-BREEDING. Partial: February to early April. Head, underparts, variable amount of mantle and scapulars, tail (not always t4–t5), and often inner primaries. Primary moult arrested with inner 3–4 (2–5) primaries new. POST-JUVENILE AND SUBSEQUENT MOULTS. Information limited. Post-juvenile complete; by midwinter, all head and underparts and part or all mantle and scapulars new, as well as central tail-feathers and 3–4 inner primaries. All head and body and remaining juvenile of tail and wing replaced in spring and summer of 2nd calendar year, mostly moulting directly from 1st immature non-breeding to 2nd immature non-breeding; 2nd series of primary moult probably April–September (but only 2 in moult examined); subsequent pre-breeding (early 3rd calendar year) as in adult pre-breeding.

Measurements. Eastern and southern USA; skins (BMNH, RMNH, ZMA).

	♂			♀		
WING AD	266	(4·16; 15)	258–272	266	(4·14; 11)	260–272
TAIL AD	174	(9·59; 8)	160–185	170	(10·8; 8)	162–193
BILL AD	39·7	(1·28; 16)	38–42	36·7	(1·09; 11)	35–39
TARSUS	24·7	(0·79; 15)	24–26	24·1	(0·99; 12)	23–25
TOE	27·6	(1·04; 15)	26–29	27·0	(1·31; 12)	26–29

Sex differences significant for bill only. Adult tail in non-breeding 128 (115–145). Juvenile wing c. 5, juvenile tail c. 60 shorter than adult; bill not full-grown during 1st year of life.

Weights. No information.

Structure. Wing rather long and narrow, pointed. 11 primaries: p10 longest, p9 8–17 shorter, p8 28–38, p7 49–58, p6 71–80, p1 158–172; p11 minute, concealed by primary coverts. Length of adult t1 66 (61–74); depth of fork (tip of t1 to tip of t6) 95–115 in adult breeding, 40–65 in juvenile, intermediate in adult non-breeding and immatures. Bill rather similar to *S. hirundo*, but deeper at base and angle of gonys slightly more marked and closer to tip of bill. No crest on nape. Tarsus and toes rather short, slender; front toes connected by webs. Outer toe c. 89% of middle, inner c. 69%, hind c. 33%. CSR

Sterna repressa **White-cheeked Tern**

PLATES 5, 7, and 8
[between pages 134 and 135, and facing page 158]

Du. Arabische Stern Fr. Sterne à joues blanches Ge. Weisswangen-Seeschwalbe
Ru. Серобрюхая крачка Sp. Charrán cariblanco Sw. Vitkindad tärna

Sterna repressa Hartert, 1916. Synonym: *Sterna albigena*.

Monotypic

Field characters. 32–34 cm (bill 3·4–3·6, legs 1·6–1·7, tail in adult 10–14, in juvenile 7 cm); wing-span 75–83 cm. Smaller and 5–10% shorter winged than both Common Tern *S. hirundo* and Arctic Tern *S. paradisaea*, but with bill size and colour, and leg length as *S. hirundo*. Graceful sea tern, with breeding plumage not only darkest but also most uniform of all medium-sized congeners (even *S. paradisaea*). Underwing and undertail pale grey. Sexes similar; marked seasonal variation. Juvenile and immature separable.

ADULT BREEDING. Long cap jet-black, contrasting with white panel across lower face; latter also isolated by dark vinaceous ash-grey tone of underparts (much darker than in *S. paradisaea*), which extends on to neck. Whole of upperparts slate-grey, with no break in uniformity over rump to tail. Wings slate-grey above and pale grey below, lacking any obvious pattern except darker tips to primaries and paler trailing edge to other flight-feathers. Bill slender, coral-red with dusky tip. Legs red. ADULT NON-BREEDING. Forehead mottled white; underparts wholly white. Upperparts remain darker grey than *S. hirundo* and *S. paradisaea*, with grey tail even more obvious in contrast to underparts. JUVENILE. Said to resemble *S. hirundo*, showing similar dusky leading edge to inner wing but distinguished by basic dark grey tone to upperparts and tail. Bill almost wholly black; legs yellow-brown. For older age groups, see Plumages.

Distinctive in uniformly dusky breeding plumage, but less certainly so at other seasons (though fully grey rump and tail said to be diagnostic). Known to swim on sea.

Voice little described but similar to *S. hirundo*, notably in alarm- and advertising-calls. In flight, a short 'kip'; from flocks, a quiet, purring 'krrr'.

Habitat. Breeds in tropical warm waters of Indian Ocean, mainly coastal and inshore, avoiding inland waters. Nests on sandy coral-girt islands, sometimes on a bare and exposed sand-flat some 400 m in from sea, or on sand blown or washed into hollows of rock surfaces. On islands in Persian Gulf favours sparsely vegetated open ground, e.g. sand-dunes above high-water mark on beaches (L Cornwallis, D A Scott). On coast of Bahrain, on flat ground or sand hummocks up to 0·5 m high (Gallagher and Rogers 1978). As colony on Sheedvar island (Iran) declined, birds remained widely distributed over open ground but withdrew from beaches (L Cornwallis). Off Kenya, on equatorial small eroded coral islands, up to 18 m high, covered with mat of low vegetation including spiky creeping shrubs among sharp projections of coral, and subject to strong winds and heavy seas during breeding season; nests mostly on edge of a large colony of Roseate Terns *S. dougallii* (Britton and Brown 1971). Off island of Tiran (Red Sea) feeds mainly over coral reefs (S Su-Aretz).

Distribution. IRAQ. Breeds on islets off Al Faw (Ticehurst *et al.* 1926; Allouse 1953); no recent information. KUWAIT. Still breeding Kubbar island, despite much interference (PRH). EGYPT. Breeds on some islands in north of Red Sea (Borman 1929; Al-Hussaini 1939) and east coast of Sinai (YP).

Accidental. Israel, Jordan.

Population. EGYPT. About 200 pairs bred Hurghada 1981 (PLM, WCM). Sinai: *c*. 50 pairs breed east coast; site varies (YP). KUWAIT. Kubbar island: 1000–1300 nests 1958 (Sales 1959, 1965).

Movements. Migratory. Mainly a summer visitor to Red Sea, Persian Gulf, and Gulf of Oman; present there April to September or October (e.g. Trott 1947, Meinertzhagen 1954, Tuck 1974). Peak spring passage, Oman, late March to end of April; autumn peak late September to mid-October (Walker 1981*a*). Some flocks remain in southern Red Sea off Eritrea, feeding offshore in winter (Smith 1957), and small numbers overwinter off Masirah island (Oman) and Bahrain (Griffiths and Rogers 1975; Hallam 1980), but in all cases many fewer than are present in summer.

Winter quarters of emigrants generally given as East Africa and west Indian subcontinent (e.g. Meinertzhagen 1954, Vaurie 1965, Etchécopar and Hüe 1967), but actually still unresolved. Vagrant to South Africa (Sinclair 1976). Only scanty records from western India; certainly common off Pakistan between March–May, but dates indicate spring passage (towards Persian Gulf) rather than wintering (Ali and Ripley 1969*b*). In East Africa, very few records from Tanzania or further south (e.g. Harvey 1974, Britton 1977), and in Kenya flocks of 100 or more seen only March–June, maximum 1400 at Sabaki estuary in July 1975 (Britton and Britton 1976); again, this corresponds to spring passage period, and in any case such numbers scarcely account for *c*. 1000 pairs which breed (July–October) on Kiunga Islands, northern Kenya (Britton and

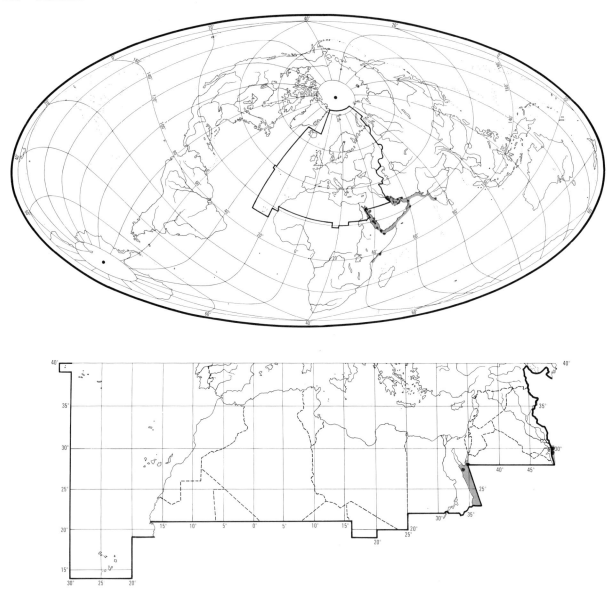

Brown 1974). Though Jackson (1938) reported it common at Mombasa in September, it is curious that neither Pakistani nor Kenyan recent observations indicate autumn passage corresponding to that noted in spring. Possibly stays well offshore in winter.

Food. Chiefly small fish and invertebrates, forced to surface by shoals of predatory fish (e.g. tuna *Thunnus thynnus*, bonito *Sarda sarda*) or disturbed by fishing boats. Forages in mixed flocks of Sternidae off Kenya, singly and in small groups which aggregate into mixed flock of usually less than 200–300 birds, constantly shifting and regrouping as fish shoals move about. Feeds mostly by dipping-to-surface; off Muscat and Masirah (Oman) always flopped on to surface from *c.* 3 m, but never submerged (Archer and Godman 1937; L H Brown, C J Feare). No specific data.

EKD

Social pattern and behaviour. Compiled largely from material supplied by M D Gallagher.

1. Mostly gregarious throughout the year, but also encountered singly or in small flocks, especially on late autumn passage (Smith 1951b, 1953; Meinertzhagen 1954; M D Gallagher). Flock of 2000–3000 recorded in September, Aden (Browne 1950); at least 70 000 at roost on Masirah island, Oman (Anon 1979). Usually migrates in small groups (less than 30, often less than 10), associating loosely at times with other species: e.g. on island of Tiran (Red Sea), breeding population of *c.* 150 pairs usually arrives in flock with Lesser Crested Terns *S. bengalensis* (S Su-Aretz). Off East African coast, forms feeding flocks

with other Sternidae, mostly Little Tern *S. albifrons*, Roseate Tern *S. dougallii*, and Bridled Tern *S. anaethetus* (L H Brown). Outside breeding season, flocks contain birds of various ages (Griffiths and Rogers 1975). BONDS. No information on nature of pair-bond. Small proportion of 1-year-olds returns to colony on Sheedvar island (Iran), most probably staying away until older (D A Scott); age of first breeding not known. Parents continue to accompany and feed offspring for unknown period after fledging (Griffiths and Rogers 1975). BREEDING DISPERSION. Colonial, typically in aggregations of 100–200 pairs (Britton and Brown 1974; T D Rogers and M D Gallagher), but sometimes much larger—estimated 300 000 pairs bred on 60 ha of the usable 78 ha of Sheedvar islands (D A Scott). Colonies smaller elsewhere, e.g. Oman: at least 50 pairs at Bandar Khaiyran, up to 150 pairs on Daimanyat islands (M D Gallagher). Off Masirah, 200 pairs on Jazirat Janzi, and 100 pairs on Jazirat Shaghaf (Griffiths and Rogers 1975). Sub-colonies may occur within large colonies, e.g. in one Bahrain colony of 2000 pairs. Existence of sub-colonies sometimes due to distribution of suitable ground, but apparently not always so: e.g. 300 pairs, Bahrain, divided into sub-colonies on more-or-less uniform sand hummocks and level sand. Scattered dispersion on Jazirat Shaghaf (1000 pairs in 4 groups) perhaps partly due to disturbance and egg-robbing (M D Gallagher). Nests can be close together (Archer and Godman 1937); in some colonies, Bahrain, neighbouring nests 46 cm apart, but also more scattered. On Jazirat Shaghaf, mean distance between 11 nests 4.6 ± 0.4 m ($3.2–7.7$) (C J Feare). On Sheedvar island, average nest density $c.$ 0.5 per m^2, maximum 2–3 per m^2 (D A Scott); on one island, Oman, average 6 per m^2 (M D Gallagher). In some places where associated with *S. bengalensis*, e.g. island of Tiran, disperses either around their dense colonies, or in neighbourhood (S Su-Aretz). Small numbers of *S. dougallii* and Swift Tern *S. bergii* nest respectively amongst and at edge of *S. repressa* sub-colonies on Daimanyat islands (M D Gallagher) and Masirah (C J Feare). Off East African coast, colonies not always well segregated from *S. dougallii*, though most nests of *S. repressa* bordered crowded colony of *S. dougallii* (Britton and Brown 1974). ROOSTING. Communally on sand-banks, rocky reefs, mud-flats in harbour (Browne 1950; M D Gallagher, S Su-Aretz), apparently also readily on sea surface (Smith 1951b). Birds often stand in water, allowing waves to wash over them (Sales 1965). May occur in large numbers typically with *S. dougallii*, *S. bergii*, *S. bengalensis*, Sandwich Tern *S. sandvicensis*, *S. anaethetus*, Sooty Gull *L. hemprichii*, and Herring Gull *L. argentatus* (C J Feare). All ages associate in roosts: flocks of up to 150 (120 adults, 30 juveniles) recorded resting at water's edge (Clapham 1964); throughout the year, Masirah, flocks often contain high proportion of non-breeding or 1st-summer birds (Griffiths and Rogers 1975).

2. FLOCK BEHAVIOUR. On arrival at breeding grounds, birds rest near colony-site on flats, rocks, etc., gathering in groups which serve as focal point for initiation of courtship behaviour (M D Gallagher). Birds loafing at tide-line perform silent, collective flights ('dreads' or 'panics') out over sea (C J Feare). In colony, first birds to detect danger give Alarm-call (see 1 in Voice) and, if on ground, take off; alerted neighbours follow suit (upflight or 'alarm') and wheel overhead, giving Alarm-calls. ANTAGONISTIC BEHAVIOUR. Outside breeding season, South Africa, single *S. repressa* in roost of 4000 Common Terns *S. hirundo* and Arctic Terns *S. paradisaea* was 'totally aggressive', vigorously chasing other birds (Sinclair 1976). HETEROSEXUAL BEHAVIOUR. No detailed information, but said to be similar to *S. hirundo* (S Su-Aretz); the following mostly from M D Gallagher. Pair-formation involves both aerial and ground behaviour at colony. Courtship activity conspicuous on arrival at colony-site (see Flock Behaviour, above). High-flights, involving 2–3 birds and version of Advertising-call (see 3 in Voice) are frequent; birds also perform Low-flight, characterized by carrying fish, flying with exaggerated ('bouncing': M D Gallagher) wing-beats, and giving Advertising-calls (see 3 in Voice). Contrast of bright red gape with dark red bill may enhance signal function of calling. On ground, ♂ circles ♀ in Erect-posture, neck and head stretched upwards. Courtship-feeding occurs, ♂ proffering small fish (C J Feare, S Su-Aretz); often a prelude to copulaion in which ♂ may mount for some time. Egg-wetting achieved by flying low over water to wet underparts (Gallagher and Rogers 1978). RELATIONS WITHIN FAMILY GROUP. At colony, parent seeks offspring with version of Kip-call (see 4 in Voice). Fledged young usually stand on shore singly, as well as in groups, awaiting return of parent with fish for which they beg by calling. Juveniles follow parents to feeding grounds, and are seen being fed well away from colonies (Griffiths and Rogers 1975). ANTI-PREDATOR RESPONSES OF YOUNG. No information. PARENTAL ANTI-PREDATOR STRATEGIES. Fairly aggressive in breeding season, but apparently much less pugnacious in defence of nest than closely related *S. hirundo* and *S. paradisaea* (P L Britton). Approach of predator, in the air or on ground, stimulates collective upflight (see Flock Behaviour, above). Birds in upflight co-operated in chasing Osprey *Pandion haliaetus* (C J Feare). Birds nesting on sand-banks, Gulf of Suez, June, flew up to chase passing Ravens *Corvus corax* (Marchant 1941). When man approaches colony, large numbers of birds join upflight (Archer and Godman 1937). Reluctant to swoop or strike (L H Brown), but does so if sufficiently provoked, often preceded by series of Anger-calls (see 2 in Voice) (M D Gallagher); known to return to nest 'within a few feet of observer' (Archer and Godman 1937). Dive-attacks also made on crabs (C J Feare).

EKD

Voice. Apparently similar to Common Tern *S. hirundo* (Mackworth-Praed and Grant 1952; Clapham 1964) and freely used at breeding grounds where both sexes have extensive repertoire. See Social Pattern and Behaviour for fuller description of contexts in which calls occur.

CALLS OF ADULTS. (1) Alarm-call. A loud and harsh 'kee-ÈRRR', 'kee-AAH' (E K Dunn, M D Gallagher), 'skeee-err' (M D Gallagher); when danger imminent, call more protracted, especially 2nd syllable, which is lower. Delivered mostly in flight, expressing alarm and fear. (2) Anger-call. A piercing 'kee-kee-kee', especially preceding dive-attack on human intruder (M D Gallagher). (3) Advertising-call. Usual version a shrill 'KIeer', 'KYah'; much briefer than call 1, with emphasis more on 1st syllable (E K Dunn). Described as 'usual call in flight'; given by advertising ♂ in Low-flight (C J Feare); commonly a more extended 'KIerri'. In High-flight, more attenuated 'keer keer keer' (E K Dunn, C J Feare). M D Gallagher renders call as a more sibilant 'skweeerr'. Nasal, complaining 'shreeark', 'sheerk', or 'seeerrk' given by adult in flight approaching young (M D Gallagher) probably the same. (4) Kip-call. A short, sharp 'kip' (E K Dunn) or 'skip' (M D Gallagher) seemingly denotes excitement; uttered often 2–3 times in rapid succession during High-flight, apparently by bird quickly answering 'keer' (see 4) of other, so overall sequence commonly 'keer

keer keer ... kip-kip ... keer' (E K Dunn, C J Feare). Related 'keep' or 'keek-keek-keek' given by parents seeking offspring. (5) Krrr-call. A quiet, purring note, heard from flock in flight or resting on shore (M D Gallagher); function not known.

CALLS OF YOUNG. Calls of downy young not described. Food-call of flying young a high-pitched 'seeeak' or 'srreea' (M D Gallagher). EKD

Breeding. SEASON. Red Sea: breeds June–August (Clapham 1964). Masirah island (Oman): eggs from 15 May to early July, chicks noted from 24 June (M D Gallagher). Bahrain: eggs from mid-May (Gallagher and Rogers 1978). SITE. On ground in the open; either on flat ground or on sand-hummocks to 0·5 m high. Nest: shallow scrape, sometimes very small (e.g. on hard ground); unlined or with a few pieces of shell and other objects, often brought during incubation (Gallagher and Rogers 1978); no nest mound made (*contra* Ticehurst 1926). Building: role of sexes unknown. EGGS. See Plate 89. Sub-elliptical, smooth and not glossy; pale buff, sometimes yellow-buff, with dark brown to blackish, also paler grey, spots, speckles, small blotches, and thin lines. 40 × 30 mm (37–45 × 27–34), sample 130; calculated weight 19 g (Schönwetter 1967). Clutch: 2(1–3). One brood. INCUBATION. Period not known. YOUNG. Precocial and semi-nidifugous. Fed by parents. FLEDGING TO MATURITY. Fledging period unknown. Not independent for at least some weeks after fledging (M D Gallagher).

Plumages. ADULT BREEDING. Forehead, upper two-thirds of lores, crown to just below eye, and nape glossy bluish-black. Rather narrow white patch from lower lores, gape, and chin to below ear-coverts and sides of nape. All upperparts except crown and nape but including back to upper tail-coverts, upper wing-coverts, and tail uniform dark ash-grey; outer web of t6 more silvery-grey (blackish when worn). Underparts dark ash-grey, only slightly paler than upperparts; paler ash-grey on under tail-coverts and on throat and cheeks where bordering white of chin and sides of head. Leading edge of wing white. Primaries uniform silvery-grey, slightly paler than upper wing-coverts when fresh, but darker when worn; inner webs duller grey with indistinct paler smoke-grey wedge on bases, wedges gradually paler towards outer primaries, nearly white on p10; innermost primaries slightly edged white at tips. Tips and edges of primaries soon blacken due to wear of those parts not protected by adjacent feathers. Secondaries and outer tertials ash-grey, narrowly fringed white on outer webs. Axillaries and under wing-coverts white with grey tinge. Slight vinaceous-red tinge to underparts at start of breeding season. ADULT NON-BREEDING. Forehead and lores white, grading to dull grey on crown; hind-crown mottled black, nape and ear-coverts black, All underparts, including cheeks, throat, and under tail-coverts white, occasionally slightly tinged grey. Remainder of body, wings, and tail as in adult breeding. DOWNY YOUNG. Upperparts buff or pale grey, patterned with blackish spots and smudges; throat to lores black; underparts white, lower belly buff (Harrison 1975). JUVENILE. Head like adult non-breeding, but mottled buff on crown, nape, and sides of head. Mantle, scapulars, and tertials ash-grey, feathers broadly fringed white, subterminally tinged buff-brown to sepia; rump to upper tail-coverts and most upper wing-coverts fringed white like scapulars, but less brown subterminally, often only slightly darker grey. Lesser upper wing-coverts along white leading edge of wing dark grey, fringed pale buff, forming dark carpal bar. Tail-feathers dull ash-grey, inner webs and tips broadly fringed white, slightly tinged sepia subterminally. Greater coverts and flight-feathers like adult, but tips narrowly fringed white when fresh and in part with faint brown subterminal dot. Underparts white. FIRST IMMATURE NON-BREEDING AND SUBSEQUENT IMMATURE PLUMAGES. In 2nd calendar year, head and body like adult non-breeding; dark grey to sepia bar across upper lesser wing-coverts; heavily worn and sepia-black juvenile outer primaries present until about July–August. At end of 2nd calendar year and during much of 3rd, faintly mottled dark wing-bar still present on lesser coverts. In spring of 3rd calendar year, similar to adult breeding, but forehead and underparts partly mottled white; outer tail-feathers often shorter than in adult breeding. Moults in 2nd and perhaps in 3rd calendar years not quite synchronous with adult; primaries often not suspended at same stage of moult as adult.

Bare parts. ADULT. Iris dark brown. Bill orange-red to deep red, much of tip black; bill darkens to blackish from about July to early spring, and red, if present, restricted to gape and to base of lower mandible. Foot orange or red; black with red tinge in autumn and winter. DOWNY YOUNG. Iris dark brown. Bill red with dark tip. Foot flesh-coloured. (Harrison 1975.) JUVENILE. Foot and base of bill pink-orange at fledging, soon darkening to adult non-breeding colour. (BMNH.)

Moults. ADULT POST-BREEDING. Complete, primaries descendant or serially descendant. Starts with inner primaries between mid-July and late August; completed with outermost late December to early February. Earlier series (started about mid-winter during pre-breeding moult and suspended during nesting in April–July with primary moult score of *c*. 15–20) either continued with next primaries (outermost completed about September–October), or not resumed, outer primaries being replaced once a year only. Body and tail August–February (forehead and parts of mantle and scapulars first). ADULT PRE-BREEDING. Partial: head, body, tail (sometimes excluding some feathers), and inner primaries. Starts late December or January with inner primaries and central tail-feathers; finishes late February to early April; primary moult suspended with innermost 3(2–5) new. POST-JUVENILE AND SUBSEQUENT MOULTS. Post-juvenile complete. Starts with head, body, tertials, and central tail-feathers from about November, followed by rest of tail, wing-coverts, and innermost primaries from January or February. Moult of primaries slow, sometimes suspended; outermost completed August–November. New series of primaries and tail-feathers starts about June, sometimes suspended later on; 1st non-breeding plumage of head and body replaced by adult non-breeding in spring and summer of 2nd calendar year. Moults from 2nd winter onwards probably as in adult.

Measurements. North and central Red Sea, April–July; skins (BMNH, RMNH).

WING AD	♂ 240	(5·72; 11)	232–248	♀ 243	(5·89; 5)	236–249
TAIL AD	142	(9·25; 9)	131–154	136	(3·30; 4)	132–140
BILL AD	35·7	(0·98; 11)	34–37	35·1	(1·34; 5)	34–37
TARSUS	18·6	(0·68; 11)	18–20	18·3	(0·54; 4)	18–19
TOE	21·1	(1·27; 11)	20–23	20·8	(0·76; 4)	20–22

Sex differences not significant. Wing of juvenile *c*. 10–25 shorter than adult, tail *c*. 30 shorter. Most tails from sample above rather worn; fresh tails mostly 140–154. Adult non-breeding tail 20 (5–30) shorter than in fresh adult breeding.

Persian Gulf, May–June; skins (BMNH).

WING AD	♂ 247	(5·57; 5)	241–252	♀ 250	(4·71; 5)	246–258
TAIL AD	151	(— ; 3)	145–157	152	(11·2; 4)	144–167
BILL AD	37·4	(1·10; 4)	36–39	35·7	(1·70; 5)	34–37
TARSUS	19·7	(0·57; 4)	19–21	18·8	(0·45; 5)	18–20

Sex differences significant for tarsus. Significantly larger than population of Red Sea, except ♀ bill and tarsus.

Weights. Laccadive Islands (India), February: ♂♂ 142 (adult), 113 (3rd calendar year) (BMNH).

Structure. Wing long and narrow, pointed. 11 primaries: p10 longest, p9 8–14 shorter, p8 25–35, p7 44–56, p6 64–74, p1 144–156; p11 minute, concealed by primary coverts. Tail deeply forked; length of central feathers 68 (65–70); in adult breeding, t2 1–3 longer than t1, t3 5–8, t4 15–20, t5 27–42, t6 60–90; in juvenile, depth of fork c. 30 mm only; immatures and adult non-breeding intermediate between juvenile and adult breeding. Feathers of nape slightly elongated in adult breeding. Shape of bill and structure of foot as in Common Tern *S. hirundo*. Outer toe c. 85% of middle, inner c. 65%, hind c. 29%.

Geographical variation. Slight. All measurements of Persian Gulf population slightly larger than those of Red Sea birds (see Measurements): only 1 out of 10 from Persian Gulf with wing below 245, only 3 of 16 from Red Sea over this.   CSR

Sterna anaethetus Bridled Tern

PLATES 9 and 10
[between pages 158 and 159]

DU. Brilstern   FR. Sterne bridée   GE. Zügelseeschwalbe
RU. Темноспинная крачка   SP. Charrán embridado   SW. Brunvingad sottärna

Sterna Anaethetus Scopoli, 1786

Polytypic. *S. a. melanoptera* Swainson, 1837, West Indies and West Africa; *antarctica* Lesson, 1831, Red Sea, Persian Gulf, and western Indian Ocean south to Maldive and Mascarene Islands. Extralimital: nominate *anaethetus* Scopoli, 1786, eastern Indian Ocean from Greater Sunda Islands and Australia east to Solomon and Palau Islands and north to Taiwan; *nelsoni* Ridgway, 1919, Pacific coast of Mexico and Central America.

Field characters. 30–32 cm (bill 3·9–4·4, legs 2 cm); wing-span 77–81 cm. Averages smaller and shorter winged than Common Tern *S. hirundo*; 10% smaller than Sooty Tern *S. fuscata*, with noticeably shorter wings. Medium-sized tern of similar structure to *S. hirundo*, except for longer, stronger bill. Plumage pattern more contrasting those of common west Palearctic *Sterna*, with black cap to head broken by white blaze from base of bill to behind eye, pale collar, grey-brown upperparts and tail-centre, black-brown primaries, and grey breast and belly. Point of underwing appears pale. Sexes similar; little seasonal variation. Juvenile and 1st-winter immature separable.

ADULT BREEDING. Broad line from bill through eye, crown, and nape jet-black, interrupted by narrow white forecrown and supercilium over eye (to point midway along upper ear-coverts). Back of lower neck white-grey, forming diffuse but obvious collar below nape. Back, wing-coverts, rump, and all but outermost tail-feathers grey-brown, contrasting with black-brown flight-feathers and white outermost tail-feathers. Pattern of upperparts somewhat suggests dirty immature *S. hirundo* or large Black Tern *Chlidonias niger* in winter plumage; much less dark and uniform than *S. fuscata*. Under wing-coverts white, contrasting with grey undersurface of secondaries and inner primaries; outermost 3 primaries show dull white inner webs, forming pale tip to underwing, unique in *Sterna*. Rest of underparts white, but strongly clouded grey on breast and belly. Bill and legs black. ADULT NON-BREEDING. Forecrown streaked white and rest of cap duller and browner; feathers of mantle with small white tips creating paler appearance than in summer. JUVENILE. Head off-white, with pale forehead and face, marked by dusky lores, obvious black panel behind eye and across nape, and black flecks on rear crown. Pale collar indistinct, since whole of back brown with feathers broadly tipped cream (and appearing paler and more mottled than in winter adult). Wing-coverts dark brown, narrowly tipped or tinged cream, appearing noticeably darker than back but paler than flight-feathers. Tail brown, with each feather tipped off-white (but lacking full white margins of adult). FIRST WINTER. Moult of head and body and wear create greater resemblance to adult, but juvenile wing and tail pattern retained; crown well streaked with white.

Darkness of upperparts distinctive, while grey-brown tone of back and pale collar allow rapid differentiation from *S. fuscata* at close and middle ranges. *S. fuscata* also shows greater contrast below, being pure white with blackish flight-feathers and tail-centre; underwing of *S. anaethetus* less contrasting, and pale grey tail-centre often not visible. Smaller, slighter form and less thrusting flight, compared with *S. fuscata*, evident at long range. When unworn, pale 'saddle' of juvenile obvious. Flight among most buoyant and graceful of genus, with noticeably elastic and exaggerated wing-beats accompanied by marked lift of body; capable of soaring slowly and accelerating suddenly. Gait and carriage as *S. hirundo*.

Less subject to extralimital vagrancy in eastern Atlantic than *S. fuscata*. Commonest call a short subdued 'kwit'; 'kee-yarr' also recorded outside breeding season.

Habitat. Tropical, subtropical, and maritime, but differs from Sooty Tern *S. fuscata* in being primarily offshore rather than pelagic. Breeds on islands and, in some areas, on mainland, nesting under bushes on sand and coral islets, but using a crevice under a ledge or the floor of a cave on limestone islands or stacks (Warham 1958). In Caribbean, nests in hollows on cliff face or under overhanging rock or broken slabs of coral (Bent 1921). In Venezuela, nests on ground near shore, usually hiding egg in crevices formed by coral debris, but also in low thick vegetation (Schauensee and Phelps 1978). Common factor appears to be preference for site offering cover from above. Although furnished with water-repellent feathers, resting on water—as once described (Murphy 1936)—not confirmed by later accounts (Smith 1951a; Warham 1958); commonly perches, however, on rocks, driftwood, harbour buoys, ships' masts, or floating wreckage (Ali and Ripley 1969b), and roosts on low salt bushes and craggy pinnacles or posts. Around nesting islands, tends to feed in deeper water away from surrounding reefs and adjacent coasts (Warham 1958). Except in display, normally keeps to lower airspace.

Distribution. IRAQ. Breeds on Dara island, near Al Faw (Ticehurst *et al.* 1926); no recent information. KUWAIT. Kubbar island: reported still nesting despite much interference (P R Haynes). EGYPT. Breeds on The Brothers and Gumrah, in northern Red Sea (Ticehurst 1924b; Al-Hussaini 1939); no recent evidence (PLM, WCM).

Accidental. Britain and Ireland: 5 found dead 1931–58, 3 alive 1979–82 (Sharrock and Sharrock 1976; Rogers *et al.* 1983); also, wing found Britain, April 1977 (Rogers *et al.* 1979).

Population. KUWAIT. Kubbar island: 2000–2500 birds in 1958 (Sales 1965). MAURITANIA. Banc d'Arguin: several hundred pairs (Naurois 1969a); 730–930 pairs (Duhautois 1974); 1480 pairs 1974, almost double 1960 population (Trotignon 1976). MOROCCO. 400 pairs Virginie island 1960 (Heim de Balsac and Mayaud 1962).

Oldest ringed bird 17 years 11 months (Rydzewski 1979a).

Movements. Little known. If not migratory out of breeding season, at least highly dispersive, leading offshore existence.

S. a. melanoptera. Population from Banc d'Arguin (Mauritania) and Morocco believed to disperse into waters offshore from Sénégal (Morel and Roux 1966). During non-breeding season (March–October) Caribbean population ranges north to Florida and south to Trinidad and Tobago (Watson 1966).

S. a. antarctica. From August, birds from Red Sea and Gulf of Aden move down east coast of Africa as far as Moçambique (Archer and Godman 1937). Regularly joins mixed flocks of Sternidae fishing off Kenya, August. Kenyan breeding population absent from coast, September–July (Britton and Brown 1974). Absent in winter on coast of Eritrea; in spring, migration seen offshore from 2nd week of April, birds moving inshore in 3rd week, and returning to colonies in May (Meinertzhagen 1954). Those breeding Persian Gulf and Makran coast (Pakistan) believed to disperse along western coast of India as far as Bombay (Archer and Godman 1937; Meinertzhagen 1954).

Little information for other races. One bird ringed Cousin Island (Seychelles) in November found 1780 km west near Pemba (Tanzania) in May (Diamond 1976). Birds vacate breeding grounds, southern Australia (e.g. Cape Leeuwin and Houtman Abrolhos) outside breeding season, presumably moving north; may be sedentary further north (Serventy *et al.* 1971). One recovery in Indonesia of bird ringed south-west Australia (Anon 1976). EKD

Food. Primarily small fish, planktonic invertebrates including crustaceans and molluscs, and the water-bug *Halobates*. Feeds chiefly from surface, mainly by hovering and dipping-to-surface. Takes prey chased to surface by schools of predatory fish (e.g. bonito *Sarda*, tuna *Thunnus*). In India, plunge-dives with complete immersion (Ali and Ripley 1969b). May be active at night; often feeds in mixed flocks of Sternidae (Bent 1921; Witherby *et al.* 1941; Mackworth-Praed and Grant 1952; Bannerman 1962; Domm and Recher 1973; Diamond 1976; LeCroy 1976; R G Brown).

Specific data not available from west Palearctic. On coasts of Ceylon, winter, fed mainly on *Halobates* (Henry 1955). Stomachs from Puerto Rico contained up to 70% fish, mostly filefish *Alutera*; 25% molluscs, both gastropods and cephalopods, especially *Spirula australis*; also one moth (Lepidoptera) and small echinoderms probably from stomachs of fish (Wetmore 1916). Regurgitated food, mostly from adults, Cousin Island (Seychelles), consisted of (by weight) 1·9% squid (Cephalopoda), 94·6% fish (4·4% Hemiramphidae, 89·8% Mullidae, 0·4% Tetraodontidae), 3·6% *Halobates*, and occasionally larval crabs (A W Diamond). On One Tree Island (Great Barrier Reef, Australia), regurgitated food from chicks consisted mainly of anchovy *Engraulis australis* 2–8 cm long and a few *Arotrolepsis filicauda* of 2–4 cm (Domm and Recher 1973); once seen to drop a balistrid fish 4–6 cm long (Hulsman

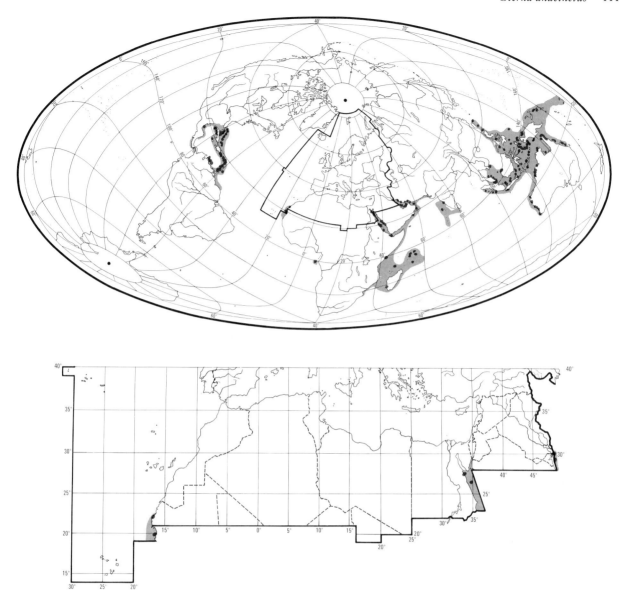

1976). Regurgitated fish fed to chicks usually already broken into small pieces, though adults frequently re-swallow large pieces and feed only smaller ones to chicks. Mean interval between 8 feeds to chicks, One Tree Island, 33 ± 211 min (K Hulsman). EKD

Social pattern and behaviour. Based largely on Australian material of J Warham from Lancelin Island (Western Australia) and K Hulsman from One Tree Island (Great Barrier Reef, Queensland).

1. Said to be gregarious at all seasons (MacDonald 1973) or solitary outside breeding season (K Hulsman). Feeds singly (Smith 1951a), or in flocks, sometimes mixed with Roseate Tern *S. dougallii*, Fairy Tern *Gygis alba*, and Black Noddy *Anous minutus* (Britton and Brown 1974; Diamond 1976; K Hulsman).

BONDS. Monogamous pair-bond, persisting for 2 or more seasons (Warham 1958; K Hulsman). Many birds arrive at nesting grounds already paired (Warham 1958; Domm and Recher 1973; Nicholls 1977). Both parents incubate and care for young, feeding them for several weeks after fledging, though not for as long as in some plunge-diving *Sterna* (Dragesco 1961b; Hulsman 1974, 1977a; Diamond 1976). Crèching does not occur (K Hulsman). BREEDING DISPERSION. Forms mostly small loose colonies (Bent 1921; Warham 1958; Naurois 1959; Diamond 1976), though compact in colony of exposed nests, East Africa (Britton and Brown 1974). Colonies off East African coast usually of less than 100 pairs (Britton and Brown 1974; Nicholls 1977), on Banc d'Arguin (Mauritania) 75–200 pairs (Naurois 1959), on Halul island (Persian Gulf) c. 2500 pairs (Branegan 1959), on Great Barrier Reef 12–326 pairs (Domm 1977; Jahnke 1977; Hulsman 1979), off coast of Western Australia 500–2000 pairs (Abbott

1978; Johnstone 1978a, b); 'hundreds of thousands' assemble to breed on islands off Zeyla, Gulf of Aden (Archer and Godman 1937; Mackworth-Praed and Grant 1952). Nests generally 1–5 m apart (Naurois 1959; K Hulsman); minimum 30 cm (Naurois 1959), maximum 20–30 m (Serventy et al. 1971). Average nest density on Banc d'Arguin c. 1 nest per 6 m² (Naurois 1959). No sub-colonies apparent within colonies (K Hulsman) and nests often evenly dispersed over available ground (Domm and Recher 1973; Domm 1977). Small nest-area territory serves for ground courtship, nesting, feeding young, and resting of off-duty bird. In East Africa, sometimes forms small, compact colonies among breeding *S. dougallii* (Britton and Brown 1974). Some birds known to return to same colony in successive seasons (Diamond 1976; K Hulsman). ROOSTING. Outside breeding season, roosts on islands, floating debris, buoys, and ships (Ticehurst 1924a; Smith 1951a; Warham 1958; Ali and Ripley 1969b; MacDonald 1973). Said to roost occasionally on sea (Witherby et al. 1941; Watson 1966), but this not corroborated (Smith 1951a). On Bird Island (Seychelles), over 5000 non-breeders roosted at night in casuarina trees *Casuarina equisetifolia* (Feare 1979a). Pre-breeding assemblages roost on sand-banks or similar 2–3 km from breeding grounds, moving to shoreline and other exposed places near nesting area when occupation imminent. On shore, flock stands at least 3 m from water's edge (Domm 1977; K Hulsman). Daily pattern of pre-breeding roosting times, One Tree Island, showed strong diurnal influence, independent of tidal cycle. Birds arrive at shore roost at c. 08.00 hrs and leave at c. 16.30–19.00 hrs; move *en masse* into nesting grounds for roosting overnight, but are absent by day in first few days of occupation (Warham 1958; Hulsman 1977a). Day-time visits begin c. 2 weeks after first arrival (Domm and Recher 1973; Hulsman 1977a), and are initially brief, birds staying only c. 1 hr at sunrise before departing. In succeeding days, progressively more of morning spent in colony before departure, and birds return earlier in afternoon until eventually pairs spend most of day at nest-sites, roosting on them at night (Warham 1958). Full occupation of colony c. 4 weeks after first arrival; during week before full occupation, periodic collective upflights occur from shore roost (Hulsman 1977a) (see Heterosexual Behaviour, below). After clutch laid, off-duty bird loafs singly in nest-territory or on habitual vantage point nearby—usually rock, branch, or post (Warham 1958; Dragesco 1961b). Non-territorial birds loaf in trees or bushes during the day (Archer and Godman 1937), at One Tree Island often with *A. minutus*. After chicks hatch, marked resurgence of collective loafing by adults on tree-tops, shoreline, etc. (K Hulsman).

2. FLOCK BEHAVIOUR. About 4 weeks after first arrival at colony, One Tree Island, large flock returning from feeding congregates over lagoon and reef flats at dusk, flying and soaring at 15–20 m before entering colony (K Hulsman). Early in breeding season, 'dreads' from colony are frequent; stimulated by Dread-call (see 8 in Voice) from bird darting swift and low out to sea, wings widely swept and body canting from side to side, making violent evasive movements. Neighbouring birds repeat call, and fly seaward in like fashion. Exodus silent after initial call, but after birds swoop low over water they rise and return, uttering Alarm-call (see 1 in Voice). Dread usually localized but may spread to affect whole colony (Warham 1958; Serventy et al. 1971; Hulsman 1977a). ANTAGONISTIC BEHAVIOUR. Generally shy and wary in breeding season when threatened by intruders, but aggressive towards conspecifics and nesting associates in defence of territory, nest, and young. (1) Ground encounters. Bird signals ownership of territory to conspecific by Head-nodding (Fig A): standing bird tilts whole body forwards at c. 45°

A

or more and nods rapidly; carpal joints held away from body and bird may give Advertising-call (see 3 in Voice) with bill wide open and tongue vibrating. In territorial dispute, opponents face one another and nod with raised hindcrown feathers; Advertising-call given especially after intruder driven off. Birds often interrupt ground-courtship and copulation between another pair, displaying ♂ usually repelling intruder as described above (Warham 1958; Serventy et al. 1971). Wandering chicks only sometimes attacked but, in Mauritania, chicks of other Sternidae and Laridae may be attacked violently, even lifted bodily (Dragesco 1961b). Fighting adults grab and tug each other by the bill, and sometimes beat each other with wings, uttering Anger-calls (see 2 in Voice); bird may be grabbed by nape, wing, or tail to prevent escape. (2) Aerial encounters. Especially in fights between territorial neighbours, participants may rise vertically from ground (Upward-flutter), half-hovering with rapid wing-beats and intermittently flying bill-first, accompanied by Anger-calls; at top of ascent, birds separate and usually retire to respective territories. Birds perched on vegetation overhanging another's nest are frequently buzzed by the nest owners (K Hulsman). HETEROSEXUAL BEHAVIOUR. (1) General. Pair-formation involves social and individual phases of aerial courtship, as well as ground behaviour at colony. (2) Aerial courtship. Collective upflights occur in week before regular day-time occupation of nesting grounds. Birds fly up together calling, resettling within c. 3 min (Hulsman 1977a; K Hulsman). Individual phase characterized by High-flight and Low-flight; former not common at colony, though may go unnoticed at sea where it also occurs (Domm and Recher 1973; K Hulsman). Usually 2 birds (rarely 3) make rapid, spiral ascent up to c. 300 m, level off, and then begin downward glide; sway from side to side—mostly in unison, though pursuer may show delayed reactions to movements of leader; birds overtake each other (Pass-ceremony) one or more times during glide (Serventy et al. 1971; K Hulsman). During High-flight, food rarely carried; some calling (not described) occurs (K Hulsman). Low-flights occur prior to egg-laying, mostly initiated by ♂♂ but a few by ♀♀ (K Hulsman). Bird picks up a stick or leaf, or sometimes regurgitates (and retains) a piece of fish, and flies about colony giving Advertising-call with slow,

B

emphasized wing-beats—'Butterfly Flight' of Warham (1958); aerial version of Bent-posture adopted, with elongated neck held slightly below horizontal, bill downward. Advertising bird pursued by up to 8 others, often at high speed. Low-flight sometimes started from a Fly-up (K Hulsman): ♂ on nest-territory arches neck with head low and flies up almost vertically; after ascending, sometimes comes straight down in same posture. Resurgence of display flights after hatching, especially High-flight, performed by unpaired birds (often arriving late in season) and failed breeders. ♂ with young may perform Fly-up and Low-flight (Hulsman 1977a; K Hulsman). (3) Ground courtship. Typically occurs on nest-territory. 2 birds assume ground Bent-posture: body tilted forwards, bill c. 30° below horizontal, carpal joints lowered and held slightly outwards, wing-tips crossing above or below uptilted tail; plumage sleeked, except for hindcrown feathers which may be raised. Holding this posture, 2 movements occur. (a) Parade. 2 birds strut with short steps side by side, heads slightly averted so that nape presented to partner (Fig B). Occasionally, one lightly touches back of other's head with bill. After moving short distance, birds turn to face each other and Bill-fencing occurs: wave lowered bills from side to side, without touching, or raise heads and wave bills at 10–15° above horizontal (Warham 1958; Serventy et al. 1971; K Hulsman). (b) Pirouette. Often follows Parade; takes 2 forms. In first version, both birds circle each other alternately as they move forwards; when one is circling, other is stationary. In second version, one bird, usually ♂, closely circles other who leans forwards, shuffling to keep roughly facing him, though both keep downtilted heads slightly averted. ♂ often reverses direction repeatedly, describing arc in front of ♀ (Warham 1958; K Hulsman); occasionally ♀ circles ♂ (K Hulsman and N P E Langham). During Parade or Pirouette, both birds sometimes pick up and mandibulate stones or bits of vegetation, and give short Kek- or Growl-calls (see 4–5 in Voice). (4) Mating. May follow Courtship-feeding or Pirouette. Before copulating, birds usually face, and make pecking movements at ground; ♀ crouches slightly and allows ♂ to mount (Warham 1958). (5) Courtship-feeding. ♂ feeds ♀ by regurgitation (Hulsman 1977a); ♀ faces or stands alongside ♂ and adopts Hunched-posture, as in other Sternidae; gives Begging-call (see 6 in Voice) and pecks repeatedly at base of ♂'s bill. ♂ arches neck to regurgitate, and ♀ takes food from bill. Courtship-feeding much less frequent than in many other Sterna (K Hulsman). (6) Nest-relief. Takes several forms. (a) Incoming bird approaches on ground, with fore-body lowered and head retracted, similar to Hunched-posture; when nest reached, bird nods and sitting bird leans forwards and nods as it stands up and moves off nest. (b) Incoming bird gives Advertising-call before landing, often eliciting same call from sitting bird which then rises, and pair perform mutual Pirouette. (c) Incoming bird gives Advertising-call and sitting bird leaves, without interaction, in different direction to that used by mate. (d) Bird continues sitting despite arrival of mate which then either regurgitates piece of fish, or picks up piece of vegetation, and flies around giving Advertising-call; sitting bird responds by moving off nest quickly and adopts ground Bent-posture; incoming bird immediately goes to nest and re-swallows fish or drops vegetation before sitting. In 97 nest-reliefs with eggs and young, One Tree Island, relieved bird left immediately (35% of cases), preened (18%), drank in flight (27%), picked up pieces of vegetation and discarded them (7%), defecated in flight (6%), or performed some other activity (7%) (K Hulsman and N P E Langham). Incubating bird occasionally flies from nest to wet belly in sea and so cool eggs (Sales 1965). RELATIONS WITHIN FAMILY GROUP. Young leave nest at c. 3 days but remain in vicinity until fledging. When begging, chick crouches slightly, retracts neck, and calls persistently, lifting bill to c. 30° above horizontal at each call; often pecks at parent's bill. Returning adult gives Advertising-call, to which older young respond by running through vegetation. If chick delayed in reaching spot, parent takes off and circles area, sometimes with slow emphatic wing-beats, uttering Advertising-call, and lands again—if necessary alighting on vegetation and dropping through to locate chick. Once fledged, young wait on shoreline or reef crest to be fed. Begin to practise dipping-to-surface (hunting method most often used by adults) c. 1 week after fledging. Fed on shore for c. 35 days, then disperse; not known if parents accompany or feed young after dispersal from colony (K Hulsman). ANTI-PREDATOR RESPONSES OF YOUNG. On hearing Alarm-call of parent, young less than 3 days old lie flat in nest with bill prone to ground. Older young leave nest and hide in rocks or vegetation; are then difficult to find (Domm 1977). If closely approached by human they run and seek new refuge. PARENTAL ANTI-PREDATOR STRATEGIES. (1) Passive measures. Parents give Alarm-calls on approach of danger (see above for responses of young). Described as timid, leaving nest furtively well before arrival of human intruder, and flying off (Warham 1958; Naurois 1959). Escape facilitated by tunnels under vegetation to clearings (Cooper 1948). (2) Active measures: against birds. Silver Gulls *Larus novaehollandiae* loitering over colony, One Tree Island, are chased, while Buff-banded Rails *Rallus phillipensis* which take eggs are buzzed vigorously while birds give Anger-calls. (3) Active measures: against man. When human remains near nest, adult sometimes returns (from escape flight) and tries to lure intruder away with broken-wing display (Cooper 1948; K Hulsman), but this less common than dive-attacks with Anger-calls. At Khubbar island (Persian Gulf), birds feigned injury by fluttering among shrubs below which nests sited; not seen in birds nesting in the open (Sales 1965). Humans not struck but may be defecated on (Hulsman 1979; K Hulsman).

(Figs A–B from photographs in Nicholls 1977.) EKD

Voice. Highly vocal at breeding grounds, with varied repertoire.

CALLS OF ADULTS. (1) Alarm-call. A staccato 'wep wep' or 'wrep wrep', likened to yapping of dog, or barking of Black-winged Stilt *Himantopus himantopus* (Serventy and Whittell 1948: Fig I, first 3 units). Most commonly heard call on disturbed breeding grounds, increasing in pitch with imminence of danger (Serventy and Whittell 1948; Warham 1958; Nicholls 1977; K Hulsman). (2) Anger-call. A guttural, tremulous 'GRRrrrrr', emphatic at the beginning, then trailing off (Fig II); used to threaten conspecifics (E K Dunn); heard intensely during swoops on ground- and aerial predators (K Hulsman), when alternates with Alarm-call (LeCroy 1976); (3) Advertising-call. A loud, tremulous 'hrrr' or 'greer', likened to whinny of horse (K Hulsman) or squeaky toy (LeCroy 1976); loud jarring 'cherr' (Mackworth-Praed and Grant 1952: Fig I, last unit); usually delivered with bill wide open and tongue vibrating. Given by advertising ♂, and by bird arriving to relieve mate or feed young; in latter, sitting mate may reciprocate (Warham 1958; K Hulsman). (4) Growl-call. A mild rendering of Advertising-call—a soft, growling 'hrrr' or 'greer'. Invites close approach, e.g. when brooding young, and during ground courtship and nest-relief

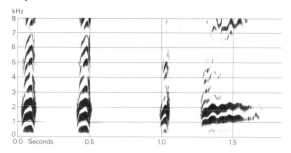

I D Turner Seychelles July 1976

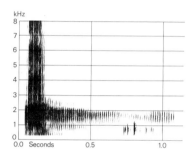

II D Turner Seychelles July 1976

(Warham 1958). (5) Kek-call. In ground courtship, 'kek' or 'kuk', apparently as contact-call (Warham 1958; K Hulsman). (6) Begging-call of ♀. No details, but perhaps a chuckling (LeCroy 1976). Given during Courtship-feeding. (7) Fishing-calls. Variety of harsh, grating cries, rendered 'krek', 'karr', 'k-rrr', 'kr-arr', 'ka-karr', 'k-ow', 'kew', and less commonly a low subdued 'kwit' (Smith 1951a; Ali and Ripley 1969b; Slater 1970); 'kee-yarr' noted outside breeding season (Nicholls 1977). (8) Dread-call. A 'mer-er-er'; given by initiator of 'dread' and taken up by others as they take flight (Warham 1958).

CALLS OF YOUNG. Small young beg with repeated high-pitched 'peep' (K Hulsman); fledged juveniles with tremulous 'prreee prreee' (E K Dunn).   EKD

Breeding. SEASON. Red Sea: eggs laid mid-June and July (Meinertzhagen 1954). Banc d'Arguin (Mauritania): nests mid-May to July (Naurois 1959). In Seychelles, breeds every $7\frac{1}{2}$–8 months (Diamond 1976). SITE. Variable: on ground in the open, or in shade of small bush or plant; in shelter of rock or boulder, even in cave or dark crevice; also well hidden in tussocks, or on top of same; also under old nest of Long-tailed Cormorant *Phalacrocorax africanus* (Warham 1958; Naurois 1959). Colonial or, more commonly, loosely colonial. Nest: shallow scrape, or eggs laid on bare rock; little or no lining. Building: no information. EGGS. See Plate 89. Sub-elliptical, smooth and slightly glossy; pale buff to cream, with fine dark brown or red-brown speckles, spots, and blotches. 44×31 mm (40–46 × 29–33), sample 108 (Schönwetter 1967). Mean weight, Seychelles, 20 g (sample 6) (Diamond 1976). Clutch: 1. One brood. INCUBATION. 28–30 days (Hulsman 1974, 1977a; K Hulsman). By both sexes (Warham 1958). In very variable spells: in Australia, 160 ± 119 min, $n = 48$ (K Hulsman); in Venezuela, apparently 24 hrs or longer (LeCroy 1976). YOUNG. Precocial and semi-nidifugous. Cared for and fed by both parents. Leaves nest at 3 or more days and hides in nearby vegetation (Diamond 1976; Hulsman 1974, 1977a). FLEDGING TO MATURITY. Fledging period 55–63 days. May become independent on leaving colony area c. 35 days after fledging (Hulsman 1974, 1977a; Diamond 1976; K Hulsman). No further information.

Plumages (*S. a. melanoptera*). ADULT BREEDING. Forehead and streak over eye to above central ear-coverts white, contrasting sharply with black crown and nape and black streak from lores through eye to nape. Hindneck and upper mantle pale grey to grey-white; lower mantle and upper sides of chest medium grey, grading to sooty-olive-brown on scapulars, tertials, and median and greater upper wing-coverts. Rump to upper tail-coverts medium grey. Chin, throat, under tail-coverts, and sides of head below black of lores and ear-coverts white; chest, belly, and vent white with faint blue-grey tinge. Tail-feathers medium grey (white towards base and white on inner edge of t2–t3); t4 usually with white edge to outer web, t5 with much of outer web white (except for some grey bordering shaft near tip), t6 completely white except for grey streak 2–5 cm long on tip of inner web. Sometimes retains some non-breeding tail-feathers, these showing more extensive grey on tips. Flight-feathers black (showing slight grey bloom when fresh); secondaries faintly fringed white at tip and with broad white margin to inner web; outer primaries have contrastingly white wedge on base, duller and less contrasting on inner. Lesser coverts and primary coverts black, contrasting slightly with sooty-olive-brown of median and greater coverts and strongly with white coverts along leading edge of wing. Under wing-coverts and axillaries white. In fresh plumage, sooty-olive-brown of upperparts and upper wing-coverts slightly tinged grey; in worn plumage, light grey on feather-tips of hindneck abraded and bleached to all-white, forming conspicuous collar; underparts all-white; white to tips of outer webs of outer tail-feathers abraded. ADULT NON-BREEDING. Forehead and streak to above eye white. Streak from lores to eye black, finely mottled white; crown black, streaked white, streaks narrower and less distinct towards centre of nape; small spot in front of eye, ear-coverts, and sides of nape almost uniformly black. Hindneck white with some black streaks or dots. Mantle pale grey with much white to feather-bases and with white fringes to tips; scapulars and tertials dark olive-grey with off-white fringes to shorter feathers and narrower edges to tips of longer ones. Feathers of back and rump medium grey with white fringes to tips, sometimes showing white of bases, especially at sides. Underparts completely white. Tail like adult breeding, but outer feathers with more grey on tip and both webs of t5–t6 usually grey for 5–6 cm of tip. New median and greater upper wing-coverts like scapulars; remainder of wing as in adult breeding. When worn, mantle and back to upper tail-coverts may look very pale. DOWNY YOUNG. Long and fine down of head, upperparts, throat, chest, wing, and thighs closely speckled dark brown, grey, and buff, sometimes almost uniform pale buffish-grey; underparts uniform dull white or buff (Witherby *et al.* 1941; Harrison

1975; K H Voous). JUVENILE. Forehead and streak over eye to above ear-coverts white. Crown streaked black and white; much white on forecrown, less towards nape. Lores white, finely speckled black. Small spot in front of eye black. Ear-coverts dark grey to black, finely streaked white. Some birds show distinct uniform black band from nape to hindcheeks, contrasting with paler crown and mantle, but others have nape same colour as hindcrown and ear-coverts. Upper mantle medium grey with much white on feather-bases, showing much white when worn. Lower mantle, scapulars, tertials, and back medium grey with broad cinnamon-buff fringes to feather-tips, these often subterminally bordered by dark sepia mark; rump and upper tail-coverts medium grey with narrower buff fringes. Underparts uniform white (slightly tinged grey when fresh); often grey patches on sides of chest, especially in birds with dark nape-band. Tail medium grey, sometimes with narrow white outer edges to outer feathers and occasionally with indistinct buff fringes and sepia subterminal mark on tip; inner webs of outer feathers white except for 3–5 cm of tip. Flight-feathers like adult breeding, but primaries without grey bloom of fresh adult and secondaries often narrowly fringed white at tip, especially innermost. Upper wing-coverts fringed buff like scapulars, fringes narrow or absent on lesser coverts. In worn plumage, crown, nape, and ear-coverts nearly uniform black, strongly contrasting with white upper mantle; cinnamon-buff fringes of upperparts and upper wing-coverts bleached to off-white and strongly abraded. Rather marked variation in upperparts: besides those described above, some nearly uniform grey above with faint buff fringes to mantle and scapulars and narrow white ones to tertials, only latter with sepia subterminal mark; others uniform sooty-black above, including mantle, except for white-streaked forecrown and white forehead and supercilium. FIRST IMMATURE NON-BREEDING. Like adult non-breeding, but heavily worn outer primaries (more pointed at tip than those of adult), inner secondaries, and some wing-coverts juvenile until 9–12 months old. When outer primaries new or nearly so, next series of primaries presumably starts and 1st non-breeding directly replaced by 2nd non-breeding; plumages and moult cycle probably similar to adult from age of c. 15 months and adult breeding attained at c. 22 months.

Bare parts. ADULT AND JUVENILE. Iris dark brown. Bill black. Foot black, in juvenile sometimes with grey tinge. DOWNY YOUNG. Iris black. Bill and foot black with slight blue tinge. (Harrison 1975; RMNH, ZMA.)

Moults. ADULT POST-BREEDING. Complete, primaries descendant. In populations with 12-month cycle, starts when feeding young or when young fledges; p1 usually shed first, but sometimes a few feathers of face, crown, or mantle before wing. When primary moult score of c. 25 reached (after c. 2 months), head and body mainly in non-breeding, but not all tertials, tail, back to upper tail-coverts, or wing-coverts. Tail starts shortly after primaries, usually all new with primary score of c. 40; sequence of replacement t1–t6–t3–t2–t4–t5, but t6 sometimes before t1, or t2 before t3. All moult completed with regrowth of p10, primary moult cycle taking 6–7 months. On Seychelles (breeding every $7\frac{1}{2}$–8 months), primary moult starts soon after hatching; average moult score of c. 25 reached $2\frac{1}{2}$ months after hatching, c. 45 after $4\frac{1}{2}$ months; completed when about to lay or during early incubation in next breeding season, c. 7 months after hatching of young of previous season (Diamond 1976). ADULT PRE-BREEDING. Partial. In birds with 12-month cycle, includes head, neck, mantle, underparts, tail (not always all feathers), and uncertain amount of scapulars, back to upper tail-coverts, and upper wing-coverts; starts with completion of p10 of previous post-breeding. In populations with cycle shorter than 12 months, pre-breeding presumably limited to small amounts of feathering of head and body and to t6(–t1); much of head, body, and wing-coverts replaced directly by breeding during post-breeding moult. A few show serially descendant primary moult (scores 1 and 48, or 18 and 49), these perhaps immatures adapting to adult moult cycle. POST-JUVENILE. Complete. Starts 3–6 months after fledging; face, crown, mantle, and t1 first, followed by remainder of head, body, and tail shortly afterwards and by p1 at 4–7 months. No further information; when wing completed at age of c. 1 year, 1st non-breeding on head and body presumably replaced directly by 2nd non-breeding at same time as adult starts post-breeding.

Measurements. ADULT.
S. a. melanoptera. Caribbean and a few from West Africa; skins (RMNH, ZMA). Tail (t1) is length of central pair.

WING	♂ 266	(6.74; 6)	255–274	♀ 260	(4.69; 4)	253–263
TAIL (t1)	72.1	(2.91; 6)	69–76	72.6	(1.55; 4)	70–74
BILL	40.7	(– ; 2)	40–42	38.4	(0.79; 4)	38–40
TARSUS	21.0	(0.97; 6)	20–22	21.0	(1.12; 4)	20–22

Sex differences significant for bill. Adult breeding fork (tip of t1 to tip of t6) 90.4 (5) 86–94.

S. a. antarctica. Red and Arabian Seas south to Madagascar and Laccadive Islands; skins (RMNH, ZMA).

WING	♂ 263	(2.93; 7)	259–266	♀ 254	(6.82; 11)	242–262
TAIL (t1)	73.0	(3.37; 7)	69–77	70.0	(3.14; 10)	67–76
BILL	43.2	(1.83; 6)	42–46	40.3	(1.41; 5)	38–41
TARSUS	21.2	(0.99; 7)	20–22	19.7	(0.70; 11)	19–21
TOE	28.4	(1.56; 7)	27–30	26.1	(1.13; 10)	25–28

Sex differences significant for wing and bill. Adult breeding fork 101, 102.

Nominate *anaethetus*. Indonesia and northern Australia; skins (RMNH, ZMA).

WING	♂ 271	(6.88; 13)	265–288	♀ 263	(10.6; 9)	251–280
TAIL (t1)	73.8	(2.69; 12)	70–78	71.8	(3.60; 7)	68–77
FORK	106	(9.71; 12)	94–120	110	(11.3; 7)	95–128
BILL	42.5	(1.30; 14)	40–44	39.9	(1.00; 9)	38–42
TARSUS	21.9	(1.12; 12)	20–23	21.2	(1.08; 9)	20–22
TOE	29.2	(0.79; 12)	28–31	28.4	(1.34; 9)	26–30

Sex differences significant for wing and bill.

Adult non-breeding fork for all races 75 (5) 65–90.

JUVENILE. Whole breeding range, combined.

WING	♂ 246	(6.36; 7)	242–258	♀ 245	(6.40; 7)	235–251
FORK	48.1	(2.73; 7)	45–52	46.1	(7.84; 7)	38–56

Juvenile tail (t1), tarsus, and toe similar to adult from fledging; juvenile bill not full-grown until at least 1 year old.

Weights. *S. a. melanoptera.* St Martin (Lesser Antilles), breeding, June: ♂ 137, ♀♀ 119, 126 (ZMA). *S. a. antarctica.* Seychelles 95.6 (6.58; 69); variation throughout breeding and moulting seasons limited (Diamond 1976). Nominate *anaethetus*. Northern Australia, breeding, November: ♂♂ 135, 150, ♀ 125 (ZMO). Off Java, adult ♀ in moult, January: 110 (ZMO). Average, Australia, c.121 (Serventy *et al.* 1971).

Structure. Wing long and narrow, pointed. 11 primaries: p10 longest, p9 3–9 shorter, p8 15–27, p7 35–48, p6 57–70, p1 148–169; p11 minute, concealed by primary coverts. Longest tertials reach to tip of p4 in folded wing. Tail deeply forked in adult breeding, less so in non-breeding and juvenile (see Measurements). Bill about equal to head length, straight and sharply tipped, heavier than in (e.g.) Common Tern *S. hirundo*; tip laterally compressed, gape wide; average depth at basal corner of nostril 8·95 and 8·15 in adult ♂ and ♀ respectively of *melanoptera*, 8·63 and 7·54 in *antarctica*, 8·75 and 7·89 in nominate *anaethetus* (measured in skins, perhaps slightly more in fresh birds). Tarsus short and slender. Toes rather long, front ones connected by deeply incised webs; outer toe *c.* 86% of middle, inner *c.* 64%, hind *c.* 30%.

Geographical variation. Rather slight. Nominate *anaethetus* from Indonesia and Australia to western tropical Pacific differs in breeding plumage from Atlantic *melanoptera* by darker upper mantle, almost the same colour as scapulars and upper wing-coverts, without pale grey to off-white collar between black nape and grey upperparts; upperparts darker sooty-greyish-brown rather than sooty-olive-brown, especially on lower mantle, back, rump, upper tail-coverts, and central tail-feathers, these parts the same colour as scapulars and upper wing-coverts rather than paler slate-grey as in *melanoptera*; white on tail more restricted, outer web of t6 usually with 4–9 cm of tip dark grey (white often extends in narrow strip along outer edge, nearly reaching tip; outer web all-white in 4 of 20 examined), tip of inner web more extensively grey, and t4–t5 completely dark (rather than t6 almost completely white, t5 with nearly all-white outer web, and t4 with narrow white outer edge as in *melanoptera*); underparts with distinct pale grey tinge from chest to vent. In worn plumage and adult non-breeding, nominate *anaethetus* often shows white collar like *melanoptera*, however, in non-breeding differing only by more extensive grey to outer web of non-breeding t6. Differences between juveniles of these races small: white on base of inner web of t6 reaches to 5–7 cm from tip in nominate *anaethetus*, 3–5 in *melanoptera*; former also more often with black nape-band and larger grey patch on side of chest. *S. a. antarctica* from Red Sea, Persian Gulf, and western Indian Ocean between *melanoptera* and nominate *anaethetus* in plumage: outer web on t6 often all-white, but tip of inner web of t6 and both webs of t5 dark for at least half of feather length, as in nominate *anaethetus*; grey tinge to underparts paler than in nominate *anaethetus*, often appearing all-white like *melanoptera*; hindneck, mantle, and back to central tail-feathers nearly as dark as scapulars and upper wing-coverts, as in nominate *anaethetus*. *S. a. antarctica* slightly smaller than nominate *anaethetus*, especially in south-west of range, but bill slightly longer though less deep at base. Bill of *melanoptera* even shorter than in *antarctica* and nominate *anaethetus*, but with deeper base. *S. a. nelsoni* from Pacific coast of Mexico and central America larger than *melanoptera* and with longer bill and greyish underparts (Ridgway 1919); hence, close to nominate *anaethetus* in these respects, but white on tail and colour of upperparts apparently as *melanoptera*. CSR

Sterna fuscata Sooty Tern

PLATES 9 and 10
[between pages 158 and 159]

Du. Bonte Stern   Fr. Sterne fuligineuse   Ge. Russseeschwalbe
Ru. Тёмная крачка   Sp. Charrán sombrío   Sw. Sottärna

Sterna fuscata Linnaeus, 1766

Polytypic. Nominate *fuscata* (Linnaeus, 1766), Caribbean and Atlantic. Extralimital: *nubilosa* Sparrman, 1788, Red Sea, Persian Gulf, and Indian Ocean, east to Greater Sunda Islands, Philippines, and Ryu Kyu Islands; *oahuensis* Bloxham, 1826, tropical northern Pacific from Marcus to Christmas Islands; *crissalis* (Lawrence, 1872), Pacific coast of Central America south to Galapagos; *serrata* Wagler, 1830, western and northern Australia through southern Pacific east to Easter Island; *kermadeci* (Mathews, 1916), Kermadec Islands; *luctuosa* Philippi and Landbeck, 1866, islands off Chilean coast.

Field characters. 33–36 cm (bill 3·9–4·8, legs 7·1–8·6 cm); wing-span 82–94 cm. Close in size to Sandwich Tern *S. sandvicensis*; up to 10% larger and bulkier than Bridled Tern *S. anaethetus*, with 15% greater wing-span. Medium-sized but noticeably long tern, with pointed but strong bill, deeply forked tail, and firm, thrusting flight action. Plumage remarkably contrasting in adult, when almost entirely black above and white below, but uniform in juvenile, almost wholly dark brown, with white flecks and spots above. Adult underwing distinctly 2-toned along whole length. Commonest call multisyllabic, loud and far-carrying. Sexes similar; some seasonal variation. Juvenile and 1st-winter immature separable.

ADULT BREEDING. Bold line from bill to eye, top of crown, and nape jet-black, contrasting with white forehead which extends in U-shape round sides of crown as far as eye. Rest of upperparts, upperwing, and all but outermost tail-feathers brown-black, lacking any obvious pattern. Outer webs of tail-feather white, sharply contrasting with rest, though may be abraded after breeding season. Underparts white, with faint grey clouding on lower flanks and vent rarely obvious in the field. Under wing-coverts white, contrasting with dark grey undersurface of flight-feathers. When worn, some may show off-white shading on lower hindneck and pale edges to tertials and central tail-feathers. Bill and legs black. ADULT NON-BREEDING. Feathers of upperparts variably fringed whitish. JUVENILE. Sooty-brown above, with buff fringes to feathers of back and smaller wing-coverts rarely visible but white fringes or tips on scapulars, tertials, larger wing-coverts, longest upper tail-coverts, and tail-feathers obvious at close range, forming spotted pattern. Grey-brown below, unmarked except for grey-white vent and grey mottling under tail. Underwing dark brown-grey with obscure mottling on longer coverts sometimes visible as pale panel (when fully lit). FIRST WINTER. Resembles juvenile until body plumage and smaller wing-coverts moulted; thereafter, forehead hoary, crown and back darker (though still with feathers fringed white), inner forewing unspotted, and underparts paler, more hoary in tone. FIRST SUMMER. Resembles adult except for variable retention of brown on underparts, dusky outer tail-feathers, and some white marks on crown. SECOND WINTER. Apparently as adult.

Adult unmistakable. Juvenile and 1st-winter bird liable to confusion with Brown Noddy *Anous stolidus* at distance, but distinguished by forked (not wedge-shaped) tail, pale grey vent, and white spots on upperparts. Flight usually with noticeable forcing of wing-beats in more direct progress than smaller *Sterna*; capable of range of aerobatics when wheeling, soaring, diving down to water, and taking food from surface. Spends long periods on the wing, rarely settling except at breeding colonies; rarely swims. Carriage on ground markedly horizontal (except for head) and low-slung. Gait tripping but free, either as walk or run.

Vagrants to west Palearctic appear off coasts and exceptionally inland. Flight call 'wide-awake' or, more accurately, 'ker-wacki-wack'. Alarm calls variable, including croaking 'krarrk' and extended whinneying 'kreeaa'.

Habitat. In low latitudes, nearly all tropical or subtropical. Always maritime and markedly aerial, being apparently capable of continuous flight over long periods in lower and upper airspace above ocean, without settling on water or on any floating or fixed perch to rest or roost (Ashmole 1963b). Breeds gregariously on oceanic, offshore, and inshore islands, where necessary on open exposed sand, coral, earth, bare rock, or lava, but preferably among grass or by or under bushes. On Bird Island (Seychelles), however, avoided nesting under planted coconuts *Cocos nucifera* and was encouraged to spread by cutting down large *Scaevola* bushes (Feare 1976). Locally, nests on narrow or wide ledges of rocky walls up to *c.* 12 m above beach, in one case under a cactus on top of the rocks (Bent 1921; Ferguson-Lees 1957). Among large colonies where nests mostly completely exposed to tropical sun are those on Pacific island of Midway (Howell and Bartholomew 1962), and Ascension in South Atlantic, where proved that area with rocks projecting above surface preferred to one flat and featureless (Ashmole 1963b). Choice of breeding islands evidently conditioned by necessity for marine environment to be unusually rich in plankton which leads to abundance of prey species.

In Seychelles, breeds usually during period of south-east trade winds, which may be necessary to bring food to surface. Often feeds 80–100 km distant (Feare 1976). Outside breeding season, ranges far from land over tropical oceans. Proven inability to withstand soaking of feathers and known powers of flight and feeding patterns point to wide-ranging movements to favourable areas of upwelling and other conditions attractive for surface fishing. Undoubtedly avoids coastal or shallow sea areas.

Distribution. Widespread breeder in tropical and subtropical zones of Atlantic, Indian, and Pacific Oceans.

MADEIRA. Possibly bred Selvagens 1982 (F Roux).

Accidental. Azores, Iceland, Britain, France, Spain, Italy, West Germany, Norway, Sweden, Tunisia.

Movements. Strongly pelagic outside breeding season, movements varying from generally dispersive to migratory; may be nocturnal as well as diurnal. Pattern of movements associated with exploitation of seasonal areas of high productivity, especially convergence zones (Ashmole and Ashmole 1967). Little information for populations of eastern Atlantic and Indian Ocean. More known from extensive ringing of birds in western Atlantic (Robertson 1969) and Pacific (e.g. Lane 1967, Gould 1974).

ATLANTIC POPULATION. Movements of east Atlantic population (breeding every 9–10 months: Ashmole 1965) little known. May be origin of sightings (2 birds in each case) in May off coast of western Sahara (Bierman and Voous 1950; Pitman 1967). Where identification made, vagrants to Europe attributed to nominate *fuscata* (e.g. Witherby *et al.* 1941), including specimen from Northampton (England), May 1980. Movements of Caribbean population (breeding every 12 months) known in detail only for colony at Dry Tortugas (Florida, USA); the following largely from Robertson (1969). Birds disperse from mid-July onwards, adults spreading into Gulf of Mexico, Straits of Florida, and (less often) Caribbean, apparently to winter in these areas. Several hurricane-related records on Atlantic coast of North America, north to Nova Scotia (American Ornithologists' Union 1957; Godfrey 1966). Juveniles initially accompany parents into winter dispersal range; during late August to early October, cross south Caribbean (M and W Robertson) and travel south (e.g. sightings in December, Surinam: Bourne and Dixon 1973) as far as northern Brazil. Thereafter, major transatlantic passage of juveniles to Gulf of Guinea, avoiding tropical cyclone belt; numerous recoveries between Sierra Leone and Congo (9°N–5°S, 14°W–10°E), maximum distance 11 300 km by supposed route. Recoveries in all months, mostly October–January; 1 bird reached southern Cameroun 28 October, 107 days after ringing. Perhaps not all juveniles cross Atlantic, for 1st-winter bird from Dry Tortugas recovered off Belize, 26 December. Not known if juveniles from colonies in south Caribbean perform such transatlantic movement, but notable absence in adjoining waters during winter suggests wide dispersal at least (R van Halewijn). First returns to colony (a few birds only) occur at 3 years old; most breed first at 6, some not until

9–10 (Robertson 1969; Harrington 1974; M and W Robertson). Bird in 5th winter the oldest recovered in West Africa (M and W Robertson).

INDIAN OCEAN POPULATION. Little information. After breeding, Seychelles birds believed to disperse widely from waters adjoining colonies (Gill 1967; Bailey 1968), and to move south with shifting belt of south-east trade winds (i.e. c. 5–20°S), October–May. Juvenile ringed Bird Island, and last seen 29 September, recovered northern Australia $3\frac{1}{2}$ months later, suggesting extensive dispersal of young (Feare 1976). Pre-breeding concentration around colonies, Seychelles, noted in May (Bailey 1968). Birds from Arabian Sea also appear to winter mostly south of equator since none encountered from Arabian Sea south to Seychelles, January–February (Gill 1967); no records from west Palearctic.

PACIFIC POPULATIONS. (1) North- and south-central Pacific. At Johnston Atoll, post-breeding dispersal June–August. Adults begin to return late November, occupying colony February–April. Many birds of all ages disperse westwards, reaching Philippines, occasionally Japan. Especially abundant in convergence area (4–5°N) of Equatorial Counter-current. (2) Eastern Pacific. In spring, most birds appear to be west of 110°W, indicating westerly dispersal; in autumn, high densities over wide area east of 110°W, consistent with timing of breeding on Galapagos and Clipperton Atoll. (King and Pyle 1957; Gould 1974.)

EKD

Voice. See Field Characters.

Plumages (nominate *fuscata*). ADULT BREEDING. Forehead and streak above lores to just above front of eye white. Streak from lores to eye, and crown and nape down to eye and ear-coverts deep black, slightly glossed blue when fresh. Central hindneck black, sides white. Upperparts and upper wing-coverts black; less deep and glossy than crown, except for lesser coverts, which contrast markedly with white leading edge of wing. Underparts, including under tail-coverts, white, slightly tinged grey on flanks and belly when fresh. Tail black like upperparts, but outer web and much of base of inner web of t6 white. Flight-feathers black with indistinct dull grey wedge to bases of inner web of primaries, slightly paler grey on inner web of outer primaries; outer web of primaries with dull grey bloom when fresh. Under wing-coverts and axillaries white; tip of greater under wing-coverts often tinged grey. In worn plumage, white feather-bases of central hindneck and mantle exposed, forming white collar; upperparts more brown-black, especially on scapulars, tertials, and median and greater upper wing-coverts; white outer web of t6 partly abraded. Some birds retain non-breeding t6, dull black except for small white tip and grey base to inner web. ADULT NON-BREEDING. Similar to breeding, but crown and lores finely speckled white, hindneck and upper mantle dull black to dark slate-grey with narrow white fringe to feather-tips and much white of feather-bases exposed; lower mantle, back, rump, and scapulars variable, from dull black with faint white feather-fringes to dark slate-grey with white fringes up to 2 mm wide; birds with latter somewhat resemble Bridled Tern *S. anaethetus*. Underparts uniform white. Tail like adult breeding, but t6 black with small white patch on tip and basally grey inner web; outer web sometimes narrowly margined or clouded grey. Upper wing-coverts either like adult breeding or a mixture of worn brown (tips often faded to pale sepia) and fresh black feathers (tips in part narrowly fringed white). Some birds do not attain non-breeding or a limited amount only, moulting almost directly from breeding to breeding. JUVENILE. Almost completely uniform sooty-black; feathers of upper mantle, back, and rump, as well as upper wing-coverts and upper tail-coverts narrowly tipped cinnamon-buff or cream-buff, lower mantle, scapulars, and tertials more widely; underparts tinged grey, especially chin, throat, and belly; vent and under tail-coverts pale grey. Tail uniform sooty-black, central feathers sometimes with buff on tips. Flight-feathers black as in adult, showing grey bloom when fresh, but secondaries and inner primaries narrowly edged white. Leading edge of wing white with grey clouding, contrasting with dark upperparts; under wing-coverts and axillaries pale grey, suffused darker grey on tips, smaller coverts almost white. In worn plumage, sooty-black of upperparts browner, and buff feather-tips of upperparts bleach to pale buff or off-white, those on scapulars showing as oblique white spots; some off-white feather-bases of hindneck, upper mantle, throat, belly, and flanks sometimes exposed; vent and shorter under tail-coverts pale grey or off-white. FIRST IMMATURE NON-BREEDING. Mainly black and hence rather similar to juvenile. Forehead, lores, and cheeks mottled dull black-brown and dark grey; crown dull black with faint dark grey-brown streaks. Nape and patch from round eye over ear-coverts uniform black, often extending down to sides of upper chest. Mantle, scapulars, and tertials dull black or dark slate grey, feather-tips fringed white; back to upper tail-coverts and upper wing-coverts uniform black, occasionally with faint narrow pale grey fringes to tips. Chin and throat pale grey, streaked and mottled dull black; chest similar, but feather-tips more extensively black, tending to form dark chest-band. Feathers of belly dark grey or slate-grey with off-white bases and paler grey tips, appearing grizzled dark grey; vent and under tail-coverts paler grey. Tail like adult non-breeding; under wing-coverts and axillaries like juvenile or slightly paler grey. Up to age of 6–9 months, outer primaries, inner secondaries, some rows of upper wing-coverts, and part of tertials and tail still juvenile, dark brown and heavily abraded. SECOND IMMATURE NON-BREEDING. At age of c. 1 year, 1st immature non-breeding replaced by 2nd, which is rather similar to 1st: mainly dark head and dark chest-band, but chin, throat, and belly to under tail-coverts paler, feathers white with limited dull clouding on tips; outer primaries not juvenile and worn, but new or in growth; 2nd moult series usually started with innermost primaries, birds at this age showing serially descendant moult. 2nd immature non-breeding followed by plumage similar to adult non-breeding, but with dark mottling on underparts and sometimes with grey feather-fringes on face and throat (Clancey 1977b). Some immatures visiting breeding colony retain part of non-breeding; 30% of 3-year-olds in breeding plumage had dark speckling on underparts, 11% of 4-year-olds, and 5% of 5-year-olds (Harrington 1974).

Bare parts. ADULT. Iris dark brown. Bill and foot black. JUVENILE. Like adult, but base of upper mandible and foot dark horn with red tinge at fledging; soon darken to adult colour (RMNH, ZMA).

Moults. ADULT POST-BREEDING AND PRE-BREEDING. Strongly variable, depending on breeding cycle of population concerned: most nest every 12 months, especially where sea temperatures above 23°C for only part of year (Ashmole 1965); elsewhere, may nest every 6 months (though probably not involving same individuals, successful breeders nesting every 12), or every 9–10 months. Nesting and moulting often mutually exclusive, and as length of nesting season roughly similar for all populations, time remaining for moult variable. Moult occurs mainly at sea, and origin and thus breeding cycle of birds taken at this time usually unknown. The following outline, based on a limited number of specimens, should be considered tentative. (1) Birds with 12-month cycle. Post-breeding starts when feeding young or shortly after leaving nesting area; p1 first, sometimes together with some feathers of face, crown, mantle, or tail. Failed breeders may moult up to 4–5 inner primaries before leaving colony (M and W Robertson). At primary moult score of $c.$ 25, head, neck, mantle, underparts, and part of scapulars, tertials, and tail mainly new; primary moult completed with regrowth of p10 $c.7\frac{1}{2}$ months after start, when all head, body, and tail also show fresh non-breeding and wing-coverts and tertials new. Pre-breeding probably follows post-breeding immediately; involves head, mantle, scapulars, underparts, and tail, and apparently a variable number of inner primaries (M and W Robertson)—birds showing some new inner primaries are not necessarily sub-adult (*contra* Ashmole 1963*a*). Approximate sequence of tail moult is t6–t1–t2–t5–t3–t4–t6–t1–t2–t5(–t3–t4), but t1 sometimes before t6; moult of first 6 feathers may be considered as post-breeding, of last 6 as pre-breeding, though moult more or less continuous. (2) Birds with shorter cycles. Start at same time as those with 12-month cycle, and still need $c.$ $7\frac{1}{2}$ months for complete flight-feather replacement, but amount of non-breeding attained limited—usually on head, neck, part of mantle, and scapulars only; old breeding on remainder of head and body and all wing-coverts replaced directly by fresh breeding. Breeding plumage complete at about same time as regrowth of p10. Fewer tail-feathers replaced, and often only t6 and t1 (or also t2 and t5) replaced twice during combined post- and pre-breeding moult. In 6-month cycle, apparently hardly any non-breeding attained and tail-feathers replaced once only, some birds nesting with dark non-breeding t6; primary moult suspended with inner (5–)7–9 primaries new (Ashmole 1963*a*), continuing in next moult cycle, when new series of inner primaries may also start. POST-JUVENILE AND SUBSEQUENT IMMATURE MOULTS. Timing in relation to hatching difficult to establish due to pelagic life. Post-juvenile starts (presumably at $c.$ 6 months old) with head and upperparts, soon followed by primaries and underparts; at primary moult score of $c.$ 25, head and body largely in 1st immature non-breeding, tail and tertials shortly afterwards. 2nd immature non-breeding starts to appear when $c.$ 1 year old, p1 often shed before previous series of primaries completed. Further moults unknown; see also Plumages.

Measurements. Nominate *fuscata* (Caribbean and West Africa) and *nubilosa* (Indian Ocean from Gulf of Aden to Indonesia), combined, as basically similar; skins (RMNH, ZMA). Tail (t1) is length of central pair; fork is tip of t1 to tip of t6; adult fork fresh breeding only.

		♂			♀		
WING	AD	294	(6·01; 16)	280–304	287	(6·75; 13)	276–297
	JUV	282	(6·05; 4)	276–290	279	(5·74; 4)	272–284
TAIL (t1)		73·4	(3·17; 21)	68–78	71·6	(3·30; 13)	68–77
FORK	AD	98·7	(13·0; 7)	85–119	111	(—; 3)	98–124
	JUV	41·0	(—; 3)	39–43	44·5	(0·76; 4)	37–52
BILL	AD	42·3	(1·98; 16)	40–46	40·2	(1·79; 9)	38–43
TARSUS		23·6	(0·85; 18)	22–25	23·1	(1·20; 12)	21–25
TOE		26·6	(1·17; 19)	24–29	25·8	(0·93; 12)	24–28

Sex differences significant for adult wing and bill. Adult non-breeding fork shorter than breeding, 70·3 (8·91; 13) 57–86. Toe rather variable, as claws long at start of breeding, blunt at end of breeding and shortly afterwards. Juvenile t1, tarsus, and toe similar to adult from fledging; juvenile bill not full-grown until at least 1 year old.

Weights. Average of $c.$ 8700 adults, spring 1971–83, Dry Tortugas (Florida, USA), 189; average annual spring weight 179–205, depending on feeding conditions (M and W Robertson). Average of 19 adults, Ascension Island (southern Atlantic), 175 (Stonehouse 1963). Average of many adults, Christmas Island (Pacific), 173 (Ashmole 1968). On day of hatching, Ascension, range in one sample 18–32, average of another sample of 25, 25·5; in season with abundant food, strong gradual increase to 1 month old when $c.$ 150 (125–170), slower later on, fledging at $c.$ 175 (145–200); in season of scarcity, average only $c.$ 72 at 1 month, starving later on (Ashmole 1963*b*). On Hawaii, 147 (24) when $c.$ 1 month old, reaching maximum of $c.$ 187 (22) on 44th day, fledging with that weight at 57 (53–66) days (Brown 1976*b*).

Structure. Wing long and narrow, pointed. 11 primaries: p10 longest, p9 4–8 shorter, p8 20–26, p7 42–48, p6 66–74, p5 91–109, p1 166–181; p11 minute, concealed by primary coverts. Longest tertials reach to tip of p4 in folded wing. Outer tail-feathers elongated, t6 especially; tail deeply forked (see Measurements). Bill rather heavy compared with other *Sterna* of same size; wide at base, laterally compressed at tip; straight with sharply pointed tip and slight angle at gonys; culmen slightly decurved at tip, cutting edges of mandibles finely serrated. Depth of bill at basal corner of nostril in *nubilosa* and nominate *fuscata* 9·12 (0·31; 16) 8·7–9·6 in adult ♂, 8·30 (0·36; 9) 7·8–8·9 in adult ♀; less deep in juveniles up to at least 1 year old. Tarsus short and slender; $c.$ 12 mm of lower tibia bare. Front toes connected by deeply incised webs; outer toe $c.$ 86% of middle, inner $c.$ 64%, hind $c.$ 29%.

Geographical variation. Rather slight, involving colour of underparts in adult breeding and size. *S. f. nubilosa* from Red Sea and Indian Ocean east to western Pacific similar in size to nominate *fuscata* from Atlantic, but belly to under tail-coverts and under wing-coverts usually extensively tinged pale grey in fresh adult breeding, rather than uniform white with faint grey wash at most as in nominate *fuscata*; in worn breeding, grey underparts of *nubilosa* often partly bleached to off-white, however. Number of races warranting recognition in remainder of Pacific uncertain; some described as possessing largely black t6, but this characteristic of non-breeding plumage and depends on breeding cycle. Recognition of *serrata* (Australia and southern Pacific) and *crissalis* (off western Central America) perhaps not warranted, as measurements similar to nominate *fuscata* and to *nubilosa*, and data on plumages at variance. Size of *oahuensis* of northern Pacific close to latter 2 races, but bill slightly heavier at base; *kermadeci* of Kermadec Islands (south-west Pacific) differs by long and heavy bill (average length 47·0, depth at basal corner of nostril 10·0), *luctuosa* of Los Desventurados and Juan Fernández Islands (south-east Pacific off Chile) by long wing.

CSR

Sterna albifrons Little Tern

PLATE 11
[between pages 158 and 159]

Du. Dwergstern   Fr. Sterne naine   Ge. Zwergseeschwalbe
Ru. Малая крачка   Sp. Charrancito   Sw. Småtärna   N. Am. Least Tern

Sterna albifrons Pallas, 1764

Polytypic. Nominate *albifrons* Pallas, 1764, Europe and North Africa, east to central Asia, south to Morocco, Egypt, Iraq, central Iran, and northern Pakistan and India; *guineae* Bannerman, 1931, West and central Africa, south from Banc d'Arguin (Mauritania). Extralimital: *sinensis* Gmelin, 1789, south-east and eastern Asia, east from Ceylon, south through Philippines and Indonesia to New Guinea and northern and eastern Australia; *antillarum* (Lesson, 1847), coasts of USA and Caribbean; *athalassos* Burleigh and Lowery, 1942, Mississippi basin; *mexicana* van Rossem and Hachisuka, 1937, Sonora (Mexico); *staebleri* Brodkorb, 1940, Chiapas (Mexico).

Field characters. 22–24 cm (bill 2·7–3·2, legs 1·5–1·8, tail of adult 4–6, of juvenile 3 cm); wing-span 48–55 cm. Only two-thirds size of Common Tern *S. hirundo*, with 40% shorter wings and 50% shorter tail (lacking obvious streamers but still noticeably forked); longer billed and larger headed than *Chlidonias* terns. Smallest of genus, with proportionately longer bill, more angular head, narrower wings, shorter tail, and more fluttering flight than medium-sized congeners. Forehead white at all ages and general appearance always whiter than *S. hirundo* and *S. paradisaea*. Except in 1st year, wings lack obvious pattern. Most of bill and legs yellow. Sexes similar; little seasonal variation. Juvenile and immature separable.

Adult Breeding. Wide forehead and streak over eye white; line across lores through eye, crown, upper ear-coverts, and long nape jet-black. Back pale blue-grey, fading on rump; tail white. Underparts pure white. Most of wing pale blue-grey, fading to white on secondaries and on bases of primaries; 2–3 outer primaries show grey-black lines, creating dark leading edge to uppersurface of outer wing (but this not visible from below). Bill almost as long as head, tapering to fine point and needle-like at distance; yellow with black tip. Legs orange-yellow in ♂, paler in ♀. Adult Non-breeding. Lores and forehead wholly white; spot before eye and band from eye to nape dull black, with faintly speckled grey crown. Grey of upperparts extends to tail-centre. Distinct dark carpal bar. When worn, primary coverts and outer primaries become dusky and contrast with rest of wing. Bill black or dark horn-brown. Legs dull yellow, grey, or mahogany. Juvenile. Smaller than adult, with blunter wing-tip. Initially, cap buff with black patch through eye and black streaks over crown, contrasting little with buff-white hindneck. Back sandy-grey, with brown-black spots and bars on mantle and particularly scapulars, becoming more uniform ash-grey over rump; tail grey in centre and white on edges, with dark brown spots on ends of feathers rarely visible except at close range. Underparts wholly white. Wing more boldy patterned than in adult breeding, with black-grey lesser coverts, primary coverts, and outer primaries combining to form dusky leading edge, contrasting with pale sandy-grey to white central coverts, and almost white secondaries and inner primaries. Later, sandy and buff tones, and sometimes grey tones, of upperparts lost, with consequent exaggeration of wing pattern and whitening of whole bird. Bill initially brown, with dull yellow on base of lower mandible; legs dull yellow. First Winter to Second Winter. After moult, October–November, like winter adult. On some, dark flecks on crown form trace of summer pattern. Bill all-black, legs orange-red. Second Summer. Most resemble summer adult but some retain darker grey lesser coverts, and dusky primary coverts and outer primaries. No '*portlandica*'-type plumage recorded in west Palearctic.

Unmistakable whether alone or alongside medium-sized congeners in west Palearctic, but subject to high risk of confusion with closely related Saunders' Little Tern *S. saundersi* in adjoining areas in south-east (see Fig I and Geographical Variation). Small size may suggest *Chlidonias* tern, but much longer and usually yellow bill, permanently grey-white plumage, and much faster wing-

Fig I Little Tern *Sterna albifrons* (top left and bottom right) with Saunders' Little Tern *S. saundersi* (top right and bottom left). Flight silhouette and action identical, but adult *S. saundersi* distinguished by lack of white eyebrow, slightly paler upperparts, more black on outer 3 primaries, and brown (not bright yellow) legs. Beware use of primaries as character when plumage worn. (From field comparison in Iran, April.)

beats prevent confusion. Flight light and active, lacking fluid grace of medium-sized congeners but showing instead rapid rate of rather stiff wing-beats; in normal flight, stroke usually shallow, but in persistent half-hover during searching for food or full hover before plunge-dive after fish, noticeably deeper; has habit of altering height of hover by short, sudden drop. Gait free, run quite rapid. Carriage markedly horizontal, with head normally sunk into shoulders.

Voice varied but lacks loud harsh tones. Commonest call 'kit-kit' or 'quit-quit', extended in excitement or self-advertisement to prolonged chittering trill; more liquid in tone than calls of other terns. Alarm call 'krüit' or 'queet', extended into scolding trill.

Habitat. Breeds in west Palearctic in middle and lower middle latitudes, both continental and oceanic, from cool temperate to steppe and Mediterranean zones. Generally confined to lowlands, although occurs exceptionally to 2000 m on mountain lake in Armeniya (Dementiev and Gladkov 1951c). Frequently coast-dwelling, more along mainland than on islands, but spreads freely up suitable reaches of major rivers and to some lakes where suitable conditions occur. Strongly prefers linear strips of bare shingle, shell-beach, or sand, only just above normal tide or flood limits and often only a few metres from shallow clear water, saline or fresh, where fish of suitable size can be caught by plunging, without necessity for extended foraging flights. In Britain and Ireland, 1967, 72% of colonies 1·5 m or less above high-water mark; 63% 18 m or less inland (Norman and Saunders 1969). Sometimes on other flat patches of sand or fine shingle, in sand-dunes, or exceptionally in growing corn or on short grass and caked mud on island in reservoir a little distance inland (Pyman and Wainwright 1952), but normally entirely clear of vegetation. In Scotland, on crumbling bases of abandoned aircraft hangers (Sharrock 1967). Consequent vulnerability of eggs and young to predation and disturbance renders many physically suitable sites untenable, especially where tourist uses coincide with breeding season, or human activities give rise to infestation by rats or feral cats. Shingle extraction, angling, and low-flying or alighting helicopters also cause trouble. In North America (especially Florida), since 1960s, habit has been growing of colonizing extensive flat roofs of buildings, covering up to several ha and up to c. 20 m high, surfaced with shell gravel or 'pea-rock'. In some cases, buildings used to occupy former natural nest-sites, but cases occur up to nearly 20 km inland, with feeding areas up to 4 km distant. Over 40 buildings occupied included warehouses, boat docks, motels, shopping centres, and public buildings such as port offices and pier entertainment. Similar habit developed simultaneously in Finland (Fisk 1978).

Outside breeding season, mainly marine, frequenting harbours in Somalia (Archer and Godman 1937), and tidal creeks, coastal lagoons, and saltpans in India (Ali and Ripley 1969b) and West Africa (E K Dunn). Birds recorded feeding (with Common Terns *S. hirundo*) from dawn to dusk at upwellings up to 600 km off Guinea (Grimes 1977). In Australia, on inshore waters near sand-dunes and shingle beaches; occasionally on rivers and lakes (Slater 1970). In New Zealand, feeds over shallows in harbours and marine inlets (Falla *et al.* 1970). Apparently does not normally rise much out of lower airspace, and does not swim.

Distribution. Colonized Finland in early 1960s, but range elsewhere has tended to contract in 20th century, especially inland in central Europe.

BRITAIN. Has occasionally bred inland (Cramp *et al.* 1974). BELGIUM. Bred in small numbers (c. 75 pairs 1937) until 1964 (Lippens and Wille 1972); attempted breeding 1973 and 1979 (PD). SWEDEN. Some change in range, with shift from south and south-west coasts further into Baltic (SOF 1978). WEST GERMANY. Formerly bred regularly inland along Elbe and Rhine and on some Bavarian lakes (Niethammer 1942; Nadler 1976). FINLAND. First bred early 1960s (OH). EAST GERMANY. Bred regularly inland in 19th century along several rivers; only occasional in 20th century (Nadler 1976; WM). POLAND. Some inland sites deserted after regulation of rivers (Tomiałojć 1976a). CZECHOSLOVAKIA. Formerly scarce breeder; presumably still regular but last proved 1955 (Hudec and Černý 1977). AUSTRIA. Formerly bred in Danube and Neusiedler See areas, last 1962 (HS). YUGOSLAVIA. Has bred, probably irregularly; last proved 1975 and 1980 (Matvejev and Vasić 1973; VFV). SYRIA. Breeds occasionally, no recent proof (Kumerloeve 1968b; HK). LEBANON. Formerly bred on coast (Kumerloeve 1968b). CYPRUS. Bred 1905, 1910, and 1946 (PRF, PFS). ALGERIA. Also said to breed along coast (Etchécopar and Hüe 1967), but no recent information (EDHJ).

Accidental. Faeroes, Norway, Lebanon, Malta, Canary Islands.

Population. Marked declines in recent years in Britain and Ireland, France, Belgium, Netherlands, West Germany, and Spain, due mainly to human pressures on beaches, etc., and habitat changes; little recent change Sweden and slight increase Finland. Limited information for other areas.

BRITAIN AND IRELAND. Decreased 19th century; increased early 20th century, perhaps reaching peak numbers in 1920s and 1930s, then decreased due to disturbance and habitat changes (Parslow 1967). Under 1600 pairs 1967 (Norman and Saunders 1969), 1814 pairs 1969–70 (Cramp *et al.* 1974); surveys up to 1979 suggest increase of c. 15% since (Thomas 1982). FRANCE. Declined due to disturbance and habitat changes; some hundreds of pairs (Yeatman 1976). Now estimated 700–900 pairs (J-YM); largest numbers in Camargue where fluctuating, e.g. 400 pairs 1956, 450 pairs 1976, 250 pairs 1979 (A Johnson),

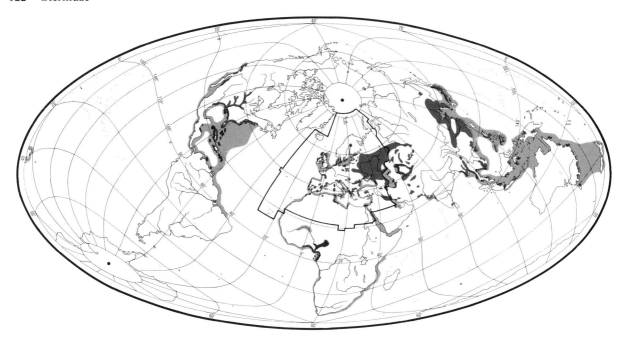

and valleys of Loire and Allier where increased from 100–150 pairs 1905 to c. 380 pairs 1980 (D Muselet). PORTUGAL. Over 100 pairs 1979 (Thomas 1982). SPAIN. Decreasing due to disturbance, habitat destruction, and pollution. Estimated 500 pairs 1976 (Thomas 1982). About 400 pairs Ebro delta 1979 (AN) and 652 pairs 1980 (A Martinez). NETHERLANDS. Declined. Some 2000–4000 pairs 1900–10, c. 1000–2500 pairs 1955 (CSR), marked decline late 1950s to c. 50 pairs 1965, then some recovery to c. 150 pairs 1971 (Rooth and Jonkers 1972), 225–275 pairs 1976–7 (R M Teixeira, CSR), and c. 320 pairs 1978 (Teixeira 1979). WEST GERMANY. Decreased; 320–380 pairs (Szijj 1977). About 500 pairs 1978–9 (Thomas 1982). Schleswig-Holstein: marked recent decline in most areas due mainly to tourist pressures; only c. 100 pairs on Baltic coast (Schmidt and Brehm 1974); over 300 pairs North Sea coast 1964–7 (Nadler 1976). DENMARK. At least 600 pairs (Mardal 1974); slight decrease, 600–800 pairs (TD). SWEDEN. 540 pairs 1973, little change since 1940s (SOF 1978). FINLAND. Increased from 2–4 pairs early 1960s to nearly 30 pairs (OH). EAST GERMANY. About 150 pairs (Nadler 1976). Mecklenburg: some decrease from 200–300 pairs early in 20th century to 100–150 pairs fluctuating 1964–75 (Klafs and Stübs 1977). POLAND. Surveys in 1973 showed 45 pairs on coast and nearly 450 pairs on rivers (Tomiałojć 1976a); c. 800 pairs 1974–9 (Thomas 1982). HUNGARY. About 10 pairs (Thomas 1982). ITALY. About 3000–3300 pairs 1982; decreasing Sardinia (G Bogliani, PB). Sardinia 430–485 pairs (Schenk 1976); Sicily c. 30 pairs (B Massa). GREECE. About 400 pairs 1978 (Thomas 1982). Decreasing (HJB, WB, GM). USSR. Estonia: c. 400 pairs (Onno 1966); c. 300 pairs 1973–7 (Thomas 1982). Latvia: 100–200 pairs (Viksne 1978).

Lithuania: c. 20 pairs decreasing (Jankevičius et al. 1981). Kuybyshev region: threatened with extinction (Titavnin 1981). TURKEY. Under 400 pairs (OSME). ISRAEL. Decreased; bred formerly Lake Huleh and coast, now c. 10 pairs (AP). EGYPT. Scores breed north Sinai coast (AP). ALGERIA. Lake Boughzoul: 8 pairs 1978 (Jacob and Jacob 1980). MAURITANIA. Fluctuating; 100–150 pairs (Naurois 1969a), 95–110 pairs 1967 (Westernhagen 1970), c. 30 pairs (Trotignon 1979).

Oldest ringed bird 21 years 1 month (Clapp et al. 1982).

Movements. Migratory. Movements of most extralimital populations not well known; status of S. albifrons and similar Saunders' Little Tern S. saundersi on east coast of Africa still in need of clarification (Clancey 1982). West European nominate albifrons winters in West Africa. Birds wintering in South Africa are all nominate albifrons, perhaps most likely from western Europe. Nominate albifrons extends north to Kenya, probably also to archipelagoes in western Indian Ocean (Clancey 1982). East European and west USSR populations winter in Red Sea and southern Arabia (and may occur in East Africa), those from Aral–Caspian region probably further east in Persian Gulf and on Makran coast as far as India. East Asian sinensis winters Malay archipelago, Indonesia, and Philippines; bird ringed Java, March 1949, recovered Denu (Ghana), December 1952 (Grimes 1978), indicating possibility of occurrence in west Palearctic. Winter quarters of Australian sinensis inadequately known (Serventy et al. 1971), but occurs Java, Lesser Sunda Islands, Moluccas, and Celebes (C S Roselaar). In Nearctic, antillarum winters on Atlantic coast south through Caribbean to northern Brazil, winter range of other races not known (Blake 1977).

Sterna albifrons 123

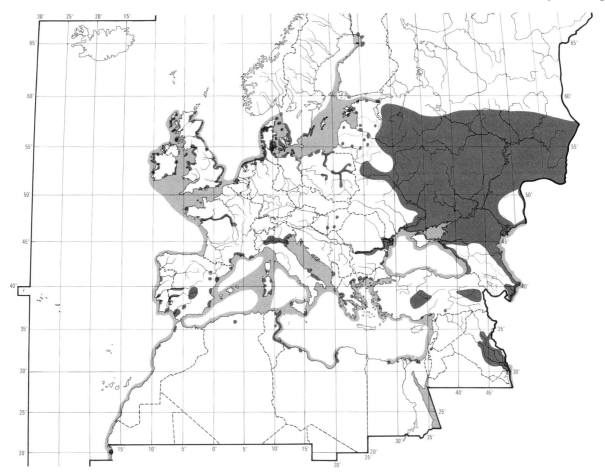

WEST EUROPEAN POPULATION. Post-breeding movement of individuals and family parties begins first half July (more often late July), gains momentum August, and continues into early October. Young birds presumably undergo dispersal before migration proper, though no evidence. Southward movement mostly along Baltic, North Sea, and Atlantic coasts. However, inland passage also occurs, e.g. nocturnal, south-westerly movement, early July to beginning of September, 75 km inland on south-west Veluwe, Netherlands (Bijlsma 1977). Most follow coast of western Europe south, and numerous recoveries from all countries, especially Portugal. Rapidity of some movements shown by bird ringed 21 August, Essex, recovered 27 August, Portugal. Some, especially inland breeders, cross central Europe via river valleys and lakes, concentrations regularly occurring north of the Alps on suitable waterbodies (Nadler 1976); thereafter, presumably follow river valleys (as does Common Tern *S. hirundo*) to Mediterranean. Common on passage Balkan peninsula. Westerly movement to Straits of Gibraltar, thence to Atlantic seaboard of Africa. Common migrant, Tunisia, beginning of August to end of October (Heim de Balsac and Mayaud 1962). Recoveries in north-west Africa restricted to Morocco where passage heavy, September to mid-October (Smith 1965). Sight records further south liable to confusion with local breeders (e.g. Dick 1975).

Paucity of recoveries in winter quarters reflects wariness of baited traps (E K Dunn), but range believed to extend from Mauritania to unknown southern limit, probably most in Gulf of Guinea. Common in Sénégal early September to November (Morel and Roux 1966). Few recoveries from presumed wintering range: British bird from Bassoul (Sénégal), April; Swedish bird from Keta (Ghana), March. Spring movements less well documented. From 13 April, heavy passage seen up Mauritanian coast (Bird 1937). Unlike some Sternidae, no evidence of inland (trans-Saharan) movement (Smith 1968b). Main passage April–May; more rapid and more strictly coastal than in autumn (Nadler 1976). In Netherlands, no inland passage, and coastal movement twice as heavy as in autumn (Bijlsma 1977).

Presumably remains in winter quarters during 1st summer; French 1-year-old found Ivory Coast, May. At 2 years old, northward movement occurs, unknown proportion reaching colonies, and, from ringing evidence, some breeding (N K Atkinson). 2-year-olds may visit regions other than natal one, e.g. British bird controlled Pisa

(Italy) in May (only recovery of British bird of any age from central Mediterranean), recovered Pas de Calais (France) 2½ months later. Extensive exchanges occur between colonies via first-time breeders: British-bred birds found breeding Denmark (2) and West Germany (3); 1 West German bird found breeding England.

EAST EUROPEAN AND WEST USSR POPULATIONS. Autumn passage from eastern Europe, Black Sea, and Asia Minor occurs late August to early October (Dementiev and Gladkov 1951; Borodulina 1960). Occasional on passage at Bosporus (Kumerloeve 1961), and small numbers recorded elsewhere in Turkey (Vittery et al. 1971). Occasional on Cyprus (Stewart and Christensen 1971). From south coast of Mediterranean, penetrates Nile and shores of Red Sea. Large passage flocks occur Egypt, September–November (Meinertzhagen 1930). Winters south to Somalia (Meinertzhagen 1930; Archer and Godman 1937), though possible confusion with S. saundersi (said to winter south to Tanzania: Britton 1980) makes limits uncertain; no evidence of wintering in East Africa (Backhurst et al. 1973). Spring passage, Egypt, mostly May; also Jordan and Syria, April–June (Nelson 1973; Macfarlane 1978). Arrives Crimea mid-May (Dementiev and Gladkov 1951c).

CENTRAL USSR POPULATION. Birds from Aral–Caspian region probably take more easterly route than those breeding further west—through Iraq and Iran (Nadler 1976). Common on southern Caspian (moving south and west), August–September (Passburg 1959; Feeny et al. 1968). Conspicuous passage through Tigris and Euphrates valleys to winter quarters in Persian Gulf and beyond. Route apparently mainly north-west of Afghanistan, where only scarce passage migrant (Nadler 1976). Spring passage recorded Iraq from April onwards (Moore and Boswell 1956a). Arrives south-west Caspian from end of April, occurring in thousands during first half of May (Borodulina 1960). For arrival times in various regions of USSR, see Dementiev and Gladkov (1951c). EKD

Food. Small fish and invertebrates, especially crustaceans and insects. Usually feeds singly, in small parties, or in widely scattered flocks; also on occasion in mixed-species flocks of Sternidae. Most commonly, works back and forth over water with quick wing-beats, head directed downward. Height of feeding flight dependent on light intensity, wind strength, and prey; on average 5–7 m (Dementiev and Gladkov 1951c; Flint 1975). Often hovers before making fast vertical or near-vertical plunge-dive with partial or complete immersion, or drops more slowly to surface with wings at 80° to horizontal, entering water vertically (Taylor 1975). Hovering height variously reported as c. 3–4.5 m (Bannerman 1931), 6 m (Ali and Ripley 1969b), and, surprisingly, 9–12 m (Flint 1975). Duration of hovering may vary from 3.5 s in calm weather to 5.1 s in windy conditions; average 4.3 s, maximum 8.0 s (Borodulina 1960); elsewhere, average 6.0 s in calm weather (Flint 1975). Strong winds depress fishing success (N K Atkinson). Usually hovers more often and for longer than other Sternidae, and probably morphologically adapted for frequent, prolonged hovering and diving into shallow water (Taylor 1975). Diving rate 1–7.3 dives per min, tending to be inversely proportional to success (H M C McLannahan); 109 per hr (Borodulina 1960). Fishing success at sea (aborted dives not included), June–July, 16.9–60%; average over 7 days was 39.7% (H M C McLannahan). Plunge-dives 4 times as often as Common Tern S. hirundo (Borodulina 1960). Carries only 1 fish at a time (Tomkins 1959). Will also feed by steeply dipping-to-surface to skim prey off water with backward flick of bill (Ali and Ripley 1969b; Buckley and Hailman 1970). Takes insects from ground as well as water (Dementiev and Gladkov 1951c), and at Lake Baykal (USSR) insects plucked in flight off vegetation (Polivanova 1971). Takes insects in aerial-pursuit (Witherby et al. 1941); catches gnats by zigzagging nearly level with water to pick them up like marsh tern Chlidonias with forward-back flick of bill (Besson 1969b). Most hunting close to colony: maximum 4.9 km (Tomkins 1959); maximum 6 km, but not more than 1.5 km offshore (N K Atkinson).

Fish include sand-eel *Ammodytes*, roach *Rutilus rutilus*, rudd *Scardinius erythrophthalmus*, carp *Cyprinus carpio*, perch *Perca fluviatilis*, pike *Esox lucius*, bleak *Alburnus alburnus*, white bream *Blicca bjoerkna*, ruffe *Gymnocephalus acerina*, pipefish *Syngnathus*, smelt *Osmerus eperlanus*, gobies (e.g. *Gobius niger*), gudgeon *G. gobio*, 'whitebait' (*Clupea harengus*, *Sprattus*), unidentified Cyprinidae, Atherinidae, and small flatfish (Pleuronectiformes). Crustaceans include shrimps (*Crangon*, *Squilla*), prawns *Leander*, unidentified crabs, Mysidacea, and Isopoda. Insects include grasshoppers (Orthoptera), adult and larval dragonflies (Odonata), flies and gnats (Diptera), beetles (Coleoptera), and ants *Formica*. Also annelids and molluscs. (Naumann 1905; Collinge 1924–7; Culemann 1928; Goethe 1932; Peters 1933; Witherby et al. 1941; Dementiev and Gladkov 1951c; Kistyakivski 1957; Borodulina 1960; Schönert 1961; Besson 1969; Taylor 1975; Nadler 1976; H M C McLannahan.)

Considerable variation in frequency of main prey. 6 stomachs, Britain, contained by volume 97% crustaceans (including Mysidacea) and annelids, 2% fish, and 1% marine molluscs (Collinge 1924–7). 4 stomachs contained only fish, 1 being full of sand-eels (Witherby et al. 1941). On Ythan estuary (Scotland), all birds in 26 observations took shrimps *Crangon vulgaris*, averaging 1.9 cm (Taylor 1975). At St Cyrus (Scotland), bulk of food brought to young was *C. harengus* (3.8–6.7 cm) and *Ammodytes* (N K Atkinson). Similar diet Norfolk, England (H M C McLannahan). 49 stomachs from eastern North Sea coast contained 95% fish, and only 5% crustaceans; 75 stomachs, summer, all contained fish, with insects in only 4–5 (Nadler 1976). In Ukraine (USSR), 28 stomachs contained mainly fish (Cyprinidae in 29%, bleak 25%, pike

11%, perch 7%, and at least 7 other species), a few insects including ants *Formica* and dragonfly larvae, and isopod crustaceans (Kistyakivski 1957). In Crimea (USSR), stomach contents and regurgitations comprised mostly fish (in 91%) and fewer shrimps and crabs (27%); in contrast in Sivash, insects occurred in 99% of regurgitations, fish in 54%, and crustaceans in only 2·1%; on lower Volga, diet exclusively fish (gobies and roach of 4–5 cm); at fish farms, Akhtarsk (Sea of Azov), material from 25 stomachs and 19 regurgitations comprised almost entirely fish but mainly non-commercial species, e.g. atherinids and gobies; at Chernomorski reserve, dominant food was young gobies (Borodulina 1960). Observations, June–August, Yagorlitskiy Gulf (Black Sea) and Rybinsk reservoir, indicated fish main diet (Flint 1975).

At Gibraltar Point (England), probably over 90% of food brought to young was crustaceans, especially prawns. Fed most often early in morning and in evening; 2 young received 2–35 feeds per 2 hrs (Davies 1981; see also for effect of tide). Tends to feed young more frequently than do other Sternidae (Tomkins 1959). At Chernomorski (USSR), young (number in brood not given) fed small fish 51–63 times per 24 hrs; average 3·6 feeds per hr, maximum 14 per hr (Borodulina 1960). Young 1–5 days old fed 1·85 times per hr, based on 122 feeds in 66 hrs to brood of 3 (Nadler 1976). Feeding occurs in 'bouts', frequency increasing as young age: mean frequency (items per chick per hr) 2·7 at 1–5 days old, 4·1 at 6–10 days, 9·0 at 11–15 days, 10·4 at 16–20 days. According to Hardy (1957) and Nadler (1976), parents bring smaller items to smaller young, though young 2–3 days old are brought 7-cm sandeels. However, H M C McLannahan found size of items apparently not dependent on age of young though young rejected unmanageable fish: 1-day-old young did not eat fish over ·5 cm, and even at 2–3 days old young had difficulty with some fish over this length; 2 young, *c.* 4 days old, found dead with fish stuck in gullet, and on 2 occasions adults twice seen to pull fish from gullets of 3-day-old young. Length of all prey fed to young of all ages ranged from less than 1 to 9 cm, mean 5·1 (H M C McLannahan; see also Davies 1981). In one area, young fed small crabs till 3rd or 4th day, and only afterwards fish (Culemann 1928).
EKD

Social pattern and behaviour. Includes material from USA: on *antillarum* from California (B W Massey) and north-east USA (I C T Nisbet), and on *athalassos* (Hardy 1957).

1. Generally gregarious throughout the year, especially when breeding and roosting, though does not form dense feeding flocks like some Sternidae, and solitary birds often occur. Spring migrant flocks, southern England, average 3·1 birds (M J Rogers). In spring, inland USA, *athalassos* arrives gradually in small loose flocks of 5–20 (Hardy 1957). After breeding, family parties gather on shore prior to migrating in small flocks; family parties also migrate alone, Finland, mean 1·8 young with single parent (Soikkeli 1962). Feeding groups in breeding season typically loose flocks of 2–4 (Hardy 1957; H M C McLannahan). BONDS. Essentially monogamous. In Massachusetts (USA), ♂♂ of *antillarum* (probably mostly unpaired) often feed and copulate with more than one ♀ (I C T Nisbet). In *antillarum*, pair-bond stable in some cases for 2 or more years, but mate-changes not uncommon (B W Massey). In nominate *albifrons*, Nadler (1976) considered changes of mate more frequent (especially in more disturbed shore colonies) than in some other Sternidae and Laridae. In nominate *albifrons*, average age of first breeding 3 years, $n=44$ (Schmidt and Siefke 1981), but sometimes breeds at 2 years (Schönert 1961; N K Atkinson), and this not uncommon in *antillarum*; some 2-year-old *antillarum* in adult plumage visit colony too late in season to breed (Massey and Atwood 1978; B W Massey). Both sexes incubate and care for young, though ♂ *antillarum*, Massachusetts, typically performed 60% of incubation, and brought 60–70% of food to young (I C T Nisbet). ♀ nominate *albifrons* generally takes greater share of incubation, especially in first few days; at hatching, ♀ stays to brood while ♂ does most fishing; both sometimes forage after a few days (Schönert 1961; Davies 1981). Juveniles fed by parents for several weeks (2–3 months in *athalassos*: Hardy 1957), and many at least accompany parents on migration. Parents often divide fledged offspring between them, or only one parent may be seen with all (Schönert 1961; Soikkeli 1962; Nadler 1976). BREEDING DISPERSION. Nests in small loose colonies or solitarily; commonly 5–15 pairs, seldom more than 150; *c.* 42% of colonies, east coast of North Sea, have up to 5 pairs, 60% up to 10, 95% up to 30–40 (Nadler 1976). In Britain and Ireland, 1973, colonies of 2–140 pairs (Lloyd *et al.* 1975); in 1969, 50% had up to 5 pairs, 85% up to 25, 3% more than 50, and only 1 colony exceeded 150 (Norman and Saunders 1969). For colony sizes of *antillarum*, Massachusetts, see Nisbet (1973b); some colonies, USA, recently of 500 or more pairs (I C T Nisbet). Birds which participate in aerial courtship together tend to form discrete nesting groups producing scattered sub-colonies (R Chestney). Distance between nests varies, often with space available, but usually not less than 2 m (Nadler 1976). At one colony, Smalenyi Island (southern USSR), only 0·36–0·85 m (Borodulina 1960), and at another, Black Sea, mean 1·3 m, closest 0·5 m (Zubakin and Kostin 1977). At one colony of 5 nests, West Germany, 5–34 m, mean 18·9; at another of 5 nests, 0·6–2·1 m, mean 1·14 (Schönert 1961). At 3 colonies, Norfolk (England), means of 11, 13, and 19 m (H M C McLannahan). ♀ makes (and both sexes defend) 1st scrape 3–10 days before 1st egg laid (N K Atkinson). In *antillarum*, small nest-area territory established by ♀ 3–4 days before laying—in loose colonies more by avoidance than aggression (I C T Nisbet); used mainly for nesting and care of young up to 2–3 days old. May defend feeding territory near colony, and some evidence that ♂ more aggressive in its defence (N K Atkinson, R Chestney). Generally returns to colony of previous breeding season, but probably not to previous nest-site (Schönert 1961). First-time breeding of *antillarum* often occurs at colony other than natal one (B W Massey). Older birds generally shift colony-site only in response to habitat alteration (I C T Nisbet). Replacement clutch laid up to 150 km from original nest-site (Schmidt and Siefke 1981). Older, experienced *antillarum* arrive at colony earlier than younger ones (Massey and Atwood 1978). In USSR, forms mixed colonies with Common Tern *S. hirundo* and Little Ringed Plover *Charadrius dubius* (Borodulina 1960; Blinov and Blinova 1980). ROOSTING. Readily forms communal roost outside breeding season, less so otherwise. Near colony, California (USA), night-roost persists throughout season (B W Massey). Prefers open ground such as shingle, sand, or mudflats near high-water mark; also saltpans. Up to 20 birds on spring migration, West Germany, formed night-roost, probably arriving after sunset (Stiefel 1966). Pre-breeding roost forms on shore near, but outside colony-site (Lewis 1920b; Tomkins 1959;

Schönert 1961). Returning off-duty bird sometimes loafs near nest (at 25–50 m) before relieving mate (Schönert 1961). Fledged young loaf on fixed spot on shore where parents can locate them. Post-breeding loafing flocks of juveniles and adults formed outside colony, St Cyrus (Scotland) by end of June (N K Atkinson); sometimes gather with other Sternidae (e.g. Browne 1950).

2. FLOCK BEHAVIOUR. In colony, first bird on ground to detect danger gives version of Alarm-call (see 1a in Voice) and takes off (Massey 1976). Alerted non-incubating and then sitting birds follow suit in 'upflight' or 'alarm' (Schönert 1961) and give more intense version of Alarm-call (see 1b in Voice). Injured conspecific elicits similar reaction to aerial intruder on ground (see Parental Anti-predator Strategies, below). 'Dreads' occur as in *S. hirundo*, sometimes in response to sudden danger, e.g. emergence of observer from hide; flock initially loose, coalescing over water prior to reoccupying colony (Hardy 1957). Frequency of occurrence decreases through season; rare in late incubation (Nadler 1976). ANTAGONISTIC BEHAVIOUR. Depending on nest-spacing, territorial defence more or less pronounced (e.g. Schönert 1961, Wolk 1974, B W Massey, I C T Nisbet). In dense colony, California, defence prominent, especially in early part of incubation period (B W Massey). (1) Ground encounters. Strange adults (usually non-breeders) intruding on territory are chased away, usually only by ♀ in *antillarum* (I C T Nisbet), but often by both sexes in dense colony (B W Massey). Wandering chicks may be met with invariable, sometimes severe, attack (Davies 1981), or indifference (B W Massey). Initially, territory-owner usually approaches adult intruder on foot, adopting Aggressive-upright, intensity varying with gravity of threat: forebody held high, neck craned upwards and a little forwards towards opponent, carpal joints slightly free of body, and hind-crown feathers slightly raised. Bill opened, either silently or to give Gakkering-call (Schönert 1961; Nadler 1976: see 2 in Voice). Aggressive-upright not recorded in *antillarum* in which intention to attack signalled by raising outspread wings; in physical attack which may follow, bird flutters along ground towards opponent with wings raised (I C T Nisbet). Incubating bird jabs with bill at intruder passing too close (Schönert 1961; H M C McLannahan). ♀ *antillarum* strongly dominant over ♂ which may be soundly pecked, even by well-established mate, if approach too sudden (I C T Nisbet). (2) Aerial encounters. If intruder does not retreat, territorial bird may make flying attack, usually horizontal, but sometimes proceeding to dive-attacks accompanied by Alarm-calls (see 1c in Voice). Ground attack in colony, California, commonly led to Upward-flutter, as in *S. hirundo*; rarer in nominate *albifrons*, but sometimes seen in disputes over feeding territories (E Sutton). HETEROSEXUAL BEHAVIOUR. (1) General. Pair-formation involves both aerial and ground behaviour at colony. Initially rather silent on first arrival at breeding grounds, resting for long periods at high-water mark near colony-site; some begin ground courtship soon after arrival (Hardy 1957; Schönert 1961), though usually later than aerial courtship, and up to 16 days before start of egg-laying (N K Atkinson). (2) Aerial courtship. Starts at breeding grounds as soon as birds arrive, and occurs until end of breeding season (N K Atkinson). As in 'crested' terns (e.g. Sandwich Tern *S. sandvicensis*), early courtship focused on pre-breeding roost, relatively little occurring on nest-territory. ♂ first attracts attention of ♀ by advertising flight: usually carries fish, and flies towards pre-breeding roost giving Advertising-call (see 3 in Voice). Typically chased by 1–2 (B W Massey), or up to 6 birds (Schönert 1961) which also give Advertising-call. Chases may last up to 15 min (Schönert 1961), and may proceed to High-flight, which usually involves few pursuers (commonly 2) or even none. High-flight of 2 birds described most fully in *antillarum* though apparently similar in nominate *albifrons*. ♂, almost always carrying fish, leads ascent giving Advertising-call while other, usually ♀, follows, often initially uttering Advertising-call, at variable distance behind; with rapid wing-beats, birds ascend in steep broad spiral to 100 m or more (Nadler 1976; I C T Nisbet); leading ♂ suddenly begins downward glide with wings in V and ♀ follows suit. Descent broadly spiral, faster and more erratic than in larger *Sterna*, with pair zigzagging and repeatedly closing and parting again, giving Advertising-call (Edwards and Woodfall 1979), though glide silent in *antillarum* (Wolk 1974). ♀ may overtake ♂ (Pass-ceremony), more often when 3 birds involved than 2; pull out only when close to ground (Wolk 1974; Moseley 1979; B W Massey, I C T Nisbet). Glide may be interspersed with slow, exaggerated wing-beats (Marples and Marples 1934; Edwards and Woodfall 1979). Up to 4 High-flights may follow consecutively (Schönert 1961; E Sutton). (3) Ground courtship. May precede or follow aerial courtship. Mostly at pre-breeding roost (Lewis 1920b; Tomkins 1959), sometimes on elevated ground. Courtship-feeding (see below) generally integral part of display; ♂ arrives at roost with fish, uttering Advertising-call, and lands close to another ♂ or ♀. ♂ typically rebuffs approach, or attempts to solicit fish in Hunched-posture, as in other Sternidae. If unfamiliar ♀ approached, both birds may initially adopt Forward-erect: forebody held high, bill about horizontal. Receptive ♀ adopts Hunched-posture, sometimes quivering wings, and gives Begging-call (see 5 in Voice). ♂ then usually switches to Bent-posture, as in *S. hirundo*, and circles ♀ closely ('parade') with characteristic shuffling or strutting gait, before transferring fish. After presentation, ♀ adopts Erect-posture: neck craned, head tilted slightly away from mate, bill 45° above horizontal or steeper, carpal joints free of body, and wings sometimes outspread or raised (Fig A). ♂ gives Advertising-call and relaxes to Forward-erect, ♀ likewise. (4) Copulation. Precopulatory display prolonged in *antillarum* (up to 8 min), and reflects dominance of ♀ throughout courtship (I C T Nisbet); apparently similar in nominate *albifrons* (e.g. Lewis 1920b). ♂ without fish sometimes mounts ♀, but fish almost certainly prerequisite for successful copulation (B W Massey and I C T Nisbet, who provide following account; see also Wolk 1974). Fish-bearing ♂ approaches ♀ very slowly and obliquely from rear until behind and to one side of her. ♀ gradually adopts Hunched-posture with Begging-call (may remain silent in *antillarum*: B W Massey), and wags head slowly from side to side (Head-turning); also half-spreads drooped wings, fluttering them open and shut until nudging ♂'s breast. May thus move away from ♂ who follows with breast puffed out, neck craned, and bill horizontal, Head-turning faster than ♀. ♂ starts to lift wings, opening and closing them, gradually (up to 6 min) raising them until quivering over back; periodically flicks bill upwards, as if drawing attention to fish. ♂'s movements accelerate and he mounts; sometimes stands, fluttering wings and Head-turning, before lowering tail. ♀ turns bill up towards ♂, stops calling, and copulation occurs. ♀♀ of *antillarum* take fish at precise moment of cloacal contact; no contact occurs if fish denied. Sometimes ♀ grabs fish and shakes ♂ off, but after successful copulation ♂ often remains mounted and copulates several times, for up to 3 min in nominate *albifrons* (Schönert 1961). Throughout copulation, ♂ *antillarum* utters low, rasping call (B W Massey). When ♂ steps down or is dislodged by ♀ running forwards, both adopt Erect-posture, heads slightly averted; one or both may preen and/or go to bathe after which further mating may occur (Schönert 1961; Nadler 1976). (5) Courtship-feeding. In *antillarum*, ♀ occupies territory 3–4 days before egg-laying and waits there for ♂ to feed her. In *antillarum*, several ♂♂ may visit and feed same ♀. In early courtship, ♀ commonly adopts Hunched-posture and gives

Begging-call, or stands up and gives Advertising-call, on approach of ♂. Feeding rate increases towards laying, remaining high in early incubation when ♂ may feed ♀ at intervals of 5–10 min (Lewis 1920b). Rate generally drops markedly after 6th day of incubation when ♂ begins to incubate more (Schönert 1961). In Norfolk, mean of 3·6 feeds per hr brought in days 1–5 of incubation (after clutch completion), 0·8 for days 6–10, 0·1 for days 11–15, 0·2 for days 16–20 (H M C McLannahan). Courtship-feeding may lead to scrape-making by ♀ (Schönert 1961; N K Atkinson, B W Massey), though both sexes said to participate (Wolk 1974; Nadler 1976). Up to 7 scrapes made before one chosen (N K Atkinson). As ♀ scrapes, ♂ circles her in Bent-posture, giving Cooing-call (see 6 in Voice). (6) Nest-relief. Early in incubation, precipitated by ♂ arriving with fish, giving Advertising-call in flight. On alighting, sequence of display initially as in Courtship-feeding. ♂ approaches ♀ in Hunched-posture, similar to that of ♀ (Fig B, left). Just before fish transferred, both birds sometimes Head-turn slowly (Lewis 1920b). ♂ feeds ♀ either on or near nest. After 7–8 days of incubation, fish less often brought, and only mutual Erect-posture, or Forward-erect, precedes nest-relief. Thereafter, sitting bird increasingly (180 of 200 exchanges) flies off nest at aerial approach of mate who continues Advertising-call while alighting on nest (Schönert 1961; Edwards and Woodfall 1979). Arriving ♂ may land 25–50 m from nest and preen before relieving mate. When eggs hatch, sitting bird waits until mate is standing alongside before departing (Schönert 1961). Settling bird often kicks sand out of scrape, sometimes to uncover partially buried eggs; may bill pebbles towards rim of nest (sideways-building), especially on approach of mate (Lewis 1920b). Before flying away, relieved bird may walk away from nest and perform behaviour equivalent to side-throwing of other Sternidae; directs pebble, shell, etc., backwards towards nest; according to Lewis (1920b), this performed only by ♀, and only up to 4th day after clutch completion. RELATIONS WITHIN FAMILY GROUP. Incubating or brooding bird, southern USA, known to dip in water and shake drops from breast on to eggs or young, presumably to cool them (Tomkins 1942; Hardy 1957). Chicks stimulate parents to feed

them, from a few hours old, by pecking at bill-tip and uttering food-call. ♂ sometimes passes food to brooding ♀ for transfer to young (Smith 1921; Schönert 1961). Parents retrieve, and sometimes wash, fish dropped by small chicks. From 3–6 days, young recognize Advertising-call of approaching parent and run towards it (Schönert 1961; Nadler 1976). Run from under ♀, calling and flapping (Davies 1981). Young learn parents' calls on 1st day (Massey 1976; Davies 1981). In nominate albifrons, parents apparently recognize offspring from c. 3 days. From 1–2 days, young highly mobile and make short excursions from nest, returning in response to Cooing-call. If undisturbed, brooded and fed at or near nest for first 3 days. Once attachment to nest ceases, brood disperses and young often become widely separated; one family moved at least 1 km (Davies 1981). At night, parents stay with small chicks and make scrape to brood them (Schönert 1961). By day, ♀ digs brooding scrapes for young (Davies 1981), though chicks also make own scrapes (Tomkins 1959). In bad weather, brooded up to 20 days old (Davies 1981). At c. 20 days, young make first short flights and, after c. 23 days, practise diving, sometimes retrieving floating leaves, etc., though not catching fish (Tomkins 1959; Nadler 1976; I C T Nisbet). Follow parents to fishing grounds and may be fed on water surface (Tomkins 1959). ANTI-PREDATOR RESPONSES OF YOUNG. On approach of danger, downy young crouch prone and freeze, usually alerted by parent's Alarm-call (1a or b, see Voice); older young may crouch where they are or run to refuge and crouch; rarely swim (Schönert 1961; Davies 1981). Older young of antillarum tend to hide, rather than simply flattening on ground, and often run in zigzag before doing so. PARENTAL ANTI-PREDATOR STRATEGIES. (1) Passive measures. At approach of danger, adult leaves nest, and flies low for 3–6 m before gaining height and giving Alarm-call (1a or b, see Voice) (Massey 1974). (2) Active measures: against birds. Predator on ground attracts small flock hovering overhead, from which individuals make dive-attacks. Approach of flying predator stimulates upflight (see Flock Behaviour, above). Whole colony takes part in attack on Accipitriformes, Strigiformes, Laridae, and Corvidae, reacting to, and starting to mob predator before it reaches colony (Hardy 1957; H M C McLannahan, I C T Nisbet). Laridae comprised 66% of birds attacked at colony, Northumberland, England (Edwards and Woodfall 1979). Birds giving Alarm-calls (see 1c in Voice) gather above intruder in flock from which individuals or groups of 2–3 mount continuous barrage of swoops, though rarely striking; intruder often repelled. S. a. antillarum attacks bird predators especially strongly, and will strike those as large as Great Black-backed Gull Larus marinus (I C T Nisbet). If predator lands in colony, only those with nests or young nearby continue to mob (H M C McLannahan). If predator retires to colony edge, attack usually ceases there (Schönert 1961). (3) Active measures: against man. Dive-attacks usually withheld until intruder within colony, then proceed as above—though solitary pairs (those with eggs at least) do not make dive-attacks (D J Brooks). More likely to attack when young still in nest. Rarely strikes intruder, but occasionally defecates on him (Wolk 1974; Massey 1976; Davies 1981). Last 3 m of dive typically made with wings in V and feet lowered (Marples and Marples 1934; Wolk 1974). In antillarum, if intruder remains near young, adult may fly low, away from intruder, giving Alarm-call (see 1b in Voice), and land nearby, still calling; may continue walking away from (but watching) intruder, take off, and land again further off, as if trying to lure it from young (Wolk 1974). (4) Active measures: against other predators. Attacks other mammalian predators vigorously, notably foxes and dogs (Canidae) and stoats and weasels (Mustelidae) (Schönert 1961; Davies 1981). S. a. antillarum swoops at, but does not strike, dogs; response generally less vigorous than with aerial predators (I C T Nisbet). For predators, North America, and response to them, see Massey (1974) and Wolk (1974).

(Fig A from photograph by B Hawkes; Fig B from photograph in Lewis 1920b.)   EKD

Voice. Most vocal at breeding grounds. Except where indicated, calls attributable to nominate *albifrons*. For sonagrams of calls 1a, 1b, 5, and 6, and for comparison with voice of *antillarum*, see Massey (1974, 1976).

CALLS OF ADULTS. (1) Alarm-calls. Massey (1976) recognized 3 calls, representing various levels of anxiety. (a) At lowest intensity, 'wiik' (Massey 1976) or 'krüit-krüit' (Nadler 1976), less clipped than call 1b; given (e.g.) if source of danger distant from colony. (b) At medium intensity, a clipped, staccato 'kit-kit-kit' (Massey 1974, 1976); 'quit' or 'queet' (P J Sellar); 'witt-witt-witt' (Schönert 1961); 'titt-titt-titt' (Nadler 1976). (c) In maximum alarm and hostility (e.g. when human handles eggs or young) while mobbing, a chattering 'tittittittrittittritt' (Schönert 1961) and, at lowest point of swoop, harsh 'bzzz' (Massey 1976); latter rendered 'chäk', 'schäk', or—at most intense—'kekechäk' (Schönert 1961). (2) Gakkering-call. 'Rä-gä-gä-rä-gä-gä', given by ♂ with open bill in Aggressive-upright towards conspecifics or other bird intruding on ground into nest-territory (Nadler 1976). Harsh, deep, scolding call of *antillarum*, California (USA), in territorial disputes probably the same (B W Massey). (3) Advertising-call. A rapid, warbling, 'churr-er-te-tet churr-er-te-tet-ee' (J Hall-Craggs); 'purrittittitTIT' (Schönert 1961); 'wididiIT' (Massey 1976). Highly variable; can have 2–5 notes, often 4 (Figs I–III show 2-, 4-, and 5-note calls respectively); of 10 calls, Norfolk (England), duration 0·275–0·425 s, mean 0·335 s (Massey 1976). Typically given while gliding down to feed or relieve mate or to feed or brood offspring; quickens to a trill on close approach (Edwards and Woodfall 1979). In *antillarum*, 'purrit-' (homologous with Fig I) identifies caller, '-titTIT' signals intention to approach without aggression (Moseley 1979). In nominate *albifrons*, part of the call, 'purrIT', given by ♂ after catching fish; expanded to longer Advertising-call on approaching ♀, who reciprocates (Massey 1976). (4) Fishing-call. A 'tschik', likened to call 1a by Massey (1976), occasionally given by ♂ while searching for fish. According to Schönert (1961), serves as contact call throughout year. (5) Begging-call. 'Wu-du-du' (Massey 1976); 'dirrdirrdirr' or 'bürrbürrbürr' (Schönert 1961) given by ♀ in Hunched-posture; according to Massey (1976), rarely used. (6) Cooing-call. A soft, clucking 'rog-rog' or 'gog-gog' (Schönert 1961); 'yupp yupp' (Smith 1921). Heard in all contexts involving nest (I C T Nisbet): nest-site selection, scrape-making, incubation, hatching, and brooding; summons young outside nest for brooding (Hardy 1957; Schönert 1961; Massey 1976; I C T Nisbet). Elements possibly derived from call 3 (Massey 1974, 1976), and rendered 'pit' by Nadler (1976), also given by parents with small young. (7) Exit-call. In

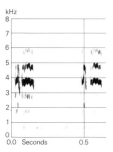

I B W Massey England May 1973

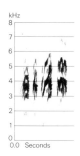

II B W Massey England May 1973

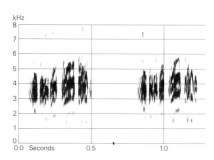

III J Kirby (1973) England May 1963

antillarum, a chittering 'chi-ti-ti-ti-ti-tik', sometimes shortened to 'chi-tik', usually given by ♂ as he flies off after feeding ♀ (Massey 1976; I C T Nisbet); ♀ usually reciprocates with disyllabic call, rarely multisyllabic one (I C T Nisbet). Not yet established whether nominate *albifrons* has homologous call.

CALLS OF YOUNG. Food-call, 'ppeppieppiepp', initially like cheeping of domestic chick *Gallus*, becoming deeper c. 1 week after hatching (Schönert 1961). After 2 weeks, changes to 'wwrätt' (Nadler 1976).

Breeding. SEASON. North Sea area: see diagram. Mediterranean region: laying begins 2–3 weeks earlier. Crimea (USSR): main laying period from second half of May (Borodulina 1960). SITE. On ground in the open. Colonial (often loosely) or solitary. Nest: shallow scrape, unlined or with pieces of available vegetation, small stones, and debris. Average diameter 10·5 cm, depth 2·5 cm, no sam-

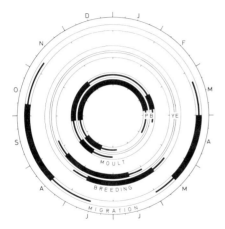

ple size given (Borodulina 1960). Building: by ♀ (Schönert 1961; Massey 1974; N K Atkinson); by both sexes according to Nadler (1976). EGGS. See Plate 90. Sub-elliptical, smooth and not glossy; very pale cream to buff (or greenish or bluish), with black, dark brown, or pale grey spots, blotches, and occasional streaks; rather variable. 32 × 24 mm (30–37 × 21–26), sample 400 (Schönwetter 1967). Mean weight 10 g (8–12), sample 270 (Nadler 1976). Clutch 1–3; 4–5 rarely recorded, at least sometimes probably misidentified Ringed Plover *Charadrius hiaticula* (Schönert 1961; N K Atkinson). Mean 2·05, sample 64 (Marples and Marples 1934); 2·23, from 7 annual means, Fife, Scotland (N K Atkinson). 4 annual means, Norfolk (England), 2·20–2·43 (N Sills). Of 527 clutches, Hampshire (England): 1 egg, 21%; 2, 58%; 3, 20%; 4, <1%; 5, <1%; mean 2·0 (N Ost, E J Wiseman). Of clutches recorded Crimea (no sample size); 1 egg, 15%; 2, 15%; 3, 69%; 4, 1%; mean 2·56 (Borodulina 1960). One brood. Replacements laid after egg loss; always smaller than 1st clutch (Borodulina 1960). Laying interval 1–2 days; 3-egg clutch laid in 4–5 days (Massey 1974; Nadler 1976; N K Atkinson). INCUBATION. 18–22 days, mean 21·5 (N K Atkinson). In *antillarum*, California (USA), 20–25 days, mean 22 (Massey 1974). Mainly by ♀, especially early on, but becoming more equal (Borodulina 1960; Nadler 1976); in *antillarum*, mainly by ♂, mean 60% on days 6–21 (I C T Nisbet). Shifts vary from a few minutes to over 1 hr, commonly c. 1 hr (Nadler 1976); 40 min to 4 hrs (N K Atkinson); mean 28 min (Schönert 1961). Begins with 1st egg but often left for long intervals; not continuous until 3rd laid; hatching near synchronous (Schönert 1961; Massey 1974; Nadler 1976). YOUNG. Precocial and semi-nidifugous. Cared for and fed by both parents; brooded while small, mostly by ♀ (I C T Nisbet). Leave nest when a few days old and hide in shingle or vegetation (Schönert 1961). FLEDGING TO MATURITY. Fledging period 19–20 days, exceptionally 15–17 (Schönert 1961; Nadler 1976; N K Atkinson). Age of independence not known, but in Mississippi (USA), *athalassos* seen being fed by adults 2–3 months after fledging (Hardy 1957), and family groups occur on migration (Soikkeli 1962). Age of first breeding sometimes 2 years, usually older (see Schönert 1961); 2 years not unusual in *antillarum*, California (Massey and Atwood 1978; B W Massey). BREEDING SUCCESS. Much affected by disturbance, especially human, but other predators, high tides, and wind-blown sand also locally important. In wardened areas, England, 154 pairs raised 177 young compared with 87 pairs raising 9 young before wardening (compiled from Lloyd *et al.* 1975). At wardened Welsh colonies, 40 pairs raised 56 young, 1·4 per pair (Thomas and Richards 1977). Fledging success at 4 colonies, Norfolk, 0–2·5 young per pair; overall, 116 pairs raised estimated 189 young, 1·63 per pair; of 116 clutches started, 37 lost to high tides, 21 to predators (H M C McLannahan). At colony, Fife, hatching success over 7 years 15% to c. 76%, fledging success 5% to c. 46%; 0·03–0·40 young fledged per pair (N K Atkinson). At colony, Norfolk, hatching success over 4 years 26–71%; main losses to high tides, and predation by Oystercatcher *Haematopus ostralegus* (N Sills). Fox *Vulpes vulpes* and Kestrel *Falco tinnunculus* also locally important predators of eggs, young, and adults (Norman and Saunders 1969; Thomas and Richards 1977). At Hiddensee (East Germany), most losses of eggs and young due to foxes and gulls *Larus* (Schmidt and Siefke 1981).

Plumages (nominate *albifrons*). ADULT BREEDING. Forehead and streak above lores reaching to above middle of eye white. Streak from lores to eye, crown, nape, and ear-coverts from behind eye backwards black. Mantle, scapulars, tertials, back, and all upper wing-coverts light grey, slightly bluish when fresh, similar to Common Tern *S. hirundo*; often slightly paler at border with black nape and towards sides of chest, longer scapulars and tertials fringed pale grey to off-white when fresh. Rump, upper tail-coverts, and tail variable: in some, appears all-white (especially when worn) though usually with faint pale grey tinge present on rump, central upper tail-coverts, t1, and outer webs of t2–t3; others have all rump to tail pale grey (paler and less bluish than remainder of upperparts) with white restricted to fringes along outer webs of t4–t5 and to outer web and basal $\frac{3}{4}$ of inner web of t6. Cheeks, sides of neck, and underparts white, often with grey tinge to sides of chest, occasionally faintly tinged pale grey from chest to vent, especially in birds with greyish rump to tail. Colour of primaries dependent on moult: 3 different moult series present, arrested when nesting, each differing in wear. Fresh primaries pale grey (except for broad white border along base of inner web), distinctly tinged silvery-grey; older ones duller grey, hardly silvery; oldest ones black. In Britain, Netherlands, Sweden, and Germany, 1 (7% of 50 birds examined), 2 (42%), or 3 (51%) outer primaries of 1st post-breeding series retained; in Mediterranean area east to Iraq, 1 (11% of 50 examined), 2 (67%), or 3 (22%). These old outer feathers black (except for white border to basal inner web), shaft of p10 white to pale horn, of p9(–p8) dark horn to black, uncommonly white or pale horn. (p2–)p3–p5(–p6) to p7–p8(–p9) are from fresher 2nd series; outer feathers light grey with silvery tinge, contrasting with blackish neighbouring outermost feathers of 1st series; inner feathers duller grey or brown-grey with frayed tips, shafts white. Innermost primaries are from 3rd series, all fresh and light grey, this series arrested with up to p1 (4% of 50 birds examined), p2 (30%), p3 (48%), p4 (12%), p5 (2%), or p6 (4%) new in western and central Europe, 0%, 18%, 26%, 44%, 12%, and

0%, respectively, in 50 birds from Mediterranean area to Iraq; 2 out of 50 from western and central Europe and 8 out of 50 from Mediterranean to Iraq had started a 4th series on innermost primaries. Secondaries light grey, tips and inner borders white. Leading edge of wing, under wing-coverts, and axillaries white. In worn plumage, some grey of feather-bases of crown and off-white of those of mantle visible; primaries of 2nd series duller, more brownish-grey, less silvery, shafts pale horn. Sexes similar, but black loral line in ♂ wider in front of eye, narrowing more abruptly near base of mandible; in ♀, narrows gradually from eye to bill; useful only for sexing of known pairs (Nadler 1976); see also Bare Parts. ADULT NON-BREEDING. Lores, forehead, and crown white, often with grey tinge and dark speckling on crown, contrasting with black patch in front of eye and black streak backwards from eye, widening towards black nape. Mantle, scapulars, tertials, back, median and greater upper wing-coverts, and secondaries as in adult breeding, but duller grey, more bluish, or mixed dull and bluish depending on moult and wear. Rump, upper tail-coverts, and tail darker grey than in adult breeding, same colour as remainder of upperparts or slightly paler; only t6(–t5) white. Underparts uniform white. Shorter lesser upper wing-coverts slate-grey, forming dark carpal bar from shoulder to base of primaries, contrasting with white leading edge and with light grey of remaining coverts. Colour of primaries depends on moult. Up to September–October, outer primaries old and with contrast between blackish outer 2–3 and grey-brown neighbouring inner usually still visible. During November–January, outer primaries new or growing, grey with silvery tinge; 2nd series actively moulting on inner primaries. From February to early April, outer primaries dark grey to blackish, though usually still with silvery-grey bloom; 3rd moult series active on innermost primaries, 2nd on p5–p8. In fresh plumage, black feathers on nape-patch narrowly fringed white, central nape occasionally appearing largely white. DOWNY YOUNG. Down of upperparts short, sometimes slightly spiny; sandy-buff to pale cream, finely speckled black. Speckles either scattered and diffuse or partly joining to form 3 broken lines on crown and 2 on back. Underparts uniform white or cream, without black or brown throat usually present in *S. hirundo* and Arctic Tern *S. paradisaea*. (Fjeldså 1977; RMNH, ZMA.) JUVENILE. Lores and forehead grey-buff, crown streaked grey-buff and black. Patch in front of eye and streak from behind eye across nape black, feathers narrowly fringed buff. Often narrow white collar on hindneck. Mantle, scapulars, and tertials pale buff, each feather with distinct black subterminal arc and broad off-white fringe. Back, rump, upper tail-coverts, and tail light grey, feathers fringed white, widest at tips (tail-coverts especially appearing white); feathers of back and central tail often with dark subterminal arc or dot, t6(–t5) usually white and unmarked. Underparts uniform white (slightly buff when fresh), sides of chest sometimes buff with dark arcs. Primaries rather different from adult breeding, without contrast caused by arrested moult: inner primaries light grey, outer gradually darker, p10 almost black; uppersurface of shafts white on inner, black or dark horn on outer; outer webs and tips with faint grey bloom only, not as silvery as adult; inner webs white with broad grey streak along shafts, like adult. Tips of secondaries and inner primaries more widely margined white than in adult, narrow white fringes present up to tip of p7–p8. Upper wing-coverts light grey (duller than in adult), tips broadly bordered pale buff to off-white, inner coverts and sometimes longer lesser coverts with dark subterminal arc; shorter lesser coverts and primary coverts greyish-black, narrowly fringed grey, forming distinct carpal bar as in adult non-breeding. In worn plumage, lores, forehead, and crown bleach to off-white, and buff tinge and brown arcs on upperparts and upper wing-coverts disappear, partly by fading, partly due to early start of post-juvenile moult; may then resemble adult non-breeding, but primaries, tail, and upper wing-coverts still distinctly juvenile. FIRST IMMATURE NON-BREEDING. Like adult non-breeding; difficult to distinguish by plumage alone when last juvenile feathers of tail, tertials, and wing-coverts lost by October–November, but differs from adult by primary moult and wear. By October of 1st autumn, juvenile outer primaries rather fresh and grey, not old or showing contrast caused by arrested moult as in adult. By November–January, outer primaries worn, dark brown (in adult, new or growing), and inner moulting. In February–April, outer primaries new or growing, 2nd series started on inner, scoring 2–25 (in adult, outer rather new, 2nd series scoring 24–45, 3rd series started, scoring 2–20). Non-breeding retained throughout 2nd calendar year, except sometimes for a few black feathers on crown or grey feathers in dark carpal bar. Primary moult either continued during summer, or arrested with scores 5–15 (3rd series) and 25–35 (2nd series), instead of 10–20 and 35–40 as in adult. SECOND IMMATURE NON-BREEDING. Like adult non-breeding, differing only by relatively newer outer primaries; sometimes shows irregularities in arrested primary moult (scoring different from adult), and may also start new series on inner primaries earlier than adult. FIRST BREEDING. In 2nd spring, usually similar to adult breeding, but occasionally some white non-breeding on crown and some black of non-breeding carpal bar retained.

Bare parts. ADULT. Iris dark brown. Bill bright yellow with 4–8 mm of tip contrastingly black, ♂ often showing more black than ♀ (Nadler 1976); black absent in 3 of 20 west Palearctic spring birds examined and only just present in 2 others. During summer, black tends to fade and by late July and August tip often appears largely yellow; at same time, dusky bluish-black of non-breeding starts to spread forwards from base of culmen and in both directions from halfway between gonys-angle and bill-tip; bill often completely horn-black by late September, all-black during winter. Yellow starts to spread again from base when breeding feathers appear on head. In breeding, foot deep chrome-yellow to reddish-orange (paler in ♀: N K Atkinson), claws black; in non-breeding (about September–March), foot dull yellow-brown, dusky grey with yellow tinge, or mahogany, duskiness spreading in early autumn from joints and tibia; rear of tarsus and soles yellow. DOWNY YOUNG. Iris dark brown. Bill pinkish or greyish-flesh, tip dusky brown or black. Foot pale pink-flesh. JUVENILE AND IMMATURES. Iris dark brown. Bill dark horn-brown to nearly black, basal cutting edges of upper mandible and base of lower mandible pink-yellow to dull yellow. Foot grey-flesh to yellow-brown. Adult non-breeding colours obtained from September onwards and usually retained during 2nd calendar year, but a few birds show yellow on base of bill June–July and occasionally basal half yellow in August. Birds with only basal half of bill yellow, culmen dark, and foot yellow-brown (although otherwise in breeding plumage) are perhaps 3rd calendar year. (RMNH, ZMA.)

Moults. Unusual among west Palearctic Sternidae in showing 3 successive moult series in primaries, each starting with p1 and arrested with different moult score at start of nesting season; Roseate Tern *S. dougallii* and some marsh terns *Chlidonias* show similar moult, as do other members of the *S. albifrons* superspecies (see Geographical Variation). For convenience, only average data for moult scores mentioned below, although attainment of each may vary by *c.* 1 month more or less. ADULT POST-BREEDING. Complete; primaries descendant, inner primaries twice. 1st series starts with p1 mainly when feeding young late June to late

August. Average primary moult score of 15 reached mid-August, 25 in last days of August, 30 by mid-September, 38 mid-October, 42 mid-November, completed mid-December. This primary moult arrested during migration with score 10–35; completed in winter quarters. Shortly after p1 (starting from primary score 4–20), moult of crown and upperparts starts; during migration, head and body arrested with about half feathering of crown, mantle, scapulars, sides of breast, tertials, and back to upper tail-coverts new, as well as some median upper wing-coverts. Tail mostly starts when p1 full-grown; centrifugal, but t6 with t4 and hence before t5. 2nd series starts at primary moult score of 35–40, when head, body, tail, and wing-coverts in fresh non-breeding; p1 shed mainly mid-October to mid-December, occasionally earlier. Average score 14 in last days of November, 28 mid-January, 33 mid-February, 36 mid-March; arrested at onset of breeding with score 35–40(–45), or up to p7–p8(–p9) new. In some at least (perhaps all), tail also starts for 2nd time with t1. ADULT PRE-BREEDING. Late February to early April; involves head, body, wing-coverts, and tail, latter probably in part a continuation of 2nd moult started in post-breeding. Inner primaries start for 3rd time, mainly February to mid-March; arrested shortly before start of incubation with score (5–)10–20(–25), or up to (p1–)p2–p4(–p5) new. POST-JUVENILE. Complete. Starts early August to late September; mantle and scapulars first, followed by head, tertials, remainder of upperparts, some wing-coverts, and underparts. Largely in first non-breeding (except flight-feathers, tail, and part of wing-coverts) late August to early November (usually early October). Flight-feathers and tail start with p1 and t1 respectively between late September and early December; average primary moult score 20 by mid-December, 33 by mid-February, 1st series completed April–May. 2nd series starts February–March, reaching score of c. 25 by mid-May. FIRST IMMATURE PRE-BREEDING. Limited. Some birds start 3rd series of primaries in May; sometimes also attain scattered breeding feathers on crown, mantle, scapulars, and wing-coverts, and arrest primary moult late May and June with scores 25–35 (2nd series) and 5–15 (3rd series). Further data confusing: birds arresting 2nd and 3rd series probably start new series in July without continuing previous series, hence moult similar to adult; more retarded birds change 1st non-breeding directly for 2nd non-breeding from July without arresting 2nd series of previous post-juvenile and at same time starting new series. Some 1-year-olds show 4 primary moult series with (e.g.) 1st just completed with p10, 2nd arrested with p7, 3rd arrested with p4, and 4th actively moulting p1–p2, or show arrested moult with p1, p4, p6, and p9 new.

Measurements. Nominate *albifrons*. Netherlands, West and East Germany, and southern Sweden; adults April–August, juveniles August–September; skins (BMNH, RMNH, ZFMK, ZMA). Fork is tip of t1 to tip of t6.

		♂			♀	
WING	AD	181	(3·27; 16)	176–187	175	(3·31; 19)	167–180
	JUV	168	(3·35; 4)	164–172	171	(3·76; 11)	166–177
TAIL (t1)		44·5	(2·35; 10)	42–48	43·4	(2·20; 14)	39–46
FORK	AD	42·8	(4·90; 6)	36–49	36·1	(3·15; 10)	29–41
	JUV	14·3	(1·96; 5)	12–17	15·4	(1·56; 11)	13–18
BILL	AD	30·2	(1·72; 13)	27·8–33·1	28·7	(1·13; 17)	26·7–30·8
TARSUS		16·8	(0·67; 13)	15·6–17·8	16·6	(0·72; 18)	15·1–17·8
TOE		18·0	(0·87; 13)	16·9–19·3	17·4	(1·10; 17)	15·8–18·9

Southern USSR, Western Sinkiang (China), and Afghanistan.
WING AD ♂ 180 (4·13; 6) 176–186 ♀ 176 (— ; 2) 174–178
BILL AD 30·8 (1·32; 6) 28·4–31·9 28·4 (— ; 2) 27·2–29·5

Southern Portugal, Morocco, Algeria, and Tunisia.
WING AD ♂ 176 (2·62; 7) 172–180 ♀ 175 (1·86; 10) 172–178
BILL AD 29·7 (1·28; 6) 28·3–31·9 29·4 (1·23; 10) 27·8–31·4

Balkans, Turkey, northern Egypt, and inland Iraq.
WING AD ♂ 174 (3·28; 8) 170–179 ♀ 171 (— ; 2) 170–172
BILL AD 29·4 (1·61; 8) 26·8–32·2 27·8 (— ; 2) 26·9–28·8

Head of Persian Gulf: Al Faw area (Iraq) and Dara and Buneh islands (Iran).
WING AD ♂ 173 (3·44; 9) 167–177 ♀ 169 (3·89; 8) 163–174
BILL AD 29·4 (1·58; 8) 27·4–31·8 28·0 (1·19; 8) 26·3–29·9
Sex differences significant for wing, fork, and bill of west and central European adult only.

S. a. guineae. Banc d'Arguin (Mauritania), 6 adult ♀♀: wing 159 (155–168), bill 28·6 (26·5–32), tarsus 16·1 (15·5–17) (Naurois and Roux 1974).

S. a. sinensis. Mainly Japan and Taiwan, adult summer, sexes combined: wing 183 (24) 178–191, fork 59·9 (9) 47–72, bill 30·4 (14) 29–32, tarsus 17·7 (13) 17–19 (RMNH, ZMA).

S. a. antillarum. South Caribbean, breeding, sexes combined: wing 163 (11) 156–170, fork 40·7 (7) 34–50, bill 28·0 (9) 26–30, tarsus 15·3 (11) 15–16 (RMNH, ZMA).

Saunders' Little Tern *S. saundersi*. Adult breeding; skins (BMNH, RMNH). Karachi (Pakistan); includes a few birds intermediate in plumage between *S. saundersi* and nominate *albifrons*.
WING AD ♂ 170 (2·50; 7) 166–173 ♀ 169 (3·21; 5) 166–172
BILL AD 29·5 (0·79; 7) 28·0–30·5 27·5 (0·36; 5) 27·0–28·0

Bahrain, Oman, and Hadhramawt (South Yemen).
WING AD ♂ 168 (— ; 2) 165–172 ♀ 165 (2·69; 4) 162–168
BILL AD 30·1 (— ; 2) 28·7–31·5 27·5 (1·49; 4) 25·9–29·5

Red Sea
WING AD ♂ 165 (— ; 3) 164–168 ♀ 162 (4·45; 8) 157–167
BILL AD 28·2 (— ; 3) 28·0–28·6 26·6 (1·49; 7) 23·5–27·9
Sex differences not significant, except Karachi bill. T1 for all *S. saundersi* combined 37·0 (13) 35–40, fork 28·6 (13) 22–35.

Weights. Nominate *albifrons*. Scotland, summer, adult 57 (30) 50–63; Morocco, September, adult 41·9 (3·33; 7), juvenile 38·4 (8) (Pienkowski 1975). Netherlands, summer: adult ♀♀ 50, 50, 55; juveniles 42, 50, 54, 58, 60; exhausted adult 38 (A L Pieters, RMNH, ZMA). Adult ♀, Makedhonia (Greece), 49 (Makatsch 1950).

S. a. sinensis. Hopeh (China): ♂ 60 (26) 45–108, ♀ 55 (23) 44–68 (Shaw 1936). Taiwan: June, ♂ 52, ♀♀ 50, 58; August, ♂ 52, ♀♀ 50, 50 (RMNH).

S. a. antillarum. St Martin (Lesser Antilles): June, ♂ 39, ♀ 39; September, ♂ 70, ♀ 52 (ZMA). Surinam, May–September 2nd calendar year: ♂ 39; ♀♀ 46, 49; unsexed 46, 49 (RMNH).

Structure. Wing long and narrow, pointed. 11 primaries: p10 longest, p9 4–14 shorter, p8 15–26, p7 31–41, p6 46–58, p5 59–72, p1 101–124; p11 minute, concealed by primary coverts. Longest tertials reach to tip of p4 in closed wing. Tail rather short, rather shallowly forked (more deeply in *sinensis*—see Measurements). Bill about equal to head length, straight and acutely pointed; culmen very gently sloping to tip, gonys nearly half of bill length, without marked angle. Depth of bill at basal corner of nostril in west European adult ♂ 6·48 (0·17; 16) 6·2–6·8, in ♀ 5·99 (0·28; 18) 5·3–6·3. Foot small and slender. Front toes fully webbed; outer toe c. 81% of middle, inner c. 66%.

Geographical variation. Rather slight and gradual. Within nominate *albifrons*, involves extent of primary moult and size.

Birds from Britain and central Europe east to USSR, Afghanistan, and Sinkiang (China) larger, those from Mediterranean and inland Turkey and Iraq slightly smaller, bill more slender; breeders at head of Persian Gulf almost similar to east Mediterranean birds—see Measurements. Range of 36 wings northern Pakistan and India 163–180 (Junge 1948) and size of these birds hence apparently similar to those of northern Persian Gulf. 2nd moult series on primaries about equally often arrested with either p7 or p8 in breeders from western and central Europe and western Mediterranean (hence 2–3 outer primaries old and blackish in summer), 3rd series with p2 or p3; 2nd series most often arrested with p8, and 3rd with p3–p5 in populations of southern USSR, western China, eastern Mediterranean, Turkey, and both inland and coastal Iraq. *S. a. guineae* from West and central Africa (Mauritania to Sudan, Zaïre, and probably Lake Turkana) smaller than nominate *albifrons* (see Measurements) and bill usually completely yellow. *S. a. sinensis* from eastern Asia slightly larger than nominate *albifrons* (see Measurements), t6 especially longer (up to 140 in ♂), but t1 similar to nominate *albifrons*, hence tail more deeply forked. Bill longer, heavier at base. Shafts of outer primaries white, that of (p8–)p9 rarely pale horn (mainly in juvenile), while shaft of p9 in most adults and all juveniles of nominate *albifrons* is dark horn to black. Arrested primary moult approximately similar to eastern populations of nominate *albifrons*: in *sinensis*, 90% of 30 examined had 2 outer primaries old, 10% p10 only; in 30 of eastern *albifrons*, 17% had old p10 only, 70% p9–p10, 13% p8–p10. Rump to tail of breeding adult pale grey or white, similar to European nominate *albifrons*. Size, plumage characters, and moults of populations from New Guinea and Australia as in east Asiatic *sinensis*. Birds of Philippines and Sunda Islands (sometimes separated as *pusilla* Temminck, 1840) also similar to *sinensis*—e.g. wing of breeding adults 183 (14) 172–192—but rump and tail-coverts often darker grey, almost same colour as upperparts. Breeders from Ceylon have primary-shafts white and wing 173–188, hence also belong to *sinensis* (Ali and Ripley 1969). *S. a. antillarum* from Caribbean and both coasts of USA distinctly smaller than nominate *albifrons*, but tail relatively deeply forked as in *sinensis*; breeders from coastal USA slightly larger than those of south Caribbean cited in Measurements. Rump, upper tail-coverts, and t1–t5 grey, similar to upperparts; underparts sometimes tinged grey. No specimens examined of 3 races described from inland USA and Pacific coast of Mexico, and status of these uncertain; for details, see Brodkorb (1940) and Burleigh and Lowery (1942); for variation within USA and status of *antillarum*, see Massey (1976).

Saunders' Little Tern *S. saundersi*, breeding Persian Gulf and Red and Arabian Seas, treated here as separate species, following Vaurie (1965) and Voous (1977a); however, reasons for not treating it as a race of *S. albifrons* seem poor. Characterized in adult breeding by (1) small size, (2) deep black outer primaries with faint grey bloom only, (3) black outer-primary-shafts, (4) moult of 2nd series of primaries arrested with p7 (hence 3 outer primaries dark), (5) more white on forehead and less above eye (forehead patch appearing squarer), (6) darker grey on rump to central tail-feathers, (7) olive or brown foot, yellow only on rear tarsus and soles, (8) preference for salt water. These characters hold true for Red Sea population (though specimen labels for foot colour in BMNH read yellow and yellow-green), but not for all birds of other populations: 1 out of 6 from northern and eastern Arabia and 3 out of 12 Karachi breeders had colour of outer primaries and shafts intermediate between *saundersi* and nominate *albifrons*. Occurrence of intermediate birds and increase in size northward (Karachi birds almost similar in size to breeders of head of Persian Gulf near Al Faw, which are pure nominate *albifrons*) suggest that *saundersi* and *albifrons* intergrade. Of points mentioned above: (1) small size only refers to Red Sea and Arabian birds, not to Pakistan *saundersi*; (2, 3) mainly black outer primaries and shafts are not valid for all *saundersi*, as a few nominate *albifrons* from North Africa, Middle East, and northern India are dark too; (4, 5, 6) 3 dark outer primaries, square forehead patch, and grey rump to tail frequently occur also in nominate *albifrons*; (7) brownish legs apparently occur in Karachi birds only (yellowish-brown and brownish-yellow on BMNH labels) and not elsewhere (e.g. Bahrain bird had legs dark yellow); note that legs of *S. albifrons* also brownish in non-breeding; (8) preference for salt water shared by many *S. albifrons*, e.g. by those breeding at head of Persian Gulf and by many Eurasian breeders. *S. a. albifrons* and *saundersi* are not separable in non-breeding plumage except for Red Sea birds which are smaller (and hence similar then to non-breeding *S. a. guineae*). See also Clancey (1982).

S. albifrons forms superspecies with *S. saundersi*, Peruvian Tern *S. lorata* (Pacific coast of Ecuador to northern Chile), Yellow-billed Tern *S. superciliaris* (inland South America from Venezuela and Brazil to northern Argentina), and Fairy Tern *S. nereis* (Western and southern Australia, Tasmania, New Zealand, and New Caledonia), each differing only in minor plumage details from *S. albifrons* and hardly in size and moults; breeding ranges show no or only marginal overlap. CSR

Sterna saundersi Hume, 1877 Saunders' Little Tern

FR. Sterne de Saunders   GE. Orientseeschwalbe

Some uncertainty about this species in west Palearctic. Known to breed in Iran and on Gulf coast of Saudi Arabia, but published record of breeding in Kuwait (Jennings 1981c) is incorrect (M C Jennings, P R Haynes) and there are no recent records of any kind within west Palearctic. However, C J O Harrison, after examining specimens in the British Museum (Natural History) considered that 3 specimens obtained at Fao (Iraq) in 19th century (2 in 1886, 1 undated) were *S. saundersi*, and that from 1 from Amara (Iraq) in 1918 apparently belonged to this species. Ticehurst *et al.* (1922) examined these specimens and considered them all to be Little Tern *S. albifrons*; he did not admit *S. saundersi* to Iraq list. C S Roselaar, who also examined these specimens, was of same opinion.

For descriptions, see Field Characters and Geographical Variation of *S. albifrons*.

Chlidonias hybridus Whiskered Tern

PLATES 12, 13, and 14
[between pages 158 and 159, and facing page 206]

DU. Witwangstern   FR. Guifette moustac   GE. Weissbartseeschwalbe
RU. Белощекая крачка   SP. Fumarel cariblanco   SW. Skäggtärna

Sterna hybrida Pallas, 1811. Synonym: *Hydrochelidon leucopareia*.

Polytypic. Nominate *hybridus* (Pallas, 1811), North Africa, Europe, Middle East, northern India and southern Siberia, east to Ussuriland (USSR), China, and Assam (India). Extralimital: *delalandii* (Mathews, 1912) (synonym *sclateri*), southern and East Africa and Madagascar; *javanicus* (Horsfield, 1822) (synonym *fluviatilis*), Australia.

Field characters. 23–25 cm (bill 3·0–3·3, legs 2·2–2·5, tail up to 6 cm); wing-span 74–78 cm. Larger and bulkier than Black Tern *C. niger* and White-winged Black Tern *C. leucopterus*, with 10–15% longer wings, longer, heavier bill, flatter head, and almost 50% longer legs; overlaps with immatures of several medium-sized sea terns *Sterna* in size, but has shorter, less forked tail. Largest marsh tern, with more dagger-shaped bill than congeners and plumage patterns recalling *Sterna* at all ages. Breeding plumage similar in pattern to that of Arctic Tern *S. paradisaea* and (even more so) White-cheeked Tern *S. repressa* but more intensely coloured; other plumages lack distinctive pattern and thus identification at all seasons requires clear observation of structure and behaviour. Flight slower and less erratic than that of congeners. Alarm call loud and grating. Sexes similar except for bill size; marked seasonal variation. Juvenile and 1st-year immature separable.

ADULT BREEDING. Forehead, rather deep crown, and long nape jet-black, contrasting with white lower face (shading to soft grey below) which forms broad pale stripe across head (particularly when seen head-on). Hindneck and rest of upperparts dark ash-grey, almost uniform in tone and thus lacking paler rump and tail of all *Sterna* except *S. repressa* and Lesser Crested Tern *S. bengalensis*. From chin, underbody becomes slate-grey to black-grey, darkest on flanks (and there much more intense in tone than *S. paradisaea* or even *S. repressa*); vent and under tail-coverts white, but sometimes contrasting less with paler grey rear belly than with darker grey rump and tail (and not forming as obvious a character as on congeners). Upperwing ash-grey, with coverts slightly paler than back and dusky ends to primaries contrasting with more silvery bases and primary coverts. Under wing-coverts white, contrasting with greyer flight-feathers; dusky ends to outer 7–8 primaries create diffuse trailing edge to outer wing. Important to recognize that underwing pattern lacks sharp black line along trailing edge of outer wing and translucent flight-feathers of *S. paradisaea*, and uniform grey appearance of *S. repressa*. Bill dark vinaceous-red, as in *S. paradisaea*, or even darker; more than half as long as crown but, since it is relatively deeper and with more prominent gonys than in congeners, it often appears noticeably shorter; shorter and more slender in ♀. Legs red. ADULT NON-BREEDING. Plumage loses general greyness of summer. Forehead largely white, rest of crown speckled black and white; nape black with some white flecks, leaving only panel behind eye as solidly black. Hindneck grey, but wears white. Rest of upperparts and upperwing ash-grey, wearing paler, particularly in centre of upperwing; lacks uniform silvery tone of *C. leucopterus*. Outer primaries show grey lines, these later wearing to dusky panel along edge of outer wing; secondaries show grey trailing edge. Underwing and underparts uniformly white. Tail-feathers show white edges when fresh. Bill and legs darker. JUVENILE. Head and underparts as winter adult, though forehead paler; spot before and panel behind eye blacker and face washed buff when fresh. Hindneck feathers tipped dark grey. Whole of upperparts basically grey, with black-brown tips on mantle, black-brown and buff tips particularly obvious on scapulars and tertials (which appear barred), and black-brown spots on tail-feathers; all markings more diffuse and sparser than on congeners. Upperwing as winter adult, but with larger coverts spotted brown and tips of flight-feathers coloured darker grey. Bill black. FIRST WINTER. Resembles winter adult except for retention of juvenile wing pattern and outer tail-feathers, latter appearing dark-tipped. On some, central wing-coverts fade and contrast quite strongly with lesser coverts and worn dusky flight-feathers (though pattern hardly approaches that of '*portlandica*'-type plumage pattern in *Sterna* terns). FIRST SUMMER. Apparently as summer adult but at least some appear to assume paler underparts.

Summer adult subject to confusion with *S. paradisaea* and *S. repressa* due to similarity of plumage pattern, but *C. hybridus* smaller, less attenuated, shorter tailed (with depth of fork hardly more than 2 cm, compared to minimum of 9 cm in *S. paradisaea*), and slightly blunter winged. Thus character of *C. hybridus* much closer to that of congeners than to *Sterna* tern. Differences in character, flight, and behaviour even more important for distinction of *C. hybridus* at other ages from *Sterna*, since plumage patterns overlap with those of several species and differ certainly only in details of wing pattern and rump and tail colours. *C. hybridus* in 1st-winter and winter adult plumage long considered indistinguishable from *C. leucopterus* but risk of confusion now reduced: *C. hybridus* clearly indicated by (1) longer, more dagger-like bill in ♂♂, (2) more scattered head pattern, (3) lack of obvious collar, and (4) greyer rump. Flight variable: at times much as *C. niger* or *C. leucopterus*, though slower in most actions; at others, more like medium-sized *Sterna*, with rather more regular and deeper wing-beats and more direct track than congeners. Shares ability with congeners to hawk insects, and with *Sterna* to plunge for fish. Gait and carriage as congeners, but stands higher than both.

Calls harsher than *C. niger* or *C. leucopterus*, particularly when breeding. Advertising-call a rasping 'kyick' or 'cherk'; alarm-call a loud 'kerch', often repeated, and more rasping than advertising-call.

Habitat. Breeds in west Palearctic in middle latitudes, in temperate and, especially, Mediterranean and steppe zones, from continental interior to oceanic coastal regions. Generally in lowlands, but formerly by lakes in Armeniya up to 2000 m. More demanding climatically than Black Tern *C. niger* or White-winged Black Tern *C. leucopterus* requiring at least 20°C mean July temperature for breeding (Voous 1960). In USSR, by standing or flowing waters in steppe and desert zones, amid reeds *Phragmites* or water-lilies *Nymphaea*, preferably by lakes with clear water, but sometimes on marshes, making floating nests (Dementiev and Gladkov 1951c). Depth of water 15–150 cm, commonly 60–80 cm (Bourke 1956; Borodulina 1960; Swift 1960b; Kapocsy 1979); may prefer more open water than *C. niger* (Swift 1960b). In Camargue (France), breeding some 50 m from edge of freshwater marsh almost completely overgrown with such aquatic plants as bulrush *Scirpus maritimus* and water milfoil *Myriophyllum*, nests being on clumps of vegetation or floating; while reeds torn up by horses or bulls used as material such a supply not essential. There, forages mainly in ricefields, preferably at early stage of growth, and in shallow parts of freshwater marshes, less than 30 cm deep. In Dombes region (France), 79% of 84 birds fed in marshes, 18% in newly sprouted cornfields, and 39% in flight for insects (Isenmann 1976a). In Hungary, nests on water with light covering of vegetation amid open areas (Kapocsy 1979). Habitat similar in marismas of Guadalquivir, Spain (Valverde 1958). In Hungary, considered readier than *C. leucopterus* to adapt to nesting on such artefacts as fish-ponds, especially where sedge *Carex* has been mown, but unwilling to breed close to colonies of other Sternidae (Bannerman 1962). Breeds to c. 1500 m in Kashmir (India) and in Madagascar up to 1280 m (Milon 1947). Resorts to flooded paddyfields, inland jheels, and marshes, as well as coastal lagoons, tidal mudflats, and estuaries, presumably after breeding (Ali and Ripley 1969b). In Zambia, found on rivers, flood plains, lagoons, or pans (Benson *et al.* 1971).

Apart from high-flight display and migration, keeps mainly to lower airspace, and poorly adapted for swimming, walking, and perching. Declined in some areas due to drainage (Mees 1979b), but in others has benefited from changes in habitat, e.g. fish-ponds.

Distribution. Breeds sporadically outside main range: see map in Mees (1979b).

BELGIUM. Bred 1950 and 1957 (Lippens and Wille 1972; PD). NETHERLANDS. Bred in small numbers 1938, 1945, and 1965 (CSR). WEST GERMANY. Bred 1854, 1862, and 1931 (Mees 1979). POLAND. Bred in 1850s, 1968, 1970,

PLATE 1 (*facing*).
Gelochelidon nilotica Gull-billed Tern (p. 6): **1–2** ad breeding, **3** 1st imm non-breeding (1st winter).
Sterna caspia Caspian Tern (p. 17): **4–5** ad breeding, **6** 1st imm non-breeding (1st winter).
Sterna maxima Royal Tern (p. 27): **7–8** ad breeding, **9** ad non-breeding, **10** 1st imm non-breeding (1st winter).
Sterna bergii velox Swift Tern (p. 36): **11–12** ad breeding, **13** 1st imm non-breeding (1st summer, inner primaries fresh but other plumage worn), **14** 1st imm non-breeding (1st winter).
Sterna bengalensis Lesser Crested Tern (p. 42): **15–16** ad breeding, **17** 1st imm non-breeding (1st winter).
Sterna sandvicensis Sandwich Tern (p. 48): **18–19** ad breeding, **20** 1st imm non-breeding (1st winter). (NWC)

PLATE 2. *Gelochelidon nilotica* Gull-billed Tern (p. 6): **1** ad breeding, **2** ad non-breeding, **3** 2nd imm breeding (2nd summer), **4** 1st imm breeding (1st spring), **5** juv, **6** downy young.   *Sterna sandvicensis* Sandwich Tern (p. 48): **7** ad breeding, **8** ad non-breeding, **9** 1st imm non-breeding (1st spring), **10** 1st imm non-breeding (1st winter), **11** juv, **12** downy young. (NWC)

PLATE 3. *Sterna caspia* Caspian Tern (p. 17): **1** ad breeding, **2** ad non-breeding, **3** 2nd imm breeding (2nd summer), **4** 1st imm non-breeding (1st winter), **5** juv, **6** downy young.   *Sterna maxima* Royal Tern (p. 27): **7** ad breeding, **8** ad non-breeding, **9** 2nd imm breeding (2nd spring), **10** 2nd imm non-breeding (2nd winter), **11** juv, **12** downy young. (NWC)

PLATE 4. *Sterna bergii velox* Swift Tern (p. 36): **1** ad breeding, **2** ad non-breeding, **3** 1st imm non-breeding (1st summer), **4** 1st imm non-breeding (1st winter), **5** juv, **6** downy young. *Sterna bengalensis* Lesser Crested Tern (p. 42): **7** ad breeding, **8** ad non-breeding, **9** 1st imm non-breeding (1st summer), **10** 1st imm non-breeding (1st winter), **11** juv, **12** downy young. (NWC)

PLATE 5. *Sterna dougallii* Roseate Tern (p. 62): **1** ad breeding, **2** ad non-breeding, **3** downy young. *Sterna hirundo* Common Tern (p. 71): **4** ad breeding, **5** ad non-breeding, **6** downy young. *Sterna paradisaea* Arctic Tern (p. 87): **7** ad breeding, **8** ad non-breeding, **9** downy young. *Sterna forsteri* Forster's Tern (p. 102): **10** ad breeding, **11** ad non-breeding. *Sterna repressa* White-cheeked Tern (p. 105): **12** ad breeding, **13** ad non-breeding, **14** downy young. (NWC)

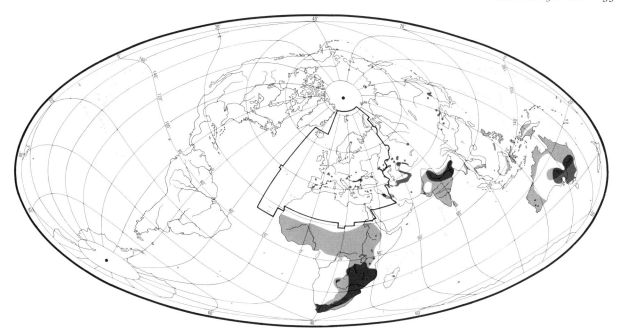

and 1972 (Tomiałojć 1976). CZECHOSLOVAKIA. Bred 1882, 1959, 1968, and 1971 (Hudec and Černý 1977); recently a few pairs have bred almost regularly in eastern Slovakia (Mees 1979b). AUSTRIA. Bred Hansag until c. 1890, and possibly Leitha marshes 1951 (HS). USSR. Bred Lithuania 1959 and 1978 and Ukraine 1942 and 1974 (Mees 1979b; HV). PORTUGAL. Bred Golega 1931 (Coverley 1932); probably nested there until 1940s (MDE); no recent evidence (NGO, RR). TURKEY. Formerly bred Amik Gölü (Mees 1977b). IRAQ. Breeding in south doubtful (Mees 1977b). SYRIA. Said to have bred, but no proof (Kumerloeve 1968b). LEBANON. Possibly breeding (Tohmé and Neuschwander 1974). TUNISIA. Said to nest in small numbers in north and centre (Heim de Balsac and Mayaud 1962) but evidence slight (Mees 1977b); only proof in 1977, when c. 60 pairs in 2 colonies (M A Czajkowski). ALGERIA. Formerly bred in some numbers (Heim de Balsac and Mayaud 1962; Mees 1977b); recently bred Lake Oubeira (EDHJ). MOROCCO. Formerly bred Ras-el-Douara and Sidi ben Mansour; now drained (Heim de Balsac and Mayaud 1962; Mees 1977b). Bred Iriki 1965–6 (Robin 1968).

Accidental. Britain, Ireland, Belgium, Luxembourg, Netherlands (almost annual), West Germany, Denmark, Norway, East Germany, Poland, Malta, Portugal, Madeira.

Population. Fluctuating numbers and erratic breeding make estimations difficult; population in Europe thought to be under 25000 pairs (Mees 1979b), but likely to be much higher if recent estimates for Spain substantiated.

FRANCE. Marked fluctuations from under 1000 pairs to c. 1500 pairs (Yeatman 1976; Mees 1979b). SPAIN. Estimated 3600 pairs, decreased (Mees 1979b); however, thought to be 50000 birds in marismas of Guadalquivir (Valverde 1960) and probably now over 100000 pairs in Andalucia, with smaller numbers elsewhere (AN). HUNGARY. Fluctuating, 200–250 pairs; declined 1900–46, increased since (Kapocsy 1979; Mees 1979b). ITALY. 100–150 pairs, perhaps other colonies elsewhere (Mees 1977b, 1979b); 200–250 pairs (Spina 1982). YUGOSLAVIA. At least 2000 pairs (Mees 1979b). BULGARIA. At most 15–20 pairs (JLR). RUMANIA. Total population unknown, perhaps 4000 pairs (Mees 1979b). GREECE. Nearly 150 pairs (Mees 1979b); c. 300 pairs (B Hallmann). USSR. Estimated 6000 pairs, but perhaps more (Mees 1979b). TURKEY. Under 1000 pairs (OSME). IRAQ. Common (Allouse 1953); no recent information.

Movements. Migratory. The following based largely on study by Mees (1977b, 1979b). Nominate *hybridus* from south-west Europe winters mostly in tropical West Africa, probably south to northern Zaïre. East European birds probably winter in Iran, Pakistan, Sudan, and Ethiopia, south to Kenya and possibly beyond. Populations from Caspian and Turkestan winter mainly Iran and Pakistan,

PLATE 6 (*facing*).
Sterna dougallii Roseate Tern (p. 62): **1–2** ad breeding, **3–4** 1st imm non-breeding with juv lesser coverts (1st winter).
Sterna hirundo Common Tern (p. 71): **5–6** ad breeding, **7** 1st imm non-breeding (1st summer, inner primaries fresh), **8–9** 1st imm non-breeding (1st winter).
Sterna paradisaea Arctic Tern (p. 87): **10–11** ad breeding, **12** 1st imm non-breeding (1st summer), **13–14** 1st imm non-breeding (1st winter). (NWC)

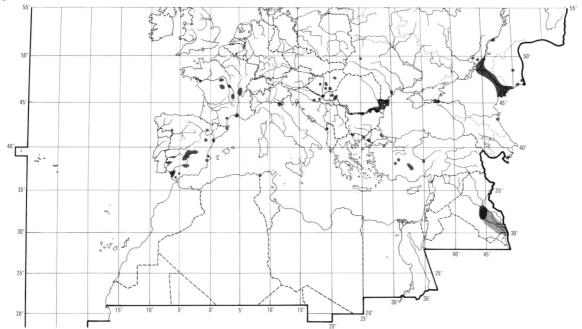

probably also reaching East Africa, India, and Ceylon (see Mees 1979b). For times of passage, arrival, and departure in Aral–Caspian–Turkestan region, see Dementiev and Gladkov (1951c), Borodulina (1960), and Feeny et al. (1968). Birds from China and eastern Asia winter south to Ceylon, south-east Asia, Greater Sunda Islands, and Philippines. Australian birds move north to winter New Guinea, Moluccas, Celebes, Java, and Borneo. C. h. delalandii of Afrotropical region strongly nomadic, but movements not well known (see Britton and Brown 1974).

In west Palearctic, spring migration March–May, mostly late April, autumn migration late July to early October (for later sightings, see Mees 1977b). Ringing shows post-breeding movement of west European birds to be initially south-west, and predominantly inland along major river courses and associated wetlands, e.g. birds ringed France recovered in autumn at Ebro delta and upper Guadalquivir (Spain). Winters regularly (in small numbers) in Mediterranean, especially Spain (Erard and Vielliard 1966), Algeria (Jacob 1979), and Tunisia (Isenmann 1972d); also recorded from Camargue (France). Though most west European birds winter in West Africa, notably Ghana (according to ringing recoveries), migration route not clear. Record of at least 500 birds near coast, Sénégal, December–January (Smet and Gompel 1980), and sightings along Atlantic seaboard in autumn and spring indicate at least some coastal movement. Passage through Straits of Gibraltar occurs August and April (Irby 1875), and 2 recoveries of French birds in Morocco, west of Gibraltar. Single recovery (April) on Sénégal coast (where appreciable numbers winter) of bird ringed Camargue (Morel and Roux 1966). Wintering in Niger inundation zone and Chad suggests trans-Saharan migration (Duhart and Descamps 1963; Mees 1977b). Arrives central Chad from about mid-August, and numerous by early September (Newby 1979). Trans-Saharan spring migration indicated by large numbers in south-east Morocco, mid-April (Smith 1968b). Spring passage through Coto Doñana (Spain) April–May.

In south-east of west Palearctic, a few winter on Rumanian coast (Lebret and Ouweneel 1976; Mees 1977b), occasionally in Turkey (Beaman et al. 1975), and regularly in Iraq (Moore and Boswell 1946); also in Iran (L Cornwallis). Winters in large numbers in Nile delta region: in 1978–9 and 1979–80, over 25000 birds on Lakes Manzala and Burullus (P L Meininger and W C Mullié), and 6500 along 25 km of coast, Matarîya–Port Said, February (see Mees 1979b). Such numbers compatible with possible wintering by large proportion of population from eastern Europe and Turkey (Mees 1979b). Large numbers, thought to be of Balkan origin (Mees 1979b), also pass through Nile valley in spring (April–May) and autumn (August to early October) (Meinertzhagen 1930, 1954). Heavy passage recorded Khartoum, late April (Dementiev and Gladkov 1951c).

Birds probably remain in winter quarters in 1st summer, returning to breed at 2 years old (Mees 1977b). Long-distance inter-colony exchanges occur, presumably more so in first-time breeders: e.g. bird hatched at Alicante (Spain) recovered 2 years later in May in Hungary (Schmidt 1978). Recorded once in Barbados, West Indies (Bond 1971). EKD

Food. Mainly insects and their larvae, and small fish, and amphibians. Forages over wetlands, and, on occasion, over dry areas. Sometimes singly but usually in small flocks, and at times with other species: e.g. in breeding area, Hungary, with Black-headed Gulls *Larus ridibundus* (Máté

1962), and on migration and in winter quarters often in large mixed flocks with other Sternidae, e.g. in East Africa (Britton 1977) and Zaïre (Ruwet 1964). In breeding season, Forez (France), flocks of up to 18 fly line-abreast up and down fields, hunting for beetles (Coleoptera) (E K Dunn). Normally uses slow, low (c. 2–4 m) searching flight into wind, periodically descending to take prey then turning to fly high and fast downwind before turning again to repeat (Ticehurst 1924a; Swift 1960b; Britton 1977; L H Brown, E K Dunn). 4 techniques. (1) Surface-plunging (most common). Usually dives from c. 2–3 m, often preceded by hovering (Schifferli 1955; Borodulina 1960; Swift 1960b; Little 1970; Kapocsy 1979; E K Dunn). (2) Dipping-to-surface. Takes prey from water (non-contact and contact) or vegetation (Ticehurst 1924a; Schifferli 1955; Bannerman 1962; Craig 1974); in breeding season, France, 0·23 dips per s (E K Dunn). (3) Drops vertically (not head first) into grassy fields for beetles sometimes hovering briefly beforehand; during period of courtship-feeding this is the most frequent feeding method, with maximum activity usually during 08.30–11.30 hrs and 16.00–19.00 hrs (E K Dunn). (4) Aerial-pursuit of flying insects (Dementiev and Gladkov 1951c; Herberigs 1958; Ferry and Dufour 1959; Crawford 1977)—usually caught low over marsh (c. 3 m), though in midday heat when insects rise, up to 15 m (E K Dunn). Prey sometimes washed by dipping into water whilst flying or whilst perched on edge of nest (Schifferli 1955; Swift 1960b). Maximum foraging distances from colonies c. 1 km (Little 1970), c. 2·5–4 km (Swift 1960b), c. 6 km (Tarboton et al. 1975), 9 km (Ferry and Dufour 1959).

Chiefly insects and their larvae; beetles, especially water-beetles (including *Dytiscus* and *Cybister*), dragonflies (including *Libellula depressa*, *Crocothemis erythraea*, *Agrion*, *Diplax*), water-bugs (*Notonecta glauca*), flies *Stratiomys*, grasshoppers and crickets (Orthoptera), and ants (Formicidae); also small fish (including carp *Cyprinus carpio*, bleak *Alburnus alburnus*, roach *Rutilus rutilus*, sun perch *Lepomis gibbosus*) and amphibians (including newts *Triturus*, frogs *Rana esculenta*, *R. ridibunda*) and their tadpoles; occasionally shrimps *Gammarus*, spiders (Araneae), and worms (Annelida) (Witherby et al. 1941; Dementiev and Gladkov 1951c; Schifferli 1955; Ferry and Dufour 1959; Borodulina 1960; Swift 1960b; Keve 1962a; Papadopol 1963; Szabó 1965; Isenmann 1976a; Kapocsy 1979; E K Dunn).

In Camargue (France), observations on 7 nests indicated birds fed mainly in ricefields and marshes; in 253 feeding flights, 07.00–11.45 hrs on 2 days, July, birds took 37 frogs, 12 tadpoles, 9 large larvae, 1 large grasshopper, 21 dragonflies, 6 flying insects, 159 small prey (possibly crustaceans), and 7 unidentified items (Schifferli 1955). Diet in same area given as chiefly larvae and adults of water insects, frogs, and tadpoles, and small carp (Swift 1960b). Food brought by adults (for mates and young?), Côte d'Or (France), mainly fish (including bleak and sun perch); also insects caught in flight (Ferry and Dufour 1959). Main food at French colony, small black beetles c. 2 cm long (c. 95% or more of items brought to mates at nest during laying); also a few fish (c. 5 cm long) and frogs, especially *R. esculenta* (E K Dunn). In Hungary, nesting birds, Hortobagy, fed chiefly on larvae of *Dytiscus* (Szabó 1965); on Lake Balaton, May, regularly caught larvae of dragonflies and tiny fish (Keve 1962a). At Prundu (Rumania), 5 stomachs, June, contained 4 tadpoles (in 2), 5 dragonflies (2), 14 larvae of *Stratiomys* (1), 3 larvae of *Dytiscus*, and much chitin (Papadopol 1963). On USSR steppes, chiefly insects (mainly beetles, grasshoppers, crickets, ants) and also spiders. Frogs and tadpoles important near fisheries on Volga delta, in places occurring in 92·7% of stomachs, and forming 68·3% of total food weight, with fish 20·4% by weight (Borodulina 1960). In Ghana, February, took small fish (Pomadasyidae) (E K Dunn). In Zaïre, August–November, fished for fry of cichlid fish: in June–July, hunted insects over marsh vegetation (Ruwet 1964). 4 stomachs, Darwin area (Australia), from birds feeding over fallow rice-paddy bays, December, contained larvae of water-beetles, weevils (Curculionidae), ants, earwigs (Dermaptera), lepidopteran larvae, spiders, crustaceans, and a frog; 1 stomach, November, contained over 100 winged ants; in September, bird seen to catch dragonfly in flight (Crawford 1977). In south-west Australia, September, fed with Gull-billed Terns *Gelochelidon nilotica* on mice *Mus musculus* present in plague proportions (Hobbs 1976). In Cape Province (South Africa), seen to pluck arum frogs *Hyperolius horstocki* from within flowers (Craig 1974).

In Camargue: stomach of chick c. 1–2 days old contained a small dragonfly adult and larvae, and pieces of waterplant; chick c. 5 days old contained part of frog *R. ridibunda* and 1 water-beetle larva; chick c. 5 days old contained 2 larvae of *Crocothemis erythraea*; 3-week-old contained beetle larve (possibly *Cybister lateralimarginalis*) and an amphibian bone (Schifferli 1955). In South Africa, food brought to young included small frogs and fish; to very young, insects and slime-worms (Milon 1947). In Australia, food brought to young included dragonfly nymphs, crustaceans (*Lepidurus*, crayfish), frogs and tadpoles, small fish, lepidopteran larvae (Noctuidae), and plague locusts (large number of pest insects taken over dry land); in 1 hr, 11 feeding trips made to 4-day-old brood of 3 (Bourke 1956). EKD

Social pattern and behaviour.
1. Mostly gregarious throughout the year; flocks of 50–100 commonly encountered in winter, Java (Hoogerwerf and Rengers Hora Siccama 1937). In tropical and sub-tropical latitudes, nomadic flocks, e.g. of 60 birds, may converge in thousands where suitable conditions obtain, e.g. seasonal flooding (Bourke 1956; Mees 1977b). Typically hunts for fish in groups of 2–3 (Ruwet 1964). Initial post-breeding flocks contain family parties, but all-juvenile flocks also occur (Smith and O'Connor 1955; Bourke 1956). On migration, travels alone or, more commonly,

in parties of up to 70 (Feeny et al. 1968). Both during and outside breeding season, often feeds with other Sternidae and gulls (Laridae), notably Black Tern *C. niger*, White-winged Black Tern *C. leucopterus*, and, less often, Common Tern *S. hirundo* (Moore and Boswell 1956a; Herberigs 1958; Máté 1962; Béldi 1963; Ruwet 1964; Halle 1968). BONDS. Pair-bond probably monogamous, persisting from year to year, though no evidence. Age at which bonds form and breeding begins unknown. Pair share nest-duties and care for young until shortly after fledging (Donahue and Ganguli 1965), though ♀ sometimes monopolizes incubation (Svoboda 1968) and brooding (Bourke 1956). Brooding parent sometimes fed by mate (Bourke 1956). BREEDING DISPERSION. Colonial, but without dense nesting of many ground-nesting Sternidae. Weak fidelity between years to specific colony area, and even less to nest-site, especially in tropics where species an opportunist breeder with extended season (Mees 1977b). Colonies in temperate latitudes usually of not more than 150 pairs, frequently less than 50. Rarely, Europe, several hundred pairs (Mountfort and Ferguson-Lees 1961; Kapocsy 1979). Typical colony, Hungary, 10–16 pairs (see Bannerman 1962); mean 13 pairs (17 colonies in natural waterbodies), 26 pairs (9 colonies in artificial fish-ponds) (Kapocsy 1979). Colonies especially small near edge of range: e.g. 2–11 pairs, Netherlands and Belgium (Mees 1977b); 11 pairs, Lithuania, USSR (Ivanauskas 1961). Extralimital colony size equally variable, but breeding concentrations, especially of nomadic populations, sometimes large; 1000 or more pairs, southern Asia (Voous 1960); up to 2000 pairs, eastern Australia (Bourke 1956). Colonies of more than a few pairs usually divided into sub-colonies. Groups exhibit highly synchronized laying (Swift 1960b). In colony, Rhone delta, first nests closely adjacent; subsequent ones in concentric rings till colony had c. 180 m radius (no sub-colonial structure mentioned) (Yeates 1948). Distance between nests variable, 1–50 m. In colony, Forez (France), mean of 18 nests 2·63 m (1·52–8·23) (E K Dunn). In Hungary, mean of 2 sub-colonies 2·5 and 8 m (Radetzky 1962), but other colonies 10–50 m (Nadler 1967; Bod and Molnár 1976). In Camargue (France), mean 2·2 m (11 nests), minimum 1 m, below which nests apparently suffer reduced breeding success from constant squabbling between neighbouring adults (Swift 1960b). Poorer visibility of neighbours in denser cover may thus allow closer spacing; of 2 colonies of 8 pairs each, South Africa, nests 10 m apart in completely grassed-over 'pan', compared with 20 m in nearby partly grassed-over pan (Tarboton et al. 1975). See Kapocsy (1979) for other measurements, Hungary; extralimital measurements in Bourke (1956), Fuggles-Couchman (1962), and Steyn (1966). Small nest-area territory defended; used for ground phases of courtship and pair-formation, nesting, and resting of off-duty bird. Colonies often established near, but generally discrete from those of *C. niger*, *C. leucopterus*, and Black-headed Gull *Larus ridibundus* (Radetzky 1962; Nadler 1967; Sóvágó 1968). *S. hirundo* occasionally nests in small numbers among *C. hybridus* (see Bannerman 1962). Commonly nests in close association with grebes (Podicipedidae)—minimum distance 1–1·5 m, Lake Csaj (Hungary); mixed clutches sometimes occur (Szlivka 1965a; Bod and Molnár 1976). In some colonies, Hungary, Little Grebe *Tachybaptus ruficollis*, Great Crested Grebe *Podiceps cristatus*, Black-necked Grebe *P. nigricollis*, and Red-necked Grebe *P. grisegena* all nest close to *C. hybridus* (Csornai et al. 1958; Bod and Molnár 1976). Evidence that breeding *C. hybridus* may actually attract and encourage *P. nigricollis* to nest in greater numbers than otherwise (Bod and Molnár 1976). ROOSTING. In breeding season, may use both temporary and more permanent collective sites—sometimes outside colony area, up to 2 km away—for preening, sleeping, and bathing (Swift 1960b). Flock of c. 200 perched (loafing) at up

A

to 18 m on branches of dead trees emergent from lake (Bourke 1956). Off-duty birds most often rest at edge of nest platform, alongside incubating or brooding mate, or on vantage point such as branch or pole up to 20 m from nest (Steyn 1960; J-D Lebreton). ♂ roosted at night on nest-platform with ♀, or in dense reedbed, small dead trees (not green ones), or bank of nearby dam, never on water (Bourke 1956). In winter quarters, loafs by day and roosts at night, often with other Sternidae, on flats, saltpans, and other open areas (E K Dunn). Large flocks of juveniles roost at suitable sites, often 1–2 km from nesting area (Smith and O'Connor 1955; Bourke 1956).

2. FLOCK BEHAVIOUR. Arrival on spring migration sometimes involves sudden appearance of flock: group of c. 60 'tumbled down' from great height in sky towards flood-waters, and flew up and down over surface in compact flock (Bourke 1956). In colony, first bird to detect danger gives Alarm-call (see 1 in Voice), and often raises wings momentarily before taking off. Alerted neighbours then perform 'upflight' giving Alarm-calls, or, especially during early incubation, 'dread' in which birds fly low and silently for some distance, swerving from side to side; shortly start calling again (Advertising-call: see 5 in Voice), disperse and return to nests (Swift 1960b). In colony adjoining one of *L. ridibundus*, however, no 'dreads' seen, and upflights occurred only when gulls performed one first, *C. hybridus* lagging momentarily behind. Small compact flock hovers low, usually without calling, over disturbance which poses no threat, such as injured conspecific (E K Dunn). For flock reaction to predators, see Parental Anti-predator Strategies (below). Individuals foraging in flock periodically give Advertising-call (E K Dunn). ANTAGONISTIC BEHAVIOUR. Well developed, notably in breeding season. The following adapted mostly from Swift (1960b). (1) Ground encounters (usually only 1 of contestants on ground). Bird on nest threatens conspecific intruder, flying up to 5 m away, by facing it head-on, either sitting down or standing in Aggressive-upright posture (Fig A); neck stretched upwards and forwards (sometimes retracted), with bill at any angle from just below horizontal to c. 50° above; carpal joints held slightly forwards and out from body; feathers of crown, cheeks, nape, and breast, and occasionally wing-coverts ruffled. Ground Threat-call (see 4 in Voice) may be uttered, or bird threatens with silent gape, showing crimson lining. Bickering between neighbours intensifies after hatching. Wandering chicks over c. 3 days recognized as strangers, driven off with blows to head, and occasionally killed (Schifferli 1955). (2) Aerial encounters. Chasing common and may develop into fight. Flying bird

B

threatening other in the air stretches neck and head upwards and gives Aerial Threat-call (see 3 in Voice). Upward-flutter (see *S. hirundo*) rare, but one seen up to 30 m (E K Dunn). HETEROSEXUAL BEHAVIOUR. (1) General. Pair-formation involves both aerial and ground behaviour at colony. (2) Aerial courtship. Usually begins at colony, sometimes at feeding grounds nearby; also seen in winter quarters, Ghana, February–March (A J M Smith). Following description by E K Dunn: advertising ♂ flies low, usually over colony, with shallow wing-beats, most often without food in bill, and giving Advertising-call. If 1 or more birds attracted to follow, advertising ♂ usually leads them into High-flight. Leader rarely (only 7% of 121 occasions) carries food in High-flight. With pursuer(s) 15–50 m (typically *c*. 30 m) behind ♂, birds make steep jerk-flying ascent with rapid, deep wing-beats, periodically (alternately when 2 birds) giving Advertising-calls. Ascent broadly spiral, with participants—where only 2—often diametrically opposite as they circle. At peak of ascent (30–150 m), level out and often fly around horizontally with slower wing-beats. After a few moments, leader initiates downward glide with wings upraised in V; pursuer(s) follow suit. At very start of glide, less commonly thereafter, bird (not known if leader or pursuer) gives Kiririk-call (see 6 in Voice); unless ascent short, calling rare during rest of glide, though Begging-call (see 7 in Voice) occasionally given by pursuer. Pursuer rarely overtakes leader in glide or any other stage of High-flight. If ascent not high, e.g. *c*. 30 m, only 1 glide—sufficient to return to ground level—occurs, but descent from greater heights made in series of glides, up to 10, depending on starting height (median 3, each *c*. 10 s duration, sample 44 High-flights). Between each glide, leader, followed in like manner by pursuer(s), swoops up, levels out, or circles for 5–10 s with flapping flight, before initiating next glide. Direction of each glide usually different, so descent broadly zigzag or spiral. Occasionally, 2nd ascent starts at end of glide, whether at ground level or during descent. Final glide often ends near, or over colony, in which case leading ♂ typically makes hovering descent from 1–10 m, presumably over proposed nest-site. Pursuer may land nearby, but rarely on site. If no hovering descent, ♂ flies around colony with shallow wing-beats, often for *c*. 2 min, before going off to feed, or initiating new High-flight. 1–5 birds participate in High-flight (mean 2·5, sample 121). Where more than 2 birds, pursuers often lag behind and drop out on ascent; often, starting group of 4 splits up into 2 High-flights, each of 2 birds. In common variation, birds fly horizontally at peak of ascent for up to several km, and may not return to colony for 10–15 min; especially common when several birds involved. Where display confined to vicinity of colony, mean duration of complete High-flight 2 min 27 s (range 50 s to 5 min 20 s, sample 31). Frequency of High-flights varies with several factors. Follows diurnal pattern in laying period, with maximum 4·7 per hr from 07.00 to 08.00 hrs, declining to 0·5 per hr by 13.00–14.00 hrs, increasing thereafter to 2·8 per hr at 17.00 to 18.00 hrs; frequency higher in warm, sunny weather (E K Dunn). Frequency declines after laying, though High-flights occur throughout breeding season, with young non-breeding birds perhaps involved later on. Swift (1960b) observed almost entire colony of *c*. 300 birds taking part in High-flight as late as 11 July. Such mass participation not seen in 121 High-flights in May at colony in Forez (E K Dunn); large groups apparently much less common than in *C. niger*. Low-flight, described for *C. niger*, not well developed in *C. hybridus*. (3) Ground courtship (see also below). Most prevalent once pair-bond established, and usually involves only 2 birds. Begins, however, before colony occupation, e.g. on emergent posts on lake where no colony yet established (E K Dunn). Once nest-site established, and during laying, arrival of ♂, often with food and giving Advertising-call, initiates following sequence: on ♂'s aerial approach, ♀ commonly stretches up to face him, uttering Advertising-call in answer to his calls (see 5 in Voice); ♂ typically hovers briefly before alighting, and once on ground may keep wings raised for several seconds, more briefly if delivering food. Before or after transferring food, ♂, facing ♀, stretches neck upwards and forwards, with bill approximately horizontal (Forward-erect: Fig B, left) to (less commonly) *c*. 50° above horizontal (Erect-posture); carpal joints held clear of body. ♀ typically faces ♂ in Stoop-posture (Fig B, right), up-ending body so that bill almost touches nest-platform (Swift 1960b); this posture (by ♂ or ♀) often accompanied by Cooing-call (see 7 in Voice). Then both birds relax, or ♂ also Stoops (so-called 'mutual stoop'), so that both face each other, often apparently pointing at centre of nest-platform (E K Dunn). If sitting on nest, ♀ may start Stoop-posture and call while ♂ still hovering overhead. If both birds alight together, both adopt Forward-erect or Erect-posture, seldom face to face (Swift 1960b). After ground courtship outside colony area, departing ♂ flew off in Head-down posture with back arched and bill *c*. 45° below horizontal (E K Dunn). (4) Mating. Occurs mainly on nest, but also at pre-laying roost outside colony (Kapocsy 1979). ♀ in Stoop-posture often solicits copulation on newly completed nest (Steyn 1960). ♂ often adopts Forward-erect before mounting, with ♀ in Hunched-posture, as in other Sternidae. While copulating, ♂ gives version of Cooing-call (J-D Lebreton). (5) Courtship-feeding. At *c*. 100–200 m from nest, approaching ♂ flies with shallow wing-beats. At *c*. 10–20 m from nest, sets wings in V and glides, making steep, swift descent to nest-site (E K Dunn). Usually lands, but also recorded passing large prey on the wing, or dropping it on nest from the air. For postures on alighting, see above. After fish transferred, birds very excited, holding tails erect and rubbing bills (Herberigs 1959). Courtship-feeding frequent during laying, declining thereafter (E K Dunn). For frequency and duration of feeds, see Kapocsy (1979). (6) Nest-relief. Arriving bird gives Advertising-call in flight. Though J-D Lebreton found Stoop-posture rare after laying, Swift (1960b) recorded it (less intensely) during incubation when nest-relief sequence can be as follows. Relieving bird (of either sex) lands and stands parallel, or at right-angles, to bird in Forward-erect or Erect-posture; relieved bird does likewise; then both perform 'mutual stoop'; this probably in early incubation, and as incubation proceeds, relief generally occurs with much less ceremony (E K Dunn). Nest-material occasionally brought by relieving bird. Prior to departure, relieved bird often pecks at, or rearranges rim of nest, and may pick up and throw material back on to nest with sideways head-jerk (side-throwing). One incubating bird made excursions to edge of nest to paddle in water before returning to eggs—apparently deliberate egg-wetting behaviour (Tarboton *et al.* 1975). RELATIONS WITHIN FAMILY GROUP. As soon as dry, chicks may walk to water's edge to drink (Smith and O'Connor 1955). Chick first fed 80–125 min after hatching (3 chicks watched) (Bourke 1956). Chick elicits parental feeding response by pecking at bill and giving food-call. Chicks recognize parents, probably by Advertising-call, at less than 4 days (Bourke 1956); parents recognize offspring from 4–5 days (Schifferli 1955; Swift 1960b). According to Bourke (1956), only ♂ feeds young for first 2 days, alighting on nest platform; thereafter, increasing amounts delivered by hovering in order to pass food to young (Steyn 1960) or by dropping it to them on nest-platform. Bourke (1950) found all food dropped to chicks 6–8 days old. Large items may be washed by adults dipping to water surface, or when perched at edge of nest (Steyn 1960). Chicks can swim to surrounding vegetation from nest when a few hours old (see Anti-predator Responses of Young, below); may be fed away from nest at 48 hrs

(Smith and O'Connor 1955), though return to nest in response to Cooing-call (see 7 in Voice), Stoop-posture, or both (Bourke 1956). Young leave nest at 4–10 days (Svoboda 1968); no longer return to nest after c. 1 week when parents may build separate feeding platform (Schifferli 1955; Bourke 1956); otherwise, move around colony until fledging (Swift 1960b). When fledged, young usually retire to mudbank or similar where both parents continue to feed them, alighting to pass items directly (Little 1970); also fed on water surface (Bourke 1956). Fledglings beg persistently; one seen calling and stamping feet when parent approached with fish (Donahue and Ganguli 1965). Fledglings develop hovering skills by retrieving inanimate objects from water (Little 1970). Juvenile may follow one parent, occasionally alighting in water to be fed, but in little more than 1 week after fledging may fly several km to hunt for itself (Swift 1960b). Shortly after fledging, juveniles seen without parents, but mixed with *C. niger* (Svoboda 1968). ANTI-PREDATOR RESPONSES OF YOUNG. On hearing Alarm-call of parent, chick seeks refuge from danger by swimming from nest to hide in vegetation, usually c. 3–5 m away, within 1–2 hrs of hatching (Smith and O'Connor 1955; Bourke 1956; Steyn 1960, 1966; Fuggles-Couchman 1962; Tarboton *et al.* 1975). At 3 days, young capable of swimming up to 20 m from nest (Schifferli 1955). Loafing young near fledging age, South Africa, made threatening lunges at Crested Coot *Fulica cristata* (Little 1970). PARENTAL ANTI-PREDATOR STRATEGIES. (1) Passive measures. At approach of danger, adults give Alarm-call (see above for response of young). (2) Active measures: against birds. Especially in single-species colonies, bold in aerial defence of nest and young. When predator approaches edge of colony, first bird to detect it gives Alarm-call, eliciting collective upflight (see Flock Behaviour, above); birds mob intruder in tight flock, sometimes circling round it in so-called 'whirl-flight' (Swift 1960b); this response seen to Black Kite *Milvus migrans*, Herring Gull *L. argentatus* (once), and Marsh Harrier *Circus aeruginosus* (main predator in southern Europe), though usually ineffective in deterring latter (Swift 1960b). ♂♂, Hungary, strongly attacked *S. hirundo*, *L. ridibundus*, *C. aeruginosus*, buzzards *Buteo*, and herons *Ardea*; incubating ♀♀ joined ♂♂ in mobbing only when danger extreme (Svoboda 1968). At mixed colony, Forez, breeding *L. ridibundus* attacked in the air and successfully repelled, but challenged only when it came very close to nest (E K Dunn). (3) Active measures: against man. Mobs vigorously. Individuals swoop at head of intruder from flock wheeling overhead; at low point of swoop, may utter Anger-call (see 2 in Voice), and sometimes defecates (Tarboton *et al.* 1975); rarely strikes, but may draw blood (Steyn 1960). Pugnacity greater around hatching time, and apparently greater in colonies where not associated with breeding *L. ridibundus* (E K Dunn).

(Figs A–B from photographs by E K Dunn.) EKD

Voice. Highly vocal, especially at breeding grounds. Calls generally louder, more croaking and cawing, and longer than those of congeners (Nadler 1967).

CALLS OF ADULTS. (1) Alarm-call. A hoarse 'kerch' or 'kechk' (Witherby *et al.* 1941: Fig I, left), like rusty nail being drawn from plank (Steyn 1966); more emphatic, rattling 'krrrerch', 'krrrek', or 'kerrrk' on close approach, when bird starts mobbing intruder. Also heard in aerial pursuit of Black-headed Gull *Larus ridibundus* (E K Dunn). See Donahue and Ganguli (1965) and Kapocsy (1979) for further renderings. (2) Anger-call. A very hoarse 'kzek' delivered up to 3 times during swoops at intruders, especially at low point of swoop (E K Dunn). May be pre-

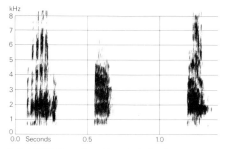

I P A D Hollom Israel (captive) April 1979

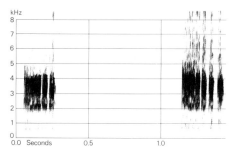

II J-C Roché (1966) Spain May 1965

ceded by shorter notes, thus: 'ke-ke-ke-KEEK' (Svoboda 1968). (3) Aerial Threat-call. A loud, high-pitched 'kwee', sometimes given by one flying bird to another (Swift 1960b); perhaps same as 'krreee' (see 6). (4) Ground Threat-call. A hard, sharp 'ku-ku' given by one bird on ground to another on ground or in the air; may also be given just before alighting on nest (Swift 1960b). (5) Advertising-call. Brief 'cherk' or 'kerk', or longer 'keerk' (E K Dunn); slightly rising 'kreek' (Donahue and Ganguli 1965); 'keerr', 'schkerr' (Kapocsy 1979). Given by both sexes when approaching nest, especially with food. Also heard during High-flight ascent; occasional 'kek' by pursuer in gliding phase of High-flight probably a related call (E K Dunn). (6) Kiririk-call. See Fig II. Sometimes heard at start of glide in High-flight (Swift 1960b; E K Dunn); sometimes more continuous 'krreee' (E K Dunn); possibly related to rattling version of call 1, and to call 3, since origins of High-flight lie probably in hostile pursuit (Cullen 1960a). (7) Cooing-call. A low, crooning, 'kura-kura-kura' or 'kura-kura-kiu' (Swift 1960b); 'kek-kek-kek' (E K Dunn); delivered in Stoop-posture, invites close approach of mate or offspring for feeding or brooding; low 'kuk-kuk-kuk' by ♂ during copulation probably a related call (E K Dunn).

CALLS OF YOUNG. Young highly vocal. Food-call of downy young a shrill wheezing (Steyn 1966); a loud squeaking (Swift 1960b). Svoboda (1968) described 'bsi bsi' inside egg near to hatching, a sharper 'bsit bsit' up to 3 days after hatching, and a louder, penetrating 'chee

chee' when parents approach with food (Donahue and Ganguli 1965; E K Dunn). EKD

Breeding. SEASON. East-central Europe: see diagram. South-west Europe and North Africa: up to 2 weeks earlier. Volga (USSR): breeding begins at end of May (Borodulina 1960). SITE. On floating vegetation, including water lily *Nymphaea alba*. Nest: raft of vegetation anchored to emergent or submerged plants; often cone-shaped, up to 75 cm across at base tapering to 20–30 cm at water-level; up to 12 cm above water with cup 2 cm deep (Bourke 1956; Fuggles-Couchman 1962; Radetzky 1962). On water-lily leaves only slight structure built (Borodulina 1960). Can also be resting on bottom (Radetzky 1962). Building: by both sexes, using tossing and building movements. EGGS. See Plate 90. Oval, smooth and slightly glossy; pale blue, buff, or grey, with black or brown spots and blotches. 39 × 28 mm (35–44 × 26–30), sample 168; calculated weight 16 g (Schönwetter 1967). Clutch: 2–3; occasional clutches of 4 or even 5 presumably by 2 ♀♀ (E K Dunn). 71% of clutches, Volga (USSR), had 2 eggs, 29% had 3, sample not given (Borodulina 1960). Mean of 81 nests, France, 2·74, with 4 of 4 eggs (E K Dunn). One brood. Replacements probable after egg loss. INCUBATION. 18–20 days. By both sexes; in approximately equal turns of *c*. 25 min (Swift 1960b), but 10–20 min, Tanzania (Fuggles-Couchman 1962). In Hungary, ♀ noted as incubating almost exclusively (Svoboda 1961). Begins fully with last egg; hatching approximately synchronous. YOUNG. Precocial and semi-nidifugous. Cared for and fed by both parents. Leave nest at 4–10 days old and seek refuges away from it, moving further if disturbed (Swift 1960b; Svoboda 1961); before 4 days old will leave nest temporarily if disturbed. FLEDGING TO MATURITY. Fledging period 23 days (Svoboda 1961). Age of independence not known, though young fed by parents for short time after fledging (Little 1970) but may migrate separately (Svoboda 1961). Age of first breeding not known. BREEDING SUCCESS. No data for west Palearctic. In Australia, estimated hatching success at one colony 66% (Smith and O'Connor 1955).

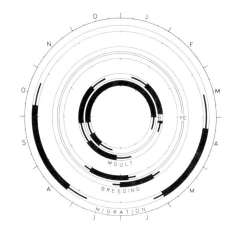

Plumages (nominate *hybridus*). ADULT BREEDING. Forehead, lores, crown down to eye, and hindneck black. Streak from chin and gape extending below eye to sides of neck white. Mantle and back medium grey; scapulars, tertials, rump, and upper tail-coverts slightly paler, light grey, whole upperparts appearing uniform grey. Upperparts distinctly paler than in Black Tern *C. niger*, slightly paler than White-winged Black Tern *C. leucopterus*, but darker than Common Tern *S. hirundo*. Chest medium grey like sides of neck and mantle, medium grey shading to light grey on throat and to dark grey or blackish-grey on breast and belly. Flanks and vent light grey, under tail-coverts white. Tail light grey, outer web of t6 white except for slightly greyish tip. Outer web of p1–p9 pale silvery-grey, outer web of p10 medium grey, slightly tinged silvery when fresh. Inner webs of primaries white, except for broad medium grey streak alongside shaft, medium grey tip (both slightly tinged silvery when fresh), and narrow dull black streak along outer edge of tip, latter forming narrow black rim along trailing edge of wing when seen from below. Worn primaries, especially oldest ones of 1st and 2nd moulting series (mainly p6–p7 and p9–p10, respectively) fade to dark grey or blackish with grey bloom hardly apparent. Fresh inner primaries narrowly bordered white. Secondaries light grey with slightly paler inner border and narrow white tips. Upper wing-coverts light grey like scapulars, slightly paler and more silvery towards wing-bend and on greater primaries coverts. Axillaries pale grey, under wing-coverts and leading edge of wing white. In fresh plumage, grey of upperparts slightly tinged bluish, light grey of throat contrasting with white streak below eye; in worn plumage, grey of underparts less deep and some white of feather-bases visible, throat white, and primaries less silvery, tips especially dull grey or blackish. Sexes similar, but underparts of ♀ sometimes slightly paler. ADULT NON-BREEDING. Forehead and lores white. Crown white, variably streaked black: in some, almost uniform white with a few dark shaft-streaks only, a few others black with rather narrow white feather-fringes, sometimes even forehead black. Small spot in front of eye and streak from behind eye over ear-coverts to across nape black, feathers narrowly edged white when fresh, especially on nape. Mantle, scapulars, tertials, and back to upper tail-coverts uniform pale grey, no white collar on hindneck (white feather-bases sometimes visible through wear, but this does not form clear-cut collar). Underparts white, often with narrow pale grey line extending down from upper mantle to upper sides of chest. Tail pale grey, outer web of t6 nearly white. Flight-feathers like adult breeding, contrast between older dark feathers and silvery-grey new feathers variable, depending on wear and moult. Upper wing-coverts pale grey, lesser coverts with slightly darker centres; no pronounced dark carpal bar, but indistinct one sometimes formed by worn dark grey lesser coverts. Leading edge of wing, axillaries, and under wing-coverts white. Resembles *C. leucopterus* in usually having rather pale crown and black ear-coverts and nape-patch, but ear-coverts and nape connected by single black streak, while in both *C. niger* and *C. leucopterus* ear-coverts are separated from black central nape by white wedge from sides of neck to above ear-coverts; besides this, crown of *C. niger* black, contrasting sharply with white forehead. No white collar on hindneck, in contrast to *C. leucopterus*; in latter species, collar more marked—bordered behind by black band across mantle, which is lacking in *C. hybridus*; faint collar often present in adult non-breeding *C. niger*, more distinctly in juvenile of latter. Upperparts and upper wing-coverts distinctly paler than in *C. niger*, slightly paler than *C. leucopterus*, and usually without distinct carpal bar which is present in many *C. leucopterus* and indicated in some *C. niger*. Narrow grey bar extending to sides of chest in *C. hybridus* much smaller than distinct patches of *C. niger*; *C. leucopterus* lacks grey at sides of chest. Non-breeding *C. hybridus* also rather similar to non-breeding and juvenile *S. hirundo* and Arctic Tern *S. paradisaea*, but these have carpal bar distinct, upperparts paler grey, rump to tail paler grey or white with outer web of t6 darker than inner web instead of inner darker than outer, tail more deeply forked, and no dark on sides of chest. DOWNY YOUNG. Closely similar to both *C. niger* and *C. leucopterus*, but hardly any white round eye, less white on wing-tip, black spot on forehead extending to gape, and all underparts white except dark brown chest and buff flanks (Fjeldså 1977). JUVENILE. Like adult non-breeding, but forehead and lores tinged buff when fresh, crown mainly black, feathers of crown and nape narrowly fringed buff to off-white when fresh. No pale collar on hindneck. Feathers of mantle and back, scapulars, tertials, inner upper wing-coverts, and tail-feathers pale grey, with distinct sepia-black subterminal bar and narrow buff-white to off-white fringes, quite distinct from adult; feathers of rump, upper tail-coverts, and remainder of upper wing-coverts pale grey with ill-defined pale buff or off-white fringes, sometimes slightly darker grey subterminally. Underparts white; narrow light grey bar extending down from sides of mantle to sides of chest as in adult non-breeding, consisting of a few dark-tipped feathers only; as in adult, bar may become indistinct through bleaching and wear. Flight-feathers like adult, but slightly duller grey and without contrast caused by partial moult. In worn plumage, buff tinges fade to off-white, but sepia-black marks on upperparts and largely dark crown and hindneck still prominent. Differs from juveniles of *C. leucopterus* and *C. niger* by absence of white wedge between rear of ear-coverts and nape; absence of white collar on hindneck; heavily variegated pale grey and sepia-black mantle, scapulars, and back, rather than nearly uniform black-brown (*C. leucopterus*) or less contrastingly marked dark grey and deep brown (*C. niger*); absence of contrastingly white rump, unlike *C. leucopterus*; no carpal bar or a faint one only; narrow grey bar on sides of chest. FIRST IMMATURE NON-BREEDING. Like adult non-breeding, and separable with difficulty after loss of last juvenile feathers with characteristic dark subterminal marks by October–January; primary moult in post-juvenile starts *c.* 2 months later than in adult post-breeding, juveniles showing lower scores than adults at same time, but some start at same time, differing only by outer primaries being uniformly worn and coloured rather than with strong contrast between some neighbouring feathers as in adult. For differences from other species, see adult non-breeding. SUBSEQUENT PLUMAGES. Although hardly any breeding obtained at 1 year old in other *Chlidonias*, only a few *C. hybridus* in non-breeding plumage found among many summer specimens examined, in contrast to *C. niger* and *C. leucopterus*; thus perhaps at least some 1-year-olds assume adult breeding plumage (see also Moults).

Bare parts. ADULT. Iris dark brown or reddish-brown. Bill dark crimson or dark blood-red in breeding; black in non-breeding, tinged red-brown, especially on base. Foot crimson in breeding (paler than bill); dull carmine or dirty red with variable grey, brown, or black tinge in non-breeding. DOWNY YOUNG. Iris dark brown. Bill dark grey with deep vinaceous base. Foot greyish-pink (Fjeldså 1977). JUVENILE. Iris dark brown. Bill brown-red with black tip. Foot red-brown or dirty brownish-red. (Kapocsy 1979; RMNH, ZMA.)

Moults. More variable than in any other Palearctic Sternidae; this species apparently highly opportunistic, suspending moult sometimes for considerable time in winter quarters when conditions unfavourable, continuing with sometimes up to 5 primaries growing at same time when conditions good (in other Sternidae, rarely more than 3). In Sternidae generally, moult of arrested pre-breeding primary moult series not continued after breeding season, adult post-breeding starting with p1 irrespective of point at which older series arrested, but *C. hybridus* frequently continues with some feathers of arrested pre-breeding series, as well as starting again with p1. Of 63 adult specimens from winter quarters, 17% showed one actively moulting or arrested series, 32% 2 series, 41% 3 series, and 10% 4 series. The following a generalized survey of complex situation. ADULT POST-BREEDING. Complete, primaries descendant, inner primaries and central pairs of tail-feathers often moulted twice. Starts in June–July on breeding grounds; crown, mantle, scapulars, and lores first, soon followed by forehead, upper tail-coverts, underparts, and tail (sequence t1–t2–t3–t6–t4–t5). Primaries start with p1 between late July and late September, occasionally also a few outer feathers replaced. Leaves breeding grounds with moult arrested, up to 3–7 inner primaries new, and head, body, wing-coverts, and tail a mixture of breeding and non-breeding. Moult continued in winter quarters; scattered breeding feathers retained longest on crown, rump, scapulars, belly, and upper wing-coverts. First specimens in full non-breeding by mid-October, most by late November, but a few retain much breeding until December–January. 1st series of primaries completed between January and early March. 2nd primary moult series starts late October to early February; birds starting early do not necessarily reach high scores, as moult often temporarily suspended. Birds starting as late as January–February do not usually start a further (3rd) series in pre-breeding. ADULT PRE-BREEDING. Partial, mainly February–March. Involves head, body, wing-coverts, tail, and usually inner primaries. Starts with some wing-coverts, scapulars, and t1 from late January; some birds completely in breeding by mid-February, but usually not so until early April, when some still growing parts of mantle, scapulars, and belly. 3rd series of primaries starts between late December and early March, but not if 2nd series started late. At start of breeding season, 1st primary moult series completed, 2nd arrested with up to p2 (4% of 51 examined), p3 (2%), p4 (2%), p5 (10%), p6 (41%), p7 (29%), p8 (10%), or p9 (2%) new; 3rd series not started (34% of 46 examined) or arrested with up to p1 (29%), p2 (16%), p3 (9%), or p4 (12%) new. POST-JUVENILE. Complete. Starts with some feathers of mantle and scapulars, t1, and p1 soon after fledging, in early birds from late June or July, but more often October to early November; 1st immature non-breeding complete by (August–)November–January except for secondaries and outer primaries. 2nd primary moult series starts January–April, when 1st flight-feather moult series completing. Further data scanty, perhaps because advanced specimens inseparable from adults and, like them, start pre-breeding from February; a few immatures examined changed 1st immature non-breeding directly for 2nd, these perhaps representing birds starting post-juvenile late.

Measurements. Nominate *hybridus*. Europe and North Africa, May–August; skins (RMNH, ZMA). Tail is to tip of t1, fork is tip of t1 to tip of t6; latter includes data from some other populations (basically similar).

WING AD	♂ 242	(7.06; 8)	231–250	♀ 232	(3.45; 12)	228–238
TAIL (t1)	68.1	(3.07; 8)	66–72	64.7	(2.61; 11)	62–70
FORK AD	16.4	(2.22; 14)	14–21	16.7	(2.54; 25)	13–22
JUV	12.0	(—; 3)	11–13	11.6	(1.52; 5)	10–14
BILL AD	31.6	(1.09; 8)	30–33	28.5	(1.53; 11)	26–30
TARSUS	23.3	(0.58; 9)	23–24	22.6	(1.40; 11)	21–24
TOE	28.3	(1.10; 8)	27–30	27.4	(0.91; 9)	26–29

Sex differences significant for wing, tail, and bill. Adult non-breeding and breeding fork similar, combined. Juvenile wing on average 8 shorter than adult; t1, tarsus, and toe similar from fledging; juvenile bill not full-grown until c. 6 months old.

Adults from (1) India and Ceylon, and (2) eastern Asia, wintering in Greater Sunda Islands; both October–April, skins (RMNH, ZMA).

WING (1) ♂ 233 (— ; 3) 230–236 ♀ 226 (2·82; 10) 221–231
     (2) 226 (4·80; 4) 220–231 224 (6·21; 16) 215–238
BILL (1) 31·8 (— ; 2) 31–33 27·9 (1·03; 14) 26–29
     (2) 30·8 (1·09; 5) 30–32 27·4 (0·93; 16) 26–29

Sex differences significant, except wing (2).

Weights. Nominate *hybridus*. North-west Iran: adult ♂, May, 88·4 (3·36; 7) 83–92 (Schüz 1959). Hopeh (China): ♂ 90 (6) 84–94, ♀ 86 (5) 79–92 (Shaw 1936). Taiwan: juvenile ♂, June, 86 (RMNH).
  C. h. delalandii. South Africa: adult ♀, October, 100 (ZMA).
  C. h. javanicus. Southern New Guinea, adults: ♂♂ 79, 85, 90 (June); ♀♀ 71 (May), 79, 90 (June) (RMNH). Australia: 85–100 (Serventy *et al.* 1971).

Structure. Wing rather long and narrow, pointed. 11 primaries: p10 longest, p9 5–14 shorter, p8 20–30, p7 38–50, p6 57–68, p1 125–150; p11 minute, concealed by primary coverts. Longest tertials reach to tip of p4 in closed wing. Tail rather short, shallowly forked (see Measurements); 12 feathers. Bill straight, slightly shorter than head (especially short in ♀); rather heavy and wide at base, as in *C. leucopterus*, but relatively longer than this species, not as slender as in *C. niger*; relatively shorter than in *S. hirundo*, with tip of culmen more strongly decurved and gonys angle less marked. Depth of bill at basal corner of nostril 7·86 (0·39; 8) 7·5–8·5 in adult ♂, 7·33 (0·32; 9) 6·8–7·7 in adult ♀. Tarsus relatively long and heavy when compared with (e.g.) *S. hirundo*. Toes long, webs deeply incised as in *C. leucopterus*; outer toe c. 91% of middle, inner c. 69%, hind c. 34%.

Geographical variation. Slight in size, more marked in colour. Nominate *hybridus* constant in colour over whole Eurasian range. *C. h. delalandii* from southern and East Africa distinctly darker in adult breeding, chest, sides of neck, mantle, scapulars, and back dark slate-grey, belly black; chin, rump to tail, flight-feathers, and wing-coverts slightly paler than chest and scapulars but distinctly darker than in nominate *hybridus*; white streak below eye narrower; non-breeding and juvenile inseparable from nominate *hybridus*. Australian race *javanicus* distinctly paler than nominate *hybridus*; in breeding, chest and all upperparts and upper wing-coverts light grey (similar to *S. hirundo*), throat pale grey, indistinctly defined from broad white streak below eye, and belly grades from medium grey near chest to dark slate-grey near vent; non-breeding and juvenile plumages paler grey than nominate *hybridus*, sometimes close in tinge to Roseate Tern *S. dougallii*. West European and North African birds large, gradually smaller through Kazakhstan (USSR) and Indian sub-continent to eastern Asia (see Measurements); difference most marked in wing and tail, less in bill, tarsus, and toe. This rather slight clinal variation does not warrant recognition of separate eastern race *indicus* (Stephens, 1826). Afrotropical *delalandii* as large as west European birds. Australian *javanicus* similar in size to east Asian birds: e.g. adults, wintering Java and Celebes, wing ♂ 227 (6·11; 13) 219–235, ♀ 226 (6·46; 13) 216–230; bill ♂ 30·6 (1·37; 16) 28–32, ♀ 28·9 (1·70; 11) 27–31 (RMNH). Black-fronted Tern *Sterna albostriata* from New Zealand sometimes considered a race of *C. hybridus* (e.g. Vaurie 1965); shows superficial resemblance in bill and in colour of adult breeding, but rump white, underparts uniform grey, tail well-forked, non-breeding different, foot small and slender, toes fully webbed, and ecology different—all closer to typical *Sterna* than to *Chlidonias*. (Mees 1977b; Lalas and Heather 1980; C S Roselaar.) CSR

Chlidonias niger **Black Tern**

PLATES 12, 13, and 14
[between pages 158 and 159, and facing page 206]

Du. Zwarte Stern   Fr. Guifette noire   Ge. Trauerseeschwalbe
Ru. Чёрная крачка   Sp. Fumarel común   Sw. Svarttärna

Sterna nigra Linnaeus, 1758

Polytypic. Nominate *niger* (Linnaeus, 1758), Eurasia; *surinamensis* (Gmelin, 1789), North America.

Field characters. 22–24 cm (bill 2·7–2·8, legs 1·5–1·7, up to 6 cm); wing-span 64–68 cm. Distinctly smaller than all sea terns *Sterna* except Little Tern *S. albifrons*, and with shorter, less sharply forked tail than usual in that genus. Slightly smaller and slimmer than Whiskered Tern *C. hybridus*, with 15% shorter wings and finer bill; close in size to White-winged Black Tern *C. leucopterus* but less compact in appearance, with rather narrower wings, longer and more forked tail, and longer, finer bill. Small, slim marsh tern, often dipping in flight. Breeding plumage dark slate-grey rather than black, relieved only by pale grey underwing and white vent. Juvenile and non-breeding plumage patterned more like *Sterna* but usually darker above, with characteristic dark smudge at shoulder contrasting with white underparts. During spring or autumn moult between winter and summer plumage, adult noticeably blotched below. All calls quieter (and less frequently uttered) than in *Sterna*. Sexes almost similar; marked seasonal variation. Juvenile and 1st-year immature separable.

ADULT MALE BREEDING. Head black, becoming dark

slate-grey over neck and body and slightly paler slate-grey on wings and tail. At rest, plumage may appear uniformly coloured, but in flight, white vent and under tail-coverts and pale ash-grey underwing contrast markedly (from below). Bill almost as long as head, narrow and tapering to fine and slightly drooping tip; black, with red gape invisible except at close range. Legs dusky red-brown. ADULT FEMALE BREEDING. When compared at close range, separable from ♂ on greyer appearance, with dull black on head restricted to crown and nape. ADULT NON-BREEDING. With moult starting on face and spreading quickly to neck and lower body, white feathers replace black or grey ones and bird passes through remarkably blotched appearance. When moult complete, head noticeably capped with black crown, ear-coverts, and upper nape (with downward extension of black on cheeks often pronounced); small black spot in front of eye, and rest of head, collar, and underbody white. In flight, variable but usually striking dusky patch at or just below shoulder stands out against pale grey underwing and white chest. Back, upperwing, and tail dusky-grey, becoming slightly paler overall than in summer and less dusky over rump and tail. In flight, upperparts show stronger pattern, with darker, more slaty upper mantle and lesser wing-coverts forming dark band across leading edge of inner wing and behind white collar. With wear and fading, larger wing-coverts may become paler grey and old outer primaries duskier (on both surfaces). With full wear and bleaching (in Africa), many show noticeably pale grey upperparts in late winter but even these retain duskier back, lesser wing-coverts, rump, and tail than adult *C. leucopterus*. Bill appears longer, due to contrast with white fore-face. JUVENILE. Plumage pattern basically as winter adult. When fresh, forehead and back of neck washed pale brown, and dusky back mottled brown and grey, with almost white tips to longest scapulars. Back distinctly and more uniformly dark than that of adult; lacks discrete saddle-like appearance of juvenile *C. leucopterus*, being extended across inner wings by dark dusky inner median coverts and wholly slate lesser coverts to form T or shallow Y pattern. Rest of upperwing less uniform grey than in adult with hoary tips to outer median and greater coverts visible as shade across inner wing, but this much less extensive than broad and striking silvery central panel on upperwing of juvenile *C. leucopterus*. Underwing coverts white, and contrasting slightly with greyer undersurface of flight-feathers. Cap less compact than on adult, with downward extension of cheek-patch obvious and (usually) a notch between it and rear crown. Shoulder-patch often joined to upper mantle, at first more extensive than in winter adult but often breaking up with wear. Initially, base of bill and legs tinged yellow. FIRST WINTER. Change of plumage inadequately studied but certainly erratic in West Africa, with many retaining juvenile appearance and others more like winter adult. On former, wing pattern frequently patchy with either pale grey or dirty white central wing-coverts forming striking panel between dusky lesser coverts and dark grey secondaries, and worn, dusky primaries creating darker outerwing; these changes approach those associated with '*portlandica*'-type plumage pattern shown by several *Sterna* terns. In addition to changes in wing pattern, birds may show paler rump and tail but confusion with *C. leucopterus* still ruled out by retention of dusky leading edge to inner wing. On those assuming more adult plumage pattern, wings more uniform but usually still show dusky leading edge and worn, darker primaries. Black patch at shoulder often shrinks to spot, becoming indistinct at long range. SUBSEQUENT PLUMAGES. Moult erratic, with many retaining vagaries of winter plumage through 1st summer. Summer birds in breeding plumage with paler or white-patched underparts are probably 2 years old (see Plumages and Moults).

Summer adult unmistakable and bird in any other plumage safely identifiable on (1) attenuated structure, particularly long, fine bill, (2) dusky band across leading edge of wing and back, forming, with wholly dark back, T-form in juvenile, (3) usually prominent smudge at shoulder, and (4) usually uniform grey rump and tail. Juvenile formerly considered inseparable from *C. leucopterus*, but this corrected by Williamson (1960). In west Palearctic, general behaviour of *C. niger* unlike that of sea terns, with lazy inland passage flight of migrants and swooping, erratic flight of breeding birds over marshes quite distinct from more purposeful and usually quicker actions of *Sterna*. However, in coastal winter haunts, flight actions of *C. niger* much modified and closer to those of *Sterna*, with much more frequent incidence of hovering and plunging associated with faster wing-beats and more direct track. Gait tripping, with noticeably short steps. Carriage low and usually horizontal. Swims readily with wing-tips and tail cocked high.

Far less vociferous than *Sterna* terns, but birds on spring passage and in breeding haunts regularly utter a quiet 'kik-kik' in flight and sharper 'teek-teek' or 'teeuw' in alarm. Advertising-call 'kreert' or 'krierik'.

Habitat. Breeds in west Palearctic in continental and maritime middle latitudes, mainly in well-watered lowlands but locally at mountain lakes up to 2000 m (Dementiev and Gladkov 1951c). In temperate, steppe, and Mediterranean climatic zones, in cool or warm, relatively dry seasonal conditions, normally free of snow or frost. Prefers fresh or brackish waters c. 1–2 m deep, permanent or sometimes temporary, neither too extensive nor too narrow nor circumscribed by steep or forested terrain, and rich in low marginal and floating or emergent aquatic vegetation including strewn stems due to cutting. Water-soldier *Stratiotes aloides* appears especially favoured. Nesting sites include small pools, lakes, ditches, overgrown canals, quiet reaches of rivers, oxbows, marshes, and swampy or relatively dry meadows or pastures, including

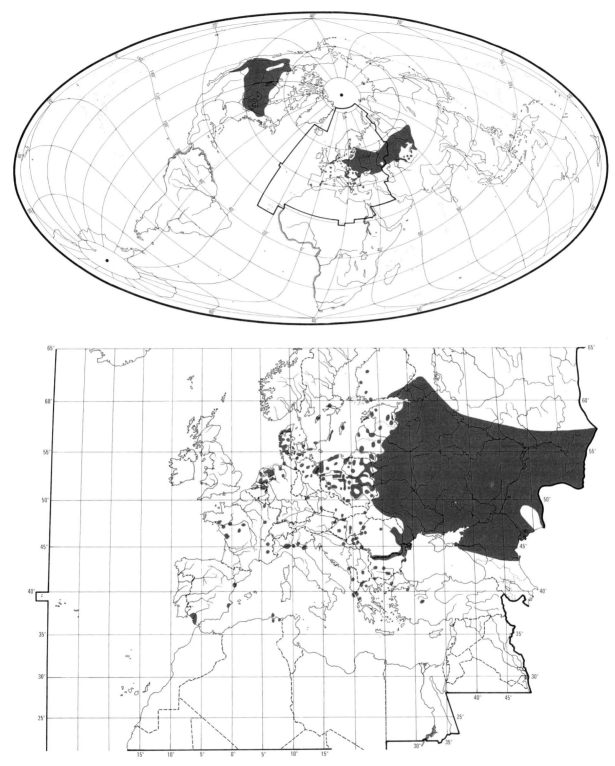

inundated areas (Niethammer 1942; IJzendoorn 1950; Haverschmidt 1978a). Reedbeds often flank or partly invade favoured waters; in North America particularly, these, when not too dense, often used for nesting, as are old muskrat *Ondatra zibethicus* houses and nests of grebes *Podiceps* or coots *Fulica*, and occasionally driftwood or boards (Bent 1921). While foraging, flights are largely over still or gently flowing water, but reedbeds, grasslands, and other drier terrain not neglected. On occasion, follows the plough (Pough 1951).

After breeding season, shifts to reservoirs, lakes, sewage farms, and eventually to salt-marshes, estuaries, bays, and coastlines, migrating both over land and sea (Bannerman 1962). In West African winter quarters, mainly frequents coasts, estuaries, and coastal lagoons, few penetrating inland (Serle and Morel 1977).

Flies freely, mainly in lowest airspace, and often rests standing or perching on posts, fishing stakes, buoys, or even bushes. Adaptable to some man-made changes in environment, but decreased in Europe due to habitat changes, especially drainage.

Distribution. Range decreased in several areas, probably mainly due to habitat changes but colonized Finland from 1960 and recently bred irregularly Britain and Ireland.

BRITAIN. Formerly bred regularly in some areas of eastern and south-east England, last certain record 1858 (Sharrock 1976); bred 1966, 1969–70, 1975, and 1978 (Sharrock et al. 1980). IRELAND. Single pairs bred 1967 and 1975 (CDH). FRANCE. Marked decrease in range, now local (Yeatman 1976). WEST GERMANY. Formerly bred Baden, Bayern, and Württemberg (Niethammer 1942). FINLAND. First bred 1960 (Haartman 1973). AUSTRIA. Bred regularly in east until c. 1890, then irregularly Neusiedlersee and Waldviertel until 1966 (Festetics 1967). GREECE. Formerly bred Lake Karla; probably breeds Luros delta (HJB, WB, GM). BULGARIA. Breeding Burgas area early 1960s; now absent (JLR). ISRAEL. Bred Lake Huleh until 1950s (HM). USSR. Formerly bred Armeniya (Dementiev and Gladkov 1951c); no recent evidence (VF).

Accidental. Iceland, Faeroes, Norway, Iraq, Kuwait, Syria, Lebanon, Jordan, Madeira.

Population. Decreased over much of European range, probably mainly due to habitat changes.

FRANCE. Decreased; under 10000 pairs (Yeatman 1976). Brittany: 60–100 pairs (Guermeur and Monnat 1980). BELGIUM. Decreased, c. 75 pairs until 1956, c. 20 pairs (Lippens and Wille 1972); c. 30 pairs (PD). NETHERLANDS. Decreased, 7500–10000 pairs 1945–50, 2700–4200 pairs 1970 (CSR), 1800–3500 pairs 1976–7 (R M Teixeira, CSR), 2000–3000 pairs 1978 (Teixeira 1979). WEST GERMANY. Schleswig-Holstein: marked decline, over 50% in last 25 years; maximum 800 pairs 1969. Hamburg: steady decline. Niedersachsen: decreasing, especially since 1960. Westfalen: declining, perhaps extinct. Rheinland: declining, now only a few pairs. (Haverschmidt 1978a.) Now probably under 200 pairs (Rheinwald 1982). DENMARK. Decreased; c. 700 pairs 1950, 400 pairs 1963–5, c. 200 pairs 1970s (Dybbro 1976). SWEDEN. 145–170 pairs 1975 (SOF 1978). FINLAND. Very rare (Haartman 1973). EAST GERMANY. Perhaps 50–180 pairs, fluctuating (Haverschmidt 1978a). Mecklenburg: declined since 19th century; probably 350–400 pairs (Klafs and Stübs 1977). POLAND.

Biebrza marshes: 537 pairs 1979 (AD). CZECHOSLOVAKIA. Decreasing (KH); largest colonies, up to 600 pairs, in eastern Slovakia (Hudec and Černý 1977). HUNGARY. Very common (LH). ITALY. About 80 pairs (S Allavena). GREECE. Decreased, now rare (HJB, WB, GM). TURKEY. Not common, perhaps c. 100 pairs or rather more (OSME). USSR. Estonia: c. 1000 pairs (Onno 1966). Latvia: 1000–2000 pairs, marked fluctuations (Viksne 1978). Ukraine: marked decrease in north-west and south-west due to drainage (Srebrodolskaya 1975; Talposh 1975b).

Oldest ringed bird 17 years 2 months (Rydzewski 1978).

Movements. Migratory throughout west Palearctic. Movements of Eurasian and Nearctic populations reviewed by Haverschmidt (1978a) and of USSR populations by Tamantseva (1955). Main wintering area for Eurasian birds is tropical West Africa where largely coastal. A few USSR birds winter on Black Sea (Strokov 1974) and Caspian Sea (Samorodov and Samorodov 1972). West Nearctic populations winter south through Panama to Peru; eastern populations south through Panama to Venezuela, rarely West Indies (Schauensee 1970); *surinamensis* vagrant to Iceland, usually in summer (A Petersen).

Autumn migration in west Palearctic begins last third of June, sometimes earlier, immature non-breeders probably moving first. Adults begin earlier than juveniles, perhaps by c. 1 month. Observations suggest nocturnal, as well as diurnal migration, sometimes at considerable height (Baggerman et al. 1956). Initially, juveniles may perform dispersive movements, sometimes to north, e.g. bird ringed Latvia (USSR) recovered Estonia in August (Tamantseva 1955). In Netherlands, peak passage July–August, continuing into September (by which time adults rare), latest 26 November. Typically assembles in huge numbers in traditional feeding areas *en route*: e.g. frequently more than 10000 in Elbe estuary and up to 80000 in IJsselmeer area, early to mid-August (Haverschmidt 1978a; Mulder and Tanger 1980; C J G Scharringa; see also Henss 1963). Netherlands thus probably main autumn staging post for populations of all northern and eastern Europe and possibly also western Siberia; these birds continue south mainly through France and Iberia (Haverschmidt 1978a; C S Roselaar). 1 bird ringed 21 August in Arnhem (Netherlands) recovered 4 years later (July) at Gomel (Byelorussiya). In some years, west and south-west passage of east European birds very heavy, e.g. through Baltic, Britain, and Switzerland in 1960 (Stromberg 1964). Passage through southern Bayern (West Germany) and Czechoslovakia from end of July to end September (Bezzel and Reichholf 1965; Fiala 1974). In August–September, birds crossing from upper Rhine occur Lac Léman (Halle 1967); those crossing central Europe reach Mediterranean coast. Recoveries in Spain are of birds ringed in Czechoslovakia, East Germany, and near Moscow (USSR); in southern France, of bird ringed Hungary. Numerous on passage, mid-August to early September, between southern France and Corsica (Ash 1969a). 1 Lithuanian bird recovered September in Italy, 1 Russian bird at Tunis (Tunisia) in August (Bernis 1967). Birds breeding east as far as south-west Siberia thought to move west on broad front to reach north-west coast of Italy (Tamantseva 1955). Up to 220 per day recorded at Bosporus (Turkey), August–September 1959 (Ballance and Lee 1961).

From eastern Mediterranean, some follow Nile and Rift valley south; common in Egypt, 24 August–10 November (Meinertzhagen 1954). Winters from southern Egypt to Sudan where fairly common October–March (Meinertzhagen 1930; Cave and MacDonald 1955; P L Meininger and W C Mullié); uncommon further south in East Africa, though recorded in late October at Lake Tanganyika, Burundi (Gaugris 1979), and at Lake Victoria and Kisumu, Kenya (Britton and Brown 1974). Comparative rarity in East Africa and Middle East suggests records from further south (see below) may derive from Atlantic coast passage. In autumn, considerable passage from Egypt along south coast of Mediterranean; often recorded as commonest migrant in Tunisia, Algeria, and Morocco, occurring in thousands from end of July to September (e.g. Brosset 1959); occurs mostly on coast, but also inland. Common in August at Straits of Gibraltar and hinterland (Henty 1961; Lathbury 1970). Trans-Saharan migration proposed by Heim de Balsac and Mayaud (1962) but this suggestion criticized by Moreau (1967) since sightings restricted to oases in northern Sahara in September–October, with no observations from central and southern Sahara at same time. A few sightings in Chad (Newby 1979), however, indicate some trans-Saharan movement, since winter distribution otherwise largely coastal. Majority follow Atlantic coast of Africa from western Mediterranean. Abundant on passage, September–October, from Morocco to Abidjan, Ivory Coast (Douaud 1953; Serle 1965; Smith 1965); route mainly coastal or offshore, some inland (Smith 1965). At beginning of September, estimated 100000 Banc d'Arguin, Mauritania (Gandrille and Trotignon 1973). This is thus major staging post, and some may winter; 1 Italian-ringed bird recovered there in November. Of birds ringed Banc d'Arguin (October–November), 2 recovered in Ghana the following spring, and 1 in Poland the following July (Dick 1975). Numerous in Sénégal December–January (Smet and Gompel 1980). Most, however, continue south to winter in Gulf of Guinea (notably Ghana to Nigeria) and beyond. On 12 December, flock of thousands, extending for miles, seen from ship crossing equator at 10°W (Hammond 1958). Recoveries in Ghana (most numerous), Ivory Coast, Nigeria, and Angola are of birds ringed Netherlands, West Germany, East Germany, and Italy (Haverschmidt 1978a; Brichetti and Martignoni 1981). Winter range extends along entire western seaboard of Africa, with records from Angola, Namibia, and South Africa, east to Natal and Transvaal (e.g. Jensen and Berry 1972, Schmidt et al. 1973, Tree 1974). 1 bird ringed Namibia recovered in July at Denisoka, USSR (52°26′N 61°41′E).

Spring migration of adults begins late March. Route lies mostly along West African coast or slightly inland, with passage heavy in Mauritania April–May (Bird 1937). Some may cross Sahara: numerous sightings in late April at oases in south-east Morocco (Smith 1968b) and in April–May at most northerly oases in Algeria and Libya (Moreau 1967; Haverschmidt 1978a). Passage through Nile valley and Egypt apparently lighter than in autumn, though large flocks occur Faiyum and Lake Menzala, late May (Meinertzhagen 1930, 1954). Many pass through Straits of Gibraltar from north-west Africa into Mediterranean, reaching southern coast of Spain (Nisbet et al. 1961). Some may follow Atlantic coast north to reach Britain. 80% of recoveries of birds ringed Tuscany (Italy) are in USSR, with movement largely north-east (Brichetti and Martignoni 1981). In Netherlands, passage strong in May (though numbers much smaller than in autumn) at favoured sites. Some reach Scotland, Ireland, Norway, Iceland, and Faeroes (Haverschmidt 1978a; A Petersen).

From Straits of Gibraltar, others move east along south coast of Mediterranean; common in Algeria and Tunisia (Dupuy 1969), though less so than in autumn in eastern Mediterranean. Notable passage May–June, probably of both west and east Mediterranean origin, up west coast of Italy (Waller 1955) and through Balkans. Ringing at Bologna shows that birds migrating through Italy fan out widely over central Europe and USSR as far east as Stalinsk, central Siberia (53°45′N 87°12′E): e.g. recoveries in Hungary (1 only 2 days after ringing), Bulgaria (9 days after ringing), Ukraine, Black Sea, Caspian, and Lake Chany (Novosibirsk).

Central and northern Europe reached via Rhône valley and Swiss lakes, thence to the Rhine (Halle 1967). Passage through Switzerland, Czechoslovakia, and southern Bayern (West Germany), larger in spring than autumn (Bezzel and Reichholf 1965; Fiala 1974), suggesting loop migration may occur with northern birds in autumn initially migrating west towards coast then south, while in spring they travel mainly south-east to north-west (in Europe) but along a more easterly path (Haverschmidt 1978a).

1-year-old nominate *niger* remain in winter quarters or move north, without reaching breeding grounds (Tamantseva 1955), though 1-year-olds of *surinamensis* apparently visit colonies (Tordoff 1962). Flocks of immatures recorded summering in West Africa. At 2–3 years old, many visit breeding grounds throughout range, but do not always breed. First-time breeders commonly settle at colony other than natal one (Bernis 1967; Haverschmidt 1978a).
EKD

Food. When breeding, mainly insects and other invertebrates; sometimes also fish and amphibians. On migration and in winter, mainly marine fish; sometimes also insects and crustaceans. Forages in groups (2–20), occasionally singly, flying into wind (if any), at height of c. 2 m (0·5–4·6 m, depending on light, wind, and prey) (Douglas 1950; Borodulina 1960; Bower 1970; Flint 1975). On migration will also forage in flocks of several thousand, following shoals of fish (Peters 1933; Bird 1937; Murphy 1938; Bannerman 1962; Grimes 1977), often forced to surface by dolphins (Odontoceti) and predatory fish, e.g. tunny *Thunnus* (Ash 1969b). 3 feeding methods. (1) Most common, dipping-to-surface (water, mudflat, plant stems or leaves, etc.); rhythmically rises and falls (little or no hovering between dips) to snatch up prey, sometimes barely touching surface. (2) Far less common, surface-plunging; sometimes immerses head and bill, occasionally submerging up to wing-tips. (Witherby *et al.* 1941; Douglas 1950; Dementiev and Gladkov 1951c; Cuthbert 1954; Baggerman *et al.* 1956; Borodulina 1960; Eddy 1961; Halle 1968; Bundy 1971; Vernon 1971c; Isenmann 1976b; Haverschmidt 1978a.) Prey usually swallowed in mid-air, and this often followed by swoop to drink (Baggerman *et al.* 1956). (3) Aerial-pursuit of insects (hawking), often in flight like swallow *Hirundo* or in zigzag fashion (Bent 1921; Dementiev and Gladkov 1951c; Cuthbert 1954; Borodulina 1960; Bower 1970). Though most feeding inland or inshore is in shallow freshwater or brackish-water areas, will occasionally feed over dry land (Bundy 1971; Scott 1971; Vernon 1971c), sometimes following the plough (Naumann 1840; Pittman 1927; Brewer 1969; Goethe 1970). At nesting time, USSR, feeds from dawn to dusk, stopping briefly at midday (Borodulina 1960). On equator, mid-December, flocks of thousands seen feeding c. 600 km offshore (Grimes 1977). Most food for young gathered within 22 m of nest (Baggerman *et al.* 1956).

Insects and their larvae, including: beetles (*Carabus, Poecilus, Harpalus, Pterostichus, Amara, Ilybius, Rhantus, Cybister, Hydrophilus, Laccobius, Enochrus, Berosus, Aphodius, Melolontha melolontha, Bembidion, Onthophagus, Anisoplia, Anomala, Agriotes, Donacia, Lixus, Chlorophanus viridis*, Staphylinidae), flies (Tipulidae, *Simulium, Culex*, Chironomidae, Ephydridae, Tachinidae, Stratiomyidae, *Tabanus, Helophilus*), dragonflies and damsel flies (Agrionidae, *Gomphus, Libellula, Diplax*), caddisflies (Phryganeidae, *Limnephilus griseus*), bugs (Jassidae, *Notonecta glauca, Ilyocoris cimicoides, Corixa, Gerris*, Lygaeidae, *Hydrometra, Tropicoris rufipes*), moths and butterflies (Lepidoptera), mayflies (Ephemeroptera), earwig *Forficula tomis*, grasshoppers, crickets and locusts (*Gomphocerus lineola, Myrmeleotettix, Oedipoda*, Gryllidae, Acrididae), cicadas (Cicadidae), hymenopterans (*Formica rufa, F. fusca*, Ichneumonidae), lacewings (Neuroptera), termites (Isoptera). Also spiders (Araneae), earthworms (Lumbricidae), leeches (Hirudinea), molluscs (*Planorbis*), crustaceans (*Artemia salina, Triops, Branchipus*), small frogs and tadpoles (*Rana, Hyla arborea*), and fish (*Esox lucius, Perca fluviatilis, Gasterosteus aculeatus, Cyprinus, Osmerus eperlanus, Engraulis guineensis, Brachydeuterus auritus, Sardinella eba, Rutilus rutilus*). (Peters 1933; Witherby *et al.* 1941; Dementiev and Gladkov 1951c; Ruthke 1951; Schmidt 1953; Maclaren 1954; Baggerman *et al.* 1956; Kistyakivski 1957; Borodulina 1960; Cvitanić and Novak 1966; Tima 1968; Dunn 1972a; Gandrille and Trotignon 1973; Isenmann 1976b). Data from Nearctic excluded.

Observations at nesting colony in Oder delta (East Germany/Poland) indicated diet mainly aquatic insects and their larvae (especially dragonflies, mayflies, water-beetles *Hydrophilus* and *Laccobius*, small lepidopterans, and reed-beetle *Donacia*), fish (including bleak *A. lucidus*, smelt *O. eperlanus*, roach *Rutilus rutilus*), and swarming ants (Ruthke 1951). In Netherlands, breeding birds seen to take chiefly fish (58% of 50 observations) c. 2·5–5 cm long, and large dragonflies (18%) (Baggerman *et al.* 1956). Breeding birds, USSR, fed largely on water-beetles and their larvae (67% frequency); also terrestrial beetles (74%), cockchafers *Melolontha* (25%), moths (10%), fish (7%), some frogs and tadpoles, and at times swarming ants and locusts. At fish-farms, mainly frogs (41·3% by weight; one bird ate average 12·5 g or 6·5 frogs per day) and fish (21·3%, predominantly wild carp *Cyprinus*); usually hunted in canals below sluice through which young fish released; fish taken mostly dead (Borodulina 1960). In 95 stomachs, Ukraine (USSR), insects and their larvae predominated; beetles (at least 16 species, especially water-beetles, and in 40% *Donacia*), flies (at least 6 species, including in 27% larvae of Stratiomyidae), dragonflies (including Agrionidae in 21%), water-bugs (at least 5 species, including *Ilyocoris cimicoides* in 26%), caddisflies, mayflies, grasshoppers, and ants; also spiders (in 12%), 6 frogs (*Rana, H. arborea*), and 18 fish (including pike *E. lucius*, perch *P. fluviatilis*, stickleback *G. aculeatus*, Cyprinidae, Acipenseridae) (Kistyakivski 1957). Of 3 stomachs, Yugoslavia, May, 2 contained flies, and 1 contained beetles *Bembidion* and lacewings (Cvitanić and Novak 1966). In Camargue (France) diet of migrants related to seasonal differences in habitat use: in spring fed mainly in freshwater areas (probably chiefly on insects though diet not fully determined), and also in ricefields (chiefly on chironomids, and crustaceans *Triops* and *Branchipus*); in summer fed mainly in (1) brackish lagoons (catching sand-smelt *Atherina* near water pumps and in areas drying out, also chironomids, sometimes amphipods *Gammarus*, and, in the air, winged ants), and (2) in ricefields (mainly on adult dragonflies taken from leaves); in summer and autumn fed also in salines (mainly on recently emerged chironomid and ephydrid flies, and earlier in season brine shrimp *Artemia salina*) (Isenmann 1976b). Autumn migrants, IJselmeer (Netherlands) and Elbe estuary (West Germany), took young smelt *O. eperlanus*

(Peters 1933; C J G Scharringa). In West Africa and Namibia took waste thrown from fishing boats and from fish factories (Pilaski 1967; Jensen and Berry 1972; Sinclair 1974). In Ghana and Sierra Leone frequented sewage outlets; also took small fish from seine nets as they were drawn inshore, and insects from saltpans and lagoons, and those blown out to sea (E K Dunn). In West Africa fed mainly on tide-blown flotsam, and occasionally on flying insects, including termites (Isoptera) (Maclaren 1954). 2 birds from winter quarters in Sierra Leone contained sardines *S. eba*, and unidentified insects; 6 birds, Ghana, contained fish (*B. auritus* and *E. guineensis*), and insects (Formicidae and Gerridae) (Dunn 1972a). In Netherlands, young fed mainly insects and other invertebrates, ranging from small insects or snails to large dragonflies which even very small chicks swallowed whole with wings (Baggerman *et al.* 1956). In East Germany, young fed mostly on dragonfly larvae, damsel flies, mayflies, and caddis flies; rarely fish, though usually given for first few feeds (Spillner 1975). In Michigan (USA), fed mainly insects (93·6% of 602 feedings brought to chicks 0–8 days old, of which 78·1% unidentified insects, 10·3% damsel flies, 2·7% dragonflies, 2·5% mayflies, 0·02% cicadas, and 4·9% fish (including *Notropis*), and 1·5% unidentified; number of feeds increased from 1·2 to 16·8 per young per hr over first 8 days (Cuthbert 1954). Single 5-day-old chick fed more often than older chicks in other nests (5·4 times per hr compared with 3·6 per hr for single 15-day-old chick and 1·7 per hr for each of 2 19-day-old chicks) with apparent increase in feeding rate over first few days, declining after about 10th day. Of 56 items fed, 13% were minnows, 6% dragonflies, and remainder small items including many insect larvae; one parent brought mainly fish, the other mainly small items in rapid succession (Dunn 1979a). EKD

Social pattern and behaviour.
1. Mostly gregarious throughout the year. Winter flocks often very large; thousands 'extending for miles' in December, Gulf of Guinea (Hammond 1958). In breeding season, typically feeds in groups of 2–20, rarely singly (Baggerman *et al.* 1956). Post-breeding flocks consist of family parties, sometimes all juveniles (Zimmerman 1931). Autumn migrant flocks, West Germany, usually less than 10, exceptionally more than 200 (Bezzel and Reichholf 1965; see also Henss 1963); 1–100, Italy (Waller 1955); up to 150, Caspian Sea (Feeny *et al.* 1968); up to 220, Bosporus (Ballance and Lee 1961); generally 10–70, but up to 3000, south-west Spain (Henty 1961; Nisbet *et al.* 1961). Spring migratory flocks up to 600, south-west Spain (Mountfort and Ferguson-Lees 1961a), where migrating flocks often described as continuous stream of thousands. See also Roosting. Both during and outside breeding season, often associates with other Sternidae, less often with gulls (Laridae). Migrant flocks, autumn, mixed with Whiskered Tern *C. hybridus* and White-winged Black Tern *C. leucopterus* (Wadley 1951; Béldi 1963); in spring, Mauritania, with Roseate Tern *S. dougallii* (Bird 1937). BONDS. Monogamous pair-bond. No information on length of bond. Some 1-year-olds visit North American colonies late in season, and breeding (if any) must be very rare. Age of first breeding presumably 2 years (Tordoff 1962; see Borodulina 1960). Pair share nest duties and care for young until a 'long time' after fledging (Baggerman *et al.* 1956; Eddy 1961), though sometimes apparently only one parent cares for fledged young (Saunders 1926; Cuthbert 1954). BREEDING DISPERSION. Colonial, though solitary nesting not uncommon. Colonies generally of less than 30 pairs, rarely over 100; typically 15–20 pairs, Netherlands and France (Morsier 1947; Baggerman *et al.* 1956); 2–150 pairs, Poland (Bocheński 1966); 10–30 pairs, maximum 300, Hungary (Csornai 1957; Csornai *et al.* 1958; Sóvágó 1968). Large colonies usually divided into sub-colonies, interspersed with isolated nests (see Bocheński 1966); latter especially likely on small ponds (Provost 1947). Distance between nests highly variable: 0·6 m to tens of metres (Baggerman *et al.* 1956); 2·5–3 m in clumps of rushes, 2–7 m on floating vegetation (Spillner 1975). In Michigan (USA), closest nests 9·2 m apart (Cuthbert 1954). Defends nest-area territory of *c.* 3 m radius around nest (Eddy 1961; Spillner 1975); serves for ground courtship, nesting, and resting of off-duty bird. Often returns to breed close to nest-site of previous year, if conditions permit; of 5 birds re-trapped at nest, Ontario (Canada), furthest distance moved from previous year was *c.* 75–100 m (Dunn 1979a). Young birds often return to natal area to bred (Haverschmidt 1978a). Commonly nests close to, though not mixed with, *C. hybridus*, *C. leucopterus*, and Black-headed Gull *L. ridibundus*; also grebes (Podicipediformes) and, less often, Common Tern *S. hirundo* (Csornai 1957; Csornai *et al.* 1958; Máté 1962; Sóvágó 1968). When young fledge or possibly earlier, parents, aided by offspring, defend temporary feeding territories (one held for 5 days) against conspecifics, other Sternidae, and gulls (Laridae), generally by chasing them off (Cuthbert 1954). ROOSTING. On autumn migration and in winter, loafs by day and roosts by night, often with other Sternidae, on mudflats, beaches, saltpans, and other open areas; large flocks recorded, e.g. 40000 in night roost, Netherlands (C J G Scharringa). Between feeding bouts, Ghana, loafed singly offshore on driftwood; flock of 150 seen resting on sea. By day and night, thousands occupied walls between ponds at saltpans; numbers increased at night, arriving to roost at dusk often from great height, swooping rapidly and erratically to land; departed again before dawn (E K Dunn). Gathering at roost after dusk also recorded during migration (see Haverschmidt 1978a). In breeding season, before laying, communal roost established outside colony, up to 1·6 km away (Baggerman *et al.* 1956), on floating vegetation, reeds, mud, poles, bushes, and emergent dead branches; used for loafing, preening, courtship, copulation, and overnight roosting (Baggerman *et al.* 1956; Spillner 1975). Both members of pair use pre-laying roost until 1st egg laid; thereafter off-duty bird may roost on nest-platform or on vegetation up to several metres away (Baggerman *et al.* 1956; Haverschmidt 1978a). Between bouts of feeding over land, Manitoba (USA), rested on fence-posts and wires (Pittman 1927). While waiting to be fed, fledged young loaf at various sites, often in parental feeding territory (Cuthbert 1954).

2. FLOCK BEHAVIOUR. Arrives on spring migration in small flocks which then explore potential colony-sites for 10–14 days (Spillner 1975). After choosing site, flocks focus attention there, flying around it but not alighting till laying imminent. First landings brief; usually at least 3 days elapse between occupation and start of laying (Baggerman *et al.* 1956; Spillner 1975; Haverschmidt 1978a). Behaviour—preening, sleeping, bathing—in pre-breeding flocks typically synchronized (Spillner 1975); up to 106 seen bathing together, Somerset (England) (Bennett and King 1959). In breeding colony, first bird to detect danger gives Alarm-call (see 1 in Voice). Alerted birds standing or incubating in colony may perform 'dread': quit colony in silence, swerving in flight from side to side for *c.* 50–70 m, after which they resume

calling and return to nests. Dreads especially common early in breeding season, brief ones occurring among small flocks hunting over water prior to colony occupation; stimulus usually not obvious (Baggerman *et al.* 1956). Birds in winter quarters attacked by falcon (Falconiformes) bunched into several flocks of 300–400 and rapidly rose to *c.* 300 m (MacLaren 1954). ANTAGONISTIC BEHAVIOUR. Mainly after Baggerman *et al.* (1956). Vigorous in defence of nest-territory, especially ♂; ♀ drives off conspecifics of both sexes, but foregoes physical contact (Spillner 1975). Disputes especially common during early occupation. (1) Ground encounters. Bird on territory adopts Aggressive-upright towards intruders: body about horizontal, legs bent; carpal joints held slightly free of body, wings sometimes raised, and body plumage slightly ruffled; neck upright and bill directed at ground intruder, though neck retracted with bill between horizontal and vertical if intruder overhead (Fig A). Utters Threat-call (see 4 in Voice) with bill gaping. Growl-calls (see 3 in Voice) given in physical encounter (Baggerman *et al.* 1956; Spillner 1975). (2) Aerial encounters. Chases may result from ground encounter, threatening bird (or nesting pair) flying purposefully at intruder, with Threat-, Anger-, and Growl-calls (see 2 in Voice). Attacks on conspecifics may develop into prolonged physical encounters in which 2 combatants hover beside one another, trying to peck and strike with feet, ascending into air (Upward-flutter), and sometimes tumbling into water (Baggerman *et al.* 1956). Drives chicks, swimming in the open, into cover with swoops and pecks at head (Hoffman 1927). HETEROSEXUAL BEHAVIOUR. (1) General. Pair-formation involves both aerial and ground behaviour at colony (see also Flock Behaviour, above). Descriptions below mainly from Baggerman *et al.* (1956). (2) Aerial courtship. Usually begins near colony, at feeding grounds; probably also on spring migration (see Witherby *et al.* 1941; Wootton 1950). High-flights seen soon after arrival at breeding area, especially on calm sunny days (Spillner 1975), continuing until 1st week of incubation. Mild resurgence occurs at hatching time (Cuthbert 1954), and similar flights, involving most colony members, even juveniles, seen up to August, North America, notably at dusk (Trautman 1939; Salter 1948; Eddy 1961). Early in season, sequence as follows. Advertising-call (see 5 in Voice) given by one bird and rapidly taken up by others signals interruption of hunting and start of High-flight in which 2–20 birds, rarely any carrying food, ascend spirally in jerk-flying manner, rapidly-beating wings almost meeting over back; mean wing-beat rate 3·6 per s, compared with 2·8 per s in normal flight (Baggerman *et al.* 1956). Sub-groups of 2–5 birds, each with their own leader, may form within main party. Pursuers, uttering occasional Threat-calls, appear to chase leader which may swerve evasively and be overtaken. Some individuals drop out during ascent which can reach 200 m; sub-groups may level out independently at different heights (Spillner 1975). With 1 or more acceleratory wing-beats, group leader eventually initiates downward glide, an oblique or broadly spiral descent, carpal joints strongly bent. Usually silent during glide (Cuthbert 1954). The leading 2 birds, probably prospective pair, usually stay close together, others following at a distance and often interspersing glide with wing-beats

(Baggerman *et al.* 1956). When near ground or water, swoop up to *c.* 15 m silently, veering from side to side in unison and still gliding on upstretched wings (Cuthbert 1954; Haverschmidt 1978a). Early in season, gliding phase commonly involves 3–4 birds, later on typically 2 (♂ and ♀). Early-season High-flights also much longer than later—up to 40 min compared with 40 s to 4 min (Cuthbert 1954). For variations in High-flight, see Baggerman *et al.* (1956). Low-flight occurs from early in occupation of colony area until end of 1st week of incubation, but independently of High-flight. Begins when ♂ catches prey item; retaining prey in bill and holding head down to render prey conspicuous, flies around, 10–15 m above colony (B Wetton), with shallow wing-beats, uttering extended version of Advertising-call (see 5 in Voice) at regular intervals. If ♀ attracted, she follows up to tens of metres behind, usually in silence or with subdued Begging-call (see 6 in Voice); sometimes overtakes ♂ (Pass-ceremony) and leads briefly. Eventually ♀ lands, almost always outside colony area before nest-site occupation. ♂ follows and alights alongside to transfer food. Rarely, ♂ may attempt to transfer food in flight. 2–3 birds, including ♂♂, may pursue food-bearing ♂; pursuing ♂ may try to dispossess food-bearer. 2 pursuing ♀♀ may chase one another, uttering Threat- and Anger-calls (Baggerman *et al.* 1956; Haverschmidt 1978a). (3) Ground courtship. May occur at feeding grounds before colony occupation, but most often at nest-site when pair-bond better developed. When pair alight at nest-site together, both assume Erect-posture (Fig B): fore-body raised, bill 40–70° above horizontal, wings lifted free of body feathers, sometimes half-open; each may give Advertising-call and turn head away from the other (Baggerman *et al.* 1956); then both often relax to Forward-erect, with bill pointing horizontally forward. Courtship-feeding (see below) integral part of ground courtship, which often follows Low-flight. On approach of ♂ in flight, ♀ stretches up to face him, uttering Begging-call with bill continuously open. When ♂ closer, ♀ may lower body to Hunched-posture (Spillner 1975), or adopt Stoop-posture, bowing body forwards, with straight back at *c.* 30° to horizontal (Fig C), accompanied by Cooing-call (see 7 in Voice); of 60 instances when ♂ alighted with food, ♀ responded with Erect-posture on 92% of occasions, Stoop-posture 5%, Erect-posture then Stoop-posture 3%. Of 24 instances when ♂ returned without food, frequencies were 13%, 83%, and 4% respectively (Baggerman *et al.* 1956). In such interactions, less intense Forward-erect may take place of Erect-posture. On landing, ♂ adopts Erect-posture

and, if bringing food, lowers head and fore-body appeasingly to transfer food, sometimes with Cooing-call. After food passed, both birds revert to Erect-posture. If ♀ adopts Stoop-posture, she often rotates, and ♂ circles with short, rapid steps to keep in front of her, repeatedly giving Cooing-call. Before egg-laying, ♂ commonly adopts steeply angled Stoop-posture and ♀ follows suit, but at shallower angle ('mutual stoop') (Fig C). Both may turn from left to right, with short steps, but mostly face each other, cooing. ♂ may sit briefly, and both may peck at weed. (4) Mating. Occurs mostly on nest-platform and also on resting sites; often, but not always, immediately preceded by courtship-feeding. In Forward- or Down-erect, with head level or slightly downtilted respectively, ♂ initially stands at *c.* 90° to ♀, and may give Lure-call (see 8 in Voice). ♀ assumes Hunched-posture, giving Begging-calls, and ♂ mounts. ♀ lifts tail as signal to ♂ to commence copulation, during which ♀ continues Begging-call, or both give Cooing-call (Baggerman *et al.* 1956; Haverschmidt 1978a). ♂ may grasp ♀'s neck in bill. Mating may last 3-4 min; several cloacal contacts may occur (Spillner 1975). After dismounting, ♂ commonly adopts Erect-posture, departs to find food for ♀, or both preen (Baggerman *et al.* 1956; Spillner 1975). (5) Courtship-feeding. Integral part of ground courtship (see above). Continues until laying of 2nd egg (Spillner 1975), or until clutch completion, diminishing thereafter (Baggerman *et al.* 1956). ♂ feeds ♀ in various circumstances: after Low-flight; when ♀ in the air sees ♂ alighting with food and joins him; when ♀ joins ♂ (without food) at nest and performs sequence of Erect-posture–Stoop-posture–Erect-posture, ♂ then departing to find food. 5 feeds recorded in 12 min (Spillner 1975); 1 feed per min for 10 min, though usually less frequently. ♀ continues to feed herself during period of Courtship-feeding (Baggerman *et al.* 1956). (6) Nest-relief. Arriving bird gives Advertising-call in flight, more so during laying than thereafter. Sitting bird, of either sex, starts Stoop-posture on approach of mate. Alighting bird initially assumes Erect-posture, less so as incubation proceeds. If sitting bird does not rise promptly, mate adopts Stoop-posture, gives Cooing-call, and may peck at nest-material. Sitting bird usually reciprocates, bowing forwards while sitting; this display persists until hatching (Baggerman *et al.* 1956; Spillner 1975). Relieved bird often pulls small piece of material out of nest rim, or from water nearby, and tosses it backwards towards nest (side-throwing). Birds may pick up material in flight up to 20 m from nest and toss it haphazardly (Spillner 1975). RELATIONS WITHIN FAMILY GROUP. When marsh dried out, parent made 2 flights, each of at least 3 km, to carry newly-hatched young to vegetated islet (Guérin 1924). ♀ often does most brooding (e.g. Turner 1920). Both parents feed young, alighting to deliver food when chicks downy; food sometimes passed to brooding bird for transfer to young. Young brooded regularly up to 4 days, rarely after 9; parent summons them for brooding with Stoop-posture and Cooing-call. Parent with food coaxes young out from beneath brooding bird by soft Advertising-call, Cooing-call, or (in ♂) Lure-call. Small young give food-call, approach adult, and peck at bill-tip. At 1 week old, young take greater initiative, running vociferously towards adult. Young seem not to recognize parents by voice at 2-3 days, but recognition improves over 1st week. From 2 days, young make sorties of increasing distance from nest, usually leaving it permanently in 2nd week to seek refuge in vegetation. Young ranged up to 6 m from nest at 5 days, 13 m at 8 days, 25 m at 9-25 days (Cuthbert 1954). Spillner (1975) found young left nest early, and were fed and brooded at constantly changing sites. Young over *c.* 5 days fed by parent passing food in flight after gliding approach (Cuthbert 1954). Food dropped in flight to well-grown young which pick it up from vegetation (Saunders 1926). Fledged young loaf at variety of sites in feeding territory (see Roosting, above) where parents feed them; young also alight on water to be fed by hovering parent (Du Bois 1931). Brood members may be 6-10 m apart, sometimes fed by one parent only; some evidence that other took no part (Saunders 1926; Cuthbert 1954). Fledglings practise hunting skills (dipping-to-surface, plunge-diving, sometimes to pick up floating material) and make flights lasting 5-15 min, but initially take little food for themselves. 7-13 days after fledging, however, feed competently on insects plucked from stems while hovering (Cuthbert 1954; Borodulina 1960). ANTI-PREDATOR RESPONSES OF YOUNG. In response to slowly repeated Alarm-call of parent, young up to 2 days old usually crouch motionless in nest, but, if call rapidly repeated, will swim, even at 3 hrs old, to hide in nearby vegetation, returning when Alarm-calls more slowly repeated, or cease (Baggerman *et al.* 1956). When parents' Alarm-call alerted young to Marsh Harrier *Circus aeruginosus*, young remained in nest, but left when Alarm-call warned of human intruder (Spillner 1975). PARENTAL ANTI-PREDATOR STRATEGIES. (1) Passive measures. No information. (2) Active measures: against birds. First bird to spot intruder gives Alarm-call, eliciting upflight of colony members. Will fly to intercept approaching flying predator up to 100 m from colony, then pursue it in tight flock, while individuals periodically make swoops at it until driven out of colony area (Baggerman *et al.* 1956; Spillner 1975; B Wetton). Chasing flock may swerve erratically around bird of prey, heron (Ardeidae), or White Stork *Ciconia ciconia* in so-called 'whirl-flight' (Baggerman *et al.* 1956). Co-operates with *S. hirundo* in joint attacks (B Wetton). Single nesting pair, Attenborough (England) also attacked wide variety of ducks and geese (Anseriformes) and passerines (Passeriformes), striking swimming Canada Goose *Branta canadensis* on the back. Waders (Charadrii) also commonly driven off (Turner 1920; B Wetton). (2) Active measures: against man. Flies to confront intruder up to 60 m from colony (Provost 1947). Flock gathers overhead, from which individuals persistently swoop to rear of intruder, pecking at head, defecating, and giving Anger-calls. Dive-attacks especially fierce from late incubation onwards (Baggerman *et al.* 1956; Eddy 1961; Haverschmidt 1975). (4) Active measures: against other predators. Makes dive-attacks on dogs and hares *Lepus* (Haverschmidt 1975). For predators in North America, and reactions to them, see Cuthbert (1954).

(Figs A-C from photographs in Baggerman *et al.* 1956.)   EKD

Voice. Highly vocal, especially at breeding grounds.

CALLS OF ADULTS (1) Alarm-call. A brief, high-pitched 'teek-teek-teek' (Baggerman *et al.* 1956); 'keek-keek-keek' (Witherby *et al.* 1941), changing to sharper, longer 'tee' or 'teeuw' as danger becomes more imminent. (2) Anger-call. A harsh 'kek' (Fig I); 'gäg' or 'kjip' (Meyer 1965); intensified version of call 1, given in extreme alarm, especially in dive-attacks, or in aerial chases, whether expelling attack or courtship. (3) Growl-call. A harsh, low 'krrr', given as combatants try to peck one another in ground and aerial disputes (Baggerman *et al.* 1956). (4) Threat-call. A shrill, often rapidly repeated 'kreea' (Fig II); 'kyeh' (Baggerman *et al.* 1956); 'kriiah' (Meyer 1965; Kapocsy 1979); staccato 'kjiää-kjia-kjaa', sometimes 'kjä-kjä-kä-kä-kä (Spillner 1975). Directed by bird on nest-territory at aerial intruder; also given by chasing birds, e.g. in Low-flight (Baggerman *et al.* 1956). Heard in disputes in winter quarters (Bromley 1952). (5) Advertising-call. Usually loud 'kreerk' or 'krierk' (Baggerman *et al.* 1956); 'kiap-

kiap-kiap' (Spillner 1975), given by ♂ in High- and Low-flights, and by both sexes approaching nest, probably serving to identify caller to mate and offspring; also given in Erect-posture and Forward-erect. More drawn-out version, 'kree-e-rick', 'kree-e-rickick', or 'kree-e-rick-kickick' (Baggerman *et al.* 1956), or 'ki-e-rek' (Meyer 1965), given when ♂ carries food in Low-flight or to nest; very rarely by ♀ carrying food. (6) Begging-call. High-pitched 'eew' or 'keew' (Baggerman *et al.* 1956); 'kieuw' or 'wie-ieeuw' (Haverschmidt 1978*a*). Given by ♀ soliciting food from ♂; during copulation, soft and rapidly repeated; occasionally uttered by ♀ in Low-flight. (7) Cooing-call. Soft, low, rapidly repeated 'krrr' (Baggerman *et al.* 1956); crooning, purring 'keerr' (see Witherby *et al.* 1941; Kapocsy 1979); 'pierr-pierr-pierr' (Spillner 1975), in association with Stoop-posture. Also by both sexes during copulation (Haverschmidt 1978*a*). (8) Lure-call. ♂ invites close approach of brooded offspring for feeding with 'kreeuw' (Baggerman *et al.* 1956); 'pjirru-pjirrur-pjirru' given by ♂ just prior to mounting ♀ possibly the same (Spillner 1975). (9) Kip-call. Brief 'kip' maintains contact between flock-members during hunting, when flying around colony, or on migration. Individual calling 'keep' apparently brought dispersed birds together in fishing flock (Bower 1970). Also rendered 'keek', 'gik', 'kik', 'kit' (Baggerman *et al.* 1956; Meyer 1965; Bower 1970; Kapocsy 1979).

CALLS OF YOUNG. Food-call of chicks, 'chirr', given within 30 min of hatching (Cuthbert 1954). Fledglings beg with shrill 'br(i)rrr' (Meyer 1965); 'tri' (Heinroth and Heinroth 1927–8) which increases in volume as parent approaches with food.

EKD

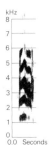

I S Palmér/Sveriges Radio (1972) Sweden May 1959

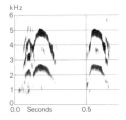

II S Palmér/Sveriges Radio (1972) Sweden May 1959

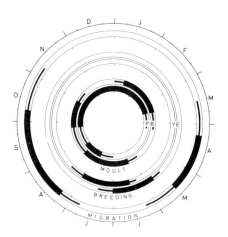

Breeding. SEASON. North-west Europe: see diagram. Southern Europe: up to 2 weeks earlier. Volga region (USSR): breeding begins late May (Borodulina 1960). SITE. On mat of floating vegetation, or in very shallow water, or on ground among marsh vegetation; rarely on dry land. Loosely colonial. Nest: low heap of water weeds, compressed and with shallow depression in top; on drier ground, a shallow scrape lined with vegetation. Outer diameter 15 cm (11–21), sample 45; internal diameter 8 cm (6–9), sample 43; cup depth 1·7 cm (0·5–3·0), sample 45 (Bocheński 1966). At laying time, nest often flat-trodden heap of weeds, but built up by incubating bird and at time of nest-relief to more substantial nest with rim and cup (Baggerman *et al.* 1956). Building: by both parents. EGGS. See Plate 90. Oval, smooth and slightly glossy; cream to pale buff, with fairly large spots and blotches of dark brown and black. 35 × 25 mm (31–40 × 23–27), sample 400; calculated weight 11 g (Schönwetter 1967). Clutch: 2–4. Of 60 nests, Volga (USSR): 2 eggs, 16; 3, 41; 4, 3; mean 2·83 (Borodulina 1960). Of 101 nests, East Germany: 1 egg, 3%; 2, 12%; 3, 77%; 4, 8%; mean 2·91 (Spillner 1975). Mean of 151 clutches, Iowa (USA), 2·6 (Bergman *et al.* 1970). One brood. Replacements laid after egg loss. Laying interval *c*. 1 day. INCUBATION. 21–22 days; mean of 28 nests 21·4 days (Bergman *et al.* 1970); 22 days (Spillner 1975); mean of 6 marked eggs 21 days (Cuthbert 1954). Much shorter periods, quoted in Witherby *et al.* (1941) and many times subsequently, unrealistically short. Stated by Haverschmidt (1945) and Spillner (1975) that incubation begins after laying of 1st egg, though perhaps not sitting tightly at first; confirmed by Cuthbert (1954) who found hatching interval of 25–26 hrs; Spillner (1975) recorded eggs hatching at shorter intervals than laying interval, sometimes as little as 4 hrs after 22 days incubation; Baggerman *et al.* (1956) reported hatching at intervals of 1–2 days, but stated that incubation does not begin until clutch complete, which seems unlikely. Incubation spells of ♂ averaged 23·3 min, and of ♀ 27·7 min, sample 230 (Baggerman *et al.* 1956); at one nest, mean spell for each bird (sexes unknown) 36·5 and 43 min, total sample 69 (B Wetton); up to 4–5 hrs by one bird at night (Spillner 1975); spells of 1·5–2·0 hrs reported from Volga, with ♀ taking larger share (Borodulina 1960). YOUNG. Precocial and semi-nidifugous. Cared for and fed by both parents. Brooded for most of 1st week while other parent collects food; fed 65–89 times per day (Borodulina 1960). Stay in nest for first 2–3 days, then wander into nearby vegetation, especially if disturbed, and do not leave nest vicinity finally until 2nd week, then staying in available vegetation cover (Cuthbert 1954; Baggerman *et al.* 1956). FLEDGING TO MATURITY. Fledging period from 19 days (Borodulina 1960; Dunn 1979a), but up to 25 days for full flight (Cuthbert 1954). Feed independently within a few days of fledging, but seen being fed 13 days after fledging (Cuthbert 1954). Age of first breeding not known but probably at least 2 years. BREEDING SUCCESS. Of 89 eggs laid, East Germany, 86·5% hatched (Spillner 1975). In USA: at least 1 egg hatched in 29% of 192 nests (Bergman *et al.* 1970); all eggs hatched in 27% of 55 nests (Dunn 1979a).

Plumages (nominate *niger*). ADULT MALE BREEDING. Head and neck black, deepest and with faint green gloss on forehead, crown, and nape, slightly duller on chin and throat. Mantle grading from dull black at border with nape to slate-grey at border with scapulars. Scapulars, tertials, back, rump, and upper wing-coverts uniform slate-grey; upper tail-coverts and tail similar or slightly paler, medium grey; outer web of t6 sometimes light grey. Chest, flanks, and belly dull black to blackish-slate. Vent and under tail-coverts white. Flight-feathers slate-grey, similar in tinge to upperparts but slightly more silvery, especially fresh inner primaries; inner webs of primaries darker grey with ill-defined dark smoke-grey wedge towards base. Leading edge of wing light grey, not contrasting with slate-grey lesser upper wing-coverts; under wing-coverts and axillaries pale grey, occasionally almost white. In fresh plumage, slate-grey of upperparts and upper wing-coverts more bluish-grey on feather-fringes. In worn plumage, flight-feathers duller, less silvery, 4–7 outer primaries nearly black; underparts with slate-grey of feather-bases showing through, less uniform blackish-slate, throat and belly mottled grey especially. ADULT FEMALE BREEDING. Like adult ♂ breeding, but mantle and underparts slate-grey, similar in tinge to scapulars and upper wing-coverts instead of distinctly darker, wing-bend of resting bird not contrasting with sides of body; black on head restricted to forehead, crown, and nape, reaching down to gape and ear-coverts, and often contrasting sharply with grey of throat, neck, and mantle, instead of all head being same colour as underparts; chin and throat often light grey rather than slate-grey; outer web of t6 rarely light grey. A few ♂♂ have chest and belly as pale as ♀, but sides of head usually blacker and without contrastingly pale chin and throat. Underparts of ♂ in worn breeding sometimes as pale as fresh ♀ breeding. ADULT NON-BREEDING. Sexes similar. Forehead, lores, eye-ring, and cheeks white. Small patch in front of eye black. Crown down to ear-coverts and centre of nape black, feathers slightly fringed off-white when fresh. Hindneck white, some feathers tipped grey, forming narrow collar round neck. Upperparts, including upper tail-coverts and tail, and all upper wing-coverts uniform medium grey, mantle, scapulars, back, and wing-coverts slightly paler and less bluish-slate than in breeding; fresh feathers narrowly fringed pale grey or off-white on tips. Shorter lesser upper wing-coverts slate-grey, hardly contrasting with other coverts when fresh, but forming rather indistinct and narrow black carpal bar when worn. Underparts white, except for distinct medium grey patches extending down from mantle to sides of chest. Under wing-coverts and axillaries greyish-white to white. Colour of flight-feathers dependent on wear and moult: in autumn, dusky grey to black, except for slate-grey inner primaries; in winter, all slate-grey with slight silvery bloom; in spring, as in winter but outer primaries slightly duller slate and less silvery than 3–6 inner. DOWNY YOUNG. Upperparts cinnamon-buff with large ill-defined dull black patches on hindcrown, nape, sides of back, upperwing, and rump, sometimes joining to form broad irregular streaks. Area round eye, lores, chin, and often wing-tip white. Lower cheeks and throat dull black to pale rufous-brown, flanks buff-brown grading to vinous-grey on chest and sides of belly and to off-white on central belly. JUVENILE. Lores and forehead pale grey, vinous-grey, or pale buff, soon fading to white. Small patch in front of eye black, narrow ring round eye (except in front) white. Crown and central nape black, extending down from crown behind eye to lower ear-coverts; feathers of crown and nape often narrowly fringed or streaked pale grey or buff when fresh. Mantle, scapulars, and tertials rather variable: feathers usually medium grey grading to dull black towards tip and fringed black-brown, deep-brown, or buff; others mainly grey with narrow fringes only; mantle often nearly uniform black. Back, rump, upper tail-coverts, and tail medium-grey to light grey, feather-tips narrowly fringed brown or buff on back, narrowly white on tail, broadly white on lower rump and upper tail-coverts, latter appearing occasionally almost uniform white; tail-feathers often dull black subterminally. Underparts white except for dark grey patches extending down from mantle to sides of chest. Flight-feathers as in adult breeding, but without contrast between fresher inner and older outer primaries caused by arrested moult; secondaries and up to 6–8 inner primaries narrowly fringed white on tips when fresh. Shorter lesser upper wing-coverts dull black, forming distinct carpal bar; remainder of upper wing-coverts medium grey with off-white, pale brown, or deep brown tips, appearing scaled. Leading edge of wing, axillaries, and under wing-coverts white. In worn plumage, rather similar to adult non-breeding, but differs in rather broad and frayed off-white fringes to scapulars and tertials and narrower fringes to upper tail-coverts; dark brown or dull grey subterminal marks to lower scapulars, tail-feathers, or inner upper wing-coverts; paler grey rump and upper tail-coverts with broad white fringes to tips; no active or arrested flight-feather moult. FIRST IMMATURE NON-BREEDING. Like adult non-breeding and plumage indistinguishable after loss of last juvenile lower scapulars, tertials, tail-feathers, or upper wing-coverts, but differs in flight-feather moult: in general, adult post-breeding primary moult late July to late January, post-juvenile primary moult early January to early July (see Moults). SECOND IMMATURE NON-BREEDING AND SUBSEQUENT PLUMAGES. When *c.* 1 year old, 1st immature non-breeding probably replaced directly by 2nd and no dark feathers of adult breeding obtained; 1st full breeding obtained in early spring of 3rd calendar year. Some birds visit western Europe with head and part of neck and chest in non-breeding, but remainder of underparts dark as in breeding; these are probably retarded 2-year-olds or perhaps advanced 1-year-olds (see Moults); among specimens from Netherlands, April–June, only 2 out of 60 birds in such a plumage (both late June), but 7 of 20 July–August; none of these taken in breeding colonies. As adult post-breeding moult advanced in July–August, these sub-adults hard to distinguish then, except for difference in wear of non-breeding head.

Bare parts. ADULT BREEDING AND NON-BREEDING. Iris dark brown. Bill black, basal cutting edges red, gape lavender-pink to orange. Foot dark red-brown, dusky livid-purple, or black with dark blood-red tinge of variable intensity. DOWNY YOUNG. Iris dark brown. Bill dusky, vinaceous at gape. Foot vinaceous. JUVENILE. Iris dark brown. Bill dark horn-brown to black, base of lower mandible bluish-pink to yellow-horn. Foot pink-brown or yellow-brown. Similar to adult from late autumn. (Ridgway 1919; Witherby *et al.* 1941; Fjeldså 1977; RMNH, ZMA.)

Moults. ADULT POST-BREEDING. Complete; primaries descendant. No double moult of inner primaries (unlike other *Chlidonias*); wing moult similar to (e.g.) Common Tern *Sterna hirundo*. Starts from late May or early June with feathering at gape and in front of eye and with eye-ring, soon followed by lores, chin, forehead, and throat. By mid-July most birds show much non-breeding on head, neck, and chest, as well as scattered feathers on underparts, mantle, and scapulars, but a few (mainly ♂♂) have just started with feathers at gape. By mid-August, head, body, t1–t2(–t3), and many median and some lesser upper wing-coverts in non-breeding, but usually not all tertials and back nor some scattered feathers on underparts; starts migration at this stage. Moult of flight-feathers starts with p1 early July to mid-August; moult arrested during autumn migration with up to p2–p3(–p4) new. Completed in winter quarters; sequence of tail t1–t2–t3–t6–t4–t5; flight-feathers completed with p10 January–February. ADULT PRE-BREEDING. Partial: head, body, tail, inner primaries, and wing-coverts except occasionally for t4–t5 and part of median and lesser upper wing-coverts. 2nd series of inner primaries starts with p1 between late November and early February, often before 1st (post-breeding) series completed; moult arrested March with up to p2 (1% of 83 examined), p3 (9%), p4 (31%), p5 (48%), or p6 (11%) new; no 3rd primary moult series started in early spring in nominate *niger*, but 1 of 8 adult breeding *surinamensis* examined had p1 new (2nd series had reached p6). Moult of body, tail, and wing-coverts mainly January–February, head, neck, chest, median upper wing-coverts, and t4–t5 last; pre-breeding completed mid-March to early April. POST-JUVENILE AND SUBSEQUENT MOULTS. Post-juvenile complete. Usually no moult before arrival in winter quarters. Mainly starts October–November with head, mantle, scapulars, and t1; p1 shed December–January. By early March, head, body, tail, and upper wing-coverts in 1st immature non-breeding, often except for some tertials, part of lesser upper wing-coverts, and greater coverts. All flight-feathers new by June–August of 2nd calendar year; some birds start 2nd primary moult-series May–June. 1st immature non-breeding directly replaced by 2nd immature non-breeding at about same time as adult post-breeding; primaries start about July. Birds with breast and belly in breeding, head, neck, and chest in non-breeding, and arrested primary moult scores as in adult visit western Europe from late June (see Plumages—wing-coverts, upperparts, and tail worn, probably non-breeding, these feathers not certainly separable from worn breeding); these are perhaps advanced 1-year-olds, as suggested by Van Rossem (1923) and Tordoff (1962), but captive birds kept by Heinroth and Heinroth (1927–8) did not obtain breeding at this age and West African 1-year-olds examined (though perhaps retarded) had not yet completed post-juvenile moult on outer primaries by late July and August and were in rather worn 1st immature non-breeding, starting moult to 2nd non-breeding with mantle, scapulars, and wing-coverts; hence, sub-adults visiting western Europe are probably 2-year-olds, perhaps retarded.

Measurements. Netherlands and Germany, spring and summer; skins (RMNH, ZFMK, ZMA). Tail is to tip of t1, fork is tip of t1 to tip of t6.

		♂			♀	
WING	AD	218	(4·03; 56)	210–226	213	(5·17; 38) 204–224
	JUV	208	(4·17; 12)	202–215	206	(3·36; 18) 201–211
TAIL	AD	64·2	(2·30; 39)	60–69	62·2	(2·46; 31) 58–66
	JUV	60·4	(3·09; 14)	57–65	60·8	(2·78; 11) 58–65
FORK	AD	18·8	(2·18; 38)	15–25	17·2	(2·45; 31) 13–21
	JUV	13·2	(1·31; 14)	11–16	12·2	(1·17; 11) 10–14
BILL	AD	27·8	(1·21; 47)	26–30	26·5	(0·82; 36) 25–28
TARSUS		16·4	(0·60; 39)	15–18	16·3	(0·51; 38) 15–17
TOE		21·8	(1·03; 41)	20–24	21·1	(0·86; 31) 20–23

Sex differences significant for adult wing, tail, fork, and bill, and for toe. Juvenile tarsus and toe full-grown at fledging, juvenile bill at age of *c.* 1 year.

Migrants, Tunisia, late April; skins (ZFMK).
WING AD ♂ 221 (4·37; 11) 216–226   ♀ 220 (3·85; 7) 215–225

Weights. Nominate *niger*. Netherlands: August, adult 73·5 (7·11; 28) 60–86; juvenile 71·9 (7·81; 50) 56–88 (A L Pieters); June–July, adult ♂♂ 59, 61, 63, 64, adult ♀♀ 54, 60, 62; exhausted adult 47 (ZMA). Hamburg (West Germany): early August, mainly adult, ♂ 73 (9) 63–88, ♀ 72 (4) 71–73 (Peters 1933). Morocco, September: adult 47·5 (2·0; 6), juvenile 45·8 (3·5; 75) (Pienkowski 1975). Kazakhstan (USSR): ♂♂ 63 (May), 55, 59 (June), 68, 71 (August), 66 (October); ♀♀ 63 (June), 60, 73, 83 (August) (Gavrin *et al.* 1962). South Africa: January, unsexed, 59, 73 (Schmitt *et al.* 1973).

C. n. surinamensis. Panama: February, 37, 39 (Burton 1973*b*). USA: July, ♂♂ 51, 64; ♀ 52 (Grinnell *et al.* 1930; Norris and Johnston 1958). Surinam: October, adult ♀ 56; December, juvenile 62 (RMNH).

Structure. Wing long and rather narrow, pointed. 11 primaries: p10 longest, p9 7–15 shorter in adult, 4–9 in juvenile; p8 20–28 shorter, p7 36–48, p6 54–61, p5 70–78, p1 115–134; p11 minute, concealed by primary coverts. Tip of longest tertial reaches to tip of p3–p4 in closed wing. Tail rather short, shallowly forked; 12 feathers, tip of t6 slightly pointed, t6 on average 18 longer than t1 in breeding, 16 in adult and immature non-breeding, 12·5 in juvenile. Bill slender, straight, compressed laterally, about equal to head length; culmen slightly decurved at tip, without marked angle at gonys. Depth at basal corner of nostril 5·98 (0·23; 38) 5·7–6·5 in adult ♂, 5·53 (0·18; 31) 5·2–5·9 in adult ♀, variably less in juvenile, depending on age. Tarsus short and slender. Toes slender, relatively long when compared with *Sterna* terns. Front toes connected by incised webs, but incisions less deep than other *Chlidonias*: in outer web, reach to 2nd–3rd phalange, in inner web, to 1st–2nd phalange. Outer toe *c.* 92% of middle, inner *c.* 71%, hind *c.* 34%.

Geographical variation. Slight. North American *surinamensis* slightly smaller and darker: wing of adult ♂ 212 (12) 201–222, ♀ 209 (18) 199–220; head, chest, and belly of adult ♂ breeding deep black (similar to White-winged Black Tern *C. leucopterus*), mantle and scapulars darker grey than adult ♂ breeding of nominate *niger*; adult ♀ *surinamensis* about as dark as adult ♂ nominate *niger*. Leading edge of wing white, broader and contrasting more with upperwing than in nominate *niger*. Upperparts of non-breeding *surinamensis* slightly darker grey than nominate *niger*; juveniles similar but dark grey patches on sides of chest in *surinamensis* often extend to flanks, with underparts white only on throat, centre of chest and belly, vent, and under tail-coverts.

CSR

Chlidonias leucopterus White-winged Black Tern

PLATES 12, 13, and 14
[between pages 158 and 159, and facing page 206]

Du. Witvleugelstern   Fr. Guifette leucoptère   Ge. Weissflügelseeschwalbe
Ru. Белокрылая крачка   Sp. Fumarel aliblanco   Sw. Vitvingad tärna   N. Am. White-winged Tern

Sterna leucoptera Temminck, 1815

Monotypic

Field characters. 20–23 cm (bill 2·3–2·4, legs 1·9–2·2, tail up to 5 cm); wing-span 63–67 cm. Close in size to Black Tern *C. niger* but with shorter, stubbier bill, tubbier body, blunter wing-tips, and shorter, even less forked tail; at least 12% smaller but proportionately less slim than Whiskered Tern *C. hybridus*. Small and rather compact marsh tern, often recalling Little Gull *Larus minutus*. Breeding plumage mainly black, strikingly relieved above by silver-white wing-coverts and white rump, vent, and tail; wing pattern reversed on undersurface, where black wing-coverts contrast with pale grey undersurface to flight-feathers. Juvenile plumage of black-brown saddle, contrasting with silvery wings, broad white collar, and white rump, diagnostic. Immature and winter adult less distinctive but both paler above than *C. niger*, lacking any mark at shoulder. Sexes almost similar; marked seasonal variation. Juvenile and 1st-year immature separable.

ADULT BREEDING. Head, back, body, and axillaries black, glossed blue on head and mantle. Rump, vent, and tail white. At rest, folded wings show bold white carpal area and silvery central coverts contained within dark grey greater coverts and secondaries, and silver-grey edges to folded primaries. In flight, pattern of both surfaces of wing striking: above, silver-white panel on inner forewing contrasts with darker surround; below, black wing-coverts are isolated against paler, grey flight-feathers. Bill obviously shorter than head, relatively deeper and less tapering than in *C. niger* and appearing stubby; black, tinged crimson. Legs bright red. ADULT NON-BREEDING. Change in plumage pattern and colour even more pronounced than in *C. niger*; relatively darker bird becomes relatively paler, passing through period of pronounced piebald appearance on head, body, back, and under wing-coverts. When moult complete, shows (1) less black on head than *C. niger*, with black spot before eye and patch behind eye both smaller, black crown shorter and speckled white overall (only looking uniform at longer range), and narrow white extension of forehead above and behind eye (separating speckles of crown from black patch behind eye), (2) wide white collar, not invaded by black mark at shoulder as in *C. niger*, (3) pale grey back and upper wing-coverts, not uniform with duskier flight-feathers nor showing dusky leading edge as in *C. niger*, (4) almost white rump, contrasting with back and greyer tail and not uniform as in *C. niger*, and (5) pure white under wing-coverts, not pale grey as in adult *C. niger*. With wear, outer primaries may show dusky lines, and tail fades to almost white. Bill black; legs dark red. JUVENILE. Unlike *C. niger*, does not closely resemble winter adult, having striking, variegated upperpart pattern. Head and body as adult but head with rear crown uniformly black, and buff wash overall when fresh. Back shows as rich black-brown saddle at distance, but at close range its colours appear in separate bands. At distance, uppersurface of wing appears almost entirely silvery, but at close range shows (1) dusky outer 3 primaries, creating dark leading edge to most of outer wing, (2) dusky band along trailing edge of innermost primaries and outer secondaries, and (3) narrow faint brown panel across inner lesser wing-coverts, not forming obvious dark leading edge as in *C. niger*. On most, effect of these 3 characters is to contain silvery area to broad central panel, but on some, 2nd character not obvious and most of wing appears uniform silvery. Underwing white, with dusky outer primaries and dark trailing band to inner flight-feathers obvious in some lights. Striking contrast between 'saddle' and wings diagnostic and darkness of former also makes collar appear broader than in *C. niger*, often visible from behind. Rump and tail white, with dusky lines or tips on latter. Bill initially with yellow base; legs dull red. FIRST WINTER. Plumage change inadequately studied but some moult erratically or suffer unusual wear and fading. When body plumage renewed, dark 'saddle' completely lost and head on some noticeably pale, but separation from adult still allowed by retention of worn juvenile wing-feathers (often showing complete dusky rim from primary coverts round wing-tip to tertials, and paler greater coverts) and duller grey, virtually square tail. No records of '*portlandica*'-type plumage pattern—this clearly related to initial paleness of upperwing. SUBSEQUENT PLUMAGES. In 1st summer and 2nd winter (spent in Africa), resembles winter adult; may show small incidence of breeding plumage (e.g. partially black under wing-coverts) in 1st summer, but not fully acquired until 2nd spring. Birds visiting west Palearctic which differ from full adult only in darker outer primaries, less complete silver-white carpal panel, and duller black body thus at least 2 years old.

Summer adult and juvenile unmistakable. 1st-year bird and winter adult liable to confusion with *C. niger*, but in addition to characters discussed above, flight, structure, and habits differ (see below). Birds of same age also show similar plumage pattern to *C. hybridus* but are smaller, less rakish in form, shorter billed, and less dashing in flight; see *C. hybridus*. Flight of similar style to *C. niger* but shallower wing-beats produce rather steadier track; at times, less erratic flight, rather blunt wing-tip, tubby body, and (from some angles) plumage pattern may combine to recall Little Gull *L. minutus*. In addition to blunter wing-tip and tubbier body, stubbier bill and shorter and squarer tail combine in less attenuated configuration than that shown by *C. niger*. Gait and carriage as *C. niger* but stands slightly higher. Retains preference for inland marshes and waters throughout; infrequent on sea-coasts.

Voice more churring and slurred than *C. niger*, and thus less piercing. At all times of year, a loud 'kerr', 'krek', or 'kreek' denotes excitement or alarm.

Habitat. Breeds in west Palearctic in continental interior middle latitudes from boreal through temperate and steppe to Mediterranean zones, mainly in lowlands but locally to 2000 m in Armeniya (Dementiev and Gladkov 1951c). Avoids mountains, deserts, and forests, and among wetlands is largely confined to natural shallow flooded grasslands or swampy standing water, often bordering large rivers, or lakes which may be freshwater or alkaline, with open areas bordered by stands of reed *Phragmites*, sedge *Carex*, and other aquatic plants. In Hungary, misses breeding in drought years, compensating in succeeding wet springs, especially when heavy rainfall follows snowy winters; prefers there rather marshy alkaline habitats, and, unlike Black Tern *C. niger*, does not occupy fish-ponds, ricefields, or ornamental waters. Also distinguished ecologically by alighting on water more frequently, and wading in search of food (see Bannerman 1962; Kapocsy 1979). Accounts differ as to frequency of sharing breeding colonies and foraging areas with *C. niger*, but appears less linked with open water fringed by reedbeds and more often to occupy transitional or fluctuating marginal inundations, small pools, and swamps than deeper stable lakes, pools, or lagoons. Such distinctions, however, not clear-cut.

Outside breeding season, occurs also along coasts on lagoons and in mangrove swamps, but mainly along rivers, on floodlands, and by lakes, foraging as well for grasshoppers and other insects on land; over cornfields and stubbles in Iraq. Noted above 2500 m in Ethiopia (Bannerman 1962). May be found far from water over dry plains, and has been recorded following the plough (Benson *et al.* 1971). Sometimes rises to higher airspace, although normally flying within *c.* 6 m of water or ground.

Distribution. FRANCE. May have nested (Yeatman 1976). BELGIUM. Bred 1937 (Lippens and Wille 1972; PD). WEST GERMANY. Bred 1936, and earlier in Bayern (Niethammer 1942). SWEDEN. Bird sitting on eggs Öland 1978; no mate seen and unsuccessful (LS). CZECHOSLOVAKIA. May have nested Slovakia 1949 and 1967, but no proof (Hudec and Černý 1977). ITALY. Pair nested Piemonte 1978, 4 pairs 1982 (C Pulcher). AUSTRIA. Formerly bred in east, last 1892 (MS). YUGOSLAVIA. Bred Vojvodina 19th century and perhaps 1963–7; bred Lake Skadar 1975. BULGARIA. Bred Lake Sreburna 1955; may nest occasionally on coast (SD). USSR. Breeds irregularly in Estonia: 1 pair 1982 (HV); nests occasionally Latvia and Lithuania (Valius *et al.* 1977; Viksne 1978). Northward extension in Yaroslavl' region since *c.* 1970 (Belousov and Makkoveeva 1981). TURKEY. Has possibly bred, but no proof (OSME). SYRIA. No proof of breeding (Kumerloeve 1968b).

Accidental. Iceland, Faeroes, Ireland, Britain (now annual), Belgium, Norway, Sweden, Finland, East Germany, Lebanon, Portugal, Madeira.

Population. POLAND. Fairly numerous in 19th century in Lublin region, now extremely scarce (Tomiałojć 1976a). Bred 1938 and since 1966; main area Biebrza marshes where 20–25 pairs 1966–8, *c.* 90 pairs 1971, 200–250 pairs 1978, and 475 pairs 1979, with smaller colonies elsewhere (Tomiałojć 1976a; AD). HUNGARY. Breeds in small numbers; for details of colonies see Kapocsy (1979). RUMANIA. Rare (MT); decreasing due to drainage (VC); nests irregularly elsewhere (Vasiliu 1968). IRAQ. Common (Allouse 1953).

Movements. Migratory. West Palearctic populations winter widely in inland waterbodies of Afrotropical region; large numbers in Africa suggest many birds from central USSR may also winter there. Origins of birds wintering Persian Gulf, Pakistan, and India south to Ceylon not certain; those wintering Burma and Malay peninsula are probably from central and east Asian population, which also winters southern China, Indonesia, and Australia.

Autumn migration begins mid-August in northern USSR, late August to early October further south, though birds may leave breeding grounds in Volga region late July (Borodulina 1960; Kapocsy 1979). Passage slight in northern and western Europe (Bernis 1967; Kapocsy 1979).

Chlidonias leucopterus 157

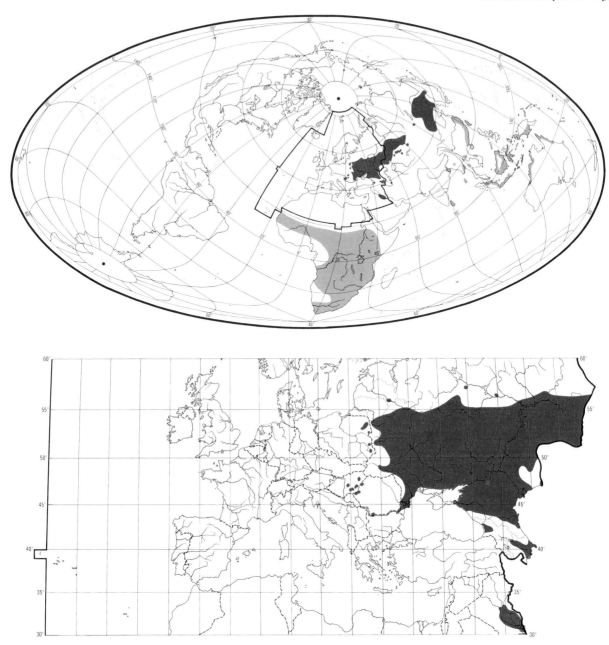

Adults rare but immatures occur annually in Britain and Ireland, mainly August–September; see Sharrock (1974) for discussion of origins of these birds. In central and south-east Europe, much less common in autumn than spring, e.g. southern Bayern (Bezzel and Reichholf 1965), Switzerland (Jacoby *et al.* 1970); no autumn passage recorded in western Czechoslovakia (Fiala 1974). Significant passage through Balkan peninsula and eastern Mediterranean; occurs Italy during August–September (Brichetti 1976), and Greece mainly from 1st week of August to mid-September (Bannerman and Bannerman 1958). Common on passage August–September at Azraq (Jordan) and in Israel, Iraq, and Iran (Moore and Boswell 1956; Passburg 1959; Nelson 1973; L Cornwallis), though much less common at Elat (Israel) than in spring (Safriel 1968). A few winter on eastern shore of Black Sea (Strokov 1974). Huge flocks occur on eastern coast of Sea of Azov in late August (Borodulina 1960). Slight passage recorded Bosporus, Turkey (Ballance and Lee 1961). In western Mediterranean, negligible movement through Straits of Gibraltar, rare in Morocco, but more common (though never abundant) Algeria and Tunisia (Bernis 1967; Bundy and Morgan 1969).

Birds wintering West Africa may travel south-west

across Sahara since no coastal movement, and evidence of inland movement in spring (see below). Numerous in Sénégal from beginning of September (Morel and Roux 1966; Smet and Gompel 1980). Movement across Sahara perhaps also accounts for birds wintering in upper Niger inundation zone (Mali), where abundant (Duhart and Descamps 1963), and Chad, where commonest of Sternidae, arriving mainly mid-August and September, earliest 29 July (Newby 1979). Elsewhere in West Africa, probably mostly a passage migrant. Access to East, central, and southern Africa initially mostly via Nile. Heavy passage through Egypt, 16 August–18 October, though rare in Sinai (Meinertzhagen 1930). Spreads thus into Sudan, occurring on all main rivers, and numerous wherever open water available south to Cape Province; particularly numerous in Rift Valley lakes of Uganda, Kenya, and Tanzania (Gaugris 1979). Distribution extends to western seaboard; numerous in Zaïre from October (Ruwet 1964). In contrast to Black Tern *C. niger*, essentially an inland species, occurring rarely on coast. Within Africa, as in Australia, may move long distances in response to changes in local feeding conditions (e.g. Ruwet 1964).

In spring, mass concentrations occur on passage in Rift Valley lakes, March–May (Britton and Brown 1974). Numbers in Zaïre build up February–April (Chapin 1939). Spring migration routes in Africa probably reverse of autumn. Recorded resting at 2580 m on Ethiopian plateau (Bannerman 1962). About 1000 recorded at confluence of Blue and White Niles (see Bernis 1967), and common in Egypt (Meinertzhagen 1930). Trans-Saharan passage indicated by northward movement 500 km inland in Sénégal, April (Morel and Roux 1966), and by appreciable numbers in south-east Morocco, late April (Smith 1968b). Usually scarce in Morocco, Algeria, and Tunisia, though more common in north-west Africa in spring than autumn (Etchécopar and Hüe 1967), and hundreds once seen in Algeria in early May (Makatsch 1957). Slight passage through Portugal, Ebro delta, and occasionally Guadalquivir (Spain), but rare in rest of Iberia (Kapocsy 1979). Also passes regularly in small numbers, often with *C. niger*, through Camargue (France) and Malta (Guichard 1956a; Isenmann 1975b; Sultana *et al.* 1975; Kapocsy 1979). Regular occurrence, May–June, in Switzerland, southern Bayern, Rhine valley, and north-west West Germany may originate from western Mediterranean (Wehner 1966; Müller 1967). Less frequent in Britain in spring (chiefly May–June) than autumn (Sharrock 1974). However, most enter eastern Mediterranean from Egypt. Some cross Mediterranean, reaching Balkan peninsula via Crete; others travel through Jordan, Israel (e.g. flocks of hundreds recorded at Elat: Safriel 1968), and Syria (Hollom 1959; Nelson 1973), after which routes probably diverge, some birds going through Turkey to eastern Europe and European USSR, others turning north-east towards region between Black and Caspian Seas (Kapocsy 1979). Arrives south-west USSR first half of May, 15–20 May in north (Dementiev and Gladkov 1951c; Borodulina 1960).

Known to summer in tropical Africa, e.g. Rift valley lakes, Zaïre, Sénégal (Ruwet 1964; Morel and Roux 1966). Plumage of African specimens indicates birds stay in winter quarters throughout 2nd calendar year (Williamson 1960; Lippens and Wille 1976; Kapocsy 1979). One Nearctic record in May in New Brunswick, Canada (Kapocsy 1979), another in October in Barbados, West Indies (Watson 1966; R van Halewijn). EKD

Food. Mainly aquatic and terrestrial invertebrates (mostly insects); sometimes fish, and less frequently amphibians. Prefers freshwater feeding habitat at all times of year, especially lakes and rivers; in Kenya also feeds far from water (Britton and Brown 1974); in USSR, takes insects in aerial-pursuit over waterless steppes (Borodulina 1960), and in Mali seen catching grasshoppers (Orthoptera) disturbed by grass fires (Curry and Sayer 1979). Swifter and more agile in flight, with less hovering whilst foraging, than Black Tern *C. niger* with which it often feeds (Witherby *et al.* 1941; Borodulina 1960; Halle 1968). Migrants often feed in large flocks (Borodulina 1960; Bannerman 1962; Ruwet 1964; Britton and Brown 1974). Skims insects from water surface or takes insects from vegetation, or hovers and catches prey by dipping-to-surface (Moore and Boswell 1956a; Amiet 1957; Borodulina 1960; Begg 1973; Kapocsy 1979); on Lac Léman (Switzerland) when wave action prevents skimming, forced to feed by dipping into water for each capture (Halle 1968). In New Zealand, recorded walking along muddy fringe of pools snapping up insects (Sibson 1954). Follows predatory fish and porpoises *Phocoena* to take small fish driven to surface (Simmons 1952; Ridley 1954).

Insects include: dragonflies (Odonata), beetles (Carabidae, Aphodiinae, Elateridae, Bruchidae, Cerambycidae, Chrysomelidae, Curculionidae, Dytiscidae), bugs (Cydnidae, Coreidae, Lygaeidae), grasshoppers and crickets (Orthoptera), flies (Diptera), ants (Formicidae), mayflies *Emphemera vulgata*, caddis flies (Trichoptera), termites (Isoptera), and caterpillars of moths (Noctuidae); other invertebrates include spiders (Araneae), molluscs, and crustaceans. Fish include carp *Cyprinus*, sticklebacks (Gasterosteidae), Tanzanian sardine *Limnothrissa miodon*, *Haplochromis*, and *Synodontis*; amphibians include adult frogs, and tadpoles. (Witherby *et al.* 1941; Dementiev and Gladkov 1951c; Borodulina 1960; Cvitanić and Novak 1966; Polivanova 1971; Serventy *et al.* 1971; Begg 1973; Crawford 1977; Kapocsy 1979.)

In USSR, chiefly aquatic insects: tends to take more adults, especially of dragonflies, but fewer water-beetles, than *C. niger*; stomachs of birds feeding on overgrown pools and swamps in Novosibirsk contained mainly dragonflies (52%) and larvae of water-beetles (39%); in Volga delta fed mostly on longhorn- and leaf-beetles, with some ants, dragonflies, and spiders; those foraging on fish-farms

PLATE 7. *Sterna dougallii* Roseate Tern (p. 62): **1** 1st imm non-breeding (1st winter), **2** juv. *Sterna hirundo* Common Tern (p. 71): **3** 1st imm non-breeding (1st winter), **4** juv. *Sterna paradisaea* Arctic Tern (p. 87): **5** 1st imm non-breeding (1st winter), **6** juv. *Sterna forsteri* Forster's Tern (p. 102): **7** 1st imm non-breeding (1st winter). *Sterna repressa* White-cheeked Tern (p. 105): **8** 2nd imm breeding (2nd summer), **9** juv. (NWC)

PLATE 8. *Sterna hirundo* Common Tern (p. 71): **1** ad breeding, **2** 1st imm non-breeding (1st winter). *Sterna forsteri* Forster's Tern (p. 102): **3–4** ad breeding, **5** 1st imm non-breeding (1st winter). *Sterna repressa* White-cheeked Tern (p. 105): **6–7** ad breeding, **8** 1st imm non-breeding (1st winter). (NWC)

PLATE 9. *Sterna anaethetus* Bridled Tern (p. 109): **1** ad breeding, **2** ad non-breeding, **3** 1st imm non-breeding, **4** juv, **5** downy young. *Sterna fuscata* Sooty Tern (p. 116): **6** ad breeding, **7** juv (worn plumage). *Anous stolidus* Brown Noddy (p. 163): **8** ad, **9** juv (worn plumage). (NWC)

PLATE 10. *Sterna anaethetus* Bridled Tern (p. 109): **1–2** ad breeding, **3–4** juv. *Sterna fuscata* Sooty Tern (p. 116): **5–6** ad breeding, **7–8** juv. *Anous stolidus* Brown Noddy (p. 163): **9–10** ad, **11** juv (worn plumage). (NWC)

PLATE 11. *Sterna albifrons* Little Tern (p. 120): **1–2** ad breeding, **3** ad non-breeding, **4–5** juv moulting into 1st imm non-breeding (1st autumn), **6** juv, **7** downy young. (NWC)

PLATE 12. *Chlidonias hybridus* Whiskered Tern (p. 133): **1** ad breeding, **2** 1st imm breeding (1st summer), **3** downy young. *Chlidonias niger* Black Tern (p. 143): **4** ad breeding, **5** ad breeding moulting into non-breeding, **6** downy young. *Chlidonias leucopterus* White-winged Black Tern (p. 155): **7** ad breeding, **8** 2nd imm breeding (2nd summer), **9** downy young. (NWC)

also took small fish (including carp), and small frogs; in July, flocks gathering in Volga delta after breeding take mainly sticklebacks many of which probably already dead; recorded 'exterminating' plague of caterpillars of army worm (Noctuidae) (Borodulina 1960). In Yugoslavia, 6 stomachs, April–May, contained the crustacean *Upogebia litoralis* (from sea shallows), dragonfly larvae and beetles (from river shores), and flies and beetles taken in flight (Cvitanić and Novak 1966).

In winter quarters, Kenya, several thousand fed on concentrations of army worm caterpillars *Laphygma exempta*; in Lake Victoria (Uganda), fed on abundant whitebait *Mulcene* (Britton and Brown 1974). Seasonal variation noted in West Africa: in November hunts for insects over plains regenerated by first rains, in December–January catches fry of cichlid fish trapped in residual pools, and in February–March hunts over newly flooded plains, often taking insects from tips of stems, leaves, and flowers of *Oryza* and *Cyperus* (Ruwet 1964). 33 stomachs, Lake Kariba (Zimbabwe), August–January, contained mainly fish, especially *L. miodon*, and less *Haplochromis* and *Synodontis*; in marginal areas takes mainly aquatic insects (particularly dragonflies), and terrestrial insects (especially grasshoppers, bugs, beetles, and termites) found drifting on mats of *Salvinia*—also takes them by hawking or skimming from surface (Begg 1973). In Australia, feeds at sewage outlets, over land, and at sea on shoals of small fish drawn to surface by tunny *Thunnus* (Serventy *et al.* 1971). From Darwin (Australia), stomachs of birds feeding over fallow ricefields, December, contained only insects, including water-beetle larvae, weevils, ants *Oecophylla virescens*, grasshoppers, and spiders (Crawford 1977). In winter quarters, Western Australia, small parties (*c.* 50) hovered over shore catching dragonflies *Hemianax papuensis* (Alexander 1917).

Chicks at Lake Baykal (USSR), fed initially on insects, primarily dragonfly larvae, from which young often extracted and ate only soft viscera; also fish 0·8–1·5 cm long (Mel'nikov 1977).

Social pattern and behaviour. Data sparse from west Palearctic. Compiled largely from work of Mel'nikov (1977) at Lake Baykal (eastern USSR) and observations of birds on migration and in winter quarters; some material from Australia.

1. Markedly gregarious. Throughout the year, often hunts in flocks, usually small but sometimes of several hundred; especially large assemblages, up to tens of thousands, found in winter quarters (Bannerman 1962; Britton and Brown 1974; Gaugris

PLATE 13 (*facing*).
Chlidonias hybridus Whiskered Tern (p. 133): **1–2** ad breeding, **3** ad non-breeding, **4** juv.
Chlidonias niger Black Tern (p. 143): **5–6** ad breeding, **7** ad non-breeding (winter), **8** juv (worn plumage).
Chlidonias leucopterus White-winged Black Tern (p. 155): **9–10** ad breeding, **11** ad non-breeding (autumn, fresh plumage), **12–13** juv moulting into 1st imm non-breeding (1st autumn or winter), **14** juv. (NWC)

1979). Feeding flocks in breeding season 3–15 birds. On migration, flocks vary in size: usually less than 10 on Caspian Sea in autumn (Feeny *et al.* 1968); 20–30, northern Iran, May (Passburg 1959); up to 50, Iraq (Chapman and McGeoch 1956); hundreds or thousands (Britton and Brown 1974). 2 migrating flocks, southern Bayern (West Germany), May, were of 40–50 and 110 birds (Bezzel and Reichholf 1965); autumn flocks, Lake Baykal, 30–1000 (Mel'nikov 1977). Flock composition varies through the year. Outside breeding season, immatures associate with older birds (e.g. Alexander 1917). At Lake Baykal, late migrant flocks comprised 10–64% juveniles (Mel'nikov 1977). Flock of up to 100 immatures, with no adults, recorded Nigeria (Wallace 1973). Commonly mixes with Black Tern *C. niger* and Whiskered Tern *C. hybridus* on migration, and for feeding (Wadley 1951; Feeny *et al.* 1968). BONDS. Monogamous pair-bond. No information on length of bond or age at which it develops. No breeding until at least 2 years old (see Plumages). Both sexes incubate, and care for young (Witherby *et al.* 1941; Bannerman 1962). Typically, 1 parent (usually ♂) accompanies fledged young, bond probably persisting for a few weeks; at Lake Baykal, 128 family parties comprised 1 adult and 2 young, 3 had 1 adult and 3 young, and 1 had 1 adult and 4 young (Mel'nikov 1977). BREEDING DISPERSION. Forms colonies, usually small; sometimes in band near shore, sometimes more scattered (Mel'nikov 1977). Of the 26 Hungarian colonies, 1951–72, 17 had 1–5 pairs, 7 had 10–20 pairs (Kapocsy 1979). According to L Horváth (see Bannerman 1962), breeds in Hungary in large numbers (colonies of 20–40 pairs, exceptionally 100–120) only in unusually wet springs following several years of drought; only sparsely (colonies of 4–5 pairs) in wet springs following a single dry year. In USSR, colonies mostly of up to 40 pairs (Dementiev and Gladkov 1951c), but over 1000 pairs at Lake Evoron, eastern USSR (Roslyakov 1979). Colonies usually divided into sub-colonies, e.g. 90 pairs in 2 groups, Lake Nagyiuan, Hungary (Cramp 1968); at Lake Baykal, sub-colonies usually of 2–10 pairs (Mel'nikov 1977). Distance between nests variable, 2·5–30 m (Nadler 1967; Kapocsy 1979). Colony of 16 pairs, Hungary, comprised 2 groups of 5 and 6 nests each, remaining 5 more isolated from one another. 4–10 m between nests within groups, compared with 4–25 m for whole colony; most frequently recorded distance 7 m, mean 10·2 (Kapocsy 1979). On large lakes, density 0·82–0·90 nests per 100 m^2, on small lakes up to 14 nests per 100 m^2 (Mel'nikov 1977). Small nest-area territory defended (Pulcher 1979); serves for ground courtship, nesting, and resting of off-duty bird. Opportunistic in choice of when and where to breed, showing weak fidelity between years to colony area, and even less to specific nest-site. If colony partially flooded, nests may be rebuilt; if all nests lost, colony-site usually abandoned for that season and, if sufficient time, all pairs re-lay elsewhere (Mel'nikov 1977). In Hungary, usually breeds near Black Tern *C. niger*; also with Whiskered Tern *C. hybridus* and Black-headed Gull *L. ridibundus*. Rarely mixed with these, though 1–2 occasionally among *C. niger* (Tarjan 1942; Béldi 1963; Nadler 1967; Sóvágó 1968; Kapocsy 1979; Pulcher 1979). Grebes (Podicipediformes) also close nesting associates (Bod and Molnár 1976). ROOSTING. At least early in breeding season, communal resting place may be established outside colony area. During incubation, off-duty bird loafs at edge of nest-platform or on suitable vantage point nearby, e.g. floating vegetation, driftwood, post. Post-breeding flocks of 50–1000 birds sometimes assemble near colony (Polivanova 1971; Mel'nikov 1977). Outside breeding season, loafs on piers, piles, mud- or sand-banks, or saltpans (Amiet 1957; Borodulina 1960; Bannerman 1962; Halle 1968), rarely on sea-shore (but see Bannerman 1962). Overnight roost, Australia, in reedbeds (Alexander 1917). On spring migration,

Ethiopian plateau, 4 seen roosting at 2580 m on ground (see Bannerman 1962). Roost size usually varies from a few to hundreds (Alexander 1917; Amiet 1957; Bannerman 1962; Ruwet 1964), but thousands seen on sand-banks, Volga delta, July (Borodulina 1960). Temporary loafing groups may form between feeding bouts, Australia, especially at high tide; the higher the tide, the bigger the roost (Hamilton 1957). At dusk, immature birds (presumably due to less efficient feeding) may join roost later than older birds (Alexander 1917). Often roosts with other Sternidae, especially *C. hybridus*, in winter quarters; also with gulls (Laridae) and waders (Charadrii), though preferring conspecifics as neighbours (e.g. Sibson 1954, Amiet 1957, Hamilton 1957, Tree 1978).

2. FLOCK BEHAVIOUR. Communal bathing recorded on autumn migration (Halle 1968). ANTAGONISTIC BEHAVIOUR. Especially at start of breeding season, territory-owner threatens conspecific intruder with version of Alarm-call (see 1 in Voice), delivered with gaping bill, while watching every movement (Kapocsy 1979). HETEROSEXUAL BEHAVIOUR. Little information. (1) General. Pair-formation involves both aerial and ground courtship at colony. (2) Aerial courtship. Begins *c.* 5–6 days after arrival at colony-site. At Lake Baykal in late May, groups of *c.* 100–150 birds make circling and weaving flights over marshy areas close to prospective colony-site. Individuals may transfer between adjacent flocks. By June, displaying flocks 20–25, maximum *c.* 50 birds (Mel'nikov 1977). High-flights presumably develop from flock displays. High-flight usually involves several birds—as in *C. niger*, to which generally similar (Kapocsy 1979). (3) Mating. Occurs at nest, display areas, feeding grounds, and roosts, mostly in afternoon (Mel'nikov 1977; Kapocsy 1979). Before mounting, ♂ adopts Forward-erect, as in *C. niger* (Kapocsy 1979). (4) Courtship-feeding. ♂ feeds ♀ at nest-site before and during laying, and for several days after start of incubation. ♂ hovers to transfer food to ♀ on nest, without alighting (Kapocsy 1979; Pulcher 1979). (5) Nest-relief. Incubating ♀ answers Advertising-call (see 3 in Voice) of ♂ flying overhead with Cooing-call or other call (see 4–5 in Voice); call 5 of ♀ appears to precipitate nest-relief. Relieving bird typically raises wings on alighting (Kapocsy 1979). RELATIONS WITHIN FAMILY GROUP. From Mel'nikov (1977). Parents may peck at shell to help chick out of egg, especially if debilitated by cold. Both parents feed young from 1½–2 hrs after hatching. During first 2 days after 1st chick hatches, parent broods almost constantly; if it leaves, does so briefly around midday. From 4–5 days old, young left alone for most of day. Brood seeks cover near nest 2–3 days after 1st chick hatches, returning to nest on arrival of parent. About 30% of 3-day-old young move 50–80 m from nest, at 4–5 days 150–200 m; chicks especially mobile in small colonies. On returning to colony, parent seeks young where last fed, and when chick calls, parent lands nearby. Siblings usually stay together. If one gets separated, parent locates it by voice, and calls to coax it back. If vegetation thick parent may lure chick to sparser vegetation for feeding. If small chick gets wet, parent may build platform of grasses, and brood chick. Young finally led up to 700 m from nest to near open water *c.* 30–70 cm deep; sometimes little dispersal from nest, however (probably varies with disturbance). Broods thus become scattered through colony area. After fledging, remain with parents in colony for 3–5 days, then disperse into adjoining areas where young alight on vegetation to be fed; this frequent for *c.* 10 days after fledging. Young maintain contact with parent by calling. As young become more skilled, increasingly hunt separately from parents in parties of 3–5 juveniles (Mel'nikov 1977). ANTI-PREDATOR RESPONSES OF YOUNG. In response to parental Alarm-call, small young swim from nest to seek refuge in vegetation, usually returning to nest when danger passes, or in response to undescribed call of parent (Mel'nikov 1977; Kapocsy 1979). PARENTAL ANTI-PREDATOR STRATEGIES. At approach of danger, gives Alarm-call (see 1 in Voice); see above for response of young. Parent coaxing stray chick back to join brood makes dive-attacks at conspecifics which threaten it. Nesting birds swoop at Common Gull *L. canus* from above and behind, striking with bill; defence more effective in larger colonies (Mel'nikov 1977). Swooped at human intruder to within 2 m of head, uttering Alarm-call (Nadler 1967); also gives Anger-call (see 2 in Voice) when flying overhead (Meyer 1965).   EKD

Voice. Vocal at breeding grounds, rather silent on migration (Meyer 1965). Generally less powerful than hoarse wheeze of Whiskered Tern *C. hybridus* (Moore and Boswell 1956*a*; Kapocsy 1979); similar to Black Tern *C. niger* but more churring and slurred, and thus less piercing (see Kapocsy 1979; E K Dunn).

CALLS OF ADULTS. (1) Alarm-call. A loud, rattling or churring 'kerr', more like *C. hybridus* than *C. niger*, but distinct from either (Witherby *et al.* 1941); 'karr' (Dementiev and Gladkov 1951*c*); 'chorr' (Meyer 1965); 'scher scher' (Kapocsy 1979); harsh, creaking 'kreh' in confrontation with conspecifics in winter quarters (Bromley 1952); 'krek krik krek', with 2nd unit slightly higher (J Hall-Craggs: Fig I); 'kreek-kreek' (Anon 1976) and 'kreet' (Hollom 1959), both heard in winter quarters, probably the same. (2) Anger-call. A single, short 'gäg' or 'güg' given by adults circling overhead when human near fledged young and interpreted as being analogous to call 2 of *C. niger*; also heard from flock shortly before take-off, perhaps in response to unseen danger (Meyer 1965). Related call (no context given) may be that rendered 'kek-kek-kek' or 'chik-a-tik-tik', and interpreted as threat-call (Anon 1976). (3) Advertising-call. A sharp 'keerr' or 'shkerr' given in flight by ♂ arriving at nest to relieve mate (Kapocsy 1979), and therefore assumed here to be Advertising-call. High-pitched 'keerrr' also accompanies (undescribed) ground threat-call (Kapocsy 1979). (4) Cooing-call. ♀ on nest responds to Advertising-call of ♂ with low, clucking call, repeated 2–3, rarely 5 times (Kapocsy 1979). Function unknown but, by analogy with *C. niger*, probably Cooing-call associated with Stoop-posture, inviting close approach. (5) ♀ being relieved on nest may also answer ♂ with call sharper and higher pitched than ♂'s Advertising-call (Kapocsy 1979). May be ♀ Advertising-call or Begging-call.

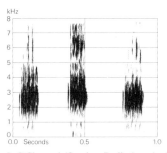

I C Chappuis/Sveriges Radio (1972) Rumania May 1967

CALLS OF YOUNG. Food-call of fledged young indistinguishable from that of *C. niger* (Meyer 1965). EKD

Breeding. SEASON. East-central Europe: see diagram. Novosibirsk (USSR): breeding begins first 2 weeks June (Borodulina 1960). Lake Baykal (USSR): arrives from mid-May; laying begins 16–20 days later; total spread of laying 40–44 days; last chick hatches c. 5 August (Mel'nikov 1977). SITE. On mat of floating vegetation (at Lake Baykal, in water 0·3–1·2 m deep); less commonly on shore (Mel'nikov 1977). Colonial. Nest: heap of waterweed, partially submerged with shallow depression on top. Average external diameter 15 cm; internal diameter 9 cm, cup 1·5 cm deep, height above water 4–5 cm (Borodulina 1960); on shore, unlined scrape. Building: probably by both sexes. EGGS. See Plate 90. Oval, smooth and slightly glossy; cream to pale buff, with large black or dark brown spots and blotches. 35×25 mm (29–37×23–27), sample 100 (Schönwetter 1967). Weight 10–11 g (Bannerman 1962). Clutch: 2–3(–4); clutches of 6, Lake Baykal (Mel'nikov 1977), probably by 2 ♀♀; laying interval of 18–24 hrs or longer (Mel'nikov 1977). Of 21 clutches, Hungary: 2 eggs, 4; 3, 17; mean 2·8 (Bannerman 1962). Of 35 clutches, Novosibirsk: 1 egg, 1; 2, 6; 3, 26; 4, 2; mean 2·83 (Borodulina 1960). One brood. Up to 2 replacement clutches, Lake Baykal; 1st clutch averaged 2·78 eggs, 2nd 2·56, 3rd 1·47 (Mel'nikov 1977). INCUBATION. 18–22 days, mostly 18–20 (sample 35) (Mel'nikov 1977); 19–20·5 days per egg (sample 5) (Kapocsy 1979). By both sexes, ♀ doing slightly more. In 58 nests of 3 eggs, Lake Baykal, began with 1st egg in 46·6%, 2nd egg in 44·8%, 3rd in 8·6%. Spread of hatching 1–4 days, usually 2–3 (Mel'nikov 1977). At start of incubation, ♀ has longer spells (19–30 min) than ♂ (10–13 min), but spells more equal (♀ 26 min, ♂ 23 min) close to hatching (Kapocsy 1979). YOUNG. Precocial and semi-nidifugous. Cared for and fed by both parents. Leave nest at a few days old and hide in surrounding vegetation (Kapocsy 1979). FLEDGING TO MATURITY. Fledging period 24–25 days (Mel'nikov 1977); 19 days (Borodulina 1960) probably minimum. Captive-reared young flew at 20–22 days (Kapocsy 1979). Age of independence not known, though juveniles acquire feeding skills within a few weeks (Mel'nikov 1977). Age of first breeding at least 2 years. BREEDING SUCCESS. At Lake Baykal, hatching success in small colonies 69–79%, in large colonies 38–62%; clutches of 3 eggs had 64·5% hatching success, 2 eggs 45·0%, 1 egg 7·1%. Hatching failure highest in 3rd eggs, and in 3rd replacement clutches. Chick mortality in first 3–4 days 5·3–20·0%; in some years, c. 25–35% fledge. Rarely more than 2 young fledge from 3-egg clutches. Main cause of mortality wind, rain, and subsequent flooding of nests; at Lake Baykal, also predation of eggs by Common Gull *L. canus*, especially in small colonies (Mel'nikov 1977).

Plumages. ADULT BREEDING. Head, neck, mantle, and underparts to vent and flanks black, slightly glossed blue-green on crown, sides of head, and neck. Scapulars and back dull black to dark slate-grey; rump, upper tail-coverts, and tail contrastingly white. Lower vent and under tail-coverts white, contrasting with black belly and flanks. Colour of primaries dependent on moult and wear: as in Little Tern *S. albifrons*, primaries show 3 successive moulting series in non-breeding season, arrested during breeding; 1st series oldest, restricted to outer 2–3(1–4) primaries, blackish, slightly tinged silvery-grey in spring, brown-black in autumn; 2nd series, (p1–)p2–p4 to (p6–)p7–p8(–p9), pale grey on inner webs, pale silvery-grey or nearly white on outer webs, contrasting with black 1st series, inner feathers slightly duller grey; 3rd series, extending to p1–p3 (sometimes none), pale silver-grey. Inner webs of all primaries with white wedge at base. Narrow black or dark grey border beneath trailing edge of outer primaries. Secondaries and tertials medium grey, inner secondaries often slightly duller. Inner greater upper wing-coverts medium grey, grading to pale grey on outer greater and on median coverts and to uniform white on lesser coverts. Under wing-coverts and axillaries black, except for grey greater coverts and white coverts along leading edge. Sexes closely similar, but ♀ often less glossy on head; scapulars and underparts slightly tinged slate-grey, especially when worn; tail partly tinged grey in about 40% of those examined, varying from slight grey wash on tips of tail-feathers to all tail light grey except for outer web of t6; exceptionally, ♂ also shows slight grey tinge to tail-tip. ADULT NON-BREEDING. Forehead and forecrown white; ring round eye white, broken by small black patch in front of eye. Hindcrown and square patch on centre of nape black (hindcrown occasionally light grey), feathers fringed white; ear-coverts black. Collar across upper mantle white, bordered behind by transverse black of dark slate band. Slate of mantle not extending down to sides of chest as in Black Tern *C. niger*. Lower mantle, scapulars, and back light grey, rump and upper tail-coverts slightly paler; tail light grey with white outer web to (t5–)t6. Underparts, including under wing-coverts and axillaries, uniform white. Colour of flight-feathers depends on state of wear and moult; in autumn (September–December), secondaries and outer primaries black (sometimes with slight grey bloom), inner primaries light grey; during December–February, outermost primaries medium grey with silver-grey bloom, inner 1–5 and all secondaries pale grey; from March, similar to adult breeding. Upper wing-coverts light grey or pale grey, occasionally with slightly darker centres; shorter lesser coverts slate-grey, forming rather indistinct carpal bar when fresh, September–October,

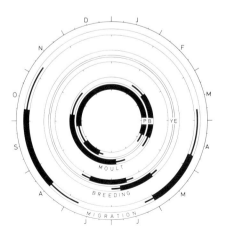

darkening to dark slate or black bar c. 1 cm wide during winter and spring, markedly contrasting with other coverts, and pied in spring when white breeding lesser coverts are growing in. Closely similar to non-breeding *C. niger*, but forecrown white and hindcrown paler, without sharp contrast between white forehead and uniform black crown of *C. niger*; black of central nape extends further down hindneck; dark bar across mantle more contrasting; grey of mantle, scapulars, upper wing-coverts, and back to tail slightly paler; dark carpal bar (when present) wider and more contrasting than in adult non-breeding *C. niger* (that of juvenile *C. niger* about similar, however); underparts uniform white without distinct dark patches at sides of chest (small patches at upper sides of chest sometimes present); a few black or dark grey under wing-coverts occasionally present up to December, sometimes reappearing from mid-February. DOWNY YOUNG. Closely similar to *C. niger*, but throat more often black-brown than in latter and usually more extensively so, more strongly contrasting with white area round eye; forehead and underparts often darker brown; less white to central chest and belly (Fjeldså 1977; RMNH). JUVENILE. Similar to adult non-breeding, but forehead and crown tinged buff-brown, feathers of mantle and back deep brown to brown-black; scapulars, tertials, back, tail, and upper wing-coverts light grey with brown feather-tips (often darker sub-terminally); rump greyish-white, markedly contrasting with back; upper tail-coverts pale grey; flight-feathers grey with slight silvery-grey bloom, without marked contrast due to moult as in adult; no black on under wing-coverts, unlike some non-breeding adults. In worn plumage, brown of feather-tips tends to bleach and abrade; then resembles adult more closely, but upper wing-coverts less uniform pale grey and with traces of darker subterminal marks on innermost. Closely similar to juvenile *C. niger*, but grey of scapulars, tertials, tail-, flight-feathers, and upper wing-coverts paler, rump nearly white (matched by some *C. niger*, but these lack contrast with dark back); crown more extensively white, often isolating black ear-patch from black central nape; mantle, back, and part of scapulars uniformly dark, forming dark saddle, contrasting with pale upper wing-coverts (some *C. niger* show similar though less contrasting saddle); side of chest lacks dark patch or shows a trace at upper side only. FIRST AND SECOND IMMATURE NON-BREEDING. Like adult non-breeding and difficult to distinguish by plumage alone when last juvenile tertials, tail-feathers, and upper wing-coverts shed December–March, but primary moult different: during December–February, many outer primaries juvenile, black-brown and heavily worn, and 1–5 inner primaries new or moulting (in adult, 1st series actively moulting outer primaries or these just completed, all new and grey, 2nd series started with inner primaries, 1–5 new or moulting); during March–June, outermost primaries still juvenile, fresh 1st series reaching to p6–p8, 2nd to p1–p3 (in adult, 1st series reached p10, outer blackish but fresh, 2nd to p6–p9, 3rd to p1–p3, and all plumage in breeding or nearly so). During summer, 1st non-breeding replaced directly by 2nd non-breeding; attains only a limited aount amount of breeding or none at all—restricted to scattered feathering of scapulars, belly, or lesser coverts. 2nd immature non-breeding like adult non-breeding, but former has outer primaries rather fresh in autumn and early winter, rather than distinctly worn. Non-breeding plumages differ from those of *C. niger* in same respects as adult non-breeding. Breeding plumage first attained in spring of 3rd calendar year; birds retaining scattered non-breeding feathers on head, underparts, tail, and wing-coverts are presumably retarded 2-year-olds.

Bare parts. ADULT. Iris dark brown. Bill dark red or black with crimson tinge in breeding, all-black in non-breeding, occasionally with some red at gape. Foot coral-red or orange-red in breeding, dark red to orange-red in non-breeding, in latter with grey or black shade variable in extent; claws black. DOWNY YOUNG. Iris dark brown. Bill dark grey, vinous at gape. Foot greyish-pink. JUVENILE. Iris dark brown. Bill black-brown with some pink-red or orange-red at gape, foot red to red-brown, both soon shading to adult non-breeding colour. Non-breeding colour retained throughout 2nd calendar year. (Witherby *et al.* 1941; Fjeldså 1977; Kapocsy 1979; RMNH.)

Moults. ADULT POST-BREEDING. Complete; primaries descendant. Starts from June; feathering at gape and eye-ring first, soon followed by lores and chin. When face and throat largely white, chest, sides of neck, crown, and inner primaries follow from about July. Leaves breeding area with head, neck, chest, and part of mantle, scapulars, breast, and belly in non-breeding; primaries arrested with up to p1–p3 new, t1 often new. Only a few reach winter quarters in this plumage; apparently much moult during stops on migration, and on arrival in October plumage completely non-breeding, including tail (sequence t1–t2–t3–t6–t4–t5), but sometimes excepting some old breeding on underwing, tertials, lesser upper wing-coverts, or t4–t5; primaries arrested with up to p5 or p8 new; variable number of outer secondaries new. 1st series of flight-feathers completed by December or March. Some birds retain some old black feathers on underwing and underparts up to December. A 2nd series of primaries starts late November to early January, arrested late April or early May with up to p6 (17% of 23 examined), p7 (39%), p8 (35%), or p9 (9%) new; 2nd series of tail starts at same time, but not known whether moult complete; 2nd moult of outer secondaries follows slightly later. ADULT PRE-BREEDING. Partial: all head, body, tail, wing-coverts, and inner primaries. Starts early February to early March; lesser and median upper wing-coverts, tertials, and longer scapulars first, soon followed by under wing-coverts and belly. By early April, wing, underparts, tail, and much of upperparts in full breeding, but head and chest largely non-breeding; full breeding attained late April or early May. 3rd series of inner primaries starts between early February and mid-April; moult arrested May with up to p1 (24% of 25 examined), p2 (30%), p3 (22%), or p4 (4%) new; 20% showed no new inner primaries. POST-JUVENILE. Complete. Timing rather variable: some start September and head, body, central tail-feathers, and part of upper wing-coverts then new by late November; others start November and do not reach similar stage until late February. Usually in complete 1st non-breeding by March, but a few retain some juvenile feathers of mantle or wing-coverts until early May. Flight-feathers start with p1 late November to early February, completed June–August. 2nd series of primaries starts from May–July of 2nd calendar year; at same time, 1st non-breeding of head, body, tail, and wing-coverts replaced by 2nd non-breeding, and no breeding attained. Subsequent moults like adult.

Measurements. Whole geographical range, no difference between eastern and western populations; skins (RMNH, ZMA). Tail is to tip of t1, fork is tip of t1 to tip of t6.

		♂			♀		
WING	AD	215	(5·68; 24)	208–221	212	(5·72; 31)	203–221
	JUV	210	(4·59; 6)	204–216	203	(5·46; 5)	198–209
TAIL		62·0	(3·14; 24)	58–66	60·7	(1·98; 29)	58–65
FORK		8·3	(2·52; 26)	5–11	8·7	(2·27; 30)	6–13
BILL	AD	25·9	(1·09; 28)	25–28	24·7	(0·98; 34)	23–26
TARSUS		20·0	(0·81; 21)	19–22	19·6	(0·70; 28)	18–21
TOE		23·9	(1·12; 21)	22–26	23·7	(0·79; 28)	22–25

Sex differences significant for wing and bill. One exceptionally large adult ♂ (wing 234, tail 68, and fork 14) and one small adult ♀ (wing 199, tail 56) excluded from table. Tail and fork of adult breeding, non-breeding, and juvenile similar. Juvenile tarsus and toe similar to adult from shortly after fledging, bill from *c.* 1 year old.

Weights. North-west Iran: May, ♂ 75 (Schüz 1959). Kazakhstan (USSR): ♂♂ 66 (May); 60, 62, 64, 75 (July) (Gavrin *et al.* 1962). Mongolia, early June: ♂♂ 66, 70, 70, 72, 80; ♀ 75 (Piechocki 1968c). Zaïre: August–November, ♂♂ 49, 50, ♀♀ 55, 55, 58; December, ♂ 63·3 (3) 56–71, ♀ 56·3 (4·80; 11) 50–65 (Verheyen 1953). South Africa: October, adult ♀, 70 (ZMA); January, unsexed, 54·2 (4·91; 110) 42–66 (Schmitt *et al.* 1973). Hopeh (China): ♂ 76 (6) 62–80, ♀ 69 (11) 63–77 (Shaw 1936). Southern New Guinea: April, adult ♀ 69, juvenile ♂ 69 (RMNH). Malaysia: early May, adult ♀ 80 (ZMO). 2 captive juveniles 71 and 77 at *c.* 2 weeks old; rather stable until starting to fly at *c.* 3 weeks when weight lost rapidly for some days and then maintained at lower level (Kapocsy 1979). 2 captive adults both varied between 60 and 70 over long period (Heinroth and Heinroth 1927–8).

Structure. Wing rather long and pointed, relatively slightly broader than in *C. niger*. 11 primaries: p10 longest, p9 5–12 shorter, p8 20–25, p7 34–44, p6 50–61, p5 65–77, p1 117–127; p11 minute, concealed by primary coverts. Tail relatively shorter and more shallowly forked than in *C. niger*; tips of outer feathers rounded rather than pointed; t6 sometimes slightly shorter than t5. Bill shorter than in *C. niger*, but deeper and wider at base, culmen more strongly decurved at tip, gonys relatively short; depth of bill at basal corner of nostril 6·41 (0·23; 27) 6·1–6·8 in adult ♂, 5·98 (0·23; 33) 5·5–6·3 in adult ♀ (5·98 and 5·53 respectively in *C. niger*). Tarsus longer and relatively heavier than in *C. niger*, toes longer. Front toes connected by small webs: curve of web between outer and middle toe reaches to joint between 1st and 2nd phalange; between inner and middle toe, reaches to halfway along 1st phalange (in *C. niger*, to 3rd phalange and to joint between 1st and 2nd phalange respectively). Outer toe *c.* 91%, of middle, inner *c.* 71%, hind *c.* 37%. CSR

Anous stolidus Brown Noddy

PLATES 9 and 10
[between pages 158 and 159]

Du. Noddy   Fr. Noddi niais   Ge. Noddiseeschwalbe
Ru. Нодди   Sp. Charrán pardelo   Sw. Noddy

Sterna stolida Linnaeus, 1758

Polytypic. Nominate *stolidus* (Linnaeus, 1758), Caribbean and Atlantic south to Gough Island. Extralimital: *pileatus* (Scopoli, 1786), Red Sea and Indian and Pacific Oceans east to Hawaii, Marquesas, Easter, and Los Desventurados Islands; *ridgwayi* Anthony, 1898, Tres Marias, Revilla Gigedo, Clipperton, and Cocos Islands, off western Central America; *galapagensis* Sharpe, 1879, Galapagos Islands.

Field characters. 38–40 cm (bill 5·3, legs 2·5 cm); wingspan 77–85 cm. Slightly smaller than Sandwich Tern *S. sandvicensis* and Sooty Tern *S. fuscata*, but bill and tail longer. Medium-sized tern, with almost all-dark plumage except for pale cap. Wedge-shaped tail unique in terns occurring in west Palearctic. Sexes similar; no seasonal variation. Juvenile separable.
ADULT. Forehead and fore-crown grey-white, becoming pale lavender-grey on rear crown and nape; this cap and white crescent below eye contrast markedly with black lores and patch below eye and lead-grey throat and sides of head. Rest of body, coverts of inner wing, and tail sooty-brown, contrasting little with brown-black flight-feathers and tail. Underparts more sepia below and generally paler when worn. Bill black and noticeably long, approaching length of head and slightly drooping over whole length, with marked gonys. Legs dark red-brown. JUVENILE. Wholly dark brown, lacking sooty tone of adult and with pale grey-brown cap indistinct; narrow white line above eye.
Adult showing remarkable pale cap unmistakable but, at distance, both adult and immature could be taken for small dark skua *Stercorarius*. However, normal lethargic slow flight quite unlike that of *Stercorarius* or *Sterna* terns, and behaviour much more like that of idly foraging gull with characteristic frequent settling on reefs, flotsam, or beach. Immature may be confused with juvenile *S. fuscata*, which shows similar brown plumage at distance; however, *S. fuscata* always shows clearly forked tail (not wedge-shaped or ragged as in *A. stolidus*), almost white vent, and typical dashing flight of genus. Gait rather restricted and pattering. Swims only reluctantly, riding buoyantly like small gull but showing rather shaggy end to silhouette.
Vagrants usually silent.

Habitat. In tropical and subtropical low latitudes, nesting on marine islands and islets of varied character. Off Somalia, prefers most inaccessible islands and coral reefs, exposed in breeding period to stormy conditions, preferring to nest on cliffs or in clefts and crannies among boulders, and, in contrast to its usual habits, rising to almost 500 m (Archer and Godman 1937). Farther east in Indian Ocean, nests placed not only on rock-shelves and on bare shingle but in the head of a coconut palm *Cocos nucifera* or on a *Pandanus* or other bush (Ali and Ripley 1969b). In the Atlantic, nests on Ascension Island in niches and large crevices in rock, on small stacks round the main island; here there are the required broad ledges and immunity from predation by cats *Felis*, though severe losses may be suffered from heavy rollers washing nests away (Dorward and Ashmole 1963). In Caribbean, characteristic habitat is small low-lying coral island, relatively sheltered from tropical storms, with temperature of coral sand surface ranging up to 50–60°C or more. Nests mainly in mangroves *Rhizophora* and low shrubs, sometimes on bare rock in southern Caribbean (R van Halewijn). Elsewhere, however, breeding habitat centres on shallow cavity in rock or crevice of a cliff, thus avoiding such frequent conflicts as occur when frigatebirds *Fregata* attempt to rest, sun, or roost on bushes bearing nests (Bent 1921).

Forages in lower airspace up to 50 km from breeding colony, often drinking, bathing, or picking up surface food in flight, by day or by moonlight. Does not plunge or dive but in some areas settles on sea, or rests on buoy, ship's rigging, fishing stake, flotsam, or even the head of a Brown Pelican *Pelecanus occidentalis* (Bent 1921; Murphy 1936). Outside breeding season lives mainly at sea.

Distribution. Widespread breeder in tropical and sub-tropical zones of Atlantic, Indian, and Pacific Oceans.

Accidental. West Germany: 1 killed in Schleswig-Holstein, October 1912 (Niethammer *et al.* 1964). Norway: 1, August 1974 (GL, VR).

Movements. Poorly known. Perhaps less pelagic than usually assumed, though seldom reported from mainland coasts. Only Atlantic race, nominate *stolidus*, considered here.

Migratory at Tristan da Cunha, where absent April–August (Elliott 1957); also absent from Tortugas and most West Indian colonies, from September or October to March (Watson 1966). However, at tropical Atlantic colonies (e.g. Ilha da Trinidade, Ascension Island, St Helena) breeding birds have been found over much of year, though with egg-laying peaks November–April; fewest birds (or even none) present inshore July to early October, and suspected that all birds leave colonies for part of year (Murphy 1936; Dorward and Ashmole 1963). Non-breeding range assumed to lie in warm latitudes, between Tropics of Cancer and Capricorn, necessitating northward movements by Tristan da Cunha and Gough Island birds.

Very few pelagic sightings, however, the most significant being (a) in Atlantic, large numbers at 7°18′N 29°50′W (between St Paul's Rocks and Cape Verde Islands) in July and 40 at 4°S 32°30′W in December (Bourne and Dixon 1973); (b) in Caribbean, 1000 at 17°45′N 65°15′W on 11 September after hurricane (Bourne 1967). In south-east Caribbean, present all year within 130 km (mainly within 80 km) of land, for even outside breeding season birds prefer to roost ashore (on islands) by night; seldom seen in western Atlantic outside Windward Islands or off Guyanas and Surinam (R van Halewijn). Adult ringed Trinidad shot St Lucia in February, and bird hatched Trinidad found one year later (in June) at Aruba (Netherlands Antilles) 1000 km west (ffrench 1973). Extent of movement during immaturity unknown; hundreds off Bahia (Brazil) in April were mainly this age (Murphy 1936), and this seems to be the only region (outside Caribbean) where large numbers approach continental coasts (Watson 1966). Enters Gulf of Mexico irregularly, under influence of hurricanes (Duncan and Havard 1980), and there are August–September hurricane-related records on eastern USA seaboard north to New England (Mason and Robertson 1965). Only a vagrant to West African mainland (Ghana to Cameroun), despite proximity to Gulf of Guinea colonies (Dekeyser and Derivot 1966; Wallace 1973).

Voice. See Field Characters.

Plumages (nominate *stolidus*). ADULT. Forehead pale grey to greyish-white, crown and nape slightly darker, more ash-grey, bordered below by narrow white line from base of culmen to above eye. Lores and narrow line just above and behind eye contrastingly black; lower eyelid and centre of upper white, remainder of upper black. Remainder of head and all body dark sooty-brown, distinctly tinged medium grey on ear-coverts, cheeks, chin, throat, neck, chest, and under tail-coverts in fresh plumage. Tail black-brown, faintly tinged grey when fresh, slightly darker than tail-coverts. Upper wing-coverts mostly dark sooty-brown, similar to upperparts; lesser coverts plumbeous-black, contrasting slightly with other coverts; flight-feathers, greater primary coverts, and bastard-wing deep black, more distinctly contrasting. Under wing-coverts and axillaries dark sooty-brown. In fresh plumage, grey tinge to cheeks, throat, and chest distinct, appearing dark grey in some lights. In worn plumage, all body uniform dark brown except for milky-white forehead and dull ash-grey crown; white line from base of culmen to above eye usually just discernible; upper wing-coverts slightly paler brown than body, especially longer lesser and median coverts, distinctly contrasting with blackish coverts along leading edge and with flight-feathers. JUVENILE. Closely similar to adult, in fresh plumage differing by slightly darker body with less grey tinge, by medium grey forehead (feather-fringes slightly paler) grading to dark plumbeous-brown on crown, separated from black lores by dotted white and grey line, and by fringes of feathers of mantle, scapulars, and upper wing-coverts being paler than centres, buff-brown or dark grey-brown. In rather worn plumage, crown almost same colour as upperparts, only forehead still greyish; feather-fringes on upperparts and wing-coverts bleached to pale sepia. In heavily worn plumage, no traces of grey cap, pale line from bill to eye indistinct, and feather-fringes on all body bleached; longer lesser and median upper wing-coverts strongly bleached and heavily abraded. Further ageing of immatures sometimes possible by moult (see Moults).

Bare parts. ADULT AND JUVENILE. Iris dark brown. Bill black. Foot black with red-brown to dark brown-grey tinge to tarsus and toes, sometimes ochre on webs.

Moults. ADULT POST-BREEDING. Complete; primaries descendant. Usually starts with p1 when adults feeding young or shortly after fledging; on average, p1 shed 2 months after hatching. All primaries new $6\frac{1}{2}$–7 months after loss of p1. Tail mainly moulted between primary moult scores 10 and 40; sequence of replacement approximately t1–t6–t3–t2–t4–t5, often asymmetrical and sometimes seemingly irregular. Head, body, and wing-coverts mainly moult between primary scores of 20 and 40; sides of head and scattered feathers of mantle and scapulars first, followed by throat and chest, shorter lesser wing-coverts, rest of head (nape often late), remainder of upperparts, breast to vent, and other wing-coverts. On Ascension (South Atlantic), situation unusual as adults moult primaries during breeding season; start 2–4 months before laying and finish when young about to fledge or shortly afterwards, sometimes with 2nd series of primaries starting c. 2 months after 1st (Dorward and Ashmole 1963); no details for body and tail. ADULT PRE-BREEDING. Partial; extent unknown, but probably varies with breeding cycle of population concerned. In birds with 12-month cycle (e.g. in Caribbean), involves at least head and apparently all upperparts, throat, chest, and some or all tail-feathers. Occasionally, a few birds replace inner primaries; up to p1 and up to p3 recorded in Caribbean. POST-JUVENILE AND SUBSEQUENT MOULTS. Post-juvenile complete. Starts with scattered feathers of upperparts and head, followed by p1 7–11 months after hatching. Head, body, tail, secondaries (s1–s2 last), and wing-coverts in heavy moult from primary score 10–30; all new from age of 11–14 months. 2nd moult series usually starts at same time as adult post-breeding, when 12–15 months old; as 1st series then not yet completed, immatures of this age show serially descendant moult, in contrast to adult. Head and body still rather new when 2nd series of primaries starts; sides of head, throat, chest, mantle, and scapulars start to moult when 2nd series scores c. 20 and 1st series completed or nearly so.

Measurements. ADULT. Nominate *stolidus*. Caribbean and Atlantic (mainly Lesser Antilles and winter birds off northern South America); skins (RMNH, ZMA). Tail to longest feather.

WING	♂	272	(8·93; 9)	262–285	♀ 267	(5·60; 15) 259–276
TAIL		142	(10·9; 9)	134–151	136	(4·96; 15) 129–146
BILL		44·3	(1·59; 9)	42–48	41·0	(1·33; 14) 39–43
TARSUS		25·7	(0·81; 9)	25–27	24·4	(0·93; 15) 23–26
TOE		37·7	(1·52; 9)	36–40	35·4	(1·42; 15) 34–38

Sex differences significant for bill, tarsus, and toe.

A. s. pileatus. Red Sea, Gulf of Aden, and western Indian Ocean; skins (RMNH, ZMA).

WING	♂	282	(7·72; 6)	273–294	♀ 274	(7·05; 5) 266–281
TAIL		152	(4·12; 6)	144–157	142	(8·44; 5) 134–152
BILL		44·2	(2·09; 5)	42–48	40·7	(1·51; 4) 39–43
TARSUS		25·5	(0·87; 6)	25–27	25·6	(0·52; 5) 25–26
TOE		38·7	(1·63; 6)	36–41	38·4	(0·74; 5) 37–39

Sex differences significant for tail and bill.

JUVENILE. Average wing c. 6 less than adult, tail c. 11. Tarsus and toe similar to adult from fledging. Adult bill size reached when c. 1 year old.

Weights. Nominate *stolidus*. Ascension (South Atlantic), breeding adult 186 (15·3; 10) 160–205 (Dorward and Ashmole 1963). St Martin (Lesser Antilles), September, in moult, ♂ 171, ♀ 161 (ZMA). Gough Island (South Atlantic), 145 (Swales 1965). Belize (Central America), April: ♂♂ 164, 167; ♀♀ 186, 186, 203 (Russell 1964). On Ascension, juveniles gradually gain weight until c. 1 month old, when average c. 215–220 (range 180–260); this approximately maintained until fledging at 2–2$\frac{1}{2}$ months (Dorward and Ashmole 1963). On Dry Tortugas (Florida), maximum of c. 160 reached when c. 40 days old (Ricklefs and White-Schuler 1978). Exhausted ♀♀ off Venezuela and Surinam, November–April, 102, 103, 118 (juveniles), 110 (adult) (RMNH, ZMA).

A. s. pileatus. Marianas (western Pacific), ♂ 197 (4) 187–204, ♀ 189 (3) 177–203 (Baker 1951). Australia, 170–227 (Serventy *et al.* 1971). Tuamotu and Gambier islands (south-east Pacific), ♂ 200 (14) 155–230, ♀ 180 (6) 160–210 (Lacan and Mougin 1974). Christmas Island (central Pacific), average of large number, 173 (Ashmole 1968). Christmas Island (Indian Ocean), 180, 185, 215 (Voous 1964).

Structure. Wing long and narrow, pointed. 11 primaries: p10 longest, p9 1–6 shorter, p8 13–20, p7 32–41, p6 51–62, p1 140–159; p11 minute, concealed by primary coverts. Longest tertials reach to tip of p4–p5 in closed wing. Tail long, more or less graduated, 12 feathers: t3 longest, t2 0–4 shorter, t1 6–14, t4 3–10, t5 19–26, t6 43–58. Bill rather stout, slightly longer than head, cutting edges and culmen straight for basal half, slightly decurved at tip; gonys with rather indistinct and rounded angle. Bill depth at basal corner of nostril in adult nominate *stolidus*: ♂ 8·76 (0·31; 7) 8·5–9·2, ♀ 8·03 (0·39; 13) 7·5–8·6. Tarsus short and slender. Toes relatively long; front toes connected by large hardly incurved webs; outer toe c. 92% of middle, inner c. 74%, hind c. 24%.

Geographical variation. Slight. Involves colour of head and body, and size. Colour strongly subject to bleaching and wear, making identification of single vagrants virtually impossible. Fresh adult *pileatus* from Red Sea, Indian Ocean, and much of Pacific differs from Atlantic nominate *stolidus* by slightly deeper bluish-grey forehead and crown, less ashy-white; more extensively black lores; plumbeous rather than medium grey tinge to fresh plumage of ear-coverts, cheeks, throat, neck, and chest; slightly darker and more plumbeous body. *A. s. ridgwayi* from islands off western Central America described as slightly darker than nominate *stolidus* and with darker grey crown (Ridgway 1919), hence perhaps not differing from *pileatus*; *galapagensis* from Galapagos distinctly darker than others, body plumbeous-black, crown dark grey, bordered at culmen and on sides by indistinct line of pale grey rather than white. Within *pileatus*, colour not uniform, as populations in northern Indian Ocean and on Hawaii have body paler, browner, less plumbeous, forehead more chalky-white (hence apparently close to nominate *stolidus*); in Indian Ocean, gradually darker towards Mascarene Islands; in Pacific, darker towards Malaysia, Australia, and southern Polynesia (Baker 1951). Bill of adult *pileatus* relatively heavier at base than that of nominate *stolidus*: ratio of bill length/bill depth of latter over 4·70, but 30% misidentified by this. Wing length approximately constant throughout range for nominate *stolidus* (average 269), *ridgwayi* (278), and *galapagensis* (277), but clinal in *pileatus*—longest (average *c.* 290) in southern Polynesia (Cook and Tuamotu to Los Desventurados Islands), *c.* 284 in central Polynesia (Fiji, Tonga, and Ellice to Christmas and Marquesas Islands), *c.* 280 for populations of Java Sea, Pelew, Micronesia, Wake, and Hawaii, *c.* 277 north-east Australia, Lord Howe, Norfolk, Kermadec, New Caledonia, New Hebrides, and Solomon Islands, *c.* 270 for China Seas, Philippines, Ryukyu, Volcano, and Bonin Islands, 284 for Mascarene Islands, and *c.* 276 for northern and north-east Indian Ocean (Baker 1951; RMNH, ZMA). Variation in tail and bill length similar, though less marked. Birds of Western Australia north to Savu Sea (Indonesia) average 261 only; plumage blacker than elsewhere in Indian and Pacific Ocean (except *galapagensis*) and bill more slender, perhaps warranting sub-specific recognition, but no name apparently available yet.

CSR

Family RYNCHOPIDAE skimmers

Medium-sized, tern-like charadriiform seabirds (Lari) with 'intermediate' type plumages (in sense of Simmons 1972). 3 closely related species in single genus *Rynchops*: Black Skimmer *R. nigra* of America, Indian Skimmer *R. albicollis* of India and south-east Asia, and African Skimmer *R. flavirostris* (accidental in west Palearctic); best regarded as comprising a superspecies rather than a single, polymorphic species. Confined to tropical and sub-tropical inland and coastal waters of Old and New Worlds.

Body shape like that of robust tern (Sternidae). ♂♂ slightly larger than ♀♀. Necks stout, anterior cervical vertebrae being elongated and modified for attachment of hypertrophied muscles involved in aerial-skimming feeding method—in which bird flies close to surface with tip of much longer lower mandible (knife-thin and flexible, with minute oblique ridges on side) in water, snapping shorter upper mandible down when prey detected to close bill on it while simultaneously swinging head down and back. Wings very long and narrow (flight fast and powerful, resembling that of large *Sterna* tern); tails short, forked; bill-shape unique, laterally compressed and specialized for aerial-skimming (see above); tarsi short, weak, scutellated; feet small, front toes partially webbed; hind toe raised, vestigial (see *R. flavirostris* for further details of external structure). No information on oil-gland. Caeca rudimentary. Down on both pterylae and apteria. Supra-orbital salt-glands well developed. Cat-like pupil of eye able to contract to vertical slit, unique in birds; even when dilated, never completely circular (Wetmore 1919). Highly specialized Rynchopidae differ from other Lari in a number of features linked with unique feeding method, justifying family rank (Zusi 1962). Affinities of feather-lice (Mallophaga) indicate close relationship with Sternidae (Timmermann 1957) as does general appearance and habits; according to Sears *et al.* (1976), however, behaviour suggests that skimmers diverged from ancestral larid line at least as early as separation of gulls (Laridae) and Sternidae, and possibly as early as skuas (Stercorariidae).

Plumages black or dark brown above (including crown), white beneath. Sexes alike. Feathers rather long and dense on underparts. Bills red, orange, or yellow; tipped with black only in *R. nigra*. Post-breeding moult complete, pre-breeding partial (extent not fully known). Young precocial and, if undisturbed, semi-nidifugous; food-dependent on parents (fed by regurgitation on to ground at first, later fish brought in bill). Down soft and thick, cryptic: buff above, with sparse dark mottling (least on head); white below. Juveniles heavily mottled above and below in *R. nigra*; more similar to adults in Old World species, but with slight mottling or streaking on head.

Rynchops flavirostris African Skimmer

PLATE 15
[facing page 206]

Du. Afrikaanse Schaarbek   Fr. Bec-en-ciseaux d'Afrique   Ge. Braunmantelscherenschnabel
Ru. Африканский водорез   Sp. Rayador africán   Sw. Afrikansk saxnäbb

Rhyncops flavirostris Vieillot, 1816

Monotypic

Field characters. 38–40 cm (bill 5·6–8·5, legs 2·3–2·6 cm); wing-span 125–135 cm. Size close to that of Royal Tern *Sterna maxima* and Swift Tern *S. bergii*, but with remarkably long and vertically narrow bill (with lower mandible extending well beyond upper), proportionately large head, long wings, and short forked tail. River and lake bird, combining mainly black upperparts with white face and underparts and yellow or red bill and legs. Juvenile distinguished by buff tips to upperparts and dusky bill tip. Sexes similar; little seasonal variation. Juvenile separable.

ADULT BREEDING. Crown from just before eye, hindneck, back, and rump dark brown-black; wings and tail-centre similarly coloured, but wings with broad white trailing edge to inner half and tail with boldly white outer feathers. Forehead, face, and rest of underparts white, with grey tone across face and smoky-brown under wing-coverts. Bill orange or red on upper mandible, yellower on lower. Legs red. ADULT NON-BREEDING. Hindneck becomes mottled white. JUVENILE. Lacks strongly contrasting black and white appearance of adult, with head and upperparts (other than outer wing) tipped buff. Bare parts paler, with bill mandibles tipped dusky.

For distinctions from extralimital congeners, see Geographical Variation. Any possible confusion with Sooty Tern *S. fuscata* instantly removed by sight of bill and head of *R. flavirostris*, which extend both forward of and well below body in flight. Flight action mechanical, with most of wing-beat accomplished above body line and characteristic bucking of shoulders on each stroke (without disturbance of head position). To feed, skims over water with bill held open and lower mandible immersed. Gait awkward, a shuffling walk. Carriage noticeably low, with drooping bill and long wing-tips creating untidy outline.

Gregarious. Commonest call a loud harsh 'kip' or 'kik'.

Habitat. In low tropical and subtropical latitudes from savanna belt southwards, generally avoiding sea and coastlines, but favouring shallow extensive open waters of fresh or saline lagoons, rivers, lakes, and open marshes outside flood season. Requires wet or dry sand-banks with good field of view for quiet resting during heat of day and for breeding. Mainly lowland, but occurs up to 500 m or more on major lakes. Found on relatively small pools during migration (Bannerman 1931; Benson *et al.* 1971; Bates 1934*a*; McLachlan and Liversidge 1970; Prozesky 1970; Serle *et al.* 1977). Flies in lowest airspace when foraging, and apparently does not swim, perch, run, or even walk except short distances. Frequently shifts habitat to avoid rising waters.

Distribution. Africa south of Sahara, from Sénégal to Sudan and south to Angola, the Zambezi and its tributaries, northern Botswana, and northern Natal.

EGYPT. Formerly occurred, mainly in Upper Egypt, and perhaps bred (Meinertzhagen 1930); not recorded between *c.* 1924 (Flower 1933) and October 1979 when *c.* 40 at Kom Ombo; 2 at Aswan, July 1981 (C G R Bowden, G A Tyler, and M D Linsley); 1 at Hurghada, May 1982 (Wimpfheimer *et al.* 1983).

Accidental. Israel: 2 collected Tel-Aviv *c.* 1934 (HM).

Movements. Dispersive and migratory. Governed by rainfall, with birds breeding in regionally variable dry seasons and moving away from rivers when sand-banks covered in wet-season floods; extent of movement varies between regions.

Longest recorded movements in East Africa. Those breeding Zambia and Tanzania (May–November) spend local wet season on western Rift Valley lakes (1500 reported Lake Rukwe in January), some crossing equator as far as Murchison Falls where regularly 300 December–April; those breeding Lake Rudolf (March–September) disperse widely over Kenya, occasionally reaching coast (Tree 1969; Britton and Brown 1974). In Nigeria, (breeding March–June), some move downriver towards coast as river water-levels rise in September, returning upstream in November–December as sand-banks are uncovered again, and also wander to various reservoirs and lakes; present Lake Chad from April to November (mainly July–September), but leaves with onset of regular strong north-east winds which make fishing difficult and also cause water to rise over roost sites (Elgood *et al.* 1973; Britton and Brown 1974). Present all year in southern Chad and in Khartoum area of Sudan, though in both regions numbers are increased in wet season (Macleay 1960; Salvan 1968*b*), which indicates northward movement from higher rainfall breeding areas further south. Formerly occurred regularly in Egyptian Nile valley, records spanning March to September (Shelley 1872; Gurney 1876; Flower 1933), but never established whether these were migrant breeders or wet-season emigrants from a southern breeding area; see also Distribution.

Voice. See Field Characters.

Plumages. ADULT BREEDING. Crown down to ear-coverts and hindneck blackish-brown. Forehead, lores, sides of head below eye, and sides of neck white. Mantle, scapulars, tertials, centre of back and rump, and central upper tail-coverts blackish-brown or dark chocolate-brown; sides of back and rump and lateral tail-coverts white. Underparts uniform white. T1 dark brown with white edge to outer web, t2 white with broad brown shaft-streak; outer feathers similar but shaft-streak gradually narrower towards t5, and t6 often completely white except for brown or pale horn shaft and slight brown suffusion to tip; brown of fresh tail has slight grey bloom. Primaries black, tinged purple-bronze when fresh, browner when worn; inner webs greyish-brown, p1–p4 paler than others and fringed white on tip and inner webs (widest on p1). Secondaries black-brown with ill-defined broad white tips; gradually less white on tips towards s1 and on tertials. Upper wing-coverts blackish-brown like mantle and scapulars, darkest on lesser and primary coverts. Underwing pale smoke-grey; coverts on leading edge of wing dark brown or dark grey-brown; axillaries white. In fresh plumage, blackish-brown of head and upperparts slightly plumbeous; when worn, feather-fringes bleached to pale brown, especially on scapulars, tertials, and upper wing-coverts. ADULT NON-BREEDING. Like adult breeding, but feathers of forehead, lores, sides of head, and hindneck grey-brown with broad white fringes, hindneck showing ill-defined pale collar, and face and cheeks appearing slightly mottled brown when worn. JUVENILE. Crown down to ear-coverts and hindneck dull black, feathers broadly fringed warm buff; forehead, lores, sides of head below eye, and sides of neck buff speckled grey, darkest on lower sides of neck. Mantle, scapulars, centre of back and rump, central upper tail-coverts, and all upper wing-coverts dull black, feathers fringed buff, except for greater coverts; boundary between black centre and buff fringe rather wavy on scapulars. Tertials dull black with buff spots or notches along edges. Tail grey (white towards base), feathers with white fringes separated from grey centres by wavy dull black streak (widest subterminally) tending to form notches on sides of feather-tips. Flight-feathers like adult, but narrow buff fringes present up to p7–p8 and white tips to secondaries narrower and more poorly defined. Underparts and underwing like adult, but leading edge of wing white (black-brown in adult). Buff fringes bleach soon and disappear by wear, tertials and sides of tail-tip appearing indented. Worn juvenile differs from adult by having upperparts dull greyish-black rather than deep brownish-black; forehead and lores mottled grey and brown, almost similar in colour to crown; tail mostly dark instead of showing much white on sides; leading edge of wing white; traces of off-white fringes still present on upper wing-coverts. FIRST IMMATURE NON-BREEDING. Like adult non-breeding; indistinguishable when last juvenile tertials, tail-feathers, or upper wing-coverts shed. Not known at what age adult breeding attained.

Bare parts. ADULT. Iris dark brown. Upper mandible vermilion or deep orange-red, slightly paler on tip; lower mandible orange-yellow, vermilion restricted to base, distal third yellow. Foot vermilion. In non-breeding, bill paler, more yellowish; foot yellow-red. JUVENILE. Iris dark brown. Bill brown-black with some yellow at base at fledging, yellow becoming more extensive with age. Foot brownish-yellow. (Bannerman 1931; Chapin 1939; RMNH.)

Moults. ADULT. Post-breeding complete, starting shortly after nesting; flight-feather and pre-breeding moults completed shortly before next nesting season. Extent of pre-breeding not fully known: involves at least head, neck, mantle, and scapulars, but not flight-feathers or tail. Post-breeding body moult starts at primary moult score $c.$ 20, completed at score $c.$ 40. Pre-breeding body moult starts when p10 about full-grown. In Egypt and Sudan, 6 birds showed fresh breeding February–April; one, November, had primary moult score 47 and pre-breeding started on body. However, 3 others February–April scored 27, 33, and 37, and body in active moult to non-breeding. JUVENILE. No information.

Measurements. ADULT. Mainly Egypt and Sudan, mainly February–April; skins (BMNH, RMNH). Bill is to tip of upper mandible, tail to t6, fork is tip of t6 to tip of t1.

WING	♂ 363	(11·3 ; 11)	346–378	♀ 349	(13·0 ; 10)	332–371
TAIL	120	(7·25; 11)	108–134	117	(4·71; 10)	111–123
FORK	31·3	(4·50; 10)	23–36	33·8	(4·39; 10)	26–40
BILL	65·8	(3·04; 11)	61–70	54·5	(3·10; 10)	51–62
TARSUS	28·7	(1·82; 11)	26–31	27·1	(1·15; 10)	26–29
TOE	26·2	(1·73; 11)	24–29	24·9	(0·83; 10)	24–27

Sex differences significant for wing, bill, and tarsus.

JUVENILE. Wing on average 56 shorter than adult, t6 and fork 27. Shape of bill at fledging similar to adult, but length distinctly less; time taken to reach adult length not known.

Weights. Zaïre, November–December, probably immature: ♂ 155; ♀ 111, 140; sex unknown 170 (Verheyen 1953). Kenya, August, immature: sex unknown, 162, 175, 177, 200, 204 (P L Britton).

Structure. Wing long and narrow, sharply pointed. 11 primaries: p10 longest; in adult, p9 18–32 shorter, p8 44–60, p7 69–88, p6 94–115, p1 196–235; in juvenile, p9 5–12 shorter, p8 24–34, p7 50–62, p6 76–94, p1 174–192; p11 minute, concealed by primary coverts. Tail rather short, shallowly forked; 12 feathers, t1 23–40 shorter than t6 in adult, 18–28 in juvenile. Longest tertials reach to about tip of p4 in closed wing. Bill straight, much longer than head, lower mandible longer than upper, both strongly compressed to form knife-like blades. Upper mandible swollen at base, gape wide, culmen and cutting edges slightly decurved towards sharply pointed tip; distal $\frac{3}{4}$ strongly compressed laterally, sharply keeled above and below and with narrow but deep groove along underside. Lower mandible strongly compressed over almost entire length; wide at base but from below nostril abruptly constricted and deepening to 10–13 mm; from there on, blade-like up to bluntly rounded tip 5–7 mm deep; both upper and lower edge of lower mandible sharp, sides with row of shallow oblique grooves, longest and most distinct near base of bill. Tip of lower mandible extends 18–24 mm beyond upper in adult ♂, 9–15 in adult ♀, less in juveniles; mandibles equal in length in small downy young, when bill like tern (Sternidae), though with lower mandible noticeably deep at base. Tarsus short but rather stout; 12–16 mm of lower tibia bare. Toes rather short, slender, front toes connected by incised webs; outer toe $c.$ 81%, of middle, inner $c.$ 77%, hind $c.$ 34%.

Geographical variation. None described.

Forms superspecies with Black Skimmer *R. nigra* from North, Central, and South America and Indian Skimmer *R. albicollis* from Pakistan and India east to south-east Asia, being intermediate in plumage between these. *R. albicollis* like *R. flavirostris* but more extensively white: hindneck white in breeding, back to upper tail-coverts and tail almost completely white, scapulars and tertials with white tips, secondaries with distal halves and inner webs white, and underwing white. *R. nigra* deep black above, including hindneck in breeding, and almost all back to upper tail-coverts; secondaries more contrastingly tipped white; amount of white on secondary tips and on sides of black tail, and colour of underwing varies between races. Bill of *R. albicollis* paler orange than *R. flavirostris*, that of *R. nigra* scarlet with black tip. *R. albicollis* slightly larger than *R. flavirostris*, *R. nigra* distinctly so. CSR

Family ALCIDAE auks

Small to moderately large, highly aquatic, diving, charadriiform seabirds (suborder Alcae) with mostly 'intermediate' plumages (in sense of Simmons 1972). Exclusively marine. 22 living species in 13 genera: (1) *Uria* (guillemots, 2 species); (2) *Alca* (single species—Razorbill *A. torda*); (3) *Cepphus* (black guillemots, 3 species); (4) *Brachyramphus* (murrelets, 2 species); (5) *Endomychura* (murrelets, 2 species); (6) *Synthliboramphus* (murrelets, 2 species); (7) *Alle* (single species—Little Auk *A. alle*); (8) *Ptychoramphus* (single species—Cassin's Auklet *P. aleuti-*

cus); (9) *Aethia* (auklets, 3 species); (10) *Cyclorrhynchus* (single species—Parakeet Auklet *C. psittacula*); (11) *Cerorhinca* (single species—Rhinoceros Auklet *C. monocerata*); (12) *Fratercula* (puffins, 2 species); (13) *Lunda* (single species—Tufted Puffin *L. cirrhata*). One species, Great Auk *Pinguinus impennis*, recently extinct. In west Palearctic, 6 species breeding (2 *Uria*, *Alca torda*, 1 *Cepphus*, *Alle alle*, 1 *Fratercula*), and 1 (*Aethia*) accidental. Distribution Holarctic: restricted to seas and coasts of northern hemisphere; mainly in arctic and subarctic, breeding range of some species extending north almost as far as there is any land, of a few others south to warm temperate or (in eastern Pacific) subtropical coasts. Most species (19) found in North Pacific where ecological variety of forms great; of 11 genera involved, 8 endemic. Genera *Uria*, *Cepphus*, and *Fratercula* of circumpolar distribution. For zoogeographical survey, see (e.g.) Udvardy (1963).

Bodies elongated; variable, depending on shape of thorax and length of sternum and particularly of pelvis which is most elongated in *Uria*. Sexes similar in size. Necks short. Wings short and narrow, with humerus longer than ulna (wing-tip long and pointed); used for both flight and swimming. Flight direct and strong, with rapid, whirring wing-beats; *P. impennis* was flightless. 11 primaries, p11 minute. 16–21 secondaries. Tails mostly short, rounded to wedge-shaped; 12–16 feathers (18 in *Cerorhinca*). Bills highly variable: short and stubby in *Alle* and the numerous Pacific murrelets and auklets of genera *Brachyramphus*, *Aethia*, etc.; long and pointed in *Cepphus* and *Uria*; long and laterally compressed in *Alca* (as also in *Pinguinus*); reach extreme in deep, laterally compressed bills of *Fratercula* and *Lunda* which are proximally covered in brightly coloured horny plates during breeding season. Rhamphotheca simple in west Palearctic species other than *Aethia* and *Fratercula*. Tarsi rather short, laterally flattened; fully scutellate, front covered with transverse scales. Toes strong, 3 front ones fully webbed; hind toe absent or vestigial. Long and narrow synsacrum and long pre-acetabular part of pelvis give legs position far to rear of body; for adaptive characters of hind limbs, see Storer (1945). Legs used for swimming and limited terrestrial movements; under water, used for steering. Characteristically awkward on land, upright body posture and waddling gait recalling penguins (Spheniscidae). However, *Fratercula*, *Lunda*, *Cepphus*, and *Aethia* have post-acetabular part of pelvis broad and are able to run and to dig nesting burrows with feet. In *Cepphus*, large pectoral muscle attached further forward than in other Alcidae facilitating terrestrial locomotion and freer flight. Oil-gland with at least 3 apertures on either side; tufted. Caeca vestigial in most species but long and functional in *Cepphus*. Down on both pterylae and apteria. Supra-orbital salt-glands well developed. Gular and sublingual pouches found in (e.g.) *Alle*, *Ptychoramphus*, and *Aethia*; oesophageal crops in *Fratercula* and *Lunda*. Skeletal and feather structures protect eardrum against entry and pressure of water (Kartashev and Ilyichev 1964). For fuller technical diagnosis, see Ridgway (1919), Stresemann (1927–34), and Witherby *et al.* (1941).

Alcidae as a whole constitute well-defined family, sharing many characters, e.g. elongated hypapophyses to anterior thoracic vertebrae for enlarged attachment of long neck muscles necessary for manoeuvrability under water and, as adaptation to swimming with wings, reduced development of external or major pectoral muscles in favour of minor pectoral muscles, former being 3–4 times as heavy as latter in auks—as against 10 times in gulls (Laridae) and 18 times in foot-propelled divers (Gaviidae) (Kartashev 1960). Thoracic vertebrae not so fused as those of landbirds, facilitating body movement under water. Preliminary character-compatibility analysis indicated 3 main lines of development in Alcidae: one leading to *Fratercula*, one to *Aethia*, and one to *Alca* and *Uria*, with *Cepphus* as an early offshoot (J G Strauch).

Plumages typically dark (brown or black) above and pale (white or whitish) below to greater or lesser extent with a few exceptions, mainly in Pacific. *Cepphus* largely dark in breeding plumage, predominantly white in non-breeding. Seasonal changes generally involve increase in white areas to greater or lesser extent. Sexes alike. Feathers dense with numerous feather tracts and down covering whole body; lores fully feathered to nostrils. Number of feathers per unit surface area twice that found in Laridae but total weight of plumage half due to remarkable shortness of feathers. Bare-part colours variable: usually, irises brown and bills black, but sometimes brightly coloured. Structure and colour of bill give some species a most bizarre appearance, particularly when long and curved feathers on head present, as in some *Aethia* and in *Lunda*. Post-breeding moult complete, flight-feathers simultaneous (descendant in some small extralimital auklets); pre-breeding moult slight, involving small feathers only and often restricted to head, neck, and chest (see Stresemann and Stresemann 1966). Rhamphotheca shed 1–2 times per year. Young precocial but nidicolous; food-dependent on parents at least up to fledging, with long period of post-fledging care in some. Chicks of species nesting in exposed cliff-sites leave for sea early, well before full fledging and while still partly covered in down. Down pattern highly variable, both between and within species; often dark above and pale below to greater or lesser extent. Juvenile plumages resemble adult breeding in general colour pattern; usually (not in, e.g., *Fratercula arctica*) consist of rather down-like feathers worn for short period only. Adult plumage acquired at 3–15 months, but adult-sized rhamphotheca (in, e.g., *Aethia*, *Fratercula*) not acquired until age of 2–4 years.

Alcae, though treated as separate order by some authorities (e.g. Stresemann 1959, Verheyen 1961), now widely considered a specialized offshoot of the wader (Charadrii) and gull–skua–tern–skimmer (Lari) assemblages. Share a number of adaptative characters with divers (Gaviidae), with which they were once classified, but Storer's (1960) opinion that this indicates close relationship appears no longer tenable in light of new evidence (e.g. Prager and Wilson 1980, Cracraft 1982). Biochemical characters examined by Sibley and Ahlquist (1972) also indicate greater affinity between Alcae and Lari than between Alcae and Gaviidae, and Lari and Alcae alone among Charadriiformes share sesamoid bone in patagial fan (see Strauch 1978). In appearance and body posture, Alcae resemble totally unrelated Spheniscidae of southern hemisphere of which they are ecological counterpart in northern; lack scale-like feathers of penguins, however, and differ greatly in skeletal characters and anatomy (see Kuroda 1954, Kozlova 1957, Kartashev 1960).

Uria aalge Guillemot

PLATES 16, 22, and 23
[facing page 207, and between pages 278 and 279]

Du. Zeekoet   Fr. Guillemot de Troïl   Ge. Trottellumme
Ru. Тонкоклювая кайра   Sp. Arao común   Sw. Sillgrissla   N. Am. Common Murre

Colymbus Aalge Pontoppidan, 1763

Polytypic. Nominate *aalge* (Pontoppidan, 1763), eastern Canada, Greenland, Iceland, Faeroes, Scotland north of c. 55°38′N, Baltic, and Norway north to c. 69°N; *albionis* Witherby, 1923, Britain south of c. 55°38′N, Ireland, Helgoland, Brittany, and western Iberia; *hyperborea* Salomonsen, 1932, Norway north of c. 69°N, coast of Murmansk, Bear Island, Spitsbergen, and Novaya Zemlya. Extralimital: *californica* Bryant, 1861, California; *inornata* Salomonsen, 1932, North Pacific.

Field characters. 38–41 cm (34–37 cm excluding tail); wing-span 64–70 cm. Averages slightly larger, deeper chested, and longer bodied than Razorbill *Alca torda* but with shorter tail; averages slightly smaller, less bulky, and shorter winged than Brünnich's Guillemot *U. lomvia*, with longer tapering bill; however, overlaps occur between all 3 species. Commonest large auk in west Palearctic with form markedly adapted to marine existence and fish-catching, as in *A. torda* and *U. lomvia*. Head, neck, and upperparts dark brown, underparts largely white. *U. aalge* most easily distinguished by long, tapering, and unmarked bill, streaked flanks (in adult) and indented furrow showing as dark line behind eye (in all plumages, but particularly evident across white rear of head in immature and winter adult). Sexes similar; marked seasonal variation. Juvenile separable at closer ranges. 3 races in west Palearctic, 1 distinguishable from other 2 in the field.

(1) South and west European race, *albionis*. ADULT BREEDING. Head and neck warm dull brown, with faint grey cast overall and black suffusion on crown-centre and hindneck, but these parts generally paler than otherwise dark grey-brown upperparts. 'Bridled' morph (up to 25% of birds, with incidence increasing from south to north) shows narrow but striking white spectacle round eye and white line along indented furrow behind eye. Most birds, however, show just dark indented furrow, with upperparts relieved only by conspicuous white tips to secondaries (forming bar across folded wing or trailing edge to inner wing in flight). Foreneck below throat and rest of underbody white, except for usually copious brown streaks along upper flanks. In flight, flank-streaking combines with mostly brown axillaries to form rather dark wing-pit, and grey-brown mottling on larger wing-coverts and almost brown greater coverts sully white wing-lining, which looks less clean than in *A. torda*. Border between brown and white on foreneck ∩-shaped. Bill brown-black and evenly tapering, with long straight culmen and little-angled gonys extending narrow triangular shape of head; on a few, basal cutting edge of upper mandible horn-coloured (resembling *U. lomvia*). ADULT WINTER. Pattern of head and neck much changed, with chin, throat, rear cheeks, and sides of rear crown and nape becoming mottled white, then hoary and finally white, but retaining dark indented furrow behind eye and around upper edge of cheeks (appearing as sharp dark line at distance) and indication of collar on lower throat. In bridled morph, spectacle but not eye-line remains visible at close range. Rest of plumage similar to breeding adult but wings paler when worn. JUVENILE. Resembles winter adult but distinguished in comparison by much shorter bill (by 50% initially), smaller size, ungrown flight- and tail-feathers, and looser plumage. At close range, separable by unstreaked flanks and black margins to back feathers. FIRST WINTER. With growth of flight-feathers and tail completed, bird resembles winter adult except for much bolder area of white on nape and hindneck, less-streaked flanks, and (still) smaller bill. FIRST SUMMER. Assumes almost complete breeding plumage, but later than adult; distinguished by worn, pale brown wings and (often) white flecks on chin and throat. At all ages, legs ochre, with blacker joints and webs. (2) North-western and northern races, nominate *aalge* and *hyperborea*. At all ages, upperparts darker brown, even appearing black, and flanks more heavily streaked than in *albionis*; on most, underwing also more heavily spotted. In winter, dark col-

lar on lower throat pronounced, often appearing to form complete band round neck. Larger individuals, noticeably bulkier than most *albionis*, overlap in size with *U. lomvia*.

Unmistakable at close range, with combination of unmarked, long tapering bill, shape of throat pattern, and streaked flanks obvious on breeding adult, and sharp, dark furrow behind eye (etched against white sides of crown and cheeks) diagnostic at all other ages. All other characters subject to overlap with *A. torda* and *U. lomvia*. Much more difficult to identify at middle ranges, but in breeding plumage and summer light, *albionis* noticeably paler than nominate *aalge*, *A. torda*, and *U. lomvia*, while long narrow triangle of bill and head (with concave upper edge) usually evident. In immature and winter adult, head pattern easiest character to see. Identification at long range often impracticable (though pale rear head of non-breeding plumage, whitest of all auks, may catch eye in good light). Flight action and behaviour similar to *A. torda* but since wing-loading less high, wing-beats slightly slower so that progress often appears freer. In flight, bill often up-tilted and end of body appears blunter than in *A. torda* (due to shorter tail). Swimming, diving, and gait essentially as *A. torda* but differences in structure (most evident in narrower bill, head, and neck, and shorter tail) create different silhouette, evident when swimming, sitting, standing, or flying. When fully extended, neck noticeably longer than in *A. torda* and *U. lomvia* (and its length may prompt confusion with divers *Gavia* and grebes *Podiceps*).

Markedly gregarious at all seasons, undertaking mass movements off coasts. Most common note of breeding adult a prolonged, angry-sounding crow, 'arrr', with slightly variable pitch; sounds loud in cliff chorus. Juvenile utters far-carrying, plaintive, musical 'PLEEo'.

Habitat. In and by marine offshore and inshore waters, mainly boreal and low Arctic but overlapping temperate and high Arctic zones. Tolerates windy, rainy, and chilly climates, but avoids ice. Mainly in waters with August temperatures of 5–20°C, and salinity above 34 parts per thousand, except for small Baltic population at much lower salinity, as also locally in certain inshore and offshore waters. Brackish or very shallow waters normally avoided. Differs from other west Palearctic Alcidae (except Brünnich's Guillemot *U. lomvia*) in using open nest-site. For breeding, prefers rocky cliffs, stacks, or islets with broad or medium ledges facing sea, or fissures and crevices between or beneath boulders. Also often on extensive bare rock surfaces, on Bear Island using slopes of 6° maximum (Williams 1972). On Seven Islands (Murmansk, USSR), majority nest on ledges covered with peat layer (Tuck 1961a); detritus also used. Height above sea-level up to 150 m (Kay 1947). After breeding over, stays inshore near colony more commonly than does Brünnich's Guillemot *U. lomvia*, and is less given to roaming far out in offshore zone. Can dive to *c*. 60 m (see Food); most fishing probably within *c*. 20 m of surface or less. Flies mainly in lower airspace, and probably only exceptionally above *c*. 200 m.

Human exploitation no longer serious at most breeding sites (Nørrevang 1978). Indiscriminate shooting for enjoyment was successfully restrained by earliest bird protection legislation in Britain (1869). A greater modern threat is marine pollution by oil and chemicals, often coinciding with favoured fishing waters, especially near the more southerly colonies. In the Irish Sea disaster of autumn 1969 (of which the vast majority of victims were *U. aalge*) suspicion finally focused on polychlorinated biphenyls derived from industry (Cramp *et al.* 1974).

Distribution. BRITAIN. Ceased to nest Sussex *c*. 1878 and Kent *c*. 1910 owing to cliff-falls. DENMARK. Bred Bornholm until 1880s and Christiansø from 1929 (TD). NORWAY. Some southern colonies no longer exist (Haftorn 1971). FINLAND. First bred 1957 (OH). JAN MAYEN. First bred 1983 (J A van Franeker and C J Camphuysen).

Accidental. East Germany, Czechoslovakia, Switzerland, Italy, Bulgaria, Rumania, Malta, Morocco, Azores.

Population. Censusing difficult, especially of large colonies, so all estimates tentative. Populations have declined, often markedly, in Faeroes, Norway, southern Britain, France, Portugal, and Spain, though some increase in Baltic.

SPITSBERGEN. Only one known colony, with 50–100 birds 1966 (Norderhaug 1974). BEAR ISLAND. Estimated *c*. 1 000 000 pairs 1980 (Luttik 1982). JAN MAYEN. Breeding 1983; at least 200 birds on nests (J A van Franeker and C J Camphuysen). ICELAND. Perhaps *c*. 1 500 000 pairs (Einarsson 1979); no known changes in 20th century (AP). FAEROES. Estimated 390 000 pairs 1972, possible decline of *c*. 20% since 1961 (Dyck and Meltofte 1975); sample count 1980 suggested decline continuing (BO). BRITAIN AND IRELAND. Perhaps nearly 577 000 pairs 1969–70 (only tentative estimates for some colonies), of which *c*. 78 000 pairs in Ireland; considerable decreases, mainly in last 30 years, at most colonies in southern England and Wales (Cramp *et al.* 1974). Counts 1971–9 suggested significant increases at most study plots in northern and eastern Britain (Stowe 1982). PORTUGAL. About 6000 pairs (Lockley 1942); *c*. 75 pairs 1982 (F Bárcena). SPAIN. Over 2180 pairs in *c*. 1960, 38 pairs 1982 (F Bárcena). FRANCE. About 300 pairs; apparent decrease from mid 19th century, increased to 1930s–1940s, sharp decline since (Yeatman 1976; Guermeur and Monnat 1980). WEST GERMANY. Helgoland: *c*. 3500 pairs 1880, decreased to *c*. 1000 pairs 1950s, increased to *c*. 2000 pairs 1981 (Vauk-Henzelt 1982; GR). DENMARK. Single colony increased to 1400–1500 pairs (TD). NORWAY. Estimated 120 000–160 000 pairs (Brun 1969a); *c*. 155 000 pairs 1964 and *c*. 80 000 pairs 1982) (GL, VR). Declines attributed to human predation on birds and eggs, oil pollution, and deaths in fishing-nets (Norderhaug *et al.* 1977). SWEDEN. Marked decline 19th century, increased after protection; *c*. 8000 pairs 1974,

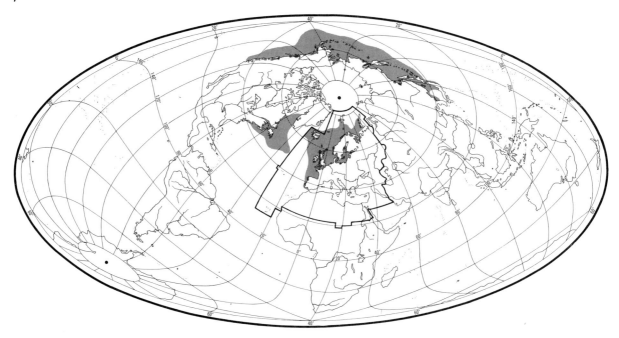

probably still increasing (Hedgren 1975; S Hedgren); c. 7800 pairs 1983 (P E Jónsson). FINLAND. First bred 1957; slow increase since at single colony, 14 pairs 1974, c. 20 pairs 1978, and 40–45 1983 (Hedgren 1975; OH, M Kilpi). USSR. Murmansk coast: c. 16000 pairs (Gerasimova 1962).

Survival. Annual adult survival: Britain 87·9% (Birkhead 1974), 93·7% (Mead 1974); Skomer Island (Wales) 91·5% (Birkhead and Hudson 1977); Helgoland (West Germany) 87·1% (Mead 1974). Survival to 5th year: southern Britain 27·0%; Helgoland 31·3% 1933–43, 30·2% 1955–66; Norway 28·6%; Canada, 4 areas, 29·5%, 36·0%, 37·1%, 41·1% (Mead 1974; Birkhead and Hudson 1977). Oldest ringed bird 32 years 1 month (Rydzewski 1978).

Movements. Dispersive. Many birds (especially adults) present all year in seas close to colonies (even northern ones), while others (especially 1st-years) disperse over long distances. Normal marine range extends south (in Pacific) to latitudes of Japan and central California, and (in Atlantic) to New England and Portugal; only stragglers further south (or in Mediterranean), and accidental inland. Marine range chiefly over continental shelves; avoids deep oceanic water.

Young birds stay at sea all year, with maximum dispersion for each population (as shown by recoveries) reached in late autumn and winter (Holgersen 1961; Birkhead 1974a; Mead 1974). Immatures start visiting colonies at 2 years old, and breed at 4–5 (see Social Pattern and Behaviour). Adults vacate colonies late July to early August, and return February–May according to latitude.

However, in Britain evidence that since 1960s adults are tending to return much earlier, with thousands ashore at some colonies (in fine weather) from October, so that such birds are absent only for moult period when flightless (Taylor and Reid 1981). Such winter attendance irregular, and at some sites has involved larger peak numbers than subsequently bred there (Bourne 1981).

Dispersal studied through ringing recoveries, but these give biased picture, under-representing marine range while over-emphasizing areas where Alcidae hunted for food or sport (e.g. Faeroes, Norway). Recoveries also over-emphasize immature birds, partly because of their greater mortality and partly because ring wear artificially reduces adult samples. For Canadian results, see Tuck (1961a); remainder of account refers to west Palearctic populations, incorporating data from Baillie (1982) and S R Baillie.

NORTH NORWAY AND USSR COLONIES (north of Arctic Circle). Main wintering areas lie off Murmansk and northern Norway; even in midwinter, majority of recoveries are from Finnmark, Troms, and Nordland provinces. Birds from Novaya Zemlya and Murmansk colonies (*hyperborea*) recovered in Norway found mainly Lofoten Islands northwards, though this race known to reach northern North Sea and Skagerrak in small numbers in winter (Holgersen 1952, 1961; Bourne 1968). Of birds ringed northern Norway, only c. 20% found south of Arctic Circle; all immature birds, and all but 5 in winter (November–April).

SOUTH NORWAY COLONIES (south of 64°N). Recoveries spread along Norwegian coast, both north and south of colonies: found from Troms (exceptionally Finnmark) to Skagerrak. 1st-years predominate among those found north of Arctic Circle. Also 3 recoveries from southern

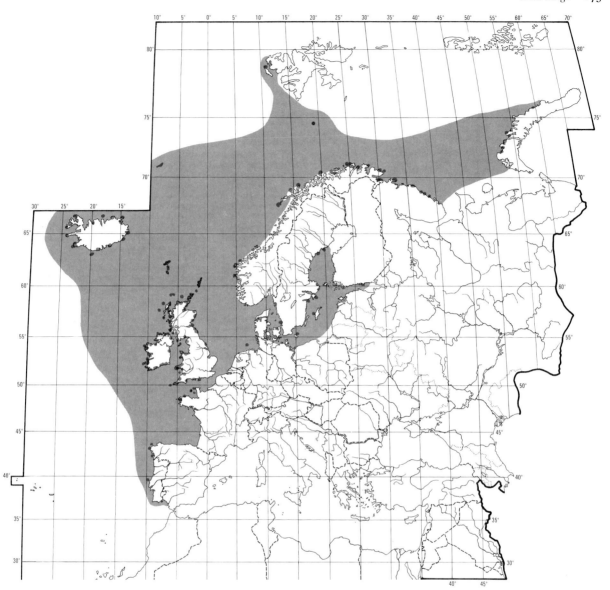

North Sea (West Germany, Netherlands), of which 2 were winter adults.

BALTIC COLONIES. Marine range almost entirely confined to Baltic. Immatures in summer and early autumn, and juveniles in post-fledging dispersal, penetrate north and east into Gulfs of Bothnia and Finland; others present then in southern Baltic, which recoveries indicate to be principal wintering region (few winter recoveries north of Gotland). A good many birds penetrate through Danish archipelago to east coast of Jutland, but few beyond: of 1409 recoveries, only 23 in Kattegat and Skagerrak and only 9 in Norwegian Sea and North Sea (all but 1 of these 32 being immatures).

HELGOLAND COLONY. Great majority of recoveries are in eastern North Sea, from south-west Norway southwards, with a few in south-east England and English Channel and an exceptional 1st-year bird in Brittany. Many birds enter Skagerrak, but few penetrate into Kattegat or Danish archipelago, and none yet found in Baltic proper. See Schloss (1969).

FAEROES COLONIES. Most recoveries from summer shooting in Faeroes, and only 144 in other categories. Birds recovered from southern Iceland to Norway (Troms to Skagerrak), and through Shetland/Orkney to eastern Britain. Of 32 recoveries in period November–January, 26 along Norwegian coast.

SHETLAND, ORKNEY, AND EAST BRITAIN COLONIES. Colony attendance counts in winter show many adults must spend most of year within reach of colony, though some disperse alongside younger birds. Recoveries spread along coast of Norway and into Skagerrak, and along both sides of North Sea. Maximum dispersion November–April.

Orkney and Shetland birds found on average further north in Norway (reaching Lofoten Islands) and these include adults as well as immatures. Birds from east Britain colonies found on average further south in North Sea (some reaching English Channel) and these show more marked separation of age groups in winter: most Norwegian recoveries are 1st-years, while most adults found in western North Sea. Orkney and Shetland birds also recovered in Faeroes (3 adults in summer) and Iceland (1 1st-year in winter).

WEST SCOTLAND, IRELAND, AND IRISH SEA COLONIES. Recoveries of those deserting colony vicinity show wide dispersal: southern Norway, both sides of North Sea, Irish Sea, English Channel, Bay of Biscay, and more infrequently western Iberia (also 1 exceptional bird on Mediterranean coast of Spain). Birds from west Scotland colonies have greater tendency to pass via northern Scotland into North Sea, and account for more recoveries there and in Norway than do birds from Ireland and Irish Sea colonies. Latter produce significantly more recoveries in western English Channel, Brittany, and Bay of Biscay. Recoveries to south of Britain (and the few Irish Sea birds found in North Sea) were all immatures.

Food. Chiefly fish, supplemented by some invertebrate food. May spot prey by dipping head repeatedly into water before surface-diving (Nørrevang 1958; Stettenheim 1959, which see for details of underwater propulsion). May also 'crash-land' over fish-shoal, and dive almost immediately (Bourne 1976a). Maximum depth of dive c. 20 m (Belopol'ski 1957), c. 55 m (Gurney 1913); bottom-dwelling fish taken from depths of up to 60 m (Scott 1973). Average submersion time near littoral zone c. 45 s, maximum 68 s (see Dewar 1924); maximum 71·1 s (Cody 1973). Mean submersion time $71·3 \pm SD\ 38·7$ s ($n = 228$), maximum 152·2 s (Scott 1973). Duration of dive closely related to depth of water; rest period between dives increased markedly (and apparently linearly) with depth of dive (Scott 1973, which see for details). Often feeds in loose flocks (e.g. Scott 1973; see Social Pattern and Behaviour). Off Skomer Island (Wales), 78% of birds fed in flocks of 2 or more, often in company with other seabirds, notably gulls (Laridae) (Birkhead 1976). Birds often feed swimming in lines, but occasionally encircle and herd shoal, catching fish at periphery (Bent 1919; Storer 1952; Stettenheim 1959). Food swallowed before surfacing (Stettenheim 1959; Oberholzer and Tschanz 1969), unless intended for chick, and then carried head first and in line with bill. Feeding rate lower, and length of trip longer, in rougher than in calm seas; sea conditions accounted for 32% of variation in feeding rates of young (Birkhead 1976). For tidal influence on feeding, see Slater (1976). Birds from different colonies on island tended to feed in different areas (Birkhead 1976). At Fair Isle (Scotland), most birds occurred within 6 km of colony (P Hope-Jones). Seen carrying food up to 16 km from colony (Storer 1952); up to 20–50(–80) km (Belopol'ski 1957). In pre-laying period, recoveries suggest birds may travel much further (in 2 cases c. 200 km) from colony to feed (Birkhead 1976).

Fish include Clupeidae—herring *Clupea harengus*, sprat *Sprattus sprattus*, capelin *Mallotus villosus*; Ammodytidae—sand-eels *Ammodytes marinus*, *A. tobianus*, *Hyperoplus lanceolatus*; Gadidae—cod *Gadus morhua*, arctic cod *Boreogadus saida*, haddock *Melanogrammus aeglefinus*, whiting *Merlangius merlangus*, blue whiting *Micromesistius poutassou*, pollack *Pollachius pollachius*, saithe *P. virens*, Norway pout *Trisopterus esmarkii*, bib *T. luscus*, poor-cod *T. minutus*, five-bearded rockling *Ciliata mustela*; Gobiidae—two-spot goby *Gobiusculus flavescens*, crystal goby *Crystallogobius linearis*; Blennidae—butterfish *Pholis gunnellus*, viviparous blenny *Zoarces viviparus*; sculpins (Cottidae), mackerel *Scomber scombrus*, three-spined stickleback *Gasterosteus aculeatus*, pearl-side *Mauriolicus muelleri*, crucian carp *Carussius carussius*. Crustaceans include Malacostraca—crabs; Euphausiacea—*Meganyctiphanes norvegica*, *Thysanoessa*; Amphipoda—*Parathemisto abyssorum*; Copepoda; Isopoda. Also polychaete worms *Nereis*, squid, and bivalves, though latter perhaps taken only because present in fish stomachs. (Collett and Olsen 1921; Collinge 1924–7; Salomonsen 1935; Kaftanovski 1938; Timmermann 1938–49; Perry 1940; Uspenski 1956; Belopol'ski 1957; Kozlova 1957; Madsen 1957; Pearson 1968; Harris 1970; Evans 1975; Birkhead 1976; Hedgren 1976; Hedgren and Linman 1979; Slater 1980; B F Blake.) For North America, see Tuck (1961a), Swartz (1966), Cody (1973), and Scott (1973); for Bering Sea, see Ogi and Tsujita (1973, 1977).

In breeding season, Novaya Zemlya, stomachs of 20 adults yielded fish in 17, including arctic cod in 8, cod in 2, sand-eels in 3, capelin in 2, sculpins in 2, remains of bivalves in 1 (Uspenski 1956). In Barents Sea, 159 stomachs gave 95% (by numbers) fish, 1·9% crustaceans, 0·6% each polychaetes, molluscs, and insects, 1·3% plant matter. In eastern Murmansk (USSR), 158 stomachs, collected over 4 years, contained herring and capelin, each 13·6–41·7%, cod 10·0–52·3%, sand-eels 4·1–32·5% (Belopol'ski 1957, which see for sexual and seasonal variation). In eastern Murmansk diet in spring mainly cod, plus c. 10% crustaceans and polychaetes; in summer and autumn exclusively fish, especially herring (Kozlova 1957). In Faeroes, sprats main food of adults and young (Salomonsen 1925). In Iceland, haddock, whiting, sand-eels, capelin, and herring (Timmermann 1938–49). On Fair Isle, June, 42 stomachs included 36 with fish remains: sand-eels *Ammodytes marinus* in 36, Gadidae in 4, gobies in 1; summer preponderance of sand-eels gives way to mixed sand-eels and sprat diet in early autumn; sprats, gobies, and Gadidae (mainly *Trisopterus*) in winter, sand-eels again in spring (B F Blake). In early March, Firth of Forth (Scotland), 44 stomachs all contained remains of fish: *Ammodytes marinus* in 43, Gadidae in 8, gobies in

8; polychaetes, which may be important in winter, occurred in 24 (B F Blake).

Most information on winter diet from Europe. Of the birds collected Kattegat and Belt Sea (both Denmark), November–February, 10 had eaten fish of 1 kind, 4 of 2 kinds; herring had been eaten by 9, gobies by 2, sticklebacks by 3; also single individuals of viviparous blenny, mackerel, and cod (Madsen 1957). In 4 areas of Skagerrak (Norway–Sweden), January, 57–79% of 684 stomachs contained gobies, 25–72% contained Clupeidae (mainly sprats), 34–40% Gadidae (mainly *Trisopterus*, also whiting, pollack, cod), 1–21% sand-eels (B F Blake).

Diet of young widely studied. On Kharlov Island, eastern Murmansk, 198 items brought to young comprised 80·3% *Ammodytes tobianus*, 16·2% herring, 3·0% capelin, 0·5% cod (species not given) (Kaftanovski 1938); thus apparently many more sand-eels than in adult diet in same region. On Stora Karlsö (Swedish Baltic), 176 items comprised 91·5% sprats, 5·1% herring, 2·3% *Hyperoplus lanceolatus* and unidentified sand-eels, 0·6% viviparous blennies and cod (Hedgren 1976; see also Hedgren and Linman 1979). In Orkney and Fair Isle, mainly sand-eels (Slater 1980; B F Blake). On Farne Islands (England), 83 items comprised 49% by number (57% by weight) *Ammodytes tobianus*, *A. marinus*, and *Hyperoplus lanceolatus*, 42% (30%) butterfish (Pearson 1968). 1190 items over 3 years, Skomer, comprised 95·4% Clupeidae (mainly *C. sprattus*), 0·7% sand-eels, 3·9% unidentified (Birkhead 1976; see also Harris 1970).

Few data on size of prey taken by adults. Estimated length up to 20 cm (Ogi and Tsujita 1977), up to 28 cm (Collett and Olsen 1921). Limiting factor not length but depth of body, captive birds taking fish with body up to 40 mm deep (Swennen and Duiven 1977). Fish presented to mates, Skomer, smaller than those fed to young: e.g. most sand-eels offered to mates 5–7 cm ($n=16$), while most offered to young on same day larger, except those given to young less than 3 days old. Mean length of Clupeidae brought to young and collected on ledges 10·4 cm (8–13, $n=46$), mean weight 8·8 g (Birkhead 1976). For other data, Skokholm and Skomer, see Harris (1970). Mean length of all fish brought to young, Farne Islands, 100–125 mm, mean weight 8 g ($n=83$) (Pearson 1968). At Stora Karlsö, mean length of sprats $135 \pm 10\cdot7$ mm (95–124), mean weight 13·3 g (Hedgren 1976). On Skomer, mean rate of feeding before noon $1\cdot03 \pm SD\ 0\cdot20$ fish per chick per 4 hrs ($n=8$), after noon $0\cdot77 \pm 0\cdot11$ ($n=12$); feeding rate decreased with age of chick, especially after c. 10–12 days. Food consumption of 60 chicks (average 10 days old) during hours of daylight $3\cdot23 \pm SD\ 1\cdot42$ fish per chick per day; average daily food intake of young of all ages 28·4 g wet weight (Birkhead 1976, 1977b; see also Sanford and Harris 1967). On Fair Isle, mean 0·21 feeds per chick per hr (6 pairs, 30–40 hrs); thus 4·2–5·8 fish per 20-hr day (Slater 1980). At Stora Karlsö, mean feeding rate 2·4 fish per day, equal to c. 32 g wet weight (Hedgren 1976). On Skomer, parents showed small peak of feeding soon after dawn, and provisioning rate higher in morning than afternoon (see above: Birkhead 1976). Based partly on estimated 2 feeds per day to young (Tuck 1961a), parents estimated to spend 16% of daylight hours fishing (Pearson 1968). EKD

Social pattern and behaviour. Based on outline supplied by T R Birkhead, also studies by Birkhead (1976, 1977a, 1978a, b), Tschanz (1968), and Williams (1972).

1. Gregarious in breeding season, and probably in rest of year. After breeding season, said to occur initially in family groups of 2; young accompanied by only 1 parent (see below) from outset (Tuck 1961a; Williams 1972; Scott 1973; Birkhead 1976). Where young accompanied by 2 adults, 1 only temporarily associated (Scott 1973; see also Storer 1952). In winter, feeds in large flocks especially in north of range where populations large (Storer 1952; Tuck 1961a); off Oregon (USA), flock of 900–1200 birds recorded (Scott 1973). Further south, adults mostly encountered singly after young independent (Storer 1952). By late spring, adults and immatures largely in separate flocks (Tuck 1961a). BONDS. Monogamous mating system; pair-bond maintained from year to year, but not, apparently, outside breeding season (Storer 1952; Tuck 1961a). Birds whose mates have been absent from colony area for 2 or more days frequently promiscuous; ♀ once obtained new mate within 5 days (Williams 1972). For hybrid with Brünnich's Guillemot *U. lomvia*, see Tschanz and Wehrlin (1968). Some evidence that ♂♂ first to arrive close inshore in spring (Tuck 1961a), and, according to Williams (1972), establish territories. Most birds breed first at 5 years, a few at 4 (Birkhead 1976; Birkhead and Hudson 1977; Hudson 1979a; Hedgren 1980a). Of 22 4-year-olds, Skomer Island (Wales), only 2 bred, both unsuccessfully; no 1-year-olds present, and 2-year-olds spent little time there; 3- and 4-year-olds, which arrived earlier in season than younger birds, spent more time, mainly in 'clubs' on tidal rocks near colony area, but, late in season, increasingly began to visit breeding ledges (Birkhead 1976; Birkhead and Hudson 1977; Hudson 1979). Ledges visited by immatures are alongside, but often quite distinct from, those of breeders (Hudson 1979a). Both sexes care for young at nest-site, but usually ♂ alone accompanies young to sea; young dependent for up to c. 12 weeks after fledging (Birkhead 1976; see also Varoujean *et al.* 1976). BREEDING DISPERSION. Forms large dense colonies of up to 100 000 pairs; up to 500 000 pairs Funk Island, Canada (Tuck 1961a). In Britain, 1969–70, 37 colonies of 3–30 790 pairs, mean 3320 (see Cramp *et al.* 1974). Colonies divided into sub-colonies (Johnson 1941; Tuck 1961a; Birkhead 1976). Average density on flat rocky surfaces, Funk Island, c. 20 pairs per m^2, up to 70 on gravel surfaces (T R Birkhead and D N Nettleship; see also Tuck 1961a). Distance between neighbours 0–50 cm (Tschanz 1968). Bird may be in physical contact with several neighbours; most less than 5 cm apart, but a few more than 30 cm, especially in 'sparse' colonies (Birkhead 1976, 1977a, 1978b). Thus defends smallest known nest-area territory of any bird, c. 0·05 m^2 (T R Birkhead): nest-site plus flexible adjacent area on which off-duty bird may loaf (Williams 1972). Territory serves for part of courtship, copulation, and raising of brood to 'fledging' (here defined as time when young first leave nest-territory to enter sea, though unable yet to fly). Territory moves somewhat with egg (see Nørrevang 1958 and Tschanz 1959 for egg-rolling). Both members of pair defend territory. Close physical contact tolerated with incubating neighbours, but non-incubating birds rebuffed at c. 15 cm (Nørrevang 1958). Birds at edge

of breeding group may also drive off intruders from adjacent neutral ground (Williams 1972). Immatures may establish temporary territories in club (Birkhead 1976). Forms colonies with other Alcidae, especially *U. lomvia*, which may nest as close to *U. aalge* as conspecific birds do; also Razorbill *Alca torda* and Kittiwake *Rissa tridactyla*. Apparently dominant over *U. lomvia*, and may usurp its nest sites (Belopol'ski 1957; Kartashev 1960; Brun 1965; Spring 1971; Williams 1972, 1974). High site-fidelity between years: over 95% of 74 birds, Skomer, occupied same site in successive years; of only 3 that moved, 2 shifted less than 0·5 m, 3rd to another colony 200 m away (Birkhead 1976, 1977a; see also Southern *et al.* 1965, Hedgren 1980a). Unsuccessful breeders more likely to move (Johnson 1938; Hedgren 1980a), and large-scale disturbance occasionally causes whole colony to shift (Johnson 1938; Birkhead 1976). ROOSTING AND COLONY-ATTENDANCE PATTERNS. Off-duty birds may loaf near incubating or brooding bird, but roost on sea away from colony. Incubating or brooding birds often doze or sleep; on cliff, relaxed bird faces inwards, body inclined slightly forwards, head hunched into body, bill somewhat raised; also frequently squats with breast against rock; may turn head slowly from side to side, and blink eyes; may give Contentment-call (see 7 in Voice); in sleeping posture, bill-tip inserted into scapulars (Nørrevang 1958; Williams 1972). Club birds loaf on tidal rocks; this presumably the nature of collective loafing area on flat ground between colony area and sea, Funk Island; there, birds always loafed standing up, and facing sea, and numbers increased as season advanced (Tuck 1961a). Loafing birds perform usual comfort behaviour, including bathing at sea; bathing bird typically rolls over on side. Pattern of attendance at breeding areas varies with latitude and season. In Britain, birds may return October or early November (see Movements), in eastern Murmansk (USSR), not until March (Bianki 1967); then roost at sea overnight, returning early morning to remain either for only a few hours, or until dusk, this varying between days; length of occupation increases as pre-laying period progresses. In pre-laying period (April–May), Skomer, attendance peaked every 5–6 days, birds staying from dawn until just before dusk; on 3rd–4th day, numbers decreased relatively early in day. Numbers highest in calm weather—at peak, about twice number that eventually bred; cyclical pattern ceased once incubation began (Birkhead 1976, 1978a; see also Corkhill 1971, Lloyd 1973). For tidal effect on attendance, see Slater (1976). During incubation and chick-rearing periods, all off-duty birds and immatures left ledges before dusk to roost at sea (Birkhead 1976, 1978a; Hudson 1979a). On Skomer, 26 May, off-duty birds spent on average 30% of day at colony, on 15 June 42%, during chick-rearing 36%; attendance peaked morning and evening, coinciding with nest-relief, morning exchange usually when male returned from roosting (Birkhead 1976, 1978a; see also Gibson 1950, Lloyd 1973, Hedgren 1975, Slater 1980).

2. Variety of behaviour expresses alarm. In response to disturbance (e.g. falling rock), or approaching intruder, bird adopts Alert-posture: stands upright, neck extended, head and neck plumage sleeked, eyes wide open, and carpal joints exposed (Williams 1972). Alarm-bowing often follows (Conder 1950): moves neck rapidly down through arc of 90° (while bill remains horizontal or pointing down) then quickly up again; each bow takes $\frac{1}{4}$–$\frac{1}{2}$ s (Tschanz 1968). At low intensity, bill may not reach below breast, producing bobbing motion. Alarm-bowing may be performed at any time, on breeding ledge or in club, by solitary or densely packed birds; especially when in groups, usually accompanied by Alarm-bowing call (Tschanz 1968; Birkhead 1976; T R Birkhead: see 1 in Voice). Birds nesting sparsely spent more time Alarm-bowing and less time sleeping during incubation than in dense groups (Birkhead 1977a). In extreme alarm, gives Distress-call (see 2 in Voice); may take flight and splash-dive into sea. Wing-flapping, in which bird stands up on tarsi and flaps wings briefly, is settling movement after flight, and also marks transition from one behaviour to another; often ends with slight rotation of head (Williams 1972); Wing-flapping often stimulates response, e.g. Footlooking, Mutual Fencing, or Allopreening (for all of which, see below) in near neighbours (Williams 1972). Birds frightened on sea, even by approaching conspecifics, often Bill-dip: immerse bill for 1–2 s, repeatedly at intervals of 5–10 s (Forssgren and Sjölander 1978; see also Conder 1950). When danger imminent, birds dive (Nørrevang 1958) often preceded by 'rushing' apart, breast, neck, and bill raised at *c.* 45° (Forssgren and Sjölander 1978). FLOCK BEHAVIOUR. Forms flocks for feeding, migration, and pre-laying behaviour; no pair-formation thought to occur in waterborne flocks (Forssgren and Sjölander 1978, which see for details of behaviour). In latter, small rafts (sometimes mixed with other Alcidae: Conder 1950) gather on water near colony area and begin to swim rapidly around; also skitter over sea, sometimes gaining great speed with aid of wings, but never actually fly. Birds may make shallow dives simultaneously, and chase under water; often cause others to dive by surfacing directly beneath and hitting them (Forssgren and Sjölander 1978; J Cayford). After simultaneous dive, many surface with upstretched neck, bill vertical and open; probably version of Head-vertical posture (Conder 1950; see Heterosexual Behaviour, below). Within flocks, birds never less than 1 body-length apart. For swimming formations, especially line-abreast, see Paludan (1947), Conder (1950), and Forssgren and Sjölander (1978). Rafting and aerial behaviour much less developed than in *U. lomvia* (T R Birkhead). ANTAGONISTIC BEHAVIOUR. (1) General. Following mostly from outline supplied by T R Birkhead, and from Birkhead (1976, 1977a, 1978b) and Williams (1972); see also Nørrevang (1958). Aggression stronger and more frequent in dense than in sparse nesting groups; typically accompanied by variant of Crowing-call (see 3a in Voice). In dense groups, birds have aggressive interaction on average every 20 min, each lasting *c.* 4·5 s (Birkhead 1978b). In Britain, aggression highest in winter during site re-establishment (defence against trespassing neighbours, also occasional conflict between pair members, more often ♂ against ♀: Greenwood 1972; Williams 1972); aggression lowest in pre-laying period, higher again during incubation and chick-rearing periods (defending sites from prospecting immatures). On water, ♂ defends ♀ mate from birds intruding within 2–3 body-lengths (Conder 1949). Birds attempting to establish nest-site for first time (usually 3-year-olds) markedly non-aggressive, but elicit aggression by their persistent presence at site (Birkhead 1976; T R Birkhead). Appeasement behaviour highly developed as adaptation to high-density nesting; most interactions comprise bird threatening another which usually signals appeasement, but may retaliate briefly first. (2) Threat and fighting. Bird threatens by pointing or stabbing

A

B

C

with bill. Especially on narrow ledges, confronts rival in Alert-posture, bill sometimes initially angled slightly down over breast (Williams 1972); raises bill c. 5–10(–45)° above horizontal and points it at opponent. As intensity increases, challenger may stand up off tarsi, lift wings slightly, and confront opponent almost bill-to-bill. Fight may follow (Fig A), rivals pivoting on well spread tarsi, jabbing and grappling with bills, wing-cuffing and calling (see 3a in Voice). Fights uncommon and usually less than 1 min; longest observed 10·5 min (Birkhead 1976). Rarely, injury results. Parents and chicks moving to edge of ledge or colony at fledging time often severely pecked *en route* (Greenwood 1974). Rarely fight on water (Storer 1952). Various behavioural sequences interpreted as displacement activities (Birkhead 1976) often follow fighting. (a) Footlooking: bird bends head and neck forward so that bill almost touches ground, maintaining posture for a few seconds, and may bite or nibble feet, or just above; especially common after egg-loss. (b) Similar action to Footlooking, but bird picks up (e.g.) small piece of stone and drops it near its feet; then raises bill to normal angle, but may repeat the action; often followed by Head-shaking. (c) Wing-flapping (see above) (Williams 1972; Birkhead 1976, 1978b). (3) Appeasement behaviour. Threat and fighting incorporate appeasement behaviour: Side-preening, Turning-away, Stretching-away. Side-preening bird turns head for 1–3 s, or longer, and may or may not actually preen; performed by non-incubating and non-breeding birds in several contexts (Birkhead 1976, 1978b). (a) After or during aggressive encounter, often by loser. Side-preening (Fig B) often occurs during site disputes in post-fledging period when only 1 member of pair on nest-site (Williams 1972). (b) After alighting (see below), especially if partner absent and neighbours within 0·5 m; alighting bird Side-preened on 46% of occasions when partner absent (n = 258), 5% when present (n = 220) (Birkhead 1976, 1978b). Bird performing Turning-away (Fig C, left) swivels head 90–150° away from aggressor, either during fight or in response to neighbours fighting. According to Nørrevang (1948), Turning-away followed by Head-shaking always stops a fight—performed by submitting bird, sometimes also by victor (also Birkhead 1976). Bird performing Stretching-away (Fig C, right) makes rapid out–in movement of neck, usually away from aggressor; in extreme cases, neck outstretched for some time ('prolonged stretch-away': Birkhead 1976). Stretching-away performed mainly by incubating birds, and in response to threat from conspecific, or from conspecifics moving or fighting nearby. Prolonged Stretching-away performed occasionally by rivals at end of a fight and interpreted as 'I submit but am staying here' (Birkhead 1976, 1978b). Alighting birds commonly adopt postures which may indicate appeasement or site-ownership (Birkhead 1976, 1978b). Thus, in Landing-posture (Fig D, right) assumed immediately but briefly (for 1–4 s) upon alighting, bird stretches head and neck upwards at c. 60°, sometimes raises wings over back, and may stand up off tarsi (Birkhead 1976, 1978b; also Storer 1952). Site-owners adopt Landing-posture on 80% of landings, most often if conspecifics nearby; non-breeders perform less frequently (8%). Wings-high walking (Fig D, left) is version when newly alighted bird is moving: walks in posture similar to raised-wings version of Landing-posture, but angle of head varies with proximity of conspecific birds; if walking through group of conspecifics, head may be held up at 70–90°; if alongside (within 0·5 m) of conspecifics, held down at c. 45°. Both postures thought to signal 'I am moving but have no aggressive intentions'. If no conspecific birds nearby, walking bird adopts neither posture (Birkhead 1976, 1978b). HETEROSEXUAL BEHAVIOUR. Following account mostly from Williams (1972), Birkhead (1976), and T R Birkhead; earlier accounts by Johnson (1941) and Storer (1952). (1) Club areas and pair-bonding behaviour. Preliminary mate selection and pairing probably occur in clubs. By the time birds resort to ledges for breeding, pair-bond probably well-established (Hudson 1979a). Breeding birds, Skomer, very rarely visited clubs, and usually only after breeding failed (Birkhead 1976). On Skomer, club consisted mainly of 3-year-olds (63·5%), fewer 4-year-olds (18·7%) and 2-year-olds; probably also very small proportion aged 5–6 years (Hudson 1979a). Main behaviour (see below) Head-vertical posture ('skypointing'; Williams 1972), Bowing, Mutual Fencing, attempted copulation, and various forms of aggression and its associated behaviours, notably displacement activities (see above) (Williams 1972; Birkhead 1976, 1978b; Forssgren and Sjölander 1978). No definite information on sequence of behaviour in pair-formation. ♂♂ (rarely ♀♀: Birkhead 1976) in clubs adopt Head-vertical posture, usually before copulation attempt (see 3d in Voice for associated call); posture rarely seen on breeding ledges, and interpreted as advertising-display—head and neck held vertically up for 2–5 s, and hyoid may be depressed, producing characteristic outline to neck (Williams 1972; Birkhead 1976). Following account (Wil-

D

liams 1972) applies to (less common) performance on 'neutral ground adjacent to breeding area': bird with established site rushes towards alighting bird; adopts Head-vertical posture, either when rushing or once stopped, whereupon usually gives Pre-copulation call (see 3d in Voice); ♀ usually Footlooks during ♂'s approach, and ♂ adopts Head-vertical posture only if ♀ raises bill during his approach; posture then held until ♀ lowers bill again. In clubs, Head-vertical posture assumed usually as another bird (of unspecified sex) approached or walked past the performer; latter usually remained stationary, apparently on temporary territory, and waited for other birds to approach (Birkhead 1976). (2) Site-ownership displays (see also above). In pre-laying period (only), birds frequently Bow (for calls, see 9 in Voice), probably to denote site-ownership. Birds which had to make small shifts in nest-site Bowed 3 times as often as those that did not move (Birkhead 1976). One or (more usually) both members of pair may Bow together (mutual Bowing); in either case, bird leans forward and places head under breast, often for some seconds (compare rapid action of Alarm-bowing, above); wings may be slightly drooped. In mutual Bowing, birds may grasp bills in bowed position. In pre-laying period, ♂ may nibble ♀'s bill-tip, proceeding to base (Williams 1972). During mutual Bowing, site-owner may pick up material (Picking), and occasionally exchange it with mate (Williams 1972). Between bouts of mutual Bowing, birds raise bill to normal angle, but may repeat Bowing action several times (Williams 1972). Prospective breeders during pre-laying period, and mainly failed breeders during chick-rearing period, perform Fish-presentation ceremony; possibly intense site-ownership display. Fish-bearer nearly always ♀ (Perry 1940; Birkhead 1976). Bird returns to nest-site carrying single fish in bill; if mate absent, ♀ stands holding fish for up to 4 hrs (Johnson 1941; Nørrevang 1958; Birkhead 1976). If mate present, or when he arrives, both birds Bow, and fish exchanged at least once under the breast or between the feet of one bird. No courtship-feeding function attributable—fish eaten (by either bird) on only 30% of occasions, otherwise discarded (Birkhead 1974b, 1976). On Skomer, Fish-presentation increased in frequency from 0·01 ceremonies per bird per hr in pre-laying period to 0·06 in chick-rearing period; most frequent in early morning when mates returned from roosting. Ceremony rare among club birds so unlikely to be significant in pairing. (3) Meeting-ceremony. Established pairs, on meeting, perform mutual Fencing-display (Conder 1950; Storer 1952; Williams 1972; Birkhead 1976; T R Birkhead). Participants clash open bills together, with grasping movements (Fig E), and both give Crowing-calls and then Barking-calls (see 3b, 4 in Voice). Mutual Bowing usually follows (Birkhead 1976). Pairs also Fence when 1 of them or a neighbour, has been involved in confrontation, winner calling more than loser; also when neighbour returns with fish to feed chick (Birkhead 1976). As season progresses, Fencing involves less grasping and more touching of crossed bills (Williams 1972). (4) Allopreening. When birds meet, mutual Fencing, and especially mutual Bowing, often precede Allopreening,

F

the most frequent behaviour throughout season between pair members (Storer 1952; Nørrevang 1958; Birkhead 1976, 1978b): bird preens head and neck of mate and, if latter incubating, also back (Fig F); head feathers of recipient, and sometimes preener, erected. Bird being preened on chin and throat often raises bill almost vertically (Williams 1972; Birkhead 1976). After mutual Fencing, preening is usually of ♀ by ♂ (Williams 1972); occasionally 2 birds Allopreen simultaneously (Nørrevang 1958). Over entire season, Skomer, mean frequency of Allopreening bouts 1·14 ± SD 0·35 per pair per min; 1·36 before chick-rearing, 0·72 during chick-rearing; duration of bouts very variable, mean c. 7 s during winter, pre-laying, and chick-rearing, but longer (14·3 s) during incubation; neighbouring conspecifics, especially incubating birds, are also Allopreened, but less often (0·09 ± SD 0·04 bouts per bird per min) and for longer (28·3 ± SD 30·2) (Birkhead 1976, 1978b; T R Birkhead). Mutual Allopreening with *U. lomvia* may occur (Williams 1972). (5) Mating. The following from Williams (1972) and Birkhead (1976). Usually ♀ initiates mating (41 of 48 cases: Birkhead 1976, 1978b)—falls forward on to breast and, remaining prone, utters ♀ Copulation-call (see 6 in Voice), tossing head back (Birkhead 1976); ♂ sometimes initiates—by giving ♂ Pre-copulation call. In either case, ♂ approaches in an upright posture, and mounts from side; as he does so, call changes (see 5 in Voice). With flapping or drooped wings, treads ♀'s back, still calling, and moves towards rear. During cloacal contact, ♂ rears up, head and neck plumage sleeked, wings spread and braced against ground or along ♀'s sides. If first attempt fails, ♂ usually moves forward and ♀ begins to call until he moves back again. Up to 14 cloacal contacts, of mean 3·5 s (1·9–5·4, n = 18) in one mounting; overall duration 17 s (4–40, n = 22) (Birkhead 1976). Usually terminated by ♀ standing up. Both birds then adopt Alert-posture (Nørrevang 1958; Williams 1972; Birkhead 1976) and often Bow. Copulation occurs throughout breeding season, in Britain from October–November onwards (Greenwood 1972; T R Birkhead); stops once egg laid, but failed and non-breeders continue to end of season (Williams 1972; T R Birkhead). High frequency of brief, 'token' copulations in individual pairs thought by Williams (1972) to reduce ♂ aggression, but may function to disguise ♀'s fertile period and so avert cuckoldry (Birkhead 1978b). On Skomer, peaks in first few hours of daylight, coinciding with return of roosting birds; at nest-site, copulation occurred within 2 min of arrival of mate in 22 of 24 cases (Birkhead 1976); during c. 4 weeks prior to egg-laying pairs copulated in early morning, and up to 4 times per day overall (Birkhead 1976). In club, copulation less frequent and rarely successful (Williams 1972). Attempted copulation between birds not paired to each other ('rape') frequent among breeders; protagonist, often unpaired, approaches silent ♀, sometimes in Head-vertical posture, calls, thrusts neck around ♀'s, and attempts to mount (Williams 1972); sometimes tries 'fly-on' mount (Birkhead 1976); ♀ usually starts, moves away, and aggressive encounter may follow (Williams 1972). Occasionally mass rape attempt occurs with 10 or more birds trying simultaneously to mount one ♀ (though sex rarely determined); rape attempts, by whatever number, probably rarely successful

E

('T R Birkhead). One ♀ mounted by 3 different ♂♂ in 20 min (Nørrevang 1958). (6) Behaviour at nest. Descriptions from Williams (1971, 1972), Birkhead (1976), and T R Birkhead; see also Johnson (1941), Paludan (1947), and Nørrevang (1958). Scraping associated with nest-building. Bird (usually ♀), facing cliff, crouches forward on breast, wings hanging loosely (sometimes partly supporting bird), and scrapes feet backwards; no pivoting occurs; if substrate soft, may form shallow depression. Bird may pick up small stone or other available object before or during Scraping, and drop or throw it between feet (82% of occasions) or near nest-site (18%) (Birkhead 1976). Between bouts of Scraping, new object may be sought; if none available, may bite at cliff face or faeces. Scraping bouts usually last only a few seconds, but may be repeated several times; one bird scraped for 6 hrs before laying (Williams 1971, 1972). Occurs mainly in the hours immediately preceding laying, and sometimes after egg-loss. ♂ usually present at laying and shows strong interest in egg (Williams 1972). At nest-relief during incubation, arriving bird Bows towards egg, and mutual Bowing follows before change-over (Nørrevang 1958). Incubating birds reach for material which they drop in front or slightly to one side (side-building: Johnson 1941; Williams 1971, 1972). After rising to preen or defecate, incubating bird may pick up objects up to 1 m away, and place them on or against its feet (homologous with side-throwing) or may wedge them against egg, before re-settling; occasionally birds thus accumulate small pile of stones near nest-site (Williams 1972). RELATIONS WITHIN FAMILY GROUP. Parents respond to calling of chick in egg with variant of Lure-call (see 8c in Voice) which chick learns, and responds to specifically after hatching (Tschanz 1968). Later, visual recognition (of parents by chick) reinforces vocal cues (Tschanz 1968; Schommer and Tschanz 1975; Wehrlin 1977). Chicks brooded closely against broodpatch for most of first few days; after 4th–5th day brooded under wing; brooded for 96% of first 2 days and for 75% when approaching fledging (Wehrlin 1977); young thermoregulate by 9–10 days (Johnson and West 1975) but older young stand beside parent only in good weather and in absence of predatory gulls (Laridae) (Wehrlin 1977, which see for response of small young to various stimuli). Until c. 4–5 days, when effective parent-chick recognition established, young may approach strange adults and be brooded by them (Johnson 1941; Wagner et al. 1957; Nørrevang 1958). If parent absent, more than 1 young may seek refuge under 1 adult (Perry 1944; Tuck 1961); continues up to 14–18 days (Kartashev 1960). Up to 3 days old, young's desire for brooding so strong that it will remain under strange adult despite hearing parent's Lure-call; however no permanent adoption known (Tschanz 1968). Chicks fed within 24 hrs of hatching, and right up to day of fledging. On hearing Lure-call or Crowing-call of parent, chick pecks vigorously at its bill, and gives Food-call. Parent arriving with fish gives variant of Lure-call, and chick emerges from brooding parent. Keeping chick between them, both parents lift wings protectively; fish-bearer bows deeply to bring fish close to chick. Chick takes fish sideways in bill from parent's bill, then turns it to swallow it head first (Tschanz 1959; Oberholzer and Tschanz 1968; Tschanz and Hirsbrunner-Scharf 1975). Parent sometimes drops fish for chick to retrieve (Belopol'ski 1957; Nørrevang 1958). Despite suggestions to contrary (e.g. Nørrevang 1958, Tuck 1961), young fed only by own parents (Tschanz 1959). Adult often preens head of approaching offspring (Nørrevang 1958). From 5–6 days, chicks pick up and drop stones and other material (Williams 1972). Fledging studied in detail (Kaftanovski 1938; Perry 1940, 1948; Kay 1947; Storer 1952; Nørrevang 1958; Tschanz 1959; Tuck 1961; Greenwood 1964; Williams 1972, 1975; Birkhead 1976). Majority fledge on calm evenings, just before dusk, e.g.

none outside 20.45–22.00 hrs (Greenwood 1964); most 22.00–24.00 hrs, Gotland, Sweden (Hedgren 1979). Following account of cliff-fledging mainly after Greenwood (1964). Active pre-fledging period may be less than $\frac{1}{2}$–2 hrs. Young, which had formerly hugged back of ledge, begins to walk around, preen, and flap wings, sometimes climbing small eminences. Bobs whole body, but especially neck, with growing intensity (Greenwood 1964). Though both parents occasionally present on ledge before chick departs, only ♂ attends young at fledging, and accompanies it to sea (Birkhead 1976). Parent starts Bowing slowly (and giving Leap-call: see 8d in Voice) towards sea when chick is near edge of ledge. Much Billing and mutual Allopreening occur. Parent or chick may take initiative in moving to distant launching-off point. When parent and chick both move to edge of ledge, Bobbing and Bowing intensify (adult's Bows deeper and longer). Finally chick launches off, adult just behind (rarely in front). Parent provides no physical assistance, contra some earlier reports. Fluttering wings and large surface area of feet reduce chick's rate of descent and it lands on belly (see also Hedgren 1980b). Chick's descent usually almost vertical, sometimes at c. 45°, increasing likelihood of landing on water; landing substrate varies with colony, e.g. at Gotland, most landed on beach, only c. 15% in sea (Hedgren 1979). On low, flat-topped islands (e.g. Funk), parent and chick walk for long distances along traditional routes to sea (Tuck 1961a). If parent in water first, or if chick impeded on descent, parent stays near water's edge, giving Leap-call, and awaiting chick (Tschanz 1959); however, 'calling down' chick (Perry 1940) not normal pattern. Once on water, both parent and chick call continuously until reunited, and with chick close by parent's side, both swim out to sea, diving together several times on parent's initiative; on one occasion performed mutual Billing (Forssgren and Sjölander 1978). Sea-going chicks mobbed by raft of birds bordering colony; great excitement pervades rafting birds which dive synchronously and chase one another under water; may surface beneath isolated chick (Nørrevang 1958; Forssgren and Sjölander 1978). Chicks which lose contact with own parent may approach strange adults; these show interest, but adoption unlikely, and orphaned chicks probably doomed (Greenwood 1964). However, strange adults occasionally accompany parent and chick to sea, at least initially. Of 145 cases of chicks at sea within c. 0.8 km of colony, 137 (95%) accompanied by single adult, 6 (4%) by 2 adults, 2 (1%) by 3 (Birkhead 1976; see also Scott 1973). In 6 cases of marked birds, guardian was ♂ parent (Birkhead 1976); in 17 cases, always ♂ (Scott 1973). ♂ parent cares for offspring for up to 12 weeks at sea (Birkhead 1976; see also Varoujean et al (1976). Food-call of young heard October–November (Keighley and Lockley 1947; King 1980). ♀ stays at nest-site for up to 3 weeks after fledging (Birkhead 1976). ANTI-PREDATOR RESPONSES OF YOUNG. Alerted by Alarm-bowing calls, unattended chick on nest-site seeks shelter in rocks, or brooding by other adults (Tschanz 1959; see above). Young swimming from colony area with parent dives to evade swooping gulls (Laridae) (Kay 1947; Greenwood 1964). PARENTAL ANTI-PREDATOR STRATEGIES. (1) Passive measures. Compared with some other Alcidae, incubation shifts lengthy and more continuous, and nest-relief rapid, resulting in high incubation efficiency, and thus protection of egg (Ingold 1980)—likewise in brooding of small young (Tschanz and Hirsbrunner-Scharf 1975). At change-over, both parents shield chick in 'tent-like' manner with wings (Storer 1952). When human approaches colony, birds adopt Alert-posture and, on closer approach, begin Alarm-bowing and calling (see 1–2 in Voice). Non-incubating and non-brooding birds move to take-off point, and are first to flee; sitting birds less likely to quit nest-sites. In sparse colonies, approach of man more often causes temporary abandonment of nesting area; in dense colonies, birds tamer and more approachable (Bent 1919; Johnson 1938; Birkhead 1976, 1977a, 1978b). (2) Active measures. Birds on eggs, Skomer, threatened and lunged at approaching gulls. Birds in denser groups thus better protected (Birkhead 1976, 1977a).

(Figs A–F from photographs in Birkhead 1978b and from drawings by T R Birkhead.) EKD

Voice. In breeding season, highly vocal on land, less so on water and in flight. Calls difficult to render phonetically, and, as no complete spectrographic analysis available, authors differ in number of calls interpreted. Following scheme based on Tschanz (1968); also Birkhead (1976) and notes by T R Birkhead. No information on adult calls outside breeding season.

CALLS OF ADULTS. (1) Alarm-bowing call ('nodding call': Tschanz 1968). Recording (Fig I) suggests nasal 'ARr' (J Hall-Craggs) or 'HADa' (P J Sellar); average 0·32 s long (Tschanz 1968); low amplitude and low frequency (Birkhead 1976). Uttered during Alarm-bowing. (2) Distress-call ('excitement call': Tschanz 1968). Usually loud, prolonged (0·37–0·48 s) 'aaargh' given by incubating or brooding bird on approach of observer, and expressing great agitation; for sonagram see Tschanz (1968), Fig 11E. (3) Crowing-call ('Bill-arring call': Birkhead 1976). Several variants occur, 4 notable. (a) 'Battle-crowing' (Tschanz 1968). A lengthy (over 5 s), gurgling 'arrr'; 'gwōō-err' (Witherby et al. 1941); directed by site-owners at intruders, e.g. birds moving along ledge (Wings-high walking), straying chicks, human intruder; also given in fights (Tschanz 1968; Birkhead 1976). For low-intensity variant, delivered as single, continuous syllable ('disapproval call'), directed at intruding conspecifics, strange chicks trying to be brooded, or emergence of own chick from brooding, see Tschanz (1968). (b) 'Reception-crowing' (Tschanz 1968), 'greeting call' (Birkhead 1976). Lasts up to 10 s (Birkhead 1976), mean 2·18 s (Tschanz 1968); consists of initially low, throaty sound (up to 4 s: Birkhead 1976), followed by rattling crescendo, pitch tending to fall at end (Fig II); whole resembles reverberating, rolling 'aargh' (Williams 1972). May be preceded by 1–2 sharp clucking sounds (P J Sellar). Given by bird arriving without fish, and by mate at nest-site; associated especially with mutual Fencing. For variant, given to young, see 8d. (c) 'Contact-crowing'. Very loud, clear, polysyllabic call, given by meeting pair members when one is sitting (Tschanz 1968). (d) Pre-copulation call of ♂. A loud 'graaa' or 'waaa', similar to 3d, invariably given (with open bill and vibrating tongue) prior to mounting (Williams 1972). ♂ approaching ♀ with intent to rape utters long, drawn-out 'Arr' (Birkhead 1976). (4) Barking-call. Far-carrying call (duration 0·12–0·16 s: Tschanz 1968), similar to barking of large dog, given in bouts of 2–4; this probably the 'ga gow gow' of Williams (1972). Uttered by both members of pair, immediately after call 3b in Meeting-ceremony (Tschanz 1968; Birkhead 1976). (5) ♂ Copulation-call. Once mounted, a continuous 'ah ah ah ah' with about 4 'ah' units to every 1 of ♀'s (call 6) (Birkhead 1976); 'wug-wug-wōōr-wug-wug-wug' (Witherby et al. 1941). Similar to call 4, but higher pitched and delivered 2–3 times per s (Tschanz 1968). (6) ♀ Copulation-call. Loud, disyllabic 'uragh' (Williams 1972); 'gurt-er' (Witherby et al. 1941); repeated at intervals of c. 1 s, with head thrown back; given by ♀ as she squats to invite copulation (Birkhead 1976; see also Nørrevang 1958). (7) Contentment-call. Soft, short 'm', uttered when bird is dozing or quietly brooding (Tschanz 1968). (8) Calls to young. (a) More or less muffled 'achrachr', 1st part open, 2nd part lacking clear tones, mean duration 0·66 s; given by parent to chick emerging from brooding or standing nearby; see Fig 12K in Tschanz (1968). (b) Alarm-contact call. Short, guttural, disyllabic 'aha', like call 6, but highly variable in volume, pitch, and rate of delivery; given by anxious parent to (e.g.) chick refusing to leave refuge, chick that has leaped off ledge but failed to reach water, chick approaching or approached by neighbour, or chick mishandling fish (Tschanz 1968). (c) Lure-call ('cooing': Williams 1972; 'crooning': Birkhead 1976). Melodious sound (1·2–1·46 s), beginning with short 'a', changing to soft 'r'; of variable pitch, but ends with descending, softer, clearly separated 'r' sounds; described as soft and drawn-out (Williams 1972; Birkhead 1976). Given by adult Bowing to approaching chick (present, or calling from refuge), inviting chick to be fed, inspecting egg at nest-relief, or responding to calls of chick in pipping egg (see Figs 9–10 in Tschanz 1968; Tschanz and Schommer 1975). (d) Leap-call ('leap-sound': Nørrevang 1958; 'crowing on water': Tschanz 1968). Distinctive crowing call (mean duration 1·85 s), like call 3b, but less marked fall in pitch

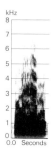

I E Simms/BBC (1971) England July 1955

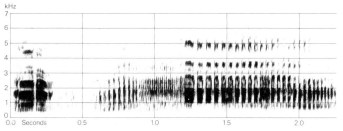

II P J Sellar Scotland June 1975

at end; distinctive, prolonged, and shrill (Nørrevang 1958); long, high-pitched growl (Greenwood 1964). Given by adult to young on verge of fledging, until reunited on water; chick distinguishes parental call (Fig 13 in Tschanz 1968). (9) Other calls. Various quiet calls associated with Bowing, e.g. 'ah ah ah' and 'arring call' (Birkhead 1976).

CALLS OF YOUNG. 9 different calls, according to study of captive young by Tschanz and Schommer (1975), which see for sonagrams and details of pitch, timbre, development with age, and individual variation. Calls intergrade considerably, often differing chiefly in intensity, rate of delivery, etc., and phonetic distinction difficult. Young in pipping egg, and hatched young up to c. 1–2 days, give brief peeping ('Small chick' sound); stimulates parent to rise frequently, roll egg, utter call 8c, and make feeding intention movements towards egg (Tschanz 1968). Hatched young give 'Pecking call' when pecking at fish, stone, own body, etc.; in the wild, this presumably Food-call of hungry chick (E K Dunn). Other calls longer, and more complex. Relaxed chick, when 'nuzzling' (inviting parent to brood), sometimes when preening, gives 'Contact call' with slight toss of head. In unfamiliar surroundings or (if parent slow to brood) when agitated, young gives 'Contact-distress call' or, alternatively, 'Distress call'. Any sudden movement may elicit 'Crying (or Screaming) call', along with escape reaction; more immediately threatened chick crouches, and lunges with 'Threat call'. After eating over-sized fish, chick may utter 'Pain call'. From c. 3 weeks old, and especially in evening, young give 'Water call' in response to parental Leap-call. In recording (Fig III) on nesting ledge, call a persistent 'PLEEo PLEEo' (P J Sellar). Calls of individuals vary markedly in pitch, facilitating recognition and contact between parent and offspring after fledging (Tschanz 1968; Tschanz and Schommer 1975). On hearing Leap-call, chick approached parent, began to give Water-call, and tried to bill with parent (Nørrevang 1958).   EKD

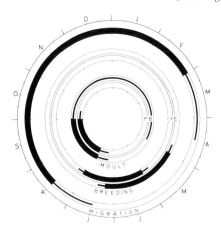

III P J Sellar Scotland June 1975

Breeding. SEASON. See diagram for Britain and Ireland; southern Iceland and Baltic up to 1 week later; northern Norway up to 2 weeks later; Novaya Zemlya c. 3 weeks later (Dementiev and Gladkov 1951b; Makatsch 1974). SITE. On cliff-ledge or flat rocky surface; occasionally in crevice. Colonial. Strong site tenacity (see Social Pattern and Behaviour). Nest: normally none; eggs laid in gravel-bottomed crevices, Canada, which could flood to 3–5 cm deep, were raised on small 'islands' of stones (Johnson 1941). Building: incubating bird in gravel-bottomed crevice, Canada, reached out for small stones and dropped them near its breast (Johnson 1941); Nørrevang (1958) suggested excreta may sometimes be important in holding egg on ledge. EGGS. See Plate 91. Pyriform, roughened, not glossy; very variable in ground-colour and markings, from blue to green, reddish, or ochre, through pastel shades to cream or white; completely unmarked to heavily blotched, spotted, or scribbled, or with combination of these, in brown, yellowish-brown, reddish, or black; can be overall dark, or with pale zones at middle or ends. 82×50 mm $(75–89 \times 45–56)$, $n = 100$; calculated weight 108 g (Schönwetter 1967); no apparent variation in size between nominate *aalge* and *albionis*. Clutch: 1. One brood. Replacements laid after egg loss, sometimes twice (Nørrevang 1958); on Skomer Island (Wales), 52% of lost eggs replaced, sample not given; re-laying interval $14.8 \pm$ SD 1.2 days, $n = 9$ (Birkhead and Hudson 1977); in arctic USSR, interval 15–22 days (Uspenski 1956); 16–18 days but varies according to when 1st egg lost (Belopol'ski 1957); 2nd replacement laid 14 days after loss (Nørrevang 1958); experienced breeders more likely to re-lay than inexperienced (Hedgren 1980a). Replacement eggs from 11 birds averaged 6.4% smaller by volume than 1st eggs (Hedgren and Linnman 1979). Overall laying period, Skomer, 1975, 1 May–5 June, median 15 May. Considerable synchrony of laying in dense colonies or sub-colonies; within such groups, 61–74% of eggs laid in 10 days from start of laying, compared with 13% in 1st 10 days on island as a whole (Birkhead 1980). INCUBATION. 28–37 days (Witherby *et al.* 1941); 34–49 days (Perry 1940); $32.4 \pm$ SD 0.2 days, $n = 21$ (Hedgren and Linnman 1979). By both sexes; in roughly equal shares; in spells of up to 16–24 hrs (Johnson 1941). Egg rested on tarsi, pointed end towards tail (Nørrevang 1958). Begins on laying. YOUNG. Semi-altricial and nidicolous. Cared for and fed by both parents, and brooded until near fledging. FLEDGING TO MATURITY.

Young leaves ledge for sea before true fledging; formerly stated to leave at c. 14 days (Witherby et al. 1941), or 15–17 days (Keighley and Lockley 1947), but at $19.3 \pm$ SD 1.5 days, $n = 15$, Baltic (Hedgren and Linnman 1979); 20–25 days, subarctic and Arctic (Greenwood 1964); $21.21 \pm$ SD 2.6 days, Skomer (Birkhead 1976); see also Social Pattern and Behaviour. Capable of flying at c. 50–70 days (Belopol'ski 1957). Fed by parents for first few weeks at sea (Hedgren and Linnman 1979); age of full independence not known. Age of first breeding mostly 5 years, a few at 4 (Birkhead and Hudson 1977). BREEDING SUCCESS. Of 486 pairs laying, Skomer, 1973–5, 80.7% eggs hatched and 89.0% of chicks survived to leave ledges; over 3 years, overall success 71.8% (61.7–74.2%), or 0.7 chicks per pair. Of 123 1st layings, 93 (75.6%) chicks left ledges, while of 12 replacement layings, 7 (58%) did so, difference not significant (Birkhead and Hudson 1977). Main egg losses due to rolling off ledges, and to predation by gulls *Larus*—chicks also taken from ledges and whilst leaving them, and some chicks fail to reach sea (Greenwood 1964). Breeding success dependent on nest density and related to synchrony of laying within groups. Birkhead (1977a) found that dense groups had shorter spread of laying than sparsely nesting birds, and so minimized number of birds out of phase at start or end of season; birds out of phase much more vulnerable to predation by gulls as they lacked neighbours who would join in defence. In colony areas classified as dense, 87.5% of 144 pairs reared chicks to leaving ledge; 73.9% of 169 pairs in moderately dense areas, 30.8% of 26 pairs in sparse areas. Of 1475 1st eggs laid, Baltic, 81.5% hatched and 96.2% of chicks left ledges; of 94 replacement eggs, 44.7% hatched and 67% of chicks left ledges, difference highly significant; experienced pairs more successful than inexperienced, pairs with 2 or more years' experience having 0.84 chicks leaving ledge per pair, with 1 year's experience 0.77, and first-time breeders 0.51 (Hedgren 1980a).

Plumages (nominate *aalge*). ADULT BREEDING. Head, neck, and throat chocolate-brown, slightly blacker on top of head and neck. Narrow furrow in feathers running backwards from posterior edge of eye. In 'bridled' morph, narrow white eye-ring and white line extending backwards along lower edge of furrow. Upperparts black or brownish-black. Underparts white, flanks and thighs streaked brownish-black, most streaks concealed under folded wing. Tail, secondaries, and primaries as upperparts, primaries with pale brown inner webs becoming whitish towards edge and base, and increasingly blackish towards tip. Shafts of outermost primaries mostly horn-coloured, becoming pale brown towards tip; shafts of inner primaries increasingly brown-black, shafts of secondaries and tail-feathers black. Secondaries broadly tipped white (not all outer ones when worn), inner webs pale brown. Upper wing-coverts as upperparts. Primaries and secondaries dark silver-grey below; primaries darker towards tips, outer webs brown or pale brown, inner webs with pale brown shaft-streaks. Greater under wing-coverts dark silver-grey, some tipped or edged white. Rest of under wing-coverts including primary coverts, with variable amount of pale brown spots and shaft-streaks. Bases of median and lesser coverts pale brown. Axillaries brown. Partial or complete melanism and albinism occur occasionally (Storer 1952; Tuck 1961a; RMNH). ADULT NON-BREEDING. Upperparts black; through wear, white feather-bases often form narrow white band across hindneck. Sides of head white, with black lores and patch below eye, and tapering black streak running backwards along lower edge of furrow. In bridled morph, narrow white line along upper margin of black streak joining narrow white eye-ring. Underparts white, feathers of lower throat and sides of neck often with blackish tips and dark shafts at border with upper neck, forming mottled black and white band across throat. Rest of plumage as adult breeding. DOWNY YOUNG. Head and neck closely covered with black-brown down, streaked by white sheaths partially enclosing several filaments. Lores, sides of head, and throat irregularly streaked white. Rest of upperparts sooty-brown to blackish slate. Underparts white. In *hyperborea*, down replaced by juvenile plumage c. 14–19 days after hatching (Kartashev 1960). JUVENILE. Upperparts, lores, patch below eye, and sides of neck blackish-brown; through wear, grey feather-bases partly visible, giving mottled appearance. Some white feathers behind eye at upper edge of furrow and across hindneck, or sides of head white with narrowing brown line backwards from eye along lower edge of furrow. Underparts white, no streaks on flanks. All wing-coverts as upperparts. Tail, primaries, and secondaries only start to grow with moult to next plumage. Feathers of loose structure, often with some hairy filaments of natal down at tips. Natal down retained longer on throat, and upper side of head and neck. FIRST IMMATURE NON-BREEDING. Like adult non-breeding but stretched wing shows contrasting boundary between moulted and unmoulted series of greater upper wing-coverts. Inner series of coverts new, feathers blackish; in outer portion, greater coverts remain unmoulted, brownish and worn. Greater under wing-coverts tipped white (Kuschert et al. 1981; ZMA). Bill smaller than in adult. FIRST IMMATURE BREEDING. Like adult breeding, usually retaining some white feathers on throat and sides of head; flight-feathers and tail contrastingly worn. Occasionally without brown throat (Verwey 1927; Witherby et al. 1941).

Bare parts. ADULT. Iris blackish-brown. Bill black, changing to slate-black in non-breeding, more brownish in 1-year-olds; occasionally, horn-coloured cutting edge to basal half of upper mandible (Verwey 1922; Salomonsen 1944; Grandjean 1972; ZMA). Leg and foot dark grey, brownish-black, or (rarely) black, anterior parts often greyish-ochre to ochre-yellow. DOWNY YOUNG. Iris blackish-brown. Bill bluish-grey, tip blackish. Leg and foot as adult (Harrison 1975; Fjeldså 1977). JUVENILE. Bill pale brown, darker towards tip. Leg and foot as adult.

Moults. ADULT POST-BREEDING. Complete. Starts with head and body in late July, completed November. Primaries late July to September; shed within 1–3 days. Secondaries late August to late September; shed simultaneously but not until primaries about half-grown. Tail-feathers shed simultaneously late August to late October, not until primaries half-grown. Flightless for c. 45–50 days, until primaries 70–80% grown. Greater upper wing-coverts and lower row of median upper wing-coverts shed simultaneously at same time as primaries and secondaries. Northern races moult later than southern races. ADULT PRE-BREEDING. Partial, November to end of March; involves head, throat, and neck, in 1- and 2-year-olds also feathers of underparts and mantle. Moult later in *hyperborea*, mid-April to late May. In birds of 1–2 years old, post-breeding moult starts earlier and pre-breeding later than in older birds. POST-JUVENILE. Partial: body and wing-coverts; variable number of outer juvenile greater upper wing-coverts and all primary coverts retained, August to mid-October. Flight-feathers and tail grow late July to September. Immature pre-breeding body moult later than in adult—second half February to late May. (Verwey 1922, 1924, 1927; Witherby et al. 1941; Salomonsen 1944; Dementiev and Gladkov 1951b; Birkhead and Taylor 1977; Swennen 1977; Kuschert et al. 1981; C Swennen.)

Measurements. Depth of bill measured at angle of gonys (sheath moulted after breeding season, resulting in reduced bill depth in non-breeding plumage). ADULT. Nominate *aalge*. Faeroes, breeding; skins (W J R de Wijs); data on tail from Dutch winter specimens (ZMA).

WING	♂ 204	(3·47; 16)	194–209	♀ 206	(4·94; 20)	198–218
TAIL	45·0	(2·90; 11)	40–50	46·6	(4·08; 7)	43–50
BILL	47·7	(2·14; 13)	44–51	46·0	(2·14; 20)	43–51
DEPTH	13·2	(0·47; 16)	12·2–13·9	12.8	(0·51; 19)	12–13·7
GONYS	30·1	(2·03; 15)	26–34	29·5	(1·69; 20)	26·5–33·5
TARSUS	38·4	(1·09; 16)	36·5–41·0	38·2	(1·38; 20)	36–42
TOE	54·2	(1·25; 15)	52–56	54·5	(2·30; 19)	51·5–58·0

U. a. albionis. Britain and Ireland, breeding; skins (W J R de Wijs); additional data on tail from Dutch winter specimens (ZMA).

WING	♂ 196	(4·63; 8)	191–203	♀ 195	(3·88; 6)	190–200
TAIL	43·6	(2·21; 11)	40–47	44·7	(4·06; 9)	38–50
BILL	47·0	(1·83; 8)	44–50	46·3	(1·63; 6)	44–48
DEPTH	12·9	(0·54; 8)	12·0–13·8	12·7	(0·67; 7)	12·1–14·0
GONYS	30·4	(2·48; 8)	27·5–33·5	30·1	(1·54; 7)	28·0–32·5
TARSUS	37·4	(0·52; 8)	37–38	37·4	(0·79; 7)	37–39
TOE	52·3	(1·73; 8)	50·0–54·5	51·4	(0·99; 7)	50·0–52·5

U. a. hyperborea. Bear Island, breeding; skins (W J R de Wijs, ZMA).

WING	♂ 212	(3·47; 11)	208–219	♀ 216	(4·62; 10)	210–225
TAIL	48·5	(3·21; 10)	44–53	48·7	(2·35; 9)	45–52
BILL	49·4	(1·79; 12)	46–52	46·6	(2·13; 10)	43–50
DEPTH	13·9	(0·40; 12)	13·4–14·4	13·5	(0·61; 10)	12·8–14·7
GONYS	30·7	(1·83; 12)	28–35	29·1	(2·40; 10)	24·0–31·5
TARSUS	41·4	(0·95; 12)	39·5–43·0	40·9	(1·0; 10)	39–42
TOE	58·0	(1·55; 12)	56–61	57·2	(2·17; 10)	52–60

Sex difference significant for bill length and depth of nominate *aalge* and for bill length of *hyperborea*.

Weights. ADULT. Nominate *aalge*: Dutch winter specimens: ♂ 649 (95·6; 27) 490–844, ♀ 670 (86·1; 16) 561–863 (ZMA); Labrador (Canada), summer, 965 (89) 815–1150 (Johnson 1944).

U. a. albionis: Skomer Island (Wales), breeding, ♂ 853 (27·8; 11), ♀ 870 (42·4; 13) (Birkhead 1976); Britain and Ireland, ♂ 634 (3) 618–659, ♀ 652–798 (W J R de Wijs).

U. a. hyperborea: Bear Island, August, ♂ 1007 (31·1; 12) 930–1040, ♀ 994 (51·2; 10) 930–1070 (ZMA); eastern Murmansk (USSR) April–August, ♂ 1066 (211) 869–1285, ♀ 1040 (164) 825–1260; Novaya Zemlya, ♂ 1142 (20) 1032–1270, ♀ 1153 (25) 1045–1271 (Belopol'ski 1957).

YOUNG. At hatching: Quebec (Canada) 77·0 (3·46; 6) 72–80 (Johnson 1944); England, average 67·3 (Birkhead 1977b); eastern Murmansk, average 74 (Belopol'ski 1957). On leaving nest-ledge: 254 (30·6; 14511) (Hedgren 1979); average 215 (Birkhead 1977b); average 263 (Kartashev 1960). Chicks hatching late, after about 1 July lose weight shortly before leaving ledge (Hedgren and Linnman 1979).

Structure. Wing short and pointed. 11 primaries: p10 longest, p9 3–6 mm shorter, p8 8–11, p7 18–21, p6 30–33, p5 38–45, p1 82–92; p11 minute, concealed by primary coverts. Tail short and slightly rounded; 12 feathers, obtusely pointed. Bill elongated, laterally compressed; lower mandible with prominent gonydeal angle. Nostrils narrow and covered with feathers. At gape, feathering reaches cutting edge of upper mandible. Tarsus laterally compressed. Outer toe *c.* 94% of middle, inner *c.* 75%; no hind toe.

Geographical variation. Bridled morph confined to Atlantic and arctic populations, percentage increasing gradually from south to north: western Iberia 0%, England 1–5%, Scotland 6–17%, Iceland 7–53% (Southern 1962), southern Norway 12·5%, northern Norway 19·4–24·6% (Brun 1970b), Bear Island 57·3% (J A van Franeker), Novaya Zemlya 36–50% (Southern 1962). Colour variation of upperparts, head, and neck more or less clinal. *U. a. albionis* paler, less blackish, than nominate *aalge*; Bear Island birds dark. In Pacific, northern birds palest, southern birds darkest. Birds from Bear Island, Faeroes, and Shetland mostly heavily streaked on flanks (those of Faeroes for this reason sometimes separated as *spiloptera* Salomonsen, 1932), those of Baltic and southern Britain with few streaks, birds from Iceland and Scotland intermediate; in Pacific, few or no streaks. Heavy spotting on under wing-coverts in birds from Bear Island, Faeroes, and Shetland, intermediate in Scotland and Iceland, few or none in rest of Britain, Baltic, eastern North America, and Pacific; variable in Norway (Pethon 1967). Degree of spotting not invariable within populations: heavy or intermediate spotting also in small percentage of birds from Britain and Baltic, and specimens with few or no spots may be found in populations which are mainly heavily spotted. Variation in size more or less clinal or irregular. In both Atlantic and Pacific, wing and tarsus increase with latitude; increase in tarsus less pronounced. Variation in average bill length irregular: Bear Island, Shetland, and Baltic characterized by rather large averages, value for eastern North America rather small. Bill depth and length of middle toe show no distinct trends; in Atlantic, Bear Island shows largest averages in both characters. Races poorly differentiated, differences being part of clinal or irregular variation; conservative view followed here (Dementiev and Gladkov 1951b; Storer 1952; W J R de Wijs).

RS

Uria lomvia Brünnich's Guillemot

PLATES 17, 22, and 23
[facing page 207, and between pages 278 and 279]

Du. Dikbekzeekoet   Fr. Guillemot de Brünnich   Ge. Dickschnabellumme
Ru. Толстоклювая кайра   Sp. Arao de Brünnich   Sw. Spetsbergsgrissla   N. Am. Thick-billed Murre

Alca Lomvia Linnaeus, 1758

Polytypic. Nominate *lomvia* (Linnaeus, 1758), arctic North Atlantic from eastern Canada east to Franz Josef Land and Novaya Zemlya. Extralimital: *eleonorae* Portenko, 1937, eastern Taymyr peninsula, east to Novosibirskiye Ostrova; *heckeri* Portenko, 1944, Wrangel and Herald Islands and northern coast of Chukotskiy peninsula; *arra* Pallas, 1811, North Pacific.

Field characters. 39–43 cm (36–38 cm excluding tail); wing-span 65–73 cm. Averages very slightly larger than Guillemot *U. aalge*, with shorter, thicker bill (in adult), rather thicker head and neck, slightly longer wings, and bulkier body; also averages larger and longer winged than most Razorbills *Alca torda* (but northern populations of all 3 species are larger and longer winged than southern ones). Auk of similar appearance and behaviour to *U. aalge*, with field characters subject to confusing and little-studied overlap with those of that species and *A. torda*. Breeding adult has short, thick bill (usually with whitish line along basal edge of upper mandible) and very dark upperparts with pointed extension of white from upper neck into lower throat. Adult at other seasons and immature show more black on rear of head than *U. aalge* and *A. torda*. Flanks of adult unstreaked; underwing more strongly marked than in most *U. aalge* and all *A. torda*. Outline of culmen, forehead, and forecrown markedly level (even slightly convex on forehead), unlike *U. aalge* (markedly concave on forehead). Sexes similar; some seasonal variation. Juvenile separable only at close range.

Adult Breeding. Plumage similar to *U. aalge* but differs in: (1) intense brown-black tone of upperparts, with only faint grey suffusion on forehead, crown, hindneck, and upper mantle, and paler rich brown cast to lower head, throat, and sides of neck (overall tone thus between that of *U. aalge aalge* and *A. torda*); (2) longer white tips to secondaries, forming not only wing-bar across folded wing but also bold patch on bunched inner feathers (patch most obvious from behind); (3) unstreaked flanks (unlike adult *U. aalge*); (4) sharply pointed extension of white up on to lower throat (unlike both *U. aalge* and *A. torda*); (5) shorter, stubbier, and more decurved black bill, with distinctive, horn-white streak on basal cutting edge of upper mandible, and more prominent gonys. In flight, upperwing pattern appears similar but (in ideal circumstances) bird shows unstreaked flanks and largely white axillaries (not creating dark wing-pit), and grey-mottled and dark-furrowed under wing-converts, more heavily marked than in most *U. aalge* and all *A. torda*, particularly on leading edge and larger feathers. Adult Winter. Becomes more distinct from *U. aalge*, differing in: (1) almost black upperparts; (2) much greater retention of black on head, particularly on chin, lower lores, cheeks under eye, and sides of crown and hindneck (thus head evenly divided black above and white below, lacking prominent invasion of rear part by white, so obvious in *U. aalge* at all ages and *A. torda* in adult and 1st winter from at least November); (3) less-streaked flanks; (4) greater incidence of black flecking on lower head and across lower throat (where often sufficient to form band, as in *U. aalge aalge*). Dark stubby bill and fully black upper head combine to create noticeably capped appearance, lacking in both *U. aalge* and *A. torda*. Bill line duller than in summer and quite often so indistinct as to be invisible in the field. Juvenile. Smaller than adult, with flight- and tail-feathers ungrown. Plumage pattern, colours, and bill shape closely similar to juvenile *A. torda*, with initially uniformly coloured head and throat; at close range, separable by black-margined, brown-black back feathers and wing-coverts. Conversely, lack of white on rear head and dark furrow behind eye allow prompt distinction from *U. aalge*. Bill line indistinct. First Winter. With growth complete, resembles winter adult (though bill still smaller and throat usually more mottled black). First Summer. Resembles breeding adult but easily distinguished by white flecks on throat, and (when worn) brown back and wings.

At close range, adult (at all times) and immature (from November) quickly separable from *U. aalge* and *A. torda* on bill shape and pattern, outline of border to throat (in summer), head pattern (in winter), and also from *U. aalge* by unstreaked flanks (in summer). At middle and long ranges, differentiation extremely difficult, with only full deep head-cap serving as ready means of diagnosis of adults and immatures (from November). Separation of distant juvenile and immature *U. lomvia* and *A. torda* in late summer and early autumn impracticable, since bill size and shape similar and head pattern not certainly divergent until October. No proof obtained of differences in flight action between *U. lomvia* and *U. aalge*, but rather longer wings (with slightly fuller fan of primaries than in *U. aalge*) probably produce looser and slower wing-beats; differences from smaller *A. torda* likely to be even more marked. Other actions and behaviour said to be similar to *U. aalge*, but *U. lomvia* far less studied.

Distribution in winter obscured by difficulty of field identification but clearly more pelagic than *U. aalge* and *A. torda*; strays overland (in Nearctic and to central

Europe) and may well be commoner in offshore waters of North Sea and Atlantic than the few records suggest.

Habitat. Differs from Guillemot *U. aalge* in being concentrated within high Arctic zone of lower salinity (below 34 parts per thousand) and only secondarily overlapping low Arctic, where purer polar waters mingle at surface with boreal waters from south at August sea temperatures of 5–10°C—exceptionally, in Gulf of St Lawrence, up to 15°C (Storer 1952). Avoids extensive permanent pack-ice, breeding only in areas where vigorous vertical circulation creates open water supporting rich planktonic biomass, near cliffed coasts where steep stratified faces, weathered pinnacles, or rocky islands afford protection from predators such as arctic fox *Alopex lagopus*. Occupied ledges may be up to 200–450 m above sea-level. Where sites shared with *U. aalge*, that species occupies broader ledges and flat open rock surfaces, while *U. lomvia* (having enlarged pectoral muscles for deeper diving and more prolonged offshore flying) needs support of neighbouring rock face during incubation (Spring 1971). Tolerates steeper (up to 10°: Williams 1972), smaller, and narrower ledges than *U. aalge*, even when too small for both of pair to be at nest together. Such situations often occur on arctic sea-cliffs due to rock-falls resulting from frost-shattering or erosion, sometimes destroying established breeding sites but otherwise transforming them to advantage of *U. lomvia*, and leading to partial segregation from *U. aalge* which needs more standing room and space to alight (Williams 1974). Never breeds in groups as dense as *U. aalge* whose exceptionally high tolerance of bodily contact permits breeding densities greater than for any other bird species. Most chosen sites face directly on to sea, enabling fledglings to glide safely down to water (Tuck 1961a; Gaston and Nettleship 1981). Sites can be 1 km from sea.

Even winter movements generally contained within cold currents of Arctic origin. Excellent underwater swimming capability enables fishing to depths of *c*. 75 m, as evidenced by catches in trawls, although as a rule much nearer the surface, often aided by upwellings (bringing up prey). Heavy wing-loading leads to sparing use of middle and upper airspace except in neighbourhood of breeding colonies. Heavy losses in salmon nets off western Greenland (including wintering birds from Spitsbergen) a serious threat (see Population).

Distribution. NORWAY. In 19th century, several unconfirmed reports of breeding; first proved 1964 (Brun 1965; Ingold and Vogel 1965).

Accidental. Ireland, Britain, France, Belgium, Netherlands, West Germany, Denmark, Sweden, Finland, East Germany, Poland, Austria, Azores.

Population. No complete counts in west Palearctic and population trends unknown. However, except for temporary decline in Novaya Zemlya (USSR) in early 20th century due to human predation, no suggestions of massive declines that have been reported in western Atlantic, e.g. decline of over 50% to *c*. 4·6 million 1961–80 in eastern Canada (Gaston 1980) and annual estimated kill 1969–71 in western Greenland of $\frac{1}{2}$ million birds in salmon nets which, with some $\frac{3}{4}$ million harvested for food there, together almost equal estimated production of young from all colonies in western Greenland and east Canadian Arctic (Tull *et al.* 1972; see also Evans and Waterston 1976).

USSR. Franz Josef Land: breeds in vast numbers (Tuck 1961a); no counts, but colonies listed by Norderhaug *et al.* (1977). Novaya Zemlya: estimated $2\frac{1}{4}$ million birds 1942 and later at least 1·9 million birds (Uspenski 1956). Heavily exploited by man in first half 20th century and decreased; protected from 1947, since when northern populations have recovered and southern populations increased (Uspenski 1956; Tuck 1961a). Kola peninsula: some 4000 pairs (Gerasimova 1962). SPITSBERGEN. Very large numbers (Tuck 1961a; Løvenskiold 1964), but no counts; see Norderhaug *et al.* (1977) for list of colonies. BEAR ISLAND. 1 200 000–1 800 000 pairs 1980 (Luttik 1982). JAN MAYEN. Breeds in large numbers (Bird and Bird 1935; Seligman and Willcox 1940); perhaps 10 000 adults 1983 (J A van Franeker and C J Camphuysen). ICELAND. Estimated *c*. 1 950 000 pairs at 19 breeding stations (4 small colonies not counted) 1937–8 (Einarsson 1979); no known changes in breeding population in 20th century (AP). NORWAY. About 1057 pairs (Norderhaugh *et al.* 1977); *c*. 800 pairs 1982 (GL, VR).

Survival. Immature survival to 5th year: 34·5% and 52·9% in 2 colonies, Canada; 33·0% 1958–62 and 19·0% 1965–9, Greenland. Annual adult survival rate in northeast Canada 91·0% (Birkhead 1976; Birkhead and Hudson 1977). Oldest ringed bird 22 years 8 months (Clapp *et al.* 1982).

Movements. More or less migratory in breeding areas where seas largely closed by ice in winter (northern Siberia, northern Bering Sea, arctic Canada, Baffin Bay). In low-arctic Greenland and in European sector of polar basin part of population present all year, though some birds (notably immatures) move away in autumn; hence dispersive there, with movements more comparable to those of Guillemot *U. aalge*.

Marine range lies over continental shelves, with oceanic records rare; south to latitudes of northern Japan and Alaskan peninsula in Pacific, northern New England and Norwegian Sea in Atlantic area, but casually further south (e.g. Ireland, Britain, North Sea). Sometimes forced inshore by bad ice conditions, and onshore winds then may cause rare wrecks, e.g. to Great Lakes region of North America (Snyder 1957; Tuck 1961a) and Finland (Merikallio 1958). Immature birds at sea all year, some staying even as far south as Newfoundland Banks (Salomonsen 1967; Brown *et al.* 1975). Ringing returns available from eastern Canada, Greenland, Spitsbergen, and northern

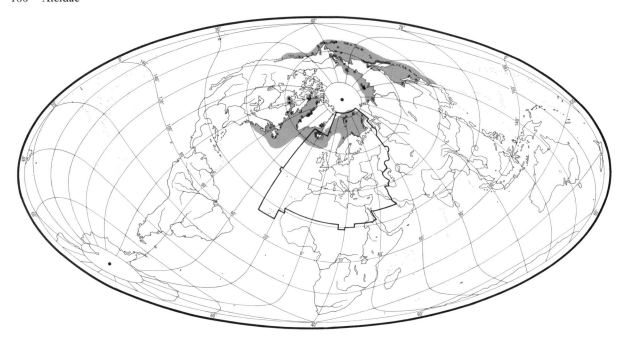

USSR; these reviewed by Salomonsen (1967a, 1971) and Tuck (1971), on which the following based.

EAST CANADIAN POPULATIONS. Ringing restricted to a few colonies, so full picture unclear. Some birds from Hudson Bay and Hudson Strait (c. 62°30′N) may winter within Hudson Bay and off Labrador, but most recoveries from Newfoundland region. Autumn exodus gradual; do not reach Newfoundland Banks in large numbers until November and numbers peak there January–February, or into April in bad ice years. In contrast, ringing on Bylot Island (c. 73°30′N) produced recoveries mostly from middle part of Greenland side of Davis Strait, where such birds largely replace local population that has mostly moved further south (see below); winter recoveries mainly from Egedesminde and Holsteinborg Districts, and West Greenland Shelf doubtless principal wintering region for very large numbers which breed in Lancaster Sound area, though some from there known to reach Labrador and Newfoundland. Birds arrive Greenland seas in October; slow spring return, March–May, follows break-up of ice.

GREENLAND POPULATIONS. East Greenland birds follow drift-ice south to winter off south-east coasts and in Angmagssalik District, with some reaching south-west Greenland; may be forced to desert East Greenland seas completely in winters when ice conditions bad. In West Greenland, colonies deserted in August and (as elsewhere) initial dispersal by swimming since young still flightless and adults beginning to moult. Recoveries indicate that part of population, mainly adults, winters within southern Davis Strait (where outnumbered by winter visitors from elsewhere), but biggest portion follows Labrador current to major wintering areas on Newfoundland Grand Banks where (following gradual nature of autumn movement) numbers peak December–January. Return movement March–May; low-arctic colonies reoccupied second half of April, high-arctic ones in second half of May. Some immature birds move north in summer to latitudes of natal colony; in autumn these return south 1–2 months earlier than adults.

SPITSBERGEN POPULATION. In spring, birds appear inshore in March and occupy colonies in April; return dependent on ledges being snow-free as well as on sea-ice conditions. Departures in late July and August. Inshore records exist for winter period, but birds involved often in poor condition or dead (Løvenskiold 1964). 1 recovery in Newfoundland (January), and all 32 others in Greenland; hence no evidence that this population represented among birds that winter off northern Norway (see below). Main movement seems to be towards south-west along outer pack-ice belt, to winter off southern and south-west Greenland; in Davis Strait penetrates north to Sukkertoppen only, so little contact with Canadian birds coming down from north. Arrives Greenland waters in November, and return under way by end of February.

NORTH RUSSIAN POPULATIONS. Several winter ringing recoveries close to colonies in Murmansk and Novaya Zemlya, and birds seen at sea in winter even around northern Novaya Zemlya (Dementiev and Gladkov 1951b; Uspenski 1956). Main wintering areas for these populations appear to be among pack-ice of Barents Sea and westwards to northern Norway (Finnmark and Troms) (Salomonsen 1967a; Haftorn 1971). Latter probably also the wintering area for (unringed) Norwegian breeding birds. 11 Murmansk and 5 Novaya Zemlya birds recovered in south-west Greenland, but these numbers small in relation to numbers ringed at colonies concerned; presumably only

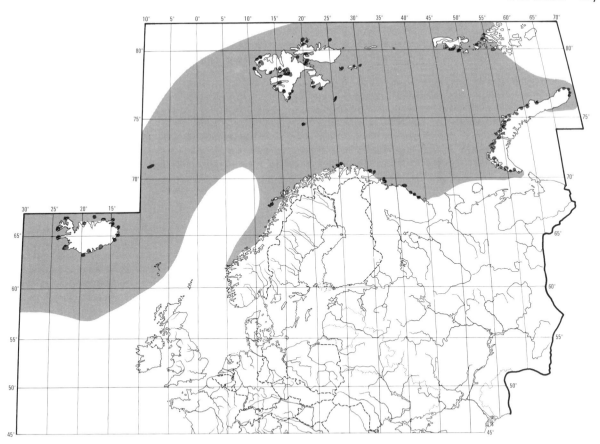

minority reach Greenland seas, and 13 of these 16 were of pre-breeding age at death.

Food. Chiefly fish, supplemented by invertebrates. Prey caught by surface-diving. Frequently swims with head dipped in water, searching for prey; dives at shallow angle with single, forceful beat of wings, propelling itself underwater with half-open wings, as in other Alcidae. Thus covers up to 30–40 m underwater during dive (Kartashev 1960); depth of dive up to 30 m (Scott 1973), 54 m (Gurney 1913), and, in Newfoundland, apparently sometimes caught in trawls at c. 75 m; obtains most food in relatively shallow water (Tuck and Squires 1955; Tuck 1961a). May stay submerged for 40–50 s, sometimes much longer (Uspenski 1956). Usually solitary feeder, occurring at most in well-dispersed flocks (Tuck 1961a; Løvenskiold 1964; A J Gaston). For mixed feeding assemblages, see Uspenski (1956). Flocks of 20–200 sometimes form sinuous moving line on surface corresponding to outline of shoal (Kartashev 1960). Only 1 fish at a time transported to young; carried head first, in line with bill (Krasovski 1937; Uspenski 1956; Gaston and Nettleship 1981); rarely, food regurgitated for young (P G H Evans). Little known about factors affecting feeding success: Tuck and Squires (1955) found feeding of young stopped under stormy conditions, but no relationship between weather and rate of feeding young found by Gaston and Nettleship (1981). Frequents edges of sea troughs (Tuck 1961a), leads in pack-ice (Løvenskiold 1964; see also Gaston and Nettleship 1981), and, at least early in breeding season, edge of ice shelves (Uspenski 1956; Bradstreet 1980; Gaston and Nettleship 1981). See Uspenski (1956) for other feeding sites close inshore. In breeding season, feeding range varies widely with local conditions, but can be considerable; 0·5–15 km (Kartashev 1960); at Akpatok Island (Canada), mostly within 1·5–8 km of colony, rarely exceeding 16 km (Tuck and Squires 1955; Tuck 1961a), though this considered an unusually small radius by Gaston and Nettleship (1981). At Cape Thompson (Alaska), birds concentrated 32–64 km offshore (Swartz 1966). At Novaya Zemlya, up to 100 km (see Uspenski 1956). At Prince Leopold Island (Canada), birds fed up to 150(–175) km (mean c. 80 km) from colony, groups from different parts of colony consistently using different areas (Nettleship and Gaston 1978; Gaston and Nettleship 1981, which see for means by which outbound flocks thought to locate productive feeding areas). Said to feed up to c. 200 km from colony, Spitsbergen, when incubating and chick-rearing (see Løvenskiold 1964).

In breeding season, fish include Clupeidae—herring *Clupea harengus*, capelin *Mallotus villosus*; Ammodytidae—sand-eel *Ammodytes tobianus*; Gadidae—arctic cod

Boreogadus saida, cod *Gadus morhua*, haddock *Melanogrammus aeglefinus*; Cottidae—bull rout *Myoxocephalus scorpius*, *M. quadricornis*, sea scorpion *Icelus bicornis*, sculpin *Gymnocanthus tricuspis*; Blennidae—blenny *Leptoclinus maculatus*; Agonidae—*Aspidophoroides olrikii*; Zoarcidae—*Gymnelis viridis*, *Lycodes* (probably *pallidus*); flatfish (Pleuronectiformes). Crustaceans include shrimps *Hippolyte*, *Sclerocrangon*, prawns *Spirontocaris gaimardii*, euphausiid *Thysanoessa inermis*; Amphipoda—*Gammarus locusta*, *Themisto*, *Euthemisto*; Mysidacea; *Calanus* (Copepoda); *Mesidotea* (Isopoda). Also polychaete worms *Nereis*; Mollusca—pteropods and cephalopods, less commonly bivalves and gastropods (including *Margarites*) (Demme 1934; Hartley and Fisher 1936; Krasovski 1937; Kaftanovski 1938; Witherby *et al.* 1941; Duffey and Sergeant 1950; Uspenski 1956, 1972; Belopol'ski 1957; Kozlova 1957; Kartashev 1960; Løvenskiold 1964, which see for summary of studies on Spitsbergen). For diet in arctic Canada, see Tuck and Squires (1955), Tuck (1961*a*), Bradstreet (1980), Birkhead and Nettleship (1981), and Gaston and Nettleship (1981); see latter also for valuable discussion of stomach analysis in relation to rate of digestion; for Alaska, see Swartz (1966), in which most invertebrates—annelids, molluscs, crustaceans—were bottom-dwelling forms; for eastern Bering Sea, see Ogi and Tsujita (1973, 1977).

Adult diet in breeding season 90–95% fish (Kartashev 1960); of length 5–16 cm, weight 5–30 g (Uspenski 1956). Fish remains (especially arctic cod) in 96% of stomachs, Prince Leopold Island; crustaceans (especially amphipods) in 21%; length of prey ranged from 5 mm (crustaceans) to 20 cm (fish), most fish 6–18 cm; adults probably consume smaller prey than fed to young (Gaston and Nettleship 1981). In Barents Sea (USSR), 364 stomachs yielded (by number) 95·6% fish, 2·0% crustaceans, 0·8% polychaetes, 0·5% molluscs; remainder insects and plant matter (Belopol'ski 1957). In eastern Murmansk, fish 89·6% of diet, remainder crustaceans and polychaetes (Kozlova 1957). On Kharlov Island (eastern Murmansk), fish in 111 stomachs comprised 38·5% herring, 23·1% sand-eels, 20·5% cod (*G. morhua*), and 17·9% capelin (Belopol'ski 1957). At Bezymyannaya Inlet, Novaya Zemlya, 314 stomachs yielded 51·3% arctic cod, 44·3% cod, 1·7% capelin, 0·9% sand-eels, 0·9% *Gymnelis viridis*, 0·6% herring, and 0·3% bull rout; other fish species in small amounts (usually less than 5%), May–August, were haddock, *Icelus bicornis*, *Leptoclinus maculatus*, *Gymnocanthus tricuspis*, *Aspidophoroides olrikii*, *Lycodes*, and flatfish (Belopol'ski 1957). In breeding season, Novaya Zemlya, 683 stomachs yielded 896 fish and 35 crustaceans; of fish, 47% arctic cod, 20% cod, less than 5% each haddock, sand-eels, sea scorpions, sculpins, blennies, herring, capelin, *Myoxocephalus quadricornis*, *Gymnelis*, *Lycodes*, and flatfish; crustaceans included small numbers of Decapoda, Amphipoda, Mysidacea, Copepoda; also small amounts of polychaetes and molluscs (Uspenski 1956). On Franz Josef Land, crustaceans (44·4% of prey numbers) and polychaetes (11·4%) more important (Demme 1934). On Spitsbergen, varied diet, but mostly arctic cod, and as available, large quantities of *Gammarus locusta*, *Thysanoessa inermis*, pteropods, and cephalopods; in one stomach, 228 *Gammarus* (see Løvenskiold 1964). In June–September, Bering Sea, fish up to 33 cm ingested (Ogi and Tsujita 1977).

Little information on diet in winter, but evidence suggests much higher proportion of invertebrate prey than in summer. On Franz Josef Land, April–May, 17 stomachs yielded crustaceans in 8, fish in 7, polychaetes in 2, and molluscs in 1. Small invertebrates only secondary prey in summer when small fish more abundant (Demme 1934; see Ogi and Tsujita 1973).

Young fed almost exclusively on fish (e.g. Krasovski 1937, Uspenski 1956, Løvenskiold 1964, Gaston and Nettleship 1981), on Spitsbergen notably arctic cod (Løvenskiold 1964; see also Duffey and Sergeant 1950). On coast of Murmansk, 36 uneaten fish collected on ledges comprised 86·1% *Ammodytes tobianus*, 8·3% herring, 2·8% cod, and 2·8% sculpin *Gymnocanthus tricuspis* (Kaftanovski 1938). In Novaya Zemlya, 77% cod, 18% capelin, 5% *Aspidophoroides olrikii* (Krasovski 1937). See Uspenski (1956) for other samples from Novaya Zemlya, in which arctic cod and haddock predominated. On Prince Leopold Island, 75–82% fish in diet over 3 years ($n=181$), mostly arctic cod; length of items, including crustaceans, 28–193 mm; older young receive bigger items than smaller young (Gaston and Nettleship 1981). Even before 1 week old, young fed on fish up to 15 cm long (Tuck 1961*a*). Weight of individual fish fed to young: *c.* 15–25 g (Uspenski 1956), *c.* 25 g (Tuck and Squires 1955), 5–29·5 g, mean 13·4 (A J Gaston). At Novaya Zemlya, young 3–8 days old fed once per day, 2–3 times per day thereafter; daily food intake *c.* 30–45 g (Uspenski 1956; see also Tuck 1961*a*). At Digges Island (Canada), young fed about twice per day (A J Gaston). At Prince Leopold, young of all ages (3–22 days) received mean 4·62 meals per day; *c.* 3·5 per day at beginning of season, *c.* 6–8 at end; young 1 day old received *c.* 26 g per day, at 3–20 days *c.* 58 g; over entire fledging period, each chick received *c.* 53 g of fish per day, compared with adult requirement of *c.* 250 g per day (Gaston and Nettleship 1981). Average feeding rate (35 chicks), Coburg Island and Cape Hay (arctic Canada), 0·75 and 0·81 meals per chick per 4 hr, average duration of feeding trips 99·5 and 70·7 min ($n=95$) (Birkhead and Nettleship 1981). At Novaya Zemlya, no diurnal pattern of feeding evident (Uspenski 1956). At Prince Leopold, no clear pattern until mid-August, after which consistent pattern emerged, with maxima at 00.00–04.00 hrs and 14.00–16.00 hrs, and minimum at 06.00 hrs (Gaston and Nettleship 1981). EKD

Social pattern and behaviour. Based mainly on Williams (1972) and, for arctic Canada, Gaston and Nettleship (1981). Also on notes compiled by T R Birkhead and P G H Evans.

1. Gregarious in breeding season, less so in winter. At end of breeding season, dispersing family groups typically comprise ♂ parent and offspring, ♀♀ dispersing later (Bradstreet 1976; P G H Evans, A J Gaston; see also Løvenskiold 1964, and Bonds and Relations within Family Group, below). According to Uspenski (1956), adults and young dispersing from Novaya Zemlya (USSR) do so in small flocks usually comprising ⅔ adults, ⅓ young; however no support from other studies that family parties include both parents. In winter, arctic Canada, solitary or in loose flocks of up to 500 (A J Gaston); older birds further offshore than immatures (Salomonsen 1967a). In severe ice conditions, large flocks assemble and may move long distances. In spring, arctic Canada, density higher along land–sea and ice–sea boundaries than in areas of open water (Bradstreet 1979). By late spring, adults and immatures mostly in separate flocks (Tuck 1961a). BONDS. Monogamous mating system. Marked mate fidelity between years (Uspenski 1956; Kozlova 1957; Gaston and Nettleship 1981; A J Gaston). However, pairs unlikely to consort outside breeding season, given dispersal pattern, above (Tuck 1961a; A J Gaston). For hybrid with Guillemot *U. aalge*, see Tschanz and Wehrlin (1968). Age of first breeding not known. Uspenski's (1956) estimate of 3 years probably too low. Both sexes care for young at nest-site but usually ♂ alone attends young at sea. Young thought to be cared for at sea for several weeks (Bradstreet 1976), 1–2 months (Uspenski 1956); no evidence that parent feeds young, though this not impossible (see *U. aalge*). BREEDING DISPERSION. Strongly colonial and highly gregarious. Colonies of less than 100 pairs uncommon, except at edge of range (e.g. Norway, Brun 1965; Iceland, Björnsson 1976). In Novaya Zemlya, colonies of up to 300000 pairs (Belopol'ski 1957), with maximum density 32 pairs per m² (Uspenski 1956), 37 pairs per m² (Krasovski 1937). Thus very small nest-area territory defended, apparently equally by ♂ and ♀ in incubation and chick-rearing periods (Gaston and Nettleship 1981); used for courtship, copulation, and raising chick until it leaves cliff (hereafter 'fledging'). For nesting association with *U. aalge* (which can apparently oust it from ledges when space limited), see Belopol'ski (1957), Spring (1971), and Williams (1972, 1974). May also compete for space with Kittiwake *Rissa tridactyla*, though not considered serious competitor (see Tuck 1961a, Løvenskiold 1964). In some colonies, shares ledges with Fulmar *Fulmaris glacialis*. Nest-site fidelity high between years (Storer 1952; Uspenski 1956; Kozlova 1957; Kartashev 1960). At Prince Leopold Island (Canada), 10 out of 12 birds occupied same site in 2 successive years; at Digges Sound (Canada), 20 out of 29; shift of site sometimes associated with breeding failure (Gaston and Nettleship 1981; A J Gaston). ROOSTING AND COLONY-ATTENDANCE PATTERNS. For roosting and loafing sites and postures, see *U. aalge* (also 7 in Voice). Loafing non-breeders preen, Pick (flick stones, etc., between their feet: Pennycuick 1956), yawn, wing-stretch, and wing-flap; non-breeding pair described as 'sleeping pressed close together' on rock (Pennycuick 1956). Unlike *U. aalge*, immatures do not assemble in 'clubs' on tidal rocks (Pennycuick 1956; Gaston and Nettleship 1981; T R Birkhead), but see Heterosexual Behaviour (below); breeders roost at sea, especially in pre-laying period (Tuck 1961a) when all nest-sites vacated at night. Birds (all ages) may loaf on ice-floes, standing upright (Løvenskiold 1964; Gaston and Nettleship 1981). On return from winter quarters, Novaya Zemlya, adults initially remain far offshore but start to occupy cliffs 5–17 days after first arrival in vicinity (Uspenski 1956). Egg-laying, Novaya Zemlya, begins 7–10 days after occupying nest-sites; in eastern Murmansk (USSR), 4 weeks after occupation (Kozlova 1957); for occupation dates, Prince Leopold, see Gaston and Nettleship (1981). No detailed information on pattern of attendance for west Palearctic; varies especially with latitude, season, and probably proximity of feeding grounds. At Prince Leopold, pre-laying period typified by alternating periods of presence and absence, which, with approach of egg-laying, become respectively longer and shorter (Gaston and Nettleship 1981); thus, in 1975, from 29 May onwards (perhaps 2–3 weeks after initial arrival), intervals between successive peaks 7, 6, 3, and 2 days (A J Gaston). Build-up from low to peak usually c. 2 days, departure more rapid and highly synchronized, numbers often dropping within 12 hrs from several hundred birds to only a few. During peaks, number of birds on study sites 50–82% of their total breeding population, maximum when laying imminent; then also large numbers on sea up to 1 km offshore. Some evidence that during peaks most of arriving birds ♂♂ whereas most already on cliffs ♀♀. After laying, fluctuations on cliffs smaller. From beginning of egg-laying to start of fledging, most sites permanently occupied by 1 member of each breeding pair. Attendance on non-breeding ledges built up throughout season. Immatures visited ledges late in season, at first exploring often peripheral sites, later landing apparently closer to occupied sites. Marked diurnal cycle of attendance involving both breeding and non-breeding birds. During incubation period, numbers higher at 08.00 hrs, lower at 22.00–24.00 hrs; attendance of adjacent nesting groups more synchronized than widely separated ones. After hatching, peak numbers at c. 18.00 hrs (Gaston and Nettleship 1981; A J Gaston). In north-west Greenland, peak at 12.00–14.00 hrs during incubation, at 02.00 and 10.00–12.00 hrs during chick-rearing (P G H Evans). At Jan Mayen, 30 July–1 August, and at Spitsbergen during chick-rearing, apparently no marked diurnal patterns of attendance (Cullen 1954; Pennycuick 1956).

2. Response to disturbance similar to *U. aalge* but noticeably less responsive to sudden movements of other birds nearby (Williams 1972). Alarmed bird adopts Alert-posture (see *U. aalge*, Fig A); this also adopted by immatures landing on occupied ledges. If not driven off at once, immature stands with neck upstretched, bill obliquely raised, plumage sleeked, carpal joints slightly raised, and turns head from side to side; jumps whenever nearby bird moves (Pennycuick 1956; Gaston and Nettleship 1981). Site-owners also show alarm by Alarm-bowing (for calls see 1–2 in Voice): make shallower forward bow than *U. aalge*, so that action reduced to jerky nodding or bobbing of head, forward to horizontal; however on broad, open sites, may bend right over (Pennycuick 1956; Williams 1972; Gaston and Nettleship 1981; see also Foster *et al.* 1951). Wing-flapping as in *U. aalge* (see that species); compared with *U. aalge*, however, evokes less response from near neighbours (Williams 1972). FLOCK BEHAVIOUR. Forms flocks for feeding, dispersal, and pre-laying behaviour. Flocks on feeding trips may fly in no distinct formation, or in long lines or echelons (Løvenskiold 1964); discrete flocks of up to 300 birds fly to feeding grounds, but well dispersed when feeding (A J Gaston). Pre-laying behaviour includes skittering over surface, synchronous diving, and underwater chasing, as in *U. aalge* but more marked (Tuck 1961a; T R Birkhead). Often 2 birds chasing just above sea follow circular or figure-of-eight path, keeping 2–3 body-lengths apart, before continuing pursuit under water; occurred frequently, Prince Leopold, in pre-laying period, and irregularly throughout season, especially when sea calm (Gaston and Nettleship 1981). Rafting behaviour often develops into aerial display of sort unknown in *U. aalge* (T R Birkhead): often initiated by 2–3 birds which attract others; flock alternately soars close to cliffs, and swoops low over sea; after a few days, birds break away from flock as it skirts cliffs and make brief landings, these increasing in frequency and duration as days pass (Tuck 1961a). At Prince Leopold, mainly individuals but also small groups (5–10 birds) leave rafts and make sorties

to cliffs, some landing but most circling and leaving again (Gaston and Nettleship 1981). ANTAGONISTIC BEHAVIOUR. (1) Threat and fighting. Said to be more pugnacious than *U. aalge* (e.g. Foster *et al.* 1951, Nørrevang 1958, Tuck 1961a), though fighting apparently not 'fiercer' than in that species (Williams 1972); little aggression shown by incubating and brooding birds to their neighbours (Pennycuick 1956), but site-holders often attack birds prospecting late in season; at Prince Leopold, 25% of 57 prospectors that landed left after such a confrontation (Gaston and Nettleship 1981). Parents and chicks making way to edge of ledge at fledging often soundly pecked (Gaston and Nettleship 1981). Fights common but usually brief, exceptionally up to 1 hr or more (Tuck 1961a; T R Birkhead); most start when incubating or brooding bird threatened by non-breeder. 2 birds on verge of fighting stretch necks up and sleek plumage (Pennycuick 1956; see Upright-threat posture in *U. aalge*). As in *U. aalge*, fighting involves jabbing (mostly at head) and, at higher intensity, gripping with bill and cuffing with wings (Gaston and Nettleship 1981). During fight, neighbours give variant of Crowing- and Uggah-calls (Pennycuick 1956: see 3a, also 2, in Voice). Fighting often ends with 1 or both birds falling off cliff (Pennycuick 1956; Gaston and Nettleship 1981). If both topple off, may lead to pursuit-flight in which pursuer—with variant of Crowing-call (3a in Voice)—tries to touch leader (Pennycuick 1956; Williams 1972); birds dislodged from cliff often land in water with bills still locked, there to continue fight (Tuck 1961a; Williams 1972; Gaston and Nettleship 1981). On sea, birds circle each other with head and bill raised, feinting with wings, then suddenly grip bills; fighting often under water; after fight in water, participants often splash along surface with neck outstretched, flailing water with wings (Pennycuick 1956). Incomplete information on repertoire of behaviour to which 'displacement' function attributed in *U. aalge*. Unlike that species, solitary *U. lomvia* does not Footlook though Williams (1972) suggests brief breast-touching (see below), performed in situations where *U. aalge* would Footlook, may be derived from tendency to look down. Fencing performed by birds after they, or their neighbours, involved in aggressive interaction (T R Birkhead)—suggests displacement activity or appeasement function. (2) Appeasement behaviour. Performs same range of appeasement behaviour as *U. aalge* (Side-preening, Turning-away, Stretching-away), and in similar contexts (T R Birkhead). To break off milder aggression on ledges, may slowly but briefly turn head from side to side, and finally bird may touch own breast (Williams 1972). Where birds do not fall off cliff during encounter, serious aggression ended by Side-preening, as in *U. aalge*. Alighting bird often Side-preens, and much more commonly (50% of occasions) when partner present, than in *U. aalge* (T R Birkhead). Although *U. aalge* can apparently displace *U. lomvia* from nest-site, on Bear Island *U. lomvia* often dominant in encounters (Williams 1972); these mostly ended by 1 bird Side-preening, but after hostilities *U. lomvia* waited longer than opponent before Side-preening and was thus ready for any suddenly renewed attack. *U. lomvia* Allopreened *U. aalge* at about same frequency as it did conspecifics (Williams 1972). Alighting birds often adopt postures which may indicate appeasement or site-ownership. At Bear Island, birds able to land directly at nest-site did so in Landing-posture—wings raised over back, head bent so that bill over breast (Williams 1972). At Prince Leopold, prospecting birds stood with wings raised after landing, often continuing to flap for several seconds, before adopting Alert-posture and Side-preening wing-coverts; rarely sat down (Gaston and Nettleship 1981). Wings-high Walking—with head bent down (as in Landing-posture)—occurs, but apparently only prior to take-off, not on landing; 'head-up' variant of *U. aalge* does not occur (T R Birkhead). HETEROSEXUAL BEHAVIOUR. (1) Pair-bonding behaviour. Compared with *U. aalge*, no Head-vertical posture ('skypointing': Williams 1972), and very little mutual Fencing, though copulation much more frequent (Williams 1972; T R Birkhead). Instead of forming clubs, immatures fly along breeding ledges and alight at edge of area occupied by nesting birds (Pennycuick 1956). At Prince Leopold, July, large flocks flew back and forth along cliffs and landed on broad, sparsely occupied ledges (Gaston and Nettleship 1981, which see for subsequent behaviour). More courtship therefore presumed to occur on ledges than in *U. aalge*. Copulation (see below) began on day of arrival at cliffs, Alaska (USA), suggesting some pairing away from colony area (Swartz 1966). No definite information on sequence of behaviour in pair-formation. Mutual Footlooking, picking up and exchanging stones (and other material) occur on broader ledges (Williams 1972). See also sections 2 and 4, below. (2) Site-ownership displays. Bowing, in which bird arches neck, lowering bill (Gaston and Nettleship 1981), less common than in *U. aalge*, and Fish-presentation very rare (T R Birkhead). So-called 'hawing' occurs—apparently not performed by *U. aalge*, but posture similar to Turning-away (Williams 1972): bird arches neck slightly, usually to one side, and gives Hawing-call (see 3d in Voice) which shakes whole body. After calling, bird frequently bends head to touch cliff or material on ledge; debatable whether this a distinct display (T R Birkhead). (3) Meeting-ceremony. In pairs 'which have no chick' (Pennycuick 1956), mutual Bowing sometimes performed when one of them arrives at nest-site. Bird alights within a few cm of sitting mate which instantly stands and Bows at site, giving Crowing-call (see 2d in Voice), arriving bird then thrusts head forward beneath breast of mate; both stand up and perform mutual Billing, nibbling at each other's bill for a few seconds (Gaston and Nettleship 1981). Unlike *U. aalge*, in which both birds call and vigorously Fence for several seconds, only one bird calls (see 2b and 3 in Voice), for a shorter time, and there is no Fencing (T R Birkhead). In pre-laying period, Meeting-ceremony often followed by copulation (Gaston and Nettleship 1981: see below). Descriptions of greeting during chick-rearing vary; brooding bird greets mate with fish by rising (allowing chick to emerge), sleeking plumage, stretching head up, and calling (see 3b in Voice); call then taken up by neighbours, and loud chorus (2 and 3b in Voice) results (Pennycuick 1956). According to Williams (1972), however, arrival of bird with fish causes brooding mate to bow and call (see also Gaston and Nettleship 1981). If site empty when bird arrives, turns and Side-preens (Williams 1972). (4) Allopreening. Significantly less frequent than in *U. aalge*, though mean length of bouts similar (T R Birkhead). According to Williams (1972), frequency similar to that of *U. aalge*. For Allopreening of *U. aalge* by *U. lomvia*, see Antagonistic Behaviour (above). (5) Mating. Begins on first arrival at cliffs (Swartz 1966). Much more frequent than in *U. aalge*, thus possibly serving to reduce ♂'s aggression (Williams 1972) and/or disguise ♀'s fertile period from neighbours (Birkhead 1978b). Mating often preceded by Allopreening (Gaston and Nettleship 1981). Mounting always initiated by ♂ giving Pre-copulation call (see 3c in Voice) close to ♀'s nape. Receptive ♀ immediately squats, giving Copulation-call (Williams 1972: see 5 in Voice). Once mounted, ♂ gives Copulation-call (see 4 in Voice); flaps wings more than does *U. aalge*, this possibly related to usually narrower ledges; occasionally ♂ fell off cliff during or immediately after copulation (Williams 1972). ♂ may grasp ♀'s crown feathers; cloacal contact usually lasts 3–10 s (Gaston and Nettleship 1981). Behaviour after copulation varies: birds may Footlook (Williams 1972) or Side-preen (Williams 1972; Gaston and Nettleship 1981). At Prince Leopold, copulation without cloacal contact most common in pre-laying period; immediately after landing, 14% of 160 copulations unsuccessful; failure resulted from non-cooperation of mounted bird which stood up and pecked at ♂; when severe fighting broke out, mounted bird possibly ♂ (Gaston and Nettleship 1981; A J Gaston). Strong association between landing and incidence of copulation, especially during pre-laying period (Williams 1972; Gaston and Nettleship 1981). In 58 landings by one member of pair at nest-site, copulation followed in 28 (Williams 1972). Copulation followed landing in 51% of cases at one site, Prince Leopold, mid-June (A J Gaston); most often (80%, $n = 138$) when alighting bird was ♂ (Gaston and Nettleship 1981). Copulation also occurs when both birds return to site after absence (Williams 1977). Successful copulation never observed in incubating pair; ♂ attempting to copulate immediately after egg laid usually rejected by ♀ (Gaston and Nettleship 1981). According to Williams (1972), 'token copulations' occurred after egg laid, ♂ merely stepping on to and off ♀'s back as she called. (6) Behaviour at nest. For scraping behaviour, see *U. aalge*. May be prolonged. One bird scraped for 8 hrs in 126 bouts of 3–8 s each, latterly at 1-min intervals (Williams 1971, 1972). Some birds, Prince Leopold, shifted stones about in such a way that prospective nest-site might be improved, e.g. by dropping large stones into crevices where eggs might otherwise have become lodged (A J Gaston). Usually, however, pebbles flicked between feet, or picked up again, or abandoned for another object a few cm away (Pennycuick 1956). After feeding chick, fish-bearer takes over brooding, either at once or within $c. \frac{1}{2}$ hr; if brooding bird reluctant to leave, may be shouldered off; relieved bird may remain on ledge for a while before flying off (Pennycuick 1956). Site-owner often raises wings high for several seconds before flying off; within 10 s or so of leaving, performs Butterfly-flight, not seen in prospecting birds (Gaston and Nettleship 1981) nor in *U. aalge* (T R Birkhead). RELATIONS WITHIN FAMILY GROUP. Chick brooded under brood-patch just after hatching, thereafter under wing (Gaston and Nettleship 1981), right up until fledging (Pennycuick 1956; Uspenski 1956), sometimes (presumably when small) by neighbouring adults (Uspenski 1956). Young fed from within 24 hrs of hatching (*contra* Tuck 1961*a*) to within 24 hrs of fledging (Pennycuick 1956, *contra* report (Uspenski 1956) that parents starve young some days before fledging. Young thermoregulate from $c.$ 3–4 days (Rolnik 1948; Kartashev 1960), completely by 9–10 days (Johnson and West 1975). Food often brought while egg still pipping, in which case usually deposited on ledge close to egg. Chick takes fish from incoming adult which bows and thrusts bill towards it (Williams 1972; Gaston and Nettleship 1981). For other details of fish transfer, see *U. aalge*. Sometimes fish dropped by parent or chick but immediately retrieved either by parent for re-presentation, or by (at least older) young (Williams 1972; T R Birkhead). Fledging procedure studied in detail (Pennycuick 1956; Kartashev 1960; Williams 1972, 1975; Gaston and Nettleship 1981). At Bear Island, where little variation in light intensity, young fledge at any time of day (Pennycuick 1956), though Løvenskiold (1954) implied most birds, Spitsbergen, fledged in evening and night hours when light intensity lowest; elsewhere, fledging mainly in evening up to midnight (e.g. Salomonsen 1950, Cullen 1954, Swartz 1966, Williams 1975, Daan and Tinbergen 1979). Most fledge on calm nights (Swartz 1966; Birkhead and Nettleship 1981). When fledging imminent, parent periodically stands up from brooding, allowing chick to walk to edge of ledge; where ledge unsuitable for fledging, parent shepherds chick sometimes several metres to more suitable spot. There, chick highly active, exercising wings, and bowing at edge of ledge (Gaston and Nettleship 1981). Much vocal contact between parent (see 8 in Voice) and chick (Pennycuick 1956). Usually only 1 parent present (Roelke and Hunt 1978; Gaston and Nettleship 1981). However, both parents said to be present, Bear Island, both periodically taking off and flying around for a few minutes. Accounts vary as to whether parent or chick leads in jumping off. At Prince Leopold, chick jumped first, flapping vigorously, then gliding when clear of cliff; parent followed, gliding $c.$ 5–10 cm behind (Gaston and Nettleship 1981). At Bear Island, parent flew slowly down to water, feet and tail spread, body rolling to left and right, and chick followed, feet spread, wings fluttering (Pennycuick 1956); both silent during descent (Kartashev 1960). At Spitsbergen, 2 adults, less often 1, followed young on descent, calling all the way (Løvenskiold 1954). Angle of descent often $c.\ 45°$ (e.g. Løvenskiold 1954), though young on cliffs up to 190 m high, Akpatok Island (Canada), said to land on sea up to $c.$ 800 m from cliff (Tuck 1961*a*). Young levels out just before reaching water (Gaston and Nettleship 1981). On landing, young immediately starts swimming vigorously and erratically, giving loud contact-calls, sometimes briefly diving once or twice. Parent and offspring usually rush to meet each other. On finding parent, chick presses close up beside it and nibbles its bill; both stop calling and swim slowly out to sea, chick staying close (Pennycuick 1956). When chick landed on ledge and failed to reach water, parent usually made no attempt to find it, but returned to nest-site (Gaston and Nettleship 1981), though adults sometimes land on ground beside young which fall short of sea (Daan and Tinbergen 1979). At Bear Island, as chick fledged, chorus of crowing arose from neighbouring adults, a dozen or more of which flew down and joined chick; swam around it giving Crowing-calls; these and others attracted by parent–chick calls mobbed chick sometimes for hours, pecking gently at its back and tail, and diving (together) around it; surfacing bird sometimes jumped on chick's back, pushing it under water (Pennycuick 1956; P G H Evans). Up to 10 birds (prospectors or site-holders without young: Gaston and Nettleship 1981) commonly accompany young on downward flight, usually in echelon behind it, but after much jostling on water, usually only 1 adult (parent) accompanies young to sea (Løvenskiold 1954; Tuck 1961*a*; Gaston and Nettleship 1981; M A Ogilvie); however calls of chick, released 32 km out to sea, very quickly attracted several adults to it (Tuck 1961*a*), but no evidence that such associations more than temporary (T R Birkhead). In numerous sightings, Spitsbergen, single adult accompanied young (Løvenskiold 1964). In 47 adult–young 'pairs' collected, arctic Canada, 46 adults were ♂ (Bradstreet 1976). At Prince Leopold, parent (♀) absent at fledging, continued to attend nest-site for up to 20 days; at this time site defence important, and fights frequent; thereafter attendance sporadic; both parents sometimes attended, indicating chick lost during or after fledging (Gaston and Nettleship 1981). ANTI-PREDATOR RESPONSES OF YOUNG. Fledging chicks which land on sea dive at approach of gull *Larus* (Gaston and Nettleship 1981) or skua *Stercorarius* (P G H Evans). Chicks harassed by man may jump from cliff prematurely (Uspenski 1956). See *U. aalge* for other responses. PARENTAL ANTI-PREDATOR STRATEGIES. (1) Passive measures. See *U. aalge*. Sits more tightly than *U. aalge*, especially after 5–8 days of incubation (Kaftanovski 1938; Kartashev 1960; T R Birkhead). At approach of gull, either on foot or in flight, birds gave Uggah-call while Alarm-bowing; birds thus alerted adopted Alert-posture (Gaston and Nettleship 1981). In Alarm-bowing during human intrusion, incubating or brooding bird shuffles backwards from cliff-face, thus gently expelling egg or chick and expediting own escape flight (Pennycuick 1956). (2) Active measures: against birds. Incubating and brooding adults pointed bills silently at nearby gull; occasionally jabbed at and grappled with gulls attempting to dislodge them (by grabbing at wing or tail); sometimes adults without chicks lunged at gull. Liability to attack from gulls decreased with increasing nest density, and with distance of egg from edge of cliff (Gaston and Nettleship 1981). When Glaucous Gull *L. hyperboreus* threatened chick that had landed on beach, parent on water rushed up beach between chick and gull, giving Crowing-calls (see 2a in Voice) and waving wings (Pennycuick 1956). (3) Active measures: against other animals. Collective aggression (jabbing, etc.) of birds on ledges thought capable of intimidating arctic foxes *Alopex* (Pennycuick 1956). EKD

Voice. Highly vocal in breeding season. No information on calls outside breeding season. Repertoire apparently similar to Guillemot *U. aalge*, on which this scheme based.

CALLS OF ADULTS. (1) Alarm-bowing call. A low, short (*c*. 0·25 s), sharp 'grrr', given in response to approach of any potential predator; quite different in tone from call 4 (Pennycuick 1956). (2) Uggah-call. Sharp, disyllabic 'uggah' (*c*. 0·2 s: Fig I); rendered 'owka' (Tuck 1961*a*), 'ger-ou', or 'jer-ow' (Williams 1972). Interspersed with call 1, and given in same situations (Pennycuick 1956), e.g. while Alarm-bowing during aerial or walking approach of gull *Larus* (Gaston and Nettleship 1981). Adult called 'owka-owka-owka' when chick persisted in swimming to shore (Tuck 1961*a*). (3) Crowing-call. A loud, hoarse 'arrrr-rrr-rr-r' of *c*. 0·75 s duration and of variable pitch; given with bill wide open (Pennycuick 1956). A guttural 'krrr' (Williams 1972). Several variants occur; following names follow *U. aalge* (see Tschanz 1968): (a) 'Battle-crowing'. In recording (Fig II), call has especially gruff, growling quality. Given when fighting, or when fight in progress nearby; also in distress (e.g. when chick approached by human intruder, and then given just before take-off) in flight—and on water (Pennycuick 1956); probably homologous with Distress-call of *U. aalge*. (b) 'Reception-crowing'. Call rises rapidly in pitch to a squeak, then falls (Pennycuick 1956). In recording (Fig III), 'rrrrraw-rraw-rraw arrrr-arrr', 1st unit rising in pitch, though not markedly so (E K Dunn). Given by bird when its mate, or neighbour, arrives on ledge with fish, apparently expressing great excitement (Pennycuick 1956). (c) Pre-copulation call (♂). Growling 'grawu', given with gaping bill close to ♀'s nape, serves as 'demand' for copulation (Williams 1972). (d) Hawing-call. Although designated a separate call by Williams (1972), 'rrr-haw-haw-haw-haw-haw-haw' (Pennycuick 1956) is a variant of Crowing-call; laughing, convulsive sound (Williams 1972), varing in pitch, resonance, number of units, and rate of delivery; often lasts *c*. 2 s (Pennycuick 1956). Typically given in chorus on arrival of nearby bird with fish, or when fight begins nearby (Pennycuick 1956); given in 'hawing' posture (Williams 1972). (4) ♂ Copulation-call. Gruff 'ha-ha-ha' (T R Birkhead, P J Sellar), likened to barking of dog (Uspenski 1956), and given by ♂ once

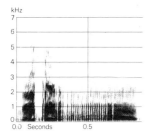

I P J Sellar Iceland June 1968

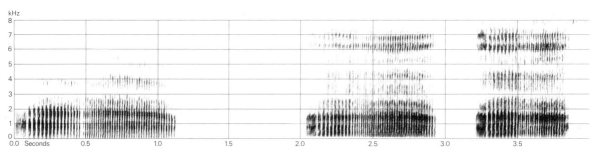

II P J Sellar Iceland May 1967

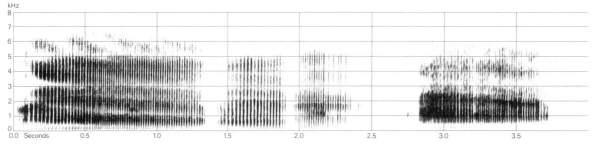

III P J Sellar Iceland June 1968

mounted. (5) ♀ Copulation-call. ♀ calls 'arauuw' on squatting (Williams 1972); loud, disyllabic 'ca-cauk' given once or twice per s (Gaston and Nettleship 1981). (6) Aowk-call. Rather wheezy 'aowk' given by bird during slow Butterfly-flight, once per wing-beat (Tuck 1961a). (7) Contentment-call. A low, throaty, gargling 'rrr', resembling first unit of call 3a, given by 'contented' loafing bird, with head drawn in, feathers ruffled, eyes half-closed, bill upwards (Pennycuick 1956). (8) Calls to young. Leap-call. Similar to call 2, given by parent encouraging chick to leap off ledge, or when approaching newly fledged chick on water; thus serves as contact between parent and offspring at fledging time, and invites close approach (Pennycuick 1956; Gaston and Nettleship 1981). Increases in volume and frequency of delivery immediately prior to fledging (Gaston and Nettleship 1981).

CALLS OF YOUNG. From c. 1 day before hatching to c. 5–6 days old, chick gives fretful cheep. From c. 7 days, Food-call a thin, wavering 'weeoo weeoo' (each unit c. 0·1 s); given intermittently when awake, but especially to solicit food or brooding (Pennycuick 1956; Tuck 1961a); described as a whimpering, disyllabic pipe (Salmonsen 1967a). During last few nights before fledging, and persistently until reunited with parent in water, chicks gave high-pitched, squeaking 'wee-wee-wee', each unit c. 0·5 s (Pennycuick 1956; Tuck 1961a; P J Sellar); rendered 'piu-piu' (Gaston and Nettleship 1981). Same or similar call, particularly loud and shrill, given by even quite small chicks in extreme distress (Pennycuick 1956); described as wheezy, cat-like sound (Tuck 1961a). EKD

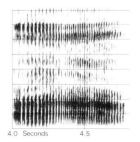

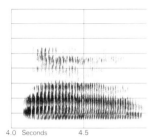

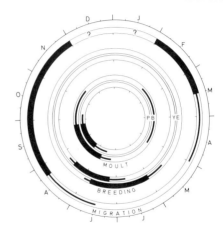

Breeding. SEASON. Spitsbergen: see diagram. Iceland: up to 1 week earlier (Hantzsch 1905). Novaya Zemlya: similar to Spitsbergen but varies between colonies and season, and affected by ice conditions (Belopol'ski 1957). SITE. On cliff-ledge; on Bear Island not found on flat rocky tops used by Guillemot *U. aalge* (Williams 1974). Colonial. Nest: none, but see Social Pattern and Behaviour for scraping behaviour. EGGS. See Plate 92. Pyriform, roughened, not glossy; very variable, as *U. aalge* (see that species) though reddish markings less usual (Witherby *et al.* 1941). On Novaya Zemlya 79 × 51 mm (70–96 × 46–59), $n = 430$ (Uspenski 1956); Spitsbergen similar (Makatsch 1974); on Iceland possibly smaller—77 × 49 mm (64–83 × 43–52), $n = 13$ (Timmermann 1938–49). Calculated weight, Novaya Zemlya, 105 g; average weight of 54 fresh eggs, Canada, $98·5 \pm SE$ 1·1 g (Gaston and Nettleship 1981). Clutch: 1. One brood. Replacements laid after egg loss; 30% of birds losing eggs re-laid once, 11% twice, and overall 67–80% of birds losing eggs laid at least 1 replacement (Tuck 1961a). Re-laying interval 10–16 days (Tuck 1961a); 15–22 days in northern USSR (Uspenski 1956); once, 2 replacements laid in 26 days (Tuck 1961a). INCUBATION. 30–35 days. Average 33·5 days (32–34), $n = 100$ (Tuck 1961a). By both sexes. Egg rested on tarsi. Begins at laying. Sexes change over at any time of day; average 0·9 times per day early in incubation, increasing to 2·0 times (Tuck 1961a). YOUNG. Semi-altricial and nidicolous. Cared for and fed by both parents, brooded while small. FLEDGING TO MATURITY. Young leaves ledge before true fledging, at 15–30 days (Gaston and Nettleship 1981); average 21·2 days, $n = 131$ (Gaston and Nettleship 1981). Accompanied on glide to sea by ♂ parent (Roelke and Hunt 1978; Gaston and Nettleship 1981), but also frequently by 1–10 other adults (M A Ogilvie). Period until fully fledged not recorded. Cared for by parent for several weeks after leaving ledge (Birkhead and Nettleship 1981). Age of first breeding not known. BREEDING SUCCESS. At one colony, Canada, 46% of eggs laid lost before hatching; at another, 20% lost (no sample sizes); greatest losses due to frost-loosened rocks which broke eggs and scared adults, allowing gulls *Larus* to take eggs; losses to gulls mainly of smaller eggs, and so from younger or poorer-quality birds; earlier breeders more productive than later (Birkhead and Nettleship 1981). Of 885 eggs, Canada, 84·0% hatched and 0·80 chicks left ledge per breeding pair (0·73–0·89); re-layings only half as successful as 1st eggs, with 40% of eggs yielding young which left ledges, mostly due to low success of hatched eggs (52%); greatest cause of failure was egg-robbing by Glaucous Gull *L. hyperboreus*; less often taking of chicks (Gaston and Nettleship 1981). In northern USSR, egg loss *c.* 50% (highest on narrow sloping ledges); on Sem'ostrovov, most eggs taken by Great Black-backed Gull *L. marinus* and Herring Gull *L. argentatus*, and on Novaya Zemlya by *L. hyperboreus*; arctic fox *Alopex lagopus* also responsible; losses of young averaged 57·3%, mostly due to low temperatures (Uspenski 1956). During fledging, Spitsbergen, young taken in flight by *L. hyperboreus* and also by gulls and arctic foxes when landing short of sea (Daan and Tinbergen 1979; M A Ogilvie).

Plumages (nominate *lomvia*). ADULT BREEDING. Upperparts black with velvet gloss on crown and nape. Lores, sides of head to upper edge of eye, sides of neck, and throat dark chocolate-brown. Narrow furrow in feathers extending backwards from posterior edge of eye. Underparts white, extending in ∧-form up to middle of brown throat. Lower part of flanks with some brownish-black streaks. Tail brownish black. Primaries and secondaries as upperparts, inner webs pale brown grading to whitish on edge and base. Shafts brownish-black, dark horn at base. Secondaries (except outermost) broadly tipped white. Upper wing-coverts as upperparts. Underside of flight-feathers dark silver-grey; outer web of p10 black, inner web with brown shaft-streak. Greater under wing-coverts dark silver-grey, some tipped or edged white; median and lesser coverts white with brown shafts and bases; brown on median under primary coverts sometimes extending on one web to tip of feather. Axillaries with variable pattern of brown and white. Due to abrasion, flight-feathers, tail, mantle feathers, and upper wing-coverts become pale brown, especially at tips, about mid-August. Melanism rarer than in Guillemot *U. aalge*; albinism probably more frequent. In partial albinism, plumage pattern retained but black feathers are light grey or grey-brown, or some normally black flight-feathers, tail-feathers, and mantle feathers are white (Storer 1952; Tuck 1961*a*; RMNH). ADULT NON-BREEDING. Upperparts brownish-black. Black of crown extends behind and below eye to level of gape and to apex of chin. Sides of neck black; adjoining feathers on lower throat tipped black, forming variable band across throat. Underparts white, some feathers on throat and lower ear-coverts tipped greyish-black. Rest of plumage as adult breeding. DOWNY YOUNG. As in *U. aalge* but head more densely streaked white and upperparts more brownish. Throat varies from blackish to greyish-white (Harrison 1975; Fjeldså 1977). Down replaced by juvenile feathers *c.* 16–20 days after hatching (Kartashev 1960). JUVENILE. 2 plumage types, one resembling adult breeding, the other adult non-breeding. In adult-breeding type, forehead, crown, and nape deep black; chin, upper foreneck, cheeks, sides of neck, and hindneck black with brownish hue. Upperparts black. Underparts white, narrow blackish feather tips to sides of chest and breast, slightly sooty feather-tips on lower flanks and thighs. White of foreneck tapering in ∧-form to throat. In adult-non-breeding type, chin, throat, and cheek white. Flight-feathers and tail grow with moult to next plumage. Feathers of loose structure, last down retained on nape, rump, lower flanks, and thighs. FIRST IMMATURE NON-BREEDING. Like adult non-breeding (but bill smaller). FIRST IMMATURE BREEDING. Like adult breeding, but upperparts and flight-feathers browner; throat often mottled white (Witherby *et al.* 1941; Dementiev and Gladkov 1951*b*).

Bare parts. ADULT. Iris dark brown. Bill black, tips of both mandibles and (in winter) gonydeal angle dark horn-coloured. Streak along basal half of cutting edge of upper mandible bluish-grey, slate-coloured, or whitish, in winter (after shedding white sheath) yellow-brown. Leg and foot anteriorly yellow-brown with dark bands across joints; rear and webs black. DOWNY YOUNG. As *U. aalge* but bill deep grey (Fjeldså 1977). JUVENILE. No information.

Moults. Poorly known. ADULT POST-BREEDING. Complete. Small feathers of head and body from late July to late September, perhaps (as in *U. aalge*) extending to November. Flight-feathers shed from late August to early September (Salomonsen 1979). ADULT PRE-BREEDING. Partial: head and neck. Full plumage acquired late April or early May; in 1-year-olds, not until early to late June (Salomonsen 1944; Dementiev and Gladkov 1951*b*). POST-JUVENILE. Growth of flight-feathers completed by September–October (Salomonsen 1979).

Measurements. Depth of bill measured at angle of gonys. ADULT. Nominate *lomvia*. Eastern Greenland, Bear Island, Iceland, and Spitsbergen, breeding; skins (RMNH, ZMA).

	♂			♀		
WING	216	(4·4; 11)	208–221	218	(4·3; 11)	208–223
TAIL	46·6	(2·5; 10)	43–50	47·8	(3·6; 11)	44–53
BILL	40·5	(2·3; 11)	36–44	38·6	(2·0; 11)	35–42
DEPTH	14·4	(0·7; 9)	13·0–15·0	13·8	(0·7; 8)	12·9–15·2
GONYS	22·0	(0·7; 10)	20·6–22·9	21·1	(1·0; 9)	20·0–22·6
TARSUS	37·9	(1·7; 12)	35–40	37·3	(1·3; 11)	36–40
TOE	52·5	(1·7; 12)	50–55	52·2	(1·2; 11)	49·7–54·2

Sex difference significant for gonys.

Bear Island (Glutz and Bauer 1982).

	♂			♀		
WING	219	(38)	205·5–229	214	(36)	202–227
BILL	53·5	(38)	51–62·5	54·5	(36)	50–60

Spitsbergen (Glutz and Bauer 1982).

	♂			♀		
WING	218	(20)	208–226	217	(15)	211–226
BILL	54·9	(20)	51–58	54·4	(15)	50–58

Weights. ADULT. Nominate *lomvia*. Eastern Greenland, Bear Island, and Spitsbergen, May–August, ♂ 957 (80·9; 12) 810–1080, ♀ 882 (113·4; 7) 730–1050, sex difference not significant (ZMA). Sexes combined, Northwest Territories (Canada), July, 861 (53·3; 57) (Birkhead and Nettleship 1981); Prince Leopold Islands (Canada), pre-laying 907 (70; 10), incubation 909 (57; 9), chick-rearing 869 (54; 17) (Gaston and Nettleship 1981). Eastern Murmansk (USSR), May–August, ♂ 989 (151) 760–1160, ♀ 999 (83) 840–1200; Novaya Zemlya, ♂ 1027 (261) 827–1206, ♀ 996 (224) 835–1206 (Belopol'ski 1957).

YOUNG. At 1–2 days: Northwest Territories 67·2 (8·8; 106) (Birkhead and Nettleship 1981); Barents Sea 70·6 (50·0–98·0) (Kartashev 1960). At 14 days, Northwest Territories, 188 (59) (Birkhead and Nettleship 1981). On leaving nest-ledge: Hudson Bay (Canada) 200–275 (Tuck 1961*a*); Barents Sea average 250; Novaya Zemlya 120–150 (Kartashev 1960); Northwest Territories, 200 (25·3; 60) (Birkhead and Nettleship 1981). Weight of fledglings differs between localities (Tuck 1961a; Johnson and West 1975). Shortly before leaving ledge, birds lose weight (Tuck 1961a; Sealy 1973; Johnson and West 1975).

Structure. Wing short and pointed. 11 primaries: p10 longest, p9 2–7 shorter, p8 8–13, p7 20–28, p6 28–40, p5 40–53, p1 85–97; p11 minute, concealed by primary coverts. Tail short, slightly rounded, 12 feathers, each obtusely pointed. Bill shorter and deeper than in *U. aalge* and with more prominent gonydeal angle. Nostrils narrow, concealed by feathering, which does not reach cutting edges of upper mandible. Outer toe *c*. 94% of middle, inner *c*. 77%; no hind toe.

Geographical variation. Colour variation difficult to evaluate. *U. l. eleonorae* and *heckeri* said to average paler and *arra* darker than nominate *lomvia* (Salomonsen 1944; Dementiev and Gladkov 1951b; Vaurie 1965). Variation in length of wing, culmen, and tarsus and depth of bill more or less clinal, with larger averages occurring in eastern part of range (see Glutz and Bauer 1982). *U. l. heckeri* and *eleonorae* also poorly differentiated on size and probably better assigned to *arra*.

RS

Alca torda Razorbill

Du. Alk   Fr. Petit Pingouin   Ge. Tordalk
Ru. Гагарка   Sp. Alca común   Sw. Tordmule

PLATES 18, 22, and 23
[facing page 230, and between pages 278 and 279]

Alca torda Linnaeus, 1758

Polytypic. Nominate *torda* Linnaeus, 1758, North America, Greenland, Bear Island, Denmark, Norway, Murmansk to White Sea, and Baltic region; *islandica* C L Brehm, 1831, Iceland, Faeroes, Britain, Ireland, and Brittany.

Field characters. 37–39 cm (31–32 cm excluding tail); wing-span 63–68 cm. Has larger bill and head than guillemots *Uria* but body and wings average slightly smaller, particularly in southern populations. Auk of similar appearance to *Uria* but with deeper, more rectangular bill (contributing to distinctly parallel-sided silhouette to deep head), thicker neck, smaller wing area, and long pointed tail-feathers. Upperparts blacker than those of southern Guillemot *U. aalge* and (less obviously) Brünnich's Guillemot *U. lomvia*. Bill of adult shows broken white line across middle; narrow white line from eye to base of culmen obvious in breeding plumage. Position of diffused black and white border on rear head important in identification of immature and winter adult. Sexes similar; marked seasonal variation. Juvenile separable.

ADULT BREEDING. Head, neck, and upperparts velvet-black, with brown tones on lower head and sides of neck obvious only in sunlight. Black upperparts relieved only by straight white line in front of eye and white tips to secondaries (forming narrow bar across folded wing or conspicuous trailing edge to inner wing in flight at close range). Lower foreneck and underbody white; under wing-coverts also white (longest feathers suffused grey-brown), contrasting sharply with dark undersurface to flight-feathers. Rounded ∧ border between white foreneck and black-brown throat and neck-sides. Bill intensely black; at close range, shows 3 vertical ridges with vertical but broken white line on innermost. Legs and feet black.
ADULT WINTER. As in *Uria*, head and neck pattern much changed with chin, throat, most of rear cheeks, variable patch on sides of nape, and neck except hind-band becoming mottled white, then hoary, then mostly white. White line in front of eye becomes less distinct; vertical line on bill turns grey. Rest of plumage less glossy and slightly browner, with paler wings (when worn). JUVENILE. Distinctly smaller and browner than adult, with small, slightly bulbous, not razor-shaped bill (lacking transverse ridges and pale line), initially no visible flight- or tail-feathers, and looser plumage. Unlike *U. aalge* and most *U. lomvia* at same age, whole head, chin, and upper throat uniformly dark brown; thus pattern recalls breeding (not winter) adult and may be retained until mid-autumn. FIRST WINTER. By November, when dark lower head fully moulted, immature resembles winter adult except for less diffuse white patch on rear head, usually a white band extending forward from there to behind eye (recalling similar but much more pronounced character in *U. aalge*), and thinner white tips to secondaries. Bill remains less deep and less rectangular in outline. FIRST SUMMER. Easily distinguished from adult by worn, pale wings and (on some) incomplete head pattern.

At close range, adult and (from late in 1st winter) immature unmistakable on sea or on rocks; shape and pattern of bill and head prevent confusion. When not close and particularly with flying bird in rough weather, separation from *Uria* difficult and best made on: (1) head shape and rectangular, not pointed, bill (making front of bird look heavy); (2) lack of sharply delineated white area or distinct black furrow behind eye; (3) whiter underbody and under wing-coverts; (4) lack of prominent black half-collar on neck (particularly obvious in northern populations of *U. aalge*). Important to check presence in *U. aalge* or absence in *A. torda* of dark furrow behind eye and sharper edge to border of black and white areas. For distinction from *U. lomvia*, see that species. Flight sustained only by constant and rapid whirring beats of small and

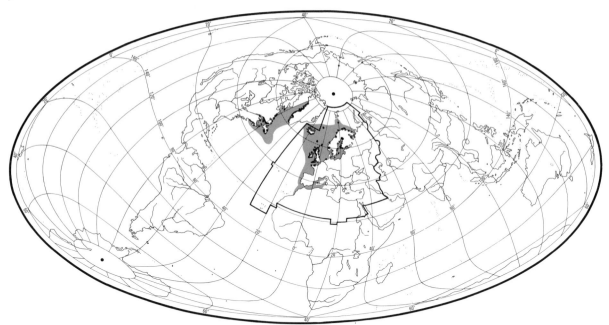

rather narrow wings (fastest of family) and apparently additional lift acquired through spreading of rear body plumage, tail, and feet; thus in flight appears bulky behind wings, with slight up-tilt to body line. Small wing area relative to total mass more clearly evident than on *Uria*, but flight fast in spite of this (even into gales). Lacks close manoeuvrability around breeding colonies; several fluttering attempts may precede landing. When departing from high cliff, regularly uses curiously halting and slow wing-strokes and conspicuous spreading of rear body and tail to extend glide to sea-level. Landing impact taken by bill, head, and drawn-in neck. Takes off freely, using kicks of feet. Dives with obvious flick of feet and quarter-opened wings, showing characteristic submerging of 3-point profile—both sets of primaries and long tail. Underwater, mostly swims with half-opened wings, accompanied by tilts of body and foot-steering. Gait restricted; usually shuffles or waddles on whole of foot and tarsus but occasionally walks or runs on foot, flapping wings. Stance rather upright; outline of bird on cliff (or on sea) less swollen-chested but thicker-necked than *U. aalge*.

Undertakes both swimming and flying migrations but usually only within inshore and offshore waters. Less gregarious than *U. aalge*. Most common calls of adult are a quiet, tremulous snore and a prolonged growl; of juvenile, a plaintive whistle, less loud and insistent than that of young *U. aalge*.

Habitat. Breeds on temperate and boreal coasts, feeding in inshore and offshore waters of high salinity, and in low-arctic seas, at water temperatures of 4–15°C; small relict populations, however, survive in brackish Baltic Sea. Prefers continental-shelf waters well clear of pack-ice. Accepts wet, chilly, and windy climates and stormy seas, but not typically pelagic and stays based on inshore waters throughout most of first half of year. In one case, regular fishing grounds not directly under breeding cliffs but some 500–700 m out to sea (Cayford 1981). Although often mixing with Guillemot *Uria aalge* on breeding cliffs, more commonly selects wider ledge beneath rock overhang, or a crevice; equally at home amongst talus and boulders of undercliff, and to lesser extent on boulder-strewn shores with no cliffs at all. On Skomer Island (Wales), uses 2 basic types of nest-site: (1) crevice on ledge with walls but no roof; (2) within *c*. 1 m of entry, along earth tunnel usually burrowed by rabbit *Oryctolagus cuniculus* or Puffin *Fratercula arctica*, or occurring among fallen boulders. (Witherby *et al.* 1941; Hudson 1979a.) Although nesting usual near water's edge, nests up to 300 m inland in Greenland (Salomonsen 1950–1). Unlike *U. aalge*, content with nest-sites not permitting high-density occupancy and not open. Also differs in wintering down to warmer and often shallower waters in warm temperate and Mediterranean zones. Although a fast and skilful underwater swimmer, normally dives only to *c*. 6 m. Roosts on water and usually flies in lower airspace above it; rarely high or over land. Despite relatively inaccessible haunts, vulnerable to human impacts through becoming trapped in fishermen's nets (Whilde 1979) or by spilt oil (Bourne 1976b), and locally by shooting or exploitation at breeding colonies.

Distribution. SPITSBERGEN. Breeding reported 1906, probably still nesting (Løvenskiold 1964); no recent evidence (Norderhaug *et al.* 1977). NORWAY. Some colonies extinct in south (Haftorn 1971). USSR. Reports of former breeding Lake Ladoga improbable (Løppenthin 1963).

Accidental. Jan Mayen, East Germany, Poland, Cze-

Alca torda 197

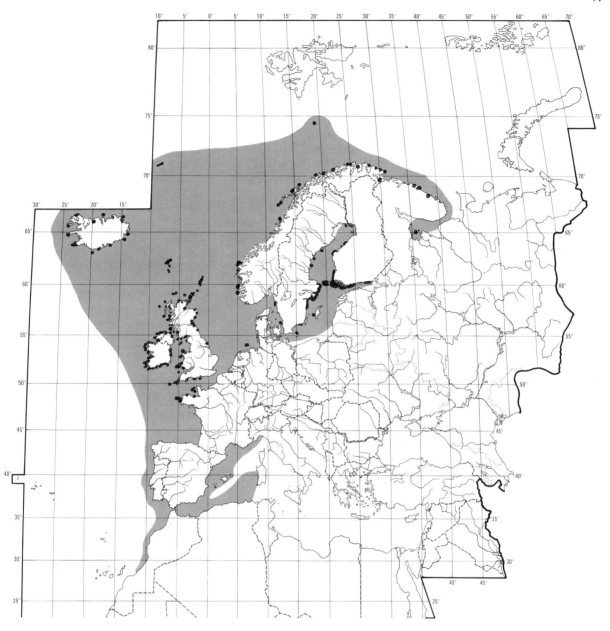

choslovakia, Hungary, Switzerland, Italy, Yugoslavia, Egypt, Malta, Azores, Canary Islands.

Population. Accurate counts of breeding populations extremely difficult; tentative estimate of world population c. 208 000, of which some 187 000 in west Palearctic (Lloyd 1976b). Impossible to give any current west Palearctic total, but likely to be higher (see, e.g., Iceland).

BEAR ISLAND. Small numbers breed regularly (Bertram and Lack 1933); none seen 1948 and 1958 (Løvenskiold 1964); at least 8 pairs bred 1970 (Brun 1970a); possibly breed 1980 (Luttik 1982). JAN MAYEN. Breeding 1983; 100–200 pairs (J A van Franeker and C J Camphuysen).

ICELAND. Tentative estimate of 5000 pairs (Lloyd 1976b) much too low, though no complete census (AP); no known changes in numbers in 20th century (AP). FAEROES. Estimated 5000 pairs 1974 (Lloyd 1976b); no censuses, but numbers thought to have declined (BO). BRITAIN AND IRELAND. Perhaps c. 144 000 pairs 1969–70; limited data suggest some decrease, mostly in recent years (Cramp et al. 1974). Counts 1971–9 suggested increases in most study plots in northern and eastern Britain (Stowe 1982). FRANCE. Under 100 pairs 1975; apparently some increase in 1950s followed by recent declines due to oil-spills (Guermeur and Monnat 1980). WEST GERMANY. Breeds only Helgoland: 4–18 pairs 1924–40, none 1960–74, 5 pairs 1981 (Vauk-Henzelt 1982; GR). DENMARK. 10 pairs 1925, over 300 pairs 1939, 60 pairs 1942, 100–200 pairs since 1970 (Dybbro 1976; TD). NORWAY. Estimated 26 600 pairs in north; declined some areas (Norderhaug *et al.* 1977); some 12 000 pairs further south, decreased (Haftorn 1971). SWEDEN. About 400 pairs 1975 and 1980 (Lloyd 1976*b*; S Hedgren); increased rapidly, from low level in mid 19th century, after protection, with slow increase in recent years (LR). Estimated 20 000–25 000 pairs 1982; reduced by *c.* 10% in last 20 years (GL, VR); 43 000 pairs 1983 (P E Jónsson). FINLAND. Numbers markedly affected by severe winters, especially 1939–41; thus perhaps 1500–2000 pairs 1930s now reduced to half (Merikallio 1958); slow recovery since to some 2000 pairs (Hildén 1978; OH); perhaps 4000 pairs 1983 (M Kilpi). USSR. Murmansk coast *c.* 300 pairs (Gerasimova 1962); no counts of small colonies in White Sea (Norderhaug *et al.* 1977).

Survival. Annual adult survival: Britain and Ireland 89·0% (Lloyd 1974), 91·4% (Mead 1974); Wales (Skokholm) 90–92% (Lloyd and Perrins 1977); Scotland (Shiant Islands) *c.* 92% (Steventon 1979). Survival from fledging to breeding age: Wales (Skokholm) *c.* 18% (Lloyd and Perrins 1977); Scotland (Shiant Islands) *c.* 11–14% (Steventon 1982). Oldest ringed bird 20 years 3 months (Rydzewski 1978).

Movements. Northern populations mainly migratory. In southern populations, young (especially 1st-year) birds make long displacements of migratory nature, though adults stay nearer colonies and are dispersive only. Overall marine range (marine inshore waters) extends south to New England in western Atlantic and to Morocco and Algeria in east.

Young leave colonies second half of July, while still flightless, and dispersal initially by swimming. Immatures stay at sea until they begin visiting colonies as pre-breeders when 2–3 years old, such birds coming ashore later in summer than adults (Lloyd 1974). Adults reoccupy colonies late February to May, according to latitude. However, in Scotland evidence of progressively earlier return since 1960s, with hundreds ashore at some sites from October onwards in recent years, though winter colony attendance not so pronounced as in Guillemot *Uria aalge* (Taylor and Reid 1981).

Significant amounts of ringing data only from Ireland, Britain, Norway, and Baltic. Recoveries subject to same biases as are those of other Alcidae: marine range under-represented, while over-emphasis on areas where shot for sport or accidentally netted in intensive sea fisheries.

RUSSIAN AND NORWEGIAN POPULATION. In contrast to Guillemot *U. aalge*, only a few winter off Murmansk and northern Norway, since most birds of all age-groups move further south (Dementiev and Gladkov 1951*b*; Holgersen 1952). The few winter recoveries in northern Norway concern juveniles and adults, but main wintering region is off south-west Norway and into Skagerrak, where birds arrive second half of September onwards. Hence birds from Norwegian colonies south of Arctic Circle disperse over relatively short distances only. Some Russian and Norwegian birds enter Kattegat in winter, with recoveries on Swedish and Danish coasts, and one Norwegian juvenile found November in Rostock (East Germany). For these populations, recoveries south or south-west of Skagerrak infrequent, though Norwegian birds found Netherlands, Pas de Calais (France), and Wicklow (Ireland), and 4 Russian birds found in Britain (east and west coasts); all recovered November–March, and only 2 were of breeding age. However, northern race (nominate *torda*) known to reach Scotland regularly in winter, based on evidence from beached corpses (Bourne 1968).

BALTIC POPULATIONS. Finnish, Swedish, and Danish (Bornholm) ringing results show these populations for most part remain within Baltic all year, shifting south and south-west for winter. Danish and Swedish birds range north and east in autumn (and as pre-breeders) into Gulfs of Bothnia and Finland, but these virtually deserted in winter, when most recoveries from southern Sweden and among Danish archipelago, though minority east to Latvia (Paludan 1947, and subsequent data). Some penetrate through Kattegat into Skagerrak, but rarely further: 2 Swedish immatures found in summer in western Norway (Möre og Romsdal, Nordland), and 2 Finnish immatures in Hordaland (Norway), October, and eastern England, February.

SCOTTISH POPULATION. Autumn movement around northern Scotland (for west coast birds) and into North Sea, with all age classes being recovered regularly around south-west Norway August–November; these Norwegian seas important as feeding and moulting area for birds from northern and eastern Britain. In October–November, 1st-year birds move southwards through North Sea and Straits of Dover; Bay of Biscay forms important wintering area for them, while others continue south to Morocco and enter western Mediterranean. Immature birds (2–4 years old) also move south in autumn, but smaller proportion (than in 1st year) penetrates south of Biscay. Adults recovered in winter mainly in North Sea and English Channel (only exceptionally as far south as Mediterranean), consistent with winter colony attendance (see above) and much reduced tendency to disperse far. Immatures tend to return north in spring (few Iberian or Mediterranean recoveries, April–July), even if they do not visit colonies; several summer recoveries from Faeroes and Norway. See Lloyd (1974) and Mead (1974).

IRISH AND IRISH SEA POPULATIONS. In contrast to Scottish population, few juveniles reach Norwegian seas. Instead, these remain near Irish and Irish Sea coasts in August; most move south to English Channel and Bay of Biscay by September and to Iberian coasts by October, and those which continue further reach Morocco and western Mediterranean in November (numerous there in

December). Few 1st-year birds winter as far north as southern Britain and southern North Sea. Many immature birds (especially 2–3 years old) migrate as far south as juveniles, but have a later mean recovery date in Mediterranean (1 January for 1st year, 6 February for 2nd–4th years), due to delay of having moulted further north. Most immatures return northwards in summer; some stay in Bay of Biscay, but others reach English Channel, Irish Sea, and North Sea, and it is this age group which is recovered in Norway in autumn (meeting Scottish population there) before migrating south again. In winter, adults recovered mainly from Irish Sea to northern Bay of Biscay, though minority penetrate as far as Mediterranean. However, winter recoveries of adults average much nearer colonies than do those of pre-breeders. See Lloyd (1974), Mead (1974), and North (1980).

Food. Chiefly fish, with some invertebrates. Prey caught mainly by surface-diving. Before diving, birds dip head into water, sometimes several times, while swimming around, apparently to spot prey (Nørrevang 1958; Bédard 1969; similar to Bill-dipping—see Social Pattern and Behaviour). May 'crash-land' over fish shoals and dive immediately (Bourne 1976a). Preferred depth of diving 2–3(–5) m (Madsen 1957); maximum 5–7 m (Kozlova 1957), possibly 10 m (Bianki 1967; see also Dewar 1924). According to Paludan (1960), dives to 5 m, staying submerged for c. 45 s. Usual submersion not more than 22 s, maximum 40–52 (Kozlova 1957; see also Dewar 1924). May catch several fish in one dive (Lloyd 1976a) or foraging trip, and, if returning to feed young, holds fish crosswise in bill. Up to 20 sand-eels (Ammodytidae) may be carried thus (Lloyd 1976a), but usually fewer: e.g. at Lundy 1–9, usually 5–6 (of which chick rarely ate more than 4, at least at outset: Perry 1940); 1–4 (Bédard 1969). Often forages in loose flocks (Bianki 1967). Mean number in mixed feeding flocks, Skomer Island (Wales), 15 (4–40); occurred in 90% of mixed assemblages, of which various gulls (Laridae) usually the nucleus (Birkhead 1976). Feeding range in breeding season usually 15–20 km from colony (Kaftanovski 1951; Kartashev 1960). At Skokholm Island (Wales), most birds, presumed feeding, concentrated 9–13 km from colony (Lloyd 1976a). At some colonies, kleptoparasitizes other Alcidae. At Vedøy (Lofoten Islands, Norway), robs Puffins *Fratercula arctica* by flying under them and taking fish from their bill, thus achieving abnormally high feeding rate; thought to provide significant proportion of diet for a few specialist individuals (Ingold and Tschanz 1970, which see for details). In Wales, also successfully steals fish from food-bearing *F. arctica*; chases take place underwater and (for up to 1·5 km) in the air (Lloyd 1976a; Warman *et al.* 1983; M P Harris); may surface beneath food-bearing bird in water, strike it from below, and collect fish dropped (Corkhill 1968).

Fish in diet varies regionally. Mainly sand-eels Ammodytidae—*Ammodytes marinus*, *A. lancea*, *Hyperoplus lanceolatus*; Clupeidae—sprat *Sprattus sprattus*, herring *Clupea harengus*, capelin *Mallotus villosus*, sardine *Sardina pilchardus*, anchovy *Engraulis encrasicolus*; Gasterosteidae—three-spined stickleback *Gasterosteus aculeatus*, ten-spined stickleback *G. pungitius*. Also Gobiidae—white goby *Aphya pellucida*, two-spot goby *Gobiusculus flavescens*; Gadidae—cod *Gadus morhua*, poor-cod *Trisopterus minutus*, five-bearded rockling *Ciliata mustela*; garfish *Belone belone*; flatfish (Pleuronectiformes). Crustacea include *Gammarus* and Mysidacea. Also polychaete worms and molluscs. (Collinge 1924–7; Perry 1940; Witherby *et al.* 1941; Paludan 1947, 1960; Kaftanovski 1951; Belopol'ski 1957; Madsen 1957; Kartashev 1960; Bianki 1967; Salomonsen 1967a; Harris 1970; Corkhill 1973; Evans 1975; Lloyd 1976a.)

In breeding season, Faeroes, adult diet almost exclusively sand-eels (Madsen 1957). In Barents Sea (USSR), 75 stomachs contained (by number) 92% fish, 1·3% crustaceans, 1·3% polychaetes, and 5·4% plant material; of fish, mostly herring (36·1%) and sand-eels (31·9%) (Belopol'ski 1957, which see for seasonal changes). At Kandalaksha Bay (USSR), adults supplement fish diet (for species, see final paragraph) with small amounts of crustaceans and polychaetes (Bianki 1967). In Greenland, mainly capelin (as elsewhere in arctic: Madsen 1957); also some crustaceans. In Europe, little direct study of adult diet, though assumed to be mostly sand-eels and sprats, as for young (see below). Of 49 stomachs (22 ♂♂, 27 ♀♀) collected throughout the year, Britain, fish remains in 43; overall 46·91% fish, 42·23% crustaceans and annelids, 10·45% molluscs (Collinge 1924–7); however, not known if molluscs contained only in stomachs of fish taken.

In winter, diet little known except for detailed study in Denmark (Madsen 1957). Of 120 birds collected November–February, food found in 71, of which 97% contained fish, 83% exclusively so. One contained only crustaceans, another only polychaetes. 28 birds (40%) had fed mainly on sticklebacks, especially three-spined stickleback and to lesser extent ten-spined stickleback. 1 bird had in its crop 48 ten-spined sticklebacks, each 2·5–3 cm long; another contained 49 three-spined sticklebacks, each c. 5 cm. 24 (34%) had fed on herring up to 15 cm long, and though less frequent numerically, herring formed greater proportion by weight (30%) than did sticklebacks (25%). 23 (32%) had fed on gobies, notably white goby, and, in 1 case, two-spot goby; 7 (10%) contained cod and poor-cod, 3 (4%) garfish, and 2 (3%) flatfish. 6 birds had fed on 2 kinds of fish, 5 on 3, 1 on 4. 11 (16%) had remains of crustaceans (Mysidacea, *Gammarus*). 1 bird contained 1400 *Mysis*, 2 *Gammarus*, 1 ten-spined, and 2 three-spined sticklebacks (Madsen 1957). Diet in Mediterranean in winter includes sardines and anchovies (see Witherby *et al.* 1941). Captive birds preferred smaller fish, and factor limiting ingestion was not length but depth of body, birds taking fish up to 23 mm deep (Swennen and Duiven 1977).

Diet of young throughout range comprises fish. In Europe, chiefly *Ammodytes* and *Clupea* (Perry 1940; Paludan 1947, 1960; Harris 1970; Corkhill 1973; Evans 1975; Lloyd 1976a). At Skokholm, sand-eels chiefly *A. marinus*, fewer *A. lancea* (Lloyd 1976a). On Graesholm (Denmark) young fed exclusively on small herring (Paludan 1947). In Barents Sea, more sand-eels taken by young than by adults (Belopol'ski 1957). At Kandalaksha Bay, predominantly sand-eels, rarely herring; capelin of secondary importance at Onega Bay (Bianki 1967). At Sainte-Marie islands (Gulf of St Lawrence, Canada), young fed chiefly on sand-eels, but prey changed as others became more abundant (Bédard 1969). Numerous studies on length of prey brought to young. For sand-eels, mean length in 2 seasons, Skokholm and Skomer, $53.1 \pm SD$ 20.1 mm ($n=168$), and $79.0 \pm SD$ 25.5 mm ($n=11$); total range (both seasons) 20–158 mm. In 1 season, Skokholm, mean length of 21 Clupeidae $53.8 \pm SD$ 21.7 mm (25–110) (Harris 1970; Lloyd 1976a, which see for summary of other studies). See also Corkhill (1973) and Evans (1975) for similar findings. Young fed by both parents until day of fledging (Paludan 1947, 1960; Bédard 1969; Lloyd 1976a). No difference in weight of feeds, or length of fish, brought to young of different ages; some parents consistently brought many small fish, others a few big ones (Lloyd 1976a). At several colonies, peak of feeding occurs early in morning (e.g. Birkhead 1976, Lloyd 1976a). On Lundy, apparently no feeding early in morning, activity peaking 10.00–11.00 hrs; there, 2 feeds per chick per day up to 12–13 days, many more thereafter (Perry 1940), though this may have reflected food supply rather than age of young (Bédard 1969). On Skokholm, 3–5 feeds per day (Lloyd 1976a). For feeding rates at colony where some birds are kleptoparasites, see Ingold and Tschanz (1970).

EKD

Social pattern and behaviour. Based largely on studies by Paludan (1947, 1960), Bédard (1969), Lloyd (1976a), and Birkhead (1976), and on notes supplied by T R Birkhead.

1. Gregarious in breeding season—less so in winter. Birds from one colony or several adjacent ones believed to remain together for some time after chicks fledge; flocks of up to 100 apparently flightless birds seen near Skokholm and Skomer Islands (Wales) after breeding season, and similar flock seen, autumn, elsewhere in Irish Sea (Lloyd 1976a). Early post-breeding flocks presumably comprise ♂ parent and dependent offspring, along with other adults (see Bonds, and Relations within Family Group, below). In winter, occurs in small flocks (Kozlova 1957), these apparently aggregating in spring (e.g. Bédard 1969). BONDS. Monogamous mating system; pair-bond maintained from year to year (Paludan 1947, 1960; Lloyd 1976a), but close contact unlikely outside breeding season, since ♂ parent leaves colony with chick before ♀ (Paludan 1947, 1960; see also Plumb 1965, Lloyd 1976b). On Skokholm, 72% of birds retained same mate between breeding seasons; of 58 birds whose mates were identified in 2 or more years, 40 pairs remained intact for 2 successive years, and 18 for at least 3 years; the rest changed mates once ($n=11$) or twice ($n=7$) in 3 seasons; only 2 pairs broke up when both birds were still alive (Lloyd 1976a). On Skokholm, 35% bred first at 4 years old, 60% at 5, 5% at 6 ($n=20$) (Lloyd 1976a; Lloyd and Perrins 1977). At Clo Mor (Scotland), some bred at 3 (J L F Parslow). At Skokholm, no 1-year-olds recorded on land or in rafts (Lloyd 1974); 2-year-olds also scarce but progressively more returned from 3 to 5 years, and gradually earlier in season. 3 2-year-olds stayed on land average $12.6 \pm SD$ 5.5 days, for 4 non-breeding 5-year-olds average $33 \pm SD$ 7.3 days (see Lloyd and Perrins 1977). At Sainte-Marie islands (Canada), a few 1-year-olds (identified on bill and plumage characters) visited 'clubs' (Bédard 1969: see below). According to Paludan (1947), birds did not return to Graesholm (Denmark) until 3rd year, but younger birds may have been overlooked. Both sexes care equally for young at nest-site, but usually ♂ alone accompanies young to sea. Reports that parent continues to feed young at sea up until independence (e.g. Belopol'ski 1957) not substantiated, but may well provision young at least initially; after departing to sea, Skokholm, adults with offspring seen surfacing with fish in bill and assumed to be feeding young (Lloyd 1976a). Duration of dependence not known but, as in Guillemot *Uria aalge*, probably several weeks. BREEDING DISPERSION. Loosely colonial; contiguous nesting rare—nests usually several metres apart. In Britain, 1969–70, mean colony size 799 pairs (1–8370, $n=37$) (see Cramp et al. 1974). Of 26 'colonies', Norway, 6 contained up to 10 pairs, 9 up to 100, 3 up to 1000, 7 up to 10000, 1 with 11000 though these larger assemblages presumably comprised smaller colonies and sub-colonies (Brun 1969a). Colonies (or sub-colonies), Sainte-Marie, 2–37 pairs, mean (including solitary pairs) 9.7 ± 6.5 pairs ($n=38$); sub-colonies 0–10 to 71–80 m apart, most 20–50; within groups, never more than 4 pairs per m². Both members of pair defend nest-area territory against both neighbours and strangers; minimum radius 20–30 cm (Bédard 1969). None of 460 sites, Skomer, was less than 15 cm apart (Birkhead 1976). Nest-territory consists of nest-site and small surrounding area; plays only minor part in pre-breeding activities, serving mainly for raising young and providing refuge. 20–30 pairs, e.g. along extended crevice, may share common landing strip (Bédard 1969). Commonly associates with *U. aalge*, and Puffin *Fratercula arctica*; 24 out of 38 colonies, Sainte-Marie, associated with one or other of these (Bédard 1969). For suggestion of active attraction to gull (Laridae) colonies, see Nordberg (1950). Shows strong fidelity to both colony and nest-site (e.g. Paludan 1947, Bianki 1967). Of 245 re-trapped birds originally ringed as adults breeding on Skokholm, 95% re-trapped breeding in colony (of which several on island) of ringing, 5% in another colony on island. Only 11 adults (8.5%) moved from nest-site in which first caught, but none changed colony. In these cases, change of mate, associated with death or breeding failure, known or suspected; of the 11 which moved, 7 of the 8 sexed were ♀♀, suggesting ♂ selects site; ♂ which moved belonged to pair that failed to hatch an egg in 2 successive years (Lloyd 1976a). ROOSTING AND COLONY-ATTENDANCE PATTERNS. In contrast to *U. aalge* in which adults rarely join clubs, both adult and immature *A. torda* use club for loafing (Bédard 1969). Club situated near breeding area but usually above height of same in *U. aalge* (Birkhead 1976); often a large flat rock outcrop, boulder, or ledge. Each colony or sub-colony usually has its own loafing area, but birds from 2 or 3 nesting groups may loaf together at 1 or 2 areas (Bédard 1969). Loafing areas especially well attended by burrow- (as opposed to ledge-) nesters (Hudson 1979a). Apart from nesting and comfort activity, loafing area used for some copulating (Bédard 1969; Lloyd 1976a; Hudson 1979a). Loafing bird often performs Footlooking: downward stare maintained for 1–4 s, during which bird inspects feet or ground; sometimes picks up small object with tip of bill and may apparently swallow it with strong sideways shake of bill, perhaps thus ridding the

Alca torda 201

A

webs of ectoparasites (Bédard 1969). For further uses of loafing area, see Antagonistic Behaviour and Heterosexual Behaviour, below. Unlike *F. arctica*, few birds sleep in rafts by day with head tucked under wings, although periods of inactivity occur with head and neck withdrawn (Lloyd 1976a). Early in season, both members of pair spend much time loafing at nest-site; leave colony to roost at sea overnight, and during incubation and chick-rearing off-duty bird roosts at sea; by day, off-duty bird either sits beside mate at nest-site or leaves colony (Lloyd 1976a). If it leaves, usually settles on sea away from rafts already present, and often bathes and preens for 5 min or more before returning to loafing area (Bédard 1969). In bathing, rarely rolls over on side as *U. aalge* does (Lloyd 1976a). During incubation on Skokholm, off-duty birds spent, on average, 46% of day (04.30–18.30 hrs) at nest-site; present more in morning than in afternoon; off-duty bird often departed in afternoon, few remaining after 17.30 hrs, but in chick-rearing period spent more time on land in afternoon than in morning and evening (Lloyd 1976a). Daily activity rhythm varies with season and colony, depending on (e.g.) proximity of food supply. At Kandalaksha Bay (USSR), most active 05.00–10.00, 13.00–14.00, and 18.00–19.00 hrs, with periods of rest between; activity declined 22.00–02.00 (–05.00) hrs, when majority resorted to sea (Bianki 1967). Pre-breeding attendance at breeding areas usually begins later than in *U. aalge*—usually spring, but at some colonies from October (for review, Britain, see Taylor and Reid 1981). Attendance initially irregular, with few landfalls, birds mostly gathering in rafts up to 2 km offshore (Lloyd 1973, 1976a; see also Flock Behaviour, below). First landings on shore during the few hours after dawn, birds returning to sea thereafter (Perry 1940; Paludan 1947; Keighley 1950; Bédard 1969; Corkhill 1971). As pre-laying period progresses, number of birds and duration of stay increase, peak occurring later in morning; birds now assemble more in traditional loafing areas, finally prospecting sites (Lloyd 1976a). Shortly before laying, most left colony area, Skokholm, at 16.00–18.00 hrs, all by 20.30 hrs; early in incubation, peak attendance during hour after dawn when many birds returned to relieve sitting mate; later in incubation, and in chick-rearing period, many returned *c.* 06.00–08.00 hrs and remained throughout

day (Lloyd 1973, 1976a). At mixed colonies, usually high correlation in pre-laying attendance with other Alcidae, especially *U. aalge*. As in that species, pre-laying attendance (especially April) often cyclic. At Skomer, peaks every 3–7 days, mean 5·2 (Corkhill 1971); at Skokholm, *c.* 5 days (Lloyd 1973, 1976a). At Lundy, attendance high for 2–3 days (good weather) separated by periods of 1–2 days when attendance almost nil (Perry 1940). Attendance nil in gales, but moderate winds no deterrent (Corkhill 1971; see also Plumb 1965, Lloyd 1973, 1976a). At Skokholm, non-breeders visited little during laying period, and rarely before 07.00 hr; at start of incubation period, arrived early morning in company with breeding birds relieving mates, peaking *c.* 09.00 hrs; few remained after 14.00 hrs, and all had left by 18.00 hrs. Later in incubation, similar pattern occurs, but birds stay until late afternoon before again departing for night (Lloyd 1973, 1976a).

2. Especially at start of season, birds very nervous, leaving cliff when disturbed by gull *Larus* or human (Lloyd 1976a). Different postures express varying degrees of alarm. Alert-posture similar to *U. aalge* (Birkhead 1976). At first hint of disturbance, turns head rapidly, seeking source, and may give Alarm-call (see 1 in Voice); at higher intensity, wings raised slightly (Fig A), and bird may eventually take off in silence, from cliff-site typically with Butterfly-flight (Conder 1950; Bédard 1969). May land in sea and make swimming approach to land (Bianki 1967). On sea in rafts, birds occasionally Bill-dip (Perry 1940; Nørrevang 1958; Bédard 1969), perhaps expressing mild alarm, as in other Alcidae; dives when threat imminent. For other alarm reactions, see Parental Anti-predator Strategies, below. FLOCK BEHAVIOUR. Flocks ('rafts') form for feeding, dispersal, and pairing behaviour. At Skokholm, rafts of 2–25 birds, mean 13 (Lloyd 1976a). Usually associates closely with *U. aalge*, less so with *F. arctica*, which tend to be chased off (Paludan 1947). Mixed rafts, Sainte-Marie, up to 300 birds, including 15–75 *A. torda*, depending on size of prospective colony; during pre-laying period, rafts approached ever closer to shore, and adjacent to colony area 15 days before egg-laying (Bédard 1969); in any raft, mostly same birds present each day (Lloyd 1976a). Many birds appear paired from outset (Conder 1950; Bédard 1969), proportion increasing during pre-laying period (Lloyd 1976a). Rafting behaviour complex. Communal bathing common (Lloyd 1976a). In early stages of rafting, birds (up to 50: Paludan 1947) adopt straggling line-abreast formation (Conder 1950); thus float or swim in same direction for several minutes, turning together, and finally coalescing into compact group of milling birds (Paludan 1947; Conder 1950; Gordon 1951; Lloyd 1976a). In so-called 'swimming together', turn heads briskly from side to side ('Head-shaking': Lloyd 1976a), as if alert, and swim rapidly around within flock as if trying to avoid one another. Suddenly most dive, and may surface holding Head-vertical posture for 3–4 s: head and neck vertical, bill forward or upward and slightly open, tail cocked exposing white under tail-coverts; in extreme form, wings slightly open (Darling 1938). Head-vertical posture common in larger rafts; may serve to attract potential mate, since performed mainly by ♂ (at least on land: Birkhead 1976; see also Paludan 1947, Lloyd 1976a). Sometimes one bird swims vigorously ahead of another, leading to chase, both birds splashing over water with wings and feet. Usually, however, leader not followed, but after swimming away for some distance, turns and swims quickly back to partner in Hunched-posture: tail low, neck and head stretched forward (Lloyd 1976a). On meeting, the 2 birds face each other and perform meeting-cermony (Bédard 1969): hunched bird points bill up at partner and Billing follows—each nibbles and grasps other's bill for a few seconds to 1 min or more (Birkhead 1976; Lloyd 1976a); head plumage often erected and Billing-

B

C

calls (see 4 in Voice) given (Birkhead 1976a). Billing said to be contagious in flock (Conder 1950). In variant of chasing, bird, with neck and tail markedly lowered (possibly aerial version of Hunched-posture) flew 5–6 m from its partner and landed, swam back, and Billed with it (Bédard 1969); this probably Butterfly-flight (Lloyd 1976a); see also Paige (1948). Sometimes both birds flew thus from flock, landed together some distance away, and Billed (Bédard 1969). Billing almost always leads to bout of Allopreening which occurs as frequently on sea (unlike *U. aalge*) in pre-breeding period as on land (Birkhead 1976); behaviour as in *U. aalge* on land, but erection of head plumage more pronounced (Birkhead 1976). Frequency of Allopreening, Skomer, 0·96–1·88 bouts per bird per min, mean 1·31 (Birkhead 1976). For case of Bill-dipping with seaweed, and apparent Head-tossing (different from movements in copulation), see Bédard (1969). Birds in raft periodically Wing-flap: bird raises itself from water, exposing fore-body, and flaps wings vigorously; often occurs after confrontation, or preceding Billing (Lloyd 1976a); function not clear. For Antagonistic Behaviour in rafts, see below. Rafting behaviour continues until end of breeding season, but after egg-laying, only unpaired birds participate (Bédard 1969, which see for diving behaviour late in season; see also Gordon 1951). ANTAGONISTIC BEHAVIOUR. Less frequent than in *U. aalge*, owing to wider spacing in colony (Birkhead 1976, which see for frequencies of aggressive encounters). As common in rafts as on land. On water, ♂ challenges intruder approaching within 2–3 body lengths of mate (Conder 1949); on land, ♂ defends small area around ♀ at start of season (Plumb 1965). Threat much more common than fighting (Birkhead 1970). On land or on water, threatens by Bill-gaping which exposes bright yellow lining; sometimes accompanied by growling (see 2 in Voice); this usually sufficient to deter intruder (Bédard 1969). In stronger challenge, bird may 'rush' opponent (Bédard 1969), usually with wings raised and even outspread, and head plumage often erected; on water, swims rapidly towards rival. Stretches towards rival and may jab with gaping bill, sometimes making contact, while loud calls given (Bédard 1969). Rivals threaten face to face (side by side in *U. aalge*). In fierce encounter, combatants grasp and twist with bills and cuff with wings (Birkhead 1976). Interspecific encounters with *U. aalge* in rafts and loafing groups likewise include Bill-gaping and rushing (Bédard 1969). At sea, bird engaged in fight often dives to escape, and is pursued under water (Bédard 1969). Encounters most frequent among ♂♂ contesting for ♀♀ at start of season (Bédard 1969; Lloyd 1976a), and in chick-rearing period when other birds, possibly prospecting immatures, intrude on established pairs (Birkhead 1976). During site occupation, bird often attacked arriving mate (Lloyd 1976a). Before egg-laying, significantly more threat and fighting among immatures than adults (Hudson 1979a). Frequency and duration of fights vary with nest-density, but usually rare and brief (Paludan 1960; Bédard 1969; Birkhead 1976). One fight, however, lasted more than 1 hr (Hudson 1979a). No ritualized appeasement postures; after 6 out of 14 fights, however, bird began preening (Birkhead 1976: see Side-preening of *U. aalge*). Bird sometimes performs 'Turning-away' which may be appeasement gesture (Lloyd 1976a, which see also for 'spinning' on water). Alighting bird sometimes adopts Landing-posture (Fig B): with body tilted forwards and bill and wings raised, walks to nest-site; possibly denotes appeasement and/or site-ownership; much less frequent than in *U. aalge* (Birkhead 1976, which see for details). Landing-posture adopted in 88% of 119 cases where bird alighted beside mate, 62% of 266 when mate absent. Unlike *U. aalge*, likelihood of display not influenced by proximity of neighbours (Birkhead 1976). HETEROSEXUAL BEHAVIOUR. (1) Pair-bonding behaviour and meeting-ceremonies. Mate selection and pairing thought to occur in rafts and loafing area, which also serve as clubs for immatures (see Roosting and Flock Behaviour, above). Spends more time displaying in rafts than *U. aalge* (Birkhead 1976). By the time pair occupy nest-site, bond probably well established or re-established. No definite information on sequence of pairing behaviour, but includes Head-vertical display (below), Billing, Allopreening, etc., all of which occur on land as well as on water. During pre-laying period, partners sometimes perform mutual Bowing; function not clear, but possibly homologous with same of *U. aalge* (Birkhead 1976). ♂ probably advertises by Head-vertical (so-called 'ecstatic') display; complete sequence occurs apparently only on land (Bédard 1969). Bird, usually ♂ (see below), begins by pointing head forwards (Forward-posture) and calling (Bédard 1969: see 3 in Voice for entire sequence of calls). Then lifts head vertically (Head-vertical posture: Fig C, left) and sometimes bends it markedly backwards; during this phase, growls while mandibles vibrated noisily (Birkhead 1976: see 3 in Voice). Finally, bird stands upright and sweeps head down against breast with bill open (Tucked-posture: Fig C, right), as growling dissipates (Bédard 1969). Usually, performer is lone ♂ at nest-site, though ♀ also performs occasionally (Perry 1940), sometimes in absence of partner; in 2 pairs, presumed ♂ performed display 10 and 14 times, ♀ 3 and 2 times respectively in given period (Bédard 1969). Head-vertical display interspersed with other courtship behaviour in 5 out of 51 cases: e.g. after Allopreening, members of pair performed display simultaneously and Tucked-posture led to Billing (Bédard 1969). Mutual display much less common (2 out of 21 cases) than solitary display, but always followed by Billing (Birkhead 1976). Head-vertical display often associated with return of ♀, when may serve as meeting-ceremony; also used near breeding sites late in season, apparently directed at other birds flying or settling nearby (Birkhead 1976). Especially frequent among parents with young about to fledge (Bédard 1969). Simpler meeting-ceremony nearly always used when partners reunite at nest-site: face each other, Bill concertedly while calling (see 4 in Voice), then simultaneously point bills at ground. This meeting-ceremony performed in 94% of 171 cases where birds

D

alighted beside mate; no seasonal variation in frequency (Birkhead 1976). (2) Allopreening. Members of pair Allopreen alternately, sometimes simultaneously, with nibbling movement, bill sometimes opening and closing rapidly. More than 98% of preening is of head and neck (Hudson 1979a). Though apparently ritualized, may also serve to remove ectoparasites (see Bédard 1969). Usually, birds lie side by side, touching at breast (Fig D); if further apart, Allopreen with sideways stretch of neck. Sitting bird may thus preen newly arrived mate. Also Allopreen while standing up facing; then especially vigorous, and recipient may bend head over back (Fig E). Participants usually silent but intense Allopreening accompanied by calling (Paludan 1947, 1960; Bédard 1969: see 4 in Voice). Duration of bout variable, up to 12 min (Bédard 1969). Most frequent in incubation; mean frequency, Skomer (all season), 0·95 bouts per pair per min, mean duration 7·24 s ($n=438$) (Birkhead 1976); c. 0·5 bouts per pair per min, duration 6–7 s, same as frequency in rafts (Hudson 1979a). At nest-site bird preens only mate (Birkhead 1976). Copulation and chick-feeding usually followed by Allopreening (see below); sometimes degenerates into brief confrontation between partners (Bédard 1969). (3) Mating. Unlike in *U. aalge*, ♂ initiates (in all of 22 cases), and mating much less synchronized with arrival of mate at nest-site than in *U. aalge*; followed within 2 min of return in only 4 out of 36 cases (Birkhead 1976). The following mainly from Perry (1940) and Bédard (1969). As in *F. arctica*, no preliminary to mounting, except sometimes meeting-ceremony. ♂ usually mounts from side, and may tread for some time before copulating, meanwhile sometimes beating wings or resting wing-tips on ground (Fig F). During cloacal contact, ♂ stands almost upright, and ♀ begins to toss head up at intervals of c. 1–2 s, opening bill wide with each toss and uttering ♀ Copulation-call (Paludan 1947, 1960; Birkhead 1976: see 6 in Voice). ♂ sometime calls during mating, sometimes not (Bédard 1969: see 5 in Voice) or said to call continuously throughout (Paludan 1947; Birkhead 1976). After cloacal contact, ♂ may descend voluntarily, and pair then lie side by side, mutual Allopreening usually occurring; if ♀ shrugs ♂ off, vigorous Billing usually follows (Bédard 1969). No records of mating attempts between birds not paired to each other (Bédard 1969; Birkhead 1976). Mating always occurs on land, usually among groups in loafing areas, at edge of nesting area, or, if space available, at nest-site (Paludan 1947; Bédard 1969; Lloyd 1976a). Mating begins 2–3 weeks before laying, ceasing after laying (Paludan 1947). During calm weather in pre-laying period, Skokholm, birds often sat and mated on rocks exposed at low tide (Lloyd 1976a). Around egg-laying, mating frequent in early morning when birds returned to cliffs from roosting (Lloyd 1976a). At Graesholm, peak 08.00–10.00 and 16.00–18.00 hrs; pairs sometimes copulated several times per day, once 4 times in $3\frac{3}{4}$ hrs (Paludan 1947). On Skokholm, breeding pairs rarely copulated after onset of incubation, whereas failed or non-breeders copulated throughout season (Lloyd 1976a). (4) Behaviour at nest. Scraping action as in *U. aalge* (Paludan 1947; Nordberg 1950; Plumb 1965; Bédard 1969; Birkhead 1976). Scrapes with one

F

E

side of body turned to cliff, moving foot nearest cliff (Williams 1971). Picks up small stones or other available debris at or near nest-site (usually within 0·5 m), and brings them below breast with sideways movements of bill (side-throwing, or, if sitting, side-building); described as repeatedly picking up fragments and dropping them immediately in front of itself (Bédard 1969; Lloyd 1976a). Behaviour frequent before laying, and persists—especially in bird departing at nest-relief—up to chick-rearing (for details, see Bédard 1969; see also Perry 1940), and occurs among immatures prospecting late in season (Bédard 1969). At nest-relief, arriving bird settles on egg, sometimes also on small chick, with Settling-call (see 7 in Voice). Relieved bird—or in pre-laying period both birds—often leaves cliff site with Butterfly-flight (Conder 1950), though function debatable. On landing in water, may Head-toss; sitting bird may also Head-toss (Bédard 1969). RELATIONS WITHIN FAMILY GROUP. For calls of young, and identification cues used by adults, see Ingold (1973), and Voice. Within 24 hrs of hatching, chick mobile and solicits brooding, first under brood-patch, later standing on parent's webs under wing (Bédard 1969; see also Kartashev 1960). Parent returning after panic summons young with Lure-call (see 8 in Voice) and gives Settling-call just before brooding; may also Wing-flap, shift nest-material, and Head-shake (Paludan 1947). Brooding bird sometimes manoeuvres small chick with bill (Bédard 1969). Young brooded almost continuously until they begin to thermoregulate at 3–4 days (Belopol'ski 1957) or 4–6 days (Kaftanovski 1951), then brooded more intermittently (Brun 1959; Bédard 1969; Ingold 1973): e.g. in the case of 2 10-day-old chicks, parents spent only 41 min out of 430 min in nest (Kaftanovski 1951), though usually brooded intermittently right up to fledging (Bédard 1969; Lloyd 1976a; see also Perry 1940, Paludan 1947). Usually guarded by at least one parent throughout nestling period, but in hidden sites (e.g. in boulders) young more than 1 week old often left alone (Bédard 1969; Lloyd 1976a). At only a few hours old, chick pecks at white stripe on adult's bill, presumably to beg food (Paludan 1947). Young fed by both parents from day of hatching until fledging (Paludan 1947; Bédard 1969; Lloyd 1976a). Unlike *U. aalge*, parent arriving with food gives no special call; leans towards young from which brooding parent stands up; incoming parent, with head slightly bowed, offers fish which is taken head first (Tschanz and Hirsbrunner-Scharf 1975, which see for differences from *Uria*), but sometimes guarding adult takes fish from mate and presents it to young (Ingold and Tschanz 1970). If more than one fish is carried, these taken by the chick one by one (Lloyd 1976a). If fish fall to ground, one parent retrieves and offers them one at a time; once, a fish thus presented 5 times (Bédard 1969). Older young also retrieve dropped fish (Bédard 1969; Lloyd 1976a). Refused fish may be eaten by parents or discarded (Lloyd 1976a). After feeding chick, parents usually perform meeting-ceremony and Allopreen (Bédard 1969). In 92% of cases ($n=60$), parent bringing fish to young took over brooding and guard duty (Ingold 1973). Up until day of fledging, or even a few hours before, chick does not move more than *c*. 30 cm from nest. Fledging behaviour closely studied (Perry 1940; Keighley and Lockley 1947; Paludan 1947, 1960; Greenwood 1964; Bédard 1969; Lloyd 1972, 1976*a*; Lawman 1975; Williams 1975). Young almost invariably fledge in late evening, after sunset, usually between 21.00 and 23.00 hrs (e.g. MacGinley 1913, Ussher 1913, Greenwood 1964, Bédard 1969, Ingold 1973, 1974), sometimes earlier if disturbed (Lawman 1975). Of 21 fledgings, Skokholm, all occurred in late evening. High winds and rough seas curtailed fledging (young brooded instead), but rain and fog had no effect (Lawman 1975). Fledging procedure differs for open (ledge) and enclosed (boulder, burrow, etc.) sites, and for vertical and more sloping site. Following account from Lawman (1975) for cliff-ledge sites, Skokholm. In evening, young notably active up to 3–4 days before fledging—move around, exercising wings and preening vigorously. On evening of fledging, active from a few minutes to 2 hrs before departure. At this time, guarded by one parent; other departs to sea, as usual, well before sunset (Bédard 1969; Ingold 1973; Lawman 1975; Lloyd 1976*a*; see also below). When fledging imminent, adult and young evidently excited. Activity of chick begins away from edge of ledge, parent standing away from chick, occasionally exercising wings and calling (see 8 in Voice). Chick then ventures towards edge, faces sea, and Bobs repeatedly, alternately stretching and retracting neck, whereupon parent joins it and also Bobs, calling continuously. If chick then retreats, parent usually follows it, and then returns to edge with chick behind, this often happening repeatedly (Bédard 1969; Lawman 1975). Chick eventually jumps off and descends with whirring wings, sometimes almost vertically but, if from higher than *c*. 10 m, gradually pulling out to enter water at *c*. 45°. Parent follows a few seconds later, and dives if chick submerged. The two often surface together, but if chick does so first, it calls and is then located by parent who also calls until reunited. Both then swim out to sea in silence, chick closely beside and just behind parent. Often both dive and swim under water for short distance. If chick, on fledging, fails to reach water, both parent and chick call incessantly, but parent eventually abandons trapped chick; if chick has to negotiate boulders near shore, parent takes lead (Lawman 1975). Once, adult tried to coax chick, which persistently swam towards cliff, by alternately prodding it several times with bill, and swimming 1–2 m towards open sea; chick failed to follow and was eventually eaten by Great Black-backed Gull *Larus marinus* (Greenwood 1964). Once, young separated from parent at sea spent all night calling and swimming up and down below colony; in morning usually repeatedly approached groups of adult conspecifics and *U. aalge* on water, only to be driven off (Lloyd 1976*a*). Of 21 fledgings from cliff sites, 19 accompanied by 1 adult (parent), but in 2 cases parent and chick joined by strange adult which flew down from cliffs; in one case stranger joined only briefly; in the other, bird swam out to sea with parent and chick, despite attempts to drive it off (Lawman 1975). However, young never mobbed, as in *Uria* (Bédard 1969). In enclosed nest-sites, chick may perform all pre-fledging exercise inside site, finally emerging, with parental encouragement, on to nearby boulder to continue Bobbing, etc. If angle of slope allows, chick scrambles down ahead of parent, but if hampered, parent leads (Lawman 1975). Parent provides no physical assistance, from any kind of site, *contra* some earlier reports (Lawman 1975). Parent accompanying young almost invariably ♂ (Bédard 1969; Lloyd 1976*a*). Other (♀) parent sometimes brought food to site early in morning after fledging; in 11 out of 12 cases, bird thus returning to site was ♀ parent. ♀ (or, if chick lost, both parents) attended site for considerable time each day up to 1 week after fledging (Lloyd 1976*a*; see also Perry 1940). Fledging notably synchronous, Kandalaksha Bay, and mass exodus from colony followed, whereupon some birds still incubating abandoned clutches (Bianki 1967). ANTI-PREDATOR RESPONSES OF YOUNG. Escape reaction develops at 3 days (Paludan 1947; Bédard 1969), threatened chick seeking refuge, head first, in crevice, or pressed against rock, or under brooding adult. At sea, young harassed by gulls evaded capture by persistent diving (Greenwood 1964; Lloyd 1976*a*). PARENTAL ANTI-PREDATOR STRATEGIES. (1) Passive measures. If off-duty bird suddenly leaves sitting mate at nest, or if human intruder suddenly corners sitting bird, latter gives Alarm-call (Bédard 1969). On close approach of predator, sitting bird may move head slowly from side to side and Bill-click (Hudson 1979*a*, which see for other contexts in which Bill-clicking occurs); sits tight and flies off only if danger acute (Bianki 1967). At exposed site, one parent guards young constantly (see above), squatting at edge of nest-site or nearby (Bédard 1969). Alarm-call occasionally given in flight, e.g. when chick below being handled (Paludan 1947, 1960; Bédard 1969). (2) Active measures. When *L. marinus* nearby, birds (presumably sitting) gave Attack-calls (2 in Voice) and struck gull if it came too close (Bédard 1969). Once, when gull alighted near chick, adult rushed it, Bill-gaping and calling, with wings slightly raised, until gull driven off (Lloyd 1976*a*). Similar response to close approach of human; if handled, nips hard with bill (e.g. Bianki 1967). Parent swimming out to sea with chick encountered flock of newly fledged Herring Gulls *L. argentatus* which swam toward chick. At first parent tried to drive gulls off, but then swam and dived with chick, surfacing *c*. 5 m away (Lawman 1975).

(Figs A, D, and E from photographs in Bédard 1969: Fig B from drawing in Birkhead 1978*b*; Figs C and F from photographs in Paludan 1947.)   EKD

Voice. Various calls associated with breeding, mostly given on ground. Little information for winter, but see King (1980*b*). All calls are variations on a growling sound (e.g. 'knorrrr' or 'narrrrv': Paludan 1947), and difficult to distinguish phonetically. Summarized by Bédard (1969) as a series of unmelodious sounds of wide fundamental frequency range (0·5–1·7 kHz), commonly *c*. 1·2 kHz. Harmonics are superimposed, usually one at *c*. 2·4 kHz, sometimes another at *c*. 3·5–4 kHz. Number of sound pulses ranges from 31 to 146 per s, lending voice its distinctive vibratory quality (aptly likened to whirring clockworks: P J Sellar). Calls differ mainly in temporal patterning of units, less so in pitch or tonality. Quality also influenced by opening of bill (Bédard 1969). Following account is based chiefly on descriptions by Bédard (1969) at Sainte-Marie islands (Canada).

CALLS OF ADULTS. (1) Alarm-call. Short growl, usually not longer than 0·5 s, given once, or frequently twice (total duration *c*. 1·5 s), or repeated in more-or-less regular rhythm. Given by bird in state varying from mild to extreme agitation. Apparently the only call given in full flight, e.g. when intruder at nest. (2) Attack-call. Protracted growl (*c*. 1·5 s), otherwise similar to call 1, given during 'rushing'; call given during defensive Bill-gaping possibly the same. (3) Ecstatic-growl ('nest-song': Paludan 1947). Consists of 3 parts, associated with the 3 phases of Head-vertical display (Bédard 1969). (a) 2–5 growling sounds

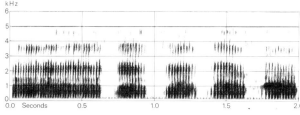

I J-C Roché (1970) France May 1978

II P J Sellar Scotland August 1970

of increasing rapidity, finally blending into a continuous growl, given in Forward-posture (Bédard 1969). (b) Long growl (2–3 s) given in Head-vertical posture (Bédard 1969); also, lower mandible vibrated rapidly against upper one, producing distinctive rattle (Birkhead 1976), likened to castanets (Perry 1940). (c) 4–5 feeble, relatively well-spaced growls of steadily diminishing intensity, given in Tucked-posture (Bédard 1969). (4) Billing-call. First growl sound up to $c.$ 2 s duration, dissipating in shorter growl units—total duration up to $c.$ 3 s (Fig I). For sonagram of Billing-calls by 2 birds (simultaneously), see Fig 21 in Bédard (1969). Given during meeting-ceremony (on land or water), sometimes during the bout of Allopreening which follows (Bédard 1969). (5) ♂ Copulation-call. Sometimes given during copulation. No description given (Bédard 1969). ♂ may call continuously throughout copulation (Birkhead 1976). (6) ♀ Copulation-call. Short growl, coinciding with each toss of head during copulation (Bédard 1969); described as disyllabic growl at regular intervals of 1–2 s (Birkhead 1976). (7) Settling-call. Succession of growl sounds of decreasing length, first $c.$ 1 s or more, given by both sexes when settling on egg or (especially when returning after panic flight) on small chick (Bédard 1969). (8) Other calls to young. Lure-call. Single growl given at intervals of 1·5–2 s, or disyllabic growl at intervals of 4–5 s (for sonagrams, see Fig 19, left, and Fig 20, left, in Bédard 1969). Invites close approach, and thus especially conspicuous around fledging time, when may coax young to quit nest-site and facilitate contact at sea (Bédard 1969; Ingold 1973).

CALLS OF YOUNG. 4 calls distinguished; after first 1–2 days of life, mostly penetrating whistles which intergrade but differ in pitch (and length), more so as chicks age (Ingold 1973, which see for details). Newly hatched, and very young chicks give soft cheeping 'dü di dü dü' or 'di di dü . . .' ('Young chick call') to solicit brooding. Young (any age) give loud 'düi' or 'düiüiüi . . . ' at intervals of 2–3 s ('Contact call') when separated from parent, often stimulating brooding response (Ingold 1973). If parent slow to respond, chick may give loud, wavering 'düiiie' or 'düiüiüiie . . .' ('Distress call': Ingold 1973), often at regular intervals, to elicit brooding. Distress-call ($c.$ 0·9 s long), e.g. when handled, described by Bédard (1969) probably similar, though more intense. From 1st week, but especially at fledging, chick gives long (up to $c.$ 1 s), wavering whistle, repeated at intervals of $c.$ 0–35 s: 'düieee düieee' or 'düiüiüieee düiüiüieee' ('Leap call': Ingold 1973); 'psee-ee psee-ee' or 'psee-ee-ee' (Perry 1940). According to Bédard (1969), call $c.$ 0·5 s long, between 3·9 and 4·8 kHz, dropping in pitch in middle, and not distinguishable (by sonagram analysis: see his Figs 19, 20) from call given, when younger, to solicit food or brooding (presumably 'Contact call', above). In recording (Fig II) of young on sea, 2 calls given in rapid succession, then gap of $c.$ 18 s before another 2 (E K Dunn, P J Sellar). Call is loud and resonant, and, relative to others, markedly variable between individuals. Thus facilitates identification and location of fledged young by parents, which responded specifically to own offspring, calling thus, 10 days after hatching, but not before. Leap-call stimulates, in parent, Lure-call and Billing response (Ingold 1973). EKD

Breeding. SEASON. Britain and Ireland: see diagram. Northern Scandinavia and Iceland: up to 3 weeks later (Makatsch 1974). SITE. In cliff crevice or cavity, on sheltered ledge, under boulders, rarely on open ledge; may use old, or even occupied, burrow of Puffin *Fratercula arctica* or rabbit *Oryctolagus cuniculus*. Of 1688 nests, Skomer Island (Wales), 87% on ledge or in crevice, 13% in burrow or under boulders (Hudson 1979a). In White Sea (no sample size), 65% under rock fragments or boulders, 30% in vertical or slightly sloping crevice, and 5% in cave; in western White Sea, also nests in piles of driftwood, and in cavities and burrows in peat (Bianki 1967). In cavities and burrows, 12–20% of eggs lie 0·5–1·0 m from entrance (Bianki 1967). Colonial. Nest: usually none, but tiny stones sometimes accumulated round egg, rarely also plant fragments (Paludan 1947; Plumb 1965). Of 67 nest-sites,

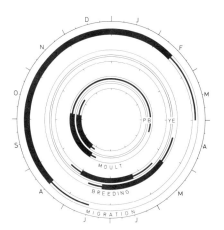

Canada, 41 marked by accumulation of roots and dried grass, 55 had gravel or loose stones collected up, 6 were on humus, and only 6 on bare rock (Bédard 1969). Building: some material may be brought around egg from within reach of sitting bird. EGGS. See Plate 92. Oval or sub-elliptical, roughened, not glossy; brown or pale brown to white, occasionally greenish, very variably spotted, blotched, or scribbled with dark brown, sometimes zoned. Nominate *torda*: 75×48 mm ($63-84 \times 42-52$), $n=250$; calculated weight 90 g (Schönwetter 1967). British population of *islandica*: 72×47 mm ($67-80 \times 42-51$), $n=186$; weight 92 g ($80-107$), $n=47$ (Plumb 1965). Icelandic population of *islandica* as nominate *torda* (Makatsch 1974). Clutch: 1, very rarely 2, and then perhaps always by 2 ♀♀, though 2nd egg of 1 clutch of 2, Skokholm (Wales), laid 17 days after 1st, similar to period between loss and re-laying (Plumb 1965). One brood. Replacements laid after egg loss; on Skokholm 25% of lost eggs replaced, $14 \cdot 1 \pm$ SD $1 \cdot 5$ days ($n=14$) after loss; eggs lost late in season not replaced; replacement eggs always smaller than original (Lloyd 1976a). In northern USSR, 35% of lost eggs replaced (Belopol'ski 1957). Re-laying only took place when egg lost less than 14 days after laying; re-laying interval 13–18 days, $n=7$ (Paludan 1947). INCUBATION. 36 days (32–39), $n=29$ (Plumb 1965). 25–35 days (Belopol'ski 1957). $35 \cdot 1 \pm$ SD $2 \cdot 2$ days, $n=239$ (Lloyd 1979). By both sexes. Begins on laying of egg. ♀ may incubate for 48 hrs after laying before first nest-relief; average 2 change-overs per 24-hr period, both ♂ and ♀ incubating at night (Lloyd 1976a). YOUNG. Semi-altricial and nidicolous. Cared for and fed by both parents, brooded while small. FLEDGING TO MATURITY. Fledging period 18·5 days (14–24), $n=34$ (Plumb 1965). $17 \cdot 2 \pm$ SD $2 \cdot 2$ days, $n=163$ (Lloyd 1976a). At Kandalaksha (USSR), 15–20 days, $n=25$ (Bianki 1967), at Sem' ostrovov 19–22 days (Kaftanovski 1951). Cared for by parents after fledging but period unknown. Age of first breeding mostly 4–5 years. BREEDING SUCCESS. Of 735 1st eggs, Skokholm, 71% hatched and 94% chicks fledged; overall, 0·71 chicks per breeding pair; replacement eggs more successful than 1st eggs laid at same time, suggesting that re-laying birds still better parents than later layers; egg losses mainly to Jackdaw *Corvus monedula* and Herring Gull *Larus argentatus*; breeding success improves with age of parents (Lloyd 1976a). Of 170 eggs laid, northern USSR, 84% hatched and 96·5% of chicks fledged; 0·81 chicks per pair (Bianki 1967). On Skomer, nests in burrows and crevices produced 0·70 chicks per pair; on ledges (more vulnerable to bird predation), 0·55 chicks per pair (Hudson 1979a).

Plumages. ADULT BREEDING. Upperparts black. Chin, throat, and sides of neck dark chocolate-brown. Conspicuous narrow white line from front of eye to frontal feathering of bill. Rest of underparts white, extending on to throat in Λ-form. Tail and flight-feathers black, inner web of primaries grading to dull grey on edge and base; secondaries with conspicuous white tip. Shafts of flight-feathers black or brown-black, pale horn-coloured towards base; shafts of tail-feathers entirely black. Upper wing-coverts black. Underside of wing dark silvery-grey, except for white median and lesser coverts in which small brown spot at base often present but usually concealed. Axillaries white. ADULT NON-BREEDING. Upperparts black, extending down to sides of breast; often narrow white mottled band across back of neck. Lores and streak under eyes brown-black, extending backwards from eye as speckled streak. Faint narrow white line from front of eye towards base of upper mandible; sometimes absent. Chin, throat, and rest of underparts white. Wing and tail as adult breeding. DOWNY YOUNG. Lores, forehead, crown, and nape white. Down on head very short. Neck and throat mottled brown and white due to brown feather-bases and white tips. Mantle mottled brown and buff due to buff tips to dark brown down feathers. Rest of upperparts dark brown with faint buff feather-tips. Mottling on wing pads and flanks more pronounced because of large pale buff tips to brown down feathers. Breast and belly white. In nominate *torda*, down replaced by juvenile plumage *c*. 15 days after hatching. (Bent 1919; Dementiev and Gladkov 1951b; Kartashev 1960). JUVENILE. Head and neck dark chocolate-brown; chin and throat sometimes white or mottled brown and white (Witherby *et al.* 1941; Fjeldså 1977); narrow white line from front of eye towards culmen often conspicuous but sometimes only faint. Feathers of upperparts dark brown with black-brown margins. Underparts white. Upper wing-coverts like upperparts. Flight-feathers and tail grow during moult to next plumage. FIRST IMMATURE NON-BREEDING. As adult non-breeding, but white tips to secondaries narrower. FIRST IMMATURE BREEDING. As adult breeding, but white tips to secondaries narrower; some white feathers may be retained on throat. Wing and tail very worn (Witherby *et al.* 1941; Salomonsen 1944).

Bare parts. ADULT BREEDING. Iris dark brown (Hartert 1921–2; Witherby *et al.* 1941). Bill black with slightly curved prominent white line across upper mandible, continuing on lower mandible as transverse white line. Inside of mouth yellow to deep lemon (Ridgway 1919; Witherby *et al.* 1941). Leg and foot black. ADULT NON-BREEDING. Bill and white line duller. IMMATURE. In 1st winter and summer, white line absent; present but narrow and dull in 2nd winter and summer; for development of grooves, see Structure. Leg and foot black. DOWNY YOUNG. Bill, leg, and foot black. Tip of upper mandible has white patch with egg-tooth. JUVENILE. White patch and egg-tooth at tip of upper mandible still present.

Moults. ADULT POST-BREEDING. Complete; August–October. Starts with body feathers and lesser wing-coverts. Flight-feathers, greater wing-coverts, primary coverts, and tail-feathers all moulted simultaneously. ADULT PRE-BREEDING. Partial; January–April. Often head and neck only, but sometimes also whole body. POST-JUVENILE. Partial: body and wing-coverts; variable number of juvenile upper wing-coverts and primary coverts retained (Kuschert *et al.* 1981). July–October, extending to November. Flight-feathers and tail-feathers grow from *c*. 5 weeks after hatching (Verwey 1922; Heinroth and Heinroth 1931–3; Witherby *et al.* 1941; Salomonsen 1944; Kozlova 1957; Swennen 1977). IMMATURE PRE-BREEDING. Moult into 1st immature breeding probably later than in adult: March–May. IMMATURE POST-BREEDING. Moult into 2nd winter plumage much earlier than in adult: middle of June to September.

Measurements. ADULT. Nominate *torda*. Summer, Sweden, Norway, Greenland; skins (Salomonsen 1944). Depth of bill measured at gonydeal angle.

PLATE 14. *Chlidonias hybridus* Whiskered Tern (p. 133): **1** ad non-breeding, **2** 1st imm non-breeding (1st winter), **3** juv. *Chlidonias niger* Black Tern (p. 143): **4** ad non-breeding, **5** 1st imm non-breeding (1st winter), **6** juv. *Chlidonias leucopterus* White-winged Black Tern (p. 155): **7** ad non-breeding, **8** juv moulting into 1st imm non-breeding (1st autumn or winter), **9** juv. (NWC)

PLATE 15. *Sterna aleutica* Aleutian Tern (p. 100): **1–3** ad breeding, **4** ad non-breeding, **5–6** juv. *Rynchops flavirostris* African Skimmer (p. 166): **7–8** ad breeding, **9** ad non-breeding, **10–11** juv. (DIMW)

PLATE 16. *Uria aalge* Guillemot (p. 170). Nominate *aalge*: **1** ad breeding, **2** ad breeding bridled morph, **3** ad non-breeding, **4** 1st imm non-breeding (1st summer), **5** juv, **6** downy young. *U. a. albionis*: **7** ad breeding. *U. a. hyperborea*: **8** ad breeding. (RG)

PLATE 17. *Uria lomvia* Brünnich's Guillemot (p. 184): **1** ad breeding, **2** ad non-breeding, **3** 1st imm breeding (1st summer), **4** juv ad-non-breeding type, **5** downy young. (RG)

WING   ♂ 209·7 (5·05; 21) 201–216   ♀ 207·6 (4·33; 24) 201–216
BILL   35·7 (1·65; 26) 32–39   34·5 (1·42; 26) 32–37
DEPTH   24·7 (1·15; 25) 23–27   23·7 (0·78; 26) 22–25

Sex difference significant for bill and depth.

A. t. islandica. ♂, Britain, Ireland, Faeroes, Iceland, breeding; ♀, Dutch coast, summer; skins (RMNH, ZFMK, ZMA).

WING   ♂ 193·6 (4·48; 13) 187–200   ♀ 196·8 (2·39; 5) 194–200
TAIL   75·7 (5·70; 13) 66–86   78·4 (2·79; 5) 76–83
BILL   34·2 (1·79; 13) 31·3–38·5   32·3 (1·18; 5) 30·7–33·6
DEPTH   20·7 (0·96; 11) 19·1–22·4   20·2 (0·78; 5) 19–21
TARSUS 30·7 (1·44; 13) 28–33·2   29·5 (1·60; 5) 27·6–31·2
TOE   45·3 (1·47; 12) 44–49   45·4 (1·30; 5) 44–47

Sex difference significant for bill

Weights. ADULT. Nominate *torda*: (1) Murmansk (Dementiev and Gladkov 1951b); (2) summer, eastern Murmansk (Belopol'ski 1957). (3) *A. t. islandica*, December–March, Dutch coast (ZMA).

(1) ♂ 727   (11)   662–800   ♀ 717 (8)   631–810
(2)   734   (81)   524–890   700 (61)   620–800
(3)   485   (77·9; 11) 391–645   453 (57·2; 7) 372–518

Unsexed, Outer Hebrides, June: 634 (45·9; 7) 585–730 (D J Steventon, BTO).

YOUNG. Nominate *torda*: at hatching, mean c. 63 (53·9–74) (Dementiev and Gladkov 1951b), mean 64 (52·5–75) (Kartashev 1960); at 1 day old, 50–70 (5) (Bianki 1967); at fledging, mean c. 250 (Dementiev and Gladkov 1951b; Kartashev 1960, 170–190 (25) (Bianki 1967). *A. t. islandica*: at hatching, mean 58 (Brun 1959). Chicks lose weight before fledging (Brun 1959; Bianki 1967; Sealy 1973) and leave nest at 27·7–35% of adult weight (Brun 1959; Kartashev 1960; Plumb 1965).

Structure. ADULT. Wing narrow and pointed. 11 primaries: p10 longest, p9 3–6 shorter, p8 8–13, p7 20–26, p6 31–38, p5 43–51, p1 90–101; p11 minute, concealed by primary coverts. Tail relatively long, wedge-shaped; 12 feathers, pointed, t1 longest. In adult breeding, bill laterally compressed, upper mandible prominently arched downwards with almost vertically directed tip enclosing tip of lower mandible. On lower edge of lower mandible, distinct rounded angle in front of which lower edge gradually slopes upwards. Base of culmen with horny ridge from above nostril to frontal feathering; narrow furrow in front of it. Broad, curved, white groove on upper mandible, extending on lower mandible as vertical white groove; 2–3 uncoloured curved grooves in front of it. Base of bill feathered except for horny ridges along cutting edges of mandibles. Nostrils narrow, situated just above cutting edge and partly covered by point of feathering extending from lores. Rictal ridge absent in adult non-breeding; rest of bill-sheath also shed, resulting in smaller bill depth (difference between summer and winter specimens statistically significant) and slightly smaller bill length (measured from nostril). Immatures in 1st–2nd winter have bill less deep than in adult and without grooves or rictal ridge. In summer of 3rd calendar year, bill still less arched than in adult, but with white groove. In subsequent years, distal grooves develop; in summer of 4th calendar year, bill probably like adult. In downy young, gonydeal angle already prominent. Outer toe c. 95% of middle, inner c. 76%; no hind toe.

Geographical variation. Nominate *torda* larger than *islandica* (see Measurements). *A. t. pica* Salomonsen, 1944, sometimes recognized, comprising breeding birds of coast of Norway, Murmansk, western Greenland, and eastern Canada; based on presence of 3rd uncoloured furrow in more than half of the number of specimens, but feature less common than that and because of overlap with other races recognition not warranted.   RS

Pinguinus impennis (Linnaeus, 1758) **Great Auk**

FR. Grand Pingouin   GE. Riesenalk

Extinct since June 1844, when last known pair killed on Eldey (Iceland). Formerly occurred over continental shelves on both sides of North Atlantic in boreal to subarctic latitudes; breeding stations known Newfoundland (Funk Island and possibly elsewhere), Iceland (south-western skerries), Scotland (St Kilda, possibly Orkney), and probably Faeroes. Scanty records from elsewhere indicate that—like its nearest relative, Razorbill *Alca torda*—it was widely dispersed at sea when not breeding; immatures occurred north of colonies in summer and autumn (e.g. in Greenland seas), while birds also ranged from Labrador to Carolinas, and (in Europe) in North Sea as well as Atlantic seaboard. Prehistoric remains also found Florida and west Mediterranean sites. For detailed accounts, see Newton (1861), Grieve (1885), and Greenway (1958).

A flightless species which bred colonially on small rocky islands with shelving access to sea. For egg, see Plate 93. Like other Alcidae, young taken to sea while still small and (presumably) dependent. For recent attempt to reconstruct feeding ecology, through fish remains in Funk Island soil samples, see Olson *et al.* (1979), but results should be treated cautiously since samples possibly included items imported on nets of fisherman-hunters (R G B Brown). Species declined due to excessive human predation at colonies: killed for food and oil, and later (as numbers dwindled) for collection specimens also.

Cepphus grylle Black Guillemot

PLATES 19, 22, and 23
[facing page 230, and between pages 278 and 279]

Du. Zwarte Zeekoet Fr. Guillemot à miroir blanc Ge. Gryllteiste
Ru. Чистик Sp. Arao aliblanco Sw. Tobisgrissla

Alca Grylle Linnaeus, 1758

Polytypic. Nominate *grylle* (Linnaeus, 1758), Baltic Sea; *mandtii* (Mandt, 1822), arctic North America, Hudson Bay, James Bay, northern Newfoundland, Labrador south to *c*. 58°N, western Greenland south to *c*. 72°N, eastern Greenland south to *c*. 69°N, Jan Mayen, Bear Island, and Spitsbergen, east to eastern Siberia and northern Alaska; *arcticus* (Brehm, 1924), North America, southern Greenland (south of *mandtii*), Britain, Ireland, western Sweden, Denmark, Norway, Murmansk, and White Sea; *faeroeensis* Brehm, 1831, Faeroes; *islandicus* Hørring, 1937, Iceland.

Field characters. 30–32 cm; wing-span 52–58 cm. About 20% smaller than Guillemot *Uria aalge*; *c*. 10% larger than Puffin *Fratercula arctica*. Medium-sized auk, with relatively small, fine bill, round head, noticeably tubby body, oval-shaped wings, and striking difference between summer and winter plumages. Broad white oval patch on wing-coverts at all seasons; rest of plumage uniformly brown-black in breeding season, dark-saddled piebald during moults, and black-speckled white in winter. Legs and feet bright red. Sexes similar; marked seasonal variation. Immature separable at close range. 5 races in west Palearctic, with complex differences (see Geographical Variation).

ADULT BREEDING. Plumage brown-black (with least gloss and browner tone on underparts), relieved only by large oval white panel on all but innermost median and greater wing-coverts. Underwing shows larger white patch than upperwing, extending on to primary coverts and axillaries. Rare variant has no wing-panel; Icelandic race *islandicus* shows broken brown line across lower edge of white panel. ADULT NON-BREEDING. Dark uniformity of summer plumage lost from late July, as predominantly white plumage invades underparts and white-fringed black feathers cover upperparts—white fringes widest on head, lower rump, and longest scapulars; white flanks retain obvious dusky marks; bird thus appears distinctly pale-headed and hoary at distance and strikingly close-barred over back at close range (both patterns absent in other auks occurring in west Palearctic). White wing-panel less obvious on swimming bird but still striking in flight; retention of uniformly dark flight-feathers and tail most obvious from behind or when diving. Rare variant retains black or partly black wing-panel; Icelandic race, *islandicus*, retains brown-black back; Arctic race, *mandtii*, shows whiter scapulars and body. JUVENILE. Resembles winter adult but crown, mantle, lower back, and rump less hoary, mostly brown-black. At close range, white wing-panel shows rows of dusky spots. FIRST WINTER. Closely resembles winter adult but upperparts still less hoary. Wing-panel retains dusky spots. FIRST SUMMER. Duller than breeding adult, with white flecks on underparts, worn brown wings, and mottled wing-panel. Bill black in adult but with strong orange tone in immature. Legs and feet coral-red, with brown or dusky tone in immature.

Unmistakable. All other auks are wholly dark above at all seasons and none occurring in west Palearctic shows similar white wing-panel. At long range, birds in non-breeding plumage may suggest small grebe *Podiceps* or even ♂ Smew *Mergus albellus* but oval shape of wings combine with small head and rounded body to give most distinctive silhouette, while fluttering action much exaggerated by flickering of white panels on both wing surfaces. Swimming ability marked, with greater use of feet than in *Uria* and noticeably sudden dive. Walks more easily than *Uria*, being able to walk and stand on raised tarsi. Like other auks, forms lines on water (often remarkably straight).

Markedly less social than other auks, not forming large offshore communities and wintering in small groups in inshore waters. Quite exceptional inland. Voice feeble.

Habitat. From beyond 88°N on polar ice through arctic, subarctic, and boreal to north temperate coastal zones in narrow ribbons of inshore, inlet, and fjord waters and their immediately bordering land frontages. Feeds regularly in unfrozen leads in drift-ice and by advancing thaw along edge of pack-ice. Exclusively marine except for relict stocks on archipelagos and coasts of relatively warm brackish Baltic (Dementiev and Gladkov 1951b). As a bottom-feeder, however, does not fish beyond shallow water, although covering long distances during dives. Breeds at varying heights, mainly near sea-level but up to 150 m in Greenland (Salomonsen 1950–1), and to 600 m on Spitsbergen, where nests may be found up to 2–3 km inland (Birulya 1910), although normally 1 km exceptional. Remarkably adaptable in choice of nest-sites, although these typically in natural holes and crevices, caves, blow-holes, under fallen slabs or boulders on storm-beaches, or in scree or talus (Cairns 1978; Asbirk 1979a; Petersen 1981). In some regions, often on skerries or inshore islands.

In winter, remains as far north as unfrozen water can be found, even up to high Arctic (Salomonsen 1950–1). Normally flies in lower airspace, and is inhibited from crossing land even for access to favourable sites, but will rise to some hundreds of metres when prospecting for clear water (Salomonsen 1950–1). Pattern of dispersal and avoidance of exposed nest-sites minimize serious human exploitation, but local indirect impacts result from introduced mink *Mustela vison*, and from oil-spills and fishing nets (Heubeck and Richardson 1980; Petersen 1981).

Distribution. ICELAND. Some local extinctions on mainland due to introduced mink *Mustela vison*, with increases on mink-free islands (Petersen 1979). BRITAIN. Extinct as breeder in south-east Scotland in 19th century and Yorkshire since 1948, but some recent range expansion in Irish Sea area (Parslow 1967; Sharrock 1976). USSR. Reports of former breeding Lake Ladoga improbable (Løppenthin 1963).

Accidental. France, Belgium, Netherlands (almost annual in recent years), West Germany, East Germany, Czechoslovakia, Yugoslavia.

Population. Few reliable counts; some local fluctuations, but overall trends unknown.

USSR. Franz Josef Land: very common, sites listed (Norderhaug *et al.* 1977). Novaya Zemlya: no counts, sites listed (Norderhaug *et al.* 1977). Kola peninsula: *c.* 1200 pairs (Norderhaug *et al.* 1977). SPITSBERGEN. Relatively abundant, but far less numerous than other auks (Løvenskiold 1964); no counts, but major sites listed by Norderhaug *et al.* (1977). BEAR ISLAND. At least 300 pairs 1980 (Luttik 1982). JAN MAYEN. Comparatively small numbers (Bird and Bird 1935); a few seen, no evidence of breeding (Seligman and Willcox 1940). ICELAND. No counts, but probably number tens of thousands or low hundred thousand birds (AP). BRITAIN AND IRELAND. No exact counts; in 1969–70 estimated *c.* 8340 pairs, of which *c.* 740 in Ireland (Cramp *et al.* 1974). DENMARK. Increased from *c.* 100 pairs in 1928 to *c.* 450 pairs in 1977 (Asbirk 1978). NORWAY. Some decrease in south (Haftorn 1971); *c.* 11000–12000 pairs 1983 (P E Jónsson). SWEDEN. Estimated 11000 pairs 1980 (S Hedgren). Marked decline in archipelago off east coast, 1958–72, due to introduced mink *Mustela vison* (Olsson 1974). FINLAND. Estimated 1500 pairs, after decline from peak *c.* 4000 pairs earlier in 20th century, due to cold winters and former human predation of eggs and young (Merikallio 1958). Has increased in some areas, but little change in others (OH); at least 5000–6000 pairs (M Kilpi). In Valassaaret islands (Gulf of Bothnia), fairly stable numbers 1954–60, then slight increase to 1965 (Hildén 1966).

Survival. Pre-breeding survival 16% Norway, 21% Sweden, 25% Denmark, 32% Finland, and 27% Greenland (Petersen 1981). Annual adult survival 85% Denmark (Asbirk 1979a), at least 87% Iceland (Petersen 1981), and 81% eastern Canada (Preston 1968). Oldest ringed bird 20 years (BTO).

Movements. Resident and dispersive. Shows less movement than other west Palearctic Alcidae, and less often encountered outside breeding range or in offshore waters. Dispersal most marked in high latitudes, but no breeding area deserted completely in winter. Some individuals (chiefly young birds) will disperse southwards in temperate zones also, but occurs only as a straggler south of New England (in western Atlantic) and North Sea (in Europe). Young leave cliffs before capable of flight, hence initial movements by swimming; indications from Greenland and southern Norway of significant post-fledging movement northwards, recalling gulls *Larus* (Salomonsen 1967a; Myrberget 1973a).

ARCTIC POPULATIONS (*mandtii*). Scale of dispersal not yet clarified, since few ringed except in western Greenland. In latter, birds winter as far north as open water exists, using (e.g.) tide rips and leads through fast ice; remains common in Disko Bay, but numbers further north are reduced in midwinter, following southward movement into open-water areas of Davis Strait. In eastern Greenland, remains fairly common during winter at ice edge off Scoresbysund (Salomonsen 1967a). Many winter around Spitsbergen in seasons when seas remain open, but these infrequent; in most winters, majority of birds forced to move out, though perhaps go no further than edge of pack-ice. Such exodus from Svalbard occurs September–October, with return movement from mid-March (exceptionally late February), depending on ice conditions; one late March report of thousands of *C. grylle* and Little Auks *Alle alle* seen flying past Nordaustlandet towards Franz Josef Land (Løvenskiold 1964). Further east, frequently encountered in winter in White Sea and around Novaya Zemlya; in good ice years, vanguard returns inshore from late February to Franz Josef Land and Novaya Zemlya (Kozlova 1957). One specimen of high-arctic race *mandtii* from eastern Iceland, a juvenile in late October (Petersen 1977).

ICELANDIC POPULATION (*islandicus*). Based on data from A Petersen; see also Petersen (1977). Considered sedentary by Timmermann (1938–49) and Gudmundsson (1953b), and this may be broadly true for adults, though one 6-year-old found in October 200 km (by sea) from colony. Juveniles and other pre-breeders disperse more widely in Icelandic seas and even further afield: of 15 September–April recoveries, 11 were in Iceland (up to 663 km from colony) and 4 in south or south-west Greenland (October–February, 3 1st-years and 1 2nd-year). Some immatures stay well away from colony in summer (especially 1st summer), but others visit nesting places where some arrive well before adult egg-laying, this being unusual in Atlantic Alcidae and perhaps related to earlier sexual maturity in this species. Only small numbers of 1-year-olds visit colonies, but more at 2–3 years old. As in other Alcidae, first-time breeders may join a colony other than their natal one, but are site-faithful thereafter.

MURMANSK AND NORWEGIAN POPULATIONS. Common all winter in White Sea–Murmansk region, especially in Onega, Kandalaksha, and Kola Bays (Kozlova 1957). Likewise in Norway, birds winter along entire coastline

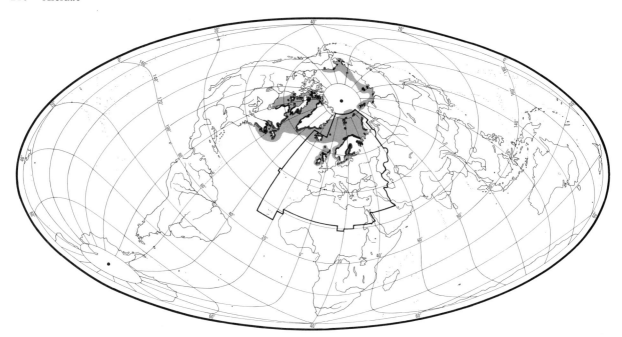

round to Varangerfjord and Russian frontier (Haftorn 1971). However, some individuals from northern Norway move south in autumn, with extreme displacements of 1000–1500 km for those reaching as far as Skagerrak; long-distance recoveries mainly immature birds, though this possibly biased by higher mortality of younger age groups (Myrberget 1973a). Birds from west Norway colonies found in winter both north and south of natal colony, though longer movements are southwards (up to 500 km). Some Norwegian birds, including those from Skagerrak colonies, penetrate into Kattegat, with recoveries on Danish and Swedish sides, whereby even birds from south Norway colonies can show displacements of up to 400 km (again, longer movements more characteristic of younger birds). One exceptional juvenile from Vestfold found in Côtes-du-Nord (France) in October.

BALTIC AND KATTEGAT POPULATIONS (nominate *grylle*). 2 groupings discernible: birds from Kattegat colonies (Denmark, south-west Sweden) winter there, and have minimal contact with Finland and east Sweden population which winters mainly along Swedish Baltic Coast (Rosendahl and Skovgaard 1968; Andersen-Harild 1969). Kattegat birds disperse from July onwards, with winter recoveries mostly from southern Kattegat (displacements not normally over 200 km); only stragglers south of Lille and Store Baelts (Denmark), or northwards into or beyond Skagerrak, though 2 exceptional Swedish juveniles (from Halland province) recovered eastern England in July and February. For Baltic population (Finland, eastern Sweden), principal wintering area is along Swedish coast between Stockholm and Bornholm, since Gulfs of Bothnia and Finland largely closed by ice; also small numbers of recoveries along southern Baltic (Poland, East and West Germany) and westwards to Danish islands, but not from Skagerrak or North Sea.

BRITISH AND IRISH POPULATIONS. The most sedentary of British and Irish seabirds, few moving far even in mid-winter (Cramp *et al.* 1974). Recoveries in Irish Sea, western Scotland, and Northern Isles usually involve displacements of under 50 km, though some autumn–winter movement between Shetland and Orkney islands of up to 100 km. Regular sightings off eastern Scotland indicate that some northern birds must penetrate into North Sea, though little ringing evidence. Longest movements by 2 Fair Isle (Shetland) birds found in winter: in North Yorkshire (as adult, 590 km) and Essex (as juvenile, 880 km).

Food. Chiefly marine fish in south of range, more crustaceans in Arctic regions. Opportunistic, switching rapidly between prey types as availability changes. Considerable variation in diet between colonies, seasons, years, and individuals (Belopol'ski 1957; Slater and Slater 1972a; Cairns 1978; see also Petersen 1981). Fish caught by surface-diving, mostly in depths of 1–8 m (Madsen 1957), 5–20 m (Bergman 1971a), *c.* 10 m (Belopol'ski 1957); mean submersion time 45·3 s (30–75, $n=20$), mean interval on surface between dives 15·1 s (4–45, $n=14$) (Nicholson 1930); submersion usually 30–35 s (25–40), interval (3–)7–10 s (Bianki 1967); *c.* 30–60 s, interval *c.* 15 s (Hyde 1937); maximum submersion 50–60 s, during which bird travels up to 75 m or more (see Kozlova 1957). Typically hunts alone but said sometimes to fish shoals co-operatively in line or semi-circle (Kaftanovski 1951). Occasionally robs conspecifics returning to colony with food (Slater and Slater 1972b; Petersen 1981). Fish for young brought singly to surface and carried crosswise in bill, usually just

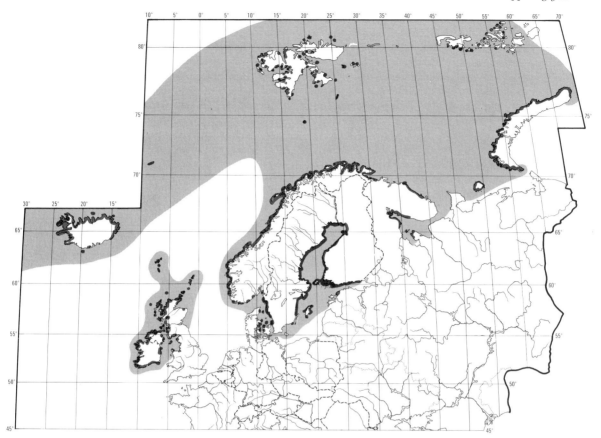

behind the gills. At surface, captured prey may be shaken, dropped, and retrieved—sometimes with shallow dive (Storer 1952; Kozlova 1957). Prey usually killed outright but butterfish *Pholis gunnellus* commonly brought alive to young (Petersen 1981). Hermit crabs *Eupagurus* are shelled, and claws removed shortly after capture (Petersen 1981). Feeding rate (to young) depressed by wind (Slater and Slater 1972a; Petersen 1981). Tidal stage may affect feeding rate, area used, and thus prey taken (Preston 1968). High tide appears to facilitate capture of littoral prey, especially butterfish *Pholis* (Preston 1968; Petersen 1981). Different nesting groups in colony may use different feeding grounds (Petersen 1981). Mainly inshore feeder in breeding season, frequenting especially littoral–sublittoral boundary; more offshore in winter. In Barents Sea, USSR, hunts mostly at ice edge during breeding season, entering fissures and even 'crawling' under stones (Uspenski 1956). Similar behaviour noted Spitsbergen where fed closer to icebergs and glacier than any other Alcidae (Hartley and Fisher 1936; see also Bradstreet 1979); also feeds in freshwater lakes (Summerhayes and Elton 1928). In winter, birds in Arctic feed at edge of pack-ice in wide range of water depths (see Storer 1952). In breeding season, foraging range 2 km or more (Bianki 1967), c. 0·5–4 km (Asbirk 1979a), 1·5–4 km (Bergman 1971), 2–4(–7 or more) km (Petersen 1981).

Especially in breeding season in lower latitudes, mainly benthic inshore fish, notably Blenniidae—butterfish *Pholis gunnellus*, viviparous blenny *Zoarces viviparus*, snake blenny *Lumpenus lampetraeformis*, *Leptoclinus maculatus*; Cyclopteridae—lumpsucker *Cyclopterus lumpus*; Cottidae—bull rout *Myoxocephalus scorpius*, sculpins *Triglops murrayi* and *Gymnocanthus ventralis*, sea scorpion *Taurulus bubalis*, *Icelus bicornis*; gobies (Gobiidae); sand-eels (Ammodytidae)—*Hyperoplus lanceolatus*, *Ammodytes tobianus*; Gadidae—cod *Boreogadus saida*, *Gadus morhua*, saithe *Pollachius virens*, haddock *Melanogrammus aeglefinus*; Clupeidae— herring *Clupea harengus*, capelin *Mallotus villosus*; garfish *Belone belone*, pipefish *Nerophis*, wrasse *Ctenolabrus rupestris*, redfish *Sebastes*, sea snail *Liparis montagui*, lamprey *Petromyzon*, ten-spined stickleback *Spinachia spinachia*, eel *Anguilla anguilla*, dab *Limanda limanda*, sole *Microstomus kitt*, arctic char *Salvelinus alpinus*. Crustaceans include Malacostraca—crabs (*Carcinus maenas*, *Hyas*, *Porcellana longicornis*, *Portunus*, *Eupagurus*), squat lobster *Galathea*, prawns and shrimps (*Crangon crangon*, *Pandalus*, *Palaemon*, *Spirontocaris*, *Sclerocrangon boreas*); *Mysis* (Mysidacea), especially *M. oculata*; Amphipoda—*Gammarus* (especially *G. locusta*), *Gammarellus homari*, *Euthemisto*, *Themisto*, *Atylus carinatus*, *Paramphithoe hystrix*, *Acanthostepheia behringensis*, *Anonyx nugax*, *Caprella*, *Jassa*; *Thysanoessa inermis* (Euphausiacea); Isopoda—*Idotea, Mesidotea entomon*; Copepoda—*Calanus*. Molluscs taken in small amounts; include Gastropoda—*Littorina, Spirorbis planorbis, Lacuna, Natica, Lunatia pallida, Onoba, Margarita*, whelk *Buccinum lapillus, Bela, Admete viridula, Amauropsis groenlandicus*; Lamellibranchia—*Nucula tenuis*, mussels *Mytilus edulis, Modiolus modiolus*; occasionally squid (Cephalopoda), *Clio* (Pteropoda). Also in small numbers, polychaete worms *Nereis, Harmothoe imbricata, Polynoe, Onuphis conchyleca*, and *Arenicola*; rarely sponges (Bryozoa), coelenterates (Hydrozoa), *Beroe* (Ctenophora), dead craneflies (Tipulidae) floating on surface, and plant matter (Naumann 1844, 1903; Römer and Schaudin 1900; Patten 1906; Collett and Olsen 1921; Hartert 1921–2; Montague 1926; Summerhayes and Elton 1928; Gorbunov 1932; Demme 1934; Bird and Bird 1935; Hartley and Fisher 1936; Witherby *et al*. 1941, Duffey and Sergeant 1950; Dementiev and Gladkov 1951b; Kaftanovski 1951; Gudmundsson 1953b; Uspenski 1956, 1972; Belopol'ski 1957; Kozlova 1957; Madsen 1957; Kartashev 1960; Bergman 1971a; Slater and Slater 1972a; Asbirk 1979a; Petersen 1981; see also Storer 1952, Løvenskiold 1964). For prey sizes (of items brought to young), see final paragraph.

Adult diet in breeding season mainly fish, but includes more invertebrates than brought to young, and more crustaceans in higher latitudes. In Danish waters, adult diet (26 stomachs) c. $\frac{2}{3}$ fish, mostly gobies (in 19); also viviparous blenny (4), ten-spined stickleback (4), butterfish (3), and fewer cod, herring, garfish, eel, pipefish, wrasse; apart from a few polychaetes (*Nereis, Polynoe*), remaining $\frac{1}{3}$ crustaceans—crabs (9 stomachs), shrimps and prawns (8), *Idotea* (6), squat lobster (1) (Madsen 1957). In breeding season, Flatey Island (Iceland), adults took 9 species of fish, forming c. $\frac{1}{3}$–$\frac{1}{2}$ diet in different years. In stomachs of shot (38) and drowned (31) birds, butterfish occurred in 37% and 19% respectively, sand-eels in 8% and 39%, bull rout in 5% and 13%, snake blennies in 11% and 0; rest mostly herring, redfish, cod, and lumpsucker. 27 species of invertebrates taken (up to 88% by number of items), notably crabs *Hyas*, shrimps *Pandalus* and *Spirontocaris*, amphipods *Anonyx nugax*, molluscs (especially *Onoba striata*), and polychaetes (especially *Nereis* which occurred in 39% of shot and 68% of drowned birds) (Petersen 1981). In breeding season, Barents Sea, stomachs of 88 adults contained (by number) 73·3% fish, 20·7% crustaceans, 3·0% molluscs, 1·5% polychaetes, and 1·5% insects; of fish, 31·9% cod, 21·7% sand-eels, 16·0% herring, 20·3% benthic species, notably butterfish and bullheads (Cottidae), 10·1% capelin (Belopol'ski 1957, which see also for seasonal and sexual variations in diet). In Spitsbergen, 21 adults contained many more crustaceans (61·2% by number) and molluscs (19·4%), and fewer fish (19·4%); snake blenny *Leptoclinus* found in 3; of crustaceans, *Thysanoessa inermis* in 6, *Euthemisto libellula* in 4, *Mysis oculata* in 5, *Gammarus locusta* in 1, *Spirontocaris gaimardii* in 3, *Sclerocrangon boreas* in 1, *Hyas araneas* in 1, *Eupagurus pubescens* in 1; of molluscs, gastropods in 5, lamellibranchs in 1; polychaete *Harmothoe imbricata* in 1 (Hartley and Fisher 1936). Birds feeding in freshwater lakes, Spitsbergen, believed to take arctic char (Summerhayes and Elton 1928). In Franz Josef Land (USSR), March–May, 28 stomachs contained (by number) 43% polychaetes, 37% amphipods, 20% fish (Demme 1934); 3 stomachs, April, contained cod *Boreogadus saida*, crab, and *Calanus* (Uspenski 1972). In Novaya Zemlya, of 34 stomachs, 17 contained fish, 17 crustaceans, 2 polychaetes, 1 molluscs, 1 sponges, and 3 algae, including *Phaeophyta* (Uspenski 1956).

Adult diet in winter less well known. In Norway, 11 stomachs contained (by number) 76·9% crustaceans, 15·4% fish, 7·7% molluscs (Naumann 1903). At Sem' ostrovov (White Sea, USSR), December, 2 birds contained *Gammarus* and *Littorina*, with fish remains also in 1; summer diet mostly sand-eels (in 4 of 7 stomachs), cod (3), and butterfish (1), with (in 2) mussels *Mytilus edulis* (Kaftanovski 1951).

Throughout range, diet of young dominated by fish, especially butterfish and, in Baltic, viviparous blenny. On Fair Isle (Scotland), entirely fish: of 544 items brought in by 13 pairs, 46·9% butterfish, 17·4% sand-eels, 18·4% Gadidae, 8·1% sea scorpions, 7·0% flatfish, 2·2% unidentified; butterfish comprised 12–79% (by number of items) of total diet of individual chicks (Slater and Slater 1972a; P G H Evans). In Denmark, young fed on 98–99% fish (by numbers), of which 67% butterfish, 22% sand-eels, 7% wrasse, 4% viviparous blenny; rest (1–2%) crustaceans (Asbirk 1979a). In Finland, 95% viviparous blennies, 5% sand-eels (Bergman 1971a). In Norway, 41% butterfish (R T Barrett). At Flatey, 94% fish (12 species), mainly butterfish ($\frac{1}{3}$–$\frac{1}{2}$ by number and weight), bull rout, and sand-eels; also a few hermit crabs and shrimps. Young chicks fed proportionately more sand-eels than older chicks (Petersen 1981). In Barents Sea, diet of young (55 stomachs) did not differ markedly from adults (see above), except in containing relatively more cod (57·6%), and relatively less butterfish and other littoral–benthic species (5·1%), herring (10·2%), and capelin (3·4%); chicks fed by parents exclusively on crustaceans starved to death (Belopol'ski 1957). At Sem' ostrovov July–August, 503 items for young comprised 81·8% fish, of which 45·3% *Ammodytes tobianus*, 42·4% cod *Gadus*, 1·4% bullheads, 0·2% butterfish, 10·7% unidentified; of 22 stomachs, 4 contained prawns, 18 cod, haddock, sand-eels, bullheads, and butterfish (see Kaftanovski 1951). At Kandalaksha Bay (White Sea, USSR), exclusively fish: 356 collar samples gave 52·2% gobies, 34·3% lumpsuckers, 6·5% cod, 3·4% herring, 0·8% sand-eels, 1·1% butterfish, 1·7% unidentified (Bianki 1967). For diet in Nearctic, see especially Preston (1968) and Cairns (1978). Prey size increases with age of young (Hyde 1937): chicks less than 10 days old have difficulty ingesting deeper-bodied fish of

10–16 cm; mean prey size at 1 day c. 10 cm, at 30 days c. 15·3 cm (Asbirk 1979a); mean c. 3 cm at 0–4 days, c. 10 cm at 35–39 days (Petersen 1981). Prey size varies with species: at Flatey, butterfish 7·5–23 cm ($n = 334$), bull rout 5–12·5 ($n = 100$), sand-eels 9–19 ($n = 88$), blennies 12–29 ($n = 6$) (Petersen 1981). At Kandalaksha Bay, butterfish 14·5 cm (13–16), sand-eels 8 (7·5–8·3), gobies 9·3 (6·0–12·3), cod 10·7 (10–12), herring 13 (11·2–14·5), lumpsuckers 13 (6–23·4) (Bianki 1967). At Sem' ostrovov, butterfish up to 17 cm, sand-eels up to 16 cm, cod 10–16 cm (Kaftanovski 1951). At Flatey, broods of 1 received 13·9 feeds per 24 hrs, broods of 2 received 18·8; feeding rate increased with age up to day 25–29, then declined (Petersen 1981). At Kandalaksha Bay, (7–)11–15 fish per chick per 24 hrs, depending on day-length and age of chick (see Bianki 1967). At colony, Denmark, 0·57 feeds per hr to broods of 1, 0·44 per chick per hr to 2, 0·46 to 3; rate for broods of 2 constant at 16·5 feeds per day at all ages, yielding c. 85 g per day for 5-day-old young, c. 165 at 30 days, c. 265 at 40 days, greater input with age resulting from increasing prey size (Asbirk 1979a). For other feeding rates (as low as 5·3 feeds per nest per 24 hrs in years of food scarcity), see Bergman (1971a). ♂ generally takes greater share in feeding young: at unspecified number of nests, ♂♂ brought 134 items, ♀♀ 31 (Preston 1968); ♂♂ 75, ♀♀ 55 (Asbirk 1979a). Feeding peaks at 05.00–06.00 and 19.00–22.00 hrs; in July, Iceland, continuous throughout night, though by August not in middle of night (Petersen 1981). EKD

Social pattern and behaviour. Based on studies by Asbirk (1979a, c), Petersen (1981), and notes compiled by A Petersen, and, for Nearctic, Preston (1968) and Cairns (1978). For closely related Pigeon Guillemot *C. columba*, see Thoreson and Booth (1958) and Drent (1965); also Storer (1952).

1. Only moderately gregarious compared with other Alcidae. Little information on dispersion outside breeding season. Encountered singly, in twos, and in small flocks at end of February, larger flocks March–April (Darling 1938). In winter often found singly (Kaftanovski 1951; Kozlova 1957), but, according to Uspenski (1956), remains in pairs or small flocks throughout the year. After fledging, young disperse independently of parents (Kaftanovski 1951; Uspenski 1956; Bianki 1967; Salomonsen 1967a; Asbirk 1979a; Petersen 1981), single juveniles sometimes accompanied by adult at sea—but kinship not known (P G H Evans). After breeding, moulting (flightless) adults, and possibly juveniles, congregate in small flocks at sea; on 26 August, in one small area c. 25 km south of breeding area Flatey Island (Iceland), 4 flocks of 20–30 birds (Petersen 1981). BONDS. Monogamous mating system, pair-bond maintained from year to year. Annual divorce rate 5% (Petersen 1981), 7% (Asbirk 1979a). Among 16 pairs, 2 paired for at least 3 years, 4 for 4 years (Petersen 1981). Of 28 pairs, 18 remained intact over 2 years, and of the 10 changed pairs, 4 suffered loss of mate, 2 clearly divorced. 5 pairs intact for at least 3 years, 1 pair for at least 4 years (Asbirk 1979a). No reliable information on bonds outside breeding season, though some pairs perhaps in contact at all times (see above). First breeding at (2–)4 years; in 6 years, Flatey, only one 2-year-old known to breed (Petersen 1981). Reports (e.g. Bianki 1967) of regular breeding at 1–2 debatable without evidence of nesting. Average age at first breeding may change with level of local population (Petersen 1981). Small numbers of 1-year-olds attend colonies (Asbirk 1979a). At Flatey, c. 17% of site-holders were immature non-breeders, mostly 2 years old; over 3 years, 7·8–18·2% of maximum numbers at colony were non-breeders (Petersen 1981). Both sexes care for young in nest, though ♂ perhaps does more feeding than ♀ (Preston 1968; Asbirk 1979a); ♀ thus thought to do more brooding, though ♂ found to do most on Fair Isle, Scotland (R Broad). Young usually independent on leaving colony. BREEDING DISPERSION. Typically in small, loose colonies, though solitary nesting occurs. Occupancy rate of available sites suggests gregarious nesting tendency (Cairns 1980). On criterion that birds nesting 10 m or more apart are solitary, 37% of pairs solitary, 63% in colonies of 2–28 ($n = 411$) (Asbirk 1979a); colonies, Denmark, up to c. 170 pairs, most less than 20 (Asbirk 1978). Mean colony size, Flatey, 1977, 28 pairs (4–133, $n = 15$) (from Petersen 1981). Larger aggregations reported extralimitally: c. 2000 pairs, Greenland (Salomonsen 1950–1), c. 2000–10000 pairs, arctic Canada (Brown *et al.* 1975)—invariably comprise smaller colonies. Distance to nearest neighbour, Kent Island (Canada), c. 1 m to over 150 m (Preston 1968). Traditional nest-sites used annually, and site tenacity marked (Kaftanovski 1951; Bianki 1967). About half the sites used in each of 2 areas were occupied every year; 68% of adults returned to nest-site of previous year (Preston 1968). Of 39 birds, 22 (56%) held same site in 2 years; 48 (79%) out of 61 in another 2 years; 15 held site for 3 years, 3 for 4, 1 for 5 (Asbirk 1979a). Of 91 bird-years, Flatey, only 9 (10%) involved site change, 1 bird moving twice; mean distance moved 12·4 m (2·1–28·4), always within same nesting group (Petersen 1981). Distance moved 0·5–150 m ($n = 34$) but in only 4 cases exceeded 30 m (Asbirk 1979a, which see for nearest-neighbour distances). Successful pairs move less than unsuccessful ones (Petersen 1981). Parents defend nest from conspecific birds to radius of several tens of cm (Bianki 1967), usually embracing habitual perch-site (Preston 1968; see Roosting, below). Much courtship and all other breeding activities confined to nest-area territory (Preston 1968). Site-holders losing eggs, or not laying, continue to defend nest-site throughout season, but not as continuously present as successful breeders (Preston 1968). Nesting with gulls or terns (Lari) found in 7 out of 8 skerries investigated; birds thought to benefit from greater pugnacity of Lari towards predators (Nordberg 1950). ROOSTING AND COLONY-ATTENDANCE PATTERNS. During breeding season, communal roosts and loafing groups of adults and older non-breeders form at edge of colony, on land or water, especially early in season, during chick-rearing period, and at end of season; location of roost shifts with tide; throughout season, both members of pair also loaf on perch-site near nest (Bianki 1967; Preston 1968; Slater and Slater 1972a; Asbirk 1979a; Petersen 1981). On land, behaviour in flock mainly resting, comfort activity, courtship, and copulation (Preston 1968). Resting bird typically raises feet from ground, inserting them into belly feathers. 1-year-olds more nervous, spending most time in water, landing only when other birds present; adopt resting posture less often and typically sit apart from others (Preston 1968). Early in season, birds came ashore early in morning and assembled on shore in groups of 10–20 (Kaftanovski 1951); initial landings coincided with high tide (Preston 1968). In early stages, most birds leave colony area at night to roost on water, thus abandoning eggs during laying (Bergman 1971a; Petersen 1981); return before sunrise to copulate and prospect sites (Asbirk 1979a). As laying proceeds, increasing numbers of birds spend night on foreshore, and roosting at sea then presumably mostly by non-breeders (Petersen 1981). Once nesting started, attendance typically peaks early in morning, decreasing

after noon; remains low in afternoon, building up in evening to peak around sunset; most off-duty birds and non-breeders continue to roost on sea (Uspenski 1956; Preston 1968; Bergman 1971a; Slater and Slater 1972a; Ramsay 1976; Asbirk 1979a; Petersen 1981). For details of nest-site attendance in pre-laying, laying, and incubation periods, see Bianki (1967). Adults with older young in nest roost at sea, returning early morning (Asbirk 1979a). Level of attendance inversely proportional to temperature, wind speed, and length of low-tide period (Petersen 1981). Peak attendance in incubation period on calm, sunny days (Bergman 1971a). Attendance increased at high tide (Preston 1968; Cairns 1978), and feeding rate in some colonies also highest around high tide (Preston 1968).

2. Alarmed or excited birds on water often Bill-dip, immersing head to above eyes. Birds, often with fish, threatened by gulls or crows (Corvidae), or those alerted by intense Alarm-call (see 1 in Voice) of another, escape by flying, diving, or both; often 'skitter' across water, using wings (Bianki 1967; Preston 1968; Asbirk 1979c; Petersen 1981). When Peregrine *Falco peregrinus* flew through flock, birds plunged into sea from flight (Darling 1938). FLOCK BEHAVIOUR. At beginning of season until laying (at Flatey, from 9–10 weeks before onset of laying: Petersen 1981), flocks (rafts) gather at sea off colony area. Unlike other Alcidae in which attendance cyclical, rafts at Flatey well attended every day unless weather bad; contained (in different years) 1·1–3·4% 1-year-olds, estimated 5·4–14·8% 2 years old and older immatures, rest prospective breeders (Petersen 1981). At Kent Island, initially assembled only for a few hours around high tide before dispersing, and did not approach shore within c. 30 m. First landings brief (c. 2 hrs), communal, and ended by mass flights to water (Preston 1968). During pre-laying period, Gulf of St Lawrence, up to several hundred birds periodically engaged in mass flights lasting several minutes; rose in tight formation high above water (Cairns 1978); at Fair Isle, flock usually landed a little way out and gradually came ashore (Slater and Slater 1972a). Such flights said to indicate nervousness (Preston 1968); homologous with mass flights of guillemots *Uria* (e.g. Tuck 1961a), and possibly with 'dreads' of Lari. Antagonistic encounters and courtship occur in flock from outset (see below). Several birds (10–30: A Petersen) may float or swim line-abreast; formation-swimming considered distinct display by Darling (1938) but chance event by Preston (1968) and A Petersen. ANTAGONISTIC BEHAVIOUR. (1) General. Wide variety incorporates different levels of intensity. Used in defence of individual distance or fixed site; especially evident during nest-prospecting when established pairs rebuff intruders, or pairs contest same site. Birds in communal roost or raft frequently fly to nest or perch to drive off intruder. Though ♂♂ much more aggressive than ♀♀, no differences between sexes in relative frequency of different displays; nor do breeders differ from non-breeders (Preston 1968). Descriptions mainly from Preston (1968) and Asbirk (1979c). (2) Threat. Pair or single bird may confront intruder in Hunched-posture while Head-tossing and calling (see 2 in Voice): bird usually lies on tarsi, tail cocked, wings close to body, neck withdrawn, head often pointed at intruder and repeatedly tossed upwards and backwards, bill closing on downswing and opening on upswing (Fig A), exposing red gape. Typically, head moved through progressively smaller arc so that gape eventually

directed steadily upwards at c. 30°. Bird may also Head-toss at mate; this the only form occurring both between mates and non-mates (Preston 1968; Asbirk 1979c). Head-tossing expresses mild threat (Preston 1968); homologous with 'Hunch-whistle' (Drent 1965) of *C. columba*. In higher-intensity threat (Preston 1968), adopts Oblique Hunched-posture (Fig B), either standing or walking towards intruder. Neck withdrawn and opened bill pointed towards opponent, but angled 30–45° below horizontal (Asbirk 1979c). Bill typically open but no call given; feathers of crown and nape raised slightly, carpal joints lifted, secondaries raised and spread, and wing-tips sometimes trailed along ground. When performed on water, bird adopts a hunched posture with raised carpal joints and spread secondaries (Preston 1968; Asbirk 1979c). 1-year-olds on land may adopt Oblique Hunched-posture when approached by others of same age (Preston 1968). Incubating bird may warn off approaching intruder in a hunched posture, neck withdrawn, wings lowered, and head pointed slightly upwards (eyes tracking intruder), uttering piping notes (Asbirk 1979c); similar to Squat-peeping posture and call (see 3 in Voice) which is, however, not exclusive to nest. Squat-peeping bird sits horizontally on rock, with nape, crown, and secondary feathers raised, tail often somewhat depressed and apparently bobbing in time with calling; function obscure but typically performed after retreat of rival (Preston 1968). In confrontations, head may periodically be turned to rest on scapulars (Head-turning, likened to Head-flagging in Laridae: Preston 1968). In highest-intensity threat, bird stands on tarsi in Oblique-upright posture ('Twitter-waggle': Drent 1965: Fig C)—neck stretched forward, bill angled down; wings lifted from sides of body and may be held stiffly over back, exposing white undersides; back plumage ruffled and tail cocked. Gives Twittering-call (see 4 in Voice) given with bill open. Bird Head-shakes vigorously (Preston 1968; Asbirk 1979c)—more frequently on alien than on own territory, suggesting appeasement (Preston 1968); occasionally performed on water, body horizontal. If bird adopts Oblique-upright posture with back to attacker, usually appeases latter (Asbirk 1979c). (3) Attack and fighting. In outright attack, bird lunges with neck outstretched, head low, bill open, giving Alarm-call. Wings lifted from sides of body and sometimes upraised (Fig D). Intruder may be grasped and bitten, usually on head or neck; attacking bird also beats its wings and tries to scratch with claws. Victim

A

C

D

E

typically retaliates and birds often tumble about. If attack made on water, victim often dives and may be pursued under (Asbirk 1979c), stimulating others to dive (Armstrong 1947). Fighting birds may leap up from water. Single or paired adults often attack 1-year-olds attempting to approach them (Preston 1968). Lunge may culminate in flying attack; victim often pursued over long distance ('Duet Flight': Drent 1965) on wide circular path, usually higher than normal flight; birds twist, turn, and swoop in unison, sometimes switching lead, pursuer landing on water when other does (Preston 1968). In behaviour of debatable function, but considered antagonistic (participants usually not mates) by Preston (1968), bird on water Leap-frogs another, just in front. Trailing bird takes off with whirring wings, neck extended, head angled upwards, and legs trailing; raising wings stiffly over back, then glides over, or parallel to other bird, splashing down in front. Wings typically held up for a few seconds after landing. Though bill kept open during Leap-frog, no call apparently given. Often occurs within flocks, especially during chasing (see Heterosexual Behaviour, below) (Preston 1968; Asbirk 1979c). 2 birds Leap-frogged each other repeatedly, Bill-dipping after bout over (Williamson 1951). Leap-frogging occurs early in season during interactions between members of pair; when 1 member of apparent pair began to Head-toss at partner, latter Leap-frogged in c. 25% of cases. However, most often occurs during Strutting-display on water (see Heterosexual Behaviour, below) in which participants usually not paired. Leap-frogging sometimes performed by birds moving from one rock to another during prolonged bout of display (Preston 1968). HETEROSEXUAL BEHAVIOUR. Mostly from Preston (1968) and Asbirk (1979c). (1) Communal display and pair-bonding behaviour. Though communal display most common early in season (on water), occurs throughout season (Asbirk 1979c), and simple courtship function not easily ascribed. Frequent interchange of partners and roles. Normally only one member of pair takes part, and if both participate they appear to behave as individuals (Asbirk 1979c). No 1-year-olds involved, and functions of assemblage may thus include mate selection and pairing among birds of breeding age (Preston 1968). May be prolonged, aquatic bouts lasting up to 15 min. On land, main feature is Strutting-display (Fig E). Participants usually not paired to any bird (Preston 1968). Strutting bird walks on toes with body upright, using distinctive high-stepping gait. Neck upstretched and slightly bowed, head directed forward or towards another bird (which may be an aggressor); carpal joints may be lifted or wings raised stiffly over back, tail cocked, mouth held open (Asbirk 1979c), or opened and closed rapidly (Preston 1968) as Strutting-call (see 5 in Voice) repeated, breast heaving with calling. 2 birds may thus Strut, one apparently pursuing other; pursuer tends to expose more of white wing-patch than pursued bird (A Petersen). Participants may stand, upstretched and breast-to-breast for minutes at a time, calling continuously. On water, Strutting-display basically similar; as bird rears up, hind-body may submerge so that bird appears to plough through water; early in season, sometimes a response to Head-tossing (Preston 1968). On land, Strutting may be initiated by 2 birds from adjacent territories, also by non-breeding bird (Asbirk 1979c). 2 birds close to each other often Strut when a 3rd approaches (A Petersen). Strutting highly contagious: birds may fly up to c. 75 m to join in. 15–20 birds frequently perform together in communal roost (Asbirk 1979c). Early in season, 84% of performances involved 2–6 birds, rest 7–11 or more. Also occurs later in season when birds on shore, or following alarm; apparently usually associated with crowding (Preston 1968). If one bird Struts in water towards another, latter assumes like posture and often swims ahead, looking back (Fig F); may turn and swim towards other bird; when they meet, both Strut in parallel, or one may dive or lead chase, often under water, stimulating widespread chasing in flock. Birds swimming on surface dive rapidly when approached from under water. As pairing advances, members of pair—especially after brief separation—indulge in frequent mutual Head-bobbing ('Billing': Drent 1965): lying, standing, or afloat, partners face each other closely and bob heads up and down irregularly but continuously; bills occasionally interlock (Witherby et al. 1941); if bird standing, tail cocked. Bill may be opened and closed with Staccato-call (see 6 in Voice) or may remain slightly open throughout; breast heaves with calling. Variations in intensity of Head-bobbing conspicuous. After copulations cease, Head-bobbing is the only display between members of pair. Occasionally performed by 1-year-olds (Preston 1968). Bird of either sex may, while Head-bobbing, or holding head steady, begin to move round its partner ('Circling': Drent 1965), giving call similar to Staccato-call (Preston 1968; Asbirk 1979c). (2) Mating. Always on land, usually after bout of Head-bobbing. Prior to copulation, ♂ stands upright with neck erect and bill downwards, giving Staccato-call. Thus begins to Strut and circle around ♀ who may also call (not described). ♀ responds by rising and circling ♂ with body hunched (Fig G). If rebuffing ♂, ♀ soon lies down again ('Settling': Preston 1968) with upright neck and bill pointing towards ♂. If receptive, ♀, after longer circling, settles with neck stretched along ground, tail cocked; ♂ may push against ♀'s side with his breast. ♀ ceases calling; ♂ may cease calling before or as he begins to mount. ♂ mounts and tramples; may raise or beat wings, but not while treading or making cloacal contact; however may lift carpal joints and droop wing-tips. ♀ may raise head to normal resting position

or direct bill upwards toward ♂ (Preston 1968; Asbirk 1979c). Occasionally ♂ grasps or pecks back of ♀'s head. After cloacal contact, ♂ may move forward and resume treading; ♀ may shrug ♂ off, sometimes when ♂ pecks back of her head. Mounting lasts from a few seconds (no cloacal contact) to over 2 min (6 or more cloacal contacts). ♂ deliberately dismounted in 12% of matings, these usually over 2 min. After mating, pair lie down near each other and preen, Head-bobbing occasionally (Preston 1968; Asbirk 1979c). First copulations recorded before birds had definitely acquired nest-sites, and 3–4 weeks before laying (Preston 1968); on Kharlov Island (USSR), began 2 weeks before laying (Belopol'ski 1957); at Flatey, 3 weeks before (A Petersen). Most frequent mid-morning and mid-afternoon (Winn 1950). Decreases rapidly in frequency after laying, when ♀ more likely to refuse ♂; with 2-egg clutches, not seen after laying of 1st egg, but up to 12 days after laying single egg (Preston 1968). (3) Behaviour at nest. Single and paired birds prospecting nest-sites often Head-toss inside cavity (Preston 1968). On 4 occasions, side-throwing movements seen outside nest—straws taken in bill and passed along sides of body, as in Lari (Asbirk 1979c). At nest relief, especially early in laying period, birds Head-bob; begin bout with bills open (high intensity), but eventually closed (Preston 1968). RELATIONS WITHIN FAMILY GROUP. Young brooded, mostly at night, for 3–6 days (Kaftanovski 1951; Bianki 1967); receive food during 24 hrs after hatching (Bianki 1967; Asbirk 1979c; Petersen 1981; contra, e.g., Kartashev 1960, Preston 1968). Young start calling on hearing approach of any adult (Slater and Slater 1972a; Petersen 1981). Parent usually passes fish directly (head first) to chick, but sometimes (possibly when disturbed: A Petersen) dropped into nest-cavity (Slater and Slater 1972a). Mate may try to seize food. Older chicks come to nest entrance as adult approaches (Petersen 1981). Young fed right up to departure; latterly make numerous short excursions from nest to exercise wings, usually after sunset (Cairns 1978), around midnight (Petersen 1981). Departing young (unable to fly) scramble over rocks to sea, or drop from cliff; not accompanied and, contra Winn (1950), no evidence that parents lure chick to water with fish (Cairns 1978). Return of parents (with food) to nest, whether after departure of single chick, or to attend unfledged 2nd chick, further suggests independence of young upon leaving nest (Bianki 1967; Slater and Slater 1972a;

Asbirk 1979a; Petersen 1981). Parents quit site shortly after brood (Cairns 1978), c. 1 week after (A Petersen). ANTI-PREDATOR RESPONSES OF YOUNG. On approach of human, young in nest retreat into corner. Just before ready to leave nest, chick avoiding capture tries to escape over rocks rather than seek refuge; once in sea, dives if pursued (Kaftanovski 1951; Bianki 1967). PARENTAL ANTI-PREDATOR STRATEGIES. (1) Passive measures. Alert-posture often associated with presence of gulls (Laridae): bird stands upright with neck extended, plumage sleeked, wings close to body; makes quick sideways turns of head, with pauses between turns (Preston 1968; Asbirk 1979c). When human approaches colony, birds leave nests, sitting and turning nervously on rocks, or swimming nearby. Alarm-call given with body horizontal, but neck typically upstretched; in more serious disturbance, may adopt Alert-posture then take flight (Kaftanovski 1951; Asbirk 1979c). At beginning of incubation, easily flushed off nest but gradually becomes more bold and may allow itself to be touched (Kaftanovski 1951). (2) Active measures. Adults disturbed at nest said to give Hissing-call (Bent 1919: see 7 in Voice).

(Fig F from Preston 1968; others from Asbirk 1979c.) EKD

Voice. In breeding season, vocal on land, water, and, less often, in flight. Difficult to render phonetically; variants of 'peeeeee', described as shrill and complaining (Nicholson 1930), high-pitched and plaintive (Darling 1938), or shrill and feeble (Witherby et al. 1941). Differences between calls based mainly on variations in temporal patterning and duration of individual notes, consisting of thin peeps or whistles (Preston 1968). Following scheme thus tentative. No information on calls outside breeding season.

CALLS OF ADULTS. (1) Alarm-call. A high-pitched, prolonged piping given with gaping bill, from the ground, sometimes in flight (Asbirk 1977c: Fig I); a long shrill, and sometimes wavering whistle (Preston 1968); a long, mewing call (Slater and Slater 1972a). Given especially when suddenly scared by large gull Larus or crow Corvus, or in fight with conspecific bird (Asbirk 1979c). (2) Head-

I J-C Roché (1970) Finland May 1964

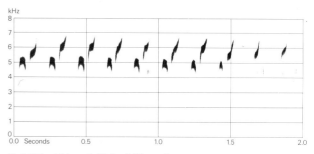

II J-C Roché (1970) Finland May 1964

tossing call. A rising and falling piping, given during Head-tossing (Asbirk 1979c); a single peep given in time with upward swing of head, and becoming progressively longer (Preston 1968); in bout of Head-tossing, effect is a rhythmical series of short mewing sounds (Slater and Slater 1972a). (3) Squat-peeping call. A series of short thin peeps uttered with bill open in Squat-peeping posture (Preston 1968). (4) Twittering-call. Whistle repeated so as to become a twitter (Witherby et al. 1941: Fig II). Given with open bill in Oblique-upright posture (Asbirk 1979c). May be same as call 3 or, at most, slight variant. (5) Strutting-call. A loud, disyllabic peeping, directed by one bird at another in communal display, especially during Strutting-display. Function obscure (see Social Pattern and Behaviour) (Preston 1968; Asbirk 1979c). (6) Staccato-call. A series of quick staccato sounds given by both sexes during Head-bobbing, and by ♂ as he Struts round and circles around ♀ prior to copulation (Preston 1968; Asbirk 1979c); probably the 'i-i-i' of Demme (1934). (7) Hissing-call. A hissing sound of protest when disturbed at nest (Bent 1919); not described by other sources.

CALLS OF YOUNG. Food-call similar to high-pitched whistle of adult (Nicholson 1930); shriller, but less clear, than adult whistle (Slater and Slater 1972a). EKD

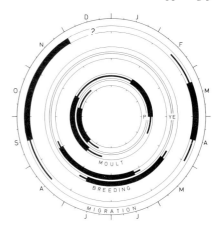

Breeding. SEASON. See diagram for Iceland and Scandinavia; up to 2 weeks later in Spitsbergen and Novaya Zemlya (Uspenski 1956; Løvenskiold 1964; Makatsch 1974). SITE. Boulder areas and screes, and crevices in low cliffs; also wide range of natural and artificial holes (see Petersen 1981). Of 411 nests, Denmark, 57% under or between stones, under fish-boxes and driftwood. Colonial to solitary. Nests re-used in subsequent years (Asbirk 1979a). Usually near sea but can be 2–3 km inland (Birulya 1910). Nest: slight hollow, or none, sometimes with apparent lining of debris, but this probably not constructed; all 212 nests in 3 colonies, Canada, had eggs laid on soil, small stones or organic matter, and none on bare rock. Average diameter of 212 nests in 3 colonies, Canada, 13·4–15·2 cm, depth 7·5–10·2 cm (Cairns 1980). EGGS. See Plate 92. Oval or pointed oval, smooth, not glossy; white, often tinged cream, buff, or blue-green, spotted and blotched black, red-brown, and grey. Nominate *grylle*: 59×40 mm ($53-68 \times 35-43$), $n = 150$; calculated weight 50 g (Schönwetter 1967); no significant variation in *mandtii* or *arcticus*. Iceland: 50 g (42–60), $n = 98$ (Petersen 1981). Denmark: 48 g (39–57), $n = 106$; weight losses during incubation averaged $10·9 \pm$ SD $3·1\%$, $n = 55$ (Asbirk 1979a). 1st egg 2·2% heavier than 2nd egg (Petersen 1981). Clutch: 1–2, rarely 3–4 and then almost certainly by 2 ♀♀. Of 935 clutches, Shetland, average 1·43 (R Broad); of 194 clutches, USSR, 2·0 (Bianki 1967). Clutches of first-time breeders average smaller than older birds (Petersen 1981). Laying interval 3 days (Preston 1968); 2–5 days (Hyde 1937). Replacements laid after early egg loss, but not if incubation of 1st clutch advanced (Petersen 1981); average re-laying interval 14·7 days (12–23), $n = 28$ (Asbirk 1979a). Replacements of 2 eggs laid. INCUBATION. 23–40 days, varying for 1st and 2nd egg. For 1st egg, $31·1 \pm$ SD 2·9 days (23–39), $n = 38$; for 2nd egg, $28·5 \pm$ SD 2·3 days (25–36), $n = 41$ (Asbirk 1979a). By both sexes in approximately equal shares; average length of shift in 4 nests 16·0–48·5 min. Proportion of time for which egg covered lower (73%) in 1-egg clutches than in 2 (84%) (Preston 1968). Full incubation does not start until after 2nd egg laid; in nests with 1 egg, starts 4–5 days after laying; eggs usually hatch less than 24 hrs apart (Preston 1968). YOUNG. Semi-altricial and nidicolous. Cared for and fed by both parents, and brooded while small. FLEDGING TO MATURITY. Fledging period $39·5 \pm$ SD 3·9 days (31–51), $n = 37$; usually 1–2 days between chicks, but up to 8 days recorded (Asbirk 1979a). First breeding at (2–)4 years. BREEDING SUCCESS. Of 683 eggs laid, Denmark, 59·0% hatched, and 54·8% of chicks raised to fledging; main identifiable losses of eggs are predation and flooding by tide; of chicks, predation and starvation. Overall breeding success (chicks fledged/eggs laid) increased from 30% for birds in 1st year of study, to 36% for same birds 1 year later, to 42% for same birds 2–5 years later; fledging and overall success higher in colonially breeding birds than in solitary pairs, though hatching rate similar (Asbirk 1979a). Of 935 eggs laid, Iceland, 79·5% hatched and 89·2% of chicks raised to fledging; success greater from larger eggs, and therefore from older birds; flooding probably most important 'independent' factor in breeding success, but mink *Mustela vison* and rats *Rattus* major causes of declines and colony desertions (Petersen 1981). Preston (1968) found breeding success lower in areas of high nest density, concluding that higher aggression rates were the cause, but Cairns (1980) found no link between density and breeding success, though aggression was more frequent in high density areas.

Plumages (*C. g. mandtii*). ADULT BREEDING. Brownish-black above and below; in fresh plumage, upperparts, sides of neck, and breast with greenish gloss. Tail black. Outer web of flight-feathers black, distal half of inner web black, tapering along shaft and margin; remaining part of inner web white. Occasionally, a few inner secondaries narrowly tipped white. Shafts of flight-feathers black, becoming horn-coloured towards base. Greater upper primary coverts black, innermost frequently with variable amount of white on inner, outer, or both webs; median upper primary coverts black, but innermost show varying amount of white. Lesser and innermost median upper wing-coverts black; greater and remainder of median coverts white, forming white patch on middle of wing. Under wing-coverts entirely white, but marginal coverts brown-black, and sometimes a few outermost primary coverts with brown spot at tip of outer web. Axillaries white. In 'motzfeldi' variant, wing-patch black instead of white, or only slightly paler than other black feathers; recorded from northern Norway, Outer Hebrides (Scotland), western Greenland, Iceland, and Spitsbergen (Salomonsen 1944; RMNH). In winter plumage, this variant shows considerable variation, some specimens only differing from normal non-breeding plumage by their black wing-patches, others being entirely black or showing partly melanistic non-breeding plumage (Salomonsen 1941). Entirely albinistic specimen collected Greenland (RMNH), and another recorded from Dutch coast (Voous 1955). ADULT NON-BREEDING. Head and neck pure white; due to wear of tips, brown-black feather-bases may become visible. Mantle mottled black and white, due to large white tips on black-brown feathers; in fresh plumage, white tips almost conceal dark feather-bases. Rump pure white. Upper scapulars entirely white with dark shafts or with variable amount of black along shaft; lower scapulars mainly black, becoming white at base. Underparts entirely white. Tail and wings as in adult breeding. DOWNY YOUNG (*C. g. arcticus*). Black-brown, underparts slightly paler. Natal down replaced by juvenile feathers July–August, extending to early September (Salomonsen 1944); starts at *c.* 10 days old and completed at *c.* 25–35 days (Winn 1950; Kartashev 1960; Bianki 1967). JUVENILE. Forehead, lores, and mantle black-brown. Head, neck, back, and rump variably mottled black and white. Scapulars black, upper ones with grey-brown base, white middle part, and black tips, giving mottled appearance. Upper tail-coverts white with narrow black tips. Underparts mottled black and white due to narrow black tips to white feathers; throat and chin almost white. Tail and wing as in adult, but secondaries and greater primary coverts tipped white to varying extent (tips absent in other races—see Geographical Variation); feathers of wing-patch white with brown-black tips, patch appearing barred or mottled. Under wing-coverts white, some faintly tipped brown. Axillaries white. FIRST IMMATURE NON-BREEDING. Like adult non-breeding, but wing still juvenile, feathers of wing-patch white with black-brown tips. A good many black-tipped juvenile feathers sometimes retained on underparts, and sometimes also some upper tail-coverts. FIRST IMMATURE BREEDING. Like adult, but wing still juvenile, wing-patch as in 1st immature non-breeding, though black-brown tips more abraded. Juvenile flight-feathers heavily worn and faded, white tips to secondaries much abraded.

Bare parts. ADULT. Iris brown. Bill black, mouth vermilion. Leg and foot bright red. DOWNY YOUNG. Iris brown. Bill blackish, tinged red at gape and base of lower mandible; mouth pink. Leg and foot dark brown. JUVENILE AND FIRST IMMATURE NON-BREEDING. Mouth orange. Leg and foot reddish-brown. (Ridgway 1919; Witherby *et al.* 1941; Harrison 1975; Fjeldså 1977.)

Moults. ADULT POST-BREEDING. Complete. Early August to late November, sometimes as early as July. Flight-feathers shed simultaneously, some time after onset of body moult; tail-feathers also, but later than flight-feathers. ADULT PRE-BREEDING. Partial: all body, but not wing and tail; some white winter feathers sometimes retained on underparts. In *arcticus*, moult mid-December to late February(–March); in nominate *grylle* and *mandtii*, early March to late April. POST-JUVENILE. Partial: body only. Moult to immature non-breeding late August or early September to first half December. FIRST IMMATURE PRE-BREEDING. Partial: body only; white winter feathers sometimes retained on underparts. Later than adult pre-breeding: early March to late June (no information on possible racial differences). (Witherby *et al.* 1941; Salomonsen 1944; Dementiev and Gladkov 1951b; Asbirk 1979b). Subsequent moults as in adult.

Measurements. ADULT. Bill depth measured at frontal feathering. Except for toe, data from M J Leloup; nominate *grylle* after Salomonsen (1944).
Nominate *grylle*. Baltic Sea.

WING	♂ 174	(3·4; 16)	169–180	♀ 177	(4·8; 6)	169–182
BILL	32·8	(1·1; 16)	30–35	33·3	(1·2; 6)	32–35
DEPTH	11·5	(0·5; 17)	11–12	11·3	(0·5; 6)	10·5–12·0

C. g. mandtii. Spitsbergen, breeding; skins.

WING	♂ 171	(3·5; 21)	168–177	♀ 169	(3·6; 9)	164–174
TAIL	49·2	(2·2; 20)	46–55	48·9	(1·6; 7)	47–51
BILL	32·1	(1·6; 18)	29·7–35·9	31·3	(1·3; 11)	29·2–33·6
DEPTH	9·7	(0·5; 13)	9·0–10·8	9·2	(0·3; 9)	9·0–9·8
TARSUS	30·6	(0·7; 21)	28·9–31·7	30·5	(1·4; 11)	27·6–32·0
TOE	41·4	(2·8; 7)	37·5–46·1	41·1	(1·1; 7)	40·0–42·8

C. g. arcticus. Britain, Ireland, breeding; skins.

WING	♂ 164	(3·8; 13)	157–171	♀ 163	(2·8; 7)	159–166
TAIL	49·2	(2·0; 13)	45–52	49·2	(0·8; 5)	48–50
BILL	31·9	(1·5; 13)	29·7–34·5	30·9	(1·6; 6)	29·1–33·6
DEPTH	10·8	(0·6; 12)	10·0–11·8	10·3	(0·4; 6)	9·8–10·8
TARSUS	31·3	(0·6; 13)	30·1–32·6	30·4	(1·0; 7)	28·7–31·8

C. g. faeroeensis. Faeroes, breeding; skins.

WING	♂ 158	(3·2; 9)	153–162	♀ 159	(2·4; 8)	155–162
TAIL	47·3	(1·9; 6)	45–50	47·4	(1·4; 8)	46–50
BILL	31·3	(1·7; 9)	29·0–33·7	31·7	(2·1; 8)	27·5–34·5
DEPTH	10·6	(0·5; 8)	9·8–11·2	10·4	(0·5; 6)	9·6–11·0
TARSUS	31·5	(0·8; 9)	29·9–32·8	31·4	(1·1; 8)	30·1–32·8

C. g. islandicus. Iceland, breeding; skins.

WING	♂ 160	(4·4; 25)	152–168	♀ 160	(3·2; 21)	153–165
TAIL	48	(2·5; 25)	43–53	48·4	(2·2; 21)	45–52
BILL	29·8	(1·2; 22)	28·0–32·4	29·8	(1·4; 18)	26·4–31·2
DEPTH	10·2	(0·7; 23)	8·9–11·5	9·9	(0·7; 17)	8·5–11·2
TARSUS	31·1	(1·2; 24)	29·3–32·9	31·3	(0·9; 18)	29·7–32·3
TOE	—			41·1	(1·7; 4)	39·2–42·9

Sex difference significant for bill depth of *mandtii* and tarsus of *arcticus*.

Weights. ADULT. *C. g. mandtii*. (1) Spitsbergen, April–June (ZMA); (2) Novaya Zemlya, summer (Dementiev and Gladkov 1951b); (3) Novaya Zemlya, summer (Belopol'ski 1957).

(1)	♂ 382	(23·4; 10)	335–426	♀ 397	(15·1; 7)	382–418
(2)	385	(—; 11)	342–424	402	(—; 4)	378–466
(3)	391	(—; 16)	304–460	412	(—; 12)	377–446

C. g. arcticus. (4) Eastern Murmansk, April–August (Belopol'ski 1957); (5) Denmark, breeding season (Asbirk 1979a); (6) Jameson Land (eastern Greenland), May–August (ZMA); (7) White Sea, summer, sexes combined (Bianki 1967).

(4)	♂ 431	(—; 78)	347–534	♀ 434	(—; 42)	372–500
(5)	376	(23·8; 40)	—	380	(18·2; 24)	—
(6)	408	(26·3; 17)	355–465	400	(25·4; 9)	341–429
(7)	♂♀ 348	(42)	298–500			

YOUNG. At hatching: *arcticus* 34·3 (3) (Johnson 1944), 31 (2) (Winn 1950), *c*. 33 (Asbirk 1979*a*); race unknown 34 (28–41) (Kartashev 1960). On leaving nest: *arcticus*, New Brunswick (Canada), 35 days old, 382 (2) (Winn 1950); Denmark 366 (49·6; 108) (Asbirk 1979*a*). 1 day old: *arcticus* 34 (14) 28·5–38 (Bianki 1967); race unknown 31–33 days old, 320–360 (Kartashev 1960). In some years, maximum reached before leaving nest higher than average adult weight; leave at average *c*. 6% lower than maximum, or about adult weight (Asbirk 1979*a*; Bianki 1967).

Structure. Wing short and pointed. 11 primaries: p10 longest, p9 3–5 shorter, p8 8–10, p7 16–21, p6 21–29, p5 32–37, p1 63–70; p11 minute, concealed by primary coverts. Tail long compared with guillemots *Uria*, slightly rounded, feathers obtusely pointed; usually 12 feathers, occasionally 13–14 (Salomonsen 1944). Bill almost straight, no prominent gonydeal angle. Nostrils very narrow, only posterior end covered by feathers. Feathering on upper mandible extends to deep groove above nostrils; on side of lower mandible, feathering extends to much smaller groove near gape; on ventral surface of lower mandible feathering does not reach gonydeal angle. Outer toe *c*. 95% of middle, inner *c*. 78%; no hind toe.

Geographical variation. Complex; widely differing opinions concerning subspecific assignment of local populations. In adult non-breeding plumage, *arcticus* and nominate *grylle* have much less white on upperparts than *mandtii*; feathers of head and neck with large black bases and small white tips, resulting in pronounced mottled appearance; black patch in front of and below eye, sometimes also behind; mantle blacker due to smaller white feather-tips; amount of white on tips of rump feathers variable, resulting in rump either mainly white or mottled black and white; upper scapulars black with broad white tips, lower scapulars entirely black. In juveniles of nominate *grylle*, *arcticus*, *faeroeensis*, and *islandicus*, upperparts entirely black or with some mottling on neck and rump; upper scapulars with some small white spots, and feathers of wing-patch with larger black tips; crown, lores, and patch below and behind eye black. Compared with other populations, birds from Iceland and Faeroes are short-winged, those from Baltic long-winged. In high Arctic (north of 70°N), weak clinal increase in wing length from west to east. Birds from Baltic, however, on average have even longer wings than those from Wrangel Island (eastern Siberia). In western Greenland, cline of increasing wing length from north to south. Hardly any variation in wing length along western coast of Atlantic, including Hudson Bay, and Baffin and Ellesmere Islands. Clines of increasing bill length run along west coast of North Atlantic from Hudson Bay to Maine and from north-west to south-west Greenland. Large mean bill length in Baltic, rather short in Iceland. Variation in bill depth (at base) slight and irregular; averages rather small in Canada, eastern Greenland, and Spitsbergen, largest in Baltic. Tarsus increases clinally from northern Baffin Island to Maine; increases from north to south in western Greenland; largest in Baltic. Variation exists in amount of white on greater wing-coverts in wing-patch and extent of white on inner web of primaries. On eastern coast of North America, western Greenland, and eastern side of Atlantic, clines of increasing amount of white in wing-coverts run northward, birds from high Arctic having no or almost no black at base of feathers. In Iceland, however, black base of greater coverts is rather large and extends along margin of outer web as thin streak. Of other populations, birds from Faeroes show greatest extension of black. Northern birds average less white on p10 than birds from south; Faeroes birds are extreme in showing no white on p10. Also variation in non-breeding and juvenile plumage, birds from high Arctic in both plumages being whiter on upperparts than birds from south. Thus, geographical variation probably best expressed by recognition of 5 races: *C. g. grylle*, *mandtii*, *arcticus*, *faeroeensis*, and *islandicus*. Birds of arctic North America and north-western Greenland showing slightly more white in wing-patch in non-breeding plumage and having shorter bill sometimes separated as *ultimus* Salomonsen, 1944, but included here in *mandtii* as juvenile plumage similar, especially with respect to white tips on secondaries and primary coverts (Salomonsen 1944; Storer 1952; Vaurie 1965; Asbirk 1979*b*; M J Leloup, R Sluys).   RS

Alle alle Little Auk

PLATES 20, 22, and 23
[facing pages 231, 278, and 279]

Du. Kleine Alk   Fr. Mergule nain   Ge. Krabbentaucher
Ru. Люрик   Sp. Mérgulo marino   Sw. Alkekung   N. Am. Dovekie

Alca Alle Linnaeus, 1758

Polytypic. Nominate *alle* (Linnaeus, 1758), Ellesmere Island (Canada), Greenland, Grimsey (Iceland), Jan Mayen, Bear Island, Spitsbergen, and Novaya Zemlya; *polaris* Stenhouse, 1930, Franz Josef Land. Not certainly known which race breeds on Severnaya Zemlya and in North Pacific.

Field characters. 17–19 cm (bill only 1·5 cm); wing-span 40–48 cm. Actually $\frac{2}{3}$ length of southern race of Puffin *Fractercula arctica grabae*, but appears even smaller; has insignificant bill, smaller head, shorter tail, and much shorter wings (by 20%); close in size to Starling *Sturnus vulgaris*. Distinctive small snub-nosed auk, with compact build, and black-and-white plumage. Except in alarm or display, appears neckless; small bill invisible except at close range. Sexes similar; marked seasonal variation. Juvenile separable.

ADULT BREEDING. Head, neck, chest, upperparts, and tail glossy black, with brown tone on face and chest, and relieved only by tiny white fleck above eye, white fringes to scapulars (forming lines along sides of back), and white tips to secondaries (forming narrow but conspicuous bar across folded wing). Underparts below chest white. Underwing rather dark, but white patch in centre of wing-pit can combine with silvery undersurfaces to flight-feathers and coverts to give pale appearance in full light. ADULT WINTER. Brown-black feathers of chin, throat, lower cheeks, sides of neck, and chest become white, leaving ear-coverts as bulging dark patch below eye. Nape variably invaded with white feathers, only rarely forming narrow collar. JUVENILE. Resembles breeding adult (as in Razorbill *Alca torda*), but, at close range, paler throat, lack of distinct eye-fleck, and browner, black-margined feathers of upperparts allow separation. Underwing probably duller, with white patch in wing-pit smaller and under primary coverts mottled grey-brown. FIRST WINTER AND SUMMER. Upperparts browner, less glossed than in adult winter; white feathers on back of neck usually form narrow, rough collar. Underwing as juvenile and both flight-feathers and tail noticeably brown (especially when worn). At all ages, black bill small and stout (recalling that of Bullfinch *Pyrrhula pyrrhula*). Legs slate-grey to dull brown, with black webs.

Unmistakable at closer range but subject to confusion with other auks, particularly *F. arctica*, small waders (Charadrii), and even passerines at longer range or in poor light. Most frequent trap for unwary observer is juvenile *F. arctica*, which may be 10–15% smaller than parent, lacks deep bill, and has dark face, and can therefore suggest *A. alle*. Even at long range, however, *A. alle* lacks following characters of *F. arctica*: (1) relatively long body line, with heavy head; (2) darker underwing; (3) broader outer wing; (4) more forward-flapping, less backwards-flicking wing action. Important to obtain clear size comparison in identification of small auks (and this usually practical, since *A. alle* rarely occurs south of regular wintering range except during general displacement of seabirds). *A. alle* appears close in size to Dunlin *Calidris alpina*, whereas *F. arctica* is at least 50% larger. Flight has some of the character of other auks but is altogether more free, even than *F. arctica*. Rises directly from water, with none of the fluttering, splashing take-off obvious in *Uria* or *Alca*, and shows much more manoeuvrability over sea or at breeding colonies, with noticeably fast, flicked, yet fluid wing-beats, frequent changes in direction, and even steep dives all contributing to performance exceptional within Alcidae. Also much more agile amongst rocks and ledges, often loafing there; able to stand and walk freely on raised tarsi. Swims buoyantly, bobbing about on choppy water; when surprised or displaying, raises head and tail in more typically auk-like silhouette.

In western Europe, known mostly as erratic winter visitor to coastal waters and occasionally occurring in extensive wrecks, during which some overfly land and may appear anywhere. Markedly gregarious in north and during major displacements; most southern records are of widely scattered individuals. Chatters noisily at breeding colonies; usually silent in winter except when settled in sheltered waters.

Habitat. Most arctic of the Alcidae. Mainly in high Arctic in breeding season, where at any height from sea-level to *c*. 500 m, nesting in fissured precipices which may remain snow-covered for some time after laying, and even until hatching. Most colonies face open sea, but some are up fjords or even in valleys running inland. Majority, however, sheltered in loose talus rocks or scree broken from cliffs by frost action. Lower edge of colonies flanked by dense green mats of thick moss (e.g. *Dicranum*) and lichen (e.g. *Caloplaca elegans*), also lush growth of higher plants such as chickweed *Cerastium alpinum* and grass *Alopecurus alpinus* (Evans 1981), attracting Ptarmigan *Lagopus mutus* and other herbivorous animals (Bruemmer 1972). Breeding in high Arctic occurs in crevices in rock walls and much more commonly in unvegetated talus, usually spread as scree on slopes, but sometimes in eroded rubble at foot of cliffs. Commonly on slopes of 25–35° allowing easy take-off without involving site instability. Sites close to sea; at Hornsund (Spitsbergen), usually within 3 km and at most 6 km. Prefers areas with early snowmelt and relatively mild climate, with shelter from wind and absence of moisture (Stempniewicz 1981). On upper sides, colonies often bordered by nests of Glaucous Gull *Larus hyperboreus* which, along with arctic fox *Alopex lagopus*, is regular predator of eggs, adults, and young. Nests sited underneath boulders which are used also as vantage points. In low Arctic, colonies often very small, and located in outlying skerries where nests placed in talus near shore or in crevices in firm rock. Where large enough, sometimes shared with other Alcidae forming loose mixed colonies, whereas the much larger high-arctic colonies are very rarely mixed (Salomonsen 1950–1; P G H Evans).

Commonly loafs on ice-hummocks on border of pack-ice, covering them with reddish droppings. In spring, congregates offshore (where ice cover may range from *c*. 51–90%), generally avoiding totally iced-up and ice-free areas (Renaud *et al.* 1982). Way of life made possible only by enormously rich biomass of high-arctic coastal waters near breeding cliffs. Onset of winter gradually forces movement from open waters with dwindling food, also entirely iced-up waters, in which, however, late leavers are not infrequently trapped. Even in winter, follows cold currents and avoids warmer seas influenced by Gulf Stream. At sea, generally flies in lower airspace. While regularly and intensively exploited for food and clothing in some regions where breeding colonies are near primitive human settlements (see Population), finds relatively light competition in harsh but rich ecological niche. (Salomonsen 1950–1.)

Distribution. Some evidence that small colonies in south abandoned in 20th century, presumably due to climatic amelioration, e.g. in Greenland (Salomonsen 1967a) and Iceland, where apparently formerly bred Langanes, in

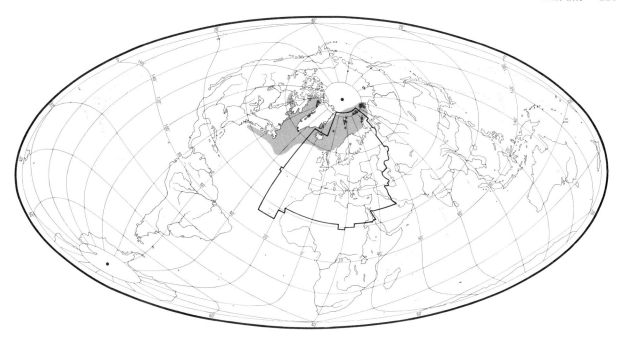
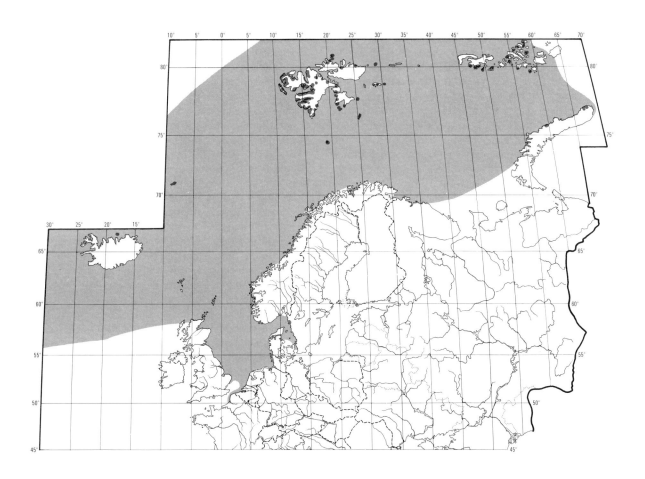

north-east (Timmerman 1938–49).

NORWAY. Possibly bred Finnmark 1981–2 (GL, VR).

Accidental. Finland, East Germany, Poland, Czechoslovakia, Italy, Malta, Spain, Portugal, Azores, Madeira.

Population. No accurate counts, but said to be many millions in more northern areas. Trends unknown, but massive human predation in Greenland thought unlikely to affect population (Salomonsen 1967a).
USSR. Franz Josef Land: said to breed in millions, with some colonies of over 100 000 birds (Norderhaug et al. 1977). Novaya Zemlya: only 3 colonies (Norderhaug et al. 1977); perhaps 5000 pairs 1969 (W R P Bourne). SPITSBERGEN. For details of colonies, see Løvenskiold (1964) and Norderhaug et al. (1977); no detailed counts, but former estimated over 1–2 million birds in Hornsund area, and latter gave order of abundance for many colonies. BEAR ISLAND. Breeds in small numbers (Løvenskiold 1964); at least 10 000 pairs 1980 (Luttik 1982). JAN MAYEN. Numerous (Clarke 1890); considerable numbers (Seligman and Willcox 1940); over 100 000 birds 1983 (J A van Franeker and C J Camphuysen). ICELAND. Nests only on Grimsey Island, where has declined since c. 1900, when 150–200 pairs (Hantsch 1905), to 5–10 pairs in recent years (AP).

Movements. Migratory and dispersive. Details (especially in relation to age) less well known than for other Atlantic Alcidae, since few ringing data available.
Important wintering areas among broken pack-ice from Barents Sea westwards into Norwegian Sea, through Denmark Strait to Davis Strait, and south to Grand Banks of Newfoundland. Latter especially important (for Baffin Bay population), and birds numerous offshore, notably north-eastwards, to 45°W; however, show definite tendency to avoid warmer water of Gulf Stream (Rankin and Duffey 1948). Also more or less regular, normally small numbers only, south to latitudes of Nova Scotia and northern New England, and to Scotland and Skagerrak, but will occur even further south in wrecks (see below), exceptionally to Caribbean and west Mediterranean. Few immatures summer on Newfoundland Banks or in low-arctic zone generally; majority move north into summer drift-ice of Baffin Bay and Norwegian and Barents Seas.
Juveniles and adults leave colonies during August, and most soon move offshore; autumn passage gradual, however, with numbers in wintering areas (e.g. Davis Strait and Newfoundland Banks) building up during October–December. Main return movement in April, with colony reoccupation in first half of May (later in bad ice years). But where part of population winters as close to colonies as ice conditions permit, vanguard arrives earlier; at Franz Josef Land and Spitsbergen, first birds occur inshore in late February and March (Gorbunov 1932; Løvenskiold 1964).
Birds from Baffin Bay colonies thought to winter especially off Labrador and Newfoundland, after moving south in Labrador Current (one Newfoundland recovery of a western Greenland bird); while species winters south-west Greenland (Egedesminde southwards), no Greenland-ringed bird found there in winter, though 13 from Spitsbergen (Salomonsen 1967a). Svalbard population evidently follows pack-ice belt of East Greenland Current; presumably winter in part in Norwegian Sea, but others pass through Denmark Strait (one Icelandic recovery) to south-west Greenland, where recoveries November–February (Norderhaug 1967; Salomonsen 1967a). Presumably east Greenland and Jan Mayen birds (possibly even some from USSR) are included among winter visitors to seas of southern Greenland and Iceland. Origins of birds wintering off Norwegian coasts, especially Finnmark, are unconfirmed, but probably from USSR sector of Polar Basin (Haftorn 1971); large westerly passage past Pechenga (Murmansk) in October, returning in April (Meinertzhagen 1938). Winter range of *polaris* (breeding Franz Josef Land and, perhaps this race, Severnaya Zemlya) unknown, though suggested by Vaurie (1965) to lie among pack-ice of Kara and Barents Seas; 2 long-winged January specimens from Shetland (Scotland) attributed to this population (*Bull. Br. Orn. Club* 1956, **76**, 107).
Most prone of Atlantic Alcidae to being blown inshore and even inland in gales, such wrecks being typically a feature of southern parts of winter range. These occurrences fairly frequent on small scale, and large wrecks occur occasionally (on both sides of Atlantic) involving hundreds or thousands of birds. While able to withstand storms at sea under normal conditions, prolonged gales may cause planktonic food to descend beyond reach; weakened birds then drift downwind, further south than normal, and become wrecked on meeting leeward coastline in continuing storm. See Murphy and Vogt (1933), Sergeant (1952), and Bateson (1961).

Food. Almost entirely planktonic crustaceans, especially for young in nest. Adult diet supplemented by small amounts of fish fry, annelid worms, and molluscs. Prey caught by surface-diving, especially in leads in ice (Hagerup 1926; Salomonsen 1950–1, 1967a; Roby et al. 1981); also at glacier faces (Hartley and Fisher 1936; Roby et al. 1981), and at convergence fronts (Brown 1976a); swims just below surface with aid of feet and wings (Bruemmer 1972). Dives for c. 30 s, and may travel considerable distance underwater (Bateson 1961). Dives for 25–40 s, with 10–20 s between dives (Kartashev 1960). For shorter dives in shallow water, see Dewar (1924). Recorded feeding (without diving) on waste from trawler, perhaps taking small crustaceans from fish stomachs (Rees 1983). Not gregarious when feeding (see Montague 1926). In Spitsbergen, precipitation and reduced visibility appeared to reduce rate of provisioning young (Norderhaug 1980). Food for young (and also for self-maintenance: see Montague 1926) transported in distensible gular pouch. Often feeds young largely at night when plankton nearest surface

(Bateson 1961; Evans 1981). In Iceland, feeds most actively 22.00–04.00 hrs (Foster et al. 1951). On Horse Head Island (Greenland), feeding visits to young at maximum c. 18.00–06.00 hrs, minimum 12.00–16.00 hrs (Evans 1981). In Spitsbergen, however, feeding rate at maximum 15.00–21.00 hrs, minimum 24.00–03.00 hrs, increasing thereafter through morning and afternoon, suggesting factors other than planktonic movements probably involved (Norderhaug 1980). Unable to feed in total darkness (Gorbunov 1932). At Horse Head Island, birds fed mainly within 2·5 km of colony, but once up to 2000 birds seen feeding or flying 32 km away (Evans 1981). According to Cody (1973), feeding range 12–16 km in north-east Iceland. Often over 50 km, western Spitsbergen (Rüppell 1969). In breeding season, Greenland, birds found feeding 100–150 km south of nearest colonies; not known if these were breeding, but this range consistent with known feeding rates and flight energetics (Brown 1976a; see also Kolthoff 1903).

Little information on diet outside breeding season. In breeding season, data largely from prey brought to young. Mainly crustaceans (*Calanus finmarchicus*, *Mysis*, *Eualus gaimardii*, *Sabinea septemcarinata*, *Gammarus*, *Euthemisto*, *Parathemisto oblivia*, *Atylus carinatus*, *Anonyx nugax*, *Acanthostepheia behringensis*, *Paramphithoe hystrix*, *Thysanoessa*, *Mesidotea*). Also molluscs (*Argonauta arctica*, Pteropoda) and, rarely, annelids and small fish. (Birulya 1910; Montague 1926; Collinge 1924–7; Gorbunov 1932; Demme 1934; Bird and Bird 1935; Hartley and Fisher 1936; Uspenski 1956, 1972; Kartashev 1960; Løvenskiold 1964; Norderhaug 1970, 1980; Bruemmer 1972; Evans 1981.)

Adult diet in breeding season mainly crustaceans. In Spitsbergen, principally *Mysis*, caught in deeper waters (Montague 1926). Of 3 adult stomachs, Spitsbergen, 2 contained *Thysanoessa*, 1 *Euthemisto* (Hartley and Fisher 1936). On Franz Josef Land, adult stomachs (no sample size given) contained *Calanus finmarchicus*, *Atylus carinatus*, *Paramphithoe hystrix*, *Anonyx nugax*, *Acanthostepheia behringensis*, and unidentified small fish (see Uspenski 1956). On Jan Mayen, shot birds (no sample size) contained *Mysis relicta* and *Calanus* (Bird and Bird 1935). 29 stomachs (no further details, but possibly year-round sample) contained 97·8% animal food (remainder algal fragments); crustaceans 80% of total, fish 1·2%, annelids 9·4%, larval molluscs 7·3% (Collinge 1924–7). In Greenland, pteropods said to be taken (Bruemmer 1972).

Food of young, Spitsbergen, at least 95% planktonic crustaceans. In 116 samples collected from adults on way to feed young, fish larvae recorded only once. *Calanus finmarchicus* by far the most important species, but also *Mysis*, *Parathemisto oblivia*, *Eualus gaimardii*, and *Sabinea septemcarinata* (Norderhaug 1980). At Horse Head Island, 460 dropped meals, 20 July–10 August, comprised (by number of items), 94% *Calanus finmarchicus* and 6% amphipods (Evans 1981, which see for species of amphipods). At Cape Atholl (Thule District, Greenland) 204 meals mostly copepods (c. 90% of items); remainder amphipods, and a few decapod larvae; characterized by low species diversity and small size variation of items (Roby et al. 1981, which see for species). In Spitsbergen, mean feeding rate 8·5 feeds per chick per 24 hrs (4–14, 320 feeds, 880 hrs) with no evidence for change in rate with age of young. However, average wet weight of feeds increased from mean 2·3 g ($n=24$) in 1st week after hatching, to mean 3·4 g ($n=33$) in 4th week; overall range 0·68–6·8 g ($n=116$). (Norderhaug 1970, 1980). Up to fledging, each chick consumed estimated 690 g plankton (calculated from Norderhaug 1970). On Horse Head Island, chicks received mean $5·25 \pm SE$ 0·12 feeds per 24 hrs (18 chicks, 88 hrs) over entire nestling period; rate lower from 21 days up until fledging (at 28 days) (Evans 1981). At Cape Atholl, mean wet weight of meals 3·48 g (1·0–6·5, $n=204$), and no difference in weights brought by ♂ and ♀ parents (Roby et al. 1981). At Spitsbergen, a low feeding rate by 1 parent was partly compensated for by the other (Norderhaug 1980). EKD

Social pattern and behaviour. Most information for west Palearctic from Stempniewicz (1981). Otherwise heavily based on studies in Greenland, especially Thule District (Ferdinand 1969; Evans 1981; Roby et al. 1981), and partly on notes supplied by P G H Evans.

1. Gregarious throughout the year. Outside breeding season, occurs in loose aggregations of small flocks, producing large overall concentrations in chief wintering grounds (Nichols 1913; Wynne-Edwards 1935; Rankin and Duffey 1948): e.g. in late February, Newfoundland, rafts composed of flocks of up to 15 birds each (McKittrick 1929). At least initially, flocks probably contain birds of all ages; at end of breeding season, adults and young may leave colony together (Løvenskiold 1964). For flock sizes in winter wrecks, see (e.g.) Snyder (1953) and Bannerman (1963a). Just before breeding, large rafts assemble offshore from colony area (e.g. Ferdinand 1969, Uspenski 1972), and enter breeding grounds en masse (Kristoffersen 1926; Salomonsen 1950–1; Ferdinand 1969: see Flock Behaviour, below), though at first make periodic, tentative landings (P G H Evans). In summer, Greenland, flocks along coast mostly of 1-year-olds (Salomonsen 1967a); feeding flocks in ice-leads of 8–10 birds (Salomonsen 1950–1), up to 100 (Roby et al. 1981). In breeding season, western Spitsbergen, flocks of 5–30 birds returned from feeding (Rüppell 1969). At end of breeding season, Spitsbergen, large noisy flocks assemble on sea off colony area (see Montague 1926, Løvenskiold 1964); 17 such birds shot all ♂♂ (Montague 1926). BONDS. Monogamous mating system. Marked mate fidelity between years. Of 10 pairs, 5 retained same mate over 3 successive years, 4 over 4 years (Norderhaug 1968). Age of first breeding not known. In late incubation and early chick-rearing, Cape Atholl (Greenland), colony visited by large numbers of non-breeders, of which high proportion 1-year-olds; others thought to be 2- and 3-year-olds (Roby et al. 1981; see also Roosting and Attendance Patterns, below). Both sexes care equally for young (Stempniewicz 1981), though ♂ may do more feeding of older young (Roby et al. 1981). Despite report that young independent on leaving nest (Uspenski 1956), not known if parental care ceases after fledging (see Relations within Family Group, below). BREEDING DISPERSION. Strongly colonial and highly gregarious, forming largest colonies of any Alcidae. In

low Arctic, colonies sometimes c. 50 pairs or less (Løvenskiold 1964; Evans 1981); in high Arctic, up to several hundred thousand pairs (Løvenskiold 1964; Norderhaug 1970) and sometimes well over 1 million pairs in Greenland (Salomonsen 1950–1; Bateson 1961). Colonies divided into discrete sub-colonies (Ferdinand 1969; Evans 1981; Stempniewicz 1981). Sub-colonies, Cape Atholl, each c. 200–250 pairs (Ferdinand 1969), but some less than 10 pairs on Horse Head Island, north-west Greenland (PGH Evans). Most sub-colonies at Hornsund (Spitsbergen) 'several score' to several hundred metres apart (Stempniewicz 1981); distance between neighbouring sub-colonies, Horse Head Island, c. 100–200 m ($n=4$) (see Evans 1981). Nest density, Spitsbergen, variable, but more than 1 pair per m² in some colonies (Norderhaug 1970, 1980), 1 pair per 1·5–2·0 m² (Stempniewicz 1981); 1 pair per 2·0 m², Horse Head Island (PGH Evans). Defends nest-area territory, comprising nesting cavity and (usually) stone near entrance which serves as platform for courtship, copulation, take-off, and landing, and as vantage point, loafing site, and exercising ground for young prior to fledging (Stempniewicz 1981); if no stone beside nest area, more distant one often used for courtship and copulation (PGH Evans). In high Arctic most colonies pure *A. alle*, but some close to other Alcidae, Kittiwakes *Rissa tridactyla*, and Fulmars *Fulmarus glacialis* (Kartashev 1960). Except in rare cases, same nest-site used by same pair each year (for details, see Norderhaug 1968). ROOSTING AND COLONY-ATTENDANCE PATTERNS. Off-duty bird may loaf just outside nest-entrance on vantage point if available or in 'club' (usually rock outcrop) associated with each sub-colony (Bateson 1961; Ferdinand 1969; Rüppell 1969; Evans 1981), or resort to sea for bathing and roosting. In club, Horse Head Island, birds spend 70–80% of time either loafing with body hunched low, or sitting on tarsi but with body upright (Evans 1981); sometimes gave Clucking-calls (Ferdinand 1969: see 4 in Voice). Commonly preened, most often 18.00–02.00 hrs, following main return to colony from sea; also yawned and Wing-flapped; on land, seldom slept (and then with eyes closed or head tucked over back), and birds probably slept mostly in rafts at sea (Evans 1981; see also Ferdinand 1969, Stempniewicz 1981, and below). After feeding, often assembles in large numbers on ice-floes to rest (Hagerup 1926; Salomonsen 1950–1). In late August, will rest, sleep, etc., in compact rafts (Montague 1926; Løvenskiold 1964); much calling in raft (Montague 1926; CS Elton). At start of breeding season, first arrivals at colony area, Cape Atholl, formed circling flocks at c. 24.00 hrs, and made first landings at c. 04.00 hrs; after 08.00 hrs, most birds returned to sea (Ferdinand 1969), more landing the following night. Attendance in colony at first erratic or cyclic, with large numbers present on some days (especially calm, sunny weather), after which all birds depart for a few days (Salomonsen 1950–1; Kartashev 1960; Bateson 1961; Ferdinand 1969). Eventually birds stay every night, leaving only by day, finally settling permanently (Salomonsen 1950–1). During laying and incubation, Cape Atholl, birds came ashore c. 01.00–02.00 hrs, and activity high until late morning, when it began to subside, birds leaving colony area in flocks 11.00–13.00 hrs, after which all visible birds had left; scarcely any birds left colony 13.00–01.00 hrs (Ferdinand 1969). Pattern of diurnal activity comparable during incubation, Horse Head Island; from large rafts, which gathered just offshore 21.00–24.00 hrs, mass flights made over colony area, from which landings at maximum c. 06.00 hrs; from 14.00 to 18.00 hrs, when nest-relief widespread, most off-duty birds left for sea. During chick-rearing, rafting peaked at 24.00, 06.00, and 18.00 hrs, but many birds remained ashore from 04.00–14.00 hrs. As chick-rearing progressed, rafting more erratic, with peaks earlier (04.00 hrs) and later (20.00 hrs), associated with decreasing time spent on land. During late chick-rearing, colony almost deserted 15.00–19.00 hrs (Evans 1981). During incubation and chick-rearing, Hornsund, apparently no differences in numbers of birds in colony during day (Stempniewicz 1981). In late incubation, Cape Atholl, 46% of birds in colony area 1-year-olds, remainder c. ½ each older non-breeders and breeders. After peak hatching, 1-year-olds much scarcer, and by late chick-rearing, few non-breeders of any age present. Non-breeders attended colony mainly 08.00–20.00 hrs, otherwise at sea (Roby *et al.* 1981).

2. When alarmed, especially by aerial predators, sitting birds adopt Horizontal-posture: sit low on tarsi facing seawards, head and neck more or less in line with body, neck sometimes upstretched; give Alarm-call (see 1 in Voice), and may flee (Ferdinand 1969). When closely approached by human, bird wintering on inland water, Suffolk (England), stretched neck upwards, gave Alarm-call, and swam rapidly off, eventually diving (King 1972b), as birds do if threatened at sea (Bateson 1961). FLOCK BEHAVIOUR. Aerial predators in colony constantly cause mass upflights (Løvenskiold 1954, 1964; Rüppell 1969; Bruemmer 1972; see also Parental Anti-predator Strategies, below). Upflights similarly caused by any sudden disturbance, e.g. break-up of glaciers or icebergs (Roby *et al.* 1981). On first arrival at breeding grounds, large proportion of population circles over colony area for hours, giving Trilling-calls (see 3 in Voice). At Cape Atholl, just after midnight, started circling in thousands over sea near colony, increasing to tens of thousands over 1–2 hrs. Entered colony area as a few large flocks which divided into smaller ones consistent with sub-colony groupings (Ferdinand 1969); latterly, usually 100–200 birds per flock (Salomonsen 1950–1), sometimes only 15–30 (PGH Evans). Once over colony, flocks flew at c. 200–300 m above ground, in circles of c. 400–500 m diameter moving faster and calling more loudly than previously; some birds rolled over as they circled, underparts upwards (Ferdinand 1969). Rapid, high flight (up to 500–600 m: Salomonsen 1950–1) reported from other colonies (Kristoffersen 1926; Salomonsen 1950–1). Similar mass flights, with much calling, arising from rafts of c. 1000–3000 birds, characteristic of incubation period. Within mass flights of several thousand birds, sub-flocks of several hundred evident. Smaller flocks rarely exceeding 50 birds, and circling in silence ('rushing flight': Evans 1981), arose from distinct sub-colonies, to which they returned once flight over (Evans 1981; see also Ferdinand 1969, Stempniewicz 1981). At this time, rafts probably represent sub-colony populations (Norderhaug 1964; Ferdinand 1969). Each sub-colony behaves independently, its circling and feeding flights not mixing with those of other sub-colonies (Stempniewicz 1981). ANTAGONISTIC BEHAVIOUR. Most intense early in breeding season, during nest-site establishment (Stempniewicz 1981). In mildest form, a mere intention of jabbing at another bird, though this probably most often between members of prospective pair (Evans 1981). In club, bird may push others away with breast (Rüppell 1969). In club, Horse Head Island, aggression rare—only 4% of interactions antagonistic, and of intention-jabbing sort (Evans 1981). Aggression more intense, and frequent, during

A

B

site establishment. For Head-vertical posture (Fig A), probably expressing site-ownership, and aggressive reactions to it (Evans 1981), see Heterosexual Behaviour, below. Disputes over nest-sites may lead to fighting (Stempniewicz 1981). Rivals, with bills locked, often rolled 'several score metres' down slope, eliciting aggressive responses from birds through whose territories they passed. Fight sometimes developed into pursuit flight, after which participants always returned to alight on vantage point (stone) inside own nest-territories; occasionally, if no such site available, bird landed on neighbour's and immediately scurried into its own nest-area (Stempniewicz 1981). Towards end of chick-rearing, apparently unpaired, prospecting birds often chased from occupied burrows, which they entered carrying small stones in bill (Evans 1981; see Heterosexual Behaviour, below). Appeasement behaviour much less marked than in cliff-nesting Alcidae. Bird landing in club invariably adopts Landing-posture (Fig B)—walks slowly and in silence, with body upright and bill and tail pointing at ground; sometimes droops wings when passing near to other birds. Landing-posture apparently signals appeasement, usually eliciting no aggressive response (Evans 1981). Birds already settled in club also adopt Landing-posture when approaching other birds (often as prelude to Head-wagging: Evans 1981; see below); then called Rolling-walk display (Ferdinand 1969), and said to elicit same (though with tail cocked rather than down) in one or more birds nearby (Ferdinand 1969; Rüppell 1969). Bird walks with 3–4 steps per s (normally 4–6), raising foot high at each step so that body tilts (Fig C: Rüppell 1969). Apparently no calls given (Ferdinand 1969) or only soft Trilling-call (Rüppell 1969). Heterosexual Behaviour. (1) General. No courtship or copulation seen on first arrival (3 June) at colony, Cape Atholl. Birds initially sat individually. By 20 June (incubation), more than half the visible birds sat in pairs, either on snow or in club (Ferdinand 1969). (2) Club-areas and pair-bonding behaviour. From outset, club is focus of social interaction in sub-colony. Although wide variety of behaviour reported, significance of displays not well understood, and sequence in pair-formation not known. At start of breeding season, first birds to arrive in colony, Cape Atholl, sat on lower slopes, facing seawards, necks upstretched and bill at $c.$ 45° upwards, jerking head rapidly from side to side; this probably related to Head-wagging (below). Thus settled, birds gave Trilling-calls, apparently directed at others circling overhead, which called in like manner (Ferdinand 1969). Trilling-call predominant at beginning of season, given by birds in clubs, rafts, or in the air (Ferdinand 1969). Trilling often initiated by 1 bird in club, and stimulates same in others in own and nearby clubs, also those overhead, chorus often continuing for several minutes ('Mass song': Rüppell 1969). Rolling-walk display and Head-wagging (below) may follow. Up to $c.$ 30 July (incubation, Horse Head Island), Head-bowing frequent (Evans 1981). 2 birds face each other closely and bow their heads several times in quick succession, sometimes for up to 1 min, though usually interrupted after $c.$ ½ min by brief rest. Head-bowing rare after hatching, and apparently replaced, in established pairs, by Head-wagging (Fig D: Evans 1981); like Head-bowing, seems to invite close approach. 2 birds, sitting or standing close together with tails cocked, wings sometimes drooped and fluttering, turn to face each other and make rapid sideways movements of head for $c.$ 30 s. Partners may lightly touch bills (Billing) as they Head-wag, and give Billing-calls (see 5 in Voice: Ferdinand 1969; Evans 1981; see also Rüppell 1969). May Head-wag with bill up, down, or horizontal, depending partly on whether 1 bird situated above the other, or both on level ground (see figures in Rüppell 1969 and in Evans 1981). However, Head-waggings, with bill up, elicited aggressive response more often than did other bill attitudes (Evans 1981, which see for range of responses). Before and after Head-wagging, bird often Head-shakes—a much more rapid and briefer action (lasting $c.$ 2 s: Evans 1981) than Head-wagging; with bill lowered, bird thus scatters the snow in front of it (Ferdinand 1969). Head-shaking seen only occasionally on Horse Head Island, and elicited no response in neighbours (Evans 1981). Movement (of any kind) by bird often elicited Head-cocking (rapid sideways tilt of head) in one nearby (Evans 1981). Sometimes, bird approached another in Head-vertical posture ('skypointing': Evans 1981): body upright, neck upstretched, bill obliquely upwards (Fig A); often accompanied by calling (not described). Posture also adopted by stationary birds. In 4 cases observed, 2 elicited aggression (Evans 1981). Head-vertical posture possibly advertising-display, as interpreted in (e.g.) guillemots *Uria* and Razorbill *Alca torda*. One or both members of established pair may depart downhill from club with Butterfly-flight (Rüppell 1969; Bruemmer 1972; Evans 1981), and usually with call similar to Clucking-call (Ferdinand 1969; P G H Evans). Before taking off, bird often stretches wings for ½–1 s. In Butterfly-flight, typically 3–5 wing-beats per s (normally 12–18), and body tilted up at $c.$ 30–45°. Flies thus up to 500 m or more, then assumes normal flight (Rüppell 1969). If both birds leave club, usually return together to same place shortly after. Mutual Butterfly-flight sometimes preceded, and immediately followed, by intense bout of Head-wagging. On Horse Island, Butterfly-flight most common early in day, up until 16.00 hrs; virtually absent 18.00–20.00 hrs. Late in fledging period, up to 90% of birds flying from colony in early afternoon performed Butterfly-flight. Function not clear, but considered homologous with similar flight of Brünnich's Guillemot *Uria lomvia* and *A. torda* (Rüppell 1969; Evans 1981). (3) Mating. At Cape Atholl, usually occurred on snow, occasionally on stones lying on it. Always preceded by 2–3 bouts of Head-wagging and Billing (Ferdinand 1969; see also Rüppell 1969). Following account from observations at Cape Atholl (Ferdinand 1969). ♂

C

slowly mounts from behind or obliquely behind ♀. When mounted, droops wings along ♀'s sides; once, flapped wings upwards for some seconds to touch ♀'s down-tilted bill. ♂ remains mounted for *c*. 30 s. After dismounting, 2–3 bouts of Head-wagging and Billing follow, then pair walk about briefly before preening. Not known if any calls given. During study, Cape Atholl, 10 copulations successful, 10 unsuccessful (Ferdinand 1969). During chick-rearing period, Horse Head Island, copulation attempts usually evoked aggression from ♀. Promiscuous mating attempts occasional (Evans 1981). (4) Behaviour at nest. The following based on Ferdinand (1969). On 5–6 occasions, birds presented stone to partner. Once, after pair performed 2–3 bouts of Head-wagging and Billing, one bird stepped down from rock with Rolling-walk display, picked up stone, returned to partner as if to present it, then dropped it; sequence repeated a few minutes later. Stone-carrying apparently site-prospecting behaviour. One prospecting bird retained stone for 5–10 min. After stone-carrying, bird lay on ground with plumage of underparts spread, quivering whole body ('as if it had settled on an egg') while giving weak Snarling-call (see 6 in Voice); other bird stood alongside. Immediately after, pair disappeared among rocks. RELATIONS WITHIN FAMILY GROUP. During hatching and for 24 hrs after, one parent remains on nest almost constantly, brooding partly under wing (Stempniewicz 1981). Effective thermoregulation achieved by 3–4 days; from 5th day, adults observed in only $\frac{1}{3}$ of 50 nests (Norderhaug 1980). Up to 3rd day, parent present for estimated 70% of time, 30% up to 5th day, 10% up to 8th (Stempniewicz 1981). Thereafter, young brooded irregularly up to 20 days (Norderhaug 1980; Stempniewicz 1981). Young beg with Food-call (see Voice) audible throughout colony in late chick-rearing (Løvenskiold 1964). Parents regurgitate from gular pouch, and chick pecks food from mouth. Fed in nest up to 15th day, emerging thereafter to exercise wings beside parents near nest-entrance, and then may be fed in open (Stempniewicz 1981). At Horse Head Island, however, fed in nest until fledging (PGH Evans). Duration of exercise increases up to fledging (Stempniewicz 1981). Fed until fledging (Norderhaug 1980; Evans 1981); for suggestion that parents may stop feeding young shortly before fledging, see Løvenskiold (1964). In Spitsbergen, most fledge afternoon and night (Løvenskiold 1954); 21.00 hrs until early morning (Norderhaug 1980). Young fly out to sea singly, in small flocks with others, or mixed with adults, though not known if these are parents (Stempniewicz 1981). Young said to fledge with parents or alone (Løvenskiold 1964); with 1 or both parents (Norderhaug 1980). Young may fly falteringly (Bruemmer 1972), but presumed not to alight on sea until well away from colony, as never seen to alight within sight of land-based observer (Norderhaug 1980; Stempniewicz 1981). Young that left with 'parents' did not return (in short term) to colony area, while those that left alone occasionally returned if first flight short (Løvenskiold 1954, 1964; see also Norderhaug 1980). Young said to be independent on fledging (Uspenski 1956), but no evidence known. At colony in Spitsbergen, fledging spread over 2 weeks (Norderhaug 1980). At Hornsund, fledging highly synchronized (most leaving within 2–3 days), more so than hatching (Stempniewicz 1981). Breeders and non-breeders quit colony at same time as young, so that exodus

rapid, occupying last 10 days of August at Hornsund (Stempniewicz 1981; see also Kartashev 1960). ANTI-PREDATOR RESPONSES OF YOUNG. Young able to run to safety inside nest-burrow from 3rd day (Stempniewicz 1981). Older young outside burrow run inside it, or sometimes into nearest other available crevice, if threatened; often thus respond to upflight and Alarm-calls of adults (Løvenskiold 1954; Bateson 1961; Stempniewicz 1981). Some newly fledged young show considerable evasive capability in the air, e.g. swooping or diving into the sea if harried by gull *Larus* (Stempniewicz 1981). PARENTAL ANTI-PREDATOR STRATEGIES. (1) Passive measures. At nest-site, sentinel bird alerts sitting mate with Alarm-call (e.g. Rüppell 1969). In response to excavation of nests by fox *Alopex*, birds move (under wing) eggs and young deeper into nesting burrow (Stempniewicz 1981). Flushing distances for ground predators (and man) often only a few metres (see Roby *et al*. 1981). (2) Active measures: against birds. Mass upflights presumably distract aerial predator, especially Glaucous Gull *L. hyperboreus* by density, mobility, and cacophony of Warning- and Alarm-calls (Stempniewicz 1981: see 2 in Voice). At Savigsivik (Greenland), threatened flocks rush downhill at speed, wheel, and usually swoop back and forth across slope (Bruemmer 1972). At Cape Atholl, birds flew out over water, remaining there for more than $\frac{1}{2}$ hr when gulls made prolonged patrols (Roby *et al*. 1981). (3) Active measures: against man and other animals. Likewise performs mass upflights if disturbed. Foxes *Alopex* elicit stronger reaction than gulls, and birds reluctant to land long after predator's departure (Løvenskiold 1964; Stempniewicz 1981).

(Fig C from drawings in Rüppell 1969; others from drawings in Evans 1981.) EKD

Voice. Calls contain essentially 2 kinds of units ('glissando' and 'serrated'), combined in varying rhythm and number (Ferdinand 1969, which see for further sonagrams, details of harmonic structure, etc.).

CALLS OF ADULTS. (1) Alarm-call. Hoarse, unmelodious sound, consisting of single glissando unit of 0.12–0.15 s duration (2 calls, same bird). Frequency and structure resemble units in 3rd part of call 3. Given by restless bird separated from raft (Ferdinand 1969). This presumably the single 'wep', or sometimes 'weep', similar in quality to Coot *Fulica atra*, given by birds on water during breeding season, and in colonies at end of season (PJ Sellar). Alarmed bird wintering inland in Britain gave short, sharp 'wow' as it swam away from approaching human (King 1972*b*). (2) Warning-call. Clipped, metallic 'dudududu' (Rüppell 1969). Described as whinnying 'whuwhuwhu', like beating of wings, and given by birds wheeling over human intruder in their sub-colony (Fig I); heard only as they pass over intruder, not elsewhere (PJ Sellar). Also 'ha- ha- ha-' calls (J Hall-Craggs: Fig II) as flock circles over human intruder (PJ Sellar), sounding like Warning-call with units more spaced, or else repeated Alarm-call, and suggesting gradient occurs between calls 1 and 2 (EK Dunn). Birds may give call whether or not gular pouch filled with food (Stempniewicz 1981; see also Norderhaug 1970). (3) Trilling-call ('Song': Rüppell 1969). Most characteristic sound at the start of breeding season. A twittering, chattering call of 3 parts: 1st 'kreea', 2nd a rippling sound, 3rd a laughing 'hahaha ...' (PJ Sellar: Fig III); rendered 'kri-ri-ri-kikiki-ki-ki' (Birulya 1910); 'krääk ääk

D

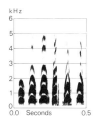
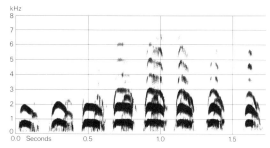

I P J Sellar Spitsbergen August 1981   II P J Sellar Spitsbergen August 1981

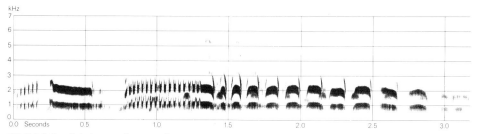

III P J Sellar Spitsbergen August 1981

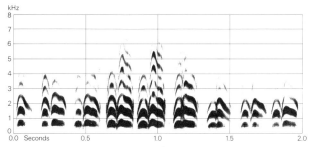
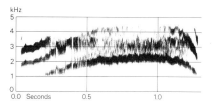

IV L Ferdinand Greenland June 1964   V J P Varin/Sveriges Radio (1972) Iceland June 1967

ak ak ak ak ak ak' (see Witherby *et al.* 1941); 'drä dräää däd däd däd . . .', last units uttered with increasing rapidity, but decreasing volume; 'ä' may be modified to 'i' or 'u' (Rüppell 1969). Following details from Ferdinand (1969). Call lasts 1·15–3 s ($n=8$). 1st part shortest (0·18–0·32 s, $n=6$), consisting of a number of units of rapidly rising and falling frequency, usually glissando at beginning and end. 2nd part, 0·08–0·2 s later, is loud, snarling sound (0·55–0·65 s) whose units resemble structure, frequency, and rhythm of 1st part, occurring regularly at *c*. 19 per 0·5 s. 3rd part comprises 3–13 glissando units, each 0·01–0·1 s ($n=5$) at 0·04–0·05 s intervals. Call highly variable between individuals in duration and structure of units. Given by birds settled both on land or water, and in flight (i.e. mass flights) (Ferdinand 1969; Rüppell 1969; P J Sellar), also sometimes by incubating bird (Ferdinand 1969, which see for details of calls in mass flights). In recording by P J Sellar, stuttering variant given in flight. (4) Clucking-call. Soft, clucking sound, audible only at short distance, comprising units similar to those of call 1. Given in duet by birds sitting quietly on rocks, the participants alternating calls. Similar call given during take-off, and during flight in small flocks (Ferdinand 1969). According to Evans (1981) same call given during Head-wagging and Butterfly-flight. This probably the soft, clipped, high-pitched 'ud ud ud . . .' or 'drud drud drud . . .', or 'au au au . . .', described as a persistent, chattering, conversational call of low excitement from birds settled in colony. When it precedes call 3, many birds stand up, and sound changes to hard, short 'dä dä dä . . .' (Rüppell 1969). (5) Billing-call. A hoarse, unmelodious call, given at shorter intervals at beginning and end—than in middle of bout. Bout length 2·1–3·7 s ($n=5$), consisting of 20–30 units, resembling those of call 1, and audible at up to 100–200 m (Ferdinand 1969). In typical performance, initiated by 1 bird, develops into duet, the participants alternating rapidly, and ends with 1 bird again (Ferdinand 1969). Apparent duet in recording by L Ferdinand (Fig IV) a cackling crescendo (P J Sellar). Compressed, high-pitched 'drrrrrru' or 'drrrrrr' (Rüppell 1969, whose rendering, with no mention of duet, may possibly represent 2 birds). Often heard from laying onwards among paired birds whilst Head-wagging and Billing (Ferdinand 1969; Rüppell 1969). (6) Snarling-call. Rapid, rhythmic series of units (26–28 per s), producing snarling sound, given by birds prospecting sites (Ferdinand 1969).

CALLS OF YOUNG. In recording of one chick in nest, variety of hoarse, wheezy whistles given, some short and chirping, others longer (Fig V) and rising in pitch, with slight fall at end (E K Dunn). Food-call of older young 'tick tick tick tick' (Løvenskiold 1954, 1964). Fledged young flying around colony in flocks had lighter call than adults, sounding intermediate between Food-call and 'full voice' of adult (Løvenskiold 1954). EKD

Breeding. SEASON. Spitsbergen and Novaya Zemlya: eggs laid second half June and early July (Dementiev and Gladkov 1951b; Løvenskiold 1964; Stempniewicz 1981). Iceland: eggs laid late May and first half June (Hantzsch 1905). Laying time related to snow cover and depth, and to air temperature (Norderhaug 1980). SITE. Well hidden in boulder scree; also in crevice in rock or cliff. Usually 0·5 m under surface, but can be up to 1·0 m (Norderhaug 1980; Stempniewicz 1981). Colonial. Nests re-used in subsequent years (Norderhaug 1967). Nest: shallow layer of pebbles, 2–4 cm in size, plus occasional fragments of lichen or straw (Norderhaug 1980). Building: pebbles brought in or gathered in nesting crevice (Norderhaug 1980; Stempniewicz 1981). EGGS. See Plate 92. Sub-elliptical, smooth, not glossy; pale greenish-blue, rarely darker, unmarked or with buff to brown, rarely dark green or pink, spots and scribbles usually concentrated at broad end, but occasionally uniformly distributed. 49 × 34 mm (45–56 × 31–39), $n = 195$ (Norderhaug 1980); see also Stempniewicz (1981) for seasonal variations. Calculated weight 28 g (Schönwetter 1967). Clutch: 1; 2 reported (Norderhaug 1980) but almost certainly by 2 ♀♀. One brood. Replacements laid after egg loss, Iceland (Hantzsch 1905), but not reported further north. INCUBATION. Widely quoted 24 days (Faber 1825–6) possibly wrong; 29 ± SD 0·8 days (28–31), $n = 38$ (Stempniewicz 1981). By both sexes. Begins on laying of egg. On average, ♂ and ♀ alternated 4 times in 24 hrs (Stempniewicz 1981), though ♀ said to do most by day, ♂ most by night (Hantzsch 1905). YOUNG. Semi-altricial and nidicolous. Cared for and fed by both parents, brooded continuously for first 2–4 days (Norderhaug 1970) and, as with egg, partly under wing (Stempniewicz 1981). FLEDGING TO MATURITY. Fledging period 27·0 ± SD 1·81 days (23–30), $n = 33$ (Stempniewicz 1981). Mean 28·3 days, western Greenland (Evans 1981). Age of independence not known. Age of first breeding not known. BREEDING SUCCESS. Of 98 eggs laid, Spitsbergen, 65·3% hatched and at least 80% of chicks fledged (Stempniewicz 1981); often heavy predation on young by Glaucous Gull *Larus hyperboreus* and arctic fox *Alopex lagopus* (e.g. Løvenskiold 1964). Of 20 eggs laid, western Greenland, 65% hatched and 77% of chicks fledged (Evans 1981).

Plumages (nominate *alle*). ADULT BREEDING. Lores, sides of head, neck, chin, throat, and breast dark chocolate-brown. Small white patch at upper margin of eye. Upperparts black, tips of scapulars with narrow white edges. Underparts white, but long feathers on flanks with variable pattern of black and white. Tail-feathers black; tip of outermost ones often with white spot or edges, others occasionally so. Primaries black-brown on outer web and tip, paler on inner web and towards base. Secondaries similar but with prominent white tip. Shafts of flight-feathers mainly black, but horn-coloured towards base. Upper wing-coverts black. Greater under wing-coverts and greater under primary coverts dark silvery-grey; remainder of underwing pale brown. Axillaries brown-black. ADULT NON-BREEDING. Upperparts glossy black often with narrow partial or complete white speckled band across hindneck. Lores and just below and behind eye brown-black; remainder of sides of head and neck white, speckled black-brown. Upper chin brown-black, rest of chin and throat white or slightly mottled brown. Feathers of breast grey-brown with white tips, usually forming broad mottled band across breast. Rest of underparts white. Wing and tail as in adult breeding. DOWNY YOUNG. Upperparts and throat blackish-brown. Underparts paler, centre of belly varying between brown, grey, and nearly white (Witherby *et al.* 1941; Harrison 1975; Fjeldså 1977). Down replaced by juvenile plumage *c.* 4 weeks after hatching (Bent 1919; Kartashev 1960). JUVENILE. Like adult breeding, somewhat browner on upperparts, especially on throat and breast. Primaries still growing after leaving nest. White spot at upper margin of eye very small. Subsequent plumages as in adult.

Bare parts. ADULT. Iris dark brown to black. Bill black. Leg and toes black to light grey, joints darker; webs black or slate-black. Mouth pale flesh. DOWNY YOUNG. Bill blackish. Leg and foot dusky (Fjeldså 1977). FIRST IMMATURE NON-BREEDING. Bill brown. (Salomonsen 1944; Kartashev 1960.)

Moults. Poorly known. ADULT POST-BREEDING. Complete. Body feathers May–September. Flight-feathers and tail September–October. ADULT PRE-BREEDING. Partial: body only. December–March. POST-JUVENILE. No exact data; body feathers about September–October; in full 1st non-breeding by early November.

Measurements. ADULT. Nominate *alle*. Spitsbergen and eastern Greenland, breeding; skins (ZMA).

WING	♂ 123·3 (2·98; 16) 116–127	♀ 120·7 (3·09; 10) 116–125
TAIL	32·3 (2·56; 17) 27–36	32·3 (1·87; 10) 29–35
BILL	16·1 (0·56; 15) 15–17	15·5 (1·22; 9) 13–17
TARSUS	20·6 (0·81; 18) 19·3–22·0	20·2 (0·87; 10) 19–22
TOE	29·0 (1·41; 13) 26·7–32·1	28·6 (1·03; 9) 26·3–29·7

Sex difference significant for wing.

Spitsbergen, breeding; live, sexes combined (Stempniewicz 1981).

WING	♂♀ 120·5 (3·2; 97)	114–127·5
TAIL	33·5 (2·4; 97)	27–38
BILL	15·8 (0·7; 91)	14·0–17·5
TARSUS	21·8 (0·8; 97)	20–23·5

A. a. polaris Franz Josef Land; wing from Vaurie (1965); tarsus and bill from Stenhouse (1930).

WING	♂ 131·9 (—; 27) 124–138	♀ 131·7 (—; 7) 129–137
BILL	♂♀ 17·0 (0·75; 10) 15·0–17·5	
TARSUS	♂♀ 22·7 (1·03; 9) 21–24	

Weights. ADULT. Nominate *alle*. Eastern Greenland and Spitsbergen, May–August: ♂ 154 (9·85; 10) 140–166, ♀ 151 (6·22; 10) 143–161 (ZMA). Spitsbergen, June–August, sexes combined: 163 (12·05; 86) 134–192 (Stempniewicz 1981). Fair Isle (Scotland): November 115 (13·6; 4) 104–134; January 150 (BTO). Dutch coast, November–March: ♂ 123 (26·96; 6)

95–173, ♀ 108 (18·91; 8) 87–107 (ZMA). Quebec (Canada), winter: 163 (8) 127–188 (Johnson 1944). New York (USA), November, sexes combined: 94·9 (19·58; 31) 70·9–141·7 (Murphy and Vogt 1933).

JUVENILE. On leaving nest, Spitsbergen, 108 (39) 88–135 (Glutz and Bauer 1982).

Structure. Wing narrow and pointed. 11 primaries: p10 longest, p9 1–4 shorter, p8 5–8, p7 11–16, p6 17–22, p5 24–29, p1 50–60; p11 minute, concealed by primary coverts. Tail short, slightly rounded; 12 feathers, tips rounded to obtusely pointed. Bill stubby, broad at base; upper mandible curved downwards, tip extending over lower mandible; both mandibles notched behind tip. During summer, slightly thickened ridge in front of nostril extends backwards to frontal feathering and ridge on lower mandible runs from base to gonydeal angle. Nostril oval, frontal and dorsal part covered with horny operculum. Outer toe $c.$ 93% of middle, inner $c.$ 76%; no hind toe.

Geographical variation. Involves size only; see Measurements. Measurements of birds from Severnaya Zemlya and North Pacific unknown, hence subspecific status undecided.   RS

Aethia cristatella Crested Auklet

PLATES 20, 22, and 23
[facing page 231, and between pages 278 and 279]

Du. Kuifalkje   Fr. Starique à crête   Ge. Schopfalk
Ru. Большая конюга   Sp. Mérgulo crestado   Sw. Tofsalka

Alca cristatella Pallas, 1769

Monotypic

Field characters. 18–20 cm; wing-span 40–47 cm. Closest in size to Little Auk *Alle alle* but with wing length approaching that of Puffin *Fratercula arctica*. Auklet with similar form and behaviour to *A. alle* but differing in rather uniform and dusky plumage. Adult has forward-drooping or coiled crest and bright red or yellow bill. Immature less distinctive, lacking crest and having dull bill. Sexes similar; little seasonal variation. Immature separable.

ADULT BREEDING. Most of head black, with long, forward-curved crest, white eye, and drooping white plumes behind eye as striking as white-tipped, orange-red bill with basal tubercles. Upperparts slate, lacking any marks except (when worn) grey mottling. Chest dusky grey, becoming brown-grey on lower body and vent. Underwing has slate-grey coverts and axillaries, and silvery-grey flight-feathers; should thus appear even darker than those of *Fratercula arctica* and (especially) *A. alle*. ADULT WINTER. Appearance little changed except in shorter crest (often appearing as a spray), fewer (or no) eye-plumes, and less swollen, pale orange bill. IMMATURE. Differs from adult in having sootier underparts and dull ochre-brown bill, and lacking crest, bright white eye, and head-plumes. Legs grey at all ages, with blue or violet tone in adult.

Unmistakable in west Palearctic, with almost uniformly dark appearance particularly obvious at distance (and small size and lack of white wing-panels rule out confusion with Black Guillemot *Cepphus grylle*). Since records of *A. cristatella* and sympatric Parakeet Auklet *Cyclorrhynchus psittacula* represent astonishing dispersal by North Pacific seabirds, important not to totally discount chance of other auklets of similar appearance. *C. psittacula* black above, with bill red in summer (but black in winter), and white eye and eye-plume; white below, unlike *A. cristatella*. For other north Pacific auklets, see Robbins *et al.* (1983). Beware oiled *A. alle* which could resemble this species. Flight, gait, and behaviour similar to *A. alle*.

No records of voice outside breeding season.

Habitat. In boreal Pacific ocean, marine and coastal, in cool or cold but usually ice-free waters, capable of surviving severe storms, and of remaining near breeding habitat throughout year. Breeds deep in crevices among boulders on beach, in cavities in cliffs, and among jumbled lava rock on high slopes of mountainous volcanic islands, especially those with steep cliffs and spires. Where sharing site with Least Auklet *Aethia pusilla* tends to occupy lower levels near water, sitting in rows along tops of big boulders near nest-sites. Forages in sea, often in tide-rips between islands, up to at least 16 km from nest, and has been found eaten by cod *Gadus*, having probably been caught some 50 m below surface. Flies mostly in lowest airspace, but in mass evolutions at breeding colonies will rise to considerable heights. (Gabrielson and Lincoln 1959; Murie 1959.)

Distribution. Breeds in Siberia (Chukotskiy peninsula), islands of Bering Sea, central Kuril Islands, and in Alaska on Aleutians, Shumagin Islands, and Kodiak Island; has occurred more widely in nesting season, e.g. in Anadyrland and arctic coast of Alaska. Winters offshore and in open seas within breeding range, south to northern Japan.

Accidental. Iceland: 1 collected in waters to north, August 1912 (Hörring 1933).

Movements. Strongly dispersive (pelagic). Colonies vacated in September; some birds ashore from mid-March but main re-occupation in May.

Marine range not well documented, but lies within Bering Sea and adjacent North Pacific. Many birds present all year offshore from southern colonies, e.g. Aleutian and Kuril Islands, probably as far north as birds can find open water (Bent 1919; Kozlova 1957). On Asiatic side, occurs in winter in Japanese seas south to Hokkaido and northern Honshu (Japanese Ornithological Society 1958), but on American side not known to penetrate south of Alaska.

In wreck off Kodiak (Alaska), 18 January 1977, c. 6000 came aboard a fishing boat at night, attracted to lights (Dick and Donaldson 1978); 100 examined ashore were all 2nd-winter or older, suggesting these age groups stay closer to colonies than 1st-winter birds.

Voice. See Field Characters.

Plumages. ADULT BREEDING. Upperparts, including recurved frontal crest, entirely black; wear may expose grey feather-bases giving mottled black and grey appearance. Several narrow white plumes behind eye, extending along sides of neck. Forehead, sides of head behind eyes, lower cheeks, sides of neck in front of white plumes, and underparts slate-grey; slightly paler towards vent and under tail-coverts. Outer webs of flight-feathers and tips of primaries black; inner webs dull brownish-grey. Upper wing-coverts and tail-feathers black. Flight-feathers silvery-grey below. Under wing-coverts and axillaries dark slate-grey. ADULT NON-BREEDING. As adult breeding, but frontal crest slightly smaller and white plumes behind eye less conspicuous or largely absent. See also Structure. JUVENILE. As adult breeding but frontal crest and white plumes behind eye entirely absent. This plumage retained during winter up to moult to 1st immature breeding (Bent 1919). See also Structure. FIRST IMMATURE BREEDING. Like adult breeding, but juvenile flight-feathers, wing-coverts, tail, and scattered feathers of back, rump, and upper tail-coverts retained, very worn. Frontal crest shorter, feathers not as attenuated as adult; plumes behind eye shorter and fewer.

Bare parts. ADULT. Iris white or yellowish-white. In adult breeding, bill bright orange-red with pale horn-yellow or whitish tip; in adult non-breeding, bill brownish. Leg and toes bright slate-blue or violet-grey; joints darker; rear sides of tarsus and soles blackish; webs black. JUVENILE. Bill brownish. FIRST IMMATURE BREEDING. As adult breeding. (Ridgway 1919; Hartert 1921–2, BMNH.)

Moults. ADULT POST-BREEDING. Complete; primaries descendant, flight capability retained. Starts during end of breeding cycle or shortly afterwards, August; completing about October–November (Bent 1919; Stresemann and Stresemann 1966). ADULT PRE-BREEDING. Partial, body only. Late winter or early spring (Bent 1919; Hartert 1921–2). POST-JUVENILE. Partial, body only. In spring, moult directly to 1st immature breeding (Bent 1919).

Measurements. ADULT. North Pacific.
Skins (BMNH); bill is culmen length from basal ridge on breeding adult.

WING	♂	144·6 (4·56; 5)	140–152	♀ 141·3 (1·89; 4)	140–144
TAIL		37·8 (1·82; 5)	35·5–40·0	35·8 (1·19; 4)	34·5–37·0
BILL		12·4 (0·42; 4)	11·9–12·9	11·9 (0·40; 3)	11·4–12·1
TARSUS		27·7 (1·29; 5)	26·2–29·7	27·4 (0·85; 4)	26·7–28·6
TOE		39·0 (3·02; 5)	35·3–43·6	39·2 (0·91; 4)	38·9–40·4

Data from Ridgway (1919); depth is depth of bill at gonydeal angle.

WING	♂ 134·8(10)	125–143	♀ 134 (7)	131–137
BILL	11·2(10)	10–12	10·9 (7)	10·5–11·5
DEPTH	11·1(10)	10–12	10·5 (7)	9·5–12

Weights. Approximate average weight 250 (Fay and Cade 1959). January, Alaska: 249 (16; 11) 221–273 (Dick and Donaldson 1978). Chukotskiy peninsula (USSR), November: ♂ 260 (16·1; 6) 230–275, ♀ 212 (3) 190–245 (Portenko 1973).

Structure. Wing narrow and pointed. 11 primaries: p10 longest, p9 2–5 shorter, p8 5–8, p7 12–13, p6 18–22, p5 26–28, p1 58–62; p11 minute, concealed by primary coverts. Tail short, tip square or inner tail-feathers slightly shorter; 14 feathers, tips rounded. Recurved crest on forehead: of 14–20 narrow, elongated feathers; in adult breeding, largest feathers c. 40 mm long, but several anterior ones much shorter. White plume-feathers behind eye very narrow; those directly behind eye short, but posterior ones, implanted below ear-coverts, up to c. 30 mm long. In adult breeding, bill short and stout; upper mandible arched, tip extending over tip of lower mandible; lower mandible bent upward with pronounced gonydeal angle, tip blunt; gonys straight or slightly convex. Cutting edges of upper mandible shallowly ∪-shaped, of lower mandible straight except for upturned portion just before tip. Nostril longitudinal, narrowly ovate. In adult breeding, bill covered by horny plates: (1) nasal plate, from loral feathering to nostril, with transverse ridge and deep furrow from cutting edge to frontal feathering in front of it; (2) narrow sub-nasal plate along cutting edge of upper mandible, extending backwards from below nostril, basal part forming pronounced horizontal ridge; (3) conspicuous semi-circular plate at upper side of gape; (4) large elongated plate enclosing basal part of lower mandible, at gape developed into horizontally protruding ridge. Horny plates shed during post-breeding moult, resulting in small triangular bill in adult non-breeding, but transverse groove in front of nostril and pronounced gonydeal angle still present. In juvenile, frontal crest and plumes behind eye absent, bill as in adult non-breeding, but smaller still, culmen and cutting edges almost straight, gonydeal angle less pronounced. In 1st immature breeding, frontal crest small, few plumes behind eye; bill as in adult breeding, but less deep at base and plates less developed; that at gape especially small. Outer toe c. 96% of middle, inner c. 77%; no hind toe. RS

Cyclorrhynchus psittacula (Pallas, 1811) Parakeet Auklet

FR. Starique perroquet   GE. Rotschnabelalk

Breeds Ayan area on Sea of Okhotsk, but mainly in Bering Strait and Bering Sea: Chukotskiy peninsula (USSR), Seward peninsula (Alaska), and islands south of Komandorskiye and Aleutian Islands; also Sernidi, Chirikof, and Kodiak Islands in Gulk of Alaska. Penetrates north into Chukchi Sea, as far as Point Barrow, in summer–autumn; winters in North Pacific from Aleutians to Kuril Islands and Sakhalin (in west), and to mid-California (in east). See Kozlova (1957) and Vaurie (1965).

Sole west Palearctic occurrence is 1 collected Lake Vät-

PLATE 18. *Alca torda* Razorbill (p. 195): **1** ad breeding, **2** ad non-breeding, **3** 1st imm breeding (1st summer), **4** 1st imm non-breeding (1st winter), **5** juv, **6** downy young. (RG)

PLATE 19. *Cepphus grylle* Black Guillemot (p. 208): *C. g. arcticus* (p. 208): **1** ad breeding, **2** ad non-breeding, **3** 1st imm non-breeding (1st summer), **4** juv, **5** downy young. *C. g. islandicus*: **6** ad breeding. (RG)

PLATE 20. *Alle alle* Little Auk (p. 219): **1** ad breeding, **2** ad non-breeding, **3** ad moulting, **4** juv, **5** downy young. *Aethia cristatella* Crested Auklet (p. 229): **6** ad breeding, **7** ad non-breeding, **8** juv (retained through 1st winter). (RG)

PLATE 21. *Fratercula arctica* Puffin (p. 231). *F. a. grabae*: **1** ad breeding, **2** ad non-breeding, **3** juv in 1st winter, **4** young juv, **5** downy young. Nominate *arctica*: **6** ad breeding. (RG)

tern (Sweden), December 1860. A puzzling record, but which bears similarity to Crested Auklet *Aethia cristatella* in Iceland (see that species). Moreover, 3 other Bering Sea/North Pacific auks (Marbled Murrelet *Brachyramphus marmoratus*, Ancient Murrelet *Synthliboramphus antiquus*, Tufted Puffin *Lunda cirrhata*) have straggled to eastern USA and Canada (Godfrey 1966; Sealy *et al.*1982).

Fratercula arctica Puffin

PLATES 21, 22, and 23
[facing page 231, and between pages 278 and 279]

Du. Papegaaiduiker   Fr. Macareux moine   Ge. Papageitaucher
Ru. Тупик   Sp. Frailecillo   Sw. Lunnefågel   N. Am. Atlantic Puffin

Alca arctica Linnaeus, 1758

Polytypic. Nominate *arctica* (Linnaeus, 1758), Iceland, central and northern Norway, Bear Island, southern Novaya Zemlya, south-west Greenland, and eastern North America; *grabae* Brehm, 1831, Britain, Ireland, Faeroes, Channel Islands, France, and southern Norway; *naumanni* Norton, 1901, north-west and eastern Greenland, Spitsbergen, and northern Novaya Zemlya.

Field characters. 26–29 cm (bill 3–4 cm long, 3–4 cm deep); wing-span 47–63 cm. 30% smaller than Guillemot *Uria aalge*, but with relatively larger head enhanced by deep bill; 40–50% larger than Little Auk *Alle alle*, with much larger bill and head, broader outer wing, and larger tail. Small compact auk, with large, colourful bill (shaped like rounded triangle) set on grey face, smart black and white plumage, orange feet, and comical, endearing character. In winter, bare parts duller and face darker. Immature lacks fully developed bill but otherwise resembles adult. Sexes similar; little seasonal variation. Juvenile and immature separable at close range.

ADULT BREEDING. Both mandibles of bill basally blue-grey and distally bright red, these colours partly outlined and fully separated by yellow; distal part of bill symmetrically ridged; gape made obvious by yellow rictal tubercles. Round face ash-grey, appearing grey-white in strong light and setting off red eye with blue-horn appendages above and below. Crown, narrow nape, collar, and rest of upperparts black, glossed and with grey tones obvious at close range. Wing uniformly grey-black above and strikingly dark (in most lights) below; pattern of dull silver-grey under wing-coverts and drab brown-black undersurface to flight-feathers visible only at close range and when fully lit. Underparts white. Legs and feet bright orange-red. ADULT WINTER. Basic patterns of bill and plumage unchanged, but bill becomes less triangular and duller (losing yellow surround to base of upper mandible, lower edge of lower mandible, and swollen gape), orbital appendages lost, and legs become yellow. Front of face darkens, giving bird depressed expression. Rest of plumage unchanged, except sometimes for grey band across nape. JUVENILE. Apart from even darker face and dusky smudges on rear flanks, plumage like winter adult. Most obvious distinction is much smaller, less deep, and thus more pointed bill, lacking ridges and coloured dark grey-brown basally and red-brown distally. Legs flesh. FIRST WINTER AND SUMMER. Resembles juvenile, retaining dull but growing bill (with only faint indications of ridges visible in 1st summer), dark face, and dull legs.

Unmistakable at all ranges when bill structure and plumage pattern visible. More difficult to identify at distance and in poor light. However, no common larger or smaller auk has as blocked an outline to the front of its silhouette, and none shows typically as dark an underwing. Identification at long range also assisted by often markedly distinct flight, with actions including not just fast whirring of wings as in other species but also noticeably looser flapping of broad-ended wings, accompanied by frequent shifts in body angle not normally shown by guillemots *Uria* or Razorbill *Alca torda*. Like *A. alle*, more agile than larger auks, with easier take-off, more sudden landing, accomplished but rolling walk, pattering run (with wing flapping), and persistently upright stance (on raised tarsi).

Highly gregarious. Unlike *U. aalge* rarely visits breeding stations during winter and reappears suddenly in spring. Only accidental inland. With *U. aalge*, most vocal of west Palearctic auks: commonest call a growling trisyllabic 'kaa-arr-arr'; short growl when startled.

Habitat. Breeds on coasts and islands facing ocean from margin of high-arctic through low-arctic and boreal to temperate zones, fishing from inshore across offshore to pelagic waters at temperatures ranging from 0 to 15°C. In and near high Arctic, excavation of burrows in turf normally prevented by frozen ground; accordingly, there resorts to steep and inaccessible cliffs of large islands or mainland, breeding in crevices and crannies, sometimes near colonies of Little Auk *Alle alle* or Brünnich's Guillemot *Uria lomvia*, up to considerable altitudes (Salomonsen 1950–1). In low Arctic, only rarely uses cracks in steep sea-cliffs up to *c.* 80 m, sharing with Razorbill *Alca torda* and Black Guillemot *Cepphus grylle*, with which it more frequently lives as neighbour in rough frost-shattered talus or in rock crevices near sea-level. Prefers, however, grass-covered peaty turf, seasonally frost-free, on medium-sized

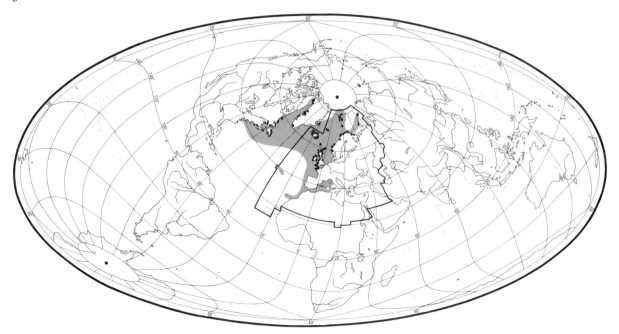

or small islands and skerries, on low, gently sloping coastal cliffs, or higher cliff terraces or level ground, where its manuring develops a tall and luxuriant growth of grasses and small erect plants (Salomonsen 1950–1; Evans 1975). In temperate zone, this preference leads to fuller ecological separation from other Alcidae, and to compensation for greater vulnerability of nest-sites by choice of relatively inaccessible islands, free of human disturbance and rats *Rattus*. Here large colonies in burrows stimulate growth of plants such as grasses (e.g. *Festuca*, *Poa*), thrift *Armeria maritima*, and sea campion *Silene maritima* (Evans 1975). Gregarious breeding and persistent interference with often thin and unstable soil mantle accordingly result in significant and enduring modifications to local ecology.

Wider-ranging at sea than other Alcidae, mainly in offshore and pelagic zones, from pack-ice to Mediterranean and subtropical waters. Swims and dives expertly, and walks relatively well, with special skill in hard burrowing. Often flies higher than most other Alcidae. Owing to more pelagic post-breeding dispersal, less often suffers from oiling, although vulnerable when close to coast. Sometimes killed in fishing nets; also suffers from impacts of human over-fishing, e.g. off Norway where collapse of local herring *Clupea* fishery led to breeding failure at Røst colony (Lid 1981).

Distribution. ICELAND. Some changes in distribution in 20th century, due to overexploitation by humans and presence of mink *Mustela vison* (AP). BRITAIN AND IRELAND. General distribution pattern unchanged, but for loss of various colonies (mostly small), see Parslow (1973), Cramp *et al.* (1974), and Harris (1976*b*). FRANCE. Some small colonies abandoned in Normandy and Brittany (Guermeur and Monnat 1980). WEST GERMANY. Helgoland: last bred 1830 (GR). SWEDEN. Bred annually until 1959, then again 1967–8 and 1970; *c.* 50 pairs in early 20th century, declining to 5–15 pairs in 1940s and 1950s (LR).

Accidental. East Germany, Finland, Poland, Hungary, Austria, Yugoslavia, Malta.

Population. Counts of breeding colonies extremely difficult; at large colonies only burrow counts, where possible, reliable (Harris 1976*b*). Estimation of trends also impossible, but in 20th century declines thought to have occurred in Britain and Ireland (mainly in south), France, and southern Norway. Various local factors involved at specific colonies, but overall declines probably due to climatic changes affecting food supplies, as contamination by pollutants does not appear to be an important cause (Harris 1976*b*; Harris and Osborn 1981).

SPITSBERGEN. Nowhere very numerous (Løvenskiold 1964). BEAR ISLAND. At least 600 pairs 1980 (Luttik 1982). JAN MAYEN. Fairly common (Bird and Bird 1935); probably a few hundred pairs (Harris 1976*b*); over 1000 pairs 1983 (J A van Franeker and C J Camphuysen). ICELAND. Perhaps 8–10 million birds; major decrease *c.* 1850 to *c.* 1870, then largely static due to change in netting methods; no general decline in recent years, perhaps even some increase (Harris 1976*b*). FAEROES. Large declines early 20th century, numbers since generally stable at perhaps 400000 to 1 million pairs (Harris 1976*b*). BRITAIN AND IRELAND. First complete census attempted 1969–70; due to census problems, impossible to suggest more than a tentative estimate of some 490000 pairs (Cramp *et al.* 1974); perhaps *c.* 550000 pairs, based on more accurate counts at some colonies (Harris 1976*b*). Dramatic declines

Fratercula arctica

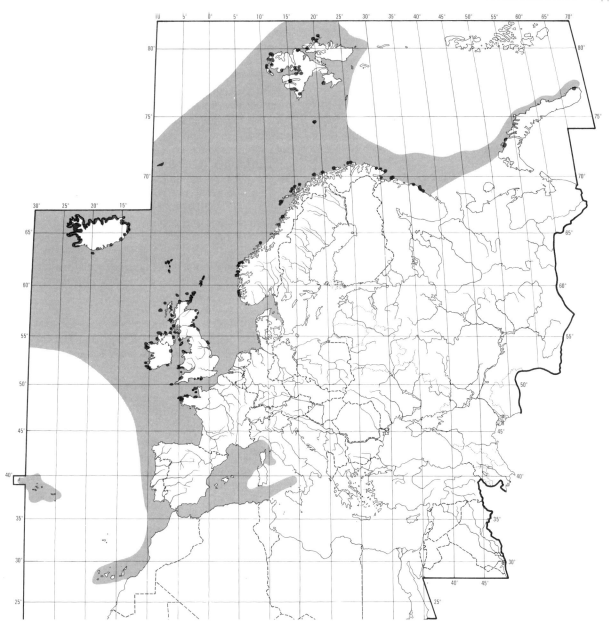

at some colonies in 20th century, mainly in south but also in Scotland, e.g. St Kilda (Flegg 1972), but overall decline apparently halted (Harris 1976b); this supported by detailed counting of burrows at 7 Scottish colonies (Harris and Murray 1981). Many possible causes suggested for earlier overall declines, but oil and other pollutants not apparently important factors and climatic changes affecting food supplies more likely (Harris 1976b). FRANCE. Perhaps 400–450 pairs (Harris 1976b); marked decline 20th century, probably due to oil pollution and climatic changes (Guermeur and Monnat 1980; Pénicaud 1980). NORWAY. Perhaps 1½–2 million pairs (Brun 1966). The most recent estimates are *c.* 1 156 000 pairs in northern colonies, where numbers maintained relatively well (Norderhaug *et al.* 1977), and *c.* 33 000 pairs in south where some decrease (Brun 1971b). However, although few chicks raised 1969–73 and 1975–80 at largest colony, Røst, apparently due to food shortage, population shows no sign of declining (Lid 1981). USSR. Novaya Zemlya: probably under 100 pairs (Norderhaug *et al.* 1977). Murmansk coast: *c.* 16 200 pairs (Skokova 1962).

Survival. Britain: annual adult survival *c.* 95·5% (Mead 1974), *c.* 95% (Ashcroft 1979), *c.* 96% Isle of May (M P Harris); 1st-year survival probably *c.* 33% (Mead 1974);

survival to 4 years old 10·3–15·6% in years 1971–3 (Ashcroft 1979); probably too low as no allowance for emigration—on Isle of May c. 38% survive to 4 years old (M P Harris). Oldest ringed bird still alive at 21 years 3 months (BTO).

Movements. Strongly dispersive. Essentially marine outside breeding season, keeping well offshore. Mid-ocean occurrences not rare (e.g. Rankin and Duffey 1948), due to transatlantic movements by some immature birds (see below).

Southern colonies occupied late March to mid-August, northern ones May to early September. On fledging, juveniles make their own way to sea and stay there until they begin to visit colonies again as pre-breeders. Pre-breeders of 4–5 years of age attend colonies again from April onwards, with progressively younger birds arriving as summer advances; most birds do not visit colony until 2 years old, and those 1-year-olds which visit do not appear until July–August, since they undergo wing moult in spring and early summer (M P Harris). Immatures frequently visit colonies other than their natal one, and some 20–25% of 1st-time breeders settle in a new colony; highly site-faithful thereafter (M P Harris; see also Myrberget 1973b, Harris 1976c, Petersen 1976a). Such inter-colony movements normally between adjacent groups of colonies, and emigration highest from colonies lacking room for expansion (M P Harris).

Winter distributions and summer ranges of immatures are poorly known, since relatively few ringing recoveries, especially from offshore zones. Occurs south to New Jersey in western Atlantic and to Morocco in eastern, with southern breeding populations being those which disperse furthest south in winter.

HIGH-ARCTIC POPULATIONS (*naumanni*). Non-breeding distribution virtually unknown. Southernmost record concerns a November specimen from Vesterålen, Norway (Haftorn 1971). Hence probably remains in East Greenland Sea and adjoining waters (probably extending to northern Norway) where seas remain unfrozen. However, Thule (north-west Greenland) populations must move further south, since no wintering occurs north of Sukkertoppen in southern Davis Strait (Salomonsen 1967a).

LOW-ARCTIC POPULATIONS (nominate *arctica*). Winter range extends south to c. 55°N in western Atlantic where Newfoundland seas are believed important for Canadian and west Greenland populations (Salomonsen 1967a; Tuck 1971). From extensive ringing in Norway (mainly Nordland), most recoveries in Norwegian waters and especially in south-west and south (Möre og Romsdal to Skagerrak). This distribution probably reflects regional trend in shooting of Alcidae, of which there is little in northern Norway; higher proportion of winter recoveries for 1st-year birds than for older ones may indicate greater vulnerability of former, or that older birds winter further offshore (Myrberget 1962c). Ringing at Lovunden colony (66°20′N, just south of Arctic Circle) showed birds having south-westerly trend towards Skagerrak, with minority penetrating into North Sea (Scotland, Helgoland Bight). In contrast, birds ringed only 250 km further north (Røst in Lofotens) produced fewer recoveries in southern Norway, but November–February recoveries of immatures far to the west, in Faeroes, Iceland, south-west Greenland, and 1 in Newfoundland (Myrberget 1973b). Data from other Norwegian colonies sparse, but support indication that Lofoten and Vesterålen birds are more likely to move out into Atlantic than birds from colonies south of Arctic Circle. Of 21 Newfoundland recoveries listed by Tuck (1971), 18 were Icelandic birds, all 1st-winter and found 8 October–11 March; no other data on dispersal of Icelandic population (no recoveries in Greenland listed by Salomonsen 1971).

SOUTHERN POPULATIONS (*grabae*). Ranges in winter south to c. 30°N and penetrates western Mediterranean (east to Italy). Most ringing data from British and Irish colonies, with 175 foreign recoveries to 1982. Birds ringed at east coast colonies (Fife, Northumberland) remained in North Sea all year, including birds found southern Norway in autumn and winter (Mead 1974; M P Harris). Birds ringed northern Scotland and Irish Sea also produced a small number of Norwegian recoveries in autumn and (later) further south in North Sea. Otherwise recoveries for these northern and western colonies indicated southward Atlantic movement by adults and immatures to Bay of Biscay, Atlantic coast of Iberia, and western Mediterranean; also recoveries in Morocco (7), Madeira (1), and Canary Islands (2). Immature birds found in summer from Bay of Biscay northwards to Faeroes and Iceland; many caught then in fishing nets, reflecting not only their inexperience but also their presence in areas of high marine productivity where also many fishermen, e.g. off Atlantic coast of Ireland (M P Harris). Some young birds move further west or northwest in Atlantic in winter: 4 Scottish 1st-winter birds found Newfoundland in December–January, and 2 Greenland recoveries in November (Scottish juvenile) and December (Welsh 4-year-old).

Recovery rates from ringing are particularly low in this species compared with other Alcidae, probably due to more offshore (pelagic) distribution. In British birds: *Uria aalge* 2·7%, *Alca torda* 3·0%, *F. arctica* 0·7%.

Food. Chiefly marine fish; in adult diet, more invertebrates, especially crustaceans, in arctic regions. Birds searching for prey commonly dip head in water, surface-diving once prey spotted, and swimming underwater mainly with wings, as in other Alcidae. Most prey caught near surface, dives probably not exceeding 15 m (Harris and Hislop 1978). On Jan Mayen, horizontal movements underwater of 30–50 m reported, taking 5–10 s (Bird and Bird 1935). Crustaceans taken by deep-diving, not at very surface (Bird and Bird 1935). Fish caught apparently only by day (Harris and Hislop 1978). For surface-diving after 'crash-landing' approach over shoal, see Bourne (1976a). In breeding season, often feeds in loose flocks, of which gulls (Laridae) usually the nucleus; mean number of *F. arctica* in mixed flocks, Irish Sea, 15 (2-120) (Birkhead 1976). For suggestion that synchronous 'splash diving' may be signal to initiate collective capture attempt, see Taylor (1978). Several fish may be caught during single dive, and, if intended for young in nest, stacked crosswise in bill (for details, see Perry 1940). Birds may bring back a few large fish or several small ones (Ashcroft 1976). Usual load *c*. 5-10 fish, seldom more than 10-12 unless feeding on larval sand-eels (Harris and Hislop 1978; see also Myrberget 1962a, Corkhill 1973). Maximum recorded, St Kilda (Scotland), 61 sand-eels (Ammodytidae) and 1 three-bearded rockling *Gaidropsarus vulgaris* (Harris and Hislop 1978; E K Dunn). Despite suggestions to the contrary (Perry 1940; Nash 1975), no evidence that fish caught with regular swing of head, alternately to left and right; orientation of individual fish held in the bill is random (Myrberget 1962a). Prey usually killed by capture, but sometimes still alive when brought to young (Harris and Hislop 1978). Rough sea conditions probably reduce feeding rate, though no direct evidence. Suggestion by Lockley (1934b) of tidal influence on feeding times at Skomer Island (Wales), disputed by Corkhill (1973), though possible tidal influence detected at Sept-Iles, Brittany, France (Pénicaud 1978). In breeding season most birds feed close to colony, probably *c*. 2-10 km away (Harris and Hislop 1978). At Skomer, most apparently within 8 km (Ashcroft 1976); 85% within 3 km of island, maximum *c*. 37 km (Corkhill 1973). Similar range in Co Kerry, Ireland (P G H Evans). Birds feed close inshore, Barents Sea (USSR) but ♂♂, Ainov Islands (USSR), may fly 15-25 km (Belopol'ski 1957).

Especially in breeding season, and at lower latitudes, mainly small pelagic fish, notably Ammodytidae—sand-eels *Ammodytes marinus*, *A. lancea*, *A. tobianus*, *Hyperoplus lanceolatus*; Clupeidae—sprat *Sprattus sprattus*, herring *C. harengus*, capelin *Mallotus villosus*; Scombridae—mackerel *Scomber scombrus*; Gadidae— arctic cod *Boreogadus saida*, cod *Gadus morhua*, whiting *Merlangius merlangus*, haddock *Melanogrammus aeglefinus*, saithe *Pollachius virens*, pollack *Pollachius pollachius*, three-bearded rockling *Gaidropsarus vulgaris*, five-bearded rockling *Ciliata mustela*, northern rockling *C. septentrionalis*. Less commonly taken fish include Gadidae—Norway pout *Trisopterus esmarkii*, blue whiting *Micromesistius poutassou*; snake blenny *Leptoclinus*; crystal goby *Crystallogobius nilssoni*; ten-spined stickleback *Spinachia spinachia*; lesser weaver *Trachinus vivipera*; gurnard (Triglidae). Crustaceans all shrimp-like or planktonic: include, especially, *Thysanoessa inermis*, *Mysis*, *Euthemisto*, and Copepoda. Molluscs include squid *Alloteuthis subulata* and pteropods *Limacina helicina* and *Clio borealis*. Also a few pelagic polychaete worms (Nereidae). (Bird and Bird 1935; Hartley and Fisher 1936; Perry 1940; Gudmundsson 1953a; Lockley 1953; Uspenski 1956; Belopol'ski 1957; Kartashev 1960; Myrberget 1962a; Løvenskiold 1964; Pearson 1968; Harris 1970; Nettleship 1972; Corkhill 1973; Evans 1975; Ashcroft 1976; Harris 1978; Harris and Hislop 1978; Hudson 1979a; Lid 1981; M de L Brooke.)

Over most of range, little known about diet of adults, in or out of breeding season. Of 37 birds killed at colonies, Britain, 20 had empty stomachs, 11 contained remains of small fish (some sand-eels), 3 contained Nereidae, 1 was full of Euphausiacea (Harris and Hislop 1978). Early in breeding season, Isle of May (Scotland), planktonic crustaceans taken (Harris and Hislop 1978), as in Barents Sea (Belopol'ski 1957). 124 stomachs, Barents Sea, yielded 67·0% (by number) fish, 4·0% crustaceans, 4·0% polychaetes, 0·8% molluscs; remaining 24% sponges (Porifera), insects, and plant matter, presumed taken accidentally. At Sem' ostrovov (east Murmansk, USSR), 100 stomachs contained 56·8% sand-eels, 21·0% capelin, 19·7% herring, 2·5% cod. At Ainov Islands, 39 stomachs contained 50% herring, 43·7% capelin, 6·3% sand-eels; only ♂♂ appeared to take sand-eels, ♀♀ taking proportionately more capelin (Belopol'ski 1957, which see for other sexual, and seasonal, variations). In Spitsbergen, various studies, summarized by Løvenskiold (1964), suggested adult diet in breeding season mainly crustaceans and fish, supplemented by molluscs and polychaetes. Of 10 stomachs, *Thysanoessa inermis* in all, *Euthemisto* in 3, *Mysis* in 2, fish remains (including arctic cod and possibly snake blennies) in 7. In other studies, diet included gammarids and pteropods (see Løvenskiold 1964). On Jan Mayen, crops of adults contained large numbers of *Mysis relicta* and copepods (probably *Calanus finmarchicus*), but only 1 fish specimen (Bird and Bird 1935). Winter diet not known, though likely to include more crustaceans than in summer (see Belopol'ski 1957, Harris and Hislop 1978).

Diet of young widely studied. Almost entirely small fish. In Europe, bulk is sprats, sand-eels, whiting, and rockling. For full list of species reported from British coastal waters, see Harris and Hislop (1978). Marked variations may occur between even quite close colonies, and between years; also within breeding season at given colony. Thus at St Kilda, sprats in diet varied from less than 1% ($n = 359$ fish) to 76% ($n = 992$) in different years; at Isle of May, 3-73%. On St Kilda, rockling 50-60% of diet late June (49 meals, 282 fish) but less than 10% by late July or early August (65 meals, 910 fish) (Harris and Hislop 1978, which see for other seasonal fluctuations; see also Corkhill 1973). On Rona (Scotland), almost equal proportions of sand-eels and five-bearded rockling (Evans 1975). Whiting important at times on St Kilda and Shiants, but not further south in Britain (Harris and Hislop 1978). In southern and eastern Britain, sand-eels and sprats predominate; sand-eels chiefly at beginning of chick-rearing period, sprats later on (Pearson 1968; Corkhill 1973; Ashcroft 1976). Of 1387 items brought to young, Skomer, 53·6% *Ammodytes marinus*, 42·0% sprats; herring much less common

(3·26%), and 0·2% each Gadidae and *Hyperoplus lanceolatus* (Corkhill 1973; see also Ashcroft 1976, Hudson 1979a for similar results, same island). At Farne Islands (England), sand-eels chief prey—85% by number, 80% by weight ($n = 219$) (Pearson 1968). In July at Mykines (Faeroes), entirely sand-eels (M de L Brooke). At Lovunden (Norway), 73·3% *Ammodytes lancea*, 22·2% herring, 3·9% saithe, 0·6% cod (Myrberget 1962a). At Røst (Norway), commonest prey mackerel, then herring; also some sand-eels and Gadidae (Lid 1981). Of fish fed to young, Barents Sea, sand-eels more frequent than in adult diet (Belopol'ski 1957). Size of fish taken by captive adults determined more by body-depth than length of fish, limit 23 mm depth (Swennen and Duiven 1977). Typical length of prey fed to young, Lovunden, 6–10 cm (Myrberget 1962a). Numerous studies of prey size, Britain, show most fish for young *c.* 5–8 cm; from large samples, derived from several colonies and years, sand-eels *c.* 40–120 mm, sprats *c.* 40–70 mm, rockling *c.* 30–37 mm, whiting *c.* 40–63 mm; largest fish of any kind a sand-eel of 207 mm (25 g) (Harris and Hislop 1978; see also Myrberget 1962a, Pearson 1968, Harris 1970, Corkhill 1973, Evans 1975, Ashcroft 1976). Herring fed to young, Lovunden, 39–137 mm ($n = 102$, Myrberget 1962a). Mackerel fed to young, Røst, 35–90 mm (Lid 1981). Weight of fish-loads brought to young varied from 0·2–29 g (Harris and Hislop 1978, which see for calorific values, and differences between colonies and years; see also Harris 1978). At Lovunden, mean weight 10·3 g (1·6–19·9, $n = 89$) (Myrberget 1962a). Mean weight, Skomer, $8·0 \pm 2·9$ g (1·7–18·5, $n = 96$) (Ashcroft 1976; see also Corkhill 1973). At Sept-Iles, at least 4·8 feeds per chick per day (Pénicaud 1978); at Skomer, 6·9 feeds per chick per day (2 days; 10, 12 chicks respectively: Corkhill 1973). At colonies, Scotland, 3·8–15·7 feeds per day, maximum 24 (Harris and Hislop 1978). Numerous statements (Perry 1940; Lockley 1953; Myrberget 1962a; Pearson 1968; Brun 1971a) that young fed only 2–3 times per day probably, or certainly, underestimates (Harris and Hislop 1978). See also Nettleship (1972). Feeding rate varies with age of young. At Skomer, rate increased from *c.* 2·5 times per day (20 g intake) at 0–4 days, to *c.* 10 times per day (74 g) at 24–28 days, declining to less than 4 times per day (25 g) between 36 days and fledging (*c.* 40 days) (Ashcroft 1976, 1979; for similar results see Myrberget 1962a, Harris 1976a, Harris and Hislop 1978, Hudson 1979a). For other estimates of daily food intake (based on lower assumed, or actual, feeding rates), see Myrberget (1962a), Nettleship (1972), and Tschanz (1979). After reaching maximum weight (70–80% of adult weight) at 28–30 days, young lose 8–10% of this during 6–10 days before fledging (M P Harris). Young said to be deserted latterly in nest (Lockley 1934b; Myrberget 1962a), but may also be fed up until day of fledging (Ashcroft 1976, 1979; Harris 1976a; Hudson 1979b). Variation possibly due to differences in timing of breeding season (Tatarinkova 1982). For regular fluctuations of food input, synchronized within local groups of burrows, see Ashcroft (1976) and Hudson (1979a). Captive young voluntarily reduced food intake after *c.* 30 days (Harris 1978), and apparently influence parental feeding rate (M P Harris). For evidence that parents determine quantity of food for offspring, see Hudson (1979b). Most food delivered early morning, some in afternoon, and slight resurgence in evening, exact timing varying with colony (Lockley 1934b, 1953; Myrberget 1962a; Roberts *et al.* 1963; Corkhill 1973; Harris and Hislop 1978; Pénicaud 1978). Adults thought to begin foraging at dawn, fulfil their own requirements, and then collect food for young. Rarely feed young when too dark (for human) to see (Harris and Hislop 1978). EKD

Social pattern and behaviour. Compiled partly from notes and material supplied by M P Harris and K Taylor.

1. Gregarious when breeding, otherwise usually occurs in ones and twos (M P Harris). In spring, birds assemble in rafts close inshore near breeding areas (see Flock Behaviour, below). During breeding season, near Skomer Island (Wales), rarely in large flocks when feeding, except in mixed assemblages with other seabirds. Usually in small loose flocks of 2–7 on sea, 69% of which comprise 2 or more birds, 53% 3 or more (Ashcroft 1976). 96·5% of birds fishing off eastern Scotland in groups of 2–8, median 2 (Taylor 1978). At Skomer, 90% of birds leaving colony, and returning, did so in groups of 2–3, rest 4–6 (Ashcroft 1976). At Vik (Iceland), departing flocks usually of 2–4 birds, returning flocks larger: mean $9·34 \pm 0·43$ ($n = 50$), sometimes exceeding 20 (Taylor 1982). BONDS. Monogamous mating system. Pair-bond maintained from year to year. No information on bonds outside breeding season. Most immatures do not return to colony until 2nd year, then visiting it during chick-rearing period. In following years, return progressively earlier (Ashcroft 1976, 1979; Hudson 1979a; Harris 1983a). Small proportion of 1-year-olds visit some colonies late in season, most remaining in rafts (Myrberget 1962c; Corkhill 1972; M P Harris; see also Flock Behaviour, below, and Movements). Age of first breeding seems to vary between colonies, but most often 5–6 years. At expanding colony on Isle of May (Scotland), breeds rarely at 3 (1 out of 54), usually 4–5 (Harris 1983a); 93% of 46 birds of 6 years or older caught in burrows had eggs or young (Harris 1983a). At Farne Islands (England), no records of breeding at 3–4 ($n = 42$) (Harris 1981), and on Skomer none attempted to breed before 4 (Ashcroft 1976, 1979; Hudson 1979a). At Vestmannaeyjar (Iceland), most start to breed when 5–6, ♂♂ possibly earlier than ♀♀ (Petersen 1976a). For age of first breeding, Egg Rock (Maine, USA), see Kress (1981). In 142 pair-years, Skomer, 7·8% of pairs divorced per year (Ashcroft 1979). Replacement birds usually heavier than predecessor, suggesting forcible eviction by former (Ashcroft 1976; see also Nettleship 1972). Temporary changes also occurred, in each of 3 cases the ♀ being replaced by another for 1 year only; apart from these, more birds (3 out of 4) changed mate after breeding failure than after success (Ashcroft 1976). Both sexes incubate and care for young until fledging: suggestion that ♀ may incubate more (Lockley 1934b, 1953) unlikely (M P Harris). BREEDING DISPERSION. Nests in colonies, mostly large and dense. Solitary nesting rare, and then usually in cliffs (M P Harris). Colonies of over 50000 pairs not uncommon in main breeding range (Gudmundsson 1953; Myrberget 1959a; Harris 1976b; Harris and Murray 1977). For colony sizes in west Palearctic, see Harris (1976b). Colonies divided into sometimes contiguous sub-colonies which are to some extent independent

units (see Flock Behaviour, below). Density of occupied burrows varies, but often 0·2–0·6 per m² (Harris 1976b, which see for review). On St Kilda (Scotland), 0·33 per m² (Harris and Murray 1977), mean in densest part 0·502 per m², in sparsest 0·16 per m² (Harris 1980; see also Harris 1976b). Comparable densities elsewhere in Scotland (Harris 1976b) and in Norway (Myrberget 1959a; Brun 1966), but higher in Ainov Islands, USSR (average 0·5–1·5 per m²: Skokova 1962, 1967), Lambi, Faeroes (average 1·74–1·84: Watson 1969), and said to be considerably higher on Vestmannaeyjar (Fridriksson 1975). In colonies where gulls (Laridae) and skuas (Stercorariidae) habitually rob *F. arctica* of food, nest density often highest near cliff edge, lower inland (see Grant and Nettleship 1971 and Nettleship 1972 for details). Nest-area territory defended, apparently mostly by ♂ (Lockley 1953; Nettleship 1972); confined to burrow and entrance (Myrberget 1961; Nettleship 1972), and serves only for egg-laying and chick-rearing. Adults show marked fidelity to previous nest-site (e.g. Skokova 1967, Nettleship 1972). Following account for Skomer from Ashcroft (1976). 95% of birds retained same site between seasons. Neither sex more likely to move than the other; 5·3% of ♂♂ moved (n = 169 bird-years) and 5·2% of ♀♀ (n = 232 bird-years). No bird changed site more than once in 4 years of study. Of 21 moves, about half due to eviction, others voluntary. One or both members of pair may be evicted (Ashcroft 1976). Move more likely after breeding failure (Myrberget 1961). 17% of 71 unsuccessful birds moved compared with 6% of 304 successful ones (Ashcroft 1979). Both sexes 'own' the burrow (Ashcroft 1976), and birds of either sex whose mates disappeared showed strong tendency (10 out of 13 birds) to stay in same burrow, there to become paired with new bird and breed the following season (Ashcroft 1979). Only 1 out of 5 ♀♀ and 0 out of 3 ♂♂ moved after death of mate (Ashcroft 1976). In a few areas, breeds in close association with Manx Shearwater *Puffinus puffinus*, some burrow systems accommodating both species; each, however, has slight advantage in its optimum habitat (Ashcroft 1976, which see for interspecific competition). Often associated with Leach's Petrel *Oceanodroma leucorhoa*, sometimes with Razorbill *Alca torda* and Black Guillemot *Cepphus grylle* (Brun 1966; M P Harris). ROOSTING AND COLONY-ATTENDANCE PATTERNS. Roosts largely at sea, at all times of year; in breeding season also sometimes at edge of colony. Loafs on land or sea, on land often with others in suitable part of colony, usually flat or sloping surface near cliff-top, or rocks and large tussocks within colony (K Taylor). Loafs and sleeps sitting on tarsi, breast touching ground; on land or water, rests with eyes half-closed, head withdrawn with bill forwards, or else bill tucked into scapulars. Loafing birds commonly yawn, preen, and bathe. On land or water, often raise fore-body to Wing-flap and shake tail, with plumage ruffled and eyes partially closed; Wing-flapping interpreted as comfort and/or displacement behaviour, expressing both readiness and reluctance to fly (Taylor 1976). Bathing and preening commonly performed by off-duty bird immediately after nest-relief (Lockley 1953). In pre-laying period, birds arriving to join raft initially preened and swam around, subsequently spending much time sleeping (Taylor 1976). On first arrival at Skokholm Island (Wales), birds remain offshore for c. 1 week (Harris 1982b). Rafts form inshore from c. 07.00 hrs, increasing in size up to 10.00 hrs, then dwindling and finally disappearing by c. 17.00 hrs as birds depart to roost further offshore. After 5–10 days of rafting, interspersed with retreats further offshore (see below), first flights made over colony area (Lockley 1953). At Isle of May, rafts initially of less than 10 birds each, 100 m or more offshore, from mid-morning. First landings rare before midday (Lockley 1953; Myrberget 1959b; Corkhill 1971; Pénicaud 1978); at Isle of May, concentrated at c. 10.00–21.00 hrs (Taylor 1982). First landings lasted c. 1–2 hrs, increasing in length during successive days as birds established sites (Lockley 1953). On Lundy (England), birds entered burrows from first day of landing (Perry 1940). Attendance at colony in pre-laying period characterized by 'quasi-cyclical' pattern of high numbers and absence (or low numbers) (Lockley 1934b, 1953; Perry 1940; Myrberget 1959a; Corkhill 1971; Nettleship 1972; Ashcroft 1976, 1979). Thereafter, attendance continues to vary, though in less marked and regular fashion. At Skomer, pre-laying peaks are 3–7 days apart, mean 5·2 (Corkhill 1971; see also Ashcroft 1976); at Isle of May, peaks every 4–5 days in both pre-laying and incubation periods; at St Kilda, peaks every 2–4 days during both incubation and chick-rearing (Taylor 1982); at Lovunden (Norway) peaks every 4–11 days (Myrberget 1959b). During chick-rearing period, diurnal rhythm of activity roughly similar in all colonies studied, with peak of feeding young between dawn and mid-morning, during which time colony almost devoid of loafing birds; thereafter, as feeding subsides, attendance at colony and just offshore builds up, with most birds present 1–2 hrs before sunset. At dusk, off-duty and immature birds disperse to roost, mostly at sea (Lockley 1953; Ashcroft 1976; Harris and Hislop 1978). Despite numerous studies, no predictable relationship between attendance and weather, though often higher in low wind speeds and may be nil in gales (e.g. Lockley 1953, Myrberget 1959b, Corkhill 1971, Brooke 1972a, Dott 1974, Ashcroft 1976). Numbers usually increase through season as immatures return to colony, reaching maximum during chick-rearing. At start of season, Isle of May, first birds ashore mostly 5 years old or older; progressively younger birds as season advanced, finally a few 1-year-olds in late July–August; successful and unsuccessful breeders, and immatures of all ages, remained until end of season and departed within a few days (Harris 1982c, 1983a; see also Myrberget 1962c and Ashcroft 1976).

2. At start of season, birds especially nervous and flighty (see Flock Behaviour, below). Large gulls *Larus* or Jackdaw *Corvus monedula* flying over may cause birds to fly off *en masse* or bolt down burrows (Perry 1940; Lockley 1953; Corkhill 1971). Birds in the air bunch if threatened by Great Black-backed Gull *Larus marinus* (Taylor 1982). For avoidance responses to Arctic Skuas *Stercorarius parasiticus*, see Grant (1971). As in other Alcidae, alarmed birds in water Head-dip: a brief submersion of head, often repeated rapidly many times (Taylor 1976). FLOCK BEHAVIOUR. Birds in early-season rafts near colony constantly shift from one group to another. Birds mostly c. 1 body length apart, though others, presumably pairs or prospective pairs, closer (Conder 1949). Much behaviour in raft associated with courtship and mating (see Heterosexual Behaviour, below). Birds within a group may dive synchronously (Lockley 1953; Taylor 1978). Following account of rafting and Wheeling-flight based on Taylor (1976, 1982). Rafts periodically coalesce. Flock members highly active at this time (none sleeping), swimming erratically to-and-fro (Taylor 1976) through one another, and Wing-flapping frequently; such birds hold head high, but keep body sunk low (Myrberget 1962b). Active groups typically of 4–6 birds, and most common at dusk when c. $\frac{1}{5}$ of raft may be mobile at any one time (Myrberget 1962b). Perhaps indicates pre-flight excitement before going offshore to roost (Taylor 1982). By day, rafting actively followed by mass take-off, birds making low, erratically circling passes over rafting area. Such flights usually brief, but if many birds join in, path may take flock also over colony area (Wheeling-flight) allowing reconnaissance (Taylor 1982, which see for discussion of the function of 'Wheels'). In pre-laying period, Isle of May, Wheeling-flights usually begin late morning; if they continue, raft moves closer inshore, leading to mass landings and eventually to establishment of nest-sites (Lockley 1953;

Corkhill 1971; Taylor 1982). Seasonal and diurnal timing of rafting behaviour differs between sub-colonies, each of which forms discrete raft and Wheeling flock. Wheeling-flights continue throughout season, sub-colonies still performing separately (Taylor 1982); may contain thousands of birds and continue for hours (Skokova 1962, 1967; Harris 1980). High proportion of birds in Wheeling flock immature and adult non-breeders (56% of 101 caught, compared with 2% of 137 caught flying to and from burrows) (Ashcroft 1976; M P Harris). Each Wheeling flock occupies more-or-less defined area, coinciding with sub-colony below it (Taylor 1982), birds following path over colony area and adjacent sea (Lockley 1953; Myrberget 1962b; Skokova 1962, 1967; Nørrevang 1977; Taylor 1982). Turning points are often features such as promontories, gullies, etc. Birds fly downwind on outer (sea) track, upwind on inner track. In calm conditions, path circular–elliptical and birds at different heights may fly in opposite directions. In moderate winds, path more often a figure-of-eight; in high winds, birds may close ranks and fly zig-zag course back and forth, overshooting boundaries of normal Wheeling-flight. In any weather, birds often deviate, making outward loop before rejoining near point of leaving; looping behaviour more frequent in higher winds and sparser flocks (see Taylor 1982 for details; also Myrberget 1962b, Skokova 1962 for variants of Wheeling-flight). Individuals averaged $1·56 \pm 0·15$ circuits ($n = 168$) before leaving Wheeling flock; usually land in colony, often in groups (Ashcroft 1976), but 5% landed in sea immediately adjacent. Fish-carriers made fewer circuits than birds without fish. All birds made more circuits, more loops, and flew closer together when large gulls *Larus* (food-pirates and direct predators of *F. arctica*) nearby (Taylor 1982). At Vik, birds appeared to leave colony at random during day (except at dawn) but formed groups of increasing size as they headed out to sea. Some incoming birds likewise formed groups by circling 0·5–1 km out from colony (Grant 1971; Taylor 1982; see also Skokova 1962, Ashcroft 1976). Apparently unlike other Alcidae, *F. arctica* shows behaviour possibly associated with synchronization of roosting movements (Taylor 1976, 1982). In Moth-flight (Fig A), bird of either sex flies with wings angled stiffly upwards, wing-tips fluttering rapidly in shallow arc, greatly reducing normal flight speed; body arched, head angled down, and feet often crossed (Myrberget 1962b; Taylor 1976, 1982). Moth-flight used on take-off or landing, when alone or in groups (Lockley 1953). Often seen on initial take-off from raft where birds displaying (Myrberget 1962b). Especially frequent, however, around dusk when repeatedly circle thus over colony until joined by others, whereupon group heads offshore together in close formation, evidently to roost. Moth-flight may signal intention to depart, rather than roost, as also seen at sea by day before leaving feeding flock (M Tasker). Probably not homologous with Butterfly-flight of *Alca* and guillemots *Uria*, which appears to have stronger territorial function. Also, as dusk approaches, birds of both sexes begin to jerk head upwards, while uttering variant of Grunting-call (see 3 in Voice). Head movement, sometimes slightly sideways, is at shallower angle than in Head-flicking (see below) and at rate of up to one toss per second. Movement, performed by stationary or moving bird, often engaged in Low-profile-walking (see below) stimulates same in others and thought to aid synchronized departure for roosting at dusk (Taylor 1976) when commonly performed simultaneously by almost every bird in given part of colony (Taylor 1976; P G H Evans). May be interspersed with brief bouts of Bill-gaping (Taylor 1976: see below). ANTAGONISTIC BEHAVIOUR. (1) Threat and fighting. In rafts, ♂ defends ♀ when conspecific intruder approaches within *c.* 2–3 body lengths (Conder 1949). ♂ swims between them, warding off or chasing intruder with flailing wings and lunging bill (Lockley 1934b, 1953). Intruders threatened by Bill-gaping (Fig B); if not repulsed, may be chased over water (Lockley 1934b, 1953; Myrberget 1962b). On land, Bill-gaping (by either sex) often associated with Forward-threat posture (Fig B) in which bird stretches low towards rival (Perry 1940; Myrberget 1962b; Taylor 1976) and may approach it, tail cocked (Lockley 1953); often turns head side on to rival (D N Nettleship). Bill opened widest in strong threat; tongue may be raised and head and neck plumage ruffled. Protracted Bill-gaping ('threat yawn' in Lockley 1953) accompanied by Creaking-call (Lockley 1953: see 1 in Voice). Bird landing in group often elicits brief Bill-gaping from nearby birds which often bite landing bird. Bill-gaping may be reciprocated (Gape-contest), the 2 birds facing and slowly sweeping their heads to left and right in unison, i.e. bill-tips kept pointed at each other (Taylor 1976). If one bird does not retreat, fight may follow: rivals face each other with plumage ruffled, and wings outstretched to cuff opponent or press on ground for balance; grapple and twist with bills, uttering Growling-calls (see 2 in Voice), and claw upwards with feet. Interlocking rivals may tumble down slope for several metres, sometimes toppling over cliff (Lockley 1934b, 1953; Perry 1940; Taylor 1976). Threat and fighting usually associated with defence of burrow (Perry 1940; Nettleship 1972), or of suitable ground (Taylor 1976). On Great Island (Newfoundland, Canada), frequency of fighting at maximum during egg-laying, declining thereafter (Nettleship 1972). (2) Appeasement behaviour. Various postures used for signalling lack of aggressive intent when landing or walking through colony. Bird alighting less than 2·5 m from others adopts Landing-posture (Fig C, right): partly bends legs and usually places one foot prominently in front of the other; bends forwards so that fore-body and bill tilted not more than *c.* 20° above horizontal; raises wings high over back, and usually cocks tail; slightly sleeks plumage (Danchin 1983; see also Taylor 1976). Bird maintains posture for up to *c.* 4 s before relaxing and folding wings (Danchin 1983)—the nearer to another bird the longer is the posture usually held (Taylor 1976); some birds follow posture by raising fore-body and bill amost vertically (Fig C, left). Landing-posture adopted by 95% of 277 birds, without fish, landing less than 2·5 m from another; thus almost invariably performed by immatures landing among others in 'club' area (see below). Infrequently performed (1 out of 29) by fish-bearing birds landing less than 2·5 m from another bird (Danchin 1983). Bird moving outside its own nest-territory, and near others, typically performs related Low-profile-walking display (Fig D): with

B

A

Fratercula arctica 239

C

D

body horizontal, close to ground, head in line, and carpal joints raised, bird usually runs quickly, stopping frequently to rise up and look round; performed most often near dusk when most nest-prospecting occurs (Taylor 1976). HETEROSEXUAL BEHAVIOUR. (1) Pair-bonding behaviour. Not well understood, except in re-formation of bond (strongly linked with possession of nest-site, see above) in experienced pair. In these, courtship and copulation occurs in rafts, so that birds may enter colony almost ready to lay (see Corkhill 1973). In 3 cases where ♀ late in returning at start of season, ♂ paired temporarily with new ♀ until former mate appeared (Ashcroft 1976). Birds without burrows (at Skomer 20–27% of individuals present: Ashcroft 1976) not usually paired. Adults losing burrows also appeared to forfeit mate; 4-year-olds capable of holding burrow even temporarily were paired (Ashcroft 1976; see below). At Isle of May, 2-year-olds move around colony, sometimes displaying briefly with other birds, but rarely enter burrows; 3-year-olds often defend burrows, but soon evicted by older birds (Harris 1983a). At Skomer, 2- or 3-year-olds often visit burrows but never attempt to defend them; none paired, though individuals sometimes attempted Billing (see below); paired 4-year-olds often try to occupy burrows, and do so until displaced by returning owners. Older immatures spend at least c. 2 hrs on land each evening, engaged in prospecting, clearing, and repairing burrows, carrying nest-material, courtship, loafing, and visiting neighbouring burrows and birds (Ashcroft 1976). Late in season, clubs, mainly of 2- and 3-year-olds (no adults), assemble on tidal rocks, Skomer (Hudson 1979a), though club-formation much less marked than in *Alca/Uria* (M P Harris). Sequence of behaviour in pair-formation not known, though components well described. ♂ may attract ♀ by sharp upward Head-flicking past vertical; repeated continuously for up to 10 min, at c. 1 Head-flick per s (Myrberget 1962b; Hudson 1979b; Taylor 1982). With each flick, bird often gives Grunting-call (Lockley 1953; Myrberget 1962b). For suggestion that ♀ also Head-flicks, see Lockley (1953). Head-flicking most often performed by ♂ pursuing ♀ in raft as pre-copulatory display. As Head-flicking ♂ comes close to ♀, he raises breast and begins Wing-fluttering in short bursts; sometimes sustained for several seconds and thus clearly distinguishable from Wing-flapping (Taylor 1976); according to Myrberget (1962b), 1 beat of wings per Head-flick. One ♂ displayed thus to 7 different ♀♀ in c. 20 min (Myrberget 1962b). Occasionally performed by ♂ on land attempting to entice ♀ to burrow; appears to softly buffet her towards entrance (K Taylor). 4–5(–12) immatures may gather on land (usually in areas of high burrow density) and Head-flick together ('Social nodding': Myrberget 1962b), though less intensely than otherwise, bill flicking up to only c. 45–60° (Myrberget 1962b). On land or water, Billing a conspicuous and common display between 2 birds, usually already paired (Myrberget 1962b). At start of bout, bird of either sex approaches another with Low-profile-walking display, swinging bill from side to side; may then nuzzle or nibble under other's bill. Birds begin to clash bills noisily broadside, bouts lasting from a few seconds to c. 2–3 min. One bird often stands upright with head and neck plumage ruffled and bill angled down towards partner who keeps lower with more sleeked plumage. Both birds may cock tails, and pad slowly around on the spot with outspread webs (Taylor 1976). According to Lockley (1953), ♂ may Bill so vigorously as to push ♀ down slope. On water, Billing birds tend to pirouette as their bills clash near water surface. At end of bout in or out of water, one bird often keeps swinging its head for several seconds, and some nibbling of partner's fleshy gape may occur (Taylor 1976). Mandibles usually opened and closed a few times ('sippering') at end of bout (Perry 1940). Billing may start (e.g.) after 1 of the participants has just landed (though does not serve as ritualized meeting-ceremony: T R Birkhead), or when a 3rd bird passes close by, or when fight starts nearby; see also Myrberget (1962b). Several nearby birds often encircle Billing pair; spectators may attempt to join in and are usually chased off, but sometimes an initiator starts Billing with a spectator. Billing also commonly induces like behaviour in neighbours (Taylor 1976). Allopreening rare, other than nuzzling and nibbling at base of bill (Lockley 1953; Hudson 1979a; T R Birkhead). For account of ♂ nibbling eye-ring of neighbouring ♀ (not mate), see Lockley (1953). Billing sometimes preceded by adoption of invitatory Bowing-posture in which bird stands with head bent to ground, sometimes touching it; head turned slowly from side to side (for up to 1 min). Plumage of head and neck ruffled (Taylor 1976). On the sea, bill bowed to water surface. Bowing not common, but mostly occurs at sea as part of ♂'s pre-copulatory display, especially if ♀ reluctant. Sometimes performed by ♀; once, ♀ stood in strong Bowing-posture, and ♂ rushed over from 3 m away to initiate Billing (Myrberget 1962b, which see for other contexts). (2) Site-ownership displays. After it has landed, or when another bird (not mate) passes or lands nearby, bird performs Pelican-walk around own burrow entrance (Fig E): stiffly erect with head lowered to touch puffed-out breast, and tail often cocked; feet raised and lowered slowly in exaggerated manner. Pelican-walk also sometimes performed by bird approaching another prior to Billing. In similar contexts, bird may signal ownership of burrow or place in assembly area by Spot-stomping: stands still, raising

E

feet alternately with webs markedly outspread; a frequent response to bird landing nearby. Rapid lateral Head-shaking, used to dispel water from bill, also probably functions as site-ownership signal; more birds landing at own burrow Head-shake than do those landing elsewhere; paired birds Head-shake more than unpaired ones. (Taylor 1976.) (3) Mating. Copulation almost invariably on sea. ♂ may solicit ♀ on land with Head-flicking and Wing-fluttering displays, and mount her briefly, but not usually successfully (Myrberget 1962b; Taylor 1976). At sea (if ♀ does not swim off or dive: Perry 1940), ♂'s Wing-fluttering, hitherto in short bursts, becomes continuous (Taylor 1976). ♀ initially swims low in water, with neck retracted; if receptive, holds head higher, keeping rear of body low. Pair begin to swim slowly round each other (Myrberget 1962b) and may Bill together before ♂, flapping for balance, mounts for 5–6 s (Lockley 1953; Taylor 1976). Once mounted, ♀ continues to keep head and breast high, though body almost totally submerged; may dislodge ♂ by diving (Perry 1940). After dismounting, both birds often preen vigorously (Lockley 1953). One ♂ solicited 5 ♀♀ in succession, attempting to mate with 3 within 15 min; most such attempts probably unsuccessful (Myrberget 1962b). Copulation apparently less frequent than in *Uria/Alca*; occurs only a few times before egg-laying, and ceases thereafter (Lockley 1953; Hudson 1979a; M P Harris). (4) Behaviour at nest. No special meeting-ceremony occurs at nest-relief, though sitting bird may greet mate with Moaning-call (see 4 in Voice); this continues after hatching (Lockley 1953). Nest-relief lengthy, average 14·1 min (4–41, $n=9$). Incubating bird occasionally leaves burrow to defecate, sometimes to fly around; mean duration of absence 7·4 min (1–22, $n=5$) (Myrberget 1962a, which see for total duration of absence during day). When ♀ leaves burrow, ♂ may 'drive and lead' her, as if pressing her to resume incubation (Lockley 1953). RELATIONS WITHIN FAMILY GROUP. Parents apparently do not recognize offspring of any age since successful cross-fostering achieved at nest up to fledging (Myrberget 1962a; M P Harris). Young brooded, sometimes regularly, until 9 days old (Hudson 1979a). Thermoregulate efficiently after c. 5–7 days (see Tschanz 1979). Begging chick said to outstretch neck towards adult, gaping and fluttering wings (Myrberget 1962a), though this not seen by Corkhill (1973) in small young. Parent returning with fish for young less than 4 days old stood in middle of nest chamber, giving Lure-call (see 5 in Voice); chick, uttering food-call almost continuously, took first 3–4 fish from bill, then parent dropped rest for chick to retrieve from ground (Corkhill 1973). Small young periodically pick up pieces of nest-material from burrow floor (Myrberget 1962a). As chick grows older and more agile, parent often drops food just inside burrow entrance (Lockley 1934b, 1953; Corkhill 1973; Hudson 1979a; M P Harris). At 3 weeks old, chick walks towards entrance to meet arriving parent (Lockley 1953). Older young, especially from c. 28–32 days, increasingly exercise wings at burrow entrance, mostly at night (Myrberget 1962a; Ashcroft 1976; Lid 1981), sometimes also during daytime social gatherings (Ashcroft 1976). Chicks in spacious burrows may exercise inside (Myrberget 1962a). Despite reports to contrary (e.g. Lockley 1953, Myrberget 1962a, Skokova 1967), young in nest not invariably deserted near fledging (Harris 1976a, 1983b; see also Tatarinkova 1982). Young quit nest for sea at night, exceptionally by day (Harris 1982a; K Taylor); may walk to suitable take-off point if one available, otherwise fly straight out to sea (Lockley 1934b, 1953; Perry 1940; Myrberget 1962a; Ashcroft 1976; Lid 1981). Young independent at fledging. For behaviour of released young on reaching sea, see Lockley (1934b). On Isle of May, parents frequent burrows for up to 3 weeks after young fledged; many, perhaps most, adults remained at colony for several weeks after most young had left (Harris 1976a, 1982b). ANTI-PREDATOR RESPONSES OF YOUNG. Scurry into rear of burrow from 1 week old (Lockley 1953). Young capable of diving immediately upon fledging (Lockley 1934b, 1953). PARENTAL ANTI-PREDATOR STRATEGIES. (1) Passive measures. Often leaves egg or young at slightest disturbance (M P Harris). (2) Active measures. If threatened in burrow by human, adult usually crouches, facing away from intruder, at back of nest chamber (M P Harris). Said to repulse burrow-intruders by lunging and biting (Lockley 1953).

(Fig A from photograph in Urry and Urry 1970; Figs B, D, and E from drawings in Taylor 1976; Fig C from drawing in Danchin 1983.)   EKD

Voice. Less vocal in breeding season than other west Palearctic Alcidae, and with smaller repertoire. No information on calls outside breeding season. No detailed study; following scheme provisional.

CALLS OF ADULTS. (1) Creaking-call. Creaking or gurgling sound given during antagonistic Bill-gaping (Lockley 1953). (2) Growling-call. Given when fighting: a deep, loud 'arr' (Fig I), similar to call 4, but more emphatic and clipped (Lockley 1934b; Taylor 1976); a short, harsh 'urrr' (Perry 1940). (3) Grunting-call of ♂. During Head-flicking, 'arr' sound given with each movement (Myrberget 1962b); a faint porcine grunt (Lockley 1953). Probably serves to

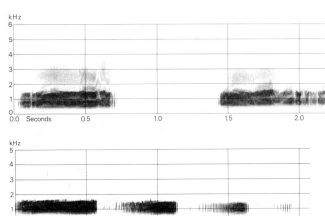

I P J Sellar Iceland June 1968

II L Shove/J-C Roché (1970) Wales June 1965

appease ♀, inviting close approach in pre-copulatory display. Similar call, 'uh', given, presumably by both sexes, with each movement during variant of Head-flicking associated with roosting (Taylor 1976, 1982). (4) Moaning-call. Low, growling or moaning 'arr', sometimes uttered singly, but often 3 times in slow succession, each syllable lower pitched than preceding one (Lockley 1934b). Rendered 'co-o-or-aa', 'haa-aa', or 'aa-aa-aa', descending in pitch (Perry 1940). Recording (Fig II) suggests a nasal, moaning 'AWW-aaah Aaah-aaah-aaah-aah-ah'. 1st syllable low, 2nd relatively high, subsequent ones steadily descending in pitch. Distinct gaps between syllables produce lazy, even laboured quality (EK Dunn). Mostly given underground, and may greet arrival of mate (Lockley 1953). Especially frequent during site-establishment (Perry 1940). Also heard from birds in loafing group (EK Dunn). (5) Lure-call. A repeated, soft, clicking sound, given by parent entering chamber with fish to feed young (Corkhill 1973). (6) Other calls. Variety of low grunts, growls, and groans, perhaps differing from those above.

CALLS OF YOUNG. Small young give a rather sharp, repeated 'chip-chip-chip' when being fed in nest (Lockley 1934b; see also Corkhill 1973). Towards fledging, young give a petulant, whistling, gosling-like 'whee-er-er whee-er-er' (Lockley 1953), whether parents present or not (Lockley 1934b).   EKD

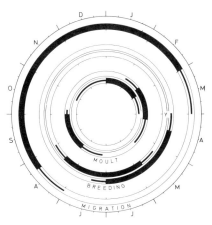

Breeding. SEASON. North Sea: see diagram. Colonies in west of Britain 1–3 weeks later (Harris 1982b). Iceland and Norway: laying from middle of May (Hantzsch 1905; Myrberget 1962a). Birds returning to breed at east Scottish colonies now arrive earlier than formerly (Harris 1982b). SITE. Shallow burrow, excavated by bird itself, or by Manx Shearwater *Puffinus puffinus* or rabbit *Oryctolagus cuniculus*; sometimes in natural crevice or under boulder. Burrow length, Iceland, usually 1–2 m, average just under 1·5 m (Gudmundsson 1953); individual burrows, Kandalaksha (USSR), usually curved and c. 2 m long, but many burrows interlinking with common entrances leading to complex of passages and chambers (Skokova 1967). Re-used in subsequent years, particularly if successful (Ashcroft 1979). Colonial. Nest: shallow hollow at end of burrow, or 1–1·5 m from entrance in longer burrow (Lockley 1934). Feathers, grass stems, leaves, etc., often taken into burrow but not arranged as lining. Building: burrow dug by ♂ and ♀. EGGS. See Plate 92. Sub-elliptical, smooth, not glossy; white, but faint brown or purplish shell markings often show; occasionally well-marked brown or purplish, often at broad end. Nominate *arctica* (*naumanni* similar): 63×44 mm ($57–69 \times 39–48$), $n = 100$; calculated weight 64 g. *F. a. grabae*: 61×42 mm ($56–67 \times 39–45$), $n = 88$; calculated weight 60 g (Schönwetter 1967). On Skomer Island (Wales), egg size found to decline through season (Ashcroft 1976). Clutch: 1, rarely 2, and then probably by 2 ♀♀. One brood. Replacements laid after early egg loss; of 75 pairs losing eggs, 7 (9·3%) laid replacement, 13–23 days later (Ashcroft 1979). INCUBATION. 39 days (36–43), Skomer (Ashcroft 1976). 41·8 days (40–45), $n = 42$, Lovunden, Norway (Myrberget 1962a). Mainly by ♀ (Lockley 1934), or more equally by both sexes (Myrberget 1962a). Begins on laying. At 11 nests, average incubation spell 32·5 hrs (23·4–69.8), $n = 47$, with changeovers slightly more frequent at night (Myrberget 1962a). YOUNG. Semi-altricial and nidicolous. Cared for and fed by both parents. FLEDGING TO MATURITY. Fledging period variable: 38 days (34–44), $n = 241$, Skomer (Ashcroft 1976, 1979); 47·7 days (43–52), $n = 32$, Lovunden (Myrberget 1962a); 34–50 days, $n = 312$, Isle of May (Scotland); 35–60 days, $n = 214$, St Kilda, Scotland (MP Harris); 39–83 days, $n = 180$, Great Island, Newfoundland, Canada (Nettleship 1972). Become independent just before fledging. Age of first breeding normally 5 years, occasionally 3–4 (Harris 1981). BREEDING SUCCESS. Of 239 eggs laid, Skokholm Island (Wales), 1973–5: 77·8% hatched and 66·1% raised to fledging; main egg losses from desertion, especially after interference from Manx Shearwater, and predation by Jackdaws *Corvus monedula*; chicks, particularly hungry ones coming to mouth of burrow, lost to gulls *Larus* (Ashcroft 1976, 1979; Harris 1980); change of mate or burrow had no effect on success, but estimated 2·2% of birds 4 years or older did not breed in one of the years 1973–5; birds breeding for first time at 5 years more successful than those doing so at 4 (Ashcroft 1976, 1979). Because of differential predation and disturbance by gulls *Larus*, birds breeding in dense areas of colony more successful than others: of 185–192 burrows in dense areas, St Kilda (Scotland), 1974–7: 78·6–83·4% had eggs, 76·4–88·7% hatched, 69·0–83·8% eggs produced fledged young; of 122–145 burrows in sparse areas, 49·2–55·2% had eggs, 50·6–84·7% hatched, 33·7–64·8% eggs produced fledged young; early layers more successful than later (Harris 1980). On Røst (Lofoten Islands, Norway), unusually high chick mortality associated with food shortage (see Lid 1981). On Great Island, more eggs either failed to hatch or disappeared on level ground than on slopes; fledging success also higher on slopes, with fewer deaths or disappearances of chicks; on level ground, more panic flights and so more eggs kicked out and hungry chicks at mouths of burrows more easily taken by gulls (Nettleship 1972). Number of young fledged per breeding pair at 3 British colonies: $0.80 \pm SE$ 0.035 (Isle of May, 5 seasons); $0.77 \pm SE$ 0.03 (St Kilda, 5 seasons); $0.64 \pm SE$ 0.02 (Skomer, 3 seasons) (Ashcroft 1979; Harris 1980; MP Harris).

Plumages. ADULT BREEDING. Crown and nape greyish-black, slightly paler towards forehead. Lores and side of head below and above eye dull white, grading to light mouse-grey on side of head, neck, and upper cheek. Narrow furrow in feathers extends backwards from posterior edge of eye. Chin and throat light mouse-grey; streak on lower cheek dark mouse-grey. Upperparts black, separated from crown by narrow greyish band across upper hindneck. Black of lower hindneck extends as broad band across foreneck, dull black on centre of throat. Underparts white, but long feathers of flank with variable brown-black pattern. Tail and upperwing black; inner web of flight-feathers grey-brown, shaft dark horn, grading into brown towards tip; marginal upper wing-coverts white. Undersurface of flight-feathers dark silvery-grey, outer web of p10 black. Greater under wing-coverts dark silvery-grey; median under wing-coverts glossy white to pale silvery-grey; lesser under wing-coverts dark silvery-grey with concealed brown bases. Occasionally, outermost median under primary coverts with brown-black outer web. Axillaries greyish-brown. ADULT NON-BREEDING. Like adult breeding except for side of head. Lores and patch above and below eye dull black. Side of head, cheek, chin, and throat mouse-grey; faintly developed dark mouse-grey streak on lower cheek. DOWNY YOUNG. Down very long and soft, shorter on belly and head; black-brown except for white centre of belly. Natal down replaced by juvenile feathers at $c.$ 20–22 days, completed $c.$ 30–32 days after hatching (Kartashev 1960), i.e. late June to early September (Salomonsen 1944; Dementiev and Gladkov 1951b). Down retained longest on neck, rump, and flanks (Bent 1919; Kartashev 1960; RMNH). Wing- and tail-feathers develop together with juvenile body plumage (Witherby et al. 1941; Salomonsen 1944; RMNH). JUVENILE. Like adult non-breeding, but black of lores and orbital region more intense, band across throat browner. This plumage retained during winter until moult into 1st immature breeding (Bent 1919; Witherby et al. 1941; Salomonsen 1944). FIRST IMMATURE BREEDING. As adult breeding, but some birds have dusky patches on face (Harris 1981).

Bare parts. ADULT BREEDING. Iris light blue, grey-brown, or hazel-brown (Ridgway 1919; Hartert 1921–2; Witherby et al. 1941), orbital ring vermilion; horny appendages above and below eye blue-grey. Swollen, horny rim at base of upper mandible and first diagonally-running ridge across both mandibles dull yellow. Proximal part of bill deep blue-grey, distal part scarlet, sometimes with white chalk-like substance at tip and in distal groove. Fleshy rosette at gape bright red. Leg and foot orange-red. ADULT NON-BREEDING. No horny appendages above and below eye. Orbital ring red or purplish-red (Ridgway 1919; Witherby et al. 1941). Proximal portion of bill brown, distal part orange-red. Rosette at gape shrunken and dull purplish-red (Ridgway 1919). Leg and foot yellow, webs darker (Witherby et al. 1941). DOWNY YOUNG. Iris black-brown. Bare skin round eye black (Kartashev 1960). Bill black to dark reddish-grey, lower mandible paler; mouth flesh. Leg and foot black, grey, or dusky (Witherby et al. 1941; Harrison 1975; Fjeldså 1977). JUVENILE. Proximal part of bill dark grey-brown, distal portion reddish-brown. Leg and foot flesh (Witherby et al. 1941).

IMMATURE BREEDING. Orbital ring red-brown and outer part of bill duller red in 1st and 2nd years. Leg and foot yellowish (MP Harris). IMMATURE NON-BREEDING. Probably as in immature breeding.

Moults. ADULT POST-BREEDING. Head, body, bill and lesser and median wing-coverts late July to September, after breeding (Witherby et al. 1941; ZMA). Flight- and tail-feathers and greater wing-coverts much later than in other Atlantic Alcidae: October–April, most often January–February; many individuals do not shed flight-feathers and tail-feathers until pre-breeding moult. Flight-feathers shed simultaneously, but very rarely one or more primaries not replaced (Salomonsen 1944; Harris and Yule 1977). ADULT PRE-BREEDING. Head, body, and bill late February to last third of May; completed just before return to colony (Salomonsen 1944; Dementiev and Gladkov 1951b; Harris and Yule 1977). POST-JUVENILE. Juvenile plumage retained during winter, replaced early March to July (Stresemann and Stresemann 1966; Harris and Yule 1977). IMMATURE POST-BREEDING. Scant information; probably as in adult. Body and upper and lower tail-coverts start late July or early August (ZMA). Primaries replaced July–March (Harris and Yule 1977). IMMATURE PRE-BREEDING. As in adult.

Measurements. Breeding adults. Bill (1) includes horny rim at base, bill (2) excludes it; bill (1) 1.5 (0.6; 40) 0.7–2.6 longer than bill (2), independent of race (ZMA). Depth is maximum bill depth at base.

F. a. grabae
Skomer Island (Wales); live (Corkhill 1972).

	♂			♀		
WING	159.7	(2.0; 88)	147–170	156.9	(2.0; 70)	146–168
DEPTH	36.4	(1.1; 88)	33.5–38.7	33.6	(1.6; 70)	28.2–35.6

Isle of May (Scotland); live (Harris 1979); range theoreticcal.

	♂			♀		
WING	161.8	(3.44; 29)	152–172	161.7	(2.66; 19)	154–170
BILL (2)	46.5	(1.42; 30)	42.2–50.8	44.2	(1.41; 22)	40.0–48.5
DEPTH	37.1	(1.15; 30)	33.7–40.5	34.6	(1.17; 22)	31.1–38.1

St Kilda (Scotland); live (Harris 1979); range theoretical.

	♂			♀		
WING	157.1	(5.77; 20)	140–174	156.0	(3.04; 20)	147–165
BILL (2)	44.2	(1.14; 21)	40.8–47.6	42.1	(1.34; 20)	38.0–46.1
DEPTH	35.3	(1.42; 21)	31.0–39.6	32.2	(1.65; 20)	27.2–37.2

British Isles; skins (Vaurie 1965).

	♂			♀		
WING	158.7	(—; 10)	152–164	156	(—; 13)	149–166
BILL (2)	46.3	(—; 10)	44–49	44	(—; 13)	41–46
DEPTH	36.1	(—; 10)	33–38	34	(—; 13)	30–37

Faeroes; skins; ♂ from Petersen (1976b), ♀ from Vaurie (1965).

	♂			♀		
WING	161.6	(3.8; 5)	156–166	157.2	(—; 20)	152–165
BILL (2)	44.1	(1.6; 5)	41.5–45.8	40.8	(—; 20)	40–47
DEPTH	36.0	(0.7; 5)	35.0–37.0	34.2	(—; 20)	29–38

Nominate *arctica*.
Southern Iceland; skins (Petersen 1976b; ZMA).

	♂			♀		
WING	165.0	(3.9; 19)	159–174	161.4	(3.6; 19)	154–166
TAIL	50.6	(2.4; 10)	47–55	49.9	(2.5; 7)	47–54
BILL (2)	46.1	(1.7; 19)	41.8–49.0	43.4	(1.7; 18)	39.7–45.6
DEPTH	37.7	(1.3; 10)	35.4–39.3	35.3	(1.4; 11)	32.6–37.5
TARSUS	27.9	(1.1; 10)	26.3–29.7	27.1	(0.8; 7)	26.0–28.4
TOE	45.1	(1.9; 10)	42.0–47.5	42.9	(1.5; 7)	41.0–44.9

Northern Iceland; skins (Petersen 1976b; ZMA).

	♂			♀		
WING	169.8	(4.0; 9)	164–176	167.8	(4.5; 12)	159–174
BILL (2)	49.7	(1.4; 9)	48.1–51.8	47.3	(1.5; 12)	44.7–50.5
DEPTH	40.6	(1.3; 5)	39.0–40.6	37.5	(0.8; 8)	36.5–38.8

Lovunden, Nordland, Norway; freshly dead (Myrberget 1963).

WING	♂ 168·9 (3·71; 96) 160–178	♀ 164·4 (3·60; 94) 158–175
BILL (I)	48·2 (1·34; 96) 45·3–52·2	46·0 (1·27; 94) 42·6–48·8
DEPTH	38·8 (1·13; 96) 35·9–42·0	37·1 (1·28; 94) 33·0–40·4
TARSUS	28·0 (0·66; 96) 26·3–29·5	27·2 (0·79; 94) 25·4–28·9

Bear Island; skins (Salomonsen 1944).

WING	♂ 168 (6·25; 3) 161–173	♀ 160·3 (6·95; 4) 150–165
BILL (I)	47·3 (1·15; 3) 46–48	45·0 (2·58; 4) 42–48
DEPTH	37·3 (3·51; 3) 34–41	36·4 (3·40; 4) 33–41

F. a. naumanni. Spitsbergen, skins (Salomonsen 1944; ZMA).

WING	♂ 183·9 (8·35; 13) 159–193	♀ 180·6 (10·1; 10) 156–191
BILL (I)	53·9 (3·32; 12) 45–57	53·4 (4·87; 10) 43–60
DEPTH	43·7 (3·64; 9) 35–48	43·3 (4·00; 9) 35–49

Weights. ADULT. Skomer, summer: ♂ 398 (25·0; 88) 345–488, ♀ 368 (22·0; 70) 310–445, sex difference significant (Corkhill 1972). St Kilda, June–August, sexes combined, 368 (737) (Harris 1979). Isle of May, July–August, sexes combined, 387 (620) (Harris 1981). Outer Hebrides (Scotland), June, sexes combined: 385 (28·5; 46) 320–445 (BTO, D J Steventon). Kerry (Ireland), July: 379 (30·1; 4) 340–405 (BTO, R J Kennedy). Iceland: ♂ 512 (10) 430–588, ♀ 456 (8) 420–525 (Timmermann 1938–49). Lofoten Islands (Norway), sexes combined: April 484 (38); May–June 459 (104); July 441 (248); August 399 (23) (Lid 1981). Lovunden, summer: ♂ 457 (25·0; 85), ♀ 426 (25·4; 79) (Myrberget 1963). Eastern Murmansk (USSR), April–August: ♂ 510 (70) 305–675, ♀ 472 (72) 308–565 (Belopol'ski 1957). Newfoundland, May–July: ♂ 481 (33·7; 91) 384–562, ♀ 442 (27·6; 134) 372–511 (Nettleship 1972).

DOWNY YOUNG. Skomer 39·7 (3·9; 20) (Ashcroft 1979); Labrador 42 (1) (Johnson 1944); USSR 50 (40–60) (Kartashev 1960); 40–54 (Fjeldså 1977); at 1 day old *c.* 45 (32) (Tschanz 1979).

JUVENILE. On Skomer, peak weight of nestling 317 (24·3; 241) (Ashcroft 1979), on Lovunden 330 (19) 270–360 (Myrberget 1962a). Juveniles experience reduced feeding frequency before leaving nest (see Food) and lose 8·1% of peak weight (Ashcroft 1979), fledging at 69–78% of adult weight (Ashcroft 1979; Harris 1982a). Weight at departure from burrow: Skomer 291 (23·2; 36) 255–375 (Corkhill 1972), 291 (24·7; 241) (Ashcroft 1979); Isle of May 291 (142) (Harris 1981); St Kilda 252 (4403) (Harris 1982a); northern Norway 276 (19) 240–320 (Myrberget 1962a); Newfoundland 258 (38·8; 179) 137–330 (Nettleship 1972). Mean fledging weight varies with calorific value of food, habitat, and time of year (Nettleship 1972; Harris 1982a).

Structure. Wing narrow and pointed. 11 primaries: p10 longest, p9 0–5 shorter, p8 3–9, p7 11–15, p6 20–25, p5 31–35, p1 69–79; p11 minute, concealed by primary coverts. Inner and outer secondaries about equal and shorter than p1, middle ones longer; thus trailing edge of inner wing rounded. Tail slightly rounded, but t1 shorter than adjacent feathers; (14–)16 feathers, tips rounded. Shape of bill differs with age and season. Bill laterally much compressed. In adult breeding, very deep, almost triangular in shape. Culmen arched, tip decurved; gonys proximally slightly concave, distally straight or slightly convex. Cutting edges of both mandibles straight, notched at tip. Swollen horny rim with aborted feather growth along basal margin of upper mandible; very narrow, scarcely raised, smooth, horny rim at base of lower mandible. Proximal part of both mandibles smooth, distal portion with 2 or more curved, oblique grooves, and corresponding ridges. Nostril narrow and slit-like, situated along cutting edge of smooth basal part of upper mandible. Fleshy rosette at gape consists of rounded patch of soft, swollen wrinkled skin; smaller in immature. Small triangular horny ornament above eye, rectangular one below; smaller in 1st year. Bill in adult non-breeding smaller and of different shape from adult breeding due to shedding of sheath during winter: lower outline of lower mandible with prominent angle; triangular basal part of bill replaced by brown lamina with narrow ridge along proximal margin; grooves and ridges in distal portion of bill less distinct; no horny rim at base of upper mandible; rosette at gape shrunken; also, horny appendages above and below eye absent. Bill features distinguish immature and adult breeding (although large individual variation makes definite ageing impossible): in 1- and 2-year-olds, bill distinctly triangular, with 1 or less than 1 groove on upper mandible; in 3-year-olds, bill more convex with 1 distinct groove and 1 shallow groove in front of it; in 4- and 5-year-olds, 2 deep and narrow grooves and 1 indistinct one (filled with chalk-like substance) in front of them; in 5- and 6-year-olds, 3 deep and narrow grooves. Some individuals never develop more than 2 grooves (Petersen 1976b; Harris 1981). Bill in downy young laterally compressed; gonydeal angle prominent; proximal part of bill darker than distal part and also wrinkled. Bill in juvenile and 1st immature non-breeding relatively small and shallow; upper mandible with indistinct and irregular ridge along distal margin of dark basal lamina; culmen gradually sloping; proximal and distal portions of lower edge of lower mandible straight with prominent gonydeal angle about half way along; no grooves in distal part of bill; rosette at gape very small; no eye ornaments. Claw of outer and middle toe curved, of inner toe sickle-shaped; claw of middle toe largest, innermost margin broadened laterally. Outer toe *c.* 94% of middle, inner *c.* 72%; no hind toe.

Geographical variation. Involves size only; see Measurements. Traditional division into races followed here but size increases clinally from south to north, making delimitation arbitrary because adjacent populations show much overlap (see Measurements). Variation in Norway clinal: in southern Norway, average wing of 16 skins (sexes combined) 162·7, bill (excluding horny rim) 44·6, bill depth 36·3, and hence similar to *grabae*; in Finnmark (nominate *arctica*), wing 173·7 (8), bill 48·0 (8), depth 38·2 (8) (Pethon 1967; see also Measurements). Birds of Jan Mayen and Murmansk coast considered intermediate between nominate *arctica* and *naumanni* (Salomonsen 1944; Vaurie 1965). Delimitation of races further complicated by birds from northern Iceland being significantly larger than those from southern Iceland (Petersen 1976b); Harris (1979) reported significant differences in size between birds from north-west and south-east Scotland.

RS

Order PTEROCLIDIFORMES
Family PTEROCLIDIDAE sandgrouse

Medium-sized, mainly seed-eating, terrestrial birds of arid habitats. A well-defined, homogeneous group of uncertain affinity (see further, below). Highly specialized for desert life. 2 genera: *Pterocles* (14 species) and *Syrrhaptes* (2 species). Represented in west Palearctic by 6 *Pterocles* (breeding) and 1 *Syrrhaptes* (irregular visitor and occasional breeder). Family of African–Eurasian distribution, range extending to Iberia, southern France, Kazakhstan, and Mongolia in the north, to Gobi Desert and central India in the east, and to Madagascar and southern Africa in the south. Most *Pterocles* found in Africa; *Syrrhaptes* confined to central Asia. For survey, see Hüe and Etchécopar (1957).

Bodies compact, elongated, and streamlined; heads small. Sexes similar in size. Necks rather short. 15–16 cervical vertebrae. Wings long and pointed. Flight direct and rapid. 11 primaries; p1 minute. 17–18 secondaries; diastataxic. Tails wedge-shaped, central feathers (t1) greatly elongated in some species; 14–18 feathers. Bills short, resembling those of certain gamebirds (Galliformes) and pigeons (Columbiformes); no cere. Tarsi short; densely feathered. Front toes strong, with thick soles covered with small scutes; fused and densely feathered on upperside in *Syrrhaptes*, forming kind of 'sand-shoe'. Hind toe small and raised in *Pterocles*, absent in *Syrrhaptes*. Gait a normal walk. Oil-gland small; naked. Caeca long and functional. Oesophagus with large, single-lobed crop. Syrinx different from Columbiformes with symmetrical sterno-tracheal muscles and 2 pairs of tracheo-bronchial muscles (Stresemann 1927–34). Gall bladder present. No supra-orbital salt-glands. Feathers with small aftershaft. As in pigeons, plumage dense with feathers loosely set in skin; structure of feather-shafts also pigeon-like (see Columbidae). Down developed on apteria, but only sparsely over greater part of plumage; modified to some extent on underparts and flanks as powder-down. Down tracts few but those on abdomen, together with downy parts of feathers there, have highly specialized structure facilitating transport of water to young (Cade and Maclean 1967). For fuller technical diagnosis, see Stresemann (1927–34), Witherby *et al.* (1940), and Fjeldså (1976).

Plumages cryptic; intricately variegated with spots and bars on brown or buff ground-colour, often with striking pattern on head or with dark area on breast or belly. Sexually dimorphic. No seasonal changes except in Pin-tailed Sandgrouse *P. alchata*. Irises red-brown to dark brown; bare skin round eye often yellow or blue. Bills grey or blue-grey, sometimes brown. Post-breeding moult complete; slow, often taking 6 months. No pre-breeding moult except in *P. alchata* (partial, starting in autumn and interrupted in winter). Downy young precocial; seminidifugous. Self-feeding, but brought water by ♂ parent who transports it in soaked belly feathers, often over long distances (Cade and Maclean 1967; Maclean 1968; George 1969). Natal down dense. Structure on dorsal surface complex, each down feather resembling a circular spray of miniature feather vanes—rather similar to condition found in coursers (Cursorinae); that on ventral surface straight and looser (Fjeldså 1976, 1977). Light brown above with intricate markings (though these faint in some species living in bare desert, e.g. Lichtenstein's Sandgrouse *P. lichtensteinii*); wader-like (see Charadriiformes, Volume III), recalling highly modified pebbly-pattern type found in coursers but even more elaborate—with black tips to each down spray and lattice-like pattern of white bands, bordered black. ♀-like juvenile plumage develops shortly after hatching, growth of wing-feathers allowing a degree of precocious flight, as in gamebirds—by 2 weeks in *Syrrhaptes* and 4 in *Pterocles*; early- and late-developing feathers may be of different structure (see, e.g., Black-bellied Sandgrouse *P. orientalis*). Adult plumage acquired some months after hatching.

Sandgrouse once usually considered to form suborder Pterocletes of order Columbiformes (e.g. Peters 1937, Mayr and Amadon 1951). Recently, Maclean (1969) and Fjeldså (1976) argued strongly in favour of closer relationship to waders (Charadrii), latter author even indicating Glareolidae (pratincoles and coursers) as nearest relatives. Stegmann (1968, 1969), however, maintained that sandgrouse form specialized offshoot of Columbiformes, having become secondarily terrestrial. Strauch (1979) criticized Fjeldså's (1976) conclusions while indicating that Charadriiformes, Pteroclidiformes, and Columbiformes may belong close together. This also tentative conclusion of Cracraft (1981). Obviously, relationship of these groups in need of further study. For present then, best to treat sandgrouse as separate order (as already done by Stresemann 1927–34). See also Columbiformes.

Pterocles lichtensteinii Lichtenstein's Sandgrouse

PLATES 24 and 26
[facing pages 302 and 326]

Du. Lichtensteins Zandhoen Fr. Ganga de Lichtenstein Ge. Wellenflughuhn
Ru. Чернолобый рябок Sp. Ortega de Lichtenstein Sw. Strimflyghöna

Pterocles lichtensteinii Temminck, 1825

Polytypic. Nominate *lichtensteinii* Temminck, 1825, Israel, Jordan, and Sinai, and from south-east Egypt south to northern Kenya, Arabian peninsula (except Hadhramawt), and Socotra; *targius* Geyr von Schweppenburg, 1916, western, southern, and central Sahara. Extralimital: *ingramsi* Bates and Kinnear, 1937, Hadhramawt; *sukensis* Neumann, 1909, south-east Sudan, northern Uganda, and western and central Kenya.

Field characters. 24–26 cm; wing-span 48–52 cm. Head and body similar in size to those of Turtle Dove *Streptopelia turtur* but tail rather short and graduated. Smallest and shortest-tailed west Palearctic sandgrouse, with rather dark, close-barred plumage and compact appearance. ♂ shows white forehead with bold transverse black bar, but otherwise plumage of both sexes lacks obvious marks. Sexes rather similar, unlike west Palearctic congeners; no seasonal variation. Immature doubtfully separable from ♀. 2 races in west Palearctic; see Geographical Variation for differences.

ADULT MALE. Ground-colour of plumage yellow-buff on back, rump, and wing-coverts, and off-white on head, neck, and most of underparts; densely barred black overall except on spotted and streaked crown, cheeks, and lower throat. Most obvious characters are: (1) bold white supercilium and large forehead, splashed with black above eye and boldly barred black in front of eye; (2) pale gold-buff bars across folded wing (of which at least 6 usually visible); (3) similarly coloured patches or spots on scapulars, tertials, and longer upper tail-coverts; (4) when seen head-on, broad yellow-buff chest-band, with narrow black line across centre and along lower edge. At certain angles (particularly from behind) and at longer range, bars on upperparts merge to give dark back. In flight, wing shows noticeably dark outer half (except for pale bases to primaries) and pale buff-grey under wing-coverts. Bill dull red; orbital ring strongly yellow, unlike any other sandgrouse in west Palearctic. Feet orange-yellow. ADULT FEMALE. General plumage pattern and colours only a little duller than adult ♂, but forehead and face yellow-buff, crown and neck more strongly spotted, and chest-band absent or restricted to narrow yellow-buff line. JUVENILE. Resembles adult ♀ but upperpart markings little more than vermiculations. Flight-feathers grey.

When size clear and short tail visible, confusion likely only with ♀ and immature Crowned Sandgrouse *P. coronatus* but that species has throat strongly yellow (white or pale buff and streaked in *P. lichtensteinii*) and barring indistinct below chest with no obvious buff bands across wing. Confusion also possible with ♀ and immature Chestnut-bellied Sandgrouse *P. exustus*, but that species has long tail-streamers, heavily spotted upper breast, noticeably yellow face, and dark-barred brown belly (thus lacking uniformity of plumage pattern so characteristic of *P. lichtensteinii*). Feeding behaviour and general appearance most galliform of genus, with short tail often held clear of ground and waddling gait reminiscent of bantam hen. Flight rapid, but silhouette much more compact than all other species except *P. coronatus*. Escape flight often short (bird landing within 200 m).

Least gregarious of sandgrouse, apparently more sedentary than any other species and markedly localized in some regions. Most characteristic call 'quwheeto' given twice, to some recalling whistle of Wigeon *Anas penelope*.

Habitat. In lowest latitudes, tropical and subtropical. Distinguished from congeners by extent of crepuscular or even nocturnal activity and by avoidance of open desert, semi-desert, or arid cultivated areas. Inhabits wadi beds, wooded or with bush-covered rocky slopes, narrow rocky valleys, or scree and boulders. In Africa, in clearings in thornbush or scrub, or open patches of *Acacia*, keeping under shade and moving from tree to tree, searching for seeds. In Sahara, almost confined to areas with *Acacia sayal* (Heim de Balsac and Mayaud 1962). Found nesting on low stony or rocky ridges covered by thick scrub, scattered thorn, or dense herbage, and near river or nullah containing drinking pools. Like congeners, requires daily access to water, flying considerable distances to reach it, or even descending into deep wells for it. Flies strongly and is mobile in the air, but on preferred rough terrain short legs restrict movement on ground. Association with cover, and localized dispersed pattern of living reduce vulnerability to serious human interference. (Geyr von Schweppenburg 1918; Jackson 1926; Archer and Godman 1937; Cave and MacDonald 1955; Hüe and Etchécopar 1957; Thomas and Robin 1977.)

Distribution. Imperfectly known. Sporadic and irregular in Sahara (Heim de Balsac and Mayaud 1962).

EGYPT. May breed eastern Sinai (PLM, WCM).
MOROCCO. Does not breed in very dry years (P Robin).
Accidental. Iraq.

Population. JORDAN. Scarce resident (DIMW).

Movements. Very little information but presumed that, like other *Pterocles*, to some extent nomadic in dry season. Small numbers seen in Mali (notably Adrar des Iforas

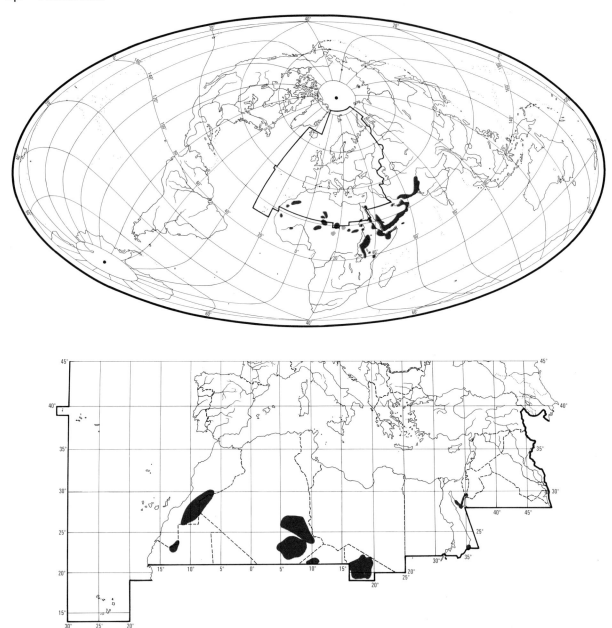

region) were mainly in April, these perhaps visitors from elsewhere (Lamarche 1980). From central Chad birds move north into Erdi and Tibesti for hottest months (Newby 1979). Even less known about Middle East population, except that the species has been reported in the past as rare winter visitor to southern Iraq (Ticehurst *et al.* 1922; Allouse 1953).

Food. Almost exclusively vegetable matter, predominantly seeds. Reported to feed in morning and towards evening (Heuglin 1873), keeping mainly in shade during day (Meinertzhagen 1954).

Seeds recorded include those of acacias (e.g. *Acacia sayal*), asphodel *Asphodelus tenuifolius*, kandi *Prosopis spicigera*, and *Cassia* (Butler 1909; Geyr von Schweppenburg 1918; Bowen 1928; Meinertzhagen 1954; Ali and Ripley 1969; Gallagher 1977; Thomas and Robin 1977). Only one report of animal food: small beetles (Coleoptera), ants (Formicidae), and larvae of ant-lion *Myrmeleon*, north-east Africa (Hartmann 1863).

In Morocco, takes mainly seeds of *Acacia sayal* in breeding season; 2 ♂♂ contained only seeds of *Asphodelus tenuifolius* and *Acacia sayal* (Thomas and Robin 1977). In Saudi Arabia, seeds of *Acacia* favoured, but turns to *Cassia* and *Asphodelus* when this source exhausted; 10 birds yielded 3674 seeds of *Acacia* (Meinertzhagen 1954). ♂ and ♀, South Yemen, December, contained seeds of *Salsola* and baubul (Yerbury 1886). ♀ Oman, April, contained seeds of *A. tenuifolius* (Gallagher 1977). See also Butler (1909) and Bowen (1928) for Sudan, and Geyr von Schweppenburg (1918) for Tuareg mountains area (Mali).

No information on food of young. MGW

Social pattern and behaviour. Most aspects poorly known.

1. Less gregarious than some congeners; pairs or small scattered parties the rule during day at all seasons. Largest concentrations reported during hottest time of day, north-east Africa (Heuglin 1873), but large aggregations associated primarily with crepuscular watering flights (Blanford 1870; Hume and Marshall 1878; Yerbury 1886; Baker 1914a; Jackson 1926; Moltoni 1945; Meinertzhagen 1954; see also Flock Behaviour, below). BONDS. No information on mating system. 'Pairs' observed in winter flocks, Pakistan (Baker 1914a). At Elat (Israel), March, sex-ratio 1♂:2♀♀ (K M Olsen). BREEDING DISPERSION. No information. ROOSTING. Roosts communally on ground during day, mainly at hottest time. May not roost nocturnally at all, particularly on moonlit nights; calling, display, flying, and drinking all recorded during night. (Heuglin 1873; Marshall *et al.* 1911; Hartert 1912–21; Bates 1937b.)

2. Birds generally crouch low and rely for protection on cryptic colouration (Hartmann 1863; Bates 1937b). At watering place, Elat, not disturbed by car lights 6–10 m away (K M Olsen). FLOCK BEHAVIOUR. Flock tended to occupy rather restricted area, Sudan, flushed birds usually landing within a few hundred metres (King 1921); see also Ticehurst (1923c) for Pakistan. In Saudi Arabia, made skilful use of thorn-bush cover and sudden last-minute noisy eruptions, followed quickly by renewed landing, to deter hunting Pallid Harrier *Circus macrourus* (Meinertzhagen 1954, which see also for behaviour in presence of wild cat *Felis*). Movement to water usually after sunset (small numbers regularly at 18.15 hrs, March, Elat: K M Olsen) when birds have fed and, in some cases, again at daybreak, before doing so; may continue well into darkness and drinking at night also recorded. Crepuscular drinking habits apparently general throughout range (Wise 1876; Butler 1909; Hartert 1912–21; Baker 1914a; Geyr von Schweppenburg 1918; Jackson 1926; Bowen 1927; Meinertzhagen 1934, 1954; Bates 1937b; Ali and Ripley 1969; Érard and Etchécopar 1970). Arrives singly or in flocks of 2–6, flying silently, fast and low (Wise 1876; Baker 1914a; Jackson 1926), though may call upon arrival and whilst landing (Geyr von Schweppenburg 1918; Ticehurst 1923c: see 1 in Voice). After arrival, birds fly around, gradually forming larger pack, then land *c.* 30 m from water to remain stationary and alert for *c.* 1 min before moving to water (Meinertzhagen 1954) or remain for quite long period and leave gradually and mostly in pairs, with calls being given on ground prior to departure, on rising, and thereafter in flight (Geyr von Schweppenburg 1918: see 1, 3d in Voice). Some birds in groups of 12 watering in darkness, Iran, crouched at approach of vehicle, others ran, then took off, moving *c.* 10 m upwards in confused and noisy eruption, only then gathering speed in horizontal flight, when calls also given (Érard and Etchécopar 1970: see 2 in Voice). ANTAGONISTIC BEHAVIOUR. Only report of display refers to nocturnal activity: birds had wings drooped, tails raised and fanned. Captive birds pouted and strutted, made turning and hopping movements, and raising of rear end and of wings was prelude to attack; movements accompanied by calls and wing-clapping (Heuglin 1873: see 3e in Voice). PARENTAL ANTI-PREDATOR STRATEGIES. Said to perform distraction-lure display of disablement type, fluttering off nest 'like huge moth' before taking off (Jourdain 1936b). No further information. MGW

Voice. Calls frequently during hours of dusk and darkness (e.g. Moltoni 1945, Gallagher 1977), but perhaps mainly quiet during day (S Cramp). At watering places, wing-noise from arriving flock may be heard before birds visible (Baker 1914a), and recording (M E W North) suggests much wing-noise also created at take-off. For wing-clapping during display, see Social Pattern and Behaviour.

CALLS OF ADULTS. (1) Main calls. A sucking, whistling 'quwheeto' (P J Sellar: Fig I) given on ground and in flight. Also described as a whistling '(see)chee-weeup' given at intervals; first sibilant audible only at close range, other sub-units liquid, clear, and musical (Gallagher 1977; Gallagher and Woodcock 1980). A 'huiiu' or more distinct 'witch-ouuu' lasting *c.* 1·5 s, with 2nd sub-unit descending in pitch; like Whistle-call of ♂ Wigeon *Anas penelope*, and rather feeble so that probably audible only up to *c.* 50 m (K M Olsen). See also Heuglin (1873), Butler (1909), Lynes (1925a), Bannerman (1931), Archer and Godman (1937), Meinertzhagen (1954), and Hüe and Etchécopar (1957). Various 'queep-queep' (Baker 1914a), 'kweep kweep eep' (Meinertzhagen 1930), and short, double 'kwieb kwieb' (Hartert 1912–21) sounds, also a 'tluit-tluit' (from ♂♂) like Little Owl *Athene noctua* (Geyr von Schweppenburg 1918), and a 'chirrup' like sparrow *Passer* (Ticehurst 1923), invariably associated with watering and unlikely to be distinct. May also give a shorter 'kwit' or 'kiti' (Gallagher and Woodcock 1980). The sonorous 'külü

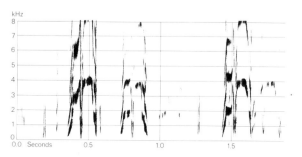

I M E W North Kenya February 1955

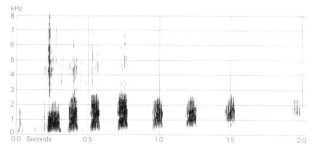

II M E W North Kenya February 1955

klü klü ör' of Marshall *et al.* (1911) presumably involves elements from calls 1 and 2 (see below). (2) Contact-alarm calls. Sharp and cackling (Hume and Marshall 1878). Harsh or hoarse 'arr-arr' given invariably when flushed (Geyr von Schweppenburg 1918). Recording (Fig II) suggests a harsh, nasal cackling 'qua qua qua' with duck-like or frog-like quality. When flushed during day, a loud hard rasping 'quack' given rapidly 5–25 times (King 1978). Also rendered as a guttural 'krerwer wer wer' (Gallagher and Woodcock 1980). Birds flushed at night, Iran, gave rapid and raucous 'kerrek-kerrek-kerrek'; on another occasion 'krrr-krrr-krrr-kru-kru' on taking off, then a number of disyllabic whistling sounds (Érard and Etchécopar 1970: see also 1, above). (3) Other calls. (a) Persistently repeated 'coquì-coquà' from flocks during evening watering session (Moltoni 1945). (b) Whilst drinking, a constant musical chatter termed 'creening' (Archer and Godman 1937; Meinertzhagen 1954). (c) A kind of musical mewing, context not given (Hüe and Etchécopar 1957). (d) Both sexes gave a pleasant, soft cooing like doves (Columbidae) whilst resting after drinking (Geyr von Schweppenburg 1918). (e) Displaying captive birds gave grumbling or growling sounds (Heuglin 1873), termed 'croaking' (Hume and Marshall 1878). A double croak thought possibly to be given only by ♀ or only in breeding season (Lynes 1925a).

Calls of Young. No information. MGW

Breeding. Season. Somalia: egg found July (Archer and Godman 1937). Site. On ground in the open, or in shelter of low shrub. Nest: shallow depression, unlined. Building: no information. Eggs. Elliptical, smooth, and quite glossy; pale buff to pinkish, lightly speckled and spotted reddish-brown, with faint blotches of purple or purplish-grey. Sudan: 42×26 mm ($40–45 \times 26–27$), $n=2$; calculated weight 16 g (Schönwetter 1967). No further information.

Plumages (nominate *lichtensteinii*). Adult Male. Forehead white crossed by broad black band. Crown and nape cinnamon-rufous (much paler in bleached plumage), densely streaked black-brown. Supercilium white, interrupted by black spot over eye. Rest of upperparts rufous-tawny to buff (depending on wear), densely and narrowly barred black-brown; scapulars and long upper tail-coverts broadly tipped golden-ochre. Sides of face buff spotted black, ear-coverts with narrow black shaft-streaks. Chin buff, spotted dark brown at sides. Throat and chest buff, barred black-brown. Breast golden-ochre with dark brown band across centre of golden area and black band separating breast from belly. Belly, flanks, and under tail-coverts buff or off-white, barred black-brown. Feathers on front of tarsus pale buff. Tail-feathers buff, barred black like upperparts and with wide golden-brown tips; subterminal black bar heavy, extended into sharp black point along shaft. Primaries brown, tipped off-white when fresh, outer webs at base pale buff. Primary coverts dark brown, contrasting with pale bases of primaries. Secondaries dark brown, dirty white or buff at base; s8–s9 with outer web barred black-brown and pale buff; tertials (from s10 inwards) entirely barred and with golden-ochre tip, similar to longer scapulars. Greater and median upper wing-coverts pale buff or off-white, broadly barred black-brown and widely tipped golden-ochre; lesser upper wing-coverts like rest of upperparts but narrowly tipped golden-ochre. Underside of flight-feathers grey. Under wing-coverts and axillaries pale buff-grey, marginal coverts mottled darker grey. Adult Female. Ground-colour pale buff. Crown streaked, chin and nape spotted, rest of body and upperwing narrowly barred dark brown. Scapulars and greater upper wing-coverts broadly tipped buff. Greater coverts less densely barred than rest of upperparts. Tail more broadly barred brown than in ♂, tips of feathers not as brightly tawny-ochre, and subterminal bar narrower. Primaries, secondaries, and under wing-coverts as ♂. Downy Young. Crown, nape, rest of upperparts, and wing warm light brown; underparts, thighs, and feathers on tarsus slightly paler. Sides of head slightly darker with stripe on side of crown, spot behind eye, and moustachial streak cream-buff. Unlike most other species of Pteroclididae, no pattern of pale stripes on upperparts (Thomas and Robin 1983). Juvenile. Head buff, strongly speckled grey. Upperparts buff to rufous-buff narrowly and densely vermiculated grey, feather-tips apparently sometimes unmarked (ZMA). Underparts like adult ♀. Tail buff, barred dark grey, subterminal bar narrower than in adult ♀. Primaries grey with buff margin to outer web and broad buff tip, latter speckled grey; buff tip smaller on p9–p10. Secondaries dark grey, margined buff and with buff base mottled grey; tertials like rest of upperparts. Greater upper wing-coverts buff, vermiculated grey; broadly tipped plain buff like adult ♀. Rest of upper wing-coverts like upperparts. Under wing-coverts pale buff-grey, indistinctly vermiculated darker grey. Subsequent Plumages. As adult, separable only while juvenile outer primaries retained.

Bare parts. Adults and Juvenile. Iris dark brown; bare skin round eye yellow. Bill brown-red. Bare portion of leg orange-yellow. Downy Young. Eye-ring, bill, and bare portion of leg light grey (Thomas and Robin 1983).

Moults. Adult Post-breeding. Complete, primaries descendant. Timing poorly known, but June specimens from southern Sahara actively moulting, having replaced 1–4 primaries (ZMA); January and March specimens not moulting (BMNH, RMNH). Progress of primary moult slow. Post-juvenile. Almost complete, but outer primaries retained.

Measurements. Adult, races combined; skins (BMNH, RMNH, ZMA).

WING	♂ 191 (2·86; 5) 187–195	♀ 176, 179, 187
TAIL	75·7 (3·55; 7) 70–81	66, 72, 72
BILL	14·9 (1·31; 7) 13·3–17·1	13·7, 13·8, 14·6
TARSUS	26, 27, 28	25, 25
TOE	25, 26, 26	23, 25

Weights. Adult ♂, Morocco, 240, 255 (Thomas and Robin 1977).

Structure. Wing narrow, rather pointed. 11 primaries: p10 longest, p9 0–4 shorter, p8 5–10, p7 15–22, p6 27–34, p5 40–46, p4 51–56, p1 83–91; p11 minute. Differences between p10 and other primaries smaller in birds having retained juvenile outer primaries. About 17 secondaries; s13 longest. Tail strongly rounded, 12 feathers; t6 *c.* 20–25 shorter than t1. Longest under tail-coverts much longer than t6. Bill slender, elongated; nostril a horizontal slit in a depression. Legs of medium length, tarsus densely feathered in front. Middle toe *c.* 95% of tarsus; inner *c.* 67% of middle, outer *c.* 67%, hind *c.* 20%.

Geographical variation. Involves mainly tinge of plumage and relative width of light and dark bars. Obscured by individual variation. Many more races have been named than recognized here (Meinertzhagen 1954; Vaurie 1965). *P. l. ingramsi* palest race with strongly reduced barring (Meinertzhagen 1954); *targius* also pale, entirely barred, but dark bars much narrower than light, particularly on underparts of ♀; ground-colour pale sandy-buff. Nominate *lichtensteinii* darker, ground-colour more rufous-buff; dark and light bars about equal in width; *sukensis* dark, with dark bars wider than pale ones, in places almost fused to dark brown patches, particularly on long scapulars (Neumann 1909; Friedmann 1930).

JW

Pterocles coronatus Crowned Sandgrouse

PLATES 24 and 26
[facing pages 302 and 326]

Du. Kroonzandhoen   Fr. Ganga couronné   Ge. Kronenflughuhn
Ru. Светлобрюхий рябок   Sp. Ganga coronada   Sw. Kronflyghöna

Pterocles coronatus Lichtenstein, 1823

Polytypic. Nominate *coronatus* Lichtenstein, 1823, Sahara from Morocco to Egypt; *vastitas* Meinertzhagen, 1928, Sinai peninsula and deserts of southern Israel and Jordan; *atratus* Hartert, 1902, deserts of Arabian peninsula and from eastern Iran east to Afghanistan. Extralimital: *saturatus* Kinnear, 1927, mountains of interior Oman; *ladas* Koelz, 1954, Pakistan.

Field characters. 27–29 cm; wing-span 52–63 cm. Averages 10% larger than Lichtenstein's Sandgrouse *P. lichtensteinii* but *c.* 20% smaller than all other sandgrouse. Second smallest of short-tailed sandgrouse. Plumage pattern of mainly pink-buff ♂ recalls Spotted Sandgrouse *P. senegallus*; close-barred ♀ and immature recall *P. lichtensteinii*. At close range, combination of dull red and blue crown and black facial marks on ♂, and combination of yellow throat and close-barred foreneck on ♀ diagnostic. Commonest call trisyllabic and rising in pitch, unlike congeners. Sexes dissimilar; no seasonal variation. Juvenile resembles ♀. 3 races in west Palearctic; nominate *coronatus* described here, others differing only slightly (see Geographical variation).

ADULT MALE. Body and wing-coverts essentially pink-isabelline, with delicate brown marbling and off-white streaks and spots indistinct on back but obvious on scapulars and wing-coverts; tail brown-isabelline with white tips to outer feathers. Dark brown flight-feathers and beautifully patterned head contrast markedly at close range, with head showing: (1) pale red-brown crown with pale blue circlet; (2) pale buff area round eye and on forehead, with black mask on sides of forehead and chin; (3) golden-yellow lower face and lower throat, extending back to form band under blue nape. In flight, underwing pattern plainest of all ♂ sandgrouse, with pale buff underbody and under wing-coverts relieved by dark brown undersurface to flight-feathers. Bill blue-grey, orbital ring dull yellow. Feet grey. ADULT FEMALE. Distinct from adult ♂: ground-colour paler (especially below) and most of plumage flecked and close-barred black or dark brown. Resembles ♀ *P. lichtensteinii* on ground and in flight but distinguished by: (1) yellow-ochre and only faintly streaked face and throat; (2) grey, not pale yellow bill; (3) more open and broader barring above and more open and obsolete barring below; (4) lack of discrete buff bands over wing-coverts. JUVENILE. Closely resembles adult ♀ but throat whiter and barring of upperparts less pronounced.

♂ unmistakable at close range but open to confusion with *P. senegallus* at a distance; *P. senegallus* larger, long-tailed, and has black centre to belly and paler uppersurface to flight-feathers. ♀ much less distinctive; for separation from *P. lichtensteinii*, see above. Flight and gait as *P. lichtensteinii*.

Flight-call a loud, hoarse 'ch-ga ch-gar-ra'.

Habitat. In low latitudes, in hot arid Mediterranean and subtropical climatic zones, on level and montane stony desert and tracts of very sparse grasses, generally avoiding sand. Tolerant of saline water (Harrison 1982). In northwest Africa, breeds among stones on dark red-brown sandstone matching plumage colour, and in desert or on sandy plains away from mountains (Heim de Balsac 1925; George 1969). In Tibesti, common in mountainous zone (Heim de Balsac and Mayaud 1962). In Oman, in stony areas and ravines—a bird of absolute desert (Meinertzhagen 1954). In Pakistan, affects barest desert with scraggy grass, and stony wastes, up to *c.* 1800 m; breeds on barren windswept sand-dunes. Somewhat crepuscular (Ali and Ripley 1969). Can replace or co-exist with Spotted Sandgrouse *P. senegallus* in Algerian Sahara where both common (Dupuy 1969).

Distribution. Imperfectly known. Varies with climatic conditions before period of sexual activity (Heim de Balsac and Mayaud 1962), though in very dry year, north-west Sahara, breeding apparently unaffected (George 1970).

IRAQ. According to Hüe and Etchécopar (1970) breeds in southern deserts, but no proof even of occurrence (Ticehurst *et al.* 1922; Allouse 1953).

Population. No precise information.

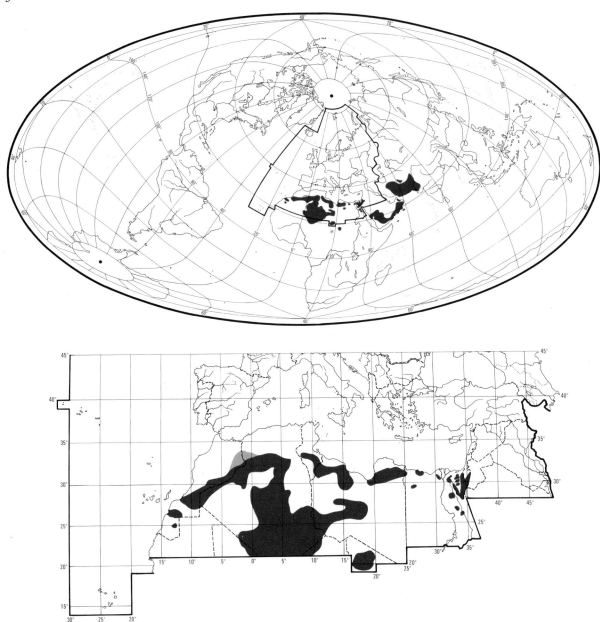

MOROCCO. Not common (JDRV); in former Spanish Sahara much less common than Spotted Sandgrouse *P. senegallus* (Valverde 1957). LIBYA. Tripolitania: least numerous of Pteroclididae (Bundy 1976). ISRAEL. Not as common as *P. senegallus* (HM).

Movements. Little known, but to some extent nomadic in dry season, like other *Pterocles*. Recorded in various localities in Morocco, Algeria, and northern Mauritania where not known to breed (Heim de Balsac and Mayaud 1962); common in latitude of Berguent (north-east Morocco) from late summer to March, with 43 of 47 collected being ♂♂ (Brosset 1956a). Scarce north of 32°N in Libya, where (at least in Wadi Kaam, Tripolitania) it is most frequent in dry season, June–September (Bundy and Morgan 1969; Bundy 1976). Central Chad vacated in hottest months, when birds move north into Erdi and Tibesti (Newby 1980). North African nominate *coronatus* also recorded exceptionally near Red Sea coast of Egypt in May and December (Goodman and Watson 1983).

Even less known about movements in Middle East. The few records for United Arab Emirates are all November–December, but nevertheless may be resident in foothills (Bundy and Warr 1980). Resident in Pakistan, though numbers augmented through immigration in winter (Ali and Ripley 1969).

Food. Vegetable matter. Reported to be more active on ground than congeners, running about and picking up seeds like Partridge *Perdix perdix* (St John 1889).

Takes primarily minute hard seeds and also shoots of various desert grasses and other plants (Meinertzhagen 1954; Ali and Ripley 1969). Insects may be taken occasionally but confirmation required (Bannerman 1931). In Sudan, favoured amioga *Tephrosia apollinea*, Sudan (King 1921). In Tunisia, stomachs contained only seeds and vegetable matter (Whitaker 1905). 5 collected at watering site in Spanish Sahara produced similar result to that for Spotted Sandgrouse *P. senegallus* (Valverde 1957; see that species).

No information on food of young. MGW

Social pattern and behaviour. Poorly known.
1. Gregarious outside breeding season, especially for watering: e.g. flocks of up to 50, Tunisia (Whitaker 1905); 80–175 assembled at water, Israel (S Cramp, K M Olsen); occurs in hundreds at water in Pakistan (Ticehurst 1923b) and Oman (Walker 1981b). Mainly in pairs, June, Morocco (George 1970). BONDS. No information. BREEDING DISPERSION. In Morocco, 2 pairs probably breeding 'close' to pair with nest at c. 1 km from nearest waterhole; similar group of 3 pairs 2 km away and c. 3 km from water (George 1970). ROOSTING. No information.
2. Described as tame and easily approached, in contrast to (e.g.) Black-bellied Sandgrouse *P. orientalis* (Baker 1913, 1928; Meinertzhagen 1954; Ali and Ripley 1969); allowed approach to c. 10 m in car, Sinai (S Cramp); but see Whitaker (1905) and Flock Behaviour, below. Said to have curious fluttering flight, seeming to hover, especially before landing (Baker 1913). FLOCK BEHAVIOUR. Flocks regularly perform various aerial evolutions (Zedlitz 1909). Birds flying at moderate height, southern Sahara, suddenly plummeted down, apparently showing interest in moving vehicle; crouched motionless after landing and allowed vehicle to approach but rose up when it stopped (Blondel 1962). Surprised in the open by Golden Eagle *Aquila chrysaetos*, birds at first crouched; while 2nd *A. chrysaetos* stooped, they allowed 1st to approach to c. 2 m, then flew up noisily and suddenly (Meinertzhagen 1930). Watering flight performed only in late morning, Tuareg mountains area (Mali) and Ahaggar plateau of Algeria (Geyr von Schweppenburg 1918; Meinertzhagen 1934); at c. 08.00 hrs, March, and from 06.45 to c. 07.45 hrs, April, Sinai (S Cramp, K M Olsen); 07.00–10.00 hrs, early March, Tunisia (Whitaker 1905); 3 pairs arrived 08.30 hrs, June, eastern Morocco (George 1970); early morning, continuing until c. 2 hrs after sunrise, but also again in late evening, India and Pakistan (Ali and Ripley 1969). No evening watering, Arabia, but birds generally did so from 08.00 to 10.00 hrs, depending on weather (Meinertzhagen 1954). See also Hüe and Etchécopar 1957) and Walker (1981b). On arrival, birds gather c. 50–100 m from water; may feed desultorily or sit about before approaching water (Ali and Ripley 1969). 3 pairs, Morocco, landed c. 20 m from water; first crouched, then moved rapidly towards it; afterwards ran c. 5 m back into stony area (George 1970). In Sinai, flocks of c. 15 usually landed in dunes and after c. 5 min, each party drank for c. 10 min (K M Olsen); left in small groups (S Cramp). Birds disturbed at water may stay nearby, tending rather to return constantly until normal departure time (Whitaker 1905). Unlike Spotted Sandgrouse *P. senegallus*, no reconnoitring behaviour noted, but on ground birds gave Warning-calls at approach of human or other predator (George 1970: see 2 in Voice). For observations on watering birds, Spanish Sahara, see Valverde (1957). ANTAGONISTIC BEHAVIOUR. No information. HETEROSEXUAL BEHAVIOUR. In eastern Morocco, moves on to breeding grounds proper at least 16 days before laying (George 1970). No information on pair-bonding behaviour. Both sexes incubate (Ticehurst 1923b). ♂ with wet breast feathers shot just after relieving ♀ at nest containing fresh clutch suggests egg-wetting occurs (Thomas and Robin 1977). RELATIONS WITHIN FAMILY GROUP. Behaviour of ♀ at water, eastern Morocco, indicated young watered in same fashion as (e.g.) *P. senegallus* (George 1970); in Sind, June, mostly ♂♂ at water and some soaked belly feathers (see Ticehurst 1923b). ANTI-PREDATOR RESPONSES OF YOUNG. Young swift-running and adept at hiding (Heuglin 1873). PARENTAL ANTI-PREDATOR STRATEGIES. ♀ sitting on well-incubated eggs, southern Morocco, allowed close approach (Robin 1966). At another locality, Morocco, ♀ left nest apparently alarmed by observer's hide and ♂ took over incubation immediately (George 1970). MGW

Voice. Described as less vocal than other Pteroclididae (e.g. Meinertzhagen 1930, 1954, Ali and Ripley 1969) but, in Tunisia at least, noisy when coming to drink (Whitaker 1905).

CALLS OF ADULTS. (1) Contact- and flight-calls. (a) Normal flight-call. A loud 'chgá chagarrá', with drawn-out 'r' in 2nd unit (Valverde 1957); recording (Fig I) similarly suggests 'ch-ga ch-gar-ra' (J Hall-Craggs) or 'que quet querrooo' (MG Wilson) from flock at take-off; hoarse quality (George 1970). Also rendered 'chee-wuk chewukeroo' or 'gatut-gadidada' or 'chiruk-chirugaga' (Gallagher and Woodcock 1980). Commonest call rendered 'not a warple'; also 'wok tu wok kock' (Walker 1981b). (b) A whistling '(wh)eee(k)' (Fig I) descending in pitch, rather like petulant begging call of young gull *Larus*, also given in flight and may be mixed with call 1a; less sonorous and higher pitched than mewing call 1b of Black-bellied Sandgrouse *P. orientalis* (J Hall-Craggs, P J Sellar). (c) Various rather soft and feeble 'ka' or 'kla' sounds given in flight or on ground (Meinertzhagen 1930, 1954; see also Hüe and Etchécopar 1957, Ali and Ripley 1969); also described as loud and reminiscent of clucking hen (Whitaker 1905; see also Hartert 1912–21). Recording by J-C Roché, March, of birds on ground suggests both monosyllabic 'kla' or 'quet', and disyllabic 'kla-kla' and 'quet-a' sounds. (2) Warning-call. A soft 'hu hu' given 2–3 times by alarmed birds (George 1970). (3) Twittering call mentioned by Baker (1913).

CALLS OF YOUNG. No information. MGW

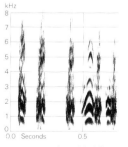 

I J-C Roché (1967) Morocco March 1966

II J-C Roché (1967) Morocco March 1966

Breeding. SEASON. Western Sahara: found breeding 10 April–20 June (Heim de Balsac and Mayaud 1962). Southern Morocco: nest found early June (Robin 1966). Eastern Morocco: nest found late May and June (George 1970). Southern Algeria: breeding from about mid-April and throughout May (Heim de Balsac and Mayaud 1962); nest found 11 June (Robin 1966). Northern Chad: chicks found 12 April (Heim de Balsac and Mayaud 1962). SITE. On ground in the open. Nest: shallow depression, without lining but can be within circle of stones; small stones present in nest at start of laying removed to nest-rim by the time clutch completed (George 1970), sand and dust also removed. Building: no information on role of sexes. EGGS. See Plate 94. Elliptical, smooth and fairly glossy; cream, spotted and blotched pale brown and purplish-grey. 39×28 mm (37–41×27–29), $n = 5$; calculated weight 17 g (Schönwetter 1967); average of 9 eggs, 41×27 mm (Etchécopar and Hüe 1967). Clutch: 2–3. INCUBATION. Period unknown. By both sexes; ♀ found on eggs at c. 19.00 hrs and 17.00 hrs (George 1970). YOUNG. Precocial and nidifugous. Cared for by both parents. No further information.

Plumages (nominate *coronatus*). ADULT MALE. Centre of forehead buff. Sides of forehead, frontal part of lores, and chin black, forming striking mask on face. Crown rufous-cinnamon, entirely surrounded by blue-grey. Hindneck with golden-ochre band immediately below blue-grey nape, contiguous with golden ochre sides of face. Rest of upperparts rufous-buff, variably clouded grey (by mainly concealed dark grey areas on feathers). Long scapulars darker, almost chestnut, marked with pale buff wedge near tip. Upper tail-coverts lighter and more ochre than rest of upperparts. Lores cream-white. Sides of head and upper throat bright golden-ochre. Lower throat and chest buff, suffused with light grey. Lower breast, abdomen, and feathers on leg pale ochre-buff; flanks and under tail-coverts cream-white. Central tail-feathers (t1) rufous-buff; t2 similar but with off-white tip and some vague black markings subterminally; t3–t8 with bright white tip and pronounced black subterminal bar, extending distally in narrow point along shaft of feather. Primaries dark grey-brown, shaft of p9–p10 mainly white, p6 with narrow buff margin distally on inner web, p5 with wider buff margin, p1–p4 with inner web broadly tipped buff, shading to white at edge. Primary coverts like primaries. Secondaries dark grey-brown, s1 with small white spot at tip, s9–s10 with outer web rufous-brown. Tertials (from s11 inwards) rufous-buff with paler buff tip. Greater upper wing-coverts rufous-buff with pale brown inner webs, inner pale buff at tip. Median and lesser coverts rufous-buff with pale buff wedge at tip bordered dark grey. Greater under wing-coverts pale brown, lesser pale buff; axillaries off-white. ADULT FEMALE. Crown rufous-buff, streaked and spotted dark grey. Hindneck more boldly spotted. Rest of upperparts, including upper wing-coverts and upper tail-coverts buff to rufous-buff, barred dark grey. Chin and upper throat golden-ochre, minutely speckled dark grey. Rest of underparts pale buff or cream-white, barred grey; bars narrower than on upperparts; on abdomen, less prominent, tinged brown. Under tail-coverts broadly tipped white, subterminal dark bar projecting like arrowhead along shaft. Central tail-feathers (t1) barred buff and grey, t2 similar; t3 similar, but tipped white; t4–t8 tipped white, with bars progressively more restricted to distal part of outer web, and inner web and base of outer web grey-brown. Primaries and secondaries as ♂, tertials barred buff and grey like rest of upperparts. DOWNY YOUNG. Apparently resembles downy young of Lichtenstein's Sandgrouse *P. lichtensteinii*, being almost uniform pale brown with darker sides of head (George 1978; Thomas and Robin 1983), but possibly with indistinct pattern of pale lines similar to other *Pterocles* (Fjeldså 1976). JUVENILE. Like adult ♀, but more vaguely barred, with tips of feathers on upperparts speckled and vermiculated grey-brown. Chin and upper throat cream-white, not golden-ochre. Primaries: p1–p8 with buff tips, speckled and vermiculated dark brown. Slightly older birds recognizable as long as some juvenile primaries retained.

Bare parts. ADULT. Iris dark brown to black, bare skin near eye dull yellow, narrow ring round eye pale blue. Bill pale blue-grey or blue, darker grey at tip. Foot pale grey, claws blue-grey. (Baker 1928; Meinertzhagen 1954; Stanford 1954; J Swaab). DOWNY YOUNG, JUVENILE. No information.

Moults. Few data. ADULT POST-BREEDING. Complete; primaries descendant. Birds in 2nd calendar year perhaps moult primaries in 2 series, replacing outer (retained from juvenile) at same time as inner, as in Pin-tailed Sandgrouse *P. alchata*. Timing apparently variable or moult arrested—specimens with some old and some renewed outer primaries collected June and December, Algeria. Paludan (1959) recorded moult in June, Afghanistan. POST-JUVENILE. Almost complete, some outer primaries retained. Primaries descendant. Bird from Israel, August, some months after hatching, had renewed p1–p3 (ZMA).

Measurements. Races combined. Adult: skins (RMNH, ZMA).

WING	♂ 203 (3·19; 5) 199–204	♀	181, 188
TAIL	83·0 (2·45; 4) 80–85		78, 81
BILL	12·1, 13·0, 13·4		11·2, 11·9
TARSUS	24, 24, 25		c. 23
TOE	24, 24, 24		c. 21

Weights. Adult ♂, Morocco, 300 (Thomas and Robin 1977).

Structure. Wing pointed. 11 primaries: p10 longest, p9 0–7 shorter, p8 9–14, p7 18–26, p6 32–40, p5 45–54, p4 58–66, p1 99–116; p11 minute, concealed by primary coverts. 18 secondaries, s11–s18 form tertials. Tail wedge-shaped, 16 feathers, t1 30–35 longer than t8. Bill short. Middle toe about as long as tarsus; outer and inner c. 70% of middle; hind toe raised, consisting of little more than small claw, less than 20% of middle.

Geographical variation. Involves ground-colour of plumage and amount of dark grey or black markings. Variation occurs within, as well as between races; may be linked to colour of soil in different areas. Nominate *coronatus* generally with rufous ground-colour of plumage and with palest grey markings, locally (e.g. in Ahaggar, central Sahara) somewhat darker and approaching *vastitas*. *P. c. vastitas* slightly less rufous on upperparts in ♂, more heavily spotted dark grey; ♀ more densely and darkly barred; neither sex as dark as *atratus*, which is considerably more darkly spotted in ♂ and has black bars wider than pale ones in ♀. *P. c. saturatus* darker still (Meinertzhagen 1954). *P. c. ladas* paler than *atratus* in both sexes; ♂ less rufous, more grey in ground-colour, less strongly marked black; ♀ less strongly barred, about similar to *vastitas*, but paler, less rufous in ground-colour. (Koelz 1954; Vaurie 1961c, 1965.) JW

Pterocles senegallus Spotted Sandgrouse

PLATES 24 and 26
[facing pages 302 and 326]

Du. Woestijnzandoen   Fr. Ganga tacheté   Ge. Wüstenflughuhn
Ru. Сенегальский рябок   Sp. Ganga moteada   Sw. Ökenflyghöna

Tetrao senegallus Linnaeus, 1771

Monotypic

Field characters. 30–35 cm; wing-span 53–65 cm. Little longer bodied but much bulkier overall than Crowned Sandgrouse *P. coronatus*, with long tail-streamers; slightly smaller, with shorter tail, than Pin-tailed Sandgrouse *P. alchata*. Quite large sandgrouse, both sexes showing at close range diagnostic long black division to plain, pale lower belly. Contrary to English name, pale spots on sides of ♂'s back and black spots of ♀ indistinct at distance, upperparts appearing olive-toned. Calls more liquid and musical than those of congeners. Sexes dissimilar; no seasonal variation. Juvenile separable at close range.

ADULT MALE. Body and tail mainly sandy-isabelline, with grey marbling on scapulars, grey cast on upper breast and neck, and brown bases to larger wing-coverts. At close range, yellow-buff tips to scapulars and larger upper wing-coverts create obvious pale spots but at distance, these merge with other colours to give olive cast to back. Head yellow-ochre, with broad pale grey supercilium and pink-brown crown visible only at close range; pattern recalls that of washed-out Partridge *Perdix perdix* and lacks decorative black marks of *P. coronatus*. Pale buff edges to grey-brown primaries decrease their contrast with rest of folded wing. Centre of belly black, contrasting with isabelline surround, but obvious only in flight at close range. In flight, underwing pattern lacks clear white centre, being pale buff with greyer flight-feathers. Bill blue-grey, orbital ring yellow. Feet pale grey. ADULT FEMALE. Ground-colour of plumage paler below and pinker above than ♂, with obvious dark streaks over crown and hindneck and both dull and intense black spots fully but not densely scattered over breast, mantle, back, rump, tail, and wing-coverts; spotting boldest on sides of breast, scapulars, and larger wing-coverts. Lower face yellow-white to ochre (above breast). Wing pattern and bare parts as ♂. JUVENILE. Tail not elongated. Plumage as ♀ but with sandier ground and wavier feather-marks. Chin and throat white; centre of belly dark brown.

Unmistakable in flight overhead or when black belly-divide visible on ground. ♂ also distinguished from *P. coronatus* by head pattern. ♂ may suggest ♀♀ of Chestnut-bellied Sandgrouse *P. exustus* and even Black-bellied Sandgrouse *P. orientalis*, but upperparts lack yellow ground of those species, black on belly never covers whole width of lower body, and no black line across lower breast. Flight action and speed as congeners. Feeding and drinking behaviour typical of genus. Seldom forms large flocks.

Commonest call distinctive, with liquid musical quality, 'quitoo quitoo'.

Habitat. With Crowned Sandgrouse *P. coronatus*, ranks as most desert-dwelling of west Palearctic Pteroclididae, inhabiting same low latitudes, in hot (up to 48°C) and arid Mediterranean and subtropical climates. Generally avoids trees and scrub, cultivation, and any but the sparsest vegetation; in East Africa, however, occurs in dry sandy and stony spots dotted with small acacias on plains (Jackson 1926). Dependent on water holes, but, unlike *P. coronatus*, will not drink highly saline water (Etchécopar and Hüe 1967). Inhabits clay or loamy soils in open desert, sandy wastes, bare stony slopes, salt-encrusted expanses, mudflats, and sand-dunes, mainly in lowlands and plains, but up to *c.* 600–900 m when wintering in Somalia (Archer and Godman 1937). Detailed study in eastern Morocco at *c.* 500–800 m found preference for extensive stony desert (hammada) terrain with larger stones. Shifts with sparse seasonal plant cover, rarely attempting to breed in true desert, as pockets of vegetation of desert steppe type are essential. Adversely affected by exceptionally dry seasons and unable to breed when humidity falls too low. Prefers to nest on pale brown substrate with blackish stones *c.* 2–4 cm in diameter (matching colour of sitting ♀'s upperparts) among body-size stones considered to assist in concealing bird and in giving protection against attack by low-flying falcons *Falco*. During pre-breeding season favoured area with growth of *Euphorbia guyoniana* *c.* 50 km from breeding area. Avoidance of high plateau related to inability to compete with Pin-tailed Sandgrouse *P. alchata* present there in large numbers (George 1969, 1970). In Algeria, sometimes coexists with Crowned Sandgrouse *P. coronatus* and sometimes replaced by it; nests average *c.* 10 km from water which is brought to young by parent (Dupuy 1969).

Distribution. Imperfectly known. Varies with climatic conditions before the period of sexual activity (Heim de Balsac and Mayaud 1962); in an area of north-west Sahara, large population in 1968 when ample rainfall but in very dry year 1969 only a few in small population nested at all (George 1970).

SYRIA. Found in south and centre in past, but no proof of nesting (Kumerloeve 1968b; HK).

Accidental. Italy.

Population. No precise information.

IRAQ. Common, but rather local (Ticehurst *et al.* 1922; Allouse 1953).

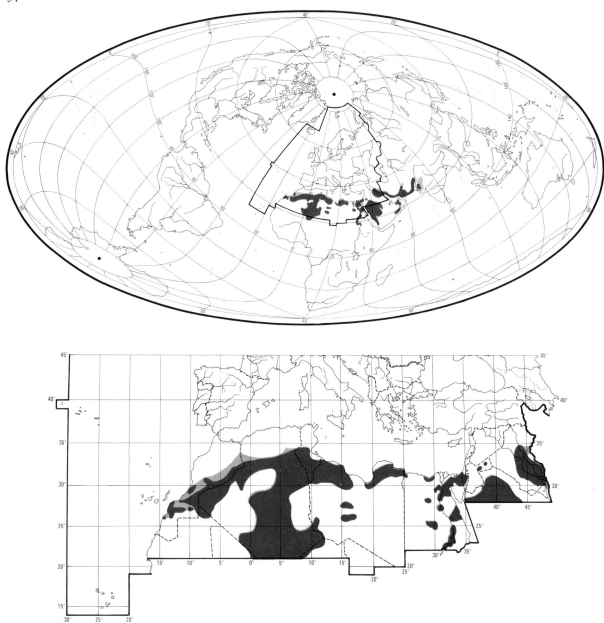

Movements. Few data, but clearly nomadic during dry season. Extent of wandering not readily defined since breeding range also imperfectly known and probably inconstant between years (see Distribution). In and around Sahara, general dispersal related to seasonal aridity begins about July, birds moving north and south; numbers (sometimes large) noted Berguent (Morocco), Biskra, and Colomb-Bechar (Algeria) October–March (Heim de Balsac and Mayaud 1962). Even reaches coastal areas: e.g. present Cape Juby (Western Sahara) May–June (Heim de Balsac and Mayaud 1962), and especially frequent in Wadi Kaam (Libya) in driest months, June–September (Bundy and Morgan 1969). In Darfur (western Sudan), occasional visitor at end of dry season (April), probably searching for water (Lynes 1925a).

In addition to characteristic nomadism (e.g. Ticehurst *et al.* 1922 for Iraq), definite seasonal displacements occur in east of range. Small numbers resident in Pakistan and presumably breed, though this unproved; but much larger numbers present (and extend also into north-west India) in winter, with thousands congregating where and when surface water restricted. However, somewhat irregular between years in abundance and timing, with winter visitors sometimes arriving as early as August–September (Ali and Ripley 1969).

Food. Predominantly plant material, mostly seeds, but considered by some authors less purely vegetarian than (e.g.) Chestnut-bellied Sandgrouse *P. exustus* (Hume and Marshall 1878; see also Heuglin 1873). Generally trots about, picking up seeds from ground (Baker 1914a).

Plant material includes seeds and other parts of various Cruciferae, seeds of *Euphorbia guyoniana*, various parts of *Salicornia*, and amioga *Tephrosia apollinea*; also grain including that from animal droppings (Heuglin 1873; Koenig 1896; King 1921; Meinertzhagen 1930; Mackworth-Praed and Grant 1952; Marchant 1961a; George 1969).

In Algeria, like Black-bellied Sandgrouse *P. orientalis*, particular preference for various parts of crucifer *Sisymbrium cinereum*; ♂, early March, contained *Sisymbrium*, *Salicornia* (apparently), and many grass seeds and grains of quartz (Koenig 1896). Of 41 stomachs collected when birds assembled at water, Spanish Sahara, 50% empty, 35% contained some food, and only 15% relatively full; birds had taken seeds only, of *c*. 6 plants, 2 of which were abundant although none specifically identified (Valverde 1957). In eastern Morocco, main food was seeds of *Euphorbia guyoniana* in pre-breeding season (George 1969). More than 50 specimens taken Iraq, Sahara, and Saudi Arabia, contained only hard seeds; birds fed early and watered on full stomach (Meinertzhagen 1954). In contrast, ♂ and ♀ shot Iraq, October, in or at edge of cultivation, had crops full of barley (Marchant 1961a).

In Morocco, young took mainly seeds and small ants (Formicidae) (George 1969). MGW

Social pattern and behaviour. Most detailed study by George (1969, 1970) in Morocco; see also Marchant (1961a) for Iraq.

1. Gregarious, at least outside breeding season; varying reports on flock size and aggregations compared with other Pteroclididae presumably to some extent reflect population changes. In Saudi Arabia, small flocks concentrate in large packs for watering (Meinertzhagen 1954); in Iraq, unlike Pin-tailed Sandgrouse *P. alchata*, tends not to form very large aggregations (Marchant 1962b); generally less gregarious than Chestnut-bellied Sandgrouse *P. exustus*, north-east Africa, but watering flocks still large (Heuglin 1873). Flocks of up to 20 recorded flying to water, Egypt (Meinertzhagen 1930), and old report of thousands sometimes assembling at water, India (Hume and Marshall 1878); for further details and comparisons, see Baker (1914a) and Ali and Ripley (1969). Non-breeders remain in flocks during breeding season, Morocco, e.g. up to 80 in exceptionally dry year (George 1970). Winter flocks of 5–50, India, said to be often of one sex (Hume and Marshall 1878). Not seen in mixed-species flocks, Morocco (George 1969), and various reports of association with congeners (e.g. Koenig 1896, Ticehurst *et al.* 1922, Moltoni 1945, Robin 1966) probably refer in the main merely to simultaneous watering flights and drinking (see also Archer and Godman 1937 for India). BONDS. Observations by George (1969, 1970) indicate monogamous mating system. Both parents care for young until 7–10 days old, India (Bolster 1922). BREEDING DISPERSION. Scanty data indicate that breeding sometimes loosely colonial. In eastern Morocco, 12 nests on 1–1.5 km² gave impression of colony, none otherwise in vicinity; 8 others widely dispersed over *c*. 100 km². At another site, 5 pairs in 'island' of vegetation of *c*. 2 km². Occupied area may be influenced by extent of vegetation cover which varies with rainfall (George 1969, 1970). 'Large colony' between Museyib and Baghdad, Iraq (Ticehurst *et al.* 1922), but no details. 2 nests *c*. 400 m apart, Iraq (S Marchant). Reported to nest always at least 50 m from nearest dry wadi, Morocco. Shortest recorded distance between nest and waterhole 3 km, maximum 8 km; most at *c*. 4 km (George 1969, 1970). Nests generally in radius of 30–50 km from water (Heim de Balsac and Mayaud 1962). In Iraq, found nesting close to Pin-tailed Sandgrouse *P. alchata* (Marchant 1961a). ROOSTING. In pre-breeding season, Morocco, communal and nocturnal, non-breeders (in compact group) continuing thus also during breeding season, and family parties may also be associated with such assemblies (but see further below). Generally in stony areas between wadis; quits wadis after evening foraging session, probably to avoid predators. With rubbing movements of body against ground, birds make fresh hollows (*c*. 5 cm deep) each evening and roost there crouched low, such hollows *c*. 1 m apart (George 1969, 1970). Reported to roost in groups like partridges (Phasianidae), heads outwards, tails inwards (Bolster 1922). Up to 50 which had foraged together during day also formed roosting assembly (George 1969). In one case, ♀ brooded 1-day-old chicks at night while ♂ roosted *c*. 30 m away; chick first attempted to make hollow at *c*. 10 days old but still roosted under ♀. 2 families roosted in some area as, but at least *c*. 30 m away from, *c*. 30 non-breeders, some of which made evening watering flight (George 1969, 1970; see also Flock Behaviour and Antagonistic Behaviour, below). Outside breeding season, birds generally leave roosts in early morning for communal foraging (George 1969). Roosts during day in shade of bushes, etc., when very hot; family did so from 11.00 to *c*. 16.00 hrs, May, Morocco (George 1969); may also loaf in early morning and after drinking (Hume and Marshall 1878; Baker 1914a).

2. Reportedly less wary than other Pteroclididae if unmolested (e.g. Hume and Marshall 1878, Archer and Godman 1937), especially when only 2–4 together (Heuglin 1873). Disturbed birds crouch low; can generally outfly Lanner *Falco biarmicus* and Barbary Falcon *F. pelegrinoides* in level flight (George 1969, 1970). See also Maclean (1976). FLOCK BEHAVIOUR. Watering flights normally in morning and, particularly if hot, sometimes also in evening (e.g. Hume and Marshall 1878, Baker 1914a, Meinertzhagen 1954, George 1969, 1970). In eastern Morocco, birds moved out of roost in early morning, then foraged; this followed by increasing number flying up and calling, with vocal response from birds on ground (see 1 in Voice). Flew to water (at *c*. 3 km from feeding grounds) in flocks of up to 60, calling constantly; birds passed rapidly several times over watering place before landing, calling loudly (see 1 in Voice), several metres from water. Normally waited 1–2 min and were very wary; then moved in jostling pack to water, some wading in a short distance. After *c*. 15 s returned to landing spot before flying off. If prevented from drinking by presence of (e.g.) nomads, birds waited at 200–1000 m; in long delay, foraged nearby. When resting on ground in such a situation, not usually able to see water-hole: at irregular intervals, single bird would fly up in reconnoitring flight at height of 50–80 m, moving very slowly several times over hole; rapid but shallow wing-beats recalled Skylark *Alauda arvensis*, and calls uttered constantly (see 2 in Voice). Bird then rejoined flock, even when danger had passed, and flew with it to water. Behaviour as described typical of non-breeders; pairs bolder, repeatedly rising up and attempting to reach water-hole separately, not tolerating delay of more than 1 hr and thus even landing when men and animals by water (George 1969, 1970). See also Archer and Godman (1937), Meinertzhagen (1954), Hüe

and Etchécopar (1957), and Valverde (1957). ANTAGONISTIC BEHAVIOUR. According to Hume and Marshall (1878), ♂♂ generally less quarrelsome than those of *P. orientalis*. However, vigorous ground fights with *P. alchata* occur, north-west Africa (George 1970). When pair approached by 2 ♀♀, eastern Morocco, these driven off by ♀ of pair, perhaps in defence of pairing territory: head held far forward on extended neck, attack launched, and calls given (George 1969: see 3 in Voice); see also Relations within Family Group, below. In Iraq, pair whose chicks not visible (probably crouching) foraged and preened; eventually joined by 2nd pair, 1 bird of which was attacked ('torpedo'-style: thus probably as just described) after quite long period of relative inactivity by both parents who then moved rapidly away; not clear if direct defence of young involved (S Marchant). At roost, ♂ drove off any non-breeders which attempted to roost near family party (George 1970). HETEROSEXUAL BEHAVIOUR. In eastern Morocco, pair-formation took place in pre-breeding areas where flocks quickly (often over only a few days) broke up into pairs prior to departure for breeding grounds proper. No descriptions of behaviour leading to pair-formation but once, early May, 6 birds (thought to be 5 following ♀, but not always clear which pursuer and which pursued) observed in aerial chase, sometimes at ground level, sometimes up to *c.* 100 m high; for behaviour of extralimital Namaqua Sandgrouse *P. namaqua*, see Maclean (1968). In nest-site selection, pair wandered around, ♀ seeking out collections of larger stones: in 1 hr, ♀ made 5 scrapes near or in such accumulations, while ♂ stood by, alert (George 1969, 1970). Until clutch complete, incubation mostly by ♀, at least during daylight, with ♂ often in close attendance; in Iraq, later pattern involved ♀ incubating until *c.* 19.00 hrs, ♂ through night until 09.30–10.00 hrs, i.e. similar to *P. alchata*; ♂ took over in 2 cases after 18.00 hrs, Morocco (Marchant 1961a; George 1969, 1970) but see also below. Claim in Whitaker (1905) that ♂ brings water to incubating ♀, transferring it by regurgitation, considered erroneous by Marchant (1961a). In Morocco, ♂ flew off sitting ♀ between 06.30 and 08.00 hrs, calling loudly (not described), both then flying off to drink. On return, both landed at some distance from nest but ♀ made for it immediately, while ♂ remained in vicinity. When ♀ left nest to feed for *c.* 1 hr in afternoon, ♂ took over; on return of ♀, ♂ side-threw small pebbles from nest-rim before change-over (George 1969). At one nest, Iraq, ♂ waited until ♀ alongside, stepped off, paused at *c.* 2 m while ♀ settled, then flew off out of sight. At another site, ♂ foraged at *c.* 200 m but alert and moved towards nest; ♀ flew off calling (not described) when ♂ at *c.* 80 m, landed *c.* 150 m from nest, calling intermittently, then made final departure roughly as ♂ settled (Marchant 1961a; S Marchant); see also Ticehurst *et al.* (1922). RELATIONS WITHIN FAMILY GROUP. Chicks call from inside egg as early as 2 days before hatching. Leave nest after last one hatched—on following morning if this occurs during night. In one case, ♀ and chicks left nest when youngest *c.* 4 hrs old, joined immediately by ♂; covered *c.* 50 m in 1 hr, then roosted. On leaving roost, family moves to wadi, chicks constantly seeking shade from parents, especially during first few days. Maximum distance covered per day usually *c.* 200–300 m. When family moved purposefully to wadi 1 km away, journey took 2 days: ♀ in front of chicks, ♂ behind; ♂ called (see 3 in Voice) with neck outstretched, often pecking at last chick; only short foraging breaks *en route*. In family party, Iraq, ♀ apparently acted as sentinel, following ♂ and chick at *c.* 50–20 m (S Marchant). Young self-feeding from early age (Ticehurst 1926), but parents peck vigorously at ground and often turn it over. ♂ brooded chicks in conditions of swirling sand. At night (see also Roosting), young brooded by ♀ alone, ♂ taking over from 06.00 to 07.00 hrs while ♀, who resumed on return, left to drink. Once

A

wadi reached, family remained there throughout day and moved only short distance within it (probably because of above-average temperatures) and tended to enter and leave at same spot. Mostly avoided broader parts of wadi where vegetation too high (presumably greater danger from predators), and normally moved directly to next belt of vegetation. (George 1969, 1970.) In family with 2 chicks, Iraq, each kept close to one parent; a single chick moved constantly between ♂ and ♀ (S Marchant). Earlier reports (e.g. Whitaker 1905) indicated both adults brought water to young and fed this to them by regurgitation. More recent studies have shown that only ♂ brings water, carrying it in absorbent plumage of underparts (Marchant 1961a; George 1969, 1970). ♂ flies to water alone when chicks still small and tended by ♀. In one case, first occurred immediately after last chick hatched. Later (at *c.* 4 days), chicks left and procedure as follows: parents call in response to other adults flying over to water (see 1 in Voice); on hearing calls, chicks crouch motionless. Parents stretch wings upwards several times, then fly off. At watering place ♂ wades in (after both have drunk) with lateral body feathers spread wide, though only central belly feathers soak up water (Fig A). (See, e.g., Cade and Maclean (1967) for water-holding ability of feathers, etc., in 4 extralimital *Pterocles*.) In case of young chicks brooded by ♀ in roosting hollow, these emerge first when ♂ returns. Otherwise, both parents come back after *c.* 10–30 min and land, holding wings high for a moment, *c.* 20–30 m from young. Chicks respond only to constant calls of own parents (see Voice). ♀ drives off any non-breeders which may have returned with pair; otherwise runs past chicks and may begin foraging. According to Marchant (1961a), ♂'s flight and gait both appeared laboured, and he landed by chick he had previously tended. At 1–2 m from chicks, ♂ assumes erect Watering-posture (Fig B) and ruffles lateral belly feathers and those on neck and throat. Young run up and, when small, sit in line to take water from feathers (in *P. namaqua*, chicks draw feathers through bill in 'stripping' action: Cade and Maclean 1967). When chicks older, lack of space produces attempts at drinking from side. ♂ may remain motionless for 10 min while chicks drink; when chicks move, holds posture for several minutes more, then leads young to sandy spot where all sand-bathe (see George 1969, 1970). In Iraq, ♀ stood by at *c.* 15 m throughout watering and was eventually joined by 1 chick; afterwards, whole family foraged (Marchant 1961a). ♂ normally goes for water in early morning but timing may be irregular: 2 young *c.* 7 days old constantly attended by both parents for *c.* 3½ hrs, morning, Iraq (Marchant

B

1961a). ♂ held 24·5 g of water in belly feathers after flight of c. 10 km (Kainady 1977). ANTI-PREDATOR RESPONSES OF YOUNG. According to George (1970), chicks 1–4 days old and with grey-brown down tend to hide among stones of matching colour; between 4th and 6th day become increasingly sandy-yellow (possibly due to bleaching: Maclean 1976) and seek patches of wind-blown sand by small bushes, immediately digging themselves in, covering c. 50% of body in short time. Crouching most common response in presence of avian or ground predator or on hearing warning call (not described) of ♂ (George 1969, 1970). 2 young in India remained motionless until about to be seized; larger chick then ran off flapping wings (Bolster 1922). May feign death (Ticehurst 1926), but no details. PARENTAL ANTI-PREDATOR STRATEGIES. (1) Passive measures. Tends to nest in stony areas where plumage of ♀ matches substrate (George 1969, 1970). Incubating birds allow varyingly close approach by man: one ♀ Morocco, remained unusually long on nest when eggs hatching (George 1970); another Iraq, could almost be touched; one ♂ tended to remain motionless, while another ran off, head held high, when men and camels at c. 150–200 m (S Marchant). Tend to crouch lower at approach of grazing animals (George 1969; see also below). While observer nearby, ♀ spent 5–20 min in approach to nest, sometimes making short flights to and fro (Marchant 1961a). In one case, Morocco, ♀ removed egg-shell to c. 20 m, after departure of human intruder (George 1970). Family heading for large wadi, Morocco, crouched when man in path, then attempted wide detour (George 1970). (2) Active measures: general. ♂ remains nearby when ♀ incubating; gives warning calls (not described) at approach of danger (George 1969). Both sexes may perform distraction-display of injury-feigning type (Meinertzhagen 1954). (3) Active measures: against birds. When kite *Milvus* overhead, Iraq, parents flew about nervously, landed by young, flew again, walked past young, etc. (S Marchant). In Morocco, when large falcon *Falco* over breeding area where adults had returned to water young, contact calls ceased and birds crouched. Once, when family attacked by low-flying *F. pelegrinoides*, parents flew a few metres ahead of predator at about half maximum speed, apparently as distraction-lure display; after some time, flew steeply upwards, accelerating, and flew off. Great Grey Shrike *Lanius excubitor* swooping on to insect near family party at once attacked by ♂ (George 1970). (4) Active measures: against man. Birds generally perform distraction-lure display of injury-feigning type, but no details; may call (see 4 in Voice). One ♀, Iraq, feigned injury only after flying c. 20 m and then running another c. 80 m (S Marchant). Another allowed approach to c. 20 m, then crept off nest, trailing wings, and remained nearby, flying up only when approached very close, and further brief descriptions indicate furtive departure and possible mock brooding (Ticehurst *et al.* 1922); one ♀ flew off nest and fluttered on ground (Ticehurst 1926). Parents in another case flew around persistently until chicks taken away (Bolster 1922). 10 min after parents had departed, apparently on watering flight, they returned within 1 min when observer approached chicks, and were assumed to have been watching from air: landed c. 50 m from intruder and beat wings 2–3 times, making little leaps (George 1970). (5) Active measures: against other animals. In Morocco, family crouched as jackal *Canis* passed at c. 20 m; when predator at c. 200 m (out of sight), parents flew up and followed it in slow flight at height of c. 10 m (see also Flock Behaviour, above). ♀ on nest approached closely by camel, abruptly raised and fanned tail exposing white spots; whole herd made detour (George 1969, 1970).

(Fig A based on photograph of *P. exustus* by C J F Coombs; Fig B from photograph in George 1969.) MGW

Voice. No major study; most comprehensive description in George (1969, 1970).

CALLS OF ADULTS. (1) Contact- and flight- calls. A 'quito quito' (George 1969). Recording (Fig I) suggests 'quitoo', 'whitoo' or 'wicko wicko'. Also rendered as: a distinctive and far-carrying 'wakhu' ('waqu-waqu'), 'wey-hu wey-hu', 'whit-hu whit-hu', 'gūtū-gūtū', or sometimes 'wheet-wheet whit-hu' (Bolster 1922; Bannerman 1931; Archer and Godman 1937; Ali and Ripley 1969); a loud 'wittow-wittoo' or 'crakow-crakow' (Gallagher and Woodcock 1980). See also Valverde (1957) where further described as clearer and more musical than Crowned Sandgrouse *P. coronatus*, though according to Whitaker (1905) and Heim de Balsac (1925), may be difficult to separate from that species. In flight, 'hi hü-kawá-kawá-hi hükawá kawá kawá' (Koenig 1896; see also Hartert 1912–21). A musical 'cuddle cuddle cuddle' (Meinertzhagen 1954; Hüe and Etchécopar 1957); gurgling 'quidle quidle quidle', like blowing through tube with one end under water, and recalling Chestnut-bellied Sandgrouse *P. exustus*, but less harsh (Hume and Marshall 1878; see also Magrath 1919, Ticehurst *et al.* 1922). Flocks give a musical gabbling (Gallagher and Woodcock 1980); in distance, like yapping of small dogs (P J Sellar); when approaching but still distant, sounds like hookah or 'hubble-bubble' being smoked (Ali and Ripley 1969). These calls given by circling birds and those on ground prior to watering flight, as birds land, in interactions with young, etc. (George 1969, 1970). (2) Reconnoitring-call. A distinct, penetrating 'quit-o quit-o'; first sub-unit short and sharp, long-drawn 'o' following after c. 2 s; sometimes only 'quit' given. Uttered during survey flight over water (George 1970). (3) Rattling sounds given by ♀ in antagonistic encounter with other ♀♀, and also by ♂ driving young to new feeding grounds (George 1969, 1970). (4) Distraction-call. Plaintive sounds given by ♀ feigning injury near nest (George 1970).

CALLS OF YOUNG. Cheeping of young audible 2 days before hatching. Later, 'quito quito' calls indistinguishable from those of adults (George 1969), e.g. by chick c. 7–10 days old (Bolster 1922). MGW

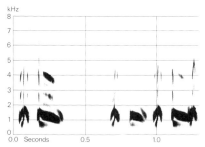

I J-C Roché (1967) Morocco March 1966

Breeding. SEASON. North Africa: eggs laid early April to mid-July (Heim de Balsac and Mayaud 1962). Eastern Morocco: eggs found from last third April to early June (George 1969, 1970). Iraq: similar to North Africa (Marchant 1961a). SITE. On ground in the open. In eastern

Morocco always near conspicuous stone, either larger than neighbours, or of conspicuous colour (George 1969). Nest: shallow scrape or, in Iraq, more usually natural hollow such as animal footprint (Marchant 1961a). Building: by ♀, with rocking movements of body (George 1969, 1970); sometimes, probably no building (S Marchant). EGGS. See Plate 94. Elliptical, smooth and fairly glossy; pale buff, lightly speckled, spotted, and blotched light brown and pale purplish-grey. 41 × 28 mm (36–44 × 26–30), $n = 33$; calculated weight 18 g (Schönwetter 1967). Clutch: 2–3, always 3 in eastern Morocco (George 1969). 8 clutches of 3 found in Iraq (Marchant 1963c). Of 11 clutches, North Africa: 5 of 2, 6 of 3 (Etchécopar and Hüe 1967). Laying interval 24–48 hrs (Marchant 1961a; George 1969). INCUBATION. 29–31 days (George 1969). Starts with first egg; hatching asynchronous. By both sexes. In Iraq, ♀ during day, and ♂ at night; change-overs recorded at 09.27, 10.18, and 19.00 hrs (Marchant 1961a). In eastern Morocco, almost exclusively by ♀ at some nests, but ♂ twice took over after 18.00 hrs (George 1969, 1970). Suggestion by Whitaker (1905) that ♂ supplies incubating ♀ with water by regurgitation strongly disputed by Marchant (1961a). YOUNG. Precocial and nidifugous. Cared for by both parents, but each parent may look after one young in brood of 2 (Marchant 1961a). Self-feeding from outset, with some assistance from parents (George 1969); see Social Pattern and Behaviour. No further information.

Plumages. ADULT MALE. Forehead and crown cinnamon-buff, bordered at sides by wide silver-grey stripes over and behind eye. Hindneck buff, suffused grey. Mantle cinnamon-buff, paler in worn plumage. Feathers of lower mantle suffused golden-ochre near tip. Scapulars brown-grey with extensive yellowish-buff tips. Back cinnamon-buff; upper tail-coverts yellowish-buff. Lores pale grey; sides of head, chin, and throat bright golden-ochre; lower foreneck and chest buff washed grey. Rest of underparts buff with longitudinal black area on centre of belly. Under tail-coverts distally white, basally dark brown. Central tail-feathers (t1) grey, distally grading into ochre-buff and with strongly elongated narrow black tip, t2–t8 grey with subterminal black area and broad white tip; outer web of t8 white. Outer webs of p1–p9 buff, grading into dark brown-grey near tip; inner webs grey. P10 with narrow outer web and shaft brown; p1–p5 distinctly tipped buff, p6–p7 more vaguely margined buff. Outer webs of upper primary coverts buff, inner grey-brown. Secondaries grey-brown, outer web basally buff, s1–s7 narrowly margined off-white at tip; tertials (from s11 inwards) grey with extensive bright golden-ochre area at tip, similar to scapulars but brighter. Upper wing-coverts violet-brown with buff spot at tip, upper rows of lesser coverts entirely buff. Greater under wing-coverts grey-brown, tipped pale buff; rest of under wing-coverts and axillaries pale buff. ADULT FEMALE. Forehead, crown, and nape cinnamon-buff, finely streaked dark brown. Lores and streak over and behind eye off-white, similarly streaked. Upperparts, including lesser upper wing-coverts, buff, conspicuously marked with rounded dark brown spots. Ground-colour of upper tail-coverts strongly washed ochre. Sides of head, chin, and upper throat golden-ochre, not as bright as in ♂. Lower throat and chest pale buff marked with spots similar to those on upperparts. Rest of underparts pale buff with longitudinal black area on centre of belly; lower flanks off-white; under tail-coverts white, basally dark brown. Tail like ♂, but t1 vaguely barred dark grey and with less strongly elongated tip. Primaries and secondaries like ♂, tertials buff, distally washed golden-ochre and spotted like upperparts. Greater and median upper wing-coverts rufous-buff with pale buff tip; spotted dark brown. Outer webs of primary coverts buff, clouded grey. Under wing-coverts and axillaries like ♂. DOWNY YOUNG. Similar to other *Pterocles*, but ground-colour pale. Upperparts mottled mixture of pale sandy-buff and brown, with black tips of down. Pattern of pale lines resembling other Pteroclididae. Underparts pale buff. (Ticehurst *et al.* 1922; George 1969; Fjeldså 1976.) JUVENILE. Upperparts sandy-buff with wavy pale grey-brown bars and vermiculations. Chin and throat white. Chest pale buff, wavily barred like upperparts. Centre of belly dark brown. Central tail-feathers buff, barred brown, not elongated; rest of tail-feathers with brown inner web and barred outer web; outer web of t7–t8 white. Primaries with pale buff outer web, brown inner web, and buff tip mottled brown. P10 like adult. Secondaries like adult. 1st adult plumage develops soon after juvenile, specimens recognizable as long as some juvenile primaries retained.

Bare parts. ADULT. Iris brown; bare skin round eye yellow, but close to eye grey. Bill blue-grey, darker at tip. Feet pale grey, claws dark grey. (Hartert 1912–21; Baker 1928; BMNH.)

Moults. ADULT POST-BREEDING. Complete; primaries descendant. Timing poorly recorded, probably about May–December. Hartert (1924) recorded specimens in fresh plumage October. POST-JUVENILE. Almost complete, some outer primaries retained. Summer or early autumn of 1st calendar year.

Measurements. Adult; skins (BMNH, RMNH, ZMA).

WING	♂	206	(3.67; 5)	203–212	♀ 196	(4.97; 6)	189–203
TAIL (t1)		132	(9.66; 5)	120–146	109	(7.71; 9)	99–120
(t2)		88	(1.41; 4)	86–89	80.4	(4.53; 8)	76–89
BILL		12.0	(0.79; 6)	11.1–13.3	11.7	(0.80; 9)	10.7–13.3
TARSUS		21, 23, 24			21, 21, 22		
TOE		21, 23, 24			21, 22, 22		

Sex differences significant for wing and tail.

Weights. Adult. Morocco, ♂ 280, ♀ 255 (Thomas and Robin 1977); Iraq, ♂ 264 (Kainady 1977).

Structure. Wing pointed. 11 primaries: p10 longest, p9 3–11 shorter, p8 12–22, p7 25–36, p6 38–47, p5 50–61, p4 64–73, p1 104–114; p11 minute, concealed by primary coverts. 18 secondaries, s11–s18 tertials. Tail wedge-shaped with strongly elongated central feathers, t1 *c.* 45 longer than t2 in ♂, *c.* 35 in ♀. Bill short, stubby. Middle toe about as long as tarsus; outer toe *c.* 67% of middle, inner *c.* 63%; hind toe reduced to small nail, raised, at side of tarsus, *c.* 14% of middle toe.

Geographical variation. Birds from Kutch (India) have been described as darker than other populations and with grey on crown (Neumann 1924) and named *remotus* but, in view of general variability in colour, recognition not warranted (Meinertzhagen 1954; Vaurie 1965).

JW

Pterocles exustus Chestnut-bellied Sandgrouse

PLATES 24 and 26
[facing pages 302 and 326]

Du. Kastanjebuikzandhoen Fr. Ganga à ventre brun Ge. Braunbauchflughuhn
Ru. Краснобрюхий рябок Sp. Ganga moruna Sw. Brunbukig flyghöna

Pterocles exustus Temminck, 1825

Polytypic. *P. e. floweri* Nicoll, 1921, northern Egypt. Accidental: *erlangeri* (Neumann, 1909) western and southern parts of Arabian peninsula. Extralimital: nominate *exustus* Temminck, 1825, south of Sahara from Sénégal to Sudan; *ellioti* Bogdanow, 1881, south-east Sudan, northern Ethiopia, and Somalia; *olivascens* (Hartert, 1909), southern Ethiopia south to Tanzania; *hindustan* Meinertzhagen, 1923, south-east Iran, Pakistan, and India.

Field characters. 31–33 cm; wing-span 48–51 cm. Bulk as smaller, short-tailed congeners, but with long tail-streamers. Small sandgrouse with distinctive dusky and yellow-toned plumage, streaked and barred in ♀. Broad dark belly shared only by much larger Black-bellied Sandgrouse *P. orientalis*. Upperparts of both sexes show pale spots. Wholly dark underwing diagnostic. Sexes dissimilar; juvenile separable at close range.

ADULT MALE. Most of forebody and back dusky isabelline with marked olive wash in some lights, relieved below by narrow black line round lower chest and increasingly dark brown to black belly and vent, and above by pale yellow black-edged spots on larger scapulars and centre of back. Most of head yellow-ochre, contrasting with dusky crown and neck. Tertials and wing-coverts patterned as scapulars, with bold black-edged yellow spots obvious at close range; whole area forms pale dusky-yellow patch at distance. Primaries noticeably black, inner ones tipped white. Underwing unique in genus, with dusky brown under wing-coverts appearing uniform with black flight-feathers. Bill pale grey-horn, orbital ring off-white (both obvious at close range). ADULT FEMALE. Basic plumage pattern close to that of ♂, but crown and hindneck streaked, while breast spotted and outlined black. Upperparts narrowly barred black, with pale cream spots on back more obvious than on ♂ and extending to form at least 5 bands over wing-coverts; dark belly indistinctly barred buff. Wing pattern as ♂. JUVENILE. Closely resembles ♀ but duller and more narrowly barred above; tail not elongated.

Unmistakable in flight, when underwing visible. On ground, may be confused with *P. orientalis* and Spotted Sandgrouse *P. senegallus*. Unlike those species, ♂ lacks grey on foreparts and has black-margined feathers on upperparts; lack of black throat mark a further distinction from *P. orientalis*. Separation of ♀ and immature from *P. senegallus* more difficult, since they share spotted and barred upperparts. Best distinction is extensively dark belly of *P. exustus*. Distinction of ♀ and immature from *P. orientalis* also difficult, particularly when size obscured, but pin-tail, generally yellower plumage tone (particularly on foreparts), back spots, wing-bands, and dark belly diagnostic. Behaviour typical of genus.

Most gregarious of genus. Commonest call suggests Willow Grouse *Lagopus lagopus*—guttural chuckling 'whit kt-arr . . .', etc.

Habitat. In hot arid subtropical and tropical zones, on dry barren plains and plateaux up to *c*. 1500 m or more. Prefers extensive level, rolling, open tracts or slopes, of steppe or semi-desert, with loose silty or dusty soil, or stony or rocky surface, but usually unattracted to sand. Also frequents waste patches surrounded by dry cultivation and fallows, dry stubbles, and sun-baked ploughlands. Associated with sparse vegetation but is attracted by leguminous herbs, grasses, and weed plants, and even young cereal crops. Sometimes among scattered thorny shrubs or trees such as *Acacia*, but avoids dense cover, and wooded, rocky, or broken terrain as well as wetlands and areas of more than light precipitation. In Somalia, said to contrast with Lichtenstein's Sandgrouse *P. lichtensteinii* in inhabiting open bare semi-desert country and native cultivation, devoid of trees and shrubs to which *P. lichtensteinii* is attached. Here, exceptionally, it extends to coast and inshore islands (Archer and Godman 1937). Flies regularly to drink at wells or pools, often through upper airspace. Otherwise terrestrial. (Koenig 1926; Christensen and Bohl 1964; Ali and Ripley 1969.)

Distribution. EGYPT. Formerly resident in Faiyum and along border of the cultivated Nile valley (Meinertzhagen 1930). No recent records (PLM, WCM). SYRIA. No skins or observations (Kumerloeve 1968*b*).

Accidental. Hungary: ♀ obtained from flock of Pallas's Sandgrouse *Syrrhaptes paradoxus*, August 1863 (Keve 1960*a*).

Population. EGYPT. In latter half of 19th century abundant near Suez and said to be commonest sandgrouse in Egypt; scarce by 1929 (Meinertzhagen 1930). No recent records (PLM, WCM).

Movements. Resident and nomadic.

On southern edge of Sahara, big concentrations occur in dry season when surface water scarce and localized, and scale of these suggest extensive nomadism. In Mali, 50 000 by Lake Kabara in late March, and comparable numbers observed (December–April) in many parts of Sahel zone,

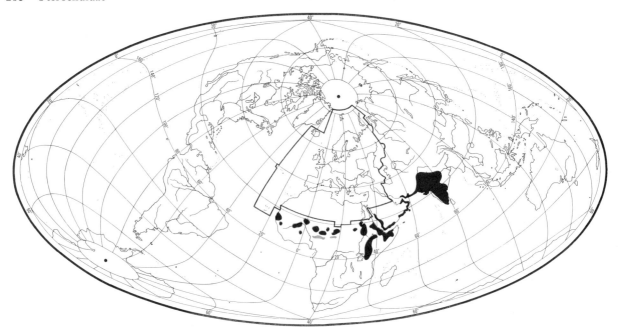

including 40 000 by pool of 0·5 ha at Tahabanet (Lamarche 1980). Non-breeding visitor to Oudi Rime Reserve (central Chad) at beginning and end of wet season, small bands remaining until December or January after heavy rains; withdraws to north for hottest months, April–June (Newby 1980). Approaches status of rains migrant in northern Nigeria, where non-breeding visitor in dry season (October–May), returning northwards into Niger during rains (Elgood *et al.* 1973; Elgood 1982). Also occasional visitor to eastern Gambia, season not stated (Gore 1981).

Few data for Middle East. South-easterly movement detected along coastal dunes at Awamir (northern Oman) in August–September (Walker 1981*a*); water-related dispersals noted Dhofar, mid-May to September, birds remaining near water source until return to breeding areas in March (Walker 1981*b*). Comparable nomadic movements recorded in Pakistan and India (Ali and Ripley 1969).

Food. Almost exclusively vegetable matter, mainly seeds. Feeds by picking up seeds from ground; may also pluck off parts of growing plants, like Black-bellied Sandgrouse *P. orientalis* (Glutz *et al.* 1977). Reported to scratch like partridge (Phasianidae) when foraging (Heuglin 1873) and, according to Reichenow (1900–1), favours areas where antelopes (Antilopinae) have been lying, scratching there in dry dung. Feeds before and after drinking, i.e. in very early morning and then, normally at different site, in more desultory fashion later; may also feed after midday roosting, sometimes in vicinity (Hume and Marshall 1878; see also Social Pattern and Behaviour).

Seeds taken include *Alysicarpus*, amaranth *Amaranthus*, rattlewort *Crotalaria*, *Cyamopsis psoralioides*, beggarweed *Desmodium*, spurge *Euphorbia*, *Gynandropsis gynandra*, heliotrope *Heliotropium strigosum*, various species of indigo *Indigofera*, grasses *Panicum*, pulse *Phaseolus radiatus*, *Tephrosia*, possibly also *Acacia tortilis*, *Cyperus conglomeratus*, *Chrozophora*, *Eleusine aristata*, *Pseudanthispteria hispida*, and thistle (Compositae). Blades of grass and leaves of mustard (Cruciferae) also recorded (Adams 1864; Baker 1914*c*; King 1921; Bowen 1928; Bates 1934*b*; Faruqi *et al.* 1960; Christensen and Bohl 1964; Ali and Ripley 1969).

Analysis of 47 crops from Pakistan and north-west India showed birds had fed almost entirely on seeds, primarily large numbers of the very small leguminous Indigoferae, though larger seeds (e.g. *Tephrosia*) also taken. Seeds of cultivated plants (*Phaseolus*, *Cyamopsis*) taken in only minor quantities. Nearly 50% of crops contained grit; picks up sand and gravel at watering place (Baker 1914*c*), and also reported by Bannerman (1931) to eat salt. (Faruqi *et al.* 1960, Christensen and Bohl 1964, where further details including seasonal analysis.) 4 crops from southern Sahara contained seeds of *Panicum turgidum* (most abundantly), *Cyperus conglomeratus*, *Chrosophora* and 2 species of Leguminosae, the larger probably *Acacia tortilis* (Bates 1934*b*). According to Faruqi *et al.* (1960), seeds of thistle taken preferentially in some areas; reported to feed extensively on durra, Egypt and Sudan (Adams 1864). In Sudan, favoured amogia *Tephrosia apollinea* (King 1921; Bowen 1928). At Ouadi Achim (Chad), mainly small to medium-sized seeds but also seen to take black ant (Formicidae) (Newby 1979). One crop, India, contained ants (Formicidae); another, small beetles (Coleoptera) (Hume and Marshall 1878), but no other reports of animal matter being taken.

No information on food of young. MGW

Social pattern and behaviour. Fullest study, based on observations in India and Pakistan, and review of literature, by Christensen and Bohl (1964). Many aspects still inadequately known.

1. Gregarious to highly gregarious; nearly always in small flocks, Egypt (Meinertzhagen 1930) and southern Sahara (Bates 1934b); up to 200–300 at water, Saudi Arabia (Meinertzhagen 1954). In Thar desert (India), 5–40 frequently recorded, and 30–40 regular during monsoon. Largest aggregations (up to 50000: Lamarche 1980) occur in extralimital parts of range when watering sites restricted (Christensen and Bohl 1964); see also Heuglin (1873), Hume and Marshall (1878), Baker (1914c), and Ali and Ripley (1969). BONDS. No information on mating system or duration of pair-bond. Sex-ratio apparently 1:1 in India and Pakistan (Christensen and Bohl 1964). First breeding at 1 year according to Christensen and Bohl (1964). However, in Sénégal, 1971, birds which hatched at beginning of breeding season (February to August or September) bred for first time at end of same season (when moulting out of juvenile plumage); in following year, all birds bred from beginning of new season, thus including those only c. 6 months old (Morel 1973). Care of young by both parents (Meade-Waldo 1922), probably at least until fledging; in one case apparently beyond (Aldrich 1943). BREEDING DISPERSION. In India, 2 eggs allegedly belonging to different pair found only a few cm from nest containing normal clutch of 3 (Hume and Marshall 1878). No further information. ROOSTING. Nocturnal, in compact assembly on ground; apparent preference for open country with minimum of cover. Also during day; particularly at hottest time, in shade of shrubs, small hollows, etc.; tends to be more dispersed than at night. Birds may loaf, e.g. after drinking (Hume and Marshall 1878; Reichenow 1900–1; Baker 1914c; Christensen and Bohl 1964). Sometimes active on moonlit nights: birds foraged by water, Lake Chad, remaining there until dawn (Bannerman 1931).

2. Generally alert, flying readily and, when flushed, almost vertically upwards (Hume and Marshall 1878; Baker 1914c; Christensen and Bohl 1964). May also crouch or run off, depending on season (Heuglin 1873). FLOCK BEHAVIOUR. Disturbed flocks may execute several rapid turns and fly in great arc before landing (Heuglin 1873). When foraging, birds often move gradually away from afternoon roosting site, those at rear flying to front, etc. (Hume and Marshall 1878). In Dhofar (Oman), winter, birds water 08.00–09.00 hrs; in hot weather, 07.00–09.30 and 16.30–17.30 hrs (Walker 1981b). In Saudi Arabia, watering flights normally in morning, from 08.00 hrs on fine, hot days, not until c. 10.00 hrs in cold weather; a few may water again in evening before sunset in hot weather (Meinertzhagen 1954); such evening drinking regular in some areas in hot, dry conditions, but otherwise normally much reduced or absent (Baker 1914c; Hüe and Etchécopar 1957; Christensen and Bohl 1964). Smaller flocks band together as they fly in; land c. 400 m from water and wait c. 10 min before flying to drink (Meinertzhagen 1954). In Dhofar, land directly in water (Walker 1981b). In Thar desert, birds assemble on nearby flat; eventually rise as flock, circle water-hole and, if no disturbance, land quickly at water's edge, sometimes actually in water, drink, and depart (Christensen and Bohl 1964). In other accounts, reported to feed or loaf before departure (Hume and Marshall 1878; Baker 1914c; Bowen 1928; Bates 1937b; Meinertzhagen 1954). May fly up to 16–80 km to water and drink regularly at same site despite disturbance, though if this becomes excessive, shift to another (distant) locality (Butler 1905; Christensen and Bohl 1964). ANTAGONISTIC BEHAVIOUR. ♂♂ reportedly pugnacious; in fights, call (see 3 in Voice), strike with wings, and, in general, resemble fighting pigeons (Columbidae) (Heuglin 1873). HETEROSEXUAL BEHAVIOUR. No information on pair-formation or courtship. In India, ♀ seen in scrape before laying with ♂ standing nearby (Aldrich 1943). Incomplete clutch covered by ♂ throughout day; after completion, ♀ incubates by day, ♂ at night (J F Reynolds). ♂ may relieve ♀ briefly during day when ♀ away feeding (St Quintin 1905). RELATIONS WITHIN FAMILY GROUP. Chicks leave nest shortly after hatching (Meinertzhagen 1954). In captivity, self-feeding from outset (Meade-Waldo 1922); brooded by both parents (St Quintin 1905). Further observations on captive birds showed watering of young to be as in (e.g.) Spotted Sandgrouse *P. senegallus* (see St Quintin 1905, Meade-Waldo 1922). See also Stegmann (1969) and, especially, Maclean (1976) for discussion of watering in Pteroclididae. ANTI-PREDATOR RESPONSES OF YOUNG. In India, family attempted to make inconspicuous escape from observer in car, chicks constantly seeking to hide under tail or close to body of parent (Aldrich 1943, which see for photograph). Captive chicks crouched low at approach of observer (St Quintin 1905). PARENTAL ANTI-PREDATOR STRATEGIES. Often leaves nest only when intruder at c. 1 m (Ali and Ripley 1969); once ♀ crouched almost flat on nest, feathers spread all round her (Beadon 1915). Bird once flew straight up off nest and out of sight at approach of man; brief description apparently indicates furtive run followed by mock-brooding (Hume and Marshall 1878). MGW

Voice. No major studies. Described as noisy during flight to water (Bannerman 1931).

CALLS OF ADULTS. (1) Contact- and flight-calls. Guttural chuckling sounds recalling Red Grouse *Lagopus lagopus*: 'whit kt-arr wit wit-ee-er kt-arrr-arr (J Hall-Craggs, P J Sellar: Fig I). Also described as a penetrating 'kut-ro', not particularly loud, but far-carrying (Ali and Ripley 1969); loud, musical 'gutter gutter' (Meinertzhagen 1930, 1954); sonorous and melodious gurgling 'gatta(r) gatta(r)' or 'gutta gutta...' (Glutz *et al.* 1977); sharp, clear 'gitt-ah gitt-ah' or 'qitt-ah qitt-ah' on taking off or as flight-call (Heuglin 1873; Hartert 1912–21); a deep 'kurra-kurra' or 'gata-gata' (Gallagher and Woodcock 1980). Commonest call 'chocka chocka'; 'quock ca-car' and 'wot worp wot worp' also given (Walker 1981b). Like continual gurgling (Adams 1864) or creak from movement of rusty springs (Bannerman 1931); see also Hüe and Etchécopar (1957). Recording of flock in flight suggests harsh gobbling with combination of twangy and harsh rasping sounds. (2) Soft twittering calls (Serle *et al.* 1977), e.g. from ♀ flying around after mate shot (Bates 1937b) presumably distinct; however, low soft twittering or 'cluttering' apparently treated (confusingly) as normal flight-call by Bates (1934b). (3) Grumbling sounds given by fighting ♂♂ (Heuglin 1873). (4) Continuous low murmuring or babbling when

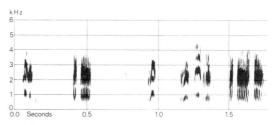

I C Chappuis/Sveriges Radio (1973) Mauritania January 1969

drinking undisturbed (Archer and Godman 1937); a musical 'creen' (Meinertzhagen 1954), though, according to Hüe and Etchécopar (1957), such sounds may also be uttered during flight to water. Crooning from captive birds thought to express contentment (Barnby Smith 1910), presumably also belongs here.

CALLS OF YOUNG. No information. MGW

Breeding. Little information from west Palearctic. SEASON. Somalia and Saudi Arabia: eggs laid from end of April to mid-June (Archer and Godman 1937; Meinertzhagen 1954). North-west India: primary nesting period February–April, but also (e.g.) November, December (Christensen and Bohl 1964; Ali and Ripley 1969). SITE. On ground in the open. Nest: shallow depression, unlined. Building: no information. EGGS. See Plate 94. Elliptical, smooth and slightly glossy; cream, with brown speckles and spots, and some purplish-grey markings. Nominate *exustus*: 36×25 mm ($34-39 \times 24-26$), $n=10$; calculated weight 13 g. *P. e. floweri*: 38×26 mm ($36-39 \times 25-27$), $n=4$; calculated weight 14 g (Schönwetter 1967). Clutch: 2–3. Strong indication of 2 broods in northern India (Christensen and Bohl 1964). Replacements laid after egg loss. INCUBATION. About 23 days in captivity (Meade-Waldo 1922). By both sexes, ♀ during day, ♂ at night (St Quintin 1905). YOUNG. Precocial and nidifugous. Cared for by both parents. FLEDGING TO MATURITY. First breeding at 1 year or less. No further information.

Plumages (nominate *exustus*). ADULT MALE. Forehead, crown, and upper mantle ochre-grey; hindneck faintly tinged yellow. Feathers of lower mantle and scapulars darker, olive-brown, with ochre spot near tip; scapulars narrowly tipped brown. Rump and upper tail-coverts olive-brown. Sides of head and chin golden-ochre; throat and upper breast buff with vinous hue. Black line across breast, narrowly margined white. Lower breast ochre-buff; belly dark chestnut to black-brown; flanks dark chestnut indistinctly barred black. Under tail-coverts pale yellow-ochre. Feathers on leg pale cream. Central tail-feathers (t1) elongated, black, basally dark ochre-grey. Other tail-feathers with broad white tip and black subterminal area; inner web basally grey, progressively darker towards outer feathers; outer web barred black and buff, but outer web of outermost (t8) black with white margin. Primaries: p10–p7 dark brown, p6 with narrow white margin to inner web, p5 with broader white margin and narrow white tip, p4–p1 broadly white at tip and inner web, basally dark brown. Underside of primaries dark brown, but shafts of p10 and p9 mainly white. Primary coverts dark brown. Secondaries (s1–s10) dark brown; s1–s2 narrowly tipped white; tertials (from s11 inwards) distally ochre-grey, innermost with ochre-yellow area and narrow brown apical line, like scapulars. Upper wing-coverts golden-ochre, basally buff; inner greater, median, and lower row of lesser coverts with brown apical margins, forming narrow bars on upperwing. Inner greater and median often with subapical white crescent. Most lesser coverts with apical brown spot; marginal coverts ochre-grey. Under wing-coverts and axillaries dark brown. ADULT FEMALE. Crown, nape, and upper mantle buff, boldly streaked brown. Lower mantle, scapulars, rump, and upper tail-coverts barred buff and brown; scapulars and some feathers of lower mantle with pale ochre subterminal area; scapulars narrowly tipped brown. Sides of head and chin golden-ochre; throat and upper breast buff spotted brown, separated by narrow black line from pale buff lower breast. Belly dark brown, indistinctly barred rufous-buff; flanks barred buff and brown; under tail-coverts and feathers on leg cream. Central tail-feathers (t1) barred buff and brown, elongated tip dark grey; t2–t8 broadly tipped pale ochre; t2–t3 barred buff and brown on both webs, inner webs of t4–t8 progressively darker grey and bars less distinct. Primaries and primary coverts like ♂ but paler. Secondaries brown, s1–s2 tipped white, s3–s4 narrowly tipped white, s10 with outer web distally pale ochre. Tertials barred buff and brown, distally pale ochre; inner margined brown at tip like scapulars. Greater and median upper wing-coverts pale golden-ochre, basally buff; lesser barred buff and dark brown, with warmer brown spots at tips; marginal buff spotted brown. Under wing-coverts and axillaries grey-brown. DOWNY YOUNG. Upperparts pale brown mottled dark brown and cream, with pattern of pale streaks forming figure of 8 (Baker 1928). White bar across forehead runs down over lores, joining white rictal stripe. White line across crown from eye to eye. Nape with pronounced central white streak and narrower lateral ones. Throat and chest sandy-buff, rest of underside off-white (BMNH). JUVENILE. Feathers developing shortly after hatching loosely structured and indistinctly barred, later ones more like adult; this occasionally interpreted as double juvenile plumage, but no clear distinction present between these. Upperparts of early stages of juvenile rufous-buff, narrowly barred and vermiculated dark brown; scapulars tipped pale buff. Throat and breast similar to upperparts, but bars narrower, more V-shaped. Belly dark brown in centre, barred rufous-brown in front and at sides. Under tail-coverts buff, irregularly spotted brown. Tail barred, feathers tipped rufous; t1 not elongated. Primaries brown, distally rufous-buff, peppered dark brown, tipped with narrow dark brown line and cream margin. Outer secondaries broadly tipped white, inner narrowly; tertials buff, irregularly barred. Greater upper wing-coverts ochre-buff, median and lesser rufous-buff with narrow brown subterminal V and wide pale buff edge. In later stages, ♂ and ♀ separable. ♂ has upperparts like adult, but ochre-grey feathers subterminally vermiculated dark brown and margined pale buff; breast vinous-buff with indistinct wavy bars, without black band; belly like adult; p10 dark brown, different from juvenile primaries; tertials buff, barred brown; upper wing-coverts margined warm brown. ♀ mainly like adult, but feathers of lower mantle and scapulars with pale ochre area at tip and brown spots on breast crescent-shaped. (Kalchreuter 1979, on *olivascens*; BMNH, RMNH.) SUBSEQUENT PLUMAGES. Like adult, but recognizable when some juvenile outer primaries retained; according to Kalchreuter (1979) also by regular wear and bleaching of secondaries which are irregularly worn in adults due to irregular moult schedule.

Bare parts (various races). ADULT. Iris dark brown; bare skin round eye yellow to greenish-yellow. Bill pale blue or pale grey, also recorded as yellow-white changing to grey after death (Sassi 1906). Foot pale yellow-grey to pale blue-grey, back of tarsus and sole of foot dark brown (Kalchreuter 1979), claw recorded as grey or blackish. (Baker 1928; Friedmann 1930; BMNH.) DOWNY YOUNG, JUVENILE. No information.

Moults. ADULT POST-BREEDING. Complete; primaries descendant. North of equator apparently from early May to late November (BMNH), south of equator mainly from September to May. Progress of primary moult slow, each feather falling when preceding about full-grown. Secondaries and tail usually start when primaries about half renewed, moulting in irregular sequence. Tertials actively moulting during entire primary moult and beyond, possibly renewed twice in course of annual cycle. Same may apply to body moult. POST-JUVENILE. Almost complete, but at least juvenile p10 retained; starts at age of a few weeks. Moult of secondaries starts at 3 moult-centres when primary moult has reached p5–p7 and proceeds regularly descendantly and ascendantly. Tail moults at same time. Body and tertials start simultaneously with primaries, many feathers possibly renewed twice before adult plumage acquired. (Kalchreuter 1979.)

Measurements. Adult, races combined; skins (BMNH, RMNH, ZMA).

WING	♂ 182	(4·36; 15)	175–190	♀ 175	(7·10; 11)	160–184
TAIL (t1)	124	(9·06; 15)	112–136	99·8	(9·24; 13)	84–119
(t2)	74·6	(4·34; 16)	66–80	70·9	(4·30; 13)	65–78
BILL	11·9	(0·78; 14)	10·8–13·4	11·5	(1·03; 11)	10·3–13·8
TARSUS	23·0	(0·52; 14)	22–24	21·8	(0·92; 8)	20–23
TOE	22·4	(0·96; 14)	21–24	21·2	(0·75; 8)	20–23

Sex differences significant except for bill.

Weights. *P. e. olivascens.* Kenya: ♂ 200; ♀♀ 170, 175, 200 (Britton 1970).

P. e. hindustan. ♂ 227–284, ♀ 213–241 (Baker 1928); Thar desert (India) ♂ 184–234, ♀ 170–213 (Christensen and Bohl 1964).

Structure. Wing pointed. 11 primaries: p10 (usually) or p9 longest or either one 0–1 shorter than other; p8 5–7 shorter than longest, p7 15–16, p6 25–27, p5 37–39, p4 49–50, p1 83–87; p11 minute, concealed by primary coverts. 18 secondaries, s11–s18 form tertials. Tail strongly wedge-shaped, with elongated central feathers, t1 *c.* 50 mm longer than t2 in ♂, *c.* 30 in ♀; t2 *c.* 25 longer than t8. 16 (14–18) tail-feathers (Kalchreuter 1979). Bill short and slender. Middle toe about as long as tarsus; outer and inner *c.* 65–70% of middle, hind toe raised, *c.* 25% of middle.

Geographical variation. Concerns mainly tinge of colouration of upperparts and upper wing-coverts; differences of opinion in literature may be due to local and individual variation. Nominate *exustus* pale, but *ellioti* paler still, with breast in ♂ more rufous, warmer grey, upperparts in ♀ more rufous. *P. e. olivascens* darker, with upperparts dark olive-grey in ♂ and more heavily barred in ♀; throat and chest darker; upper wing-coverts yellow-ochre rather than golden-yellow. *P. e. floweri* lighter than *olivascens*, darker and greyer than nominate *exustus*; aberrant ♂ specimen in BMNH extremely dark (see Meinertzhagen 1930, plate XXIV). *P. e. hindustan* greyer on upperparts than nominate *exustus*, similar to *floweri*, but duller; upper wing-coverts and subterminal spots on scapulars more ochre, less yellow. *P. e. erlangeri* paler than *hindustan* and nominate *exustus*; scapulars and upper wing-coverts vinous-buff. (Vaurie 1961.) JW

Pterocles orientalis Black-bellied Sandgrouse

PLATES 25 and 26
[facing pages 303 and 326]

Du. Zwartbuikzandhoen   Fr. Ganga unibande   Ge. Sandflughuhn
Ru. Чернобрюхий рябок   Sp. Ortega   Sw. Svartbukig flyghöna

Tetrao orientalis Linnaeus, 1758

Polytypic. Nominate *orientalis* (Linnaeus, 1758) Iberian peninsula, Fuerteventura (Canary Islands), North Africa, Cyprus, and Asia Minor, east to about the Turkey–Iran border; *arenarius* (Pallas, 1775) from the lower Volga east to Chinese Turkestan, south to Iran and Afghanistan (Vaurie 1965).

Field characters. 33–35 cm; wing-span 70–73 cm. Bulkier than any other west Palearctic sandgrouse but lacking tail-streamers. Large heavy-bodied sandgrouse with relatively short tail and rather broad wings. Combination of long and broad black belly with white under wing-coverts diagnostic. Mainly grey head and chest of ♂ interrupted by black and chestnut throat and terminated by bold black transverse line. ♀ spotted on chest and irregularly barred above. Sexes dissimilar; no seasonal variation. Juvenile scarcely separable. 2 races in west Palearctic, not separable in the field.

ADULT MALE. Plumage most strongly patterned of west Palearctic sandgrouse. Head, neck, and fore-body grey, palest on lower chest and relieved by chestnut and black half-collar on lower throat and distinct black line across lower breast, and contrasting sharply with wholly black lower body and vent. From hindneck to rump and over all tertials and wing-coverts, plumage basically yellow-ochre, but olive-black centres obvious on all larger feathers except greater coverts. In flight, grey-black primaries and dusky primary coverts contrast with grey-spotted yellow inner wing (on which yellowest greater coverts may appear as pale central panel). On underwing, black flight-feathers contrast sharply with white coverts. Pointed but short tail barred yellow and black. Bill black-horn, orbital ring off-white. Feet black. ADULT FEMALE. Pattern as ♂, but lacks chestnut half-collar and grey foreparts; head and chest pale buff, streaked and spotted dark black-brown (with chest-marks contained by black line positioned as in ♂), and whole of upperparts yellow-ochre, closely barred and marbled dark brown or black. JUVENILE. Closely resembles ♀.

Largest sandgrouse of west Palearctic and unmistakable when seen well. Confusion with Chestnut-bellied Sandgrouse *P. exustus* and Pallas's Sandgrouse *Syrrhaptes paradoxus* only likely in distant observation of silent birds and even then instantly removed by sight of underwing and underbody pattern—strikingly black and grey-white in present species, almost wholly dark in *P. exustus*, and wholly pale with discrete black patch on mid-underbody in *S. paradoxus*. Flight action slower and more powerful than congeners, with slower take-off sometimes obvious. Noticeably wilder than most other sandgrouse.

Most often seen in small groups or pairs; occasionally more gregarious. Commonest call a rather gruff, yet soft, gurgling or bubbling suggesting Black Grouse *Tetrao tetrix*.

Habitat. Extends over lower middle latitudes from warm temperate and Mediterranean to arid or semi-arid steppe and subtropical climatic zones, summer range being between 21°C and 29°C July isotherms; winter range in Asia between *c*. 5°C and 18°C January isotherms, in North Africa and Iberia between 10°C and 18°C, with precipitation normally 250–500 mm. Accordingly adapted to somewhat cooler and less arid conditions than most congeners and to greater climatic variety. Differs from nearest geographic counterpart Pin-tailed Sandgrouse *P. alchata* in ascending to higher altitudes and in greater attraction to vegetation cover, and from Lichtenstein's Sandgrouse *P. lichtensteinii* in dislike of trees and scrub, and generally of rough rocky terrain. Chiefly on flat plains, saltflats, and sandy, loamy, clay, or gravelly soils or dusty patches and tracks, sometimes covered with stones or varied by hummocks, hillocks, eroded slopes, or worn-down rocky outcrops. Saline or alkaline flats with scattered patches of vegetation are attractive. Occasionally occupies artificial sites such as airfields. In some regions ascends to mountain valleys and upland plateaux with grassy steppe vegetation and grazed arid rangelands, up to more than 3000 m. In different parts of range associated with wide variety of grasses and flowering plants, especially after rains; also with cultivated crops under dry farming or irrigation and accompanying fallows and stubbles, on which spilt grain sometimes eaten. Tolerates presence of occasional low shrubs and even trees such as mesquite *Prosopis* where there are expanses of bare ground with sparse patches of seasonal vegetation. Wintering habitat similar. Access to water essential, as for congeners, even when flight of up to *c*. 60 km needed to reach it, and even when it is brackish or salt. Flock recorded alighting in deep water in River Tigris (Meinertzhagen 1954). Moves freely on unobstructed terrain, and flies strongly at varying heights.

Dispersive habits over wide suitable habitat, and speed, altitude, and wariness in flight have enabled population to maintain itself fairly well in face of occasional heavy hunting losses (Ali and Ripley 1969). For detailed study of plant species typical of particular parts of range, see Bump and Bohl (1964).

Distribution. TURKEY. Some range reduction in west of central plateau due to increased cultivation (Kumerloeve 1961). SYRIA. May breed in north but no proof (Kumerloeve 1968b; HK).

Accidental. Belgium, East Germany, Malta, Kuwait.

Population. PORTUGAL. Believed to be decreasing (RR). SPAIN. Rather numerous (AN). CANARY ISLANDS. Breeds only on Fuerteventura, and there becoming very scarce (Bannerman 1963b; KWE). LIBYA. Not scarce (Bundy 1976). CYPRUS. Now scarce (PF, PS). TUNISIA. Common (MS).

Movements. Resident in Iberia and North Africa, migratory to partially migratory in southern Asia.

Present all year within Iberian and North African breeding range, and less evidence than for other *Pterocles* (e.g. Pin-tailed Sandgrouse *P. alchata*) of extensive nomadism there. In north-west Africa, southernmost birds move northwards away from most arid zone about July, as breeding finishes (Heim de Balsac and Mayaud 1962). In Libya, commonest in Wadi Kaam (northern Tripolitania) in dry period June–September, and recorded occasionally in Cyrenaica (Bardia, Gazala, Tobruk) in winter (Bundy and Morgan 1969; Bundy 1976). Vagrants to Europe, including flock of 20 in Belgium (August 1917), attributed to western race, which also breeds east to Transcaucasia (Vaurie 1965).

Present all year on central plateau of Turkey, but perhaps partially migratory since small numbers occur regularly on passage in southern Turkey (Beaman *et al.* 1975); passage through Cyprus in April and September–November, though numbers now much smaller than formerly (Stewart and Christensen 1971). Also partially migratory in Transcaucasia, with passage birds noted near Turkish border in September (Dementiev and Gladkov 1951b). Presumably such emigrants winter in Middle East, south to Sinai where scarce but regular, though only straggler to Arabian peninsula (Jennings 1981). Winter visitor, October–April, to Iraq, though these may come from Iranian plateau (Allouse 1953; Moore and Boswell 1956b). Also scarce or irregular winter visitor (mid-October to mid-April) to Kuwait, and has occurred Bahrain in January (Bundy and Warr 1980).

Largely migratory in Asiatic USSR (*arenarius*). A few winter in Tadzhikistan, and more consistently in Turkmeniya where apt to concentrate on fringes of cultivated land in cold weather, but wholly migratory further north in central Asia (Dementiev and Gladkov 1951b). There, flocks formed in August and major flights over Turkmeniya occur September–October, tailing off in November; spring return begins early March, and even northern breeding areas reoccupied by mid-April. Presumably these migrants use the important wintering areas in Pakistan (where also a small resident population in Baluchistan) and in north-west India, where birds occur locally in enormous numbers from September–October to February–March; however, winter distribution somewhat erratic, both

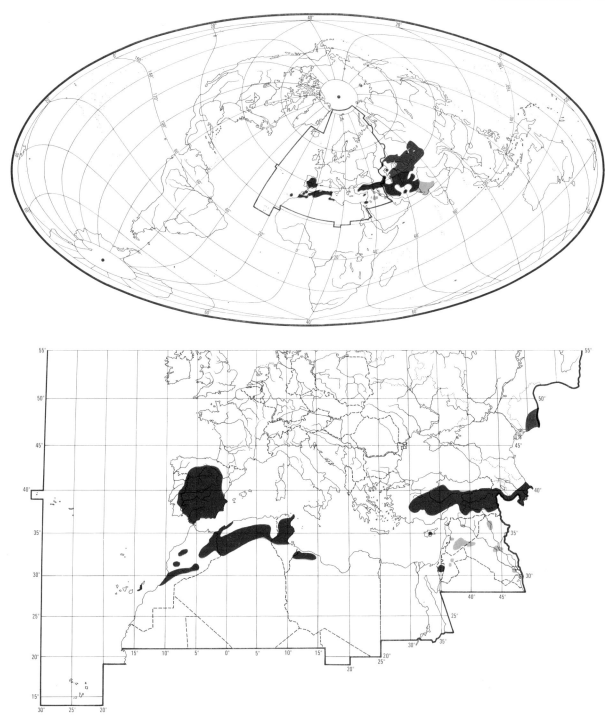

locally and between years, and birds may forsake long-established haunt after even a short run of aberrant seasons (Ali and Ripley 1969).

Food. Mostly seeds. Forages in flocks of 3–25 birds; larger flocks sometimes in winter (Bump and Bohl 1964). Food mainly picked up from ground but will also pluck off vegetative parts and flower heads from growing herbs or bushes (Glutz *et al.* 1977).

Seeds recorded include melilot *Melilotus*, milk vetch *Astragalus*, sainfoin *Onobrychis*, indigo *Indigofera linifolia* and *I. cordifolia*, pulse *Phaseolus radiatus* and *P. acontifolius*, *Tephrosia purpurea*, *Cyamopsis psoralioides*, panic grass *Panicum* (also, rarely, leaves), heliotrope *Heliotropium strigosum*, *Gynandropsis gynandra*, glasswort *Salicornia*, *Sisymbrium cinereum*, *Ammodendron*, *Polygonum*; seeds and shoots of wormwood *Artemisia*, camel thorn *Alhagi camelorum*, and thistle *Salsala*; grain, including barley *Hordeum* and wheat *Triticum*; berries also recorded. Fewer insects (of only very sporadic occurrence and probably taken involuntarily: Hume and Marshall 1878): white ant (Hymenoptera), beetles (Coleoptera), and various larvae. Grit (e.g. quartz granules) invariably present in birds examined. (Hume and Marshall 1878; Koenig 1896; Hartert 1912–21; Baker 1913a, 1928; Sladen 1919; Ticehurst *et al.* 1922; Dementiev and Gladkov 1951b; Rustamov 1954; Gavrin *et al.* 1962; Bump and Bohl 1964; Ali and Ripley 1969.)

Like congeners, often takes very large numbers of smallest seeds (see below), but will also take far more large seeds, including waste grain, than other Asian Pteroclididae (Bump and Bohl 1964).

1 collected Iraq, contained *c.* 30 000 seeds of *Melilotus* and *Astragalus* (Ticehurst *et al.*1922). Of 20, central Turkey, July, 13 contained only weed seeds, 4 weed seeds and wheat or cultivated legumes, 3 were empty; most birds had taken grit (a few small stones); 3, May, contained only weed seeds. Of 2, Iran, October, 1 held curled tips of an unidentified desert plant, the other *c.* 70% weed seeds and *c.* 30% wheat (Bump and Bohl 1964). In Algeria, March–April, preferentially took seeds and other parts of *Sisymbrium cinereum*; birds flew to water when skin folds at either side of crop bulging (Koenig 1896). In central Morocco, a little green food taken with seeds (Meinertzhagen 1940). In Israel, August–September, birds held barley and wheat grains, and other small seeds (Sladen 1919). Of 2 from Kara-Kumy desert, Turkmeniya (USSR), June, 1 held 80% (by volume) seeds of *Polygonum polycnemoides* and 20% seeds of *Astragalus*, the other similar but also some of *Onobrychis*; 1, April, held seeds of *Astragalus* and *Ammodendron* (Rustamov 1954). For data from Kirgiziya (USSR), see Yanushevich *et al.* (1959); from Thar desert (India), Bump and Bohl (1964).

Young birds possibly take more insects than adults but no details (Gavrin *et al.* 1962). According to Dementiev (1952), chicks given mushy mixture of seeds and water from crop.   MGW

Social pattern and behaviour. For fullest study and literature review, see Bump and Bohl (1964). Much of behaviour largely unknown.

1. Gregarious but less so than Pin-tailed Sandgrouse *P. alchata*. Flocks of a few up to several hundred in winter, Iberia and Middle East; main concentrations, as in other Pteroclididae, at watering places, particularly on wintering grounds, India: e.g. once *c.* 8000 at Gajner tank, Bikaner. Otherwise, more often in pairs or smaller parties of 3–25(–50) for foraging, etc. (Bump and Bohl 1964). In Kazakhstan (USSR), ♂♂ form separate flocks for foraging while ♀♀ on eggs (Mitropol'ski 1977). Post-fledging flocks of hundreds or at times thousands, Kazakhstan, USSR (Gavrin *et al.* 1962). On spring migration, USSR, in pairs or small flocks (8–10); in autumn, sometimes several hundred together (Dementiev 1952; Yanushevich *et al.* 1959; Gavrin *et al.* 1962). Autumn–winter flocks in India said to be often single-sex initially, ratio about equal by departure (Hume and Marshall 1878; Baker 1913a); pattern not confirmed by Bump and Bohl (1964). BONDS. Monogamous mating system (Whitaker 1905), pair-bond possibly lasting for several years (Mitropol'ski 1977). Pair-formation occurs during spring migration or shortly after arrival on breeding grounds, Turkey and USSR (Dementiev and Gladkov 1951b; Gavrin *et al.* 1962; Bump and Bohl 1964); birds thought to have maintained pair-bond through winter, Tunisia (Zedlitz 1909). In eastern migratory *arenarius*, birds arrive paired in March (Dementiev and Gladkov 1951b), though groups of apparently unpaired birds also noted by Mitropol'ski (1977). Care of young by both parents (Meade-Waldo 1922; Gavrin *et al.* 1962), but duration not known. Age of first breeding 1 year (Bump and Bohl 1964). BREEDING DISPERSION. Generally solitary though nests occasionally close together (Gavrin *et al.* 1962). In wormwood *Artemisia* steppe, central Anatolia (Turkey), *c.* 3 pairs on 1 km² (Lehmann 1971); on high-mountain steppes of eastern Anatolia, 2·5–5 pairs per 100 ha (Bump and Bohl 1964). On middle reaches of Zeravshan river, Tadzhikistan (USSR), *c.* 16 pairs per 100 ha of dry clay pans, 6 pairs in stony desert, and 0·5 pairs on feather grass *Stipa* steppe (Dementiev and Gladkov 1951b). ROOSTING. Communal (possibly in small dispersed groups: Bump and Bohl 1964) and nocturnal; on open plains or fallow fields, in young crops, on dusty tracks, or possibly on sandy or stony hummocks; sites may be traditional, birds flying there after evening foraging or watering (Whitaker 1905; Dupond 1942; Whistler 1949; Bump and Bohl 1964). Dust-bathing and basking frequent in early morning, on sites as above. Birds also roost around midday when sun hot; not all members of flock sleep at same time—some move around calling softly (Hume and Marshall 1878; Baker 1913a: see 1a in Voice).

2. Often described as wildest and wariest of west Palearctic Pteroclididae (e.g. Bates 1937b, Meinertzhagen 1954, Ali and Ripley 1969); in Semirech'e (USSR), said to be less so when in pairs (Shnitnikov 1949). Tends to haunt areas where small bushes or stones roughly match body size, and there may crouch in danger (Meinertzhagen 1940). When flushed, habitually springs up *c.* 2–3 m before flying off (Bump and Bohl 1964). FLOCK BEHAVIOUR. When threatened, birds tend to fly great distance (Whitaker 1905; Bates 1937b); alarm-call (not described) may be given at approach of man (Baker 1913a) and birds sometimes run rapidly away from such danger (Koenig 1896). Early-morning loafing flocks often close-packed (Hume and Marshall 1878), also at times for foraging (Whistler 1941), but generally less compact, both in air and on ground, than *P. alchata*, feeding birds tending to be in line (Meinertzhagen 1940) but may straggle widely (Hume and Marshall 1878). At midday roost, usually more dispersed (Hume and Marshall 1878; Baker 1913a.) Several flocks may unite for early-morning basking, etc., and birds then generally tolerant of each other; nearer breeding season much skirmishing (brief description indicates probably largely individual-distance disputes) between ♂♂ of a flock in early morning, and at water (Hume and Marshall 1878; Baker 1913a; see also further, below). Regular and punctual in watering habits (Baker 1913a) and, according to Thomas and Robin (1977), high degree of intra- and interspecific (with *P. alchata*) synchrony evident in flight times and vocal activity, Morocco. Flies up to *c.* 60 km to water (e.g. Gavrin *et al.* 1962; see also Bump and Bohl 1964). Morning watering the rule in most parts; evening flight may be restricted or absent, particularly in colder weather (e.g. Whitaker 1905, Meinertzhagen 1954, Ivanov 1969, Gavrin *et al.* 1962). In north-west India, arrived before Chestnut-bellied Sandgrouse *P. exustus*, i.e. shortly after sunrise (Bump and Bohl

1964). On arrival, flocks circle 1–2 times and fly down steeply to land near water where may wait up to 30 min. At Bikaner, regularly assembled (sometimes in thousands) on barren flats near water (birds generally favour watering sites clear of vegetation), then circled tank *en masse* before landing to drink quickly and depart. May at times alight on water (see also Meinertzhagen 1954) and drink. Much jostling and squabbling typical, especially when flocks arrive together, but also between ♂♂ of same flock (Baker 1913a). In Turkey, summer, birds left in flocks of up to 50 which broke up into smaller groups at distance. May associate with congeners at water; in central Turkey, December, small flocks apparently integrated with huge aggregation of *P. alchata* (Bump and Bohl 1964). ANTAGONISTIC BEHAVIOUR. See above; no further information. HETEROSEXUAL BEHAVIOUR. Courtship begins some time after arrival on breeding grounds (see also Bonds, above): ♂ flies after ♀, calling (not described), then circles her thus on ground (Dementiev and Gladkov 1951b; Gavrin et al. 1962). Once, Kazakhstan, late July, ♂ displayed near ♀ with tail raised and wings drooped; perhaps preliminary to 2nd brood (Mitropol'ski 1977). Nest-scrape made by both sexes (Glutz et al. 1977). Both sexes incubate: ♀ by day, ♂ at night. ♀ once recorded shading eggs with half-open wings, Turkey, August (Dementiev and Gladkov 1951b; Bump and Bohl 1964). RELATIONS WITHIN FAMILY GROUP. Behaviour of captive birds as in *P. alchata* (Meade-Waldo 1922; see *P. alchata*). 2 adults seen tending brood of 3 (*c.* ⅓ grown), northern Afghanistan (Bump and Bohl 1964); see also Mitropol'ski (1977). Young watered by ♂ (Meade-Waldo 1906, 1922; Glutz et al. 1977), exactly as in Spotted Sandgrouse *P. senegallus* (see that species). ANTI-PREDATOR RESPONSES OF YOUNG. In Karakumy desert (USSR), chicks (age not given) attempted to flee at approach of man (Rustamov 1954). PARENTAL ANTI-PREDATOR STRATEGIES. In later stages of incubation, tends to sit tight and allow close approach (Dementiev and Gladkov 1951b). In Semirech'e (USSR), ♀ finally moved off nest and performed apparent impeded-flight low over ground (as if injured), then dropped down and, whilst calling (see 3 in Voice), feigned dying (Shnitnikov 1949). While chicks fled (see above) ♀ performed distraction-lure display of disablement type and ♂ stood not far off (Rustamov 1954). MGW

Voice. No detailed studies. In flight, wings make soft swishing, almost a whistle at close range (Baker 1913a).
CALLS OF ADULTS. (1) Contact- and flight-calls. (a) A slightly gruff, yet soft, rather low-pitched gurgling or bubbling suggesting Rookooing-call of Black Grouse *Tetrao tetrix*, usually given 2–3 times; far-carrying. Similar but quieter, normally single calls given when foraging or drinking (Glutz et al. 1977; D J Brooks). Recording (Fig I) suggests 'tchowrrr rerr-rerr', the last 2 sub-units having a slightly more petulant timbre (M G Wilson); a hard rattling 'char' followed by a rippling sound and ending with a low-pitched musical tone (J Hall-Craggs). Other descriptions of main call: fairly loud, clucking and musical 'churr-churr-rur', given in flight, and occasionally by birds on ground (which otherwise generally silent) to passing flock (Bump and Bohl 1964); a brief tremulous trill, slightly descending in pitch (Shnitnikov 1949). Markedly different (lower pitched and throatier: Meinertzhagen 1940) from Pin-tailed Sandgrouse *P. alchata* or Pallas's Sandgrouse *Syrrhaptes paradoxus* (Gavrin et al. 1962). See also Whitaker (1905), Baker (1913a), Meinertzhagen

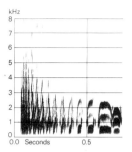
I J-C Roché (1966) Morocco March 1966

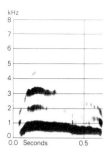
II J-C Roché (1966) Morocco March 1966

(1954), and Ali and Ripley (1969). (b) A wild mewing recalling Buzzard *Buteo buteo* but lower pitched and with weaker harmonics; duration *c.* 0·6 s compared with *c.* 1·0 s of *B. buteo* (J Hall-Craggs, P J Sellar: Fig II), interspersed with call 1a; allegedly given only when bird rises of its own volition, and described as a sonorous 'hou' (J-C Roché). (2) Alarm-call. On flushing, ♂ may give rattling sound (Zedlitz 1909); see also Whitaker (1905). (3) Distraction-call. A feeble squeaking sound (Shnitnikov 1949; Gavrin et al. 1962).
CALLS OF YOUNG. No information. MGW

Breeding. SEASON. Canary Islands: eggs laid from mid-March, with most in April, but also found June (Bannerman 1963b). Spain: eggs found mid-May to late June (Makatsch 1976). Algeria and Tunisia: eggs found early April to mid-July (Heim de Balsac and Mayaud 1962). Cyprus: eggs laid May (Bannerman and Bannerman 1958). Turkey and Afghanistan: eggs found May–June, but in central Turkey also young chicks and slightly incubated eggs in late August (Bump and Bohl 1964). SITE. On ground in the open. Nest: shallow depression, unlined or with a few pieces of dried grass; sometimes in circle of small stones (Bannerman 1963b). Building: by both sexes. EGGS. See Plate 94. Elliptical, smooth and glossy; pale to creamy-buff, or greenish-grey, heavily spotted, blotched, and scrawled brown or buff-brown, and lavender and pale purplish-grey. 48×32 mm ($44-53 \times 30-37$), $n = 109$; calculated weight 28 g (Schönwetter 1967). Clutch: 2–3; of 17 clutches, North Africa, 7 of 2, 10 of 3 (Etchécopar and Hüe 1967); on Canary Islands, 2 (Bannerman 1963b); 12 clutches of 3, Anatolia, Turkey (Lehmann 1971). 2 broods probably, but unproven (Bump and Bohl 1964). Replacements laid after eggs lost. INCUBATION. 23–28 days in captivity (Meade-Waldo 1922). Begins with 1st egg (Dementiev and Gladkov 1951b). By both sexes, ♀ by day, ♂ at night (Austin 1961). YOUNG. Precocial and nidifugous. Cared for by both parents; self-feeding from outset (Meade-Waldo 1922). FLEDGING TO MATURITY. Fledging period unknown. Age of first breeding 1 year.

Plumages (nominate *orientalis*). ADULT MALE. Forehead, crown, nape, centre of hindneck, and upper mantle buff-grey, sides of head paler. Chin, lower sides of face, and sides of neck orange-chestnut. Triangular black spot on throat. Mantle and scapulars spotted: feathers with rounded golden-yellow apical spot and broad dark grey subterminal band; concealed basal part of feathers rufous-cinnamon with grey specks and vermiculations. Back and rump like mantle, but golden feather-tips paler, dark subterminal area narrower, and bases showing through. Upper tail-coverts mainly rufous-cinnamon (pale buff when bleached), narrowly barred grey; tip grey, often with elongated golden spot along shaft, much paler and strongly worn in older feathers. Foreneck and upper breast delicately buff-grey like crown, separated from pink-buff lower breast by narrow black band with indistinct white margins. Belly and flanks black. Feathers on legs off-white. Under tail-coverts white with dark grey-brown bases. Central tail-feathers (t1) cinnamon-rufous (pale buff when bleached) barred grey; tip grey, sometimes tinged golden. T2 like t1, but with distinct white tip and black subterminal bar, both broader on inner than on outer web. Rest of tail-feathers progressively greyer and less distinctly barred, all with broad white tip. Primaries blue-grey (browner when worn) with dark brown shaft, sometimes with narrow white margin to inner web near tip, outer web of p10 dark brown. Primary coverts like primaries, marginal ones cream mottled grey. Secondaries with distal half grey, basal half white to pale ochre; undersurface of flight-feathers grey-black. Tertials (from about s10 inwards) with golden-yellow streak near tip of outer web; innermost with golden-yellow spot at tip, grey subterminal band, and rufous-cinnamon base barred grey, like longer scapulars. Greater upper wing-coverts golden-yellow, inner web paler with some grey blotches or bars. Rest of upper wing-coverts grey with golden spots like mantle. Under wing-coverts white, greater with grey-brown outer web. Axillaries white. ADULT FEMALE. Forehead, crown, hindneck, and upper mantle cinnamon-buff heavily streaked black-brown. Rest of upperparts cinnamon-rufous (paler when worn), spotted and barred black-brown. Sometimes some feathers in centre of mantle with indistinct golden tip. Upper tail-coverts cinnamon rufous, barred dark brown; bars tending to Vs. Sides of head and neck cream, narrowly streaked black, slightly tinged tawny at sides of neck; chin and upper throat cream. Short, narrow black bar across lower throat, bordered below with indistinct grey area (sometimes hardly noticeable). Foreneck and upper breast pale cinnamon-rufous, densely spotted with rounded black-brown spots. Narrow black bar across breast as in ♂. Rest of underparts as ♂. Tail like ♂, but t1 more rufous and all feathers more boldly barred. Primaries brown-grey, inner with distinct white margin to tip and distal half of inner web. Primary coverts grey, narrowly margined white; marginal coverts buff with dark brown spots. Secondaries brown-grey with off-white or buff base; outer often tipped white; undersurface of flight-feathers grey-black. Tertials with base more rufous, often marked with dark brown bars or spots; innermost cinnamon-rufous barred black-brown like scapulars. Upper wing-coverts cinnamon-rufous, barred and spotted black-brown, bars at tips of feathers V-shaped; outer webs of greater coverts golden-yellow. Under wing-coverts and axillaries like ♂. DOWNY YOUNG (data for both races). Upperparts tawny-buff densely spotted black and with poorly defined pattern of off-white lines, less contrastingly patterned than Pin-tailed Sandgrouse *P. alchata*. Throat and foreneck off-white, sometimes with pink hue; rest of underparts pale buff, flanks faintly barred grey. Leg with buff down in front. (Zaletaev 1965; Fjeldså 1977.) JUVENILE. Like adult ♀, but apparently less distinctly patterned. Feathers developing shortly after hatching are of loose structure, later ones more like adult; this occasionally interpreted as double juvenile plumage (e.g. Dementiev and Gladkov 1951b). Early feathers on upperparts cinnamon-pink with narrow concentric dusky lines; chin, throat, and breast similar; belly black, under tail-coverts white. Primaries grey with diagnostic rufous-buff edges vermiculated black; p10 more like adult p10. (Hartert 1912–21; Baker 1928; Fjeldså 1977; Glutz et al. 1977.) SUBSEQUENT PLUMAGES. Like adult, but recognizable when some juvenile outer primaries retained.

Bare parts. ADULT. Iris brown; edge of eyelid lemon-yellow. Bill grey, darker at tip. Leg and foot grey to brown-grey, claws darker (Hartert 1912–21; Baker 1928). DOWNY YOUNG. Iris brown. Bill grey. Leg and foot dirty grey or buff-pink (Zaletaev 1965; Fjeldså 1977).

Moults. ADULT POST-BREEDING. Complete; primaries descendant. Primaries start May, sometimes April (♂, 4 May, southern Algeria, had p1–p2 renewed, p3–p4 growing), apparently dependent on latitude of breeding area; finish October. Tail and body start when primaries less than half renewed, also finish October. POST-JUVENILE. Almost complete, but some juvenile outer primaries retained; autumn of 1st calendar year, starting August. Baker (1928) recorded ♂ in juvenile plumage October. (Dementiev and Gladkov 1951b; RMNH, ZMA.)

Measurements. Adults, data for both races combined; skins (RMNH, ZMA).

WING	♂ 237	(6·00; 8)	227–244	♀ 228	(11·8; 3)	221–242
TAIL	94·4	(4·27; 8)	88–101	91·3	(8·74; 3)	84–101
BILL	13·4	(1·25; 8)	12·0–15·6	13·4	(1·00; 3)	12·4–14·4
TARSUS	30·3	(2·50; 7)	26–34	31·0	(4·58; 3)	26–35
TOE	23·4	(1·40; 7)	22–26	24·7	(0·58; 3)	24–25

Weights (*P. o. arenarius*). September, Kazakhstan, ♂ 428 (9) 400–460; ♀ 383 (11) 300–420; juvenile, both sexes 395 (4) 350–450 (Gavrin et al. 1962). ♂♂ 480, 507, 520, 550; ♀♀ 410, 422, 440, 465 (Dementiev and Gladkov 1951b).

Structure. Wing narrow and pointed. 11 primaries: p10 longest, p9 2–6 shorter, p8 10–15, p7 24–30, p6 37–45, p5 50–58, p4 64–75, p1 113–127; p11 minute, concealed by primary coverts. Tail strongly wedge-shaped; 14 feathers, t1 slightly protruding when fresh; t7 *c*. 30–40 shorter than t1. Middle toe *c*. 80% of tarsus; outer toe *c*. 68% of middle, inner *c*. 75%; hind toe vestigial, *c*. 10% of middle. Rest of structure as *P. alchata*.

Geographical variation. Slight. 2 races recognized here, following Vaurie (1965), though treatment in need of revision. According to Vaurie (1961c), ♂ *arenarius* has upper wing-coverts and tips of feathers of upperparts yellower, less golden than nominate *orientalis*; upper breast greyer, less buff; ♀ paler, lower breast whitish, not pink-buff. Individual specimens may be inseparable, however, and plumage colour sensitive to wear and bleaching. Moreover, occasional adult ♂♂ of both races are dark on central mantle with feather-tips suffused tawny, obscuring golden-yellow.

JW

Pterocles alchata **Pin-tailed Sandgrouse**

PLATES 25 and 26
[facing pages 303 and 326]

Du. Witbuikzandhoen   Fr. Ganga cata   Ge. Spiessflughuhn
Ru. Белобрюхий рябок   Sp. Ganga común   Sw. Vitbukig flyghöna

Tetrao Alchata Linnaeus, 1766

Polytypic. Nominate *alchata* (Linnaeus, 1766), Iberian peninsula and southern France; *caudacutus* (Gmelin, 1774), North Africa and south-west Asia.

Field characters. 31–39 cm; wing-span 54–65 cm. Less bulky and 15% shorter winged than Black-bellied Sandgrouse *P. orientalis* but with noticeably long tail-streamers; *c.* 10% larger than Spotted Sandgrouse *P. senegallus*. Rather large sandgrouse, with strikingly white underbody and under wing-coverts and most intricately patterned and colourful fore- and upperparts in genus. ♂ has chestnut face, black throat, black-edged chestnut chest-band, and distinctly green upperparts. ♀ has yellow-buff face and breast, distinctly lined black through eye, twice round mid-neck, and across lower breast, and strongly marbled back and inner wing. Call distinctive. Sexes dissimilar; some seasonal variation in ♂. Juvenile separable.

ADULT MALE BREEDING. Plumage changes slightly during season (see Plumages). Crown, most of neck, back, and fore- and inner wing-coverts bright oily-green, with yellow spots (in spring only) obvious on scapulars and centre of back and with black edges to wing-coverts. Face chestnut-yellow, with black line through eye; black centre to throat. Broad chestnut-buff chest-band outlined black. Outer wing-coverts vinaceous-chestnut, outlined white and black (and conspicuous on standing bird). Primaries grey-black, with pale bloom and edges so obvious that in flight upperwing shows black trailing edge only on inner primaries and secondaries. Unlike any other sandgrouse, pure white underbody and mainly white underwing combine to form marked contrast with black outer primaries and dusky trailing and leading edges of wing. Rump and tail boldly barred yellow and black, except for dusky tail-streamers. ADULT MALE NON-BREEDING. When moult from breeding plumage complete, upperparts (including crown and upper face) patterned as rump and tail but become buff. Chin white or partly so. Bill brown- or grey-horn; orbital ring blue-grey. Feet grey-black. ADULT FEMALE BREEDING. All upperparts barred and splashed black and pearl-grey on buff and yellow, these colours spreading on to wing-coverts and appearing as widening bars from lesser to greater coverts, ending in sharply etched black and white outer feathers along lower edge of folded wing. Chin white, but face, neck, and chest golden-buff, with black line through eye, broad black bar across lower throat, narrow black line across neck (these two appearing as double neck-ring), and broad black bar below chest. Rest of plumage as ♂ breeding but rump and tail less distinctly barred. Central tail-feathers shorter than in ♂. ADULT FEMALE NON-BREEDING. Loses pearl-grey bars on upperparts; throat becomes spotted. JUVENILE. Resembles ♀ but tail not elongated; upperparts lack grey, and markings of foreparts are subdued, with no black division between chest and underbody. Almost white rear supercilium and pale buff tips to greater and primary coverts are eye-catching, latter forming wing-bar across hindwing.

Unmistakable; no other west Palearctic sandgrouse shows pure white underbody below chest. Flight action free and rapid, reminiscent of both pigeon *Columba* and plover *Pluvialis*; deep, regular wing-beats alternate with occasional glides or tumbling descent. Like all *Pterocles*, wing-beats produce slight whistling sound. Gait restricted by marked shortness of legs; walk waddling and run pattering. Feeding actions recall those of *Columba* and small gamebird *Coturnix* or *Perdix*. Drinking behaviour characterized by mass assemblies of chattering birds at water sources, during which they show excited flight evolutions, cautious landings, and approaches to edge, then rapid drinking from edge or in shallows and sudden withdrawal. Usually shy, performing long escape flights.

Predominant sandgrouse of less arid desert and steppe of west Palearctic, usually outnumbering all congeners. Commonest call a loud, far-carrying, harsh but ringing 'catar catar' suggesting Jackdaw *Corvus monedula* at distance; in chorus, has strong chattering tone.

Habitat. On warm arid steppe and in Mediterranean climatic zone. Mainly on lowland plains, including baked mud on dried-out marshes by tidal water on Spanish marismas, stony hammada on desert edge, and bare clay expanses alternating with sand, as well as dusty or sunbaked flats and sand-dunes. Often on infertile areas of dry cultivation, or near irrigation ditches. Inhabits areas of scattered tamarisk *Tamarix*, *Retama*, and other bushes, but avoids trees and scrubland, and rocky broken terrain. Except in certain areas, such as Spanish marismas, appears less attracted to vegetation cover than *P. orientalis*, but statements that it is more of a desert bird probably go too far. In central Spain, feeds on agricultural land in early summer, moving to more natural habitat later (Casado *et al.* 1983). Prefers more humid habitat (up to 400 mm isohyet) when wintering in Chad; influenced by extent of southward spread of humid fronts, varying with season (Dupuy 1969). Apparently less tolerant of rolling country or uplands although in Morocco on high plateaux. In the Crau (southern France), inhabits stony waste with stones

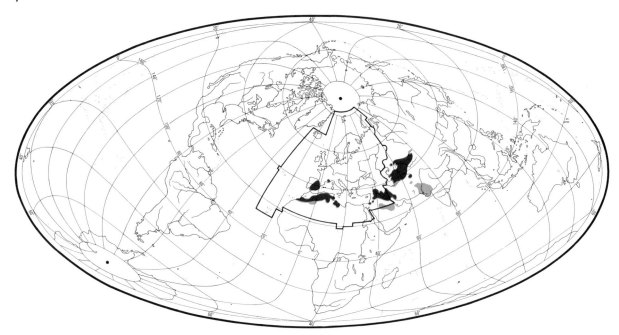

of various sizes and sparse vegetation, especially *Asphodelus*; apparently also associated with vehicle tracks. Generally, however, more attracted to sand and less to stony or grassy terrain than *P. orientalis*. Unable to exist far from water, essential not only for adults but for carrying in soaked belly feathers to chicks, in regular social flights. Will drink brackish water in absence of fresh, and will not only wade but alight on river far from shore, floating high like gull *Larus* and taking off without difficulty. Flies freely in upper as well as lower airspace. Agriculturally marginal and climatically unfavourable character of habitat affords safeguards against human land use and disturbance. (Dementiev and Gladkov 1951b; Paludan 1959; Frisch 1965, 1969a; Ali and Ripley 1969; Ferguson-Lees 1969a; George 1970.)

Distribution. PORTUGAL. Bred formerly; only one recent sighting of a pair, spring 1981 (RR). LIBYA. Southern limits obscure; probably breeds in Fezzan. KUWAIT. Last bred c. 1942 (PRH). ISRAEL. In 1938 and 1961 mass nesting in Huleh area and Beth Shean valley, c. 200 km north of normal range (HM).

Accidental. Italy, Malta, USSR, Kuwait, Lebanon, Cyprus, Egypt.

Population. FRANCE. About 120 pairs (Cheylan 1975). SPAIN. Not scarce (AN). MOROCCO AND ALGERIA. No estimate of breeding populations but in 1964 c. 200 000 on 400 km² of Haut Plateau (George 1970). TUNISIA. Decreasing (Blanchet 1955). IRAQ. Abundant in most parts of range (Ticehurst *et al.* 1922; Marchant 1961a).

Movements. Resident and nomadic in southern Europe, North Africa, and Middle East; largely migratory in Soviet central Asia.

Rather little movement occurs in southern Europe, dispersal being restricted; Iberian birds occur irregularly outside breeding range, e.g. in Guadalquivir delta, and French birds (from Crau population) are erratic visitors to Gard and Provence. Nomadic movements more marked in North Africa, where numbers vary regionally (especially late summer to March) as flocks settle in areas of local rainfall and penetrate further into desert after heavy rains (Heim de Balsac and Mayaud 1962; Smith 1965; Erard 1970). In Libya, becomes more frequent in coastal wadis in near-rainless months of June–September (Bundy 1976).

In Middle East, extensive dispersals reported from Iraq especially; birds will push out further into desert areas during autumn rains, though, in winter, flocks are conspicuous in river valleys before spring return to breeding areas (Allouse 1953; Moore and Boswell 1956a). Irregular visitor to adjacent Kuwait, sometimes in flocks (Bundy and Warr 1980). Resident in northern and central Saudi Arabia, where numbers increase in winter and nomadism apparent (Jennings 1981a). In Turkey, where it occurs mainly in south-east, records are concentrated April–September so possibly summer visitor there (Beaman *et al.* 1975); however, also an earlier report of 50 000 at watering place in December (Bump and Bohl 1964). In North Africa and Middle East, nomadism results in occasional breeding beyond normal limits.

Largely migratory in Asiatic USSR, except close to southern frontiers where it winters regularly. Big flocks form in August, and are seen in Uzbekistan September–October. In Turkestan, main exodus occurs second half of October and first half of November, migrant flocks

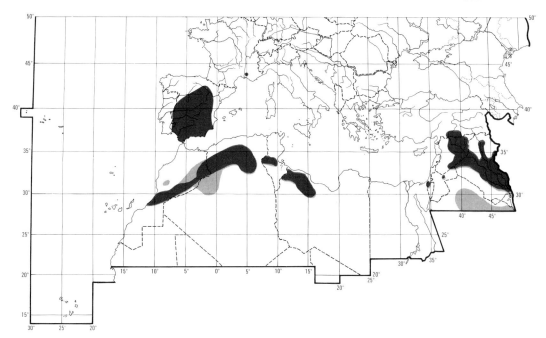

circuiting west of Kopet Dag mountains; a few linger there to late November, and may overwinter in mild seasons. Spring return through Turkestan in March and early April (Dementiev and Gladkov 1951b). These emigrants from Soviet central Asia assumed to comprise (or be included amongst) large numbers of winter visitors that occur in Pakistan (especially Baluchistan and Sind) and to a lesser extent in north-west India (Ali and Ripley 1969). Stragglers to Volga-Ural steppes, Transcaucasia, and Ukraine probably originate from this migratory population.

Food. Almost exclusively vegetable matter (but see below): mainly seeds, fewer shoots and green leaves (Ferguson-Lees 1969a; Cheylan 1975). Seeds normally picked up from ground but, like other parts, may at times be plucked off growing plants. Captive birds flicked sand and soil away with lateral movements of bill, but did not scratch with feet (Frisch 1969a). According to Moore and Boswell (1956b), may take food from camel droppings, Iraq, but Marchant (1962b) never saw birds show any interest in this source. Sometimes feeds in huge flocks, moving steadily across (e.g.) newly sown field picking up grains (Bump and Bohl 1964).

Takes primarily seeds of chenopods, grasses (Gramineae), and other wild plants, also grain. Seeds recorded include pulse *Phaseolus*, seeds and more rarely shoots of glasswort *Salicornia*, wormwood *Artemisia*, camel thorn *Alhagi camelorum*, *Arthraterum pungens*, seeds and leaves of asphodel *Asphodelus fistulosus*, *A. alba*, vetch (Leguminosae), *Calligonum*, bistorts *Polygonum*, *Fagopyrum*, green leaves of (e.g.) clover *Trifolium*, leaves or flowers of *Mesembryanthemum nodiflorum*. Of animal matter only beetles (Coleoptera) mentioned specifically (see below). Much grit, quartz, and sand also regularly taken. (Dresser 1871–81; Hume and Marshall 1878; Baker 1914b, 1928; Ticehurst *et al.* 1922; Dementiev and Gladkov 1951b; Ivanov *et al.* 1953; Guichard 1961; Gavrin *et al.* 1962; Ferguson-Lees 1969a; Frisch 1969a; Thomas and Robin 1977.)

In the Crau (southern France), pellets contained plant debris and chitinous parts of insects: birds considered to have taken *c.* 50% plant food (mainly seeds, also some leaves of *Asphodelus fistulosus*; seeds of *A. alba* also favoured: Frisch 1969a) and 50% insects, primarily small beetles (Guichard 1961). However, almost entirely vegetarian in other parts of range (Cheylan 1975), and insects otherwise rarely mentioned or thought to be taken accidentally (see, e.g., Hartert 1912–21, Dementiev and Gladkov 1951b, Ivanov *et al.* 1953). Neither adult nor young captive birds ever took any animal food (Frisch 1970). In central Spain, 48 birds contained seeds (see Casado *et al.* 1983). In Syria, stomachs contained tiny seeds of small grass-like shrub, cereal grains, and green leaf fragments; 1 bird which had been feeding on newly sown wheat and barley held *c.* 58 cereal grains (Ticehurst *et al.* 1922). Stomachs collected winter, India, contained green leaves, seeds of pulse *Phaseolus*, and various kinds of grain; gizzards held small stones (Baker 1914b). Grit, etc., may constitute up to $\frac{1}{3}$ (by weight) of stomach contents (Meinertzhagen 1964). In Turkmeniya (USSR), birds on autumn passage favoured camel thorn seeds (Dementiev 1952). In Morocco, a ♀ fed preferentially on the exceptionally salty *Mesembryanthemum nodiflorum* (Thomas and Robin 1977).

Young feed entirely on minute seeds (Ferguson-Lees 1969a). See also Social Pattern and Behaviour. MGW

Social pattern and behaviour. Best studied of west Palearctic Pteroclididae: see Frisch (1969a, 1970) for study of captive birds and observations in the Crau (southern France), also Guichard (1961) and Cheylan (1975); for data from Iraq, see Marchant (1961a), and, for general review, Ferguson-Lees (1969a).

1. Strongly gregarious, particularly in autumn and winter and for watering (Ferguson-Lees 1969a): e.g. at least 50 000 at watering place near Mardin (Turkey), December (Bump and Bohl 1964). Yearly flocking pattern well illustrated in study of small population in the Crau: in winter in groups of c. 100 or virtually whole population (c. 400) together; flocks of c. 50–80, March, reduced to c. 10–20 by April and gradual break-up into pairs takes place from second half of March to May (similar pattern in Iraq: Marchant 1963c); increasing tendency for families to join together and adults to flock in fledging period, and up to 120 together again from September (Cheylan 1975; also Ern 1960, Ferguson-Lees 1969a); according to Guichard (1961), gregarious habit re-established from early June, particularly for evening drinking. In Karakumy desert (USSR), spring passage flocks generally small, c. 20–30(–150); in Turkmeniya (USSR), thousands at times on autumn migration (Dementiev 1952). In the Crau, may associate with Little Bustard *Tetrax tetrax* (Cheylan 1975; see also Roosting, below). For further details, see (e.g.) Hume and Marshall (1878), Whitaker (1905), Baker (1914b), Ticehurst et al. (1922), Meinertzhagen (1954), and Kumerloeve (1968a). BONDS. Observations on wild and captive birds indicate essentially monogamous mating system (Dresser 1871–8; Marshall et al. 1911)—but see also below and part 2. In Aral Sea area (USSR), flocks break up and pair-formation takes place in late April (Dementiev and Gladkov 1951b). Structure of flocks, the Crau, suggests pair-fidelity possibly extends beyond breeding season and bond may in fact be life-long (Cheylan 1975), which apparently confirmed by behaviour of captive ♀ (Meade-Waldo 1906). Pairs observed close together in flocks on ground and in the air, Algeria, March–May (Heim de Balsac 1926). Trios (2♂♂, 1♀) not uncommon in breeding season in the Crau, and rape and casual sexual encounters probably also not infrequent; superfluous captive ♀ maintained at least tenuous contact with a pair despite attempts to expel her (Frisch 1969a, 1970; see also below). No pairing behaviour noted in captive birds until c. 2 years old (Frisch 1970), but according to Dementiev and Gladkov (1951b), birds attain sexual maturity at 1 year. Both parents tend young until fledging (Guichard 1961); see also Relations within Family Group, below. BREEDING DISPERSION. Often in groups, but many nests more isolated (Ferguson-Lees 1969a). Colonial or semi-colonial in USSR, probably determined by availability of suitable habitat; nests at times c. 20 m apart, usually less dense (Dementiev and Gladkov 1951b); 10–30 m apart in Kazakhstan (USSR), and more colonial than Black-bellied Sandgrouse *P. orientalis* or Pallas's Sandgrouse *Syrrhaptes paradoxus* (Gavrin et al. 1962); see also Dementiev (1952) and Rustamov (1954). In the Crau, c. 60–80 pairs on 60 km²; after break-up of small flocks into pairs (late May), each pair remains roughly within c. 1-km radius (Guichard 1961; Cheylan 1975, which see for further details, including diagrams); 3 pairs along 3-km stretch, 700–1000 m apart (Guichard 1961). In Sheik-Saad area (Syria), thousands of pairs on a few km² (Ticehurst et al. 1922; see also Baker 1928). ROOSTING. Communal and nocturnal, on ground (for diurnal roosting, see below). In the Crau, 76 roosted at night as compact group within larger roosting assembly of 800 *T. tetrax*, each bird in its own self-made scrape (Cheylan 1975). Birds sleep on belly, head held forward and neck raised (Frisch 1969a). Some spasmodic flying and vocal activity throughout night, Iraq, mid-June (S Marchant). Generally also roosts in close-packed groups during heat of day (Ticehurst et al. 1922). In the Crau, birds showed fondness for dusting and basking, sometimes remaining virtually immobile for hours (Guichard 1961). Dusting behaviour of wild and captive birds including chicks thought possibly unique: slide along short distance on back, head and neck twisted, eyes closed legs pointed skyward; birds sand-bathed also 'more normally' on belly (Frisch 1969a).

2. Some variation in descriptions of relative wariness: probably due at least in some measure to level of disturbance. In the Crau, birds generally wary, flying at c. 100–200 m from human intruder: rise almost vertically, calling then and also afterwards in forward flight (see 1 in Voice); rapidly attain great height and may describe huge circles before disappearing from view (Guichard 1961; see also Ern 1960). Birds allow close approach in car, then fly up suddenly with loud calls. If observer further away, freeze on hearing alarm-call (see 3 in Voice); remain crouched for several minutes before moving slowly away with body held low (Frisch 1969a). In Algeria, birds considered less wild (and pairs less so than flocks: Heim de Balsac 1926) than in southern Europe: usually escape by running, and fly only when pursued (Germain 1965). Up and down Head-bobbing commonly given prior to running or flying (Ferguson-Lees 1969a): not so according to Frisch (1969a) where, however, reported that captive birds raised and fanned tail like Snipe *Gallinago gallinago* in sudden alarm (see account for *P. senegallus*). See also Hume and Marshall (1878), Whitaker (1905), Bates (1937b), Gavrin et al. (1962), and Kumerloeve (1968a). FLOCK BEHAVIOUR. Feeding flocks usually compact (Meinertzhagen 1940); in Iraq, c. 2000–3000 spread over large area but tended to fly together (S Marchant). Movements of feeding flock like Starlings *Sturnus vulgaris* at roosting time; in alarm, birds at front fly up first (Moore and Boswell 1956b). For behaviour of flock in relation to grazing animals, the Crau, see Guichard (1961). Flocks of 7–17 birds (predominantly ♂♂ and thus probably non-breeders) performed repeated aerial evolutions daily in late afternoon/evening in the Crau: flew up suddenly, indulged in rapid changes of height and direction, and landed after c. 1 min (Frisch 1969a). 2 pairs in the Crau, late April, showed remarkably co-ordinated behaviour: foraged 15.00–15.50 hrs, then preened and dust-bathed, foraged again until 16.25 hrs, then came together to rest, remaining immobile until 16.39 hrs, when all stretched suddenly and simultaneously, then resumed feeding (Cheylan 1975). Watering flight occurs 07.00–09.00 hrs in Arabia; birds noisy in flight and when watering (Meinertzhagen 1954: see 1 in Voice). May wheel and circle over water before landing (Bump and Bohl 1964); see also Gavrin et al. (1962) for description of such performance in Kazakhstan (USSR). In Syria, June, generally watered 06.00–08.00 hrs and again 17.00–18.30 hrs, though sporadic drinking took place at various times of day; when cold, c. 08.00–12.00 hrs. In July, arrived from c. 05.00 hrs. After flocks had gathered in area of bare earth, all rose up and moved to water; usually waded in and, after drinking, departed singly or as re-formed flocks (Ticehurst et al. 1922). In Tunisia, unlike congeners, birds tend to alight directly in shallows (Zedlitz 1909); if disturbed, usually make 1–2 attempts to return before finally quitting (Whitaker 1905). Pair recorded alighting on water and drinking, Iraq (Magrath 1917). In Morocco, where pronounced synchrony with activity of *P. orientalis* (see that account), participation in, and timing of, watering flights temperature-sensitive: birds apparently anticipate hotter conditions by making earlier, briefer flights to drink (Thomas and Robin 1977). ANTAGONISTIC BEHAVIOUR. Captive birds generally amicable, but in breeding season ♂♂ become vocal and engage in persistent pecking bouts (Koenig 1896). Vigorous ground fights occur with *P. senegallus*, north-west Africa (George 1970). In breeding season, ♂ liable to attack any other ♂, or other bird (e.g. lark Alaudidae), that approaches

his mate: calls given (see 6 in Voice) and warning posture adopted with head lowered and wings arched. May expel intruder in rapid run with head held low and forward, tail raised, wings close to body, and feathers sleeked; intruder usually retreats immediately (Marshall *et al.* 1911). In larger feeding flocks and associations of several pairs or families, ♀♀ often drive off ♂ coming too close: ♀ runs at ♂, neck extended, calling (see 6 in Voice); in trios (see Bonds, above), ♀ may even peck at strange ♂, but no serious fights develop. Captive ♀♀ particularly aggressive towards one another when they approach closer than 1 m; superfluous ♀ was also driven off (see Bonds, above) and, during a copulation attempt, ♂ of another pair intervened. In the Crau, late June, ♀ of pair copulated with strange ♂; after initial hesitation, her own ♂ tugged at intruder's back feathers; intruder then moved off with own ♀ who had just approached, and soon both pairs foraged together peacefully (Frisch 1969a, 1970). A captive pair evicted conspecifics (as described above) from vicinity of nest (Frisch 1970). In Iraq, ♂ foraging near incubating ♀ crouched and 'pointed' (presumably posture as indicated for expelling attack, above) from *c.* 4–5 m apparently causing 2nd ♂ to walk past. Pairs tending young exhibited varying degrees of tolerance and aggression towards intruding conspecifics of various ages (S Marchant). HETEROSEXUAL BEHAVIOUR. (1) Pair-bonding behaviour. During pair-formation, birds more active than normal; make more spontaneous flights, also perform courtship flights similar to those of Garganey *Anas querquedula*, but at great height. Small group flies up and wild pursuit follows; joined by other birds, all then performing kind of high-speed sky-dance, calling constantly and moving rapidly away (Guichard 1961: see 1 in Voice). Little information on display postures. ♂ twice followed ♀, both birds moving with rather deliberate gait, heads lowered, and tails raised and partly fanned (Ferguson-Lees 1969a): possibly similar to 'strutting' of Namaqua Sandgrouse *P. namaqua*, possibly connected with pair-formation (Maclean 1968). According to Marshall *et al.* (1911), ♂ may move round ♀ on ground like pigeon (Columbidae), with feathers ruffled, wings raised or arched, and tail sometimes fanned. (2) Mating. In captivity, ♂ followed ♀ in stiff-legged gait, tail depressed and fanned, mounted after a few metres, and copulated; no invitatory behaviour by ♀ (Frisch 1970) (3) Nest-site selection. According to Guichard (1961), in the Crau ♀ chooses nest-site about mid-May. Once, ♂ made rotating movements with wings slightly raised, whereupon ♀ ran up to inspect site. Captive ♂ led ♀ to particular spot for scrape, calling quietly and persistently (see 5 in Voice). Both birds scrape with feet (but not in manner of foraging hens); also peck around in scrape and remove any small stones, faeces, or feathers (Frisch 1969a, 1970). (4) Behaviour at nest. Both sexes incubate. During laying, eggs covered by ♀ during daylight; ♂ often in attendance (in the Crau at *c.* 100 m, and also accompanied ♀ when she left to feed or drink: Guichard 1961), but not after clutch complete (Marchant 1963c), when ♂ only rarely comes near (Ferguson-Lees 1969a). When 1st egg laid by captive ♀, pair maintained contact vocally when out of sight of each other, then ran together (Frisch 1970: see 2 in Voice). In Iraq, ♀ rarely, if ever, left eggs voluntarily and change-over took place twice a day, apparently without ceremony: e.g. incoming ♂ landed up to 300 m away from nest and ♀ flew away directly while ♂ still at *c.* 30–60 m; ♂ took over at *c.* 18.00 hrs and then incubated all night. ♂ once called (not described) to approaching ♀ only *c.* 10 min after settling, then flew off without ceremony when ♀ at *c.* 2 m. Early-morning nest-relief similar: on 3 occasions, ♂ remained until ♀ alongside, then stepped off and waited at *c.* 1 m until ♀ settled, then flew to *c.* 500–600 m or out of sight; ♀ once sat quite erect for a while before settling right down and once raised herself up on nest as ♂ flew over; on another occasion left with a ♂ who had accompanied her own mate on arrival. Incubating bird generally motionless (Marchant 1961a, 1963c; S Marchant.) ♀, Morocco, responded to rising temperature with raising of dorsal feathers, gular fluttering, and bill gaping (see Thomas and Robin 1977). ♀, Iraq, once attended at nest by 4 ♂♂ who kept walking round her; after departure of 3 of these, single ♂ continued thus, ♀ on nest following his movements by rotating, at times pointing at him with neck on ground and occasionally peering at him over shoulder; similar behaviour noted when ♀ on nest and ♂ foraging nearby (Marchant 1961a; S Marchant). RELATIONS WITHIN FAMILY GROUP. Young leave nest as soon as dry, i.e. within 12–24 hrs; self-feeding from this point (Ferguson-Lees 1969a). Captive chicks took food independently from 2nd day, but, initially, ♀ showed food (grains) to young by first picking it up or tossing it in front of them; in warm weather, ♀ pecked away almost uninterruptedly to show chicks food (Frisch 1970). In another case, captive adults 'broke up' food for chicks which, however, independent enough by 3rd day to forage for themselves (Koenig 1896). Chicks seek shade of shrubs, etc., during heat of day, but otherwise follow parents closely; in brood of 2, each chick tended by 1 parent; in association of 2 pairs and 3 young, 1 chick ran indiscriminately between 2 ♀♀, other 2 chicks eventually brooded by 1 ♀ (Marchant 1961a; Ferguson-Lees 1969a; S Marchant). Last part of incubation and brooding in nest by ♀ alone, including at night (uninterruptedly for first 2 days: Koenig 1896). Away from nest, ♂ also broods but less often than ♀. Chicks led back to nest-scrape for brooding at night or in rain (Frisch 1970). Constant vocal contact maintained with young (Koenig 1896: see 2 in Voice). Captive chicks roosted independently of adults at 10 days, and as far apart as possible (Meade-Waldo 1896, 1906). Fly quite strongly at *c.* ⅓–½ grown; chick of this size apparently behaved quite independently in small group of adults (Marchant 1961a). Probably fledge at just under 4 weeks, but 1 fledged chick, early July, still followed ♂ closely (Frisch 1969a): presumably still dependent on ♂ for water (Ferguson-Lees 1969a; see also below). Quarrelling captive chicks rushed at one another, calling (see Voice); adopted an erect posture, made lunging movements with bill, and weaker bird finally thrown on to back where it crouched to indicate submission (Frisch 1969a). Watering of young largely as in *P. senegallus* (see that species; also Meade-Waldo 1906 and, especially, Marchant 1962a). In captive birds, began with hatching of 1st chick: ♂ apparently stimulated to wet belly feathers by sight of empty eggshells or, more likely, call of freshly hatched young. Thirsty chicks made thrusting movements into ♂'s belly feathers, but 1-day-old chick also drank from bowl (Frisch 1970). Captive ♂ rubbed breast against ground prior to wetting feathers (possibly not normal behaviour: Frisch 1970); performed Head-bobbing as if about to fly up (see above), then ran to ♀, and chicks emerged to drink (Meade-Waldo 1896). In Iraq, ♀ stood by, not participating; in one case (young well-grown), birds moved apart, after only *c.* 1 min watering to begin foraging (Marchant 1962a). Large numbers coming to water at Sarysu river, Kazakhstan (USSR), late July, reported to be almost exclusively ♂♂, but no explanation given (Gavrin *et al.* 1962). ANTI-PREDATOR RESPONSES OF YOUNG. Crouch motionless on hearing warning-call of parent (see 3 in Voice) or (if parents absent and human nearby or raptor sighted) equivalent from another chick. Crouch with body low, but neck extended and head held forward, unlike most waders (Charadrii); get up only when adult close. Fledged chick first crouched, then flew up when vehicle near. Need to seek shade made otherwise shy chicks 1 to several days old run towards man and sit down in his shadow. When being led away, chicks walk behind and to side of parent (Frisch 1969a, 1970). PARENTAL ANTI-PREDATOR STRATEGIES. (1) Passive measures.

Exceedingly wary up to hatching (Guichard 1961). ♀ incubated with head low and body flattened; head and neck raised if suspicious. In desert areas at least, does not normally leave nest during day, unless disturbed. At approach of man, usually performs Head-bobbing (see above), then walks or runs from nest while intruder at c. 100–200 m; may sit tight when eggs near hatching and approach possible in car to a few metres (Ferguson-Lees 1969a). Captive ♀ raised mantle feathers at sight of distant raptor (Koenig 1896). In the Crau, ♀ departs stealthily, moves far off before flying; returns with equal caution (after c. ½ hr, if no longer anxious): lands some way off and moves slowly in with back hunched and head held forward and low (raised only rarely), making use of available cover. ♂ may stand guard at c. 100 m and birds depart in different directions if disturbed (Guichard 1961). In Iraq, ♀ leaving nest was joined by ♂ who eventually led her cautiously back; regular return route established at one site (S Marchant). Parents generally remain inconspicuously near young during day; if disturbed, fly up at some distance, circle, and land in vicinity (Guichard 1961). Return may be delayed by 1–2 hrs and birds may use more circuitous route; ♂ invariably returns first. In captive family, ♂ always led and gave warning-call at any sign of danger (Frisch 1969a, 1970: see 3 in Voice). When fledged chick took flight, the Crau, July, ♂ at once joined it and called (not described); both landed together (Frisch 1969a). (2) Active measures. Information only for human intruders. Distraction-lure display of disablement type usual when incubating bird disturbed (Meinertzhagen 1954). According to Ticehurst et al. (1922), performance more vigorous and elaborate after hatching. On day of hatching, ♂ feigned broken wing not far from nest; ♀ left at last moment, then staggered, rolled over, and moved on with tail raised, wings trailing, head low, neck outstretched, and plumage (especially of underparts) ruffled; finally rose and flew off. May run off dragging wings with tail fanned and give Distraction-call (see 4 in Voice); or fly up noisily from nest (Ferguson-Lees 1969a). ♂, Iraq, once flew off nest (with 2 eggs) and performed distraction-lure display at c. 100 m; one ♀ moved c. 150 m off nest and performed ground-pecking. Another ♀ flew over area calling (see 3 in Voice), and one ♀ flew to and fro on 3 occasions before coming to relieve ♂. ♀ who had left nest at last moment, flew around with 2 ♂♂ and called (not described) at c. 200 m, until intruder had departed (Marchant 1961a; S Marchant). MGW

Voice. For most comprehensive treatment (based partly on study of captive birds), see Frisch (1969a, 1970). In the Crau (southern France), normally silent on ground, but noisy in flight (Guichard 1961); in Algeria, especially noisy during morning watering flight (Koenig 1896; see Social Pattern and Behaviour). Described as generally vocal, India (Hume and Marshall 1878; Baker 1914b). All softer calls and beginning of louder ones given with bill closed (Frisch 1969a). Several calls may be duplicated below.

CALLS OF ADULTS. (1) Flight-calls. A loud, ringing and distinctive 'catar-catar' or 'guettarr' (Whitaker 1905; Ferguson-Lees 1969a); see also Hartert (1912–21) and Hüe and Etchécopar (1957). A nasal 'ga-ga-ga' (Frisch 1969a), 'ga-ga' or 'ga-ng' (Dementiev and Gladkov 1951b); 'ha han ... ha han' given as long series, the 2 units almost the same, raucous and with mournful quality, low-pitched but far-carrying (Guichard 1961). Abrupt, guttural, somewhat nasal 'gang gang', reminiscent of Jackdaw *Corvus*

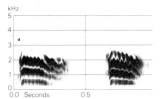

I E D H Johnson/Sveriges Radio (1972) Morocco April 1963

monedula when heard at distance (Gavrin et al. 1962); similar comparison in Kumerloeve (1968b) where rendered as 'kiaou-kiaou' and said to be given typically when flying to water. Recording (Fig I) supports comparison with *C. monedula*; suggests, however, that a rapidly repeated series, or at times only 2, rather low-pitched and nasal 'oing' or 'han(g)' sounds are followed by an emphatic harsh 'ärrrr' or 'arrrr' though these latter sounds may also be given in isolation. Further confirmation presented in Frisch (1969a) that a sharp 'ärr-ärr' and also a loud 'rau rau' or 'crau' may be given in flight; 'crau-crau' uttered typically after flying up and before landing, sometimes just before taking off. For further details and descriptions, see Koenig (1896), Magrath (1919), Ticehurst et al. (1922), Meinertzhagen (1954), and Germain (1965). In Iraq, birds normally gave loud, single, well-spaced 'ghirr-ghirr-ghirr' sounds; also a more rapid 'dōit-dōit-dōit-dōit' or 'whoick' and variants such as 'ik-o'; 3 birds flying near nest of one of them gave 'whack-kik-i-kik' calls (S Marchant). (2) Contact-calls. ♂ gives a loud 'arrr-arrr', ♀ a slightly higher pitched 'ärrr-ärrr' or a soft 'goggering' sound (Frisch 1970); a soft 'kokkok-kokkok' (Homeyer 1864). In interactions with young, a 'kück-kück' (Koenig 1896). (3) Warning- and alarm-calls. A soft 'grü-grrü', or 'rrü-rrü' (Frisch 1969a); both ♂ and ♀ may give 'twoi twoi twoi' sounds (Ticehurst et al. 1922). A loud 'crau-crau' given by ♂ as warning to young (Frisch 1970). Alarmed ♀ flying near nest called with a strangled 'choick' (S Marchant). (4) Distraction call. A high-pitched croaking churr (Ferguson-Lees 1969a). (5) Cooing-call. A soft cooing, rather like distant Black Grouse *Tetrao tetrix*, given by ♂ leading ♀ to potential nest-site (Frisch 1970). (6) Aggression calls. 'Gaggering' sounds given by ♀ driving off another ♀ or ♂ (Frisch 1970). ♂ driving off rival may give 'drohd droh drah dräh' sounds (Marshall et al. 1911). Both ♂ and ♀ (captives) gave 'og-og-og-og-og-gerrrrrr' sounds when expelling conspecifics from vicinity of nest (Frisch 1970). (7) Other calls. Birds emitted 'gaggering' sounds, as if to themselves, whilst foraging (Frisch 1970).

CALLS OF YOUNG. Contentment expressed by a soft whispering 'wi-wü' or 'wü-wö', interspersed with a soft gaggering 'gugglug'. Chicks give a similarly soft 'örr-örr' as a warning, this corresponding to the warning 'grü-rü' of adults (Frisch 1969a). According to Koenig (1896), respond with cheeping sounds to contact note of adults. Contact call of chicks otherwise described as a loud 'ääg' or disyllabic 'quä-äg', which changes to 'crau' or 'raw' around fledging. Full loud 'crau' and sharp 'ärr' of adults given by chick at c. 4 months. Younger fighting chicks emit a loud rattling 'gerrerreg-greggerreg'; at c. 4 months, fights accompanied by gaggering 'quä-quä-quä' sounds or a low-pitched 'gogogogogog', though this may change to the rattling 'gerrereg' of younger birds. Squeaking and compressed sounds also heard, but no context given (Frisch 1969a). MGW

Breeding. SEASON. North Africa, Spain, France, and Middle East: normal peak of laying from 2nd week May to late June, but early clutches from mid-April, and laying can continue into August (Meinertzhagen 1954; Guichard 1961; Marchant 1961a, 1963c; Ferguson-Lees 1969a; Makatsch 1976). SITE. On ground in the open, or sometimes by small tuft of vegetation or low scrub. Nest: shallow depression 10–12 cm in diameter, 1–4 cm in depth, unlined. In Iraq, 15 of 23 nests in unaltered footprints, c. 8 cm below general surface (Marchant 1961a). Building: only observation involved scraping by ♂ (Frisch 1969a); in captivity, both birds made scrape (Frisch 1970); sometimes probably no building (S Marchant). EGGS. See Plate 94. Elliptical, smooth and glossy; buff, with brown and pale grey blotches, spots, and speckles. Nominate *alchata*: 47×31 mm ($43–51 \times 29–33$), $n=60$; calculated weight 25 g. *P. a. caudacutus*: 45×31 mm ($40–50 \times 28–34$), $n=116$; calculated weight 24 g (Schönwetter 1967). Clutch: 2–3. Of 24 clutches, Iraq, 1 of 2, 23 of 3 (Marchant 1963c). No proof of 2nd brood or replacements, but length of laying season suggests either or both probable. Laying interval 24–48 hrs (Meade-Waldo 1897; Marchant 1961a). INCUBATION. Possibly variable. 19–20 days, Iraq (Marchant 1961a); 23 days, France (Guichard 1961); in captivity, 21–23 days (Meade-Waldo 1906) or 25 days (Koenig 1896). By both sexes, ♀ by day, ♂ at night; change-overs seen 08.00–08.30 hrs and c. 18.00 hrs (Marchant 1961a). Full incubation begins with last egg, though ♀ will cover earlier eggs during day to protect them from sun (Marchant 1961a). YOUNG. Precocial and nidifugous. Cared for by both parents, who may split brood between them (Marchant 1961a). FLEDGING TO MATURITY. Fledging period c. 4 weeks. May become independent as early as 10 days old (Meade-Waldo 1906; Marchant 1961a; Frisch 1969a). No further information.

Plumages (nominate *alchata*). In course of annual cycle (in both races), 2 types of breeding plumage develop: early phase in autumn during latter part of post-breeding moult, late phase in spring during pre-breeding moult. In ♂, late phase differs from early phase by presence of golden-yellow spots on greenish mantle and scapulars (Hartert 1912–21; Stresemann and Stresemann 1966; BMNH). In ♀, differences confined to type of barring. ADULT MALE BREEDING, LATE PHASE. Crown and hindneck greenish-ochre, sometimes with remains of some barred feathers showing through. Mantle and scapulars grey-green, most feathers with big subterminal yellow spot and dark margin. Back, rump, and upper tail-coverts buff, densely barred brown. Sides of face orange with black streak behind eye. Centre of chin and upper throat black, broadly bordered deep orange. Lower throat pale greenish-ochre. Breast bay-brown, bordered in front and behind by black bar. Rest of underparts white; under tail-coverts basally barred cream and brown. Feathering on tarsus and tibia white, occasionally barred pale grey-brown. Central tail-feathers (t1) elongated, black but basally barred buff and brown. Other tail-feathers grey-brown, broadly tipped white; outer webs basally barred buff, progressively paler towards outer feathers, outer web of t8 distally white. Primaries: p10 with outer web dark grey, inner brown-grey, but blue-grey along black-brown shaft. Outer web of p1–p9 pale grey; inner web brown, narrowly margined white and with pale area near shaft. White margin prominent on p1–p6, extending to tip on p1–p2. Secondaries dark brown, tip and outer web edged white, inner web basally white. Inner secondaries from s10 inwards with increasingly large patch of yellow-olive on outer web, inner web pale brown-grey; innermost like longer scapulars with yellow spot at tip. Outer web of greater upper wing-coverts deep brown with yellow line along margin, extreme margin brown, separated from yellow by very fine black line; inner web and base of outer web grey with faint brown hue. Median and lesser coverts similar, but brown bar and yellow line shifted progressively towards subterminal position on both webs. Coverts on inner wing olive-yellow with black margin; marginal coverts and upper rows of lesser dark olive-brown. Under wing-coverts and axillaries white, those in wide zone along margin of wing dark grey; greater under primary coverts paler grey. ADULT MALE BREEDING, EARLY PHASE. Crown, sides of head, chin, and upper throat either like late-phase breeding or like non-breeding or showing mixture of both. Mantle and scapulars predominantly dark grey-green without yellow spots, mixed with some barred feathers retained from non-breeding. Rest of plumage like late-phase breeding. In birds that are actively moulting brown area on breast, vague bars on pale feather-bases show through. ADULT MALE NON-BREEDING. Crown and hindneck dark rufous-buff heavily barred dark brown. Sides of face similar, spotted dark brown. Black streak behind eye mixed with white feathers or almost absent. Centre of chin and upper throat white; lower throat showing bars and spots. Mantle and scapulars barred buff and black, mixed with unbarred feathers (new feathers of early-phase breeding or those retained from late-phase breeding). Some innermost secondaries also barred. Rest of plumage like breeding. ADULT FEMALE BREEDING, LATE PHASE. Forehead, crown, and hindneck rufous-buff, barred black-brown; supercilium and sides of face orange-buff; black streak behind eye. Mantle and scapulars barred golden-yellow, silver-grey, and black; feathers with narrow black tip, wide golden subterminal bar, silvery grey bar narrowly outlined in black, and several rufous-buff and dark brown bars. Occasional retained non-breeding feathers have pale bars more rufous and lack silver-grey. Chin and upper throat white. Lower throat orange-buff crossed by black band; breast bay-brown, bordered black; thus, ♀ shows 2 black bands on lower throat and upper breast (♂ only 1). Rest of body, tail, primaries, and secondaries like ♂, but innermost secondaries barred, resembling long scapulars. Pattern of upper wing-coverts different from ♂. Outer greater coverts pale grey, outer web golden-yellow or bay-brown near edge, often narrowly margined black; inner progressively more barred and resembling scapulars. Median and lower rows of lesser coverts deep golden-yellow or orange-brown, boldly outlined by black crescents on feather-edges; basally grey-brown on outer wing, barred buff and brown on inner. Marginal and upper rows of lesser coverts densely barred rufous-buff and grey-brown. Underwing like ♂. ADULT FEMALE BREEDING, EARLY PHASE. As late phase, but mantle and scapulars barred rufous-brown and black with grey to dark grey subterminal bar, far less beautiful than late phase (Stresemann and Stresemann 1966). ADULT FEMALE NON-BREEDING. Differs from breeding in spotted sides of head and lack of pale silver-grey bars on upperparts. Crown and hindneck more densely barred and spotted. Sides of face rufous-buff with rounded black-brown spots. Lower throat densely spotted, and with only a suggestion of black band. Mantle feathers and scapulars barred rufous-buff and black. Rest of plumage like breeding. DOWNY YOUNG. Upperparts and sides of head ochraceous-brown, mottled by black tips of down feathers. Crown with central cream streak and cream mottling at sides; sides of head with 2 irregular streaks behind eye and 1 from corner of mouth. Upperside of body with central and lateral cream streaks and 2 bands across back. Chin and throat pale rufous-buff; pale buff crescent at side of neck. Rest of underparts pinkish-buff to sandy-ochre. Wings like back, distally margined cream (Fjeldså 1977; BMNH). Feathers of juvenile develop some days after hatching, down adhering to tips of growing feathers. JUVENILE. Crown and hindneck rufous-buff, marked with pale brown Vs or wavy bars. Sides of head finely speckled pale brown on rufous-buff ground; cream-white streak behind eye (Hartert 1912–21). Mantle, scapulars, and upper wing-coverts rufous-buff, feathers broadly edged pale cream and marked with dark brown bars of elongated V-shape. Upper tail-coverts buff, densely barred and marked dark brown. Chin white; wide band across throat and chest buff with pale brown bars, parallel to outline of feathers. Rest of underparts white, blotched brown-grey on lower breast and belly. Central tail-feathers not elongated, barred buff and brown; t2 also barred on both webs; t3–t7 barred on outer web only, inner web brown-grey tip white; t8 with outer web distally white as in adult. P1–p9 brown-grey, broadly and often irregularly tipped pale buff to off-white; p10 grey-black, curiously pointed, developing much later than other primaries. Secondaries like adult, but margins buff rather than white; inner like scapulars. Greater under wing-coverts brown-grey; marginal dark brown; rest of coverts and axillaries white. IMMATURE AND SUBSEQUENT PLUMAGES. Juvenile plumage worn for c. 2 months only, immature developing in late summer of 1st calendar year. Like adult non-breeding; ♂ and ♀ separable by type of upper wing-coverts. In early stage of immature plumage, brown band across breast often mottled with dark grey-brown buff-margined feathers. Inner primaries (p1 to p4–p7) of adult type: outer still juvenile, strongly worn, with black-grey pointed p10 diagnostic of immature until summer of 2nd calendar year when 1st breeding plumage present; otherwise resembles adult breeding.

Bare parts. ADULT. Iris dark brown; eye-ring blue-grey. Bill horn-grey, tinged bluish, occasionally brownish. Back of tarsus black-brown; foot pale grey, sometimes with green tinge; claws black (Hartert 1912–21; Dementiev and Gladkov 1951b; BMNH). DOWNY YOUNG. Iris dark brown. Bill horn. Foot buff-pink (Fjeldså 1977). JUVENILE. No information.

Moults (data for both races combined). ADULT POST-BREEDING. Complete, primaries descendant. Primaries start May (rarely April) finishing October; moult slowly, each feather shed when preceding almost full-grown. In some cases, moult apparently starts much later and is suspended in November or some feathers are skipped, leading to irregular pattern of wear (BMNH); further study needed. Tail moults quickly when primaries almost finished. Mantle and scapulars start June, slowly replacing late-phase breeding by non-breeding; later, from September onward, early-phase breeding feathers appear; moult stops late October (Stresemann and Stresemann 1966). Head and underparts more quickly moulted August–September. ADULT PRE-BREEDING. Partial: head and upperparts. Head and throat November–March; in ♂, throat often entirely black by midwinter. Mantle and scapulars early March to late April, replacing much of early-phase breeding by late phase (Stresemann and Stresemann 1966); occasional late-phase feathers acquired before midwinter. POST-JUVENILE. Almost complete, but some outer primaries not replaced. P1 shed before juvenile p10 fully grown. Primary moult suspended September–October at p4–p7. Body starts c. 8 weeks after hatching, finishing late autumn. Elongated central tail-feathers acquired early autumn. SUBSEQUENT MOULTS. In spring of 2nd calendar year, 1st pre-breeding moult as adult. Primary moult in 2nd calendar year serial, starting with p1 April–May; also (at least in some birds) continues with p5–p8 from where suspended in previous autumn (RMNH, ZMA). Not known whether outer primaries replaced twice in single season; retention of some may explain irregularities mentioned under adult post-breeding.

Measurements. Nominate *alchata*. Adult; Iberian peninsula and southern France, all year; skins (BMNH, RMNH). Length of t2 given as indication of tail length, elongated t1 being very variable.

WING	♂ 209	(2·62; 10)	205–214	♀ 205	(3·77; 7)	201–211
TAIL (t1)	157	(9·04; 8)	147–173	138	(6·62; 7)	129–146
(t2)	85·0	(3·51; 8)	79–89	86·3	(3·55; 7)	81–91
BILL	13·6	(0·38; 9)	13·1–14·3	13·8	(0·57; 7)	13·0–14·5
TARSUS	27·9	(0·93; 9)	27–30	26·0	(0·29; 7)	25–27
TOE	25·3	(1·15; 9)	24–28	24·3	(0·95; 7)	23–26

P. a. caudacutus. Whole geographical range, all year; skins (BMNH, RMNH, ZMA).

WING	♂ 214	(4·79; 22)	205–224	♀ 210	(5·31; 23)	200–217
BILL	13·2	(0·70; 23)	12·0–14·3	13·3	(0·54; 22)	12·1–14·3

Sex difference significant for wing, tail, and tarsus. In nominate *alchata*, wing of adult significantly longer than that of immature: 195 (4·00; 7) 190–201, sexes combined. Bill of adult slightly longer than that of immature: 13·3 (0·45; 7) 12·8–13·9. Wing of *caudacutus* significantly longer than nominate *alchata*, but more variable; bill significantly shorter in adult ♀.

Weights. *P. a. caudacutus*. ♂ c. 250, maximum 290; ♀ c. 225, maximum 230 (Dementiev and Gladkov 1951b). Kazakhstan: ♂ 250 (June), ♀ 245 (May) (Gavrin et al. 1962).

Structure. Wing narrow and pointed. 11 primaries: p10 longest (rarely 1 mm shorter than p9), p9 0–8 shorter, p8 11–18, p7 21–31, p6 33–45, p5 46–57, p4 58–71, p1 96–103 in ♀, 104–110 in ♂; p11 minute, concealed by primary coverts. About 18 secondaries, s13 or s14 longest. Tail strongly pointed; 16 feathers, occasionally 18, t1 elongated, 35–94 longer than t2, protruding part of feathers narrowly lanceolate; t8 c. 25–35 shorter than t2. Plumage dense. Bill short and sturdy, chicken-like, upper mandible strongly decurved, slightly longer than lower; nostril oblique, wide, partly hidden under dense short feathers. Legs of medium length, tarsus feathered in front. 4 toes, hind-toe small, raised; middle toe c. 92% of tarsus, inner c. 66% of middle, outer c. 66%, hind c. 21%. Claws broad.

Geographical variation. *P. a. caudacutus* longer winged and paler than nominate *alchata*, most obviously in brown on breast (particularly ♀) and on upper wing-coverts; upperparts also paler on average. In ♂, pale line in brown area on upper wing-coverts white, not yellow; in ♀, outer greater and median upper wing-coverts white to pale ochre inside black rim, not deep ochre or tawny.

JW

Syrrhaptes paradoxus Pallas's Sandgrouse

PLATES 25 and 26
[facing pages 303 and 326]

Du. Steppenhoen Fr. Syrrhapte paradoxal Ge. Steppenhuhn
Ru. Саджа Sp. Ganga de Pallas Sw. Stëpphöna

Tetrao paradoxus Pallas, 1773

Monotypic

Field characters. 30–41 cm; wing-span 63–78 cm, 6–10 cm of which is elongated primaries. Less bulky than Pin-tailed Sandgrouse *Pterocles alchata* and Black-bellied Sandgrouse *P. orientalis* but with even longer pins on central tail-feathers and (uniquely) on longest primary. Structure differs from other sandgrouse: in flight, rather smaller head and long tail make wings appear set further forward. Quite large, relatively attenuated sandgrouse, with diagnostic combination of wholly pale underwing and underbody except for bold black patch across rear belly. ♂ up to 10% larger than ♀, with orange-ochre face and pale buff inner wing. ♀ and immature speckled on back and wing. All show vinaceous-chestnut bar across greater wing-coverts. Sexes not markedly dissimilar; no seasonal variation. Juvenile separable at close range.

ADULT MALE. Ground-colour of back, all wing-coverts, secondaries, rump, and tail sandy-buff, boldly chevroned black on back and scapulars and spotted or lined black or black-brown on larger inner wing-coverts, inner secondaries, primary coverts, and inner primaries. Primaries noticeably pale dove-grey above. Head orange-ochre, with grey on rear crown, from eye to neck, and around breast and chest; chest delimited by transverse lines of small black bars. Underparts below chest buff-white with bold, broad black band across belly and long white vent. In flight, upperwing little patterned except for chestnut bar across inner wing and dark trailing edge of secondaries and inner primaries which extends forward at their union. Underwing virtually unpatterned, with black-spotted, pale sand-buff axillaries and wing-coverts contrasting little with pale ash-grey undersurface of flight-feathers; whole wing only narrowly bordered dusky. Shorter tail-feathers show pale outer margins. Bill dull blue; eye-ring dark blue-grey. Legs and feet fully feathered (white), unlike other sandgrouse. ADULT FEMALE. Resembles ♂ but duller, with speckles and spots on crown, rear ear-coverts, hind-neck, and all wing-coverts; belly-patch purple-brown. Short black bar under buff throat but no deep grey chest. JUVENILE. Resembles adult ♀ but lacks ochre on head; barring on upperparts less intense and belly-patch distinctly brown. Lacks bar on greater wing-coverts until 1st winter. Outer primaries shorter than adult, tail not elongated.

Unmistakable on ground, strongly recalling Partridge *Perdix perdix*. Flight pattern unique among sandgrouse, showing wholly pale underwing, while dark belly-patch much smaller than on *P. orientalis* and Chestnut-bellied Sandgrouse *P. exustus*. Flight action resembles that of *Pterocles* but probably more powerful and rapid, with noise produced by regular wing-beats louder and noticeably sibilant. Flight usually direct but occasionally circular and undulating. Gait as *Pterocles* but carriage noticeably low, with body held almost horizontal.

Westward vagrancy much decreased since 1908, but a few still appear over or on open cultivation, marshes, and islands. Commonest call a fairly low-pitched 'cu-ruu cu-ruu cu-ou-ruu'.

Habitat. In continental middle latitudes, on arid open plains and uplands in steppe climate and vegetation zone, subject to wide changes in temperature, and to occasional closure under ice-coated snow, compelling mass evacuation. Avoids drifting sands, extensive waterless desert, mountains, forests, wetlands and settled areas. Prefers clay or loams to sandy, rocky, or stony soils; favours low, fairly sparse, semi-desert, steppe vegetation cover, such as wormwood *Artemisia* and *Agriophyllum gobicum* and steppe grasses, on both flat and hilly terrain. Where suitable valleys provide such conditions will occupy them up to altitudes ranging from 1300 m (Sinkiang) to, exceptionally, 3200 m (Naryn basin). Likes deep dust for wallowing. Availability of surface water essential within daily flying range, but this can extend to several dozen km, except when rearing young. Climatic constraints lead to regular migrations from north of range, and historically a number of mass eruptive movements have been triggered off, usually in winter, in a westerly or easterly direction, sometimes leading into temperate, boreal, or Mediterranean zones and exceptionally into oceanic climates where breeding has been attempted, e.g. on grassy coastal sand-dunes (see Distribution). (Dementiev and Gladkov 1951*b*; Yanushevich *et al.* 1959; Bannerman 1959.)

Distribution. Boundaries of range apparently fluctuate. In some years, large emigrations to both east and west of main range (Dementiev and Gladkov 1951*b*). Major irruptions into Europe in 1863, 1888, and 1908 (see Movements), following which sporadic breeding occurred in west Palearctic.

BRITAIN. Nested Yorkshire (England) 1888 (2 clutches), Elgin (Scotland) 1888–9 (young seen) (Witherby *et al.* 1940), and probably Suffolk (England) 1888 (Payn 1978). BELGIUM. Breeding attempted 1889 (Glutz *et al.* 1977). NETHERLANDS. Single clutches found 1863 and 1888 (CSR). WEST GERMANY. Nested Niedersachsen and Schleswig-Holstein 1888 (Glutz *et al.* 1977). DENMARK. Many pairs bred Jutland 1888, some possibly nested 1889

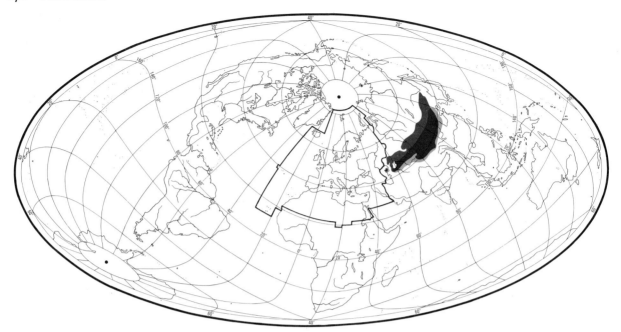

(TD). SWEDEN. 2 pairs bred 1888 (LR). AUSTRIA. Possibly bred 1864 (Glutz et al. 1977). USSR. Nested in area of lower Volga and Volga-Ural steppes in 19th century, and near Moscow, Crimea, and Ukraine in 1908 (Sudilovskaya 1935; Dementiev and Gladkov 1951b).

Accidental. Faeroes, Britain, Ireland, Spain, France, Belgium, Netherlands, West Germany, Denmark, Norway, Sweden, Finland, East Germany, Poland, Czechoslovakia, Switzerland, Austria, Hungary, Italy, Yugoslavia, Greece, Bulgaria, Rumania, USSR, Turkey.

Population. Numbers fluctuate markedly (Dementiev and Gladkov 1951b). Reasons for irruptions in west Palearctic in second half of 19th century and early 20th century not understood (see Movements).

Movements. Partially migratory, with extent of movement varying between years; also eruptive at times.

Not known to winter regularly in northernmost parts of breeding range, e.g. above c. 45°N in Kazakhstan and Mongolia, but extent of southward movement variable and dependent on snow fall: deep snow, especially with ice crust, inhibits feeding, and this (rather than high population level) is probable cause of large-scale autumn/winter movement (Dementiev and Gladkov 1951b). In normal winters, displacements relatively short-distance, being into southern breeding range or just beyond, e.g. into Turkestan (USSR), Sinkiang, and Inner Mongolia (northern China), but large numbers will penetrate further south in occasional very cold winters, affecting eastern part of range more than western. Then, reaches Soviet Maritime Territories, Manchuria, Hopeh, and (to lesser extent) Transcaspia. Autumn movement in Soviet Central Asia, Mongolia, and northern China occurs October–November, continuing later in cold-weather exodus; spring return noted February in north-east China, early March in Mongolia and Turkestan, and even northern breeding areas reoccupied by mid-April. See Grote (1936), Dementiev and Gladkov (1951b), Hemmingsen (1959), Cheng (1963), Hemmingsen and Guildal (1968), Kuleshova (1968), Potorocha (1968), Shibaev (1968), and Yakhontov (1968).

Well known for remarkable (though infrequent) eruptions, in which the species has occurred across full width of Palearctic, leading to temporary expansions of breeding range and isolated nesting records far outside normal distribution (see Distribution). Such eruptions not cyclic, as formerly supposed (Lack 1954), and suggested relationship with sunspot activity cannot be upheld (Hemmingsen 1959). Irruptions into Europe and eastern Asia occurred in different years, indicating that local factors affected different segments of population (Dementiev and Gladkov 1951b). In eastern Asia, irruptive wandering in spring

PLATE 22 (facing).
Uria aalge aalge Guillemot (p. 170): **1** ad breeding, **2** 1st imm non-breeding (1st winter).
Uria lomvia Brünnich's Guillemot (p. 184): **3** ad breeding, **4** 1st imm non-breeding (1st winter).
Alca torda Razorbill (p. 195): **5** ad breeding, **6** 1st imm non-breeding (1st winter).
Cepphus grylle arcticus Black Guillemot (p. 208): **7** ad breeding, **8** 1st imm non-breeding (1st winter).
Alle alle Little Auk (p. 219): **9** ad breeding, **10** ad and imm non-breeding.
Aethia cristatella Crested Auklet (p. 229): **11** ad breeding, **12** juv (retained through 1st winter).
Fratercula arctica Puffin (p. 231): **13** ad breeding, **14** juv in 1st winter. (RG)

Robert Gillmor

seems to develop from abnormal displacement in preceding cold winter (Hemmingsen 1959), but different causes thought to be involved in Soviet Central Asia, where cold weather migration much less marked, and eruptions can be preceded by irregular mass movements within normal range 1 or even 2 years earlier—e.g. disappeared from Zaisan basin in autumn 1907, ahead of major eruption (that reached western Europe) in spring 1908 (Sudilovskaya 1935). Causes not certainly known, though food supply almost certainly implicated (Lack 1954). Mass movements prior to eruption suggest poor seed crops in such years. Early spring migration brings birds back into northern breeding areas while weather still very cold, and perhaps food sometimes still inaccessible beneath snow; flooding after sudden thaw may make habitat temporarily unsuitable (Sushkin 1908; Sudilovskaya 1935; Dementiev and Gladkov 1951b).

Irruptions into Europe included spectacular ones in spring–summer of 1863, 1888–9, and 1908, during which recorded north to Arkhangel'sk and Fenno-Scandia, and west to Faeroes, Ireland, and Spain. Flocks began reaching eastern Europe in second half of April, spreading west during May–June; in 1888, birds moving in European USSR in March, and vanguard reached North Sea in late April. Some easterly return movement in late summer and autumn noted on Continent, but on much smaller scale than initial irruption (Newton 1864; Tschusi zu Schmidhoffen 1909; Sudilovskaya 1935). Since 1908, irruptions into Europe rarer, with only a few birds each time outside USSR. Reason for this curtailment uncertain; possibly linked to contraction of west Siberian range, whose normal limits inadequately defined—may have expanded temporarily (but breeding perhaps sporadic) during second half of 19th century.

Food. Mainly seeds; also green shoots. Picks up seeds from ground but will also pluck off food from growing plants (e.g. Reichenow 1889); vagrants in Kiel area (West Germany) bent down stems to reach heads of lyme grass *Elymus arenarius* (Werner 1889). According to Dementiev and Gladkov (1951b), regularly digs for food like chicken; will scratch out seeds of buckwheat *Fagopyrum* (Meier 1865). Vagrant in Shetland made continual sideways movements with bill whilst foraging (Holbourn and Gear 1970). Feeds with head bowed low and tail raised high (Holtz 1863). In Pamirs (USSR), winter, tends to forage close to herds of sheep, presumably benefiting from grazing activity (Scheifler 1979). Birds foraging on newly sown land kept more or less in line, moved in one direction, and rapidly picked up grain like pigeons (Columbidae) (Evans 1889).

Wide range of seeds recorded include those of legumes (Leguminosae), docks (Polygonaceae), pinks (Caryophyllaceae), goosefoots (Chenopodiaceae), grasses (Gramineae), sedges *Carex*, wormwood *Artemisia*, cranesbill *Geranium*, nettle *Urtica*, crucifers (Cruciferae), Cannabaceae, eyebright *Euphrasia*, dill *Anethum*, buttercup *Ranunculus*, bugloss *Lycopsis*, *Allium*. Shoots of Leguminosae, sorrel (Polygonaceae), stonecrop *Sedum*, glasswort *Salicornia*, sea-blite *Suaeda*, sea sandwort *Honkenya peploides*, bog myrtle *Myrica gale*, cereals, cabbage (Cruciferae), chickweed *Stellaria*, and a moss (Musci); fruits of *Nitrana schoeberi*, berries, and heads of grass *Poa*. Animal food exceptional: includes pupae of small moths (Lepidoptera), beetles (Coleoptera), and pupae of small flies (Diptera); in Essex (England), birds reported to have run about picking up insects (Bree 1863), but no details. Sand, small stones, etc., taken regularly as grit, and winkles *Littorina*, Helgoland (West Germany), presumably served same purpose. (Moore 1860; Altum 1863a,b; Radde 1863; Meier 1865; Southwell 1888; Tegetmeier 1888; Evans 1889; Howard 1889; Macpherson 1889a; Rohweder 1889; Werner 1889; Gätke 1900; Sim 1903; Dementiev and Gladkov 1951b; Yanuschevich *et al.* 1959; Dubrovski 1961; Pek and Fedyanina 1961; Gavrin *et al.* 1962; Glutz *et al.* 1977.)

19 birds from Kirgiziya (USSR), summer, contained mainly seeds and leaves of Leguminosae (11 stomachs) and Cruciferae (8); next in order of importance came seeds of *Corispermum* (Chenopodiaceae) in 6, and *Lycopsis* in 5; 2 each contained seeds of *Allium* and wheat. In winter, birds stay mainly in areas free of snow and take seeds of desert grasses and shoots, also grain from harvested fields (Yanushevich *et al.* 1959); for further analysis from Kirgiziya, see Pek and Fedyanina (1961). In north-west Kazakhstan (USSR), birds took wheat and millet *Panicum* on roads; crops contained also various grass seeds and once leaves of orache *Atriplex* (Dubrovski 1961); see also Gavrin *et al.* (1962) and Krivitski (1977). In Transbaykalia (USSR), spring, contained mainly *Salsola* seeds, birds showing fondness for young *Salicornia* shoots when these became available (Radde 1863). Birds from Chuskaya steppe and south-east Altay (USSR) contained flower heads of *Allium tenuissimum* and seeds of *Astragalus* and *Elymus dasystachys*. In Mongolia, birds took mainly *Agriophyllum gobicum* and seeds of *Artemisia* and Leguminosae (Dementiev and Gladkov 1951b). Another study, Mongolia, showed main food as flower heads and seeds, more rarely also leaves, of *Chenopodium acuminatum*, *Allium polyrrhizum*, *A. mongolicum*, and *A. tenuissimum*, as well as of various *Astragalus* (Glutz *et al.* 1977).

PLATE 23 (*facing*).
Uria aalge aalge Guillemot (p. 170): 1–2 ad breeding, 3 ad non-breeding.
Uria lomvia Brünnich's Guillemot (p. 184): 4–5 ad breeding, 6 ad non-breeding.
Alca torda Razorbill (p. 195): 7–8 ad breeding, 9 ad non-breeding.
Cepphus grylle arcticus Black Guillemot (p. 208): 10–11 ad breeding, 12 ad non-breeding.
Alle alle Little Auk (p. 219): 13–14 ad breeding, 15 ad non-breeding.
Aethia cristatella Crested Auklet (p. 229): 16 ad non-breeding.
Fratercula arctica Puffin (p. 231): 17–18 ad breeding, 19 ad non-breeding. (RG)

Studies outside normal range. Individual birds from Norfolk (England) contained seeds as follows: 80 knotgrass *Polygonum aviculare*, 49 bird's foot *Ornithopus*, 17 clover *Trifolium*; *Trifolium procumbens* and sainfoin *Onobrychis*; mostly chickweed *Stellaria media* and an unidentified grass; mainly *P. aviculare* and a few white clover *T. repens* and bent *Agrostis*; dove's-foot cranesbill *Geranium molle* (mostly), nettle *Urtica dioica*, common sorrel *Rumex acetosa*, *Stellaria media*, and a grass; birds also favoured seeds of campion *Silene* and *Suaeda maritima*, and leaves and shoots of latter (Southwell 1888). Most birds from Scotland contained clover (predominantly) and grass seeds, with some grain—barley or oats; birds from Shetland, particularly, took green material. Birds from eastern Scotland, January, held only seeds of *Atriplex* and *Polygonum*. 2 crops contained dipteran pupae (Evans 1889). 7 birds from Lancashire (England) contained mainly seeds of clover, grass *Lolium multiflorum*, knotgrass *Polygonum*, fat-hen *Chenopodium album*, and mouse-ear *Cerastium*; gizzards held from 50 to over 80% (by volume) quartz fragments (Howard 1889). For further British data, see (e.g.) Tegetmeier (1888), Macpherson (1889a), and Sim (1903). Long-staying birds on Borkum island (West Germany) took chiefly seeds (also leaf buds and leaves) of *Suaeda maritima*; several crops contained only seeds of spurrey *Spergularia marina*; others held fruits of a grass and *S. marina*; in late August, bird's-foot trefoil *Lotus corniculatus* predominated (Altum 1863a,b). Several authors mention preference for buckwheat *Fagopyrum* (Leverkühn 1889). For analysis of bird (♂) obtained Franken (West Germany), see Glutz *et al.* (1977). Captive birds showed preference for small seeds over larger cereal grain; animal food ignored (Altum 1863b).

Crop of chick *c.* 2 days old, Scotland, contained 45 seeds including grasses *Lolium perenne*, *Deschampsia caespitosa*, *Poa annua*, broom *Sarothamnus scoparius*, and redshank *Polygonum persicaria* (Newton 1890). MGW

Social pattern and behaviour. No detailed study on regular (extralimital) breeding grounds. Most west Palearctic data refer to brief observations of vagrants during irruptions; for fuller study, Borkum (West Germany), see Altum (1863a,b, 1864). Poorly known.

1. Gregarious throughout year, more so outside breeding season: e.g. 2 flocks each of *c.* 1000, Kirgiziya (USSR), October (Yanushevich *et al.* 1959), and reportedly in flocks of several thousand, Mongolia, winter (Dresser 1871–81; see also Piechocki *et al.* 1981). In Kazakhstan (USSR), not in such large (autumn) flocks as Pin-tailed Sandgrouse *Pterocles alchata* (Gavrin *et al.* 1962). In pairs or smaller flocks during breeding season (Grote 1936; Yanushevich *et al.* 1959), though even at this time larger flocks may be formed for watering (Tegetmeier 1888). Non-breeders may remain in flocks during breeding season (e.g. Krivitski 1965), though large aggregations noted by Radde (1863), Transbaykalia (USSR), May, presumably post-breeding assemblies. Migrant flocks of up to 400, western USSR (Sudilovskaya 1935). Vagrant, Norfolk (England), June, associated with Grey Plover *Pluvialis squatarola* (Stevenson 1866), and one flushed together with Partridge *Perdix perdix*, Denmark, October (Fog 1970). Flocks in Norfolk always remained apart from other species (Southwell 1888). BONDS. Monogamous mating system (Dementiev and Gladkov 1951b). Age of first breeding not known, but non-breeders in flocks, Kazakhstan, thought to be 1 year old (Krivitski 1965). Sex-ratio in vagrant flocks, Norfolk, apparently about equal (Southwell 1888). In Transbaykalia, birds arrived already paired, but initially remained in flocks (Radde 1863); pair-formation after arrival on breeding grounds, Kazakhstan (Gavrin *et al.* 1962); in Kirgiziya, arrived mid-April, but pairs not formed until first half of May (Grote 1936). In Scotland, birds formed pairs shortly after arrival in mid-May, but most then reunited in flock after *c.* 7–10 days (Evans 1889). Close association of pair apparent even in invasion flocks (e.g. Dubrovski 1961). ♀ seen with chicks, Transbaykalia, late April (Radde 1863); both parents attending young *c.* 3–4 days old, Scotland (Newton 1890). No further information on care of young. BREEDING DISPERSION. Generally in small, scattered colonies, USSR (e.g. Radde 1863, Seebohm 1884, Gavrin *et al.* 1962, Belik 1977). In Aral Sea area, estimated 50000 birds on 5000 km² based on census of watering flocks; breeding density in typical habitat probably considerably higher (Sarzhinski 1977). In stony desert, Kirgiziya, no compact colonies, presumed to be because of poor food supply, and nests mostly 800–1000 m apart (Yanushevich *et al.* 1959); elsewhere in USSR, not uncommonly 5–6 m apart (Dementiev and Gladkov 1951b). Two nests, Denmark, *c.* 1000 m apart (Leverkühn 1889). ROOSTING. Few observations refer to nocturnal roosting, but Schwaitzer (1865) deduced from circumstantial evidence that 4 in Poland roosted together in hollows; others on Borkum (West Germany) also evidently roosted communally and close together on ground (Altum 1863a). In Rumania, various kinds of fields used (Tschusi zu Schmidhoffen 1909). West German vagrants established traditional flight-path from feeding grounds to (unusual) roost in oak thicket (Rohweder 1889). Flocks in Transbaykaliya which evidently migrated on during night, still vocal on ground after dusk (Radde 1863). Leaves roost at sunrise, Mongolia, flying into desert to feed (Tegetmeier 1888). On Borkum, birds remained at roost (where water also present) until *c.* 09.00 hrs, then later flew to feeding grounds on flats (Altum 1863a). Diurnal roosting similarly communal and on ground, in shallow excavated pits; in Transbaykalia, pits on slight grassy elevations in saltpans. Birds seek to get well down into pit but feathers ruffled; whilst some sleeping, others may forage (Radde 1863; Newton 1864). In Kiel area (West Germany), birds formed randomly dispersed assembly (Werner 1889). Often loafs and dust-bathes at watering place (e.g. Gavrin *et al.* 1962). 5 captive birds roosted together in summer, but maintained at least a certain individual distance; in snow and cold, formed close-packed group in alternate head-to-tail order (Marshall *et al.* 1911).

2. Varying reports of relative wariness of European vagrants probably attributable to level of disturbance which clearly excessive at times. Birds tend to haunt more open areas and to rely on cryptic colouration for protection (Altum 1863a). Flocks wary, Transbaykalia, May (Radde 1863). On Borkum, generally allowed approach to *c.* 200 m; a solitary bird was less shy (Altum 1863a). On Helgoland (West Germany), tolerated man at 20–30 m (Gätke 1900), and migrants normally allowed close approach, USSR (Sudilovskaya 1935). May crouch at first in danger, then usually flies up with noisy beating of wings, like pigeon (Columbidae), and may also call (Altum 1863a; Tegetmeier 1888: see 2 in Voice). Flushed birds tend to fly higher than usual (Altum 1863a). May also run away (Macpherson 1889b). One of several captive birds raised and fanned tail at approach of observer (Rohweder 1889); a captive ♀ performed similarly, also extended neck upwards, called, and lunged with bill

(Holtz 1864: see 3 in Voice). FLOCK BEHAVIOUR. High degree of 'unanimity' noted (Macpherson 1889b). In Mongolia, small flocks occasionally rose high, individuals swooping down at intervals, then ascending again (Tegetmeier 1888). Large flock, Borkum, flew in gentle undulations, like pigeons returning from feeding grounds. Attacked by Marsh Harrier *Circus aeruginosus*, flock split up allowing raptor to pass through (Altum 1863a), Birds circle before settling to feed (Tegetmeier 1888); see also below. On Borkum, waited motionless for *c*. 20 min after landing, then moved forward, foraging on broad front; at approach of man, one bird moved on to small mound, stretched up and called (see 2 in Voice); flock compacted and birds crouched (Altum 1863a). Other reports refer to birds running together, then flying up (e.g. Bolam 1889, Evans 1889). In Transbaykalia, disturbed birds, some reacting to alarm-calls (not described), formed large flock in flight, then split up into smaller flocks again later (Radde 1863). Tend to circle when alarmed (Bolle 1863a), though often land close to where disturbed (Southwell 1888). Like other Pteroclididae, uses traditional watering sites. In Mongolia, large numbers remained throughout day at such a site despite water being frozen (Piechocki 1968c). Birds reported coming to water up to *c*. 09.00 hrs, April (Radde 1863), after morning foraging session (Tegetmeier 1888); from *c*. 07.00 hrs, early June (G Rinnhofer); in Kazakhstan, generally 08.00–11.00 hrs (Gavrin *et al.* 1962; Scheifler 1972); in Kirgiziya, 10.00–16.00 hrs (Grote 1936); in Semirech'e (USSR), at 06.00 and 11.00 hrs, May (Shnitnikov 1949). Evening watering may occur around sunset (e.g. Grote 1936); in Kazakhstan, not all birds participate (Gavrin *et al.* 1962), but see also Krivitski (1977). No flight following rain (Gavrin *et al.* 1962). In Mongolia, recorded flying to salt-marsh, probably to drink dew (G Rinnhofer). In Kiel area, birds drank sea water close to day-time roost (Werner 1889), and on Borkum drank from rainwater pool at nocturnal roost (Altum 1863a); see also Roosting, above. May fly to water in pairs (breeding season) or larger flocks (Seebohm 1884; Tegetmeier 1888); in Mongolia, early June, up to 25 (G Rinnhofer). In Transbaykalia, April, pairs called on arrival, birds on ground responding; drank fairly hastily and departed for foraging grounds (Radde 1863). Flocks invariably circle watering place before landing (Tegetmeier 1888). May spend several hours at water, alternately loafing, feeding, and drinking (Grote 1936). ANTAGONISTIC BEHAVIOUR. Described as intolerant and quarrelsome in feeding flock (Schwaitzer 1865); no antagonism seen, Kiel area (Werner 1889). In individual-distance disputes, birds faced one another, wings raised, head withdrawn, calling (see 3 in Voice); also briefly leaped up at one another (Rohweder 1889). Captive ♂♂ generally amicable, but disputes developed with onset of courtship, involving aggressive posture markedly different from that of *Pterocles* sandgrouse: front part of body raised, feathers on neck, breast, and upper back ruffled, and wings slightly raised; blows delivered with bill and fights vigorous but not damaging (Marshall *et al.* 1911). High-speed pursuit-flights of ♂♂ mentioned by Gavrin *et al.* (1962), but no details. Displaying ♂ reported to chase off other ♂♂ (Dementiev and Gladkov 1951b). HETEROSEXUAL BEHAVIOUR. Few details on pair-formation and courtship. ♂ reported to call (not described, but see 5 in Voice), and to run round ♀ like pigeon, although neck not inflated and no bowing movements made; sometimes flies up and circles ♀ (Dementiev and Gladkov 1951b). ♀ may fly up, and is then pursued by ♂ (Gavrin *et al.* 1962); such pursuit-flights accompanied by loud calls (Piechocki 1968c). 'Pairs' sometimes broke away from migrating flocks, these birds alternately performing rapid stoops at one another (Glutz *et al.* 1977). Empty scrape often *c*. 1–2 m from occupied nest, south-east Transbaykalia (Belik 1977). Incubation by both sexes. In Kirgiziya, ♀ seen on nest in morning and in afternoon around 16.00–17.00 hrs (Glutz *et al.* 1977); in Denmark, ♂ on nest in morning, ♀ in afternoon (Leverkühn 1889). While ♀ incubating, ♂ usually stands nearby (Grote 1936). RELATIONS WITHIN FAMILY GROUP. Young reported not to open eyes until 2nd day; hide under bushes to escape sun (Yanushevich *et al.* 1959). Watering of young not studied but unlikely to be different from (e.g.) Spotted Sandgrouse *P. senegallus* (S Marchant). ANTI-PREDATOR RESPONSES OF YOUNG. 2 chicks *c*. 3–4 days old crouched *c*. 10 m apart, also after release (Newton 1890). PARENTAL ANTI-PREDATOR STRATEGIES. Incubating bird usually sits tight (e.g. Moore 1860, Leverkühn 1889, Rohweder 1889, Yanushevich *et al.* 1959). In Denmark, crouched low on nest, then performed distraction-lure display of injury-feigning type on departure: fluttered along as if shot, repeatedly making as if to crouch and hide (Leverkühn 1889). In Schleswig-Holstein, flew off when man at *c*. 3 m—more slowly than usual and on low, rather zigzag flight-path; landed *c*. 100 m from nest and disappeared (Rohweder 1889). According to Yanushevich *et al.* (1959), bird first runs from nest, then flies up, but no distraction-display recorded. Generally wary when with young (Newton 1890); performs distraction-display (Gavrin *et al.* 1962), but no details. MGW

Voice. Generally vocal, both on ground and in the air; sometimes described as quieter in autumn (Hartert 1912–21). Wings make far-carrying humming or whistling sound; flock may sound like sighing of wind (Tegetmeier 1888); for wing-noise at take-off, see Social Pattern and Behaviour. Function of most calls poorly understood and several may be duplicated below. Full study required.

CALLS OF ADULTS. (1) Flight- and contact-calls. A fairly low-pitched 'cu-ruu cu-ruu cu-ou-ruu' (D I M Wallace); trisyllabic, clear but not shrill, rapidly repeated 'köckerík-köckerík . . .', carrying at least 400 m (Altum 1863b, 1864; Southwell 1888). Recording (Fig I) suggests rapid musical chirruping or trilling recalling chorus of waders (Charadrii); number of syllables apparently not constant. Difficult to render satisfactorily but at times roughly 'cherrcherri-chicherr' or 'kwepkwekerikikerr'; such calls and, more particularly, single explosive 'tchep' or 'kep' sounds (Fig II) which frequently interspersed with them, sometimes reminiscent of Moorhen *Gallinula chloropus* (M G Wilson). Otherwise described as having soft, pleasant-sounding, half gurgling, half piping, chattering or chuckling quality (Moore 1860; Reinhardt 1864; Macpherson 1889a; Rohweder 1889). For attempts to render various di- and monosyllabic sounds (sometimes given in rapid series: e.g. Altum 1863a, Gätke 1900, Kate 1966), including comparisons with miscellaneous Charadrii and also Black Grouse *Tetrao tetrix*, see (e.g.) Bolle (1863b), Holtz (1863), Bolam (1889), Howard (1889), Macpherson (1889a), Rohweder (1889), Gavrin *et al.* (1962), and Hüe and Etchécopar (1978). (2) Calls given on taking off, also possibly expressing alarm. Often much as above, e.g. 'kök-kerik' or loud 'köckerick köckerick' (Altum 1863a). According to Macpherson (1889b), harder and more guttural: a rapid 'kriktikrik' (Etchécopar and Hüe 1978), 'tick-a-rick' (Bolam 1889); but described by Meier (1865) as penetrating, loud, and often fluting. Apart from a low,

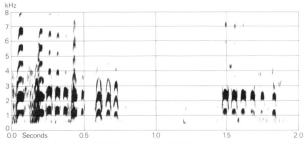

I B N Veprintsev and V V Leonovitch USSR June 1975

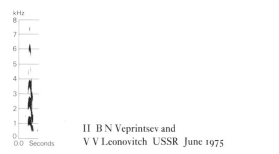

II B N Veprintsev and V V Leonovitch USSR June 1975

sharp 'tuck' (see call 1), a hurried 'purr-t' like Curlew Sandpiper *Calidris ferruginea* (Evans 1889). Lone vagrant, Netherlands, gave a quite strong 'kiierk' (with distinct 'ie' sound), 'kriek-kriek', 'krúiek', 'krwiek', and also 'kèjar-kejarjar' sounds (Kate 1966). (3) Threat- or attack-calls. A 'kriktikrik' (Altum 1863a); see also call 2. Captive ♀ gave a low-pitched 'guck', 1–3 times in rapid sequence; at closer approach of observer, this developed into a 'gurrrrrrr', with rising pitch: considered to express disapproval (Holtz 1864). (4) Distress-call. When seized, captive bird uttered a screeching 'krwä', like hare *Lepus* in similar situation (Rohweder 1889). (5) Other calls. Whilst foraging, and otherwise moving about on ground, birds gave a soft 'kök kök' (Altum 1863a; Bolle 1863b); presumably the same as guttural 'kok kok' of Etchécopar and Hüe (1978). At rest, birds give a variant of the normal chuckling 'truck-turuk', more like Partridge *Perdix perdix*. Captive (and injured) ♀ gave a 'coo' like pigeon (Columbidae) when frightened or disturbed; 'cuck-cuck' sound considered to express contentment (Macpherson 1889b). Captive pair also gave low cooing sound. Whilst circling ♂, ♀ uttered a rather cat-like 'purr', and this also taken up by ♂ (Southwell 1888).

CALLS OF YOUNG. Captive-bred chick gave gurgling 'gilik' like adult but higher pitched (Winge 1892). MGW

Breeding. SEASON. Southern USSR: eggs laid mid-April to June (Johansen 1959). Kirgiziya (USSR): laying begins late March or early April; eggs found late May, and young chicks early August (Yanushevich *et al.* 1959). Probably everywhere very extended. SITE. On ground in the open or sheltered by small tuft of vegetation. Nest: shallow scrape in rubble-covered substrate, usually without lining, but sometimes thin layer of dry *Ephedra* twigs; internal diameter 12–13 cm, depth 3–4 cm (Yanushevich *et al.* 1959). Building: no information. EGGS. See Plate 94. Long elliptical, smooth and glossy; pale buff to cream, with brown spots, blotches, and speckles. 43×30 mm (39–47×27–32), $n = 100$; calculated weight 21 g (Schönwetter 1967). Weight 16–20 g (Yanushevich *et al.* 1959). Clutch: 2–3(–4). 2 (possibly 3) broods (Witherby *et al.* 1940). Laying interval 24–48(–72) hrs (Glutz *et al.* 1977). INCUBATION. About 28 days (Yanushevich *et al.* 1959). 23–24(22–27) days under domestic hens; 28 days in incubator (Witherby *et al.* 1940). By both sexes; ♂ with brood-patch shot by nest (Yanushevich *et al.* 1959). Eggs may be left for long periods in day while adults fly off to drink (Grote 1936). Begins with or soon after laying of 1st egg (Glutz *et al.* 1977). Hatching asynchronous (Yanushevich *et al.* 1959). YOUNG. Precocial and nidifugous. Young left for long periods during day (Grote 1936). No further information.

Plumages. ADULT MALE. Forehead, front half of crown, collar across hindneck, lores, and chin grey, strongly tinged ochre; supercilium deeper ochre extending to orange-ochre spot on side of neck. Malar stripe and gorget on upper throat orange-ochre, separated from spot on side of neck by grey crescent running toward ear-coverts. Nape and lower hindneck grey. Mantle and scapulars cinnamon-buff, barred black; scapulars with broad buff tip, subterminal black bar and purple-brown spot, in longest feathers on both webs, in shorter on outer only. Back and rump like mantle, but barring more irregular. Upper tail-coverts elongated, buff, with V-shaped pattern of dark grey. Chest grey, joined to grey of lower hindneck. Band of silver-white feathers with narrow subterminal black bar across upper breast. Rest of breast buff with pink hue; belly black, mottled buff by feather-bases showing through. Vent, lower flanks, and under tail-coverts white; under tail-coverts buff at base barred dark brown, separated from white edge and tips of feather by elongated dark brown V. Feathering on legs off-white. Central tail-feathers (t1) buff, barred (sometimes streaked and blotched) dark grey, shaft black; tip strongly elongated, black. Rest of tail-feathers tipped white with outer web and inner near shaft grey, edged and streaked buff, and rest of inner barred dark grey and buff; outer web of outermost (t8) white. Primaries grey with black shaft. P10 with narrow black outer web and elongated tip (see Structure), others with outer web narrowly edged off-white. Inner web of p4–p6 subterminally dark grey, margined buff; p2–p3 with pronounced subterminal black area and bright buff zone along distal edge of inner web. Underside of primaries silver-white. Primary coverts buff with broad black shaft-streak. Secondaries buff, more deeply tinged towards tip and with distal part of outer web black, margined buff; on s1–s2, subterminal black spot extends to inner web. Tertials (from s11 inwards) buff, rufous buff on inner web; s11–s12 with broad grey shaft-streak, others irregularly barred and streaked dark grey to dark brown. Upper wing-coverts buff; greater with outer web distally bay, forming panel on wing; marginal and inner median with rounded black spot near tip. Under wing-coverts and axillaries silver-white, latter with big black spot near tip. ADULT FEMALE. Forehead and crown buff, tinged ochre and streaked black. Supercilium deeper ochre, extending to golden-ochre spot at side of neck. Chin off-white; malar streak and gorget golden-ochre, separated from spot on side of neck by buff crescent spotted black (finely streaked on ear-coverts), and from lower throat by narrow black border.

Nape, hindneck, and upper mantle buff, spotted black. Rest of upperparts buff, barred and spotted black; ground-colour less cinnamon than ♂, barring, narrower denser, and more irregular, scapulars without extensive buff tip and with only indistinct purplish spot subterminally. Upper tail-coverts basally barred, near tip with black shaft-streak or elongated V-mark. Lower throat, chest, and upper breast grey, without cross-band; sides of chest buff with rounded black spots. Lower breast off-white. Belly dark chestnut to black. Lower flanks, vent, under tail-coverts, and feathering on legs off-white; latter slightly tinged brown or buff; under tail-coverts buff, heavily streaked and barred black on hidden feather-bases. Tail buff, barred grey; tip of t1 elongated, dark grey with black shaft; tip of t2–t8 white; outer web of t8 white, basally grey. Primaries, primary coverts, and secondaries coloured as in ♂ (but tip less elongated—see Structure); tertials (from s11 inwards) buff, barred and spotted black on outer web, brown on inner. Greater upper wing-coverts as in ♂; median and lesser cinnamon-buff, inner median and all lesser with rounded black spot near tip. Underside of wing and axillaries as in ♂. DOWNY YOUNG. Upperparts tawny-buff, mottled with fine specks of black and marked with pattern of white lines (often bordered black) running along centre and sides of crown, mantle, and back, connected by bars. Sides of head tawny-buff with white lines over, through, and below eye. Underparts off-white, chin and throat finely streaked black. Tarsus and upperside of toes clad in off-white down (toes bare in other Pteroclididae). (Fjeldså 1976, 1977; ZMA.) JUVENILE. Feathers developing shortly after hatching more loosely structured and with paler marking than those developing later; this interpreted as double juvenile plumage by Witherby *et al.* (1940) and Dementiev and Gladkov (1951*b*), but no clear distinction present (Glutz *et al.* 1977; see also Chestnut-bellied Sandgrouse *Pterocles exustus* and Black-bellied Sandgrouse *P. orientalis*). Pattern of plumage resembles adult ♀. No ochre on head; bars on upperparts browner, less black, subterminal more crescent-shaped; scapulars without purplish spots. Chin, throat, and upper breast buff; lower throat and upper breast streaked and mottled black. Dark area on belly smaller. All tail-feathers broadly barred brown and tipped buff; t1 without long attenuated tip. P1–p9 browner; tips buff, with brown lines and frecklings. Primary coverts with irregular brown markings at tip; greater upper wing-coverts lacking bay outer webs; rest of coverts with brown subterminal crescents instead of round black spots. (Witherby *et al.* 1940.) SUBSEQUENT PLUMAGES. Like adult, but recognizable as long as some juvenile outer primaries retained.

Bare parts. ADULT. Iris dark yellow-brown; bare skin round eye dark blue-grey. Bill blue-grey, paler at tip. Sole of foot brown, claws blackish. DOWNY YOUNG. Bill blackish, claws pale grey. (Glutz *et al.* 1977.)

Moults. ADULT POST-BREEDING. Complete; primaries descendant. Primaries late May to September or early October; moults more quickly than *Pterocles* with usually 2 adjacent feathers growing at same time. Tail starts (usually with t1) late July when about half of primaries replaced. Body starts on back about middle of June (though some feathers on back may be among latest to be replaced), continuing with head late June to early July; moult most intense late July; finished September–October, occasionally December (Roding 1973) or even January–February (Witherby *et al.* 1940). POST-JUVENILE. Almost complete, but some outer primaries retained; from summer to early autumn, about 2 months after hatching. (Piechocki 1968*c*; Glutz *et al.* 1977; ZMA.)

Measurements. Adult; skins (RMNH, ZMA). Tarsus and toes densely feathered in front, and thus measured on rear; tarsus from tarsal joint to border of feathers, toe along bare sole of foot.

	♂			♀		
WING	253	(5·77; 14)	243–259	222	(5·81; 12)	214–235
TAIL(t1)	190	(20·3; 11)	165–228	144	(8·24; 11)	131–160
(t2)	106	(4·73; 12)	99–117	98·4	(4·99; 11)	88–106
BILL	10·0	(0·69; 15)	8·9–11·5	9·6	(0·58; 15)	8·8–10·7
TARSUS	21·6	(0·98; 7)	20–23	20·9	(1·25; 8)	19–23
TOE	23·0	(1·15; 7)	21–24	22·4	(1·40; 7)	21–25

Sex differences significant for wing and tail.

Weights. Mongolia, summer, adult ♂ 274 (16·5; 8) 255–300; ♀ 252 (14·5; 5) 235–270; juvenile ♀ 180 (Piechocki 1968). Kazakhstan, adult ♂♂, June 264, 265, 305, September 264 (10) 250–280; adult ♀, June 272, September 220 (6) 200–250; juvenile 248 (6) 220–260 (Gavrin *et al.* 1962). Averages during irruption 1888, Scotland, May ♂ 247, ♀ 238; June, ♂ 278; September, ♀ 260 (Glutz *et al.* 1977). Netherlands, December, ♀ 268 (Roding 1973).

Structure. Wing long, strongly pointed; p10 with characteristic narrowly lanceolate elongated tip, much longer in ♂ than in ♀. 11 primaries: p10 longest; in ♂, p9 35–46 shorter, p8 51–65, p7 67–73, p6 82–89, p5 95–101, p4 110–116, p1 151–158; in ♀, p9 16–21 shorter, p8 30–35, p7 43–48, p6 55–63, p5 67–76, p4 81–90, p1 122–130; p11 minute, concealed by primary coverts. Tail strongly wedge-shaped with elongated tip of t1 protruding *c.* 80 in ♂, *c.* 45 in ♀; in juvenile, not elongated or pointed, protruding only a little; 16 feathers (occasionally 18), t2 *c.* 30–40 longer than t8. Bill short, nostrils covered by dense feathers. Leg very short, densely feathered, except on sole of foot. Toes short, joined up to base of claws; middle toe about as long as tarsus, inner and outer toe *c.* 70% of middle; no hind toe.

Geographical variation. None.

JW

Order COLUMBIFORMES
Family COLUMBIDAE pigeons

Small to large arboreal and terrestrial birds. Now sole family in order, related dodos (Raphidae) and solitaires (Pezophapidae) being extinct—though these considered to be flightless rails (Rallidae) by Lüttschwager (1961). A well-defined, natural assemblage with no known close relatives (see further, below). Ecological differentiation considerable, but strictly arboreal, mainly frugivorous and granivorous forms predominate; strictly terrestrial forms occur in humid tropics and Australia. About 300 species in *c.* 42 genera (including 17 monotypic ones and 12 of less than 5 species); 295 species recognized by Goodwin (1970), including 2–3 recently extinct, 303 by Bock and Farrand (1980). Main genera include: (1) *Columba* (typical pigeons, 51 species); (2) *Streptopelia* (turtle-doves, 15 species); (3) *Geotrygon* (American quail-doves, 13 species); (4) *Gallicolumba* (Old World quail-doves, 18 species); (5) *Treron* (green pigeons, 23 species); (6) *Ptilinopus* (fruit-doves, 49 species); (7) *Ducula* (imperial pigeons, 36 species). Numerous attempts at establishing formal sub-divisions within family have proved unsatisfactory, and none recognized here; for informal species-groups, see Goodwin (1970). 13 species of 3 genera represented in west Palearctic, 11 breeding—6 *Columba*, 4 *Streptopelia*, and Namaqua Dove *Oena capensis*. Family of cosmopolitan distribution except for Arctic and Antarctic; greatest diversity found in tropical Central America (30 species), tropical south-east Asia (62), and Australia (21). NB Names 'pigeon' and 'dove' are synonymous; 'pigeon' popularly used for larger species, 'dove' for small—but exceptions numerous.

Bodies plump and compact, heads rather small. Sexes similar in size. Necks short. 37–39 vertebrae (in west Palearctic species at least), including fused pelvis and pygostyl (H Sandee). Wings long and broad in many species but short in others; tips rounded. Flight strong and direct in majority of species; no capacity for soaring flight though gliding widespread, especially in display. 11 primaries; p1 reduced. 10–15 secondaries (including tertials); diastataxic in most species but some eutaxic. Flight-feathers rigid, causing loud and characteristic clapping sound when bird flies away (and in display). Tails usually fairly long and broad with tip square or slightly rounded, but very long and pointed in some species; 12–14 feathers (16–18 in crowned pigeons *Goura* and pheasant-pigeons *Otidiphaps*). Bills short, weak, and plover-like except in large arboreal, fruit-eating species; tip hard (sometimes hooked), base soft. Nostrils narrow, obliquely placed in cere at base of bill, and covered by narrow scale or fleshy operculum. Ceres usually small, but greatly enlarged in a few species by caruncles and other protuberances. Tarsi usually short, particularly in arboreal species; covered laterally and behind with small hexagonal or rounded scales. Feet of perching type, with 3 front toes and large and functional hind toe; basal phalanges long. Gait deliberate, with characteristic bobbing movements of head—head remaining at constant level in relation to moving body. Oil-gland absent in many species, small and unfeathered if present; preen-oil not used to dress feathers during preening, being replaced by powder-down (see below) which permeates plumage (Goodwin 1970). Caeca small and non-functional in some species, absent in others. Crop large, with 2 lobes—causing asymmetry of extrinsic syringeal muscles; during relevant stage of breeding cycle, produces a nutritious secretion ('crop milk') from lining for feeding small young. Gizzard heavily muscled. No gall bladder. No supra-orbital salt-glands. Syrinx tracheo-bronchial. Eyes appear small but have very large orbits; orbital skin bare. No aftershaft. Plumage dense, with feathers very loosely set in a thin skin; feather-shafts strong and broad, tapering abuptly in thin point. No down tracts, these being replaced by downy barbs at base and basal lateral edges of all body feathers; those on flanks modified as powder-down. For fuller technical diagnosis, see Ridgway (1916).

Plumages of many species soft shades of brown, grey, or vinous but highly colourful in others (e.g. vivid green, orange); often bold in pattern, with iridescent, white, or black patches on sides of neck and breast, on wing, or on tail; a few species crested. In majority of family, sexes differ only slightly in appearance—with ♂ brighter in pattern; in a few others, sexes exactly alike or differ strikingly. Bare parts of many colours: irises (e.g.) red, orange, yellow, brown, and green; orbital skin (e.g.) greyish, red, pink, and purple; bills (e.g.) black, brown, yellow, white, grey, green, and blue—tip and base often of different colours from one another and from rest of bill; ceres (e.g.) white, grey, black, red, and green; legs and feet (e.g.) red, pink, and purple. Post-breeding moult complete; primaries descendant, occasionally serially descendant; very slow, taking up to 10 months—not, or usually not, suspended during breeding. No pre-breeding moult. Young altricial, nidicolous, and wholly dependent on parents for food; blind at hatching, with sparse, coarse, hair-like, loose, white or rufous down mainly on upperparts. Juvenile plumages usually distinct from adults', duller with dark subterminal band and light edging on contour feathers. Adult plumages attained in complete post-juvenile moult starting 1–3 months after hatching; on head, body, and wing-coverts from 3–6 months, but replacement of flight-feathers and tail slow, taking 6–14 months.

Columbidae share a number of characters with Charadriiformes (waders, gulls, etc.) including structure of palate and nares, presence of small basipterygoid processes, type of syrinx, configuration of flexor tendons of toes, and mainly diastataxic structure of wing. Together with Pteroclididae (sandgrouse), pigeons have been united in one order with waders and allies by Gadow (1893), Seebohm (1895), Ridgway (1901), and Fjeldså (1976). They differ from Charadriiformes (as here classified) in having a more rigid structure of vertebral column, presence of a large and functional hind toe, and a number of other structural features (see above), as well as in their biology, and behaviour and type of young. Any relationship with Charadriiformes seems distant at best, as does that with Pteroclididae (see that family) and Psittaciformes (parrots). Egg-white proteins differ from those of all other families (Sibley and Ahlquist 1972), as does chemical composition of lipid oil-gland secretion (Jacob 1978).

Columba livia Rock Dove

PLATES 27 and 29
[facing pages 327 and 350]

Du. Rotsduif   Fr. Pigeon biset   Ge. Felsentaube
Ru. Сизый голубь   Sp. Paloma bravía   Sw. Klippduva

Columba livia Gmelin, 1789

Polytypic. Nominate *livia* (Gmelin, 1789), western and southern Europe, Canary Islands, Maghreb, northern Libya, and north-west Egypt, east to southern Urals, western Kazakhstan, northern slopes of Caucasus, Georgian SSR, Cyprus, Turkey, and Iraq; *palaestinae* Zedlitz, 1912, Syria to Sinai and Arabia; *gaddi* Zarudny and Loudon, 1906, Azerbaydzhan (USSR), Iran, western and northern Afghanistan, Turkmeniya, Ustyurt Plateau in Transcaspia, and Uzbekistan; *targia* Geyr von Schweppenburg, 1916, Sahara from Mali to Sudan; *dakhlae* Meinertzhagen, 1928, Dakhla and Kharga oases (central Egypt); *schimperi* Bonaparte, 1854, Nile delta and valley in Egypt and northern Sudan. Extralimital: *neglecta* Hume, 1873, mountain ranges of Central Asia from Tarbagatay, Tien Shan, and Pamir to eastern Himalaya, grading into *gaddi* in western Kirgiziya and Tadzhikistan and in Hindu Kush (Afghanistan); *intermedia* Strickland, 1844, peninsular India and Ceylon; *gymnocyclus* Gray, 1856, Sénégal and Guinea to Ghana and Nigeria.

Field characters. 31–34 cm; wing-span 63–70 cm. Wild birds noticeably smaller than Woodpigeon *Columba palumbus* but bulkier than Stock Dove *C. oenas*; feral birds often larger when interbred with larger escaped domesticated forms. Medium-sized and compact, blue-grey pigeon, with 2 obvious black bars across rear half of inner wing; in flight, shows white under wing-coverts and white patch on lower back (though latter indistinct or absent in populations of Middle East and Canary Islands). In domesticated and feral birds, plumage extensively variable from all black to all white through various piebald, blue-grey, and red-brown variants; some show double wing-bar of wild birds. Flight confident and fast. In wild birds, ♀ slightly duller than ♂, with less intense gloss on neck; no seasonal variation. Juvenile separable. 6 races in west Palearctic, 3 geographically isolated, the others clinal but recognizable in some areas.

Adult. (1) European, Canary Islands, and North African race, nominate *livia*. Head, underbody, and rump blue-grey, but lower back and rump pure white (sometimes greyish on Canary Islands), forming bold patch. Nape, neck, and upper breast glossy green and purple (mostly purple around breast, mostly green higher up). Mantle, scapulars, and inner wing ash-grey, distinctly paler than head and body; inner wing marked black on tertials, and broadly barred black on greater coverts and inner secondaries. Outer primaries and tips of secondaries and inner primaries dusky (but not contrasting boldly with rest of upperwing). Axillaries and under wing-coverts white, contrasting strongly with grey underbody and dusky undersurface of flight-feathers (and forming bold underwing pattern shared only by Yellow-eyed Stock Dove *C. eversmanni*). Tail blue-grey, with quite broad, brown-black terminal band. Bill lead-coloured, with mealy white cere. Eyes orange, legs and feet dull to bright red. (2) Egyptian desert race, *dakhlae*. Markedly paler than other races, with mantle and underbody almost as white as back; looks dark-headed and -chested. (3) East Mediterranean race *palaestinae*, south-west Asian race *gaddi*, Saharan montane race *targia*, and Nile valley race *schimperi*. Paler than nominate *livia*, but white back usually obscure or replaced by grey. Size of bird decreases markedly to south. Juvenile. All races. Distinctly duller and darker than adult, with all characters (except white back) less clear. Underwing pale grey. Gloss on neck feathers minimal. Throughout range, plumage pattern subject to dilution or distortion through widespread interbreeding with domestic and feral birds. Many feral birds show grey back and under wing-coverts, which invite confusion with *C. oenas*.

Combination of white underwing with black bars across inner upperwing diagnostic. Possibility of confusion with *C. eversmanni* not studied: it too shows white underwing and white on back; lacks, however, almost-black flight-feathers, bold inner wing-bars, and sharp black tail-band, and has duller head uniform with mantle. Confusion of wild birds with *C. oenas* unlikely in flight, since that species lacks white underwing and white on back, and shows broad pale centre to upperwing. Separable from *C. oenas* on ground by close observation of inner wing pattern (boldly barred black in *C. livia*, showing only short black lines or spots on innermost feathers in *C. oenas*), mantle colour (paler than head in *C. livia*, uniform with head in *C. oenas*), and neck-surround colour (intense in *C. livia*, light and essentially green in *C. oenas*). Flight most dashing of genus, with rapid, much angled beats of wings producing great speed; has also marked gliding and wheeling ability. Take-off explosive, with clatter of wings; landing often preceded by long glide, but accomplished with final flutter. Gait freer than congeners, with walk often becoming loping half-run. Settles freely on rocks, cliffs, and open ground, but wild birds usually shun perching on branches except where close to cliffs.

In display, gives a polysyllabic coo; otherwise rather silent.

Habitat. In middle and lower latitudes of west Palearctic, mainly in continental temperate, Mediterranean, steppe, desert, and subtropical zones, but also on oceanic coasts and boreal to subtropical islands. Apart from natural range, has through human intervention developed feral populations which have originated from captive birds, some joining and interbreeding with wild populations, while others have formed concentrated colonies commensal with man on and around buildings, often in city centres. These artificially based populations hardly overlap with those persisting within or reverting to natural fringe of range. Some have recently penetrated to Murmansk in Russian Arctic (Glutz and Bauer 1980). Currently domesticated birds of various breeds may be based on dovecotes situated in farmlands and other intermediate habitats, thus sharing ecological niche with other species such as Stock Dove *C. oenas*. Most recently, situation further complicated by great expansion of Collared Dove *Streptopelia decaocto*, also commensal with man over much of Europe, but hitherto not normally penetrating into main inner city strongholds of feral *C. livia*.

Natural habitat linked with nest-site on rock-faces, especially coastal: according to Scottish records, predominantly in clefts or ledges in caves within sea-cliffs, used also for roosting; clefts in other rocks and ruined buildings less frequently chosen (Baxter and Rintoul 1953). Mostly frequents narrow belt of often marginally cultivated land behind cliff-tops, where weed plants abound and grain can be picked up, with access also to drinking water. Commonly treeless, such habitats discourage competitive tree-based congeners and favour wariness against possible predators, as well as permitting breeding during most of year. In USSR and elsewhere in Eurasia, similar rocky habitats locally occupied deep inland (Dementiev and Gladkov 1951b), but over most of Europe only feral form attached to human settlements occurs (Glutz and Bauer 1980). In Algerian Sahara, common wherever rocks, some vegetation, and permanent water occur, even in desert where desert melon *Coloquintus vulgaris* provides food and moisture (Dupuy 1969). In Tunisia, often nests in deep wells (Jarry 1969). Avoids dense and tall vegetation, from grassland to woodland, and also broken or enclosed terrain, perching rarely. Occasionally alights on water, even on sea, apparently to drink.

Feral form attached to human settlements, especially city centres; also prefers to breed in groups, nesting under cover in church towers, steeples, and large institutional buildings such as public offices, museums, railway stations, factories, warehouses, and architecturally ornamented structures; also large gabled houses, especially with turrets, cupolas, angled rain-pipes, ventilators, air-brick spaces, eaves, lofts, and attics. Most nest-sites fairly high above ground. Secure loafing places such as roofs, gutters, ledges, window-sills, etc., almost essential, as is ready access to drinking water and bathing places. Although tolerant of close human approach on ground, prefers abandoned structures or parts of structures for breeding; in a deserted seaside hotel every room occupied by birds flying in and out through broken or open windows. Suitable structures often have to be wired against entry, or are deserted by birds when repair or maintenance work is undertaken. Prefers buildings separated only by bare ground, paving, or shortest or sparsest plant cover, or facing margins of water. Where chosen sites are fronted by trees, will sometimes perch on larger horizontal branches, or even nest in holes in them. In city parks, gardens, allotments, and waste ground may compete for food with town-dwelling Woodpigeons *C. palumbus*, but (apart from House Sparrow *Passer domesticus*) usually has monopoly of artificially provided food in paved squares and at markets and railway stations. (Based largely on Simms 1975; see also Goodwin 1954a, 1970, and Gompertz 1957.) While wild birds in natural habitats show conservative and stable life-style, feral stocks have made rapid and far-reaching adaptations.

Distribution. Original distribution obscure because of long history of domestication by man, e.g. for food (dovecotes and special platforms in breeding caves) and for breeding (as racing pigeons or for producing fancy varieties). Many became feral, especially in urban areas, and are still being reinforced by escaped birds (Goodwin 1970); also suggested by Glutz and Bauer (1980) that wild birds may have joined feral colonies. Present distribution inadequately known, as impossible in many countries to distinguish wholly wild colonies (see, e.g., Murton and Clarke 1968). In west Palearctic, impossible to construct any detailed distribution map, as until recently largely neglected by ornithologists and some recent atlas surveys have not included feral forms. An attempt has been made, however, to indicate present world distribution, but for reasons given by Long (1981a) this must be regarded as extremely tentative.

Population. Limited information. Scattered evidence of increase in feral birds mostly in late 19th and 20th centuries, still continuing.

BRITAIN AND IRELAND. Breeding population could exceed 100 000 pairs (Sharrock 1976); still increasing inner London (Cramp and Tomlins 1966). NETHERLANDS. Probably marked increase in feral populations 1950–70 (CSR). FRANCE. Increase of feral birds relatively recent, e.g. in Paris from early 20th century (Yeatman 1976).

Survival. England (Salford): mortality in 1st year of life $43 \pm 7\cdot3\%$, annual adult mortality $33\cdot5 \pm 4\cdot9\%$ (Murton *et al.* 1972b). England (Flamborough Head): annual adult mortality from shot sample *c.* 30% (Murton and Clarke 1968). Oldest ringed bird 6 years 4 months (Rydzewski 1978).

Movements. Despite pronounced homing ability over long distances shown by domesticated birds often used as subjects for orientation experiments, free-living populations (wild and feral) are resident or even sedentary. Local

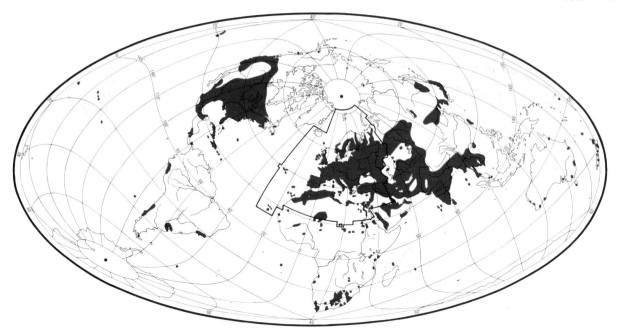

movements more marked in seasonally arid regions: e.g. present on coastal plains of northern Oman from June/July to early March, then move to inland wadis and hills to breed (Walker 1981a); big flocks congregate in Niger inundation zone, Mali, November–May (Lamarche 1980). On Fair Isle (Scotland), small passages detected late September–December and (fewer) February–April (Williamson 1965); these presumably short-distance movements between Northern Isles.

Recoveries ($n = 13$) for wild stock in Britain moved 0–5 km (8), 6–8 km (3), 16 km (1), and 28 km (1) (BTO). Recoveries ($n = 1269$) for feral birds in Manchester (England), showed these remain all year within urban habitat and even within same building (juveniles showing slightly more tendency to disperse); 86% found within 90 m of ringing site and nearly all within 1 km; only 5 birds exceeded 1·3 km, with longest at 4 km, 4·8 km, and 6·4 km (Murton et al. 1972b, 1974a). A feral population in Brno (Czechoslovakia), roosted in town and fed by day on surrounding farmland (including grain stores); movements mainly 5–8 km and rarely over 18 km, but exceptionally up to 30 km, latter birds probably dispersing away from Brno population (Havlín 1979).

Food. Chiefly seeds of cereals, legumes, and weeds; occasionally green leaves or buds, and invertebrates. Feral birds also take wide variety of artificial foods. Food almost always taken on ground (usually in the open) by walking and pecking. Rarely, feral birds feed on ground in woodland, or in trees, vines, etc.; not recorded for wild form. Appears quite inexpert when feeding in fine tree branches, overbalancing almost constantly. On ground, often feeds in flocks. Dominance hierarchy established in feral flocks, birds at centre obtaining more food per min and per area searched; peripheral, apparently subordinate, birds obtain less food and are lighter than central birds. At Salford Docks (England) in summer, feral birds spent 6 hrs (46% of daylight) in vicinity of food sources, mainly in early morning and late afternoon, but fed c. 10 min per hr; in winter, birds at food c. 7 hrs per day (88% of daylight). (Murton et al. 1972b.) Feeding rates of 33 pecks per 15 s (Murton et al. 1972b) and 25 pecks per 15 s (Coombs et al. 1981) measured for feral flocks in Cheshire (England). For details of morphology of feeding apparatus, see Zweers (1982). Drinks from edge of, or in shallows of, water; sometimes alights on surface to drink (see Yarrell and Saunders 1882–4; Butler 1930; Ingram 1978) or hovers above with bill immersed (Blackburn 1895). Pecks at frost for water (Simms 1979).

Adult diet of both wild and feral birds poorly studied considering their range and abundance. Wild birds take grains of wheat *Triticum*, barley *Hordeum*, oats *Avena*, rye *Secale*, maize *Zea*, and millet *Panicum*; seeds of cultivated and weed species such as docks *Rumex*, knotgrass *Polygonum*, goosefoots (*Chenopodium*, *Atriplex*), chickweed *Stellaria*, buttercup *Ranunculus*, fumitory *Fumaria*, cruciferous (*Raphanus*, *Capsella*, *Sinapis*, *Brassica*), legumes (*Pisum*, *Vicia*, *Phaseolus*, *Melilotus*), cleavers *Galium*, pansy *Viola*, speedwell *Veronica*, plantain *Plantago*, and grasses (Gramineae). Also fruits and some green leaves (e.g. *Ranunculus*, *Stellaria*), and pieces of potato. Animal matter occasionally taken, especially cocoons of earthworms (Oligochaeta) and small snails (*Helix*, *Helicella*, *Trichia*, *Bulimus*), but also earthworms, slugs (*Arion*, *Agriolimax*), arachnids, and insects. Invertebrates sometimes taken in large numbers, e.g. crop of one bird crammed with moth (Geometridae) larvae. (Macgillivray 1837; Hodgkinson 1844; Naumann 1905; Murton and Westwood 1966;

Murton and Clarke 1968; Simms 1979.) In addition, feral birds take seeds of garden and agricultural plants and weeds such as crucifers (*Barbarea*, *Sisymbrium*, *Erysimum*), corn spurrey *Spergula*, clover *Trifolium*, tomato, cucumber, spurge *Euphorbia*, hemp-nettle *Galeopsis*, composites (e.g. *Taraxacum*, *Helianthus*) and garlic *Allium*. Fruits and seeds of oak *Quercus* (usually broken or rotten), plane *Platanus*, *Prunus*, ivy *Hedera*, *Viburnum*, poison ivy *Rhus*, privet *Ligustrum*, and nightshade *Solanum*; green leaves of sea-stock *Matthiola*, primrose *Primula*, and tulip *Tulipa*; roots of couch grass *Agropyron*, star-of-Bethlehem *Ornithogalum*, saxifrage *Saxifraga*, *Ficaria*; tree buds; flowers of *Ulmus*, sycamore *Acer*, and daisy (Compositae); spangle galls from *Quercus* leaves, and seaweed. Much artificial food and domestic scraps eaten: e.g. bread, biscuits, processed cattle feed, peanuts, banana, apple peel, cooked potato, cheese, fish, meat, fat, chocolate, and ice cream. (Blackburn 1895; Naumann 1905; Townsend 1915; Goodwin 1954a, 1960b; Gompertz 1957; Reinke 1959; Murton and Westwood 1966; Dilks 1975a; Moeed 1975; Pierson et al. 1976; Potvin et al. 1976; Kotov 1978; Havlín 1979; Simms 1979.) Small stones and grit collected from roads and pavement in early morning; also recorded eating mortar from buildings, presumably for calcium content (Goodwin 1954).

Diet diverse, but 2–3 foods usually dominate; however, considerable variation with season and locality. In western Siberia (USSR), 7·6 food types per bird ($n=198$); 7 food types occurred in 75% or more of birds (Kotov 1978). 18 birds, Christchurch (New Zealand), averaged 3·4 food types per bird (Moeed 1975). In Virginia (USA), 2·8 food types per bird ($n=144$), with only maize occurring in 75% or more of crops (Pierson et al. 1976). Experiments in captivity showed food preferences of feral birds (in descending order): peas, grain (wheat, oats, barley), millet, maize (Brown 1969).

Contents of 86 crops of wild birds, Flamborough (England), throughout the year (except April), included 73% (dry weight) grain including 57% barley (90–99% of November–February totals) and 16% wheat or oats, 10% cultivated legume seeds, 5% *Stellaria*, 4% grass seed, 2% *Ranunculus*, 1·5% *Fumaria*, 1% *Polygonum*, and 1% animal matter (Murton and Westwood 1966).

244 crops of feral birds, Leeds (England), throughout the year (except December), contained 59% (dry weight) artificial food, including 55% bread or cake, 2% exotic seeds, and 2% other domestic scraps; 19% barley, 14% wheat, 5% maize, 2% wild seeds, and 1% peas or beans (Murton and Westwood 1966). In western Siberia (USSR), April–September, 100% of 198 birds contained wheat and barley, 91% peas, 88% oats, 88% sunflowers, 85% green parts of plants, 75% mollusc shell fragments, 70% complete snails, 30% maize, 24% wild hemp, 23% *Barbarea* seeds, 19% *Atriplex*, 13% *Polygonum*, 13% *Populus*, 8% *Vicia*, 6% earthworms, and 4% slugs (Kotov 1978). Studies in New Zealand (Dilks 1975a) and USA (Pierson et al. 1976) showed similar results.

Captive birds from feral stock consumed c. 30 g of cereal per day (Murton et al. 1972b); wild birds ate 17·5–43 g in one feeding bout (2 bouts per day) (Kotov 1978). Average volume of food in crop varies throughout year in New Zealand from 25·3 cm^3 in June (winter) to 50·2 cm^3 in January (summer) (Dilks 1975a). In Virginia (USA), average 18·5 cm^3 of food in crop in December (Pierson et al. 1976). Extreme examples of number of items in crop: 7424 rape seeds plus 65 corn and barley; 3950 clover seeds; 1309 barley grains; 1089 rye grains; 328 black nightshade berries; 199 peas; 128 maize grains (Nelson 1907; Witherby et al. 1940; Dilks 1975a).

At Flamborough, diet of wild nestlings similar to that of adults, but with higher proportion of animal material. Crop milk very important; sole food initially. Increasing percentage of seeds and other food given from c. 5th day until only traces of crop milk by fledging. Crop milk composed of 64–82% water, 10–19% protein, and 7–13% fat; contains amylase, saccharases, rennet, and vitamins A, B, C, D, and E (Levi 1945; Murton and Clarke 1968; Kotov 1978). 147 crops of young birds, throughout the year, contained 52% (dry weight, excluding crop milk) barley, 37% wheat, 8% oats, 2% weed seeds and leaves, and 1% animals, including earthworm cocoons (in 50), snails (9), and slugs (9) (Murton and Clarke 1968). Nestlings first fed 4–6 hrs after hatching; fed 3–4 times per day up until 5–7 days, thereafter twice a day (morning and evening) (Kotov 1978).

HAR

Social pattern and behaviour. Few detailed studies of truly wild, mainly cliff-dwelling *C. livia*, but some aspects covered by Lees (1946) for Scotland, and Petersen and Williamson (1949) for Faeroes; see also comments by Murton and Clarke (1968) for Flamborough Head (England). Generally considered, however, that no essential differences exist between wild birds and birds of urban and other habitats (e.g. Heinroth and Heinroth 1949, Goodwin 1970), but see part 2. Following based mainly on Goodwin (1970), and on Heinroth and Heinroth (1949) which deals with captive birds of wild stock, domesticated homing pigeons, and progeny of mixed pairs. See also Gompertz (1957) for birds in London (England), Murton et al. (1972a, b, 1974a) for Salford (England), and Spiteri (1975) for study on homing pigeons.

1. Gregarious, especially when roosting, feeding, flying some distance, or loafing; commonly in pairs or small flocks (Goodwin 1970). In Faeroes, mainly in pairs during summer, flocks forming at other times (Petersen and Williamson 1949). Considerable numbers may gather where food abundant, e.g. up to 6000 at grain silo, Brno (Czechoslovakia), and hundreds or thousands in city squares where birds fed by public (Bruns 1959; Goodwin 1970; Brion and Vacher 1970; Havlín 1979; Simms 1979). For further data on flock sizes, see Cramp et al. (1964), Murton et al. (1972a, b), Havlín (1979), and Simms (1979). In feeding flocks, Berlin, sex-ratio of young birds about equal (Becker 1963, 1964; see also Flock Behaviour, below). In Salford, flocks (though controlled) stable in structure for long period (Murton et al. 1972b, which see for juvenile–adult ratio). In London, juveniles hung on longest when (artificial) food supply terminated; same juveniles formed core of group eventually returning when supply restored (Gompertz 1957). In southern Urals and Kemerovo region (USSR), 98 1st-laid eggs produced 69 ♂♂ and 29 ♀♀, 88 2nd-laid 34 ♂♂ and 54 ♀♀ (Kotov 1978). BONDS. Most commonly mono-

gamous mating system with life-long pair-bond (Heinroth and Heinroth 1949; Bruns 1959; Goodwin 1970). In case of feral and domestic birds, length of bond probably due to birds being sedentary or caged, and almost constantly together and in breeding condition (Goodwin 1970). In USSR, some pair-bonds last 10–12 years but duration for single season also not unusual: of 26 pairs studied 1965-9, 4 ♂♂ and 5 ♀♀ took new mates each year, others showing fidelity throughout (Kotov 1978); see also Murton et al. (1974a), and below. In West Bengal (India), however, study of ringed birds indicated partners changed each season (Sengupta 1976). For relative strength of individual pair-bonds, see (e.g.) Gompertz (1957) and also Heterosexual Behaviour (below). For colour preferences for mating in polymorphic domestic or feral birds, see Goodwin (1958a, 1970), Mainardi (1964), Murton and Clarke (1968), Murton et al. (1973b, 1974a), and Kotov (1978). Relatively little interbreeding between feral, dovecote, and wild birds on one hand and domestic birds on other; barrier probably formed by ecological differences reinforced by mating preferences. Also apparently few records of mixed wild and feral pairs (Goodwin 1954a, which see for definitions of above categories). Promiscuous copulations by paired birds rare, particularly in wild and feral ones; where ♂ in loft has such encounters, other ♀ not tolerated near nest. If ♂ becomes less sexually active shortly before laying, ♀ may copulate with another ♂, but this does not lead to divorce (Heinroth and Heinroth 1949, which see also for polygyny). In another study, a few ♀♀ copulated with other ♂♂ during breaks from incubation; similarly did not lead to break-up of pairs (Kotov 1978). For cases of pairing and other heterosexual interactions between parent and its (in some cases) not yet sexually mature young, see (e.g.) Goodwin (1947a, 1958a) and Heinroth and Heinroth (1949). Bisexual tendencies strong in young birds (Goodwin 1958a). In USSR, several records of ♂-♂ and ♀-♀ pairs; unlike ♀-♀ pairs, ♂-♂ pairs formed only from young birds (Kotov 1978). Pair-bond maintained and nest-territory defended in most cases throughout the year, even when birds not breeding (Heinroth and Heinroth 1949; Murton et al. 1974a; see also below). In feral birds, earliest reported age of first breeding c. 6 months, e.g. many young fledging in January bred in summer of same year. 2 ♂♂ losing mates re-paired with juveniles from nearby nests; these ♀♀ laid at 6–7 months old (Murton et al. 1972b, 1974a). According to Abs (1975) sexual maturity attained at c. 4–6 months. In feral population, Tampere (Finland), ♂♂ first bred on average when slightly younger than ♀♀ (Häkkinen et al. 1973). Only c. $\frac{1}{3}$ of Salford feral population bred in any one year (perhaps ever), though all birds apparently sexually mature (Murton et al. 1972b). Both parents care for young (e.g. Bruns 1959, Kotov 1978). Where 2nd clutch laid, 1st brood tended mainly by ♂ (may include some post-fledging care); usually expelled from territory by ♀ when 2nd hatches (Heinroth and Heinroth 1949; Petersen and Williamson 1949; Murton and Clarke 1968; Goodwin 1970; Kotov 1978). BREEDING DISPERSION. Both wild and feral birds generally loosely colonial, though many pairs more isolated (Fitter 1949; Simms 1979). In USSR, 'colonies' of 80–200 pairs in various town buildings; 40–80 pairs in (!) country churches (Kotov 1978). About 30 pairs in one large cave, Scotland (Lees 1946); on Islay (Scotland), small scattered colonies quite usual (Gompertz 1957). At breeding site in mill, Scotland, c. 50 pairs of feral birds thought to represent true colony; over 9 years persisted in nesting close together or in groups when wider spacing possible (Riddle 1971). However, according to Gompertz (1957), close-knit colonies not typical of either wild or feral birds. In large colonies of feral birds, nests may be 0·5–1 m apart, more dispersed elsewhere (Kotov 1978). Not usually close together, Tampere (Häkkinen et al. 1973). Few detailed reports of population density. In central London, c. 10000–20000 pairs on 10360 ha (Cramp and Tomlins 1966). In Salford, c. 1500 birds present in study areas of c. 2243 ha, April, but only 256 nests located (Murton et al. 1972b; see also Bonds). In Hamburg (West Germany), suburban garden area held 1·0 pairs per 10 ha, dockland and industrial area 6·7, old residential area 15·6, city centre 25 (Glutz and Bauer 1980). In Brno, 2 birds per 10 ha in suburbs, 6·3 in industrial area, 8 in old residential area, 109 in centre (Hudec 1977). ♂ (less so ♀ mate) defends nesting territory (Murton et al. 1974a) of variable size in which ♀ may use habitual perch (possibly because ♀ subordinate to ♂♂); ♂s perch often outside nesting territory which is used for some courtship and rearing of young (Heinroth and Heinroth 1949 on study of loft birds; Goodwin 1954a). In lofts at least, young also defend parental territory. Most pairs use 2 nest-sites alternately, not normally the same for successive broods (Goodwin 1960). In Salford, all-year breeders occupied same nesting territories; majority of pairs used same or adjacent site over several years (Murton et al. 1972b, 1974a). In Tampere colony, same number of pairs bred each year independently of fluctuating population (Häkkinen et al. 1973). In Paris (France), birds range c. 600 m from nest-site (Brion and Vacher 1970). In Vienna (Austria), c. 500–600 m; in Hamburg, 4 flocks moved up to 500 m between roost and feeding sites but kept generally within c. 150-m radius of feeding sites during day (Bruns 1959; Reinke 1959). In Salford, flocks similarly sedentary (Murton et al. 1972b). At Augsburg Zoo (West Germany), flocks based on 4 dovecotes maintained remarkable mutual isolation (closest approach c. 10–15 m); home range varied with size of flock (Steinbacher 1959b). Varying reports on size of nesting territory: 'quite extensive' if alternative sites available (Gompertz 1957); a few m^2 (Häkkinen et al. 1973); defends only immediate vicinity of nest (Riddle 1971). Area depends mainly on strength of opposition (D Goodwin). Where site offers space for more than one pair but access restricted, occupying ♂ repels attempts to enter by other prospecting pairs; if colony has several entrances, territorial ♂ shows aggression towards others trying to use 'his' entrance (Gompertz 1957). For discussion of territory in domestic pigeon lofts, see Heinroth and Heinroth (1949). Considerable numbers of non-breeders present in Tampere (Häkkinen et al. 1973), and in USSR unpaired ♀♀ may lay in nests of breeding pairs (Kotov 1978). In Salford, non-breeders remain in flocks in general area of colony (see Murton et al. 1972b). ROOSTING. Mainly communal and nocturnal. Feral birds often markedly gregarious for roosting but tendency less pronounced among those breeding. In feral birds, roost and colony may be same (e.g. Schein 1954, Bruns 1959). Some pairs or single birds conspicuously unsocial, habitually roosting alone (Gompertz 1957). Wild birds normally roost on ledges in caves, in rock holes and crevices, etc. (Petersen and Williamson 1949; Goodwin 1954a; Simms 1979). Sites chosen by feral birds often nearest artificial equivalent; apparent preference for all-round cover so that sites similar to those used for nesting may be hidden in ruins, railway stations, etc., or more exposed (to weather, and artifical lighting) on ledges, though latter particularly favoured when broad overhang present (Townsend 1915; Goodwin 1954a; Gompertz 1957; Simms 1979). In London, often on ledges at height of c. 10 m or higher, a few lower (Cramp et al. 1964; S Cramp). Pairs or single birds may discover good site where only room for one or two and then often use it throughout year (Goodwin 1970). Some birds have several alternative sites and may use that closest to where they last fed, or choice possibly dictated by weather (Gompertz 1957). Roosts often used by birds from different feeding flocks and perhaps function partly as information centres (Murton et al. 1972a; see also Zahavi 1971). Disturbance at roost by man or other predator usually causes (temporary) desertion (Goodwin 1954a; Gompertz 1957). In Brno, exodus from roost or nests shows peak at 04.00–05.00 hrs in spring and mainly involves ♂♂, also non-breeders and some juveniles; less conspicuous peak c. 11.00–12.00 hrs mainly of ♀♀ flying out after nest-relief. In winter, birds mostly leave 09.00–11.00 hrs, returning c. 12.00–14.00 hrs. Some activity evident in streets before dawn when still completely dark. Activity may continue throughout night, at (e.g.) brightly lit railway station. Dispersal to feeding grounds at Brno (mostly up to 6–8 km away) takes place in small, widely separated flocks or long strings; some break-up or compacting takes place en route, and very large flocks not stable. Flight altitude primarily weather-dependent. (Havlín 1979.) Bad weather may delay flight from roost which is often also occupied during day in gloomy or stormy conditions (see Townsend 1915, Bruns 1959, Reinke 1959). Paired birds showing some courtship activity may still use communal roost well away from nest-site before laying (Goodwin 1960). In West Bengal, ♀ begins roosting on nest 1 day before laying (Sengupta 1976). Off-duty ♂♂ used communal roost during incubation, Hamburg (Reinke 1959); domesticated ♂ may roost close to nest before 1st egg laid, but afterwards usually well away from incubating ♀ (Goodwin 1970). After c. 20 days, unfledged young leave nest during day but return to roost there at night up to 28th day (e.g. Kotov 1978). In feral birds, communal loafing, preening, and sunning frequent and typical of cold but sunny, early mornings (Goodwin 1970; Havlín 1979). Perching in trees regular in some areas (Goodwin 1970); in London, sometimes used for resting during day but not for nocturnal roosting (Cramp et al. 1964). In Salford, birds tend to use exposed vantage points for long periods whilst surveying area and waiting for suitable feeding opportunities, though return to sheltered nocturnal roost-sites for resting proper (Murton et al. 1972a). Vantage points and nocturnal roosts may be the same (Simms 1979). Birds exceptionally alight briefly on water to drink (Goodwin 1970; see also Flock Behaviour, below). Birds sleep with head drawn in, bill resting on fore-neck; feathers often ruffled and one leg usually drawn up. During day, may also squat, or lie down, as when sunning or after bathing (Heinroth and Heinroth 1949; Goodwin 1970).

2. Relative timidity of wild and feral birds towards man mainly dependent on past treatment by him (D Goodwin), though behaviour otherwise essentially similar (Heinroth and Heinroth 1949; Gompertz 1957). In cities, often tame (eating out of hand, etc.), but where trapping, etc., undertaken, birds likely to avoid area, or at least to become more cautious for some time (Bruns 1959; Goodwin 1970; Simms 1979). At sight of man or other ground predator, bird usually freezes in erect Alarm-posture (Fig A), with neck extended and feathers sleeked (shaking of feathers invariably following if alarm subsides: Heinroth and Heinroth 1949); Distress-call (see 3 in Voice) usually given (also frequently when handled). Similar but more crouched posture typically adopted in presence of avian predator. Low-intensity flight-intention movement involves sleeking of feathers and sudden lowering of tail (then returned to normal position); thrusting or throwing forward of head usually a more definite indication of impending escape (Goodwin 1970). Loud and sharp Wing-clapping characteristic of birds flying up in alarm (Townsend 1915), or intending to fly some distance (D Goodwin); otherwise normally fly up silently (Townsend 1915). Sudden take-off or Distress-calls induce alarm and alertness in other birds; panic may result at roost. Observations on wild birds, Scotland, showed that during day they apparently need to see source of disturbance before fleeing in alarm, while others on ledges equally near but apparently less anxious because of cover, tend to adopt Alarm-posture (Goodwin 1970; see also for reactions of feral birds in towns). Wild birds, Islay, reported to leave caves on hearing calls from Oystercatcher Haematopus ostralegus (Gompertz 1957). Usually flies close to cliff or hillside, to avoid predation by falcons Falco (see also Selous 1905b for behaviour in presence of Golden Eagle Aquila chrysaetos, and Maser 1975 for account of birds threatened by Raven Corvus corax). Pursued by falcon, birds fly low and fast, then jink sideways or down at last moment to evade seizure; make for cover wherever possible; may also attempt to circle upwards (Goodwin 1954a, 1970; Dilks 1975b). Similar tactics employed when molested by sparrows Passer, but does not head for cover (Goodwin 1954a). According to Lees (1946), feral birds among colony of wild ones more likely to be taken by Sparrowhawk Accipiter nisus; see also Mainardi (1964) and, especially, Murton and Clarke (1968). In presence of predatory Glaucous-winged Gull Larus glaucescens, birds feeding on grain often only flew or ran in short arc around it (Jyrkkanen 1975). Escape from cat or car often in last-minute upward flight (Bruns 1959; Goodwin 1970). Circling often precedes landing, particularly at unfamiliar feeding place. Sometimes briefly alert before beginning to forage (Townsend 1915; Goodwin 1954a). Mobbing of predatory birds apparently rare in Columbidae (Goodwin 1970); once, bird repeatedly dive-bombed 2 flying Black-headed Gulls L. ridibundus (Azzopardi 1979). Compared with many other Columbidae, Head-shaking (rapid, mainly lateral, shaking or jerking movements) rather rare; may be expression of general discomfort (Goodwin 1956a, 1970); according to Simms (1979), regular in feral birds after drinking. FLOCK BEHAVIOUR. Flocks of varying size commonly formed for virtually all activities away from nesting territory (Glutz and Bauer 1980). Flocks compact, particularly when crossing large tracts of open country (Goodwin 1970; see also above). If flock of domesticated birds splits in flight, young birds often perform low-intensity Display-flight (see Antagonistic Behaviour, below) when momentarily undecided about which group to follow. Lost or immature birds in particular show marked 'company-seeking' tendency, joining or following others (Goodwin 1970). In Ruhrgebiet (West Germany) domesticated birds arriving at fields to feed normally circle, gain height, and circle again before landing, generally in open area with all-round view; such an approach typical of feral birds wary because no others already present (D Goodwin). On ground, tend to avoid hollows. Preening, sunning, and some courtship indulged in, but birds frequently forage all at once, some from rear of flock flying to front, etc. Birds often fly up and land again at virtually same spot. Raising of wings or short flutter off ground usual alarm signal. (Eber 1962.) For experimental study of reactions in other flock members to escape-intention movements and abrupt take-off, see Davis (1975). Attacked by falcon, notably Peregrine F. peregrinus, flock of domesticated or feral birds tends to scatter; wild birds (e.g. in Scotland) reported to form compact flocks and to perform unified manoeuvres (Goodwin 1970; Simms 1979). Circling Buzzard Buteo buteo attacked 7 times in succession by small flock of domesticated birds (see Kytzia 1954); similar be-

A

haviour noted towards North American *Buteo*, and *Accipiter* and Kestrel *F. tinnunculus* (Goodwin 1970; D Goodwin). In Salford, flocks more or less discrete units attached to specific resting, roosting, and feeding areas. Flocks formed at vantage points where central position possibly advantageous, as certainly so in feeding flocks (see below) (Murton *et al.* 1972*a*). Feral birds (in London) said to lead individual existence for feeding and roosting, flocks being rather chance aggregations (Gompertz 1957); also evident when feral birds fly out to feed on fields (D Goodwin). In Salford, availability of food at predictable time during day allows most birds to disperse to night roosts when not feeding. Where supply unpredictable, birds tend to stay together and watch; remain alert and quick to follow when one sees opportunity to feed (Murton *et al.* 1972*a*; see also for social hierarchy in flocks). Between feeding bouts, birds may form compact groups, squatting with tarsi and often breast flat on ground (Townsend 1915). For bathing and swimming of small flock, see Cottam (1949). ANTAGONISTIC BEHAVIOUR. (1) General. Birds generally tolerant at colonies as long as each pair has nest-site and perch (Heinroth and Heinroth 1926–7; Riddle 1971). For observations on birds in artificially crowded conditions of loft (thus perhaps not typical of wild or feral populations), see Heinroth and Heinroth (1949). (2) Display-flight. Sexually active ♂ flies out from cliff or building, flapping flight with exaggeratedly slow and deep wing beats being followed by several loud Wing-claps and Gliding with wings in V and tail spread; final descent may be made with slight rocking motion. Version given by wild birds less intense. Less frequent and emphatic in unpaired ♀; paired ♀ may follow mate in Display-flight with less vigorous Wing-clapping if he initiates performance. Often given over neutral ground; however, stated by Glutz and Bauer (1980) that, apart from heterosexual function, serves to demarcate territory. (Based mainly on Goodwin 1970, which see for contexts in which Display-flight given; also Heinroth and Heinroth 1949, Simms 1979.) (3) Threat and fighting. Frequent prelude to fight is mutual threat through Bowing-display: birds first in an erect posture and throat vibrates; crown, neck, and back feathers ruffled; conspicuously swollen look to front of body results from ruffling of feathers and inflation of cervical air sacs. As head lowered, first 2 units of Display-call given then final unit as head raised slightly. 2 ♂♂ may perform Bowing-display sideways-on or frontally; normally given while walking in circle, alternating clockwise and anti-clockwise (Spiteri 1975; Glutz and Bauer 1980: see 2 in Voice). According to Goodwin (1970) pattern more complex and variants of Bowing-display may be distinguished (see also Heterosexual Behaviour, below). In self-assertive form, given by ♂ in breeding condition, landing in or near territory, at approach of stranger, known rival ♂, etc., head rarely lifted, and tail not spread. In defensive form, head simply lowered and Display-call given. Both variants given rarely by ♀ at own roost or nest-site; usually executes only half-turn, then moves back through same arc. For cliff-ledge dispute in wild birds, Scotland, see Simms (1979). According to Murton *et al.* (1974*a*), bird may also threaten with neck extended forward, body held low but feathers sleeked. Appeasement indicated by slight trembling of tail and wing-tips; also performed by ♀♀ and young birds, least often by strong ♂♂ (Heinroth and Heinroth 1949). Nodding involves downward movements of head with bill at lowest point *c.* 85–90° to horizontal. Given in variety of contexts; variant associated with threat usually faster and more emphatic though more often incomplete. Jumping towards or away from another bird similarly indicates either sexual or aggressive excitement; particularly linked with Bowing-display (Goodwin 1956*b*, 1963, 1970). Defensive variant of Bowing-display normally indicates unwillingness to yield and may be accompanied and further

B

strengthened by short bout of Wing-twitching: abrupt movements of folded wings (often only one, held close to body) given at intervals of 0·25–1 s. Wing-twitching may also occur in heterosexual contexts and parent–young interactions (see below). Defensive-threat display (Fig B) indicative of attack–escape conflict: plumage ruffled, tail spread, one or both wings raised (if both, tendency to flee predominant), and head drawn in and pointed at assailant. Bird may leap into the air with quick downstroke of wings (Goodwin 1970). Initial part of fight which frequently ensues consists of sparring with wings (Wing-cuffing); Wing-twitching probably an intention movement to Wing-cuff (Heinroth and Heinroth 1949); powerful and swift blows dealt at opponent's head and neck with carpal joint of nearer wing, while other raised for balance. According to Spiteri (1975), dominant bird more likely to employ downward flick of wing, while upward or outward movement characteristically used in defence; see also Townsend (1915). Usually followed by more damaging sequence with pecking (mainly at opponent's face), attempts to grip other's head or neck, shaking, interlocking of bills, buffeting and, at high intensity, treading on other's tail; blood may be drawn and feathers fly. Dominant bird may clamber on to other's back, continuing pecks at head. Retreat by opponent to territory limit may mark end of fight but Bowing-display and a crouched threat posture much in evidence on limit (Townsend 1915; Heinroth and Heinroth 1949). For chasing by dominant ♂, see Spiteri (1975). Fight may last several minutes and during its course ♂ may perform Display-flight; on his return, Nodding suggests fresh attack imminent. ♀ (if present) remains impassively by nest; ♂ usually directs attention at her (strutting and cooing) once intruder expelled (Townsend 1915; Sengupta 1976). According to Murton *et al.* (1974*a*), mate often returns to assist in territory defence, when only one bird in attendance. ♀ exceptionally performs variant of Bowing-display to ♂ after territorial dispute with neighbours (Goodwin 1970). Fight may result in weaker bird being forced off ledge, being completely exhausted, or dying (Townsend 1915). 3rd ♂ never intervenes (Heinroth and Heinroth 1949). Advertising-call (see 1 in Voice) usually announces presence of ♂ (exceptionally ♀) to potential mates. Commonly given after hostilities by dominant bird, though mutual challenging may develop once intruder has retreated into own territory (Goodwin 1970). Young ♂ attempting to establish territory gives Bowing-display (see above) and Display-call; fighting frequent, but younger bird may benefit eventually from greater tolerance shown by old ♂ to own ♂ progeny. At *c.* 40 days, young defend parental territory in simpler fashion: call given (see Voice) and attack follows immediately; method sometimes employed by older but weak birds in defence of territory against higher-ranking individuals (Heinroth and Heinroth 1949). Interference with copulation common, at least in domestic and feral birds. Attacks performed mainly by ♂♂ in full breeding condition (Heinroth and Heinroth 1949).

According to Goodwin (1970), shows strong link with Driving (see below), but not so according to Brown (1968). Interfering bird(s) may show alertness at pre-copulatory behaviour of pair; then runs or flies in, dislodges ♂, and attacks ♀. Rarely, copulates with ♀ if she remains crouched; disturber usually departs rapidly though chasing and further pecking may occur (see Heinroth and Heinroth 1949, Goodwin 1956a, 1970). HETEROSEXUAL BEHAVIOUR. (1) General. Maximum activity over c. 10 days before and during Driving (Heinroth and Heinroth 1949; see subsection 3, below). At Flamborough Head, pair-formation and pre-incubation behaviour lasts c. 8–14 days (Murton and Clarke 1968). Speed of pair-formation depends mainly on birds' age and condition (see Goodwin 1970; also Glutz and Bauer 1980). For behaviour of homosexual pairs, see (e.g.) Kotov (1978); for bigamous ♂, see Heinroth and Heinroth (1949). Some pairs remain together more or less throughout year, roost in close physical contact, and show more 'affection' (Heinroth and Heinroth 1949; Petersen and Williamson 1949; Gompertz 1957; Goodwin 1960, 1970). When sexually excited, bird may perform Wing-lifting, with primaries slightly spread and folded wings often partly open, especially on landing; usually given in response to conspecific bird approaching from above or behind and often linked with Parading (Goodwin 1956b, 1970; see also below). (2) Pair-bonding behaviour. In feral birds, many pairs formed before search for nest-site undertaken (D Goodwin). For claim that pair-formation begins only when ♂ has occupied nest-site and gives Advertising-call from it to attract ♀, see (e.g.) Heinroth and Heinroth (1949), Riddle (1971), and Kotov (1978). According to Goodwin (1970), in free-flying birds at least, ♀ more likely to become attracted first by Display-flight or sexual variant of Bowing-display. ♂ flies towards ♀ with Wing-clapping (probably abbreviated Display-flight), also during closer approach when loud Display-call (see 2 in Voice) uttered; neck swollen, and feathers on neck, rump, lower back, and belly ruffled. When near ♀, ♂ lifts head high but bill still pointed slightly down. Bowing-display (Fig C) accompanied by fanning, closing, and depressing tail (even scraping it on ground); movements often in synchrony with raising of head. Feet raised higher than usual; bird rotates quickly, Bows low, and gives (elements of) Display-call, then raises head and turns away. May make occasional short hops towards ♀ and may Nod briefly during pauses. ♂ often runs in front of, then around ♀, if she moves away (Townsend 1915; Goodwin 1970; also Heinroth and Heinroth 1949, Fabricius and Jansson 1963). ♀ may respond by Nodding and Parading: head held high, wings slightly lowered and primaries spread a little; feathers on back and rump usually ruffled. Moves in characteristically high-stepping gait (Heinroth and Heinroth 1949, where, however, stated that it occurs typically after copulation; see below). ♂ then performs Mock-preening: head turned and bill thrust between scapulars and body, then quickly withdrawn to face forward again (cracking sound audible at close quarters: Fabricius and Jansson 1963); may also be directed at sides of breast (usually low intensity). Performed less commonly by ♀. (Based mainly on Goodwin 1970, which see for more interpret-

D

ative detail.) ♀ normally Allopreens ♂: usually gentle nibbling movements directed mainly at partner's head and nape; may serve to remove ectoparasites, though clearly also strong sexual stimulus. Often indicates partner not yet ready to mate (Goodwin 1970). (3) Driving. Directed by ♂ towards mate; primary aim (contra Heinroth and Heinroth 1949) not to Drive ♀ to nest but rather to remove her from possible rivals and copulate undisturbed (but see also Hollander 1945 and Spiteri 1975). Performed frequently by wild and feral birds where numbers gathered at feeding grounds (Goodwin 1970); ceases with laying (Heinroth and Heinroth 1949). ♂ Drives ♀ by chasing her with long, rapid steps (Fig D); in more intense form, follows her almost everywhere, treading on her tail and pecking at her head. May also Drive in flight, following close behind and above; normally ceases when c. 30 m from conspecific birds. No call given during Driving (Heinroth and Heinroth 1949; D Goodwin). For further details and discussion, see (e.g.) Whitman (1919), Diebschlag (1941), Fabricius and Jansson (1963), Goodwin (1963), Spiteri (1975), and Simms (1979). (4) Courtship-feeding (Billing). According to Heinroth and Heinroth (1949), solicited by ♀ with gentle pecking movements at base of ♂'s bill (Begging). Otherwise, Allopreening followed by ♀ inserting bill into that of ♂; both then perform 'regurgitating' movements, more pronounced in ♂ (Fabricius and Jansson 1963; Spiteri 1975). Act more symbolic as food passed only rarely or in small amounts (Goodwin 1946, 1970; Heinroth and Heinroth 1949; Murton et al. 1969). (5) Mating. May occur when ♀ Driven by ♂ suddenly stops, turns, and Begs from him (Heinroth and Heinroth 1949). Birds may run behind one another in circle, ♂ Mock-preening intermittently. ♀ adopts pre-copulatory posture, crouching with wings closed but slightly raised. ♂ leaps up from side and ♀ cocks tail, base of which ♂ grips with bent leg, after flapping wings and wagging tail to achieve cloacal contact. Afterwards, ♂ dismounts to one side (Heinroth and Heinroth 1949). According to Spiteri (1975), overbalances from thrusting movements and may peck at head or neck of still crouching ♀. Otherwise, ♀ straightens up in one movement and Parades (Heinroth and Heinroth 1949); ♂ often also (Goodwin 1970). ♂ then commonly flies off in Display-flight, ♀ following ; ♂ rarely gives Bowing-display immediately after dismounting. No calls during mating. Shortly before laying, ♂ often crouches apparently for reversed copulation after dismounting; may be elicited by ♀ Allopreening him or pecking at his back. Mock-preening and Courtship-feeding generally absent but claimed by Heinroth and Heinroth (1949) that reversed copulation does involve insemination. Pair disturbed during Mating slow to initiate new sequence. Copulation may take place at certain favoured sites but never in nest (Heinroth and Heinroth 1949; also Goodwin 1970; for further detailed description, see Spiteri 1975). According to Kotov (1978), mating occurs 5–7 times per day during nest-building, 2–3 times per day 1–2 days before laying, 1–2 times between 1st and 2nd egg. See also Sengupta (1976). (6) Nest-site selection. ♂ may have occupied site before courtship started. Pair may otherwise search together, though ♂ normally takes initiative (Goodwin 1970). (7) Behaviour at nest. In Nest-calling (most common during early stages of pre-incubation phase: Spiteri 1975), ♂ at nest-site adopts a crouched posture, tail and rump slightly raised, head

C

lowered, and gives Advertising-call (Goodwin 1970). Swollen throat moves rhythmically with call (Heinroth and Heinroth 1949). ♀ usually hops or runs in with tail fanned; posturing similar to that of ♂ in sexual variant of Bowing-display. ♂ may Nod and Wing-twitch: both wings usually moved spasmodically if ♀ approaches directly from front. Wing-twitching less frequent and less intense in ♀; given more by ♀ later in cycle (Spiteri 1975). ♀ may be driven away initially, particularly by younger ♂, using Wing-twitching and lunging with bill. ♀ may also Allopreen ♂ or peck more violently at his head (Fig E); in latter case, ♂ pushes head under ♀'s breast, and role may be reversed but head always focal point of attack and thus concealed. Both birds may then sit close together in nest; give Advertising-call ('Nest-call': see 1 in Voice), or (mainly ♀) variant (see 8 in Voice); much mutual Allopreening typical at this time, and Nodding and Wing-twitching also given. Allopreening and Nest-calling held to play important role in cementing pair-bond. Nest-calling indicates site shown to mate for approval; given most intensely by newly paired or pairing birds at newly discovered site, much less by well-established pairs starting 2nd or subsequent brood (Goodwin 1970). Scraping, scratching, and rotating movements often executed by ♂ in early stages, and both may initially bring in material. Building more systematic once ♀ has squatted at selected site; she may urge ♂ off nest by pushing head under his breast. Shortly before laying, ♂ brings material; for presentation to ♀, see Fabricius and Jansson (1963). ♀ does most building and often Nest-calls and Nods. According to Kotov (1978), ♂ attends ♀ closely during 3-4 days before laying. ♀ usually remains on nest for some time before laying. Before 2nd egg laid, birds stand alternately over 1st without incubating (Heinroth and Heinroth 1949). ♂ then usually incubates from about midmorning until c. $1\frac{1}{2}-2\frac{1}{2}$ hrs. before dusk, ♀ at other times (D Goodwin); for simultaneous incubation at 2 nests by birds of a pair, see Simms (1979). Nest-relief without ceremony; incoming bird may gently urge partner off nest. Particularly at hatching, bird (usually ♀) reluctant to leave; ♂ may bring nest-material, nibble at ♀'s bill, or push. ♀ may call (see 4 in Voice), Wing-twitch, and lunge at ♂ (Heinroth and Heinroth 1949; Fabricius and Jansson 1963; Goodwin 1970). According to Sengupta (1976), off-duty bird usually remains close to nest; but see Roosting (above). For behaviour at start of a subsequent brood, see Heinroth and Heinroth (1949). RELATIONS WITHIN FAMILY GROUP. See also Bonds (above). Adult does not assist with hatching (Heinroth and Heinroth 1949). Young usually fed within 1-2 hrs of hatching (Goodwin 1970); 1st chick after c. 4-6 hrs, 2nd after c. 24 hrs (Kotov 1978; see also Spiteri 1975). Initially, parent may make shaking movements with bowed head and lightly touch chick's bill, back, or neck to stimulate response; also gives Feeding-call (Kotov 1978: see 6 in Voice). When chick raises head, adult gently takes its bill and regurgitates crop milk; other food mixed in from c. 5-7 days (Heinroth and Heinroth 1949; Goodwin 1970; Kotov 1978). Striking colour pattern of chick's bill may provide stimulus in poor light. During feeding, chick 1-3 days old only gives food-call; from 3rd day, also pushes bill up more vigorously towards parent's and may shake head (Kotov 1978). From 7th day, begs persistently and loudly, stretching up and pushing towards arriving adult which gives Announcement-call (see 5 in Voice). Older young flap wings against parent's back, apparently to stimulate regurgitation. Young also given water from crop (Heinroth and Heinroth 1949; Goodwin 1954a, 1970). If one nestling weak or noticeably smaller, other may open wings and swell up to prevent it from feeding (Goodwin 1954a, 1970; Kotov 1978). 2 young chicks may be fed simultaneously; parent usually not able to cope with this later. Wing-twitching of adult may be given in response to over-zealous begging of young (Heinroth and Heinroth 1949). Normally brooded constantly for c. 7 days; not during day from 7th or 8th day if warm. Able to thermoregulate at c. 9 days, but usually sit in close physical contact. No resumption of brooding, once it has ceased, if weather turns cold. Adults generally stay away from nest when not feeding (Kotov 1978). Adults first recognize own young at about fledging time. Feral birds often ferociously attack trespassing juveniles, while domesticated birds readily adopt nestlings of roughly same age and appearance as their own, but usually hostile towards fledglings (Goodwin 1970). According to Heinroth and Heinroth (1949), young able to recognize parents when about half grown. In 5 cases, USSR, where nests only c. 1-1.5 m apart, 2 young, when mobile moved into neighbouring nest where all 4 fed by both sets of parents in turn (Kotov 1978). Allopreening may occur between siblings still in nest; may follow bout of aggression. Rudimentary heterosexual behaviour also apparent. Adults may also allopreen young; more typically done by ♂ with fledgling (Goodwin 1947a, 1970; Heinroth and Heinroth 1949). Young of 1st brood normally evicted when 2nd brood hatches (♀ less tolerant than ♂); sight of newly hatched young probably arouses hostility towards 1st brood, so that poorly-lit site may allow them to remain near nest (Goodwin 1970; also Heinroth and Heinroth 1949, which see for further interactions between ♂ and young of 1st brood). Where conditions allow, leave nest at c. 25-26 days but return there to roost and are often fed on or near it (Riddle 1971; Kotov 1978). In Faeroes, finally leave colony at c. 5 weeks (Petersen and Williamson 1949). According to Gompertz (1957), c. 2-3 weeks after leaving nest, young feral birds tend to follow ♂ to his feeding grounds. May first attempt to take objects from adult's bill whilst begging; quickly learn to feed properly (Goodwin 1954a; see also Riddle 1971). Young initially stay in or near natal area (Bruns 1959). ANTI-PREDATOR RESPONSES OF YOUNG. Small chicks (c. 4-8 days old) tend to crouch low in nest (Kotov 1978). From c. 9th-13th day, when threatened from front, bird rears up with feathers ruffled, especially on neck; neck inflated and deflated; accompanied by Bill-snapping, and puffing or hissing sounds. May also lunge and peck at intruder and attempt to strike with wings; ♂♂ use such tactics earlier and more readily than ♀♀ (Spiteri 1975; Kotov 1978). At c. 20 days, sometimes leap out of nest if disturbed. At time of fledging, exceedingly timid, fleeing whenever threatened; call loudly when seized (Heinroth and Heinroth 1926-7, 1949; Goodwin 1970; Kotov 1978; see also Voice). On hearing warning call (not described, but see 3 in Voice) of parent during feeding, nestlings fall silent temporarily, then resume begging (Kotov 1978). PARENTAL ANTI-PREDATOR STRATEGIES. (1) Passive measures. Feral birds wary at nest-site only if persecuted by man (D Goodwin); may flee even at night when most Columbidae loath to fly (Bruns 1959; Reinke 1959), though apparently then unable to find their way back (Heinroth and Heinroth 1949). In Hamburg, small colony deserted when pair of Kestrels *F. tinnunculus* settled to breed nearby (Volkmann 1959; see also Bruns 1959). (2) Active measures: against man. Generally sits very tight c. 10 days either side of hatching (Lees 1946). At close approach of intruder, may initially assume slightly squatting posture with ruffled feathers (Spiteri 1975); otherwise, performs highest-intensity Defensive-threat display and may call, before attacking with bill and carpal joints (Heinroth and Heinroth 1949; Fabricius and Jansson 1963; Goodwin 1970). Bird flushed from hatching eggs, South Uist (Scotland), performed low-intensity Display-flight at c. 50 m from man; at colony, London, where men working, birds attempting to return to it reacted similarly on or immediately after veering away (Goodwin 1970). During feeding, parent may periodically give warning call (Kotov 1978; see above).

(Figs A and B based on drawings in Goodwin 1970; Figs C and D from photographs in Simms 1979; Fig E from photograph in Fabricius and Jansson 1963.)   MGW

Voice. Calls of wild birds of Egypt and Libya more musical and higher pitched, less deep and gruff than large domestic breeds (Goodwin 1955). ♀ generally calls less often and less loudly than ♂; does not normally give full intensity sexual form of Display-call (Goodwin 1956b; see also below). No contact or flight calls (Heinroth and Heinroth 1926–7, 1949). For Wing-clapping during Display-flight and at other times, and for sounds made by young threatening predators, see Social Pattern and Behaviour.

CALLS OF ADULTS. (1) Advertising-call (Song or Nest-call). Characteristically given by territorial ♂. A moaning 'ōōrh' or 'oh-oo-oor', subject to variation. Also used by both sexes as Nest-call (Goodwin 1970; see also below). Slight rise in pitch in middle of moan; audible above sound of surf or city traffic. May also be heard outside territory (Simms 1979). A slightly hoarser, softer, rattling variant, lacking moaning quality, given occasionally by sexually excited ♀ in nest (Heinroth and Heinroth 1949; also Fabricius and Jansson 1963). Described by Spiteri (1975) as a simply structured 'aoo', though can be rich in harmonics; sequence may last c. 1 min and temporal intervals between calls more or less constant. In recording of 2 birds calling together at nest, presumed ♂ gives moaning 'oorh...' at c. 1 call per s (Fig Ia) while presumed ♀ gives hoarser, more clipped 'curh...' or 'urh...' sounds (Fig Ib), almost twice as fast (J Hall-Craggs). (2) Display-call. A rather hurried-sounding 'oo-rŏŏ-cŏŏ t'cōō'; may become blurred, with final unit emphasized and more drawn-out, in intense sexual variant of ♂'s Bowing-display (Goodwin 1954b, 1956b, 1970); 'oo-r-ooooo-t-ooo' (J Hall-Craggs: Fig II). Given at intervals of c. 1 s, more excited variant having rolling guttural 'rrr' in middle (Simms 1979). Given in heterosexual and antagonistic contexts (also by captive ♀♀ of wild stock); see Heinroth and Heinroth (1926–7, 1949). See also Riggenbach (1961–2) and, for different classification and nomenclature, Spiteri (1975). (3) Distress-call and related vocalizations. A gasping, grunted 'ōōrh'. Drawn-out, loud, panting variants given by timid domestic or captive birds in nest defence; extreme variant almost a high-pitched scream, e.g. a harsh 'hinnnh', usually from ♀ (Spiteri 1975; see also below). Also given as warning call. Low-pitched grunting 'coo' expressing fear or annoyance; given (e.g.) when flying raptor sighted. Apparently, no specific ground-predator alarm call (Goodwin 1954b, 1970). In fright rather than as warning, 'wao' (Fabricius and Jansson 1963) or short emphatic 'ru' like

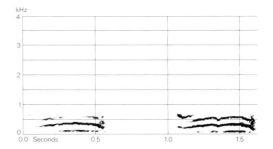

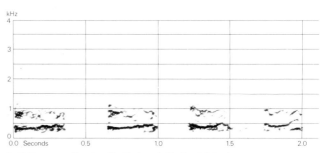

I T Gompertz/Sveriges Radio (1972) England March 1955

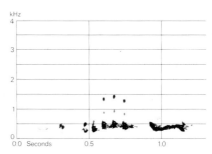

II T Gompertz/Sveriges Radio (1972) England March 1955

Woodpigeon *C. palumbus* (Heinroth and Heinroth 1926–7, 1949). (4) Defence-call of ♀. A sharp 'rúckuh' (Heinroth and Heinroth 1949; see also below). (5) Announcement-call. A grumbling sound, audible only at close quarters. Given by adult arriving to feed older young (Heinroth and Heinroth 1949). (6) Feeding-call. A hoarse, soft 'au' from adult ready to feed young (Heinroth and Heinroth 1949). A tender, inviting 'coo', given only in this context (Kotov 1978). According to Goodwin (1956b), adult gives Advertising-call (call 1) when ready to feed older young. (7) A low fretful murmuring sound, audible only at very close range and given apparently when bird upset or angry (Goodwin 1970). (8) In appeasing contexts, ♀ may give form intermediate between calls 1 and 3 towards ♂; more rarely vice versa (Goodwin 1970).

CALLS OF YOUNG. A thin, squeaky, sibilant wheezing 'weeeee erweeeee' as food- or appeasement-call (Goodwin 1970; Simms 1979). Variably loud, depending on degree of excitement but rather loud and persistent compared with other Columbidae (Heinroth and Heinroth 1949). Barely audible 'cheep' sounds gain in strength from c. 4–7 days. Clicking sound made when bill opened at beginning of each call; disappears at c. 8 days. Calls have remarkably large frequency range though this is reduced with age. Quiet 'cluck' sounds may accompany threat behaviour at c. 10–11 days (Spiteri 1975). First attempts at adult calls given at or just before fledging; wheezing sounds uttered in same rhythm as 'coos' of adult (Goodwin 1970). At c. 7–8 weeks, voice breaks and birds practically mute for c. 1 week, although occasionally utter high- and low-frequency sounds in succession; gradually change to typical adult low-frequency sounds (see adult call 3) then, at c. 9 weeks, rhythm is characteristic of adult (courtship) calls (Abs 1975, which see also for hormonal influence). According to Goodwin (1970), wheezing sounds still given as food- or appeasement-call when bird calls more or less like adult, but proper sound and intonation not yet established. At c. 40 days, may call 'rúckuh' in defence of parental territory (Heinroth and Heinroth 1949: see adult call 4). According to Spiteri (1975), 'Bowing-calls' (see adult call 2) of young ♂♂ have softer warbling quality at c. 3–4 months; at c. 3–5 months old (i.e. before sexual maturity), ♀♀ give irregular variant, and more often than adult ♀♀ (see above).   MGW

Breeding. SEASON. Britain and Ireland: see diagram. Breeds throughout the year in Britain: in Scotland, peak in April and minimum in July; in Humberside, peaks April–June and August–September, and only about one quarter of population breeds in winter (Lees 1946; Murton and Clarke 1968). Faeroes: late March to early October, but most mid-April to July (Petersen and Williamson 1949). Mediterranean: eggs laid March–July (Makatsch 1976). Cyprus: eggs laid 20 March–7 May (Bannerman and Bannerman 1958). Canary Islands: breeds end of April to mid-July. Madeira: breeds March–July. Cape Verde Islands: known to breed January–April but probably in other months too (Bannerman 1963b; Bannerman and Bannerman 1965, 1968). SITE. On natural or artificial ledge, or in hole inside cave or deep crevice of cliff. Colonial. Nest: loosely constructed cup of roots, stems, and leaves, small pieces of driftwood, seaweed, and feathers; no true lining. Building: by both sexes, but mainly by ♀ (Petersen and Williamson 1949). EGGS. See Plate 95. Sub-elliptical, smooth and slightly glossy; white. 39×29 mm (36–43×27–32), $n = 80$; calculated weight 18 g (Schönwetter 1967). Clutch: 2, occasionally 1. Of 392 clutches, England: 1 egg, 4%; 2 eggs, 96%; average 1·93 (Murton and Clarke 1968). Probably at least 5 broods (Murton and Clarke), perhaps more (Lees 1946). Laying interval 48 hrs (Kotov 1978). INCUBATION. 16–19 days. By both sexes, beginning with 1st egg; hatching asynchronous. YOUNG. Semi-altricial and nidicolous. Cared for and fed by both parents; crop milk given to nestlings at first, but reduced as young grow and natural food brought in (Murton and Clarke 1968). FLEDGING TO MATURITY. Fledging period 35–37 days (Lees 1946; Petersen and Williamson 1949); c. 25 days (Murton and Clarke 1968). When parents lay further clutch, young become independent at fledging or soon after; otherwise tended by parents for c. 10 days after fledging (Lees 1946; Murton and Clarke 1968). Age of first breeding 1 year for wild birds, but feral birds may breed at c. 6 months (Murton et al. 1974). BREEDING SUCCESS. Of 812 eggs laid, Humberside, 1965–6, 541 (66·6%) hatched and 384 young (47·3% overall) fledged; egg loss due to predation (18·7%), desertion (9·2%), and infertility (5·3%); young lost through predation or disappearance (8·1%), dying in nest (perhaps mainly from starvation), and flooding (20·9%); an average pair laid 5 clutches per year, hatched 4 of them, and reared 5 young to flying; of 29 clutches of 1 egg, 38% hatched and 48% were predated before hatching, while of 363 clutches of 2 eggs, 67% of eggs hatched and 19% were predated; hatching success showed little seasonal variation but fledging success, and therefore overall rearing success, best May–October, when twice as good as in winter months, with losses of young mainly from starvation at times of food shortage (Murton and Clarke 1968). Of 327 eggs laid Manchester (England), 78% hatched, 12% infertile, and 10% lost or predated; eggs laid by pairs of dissimilar plumage were significantly more likely to hatch than eggs of pairs of similar plumage (Murton et al. 1973b).

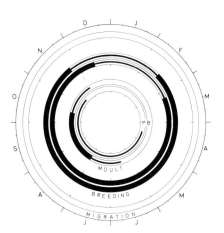

Plumages (nominate *livia*). ADULT. Head and nape plumbeous-grey, slightly glossed green or purple on ear-coverts. Broad glossy ring round neck, from lower nape, hindpart of cheeks, and lower throat down to upper mantle and chest; strongly glossed green or deep purple (depending on angle of light), usually green on hindneck and upper foreneck, usually purple on upper mantle and chest. Lower mantle, scapulars, and lesser and median upper wing-coverts light bluish-grey or ash-grey. Upper back pale grey, lower back and upper rump white, lower rump and upper tail-coverts dark bluish-grey or plumbeous-grey. Underparts below chest medium bluish-grey, under tail-coverts dark bluish-grey. Tail medium grey, c. $2\frac{1}{2}$–3 cm of feather-tips black, extreme tips often slightly grey; middle portion of outer web of t6 white, extreme base and narrow outer edge grey. From below, tail black, except for light grey middle portion of outer web of t6. Primaries, greater upper primary coverts, and bastard wing medium grey, slightly darker towards tips, slightly paler towards bases; outer webs tinged light silvery-grey when fresh. Both webs of tertials and outer webs of secondaries (inward from about s6) black, tips ash-grey, black forming band c. $1\frac{1}{2}$–2 cm wide across spread wing; outer secondaries medium grey with black or greyish-black tips c. 2–$2\frac{1}{2}$ cm wide. Greater upper wing-coverts like inner secondaries; black bases visible as band c. 1 cm wide across spread wing, this band about double the length of broader one across tertials and inner secondaries. 1–3 outer greater coverts variable: uniform ash-grey, or ash-grey with solid black patch or black freckling on basal outer web. Lesser upper primary coverts light bluish-grey or ash-grey, like median and lesser upper wing-coverts. Under wing-coverts and axillaries white, lesser coverts along leading edge of wing light or pale grey, greater under primary coverts and undersurface of flight-feathers pale silvery-grey, darker grey on tips of flight-feathers and on outer web of p10. Sexes similar. DOWNY YOUNG. Down sparse and coarse; yellowish with reddish tinge. JUVENILE. Head, neck, upper mantle, and chest slate-grey with brown tinge, not or hardly glossy, sometimes with rufous feather-edges; lower mantle, scapulars, and upper wing-coverts paler grey than adult and with brown tinge, often narrowly fringed white or pale buff; lower rump feathers and upper tail-coverts duller grey, faintly fringed brown when fresh; underparts duller and browner grey than adult, feathers sometimes narrowly fringed white. White of outer web of t6 usually extensively tinged grey. Black of tail-tip and (especially) bands across greater upper wing-coverts, tertials, and inner secondaries duller and browner, less sharply defined; ash-grey tips to greater tertial coverts and tertials usually greyish-brown and partly freckled dark grey rather than pure ash-grey as in adult. T6 21–27 mm wide at tip (26–33 mm in adult). Flight-feathers slightly browner than adult, tips narrowly fringed white. Forehead, gape, and chin bare. Feathers appearing last (face, part of head, hindneck, mantle, outer scapulars, and some rows of longer lesser upper wing-coverts) intermediate in shape and colour between adult and juvenile: purer grey and less brownish than in juvenile, but not as long and broad as in adult; neck feathers usually glossed green or slightly purple on tips; these feathers mostly replaced again later on in post-juvenile moult; not known whether this should be considered late juvenile, early full adult, or a separate immature plumage. FIRST ADULT. Like adult, but variable amount of juvenile feathers retained. Not recognizable after last tail-feathers, secondaries, and p10 replaced. In least advanced birds, many narrow and short brownish-grey juvenile feathers contrast with wider and longer purer bluish-grey adult feathers, especially on upper wing-coverts; black pattern of inner greater upper wing-coverts, tertials, and inner secondaries brown and not sharply demarcated from narrow brownish feather-tips, contrasting with wider and clear-patterned bluish-grey and black feathers; some tail-feathers still juvenile, shorter, narrower, and more worn than fresh 1st adult ones. More advanced birds retain some juvenile flight-feathers only; outer primaries browner and more sharply pointed than blacker and more rounded new inner ones; when not too worn, juvenile p8 6–12 mm shorter than longest (p9), p7 22–30 (in adult, p8 mainly 2–7 and p7 17–23). Last juvenile feathers to be lost are some secondaries; colour not reliable for ageing, as adult may also retain old faded secondaries, but in at least some birds retained juvenile secondaries are recognizable by shape—shorter, narrower, and more rounded or sharply pointed at tip than adult, not all equal in length, width, and shape.

Bare parts. ADULT. Iris orange, red, or golden-orange with paler yellow inner ring. Skin round eye bluish-grey, similar or slightly paler than surrounding feathers (pink in some feral birds). Bill black or greyish-black, cere with white powdering. Leg and foot dark pink-red to purple-red, claws blackish. DOWNY YOUNG. Iris reddish-grey. Bill laden-grey, tipped light flesh-colour. Leg and foot leaden-flesh or greyish-pink. JUVENILE. Iris pale grey, dull brown, ochre, or yellow. Skin round eye pale or dark grey. Bill leaden-grey, flesh-colour of tip gradually disappearing and white powdering of leaden-grey cere gradually developing with post-juvenile moult. Leg and foot partly grey or fully reddish with brown, pink, or orange tinge. (Witherby *et al.* 1940; Goodwin 1970; Harrison 1975; Glutz and Bauer 1980; BMNH.)

Moults. ADULT POST-BREEDING. Closely similar to domesticated birds and to Eastern Rock Dove *C. rupestris* according to Kotov and Noskov (1978), which see for details. Complete; primaries descendant. Start strongly variable, individually as well as with locality. Starts with p1, in most birds May–June, but some from January (a few from Ahaggar and Canary Islands), others not until July or early August (some from western Europe, Caucasus, and Turkey). Moult completed with p10, mid-October to December(–February); total duration of primary moult $5\frac{1}{2}$–6 months (Kotov and Noskov 1978) or 7–8 months (Stresemann and Stresemann 1966). Of 64 moulting adults examined, 81% started primary moult April–May and completed mid-October to early December. Secondaries, tail, and body start mainly with loss or regrowth of p5, completed at about same time as p10; most intense August–September (Kotov and Noskov 1978) or September–October (Dementiev and Gladkov 1951b), and breeding cycle then usually interrupted. Tail starts with t1 or t2, completed with t4 or t5. Secondaries start with innermost. In feral birds, primaries may start from April–May, moult slow or suspended during peak of breeding season, resumed July and all birds examined had started by August; moult August–November more rapid than in spring, with p10 new (September–)November–December, or (after suspension in midwinter) February–March (Murton *et al.* 1972b; Glutz and Bauer 1980). POST-JUVENILE. Even more variable than adult. Starts with face, mantle, outer scapulars, and longer lesser upper wing-coverts (see Plumages), usually soon followed by p1 at age of c. 50 days (Heinroth and Heinroth 1926–7); however, some birds show largely 1st adult body plumage with all primaries still juvenile, especially in winter. In general, plumage 1st adult when primary moult score of 30 reached, except for some inner tertials, outer or middle secondaries, and some tail-feathers which are usually replaced at same time as outer primaries. Depending on hatching date and locality, may start in any month, except perhaps November–December. Thus, Canary Island birds examined had lost p1 between late January and September and had all primaries new October–June; in more temperate areas like western Europe, starts May–October, and timing of early-hatched birds similar to adult post-breeding, but late-hatched juveniles delayed until spring. Primary moult usually suspended at any stage during (October–)November–February(–March), resuming in spring. As in other *Columba*, serially descendant moult (next series of primaries starting with p1 when post-juvenile moult still active in outer primaries) occasionally recorded (Kotov and Noskov 1978).

Measurements. All year; skins (BMNH, RMNH, ZMA, ZFMK, ZMM).
    Nominate *livia*
Adult wing of (1) Britain, (2) southern Europe (Iberia, Balearic Islands, Sardinia, Italy, Yugoslavia, and Greece); other measurements combined.

WING (1) ♂ 228 (4·57; 16) 220–235   ♀ 219 (3·57; 15) 214–227
     (2)   226 (5·37; 11) 219–234   219 (4·03; 6)   213–223

JUV		223	(5·98; 11)	215–232	216 (3·52; 10)	211–223
TAIL AD		114	(4·21; 9)	108–122	111 (3·65; 5)	106–116
BILL		19·0	(0·75; 13)	18·0–20·5	18·6 (0·63; 12)	17·5–19·4
TARSUS		30·6	(0·82; 13)	29·7–32·2	29·5 (1·40; 12)	27·8–32·0
TOE		34·4	(0·75; 10)	32·8–35·7	32·8 (1·40; 11)	31·0–35·0

Wing of adults of other populations provisionally included in nominate *livia* (see Geographical Variation): (3) Canary Islands, (4) northern Algeria and Tunisia, (5) Cyrenaica (north-east Libya) and Libyan plateau in north-west Egypt, (6) Asia Minor and Georgian SSR (USSR), (7) northern Caucasus (Krasnodar to Dagestan), (8) Iraq.

WING (3)	♂	224	(5·87; 11)	216–232	♀ 216 (3·68; 12)	211–220
(4)		224	(6·92; 6)	218–233	219 (5·78; 7)	212–226
(5)		221	(5·11; 5)	216–227	216 (5·32; 8)	209–224
(6)		234	(6·33; 8)	226–241	221 (7·47; 6)	216–230
(7)		234	(2·55; 5)	231–238	222 (3·29; 8)	218–226
(8)		220	(2·48; 6)	217–222	218 (2·22; 4)	216–221

C. l. palaestinae. Syria, Jordan, and Israel.

WING AD ♂ 221 (3·78; 11) 216–225 ♀ 217 (7·15; 6) 210–224

C. l. schimperi. Unsexed birds, Nile delta and El Faiyum, adult wing 208 (4·45; 6) 200–213. Others, Egypt (Vaurie 1965):

WING ♂ 206 (— ; 8) 194–217 ♀ 200 (— ; 10) 195–209

C. l. dakhlae. Dakhla oasis (Egypt): wing adult ♀, 203 (4·86; 5) 197–207

C. l. targia. Ahaggar (Algeria), Tibesti, and Ennedi (Chad).

WING AD ♂ 214 (4·61; 12) 209–219 ♀ 207 (3·47; 12) 202–213

C. l. gymnocyclus. Sénégambia, Mali, and Ghana.

WING AD ♂ 216 (3·83; 7) 211–220 ♀ 203 (— ; 3) 202–203

Weights. Data on wild birds limited.

Nominate *livia.* Shetland (Scotland): 2 unsexed, 302 (adult, April), 238 (1st year, November) (BTO). Turkey: June, ♂ 370 (Kumerloeve 1969); July, ♂ 355, ♀♀ 285, 325 (ZFMK). Canary Islands: October–December, ♂ 233, ♀ 200 (ZFMK).

C. l. targia. Ahaggar, December: ♂ 250, ♀ 235 (ZFMK). Ahaggar, ♂ 248–250 (3), ♀ 202–248 (7); Ennedi, ♂ 180–220 (5) (Niethammer 1955).

C. l. gaddi. Kazakhstan (USSR), June–July: ♂ 280 (5) 258–300, ♀ 314 (6) 243–360 (Gavrin et al. 1962).

C. l. intermedia. Hopeh (China): ♂ 294 (19) 193–347, ♀ 267 (14) 195–320 (Shaw 1936).

Structure. Wing rather long, pointed; relatively broad at base. 11 primaries: p9 longest; in adult, p10 1–6 shorter, p8 2–7, p7 17–23, p6 31–42, p5 44–55, p1 86–102; in juvenile, p10 2–6 shorter, p8 6–12, p7 22–30, p6 34–49, p5 51–66, p1 92–106; p11 minute, concealed by primary coverts. Tail rather short, slightly rounded; 12 feathers, t6 6–14 shorter than t1. In adult, tail-feathers slightly widened at tip, maximum width of t1 near tip 31–42, t6 26–33; in 1st adult (with new tail), t1 30–35, t6 27–31; in juvenile, sides of feathers parallel, not widened at tip, width of t1 at tip 26–31, t6 21–27. Bill slender, shorter than head, rather soft except for nail and cutting edges; base of upper mandible with a soft cere, appearing powdered white in adult, partly covering slit-like nostrils. Crown of ♂ higher than in ♀ (Goodwin 1954a). Tarsus stout and short, upper front feathered. Toes rather stout and short; outer and inner toe *c.* 78% of middle, hind toe *c.* 56%. Claws strong, short and curved.

Geographical variation. Marked; involves size, depth of grey on body, and (independent of body colour) colour of rump. Depth of grey depends on wear (paler in fresh plumage, duller when worn) and only equally fresh birds should be compared. Typical nominate *livia* intermediate in size and combines dark body with white rump; occurs Britain and Ireland, France, southern Europe (Iberia to Balkans and Greece, including islands in western Mediterranean), Ukraine, Crimea, Rostov, and western Kazakhstan north to southern Urals. Certain other populations here provisionally included in nominate *livia*, though quite heterogeneous; perhaps better split into several races, though no names available. These comprise (1) population inhabiting northern Caucasus, with colour similar to typical nominate *livia* but size larger (see Measurements); (2) birds of Asia Minor and Georgian SSR, with large size similar to Caucasus birds, but body slightly paler grey and rump white, rarely pale grey, grading into smaller and paler birds of south-central Turkey and into large and pale birds (*gaddi*) of extreme eastern Turkey, Armeniya, and Azerbaydzhan; (3) birds from Canary Islands and (apparently) inland Madeira (Bannerman and Bannerman 1965) with colour and size close to typical nominate *livia*, slightly narrower rump-patch sometimes partly suffused pale grey, and perhaps slightly darker underparts sometimes separated as *canariensis* Bannerman, 1914, but differences very small; (4) populations of northern Africa from Morocco and northern Algeria east to Libyan plateau in north-west Egypt, with size similar to or slightly smaller than typical nominate *livia*, body slightly paler grey both above and below, like birds of Asia Minor (darkest birds in these populations similar to palest ones of typical nominate *livia*), and rump white, but frequently (6 out of 10) pale grey or greyish-white in Tunisia; (5) birds inhabiting Cyclades, Crete, Rhodes, and Cyprus, with size and body colour similar to North African birds, and rump white. Population of Iraq also provisionally included in nominate *livia*, but slightly smaller and body slightly paler grey than typical nominate *livia*; in both these characters similar to birds of Cyrenaica (Libya) but rump greyish-white to pale grey, sometimes similar to colour of mantle, only rarely (1 out of 10) white, unlike birds of Cyrenaica. *C. l. palaestinae* of Syria, Lebanon, Israel, and Jordan paler than any of previous populations; about similar in colour to *gaddi* of Iran eastward and to *schimperi* of Nile valley; mantle, scapulars, and upper wing-coverts pale grey rather than the light bluish-grey or ash-grey of typical nominate *livia*, underparts light bluish-grey rather than medium bluish-grey; rump usually pale grey, similar to colour of mantle, occasionally (5 out of 17) white or greyish-white, size rather small, clinally decreasing from northern Syria to Negev—average wing of ♂ in Syria 223, ♀ 221, Jordan and Israel ♂ 220, ♀ 215. Birds of western and southern Arabia close to *palaestinae* of Jordan and Israel in size and colour, those of eastern Arabia nearer *gaddi*. *C. l. gaddi* from Azerbaydzhan, Iran, Afghanistan, and Transcaspia east to Uzbekistan similar in colour of body to *palaestinae*, rump variable (white, greyish-white, or pale grey), but size large: average wing in western Iran ♂ 229, ♀ 225; further east, larger still, average wing for eastern Iran ♂ 234, ♀ 227½ (Vaurie 1965). *C. l. schimperi* of Nile delta and valley, Suez Canal area, El Faiyum, and north-east Sudan about as pale as *palaestinae* and *gaddi*, or even paler, especially below; rump nearly always pale grey (similar to mantle), rarely greyish-white; size much smaller than *palaestinae*. Sinai birds intermediate in size and colour between *palaestinae* and *schimperi*; wing single adult ♂ 214, ♀ 212. *C. l. dakhlae*, inhabiting Dakhla and Kharga oases in central Egypt, is palest race; mantle, scapulars, upper wing-coverts, and belly whitish-grey, rump white, upper tail-coverts pale grey, broad purple-green collar round neck and chest subdued by greyish feather-tips; size as *schimperi*. In contrast to this, another isolated race, *targia* from Saharan hill tracts, is rather dark: upperparts similar to typical nominate *livia*, underparts similar to upperparts (hence paler than in nominate *livia*, more like *gaddi*), rump grey like mantle, purple-green collar slightly duller; slightly larger than *schimperi* and *dakhlae*, but distinctly smaller than any population of nominate *livia*. Of extralimital races, *neglecta* similar in colour to typical nominate *livia*, but larger (average wing of ♂ 236, ♀ 228: Vaurie 1965) and rump pale grey to slate-grey, exceptionally white; *intermedia* has upperparts darker than nominate *livia*, rump dark bluish-grey, size as European nominate *livia*; *gymnocyclus* (synonym: *lividior* Bates, 1932) has body dark slate-grey (blacker in worn plumage, paler bluish in fresh), head slate-black with contrastingly red bare patch round eye, rump white or greyish-white, glossy collar round neck wider, green, extending up to throat, cheeks, and nape and down to upper flanks and faintly over breast. Position of birds separated as *atlantis* Bannerman, 1931, from Azores, Madeira, and Cape Verde Islands problematical: about similar to melanistic 'blue-chequer' or 'velvet' variant of feral birds with many black spots on upperparts (obscuring black wing-bars) and with white or grey rump, or completely black on mantle and upperwing; derived either from feral birds, or melanistic mutant of Canary Islands birds, to which it is of similar size. Similarly dark birds occur in apparently wild populations of other Atlantic islands like Faeroes and Islay, Scotland (Petersen and Williamson 1949). Feral birds strongly variable; some populations, descended from dovecote birds with hardly any influence on breeding by man, rather small in size (like nominate *livia*) and colour mainly like wild or blue-chequer birds; others are descended from homing and domestic birds and these generally larger in size (close to *gaddi* and *neglecta*, but bill heavier at base, cere more strongly developed, and large bare patch round eye) and strongly variable in colour. For variation in feral birds, see Goodwin (1954a); for occurrence of blue-chequer variants in feral population, see Murton and Clarke (1968), Parkin (1970), and Murton (1970). CSR

Columba oenas Stock Dove

PLATES 28 and 29
[facing pages 327 and 350]

Du. Holenduif Fr. Pigeon columbin Ge. Hohltaube
Ru. Клинтух Sp. Paloma zurita Sw. Skogsduva

Columba Oenas Linnaeus, 1758

Polytypic. Nominate *oenas* Linnaeus, 1758, Europe and North Africa east to northern Iran, Caspian Sea, and through south-west Siberia and northern Kazakhstan to Semipalatinsk. Extralimital: *yarkandensis* Buturlin, 1909, eastern Uzbekistan, Kirgiziya, and Tadzhikistan east through Tien Shan and Tarim basin (Sinkiang) to Lop Nor.

Field characters. 32–34 cm; wing-span 63–69 cm. About 25% smaller and slighter than Woodpigeon *C. palumbus* with much shorter tail; also slighter than Rock Dove *C. livia*. Medium-sized, compact but rather delicate, blue-grey pigeon, lacking obvious plumage characters but in flight showing diagnostic pale-centred, dark-rimmed wing. At all ages, lacks any striking white area of plumage. Flight confident but not fast. Sexes mostly similar; no seasonal variation. Juvenile separable.

ADULT. Head and most of upperparts and underparts grey, with blue tone on head and underbody and slate tinge on mantle and scapulars. Sides of hindneck glossy green (with amethyst sheen); lower throat and upper breast mauve-pink (lacking sheen), less intense than on congeners. Back, rump, and most of upperwing paler grey-blue, and unmarked except for 2 short black bars (and often a trace of a 3rd) on tertials and inner greater coverts; in flight, shows striking pale panel on upperwing, since pale grey centre surrounded by slate lesser coverts and dusky to black primary coverts, outer primaries, and broad tips to inner primaries and secondaries. Underwing wholly blue-grey, becoming dusky on trailing edge and tip. Tail grey with broad black terminal band. Bill yellow-horn, pink at base below mealy white cere; eye brown; orbital ring grey. Legs bright pink-red, duskier on some ♀♀. JUVENILE. Noticeably duller and browner than adult, lacking obvious green sheen on neck; rusty tone on breast. Small black bars on inner wing even more restricted than on adult.

Unmistakable in flight, when pale-centred, dark-rimmed wings and pale rump create subtle but distinctive pattern. Beware, however, some feral Rock Doves *C. livia* which may show rather similar appearance, including grey back and underwing. With bird on ground, confusion with *C. livia* (particularly juvenile) possible but pale bill, difference in inner-wing marks, and more delicate character and carriage quickly recognizable with experience. At all times, lack of white wing-crescents rules out confusion with *C. palumbus*. Flight actions close to those of *C. livia* but generally less dashing, not so fast and including characteristic soft flutter of wings (both on take-off and landing). Wings usually appear less tapering and less angled than in *C. livia*, and bursts of wing-beats recall *Streptopelia* doves. Gait less rapid on open ground than *C. livia*; noticeably agile on branches. Carriage less horizontal than *C. livia* or *C. palumbus*, with (when searching for food) characteristic half-upright stance—small head often held high and low belly close to ground.

Most characteristic call rather gruff, deep, and emphatic: 'ooo-wu', given 4–12 times.

Habitat. In upper and lower middle latitudes, continental and oceanic, centred on temperate zone but extending into boreal and Mediterranean, and marginally into steppe. Mainly lowland, but ranges freely up to *c*. 500 m and in places sparsely up to 1000 m or more. Flourishes best along the often narrow and shifting border between forest and open country, where well-grown trees with hollows or holes for breeding are within easy access of fields rich in weed species or crop seeds lying on preferably bare ground, and also conveniently near drinking water. Where ample forage and water are available at a distance from trees will adapt freely to breeding in rock crevices, rabbit *Oryctolagus cuniculus* burrows, or holes in buildings and ruins, or other artefacts. Conversely, avoids even suitable tree stands cut off from feeding range by rocky or mountainous terrain or by unbroken forest or wetlands. In USSR, prefers open forests or those interspersed with open stretches and glades, and has suffered through deforestation, although will use for breeding hollow trees of almost any species, broad-leaved or coniferous, but these must be close to suitable area for collecting food from ground (Dementiev and Gladkov 1951b). Although occasionally frequenting beaches, normally replaced on coasts by Rock Dove *C. livia* except where cliffs absent and sand-dunes or light soils afford it advantage for breeding sites. Preferred inland sites in western Europe are ancient parklands or avenues, spinneys and copses, riverside and hedgerow trees, and even city parks, especially where trees over-mature. Within the range of Black Woodpecker *Dryocopus martius*, its old large nest-holes provide additional opportunities. Readily accepts nest-boxes. Spread of agricultural weeds with expansion of farming in 19th and early 20th centuries enabled colonization of much additional territory, but widespread use of herbicides after *c.* 1950 led to serious reverses in parts of range (Murton 1971a).

Distribution. Expansion of range in parts of western Europe; in Britain, Ireland, Netherlands, and France from latter half of 19th century, and more recently in Belgium and Luxembourg. Range decreased in Finland and Bulgaria.

BRITAIN AND IRELAND. Marked expansion. In early 19th century confined to southern and eastern England; north and south-west England colonized 1870s, first known breeding Scotland *c.* 1856 and Ireland *c.* 1877, spread continuing at reduced rate until *c.* 1930s in England and in 1950s in Ireland. In late 1950s, marked decline (see Population) led to local extinctions, especially in much of eastern England; since recovered, but pre-1950 distribution not fully regained (Parslow 1967; Sharrock 1976; R J O'Connor and C J Mead). Occasionally breeds Channel Islands (Long 1981b). FRANCE. Some expansion—e.g. Marne *c.* 1870, Normandy 1925, Languedoc 1936, Saône-et-Loire 1947, and Finistère 1964 (Yeatman 1976). BELGIUM. Spread early 20th century (Verheyen and Damme 1967). LUXEMBOURG. Spread from *c.* 1939 (Glutz and Bauer 1980). NETHERLANDS. First bred 1880; in early 20th century, still local in centre and east; from 1930–8 colonized north and south (CSR). FINLAND. Marked contraction of range in recent decades (OH). BULGARIA. Range decreased (Z Spiridonov). ALGERIA. May breed Djebel Babor, but no proof (EDHJ). TUNISIA. No proof of breeding (MS).

Accidental. Faeroes, Lebanon, Cyprus, Malta, Tunisia.

Population. Marked increases in Britain, Ireland, and Netherlands (interrupted in Britain and Netherlands by temporary declines due to toxic seed-dressings). Declines in West Germany, Norway, Sweden, Finland, Poland, Switzerland, and USSR, and at least locally in France.

BRITAIN AND IRELAND. Marked increase, associated with range expansion, from late 19th century until 1930s (Britain) and 1950s (Ireland). Sharp decline, especially eastern England, late 1950s and early 1960s caused by seed-dressings; numbers largely recovered since (Parslow 1967; Sharrock 1976; R J O'Connor and C J Mead). Estimated more than 100000 pairs (Sharrock 1976). FRANCE. Under 10000 pairs (Yeatman 1976); some recent decrease, especially Alsace (A Le Toquin). BELGIUM. Rare *c.* 1900, now *c.* 16000 pairs (Lippens and Wille 1972), but 2500–10000 pairs 1967 (Verheyen and Damme 1967). LUXEMBOURG. About 180 pairs (Lippens and Wille 1972). NETHERLANDS. Increase continued from colonization in 1880 to 1940–3; stable 1943–50; declined by *c.* 80% 1950–5 to 1965–70; increased after ban on toxic seed dressings, and estimated 13000–17000 pairs 1975–7 (Teixeira 1979). WEST GERMANY. Indications of general decline during *c.* 1900–30, and again 1950–65 (Glutz and Bauer 1980). For local counts, see Bauer and Thielcke (1982). Now perhaps 11000 pairs (Rheinwald 1982). DENMARK. About 100 pairs 1974, decreased (Dybbro 1976; TD); 100–200 pairs 1979 (I Clausager). NORWAY. Decreased over last 50 years (Haftorn 1958; GL, VR). SWEDEN. About 50000 pairs (Ulfstrand and Högstedt 1976); decreased since 1940s, with passage at Falsterbo dropped by *c.* 75% since then (LR). FINLAND. About 800 pairs (Merikallio 1958); marked decrease recent decades, causes unknown (OH). EAST GERMANY. Karl-Marx-Stadt province: marked decline since *c.* 1950 (Möckel 1981). POLAND. Scarce or very scarce; declined since 19th century (Tomiałojć 1976a). CZECHOSLOVAKIA. Marked decline, still continuing (Hudec and Černý 1977). AUSTRIA. Scarce except in Danube flood-plain and edge of Wienerwald, where numbers unchanged in last 30 years (Glutz and Bauer 1980). SWITZERLAND. Decreased since 1950, and especially since 1972, reasons unknown (Schifferli *et al.* 1970). HUNGARY. Numbers relatively stable (LH). BULGARIA. Declined; formerly numerous (Z Spiridonov). USSR. Rare, decreased since 19th century (Dementiev and Gladkov 1951b). Estonia: now scarce; recent marked decrease (HV). Latvia: fairly rare, decreasing since 1930s (Baumanis and Blūms 1972). Lithuania: numbers small, decreased from *c.* 1930, owing to shortage of nest-sites (Ivanauskas 1964).

Survival. Britain: 1st-year survival *c.* 40%, adult survival *c.* 53·7%, but lower in late 1950s and early 1960s (R J O'Connor and C J Mead). Finland: 1st-year mortality 57·5%, adult mortality 44·5% (Saari 1979b). Oldest ringed bird 12 years 7 months (Jacquat 1975).

Movements. Resident (dispersive) in temperate regions, migratory to partially migratory elsewhere.

300 Columbidae

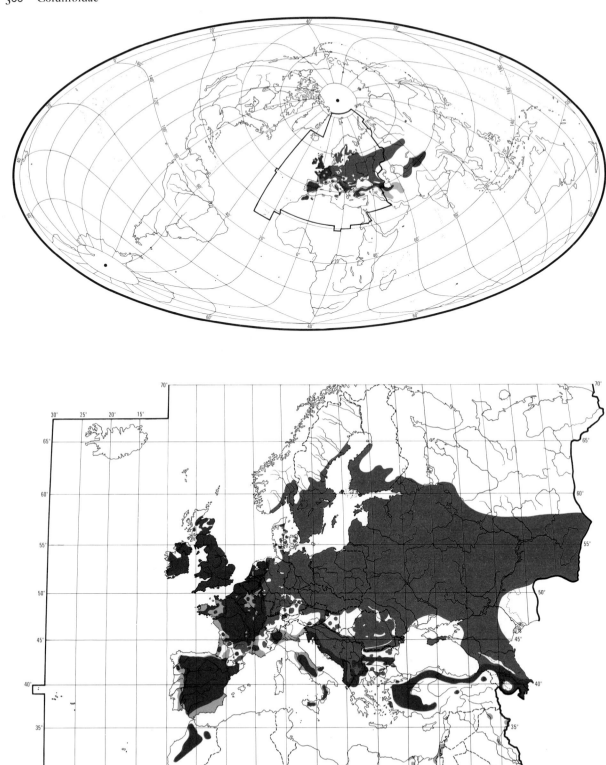

Almost entirely migratory in Fenno-Scandia and eastern Europe, and rather few remain for winter in northern provinces of East and West Germany. Becomes progressively less migratory further south and west (under less severe winter weather conditions), until mainly resident in southern Europe and Asia Minor, and (under influence of maritime climate) in western seaboard regions north to Britain (e.g. Murton 1966, Bernis 1967). Migrants winter especially in southern half of France and in Iberia, and (further east) along northern Mediterranean basin; winter ringing recoveries in Europe mainly west and south of 5°C January isotherm (Glutz and Bauer 1980). Within USSR, main wintering region for nominate *oenas* is Transcaucasia (Dementiev and Gladkov 1951b).

Local movements noted north-west Africa, November–March, but not known whether these solely Maghreb breeders or include winter visitors from Europe (Heim de Balsac and Mayaud 1962); 1 migrant noted Cap Spartel in April (Pineau and Giraud-Audine 1974). Winter immigration there probably irregular: only 2 modern records at Gibraltar, though included flock of 28 in October 1979 (Cortes *et al.* 1980). In Turkey, most common in winter in Anatolia and south coast regions (Beaman *et al.* 1975); probably these include immigrants from eastern Europe, as well as Turkish population moving seasonally into less severe western areas. Irregular winter visitor to Cyprus, though quite common (late October to late February) in some years, probably related to severity of Turkish winter (Bannerman and Bannerman 1971; Stewart and Christensen 1971). Rare large-scale winter incursions reported as far south as Sinai (Meinertzhagen 1930). No certain records from Arabia or Afrotropics; one report of numbers in Niger, December (Fairon 1971), probably refers to *C. livia*.

Ringing recoveries summarized by Rendahl (1965) and Glutz and Bauer (1980). Birds flock after breeding and local dispersal may begin soon after: one Swiss nestling found 78 km north 3 weeks later. Exceptionally, Belgian-ringed nestling (May) found Estonia in September of same year. Autumn migration in Fenno-Scandia at height in second half of September and October, with conspicuous passage flocks crossing coasts by day. Migration oriented towards south-west (with Finnish birds recovered south-west Sweden and south-east Norway); winter recoveries from Netherlands to Iberia, but majority in south-west France and north-east Spain (Gironde-Pyrénées region). Atypically, 2 Finnish birds found in Austria November–December (Rendahl 1965). Under overcast conditions, continental migrants occasionally drifted to eastern England (e.g. Murton and Ridpath 1962), and 1 Finnish bird found Norfolk in March; no known regular winter immigration, however. Mainly migratory populations from West Germany and Switzerland show same south-west orientation, with same grouping of winter recoveries in south-west France and Spanish Pyrénées (mapped in Glutz and Bauer 1980). Fewer recoveries for east European populations: birds from Baltic states and Czechoslovakia show same trend, though also recoveries in Italy; one ringed Italy in October was found Poland in June. One Russian bird (from Ukraine) found Greece in December. Return migration begins February, increases in scale through March, and northern breeding areas reoccupied second half of March and April.

In contrast, breeding population of Low Countries shows less movement. Of 44 recoveries for Dutch-bred birds, only 5 abroad: in West Germany, England, Belgium (2), and France (Speek 1973). Estimated *c.* 50% resident in Belgium (Lippens and Wille 1972); emigrants recovered predominantly in France (only 1 in Spain). Essentially resident in Britain, with juveniles only slightly more prone than adults to disperse, and no southerly emphasis to movement (Murton 1966): of 269 recoveries, only 12·6% had moved more than 25 km—10·9% of birds ringed as adults, 13·2% of juveniles; only 2 of these (both juveniles) found abroad, in south-west France (November) and north-east Spain (October) (O'Connor and Mead 1981).

Central Asian race (*yarkandensis*) may be dispersive only, since sole record away from breeding range is 1 skin from Afghanistan (Vaurie 1965). However, some breeding areas certainly vacated in winter, and extent of movement still unclear (Dementiev and Gladkov 1951b).

Food. Chiefly plant material: seeds, green leaves, buds, and flowers, occasionally invertebrates. Food taken mainly on ground, by walking and pecking; sometimes feeds in trees or amongst shrubs. On ground often feeds in flocks, but in trees usually solitary. Feeding peaks in early morning, near midday, and late afternoon; intermediate periods spent resting, usually in woodland. Spends *c.* 30% of daylight hours feeding (Kotov 1974b; H A Robertson).

Diet includes grain of wheat *Triticum*, barley *Hordeum*, oats *Avena*, rye *Secale*, maize *Zea*, and millet *Panicum*; seeds of cultivated and weed species such as goosefoots (*Chenopodium*, *Atriplex*), rape and turnip *Brassica*, charlock and mustard *Sinapis*, legumes (*Pisum*, *Phaseolus*, *Lathyrus*, *Vicia*, *Melilotus*, *Medicago*, *Trifolium*), corn spurrey *Spergula*, campion *Silene*, spurge *Euphorbia*, knotgrass *Polygonum*, buckwheat *Fagopyrum*, buttercup *Ranunculus*, bedstraw *Galium*, pansy *Viola*, milkwort *Polygala*, cowherb *Vaccaria*, dock *Rumex*, scarlet pimpernel *Anagallis*, fluellen *Kickxia*, speedwell *Veronica*, field madder *Sheradia arvensis*, *Helianthus*, *Centaurea*, grasses (*Avena*, *Setaria*), sedge *Carex*, *Schoenoplectus*, and fumitory *Fumaria*; also acorns *Quercus*, beech-mast *Fagus*, seeds of pine *Pinus* and hornbeam *Carpinus*, currants and gooseberries *Ribes*, and bilberries *Vaccinium*. Leaves of swede, kale, cabbage, brussels sprouts, and turnip *Brassica*, clover *Trifolium*, and various weeds like *Ranunculus* and dandelion *Taraxacum*, and succulent parts and seeds of glasswort *Salicornia*, flower buds of mustard *Sinapis*, and flowers of ash *Fraxinus*. Animal food includes land snails (*Pupilla*, *Succinea*, *Vallonia*, *Limax*, *Monacha*,

Cochlicopa, Fruticicola, Hyalina), aquatic molluscs (*Valvata, Planorbis, Sphaerium*), slugs, cocoons of earthworms (Oligochaeta), centipedes (Chilopoda), and insects such as larvae of gall-midge *Oligotrophus*, weevils (Curculionidae), and potato beetles *Leptinotarsa*. (Bolam 1912; Collinge 1924–7; Ticehurst 1932; Witherby *et al.* 1940; Likhachev 1954; Niethammer and Przygodda 1954; Klinz 1955; Murton *et al.* 1964*a*, 1965; Goodwin 1970; Musil 1970; Glutz and Bauer 1980; Lang 1981.) In urban areas, occasionally takes bread (Goodwin 1960*b*). Drinks fresh water from edge of, or in shallows of, ponds, streams, ditches, puddles, or drinking troughs; sometimes alights on surface of water, and also drinks from water-lily leaves (Jäckel 1891). Occasionally drinks brackish or salt water (Bolam 1912); recorded drinking ooze from manure heaps (Murton *et al.* 1964*a*). Small stones and grit often collected; mean 81 stones (0·63 g) per nestling (*n* = 43), maximum 541 (Glutz and Bauer 1980).

Seasonal variation in diet mainly due to differences in food availability, but diet more varied than in Woodpigeon *C. palumbus*; individuals often take seeds of wide variety of weed species in addition to other foods. 166 crops collected throughout the year, Cambridgeshire (England), showed seasonal variation in content: 37% (of items) cereals (up to 69% November–December), 23% *Sinapis/Brassica* seeds (up to 45% September–October), 28% weed seeds (including 5% *Stellaria*, 4·5% knotgrass *Polygonum*, and 2% wild oat *Avena ludoviciana*); also important were *Chenopodium album* (January–February), *Melilotus officinalis* (March–April), and *Euphorbia exigua* (September–December), 4% *Sinapis* flower buds (May–June only), 4% peas and beans (Mainly July–August), 2% weed and clover leaves, and 2% animal matter (Murton *et al.* 1964*a*). For details of 6 crops and stomachs, February–March, West Germany, see Niethammer and Przygodda (1954). 22 crops, Danube delta (Rumania), contained (in addition to green parts of plants) 58% (of items) charlock *Sinapis arvensis*, 27% sunflower *Helianthus*, 6% peas, 4% *Vicia*, 3% *Trifolium*, and 9 other species (Glutz and Bauer 1980). Of 85 crops collected throughout the year, Caucasus (USSR), 65% contained *Vicia hirsuta*, 50% cultivated *V. sativa*, 50% wheat, 35% *V. angustifolia*, 30% *Fagopyrum esculentum*, 30% *Euphorbia virgata*, 26% millet, 26% *Vaccaria parviflora*, 15% peas, and 14% rye; 27 further plant species also occurred. 44% of crops contained molluscs (29 species): most frequently, *Pupilla muscorum* (in 16·5% of crops), *Vallonia pulchella* (13%), *Succinea putris* (8%), and the freshwater species *Valvata piscinalis* (8%) (Likhachev 1954; Murton *et al.* 1965).

Crop weight of 4 birds killed late afternoon, February, West Germany, 32–45 g; gizzard weight of 2 birds, March, 26 and 40 g (Niethammer and Przygodda 1954). Extreme examples of capacity of crop include: 3745 seeds including 3631 *Vicia*; 7942 *Ranunculus* and 2665 *Vicia* seeds; 1260 *Galium aparine* seeds and 2 barley grains; 321 *Helianthus* seeds; 451 *Lathyrus* seeds; 2216 *Sinapis arvensis* seeds (Muirhead 1895; Glutz and Bauer 1980).

Food of nestlings similar to that of adults (Murton *et al.* 1964*a*). Crop milk very important, initially forming sole food. Increasing percentage of small seeds given from 3–4 days, so that by 12th day 50% of total content (by volume) is crop milk; 30% by 16th day, and only traces at 24th day (Kotov 1974*b*; Heinen and Margrewitz 1981; Lang 1981; H A Robertson). In 14 crops, July–August, Cambridgeshire, 49% (of items, excluding crop milk) *Sinapis* or *Brassica* seeds, 23% cultivated *Vicia*, 9% cereal, 8% grass seeds, 5% other weed seeds, 2% each peas, weed leaves, and animals (earthworm cocoons and snails). 46 crops and gizzards, Nordrhein-Westfalen (West Germany), contained 71% (of items) wild vetch *Vicia*, 18% *Ranunculus*, 6% *Brassica*, remainder equally divided among cereals, chickweed *Stellaria*, and cornflower *Centaurea cyanus* (Glutz and Bauer 1980). Near Stuttgart (West Germany), seasonal variation in nestling diet reflected availability: cereals dominant April–May, but virtually absent June and early July; unripe barley taken late July, and cereals dominant again August–October (Lang 1981). Nestlings fed 3–4 times per day until 5 days old, thereafter twice a day, morning and evening (Kotov 1974*b*). 46 nestlings (West Germany) averaged 2314 seeds, and 14 nestlings (England) averaged 553 seeds per crop (Murton *et al.* 1964*a*; Glutz and Bauer 1980). Crop contents of nestling weighed 46 g (Glutz and Bauer 1980). HAR

Social pattern and behaviour. Less well studied than either Rock Dove *C. livia* or Woodpigeon *C. palumbus* but many aspects similar to one or both of these (e.g. Goodwin 1970; see below). For major study at Hainaut (Belgium), see Delmée (1954); for study in 2 regions of USSR, see Kotov (1974*b*); see also Glutz and Bauer (1980) for recent summary.

1. Often gregarious, though less so than *C. palumbus*; largest flocks (of up to 500) on feeding grounds in winter, though substantial numbers may also gather to feed in breeding season (Bannerman 1959; Glutz and Bauer 1980). Depending on size of local population, flocks of 100 or more juveniles may form during summer (Glutz and Bauer 1980). For further details on post-breeding-season flocking patterns, see Steinfatt (1941*a*), Geyr von Schweppenburg (1942), Delmée (1954), and Kotov (1974*b*). Migrant flocks generally small (Fischer-Sigwart 1914; Haller 1934; Labitte 1955; Kotov 1974*b*). Winter roosting assemblies near Bonn (West Germany) mainly ♂♂ (Niethammer and Przygodda 1954). Slight ♂ surplus throughout breeding season, Hainaut (Delmée 1954). Readily associates with *C. palumbus* on migration, integrating completely (Gatter and Penski 1978) or not (Steinfatt 1941*a*). Loose feeding association with *C. palumbus*, also

PLATE 24 (*facing*).
Pterocles lichtensteinii Lichtenstein's Sandgrouse (p. 245): 1 ad ♂, 2 ad ♀, 3 juv moulting (wing-converts and underbody partly ad ♂), 4 downy young.
Pterocles coronatus coronatus Crowned Sandgrouse (p. 249): 5 ad ♂, 6 ad ♀, 7 downy young.
Pterocles senegallus Spotted Sandgrouse (p. 253): 8 ad ♂, 9 ad ♀, 10 juv ♂, 11 downy young.
Pterocles exustus floweri Chestnut-bellied Sandgrouse (p. 259): 12 ad ♂, 13 ad ♀, 14 late juv ♂, 15 early juv. (CJFC)

with *C. livia*, similarly occurs (Fischer-Sigwart 1914; Haller 1934; Labitte 1955; Kotov 1974b; D Goodwin; see also Roosting and Flock Behaviour, below). BONDS. Monogamous mating system with single-season or, at least in some populations, long-term (up to 4 seasons recorded) pair-bond promoted by site-fidelity. Birds of a pair may sometimes remain together on passage and in winter flocks where low-intensity courtship also observed (Steinfatt 1941a; Delmée 1954; see also Doucet 1969a, Glutz and Bauer 1980, and Heterosexual Behaviour, below). In town parks, pairs may remain as such and stay in general breeding area (Niethammer and Przygodda 1954). For case of ♂ rapidly re-pairing after loss of ♀, and successful raising of young, see Hewett (1881). In Rominter Heide (Poland), all birds paired by c. mid-April (Steinfatt 1941a). Exact age of first breeding not known, but most probably 1 year (see Lofts *et al*. 1966, Murton 1966); hand-reared birds of both sexes taken from nests in the wild first bred at 9–10 months (D Goodwin). Care of young by both sexes up to fledging and, to some extent, beyond (e.g. Kotov 1974b). BREEDING DISPERSION. Largely dependent upon availability of suitable nest-sites (e.g. Goodwin 1955, Alpers 1973, Glutz and Bauer 1980); in favourable areas (e.g. quarries, stand of trees with many natural cavities, or where artificial sites provided) small neighbourhood groups established (Delmée 1954; Goodwin 1955); on Texel (Netherlands), birds nesting in burrows tend to form clumps of 4–5 pairs, despite surplus of burrows (Dijksen and Dijksen 1977). According to Steinfatt (1941a) no clearly demarcated territory defended against conspecifics, unlike (e.g.) *C. palumbus* and Turtle Dove *Streptopelia turtur*; see also Delmée (1954). However, in Lüneburger Heide (West Germany), 'nesting territory' said to be definitely defended and average size calculated to be 9·2 ha (Alpers 1973); not clear how much of this actually defended (see below). In Rothrist (Switzerland), c. 60 pairs on c. 20 km², with 18–20 pairs on island of c. 240 m² (Haller 1934; Glutz 1962). In Rominter Heide, c. 1 pair on 250 ha (Steinfatt 1941a). In Lüneburger Heide, largely dependent on holes of Black Woodpecker *Dryocopus martius* in ancient beeches *Fagus*; where 14 pairs on 1·3 km² (of a 235-km² study area), average 1·10 pairs per 10 ha (Alpers 1973). For Karl-Marx-Stadt (East Germany), see also Möckel (1981). Exceptionally high densities in central Europe (Glutz and Bauer 1980): 40 pairs on 2·35 km², south of Vienna (Austria), and 36 pairs on 4 km², Prater park in Vienna; in West Germany, 25–30 pairs on 30 km², Unterfranken, and 100 pairs on 53 km² near Bamberg; densities also high on some islands where birds nest mostly in burrows: 30–60 pairs on 20 km², Texel (Dijksen and Dijksen 1977). Up to 24 pairs (mainly in boxes) on 13·3 ha, Wesel, West Germany (Heinen and Margrewitz 1981). In southern England, 3 occupied nest-boxes in garden of c. 0·4 ha (P A D Hollom). In Orenburg (USSR), 2–4 birds per 10 km in June; in Kemerovo, (USSR), 4–5 (Kotov 1974b). Several pairs sometimes nest relatively close together (e.g. Lippens 1935, Alpers 1973, Kotov 1974b). Nests not infrequently less than 50 m apart; may include 2 pairs in same tree (Glutz and Bauer 1980). High proportion of non-breeders a feature of some populations. In Hainaut, 20 birds ringed 1946–50 were breeders; of 83 others, 54 were roving (mostly 1st-years and unpaired ♀♀) and 29 others attached to local group though unable to find nest-site (Delmée 1954); near Bonn (West Germany), number of breeding pairs increased from c. 7–8 to c. 11–12 during season as more *D. martius* holes vacated (Krambrich 1953); see also Möckel and Kunz (1981). Most juveniles return to general area of birth. Adults show marked site-fidelity: e.g. one ♂ bred for 3 years at same site but paired with 3 different ♀♀ (Delmée 1954); see also Doucet (1969a). Same site often used for 2nd and subsequent broods (e.g. Steinfatt 1941a, Campbell 1951, Bartholomew 1953, Delmée 1954); where not, breeding failure and brood-overlap likely factors (H A Robertson; see also Musil 1970 and Relations within Family Group, below). Activity radius during breeding season, Belgium, 15–20 km (Lippens 1935; Delmée 1954). Reported to nest amicably alongside *C. palumbus*, southern France (Hüe 1970); but see also Antagonistic Behaviour (below). For other nesting associates, see Glutz and Bauer (1980). ROOSTING. Communal and nocturnal, at least outside breeding season (e.g. Delmée 1954, Labitte 1955, Glutz and Bauer 1980). Non-breeders continue to roost communally during breeding season (Huber 1954). Said mainly (perhaps exclusively) to use holes in breeding season, Eicks (West Germany); at other times trees, particularly older pines *Pinus* (Geyr von Schweppenburg 1942); see also Huber (1954). In Hainaut, artificial sites not used outside breeding season, birds roosting rather in small groups in bare trees, though moved to thicket of holly *Ilex* in colder weather (Delmée 1954). At Kurgal'dzhin (USSR), winter, birds roosted in ricks, making hollows in straw; some also used fissures in river bank (Krivitski 1962). Cracks and holes in cliffs favoured, Derbyshire, England (Swaine 1945). Roosts generally at edge of open feeding areas; may be traditional over several years (Glutz and Bauer 1980). In Eure-et-Loir (France), c. 12 moved to roosting tree in middle of copse at c. 15.30 hrs, rested for c. 30 min, then each bird assumed perch for roosting proper—apart from *C. palumbus* (Labitte 1955). ♀ roosts in nest when incubating, or brooding (Kotov 1974b; see also Relations within Family Group, below); ♂ in tree or some kind of cavity nearby (Steinfatt 1941a; Swaine 1945; Delmée 1954). Fledged young roost in nest, at least for a time; one (or two) did so with well-grown chicks of another brood (Campbell 1951). In early morning, birds also spend much time loafing high in favoured trees (Naumann 1833). Nocturnal roost-sites also regularly visited during day (Glutz and Bauer 1980), e.g. during bad weather in early breeding season (Kotov 1974b, which see also for loafing around midday). In period preceding autumn emigration, birds spend c. 30% of day foraging, otherwise loaf and roost (Kotov 1974b).

2. Occupies intermediate position between *C. livia* and *C. palumbus* with regard to certain displays and also some aspects of plumage colour and pattern (D Goodwin). Generally wary of man, and ♀ more so than ♂ (Delmée 1954). Often described as less wary than *C. palumbus* (e.g. Naumann 1833, Lippens 1935), particularly in woods, and when not persecuted (Steinfatt 1941a; see also Labitte 1955). Shyer than *C. palumbus* nearer human habitations (Geyr von Schweppenburg 1942). Birds trapped and unable to escape will attack with blows of carpal joint (Doucet 1969a; see also below). Flight agile and often accompanied by whistling sound. Very little if any Wing-clapping at normal take-off, unlike *C. palumbus* (Naumann 1833; Bussmann 1925; Heinroth and Heinroth 1926–7; see also Hagen 1917 and Parental Anti-predator Strategies, below). Like *C. livia*, birds make sudden upward and downward movement in flight, in attempt to escape from (e.g.) Goshawk *Accipiter gentilis* (Klinz 1955). For alighting on water, see (e.g.) Barclay (1935), Cornish (1947),

PLATE 25 (*facing*).
Pterocles orientalis Black-bellied Sandgrouse (p. 263): **1** ad ♂, **2** ad ♀, **3** juv moulting into ad ♂, **4** downy young.
Pterocles alchata Pin-tailed Sandgrouse (p. 269). *P. a. caudacutus*: **5** ad ♂ breeding late phase (early summer), **6** ad ♀ breeding late phase (early summer), **7** juv moulting into ad ♂ non-breeding. Nominate *alchata*: **8** ad ♀ breeding late phase (early summer), **9** downy young.
Syrrhaptes paradoxus Pallas's Sandgrouse (p. 277): **10** ad ♂, **11** ad ♀, **12** juv, **13** downy young. (CJFC)

Höhn (1947), and White (1947). FLOCK BEHAVIOUR. Winter feeding flocks generally more compact than in *C. palumbus* (Goodwin 1960*b*); in France less so than *C. palumbus* when flying on migration (Labitte 1955). Generally arrive at and depart from feeding grounds separately from other Columbidae (D Goodwin). As in *C. livia*, flock may take off suddenly, usually soon landing again (Geyr von Schweppenburg 1942); such upflights usually due to momentarily 'hawk-like' appearance of newly arriving pigeon (Columbidae) or less often, thrush *Turdus* or crow *Corvus* (D Goodwin). In Yorkshire (England), *c*. 40 hovered and trod water for *c*. 2–3 s; head pointed up, wings held back, tail fanned and depressed; performance given several times (Crackles 1948). ANTAGONISTIC BEHAVIOUR. (1) General. Some descriptions may refer to interactions between ♂ and ♀ (see, e.g., Selous 1901, also Heterosexual Behaviour, below). Advertising-call given by ♂ from site near nest (Kotov 1974*b*: see 1 in Voice); also in area where no suitable sites (Steinfatt 1941*a*), perhaps by first-time breeders (Delmée 1954). ♂ sits erect, with iridescent neck feathers ruffled and/or cervical air-sac inflated (Heinroth and Heinroth 1926–7; Musil 1970; Glutz and Bauer 1980), though posture described elsewhere as rather stiff and bowed over (Steinfatt 1941*a*). Head may be moved slightly with rhythm of call (Kotov 1974*b*). At peak of season; given almost throughout day; highest intensity in early morning, marked decline after 09.00 hrs and birds generally silent around midday and early afternoon; less intense calling later (Steinfatt 1941*a*; see also Selous 1901, Kotov 1974*b*). (2) Display-flight. Similar to that of *C. livia* (Goodwin 1970); frequently performed between bouts of vocal activity (Delmée 1954; Kotov 1974*b*). In full-intensity version, ♂ flies on more or less horizontal plane (drops *c*. 0·5–1 m immediately after take-off: Kotov 1974*b*) and gives rather slow, deep wing-beats, primaries meeting over back to produce Wing-clapping (less loud and emphatic than in *C. livia*). Followed by generally horizontal glide with wings held *c*. 15° above horizontal and tail spread; crop region appears swollen (Steinfatt 1941*a*; Geyr von Schweppenburg 1942; Goodwin 1956*b*, 1970; Kotov 1974*b*). For possible function of tail markings, including comparisons with other Columbidae, see Goodwin (1955). Bird may glide thus with turns and no great loss of height, but additional wing-beats given when necessary, and at end of flight usually flies up to perch where more Advertising-calls may be given. (Rarer) low-intensity variant involves only slow, emphatic wing-beats—thus not much different from ordinary flight, which anyway sometimes interrupted for performance of Display-flight (Steinfatt 1941*a*; Geyr von Schweppenburg 1942; Kotov 1974*b*). Short version of Display-flight may be given to ♀ on ground (usually walking): ♂ flies up with deep wing-beats and drops or glides down to alight by ♀ and perform Bowing-display (D Goodwin; see also Heterosexual Behaviour, below). ♂ (paired with *C. livia*) flew up steeply with loud Wing-clapping, then glided down with wings held up at acute angle; continued for long period with pauses of 3–7 min (Musil 1970). Such variants occur when object of display walking about on ground (D Goodwin), and thus perhaps more widespread in ground-nesters (see Lippens 1935, Glutz and Bauer 1980). Display-flight often between trees *c*. 50–800 m apart, though sometimes over *c*. 2–3 km (Geyr von Schweppenburg 1952; Kotov 1974); circular flights also common (Steinfatt 1941*a*; Delmée 1954). Generally more confined to nesting area than in *C. palumbus* (Murton and Isaacson 1962). Repetitions frequent and several birds may participate at same time; up to 4 did so on criss-cross paths over clearing with nest-sites at edge. ♀ may possibly perform low-intensity Display-flight over short distances (Geyr von Schweppenburg 1942; see also Heterosexual Behaviour, below). Display-flight given from establishment of 'territory' to hatching, though some ♂♂ cease completely at start of incubation (Kotov 1974*b*). (3) Threat and fighting. In Hainaut, fights between ♂♂ rare once nest-sites occupied as long as birds do not come too close (Delmée 1954); similarly, in Rominter Heide, after initial disputes pairs settled to breed as amicable neighbours (Steinfatt 1941*a*). However, many reports indicate marked aggression shown in defence of nest-sites against conspecifics, and most bitter and prolonged disputes arise when these contested (e.g. Goodwin 1955, 1970). Self-assertive Bowing-display may be given by site-owner when conspecific lands nearby (Goodwin 1956*b*). Opponents may stand head to head, or adjacent and parallel to perform Bowing-display (Selous 1901); when given sideways on, indicates tendency to flee. In Bowing-display (for full description of highest-intensity variant, see Heterosexual Behaviour, below), head raised with neck feathers ruffled, then lowered and Display-call given (see 2 in Voice); tail fanned and raised. Bowing-display may be given before and during fight. Nodding (as in *C. livia*, or more particularly *C. palumbus* and typically given in same situations) between attacks indicates refusal to yield; head position in fairly intense Nod more or less as in Bowing-display. Mutual threat thus largely similar to that of *C. livia* (Goodwin 1955, 1956*a,b*, 1963, 1970; Glutz and Bauer 1980). Birds fight with blows from carpal joint and pecking; one may attempt to jump on to other's back. Often takes place on branch and dislodging attempts thus frequent, but often continues on ground and exhaustion of participants may result (Selous 1901; Goodwin 1955; Glutz and Bauer 1980). Aerial continuation of combat involving one bird flying just above the other and delivering blows with wings (see Selous 1901), pos-'sibly a heterosexual interaction (see below); in another case, however, mid-April, 2 presumed ♂♂ held bodies vertical, fanned tails, and buffeted with breasts (Hagen 1917). Apparently no interest shown in fight by other nearby birds (Selous 1901), but when 2 pairs involved in combat, all birds may fight each other (see Goodwin 1955). According to Lippens (1935), 3–4 ♂♂ may chase one ♀ and fight amongst themselves, breaking off only in attempt to court her. In Rominter Heide, fights quite common after spring arrival; one of several minutes' duration ensued when ♂ arrived in response to Advertising-call of another; afterwards, both flew off and wings made rushing sound; Display-flight (see below) performed by one bird on his return (Steinfatt 1941*a*). According to Kotov (1974*b*), most intruders into occupied sites are unpaired ♂♂ seeking to court off-duty ♀♀; evicted by site-owner using threat or attack as described. Similarly, in Hainaut, most intruders quickly flee if site owner makes as if to attack. 2nd ♂ attempting to intervene when ♂ courting ♀ is charged and expelled in most cases but may continue to follow pair around (Delmée 1954). Interference with copulation probably as in *C. livia* (Heinroth and Heinroth 1926–7, also Glutz and Bauer 1980 but see also section 4 in Heterosexual Behaviour, below). Incubating ♀ sometimes allows stranger to enter nest-box and to lay in cavity; at other times such birds evicted with blows from carpal joint (Delmée 1954). Four captive and fledged young maintained individual distance of *c*. 20 cm; threat indicated by pecking intention-movements, sometimes reciprocated (Steinfatt 1941*a*). One captive ♂ chased another away from food, running after him with tail spread and brushing ground; no call given (Heinroth and Heinroth 1926–7). Although does not normally compete for food with *C. palumbus*, interspecific supplanting attacks and actual fights occur: *C. palumbus* always dominant (Murton and Isaacson 1962). Conspicuously aggressive towards Jackdaw *Corvus monedula*, pair sometimes working together to evict such an intruder (Delmée 1954). However, *C. oenas* probably subordinate in most such instances of competition with the other species (Goodwin 1980). Once, *C. monedula* driven away in hovering flight with rapid winnowing wing-beats (Dickens 1953, which see also for exceptional attacks leading to death of Little Owl *Athene noctua*). HETEROSEXUAL BEHAVIOUR. (1) General. In sedentary population, Oxford (England), birds inspected and flew around 'colony' through autumn and winter; display (no details) also noted (Campbell 1951). In Hainaut, Advertising-call and pursuit-flights (not described) also occur in winter (Delmée 1954; see also Bonds, above). For further details of early display activity, see (e.g.) Fischer-Sigwart (1914), Haller (1934), Labitte (1955), and Kotov (1974b). From arrival, older ♂♂ without returning mate (and probably younger unpaired birds) occupy nest-site and defend this against intruders (see Antagonistic Behaviour, above); at same time attempt to attract nest-seeking ♀ (see below). ♂ may initially remain even if ♀ unresponsive but finally quits area if unsuccessful; may also join a ♀ in her search for a nest-site (see below). Older ♀♀ may take possession of a nest-site whilst awaiting return of ♂ (Delmée 1954). In Karl-Marx-Stadt, laying begins c. 2 weeks after arrival (R Möckel). For courtship display of hybrid *C. oenas* × *C. palumbus*, see Creutz (1961). (2) Pair-bonding behaviour. Essentially as in *C. palumbus* and *S. turtur* (Goodwin 1970). Advertising-call most conspicuous feature in recrudescent courtship activity at start of each brood (Steinfatt 1941a), but overall courtship behaviour much abbreviated for 2nd and subsequent broods (Delmée 1954), though pairs vary individually (Glutz and Bauer 1980). Pair-formation normally concluded (at latest) when ♀ accepts site offered by ♂ (Glutz and Bauer 1980; see also below). Attracted by Advertising-call and/or Display-flight of ♂, a ♀ may sit nearby, whereupon ♂ approaches with tail partly spread and lowered to perform high-intensity variant (see also Antagonistic Behaviour, above) of Bowing-display. Movements rather slow and deliberate; performance similar to that of *C. palumbus*. When near ♀, ♂ raises head with neck feathers ruffled (much as in *C. livia*), then low Bow given, while tail at same time fanned and raised to c. 80°, though normally closed again before reaching highest point; Display-call uttered. Unlike *C. livia*, no pirouetting or turning (Goodwin 1955, 1956b; see also Musil 1970). According to Selous (1901), tail more arched than fanned (for possible signal effect of whitish-grey edges, see Glutz and Bauer 1980), and Bow usually held for several seconds before head raised for repeat (3–4 performances typically given); when tail lowered, may be spread again and then brush ground. ♂ frequently pecks ♀ or strikes her with wing after Bowing-display; ♀ normally responds to Bowing-display by Parading and Wing-lifting; latter display particularly well developed in *C. oenas* (Goodwin 1956b, 1970; Kotov 1974b; see also *C. livia*). In early part of courtship, where ♂ persistent, ♀ may peck him, strike with wing, retreat along branch, or fly about; brief scuffles, or even fierce fights also occur. ♀ sometimes also flies off Wing-clapping and then pursued by ♂, so that whole process repeated later (Selous 1901; Steinfatt 1941a; Delmée 1954; Goodwin 1956a; Kotov 1974b). Bowing-display given by ♀ more or less as in *C. livia*, though rather less freely (Goodwin 1956b); barely audible Display-call uttered (D Goodwin). According to Kotov (1974b), ♀ Bows silently in response to Bowing-display of ♂; also given after short fight when ♂ retreating (Selous 1901). ♂ may also fly off to perform Display-flight (Selous 1901; see above); once, ♂ entered box and gave Nest-call (see 1 in Voice) for c. 1 min while ♀ Mock-preened nearby (R Möckel). (3) Driving. Highly developed, resembling that of *C. livia*. Normally performed, as in that species, shortly before laying, i.e. by pairs which have already mated and started nesting. Once, in 'colony' area, ♂ Drove ♀ away from another ♂ on ground (Goodwin 1955, 1956a, 1963; Murton and Isaacson 1962). If ♀ flies to ground after retreating from ♂, he may Drive her there very persistently (Delmée 1954). ♂ Driving on ground follows ♀ in rather high-stepping gait, making gentle pecking

A

movements at her head (Steinfatt 1941a). Said to strike ♀ fiercely with wings (Heinroth and Heinroth 1926–7), though this more likely an aggressive encounter at early stage of courtship (Goodwin 1955; see also above); for such attacks later in cycle, when ♀ outside nest-hole, see Musil (1970), where ♂ described as generally more vigorous than *C. livia* in relationship with ♀. In Aerial Driving (Fig A), birds fly in hesitant manner, ♂ positioned directly above and behind ♀ who apparently makes no attempt to escape (Murton and Isaacson 1962; Goodwin 1963). Aerial pursuits described by Geyr von Schweppenburg (1942) thought to be continuation of courtship in trees (see above); as 3 birds (probably 2 ♂♂ and a ♀) occasionally participated, not attributable to Driving (D Goodwin). (4) Courtship-feeding (Billing). Similar to that of *C. livia* including fact that often no food passed (Goodwin 1948a). Less prolonged than in *C. livia* or *C. palumbus* (Kotov 1974b). Usually preceded by Wing-trembling of ♀, in manner of food-begging chick (Goodwin 1948a; see also below). (5) Mating. According to Kotov (1974b), frequently solicited by ♀ when ♂ at most active in Bowing-display (see above); however, normally preceded by mutual Allopreening of head and neck (Fig B), Courtship-feeding, and Mock-preening (Haverschmidt 1932; Kotov 1974b; see *C. livia* for fuller details). ♀ generally passive (Delmée 1954). ♂ Parades with wings held out horizontally and more widely spread at shoulder than *C. livia* (Goodwin 1948a); supports himself on ♀'s raised wings after mounting (Haverschmidt 1932). According to Klinz (1955), call given by ♂ as he slips off ♀ (see 4b in Voice). After dismounting, ♂ gives (single) Bowing-display and Display-call; mating otherwise much like that of *C. palumbus* (Goodwin 1948a, which see also for study of captive birds; also Goodwin 1970). Elsewhere, both reported to Nod rapidly after copulation, feathers on back and rump being ruffled. During nest-building period, birds copulate 4–6 times per day; 2–3 times per day shortly before laying; continues up to start of incubation and may take place around midday when birds otherwise loafing (Kotov 1974b). Pair once mated (after Allopreening and Courtship-feeding) on edge of flock, without interference (Murton and Isaacson 1962). (5) Nest-site selection. Final choice made by ♀ who may accept site offered by ♂ (see above). ♀ first-time breeders usually move around alone in search for site (Delmée 1954). Otherwise, pair may search together, ♀ again the more active. ♂ may sit nearby and call (not described, but presumably 1 in Voice), watching ♀ alertly; may also perform low-intensity Display-flight. ♀ flies off with Wing-clapping, followed by ♂, if search fruitless in particular area (Steinfatt 1941a). According to Kotov (1974b), ♂ sits by a hole

B

giving Nest-call (see 1 in Voice) and Wing-twitching (see *C. livia*); both then inspect site alternately. (6) Behaviour at nest. ♂ gives up to 10–15 successive Nest-calls from nest; changes to Affection-call (see 4a in Voice) if ♀ enters hole (Steinfatt 1941a). ♂ may also scratch with feet while Nest-calling (R Möckel). At this time, frequent alternation of Nest-call from nest, Advertising-call from perch nearby, and Display-flight with ♀ following (Steinfatt 1941a). Gradually more frequent and more prolonged visits made by pair to nest (Kotov 1974b). Even at early stage, pair may spend long period (once c. 3 hrs) together in nest, both giving persistent Nest-calls (P A D Hollom). Nest-call given in same posture as *C. livia*. When pair together at nest, Wing-twitching and Nodding (no twig-fixing movement, unlike in *C. palumbus*) given mainly by ♂, but behaviour essentially as in *C. livia* (Goodwin 1955, 1970; D Goodwin). ♂ collects material on ground, most building done by ♀ (Kotov 1974b); like *C. livia*, ♂ shows strong tendency to present material to ♀ from behind (Goodwin 1955). ♀ may give Nest-call when ♂ flying in. ♂ may sit nearby and call (not described, but presumably Advertising-call) whilst ♀ incubating (Steinfatt 1941a). According to Delmée (1954), ♂ sometimes attempts to attract another ♀ to different site. For strange ♀♀ entering nest-hole, see Antagonistic Behaviour (above). Nest-reliefs often difficult to observe. Said to occur twice a day; both birds may be briefly in nest-hole together, but only during building and in early nestling period (Kotov 1974b). ♀ once flew to tree, looked inside hole and ♂ left; at another site, ♀ once left briefly to fly around before returning (Steinfatt 1941a; see also Parental Anti-predator Strategies, below). RELATIONS WITHIN FAMILY GROUP. Young blind and helpless at birth, eyes opening at c. 4–6 days (H A Robertson). Growth and development much as in *C. livia*. During early period, nest-reliefs as during incubation (see above). Nestlings brooded constantly for 5–6 days (H A Robertson) or c. 7–8 days, then no longer during day (up to c. 10 days if cold: H A Robertson), though ♀ always present in nest at night until young 13–15 days old (Kotov 1974b; see also Roosting, above). According to Doucet (1969a), no longer brooded at night after c. 8 days. Young beg by calling (see Voice) and trembling or flapping wings. Fed by both parents who call (P A D Hollom: see 4c in Voice); initially, only after nest-relief, but more frequently once day-time brooding ceases. Normally defecate in nest and droppings not removed (Steinfatt 1941a); rarely, older birds defecate through entrance-hole (Kotov 1974b). Tend to sit facing away from entrance-hole (Kotov 1974b) or often head to tail (H A Robertson). Self-preen actively from c. 10–12 days (Kotov 1974b); mutual Allopreening by siblings and preening of young by parents presumably as in *C. livia* (see Glutz and Bauer 1980). Some sites allow young limited excursions from c. 10 days (Doucet 1969a). Young nestlings may give food-calls while man present at nest; intertwining of necks suggests inter-sibling strife may occur (Campbell 1951). Despite lack of nest sanitation, same site often used for subsequent brood, and ♀ may lay whilst young of previous brood still present, so that ♂ probably assumes more responsibility for these nestlings (Lippens 1935; Steinfatt 1941a; Campbell 1951; Bartholomew 1952, 1953; Paulussen 1953; Delmée 1954; Varga 1977b), feeding them from crop for c. 6–8 days after fledging (Kotov 1974b), at ever-increasing intervals (Delmée 1954, where also reported that adults accompany fledged young to feeding grounds). After fledging at c. 28–30 days, young remain c. 7–12 days in general area, becoming fully independent at c. 37–40 days, when small nomadic flocks may be formed (Kotov 1974b). According to Huber (1954), young rarely return to nests after fledging (but see also Roosting, above). ANTI-PREDATOR RESPONSES OF YOUNG. According to Kotov (1974b), nestlings fall silent on hearing Warning-call of adult (see 3 in Voice). Very small young show no fear of man; tend rather to move heads about and give Food-call (Campbell 1951; see also Relations within Family Group, above). At c. 7–8 days (also when older if nest visited at dusk), may crouch low in nest and freeze, but if attempt made to touch them, perform threat-display, rearing up with feathers or down ruffled, Bill-snapping (see *C. livia*) and making puffing (wheezing) or hissing sounds; may strike with bill or wings. Recorded striking dormouse *Muscardinus avellanarius* (Delmée 1954). Well-grown young try to retreat further into cavity, though often still perform threat-display, and normally attack if handled (Campbell 1951; Kotov 1974b; H A Robertson). For performance of threat-display by captive (and injured) adult, see Goodwin (1956b). PARENTAL ANTI-PREDATOR STRATEGIES. (1) Passive measures. Generally sits tight (on well-set eggs or small chicks: H A Robertson) though some birds may leave when man still at some distance. Generally quick to return after disturbance (Naumann 1833; Geyr von Schweppenburg 1924; Bussmann 1925; Heinroth and Heinroth 1926–7; Steinfatt 1941a; Haller 1934; Klinz 1955). (2) Active measures. Pair recorded performing apparent simple aerial demonstration when man present (Oeser 1971). Uses pecking and blows with carpal joint to expel *D. martius*, and squirrel *Sciurus* (Steinfatt 1941a); see also Antagonistic Behaviour, above. If present, sentinel adult gives loud Warning-call (see 3 in Voice) at approach of danger and departs with Wing-clapping; mate then also leaves; pattern apparently same at all stages of breeding cycle (Kotov 1974b). Bird leaving nest at approach of man gave 'wiwiwi. . .' or 'ikikikik' sounds with wings, not heard at other times (Hagen 1917). Adult may give Warning-call periodically while feeding young (see above). No distraction-displays recorded (Kotov 1974b).

(Fig A from Goodwin 1955; Fig B from photograph in Heinen and Margrewitz 1981.) MGW

Voice. Repertoire restricted; some calls may be duplicated below. Captive birds may call at night (see Goodwin 1969). For Wing-clapping and other wing noises, also for Bill-snapping and puffing (wheezing) sounds of young when threatened, see Social Pattern and Behaviour.

CALLS OF ADULTS. (1) Advertising-call (Song). Recording (Fig I) suggests 'ooo-uh' or 'ooo-er' given 10 times in an accelerating sequence; first c. 5 calls sound gruffer or more strained and, at times, trisyllabic; later calls in sequence clearer and purer in tone—more like 'ooo-(w)up' (M G Wilson). Each call normally with 2 clearly audible syllables: 1st long-drawn (an unstressed but also long-drawn syllable is linked with 1st but difficult to distinguish); 2nd short but with equal emphasis (Geyr von Schweppenburg 1942). Also rendered as 'huuwútt' (Steinfatt 1941a), 'hou-woup' with stress on either syllable (Delmée 1954), a soft, musical but emphatic 'oo-er-oo', variable in tone and modulation (Goodwin 1955), a rather hurried-sounding 'oo-ŏŏ-ŏŏ' or 'cōō-ōō-rŏŏ' (Goodwin 1956b), a deep 'ōō-er-ōō' or 'cōō-ōō' (Goodwin 1970). Rhythm variable; intensity may be increased with ever shorter intervals, perhaps when ♀ present (Bettmann 1959; see also Runte 1959). Given 4–6 times in succession, at highest intensity up to 15–20 times (Steinfatt 1941a); c. 18–20 calls per 8 s (Simms 1971). Slightly hoarse quality; pauses of 1–2 min between bouts (Kotov 1974b, which see for further details). Nest-call similar (Goodwin 1955). Also, growling crooning sounds given by both birds of a pair when together at nest (P A D Hollom). A muffled, moaning sound given by ♀ (Kotov 1974b); presumably ♀ (variant of) Nest-call. (2) Display-call. Given by ♂ in Bowing-display to ♀. Considered a variant of call 1; not so emphatic, more grumbling and long-drawn— 'huuhwurr huuhwurr' (Steinfatt 1941a). A soft, low-pitched droning 'ōōh-ōōh' or 'cōō-ōō cōō-ōōh', audible only at close quarters. Soft snapping or clicking sound, apparently from mandibles, but possibly vocal, given in middle of call (Goodwin 1947b, 1955, 1956b, 1970). According to Kotov (1974b), clicking sounds soft and guttural and given in middle and at end of call. (3) Distress- and Warning-call. As in *C. livia* but perhaps slightly higher pitched; given in same situations (Goodwin 1956b); 'guuu guuu guuu', or simple 'guuu' (Kotov 1974b); a short 'ru' (Glutz and Bauer 1980). (4) Other calls. (a) Monosyllabic 'gurrr' with upward inflexion, softer and more gentle-sounding than call 1. Given by ♂ when ♀ has entered nest-hole ('Affection call': Steinfatt 1941a). (b) Immediately after copulation, ♂ calls 'ru'u' (Klinz 1955); possibly part of call 2. (c) Various 'whoop', 'h'woop', or 'oop' sounds, with variation in pitch and volume, given by parent feeding young (P A D Hollom).

CALLS OF YOUNG. Food-call a long-drawn, faint, seeping whistle of *c*. 5–6 kHz (Fig II, lower-frequency trace represents regurgitation sounds), much as in *C. livia* (Naumann 1833; Steinfatt 1941a; Campbell 1951). Weak at *c*. 2–3 days old, in full strength only from *c*. 5–6 days (Kotov 1974b), penetrating and shrill near fledging (P A D

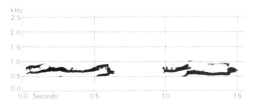

I P A D Hollom England July 1977

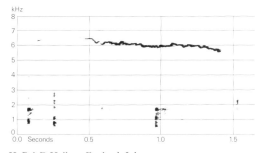

II P A D Hollom England July 1977

Hollom), and lost at independence (Steinfatt 1941a). Also given by young *c*. 8–15 days old when removed from nest; possibly has contact function (H A Robertson). Advertising-call (adult call 1) given (not full version) by captive birds at *c*. 8–9 weeks (Steinfatt 1941a); according to Musil (1970), at *c*. 3 months.   MGW

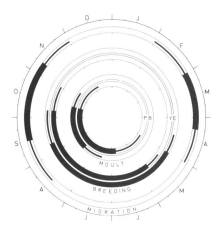

Breeding. SEASON. Britain and Europe: see diagram. Little variation across Europe but laying begins late February to early March in west of range including Britain and Ireland (Makatsch 1976). SITE. Hole in tree, building, or cliff; occasionally in rabbit *Oryctolagus cuniculus* burrow, in dense clump of twigs on tree, or even under bush. Locally, in old nest of bird or drey of squirrel *Sciurus* (C S Roselaar). Solitary to semi-colonial. Sites reused in subsequent years. Nest: often none, but small amounts of grass, twigs, and leaves sometimes added; rarely, considerable lining constructed. Building: by both sexes (Delmée 1954). EGGS. See Plate 95. Elliptical, smooth and slightly glossy; white, tinged creamy-white. 38 × 29 mm (33–41 × 26–31), *n* = 80; calculated weight 17 g (Schönwetter 1967). Clutch: 2, occasionally 1; 3–6 recorded but probably always by 2 ♀♀ (Campbell 1951). Of 156 clutches laid by 42 pairs, Belgium: 1 egg, 4%; 2, 81%; 3, 9%; 4, 3%; 5, 1%; 6, 2%; average 2·3 (Delmée 1954). Multi-brooded; of 42 pairs, Belgium, 3 pairs laid 2 clutches; 9, 3; 23, 4; 7, 5; average 3·8 (Delmée 1954). Of 50 pairs, East Germany, average 2·46 (1–4) broods per pair (Möckel and Kunz 1981). Laying interval 48 hrs (Delmée 1954); can be 24 hrs (Freeman and Bates 1937). Replacement clutch laid after loss, 3–12 days later (Huber 1954). INCUBATION. 16–18 days. By both sexes, with ♀ during night and for much of day (Delmée 1954). May begin with 1st egg and hatching asynchronous (Delmée 1954). YOUNG. Semi-altricial and nidicolous. Cared for and fed by both parents, brooded while small. Initially fed exclusively on crop milk, later on mixture of milk and natural food. FLEDGING TO MATURITY. Fledging period variable, from 20–30 days (Delmée 1954). Become independent shortly after. Age of first breeding probably 1 year. BREEDING SUCCESS. Of 360 eggs laid, Belgium, 193 (54%) hatched, and 136 young

(37·8% overall) fledged; failure to hatch due to desertion (21·2% of eggs laid), predation (by corvids and dormouse *Muscardinus avellanarius*) (19%), and infertility (5·8%); losses of young due to dying in nest (19·7%, probably mainly starvation) and predation (9·8%); of 113 young fledging in 76 nests, 38 in broods of 1, 72 in broods of 2, and 3 in broods of 3; average 1·5; overall, 62 pairs reared 1–7 young each, average 3·06 (Delmée 1954). Of 152 eggs laid, England, 101 (66·4%) hatched, and 61 young (40·1% overall) fledged, with greatest success from eggs laid in July (Campbell 1951). In East Germany, average 1·78 young reared per successful brood in 399 nests, or 1·3 young per nest started; on average, each pair reared 3·25 (1–8) young per year to fledging; 38·6% mortality of eggs and young due to desertion (including death of young due to bad weather and predation of adults), 16·8% due to flooding of nest-holes, 7% to predation on young, remainder unknown (Möckel and Kunz 1981).

Plumages (nominate *oenas*). ADULT MALE. Forehead, crown, nape, sides of head, chin, and throat medium bluish-grey. Patch from upper sides of neck widening downwards and extending across lower hindneck glossy green or purple-red (depending on light). Mantle, scapulars, and lesser upper wing-coverts rather dark bluish-grey; back and rump light bluish-grey, slightly darker grey towards longest upper tail-coverts. Foreneck and chest deep greyish-vinaceous, not glossy as in Rock Dove *C. livia* or Yellow-eyed Stock Dove *C. eversmanni*, vinaceous tinge faintly extending to grey upper breast and sides of breast. Remaining underparts light bluish-grey or ash-grey; under tail-coverts medium bluish-grey. Tail light bluish-grey; tip black (tinged grey when fresh), $3\frac{1}{2}$–4 cm wide on t1, 2–$2\frac{1}{2}$ on t6; black tip proximally bordered by pale grey band *c*. 1 cm wide (most pronounced on t6) which is faintly bordered proximally by another dark grey band; basal and middle portion of outer web of t6 white or greyish-white. Undersurface of tail mainly glossy greyish-black with rather faint light grey subterminal bar and grey on middle of outer web of t6. Outer primaries brownish-black (tinged slightly grey when fresh), outer web of p6–p9 with narrow off-white outer border; inner primaries and secondaries light bluish-grey with black tip $2\frac{1}{2}$–3 cm wide; basal and middle portions of inner web of primaries dark grey; light bluish-grey of middle portions of outer web extending to p5(–p7). Inner 3–4 tertials and inner 3–6 greater and median upper wing-coverts medium grey with solid black patch on middle portion of outer web, latter forming 3 short bars on inner wing; those on tertials and median coverts sometimes small or partly missing, on median coverts often largely concealed by overlaying tips of lesser coverts. Outer greater primary coverts brown-black, inner light grey, sometimes with large black dot on tip of outer web. Middle and outer greater coverts light bluish-grey, together with light inner greater primary coverts and light middle portions of outer secondaries and inner primaries forming conspicuous light grey panel on upperwing, markedly contrasting with black flight-feather tips and brown-black outer primaries and outer greater primary coverts, and slightly contrasting with medium bluish-grey median coverts and outer lesser coverts. Under wing-coverts and axillaries light bluish-grey, not white or greyish-white as in *C. livia* and *C. eversmanni*; undersurface of flight-feathers darker and feather-tips more broadly black than in those species. In worn plumage (May–September), feather-fringes of mantle paler grey, scapulars and tertials tinged brown-grey, and deep vinaceous of neck and upper breast purer and more extensive. ADULT FEMALE. Similar to adult ♂, but mantle, scapulars, and especially tertials brownish-grey rather than bluish-grey (but ♂ in worn plumage similar); width of glossy collar at sides of neck about similar to that of hindneck, not as broad on lower sides of neck as in ♂, in which patch extends up to just behind ear-coverts and further down to sides of chest; some ♀♀ have grey of feather-bases visible in glossy neck-patch; vinaceous of chest less deep and more greyish in ♀, not extending to upper breast, but difference hardly visible when plumage fresh, marked only April–September, when ♂♂ have chest and upper breast uniform deep vinaceous, ♀♀ paler vinaceous on chest only and with some grey of feather-bases visible. Belly of ♀ during summer more subject to brown soiling than ♂. DOWNY YOUNG. Down coarse, hairy, and sparse, absent around eyes and chin; warm yellowish-buff, more deeply coloured than in Woodpigeon *C. palumbus* (Witherby *et al.* 1940; Harrison 1975). JUVENILE. Rather like adult, differing mainly by lack of gloss on neck and by feather structure. Feathers softer, shorter and narrower than in adult. Face bare. Head, throat, hindneck, sides of neck, mantle, scapulars, tertials, and lesser and median upper wing-coverts pale ash-grey with brown tinge, neck without glossy feather-tips; black marks on inner tertials and inner greater and median coverts smaller, less distinctly defined, often lacking on some feathers and occasionally completely absent. Vinaceous tinge on underparts restricted to chest; faint, much mixed and tinged with grey, and sometimes hardly visible, especially in ♀. Feathers appearing on face, head, neck, outer mantle, and longer lesser coverts shortly before and just after fledging purer bluish-grey, like those of adult but narrower and shorter, with some feather-tips on sides of neck glossed green and purple; not known whether these feathers are late juvenile or early 1st adult, but often partly replaced later during post-juvenile moult by full adult. See also Structure. FIRST ADULT. Like adult, but strongly variable amount of juvenile retained; not distinguishable after last juvenile feathers lost. When juvenile feathers still present on body or in upper wing-coverts, ageing easy as pale brownish-grey juvenile contrasts markedly with bluish-grey adult, juvenile feathers shorter and narrower, and some feathers of neck-sides not yet glossy. When only outer primaries and some tail-feathers remain, ageing more difficult (see Structure); these immatures usually still show a few juvenile tertials. Last juvenile feathers to be replaced are central secondaries (see Moults); however, some full adults may also retain some old secondaries, and wear and brown colour of these old feathers similar in juvenile and adult; hence, retained juvenile secondaries only recognizable when narrower or shorter than new neighbouring ones, or with differently shaped tip. Birds showing several alternate groups of new and old secondaries are adult. Body colour of 1st adult in complete plumage is similar to adult, but some show less extensive vinaceous on chest and less extensive glossy green on sides of neck: some 1st adult ♂♂ resemble adult ♀, some 1st adult ♀♀ have limited gloss on sides of neck and vinaceous restricted to sides of chest with central chest almost grey.

Bare parts. ADULT. Iris dark brown to dark red-brown. Narrow ring of bare skin round eye blue-grey. Base of bill and cere bright red, purple-red, or pink-red, cere all or partly powdered white; tip yellow to ivory-white, in ♀ often with dusky suffusion. Leg and foot bright coral-red (in ♀ often with dusky tinge), paler red to yellowish-flesh on back of tarsus and soles; claws horn-brown to black. DOWNY YOUNG. Iris dark brown. Bill dark brown-grey or horn-brown with white tip and pale flesh-coloured flanges at base. Leg and foot dull flesh-coloured grey. JUVENILE. Like downy young, attaining adult colours usually during post-juvenile moult; some ♂♂ show adult colour from c. 3 weeks. (Hartert 1912–21; Goodwin 1970; Harrison 1975; RMNH.)

Moults. Based mainly on sample of 120 birds, Netherlands (RMNH, ZMA). ADULT POST-BREEDING. Complete, primaries descendant. Starts with p1 between early May and mid-July, ♂ slightly earlier than ♀; primary moult scores of birds collected 25 July–4 August, Netherlands, ♂ 20 (2·76; 8) 16–24, ♀ 15 (4·31; 7) 9–22; all primaries new from mid-October to early December. Only 1 out of 40 birds had started about February, suspending with score 10 in April. Body starts with scattered feathers of upperparts or belly from late July, and moult heavy late August to early October. Tail moulted September–November during last stages of primary moult, sequence approximately t1–2–3–4–6–5 (t6 sometimes with t3 or last). Secondaries late like tail; probably not all replaced every year, scattered feathers or 1–2 small groups of feathers retained. Body and tail completed with regrowth of p10. POST-JUVENILE. Complete; timing highly variable, depending on hatching. Starts with head, hindneck, mantle, and median upper wing-coverts slightly before or after fledging, p1 soon following at age of 55 days (Heinroth and Heinroth 1926–7); speed of primary moult similar to adult (lasting 5–6 months), head and body largely in 1st adult at primary score of 30–40. Tail moulted in later stages of primary moult, as in adult, and usually completed with regrowth of p10; t5 and t6 sometimes much later than others. Central secondaries replaced last, mainly s3–s8, s4–s7, or s3–s5, up to 6 months (or perhaps more) later than p10. Early-hatched juveniles have moult similar to adult post-breeding, completing primaries before winter; later juveniles suspend moult from November or December, resuming late February or March. Last-hatched autumn juveniles do not start primaries until February or later. Of 28 winter immatures, Netherlands, 25% had all primaries new, 28% had not yet started, 5% suspended with only p1 new, 5% up to p2, 4% up to p3, 4% up to p6, 11% up to p7, and 18% up to p8 (absence of birds suspending with up to p4–p5 new probably caused by low breeding productivity of adult during main moult period about September). If post-juvenile primary moult not completed by April, moult serially descendant, with 1st adult post-breeding starting from May with innermost primaries and post-juvenile still active on outer. Of 47 December–March birds, Netherlands, 47% showed some retained juvenile secondaries and hence were 1st adult, 53% not and probably were adult, but of 66 April–August birds of 2nd calendar year or older, only 24% had some secondaries juvenile; this lower proportion suggests that some 1st adults replace juvenile secondaries in spring and thus increasingly fewer birds are separable from full adult.

Measurements. Netherlands, Sweden, and central Europe, all year; skins (RMNH, ZMA). 1st adult wing is from immature with replaced outer primaries.

		♂			♀	
WING	AD	222	(3·15; 30)	216–228	216 (4·68; 20)	208–223
	1ST AD	221	(6·97; 14)	212–229	209 (4·17; 6)	205–215
	JUV	213	(3·85; 12)	207–219	211 (3·33; 10)	206–217
TAIL	AD	118	(4·94; 18)	110–126	114 (3·75; 13)	106–119
	JUV	110	(7·04; 4)	102–118	110 (3·20; 4)	105–112
BILL	AD	19·6	(0·95; 23)	18·0–21·2	19·1 (0·72; 16)	17·8–20·0
TARSUS		29·6	(0·96; 25)	28·0–31·2	29·2 (1·11; 16)	27·8–31·4
TOE		32·2	(1·63; 22)	29·6–34·5	31·7 (1·22; 14)	29·8–33·5

Sex differences significant for adult and 1st adult wing and adult tail. Juvenile wing and tail significantly shorter than adult. 1st adult tail on average 2 mm shorter than adult tail. Tarsus and toe of juvenile similar to adult from fledging, bill similar once face feathered. One 1st adult ♂ with wing 237 excluded from table (Netherlands, May).

Weights. Mainly Netherlands, adult (ZMA); others from West Germany (Niethammer and Przygodda 1954), northern Iran (Schüz 1959), Kazakhstan, USSR (Gavrin et al. 1962), and Asia Minor (Rokitansky and Schifter 1971), combined.

	♂			♀		
Dec–Jan	318	(— ; 3)	300–330	299	(— ; 2)	242–355
Feb–Mar	337	(26·7; 7)	297–365	292	(12·6; 4)	281–310
Apr–May	304	(— ; 3)	242–340	298	(— ;3)	279–320
Jun–Jul	302	(18·8; 12)	280–334	280	(16·2; 7)	254–302
Aug–Sep	305	(— ; 3)	299–317	263	(— ; 3)	245–273

Minimum: exhausted ♂, January, Netherlands, 188 (ZMA).

Structure. Wing rather long, tip slightly more rounded than in *C. livia*, base relatively broad. 11 primaries: in adult, p9 usually longest and p8 1–4 shorter, but sometimes equal or p8 longest and p9 1–2 shorter; p10 4–10 shorter (5–17 in 1st adult), p7 11–18, p6 26–34, p5 42–52, p1 84–94; in juvenile, p9 longest, p10 6–12 shorter, p8 3–7, p7 16–24, p6 33–43, p5 52–61, p1 91–98; p11 minute, concealed by primary coverts. Inner web of p10 emarginated, sides of distal half of p10 mostly parallel (gradually tapering to point in *C. livia*). Tail rather short and broad, tip slightly rounded, 12 feathers; width of t1 near tip in adult 33–41, in 1st adult 31–38, in juvenile 25–30; width of t6 near tip in adult and 1st adult 26–34, in juvenile 19–27. Bill as in *C. livia*, but cere in adults less thickly powdered, with thin mealy layer at most. Tarsus rather short and strong, upper front side feathered. Toes rather long and slender; outer and inner toe both c. 78% of middle, hind c. 61%. Claws short but heavy, strongly decurved, sharp.

Geographical variation. Slight. Isolated *yarkandensis* of Central Asia averages slightly paler on head, rump, and underparts (light ash-grey rather than medium bluish-grey), and wing averages 7 mm longer (Hartert 1912–21; Dementiev and Gladkov 1951b; Vaurie 1965; BMNH). Part of south-eastern and eastern populations (Turkmeniya, Kuznetskiy Alatau) of nominate *oenas* as pale as *yarkandensis*, however (ZMM), and size also large. CSR

Columba eversmanni Yellow-eyed Stock Dove

PLATES 28 and 29
[facing pages 327 and 350]

Du. Oosterse Holenduif Fr. Pigeon d'Eversmann Ge. Kleine Hohltaube
Ru. Бурый голубь Sp. Paloma zurita oriental Sw. Turkistanduve

Columba eversmanni Bonaparte, 1856

Monotypic

Field characters. 29–31 cm; wing-span 60–62 cm. About 10% smaller than Stock Dove *C. oenas* but with proportionately longer and more pointed wings; smallest *Columba* in west Palearctic. Rather small, dull-grey pigeon, with more delicate and slightly paler appearance than *C. oenas* and differing distinctly in almost white lower back and under wing-coverts, and yellowish bare parts. Flight pattern strongly recalls wild Rock Dove *C. livia*, but separated by lack of sharp black terminal tail-band and absence of bold black bars over inner wing. Sexes similar; no seasonal variation. Juvenile separable.

ADULT. On ground, plumage pattern not dissimilar to *C. oenas*, but head strongly tinged vinous (in ♂), sheen on green neck feathers and across lower throat purple to chestnut, bars on inner folded wing usually restricted to one on greater coverts, primaries duller, and rear belly paler. In flight, instantly distinguished by pale grey or white under wing-coverts, extensive whitish patch on back, dull grey rump, and tail without clear-cut terminal band; head and neck colours prompt confusion with *C. livia* but tail pattern (and lack of bold wing-bars) prevent misidentification. Upperwing lacks pale central panel of *C. oenas*, and is duller due to less black surround. Bill yellow-green, slate at base, with mealy cere. Eye deep yellow; orbital ring yellow. Legs yellow-flesh to pale red. JUVENILE. Differs little from adult. Head usually lacks vinous, and neck sheen minimal.

Little studied in the field but clearly easily confused with *C. livia* in flight (particularly at long range) and with immature *C. oenas* on ground. When tail visible, lack of sharp black band diagnostic. Flight and behaviour apparently as *C. oenas*.

Silent except during breeding season. Song very similar to *C. oenas*; a subdued 'oo-oo-oo' (Dementiev and Gladkov 1951*b*).

Habitat. In lower middle latitudes, mainly in steppe and neighbouring lowlands, and valleys issuing from mountains up to *c*. 1500 m, generally near a stream or water source. Breeds either in clefts in high loess cliffs, burrows excavated by rollers *Coracias*, in high old gnarled and hollow poplars *Populus*, elms *Ulmus*, planes *Platanus*, or in orchards or ruined buildings. Rests perching in dense foliage in crowns of trees, and will fly up to 5–10 km to forage for food.

Wintering birds in India favour groves of trees in cultivated country, feeding on ripening mulberries *Morus* plucked from trees, and on maize and other cereals gleaned in stubbles after harvest or dug up in freshly sown fields. Roosts in groves of poplars and other trees. (Dementiev and Gladkov 1951*b*; Ali and Ripley 1969.)

Distribution. Breeds in Asia, from north-east Iran north to southern Aral Sea, then east to southern Lake Balkhash and valley of Kara Irtysh and south through Tien Shan and Hindu Kush to northern Afghanistan. Winters in more southerly parts of breeding range and south to Pakistan and north-west India.

Accidental. USSR.

Movements. Mainly migratory. A few birds present all year in southern Turkmeniya (e.g. in Murgab and Tedzhen valleys), and others in northern Iran and Afghanistan, but majority migrate to winter quarters in Iranian Baluchistan, Afghanistan, Pakistan (especially), and north-west India (Dementiev and Gladkov 1951*b*; Vaurie 1965; Ali and Ripley 1969). In northern Pakistan, migrant flocks pass through Kohat and Kurram valley in spring (Whitehead 1911). Principal passages believed to circuit west of main Himalayan–Pamir ranges, for migrants rarely encountered in high mountains (Dementiev and Gladkov 1951*b*) and only one specimen known from Tibet, October (Vaurie 1972).

Autumn passage October-November; return movement mainly April, though early migrants reach Turkmeniya in late March. Stragglers within USSR include 3 in European Russia: Orenburg, Kuybyshev, Chuvash (Dementiev and Gladkov 1951*b*).

Voice. See Field Characters.

Plumages. ADULT. Rather similar to Stock Dove *C. oenas*, but head slightly paler, light rather than medium bluish-grey, crown tinged vinous (most extensive in ♂), green and purple gloss at sides of neck more narrowly extending to hindneck, but, in contrast to *C. oenas*, also in narrow band across lower throat (gloss on central hindneck and in central lower throat almost absent in some ♀♀); mantle, scapulars, and upper wing-coverts slightly paler, ash-grey, less bluish-grey, feather-tips narrowly bordered grey-brown, especially when worn; back and upper rump extensively white, more so than rather narrow patch of Rock Dove *C. livia*. Chest vinous as in *C. oenas*, but often slightly less extensive (especially hardly visible in some ♀♀) and some grey of feather-bases shining through; breast to vent pale rather than light bluish-grey, but under tail-coverts equally dark as in *C. oenas*, hence more contrasting with belly. Tail-tip black or greyish-black for 3 (t6) to 4 cm (t1), remainder light bluish-grey; (t2–)t3–t6 with a light 5–10 mm wide bar *c*. 15 mm from tip; basal half of outer web of t6 mainly white; from below, tail glossy dark grey with faint paler grey subterminal bar, narrower in width and closer to tip than in *C. oenas*. Flight-feathers as in *C. oenas*, but bases markedly paler, silvery-grey, tips slightly paler; dark tips of secondaries narrower, ½–2 cm, poorly defined from middle and basal portions, no dark tips to innermost secondaries; wing without sharp contrast between pale patch on centre of wing (formed by light bluish-grey greater and median upper wing-coverts and bases of outer secondaries and inner primaries) and broad black tips of secondaries and inner primaries as in *C. oenas*, whole upperwing rather pale instead, except for slightly darker lesser coverts and ill-defined greyish-black tips to middle and outer secondaries; short black bars on tertials, greater tertial-coverts, and innermost median coverts as in *C. oenas*, but more reduced in extent and frequently absent on median coverts or tertials. Under wing-coverts pale grey, median under wing-coverts and axillaries greyish-white or almost white. JUVENILE. Like adult, but (as in *C. oenas*) vinous tinge more restricted, hardly present in ♀ and soon lost by wear in both sexes; feathers of chest sometimes faintly edged brown; all neck dull ash-grey, virtually without green and purple gloss as in adult; slight brownish tinge to grey of scapulars and inner upper wing-coverts, short black bars on inner wing tinged brown and less sharply defined; tail-feathers and secondaries narrower than in adult. FIRST ADULT. Like adult; during 1st winter, early-hatched immatures probably fully adult and not distinguishable, but usually at least some shorter and narrower juvenile secondaries retained (mainly s2–s8), showing more rounded tip than longer and square-tipped fresh neighbouring ones; late-hatched immatures usually retain juvenile outer primaries (often more sharply pointed than in adult, p8 relatively shorter), outer tail-feathers (especially t5; narrower, shorter, and more heavily worn than neighbouring fresh ones), and sometimes part or much of juvenile body. By April–May, remaining juvenile feathers usually replaced by adult ones.

Bare parts. ADULT. Iris yellow, dull yellow, yellowish-brown, or buff. Bare skin round eye pale yellow, yellow, or cream. Cere and basal $\frac{1}{2}$–$\frac{2}{3}$ of bill slate-grey, cere slightly mealy white; bill-tip pale green to very pale greenish-yellow. Leg and foot yellowish-flesh, pale flesh, yellowish-pink, or pale strawberry-red; claws horn-brown to black. JUVENILE. Iris, bare skin round eye, and bill as adult, but iris sometimes tinged brown and cere and base of bill dark brownish-grey. Leg and foot pale pinkish-brown. (Dementiev and Gladkov 1951b; Paludan 1959; Goodwin 1970; BMNH, RMNH.)

Moults. ADULT POST-BREEDING. Complete, primaries descendant. Closely similar to *C. oenas*. Starts with p1 late June to late July, all primaries new late October to late December (–January); not known whether moult suspended during migration. Heavy moult of body and tail late August and early September (Dementiev and Gladkov 1951b), complete before p8 fully grown; hence, apparently slightly earlier than in *C. oenas*. POST-JUVENILE. As in *C. oenas*, and highly variable also. Early-hatched juveniles moult primaries at same time as adult, completing from early November, but late ones complete up to July. In contrast to *C. oenas*, primary moult not suspended in winter. Head and body start slightly before loss of p1, completed with primary moult score of $c.$ 35.

Measurements. Whole geographical range, all year; skins (BMNH, RMNH, ZFMK).

WING AD	♂ 207	(4·80; 13)	198–215	♀ 206 (3·83; 11)	200–214
TAIL AD	103	(4·23; 11)	97–110	101 (2·41; 10)	98–106
BILL	17·4	(0·62; 11)	16·4–18·2	17·5 (0·86; 10)	16·7–18·8
TARSUS	26·3	(0·75; 11)	25·0–27·5	25·5 (0·86; 11)	23·8–26·7
TOE	30·2	(1·17; 11)	28·8–31·3	30·2 (1·38; 10)	28·5–32·1

Sex differences not significant. Juvenile wing and tail both average 10 mm shorter than adult; juvenile tarsus and toe similar from fledging, bill similar when 1st adult plumage obtained.

Weights. Afghanistan, July: ♂ 204 (21·0; 7) 183–234, ♀ 183 (Paludan 1959). North-west India, November: ♂♂ 191, 209; ♀♀ 198, 220 (BMNH).

Structure. Wing shape rather as in *C. livia*. 11 primaries: p9 longest; in adult, p10 2–5 shorter, p8 3–6, p7 17–21, p6 30–40, p1 80–96; juvenile similar, but p8 5–9 shorter; inner web of p10 emarginated as in *C. oenas*. Structure otherwise close to *C. oenas*, but tail and tarsus relatively slightly shorter, average width of tail-feathers $c.$ 4 mm less (within each age-class), outer and inner toe both $c.$ 76% of middle, and large bare patch round eye. CSR

Columba palumbus Woodpigeon

PLATES 29 and 30
[between pages 350 and 351]

Du. Houtduif   Fr. Pigeon ramier   Ge. Ringeltaube
Ru. Вяхирь   Sp. Paloma torcaz   Sw. Ringduva

Columba palumbus Linnaeus, 1758

Polytypic. Nominate *palumbus* Linnaeus, 1758, Europe and North Africa east to western Siberia, eastern Turkey, and Iraq; *iranica* Zarudny, 1910, Iran and southern Turkmeniya (USSR), grading into nominate *palumbus* in eastern Turkey and Transcaucasia; *maderensis* Tschusi, 1904, Madeira (formerly); *azorica* Hartert, 1905, Azores. Extralimital: *casiotis* Bonaparte, 1854, mountains of western central Asia, west to Afghanistan and south-east Iran.

Field characters. 40–42 cm; wing-span 75–80 cm; tail 11–13 cm. Largest of Columbidae in west Palearctic, though closely approached by Long-toed Pigeon *C. trocaz* of Madeira; much larger, broader winged, and longer tailed than smaller congeners and twice as large as *Streptopelia* doves. Large, heavy-chested, long-tailed blue-grey pigeon, with white neck-patch in adult and white wing-crescents and broad black tail-band obvious at all ages. Flight powerful. Sexes similar; no seasonal variation. Juvenile separable. 3 races in west Palearctic, 2 described here (see also Geographical Variation).

ADULT. (1) European, north-west African, and west Asian race, nominate *palumbus*. Head, back, rump, and tail-base blue-grey, contrasting (in most lights) with grey mantle, scapulars, and wing-coverts. Sides and back of upper neck glossy green (with purple sheen); large white patch on sides of neck below this area; sides of lower neck glossy purple, merging into mauve-pink breast. Underbody and rump blue-grey, becoming grey-white under tail. On upperwing, bold white crescent curves round outer feathers of coverts of inner half, forming band obvious in flight; primary coverts grey-black, primaries and tips of secondaries dusky to almost black; white outer fringes to primaries obvious at close range on ground (and even at distance in flight, when they make basal $\frac{2}{3}$ of primaries appear pale). Underwing blue-grey, with darker tips to flight-feathers not obviously contrasting. Tail distinctly patterned, with (above) broad, black terminal band and (below) white central band obvious between dusky base and black end. Bill yellow, with base red-pink under

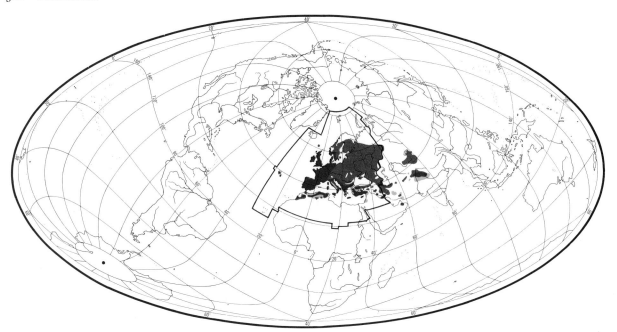

mealy white cere; eye pale straw to lemon-yellow (obvious at close range). Legs mauve-pink. (2) Azores race, *azorica*. Darker and more richly coloured than nominate *palumbus*, with more vinaceous breast and slatier rump; terminal band on tail less intensely coloured, more brown than black. JUVENILE. All races. Much duller than adult, with plumage contrasts suppressed; instantly separable by lack of more than vestigial gloss on neck and total absence of white patch there. Scapulars and wing-coverts show light buff margins at close range.

No other pigeon or dove in west Palearctic has white crescents curving back over mid-wing. Flight powerful, fast, and direct, with constant wing-beats; also includes accomplished escape tactics (with bursts of wing-beats, side-slips, and rapid turns), long glides (including tumbling falls), rapid take-off (with characteristic loud clatter of wings), and precise landing (with heavy flutter). Has almost invariable habit of raising and then lowering tail after landing. On ground, body carriage markedly horizontal; gait less free than smaller *Columba*—often a slow creep when feeding, and with head bobbing in time with steps. Perches freely, and more agile in tree cover than Stock Dove *C. oenas*, being frequently an arboreal feeder.

Wary of man where persecuted but tame in suburban and urban areas. Most characteristic call has well-marked rhythm and sonorous, lazy tone: '(c)oo COOO coo coo-coo'.

Habitat. In upper and lower middle latitudes, continental and oceanic, especially temperate but recently extending through boreal to low-arctic; also breeds marginally in steppe and Mediterranean zones where large-scale wintering occurs. Able to withstand chilly, cloudy, rainy, or misty conditions, but thinning out markedly in face of torrid, arid, frosty, or snowy climates. At once an arboreal, ground, and aerial species; ecologically dynamic and outstandingly adaptable to opportunities arising from changing land-use or climate. Avoids bare rocky mountain regions, densely vegetated wetlands, treeless plains or uplands, and exposed seacoasts. Core of habitat apparently formed by woodland, deciduous or (in parts of range) coniferous, fringing or penetrated by well-vegetated open spaces, especially in lowlands; accordingly by nature an edge-loving (ecotone) species. In Alps, breeds regularly up to 1500–1600 m, even to treeline in places (Glutz and Bauer 1980); in Himalayas in summer, chiefly at *c*. 1500–3000 m (Ali and Ripley 1969). In Britain, nests in woods of ash *Fraxinus* up to *c*. 370 m, but rare in higher woods of birch *Betula* or oak *Quercus*; in lowlands, plentiful in woods of beech *Fagus*, oak and ash, and locally of yew *Taxus*, as well as in young conifer plantations, especially where these neighbour cropland with abundance of food plants (Yapp 1962). Most favoured plantations in Britain are of sitka spruce *Picea sitchensis* and Douglas fir *Pseudotsuga menziesii*, especially those of less than 2 ha and 5–10 m high (Colquhoun 1951). As a grazer on farmland, has succeeded better than Partridge *Perdix perdix*, owing to ability also to forage in trees and shrubs and to greater flexibility of breeding period arising from arboreal nesting and wider daily range (Murton 1971*a*). Where agricultural improvement has broken up forests, and has been associated with scattered pattern of copses, spinneys, avenues, and roadside or hedgerow trees, among field crops of high protein content, basic arboreal habit has been reduced almost to secondary role, at least for feeding. In Finland, before spread of agriculture, original breeding habitat was prob-

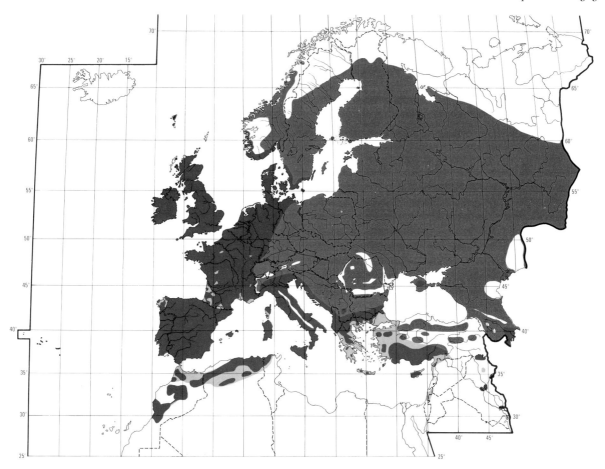

ably forest edges by natural meadows and heaths; most nests in study area, 1971–5, within 50 m of field or other opening, although nesting reported as common in Scandinavia in deep forests, those of Norway spruce *Picea abies* being preferred, followed by Scots pine *Pinus sylvestris*. This preference maintained even where, as in southern Sweden, broad-leaved trees dominant; later growth of foliage may be a factor here, as in England (Saari 1979a).

In central Europe, original nesting habitat was edge of old pine–oak forests, sometimes with spruce also, and foraging was directed to their seed crops, being facilitated by normal foraging range of up to 15 km or more from nests. From early 19th century, however, colonization of towns began, notably Paris, followed by others in lowland deforested regions between Netherlands and Poland and in Lombardy (Italy), extending in 20th century to towns in more wooded landscapes, and to Scandinavia (Tomiałojć 1976b). In some cases, breeding continued in long-occupied areas which became engulfed by expanding cities, while in others colonization occurred from outside, first to parks and cemeteries, then to roadside and garden trees, and eventually even to nest-sites on drainpipes and buildings (Schulze 1951; Simms 1975). In such surroundings, much less wary, but especially safe sites such as tree-clad islands in ornamental waters often sought for roosting. Full use continues to be made also of aerial mobility, in upper as well as lower airspace.

In Iberian winter quarters, readily finds adequate food supplies in extensive managed open woodlands of oaks often underplanted with crops or used for pasturing pigs and other livestock. By local shifts when necessary, normally possible for almost entire wintering population to remain well fed throughout the season from acorns largely collected on the ground, at the same time obtaining secure shelter. (Ortuno and Caballos 1977; F J Purroy, M Rodero, and L Tomiałojć.)

Distribution. Marked range expansion from 19th century or earlier, in Britain and Ireland; in 20th century, northward spread in Fenno-Scandia, and recently to Faeroes, with occasional breeding in Iceland. No longer breeds Madeira.

ICELAND. Bred 1964 and twice since (AP). FAEROES. First bred 1969 (Reinart 1972). BRITAIN AND IRELAND. Marked expansion began in 19th century (and possibly earlier) especially in Scotland; more gradual expansion followed, slower in Ireland; has bred Shetland (Murton 1965a; Parslow 1967; Sharrock 1976). FRANCE. Slight expansion in range since 1936 (Yeatman 1976). NORWAY. Spread in west (Haftorn 1958). SWEDEN. Marked range expansion 20th century, north along Gulf of Bothnia and inland in north (LR). FINLAND. Expanded range in recent decades (OH). LEBANON AND SYRIA. May have bred formerly but no proof (Kumerloeve 1968b; HK). MADEIRA. Extinct, no observations in last 40 years (Bannerman and Bannerman 1965; AZ).

Accidental. Spitsbergen.

Population. Details limited, but range expansion (see Distribution) and other evidence suggest increases from at least mid-19th century in much of Europe (though declined in USSR). Population increase also suggested as important factor in spread of urban breeding (Tomiałojć 1976b). For detailed studies of population dynamics of farmland population in eastern England, see Murton et al. (1964b), Murton (1965a, b), and Murton et al. (1974c).
FAEROES. 6 pairs 1981 (DB). BRITAIN AND IRELAND. Probably 3–5 million pairs 1976. Britain: not less than 5 million birds in July (Murton 1965a). No earlier estimates, but marked increase 19th century, followed by more gradual expansion, mainly linked with increased cultivation (Parslow 1967). Numbers of birds shot in Britain suggest decline after 1962–3 hard winter, rise in late 1960s, followed by some decline in 1970s (Potts 1981). FRANCE. Up to, or perhaps over, 1 million pairs (Yeatman 1976). BELGIUM. About 116000 pairs (Lippens and Wille 1972). LUXEMBOURG. About 12000 pairs (Lippens and Wille 1972). NETHERLANDS. Increased, probably from c. 1950 and still continuing (CSR); fluctuating, estimated 375000–500000 pairs 1959–61 (Doude van Troostwijk 1964a); estimated 425000–500000 pairs 1975–7 (Teixeira 1979). WEST GERMANY. Perhaps 1 million pairs (GR). SWEDEN. About 800000 pairs (Ulfstrand and Högstedt 1976); increased in 20th century (LR). FINLAND. About 164000 pairs (Merikallio 1958). POLAND. Fairly numerous, increasing in west (Tomiałojć 1976a). USSR. Considerable decreases, due to habitat changes and perhaps shooting (Dementiev and Gladkov 1951b). IRAQ. Fairly common, though local (Allouse 1953). CYPRUS. Common in forest areas (PRF, PS). ALGERIA. Perhaps some decrease (EDHJ). AZORES. Decreased, now rare (Bannerman and Bannerman 1966); recent observations suggest more numerous than previously thought (Brien et al. 1982).

Survival. Britain: juvenile mortality c. 74%, annual adult mortality c. 36% (Murton 1965b). Netherlands: 1911–53 (when no bounty paid for shooting) 1st-year mortality 49%, annual adult mortality 50%; 1959–62 (bounty) 1st-year 55%, adult 61% (Doude van Troostwijk 1964a); 1911–81 1st-year and adult both 46% (Glutz and Bauer 1980). Denmark: 1st-year mortality 54·7%, adult mortality 41·3% (Søndergaard 1983). Finland: 1st-year mortality 41·7%, 2nd-year mortality 47·6%, mortality in later years 30·5% (Saari 1979b). Oldest ringed bird 16 years 4 months (Barriéty 1971).

Movements. Mainly migratory in northern and eastern Europe and western Siberia. Partially migratory elsewhere in Europe, with degree of movement declining progressively towards south and west, until largely resident in maritime regions from Britain and Ireland to Morocco, and through Mediterranean basin to Near East. Northern and eastern boundaries of normal winter distribution lie close to 0°C January isotherm (Glutz and Bauer 1980).

Small numbers or scattered individuals remain in Scandinavia and eastern Europe. Chief winter range is from Scotland and Denmark to France, Iberia, and Morocco, then east to Iraq (plus extralimital races *iranica* and *casiotis* from Soviet Central Asia and Iran to northern India). Possibly some European migrants reach north-west Africa (Heim de Balsac and Mayaud 1962). This suggested by very small though annual passages on Malta, September–November and March–April (Sultana and Gauci 1982); rarely seen on Gibraltar, however (Cortes et al. 1980). Southernmost record, Mauritanian coast in April 1981 (Browne 1981), possibly a Moroccan bird.

MIGRATORY POPULATIONS of Fenno-Scandia and eastern Europe depart towards south-west mid-September to early November, at peak in first half October. Reluctant to cross wide seas, and readily deflected along coastlines. Hence 2 distinct migratory streams through Baltic region: (1) across Kattegat and Danish islands, (2) around and across Gulf of Finland. Biggest departures from southern Sweden follow passing of cold front, with north-west to north-east tailwinds, and movement continued in fog when landmarks obscured, though then some westward deviation from normal narrow-front crossing of extreme western Baltic (Alerstam and Ulfstrand 1974; Alerstam 1977). For correlations between migration and weather parameters in Württemberg (West Germany), see Gatter and Penski (1978). Numerous ringing recoveries from Scandinavian population show these winter mainly south-west France and (especially) Iberia, and passage recoveries across Europe concentrated in relatively narrow, direct corridor (Denmark, West Germany, eastern Belgium, Paris basin to Pyrénées) (Rendahl 1965; Lebreton 1969). Radar studies in southern Sweden confirmed migrating flocks set off along this track, with only small deviations caused by local weather (Alerstam and Ulfstrand 1974). Irregular autumn occurrences of Continental birds in (mainly eastern) Britain believed attributable to drift movements in overcast conditions over North Sea (Murton and Ridpath 1962; Murton 1965a); such birds probably originate western Scandinavia (single recoveries from Norway and Denmark) and pass through quickly, so that no noticeable winter augmentation. However, 4 foreign-ringed recoveries in Britain and Ireland in February: Danish-bred bird to Co Leix. Dutch-bred birds in Co Cork and Essex, and Dutch winter-ringed adult in Suffolk.

Finnish and east European birds (data from Hungary, East Germany, Poland, Soviet Baltic States) have broadly same wintering areas as Scandinavian population (see above), though also a few recoveries in Italy (including Sardinia) and Corsica; one ringed Italy in March recovered Ukraine (USSR) in subsequent spring. Passage birds ringed south-west France found in later years in Italy (October, November), Austria, Czechoslovakia, Poland, East and West Germany, Netherlands, Denmark, Norway, Sweden, Finland, and Estonia, USSR (Barriéty 1963, 1964). Some Finnish birds pass south-west in autumn across Sweden, but, in general, north-eastern populations follow coastlines of east and south-east Baltic (some crossing Gulf of Finland) round to Kaliningrad before striking off overland towards France (Kumari 1961; Podkovyrkin 1977). This movement, joined by birds from eastern and central Europe, crosses southern West Germany and Switzerland, and follows a line Jura–Massif Central–Pyrénées across France (Lebreton 1969). In concentrated passage through Jura in 1974, 620 000 passed over Fort de l'Ecluse on one day, 13 October (Géroudet 1976). On this route, peak movement across France occurs 10–20 October, slightly earlier than for Atlantic Zone route where maxima 1–10 November (Lebreton 1969). Less conspicuous return passage across Europe occurs mainly March–April. Longest ringing movement 3675 km: Oulu (Finland) to Lisbon (Portugal).

PARTIALLY MIGRATORY TO RESIDENT POPULATIONS of central, western, and southern Europe have been studied less, except in Britain. Majority leave Alpine regions (e.g. Switzerland) for winter, while proportions of winter recoveries abroad suggest 45–70% are resident in north-west Europe (Glutz and Bauer 1980); c. 57% in Belgium (Lippens and Wille 1972), c. 65% in Netherlands (B J Speek). Degree of emigration probably varies annually with weather and food supply. December–February recoveries of birds ringed West Germany, Netherlands, and Belgium showed many remained within 50 km of ringing site, but minority passed S–SSW as far as French Pyrénées; did not reach Iberia, however, in contrast to some Swiss birds. In general, transient birds from partially migratory populations show same south-westerly tendency as migratory populations (from further north and east), but do not penetrate so far in that direction.

In contrast, British population basically resident, even sedentary, with reports of large-scale movements in autumn–winter often due to misinterpretation of flights to/from roosts (Murton and Ridpath 1962; Murton 1965a). Analysis of 205 November–March recoveries found that 87·5% of adults and 65·5% of 1st-year birds reported within 40 km of ringing site; also evidence for regional variation in degree of dispersal, this least in Scotland where only 20% (all ages) had moved more than 8 km (Murton and Ridpath 1962). No later analysis, but simple breakdown of 1970–81 recoveries ($n = 544$) found: 0–9 km 72·5%, 10–100 km 25·3%, over 100 km 2·2% (BTO). Autumn movement by British-bred birds more marked in some years, e.g. 1959, 1961, 1975; these atypical, and linked to reduced food supply due to early ploughing of stubble fields and failure of acorn crop (Murton 1965a; Boddy 1978b). Then, small numbers cross into France (especially Normandy and Brittany), but only 25 recoveries (1·2%) overall.

Few data on movement between breeding seasons. April–July recoveries for Dutch birds ($n = 87$) were: local 64%, under 10 km 22%, 20–60 km 13%, over 60 km 1%. In study in Hamburg (West Germany), few 1-year-olds bred in natal area (Mulsow 1979). One August migrant, Helgoland (West Germany), shot on nest in Norfolk (England) 5 years later.

Food. Chiefly plant material: green leaves, seeds, berries, buds, flowers, and root crops; occasionally invertebrates. Food taken principally on ground by walking and pecking. Also feeds in trees where remarkably agile, clambering among small branches, and will even hang upside down to reach food. On ground, often feeds in flocks in which feeding hierarchy established; subordinates at front feed slower than birds in middle or rear (Murton *et al.* 1966). In winter, pecking rate on pastures increased from 70 per min in early morning to 103 per min prior to roosting; at same time, area searched decreased (Murton *et al.* 1963a). Most food taken in late afternoon (Gibb and Hartley 1957; Murton *et al.* 1963a; Mathiasson 1967). Percentage food intake of free-living Swedish birds throughout day: 06.00–09.00 hrs 8·5%, 09.00–12.00 hrs 12·7%, 12.00–15.00 hrs 8·8%, 15.00–18.00 hrs 70·0%; for captive bird, 11·2%, 16·5%, 8·8%, and 63·5% respectively (Mathiasson 1967; see also Murton *et al.* 1963a). In Denmark during breeding season, 2 feeding peaks noted: early morning (mostly ♂♂ involved) and late evening (mostly ♀♀) (Jensen and Haarløv 1963). In winter, 95% of daylight hours spent searching for food on pasture, and 64% on clover leys. On ripe cereal or stubble in summer and autumn, only 5–10% of daylight hours spent feeding, usually early morning and late evening (see Murton *et al.* 1963a). In some high-density rural areas, birds tend to forage widely: e.g. up to 15–20 km from nest (Cramp 1958; also Doude van Troostwijk 1964a, Tomiałojć 1976b); in Hamburg area, up to c. 10 km (Mulsow 1979); on Waddenzee islands (Netherlands), up to 20 km to mainland (C S Roselaar).

Adult diet well studied. Includes grain of wheat *Triticum*, barley *Hordeum*, oats *Avena*, maize *Zea*, and rye *Secale*; green leaves of clover *Trifolium*, rape, cabbage, brussels sprouts, turnip, kale, and cauliflower *Brassica*, lettuce *Lactuca*, ash *Fraxinus*, ivy *Hedera*, peas *Pisum*, sugar beet *Beta*, lucerne and medick *Medicago*, mustard *Sinapis*, spurry *Spergula*, radish *Raphanus*, buttercup and celandine *Ranunculus*, chickweeds *Stellaria*, mouse-ears *Cerastium*, sainfoin *Onobrychis*, speedwell *Veronica*, plantain *Plantago*, dandelion *Taraxacum*, nettle *Urtica*, mallow *Malva*, and many other weeds; fruits and seeds of oak *Quercus*, beech *Fagus*, elder *Sambucus*, hawthorn *Crataegus*, hazel *Corylus*, ivy *Hedera*, *Viburnum*, privet *Ligustrum*, holly *Ilex*, blackthorn, plums, etc. *Prunus*, pip fruits (e.g. apples *Malus*), buckthorn *Rhamnus*, dogwood *Cornus*, spindle *Euonymus*, yew *Taxus*, pine *Pinus*, maple *Acer*, rose *Rosa*, crowberry *Empetrum*, bilberry, cowberry, and cranberry *Vaccinium*, blackberry and raspberry *Rubus*, gooseberry and currant *Ribes*, mulberry *Morus*, mistletoe *Viscum*, rape, linseed *Linum*, peas *Pisum* and *Lathyrus*, beans *Vicia* and *Phaseolus*, mustard, *Asparagus*, and of weeds such as chickweeds, vetch *Vicia*, *Ranunculus*, knotgrass *Polygonum*, pansy *Viola*, fat hen *Chenopodium*, wild strawberry *Fragaria*, buckwheat *Fagopyrum*, and grasses (Gramineae); roots, rhizomes, or bulbs of potato *Solanum*, sugar beet, turnip, silverweed *Potentilla*, *Oxalis*, figwort *Scrophularia*, wood anemone *Anemone*, pignut *Conopodium* and *Dentaria*; flowers and buds of ash, beech, cherry, plum, elder, elm *Ulmus*, plane *Platanus*, hawthorn, hazel, oak, walnut *Juglans*, willow *Salix*, and pines; also oak galls, fungus, and moss. Animal food includes: earthworms (Annelida) and their cocoons; small insects including gall wasps (Cynipidae), beetles (Coleoptera), larvae (e.g. *Hybernia*) and pupae of Lepidoptera, scale insects (Coccoidea), and feather lice (Mallophaga); snails (especially *Succinea*, *Helix*, *Nassa*) and slugs (e.g. *Limax*, *Arion*); spiders (Araneae). (Gray 1871; Yarrell and Saunders 1882–4; Witherby 1910; Marriage 1914; Collinge 1924–7; Davies 1930; Harthan 1942; Colquhoun 1951; Havre 1951; Niethammer and Przygodda 1954; Goodwin 1960b; Lohmann 1960; Gasow 1962; Simms 1962; Lebeurier 1963; Doude van Troostwijk 1964b; Murton *et al.* 1964a; Akkermann 1965; Bettmann 1965, 1966, 1981; Murton 1965a; Bannerman and Bannerman 1966; Mathiasson 1967; Ljunggren 1968; Cramp 1972; Schnock and Seutin 1973; Podkovyrkin 1977; Tebbutt 1977; Boddy 1978a; Glutz and Bauer 1980; Bub 1981.) In urban areas, also takes bread, cakes, biscuits, and other scraps (Goodwin 1960b; Ljunggren 1968; Cramp 1972). Recorded eating seaweed at low tide (Yarrell and Saunders 1882–4), possibly for salt, as it also pecks at salt-licks for cattle (Murton *et al.* 1964a) and drinks brackish or salt water occasionally (Yarrell and Saunders 1882–4; Lippert 1975). Usually drinks fresh water from edge of, or in shallows of, ponds, streams, ditches, puddles, or drinking-troughs (Murton *et al.* 1964a), but also recorded drinking ooze from manure heap (Murton *et al.* 1964a; H A Robertson), and sometimes alights on surface of water (MacPherson 1908; Stainton 1978), possibly to drink (see Lippert 1975). Small stones and grit, often collected from roads in early morning, occur most often in stomach when hard grain taken, presumably to break it down (Murton *et al.* 1964a; Bettmann 1966); however, as soil sometimes also taken, may have nutritional value (Warncke 1962).

Diet usually dominated by 1–2 kinds of food, but considerable variation with season and locality. Diet less varied in winter, and when cereals are ripe or have been recently sown or harvested: e.g. 44 food types per 100 crops during cereal growth phase, 28 at sowing, and 17 at ripe and harvesting stages (Schnock and Seutin 1973). Of 30 crops, West Germany, July–November, 60% contained 1 food species and only 1 crop contained more than 2 species, whereas in the same months in Sweden 31% of 107 crops contained 1 species and 46% contained more than 2 species (maximum 7) (Bettmann 1965; Mathiasson 1967).

382 crops collected throughout the year near Oxford (England), showed seasonal variation in content: 41% of fresh weight cereals, mainly wheat and barley, at maximum July–November (up to 85%), but also barley and oats from March–April sowings (up to 53%); 22% tree fruits and flowers, especially hawthorn and ivy fruits, maximum November–January (up to 77%); 20% weeds, mainly November–August, maximum 92% June (mainly *Ranunculus*); 8% peas and beans, important July–September (up to 35%); 6% clover leaves, peaking February (41%); cultivated *Brassica* leaves taken January–March (up to 8%), but winter mild; animal food taken in summer, particularly August (5%). In Herefordshire, 109 crops, June–March contained 47% tree fruits, especially ivy (97% of fruits), 28% clover leaves, 13% turnip leaves, and 12% weeds, of which 59% *Ranunculus*. In Northamptonshire, 38 crops, March–April, contained 35% peas, 13% beans, 13% ivy berries, 12% barley, 8% dandelions, 7% ash flowers, and 5% linseed. 115 crops, October–April, Aberdeenshire, Scotland, contained (% fresh weight) 36% oats (100% of diet of 23 birds in December and April), 33% potatoes (only January–March, 87% in March), 17% beech mast (38% October, 62% November), 5% barley, and 5% turnip leaves (both January–March only). (Colquhoun 1951.) In Cambridgeshire (England), 629 crops collected throughout the year contained (% of items) 40% cereals (39% sown cereals in December, 91% cereal from standing crop or stubble in September), 20% clover leaves, principally January–July (up to 52%); 13% legume or *Brassica* leaves, 8% weed leaves, 6% tree fruits, 3% each of weed seeds, peas and beans, and silverweed rhizomes or sugarbeet fragments, 2% tree flowers, 1% tree leaves and buds, and small numbers of animals (Murton *et al.* 1964a).

In south-west Sweden, of 782 crops (some empty) from rural population throughout the year, 49% contained cereals, 21% rape leaves (mainly January–April), 9% seeds of rape or mustard (mainly in summer), and 8% peas; seasonal variation linked to availability (Ljunggren 1968). In same region, of 138 crops, August–October, 90% contained cereals (64% barley, 50% wheat, 23% oats), 34% clover leaves, and 19% chickweed *Stellaria media*. Other items occurred in less than 10% of crops (Mathiasson 1967).

In Netherlands, March–September, 898 crops contained (% dry weight) 61% cereals (13–83%) and 20% peas (1–34%) (Doude van Troostwijk 1964b). 673 crops from Belgium showed seasonal variation reflecting changes in food availability. During sowing and germination of spring crops (April to mid-May), 91% (fresh weight) cereals or legumes, 4% leaves of clover, buttercup, and dandelion, 3% flower buds of beech, and 2% weed seeds. In cereal-growing phase (mid-May to mid-July), diet extremely varied; only 6% cereals or legumes, 33% leaves of clover, lucerne, and ash, 45% fruits and seeds (mainly *Ranunculus* and chickweeds), 10% rhizomes of *Anemone* and bulbs of *Ficaria*, 4% flower buds, and 3% animal matter. When cereals ripe, and harvested (mid-July to mid-October); 97% cultivated cereals or legumes (including 53% wheat, 29% peas, and 11% barley), 2% seeds of wild vetch, and 1% animal matter. During remainder of year, 36% cereals or legumes (but up to 83% when snow lying), 19% green leaves, and 45% fruits and seeds, especially acorns and beechmast (Schnock and Seutin 1973).

In southern France, October, 116 crops contained (% fresh weight) 60% acorns, 27% beechmast, 11% cereals, and a few green leaves, weed seeds, and insects (Barriéty 1957); for northern France, September–April, see Lebeurier (1963). Of 380 crops from lower Rhine valley (West Germany), December–April, 51% contained *Brassica* leaves (0 in April to 92% January), mostly of brussels sprouts (49% of total number of items); 25% cereals (7% February to 53% March), 8% acorns (22% December and 17% January only), 7% clover, 6% peas (28% March only), 5% vetch (24% December), and 2% beechmast (calculated from Bettmann 1965); for similar study, see Niethammer and Przygodda (1954). In West Germany, 350 crops from all times of year held mostly cereals (56% of crops) and seeds of pasture plants (46%), especially chickweed *Stellaria* (30%); vegetables, mainly *Brassica* (31%), most important (60–80%) in winter (December–February); tree buds, fruits, and flowers in 18%, root crops 11%, grasses 4%, and animal matter 2%; beans and peas recorded only once (Bettmann 1966). For further West German study, see Gasow (1962).

Of 12 crops and stomachs, early September, Karel'skaya ASSR (USSR), 8 contained peas, 6 rye, 2 oats, 2 seedpods and seeds of cow-wheat *Melampyrum pratense*, 1 vetch, 1 bilberries, 1 cherries, and 1 snail *Chochliocopa librica*; bivalve mollusc, grit, and small lump of earth (6·4 g) also found (Podkovyrkin 1977).

Daily food requirements of captive adults 42·5 g wheat per day, and calculated to be at least 50 g per day for free-living birds (Colquhoun 1951); 84 g per day on varied diet or 88 g per day on pure grain (Mathiasson 1967). In winter, daily intake averages 60 g dry weight in clover field, 47 g in pasture (Murton *et al.* 1963a). Observed nutritional value of diet varies seasonally; highest when feeding on cereals and beech-nuts, lowest in winter when, in Aberdeenshire (Scotland), potatoes formed up to 96% of diet (Colquhoun 1951). Starch equivalent (approximates nutritional value) and proportion of digestible protein of main food types both positively associated with food preferences—in descending order: beech-nuts, wheat, barley, oats, clover, kale, turnip (Colquhoun 1951). Rape seed preferred to barley, acorns, or peas in captive bird (Mathiasson 1967). In Vale of Evesham (England), brussels sprouts preferred to cabbages (Jones 1974). Seeds fed to hand-reared birds ranked: (1) wheat, maple peas, peanuts; (2) green peas; (3) hemp, maize; (4) millet, rice, rape, mustard, linseed, sunflower (Murton and Westwood 1964). Crop often weighs more than 70 g; maximum recorded 155 g included 2074 leaf fragments of *Ranunculus* and 725 ivy berries (Davies 1930); 121 g of hawthorn berries (Colquhoun 1951), and 105 g of brussels sprouts (Bettmann 1965) also recorded. Other extreme examples of number of items in crop: 30 cherries, 72 clover leaf fragments, and 10 scale insects; 1087 barley grains; 758 wheat grains; 946 oat grains; 272 beech-nuts; 59 pin acorns *Quercus palustris* and 92 elderberries; 38 acorns of *Q. robur*; 37 acorns, 6 beech-nuts, and 2 cherries; 82 cherries; 363 elderberries; 136 hawthorn fruit and 1373 cabbage leaf fragments; 3755 clover and 183 cabbage leaf fragments; 144 peas and 7 beans; 20 potatoes; 77 snails (Gray 1871; Gladstone 1910; Scone 1925; Hale 1928; Gasow 1962; Bettmann 1966; Mathiasson 1967).

Nestling food well studied in eastern England, where diet similar to that of adults except that animal food, especially snails, and weed seeds more frequent. In July, items in 82 crops comprised 43% (by number) cereals (wheat 34%, oats 6%, barley 3%), 38% mustard *Sinapis* seeds, 11% weed seeds (6% *Ranunculus*, 2% *Cerastium*), 9% peas, 3% snails, 3% beans, and 1% clover. From same area, August–October, 106 crops contained 77% cereals (wheat 66%, barley 8%, rye 3%), 6% weed seeds, 5% beans, 5% *Brassica* seeds, 3% snails, 2% clover, and 1% peas. In Cambridgeshire, August–September, 28 crops contained 67% cereals (27% wheat, 40% barley), 16% *Sinapis* seeds, 9% weed seeds, 3% weed leaves, 1% beech-nuts, and 1% peas. Crop milk also important. Proportion of total dry weight of food intake formed by crop milk decreases as chicks age: 1–3 days, 92%; 3–5 days, 86%; 7–9 days, 49%; 10–14 days, 33%; over 15 days, 21% ($n=86$). (Murton *et al.* 1963b.) Crop milk 65–81% water, 13–19% protein, 7–13% fat, and 1·5% ash; contains amylase and invertase, rich in vitamins A, B, and B_2, but lacks carbohydrates and low in calcium (Needham 1942).

During first days while continuous brooding occurs, young fed at *c.* 1-hr intervals. Gradually fed less frequently; by 8–10 days fed twice per day by each parent, morning and evening, occasionally near midday. Sometimes near end of fledging period fed by only one adult twice per day. (Murton and Isaacson 1962; Erdmann 1971; Harmuth 1971; Kotov 1976a.)   HAR

Social pattern and behaviour. Based mainly on major studies by Cramp (1958) in central London (England), Murton and Isaacson (1962) at Carlton (Cambridgeshire, England), and important summary by Murton (1965a). See also Akkermann (1966) for Oldenburg (West Germany).

1. Often markedly gregarious, especially for feeding and roosting outside breeding season (Colquhoun 1951; Murton 1965a;

Murton *et al.* 1966). Largest flocks reported, up to *c*. 100000, Osnabrück (West Germany), early October/mid-November (Drost and Schüz 1936; see also Akkermann 1966). Sometimes thousands at winter roosts (e.g. Niethammer and Przygodda 1954, Cramp *et al.* 1964). At Carlton, mean flock sizes 57, January–February; 35, March–April; 17, May–June; 11, July–August; 29, September–October; 34, November–December; up to *c*. 1000 at times, January–April (Murton *et al.* 1964*b*). At Oldenburg, stay together as pairs in winter, several pairs sometimes feeding together (Akkermann 1966). For flocking within breeding season, see Murton (1958, 1965*a*), Dittberner (1966), Tomiałojć (1979), and Wittenberg (1980). For flocks of independent juveniles, see Zhelnin (1959), Bettmann (1970), Kotov (1976*a*), Tomiałojć (1976*b*), and Mulsow (1979). 5 summer and autumn flocks, England, contained 0–21% juveniles (Colquhoun 1951, which see also for juvenile–adult ratios in shot samples; see also Murton and Isaacson 1964); at autumn roosts, Berlin, *c*. 50% juveniles (Anders 1975). Regularly associates on passage only with Stock Dove *C. oenas* (Gatter and Penski 1978; also Bergman 1953*b*). BONDS. Monogamous mating system with pair-bond lasting for single season (Bettmann 1970), though longer (possibly for life) in sedentary urban populations (see Cramp 1958, Akkermann 1966, Goodwin 1970, Kotov 1976*a*). For birds forming new pair-bond, timing of pair-formation variable; may begin earlier in towns (Cramp 1958; Murton 1958). Birds thought to be associating as pairs in autumn flocks, USSR (Kotov 1976*a*), and in winter roosts, France (Labitte 1956*b*; but see above). For successful breeding of ♀ *C. palumbus* with ♂ *C. oenas*, see Creutz (1961). First breeding normally at 1 year old, with 1st-years usually *c*. 1–2 months behind adults, and ♀♀ slower than ♂♂ (Murton 1960*a*, 1961, 1965*a*, 1966; Lofts *et al.* 1966). For factors influencing timing of sexual cycle, see Lofts and Murton (1966); see also Lofts *et al.* (1967) for experimental study of photoperiodicity and sexual maturation. Young cared for by both sexes up to and (mainly by ♂) for some period beyond fledging (Piechocki 1956; Bettmann 1970; Erdmann 1971). BREEDING DISPERSION. Solitary and territorial, but several pairs sometimes nest close together. In Białowieża, nests usually *c*. 200–300 m apart; in urban parks, Poland, distance much reduced, sometimes only *c*. 1 m, with several cases of 2 occupied nests, and exceptionally up to 7, in one tree (Tomiałojć 1976*b*, 1979). Otherwise, usually *c*. 15–20 m apart in urban habitats (e.g. Prölss 1956, Gnielka 1978). For many data, including detailed comparisons of density patterns in Europe, see Colquhoun (1951) and Tomiałojć (1976*b*, 1979); also Glutz and Bauer (1980). Highest recorded density in central Europe up to *c*. 460–470 pairs on 36 ha, Centralny Park (Legnica, Poland), of which 330 pairs on 15 ha (Tomiałojć 1979). For further, primarily urban, data from West Germany, including some breakdown by habitat, see (e.g.) Riese (1954), Prölss (1956), Link (1958), Akkermann (1966), and Mulsow (1979). On Spiekeroog island (West Germany), *c*. 20 pairs in copse of *c*. 0·4 ha; habitat destroyed by floods, and birds became ground-nesters with 13 pairs on 1 ha of shore vegetation (Meyer-Deepen 1975). In various English counties, highest density in hedgerows, lowest in mature deciduous woodland (Murton 1960*a*, 1965*a*); see also Colquhoun (1951). High densities also typical of urban parks, cemeteries, zoos, etc. (e.g. Tomiałojć 1976*b*, 1979). Higher densities occur in relatively predator-free areas (Tomiałojć 1979); see also Akkermann (1966). In park, Legnica, local clusters in different places each year. Comparison between rural and urban habitat indicates that in urban parks, young birds attracted to high-density areas (Tomiałojć 1979). Territory established by ♂ and defended mainly by him after break-up of winter flocks or after arrival on breeding grounds (Kotov 1976*a*; Glutz and Bauer 1980); as early as November–December in London, where many territories maintained for most of year and occupied (perhaps not by same birds) for several years in succession (Cramp 1958). Some migratory or wandering populations may also exhibit territorial fidelity (e.g. Murton and Ridpath 1962, Goodwin 1970), but study of ringed birds, Hamburg, showed few 1st-years returned to breed in natal area (Mulsow 1979). Territory used for much courtship behaviour and copulation (latter mainly within territory but away from nest) and rearing of young (Murton and Isaacson 1962). For function of territory (in rural habitat, probably primarily to reduce density and thus nest-predation), see Murton (1958, 1965*a*) and Murton and Isaacson (1962); see also Cramp (1958) and Tomiałojć (1976*b*, 1979). Maintenance of territory may help to ensure sufficient nest-sites for a characteristically multi-brooded species (Cramp 1958; see below). Some vantage-point a normal requirement of territory (Colquhoun 1951). In London, most territories contain several trees; in a square of *c*. 1350 m², one pair defended 2 trees, also adjoining buildings and ground. In larger squares and parks, territory defence largely confined to trees (usually several, particularly those with good nest-sites); territorial bird probably defends (in some cases for much of year) at least one tree regardless of density of birds (Cramp 1958). Territorial bird otherwise reported to defend mainly nest or its immediate vicinity, though size of territory may vary with population density from year to year. Territory size compressible to marked extent so that nests may be only short distance apart (see above; see also Colquhoun 1951); exceptional and excessive compression however leads to territorial disputes and desertion (Murton 1965*a*; also Murton and Isaacson 1962). Usual size of territory quite unrelated to food requirements (Murton and Isaacson 1962). In larger London squares and parks, much of ground used as neutral feeding area, though some ground fights probably territorial. Birds rarely feed in territory, even when it includes ground; trees defended in most territories (see above) provide food only during restricted period (Cramp 1958). Where territorial birds feed to some extent in upper canopy of their territory, intruding ♂♂ usually tolerated there (Murton and Isaacson 1962). Territories usually contain several nests remaining from previous years; an old nest may be used for a repeat or a new one built, usually only a few metres away, exceptionally up to *c*. 20 m (Murton 1960*a*, 1965*a*; see also Tomiałojć 1979). Same nest may be used for up to 4 broods in one season or birds may alternate (see, e.g., Akkermann 1966, Gnielka 1978). Once, nest built early in season but not used until 3rd brood (Riese 1968). Previous year's nest often used (perhaps not by same pair); once for 5 years in succession (Akkermann 1966); exceptionally over 17 years and several times per season (Menzel 1975); see also Harmuth (1971) for use of old nests. Association with owls (Strigidae) and diurnal raptors (Accipitridae) not infrequent (Tomiałojć 1979). In Coto Doñana (Spain), dense clusters of nests sited around those of Black Kite *Milvus migrans* (Cain *et al.* 1982; also Cain and Hillgarth 1974). Similar clusters around nests of Kestrel *Falco tinnunculus* (Wittenberg 1958). For nesting association and other interactions with Hobby *F. subbuteo*, see Collar (1978), and for association with various raptors in Rostov region (USSR), see Kazakov (1976). ROOSTING. Nocturnal, and, particularly outside breeding season, communal, though gregarious tendency much less pronounced in breeding birds (Cramp 1958; Goodwin 1960*b*). Mainly in tall trees (e.g. Alexander 1940). Winter flocks, Eure-et-Loir (France), favoured those with cover of ivy *Hedera*, or conifers, and in deciduous trees used firm leafy branches at *c*. 6–16 m; site used nightly unless disturbed (Labitte 1956*b*). Always in trees, London, particularly on island in lake (Low 1923, 1924; Cramp *et al.* 1964). In Oldenburg area, birds roost in groups of *c*. 10 in larches *Larix* and poplars *Populus*;

in evergreens, more dispersed (Akkermann 1966). In England, larger woods (where many breed) used from about late October, birds from other breeding habitats moving in (Murton and Isaacson 1962). Perches on buildings in London, but no nocturnal roosting recorded there (Cramp et al. 1964). May fly up to c. 65 km between feeding grounds and roost (Colquhoun 1951). In winter, sometimes roosts near temporary feeding grounds rather than returning to traditional site (Murton et al. 1964b). Urban populations, Poland, may forage in fields and return to towns for roosting (Tomiałojć 1976b). Timing of roosting flight markedly light-dependent, late summer and autumn (e.g. Anders 1975). In Tiergarten (West Berlin), traditional flight-paths used, though new pattern established in late September. Birds may go to pre-roost assembly area or direct to roost proper. After perching in tree-tops first move, some time before sunset, to traditional sites for prolonged bathing and drinking. Afterwards flutter up to roost sites singly or in groups; branch-by-branch approach typical. Morning exodus much more rapid than evening arrival (see Anders 1975). According to Lack and Ridpath (1955), weather affects rapidity of departure and size of flocks. Normally all birds leave soon after daybreak (Murton and Ridpath 1962; Murton et al. 1963a). In spring, when some birds beginning to acquire territories, mass exodus still takes place but commuting between roost and feeding grounds occurs throughout day; even more pronounced in summer when food abundant. During breeding season, sexual division of roosting flights reflects pattern of nest-reliefs and territorial activity (Murton and Ridpath 1962; Murton et al. 1963a, 1964b; see also Niethammer and Przygodda 1954 for Bonn area). In early part of breeding season, when territories established, birds may still roost communally in small groups (Goodwin 1960b, 1970; Murton 1965a). ♂ and ♀ perhaps then use separate roosts (Cramp 1958). Often roost in close physical contact once pair-formation completed; when nest-building begins, use nest-tree or nearby site in territory. ♂ normally ceases to roost near ♀ once incubation begun (Goodwin 1960b, 1970). ♀ usually roosts with young until these c. 13–14 days old (Kotov 1976a). Fledged juveniles may initially return to nest to roost; in one case, first roosted away from nest at 27–28 days when parents also not roosting in vicinity (Ryves 1931; also Murton 1965a). Birds often loaf between feeding bouts, in midwinter particularly around midday. May remain on fields, pausing and standing motionless, usually adopting posture with feathers ruffled and head drawn in, often preceded by rhythmic raising and lowering of head (Murton and Isaacson 1962; Murton et al. 1963a, 1964a; Kenward and Sibly 1977, 1978). From October, c. 50% of day spent resting, reduced to c. 36% by late November and early December, and only c. 5% of a potential 8-hr activity period deeper into winter (Murton and Isaacson 1962; Murton et al. 1963a, 1964a; Murton 1965a). See also Mathiasson (1967), Cramp (1972), Kotov (1976a), and Kenward and Sibly (1977, 1978). Generally avoids roosts used by corvids (Niethammer and Przygodda 1954; Akkermann 1966); at one London roost, shared tree with Carrion Crows *Corvus corone* (Cramp et al. 1964).

2. Shows extreme fear of man in most areas (Naumann 1833; Cramp 1958), but tame in many towns (e.g. Cramp 1958). Juveniles may (e.g. Zhelnin 1959, Gnielka 1978) or may not (Tomiałojć 1976b) be less timid. Urban juveniles do not follow adults to feeding grounds to 'copy' them to same extent as Rock Dove *C. livia*; slower than *C. livia* in learning to take food from humans (Goodwin 1955, 1960b). When feeding and partially alarmed, adopts Alarm-posture (similar to Threat-posture, see below): body erect, feathers sleeked, neck extended with white marks conspicuous, and tail depressed (see, e.g., Murton and Isaacson 1962; also Flock Behaviour, below). When disturbed, foraging birds usually make for trees (Goodwin 1955). Healthy birds can outfly Goshawk *Accipiter gentilis* (Kenward 1978a, b, which see for responses to attacks by trained *A. gentilis*). A few seconds after landing (including sometimes on ground), tail almost always (slightly) spread and raised, then slowly lowered to normal position; indicates bird relaxed and intending to stay (Heinroth and Heinroth 1926–7; Goodwin 1956b, 1970; D Goodwin). For alighting on water, see (e.g.) Macpherson (1908); for aerial 'bathing', see Berndt (1962) and Erdmann (1965). For bathing and drinking by flock of c. 500–600 after alighting on water, Black Sea (USSR), see Lippert (1975). FLOCK BEHAVIOUR. Most migrants fly in discrete, shallow, horizontally extended flocks (Gatter and Penski 1978); see also Podkovyrkin (1977) for study of autumn migrants in north-west USSR. Migrant flocks, England, high flying and tightly bunched (Lack and Ridpath 1955). Sudden change of direction and downward plunge elicited by raptor (Gatter and Penski 1978). When high-flying migrant flock attacked by Peregrine *Falco peregrinus*, birds scattered and dived down, heading for cover (Geyr von Schweppenburg 1952; Goodwin 1970). Isolated bird attacked by *A. gentilis*, rejoined compact flock which apparently discouraged raptor (Bettmann 1978). See also Geyr von Schweppenburg (1942) and Kenward (1978a, b). In more general alarm, some birds adopt Alarm-posture; highly infectious, may lead to departure. In winter when food scarcer, birds less prone to take alarm. On stubble or tall clover, viewing conditions less good, leading to more frequent use of Alarm-posture, general restlessness, and leap-frogging flights by birds at rear (Murton and Isaacson 1962). When some on ground and others in trees, panic-type escape at approach of man progressive, not instantaneous; some remain, alert, peering about after bulk has gone; do not flee until danger visible (Goodwin 1970). Altitude when approaching roost weather-dependent (Ash et al. 1956). A few may apparently reconnoitre, circling roost-trees then plummeting to perch prominently; remainder joins them normally without such manoeuvres (Bettmann 1970). For long period in midwinter, flock commuting between roost and regular feeding grounds typically maintains same composition; little interchange between flocks (Murton and Isaacson 1962; Murton 1965a). Roosting individuals normally maintain pecking distance between each other (Goodwin 1960b). For morning departure, see (e.g.) Murton and Ridpath (1962). Juveniles tend to congregate in small parties at edge of wood; apparently reluctant to leave cover, feeding as near wood as possible; join adults on feeding grounds only later in year (Murton and Isaacson 1964; Murton 1965a). Perched or feeding birds, particularly large flock, strong stimulus for others to land; white wing-markings presumed to encourage landing by others. Will also join foraging *C. oenas* (Murton 1965a). Birds more dispersed when foraging fast on grain sowings (Murton et al. 1966); individual distance of c. 30–60 cm maintained (Murton and Isaacson 1962). See also Goodwin (1960b) for comparisons with *C. livia* and *C. oenas*. Flocks most integrated early in day when all birds hungry; later break up, some returning to roost, others loafing nearby; pattern most pronounced autumn and spring (Murton 1965a; see also Roosting, above). Where undisturbed, birds of flock remain as a unit, gradually moving in same direction across field, with very few upflights or changes of course. Flock usually a circle or elongate oval. Any joining flock (see above) tend to avoid middle 80% of birds. Pecking rate of front 10% of birds much lower than others; noticeably more uneasy, constantly looking up; submissive to others. Birds leave mainly from front and rear. Normal and underweight adults dominant over juveniles. Smaller flocks contain more poor-condition birds. Front birds show marked lack of co-ordination compared with dominants. Subordinates exhibit lower weight, greater adrenal stress

and poorer survival. May attempt to establish themselves in other flocks, but if food scarce and birds then forced to feed alone, spend too much time watching for predators and are thus even less successful than subordinates in flock (see Murton *et al.* 1966, 1971). See also Kenward and Sibly (1978). In winter, antagonism mostly takes form of supplanting attacks—often subtle and difficult to detect—rather than fights proper (Murton and Isaacson 1962; Murton *et al.* 1966). Heterosexual encounters may also occur; accosted bird may run or fly off, peck or strike with wings, or courtship display (rarely mating) may develop, 2 birds departing together (Murton 1958; Murton and Isaacson 1962). Both antagonistic and heterosexual interactions may be triggered by head-down (i.e. submissive) posture of feeding bird (Murton 1965a; see also below). ANTAGONISTIC BEHAVIOUR. (1) General. Territorial behaviour begins much earlier in towns, where breeding density high; a few ♂♂ often sit quietly in territory for long periods from November (Cramp 1958). Territory otherwise marked by Advertising-call (see 1 in Voice) and Display-flight, sometimes in combination (e.g. Akkermann 1966, Harmuth 1971). In London, Advertising-call may be given from December or January to late September but mostly March–June; in early period, given 2–3 times from territory after arrival from communal roost, and before flying off to feed (Cramp 1958; D Goodwin). At Carlton, at peak of breeding season, given more or less all day; maximum at about peak laying. Same song-posts used daily (Murton and Isaacson 1962; see also Breeding Dispersion, above). In Halle (East Germany), declines by laying, also towards end of season (Gnielka 1978). Bird sits on branch (sometimes at nest-site: Goodwin 1970), often rather crouched; crop region swollen, moving markedly with call (Naumann 1833). White feathers on neck ruffled (Heinroth and Heinroth 1926–7). More rarely given on ground (e.g. Ruttledge 1974), exceptionally in flight (Naumann 1833). (2) Display-flight. See Fig A. Bird flies forward and up (*c*. 20–30 m), suddenly ascends yet more steeply and, at apex, gives 1, usually more (up to 9: Akkermann 1966) loud Wing-claps (see Voice). According to Akkermann (1966), caused by contact of wing surfaces; more probably produced by whip-like crack on down-stroke (Murton 1965a). With wings at or slightly below horizontal, bird glides forward, then gently down (*c*. 6–7 m) with tail spread. Performance given up to 5 times before landing (Akkermann 1966). Advertising-call given before take-off and after landing; then return flight made. At low intensity, no audible Wing-claps (Goodwin 1955; Kotov 1976a). For signal function of tail-markings, see Goodwin (1955). After pair-formation and during incubation, lower-intensity Display-flights performed *c*. 20–30 m from nest (Kotov 1976a). Display-flight given much less frequently than Advertising-call; most common early in year when most territorial disputes take place. Strongly infectious; may also elicit Advertising-call in territorial ♂ perched below (Cramp 1958; see also Heinroth and Heinroth 1926–7, and Breeding Dispersion, above). In London squares or where birds using isolated trees or hedgerows, Display-flight probably more confined to territories; otherwise more wide-ranging, including neutral ground (Goodwin 1956b; Cramp 1958; Murton and Isaacson 1962; Murton 1965a; Akkermann 1966). Birds may glide and Wing-clap on leaving territory (see also Parental Anti-predator Strategies, below) and when passing over feeding ground (Murton and Isaacson 1962; Murton 1965a). For Display-flight by ♀, see Heterosexual Behaviour (below). Rapid circling pursuit-flights (significance unknown), said to be performed above trees (Selous 1901). (3) Threat and fighting. Territory most vigorously defended prior to laying, less so during incubation. Territorial activity usually reduced during early fledging; recrudescence apparent only with start of new cycle (Murton 1965a). During early part of season, intruders (mainly unpaired ♂♂) usually quickly expelled (Murton 1965a; Tomiałojć 1979; see also below). Woodland birds show more marked aggression in lower canopy, where nests eventually built (Murton and Isaacson 1962). Defence not continued during night and small communal roosts may be established and centred on one territory where owners hostile but do not evict intruders until following morning. Birds may be persistent in trying to settle in occupied areas (Cramp 1958; Murton and Isaacson 1962). Intrusion by paired birds more likely to result in fierce fights, and sometimes success of such intruders. In Białowieża forest, Poland (where density low), mutual avoidance, rather than threat and fighting, characteristic (Tomiałojć 1979). In London, where density high, breeding adults fiercely aggressive with much chasing and fighting (Goodwin 1955). If intruding ♂ does not depart completely at advance of territory owner, latter usually resorts to series of supplanting attacks, chasing and harrying intruder from branch to branch; often leads to expulsion (Cramp 1958). More persistent attitude by intruder elicits Threat-posture in other: neck extended, head upright, wing-tips slightly raised, and back feathers ruffled; may be adopted prior to supplanting attack (Murton 1965a). ♀ followed by one ♂ caused retreat of 2nd ♂ performing frontal Bowing-display (see below) by assuming Threat-posture with wings half open (Cramp 1958). Threat-posture often accompanied by Nodding: downward pecking movements with wings half open given much as by *C. livia*, similarly

A

indicates unwillingness to yield (Goodwin 1956a; Cramp 1958). According to Murton (1960b) and Murton and Isaacson (1962), probably different from Nodding or Bill-lowering movements given in Nest-calling. Advertising-call additionally given as threat to close neighbours and intruders; special anger call also at high intensity (see 7a in Voice). Submissiveness in intruder may elicit Bowing-display (ruffled back feathers show aggressive tendencies dominant) from territory-owner; intruder normally retreats (Murton and Isaacson 1962; Murton 1965a). In Bowing-display, tail raised and fanned but usually closed again before it reaches highest point (culmination of Bow); pupils strongly contracted; tail raised only as forward movement stops, so that bird walking after another gives only Bow and Display-call (see 2 in Voice), not raising tail (see also subsection 2 in Heterosexual Behaviour, below). Jumping performed towards (or away from) other bird in antagonistic and heterosexual encounters. Tail not spread or depressed when approaching another bird, but ♂ may rarely Jump and Wing-clap prior to attack or display (Goodwin 1955, 1956b, 1970). Incubating ♂ may first give Advertising-call, then attack intruder. (Murton and Isaacson 1962, 1964.) Territory-owner may also threaten intruder in the air, flying up with loud Wing-clap; ♂ who finally succeeded in establishing territory despite intermittent attacks from longer-established bird twice performed floating flight with deep, slow wing-beats after driving off other ♂. Another ♂ gave Nest-call (see 3 in Voice) with vigorous up-and-down tail movements after evicting intruder (Cramp 1958). Fights may be preceded by mutual challenge, one bird then giving full Display-call prior to attack and sometimes calling following successful eviction (Akkermann 1966: see 2 and 7b in Voice). May last up to 2 hrs and at times so violent that feathers fly. Birds may face one another or stand side by side, Threat-posture (see above) and advance alternating with anxious withdrawing movements of head and retreat; striking with wings, pecking, leaps, and attempts to land or clamber on to other's back and aerial buffeting occur. Bowing-display given by dominant individual as opponent retreats (see, e.g., Murton and Isaacson 1962); also, particularly after ground fight, by defeated bird (Cramp 1958). According to Cramp (1958), raising of wings indicates aggressive rather than escape tendency dominant (but see *C. livia*). As ♀♀ also defend territories, pair-versus-pair fights frequent (Murton and Isaacson 1962); ♀ once remained on nest, Wing-twitching and Nodding until ♂ had evicted intruding pair (Akkermann 1966). No pursuit after fight and expulsion (Prölss 1956). Recently fledged juveniles who trespass may be fed by territory owners but often fiercely attacked in situations where other adults tolerated (Goodwin 1960b, 1970; Murton and Isaacson 1962). Interference with copulation by 2nd ♂ rare, and successful copulation sometimes accomplished in feeding flock (see Cramp 1958, Murton 1958, 1965a, Murton and Isaacson 1962, Goodwin 1970). HETEROSEXUAL BEHAVIOUR. (1) General. Pair-formation may apparently take place when no territory held, but mostly occurs within territory (Murton and Isaacson 1962). ♂♂ take up territories straight after arrival, Orenburg, USSR (Kotov 1976a). Birds of a pair in migratory or roving populations apparently recognize and accept each other immediately when they return (Goodwin 1970). Display more restricted early in season if food less abundant and cold spells may prolong pairing (Cramp 1958; Murton and Isaacson 1962). Marked diurnal activity-rhythm (determined by food supply) regulates timing of territory acquisition and all later stages of pair-formation and breeding cycle. Pair normally together on territory, and most vocal, *c*. 07.00–09.00 and *c*. 15.00–17.00 hrs (Murton and Isaacson 1962; Murton 1965a; see below). ♂ most active in courtship during nest-building. Unpaired ♂♂ quite often attempt to court ♀♀ on feeding grounds, at watering place, or in other territories (Kotov 1976a). Recrudescence of courtship activity may occur when previous brood still in nest (Murton and Isaacson 1962). For vocal and courtship activity (including mating) outside breeding season, see (e.g.) Bettmann (1961), Radermacher (1962), Schumann (1963), Peitzmeier (1974). (2) Pair-bonding behaviour. Unpaired ♂ may sit in tree silently and self-preen intermittently. Attracts ♀ variously with Advertising-call, Nest-call, and Display-flights (e.g. Cramp 1958, Harmuth 1971a). ♀'s approach typically indirect and cautious. ♂ usually uncertain and hostile at first, adopting Threat-posture. Use by ♂ of Nodding (see above) further indicates sexual/aggressive conflict. ♀ generally maintains submissive head-down posture; may also Nod briefly (sexually excited bird lowering head only slightly: Goodwin 1955) and retreat, or also change to Threat-posture (Cramp 1958; Murton and Isaacson 1962; Murton 1965a). Repetition of Bowing-display by ♂ (see above) has apparent appeasing effect, encouraging ♀ to stay. ♂'s neck inflated to show white patches and iridescent green; wings closed and Display-call given. Also given on ground by unpaired ♂ or paired ♂ displaying to strange ♀. ♂ may walk after ♀; at higher intensity, Jumps up to *c*. 30 cm high and forward with feet tucked under (Cramp 1958; see also Selous 1901). In abrupt leaps, given 5–10 times, tail may brush ground. Before Display-call given crop region inflated. In Bowing-display (following Jumping) eyes closed to narrow slit (Harmuth 1971; Kotov 1976a). Jumping movements comparable to Driving in ♂ *C. oenas*; bird acts as if to mount and copulate (Murton and Isaacson 1962). For further interaction on ground, see Bettmann (1978). Wing-lifting (less pronounced) and Parading performed as in *C. livia* and *C. oenas*; head held high, rump feathers ruffled, wings slightly opened, and slight side-to-side movements made (Goodwin 1956b). According to Colquhoun (1951), ♂ performing Display-flight to mate often followed by 3rd bird. Like *C. livia*, ♀ accompanying ♂ in flight often gives low-intensity variant of Display-flight if ♂ does so; ♀ exceptionally performs Display-flight apparently on own initiative (D Goodwin). ♀ often joins ♂ on territory after initial approach but at first keeps her distance (even where birds arrive apparently paired); pair formed when ♀'s fear sufficiently reduced for her to allow ♂ close approach (Cramp 1958; Murton 1965a). Nest-calling given typically when ♂ and ♀ present (and paired)—from about early January in London (Cramp 1958); when alone, ♂ more likely to give Advertising-call or, exceptionally, Nest-call, in normal posture (Goodwin 1956b, 1970). According to Harmuth (1971), ♂ changes from Advertising- to Nest-call when ♀ within 10 m. Potential nest-site advertised and ♀ (or potential mate) attracted by ♂ adopting crouched posture with head lowered and tail raised, Nodding and giving Nest-call (see 3 in Voice). At low intensity, neck lowered only slightly; at higher intensity, head lowered steadily until bill movements directed at breast or pecks made at branch or twigs of old nest; may culminate in twig-fixing movement (Goodwin 1956a). Wing-twitching and sometimes Tail-jerking performed by ♂ (Cramp 1958; Murton and Isaacson 1962). ♂ Wing-twitches with left, right, or both wings depending on ♀'s approach (Goodwin 1970). ♀ usually comes closer as soon as ♂ begins Nest-calling, and gives less vigorous Nodding and (less frequently) variant of Nest-call (Cramp 1958, Murton and Isaacson 1962: see 3b in Voice); according to Harmuth (1971), ♀ Wing-twitches more than ♂. Nest-calling may continue for up to *c*. 4 hrs or cease after a few minutes (Akkermann 1966; see also Klinz 1955). Often observed well before nesting; important for strengthening pair-bond (Cramp 1958). ♀ may peck ♂, usually gently (Cramp 1958); pecking of ♂ by ♀ frequent, as in *C. oenas* (Goodwin 1956a). Soon after mutual acceptance, ♀ more likely to Allopreen ♂, concentrating vibratory movements of bill on his ruffled neck feathers and area round base of bill and eyes; usually leads to mutual Allopreening (Murton 1965a; Harmuth 1971). Also occurs away from nest, once leading to mating by pair at edge of feeding flock (Murton and Isaacson 1962; often frequent and prolonged early in year (Cramp 1958); continues throughout nest-building, but rare after start of incubation (Murton and Isaacson 1962), though also performed outside normal courtship period by older pairs with long-term bond (Akkermann 1966). (3) Driving. Both on ground and in the air less developed than in *C. livia* and *C. oenas*; may occur on feeding ground away from territory, when ♂ Drives ♀ away from feared rival ♂, chasing and attacking her until she flies off and he follows (Cramp 1958; Goodwin 1963). (4) Courtship-feeding (Billing). Unlike in *C. livia*, not preceded by circling. Typically, ♂ flies to suitable perch where joined by ♀; ♂ then offers open bill and ♀ adopts submissive posture as in Allopreening, but her bill now closed and inserted into open bill of ♂. Food passed at least sometimes (Cramp 1958; Murton 1960b, 1965a; D Goodwin). According to Prölss (1956), bills interlocked and heads moved about for several minutes; said to be briefer, less pronounced than in *C. livia* (Harmuth 1971; Kotov 1976a). Normal prelude to mating, but often occurs early in season without leading to this (Cramp 1958). (5) Mating. Most frequent during nest-building; isolated instances at other times including in winter flocks (Cramp 1958; Murton 1958; Murton and Isaacson 1962); mostly in morning, late afternoon, or evening, particularly in cool, damp conditions (Harmuth 1971). Usually on favoured branch in territory, not on nest (Haverschmidt 1932; Cramp 1958; Murton and Isaacson 1962; Murton 1965a). Often initiated by ♀ (Akkermann 1966). Pre-copulatory behaviour typically includes Mock-preening, followed by Billing (see above), quite often alternately (Haverschmidt 1932; Cramp 1958). According to Godwin (1963), Billing not always preceded by Allopreening which anyway indicates one or both not ready to copulate; ♀ may, however, Beg (i.e. nibbling at ♂'s bill). Mating may follow directly from Bowing-display (Klinz 1955; Cramp 1958); more often than in *C. oenas* (Goodwin 1956b). ♀ may Bill and Allopreen ♂ if he is reluctant to mount (Cramp 1958). Repeated walking back and forth along branch by one bird, or long-stepping run with wings drooped (probably Parading), Jumps, and turns also occur as prelude to copulation (Haverschmidt 1932; Cramp 1958). ♀ may crouch 2–3 times between bouts of Billing; ♂ mounts from side, supporting himself on ♀'s slightly arched wings, and pushes ♀'s tail aside. Copulation lasts up to 5 s, ♂ flapping wings (Haverschmidt 1932; Harmuth 1971). If either bird loses balance, leads to aggressive posturing, Nodding, and anger call (Murton 1965a; Murton and Isaacson 1962: see 7a in Voice). In one case, birds beat wings rhythmically between acts of copulation (Neumann 1942). Post-copulatory behaviour variable (Cramp 1958). Both birds may Mock-preen and Bill (Haverschmidt 1932), mock-feed or actually feed (Murton and Isaacson 1962). ♀ may raise wings and call, head bowed on stiff neck (Haverschmidt 1932: see 4 in Voice). ♂ adopts an upright posture with neck feathers ruffled, and bill held stiffly against breast. Wings slightly spread and tail spread to touch ground; may remain thus motionless for *c.* 1 min (Prölss 1956; Kotov 1976a); call given (Klinz 1955: see 4 in Voice). Mutual Allopreening, period of relative quiet with birds sitting hunched up, and self-preening, all typical of period after copulation (Haverschmidt 1932; Prölss 1956; Cramp 1958; Harmuth 1971). Jumping by ♂ (Harmuth 1971), flying off with Wing-clapping (Klinz 1955), hopping back and forth and turning by ♀ (Haverschmidt 1932), and floating flight by both afterwards (Neumann 1942) also reported. Both birds may call (Akkermann 1966: see 4 in Voice). Reversed mounting occurs (Akkermann 1966). (6) Nest-site selection. ♂ may Nest-call at several sites, ♀ choosing one; according to Akkermann (1966), ♀'s choice indicated by Wing-twitching. (7) Behaviour at nest. Building quite often follows mutual Allopreening, ♂ leaving to collect material mainly within territory (Murton and Isaacson 1962); within (8–)40–100 m (Harmuth 1971; Kotov 1976a; Gnielka 1978). Most building done by ♀ in morning and early evening (Harmuth 1971; Kotov 1976a). Nest-building and Nest-calling often simultaneous (Murton 1965a). ♀ may Nest-call (see 3b in Voice; also Piechocki 1956), Nod, and Wing-twitch; ♂ sometimes Allopreens ♀ when he brings material (Kotov 1976a). ♂ usually jumps on to ♀'s back, presenting material to her from behind (Goodwin 1955; Harmuth 1971). ♀ may make rotating movements (Dathe 1967). When pair meet at nest, ♀ may initially assume Threat-posture and fighting may occur. Such a development prevented by Head-hiding ceremony (Fig B): ♂ lowers head and withdraws

B

neck with bill pointed down; pushes head under ♀'s body so that head and white neck-marks inconspicuous. ♀ may then lower head (aggression waning), allowing ♂ to straighten up and eventually to Allopreen her. Both may then Nod asynchronously (Fig C), often standing side by side (Murton 1960b; see also

C ♂ ♀

Goodwin 1956a, 1960a). Either bird may sit quietly on nest before laying; nest-reliefs during such False-sitting as later (Murton and Isaacson 1962; see below). If both present, ♂ may Allopreen ♀ (Cramp 1958). False-sitting normally for brief period (Tomiałojć 1979); for extreme case, see Cramp (1958). Around laying, ♂ guards nest-site and accompanies ♀ on foraging or watering flights (Kotov 1976a). ♀ said to stand over 1st egg without incubating properly until 2nd laid (Klinz 1955). ♂ usually incubates *c.* 10.00–17.00 hrs (Murton and Isaacson 1962; Harmuth 1971). Sitting bird normally leaves when mate on nearby perch or nest; sometimes gently urged off. Submissive and appeasing postures, Wing-vibrating, and (occasionally: Murton 1965a) Allopreening may occur. Either bird may call (Murton and Isaacson 1962: see 5 in Voice). Off-duty bird usually leaves territory immediately (Babin 1913; Cramp 1958). Sharp Wing-clap given on several occasions by departing bird (Ryves 1931). According to Cramp (1958), full Advertising-call normally ceases with start of incubation, but occasionally given by sitting bird. Often given by ♂, never by ♀, late in incubation (Murton and Isaacson 1962; Murton 1965a). ♂ once fed incubating ♀ (Wolff and Gehren 1951). RELATIONS WITHIN FAMILY GROUP. See also Bonds (above). Nestlings brooded continuously, with nest-reliefs as during incubation. First nestling fed c. 5–7 hrs after hatching (Kotov 1976a). Able to stretch up and give weak Food-calls (see Voice) from 2nd day (Murton and Isaacson 1962; Harmuth 1971; Kotov 1976a). Generally more or less passive during feeding for first c. 5 days (Harmuth 1971). Adult may stimulate nestling to thrust head forward by pecking and holding its bill (Murton and Isaacson 1962). In early stages (up to 5–7 days: Kotov 1976a), 2 nestlings normally face parent and are fed simultaneously (alternately later) by regurgitation, pushing bills into adult's at side (Prölss 1956; Murton 1960b; Harmuth 1971). Older nestling may squeeze out younger during feeding (e.g. Kotov 1976a). More vigorous Begging, with young calling and Wing-flapping (against adult's back and shoulders) at approach of parent, reported to coincide with gradual admixture of more solid food in diet (Erdmann 1971; Harmuth 1971; see also Colquhoun 1951). ♂ normally announces arrival with several short (otherwise undescribed) calls and makes gradual approach, while that of ♀ silent and direct (Erdmann 1971). Feeding of one nestling took c. 5 min (Piechocki 1956). At c. 6–7 days, young remain motionless in nest between feeds (Erdmann 1971), but older young may give food-calls even when parent not present (Murton 1965a), although mostly rest quietly when left alone (Piechocki 1956). Chick which fell from nest ignored by parents despite food-calls (Erdmann 1971). Able to self-preen at c. 11–13 days but also Allopreened by parents (Kotov 1976a; see also below), self-preening and wing-exercising frequent at c. 3–4 weeks (Harmuth 1971). Varied reports on brooding: day-time brooding probably ceases in most cases in 2nd week, while ♀ continues at night. Usually resumed in cold, wet weather, even when chicks almost fully fledged (e.g. Ryves 1931, Piechocki 1956, Kotov 1976a, Gnielka 1978; see also Breeding). Once, young bird brooded away from nest on day of fledging (Piechocki 1956). First short excursions from nest probably in most cases at c. 21–26 days (see Heinroth and Heinroth 1926–7, Piechocki 1956, Kotov 1976a); at c. 14 days according to Klinz (1955). Young may return to nest initially, also for roosting (see part 1), or be fed nearby, then begging only with Wing-flapping (Ryves 1931). ♂ increasingly cares for young in later stages, when ♀ re-nesting (see, e.g., Babin 1913, Ryves 1931, Erdmann 1971). Young usually remain several days in parental territory after fledging (e.g. Kotov 1976a, Gnielka 1978); for calls, see Ryves (1931) and Voice. In one case, expelled by ♀ (using pecks and blows from wings) at 33 days, but fed by ♂ 10 days later (Piechocki 1956; see also Stöcklin 1952, Goodwin 1960b, 1970). According to Gnielka (1978), then move right away; elsewhere reported to accompany adults to feeding grounds and watering place for c. 3–5 days after fledging at 34–35 days (Kotov 1976a). For report of 5 juveniles being fed by adult, mid-September, see Wavrin (1971). ANTI-PREDATOR RESPONSES OF YOUNG. During first few days, usually crouch motionless in nest during disturbance (Akkermann 1966); cease food-call temporarily on hearing warning-call of parent (Kotov 1976a: see 6 in Voice). From c. 2 weeks, or from c. 7–8 days if attempt made to touch them, young adopt Defensive-threat posture: grip twigs of nest, rear up, inflate body, at times thrust head forward violently, and give varying combinations of hissing (see Voice), bill-snapping, pecking intention-movements, and (when older) Wing-flicking; may also lie on side and vertically raise wing facing away from predator, or stand with both wings raised; spitting, or regurgitating of food also occur. (Based mainly on Akkermann 1966; also Colquhoun 1951, Murton 1958, Harmuth 1971, Kotov 1976a.) 2 well-grown young from ground nest walked away to hide in nearest tuft at approach of man (Banks 1969). Usually restless at approach of man when close to fledging (Colquhoun 1951; Tomiałojć 1979), though not in London (S Cramp). PARENTAL ANTI-PREDATOR STRATEGIES. (1) Passive measures. Generally leaves nest when disturbed in early stages and quite likely to desert or at least to be a long time (up to 8 hrs: Murton 1958) in returning (Naumann 1833; Murton et al. 1963b; Harmuth 1971; Kotov 1976a). One regularly covered eggs with small twigs before departure (Trelfa 1953). In towns, birds usually sit tight and allow much closer approach by man: average 2·9 m ($n = 90$). No marked variation with stage of nesting cycle (Wittenberg 1969), though tight-sitting said elsewhere to be more common either side of hatching (e.g. Cramp 1972, Kotov 1976a, Tomiałojć 1976b). In extreme cases, bird may move only a few metres away in nesting tree, and return as soon as nest inspection completed (Heinroth 1911), or have to be pushed aside so that (hatching) eggs could be inspected (Gnielka 1978). According to Akkermann (1966), adult generally makes silent departure from nest with young when confronted by human or other animal predator, but see also below. May make indirect return to nest at this stage of breeding cycle (Harmuth 1971). Tends to stay put also in presence of corvids (Murton and Isaacson 1964; Murton 1965a; Tomiałojć 1979); in central London, left silently and without display only at close approach of *Corvus corone* (Cramp 1972). (2) Active measures: against birds. One pursued *C. corone*, once in Display-flight (see Antagonistic Behaviour, above), and several times in normal flight (Creutz 1967). (3) Active measures: against man. Wing-clapping often elicited by human intruder. Bird may Wing-clap and glide when disturbed from nest (Murton and Isaacson 1962; Murton 1965a). Departs thus from nest during building and early incubation (Kotov 1976a). Particularly after hatching, also not infrequently late in incubation (even during False-sitting: Wittenberg 1969), may perform low- or high-intensity distraction-lure display of disablement type. Recorded in 53 out of 104 cases of bird being disturbed from nest, in ⅔ when young present (Wittenberg 1969); 3 out of 200 times (Course 1943). Adult (probably ♂ in most cases: Kotov 1976a) makes apparent free-fall ('stone-like drop': Course 1943), sometimes angled, from nest to ground (see Heydt 1968 for illustrations, also of subsequent behaviour); once flapped around first in bush near nest (Gnielka 1978). At low intensity, bird then performs impeded-flight, wings not fully open, close to ground or cover for 40–300 m, sometimes making audible brushing sound (Harmuth 1971). Less common high-intensity variant typically includes stationary or mobile flapping or twitching, with tail spread; often interrupted and may be combined with low-intensity variant in performance lasting up to 5 min, bird usually flying off eventually (Wittenberg 1969; also, e.g., Sehlbach 1908, 1920, Tebbutt 1942, Stöcklin 1952, Harmuth 1971). For report of distraction-display only after bird had flown (apparently normally) c. 60 m from nest, see Walpole-Bond (1938). For further discussion, see Heinroth and Heinroth (1926–7) and Tebbutt (1942). (4) Active measures: against other animals. One ♂ evicted squirrel *Sciurus* by flying at it and striking with wings (Dathe 1967).

(Fig A from field-sketches by C J F Coombs; Figs B–C based on photographs in Murton 1960b.) MGW

Voice. Much individual variation in pitch, timbre, duration, and rhythm of calls. ♂ more vocal than ♀ (Akkerman 1966). According to Klinz (1955), ♀'s vocalizations sporadic and much softer, though repertoire basically the same. For possible reasons determining separate development of Advertising-call and Nest-call, see Goodwin (1956b) and below. ♂ may occasionally give Advertising-call at night (Klinz 1955). Normal flight accompanied by

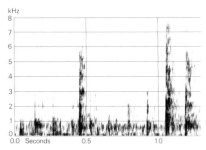

I R W Genever England July 1971

II P A D Hollom England June 1976

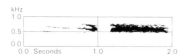

III P A D Hollom England June 1976

a sighing whistling 'wichwichwichwich' from wings (Naumann 1833); not given by juveniles (Colquhoun 1951). For Wing-clapping, see Social Pattern and Behaviour; Fig I shows 1 Wing-clap interspersed between wing-beats and 2 more presumably at the apex of the Display-flight.

CALLS OF ADULTS. (1) Advertising-call (Song). Throughout a song-bout each phrase has 5 units, except last which usually ends with a sometimes faint '(c)oo(k)'. Stress falls mainly on 2nd unit. Recording of last phrase of a bout (Fig II) suggests '(c)oo COOO coo coo-coo (c)ooo(k)', in which initial unit faint but ascending in pitch; final '(c)oo(k)' almost steady but with very slight ascent in pitch and with duration of c. 240 ms, c. 120 ms less than initial unit (J Hall-Craggs, M G Wilson). Initial unit, although faint and audible only at close quarters (sometimes more like 'woe' or 'ke': M G Wilson), is normally given (Goodwin 1956b, 1970); for claim that Advertising-call usually begins with a stressed 'cooo', see Huxley and Brown (1953). Recording (Fig II) indicates final '(c)oo(k)' to be special sound used only at end of bout and not same as 1st unit of each phrase (J Hall-Craggs); omission said to be more likely in (less usual) bout of 5–6 phrases (see Huxley and Brown 1953). For further discussion of final unit, see (e.g.) Naumann (1833), Heinroth and Heinroth (1926–7), Bettmann (1959), and Akkermann (1966). Up to 13 phrases given per bout, usually 3–5 (Bettmann 1959); 2–4 (Goodwin 1970); see also Huxley and Brown (1953). Average duration of 4-phrase bout 9–13 s (Akkermann 1966). Advertising-call shows considerable individual variation with regard to pitch, rhythm (i.e. shifting stress—see Akkermann 1966 for numerous examples), and time intervals between syllables and phrases; some syllables may be doubled (see Akkermann 1966, Goodwin 1970). Average duration of phrase increases with number of phrases; 'normal' 3-phrase call delivered more rapidly. Volume and rate of calling dependent on level of courtship activity and weather; audible over considerable distance (see, e.g., Huxley and Brown 1953, Harmuth 1971a, Kotov 1976a). A high-pitched, 3-syllable variant, likened to Advertising-call of *S. decaocto*, given regularly in Iraq (Harrison 1955); see also Goodwin (1958b). Song normally delivered with bill more or less closed, lower mandible trembling slightly (Heinroth and Heinroth 1926–7); throat puffed out (Ruttledge 1974). Advertising-call given also by ♀; regularly by captive ♀ imprinted on domestic *C. livia* (D Goodwin); for situations in which Advertising-call given by ♀ Columbidae, see Goodwin (1970); see also Akkermann (1966), and Calls of Young. (2) Display-call. A low-pitched, intense-sounding but highly variable (also in volume) 'cōō-cōō cŭ-cŭ cōō cōō' or 'cōō cu-cu-cu-cu cōō'; softer, rapid panting or laughing 'cŭ-cŭ' sounds possibly from intake of breath; given by ♂ in Bowing-display (Goodwin 1970). A variable and quiet 'coo ooo', 'coo coo coo', or 'coo ke ke coo coo' (Cramp 1958). ♂ calls 'huu-gugugu-guu' prior to launching attack in territorial dispute; a slightly attenuated 'huu' followed by rapid, cackling, lower-pitched 'gugugu' sounds, then a long-drawn, ascending and descending 'guu' (Akkermann 1966). See also Klinz (1955) and Harmuth (1971a). Drives in silence (D Goodwin). (3) Nest-call. (a) ♂ gives a low-pitched 'ŏŏ-ōōr' (Goodwin 1970: Fig III), 'oooo-aaarh' (Murton and Isaacson 1962), or a groaning, grumbling 'huh-rkuh' (Heinroth and Heinroth 1926–7). Of variable intensity, but usually softer and less loud than call 1, although strained, intense tone sometimes more apparent (Goodwin 1970; also Harmuth 1971). Also rendered 'guu-huguu(r)', in which low-pitched, fading 'guu' followed after pause (up to 2·5 s) by 'hu' which is higher pitched than beginning of 'guu'; 'guu(r)' quite often rolled, descending in pitch and becoming softer. May be given at intervals of c. 13 s over 15 min, or more sporadically during day. Full form occasionally precedes call 1 (*contra* Bettmann 1959) or Nest-call may be abbreviated to a more or less low-pitched 'gruu' (Akkermann 1966). Individual variation in tone and inflexion; distinct from calls 1 and 2 (Goodwin 1948b), though may sound like abbreviated version (i.e. 1st 2 units) of call 1 (Goodwin 1956b). (b) Much softer, lower-pitched variant given by ♀ (Murton and Isaacson 1962), a soft cooing (Kotov 1976a). (4) Calls associated with mating. After copulation, ♂ gives several high-pitched 'hyi-hyi' sounds, followed by a grumbling note (Klinz 1955). Long-drawn snoring 'ruh' (Heinroth and Heinroth 1926–7) presumably refers to 2nd of these. ♀ gives a very soft 'ruh ruh' (Haverschmidt 1932). ♂ also reported to emit a 'gu' or 'hu' similar to 7b (below), and both sexes once gave a muffled 'ru' (Akkermann 1966). (5) Nest-relief call. Either bird may give a low-pitched, soft 'oooo' like shortened version of 3 (Murton and Isaacson 1962). (6) Calls associated with alarm, fright, or warning. An abrupt 'ru' (Heinroth and Heinroth 1926–7); a grunting sound (Colquhoun 1951); resembles abrupt final unit of call 1 (Kotov 1976a; see also above). (7) Other calls. (a) A sharp 'wow' expressing anger (Murton and Isaacson 1962). (b) A groaning 'ouu-huu' sometimes given by ♂ after evicting intruder (Akkermann 1966). (c) A soft 'gu' given mainly by adult about to feed young (Akkermann 1966); a low-pitched purring 'oorh', like softer variant of call 1, given between bouts of feeding (Goodwin 1948b).

CALLS OF YOUNG. Weak, seeping calls given by very young chicks. Full food-call develops from 5–6 days (Kotov 1976a): squeaking or sibilant whistling sounds (Colquhoun 1951). Elsewhere, distinction made between long-drawn 'fiet', 'füit', or 'füt' given from 1–2 days old as a contact note (intensity higher when adult present) and an oft-repeated, higher-pitched, and more insistent 'fiet' or 'fi' given as food-call once bill-contact with parent established (Akkermann 1966). According to Harmuth (1971), long-drawn, gentle cheeping sounds given from c. 5 days have more or less steady pitch; later, calls louder, briefer, and more irregular. During Defensive-threat display a hissing 'che' sound given, as well as bill-snapping (Akkermann 1966). Low-pitched, soft cooing given by recently fledged young shortly before leaving breeding area (Ryves 1931). First 'imperfect' Advertising-call (adult call 1) given at c. 3 months (Heinroth and Heinroth 1926–7); much higher pitched, jerkier, and more abrupt than fully-developed adult call 1 (Harmuth 1971). A captive ♂ gave a low, harsh 'coo' at c. 8 weeks and a ♀ called similarly at c. 11½ weeks (Colquhoun 1951). At 83 days, a captive ♀ gave (when threatened) a nasal, whistling 'fi fi gú', softer than call 1 of which it was deemed the precursor (Akkermann 1966).   MGW

Breeding. SEASON. North-west Europe: see diagram. Considerable variation even within countries, with urban birds in Britain nesting significantly earlier (starting second half of February) than rural (starting second half of March to second half of April); peak laying period for urban birds second half of April and first half of May, and for rural first half of July to first half of September; differences dictated by food availability (Murton 1958; Cramp 1972). North Africa: eggs laid May–June (Etchécopar and Hüe 1967). Azores: eggs found May–July (Bannerman and Bannerman 1966). Central Europe: laying begins mid-April (Makatsch 1976). SITE. In fork or branches of tree, or in creepers on tree; among thinner twigs of scrub thickets; less often in thick vegetation on ground or under hedge, and on ledge of building; rarely in hole in building (Frisch 1974; Denker 1975; Glutz and Bauer 1980). Average 3–5 m above ground, most 1·5–7 m, but up to 25 m (Murton 1965a). 538 nests, Finland at 0·5–18 m, average 3·97 m: 2 m, 5·3%; 2–4 m, 47·9%; 4–6 m, 29·7%; 6–10 m, 13·3%; 10 m, 3·8% (Saari 1979a). Nest may be re-used for successive clutches and in subsequent seasons. Amount of cover is main reason for selection of tree species; early breeders choose conifers and evergreens. Nest: newly built nest a flimsy structure of twigs, up to 20 cm long, forming platform 17–23 cm across; usually lined finer twigs, sometimes with grasses and leaves; with re-use, nest becomes bulkier (Murton 1965a). Twigs 1–3 mm (0·5–4·5) in diameter; twigs in lower part of nest 2·0–3·5 mm (1·0–4·5) and in upper part 1·0–2·0 mm (0·5–3·0); of 112 twigs in lower part of nest, average diameter 2·7 mm, length 21·1 cm; of 99 twigs in upper parts of nest, average diameter 1·6 mm, length 22·7 cm (Encke 1963). Building: by both sexes, or mainly by ♀, with ♂ collecting twigs from ground or snapping them from growing plant. Of 112 twigs in lower part of nest, 78% had new breaks, and of 99 twigs in upper part of nest, 91% had new breaks (Encke 1963). Usually takes 8–12 days (Kotov 1976a); c. 2 days recorded (Murton and Isaacson 1962). EGGS. See Plate 95. Sub-elliptical, smooth and slightly glossy; white. 41 × 29 mm (37–47 × 25–32), n = 80 (Schönwetter 1967). Weight 18·5 g (16–22), n = 22 (Verheyen 1967). Weight declines from April to September (Murton et al. 1974b). No apparent subspecific variation. Clutch: 1–2, rarely 3–4, and then probably always by 2 ♀♀. Of 926 clutches, England: 1 egg, 16·2%; 2, 83·6%; 3, 0·2%; average 1·84 ± SD 0·04; significant variation through season, down to average 1·72 in May, and up to peak 1·92 in August (Murton 1958). Of 555 clutches, Finland: 1 egg, 11·4%; 2, 87·9%; 3, 0·5%; 4, 0·2%; average 1·9 (Saari 1979a). Laying interval 1–3 days. In undisturbed, unpredated conditions, could theoretically have 3 broods, but even successful rearing of 2 only achieved by small proportion of population, and true position obscured by predation followed by re-laying (Murton 1965a). Replacements laid after egg loss. INCUBATION. 17 days. By both sexes, with ♀ sitting for c. 17 hrs per day, ♂ c. 10.00–17.00 hrs. Full incubation begins with 2nd egg, but ♀ may cover 1st; hatching nearly synchronous. YOUNG. Semi-altricial and nidicolous. Cared for and fed by both parents. Brooded more or less continuously until 7–8 days old, by both parents with same diurnal routine as incubation, but hardly at all after 10

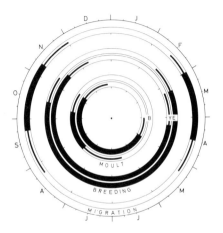

days (Murton 1965a). Fed with crop milk several times a day at first, reducing to twice a day by each parent at about 8–10 days old, when natural food also brought in, with further reduction to feeding by one parent during last days before fledging. FLEDGING TO MATURITY. Fledging period variable. 33–34 days for undisturbed birds, but range 20–35 days (Colquhoun 1951); average 28·7 days, $n = 78$ (Murton et al. 1963b); 16–20 days for some birds, and average $21·6 \pm 1·5$ days, $n = 17$ (Murton et al. 1963b). Growth rate may depend on food supply and so vary through year. Become independent at least 1 week after fledging. Age of first breeding normally 1 year (see Social Pattern and Behaviour). BREEDING SUCCESS. Of 1704 eggs laid, rural England, 42·1% hatched and 30·9% raised to fledging; breeding success highest in August, linked to availability of cereal crops providing food for young; predation accounted for 97% of egg loss—mainly by Jay *Garrulus glandarius*, Magpie *Pica pica*, Rook *Corvus frugilegus*, stoat *Mustela erminea*, brown rat *Rattus*, and grey squirrel *Sciurus carolinensis*; variation in egg loss through season caused by variation in predator activity; predation of eggs also related to nest density; some predation of nestlings but starvation, particularly May to early August, more important (Murton 1958). Of 758 eggs laid, central London, 45·5% raised to fledging, the higher success rate than in rural England being due to markedly lower predation (Cramp 1972). Of 167 nests, Finland, all with 2 eggs, 43% hatched 2 young, and 5% hatched 1 young (Saari 1979a). Of 393 clutches, East Germany, 38% lost, of which 69% destroyed or stolen, 13% deserted or addled, and 7% destroyed by weather; of 53 broods lost, 89% taken by *P. pica*, Carrion Crow *Corvus corone*, and *M. erminea* (Gnielka 1978).

Plumages (nominate *palumbus*). ADULT. Head, including nape and throat, medium bluish-grey, slightly glossed green or purple on nape and behind ear-coverts. Central hindneck and upper sides of neck strongly glossed green (purple in some lights, especially on feather-bases), bordered by conspicuous silvery-white or slightly pink-cream patch on lower hindneck. Broad band across upper mantle extending to sides of chest strongly glossed purple, glossed bronze or green in some lights on upper mantle, tinged magenta on sides of chest, and sometimes narrowly extending towards upper chest just below grey throat. Lower mantle, scapulars, and tertials dark grey with slight brown tinge; back, rump, and upper tail-coverts contrastingly medium or light bluish-grey. Chest and sides of breast deep vinaceous, grading to pale pinkish-vinaceous on breast and upper belly. Flanks and lower belly pale bluish-grey, feathers tipped pale vinaceous where bordering vinaceous of breast and upper belly; vent greyish-white or cream-white, under tail-coverts pale bluish-grey. Uppersurface of tail-feathers pale grey with ill-defined greyish-black tips and bases; dark tips $c.$ 5–6 cm wide on t1, $c.$ 3–5 cm on t6; pale centres of tail-feathers only showing as a broad ill-defined band when tail spread. Undersurface of tail with sharply demarcated greyish-white band $c.$ 4–5 cm wide and with shiny black feather-bases and -tips, latter 3–5 cm (t6) to 5–6 cm (t1) wide. Primaries, greater upper primary coverts, and bastard wing black with slight grey tinge, bases of inner primaries and outer webs of others greyish; p3–p9 with distinct white margin $c.$ 0·5–2 mm wide; margins of p1–p2 and p10 narrower and often mainly light grey; central greater primary coverts often with narrow white outer edges; in bastard wing, shortest feather has white outer web, white reduced or absent towards longest feather. Secondaries medium grey, outer slightly darker towards tips, outermost (s1) with white margin 2–3 mm wide along outer web, s2–s3(–s4) with greyish-white or light grey margin; inner secondaries bordering tertials have slight brown tinge. Outer 5–7 greater and median upper wing-coverts have much white on outer webs and mainly hidden light grey inner webs; white outer webs of these, mainly white outermost lesser coverts, white outer web of shortest bastard wing feathers, and white or pale grey margins to outer secondaries together form conspicuous white band across wing. Remaining upper wing-coverts light bluish-grey where bordering white band; elsewhere, medium or dark grey, often slightly tinged brown, like mantle and scapulars. Under wing-coverts and axillaries light or medium bluish-grey. In worn plumage, some grey of feather-bases may show on white neck-patches, glossy green tips of feathers of hindneck and upper sides of neck narrower and more purple; mantle, scapulars, tertials, and inner and shorter upper wing-coverts darker and with stronger brown tinge; chest and upper breast with deeper vinaceous tinge but lower breast and belly often paler and with more grey of feather-bases showing; flight-feathers greyish-brown instead of greyish-black. Sexes usually inseparable, except perhaps for known pairs: ♀♀ on average show slightly smaller white neck-patch, slightly narrower purple band from hindneck down to sides of chest, and, especially, a less deep vinaceous, more greyish chest. In a large series, the deepest coloured few birds were ♂♂ (adult and 1st adult), the palest few ♀♀ (mainly 1st adult); remainder (all ages and both sexes) were intermediate. Also, influence of wear on colour is marked, and thus only sexing of known pairs at similar stage of wear and of similar age seems trustworthy. DOWNY YOUNG. Down-tufts sparse, coarse, and hair-like; pale yellow; face, skin round eye, and centre of belly bare. JUVENILE. Face and throat bare. Crown, sides of head, hindneck, and sides of neck either pale brown with narrow darker brown or rusty feather-fringes (often purer grey above eye and on sides of head and neck), or rather uniform light bluish-grey with faint brown or rufous feather-fringes. No glossy feather-tips on neck or upper mantle, no white on sides of neck. Mantle pale brown or light brownish-grey; scapulars and tertials dark or pale grey-brown, feathers narrowly fringed rusty-rufous, buff-brown, pale buff, or cream-white on tips. Back and rump like adult; upper tail-coverts pale grey-brown to light bluish-grey with distinct but narrow rusty to pale buff feather-margins. Chest to belly and flanks pale vinaceous-chestnut to buff-brown with variable amount of pale grey feather-bases visible; lower flanks, thighs, vent, and under tail-coverts pale grey with ill-defined rufous to pale buff feather-tips. Tail like adult, but feathers usually shorter and narrower (see Measurements and Structure), black tip sometimes narrower and less sharply defined below. Wing as adult, but inner upper wing-coverts pale grey-brown, tips fringed rufous to off-white; fringes usually narrow but distinct, occasionally indistinct or absent; white outer coverts often with faint buff or rufous tips; primaries often browner; primaries, pri-

PLATE 26 (facing).
Pterocles lichtensteinii Lichtenstein's Sandgrouse (p. 245): 1–2 ad ♂, 3–4 ad ♀.
Pterocles coronatus coronatus Crowned Sandgrouse (p. 249): 5–6 ad ♂, 7–8 ad ♀.
Pterocles senegallus Spotted Sandgrouse (p. 253): 9–10 ad ♂, 11–12 ad ♀.

PLATE 27. *Columba livia* Rock Dove (p. 285). Nominate *livia*: **1** ad, **2** juv, **3** downy young, **4–6** feral variants. *C. l. dakhlae*: **7** ad. *C. l. schimperi*: **8** ad. (CJFC)

PLATE 28. *Columba oenas* Stock Dove (p. 298): **1** ad, **2** juv moulting into 1st ad (1st autumn or winter), **3** juv, **4** downy young. *Columba eversmanni* Yellow-eyed Stock Dove (p. 309): **5** ad, **6** 1st ad (1st winter). (CJFC)

mary coverts, and bastard wing with narrow white, pale buff, or rufous margins on tips, usually best marked and deepest rufous on greater primary coverts, but occasionally hard to distinguish or almost absent on tip of outer greater covert; under wing-coverts and axillaries with traces of buff, white, or dark grey fringes on tips. Smaller juveniles usually have fragments of pale yellow down-tufts adhering to some feather-tips of body or tail. Feathers growing late (at about fledging) purer bluish-grey, like those of adult, but usually shorter and narrower; these appear on face, throat, back, rump, outer mantle, and among longer lesser coverts. FIRST ADULT. Like adult. Once adult body plumage attained, younger birds usually still show many pale grey-brown juvenile tertials and upper wing-coverts, these with traces of white or buff fringes and contrasting with fresh uniform blue-grey feathers. When post-juvenile moult more advanced, juvenile outer primaries, outer greater and some lesser upper primary coverts, and some tail-feathers still present; outer primaries rather worn and pointed and with traces of pale fringes on tips, outer primary coverts usually with distinct rufous markings on tips, and juvenile tail-feathers shorter and narrower than neighbouring ones. When primary and tail moult completed, 1st adult sometimes still separable by retained juvenile middle secondaries, usually s5–s7, s3–s7, s3–s8, or s2–s8; these shorter, narrower, and with less broadly rounded tip than those of full adult. Adult may also show some older secondaries, but these more similar in length and shape to fresh feathers, often different in age, and often occurring in groups with newer feathers between. Any bird with moult pattern deviating from rather rigid pattern of adult (see Moults) probably 1st adult.

Bare parts. ADULT. Iris pale lemon-yellow to pale sulphur-yellow. Narrow ring of bare grey skin round eye. Base of bill and cere rose-red, purple-red, or bright red, cere powdered white except for edge just above nostril; tip bright yellow or orange-yellow, tip of nail usually white. Leg and foot deep purple-red or dark red, soles and part of rear of tarsus yellowish-flesh. DOWNY YOUNG. Iris light bluish-grey. Bill basally blue-grey or slate-grey, shading to pinkish- or yellowish-white near tip; subterminal band across nail horn-brown. Leg and foot mauve-pink or blue-grey with slight purple tinge; soles and rear of tarsus grey. Bare skin of body lead-blue with flesh tinge, flanges of bill pale flesh. JUVENILE. At fledging and prior to post-juvenile moult, iris pearl-grey with slight greenish or yellowish tinge, base of bill and cere brownish-slate with slight purple tinge, proximal half yellowish with dark horn tip, leg and foot slate-grey with purple tinge. Adult colours attained during post-juvenile moult; iris and bill similar to adult from about primary moult score of 25, but fully powdered cere and fully pale tip of bill not acquired until about completion of moult or when starting to breed (slight horn tinge on tip of bill sometimes still present in adult). Leg and foot gradually shade from light red through clear red to purple-red, latter colour attained when post-juvenile moult almost completed. (Witherby *et al.* 1940; Ljunggren 1969; RMNH, ZMA.)

Moults. Based mainly on data from Murton *et al.* (1974*b*) and Boddy (1981) for Britain, *c.* 200 skins (RMNH, ZMA) for Netherlands, Niethammer (1970) and Glutz and Bauer (1980) for West Germany, and Mathiasson (1965) for Sweden, which are all closely similar, though interpretation of some authors inaccurate, apparently caused mainly by incorrect ageing. ADULT POST-BREEDING. Complete, except for part of secondaries; primaries descendant. Starts with p1 between 1st week April and 1st week May (only some from North Africa and Iraq later in May or in early June); average primary moult score of 17 (9–22) reached mid-June, 23 (14–30) mid-July, 29 (22–37) mid-August, 38 (32–44) mid-September, 41 (32–49) mid-October, 46 (39–50) mid-November, all completed late December; only limited number of birds suspend primary moult at onset of winter (0–13% of total in various studies, but 41–54% in Boddy 1981), usually retaining old p10 only. Some suspend moult for a short period during late June to August. Body and tail mainly moulted late August to early October, between primary moult scores of 25 and 45; sequence of tail-feather replacement mainly t2–1–3–4–6–5, but occasionally t3 replaced before t1 or moult started with t1, t3, or t6. Secondaries moulted at about same time as tail and body, starting ascendantly with s1 at primary moult score of 30–35 and at same time or slightly earlier both descendantly and ascendantly from central tertials (about s9–s11); however, only a limited number of secondaries replaced each year, suspending when primary score of 45–50 reached and continuing in next moulting season with next feathers, with a result that adults often show several new secondaries or groups of secondaries separated by older ones, e.g. s1, s3, s4, s7, s8, s10, s11, and s13 new or fairly new (remainder old or rather old), or s2–s4, s6, s9–s13 new (often asymmetrically between wings). POST-JUVENILE. Strongly variable between individuals (caused by extended breeding period), but amount of variation similar all over geographical range. Starts with body and p1 at about age of 6–7 weeks (Heinroth and Heinroth 1926–7; Boddy 1981); head, mantle, and part of lesser and median upper wing-coverts first, followed by neck (white patches appearing), underparts, scapulars, upper tail-coverts, and more wing-coverts from primary moult score of 10–15 at age of 12–14 weeks. Largely in adult plumage at moult score of 25–30, except for some tertials and upper wing-coverts. Tertials mainly replaced at moult score of 25–35, tail 20–45 (sequence as in adult). Speed of primary moult as in adult. Secondaries moulted ascendantly from s1 and descendantly from s9–s10 from primary moult score 30–35, suspending from score 45–50, when (s2–)s3–s8(–s9) still juvenile, but sometimes (s3–)s4–s6(–s7) only; moult resumed from suspension point in next moulting season, but it may take 1–2 years before last juvenile s5–s6 replaced. This general outline only followed by early-hatched juveniles, which moult at same time as adults and may complete before winter. Later-hatched juveniles suspend moult, irrespective of stage reached, from November (a few) or mid-December (though body sometimes continues at slow pace). During winter in Britain, 19% of *c.* 40 birds had not yet started, 9% suspended at score 5 (only p1 new), 9% at 10 (up to p2 new), 4% 15, 9% 20, 14% 25, 16% 30, 14% 35, 4% 40, 2% completed (score 50) (Murton *et al.* 1974*b*); from much larger British sample, *c.* 14% had not started, *c.* 13% suspended at score 5, *c.* 20% at 10, *c.* 16% at 15, *c.* 18% at 20, but none higher (Boddy 1981), probably due to incorrect ageing of older immatures (see also improbably high number of adults found suspending post-breeding at high score in this study); in a West German study, the number of birds not yet started or already completed is not mentioned, but of 56 suspending birds, 5% halted at score 5, 14% at 10, 13% 15, 14% 20, 20% 25, 20% 30, 5% 35, 3% 40, 5% 45 (Niethammer 1970); of 58 birds from Netherlands, 12% had not yet started, 9% suspended at score 5, 3% at 10, 7% 15, 3% 20, 7% 25, 7% 30, 12% 35, 12% 40, 12% 45, 16% had completed. Between late February and early April, moult resumed from point of suspension; only then do late-hatched (November) birds start post-juvenile body and wing moult, early-hatched juveniles of next spring differing from them only by fresh plumage and by the fact that they do not breed. Some immatures complete post-juvenile moult with regrowth of p10 between April and November (though retaining juvenile middle secondaries), but others, although resuming moult in late winter or early spring, stop primaries for 2nd time later on, usually arresting with score 45 (i.e. retaining juvenile p10). As 1st post-breeding moult starts April–June (at same time as adult post-breeding or slightly later, but not in immatures starting post-juvenile as late as February–March), many juveniles show serially descendant moult April–October: post-juvenile active on outer primaries, post-breeding on inner. Birds arresting 1st post-breeding moult after earlier resumption about March, in June may show pattern such as: p1–2 new, p3 growing, p4–5 rather old, p6–9 new, p10 old (juvenile).

Measurements. Nominate *palumbus*. Mainly Netherlands, a few Sweden and West Germany, all year; skins (RMNH, ZMA). 1st adult wing and tail are from immature with replaced outer primaries and tail; adult bill includes 1st adult.

WING AD	♂ 252	(4·93; 44)	243–263	♀ 250	(5·98; 30)	240–260
1ST AD	251	(4·53; 26)	241–257	249	(6·32; 9)	240–256
JUV	242	(4·51; 29)	235–251	243	(5·14; 22)	234–251
TAIL AD	165	(5·61; 10)	156–174	160	(5·89; 11)	153–171
1ST AD	162	(3·74; 11)	156–168	157	(5·37; 9)	148–164
JUV	153	(5·77; 8)	142–160	150	(— ; 3)	139–156
BILL AD	21·4	(1·15; 40)	19·3–23·6	21·1	(1·25; 29)	19·3–23·4
TARSUS	32·8	(0·86; 37)	31·4–34·6	32·7	(1·28; 26)	30·2–34·5
TOE	39·9	(1·67; 39)	37·3–43·4	39·7	(1·38; 25)	37·8–42·5

Sex differences not significant. Difference between adult and 1st adult not significant, but juvenile tail and wing both significantly shorter than those of adult and 1st adult. Juvenile tarsus and toe similar to adult from fledging; juvenile bill similar once face feathered.

Rostov-on-Don and Caucasus area (USSR), north-west Iran, and Turkey (March–August), and Syria and Israel (winter); skins (BMNH, RMNH, ZMM).
WING AD ♂ 259 (6·72; 14) 250–271 ♀ 254 (7·10; 7) 246–260

Iraq, all year; skins (BMNH).
WING AD ♂ 240 (7·60; 7) 228–248 ♀ 240 (6·67; 6) 229–246

North Africa, all year; skins (Vaurie 1961*a*; BMNH, ZMA).
WING AD ♂ 258 (5·68; 17) 248–270 ♀ 253 (5·01; 6) 244–259

C. p. azorica. Azores, April–May; skins (BMNH).
WING AD ♂ 243 (5·40; 5) 237–251 ♀ 239 (6·80; 4) 232–248

Weights. ADULT AND 1ST ADULT. (1) Britain (Murton *et al.* 1974*a*); (2) Netherlands (RMNH, ZMA) and West Germany (Niethammer and Przygodda 1954; Keil and Rossbach 1979); (3) Scandinavia, without crop contents (Ljunggren 1968); smaller Swedish sample from Mathiasson (1965) similar to latter.

(1) JAN–FEB	♂ 496	(44·1; 218)	♀ 480	(39·1; 263)
MAR–APR	499	(37·4; 107)	490	(41·9; 231)
MAY–JUN	496	(31·9; 59)	482	(32·2; 80)
JUL–AUG	462	(28·4; 52)	472	(31·7; 129)
SEP–OCT	504	(32·4; 67)	475	(33·6; 70)
NOV–DEC	568	(57·0; 105)	566	(51·4; 108)
(2) JAN–FEB	544	(36·1; 29)	520	(35·1; 21)
MAR–APR	504	(39·9; 25)	500	(48·3; 13)
MAY–AUG	492	(32·1; 17)	532	(64·4; 3)
SEP–OCT	500	(— ; 14)	480	(— ; 18)
NOV–DEC	519	(— ; 18)	492	(— ; 18)
(3) JAN–MID-MAR	502	(53·5; 57)	471	(52·1; 43)
MID-MAR–MID-MAY	516	(36·1; 156)	505	(35·8; 156)
MID-MAY–MID-JUL	505	(40·2; 101)	497	(39·4; 62)
MID-JUL–MID-SEP	499	(35·1; 212)	490	(24·3; 178)
MID-SEP–DEC	526	(32·1; 114)	502	(41·5; 77)

Heaviest birds recorded 690 (range not given by all authors above); minima of exhausted adults mainly 290–350. For influence of reproductive cycle, weather, and food on weight in Belgium, see Schnock (1981); all-year sample, ♂ 494 (288) 325–614, ♀ 484 (230) 284–587.

JUVENILE. On day of hatching 16·8 (2·12; 58), on 6th day 99·0 (14·3; 55), on 16/17th day 274 (36·9; 158) (2 young in nest) or 300 (37·5; 67) (1 young in nest); increase slower after 23rd day; during 1st autumn, average *c.* 70 below adult; during 1st winter, 40–59; during April–July of 2nd calendar year, average still 16–26 below full adult (Mathiasson 1965; Murton *et al.* 1974*b*).

Structure. Wing rather long, broad, tip rounded. 11 primaries: p8 and p9 longest or either of these 0–3(–6) shorter, p10 8–17 shorter, p7 12–22, p6 31–41, p5 49–57, p1 90–101; p11 minute, concealed by primary coverts. Inner web of p10–p9 and outer p9–p8 emarginated. Tail rather long, broad, tip slightly rounded; 12 feathers. Tail-feathers wider towards bluntly rounded tip in adult (especially outermost), parallel-sided and relatively narrower and shorter in juvenile: width of t1 near tip 39·2 (35–45) in adult, 37·2 (33–44) in 1st adult (new feathers), 32·6 (29–36) in juvenile; width near tip of t6 33·4 (29–39) in adult, 32·1 (27–37) in 1st adult, 28·4 (25–32) in juvenile (for each, $n \geqslant 20$). Bill short and slender; rather strong, decurved, with slightly projecting nail to upper mandible; base of upper mandible covered by broad soft cere, bulbous above each obliquely placed narrow nostril. Cere covered with granulated or powder-like layer in adult. Tarsus short and rather thick, tibia and upper front of tarsus feathered. Toes rather long and slender, soles thick and flattened; outer toe *c.* 78% of middle, inner *c.* 77%, hind *c.* 62%. Claws rather slender and sharp, slightly curved.

Geographical variation. Slight, except for rather distinct extralimital race *casiotis*; involves size and general colour of body. Size constant over much of Europe and western Siberia, small samples of nominate *palumbus* examined from Scandinavia, Britain, and central Europe closely similar to larger sample from Netherlands cited in Measurements; single specimen examined from rare or extinct Madeiran race *maderensis* similar to these (♀, wing 250: ZMA). Birds from North Africa (sometimes separated as *excelsa* Bonaparte, 1856), eastern Turkey, Caucasus and area directly north of it, Transcaucasia, and Iran (latter involving *iranica*) slightly larger, similar in size to *casiotis* of mountains of west central Asia (east from south-east Iran to Tien Shan and western Himalayas). Birds from Iraq markedly smaller than those from Turkey, Caucasus, and Iran, tip of bill paler, and voice said to differ (Harrison 1955), but plumage similar to European nominate *palumbus* and therefore included there. Isolated *azorica* from Azores also small (see Measurements). Within range of nominate *palumbus*, some populations slightly darker and more saturated than others, e.g. those of western Scotland (sometimes separated as *kleinschmidti* Clancey, 1950) and North Africa rather dark, those of Scandinavia rather pale, but individual variation fairly large in all localities examined, in part depending on individual differences in wear and on age of skin, and recognition of further races in Europe or North Africa not warranted (see also Vaurie 1961*a*). *C. p. azorica* differs from European nominate *palumbus* by slightly darker and more dull brownish mantle, scapulars, tertials, and inner upper wing-coverts; slightly smaller white neck-patches, more often tinged pink-buff than in nominate *palumbus*, and sometimes whole neck-patch pink-buff; back, rump, and upper and under tail-coverts slightly darker grey; some ♂♂ have chest and breast slightly deeper and more greyish-vinaceous; no constant difference in tail-pattern. *C. p. maderensis* intermediate between *azorica* and nominate *palumbus*; in fact, the single specimen examined of this race was within range of variation of nominate *palumbus* and recognition of *maderensis* perhaps unwarranted. Central Asiatic *casiotis* differs markedly in colour of neck: only nape and hindneck glossed green or purple, white neck-patches restricted to narrow bar on lower sides and often completely tinged pink-buff or whitish-buff rather than white; remaining sides of neck including lower throat medium blue-grey with slight green gloss; purple band across upper mantle and immediately below buff neck-patch and grey throat distinctly narrower than in nominate *palumbus*; upperparts slightly paler brown-grey, less bluish; underparts slightly more pinkish-vinaceous, less deep vinaceous; white edges to outer webs of primaries narrower and often tinged buff. *C. p. iranica* from Kopet Dag (Turkmeniya) and much of Iran intermediate in colour between *casiotis* and nominate *palumbus*: body rather pale and size rather large, as in *casiotis*; neck similar to nominate *palumbus*, but neck-patches not as wide and often tinged buff. Rather pale birds with fairly small and occasionally buff neck-patches similar to *iranica* occur west to Transcaucasia and eastern Turkey. CSR

Columba trocaz Long-toed Pigeon

PLATES 29 and 30
[between pages 350 and 351]

Du. Trocazduif   Fr. Pigeon trocaz   Ge. Silberhalstaube
Ru. Серебристошейный голубь   Sp. Paloma torqueza   Sw. Madeiraduva

Columba Trocaz Heineken, 1829

Monotypic

Field characters. 38–40 cm; wing-span 72–76 cm. Only slightly smaller than Woodpigeon *C. palumbus*. Large, dark grey and purple pigeon, with blue-grey head, silvery patch on neck and grey subterminal band across end of almost black tail. Flight recalls *C. palumbus*, but relatively broader and more rounded wings produce hesitant action and different silhouette. Sexes similar; no seasonal variation. Juvenile separable.

ADULT. Head and foreneck blue-grey; sides of neck grey (glossed green), with bold patch of silver-tipped feathers. Hindneck and upper mantle glossy pink-purple (with green sheen). Scapulars and wing-coverts slate-grey, latter faintly relieved by paler grey crescent on greater coverts; flight-feathers black-brown. Back and rump blue-grey. Tail slate-black, with broad subterminal pale grey band. Breast vinous; rest of underparts blue-grey. Underwing dark grey. Bill mainly red; eye pale yellow; orbital ring red. Legs coral-red. JUVENILE. Duller, browner than adult, lacking glossed plumage and showing pale margins to larger body and inner wing-feathers.

Unmistakable; no other large and dark *Columba* pigeon inhabits Madeira. Beware, however, chance of confusion with large dark feral Rock Dove *C. livia* (see Recognition). Flight appearance differs from *C. palumbus*, due to more rounded wing-tips (producing less flowing action) and longer tail. Behaviour resembles *C. palumbus*.

Voice similar to *C. palumbus*.

Habitat. Only on mountainous subtropical Atlantic island of Madeira in higher forest zone covered in rain-cloud for much of year, among tall laurels *Laurus* or dense tree-heath *Erica arborea*. Favoured haunt lies along artifical watercourse at *c.* 1000 m following steep escarpment indented by ravines, with occasional large dead laurel tree and much tall tree-heath. Flies down to farms, feeding with Rock Dove *C. livia* on grain at threshing floors; also on cherries. Sometimes breeds in tree-heath but mostly in inaccessible caves on cliff faces, and in crevices on rock ledges. Shares woodland habitat with Blackbird *Turdus merula* and Chaffinch *Fringilla coelebs*. (Bannerman and Bannerman 1965.)

Distribution. Breeds only on Madeira.

Population. Probably declined in last 100 years (Bannerman and Bannerman 1965); numbers now more or less stable, at probably over 500 birds, though still no restrictions on hunting (GM, PAZ).

Movements. Resident, though descending in autumn from mountain forest to cultivated land (Bannerman and Bannerman 1965).

Food. Chiefly fruit; occasionally leaves and grain. Feeds both in tree-canopy and on ground, taking fallen fruit among dead leaves (Godman 1872).

Diet mainly fruit of bay *Laurus azoricus*, til *Ocotea foetens*, haya *Myrica faya*, and viñatigo *Persea indica* (Harcourt 1851; Godman 1872; Koenig 1890; Zino 1969b). In autumn, especially, visits cultivated areas to take grain and wide range of other foods such as cabbage *Brassica* and cherries *Prunus*; can be a pest on such crops (Meinertzhagen 1925; Sarmento 1948; Bannerman and Bannerman 1965; Zino 1969b; Berg and Wijs 1980). Also recorded feeding on grasses (Harcourt 1851).   HAR

Social pattern and behaviour. Very little information.

1. Mostly solitary or in small groups; maximum 10–20 (Sarmento 1948). Single birds and groups of 3–5 noted late January and early April. One group of 3 composed of 2 ♂♂ and one ♀. Mixes with Rock Dove *C. livia* on harvested fields and at threshing floors (Bannerman and Bannerman 1965; Zino 1969b). BONDS. No information. BREEDING DISPERSION. At least 2, possibly 3, pairs along transect of 1–2 km, mid-October; 2 disused nests *c.* 1 km and *c.* 2·5 km from occupied nest (Zino 1969b). ROOSTING. No information.

2. Very wary of man; spends much time in trees with dense foliage (Godman 1872; Schmitz 1910; R de Naurois). FLOCK

BEHAVIOUR. No information. ANTAGONISTIC BEHAVIOUR. No information. HETEROSEXUAL BEHAVIOUR. Captive birds showed Bowing-display and Nodding as in Woodpigeon *C. palumbus* (Goodwin 1970). ♂ in tree (with ♀ and another ♂), early April, performed Bowing-display: fanned tail, Bowed repeatedly, and ruffled neck feathers to expose silvery ring; walked slowly along branch after ♀ who then flew off, followed by both ♂♂ (Bannerman and Bannerman 1965). Long bout of courtship (no details), 12 October, did not lead to copulation; possibly low intensity and not indicative of imminent breeding (Zino 1969b). In captivity both sexes incubated: ♂ 10.00–16.00 hrs, ♀ at other times (Schmitz 1910). RELATIONS WITHIN FAMILY GROUP. Eyes of captive chick opened only on 7th day; after hatching 12 February, fully developed and independent by mid-April (Schmitz 1910). In the wild, brooded up to at least 10th day and fledged at 28 days (Zino 1969b). ANTI-PREDATOR RESPONSES OF YOUNG. No information. PARENTAL ANTI-PREDATOR STRATEGIES. Adult left nest containing small chick when men a few metres away; did not return for c. 45 min while men still present but hidden. On another occasion (nestling c. 1 week old), adult returned after c. 1 hr, perched c. 1 m from tree screening nest but did not make final approach. When nestling c. 10 days old, both parents returned and one eventually went to nest to brood, being apparently undisturbed by camera flash but leaving at close approach (Zino 1969b). MGW

Voice. Advertising-call of captive ♂ 'cōō–cōō cōōōō cŏŏk', given several times. Similar to that of Woodpigeon *C. palumbus* in abrupt ending and tone but less loud. Display-call like muffled version of *C. palumbus*, effect possibly caused by noise of other birds and people (Goodwin 1970); however, muffled not very sonorous quality of wild ♂'s cooing confirmed by Zino (1969b). MGW

Breeding. SEASON. Madeira: eggs found in almost every month, but mostly February–June (Zino 1969b); mostly March–May (Bannerman and Bannerman 1965); moult schedule suggests mostly January–May (C S Roselaar). SITE. In rock crevice in cliff or a few metres above flat ground or slope, or less frequently in low bush a few metres above ground, especially heaths *Erica* and laurels *Laurus*; occasionally on high horizontal branch, one 8 m above torrent (Zino 1969b). One rock crevice extended 40 cm horizontally, c. 30 cm high, and 30–40 cm wide, 5 m above steep-banked stream (Zino 1969b). Nest: untidy heap of dry twigs, mainly of *Erica*. Building: no information. EGGS. See Plate 95. Sub-elliptical; white. 47 × 32 mm (45–50 × 30–55), $n = 7$; calculated weight 26 g (Schönwetter 1967). Clutch: 1 (Bannerman and Bannerman 1965); exceptionally 2 (Zino 1969b); captive bird laid 4 eggs in 43 days, when each egg removed after laying (Schmitz 1910). INCUBATION. 19–20 days in captivity; by both sexes (Schmitz 1910). YOUNG. Semi-altricial and nidicolous. FLEDGING TO MATURITY. In one case, fledging period 28 days (Zino 1969b). No further information.

Plumages. ADULT. Head and upper hindneck medium to dark bluish-grey, slightly glossed green or purple (depending on light) on hindcrown and upper hindneck. Feathers of lower hindneck and sides of neck dark grey with broad pale silvery-grey tips, latter forming conspicuous broad and pale half-collar on neck; silvery-grey slightly glossed purple-pink or green in some lights; some feathers at lower sides of neck occasionally tipped chestnut. Broad band across upper mantle glossed deep purple and green, narrowing towards lower sides of neck; situated just below silvery-grey half-collar. Lower mantle, scapulars, and tertials dark slate-grey, back and rump slightly paler, dark bluish-grey; upper tail-coverts dark slate-grey or greyish-black, sometimes in part slightly glossed purple. Central throat and all foreneck medium bluish-grey, faintly glossed green on lower foreneck. Chest and upper sides of breast down to central belly deep vinaceous; flanks, belly, and vent medium bluish-grey, under tail-coverts dark bluish-grey; extent of vinaceous on underparts rather variable (independent of sex)—in some, down to upper breast only, in others reaches down to upper flanks and all belly. Tail black with pale subterminal band 3–3½ cm wide situated 3–4 cm from tip; this subterminal band light grey on uppersurface of tail, pale grey or greyish-white below; border of grey band sharply defined proximally, sometimes less so distally. Flight-feathers and upper primary coverts brownish-black, fresh feathers tinged dark slate-grey on outer webs; narrow margins of p5–p9 pale buff or off-white. Lesser, median, and innermost greater upper wing-coverts dark slate-grey like scapulars and tertials, but outermost lesser and median and most greater coverts paler medium grey, contrasting slightly. Under wing-coverts and axillaries dark grey. DOWNY YOUNG. Sparsely covered with coarse yellowish down. JUVENILE. Like adult, but head and neck dull brownish-grey, without or almost without green and purple gloss on nape and lower neck and without silvery-grey feather-tips on hindneck and sides of neck; mantle, scapulars, tertials, back to upper tail-coverts, and upper wing-coverts (except outermost lesser, median, and most greater, which, like secondaries, are similar to adult) dark sepia-brown, without dark slate cast of adults, feathers narrowly fringed rufous; chest and breast rufous-brown, some grey of feather-bases visible, without deep vinaceous tinge of adult; remaining underparts duller grey, less bluish. Tail-feathers narrower (see Structure). Primaries and greater primary coverts dark brown, p4–p7 with sharp but narrow pale buff margins, p8–p10 with faint rufous-brown fringes; margins of primary coverts like those of primaries. Some birds show only limited amount of rufous fringes on upperparts, upper wing-coverts, primary coverts, and primaries; these difficult to age, especially when much of juvenile body and wing-coverts replaced by 1st adult plumage; usually a faint rufous-brown edge on tip of at least p10 or on outermost greater primary covert still visible. FIRST ADULT. Almost like adult. Ageing usually not difficult when scattered juvenile feathers on body or upper wing-coverts remain; these shorter, narrower, less glossy, and more worn than fresh adult ones. More advanced birds retain juvenile outer primaries, outer greater primary coverts, and some tail-feathers only; tail-feathers shorter, narrower, and more worn than adult (see Structure); outer primaries or greater primary coverts usually show narrow rufous or buff margin at tip. Last juvenile feathers to be replaced are middle secondaries; often slightly narrower and shorter than neighbouring new feathers, but difference not always marked.

Bare parts. ADULT. Iris pale straw-yellow, orbital ring red. Bill coral-red. Leg and foot carmine-red or bright coral-red. DOWNY YOUNG. Iris black. Bill-base pinkish-grey, tip pinkish-white; broad black band across nail. Leg and foot greyish-pink with dull purple-red scutes. JUVENILE. Iris light grey. Bill blackish-grey. Leg and foot dusky red. (Schmitz 1893; Hartert 1912–21; Bannerman and Bannerman 1965; Zino 1969b; BMNH.)

Moults. Even when scanty data on *C. trocaz* (*t*), Bolle's Laurel Pigeon *C. bollii* (*b*), and Laurel Pigeon *C. junoniae* (*j*) combined, moults difficult to follow, as no July–October birds available; probably as in Woodpigeon *C. palumbus*, but as nesting season of the 3 species is even more protracted, may start in any month. ADULT POST-BREEDING. Most adults examined (21 specimens, November–May) showed fresh to rather fresh body plumage with all primaries new or fairly new (primary moult score 50). A few in active primary moult: December score 45 (*t*), March score 23 (*b*), and June score 34 (*t*); suspended primary moult found November score 30 (*t*), February score 45 (*t*), April 35 (*b*), and May score 35 (*j*); those with advanced scores may include wrongly aged immatures. POST-JUVENILE. Starts with nape, sides of neck, outer mantle, and longer lesser coverts at about fledging, followed by p1 shortly afterwards; body and wing-coverts fully adult when primary moult score of *c.* 25 reached. Tail mainly starts at score 10–20 and finishes when p10 regrown, but sometimes still fully juvenile at score 30 or fully adult at score 22; sequence t1–2–3–4–5, with t6 somewhere between shortly after t1 and just before t5, or rarely last. Secondaries and part of tertials moult last, starting from score 30–40, last (central) secondaries replaced after regrowth of p10. None in serially descendant moult examined, in contrast to *C. palumbus*. Birds examined not yet moulting primaries (score 0) were from November (*t*), January (*t*), March (*b*), April (*t*, *b*), and May (*j*). Active moult found: November, score 43 (*t*); December, scores 4 (*t*), 32 (*t*), 34 (*t*); January, scores 4 (*t*), 6 (*t*), 24 (*b*), 47 (*b*); February, score 16 (*t*); April, score 22 (*j*); May, scores 21 (*j*), 22 (*b*), 28 (*b*); and June scores 27 (*t*), and 27 (*j*). Immatures in suspended primary moult: January, scores 15 (*t*), 15 (*t*), 40 (*b*), 40 (*b*); February, scores 10 (*j*), 15 (*j*); March, scores 45 (*b*), 45 (*j*); April, scores 10 (*t*), 30 (*b*); May, scores 10 (*j*), 15 (*t*); and June score 20 (*t*). 1st adults with all primaries new (score 50): January (*b*) and May (*b*), but others probably overlooked among full adults. From combined data of adults and juveniles, May–June and November–January appear to be peak periods of active primary replacement and July–October also undoubtedly important for moulting; most adults showed new wing from November onwards; relatively many birds in suspended primary moult January–May (probably main breeding period).

Measurements. Madeira, November–June; skins (BMNH, RMNH, ZFMK, ZMA).

WING	AD	♂ 252 (3·10; 9) 248–258	♀ 244 (3·35; 9) 237–247
	JUV	243 (4·80; 8) 236–251	231 (4·60; 8) 226–240
TAIL	AD	178 (5·33; 11) 170–184	173 (7·18; 10) 165–183
	JUV	162 (—; 3) 157–168	158 (5·35; 4) 150–162
BILL	AD	21·0 (1·08; 12) 19·4–22·7	19·9 (1·32; 11) 18·1–21·6
TARSUS		36·3 (1·48; 11) 34·7–38·8	35·0 (1·12; 12) 33·8–36·8
TOE		50·9 (1·84; 12) 48·9–55·1	49·8 (2·11; 12) 47·4–53·4

Sex differences significant for wing and tarsus. Juvenile wing and tail significantly shorter than adult.

Weights. No information.

Structure. Wing rather short, broad, tip rounded; relatively broader at base and more rounded at tip than in *C. palumbus*. 11 primaries: p8 longest; in adult, p10 19–25 shorter, p9 3–8; in juvenile, p10 11–24 shorter, p9 1–5; in both, p7 2–8 shorter, p6 13–23, p5 30–43, p1 70–82; p11 minute, concealed by primary coverts. Longest tertials reach to tip of p2–p3 in closed wing. Tail long, tip square or slightly rounded; 12 feathers, t6 6–15 shorter than t1; width of t1 near tip in adult 41–52, in juvenile 37–39; of t6, 34–43 in adult, 30–34 in juvenile. Bill as in *C. palumbus*, but relatively slightly shorter and more slender. Toes relatively much longer than in *C. palumbus*, 1·42 times tarsus length rather than 1·21 (1·30 in *C. bollii* and *C. junoniae*); outer and inner toe *c.* 74% of middle, hind *c.* 60%. Claws longer than in *C. palumbus*.

Geographical variation. None.

C. trocaz (Madeira only) forms a group of closely related species with *C. bollii* and *C. junoniae*, both the latter occurring on La Palma, Gomera, and Tenerife in Canary Islands. *C. trocaz* and *C. bollii* often considered to comprise single polytypic species, but *C. junoniae* as close to *C. bollii* as *C. bollii* is to *C. trocaz* and all could easily pass for a single superspecies were it not for the complete geographical overlap of *C. bollii* and *C. junoniae*. *C. bollii* differs mainly from *C. trocaz* by smaller size, more saturated colours (more extensive glossy green on head and neck, more extensive and deeper vinaceous on underparts), and less black on tip of tail (sometimes almost no black on uppersurface of tip). *C. junoniae* is even more extensively glossy green on head, neck and chest, deeper and more extensively purple-vinaceous on underparts, and tail-tip completely pale, without black even on undersurface, and hence shows same features as in *C. bollii*, but in a more extreme form.

Recognition. Adult differs from Madeiran race of Woodpigeon *C. palumbus maderensis* by dark upperparts, more conspicuous pale band on uppersurface of tail, large grey rather than small white patch on sides of neck, grey instead of vinaceous foreneck, deeper vinaceous chest to upper belly, dark grey flanks, lower belly, vent, and under tail-coverts, and absence of white in wing; some variants of Madeiran feral Rock Dove *C. livia* 'atlantis' superficially rather similar, however, showing black mantle, scapulars, and upper wing-coverts and deep vinaceous chest. CSR

Columba bollii Bolle's Laurel Pigeon

PLATES 29 and 30
[between pages 350 and 351]

Du. Bolle's Laurierduif   Fr. Pigeon de Bolle   Ge. Kanarentaube/Bolle's Lorbeertaube
Ru. Канарский голубь   Sp. Paloma turqué   Sw. Kanarienduva

Columba bollii Godman, 1872

Monotypic

Field characters. 35–37 cm; wing-span 65–68 cm. Slightly smaller than Laurel Pigeon *C. junoniae*. Rather large, slate and dark vinous pigeon (closely related to Long-toed Pigeon *C. trocaz*), with almost black tail broadly banded pale grey. Flight more flapping than Woodpigeon *C. palumbus* but voice similar. Sexes similar;

no seasonal variation. Juvenile separable.

ADULT. Head, throat, and foreneck slate-grey; sides of neck and broad nape glossy green (with purple sheen). Mantle slate-grey (with broad glossy green band); scapulars and wing-coverts slate; flight-feathers black-brown. Back and rump slate-grey. Tail with black base and tip, and broad subterminal pale grey band; sometimes, almost no black tip, and then resembles *C. junoniae*. Breast and underbody deep vinous, but lower flanks and vent grey. Underwing dark grey. Bill mainly red; eye yellow; orbital ring red. Legs red. JUVENILE. Duller, browner than adult, lacking glossed plumage and showing narrow, pale margins to large feathers of body and inner wing. Best distinguished from juvenile *C. junoniae* by tail pattern.

Although closely related to *C. trocaz* of Madeira, confusion ruled out by geographical isolation on western Canary Islands. Distinguished from partly sympatric *C. junoniae* by tail pattern: banded grey with (usually) narrow black tip in this species, broadly tipped white in *C. junoniae*; see also *C. junoniae*. Beware chance of confusion with large dark feral Rock Dove *C. livia*. Flight appearance differs from *C. palumbus*, due to shorter, more rounded wings and longer tail, but though wing-beats more flapping, action rapid and strong. Normally flies close to tree-tops.

Calls low-pitched and muffled.

Habitat. Like Laurel Pigeon *C. junoniae*, confined to mountainous subtropical Atlantic islands, but usually segregated from it by food and other preferences linking it with dense laurel *Laurus* forests, mainly above typical range of *C. junoniae*. Also by nesting mainly in tree-heath *Erica arborea*, as well as laurel *Laurus*, *Myrica*, and *Viburnum*, sometimes at *c*. 3–6 m above ground, rather than on rocks. In heat of day, retires to deep shade of laurel and other high vegetation, but like *C. junoniae* flies down in late summer and autumn to lower levels to feed on ripe cultivated cereals and fruit. Drinks regularly at springs. (Bannerman 1963*b*.)

Distribution. Breeds only on La Palma, Gomera, and Tenerife (Canary Islands). Extinct on Gran Canaria; last records 1889 (Tristram 1889).

Population. CANARY ISLANDS. Plentiful on central and western islands of archipelago in mid-15th century; since declined markedly with destruction of laurel forests. Tenerife: considered near extinction by Bannerman (1963*b*), but now at least 200–300 birds (KWE). La Palma: relatively common in forested areas of north-east, where *c*. 100 seen August 1982 (KWE). Gomera: population larger than on Tenerife (KWE).

Movements. Resident. No detailed studies of local dispersals, but certainly include altitudinal feeding movements, especially in late summer and autumn (Bannerman 1963*b*).

Food. Chiefly fruit; also grain, and occasionally leaves, shoots, and buds. Feeds equally on ground or in tree canopy. Agile on ground, feeding with walk–stop–peck action. Clumsy in canopy with frequent wing-flapping. Shuffles or walks out along branches, stretches out neck and plucks berries from below; alternatively, lands on thin branches and lowers head to pluck berries from above (K W Emmerson). Diet strongly linked to fruiting season of forest trees; birds feed in large concentrations where there is a local abundance (K W Emmerson). Feeds most actively in early morning (Godman 1872; K W Emmerson).

On Anaga peninsula (Tenerife), summer diet consists mainly of *Persea indica*, *Rhumnus glandulosa*, *Picconia excelsa*, *Apollonias barbusana*, *Laurus azoricus*, and *Myrica faya*; in autumn and early winter, *L. azoricus* and *M. faya* predominate. *Ilex canariensis* and *R. glandulosa* taken in winter and, together with *P. indica*, in spring (K W Emmerson). Fruit of til *Ocotea foetens*, apples *Malus*, and rye *Secale* also taken; occasionally, young green shoots and buds of mocan *Visnea mocanera* (Bolle 1857; (Koenig 1890; Bannerman 1963*b*; Löhrl 1981). On Tenerife in spring, an adult ♂ had crop full of leaves of an unidentified shrub, and an adult ♀ had taken laurel berries (Reid 1887); 9 birds had crops full of viñatigo *Persea indica* (Godman 1872).

Young raised in captivity on wheat *Triticum*, rye, and other grain (see Bolle 1857). HAR

Social pattern and behaviour. Little information. Following account includes material supplied by K W Emmerson.

1. Mostly solitary or in pairs, less frequently in small flocks of 3–5; larger (*c*. 15–20) loose-knit aggregations at favourable feeding sites; exceptional flock of *c*. 100, late August (K W Emmerson). See also Laurel Pigeon *C. junoniae* for association with that species. BONDS. Monogamous mating system; duration of pair-bond uncertain (K W Emmerson). 6 birds in pairs, part of loose feeding association, late March (Löhrl 1981). Care of young by both parents until fledging (K W Emmerson; see also Koenig 1890 for study of captive birds). BREEDING DISPERSION. Solitary, though several pairs may nest in same general area. Scant data available suggest only immediate vicinity of nest defended; home ranges overlap. Pair-formation and courtship may take place away from nesting area (K W Emmerson). ROOSTING. Nocturnal and, outside breeding season, solitary or communal (12 or more together); in breeding season, pairs roost near nest. On Gomera, commonly uses tall emergent trees more or less bare of foliage; in summer, roosts situated at forest edge. Birds enter roost *c*. 2–3 hrs before nightfall, arriving in twos, threes, or up to 8 together; start to leave from first light (K W Emmerson). In early morning, fly out occasionally to forage on cornfields; drink in period 14.00–17.00 hrs (Bannerman 1963*b*). May loaf after either activity (Koenig 1890). According to Godman (1872), spends midday period deep in dark forests, not flying much unless disturbed. Rests diurnally in lower canopy of tall trees; sites often at head of ravines and small watercourses, or on ridge tops (K W Emmerson).

2. Exceedingly shy and wary (presumably due at least in some measure to persecution), flying at slightest disturbance (Reid 1887; Löhrl 1981; K W Emmerson). Flies up from ground or tree with loud Wing-clapping (K W Emmerson). ♂ more easily

approached when giving Advertising-call (Koenig 1890: see 1 in Voice). FLOCK BEHAVIOUR. No details. ANTAGONISTIC BEHAVIOUR. In Display-flight, ♂ flies out with loud Wing-clapping, glides with tail fully spread and wings held horizontally on semi-elliptical or circular flight-path, then lands near starting point (K W Emmerson). Brief earlier description indicated undulating Display-flight like that of Woodpigeon *C. palumbus* (Bannerman 1963*b*). HETEROSEXUAL BEHAVIOUR. On Tenerife, courtship activity occurs from January, with ♂ flying after ♀ above forest canopy (or noisily through trees: Koenig 1890). ♂ also gives Advertising-call (see 1 in Voice) in short bouts interspersed with self-preening; normally given while perched on lateral branch in lower canopy. One strutted up and down branch at appearance of presumed ♀ (K W Emmerson). In encounter with ♀ on branch (after pursuit flight), ♂ usually moves with tripping steps towards her and performs apparent Bowing-display, inclining front part of body, with 'crop inflated' (presumably inflated cervical air sac and ruffled feathers) and giving Display-call (see 2 in Voice); assumed to be attempt to invite copulation. If ♀ not ready, she flies off silently. ♂ may then ruffle feathers and self-preen before adopting more erect posture—'crop' again swollen, head withdrawn, and eyes closed—to give Advertising-call at regular intervals (Koenig 1890). ♂ also once performed Bowing-display apparently in manner of Rock Dove *C. livia*: circled ♀ on ground with 'crop' inflated and made bowing movements. Birds at some distance from observers and no calls detectable (K W Emmerson, A Martin). Captive ♂ reported to be extremely active in courtship of ♀ (Koenig 1890). RELATIONS WITHIN FAMILY GROUP. In captive pair, both birds present at nest near hatching. Nestling brooded 'devotedly' by both sexes for *c.* 10 days, though ♂ reported to be more solicitous in care of young (Koenig 1890). Young bird apparently remains near nest in immediate post-fledging period (K W Emmerson, A Martin). ANTI-PREDATOR RESPONSES OF YOUNG. Younger nestling motionless while observers present; nearer fledging, still little reaction—moved head and occasionally shifted in nest; no calls given (K W Emmerson, A Martin). PARENTAL ANTI-PREDATOR STRATEGIES. Tends to be slow to return to nest after disturbance (Koenig 1890); in one case, ♂ returned after *c.* 30 min, ♀ after *c.* 3 hrs (Reid 1887). Captive ♂ sat tight on egg; when nestling *c.* 10 days old, ♂ flew at intruder, pecked at his fingers, and called (Koenig 1890: see 3 in Voice). MGW

Voice. Calls 1 and 2 heard, Tenerife, December–January (H-H Bergmann). However, birds not very vocal there, November or March; definitely present but silent for hours. Possibly almost completely silent during last stage of breeding cycle. Calls are low-pitched and muffled; effect enhanced by density of laurel foliage, sound carrying only *c.* 100 m, even in still conditions (Löhrl 1981). Often noisy when foraging in tree-tops, especially when taking off or landing. In flight, wings make swishing sound with unobtrusive high-pitched component; more metallic in fast, downward flight (H-H Bergmann). For Wing-clapping, see Social Pattern and Behaviour.
CALLS OF ADULTS. (1) Advertising-call (Song) of ♂.
Recording (Fig I) suggests a 'ruor ruor(rur) ruor rup'; 2nd unit sometimes a little shorter, or less loud, than 1st and 3rd (J Hall-Craggs). A very weak guttural cooing, less 'hooting' than Woodpigeon *C. palumbus*. Song bouts made up of varying number of 4-unit phrases, stress on 1st and 3rd unit. 4th unit much shorter (H-H Bergmann). Also rendered 'wa-wa-woo-wa' and 'woo-wa-woo-wa' (K W Emmerson); 'ruk ruk gruuuk guk', 1st unit occasionally shorter than 2nd and duration of 4th unit varies individually. Song-phrases usually given in rapid succession: e.g. 52 in 60 min, with 5–6 bouts of phrases most frequent (H Löhrl). (2) Display-call. Apparently more subdued and muffled than call 1, carrying only *c.* 40 m in absolutely still conditions. A 'truú-trúuu-trŭ', 1st unit with increasing amplitude, 2nd longest, 3rd short (H-H Bergmann). Recording by H-H Bergmann, January, suggests at times a closer resemblance to call 1 with a just-audible 1st unit, though phrases separated by quite long pauses (J Hall-Craggs). See also Koenig (1890) and Goodwin (1970). (3) Captive ♂ attacking man in defence of nestling gave grunting 'jur jur' (Koenig 1890); according to Goodwin (1970), an aggressive or possibly distress call.
CALL OF YOUNG. No information. MGW

Breeding. SEASON. Gomera: January–September. Tenerife: shorter season, January–May. No information from La Palma. (K W Emmerson.) SITE. In low bush, especially tree-heath *Erica arborea* and haya *Myrica faya* (Bannerman 1963*b*). *Erica scoparia* also frequently used, as well as *Viburnum* and some others (Koenig 1890; K W Emmerson). Nest re-used in subsequent years; new nest often built on top of old (K W Emmerson). Nest: compact structure of twigs and fine branches, most frequently of *Erica arborea* and *Myrica faya*; 2 nests, 32 cm diameter and 15 cm high, and 30 × 12 cm (K W Emmerson). Unlined (K W Emmerson); lined with roots (Bannerman 1963*b*). Building: in captive pair, by both ♂ and ♀ (Koenig 1890). EGGS. See Plate 95. Sub-elliptical, smooth and not or slightly glossy; white, but may look yellowish due to yolk showing through very thin shell. 42 × 29 mm (41–44 × 28–31), $n = 17$; calculated weight 19 g (Schönwetter 1967). Clutch: 1. 2 broods. INCUBATION. In captive pair, *c.* 18–19 days (Koenig 1890). By both sexes (K W Emmerson); in captive pair, ♀ sat 19.00–09.00 hrs, ♂ rest of time (Koenig 1890). YOUNG. Semi-altricial and nidicolous. Cared for and fed by both parents (K W Emmerson). Nearly fledged young may leave nest for surrounding branches (K W Emmerson). No further information.

Plumages. ADULT. Closely similar to Long-toed Pigeon *C. trocaz*, differing as follows. Feather-tips of broad half-collar on

I H-H Bergmann Canary Islands January 1983

hindneck and sides of neck iridescent green (in some lights, purple and/or chestnut), not pale silvery-grey; green or slightly purple gloss on slightly darker grey hindcrown, nape, and foreneck more extensive; broad band across mantle glossed green (chestnut or bronze-green in some lights), not deep purple; upperparts and upper wing-coverts slightly darker, slate-black, outer lesser and median upper wing-coverts and all greater coverts not distinctly paler than others; deep vinaceous tinge on underparts extends further downwards, only lower flanks, vent, and under tail-coverts grey, but even here some feather-tips vinaceous; tinge of vinaceous underparts more reddish, less pink, glossed with purple on chest immediately below green-glossed grey foreneck. Tail as in *C. trocaz*, but black tip slightly narrower, $1\frac{1}{2}$–2 cm wide, not sharply demarcated from pale grey subterminal band 3–4 cm wide across uppersurface of tail, but sharply demarcated from white subterminal band on undersurface; on some birds, black tail-tip hardly visible, proximal 5–$5\frac{1}{2}$ cm of tail appears all pale grey from above, and white from below, strongly resembling rather similar tail pattern of Laurel Pigeon *C. junoniae*. Flight-feathers as in *C. trocaz*, but buff edges to p5–p9 narrower or almost absent. DOWNY YOUNG. Down scanty, hair-like; pale golden-yellow on upperparts, cream-yellow on underparts (K W Emmerson). JUVENILE. As in *C. trocaz*, differing from that species by darker sepia-brown upperparts, more extensive rufous-brown on underparts (extending to non-glossy sides of neck), and less or hardly any black on uppersurface of tail-tip. Most juveniles show narrower rufous margins to feathers of body and upper wing-coverts than *C. trocaz*, those on tips of primaries and greater upper primary coverts sometimes hardly visible, making ageing difficult when last juvenile tail-feathers lost (see Structure). FIRST ADULT. Differs from adult as in *C. trocaz*.

Bare parts. ADULT. Iris straw-yellow, often with red tinge. Eyelid red. Bill red, darkest on tip. Leg and foot red. DOWNY YOUNG. Bare skin of head blue-grey, of body pink; skin of central feather-tracts and wing purplish-pink. Bill black with white terminal spot. Leg and foot pink. JUVENILE. Bill black. Leg and foot dark straw-colour with reddish hue. (Koenig 1890; Hartert 1912–21; K W Emmerson, BMNH.)

Moults. See *C. trocaz*.

Measurements. Tenerife, Gomera, and La Palma (western Canary Islands); December–May; skins (BMNH, RMNH, ZFMK).

WING AD	♂ 221	(4·71; 10)	216–228	♀ 217	(4·35; 10)	208–222
TAIL AD	152	(5·75; 9)	145–159	153	(4·34; 8)	147–158
BILL AD	18·9	(0·76; 10)	18·3–20·0	19·7	(1·11; 10)	17·9–20·9
TARSUS	30·3	(1·48; 9)	28·2–32·0	29·6	(0·88; 9)	28·6–30·5
TOE	39·4	(2·36; 8)	37·0–42·3	39·5	(0·80; 8)	38·2–40·5

Sex differences not significant. Juveniles: wing, ♂ 214; ♀ 198, 208, 210; tail, ♀ 131, 132, 142. Wing of Gomera and La Palma adults perhaps slightly shorter than on Tenerife, although ranges show complete overlap: Gomera, ♂ 216, 218; ♀ 210, 218; La Palma, ♀ 217.

Weights. No information.

Structure. Wing shape as *C. trocaz*; in adult, p8 longest, p10 17–23 shorter, p9 2–6, p7 1–6, p6 10–21, p5 27–40, p1 65–76. Tail relatively slightly shorter than in *C. trocaz* and *C. junoniae*, but longer than in *C. palumbus*; t6 5–12 shorter than t1; width of t1 near tip 35–40 mm in adult, 28–32 in juvenile, of t6 27–33 in adult, 20–25 in juvenile. Middle toe longer than in Woodpigeon *C. palumbus* relative to tarsus, but not as long as in *C. trocaz*. Remainder of structure as *C. trocaz*.

Geographical variation. Probably none (but see Measurements).

Sometimes considered to be a race of *C. trocaz* (see that species). CSR

Columba junoniae Laurel Pigeon

PLATES 29 and 30
[between pages 350 and 351]

Du. Laurierduif   Fr. Pigeon des lauriers   Ge. Lorbeertaube
Ru. Белохвостый голубь   Sp. Paloma rabiche   Sw. Lagerduve

Columba junoniae Hartert, 1916. Synonym: *Columba laurivora*.

Monotypic

Field characters. 37–38 cm; wing-span 64–67 cm. Slightly smaller and more slender than Bolle's Laurel Pigeon *C. bollii*, with even shorter and more rounded wings. Rather large, slate-brown and purple pigeon, with grey rump and tail ending in broad, diffuse cream-white band. Soft, flopping flight distinctive. Sexes similar; no seasonal variation. Juvenile separable.

ADULT. Except for diagnostic tail pattern, closely resembles *C. bollii*. At close range, generally darker and browner in appearance, with much green gloss from crown to nape (making whole of head and neck shine at certain angles) and umber, not slate, tone to back. Grey back and rump fade into distinctly paler tail, which has uppersurface with grey-white centre and broad grey to cream-white tips to dusky outer feathers, and undersurface with black-brown base contrasting with broad white tip. Chest deep vinous, glossed purple-green; underparts below chest rich purple-vinous, with restricted grey patch under tail. Bill vinous at base, pink-white at tip; eye orange, orbital ring red. Legs red. JUVENILE. More closely resembles adult than in *C. bollii*, with markedly rufous tone to (and only indistinct pale margins on) body and inner wing-feathers. Best distinguished from *C. bollii* by lack of visible grey on underbody, and tail pattern.

Confusion with *C. bollii* normally prevented by difference in tail pattern: broadly tipped white in this species, grey-banded with (usually) narrow black tip in *C. bollii* (see that species). From above, contrast between brown wings and blue-grey upperparts diagnostic in good light. Flight action slowest and most flapping of all west Palearctic *Columba* (recalling slower action of Jay *Garrulus glandarius*). Less arboreal than *C. bollii*, with long, swinging gait (accompanied by raising of tail and bowing of head), and run recalling partridge *Alectoris*.

Calls little known (see Voice).

Habitat. Only on mountainous subtropical Atlantic islands, during summer mainly in moist cloudy zone immediately below high laurel forest belt. Up to *c*. 1600 m or more, on steep or precipitous wooded slopes densely clad in tree-heath *Erica arborea*, laurel *Laurus*, haya *Myrica faya*, and other thick vegetation, and in bushes at edge of crags or on ledges covered with ferns and rough herbage. Often flies in pairs or singly along mountain faces, or frequents deep gorges grown up with laurel woods. Also among pines interspersed with shrubs, especially after fire has locally destroyed preferred tall laurel forests interspersed with til tree *Ocotea foetens* and viñatigo *Persea indica*, in shade of which heat of the day is spent and much feeding occurs. Drinks at springs, often surrounded by tall laurels in which bird settles before coming down to ground. Very active on ground and can run quickly, especially when feeding in lower barley and flax fields. Often roosts near farms, sometimes living in trees outside woods, but nests on rock-ledges and in crevices where ferns abundant. Despite relative inaccessibility of haunts, has suffered much from hunting pressure, and habitat has also dwindled (Bannerman 1963*b*).

Distribution. Breeds only on La Palma, Gomera, and Tenerife (Canary Islands). Not discovered on Tenerife until *c*. 1975 but may have been overlooked earlier (Löhrl 1981; KWE).

Population. CANARY ISLANDS. La Palma: apparently still fairly common in old haunts. Gomera: although formerly reported as very rare (see, e.g., Cullen *et al*. 1952) now relatively common in restricted local range. Tenerife: extremely small population in 2 areas. (KWE.)

Movements. Resident, though makes altitudinal feeding movements between upland forest and lower agricultural land (Bannerman 1963*b*).

Food. Chiefly fruit, also grain and occasionally flowers. Feeds in trees and on the ground (e.g. Meade-Waldo 1889*a*, Cullen *et al*. 1952).

Some early reports refer to Bolle's Laurel Pigeon *C. bollii* or are otherwise not fully substantiated. Fruits of til *Ocotea foetens* and viñatigo *Persea indica* form main part of diet of *C. junoniae* (Meade-Waldo 1889*b*, 1893). On cultivated areas, takes cherries *Prunus*, flowers of flax *Linum*, barley *Hordeum*, and wheat *Triticum* (Meade-Waldo 1889*a*, *b*; Koenig 1890). Captive adult ♂ refused natural food, but preferred wheat, hempseed *Cannabis*, and rape *Brassica* (Meade-Waldo 1889*b*). HAR

Social pattern and behaviour. Very poorly known.

1. Normally in pairs, singly, or in small groups of 3–9 birds (K W Emmerson; see also, e.g., Conrad 1979*b*). More gregarious than Bolle's Laurel Pigeon *C. bollii*, with which it associates at times in flocks of up to 15–20 (Koenig 1890; Thanner 1908; Polatzek 1909). BONDS, BREEDING DISPERSION, ROOSTING. No information.

2. Not shy (e.g. Hüe and Etchécopar 1958), and generally less wary than *C. bollii* (K W Emmerson). Birds watched on Tenerife, March, remained mostly in trees, but habitat in some areas necessitated more frequent flights above canopy than normally undertaken by *C. bollii* (Löhrl 1981). Flies up with Wing-clapping like *C. palumbus* (Cullen *et al*. 1952; Hüe and Etchécopar 1958). FLOCK BEHAVIOUR, ANTAGONISTIC BEHAVIOUR. No information. HETEROSEXUAL BEHAVIOUR. In presumed courtship Display-flight, bird flies out well above canopy with soft flopping wing-beats; describes concentric circles, interrupted by periods of gliding, and descends gradually with tail full-spread (K W Emmerson). One of 2 birds rose up slowly, and gave similar performance over observers' heads (Etchécopar and Hüe 1957). In earlier brief description of apparent Display-flight, ♂ said to have described arc with tail spread, after which it landed by ♀ and soon gave Display-call (Koenig 1890: see Voice). No further information. MGW

Voice. In recording of presumed *C. junoniae* by H-H Bergmann, Tenerife, April, probable Advertising-call (Song) is a 'pu-pu-pooo', quite distinct from Advertising-call of Bolle's Laurel Pigeon *C. bollii* (H-H Bergmann, J Hall-Craggs). A series of 2-unit 'up-poooo' calls separated by quite long pauses also given; significance not known (H-H Bergmann, M G Wilson). Only other description refers to 'courtship call' (situation suggests probably Display-call: Goodwin 1970) of ♂ in display to ♀; 'kurŭh kurŭh-kurŭkĕdikŭh kurŭkĕdikŭh-kŭh-kŭh', with warbling quality in middle (Koenig 1890). MGW

Breeding. SEASON. Canary Islands: laying begins very end of April or first half of May (Bannerman 1963*b*). SITE. Bare stump or rock-ledge on side of steep overgrown precipice. Nest: apparently none. Building: no information. EGGS. See Plate 95. Sub-elliptical, smooth, slightly glossy; creamy-white. 2 eggs 40 × 29 mm and 45 × 30 mm; calculated weight 20 g (Schönwetter 1967). Clutch: 1 (Bannerman 1963*b*). No further information.

Plumages. ADULT. Closely similar to Bolle's Laurel Pigeon *C. bollii*, but differs from that species and from Long-toed Pigeon *C. trocaz* in being more deeply coloured, more extensively glossy on head and neck, and in even less black on tail-tip than *C. bollii*. Top of head and upperparts as *C. bollii*, but crown and nape grey with extensive green gloss, similar to hindneck and sides of neck; mantle, scapulars, and upper wing-coverts slightly browner, dark sepia-brown with slate tinge rather than slate-black, feather-tips slightly paler brown. Sides of head, throat, foreneck, and chest different from *C. bollii*: feathers of head-sides and throat with broad glossy green tips (partly purplish in some lights), and foreneck and chest deep vinaceous with broad glossy green or purple-green feather-tips (foreneck not grey and chest not deep purplish-vinaceous as in *C. bollii*); remainder of underparts deep vinaceous as in *C. bollii*, but vinaceous even more extensive, slightly glossed purple in some lights, only part of lower flanks, vent, thighs, and all under tail-coverts grey. Back to tail rather different from *C. bollii*: back dark grey, gradually paler grey towards upper tail-coverts and t1, with t1 pale grey or pale brownish-grey; t2–t6 dark grey with ill-defined silvery-white tips 4½–5½ cm wide, tips of outer webs of t2–t3(–t4) partly pale grey; undersurface of tail with fully white tip, contrasting rather with black-brown base; tail of *C. bollii* occasionally rather similar. Flight-feathers as in *C. bollii*, but slightly browner, p5–p10 narrowly bordered pale brown on outer webs; under wing-coverts and axillaries dark sepia-grey. DOWNY YOUNG. Down coarse, white (Harrison 1975). JUVENILE. Rather similar to adult, but general colour of upperparts warmer brown and underparts rufous-cinnamon, neck and chest not or hardly glossy. Head and neck dark grey with faint rufous feather-edges on crown and faint green or purple glossed feather-tips on hindneck and sides of neck. Mantle dark grey, feather-tips tinged brown; scapulars and upper wing-coverts like adult but appear more rufous-brown due to narrow ill-defined rufous-chestnut feather-margins (some birds without these); underparts like adult but rufous-cinnamon instead of deep purplish-vinaceous, not glossy. Tail like adult, but silvery-white tips even less sharply demarcated from grey base, t2–t6 pale grey-brown grading to silvery-white on tip. Flight-feathers as in adult, but pale edges to outer webs of outer primaries less sharply defined and tips of outer primaries more sharply pointed. Closely similar to juvenile *C. bollii*, differing mainly by tail, by more red-brown rather than black-brown scapulars, upper wing-coverts, and flight-feathers, slightly broader rufous fringes to scapulars and coverts, and by absence of grey on lower flanks and vent. FIRST ADULT. Differences from full adult as in *C. trocaz*; see also Structure.

Bare parts. ADULT. Iris orange with red outer ring. Base of bill and cere vinaceous-red, rest of bill pink or pinkish-white. Foot wax-red, claws brown. DOWNY YOUNG. No information for iris, leg, or foot; bill black. JUVENILE. No information. (Koenig 1890; Hartert 1912–21; Bannerman 1963b; Harrison 1975.)

Moults. See *C. trocaz*.

Measurements. La Palma and Gomera (western Canary Islands); February–June; skins (BMNH, ZFMK).

		♂			♀		
WING	AD	225	(3·90; 7)	221–231	214	(6·88; 5)	206–224
TAIL	AD	163	(6·71; 5)	158–171	157	(5·92; 5)	149–164
BILL	AD	18·8	(1·09; 7)	17·7–20·7	19·4	(1·48; 5)	17·7–21·3
TARSUS		33·5	(0·52; 7)	32·6–34·2	32·5	(1·02; 6)	31·0–33·6
TOE		43·3	(1·46; 7)	41·8–45·6	42·4	(2·31; 5)	38·5–44·1

Sex differences significant for wing and tail. Juveniles: wing, ♂ 212, 214, 214, 222; ♀ 205, 216; tail, ♂ 153, 157, 160; ♀ 152.

Weights. No information.

Structure. As *C. trocaz* and *C. bollii*, but wing relatively shorter and more rounded; flight-feathers softer. In adult, p7–p8 longest or each 0–3 shorter than other; p10 16–29 shorter, p9 2–6, p6 7–12, p5 22–35, p1 54–67; in juvenile, p10 13–20 shorter, p9 0–5, p6 10–14. Relative tail length similar to *C. trocaz*; t6 6–9 shorter than t1; width of tail-feathers as in *C. bollii*. Bill relatively shorter than in *C. bollii*. Relative length of middle toe similar to that of *C. bollii*.

Geographical variation. None.

Closely related to *C. trocaz* and *C. bollii* (see *C. trocaz*). CSR

Streptopelia roseogrisea African Collared Dove (Pink-headed Turtle Dove)

PLATES 31 and 34
[facing pages 374 and 398]

Du. Isabeltortel   Fr. Tourterelle rose-et-grise   Ge. Lachtaube
Ru. Африканская кольчатая горлица   Sp. Tórtola de cabeza rosa   Sw. Skrattduva

Columba roseogriseam Sundevall, 1857

Polytypic. Nominate *roseogrisea* (Sundevall, 1857), dry zone south of Sahara from central Mauritania and northern Mali east to central northern Sudan and western Ethiopia, north to Tibesti (Chad). Extralimital: *arabica* (Neumann, 1904), coasts of Red Sea in western Arabia and from northern Somalia north to Gebel Elba (north-east Sudan). Barbary Dove *S. 'risoria'* (Linnaeus, 1758) is a domesticated form of *S. roseogrisea*, locally established in feral state.

Field characters. 29–30 cm; wing-span 45–50 cm. Slightly smaller than Collared Dove *S. decaocto*, with similar structure apart from somewhat shorter tail. Rather small, vinaceous- and grey-brown dove, with character and plumage pattern close to *S. decaocto*. Usually paler than *S. decaocto* but differs distinctly only in darker rim to wing in flight. Most readily identified by voice (constant in all forms including domestic '*risoria*' variant, so-called Barbary Dove). Sexes similar: no seasonal variation. Juvenile separable. In west Palearctic, 2 wild races (not separable in the field) and 1 distinguishable domesticated form.

ADULT. (1) Wild races. Plumage pattern and colours usually paler than *S. decaocto*, differing noticeably only in stronger grey and mauve tones on body and wing-coverts. In flight, wing pattern stronger than in *S. decaocto*, with primary-tips almost black contrasting more strongly with grey upper wing-coverts and grey-white under wing-coverts than in *S. decaocto*, and secondary-tips also darker; total effect makes wing dark-rimmed, recalling

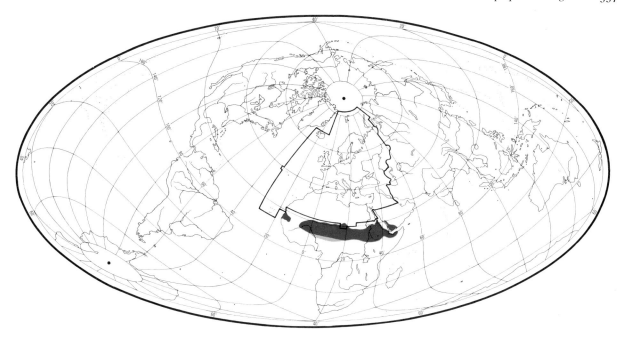

Stock Dove *Columba oenas*. (2) Domesticated variant '*risoria*'. Usually less intensely coloured than wild races—distinctly buffer, fading when worn to pale cream; grey and vinous tones suppressed or absent. Wing pattern lacks dark rim of wild bird and sometimes appears more uniform than *S. decaocto*. JUVENILE. Neck-collar indistinct; edges of upperparts margined pale. Plumage tones drabber than adult.

Separation from *S. decaocto* never easy, requiring close observation even by expert—and made more difficult by existence of domestic variant; best based on distinctive voice and circumstances of its use. In Africa, liable to confusion with 4 similarly marked congeners (see Mackworth-Praed and Grant 1952). Flight (including display), gait, and general behaviour closely similar to *S. decaocto*.

Voice distinct from that of *S. decaocto*: a soft 'ku-kruuu' or 'koo-krr-oo', with characteristic rolled 'r' in central syllable totally lacking in *S. decaocto*; in excitement, a jeering 'hek-kek-kek', suggesting trisyllabic laugh quite unlike rasping or mewing scream of *S. decaocto*, and given just after landing on ground or perch.

Habitat. Inadequately described. In tropical low latitudes, in sub-desert, but in Somalia extraordinarily plentiful at coast, where found in every garden and at every watering place, crowding every tamarisk *Tamarix* tree near overflow from wells, too tame to be driven away. Evidently, however, capable of surviving for long periods without access to water in interior; found in desert mountain terrain at Tibesti, Chad (Moreau 1966), but preferring open dry *Acacia* thorn-bush, and avoiding forest (Archer and Godman 1937). May nest *c*. 20 km from nearest watering place (Morel 1975).

Distribution and population. In west Palearctic, known to breed only in Tibesti (Chad). No information on populations or trends.

Movements. In part a rains migrant in Sahel zone; resident elsewhere, though local movements may occur.

Present all year in Sahel, though large movements apparent in Mali at beginning and end of wet season (Lamarche 1980); in central Chad, resident south of 15°N but extends further north in August–September (Newby 1980). Further south, in Nigeria, a well-defined non-breeding migrant, arriving in thorn-scrub savanna zone (Sokoto to Lake Chad) in dry season (mid-October to May), but returning north with onset of summer rains (Elgood *et al.* 1973; Elgood 1982).

Resident elsewhere but subject to local movements that are probably linked to avoidance of climatic extremes, e.g. absent from Eritrean coasts in hot summer months—probably moves inland (Smith 1957). Large numbers on Dahlac archipelago (Ethiopia), in March, with flocks flying off to north-west; not clear whether migration or local movement (Urban and Boswall 1969).

Food. Seeds of grasses and other plants. Sometimes takes spilled grain (Goodwin 1970). In southern Sahara, mainly seeds of euphorbia *Chrozophora* (Bates 1943b). Suspected to go without water for considerable periods and found in arid areas away from water (Lynes 1925a); berries can supplement requirements (Bates 1927). BDSS

Social pattern and behaviour. No detailed investigation of wild *S. roseogrisea*, though its domestic descendant, so-called Barbary Dove *S*. '*risoria*', is well studied: e.g. Craig (1909),

I C Chappuis Mali February 1969

Goodwin (1952a, which deals with free-flying birds), Miller and Miller (1958), Lehrmann (1964), and Davies (1970, 1974).

1. Gregarious at times, particularly when drinking (Lynes 1925a); also assembles where food abundant but not in compact flocks otherwise (Reichenow 1900–1). Up to 610 at watering place, northern Sénégal, commonly associating with other Columbidae (Morel 1975). BONDS, BREEDING DISPERSION. No information. ROOSTING. When drinking sites restricted (as in dry season), birds spend long periods (sometimes most of day) loafing in vicinity (Lynes 1925a; Morel 1975). Early in morning after cool night, birds perched motionless and with plumage ruffled in trees for c. 30 min before descending to drink (Morel 1975). Pairs and 'families' said to spend hottest part of day perched in trees and bushes, usually away from settlements where they often feed (Reichenow 1900–1).

2. Few details on behaviour, though said to resemble well-studied *S. 'risoria'* (Goodwin 1970), this being confirmed by observations made by D Goodwin on captive nominate *roseogrisea* at London Zoo (Goodwin). Thus probably similar to other *Streptopelia*. May be tame and confiding at watering place (Archer and Godman 1937). FLOCK BEHAVIOUR. In northern Sénégal, October–December, birds drink at puddles and pools in 2 equal peaks—morning and late afternoon. In April–June, peak numbers at 07.00 hrs, then almost complete dispersal mid-afternoon and resurgence thereafter. Arrival and departure times of first and last birds vary with month and perhaps with time of sunrise and sunset. Mean time taken to drink $11 \cdot 3$ s ($n=16$). Prefers biggest puddles and those most easily accessible. When drinking at big puddles with Laughing Dove *S. senegalensis*, sometimes climbs over their backs (Morel 1975). ANTAGONISTIC BEHAVIOUR. Bowing-display (up-and-down bobbing movements of head and shoulders) performed by one ♂ who gave persistent Display-call to another who remained silent, but no details given of any overtly antagonistic behaviour linked with this (Bates 1927: see 2 in Voice). In Display-flight by presumed ♂, bird flew up at c. 80° with Wing-clapping, then glided straight down at c. 30°, landing within c. 1 m of apparent mate (Urban and Boswall 1969). Display-flight probably longer and more pronounced than in *S. 'risoria'* (Goodwin 1970). For detailed descriptions of antagonistic interactions in *S. 'risoria'*, see Goodwin (1952a) and Miller and Miller (1958). HETEROSEXUAL BEHAVIOUR. In Bowing-display by ♂, bird faced mate in a markedly erect posture with neck feathers ruffled, then made several bowing movements. Followed by chases and (sometimes successful) copulation (Urban and Boswall 1969). In Lake Chad area, April–May, ♂♂ said to be still in state of 'sexual activity' after breeding (Bates 1927). See also Craig (1909), Goodwin (1952a), Miller and Miller (1958), and Lehrmann (1964) for *S. 'risoria'*. ANTI-PREDATOR RESPONSES OF YOUNG, PARENTAL ANTI-PREDATOR STRATEGIES. No information for *S. roseogrisea*. For *S. 'risoria'*, see (e.g.) Goodwin (1952a), also Grünefeld (1952). MGW

Voice. As in Barbary Dove *S. 'risoria'* (Goodwin 1970; Chappuis 1974; D Goodwin). For detailed discussion of repertoire in *S. 'risoria'* see (e.g.) Craig (1909), Bodenstein (1949b, where comparison with Collared Dove *S. decaocto*), Goodwin (1952a, 1970), Miller and Miller (1958), Lade and Thorpe (1964, where also extra sonagrams),

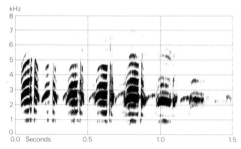

II C Chappuis Mali February 1969

Davies (1970, 1974), and Mairy (1971, 1976a, b, c, d, 1979a, b). In *arabica*, 'high-pitched instrumental sound' produced by wings in short flights (Urban and Boswall 1969); for Wing-clapping in Display-flight, see Social Pattern and Behaviour.

CALLS OF ADULTS. (1) Advertising-call (Song). Recording (Fig I) suggests 'KOOK r-r-r-r-r-r-OOooooooooooo' in which 'KOOK' begins and ends abruptly; remainder virtually a long single unit ascending then descending in pitch, with liquid 'r' sounds and marked emphasis on 'OO'; has quality of hooting owl (Strigidae) or crowing cockerel. Often followed by 'qua' sound, presumably intake of breath. (J Hall-Craggs, P J Sellar, M G Wilson.) Also rendered: 'took-tookoroo' (Lynes 1925a); 'OOH r-r-r-r-r-r-r-oo-r', with 'oo' higher pitched (Urban and Boswall 1969); hollow and sonorous 'whoo-o-o-o-o-who', or 'whoo who-o-o-o-o', the 'o-o-o-o-o' given as a rolling trill (King 1978). Less shrill or piercing than *S. 'risoria'* (Reichenow 1900–1). For suggestion of racial variation, see Urban and Boswall (1969). For additional sonagram from Mali, see Mairy (1976d). (2) Display-call. A half-purring 'coo-coo-oo' (Bates 1927); presumably similar to call 1. In *S. 'risoria'*, Advertising-call, Display-call, and Nest-call closely similar if not identical (see Goodwin 1952, 1970); observations at London Zoo suggested this also true of *S. r. roseogrisea* (D Goodwin). (3) Nest-call. For sonagram, see Mairy (1976d). (4) Excitement-call. Recording (Fig II) suggests nasal 'heh heh heh ...' with varying speed, rhythm, and timbre (M G Wilson). Excitement-calls of captive birds exactly like *S. 'risoria'* (D Goodwin). (5) Other calls. A frequently uttered 5-note call said to resemble *S. decaocto* (Urban and Boswall 1969); possibly refers to call 4 and comparison therefore refers to high-intensity variant of *S. decaocto* Excitement-call. No further information. MGW

Breeding. SEASON. No information from west Palearctic. Somalia: eggs found May–June (Archer and Godman 1937). Sénégal: breeds all year (Morel 1972), though Heim de Balsac and Mayaud (1962) stated that breeding corresponds to northern spring. Western Sudan: January–June or later (Mackworth-Praed and Grant 1952). SITE. In tree or bush, at 4–5 m. Nest: flimsy platform of twigs. Building: no information. EGGS. See Plate 95. Sub-elliptical, smooth and slightly glossy; white or whitish. 29 × 23 mm, no sample size or range given; calculated weight 8 g (Schönwetter 1967). Clutch: 2 (Mackworth-Praed and Grant 1952). No further information.

Plumages (nominate *roseogrisea*). ADULT. Rather similar to Collared Dove *S. decaocto*, but forehead, crown, and nape isabelline-buff with mauve-pink tinge of variable extent; sides of head and upper chin buff, grading to pale buff and cream on throat; colour of top and sides of head rather close to that of upperparts, not to underparts as in *S. decaocto*. Half-collar on lower hindneck and sides of neck as in *S. decaocto*. Upperparts, tertials, and most lesser and median upper wing-coverts buff-brown or pale fawn, less greyish than in *S. decaocto*; all secondaries and outer greater and median upper wing-coverts light grey with buff tinge to outer webs, not mainly uniform light bluish-grey. Chest and breast pale buff-brown or isabelline-buff with slight mauve-pink cast, flanks pale grey, belly pale buff. Vent cream-buff, grading to cream-white on under tail-coverts (all medium-grey in *S. decaocto*). Tail-feathers: t1 pale fawn-brown like back and rump, others pale grey-brown with ill-defined white tip (from *c.* 1 cm wide on t2 to 3–5 cm on t6), t2–t4 with medium grey bases to inner webs, t5–t6 with dark grey bases, outer web of t6 white; from below, tail shows greyish-black base, markedly contrasting white tip 5–6 cm wide, and mainly white outer web of t6. Primaries dark grey-brown, tips and outer webs of p6–p10 narrowly and faintly bordered cream-white. Under wing-coverts and axillaries white with buff tinge to lesser coverts along leading edge of wing and often with slight grey tinge to greater secondary and outer primary coverts. Sexes nearly similar, but ♂ has forehead, crown, and nape pale grey with slight mauve-pink cast, ♀ has forehead pale buff and crown and nape tinged sandy-isabelline; mauve-pink tinge on chest and breast slightly deeper and more extensive in ♂. Domesticated form, Barbary Dove *S. 'risoria'*, similar to *S. roseogrisea*, but colour slightly to distinctly paler (all-white in 'Java Dove' variant); in more typical specimens, head, upperparts, t1, and chest sandy-buff or cream-buff, shading to pale cream or white on chin, vent, and under tail-coverts; flanks pale grey; tail as in *S. roseogrisea*; primaries and upper primary coverts buff-brown, tinged grey when fresh, secondaries and outer upper wing-coverts pale cream-buff with grey tinge; under wing-coverts and axillaries pale grey to white; in fresh birds, head and chest tinged vinaceous-pink. Some birds are very pale creamy-isabelline, but colour of all variants strongly subject to bleaching. DOWNY YOUNG. Down coarse and hair-like; cream-buff. JUVENILE. Closely similar to adult, differing by absence of half-collar, by slightly duller colours, and by narrower and shorter feathers; fringes of feathers of scapulars and upper wing-coverts often contrastingly white or buff, but difference sometimes not marked or non-existent. See also juvenile *S. decaocto*. FIRST ADULT. Like adult, but variable number of shorter and narrower juvenile feathers retained, contrasting with fresh longer and broader feathers; sometimes still shows pale fringes on upper wing-coverts. Differs most clearly in tail-feathers (see Structure) and relatively short, narrow, and pointed outer primaries.

Bare parts. ADULT. Iris deep vermilion. Bare skin round eye grey. Bill black. Leg and foot crimson or purplish-red, claws black. In *S. 'risoria'*, iris ruby-red (exceptionally orange), bare skin round eye paler grey, and bill dark purplish-black with silvery bloom; white 'Java Dove' has iris dark or pink. DOWNY YOUNG, JUVENILE. No information for wild *S. roseogrisea*. In *S. 'risoria'*, iris changes from dull grey to yellowish and orange to obtain adult colour at completion of post-juvenile moult, bill from dull pink through pinkish-grey to purplish-black, leg and foot dull grey through pinkish-grey to red. (Mackworth-Praed and Grant 1952; Goodwin 1970; BMNH, RMNH.)

Moults. Little known (but see Morel 1983); only February–May specimens examined, mainly from Sudan. ADULT POST-BREEDING. Most adults showed fresh to rather fresh body, tail, and flight-feathers; one had moult suspended with 3 outer primaries old; 3 others (perhaps 2nd-calendar-year birds) showed evidence of suspended serially descendant moult with outer 3–4 and inner 3–4 primaries new and middle 3 old—probably, moult previously suspended with p6 or p7, starting again with p7 or p8 and p1 following at same time or shortly afterwards; one, February, had inner 2 and outer 4 primaries new with p3 growing. POST-JUVENILE. Complete, sequence and timing in relation to hatching as in *S. decaocto*. Recorded primary moult scores, Sudan, February: 29, 29, 30 (suspended), 12, 6, 6. (BMNH, ZFMK.).

Measurements. Nominate *roseogrisea*, Sudan, February–May; skins (BMNH, ZFMK).

WING AD	♂ 166	(5·24; 10)	158–174	♀ 157	(3·81; 8)	153–162
TAIL AD	115	(3·68; 7)	110–120	113	(3·90; 8)	107–117
BILL AD	15·3	(0·90; 9)	13·9–16·5	15·1	(0·53; 9)	14·3–15·8
TARSUS	22·8	(1·40; 10)	21·2–24·3	22·1	(1·45; 9)	19·9–23·9
TOE	28·4	(1·41; 10)	26·5–30·0	27·7	(1·55; 9)	25·8–29·7

S. r. arabica. Western Arabia, all year; skins (BMNH).

WING AD	♂ 159	(2·66; 11)	157–164	♀ 154	(3·90; 4)	151–160

Domesticated form *S. 'risoria'*. Skins (BMNH, RMNH).

WING AD	♂ 171	(4·24; 13)	165–179	♀ 160	(8·00; 4)	152–166
TAIL AD	124	(2·82; 12)	118–128	115	(—; 3)	113–118
BILL AD	15·2	(0·83; 13)	13·9–16·4	15·3	(0·33; 4)	14·9–15·7
TARSUS	22·9	(0·98; 7)	21·2–24·0	22·6	(0·53; 4)	21·8–23·0

In all, sex differences significant for wing only. Wing of juvenile on average *c.* 9 shorter, tail *c.* 12; both significantly shorter than adult wing and tail. Wing of single adult ♀ *arabica* from Gebel Elba (Sudan), April, 163 (BMNH).

Weights. Nominate *roseogrisea*. Ennedi (Chad), April, ♂ 135, ♀ 148; Aïr (Niger), February, ♂♂ 165, 172 (Niethammer 1955).

Structure. Wing rather short, tip slightly rounded. 11 primaries: p9 longest, p10 2–9 shorter, p8 0–3, p7 6–12, p6 18–24, p5 26–35, p1 51–62; p11 minute, concealed by primary coverts. Tail rather short, tail/wing ratio 0·70 in nominate *roseogrisea*, 0·72 in domesticated '*risoria*'; of west Palearctic *Streptopelia*, only Turtle Dove *S. turtur* (0·65) and Rufous Turtle Dove *S. orientalis* (0·68) have shorter tails relative to wing length; distinctly longer in *S. decaocto* (0·77) and Laughing Dove *S. senegalensis* (0·81). Tail slightly rounded; 12 feathers, t6 10–15 shorter than t1. In adult, width of t1 near tip 20–24 mm, of t6 17–21; juvenile tail-feathers on average 2 mm narrower and tip more pointed, less squarish. Bill short and slender, as in most other west Palearctic Columbidae. Tarsus rather short, but strong. Toes rather long and slender; outer toe *c.* 72% of middle, inner *c.* 76%, hind *c.* 62%.

Geographical variation. Slight. *S. r. arabica* from Red Sea coasts of Arabia, Sudan, Ethiopia, and Somalia differs from nominate *roseogrisea* from central and western Sudan by pale bluish-grey under wing-coverts and axillaries and by averaging slightly smaller; difference in underwing slight, however, and many nominate *roseogrisea* show greyish tinge to part of underwing at least. Populations in western part of range, east to about Lake Chad, are sometimes said to be darker in colour and separated as *bornuensis* (Bannerman 1931); however, large series examined (BMNH) did not show any constant difference from typical nominate *roseogrisea* of Sudan. Examined birds from Tibesti (Chad) showed deeper rufous-brown upperparts and paler upper wing-coverts than others, but sample too small to see whether this character constant. Domesticated *S. 'risoria'* (better named *S. roseogrisea* var. *domestica*: Nowak 1975a) similar to nominate *S. roseogrisea*, but slightly paler and more creamy, especially rump and upper tail, and wing and tail average longer.

Forms superspecies with *S. decaocto* and White-winged Collared Dove *S. reichenowi* from Giuba valley in north-east Africa (Nowak 1975a); Javanese Collared Dove *S. bitorquata* (Goodwin 1970; Snow 1978) and Red-eyed Dove *S. semitorquata* of Afrotropical region (Nowak 1975a) perhaps also rather closely related.

CSR

Streptopelia decaocto Collared Dove

PLATES 31 and 34
[facing pages 374 and 398]

Du. Turkse Tortel   Fr. Tourterelle turque   Ge. Türkentaube
Ru. Кольчатая горлица   Sp. Tórtola turca   Sw. Turkduva

Columba decaocto Frivaldszky, 1838

Polytypic. Nominate *decaocto* (Frivaldszky, 1838), Europe, Middle East, and northern Arabia, east to Afghanistan, Pakistan, northern India, Nepal, and Assam, introduced China, Korea, Japan, and elsewhere. Extralimital: *stolickzae* (Hume, 1874), Kirgiziya, south-east Kazakhstan (USSR), Sinkiang, and northern Kansu (China), grading into nominate *decaocto* in Inner Mongolia; *intercedens* (Brehm, 1855), southern India and Ceylon, grading into nominate *decaocto* in central and northern India; *xanthocyclus* Newman, 1906, north-west Burma and perhaps elsewhere in tropical south-east Asia, grading into nominate *decaocto* in Manipur (eastern India).

Field characters. 31–33 cm; wing-span 47–55 cm; tail 10–11 cm. About 10% larger than African Collared Dove *S. roseogrisea*; up to 20% larger and longer-tailed than Turtle Dove *S. turtur*; 25% larger than Laughing Dove *S. senegalensis*. Noticeably pale dove with long, rather square-ended tail. Undertail boldly marked with almost black base and broad cream-white terminal band. At close range black half-collar on hindneck of adult obvious. Song and call distinctive. Sexes similar; no seasonal variation. Juvenile separable.

ADULT. Except for wild *S. roseogrisea*, palest dove in west Palearctic with little contrast in plumage pattern. Crown pale grey, merging into pale vinous-pink (♂) or vinous-buff (♀) face. Narrow collar round hindneck mainly black, noticeably edged white above and less so below. Rest of neck, breast, and forebody vinous-buff, fading into buff-white on belly and under tail-coverts. Whole of back, scapulars, and smaller wing-coverts sandy grey-brown. Across inner secondaries and greater and primary coverts, wide panel of pale grey contrasts with dusky grey tips to outer secondaries and dusky brown primaries. Upper tail dusky brown, with cream-white tips obvious on outer feathers only, emphasizing corners. In flight, no obvious pattern on wing or tail from above, but contrast from below of almost white under wing-coverts and dusky flight-feathers often striking and basally black, distally white pattern of tail always so. White terminal band noticeably broad, shared only by *S. roseogrisea*. Bill black; eye deep red; orbital ring off-white. Legs and feet mauve-red. JUVENILE. Duller than adult; lacks half-collar and shows pale buff margins on feathers of upperparts.

Separation from wild *S. roseogrisea* difficult; see that species. Confusable with dull adult and (particularly) juvenile *S. turtur*, but that species has fan-shaped tail and is darker above, with spotted chestnut or brown scapulars and wing-coverts, darker lower back, rump, and tail, brilliant, narrow white tail-tip, and much more agile flight action. More likely confusion is with domesticated Barbary Dove *S. 'risoria'* form of *S. roseogrisea* which is slightly smaller and even paler and more uniform, with essentially sandy or white-buff plumage (lacking vinous breast) and less contrasting undertail pattern. Flight fast, with frequent bursts of clipped wing-beats and looping falls, glides, and dives; actions less energetic than those of *S. turtur* but include characteristic 'shooting' acceleration when leaving cover or in alarm. Gait free, with mincing walk (accompanied by tail movement and head-bobbing) and occasional half-run. Perches freely; spends much time on ground when searching for food, covering quite long distances on foot.

Gregarious, forming winter flocks of hundreds. Song a persistent but musical, often loud 'ku-koo-ku'. Other calls include striking, harsh swearing note.

Habitat. Has changed considerably during major range expansion. Considered under (1) original range in low latitudes of Oriental region, (2) first expansion or introduction through Asia Minor, etc., to Balkans, and (3) recent expansion through middle to upper middle latitudes of west Palearctic, including temperate and marginally boreal zones and oceanic coasts and islands.

(1) In Indian heartland of range affects open, cultivated,

essentially dry deciduous country (even semi-desert) with groves or a scattering of babool *Acacia* and similar trees. Avoids city centres and moist evergreen tracts. Occurs up to *c.* 2400 m (rarely to *c.* 3000 m) in western Himalayas in summer. Gleans in paddy stubbles, on fallows, or in newly sown millet fields, often by towns and villages. Perches freely on buildings, also boldly entering rustic dwellings, cattle sheds, and verandas, within an arm's length of the occupants, occasionally nesting on rafters in a cattle shed or outhouse but usually in a bush or small tree (Ali and Ripley 1969). Avoids heavy forest and is found in great abundance in cultivation and open country wherever trees, large bushes, and hedges provide cover (Whistler 1941). Gathers in trees and along telegraph lines (Hutson 1954). In Afghanistan only seen in summer, in willow groves, hotel gardens, poplar groves, fields, and at Herat very common in the town (Paludan 1959).

(2) In Iraq common wherever there are trees, except in Kurdistan mountains, favouring date gardens, orchards, poplar *Populus* thickets, bramble *Rubus* clumps, mulberry *Morus*, apple *Malus*, and tamarisk *Tamarix*. Nests on buildings, on an office block or under a garage roof, but not frequently, although feeding in towns in streets, gardens, and yards, often freely in fields. None found breeding more than 1 km from nearest house. Weather in Iraq is often cold and bleak for several winter weeks, with temperature falling below freezing. (Marchant 1963c.) In some parts of central Europe, making increasing use of inner cities (Bozsko 1979; Bozsko and Lajos 1981) and countryside (Kneis and Görner 1981) for breeding and feeding.

(3) Involving settlement in cooler climates, often wetter, windier, and seasonally chillier than previous environments, and demanding in some cases readiness to cross seas and to overwinter in severe frosts. Here speed and success of colonization apparently based on consistent practice of selecting sites where grain readily available through human agency either at processing plants (mills, maltings) or through spillage (as in docks) or put down as livestock feed (farms, gardens, hen-runs, rearing pens for pheasants *Phasianus*, or zoological collections). Consequently has become closely associated with suburban human habitations and activities, normally avoiding open countryside and often city centres where other Columbidae would offer strong competition (Sharrock 1976). Unlike feral Rock Dove *Columba livia*, depends largely on food provided indirectly (rather than directly) by man. In both cases, however, dependence implies sedentary existence, despite earlier migratory habit when faced with cold season.

In western Europe, favours mixed habitats with gardens, farmyards, orchards, churchyards, avenues, and similar areas, part open and part carrying trees, overhead wires with supporting poles or pylons, and other suitable perches or places for resting, roosting, or nesting. Such habitats occur particularly in suburbs or on urban fringe, or in small towns, large villages, and isolated clusters of buildings, especially where there is some bare ground such as yards or ploughland. Frequents small or medium-sized trees, pears *Pyrus* being favoured in south, with limes *Tilia*, poplar *Populus*, and other broad-leaved species, while in Britain, planted conifers, often exotic, also used (Simms 1975). Low-density residential quarters with large gardens preferred to those with smaller gardens or none. In Britain, numbers of breeding birds near centre of town inversely correlated with size of its built-up area, the hearts of major towns and cities being shunned. Further pronounced feature in Britain is coastal concentration, within *c.* 8 km of tidal waters (Hudson 1972). While in some areas density rapidly builds up to high level, other apparently suitable areas are neglected, as well as grassland, moorland, and upland, few nests in Britain being above 300 m, although in West Germany and Switzerland breeding proved up to 1000 m and suspected even higher (Glutz and Bauer 1980). Unlike *Columba* pigeons, not markedly dependent on access to water for drinking and bathing, and in western Europe infrequently nests or loafs on buildings. Flies freely, but usually for short distances and no higher than *c.* 30 m. Not seriously threatened by predation or human persecution at present, and not recognized as serious pest in most places. In Debrecen (Hungary), increasing density causes sanitary and economic problems, and displacement of other bird species (Bozsko and Lajos 1981). Elsewhere (Hungary and Czechoslovakia), may damage crops (Helešic 1981).

Distribution. Marked range expansion in Europe since *c.* 1930; initially spread from south-east mainly to north and west, in the north reaching Norway and Sweden in the 1960s and Finland by 1966, and in the west Faeroes and Iceland (not breeding regularly) by early 1970s. Since then, main expansion in USSR to east, north, and south; still continuing. Reached Spain and Portugal in 1974, and has recently begun to colonize Egypt.

Until *c.* 1930, largely confined in Europe to Balkans; earlier history there uncertain—may have been originally introduced by man (as perhaps also in Turkey and Middle East); favoured by Muslims in Ottoman Empire and may have declined in areas where their influence decreased when Balkan countries became independent. Increases in northern Yugoslavia led to start of major expansion when Hungarian plain invaded *c.* 1930 and Rumania in 1933; in next 40 years, colonized *c.* 2500000 km^2 (Glutz and Bauer 1980, with maps showing extent of spread 1932–72). Documented in detail by Fisher (1953), Stresemann and Nowak (1958), and Nowak (1965, 1975b); dates of first breeding given below; in some cases, recorded (and may have bred) earlier.

Causes of this dramatic expansion still not fully understood; suggestions include genetic mutation (Mayr 1951, 1963; Fisher 1953), change from nesting mainly in buildings to mostly in trees (Stresemann and Nowak 1958),

342 Columbidae

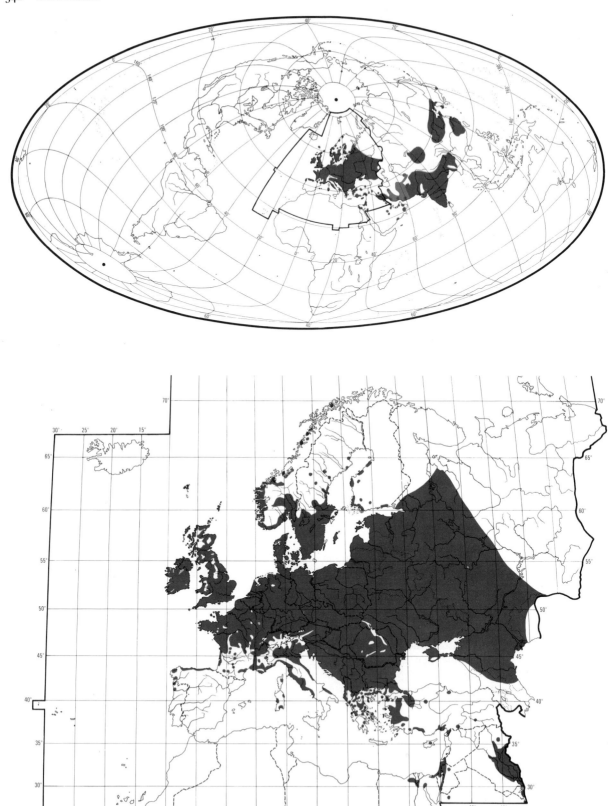

alterations in extra-specific control factors (Berndt and Dancker 1966), increase in number of broods per annum in temperate regions (Marchant 1963c; Glutz and Bauer 1980), lessened predation in suburban and urban habitats (Marchant 1963b), or combinations of these factors.

ICELAND. First bred 1971; not recorded annually since (AP). FAEROES. Breeding from early 1970s (BO). BRITAIN AND IRELAND. First bred Britain 1955, Ireland 1959, and Channel Islands 1961; rapid spread until 1965, more slowly since (maps and details in Hudson 1965, 1972, Sharrock 1976, Long 1981a). FRANCE. Bred from 1952; details of spread in Yeatman (1976). Corsica: first bred c. 1975 (J-C Thibault). PORTUGAL. Well established Oporto from 1974 (Santos Júnior 1978–9b, 1979–80); bred 50 km to north 1982 (RR). SPAIN. First bred 1974 (Morales 1974); has spread slowly along north and northwest coasts (AN). BELGIUM. First bred 1955 (Lippens and Wille 1972). LUXEMBOURG. First bred c. 1956 (Glutz and Bauer 1980). NETHERLANDS. First bred 1950; spread throughout by 1956 (CSR). WEST GERMANY. First bred 1945 or 1946 (Glutz and Bauer 1980). DENMARK. First bred 1950 (Dybbro 1976). NORWAY. First bred 1955 (Haftorn 1971). SWEDEN. First bred 1951; up to 1962 colonized some towns in south, then up to 1975 range increase, with marked northward expansion (Risberg 1978); little change since (LR). FINLAND. First bred 1966; spread since, mainly in coastal towns and villages (OH). EAST GERMANY. First bred 1946 (Makatsch 1981); now spreading outside human settlements (Kneis and Börner 1981). POLAND. First bred 1943 (Tomiałojć 1976a). CZECHOSLOVAKIA. First bred 1938; colonized whole country by 1954 (Hudec and Černý 1977). HUNGARY. Colonized from c. 1930; see Keve (1962b) for details of spread. AUSTRIA. First bred 1943 (Glutz and Bauer 1980). SWITZERLAND. First bred c. 1952 (Glutz and Bauer 1980); by c. 1970, larger towns in northern lowlands colonized; by 1978, nesting in all settled areas to c. 700 m, since spreading in valleys in Alps and Jura to c. 1600 m (RW). ITALY. First bred 1947 (Moltoni 1954b) and rapidly colonized north (SF); has spread to south and recently to Sardinia (PB, BM). YUGOSLAVIA. Breeding before 1900 in south and Belgrade area; expansion north and north-west began c. 1927 (Rucner 1952; Matvejev and Vasić 1973). GREECE. Said to have been largely exterminated after expulsion of Turks, except for some islands near Turkey (Reiser 1905), but not in eastern Makedhonia and Thrace, and spread again from there from c. 1930 (HJB, WB, GM). BULGARIA. Some spread from towns in recent years (NB, YM, PS, SS, ZS). RUMANIA. Marked expansion since 1933 (Vasiliu 1968); all except mountain areas occupied by c. 1957 (Munteanu 1970). USSR. First pairs recorded in Ukraine 1944 (Dementiev and Gladkov 1951b); spread east and north, since then more rapid colonization, reaching north to Onega (White Sea), east to Kazakhstan, and south to Krasnodar and Makhachkala, though not all intervening areas may be occupied (Bozsko 1977; Rubinshtein 1981; Varshavski 1981). TURKEY. For details of distribution, see Kumerloeve (1968b); no apparent recent change (MB, RP). SYRIA. Formerly resident Aleppo (Kumerloeve 1968b); may still breed occasionally (see Macfarlane 1978). LEBANON. Formerly bred Beirut; no recent observations (Tohmé and Neuschwander 1974). CYPRUS. Scarce resident Nicosia, probably introduced (PRF, PS). EGYPT. First recorded 1979, in Suez, and has spread to Cairo and environs (PLM, WCM, Sherif M Baha el Din).

Accidental. Morocco, Malta.

Population. Numbers in west Palearctic have increased greatly following major range expansion (see Distribution), but few population estimates.

FAEROES. A few pairs (DB). BRITAIN AND IRELAND. After colonization in 1955, estimated c. 3000 pairs 1964, 15000–25000 pairs 1970, and 30000–40000 pairs 1972; rate of increase in first 10 years c. 100% per year, dropping to under 50% (possibly only 25%) by 1970 (Hudson 1965, 1972; Sharrock 1976), and now approaching stabilization (R Hudson). FRANCE 10000–100000 pairs (Yeatman 1976). Corsica: up to 50 pairs (J-C Thibault). BELGIUM. About 40000 pairs (Lippens and Wille 1972); c. 50000 pairs (PD). NETHERLANDS. From 1950 to 1963 increased from 5 pairs to 4000–5300 pairs (Leys 1964), then to 60000–100000 pairs 1975–7 and increasing since (Teixeira 1979). WEST GERMANY. Estimated 550000–670000 pairs (GR). DENMARK. Increased, but no detailed figures; numbers fluctuate, but no apparent decrease since limited shooting allowed from 1972 (Fog 1979). NORWAY. Marked increase until 1975; since, stable or decreased in many localities (GL, VR). SWEDEN. About 1200 birds, winter 1961–2; marked increase to 1974–5 when 6000–8000 pairs (Risberg 1978); little change since (LR). FINLAND. Some hundreds of pairs (OH). EAST GERMANY. See Klafs and Stübs (1977) and Rutschke (1983) for local estimates. POLAND. Fairly numerous (Tomiałojć 1976a). GREECE. Increased after c. 1930, especially in 1950s and 1960s, but little recent change (HJB, WB, GM). EGYPT. Several hundred birds in eastern part of Nile delta, 1983 (PLM).

Survival. England: juvenile mortality 69%, annual adult mortality 39% (Coombs et al. 1981). Sweden: annual mortality 29% (Bentz 1982). Central Europe: mortality in 1st year of life 50–75%, annual adult mortality 35–55% (Glutz and Bauer 1980). Oldest ringed bird 13 years 8 months (BTO).

Movements. In southern Asia (to which formerly confined), basically resident, though in mountainous regions there can be pronounced altitudinal movements, with birds descending to lower ground and even moving away from upland areas November–March: e.g. Meinertzhagen (1920) and Ali and Ripley (1969) for Pakistan and northern India.

Among European colonists (to which rest of account

refers), breeding adults believed largely resident, as are a proportion of immatures. Other immatures make pronounced dispersals, especially westwards (see below), whereby the species has spread right across Europe this century; characterized by settlement in new region, without return movement towards natal area (dismigration).

Dispersal by young birds can begin soon after post-juvenile moult, but only small proportion occurs in autumn. In 853 arrival and passage dates for Britain and Ireland, 25% occurred August–November, 60% April–June (Hudson 1965); of 717 passage birds on Helgoland (West Germany), 1963–70, only 5% August–November, with occurrences chiefly May (60%) and June (21%) (Vauk 1972). Hence maximum dispersion occurs in late spring/early summer of 2nd calendar year. Dispersal may be triggered by density-dependent food competition in late winter (Reichholf 1976). Adults thought only occasionally to move more than 10 km (Glutz and Bauer 1980). In Britain, birds ringed as 'adults' often found to have moved 200 km or more (Coombs et al. 1981), but, due to ageing difficulties, these certainly include 2nd-calendar-year birds, and no distant recoveries for known breeding adults. 1 West German adult ♀ moved 354 km south in June, after deserting young brood hatched near her own birthplace (Lachner 1963).

European ringing data show longer displacements (more than 20 km) strongly oriented westwards (e.g. Nowak 1965, Glutz and Bauer 1980), but Helgoland migrants influenced by island's geographical position, with as many recoveries in north-east quadrant (Schleswig-Holstein, Denmark, southern Scandinavia) as west of north–south axis (Vauk 1963, 1972). In British analysis, 77% of 171 recoveries vectored SSW–N, 23% NNE–S (Coombs et al. 1981); atypical British-ringed birds found West Germany (1), Belgium (1), and France (6), and 1 atypical Helgoland bird found in Arkhangel'sk (USSR). Most dismigrants recovered within 300 km of ringing site, though displacements of 400–500 km not scarce; rarely more than 1000 km. In Continental data, 34% of 2nd year birds had moved 0–20 km, 34% 21–100 km, 32% 101–500 km (Glutz and Bauer 1980).

British and Irish ringing results indicate recent reduction in degree of long-distance dispersal, coincidental with approaching stabilization of breeding population size (BTO). Proportions of recoveries showing movement of more than 20 km: 1963–6 43% ($n=42$); 1967–70 38% ($n=101$); 1971–4 42% ($n=169$); 1975–8 24% ($n=110$); 1979–82 22% ($n=134$). Similarly, of 46 foreign-ringed birds found Britain and Ireland (originating West Germany, Netherlands, Belgium, Channel Islands), average 2·8 per year during 1962–74, 1·1 per year 1975–82. In other granivorous birds, such as finches (Fringillidae), degree of movement liable to increase at high population densities; hence opposite trend in *S. decaocto* unexpected. Original impetus for spread across Europe not fully understood (see Distribution); may have had genetic basis (Mayr 1963), but if so, this presumably now being selected against. With little or no optimum habitat remaining uncolonized, birds staying near natal area (with which they are familiar) perhaps have more chance of breeding successfully than birds moving into unfamiliar territory now almost inevitably occupied already.

Food. Cereal grain, and seeds and fruits of other herbs and grasses; occasionally, green parts of plants, invertebrates, and bread. Food taken mainly from ground. Aided by rapid wing-beats takes berries from bushes and trees (Nowak 1965; Goodwin 1970, 1978; Beretzk and Keve 1973). Will take seeds of sunflower *Helianthus* from standing plant (Kiss and Rékási 1981), and aphids and lepidopteran larvae from trees (Navasaitis 1968). Indigestible items may be ejected as pellets (Hofstetter 1954).

Plant food includes grains of wheat *Triticum*, maize *Zea*, buckwheat *Fagopyrum*, rye *Secale*, barley *Hordeum*, millet *Setaria*, and oats *Avena*; seeds of knotgrass *Polygonum*, goosefoots (Chenopodiaceae), amaranth *Amaranthus*, brassicas *Brassica*, wild mustard *Sinapis*, wintercress *Barbarea*, bedstraw *Galium*, bindweed *Convolvulus*, sunflower *Helianthus*, and grasses (Gramineae); berries of elder *Sambucus* and grapes *Vitis*. Garbage scavenged includes bread, rice, and peeled barley; green parts of plants, flowers, buds, bark, and paper occasionally taken. Small molluscs (e.g. *Planorbis*, *Bithynia*, *Anisus*) taken rarely; also beetles (Coleoptera), flies (Diptera), aphids (Aphidoidea), and lepidopteran larvae (Feriancová 1955; Nowak 1965; Rékási 1975; Rana 1976; Helešic 1981; Kiss and Rékási 1981).

In Dobrogea region (Romania), August–October, feeds on cereal grain (especially maize) or sunflower seeds. 250 stomachs contained maize (72·4%) by weight, sunflower seeds (13·1%), wheat (5·1%), grapes (3·2%), and variety of weed seeds (Kiss and Rékási 1981). In Brno area (Czechoslovakia), wheat made up 12·6% of total number of stomach items and 18·6% of total crop items; also barley (3·2%, 8·6%) and maize (2·8%, 8·4%); remainder made up of herb seeds, fruit, and tree seeds (e.g. *Tilia*, *Acer*, *Viburnum*); remains of beetles *Polydrosus* also found. Gastroliths found in 98% of birds (Helešic 1981). In Yugoslavian study, maize and wheat principal foods (Szlivka 1965). In Hungary, maize, sunflower, and millet supplemented by seeds of *Polygonum*, *Convolvulus*, and *Amaranthus*; also fruits of elder and grapes (Rékási 1975). In Lithuania (USSR), mainly grains of barley, wheat, oats, rye, and buckwheat; also peas, seeds of *Barbarea* and *Sinapis*, food scraps, aphids, and lepidopteran larvae (Navasaitis 1968). Throughout eastern Europe, maize was major feature of diet noted by Kovačević and Danon (1957) and Tutman (1960), whilst sunflower and wheat may also be favoured (Kovačević and Danon 1950–1; Bičik and Směsná 1971). Weeds such as *Amaranthus*, *Galium*, *Chenopodium*, *Atriplex*, *Setaria*, and *Fagopyrum* may also be important (Feriancová 1955; Deyl 1964; Szlivka 1965b;

Sovis and Vallo 1966; Rékási 1975; Grüll 1979). Food consumption 11–26 g per day (Feriancová 1955; Bičik and Směsná 1971; Murton *et al.* 1973*a*; Helešic 1981).

Young given crop milk up to *c*. 10 days, then also grain (Nowak 1965); older young also given berries of elder (Hofstetter 1954). Average daily consumption 7·5–8·0 g per bird (Feriancová 1955). BDSS

Social pattern and behaviour. Major studies by Bodenstein (1946*b*), Hofstetter (1952, 1954), Tomasz (1955), Lachner (1963), and Gnielka (1975).

1. Gregarious, especially outside breeding season when roosting and feeding, though concentrations persist at prime feeding sites throughout the year (Hudson 1965; Coombs *et al.* 1981). In autumn, juveniles first to flock; extent to which adults join later partly dependent on severity of weather, flocks being larger in lower temperatures (Tomasz 1955; Nowak 1965). In towns, some pairs with good feeding sites near nest-territory shun flocks, roosting and feeding alone in winter (see Breeding Dispersion, below). In spring, flock disintegrates as pairs leave to breed (Dyrcz 1956). For flock of 1400–1500 feeding on winter wheat, Nordhausen (East Germany), see Schmidt (1972*c*). In Hungary, up to 10000 recorded on 100-ha field of ripening sunflower *Helianthus* (Csernavölgyi 1975–7). For dispersing flocks of up to thousands, see Beretzk (1948–51); see also Salomonsen (1953). BONDS. Essentially monogamous mating system; pair-bond maintained throughout the year and probably tends to persist from year to year (e.g. Hofstetter 1954, Lachner 1963, Nowak 1965). Permanence of pair-bond in winter thought to be attribute only of some older birds (Hofstetter 1952). For probable bigamous ♂, see Hofstetter (1952). Independent of month of hatching, young invariably capable of breeding the following spring; sometimes also in year of hatching (see Nöhring 1965), though only if born early in year (Pomarnacki 1960). In one case, 4-month-old ♂ successfully bred in August; in another, widowed ♂ started breeding with 2½-month-old daughter (see Nöhring 1965). Both sexes care for young (e.g. Keve 1944–7). Fledged young attended at first by both parents, later only by ♂; seen begging unsuccessfully from ♂ at 29–30 days old, and most independent by 31–40 days (Hofstetter 1952, 1954, which see for ♂ feeding 44-day-old young; see also Klinz 1955). For wild hybrids with Turtle Dove *S. turtur*, see Bodenstein (1949*b*), and Voous (1963*a*); also Beretzk and Keve (1973). BREEDING DISPERSION. Solitary, forming neighbourhood groups in good habitat. In Europe, density of more than 1 pair per ha achieved only in urban areas, for which most information available. In Berlin, up to 0·29 pairs per ha (Löschau and Lenz 1967; Bruch *et al.* 1978). In Poznán (Poland), up to 3·5 pairs per ha (Górski and Górska 1979). In Basel (Switzerland) and Brno (Czechoslovakia), 1 or more pairs per ha (Willi 1972; Hudec 1976). For densities in East Germany, see Gnielka (1975) and Saemann (1968, 1975). Highest densities reported from inner city areas in Hungary, e.g. in narrow streets, Debrecen, 10·4 pairs per ha in mid-April (Bozsko and Lajos 1981, which see for comparison with other habitat); in park, Szeged, 40–50 pairs in 3 ha (Beretzk and Keve 1973). In orchard, Iraq, 11 nests in 8–10 ha (Marchant 1963*c*). In densely populated areas, distance between nests often not much more than 30 m, sometimes only 6–12 m (Hofstetter 1963; Lachner 1963); 15–20 m (Löschau and Lenz 1967). Defends territory comprising strictly defended core area in vicinity of nest (Dyrcz 1956; Hudson 1965) and weakly defended surrounding area. Thus territorial 'boundaries' of neighbouring pairs are vague, rarely contiguous, and trespassed with impunity (Hofstetter 1952; Tomasz 1955; Dyrcz 1956). Where nest-density high, neighbours virtually ignored unless they fly into nest-tree of another pair (Löschau and Lenz 1967; see also Lachner 1963). Nest-territory thought to be 0·08–0·2 ha (0·05–0·35) (Tomasz 1955). At Soest (West Germany), apparently 1·50–10·60 ha, mean 4·21, *n* = 15 (Hofstetter 1954), though estimates may represent overall density more than territory size. Territory always contains at least one (usually more) high vantage points (Tomasz 1955; Bozsko 1979). Territory of given pair said to vary with time, being largest during incubation and first 10 days of chick-rearing, smaller during 10th–16th days of chick-rearing, when ♂ most closely attached to nest. During last days of chick-rearing, ♂ apparently enlarges territory again while prospecting for new nest-site (Tomasz 1955). Territory established by ♂ and mostly defended by him, less by ♀ (Bodenstein 1949*b*; Hofstetter 1954; Tomasz 1955; Lachner 1963; Heer 1975). If winter mild, ♂ may continue to defend territory, but attachment then weak, and owner not always present (Keve 1944–7; Tomasz 1955; Lachner 1963; Löschau and Lenz 1967). For young birds attempting to defend temporary territories in 'communal' area, see Hofstetter (1954). Breeding adults may tolerate close presence of displaying and nesting *S. turtur* (Niethammer 1943*b*; Bodenstein 1949*b*; Hofstetter 1952; Tomasz 1955), or drive them from territory (Tomasz 1955; Fletcher 1979). Competition with *S. turtur* thought to be negligible (Beretzk and Keve 1973). For amicable relations with Woodpigeons *C. palumbus*, see Bettmann (1959), Löschau and Lenz (1967), and Riese (1967); for possible competition, see Nowak (1965). Numerous accounts of fierce disputes with thrushes (Turdidae) nesting or attempting to nest nearby (Keve 1944–7; Hofstetter 1954; Tomasz 1955; Forster 1957; Lachner 1963; Gilpin 1968; Hollick 1980). Marked fidelity to territory (Lachner 1963) and often nest-site (Willi 1972). One pair occupied same site 3 years in succession, another pair for 2 years (Beretzk and Keve 1973). Of 6 territory-owning ♂♂, 5 occupied same territories the following year (Hofstetter 1954). Commonly uses same nest for consecutive broods within 1 season (Niethammer 1943*b*; Klinz 1955; Gnielka 1975); once, for 6 broods (Lachner 1963). More likely to nest in same site if previous nest destroyed (Saemann 1968; Beretzk and Keve 1973). For birds invariably building new nest for each clutch, see (e.g.) Hofstetter (1952). ROOSTING. Usually roosts communally and nocturnally. Communal roosting marked outside breeding season, but some also when breeding (Lachner 1963; Gnielka 1975). In midwinter, Debrecen (Hungary), roosts of up to 1600 birds (Bozsko and Lajos 1981); for other roost-sizes see Hudson (1965), Nowak (1965), Navasaitis (1968), and Coombs *et al.* (1981). Some roost in pairs or individually in winter (Dyrcz 1956; Vlašín 1978), presumably at least sometimes in territory. Roost builds up in late summer and dwindles in spring (e.g. Hudson 1965, Bozsko and Lajos 1981, which see for variation with temperature). Juveniles first to roost communally (Rost 1957; Géroudet 1961*b*). Older birds first to break away from roost in spring (Hofstetter 1952), though some breeding birds may continue to roost communally (Lachner 1963). On break-up of flock, ♂ adopts roost-site in territory. Before laying, pair usually roost together on ♂'s site or elsewhere (Tomasz 1955); for pair roosting on nest see Hofstetter (1954). ♂ and ♀ sometimes roost separately, ♀ however always near nest (Tomasz 1955; see also Hofstetter 1952). During incubation, ♀ roosts on nest, ♂ usually nearby, often in nest-tree (Gnielka 1975). By day, off-duty bird loafs on vantage point near nest, often preening and sunbathing (Tomasz 1955; Heer 1975). When young no longer brooded, pair roosted together in canopy (Niethammer 1943*b*). Fledged young may share roost-site with parents (Tomasz 1955). Outer branches of dense foliage preferred, especially in cold weather (Bodenstein

1949b; Lachner 1963; Nowak 1965); conifers often then chosen (Béres 1948–51; Tomasz 1955; Vlašín 1978), and trees with natural defences against predators, e.g. holly *Ilex*, hawthorn *Crataegus* (Coombs *et al.* 1981); in willows *Salix* on island in lake (Hudson 1965); also thick ivy *Hedera* (R Hudson); less often in buildings (Nowak 1965). Birds often fly into cities from rural areas (up to *c*. 7 km away), seeking trees in streets, parks, and gardens (e.g. Coombs *et al.* 1981). Some individuals occupy same perch-site every night (Nowak 1965). Highly tolerant of street lighting and traffic (Vlašín 1978). Communal site not uncommonly changes (e.g. Dyrcz 1956), especially if disturbed by potential predator (Hofstetter 1952; Rost 1957). For displacement by Starlings *Sturnus vulgaris*, see Lachner (1963). Tree-sites chosen for communal roosting often in territory of still-resident pair. Latter slept close together, other birds keeping individual distance. In roosting posture, head never tucked backwards, but simply retracted; bill buried in breast feathers and feet covered by belly plumage (Rost 1957). At feeding site in severe weather, birds may spend most of day perched in roosting posture (Stolt and Risberg 1971). Rest and preen in trees between bouts of feeding (Hofstetter 1952). Loafing ♀ sunbathed with plumage raised and eyes closed (Gilpin 1962). Birds sunbathe communally and bathe on damp grass (Lachner 1963). Several reports of rain-bathing (e.g. Nagy 1959–61, Beretzk and Keve 1973, Reichling 1974b), once communally (Ringleben 1959). Birds fly to roost singly, in pairs, or in small flocks, often feeding near roost-site before retiring (Hofstetter 1952; Dyrcz 1956; Nowak 1965); often also drink (Rost 1957) and preen (Nowak 1965). In morning, moved to higher branches to preen before flying off in small flocks to feeding sites (Rost 1957). At winter roost, Poland, first single birds, then larger groups, left tree in morning, usually *c*. ½ hr after sunrise, later if cloudy (Dyrcz 1956). In winter, birds left roost shortly after sunrise, singly or in small groups, and made directly for feeding site. After feeding, and up to 14.00 hrs, some roosted at nocturnal site. Most assembled near roost at *c*. 16.00 hrs, and roosted *c*. 17.00–06.00 hrs (Nowak 1965). Usually assembled to roost *c*. 30 min before sunset, perching on bare trees before roosting in trees still in leaf. On dull days, move to roost quick and silent, but sometimes protracted and noisy with much chasing and jostling for position, while some birds circled over the area for a long time (Dyrcz 1956; see also Beretzk 1948–51).

2. May become very tame unless persecuted by man (Bodenstein 1949b; Tomasz 1955). Numerous reports of being mobbed by small passerines (Hofstetter 1952, 1954; Steinbacher 1959a; Lachner 1963; Gnielka 1975). For reviews of possible explanations, see Festetics (1973) and Schulz (1977). When mobbed, perched bird adopted apparently submissive cowering posture, with wings spread (Sharrock 1978); more likely a defensive-threat posture. Alighting bird slowly raises and lowers tail (Bodenstein 1949b). FLOCK BEHAVIOUR. In feeding flocks, individual distance 15–30 cm, less in pair or family party (see Glutz and Bauer 1980). More compact in harder weather (Hofstetter 1952). In winter flocks, evidence of continued pair-bonding—if one bird flies up (presumed) mate follows quickly (Dyrcz 1956). In winter, flocks dispersing from roost at dawn would suddenly and rapidly tower, careering to and fro high above town (Beretzk 1948–51; see also Roosting, above, for pre-roosting aerial manoeuvres). Feeding birds directly threatened by Sparrowhawk *Accipiter nisus* fly up in panic with loud Wing-clapping; others adopt Alarm-posture, standing erect with plumage sleeked, and/or give Warning-calls (see 4 in Voice), thus alerting flock. After upflight, flock breaks up into small groups which describe ever larger circles, and return later singly or in small groups (Hofstetter 1952, 1954). For other reactions to *A. nisus*, which can usually be outmanoeuvred, see Steinbacher (1959a). ANTAGONISTIC BEHAVIOUR. (1) General. ♂ guards territory by keeping vigil at vantage points, giving Advertising-call (see 1 in Voice), and mounting Display-flights (Bodenstein 1949b; Tomasz 1955; Lachner 1963). Usually starts to give Advertising-call, typically from high perch in territory, at time of break-up of flock in spring, also more softly in winter when restricted to early morning (Hofstetter 1952). In spring, calls vigorously, especially ½–2 hrs after sunrise (Dyrcz 1956; Kumerloeve 1962). When singing, ♂ sits with bill lowered, throat inflated, and plumage somewhat ruffled; bends tarsal joint slightly with each call. During incubation, ♂ defends territory early in morning while feeding, little during day while incubating, more late in afternoon when relieved by ♀ (Tomasz 1955; see also Frugis 1952). Aggressive around nest but less so elsewhere (Hofstetter 1954; Gnielka 1975). For lesser role of ♀ in territorial defence, see below. For aggressive behaviour of 1-year-olds trying to defend territories, see Hofstetter (1954). Some juveniles dominant over adults in flock (Lachner 1963). (2) Display-flight. Bird (mostly ♂) ascends at steep angle from high perch, often of exposed tree-top, with deep, rapid, Wing-clapping beats (Schüz 1948; Bodenstein 1949b). Climbs to *c*. 10 m (Rost 1953), then glides, often in spiral, with wings and tail spread, and usually accompanied by Excitement-call (see 4 in Voice) to alight on same perch or a new one (Richardson *et al.* 1957; Goodwin 1970). Not uncommonly re-ascends straight after glide (Bodenstein 1949b, which see also for soaring variant). Less likely than *S. turtur* to Wing-clap during ascent (Klinz 1955). Display-flight initiated almost invariably when conspecific bird nearby (Bowing-display if yet closer: Hofstetter 1954)—usually an intruder but also ♀ mate. Thus signals both site-advertisement and relationship to ♀ (Hofstetter 1952). Often followed by pursuit of intruder (Rost 1953; Nowak 1965). For Display-flight stimulated apparently by passerines flying nearby, see Dyrcz (1956). Display-flight occasionally performed by ♀ when searching for nest-site, and then directed at ♂ mate (Hofstetter 1954). (3) Threat and fighting. On approach of intruder, territorial bird may give Warning-call (Bodenstein 1949b). Threatens intruder by flying towards it and landing on perch or ground to perform Bowing-display near rival; as in other Columbidae, bird faces rival and alternates repeatedly between an upright and a crouched posture. With legs bent, neck region swollen (combination of inflated cervical air sacs and ruffled plumage: Glutz and Bauer 1980), and tail depressed but not fanned, bird Bows

C

forward until bill low (Fig A); then raises body, still with neck region swollen, to upright posture (Fig B); for 5 birds, mean time per Bow 1·69–1·83 s; up to 15 Bows in c. 30 s (Davies 1970). While Bowing, alternately raises and lowers feet. Accompanies display with trisyllabic Display-call (see 2 in Voice), giving 1st unit during Bow, 2nd unit at low point of Bow, 3rd when rising (Hofstetter 1952; see also Glutz and Bauer 1980). Young birds apparently Bow in silence (Glutz and Bauer 1980). For rare case of Bowing-display by ♀, see Hofstetter (1954). For Bowing-display by ♂ to attract ♀, see Heterosexual Behaviour, below. When Bowing, rivals approach each other in attacking variant of low Bowing posture: head lowered, neck region swollen, wings drooped, uttering Display-call; for related posture of perched bird about to attack perched rival, see Fig C, right. On ground, rivals thus engage each other at a run, or in 2-footed leap (Ferianc 1947; Bodenstein 1949b); described as clumsy, half-sideways, hopping motion (Hofstetter 1952). Submissive bird sleeks plumage and usually flies off instantly (Bodenstein 1949b), or runs away in sleeked, stiff posture with drooped wings (Hofstetter 1954). Crouching down, slight lowering of head and ruffling neck feathers (invitation to Allopreen) also said to denote appeasement; thus appeased, dominant bird may perform brief Bowing-display or drive intruder away (Glutz and Bauer 1980). If both hold ground, fight usually ensues—often long and fierce in territorial disputes early in year. The 2 birds sidle up to each other with backs slightly arched (Glutz and Bauer 1980), flicking carpal joints and Nodding (thought to be pecking intention-movement). Combatants begin by delivering blows with carpal joints; often then grab each other by neck and tug until 1 or both tumble over (Bodenstein 1949b). May pull out feathers (Tomasz 1955) and jump on rival's back (Hofstetter 1952; Tomasz 1955). Fighting accompanied by Excitement-calls (Klinz 1955; Dyrcz 1956). Most fights occur on roof-tops, trees, sometimes in the air, and may last up to ½ hr. Most occur between ♂♂, often ♂–♀, sometimes between ♀♀ (Tomasz 1955). In disputes with C. livia, S. decaocto often has the advantage since it attacks directly (Bodenstein 1949b). If challenged bird flies up, it is usually pursued vigorously to territory boundary; pursuer may dive on victim in flight, sometimes striking with bill (Ferianc 1947) and pulling out feathers (Bodenstein 1949b). Pursuer may also try to strike with wings in flight (Nowak 1965). During attacks and chases, both sexes give Warning-call (Bodenstein 1949b). Heterosexual Behaviour. (1) General. Most birds appear to be paired on taking up territory; young birds thought to pair in advance, probably in communal area (Hofstetter 1954). ♂ attracts ♀ by combination of Advertising-calls, Display-flight, and Bowing-display. Variant of Advertising-call (see 1 in Voice) sometimes given by unpaired ♀ (Hofstetter 1952; see also Ferianc 1947, Bodenstein 1949b). All forms of display wane in autumn (Hofstetter 1952), but may continue sporadically and in often subdued form through winter (e.g. Keve 1944–7, Reuterwall 1956, Stolt and Risberg 1971; see also Breeding). (2) Pair-bonding behaviour. In spring, unpaired ♂ attracts ♀ to territory with Advertising-call; may also Nest-call (see 3 in Voice) to any strange ♀ entering his territory during breeding season (Hofstetter 1952, 1954). ♀ responding to calls flies directly towards ♂. If ♂ already paired, drives ♀ off. If unpaired, ♂ increases calling rate. ♀ approaches ♂ in an upright posture, with bill lowered and feathers ruffled. ♂ attempts to approach ♀ and performs Bowing-display (see above). Displaying ♂ keeps changing from one foot to the other, periodically Jumping towards ♀. ♀ may approach and Allopreen neck and head feathers of ♂ who remains quietly in a crouched posture (Bodenstein 1949b; Bozsko 1979). Allopreening, however, more a feature of established pairs who spend a lot of time in close physical contact, nuzzling into each other and Billing (see Hofstetter 1954, and below). If ♀ flies off, ♂ may pursue her with Excitement-calls (see Glutz and Bauer 1980). ♂'s flight then has rather weak 'draggling' quality (Richardson et al. 1957). ♂ courting recently widowed ♀ perched just above her and turned nervously round and round, achieving close bodily contact the next day; later, ♀ paired with a new ♂ and was seen Billing with him (Dathe 1981). (3) Driving. Rare compared with (e.g.) Rock Dove C. livia. None seen despite lengthy observation (Hofstetter 1954). ♂ said to chivvy mate away from rival (Fig D, right) with apparent variant of Excitement-call (Richardson et al. 1957). (4) Courtship-feeding. Little known. Feeding of ♀ by ♂ said to occur and proceed as in C. livia; ♀ begs less vigorously than do young (Bodenstein 1949b). (5) Mating. Often preceded by Display-flight (Richardson et al. 1957). Also usually preceded by intense Bowing (♂), sometimes by aerial pursuit (of ♀ by ♂), rarely by Billing. ♀ invites copulation by crouching and slightly lifting or drooping wings (Hofstetter 1954; Géroudet 1961, which see for photograph). ♀ may crouch thus on seeing ♂ approaching in Display-flight (Hofstetter 1954). ♂ signals readiness to copulate by Mock-preening: moves head over back and under scapulars of opposite wing (Bodenstein 1949b). ♂ leaps on to ♀'s back or, after Display-flight (or in cases of undescribed 'rape'), flies on (see Glutz and Bauer 1980). During and after copulation, variant of Excitement-call may be given by ♂ (Ferianc 1947); by both sexes (Bodenstein 1949b; see also Hofstetter 1954). Copulation lasts a few seconds but may be repeated (Fiebig 1956; see also Nowak 1965); interval between successive copulations may be as little as 3 min (Hofstetter 1954). Various post-copulatory behaviour occurs. Usually pair sit close, one or both self-preen, one (usually ♂) or both fly off (Hoffstetter 1954). May Allopreen (Tomasz 1955). If ♀ flies off, ♂ may pursue; ♀ more likely to leave hurriedly if victim of 'rape' (Bozsko 1970). For ♂ attacking, then copulating with ♀ (not mate) when

D

she drooped wings, see Hofstetter (1954). ♂ sometimes resumes Bowing-display, or pair may Bill (Hofstetter 1954). Copulation occurs both in and out of territory, sometimes at nest (Hofstetter 1952, see also Glutz and Bauer 1980), usually on some high vantage point, rarely on ground. For description of mating where ♀ lay on ground, see Tomasz (1955). Copulation in any pair relatively infrequent (Tomasz 1955). Eggs laid (2–)4–8 days after copulation (Klinz 1955). Copulation especially frequent at time of nest-site selection, mainly 9–10 days before 1st egg (Hofstetter 1954); (6) Nest-site selection. ♂ shows prospective sites to ♀ (Nest-showing) with head lowered, tail raised, and neck plumage somewhat ruffled; meanwhile Wing-twitches and Nest-calls (Hofstetter 1954). May appear to Nod at prospective site (Bodenstein 1949b). ♂ alights at each potential site in turn and Nest-calls to ♀ who joins him, moving close by his side; pair may remain thus for variable period, depending on interest of ♀; may intersperse Nest-showing with Allopreening, in most intense form of mutual kind (Nagy 1959–61; Gnielka 1975). Pair sit close together, directing preening at head, especially around base of bill, neck, crop, back, and feet: Allopreening may continue for several minutes, interrupted with intervals of equal duration (Hofstetter 1954). Pair sit facing each other, inspecting prospective site. ♀ indicates acceptance by starting to Nest-call. ♂ may move aside, ♀ taking his place and Nest-showing; may Nest-call for some time after ♂ has left (Hofstetter 1952; see also Tomasz 1955). ♀ inspected one site by turning round and round on it. If site not acceptable to ♀, ♂ moves to another (Tomasz 1955). ♂, uttering Excitement-calls, brings nest-material to ♀; on arrival at nest, pair Nest-call and Bill (Hofstetter 1952, 1954). ♂ often pushes twigs directly under ♀ (Bozsko 1979). ♀ Nest-calls while building; most building in early and late morning (Hofstetter 1952, 1954). (7) Behaviour at nest. Nest effectively complete in 1–3 days (Bozsko 1979) but building continues during incubation and brooding, ♂ bringing in nest-material for ♀ just after he has been relieved by her. Change-over usually at very regular times, at least up until young c. 8 days old (Gnielka 1975). Sitting bird begins calling, which immediately summons mate perched nearby (Tomasz 1955). ♂ the more vocal on nest, usually giving Advertising-calls just before nest-relief, but on one occasion gave Nest-calls (Heer 1975; see also Niethammer 1943b). ♀ usually Nest-calls just before being relieved (Hofstetter 1952; Heer 1975). ♂ said to take initiative at nest-relief, calling to summon ♀ both on and off eggs (Richardson et al. 1957, which see for ♂ performing Bowing-display as Meeting-ceremony at nest-relief, or after any long absence from ♀). Incoming bird sometimes has to urge mate off nest; when ♂ reluctant to rise, ♀ pecked at sides of his head or throat (Gnielka 1975). RELATIONS WITHIN FAMILY GROUP. Young first call at 4–7 hrs after hatching (Rana 1973). Brooded for first 10–12 days (Hofstetter 1952; Tomasz 1955; Gnielka 1975) but at any time up until fledging if weather inclement (Niethammer 1943b; Bozsko 1979). Siblings sit in nest facing in opposite directions (Niethammer 1943b). Beg by pecking at parent's crop region; to feed, small chick inserts bill in side of parent's bill. From 6th day, both parents often preen young, which also self-preen from 11th day (Gnielka 1975). Young stand up in nest at 15–16 days (Tomasz 1955); then exercise wings and make short excursions (sometimes in faltering flight: Hofstetter 1952) on to surrounding foliage there sitting motionless for long periods (Niethammer 1943b; Hofstetter 1952; MacDonald 1967; see also Anti-predator Responses of Young, below). On arrival of parent, however, beg vigorously with Wing-shivering and food-calls (see Glutz and Bauer 1980). Initially return to nest to be fed (Gnielka 1974), also at night (Tomasz 1955; Bozsko 1979), but eventually roost outside with parents (Tomasz 1955). During 1st week after fledging, invariably accompanied by at least 1 parent (MacDonald 1967). Initially attended by both, later only by ♂ (Hofstetter 1952), at least sometimes because ♀ incubating new clutch (Tomasz 1955; Saemann 1968). For young sitting on nest alongside incubating parent, see Baege (1967) and Poos (1972). Fledged young fed for some time, though irregularly (see Bonds, above). Parents apparently do not molest independent young intruding on territory (Tomasz 1955) and contact may thus persist for several months (Glutz and Bauer 1980). ANTI-PREDATOR RESPONSES OF YOUNG. 12-day-old young reacted to human intrusion with Bill-snapping (Gnielka 1975). 'Freezing' behaviour of young on leaving nest presumably helps to avoid detection (Hofstetter 1952). 13-day-old young, responding to intruder by leaping from nest, fluttered to ground where tried to hide (Gnielka 1975). On reaching ground, threatened young make for cover with Wing-shivering, then become immobile (see Glutz and Bauer 1980). PARENTAL ANTI-PREDATOR STRATEGIES. (1) Passive measures. When disturbed on nest, adult extends neck and ruffles head plumage (Gnielka 1975); flushes readily (Hofstetter 1954). Unusually tame in some inner city areas: e.g. for bird (♀) allowing itself to be stroked on nest, see Bozsko (1979). (2) Active measures: against birds. Vigorously chases Corvidae from vicinity of nest. For expulsion of Jackdaws *Corvus monedula*, see Keve (1944–7), Tomasz (1955), Lachner (1963), Saemann (1968), and Latzel (1969), which see also for tight sitting when threatened by *C. monedula*). For attacks on Magpies *Pica pica*, see Hofstetter (1952), Riese (1967), Laferrère (1974), and Gnielka (1975). Bird approached by apparently threatening Blackbird *Turdus merula* adopted Defensive-threat posture typical of Columbidae: ruffled plumage and raised one wing at steep angle (Hofstetter 1954). (3) Active measures: against man. Incubating ♂ and ♀ delivered vigorous blows with wings to hand of intruder at nest. When young 3–7 days old, sitting bird not uncommonly fluttered down from nest, wing-action effecting distraction-lure display of disablement type (Gnielka 1975). When observer approached nest of breeding pair in captivity, bird often plunged to ground and performed variant of injury-feigning display: stretched forwards with tail closed and, with wings quite close to body, appeared to limp along, moving quite rapidly on carpal joints, with rocking motion of body (Fig E). When observer bent down to the bird, it performed apparent defensive-threat display (Fig F): brief sudden Wing- and Tail-spreading (Grünefeld 1952). If presence of intruder causes young to leap prematurely from nest to ground, parent may land nearby and creep around, Wing-shivering, until young safely under cover (see Glutz and Bauer 1980). In 3 out of c. 43 intrusions on nest, bird reacted with undescribed injury-feigning display (in 2 cases by same bird). In every case, bird had sat tight for an unusually long time, and nests were sited rather low (Hofstet-

E

F

ter 1954). With young of any age, 'broken wing' display observed several times, Herford, West Germany (Lachner 1963). Though varying from pair to pair, Oxford (England), injury-feigning not uncommon with hard-set eggs or small young, unusual at other times (H A Robertson). (3) Active measures: against other animals. Distraction-lure display of disablement type performed when squirrel *Sciurus* in nest-tree (Gnielka 1975; see above for same against man).

(Figs A–B from photographs in Davies 1970; Fig C from photograph in Ferianc 1947; Fig D from drawing in Richardson *et al.* 1957; Figs E–F from drawings in Grünefeld 1952.)   EKD

Voice. Highly vocal in breeding season, especially in early morning and evening (e.g. Frugis 1952). As incubation proceeds, calls less (Niethammer 1943*b*). For dialects and seasonal variation within local population, see Frieling (1960). For additional sonagrams, see Lade and Thorpe (1964) and Gürtler (1973). For Wing-clapping in Display-flight and alarm, see Social Pattern and Behaviour.

CALLS OF ADULTS. (1) Advertising-call. Mostly a repeated, penetrating cooing call, of basic frequency 0·5–0·55 kHz (Gürtler 1973). Variously rendered 'du-dūh-du' (Niethammer 1943*b*), 'gu-gūū-gu' (Hofstetter 1952); 'ku-koo-ku' (J Hall-Craggs: Fig I). Emphasis usually on 2nd syllable, which is 3 times as long as 1st (Gürtler 1973); 2nd syllable drops slightly in pitch towards end (Schüz 1948). Call lasts *c.* 1·2–1·3 s (Gürtler 1973). Repeated usually 3–12(–15) times (Bettmann 1959). 3rd syllable may be a sharper 'duit', or a swallowed and muffled '(g)u' (Frieling 1960). In recording by V C Lewis (Fig I), muffled sound possibly inhalation of breath (P J Sellar). Especially after long series, usually ends in disyllabic variant, e.g. 'gu-guh' (Bettmann 1959; see also Bodenstein 1949*b*, Dyrcz 1956, Frieling 1960, Kumerloeve 1962). For less common variant of 4 syllables, see Bodenstein (1949*b*), Bettmann (1959), and Kumerloeve (1962). Given mainly by ♂, single or paired, as territorial call, typically from high perch, when serves to attract ♀♀ and warn off other ♂♂. Also sometimes given by ♀ (Hofstetter 1952, 1954, which see for contexts); ♀'s call quieter (Nowak 1965), higher and clearer (Ferianc 1947). Sound ascribed to Advertising-call—given by sitting ♂ either as greeting to ♀ (Niethammer 1943*b*) or to summon her for nest-relief (Heer 1975)—is possibly call 3, though the two calls said to alternate in this, and other, contexts (Hofstetter 1954). (2) Display-call. Very similar to call 1, and often said to be identical (Bodenstein 1949*b*; see also Lade and Thorpe 1964). For slight, but apparently consistent differences in frequency from call 1, see Gürtler (1973). Given repeatedly by ♂ performing Bowing-display—to ♀ as part of courtship (e.g. Richardson *et al.* 1957, Gürtler 1973), or to rival ♂ (e.g. Hofstetter 1952); though call currently assumed identical in courtship and threat, not yet proved to be so. For division of syllables within Bowing-display, see Social Pattern and Behaviour. (3) Nest-call. Similar to call 1, but said to be lower, slower, and given singly, never in series (Dyrcz 1956). Mostly by ♂; also sometimes by ♀ (Hofstetter 1954) when quieter (Nowak 1965), higher pitched (Hofstetter 1952). Given mostly near the finished nest, also when nest-building, or before copulation (Dyrcz 1956), during search for nest-site, and to summon mate for nest-relief (Hofstetter 1954; see also Heer 1975); also sometimes given when strange bird flies overhead (Hofstetter 1952). (4) Excitement-call. See Fig II. A nasal-sounding 'gheè-gheeè' (Frugis 1952), 'chrräi' (Hofstetter 1952), 'rrräh' (Bodenstein 1949*b*; see also Bettmann 1959). Duration *c.* 0·5 s (Gürtler 1973). Given in variety of contexts where bird excited or alarmed, notably by ♂ in gliding phase of Display-flight. Also at other times in flight by both sexes, often before alighting (hence 'alighting call': May and Fisher 1953). Also in aggressive encounters on ground and in the air (Hofstetter 1952; Dyrcz 1956), when intruder near nest (Ferianc 1947), and when handled (Hofstetter 1954). Forceful angry 'kwurr kwurr' given by Driving bird (Richardson *et al.* 1957) probably the same or a variant. In high intensity, e.g. in aerial chase involving several birds (Hofstetter 1952), or during and after copulation, units run together to give loud, nasal, rattling 'grögrögrögrö' (Bodenstein 1949*b*). Low-intensity variant given by birds in minor disputes or when disturbed at feeding grounds in winter, a muffled 'ru' (Hofstetter 1952); also rendered a fairly quiet, compressed, nasal 'gu' or 'gru', expressing mild alarm or annoyance, notably on approach of conspecific intruder ('warning call': Bodenstein 1949*b*). (5) Other calls. Both sexes give 'huuuu', initially ascending, latterly descending in pitch, almost owl-like but softer; not further described (Mächler 1955*a*). So-called 'affection call' ('gu': Bodenstein 1949*b*) possibly variant of calls 3 or 4; perhaps also given when feeding young (Bodenstein 1949*b*).

CALLS OF YOUNG. Food-call a shrill, feeble 'weep', accompanied by Wing-shivering (Richardson *et al.* 1957); hoarse piping (Hofstetter 1952), similar to other Columbidae (Nowak 1965). Juvenile gave repeated, rather harsh, decending 'häää', also 'hä-hä-hä' when handled (Machler

I V C Lewis England May 1972

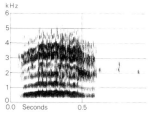

II V C Lewis England May 1972

1955*b*); probably variant of adult call 4. A 'wäh' call also presumed to be precursor of adult call 4 (Hofstetter 1954). Adult Advertising-call first heard in juvenile when 11 weeks old (Lachner 1963). Adult Nest-call first heard at 113 days (Hofstetter 1952).   EKD

Breeding. SEASON. Prolonged throughout range. Northwest Europe: see diagram. Iraq: eggs laid March–September (Marchant 1963*c*). SITE. In tree, bush, or tall hedge; occasionally on pylon, ledge of building, or in roof trough or guttering; in tree, usually near trunk, seldom on end of branch. Wide variety of tree species used, including deciduous trees in spring as well as conifers. 22 species used, West Germany, chiefly lime *Tilia* and birch *Betula* (Lachner 1963). Average height above ground of 595 nests, East Germany, 6·77 m (1·9–16), increasing through season from 5·51 m ($n = 254$) February–April, to 6·73 m ($n = 118$) May, and 8·22 m ($n = 223$) June–October (Gnielka 1974). Average height of 2139 nests, Czechoslovakia, 8·1 m (2·6–22); below 3·5 m, 3·2%; 3·5–12·5 m, 88%; 12·5–16·5 m, 7·6%; over 16·5 m, 1·2% (Pikula and Kubík 1978). Nests re-used for successive broods. Nest: usually flimsy platform of twigs, stems, and roots, occasionally bulkier, and may be added to for 2nd and later broods; wire sometimes used with or instead of natural material (e.g. Novrup 1953, Rost 1953). Average size of 25 nests, East Germany, 26 × 21 cm (24–32 × 16–25); average weight 58 g (19–280); average 142 (60–332) pieces of material (Gnielka 1975). Building: mainly by ♀ with ♂ gathering material; takes 3–4 days, usually from dawn to *c*. 10.00 hrs; material gathered from single area, not always within territory; collected from ground but also by breaking twigs from trees; ♂ carries 1 piece at a time to nest and places it there for ♀ to arrange (Tomasz 1955). EGGS. See Plate 95. Oval, smooth and slightly glossy; white. 30·5 × 23·6 mm (26–39 × 20–27), $n = 698$; of 384 1st eggs, 29·9 ± SD 1·23 × 23·6 ± SD 0·83 mm; of 314 2nd eggs, 31·5 ± SD 1·44 × 23·7 ± SD 0·85 mm (Pikula and Kubík 1978). Fresh weight 9·6 g (6·6–14·2), $n = 75$; of 40 1st eggs, 9·3 ± SD 1·3 g; of 35 2nd eggs, 9·9 ± SD 1·4 g (Pikula and Kubík 1978). Clutch: (1–)2. Occasional clutches of 3–4 by 2 ♀♀ (e.g. Frugis 1952). Of 482 clutches, East Germany: 1 egg, 3%; 2, 97%; average 1·97, with frequency of clutches of 1 declining through season (Gnielka 1975). Of 302 clutches Czechoslovakia: 1 egg, 11·9%; 2, 88·1%; average 1·88; average increased significantly from 1·79 in March to 2·0 in September (Kubík and Balát 1973). 3–6 broods per year; 4 broods in Czechoslovakia (Kubík and Balát 1973); probably only 3 broods in Iraq (Marchant 1963*c*). Replacement clutch laid after loss. Laying interval 36–48 hrs; average 38·5 hrs (37–40), $n = 7$ (Gnielka 1975). Interval between broods longer when same nest reused than if different site used (Lachner 1963); 3–7 days (Bozsko 1979). INCUBATION. 14–18 days, most probably 14–16 (Gnielka 1975). By both parents, with ♀ sitting through night and ♂ relieving her in morning for about 8 hrs. Begins with

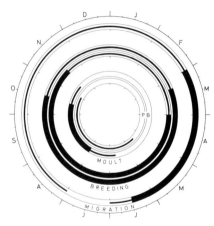

2nd egg, perhaps sometimes with 1st though then not continuous; hatching interval 12–40 hrs (Gnielka 1975). YOUNG. Semi-altricial and nidicolous. Cared for and fed by both parents; brooded continuously while small. FLEDGING TO MATURITY. Fledging period 17 days (15–19), $n = 21$ (Gnielka 1975). Longer periods reported, e.g. 24 days (Tjittes and Koersveld 1952). Young old enough to fledge at 14 days but stay in nest until 15–16 days when short flight (of only a few metres) made; may return to nest at 19–20 days (Hofstetter 1952). Become independent about 1 week later. Age of first breeding 1 year or less (see Social Pattern and Behaviour). BREEDING SUCCESS. Of 588 nests, East Germany, 65·5% hatched at least 1 egg and 48·6% fledged chicks; fledging success increased through season from 31·8% ($n = 88$) in March to 50% ($n = 200$) May–June and 70% ($n = 66$) August–October; main losses of eggs and young to predators, particularly Magpie *Pica pica* (Gnielka 1975). Of 436 eggs laid, Czechoslovakia, 86·5% hatched; of 242 hatched eggs, 79·3% young reared; overall success rate 68·6%; losses of eggs due to addling (7% of eggs laid), or destruction or ejection by flushed bird (3·7%); losses of young from being ejected (4·2%) or predated (4·2%) (Kubík and Balát 1973). Of 77 eggs laid, Iraq, 64% hatched and 35% raised to fledging (Marchant 1963*c*).

Plumages (nominate *decaocto*). ADULT MALE. Forehead, crown, nape, and sides of head and neck pale pink-grey, deepest, near light vinaceous-pink, on hindcrown, nape, and sides of neck, purer pale bluish-grey on forehead, sometimes slightly tinged buff on crown and nape. Narrow black half-collar round lower hindneck down to sides of neck, some feathers narrowly tipped white; bordered above at hindneck and at both sides on sides of neck by light grey feathers tipped white. Mantle, scapulars,

PLATE 29 (*facing*).
Columba livia livia Rock Dove (p. 285): **1–2** ad.
Columba oenas Stock Dove (p. 298): **3–4** ad.
Columba eversmanni Yellow-eyed Stock Dove (p. 309): **5–6** ad.
Columba palumbus palumbus Woodpigeon (p. 311): **7–8** ad.
Columba trocaz Long-toed Pigeon (p. 329): **9–10** ad.
Columba bollii Bolle's Laurel Pigeon (p. 331): **11–12** ad.
Columba junoniae Laurel Pigeon (p. 334): **13–14** ad. (CJFC)

tertials, back, rump, upper tail-coverts, and all upper wing-coverts except outermost and marginal ones between olive-drab and greyish-fawn, in some with slight russet tinge; some bluish-grey often visible at feather-bases on sides of back and rump and on longer uppertail-coverts. Upper chin white, grading to bright pale vinaceous-grey on throat, sides of chest, and breast. Upper flanks olive- or pinkish-drab; lower flanks, vent, and under tail-coverts medium bluish-grey, grading to pale pinkish-grey on belly. Uppersurface of central tail-feathers (t1) greyish-drab, sometimes with silvery-grey bloom along shaft; t2 pale grey with dark grey basal $\frac{3}{5}$ to inner web, broad greyish-drab border along outer edge, and grading to off-white on tip; t3–t5 have paler grey distal part, more contrastingly dark grey basal $\frac{3}{5}$, and restricted drab along outer edge; t6 has inner web black or greyish-black with straight-cut greyish-white tip 45–60 mm long, and outer web grey at base, greyish-white on tip, and black along middle portion of shaft, latter extending slightly further towards tip than on inner web; from below, tail contrastingly glossy black on basal $\frac{3}{5}$ and white on remainder, with black pattern on t6 similar to uppersurface. Primaries dark olive-brown, inner webs grading to white at base, outer webs and tips with narrow pale grey edge, and outer webs of innermost with increasing amount of pale grey. Secondaries pale grey, tips and outer webs variably washed olive-drab. 4–5 outer greater and median upper wing-coverts, outermost lesser coverts, marginal coverts, primary coverts, and bastard wing pale bluish-grey; feathers bordering fawn or drab inner coverts, longest bastard wing feathers, and tips of greater primary coverts sometimes with some drab or olive-grey suffusion. Under wing-coverts and axillaries greyish-white, almost white on central greater coverts. In worn plumage, crown and nape slightly duller vinaceous-pink with some brown tinge; remainder of upperparts paler greyish-drab (less russet-fawn); throat to breast paler pinkish, sometimes with buff tinge (less uniform vinaceous-grey). ADULT FEMALE. Like adult ♂, but crown, nape, often sides of head, and sometimes forehead tinged drab or pale olive-brown, and less pure greyish-pink; narrow feather-tips (if any) in black half-collar drab-grey rather than white; pale vinaceous-grey on sides of chest does not reach up to sides of mantle; throat and chest paler pinkish-grey, less vinaceous; breast and belly paler, pale grey with some pink suffusion, not as vinaceous-pink as adult ♂. In worn plumage, differences less marked, but crown and nape of ♂ usually still mainly vinaceous-grey, of ♀ drab- or olive-grey. DOWNY YOUNG. Down sparse, hair-like; drab yellowish-white (Harrison 1975; RMNH). JUVENILE. Rather like adult, but forehead, gape, and chin bare, forecrown and upper throat whitish, rest of head and neck fawn-brown with some pale grey of feather-bases visible and without black half-collar or with traces of it only; rest of upperparts and inner upper wing-coverts fawn as in adult, but each feather narrowly margined cream-buff to russet-brown on tip, occasionally hardly visible; throat cream-buff, chest, breast, and upper flanks pale grey or pinkish-grey with broad buff or russet-brown feather-tips; belly, vent, and lower flanks pale grey with some pale buff suffusion; under tail-coverts grey, tipped buff, and do not reach white of tail-tip (unlike adult). Tail as adult, but feathers narrower (see Structure); t1–t5 tinged buff on tip, outer web with wider drab or cinnamon edge; white on tip of t6 extending for 25–35 mm along shaft only, border with black basal $\frac{2}{3}$ incurved rather than straight as in adult, not as sharply demarcated, especially on outer web. Upperwing as in adult, but lesser and median coverts rather faintly tipped rufous, buff, or yellowish-white; greater coverts, primary coverts, feathers of bastard wing, and flight-feathers extensively tinged pale fawn or olive-drab, bordered terminally by narrow but distinct rufous-buff or yellowish-white edges (most marked on primary coverts); secondaries and outer wing-coverts not as uniform blue-grey as in adult, primaries not as uniform blackish olive-brown. Relatively late growing feathers on face, upper throat, central chest, and sides of neck more similar to adult, those on side of neck, appearing from 2–4 weeks, greyish-black with buff tips, forming small and broken neck-patches. FIRST ADULT. Like full adult, but, depending on age, variable number of juvenile feathers retained (see also Moults). Adult head and body and many wing-coverts attained within 2–3 months of hatching, complete black half-collar at 2 months (but feathers tipped buff rather than white or grey as in adult); when head and body adult, sexes separable as in adult, but some ♂♂ may retain some juvenile buff-brown feathers in crown up to 4–6 months and chest not always as bright vinous-pink as in adult ♂. Some juvenile tail-feathers present up to 5 months of age (or later, when post-juvenile moult slowed down or suspended during adverse weather); juvenile outer primaries and outer greater primary coverts up to 5–7 months; rather narrow and short middle secondaries (mainly s5–s7) with extensive drab tinge and narrow pale buff margin to tip occasionally up to 9–11 months. Strongly widened tip to t6 with straight-edged white on inner web and broad white border to outer web not attained until 4 years old (Lachner 1965).

Bare parts. ADULT. Iris deep red (rarely yellowish, perhaps only in hybrids with so-called Barbary Dove *S. 'risoria'*; see African Collared Dove *S. roseogrisea*). Narrow ring round eye greyish-white or yellowish-white. Bill greyish-black or dull black. Leg and foot deep red or purple-red. DOWNY YOUNG. Iris lead-grey, changing to dull brown when still in nest. Bill pale flesh-colour, darkening to bluish-slate before fledging. Leg and foot dark lead-grey, changing to light brownish-red at fledging. Skin between down dark. JUVENILE. At fledging, iris dull brown, bill bluish-slate, leg and foot brownish-red; all gradually change to adult colour during post-juvenile moult, iris colour deepening from dark red-brown to red, bill to black, and leg and foot from greyish-pink and brown-red to red, showing adult colour upon termination of moult. (Lachner 1965; RMNH, ZMA.)

Moults. Mainly based on data from Britain (Insley *et al.* 1980), Netherlands (*c.* 250 skins: RMNH, ZMA), and West Germany (Lachner 1965; Juckwer 1970), and some data from Makedonija, Yugoslavia (Juckwer 1970); all basically similar. ADULT POST-BREEDING. Complete; primaries descendant. Starts with p1 mainly mid-May to early July; average duration of primary moult 125–150 days (p1–p5 35–40 days, p6–p10 3–4 months), mainly completed with regrowth of p10 September–October. A few birds start from March (moult often slowing down or suspended April–May) and complete from late August; some others start as late as early August, and complete early December; only exceptionally a few adults not completed by early winter—birds showing moult or suspended moult in winter are almost always immatures. Secondaries start ascendantly from s1 and descendantly from s9, both about at loss of p6, completed

PLATE 30 (*facing*).
Columba palumbus Woodpigeon (p. 311). Nominate *palumbus*: **1** ad, **2** juv moulting into 1st ad (1st autumn or winter), **3** downy young. *C. p. azorica*: **4** ad.
Columba trocaz Long-toed Pigeon (p. 329): **5** ad, **6** juv, **7** downy young.
Columba bollii Bolle's Laurel Pigeon (p. 331): **8** ad, **9** juv, **10** downy young.
Columba junoniae Laurel Pigeon (p. 334): **11** ad, **12** juv, **13** downy young. (CJFC)

with s6 after c. $3\frac{1}{2}$ months. Tail starts from primary moult score of 25–35, usually completed at score 40–45, full replacement in c. $2\frac{1}{2}$ months; sequence approximately from t1 outwards, but t6 with t3, t5 last. Body starts from score of c. 25, scattered feathers of mantle, scapulars, and tertials first, soon followed by median coverts, upper tail-coverts, sides of neck, and underparts; head, neck, back, and rump last, completed with score of c. 45. POST-JUVENILE. Complete. Strongly variable, depending on time of hatching. Early-hatched birds start from late March and complete from August with speed of primary moult about similar to adult; birds hatched July start August and moult more rapidly, completing early December; late-hatched birds, starting September–October, slow down or suspend moult between late November and mid-March, especially in January (when 51% of a large British sample suspended, but only 19% of a sample of 21 from the Netherlands); birds hatched late autumn hardly start in winter and delay onset of intense moult to late January or February, completing up to June or early July, sometimes after suspending moult in spring. Late-hatched birds may show serially descendant primary moult in spring and early summer, with post-juvenile active on outer primaries and 1st adult post-breeding on inner (timing of latter about similar to adult post-breeding), but this relatively uncommon when compared with west Palearctic *Columba*. In all juveniles, moult starts with body from 3–4 weeks after hatching (usually before all growth of juvenile plumage completed); primaries start with p1 at 4–5 weeks, usually before black half-collar and adult iris colour developed. First adult plumage (on face, throat, centre of chest, longer lesser upper wing-coverts, and sometimes sides of neck) appears before p1 lost; at primary moult score 10–20, heavy moult on head, upperparts, and wing-coverts, with underparts largely new; all head, body, and wing-coverts new from score 20–25. Tertials moulted at 20–30, tail mainly 20–45 (sequence as in adult post-breeding; occasionally a few old feathers retained after completion of primary moult). Secondaries moulted from s1 inwards and s9 outwards, starting from primary score of c. 25, completed at c. 50, but frequently some feathers (mainly s5–s6) retained until 1st adult post-breeding.

Measurements. Netherlands, all year; skins (RMNH, ZMA).

		♂			♀		
WING	AD	182	(2·85; 32)	177–188	177	(2·64; 28)	173–182
	JUV	178	(2·43; 7)	176–182	172	(3·52; 7)	166–178
TAIL	AD	142	(3·54; 14)	136–147	136	(3·77; 12)	129–141
	JUV	130	(2·97; 5)	126–133	127	(3·52; 5)	122–132
BILL	AD	16·9	(0·71; 22)	16·0–18·1	16·6	(1·02; 18)	15·2–18·4
TARSUS		25·4	(0·74; 24)	24·1–26·7	24·7	(0·66; 19)	23·6–25·8
TOE		28·9	(1·11; 19)	27·5–30·6	28·1	(1·13; 17)	26·8–30·3

Sex differences significant, except bill and juvenile tail. Juvenile wing and tail significantly shorter than adult; tarsus and toe similar to adult from fledging, bill similar after completion of post-juvenile moult on face. First adults with wing and tail new (excluded from table), had wing on average 0·7 mm shorter than adult, tail 1·3 mm shorter; difference not significant.
Freshly dead birds, Netherlands; all year (ZMA).

		♂			♀		
WING	AD	185	(4·33; 85)	175–194	179	(4·08; 83)	170–190
TAIL	AD	142	(4·53; 73)	132–151	135	(4·87; 79)	126–146

No statistical difference in measurements between skins from Netherlands above and those from Köln (West Germany) and Makedonija (Yugoslavia) cited by Niethammer (1962). European skins measured by Nowak (1975a) averaged slightly smaller (presumably because of different measuring technique or inclusion of juveniles), but his data otherwise excellent for comparing different populations: nominate *decaocto*, (1) Germany and Balkans, (2) Iraq and Israel east to Afghanistan and north-west India, (3) Nepal, Bangladesh, and north-east India, east to eastern China, Korea, and Japan, (4) peninsular India and Ceylon; (5) *stolickzae* from central Asia; (6) *xanthocyclus* from tropical south-east Asia.

		♂			♀		
(1)		177	(5·0; 104)	162–189	174	(5·4; 85)	160–188
(2)		170	(5·6; 28)	160–184	167	(4·4; 20)	159–172
(3)		175	(4·0; 15)	168–181	166	(4·8; 13)	159–178
(4)		165	(5·7; 27)	152–178	162	(3·8; 15)	155–168
(5)		182	(7·3; 16)	169–196	179	(8·1; 9)	167–191
(6)		180	(4·4; 9)	172–185	181	(1·2; 3)	180–182

Weights. Netherlands: (1) January, shot; (2) February, shot; (3) March, shot; (4) May–August, adult, traffic and wire victims; (5) December, found dead; (6) January, frost-killed (ZMA). (7) West Germany, adult, June–September, city and rural areas combined (Lachner 1965). Portugal, adult: (8) May–July, (9) December–February (Santos Júnior 1978–9b). (10) Yugoslavia, adult, November–February (Kroneisl-Rucner 1957b). (11) Rajasthan (western India), all year (Rana 1975).

		♂			♀		
(1)		217	(6·46; 22)	171–234	200	(15·9; 13)	171–224
(2)		203	(7·35; 4)	199–214	173	(21·7; 10)	171–193
(3)		206	(14·5; 71)	173–243	197	(15·7; 67)	162–231
(4)		202	(15·4; 4)	181–218	191	(15·8; 5)	165–204
(5)		178	(22·4; 6)	146–215	148	(30·7; 12)	112–218
(6)		146	(14·0; 5)	132–168	130	(11·2; 7)	114–145
(7)		197	(—; 103)	150–242	189	(—; 83)	125–229
(8)		194	(16·4; 9)	175–225	198	(30·0; 8)	165–260
(9)		206	(11·9; 5)	190–220	196	(18·1; 7)	180–225
(10)		215	(16·7; 16)	189–249	205	(13·4; 16)	185–232
(11)		152	(11·6; 87)	115–184	146	(10·3; 80)	113–176

Maximum winter weight recorded: ♂ 264, ♀ 249 (Lachner 1965). Minimum weights of exhausted adults 130–150 (♂♂), 112–130 (♀♀) (ZMA). Juveniles fledge at 110–130; some may reach 200 within 40–60 days (maximum, ♂, 229) (Lachner 1965), but average weight still c. 20 below adult in 1st winter (ZMA).

Structure. Wing rather short, broad, tip bluntly pointed. 11 primaries: p9 usually longest; in adult, p10 4–8 shorter, p8 0–2 (rarely, p8 1–3 longer than p9), p7 7–12, p6 19–24, p5 29–33, p1 56–66; in juvenile, p10 and p8 1–6 shorter, p7 13–24, p6 23–33, p5 30–42; p10 and outer web of (p7–)p8–p9 emarginated. Tail rather long, tip slightly rounded; 12 feathers, t6 9–17 shorter than t1. Inner web of tail-feathers broadened near bluntly rounded tip in adult (except t1); sides of webs mostly parallel in juvenile, slightly attenuated towards rounded tip; difference in width between adult and juvenile often not marked, but difference in shape (and amount of white on tip) distinct: width of t1 at about middle 25–31 mm in adult, 22–28 in juvenile; width of t6 21–26 in adult, 18–23 in juvenile; 1st adult with new tail intermediate. Bill short and slender, relatively narrower than in *Columba*, with less pronounced soft flaps of cere above narrow oblique nostrils. Skin round eye bare (relatively more so than in west Palearctic *Columba*). Tarsus rather short and heavy, tibia feathered down to joint. Toe rather short and slender, with rather swollen and flattened soles; outer toe c. 76% of middle, inner c. 78%, hind c. 60%. Claws rather short, sharp, slightly decurved.

Geographical variation. Rather slight, involving size (see Measurements and Weights) and colour. Cline of gradually smaller size and slightly darker colour runs from Europe through Middle East and northern India towards Ceylon; smaller and darker birds from end of cline (named *intercedens*) here separated from larger and slightly paler European nominate *decaocto*, notwithstanding intermediate populations in India, as Ceylonese birds do not overlap in size with European nominate: wing of adult ♂ 165 (3·95; 7) 159–171, of adult ♀ 161 (2·89; 4) 158–164; wing of adult ♂ of peninsular India 173 (3·02; 6) 169–176 and thus closer to Ceylon birds than to those of Europe and therefore included in *intercedens*. *S. d. stolickzae* of central Asia is large and pale; in particular, sides of neck, chest, and belly paler than in nominate *decaocto*, head and nape vinous-pink with pale grey tinge, primaries slightly paler and with more distinct and paler edges than European birds, and perhaps more white shows on outer tail-feathers (Nowak 1975a; ZMM). Burmese race *xanthocyclus* is also large, but dark and with conspicuously yellow ring of bare skin round eye; in particular, rump dark, crown bluish, and underparts greyish; primaries black and amount of white on tail-tips rather limited (Nowak 1975a).

For relationships with other species, see African Collared Dove *S. roseogrisea*. CSR

Streptopelia turtur Turtle Dove

PLATES 33 and 34
[facing pages 375 and 398]

Du. Tortelduif Fr. Tourterelle des bois Ge. Turteltaube
Ru. Обыкновенная горлица Sp. Tórtola común Sw. Turturduva

Columba Turtur Linnaeus, 1758

Polytypic. Nominate *turtur* (Linnaeus, 1758), Canary Islands and Europe (except Balearic Islands) east through Turkey to Caspian and Irtysh river, in western Siberia south to forested steppes of Kazakhstan; *arenicola* (Hartert, 1894), Balearic Islands, north-west Africa east to Cyrenaica, and from Levant and east coast of Caspian to western Altay, Sinkiang, and Iran; *rufescens* (Brehm, 1845), Egypt (including western oases) and northern Sudan; *hoggara* (Geyr von Schweppenburg, 1916), mountains of central Sahara from Ahaggar (Algeria) to Aïr (Niger).

Field characters. 26–28 cm; wing-span 47–53 cm; tail 7–8 cm. Up to 10% larger than Laughing Dove *S. senegalensis* but 20% smaller than Rufous Turtle Dove *S. orientalis* and Collared Dove *S. decaocto*. Rather small, delicate and graceful dove, with fan-shaped tail. Adult's combination of dark-spotted, light chestnut to yellow buff wing-coverts, blue-grey diamond shaped panel on inner wing, and almost black, white-rimmed tail diagnostic. Immature much less distinctive but has adult tail pattern. Flight markedly agile, with characteristic bursts of flickering wing-beats, and white belly and under tail-coverts obvious. Purring song. Sexes rather similar; no seasonal variation. Juvenile distinct. 4 races in west Palearctic; adults of some separable in field.

ADULT. (1) European and west Asian race, nominate *turtur*. Crown and nape grey-blue; face pale vinous-brown; neck and breast mauve-pink (darker in ♂ than ♀), neck-sides with patch of black and blue- and white-tipped feathers (appearing white streaked black at distance). Mantle brown, with indistinct black centres and rufous margins to feathers; rump similar, but black centres more obvious and blue-grey margins showing particularly at sides. Scapulars, tertials, and most of wing-coverts (except greater) rich orange-brown to yellow-buff with bold black centres; whole area looks warm and strongly dappled in sunlight. Greater coverts and primary coverts blue-grey, often contrasting conspicuously with rufous inner wing in flight. Outer primary coverts and all flight-feathers brown-black. Underwing pale grey-blue, becoming dusky at tip and along trailing edge. Uppertail dark grey-brown in centre, with bases of outer feathers almost black and tips of same feathers white (forming sharp, almost complete rim to spread tail); undertail noticeably black, completely rimmed white and contrasting strongly with cream-white underbody and vent. ♀ usually paler and duller than ♂. Bill black-horn, with pale tip and mealy cere; eye golden; orbital ring purple. Legs purple-red to dark pink. (2) Balearic, North African, and Middle East race, *arenicola*. Slightly smaller and paler than nominate *turtur*, with some showing noticeably blue panel across inner wing. (3) Saharan race, *hoggara*. Crown often invaded with buff and upperparts more broadly margined with orange-buff. (4) Egyptian race, *rufescens*. See Geographical Variation. Darkest race. Lacks grey crown; strongly rufous (even fox-coloured) on upperparts and buffier below; underwing slate. JUVENILE. All races. Much duller and less colourful than adult, with all contrasts suppressed in generally drab brown appearance. Lacks grey crown, vinous breast, marked neck-patch, and strong rufous upperparts of adult, but shows partial blue panel across inner wing and diagnostic tail pattern. Except in *rufescens*, underwing paler than in adult.

Most ubiquitous of *Streptopelia* doves in west Palearctic. Confusion frequent, however, with all congeners, with adult requiring separation from *S. orientalis* and *S. senegalensis* and juvenile also from *S. decaocto*; see those species. Flight rapid, graceful, and markedly agile, with bursts of erratic, clipped, and even flickering wing-beats allowing easy changes in speed and track, and close manoeuvring in foliage; flight of long-distance migrants more regular in action, with rather looser wing-beats; glides frequently; take-off and landing accomplished with rapid flutter of

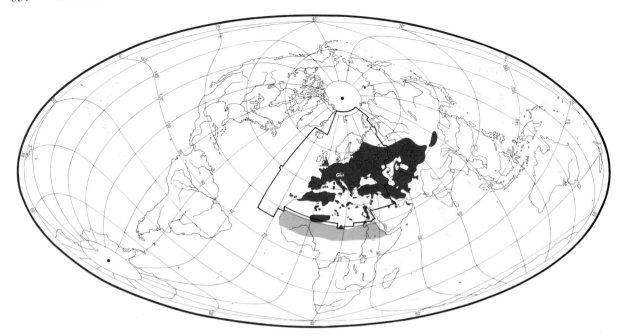

wings and spreading of tail. Gait free—as *S. decaocto* but more tripping. Perches freely, and agile on branches; fond of sitting on wires above nest cover.

Most characteristic call a deep, soporific, purred croon: 'rroorrr rrorrrr rroorrrr'.

Habitat. Across upper and lower middle latitudes of west Palearctic, breeding in temperate, Mediterranean, steppe, and semi-desert climatic zones. Avoids windy, wet, cloudy, and chilly situations, preferring fairly dry, sunny, sheltered lowlands, with accessible water and variety of cover as well as open ground. Avoids mountains, breeding in continental Europe mainly below c. 350 m and only very locally above c. 500 m (Glutz and Bauer 1980). Finds extensive unbroken forests and wetlands unsuitable, but often inhabits their borders, especially where these broken up with outliers such as groves, spinneys, copses, and open woodland, or trees along river banks, hedgerows, heaths with clumps of trees, and small plantations, orchards, large gardens, parks, and churchyards, preferably near croplands or fields with weeds (especially fumitory *Fumaria*) or deposits of seeds or corn. While tolerating human presence, dislikes breeding in or very near towns, villages, or farm settlements. A ground feeder, but otherwise largely arboreal, preferring low trees such as hawthorn *Crataegus*, especially on light soils. Aerial and mobile, flying mostly in middle or lower airspace; fond of perching on power-lines or other wires or cables. While owing much to man's cultivation and land management, not prepared to accept relationship as close as those of Collared Dove *S. decaocto*, feral Rock Dove *Columba livia*, or Woodpigeon *C. palumbus*, in this respect being even more reserved than Stock Dove *C. oenas*. Although little disturbed in summer quarters, suffers heavy losses in some areas through shooting on migration.

Distribution. Few changes in range recorded, except for northward spread in Britain; probably now breeding regularly in Denmark and perhaps in Ireland.

BRITAIN. Range extended to Wales and northern England (Cheshire and Yorkshire) before 1865 (Alexander and Lack 1944); colonized south Lancashire 1904; further slight expansion since 1940 to north-east England and south-east Scotland; recent fluctuations on periphery of range in Wales (Parslow 1967; Sharrock 1976). IRELAND. Known breeding records intermittent (Parslow 1967; Sharrock 1976); one or more pairs probably nest in most years (CDH). DENMARK. Only c. 10 breeding records, but now probably regular (Dybbro 1976; TD). SWEDEN. Over 100 birds seen annually, but no proof of breeding (LR). FINLAND. Bred 1868 (Merikallio 1958). LEBANON. Probably breeds occasionally, but no recent proof (Tohmé and Neuschwander 1974; Macfarlane 1978). MALTA. Nested unsuccessfully 1956, several attempts 1981 (Sultana and Gauci 1982). MADEIRA. Breeds occasionally (Bannerman and Bannerman 1965; GM, AZ).

Accidental. Iceland, Faeroes, Norway, Finland, Azores, Madeira, Cape Verde Islands.

Population. Fluctuating in some areas, perhaps linked with climatic changes (Glutz and Bauer 1980). Increased in Britain in 19th century and in Baltic region of USSR in 20th century; recent decreases in Portugal, Netherlands, and Italy.

BRITAIN. Increased before 1965 and probably slightly since, with some fluctuations: estimated totals—under

Streptopelia turtur 355

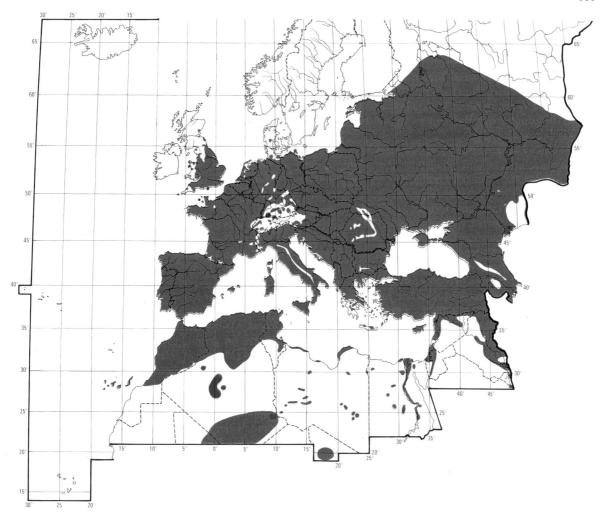

100000 pairs (Parslow 1967), over 125000 pairs (Sharrock 1976). FRANCE. Fluctuating; estimated under 1000000 pairs (Yeatman 1976). PORTUGAL. Decreased (RR). BELGIUM. About 32000 pairs (Lippens and Wille 1972); c. 25000 pairs 1982 (PD). LUXEMBOURG. About 2200 pairs (Lippens and Wille 1972). NETHERLANDS. Very common c. 1940; recent decline to 20–50%, partly due to habitat changes; in 1977, 25000–35000 pairs (Teixeira 1979; CSR). WEST GERMANY. Decreasing (GR); for local counts and fluctuations, see Glutz and Bauer (1980). EAST GERMANY. For local counts, see Glutz and Bauer (1980). ITALY. Decreased since c. 1960, mainly due to habitat changes (SF). GREECE. Common on mainland; some local decreases probably due to excessive hunting (HJB, WB, GM). USSR. Estonia. Formerly rare; marked increase since 1960s, now abundant (HV). Latvia: increased since 1960s, now fairly numerous (HV). Lithuania: increased 20th century (Valius *et al.* 1977).

Survival. Britain: estimated 1st-year mortality c. 64%, annual adult mortality c. 50% (Murton 1968). For limited data from central Europe, see Glutz and Bauer (1980). Oldest ringed bird 13 years 2 months (Rydzewski 1978).

Movements. Migratory, with possible exception of some Saharan-breeding birds (see below). Essentially a summer visitor to Palearctic, wintering in Africa; rare winter records from Europe and Kuwait (Hudson 1973; Glutz and Bauer 1980; Bundy and Warr 1980). Notable for disparity between sizes of known breeding and wintering ranges (Moreau 1972).

AFRICAN POPULATIONS. North African population of *arenicola* (breeding Morocco to northern Libya) mainly migratory, but in winter quarters below Sahara (see below) indistinguishable from migrants of same race from Asia; present all year in some oases of southern Algeria (Dupuy 1966, 1969; E D H Johnson). Egyptian race (*rufescens*) reported as migratory (Meinertzhagen 1930; Vaurie 1961b), but some winter in Dakhla and Kharga oases and on lower Nile (P L Meininger and W C Mullié); this race resident in extreme north Sudan, and museum specimens collected south to Blue Nile (Sudan) and in adjacent north-west Ethiopia (Cave and Macdonald 1955; Vaurie 1961b; Urban and Brown 1971). Central Saharan race (*hoggara*) not yet identified south of breeding range, though certainly migratory in Fezzan and Tibesti where absent mid-October to February (Guichard 1955); however, a proportion (at least) are resident in Ahaggar and in northern Niger breeding areas (Bannerman 1931a; Fairon 1971; E D H Johnson).

EURASIAN POPULATIONS. Winter quarters lie in northern Afrotropics, in semi-arid Soudanian and savanna zones from Sénégal and Guinea to Sudan and Ethiopia, and as far south as northern Ghana and northern Cameroun (Moreau 1972; Walsh and Grimes 1981). Has occurred northern Zaire; only a vagrant to East Africa (Britton 1980). Asiatic population (*arenicola*) probably occurs in eastern portion of winter range, but no adequate delineation (no ringing recoveries, and *arenicola* is also the migratory North African breeding form).

Leaves European breeding areas late July to September (adults and juveniles together), with stragglers into October; passage broad-front (with south-westerly orientation in Eurasia), but also some suggestion of partial concentration for Mediterranean crossings. Migrants from western Europe recovered especially in south-west France and Iberia (but not Galicia), and enter Africa through Morocco (e.g. Bernis 1967, Murton 1968, Pereira 1979); strongly represented in Portugal in autumn, though in spring the passage is more directly northwards across Spain. Many also cross Balkans and Italy at both seasons, to enter or leave Africa via Tunisia and Libya. On Malta, up to 20000 per day at peak spring passage (fewer in autumn) and c. 100000 shot there annually; passage birds ringed Malta recovered especially in central Europe: Czechoslovakia, East Germany, Poland (Sultana and Gauci 1982). Similarly, birds ringed central Europe (e.g. Austria, Hungary, Poland) recovered on passage in Italy and Greece. Birds of uncertain origins (2 ringed Cyprus found Ukraine in August and September) are also common migrants around eastern Mediterranean; on Cyprus (as on Malta) more numerous in spring than autumn. Marchant (1963b) noted concentrated westerly passage past Baghdad (Iraq) in autumn, with minimum of 3000000 estimated to pass along 100-km front; this passage of Asiatic birds crosses Arabia into north-east Africa. Migrant ringed Kuwait in May found Uzbekistan (72°15′E) in September.

Numerous oases records, sometimes involving large falls, show Sahara crossed at many points and probably on broad front (e.g. Moreau 1961, 1972, Smith 1968, Bundy 1976). Main movement through Sahel zone September–October, and into November in Chad at least (Lamarche 1980; Newby 1980); poorly-defined wintering areas lie further south. Reappears in Sahel February–March, when more abundant than in autumn (possibly many overfly then). Huge numbers visit ricefields in Sénégal after harvest, with 450000 in one roost at Richard-Toll, March 1970 (Morel and Roux 1966, 1973); in Niger inundation zone, Mali, mass return in February–March as waters recede from flood plains, with estimated 700000 in main area (Curry and Sayer 1979; Lamarche 1980). Rather few occur Gambia in autumn or midwinter, though very large numbers in February–March of some recent years, with over 1000000 at one roost near Kaur in early 1976; due to recent poor rainfall in Soudanian and savanna zones, birds in West Africa perhaps now tend to move west in late winter (Jensen and Kirkeby 1980; Gore 1981). One ringed Chad, November, found Greece in April; 2 ringed Ethiopia, October and December, found in southern USSR (Ukraine, Rostov) in later autumns. Only other Afrotropical recoveries involved Mali and Sénégal, and all were birds from western Europe.

Most disappear from south of Sahara in March and first half of April, though small numbers summer there (e.g. Lamarche 1980 for Chad). Main northward passage through Mediterranean in second half of April, with breeding areas reoccupied during May.

Food. Mainly seeds and fruits of weeds and cereals. Feeds mostly on ground, apparently not often from trees or hedges (Murton *et al.* 1964a).

Plant material includes seeds and fruits of Polygonaceae, Ranunculaceae, Fumariaceae, Resedaceae, Violaceae, Caryophyllaceae, Chenopodiaceae, Cruciferae, Leguminosae, Euphorbiaceae, Primulaceae, Scrophulariaceae, Rubiaceae, Compositae, Gramineae, and Scots pine *Pinus sylvestris*. Animal food includes cocoons of earthworms (Oligochaeta); occasionally, insects and small snails. (Feriancová 1955; Tutman 1960; Murton *et al.* 1964a; Goodwin 1970; Murton 1971a.)

In USSR, shows preference for seeds of oil-bearing crops when available but in early summer chiefly takes seeds of wild plants. In Soviet central Asia, may feed on mulberries *Morus* (Dementiev and Gladkov 1951b). 100 birds from USSR contained seeds (mostly of fumitory *Fumaria*, millet *Setaria*, cornflower *Centaurea*, and wheat *Triticum*) and a few molluscs (see Murton *et al.* 1965). At Dubrovnik (Yugoslavia), one stomach contained 3 seeds of vetch *Vicia*, other crushed seeds, and sand grains (Tutman 1960). In Czechoslovakia in a sample of 44 stomachs, 39% of total items were *Chenopodium* seeds, 38% *Setaria*, 12% *Triticum*, and 4% *Centaurea* (Feriancová 1955). In Hungary, in 36 birds, *Sinapis* seeds made up 38% of total items found, 27% *Chenopodium*, 12% medick *Medicago*, 10% *Setaria*, and 8% *Fumaria* (Glutz and Bauer 1980). In Romania, August–September, 42% of total seeds and fruits taken were *Setaria*, 28% goosefoots (Chenopodiaceae), 16% sunflower *Helianthus*, 7% *Sinapis*, and 3% vetch (Glutz and Bauer 1980). In England, seeds of *Fumaria* most important single item (Witherby *et al.* 1940; Murton *et al.* 1964a; Murton 1971a). In Cambridgeshire, *Fumaria* seeds and fruits of chickweed *Stellaria media* important, especially July–August; grass seeds, cereals, fruits of *Sinapis* or *Brassica*, and seeds of *Polygonum* also taken along with variety of other weed seeds. Animal food included earthworm cocoons, and occasionally insects and small snails (Murton *et al.* 1964*a*). Recorded taking seeds of Scots pine *Pinus sylvestris*. Also suspected to take insect larvae when abundant (Goodwin 1970).

Young first fed *c*. 5–7 hrs after hatching; given crop milk for first *c*. 5 days, then also other foods, until same as parents from *c*. 9 days (Kotov 1974*a*). 5 young recorded taking 2334 items of which 56% by number *Fumaria* seeds, 14% cereal seeds, 9% *Sinapis* or *Brassica* seeds, 4% *Stellaria media* seed capsules, 3% *Polygonum* seeds, 3% grass seeds, 10% other weed seeds, and 1% animal or other (Murton *et al.* 1964*a*). Parents said to bring water for young in crop (Klinz 1955; see also Goodwin 1970). BDSS

Social pattern and behaviour. No detailed studies.
1. Usually in pairs or singly; on migration mostly in small loose-knit flocks. Larger aggregations typically on favourable feeding grounds (Goodwin 1970: see also, e.g., Rékási 1979). Less commonly flocks during breeding season: e.g. up to *c*. 300, mid-June (Glutz and Bauer 1980; see also Creutz 1974). Huge roosting assemblies at times in African winter quarters: e.g. in March, Sénégal, *c*. 450000 (Morel and Morel 1979), and over 1000000 on roosting flight (Gore 1980, 1981). Family parties band together at end of breeding season (Ingram and Salmon 1924; Kölsch 1958). Small separate flocks of juveniles said to occur in Sénégal, September (Morel and Roux 1966). Not infrequently associates with other Columbidae, especially when feeding (Glutz and Bauer 1980); for associates in Africa, see (e.g.) Curry (1974). BONDS. Observations on 60 marked birds showed monogamous mating system the rule (Klinz 1955). Duration of pair-bond not known exactly but possibly extends over more than 1 season and promoted by site fidelity (Trinthammer 1859; see also Kotov 1974*a*). Pair-formation takes place on breeding grounds or in some cases possibly before; birds seen apparently in pairs on migration (Hosking 1942; Glutz and Bauer 1980); interval between arrival and laying sometimes very short (Dementiev and Gladkov 1951*b*); in Orenburg region (USSR), older birds said to arrive already paired (Kotov 1974*a*). Observations on captive birds allegedly showed that pair-bond severed outside breeding season (Naumann 1833). First breeding at 1 year (D Goodwin). Care of young by both parents; mainly by ♂ in later stages (see Relations within Family Group). Has hybridized with *S. decaocto* (e.g. Voous 1963*a*); for association between wild ♂ *S. turtur* and ♀ Barbary Dove *S. 'risoria'*, see Nichols (1907); see also Bodenstein (1949*a*). BREEDING DISPERSION. Solitary, but many pairs often nest quite close together (2 nests once only *c*. 3 m apart: Coenen 1954) when suitable habitat restricted but good feeding grounds easily accessible: e.g. in isolated field copses, or orchards in oases (Goodwin 1970; see also Hosking 1942). At Carlton (Cambridgeshire, England), 4–9 pairs on 0·65 ha (Murton 1968). In central Europe, mostly 0·4–0·6 pairs per km² (Glutz and Bauer 1980 which see for concentrations in West Germany). For Netherlands, see (e.g.) Alleyn *et al.* (1971) and Teixeira (1979), for East Germany see (e.g.) Creutz (1974), and for Switzerland see Glutz von Blotzheim (1962). At Lake Balaton (Hungary), exceptional concentration of *c*. 160–165 pairs on peninsula of *c*. 12 km² (Glutz and Bauer 1980). In northern Turkey, 8 pairs on *c*. 1 ha (Renkhoff 1972). Both members of pair (Kotov 1974*a*) defend only immediate vicinity of nest (Glutz and Bauer 1980); according to Hosking (1942), such defence embraces nest-tree or -bush. Some heterosexual behaviour takes place on or near nest (Kotov 1974*a*; see also Flock Behaviour and Heterosexual Behaviour, below). Adults feed mainly away from nest, ranging up to *c*. 3–6 km (Glutz and Bauer 1980). ROOSTING. Nocturnal and, outside breeding season, usually communal (Glutz and Bauer 1980); in 2 areas of East Germany, roosting assemblies also noted from June or late July though not clear which categories of birds involved (see Creutz 1974, Glutz and Bauer 1980). Normally in dense leafy bushes or trees, often only *c*. 2 m above ground (Naumann 1833; Glutz and Bauer 1980); in southern floodplains of Mali, birds used remote stands of *Acacia* (Curry 1974). During nest-building, pair generally roost in physical contact in crown of tree *c*. 50–200 m from nest (Kotov 1974*a*). In Sénégal, February–March, birds arrived at roost *c*. 30 min before sunset; *c*. 100 birds per tree, and roost shared with several hundred large birds of prey (Accipitridae). Communal roosting persisted until spring departure, with (e.g.) *c*. 5000–8000 still present early May (Morel and Roux 1966; see also Gore 1980). In Mali, birds roosted apart from other (resident) Columbidae (Curry 1974); independent juveniles sometimes roost with Stock Dove *Columba oenas* and Woodpigeon *C. palumbus* (Klinz 1955). On breeding grounds, birds go to roost at dusk (Naumann 1833), and active next morning when still only half light (Kotov 1974*a*). Occasionally call at night (Klinz 1955). Sites used for daytime loafing much as above; also posts and (particularly low-slung) electric wires (Freitag and Metz 1977; also Hinsche 1968); in some areas, on roofs—especially juveniles in autumn (Naumann 1833). In Mali, March–April, birds seek shelter from sun in trees and huddle under small bushes when temperature *c*. 45°C (Curry 1974). Migrants in desert areas shelter in any shade from sun and wind (Haas 1974; Haas and Beck 1979). In northern Sénégal, October–December, birds drink late morning and (higher peak) before sunset (Morel 1975; see also Morel and Roux 1966, and Flock Behaviour, below); in Mali, drink only at dusk before going to roost; in Nigeria, possibly drink also at dawn (Curry 1974). On breeding grounds, reported to drink around midday and after sunset (Naumann 1833; see also Kotov 1974*a*). At Kirchheim (West Germany), most came to loess cliff to take earth in morning and evening; some perhaps roosted there (Kölsch 1958; see also Flock Behaviour, below). Markedly fond of sun-bathing (Goodwin 1970); preceded by water-bathing in one pair, late June (Bentham 1957). For description of rain-bathing and sunning postures adopted by *Streptopelia* and other Columbidae, see Goodwin (1970).

2. Often shy and unapproachable in autumn, particularly away from woods (Naumann 1833). Wintering birds in Africa more sensitive than other Columbidae to disturbance by man (Curry 1974; G D Field). When alarmed, bird extends neck forwards (compare Alarm-posture of, e.g., Rock Dove *C. livia*) and looks around alertly, turning head (Naumann 1833; see also Heider 1953). Birds generally fly to trees when disturbed (Curry 1974; D Goodwin). May attempt to escape from aerial predators in fast, agile flight but, unlike *S. decaocto*, usually makes for cover (Naumann 1833), exceptionally entering buildings (Glutz and Bauer 1980). Head-shaking given as expression of discomfort (Goodwin 1956*a*). Wing-clapping often performed at take-off, only occasionally before landing (Tooby 1946). Peculiar slow tail-raising given after alighting (e.g. Bodenstein 1949*b*); movement exaggerated in sexual or aggressive excitement (Goodwin 1956*b*; see also *C. palumbus*). For signal function of tail pattern at take-off, see Goodwin (1970). FLOCK BEHAVIOUR. Little information. In Mali, large flocks fly in loose chevron formation to water (Curry 1974). When moving to roost birds fly low and flocks may form almost continuous stream for *c*. 30 min (Morel and

Roux 1966). To drink or bathe, birds tend to land in area devoid of vegetation, then walk to gently shelving section of shore (Naumann 1833). Occasional disputes over perches may take place in larger flocks (Freitag and Metz 1977). Courtship, including mating, also self-preening and sunbathing, similarly occur (Kölsch 1958; Glutz and Bauer 1980). For behaviour of aberrantly plumaged individual in flock of c. 7–15, see Ressl (1963). ANTAGONISTIC BEHAVIOUR. (1) Display-flight. Typically performed by ♂ after (long) bout of Advertising-calls and in presence of ♀ (Hoffman 1927; Colquhoun 1940: see 1 in Voice). Unlike S. decaocto, apparently directed only at mate (Glutz and Bauer 1980); however, according to Kotov (1974a), performed both to attract ♀ and to dissuade strange ♂♂ from approaching occupied nest-site. Display-flight invariably performed at sight of another conspecific bird flying nearby (D Goodwin). ♂ takes off from elevated perch and, with tail widely fanned, makes steeper and swifter ascent than C. palumbus to c. 20–30 m; all, or only last 2–5, of the few rapid wing-beats produce Wing-clapping—less loud than in C. palumbus (Tooby 1946). Bird glides slightly further up after ceasing wing-beats and then spirals slowly down, wings fully spread and horizontal (sometimes raised vertically: Klinz 1955); tail also fanned showing contrasting black and white pattern (see Goodwin 1955). Bird may return to starting point in wide arc or move to different perch up to c. 100–150 m away; Advertising-call invariably given after landing (Naumann 1833; Hoffmann 1927; Colquhoun 1940; Hosking 1942; Klinz 1955; Kotov 1974a; Glutz and Bauer 1980). Gliding descent may be followed directly by renewed ascent, whole performance being given up to 6 times (Kotov 1974a; Glutz and Bauer 1980); such repeats said by Colquhoun (1940) to be rare. Gliding phase may last up to c. 1–1½ min (Kotov 1974a), bird sometimes remaining high in the air for some time before landing (Goodwin 1970; see also Colquhoun 1940). ♂ observed by Hosking (1942) descended, renewed ascent, then glided round tree in which ♀ perched. Display-flight also performed by ♀ though less commonly (Glutz and Bauer 1980); however, in one study ♀♀ and ♂♂ seen regularly in Display-flight together in evenings at start of breeding season (Hosking 1942; see also Selous 1901 and Hoffmann 1927). Most Display-flights performed during pair-formation and nest-building; fairly frequent during incubation, though rarer later and normally cease with hatching of 2nd clutch (Kotov 1974a). (2) Threat and fighting. Most hostile towards conspecific neighbours c. 5–6 days before laying. If strange conspecific bird sighted during gliding phase of Display-flight, ♂ immediately breaks off and may pursue other bird up to c. 100–150 m (Kotov 1974a). If Bowing-display (see Heterosexual Behaviour, below) elicits other than appropriate sexual response, performing bird usually ceases displaying and prepares to attack: lowers head and gives Excitement-call (Goodwin 1956b: see 4 in Voice). Rather perfunctory Nodding may also be given: in Streptopelia, apparently indicates active rather than defensive aggressiveness (see Goodwin 1956a, 1963). Fullest description of antagonistic interaction refers to (not uncommon) attempt by 2 ♂♂ to court 1♀. 2 ♂♂ first gave Bowing-display with Display-calls (see 2 in Voice) either side of ♀, then each pecked her and hit her with wings, apparently trying to get her away from rival. ♀ flew off, followed by 2 ♂♂ and whole performance repeated. After disappearance of ♀, fight took place between 2 ♂♂: adopted Defensive-threat posture with feathers ruffled and wing further from opponent raised. Clicking sounds which accompanied closer approach said to be made by bill; more likely Excitement-call and thus probably vocal (D Goodwin: see 4 in Voice). Blows with wings then delivered and further 'threatening sounds' given. Such sounds said to be emitted frequently in flight when 2 ♂♂ in pursuit, or chasing ♀. In fight, most pecking directed at upper neck. At peak intensity, birds grapple with bills, hit with wings and try to push each other off perch; chase may follow. Birds also frequently threaten with head lowered and back arched, giving clicking sounds as they peck; feathers often torn out (Kotov 1974a). Other aggressive sounds (see 6a in Voice) uttered during fight described by Klinz (1955). Disputes with S. decaocto said to be rare, birds displaying in close proximity, etc. (e.g. Niethammer 1943b, Bodenstein 1949b). Territory-holding S. decaocto usually dominant, even when pair of S. turtur tries to ward off single bird of other species (see Tomasz 1955). HETEROSEXUAL BEHAVIOUR. (1) General. Courtship (exceptionally copulation) sometimes performed by passage birds (Grote 1944). ♂♂ apparently arrive on breeding grounds earlier than ♀♀ (Naumann 1833) and Advertising-call uttered until July, sporadically in August (Glutz and Bauer 1980). ♂ utters Advertising-call with head lowered and lower neck swollen (Naumann 1833); head moves slightly up and down in time with call (Kotov 1974a); given in early morning (c. 08.00–10.00 hrs), around midday, and in evening, from concealed site in bush or tree, but nearer nest often on more exposed perch (Naumann 1833). Apparently serves to bring birds of (potential) pair together, and presumably to synchronize the partners (Glutz and Bauer 1980). Pair-formation essentially as in other European Columbidae (Goodwin 1970). Most courtship said to take place in early morning and evening, with ♂ at most active during nest-building, i.e. c. 5–6 days before laying (Kotov 1974a; see also Colquhoun 1940). When pair flying to feeding grounds, ♂ usually slightly ahead (Hosking 1942); according to Kotov (1974a), ♀ always in front. On ground, birds rarely more than 1·5 m apart (Hosking 1942). For brief display directed at S. decaocto, see Bawtree (1965). (2) Pair-bonding behaviour. Unpaired ♀ most likely to be attracted initially by Display-flight of ♂ who may briefly perform with louder-than-usual Wing-clapping, apparently encouraging ♀ to land (Kotov 1974a). If she does so, ♂ performs Bowing-display which is mainly sexual in Streptopelia doves. In S. turtur, consists of a series of rapid bobbing movements (Goodwin 1970); up to 6 (Selous 1901). Mean time per complete bow 0·72 s ($n = 5$ bouts); speed and accompanying Display-calls (see below) are main distinguishing features from corresponding display of S. decaocto (Davies 1970, where also further comparison with various hybrids). ♂ rises up to full height then bows low so that belly touches branch or ground (Kotov 1974a). At culmination of each bow, ♂'s head and bill at roughly 45°, and erected plumage forms black and silver ruff either side of head (Goodwin 1970); practically all contour feathers ruffled, tail depressed, and bird changes from one foot to the other (Heinroth and Heinroth 1926–7: Fig A). During Bowing-display, ♂ gives almost incessant Display-calls (Selous 1901; Klinz 1955). Bowing-display apparently performed only by ♂ (Glutz and Bauer 1980), though ♀ possibly does so when birds in tree (Selous 1901). ♀ ready to copulate said to spread tail and wings slightly and to bow to ♂ who then gives higher-intensity Display-call (Kotov 1974a); description may refer (in part) to pre-copulatory Parading (see subsection 5, below). Bowing-display usually takes place on high branch within c. 1 m of ♀ (Colquhoun 1940); on ground ♂ additionally executed

A

'dancing step' (possibly Jumping) which caused ♀ to withdraw, then fly off (Selous 1901; see also Goodwin 1970). ♂ may also chase ♀ from branch to branch and utter smacking or rattling sounds—presumably Excitement-call (Kotov 1974a: see 4 in Voice). ♀ sometimes sits quite near ♂ when he gives Advertising-call; birds of pair also spend much time in close physical contact, when nuzzling (see *S. decaocto*) and mutual Allopreening (of area around ear coverts) typically performed; Billing (see below) apparently also, but no details (Hosking 1942; Klinz 1955). Allopreening more frequent in *Streptopelia* doves than in European *Columba* (Goodwin 1956b). Often followed by Display-flight (Hosking 1942), by either bird (Selous 1901). (3) Driving. When pair together with other conspecific birds and ♀ approached by strange ♂♂, she is normally Driven away from them by her mate who pecks lightly at her head and eventually forces her to take flight (Kotov 1974a). Description in Klinz (1955) probably refers to (part of) Bowing-display. (4) Billing. Birds apparently 'fence' with bills, ♀ not inserting hers into ♂'s (Goodwin 1970), and performance even held by Colquhoun (1940) to be same as mutual Allopreening. No detailed descriptions indicating that courtship-feeding takes place but process said by Kotov (1974a) to be certainly much briefer than in *C. livia*. (5) Mating. Often preceded by Display-flight (Hosking 1942), then brief Bowing-display lasting c. 5–10 s (Colquhoun 1940); receptive ♀ may then respond immediately by soliciting copulation, without waiting for further ceremony (Goodwin 1956b). According to Kotov (1974a), ♂ may give Bowing-display with Display-calls for c. 2–5 min before ♀ invites copulation. Unwilling or apparently uninterested ♀ may fly off, followed by ♂ (Colquhoun 1940), or be driven off by him using bill and wings (Kotov 1974a); sometimes ♂ flies short distance, initiates further mating ceremony, then joined by ♀ who gives appropriate response (Goodwin 1948a). Usual preliminary to copulation is Parading, Mock-preening (by ♂ or both), and Billing. Mock-preening as in (e.g.) *C. livia*; Parading also, except that in *Streptopelia* head lowered (Colquhoun 1940; Goodwin 1970). After copulation, ♂ briefly adopts markedly erect ('penguin-like') posture with ruffled neck feathers, as in (e.g.) Woodpigeon *C. palumbus*. ♀ Parades, or may turn round on perch in same posture (Goodwin 1948a, 1970). Elsewhere reported that both ruffle feathers, flutter wings, and sit close together for quite a long time (Naumann 1833). Call said to be given after copulation (Klinz 1955: see 6b in Voice), but not clear by which bird. Some ♂♂ said to give Bowing-display and Display-call for c. 5–10 min afterwards (Kotov 1974a). Mating takes place on nest, in tree, or on ground, including where birds forage (Naumann 1833; Colquhoun 1940; see also Flock Behaviour, above). Occurs c. 5–6 times per day during nest-building, often when birds otherwise loafing around midday (Kotov 1974a). (6) Nest-site selection. Essentially as in *S. decaocto*, with ♀ choosing site offered by ♂ (Glutz and Bauer 1980). ♂ giving Advertising-call perhaps changes to Nest-call (see 3 in Voice) when ♀ close (Goodwin 1956b). No full details of Nest-calling available (Glutz and Bauer 1980), but description in Kotov (1974a) indicates probably much as in other west Palearctic Columbidae. (7) Behaviour at nest. At chosen site, ♂ and ♀ sit close together and mutually Allopreen; sometimes give soft calls (presumably Nest-calls) and Nod. ♂ then flies off to collect material (at varying distance from site). Building usually takes c. 6–7 days; mostly done in morning and evening. May be interrupted by bad weather and then, after c. 2–4 days, same construction resumed or new one started, sometimes only c. 5–6 m away (Kotov 1974a). ♀ on half-built nest gave soft cooing (possibly Nest-call), ♂ responding (call not described) from nearby (Ingram and Salmon 1924). Incubation rhythm as in other west Palearctic Columbidae (Kotov 1974a). Nest-relief once proceeded as follows: ♀ brooding 6-day-old nestlings; ♂ called from nearby and ♀ responded immediately (neither call described); ♂ flew in and ♀ walked along branch to where he had landed c. 1·2 m from nest, briefly bill-touched with him, and flew off (Ingram and Salmon 1924). See also Hosking (1942). RELATIONS WITHIN FAMILY GROUP. Younger nestlings fed simultaneously. Feeds generally last c. 5(–10) min, parent making frequent short pauses (Ingram and Salmon 1924; Hosking 1942). Young beg by giving food-call and, from 2nd day, stretch up to adult's bill (Kotov 1974a); older nestlings flap wings and jab with bills (Hosking 1942). Up to c. 6–7 days, fed by ♀ in early morning before she leaves (Kotov 1974a); in one study, ♂ invariably departed quickly after feeding young, ♀ tended to remain (Hosking 1942). Nestlings brooded by both parents (pattern as for incubation) up to c. 7 days (Kotov 1974a), by ♀ also at night thereafter and during day in bad weather up to fledging according to Naumann (1833). According to Kotov (1974a), thermoregulate at c. 9 days, though nocturnal brooding often continued to c. 12–13 days. For report that both parents sometimes brood nestlings simultaneously, including at night, see Klinz (1955). Both parents seen to Allopreen 6-day-old nestlings (Ingram and Salmon 1924). Young usually sit side by side and head to tail in nest (Naumann 1833; Kotov 1974a), though may face same direction; possibly dependent on whether site allows parents to vary direction of approach (Schwarz 1938). At c. 5–6 days, young crawl round in nest on tarsi, also using bill for support (Kotov 1974a); at c. 6–9 days, often self-preen or head-scratch. More wing-flapping (when food-calls also given) and stretching evident from c. 12 days, when mutual bill-nibbling by siblings also noted (Ingram and Salmon 1924). Self-preening intensified at c. 13–18 days. First short excursions made from nest at c. 15–16 days (Ingram and Salmon 1924), or not until 19–20 days (Kotov 1974a). Young fly well at c. 25–30 days (Kotov 1974a), captive birds able to fly at 11–12 days (D Goodwin). Parents said to cease feeding young after c. 28–30 days, ♀ c. 4–5 days earlier than ♂, though not clear whether this refers to start of 2nd brood (see Kotov 1974a). Not known exactly when family breaks up (Glutz and Bauer 1980); for report that fledged young remain with adults for some days (or weeks) in general natal area, see Klinz (1955). For aggression of captive pair towards young of previous brood, see Stevenson (1875). ANTI-PREDATOR RESPONSES OF YOUNG. When small, crouch low in nest (Kotov 1974a). Generally quiet in presence of human intruder (Schwarz 1938). If attempt made to touch them, rear up in Defensive-threat display with ruffled down or feathers and neck swollen, bill-snapping and making hissing sounds; may also peck at fingers (Schwarz 1938; Kotov 1974a). Defensive-threat display used successfully against Jay *Garrulus glandarius* (Ingram and Salmon 1924). Older but still unfledged young leap or clamber out of nest and attempt to hide in vegetation (Naumann 1833; Kotov 1974a). PARENTAL ANTI-PREDATOR STRATEGIES. (1) Passive measures. During building or early in incubation, birds easily alarmed at approach of man; fly away, sometimes with Wing-clapping, and often desert nest completely (e.g. Kotov 1974a). Tends to sit tight later in incubation or when brooding young; one ♀ crouched lower on nestlings at slight noise from human intruders (Ingram and Salmon 1924). (2) Active measures: against birds. Captive bird flew off nest to attack approaching Magpie *Pica pica* (Goodwin 1970). (3) Active measures: against man. Most descriptions refer to distraction-lure display of disablement type, usually performed when flushed from hatching eggs or small young. Bird may perch nearby and flutter wings vigorously, or tumble to ground and feign broken wing; crawls on breast, or flaps along in apparent impeded flight with tail depressed and slightly spread, moving over or through vegetation (e.g. Hosking 1942, Goodwin 1970, Glutz and Bauer 1980, P A D Hollom). Performance once over c. 100 m (Powell 1933), though bird may fly off immediately when observer makes closer approach (Allen 1933). Distraction-lure display sometimes performed up to c. 60 m from nest after bird has made normal departure; once, performed in bracken c. 12 m from nest, bird invisible to observer (Garnett 1933; also Boyd 1933). Captive birds reported to crouch low in nest with neck extended forwards, then to rock on carpal joints; if threat more acute, may adopt Defensive-threat posture typical of Columbidae with one wing raised (Grünefeld 1952; see also *S. decaocto* and *S. roseogrisea*). 2 records of bird giving Advertising-call nearby while nest being examined (Schwarz 1938).

(Fig A from photograph in Heinroth and Heinroth 1926–7.)

MGW

Voice. Repertoire more restricted than Collared Dove *S. decaocto* (Bodenstein 1949b). Calls 1–3 all have same timbre but different rhythmic forms (Lade and Thorpe 1964). ♂ and ♀ have same repertoire, but calls 1–3 of ♀ hoarser, less musical, softer, and used less often (Goodwin 1970). Whistling noise from wings heard only rarely in flight (Naumann 1833; but see also Bodenstein 1949b and Klinz 1955). For Wing-clapping, see Social Pattern and Behaviour.

CALLS OF ADULTS. (1) Advertising-call (Song). A purring 'cōōrr-cōōrr' or 'tūrr-tūrr' (Goodwin 1970). Typically disyllabic but not uncommonly more than 2 units; usually 2–5 units in each phrase, with 3–12 phrases per bout of song (Simms 1971); each full song said to last c. 15 s (Hoffmann 1927). Recording (Fig I) suggests 'turrrr turrrr', with very quiet onset and gradual crescendo through 1st unit, while 2nd unit shows equal overall intensity (J Hall-Craggs). At times, indistinctly disyllabic—'turrurrr' or 'turrr(ug)'; may descend in pitch (see Hoffmann 1927). Duration of individual units c. 0·5–0·7 s (Lade and Thorpe 1964, which see for additional sonagrams and analysis of calls given by other *Streptopelia*, including hybrids). (2) Display-call. A faster version of call 1 (Simms 1971): a hurried and emphatic 'crŏŏr(wa) crŏŏr(wa)crŏŏr(wa)' given in time with bobbing movements of Bowing-display (Goodwin 1970). Unlike the even 'tur' of call 1, a rather more clearly phrased 'túrr turr-turr( ) túrr turr-turr', ( ) representing peculiar snapping sound (Heinroth and Heinroth 1926–7; see also call 4 below), described by Colquhoun (1940) as a wheezing gasping or grunting echo note, especially conspicuous after 1st syllable; rendered 'aarnk' by Simms (1971), where said to be uttered before 1st or between 1st and 2nd syllable. Recording of presumed Display-call (Fig II) suggests rolling, purring, sonorous 'qurrrr qurrrr (qua) qurrrrr(qua)'; 2 'qua' sounds (probably inhalation of breath) given per phrase (P J Sellar, M G Wilson). Display-call further described as 'turturturturturtur . . .' sometimes given almost without interruption; moderated, softer, and faster version of call 1 (Naumann 1833). Throbbing, and audible over some distance (Colquhoun 1940). Deep, rolling, and musical (Selous 1901). (3) Nest-call. A trisyllabic variant of call 1, with middle unit shorter: 'cōōrr-cŏŏr-cōōr'. Sometimes given as apparent alternative to call 1 (Goodwin 1970). See also Bettmann (1959). (4) Excitement-call. A short, explosive popping sound like sudden drawing of cork from bottle. Given in similar situations to homologous call of (e.g.) Barbary Dove *S. 'risoria'* (thus also expressing anger), but apparently less freely (Goodwin 1955, 1970). (5) Distress-call. A panting gasping sound, higher pitched than in *Columba*; at high intensity (e.g. when bird seized) almost a scream (Goodwin 1970). A short trumpeting 'ru' ('Fright-call'), as in many other Columbidae, is presumably the same; also given by captive bird attacking Hobby *Falco subbuteo* (Heinroth and Heinroth 1926–7). (6) Other calls. (a) Short grumbling sounds uttered by bird driving off rival (Klinz 1955). Captive pair (particularly ♂) gave cough-like croaking sounds when showing hostility towards offspring of previous year (Stevenson 1875). (b) Throttled sound emitted after copulation (Klinz 1955).

CALLS OF YOUNG. Weak squeaking or whistling sounds as food-call (Ingram and Salmon 1924; Hosking 1942). At c. 7 days, louder and like food-call of other west Palearctic *Streptopelia*. Rather hoarse wheezing sounds at c. 2 months presumably first attempt to utter adult calls. Full 'courtship cooing' (presumably adult call 1) given at c. 7–11 months (Kotov 1974a).

MGW

I E Simms/BBC England June 1952

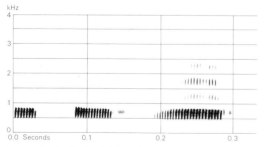

II V C Lewis England July 1969

Breeding. SEASON. North-west Europe: see diagram. Apparently similar over much of southern Europe (Makatsch 1976). Canary Islands: eggs from May onwards (Bannerman 1963b). North Africa: eggs laid April to early July (Heim de Balsac and Mayaud 1962). SITE. In tree, shrub, or hedge; of 511 nests, Britain, 43% in hawthorn *Crataegus monogyna*, 17% in elder *Sambucus niger*, 29% in other deciduous species, 5% in conifers, 6% miscellaneous (Murton 1968). Average of 411 nests, 2·4 m above ground, maximum 13 m (Murton 1968). Occasionally uses nest of other species, complete or as foundation (Glutz

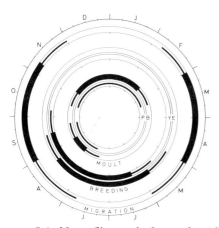

and Bauer 1980). NEST: flimsy platform of small twigs, lined finer material, including roots, grass stems, and leaves. Nest of rusty wire recorded West Germany (Merkel 1957). Building: by ♀; ♂ gathers material (Campbell and Ferguson-Lees 1971), but Klinz (1955) recorded building by both. Twigs collected from ground, not broken from tree (Glutz and Bauer 1980). EGGS. See Plate 95. Oval, smooth and slightly glossy; white. Nominate *turtur* (*arenicola* similar): 30 × 22 mm (28–33 × 20–24), $n = 70$; calculated weight 8 g. Weight 8 g (7–9·5), $n = 32$ (Kotov 1974*a*). *S. t. isabellinus*: 27 × 21 mm (26–28 × 20–22), $n = 5$; calculated weight 6 g (Schönwetter 1967). Clutch: 1–2; 3 recorded (Erard 1969). Of 277 clutches, England: 1 egg, 10·5%; 2, 89%; average 1·9 (Murton 1968). 2–3 broods; 2 broods normal in southern USSR but only in some years further north (Grote 1944). Replacement clutch laid after loss. Laying interval 39–48 hrs (Kotov 1974*a*). INCUBATION. 13–14(–16) days (Witherby *et al.* 1940; Murton 1968; Kotov 1974*a*). By both sexes; ♂ from *c.* 09.00–10.00 hrs to 15.00–16.00 hrs, ♀ remainder (Naumann 1833). Begins with 2nd egg; hatching near synchronous. YOUNG. Semi-altricial and nidicolous. Cared for and fed by both parents, brooded while small. FLEDGING TO MATURITY. Fledging period *c.* 20 days. Become independent soon after. Age of first breeding 1 year. BREEDING SUCCESS. Of 621 eggs, England, 292 (47%) hatched; losses due to predation (34%), desertion (14%), and infertility and other factors (5%), with predation greatest in June and desertion in August; of 305 young hatched, 250 (82%) fledged, giving overall success of 39%; in another sample of 134 eggs laid, 62 (46%) hatched and 50 young (37% overall) fledged. Success increased from 34% in May to 48% in July, linked with better food supply for young. Success related to clutch and brood size: 48% of 496 eggs in clutches of 2 hatched, but only 17% of 29 eggs in clutches of 1; 86% of young in 239 broods of 2 fledged but only 67% of 24 young in broods of 1. (Murton 1968.)

Plumages (nominate *turtur*). ADULT MALE. Forehead and crown medium bluish-grey, often slightly paler and tinged cream-buff on forehead, lores, and just above and behind eye. Nape, central hindneck, and upper mantle medium bluish-grey, feathers narrowly tipped olive-brown, tips wearing off during summer. Black patches on sides of neck with narrow white feather-tips in centre of patch, bluish-grey tips at edges. Mantle olive-brown, feather-centres usually dull black but occasionally not; sometimes, surround of black centres tinged bluish-grey. Scapulars, tertials, lesser and median upper wing-coverts (except outermost), and tertial coverts deep black with contrasting broad cinnamon-rufous to cinnamon-buff fringes, latter narrower and fading to pale cinnamon or pale buff when plumage worn. Back and centre of rump olive-brown with dull black feather-centres, like mantle, lateral feathers and variable number of central feathers bluish-grey; in some, back and rump mainly olive-brown, in others largely medium bluish-grey (deeper and more bluish than crown). Upper tail-coverts olive-grey or bluish-grey, tipped olive-brown. Chin and centre of throat pink-cream or cream-buff; cheeks and sides of throat to chest, sides of breast, and upper belly pale greyish-vinaceous when fresh, deeper and purer vinaceous when worn. Centre of belly cream-pink to cream-white, flanks and upper sides of thighs bluish-grey; remainder of underparts including under tail-coverts white. Central tail-feathers (t1) dull greyish olive-brown; other tail-feathers black with grey bloom, 7–15 mm (t2) to 24–34 mm (t5–t6) of tip white; tip of t2 often partly suffused bluish-grey and olive-brown; outer web of t6 white. From below, tail black with broad white tips and white outer web of t6. Flight-feathers greyish-black, tips of secondaries slightly tinged olive-brown, outer webs of secondaries and inner primaries tinged grey; fresh flight-feathers narrowly margined rufous at tip. Inner and central lesser and median upper wing-coverts and tertial coverts like scapulars; outer lesser and median coverts pale ash-grey with black or dull grey centre; greater coverts pale ash-grey with olive-brown centres and rufous-washed tip and outer edge. Upper primary coverts light grey, inner webs of greater dull black. Longest feathers of bastard wing black, shorter light grey. Under wing-coverts and axillaries light bluish-grey. ADULT FEMALE. Closely similar to adult ♂ and not always separable. In fresh plumage, forehead more buff or isabelline, paler and less grey than ♂; crown paler and duller grey; nape to centre of upper mantle olive-brown like remainder of mantle (in ♂, all nape to centre of upper mantle almost uniform medium bluish-grey); cheeks and sides of chin buff, not slightly vinaceous as in ♂; chest less deep vinaceous and with buff tinge, usually intermixed with some grey-brown or pink-grey feathers, sometimes completely greyish-buff; breast whiter; thus chest and upper breast not almost uniform deep vinaceous like ♂. In worn plumage, on average duller and more buffy than ♂, but sometimes difficult to sex. NESTLING. Down coarse and hair-like; pale straw-buff or straw-yellow in nominate *turtur*, paler yellow in *arenicola*. JUVENILE. Crown, hindneck, and mantle greyish-drab or dull drab-brown, feathers narrowly fringed buff or rufous when fresh; forehead, area just above eye, ear-coverts, and cheek paler greyish-buff; sides of neck pale ash-grey with faint buff feather-fringes, no contrastingly black-and-white patches. Scapulars, tertials, most lesser and median upper wing-coverts, and tertial coverts cinnamon-buff to rufous-brown, not sharply demarcated from greyish or dull black feather-centres; feather-tips narrowly but distinctly margined pale buff or rufous-cinnamon. Back, rump, and upper tail-coverts like crown and mantle. Chin cream-white or pale buff; throat, chest, upper breast, and upper flanks light grey with rather broad buff or rufous feather-tips, grey sometimes slightly pink but appearance much browner and never vinaceous as adult. Remainder of underparts like adult, tips of longest under tail-coverts washed rufous. Tail as in adult, but less white on tips of t2–t6 (16–24 mm on inner web of t6), some black beside shaft on outer web of t6 (not fully white as in adult), and fresh feathers with slight rufous tinge at tip. Tail-feathers slightly narrower and tips more narrowly rounded than in adult (see Structure). Flight-feathers as in adult, but outer primaries more broadly washed pale cinnamon or rufous on tip and outer web; greater upper primary coverts distinctly fringed rufous on tip. Inner and central upper wing-coverts pale cinnamon-drab or drab-brown, distinctly tipped pale buff, centres of lesser coverts blackish; outer upper wing-coverts ash-grey, slightly duller than in adult, tips of coverts with poorly defined white fringes. Remainder of wing as in adult. ♂ often distinctly darker than ♀, feather-fringes deeper rufous; ♀ often more tinged buff, fringes paler buff or cream-white. FIRST ADULT. Similar to adult when post-juvenile moult completed. A few 1-year-old birds arrive on breeding grounds with some juvenile flight-feathers retained: mainly middle secondaries (narrower and shorter than neighbouring ones), rarely 1–2 outer primaries (corresponding greater upper coverts with traces of rufous fringes).

Bare parts. ADULT. Iris orange-red, orange-yellow, golden-yellow, golden-brown, brass-brown, yellow-brown, or (in winter) bright red-brown to golden-orange. Bare skin round eye dark reddish-purple, purplish-red, raspberry-red, or (in winter) pink; a reddish-blue spot below eye. Bill and cere brownish-black, horn-black, greyish-black, or purplish-black, or with cere and base of bill with purple-red, reddish-flesh, or pink tinge. Leg and foot purple-red, crimson, or (in winter) dark pink to crimson-pink. NESTLING. Iris brown-grey. Bill plumbeous-grey with slight pink tinge to cutting edges and lower mandible. Leg and foot greyish-pink. Bare skin of body pink, of face plumbeous-grey; all plumbeous-grey from c. 5 days. JUVENILE. Iris light brown or grey-brown, changing to yellow-brown and orange-yellow during post-juvenile moult. Bare skin round eye pink. Bill greyish-black. Leg and foot pale grey-blue or pinkish-grey, gradually changing to pink (at about fledging) and red (during post-juvenile moult). (Hartert 1912–21; Heinroth and Heinroth 1926–7; Goodwin 1970; Harrison 1975; Glutz and Bauer 1980; BMNH, RMNH.)

Moults. ADULT POST-BREEDING. Some start on breeding grounds from August, departing with moult suspended; others not moulting as late as October–November, starting in winter quarters. Part of mantle, scapulars, flanks, rump, and head first, soon followed by p1. Of 8 autumn migrants, western Europe, 5 had not yet started (except for a few body feathers in some), 1 had suspended with p1 new, 2 with p1–p2 new (RMNH, ZMA); in Crete, 6 had not started, 7 suspended with p1 new, and 1 each with p1–p2, p1–p3, and p1–p4 new (Swann and Baillie 1979). Moult completed in winter quarters, January–March. Tail and secondaries replaced in winter. No information on duration of primary moult. POST-JUVENILE. Complete. Timing strongly variable, depending on date of hatching. Moult starts within c. 5 weeks of fledging, unless pre-migratory fattening or migration started. During autumn migration, moult suspended with (in birds hatched May–June) large part of body already 1st adult and with 1–4 inner primaries new; later-hatched birds start migration in full juvenile plumage, occasionally even with primaries not fully grown and remains of down still present on back and tail-coverts. Moult completed or started in winter quarters; hardly any data available, but termination date probably as variable as starting date. A few 1st adults arrive on breeding grounds with 1–2 juvenile outer primaries and greater upper primary coverts retained; more frequently, some juvenile middle secondaries retained. (Dementiev and Gladkov 1951b; Stresemann and Stresemann 1966; BMNH, RMNH, ZMA.)

Measurements. Nominate *turtur*. Western Europe, summer; skins (RMNH, ZMA).

WING AD	♂ 179	(2·53; 37)	174–185	♀ 172	(2·73; 16)	167–177
TAIL AD	116	(3·61; 35)	110–123	113	(3·01; 16)	108–118
BILL AD	16·8	(0·92; 38)	15·4–18·7	16·0	(0·68; 16)	14·9–16·6
TARSUS	23·6	(0·88; 20)	22·4–24·9	22·9	(0·62; 14)	22·1–23·8
TOE	27·4	(1·33; 20)	25·5–29·0	26·4	(0·66; 12)	25·4–27·3

Juvenile wing on average 4 mm shorter than adult, tail 11 mm; bill similar to adult when face fully feathered.

S. t. arenicola. Northern Algeria and Tunisia, summer; skins (RMNH, ZMA).

WING AD	♂ 175	(2·50; 6)	171–178	♀ 167	(3·92; 5)	162–172
TAIL AD	116	(3·27; 7)	112–121	110	(2·80; 5)	106–114
BILL AD	15·9	(0·80; 6)	15·3–17·0	16·5	(0·73; 5)	15·6–17·4
TARSUS	23·0	(0·87; 7)	21·8–24·2	22·4	(0·52; 5)	21·5–22·9
TOE	26·5	(1·04; 6)	25·2–28·0	25·9	(0·78; 5)	25·0–26·9

Jordan, Iraq, Bahrain, February–July; skins (BMNH, RMNH).

WING AD	♂ 165	(2·11; 6)	161–167	♀ 162	(2·40; 9)	157–165

Turkmeniya (USSR), summer; skins (ZMM).

WING AD	♂ 173	(3·35; 11)	169–178	♀ 166	(3·43; 6)	162–172

S. t. hoggara. Ahaggar (Algeria), Fezzan (Libya), and Tibesti (Chad), March–June; skins (BMNH, ZMA).

WING AD	♂ 173	(2·63; 6)	169–176	♀ 167	(2·57; 7)	162–169

S. t. rufescens. Nile delta and valley and Dakhla oasis (Egypt), March–July; skins (BMNH, RMNH).

WING AD	♂ 164	(3·52; 12)	160–170	♀ 163	(— ; 3)	162–165

Sex differences significant, except bill, tarsus, and toe of *arenicola* and wing of *rufescens*.

Weights. Nominate *turtur*. Camargue (France): April–May, after crossing Mediterranean, 125 (13·5; 48) 100–156; June–July, 144 (12·0; 28); August–September, adult 152 (24·3; 15) 120–208, juvenile 126 (23·9; 17) 99–170 (Glutz and Bauer 1980). Portugal: September, migrants, 132 (246) 85–170 (Santos Júnior 1978–9a). Turkey, late May to mid-July: ♂ 131(15·9; 5) 105–143, ♀ 142 (13·0; 5) 130–160 (Kumerloeve 1967, 1969, 1970). Aldabra, December: ♂ 105 (BMNH). Netherlands, May–June: ♂♂ 160, 177, 178; ♀♀ 136, 138; exhausted ♀ 95 (RMNH, ZMA).

S. t. arenicola. Iran and Afghanistan, mid-May to early August: ♂ 124 (10·0; 6) 111–140, ♀ 118 (9·11; 6) 107–131 (Paludan 1938, 1940, 1959; Desfayes and Praz 1978).

Structure. Wing rather long, slightly rounded at tip, broad at base. 11 primaries: p9 longest, p10 0–6 shorter (occasionally, p10 1–2 longer than p9); in adult, p8 2–7 shorter, p7 13–22, p6 24–34, p1 64–76; in juvenile, p8 4–9 shorter, p7 18–24, p6 28–36, p1 67–76; wing-shape independent of race. Outer web of (p8–)p9 and inner web of p10 slightly emarginated. Tail rather long, rounded or wedge-shaped at tip; 12 feathers. In adult, t1 longest, t6 14–25 shorter; in juvenile, t2–t3 longest, t1 c. 5 shorter, t6 12–20. Width of t1 at about middle 24(21–27) in adult, 21(17–23) in juvenile; t6 20·5 (18–23) in adult, 17·5(16–19) in juvenile. Bill straight and slender, with small cere, as in other *Streptopelia*. Large bare patch of skin round eye, smallest in juvenile. Tarsus and toes short and slender; outer toe c. 77% of middle, inner c. 80%, hind c. 65%.

Geographical variation. Rather slight in size, more pronounced in colour. Largest birds inhabit northern parts of range, from west and central Europe east to upper Irtysh (USSR); average wing in latter area ♂ 178, ♀ 172. South of this, size slightly smaller: Canary Islands ♂ 174, ♀ 167 (Hartert 1912–21; BMNH, ZMA), north-west Africa ♂ 175, ♀ 167 (see Measurements), Turkey ♂ 174, ♀ 170 (Kumerloeve 1967, 1969, 1970; RMNH, ZMA), Turkmeniya (USSR) ♂ 173, ♀ 166 (see Measurements), Afghanistan ♂ 171, ♀ 166 (Paludan 1959). Further south, in Egypt (*rufescens*), Levant, Iraq, and Arabian peninsula, smaller still (see Measurements), but population inhabiting mountains of central Sahara (*hoggara*) not particularly small. Variation in colour rather gradual and boundaries not sharply defined, ♀♀ especially often hard to distinguish. Northern race, nominate *turtur*, medium bluish-grey on crown, olive-brown on mantle, and black with rather bright rufous fringes on scapulars, tertials, and inner upper wing-coverts. Crown of adult ♂ *arenicola* pale bluish-grey, nape pale vinaceous-pink, crown and nape of ♀ buff with grey feather-bases visible on crown; mantle in both sexes pale rufous-cinnamon or greyish-cinnamon, not olive-brown or olive-grey as in nominate *turtur*; scapulars, inner upper wing-coverts, and especially tertials with broader and paler rufous feather-fringes, and black feather-centres smaller; central back to upper tail-coverts and t1 paler sandy- or rufous-olive, less dark olive-brown; chin slightly more extensive cream-buff, especially in ♀; chest and upper breast paler lilac-pink, on ♀ often mixed with some buff. Colours strongly subject to wear and bleaching, and worn nominate *turtur* may appear pale fringed above, while even some fresh ones (especially ♀♀) from western Europe have fringes of mantle and scapulars greyish-cinnamon, appearing as pale as *arenicola*. Typical pale *arenicola* breeds east from Levant and Caspian Sea; birds inhabiting Balearic Islands, north-west Africa, Cyrenaica (Libya), south-east Turkey, and mountains of central Asia are slightly darker, more or less intermediate with nominate *turtur*, and some individuals inseparable from latter; these provisionally all included in *arenicola*. Adult *hoggara* from central Sahara rather like *arenicola*: crown of ♂ light bluish-grey, occasionally with buff tinge, nape rufous-buff, not as uniform bluish-grey as *arenicola*; ♀ has all head buff except for pale grey tinge on forehead and forecrown; mantle and central back to upper tail-coverts deeper and more extensively rufous-cinnamon than in *arenicola*; scapulars, tertials, and inner upper wing-coverts with restricted black centres as in *arenicola*, but wide fringes slightly deeper rufous-cinnamon and less buff; chest brighter vinaceous than in *arenicola*, in ♀ with much buff tinge; tips of t1–t2 more extensively washed rufous than in *arenicola*. Situation in Egypt not fully clarified. Vaurie (1961*b*) separated 2 races, *rufescens* in Dakhla oasis (and perhaps also Kharga and Kufrah), and *isabellina* in Nile valley and delta and in Faiyum. In *rufescens*, crown rich buff-brown without grey; all mantle, scapulars, tertials, and inner upper wing-coverts bright rufous-cinnamon or fox-red, with only limited black feather-centres; limited bluish-grey on back to upper tail-coverts and on outer upper wing-coverts; chin and sides of head deep buff; chest and upper breast rather deep vinaceous; some or many rufous edges on grey under wing-coverts and flank feathers. In *isabellina*, upperparts paler buff-brown rather than rufous as in *rufescens*; some vinaceous on hindneck bordering neck-patches; underparts slightly paler than *rufescens*, throat and chest more lavender or lavender-buff; very similar to ♀ *hoggara*, but top of head usually completely buff-brown without grey on forehead, and grey on outer upper wing-coverts more restricted. These differences probably sexual, however, as all *rufescens* examined were ♂ (including singles from Nile delta and Faiyum) and all *isabellina* ♀. As no ♀♀ from Dakhla examined, situation not yet fully established. Eventually, Egyptian ♂♂ may prove to be dimorphic, like those of Laughing Dove *S. senegalensis aegyptiaca*, which has some ♂♂ bright rufous above, others from same locality olive like ♀.

Forms superspecies with Rufous Turtle Dove *S. orientalis* of Asia, Adamawa Turtle Dove *S. hypopyrrha* from highlands of Nigeria and Cameroun, and Dusky Turtle Dove *S. lugens* from highlands of north-east and East Africa (Snow 1978). CSR

Streptopelia orientalis Rufous Turtle Dove

PLATES 33 and 34
[facing pages 375 and 398]

Du. Oosterse Tortelduif Fr. Tourterelle orientale Ge. Orient-Turteltaube
Ru. Большая горлица Sp. Tórtola oriental Sw. Större Turturduva

Columba orientalis Latham, 1790

Polytypic. *S. o. meena* (Sykes, 1832), Soviet Central Asia and south-west Siberia east to Altay mountains (USSR), and south through western Himalayas to Nepal, straggling to western Europe; nominate *orientalis* (Latham, 1790), central Siberia and south-east Asia from Yenisey basin south to eastern Himalayas and northern Vietnam, straggling to western Europe. Extralimital: *agricola* (Tickell, 1833), Nepal to Bangladesh, northern and central Burma, and western Yunnan (China); *erythrocephala* (Bonaparte, 1855), peninsular India; *stimpsoni* (Stejneger, 1887), Ryukyu Islands; *orii*, Yamashina 1932, Taiwan.

Field characters. 33–35 cm; wing-span 53–60 cm; tail 8–9 cm. About 15% larger than Turtle Dove *S. turtur*, with noticeably broader wings, greater girth, and deeper belly; largest of genus in west Palearctic. Noticeably bulky *Streptopelia* with plumage similar in pattern to *S. turtur* but less bright, due particularly to larger black centres on inner wing and shoulders and vinaceous-brown rear head and hindneck. Diagnostic characters vary between races; much-vaunted grey tail-rim absent in west Asian birds. Voice distinct. Flight and gait less agile than *S. turtur*.

Sexes similar; no seasonal variation. Juvenile separable. 2 races wander to west Palearctic, separable in the field.

ADULT. (1) West Asian race, *meena*. Plumage closely similar to *S. turtur* but well marked birds differ in (a) darker grey head; (b) browner hindneck; (c) grey-blue feather-tips in black neck-patch (making patch indistinct); (d) greater coverts merely edged and tipped grey-blue; (e) outer and median coverts edged pale grey; (f) whole of primary coverts dusky (so that outer half of wing wholly dark); (g) upper tail-coverts black-brown, contrasting with bluer rump. Supposedly darker slate-blue underwing (Heinzel *et al.* 1972) disputed in this race (see Plumages). Palest race, with belly and under tail-coverts often as white as in *S. turtur*. Bill dark grey with purple base and mealy cere; eye red-gold, orbital ring grey to red. Legs and feet pink to red. (2) East Asian race, nominate *orientalis*. Differs from *meena* in darker, richer plumage, most noticeably in (a) browner crown and mantle, (b) redder margins to blacker-centred scapulars and inner wing-coverts, (c) grey-blue tips to tail-feathers, (d) vinous to grey-cream belly and under tail-coverts, and (e) darker underwing, making separation of this race from *S. turtur* simpler than for *meena*. JUVENILE. Both races. As in *S. turtur*, duller with pale fringes to wing-coverts; margins to scapulars and wing-coverts more buff than rufous.

Larger size (particularly of wings and body) than *S. turtur* obvious in comparison and visible to experienced eye at all times; also contributes to different flight action and gait. However, confusion not infrequent, and obvious plumage differences between *S. turtur* and *meena* exist only in browner rear head, hindneck, and mantle; distinction of *S. turtur* from nominate *orientalis* easier, since *orientalis* shows diagnostic grey rim to tail and also lacks the bold contrast between pale belly and black base to undertail so typical of *S. turtur*. Important to recognize that only significant difference in tail pattern between *S. orientalis* and *S. turtur* is lack of bold white edge of outermost tail-feathers on *S. orientalis*; both can show white or pale spots right across tail-tip when landing or taking off. Suggestions that *S. orientalis* lacks striking tail-tip now suspect. Flight typical of genus but much heavier and distinctly less agile than *S. turtur*, recalling Stock Dove *Columba oenas* at times; looks bulkier and shorter tailed in the air than *S. decaocto*. Walks more slowly than *S. turtur*, with tail often held up and head-bobbing less pronounced ; when searching for food, can look remarkably like small chicken. Perches freely, but takes off and lands with slower flutter than *S. turtur*.

Vagrants show no habitat preference, appearing on coasts and islands, as well as inland. Most characteristic call a hoarse drone: 'woo-woo-kak-coor', phrased like Woodpigeon *C. palumbus*.

Habitat. Shows wide diversity over far-flung range, extending into boreal zone in north and to tropics in south, and in some regions to mountains up to *c.* 4000 m in Nepal (Ali and Ripley 1969), and to 3250 m south-east of Samarkand, USSR (Dementiev and Gladkov 1951*b*). Breeds in forests of spruce *Picea*, and in Kashmir in close-growing spinneys of young firs (Bates and Lowther 1952); also in juniper stands, cedar woods, and open glades (in pinewoods mixed with birches and aspen) on plateaux and watersheds, and among low hills in poplar *Populus*, willow *Salix*, and various other broad-leaved trees on floodplains. In India in open mixed deciduous and bamboo forest, particularly in broken foothills and in vicinity of cultivation, where large trees, groves, and gardens provide shelter in easy reach of stubbles used for feeding on fallen grains and seeds. In USSR, however, feeds locally in forest on berries of *Prunus* consumed in trees, and also on pine seeds gathered underneath. Active on ground, running and walking freely, but very wary; flies strongly, but data on heights of flight lacking. After breeding, shifts largely from forests to more open and cultivated land.

Distribution. Breeds in Asia from eastern Urals to Pacific, Sakhalin, Kuril Islands, and Japan, south through China, Turkestan, and Afghanistan to India, Burma, northern Indo-China, Taiwan, and Ryukyu Islands. Resident and migratory, northern populations wintering in India, Burma, Japan, and southern China to Indo-China.

Accidental. Britain, France, West Germany, Denmark, Norway, Sweden, Italy, USSR, Greece, Kuwait.

Movements. Resident in southern Asia (*stimpsoni, orii, erythrocephala, agricola*, southern populations of nominate *orientalis*), except for some southward movement in autumn in eastern India. Migratory in Siberia and contiguous parts of Asia from Afghanistan to northern Japan (*meena*, northern populations of nominate *orientalis*). Palearctic migrants winter in south-east Asia from southern Japan to Indo-China (nominate *orientalis*), and in Indian subcontinent (*meena*) (Dementiev and Gladkov 1951*b*; Vaurie 1965; Ali and Ripley 1969). This indicates separate movements from USSR, east and west of Himalayas, and small numbers (presumably *meena*) occur eastern Iran on passage (D A Scott). Also occurs in very small numbers, or irregularly, on Arabian side of Persian Gulf (Kuwait to Oman), September to early November and more rarely May–June (Bundy and Warr 1980; Walker 1981*a, b*).

In eastern China, movements of nominate *orientalis* reported late March to June and September–October (Hemmingsen and Guildal 1968). In western Siberia, passages of *meena* evident mid-April to early June, and August to mid-October (Dementiev and Gladkov 1951*b*); passes through Kohat (northern Pakistan), mid-April to early May and September–October (Ali and Ripley 1969). 13 dated records for northern and western Europe occurred January (1), May–July (4), October–November (8).

Voice. See Field Characters.

Plumages (*S. o. meena*). ADULT. Closely similar to nominate race of Turtle Dove *S. t. turtur* and sometimes hard to distinguish, except for larger size and different wing formula. Differs in slightly darker medium bluish-grey forehead and crown, light bluish-grey rather than pale bluish-grey or white feather-tips in black neck-patches; deep reddish-vinaceous nape; darker olive-brown or vinaceous central hindneck and mantle (less olive-grey or ochre-buff than *S. turtur*); broader and duller black centres to scapulars, tertials, and inner upper wing-coverts, fringes relatively narrower, orange-rufous when fresh to grey-buff when worn; back and rump darker blue-grey, upper tail-coverts dark olive-brown; chin and upper throat warmer buff; lower throat, chest, upper breast, and sides of breast deeper vinaceous with slight ochre tinge, less pure vinaceous-pink; belly and vent cream-pink instead of white; flanks, axillaries, under wing-coverts, outer upper wing-coverts, and outer secondaries slightly darker medium bluish-grey; greater upper primary coverts black without blue-grey tinge; white of longer under tail-coverts and white of tail-tips sometimes slightly tinged pale grey; white on outer web of t6 restricted to rather narrow border (outer web largely or all white in *S. turtur*). Hardly any difference in colour of underwing (*contra* Heinzel *et al.* 1972). Sexes closely similar, but crown and chest of ♀ often tinged brown, and belly and breast more extensively cream on average. *S. o. meena* distinctly darker than southern and eastern race *arenicola* of *S. turtur*, with which it overlaps in south-west Siberia and in south-western mountain ranges of central Asia. NESTLING. Down plentiful, but absent round eye, on throat, and on mid-underparts; hair-like, pale yellow or yellowish-white (Hartert 1912–21; Harrison 1975). JUVENILE. Closely similar to juvenile *S. t. turtur* and hardly distinguishable except for larger size, different wing formula, and different pattern on t6 (see Adult). Head, neck, and mantle slightly darker and duller grey-brown (no neck-patches); scapulars, tertials, and inner upper wing-coverts darker black-brown rather than dark olive-brown or rufous-brown; fringes rufous as in *S. t. turtur*, however; chest and breast darker grey with slightly wider and browner feather-fringes. FIRST ADULT. Like adult, but some retain part of juvenile plumage: mainly secondaries (s2–)s3–s7(–s8), occasionally tail-feathers (t4–)t5, or outer primaries from p7 outwards.

Bare parts. ADULT. Iris yellowish-orange, golden-orange, light red, or orange-red. Narrow ring round eye leaden-grey, bluish-purple, dark pink, magenta, or dark red. Bill dark leaden-grey, brownish-black, or dull purple with paler horn-coloured tip and bluish-purple, purple, magenta, or dark red base and cere. Leg and foot pink, dull purple-red, magenta, purple, or red; rear of tarsus and soles paler, pink; claws horn-brown. NESTLING. in *S. turtur*; see Kotov (1976b). JUVENILE. Iris dull brown. Bill dark leaden-grey or purple-black. Leg and foot dull greyish-pink or pink-red. (Hartert 1912–21; Dementiev and Gladkov 1951b; Ali and Ripley 1969; Goodwin 1970; BMNH.)

Moults. ADULT POST-BREEDING. Complete; primaries descendant. Starts in breeding area with scattered feathers of chest, breast, scapulars, and some upper wing-coverts from early July, but apparently no primaries; moult suspended during autumn migration, and not all birds have started by then. Main moult in winter quarters, completed by February–March; duration of primary moult not known. Body and tail start from primary moult score of 20–30, completed at score 40–45, but in birds resuming primary moult after previous stop (probably 1st adults in 2nd calendar year) body moulted from higher score, e.g. in 3 with moult suspended at 30, body moult started when primary moult nearing score 50; one of these showed serially descendant moult, also moulting inner primaries. POST-JUVENILE. Complete or partial, primaries descendant. Highly variable, depending on time of hatching; suspended during autumn migration (primary moult score up to 25 recorded), but late-hatched birds not yet started then and some of these had not yet completed moult during spring migration, when moult suspended. Moult probably starts within a few weeks after fledging; some feathers of head, neck, and upper wing-coverts first, soon followed by p1. Head and body in heavy moult between primary moult scores 5 and 25; largely adult from score 30–40. Tail and secondaries start from score of *c.* 25. Early-hatched birds complete moult before spring migration, but others retain some middle secondaries and occasionally some tail-feathers during summer of 2nd calendar year; least advanced birds also retain outer primaries and outer greater primary coverts, these arriving in breeding area with moult suspended and up to 4 primaries old (see also Adult Post-breeding). (BMNH, RMNH, ZMA, ZMM.)

Measurements. Nominate *orientalis*. Sakhalin and Japan, summer, and China, all year; skins (BMNH, RMNH, ZMA).

WING AD	♂ 196	(3·33; 9)	192–201	♀ 193	(2·88; 11) 188–196
TAIL AD	133	(5·60; 10)	126–142	129	(5·34; 13) 124–138
BILL AD	17·0	(0·81; 13)	16·3–18·2	16·6	(1·10; 15) 15·3–18·3
TARSUS	27·5	(1·32; 12)	25·5–29·1	26·7	(1·34; 16) 24·7–29·3
TOE	30·8	(2·21; 7)	28·3–34·3	29·9	(1·27; 9) 28·8–32·4

Sex differences significant for wing. Juvenile wing on average 5 mm shorter than adult, tail 7 mm. Range of unsexed Japanese birds 180–207.

S. o. meena. European part of Urals and neighbouring north-west Kazakhstan (USSR), April–July; skins (ZMM).

WING AD	♂ 202	(4·09; 13)	196–210	♀ 197	(3·64; 6) 192–201

Weights. *S. o. meena*. Kazakhstan (USSR): May, ♂♂ 230, 236; ♀ 230; June, ♀ 235; July, ♂♂ 197, 210; ♀♀ 227, 238 (Gavrin *et al.* 1962). Kirgiziya (USSR), April–August: unsexed 171–270 (44) (Yanushevich *et al.* 1959). Afghanistan: May, ♂♂ 165, 221; June, ♂♂ 177, 190, 194; ♀ 204; September and early October, juvenile ♂♂ 195, 229 (Paludan 1959).

Nominate *orientalis*. Taiwan, November: ♂ 195 (RMNH). Mongolia: May, ♂♂ 245, 274; ♀ 218; June, ♂ 195, ♀ 206; July, juvenile ♂ 184 (Piechocki 1968c). South-central Siberia: at hatching, average 11·5; on 15th day 156·5, at fledging (25–26 days) *c.* 178 (Kotov 1976b).

S. o. meena or *agricola*. Nepal: ♂♂ 180, 194, 196; ♀♀ 182, 196; juvenile ♀♀ 148, 150 (Diesselhorst 1968).

Structure. Mostly as in *S. turtur*, but with relatively heavier body, broader wing with more rounded tip, and shorter tail when compared with bulk of body; more similar in size and shape to Stock Dove *Columba oenas*. 11 primaries: in adult, p8 longest, p10 5–10 shorter, p9 0–3, p7 8–16, p6 20–32, p5 33–46, p1 64–78; in juvenile, p9 longest, p10 1–8 shorter, p8 0–4, p7 13–18 (in *S. turtur* adult, p9 usually longest, p10 0–6 shorter, p8 2–8, p7 13–21; in juvenile, p9–p10 longest, about equal, and p8 4–10 shorter); p11 minute, concealed by primary coverts. Inner web of (p8–)p9–p10 and outer of (p7–)p8–p9 slightly emarginated. Differences in width of tail-feathers as in *S. turtur*, but all feathers average and range 5 mm wider; difference between longest and shortest feather similar. Outer toe *c.* 76% of middle, inner *c.* 80%, hind *c.* 61%.

Geographical variation. Marked between Siberian races *meena* and nominate *orientalis*; rather slight and clinal in more southern races. Nominate *orientalis* distinctly darker than *meena*, more markedly different from *S. turtur* (and hence probably less often overlooked than *meena* as a straggler to Europe). Forehead, lore, and area below eye pale grey-buff, crown medium bluish-grey. Nape, central hindneck, and mantle deep greyish-brown or dark ginger-brown, sometimes with slight russet tinge. Centres of scapulars, tertials, and inner upper wing-coverts extensively deep black with sharply demarcated but rather narrow rufous-chestnut (when fresh) to pale buff (when worn, especially on wing-coverts) arc-like fringes. Back to upper tail-coverts dark bluish-grey, some central feathers and longer upper tail-coverts sometimes with rufous-brown or grey feather-tips. Chin pink-buff, chest and sides of breast up to neck-patches medium vinaceous-grey with slight brown tinge (uniform dull grey when worn), breast and belly light vinaceous or deep vinaceous-pink, vent greyish-pink to cream, under tail-coverts light bluish-grey. Tail as in *meena*, but feathers tipped and outer web of t6 margined light bluish-grey instead of white. Neck-patches, outer upper wing-coverts, flight-feathers, and underwing as in *meena*. In ♀, crown averages browner than in ♂, nape and mantle less vinaceous, chest duller grey, and rufous fringes on upperparts paler. Juvenile darker than *meena* and distinctly darker than *S. turtur*; crown, hindneck, and mantle darker brown-grey with faint rufous-buff feather-edges when fresh; feather centres of scapulars, tertials, and inner upper wing-coverts brownish-black rather than olive- or rufous-brown, rather narrowly fringed buff; grey of chest darker, narrow feather-edges deeper rufous, breast to vent extensively tinged cinnamon-buff; edges of flight-feathers and greater upper primary coverts deeper rufous, more sharply defined, primaries with less rufous wash to tips; under tail-coverts and tips of tail-feathers grey as in adult. Nominate *orientalis* and *meena* intergrade in hills between upper Ob and Yenisey rivers and in east Russian Altay. *S. o. agricola* of north-east India, Bangladesh, Burma, and Yunnan (China) rather more richly coloured with reddish-rusty on upperparts than in nominate *orientalis*, crown and underparts more uniform deep vinaceous, under tail-coverts and tail-tips slate grey, and wing on average 4–5 mm shorter; intergrades with *meena* in Nepal and with nominate *orientalis* in south-west China, eastern Himalaya, and northern Burma. *S. o. erythrocephala* of peninsular India similar to *agricola*, but even more rufous on upperparts, head and neck rich pinkish-rufous, chest deep rusty-pink, and wing of ♂ on average 189, ♀ 184 (Ali and Ripley 1969; Goodwin 1970); however, much overlap in size and colour with *agricola* and recognition perhaps not warranted (e.g. Vaurie 1961*b*). *S. o. stimpsoni* of Ryukyu Islands similar in colour to nominate *orientalis* and size similar, but underparts darker and more vinaceous (Hartert 1912–21; Vaurie 1961*b*). *S. o. orii* of Taiwan close to nominate *orientalis*, but duller and more plumbeous-grey or plumbeous-brown, underparts less vinaceous, and wing slightly shorter—average 192 in 9 adult ♂♂, 189 in 14 adult ♀♀.

S. orientalis forms species-group with Dusky Turtle Dove *S. lugens* and Adamawa Turtle Dove *S. hypopyrrha* of Africa and with *S. turtur* (see latter species). CSR

Streptopelia senegalensis Laughing Dove

PLATES 33 and 34
[facing pages 375 and 398]

Du. Palmtortel   Fr. Tourterelle maillée   Ge. Palmtaube
Ru. Малая горлица   Sp. Tórtola del Senegal   Sw. Palmduva

Columba senegalensis Linnaeus, 1766

Polytypic. Nominate *senegalensis* (Linnaeus, 1766), Africa south of Sahara (except for dry south-west), north to Mauritania, Ahaggar (Algeria), Tibesti (Chad), and southern Egypt; also western Arabia, Jordan, and Israel; *phoenicophila* Hartert, 1916, north-west Libya, Tunisia, and northern Algeria; also Turkey and northern Syria (probably introduced); *aegyptiaca* (Latham, 1790), Egypt south to Aswan; *cambayensis* (Gmelin, 1789), Iran, Pakistan, southern Afghanistan, and India, occasionally central and southern Iraq; *ermanni* (Bonaparte, 1856), Transcaspia (USSR) north to Aral Sea, east to eastern Kazakhstan and south to Sinkiang (China) and northern Afghanistan, straggling to European USSR. Extralimital: *socotrae* Grant, 1914, Socotra; *divergens* Clancey, 1970, dry south-west of Africa in Namibia, Botswana, and bordering districts of South Africa and Zimbabwe.

Field characters. 25–27 cm; wing-span 40–45 cm; tail 11·5 cm. Slighter and proportionately shorter winged and longer-tailed than Turtle Dove *S. turtur*. Small but proportionately large-headed, red-brown and blue-grey dove, with gorget of black spots and tri-coloured upperwing. Tail pattern recalls *S. turtur* but tips create broader band below. Flight fluttering. Sexes rather similar; no seasonal variation. Juvenile separable. 5 races in west Palearctic, 3 described here (see also Geographical Variation).

ADULT. (1) Middle East and Afrotropical race, nominate *senegalensis*. Head, neck, and underparts to belly rich vinous-buff, interrupted on glossy rufous-copper upper breast by at least 6 transverse lines of black flecks (actually the bases of breast feathers) which form obvious gorget. Back, scapulars, and lesser and inner median coverts red-brown or chestnut, contrasting with bright grey-blue panel on primary, outer median, and all greater coverts, which in turn contrasts with dusky flight-feathers. Thus upperwing shows bold, tri-coloured pattern, with almost blue central panel often shining noticeably. Underwing blue-grey. Lower back blue-grey; rump and tail dusky-grey, latter with all but central feathers broadly tipped white (particularly below). Black base to undertail contrasts with buff-white vent. ♀ usually paler than ♂, with less red tone in plumage. Bill black; eye brown, orbital ring pink-red. Legs pink-red. (2) Saharan race, *phoenicophila*. Slightly larger than nominate *senegalensis*; duller, with browner rump. (3) Egyptian race, *aegyptiaca*. Red-brown more intense than in nominate *senegalensis*. JUVENILE. All races. Much duller than adult, lacking gorget, but diagnostic wing pattern present.

Unmistakable. No other *Streptopelia* as small or as richly coloured. Flight typical of genus but actions rather more hesitant, with wings appearing short and rather blunt-ended and wing-beats weaker and distinctly fluttering. Gait and behaviour as *S. turtur*.

Noticeably tame. Most characteristic call a quickly uttered, rising and then falling, laughing multisyllable: 'p-oooo pe-poo-ooo pup oo'.

Habitat. In low or lower middle tropical and subtropical latitudes, but to mid-latitudes in central Asia, ranging there up to *c.* 1600 m (Dementiev and Gladkov 1951*b*). In India to *c.* 1500 m (Ali and Ripley 1969) and in East Africa to *c.* 2000 m (Williams 1963). Associated with cultivation, trees but not forests, and above all with human settlements, differing from Collared Dove *S. decaocto* in preferring them at their densest. In India, restricted to dry deciduous biotope, especially in semi-desert; inhabits village environs and stony scrub-and-bush country about cultivation; particularly where interspersed with *Euphorbia* hedges and *Opuntia* thickets. Gleans with *S. decaocto* and other doves in stubble fields. Nests in open bushes or date palms *Phoenix sylvestris*, and often in inhabited dwellings (Ali and Ripley 1969). In USSR, never found far from human habitation, preferring towns with old buildings such as mud huts with flat roofs to more modern structures, calling from a smoke stack, high pole, or dead treetop branch; occasionally nests in treetops or tangles of vines, but usually in man-made structure (Dementiev and Gladkov 1951*b*). In North Africa, favours oases (Etchécopar and Hüe 1967). In Algerian Sahara, endures summer heat in palm groves in oases better than most species, although introduced here by man (Dupuy 1969). In Somalia, frequents water-holes or trickling streams in otherwise dry riverbeds, first assembling in an *Acacia* bush or tamarisk. Also perches in trees during daytime heat, and nests in thorn-bushes rather than on buildings, but spends a good deal of time on ground. Inhabits interior plateau as well as coast. (Archer and Godman 1937.) In East Africa, frequents thornbush country and cultivated areas (Williams 1963) but in West Africa mainly restricted to villages, towns, and surrounding farmland, spending much time on ground feeding on fallen seeds (Serle *et al.* 1977). In southern Africa, habitat similar: found in almost every garden (Prozesky 1970). While virtually as urban as feral Rock Dove *Columba livia*, much less dependent on being fed by man or on industrial accumulations of waste grain; type of commensalism more similar to that of *S. decaocto*. Flights normally brief and in lower airspace.

Distribution. Populations in parts of Middle East and Turkey perhaps originally introduced by man (Goodwin 1970), as in Malta (where now extinct), Western Australia, and elsewhere (Long 1981). Marked expansion in Israel and slight recent spread in Algeria, but range contracted in Lebanon.

ALGERIA. Marked spread in last 25 years; breeding Algiers from 1979 (EDHJ). ISRAEL. Marked expansion in range; in 1930s breeding only in Jerusalem, Tel-Aviv, and Jaffa, with a few isolated colonies in south, but now widespread, and since 1970 has colonized all oases and settlements in desert areas (HM). LEBANON. Local. Formerly in Beirut, Chtaura, and perhaps elsewhere when possibly spreading, but now restricted to Beirut (Macfarlane 1978; HK). MALTA. Some released birds bred for a few years in 1940s, now extinct (Sultana and Gauci 1982).

Accidental. Finland, Italy, Greece, USSR, Iraq, Cyprus, Malta, Morocco.

Population. Increased in Israel and perhaps Tunisia.

ALGERIA. Numerous in oases (EDHJ). TUNISIA. Very common in towns, as well as agricultural areas and oases; may have increased (MS). LIBYA. Tripolitania: common in coastal belt (Bundy 1976). EGYPT. Abundant (PLM, WCM). ISRAEL. Marked increase, especially 1940–60 (HM). JORDAN. Scarce (DIMW). LEBANON. Common in Beirut (Macfarlane 1978). SYRIA. In some numbers in 4 large cities and towns (Kumerloeve 1968*b*; Macfarlane 1978).

Survival. South Africa: mean annual survival rate 44% (Dean 1977), but this a gross underestimate (Dean 1980); survival rate for first 3 months 89% (Dean 1980). Oldest ringed bird 5 years 9 months (Rydzewski 1978).

Movements. Basically resident throughout range in Africa, Near East, Middle East, Indian subregion, and Soviet central Asia, but also some movement of uncertain extent, that in west Palearctic probably at individual (rather than population) level.

In high-rainfall areas of tropical Africa, largest numbers present in dry season, while in arid areas numbers increase in short rainy season (e.g. Benson *et al.* 1971, Britton 1980). Abundant resident in Nigeria, and probably sedentary though numbers fluctuate seasonally in some areas (Elgood 1982). Mainly wet-season visitor to Ouadi Rimé-Ouadi Achim Reserve (central Chad), breeding August–October (Newby 1980). In Libyan Sahara, 160 passed through Serir (among big passage of Turtle Doves *S. turtur*) in April 1969, though none noted spring 1970 (Hogg 1974), and several seen in Fezzan in March–April 1981 (Cowan 1982); if identifications correct, suggest some hitherto unsuspected trans-desert movement.

In Mediterranean, has straggled to Malta (18 records, mostly October—see Sultana and Gauci 1982), Italy, Greece, and Cyprus; while one found on ship in Bay of Biscay, April 1966 (after storms), stayed aboard until ship reached Moroccan waters (Pitman 1967). Another taken on ship in Red Sea at 15°47′N 52°25′E, August 1963, was of Middle East race, nominate *senegalensis* (Bailey 1966*b*). Iranian–Indian race *cambayensis* has provided rare winter records from Iraq and Oman (Allouse 1953; Meinertzhagen 1954), a 200-km ringing movement in March from Indian Kutch to Hyderabad, Pakistan (Ali and Ripley 1969), and ship records for Arabian Sea off Karachi in March–April (Ticehurst 1923*b*; Moreau 1938). *S. s. ermanni* resident in Soviet Central Asia, without seasonal fluctuations in numbers to indicate even local movement,

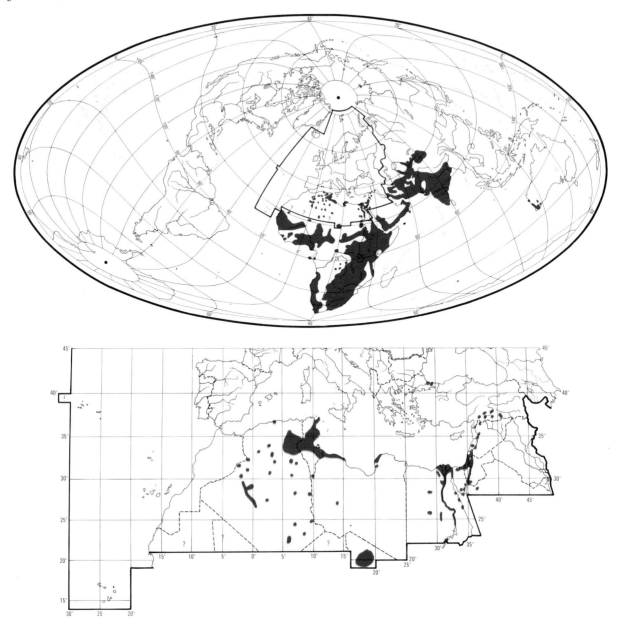

but straggled once (May 1884) to Orenburg, southern Urals (Dementiev and Gladkov 1951b).

Food. Mainly seeds, supplemented at times by insects and small snails. In Afrotropical region, forages on bare or sparsely covered ground, gravelly and sandy roads, ploughed land, etc. (Irving and Beesley 1976; Dean 1979a; D J Brooks). Feeds at grain spillages and in Australia will take scraps and garden seeds (Frith *et al.* 1976). Will stretch to pluck seeds from growing weeds up to 30–50 cm high (Ali and Ripley 1969).

At Barberspan (South Africa), takes seeds of crops including sunflower *Helianthus*, maize *Zea*, and wheat *Triticum*; also seeds of sedges and cultivated herbs, mainly those of 2 mm diameter or less, e.g. *Amaranthus*, *Cleome*, *Eleusine*, *Carex*, *Fibristylis*, and *Commelina*. Ripe and unripe grain taken. Roots and bulbs up to *c*. 6 mm diameter taken if found uprooted. Animal food (taken significantly more by ovulating ♀♀) includes larvae and pupae of house fly *Musca domestica*, harvester termite *Hodotermes mossambicus*, and small snails *Succinea striata* (Irving and Beesley 1976; Dean 1979a). In India, crop seeds also most important food, especially bajra *Pennisetum typhoides* (43% of annual consumption); also takes jowar *Sorghum vulgare* (2%), maize (0·6%), and insects (17%); grass and weed seeds taken less commonly (Rana 1976). In Perth (Australia), introduced urban population (157 birds studied) took mainly wheat (29·8% of items), bread (31·2%), flower seeds (13·4%), and grass seeds (7·1%) (Frith et al. 1976).

Young given crop milk initially; other food (seeds, also animal matter) added from c. 5 days (Irving and Beesley 1976; Kotov 1976c).
BDSS

Social pattern and behaviour. Little information from west Palearctic. Fullest studies on extralimital *ermanni* in Soviet Central Asia; see (e.g.) Annaeva (1965), Kekilova (1973), Mirzobokhodurov (1974), and, particularly, Kotov (1976c); for nominate *senegalensis* in southern Africa, see (e.g.) Robinson (1956). For study of captive birds, see Goodwin (1947c).

1. Often in small flocks though larger aggregations may occur where food plentiful, also for watering and roosting (e.g. Someren 1956, Annaeva 1965, Kotov 1976c, Gallagher and Woodcock 1980, W R J Dean). All-juvenile flocks (up to 30 birds) occur after breeding (Dementiev and Gladkov 1951b; Kotov 1976c). Sometimes associates with other Columbidae for watering and roosting (e.g. Morel 1975, Irving and Beesley 1976). BONDS. Monogamous mating system, with pair-bond lasting for several years; possibly life-long (Dharmakumarsinhji 1955; Annaeva 1965; Kotov 1976c). Said to remain in pairs throughout the year (e.g. Kekilova 1973, Mirzobokhodurov 1974). Age of first breeding not known. Both parents care for young up to and beyond fledging (Kotov 1976c; see Relations within Family Group). BREEDING DISPERSION. Usually solitary but several pairs sometimes nest close together: e.g. 2–3 in same tree, southern Africa (W R J Dean); 16 under one roof, USSR (Kotov 1976c). According to Dementiev and Gladkov (1951b), 'territory' embraces tree, roof, or yard where birds nesting. Observations by Kotov (1976c) suggest only immediate vicinity of nest defended, mainly by ♂; see also Irving and Beesley (1976), and Antagonistic Behaviour (below). For experimental study, see Stranger (1968). In Turkmeniya (USSR), 127 birds per km², Kushki valley; 175 per km², Murgab valley (Drozdov 1968). In Botswana, 0·40–0·77 pairs per ha (Irving and Beesley 1976). At Barberspan (South Africa), 21 of 51 nests in 4-ha woodland active simultaneously (Dean 1980). Same nest often used several times in a season or even in successive years (Kotov 1976c; see also Grimes 1972). May range up to c. 5 km from nest for communal foraging according to Irving and Beesley (1976), though said by Kotov (1976c) not to fly far during nestling period. ROOSTING. Nocturnal. Singly or in pairs, though sometimes communal, particularly outside breeding season (e.g. Irving and Beesley 1976). In Samarkand, autumn and winter, birds use secluded sites offering some protection from wind and cold: e.g. in lofts and ruins, less commonly in trees (Kotov 1976c; see also Mirzobokhodurov 1974). In Kazakhstan (USSR), some nocturnal roosting done in trees, but also under eaves of buildings (Gavrin et al. 1962). In Botswana, outside breeding season, rarely up to 20–30 in clump of trees with dense foliage; sites not used for more than a few days in succession (Irving and Beesley 1976). May perch high in trees for sunning prior to roosting (Someren 1956). In northern Tadzhikistan (USSR), starts feeding shortly before sunrise (Mirzobokhodurov 1974). When nest-building or rearing young (after brooding finished), pair roost nearby, ♂ also when ♀ incubating (Robinson 1956; Kotov 1976c); ♂ alleged to roost on edge of nest when ♀ incubating or brooding small young (Annaeva 1965), but this not likely to be usual (D Goodwin). Fledged young may return to nest to roost or sleep close to parents in tree (Kekilova 1973). Birds may call on moonlit nights (Dharmakumarsinhji 1955). Sites used for daytime loafing similar, usually close to feeding grounds or watering place (e.g. Kekilova 1973, Mirzobokhodurov 1974, Morel 1975); in southern Africa, also on ground in shade (Irving and Beesley 1976). In hot weather, Samarkand, mostly loafing or at water from c. 10.00–11.00 hrs to c. 17.00–18.00 hrs (Kotov 1976c; see also for details of diurnal rhythms throughout the year). Similar pattern, Botswana, where birds also forage at reduced rate during day, especially in breeding season. Usually well scattered (rarely more than 2 in same tree) for daytime loafing, though in Namibia c. 200 recorded with c. 30 Ring-necked Doves *S. capicola* in trees after drinking (Irving and Beesley 1976). See also Flock Behaviour (below).

2. Shows little fear of man where not much persecuted (Goodwin 1970); tame in some cities (e.g. Koenig 1896, Bates 1934b, Gavrin et al. 1962), though not so elsewhere (e.g. Someren 1956, Gallagher and Woodcock 1980). Generally flies up from ground almost perpendicularly and with Wing-clapping (Dharmakumarsinhji 1955). Tail-raising may be performed after alighting from Display-flight (Mundy and Cook 1974; see below). Head-shaking occurs (Goodwin 1956a). FLOCK BEHAVIOUR. In northern Sénégal, December, birds drink at puddles at any time of day though mostly around midday. May approach water by slipping through vegetation. Drinking apparently strongly contagious. At wells, several peaks during day; drinking often very rapid, birds staying only c. 4 s. In late December after cool night, behaviour as in African Collared Dove *S. roseogrisea*; see that species (Morel 1975). At Gobabeb (Namibia), birds drink mostly before 10.00 and after c. 16.00 hrs. Up to 100–200 gather in trees by water and many false starts made before first small groups fly down to land close to water; rapid drinking followed by return to trees so that steady 2-way traffic develops, even involving some aerial collisions (Cade 1965). In Botswana, birds usually perch in trees for several min (up to 30) before approaching (see Irving and Beesley 1976). For (experimental) study of feeding and drinking of flocks and alertness towards predators, see Siegfried and Underhill (1975). ANTAGONISTIC BEHAVIOUR. (1) General. Said to be pugnacious towards conspecific birds and the larger *S. capicola* (Winterbottom 1967). Advertising-call (see 1 in Voice) given up to pair-formation but also during nest-building and incubation (Dementiev and Gladkov 1951b); in Turkmeniya, more or less throughout year if winter mild (Kekilova 1973). Usually given from high perch in nesting area, mostly in morning and evening; in spring and autumn, and by unpaired birds, also quite often around midday. ♂'s crop region markedly swollen and nodding movements made in time with call (Kotov 1976c). Advertising-call sometimes used to signal aggression, e.g. in interspecific dispute (over food) with Red-eyed Dove *S. semitorquata* (M E W North). (2) Display-flight. ♂ makes rapid ascent (Dharmakumarsinhji 1955) with Wing-clapping to c. 15–20 m above trees or buildings, then gliding descent with wings and tail fully spread to c. 4–5 m (Kotov 1976c); wings bowed in glide (Gallagher and Woodcock 1980). Sometimes performed up to 5–6 times, resulting in undulating flight-path and gradual descent. Bird usually returns to starting point, or may fly to another perch up to 150 m away. Unpaired birds said to perform longer flights. Advertising-call normally uttered after landing. Display-flights frequent during nest-building (Kotov 1976c; also Mirzobokhodurov 1974). See also Heterosexual Behaviour (below). (3) Threat and fighting. ♂ usually quick to chase off any conspecific strangers approaching in flight; if such an intruder lands c. 2–3 m from nest, site-owner flies at it, dislodges it from perch, and may chase it for c. 50–70 m. Intruder may also elicit Bowing-display (as directed towards ♀: see Heterosexual Behaviour, below) from site-owner (Kotov 1976c). If intruder does not make appropriate response, site-owner lowers head and attacks, though unlike (e.g.) Turtle Dove *S. turtur*, does not give Excitement-call (Goodwin 1956b). Bird will sometimes leave eggs or young to

chase off intruder. Off-duty ♀ also defends nest, particularly around hatching and for first c. 5 days thereafter during which time ♂ also more aggressive. ♂ does not come to aid of ♀ in fight (nor vice versa), though he will chase away conspecific birds from vicinity of ♀ (Kotov 1976c). In Egypt, vigorous fights and aggressive chases noted (Koenig 1926); see also Koenig (1896) for comments on captive birds. HETEROSEXUAL BEHAVIOUR (1) General. In Soviet Central Asia, at least, various elements of courtship (including mating) occur also outside main breeding season (e.g. Annaeva 1965, Mirzobokhodurov 1974, Kotov 1976c; see also Antagonistic Behaviour). Older ♂♂ said to be less active in courtship, e.g. performing Display-flight only before 1st brood in early spring (Kotov 1976c). (2) Pair-bonding behaviour. ♀ perhaps first attracted to unpaired ♂ by his Display-flight (Dharmakumarsinhji 1955); ♂ thus displaying said to give louder Wing-clapping when ♀ near (Kotov 1976c). Display-flight performed several times while ♀ present and ♂ will follow her if she flies past (Dharmakumarsinhji 1955). At Sokoto (Nigeria), birds chased one another in rodent-like fashion over 1–2 m on ground, pursuer with feathers ruffled. ♂ also performed circular Display-flights in presence of ♀ (Mundy and Cook 1974). Closer approach by ♀ usually elicits Bowing-display in ♂ (Goodwin 1970; see also Koenig 1896): crop region swollen, fairly rapid bows executed with front part of body; at lowest point, bill only slightly below horizontal; plumage on front and sides of neck ruffled; Display-call uttered (see 3 in Voice). Mean time per complete bow 1·93 s ($n=13$) and 2·09 s ($n=5$) (see Davies 1970). Unresponsive ♀ chased away by ♂ (Kotov 1976c). (3) Driving. As in other west Palearctic Columbidae, ♂ Drives ♀ away from strange conspecific birds (Kotov 1976c). (4) Courtship-feeding (Billing). ♀ inserts bill into ♂'s and he makes slight head movement; food probably not passed (Goodwin 1970). As prelude to copulation, birds approached one another, and then rubbed and tapped bills for a few seconds (Mundy and Cook 1974). (5) Mating. Preceded by Mock-preening behind wing by one or both birds, and Billing (Goodwin 1970). May take place on roof, in tree, or on ground. ♂ said to pursue ♀, tail partly fanned and brushing ground (this doubted by D Goodwin); performs Bowing-display and gives Display-calls. ♀ runs away but intermittently turns to face ♂ and Nods several times very quickly. When ♂ at most active in pursuit, ♀ moves away, Mock-preening. ♂ finally crouches, calls (see 6a in Voice), spreads wings slightly and further invites copulation with slight Wing-twitching (movements Kotov 1976c). ♂ walks 'stiffly' round ♀ before mounting (Mundy and Cook 1974). After copulation ♂ adopts posture similar to ♂ *S. turtur* but less erect; ♀ Parades, neck and rump feathers ruffled (Goodwin 1970). ♂ said sometimes to move quickly past ♀ with wings and tail slightly spread, feathers on lower back ruffled (presumably Parading). He may also perform Bowing-display for c. 10–15 min. Birds then sit close together and mutually Allopreen head and neck. ♀ often again adopts pre-copulatory posture and calls; also Mock-preens, apparently in response to Display-call of ♂. During nest-building, birds mate 5–6 times per day, 2–3 times per day towards end of that period. ♂ quite often attempts to copulate with ♀ when she is incubating or feeding young (Kotov 1976c). Mating often takes place during evening loafing (Someren 1956; see Roosting); once on nest when ♀ building (Mundy and Cook 1974). (6) Nest-site selection. ♀ eventually chooses site offered by ♂ who gives Nest-call (see 2 in Voice) and Wing-twitches, but follows ♀ if she flies off, or he moves to another site for a repeat performance. ♀ reported also to give Nest-call to attract ♂, and 2 birds may even call from different sites simultaneously. ♂ flying to ♀ giving Nest-call settles by her, giving soft cooing sound (see 6a in Voice) and Allopreening her (Kotov 1976c; see also Goodwin 1946, Mirzobokhodurov 1974). (7) Behaviour at nest. Birds sit close together at nest-site; alternately utter soft cooing sound and Nod. ♂ then leaves to collect material (mostly nearby). With plumage slightly ruffled, ♀ Wing-twitches and periodically gives soft call (possibly Nest-call); such calls louder if ♂ slow to return. ♀ may reject some material brought by ♂. Most building done 08.00–11.00 hrs and after 16.00–18.00 hrs; takes c. 5–9 days, longer in early part of season (see Kotov 1976c), or only 1 day in some cases (Mirzobokhodurov 1974). ♂ sitting on uncompleted nest quivered tail and gave soft calls; once, when ♀ returned, ♂ Wing-twitched, then rubbed head against ♀'s throat; mutual head-rubbing then followed and ♀ eventually gently urged ♂'s departure. ♂ may continue to bring material after 1st egg laid (Robinson 1956; Mirzobokhodurov 1974). Incubation pattern as in other west Palearctic Columbidae, though change-overs more frequent in hot weather. Incoming ♂ perches on nest-rim, gives soft cooing sound, and Allopreens ♀'s head and neck; gradually moves on to nest, gently pushing ♀ off eggs. ♀ then flies off to feed and loaf and may spend some off-duty time on ground near nest. Departing ♂ gives Advertising-call before leaving and moves away (usually c. 10–15 m) in low-intensity Display-flight. ♀ said to approach unobtrusively for nest-relief (Kotov 1976c). In pair observed by Robinson (1956), nest-relief invariably accompanied by calls (see 6 in Voice) from one or both birds; incoming bird perched near nest then walked in, sometimes self-preening beforehand. For further details, see (e.g.) Mirzobokhodurov (1974). RELATIONS WITHIN FAMILY GROUP. Parent does not assist with hatching but feeds young for first time c. 1½–2 hrs afterwards. Fed by both parents, slightly more by ♂ in later period if subsequent brood started (Kotov 1976c). Captive ♂ fed young of 1st brood alone once incubation of 2nd clutch had started (Goodwin 1947c). At c. 3–5 days, young beg by giving weak food-call (see Voice) and stretching up to parent's bill (Ganguli 1975; Kotov 1976c). Parent stimulates response by touching base of nestling's bill (Ganguli 1975). At one nest, young sometimes nibbled at adult's breast feathers, while wing quivering of nestlings first noted at 7 days (Robinson 1956). Both parents may come to nest together, or one may wait nearby while other feeds young (Kotov 1976c). 2 young sometimes fed together, this taking up to c. 10 min; siblings often fight during feeding (Kotov 1976c). For prolonged conflict between young, apparently leading to death of both, see Walsh (1980). Also watered by parents; at least twice around midday when very hot (Kotov 1976c). Brooded almost continuously for first few days after hatching (W R J Dean). Normally require no daytime brooding after c. 6–7 days, but ♀ continues to brood at night (Kotov 1976c); nocturnal brooding ceases at c. 9–10 days (Robinson 1956; Mirzobokhodurov 1974). Young able to crawl around in nest at c. 5 days; thermoregulate and walk well at c. 8 days, and much wing-exercising thereafter. Young nestlings mostly sleep head to tail and parallel. Defecate in nest; when older, over edge (Kotov 1976c). Remain in nest for c. 11–14 days (W R J Dean) or 14–16 days (Kotov 1976c). Not able to fly strongly on leaving and remain in vicinity for c. 3–5 days longer; attempt to self-feed straight away, but generally successful only at c. 18–20 days, and some young may be fed by parents for c. 8–9 days after leaving nest (Kotov 1976c). Also brooded on ground or in tree; probably independent c. 7 days after leaving nest (W R J Dean). Fledged young frequently self-preen and siblings sometimes also mutually Allopreen (Kotov 1976c). In one case, both parents tended young for c. 3–4 days after fledging, but ♂, who had presumably resumed courtship of ♀, then attacked them viciously (Ganguli 1975). For exceptional case of ♀ laying in nest still containing young of 1st brood, see Bogdanov (1955). Independent young may form flocks (Dementiev and Gladkov 1951b; Kotov 1976c).

In case of last brood of season, family unit maintained for considerable time, birds staying in nesting area (Kotov 1976c). For further details see, especially, Mirzobokhodurov (1974). ANTI-PREDATOR RESPONSES OF YOUNG. Fall silent and crouch in nest if parent gives Warning-call (see 5 in Voice). At c. 6–7 days, crouch when alarmed but Bill-click also. May leave nest 1–2 days prematurely if disturbed; attempt to escape into cover (Kotov 1976c; see also Robinson 1956). PARENTAL ANTI-PREDATOR STRATEGIES. (1) Passive measures. Less timid at nest than S. turtur (Kotov 1976c); generally sits tight, showing little fear of man, and can sometimes even be seized on nest (Shnitnikov 1949; see also Goodwin 1947c for observations on captive birds). Sitting bird remains alert and, when alarmed, usually crouches in nest, head lowered and feathers sleeked (Robinson 1956; Kotov 1976c). Rarely disappears from view on leaving nest and normally returns after c. 7–10 min following disturbance (Kotov 1976c). (2) Active measures. Warning-call given at approach of danger (Kotov 1976c: see 5 in Voice). When captive birds approached closely, ♀ tended to leave nest, while ♂ would perform Defensive-threat display: plumage ruffled and tail spread, one wing raised and other used to strike (Goodwin 1947c). When man c. 2–3 m from nest, wild bird may fly up to c. 100–150 m and watch whilst raising and lowering tail agitatedly (Kotov 1976c). In one case, ♀ flew to ground and performed distraction-lure display of disablement type (trailing wings); returned to young after c. 15 min (Mirzobokhodurov 1974).   MGW

Voice. Generally much less vocal than Collared Dove S. decaocto, particularly in Display-flight (Kumerloeve 1959), though in Egypt, late October, birds gave 'flight song like S. decaocto' (F E Vuilleumier), but no further details; according to Chappuis (1974), no calls given in flight. Rather silent compared with other Columbidae of Sierra Leone (G D Field). Rhythm of calls 1–3 identical (Lade and Thorpe 1964, which see for additional sonagram). Wings said to make whistling sound at times in flight (Gallagher and Woodcock 1980); for Wing-clapping, see Social Pattern and Behaviour.
  CALLS OF ADULTS. (1) Advertising-call. Completely different from other Streptopelia (Goodwin 1970), especially in rhythm (Chappuis 1974). Recording (Fig I) suggest bubbling 'p-oooo pe-poo-ooo pup oo' lasting 1·04 s; distinct temporal pattern and considerable crescendo then decrescendo overall (J Hall-Craggs). Also described as a soft, rather musical phrase of 4–6 notes 'cŏŏ-cŏŏ cŏŏ-ōō', cŏŏrŏŏcŏŏ-cŏŏ-cŏŏcōō', or 'ŏŏ-gŏŏrŏŏrŏŏtōō' (Goodwin 1970); a hollow, rolling, laughing 'ha-ha-hoo-hoo hoo-hoo-hoo' or 'ha-ha hoo-ha' with variants (King 1978); varied but distinctive ending in rapid descending bubbling or laughing sound—'cu-coo-coo-cucuck', 'pu-poopoo-pupupooo', or 'poo poo poopidoo' (Gallagher and Woodcock 1980). In Sierra Leone, March, a 'huhu huhuhoo', or, less frequently, 'huhu huhuhuhoo' (G D Field). First 4 notes rapid, then 1–2 others long and lower pitched (Hue and Etchécopar 1970). Apparently also given by ♀ (Goodwin 1947c; see below), but no details. For further descriptions, see (e.g.) Lynes (1925a), Bannerman (1931a), Lowther (1949), Cave and Macdonald (1955), Someren (1956), and McLachlan and Liversidge (1970). (2) Nest-call. Similar to 1 (Goodwin 1970); perhaps generally softer (see Social Pattern and Behaviour). (3) Display-call. A hurried-sounding 'cōō-ōō-ŏŏ-cŏŏ-cŏŏ (Goodwin 1970); a 'coo-coorooroo' (Dharmakumarsinhji 1955); see also Koenig (1896) and Kotov (1976c). Indistinguishable from call 1 (M E W North). (4) Alarm and distress calls. A moaning gasp like Turtle Dove S. turtur (Goodwin 1947c). An almost hissing 'ker-r', perhaps containing alarm element, and possibly given by ♂ alone (Lowther 1949). The peevish 'ru-r-r' of Bannerman (1931a) presumably also belongs here. (5) Warning-call. A 'guuu guuu guuu' (Kotov 1976c). (6) Other calls. (a) A soft cooing sound uttered by ♀ adopting pre-copulatory posture; presumably the same given during various other heterosexual interactions (Kotov 1976c; see Social Pattern and Behaviour). (b) A deep cooing from one or both birds at nest-relief (Robinson 1956).
  CALLS OF YOUNG. Food-call a long-drawn peeping sound (Robinson 1956); clearly much as in other west Palearctic Columbidae. First 'squeaky' attempt to utter adult call 1 made by captive ♀ when in full post-juvenile moult (Goodwin 1947c). For Bill-clicking as anti-predator response, see Social Pattern and Behaviour.   MGW

Breeding. Little information from west Palearctic; the following based mainly on Dean (1980) and information supplied by W R J Dean for South Africa. SEASON. Tunisia and Algeria: late February to June, but occasionally starting December. Egypt: February–October (Heim de Balsac and Mayaud 1962). SITE. In tree, bush, or scrub at 1–5 m, also commonly on building—under eaves, on drainpipe, or in crack in wall. Nest re-used for successive broods, especially if previous brood successful. Of 302 sites, South Africa, 37% used twice, 15·2% used 3 times, 6–8% 4–5. Nest: frail platform of twigs, often leaf stalks, lined finer material. Building: by ♀, using material brought by ♂. EGGS. See Plate 95. Elliptical, smooth and fairly glossy; white. S. s. phoenicophila: 27 × 21 mm (26–29 × 19–22), n = 8; calculated weight 6 g. S. s. aegyptiaca: 26 × 20 mm (24–28 × 19–21), n = 40; calculated weight 5 g. (Schönwetter 1967). S. s. ermanni: 25–32 × 20–23 mm, weight 4·4–8·7 g, n = 157 (Kekilova 1973). Clutch: 2 (1–4). Clutches of 3–4 probably by 2 ♀♀ (Kotov 1976c). Of 620 clutches, South Africa: 1 egg, 4·5%, 2, 94·0%; 3, 1·3% 4, 0·2%; average 1·97. Multi-brooded; in South Africa, breeding occurs throughout the year, each pair making up to 5–6 attempts. Laying interval 24–48 hrs. INCUBATION. 12–14 days. By both sexes. Begins with or before last egg; hatching sometimes asynchronous. YOUNG. Semi-altricial and nidicolous. Cared for and fed

I M E W North Kenya December 1953

by both parents. Brooded almost continuously for first few days. May leave nest before fully fledged, and brooded and fed by parents on ground or in trees. FLEDGING TO MATURITY. Fledging period 14–17 days. Independent 4–7 days later. Age of first breeding not known. BREEDING SUCCESS. Of 84 eggs laid, south-west Turkmeniya (USSR), 66·6% hatched, 28·6% lost, 4·8% infertile; 73·2% of young fledged giving 48·8% overall fledging success (Kekilova 1973). Of 1220 eggs laid, South Africa, 573 (46·9%) hatched, and 436 young (35·7%) fledged. Predation of eggs and starvation of young main reasons for failure. 1-egg clutches produced average 0·43 young per egg; 2-egg clutches produced 0·36 young per egg; clutches of 3–4 produced no young. Overall success: 0·7 young per nesting attempt ($n = 619$), western Transvaal (Dean 1980); 0·88 young per nesting attempt ($n = 672$), Botswana (Irving and Beesley 1976).

Plumages (nominate *senegalensis*). ADULT. Forehead, crown, and hindneck dark mauve-pink, sides of head, upper sides of neck, and throat slightly paler mauve-pink, chin paler still, almost white; dull black feather-bases of hindneck may show when plumage worn. Feathers at lower sides of neck bifurcated, tips bright cinnamon-rufous, bases contrastingly black, forming large cinnamon-and-black dotted patches at sides of neck, narrowly joining at foreneck where sometimes mixed with mauve-pink without showing black. Mantle and scapulars red-brown or cinnamon-brown (deeper chestnut-red when fresh, more orange-cinnamon if worn), some olive-grey or pale olive-brown of feather-bases showing through, especially in worn plumage; upper mantle partly tinged mauve-pink. Back, rump, and upper tail-coverts medium blue-grey; longest upper tail-coverts and tips of variable number of feathers of back and rump olive-brown. Chest and sides of breast below cinnamon-and-black neck-patches dark mauve-pink, breast and belly gradually paler, lower belly and vent cream-pink or cream-white, under tail-coverts white. Flanks light bluish-grey. Central pair of tail-feathers (t1) olive-brown; t2 medium grey with slight blue tinge and olive-brown border at sides; t3 medium grey grading to light grey and greyish-white on tip with ill-defined blackish bar halfway along and slight olive-brown border to outer web; t4 with more contrast between dark grey and black basal half and light grey to white distal half; t5–t6 mainly or fully black on basal half and white on distal half, t6 also with white border along outer edge; from below, tail largely white, black basal half of feathers hidden by white under tail-coverts except when tail fully spread; tail from above with relatively darker central feathers but more extensively white lateral ones than (e.g.) Collared Dove *S. decaocto* and African Collared Dove *S. roseogrisea*. Primaries dark olive-grey, outer webs of outermost tinged silvery-grey and narrowly bordered white, outer webs of innermost partly tinged bluish-grey; greater upper primary coverts and longer bastard wing feathers black with slight grey tinge; remainder of upper-wing including secondaries, tertials, and lesser primary coverts light blue-grey, innermost wing-coverts and tertials tipped orange-cinnamon to rufous-chestnut (like scapulars) to variable extent, median and greater coverts narrowly bordered pale grey or grey-white. Under wing-coverts and axillaries light bluish-grey. Sexes mainly similar, but some ♀♀ less deep mauve-pink on head, with less black of feather-bases showing on lower sides of neck and none on foreneck; paler rufous or rufous-buff on mantle and scapulars with more olive-brown of feather-bases showing; rufous mantle more sharply demarcated from mauve-pink hindneck; underparts more extensively white or cream with paler pink and less mauvish chest and breast. NESTLING. Down sparse, hair-like; dull yellowish-buff or pale straw colour. JUVENILE. Crown, sides of head, neck, and mantle cinnamon-brown to bright rufous-cinnamon, with some grey of feather-bases visible; unlike adult, no mauve-pink tinge (except slightly on sides of neck) and no cinnamon-and-black dotted patches on neck. Scapulars rufous-brown (soon bleaching to sandy) with narrow pale buff edges to tips and much dull olive-grey on bases; tertials olive-brown or brown with faint buff edges. Underparts like adult, but chest and breast grey-brown to rufous-cinnamon (not bright mauve-pink like adult), fringed sandy-yellow when worn; belly more extensively cream or pale rufous. Back, rump, upper tail-coverts, and tail as adult, but tail-feathers differ in shape (see Structure). Primaries and greater upper primary coverts as in adult, but tips narrowly edged rufous and outer webs of primaries tinged rufous rather than grey, especially on innermost; secondaries light blue-grey with dull olive-brown tips and inner webs, tips narrowly edged rufous. Light blue-grey upper wing-coverts similar to those of adult, but slightly duller and less bluish towards tips, latter narrowly bordered off-white. FIRST ADULT. Not separable from adult when post-juvenile moult completed. In later stages of moult, when head, body, and wing-coverts in full adult, usually still recognizable by rufous edges to tips of juvenile outer primaries and (especially) outer greater primary coverts, by faint buff or rufous edges to tips of some old secondaries, and sometimes by retention of some shorter, narrower, and relatively more worn tail-feathers.

Bare parts. ADULT. Iris brown or dark brown. Narrow ring of bare skin round eye raspberry-red or grey. Bill slate-black or black. Leg and foot brick-red, deep red, raspberry-red, purple-red, or dull purplish-pink. NESTLING. Iris grey-brown. Bill pale flesh, shading to greyish-black before fledging. Leg and foot pinkish-grey. JUVENILE. Leg and food reddish-grey or dull red at fledging, gradually attaining adult colour during post-juvenile moult. Otherwise like adult. (Hartert 1912–21; Goodwin 1970; RMNH.)

Moults. ADULT POST-BREEDING. Complete; primaries descendant. In North Africa and Middle East, starts with p1 April–June, completing with regrowth of p10 October–November. Body and tail start from primary moult score 20–30, completing slightly before p10. In South Africa, primary moult takes 61–227 days, mainly (October–)November–May (southern summer and autumn); duration (computed from regression line) 120 days (Dean 1979b), from retraps *c.* 164 days (Siegfried 1971b). POST-JUVENILE. Complete; primaries descendant. Starts when *c.* 2 months old; head, body, and upper wing-coverts first, followed *c.* 2 weeks later by p1 (Hunter 1973; Dean 1979b). At primary moult score of *c.* 20, head, body, and wing-coverts largely in adult plumage and moult of tail and tertials starts. When primaries new (score 50), moult generally completed, but sometimes a few juvenile tail-feathers or secondaries (mainly s4–s6) retained. Moult may be halted at any stage under unfavourable conditions, e.g. 3 North African birds suspended primary moult during winter with scores 30, 35, and 40, while others had already completed. In those suspending, moult presumably resumed later; one bird from early breeding season had suspended or arrested with score 40 after earlier resuming from score 30 in late winter; birds like this may later show serially descendant moult, resuming post-juvenile moult on outer primaries while also starting 1st post-breeding on inner. In South Africa, no indication of suspended moult and only a few birds showed serially descendant moult (Siegfried 1971b).

Measurements. Nominate *senegalensis*. Eastern and southern Africa, all year; skins (BMNH, ZMA).

WING AD	♂ 139	(3·20; 15)	134–144	♀ 136	(2·70; 9) 131–139
TAIL AD	114	(4·37; 14)	109–124	109	(3·63; 9) 104–115
BILL AD	14·8	(0·94; 15)	12·8–16·3	14·9	(0·86; 9) 13·7–15·9
TARSUS	21·9	(0·76; 10)	21·0–23·3	20·8	(0·59; 5) 20·0–21·5
TOE	23·2	(1·33; 9)	22·0–25·9	23·1	(1·10; 5) 21·8–24·8

Sex differences significant, except bill and toe. Juvenile wing on average 5 mm shorter than adult, tail 9 mm.

Adult wing. Jordan and Israel: ♂♂ 140, 142, 143; ♀♀ 131, 139. Abu Simbel (southern Egypt): ♂♂ 138, 140, 144; ♀ 133. Tibesti (Chad): ♂ 144 (BMNH, RMNH).

S. s. phoenicophila. Algeria and Tunisia, all year; skins (BMNH, RMNH, ZMA).

WING AD	♂ 150	(2·83; 10)	147–156	♀ 147	(2·19; 6) 136–151
TAIL AD	128	(1·80; 4)	126–130	120	(— ; 2) 113–128

Turkey and northern Syria, February–May; skins (BMNH, RMNH).

WING AD	♂ 150	(4·34; 4)	143–153	♀ 151	(3·52; 4) 147–156

S. s. aegyptiaca. Nile delta and Dakhla oasis (Egypt), January–March; skins (BMNH).

WING AD	♂ 148	(1·61; 8)	145–150	♀ 142	(4·22; 6) 138–148

Weights. Nominate *senegalensis*. South Africa: (1) adult, September–April; (2) adult, May–August (southern winter) (Siegfried 1971a); (3) adult ♂, all year; (4) adult ♀, all year (Dean 1979b). Sexes combined, all year: (5) adult; (6) juvenile (Siegfried 1971a; Skead 1974; Day 1975; W R J Dean).

(1)	101 (9·83; 352) 72–139		(2)	106 (8·94; 240) 80–135
(3)	102 (6·98; 76) —		(4)	100 (9·25; 39) —
(5)	101 (9·20; 1157) 72–139		(6)	87·9 (12·6; 205) 61–134

Ennedi (Chad), April: ♀ 67 (Niethammer 1955). Nigeria, November–July: 93·2 (9·3; 5) 79–101 (Fry 1970). Kenya: ♂♂ 76, 92, 94; ♀♀ 80, 80 (Britton 1970).

S. s. ermanni. Kazakhstan (USSR): ♂ 110 (5) 104–121, ♀ 109 (5) 95–119 (Dementiev and Gladkov 1951b). On 1st day after hatching, 8·5 (0·38; 7), on 20th day 78·7 (0·49; 7), levelling off to c. 84 on 26th–30th day (Kotov 1976c).

S. s. cambayensis. Iran, May–June: ♂ 68 (Desfayes and Praz 1978). Western and south-west Afghanistan: late April, ♂ 75, ♀♀ 77, 78; late June, ♂ 76, ♀ 75 (Paludan 1959).

Structure. Wing short and broad, tip rather rounded. 11 primaries: in adult p8 longest, p10 5–12 shorter, p9 0–2 shorter, p7 2–5, p6 10–14, p5 19–23, p4 24–30, p1 36–46; in juvenile, p9 longest, p10 3–9 shorter, p8 0–2, p7 5–10; p11 minute, concealed by primary coverts. Outer web of (p7–)p8–p9 and inner web of (p9–)p10 emarginated. Tail rather long (see Structure of *S. roseogrisea*), 12 feathers; t6 16–32 shorter than t1–t2 in adult, 8–22 in juvenile. Tail-feathers rather wide and with broadly rounded tip in adult, narrower and shorter in juvenile; greatest width near tip of t1 20–28 mm and of t6 17–22 in adult, 19–24 and 15–20 respectively in 1st adult, 15–20 and 14–18 in juvenile. Bill short and slender, nail narrow, and leg and toes short and rather heavy, as in other Palearctic *Streptopelia*. Outer toe c. 73% of middle, inner c. 77%, hind c. 64%.

Geographical variation. 2 subspecies-groups: Asiatic *cambayensis* group with *cambayensis* and *ermanni* east from Iran, and nominate *senegalensis* group with remainder of races in Africa, western Arabia, Turkey, and Middle East. Groups differ distinctly in colour: nominate *senegalensis* group with broad rufous tips to olive-brown feathers of mantle and scapulars, and more extensive bluish-grey on back and rump; *cambayensis* group duller with all mantle, scapulars, tertials, back to upper tail-coverts, t1, and inner upper wing-coverts uniform olive-brown (without rufous), bluish-grey paler and restricted to outer upper wing-coverts, flanks, and sides of rump, and head and underparts paler, more pinkish, less vinaceous-mauve. However, some ♀♀ of North African and Middle East races *phoenicophila* and *aegyptiaca* as well as some of nominate *senegalensis* from Jordan, Israel, and Arabia almost or completely without rufous tips and closely similar to birds of *cambayensis* group. Within *cambayensis* group, *cambayensis* of southern Iran, southern Afghanistan, Pakistan, and India distinctly smaller, wing ♂ 130 (20) 124–135, ♀ 127 (20) 121–134; *ermanni* of USSR and western China larger, wing ♂ 142 (8) 139–146, ♀ 138 (9) 133–144; populations with intermediate size inhabit north-east Iran and northern and eastern Afghanistan (Vaurie 1965). In nominate *senegalensis* group, *socotrae* of Socotra pale and small, wing 127 (12) 122–133 (Hartert 1912–21; Vaurie 1965; BMNH); *divergens* of arid south-west Africa pale and large, wing 139 (20) 132–148 (Clancey 1970b). Nominate *senegalensis* brightly coloured but small; birds inhabiting East Africa, São Tomé, and area in and around central and West African forest block smaller, wing 135 (10) 128–144, those of southern Africa and area from Sénégal to Ethiopia, western Arabia, Jordan, and Israel slightly larger, wing 138 (15) 133–146; also, birds from Sierra Leone east to Ethiopia and south to Cape Province (except arid south-west of Africa) brighter than topotypical nominate *senegalensis* from Sénégal east to Sudan and Eritrea, and perhaps separable as *aequatorialis* (Erlanger 1904); those of western Arabia, Jordan, and Israel often duller (♂♂ with more olive-brown of feather-bases showing, ♀♀ often largely dull olive-brown above), but included in nominate *senegalensis* as size close to that race. *S. s. aegyptiaca* from Nile valley and delta south to Aswan, Suez Canal area, and Dakhla oasis slightly more brightly coloured than nominate *senegalensis*: head, neck, and chest brighter vinaceous-pink; mantle, scapulars, and tertials bright rufous-red without olive-brown of feather-bases showing; some ♂♂ and many ♀♀ hardly show any rufous to feather-tips on upperparts, however, and are markedly more uniform olive-brown than typical nominate *senegalensis*; size distinctly larger (see Measurements). *S. s. aegyptiaca* intergrades with nominate *senegalensis* in Egyptian Nile valley between Aswan and Abu Simbel. *S. s. phoenicophila* of east-central Algeria, Tunisia, and north-west Libya like nominate *senegalensis*, but mauve-pink of head, neck, and underparts slightly darker, more vinaceous; feather-tips of mantle, scapulars, tertials, and inner upper wing-coverts slightly duller and darker rufous, and much olive-brown of feather-bases showing; central back and rump, upper tail-coverts, and t1 more extensively tinged olive-brown (extent almost as in *cambayensis* group, but remainder of upperparts with rufous feather-tips, not uniform olive-brown as in that group); distinctly larger than nominate *senegalensis* and slightly larger than *aegyptiaca*, but less extensively rufous above than latter; some ♀♀ hardly separable from those of *aegyptiaca*. Birds from Turkey and northern Syria (probably introduced) completely similar to *phoenicophila*; not known whether birds of southern Syria and Lebanon are like north Syrian *phoenicophila* or like nominate *senegalensis* of Jordan and Israel.

CSR

Oena capensis Namaqua Dove

PLATES 32 and 34
[facing pages 374 and 398]

Du. Maskerduifje   Fr. Tourterelle à masque de fer   Ge. Kaptäubchen
Ru. Длиннохвостая горлица   Sp. Tórtola de Cabo   Sw. Kapduva

Columba capensis Linnaeus, 1766

Polytypic. Nominate *capensis* (Linnaeus, 1766), Africa and Middle East. Extralimital: *aliena* Bangs, 1918, Madagascar.

Field characters. 26–28 cm; wing-span 28–33 cm; tail 11–12 cm. Much smaller even than Laughing Dove *Streptopelia senegalensis*, with body size close to that of Skylark *Alauda arvensis* but diamond-shaped tail as long again. Tiny, long-and-narrow-tailed, grey, black, and white dove with markedly rufous underwing. Sexes dissimilar; no seasonal variation. Juvenile distinct.

ADULT MALE. Front of head, foreneck, and centre of breast velvet-black, contrasting with pure white underparts. Crown, sides of neck, and breast grey. Hindneck, back, scapulars, and rump ashy-brown, with 2 black bars across rump narrowly divided by grey-white. Tail grey-black, with small white tips to shorter, graduated outer feathers obvious only from below. Wing-coverts grey-brown, with dark violet-blue spots on 4 inner greater coverts and nearby tertials. Flight-feathers dusky, with black tips to primaries and rufous inner webs; latter create almost wholly rufous-chestnut underwing, which contrasts with black axillaries and white body. Bill yellow, with purple-red base; eye appears black. Legs and feet purple-red. ADULT FEMALE. As ♂ but for lack of black mask and breast, these areas being dull grey. JUVENILE. Resembles adult ♀ but feathers of upperparts subterminally barred black and broadly tipped buff-white.

Unmistakable. No other dove in west Palearctic as small or as attenuated; black, graduated tail diagnostic. Flight fast, direct, and normally close to ground, with wing-beats recalling small wader *Calidris*; lands with short flutter followed by raising and fanning of tail. Gait a mincing walk. Difficult to see on ground, with plumage often cryptic.

Usually silent but utters weak, low-pitched coo.

Habitat. In tropical and subtropical low latitudes, in open arid and semi-desert country, avoiding forest but inhabiting equatorial coastal savannas (Serle *et al.* 1977). In East Africa, also in thorn-bush country, especially in sandy areas (Williams 1963). Feeds and walks much on ground, and perches low, on fences or bushes, congregating at drinking places during the heat of the day in drier regions (Prozesky 1970). Breeds in gardens in Khartoum (Mackworth-Praed and Grant 1952). Flies strongly but mainly for short distances in lower airspace.

Distribution. ISRAEL. First recorded 1961. Now apparently breeds regularly in Arava valley, probably due to increase of irrigated agriculture there; juveniles collected, but no nests yet reported (HM). ALGERIA. Recently discovered breeding in far south; probably present since 1950s (EDHJ). 2 birds collected 1952, Tamanrasset (♂ 23 April, ♀ 27 April), both with enlarged gonads, but no proof of breeding (C S Roselaar).

Accidental. Morocco, Mauritania, Egypt, Jordan, Kuwait.

Population. ISRAEL. Not common (HM).

Movements. Partial rains migrant in sub-Saharan zones; otherwise resident, but subject to local movements.

Present all year in Chad and Mali, though in latter leaves delta area in wet season (Lamarche 1980). A well-defined rains migrant in Nigeria, where resident in Sahel zone but in dry season (September–April) spreads south to great rivers (sometimes as far as Ilorin and Gboko) and breeds there November–February (Elgood *et al.* 1973; Elgood 1982). Peak numbers pass through Borgu in October and February/early March, but timing variable, with (at Zaria) first arrivals in different years spanning 2 October–29 November, and final departures mid-April to early May. Also dry-season migrant to Gambia, late October to June (Gore 1981); late October and November records from northern Togo (*Malimbus* 1980, **2**, 115), and Sierra Leone stragglers occurred November–January (G D Field).

Local movements occur in East Africa, including concentrations at seed crops (Britton 1980), and in Zambia commoner in dry season than during rains (Benson *et al.* 1971); extent of movement unknown. Little information from Arabia, where capacity for dispersal shown by colonization of newly cultivated areas (e.g. Jennings 1981a), including Eastern Province of Saudi Arabia and Kuwait (Bundy and Warr 1980). One on ship in Red Sea, 16°N, in September 1947 (Elliott and Monk 1952).

Food. Data only from extralimital populations and few specific identifications. Almost entirely seeds; occasionally insects. Feeds on ground. Rather active when feeding, walking and running with rapid little strides; adopts hunched posture resembling rat *Rattus* or squirrel *Sciurus* from distance (Reichenow 1900–1; Archer and Godman 1937).

Where available, cereals form bulk of diet in central Chad (Newby 1980); small grains such as millet *Panicum miliaceum*, teff *Poa abyssinica*, munga *Pannisetum typhoideum*, and *Eleusine* have been noted (Bannerman 1931; Priest 1934; Chapin 1939; Bouet 1955; McLachlan and Liversidge 1970). Small wind-borne grass and weed seeds such as *Panicum laetum* and *Eragrostis pilosa* (Newby 1980)

PLATE 31. *Streptopelia roseogrisea* African Collared Dove (p. 336): **1** ad, **2** juv moulting into 1st ad (1st autumn), **3** downy young, **4** ad Barbary Dove *S. 'risoria'*, **5** juv Barbary Dove. *Streptopelia decaocto* Collared Dove (p. 340): **6** ad ♂, **7** juv moulting into 1st ad (1st autumn), **8** downy young. (CJFC)

PLATE 32. *Oena capensis* Namaqua Dove (p. 374): **1** ad ♂, **2** ad ♀, **3** ♀ with a few juv primaries and greater coverts, **4** juv, **5** downy young. (CJFC)

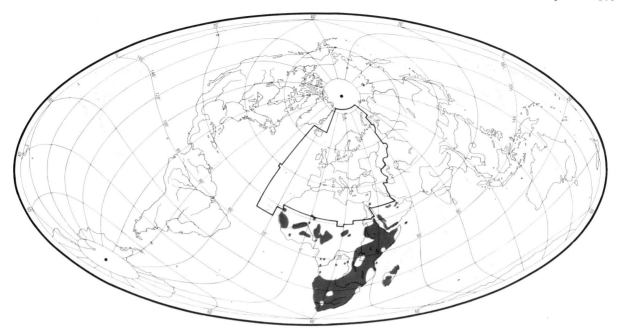

are main food in most of its range. 3 museum specimens, and birds shot in Tanzania, contained only seeds in crops (Angst 1975). Insects picked from cattle- and horse-dung (see Reichenow 1900–1); Steinbacher (1965) also mentioned insects in diet. In captivity, preferred food is very small seeds such as millet, and artificial soft-food mixtures; virtually no green matter taken (Angst 1975).

Nestlings fed crop milk for first few days, seeds thereafter (W R J Dean). HAR

Social pattern and behaviour. Little known, especially for west Palearctic.

1. Often gregarious outside breeding season, occurring in small groups or loose flocks (Heuglin 1869; Angst 1975; D J Brooks). Hundreds or thousands sometimes gather in quite a small area (I J Ferguson-Lees), notably at water-holes (see Roosting, below). At other times in pairs or family parties (Reichenow 1900–1). From November to June, Sénégal, occurs in pairs or, early in breeding season, in small parties, often 1 ♂ with 2–3 ♀♀ (Bannerman 1931). In breeding season, Sénégal, midday drinking flocks thought to consist largely of ♀♀ since ♂♂ incubating then. In drinking flocks, tolerates close company of Laughing Dove *Streptopelia senegalensis* but avoids African Collared Dove *S. roseogrisea* (Morel 1975). Feeding flocks in Namibia associated (on ground and in flight) with finch-larks *Eremopterix* and Red-billed Quelea *Quelea quelea* (Hoesch and Niethammer 1940). BONDS. No information on mating system. Both members of pair help to build nest (Someren 1956), incubate (Priest 1934;

PLATE 33 (*facing*).
Streptopelia turtur Turtle Dove (p. 353). Nominate *turtur*: **1** ad, **2** juv, **3** downy young. *S. t. rufescens*: **4** ad.
Streptopelia orientalis Rufous Turtle Dove (p. 363). Nominate *orientalis*: **5** ad, **6** juv. *S. o. meena*: **7** juv.
Streptopelia senegalensis senegalensis Laughing Dove (p. 366): **8** ad, **9** juv, **10** downy young. (CJFC)

Vincent 1946a; Someren 1956; Hoffmann 1969; McLachlan and Liversidge 1970), and feed young (W R J Dean). In captive birds, young closely attached to parents for *c.* 1 month after fledging (Angst 1975; see Roosting, below). BREEDING DISPERSION. Apparently varies from solitary to colonial; for colonial breeding in acacias, Aden, see Meinertzhagen (1954). At Barberspan (South Africa), 5 pairs in 2 ha; nearest horizontal distance between neighbours 15 m (Hoffmann 1969). ROOSTING. In trees and bushes at night; birds loaf there by day (Heuglin 1869). For ♂ 'dozing' on nest, see Someren (1956). By day, off-duty bird (mostly ♀) frequently sits several metres from nest (Hoffmann 1969). Around settlements, often settles on fences to preen between feeding bouts (Reichenow 1900–1). From *c.* 2 weeks after fledging, when powers of flight fully exercised, young of captive pair flew up to roost close by them (Angst 1975). Well known for marked activity in heat of day. Fond of sunbathing (Angst 1975); water-holes visited primarily in middle of day (Lynes 1925a; Priest 1934; Meinertzhagen 1954; Cade 1965), though morning and evening visits may be marked (Someren 1956; Newby 1980). During dry season, Sénégal, often sits around drinking-site for hours, loafing and occasionally preening (Morel 1975, which see for details of drinking behaviour).

2. Often said to be shy and nervous (Someren 1958; Angst 1975; see also Bannerman 1931), but may be tame around settlements (Reichenow 1900–1; Vincent 1946a). Birds landing at edge of river-bed sat still for a few minutes, then flew quickly and low before dropping into water-hole to drink; stayed only a few seconds (Cade 1965). For 'feigning death' when handled, see Meinertzhagen (1954). On landing, bird raises tail, fanning it, then lowers it slowly (Gifford 1941; McLachlan and Liversidge 1970; Angst 1975). FLOCK BEHAVIOUR. No details. ANTAGONISTIC BEHAVIOUR. ♂♂ said to be very jealous and constantly fighting (Heuglin 1869). HETEROSEXUAL BEHAVIOUR. Little known. In Arabia, courtship occurs throughout the year (Meinertzhagen 1954). (1) Pair-bonding behaviour. No information on sequence of events, though involves aerial and ground displays. During courtship, ♂ follows ♀ and makes small circular Display-flights, lifting tail on landing (Mundy and Cook 1974). Aerial behaviour

described as quite butterfly-like. ♂ displaying to ♀ hovered and descended with tail fanned (Gifford 1941). In Bowing-display (Gifford 1941; see also Goodwin 1970), ♂ lowers head, raises tail (not fanned) and moves both wings silently and rhythmically at c. 100 vibrations per min; meanwhile gives a cooing call (Gifford 1941: see 1 in Voice). In captive pair, ♂ started displaying again 8 days after young hatched; apparently normal in advance of further clutch (Angst 1975). High frequency of display by ♂ said to be typical (Archer and Godman 1937). (2) Courtship-feeding and mating. Before mating, ♂ feeds ♀ several times, then reaches over and presses his neck on hers. ♀ does likewise and, after this has been repeated a few times, solicits with raised tail. ♂ then mounts and copulates (Raethel et al. 1976). Once, ♂ suddenly flew up and copulated with ♀ while she was nest-building (Mundy and Cook 1974). (3) Behaviour at nest. Nest-relief occurs in silence with no special ceremony (Someren 1956). In one instance, ♂ arrived and settled on twig near nest; ♀ flew away immediately and ♂ settled on eggs. Frequently, off-duty bird sat near nest for c. 10–20 min before taking over (Hoffmann 1969). In another case, when ♀ flew over nest, incubating ♂ looked up and left nest; in a short time, ♀ dropped to ground behind nest; ♂ flew down and Billed gently with her as if encouraging her to go to nest, but she was reluctant to move; ♂ eventually flew off to preen, but after nest exposed for c. 20 min, returned to incubate and ♀ flew away; after c. 1 hr, ♂ stretched wings over head, stood up, and flew off, leaving nest unattended; ♀ found sitting later same day (Someren 1956). RELATIONS WITHIN FAMILY GROUP. After fledging, captive young joined parents on perch for sunbathing, roosting, etc. (Angst 1975). ♀ said to entice fledged young back to nest by calling (e.g. Musil 1963, 1965), but this not found by Angst (1975). ANTI-PREDATOR RESPONSES OF YOUNG. No details. PARENTAL ANTI-PREDATOR STRATEGIES. (1) Passive measures. Shy at nest. Readily flushes from nest on approach of human intruder, and stays away if intruder remains nearby (Someren 1958; Hoffmann 1969). (2) Active measures: against man. When human intruder near nest, bird may perform distraction-lure display, apparently of disablement type. On flushing from nest, often flies to nearby perch and stays for a short time fluttering wings, thereafter flying a little higher to watch intruder (Vincent 1946a). ♂, flushed from nest, with 2 small young, fluttered around, raising and lowering wings in curious shaking manner (see Priest 1934). 'Wing-beating' by perched ♀ (breeding in captivity), when observer tried to put young back in nest (Angst 1975) probably the same or a related display.   EKD

Voice. Mostly silent (Someren 1956; Williams and Arlott 1980).

CALLS OF ADULTS. (1) Advertising-call. A rather deep, low-pitched coo; usually double, sometimes single, though not known if these are different calls or variants (Goodwin 1970; Meinertzhagen 1954). A deep, mournful, booming, ventriloquial 'ho ho-ho-ho' or 'hu-hoo' (Gallagher and Woodcock 1980). A low 'twooh hoooo', 1st unit 'explosive' (McLachlan and Liversidge 1970). Other descriptions emphasize subdued quality: low, unobtrusive sounds like feeble halted beginning of call of Blue-spotted Wood Dove *Turtur afer*—a feeble cooing (Bates 1934b); a low, subdued 'hoo hoo' (Serle et al. 1977); a weak 'koo koo' (Williams 1963); a quiet 'kook-koor-ru' by ♂, ♀ often answering in like manner (Priest 1934). In recording by W J Onderstall/ Transvaal Museum, a mournful 'po-oowa' (P J Sellar) or 'hu huuoooooooooooooer' (J Hall-Craggs), 1st unit the more emphatic; drops in pitch during 'oo', rising again at 'wa'; calls each c. 1·5 s long, repeated regularly at intervals of c. 1 s (E K Dunn). In Bowing-display, gives undescribed cooing call (Display-call), perhaps different from call 1. (2) Nest-call. Single, deep gruff note by captive ♂ (see Goodwin 1970); presumably site-advertisement call, as in other Columbidae. (3) Other calls. In feeding flock, gives quiet call, described as 'pretty little conversational note' (Priest 1934). Parents, perhaps only ♀, use undescribed call to lure young towards (possibly also away from) nest (see Social Pattern and Behaviour).

CALLS OF YOUNG. No information.   EKD

Breeding. SEASON. No information from west Palearctic. Somalia: eggs laid mid-April to early June (Archer and Godman 1937). Northern Mauritania: laying begins April (Heim de Balsac and Mayaud 1962). SITE. On horizontal fork of low bush or tree, or on dry twigs of fallen trees; 0·3–2 m above ground (McLachlan and Liversidge 1970); also on top of low mound or anthill (Someren 1956). Nest: fragile platform of twigs (Archer and Godman 1937); rather solid for a dove, of twigs lined with fine rootlets and grass, etc. (Mackworth-Praed and Grant 1952; McLachlan and Liversidge 1970); c. 10 cm in diameter (W R J Dean); 2 nests 8·5 cm and 5·5 cm in diameter (Serle 1943). Building: ♂ takes large part (Heim de Balsac and Mayaud 1962). EGGS. See Plate 95. Oval, smooth and slightly glossy; cream or pale yellow. 21 × 16 mm (19–23 × 14–18), $n = 34$; calculated weight 3 g (Schönwetter 1967). Clutch: 2–3. Of 93 clutches, South Africa: 2 eggs, 92; 3, 1; average 2·01 (McLachlan and Liversidge 1970). 2–3 broods (Priest 1934; Angst 1975). Laying interval 24 hrs. INCUBATION. 13–16 days. By both sexes; ♂ by day (on average, 09.30–15.30 hrs), ♀ by night (Hoffmann 1969). YOUNG. Semi-altricial and nidicolous. Cared for and fed by both parents on crop milk for first few days, then seeds (McLachlan and Liverside 1970). FLEDGING TO MATURITY. Fledging period 11–12(–16) days. No further information.

Plumages (nominate *capensis* from Ahaggar, Algeria). ADULT MALE. Mask covering forehead, lores back to middle of eye, chin, throat, and chest black. Crown, sides of head behind eye, sides of neck, and sides of breast pale bluish-grey, gradually merging into dull greyish-drab of hindneck, mantle, scapulars, tertials, back, rump, and inner upper wing-coverts. Back marked with 2 parallel black-brown bars, separated by pale greyish-drab to off-white bar; innermost tertials with dull drab streak along inner webs; middle portions of outer webs of a number of tertials and greater and median tertial coverts with large rounded iridescent spots, glossed purple-red, magenta, purple-blue, or blue-green, varying with light and from bird to bird; number of spots rather variable and occasionally largely hidden by other wing-coverts or scapulars. Upper tail-coverts light grey, distal inner webs and tips broadly bordered black. Upper flanks pale grey; under tail-coverts black with partial pale grey or white outer webs; remainder of underparts below black chest white (occasionally stained buff or rusty from soil). Central pair of tail-feathers blackish olive-brown with variable amount of light grey bloom at base, others black with light grey base, (t3–)t4–t6 with light grey to white fringe on tip; basal outer web of t6 white; from below, tail black except for pale fringes to tips of outer feathers and for partially white outer web of t6. Rather broad tips and more narrow outer edges of p1–p9 and broad tip to p10 dark sepia to dull black, remainder of primaries contrastingly bright rufous-chestnut; tips of fresh primaries narrowly edged white, black outer edge faintly glossed blue. Broad tip and broad outer edge of secondaries dull greyish-drab, centre and inner portion chestnut on outer secondaries, dull black on inner; tips of inner secondaries paler grey. Greater primary coverts like primaries; remainder of upper wing-coverts pale bluish-grey with dusky streak along outer edge of greater (for drab innermost coverts and iridescent spots on tertial coverts, see above). Axillaries black, under wing-coverts bright rufous-chestnut, whole underwing appearing rufous except for dusky flight-feather tips and innermost secondaries. In worn plumage, crown and sides of head and neck tinged drab; upperparts and tertials duller drab, less greyish olive-drab. ADULT FEMALE. Forehead and chin buff-white, grading to buff-brown on forecrown, sides of head, and throat; narrow black line on lores. Entire upperparts, including inner upper wing-coverts, tertials, and upper tail-coverts warm buff-brown with 2 parallel dull black bars across back; dull black streak along inner tertials; black tips to upper tail-coverts; mantle, scapulars, and upper tail-coverts with slight olive-grey tinge, inner tertials and tertial coverts with iridescent spots similar to adult ♂, but usually fewer. Throat and chest warm buff-brown, restricted grey of feather-bases not visible. Remainder of underparts, tail, and all wing similar to adult ♂, but t1 tinged fawn rather than grey. Some ♀♀ (e.g. from Sudan and Ethiopia) have face whiter, forecrown pale grey, upperparts greyish olive-drab similar to ♂, and throat to chest uniform pale bluish-grey or bluish-grey with narrow buff feather-fringes. NESTLING. Down sparse and whitish (Harrison 1975). JUVENILE. Rather like adult ♀, but extensively marked with black. Forehead, sides of head, and throat pale grey-buff, each feather with faint dusky grey subterminal bar and narrow buff tip. Feathers of crown pale grey with buff tip and distinct black subterminal bar. Hindneck and sides of neck grey with faint buff feather-tips. Mantle, scapulars, and tertials grey with broad black subterminal bar and large golden-buff diamond-shaped spot on tip, tip narrowly fringed white terminally; on some longer feathers, traces of 2nd black and buff bar on middle portion; grey feather-bases largely hidden, except on tertials. Back and rump marked black and buff; 2 broad parallel black bars across back separated by grey bar. Upper tail-coverts as adult, but tip fringed buff and black less clear-cut. Throat and chest pale buff-grey with broad black feather-tips; tips with golden-buff mark, chest appearing black with buff spangles. Remainder of underparts and tail as adult, but under tail-coverts and tail-feathers tipped buff. Flight-feathers, primary coverts, and underwing as adult, but tips of primaries with rufous spot and pale buff fringe. Lesser upper wing-coverts boldly patterned, like mantle; inner median and greater coverts pale grey with ill-defined black subterminal bar and off-white and buff tip, outer black with restricted grey at base; tip off-white. FIRST ADULT. As adult when last distinctly marked juvenile feathers lost. Longest-retained juvenile feathers are outer primaries and greater primary coverts, both with rufous spot or mottling and traces of sharp pale fringes on tips, and middle secondaries, with indistinct dark subterminal bar and rufous and white mottled tip.

Bare parts. ADULT. Iris dark brown. Bare skin round eye purplish-grey or grey. In ♂, basal half of bill purplish-red, scarlet, pink-red, or orange-red, distal half orange, orange-yellow, or yellow; in ♀, bill purplish-black, dull carmine, or red-brown. Leg and foot purple, purplish-red, carmine, brick-red, or pink-red, in ♀ sometimes brown-red. NESTLING. No information. JUVENILE. At fledging, iris grey-brown, bill pinkish-grey with white tip, leg and foot greyish-pink; adult colours probably attained during post-juvenile moult. (Goodwin 1970; RMNH, ZMA.)

Moults. ADULT POST-BREEDING. Complete, primaries descendant. Several birds in moult from northern Afrotropics examined, but these mostly from different localities for which optimum breeding season (if any) not known, and hence duration of moult and relation between breeding cycle and timing of moult difficult to establish. Suspended primary moult frequent, with all scores between 10 and 40 found; as body plumage of birds showing this can be fresh or worn, suspension probably occurs during nesting, when wear generally strong. Several birds had single feathers new (sometimes asymmetrically), e.g. p7, p8, or p9, and remainder worn: probably, moult resumed after long suspension during nesting and halted again. Others had (e.g.) 4 outermost primaries new (or partly moulting) and inner old. Birds in active moult rarely showed more than 1 growing feather at a time, indicating long duration of primary replacement; these birds generally had body worn if primary score up to (10–)20–30 and fresh from 40–45; tail moulted at same time (sequence approximately t1–3–6–2–4–5, but much variation). POST-JUVENILE. Complete; primaries descendant. No data on timing (in other Columbidae, starts shortly after fledging). Part of mantle, scapulars, and median upper wing-coverts replaced first, soon followed by p1. On reaching primary moult score of 20–25, plumage mainly adult, but all greater upper wing-coverts and secondaries, some tertials and lesser coverts, and part of tail still juvenile. At score 35–40, only outer primaries, middle secondaries, and occasionally some tail-feathers juvenile, head and body often already worn adult (perhaps caused by nesting at this stage). When primary moult completed, occasionally some juvenile middle secondaries remain, mainly s3–s7 or s4–s6.

Measurements. Nominate *capensis*. West Africa, Sudan, Ethiopia, and Tanzania; all year; skins (RMNH, ZMA).

WING AD	♂ 108	(2·12; 17)	106–113	♀ 105 (1·97; 13)	103–109
TAIL AD	134	(7·26; 15)	120–147	128 (4·20; 9)	122–137
BILL AD	14·0	(0·68; 19)	12·7–15·1	13·6 (0·31; 12)	13·2–14·2
TARSUS	14·8	(0·48; 10)	14·0–15·5	14·8 (0·41; 10)	14·2–15·5
TOE	16·5	(0·47; 10)	15·9–17·2	15·6 (0·95; 10)	14·2–16·5

Sex differences significant for wing and toe. Juvenile wing on average 3½ mm shorter than adult, juvenile tail 13; juvenile bill similar to adult once face feathered. Single adult ♂ from Ahaggar (Algeria) had wing 112, tail 143; adult ♀, wing 108, tail 148 (ZMA).

Weights. Nominate *capensis*. South Africa, all year: adult ♂ 40·7 (3·35; 75) 35–47; adult ♀ 40·4 (3·75; 63) 32–54; juvenile, sexes combined, 39·0 (2·99; 22) 31–45 (Liversidge 1968; Skead 1974, 1977; Day 1975). East Africa, all year: ♂ 37·6 (7) 33–42, ♀ 35·0 (8) 32–39 (Britton 1970; Colston 1971). Nigeria, December–January: ♂ 40·9 (3·8; 5) 38–48, ♀ 36·1 (2·0; 6) 33–38 (Fry 1970). Chad: ♂♂ 32, 32, 39, 43 (Niethammer 1955).

O. c. aliena. Adults, Madagascar, June: ♂ 40, ♀ 50 (ZMA).

Structure. Wing short and rather broad, tip rounded. 11 primaries: in adult, p9 longest, p10 1–5 shorter, p8 0–2, p7 7–10, p6 14–17, p5 19–22, p1 34–42; in juvenile, p10 0–3 shorter, p8 3–5, p7 10–14; p11 minute, concealed by primary coverts. Primaries not emarginated. Tail very long, graduated; 12 feathers, t6 66–89 shorter than t1. Bill short, slender, and straight; base of upper mandible rather soft and slightly expanded, forming weakly developed cere, extending to above narrow oblique nostrils. Nail of upper mandible fine, only slightly decurved. Tarsus and toes short and rather slender; lower sides and rear of tibia bare. Outer toe $c.\ 73\%$ of middle, inner $c.\ 78\%$, hind $c.\ 70\%$.

Geographical variation. Rather slight. In southern Africa, adult ♂♂ from southern and interior eastern South Africa, Lesotho, and Swaziland have grey of crown sharply demarcated from buff-brown nape, remainder of upperparts including tertials also buff-brown; those from Namibia, Botswana, northern Transvaal, Zimbabwe, and Moçambique have grey of crown tinged with drab and remainder of upperparts drab rather than buff; in adult ♀♀, southern birds have upperparts greyish buff-brown and chest-feathers broadly fringed buff or rusty-brown, more northern birds have upperparts greyer and chest pure grey or narrowly fringed buff only. Birds similar to those of northern part of southern Africa extend north to Sahel zone and Arabia and have been named *anonyma* Oberholsern, 1904, with nominate *capensis* restricted as a breeder to southern and eastern South Africa, though migrating north to at least Zambia. (Clancey 1966.) However, birds from northern Afrotropics far from uniform either, especially ♀♀, colour varying from richly buff on upperparts and on chest of ♀ (e.g. Ahaggar, Sudan, West Africa, Tanzania) to pure grey (e.g. Ethiopia, Kenya, West Africa) independent of wear and of age of specimen. In absence of series of fresh birds from all parts of range, *anonyma* not recognized. Madagascar race *aliena* differs in adult ♂ by paler greyish-white border to black mask, pale blue-grey of crown extending almost to upper mantle, broader and blacker bars across back, slightly darker olive-grey mantle, scapulars, inner tertials, and rump, darker grey lesser upper wing-coverts; more extensive and deeper black to tip and outer web of p10; adult ♀ differs by more extensive white on forehead, front of cheeks, and chin; medium grey crown, broader and darker bars across back, darker lesser upper wing-coverts, and more black on p10 (Bangs 1918; ZMA). Differences in size slight: populations from West Africa east to Ethiopia and south to East Africa smallest (see Measurements), birds from southern Cape Province (South Africa) largest, average wing adult ♂ 113, adult ♀ 108; *aliena* intermediate, average wing adult ♂ $109\frac{1}{2}$, ♀ 108. CSR

Ectopistes migratorius (Linnaeus, 1766) **Passenger Pigeon**

FR. Colombe voyageuse   GE. Wandertaube

Extinct since September 1914, when last bird died in captivity in USA. Formerly occurred over eastern half of North America, with main breeding concentrations in New England and Great Lakes states and provinces; wintered in southern USA, straggling to Mexico in severe seasons. Bred socially, some mass nestings occupying 80 km^2; may formerly have been commonest Nearctic landbird, with total population that may have attained 3000 million birds (Schorger 1955). Rapid decline to extinction due in part to habitat destruction, but mainly to sustained excessive hunting by man; population crash pronounced by 1870, and no certain wild record after 1900. Between 1825 and 1848 at least 4 recorded Ireland, Britain, and France in circumstances consistent with vagrant origin (Alexander and Fitter 1955), but not admitted to any European list, as many imported into Europe from late 18th century onwards.

Order PSITTACIFORMES
Family PSITTACIDAE parrots

Very small to large, mainly vegetable feeders. Sole family in order. About 330 species, usually confined to tropics, but extending to higher latitudes in southern hemisphere and formerly North America. Usually divided into 6–8 subfamilies, the largest of which, Psittacinae, contains the only west Palearctic species, Ring-necked Parakeet *Psittacula krameri*, introduced into western Europe, Israel, and Egypt; other subfamilies ignored here. Body rather plump, neck short, with 14 vertebrae. Wings usually more or less rounded, but sometimes rather long and narrow. 11 primaries (11th vestigial); 11–13 secondaries, diastataxic. Tails variable, usually 12 feathers. Bill usually more or less stout, often with small cere; upper mandible attached to skull by flexible joint. Tarsus usually very short, tibia rather long covered with small, granular scales; feet zygodactyl. No caeca. Oil-gland feathered (absent in some American genera). Feathers with long aftershaft. Plumage colourful, rather sparse and usually hard and glossy; sexes normally similar. Young altricial and nidicolous. Adult plumage attained in 1st–3rd year.

Psittacula krameri **Ring-necked Parakeet (Rose-ringed Parakeet)**

PLATE 35
[facing page 399]

Du. Halsbandparkiet   Fr. Perruche à collier   Ge. Halsbandsittich
Ru. Ожереловый попугай Крамера   Sp. Periquito de collar   Sw. Halsbandsparakit

Psittacus krameri Scopoli, 1769. Synonym: *Palaeornis torquatus*.

Polytypic. Nominate *krameri* (Scopoli, 1769), Guinea and Sénégal east to western Uganda and southern Sudan, accidental in west Palearctic; *borealis* (Neumann, 1915), north-west Pakistan east through northern India to central Burma; *manillensis* (Bechstein, 1800), Ceylon and peninsular India south of 20°N (last 2 races occur in feral state in west Palearctic and elsewhere). Extralimital: *parvirostris* (Souancé, 1856), central eastern Sudan through northern Ethiopia to north-west Somalia.

Field characters. 38–42 cm; wing-span 42–48 cm; tail up to 25 cm. Body size close to that of Bee-eater *Merops apiaster*. Round-headed parakeet with long, pointed tail and almost uniform pale green plumage. Rear end exceptionally attenuated, both when perched and in flight. Flight powerful and rapid, not recalling any other wild bird in west Palearctic. Sexes distinguishable at close range; no seasonal variation. Juvenile separable. Most feral birds in west Palearctic apparently of oriental race *borealis* (described here) but some of southern Indian race *manillensis* and African nominate *krameri* also involved (see Geographical Variation for distinctions).

ADULT MALE. Generally pale yellow-green, with blue sheen or wash particularly on rear crown and tail and relieved only by duskier centres to tertials and flight-feathers. Throat, lower cheeks, and lower nape encircled by ring; this bends down from bill and up on cheeks, being black around throat and rose-red over nape. Bill and orbital ring waxy crimson. Legs olive. ADULT FEMALE. Resembles ♂ but lacks blue on head and has only indistinct emerald neck-ring. IMMATURE. Resembles ♀ but yellower, particularly on fringes of flight-feathers. Neck-ring indistinct or lacking.

Unmistakable among wild and feral birds of west Palearctic but beware also escapes of similarly coloured and shaped parakeets of various genera (see Forshaw 1978). Flight remarkably free, with rapid, flickering wing-beats propelling bull-headed but tapering bird at great speed; actions include sudden direction changes and fast dives. Perches freely and climbs expertly but waddles on ground.

Gregarious and noisy. Commonest call a shrill and harsh screech: 'kio' or 'keeak'.

Habitat. In tropical and subtropical low latitudes, mainly lowland but ascending Himalayan foothills to *c.* 1300 m and Indian peninsular uplands to *c.* 1600 m. Avoids mountains, rocky terrain, full deserts, wetlands, and forests, but widespread and abundant in moist and dry deciduous lightly wooded country including light secondary jungle, gardens, orchards, and cultivated areas in neighbourhood of human habitations; sometimes in semi-desert. Although basically arboreal, nesting, roosting, displaying, and often feeding in trees, it is locally mobile, volatile, and adaptable, exploiting not only orchard fruit but ripening fields of grain and food crops, both standing and lying on ground, which it attacks in flocks; also consumes grain spilt in transit, or bites through sacks to eat grain or ground-nuts (Whistler 1941; Ali and Ripley 1969). Switches diet seasonally as between tree and ground sources, sometimes harvesting clusters of berries in company with monkeys; often comes to pools in late afternoon, first alighting on nearby trees and then flying down to walk last steps to edge, not entering water (Hutson 1954). Also nests freely in holes in rock scarps, walls of ruined buildings, and ancient forts, and even buildings in towns, including noisy and congested bazaars. In West Africa, a troublesome major pest of drier savanna (Serle *et al.* 1977); confined to semi-arid country and thorn scrub (Mackworth-Praed and Grant 1970). In southern Sudan usually associated with woodland, wherever trees found in reasonable quantity (Cave and Macdonald 1955). Despite tropical origin, released or escaped birds have proved able to survive for long periods in temperate climates (especially in suburban parks, large gardens, and orchards), even through severe winters and in face of intense competition from native species (see Distribution). Flies strongly, but normally stays within lower airspace.

Distribution and population. In west Palearctic, introduced or escaped birds now seem to be established in Britain, Netherlands, West Germany, Israel, and Egypt, but apparently extinct in Iraq. For history of introductions elsewhere see Long (1981a); map of world distribution tentative, as often not clear whether introduced populations fully established.

BRITAIN. Bred Norfolk 1855 (Lever 1977), then no published nesting records until 1969 when bred Kent; now breeding regularly there, and in Greater London (from 1973), Greater Manchester (from 1974) and Merseyside (from 1980). Probably breeds regularly in Surrey (from 1971) and Sussex (from 1974) and has bred at least once in Berkshire, Clwyd, and West Yorkshire. Under-recorded in some areas; tentative estimate of over 1000 birds in 1982. (B Hawkes.) BELGIUM. Small feral populations Brussels (from *c.* 1970) and Antwerp (PD, CSR). NETHERLANDS. Small populations now in Amsterdam, Rotterdam, The Hague, and Zeeland. First proved breed-

380 Psittacidae

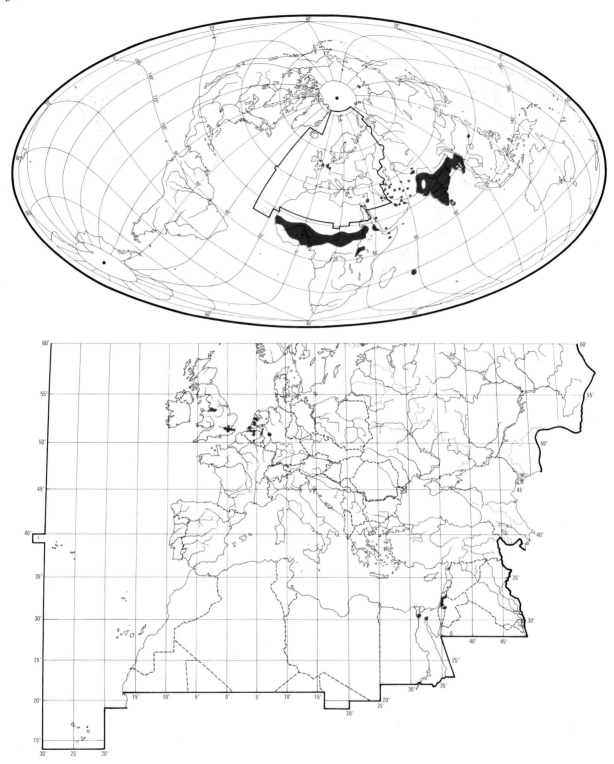

ing 1980 (but perhaps from 1968) and probably now self-supporting. About 40–50 birds 1977, probably several hundred 1981 (Teixeira 1979; Taapken 1981; CSR). WEST GERMANY. Feral breeding populations now in Köln, Brühl, and Bonn (probably also in Wiesbaden, and perhaps Leverkusen to Düsseldorf). First nested Köln 1960s, near Zoological Garden; now *c.* 50 pairs. In Brühl, 4 pairs since 1975. First bred Bonn 1979. Flock of *c.* 100

seen Düsseldorf 1979 (GR). IRAQ. Bred Baghdad in 1930s (Moore and Boswell 1956b), but apparently largely died out (Marchant and Macnab 1962). ISRAEL. About 1960, escaped from Tel-Aviv Zoo, and from aviculturists. Since increased and spread over most of coastal plain, also Galilee and Jordan valley. Now well-established breeder; nests found and juveniles seen (HM). KUWAIT. Present, especially in winter, but breeding not yet proved (PRH). EGYPT. Birds of *manillensis* escaped from Giza Zoological Gardens 1901–8; bred and increased, with 127 killed 1916–19. Now common Giza and Cairo (perhaps several hundred), with vagrants elsewhere (one as early as 1895) (Flower 1933; Goodman 1982; PLM, WCM).

Accidental. Cape Verde Islands. 1 collected 1909 (Long 1981a).

Movements. Resident in Afrotropical and Oriental regions, except that in cultivated areas there are local feeding movements governed by ripening of crops (Forshaw 1978). West Palearctic introduced populations also resident, though no studies of local dispersals.

Food. In India takes only plant food: predominantly seeds, berries, fruits, flowers, and nectar (e.g. Forshaw 1978). In Britain, feral birds omnivorous (B Hawkes). Frequently forages on ground (Harman 1974), e.g. gleans waste grain from fields and takes fallen fruit (Ganguli 1975; Olšaník 1977). Will tear open sacks of grain or ground-nuts *Arachis hypogaea* and take contents (Ali and Ripley 1969; Forshaw 1978). Rarely feeds on ground in Britain, obtaining food mainly from bird tables and garden feeders, though also in gardens and agricultural crops (B Hawkes). Often described as appearing wasteful in foraging methods, tending to gnaw at half-ripe fruits and taking only 1–3 pecks while clambering around amongst twigs (e.g. Ali and Ripley 1969, Ramzan and Toor 1973, B Hawkes). Generally considered the most serious avian pest to agriculture and horticulture in India (e.g. Shivanarayan *et al.* 1981). When feeding on cereal crops, birds may fly down to snatch a seed-head in brief hover, then return to perch; sometimes settle on stalks, biting into seed-head or severing it completely as stem bends—always seized by stalk end (Hutson 1954; Ali and Ripley 1969). Seed-heads and small fruits raised to bill in foot; seed-head may be nibbled at, then dropped (Ali and Ripley 1969; B Hawkes). 2 captive birds recorded sharing food held in foot of one (Smith 1972b). To take legumes from tree, sidles along branch and stoops to seize and pluck pod at stem end with bill; eats from stem end, holding other in foot, and works steadily along until last bite given with pod held in bill. Only seeds taken, pod discarded (Hutson 1954). Fruit not swallowed whole or crushed but pericarp usually removed before final gulp (Dubale and Rawal 1965). Pecks off petals of certain flowers to get at nectar (Ali and Abdulali 1938). Birds clinging to masonry wall of culvert thought to be after salt (Hutson 1954). Food carried to young in crop and transferred by regurgitation which may occur up to *c*.2 hrs after foraging sally (Lamba 1966; also Shivanarayan *et al.* 1981). In Britain, feral birds apparently unaffected by hard weather, as food then obtained from suburban bird tables (B Hawkes). Feeds mainly 06.00–09.00 and 16.00–18.00 hrs (Shivanarayan *et al.* 1981); similar pattern in Britain (B Hawkes).

Jaw and tongue musculature adapted for holding fruit and removing pericarp. For detailed morphological study, see Dubale and Rawal (1965); also Homberger (1980).

For detailed list of foods taken in India, see especially Hutson (1954), Ali and Ripley (1969), and Ganguli (1975); also (e.g.) Dharmakumarsinhji (1955), Toor and Ramzan (1974), Rao and Shivanarayan (1981), and Shivanarayan *et al.* (1981). In Britain, food items recorded include fruits and berries of various Rosaceae: apples and crab apples *Malus*, pears *Pyrus*, rose *Rosa*, hawthorn *Crataegus*, *Cotoneaster*, cherries and plums *Prunus*, raspberries *Rubus*, strawberries *Fragaria*; also berries of holly *Ilex* and elder *Sambucus*, grapes, peeled bananas, sliced oranges, peas *Pisum*, barley *Hordeum*, maize *Zea*, peanuts *Arachis*, beech mast *Fagus*, horse chestnuts, *Aesculus hippocastanum*, and seeds of hornbeam *Carpinus betulus*, ash *Fraxinus excelsior* and pine *Pinus*; also takes bread, bacon rind, biscuits, corn, and meat (from bone on tree). Recorded taking grit (B Hawkes).

For seasonal variation in diet outside built-up area, Delhi (India), see Hutson (1954). In Upper Sind (India), favours blackwood beans *Dalbergia latifolia*, elsewhere generally seeds of babul (Ticehurst 1923a). When no cultivated fruits or other crops in season, India, birds take wild fruits (wild figs, *Zizyphus*, etc.) and seeds. In February, congregate to take nectar from flowering silk-cotton tree. 53 stomachs from Pusa and 3 from Nagpur contained entirely vegetable matter—mustard, wheat, maize, rice, *Nephelium litchi*, and wild fruits, also seeds of *Dalbergia sissu* (Fletcher and Inglis 1924). 1-year study in north-west India showed birds took seasonal crops, causing considerable damage; preferentially guavas *Psidium guajava* (Toor and Ramzan 1974; see also Shivanarayan *et al.* 1981). See also Wilson (1949) for crop damage in Sudan. In Saudi Arabia, November, birds took seeds of peacock tree *Caesalpina pulcherrima* (P Scott). In Oman, takes mainly seeds of sunflower *Helianthus*; also legumes, dates, other fruit, and grain (Gallagher and Woodcock 1980); in winter, in north-east coastal garden belt, probably largely dependent on seeds of *Acacia arabica* (Guichard and Goodwin 1952). Stomach contents from Upper Volta/Mali, December–January, included wild figs and other fruits and seeds, millet, and flowers (Bates 1934b). In West Africa, takes ripening guinea-corn, drying ground-nuts, and nuts of oil palm *Elaeis guineensis*; otherwise bush fruits (Bannerman 1931). In Eritrea, food includes figs, dates, and fruits of jujube, tamarind, and baobab; no particular fondness for cereal grain (Moltoni 1939a). In Britain, 224 sight records of feeding birds over 4 years comprised 39% involving birds taking apples on trees, 3% apples on ground, 3% crab apples, 26% peanuts in feeders, 6% parrot food, 6% unidentified berries in shrubs, 5% growing pears, 5% flowers/buds of trees and shrubs, 4% wild bird food, and 3% other items (B Hawkes).

Diet of young in India largely as in adult (see, e.g., study by Rao and Shivanarayan 1981); also in Britain, where diet of adults shows no change during nestling period (B Hawkes). Chicks first fed soon after hatching, often when only half out of shell (Smith 1979; G A Smith); then up to 3 times per day according to Lamba (1966). Scant data on British feral birds indicate chicks fed 1–2 times per hr early in day, once per hr or less later in day (B Hawkes).

MGW

Social pattern and behaviour. Following account includes material on feral birds, Britain, from B Hawkes; few data otherwise available from west Palearctic. For more detailed studies, India, see (e.g.) Fletcher and Inglis (1924), Hutson (1954), Dharmakumarsinhji (1955), MacDonald (1960), Lamba (1966), and Ganguli (1975). For study of captive birds, see Smith (1972b).

1. Usually in small parties or larger flocks, aggregations typically of many hundreds or thousands on feeding grounds and, especially, for roosting (e.g. Dilger 1954, MacDonald 1960, Ali and Ripley 1969, Forshaw 1978). In Britain, in pairs when breeding (B Hawkes). Flocks during breeding season, Delhi (India), possibly ♂♂ and non-breeders (Ganguli 1975). At Delhi, roosting flights continue during breeding season (Hutson 1954). Average flock size, Britain, 35 ($n = 10$); family parties band together at end of breeding season (B Hawkes; see also MacDonald 1960 for India). According to Smith (1972b) and Harman (1974), ♂♂ apparently always in majority; however, in flock of 45, Britain, sex-ratio of adults 1:1; 30% of same flock adjudged 1- and 2-year-olds by tail length (B Hawkes). Roosting flocks, Barrackpore (India), July, composed mainly of immatures (Beavan 1865). BONDS. Monogamous mating system, though no detailed information (Smith 1972b; B Hawkes). Bigamy recorded in captive birds (Smith 1972b); for alleged finding of 2 ♀♀ in one nest, see Oates (1890) and Lamba (1966); fierce territoriality of ♀♀ makes such occurrence unlikely (G A Smith). Pair-bond probably of long duration (B Hawkes), possibly for life (MacDonald 1960). In captive birds, pair-bond weakened after breeding (Smith 1972b; G A Smith; see also Heterosexual Behaviour, below). Care of young by both parents (predominantly ♀) up to and beyond fledging; exact duration unknown (B Hawkes), but see Relations within Family Group (below). Age of first breeding probably 3 years in most cases; sometimes, for both ♂ and ♀, at 2. Perhaps c. 50% of captive pairs where both birds 2 years old will breed. Claims of exceptional breeding at 1 year old possibly refer only to courtship activity (Smith 1972b, 1979; also Hutton 1873, Tavistock 1929, Forshaw 1978, B Hawkes). BREEDING DISPERSION. Solitary, though not infrequently in loose groups (e.g. Ali and Ripley 1969), and up to 8 pairs recorded in same tree (Shivanarayan *et al.* 1981), while 2 pairs in same tree probably frequent (MacDonald 1960). In Delhi, c. 6 pairs in line of trees, 1 pair per tree (Hutson 1954). In Britain, solitary pairs or small loose nesting groups (c. 4 pairs) more or less equally common, but few data available; sometimes 2 nests in same tree, but birds mostly well separated in urban parks, etc. (B Hawkes). Territorial defence apparently restricted to immediate vicinity of nest-hole (Lamba 1966); see Antagonistic Behaviour (below) for role played by sexes. Courtship and pair-formation usually take place in nest-tree or nearby. No feeding done in 'territory' during courtship season and birds may have habitual feeding area away from nest-site (B Hawkes; see also Roosting and Heterosexual Behaviour, below). Nest-site may be used for many years in succession: in Britain, 7 years recorded (D Goodwin, B Hawkes); in Sudan 6 years (Cunningham-van Someren 1969). ROOSTING. Outside breeding season (to some extent also during), communal and nocturnal. In India, mainly in large leafy trees, especially in town parks and gardens; a few birds may use similar sites in steppe. Also rock crevices and (mainly breeders) in cracks or holes in walls of buildings. Sites traditional (Dilger 1954; Olšaník 1977; Forshaw 1978). In Britain, uses tall trees and, rarely, buildings (B Hawkes). At Agra (India), birds arrived in flocks of varying size, within c. 1 hr before sunset (Dilger 1954). Regular flight-paths used (Harman 1974). In Britain, arrives at roost just before dark, leaving soon after first light (B Hawkes). For behaviour at roost and in flight to and from feeding grounds, see Flock Behaviour (below). Frequently roosts in association with crows (Corvidae) and mynahs (Sturnidae) (e.g. Jerdon 1877, Lowther 1949, Lamba 1966, Olšaník 1977). ♀ roosts in nest-cavity before laying and with young up to fledging (MacDonald 1960; see also Heterosexual Behaviour, below). Presumed (feral) ♀, California (USA) did so each night, mid-February to mid-June, usually retiring at c. 17.00–17.30 hrs (Hardy 1964). According to MacDonald (1960), ♂ roosts apart, at least in pre-laying period. Elsewhere, ♂ reported to spend part of day in nest and to roost there nocturnally (Smith 1975b), and nearby during incubation (Shivanarayan *et al.* 1981). Members of captive pairs initially roosted close together on perch, ♀♀ then immediately using nest-boxes once these provided, and practice resumed c. 1 week after brood lost. One ♂ roosted at night with ♀ after c. 1 week's incubation (Smith 1972b). Fledged young reported to roost communally before dispersal (Dewar 1909). Sleeping posture very erect (B Hawkes). Sites used for loafing much as described above; off-duty bird may have special site near nest (B Hawkes). At Agra, commonly loafs on electricity wires (Dilger 1954). Captive birds spend much time dozing; feathers ruffled, head often twisted on to back, eyes almost closed; occasionally self-preen. Rain-bathing always followed by vigorous self-preening (see Smith 1972b). Wild birds, India, sit on exposed branches and soak themselves in pouring rain (Ganguli 1975). Scant data from Britain indicate sunning and rain bathing also occur (B Hawkes).

2. Described by Forshaw (1978) as 'fearless' and by Baker (1927) as 'bold'. Foraging flocks and those flying in to roost wary; give warning calls at sight of man (Olšaník 1977; see 1 and 4 in Voice). In Britain, wary of man at times and flies away giving repeated Kee-ak calls (B Hawkes: see 1 in Voice). Captive birds sleek plumage on hearing Warning-calls (Smith 1972b). Normal flight usually swift and direct (MacDonald 1960; see also Flock Behaviour, below); more irregular flight, with frequent turning and circling, typically occurs high in the air during rains. Can normally outfly most Falconiformes, being vulnerable only to attack from above; large raptors mobbed from perch or in flight (Dharmakumarsinhji 1955). Captive birds sociable but normally maintain individual distance of c. 15–30 cm, the greater distance typically when birds wide awake; 6 birds probably roosted close together on perch c.1 m long (Smith 1972b). FLOCK BEHAVIOUR. Attacked by Shikra *Accipiter badius*, flock of c. 6 scattered, calling loudly (not described); fast twisting and veering flight, with zigzag manoeuvres, used in such a situation in attempt to evade pursuer (MacDonald 1960). Following attack by Laggar Falcon *Falco jugger*, birds departed as flock, shrieking (Lowther 1949). Warning-calls of pair whose nest threatened by (e.g.) mynahs (Sturnidae) or animal predator, apparently attract other *P. krameri* who fly in and mob intruder with much screaming and screeching (Ali and Ripley 1969; see also Dharmakumarsinhji

1955 and Parental Anti-predator Strategies, below). Short, branch-to-branch chases by members of small flock who uttered 'kwurrup' sounds, then Kee-ak calls thought probably aggressive (D Goodwin: see 1 and 6c in Voice). In Sudan, small flocks seen in pursuit-flights through woodland glades, birds giving chattering calls (Bannerman 1931: see 3 in Voice); behaviour difficult to classify. At Chilaw (Ceylon), birds initially straggled in to roost (from c. 17.00 hrs), then arrived in flocks, peaking at c. 17.30 hrs. While high-flying birds made sudden plummet with aerial evolutions upon arrival, others arrived in low, silent flight (Layard 1854). See also Jerdon (1877) and Hutson (1954). Birds may strip foliage from roosting trees on alighting (Lowther 1949). Give loud Kee-ak calls and fly from branch to branch, quarrelling over sites (see Olšaník 1977). Birds equally vocal during morning exodus which may involve flocks flying out in quick succession over c. 20 min (Hutson 1954). According to Harman (1974), fly well above trees and give regular Kee-ak calls *en route* to feeding grounds. In Britain, general pattern of roosting flights similar (B Hawkes). Often come to pools in late afternoon, landing in trees, flying down, and then walking last few paces (Hutson 1954). At Mhow (central India), where birds artificially fed, foraging flocks close packed, with up to 200 on c. 40 m² (Briggs 1932). ANTAGONISTIC BEHAVIOUR. Squabbling frequent (Forshaw 1978), but no details. Study of captive birds showed ♀♀ dominate ♂♂, particularly immatures. In commonest form of threat, head moved sharply upward, then lowered (Smith 1975b). At start of breeding season, when pair-bonds becoming established, supplanting attacks more frequent, with ♂♂ driving off rivals: walking approach made, then bird lunges with open bill; Eye-blazing also given (pupil constricted so that outermost edge of iris revealed: common threat in many Psittacidae; see also Heterosexual Behaviour, below). Fights not recorded between ♂♂, but in ♀♀ occasionally have fatal outcome. Once, other conspecific birds gathered round fighting ♀♀ but probably did not participate. Birds generally tolerant during courtship; interference with pair infrequent in both captive (Smith 1972b) and wild birds (MacDonald 1960; see also Heterosexual Behaviour, below). ♀ viciously pecked at strange ♂ who attempted to copulate with her after she had mated with her own ♂ (MacDonald 1960). In Britain, said to compete aggressively for nest-holes with Great Tits *Parus major*, Jackdaws *Corvus monedula*, and Starlings *Sturnus vulgaris* (B Hawkes). In case of *C. monedula*, *P. krameri* more likely to be responding aggressively towards potential nest-predator; *S. vulgaris* the only serious competitor (D Goodwin). Once territory occupied, zealously defended against conspecific intruders, with 1 bird of pair constantly on guard in vicinity (see also Breeding Dispersion, above). Other hole-nesters may be tolerated as nesting neighbours, even in same tree (Lamba 1966; also MacDonald 1960), but driven away from vicinity of nest (D Goodwin). ♂ defends (not vigorously) area around own mate and fledged young, but never defends nest-hole; only occasionally drives off another ♂ from nearby perch, usually seeming unconcerned. ♀, in contrast, tolerates no conspecific bird other than own mate less than 30 cm from nest-hole or herself. Lunging and Bill-snapping usually enough to discourage intruder but brief pursuit may occur, with owner then returning to enter nest-box. In 2 fights between ♀♀ observed when nest-boxes first available, birds tumbled to ground together but soon separated and territory-owner dominant (Smith 1972b). In captive birds, both parents defend fledged young; other ♀♀ most likely to attack when young stray near their boxes (Smith 1972b). Dominant over many other species at bird-tables, etc.; in threat, gives Kee-ak calls and Wing-flapping (B Hawkes). HETEROSEXUAL BEHAVIOUR. (1) General. Numbers at roosts markedly reduced at beginning of breeding season, as pairs leave to seek nest-sites (Forshaw 1978). Pair-formation occurs well in advance of breeding proper. Once paired, birds generally shun company of others; spend much time perched together, when courtship occurs (Lamba 1966). In Punjab (north-west India), breeding season (January–)March–June, with courtship display and mating observed mid-February to 1st week of March (Forshaw 1978). Captive juveniles may show courtship behaviour at any time of year (Smith 1972b). ♀ dominant, once breeding season approaches; courtship display (see below) repeated many times per day and over 2–3 days or more (Smith 1972b; B Hawkes). In captive birds, close physical contact avoided: probably due to ♂'s fear of dominant ♀. Pair losing chicks showed no resurgence of courtship behaviour apart from ♂ feeding ♀, which also ceased after c. 14 days, although birds often perched close together; few signs of association after c. 6 weeks (Smith 1972b). (2) Pair-bonding behaviour. Courtship display elaborate (Forshaw 1978). Observations on captive birds showed that, in late autumn, ♂♂ sometimes Parade ('Stately-walk'): walk slowly and sedately, head high, and Eye-blaze. Bowing typical of early stage and, like Parading, also performed when ♀ not present: ♂ 'fiddles with' perch near feet in Bow, then raises head high; performed 2–3 times. ♂ may also Parade, and give stiff little Jumps if on flat ground. Captive ♀ which eventually paired with bigamous ♂ repeatedly gave variant of Kee-ak call (see 1 and 6a in Voice), at each call fanning and closing tail and Head-flicking between each bout. In associated Nodding (thought to be advertising behaviour of unpaired ♀ ready to breed), head lowered slightly, tail fanned, wings slightly drooped, and call given (see 6b in Voice), then head raised (Smith 1972b). Courtship behaviour usually takes place on tree near nest (B Hawkes). In presence of ♂, ♀ droops or spreads wings slightly (wings half spread: Hutton 1873), sways head from side to side in semicircles, rolls eyes, and gives Courtship-call (Hutton 1873; Lamba 1966; Forshaw 1978; B Hawkes: see 5b in Voice). ♂ may remain perched quietly close by (Hutton 1873). According to Smith (1972b), generally hesitant and tense in approach (see also subsection 1, above) and, particularly in early stages, performs frequent Bill-wiping or Bill-tapping against perch. Advances with small positive steps (B Hawkes), or struts towards ♀ (Lamba 1966; Forshaw 1978). ♂ then stretches ('Crane- or Stretch-preening') towards ♀ from some distance, wings held slightly away from body; Eye-blazing also performed. Allopreens ♀ in hasty movement ('pecking kiss') on nape or near carpal joint (Fig A); Allopreening of ♀'s head a later development requiring more co-operation from her (see 3, below). ♀ remains more or less completely motionless at this stage (Head-swaying, etc., of ♀ as described above, not mentioned by Smith 1972b); flinching by her may cause ♂ to fly off

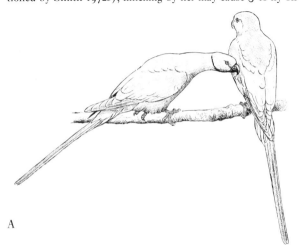

A

384 Psittacidae

B

briefly but he normally returns for resumption of courtship activity. After Crane-preening, ♂ adopts Guardsman-posture (Fig B): head drawn back and raised high, bill held tightly against neck, foot nearer ♀ lifted high (perhaps merely to assist further retreat which may be of 1–2 steps during which Bill-tapping also given). Brief pause followed by a few Perch-tapping movements with raised foot against perch or in the air before head darted forward again for further bout of Crane-preening (Smith 1972b, 1979; also Ali and Ripley 1969, Forshaw 1978). ♂ may raise feet alternately (Hutson 1954) and sway body from side to side (Dharmakumarsinhji 1955). Whole performance may be given several times and last several minutes, ♂ later Allopreening ♀ with bill-tip on side of head or crown, and Billing or Bill-rubbing (Hutton 1873; Hutson 1954; Lamba 1966; Forshaw 1978; B Hawkes). According to Adam (1873), ♂ released one foot (usually the right) to perform bowing-type movements between bouts of Billing. Each bird may turn away after repeated Billing and ♂ move to another perch, call (see 6f in Voice), and Wing-flick (no details) while ♀ remains to nibble at foliage (Hutson 1954). For report of ♂ hopping round ♀, feathers sleeked, wings slightly spread, giving bowing movements, see Harman (1974). (3) Courtship-feeding. Eventually follows from, or may be interspersed with, Allopreening or Billing (Lamba 1966; Smith 1972b, 1979; B Hawkes). Pattern essentially as in Allopreening and Billing (see subsection 2, above); ♀, however, adopts flatter posture, head withdrawn and twisted to one side. Unobtrusive regurgitatory Head-bobbing given by ♂ in barely perceptible pause preceding bill-contact. Throat slightly puffed out, neck slightly bent as food offered (Smith 1972b). ♂'s bill a little open, that of ♀ inserted to take regurgitated food (MacDonald 1960; Ali and Ripley 1969), both birds holding head high (B Hawkes). Contact broken off abruptly, ♂ Back-jerking head to reassume Guardsman-posture and watch ♀ (Smith 1972). ♂ has wings partially open at shoulders, stands c. 30 cm from ♀ (Ali and Ripley 1969). Toes of ♂'s raised foot extended towards ♀ for c. 5–25 s (B Hawkes); ♀ makes slight nibbling movements after receiving food. ♂'s foot replaced on perch for forward Head-dart to feed ♀ again (Smith 1972b). Most movements thought to be linked with strenuous process of regurgitation (MacDonald 1960). Courtship-feeding continues until after fledging but generally less during incubation and especially after hatching: done directly, ♂ having wings close to body and making only slight Head-bobbing movements. Frequently Bill-wipes between feeds. Pumping movements result primarily from Bill-jerking given by ♀. During Head-raising, foot of ♂ nearer ♀ may be shuffled slightly. Both ♂ and ♀ may sometimes (generally: B Hawkes) Back-jerk heads after food has been transferred, ♀'s movements being more or less mirror-image of ♂'s (Smith 1972b). According to Ali and Ripley (1969), ♂ may cross to give whole performance on other side of ♀, and move back and forth several times. Bigamous ♂ fed 2 ♀♀ equally (Smith 1972b). Courtship-feeding usually, but not invariably, associated with mating (MacDonald 1960). (4) Mating. May occur after only one display pattern as described, but usually follows several (B Hawkes; also Forshaw 1978). Takes place in tree, on house-top, wire, etc. (Lamba 1966). 3 pairs copulated invariably first thing in morning and about once per hr (sometimes at shorter intervals) throughout day. In one pair, regular pattern evident each morning over c. 6–7 weeks, ♂ flying to nest-hole and giving loud Kee-ak call, though this changed to gentler tone when ♀ slow to emerge. When she did so, ♂ followed; birds then copulated and ♂ fed ♀. Apparent rape attempts occurred when ♀ reluctant to mate. Mating continued after 1st egg laid but ceased about mid-March while Courtship-feeding continued (MacDonald 1960). In captive birds, ceases about half way through incubation and not resumed later (Smith 1972b). Allopreening of ♂ by ♀ never seen in captive birds, but from early/mid-winter, ♂ increasingly Allopreens ♀ from closer position than hitherto; ♀ adopts posture with back horizontal, wings held slightly away from body. Courtship-feeding, however, apparently reduced. ♀ holds head askew (as previously for Courtship-feeding); may move lowered head slightly backwards and forwards. ♂ makes many attempts to mount, and eventually Allopreens ♀'s nape rather roughly while birds stand adjacent and with bodily contact; ♀ then crouches lower. ♂ places first one then both feet on ♀'s back, droops wings, and Eye-blazes. Allopreening changes to ritualized Hammering at base of ♀'s neck (Fig C): repeatedly pounded with culmen, head raised high, bill tucked in, in rhythmical crescendo from 1 blow per 1·5 s to more than 1 per s; blows alternate between left and right of ♀'s neck as she slightly sways body and moves head from side to side (Smith 1972b, 1979; G A Smith). In pair observed by Ali (1927), ♀ squatted with neck extended forward, back flattened, and wings slightly opened and raised; ♂ mounted using bill to assist; Hammering given c. 20 times on ♀'s left side, performance then repeated on right; ♀ frequently turned head to touch ♂'s bill. In copulation attempt lasting c. 30 s, apparently unwilling ♀ pecked at ♂'s face; clumsiness by ♂ in mounting also sometimes elicits pecking. More complex pre-copulatory behaviour, with also much looking around, typical of early stages (MacDonald 1960). During more rapid Hammering, ♂ lowers tail to achieve cloacal contact (which may be repeated: B Hawkes), whole process of Hammering and copulation lasting c. 4 min (Smith 1972b). According to MacDonald (1960), decreasing amount of time spent on ♀'s back as pair-bond cemented: initially c. 3–4 min, later c. 2 min. Reported by Ali (1927) that ♂ once seen to shift to centre of ♀'s back, wings opened and drooped low either side of her, body tilted backwards to copulate (3 times in 30 min); between performances (2nd and 3rd briefer) birds moved a little apart then sidled up again. After copulation, both birds ruffle feathers and self-preen; ♂ sometimes feeds ♀ (Smith 1972b; B Hawkes). Immediately prior to laying, ♀ often

C

feeds ♂ and Allopreens his head; may even mount him, but never leads to copulation (G A Smith). (5) Nest-site selection. Apparently by both birds after prolonged inspection of various cavities, but ♀ probably makes final choice (Lamba 1966). Captive ♀♀ made excavating movements on ground before nest-boxes put up. Site inspected by ♀ Crane-peering and then withdrawing head (like ♂ Courtship-feeding ♀—see above). Selection apparently by ♀, ♂ showing little or no interest (Smith 1972b). (6) Behaviour at nest. Both sexes may excavate cavity when site occasionally requires this (Lamba 1966; also Shivanarayan et al. 1981); in Deesa (India), done by pair at least 3 months before laying (Fletcher and Inglis 1924). ♂ did not assist, Delhi; ♀ spent long periods in cavity prior to laying—also in Kent, England (D Goodwin), and in captive birds (Smith 1972b); when ♂ arrived from roost, she flew with him to feeding grounds. Birds spent long periods together near nest, mid-July to late January, and ♂ also sat nearby, accompanying ♀ if she flew off (MacDonald 1960). In early stages, ♀ always fans tail when entering box. Often joined in box by ♂, but birds usually emerge (♀ first) after c. 5–10 s, ♀ defecating and ruffling feathers before being fed by ♂ (Smith 1972b; see also subsection 3, above). Incubation entirely by ♀ (G A Smith). ♂ may visit site from time to time, Crane-peer into nest and feed ♀ (MacDonald 1960); may also spend up to c. 30 min in nest with incubating ♀ (Smith 1979). In Britain, ♂ said to remain in nest apparently alone for 1–2 hrs at this stage (B Hawkes). RELATIONS WITHIN FAMILY GROUP. Eyes of nestling begin to open at c. 10 days (Smith 1979). Adults tend to announce arrival (calls not described) when coming to feed young (Lowther 1949). According to MacDonald (1960), ♂ may pass food to ♀ by regurgitation, but rarely feeds young directly. Captive nestlings fed by both parents. In most cases, ♂ fed ♀ outside nest, then entered to feed young, often followed immediately by ♀ (Smith 1972b). Wild birds still fed (by regurgitation) by both parents at fledging stage: beg by holding heads up with open bills and give food-call. No nest-sanitation observed (B Hawkes). Captive young brooded nightly by ♀ until fledging. From c. 10 days after hatching, daytime brooding much reduced, until almost absent at c. 3 weeks. ♀ may Allopreen young (Smith 1972b). Family remains together for some days as discrete unit, before joining others; some feeding of young may take place after fledging, but fuller study required (B Hawkes). In India, young apparently independent straight after fledging (Lamba 1966). ANTI-PREDATOR RESPONSES OF YOUNG. At c. 2–3 weeks, often climb up to entrance-hole, but quickly withdraw at approach of intruder or parent; may peck if attempt made to pursue them with hand or stick (Lamba 1966). Captive nestlings gave a buzzing call as response to Warning-call of ♀ (see 4 in Voice). When handled, young give a distress call (Smith 1972b). PARENTAL ANTI-PREDATOR STRATEGIES. ♀ tends to sit tight and may even allow herself to be seized on nest, although savage pecking also occurs (Lamba 1966; B Hawkes). One ♀ had to be pushed off eggs but remained in cavity whilst eggs removed despite exit being free (Butler 1875). Does not desert easily, even after considerable disturbance at nest (Lamba 1966). Captive ♀♀ left nest-boxes on hearing Warning-calls (Smith 1972b). A threatened pair may 'summon assistance' by giving loud Kee-ak calls (Ali 1968; see also Flock Behaviour, above). In Britain, rarely shows fear of Kestrels *Falco tinnunculus*, and no alarm or hostility elicited by Magpie *Pica pica*; mobs Carrion Crow *Corvus corone* excitedly, giving Warning-calls, and may make dive-attacks, possibly involving physical contact (D Goodwin). ♀ surprised by Shikra *Accipiter badius* near nest attempted to escape by entering cavity; forced to veer away and made off in zigzag flight; returned to nest immediately once raptor had relinquished pursuit (MacDonald 1960).

(Figs A–C based on drawings in Smith 1972b.)   MGW

Voice. Extremely noisy, particularly at roost; rather less vocal during moult (July–August). Repertoire not studied in detail but thought to be restricted (Dilger 1954; Forshaw 1978; B Hawkes). Some calls may be duplicated below.

CALLS OF ADULTS. (1) Kee-ak call. A loud screeching 'kee-ak . . . kee-ak . . . kee-ak', given in flight or when perched (Forshaw 1978). Recording (Fig I) suggests 'kee-a': starting transient apparent on pronounced 2nd harmonic, but no corresponding end one; 'kee-ak' more evident in last 2 units of Fig III (J Hall-Craggs). Some 'kee-ak' sounds

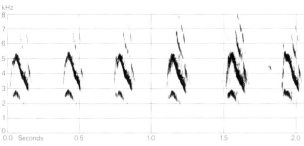

I F M Gauntlett India May 1969

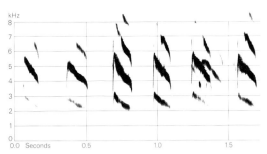

II E D H Johnson Ceylon November 1977

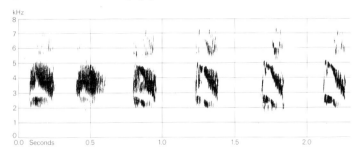

III F M Gauntlett India May 1969

shrill and piercing, others softer in tone and more musical; at times tending towards 1 rather than 2 syllables (M G Wilson). Compare Fig II which has subtle differences. See also Hutton (1873). Sometimes given in rapid succession, tempo varying with circumstances (Ali and Ripley 1969). Typically uttered by bird approaching or departing from nest area (D Goodwin). According to Smith (1972b), given most frequently when bird slightly disturbed, but not markedly alarmed; also just before roosting. Probably contains alarm element; no other specific alarm call known (B Hawkes; see also call 4). (2) A harsh grating or rasping 'errrrr' recalling Jay *Garrulus glandarius* (P J Sellar, M G Wilson: first 4 units of Fig III). Also rendered as a harsh screeching 'kreh kreh kreh' or 'kzeh kzeh kzeh'; in recording of flying flock by E D H Johnson, Ceylon, strongly resembles Alarm- and Anger-calls of Whiskered Tern *Chlidonias hybridus* (E K Dunn). Calls 1–2 merge into each other; probably little difference in function but rather in degree of motivation (J Hall-Craggs). (3) A plaintive, chattering 'chee-chee', given in flight, especially when approaching nest (Hardy 1964). Presumably same as the 'chai-chai-chai-chai' of Mackworth-Praed and Grant (1952) where, however, described as a distinctive shrill, screeching sound. Chattering screams also given in pursuit-flight (Bannerman 1931). (4) Warning-call. A series of 'ak' sounds (Smith 1972b). Rather like call 1, but shortened and accelerated, several calls given in quick succession; frenzied in quality (D Goodwin). (5) Courtship calls. (a) ♂ gives pleasant whistling notes (Smith 1972b). Also described as a pleasant gentle warble, usually associated with Allopreening and Billing (Harman 1974); see also call 6e. (b) ♀ emits soft twittering sounds (Forshaw 1978); a frequent soft chuckling (B Hawkes). (6) Other calls. (a) A repeated strident note similar to call 1, but clearer, given by 2 captive ♀♀ apparently as part of Nodding-display (Smith 1972b). (b) Call given during Nodding-display a disyllabic 'dear boy' (Smith 1972b). (c) Apparently aggressive chases accompanied by 'kwurrup' sounds, like gull *Larus* (D Goodwin). (d) When sitting quietly in tree-tops, birds utter various whistling and subdued chattering notes (Forshaw 1978; see also calls 3 and 5a). Mewing, whistling, and cheeping sounds also evident in recording of flying birds, Ceylon. (e) Immature ♂ sometimes gives soft warbling; not heard from adult ♂ (Smith 1972b), but see also call 5a. (f) ♂ gave squealing sound after moving away from ♀ following bout of courtship (Hutson 1954).

CALLS OF YOUNG. Food-call of fledglings a low-pitched, harsh sound (B Hawkes). Buzzing sounds given apparently in response to Warning-call of ♀ (adult call 4). Distress-call a harsher, further-carrying sound (Smith 1972b, 1975b).
MGW

Breeding. SEASON. Britain: eggs laid January–June (B Hawkes). SITE. In hole in tree, or sometimes in roof or wall, usually well above ground; hole of woodpecker (Picidae) often used. Site re-used in subsequent years. Solitary or colonial. Nest: layer of debris or wood dust. Building: woodpecker hole may be enlarged, but no material brought in; excavation by ♀, sometimes ♂ also (see Social Pattern and Behaviour). EGGS. See Plate 95. Broad oval, smooth, not glossy; 31 × 24 mm (29–33 × 23–25), $n = 39$ (Lamba 1966). Clutch: 3–4 (2–6) (Forshaw 1978). 2–4 in Britain (B Hawkes). Of 66 clutches, India: 2 eggs, 6%; 3, 33%; 4, 58%; 5, 3%; average 3·6 (Shivanarayan *et al.* 1981). Probably single-brooded in Britain (B Hawkes). Laying interval 1 day. INCUBATION. 22–24 (21–28) days; seldom begins until clutch near completion (Smith 1979); hatching asynchronous. By ♀ only (Shivanarayan *et al.* 1981; G A Smith). YOUNG. Altricial and nidicolous. Cared for and fed by both parents; brooded more or less continuously while small. FLEDGING TO MATURITY. Fledging period 40–50 days. Become independent some weeks later (Forshaw 1968); independent at *c.* 6 months (B Hawkes). Age of first breeding (2–)3 years. BREEDING SUCCESS. Hatching and fledging success in Britain reported as very high (B Hawkes). Of 236 eggs laid in 66 nests, India, 64·8% hatched, main losses to predation by snakes and crows; of 153 chicks hatched, 72% fledged; overall success 46·5% (Shivanarayan *et al.* 1981).

Plumages (*P. k. borealis*). ADULT MALE. Forehead, lores, ear-coverts, and cheeks bright yellowish-green with faint black line from nostril to front of eye; crown, nape, and back of ear-coverts and hindcheeks bright yellowish-green, extensively tinged sky-blue or pale lavender-blue on feather-tips, forming solid pale blue patch on hindcrown and nape narrowing to lower hindcheeks when plumage fresh, but much yellow-green of feather-bases exposed and blue almost absent when plumage worn. Band bordering base of lower mandible and along bare chin black, widening on black throat. Narrow collar round neck pink-red on hindneck, black bordered below by pink-red on sides of neck, black on foreneck where widening to throat. Upperparts including upper wing-coverts bright green, brightest on back and rump, slightly bluish just below pink-red band of hindneck, on tertial coverts, and on outer webs and tips of median and greater coverts; green of upperparts duller when plumage worn. Underparts including axillaries and lesser and median under wing-coverts yellowish-green, less bright and shiny than on face and cheeks, slightly tinged grey on chest, sides of breast, and upper flanks, more greenish-yellow on lower flanks, thighs, and under tail-coverts. Central pair of tail-feathers (t1) pale bluish-green, grading to pale turquoise on distal half or more; t2 pale bluish-green on outer web, yellowish-green on inner, t3–t6 bright yellowish-green on outer web, pale greenish-yellow on inner; all tail-feathers with contrasting horn-black shaft, narrow pale yellow fringe at tip, and with pale yellow undersurface with slight dusky distal suffusion. Flight-feathers, greater upper primary coverts, and bastard wing dark green, contrasting with paler green upper wing-coverts; shafts black, inner webs of flight-feathers broadly bordered dull black; inner web of p10 largely black and outer web dark bluish-green to cerulean-blue; tips of outer webs of secondaries, inner webs of primaries, and outer webs of (p7–)p8–p9(–p10) narrowly edged pale yellow or yellowish-white. Undersurface of primaries and greater under primary coverts greyish-black, contrasting with yellowish-green of remainder of underwing. ADULT FEMALE. Like adult ♂, but entire head and neck green, brightest and with faint dark loral line on face, slightly duller and similar to remainder of upperparts on crown, nape, and neck; some greenish-yellow of feather-bases sometimes exposed by wear. No blue on head or black on throat and no pink-red and black collar but usually a shiny emerald-green collar just visible and occasionally a faint dark green-grey bar on sides of neck, mainly caused by wear. NESTLING. Entirely naked, except for a few white hairs present at hatching, soon disappearing (Harrison 1975). JUVENILE. Like adult ♀, but green of head and body with slightly more yellowish tinge; tips of flight-feathers and greater upper primary coverts and inner webs of primaries narrowly edged yellow, these edges wider than those present in fresh adult, up to 1 mm; tail shorter than adult. FIRST ADULT. Like adult ♀ and difficult to distinguish, but usually retains some scattered worn juvenile flight-feathers with traces of rather wide yellow edges; tail on average shorter than in adult. Adult head markings in ♂ acquired after complete first post-breeding moult in 3rd calendar year.

Bare parts. ADULT. Iris pale yellow or yellowish-white. Bare skin round eye pink-yellow to orange. In *borealis*, bill coral-red, lower mandible sometimes partly marked black; in other races, lower mandible entirely black; tip of upper mandible black in *krameri* and slightly so in *parvirostris*. Leg and foot greenish-grey or greenish-slate. NESTLING. Iris bluish. Bill yellowish, changing to red before fledging. Bare skin pinkish with some faint blue patches. Leg and foot pink or whitish-pink. JUVENILE. Iris greyish-white. Bill coral-pink with pale tip. Leg and foot grey. (Ali and Ripley 1969; Harrison 1975; Forshaw 1978; BMNH.)

Moults. ADULT POST-BREEDING. Complete; primaries descendant and ascendant from p6; p6–p10 and p5–p1 moult at same time, as separate groups, or p6–p10 start earlier; sequence in p5–p1 group often irregular (Stresemann and Stresemann 1966). Flight-feather moult very slow, next feather sometimes shed when previous one already worn. In northern India, where breeding January–May(–July), moult of head, body, wing, and tail mainly May–September; feral European birds breed slightly later and moult completed up to November. Primary moult not completed in single moult season: suspended at end of season and resumed in next season from previous halting point, then also starting again with p6. Moult thus serially descendant and several series may be active on both p6–p10 and p5–p1 groups: e.g. p6, p8, and p10 growing with p7, p9, and p1–p5 still new. Secondaries serially ascendant, tail irregular. Sequence and timing of flight-feather and tail moult require confirmation, preferably in long-term study in the wild or else of captives of known age. POST-JUVENILE. Complete. Starts at age of c. 1 year; sequence as in adult, but flight-feather moult not serial; some juvenile flight-feathers retained into 3rd calendar year. FIRST POST-BREEDING. In 3rd calendar year. As in adult post-breeding.

Measurements. *P. k. borealis*. Northern Pakistan and north-west India, all year; skins (BMNH, RMNH). Bill measured to cere. Toe is outer front toe.

WING AD	♂	178	(4·36; 12)	172–187	♀	173	(3·37; 8)	168–178
TAIL AD		253	(15·2; 9)	229–279		221	(14·2; 5)	204–238
BILL AD		23·8	(0·96; 15)	23·2–26·4		22·6	(1·10; 9)	20·8–24·4
TARSUS		17·7	(0·91; 13)	16·4–18·8		18·4	(0·81; 9)	16·9–19·6
TOE		28·6	(1·35; 10)	26·9–30·7		28·1	(0·92; 5)	26·2–29·3

Sex differences significant, except tarsus and toe. Juvenile wing on average c. 10 shorter than adult, juvenile tail c. 60.

P. k. manillensis. Ceylon, all year; skins (BMNH, RMNH).

WING AD	♂	165	(3·68; 10)	160–169	♀	158	(2·65; 4)	154–160
TAIL AD		205	(20·3; 8)	182–235		178	(— ; 3)	164–188
BILL AD		24·2	(0·86; 9)	23·1–25·4		22·5	(0·27; 4)	22·2–22·8

Sex differences significant; wing and tail significantly shorter than *borealis*.

Nominate *krameri*. West Africa to western Uganda and south-west Sudan (Forshaw 1978).

WING AD	♂ 150	(— ; 12)	144–157	♀ 148	(— ; 10)	143–152
TAIL AD	231	(— ; 12)	194–278	198	(— ; 10)	177–240
BILL AD	19·6	(— ; 12)	18–21	19·8	(— ; 10)	18–21

Weights. *P. k. borealis*. Gujarat (India): ♂ 104–139 (5) (Ali and Ripley 1969). Nepal: April, ♂♂ 136, 143; March, laying ♀ 158 (Diesselhorst 1968).

Nominate *krameri*. Ennedi (Chad), April, ♂♂ 51, 85, 93; Niger, ♀ 92 (Niethammer 1955).

Structure. Wing rather long and narrow, tip pointed. 11 primaries: p9 longest, p10 and p8 0–4 shorter, p7 16–23 shorter, p6 26–35, p5 37–49, p1 70–83; p11 minute, concealed by primary coverts. Outer web of p8–p10 and inner web of p9–p10 slightly emarginated. Tail very long, graduated; 12 rather narrow and tapering feathers, t6 100–170 shorter than t1 in adult, 70–100 in juvenile. Typical parrot bill: short and broad, upper mandible strongly decurved, extending much over strongly recurved lower mandible; culmen with gently rounded ridge. Cutting edges of upper mandible sharp, indented at base of nail; tip sharp, inside flattened and with small transverse grooves; lower mandible broad and shovel-like, with straight sharp cutting edge at tip. Narrow bare cere along base of upper mandible, in which small rounded nostrils situated close to top. Tarsus short and stout, upper half covered by feathers growing from tibia. Toes rather long and strong, soles rough and flattened; outer toe directed backwards; skin at basal part of middle and inner toe fused. Outer hind toe c. 84% of outer front toe, inner front c. 69%, inner hind c. 53%. Claws strong and sharp, strongly decurved.

Geographical variation. Marked in size, slight in colour. In Indian subcontinent, variation in size clinal, in colour (except bill) negligible. Average wing (sexes combined) in Punjab (north-west India) 178, Gujarat (central western India) 173, Bombay area and Deccan (western peninsular India near 20°N boundary) 169, Madras (southern India) 166, Ceylon 161 (BMNH, RMNH). In typical *borealis*, both mandibles red, but lower mandible sometimes partly black; in typical *manillensis* lower mandible black and sometimes some black on tip of upper mandible; at 20°N in peninsular India, amount of red variable, either intermediate in extent or like typical specimens of both races (Ali and Ripley 1969). Nominate *krameri* from Sénégal east to Niger and Nigeria smaller than Asiatic races, colour paler yellow-green; upper mandible red with much black on tip, lower mandible black with limited red near base. *P. k. parvirostris* from Sennar area (on Blue Nile in eastern Sudan), Ethiopia, and northern Somalia darker green, less yellowish, pink-red collar of ♂ more prominent; wing slightly longer, average of both sexes 153, bill slightly smaller (Forshaw 1978). Birds inhabiting area between Lake Chad, northern Cameroun, Ennedi (Chad), and upper White Nile in southern Sudan and north-west Uganda like nominate *krameri* but with slightly smaller and more extensively red bill of *parvirostris*; sometimes separated as *centralis* (Neumann 1915), but better included in nominate *krameri*. Races of widely introduced feral birds not fully known; most attributed to *borealis*. Few specimens available, but Egyptian birds nearest *manillensis* (Goodman 1982) and old feral specimens examined from Netherlands had average adult wing 169 and strongly variable lower mandible colour, indicating that either a mixture of both *borealis* and *manillensis* was involved or that birds from Bombay area were introduced.

P. krameri forms superspecies with rear Mauritius Parakeet *P. echo* from Mauritius and perhaps formerly Réunion. CSR

Order CUCULIFORMES

Small to large, insect- or fruit-eaters. 2 families currently recognized (1 represented in west Palearctic): (1) Musophagidae (turacos, *c.* 19 species; Afrotropical region); (2) Cuculidae. Musophagidae sometimes confined to suborder Musophagi or even treated as separate order. Some authorities have treated the family Opisthocomidae (single species, Hoatzin *Opisthocomus hoazin* of South America) in Cuculiformes, but more usually included in Galliformes.

Family CUCULIDAE cuckoos

Small to fairly large birds, mainly insect-feeders. 6 subfamilies (3 represented in west Palearctic): (1) Cuculinae; (2) Phaenicophaeinae; (3) Crotophaginae (3 species of ani, and Guira *Guira guira*; tropical America); (4) Neomorphinae (roadrunners and ground-cuckoos, 13 species; Nearctic and south-east Asia); (5) Couinae (couas, 10 species; Madagascar and adjacent islands); (6) Centropodinae. Distinction between subfamilies not clear-cut, however, and some genera difficult to allocate.

Bodies tend to be elongated, and necks moderately long (13 cervical vertebrae). Wings variable—moderately long and rounded or long and pointed; 10 primaries, 9 secondaries, wing eutaxic. Tails may be medium length to graduated and very long; 10 feathers (8 in Crotophaginae). Bills usually moderately long and rather stout and slightly decurved, though very deep and laterally compressed in *Crotophaga* (Crotophaginae). Legs short in most arboreal species, long in those that live on ground; feet zygodactyl, with outer toe permanently reversed. Oil-gland naked. Caeca rather long. Feathers without aftershaft or with small one only. Plumages rather loose, though stiff and wiry in some species; wide variations in colour, though sexes tend to be rather similar, while juveniles may differ markedly. Post-breeding moult complete; often in winter quarters in migratory species; flight-feather moult complex. Young altricial and nidicolous. Adult plumage acquired in 1st winter. All species of Cuculinae and 3 tropical American species of Neomorphinae practise brood-parasitism.

Subfamily CUCULINAE parasitic cuckoos

42–50 species in 15–17 genera: (1) *Clamator* (4 species, Africa, southern Asia, and southern Europe); (2) *Cuculus* (12 species, Europe, Africa, and Asia to Australia); (3) *Cercococcyx* (3 species, Africa); (4) *Cacomantis* (*c.* 5 species, Oriental and Australasian regions); (5) *Chrysococcyx* (4 species, Africa); (6) *Chalcites* (4–8 species, Oriental and Australasian regions); remaining genera are monotypic, half of them restricted to Australasian region. Represented in west Palearctic by 2 species of *Clamator* (1 breeding, 1 accidental), 2 of *Cuculus* (breeding), and 1 of *Chrysococcyx* (accidental).

For general features, see Cuculidae. Nostril small, rounded, edges raised. All species brood-parasitic.

Clamator jacobinus Jacobin Cuckoo

PLATES 36 and 37
[facing pages 399 and 470]

Du. Jakobijnkoekoek   Fr. Coucou jacobin   Ge. Jakobiner Kuckuck
Ru. Двухцветная хохлатая кукушка   Sp. Críalo etiope   Sw. Svartvit gök

Cuculus Jacobinus Boddaert, 1783

Polytypic. *C. j. pica* (Hemprich and Ehrenberg, 1833), northern Afrotropics (probably including southern Arabia) south to northern Namibia, Botswana, Zambia, and Tanzania, and India south to Bombay, northern Bihar, west through Pakistan to south-east Iran. Extralimital: nominate *jacobinus* (Boddaert, 1783), Ceylon and India south of *pica*; *serratus* (Sparrman, 1786), southern and eastern South Africa north to Zimbabwe and Moçambique, south of Zambezi river.

Field characters. 31–33 cm; wing-span 45–50 cm; tail 12–14 cm. About 25% smaller than Great Spotted Cuckoo *Clamator glandarius* but with similar structure. Rather small, crested cuckoo sharing general character of *C. glandarius* but instantly distinguished by black and (usually) white plumage. Wings show bold white patch at base of primaries; tail-feathers boldly tipped white. Noisy, with laughing cackle. Sexes similar; no seasonal variation. Juvenile separable.

ADULT. Whole of upperparts and flight-feathers glossy black, relieved only by broad white bases to primaries and long white tips to tail-feathers. In central Africa, underparts usually dull white, washed yellow-buff from chin to chest and over vent; occasionally, all black or streaked black (see Geographical Variation). Bill black; legs dusky grey. JUVENILE. Upperparts dusky brown, with little gloss; underparts tinged buff or dusky. Smaller white patch on primaries than in adult.

No similar species recorded in west Palearctic, but in Africa easily confused with sympatric Levaillant's Cuckoo

C. levaillantii: no detailed criteria for separation, but *C. levaillantii* larger (*c*. 39 cm long) with (usually) throat and chest heavily streaked (never all-pale) or entire underparts black. Flight action and behaviour of *C. jacobinus* resemble those of *C. glandarius*. Frequently flirts tail. Typically inhabits dry savannas with patches of dense cover; much more skulking than Cuckoo *Cuculus canorus*.

Commonest call a harsh loud cackle, 'quer-qui-quik', with laughing quality.

Habitat. Mainly tropical and subtropical, both in lowlands and up to *c*. 2600 m in peninsular India and *c*. 1300 m in Ceylon, straggling up to *c*. 4270 m in Himalayas. In West Africa especially common in semi-arid parts of Sahel, records from forest probably being birds on passage (Serle *et al*. 1977). Often also in thick swampy bush (Mackworth-Praed and Grant 1970). Normally in mimosa and acacia savannas and thickets in thorn-scrub zone (Bannerman 1933). In southern Africa on open thornveld (Prozesky 1970). In East and central Africa frequents thorn-bush and scrub country, various types of woodland, and cultivated areas where there are scrub and trees: on spring migration common on Kenya coast (Williams 1963). In India, breeds in open well-wooded country, gardens, groves (even within precincts of towns and villages), and neighbourhood of cultivation, in plains, lowlands, and hills, in dry as well as moist deciduous regions, and in stunted jungle in semi-desert areas. Largely arboreal, but also descends into low bushes and hops about on ground in search of food (Ali and Ripley 1969). At Delhi on golf links in thickets and calling from trees and telephone wires; also in gardens, in woods, and in scrub country (Hutson 1954). In Kashmir, a main stronghold is bush-covered delta of Sind river; perches on top of a bush or creeps about lower foliage in search of insects (Bates and Lowther 1952). Flies freely, apparently keeping within lower airspace.

Distribution. Breeds in Africa south of Sahara, from Sénégal to Red Sea, south to Cape Province (except in heavily forested areas), probably in southern Arabia (see Movements), and in south-east Iran, Pakistan, India, Ceylon, and Burma. Resident and migratory (see Movements).

Accidental. Chad: specimen, 9 September 1955, Tibesti (Malbrant 1957).

Movements. Different populations dispersive or migratory; details often unclear where populations overlap during 'leap-frog' migrations.

ASIATIC POPULATIONS (breeding Iranian Baluchistan and Indian subcontinent). Northern race *pica* (Baluchistan to northern India) migratory, arriving with rains in late May to early June and leaving September–October. This race not found wintering anywhere in Asia, and presumed to migrate to Africa where not readily separable from birds of same race occupying much of Afrotropics (Whistler 1928, 1931; Friedmann 1964). Some East African specimens have high wing/tail ratio suggesting Indian *pica* (C S Roselaar). Recorded only irregularly in southern Arabia (e.g. Jennings 1981a, Cornwallis and Porter 1982) and, since 9 such birds measured were larger than those of northern India (see Measurements), some records, at least, probably relate to separate (unproven) south Arabian breeding popu-lation (C S Roselaar). Thus, passage of *pica* to Africa (if it occurs) may be direct across Arabian Sea/Indian Ocean. Nominate *jacobinus* (peninsular India) resident there yet also dispersive; Clancey (1960, 1965) attributed small specimens from south-east Africa (Malawi, Zimbabwe, Moçambique, Natal), September–April, to immigrants from this population, but wing lengths more consistent with East African origins (C S Roselaar).

AFRICAN POPULATIONS. A rains migrant in southern Africa south of *c*. 10°S (*serratus*), arriving late September and October, departing (after breeding) March–April (e.g. Clancey 1964, Benson *et al*. 1971). These migrate north towards and (to some extent) beyond equator, passage being noted especially in eastern half of continent April–May (Archer and Godman 1961; Friedmann 1964; Britton 1980). Movement probably reaches Sudan, where the species is a common non-breeding migrant March–October in south and June–September in Darfur (Lynes 1925a; Cave and Macdonald 1955); specimens of *pica* obtained from Sudan, and of *serratus* from Chad, Sudan, and Ethiopia. Complex movements in East Africa, November–January (Britton 1980), but not certain how these relate to local and southern populations and putative presence of Indian birds (see above).

Marked movements also by populations of northern Afrotropics (*pica*). A not uncommon intra-African migrant in Nigeria north of great rivers, extending northwards in rains during April–October (breeding May–July); southerly records sparse and in dry season (Elgood *et al*. 1973; Elgood 1982). A rains migrant in Mali, May–November (breeding June–September), disappearing with onset of dry season (Lamarche 1980). In central Chad, a wet-season visitor June–October, though breeding status not clarified; regularly penetrates north to Ennedi (Salvan 1968b; Newby 1980).

Voice. See Field Characters.

Plumages. Polymorphic; see Geographical Variation. Following description is of intermediate morph (of northern Afrotropics). ADULT. Forehead, crown, and crest black with, slightly glossed purple-blue or blue-green, black reaching down to lores, ear-coverts, and just below eye; crown in some lights appearing streaked by glossy black shafts. Hindneck, mantle, scapulars, tertials, upper wing-coverts, and back to upper tail-coverts black with slight green or purple-blue gloss. Underparts, including sides of neck, pale yellow-buff, deepest on chin, throat, sides of neck, and chest; throat and chest with narrow grey shaft-streaks and with some grey of feather-bases visible; upper flanks and thighs tinged grey; in worn plumage, grey of feather-bases and darker grey shaft-streaks show distinctly on throat, chest, and sides of neck; belly and vent off-white. Tail black with slight green or blue-green gloss; t1 tipped with white broken band 3–5 mm wide; increasing amount of white on tips towards outer feathers; 25–35 mm of tip white on t5. Flight-feathers black with slight green gloss, rapidly fading to blackish-brown with wear; primaries with much white at base (usually except for outer web of p10 and inner web of p1), sharply demarcated from black middle portion and tip; white extends up to c. 20 mm beyond greater primary coverts on upperwing, up to c. 30 on underwing. P10 sometimes with narrow white border along sides, not on tip. Under wing-coverts yellow-buff, longest ones and axillaries black. JUVENILE. Like adult, but upperparts and upper wing-coverts dark bronze-brown rather than black, darkest (dull greenish-black) on crown and mantle; crest feathers shorter than in adult (15–30 mm instead of 25–35), tips rounded rather than pointed; chin and sides of neck to chest light grey, with buff feather-tips and faint dark grey shaft-streaks on chest, appearing greyish and streaked when worn; remainder of underparts yellow-buff. Tail-feathers relatively narrower (15–20 mm instead of 20–30 as in adult) and with less white on tips; virtually no white on tip of t1, small spot of 3–4 mm on tip of t2 (measured along shaft), 11–15 on t3, 20–25 on t4–t5 (in adult, small spot on t1, 10–15 mm on tip of t2, gradually increasing to c. 30 on t5); often a white streak along outer web of t5. Flight-feathers slightly browner than in adult; white bases of primaries more restricted, reaching up to 10–18 mm beyond greater upper coverts, and less sharply defined from brown-black tips, showing some mottling at demarcation. Greater upper primary coverts with pale buff or white edges up to 3 mm wide on tips; primaries and sometimes greater secondary coverts narrowly tipped pale buff or white; median and longer lesser coverts often tipped buff. In worn plumage, colour differences from adult rather difficult to establish, but more limited and less clear-cut white on primary-bases and buff or off-white edges of greater primary coverts usually readily visible. FIRST ADULT. Not distinguishable from adult when post-juvenile moult completed. Last juvenile feathers to be replaced are scattered flight-feathers and greater coverts (primaries with more restricted and less sharply defined white at base than fresh neighbouring feathers, some greater upper wing-coverts tipped off-white, all strongly tinged brown and relatively narrower than fresh feathers), some tail-feathers (relatively shorter, narrower, and with less white on tips than neighbouring fresh feathers), or some occasional scapulars or median and lesser upper wing-coverts.

Bare parts. ADULT AND JUVENILE. Iris dark brown or greyish-brown. Bill black, tinged lemon-yellow or pale horn near chin; blue-grey or slightly pinkish at base shortly after fledging. Leg and foot dark grey, slate-grey, or black, soles yellow, pale brown, or pale grey, claws black. (BMNH, RMNH.)

Moults. ADULT POST-BREEDING. Complete, primary moult complex. Mainly in winter quarters, though sometimes starting in breeding area; timing not fully established due to extensive overlap between populations outside breeding season. Starts with scattered feathers of lower mantle, scapulars, ear-coverts, crown, and chest, soon followed by p6; all body new when scattered old flight- and tail-feathers still present. Sequence of primary moult not fully established: according to Stresemann and Stresemann (1969) as in Great Spotted Cuckoo *C. glandarius*, with an outer group (p5–p10) moulting in sequence of 6–9–7–10–8–5 or 6–9–7–10–5–8 and an inner group (p1–p4) moulting in irregular sequence at same time, 2 neighbouring feathers never growing simultaneously; according to specimens from BMNH and RMNH, primary moult probably starts both ascendantly and descendantly from 3 centres (p2, p6, and p9), all 3 starting at same time or that at p9 last; p6 followed by p7 and subsequently by p5 (next feather lost when previous one full-grown), while during same period p2 first followed by p1 and next by p3 (or p3 first and p1 next), and p9 by p10 and subsequently by p8; p4 either belongs to inner moult-centre (falling after p3 or p1) or to middle centre (falling after p5), and may grow simultaneously with p5 or p3. Secondary moult apparently irregular, but neighbouring feathers do not grow at same time; moult of tail irregular, but t1–t3 never all grow at same time, 1 of these pairs being retained until other 2 full-grown. POST-JUVENILE. Complete. As adult post-breeding. May start shortly after fledging when still near breeding area, or not until arrival in winter quarters.

Measurements. *C. j. pica.* Nigeria, Sudan, Ethiopia, and Somalia, March–September; skins (BMNH, RMNH). Bill (F) to forehead, (N) to nostrils. Toe is outer front toe.

WING AD	♂ 153	(3·33; 10)	148–157	♀ 156	(5·25; 9)	151–164
TAIL	176	(5·87; 8)	170–186	181	(8·34; 7)	171–195
BILL (F)	25·5	(0·85; 8)	24·6–26·8	25·8	(0·69; 5)	24·8–26·7
(N)	16·2	(0·49; 7)	15·5–16·7	16·7	(0·89; 8)	15·5–17·6
TARSUS	27·4	(0·91; 8)	26·1–28·5	27·3	(1·71; 6)	25·6–29·3
TOE	26·5	(1·16; 6)	25·5–27·7	27·0	(1·24; 4)	25·2–28·1

Sex differences not significant. Juvenile wing and tail both on average 5·5 shorter than adult.

Adults, about equal number of both sexes combined. *C. j. pica*: (1) Kenya and Tanzania, January–April (moulting birds excluded); (2) Arabia, March–June; (3) northern India and Burma, May–September. (4) Nominate *jacobinus*: Ceylon. *C. j. serratus*, southern Africa, November–March: (5) grey morph, (6) black morph. (BMNH, RMNH, ZMA.) Wing of ♀ averages 1–3 longer than ♂, tail 3–5, but sex differences not significant. Bill measured to nostril.

WING (1)	147	(2·60; 10)	143–150	(4) 138	(2·94; 14)	133–143
TAIL	168	(6·27; 10)	159–176	155	(5·41; 13)	148–164
BILL	16·4	(0·51; 10)	15·5–17·3	15·7	(0·60; 10)	14·9–16·4
WING (2)	160	(2·37; 9)	156–163	(5) 159	(3·78; 11)	154–165
TAIL	185	(9·14; 9)	172–195	180	(5·20; 9)	174–188
BILL	17·1	(0·94; 8)	15·8–18·1	16·8	(0·74; 8)	15·8–17·8
WING (3)	152	(2·78; 20)	148–158	(6) 154	(3·51; 15)	149–162
TAIL	168	(4·81; 20)	161–176	175	(5·07; 15)	168–185
BILL	16·4	(0·56; 19)	15·7–17·4	16·8	(0·56; 15)	16·0–17·7

Weights. Nominate *pica* (but may include some *serratus*). Nigeria, April–June: 66, 74, 83 (Fry 1970). Ethiopia, June: ♀♀ 80, 84 (Britton 1970). East Africa: ♂♂ 58·5, 70, 71; ♀ 82·2; immature ♂ 62, immature ♀ 50·5, sex unknown 68, 70, 86 (Meise 1937; Moreau 1944; Britton 1970; Colston 1971). Zaïre, January–February: ♂♂ 56, 69; ♀ 76 (Verheyen 1953). India 61–74 (4); very fat by September–October (Ali and Ripley 1969).

Structure. Wing rather long, base broad, tip rounded. 10 primaries: p7 longest; in adult, p8 0–2 shorter, p9 11–16, p10 50–56, p6 2–25, p5 13–17, p4 25–29, p1 44–49; in juvenile, p8 4–6 shorter, p9 18–23, p10 54–62. Tail long, graduated, 10 feathers; t5 60–70 shorter than t1. Bill as in *C. glandarius*, but nostril slightly shorter and more rounded. Leg and foot as in *C. glandarius*.

Geographical variation. Marked, in size and in occurrence of colour morphs. (1) Pied morph. Underparts uniform buff (when fresh) or off-white (when worn), throat and chest without grey feather-bases showing and without darker grey shaft-streaks, or latter only showing very faintly when plumage worn; throat and chest of juvenile buff, not uniform grey. (2) Intermediate morph (as described in Plumages). As pied morph but with grey feather-bases and grey shaft-streaks on throat and chest; all stages of transition between pied and grey morphs occur. (3) Grey morph. Like intermediate morph, but streaks on throat and chest heavier; grey on chest, upper breast, flanks, and thighs more extensive (occasionally, whole underparts grey); streaks on sides of breast often joining to form black patches; tail-tips often with less white. (4) Black morph. Entirely black (including tail-tips) except for white primary-patch. All birds of Indian subcontinent and most found in Arabia uniform buff below, as in pied morph, showing faint grey streaks on throat at most; of these, birds of Ceylon and southern India smaller, separated as nominate *jacobinus*; those of northern India, Burma, Pakistan, and south-east Iran larger, named *pica* (though paler than typical *pica* from Sudan); those found in Arabia largest, provisionally included in *pica* also. In northern Afrotropics, from West Africa east to Ethiopia and Somalia, typical *pica* occurs; most are intermediate morph and hence are darker than Indian *pica*, but black morph occurs exceptionally; wing length similar to Indian *pica* but tail averages longer. In Africa south of Zambezi, all birds are exclusively grey or black morphs, and these separated as *serratus*; large in size (especially grey morph), approaching Arabian birds. Situation in East Africa confusing, as birds from both northern and southern Afrotropics and probably Indian birds occur there in non-breeding season; local breeders similar in colour to those of northern Afrotropics, but slightly smaller; migrants show all colour morphs and have large size range. For details of distribution of morphs and migration in southern Africa, see Clancey (1960, 1965) and Liversidge (1970); *serratus* differs from other races in preferred egg hosts (Snow 1980) and in egg colour (Harrison 1971). CSR

Clamator glandarius Great Spotted Cuckoo

PLATES 36 and 37
[facing pages 399 and 470]

Du. Kuifkoekoek Fr. Coucou-geai Ge. Häherkuckuck
Ru. Хохлатая кукушка Sp. Críalo Sw. Skatgök

Cuculus glandarius Linnaeus, 1758

Polytypic. Nominate *glandarius* (Linnaeus, 1758), southern Europe, Middle East, and Africa north of tropical forests and northern Tanzania. Extralimital: *choragium* Clancey, 1951, southern and eastern Africa north to about Zambia and southern Tanzania.

Field characters. 38–40 cm; wing-span 58–61 cm; tail 14–18 cm. About 15% larger than Cuckoo *Cuculus canorus*, with broader wings and fuller, more wedge-shaped tail. Quite large, crested cuckoo, with dark upperparts and pale underparts. Wing-feathers boldly spotted off-white. Bold and noisy during breeding season but not otherwise. Sexes similar; no seasonal variation. Juvenile separable.
  ADULT. Long crested crown blue-grey, with dusky streaks and small cream spots in fresh plumage. Lores, upper cheeks, and hindneck black-brown; rest of upperparts dusky brown and distinctly spotted or fringed dull white on scapulars, inner wing, longest upper tail-coverts, and tips of tail-feathers. Boldest spotting on inner wing visible in flight. Underparts cream-white with pale orange wash on lower face and in front of shoulders. Under wing-coverts cream. Bill grey-black; eye-ring bright red. Legs brown-grey. JUVENILE. Resembles adult but easily distinguished by shorter crest on black crown, darker upperparts with smaller cream (not white) spots and fringes, and chestnut dark-tipped primaries. Chestnut primaries conspicuous in flight. Underparts distinctly warmer-toned than adult, with orange wash strong on lower face and sides of neck and chest. FIRST YEAR. Resembles adult, but retains chestnut-brown on bases of primaries and dull spots on wing-coverts.
  Unmistakable in most circumstances; basic plumage pattern similar to Jacobin Cuckoo *Clamator jacobinus* but lacks white patch at base of primaries. Flight action stronger than that of *Cuculus canorus*, with rather slower wing-beats still creating quite rapid progress; track fairly direct, occasionally interspersed with jinks and dips. Tail trails noticeably and looks bushy during landing. Perches freely and conspicuously. Parasitic on corvids; during breeding season, mobbing Magpies *Pica pica* frequently demonstrate presence of otherwise hidden bird.
  Noisy during breeding season; song a rasping chatter, turning to a gobble: 'kittera kittera ... kee-ow kee-ow ... wow wow wuh'. In alarm, a characteristic crow-like bark.

Habitat. In Mediterranean, semi-arid, subtropical, and tropical zones, generally in warm lowlands, unusual above *c*. 500 m and generally avoiding mountains, forests (especially rain-forest), and wetlands, although tolerating oceanic climate on Iberian Atlantic coast. Population in Coto Doñana (Spain) concentrated along heathlands, interspersed with cork oaks *Quercus suber*, and with thickets of bramble *Rubus*, tree-heath *Erica arborea*, juniper *Juniperus*, *Pistacia lentiscus*, and other shrubs used for nesting by many Magpies *Pica pica* on which this species is locally parasitic; to lesser extent, in open woodland of stone pines *Pinus pinea* fringing the heath, but neighbouring mobile dunes and extensive marshes avoided, although not freshwater *lagunas* on sandy heaths (Mountfort 1958; Valverde 1958). Also favours cultivated areas such as olive or almond groves and adjoining parkland or open oak and pine woodland, or willow *Salix* spinneys along stream, occasionally penetrating into towns or villages. Perches, calls, and often feeds in trees, but also descends freely to hop or run on ground, moving restlessly over wide

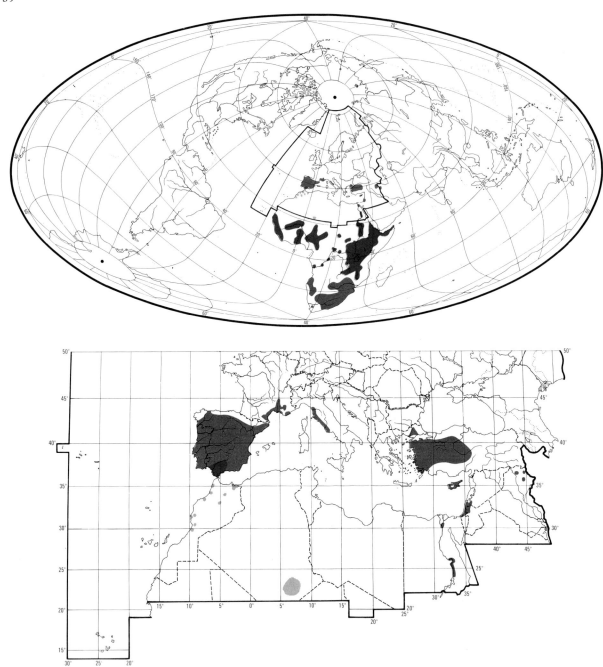

area of varied territory, and using any convenient look-out post, including artefacts. In West Africa, frequents principally arid thorn-scrub and grass woodland, but not forested country, favouring clearings and vicinity of villages, and areas where fig *Ficus* trees plentiful (Bannerman 1955). In southern Africa, also prefers fairly open wooded country (Prozesky 1970). In Somalia, ranges from sea-level to plateau at *c*. 600–1600 m, habitat varying from coastal date groves, clumps of palm, and native gardens to open thorn-bush and heavily foliaged wild sycamore-fig trees marking wells in uplands (Archer and Godman 1961). Choice between wild and man-made habitats subordinate to overriding needs for ample accessible supply of suitable food and availability of nests of preferred host species. Although a strong flier, often prefers to travel close to ground.

Distribution. Spread in France, but apparently no longer breeding in North Africa.

FRANCE. Marked spread. Before 1943, bred only 1885

and 1924 (Mayaud 1936b); from 1943 bred Hérault (Hüe 1945), 1945 Tarn (Lévêque 1968), 1947 Aude (Hüe 1947a), 1952 Bouches-du-Rhône (Hüe 1952), 1964 Gard (Lévêque 1968), and 1970 Alpes-de-Haute-Provence (Crocq 1975). ITALY. Spread south and east (SF); breeds irregularly Sardinia (PB, BM). YUGOSLAVIA. Breeding attempted or suspected in 8 years between 1889 and 1974 (Matvejev and Vasić 1973; VFV). GREECE. First proved breeding 1978 (2 pairs); since occurred at 10 sites, but breeding not proved (HJB, WB, GM). LEBANON. Bred 1954; no recent records (HK). SYRIA. Probably breeds in small numbers in north and north-west, but no proof (Kumerloeve 1968b; HK). JORDAN. May breed in north of Jordan valley (DIMW). TUNISIA. Has bred (Heim de Balsac and Mayaud 1962); no recent evidence (MS). ALGERIA. Bred 19th century (Heim de Balsac and Mayaud 1962); no recent breeding records (EDHJ). MOROCCO. No proof of nesting (M Thévenot, JDRV).

Accidental. Britain, Ireland, Belgium, Netherlands, West Germany, Denmark, Norway, Sweden, Finland, East Germany, Austria, Switzerland, Yugoslavia, Albania, Bulgaria, USSR, Kuwait, Jordan, Libya, Madeira.

Population. Increased in northern Spain, France, and Italy.

FRANCE. Increased since 1943; 100–1000 pairs (Yeatman 1976). SPAIN. Numerous in southern half (AN); increased in north (Lama 1959). ITALY. Some recent increase (SF, PB). IRAQ. Probably common in north-west; first proved breeding 1962 (Marchant and Macnab 1962; Marchant 1963c). EGYPT. Little recent information, but breeding perhaps less common than earlier in 20th century (PGM, WM). CYPRUS. Scarce (Stewart and Christensen 1971; PRF, PS).

Movements. Migratory at northern and southern ends of breeding range (southern Europe and Middle East; southern Africa), but in between is resident or makes short-distance movements to avoid excesses of drought and humidity.

AFROTROPICAL POPULATIONS. Migratory population *choragium* present in South African breeding areas late September to March; winters further north, into central Africa, but range undefined (McLachlan and Liversidge 1970; Clancey 1965). Kynes (1925a) saw marked influx of non-breeding adults and juveniles in Darfur (Sudan), May–July; hence of Afrotropical rather than Palearctic origin, but not known whether *choragium* involved this far north. In Chad and Nigeria there are indications of northwards post-breeding movement into Sahel zone for autumn rainy season (when Palearctic birds might occur also) (Salvan 1968b; Elgood *et al.* 1973).

PALEARCTIC POPULATIONS. Migratory. Small numbers winter in western Morocco (Smith 1965), in Nile valley, Egypt (Etchécopar and Hüe 1967), and at least occasionally in Arabia (Bundy and Warr 1980), but many European and Asiatic birds believed to winter in Africa below Sahara where, however, range very poorly known due to impossibility of distinguishing northern migrants from birds of same race resident there (no African ringing data). Known to migrate in numbers (seen especially in spring) through north-west Africa, Egypt, and western Arabia (Meinertzhagen 1930, 1954; Heim de Balsac and Mayaud 1962), though not Libya (Bundy 1976), which may reflect gap in European breeding range (i.e. rarity in Italy and Balkans). On present knowledge (certainly incomplete), may also be corresponding gap along southern side of Sahara (Moreau 1972). Migrants pass through Mauritania in large numbers in August, and occur Sénégambia through to February (e.g. Morel and Roux 1966, Gore 1981); in Mali, local population of Niger inundation zone clearly augmented August–March (Curry and Sayer 1979; Lamarche 1980), and circumstantial evidence indicates that winter visitors (November–March) in Sudan and Ethiopia are or include Palearctic migrants (Lynes 1925a; Moreau 1972). However, no discernible winter influx into Nigeria or Chad, where occurrence of Palearctic birds still problematic (Salvan 1968b; Elgood 1982). No good evidence at present that northern birds move further south than c. 10°N (Moreau 1972). However, transient birds in Kenya and Tanzania occur mainly September–March; status obscured by East African breeders and intra-African migrants from south, but temporal pattern consistent with (unproven) involvement of northern birds also (Britton 1980).

Adults start leaving territories in June; exodus from Europe begins July, peaking in August. Juvenile passage averages later, some remaining in Mediterranean Europe into November. Arrivals south of Sahara from August, and some late-July dates possibly Palearctic birds. Intraseasonal movement evident within winter range: leaves Sénégal in December as aerial insect numbers decline; southerly movement noted Mali and Sudan in December; in Eritrean coastal plain, winter visitors arrive as late as December (Morel and Roux 1966; Moreau 1972; Lamarche 1980). Return movement in North Africa and southern Spain from early February; main spring passage March to mid-April. With late departure of some juveniles and early appearance of spring vanguard, may be absent from Mediterranean area for only a few weeks around midwinter. 2 nestlings ringed Bouches du Rhône (France) found the following spring (17 March, 9 April) in Salerno and Rome (south-west Italy).

Food. Chiefly hairy caterpillars. Prey taken in trees and on ground, then 'dressed' to remove hairs before being eaten (e.g. Mestre Raventós 1969). In Camargue, hops along ground, with tail raised, searching for prey, and also feeds in bushes and trees (Frisch and Frisch 1967). When hunting on ground, periodically makes long bounds (Nicolau-Guillaumet and Spitz 1958). In Sierra Leone, also catches termites (Isoptera) by vertical upward flights

(G D Field). Not known if ♀ eats any of the eggs she may occasionally remove from brood-hosts' nests when laying her own.

In breeding season, recorded prey comprises hairy caterpillars (Lepidoptera), e.g. larvae of processionary moths *Thaumetopoea pityocampa* and *T. processionea*; grasshoppers, crickets, and bush-crickets (Orthoptera), dragonflies (Libellulidae), ants (Formicidae), beetles (Coleoptera), and lizards (Hüe 1952; Valverde 1953a; Laferrère 1956; Nicolau-Guillaumet and Spitz 1958; Mestre Raventós 1969; Carlo 1971).

In pine *Pinus* and juniper *Juniperus*, Spain, March to April, 11 stomachs contained exclusively *T. pityocampa* caterpillars, mean 32·8 (22–41) per stomach (see Valverde 1953a). For similar diet when caterpillars emerge, southern France, see Laferrère (1956). In Italy, almost exclusively *T. pityocampa* and *T. processionea* for as long as these available (Carlo 1971). No definite information for west Palearctic birds in winter quarters, but diet of *C. glandarius* in West Africa includes hairy caterpillars, grasshoppers and locusts (Acrididae), ants, termites, and other insects (Bannerman 1933; G D Field).

Diet of young apparently only insects, probably because they refuse other sorts of food (e.g. grain, vertebrate prey) brought by hosts (Valverde 1953a; Carlo 1971). Most frequent items in diet, Italy (where host Magpie *Pica pica*), locusts, caterpillars of *Arctia villica*, beetles, ants, and insect eggs (Arrigoni degli Oddi 1929); also dragonflies, snails *Helix*, spiders (Araneae), etc. (Carlo 1971). In 4 stomachs, Italy, many larvae of sawfly *Arge*, 2 large *Arctia* caterpillars, and over 30 other hairy caterpillars (see Carlo 1971). At Valladolid (Spain), stomach of young near fledging in *P. pica* nest, May, contained 1 cricket *Gryllus*, 1 grasshopper (Acrididae), 1 bug (Hemiptera), c. 30–40 beetles, 15 larvae (probably Coleoptera), and 1 snail (*Helix* or similar). Stomach of independent young contained 1 adult cockchafer (*Melolontha* or similar), 2 'Tettyngia' (possibly bush-cricket *Tettigonia*), and 1 caterpillar 55 mm long (Valverde 1953a). Stomach of young being fed by *P. pica*, le Midi (France), contained beetles; stomach of an independent juvenile contained Orthoptera, and 2 juveniles were seen chasing Orthopterans (Hüe 1952). For feeding rate at nest in Spain, see Alvarez and Arias-de-Reyna (1974).

EKD

Social pattern and behaviour. Based partly on material supplied by L Arias-de-Reyna. For behaviour in relation to other *Clamator*, see Friedmann (1964).

1. Mildly gregarious on migration. No definite information for west Palearctic birds in winter quarters. In Dronais (France), November, flocks of 5–6, presumably immatures, fed together (Labitte 1940). Shortly before migration, Spain, small flocks comprise 2–3 adults and several young (L Arias-de-Reyna). In June–July, Sudan, migrating parties of up to 5 (Lynes 1925a); in early February, Egypt, migrating parties of 10–20 (Meinertzhagen 1930). For flocks of 6–8, Namibia, see Hoesch (1934). BONDS. Mating system not well studied, though almost certainly monogamous (L Arias-de-Reyna); pair-bond lasts at least 1 breeding season (Friedmann 1948; Frisch 1969c). Possibility of polygamy (see Lévêque 1968) assumed from reports of more than 1 ♀ laying in same nest, and direct evidence lacking— multiple use of nest more likely to arise from overlap of breeding territories and paucity of brood-host nests late in season (Arias-de-Reyna *et al.* 1982; L Arias-de-Reyna; see Breeding Dispersion, below). In Coto Doñana (Spain), sex-ratio apparently 2–3 ♂♂ : 1 ♀ (Valverde 1953a). Since adults migrate to winter quarters before young, and young are not raised by own parents, no family bonds likely. For reports of adults feeding fledged young, see Relations within Family Group (below). Probably breeds first at 1 year (see Nicolau-Guillaumet and Spitz 1958). BREEDING DISPERSION. Solitary. ♂ establishes and, with help of ♀, defends breeding territory, especially at high densities (Valverde 1953a, 1971; Arias-de-Reyna *et al.* 1982). In one case, territories overlapped, and boundary disputes frequent; overlap attributed to rivalry between neighbouring pairs to annexe 'best' areas occupied by brood-host (Magpie *Pica pica*), leading to 2 ♀♀ laying in some nests (Arias-de-Reyna *et al.* 1982; see also Carlo 1971). For evidence (egg characteristics) of 3 ♀♀ laying in same nest, see Mountfort (1958). In Coto Doñana (host *P. pica*), 1 pair per km^2 (Valverde 1971); 30–35 pairs in 270 km^2 (Mountfort and Ferguson-Lees 1961). In Portugal, at least 3 pairs in 2 km^2 (N J Collar). In Ibadan (Ivory Coast), presumed breeding pair held territory of c. 36 ha (Pettet 1975). Size of territories, Sierra Morena (Spain), dependent on density of brood-host nests (L Arias-de-Reyna). Each territory contained up to 30 nests of the host *P. pica* (Arias-de-Reyna *et al.* 1981). One territory contained c. 40 pairs of *P. pica* (L Arias-de-Reyna). Several pairs which defended territories also fed communally, without aggression, in an area c. 2 km from their territories. Towards end of breeding season, when host-nests less numerous in territories, pairs went outside their territories into less suitable habitat, e.g. meadows, to parasitize *P. pica* and other species (Arias-de-Reyna *et al.* 1981). In pinewoods, Coto Doñana, 52·2% of 44 nests of *P. pica* parasitized (Valverde 1971; see also Álvarez and Arias-de-Reyna 1974). ROOSTING. Activity essentially diurnal, but occasionally calls at night in breeding season (Mestre Raventós 1969; see Voice); nothing else known about nocturnal behaviour. By day, interrupts activity with brief periods of sitting still and silent, typically in middle of a pine. Sunbathes with slightly spread wings (Laferrère 1956). Bathing known only from captive birds which jumped several times into water with slightly spread wings (Frisch and Frisch 1967).

2. Quite bold and approachable (Nicolau-Guillaumet and Spitz 1958; Frisch and Frisch 1967). When perched, wings may be held slightly open, and tail held fanned and frequently raised and lowered (Nicolau-Guillaumet and Spitz 1958; Germain 1965). Pair flushed, Portugal, mid-May, flew off silently and slowly on rounded wings, possibly to avoid provoking *P. pica* hosts nearby (N J Collar). ANTAGONISTIC BEHAVIOUR. (1) Display-flight. Bird (presumably ♂) flies from tree-top and climbs steeply, in silence, with rapid wing-beats and downtilted, somewhat fanned tail, as in Collared Dove *Streptopelia decaocto*. After ascending usually to c. 15–20 m, bird initiates downward glide with tail fanned and slightly depressed. Often makes series of level glides, with 1–2(–3) irregular changes of direction during descent, accompanied by Advertising-call (see 1 in Voice), before dropping down into tree canopy. Incidence of display seemed to be related to presence of other birds (of unknown sex) nearby (Laferrère 1956). However, not known whether Display-flight serves principally to rebuff other ♂♂ or to attract ♀♀. (2) Threat and fighting. At start of breeding season, ♂♂ follow each other around, and engage

in quite vigorous fights (Carlo 1971). In captive study of 2 ♂♂ and 2 ♀♀, 3 birds (separately) threatened 4th by fanning tails and raising crests; threatened bird reciprocated (Frisch 1969c). Fledged young also raised crown feathers when excited (Frisch and Frisch 1967). At start of breeding season, Coto Doñana, ♂♂ highly aggressive in defence of territory. Establish territory partly with Advertising-call. 2 ♂♂ often alighted alongside each other at territorial boundary and engaged in fierce fights in which feathers sometimes pulled out (Valverde 1953a, which see also for confrontations with *P. pica*; see also Arias-de-Reyna *et al.* 1982). In Portugal, late April, intensive aerial chasing and aggression observed between pair of *P. pica* and 2 or possibly more pairs of *C. glandarius* (N J Collar). In captive study, ♀ Roller *Coracias garrulus* attacked ♂ *Clamator glandarius*, allowing ♀ *C. glandarius* to lay during mêlée; aggressive reactions of ♂ included calling, chasing, and knocking *Coracias garrulus* away from its own nest-hole; at one point, when *C. garrulus* flew at ♂, both fluttered to ground (Frisch 1973). HETEROSEXUAL BEHAVIOUR. (1) Pair-bonding behaviour. Little information. ♂♂ usually first to arrive back at breeding grounds, Coto Doñana. For possible heterosexual function of Display-flight, see above. Several ♂♂ often seen chasing 1 ♀, Coto Doñana (Mountfort 1958). In alleged courtship, ♂ chased ♀ noisily among small trees (Vincent 1946a). For noisy chasing (participants of unknown sex), see also Mestre Raventós (1969). Captive pair, considered to be bonded during breeding season, accompanied each other constantly, and occasionally Allopreened each other around nape and eyes (Frisch 1969c). In July, Nigeria, pair perched side by side frequently touched bills and called softly (Mundy and Cook 1974). Pair-members highly vocal to each other (see Voice). (2) Courtship-feeding and mating. Feeding of ♀ by ♂ an invariable and immediate prelude to copulation. In Portugal, ♂ seen to mount and feed ♀, without copulating (N J Collar); in captivity also, ♂ fed ♀ at other times (Frisch 1969c). In Ibadan, ♂ repeatedly fed ♀ over short period before copulating (Pettet 1975). In south-west Spain, as ♂ carrying caterpillar approached ♀ sitting on ground, ♀ began rhythmically jerking body while keeping wings closed. ♂ hopped behind and mounted, at same time offering ♀ the prey which she immediately seized in bill but did not eat; both birds maintained grip on prey, apparently helping ♂ to balance. ♂ settled low on ♀, appearing to grip sides of her body with legs; remained thus for *c.* 2 s. ♀ appeared to break off copulation by starting to eat prey, which ♂ then released. ♂ dismounted and brought new caterpillar 2–3 times, each time stepping on ♀'s back from behind and passing food. ♀ repeated jerking action each time ♂ approached. ♂ mounted a 2nd time for *c.* 2 min, and finally, without any preliminary feeding, for *c.* 30 s. During entire sequence (*c.* 25 min), ♀ remained stationary (Channer 1976). In captive pair, food-bearing ♂ gave soft Pre-copulatory call (see 4 in Voice) as soliciting ♀ twitched tail up and down. ♂ mounted and gripped ♀'s slightly protruding carpal joints with his toes (Fig A). Food held by both birds (as above) while ♀ twitched wings alternately, and pair held tails close together; pair sometimes remained so for several minutes before copulating. ♂ released food item only after copulation.

If mating unsuccessful (e.g. ♀ unwilling), ♂ reluctant to forfeit food and either ate it himself or resisted ♀ so that she only tore off a bit of it (Frisch 1969c). EGG-LAYING BEHAVIOUR. Host in descriptions given below is *P. pica* unless stated otherwise. ♀ takes initiative in searching for hosts' nests (Carlo 1971). Before laying, ♀ thought to remain immobile and under cover for long period, studying movements of intended host (Hüe 1952). ♀ carefully and furtively searched area, tree by tree, for almost 2 hrs (Carlo 1971). Once host nest chosen, ♂ may assist in distracting host, especially if latter's clutch complete or nearly so, and thus closely attended. Various stratagems described. In one case, northern Nigeria, pair seemed to distract Pied Crow *Corvus albus* by persistent loud calling (Mundy and Cook 1977). In Coto Doñana, ♂ approached host's nest conspicuously, calling as it flew from bush to bush, while ♀ made direct, silent, low-level approach under cover of vegetation (Alvarez and Arias-de-Reyna 1974). In captive study, ♀ called up mate with Pre-laying call (see 5 in Voice); intended host (♀ Roller *Coracias garrulus*) then flew to confront ♂; after scuffle, ♂ chased host, whereupon ♀ *Clamator glandarius* slipped into host's nest and laid (Frisch 1973, which see for details of ♂'s behaviour). In 2 cases of observed laying, bird sat on nest-rim to lay egg which dropped into host's nest. From arrival to departure took 3 s (Arias-de-Reyna *et al.* 1982). In 8 out of 9 cases, ♀ inserted egg before host's clutch complete, mostly when half-complete; in the other case, insertion took place after incubation had begun (Arias-de-Reyna *et al.* 1982; see also Hüe 1953). In captive study, ♀, within minutes after copulation, started making repeated visits to provided nest, and there performed turning, hollowing movements with backwards action of feet (Frisch 1969c; see also Frisch 1973 for similar behaviour of ♂). Visits intensified when laying imminent, ♀ in nest trembling body, holding head well down, and making stabbing movements with bill (Frisch 1973), in one case at dummy eggs of *P. pica* (Frisch 1969b). For damage inflicted on host's eggs in the wild, see below. At time of laying each of her eggs (in the wild), ♀ said to remove 1, sometimes 2, of host's eggs, but never a previous one of her own, or one laid by another conspecific ♀ (Mountfort and Ferguson-Lees 1961). In 75% of 28 parasitized nests, usually 2 of host's eggs damaged when parasite laid (Arias-de-Reyna *et al.* 1982; L Arias-de-Reyna); 1–2(–4) damaged (Mountfort and Ferguson-Lees 1961). Parasite may damage no eggs or, occasionally, if host's clutch complete, all of them (Hüe 1953). Parasite thought to damage eggs deliberately by pecking (Bradfield 1931; Mountfort and Ferguson-Lees 1961; Valverde 1971), but direct evidence lacking and, despite intensive study, deliberate damage or removal of eggs never seen (L Arias-de-Reyna). Damage may more often arise indirectly, e.g. by vertical drop of parasite's egg at laying (Gaston 1976; see also Craib 1982); in both cases of laying from rim of nest (see above), both parasite's egg and 1 of the host's found to be damaged, presumably by impact. Damage also thought to be inflicted occasionally by parasite's feet, perhaps when laying hurriedly (L Arias-de-Reyna). Host may remove damaged eggs (Bradfield 1931; see also Brichambaut 1973). FOSTER-CARE OF YOUNG. Host in situations described below is *P. pica* unless stated otherwise. For rivalry in nest (host–parasite, parasite–parasite) where host *Corvus albus*, see Jensen and Jensen (1969). Parasite's chick usually hatches and fledges before host's young, having shorter incubation and fledging periods (Hüe 1953, 1964; Frisch and Frisch 1967; Frisch 1969c; Valverde 1971; Alvarez and Arias-de-Reyna 1974). Death of host's young certain if they hatch 3–4 days or more after parasite (Frisch 1969c; see also Alvarez and Arias-de-Reyna 1974). When 2 eggs of *C. glandarius* laid in same nest, survival of 2nd doubtful if it hatches 6 days or more after 1st (Valverde 1971). Otherwise,

A

B

2 parasites in nest appear to fledge without harming each other (e.g. Hüe 1945). Young usually cause death of host's young or of conspecific nest-mate by out-competing them for food, or by smothering them (Mountfort and Ferguson-Lees 1961). Within 30 min of hatching in incubator, young started calling (Frisch 1969c; see Voice for this, and other, food-calls). From day of hatching, parasite's young noticeably more vigorous than host's young of equivalent age (Alvarez and Arias-de-Reyna 1974; see also Frisch 1969c). Parasite's young beg more persistently, are more mobile, and point gape more accurately at incoming food; thus better able to attract attention of foster-parent and intercept food (Alvarez and Arias-de-Reyna 1974). Acceptance of parasite by adult hosts facilitated by vocal mimicry of host's young (Hüe 1945; Frisch and Frisch 1967; Mundy 1973; Mundy and Cook 1974). Preferential distribution of food to parasite thought to be stimulated partly by larger palatal papillae of parasite (Fig B), acting as stronger releaser than gape of host's own young (Mundy and Cook 1977). From 11th day, young start to quiver wings and, at same time, becoming increasingly aggressive to other nestlings, tugging at them and pecking hard at their heads; may inflict appreciable damage if host's offspring significantly younger than parasite (Alvarez and Arias-de-Reyna 1974). However, no evidence that parasite attempts to evict host's eggs or young (Hüe 1952, 1953). From 12th day (in captivity), young able to stand up and raise crest-feathers (Frisch and Frisch 1967). Sat on nest-rim from 14th day, on branches near nest from 16-18 days. After parasite fledged, hosts less attentive to nest, devoting most care to parasite, thus hastening demise of own offspring (Alvarez and Arias-de-Reyna 1974; see also Mundy and Cook 1974, 1977). In one case, where 23-day-old young of host first to fledge, 15-day-old parasite followed suit, and both were fed near nest (Alvarez and Arias-de-Reyna 1974). Young continue to be fed for some time after fledging, at first calling (see Voice) almost ceaselessly (both when perched and in flight) when foster-parents appear, and chasing after them. Meanwhile, partly self-feed (Hüe 1945). Post-fledging dependence at least 2 weeks in case of Hoopoe *Upupa epops* as host in Africa (Jensen and Jensen 1969). For adult *C. glandarius* twice feeding juvenile, September, Nigeria, see Mundy and Cook (1974, 1977). ANTI-PREDATOR RESPONSES OF YOUNG. At 14 days, threatened captive young ruffled feathers and gaped widely (Frisch 1969c); at 18 days, crouched down in nest when approached by dog or human (Frisch and Frisch 1967).

(Fig A from drawing in Frisch 1969c; Fig B from photograph in Frisch and Frisch 1967.)   EKD

Voice. Loud, harsh, and varied (Frisch and Frisch 1967), especially in breeding season (Mestre Raventós 1969). Silent outside breeding season, West Africa (Serle *et al.* 1977). No complete study, and the following scheme provisional.

CALLS OF ADULTS. (1) Advertising-calls of ♂. (a) Rasping 'keeow-keeow' (Witherby *et*. 1938); accelerating 'keyar-keyar-keyar-keyar' (Pettet 1975); very rapid 'kowkowkow-kowkowkowkow' and a slower 'kirrow-kirrow-kirrow-kirrow' (N J Collar). Other renderings in Stadler (1951) and Carlo (1971). In recording (Fig I), a clear, raptor-like 'kleeok', with up to 1 s between calls (P J Sellar). Commonest call; given by perched ♂ to attract ♀♀, also sometimes at night (Carlo 1971), when rendered a more subdued 'kioc-kioc' (Mestre Raventós 1969). Also given when ♂ approaches or meets ♀ (Nicolau-Guillaumet and Spitz 1958; Pettet 1975), and thus possibly also a greeting-call (see also call 6b). Heard less after onset of laying (Carlo 1971). (b) During Display-flight, ♂ begins with sound like grinding clockwork, followed by descending series of guttural sounds, rendered 'kianm-kianm-kian-kiacq-kkiau ...', the last sounds very muffled (Laferrère 1956). Probably the same call described by Frisch (1969c) as chattering sound preceded by rolling noise, thus 'gegerrrrrrrrrrr-wäge-wäge-wäge-wäge-wäg', given by 4 captive birds, usually with tail fanned, throughout breeding season. In recording (Fig II) bird chasing conspecific gave a quiet, rasping, staccato 'kzek', followed by bubbling 'gegegege-gerrrrr ...', finally '... keera keera keera keera keera keera' (E K Dunn). (2) Bubbling-call of ♀. Resonant, laughing 'gi-gi-gi ...' or 'ku-ku-ku ...', rich in harmonics, and given in continuous series, also sometimes more slowly. In recording (Fig III), 'gi-gi-gi-gi-gi-gi-gi-gi-ku-ku-ku-ku-ku', the change from 'gi-' to 'ku-' sounds occurring gradually (J Hall-Craggs); likened to call of Green Woodpecker *Picus viridis* or Bubbling-call of ♀ Cuckoo *Cuculus canorus* (Stadler 1951), though less bubbling and more like tern *Sterna* (P J Sellar); prolonged raptor-like 'kh-kh-kh ...' (G D Field) possibly the same. Also 'BUbuBUbuBUbuBU ...' 'ullupp ullupp ullupp ...', given in smooth continuous series (Stadler 1951). (3) Alarm-call. Short, harsh 'cark cark' (Witherby *et al.* 1938); 'kook' (Serle *et al.*1977); 'krrêk' of ♀ in flight (Nicolau-Guillaumet and Spitz 1948) probably the same. (4) Pre-copulatory call. Rapid chirruping or chittering 'chet-chet-chet-chet ...' of constant pitch given (perhaps by ♂) throughout courtship-feeding, occasionally slows into sound resembling call 1—'chet-chet-chet-chear-keyar-keyar' (Pettet 1975). Squeaky 'ch ch ch', as though made through clenched teeth, and lower grating 'ch ch ch chayr' (G D Field) possibly the same. ♂ approaching ♀ with food prior to copulation gave soft monotonous 'äg äg äg äg' (Frisch 1969c); raucous 'aaa aaa aaa' when ♂ 'courting' ♀ (N J Collar) possibly the same. According to Carlo (1971) ♂ gives more hurried and emphatic variant of call 1a at moment of copulation. (5) Pre-laying call. A monotonous 'woig-woig-woig', given by captive ♀ near nest when intending to lay, typically at moment when host left nest temporarily unattended; in the 2 cases observed, call summoned mate to distract host while ♀ laid egg (Frisch 1973). (6) Other calls. (a) Call likened to that of

I P A D Hollom Spain March 1981

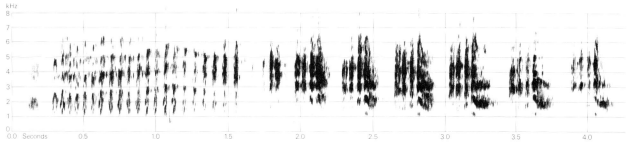

II P A D Hollom Spain March 1981

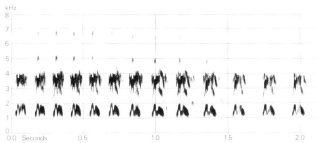

III J-C Roché (1966) Spain May 1965

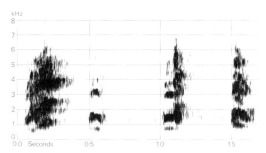

IV P A D Hollom Spain March 1981

Black-headed Gull *L. ridibundus*—'krââ-krââ-krââ' or 'kôô-kôô-kôô' (Laferrère 1956); harsh grating 'wraah' (N J Collar) possibly the same. (b) When 2 birds met, a rapid, rhythmic, vibrating trill, 30–40 s long, different from call 1 (Laferrère 1956); this possibly the high trilling 'pirrrip' described by N J Collar. (c) A hollow, deep-throated 'pyaaa pyaaa', given at dusk by a ♀ next to nest of Magpie *Pica pica* (N J Collar). (d) Various 'chrrr' and 'chow' noises uttered in flight (G D Field).

CALLS OF YOUNG. A soft 'sib sib sib', given within 30 min of hatching. For first 8 days, food-call a very faint 'i-i-i-i-i-i'. When fully feathered, a stronger 'DI-did DI-did DI-di-did . . .', ending in buzzing series of 'i-' sounds, thus '. . . dididii-i-i-i-i-iiiiiii'; in captive bird, given for as long as fed from hand. At fledging, food-call a chattering 'wä-wä-wäg', said to mimic *P. pica* (Frisch and Frisch 1967); rendered 'choc choc choc' (Hüe 1945). Calls of young raised by *P. pica* distinct from those of young raised by Azure-winged Magpie *Cyanopica cyanea* (Arias-de-Reyna and Hidalgo 1982). For evidence (supported by sonagrams) of mimicry of Pied Crow *Corvus albus* by young raised by that host, see Mundy (1973). Fledglings raised by *P. pica* also have a contact-call, 'äcip äcip'. At 26 days, a piercing 'kig', function not known. Also occasionally a soft 'gög-gög' or 'ü-üg ü-üg ügügüg', not unlike last food-call, but lower (Frisch and Frisch 1967). EKD

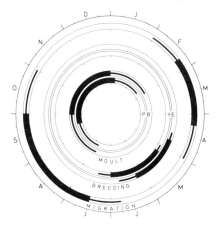

Breeding. SEASON. Spain and North Africa: see diagram. SITE. Eggs laid in nest of brood-host. In Spain, principal host Magpie *Pica pica*, with Crow *Corvus corone*, Raven *C. corax*, Azure-winged Magpie *Cyanopica cyanea*, and Jay

Garrulus glandarius also parasitized; once, Kestrel *Falco tinnunculus* using nest of *P. pica* (Valverde 1971). In northwest Africa (formerly), principally *P. pica*; in Egypt, *Corvus corone* (Heim de Balsac and Mayaud 1962). EGGS. See Plate 95. Elliptical to sub-elliptical, with blunt ends, smooth and fairly glossy; pale green-blue, thickly spotted light brown or red-brown, mimicking host species. 32×24 mm ($28-36 \times 22-27$), $n = 100$ (Schönwetter 1967). Weight 9·4 g (8–11), $n = 14$ (Mestre Raventós 1969). Egg has thick, dense shell; eggshell $c.$ 1·2–1·5 times heavier than similarly-sized egg of *P. pica* (data from Makatsch 1976). Clutch: maximum of 24 eggs thought possible (through histological examination), while most observed 18; always laid in 3 approximately equal series with 5–8 days between them; within series, eggs laid on alternate days, normally 1 per host nest, but rarely (e.g. when host nests are scarce) lays up to 3 eggs in same nest (L Arias-de-Reyna). Out of 43 parasitized nests of Corvidae in Spain, Asia Minor, and (mostly) Egypt, containing total of 54 *C. glandarius* eggs, 33 nests had 1 egg, 9 had 2, 1 had 3 (Friedmann 1964). Of 41 nests, France, with young of *C. glandarius*, 23 had 1, 15 had 2, 3 had 3 (Lévêque 1968). Rarely, up to 8 eggs in one nest (Meinertzhagen 1930; Mountfort 1958); due to more than 1 ♀ (see Social Pattern and Behaviour). Host's eggs perhaps occasionally removed (see Social Pattern and Behaviour). Laying synchronized with that of host, usually occurring before host begins incubation (Alvarez and Arias-de-Reyna 1974). INCUBATION. 12–14(11–15) days, average 12·8 ($n = 5$) (L Arias-de-Reyna). By host species only. YOUNG. Altricial and nidicolous. Cared for and fed by host. Do not eject eggs or young of host. FLEDGING TO MATURITY. Leave nest at 16–20 days (Brichambaut 1973), 19–21 days ($n = 5$) (Valverde 1971). Fledging period average 24 days (20–26), $n = 20$ (Valverde 1971). Become independent some time after fledging. Age of first breeding probably 1 year. BREEDING SUCCESS. For percentage of nests of host *P. pica* parasitized, see Social Pattern and Behaviour. Some eggs ejected or eaten by host (Alvarez and Arias-de-Reyna 1974). In 2 years, Sierra Morena (Spain), in 16 and 20 parasitized nests of *P. pica*, 25 and $c.$ 31 eggs laid respectively; 75% and $c.$ 76·9% of eggs laid were accepted; 79·0% and 91·7% of accepted eggs hatched, and 93·3% and 54·5% of hatched eggs fledged. 56·0% and 42·1% of eggs laid fledged (Arias-de-Reyna *et al.* 1982).

Plumages. ADULT. Forehead, lores, strip of feathering along lower mandible at gape, crown down to middle of eye, and crest medium grey, usually with much pale grey or silvery-white of feather-bases exposed; narrow black shaft-streaks. Ear-coverts and area from just below eye backward to nape and central hindneck slightly darker grey or brownish-black, without distinctly darker shafts and paler feather-bases. Mantle, scapulars, tertials, back, rump, and upper wing-coverts dark brown-grey; feathers of mantle, back, and rump with faint white spot or edge on tip; scapulars, tertials, and upper wing-coverts with large white triangular patch on tip, widest on greater coverts, greater primary coverts, and outer median coverts where tip fully white. Upperparts dark plumbeous-grey with white spots and tips prominent when plumage fresh, dark sepia-brown with white of tips heavily abraded when worn, sometimes traces of white only on tertials and outer wing-coverts. Upper tail-coverts dark plumbeous-grey to dark brown, longest ones tipped white, outer web of lateral ones washed white. Chin, throat, front and sides of neck, and chest straw-yellow or deep buff, dark shafts showing faintly; in worn plumage, chin to chest paler yellow to dirty white, much light grey on feather-bases showing, and dark shafts more prominent, especially on chin and throat. Remainder of underparts pale cream-yellow or white, rear of thighs grey, flanks faintly tinged grey. Tail black with greyish bloom when fresh, dark olive-brown when worn; white tips on t4–t5 (obliquely separated from dark base) $c.$ 4 cm wide, t3 $c.$ 3 cm, t2 $c.$ 2 cm, and t1 slightly less than 1 cm with rather triangular shape; white tip of t1 soon disappears through abrasion. Flight-feathers grey-brown (darker and more greyish when new, paler olive-brown when old), darkest and with faint green or bronze gloss subterminally; borders of tips white, forming crescent 2–4 mm wide. White feather-tips strongly subject to abrasion, especially on longer primaries; white sometimes completely lost, as on scapulars, tertials, and wing-coverts, but still showing on secondaries, p1, outer greater upper wing-coverts, and greater upper primary coverts. Under wing-coverts yellow-buff, longest converts and axillaries pale grey, similar to undersurface of primaries. NESTLING. Naked at hatching. Feather-sheaths appear on crown, upperparts, flanks, and wing from $c.$ 5th day; feather-tips appear on 12th–13th day, when central breast, belly, and vent still bare; appears fully feathered except for some pins on crown on 16th day; at fledging (19–22 days), fully feathered but wing and tail not full-grown; primaries full-grown at $c.$ 24th day, tail at 46th day (Frisch and Frisch 1967; Valverde 1971; Mundy and Cook 1977). JUVENILE. Rather different from adult. Forehead and crown down to gape and just below eye, central hindneck, and ear-coverts deep black, sometimes with slight plumbeous tinge; mainly dark grey in some fledglings. Mantle, scapulars, tertials, upper wing-coverts, back, and rump brownish-black with slight green gloss, distinctly darker than adult, but some grey of feather-bases sometimes exposed; scapulars and upper wing-coverts with white or cream spots on tips; spots often smaller and more rounded in shape than in adult, less straight or triangular, but much variation; tertials only narrowly tipped white (not broad as in adult); greater upper primary coverts more broadly tipped white than in adult, but boundary with brown-black base less sharp, slightly mottled. Upper tail-coverts brownish-black, outer webs of outer ones white, less greyish and freckled than adult. Chin, throat, sides of neck, and chest pale buff, not as yellowish as in adult and usually without dark grey shafts. Remainder of underparts white, but lower flanks and rear of thigh suffused dark olive-brown. Tail as in adult, but feathers blacker, less greyish, white tail-tips much reduced in extent, sometimes partly freckled or tinged brown-black. Secondaries, tips of primaries, p10, and often outer web of (p8–)p9 brownish-black, middle and basal portions of remainder of primaries bright rufous-chestnut, not sharply defined

PLATE 34 (*facing*).
Streptopelia roseogrisea African Collared Dove (p. 336): **1–2** ad, **3–4** ad Barbary Dove *S.* '*risoria*'.
Streptopelia decaocto Collared Dove (p. 340): **5–6** ad.
Streptopelia turtur turtur Turtle Dove (p. 353): **7–8** ad.
Streptopelia orientalis orientalis Rufous Turtle Dove (p. 363): **9–10** ad.
Streptopelia senegalensis senegalensis Laughing Dove (p. 366): **11–12** ad.
Oena capensis Namaqua Dove (p. 374): **13–14** ad ♂, **15–16** ad ♀.
(CJFC)

PLATE 35. *Psittacula krameri borealis* Ring-necked Parakeet (p. 379): **1–3** ad ♂, **4** ad ♀, **5** juv, **6** nestling. (CJFC)

PLATE 36. *Clamator jacobinus pica* Jacobin Cuckoo (p. 388): **1** ad intermediate morph, **2** juv intermediate morph. *Clamator glandarius* Great Spotted Cuckoo (p. 391): **3** ad (fresh plumage), **4** juv, **5** fledgling. (DIMW)

from brown-black tips. White tips of secondaries and inner primaries narrower than adult (c. 2 mm rather than 3–6 mm), those of outer primaries wider (4–6 mm instead of 1–2 mm); often some brown freckling at border of subterminal brown-black. Under wing-coverts and axillaries as in adult. FIRST ADULT. To varying degree intermediate between juvenile and adult, depending on stage of moult: feathers growing early in post-juvenile moult rather like juvenile, later ones gradually resembling adult more. Appearance rather juvenile early in moulting season (about September–November of 1st calendar year): forehead, lores, and crown black with slight grey tinge; mantle, scapulars, and upper wing-coverts similar to juvenile, new flight-feathers with strong rufous tinge on centre and inner web; some juvenile flight- and tail-feathers and occasionally scattered feathers on body retained. Later in season, more similar to adult, with crown more greyish (rarely as silvery-grey as adult), upperparts and upper wing-coverts like adult, though usually less plumbeous-grey and with slightly smaller white spots; late-grown primaries have limited rufous tinge or none at all. Not known whether 1st-adult feathers acquired early in moult and showing more juvenile characters replaced by more adult type later on; birds with juvenile-like 1st adult probably retain this appearance during 1st half of 2nd calendar year (like many Egyptian birds, where post-juvenile moult starts shortly after fledging); 1st adults with more adult appearance probably belong to late-moulting populations like those of Iberia, where post-juvenile moult postponed until arrival in winter quarters. In all 1st adults, early-grown primaries (p6, p7, and p9 especially) show variable amount of rufous tinge or mottling along shafts and on inner web. In sample of non-juvenile birds, 39% of 31 ♂♂ and 73% of 26 ♀♀ showed rufous, suggesting that ♀♀ may retain rufous beyond 2nd calendar year (unlike ♂♂) or that not all ♂♂ show rufous in 2nd calendar year. Some juvenile feathers often retained, especially among secondaries: these shorter with much narrower and shorter edge of white along tip than neighbouring fresh feathers.

Bare parts. ADULT. Iris light grey-brown, dirty yellow-brown, dark grey-brown, or dark brown. Eye-ring red or orange-red. Bill black to dark horn-brown, undersurface of lower mandible yellow; some orange-yellow at gape. Leg and foot dark leaden-grey to pale grey. NESTLING. Bare skin, including that of bill, leg, and foot, pink at hatching, darkening to bluish-grey and black from 5–6 days (upperparts, bill, leg, and foot first). Bill conspicuously edged white, inside of mouth orange-yellow; no lateral flanges (unlike Corvidae). Egg-tooth white, retained for some time after fledging. JUVENILE. Like adult, but iris apparently always dark brown. (Hartert 1912–21; Valverde 1971; Mundy and Cook 1977; Glutz and Bauer 1980; BMNH, RMNH, ZFMK.)

Moults. Based mainly on Stresemann and Stresemann (1969). ADULT POST-BREEDING. Complete; primary moult complex. Total duration probably at least c. 100 days. Probably starts July–September, directly after arrival in winter quarters; completed December–February. 2 groups of primaries moult simultaneously: p5–p10 and p1–p4; sequence of inner 4 irregular. Moult starts with loss of p6, soon followed by p9; when these both full-grown, p7 lost, soon followed by p10, when these full-grown, p8 and p5 lost almost simultaneously. In inner group, 1st feather (nearly always p1) lost between p6 and p9, 2nd (often p3) with p7, 3rd (usually p2 or p4) lost during regrowth of p7 and p10, last feather lost with loss of p5 and p8. Tertials (s8–s11) lost centrifugally from s10 or s9 during replacement of first 7 primaries; secondaries usually start late during primary moult, sequence irregular, some still old when primaries completely new; perhaps not all replaced every year. Tail rather irregular: 3 central pairs never all grow at same time, 1 pair being retained until other 2 full-grown; t5 and t1 often replaced first, t2 last; starts at same time as primaries or slightly earlier, completed at same time as primaries or slightly later. Body starts with loss of p6 or p9; head, neck, throat, chest, and upper tail-coverts first; usually all completed when p5 and p8 lost or just growing. POST-JUVENILE. Complete or partial. Timing and sequence as in adult post-breeding. Early-hatched or early-moulting juveniles already largely in 1st adult plumage (which then usually resembles juvenile) by October, late-hatched or late-moulting ones by February (when 1st adult usually resembles adult). For details, see 1st adult in Plumages. Occasionally, some juvenile flight-feathers retained, mainly s1–s5 or s3–s4, rarely p4, p5, or p8.

Measurements. Wing and tail of birds from (1) Iberia and north-west Africa, (2) Turkey and Cyprus to Iraq and Egypt; other measurements combined for these areas. Juvenile wing from Egypt only; adult wing includes fully moulted 1st adult, as these do not differ significantly. Skins (BMNH, RMNH, ZFMK, ZMA, ZMM). Bill (F) to forehead, bill (N) from nostril to tip. Toe is outer front toe.

		♂			♀	
WING AD (1)		209	(3.90; 8) 202–214		199	(3.88; 6) 194–205
	(2)	214	(5.98; 20) 205–228		203	(4.09; 19) 195–208
	JUV	206	(4.06; 5) 201–212		195	(4.41; 9) 190–201
TAIL AD (1)		210	(7.12; 8) 199–218		195	(3.85; 5) 191–199
	(2)	220	(7.25; 18) 209–236		201	(5.76; 16) 190–208
	JUV	216	(3.67; 16) 211–220		196	(13.5; 10) 180–211
BILL (F)		30.9	(1.80; 17) 27.8–34.1		30.3	(1.56; 17) 27.7–32.7
	(N)	19.5	(1.09; 19) 18.4–21.8		18.7	(0.75; 16) 17.4–19.9
TARSUS		35.4	(1.24; 14) 33.4–36.7		33.8	(2.02; 10) 31.3–36.6
TOE		33.4	(1.44; 12) 31.5–35.7		31.4	(1.18; 6) 30.2–33.0

Sex differences significant, except bill (F).

Weights. ADULT. In breeding season, Spain, ♂ 169 (14.5; 6) 153–192, ♀ 138 (Valverde 1971); in winter, lean ♂ 139 (Peréz Chiscano 1971). Average of 10 ♀♀ in breeding season, Nigeria, 130 (Mundy and Cook 1977).

NESTLING. At hatching, average of 12 c. 8; at 7 days, c. 67; on 14th–15th day, 130 (10.2; 8) 114–142; on 18th–19th day, when about to leave nest, 140 (16.0; 6) 114–160; average daily growth until 15th day (12 birds), 9–10 (Frisch and Frisch 1967; Valverde 1971; Mundy and Cook 1977).

JUVENILE. France, July–August: 140 (4) 120–158 (Glutz and Bauer 1980). Italy: September, ♂ 145, ♀ 115 (Moltoni 1965); February, ♀ 100 (Mocci Demartis 1976).

Structure. Wing rather long but wide at base, tip rounded. 10 primaries: in adult, p7–p8 longest or either feather 0–2 shorter, p9 12–19 shorter, p10 65–85, p6 10–19, p5 30–36, p4 44–52, p1 75–85; in juvenile, p8 sometimes distinctly shorter than p7, p9 17–22 shorter, p5 19–35, p4 32–48, p1 61–78. Inner webs of p7–p10 slightly emarginated. Tail long, graduated, 10 feathers; t5 75–110 shorter than t1. Crown and nape of adult with elongated and pointed feathers, forming short crest; crown feathers of juvenile shorter and more rounded, of 1st adult intermediate. Bill as in *Cuculus* cuckoos, but relatively slightly heavier, wider at base; nostril slit-like, not rounded nor with raised edges as in *Cuculus*. Leg and foot heavy, markedly scutellated, tarsus and toes not short and slender as in *Cuculus*. 2 toes directed backward; basal digit of both front toes slightly fused at base, unlike *Cuculus*; inner front toe c. 80% of outer front toe, outer hind c. 70%, inner hind c. 54%. Claws short but strong, decurved, sharply pointed.

Cuculinae

Geographical variation. Rather slight. Birds from Middle East completely similar in size to those of Egypt; birds of Spain, Portugal, and north-west Africa show slightly shorter wing and tail but otherwise similar. Populations breeding in Afrotropics from West Africa east to Somalia and Kenya also similar in size, but those breeding in southern Africa north probably to Zambia and southern Tanzania smaller in all measurements and separated as *choragium* by Clancey (1951): wing ♂ 194 (10) 187–201, ♀ 193 (10) 187–199; tail ♂ 193 (10) 182–201, ♀ 190 (10) 183–198 (Clancey 1973). Differences in colour negligible. In non-breeding season, large- and small-sized populations overlap to great extent over Afrotropics. CSR

Chrysococcyx caprius Didric Cuckoo

Du. Diederikkoekoek   Fr. Coucou didric   Ge. Goldkuckuck
Ru. Белобрюхая кукушка   Sp. Cuco cobrizo   Sw. Guldgök

PLATES 37 and 41
[between pages 470 and 471]

Cuculus caprius Boddaert, 1783

Monotypic

Field characters. 18–20 cm; wing-span 32–35 cm. Much smaller than Cuckoo *Cuculus canorus*, with head and body only $\frac{2}{3}$ size of that species but with proportionately larger bill and much shorter tail. Small, rather compact cuckoo, differing distinctly from *Cuculus* and *Coccyzus* cuckoos in less attenuated form, metallic green upperparts, well marked head, and noticeably barred wings. Behaviour recalls shrike *Lanius*. Sexes dissimilar; no seasonal variation but iridescence of adult plumage reduced by wear. Juvenile separable.

ADULT MALE. Upperparts dark metallic green, with iridescence varying noticeably (in different light angles) from coppery (particularly on nape and mantle) to silvery. Tail less green, and can appear basically black. Within dark upperparts, white streaks in front of and behind eye form broken but obvious supercilium, white spots to wing-coverts and on flight-feathers create blotches and bars, and white tips and spots on outer tail-feathers produce barred edge. Underparts essentially white; face, throat, and upper breast noticeably so, relieved only by narrow black moustache, but chest-sides, flanks, and vent thickly barred dark bronze-green, often apparently black. Underwing closely barred like flanks, with secondaries uniformly dark. In shadow at distance, looks very dark above, with broad white throat the only obvious mark. ADULT FEMALE. Pattern similar to ♂ but colour intensity and contrast reduced by duller tone to upperparts, fewer white bars on wing, and less white, less completely barred underparts. JUVENILE. In fresh plumage, lacks green upperparts of adult, being rufous-brown above. Markings more diffuse, with lower face and throat more blotched and streaked than in adult ♀. Primaries lack obvious white spots.

In Africa and southern Arabia, can be confused with Klaas's Cuckoo *C. klaas* (10–15% smaller, with ♂ almost uniform emerald-green above, whiter below, with green patches on chest-sides; ♀ and immature much less obviously different, but all plumages best distinguished by almost completely white outer tail-feathers). Plumage pattern, size, and behaviour all as likely to suggest strange shrike *Lanius* as typical cuckoo *Cuculus*. Flight actions not as loose as *Cuculus* but easy and free, with similar untidy movement of wings on take-off and landing. ♂ fond of high perch, from which it pounces on prey, as in *Lanius*.

Most characteristic call (unlikely from vagrant) a multisyllabic, mournful 'dee-dee-dee-DEE-a-dric'.

Habitat. In all types of tropical country except in heavy forest, especially in open savanna woodland, abandoned farms, cassava plantations, and swampy places, not only in lowlands but in mountainous country and on coast. Extends to more open parts of dry woodland in Sudan and even to oases of southern Sahara. Often sings from high isolated tree (Bannerman 1933, 1953). In central Africa, in papyrus swamps and commonly in most types of woodland. In East Africa, most frequent in dry thornbush and highland dry woodland (Williams 1963); common in wooded and bushy grassland and cultivation up to 2000 m (Britton 1980). In southern Africa, has adapted to exotic plantations and to built-up areas, even cities (Prozesky 1970).

Distribution. Breeds in Afrotropics from Sénégal to Ethiopia and south to South Africa; migrant in many areas (see Movements).

Accidental. Cyprus: adult ♀, Akrotiri, 27 June 1982 (Lobb 1983).

Movements. Migratory at northern and southern ends of breeding range, though present all year (perhaps with local movements) in low latitudes either side of equator.

A pronounced rains migrant in southern Africa, at least as far north as Katanga and southern Tanzania (8–10°S), birds here being present in breeding range September–October to April, a few lingering to June (White 1965;

Benson *et al.* 1971; Britton 1980; Curry-Lindahl 1981). Presumed to spend non-breeding season in equatorial belt (Clancey 1964), but unknown how far north these migrants penetrate.

Situation in northern tropics much less clear-cut: some present all year in non-arid regions, while others move northwards into savanna and Sahel zones during summer rains (May–June to September–October) (e.g. Bannerman 1953, Cave and Macdonald 1955). Movement of this type reported all across continent from Sénégal to Ethiopia. In Nigeria, for example, present all year south of great rivers, but further north arrives and breeds during rains, with abundance varying with local rainfall (Elgood *et al.* 1973; Elgood 1982); in central Chad can be numerous (early July to mid-October) in wet years or absent in dry ones (Salvan 1968*b*). Common summer visitor (mid-June to September) in Darfur (Sudan), where some bred but very large proportion did not (Lynes 1925*a*); absence of 1st-year birds suggested these not from a population that had bred earlier further south.

Voice. See Field Characters.

Plumages. ADULT MALE. Forehead, crown, hindneck, mantle, scapulars, tertials, back, rump, and upper tail-coverts green with strong metallic coppery-red lustre, colour varying between rather dull dark green and brilliantly coppery depending on angle of light; some hidden feather-bases on forehead white; lateral upper tail-coverts with white outer webs. Lores, streak in front of eye, streak below eye widening on ear-coverts and extending to sides of neck, and small feathers on rims of eyelids metallic bronze-green or rather dull brown-green; a small spot above lores in front of eye and a streak from above hindcorner of eye widening towards sides of nape white; lower cheeks white or with slight buff tinge, with a narrow and short oblique green streak from gape downwards. Underparts white, usually with a slight buff or cream tinge, especially on chin, chest, and flanks; sides of belly, flanks, and under tail-coverts evenly marked with broad metallic bronze-green bars. Central pair of tail-feathers (t1) metallic-green with coppery lustre like upperparts; other tail-feathers duller metallic-green, less glossy and with dull greenish-black inner webs and large white spot on tips; t3 with some small white spots along outer edge, t4 with larger spots, t5 with 2–3 large white spots on both webs. Tips and outer webs of p1–p7 (–p8) metallic-green; basal inner webs of p1–p7(–p8) and both webs of outermost primaries dull greyish black; p8–p10 with small elongate white spots along outer edge, basal and middle portions of all inner webs with 3–4 broad and contrasting white bars. Secondaries metallic-green, glossed coppery on tips and outer webs; outer webs with 2 rather large white spots along edge. Upper wing-coverts metallic-green, lesser and outer coverts strongly glossed coppery-red in some lights, like upperparts, inner coverts with coppery fringes; central greater upper wing-coverts with white outer webs (sometimes reduced to irregular white spots), central median and longer lesser upper wing-coverts with large white spots and broad white bars. Under wing-coverts and axillaries contrastingly barred cream-white and metallic-green (like flanks), rather similar to coarsely barred undersurface of primaries but contrasting with uniform dark undersurface of secondaries. Coppery lustre of upperparts sometimes largely lost by abrasion, appearing duller green or even greenish-black in some lights; white spots on secondaries, tail, and upper wing-coverts partly lost through wear. ADULT FEMALE. Like adult ♂, but upperparts duller, less glossed with coppery-red, in particular forehead, hindneck, mantle, and streaks on sides of head duller olive-green; white of sides of head and of chin to breast more strongly tinged buff; pattern of sides of head less contrasting than adult ♂. Feather-centres on sides of neck and breast with dusky grey marks, showing as dark triangles or irregular bars. White bars on central lesser and median upper wing-coverts narrower, white spots on greater coverts and secondaries smaller. Primaries and tail rather variable; in some, spots fully white as in adult ♂; in others spots largely rufous, only those in t5, on tips of t2–t4, and partly on inner webs of primaries more or less white. JUVENILE. Entire upperparts, sides of head down to eye, and lesser upper wing-coverts rufous-cinnamon with metallic-green feather-bases; forehead, crown, and mantle appear uniform rufous (green showing only when plumage worn); rump and upper tail-coverts mainly green; amount of rufous or green on scapulars, tertials, and lesser upper wing-coverts depends on abrasion (greener when worn); irregular supercilium white (broken above eye, as in adult), outer webs of some upper tail-coverts white. Sides of head below eye, sides of neck, and chin to chest buff or cream-white with heavy dark olive-brown blotches, tending to join to form irregular streaks; remainder of underparts like adult, but olive-green bars duller, less regular, and less sharply defined. Tail: t1 like adult, but narrowly fringed rufous and partly spotted rufous near base; t2–t4 rather different, dull green with rather close and irregular broken rufous bars, only some spots on tips white; t5 like adult but white spots smaller, less rounded, more numerous. Primaries quite different from adult, barred rufous and green, only part of bars on inner webs of outer primaries white; tips of primaries uniform dark greenish-grey, narrowly fringed rufous. Secondaries glossy dark green, spots along outer fringes rufous (white near tips of some middle secondaries), inner webs closely barred rufous. Upper wing-coverts glossy green; lesser tipped rufous, others closely barred rufous; some white spots on middle secondaries. Under wing-coverts and axillaries like adult, but dark bars less regular and less glossy green. FIRST ADULT. Like adult, but some juvenile secondaries with corresponding greater upper wing-coverts usually retained, as well as occasionally some outer coverts, closely barred rufous and green, strongly contrasting with new neighbouring wing-coverts and secondaries which are more strongly glossy green and lack any pattern except for some large white spots. Upperparts of 1st adult ♂ usually less strongly coppery-red than adult ♂, white spots on wing-coverts often smaller, and underparts deeper buff, thus more similar to adult ♀, but tail-spots always white. White spots on tail and flight-feathers on 1st adult ♀ more extensively rufous than adult ♀. Some birds perhaps indistinguishable from adult.

Bare parts. ADULT MALE. Iris red-brown, red, brilliant crimson, or carmine with yellow-white outer ring. Bare patch round eye brick-red, red, or carmine. Upper mandible and tip of lower mandible black, remainder of lower mandible dark grey, gradually paler towards base. Leg and foot dark grey-horn, bluish-black, slate, or black. ADULT FEMALE. Iris brown, red-brown, or red. Bare patch round eye brick-red. Bill, leg, and foot as in adult ♂. JUVENILE. Iris pale brown or brown. Bill pink-red, coral-red, or orange-red; darkening to black during early post-juvenile moult. Leg and foot flesh-grey, slate-grey, brown-grey, or black. (RMNH, ZMA.)

Moults. Information limited. Sequence as in Cuckoo *Cuculus canorus* (Stresemann and Stresemann 1961). Timing probably dependent on local breeding season. In birds examined from West Africa and Tanzania, plumage fresh September–December, worn March–May; single birds starting moult March and May, one completing September. Adult post-breeding primary moult may apparently be suspended at any stage. Post-juvenile not always complete; in particular, some juvenile secondaries with their coverts frequently retained.

Measurements. ADULT AND FIRST ADULT. Afrotropics, all year; skins (RMNH, ZMA). Bill (F) to forehead, bill (N) to distal corner of nostril.

	♂		♀	
WING	114 (2·66; 15) 110–118		119 (5·08; 7) 113–126	
TAIL	81·3 (3·86; 15) 77–88		81·8 (2·88; 8) 77–85	
BILL (F)	20·4 (0·74; 15) 19·3–21·6		20·5 (0·71; 8) 19·6–21·3	
BILL (N)	13·0 (0·44; 15) 12·3–13·5		13·3 (0·67; 7) 12·5–13·8	
TARSUS	17·2 (0·81; 13) 16·1–18·5		17·5 (0·64; 7) 16·8–18·4	

Sex differences significant for wing only. First adult wing and tail on average *c*. 4 mm shorter than full adult.

JUVENILE. Wing on average *c*. 8 mm shorter than full adult, tail *c*. 5 mm.

Weights. Kenya and Zambia: ♂ 28·6 (6) 27–31, ♀ 35·9 (4) 31–43, unsexed 35·5 (3) 30–38 (Dowsett 1965; Britton and Dowsett 1969; Britton 1970; Colston 1971). South Africa: ♂♂ 27·7, 33·1; ♀ 42·6 (Anon 1968b; Skead 1974; RMNH). Ghana: ♂♂ 26·2, 28·6 (Greig-Smith and Davidson 1977); ♂ 26·2 (2·2; 6) 22–29, ♀ 32·0 (0·7; 4) 31–33, unsexed 25·0 (6) 24–27 (Fry 1970).

Structure. Proportionally similar to *Cuculus canorus*, but tail relatively slightly shorter and wing-tip more rounded. P8 longest, p9 1–5 shorter, p10 33–40, p7 4–7, p6 10–18, p5 18–25, p4 23–32, p1 37–47. Tail rounded at tip, t5 11–15 shorter than t1. Outer front toe with claw 21·3 (1·4; 8) 20–23; inner front toe with claw *c*. 68% of outer front, outer hind *c*. 82%, inner hind *c*. 49%. Remainder of structure as in *C. canorus*.

Geographical variation. Slight; birds from southern Africa average larger than those from further north (Clancey 1964). For relationships, see Friedmann (1968). CSR

Cuculus canorus Cuckoo

PLATES 38 and 42
[between pages 470 and 471]

Du. Koekoek   Fr. Coucou gris   Ge. Kuckuck
Ru. Обыкновенная кукушка   Sp. Cuco   Sw. Gök   N. Am. Common Cuckoo

Cuculus canorus Linnaeus, 1758

Polytypic. Nominate *canorus* Linnaeus, 1758, Europe south to France and Italy, and east through Siberia to Kamchatka and Sakhalin, south to Turkey, Levant, Iraq, northern Elburz mountains in Iran, northern Kazakhstan, Russian Altay mountains, Transbaykalia, and Amur river; *bangsi* Oberholser, 1919, Iberia, Balearic Islands, and north-west Africa; *subtelephonus* Zarudny, 1914, Iran (except north-west and north from Elburz), Afghanistan, northern Pakistan, Kashmir, and from eastern shore of Caspian through arid parts of Kazakhstan, Turkmeniya, Mongolia, and western China east to north and central China and Japan; migrates through Middle East. Extralimital: *bakeri*, Hartert 1912, Himalayas east from Punjab, and from Assam and southern China southward.

Field characters. 32–34 cm; wing-span 55–60 cm; tail 13–15 cm. Close to size of Collared Dove *Streptopelia decaocto*, but even longer tailed. Rakish, falcon-like bird with grey plumage relieved only by darker flight-feathers and tail and barred underbody. Decurved bill, rather small head, long, narrow-ended wings, and long rounded tail typical of genus. In flight, wings not raised above horizontal. Song the well-known 'cu-coo'. Sexes similar, but rufous morph of ♀ occurs rarely; no seasonal variation. Juvenile separable.

ADULT MALE. Head, breast, and most of upperparts dull ash- to slate-grey, brightest on face, rump, and upper tail-coverts and dullest (with brown wash) on inner wing-coverts. Underparts from chest to vent white, with narrow brown-black bars forming pattern of continuous transverse lines over underbody but typically not on creamier vent and under tail-coverts. Flight-feathers dark brown-grey above. Mottled and faintly barred underwing shows quite obvious pale tract in centre. Tail-feathers almost black at ends but are tipped white and spotted or barred white particularly on inner web; chequering not striking except when seen from below or when tail spread.

ADULT FEMALE. (1) Grey morph. Common. Usually differs from ♂ in slightly browner upperparts, indistinctly barred, buff-toned breast, or additional buff-brown, black-barred band round chest and shoulders; some indistinguishable. (2) Rufous morph. Rare. Strikingly different, having upperparts rich chestnut, broadly barred black except on rump; face, throat, and breast warm buff, rest of underbody buff-white, all narrowly barred black. Flight-feathers black-brown, prominently barred chestnut. Tail-feathers barred black and chestnut, broadly tipped white. Underwing also tinged buff. Bill of adults dark horn, with pale green or yellow base to lower mandible; eye and eyelids bright yellow; legs and feet yellow. JUVENILE. Plumage variable, with 2 types common: one essentially grey and little barred above, the other rufous-brown and much barred all over. Latter occurs more frequently than rufous ♀, and only exceptionally as bright and with unbarred rump. All differ from adult in prominent white patch on nape (and occasionally some white feathers on forehead and crown), fully barred throat and breast, many narrow white fringes (but not spots) on feathers of upperparts, buffier underparts (particularly on vent and under tail), more obvious barring on flight-feathers, and larger white spots on tail. FIRST YEAR. From March, largely adult plumage assumed but both sexes usually retain some barred juvenile secondaries and greater coverts.

Generally considered unmistakable in western areas of west Palearctic though often confused with small falcons *Falco* and hawks *Accipiter* in brief glimpse when no clear impression gained of plumage pattern (which also suggests ♂♂ of several small *Accipiter*). Silhouette and flight of *C. canorus* usually distinctive in any prolonged sighting. Extremely similar to Oriental Cuckoo *C. saturatus* occurring in east of west Palearctic (and probably elsewhere on migration); see that species. Song and calls diagnostic. Flight noticeably direct and apparently hurried in action, with wings beating rapidly, deeply (not raised above horizontal), and occasionally erratically (particularly when suddenly making height or entering cover); glides frequently with wings extended or depressed, particularly during territorial patrol or before landing, which is characteristically untidy with loose flaps of wings and spreading of tail. Awkward on the ground, with shuffling walk (body and tail swivelling) and hopping. Displaying and singing birds frequently lower head, let wings droop, twist body, and raise and fan tail; silhouette then quite different from normal.

Usually solitary outside breeding season and inconspicuous as migrant, though occasionally occurring on coasts and through hill valleys. Commonest call a muted, musical but far-carrying 'cu-coo'. Also, a harsh repeated 'gowk' and (from ♀) a bubbling trill.

Habitat. Breeds in all climatic zones of west Palearctic except arctic tundra and desert. Absent from barren mountains, extensive dense forests, oceanic islands, and closely built urban areas. Ubiquitous and mobile, in both lowlands and uplands, to *c.* 2400 m in Switzerland (Glutz and Bauer 1980) and somewhat higher in USSR (Dementiev and Gladkov 1951*a*); in India, breeds at *c.* 600–4100 m and occurs exceptionally to 5250 m (Ali and Ripley 1969). Ranges indiscriminately over wide variety of habitats conventionally regarded as distinct: coniferous or broad-leaved woodland and scrub, parkland, open country with scattered trees or hedgerows including farmland of various types, wetlands with reedbeds, heaths, and coastal dunes or marshes. Restless mobility and far-carrying voice compensate for thin distribution among habitats, largely determined by choices of principal host species parasitized for rearing young, such as Meadow Pipit *Anthus pratensis* on rough grassland, Dunnock *Prunella modularis* in bushy fringes or scrub, and locally Reed Warbler *Acrocephalus scirpaceus* in reedbeds. ♀ may frequent more than one egg-laying territory, separated by up to 3 km or more from shared feeding territory where caterpillars of prey species are abundant (Wyllie 1981). In addition to adequate food supplies and populations of brood-hosts, requires plenty of well-placed commanding song-posts and look-outs (e.g. treetops, bare rocks, telegraph poles, fences, or walls) for communicating with conspecific birds and monitoring breeding stages of prospective hosts. Not apparently dependent on human land uses or artefacts, but takes full advantage of them where appropriate, e.g. disused railway lines (Wyllie 1981). In USSR, frequents fringes of taiga, thickets of willow *Salix*, forest edges, and floodland orchards; in Ukraine, in copses of birch *Betula*, in wooded steppes, in lake depressions overgrown with reeds and bushes, in ravines, and on entirely bare steppes. Most abundant in forest–steppe areas with coppices and ponds, and in mixed open forests (Dementiev and Gladkov 1951*a*). Descends freely to ground, where most feeding is done, but moves only short distances on it, and dislikes long herbage. Often in vicinity of water, without appearing dependent on it. While enjoying warm sunny conditions, shows remarkable tolerance for chilly, windy, and wet weather within limits of tolerance of prey species, and short of severe frost.

While wintering in Africa, is mobile in response to rainy seasons and fluctuating food supplies: passage observed along watercourses and in acacias, savanna, dry evergreen *Brachylaena* forest, and coconut palms (Moreau 1972). Flight normally in lowest airspace, even within centimetres of ground, but displaying birds may circle high in the air (Wyllie 1981) and on migration flies at up to *c.* 300 m (Glutz and Bauer 1980).

Distribution. IRAQ. Regarded as passage migrant only (Allouse 1953), but almost certainly breeds Kurdistan (McGeoch 1963). SYRIA. Almost certainly breeds, but no proof (Kumerloeve 1968*b*; KH). LEBANON. Probably breeds (at least occasionally) in mountain areas, but no proof (Benson 1970; Tohmé and Neuschwander 1974; Macfarlane 1978). ISRAEL. Formerly occasional breeder, but regular in last few years (HM). CYPRUS. May breed occasionally (PRF, PS).

Accidental. Iceland, Faeroes, Azores, Madeira, Canary Islands, Cape Verde Islands.

Population. Data limited, but recent decreases noted Britain, Ireland, and Finland, and possibly decreased Netherlands and West Germany. Numbers sometimes fluctuate in some areas with numbers of host species (Glutz and Bauer 1980).

BRITAIN AND IRELAND. No reliable counts. Total of under 10000 'pairs' suggested by Parslow (1967), but Sharrock (1976) thought at least 17500–35000 'pairs'. Widespread reports of decreases, especially since early 1950s but extent difficult to assess; in Ireland, general decrease in 20th century (Parslow 1967). FRANCE. Between 100000 and 1000000 'pairs' (Yeatman 1976). BELGIUM. About 14000 'pairs' (Lippens and Wille 1972); *c.* 10000 'pairs' 1982 (PD). LUXEMBOURG. About 2500 'pairs' (Lippens and Wille 1972). NETHERLANDS. About 10000 territorial ♀♀; probably some decline, but data limited

404 Cuculinae

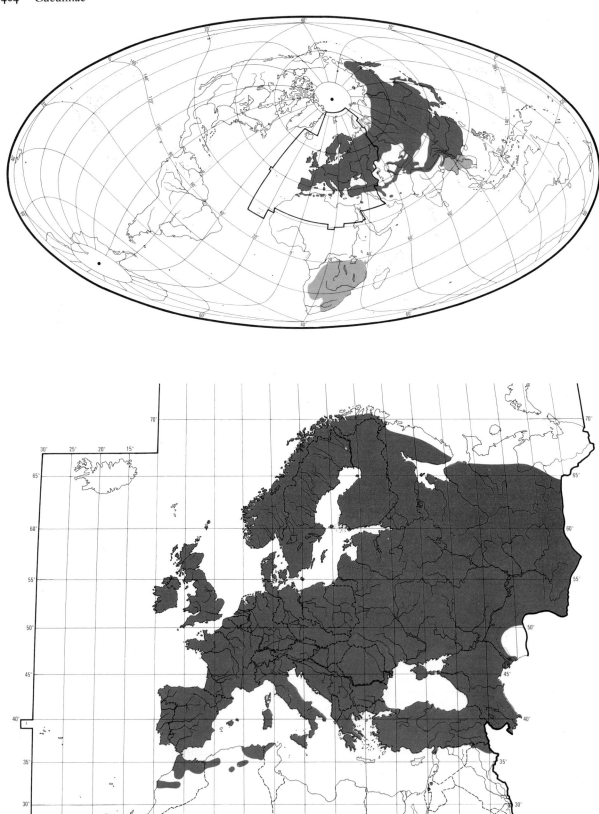

(Teixeira 1979). WEST GERMANY. Apparently declining (GR). SWEDEN. About 100000 'pairs' (Ulfstrand and Högstedt 1976); probably declining (LR). FINLAND. About 30000 'pairs' (Merikallio 1958); decreased in recent decades (OH). POLAND. Fairly numerous (Tomiałojć 1976a). CZECHOSLOVAKIA. No apparent change in numbers (KH). GREECE. Scarce (HJB, WB, GM). USSR. Abundant in forest–steppe areas of Ukraine (less so in steppe) and mixed open forests of central areas, rarer in north of range and in mountains (Dementiev and Gladkov 1951a). SPAIN. Numerous, especially in north (AN).

Movements. Migratory. All west Palearctic populations winter in Africa, as also those from large part of Asia, except east Asiatic birds which winter in south-east Asia south to Philippines; position of migratory divide not known. Though said by Ali and Ripley (1969) that nominate *canorus* (breeding eastwards across Siberia) winters regularly in peninsular India, in fact only a straggler there in winter (Desfayes 1974). Central Asian race *subtelephonus* occurs on passage across Middle East, including Iraq, Jordan, and Israel (CS Roselaar); believed to enter eastern Africa, whence specimens reported Somalia to Moçambique (Archer and Godman 1961; Clancey 1965; Britton 1980). Montane race *bakeri* (breeding Himalayas to western China) winters Burma, Thailand, and Laos; not considered further here.

Winter distribution within Afrotropical region poorly known. Only 2 ringing recoveries south of Sahara: Dutch juvenile found Togo in October, and British-ringed nestling found Cameroun in January. Birds collected western parts of West Africa (Sénégambia, Mali, Ghana) notable for short wings, corresponding to migratory race *bangsi* of Iberia and north-west Africa (Moreau 1972); this consistent with apparent scarcity of European nominate *canorus* migrating through Spain and Morocco into westernmost Africa (see below). Migrants pass through Sénégal and Mali from mid-August to October (Morel and Roux 1966; Lamarche 1980), and through Sierra Leone and Togo from October to early December (Douaud 1957; GD Field); but status often obscured by problem of field separation from African Cuckoo *C. gularis*. Regular wintering not proven for anywhere in West Africa (though casual records exist), and *bangsi* specimens collected in winter as far to south-east as Lake Tanganyika (Chapin 1939). Over most of East Africa also, occurs chiefly October–December and March–April (Britton 1980). Normal winter range (all races) considered to lie south of equator and mostly south of 10°S (Seel 1977b, 1980), reaching Namibia and South Africa.

North of equator (e.g. Sénégal, Mali, Togo, Chad, Ethiopia, Somalia), passage birds seen in autumn but hardly at all in spring; in contrast, spring records (mid-March to late April) predominate in northern Zaïre, Kenya, and Tanzania, when birds very fat. Hence inferred that birds overfly Mediterranean and Sahara in autumn (when scarce in North Africa), while in spring they fatten and initiate migration in low latitudes, overflying northern Afrotropics and Sahara in continuous flight of 4000–5000 km (Moreau 1972). Comparable strategy may apply also to Middle East, since more autumn than spring records from Arabian peninsula (e.g. Griffiths and Rogers 1975, Walker 1981b), but in Iraq spring passage more pronounced than in autumn (Marchant 1963b); data sparse, however. Seen regularly in autumn on Dahlac archipelago (Red Sea), birds sometimes arriving exhausted (Mann 1971).

Ringing recoveries, reviewed by De Smet (1967, 1970, 1972) and Seel (1977b), show directions of movement within Europe (clearly broad front), but reveal little about entry into or exodus from Africa. British birds move south-east in autumn (reverse in spring), while continental birds move south-east to south-west. Only 3 recoveries in Spain, however, from Britain (2) and Belgium (1, in December) (Bernis 1970); hence few European nominate *canorus* use Iberia–Morocco route, and (instead) cross Mediterranean from France and Algeria eastwards. Other British-ringed passage recoveries in Low Countries (8), southern West Germany (3), France (14, mainly north and east), Italy (8), Malta (2), Algeria (1), and Tunisia (1). Migrants from Low Countries found in approximately equal numbers south-west to south (in France) and south-east (towards and into Italy). Ringing in West and East Germany (recoveries from Hückler 1968) showed $c.\frac{1}{3}$ of migrants to south-west of ringing site, in Belgium and France; otherwise found south to south-east in Italy, Malta, Yugoslavia, and Greece. From eastern Baltic (including Finland), birds found south to south-east, in Mediterranean basin (Italy, Malta, Yugoslavia, Crete, coastal Egypt). Small-scale ringing in other continental countries showed same trend, with birds recovered on passage from south-west to south-east of origin, but not reaching Iberia.

For maps of recoveries by month and age-class, see Seel (1977b). Post-fledging dispersal begins July, random in direction and normally over short distances only, though some British juveniles longer: e.g. Oxford nestling (17 June) found Fyn (Denmark) 2 August, and Humberside juvenile (11 July) found Schleswig-Holstein (West Germany) 3 August. Southward autumn migration begins early August, earlier in adults than juveniles; few of any age remain in Europe beyond September, though later stragglers have included exceptional instances of overwintering. Present in Africa mid-August to April (see above); adults arrive there earlier than 1st-year birds, and depart earlier in spring (when ♂♂ precede ♀♀) (Seel 1980). Vanguard reaches southern Europe in late March, with main arrivals April to early May, though northernmost breeding areas may not be reoccupied until early June. British-ringed bird recovered northern France on 5 March of 2nd calendar year may have wintered in Europe. 1-year-old birds generally return to vicinity of natal area, arriving later than older birds, and some at least attempt breeding (Seel *et al.* 1981); timing of autumn departures does not differ between these age classes. Adults show marked fidelity to previous breeding area.

Food. Almost entirely insects, mostly caterpillars (Lepidoptera), including numerous colonial, hairy, and warningly-coloured species, largely avoided by other birds; beetles (Coleoptera) next most important prey. Relatively large prey, especially caterpillars, located mostly by perching upright and motionless on vantage point and scanning surroundings; alert to any movement within *c.* 50 m. If available prey smaller, bird uses lookouts less and searches more actively (Wyllie 1981). Once prey spotted, capture technique varies with location, type, and abundance. Bird often sits on inner branch of tree looking outwards, then picks off prey in brief fluttering flight before returning to look-out (Amann 1948; Löhrl 1950). Will pick caterpillars off bark (Link 1889) or off wall (Wyllie 1981). Also glides repeatedly from perch to seize items from herbage or ground (Löhrl 1950; Demandt 1954; Bäsecke 1955; Liedekerke 1980). Young bird repeatedly flew from fence to adjoining field, taking larvae of cinnabar moth *Callimorpha jacobaeae* from ragwort *Senecio jacobaea* without landing (Crawshaw 1963). Often hovers to seize prey (Glutz and Bauer 1980). Prey may be eaten on the spot or brought back to perch for handling (Wyllie 1981; see below). Also reported quartering heath in low, criss-cross flight-path, sometimes beating vegetation with wings, apparently to disturb insects, periodically landing to catch and eat prey (Ruthke 1949; Wolff 1949; see also Peus 1949). Where infestation or colony of caterpillars found, bird (often in company of others) clambers around picking them off (Bottomley and Bottomley 1975). 4 birds seen flapping vigorously but deliberately on tops of young pines *Pinus*, then flew to ground, presumably to retrieve larvae thus dislodged (Anon 1961). Less commonly searches on, or from, ground, e.g. leaping in the air to pluck *Abraxas grossulariata* caterpillars from low vegetation (Armitage 1978). ♂ walked up and down edge of lawn, pulling out worms (Lumbricidae) and grubs (Green 1928). In winter, Sierra Leone, usually feeds in small leafy trees, where caterpillars abundant, e.g. picking them off trunk in quick flight from branch; also feeds by hopping about on ground, searching herbage (G D Field). Large hairy caterpillars kneaded from end to end in bill, then given a few violent shakes, sometimes against surface, to expel gut contents, before being swallowed whole (Löhrl 1950; Wyllie 1981). Caterpillars mandibulated usually from one end to the other, then back again, and the head audibly crushed. Rolled up caterpillars tossed about until they unroll. After swallowing, bird often shakes head (Glutz and Bauer 1980); may also bob head, ruffle throat plumage, or wipe bill (Creutz 1970). Small hairy caterpillars swallowed whole without being mandibulated, suggesting treatment of larger items serves more to soften tissue than to remove hairs and defensive secretions (Wyllie 1981). Gizzard specially adapted for dealing with noxious hairs—mucosal lining periodically sloughed off and regurgitated, as are pellets of hairs (Stresemann 1926) and chitin (Dementiev and Gladkov 1951a). ♀ regularly eats eggs of potential and intended hosts, probably partly for food value, perhaps especially for calcium for forming own eggs; less commonly eats nestlings. Eggs nearly always swallowed whole, with backward toss of head; occasionally, if accidentally crushed, some of egg-contents eaten (Wyllie 1981; see also Social Pattern and Behaviour). Usually hunts alone, but locally abundant food may attract 2–4 birds or more; see Social Pattern and Behaviour. In breeding season, birds may travel appreciable distances to favoured feeding sites; most birds thought to have separate feeding and breeding areas although this presumably varies with host species and its dispersion. In May–June, marked ♀ fed almost entirely at site 3·2 and 2·6 km from the 2 sites where she laid all her eggs (host Reed Warbler *Acrocephalus scirpaceus*). Feeding site, an area of scrub with abundant caterpillars, shared with at least 3 other ♀♀ and *c.* 4 ♂♂. On average, marked ♂♂ sang in area of *c.* 30 ha but travelled at least 4 km away to feed (Wyllie 1981).

Insects mainly larvae (exceptionally adults, eggs, and pupae) of moths (Lepidoptera: Notodontidae, Thaumetopoeidae, Lymantriidae, Geometridae, Sphingidae, Lasiocampidae, Saturnidae, Plusiidae, Arctiidae, Caledrinidae, Hypsidae, Brephidae, Hydriomenidae, Selidosemidae, Cossidae, Zygaenidae, Tortricidae), and butterflies (Lepidoptera: Nymphalidae, Pieridae); also adult and larval beetles (Coleoptera: Cicindelidae, Carabidae, Dytiscidae, Silphidae, Staphylinidae, Elateridae, Lampyridae, Scarabaeidae, Tenebrionidae, Curculionidae, Scolytidae, Cerambycidae, Chrysomelidae, Coccinellidae). Less often dragonflies and damselflies (Odonata), mayflies (Ephemeroptera), crickets, etc. (Orthoptera: Gryllidae, Tettigoniidae, Acrididae, Gryllotalpidae), earwigs (Dermaptera), mantises (Dictyoptera), bugs (Hemiptera: Cicadidae, Cercopidae), adult and larval flies (Diptera: Tipulidae, Syrphidae), and adult and larval Hymenoptera, notably sawflies (Symphyta) and ants (Formicidae) and rarely bees (Apoidea) and wasps (Vespoidea). Non-insect food includes spiders (Araneae), centipedes (Myriapoda), earthworms (Lumbricidae), snails and slugs (Gastropoda), young frogs and toads (Anura), birds' eggs, and nestlings. Rarely vegetable matter, e.g. buds, seeds, and berries (*Rhamnus, Juniperus, Vaccinium*). (Link 1889; Denwood 1896; Bolam 1914; Collinge 1924–7; Schiermann 1926a; Chavigny and Dû 1938; Witherby *et al.* 1938; Lowe 1943; Ruthke 1949; Löhrl 1950; Dementiev and Gladkov 1951a; Neufeldt 1958b; Ganya *et al.* 1969; Korenberg *et al.* 1972; Armitage 1978; Liedekerke 1980; Wyllie 1981; Bub 1982; McDougall 1983.)

Most comprehensive, though not fully documented, study by Link (1889), summarizing 36-year study of stomach contents, mostly from Bayern (West Germany). Diet there mostly caterpillars, notably Lymantriidae (e.g. *Porthesia, Leucoma salicis, Orgyia antiqua*) and Lasiocampidae (e.g. *Malacosoma neustria, Macrothylacia rubi*); stomachs often crammed with single species, e.g. 173 *M. neustria* in one. Also taken, sometimes in large numbers when available: Pieridae, Arctiidae, Notodontidae, Caledrinidae, Selidosemidae, and Thaumetopoeidae (once 97 *Thaumetopoea processionea* in one stomach; see also below, and Great Spotted Cuckoo *Clamator glandarius*); regularly, but in small numbers, Sphingidae, Cossidae, and Plusiidae. Among beetles, most frequently Scarabaeidae (mostly adults, e.g. *Geotrupes*, but also larval cockchafer *M. melolontha*), Chrysomelidae (e.g. *Melasoma*), and Scolytidae (one stomach, April, with 60 *Ips typographus*); several records of Cicindelidae, Carabidae, Silphidae, Tenebrionidae, Cerambycidae, and Coccinellidae. Other items taken in significant amounts include various Orthoptera, damselflies, sawflies, ants, and, in Italy, mantises and cicadas. For seasonal variation, see below. Of 20 stomachs, Britain, April–August, 94% animal food: 58·5% Lepidoptera larvae, 14·5% beetles, 4·5% adult and larval sawflies, 6% Diptera larvae, 2·5% earthworms, 2% slugs and snails, 4% miscellaneous; of vegetable matter, 3·5% weed seeds (Collinge 1924–7). Of 4 stomachs, Dumfriesshire (Scotland), end of May and early June, one contained only click-beetles (Elateridae), one almost entirely larvae of *Operophtera brumata* (Geometridae), the others mainly larvae of *Erannis defoliaria* (Brephidae), c. 70 and c. 300 individuals in these 2 stomachs (Lowe 1943). Crop contents of long-dead bird, June, Cheshire (England) contained remains of mostly Tipulidae larvae and beetles (Carabidae) (D W Yalden). On first arrival, Britain, birds take many newly emerged larvae of Lasiocampidae (Lepidoptera). If spring late, or if weather wet and cold, probably more beetles, spiders, etc.; droppings of 15 or more birds feeding on ground in cold spring contained many beetle remains (Wyllie 1981). In Bayern, diet up to end of May or early June often almost entirely beetles, notably cockchafers *Melolontha* and Chrysomelidae (Link 1889). Towards end of breeding, late June to early July, colonial larvae of *Aglais urticae* and *Nymphalis io* (Lepidoptera) commonly sought in nettlebeds, while juveniles also take *Callimorpha jacobaeae* (see 1st paragraph). After most caterpillars pupated, Bayern, fields and ponds also visited, especially by juveniles, to seek grasshoppers (Acrididae), worms (Lumbricidae), damselflies and dragonflies (Odonata), etc. (Link 1889). On almost every day, 12 August–8 September, Midlothian (Scotland), juvenile visited garden to eat young amphibians, probably frogs *Rana* and toads *Bufo* (McDougall 1983). At end of breeding season, may also take plant matter, in Bayern notably berries of alder buckthorn *Rhamnus frangula* and juniper *Juniperus communis* (Link 1889). In Kirgiziya (USSR), one stomach contained stones of wild cherry *Prunus* (Pek and Fedyanina 1961, which see for analysis of 52 stomachs). Plant matter also thought to be included soon after arrival on breeding grounds, if little else available (Link 1889). Of 9 stomachs, Moldavia (USSR), only 1 contained no caterpillars; up to 20–30 caterpillars per stomach in others, mainly Lasiocampidae (46% of total items), also 14% *Euproctis similis* (Lymantriidae), 8% Geometridae, 15% *M. melolontha* larvae, 12% Curculionidae, 5% spiders (Ganya *et al.* 1969). In Kareliya (USSR), stomachs contained various beetles (Elateridae, Cerambycidae, Chrysomelidae, Staphylinidae, Carabidae, Dytiscidae), also caterpillars and ants (Neufeldt 1958*b*). 6 stomachs, Kirov (USSR), contained mainly terrestrial insects, most *M. melolontha* (Korenberg *et al.* 1972). 19 stomachs, Kalinin (USSR), contained 427 caterpillars and other larvae, 129 various beetles, and small amounts of egg remains (Dementiev and Gladkov 1951*a*). For feeding on earthworms, see Bolam (1914), Liedekerke (1980), and Bub (1982). Little studied outside breeding season, though hairy caterpillars again thought to be main prey (Wyllie 1981, which see for other seasonal changes). Little known about races other than nominate *canorus*; in Tébressa (Algeria) numerous stomachs of *bangsi* crammed with *Thaumetopoea* caterpillars (Chavigny and Dû 1938).

Young fed on whatever diet hosts would usually bring to own young, i.e. mostly soft-bodied insects. For report, disputed by Ferry and Martinet (1974), that one Great Grey Shrike *Lanius excubitor* of pair fed parasite on hairy caterpillars, while mate fed its own brood (sharing nest) on usual diet of crickets, beetles, etc., see Claudon (1951). Food brought to mixed brood of *C. canorus* and *L. excubitor* included small mammals (Ferry and Martinet 1974). Diet, however diverse, must be predominantly animal food, and no species of non-insectivorous passerine in Europe known to have successfully raised young *C. canorus* (Wyllie 1981). No evidence that hosts provision young *C. canorus* at faster rate than they would their own brood (Wyllie 1981), though in experimental study, volume of food-calls equalled those (in concert) of host Redstart *Phoenicurus phoenicurus* brood, stimulating host to feed parasite at same rate as it would whole brood (Khayutin *et al.* 1982). Fledged young fed snails and worms by casual fosterers (Bolam 1914). Once, juvenile ate dead Willow Warbler *Phylloscopus trochilus* (Bub 1943).
EKD

Social pattern and behaviour. Extensively studied, but much early work confused fact with speculation. For valuable pioneer studies, see Jenner (1788) and, especially, Rey (1892). Basis of modern understanding largely laid by Chance (1922, 1940). For important recent studies see especially Makatsch (1955), Löhrl (1979), Gärtner (1981*a, b*, 1982), and Wyllie (1975, 1981).

1. Tends to migrate singly or in small groups (De Smet 1970; Moreau 1972; Wyllie 1981). 20–100 occasionally congregate on or near breeding grounds (see Flock Behaviour, below), probably associated with locally abundant food. Similar concentration also reported, Tanzania, 30 March–2 April (Moreau 1972). Outside migration periods, habits of wintering birds little known, due to combination of retiring habits and confusion with African

Cuckoo *C. gularis* (Moreau 1972). BONDS. Mating system not well understood but, on basis of wide overlap of 'song ranges' (♂) and 'egg ranges' (♀) (see Breeding Dispersion, below), most likely promiscuous (Wyllie 1981; see also Makatsch 1955). Several ♂♂ and ♀♀ may breed in same area, and each ♂ vies with rivals to court every ♀; from marking studies, nearly every ♀ seen to be courted by more than 1 ♂ in same season, and by different ♂♂ in successive years (Wyllie 1981). ♀ copulates with several ♂♂ in laying period (Molnár 1950; Makatsch 1955; Groebbels 1957). No hard evidence for other mating systems—e.g. short or long-term monogamy (Chance 1940; Paulussen 1957; Lack 1968)—which may be falsely inferred from site fidelity (see below)—or polygamy or polyandry (e.g. Claudon 1951), polyandry often being inferred from apparent numerical bias towards more conspicuous ♂♂. Nevertheless, ♂♂ and ♀♀ often occur locally in very unequal proportions, and possibility of other mating systems cannot be dismissed. Age of first breeding of both sexes probably usually 2 years, sometimes 1 (Seel *et al.* 1981; see also Breeding Dispersion). BREEDING DISPERSION. Complex and flexible system, thought to consist essentially of home ranges rather than exclusive territories, though birds more territorial in some situations than others. (1) ♂ dispersion. If density high, ♂ said to defend 'territory' but other ♂♂ tolerated at least on periphery; at low density, ♂♂ less territorial, and dispersion maintained by song alone unless ♀♀ present, and ♂♂ then less tolerant of rivals (Löhrl 1950; Molnár 1950; Glutz and Bauer 1980). In Cambridgeshire (England), ♂♂ weakly territorial, overlapping extensively in song ranges (*c.* 30 ha per bird) and feeding areas (see Food). Dominance hierarchy likely, subordinate ♂♂ being expelled in vicinity of potential laying areas (Wyllie 1981). Many 1-year-olds thought to be nomadic and non-territorial (Glutz and Bauer 1980), though some occupy territories and may breed (Seel *et al.* 1981). Range size of ♂♂ apparently varies with habitat, host density, age of bird, and hierarchical status (Glutz and Bauer 1980). Thus, size difficult to determine, and estimates vary widely. Area of 10 ha, Hamburg (West Germany), with 80 pairs of host Marsh Warblers *Acrocephalus palustris*, held by 1 ♂ each year, but 1–2 other ♂♂ usually present on periphery (Gärtner 1981a). In various habitats, 20–50 ha per calling ♂ (Becker and Dankhoff 1973; see also Glutz and Bauer 1980). 2 ♂♂ shared *c.* 13 ha of ponds and fields, another in *c.* 94 ha of mixed open scrub and woodland; probably *c.* 27 ♂♂ in 42 km², i.e. 1·55 km² per ♂ (Melde 1982). For similar range size of single ♂♂, see Westerfrölke (1956) and Groebbels (1957). 1-year-old ♂ occupied *c.* 1·13 km² (Seel *et al.* 1981). In 'optimal habitat', Mecklenburg (East Germany), 1·17 km² per ♂ (see Glutz and Bauer 1980); in agricultural land, Niedersachsen (West Germany), 1·15 km² per ♂ (see Bögershausen 1976). (2) ♀ dispersion. Evidence better for ♀ holding laying territory, though considerable overlap between ♀♀ may occur (e.g. Simmons 1974). Dominant ♀ thought to defend area containing adequate number of host nests; in 2 cases, laying area *c.* 30 ha (Wyllie 1975, 1981). In 10 ha, occupied by *A. palustris*, 0–4 ♀♀ in each of 8 years, with 3 categories of ♀♀ occurring: (a) territory-owners, laying large series of eggs (see Breeding) in same area over several years; (b) sedentary non-territorial ♀♀ (usually 1 in area of each territorial ♀), which lay a few eggs throughout season and may eventually inherit territory of dominant ♀; (c) nomadic, non-territorial ♀♀ (perhaps 1-year-olds) which never laid more than 1 egg in owner's territory, and thought to rove over substantial area (Gärtner 1981a). In one case, dominant ♀ probably caused subordinate to lay in nests of host other than subordinate's preferred one (Wyllie 1975). Area of *c.* 150 ha, containing 54 pairs of Reed Warbler *A. scirpaceus*, held up to 6 ♀♀ and at least 2 ♂♂ (Wyllie 1975); where host *A. scirpaceus*, ♀ occupied *c.* 50 ha; greatest distance between parasitized nests 1700 m (Marbot 1959). Each year, 2–3 ♀♀ on island of 100 ha, 2–3 km from mainland, hosts Sedge Warbler *A. schoenobaenus* and *A. scirpaceus* (Ruthke 1951). Along Vezouze river (France), where host *A. scirpaceus*, 15 ♀♀ laid over average distance of 616 m (37 m–4 km) in different years (calculated from Blaise 1965). Greatest distance, 2·5 km between Pied Wagtail *Motacilla alba* nests used by same ♀ (Warncke and Wittenberg 1958). Eggs of individual ♀♀ laid in *M. alba* nests often over 4–5 km² (Hethke 1968). Few reliable studies of parasitism rates. 16% of 110 nests of *A. palustris* parasitized (Gärtner 1981a). In different areas and years, where host *A. scirpaceus*, frequency 0–38·1%; at main study site, Cambridgeshire (107–247 nests), 3·3–23·4%; higher rates tended to coincide with lower host density and/or higher density of ♀ *C. canorus* (Wyllie 1981). Site fidelity high, at least in established ♂♂ and ♀♀. For ♀♀, much evidence from egg characteristics (e.g. Evans 1922, Baker, 1922, Molnár 1950), and fidelity recorded for 7 years or more (Schiermann 1926b; Chance 1940; Labitte 1958; Blaise 1965). Based on abnormal calls, 3 ♂♂ thought to be site-faithful for 9, 10, and 10 years (Anon 1928; Westerfrölke 1956). Some evidence that home range occupied in 1st year re-occupied in subsequent years (Seel *et al.* 1981). ROOSTING. Same site typically used throughout breeding season; in one case, in one of the thickest, tallest pear *Pyrus* trees in territory (Löhrl 1950). In Cambridgeshire, usually *c.* 3–4 m up in dense scrub, especially hawthorn *Crataegus* or blackthorn *Prunus* (I Wyllie). One ♀ roosted at feeding site, *c.* 4 km from laying area (Wyllie 1981). Recorded roosting on pylon (Michaelis 1971). Sleeps with head slightly turned but bill not tucked into mantle, and reacts to slightest sounds (Löhrl 1979). Activity mainly diurnal. In midsummer, Cambridgeshire, ♂♂ started calling (see 1 in Voice) *c.* 1 hr before sunrise, and most vocal 04.00–08.00 hrs; continued to call up to 1 hr after sunset (Wyllie 1981). Sometimes call at night, even in total darkness, though rarely before midnight (Ramsbotham 1906; Löhrl 1950; Glutz and Bauer 1980). Bathing in pools not known, but readily rain-bathes with tail-feathers broadly fanned and wings raised alternately such that undersides also exposed (Bährmann 1962). In heavy rain, head held almost vertically upwards; after rain stops, remains in erect posture with tail pointed down. Sunbathes apparently only when sun quite low and shining on bird's back (Löhrl 1979).

2. Typically shy and retiring (e.g. Wyllie 1981). Frequently mobbed by passerines, especially ♀ by intended hosts, whereupon may ruffle plumage and distend throat; also gapes and gives Hissing-call (e.g. Chance 1923: see 7 in Voice). Laying ♀ attacked by Dunnock *Prunella modularis* rose with each assault, struck out with feet, and gave Gowk-call (Loyd 1912: see 2 in Voice). At approach of observer, captive birds adopt Concealing-posture (Fig A): sleek feathers and look remarkably slim; head, body, and tail form straight line, directed at source of danger with slow swaying movement (Löhrl 1979; see also section 3 in Egg-laying Behaviour, below). FLOCK BEHAVIOUR. May associate closely in large numbers during spring and autumn migration where vegetation infested by caterpillars or other suitable prey (Bau 1901; Hurrell 1980). Birds then typically excited and noisy, and move rapidly through vegetation: e.g. on 11 May, Hampshire (Eng-

A

land), c. 50 in 600 m² of woodland; sounds included apparent Gowk-calls (Christie 1979); for similar assemblage of c. 100 or more, see Cox (1980). At Dungeness (England), mid-June, feeding flocks of 8–12 adults, perhaps mainly ♂♂, commonly occur; from July, ♀♀ and juveniles better represented, and from c. 20 July, flocks entirely juveniles; birds typically silent and unobtrusive (N Riddiford). ANTAGONISTIC BEHAVIOUR. (1) From day of arrival on breeding grounds, ♂ defends area with Advertising-call (see 1 in Voice) given from almost any perch, exposed or hidden, from ground level to tree-top; also in flight. Song output usually greatest on first arrival, decreases on arrival of ♀♀, increases again during pre-laying period (mid-May), and then declines until departure (Wyllie 1981). Rival ♂♂ typically face each other when calling (Wyllie 1981). May approach each other ever closer until c. 15 m apart, calling, then withdraw (Mercier 1919). At Dungeness, territorial disputes involved noisy, hectic chases by up to 6 ♂♂ (N Riddiford). In close encounter, rivals sit a short distance apart, and threaten with open bill pointed forwards; occasionally, with slightly open wings. If Advertising-call imitated (by human) some ♂♂ flee while others approach to within a few metres and give Gowk-call and Hissing-calls (see 2 and 7 in Voice). Particularly aggressive birds 'slide' along perch and beat wings demonstratively (Glutz and Bauer 1980). Physical contact rare between ♂♂, and then usually when ♀ present (Wyllie 1981). May cuff each other with wings, seldom more physical (Molnár 1950), but see Gush (1979). Aggression between ♂♂ high early in season, waning as season progresses (Wyllie 1981); by end of season, not aggressive, e.g. will feed close together (Löhrl 1950). Physical contact also rare between ♀♀; 2 ♀♀, each trying to lay at same time in different nests c. 25 m apart, attacked each other; one flew down to buffet other as it started Gliding-flight to lay (see below), thus preventing it from laying for c. 1 hr (Chance 1940). Once, in apparent confrontation, when 3 birds approached flying, 2 others rose to challenge them, and by circling upwards to c. 100 m, gradually forced intruders away before returning to home area (Gush 1979). In similar incident, ♂ and ♀ spiralled upwards without flapping—possibly display-flight; ♂ following ♀ broke off to chase away intruding ♂ (Wyllie 1981; I Wyllie). HETEROSEXUAL BEHAVIOUR. (1) Mate attraction. Apparently ♂ or ♀ may take initiative, ♂ with Advertising-call, ♀ with Bubbling-call (see 4 in Voice). ♂ may approach ♀, calling with more animation as he nears her. ♂ alights near ♀ and performs elaborate Advertising-display, comprising head-bobbing or, at greater intensity, bowing; also opens and droops wings, and raises and fans tail; with tail fanned, may rotate body slowly about horizontal axis and sway it from side to side (e.g. MacKeith 1908, Wyllie 1981: Fig B). Without moving rest of body, bird slowly moved tail so that tip described circle, finishing with tail erect (Luton 1957); for associated call, see 3 in Voice. Body-swaying performed at rate of c. 1 movement to left or right per s, with tail raised but not fanned (Löhrl 1979). When ♂ displaying, ♀ remained silent (MacKeith 1908). Occasionally, displaying ♂ may fly down to pick up piece of grass, etc., and return to display with it as if presenting it to ♀; usually drops it after a few seconds before flying after her (Chance 1940; Matthews 1947; Walker 1949; Luton 1957; Harrison 1974; King 1974b). Though several reports of carrying material, behaviour apparently not common. Once, ♂ presented ♀ with caterpillar and, while this possibly courtship-feeding, procedure evidently very rare (Wyllie 1981). ♂ also performs short circular Display-flights, calling persistently (see Bögerhausen 1976). Whenever ♀ made slightest movement, displaying ♂ flew to her side, but remained only a few seconds before flying c. 100 m to display again on top of bush; continued thus for 15 min (Loyd 1914). Often 2 or more ♂♂ vie for 1 ♀, or 1 ♂ displays to 2 ♀♀. Rival ♂♂ frequently end up fighting, whereupon ♀ slips off secretively. Alternatively, ♂♂ may pursue ♀ in swift, darting and weaving flight (Wyllie 1981). (2) Mating. ♂ ready to copulate usually follows ♀ silently, often for lengthy period, preferring to fly to perches of moderate height (Löhrl 1979). Presentation by ♂ of piece of vegetation, etc. (see above), may directly precede copulation and probably invites it. ♀ opens wings slightly, moves tail to one side, and calls just before copulation. ♂, once mounted, leans forwards and, letting wings droop slightly, moves unopened tail slowly from side to side; ♂ said to copulate usually 2–3 times in succession before birds go separate ways (Molnár 1950). On 2 occasions, probably involving same ♂ and ♀, ♀ perched silently on branch with tail lowered, and ♂ approached in gliding flight, calling, landed directly on her back, and copulated within a few seconds. On 1st occasion, pair flew off together and ♂ called for some time after. On 2nd occasion, after mating, ♂ landed on bush and began calling vigorously, while nearby ♀ gave Bubbling-call. ♂ displayed (see above) and both flew to ground where he picked up and dropped small twigs and leaves, calling continuously; ♂ made mock attacks on ♀ who gave Bubbling-call until ♂ fell silent c. ½ hr after mating (Wyllie 1981). According to Claudon (1951), ♀ indicates sexual excitement with Gowk-call; partners then drop to ground in tangle of wings, and copulate there or on nearby stump. For mating on overhead wires, see Montfort (1947) and Molnár (1950); see also Löhrl (1979). Copulation occurs at any time until nightfall, but mostly early morning (Molnár 1950). Thought to occur 1½ days before laying egg thus fertilized, and to take place near intended host nest (Gärtner 1981a). On 3 occasions, ♀ laid previously fertilized egg on day when she also copulated (Wyllie 1981). Copulation occurs throughout laying period (e.g. Seel 1973; see also Bonds, above). EGG-LAYING BEHAVIOUR. (1) Manipulation of host nests. ♀, but never ♂ (e.g. Löhrl 1979), may take 1 or more eggs, or even small young, from host nests not used for own laying, thereby often destroying nest (e.g. Jourdain 1925, Varga 1977a). As host then usually rebuilds, potential laying span of C. canorus thus extended (Gärtner 1981a, b, which see for discussion of other possible functions, also Wyllie 1981). Such robbing perhaps more common where laying of host population highly synchronized. In some years, Hamburg, robbing from A. palustris nests reached 30%, and in 64% of cases led to desertion by host (Gärtner 1981a, b). Of 159 nests of A. scirpaceus, 48% were robbed of 158 eggs, and C. canorus laid 27% of its eggs in replacement nests (Gehringer 1979). One ♀, after laying egg in nest of M. alba which contained 3 newly hatched young, returned 2 days later to eject young (Headley 1919; see also Milburn 1915, Scholey 1924). ♀ recorded eating 4-day-old A. scirpaceus (Wyllie 1975). (2) Selection of host nests. Ovulation and laying period closely synchronized with those of host (Chance 1940; Gärtner 1981a). If nests of preferred host not available, especially early in season, occasionally lays in nest of another species (e.g. Baker 1922, Chance 1940). ♀ usually chooses nests with incomplete clutches, less commonly fresh complete ones; occasionally lays in clutches already incubated, rarely close to or after hatching (e.g. Butterfield 1915,

B

Mairlot 1919), but presumably only when no other option available (Gärtner 1981a; Wyllie 1981). Rarely, lays in unfinished or deserted nests (Butterfield 1915; Chance 1940). In 72 parasitized nests of Meadow Pipit *Anthus pratensis*, 4 empty at time of parasitizing, 17 had 1 egg, 20 had 2, 14 had 3, 13 had 4, 4 had 5 (Chance 1940). In 90 nests of *A. scirpaceus*, 4 had no eggs, 25 had 1, 33 had 2, 21 had 3, 6 had 4, 1 had 5 (Wyllie 1981). Late in season, one ♀ will rarely lay 2 eggs in same nest (e.g. Wenzel 1914, Venables and Venables 1962). More often (never commonly), 2 or more different ♀♀ lay in same nest. Of 1246 parasitized clutches, 4% contained 2 *C. canorus* eggs (Rey 1892). Of 870 eggs laid in *A. scirpaceus* nests, estimated 6–7% laid by 2 ♀♀ in same nest (Makatsch 1955). Of 312 parasitized nests of Great Reed Warbler *A. arundinaceus*, 1 *C. canorus* egg laid in 55·4%, 2 in 34·6%, 3 in 9·0%, 4 in 0·6%, 5 in 0·3% (Molnár 1950). Once, 3 eggs (2 by same ♀) in nest of Red-backed Shrike *Lanius collurio* (Schlegel 1915). (3) Location of nests. Despite some earlier reports, ♂ plays no part, though occasionally present at laying (see below). ♀ locates suitable nests by combination of prolonged watching, usually perched or less commonly in the air, and, when close, by gauging degree of agitation of hosts (see below). Most often, ♀ lies along elevated branch with commanding view of nest and up to *c.* 100 m from it in case of ground-nesting host, closer (e.g. 50 m) if nest in reedbed where visibility reduced (Chance 1940; Blaise 1965; Wyllie 1981). ♀ sits in Horizontal-posture, similar to a Nightjar *Caprimulgus europaeus*, and presumably same as or closely similar to Concealing-posture; often remains thus motionless for *c.* 1–2 hrs each day for several days during nest-construction, occasionally visiting nest-site a few days, or hours, before laying, presumably to check location or stage of development (Chance 1940; Wyllie 1981). Where host *Anthus pratensis*, ♀ never makes reconnoitring visit until host has laid at least 1 egg (Chance 1920). Perhaps more often in open country, where look-out perches scarce, ♀ thought to search for nests, or inspect chosen one, with so-called 'raptor-flight'; soars up to 60 m or more with outspread primaries, occasionally flapping wings, very like Sparrowhawk *Accipiter nisus* (Chance 1940; Ash 1965). ♀ may locate nests by combination of gliding and ground-searching; one ♀ glided around meadow, repeatedly alighting on tree or on ground which she then hurriedly searched, eventually finding nest after 7 hrs (Wyllie 1981). ♀ does not search randomly, but, by testing reaction of hosts, learns which part of host's territory is best guarded and thus where nest located (Löhrl 1950; Seppä 1969). In one case, as soon as host *A. pratensis* less agitated, searching ♀ changed direction; when close to nest, and mobbed, ♀ fluttered along with ever smaller leaps, gazing around until she spotted nest (Löhrl 1950). (4) Laying. Numerous early accounts (before *c.* 1920) erroneous, especially concerning supposition that ♀ either lays egg near nest and deposits it therein with bill or regurgitates egg into nest; these theories often thought to explain deposition of eggs in inaccessible nests, e.g. in holes, but dismissed by Chance (1940) who, in over 100 layings in nests of 8 species, never recorded any method other than direct laying by ♀ sitting in nest. This now accepted as method always practised, irrespective of nature of host nest, though details vary somewhat with host. For discussion, see Makatsch (1955), Löhrl (1979), and Wyllie (1981). Morphological adaptations to laying include stout: manoeuvrable legs, well able to straddle and grip nest; unusually extensible cloaca, facilitating laying in confined nests; thickened eggshell, resisting drop from height (Rey 1892; Chance 1940; Joy 1943; Wyllie 1981). When ready to lay in nest of *A. pratensis*, ♀ often waits until hosts absent before flying from look-out to nest in characteristic hawk-like Gliding-flight; if harassed by hosts, may be forced to return to look-out, and make several approaches

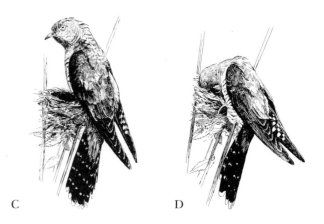

C D

before laying (Chance 1940). May be fiercely challenged by hosts at nest, and egg of *C. canorus* sometimes broken in mêlée (Rey 1892). On landing, ♀ finds nest by ground search (see above). Where host *A. scirpaceus*, ♀ lands a little way from nest, and makes final approach through reeds (Wyllie 1975; see also Gehringer 1979). ♀ intending to lay in nest of *P. modularis* alternately hopped a few paces towards nest and halted for long intervals with wings drooped and neck outstretched (Loyd 1912), presumably as in Horizontal-posture. ♂, usually excited, sometimes comes near ♀ at laying (Astley 1915; Chance 1940; Wyllie 1975, 1981); at Dungeness, ♂ often sat on top of bush in which ♀ laying, though less attentive as laying period progressed (N Riddiford). The following based mainly on Chance (1922, 1940) and Wyllie (1981). On reaching nest, ♀'s head and throat plumage often ruffled (Fig C), partly accounting for mistaken impression that egg about to be regurgitated. ♀ clings to nest-rim with feet and balances with outstretched wings and tail. Feet remain on rim, presumably to prevent damaging host eggs. ♀ then invariably picks up 1 of host's eggs in bill (Fig D) and holds it while raising herself to position cloaca over nest-cup. ♀ lays with a shudder, sometimes also with wing-quivering (Fig E), then immediately flies off, often carrying stolen egg which is swallowed whole or crushed and eaten. Sometimes 1 or more of host's eggs eaten at nest (see below), with backward toss of head, before she lays her own (Gehringer 1979; Gärtner 1981a, b; Wyllie 1981). Where host *A. palustris*, ♀ took at least 1 egg and, if at least 2 eggs present, 2 or more taken in 25% of cases. ♀ occasionally took host egg 1–2 days before laying her own, then another as she laid (Gärtner 1981a, b). If 2 ♀♀ lay in same nest, 2nd ♀ sometimes removes 1st egg laid (Molnár 1950; Wyllie 1981). In 3 seasons when 2 ♀♀ laying in same area, egg laid by 1st ♀ not infrequently taken, presumably by 2nd ♀ (Gehringer 1979, which see for details). Just after laying, ♀ typically gives Bubbling-call, to which ♂(♂) replies with Advertising-call. In case of nests in

E

holes or crevices, or domed nests (e.g. Wren *Troglodytes troglodytes*), ♀ clings vertically in order to press cloaca close to nest opening; nest of *T. troglodytes* often thus damaged (Musselwhite and Ware 1923; Chance 1940). For laying into nest-boxes from entrance, see Herberg (1960). Time from arrival at any nest to departure remarkably short, thus attracting least attention; ♀♀ spent no more than 1 min in immediate vicinity of nest, and average c. 9 s (4–16, $n=8$) laying eggs (Seel 1973; see also Molnár 1950). Typically lays in afternoon or evening (Chance 1940; Molnár 1950; Seel 1973; Wyllie 1981 which see for review, also seasonal duration). FOSTER-CARE OF YOUNG. Parasite's egg usually hatches, and chick fledges, before those of host; shorter incubation probably facilitated by partial incubation in oviduct (Perrins 1967). Parasite therefore usually hatches into host clutch, less often when some or exceptionally all of host's young already hatched. Ejection of host's eggs and young by parasite first described by Jenner (1788). The following, based on Molnár (1950) and Wyllie (1981), describes ejection of egg, though eviction of host nestlings essentially same. Newly hatched parasite still blind, initially passive and silent, but from c. 8–10 (3–36) hrs, begins to wriggle in bottom of nest until 1 of host's eggs manoeuvred against side of nest—this helped by shallow, sensitive hollow between shoulders, and by outspread wings. With head held down, almost touching belly, or braced against floor or walls of nest, and muscular legs and toes splayed and braced against sides of nest, parasite slowly works egg up side of nest. With effort of raising egg, parasite's body appears to swell, veins of neck and wing-pit markedly so; head jerks spasmodically up and down. Bouts of effort alternate with spells of rest. When egg nears nest-rim, parasite clasps rim with wing-tips (Fig F)

and egg or nestling finally ejected with brief quivering, jerking action. When sure that ejection successful (by feeling with wings, etc.) parasite drops back into nest-cup. In 2 out of 114 cases, parasite accidentally ejected itself also. Parasite took on average, c. $3\frac{1}{2}$ min to evict egg (sometimes 20 s), even from deep nest (Lancum 1929). May take only 3–4 hrs to clear several eggs or young, sometimes up to 1–3 days. Hosts may unwittingly aid process by removing eggs or young stranded on nest rim. Urge to eject nest contents dissipates after 4–5 days (Molnár 1950; Claudon 1951), at which stage hollow in back disappears (Butterfield 1916). 2-day-old parasite once ejected 7-day-old *P. modularis* (Gurney 1905). Parasite rarely thwarted in ejecting nest contents, but may be: e.g. in confined nest-site, if host young well grown, or perhaps sometimes if host's eggs too big. Rapidly growing parasite then usually kills host's young by smothering, and is raised alone. Host's young occasionally survive, and mixed broods reported with (e.g.) Robin *Erithacus rubecula* (Burton 1947) and Great Tit *Parus major* (Puhlmann 1914b; Homoki Nagy 1977). Once, 2 *C. canorus* young raised with *P. major* brood (Puhlmann 1914b). Parasite not uncommonly raised with young of Great Grey Shrike *Lanius excubitor*, 3 contained 1 parasite egg, 3 had 1 parasite plus 4 host young, 1 had 2 parasite young and 3 host young; all mixed broods in this and previous years, successfully raised (Claudon 1955). According to Claudon (1955), host's eggs often too big to eject, but this disputed by Ferry and Martinet (1974) who cite ejection of (e.g.) Blackbird *Turdus merula* eggs. In cases where 2 parasite eggs hatch in nest, 1 young usually ejects other, often after prolonged struggle; in case where *A. arundinaceus* nest contained 3 parasite eggs, 1 ejected soon after hatching, while others tried to eject each other for 4 days until both fell out of nest (Molnár 1950). Parasite brooded from hatching. After eyes open at 5–7 days (Bussmann 1947; Wyllie 1981), host broods only during bad weather and at night (Wyllie 1981); brooded until 9 days by *P. modularis* (Werth 1947). In mixed broods, parasite typically sits on top (e.g. Peltre 1931). Chick typically moves little, i.e. does not crane neck, quiver wings vigorously, or rotate excessively in nest (see also below). Relative immobility evidently to safeguard chick's bulk against damaging or dislodging nest. When fed, young cocks head back and directs gape upwards (Bussmann 1947). Young mostly silent for c. 4 days after hatching (Glutz and Bauer 1980). Colourful orange-red gape and food-calls (see Voice), which intensify from c. 7 days, serve as powerful feeding stimulus; for *T. troglodytes* feeding parasite in *P. modularis* nest, instead of its own brood nearby, see Owen (1913). Food not properly inserted is regurgitated and may have to be inserted again, sometimes repeatedly (e.g. Bussmann 1947). Food swallowed, with bill closed, only when delivery complete, thus preventing injury to host which bows deep into gape (but see Hens 1949). Young more active from 11 days, turning in nest, wing-stretching, and preening. By 16 days, turns in nest to face approaching host; quivers one outstretched wing and vibrates gape (Wyllie 1981). Before fledging, captive young reacted to caterpillar prey and dealt with them like adult, though initially some lack of success in swallowing; hungry young began to self-feed 2 days after leaving nest, but accepted food from foster-parents for 3 weeks (Löhrl 1979). Out of nest, young fed assiduously by *A. scirpaceus* foster-parents for 2–3 weeks, perhaps 4–6 weeks in some other host species (Wyllie 1981). Sometimes stands on back of young to deliver food. Young typically directs vicious peck at host just after being fed (Coward 1919). On leaving nest, young seeks cover nearby, and remains silent until fed, whereupon calls stridently. In one case, bird capable of only clumsy flight on 1st day, but by 3rd day, had moved 100 m. By 4th day, flew from branch to branch begging from any passing bird, foster-parent or not, with gaping bill and trembling wings (Fig G). By 8th day, made increasingly long flights. Last seen fed by hosts (*Hippolais* warblers) 15 days after leaving nest (Thomaz de Bossierre 1947). Young often fly after foster parents, uttering food-calls (Glutz and Bauer 1980). During dependence on hosts, young travel several hundred metres from nest-site, much further than young of host species would do before independence (Wyllie 1981). Birds other than foster-parent not uncommonly feed fledged young (e.g. Owen 1913). Recently fledged young may gather in small area; once, 3 fledged birds close together fed,

G

apparently indiscriminately, by variety of passerine species (Bolam 1914). For 2 fledged young fed by Redstart *Phoenicurus phoenicurus*, and by Rock Pipit *Anthus spinoletta*, see respectively Gautier (1968) and Thomas (1975). According to Wyllie (1981), no reliable evidence for adult *C. canorus* feeding fledged young, but for report of adult apparently feeding young 20 times in succession, see Klein (1911). ANTI-PREDATOR RESPONSES OF YOUNG. Young in nest dozed with eyes shut between feeds, but instantly alert when predators, especially crows (Corvidae), called or came close (Wilde 1974). From *c*. 7 days, when eyes open, young threaten intruders by gaping, giving food-calls (hissing quality probably intimidating), ruffling plumage, especially around head and neck, and quivering wings; also make lunging movements at approaching danger. Older young also repeatedly rise and fall in nest as if threatening attack. From *c*. 14 days, lunge quickly forwards and audibly bill-snap (Butterfield 1916; Chance 1940; Sellin 1969; Löhrl 1975; Wyllie 1981). At *c*. 16 days, give alarm-call if handled, and also excrete foul-smelling brown liquid, evidently in defence (Wyllie 1981; see also Senegal Coucal *Centropus senegalensis*). PARENTAL ANTI-PREDATOR STRATEGIES. No good evidence for any defence of offspring; once, however, bird mobbed human approaching juvenile being fed by *A. pratensis* (Bell 1965).

(Fig A from photograph in Löhrl 1979; Fig B from drawing in Astley 1915; Figs C–D from photographs in Gärtner 1981*b*; Figs E–F from photographs in Wyllie 1981; Fig G from photograph in Makatsch 1955.) EKD

Voice. Freely used during breeding season, especially by ♂. Little known outside breeding season, but see calls 4 and 5. For musical notation, see Schmitt and Stadler (1918).

CALLS OF ADULTS. (1) Advertising-call of ♂. Familiar, disyllabic 'cu-coo' (Fig I), with far-carrying, often ventriloquial quality, serving to advertise occupancy of territory (Wyllie 1981). Given at intervals of *c*. 1–1·5 s, typically in series of 10–20, with a few seconds between series. Performance of up to 300 successive calls reported; calls given in more rapid succession when excited—maximum 26 calls per 30 s (Bögershausen 1976). 1st syllable higher—especially when excited—and shorter than 2nd; pitch varies, sometimes markedly, between individuals; call may span

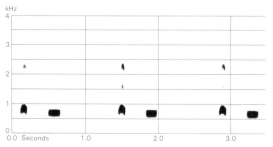

I V C Lewis England May 1970

II R W Genever England April 1968

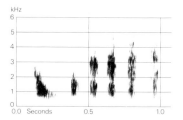

III V C Lewis England June 1979

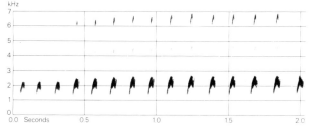

IV V C Lewis England June 1979

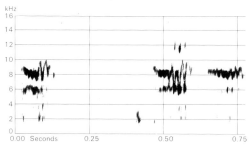

V V C Lewis England July 1959

various pitch intervals from a second to a fifth, most frequently minor third (Glutz and Bauer 1980). Delivered from any perch or in flight, usually by day, occasionally also at night (Löhrl 1950). 1st syllable given with bill open, and forward upward thrust of neck, 2nd syllable with bill closed during return to normal posture (Vollbrecht 1955; Löhrl 1979). Several variants reported. Especially when excited, commonly gives several, often 2–3, rapid 'cuck' sounds to each 'oo', or vice versa; occasionally 'cuck' alone (Witherby *et al.* 1938; Groebbels 1957; Wyllie 1981). In recording (Fig II) of bird in flight, 'cu-coo kuk kuk kuk kuk kuk kuk kuk coo' (J Hall-Craggs). Variants occur throughout breeding season (Witherby *et al.* 1938), but at beginning, and perhaps more often towards end of season, call may sound hoarse and out of tune; for seasonal frequency of calling, see Wyllie (1981). Calls given much less frequently in cold and thundery weather (Groebbels 1957; Melde 1982). (2) Gowk-call. Short, harsh 'gowk' (Fig III), not unlike Jay *Garrulus glandarius*, given in rapid series (P J Sellar). Also described as rapid, hoarse, spluttered 'kwow-wow-wow' (Witherby *et al.* 1938); short, rather gruff 'wah-wah' (Chance 1940; Wyllie 1981). For other renderings see Glutz and Bauer (1980). Given usually in flight or on alighting (Chance 1940), by both sexes, but especially by excited ♂ in association with call 1, e.g. when calling against another ♂ (Wyllie 1981), in pursuit of ♀ (Glutz and Bauer 1980), or when ♀ soliciting copulation ('crechh crechh': Claudon 1951). Also occasionally by ♀ in confrontations with rivals, when harassed by ♂, when mobbed ('cak-cak-cak-cak-cak': Wyllie 1981; also Loyd 1912); or before mating (Claudon 1951). In recording by V C Lewis of mating sequence, extended series of choked 'gowk' sounds and variants. Gruff, yet soft 'grorr-grorr-grorr' (Chance 1940), probably same or variant (Wyllie 1981). (3) Guo-call. Soft 'guo guo', audible only at close range, given by ♂ displaying to ♀, usually linked with swaying motion of tail (Löhrl 1979; see Social Pattern and Behaviour). (4) Bubbling-call of ♀. Liquid, rather musical bubbling sounds (Chance 1940); aptly described as 'water-bubbling chuckle' (Witherby et al. 1938). Consists of c. 15 units on descending scale, lasting c. 3 s; occasionally 2 or more series given in quick succession (Wyllie 1981). In recording (Fig IV), call has slight crescendo; this apparently typical (P J Sellar). Delivered in flight or from perch, particularly after egg-laying; thus most often heard in May–June (Chance 1940; Wyllie 1981). For calling on nest, see Bunyard (1926). May also be given in response to call 1, when serves to attract mate (Groebbels 1957; Melde 1982), or in response to Bubbling-call by another ♀ (Melde 1982). 'Excited bubbling calls' heard from flock on spring migration in Tanzania (Moreau 1972) possibly this call. (5) Chuckling-call of ♂. Very similar to call 4 of ♀, but deeper and harsher; occasionally given at end of series of call 1 (Wyllie 1981). Though said by Chance (1940) to be well-known and distinct from call 4, probably not common, and likely to be confused with call 4 (Wyllie 1981). Possibly given only sporadically and outside breeding season (Löhrl 1979). (6) Mewing-call. Soft, low-pitched murmuring sound, audible only at close range, said to be given by ♀ when intently watching host nests or prior to laying (Chance 1940). Not heard by Wyllie (1981), despite close observation of many ♀♀ near nests of Reed Warbler *Acrocephalus scirpaceus*. (7) Hissing-call. Short, sharp, explosive hiss, given by ♂ when chasing ♀ (Witherby et al. 1938). Also described as short, repetitive, snake-like hiss, given during chase, though not known if by ♂ or ♀ (Wyllie 1981). Said to be given after call 1, by ♂♂ chasing rivals (Groebbels 1957).

CALLS OF YOUNG. Initially, from c. 24 hrs, food-call a thin, squeaky 'seep', given only when hosts arrive with food. As bird grows, food-call gets steadily louder until, at c. 7 days, audible at several metres from nest (Wyllie 1981). Food-call of older young a persistent, piercing sibilant 'chizz-chizz-chizz' (Ziegeler and Poll 1950) or 'gigigi...' (Glutz and Bauer 1980). In recording by C Fuller of 17-day-old young begging and being fed, a high-pitched 'si-si-si-si-si-si-si-si si-si-si-si-si-si-si-si' (J Hall-Craggs). When hosts absent collecting food, young give a single chirp, repeated at intervals of a few seconds (Wyllie 1981)—'ziii ziii' or 'sriii sriii' (Glutz and Bauer 1980); given in response to contact-call of foster-parent, also especially loudly when handled (Bussmann 1947; see also below). These calls used for remainder of time young dependent on hosts (Wyllie 1981). In recording of young just after leaving nest, 'schri schri shri-shri-shri-shri'; sonagram (Fig V) shows 'schri shri-shri' part of this series (J Hall-Craggs). One fledged young appeared to imitate 'tissip' of Meadow Pipit *Anthus pratensis* (O'Connor 1962); for comparison with frequency range and volume of young Redstarts *Phoenicurus phoenicurus*, see Khayutin et al. (1982). From c. 16 days, young emit rattling, hawk-like call of alarm when handled (Wyllie 1981); this probably the low trill given by captive young as it struck out at hand bearing food (Gurney 1902).   EKD

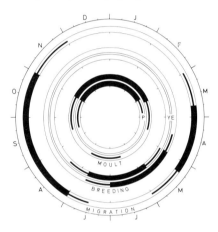

Breeding. SEASON. North-west Europe: see diagram. Laying throughout range timed to coincide with peak laying period of selected host species, so within one area individual birds have different laying periods (Wyllie 1981). SITE. Eggs laid in nest of other species; each individual lays all or most of its eggs in nests of one species. Over 100 different host species recorded in Europe; in north-west Europe, commonest are Meadow Pipit *Anthus pratensis*, Dunnock *Prunella modularis*, and Reed Warbler *Acrocephalus scirpaceus*; in central Europe, Garden Warbler *Sylvia borin*, *A. pratensis*, Pied Wagtail *Motacilla alba*, and Robin *Erithacus rubecula*; in Czechoslovakia, Redstart *Phoenicurus phoenicurus* and *E. rubecula*; in Finland, Brambling *Fringilla montifringilla* and *P. phoenicurus*; in Hungary, Great Reed Warbler *Acrocephalus arundinaceus* (Wyllie 1981). Lack (1963) listed principal species from 7 British studies, with *P. modularis* (41%), *Acrocephalus scirpaceus* (17%), and *Anthus pratensis* (10%) the commonest, but samples biased towards species whose nests are easy to find, or towards species being particularly studied (e.g. Chance 1940, Wyllie 1981), so that actual rankings not very meaningful. EGGS. See Plate 96. Sub-elliptical, smooth and fairly glossy; very variable, often resembling eggs of host species (see Baker 1923, 1942). 22 × 17 (20–26 × 15–19), n = 800 (Schönwetter 1967).

Weight 3·4 g (2·9–3·8), $n = 31$, laid in nests of *A. scirpaceus* (Wyllie 1981). Clutch: each bird lays 1–25 eggs; 1 (rarely 2) per host nest. Average 9·2 eggs (2–25), $n = 46$ (Wyllie 1981). Most eggs laid between 16.00 hrs and dusk. Eggs laid at intervals of *c.* 2 days, normally synchronized closely with host's laying period (Wyllie 1981). Egg of host usually removed; often 1–2 eaten before laying. INCUBATION. Period from laying to hatching 12·4 days (11·5–13·5), $n = 9$; from incubation starting to hatching 11·6 days (11·3–12·0), $n = 9$ (Wyllie 1981). By host species only. YOUNG. Altricial and nidicolous. Cared for and fed by host parents. At 8–36 hrs old, ejects host's eggs or young. Exceptionally, reared with host young. FLEDGING TO MATURITY. Fledging period 19 days (17–21), $n = 9$; usually leaves nest at *c.* 17 days (13–20), $n = 8$ (Wyllie 1981). When reared by *A. scirpaceus*, becomes independent at 12 days (9–25), $n = 14$, after fledging, but up to 4–6 weeks after for other hosts (Wyllie 1981). Age of first breeding (1–)2 years. BREEDING SUCCESS. Of 213 young hatched, 140 (66%) fledged (Owen 1933). Of 74 hatched, 16 (22%) fledged, 48 (65%) predated; other losses included 2 ejecting themselves from nest (Wyllie 1981). Of 189 eggs laid, West Germany, 58% hatched and 30% raised to fledging (Glutz and Bauer 1980).

Plumages (nominate *canorus*). ADULT MALE. Head, upperparts, and upper wing-coverts medium grey; distinctly tinged bluish when plumage fresh (retained longest on back to upper tail-coverts), more brownish-grey when worn (apparent on tertials and upper wing-coverts first). Fresh feathers of mantle, scapulars, and upper wing-coverts indistinctly fringed pale grey on tips. Sides of head and chin often slightly paler bluish-grey than forehead and crown; throat and chest distinctly paler, between light and pale neutral grey, less bluish than upperparts. Chest occasionally with traces of pale rufous subterminal bars. Breast, belly, and flanks white or cream-white, contrastingly marked with dark grey bars 1·4–2·8 mm wide, *c.* 3–5 mm apart. Vent and under tail-coverts cream-buff or pale cream-yellow (palest on longer under tail-coverts), marked with well-spaced (*c.* 8–10 mm) and broken dark grey bars—narrow and occasionally absent on vent and shorter under tail-coverts, 2–3 mm wide on longer coverts. Tail black with small white tip (4–8 mm wide); rows of irregular white spots along shafts and edges of feathers, occasionally almost absent on t1 and often absent along outer webs of t2–t5, those on inner webs sometimes forming short bars and occasionally joining spots along shafts to form irregular bands, especially on t4–t5 (in latter birds, black of feathers often partially invaded by grey). Flight-feathers dark grey (secondaries bluish like upper wing-coverts, outer primaries dull olive-black); tips of secondaries and p1–p6(–p8) with narrow white fringes *c.* 0·5 mm wide; inner webs of primaries with row of 6–10 large white triangular or squarish spots, extending from base to 3–5 cm from tip; spots largest and squarest near primary base. Greater upper primary coverts and bastard wing greyish-black, smaller upper primary coverts white with narrow grey bars and partially grey tips. Under wing-coverts and axillaries white or cream-yellow with grey bars (similar to breast), greater coverts uniform grey. Compared with adult ♀ and juvenile, individual variation limited; some variation in width of barring on underparts (see also Geographical Variation), in size and number of white spots on tail (larger and tending to form pale bars on outer feathers in birds with narrowly barred breast and belly), and in depth of cream-yellow of vent and under tail-coverts; some birds show slight rufous bars on chest, small pale spots on bases of outer webs of primaries, or rufous tinge to part of white spots on inner webs of primaries or on tail, but these probably 1-year-old. A rufous morph is exceptional (Ringleben 1958). ADULT FEMALE. 2 colour morphs. GREY MORPH. Similar to adult ♂ and occasionally (3 out of 36 examined) indistinguishable. Usually (in 17 out of 36) differs from adult ♂ by showing rufous or pink-buff barring on chest and sides of neck (without contrast between uniform grey chest and barred breast of adult ♂), this rufous barring frequently (12 out of 36) extending right across hindneck and slightly up to throat and crown. Unlike adult ♂, these ♀♀ usually show small rufous spots on grey median and greater upper wing-coverts, on outer webs of secondaries, and on bases of outer webs of primaries (rufous spots smaller and feathers greyer than in retained brownish juvenile feathers of 1-year-old); also, white spots on inner webs of primaries and on tail-feathers partly suffused rufous. Occasionally (4 out of 36), rufous or cream-buff barring extends over all head, mantle, many upper wing-coverts, flight-feathers, and upper tail-coverts, with uniform bluish-grey limited to scapulars, tertials, back, rump, and lesser upper wing-coverts; such birds incline towards rufous morph in appearance, though rump not rufous; closely similar to juvenile, differing in absence of white fringes along tips of flight-feathers, upperparts, and upper wing-coverts and in less rufous barring on flight-feathers and tail. RUFOUS MORPH. Crown, hindneck, mantle, scapulars, and upper wing-coverts rich cinnamon-rufous with well-spaced greyish-black bars 1–4 mm wide; back, rump, and upper tail-coverts either uniform rich cinnamon-rufous or all or some feathers (especially on back and longer coverts) with rather narrow dark subterminal mark. Sides of head and neck pale rufous-cinnamon or rufous-buff, gradually merging into pink-buff or cream-yellow chin, throat, and chest; all narrowly barred black; remainder of underparts cream-buff or white with dark bars, as in adult ♂. Tail with narrow white tip and broad black subterminal band, remainder rather evenly barred with rich rufous-cinnamon and greyish-black or black and often with white spots along shafts, but middle and basal portion of tail occasionally uniform rufous. Flight-feathers browner than in adult ♂, white fringes along tips wider (1–2 mm); outer webs of primaries with large rufous spots, white spots on inner webs partly tinged rufous; secondaries with large rufous spots along sides or completely barred rufous. Smaller upper primary coverts, under wing-coverts, and axillaries as in adult ♂, but greater under wing-coverts barred rufous and black instead of uniform grey. Proportion of rufous-morph ♀♀ in natural population not fully established: presumed to be rare, though apparently depends partly on locality. NESTLING. Naked at hatching. Closed sheaths appear on body from 3–6 days, bird appearing spiny at 1 week; sheaths open to produce feather-tips from 9–11 days, upperparts appearing feathered from 13 days, underparts from 19 days (Glutz and Bauer 1980). JUVENILE. Strongly variable, but all variations grade into each other and no morphs separable. Always distinguishable from adult by white fringes to feather-tips on upperparts, upper wing-coverts, and flight-feathers; these in part abrade soon, but are still easily visible on median upper wing-coverts, scapulars, back, rump, and flight-feathers, even by midwinter. Some white feathers on nape and occasionally on forehead and crown present up to 1–3 months after fledging. Differs from adult ♂ and grey morph of adult ♀ in blacker upperparts (less bluish-grey), absence of blue-grey on chest, and more extensively broken rufous bars on flight-feathers and tail. Some rufous juveniles rather similar to rufous morph of adult ♀, but black bars on head, mantle, scapulars, and upper wing-coverts of juvenile wider than rufous bars (not narrower); rufous of upperparts browner, less bright; black bars on sides of head, chin, and throat usually much wider; white fringes along tips of primaries extend up to p10, not to p6–p7. Darkest birds have head, neck, and chest blackish with partial off-white barring, remainder of upperparts and upper wing-coverts dark grey-brown, palest and slightly bluish on rump; rufous restricted to a few rather small spots on flight-feathers, tail, and upper wing-coverts. Other birds show more extensive rufous or cream-buff bars on nape, mantle, scapulars, upper wing-coverts, and flight-feathers, but back and rump still mainly grey, and rufous in tail limited; still others have numerous narrow and broken rufous bars on dull black upperparts and upper wing-coverts, with back to upper tail-coverts closely barred grey and rufous and tail-bars mainly rufous. The most extensively rufous birds have all upperparts and upper wing-coverts evenly barred greyish-black and rufous, including back to upper tail-coverts; only exceptionally are rump and upper tail-coverts almost uniform rufous-cinnamon as in rufous-morph adult ♀; flight-feathers similar to rufous adult ♀, but p7–p10 fringed white; tail as in rufous adult ♀, but without broad and distinct black subterminal band and remainder never uniform rufous. No correlation between colour and sex, except perhaps for exceptional ♀♀ with uniform bright rufous rump (2 of 50 examined). FIRST ADULT. Similar to adult, but part of juvenile plumage frequently retained, especially some secondaries (see Moults): these more heavily worn and often narrower and shorter than neighbouring fresh feathers; in ♂ and grey-morph ♀, old feathers also differ conspicuously in browner colour and large rufous spots or bars. In 1-year-old ♂♂, white spots on inner webs of primaries partly suffused rufous, unlike older ♂♂; not known whether those ♂♂ which show partly rufous spots but no juvenile feathers are 1-year-olds with complete post-juvenile moult or individual variants among older ♂♂.

Bare parts. ADULT. Iris yellow, rarely brownish-yellow or orange-red; in rufous morph of ♀, sometimes pearl-grey, pale yellow, or hazel. Eye-ring pale yellow, chrome-yellow, or orange-yellow. Upper mandible and tip of lower mandible dark horn-brown or horn-black, sometimes with grey tinge; basal cutting edges to below nostril on upper mandible and basal $\frac{3}{4}$ of lower mandible greenish-yellow to chrome-yellow. Corner of gape orange-yellow or orange-red, mouth orange-red or red. Leg and foot bright yellow or ochre-yellow; claws and distal sides of toes horn-brown or mixed yellow and brown. NESTLING. Iris grey, grey-brown, or brown. Bare skin, including that of bill, leg, and foot, flesh-pink; face, back, and wing dusky pinkish-grey; mouth deep orange or deep red, gape flanges yellow. JUVENILE. Iris grey, grey-brown, yellow-brown, brown-yellow, pale olive-yellow, or yellow. Eye-ring pale yellow or pinkish-yellow. Bill as in adult, but duller; upper mandible browner, base of lower mandible greyish-yellow, olive-yellow, pale yellow, or light horn-colour; red near gape. Leg and foot brownish-yellow, yellow, or orange-yellow. (Hartert 1912–21; Harrison 1975; Löhrl 1979; BMNH, RMNH, ZMA.)

Moults. Mainly based on data supplied by D C Seel, with additional information from Verheyen (1950), Stresemann and Stresemann (1961), and specimens (RMNH, ZMA). ADULT POST-BREEDING. Complete; primaries seemingly irregular, but in reality in fairly fixed order; up to 4 primaries may grow at same time, but 2 growing feathers always separated by a full-grown one, except occasionally for some inner primaries. Flight-feather moult starts while in winter quarters, (September–)October–November, exceptionally 1 feather replaced in breeding area; completed (December–)February–March, rarely later; full replacement probably in $3\frac{1}{2}$–4 months, but moult occasionally suspended (independent of stage reached) and then more time needed. Sequence of primary replacement 7–9–4–1–2–5–10–8–3–6. Tail and secondaries moult at same time as primaries; sequence of secondaries approximately 6–8–7–9–1–5–2–4–3, of tail 1–5–3–4–2 (according to Verheyen 1950) or perhaps as in Oriental Cuckoo *C. saturatus*. Scapulars and lesser and median upper wing-coverts on average moult at same time as primaries, remainder of head and body on average 15–19 days later, crown last. Some slight body moult throughout year, except August and (April–)May. POST-JUVENILE. Complete, usually except for a few flight-feathers or some other feathers. Sequence mainly as in adult post-breeding, but scapulars and lesser and median upper wing-coverts moult on average *c.* 13 days before primaries, and remainder of head and body at same time as primaries. Timing slightly later, primaries on average *c.* 1 month later than in adult post-breeding, starting September–January, completing February–July. Moult of secondaries relatively late compared with primaries, and usually not completed when moult suspended during pre-migration fattening and spring migration; in particular, juvenile s2–s4(–s5) or s3–s4(–s5) retained during summer of 2nd calendar year, together with corresponding greater coverts, occasionally only s4–s5 or s2–s3 retained; also, some feathers of bastard wing and some upper wing-coverts frequently retained, as well as rarely a few primaries (mainly p6 or p3). As in *C. saturatus*, these retained feathers probably replaced in moulting season of 2nd autumn and winter, when new moulting series starts also. A few birds still fully juvenile December and these probably the birds summering in central and East Africa, starting moult about January–February, completing about August–October.

Measurements. Nominate *canorus*. Mainly Netherlands, some Britain and central Europe, summer; skins (BMNH, RMNH, ZMA). Bill (F) is adult Bill from tip to forehead, bill (N) is adult bill from tip to distal edge of nostril. Toe is outer front toe (3rd).

WING AD	♂	221	(4.32; 52)	213–230	♀ 210	(3.86; 35)	204–216
JUV		208	(5.94; 24)	203–219	201	(3.84; 20)	195–208
TAIL AD		177	(4.14; 48)	170–186	167	(5.46; 33)	158–177
JUV		172	(6.03; 22)	164–180	163	(4.10; 18)	157–171
BILL (F)		27.7	(1.51; 44)	25.5–31.2	26.8	(1.01; 32)	25.2–28.6
BILL (N)		16.0	(0.81; 49)	14.7–17.6	15.4	(0.76; 35)	14.2–16.7
				21.3–24.2	22.2	(0.88; 25)	20.6–23.7
TOE		26.6	(1.17; 25)	25.0–28.5	26.2	(1.13; 19)	25.0–28.5

Sex differences significant, except tarsus and toe. Juvenile wing and tail not full-grown at fledging, excluded from table; once full-grown, still significantly shorter than adult. Juvenile tarsus and toe similar to adult from shortly after fledging, juvenile bill not until about midwinter: average bill (F) of fledged juvenile 3.4 shorter than adult in July, 1.9 in August, 0.9 September; bill (N) 2.3 shorter in July, 1.6 August, 1.4 September. Wing and tail of 1st adult not significantly different from adult, included in table.

South-east Europe, Turkey, European USSR, and Scandinavia, summer; skins (RMNH, ZMA, ZMM).

WING AD	♂	231	(4.19; 10)	224–236	♀ 224	(4.96; 6)	217–230
TAIL AD		179	(4.56; 9)	173–185	174	(4.51; 5)	168–179
BILL (F)		28.3	(1.55; 9)	26.9–31.4	27.3	(1.04; 4)	26.2–28.7
BILL (N)		16.4	(0.81; 9)	15.4–17.5	15.6	(0.19; 5)	15.4–15.9

Sex differences significant for wing.

C. c. bangsi. Iberia and north-west Africa, summer; skins (Vaurie 1965; BMNH, RMNH, ZFMK, ZMA).

WING AD	♂	210	(—; 14)	203–221	♀ 204	(7.39; 14)	192–213
BILL (F)		26.9	(—; 12)	24.5–29.5	27.2	(1.18; 6)	25.9–28.2

C. c. subtelephonus. Afghanistan, Turkmeniya, and Kirgiziya (USSR), and Sinkiang (China), summer; skins (BMNH, RMNH, ZMA).

WING AD	♂	212	(6·58; 7)	204–222	♀ 202	(6·95; 6)	194–212
BILL (F)		25·7	(— ; 2)	25·2–26·2	26·2	(1·06; 4)	24·7–27·0

Weights. Nominate *canorus*. ADULT. Britain (Seel 1977*a*).

APRIL	♂	129	(18; 10)	♀ 112	(14; 6)
MAY		117	(17; 84)	106	(14; 52)
JUNE		114	(23; 20)	106	(13; 14)
JULY		133	(29; 7)	112	(18; 4)

Camargue (France): after crossing Mediterranean, April, ♂ 96 (17; 15), ♀ 95 (16; 16); in May–June, ♂ 108 (14·3; 24), ♀ 98 (8·3; 15); prior to departure, July, ♂ 132 (22; 25), ♀ 113 (16; 23) (Glutz and Bauer 1980). Malta, on spring migration: 133 (22·2; 4) 100–148 (J Sultana and C Gauci). ♀♀ at end of laying period, Belgium, 95 (6) 88–98; on departure, 106 (5) 98–118 (Verheyen 1950). In late July and early August, Finland, weight of 3 birds increased 52–73% within 15–17 days, ♂ reaching 160, ♀ 149 (Hildén 1974); in captivity, then even 151, 187, and 197 reached (Löhrl 1979). In Afrotropics: during autumn migration and in winter quarters, July–March, adult ♂ 110 (21·8; 6); adult ♀ 98 (23·0; 5); juvenile 86 (14·5; 10); on spring migration in northern Afrotropics, March–May, ♂ 158 (32; 5), ♀ 107 (13; 3) (D C Seel).

JUVENILE. At hatching, 2·6 (4) 2·4–2·8; strongly increasing up to 14th day, when *c*. 75, slower until 22nd day, when fledging at 84·7 (11) 80–90 (Glutz and Bauer 1980). Britain: July 97 (29; 49), August 89 (18; 96), September–October 105 (25; 5) (Seel 1977*a*). Netherlands: July–August, ♂ 104 (11; 9) 87–110, ♀ 110 (14; 6) 94–130; September, ♂ 110 (13; 4) 100–127, ♀ 111 (3) 93–136 (RMNH, ZMA). Camargue: July 98 (36), August 97 (49), September 119 (15); total range, 67–160 (Glutz and Bauer 1980). Helgoland (West Germany), autumn: 92 (46) 68–130 (Löhrl 1979). Exhausted full-grown autumn birds, Netherlands: ♂ 65 (10; 11) 52–84, ♀ 65 (11; 14) 47–87 (RMNH, ZMA).

C. c. subtelephonus. Iran and Afghanistan, April–September: ♂♂ 81, 92, 98, 128; ♀♀ 81, 84, 91 (Paludan 1938, 1940, 1959). Hopeh (China): ♂ 96 (35) 71–127, ♀ 97 (14) 70–138 (Shaw 1936).

Structure. Wing rather long and narrow, pointed. 10 primaries. p8 longest, p9 (2–)4–10 shorter, p10 53–69, p7 8–12, p6 23–32, p5 38–52, p1 92–106. Primaries not emarginated, each slightly tapering towards rounded tip. Tail long, graduated; 10 feathers, t5 32–58 shorter than t1. Upper tail-coverts long and stiff, densely packed. Bill rather short and slender; culmen and cutting edges curved downwards towards sharply pointed tip, gonys almost straight; bill wide at gape, laterally compressed at tip. Nostrils rather small, elliptical; margin slightly raised. Tarsus short, feathered on upper half. Toes rather short, slender; outer front toe longest; inner hind toe *c*. 46% of outer front, inner front *c*. 67%, outer hind *c*. 78%.

Geographical variation. Considerable, mainly clinal; involves width of dark barring on underparts, and size; slightly also depth of grey on upperparts. Race *bangsi* of Iberia and north-west Africa small, especially wing (tail and bill less so); nominate *canorus* of Britain, central Europe, and Italy larger; in Scandinavia, south-east Europe, and European USSR larger still (see Measurements). Variation probably clinal, size in Europe increasing from south-west to north-east. (Birds of Britain, central Europe, and Italy included in nominate *canorus* only provisionally while more data of typical nominate *canorus* from Scandinavia lacking.) Populations of northern Siberia east to Kamchatka and Amur river similar in size to populations of European USSR, south-east Europe, and Scandinavia, and these included in nominate *canorus*; for details in western Siberia, see Johansen (1955). Birds of intermediate size (similar in that respect to those of Britain) occur east from Tarbagatay (USSR), eastern Sinkiang, and Kansu (China) through Mongolia and northern and central China to Japan; these included in *subtelephonus*, though typical *subtelephonus* from Transcaspia to western Sinkiang and south to Iran and Afghanistan as small as *bangsi*; former group of intermediate size perhaps separable as *telephonus* Heine, 1863, but samples examined too small and populations of east and central Asia in need of fuller investigation. Typical *subtelephonus* occurs as a migrant in Middle East (Iraq, Jordan, Israel). *C. c. bakeri* from Himalayas, Assam, and Burma similar in size to birds from Britain and central Europe; *bakeri* from southern China slightly smaller and sometimes separated as *fallax* Stresemann, 1930, but see Mees (1979*a*); *bakeri* differs from all other populations in darker grey upperparts, close to Oriental Cuckoo *C. saturatus*. In all populations, width of dark barring on underparts strongly variable, both geographically and individually. In general, bars of nominate *canorus* from Scandinavia, north European USSR, central Europe, and Britain rather broad, 1·5–2·4 mm wide on breast (mainly 1·8–2·1), locally averaging slightly wider (e.g. on Corsica and Sardinia), but within above range width of bars based on average of 5 feathers, measured on different parts of breast of single bird). width of bars of *bangsi* and *bakeri* similar to this. Birds from south-east Europe, Turkey, south European USSR, southern shores of Caspian, and much of Siberian range have bars on average slightly narrower, range 1·2–2·1 (mainly 1·5–2·0), but are included in nominate *canorus* in view of similar size. *C. c. subtelephonus* has narrow bars, 0·7–1·6 wide on breast; birds of central and north-east China and Japan show rather narrow bars also, 1·0–1·6 wide, and these (if not treated separately as *telephonus*) better included in slightly smaller-sized *subtelephonus* rather than in larger and much broader-barred nominate *canorus* (as by Vaurie 1965). Grey-morph adult ♀ *bangsi* often with markedly deep rufous on chest and round neck, resembling Red-chested Cuckoo *C. solitarius* of Afrotropics.

Forms superspecies with African Cuckoo *C. gularis* from Afrotropics, latter differing mainly in broader and more extensively yellow bill, fully white bars on outer tail-feathers with more distinct black subterminal bar in adult ♂ and grey-morph adult ♀, distinctly greyer juvenile plumage with whitish or cream-buff bars on head and neck and no rufous spots on flight-feathers, and different voice; size similar to nominate *canorus* of Britain and central Europe; width of barring on underparts mainly 1·1–1·7 mm. Often considered conspecific, but see Vaurie (1965), Kipp (1976), Payne (1977), and Snow (1978).   CSR

Cuculus saturatus Oriental Cuckoo

PLATES 39 and 42
[between pages 470 and 471]

Du. Boskoekoek   Fr. Coucou oriental   Ge. Hopfkuckuck
Ru. Глухая кукушка   Sp. Cucu siberiano   Sw. Taigagök

Cuculus saturatus Blyth, 1843. Synonym: *Cuculus optatus*.

Polytypic. *C. s. horsfieldi* Moore, 1857, European part of USSR east to Pacific, south to Japan and to about Yangtze river in China. Extralimital: nominate *saturatus* Blyth, 1843, northern Pakistan and Kashmir east through Himalayas to southern China and Taiwan; *lepidus* S Müller, 1845, Malay peninsula, Sumatra, Java, and Lesser Sunda Islands east to Timor; *insulindae* Hartert, 1912, Borneo.

Field characters. 30–32 cm; wing-span 51–57 cm; tail 12–13 cm. Slightly shorter winged and less bulky than Cuckoo *C. canorus*. Cuckoo of remarkably similar structure and appearance to *C. canorus*. Grey adult differs mostly in broader and more intense barring on underparts; rufous ♀ and juvenile differ in more strongly barred upper- and underparts. Study of field characters incomplete. Song of ♂ totally distinctive, recalling Hoopoe *Upupa epops*. Sexes similar; no seasonal variation. Juvenile separable.

ADULT MALE. A few indistinguishable from *C. canorus* but most differ in: (1) greyer chest and darker, bluer back and wings, making head appear paler, at least in direct comparison between *C. saturatus horsfieldi* and *C. canorus subtelephonus* (both of south-central Palearctic); (2) broader, wholly black barring on underparts, with bars up to ⅓ wider than those of typical *C. canorus*; (3) extension of black barring over thighs and vent; (4) heavy black blotching (not fine barring) on under tail-coverts; (5) yellower or buffier ground to underparts, particularly under tail and on under wing-coverts; (6) lack of barring on leading edge of wing, creating mainly white patch below carpal joint (particularly compared with *C. canorus canorus*). Important to realize that none of the above distinctions easy to see or free of variation. ADULT FEMALE. Occurs in both grey and rufous morphs. Grey morph differs from ♂ and ♀ *C. canorus* in coarser marks on buffier sides of neck and across chest. Rufous morph differs more obviously from ♀ *C. canorus* in broader, blacker, and hence more contrasting barring, particularly on chest, back, larger inner wing-feathers, and over rump and tail. Bare parts as in *C. canorus*, though base of lower mandible yellower and legs and feet more orange (even tinged red) in eastern birds. Beware again danger of wishful observation and variation. JUVENILE. Both grey and rufous types occur, with intergradation less complete than in juvenile *C. canorus*. Pattern of rufous juveniles resembles rufous morph ♀ and differs from rufous juvenile *C. canorus* in bolder black barring on upperparts and tail. All juveniles have bolder, more open barring on warmer underparts than juvenile *C. canorus*, with noticeably heavily marked breast-sides and chest-band.

Singing ♂ unmistakable, but silent ♂ and all other birds, whatever sex or age, subject to serious confusion with *C. canorus*. Separation from that species incompletely studied but, in addition to characters listed above, other plumage marks and slight differences in structure and flight action merit discussion. Heavier saturation of body and underwing with cream-yellow (adult) or rusty (juvenile) ground-colour distinctive in most skins; underwing colour (visible mostly on pale central tract) given as distinguishing character by Heinzel *et al.* (1972) but untrustworthy due to variation. Absence of white spot on nape of juvenile said by King and Dickinson (1975) to be usual, but, as in *C. canorus*, present at fledging and lost through abrasion a few weeks later. Thus such features true for neither *C. saturatus horsfieldi* nor nominate *saturatus* (see Plumages). More hopeful character (for acute observer) exists in form of rump- and tail-barring shown by rufous-morph ♀ *C. saturatus* which is not only bolder, coarser, and blacker, but also more even and more complete, with individual marks on feathers neither narrowing at edges nor breaking up into blotches and thus creating much more obvious pattern. Sadly, this no help with grey-morph birds and therefore important to realize that although general appearance closely similar to *C. canorus*, silhouette slighter, with more rounded crown, slightly shorter bill, and slimmer, apparently shorter tail. Thus some birds appear distinctly smaller than *C. canorus* and even more like falcon *Falco*, particularly ♂ Merlin *F. columbarius*. Flight also similar to *C. canorus*, but action seems to lack occasional looseness of that species, inviting more frequent recall of small *Falco*. Much more secretive than *C. canorus*, keeping within foliage and rarely using exposed perch. Gait as *C. canorus*.

Song of ♂ quite unlike that of *C. canorus*, consisting of series of 'bu-' notes of constant low pitch; at distance, suggests *Upupa epops*, but at close range distinguished by shorter length of each note and less sonorous quality. All calls quieter than those of *C. canorus*.

Habitat. From subarctic and boreal through temperate and steppe zones, to subtropics and (in winter) tropics. Much more a forest bird than Cuckoo *Cuculus canorus* (Voous 1960), preferring secluded and undisturbed country. In USSR, favours high coniferous forests of spruce *Picea*, pine *Pinus*, silver fir *Abies*, and mixed coniferous and broad-leaved trees, such as birch *Betula* and aspen *Populus*; sometimes in pure broad-leaved forests, steppe birch copses, riverside willows *Salix*, and thickets. Within such habitats, however, may favour more open areas: e.g. in Krasnoyarsk (USSR), most abundant in

418 Cuculinae

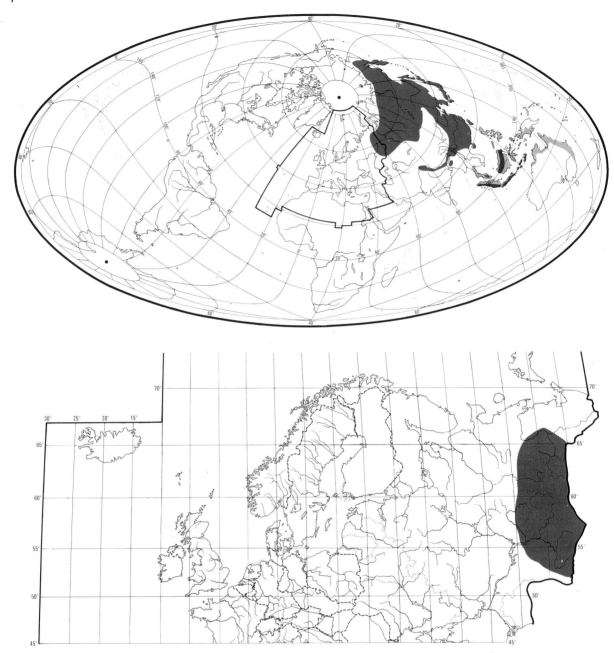

glades, along river banks, etc. (Kislenko and Naumov 1967). In Salair (USSR), *C. canorus* favours wooded steppe, *C. saturatus* commoner in secondary woodland and in depths of taiga with mature conifers (Chunikhin 1964). Inhabits northern taiga to forest edge, and riversides, ravines, fringes of wetlands, and slopes of wooded hills and mountains, but in USSR found at lower altitudes than *C. canorus* (Dementiev and Gladkov 1951a). Only rarely above 1000 m (Johansen 1955). In Himalayas, however, at 1500–3300 m, in hilly wooded country and orchards; infrequently recorded in plains (Ali and Ripley 1969). In Australian winter quarters in monsoon forest, coastal and rain forest fringes, and adjacent woodland in north; in eastern Australia in varied forests and clearings (Slater 1970).

Distribution and Population. Range poorly known due to skulking habits; there are a number of records west of mapped range, some of which may refer to breeding, possibly irregular (Ivanov 1976). In west Palearctic range, common or frequent (Dementiev and Gladkov 1951a).

Movements. Northern race *horsfieldi* (breeding USSR and northern parts of Mongolia, China, and Japan) wholly migratory, though extremely little known about its movements. Evidently winters Philippines (a few), eastern Indonesia (mainly passage migrant, but collected December–February also), Melanesia (especially New Guinea), and north and north-east Australia; also reaches Lord Howe Island fairly regularly and straggles to New Zealand (Vaurie 1965; Condon 1975; C S Roselaar). Occurs on passage through south-east Asia, but not recorded Middle East or (this race) Indian sub-region, and apparently only a straggler to south-west Soviet Asia: Turkmeniya, Uzbekistan, Tadzhikistan, and southern Kazakhstan (Dementiev and Gladkov 1951a). Hence much of migration for east European and west Asian birds must be south-east/north-west within USSR, passing north of major mountain chains of central Asia. Migrates singly or in small loose groups. Scanty data on timing: passage through south-east Asia (with nominate *saturatus*; see below) in September–November and April to early June (e.g. La Touche 1931–4, Lee 1969); obtained Indonesia late August to early May (C S Roselaar), and present northern Australia during monsoon season, November–April (Bravery 1967); in USSR, vanguard reaches lower latitudes of breeding range (even west to Urals) around mid-May, but not until early June near Arctic Circle (Dementiev and Gladkov 1951a). Stragglers reach Bering Sea islands and western Alaska (e.g. Roberson 1980).

Nominate *saturatus* (breeding Himalayas to southern China and Taiwan) mainly migratory, though some winter on plains of northern India (Ali and Ripley 1969). Otherwise, winter range poorly known; believed to overlap *horsfieldi* in Indonesia and New Guinea (Vaurie 1965), but also occurs sparingly as migrant and perhaps winters in Thailand and Malaysia (Deignan 1963; Medway and Wells 1976). Proportion of nominate *saturatus* (against *horsfieldi*) declines from west to east in Indonesia: museum skins (all October–February) 83% ($n=12$) nominate *saturatus* in Sumatra, 54% ($n=35$) in Java, 8% ($n=12$) in Celebes and Lesser Sundas, 12% ($n=16$) in northern Moluccas, 0 ($n=28$) in southern Moluccas and New Guinea (C S Roselaar). Both races occur (in about equal proportions) on passage through Borneo (Smythies 1968). Tropical races *lepidus* and *insulindae* (see Wells and Becking 1975) believed sedentary.

Food. Little known, and information mostly extralimital.
Mainly large insects and their larvae, taken from ground, shrubs, and trees (Dolgushin *et al.* 1970). In winter, Queensland (Australia), took mostly stick insects (Phasmida) by crashing into foliage of tree, flapping to maintain balance while extracting prey; then flew to bare limb where slowly prepared prey, biting it from end to end before swallowing it (Zillman 1965). In Queensland, also made sallies from vantage point to seize emerging cicadas (Cicadidae) on trunks and larger branches of trees; also took hairy caterpillars (Lymantriidae) from same sites, from foliage, and from grass below trees. On trunks and large branches, often perched sideways to extract caterpillars from under loose bark, before carrying them to vantage point for eating (Bravery 1967).

Only 1 record for west Palearctic: 4 stomachs (no further details), Volga–Kama region (USSR), contained hairy and other caterpillars (Lepidoptera), spiders (Araneae), and beetles (Melolonthidae, Cerambycidae, Carabidae, Coccinellidae) (Popov 1977). 1 stomach, Anadyr river (Siberia), contained remains of beetles (Coleoptera), wasps (Vespidae), flies (Diptera), and many hairs of caterpillars (see Dementiev and Gladkov 1951a). Ants (Formicidae) recorded in Australia (Anon 1976).

Stomach of juvenile ♂ fed by Black-throated Accentor *Prunella atrogularis*, south-east Altay (USSR), contained berries and small 'grasshopper-type' (Acrididae) insects; another juvenile ♂ contained large grasshopper-type insects, and caterpillars (Irisov 1967). EKD

Social pattern and behaviour. Little information, especially for west Palearctic. Behaviour said to be closely similar to that of Cuckoo *C. canorus* (Dementiev and Gladkov 1951a; Etchécopar and Hüe 1978).

1. Outside breeding season, Queensland (Australia), found singly, in pairs, or in small parties (Zillman 1965). May congregate in small numbers where food locally abundant; in one locality, Queensland, 8 birds on *c.* 2 ha (Bravery 1967). In Kazakhstan, migrates in loose groups of to 10 or more, sometimes accompanied by *C. canorus* (Dolgushin *et al.* 1970). BONDS. No information. BREEDING DISPERSION. Solitary. Practises brood-parasitism, apparently as in *C. canorus*. ♂ defends breeding territory (Krechmar 1966). Said to have breeding area of 1·5–2 km^2 (Popov 1977). In USSR, density varies from 1 bird per 10 km^2, Volga–Kama region (Popov 1977), to up to 5 singing ♂♂ per 100 m of transect, Krasnoyarsk (Kislenko and Naumov 1967). In pine forest, Salair (USSR), 0·5 birds per km^2, in taiga and taiga edge 1 bird per km^2; on Syukhta river (USSR), 2 birds per km^2 (Dementiev and Gladkov 1951a); on Pechora river (USSR), 2·4 birds per km^2 (Estaviev 1981). See also Reymers (1966). In Krasnoyarsk, after 1st broods of host Chiffchaff *Phylloscopus collybita* have fledged, begins parasitizing 2nd clutches and also other *Phylloscopus* (Kislenko and Naumov 1967). ROOSTING. Nothing known about roosting at night, though ♂ will call at night (Kapitonov 1962), and according to Krechmar (1966) calls mostly at night, though most often at dawn and dusk according to Dementiev and Gladkov (1951a). Outside breeding season, preening bouts last up to $\frac{1}{2}$ hr, with much alternate stretching of legs and wings (Zillman 1965). Wings stretched over head (Bravery 1967).

2. Generally shy and wary in breeding season (Dementiev and Gladkov 1951a), especially ♀ who does not advertise herself as ♂ does (Becking 1981). Less shy outside breeding season when feeding birds may permit approach to within a few metres. Apparently then tends to turn back to approaching observer, hiding more conspicuous breast plumage (Bravery 1967). FLOCK BEHAVIOUR. Whether on migration or feeding, aggregations relatively loose, birds staying well apart, e.g. 5 birds kept *c.* 20 m apart from one another while feeding (Bravery 1967). ANTAGONISTIC BEHAVIOUR. ♂ delivers Advertising-call (see 1 in Voice),

A

typically from favourite exposed vantage point in tree-top, adopting posture different from that of *C. canorus*: hunched, with tail somewhat lowered, head withdrawn and somewhat lowered, and throat markedly swollen, apparently due to inflation of cervical air sacs (Fig A). Sometimes flies slowly from perch in straight path, sometimes raising head somewhat, and may call in flight (Panov 1973). In second half of June, western Taymyr peninsula (USSR), ♂♂ regularly seen chasing others out of territories (Krechmar 1966). On one occasion at Novosibirsk, ♂ (in presence of ♀ and another ♂) adopted posture similar to that of displaying ♂ *C. canorus*: sat low and moved tail slowly up and down, after which it attempted to drive off rival ♂, gliding towards it with tail fanned. However, 2 ♂♂ in pursuit of 1 ♀ did not usually show much aggression. ♂ threatened by *Phylloscopus* fanned tail, raised wings high, and gaped widely (Fig B); when chased off

B

by *Phylloscopus*, ♂ circled in the air, uttering Kau-call (see 3 in Voice), before alighting again nearby (Panov 1973). When mobbed by other species in Queensland, *C. saturatus* raised wings above body and rapidly moved them up and down at same time adopting a 'threatening' posture. This usually sufficient to deter attacking birds, but if harassed too much, bird flew to another perch (Bravery 1967). Thought not to compete seriously with *C. canorus*, from which separated by habitat and choice of hosts (Chunikhin 1964; Kislenko and Naumov 1967), though fights recorded between ♂ *C. saturatus* and ♂ *C. canorus* near Moscow (Zykova and Ivanov 1967). HETEROSEXUAL BEHAVIOUR. Display of ♂ (see above) presumably attracts ♀♀ as well as rebuffing rival ♂♂. Courtship behaviour said to be indistinguishable in the field from that of *C. canorus* (Dementiev and Gladkov 1951a). When calling (see 2 in Voice), ♀ stretches neck forwards and upwards, and quivers half-open wings. EGG-LAYING BEHAVIOUR. One ♀ intending to lay was unusually approachable and sluggish in her movements, while nearby *Phylloscopus* became very agitated (Panov 1973). FOSTER-CARE OF YOUNG. Young eject eggs of young of host (Dementiev and Gladkov 1951a). When begging, bright orange gape of young strikingly prominent against black gape-flanges and head plumage, eliciting strong feeding response from host (Chunikhin 1964). No further information. ANTI-PREDATOR RESPONSES OF YOUNG. No information. PARENTAL ANTI-PREDATOR STRATEGIES. Presumably none.

(Figs A and B from drawings in Panov 1973.) EKD

Voice. Freely used during breeding season, relatively little at other times (Zillman 1965; Dolgushin *et al.* 1970). For comparison (including sonagrams) of *horsfieldi* with nominate *saturatus*, see Wells and Becking (1975).

CALLS OF ADULTS. (1) Advertising-call of ♂. A rapid 'bu-bu-bu-bu' (invariably 4 syllables) followed by a stifled or muted series of 6–8 disyllabic 'bu-bu' sounds (Fig I); has quality of Hoopoe *Upupa epops* (Kozlova 1930; Dementiev and Gladkov 1951a); also rendered 'ut-' or guttural 'chu-' (see Grote 1934). In Japan, 'pupu ... pupu ...' repeated up to 50 times, or a rapid 'pupupupupu ...' of 5–8 syllables (Jahn 1942), or a slower 'po po po po ...' (Austin and Kuroda 1953). Recordings show that all these variants may fall within repertoire of a single bird (Wells and Becking 1975). Bout of calling often begins with a guttural 'kkokh' or 'kkukh', not heard at other times (Panov 1973). ♂ calls day and night, sometimes from same spot for 1–2 hrs (Kapitonov 1962); heard most often at dawn and dusk (Dementiev and Gladkov 1951a); call given perched or in flight (Kozlova 1930). (2) Bubbling-call of ♀. A deep, sonorous rolling laugh (see Grote 1934), similar to that

I B N Veprintsev USSR

of Cuckoo *C. canorus*, and serving same function (Dementiev and Gladkov 1951a). (3) Kau-call. A striking 'kau' or 'khau', likened to voice of small falcon *Falco*, given by ♂ circling in the air after being mobbed by *Phylloscopus* warblers (Panov 1973). (4) Kuk-call. A laughing 'kuk kuk kuk ...' (Zillman 1965); a harsh, cackling 'gaak-gaak-gak-ak-ak-ak' (Bravery 1967), given by birds in loose feeding association outside breeding season; not loud but audible at up to *c.* 20 m; usually given just before flying from perch to capture prey. Feeding birds also uttered an occasional 'coo' (Zillman 1965). (5) Trilling-call. A ringing, mournful trill of 3 syllables, repeated several times in crescendo (Zillman 1965; see also Bravery 1967). Calls 4–5 reported outside breeding season only.

CALLS OF YOUNG. No information. EKD

Breeding. SEASON. Central USSR: eggs laid June–July (Dementiev and Gladkov 1951a). SITE. Eggs laid in nest of Arctic Warbler *Phylloscopus borealis*, Chiffchaff *P. collybita*, Willow Warbler *P. trochilus*, or other *Phylloscopus*; also Olive-backed Pipit *Anthus hodgsoni*; of 10 nests parasitized, central USSR, 9 *P. collybita*, 1 Pallas's Warbler *P. proregulus* (Chunikhin 1964). For other host species in Asia, see Makatsch (1976). EGGS. See Plate 96. Long sub-elliptical, smooth and slightly glossy; variable, apparently mimicking host, but little information; grey-blue to white, finely speckled violet, brown, or red (Dementiev and Gladkov 1951a). Eggs from all host nests:

21 × 14 mm (20–25 × 12–16), $n=45$; calculated weight 2·0 g (Schönwetter 1967). Eggs from *P. collybita* nests: 19 × 14 mm (19–20 × 13–15), $n=3$; weight 1·9 g (1·8–2·0), $n=3$ (Kislenko and Naumov 1967); weight 1·7–1·9 g, $n=4$ (Chunikhin 1964). No information on number of eggs laid, or incubation or fledging periods. In central USSR, first parasitizes early-breeding *P. collybita*; later on, as their nests become available, also other *Phylloscopus* (Kislenko and Naumov 1967). Young eject eggs or young of host species. No further information.

Plumages (*C. s. horsfieldi*). ADULT MALE. Closely similar to adult ♂ Cuckoo *C. canorus canorus* and sometimes hardly separable, all distinguishing characters for one species occurring in a few specimens of the other, no character fully reliable (see also Measurements). Grey of upperparts slightly darker and more bluish-grey, less ash-grey, though a few *C. canorus* similar; colour of upperparts in both species influenced by bleaching and wear, but difference between worn birds usually still visible on rump. Underparts usually with broader and blacker bars, 2–2·5 mm wide on breast, 3–4 mm on flanks (in *C. canorus*, mainly 1·5–2 mm on breast, up to 3 mm on flanks), broad bars extending down to vent and thighs; ground-colour of breast and belly often deeper cream-yellow (cream-white in *C. canorus*); ground-colour of vent and under tail-coverts usually deeper buff-yellow (some *C. canorus* similar on vent, but longest coverts usually paler), vent and shorter coverts often unmarked, longer under tail-coverts heavily blotched black (in *C. canorus*, under tail-coverts usually completely marked with narrow, widely spaced bars); upperside of outer upper median primary coverts almost always unmarked white or cream, except for grey tip (usually barred grey in *C. canorus*). Flight-feathers, under wing-coverts, and tail as in *C. canorus*, equally variable. Note that some races of *C. canorus* are smaller, darker above, or more heavily barred below than typical nominate *canorus*, and these may closely resemble *C. saturatus* in some characters. ADULT FEMALE. GREY MORPH. Differs from grey-morph ♀ *C. canorus canorus* in same way as adult ♂. Differs from adult ♂ by rather coarse black and pink-buff or rufous barring on lower chest and sides of neck, occasionally also on hindneck; these bars usually slightly coarser than similar bars of adult ♀ *C. canorus*. RUFOUS MORPH. Easier to distinguish from rufous-morph adult ♀ *C. canorus* than are grey-morph ♀ and adult ♂. Forehead, crown, hindneck, mantle, scapulars, tertials, and upper wing-coverts barred rufous and dark as in rufous *C. canorus*, but dark bars blacker (less greyish) and broader: *c.* 2–3 mm wide on crown, similar in width to intervening rufous bars (in *C. canorus*, width of black 1–2 mm, of intervening rufous 3–4); 3–5 mm on mantle and shorter upper wing-coverts (2 mm, rarely 3, in *C. canorus*), 4–7 mm on longer scapulars, tertials, and longer upper wing-coverts, where bars grey with black borders rather than fully black (3–5 mm in *C. canorus*). Back to upper tail-coverts marked with dark grey and rufous bars *c.* 3–5 mm wide (in *C. canorus*, uniform bright rufous or with small and short dark bars or spots only). Marked contrast between chin to chest and remainder of underparts: sides of head, chin, and throat heavily spotted black; chest yellow-buff with black bars 2–3 mm wide, sides of breast barred rufous and black; breast to vent cream-yellow or cream-buff with black bars 2–4 mm wide (in *C. canorus canorus*, sides of head, cheeks, and chin pale yellow-buff or pale cinnamon with fine black dots or short bars, distinctly paler than in *C. saturatus*; ground colour of chest yellow or pink-buff, of belly cream-white or white, all evenly marked with black bars *c.* 2 mm wide); under tail-coverts of *C. saturatus* often warmer yellow, longer coverts more heavily blotched, but some overlap with *C. canorus*. Tail of *C. saturatus* with broad black subterminal bar and white tip as in *C. canorus*, but otherwise more heavily and equally barred: unbroken black bars 5–7 mm wide, not narrowing towards sides (in *C. canorus*, narrower, often broken into blotches, narrowing towards sides of feather). NESTLING. Entirely naked. JUVENILE. More greyish and more rufous types separable, with less gradual intergradation than in juvenile *C. canorus*, but variation still large; no apparent sex-linkage. (1) Greyish birds. Upperparts as greyish juveniles of *C. canorus*, including white nape-patch, but underparts different: lores, sides of head, upper sides of neck, and sides of breast dark sepia or dark chocolate-brown with some rufous or cream-buff mottling, contrasting with throat, foreneck, lower sides of neck, and central chest, which are heavily and evenly barred cream-white and black. Almost uniform dark sides of breast and heavily barred chest contrast slightly with heavily cream-and-black-barred breast to vent (in dark-throated variants of *C. canorus*, lores, sides of head, and chin black with fine white bars, chest and sides of breast heavily marked black and cream, markedly contrasting with more widely spaced barring of breast and belly; less rufous mottling on sides of head and lores; no contrasting dark patches on sides of breast; bars on belly on average narrower; in paler variants, whole underparts cream with widely spaced dark bars, much paler than *C. saturatus*). Under tail-coverts and uppersurface of outer upper median primary coverts as in adult ♂. (2) Rufous birds. Upperparts mainly similar to rufous juvenile *C. canorus*, but back to upper tail-coverts of *C. saturatus* more contrastingly barred black or brown-black and rufous (in *C. canorus*, dark bars more greyish and less sharply defined); underparts as in greyish birds. Some closely similar to rufous-morph adult ♀, but latter lacks white edges on tips of flight-feathers. In all juveniles, white nape-patch and white feather-fringes to upperparts, upper wing-coverts, and primaries tend to disappear through abrasion. FIRST ADULT. Mainly like adult, differing from it as in *C. canorus* (see Moults).

Bare parts. ADULT. In ♂, iris yellow (occasionally red-brown, mainly in 1st adult); in ♀, yellow-grey, greyish-yellow, light yellow-brown, brown with yellow outer ring, or brown. Eye-ring lemon-yellow or clear yellow. Bill black, often with greyish, brownish, or greenish tinge; base of lower mandible and base of cutting edges of upper light greenish, olive-green, greenish-yellow, or greyish-yellow, corner of gape yellow or orange-yellow, mouth orange. Leg and foot lemon-yellow, clear yellow, butter-yellow, or orange-yellow. NESTLING. Iris dark brown. Edges of bill yellowish-white, mouth vermilion or orange. No further information. JUVENILE. Iris grey, light brown, yellow-brown, or dark brown. Eye-ring yellow. Bill dark horn-colour or black, base of lower mandible and base of cutting edges of upper pale horn, yellow, or yellow-green. Leg and foot pale, clear, or dark yellow. (RMNH, ZMA, ZMM.)

Moults. Based on wintering *horsfieldi* in Indonesia (RMNH, ZMA); nominate *saturatus* wintering in same area differs in moulting 1–2 months earlier. ADULT POST-BREEDING. Complete; primary moult complex. 2 moulting groups of primaries separable, p1–p4 and p5–p10. Sequence of outer group: 7–9–(short suspension)–5–8–(short suspension)–6–10 or –10–6. In inner group, p1 shed first, soon followed by p4 at about time of loss of p7; next p2 with p5 or p8, last p3 with p10 or p6 or slightly later; exceptionally, p6 or p3 excluded and retained for another year. First primary lost late August to late November; all completed late November to early April, sometimes after a longer suspension. Tail moulted mainly at primary moult score 15–35;

sequence 5–4–1–3–2, but often deviating slightly. Secondaries start at primary score of c. 25, completed with regrowth of last primary or slightly later, exceptionally a few old ones retained; sequence approximately 9–7–8–6–1–5–4, s2 or s3 last. Head, body, and wing-coverts moulted at primary score 0–40. In 1-year-old, retained juvenile feathers moulted first, out of normal sequence (e.g. p6, s2–s4, and t3–t4 first), starting at about loss of p1; later on, still recognizable as 2nd-calendar-year by showing these feathers slightly older at completion of moult than p1, p4, p9, t5, s7, s9, etc., instead of newer. POST-JUVENILE. Usually partial, occasionally complete. Starts between mid-September and mid-November with scattered feathers of crown, nape, mantle, chin, chest, and upper scapulars, followed by shedding of primaries (sequence as in adult) from late September to early January. Primary moult usually completed late February to early May, but birds starting from about late November or later suspend moult until next moulting season, retaining juvenile p3, p10, or p6; exceptionally, 7 primaries still juvenile in summer of 2nd calendar year. Tail moulted at primary moult score 15–45 in different sequence from adult: 1–5–2–4–3; juvenile t3(–t4) frequently retained in summer. Sequence of secondaries as in adult, starting at primary score 20–25; often still not completed when all primaries new, and juvenile s2 or s3, s2 and s3, s2–s4 (–s5), or even up to 8 secondaries retained until 1st post-breeding moult. Body and wing-coverts in 1st adult at primary score c. 40 (from January or later); occasionally, some juvenile wing-coverts retained.

Measurements. *C. s. horsfieldi*. Adult wing from: (1) European USSR, south-west Siberia, and western Kazakhstan, summer; skins (ZMM); (2) Indonesia, late August to early May. All other measurements Indonesia; skins (RMNH, ZMA). Bill (F) is adult bill from tip to forehead, bill (N) is tip to front of nostrils. Toe is outer front toe.

		♂			♀		
WING	(1)	210	(6·74; 19)	198–221	198	(7·25; 7)	191–209
	(2)	208	(5·61; 19)	197–217	195	(6·17; 23)	188–207
	JUV	201	(5·51; 24)	190–212	190	(6·29; 22)	179–201
TAIL	AD	167	(5·40; 21)	159–175	156	(4·23; 19)	150–164
	JUV	160	(5·49; 17)	152–168	150	(5·11; 19)	143–159
BILL (F)		27·7	(0·90; 27)	26·3–29·3	26·3	(1·39; 26)	23·9–28·3
BILL (N)		16·1	(0·50; 27)	15·5–17·0	15·5	(0·92; 26)	14·2–16·7
TARSUS		20·7	(0·75; 17)	19·9–22·3	20·3	(0·96; 12)	18·7–21·4
TOE		26·3	(1·29; 10)	25·0–28·5	24·9	(1·52; 18)	23·2–27·2

Sex differences significant, except tarsus and toe. Juvenile wing and tail significantly shorter than adult; juvenile bill not full-grown until December–January, excluded from table. Smaller than *C. canorus canorus*: in *C. saturatus*, adult ♂ wing usually below 215, adult ♀ 205, juvenile ♂ 207, juvenile ♀ 195; adult ♂ tail mainly below 173, adult ♀ 159, juvenile ♂ 165, juvenile ♀ 155; *C. c. canorus* mainly over these values, but 6–10% of birds wrongly identified by these criteria and the largest *horsfieldi* examined came from European USSR, making identification by size difficult in west Palearctic; also, tail of *Cuculus* difficult to measure without experience.

Nominate *saturatus*. ADULT. Taiwan, March–June; skins (RMNH).

		♂			♀		
WING		190	(3·58; 10)	184–195	177	(2·94; 8)	174–183
TAIL		161	(4·11; 9)	157–167	146	(2·71; 7)	141–150
BILL (F)		27·9	(0·95; 10)	26·7–29·2	27·0	(1·34; 8)	25·6–28·6
BILL (N)		16·8	(0·54; 10)	16·1–17·4	16·6	(0·68; 8)	15·9–17·3

Indonesia, October–February; skins (RMNH, ZMA).

		♂			♀		
WING		189	(3·44; 10)	185–195	175	(5·42; 20)	166–184
TAIL		157	(5·02; 12)	151–164	146	(5·70; 21)	135–156
BILL (F)		28·3	(0·43; 9)	27·7–28·9	25·8	(0·94; 17)	24·4–27·5
BILL (N)		16·8	(0·43; 9)	16·2–17·3	15·3	(0·76; 17)	14·2–16·4

Sex differences significant, except bill from Taiwan. Wing and tail of nominate *saturatus* significantly shorter than *horsfieldi*, but not bill. Note relatively small bill of Indonesian ♀♀.

Weights. *C. s. horsfieldi*. Western Siberia, June: ♂ 139 (ZMM). Kazakhstan (USSR), May–June: ♂♂ 91, 109, 128; ♀♀ 84, 89 (Dolgushin *et al.* 1970). South-central Siberia, spring and summer: ♂ 116 (11) 105–128, ♀ 77 (6) 75–80 (Chunikhin 1964). Kurils (USSR): July, ♂ 132; late August, juvenile ♂ 82 (Nechaev 1969). Hopeh (China), May–August: ♂ 114 (7) 98–134, ♀ 113 (6) 74–156 (Shaw 1936). Taiwan, October: juvenile ♀ 101 (RMNH). Indonesia: October, ♀ 74; April, ♀ 73 (RMNH, ZMA).

Nominate *saturatus*. Nepal, May: ♂ 105, ♀ 72 (Diesselhorst 1968). Taiwan: March–June, adult, ♂ 93·1 (9·27; 9) 75–108, ♀ 70·4 (8·04; 7) 60–80; September, immature, ♂ 80, ♀ 75 (RMNH). Indonesia: October, ♀ 83 (ZMA).

Structure. 10 primaries: p8 longest, p9 5–12 shorter, p10 50–68, p7 7–11, p6 19–28, p5 36–43, p1 85–99. 10 tail-feathers, t5 37–46 shorter than t1. Bill relatively slightly shorter than in *C. canorus*, but heavier at base. Outer hind toe c. 80% of outer front, inner front c. 67%, inner hind c. 48%. Remainder of structure as in *C. canorus*.

Geographical variation. Marked, involving size only. *C. s. horsfieldi* from European USSR east to Japan and eastern China largest; nominate *saturatus* from Himalayas to southern China and Taiwan smaller (see Measurements). Although ranges of these races meet in China, extent of mixing apparently small, as no or hardly any birds of intermediate size occur, as far as can be seen from wintering birds of both races in Indonesia (size-frequency distribution of birds of same age-group and sex shows 2 separate peaks with no overlap). *C. s. lepidus* from mountains of Malaya, Sumatra, Java, and Lesser Sunda Islands east to Timor much smaller again: wing of adult ♂ from Java 153 (26) 146–159, ♀ 143 (19) 138–147 (Becking 1975). Formerly considered to be a race of Little Cuckoo *C. poliocephalus*, but colour, egg-shell structure, breeding biology, and call like *C. saturatus* and dissimilar from *C. poliocephalus* (Becking 1975; Wells and Becking 1975): dark bluish-grey above and heavily barred below, as in nominate *saturatus*, but under tail-coverts slightly deeper pink-cinnamon. *C. s. insulindae* from mountains of Borneo similar in size to *lepidus*, but upperparts slightly darker; some Sumatran *lepidus* rather dark too, however; *insulindae* otherwise similar to nominate *saturatus* (Becking 1975; Wells 1982). In all races, 30–60% of ♀♀ are rufous morph. CSR

Subfamily PHAENICOPHAEINAE non-parasitic cuckoos

About 28 species in 11–14 genera. 2 species occur accidentally in west Palearctic, both of American genus *Coccyzus*. 2–4 other genera in New World, 1 in Africa, 8–9 in tropical Asia, and remainder between Malaysia and Philippines.

For general features, see Cuculidae. Nostrils narrow, without raised edges. Brood-parasitism not practised.

Coccyzus erythrophthalmus Black-billed Cuckoo

PLATE 40
[between pages 470 and 471]

Du. Zwartsnavelkoekoek   Fr. Coulicou à bec noir   Ge. Schwarzschnabelkuckuck
Ru. Черноклювая американская кукушка   Sp. Cucu de pico negro   Sw. Svartnäbbad regngök

Cuculus erythropthalma Wilson, 1811

Monotypic

Field characters. 27–31 cm; wing-span 38–42 cm; tail 11–13 cm. About 15% smaller than Cuckoo *Cuculus canorus*, with proportionately shorter wings and tail but retaining characteristic silhouette of family. Rather small, slim, rather dove-like Nearctic cuckoo, dark-brown above and grey-white below and lacking any obvious characters. Clear sight of wholly black bill or small white spots on brown tail-feathers essential to distinction from Yellow-billed Cuckoo *Coccyzus americanus*. Flight lighter and more graceful than *Cuculus canorus*. Sexes similar; no seasonal variation. Juvenile separable.

ADULT. Upperparts and wing-coverts uniformly bronzy-brown; flight-feathers similarly coloured but with indistinct rufous or pale buff showing occasionally on inner webs in flight (but not forming obvious splash of colour as in *C. americanus*). Tail also bronzy-brown, with small white spots and dark brown subterminal band on outer feathers. Tail pattern indistinct (and appears dull from below, with white spots contrasting little with brown-grey undersurface of feathers). Underparts white, with buff tone strong on throat and on under tail-coverts. Under wing-coverts cream-buff. Bill grey-black, with lower mandible wholly dark (unlike *Coccyzus americanus*). Eye-ring red. Legs pale grey. IMMATURE. From about September, resembles adult but upperpart feathers narrowly tipped grey-white, tail-spots buffish with no subterminal bar, and flight-feathers show more rufous than adult (inviting confusion with *C. americanus*), with secondaries and inner primaries finely tipped pale buff. Bare parts as adult but skin round eye yellow.

Smallest cuckoo occurring in west Palearctic (except for brilliantly coloured Didric Cuckoo *Chrysococcyx caprius*), with silhouette typical of family; wholly pale underparts prevent confusion with *Cuculus*, and much smaller size and lack of pronounced spotting on wings rule out confusion with Great Spotted Cuckoo *Clamator glandarius*. In brief view, distinction from *Coccyzus americanus* difficult but that species shows black and white outer tail-feathers, large rufous patch on wing, dark cheeks, and (at short range) conspicuous yellow on bill. Flight recalls *Cuculus* but even lighter, with flickering wing-beats and frequent glides and jinks in cover. More secretive than *Cuculus*, keeping to dense foliage. Feeds at lower level than *Coccyzus americanus*, walking with loping gait.

Vagrants occur on coasts and islands. Rarely vocal outside breeding season.

Habitat. In Nearctic continental mid-latitudes, breeding from fringe of boreal coniferous and mixed forest to temperate plains and uplands. Occupies somewhat more densely wooded habitat than Yellow-billed Cuckoo *C. americanus*, preferring extensive areas of upland woods with variety of trees, bushes, and vines for nest cover (Johnsgard 1979). In Canada in more open woodland, such as tangles of willow, alder, and vines, nesting in either deciduous or coniferous trees (Godfrey 1966). Fond of wet places, more often descending to ground than *C. americanus*, and much influenced by infestations of caterpillar prey (Forbush and May 1939). Found in shrubs along creeks or in wooded areas near wet ground up to c. 1700 m (Niedrach and Rockwell 1939). Also wood-borders, bushy roadsides, orchards, and cultivated grounds (Hollom 1960). On passage in Venezuela occurs in rain forest, secondary growth, semi-open woodland, and plantations, keeping low in trees. Altitudinal range 350–1100 m (Schauensee and Phelps 1978).

Distribution. Breeds in North America east of Rocky Mountains, from southern Canada south to Wyoming, Kansas, Arkansas, Tennessee, and South Carolina. Winters from north-west South America south to Peru.

Accidental. Iceland, Ireland, Britain, France, West Germany, Denmark, Italy, Azores.

Movements. Migratory. Withdraws in autumn towards winter quarters in north-western South America (Colombia, Ecuador, northern Peru); has also occurred Paraguay and adjacent northern Argentina (Misiones), and perhaps more than accidental there (Short 1972).

Common in eastern North America and average breeding distribution further north than that of Yellow-billed Cuckoo *C. americanus*, extending well into Canadian maritime provinces; however, much rarer vagrant to Europe (12 records to 1980, against 52 *C. americanus*). This is a consequence of south to south-west autumn movement from eastern North America towards restricted wintering area (see above) and away from Atlantic coasts where uncommon (e.g. Murray 1952, Stewart and Robbins 1958). Migrants cross Gulf of Mexico region (including Greater Antilles) and Central America; relative paucity of passage data from southern regions (e.g. Keast and Morton 1980) probably due to nocturnal migration with overflying of large areas. In Guatemala, 15–20 killed in one night when attracted to fire (Baepler 1962) though few other records there.

Autumn movement begins late July to early August in north, with main exodus mid-August to mid-September, but some birds linger well into October, when main movement already occurring through Panama. Return migration during April, with main influx into northern breeding areas in May and early June. Apart from one in July (France), dated European records span 23 September–6 November, most in October.

Voice. See Field Characters.

Plumages. ADULT. Forehead, lores, and narrow line up to just above middle of eye light grey with slight olive tinge, not sharply demarcated from olive-grey of crown. Crown and remainder of upperparts including ear-coverts, sides of neck, tertials, and all upper wing-coverts greyish-olive with slight bronze-green lustre, usually slightly purer olive-grey on back and rump, occasionally with faint rufous tinge on outer web of median and greater upper wing-coverts; upperparts more greyish-olive and with more bronze-green lustre than dull greyish-brown or brownish-grey of Yellow-billed Cuckoo *C. americanus*; forehead less pure greyish than *C. americanus*; no black streak from lores to lower ear-coverts (as present in adult *C. americanus*). Feathering from gape and just below eye down sides of neck, and on sides of breast greyish-cream, sometimes slightly tinged olive; chin, throat, and chest cream-yellow, often slightly suffused pale olive-grey on lower throat and chest; chin to chest of *C. americanus* usually purer greyish-white but frequently similar. Belly and vent white, sometimes with slightly olive-grey suffusion; upper flanks, thighs, and under tail-coverts greyish-cream or buff-cream, lower flanks olive-grey. Uppersurface of tail greyish-olive with slight bronze-green lustre like upperparts; undersurface pale grey (in *C. americanus*, black in adult, dark grey in juvenile); 2–4 mm of tip of t2 white, increasing to 6–8 mm on tip of t5; subterminally bordered by bar 3–5 mm wide—black on uppersurface and dark grey on undersurface, poorly defined from greyish-olive (above) or grey (below) of remainder of tail (thus, lacks broad and contrasting white tail-tips of *C. americanus*). Flight-feathers greyish-olive or slightly brownish-olive, like upperparts, but less lustrous; broad borders to base of inner web pale cream or partly cinnamon-buff, basal outer edge of inner primaries occasionally with rufous tinge, but much less than in *C. americanus*. Under wing-coverts and axillaries rather deep buff or yellow-buff, about similar in colour to undersurface of basal half of inner and middle primaries and to undersurface of basal and middle portion of secondaries. In worn autumn plumage, upperparts slightly duller and greyer olive-brown (still with bronze lustre), forehead and lores dirty olive-grey, chin, sides of neck, throat, and chest pale buff with grey of feather-bases visible, sides of neck and sides of breast less uniform grey, and chin to central chest less uniform yellow-buff. JUVENILE. Like adult, but upperparts slightly more rufous-brown, especially on nape, mantle, upper scapulars, rump, and upper wing-coverts; feathers of upperparts and upper wing-coverts narrowly fringed pale greyish-buff to off-white, appearing faintly scaled. Underparts silky-white with buff tinge on chin, throat, and chest. Tail-feathers with inconspicuous buffish-white tips *c.* 2 mm wide; no dark subterminal bar or this very faint and poorly defined; feathers narrower than in adult (10–11 mm wide at middle of t5, 13–16 in adult). Flight-feathers and greater upper wing-coverts with strong rufous tinge on basal and middle portions; narrow pale buff fringes on tips of greater primary coverts, secondaries, and inner primaries. FIRST NON-BREEDING. From about September, like adult, but juvenile flight-feathers, tail, most tertials, some upper tail-coverts, and a few or many outer upper wing-coverts retained. FIRST BREEDING. As adult; birds with slight rufous tinge to upperparts and upper wing-coverts, with some rufous suffusion on bases of primaries, and with relatively faint dark subterminal tail-bar are probably in 2nd calendar year.

Bare parts. ADULT. Iris deep brown. Eye-ring and bare skin round eye bright red. Bill dark slate or black, tinged grey-blue at base, sometimes slightly yellow on underside of lower mandible at base. Leg and foot pale blue-grey or grey-blue. JUVENILE. Like adult, but bare skin round eye and eye-ring yellow, changing to red during 1st pre-breeding moult. (Ridgway 1916; Sutton 1982; BMNH, RMNH.)

Moults. As in *C. americanus*, but 1st and adult pre-breeding perhaps slightly later, up to May (Witherby *et al.* 1938).

Measurements. ADULT. Eastern North America, May–September; skins (BMNH, RMNH, ZMA). Bill (F) from tip to forehead, bill (N) from tip to nostril. Toe is outer front toe.

WING	♂	139	(3·30; 17)	134–144	♀ 143	(2·87; 15) 139–147
TAIL		151	(5·53; 17)	144–161	157	(5·32; 15) 149–167
BILL (F)		27·7	(1·01; 9)	26·2–29·0	28·9	(0·76; 7) 28·0–29·8
BILL (N)		17·4	(0·92; 17)	16·1–19·0	17·5	(0·81; 15) 16·6–18·8
TARSUS		24·5	(0·64; 9)	23·6–25·2	24·3	(0·98; 8) 23·2–25·6
TOE		23·1	(1·18; 7)	21·2–24·4	22·7	(0·36; 5) 22·4–23·3

Sex differences significant for wing, tail, and bill (F).

JUVENILE. Like adult, but tail on average *c.* 4 shorter and bill not full-grown until about midwinter.

Weights. New Jersey (USA), mainly September: 39·1 (25) 34·7–40·0 (Murray and Jehl 1964). Florida, autumn: juvenile ♂ 67·8 (Grocki and Johnston 1974). Kentucky, June: ♂ 41·6 (Mengel 1965). Peru, April: ♂ 49·3 (Sanft 1970). Mexico, September: ♂ 42·3 (RMNH).

Structure. 10 primaries: p8 longest, p9 16–19 shorter, p10 49–55, p7 2–3, p6 10–12, p5 21–24, p1 46–52. Tail long, graduated; t5 52–65 shorter than t1, t4 20–30, t3 7–12, t2 2–6. Tail relatively longer than in *C. americanus*: usually 9–24 mm longer than wing length, exceptionally (mainly juvenile) 0–8 longer; in *C. americanus*, tail usually shorter than wing, occasionally up to 3–4(–6) longer. Bill more slender than in *C. americanus*; in particular, tip finer and culmen narrower; tip slightly less strongly decurved. Remainder of structure as in *C. americanus*. CSR

Coccyzus americanus **Yellow-billed Cuckoo**

PLATE 40
[between pages 470 and 471]

Du. Geelsnavelkoekoek   Fr. Coulicou à bec jaune   Ge. Gelbschnabelkuckuck
Ru. Желтоклювая кукушка   Sp. cucu de pico amarillo   Sw. Gulnäbbad regngök

Cuculus americanus Linnaeus, 1758

Polytypic. Nominate *americanus* (Linnaeus, 1758), North America east of Rocky Mountains, south to south-east Mexico and probably to Greater Antilles. Extralimital: *occidentalis* Ridgway, 1887, western North America from British Columbia south to north-west Mexico.

Field characters. 28–32 cm; wing-span 40–48 cm; tail 11–13 cm. Slightly smaller and slimmer than Cuckoo *Cuculus canorus*; averages larger but not longer tailed than Black-billed Cuckoo *Coccyzus erythrophthalmus*. Rather small cuckoo, with similar configuration and plumage pattern to *C. erythrophthalmus* but differing in somewhat bulkier shape, yellow base to bill, rufous patch on wing, and bold white spots on black outer tail-feathers. Sexes similar; no seasonal variation. Juvenile separable.

ADULT. Upperparts slightly bronzy-brown, with grey face and dark brown cheeks; wing-coverts, tertials, and outer webs and ends of flight-feathers brown but most of flight-feathers chestnut. Chestnut usually obvious on folded wing and eye-catching in flight. Tail-centre as upperparts but outer feathers increasingly black towards end with long, bold white spots at end (and extending along outer edge of outermost feather). Black and white pattern on tail most obvious from below or when spread. Underparts white, tinged buff-grey on throat and breast; under wing-coverts cream-buff. Bill dark horn, with striking yellow patch on most of lower mandible and base of upper mandible. Eye-ring yellow. Legs grey. IMMATURE. From about September, resembles adult, but browner above and rustier toned on whole of wing. Tail pattern less distinct than in adult. Moult of wing-feathers (from November) and tail (from midwinter) may obscure these differences.

Confusion with indigenous cuckoos unlikely but separation from vagrant *C. erythrophthalmus* needs care and best distinctions are wing pattern (in *C. americanus* colourful, chestnut; in *C. erythrophthalmus*, uniform in adult, tinged rufous in juvenile) and tail pattern (bold in *C. americanus*, indistinct in *C. erythrophthalmus*).

Vagrants occur on coasts and islands. Rarely vocal outside breeding season.

Habitat. Breeding throughout temperate Nearctic, predominantly in lowlands in zone of broad-leaved woodland; also to subtropics. Rarely in deep woodlands, preferring dense tangles of secondary growth in rural areas, moist patches, and streamside willow *Salix* thickets. Also in run-down orchards and along brush-grown country roads, nesting in dense thorny or evergreen shrubs or trees, therefore often indirectly profiting by human land use (Pough 1949). In Texas, on farms and in mesquite *Prosopis* arid chaparral country; also woodland (Peterson 1960). On great plains east of Rocky Mountains also avoids extremely dense woods, favouring secondary growth, moderately dense thickets near watercourses, brushy orchards, and deserted farmlands overgrown with shrubs and brush (Johnsgard 1979). In Louisiana, nests in trees on edge of clearing or in open grove, not infrequently by human dwelling (Lowery 1955). In southern Canada, in willow and alder *Alnus* tangles and in open woodland (Godfrey 1966). In western USA, mainly in willow and poplar *Populus* thickets by rivers, and in mesquite (Peterson 1961).

In winter in Venezuela, in open situations, open woodland, secondary growth, xerophytic areas, thicket, or paramo, usually up to 150–400 m, but exceptionally to 4200 m (Schauensee and Phelps 1978). In Guyana in November, in clearings in high rain forest (E M Nicholson). Flies low and for brief distances except on migration.

Distribution. Breeds in North America, from southern Canada (locally in British Columbia, southern Ontario, and southern New Brunswick, rarely elsewhere) south to central Mexico and the Caribbean. Winters mainly in South America from Venezuela and Colombia to Argentina.

Accidental. Iceland, Ireland, Britain, France, Belgium, Denmark, Norway, Italy, Morocco, Azores.

Movements. Migratory. Moves on broad fronts to and from winter quarters in South America, where present east of Andes and south to central Argentina; small numbers also winter in Panama (Ridgely 1976).

Growing evidence of involvement by eastern population in transoceanic autumn passage between south-east Canada/New England and West Indies/South America. Occurs fairly regularly on Bermuda and outer West Indies; radar studies on Puerto Rico in autumn showed night migrants arriving from north, with this species (among others) commonly alighting there (Richardson 1976); birds arriving Netherlands Antilles in autumn often in exhausted conditions, at 50% of winter body weight (Voous 1955; see also Weights). Overwater flight of 2000–3000 km to West Indies, 4000 km for overflying birds going on to South American mainland. This transoceanic route not used in spring, when Puerto Rico radar studies showed migratory movement (all species) oriented towards Florida (Richardson 1974).

Autumn passage in New England early August to mid-October, mainly late August and September (e.g. Stewart and Robbins 1958, Bull 1974), lasting through November in Panama and Caribbean where, however, peak passage in October (e.g. Voous 1955, ffrench 1973, Ridgely 1976). Early arrivals occur southern USA in March, but main period April–May, with immigration into northern breeding areas continuing into early June. Autumn tropical storms occasionally carry large numbers northwards into New England and Canadian maritime provinces (Godfrey 1966; Bull 1974); however, this phenomenon not thought to contribute significantly to transatlantic vagrancy. Latter believed due to migrants (especially those moving offshore) becoming disoriented in frontal zones north of Bermuda, then carried eastwards in warm sectors associated with fast-moving wave depressions (see Elkins 1979). 40 dated west European records occurred late September (3), October (25), November (11), and December (1); hence majority occurred after peak passage period in eastern North America. Over half (56%) of British and Irish records concerned birds found dead or dying, showing stress involved.

Voice. See Field Characters.

Plumages (nominate *americanus*). ADULT. Forehead, lores, and narrow line up to just above middle of eye medium grey; a narrow black line running below eye from lores to lower ear-coverts, sometimes indistinct or absent. Crown, hindneck, sides of neck, mantle, scapulars, tertials, and upper wing-coverts olive-brown with slight bronze lustre and greyish cast when fresh, duller olive-brown when worn; forecrown often all or partly tinged medium grey. Back and rump slightly greyer and less olive-bronze than mantle and scapulars, upper tail-coverts similar to mantle and scapulars. Chin, throat, and chest pale grey, usually with cream-white or cream-buff tinge on central throat and central chest; cheeks, sides of neck, and sides of breast sometimes tinged olive. Remainder of underparts pale cream or white, upper flanks, thighs, and under tail-coverts pale cream-buff, lower flanks olive-grey. Tail: t1 olive-brown with slight bronze lustre, like upperparts, tip sometimes tinged dull black; t2 black with olive-brown base and narrow buff-white tip 2–5 mm wide, but occasionally olive-brown with black distal quarter and no pale tip; t3 black with 12–20 mm of tip white and faint white border along inner and outer edge; t4 similar, but 22–30 mm of tip white; t5 black with 20–25 mm of tip white, distal half of outer web white, and faint white edge along inner web. Both from above and below, tail black with large rounded white tips to t2–t5 and olive-brown (above) or grey (below) central pair, unlike Black-billed Cuckoo *C. erythrophthalmus*. Flight-feathers and greater upper primary coverts greyish olive-brown, tips slightly darker, bases and middle portions of inner webs bright rufous (more restricted on p9–p10 and innermost secondaries). Rufous sometimes extends to basal outer webs of inner and middle primaries, but this probably depends on age and sex: of 12 examined with rufous restricted (extending to just along shaft on outer web), 10 were ♂♂ and 2 ♀♀; of 13 with rufous extending across whole base and middle portion of outer web, 8 were ♀♀ and 5 ♂♂; of latter group, at least 2 ♂♂ had a few retained juvenile feathers and were thus in 2nd calendar year; in those with rufous restricted, outer webs dark olive-grey, no rufous showing when wing closed. Under wing-coverts and axillaries cream-buff; pale bases of primaries (except p10) and to lesser extent of secondaries form pale rufous or cinnamon-rufous underwing-patch. In worn plumage, upperparts duller and browner, less grey; forehead, forecrown, and lores more olive, chin to chest less uniform grey. JUVENILE. Like adult, but all upperparts including forehead, lores, ear-coverts, and upper wing-coverts more rufous olive-brown, feathers often slightly edged with rufous or buff on tips, edges sometimes pale cream on upper wing-coverts. No dark streak from lores to lower ear-coverts. Tail rather different from adult: t2–t5 dull black or dusky grey from above, grey from below (not deep black); grey poorly defined from white tips of (t2–)t3–t5; white on tip of t3 7–13 mm wide (measured along shaft); white edges on sides of t3–t5 more extensive than in adult, especially on outer web of t4, but poorly defined; tail-feathers often narrower than in adult. Flight-feathers and upper primary coverts more extensively rufous than in adult: outer web of primaries and greater primary coverts deep rufous, except for (p8–)p9–p10 and for extreme tip of feather; often a slight olive-grey shade on distal outer edge of primaries; outer secondaries with rufous tinge to centre and base of outer web (in adult, no rufous on outer webs of primaries, outer secondaries, and greater primary coverts, except for narrow poorly-defined line along shaft; sometimes, however, rufous extends to outer edge near base—see Adult). Inner primaries and outer secondaries often with faint rufous margin at tip. FIRST NON-BREEDING. Like adult from about September, but juvenile flight-feathers, tail, upper primary coverts, most or all tertials, and a few to many outer lesser, median, and greater upper wing-coverts retained. Dark streak from lores below eye to lower ear-coverts usually absent. For flight-feathers and tail, see Juvenile; old upper wing-coverts shorter, narrower, and more rufous than in adult, often with narrow contrasting rufous-buff or cream margin at tip. FIRST BREEDING. Like adult, but apparently with more rufous on primaries and upper primary coverts (see Adult); some show much rufous on scattered primaries and less on intervening ones (rufous feathers probably grown during early part of post-juvenile moult). Occasionally, some juvenile secondaries or wing-coverts retained, heavily abraded.

Bare parts. ADULT. Iris dark brown or dark sepia. Eye-ring lemon-yellow. Small bare patch below eye grey. Upper mandible and tip of lower mandible blackish-horn to slate-black; remainder of lower mandible and basal half of cutting edges of upper lemon-yellow, bright yellow, or ochre-yellow. Leg and foot dark slate-grey to pale bluish-grey. JUVENILE. At fledging, iris black-brown; eye-ring and bare skin below eye dull grey; bill dark bluish horn colour, basal half of cutting edges of upper mandible and base and middle of lower mandible pale flesh colour; leg and foot bluish-grey (Sutton 1982). During autumn migration, September–November, iris dark brown, eye-ring yellow, bare patch below eye pale yellow; bill, leg, and foot as adult. (Ridgway 1916; RMNH, ZMA.)

Moults. ADULT POST-BREEDING. Complete. Primary moult complex, several alternate feathers (never neighbouring ones) growing at same time, but no regular sequence apparent (Stresemann and Stresemann 1961). Usually starts in breeding area, July–September, with scattered feathers of head and body, but only exceptionally a few primaries or tail-feathers replaced then. During autumn migration, moult suspended; of 15 autumn migrants examined, October and 1st week November, 3 had a few feathers of nape, mantle, scapulars, lores, and chest new, remaining 12 had 50–100% of feathers of head, mantle, scapulars, underparts, and upper tail-coverts new and rest of body, tertials, tail, and wing old; most advanced bird also had p4–p5 and t1 new. Moult resumed in winter quarters and completed March(–April).

ADULT PRE-BREEDING. Partial: head and body, but apparently not back and rump; no wing- or tail-feathers. January–April. POST-JUVENILE. Complete. Starts with scattered feathers of head, mantle, scapulars, and chest from August–September. Of 25 autumn migrants, mid-September to mid-November, 3 had only limited amount of feathers new, remainder had all head and body new (but not always all back, rump, or upper tail-coverts) as well as some or many lesser and median upper wing-coverts and inner greater coverts; occasionally, a few tertials also. Moult suspended during autumn migration, resumed in winter quarters, where flight-feathers and tail start. Moult completed up to late April. FIRST PRE-BREEDING. Extent and timing not fully known. Advanced birds probably as in adult pre-breeding, but extent probably restricted in birds which completed post-juvenile as late as March–April. In one bird examined, involved only forehead and sides of head. (Stresemann and Stresemann 1961; RMNH, ZMA.)

Measurements. Nominate *americanus*. Eastern USA, April–September, and migrants from eastern and south-east Caribbean, September–November and April; skins (RMNH, ZMA). Bill (F) from tip to forehead, bill (N) from tip to nostril, both for adult only. Toe is outer front toe.

WING AD	♂ 143	(3·05; 19)	137–148	♀ 147	(3·08; 16)	143–152
JUV	144	(4·02; 12)	139–149	147	(3·39; 19)	143–154
TAIL AD	142	(4·81; 20)	135–152	146	(2·87; 15)	142–151
JUV	143	(7·73; 11)	135–154	144	(5·46; 16)	137–155
BILL (F)	29·9	(1·28; 15)	28·2–31·6	31·4	(1·19; 13)	30·0–33·3
BILL (N)	19·0	(0·87; 16)	17·9–20·5	19·4	(0·58; 11)	18·6–20·1
TARSUS	25·9	(0·86; 16)	24·4–27·2	26·0	(0·97; 25)	24·4–27·7
TOE	24·2	(0·85; 15)	23·0–25·6	24·8	(0·84; 13)	23·6–26·1

Sex differences significant for wing, adult tail, and bill (F). Juvenile wing, tail, tarsus, and toe not significantly different from adult; by October–November, bill of juvenile ♂ similar to adult, but bill of juvenile ♀ on average still 1·5 shorter.

Weights. Georgia and South Carolina (USA): ♂, late July, 54·6; ♀, early August, 61·2 (Norris and Johnston 1958). Ohio: ♂, August, 57·8 (Baldwin and Kendeigh 1938). Kentucky, September: ♂♂ 63·2, 74·5; ♀ 61·7 (Mengel 1965). At start of autumn migration, Florida: 78·9 (10) 66·2–96·1 (Grocki and Johnston 1974). Exhausted autumn migrants after Caribbean crossing, October–November, Netherlands Antilles: adult ♂ 31·4 (2·30; 5) 28–34; adult ♀ 35·8 (3·56; 5) 32–40; juvenile ♂ 32·3 (6·19; 6) 26–40; juvenile ♀ 34·4 (6·32; 14) 29–45; most of these died afterwards (ZMA). Surinam, April: ♂♂ 45, 49·5; ♀ 60·5 (RMNH). On spring migration, Florida: 53·6 (11·4; 6) 41·7–70·2 (Grocki and Johnston 1974).

Structure. Wing rather long, broad at base, tip pointed. 10 primaries: p8 longest, p9 12–16 shorter, p10 44–58, p7 9–17, p6 20–29, p5 30–40, p1 51–60. Primaries not emarginated, outer ones gradually narrowing towards rounded tip. Tail long, graduated, 10 feathers; t1 longest, t2 about equal, t3 2–8 shorter, t4 17–30, t5 40–62; tail-feathers of juvenile on average slightly narrower than adult (width of t1 in middle 17–19 in juvenile, 18–21 in adult). Bill deep at base, wide at gape, distal half strongly compressed laterally; long compared with Cuckoo *Cuculus canorus*, tips of both mandibles more strongly decurved. Nostril narrow, edges not raised. Eye-ring and small crescent below eye bare. Feathering of rump less dense than in *Cuculus*. Tarsus rather long and slender, bare. Toes slender; outer hind toe *c*. 85% of outer front toe, inner front *c*. 65%, inner hind toe *c*. 51%. Claws short, sharp, decurved.

Geographical variation. Slight. *C. a. occidentalis* from western North America slightly larger: average of 10 adult ♂♂ and 9 adult ♀♀, California and Arizona: wing ♂ 150, ♀ 151; tail ♂ 147, ♀ 150. No difference in colour. (Ridgway 1916.)   CSR

Subfamily CENTROPODINAE coucals

Mostly large, skulking terrestrial birds with weak flight; 26–28 species in single genus *Centropus* distributed from Africa and Madagascar through southern Asia to Australasia and Solomon Islands. 1 species in west Palearctic, breeding.

For general features, see Cuculidae. Nostrils narrow, with soft flap above. Inner hind claw remarkably long, spur-like. Brood-parasitism not practised.

Centropus senegalensis Senegal Coucal

PLATES 37 and 41
[between pages 470 and 471]

Du. Senegalese Spoorkoekoek   Fr. Coucal de Sénégal   Ge. Senegal-Spornkuckuck
Ru. Шпорцевая кукушка   Sp. Cucal senegales   Sw. Senegalmaskgök

Cuculus senegalensis Linnaeus, 1766

Polytypic. *C. s. aegyptius* (Gmelin, 1788), Egypt. Extralimital: nominate *senegalensis* (Linnaeus, 1766), northern Afrotropics from Sénégal to Somalia, south to Uganda, Zaïre, and northern Angola; *flecki* Reichenow, 1893, southern Africa from Zambia and Malawi to Transvaal.

Field characters. 40–42 cm; wing-span 50–55 cm; tail 15–22 cm. Longer and much bulkier than even Great Spotted Cuckoo *Clamator glandarius*, approaching size of large crows *Corvus*. Heavy-billed, long- and broad-tailed bird, lacking form and behaviour of other cuckoos; general appearance well indicated by old name 'crow-pheasant'. Black bill and head, brown back, bright rufous wings, and black tail contrast with buff-white underparts. Rarely seen out of dense ground cover. Sexes similar; no seasonal variation. Juvenile separable.

428 Centropodinae

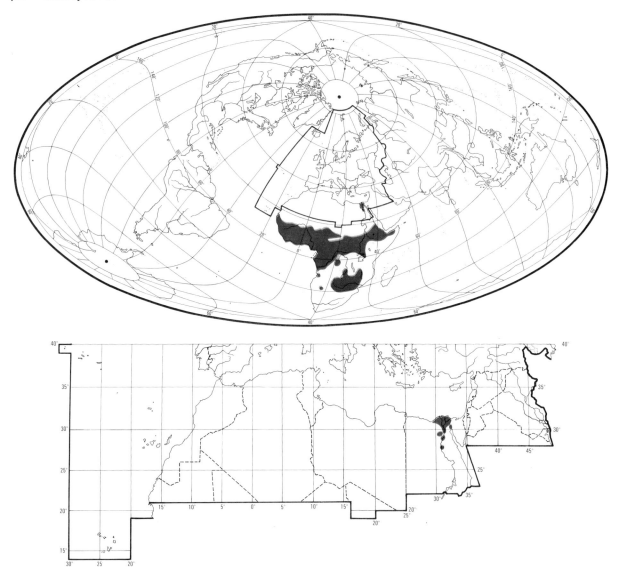

ADULT. Head, nape, and tail black, tail glossed green. Back rufous-brown; rump olive-brown. Wings bright chestnut, appearing to glow when spread in sun. Chin, throat, and sides of neck pale buff-white, body buff-white. Massive, crow-like bill black. Eye red. Legs and feet grey-black, with long claw on inner rear toe. JUVENILE. Distinguished from adult by black-barred back and strongly buff chest.

Unmistakable in west Palearctic (though subject to confusion with Blue-headed Coucal *C. monachus* south of Sahara). Flight slow and clumsy, with apparently laborious beats of rather short, much fanned wings. Gait freer than in other cuckoos, with slow stalking walk, changing to hop and run. Largely ground-feeding. Usually confined to bushes and thickets, moving clumsily among or flying laboriously between them. Rippling song like water bubbling from bottle; subdued but sweet-toned. Commonest call a persistent, monotonous 'hood-hood...' or 'hu-hu...', given most frequently around dawn and dusk.

Habitat. In tropical lowlands, avoiding forest belt and mountains, but ascending to high plateau in Sierra Leone (G D Field). Mainly in thorn-scrub and grass woodland belts, in savanna or semi-arid tree savanna, dense and light bush, grasslands, dense grass patches, fringing vegetation of streams, reedbeds, and papyrus swamps, gardens with shrubs, and cultivated areas. Often in thick cover, or creeping about among low bushes, but also feeds much on ground. In Sierra Leone, reaches forest edge and immediately colonizes felled areas (G D Field). Round edges of grassland in swamps, but not found in middle; rarely comes close to houses, though seen in hotel grounds in Faiyum, Egypt (P A D Hollom). Skulks in shrubberies and dense bushes at edge of more open areas. Most active in evening, and sometimes far into night. Flies low and rarely far. (Bannerman 1933; Prozesky 1970; Serle *et al.* 1977.)

Distribution and population. EGYPT. Locally not uncommon resident (PLM, WCM).

Movements. Non-migratory throughout Africa; however, local movements likely in the more arid areas. In Sudan, resident in Darfur though in west Kordofan not seen until onset of rains in May (Lynes 1925a); in Mali, also reported absent from the most arid parts during dry season (Lamarche 1980). Egyptian race *aegyptius* resident.

Food. Little information, and no details for west Palearctic.

Wide variety of animal food, chiefly large insects (especially Orthoptera) supplemented by small vertebrates. Feeds mostly on ground (Meinertzhagen 1930), but will often perch in tree and swoop on prey (Bannerman 1933). In Nigeria, observed chasing lizards along branches of tree; once caught half-grown ♀ *Agama agama* in bill, hammered it several times on ground, and swallowed it head first (Brotherton 1965). Also known to attack live birds caught in nets (Lynes 1925a; Brotherton 1965).

South of Sahara, diet includes grasshoppers and locusts (Acrididae), caterpillars (Lepidoptera), termites *Bellicositermes* (Isoptera), beetles (Coleoptera), bugs (Hemiptera), frogs, snakes, lizards, rodents, and nestlings of small birds (Reichenow 1902–3; Granvik 1923; Chapin 1939a; Monard 1940; Young 1946; Dekeyser 1956; Mackworth-Praed and Grant 1952; Brotherton 1965; Etchécopar and Hüe 1967; Steyn 1972; G D Field).

Stomachs of 5 adults (including 3 ♂♂, 1 ♀), Sénégal, contained Orthoptera, *Bellicositermes*, and insect debris (Dekeyser 1956). In 6 stomachs, Zaire, Coleoptera in 3, Orthoptera in 2, caterpillars in 2, hemipteran bug in 1 (Chapin 1939a).

Stomach of young bird (probably nestling), southern Rhodesia, contained 17 Acrididae and debris of other insects (Priest 1934). Stomach of newly fledged juvenile, eastern Nigeria, contained 2-cm crab, while another bird (no details) had eaten insects (Serle 1957). EKD

Social pattern and behaviour. Little known. Account based largely on extralimital material.

1. Not gregarious; no records of more than 2 birds together. Continuation of Advertising-call (see 1 in Voice) throughout year (Bannerman 1933; Chapin 1939) suggests birds possibly territorial at all times. For flocking with other species, see Flock Behaviour (below). BONDS. No information. Studies on other *Centropus* indicate monogamy cannot be assumed (see Vernon 1971b). BREEDING DISPERSION. In western Sénégal, each pair occupied breeding territory of *c.* 50 m square; trespassing rare (Moynihan 1978), but see Flock Behaviour (below). No further information. ROOSTING. Nothing known about nocturnal roosting, or loafing. Activity notably crepuscular, and, in Kenya and Uganda, birds said to emerge from hiding in late afternoon (Jackson 1938a). Vocal mostly in the evening, also often at night (Jackson 1938a; Mackworth-Praed and Grant 1952).

2. Shy and vigilant (Granvik 1923), though reluctance to fly has given rise to reports of bird being tame and approachable (e.g. Hopkinson 1913, Monard 1940). If disturbed, more readily runs than flies (Young 1946; Gore 1964); on approach of intruder, often disappears in grass, only flying when *c.* 100 m further off (Granvik 1923). Usually keeps to thick cover, where it skulks about, typically hopping from branch to branch of a bush until it reaches the top, whence it flaps heavily off and pitches forwards into the next (Mackworth-Praed and Grant 1952; G D Field). Readily mobbed by other breeding birds (G D Field). FLOCK BEHAVIOUR. Does not, apparently, flock with conspecifics but does so with other birds. In western Sénégal, pairs daily joined and followed within *c.* 1–10 m of any Red-billed Hornbills *Tockus erythrorhynchus* that entered their territories. Other species may also occur in flocks (e.g. Long-tailed Glossy Starling *Lamprotornis caudatus*, Green Wood Hoopoe *Phoeniculus purpureus*), and *C. senegalensis* has been seen to react to alarm call from *L. caudatus* (Moynihan 1978). ANTAGONISTIC BEHAVIOUR. Advertising-call probably functions partly as site-advertisement. Calling bird perches, sometimes on high exposed branch, with head bowed forward, bill vertically down, and tail also down, parallel with legs (Hopkinson 1910; Young 1946). While calling, whole body shakes violently (G D Field), and throat alternately puffs out and collapses (Hopkinson 1910). Performance of duet (see 1 in Voice) may serve as joint site-advertisement display. HETEROSEXUAL BEHAVIOUR. Advertising-call of ♂ presumably serves to attract mate as well as rebuffing other ♂♂. No further information. For pre-copulatory courtship-feeding in other *Centropus*, see Vernon (1971b) and Frith (1975). RELATIONS WITHIN FAMILY GROUP. Young leave nest before they can fly, and probably remain in thick scrub in vicinity of nest, unless disturbed (Steyn 1972). ANTI-PREDATOR RESPONSES OF YOUNG. If approached by man, young out of nest run strongly: 21-day-old chick caught only after long chase through undergrowth, and struggled vigorously when handled. From 5–6 days, handled young excrete foul-smelling brown liquid, quite different from white faecal sacs removed by parents (Steyn 1972). See also Voice. PARENTAL ANTI-PREDATOR STRATEGIES. No information. EKD

Voice. Heard throughout the year in West Africa (Bannerman 1933; Chapin 1939). However, birds much less vocal in dry season, Sierra Leone (G D Field). Account based mostly on West African material.

CALLS OF ADULTS. (1) Advertising-call. (a) Solo calling. Series of bubbling, flute-like sounds, like water being poured out of bottle (Mackworth-Praed and Grant 1952); 'hoo-hoo hu-hu-hu-hu-hu-hu-hu-hu' (Bannerman 1933). Full phrase consists of 15–20 notes, initially dropping in pitch, then rising a little before the end (Chapin 1939). In recording (Fig I), pitch rises on 2nd 'hoo' note, then drops through subsequent 11 'hu' notes (J Hall-Craggs). Almost constant accelerando through the phrase. Speed of delivery varies with individual and sex; more so than in any other *Centropus* (Chappuis 1974). Bird may call, with a few pauses, for over 1 hr (Meinertzhagen 1930). (b) Duetting. 1st phrase said to be given usually by ♂, followed instantly by call from ♀ (Rand 1951; see also Chappuis 1974), sequence being repeated several times

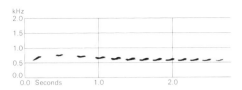

I L Grimes Ghana April 1974

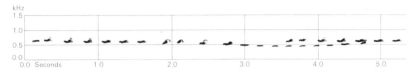

II L Grimes Ghana June 1975

(Rand 1951). In recording (Fig II), 2 birds overlap and sequence more complex: 1st bird utters 'woo WOO woo-woo-woo-woo-woo', the latter 'woo' notes (variable in number) not dropping markedly in pitch; immediately after 1st bird's last note (but between its last 2 notes in 2 other recordings of same pair), 2nd bird begins descending, then ascending, series of 13 shorter, accelerating 'wu-' notes, and 1st bird completes sequence by repeating its phrase, starting at 2nd bird's 8th note, and thus overlapping with the end of its phrase. Each of 4 sequences of duet lasts c. 5 s, with gaps of 15 s, 3·5 s, and 15 s (E K Dunn); 1st bird's contributions and timing more consistent than 2nd's (J Hall-Craggs). (2) Alarm-call. A sharp 'guk guk', given when young removed from nest by human intruder (Steyn 1972); rapid 'wūk wūk' (Young 1946) probably the same, perhaps also clacking 'tock-tock toc-toc-toc-toc-toc-toc-toc-toc' (Bannerman 1933). Cackle or mocking laugh, thought to be given in alarm (Hopkinson 1910), possibly the same. (3) Other calls. (a) Slow, vibrant 'whoo whoo whoo. . .' given at rate of 3 per 5 s, less commonly than call 1 (G D Field). This probably the call described by Chapin (1939) which, compared with call 1, comprised 'toots' of same pitch but fewer and more slowly repeated. Function not known. (b) Harsh call, very different from call 1, sometimes given (Chapin 1939).

CALLS OF YOUNG. Young 3–4 days old gave nasal, wheezing 'tjer tjer tjer', like someone with catarrh (Steyn 1972); probably equivalent to the hissing, snake-like sound interpreted as defensive reaction in other *Centropus* (Someren 1956; Vernon 1971b; Frith 1975). EKD

Breeding. SEASON. Egypt: eggs laid April (Etchécopar and Hüe 1967). SITE. In thick scrub or tall undergrowth. One nest 3·6 m above ground (Steyn 1972), another 0·3 m (Granvik 1923). Nest: bulky, more or less spherical structure with side entrance; of grass, roots, and leaves, lined fresh leaves (Granvik 1923; G D Field). Building; no information on role of sexes. EGGS. See Plate 95. Short elliptical, roughened and not glossy; white. 34×26 mm ($29–37 \times 23–38$), $n = 36$; calculated weight 12 g (Schönwetter 1967). Clutch: 3–5. INCUBATION. 18–19 days (Steyn 1972). ♂ takes part in incubation; possibly by ♂ only (Granvik 1923; Vernon 1971b). Begins with 1st egg; hatching asynchronous (Steyn 1972). YOUNG. Altricial and nidicolous. Role of parents not known. FLEDGING TO MATURITY. Fledging period 18–20 days, but young leave nest a day or two before (Steyn 1972). No further information.

Plumages (*C. s. aegyptius*). ADULT. Forehead, crown, and hindneck down to gape, just below eye, and ear-coverts black with stiff shiny black shafts and dark olive-brown lateral bases to feathers, latter shining through, giving cap brownish-black appearance. Brownish-black hindneck rather gradually merging into rufous olive-brown of mantle and scapulars; lower scapulars and tertials olive-brown, bright cinnamon-rufous of central bases largely hidden. Back rufous olive-brown, rump brownish-olive, upper tail-coverts black with olive tinge and slight green gloss. Chin, lower cheeks, sides of neck, foreneck, and chest cream-buff to cream-white, stiff shafts shiny pale cream; breast, flanks, belly, vent, and thighs pale cream-buff, under tail-coverts slightly darker cream-buff. Tail bronzy-black with slight olive-brown tinge and faint green gloss. Flight-feathers and all upper wing-coverts bright cinnamon-rufous, contrasting with olive-brown upperparts and tertials; tips of primaries dull black or dark brownish-grey for 1–1·5 cm, tips of innermost primaries and secondaries for 0·5–1 cm; dark tips not sharply demarcated from rufous. Most under wing-coverts buff or cream-buff; greater under wing-coverts and axillaries rufous, as flight-feathers. For birds with faint buff bars on upper tail-coverts and occasionally on tail-base and faint grey bars on sides of neck, flanks, and under tail-coverts, see 1st adult. NESTLING. Upperparts and flanks covered with sparse long white hairs; otherwise bare (Harrison 1975). JUVENILE. Rather like adult but, in general, feathers narrower and looser (especially on back, rump, tail-coverts, and underparts) with less stiff and shiny shafts; body, tail, and wing barred to variable extent. Crown and hindneck browner than adult; mantle, scapulars, and tertials rufous olive-brown as in adult but partly or completely marked with rather indistinct dull black bars, about equal in width to intervening olive-brown; bars most pronounced on tertials, frequently absent from mantle and scapulars. Back and rump barred olive-brown and dull grey, upper tail-coverts dark olive-brown with narrow buff bars. Underparts cream-buff, cheeks, sides of neck, chest, and upper flanks with variable amount of rufous suffusion and dark borders on feather-tips; lower flanks and under tail-coverts barred buff and grey. Tail brownish-black with slight green gloss, some or all feathers with narrow and sometimes indistinct buff bars, latter often restricted to base and middle portion of feather or to extreme tip only, sometimes hard to see or absent; some tail-feathers occasionally with pale buff edges or tip. Flight-feathers as in adult, but dark tip partly broken into bars, especially on inner primaries and secondaries; barring sometimes extends over more than distal half of feather, but occasionally restricted to some dusky freckling proximal to black tip. Upper wing-coverts rufous as in adult, but barred with black, especially on distal halves of greater coverts; dark bars on lesser and median coverts often narrower and more restricted, sometimes absent. FIRST ADULT. Like adult, but some barring similar to that of juvenile plumage present; at times, some scattered juvenile flight-feathers or occasionally a few tail-feathers retained. Sometimes traces of dusky bars on scapulars; upper tail-coverts dull black with greenish gloss, usually with faint buff bars, sides of neck, flanks, and under tail-coverts barred grey; exceptionally, tail with faint traces of buff bars towards base. Of 13 birds considered to be 1st adult by presence of new plumage on body combined with a few retained juvenile flight-feathers, most showed distinct barring on sides of neck, flanks, and tail-coverts; in some, however, barring faint, and in 4 all bars absent and plumage similar to adult; on other hand, of 24 birds considered to be adult (wing fully adult), 15 had traces of bars on sides of neck, flanks, or tail-coverts, and these either 1st adult with post-juvenile moult completed, or older with some barring still present.

Bare parts. ADULT AND FIRST ADULT. Iris crimson-red or orange-red. Bill black. Leg and foot bluish-grey, plumbeous-blue, or blackish-grey. NESTLING. Iris grey. Bill pink. Mouth pink, tongue red with black line across tip and fine, backward-pointing whitish spines. Bare skin of head and upperparts black, of underparts dark pink. Leg and foot plumbeous-black, rear of tarsus and soles pink. JUVENILE. At fledging, iris yellow-brown, olive-brown, yellow-green, or greyish-green; bill brown-black with pale horn-coloured lower mandible and cutting edges; leg and foot plumbeous-blue. Adult colours obtained during post-juvenile moult. (Chapin 1939; Mackworth-Praed and Grant 1962; Harrison 1975; BMNH, RMNH, ZMA.)

Moults. ADULT POST-BREEDING. Complete. Primary moult complex. Sequence approximately 10–8–6–4–2–9–7–5–3–1 or 9–7–5–3–1–10–8–6–4–2 (Stresemann and Stresemann 1966), up to 4 feathers growing at same time but never 2 neighbouring ones. Limited number of skins available (BMNH, RMNH, ZFMK) suggest primaries moulted in 2 groups: outer in sequence 10–8–6–9–7–5 or 9–7–5–10–8–6, inner in sequence 4–1–3–2, inner group starting at about loss of 4th–5th feather of outer group. Primary moult usually suspended during breeding, occasionally not; resumption after breeding often results in pattern difficult to explain. Secondary moult complex; several feathers grow at same time, but no neighbouring ones; sequence not established. Sequence of tail-feathers irregular, often asymmetrical; never more than 1 pair of central 3 pairs growing at same time; t2(–t3) usually last. Moult in Egypt mainly June–November, occasionally up to January; all flight-feathers, tail, and body at same time. POST-JUVENILE. Complete. Sequence of flight-feathers and tail as in adult post-breeding. Scattered feathers of head and mantle and some scapulars and upper wing-coverts first, soon followed by 1st primary, probably starting shortly after fledging; head and body largely in 1st adult when tail moult starts, and usually then with 2–3 primaries new; tail often fully 1st adult when still 2–3 juvenile primaries present. Feathers replaced last are p2, s2 (occasionally with s1, s3, or s5), and t2 (occasionally with t3 or t5), and these frequently retained if moult suspended. Primary moult in Egypt starts mainly November–December, completing February–May, but data limited; only 8 in active moult examined.

Measurements. *C. s. aegyptius*. ADULT. Egypt, mainly Faiyum, November–June; skins (BMNH, RMNH, ZFMK, ZMA; Giza Zoological Museum *per* S Goodman and P L Meininger). Bill (F) is tip to forehead, bill (N) tip to nostril; toe is outer front toe.

WING	♂ 175	(3·73; 16)	170–183	♀ 183	(2·79; 13)	177–187
TAIL	209	(11·8; 13)	188–222	227	(11·0; 13)	212–246
BILL (F)	32·4	(1·41; 9)	30·4–34·1	35·0	(1·01; 12)	33·6–36·6
BILL (N)	19·1	(0·78; 9)	18·3–20·1	19·8	(0·69; 12)	19·0–20·7
TARSUS	45·2	(1·65; 8)	42·7–46·6	46·0	(1·56; 12)	44·2–48·2
TOE	39·0	(2·00; 8)	36·0–41·2	40·2	(1·49; 12)	38·6–42·8

Sex differences significant for wing, tail, and bill.

JUVENILE. Similar to adult, but bill not full-grown until several months after fledging.

Weights. No information on *aegyptius*. Adult ♂ of slightly smaller race *flecki*, Botswana: 178 (Jackson 1969).

Structure. Wing short and broad, tip rounded. 10 primaries: p6–p7 longest or either feather 0–2 shorter, p8 7–12 shorter, p9 22–29, p10 52–62, p5 0–5, p4 2–5, p3 5–12, p2 12–20, p1 19–26; secondaries and tertials gradually shorter towards body. Tail long, graduated; 10 feathers, broad with rounded tip; t5 66–78 shorter than t1; width of t1 near tip 43–53 mm in adult, 39–48 in juvenile. Bill heavy, deep at base, strongly compressed; culmen strongly decurved; cutting edges slightly decurved, but almost straight on basal half; slight groove along basal half of sides of rounded culmen; nostrils narrow, oblique, and partly covered by soft flap above. Tarsus strong, heavily scutellated, unfeathered; toes rather long and slender. Claws rather long, decurved, that of inner hind toe long, slightly decurved, spur-like, 18–22 mm long in adult, 15–16 in fledgling. Outer hind toe *c.* 85% of outer front toe, inner front toe *c.* 68%, inner hind *c.* 91% (40% for toe proper, 51% for spur).

Geographical variation. Rather slight. Nominate *senegalensis* from northern and central Afrotropics differs from *aegyptius* in deeper black crown and hindneck with bluish or greenish gloss, and in rufous-chestnut mantle and scapulars, almost similar in tinge to upper wing-coverts and flight-feathers, only all or part of lower scapulars and tertials olive-brown; rufous upperparts contrast markedly with deep black hindneck, thus no gradual change from brownish-black crown and nape to olive-brown mantle as in *aegyptius*. Juvenile nominate *senegalensis* brighter rufous on upperparts also, dark barring more contrasting than in *aegyptius*. Nominate *senegalensis* slightly or distinctly smaller than *aegyptius*, depending on locality: in zone from Sénégal to Sudan, average wing of ♂ 162, of ♀ 171, range (both sexes combined) 158–180 (10); from Liberia to Ghana, average wing of ♂ 154, ♀ 163, range 150–167 (10). Southernmost race *flecki* even paler rufous above than nominate *senegalensis*; olive-brown (if any) restricted to tips of tertials; size slightly larger than that of nominate *senegalensis* from Sénégal to Sudan. A morph with black head and throat and rich brownish-chestnut body occurs (named '*epomidis*'); known from West Africa only (Mackworth-Praed and Grant 1970). CSR

Order STRIGIFORMES

Small to large predators; mainly nocturnal or crepuscular. 2 families, Tytonidae and Strigidae, both represented in west Palearctic. A clearly defined order, probably closest to Caprimulgiformes, although formerly sometimes linked with Falconiformes.

Features common to both families include the following. 14 cervical vertebrae. Eyes directed forwards. Bills hooked as in Falconiformes but directed more downwards. Base of bill with soft cere as in Falconiformes, but covered by bristles projecting laterally from base of bill (actively directed sidewards during eating). Outer toe reversible, but directed laterally at rest. In some species, erectile ear-like tufts of feathers on forehead. Unlike Falconiformes, crop absent and caeca large. Plumages soft.

Family TYTONIDAE barn owls and allies

Medium-sized predators, mainly nocturnal or crepuscular feeders. Differ from Strigidae mainly in having heart-shaped faces, tails usually ending in shallow V, inner and middle toes roughly equal in length, claw of middle toe with comb-like serrated edge, and wishbone and breastbone fused together. 2 subfamilies: (1) Tytoninae (barn owls); (2) Phodilinae (bay owls, 2 species, Oriental and Afrotropical regions). Phodilinae not considered further here. Tytoninae comprises single genus, *Tyto*, of *c*. 8–10 species distributed world-wide, with single representative (breeding) in west Palearctic.

Heads large and round, without ear-tufts. Heart-shaped facial discs. Eyes dark and relatively small. Ear-openings long, but small orifice covered by large flap. Wings large with rounded tip; 10 primaries; wing diastataxic. Tails rather short, 12 feathers. Feathers without aftershaft. Legs long and slender, feathered or (in grassland species) almost bare. Oil-gland virtually naked. Plumages barred or spotted with either golden- or chocolate-brown, but highly variable in Barn Owl *Tyto alba*. Sexes more or less similar. Young nidicolous. Juvenile plumage even further reduced than in Strigidae, downy and virtually unpatterned, forming a 2nd down (mesoptile), developed *c*. 1–2 weeks after 1st down (neoptile or protoptile). Post-breeding moult partial, complex; see *T. alba*. Adult plumage acquired at *c*. 3 months.

Tyto alba Barn Owl

PLATES 43 and 53
[facing pages 494 and 543]

Du. Kerkuil	Fr. Chouette effraie	Ge. Schleiereule	
Ru. Сипуха	Sp. Lechuza común	Sw. Tornuggla	N. Am. Common Barn-Owl

Strix alba Scopoli, 1769

Polytypic. Nominate *alba* (Scopoli, 1769), Britain, France, and southern Europe, south to Balearic Islands, Sicily, and Malta, east to western Yugoslavia and western Greece, also Tenerife and Gran Canaria (Canary Islands); *guttata* (Brehm, 1831), central Europe east to south-west European USSR, grading into nominate *alba* in northern and eastern France, Belgium, Netherlands, German Rhine valley, central Switzerland, and central Yugoslavia; *ernesti* (Kleinschmidt, 1901), Sardinia and Corsica; *erlangeri* Sclater, 1921, North Africa south to northern Mauritania, Ahaggar (Algeria), and Nile valley; also Crete, Cyprus, Middle East, and Arabia; *affinis* (Blyth, 1826), Afrotropical mainland, grading into *erlangeri* in southern Egypt; *schmitzi* Hartert, 1900, Madeira and Porto Santo; *gracilirostris* Hartert, 1905, Fuerteventura, Lanzarote, and Alegranza (eastern Canary Islands); *detorta* Hartert, 1913, Cape Verde Islands. Extralimital: *c*. 27 further races worldwide.

Field characters. 33–35 cm: wing-span 85–93 cm. 10% smaller but with slightly longer and narrower wings than Tawny Owl *Strix aluco*. Medium-sized, noticeably long-legged owl, with strikingly heart-shaped face, beautifully mottled light grey and buff upperparts, and white to buff underparts. Birds with white underparts ghostly in appearance; those with buff underparts also spotted below but never streaked or barred (as most other owls). Not strictly nocturnal. Flight noticeably slow when hunting, but exceptionally buoyant with characteristic wavering; legs often dangle noticeably. Commonest call a drawn-out shriek. Sexes slightly dissimilar. No seasonal variation. After leaving nest, young not separable in the field from adult. 8–9 races in west Palearctic; typical nominate *alba* and *guttata* easily separated from each other, but intergrades require care.

(1) West and south European race, nominate *alba*. ADULT MALE. Facial disc complete, consisting of oval sides

forming together remarkable heart-shape (head-on); white, with narrow buff-brown to dark brown ruff further emphasizing disc and boldly punctuated by black eyes and shadowed eye-pits, latter rusty on inner edge (and sometimes all round). Underparts silky, silver-white, appearing unmarked at most ranges but often tinged pale rufous-buff on sides of chest and sometimes finely spotted dark brown on chest and flanks. Legs densely (and tops of toes thinly) feathered white and usually free of body feathers, with tarsus and lower part of tibia clearly visible. Underwing almost entirely white, with grey-brown bars on undersurface of flight-feathers indistinct except on outer primaries. Upperparts wholly golden-buff, finely streaked, mottled, and vermiculated brown and grey (latter predominant on crown and back); fewer marks on forecrown and nape, so these areas noticeably buff. Wing-coverts coloured as upperparts but markings form patches in irregular transverse pattern. Flight-feathers also similarly coloured but markings form soft bars across primaries, secondaries, and tertials. Tail pale golden-buff, with markings faint and hardly forming bars. ADULT FEMALE. Underparts usually more numerously and heavily spotted; upperparts with greater incidence of heavier and greyer streaking and mottling, forming more obvious bars on flight-feathers and tail and denser patches elsewhere. However, individual variation occurs in both sexes; other than the most heavily marked ♀♀, only birds of known pair can be safely sexed. (2) Central European race, *guttata*. ADULT. Facial disc less white, with rust obvious around eye and black-tipped ruff even more striking than on nominate *alba*. Rest of underparts buff (♂) to rufous-buff (♀), usually spotted all over. Under wing-coverts buff, with primary coverts distinctly rimmed grey-buff. All upperparts, flight-feathers, and tail heavily marked with grey; buff tone of upperparts sometimes suppressed by grey, but grey usually merely predominant, appearing as bolder patches and more complete bars. Individual variation occurs; also, in eastern France, intergradation with nominate *alba*. (3) Other races. White-breasted birds occur in Corsica, Sardinia, and Middle East; last noticeably golden above. Buff-breasted birds occur on Madeira, eastern Canary Islands, and Cape Verde Islands; last darkest of all. JUVENILE. Plumage downy, moulted before fledging; thereafter, resembles adult.

No owl unmistakable in brief glimpse (and true appearance often changes when starkly illuminated). White-breasted *T. alba* usually distinctive, however, since only the 60% larger ♂ Snowy Owl *Nyctea scandiaca* is as white below, with broader wings and less protruding head. Buff-breasted birds less distinctive, often looking as uniformly coloured as other owls but usually showing paler and patchier plumage pattern. At close range, shape of facial disc, lack of obvious streaks or bars on underparts, and leg length diagnostic. Hunting flight usually low, slow, and wavering, but noticeably buoyant; under close observation, bird can be seen to be employing expert search tactics when hunting, with head protruding and turning laterally, wing-beats varied between slightest flicks and deeper strokes (allowing side-slips and hovering), and body and extended legs bucking slightly, as muscular effort is concentrated on holding hunting line along hedge, across stubble, etc. Flight over longer distance sometimes well above ground level, with more regular wing-beats. Carriage noticeably upright, both on perch and on ground; leg length obvious, tail length not so. Gait a rolling walk and occasional lope. Maintains regular beat round hunting territory, frequently appearing on it in late afternoon and evening and even all day in winter, particularly during periods of snow cover.

Territorial call a long eerie shriek; in alarm, a short, screaming yell.

Habitat. Breeding in west Palearctic in warm or mild middle and lower latitudes, in oceanic or moderately continental climates, with winter snow averaging less than 40 days duration and less than 7 cm depth (Glutz and Bauer 1980). Mainly in open but not treeless lowlands, especially farmland with spinneys, hedges, ditches, ponds, and banks, and some rough grass or herbage, roadside verges, and similar rough terrain where mice and other prey can be hunted in low flight. Often on fringe of villages or other small groups of buildings, offering shady or dark undisturbed daytime roosting posts as well as nest-sites. For latter prefers old churches, towers, ruins, barns, warehouses, granaries, lofts, dovecotes, windmills, chimneys, and similar artefacts offering ample room and convenient entrances and exits, free from disturbance and adjoining good hunting beats. In parts of range (e.g. Britain) also commonly nests in isolated hollow trees or in natural cavity in face of cliff or quarry or in nest-box (Sharrock 1976). Almost everywhere, use of suitable habitat limited by exceptional requirements for more floor-space and easier access than often associated with hole-sites for nests, accompanied by subdued light and dependable freedom from disturbance. Modern building design and maintenance often disqualify otherwise suitable sites, while disturbance limits acceptable daytime roosting posts which include dense foliage in trees. Fluctuations in prey numbers lead to periodic increases in requirements and local extensions of range, into marginal areas and to higher altitudes, up to or even above 800 m compared with normal limits of *c*. 600 m (Glutz and Bauer 1980). Readily nests on sea-cliffs, and even on small exposed islands. Intolerance of disturbance at nest or roost does not inhibit living closer to man than other owls and being directly or indirectly somewhat dependent on human land uses, this adaptation being eased by nocturnal habits.

Habits similar in extralimital parts of range, but in Africa is recorded as sharing large nest of Hamerkop *Scopus umbretta*, and as nesting more freely in rock crevices or caves and hollow trees far from human habitations (Archer and Godman 1961; Williams 1963; Serle *et al.*

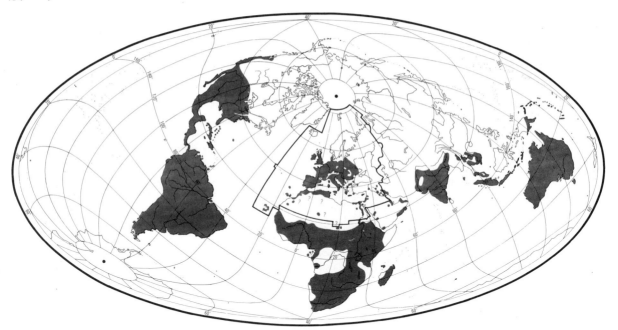

1977). In southern Africa, also in mine-shafts (Prozesky 1970). In Texas (USA), inhabits canyons and will nest in hole in bank (Peterson 1960).

Hunting methods dictate habitual flying in lowest airspace.

Distribution. Few marked changes. Colonized Latvia and Lithuania in early 1930s.

USSR. Colonized Lithuania and Latvia in early 1930s, although said to have nested also in Latvia in early 19th century (Ivanauskas 1964; Transehe 1965). TURKEY. Probable breeding area mapped, although firm proofs lacking (MB, RFP). KUWAIT. Seldom seen; no evidence of breeding (RPH). SYRIA. May breed, but no proof (Kumerloeve 1968b). LIBYA. No proof of breeding (Bundy 1976). ALGERIA. Probably breeds in Sahara (EDHJ).

Accidental. Norway, Finland, Kuwait, Azores.

Population. Numbers breeding fluctuate markedly with those of small rodent populations; also affected by severe mortality in hard winters in some areas (Glutz and Bauer 1980). General decrease, often marked, over much of west Palearctic range, variously ascribed to human persecution, habitat loss (including nest-sites), and pesticides.

BRITAIN AND IRELAND. Decreased, though few detailed figures, and may be obscured by fluctuations due to effects of hard winters and changes in small rodent populations. Estimated c. 12000 pairs in England and Wales (Blaker 1934); Britain and Ireland 1000–10000 pairs (Parslow 1967), 4500–9000 pairs (Sharrock 1976). Decline apparently began in Britain in 19th century, continued at lower rate until late 1940s, more marked after 1955; in Ireland first noticeable about 1950. Suggested causes include habitat loss, persecution, disturbance, and pesticides (Prestt 1965; Parslow 1967; Sharrock 1976; Osborne 1982). FRANCE. Decreasing, with fluctuations. Estimated a little over 10000 pairs (Yeatman 1976); this considered much too low, probably c. 60000 pairs in view of some local estimates, e.g. Baudvin 1975 (RC). SPAIN. Decreased, due to rodenticides and habitat loss. Not less than 50000 pairs 1965, not less than 30000 pairs 1975 (AN). BELGIUM. Decreased. About 1500 pairs (Lippens and Wille 1972), c. 1000 pairs 1977–82 (PD). LUXEMBOURG. About 1600 pairs (Lippens and Wille 1972). NETHERLANDS. Decreased, though fluctuating. Up to 1962, varied from 1800 pairs in poor rodent years to 3500 pairs in good years, and over 1969–74 from 230 pairs to 490 pairs (Braaksma and De Bruijn 1976; see also Honer 1963); in 1975–7, 300–600 pairs (Teixeira 1979). WEST GERMANY. Declining in several regions (Bauer and Thielcke 1982). In Bayern, 400–1000 pairs; long-term decline but some local increases (Bezzel et al. 1980). DENMARK. Decreasing. 75–100 pairs (Dybbro 1976). SWEDEN. Decreased and almost extinct. About 50 sites in 1870s, halved by early 20th century; not exceeding 18 pairs 1951–61, 5 pairs 1974, and 1 pair 1980 (Holmgren 1983; LR). POLAND. Scarce or very scarce (Tomiałojć 1976a). CZECHOSLOVAKIA. Marked decline since 1960 (KH). ALBANIA. Rare (EN). GREECE. Rare (HJB, WB, GM). MALTA. Marked decrease due to human persecution; formerly at least 10 pairs, now reduced to 1 pair (Sultana and Gauci 1982). CANARY ISLANDS. Scarce on Lanzarote and Fuerteventura; no apparent change in numbers in 20th century on Tenerife (KWE).

Mortality. East Germany: mortality in 1st year of life 72·7% (considerable annual variations), annual adult mortality 61% (Schönfeld 1974). Switzerland: mortality

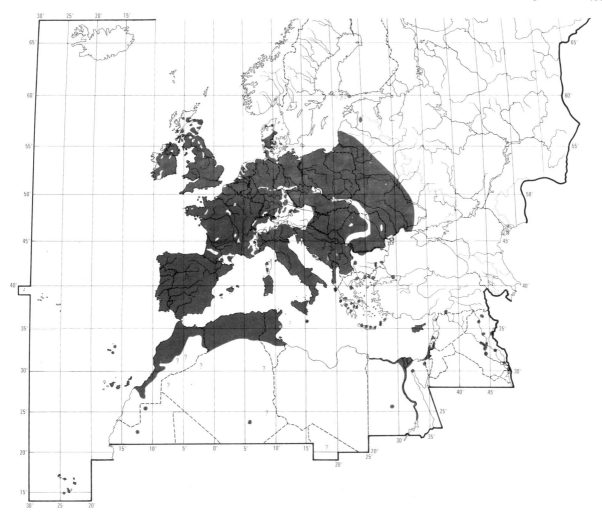

in 1st year of life 67·7%, in 2nd year 50·7%, and from 3rd year 42·9% (Glutz and Bauer 1980). For seasonal mortality distribution, see Glue (1973). Oldest ringed bird 21 years 4 months (Schifferli and Imboden 1972).

Movements. West Palearctic populations basically resident, though young birds especially make dispersals which (in Europe) are more extensive in some years than others. No clear overall directional trend to movements within Europe, except that longer movements (above 300 km) tend to be south of east–west axis. Only major topographical features (e.g. large water bodies) inhibit dispersal (Frylestam 1972; Mikkola 1983), though presence of breeding populations on many oceanic and other islands shows that water barrier often crossed successfully in the past. Nearctic population (*pratincola*) partially migratory in northern USA, with some adults as well as young birds moving up to 2000 km south in autumn, juveniles furthest (Stewart 1952; Mueller and Berger 1959); not considered further here.

In Europe, present in winter even at northern limits of breeding range—Denmark, southern Sweden (but nearly extinct there), Latvia—showing increased mortality (rather than increased emigration) in occasional severe winters, when depth of snow cover appears critical (e.g. Bühler 1964, Güttinger 1965). Instead, chief components of movement are (1) annual post-fledging dispersal of young, and (2) irregular prey-related dispersal of eruptive nature.

POST-FLEDGING DISPERSAL. Juveniles remain dependent on parents for c. 1–2 months after fledging and, after achieving independence, disperse in period September–November; based on few data, some movement probably continues throughout 1st winter. Young hatched in early summer have longer dispersal period (and on average move further) than late-hatched birds (Glutz von Blotzheim and Schwarzenbach 1979; Bunn *et al.* 1982). Dispersal directions random, where topographical features (seas, mountain ranges) allow, and nest-siblings often move to different directions and distances; see Sauter (1956) for

West Germany and Bunn *et al.* (1982) for Britain. In years of normal food supply, most juveniles settle within 20–30 km of birthplace, though (on the Continent) every year a minority move 100 km or more (e.g. Frylestam 1972, Schönfeld 1974, Glutz von Blotzheim and Schwarzenbach 1979). Overall, island populations show less long-distance movement than continental ones (see Table A for comparison of Britain and Netherlands).

Table A Dispersal in Barn Owls *Tyto alba* of Britain and Netherlands during 1st year of life (after Bunn *et al.* 1982).

Distance (km)	Britain ($n=285$)	Netherlands ($n=355$)
0–10	61·5%	25·6%
11–50	30·9%	42·3%
51–100	5·3%	15·5%
101–200	1·7%	9·6%
201–300	0·3%	2·5%
over 300	0·3%	4·5%

ERUPTIVE DISPERSAL. In British population, no annual differences in extent of dispersal have been detected (Bunn *et al.* 1982), though eruptive fluctuations known to occur in continental Europe. These happen when numbers high in autumn, following good breeding success during season of vole (Microtinae) abundance, but prey numbers then crash; the combination is needed (Sauter 1956; Honer 1963). Movement of birds then greater (both in terms of numbers and of distances moved) and their mortality increases substantially due to density-dependent effects of reduced prey availability, with cold weather (if any) only serving to heighten effect (Honer 1963). In these circumstances, *guttata* straggles to eastern Britain, whence single recoveries of birds from Netherlands and Belgium. In East Germany, 60–65% of young settled within 15 km of birthplace in years when *Microtus* densities rising, but *c.* 60% (plus some adults) moved longer distances in other years (Schönfeld 1974). In West Germany and Switzerland, *c.* 40% settled within 25 km of birthplace in vole mid-cycle years, while *c.* 15% found over 200 km distant (including *c.* 3% above 500 km) (Sauter 1956; Glutz and Bauer 1980). In Switzerland, particularly high rate of emigration in autumns 1947, 1959, 1961, and 1972 (with exceptionally high winter mortality then), while only short-distance displacements in 1948, 1955, 1963, and 1970; no related data on vole abundance, however (Glutz von Blotzheim and Schwarzenbach 1979). For local breeding densities varying with vole abundance, see (e.g.) Kaus *et al.* (1971) and Kaus (1977).

Eruptive dispersal involves chiefly 1st-year birds, though some adults can also be involved (Sauter 1956). Even in such years, most ringing recoveries within 300 km. Longest include exceptional juvenile/1st winter birds from East and West Germany, Netherlands, Switzerland, and France found in Spain (up to 1650 km south-west); also Swiss juveniles found November in Poland (1060 km north-east) and Rumania (1080 km ESE), Bavarian-bred bird in Ukraine in December of 2nd year (1180 km east), and full-grown bird from Italy (January) also found Ukraine in December (1040 km north-east).

No dispersal data for resident populations of North Africa and Middle East.

Food. Over most of west Palearctic range, small mammals, more or less as available—thus largely mice (Murinae) and small voles (Microtinae) and more shrews (Soricidae) than other medium-sized Strigiformes; also some small birds, and a few frogs and toads. Essentially nocturnal, but at all times of year commonly hunts from 1–2 hrs before sunset and also after dawn; apparently much individual variation; for study and discussion, see Bunn *et al.* (1982); also Mikkola (1983). Can locate prey by sound (if frequency over 5 kHz) in total darkness with accuracy of less than 1° in both horizontal and vertical planes; for experimental studies, see Payne (1971), Knudsen *et al.* (1979), and review by Knudsen (1981). Full binocular vision apparently also important in the wild as bird blind in one eye had great difficulty in hunting (Bunn *et al.* 1982). For sight in poor light, see Dice (1945) and Marti (1974); also Bunn (1976). Hunts mostly with searching flight, frequently at 1·5–4·5 m, but rises and falls continually—pausing, moving sideways, and hovering. In dropping to ground, may dive forward, drop vertically with wings raised, throw itself sideways, or perform apparent somersault, changing direction by 90°; occasionally gives a wing-beat to add impetus (Bunn *et al.* 1982). At last moment, withdraws head, swings feet forward, and closes eyes; spreads claws only immediately before they contact prey (but earlier in total darkness) (Payne 1971). After striking, spreads wings on to ground to steady itself. If it misses, may listen and pounce again without flying up; chase along ground may follow (bird running). In rough, open areas, may fly back and forth—slowly (hunting) into wind, then quickly downwind. Less often, hunts from perch, flying between several perches and pausing *c.* 1 min on each. Some individuals will also hunt by repeated hovering, advancing a few metres between pauses. (Bunn *et al.* 1982; see also for activity patterns.) Birds apparently taken largely from roosts, though swallow *Hirundo* once seen to be taken in aerial-pursuit in Zambia (Mitchell 1964). Will flap wings against bushes to disturb roosting birds (Bunn *et al.* 1982), and in Iraq recorded regularly taking birds from roost under floodlights by dropping down into bushes from perch above; when birds quiet again, process repeated—for up to 1 hr or more (Sage 1962); similarly in town-centre in Moldavia, USSR (Averin and Ganya 1966). Record of bird dropping vertically into water to rise with fish in feet (Stanley 1838), and also of taking good-sized roach *Rutilus rutilus* from river (Bunn *et al.* 1982). Occasionally plunders birds' nests. Recorded hunting hawk-moths *Herse* whirring around flowers (Glutz and Bauer 1980). Once seen apparently feeding on carrion (Dunsire and Dunsire 1978). Not clear if wild-caught prey killed only with feet or if bill also used; bill often used in captivity, though manner variable (see Bunn *et al.* 1982). According to Ticehurst (1935), nearly always nips through back of skull (see also Long-eared Owl *Asio otus*). Small mammals usually swallowed whole (Bunn *et al.* 1982)—sometimes, at least, up to size of young brown rat *Rattus norvegicus* (Glue 1970). Birds apparently always decapitated (head sometimes discarded); body may be eaten piecemeal (Bunn *et al.* 1982) or, according to Thiollay (1963b), swallowed whole with head and feet; larger feathers roughly plucked. Prey usually carried to nest in one foot; rarely in bill (Bunn *et al.* 1982). Recorded carrying adult Moorhen *Gallinula chloropus* in flight, apparently without difficulty (Scott 1979). Cannibalism in nest occurs occasionally, though victim has normally died of other causes. However, young 50 days old recorded being killed and eaten by siblings (Bunn *et al.* 1982). For food-caching, see Social Pattern and Behaviour.

Fresh pellets characteristically glossy black; even when dry, dark grey to blackish and smooth; harder and more compact than in *A. otus* and Tawny Owl *Strix aluco*. Appearance of fur not much changed; bones mostly unbroken (unless large), skulls (except cranium) characteristically intact with lower jaw in place. Skulls of frogs and toads usually digested. Occasionally (but only at night) ejects pellets with incompletely digested prey. Pellets may contain leaves, small stones, etc.; one pellet $\frac{3}{4}$ composed of leaves. (Thiollay 1963b; Bunn *et al.* 1982.) According to Glutz and Bauer (1980), inanimate material sometimes taken deliberately. Pellet measurements: 40–83 × 21–28 mm, $n = 165$ (Ticehurst 1935); average 30–60 × 15–35 mm (Thiollay 1963b); normally 30–70 × 18–26 mm, i.e. same as *S. aluco* (Glue 1970). Pellets of young smaller and rounded (Saint Girons 1964). Each adult pellet contains remains of 3–6 small mammals; sometimes 1, exceptionally (with shrews) 14 (Thiollay 1963b); average 2·7 vertebrates, $n = 14197$ pellets (Lange 1948). Concluded by Guérin (1928) that wild birds eject 2 pellets per day—one at night while hunting, another at roost. However, captive birds average 1·4 (Schmidt 1977; see also Bunn *et al.* 1982). One captive bird averaged 1·7 pellets per day (1·2–2·4 over 6 4-week periods); each pellet represented average 41·4 g live weight of food (Marti 1973). For discussion and review of pellet ejection (times, stimuli, physiology, etc.), see Bunn *et al.* (1982). Apparently, all vertebrate prey items eaten reappear in pellets (Glue 1967), but no thorough study. At traditional roost- and nest-sites, enormous quantities of pellets may accumulate.

For discussion of adaptive features of ears (openings set asymmetrically to permit accurate sound location—see above), eyes, and wings, see Bunn *et al.* (1982).

The following prey recorded in west Palearctic. Mammals: elephant shrew *Elephantulus rozeti*, shrews (*Sorex minutus*, *S. araneus*, *Neomys fodiens*, *N. anomalus*, *Suncus etruscus*, *Crocidura leucodon*, *C. suaveolens*, *C. russula*), mole *Talpa europaea*, blind mole *T. caeca*, bats (*Rousettus aegyptiacus*, *Taphozous nudiventris*, *Rhinolophus ferrumequinum*, *R. hipposideros*, *R. euryale*, *R. mehelyi*, *Asellia tridens*, *Myotis daubentonii*, *M. dascyneme*, *M. mystacinus*, *M. emarginatus*, *M. nattereri*, *M. bechsteini*, *M. myotis*, *M. blythii*, *M. oxygnathus*, *Otonycteris hemprichi*, *Plecotus auritus*, *P. austriacus*, *Miniopterus schreibersi*, *Barbastella barbastellus*, *Pipistrellus pipistrellus*, *P. nathusii*, *P. kuhli*, *P. savii*, *Vespertilio serotinus*, *V. nilssoni*, *V. murinus*, *Nyctalus noctula*, *N. leisleri*, *N. lasiopterus*), rabbit *Oryctolagus cuniculus*, red squirrel *Sciurus vulgaris*, European suslik *Citellus citellus*, garden dormouse *Eliomys quercinus*, forest dormouse *Dryomys nitedula*, edible dormouse *Glis glis*, dormouse *Muscardinus avellanarius*, common hamster *Cricetus cricetus*, migratory hamster *Cricetulus migratorius*, gerbils (*Gerbillus campestris*, *G. dasyurus*, *G. gerbillus*, *Pachyuromys duprasi*, *Meriones shawi*, *M. tamariscinus*), bank vole *Clethrionomys glareolus*, water vole *Arvicola terrestris*, pine voles (*Pitymys subterraneus*, *P. multiplex*, *P. savii*, *P. duodecimcostatus*), common vole *Microtus arvalis*, Cabrera's vole *M. cabrerae*, short-tailed vole *M. agrestis*, root vole *M. oeconomus*, snow vole *M. nivalis*, Günther's vole *M. guentheri*, muskrat *Ondatra zibethicus*, harvest mouse *Micromys minutus*, striped field mouse *Apodemus agrarius*, yellow-necked mouse *A. flavicollis*, wood mouse *A. sylvaticus*, zebra mouse *Lemniscomys barbarus*, black rat *Rattus rattus*, brown rat *R. norvegicus*, house mouse *Mus musculus*, spiny mouse *Acomys caharinus*, short-tailed bandicoot rat *Nesokia indica*, lesser mole-rat *Spalax leucodon*, birch mice (*Sicista betulina*, *S. subtilis*), jerboas (*Jaculus jaculus*, *J. orientalis*), stoat *Mustela erminea*, weasel *M. nivalis* (see, especially, Schmidt 1973b; also Dor 1947, Heim de Balsac 1965, Heim de Balsac and Beaufort 1966, Nader 1969, Krzanowski 1973, Herrera 1974a, Contoli and Sammuri 1978, Saint Girons and Thouy 1978, Ruprecht 1979b). Birds mostly small, communally roosting species: thrushes *Turdus*, starlings *Sturnus*, sparrows *Passer*, and finches (Fringillidae); not commonly smaller or larger than this (see, especially, Thiollay 1968, Glue 1972, Schmidt 1972a, Schnurre 1975). Reptiles: geckos (*Tarentola mauritanica*, *T. delalandei*), lizards (*Lacerta lepida*, *L. viridis*, *L. muralis*, *L. hispanica*, *Psammodromus algirus*, *Acanthodactylus erithrurus*), slow-worm *Anguis fragilis*, chameleon *Chamaeleo chamaeleon*, skink *Mabuya strangeri*, grass snake *Natrix natrix* (Moltoni 1937; Brosset 1956; Heim de Balsac 1965; Valverde 1967; Thiollay 1968; Herrera 1974a; Glutz and Bauer 1980; Bunn *et al.* 1982). Amphibians: frogs (*Discoglossus pictus*, *Pelodytes punctatus*, *Rana temporaria*, *R. ridibunda*, *R. esculenta*, *R. arvalis*, *R. perezi*, *R. delmatina*), toads (*Bufo bufo*, *Pelobates fuscus*, *P. cultripes*) (Thiollay 1968; Marián and Marián 1973; Herrera 1974a; Schnurre 1975; Veiga 1980; K W Emmerson and A Martin). Fish: trout *Salmo trutta*, roach *Leuciscus rutilus*, carp *Cyprinus carpio*, rudd *Scardinius erythrophthalmus*, perch *Perca fluviatilis* (Thiollay 1968; Bunn *et al.* 1982). Insects: largely crickets, etc. (Orthoptera: Gryllidae, Tettigoniidae, Gryllotalpidae, Acrididae)

Table B Food of Barn Owl *Tyto alba* as found from pellet analyses. Figures are percentages of live weight of vertebrates only.

	All areas of Britain (mostly) and Ireland	Galway (Ireland)	Denmark	All areas of Netherlands	Somme (N France)	West Holstein (West Germany)
	1960–71, all year	Sep 1970–Aug 1971	1932–43, mostly summer	1929–77	All year	1951–65
Mice (small Murinae)	14·5 ⎫ 68·8	62·2 ⎫ 62·2	36·5 ⎫ 62·2	11·6 ⎫ 57·8	19·7 ⎫ 76·1	2·1 ⎫ 88·2
Small voles (Microtinae)	54·3 ⎭	0² ⎭	25·7 ⎭	46·2 ⎭	56·4 ⎭	86·1 ⎭
Larger mammals[1]	12·8	24·4	9·0	5·4	5·3	0·8
Shrews (Soricidae)	16·1	6·2	22·1	33·4	15·1	10·5
Bats (Chiroptera)	+	0	0·1	+	+	0
Birds	2·1	3·2	5·8	2·7	3·5	0·1
Other	0·2 (mostly amphibians)	4·0 (frogs)	0·8 (amphibians)	0·6 (mostly amphibians)	0³	0·5 (frogs)
Total no. of vertebrates	47 865	841	38 861	92 313	16 496	11 913
Source	Glue (1974), data reanalysed	Fairley and Clark (1972)	Lange (1948), data reanalysed	Bruijn (1979), data reanalysed	Saint Girons and Martin (1973), data reanalysed	Bohnsack (1966), data reanalysed

and beetles (Carabidae, Dytiscidae, Hydrophilidae, Silphidae, Staphylinidae, Lucanidae, Geotrupidae, Scarabaeidae, Tenebrionidae, Coccinellidae, Cerambycidae, Curculionidae); also dragonflies, etc. (Odonata), cockroaches (Blattidae), mantises (Mantidae), Lepidoptera larvae and adults, earwigs (Dermaptera), ants (Formicidae). Other invertebrates include coastal isopod *Ligia*, molluscs (*Helicella*, *Rumina*, *Melanopsis*, *Euparipha*, *Limax*), and worms *Lumbricus* (see, especially, Thiollay 1968, Herrera 1974a; also Uttendörfer 1952, Bourne 1955, Haensel and Walther 1966, Valverde 1967, Buckley and Goldsmith 1972, Glutz and Bauer 1980, Veiga 1980, Amat and Soriguer 1981). For inanimate material in pellets, see 2nd paragraph.

Enormous literature of food analyses from west Palearctic, based almost wholly on pellet examination. However, diet and related aspects of ecology not yet studied with nearly the same thoroughness as done for *S. aluco*. For summary of major European studies, see Table B and Schmidt (1973b). See also the following significant studies. Britain: Brown and Twigg (1972), Buckley and Goldsmith (1972), Webster (1973), Brown (1981). Ireland: Fairley (1966). Sweden: Frylestam (1970). Netherlands: Hoekstra (1974). Belgium: Van der Straeten and Asselberg (1973), Delmée *et al.* (1979). France: Kahmann and Brotzler (1956), Heim de Balsac and Beaufort (1966), Thiollay (1966, 1968), Cuisin and Cuisin (1979), Henry (1982a). East and West Germany: Uttendörfer (1939b, 1952), Rothkopf (1970), Schnurre and Bethge (1973); see also sources in Glutz and Bauer (1980). Switzerland: Pricam and Zelenka (1964), Zelenka and Pricam (1964). Poland: Kulczycki (1964), Glutz and Bauer (1980), Goszczyński (1981). Czechoslovakia: Balát (1956), Stastnyi (1973), Viček and Vondráček (1974). Hungary: Schmidt (1966–7). USSR: Pidoplitschka (1937), Tatarinov (1965). Spain: Camacho Muñoz (1975), Sans-Coma and Kahmann (1976), Vericad *et al.* (1976), López Gordo *et al.* (1977), Campos (1978), Veiga (1980), Amat and Soriguer (1981). Portugal: Luz Madureira (1979). Italy: Contoli (1976a, b), Petretti (1977), Contoli and Sammuri (1978), Santini and Farina (1978), Contoli *et al.* (1979). Malta: Schembri and Zammit (1979). Crete (Greece): Pieper (1976). Cyprus: Glutz and Bauer (1980). In western Morocco, 1972–5, 4628 vertebrates included, by number, 69·7% *Mus*, 1·4% *Apodemus*, 0·8% *Rattus*, 8·9% *Gerbillus*, 7·9% *Crocidura*, 10·9% birds, 0·2% lizards, 0·1% amphibians (Saint Girons and Thouy 1978); at Berkane, 972 vertebrates in pellets from clock tower comprised, by number, 89·3% birds. For North Africa, see also Brosset (1956b), Valverde (1957), Heim de Balsac and Mayaud (1962), Heim de Balsac (1965), Saint Girons (1973), and Saint Girons *et al.* (1974). On Tenerife (Canary Islands), diet largely *Mus*; also *Rattus*, frogs *Rana*, geckos *Tarentola*, earwigs *Aisolabis*, crickets *Gryllus*, and sparrows *Passer* (K W Emmerson and A Martin); similar on Madeira (Bannerman and Bannerman 1965). On Cape Verde Islands, 2 pellet collections from outside towns contained only insects (Bourne 1955); on São Tiago, took *Rattus*, *Mus*, *Passer*, a gecko, and insects (Naurois 1969b); on Raso and Branco, largely reptiles and birds (largely Madeiran Storm-petrel *Oceanodroma castro* and White-faced Storm-petrel *Pelagodroma marina*) (Heim de Balsac 1965). In Palestine, 6224 vertebrates included, by number, 46·1% *Microtus guentheri*, 17·7% *Mus*, 16·3% shrews, 14·2% *Meriones*, 1·9% *Rattus*, 1·6% *Spalax*, 1·2% *Cricetulus* (Dor 1947). In Iraq, 194 vertebrates in pellets from one site included, by number, 30·4% *Mus*, 28·4% bats (roosting at same site), 19·0% birds, 17·0% *Rattus* (Nader 1969). Conclusion of Herrera (1974b), that in western Europe diet more diverse in Mediterranean than in other regions, not justified.

Diet widely assumed to reflect local availability of small-mammal prey, with no marked preferences; however, evi-

Kujawy (central Poland)	SW Spain	Tuscany (central Italy)
1964, autumn	Sep 1971–Dec 1972	1974, late summer
37·5 ⎫ 72·5 35·0 ⎭	61·7 ⎫ 74·8 13·1 ⎭	46·1⁴ ⎫ 88·9 42·8 ⎭
1·3	6·5	2·7
9·4	8·5	6·2
0·2	0·6	0
14·5	5·2	2·3
2·1 (amphibians)	3·9 (amphibians, lizards)	+ (lizards)
16944	14168	3598
Ruprecht (1979a), data reanalysed	Herrera (1974a), data reanalysed	Lovari et al. (1976), data reanalysed

1. Mostly mole *Talpa europaea*, water vole *Arvicola terrestris*, and brown rat *Rattus norvegicus*; in southern Europe, also garden dormouse *Eliomys quercinus*.
2. No voles present over most of Ireland.
3. Only mammals and birds included in this study.
4. Includes *c*. 5% dormouse *Muscardinus avellanarius* (Muscardinidae).

dence available so far (e.g. Glue 1967, Webster 1973, Bruijn 1979, Goszczyński 1981) almost wholly circumstantial. Certainly, does not show apparent aversion to shrews of *A. otus* and *S. aluco*; shrews formed 48·5% (n=683) of food by weight in Ythan valley, Scotland (Hardy 1977). Low figure recorded from Ireland (Table B) probably due to tiny *Sorex minutus* being only species present. Where voles main prey, proportions follow their population cycles: e.g. at Turew (Poland) 47·1% by weight *Microtus arvalis* when population low, 88·4% when high Gioszczyński 1981); in Netherlands, varied from 37% to 60% by weight (Bruijn 1979). In southern France, July–August, *Microtus* and *Apodemus* taken were mostly adult, despite greater abundance of young (Saint Girons 1965). 69 *Rattus norvegicus* in pellets, Britain and Ireland, mostly had live weights of 20–80 g; most studies (including, for uniformity, reanalyses made in Table B) use figure of 100 g to convert to weight percentage, so importance probably overemphasized (Morris 1979). Bats more important than in *S. aluco* and *A. otus*, but though recorded forming up to 55·3% of diet in Austria, role generally insignificant (Table B); see Bauer (1956) and Ruprecht (1979b). 2 pairs which nested by colonies of bats apparently never took them (Meinertzhagen 1959). Birds perhaps more important than usual in more built-up habitats: e.g. 19·0% by number (n=390) in suburbs, Hainaut, Belgium (Leurquin 1975); see also above for North Africa. In Spain, frogs *Rana ridibunda* taken tended to be large and ♂—presumably because calling (Calderon and Collado 1976). Fish rare in diet: 1 in 250000 prey items, central Europe (Uttendörfer 1939b). Invertebrates form only readily available prey group which appears to be avoided (Bunn *et al.* 1982). Pellets do not allow precise quantification, but, in Europe, more insect remains found in south. In Britain and Ireland, invertebrates negligible in both pellets and guts: mostly beetles—342 among 47865 vertebrates in pellets (Glue 1974). In Siena (central Italy), 34% of 53 items from gizzards of 70 birds were Orthoptera (Lovari 1974). For other gut analyses, see Lovari (1978) and Kochan (1979). Not stated in most studies whether pellets examined for chaetae of earthworms (Lumbricidae), but evidently eaten infrequently—chaetae found in 3 out of 188 pellet collections in Britain and Ireland (Glue 1974).

Long-term year-to-year studies provided only by Lange (1948) and Bohnsack (1966). Seasonal variation poorly studied; best work by Goszczyński (1981). Generally, proportion of voles decreases during about April–August, and proportion of shrews (sometimes also birds, but not larger mammals) increases correspondingly (Schnurre 1944; Glue 1967; Webster 1973; Brown 1981; Goszczyński 1981)—or sometimes proportion of small rodents increases only in autumn and early winter (Fairley and Clark 1972; Hoekstra 1974); these patterns may well be determined almost solely by seasonal fluctuations in vole populations (unlike situation in *S. aluco*), though shrews' apparent habit (in central Europe at least) of spending much of winter underground (Corbet and Southern 1977) perhaps also involved. However, Giban *et al.* (1948) found pattern reversed in Charente-Maritime (western France). Saint Girons (1963) and Saint Girons and Thouy (1978) found no regular pattern. See also Lange (1948), Pricam and Zelenka (1964), Campos (1978), and Hennache (1981). For studies discussing variation in diet with habitat, see Glue (1967), Saint Girons and Martin (1973), Herrera (1974a), Lovari *et al.* (1976), Bruijn (1979), and Hennache (1981). For comparisons with diet of other Strigiformes in same or nearby area, see *S. aluco* and *A. otus*.

Food consumption in the wild 70–104 g per day according to Schmidt (1977); 7–8 small-vole-sized animals (Bunn *et al.* 1982), i.e. *c*. 140–160 g; *c*. 100 g (Pricam and Zelenka 1964); see also Goszczyński (1976). Captive bird ate 46·4–74·0 g live weight per day, most in winter (Marti 1973). Various captive studies said to suggest 100–150 g (see Bunn *et al.* 1982); see also Glutz and Bauer (1980) and Hamilton and Neill (1981). Compare also data for (larger) *S. aluco* (said to have more efficient digestion: see Glutz and Bauer 1980). When feeding young, Kiev (USSR), caught *c*. 85–128 rodents per month, average 13 mice per ha (Averin and Ganya 1966). Pellets from one bird, northern England, suggested that severe winter weather depressed food intake by *c*. 50% (Glue and Nuttal 1971).

Food of young said to differ little from adult's, including more of the smaller species (Saint Girons 1964), but no detailed studies. Pair once recorded bringing 17 items to nest in 2 hrs (Bunn *et al.* 1982); see also sources in Glutz and Bauer (1980), as well as Hamilton and Neill (1981) and, for surplus food at nest, Baudvin (1980). For development of hunting skills in young, see Bunn *et al.* (1982).

DJB

Social pattern and behaviour. Most comprehensive accounts by Bunn and Warburton (1977) and Bunn *et al.* (1982). For review, see also Schneider (1964). Account includes material supplied by I R Taylor.

1. Solitary or in pairs outside breeding season. Pairs and individuals of either sex may hold territory throughout the year, and presumably defend it through winter. Throughout 4-year study, Lancashire (England), all known birds in 5 territories resident throughout the year (Bunn and Warburton 1977; Bunn *et al.* 1982). BONDS. Monogamous mating system the rule. ♂♂ occasionally bigamous (Baudvin 1975), once when food supply artificially supplemented (Bunn *et al.* 1982; see also Breeding Dispersion, below). In 168 breeding attempts, Scotland, 4 cases of bigamy in which both ♀♀ laid, and 2 cases where 2 ♀♀ associated with site but only 1 laid (I R Taylor). In West Germany, record of ♀ associating with 2 ♂♂ (Schönfeld and Girbig 1975). Pair-bond apparently long-term and often persists all year; some pairs maintain close contact outside breeding season, others (probably most) more loosely associated in territory (Bunn *et al.* 1982; see also Roosting). In 33 pairs of marked birds where both survived from one year to next, no divorce recorded (I R Taylor). If bird dies, mate may take new partner within the year. In several cases, both partners changed over 2 years (Baudvin 1975). No certain records of divorce. In Côte d'Or (France), frequency of *alba-guttata* pairs same as expected from random pairing between the races (Baudvin 1975). Both sexes breed first at 1 year (Schneider 1937; Stewart 1952; Bunn and Warburton 1977; Kaus 1977), though sometimes not until 2, probably depending on food supply (I R Taylor). As in other Strigiformes, ♀ takes dominant share of nest-duties: performs all incubation and brooding until young well-developed (see Relations within Family Group, below). ♂ feeds ♀ and young, ♀ also feeding young after she stops brooding. Young dependent on parents until 12–14 weeks old, exceptionally 16 weeks (Bunn *et al.* 1982). Cannibalism reported among siblings, and, though perhaps not uncommon in some areas (e.g. Baudvin 1975, 1978: see Relations within Family Group), incidence probably sometimes exaggerated, being often inferred when young disappear for other reasons (Bunn *et al.* 1982). BREEDING DISPERSION. Solitary and territorial, but occasionally loosely colonial in USA (Smith *et al.* 1974; Smith and Marti 1976). In many places, density likely to be limited by availability of sites for nesting and roosting: e.g. in lowland Scotland, maximum 23 pairs in 50 km²; in uplands (same year), where nest-sites probably fewer, 11 pairs in 100 km² (I R Taylor). In central Europe 1 pair per village the rule: e.g. 3 pairs in 3 villages in triangle of 500–600–700 m (Kaus 1977; see also Glutz and Bauer 1980). In Switzerland, 4 pairs in one area of 1 km² (see Glutz von Blotzheim 1962). In probably optimal habitat, Schleswig-Holstein (West Germany), up to 2·3 pairs per 10 km² (Ziesemer 1980). Number of breeding pairs usually fluctuates much more markedly than overall population density (Glutz and Bauer 1980). Size of territory varies according to food supply. However, boundaries often not rigidly defended and estimates of territory size likely to be approximate: e.g. in Scotland, nest area defended, but during chick-rearing neighbouring pairs overlapped extensively for hunting (I R Taylor). Where food plentiful, territory apparently *c.* 0·4–0·6 km² (Ticehurst 1935; Evans and Emlen 1947; Glutz and Bauer 1980). In area dominated by conifer plantations, average size of 5 territories, each of which shared part of boundary with the others, *c.* 2·5 km² (Bunn *et al.* 1982). In central Europe, activity radius of 800–1500 m from nest frequently reported (see Glutz and Bauer 1980). In loose colony (see above), some pairs occupied overlapping home ranges and each pair commonly defended an area only 5–10 m around its nest (Smith *et al.* 1974). Some reports of birds nesting in same building suggest bigamy involved: e.g. 3 nests 5 m apart from one another, 2 nests 10 m apart (Baudvin 1975; see also Callion 1973). However, possibility of overlapping broods of single pair (see below, and Relations within Family Group) should not be overlooked as more common explanation for some simultaneously occupied nests. Territory established by ♂; if either mate dies, other may continue to defend it (Bunn *et al.* 1982), or move out (I R Taylor; see below). Probably most hunting done within territory, which typically contains a number of favourite vantage points (Robinson 1973). Unless regularly disturbed, site fidelity marked (Hermanson and Otterhag 1963), and sites recorded in continuous use for 70 years (Bunn *et al.* 1982). In Scotland, 31 cases of site retention between years by ♂♂, only 1 change; in ♀♀, 45 stayed, 4 changed; distance moved 1–4·5 km, all following loss of mate (I R Taylor). 2 cases of same ♀♀ at nest-sites for 3 years (Kaus 1977); 1 ♀ at same site for 9 years, same ♂ at this site for 2 years (Schneider 1930). 2nd clutch of season frequently laid in same nest as 1st, sometimes even when 1st-brood young still in nest (Bunn *et al.* 1982). Occasionally nests close to Kestrel *Falco tinnunculus*, once 2·5 m away (Fellowes 1967; Baudvin 1975). ROOSTING. Diurnal. Uses regular, well-concealed site in building, etc., e.g. under roof, hole in wall, tunnel between bales of hay. Less often (though commonly in south-central Scotland: I R Taylor) in dense evergreen tree cover, notably spruce *Picea*, pine *Pinus*, and holly *Ilex* (Bunn *et al.* 1982). Once, on low stumps in reedbed (Ash 1954). For other sites, see Witherby *et al.* (1938). Also loafs for long periods on ground, as in Short-eared Owl *Asio flammeus*. Bird about to sleep or doze typically starts slight sideways rocking movement and assumes relaxed posture: with plumage loose, stands on one leg, with body leaning forward, facial disc flattened, and eyes closed. At slightest noise, opens eyes. Occasionally ruffles plumage and preens. In hot weather, frequently stamps feet rapidly (Bunn *et al.* 1982). For mutual Allopreening by roosting pair, see Heterosexual Behaviour (below). ♀ starts roosting at nest-site before laying; ♂ typically accompanies her there for 2–3 weeks until clutch complete, then resorts to roost elsewhere, sometimes up to 1·5–2 km away. When young *c.* 4–5 weeks old, ♀ also moves to separate roost-site (I R Taylor), as if to avoid close attentions of young. Outside breeding season, some pairs roost together at nest-site, or, if disturbed, elsewhere; others, probably most, roost separately in territory outside breeding season (Bunn *et al.* 1982; see also Schneider 1930). For communal roosting of several birds outside breeding season, see Smith and Marti (1976). For periods of hunting activity, see Food; also Festetics (1968) for periods of rest during night. In early morning, before roosting, and at dusk after leaving roost, captive birds regularly performed wide Head-circling movements about horizontal axis, also lateral swaying movements of body; these thought to be for orientation and focusing (Trollope 1971). Apparently sunbathes less than some other Strigiformes, but one reported lying on sunlit patch of ground and raising one wing to expose side to sun (Bentham 1962). Recorded bathing in stream on sunny afternoon (Bunn *et al.* 1982).

2. Generally nervous. At first hint of danger, roosting bird lowers leg (raised in roosting-posture), draws itself upright, and holds wings tight to body, almost wrapping them around itself, thus concealing conspicuous white underparts; closes eyes. In captivity, threatens intruder with Forward-threat posture (Fig A): bows towards source of threat and spreads wings, tilting

A

them forward; tail raised and spread. On closer approach, head lowered, showing nape to observer (Trollope 1971, which see for photographs) and swings head from side to side (aggressive Body-swaying: see below). Accompanies display with hissing sounds (see below) and Bill-snapping (Trollope 1971: see Voice). When cornered, bird performs Body-swaying display (Fig B): standing on alternate feet, sways lowered head and upper body from side to side, ruffles plumage, lowers wings, and spreads tail. Every few seconds, looks down and shakes head rapidly (Head-shaking), then looks up and continues Body-swaying. Occasionally, makes sudden lunge forwards. While glaring at source of danger, often gives Short Hissing-call with bill open (see 11 in Voice) and Bill-snaps. If more imminently threatened, bird gives Long Hissing-call (see 10 in Voice) with bill closed, this increasing in volume and vigour until it grades into Distress-call (see 9 in Voice). Combination of threat posture, hissing calls, and Bill-snapping vary more with individual than circumstances. When attacked, or closely approached, bird may raise foot and lean backwards, striking upwards and sinking claws into outstretched hand. If picked up, often feigns death, closing both eyes and lying prostrate (Bunn *et al.* 1982). See also Antipredator Responses of Young, and Parental Anti-predator Strategies (below). ANTAGONISTIC BEHAVIOUR. Threat postures and hissing calls described above not used by adults against conspecific birds. ♂♂, and apparently some ♀♀, demarcate territory by calling and Display-flight, perhaps intensely so in areas of high density. On emergence from roost or nest-site at dusk, bird repeatedly gives Advertising-call (see 1 in Voice), typically in outward flight, also when perched before departure (Bunn *et al.* 1982). Advertising-call given typically from *c.* 6–7 weeks before laying; wanes thereafter but resurgence occurs before 2nd clutch (Bühler and Epple 1980). In Display-flight, bird flies steadily over territory, repeatedly changing direction and calling every few seconds; flight brief, and bird often returns to roost after 2–3 min. In unusual flight, perhaps display, bird ascended to *c.* 50 m and made spiral descent with exaggerated Wing-clapping. Occasionally, bird strays into another's territory and may be chased by resident. Intruder not uncommonly escapes by diving for ground cover. Brief fights and angry chases sometimes occur. Most often, resident flies straight at intruder and both then make rapid almost vertical ascent, breast to breast, before trespasser retreats, sometimes hotly pursued. During ascent, aggressor screeches loudly and birds perhaps grapple with claws (Bunn *et al.* 1982). In fight over food between pair and 3 Little Owls *Athene noctua*, ♂ buffeted rivals with wings and, with aid of mate, drove them off (Lord and Ainsworth 1945). For fierce fight with *F. tinnunculus* over food, see Dunn (1979*b*). HETEROSEXUAL BEHAVIOUR. Nest-site, or prospective nest-site, the main focus for display. (1) Pair-bonding behaviour. Advertising-call, given every few seconds from roost, and Display-flight of ♂ serve to attract ♀♀ as well as repelling rivals. Unpaired birds especially vocal, and bird that lost mate increased calling rate abruptly the same evening (Bühler and Epple 1980). ♀ approached calling ♂, then flew off towards another ♂ calling *c.* 1 km away (Bunn and Warburton 1977). Initially, ♀ may respond aggressively to ♂. Pursuit-flights typical of early pairing: ♂, flying just above and behind ♀, follows her persistently, the 2 birds twisting and turning, though pursuit sometimes more leisurely; both screech frequently (see 1b in Voice). In Moth-flight, ♂ hovers for up to 5 s, legs dangling, in front of perched ♀. Wing-clapping occasionally performed by ♂, but possibly accidental, and not ritualized as in *Asio* owls (Bunn and Warburton 1977; Bunn *et al.* 1982, which see for details). (2) Nest-site selection. ♂, calling frequently (see 1 in Voice), flies repeatedly, with rapid wing-beats, in and out of potential nest-site (Nest-showing); calls to ♀ from within site and, if this fails to attract her, may also perform Display-flight (as in Antagonistic Behaviour). ♂ may crouch in site, scratching with feet, and probing with bill; may then summon ♀ with Purring-call (Bunn and Warburton 1977; Bunn *et al.* 1982: see 2 in Voice). In one sequence, presumed ♂ called (etc.) from dovecote, then nest-box *c.* 50 m away, but other bird, presumed ♀, did not follow him into either (Löhrl 1965). Suggests final choice perhaps by ♀. (3) Mating. Occurs with remarkable frequency, and not uncommonly continues well into chick-rearing; usually at or near potential or chosen nest-site. During prospecting, pair copulate every few minutes, often without prior display. Either sex may apparently solicit, ♂ sometimes crouching in front of ♀ as if inviting her to mount. Once, during chick-rearing, ♂ deposited prey beside brooding ♀ before flying a short distance from nest; ♀ followed and began to solicit by swaying body and vibrating wings; ♂ paused in mid-air and Wing-clapped before mounting. In more typical sequence, ♀ begins to solicit ♂ with increasingly rapid, quiet Snoring-calls (see 3 in Voice), and lowers body (Bunn *et al.* 1982). If ♂ ignores ♀, she may switch to Purring-call (Bühler and Epple 1980). ♂ mounts, balancing with widely spread wings and by grasping ♀'s nape with his bill. ♂ utters Copulation-call (see 6 in Voice), sometimes also a few snores. Every 2–3 s, ♂ makes cloacal contact with convulsive thrusts, accompanied by bursts of Copulation-calls. Snoring-calls of ♀ typically increase in volume up to moment of cloacal contact. Copulation lengthy (10–20 s) and, on completion, some ♂♂ give 1–2 loud screeches. After mating, ♂ often begins to doze while ♀ Allopreens him, especially on underparts and head (see also below). Copulation commonly occurs before ♂ leaves on first hunt of evening, and whenever he returns with prey. On his

B

C

arrival with prey, ♀ may fly to dark corner and solicit with Purring-call (Bunn and Warburton 1977; Bunn et al. 1982). ♂ mounts ♀ after transferring prey, ♀ typically holding it in bill during copulation (Fig C). Rate of copulation wanes during incubation and chick-rearing, but one pair with young 5–6 weeks old copulated once each night, c. 3 m from nest, always after ♂ transferred food to ♀ (Trötschel 1973, which see for photographs). (4) Courtship-feeding and food-caching. Following account after Bunn et al. (1982). ♂ starts to feed ♀ regularly from c. 1 month before laying, mostly at nest-site. At least initially, courtship-feeding invariably associated with copulation (see above). ♂ usually announces arrival with Greeting-call (see 4 in Voice), and ♀ responds with Snoring-, then Purring-calls. ♂ approaches ♀ giving Greeting- or, less often, Feeding-calls (see 5 in Voice). On taking food, bill-to-bill, ♀ crouches, inviting copulation. ♀ eats prey after mating or, if satiated, lets it drop. ♂ may feed ♀ 2–3 times each evening, but often brings surfeit which is left to rot. Surplus prey thus not effectively cached and, according to Bunn et al. (1982), true caching only sporadic, and then more at roost than nest-site. However, Baudvin (1975, 1980) believed food surplus an important safeguard against (e.g.) ♀'s need to leave nest excessively and endanger eggs. Amount of prey found at nest varies with provisioning rate of ♂ and consumption rate of ♀ and young. During incubation, when ♀ inactive and her energy needs relatively low, rarely more than 2–3 prey at nest at any time. When young hatch, ♂ brings greater quantities but, as requirements of young still quite low, prey accumulates. Sometimes 'several tens' of prey at nest, once more than 50. When young older than 21 days, consumption increases and less accumulates (Baudvin 1975, 1980). (5) Allopreening. Outside breeding season, roosting pairs regularly perform mutual Allopreening. ♀ usually more attentive, e.g. after copulation. Approaches ♂ with tremulous squeaks or whistles and preens him all over, often beginning with circling movements around bill, extending to cheeks, rest of head, neck, etc. (Fig D). Groomed bird twitters continuously (see 4 in Voice). Bill-snapping and soft snoring sounds also heard. After Allopreening, pair usually doze (Bunn and Warburton 1977; Bunn et al. 1982). For apparently exceptional Allopreening display, see Hosking and Smith (1943). For Allopreening in captivity, see Harrison (1965, 1969) and Trollope (1971). (6) Behaviour at nest. As laying approaches, ♀ becomes increasingly attached to chosen nest-site, and sits tightly there. By pulling them towards her, ♀ accumulates pellets to form layer for eggs. ♀ will break up pellets to make nest-lining (Löhrl 1962a; Schulz 1978; I R Taylor), though Schneider (1930, 1946) considered fragmentation to arise mostly from attack by larvae of moth *Trichophaga*. Once laying begins, pair more silent at nest. ♀ incubates assiduously, leaving only at lengthy intervals to dash about, preen, defecate, and collect prey left by ♂, before returning to nest after c. 5–10 min (Baudvin 1980; Bunn et al. 1982). Typically leaves nest only 2–3 times per 24 hrs, more often and for longer if food short. ♂ not thought to contribute effectively to incubation or brooding, despite rare reports of sitting on eggs or young (Baudvin 1975; Bunn and Warburton 1977; Bunn et al. 1982). For photographs and account of egg-turning, egg-rolling, and position changes during incubation, see Epple and Bühler (1981). RELATIONS WITHIN FAMILY GROUP. For development of young, see Scherzinger (1971a). Young start calling from egg 24 hrs before hatching; ♀ responds with restlessness and soft variant of Feeding-call (Bunn and Warburton 1977; Bunn et al. 1982). Hand-reared ♀ helped young to hatch, using her bill to break shell from egg and clean emerging young (Bühler 1970, which see for photographs). As soon as young dry, give discomfort-calls if uncovered (see Voice); food-calls given from 2nd day, increasing in strength and frequency with age. Some young probably not fed until 2nd day, whereupon ♂ starts bringing surplus food (see above). Large prey (e.g. rats *Rattus* and birds) usually decapitated by ♂ before arrival at nest, small prey (e.g. voles Microtinae) brought whole. ♂ leaves prey at nest for ♀ to dismember (Bunn and Warburton 1977; Bunn et al. 1982). Accompanied by Feeding-call, ♀ gives choice parts to young, eating rest herself (Baudvin 1980). ♀ feeds blind young by straddling them from rear; very patient, once taking 70 min to successfully feed reluctant chick (Bunn and Warburton 1977; Bunn et al. 1982). After 6–7 days, young begin to vomit up indigestible matter, not casting pellets until older (Delmée et al. 1978). ♀ also eats faeces of young until smallest chick c. 10 days, after which young back out of nest to defecate. ♀ now broods less closely and starts hunting for herself and for young; usually feeds young less than ♂, but this varies, some feeding young scarcely at all, others providing more than ♂. At 8–10 days, eyes of young begin to open. At 2 weeks, young able to swallow most prey whole (Trollope 1971; Bunn et al. 1982). In different years, Côte d'Or, 7·1–32·4% of young apparently eaten by older siblings; live young rarely killed, but ailing offspring may be fed by parents to young when dead or moribund (Baudvin 1975, 1978). When oldest young c. 3–4 weeks, ♀ stops brooding and soon visits nest only to feed young, arriving silently or with Feeding-call, while food-bearing ♂ gives variant of Greeting-call (see 4 in Voice). At 4 weeks, young intensely curious, exploring immediate surroundings of nest; typically make head-bobbing and rapid head-circling movements, and jockey for position when parent arrives with food. Parents stay at nest only long enough to ensure transfer (Bunn et al. 1982). Young that fall out of nest often able to climb back; those failing to regain nest may be left to starve (Bunn 1977) or continue to be fed (I R Taylor). At 5–6 weeks, young exercise wings vigorously and make walking excursions from nest. Siblings typically greet parents, also young returning from excursion, with Begging-display (Fig E): crouch, turn head upwards and weave it from side to side, and vibrate wings, turned to show undersides. Returning young approaches begging sibling, and frenzied Allopreening, head-bobbing, etc., follow. Perhaps not uncommonly, older young may then feed younger siblings with prey delivered by parents, and give associated Feeding-call of adult during presentation; such feeding apparently unique among Stri-

D

E

F

giformes (Bunn and Warburton 1977; Epple 1979; Bunn *et al.* 1982). Overlap of 1st brood and 2nd clutch at this stage not exceptional; out of 32 cases where 2nd clutch laid, 4 started before 1st brood left nest (Baudvin 1975). After oldest young finally leaves nest (at 7–8 weeks), parents continue to return to nest to feed remaining young until last fledged. When parents arrive with food, fledged young may try to return to nest (also continue to roost there); later, fed wherever they are in territory (Schneider 1964; Bunn *et al.* 1982). When all fledged, young typically line up together to be fed. Sometimes threaten one another with Forward-threat posture. At 9 weeks, fledglings regularly 'play': pounce on pellets, leaves, etc.; also swoop on and chase insects; first serious stoop from a height at 65 days, and first known capture (invertebrate prey) at 72 days; mammalian prey not caught until some time after. Young follow one another around territory, maintaining persistent chorus of excited snoring, chirruping, and Bill-snapping sounds. Parents gradually feed young less and become increasingly aggressive towards them. By 12th week, young more likely to roost away from nest. Leave territory at *c.* 10–11(–15) weeks (Bunn and Warburton 1977; Bunn *et al.* 1982). ANTI-PREDATOR RESPONSES OF YOUNG. Once eyes open, young begin to crouch, hiss, and Bill-snap when disturbed. On hearing Warning-call (see 7 in Voice) of adult, young immediately fall silent. On close approach, 26-day-old young may roll on back, hiss, and raise claws (Fig F), though seldom strike or bite until older. Others lie prostrate with eyes closed and head drooped, feigning death (Festetics 1952–5; Schneider 1964; Bunn *et al.* 1982). On approach of human intruder, young near fledging may Body-sway, then run into dark recesses of nest-site (Schneider 1964). Fledged young also seek cover, but if in the open sit glaring at intruder and frequently begin to give Mobbing-call (see 8 in Voice); may circle overhead with Mobbing-call. Fledged young typically bolder than adults; will gather around fox *Vulpes* and scream at it, also mob humans and cats. 87-day-old young adopted Forward-threat posture towards cat (Bunn *et al.* 1982); similar response reported by Hubl (1952). PARENTAL ANTI-PREDATOR STRATEGIES. (1) Passive measures. ♀ sits tightly (see above) even on close approach, not flushing readily until she stops brooding closely. (2) Active measures. Little developed, and, unlike some Strigiformes, seldom attacks, especially during incubation. Perched bird may direct Mobbing-call at ground predators. Occasionally, Long Hissing-call given by ♀ standing over young and adopting a threat posture (Schneider 1964; Bunn *et al.* 1982). Forward-threat posture, however, not known to be adopted in presence of predators. One pair regularly attacked intruders at nest, knocking their hats off and scratching their faces (Schneider 1953; see also Gerber 1960).

(Figs A–B and E from photographs in Bunn *et al.* 1982; Fig C from photograph in Trötschel 1973; Fig D from photograph in Harrison 1969; Fig F from photograph in Schneider 1964.)

EKD

Voice. Diverse repertoire in breeding season, much less vocal at other times. Several calls, mostly of screeching, wheezing, or hissing type, grade into one another, and difficult to distinguish audibly or to describe phonetically. However, major studies by Bühler and Epple (1980, which see for sonagrams and seasonal variation), by Bunn (1974) and Bunn *et al.* (1982) have greatly clarified repertoire. Following scheme based largely on Bunn *et al.* (1982), to which Bunn (1974) similar. Account includes notes and recordings supplied by P A D Hollom. Bill-snapping sound made in 2 ways: in 1st (especially common in young), tongue protruded alternately either side of bill; when withdrawn, mandibles snap together under pressure. In older birds, lower mandible forced outwards to tip of upper, or beyond, and the two clenched together, bill snaps shut when lower mandible suddenly withdrawn (Walker 1974). For Wing-clapping, see Social Pattern and Behaviour.

CALLS OF ADULTS. (1) Advertising-call. (a) ♂'s call. Best known call, popularly the 'screech', given especially in flight. A long-drawn hoarse screaming 'shrrreeeeee', with gargling or rattling 'pea-whistle' quality (Fig I, call given by perched bird); in flight, effort of wing action adds to distinct tremulous effect (Bunn *et al.* 1982). Mean length 2·03 s (1·4–2·6, $n=95$), given at intervals of 1–20 s, often 50(–250) times in series, also singly when patrolling territory. Serves to advertise territory and nest-site (Bühler and Epple 1980, which see for sonagrams of harsh and smooth variants). (b) ♀'s call. Given much less frequently than ♂'s call, perhaps by some ♀♀ only. Differs in timbre from ♂'s call; not so 'perfectly delivered', and tends to break off into less tremulous scream. Especially in aerial chases, ♀ may give lower-pitched, more broken, and caterwauling variant in noisy medley with ♂'s call (Bunn *et al.* 1982). (2) Purring-call. Subdued, wheezy screeching sound: mellow and tremulous in ♂ (Bunn *et al.* 1982: Fig II), less mellow, more sustained, and higher pitched in ♀. Given in variety of contexts, notably by ♂ during Nest-showing and by food-begging ♀ (Bunn *et al.* 1982); sometimes associated with copulation, and given by both sexes to express apparent aggression towards well-grown young (Bühler and Epple 1980). In recording at nest, regular calls (Fig II) of ♂ probably performing Nest-showing alternate in duet with Snoring-calls of ♀ (call 3, Fig III) (E K Dunn, P A D Hollom). (3) Snoring-call. Mainly by ♀, also occasionally by ♂. Highly variable in length and timbre, with hissing, rasping, wheezing, whistling, or screeching quality, depending on circumstances. Invariably, however, distinguished by persistent repetition (Bunn *et al.* 1982). According to Bühler and Epple (1980), call relatively short (*c.* 0·3–1·0 s) and, depending on intensity, 5–60 given per min. In recording at nest,

444 Tytonidae

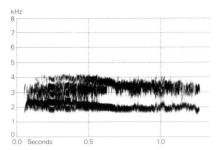

I P A D Hollom England July 1977

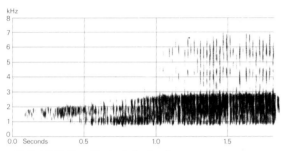

II P A D Hollom England April 1981

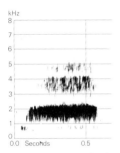

III P A D Hollom England April 1981

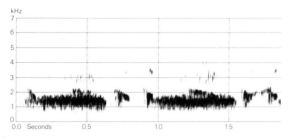

IV P A D Hollom England June 1981

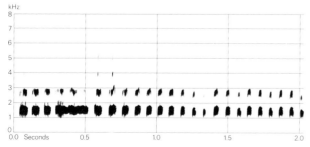

V P A D Hollom England June 1981

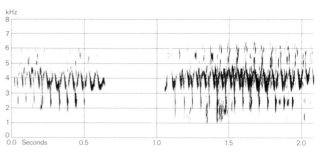

VI P A D Hollom England July 1976

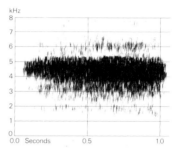

VII P A D Hollom England September 1977

Snoring-call (Fig III) alternates regularly with Purring-call of ♂ (call 2). Lower pitched and shorter than similar food-call of young (Bunn *et al.* 1982; see Calls of Young; also Bühler and Epple 1980 for comparison of sonagrams). Given apparently as a soliciting- and contact-call by incubating or brooding ♀, in response to ♂'s Greeting-call (call 4), also by ♀ when feeding young. Variants include the following. (a) During copulation, ♀ gives frenzied series of snoring 'aaahhhhhh' sounds (Fig IV: long units, interspersed with much shorter Copulation-call of ♂, call 6: J Hall-Craggs); calls get louder towards cloacal contact; described as combination of snoring and purring sounds (Bunn *et al.* 1982). (b) Soft snoring sounds, higher pitched and longer in ♂, exchanged during Allopreening (Bühler and Epple 1980; Bunn *et al.* 1982). (4) Greeting-call. Combined squeak and chirrup, given by ♂ to ♀ from before laying; more twittering variants given when he delivers food to nest, and by both sexes during Allopreening (Bühler and Epple 1980; Bunn *et al.* 1982). (5) Feeding-call. Prolonged, fast, chattering twitter, given by ♂ when passing food to ♀ early in nesting period. Also regularly by ♀ from day before hatching, and thereafter when she feeds young (Bunn *et al.* 1982); see Fig V. (6) Copulation-call of ♂. Distinctive brief squeaking sounds with plaintive quality, given repeatedly in rapid staccato manner during copulation (Bunn *et al.* 1982: Fig IV, 2nd and last short calls). Described as 'hip hip...' (J Hall-Craggs); 'gäck-gäckäck-gäck' or 'gick-gickgick', similar in timbre to Feeding-call, but sharper and at less frequent intervals (Bühler and Epple 1980). (7) Warning-call. Long-drawn, high-pitched scream or squeal, lacking tremulous or hissing quality, almost always given in flight, when nest and young threatened. Confusable with Advertising-call

(Bunn et al. 1982, which see for closely related 'Anxiety call', occasionally given by ♀ when nest threatened). (8) Mobbing-call. Described as an 'explosive yell' (Bunn et al. 1982); a very loud shrill screech (Bühler and Epple 1980). May be given in combination with, but quite distinct from, Warning-call. Directed at mammalian predators, and denotes fear and anger (Bunn et al. 1982). (9) Distress-call. A volley of loud screams with long-drawn cat-like quality (Bühler and Epple 1980; Bunn et al. 1982). Given in situations of intense agitation and fear, e.g. when seized, or in fierce fight. Often follows series of Long Hissing-calls (see below) (Bunn et al. 1982). (10) Long Hissing-call. Sustained (2–8 s) hissing sound, frequency 10–14 kHz, often ending in brief squeaking whistle, given with bill apparently closed, to intimidate predators; typically combined with threat postures, Bill-snapping, etc. (Bühler and Epple 1980; Bunn et al. 1982). Though said to be used also in confrontations with conspecific birds (Bunn et al. 1982), this denied by Bühler and Epple (1980). (11) Short Hissing-call. Brief (0·3–0·4 s) quiet hissing 'fff' sound, often given 2–3 times in succession, with bill open; homologous with Hissing-call of other Strigiformes. Given in similar context to Long Hissing-call (Bühler and Epple 1980; Bunn et al. 1982). (12) Other calls include: (a) tremulous squeak of ♀, used in response to ♂'s Greeting-call outside breeding season; (b) 'kit-kit' sound, occasionally given by bird in flight and thought to be a contact-call (Bunn 1977; see also Witherby et al. 1938).

CALLS OF YOUNG. From hatching, express discomfort and need for attention with repeated musical twittering. Timbre changes with age, and in recording (Fig VI) of ¾-grown young, a series of chittering sounds, longer and higher pitched than those of small nestlings (P A D Hollom, P J Sellar). Also heard when young quarrel; given much less once ♀ starts hunting for young (Bunn et al. 1982). From c. 2–7 days, food-call of young a hissing, snoring sound, similar to but distinguishable from adult call 3. In recording by P A D Hollom, food-calls differ in pitch between individuals in brood, presumably reflecting differences in age. From 3 weeks, food-call typically wheezy and long-drawn (Fig VII, 2 weeks after fledging). Given sporadically after 40 days. Young occasionally give a gentle pure fluting 'üüüi' sound, 0·3–1·0 s long, with slightly rising pitch, suggesting Bullfinch *Pyrrhula pyrrhula*; function not known (Bühler and Epple 1980, which see for sonagram). While still in nest, young start developing various adult calls, notably Long and Short Hissing-calls, Warning-call, Mobbing-call, and Feeding-call (Bühler and Epple 1980; Bunn et al. 1982). EKD

Breeding. SEASON. British Isles and north-west Europe: see diagram. Mediterranean and North Africa: similar start to laying season but may not be so prolonged (Makatsch 1976). SITE. In holes in tree or building, in cliff, quarry, rock outcrop. Re-uses nests for successive broods and in successive years. Nest: none made, though slight hollow may be formed in cavity. EGGS. See Plate 97. Elliptical, smooth, not glossy; white. Nominate *alba*: 40 × 32 mm (36–45 × 29–34), n = 100; weight 21 g. *T. a. guttata*: 41 × 32 mm (37–45 × 29–34), n = 57; weight 22 g (Makatsch 1976). Fresh weight 21 g (19–23), n = 21 (Verheyen 1967). Clutch: 4–7 (2–14). Of 325 1st clutches, France; 2 eggs, 1%; 3, 7%; 4, 16%; 5, 27%; 6, 20%; 7, 17%; 8, 7%; 9, 2%; 10, 2%; 11–13, 1%; average 5·7. Of 150 2nd clutches, France: 3 eggs, 1%; 4, 2%; 5, 9%; 6, 21%; 7, 21%; 8, 23%; 9, 11%; 10, 7%; 11, 4%; 14, 1%; average 7·4. Clutch size 5·6 in March (n = 36), 5·8 April–May (n = 161), 7·4 June–July (n = 158), 5·5 August (n = 29) (Baudvin 1979a). 1–2 broods. Replacements laid after loss of eggs or small young; 1 pair, USA, re-laid 6 days after loss, and 3 pairs after 13 days (Marti 1969; Smith et al. 1974). 2 broods uncommon in Britain (Bunn et al. 1982, which see also for exceptional case of 3rd brood). 2 broods in years of plentiful food (good vole years), and when 1st clutch laid early enough: e.g. in France, 2nd broods produced when 1st laid before early May (Baudvin 1979b). Interval between laying of 1st egg of 1st clutch, and 1st egg of 2nd averages 100 days (73–127), n = 83 (Baudvin 1979b). Eggs laid at intervals of 2–3 days; in USA, average 2·3 ± SD 0·21 days (1–7), n = 18 (Smith et al. 1974). INCUBATION. 30–31 days (Bunn and Warburton 1977). By ♀ only. Begins with 1st egg; hatching asynchronous, at similar intervals to laying. Eggshells pushed to side of cavity, or removed, or eaten (Bunn and Warburton 1977). YOUNG. Altricial and nidicolous. Cared for and fed by both parents; brooded more or less continuously while small, and fed by ♀ with food brought by ♂. ♀ ceases to brood when youngest chick c. 10 days old (Bunn and Warburton 1977). Able to dismember food at c. 16 days (Bunn and Warburton 1977). FLEDGING TO MATURITY. Fledging period 50–55 days. Become independent c. 3–5 weeks later. Age of first breeding 1(–2) years (see Social Pattern and Behaviour). BREEDING SUCCESS. Varies with food supply. In Netherlands, in poor vole years, average clutch size 3·6 and 2·7 young fledged per brood (average number of broods 200–300); in good vole

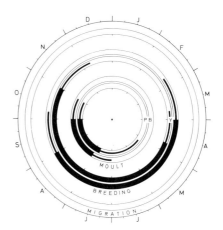

years, average clutch 4·4 and 3·5 young fledged per brood (average number of broods 300–500); overall, average 3·2 young reared from 471 broods (Braaksma and Bruijn 1976; Bruijn 1979). In France, average fledging success 65% from 475 broods, with 2975 eggs laid producing 1941 young to flying, average 4·1 young per clutch; success not dependent upon clutch size; $\frac{2}{3}$ of losses thought to be due to cannibalism, remainder to predation and infertility (Baudvin 1979a).

Plumages (nominate *alba*). ADULT MALE. Upperparts and upper wing-coverts yellow-buff to deep golden-buff, tips of feathers finely vermiculated grey and white and terminal part of shafts with some alternate black and white spots or small chevrons; crown, back, rump, and in particular hindneck usually predominantly golden-buff, vermiculated feather-tips appearing as isolated grey spots or streaks; mantle, scapulars, and upper tail-coverts with more extensively vermiculated feather-tips, grey more or less coalescent, with rather limited amount of golden feather-bases visible. Shorter lesser upper wing-coverts golden-buff, marked with small black specks or black-and-white spots, marginal coverts white; vermiculated tips of longer lesser, median, and greater upper wing-coverts rather extensive, but usually still much golden-buff of bases visible, appearing as alternate rows of golden and grey across wing. Sides of neck golden-buff with black specks and variable amount of white tinge. Distinct heart-shaped facial disc white, usually slightly tinged grey or cream; patch at inner corner of eye rufous-cinnamon, sometimes extending narrowly round eye, rarely virtually absent. A distinct ruff of rather stiff and narrow feathers round facial disc; feathers silvery-white at border of disc, golden-buff with fine black specks at border of crown and sides of head and neck, white but usually with narrow distinct black and golden-rufous tips at border of throat. Underparts, including feathering of legs and bristles on toes, white, frequently with slight pale yellow-buff wash on chest and occasionally flanks; entirely unmarked or with some dusky specks, mainly on sides of belly and flanks; occasionally, all breast, belly, flanks, and thigh with black spots up to 1–2 mm across. Ground-colour of central pair of tail-feathers (t1) pale cream-buff to golden-buff, sometimes nearly white when worn; inner webs of other tail-feathers with increasing amount of white outwards, ground-colour of both webs of t6 virtually white. T1 with 3–4 rather narrow dark grey bars showing beyond tips of upper tail-coverts, each bar partly consisting of close white and grey speckling; buff ground-colours increasingly peppered with fine grey specks towards feather-tips, tip sometimes marked with black-and-white droplet at shaft; other tail-feathers similar, but both dark bars and grey peppering reduced outwards, t6 virtually uniform except for 2(–3) short dark bars in middle and some fine specks. Some birds have tail very pale: t1 cream with traces of some dark bars on basal half and rather limited amount of fine specks on tips, other feathers virtually white except for some dusky specks or blotches. Flight-feathers golden-buff on outer webs and along shafts on inner webs with speckled grey-and-white bars, extending into uniform grey or dull black spot just across shaft on inner web; 4–6 bars show beyond upper wing-coverts on outer primaries, 3(–4) on inner primaries and secondaries; tips of flight-feathers densely speckled grey, sometimes with black-and-white droplet at shaft. Tertials rather like longer scapulars, but with traces of speckled subterminal bars. Flight-feathers occasionally very pale cream-buff or cream-white, virtually without dark bars on outer webs, pale bases contrasting markedly with golden-buff and grey upper wing-coverts; tips somewhat less speckled grey than in normal birds, but specks more contrasting. Upper wing-coverts like upperparts; greater and median upper primary coverts golden-buff with rather limited amount of grey speckling on tips, lesser cream-buff or white. Under wing-coverts and axillaries white, slightly yellowish on longer coverts in birds with yellowish chest, speckled grey or black in birds with speckled underparts; greater under primary coverts of speckled birds slightly tinged golden-buff on tips and with some dusky marks, uniform white in birds with uniform white underparts. ADULT FEMALE. Grey feather-tips on upperparts on average larger, black-and-white droplets slightly broader, underparts more often with yellow-buff wash on chest, sometimes extending to flanks and breast, underparts and under wing-coverts often more extensively marked with fine dark specks 1–2 mm across, and bars on flight-feathers and tail often more solid dark grey, less freckled grey and white; however, overlap in characters between sexes complete and sexing of single birds on these characters untrustworthy. Only in extreme cases is sexing by colour likely to be correct: if uniform white below with (at most) some tiny specks on flanks, usually ♂; if distinctly yellowish below and almost completely speckled with spots 1–2 mm across, usually ♀. DOWNY YOUNG. 1st down (neoptile) short and dense, white; patches on side of neck and rear of tarsus bare, down sparse on central belly. For development, see Bunn *et al.* (1982). JUVENILE. 2nd feathering completely downy, not intermediate in character between down and feathers as in mesoptile of Strigidae. Down long, very dense towards bases, fluffy on tips, but short on legs and forming scanty short tufts on toes. Down white, pale greyish-buff, or pale cream-buff. Starts to grow from 10th–15th day, but fine white neoptile clings to tips of 2nd down until late in 3rd week. Fully developed by 14th–20th day, gradually replaced by 1st adult feathers during 35th–60th day. 1st adult first visible on facial disc, ruff, leg, and upper wing-coverts (face showing feathers in pin from 19th–25th day); at end of 7th week, face and upper wing-coverts feathered, and feathers on scapulars and chest growing; late in 8th week, some down still on hindneck, sides of breast, belly, and thighs; last down lost and plumage fully 1st adult by 8th–9th week. Growth of flight-feathers and tail starts 12th–19th day, completed c. 70th day. (Heinroth and Heinroth 1926–7; Bussmann 1937; Schönfeld and Girbig 1975; Bunn and Warburton 1977; ZMA; see also Bunn *et al.* 1982.) FIRST ADULT. Similar to adult, distinguishable only by absence of mutual age-difference in flight-feathers. All flight-feathers equally new at fledging, all slightly worn by spring (in particular, more exposed p5–p9); worn feathers appear duller, greyer, and less clearly speckled than new feathers. When bird c. 1 year old, (p5–)p6–p7 replaced by new feathers, remainder of primaries still old, juvenile, often showing dusky grey rim along tip and pattern and extent of speckling and barring often slightly different from new feathers. In following summer, juvenile p4 (p2–p5) and p8 (p7–p9) replaced, but some other juvenile primaries and also secondaries usually retained until 4th or 5th calendar year, showing even greater heterogeneity in age between individual feathers than in 2nd winter; this heterogeneity maintained in subsequent years, though age-differences between feathers in some older individuals sometimes hard to see.

Bare parts. ADULT AND FIRST ADULT. Iris deep brown or black-brown. Eyelids flesh-brown or grey. Bill and cere pale pink-flesh, sometimes with partial blue-grey or yellow tinge. Foot strongly variable: dark grey-brown, yellow-brown, dirty yellow, or bright yellow, often dullest on 1st adult; claws dark horn-brown or black. DOWNY YOUNG AND JUVENILE. Eyes open from 8th day, pale blue at first; iris deep brown from 4th week. At hatching, bare skin (including cere, leg, and foot) pink-yellow or yellow-red, shining through down; bill and uppersurface of toes ivory-pink, sometimes partly tinged grey-horn or blue-grey. Lower mandible pink. At 35th day, cere and eyelids yellow, bill and toes slate-blue or dull black. (Heinroth and Heinroth 1926-7; Harrison 1975; Bunn and Warburton 1977; Bunn *et al.* 1982; RMNH, ZMA.)

Moults. Based on Piechocki (1961, 1974), Schönfeld and Piechocki (1974), Sutter (1975), and specimens in RMNH and ZMA, mainly referring to *guttata*, but limited number of specimens of other races examined basically similar. ADULT AND FIRST ADULT POST-BREEDING. Partial. Primaries serially descendant and ascendant from a centre at p6. Moult in 2nd calendar year mainly starts mid-July to mid-August (in a few birds, tail- and flight-feathers not moulted until 3rd calendar year). Moult in later years mainly August–September, but some tail- and flight-feathers shed May–October in ♀, (June–)July–October in ♂: breeding ♂ on average starts 6–8 weeks later than ♀, usually when young of 1st brood independent (non-breeding ♂ may start earlier). When all samples combined, 1·67(55)0–3 primaries replaced in 2nd calendar year, p6(–p7), exceptionally p5; in 3rd calendar year, 3·21(19)2–7 primaries shed, usually among p4–p5 and p7–p8, rarely (p1–)p2–p3 or p9–p10; in 4th calendar year, 3·35(13)2–6 primaries changed, mainly p3(p2–p4) and p9(–p10), and *c.* ⅓ of birds also start new moulting series with p6, replacing 0·5(13)0–5 feathers (usually p6 only). In 5th calendar year, 2·4(10)0–4 primaries of 1st series and 1·9(10)1–4 of 2nd shed, mainly last juvenile feathers (p1–p2) and p6–p7 (p5–p8) of 2nd series. In later years, 2–3 series active in primaries, each replacing only a few feathers in single season and pattern increasingly complicated; centre at p6 probably activated every other year. Secondaries moulted from 3 centres, s12, s5, and s2; 3·6(21) 0–7 secondaries replaced in 2nd calendar year, usually s12–s13 (s11–s14), occasionally s10 and s2, rarely s9 and s5; s12 shed at about same time as p6. In 3rd calendar year, 7·6(7)6–10 secondaries moulted; after this, all juvenile secondaries thus replaced, often excepting s1, s4, or s8; in next year, these last feathers shed, while a 2nd series may start with s12 (s10–s14) and occasionally s2. Tail moult irregular: t6 usually moulted first, followed by t1, t2, or t3; 5·2(13) 0–12 tail-feathers replaced in 2nd calendar year, when t6 shed shortly after p6; remaining feathers, often (t3–)t4–t5, usually replaced in following year; moults similar in later years. Rarely, all tail replaced in single season and occasionally no tail moult at all. No details for body; sequence of replacement as in post-juvenile, mainly August–September; apparently often complete, but at least occasionally partial and perhaps no moult at all in some years. POST-JUVENILE. Juvenile 2nd down completely replaced by 1st adult plumage during 35th–60th day (see Plumages), some feathers still growing by October–November; juvenile flight-feathers, tertials, greater primary coverts, and tail retained.

Measurements. Nominate *alba*. Britain, all year; skins (BMNH, RMNH, ZMA). Bill (F) to forehead (skull).

WING	♂ 289	(6·18; 18)	279–299	♀ 290	(6·19; 13)	280–300
TAIL	115	(3·64; 18)	110–122	115	(5·46; 13)	109–124
BILL (F)	30·8	(0·79; 9)	30–32	32·4	(1·05; 4)	31–33
TARSUS	57·0	(2·90; 8)	54–60	56·8	(2·45; 5)	54–60

Sex differences not significant. Bill to cere 19·0 (1·08; 13) 18–21; middle toe without claw 29·7 (1·04; 13) 28–32; middle claw 16·1 (1·15; 13) 15–18.

T. a. guttata. Netherlands (except south), all year; skins (ZMA). Bill (C) to cere.

WING	♂ 286	(5·44; 54)	275–297	♀ 287	(5·44; 66)	273–298
TAIL	115	(4·21; 23)	108–121	114	(5·20; 22)	107–123
BILL (F)	31·2	(1·73; 17)	29–33	30·8	(1·43; 18)	29–33
BILL (C)	18·8	(0·69; 35)	17–20	19·3	(0·72; 50)	18–21
TARSUS	56·6	(1·88; 49)	53–59	56·0	(1·96; 61)	52–60

Sex differences not significant, except bill (C). Single dwarf ♀ with wing 272 and tarsus 47·5 excluded. Middle toe without claw 28·7 (1·11; 17) 28–30; middle claw 15·4 (1·17; 16) 14–17.

T. a. affinis. Afrotropics, mainly Sénégambia to Sudan, all year; skins (BMNH, RMNH, ZMA).

WING	♂ 293	(8·94; 11)	283–307	♀ 294	(7·61; 7)	285–306
TAIL	114	(3·51; 11)	110–120	116	(6·57; 7)	109–123
BILL (F)	32·4	(1·13; 11)	31–34	33·3	(1·88; 8)	31–36
TARSUS	63·4	(2·58; 11)	60–68	63·8	(1·88; 8)	60–65

Sex differences not significant. Bill to cere 19·3 (1·10; 19) 18–21; middle toe without claw 32·1 (1·70; 19) 30–35; middle claw 16·9 (1·17; 17) 16–19.

Sexes combined: (1) nominate *alba*, Italy; (2) *ernesti*, Sardinia; (3) *erlangeri*, north-west Africa; (4) *erlangeri*, Egypt and Arabia; (5) *erlangeri*, Crete, Cyprus, and Middle East; (6) *schmitzi*, Madeira; (7) *gracilirostris*, eastern Canary islands; (8) *detorta*, São Tiago and Brava (Cape Verde Islands) (BMNH, RMNH, ZMA).

(1) WING	290	(7·57; 9)	277–298	TAIL	113	(3·94; 8)	106–117
(2)	293	(9·08; 8)	284–309		118	(6·48; 8)	110–126
(3)	294	(5·80; 23)	284–304		114	(5·39; 23)	107–124
(4)	300	(6·04; 13)	292–315		117	(5·87; 13)	110–126
(5)	299	(8·54; 11)	286–310		118	(3·97; 11)	112–124
(6)	280	(4·77; 9)	272–285		112	(2·99; 9)	107–116
(7)	257	(3·10; 4)	253–260		99	(3·73; 4)	94–102
(8)	295	(7·80; 5)	283–305		120	(2·99; 5)	116–124

(1) BILL (F)	32·1	(2·05; 6)	30–34	TARSUS	58·3	(1·98; 7)	55–61
(2)	30·7	(1·67; 8)	28–33		61·6	(2·12; 8)	60–66
(3)	31·6	(0·75; 10)	30–33		62·2	(2·25; 11)	59–68

Averages of some other measurements of these populations. Nominate *alba*, Italy: bill to cere 19·3, middle toe without claw 30·2, middle claw 16·3. *T. a. ernesti*: bill (C) 19·2, toe 31·1, claw 16·6. *T. a. erlangeri*, north-west Africa: bill (C) 19·6, toe 30·1, claw 16·9. *T. a. schmitzi*: bill (F) 33·6, bill (C) 20·5, tarsus 64·3, toe 30·3, claw 16·2. *T. a. gracilirostris*: bill (F) 27·2, bill (C) 16·6, tarsus 56·0, toe 27·6, claw 15·2. *T. a. detorta*, Cape Verde Islands, sexes combined: wing 290 (20) 272–299, tail 113 (16) 103–124, bill (C) 20·1 (19) 19–21 (Naurois 1983a).

Weights. ADULT AND FIRST ADULT. Nominate *alba*. Italy, October–March, excluding birds in poor condition: ♂ 271 (19·4; 10) 240–313, ♀ 298 (42·9; 6) 245–360; April–September, ♂ 258 (10·4; 5) 250–275 (Moltoni 1949). France (including some intermediates with *guttata*) 340 (95) (Bunn *et al.* 1982).

T. a. guttata (A) East Germany: (1) pre-laying period; (2) laying period; (3) during incubation and hatching of eggs; (4) feeding young; (5) during winter; (6) in spring (Schönfeld *et al.* 1977). (B) Netherlands: (7) October–December; (8) January–March; (9) starved birds, all year (ZMA).

A (1)	♂ 314	(—; 5)	290–345	♀ 351	(—; 5)	335–380
(2)				436	(—; 11)	380–480
(3)	314	(—; 9)	290–330	392	(—; 37)	330–455
(4)	311	(—; 7)	300–325	358	(—; 14)	320–405
(5)	298	(—; 14)	—	313	(—; 24)	—
(6)	304	(—; 25)	—	323	(—; 31)	—
B (7)	292	(41·5; 12)	250–400	296	(29·4; 11)	252–330
(8)	295	(20·7; 8)	265–327	309	(20·8; 8)	295–350
(9)	220	(14·9; 27)	187–242	221	(13·5; 6)	200–248

T. a. affinis. Southern Africa: 346 (21·7; 16) 295–381 (Biggs *et al.* 1979).

DOWNY YOUNG AND JUVENILE. *T. a. guttata.* 5 main periods discernible: (1) day 1–25, with daily increase 12·1 (10–14), reaching average of 310; (2) day 25–38, daily increase 5·0 (4–6), reaching 370; (3) day 38–48, when constant weight of *c.* 370 maintained; (4) day 48–65, when many feathers grow and chicks very active, causing decrease of 3·2 (3–4) daily, down to *c.* 310; (5) 65th day to fledging (day 80–90), during which lower weight level maintained. At hatching, 13·2–13·6; average on 1st day 14·6 (11), 9th 83 (9), 15th 142 (12), 22nd 264 (14), 28th 309 (12), 35th 377 (11), 42nd 369 (9), 55th–60th 339 (15), 79th 297 (3), 90th 309 (3). (Schönfeld and Girbig 1975.) Nominate *alba.* For growth of young, Rumania, see Radu (1973).

Structure. Wing rather long and narrow, tip rounded. 10 primaries: p9 longest, p10 (0–)2–8 shorter, p8 2–8, p7 18–25, p6 38–45, p5 54–65, p1 103–120. Primaries not emarginated. Outer web of p10 with rather short serrations along edge. Tail rather short, square; 12 feathers, t6 5 mm shorter to 10 mm longer than t1. Facial disc distinct, heart-shaped, surrounded by distinct ruff. Eyes rather small. Bill long, rather slender; protected above by long bristle-like feathers projecting forward from lateral base of upper mandible and from inner corner of eye, only strongly hooked tip of upper mandible and tip of shorter lower mandible not covered by bristles; tip of bill laterally compressed. Leg and toes long, slender; tibia and upper tarsus closely covered by dense woolly feathering, feathers gradually shorter and sparser towards lower tarsus; in some southern races, lower tarsus covered by scanty and short feather-tufts or thin hair-like bristles only. Toes bare except for thin hair-like bristles. For length of middle toe and claw, see Measurements. Outer toe without claw *c.* 75% of middle toe without claw, inner *c.* 93%, hind *c.* 50%. Claws long, slender, decurved; inner edge of middle claw flattened, finely serrated; outer claw *c.* 94% of middle claw, inner *c.* 116%, hind *c.* 99%.

Geographical variation. Marked; mainly clinal, except for insular populations. Concerns size (as expressed in wing and tail length), relative length and strength of bill, tarsus, and toes, proportion of tarsus feathered, and colour. 6 main aspects to colour variation. (1) Amount of grey on feather-tips of upperparts and upper wing-coverts: upperparts yellow with restricted grey on lower mantle, scapulars, and wing-coverts, or most feather-tips grey but still much yellow of feather-bases showing, or fully grey with hardly any yellow visible. (2) Size of droplets on feather-tips of upperparts: varies from small white spot bordered above and below by some black, through larger white spot with bolder black surround, to still larger white spot or small chevron with a 2nd one nearer base of each feather, both situated in long black central streak. (3) Depth of yellow and grey of upperparts and upper wing-coverts: yellow varies between pale yellow and deep golden-yellow or rusty-yellow; grey paler when vermiculation pale and sparse, or darker when vermiculation denser or when colour uniform dark bluish-grey, brown-grey, or blackish-grey. (4) Pattern and colour of flight-feathers and tail: ground-colour varies between white through yellow to rufous-brown; tips with variable amount of paler or darker grey vermiculation (sometimes no grey at all; often with black-and-white droplet near tip as on scapulars); bars on basal and middle portions of feather broad and solidly black, narrower or vermiculated grey, or virtually absent. (5) Colour of face and underparts: varies from completely silky white (usually except for black-brown or black ruff round facial disc and dark spot at eye) or white with pale yellow chest, through pale yellow with whitish wash on belly and vent, to completely pale yellow, deep yellow, or rufous-cinnamon, including variable amount of rufous on face. (6) Amount and size of black spots on feather-tips of underparts: varies from none at all through tiny specks on flanks, larger rounded dots on breast, belly, and flanks, or from chest to vent and thighs, to large blotches (up to 4 mm across) all over or with indication of chevrons; spots occasionally have 2nd spot of similar or smaller size nearer feather-base, in particular in birds with larger spots; in birds with coloured underparts, black of spots often bordered by white, forming poorly-defined drop, white occasionally accentuated by partial and narrow black borders. Most of these colour variations run about parallel: birds with paler yellow upperparts with restricted grey often also show pale underparts, flight-feathers, and tail. However, amount and size of spots frequently rather independent of remainder of colour: birds with upperparts yellow and virtually no grey may show bold black-and-white droplets above; birds with white underparts may show rather large and numerous black spots; birds with extensive and deep grey upperparts may show contrastingly pale and unmarked flight-feathers, tail, or underparts. All races fit into these colour variations, which have no readily apparent ecological basis: thus, a race from a wet temperate region can be very similar to a race from an arid tropical island far away. Also, ♀♀ average usually slightly darker or more heavily marked in all characters than ♂♂, and ♀♀ of some races may resemble ♂♂ of another; individual variation limited in some populations, extensive in others. Remainder of account refers to west Palearctic races only. In view of parallelism in colour (birds from some areas closely resemble others from a quite distant area with differently coloured ones in between), races mainly based on size, and 4 groups then separable. (1) Small *gracilirostris* of eastern Canary Islands, with rather slender bill but relatively strong tarsus and toes. (2) Rather small *schmitzi* of Madeira, with heavy tarsus and toes and lower half of tarsus and all toes bare except for heavy bristles. (3) Rather large nominate *alba* and *guttata* from Europe (except some Mediterranean islands) with rather short and slender tarsus and toes and with tarsus largely feathered. (4) Large *erlangeri* from North Africa, Arabia, Crete, Cyprus, and Middle East, *ernesti* of Corsica and Sardinia, *affinis* from Afrotropical mainland, and *detorta* from Cape Verde Islands, all with long and partly bare tarsus and rather heavy bare toes. In nominate *alba*, population from Britain palest; in ♂, upperparts mainly yellow, underparts fully white except for occasional slight yellow wash to chest or breast or fine black specks on flanks; ♀ with about equal amount of pale yellow and pale grey on upperparts, underparts usually show slight yellow tinge, often with small black spots on breast, belly, and flanks; flight-feathers and tail occasionally very pale yellowish-white with reduced barring and no grey on tips. Nominate *alba* from Italy close to those of Britain, but more often with faint yellow tinge on breast, and flight-feathers rarely as pale; Iberian birds show more spots below than those of Britain and Italy (rather similar to *erlangeri* from north-west Africa), and birds examined from Tenerife and Gran Canaria similar to these (size of latter closer to Iberian birds than to north-west African ones and hence included in nominate *alba*). Birds from western France vary in colour between British and Iberian type; those of northern and eastern France more heavily spotted, intergrading with *guttata*. *T. a. guttata* darkest of west Palearctic forms, except *detorta*; upperparts rather dark grey with limited golden-rufous of feather-bases showing (black-and-white droplets not very marked), flight-feathers and tail well-marked (only exceptionally almost uniform rufous-cinnamon), and facial disc and underparts rufous-cinnamon or rufous-yellow (occasionally paler yellow or partially white in ♂) with rather numerous black spots 1–3 mm across. Populations strongly variable in colour inhabit southern Netherlands, Belgium, German Rhine valley, central Switzerland, and central Yugoslavia, where *guttata* grades into nominate *alba*; for details, see Voous (1950). *T. a. ernesti* from Corsica and Sardinia is palest race; flight-feathers and tail white or yellowish-white with restricted bars to bases and virtually without grey on tips in ♂, slightly more yellowish and more fully barred in ♀ (latter resembling occasional pale British birds); underparts about similar to nominate *alba* from Britain, but upperparts and upper wing-coverts even paler yellow with more restricted grey or almost none at all (restricted to surroundings of droplets on lower mantle and scapulars). *T. a. erlangeri* from Crete, Cyprus, and Middle East as pale as British nominate *alba* or even paler and less marked; populations of Arabia, Egypt, and north-west Africa (south to Ahaggar in Algeria and at least occasionally to Sierra Leone) show more extensive though paler grey on upperparts; underparts in ♂ white, sometimes with yellow tinge on chest, nearly always distinctly spotted black on breast and flanks (unlike birds from Middle East), ♀ fully pale yellow (occasionally with white-tinged central belly) with black spots 2–3 mm across on chest, breast, flanks, belly, and thighs. *T. a. affinis* from Afrotropical mainland intergrades with *erlangeri* in southern Egypt; colour of *affinis* rather similar to North African populations of *erlangeri*, but upperparts darker grey, droplets on upperparts more heavily bordered black; spots on underparts larger, often double, sometimes rather triangular, underparts appearing more profusely speckled; bold dark bars across inner webs of flight-feathers and on tail. Madeiran race *schmitzi* rather like north-west African *erlangeri* in spotting, but upperparts darker grey (close to Afrotropical *affinis*; rather bluish, not as brown as in *guttata* from central Europe), underparts rather extensively yellow (like *affinis*). *T. a. gracilirostris* of eastern Canary Islands similar to *schmitzi* in spotting and colour of upperparts, but underparts (though variable) on average darker, yellowish-white to all-yellow in ♂, yellow or deep golden-yellow in ♀ (like *guttata*). Cape Verdian *detorta* darker than *guttata*, underparts on average deeper rufous-cinnamon with spots 2–4 mm across, upperparts much darker greyish-brown with large double white droplets surrounded by much black. In summary, colour gradually darker and specks heavier in sequence as follows: Sardinia and Corsica, Middle East, Britain, Italy, Iberia, north-west Africa, Egypt, Arabia, Afrotropics, Madeira, eastern Canary Islands, cental Europe, Cape Verde Islands; at the same time, ♀♀ from (e.g.) Middle East, Britain, and Italy resemble ♂♂ from Egypt and Arabia, ♀♀ from Iberia and north-west Africa resemble ♂♂ from Afrotropics and Madeira, and ♀♀ from Arabia, Afrotropics, and Madeira somewhat similar to ♂♂ of eastern Canary Islands, central Europe, and Cape Verde Islands. For extralimital races, see (e.g.) Hartert (1912–21) and Burton (1973a).

CSR

Family STRIGIDAE typical owls

Small to large predators, mostly nocturnal or crepuscular. 2 subfamilies, Buboninae and Striginae, both represented in west Palearctic. However, this division (based mainly on poorly defined differences in structure of ear—see Norberg 1977, also Buboninae, below, and Striginae, p. 525) probably not natural. Development of ear linked with latitude (hearing less important for prey location in noisy tropical environment), and (e.g.) tropical *Ciccaba* (usually placed in Buboninae) probably closely related to mainly boreal or temperate genus *Strix* (Striginae). A division of Strigidae into 3 groups based on marked differences in moult may be better founded: (1) *Otus, Glaucidium, Micrathene, Athene, Speotyto, Surnia, Rhinoptynx, Asio*; (2) *Bubo, Ketupa, Nyctea* (and perhaps *Pulsatrix*); (3) *Ciccaba, Strix, Aegolius* (C S Roselaar).

Heads round, some with ear-tufts. Breastbone and wishbone separate. Wings variable; 10 primaries; wing diastataxic. Feathers without aftershaft. Tails variable (10–)12 feathers. Bills variable, but relatively shorter than in Tytonidae, though large and powerful. Eyes relatively larger than in Tytonidae. Legs variable. Inner toe shorter than middle toe. No serrations on middle claw. Oil-gland naked or virtually so. Plumages variable, but often brown, cream, or buff; sexes usually more or less similar, though ♀♀ tend to be larger and darker. Post-breeding moult variable, either complete with primaries descendant (e.g. *Otus, Surnia, Glaucidium, Athene*) or complex and serial (e.g. *Bubo, Ketupa, Nyctea, Strix, Aegolius*). Young nidicolous. Juvenile plumages variable: feathering often reduced to long feather-like down (mesoptile), though not to same extent as in Tytonidae; patterned as in adult or with simpler pattern. Adult plumage acquired at 2–15 months.

Subfamily BUBONINAE eagle owls and allies

About 100 species in *c.* 20 genera world-wide, including the following, each with 1 (or, in *Otus*, 2) representatives in west Palearctic, all breeding: (1) *Otus* (*c.* 30 species, world-wide except northern Eurasia and Australia); (2) *Bubo* (11–12 species, world-wide except Australasia); (3) *Ketupa* (4 species, southern and eastern Asia); (4) *Nyctea* (1 species, northern Holarctic); (5) *Surnia* (1 species, northern Holarctic); (6) *Glaucidium* (12–13 species, world-wide except Australasia); (7) *Athene* (3 species, Eurasia and North Africa).

For general features, moults, etc., see Strigidae. Ears relatively smaller than in Striginae, with small symmetrical ear-flaps. Facial disc reduced.

Otus brucei Striated Scops Owl

PLATES 44 and 54
[facing pages 494 and 566]

Du. Gestreepte Dwergooruil Fr. Petit-duc de Bruce Ge. Streifenohreule
Ru. Буланая совка Sp. Autillo pálido Sw. Blek Dvärguv

Ephialtes Brucei Hume, 1873

Polytypic. *O. b. obsoletus* (Cabanis, 1875), Syria, northern Iraq, Turkmeniya, northern Afghanistan, and lowlands of Uzbekistan east to Samarkand; *exiguus* Mukherjee, 1958, Israel, central Iraq, southern Iran, southern Afghanistan, and western Pakistan north to Bannu and perhaps Kohat. Extralimital: *pamelae* Bates, 1937, Arabia (except north-east) and west of Zagros mountains in Iran at 30°N (and perhaps this race in intervening southern Iraq and in southern Israel or Sinai), intermediates with *socotranus* in Hadhramawt (Yemen) and Dhofar (Oman); *socotranus* (Grant and Forbes, 1899), Socotra; nominate *brucei* (Hume, 1873), eastern Aral Sea, Syr-Dar'ya valley, Fergana basin, and mountain valleys of northern Tadzhikistan and Kirgiziya; *semenowi* (Zarudny and Härms, 1902), Tarim basin (China) east to *c.* 88°E and mountain valleys of southern Tadzhikistan, eastern Afghanistan, and Pakistan north of 35°N.

Field characters. 20–21 cm; wing-span 54–64 cm. Averages larger than Scops Owl *O. scops*. Closely resembles commoner and more widespread *O. scops* but usually greyer and more finely streaked, appearing paler. Lacks white spots on crown and hindneck and on outer scapulars; shows paler face. Restricted to arid regions. Sexes similar; no seasonal variation. Juvenile separable.

ADULT. Differs most noticeably from *O. scops* in: (1) generally paler plumage with grey or ochre tones, lacking any rufous; (2) much paler surround to eye; (3) absence of white spots on crown and hindneck; (4) fainter and finer vermiculations and streaks; (5) less obvious paler spots on outer scapulars; (6) paler, less marked underparts; (7) if visible, pale horn-yellow base to bill. JUVENILE. Plumage less downy than in larger owls. Upperparts rather dark, grey-brown with narrow off-white barring; underparts completely barred, unlike adult and juvenile *O. scops*.

Due to wide overlap of both breeding and migratory ranges with *O. scops*, no small eared owl in Middle East identifiable without most careful and patient observation of characters noted above. Flight and behaviour as *O. scops*.

Commonest call a long series of short, soft sounds—'wup-wup-wup-. . .'—notes repeated more rapidly than in *O. scops*.

Habitat. In dry warm lower middle latitudes of west Palearctic. Breeds in USSR in riverine forest but chiefly farmlands, cultivated orchards, and towns in steppe lowlands, staying out of mountains. Abundant in foothills; nests in hollow trees, especially poplar *Populus diversifolius*, white willow *Salix*, and sometimes on leafless saxaul tree *Arthrophytum* in the open (Dementiev and Gladkov 1951*a*). Characteristic habitat in Uzbekistan is cultivated landscape with arable fields, parks, and gardens; also in thickets and woodlands along rivers, provided there are some older trees with suitable holes (Bakaev and Salimov 1973). In Iraq, nests in old nests of Magpie *Pica pica* (Marchant 1963*c*). In Pakistan in arid semi-desert and stony foothills up to *c.* 1800 m, hiding during daytime in tree holes or dense foliage, or standing bolt upright close to tree trunk, perfectly camouflaged. Will hunt by day, sitting in pistacio *Pistacia* tree or on rock ledge. Nests in hole in tree trunk or date palm (Ali and Ripley 1969). Flies in lower airspace, for short distances except on migration. Comprehensive and detailed data lacking.

Distribution. Imperfectly known. SYRIA. Breeding reported near Aleppo 1919 (Clarke 1924; Kumerloeve 1968*b*) now thought to be erroneous (HK), but probably breeds elsewhere. ISRAEL. May breed occasionally, but no nests found (HM). JORDAN. May breed but no proof (DIMW).

Population. IRAQ. Abundant, Habbaniya (Chapman and McGeoch 1956); common Hilla and Museyib (Ticehurst *et al.* 1922).

Movements. Migratory in USSR and partially so in Iraq. Inadequate information from elsewhere, though presumed resident in Arabian peninsula.

Summer visitor to USSR (Soviet Central Asia only), arriving late March and April and leaving September–October (Dementiev and Gladkov 1951*a*). Similarly in Iraq, from which December–February museum skins exist; birds normally encountered March–October only (Marchant 1963*a*). Winter ranges inadequately known. Nominate *brucei* (breeding northern Soviet Central Asia) has been collected in autumn and winter in Pakistan and in Bombay region of western India, but not elsewhere (Ali and Ripley 1969; G P Hekstra, C S Roselaar); no corresponding data for *semenowi* (breeding further east in central Asia), also suspected to be migratory.

Degree of movement by *obsoletus* and *exiguus* (breeding Turkmeniya, Uzbekistan, and Sinai to Pakistan) remains unclear due, in part, to inadequate definition of breeding ranges. Perhaps short-distance migrants or dispersive, but emigrants (e.g. from Iraq) presumably winter further south in Middle East, though known certainly only from a single specimen of *exiguus* collected Bahrain in October (Bundy and Warr 1980; G P Hekstra, C S Roselaar). Birds

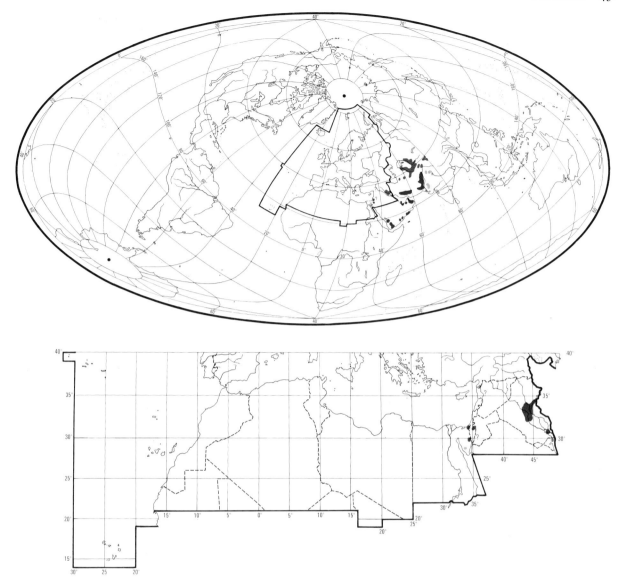

captured several times recently at Elat, southern Israel (Mikkola 1983), but these not necessarily long-distance migrants. Arabian race *pamelae* present there all year, and believed resident (e.g. Gallagher and Woodcock 1980).

Food. Insects, and small mammals, birds, and lizards. Prey taken in flight, or from ground or tree, etc. Will hunt from perch or by aerial-pursuit, to take (e.g.) moths (Lepidoptera) and bats (Chiroptera) (Ticehurst *et al.* 1922; Meinertzhagen 1959; Gavrin *et al.* 1962). Will hunt under artificial light, sometimes becoming tame (Ticehurst *et al.* 1926; Chapman and McGeoch 1956).

In Zeravshan basin (USSR), 4 stomachs contained moths (Lepidoptera), beetles (Cerambycidae, Cetoniinae, Geotrupidae, Carabidae), and locusts (Acrididae); 12 pellets from a nest contained 12–13 crickets (Gryllidae), 5–7 mole-crickets (Gryllotalpidae), 65–67 eggs of Acrididae, 9–11 Geotrupidae, and 7–8 other beetles; Gryllotalpidae taken by chicks, and stomach of fledgling contained Geotrupidae and other insects. On lower Syr-Dar'ya, remains by nests consisted of insects, small passerines, bats (Chiroptera), house mouse *Mus musculus*, and gerbil *Meriones tamariscinus*. (Bakaev and Salimov 1973; Pukinski 1977.) In Iraq, remains of House Sparrows *Passer domesticus* found in nest (Ticehurst *et al.* 1926). Said to chase bats constantly (Meinertzhagen 1959).

At nest with 2 young 18–20 days old, Zeravshan basin, adult brought food on average 3 times per hr in period 21.00–06.00 hrs; at 25–27 days, 8 times per hr (Bakaev and Salimov 1973). On lower Amu-Dar'ya, insects and small lizards *Eremias* brought to 3 chicks 29 times in 1 day (Rashkevich 1965).

DJB

Social pattern and behaviour. Little known. Following account based largely on extralimital material.

1. Nothing known about dispersion outside breeding season, but presumably solitary (see Bonds, below). BONDS. Probably monogamous, as in Scops Owl *O. scops*. ♂ highly attentive to ♀ (see Heterosexual Behaviour, below). Pair-bond thought to be 'long-lasting' (Gavrin *et al.* 1962). At start of breeding season, Zeravshan river (Uzbekistan, USSR), birds arrive singly (Bakaev and Salimov 1973), suggesting pair-bond not maintained outside breeding season. Age of first breeding not known. Only ♀ incubates, but both sexes care for young (Pukinski 1977). After leaving nest, young fed by parents for 2–4 days nearby (Bakaev and Salimov 1973). No information on duration of family bond thereafter. BREEDING DISPERSION. Solitary and territorial. Most nests apparently well dispersed. At Dzhilvan (USSR), 2 nests in 0·8 ha; at Sarazm (Uzbekistan, USSR), 2 nests in 2·5 km^2 (Bakaev and Salimov 1973, which see for other references to apparently low densities). In riverine woodland, lower Amu-Dar'ya (Uzbekistan, USSR), mean 0·3–0·8 birds per 3 km transect (Rashkevich 1965). At Tigrovaya balka (Tadzhikistan, USSR), 10–12 pairs in *c.* 400 km^2 (Potapov 1959). ROOSTING. Rarely seen by day, when hides in tree-holes or dense foliage. While ♀ incubating, ♂ sits in neighbouring tree (Kislenko 1972). In Iraq, bird called at midday from roost on branch under dense canopy of small tree (Chapman and McGeoch 1956: see 1 in Voice). Said to hunt only after sunset, Turkmeniya, USSR (Dementiev 1952), but seen feeding before dusk, Iraq (Ticehurst *et al.* 1922).

2. Tame and confiding (Pukinski 1977); in Iraq, feeds within a few feet of people (Ticehurst *et al.* 1926; Chapman and McGeoch 1956). When disturbed, adopts Sleeked-upright posture (for photograph, see Pukinski 1977): stands bolt upright, often pressed close to trunk, eyes half closed and plumage sleeked, as in *O. scops* (Gavrin *et al.* 1962; Ali and Ripley 1969). Only flies a short distance if disturbed by day (see Ali and Ripley 1969). ANTAGONISTIC BEHAVIOUR. See below. HETEROSEXUAL BEHAVIOUR. Advertising-call (see 1 in Voice) of ♂ given often for long spells at night (Pukinski 1977); presumably advertises presence to both potential mates and rival ♂♂. ♀ regularly fed (presumably during incubation) by ♂ who is very attentive to her, rarely straying far away; pair may even hunt together (Pukinski 1977), see also Roosting, above. For calls during copulation, see 2b in Voice. No further information. RELATIONS WITHIN FAMILY GROUP. Eyes of young open slightly after 3 days (Pukinski 1977, fully by 7th day. At hatching, young appear weak and unresponsive but react to disturbance by calling. Can stand on feet from 3rd day (Bakaev and Salimov 1973). Fully feathered well before leaving nest (Gavrin *et al.* 1962). For provisioning after leaving nest, see part 1. At Badkhyz (Turkmeniya, USSR), young remained close to nest in dense groves throughout summer (Pukinski 1977). ANTI-PREDATOR RESPONSES OF YOUNG. Very fierce when removed from nest and taken into captivity (Clarke 1924). PARENTAL ANTI-PREDATOR STRATEGIES. (1) Passive measures. Sits tightly (Kislenko 1979); behaved thus in all nests found, not flushing until human intruder looked into nest or inserted hand (Marchant 1963c). Sitting bird sometimes allowed eggs to be removed without flushing (Pukinski 1977). Contact-alarm call (see 2a in Voice) apparently serves to warn young in nest (Pukinski 1977), perhaps when threatened. EKD

Voice. Freely used during breeding season, especially at start. By end of season, birds quiet and unobtrusive (Pukinski 1977). Bill-snapping occurs (Ali and Ripley 1969; Gallagher and Woodcock 1980); presumably defensive as in other Strigidae.

CALLS OF ADULT. (1) Advertising-call of ♂. A long, monotonous sequence of short, soft sounds: 'wup-wup-wup-...', like sound of distant well-pump, and variable in tonal quality between individuals (Gallagher 1977); also rendered 'bup bup bup...', with 1–3 s between each long series (Gallagher and Woodcock 1980); short monotonous 'boo-boo...', resembling Stock Dove *Columba oenas* (Ticehurst *et al.* 1922), also described as 'booming' (Ticehurst *et al.* 1926). In recording by T Roberts (Baluchistan), relatively low-pitched, rhythmic 'boo-boo-boo-...', at rate of 67 units per min, quite different from less frequent (22–26 units per min) Advertising-call of Scops Owl *O. scops* (E K Dunn). Widely considered different from *O. scops*. Often given for ½ hr before stopping, almost continually on warm, still nights (Pukinski 1977); until dawn (Meinertzhagen 1954); occasionally during day (see Social Pattern and Behaviour). (2) Contact-alarm calls. (a) Call resembling sharp barking of a dog, 'au', apparently given when feeding young, and said to be an 'alarm signal' directed at them (Potapov 1959). (b) Loud, sharp 'tsirr va va vau' (Pukinski 1977); 'tsirr-va-vaa', beginning with a warble (Dementiev and Gladkov 1951a). Given in variety of contexts; sometimes precedes call 1, also sometimes given during copulation, or in response to any disturbance, e.g. man talking (Pukinski 1977). (3) Other calls. A sort of wheezing squeak, like noise produced when inflating car tyre with pump (Ali and Ripley 1969); 'mews' reported by Gallagher and Woodcock (1980) possibly the same.

CALLS OF YOUNG. Small young respond to disturbance with tremulous squeaking (Bakaev and Salimov 1973). This presumably the food-call, described as a persistent, rhythmic squealing (Pukinski 1977). Also described as a wheezing sound (Ticehurst *et al.* 1926). EKD

Breeding. SEASON. Southern USSR: first eggs laid in last third of April, complete clutches found early May (Dementiev and Gladkov 1951a); fresh eggs, Uzbekistan, last half June (Bakaev and Salimov 1973). Iraq: eggs laid mid-April to late May (Marchant 1963c). SITE. In hollow in tree, hole in river bank or in building, or in old nest of Magpie *Pica pica*; also in large nest-boxes. Of 45 nests, lower Amu-Dar'ya (USSR), 30 in *P. pica* nests, 14 in holes in trees, 1 in ventilator. Older birds occupy tree-hole sites first, young birds being forced to take suboptimal *P. pica* nests (Gavrin *et al.* 1962). Of 11 nests, Uzbekistan (USSR), 5 in willow *Salix*, 3 in mulberry *Morus*, and 3 in *P. pica* nests (Bakaev and Salimov 1973). Occasional mixed clutches of *O. brucei* and *P. pica* found, with *O. brucei* dominating, suggesting *P. pica* may sometimes be evicted from nest (Rashkevich 1965). Also recorded taking over holes of woodpeckers (Picidae) (Dementiev 1952). Height above ground 3–4·5 m (5 nests, Iraq) (Marchant 1963c); one at 6·5 m (Kislenko 1972); one at *c.* 6 m, Iraq (Ticehurst *et al.* 1926). Nest: 7 of 11 cavity nests, Uzbekistan (USSR) had linings of leaves or twigs, 4 had only wood dust; *P. pica* nests may be strengthened and relined (Bakaev and Salimov 1973). In Iraq, nests of *P. pica* not altered or added to (Marchant 1963c). Hole in date palm, Iraq, *c.* 45 cm deep (Ticehurst *et al.* 1926). Building: no information. EGGS. See Plate 97. Short elliptical, smooth and slightly glossy; white. 31 × 27 mm (29–34 × 26–29), *n* = 60 (Schönwetter 1967). Clutch: 4–5 (3–6). Replacements probable after egg loss (Marchant 1963c). Laying interval 24 hrs (Potapov 1959) or 48 hrs (Pukinski 1977). INCUBATION. About 28 days (Dementiev and Gladkov 1951a). 26–28 days (Bakaev and Salimov 1973). Begins with 1st egg, but not uncommonly begins properly when 2–3 eggs laid (Pukinski 1977); hatching asynchronous. By ♀ only (Bakaev and Salimov 1973). YOUNG. Semi-altricial and nidicolous. Cared for and fed by both parents. Leave nest-hole at 26 days, and fed by parents near entrance for 2–4 nights (Bakaev and Salimov 1973). FLEDGING TO MATURITY. Fledging period 28–30 days (Bakaev and Salimov 1973); *c.* 21–25 days (Dementiev and Gladkov 1951a; Marchant 1963c). Age of independence and of first breeding not known. BREEDING SUCCESS. Of 18 eggs found in 4 nests, Uzbekistan, 11 hatched and 6 fledged; 1 full clutch of 5 destroyed by children, 2 eggs infertile; 1 chick, last to hatch in brood, disappeared, and fate of 4 others unknown (Bakaev and Salimov 1973).

Plumages (*exiguus*). ADULT. Crown, nape, mantle, scapulars, back to upper tail-coverts, tertials, and upper wing-coverts dark grey-buff; feathers with black shaft-streaks, forming dark spots on crown and sharp streaks on rest of upperparts, sides of feathers with tiny grey speckling, specks sometimes joining to faint narrow bars. Some pink-buff of feather-bases of hindcrown visible; outer webs of outer scapulars and outer median upper wing-coverts uniform cinnamon-buff or pink-buff with black tip. Facial disc pale buff with tiny grey specks; chin almost uniform cream-white; bristles at sides of bill white, lateral ones tipped black. Feathers of ruff bordering facial disc dull black at tips, pale cinnamon-pink subterminally; this dark border poorly developed near forecrown and chest. Ground-colour of underparts pale buff, grading to cream-white on vent and to white on under tail-coverts; all feathers with narrow black shaft-streak (widest on chest), feathers of chest, breast, and flanks with tiny grey specks. Feathering on tarsus buff with fine grey spots and streaks. Tail as in Scops Owl *O. scops*, but light bars wider, less speckled at borders; on t1, *c.* 4 cream bands each *c.* 4 mm wide visible beyond upper tail-coverts; *c.* 1 cm of tail-tip freckled grey or pale buff, without traces of bands up to tips (unlike *O. scops*). Flight-feathers as in *O. scops*, but tips pale buff or grey-white freckled grey, more uniform than in *O. scops* and without broad bands; pale buff and pink-white spots on outer webs of outer primaries relatively wider, about as wide as intervening grey-speckled bands. Under wing-coverts and axillaries cream-white with indistinct grey-brown streaks, greater under primary coverts brown, forming contrasting dark patch. Some variation occurs: upperparts sometimes slightly purer grey and underparts paler cream-white than described above; black shaft-streaks sometimes broader and grey specks on sides of feathers often joining to form faint vermiculations on both upperparts and underparts. DOWNY YOUNG. 1st down (neoptile) thick, white (Bakaev and Salimov 1973). JUVENILE. 2nd down (mesoptile) softer and slightly less feather-like than in Scops Owl *O. scops*, colour not similar to adult as in that species. Mantle, scapulars, back, rump, and lesser upper wing-coverts closely barred cream or pink-buff and grey, feather-tips finely speckled dark grey; lower scapulars with tip and base white. Median upper wing-coverts irregularly and coarsely barred cream-white and dark grey or grey-brown, cream-white partly freckled grey; broad tips of coverts white. Feathering of tarsus uniform cream. Upper tail-coverts, tail, tertials, flight-feathers, and primary coverts as adult. FIRST ADULT. Not sufficiently studied; probably differs from adult in same characters as 1st adult *O. scops* differs from adult.

Bare parts. ADULT AND FIRST ADULT. Iris yellow, bright yellow, or golden-yellow. Bill horn-brown, almost black on tip, yellowish at base or middle of lower mandible. Foot olive-grey or dark grey, soles paler, claws dark horn-brown. DOWNY YOUNG AND JUVENILE. Eye covered by bluish membrane until 7 days; from then, iris greyish-yellow or yellow. Bill, foot, and claws greyish-pink at hatching; bill with distinct square white egg-tooth; on *c.* 14th day, egg-tooth lost, bill-tip, feet, and claws darkening to brownish-horn. (Hartert 1912–21; Dementiev and Gladkov 1951a; Ali and Ripley 1969; Bakaev and Salimov 1973; BMNH.)

Moults. ADULT POST-BREEDING. Complete; primaries descendant. Starts with loss of p1; in Turkmeniya (USSR), starts from late May to early July (ZMM). In Iraq, all primaries new November–December (BMNH). Of those examined, none had primary moult score of more than 10—thus no details known on duration of moult; timing and sequence of replacement of body, tail, and secondaries probably as in *O. scops*, and probably with same difference between non-migratory and migratory populations. POST-JUVENILE. Partial: juvenile flight-feathers, primary coverts, and greater coverts retained. Starts soon after fledging; juveniles found in almost complete 1st adult plumage from late July (Turkmeniya); one from Iraq still in heavy moult in September, another completed by then.

Measurements. *O. b. exiguus*. Iraq, all year; skins (G P Hekstra, BMNH). Bill (F) to forehead; to cere, on average 7·3 less.

WING	♂	153	(4·07; 7)	148–158	♀ 158	(3·27; 6)	154–162
TAIL		70·9	(4·39; 6)	66–76	73·1	(3·32; 8)	69–78
BILL (F)		19·0	(0·83; 6)	18·2–19·8	18·8	(0·71; 8)	17·9–20·1
TARSUS		30·7	(1·62; 6)	28·4–32·0	31·3	(1·58; 8)	29·4–33·5

Sex differences not significant.

Sexes combined. (A) *exiguus*: (1) southern Iran and western Pakistan; (2) Bahrain and northern Oman, (3) Israel. (B) *obsoletus*: (4) Syria; (5) Turkmeniya and Uzbekistan. (C) Nominate *brucei*: (6) Syr-Dar'ya, Fergana basin, Kirgiziya, and wintering birds India. (D) *semenowi*: (7) Tarim basin, northern Pakistan, Wakhan (Afghanistan), and Pamir. (E) *pamelae*: (8) western Arabia, (9) Hadhramawt (Yemen) and Dhofar (Oman). (F) *socotranus*: (10) Socotra. (G P Hekstra.)

		WING			TARSUS		
(A)	(1)	157	(3·61; 19)	152–163	28·5	(0·95; 16)	26–30
	(2)	153	(2·45; 4)	151–156	30·0	(0·71; 4)	29–31
	(3)	147	(6·28; 8)	139–156	29·6	(1·78; 8)	26–32
(B)	(4)	158	(5·94; 4)	150–163	29·6	(0·50; 4)	29–30
	(5)	160	(4·09; 23)	155–168	30·5	(—; 2)	30–31
(C)	(6)	164	(3·14; 38)	158–170	30·7	(2·19; 21)	28–35
(D)	(7)	165	(2·86; 12)	160–169	28·8	(1·09; 9)	28–30
(E)	(8)	144	(4·46; 8)	136–148	26·9	(1·19; 8)	26–29
	(9)	138	(4·03; 4)	134–143	26·1	(0·85; 4)	25–27
(F)	(10)	131	(3·00; 20)	125–136	23·3	(1·04; 21)	22–25

Average tail length runs parallel to wing length: from 58·9 in *socotranus* to 78·9 in *semenowi*; average bill (F) 19·5–19·9 in nominate *brucei* or *semenowi*, 19·4–19·8 in Socotra and Arabian birds, 18·8–19·1 in others.

Weights. *O. b. exiguus.* Israel, December: ♂ 100 (Museum of Tel Aviv).

O. b. obsoletus. USSR: ♂ 110 (Dementiev and Gladkov 1951a); at hatching 7·8–10·5, at 7th day average *c.* 32, at 14th *c.* 71, at 21st *c.* 88, at 26th (when fledging) *c.* 91 (Bakaev and Salimov 1973).

O. b. pamelae. Dhofar (Oman), September–October: ♀♀ 62, 71 (Gallagher and Rogers 1980; BMNH).

Structure. Wing relatively shorter broader, and with more rounded tip than *O. scops*. 10 primaries; in *exiguus* and *obsoletus*, p8 longest, p9 1–5 shorter, p10 16–24, p7 0–2, p6 4–8, p5 12–18, p4 20–27, p1 36–45(–48); tip of p10 in closed wing lies between tips of p4 and p5, rarely equal to p5 (in *O. scops*, between p5 and p6 or about equal to p6). In nominate *brucei* and *semenowi*, p8 longest, p9 1–5 shorter, p10 17–23, p7 0–2, p6 4–9, p5 15–19, p4 23–27, p1 39–46; tip of p10 between tips of p4 and p5; in *pamelae* and *socotranus*, p7 longest, p8 0–2 shorter, p9 4–9, p10 17–24, p6 2–5, p5 9–13, p4 16–22, p3 21–28, p1 (27–)30–39; tip of p10 equal to tip of p4 or between p3 and p4. Tail and tarsus relatively longer than *O. scops* (except tarsus of *socotranus*). Feathering of tarsus extends on to basal half of 1st joints of toes, sometimes covering almost all 1st joint (toes bare in *O. scops*). Middle toe without claw 17·0 (7) 15·4–18·6, middle claw 10·0 (6) 9·7–10·7 (in *exiguus* and *obsoletus*). Remainder of structure as *O. scops*.

Geographical variation. Complex. Not known whether species as treated here forms an entity or should be split into 2, and not fully understood whether *O. brucei*, Scops Owl *O. scops*, Oriental Scops Owl *O. sunia*, and African Scops Owl *O. senegalensis* should be considered as conspecific, as separate species within a superspecies, or less closely related than this. *O. brucei* abruptly replaced by *O. sunia* in Pakistan south-east of 35°N and further east, and in Afrotropics by *O. senegalensis*, pointing to single superspecies or even to single species; on the other hand, *O. scops* undoubtedly closely related also, though partly overlapping in range with *O. brucei*, and some authors treat *O. sunia* and *O. senegalensis* as races of *O. scops* rather than of *O. brucei*; see also Weijden (1973) for similarities of voice of West African *O. senegalensis* and *O. scops*, and Gallagher and Rogers (1980) for comparison of voice of *pamelae* and *O. senegalensis*. *O. brucei* as treated here separable into 4 distinct size-groups: (1) large nominate *brucei* and *semenowi* from central Asia, of which at least nominate *brucei* migrates to northern India and perhaps *semenowi* a partial migrant also; wing of 83% of 51 examined over 161; (2) middle-sized *exiguus* and *obsoletus* from Levant and north-east Arabia east to Uzbekistan and western Pakistan; wing mainly 150–161, only some Turkmeniyan *obsoletus* over this and some Israel *exiguus* below this; (3) small *pamelae* in western, southern, and central Arabia and in south-west Iran, wing mainly 136–149; (4) tiny *socotranus* from Socotra, wing mainly 125–135. *O. b. exiguus* rather greyish; ground-colour of upperparts pale cream-grey with much lateral speckling on feathers, underparts similar or purer pale grey, with rather wide but poorly defined black streaks (see Plumages). *O. b. obsoletus* more sandy-buff; ground-colour of upperparts pale buff-brown with limited lateral speckling, underparts buff or cream-buff; dark streaks on head and body narrower, but more sharply defined than in *exiguus*. Nominate *brucei* rather grey, like *exiguus*, ground-colour of upperparts sandy-grey; black shaft-streaks and limited amount of grey vermiculation finer than both *exiguus* and *obsoletus*; *semenowi* similar to nominate *brucei*, but ground-colour deep ochre-yellow and black streaks sometimes slightly broader. Arabian *pamelae* has buff-brown ground-colour; rather like *obsoletus*, but with heavier grey vermiculation and broader black shaft-streaks; sometimes rather like *O. scops*, but without rufous tinges; *socotranus* similar but purer grey on upperparts and whiter below, vermiculation and shaft-streaks heavier. Birds from north-east Yemen and southern Oman rather similar in colour to *socotranus* but nearer *pamelae* in size. Relationships between *pamelae* and *exiguus* obscure; small birds closely similar to *pamelae* occur in Iran west of Zagros mountains at *c.* 30°N, while *exiguus* occurs nearby in central Iraq and Zagros mountains; birds from Israel comprise either both small *pamelae* and larger *exiguus*, form a population intermediate in character, or are migrants (G P Hekstra, C S Roselaar.)

Recognition. Closely similar to *O. scops*, but more uniform above; no white spots on hindcrown and mantle, no rufous-brown on lower mantle, scapulars, upper wing-coverts, and chest; pale spots on outer scapulars and upper wing-coverts smaller, buff rather than cream or white, less contrasting; underparts more uniform, with narrower black shaft-streaks than *O. scops* and without fine dark bars or vermiculation and white subterminal bars of that species; facial disc more uniform, ruff round disc even less pronounced; broad tips of tail- and flight-feathers finely vermiculated (*O. scops* evenly barred up to tip), light bars on tail and primaries less accentuated by dark borders. Juvenile differs from juvenile *O. scops* in barred instead of closely freckled or vermiculated upperparts and in less closely barred and freckled underparts; differences in tail and wing as for adult. See also Structure.

GPH, CSR

Otus scops Scops Owl

PLATES 44 and 54
[facing pages 494 and 566]

Du. Dwergooruil   Fr. Hibou petit-duc   Ge. Zwergohreule
Ru. Обыкновенная сплюшка   Sp. Autillo   Sw. Dvärguv

Strix Scops Linnaeus, 1758

Polytypic. Nominate *scops* (Linnaeus, 1758), France, Sardinia, and Italy east to Volga, south to northern Greece, northern Turkey, and Transcaucasia; *mallorcae* Jordans, 1923, Iberia, Balearic Islands, and (probably this race) north-west Africa; *cycladum* (Tschusi, 1904), southern Greece (Pelopónnisos, Cyclades, Crete), east through southern Asia Minor to Adana region (Turkey) and south to Israel and Jordan; *cyprius* (Madarász, 1901), Cyprus; *turanicus* (Loudon, 1905), Iraq (and perhaps this race in south-east Turkey) east through Iran and southern Transcaspia to Afghanistan; *pulchellus* (Pallas, 1771), Volga east to Lake Baykal, south to Altay and Tien Shan mountains.

Field characters. 19–20 cm; wing-span 53–63 cm. 2nd smallest owl of west Palearctic; about 15% larger (and wider-headed) than Pygmy Owl *Glaucidium passerinum*; 10% smaller and less round-headed than Little Owl *Athene noctua*; averages smaller than Striated Scops Owl *O. brucei*. Small, but large-headed eared owl, with plumage and silhouette somewhat recalling those of *Asio* owls but only half their size. Can appear stocky on perch but looks longer winged than other small owls in flight. Plumage pattern rather uniform; colour variable, with most birds brown-grey and some rufous-brown. Expression lacks frown of *A. noctua*; more sorrowful. Apart from ear-tufts, most striking features are dark surround to eye, line of pale spots on scapulars, and pale-barred primaries. Commonest call a soft but penetrating, slowly repeated monosyllable. Sexes similar; no seasonal variation. Juvenile inseparable.

ADULT. Brown-grey to rufous-brown above, with black-brown streaks, bars, and vermiculations; sides of crown, ears, and scapulars fringed white or pale buff, with marks on scapulars enhanced by almost black inner webs and large enough to show as spots. Sides of mantle usually more rufous. Face below eyes shows incomplete grey disc enclosing dark grey-brown surround to pale yellow eye. Wing-coverts and inner flight-feathers as back; inner flight-feathers only faintly barred, outer 6 primaries broadly barred buff-white. Brown to buff-white below, with black-brown streaks (widest on flanks), bars, and vermiculations. Underwing and vent buff-white, only faintly marked. Legs noticeably thin, feathered buff-white. Bill blue-black; feet grey. JUVENILE. Plumage less downy than in larger owls. Streaks and vermiculations on underparts less obvious than on adult.

At close range, confusable only with *O. brucei*; see that species for distinctions. At distance, separation from *A. noctua* difficult in poor light but *O. scops* has more upright stance, often lengthier silhouette, less fierce appearance, and dissimilar plumage pattern with few bold marks (in *A. noctua*, pale spotted above and boldly streaked below). Distinction from *A. noctua* also assisted by much less undulating and bounding flight, with quite rapid, more regular wing-beats maintaining more even track. Relaxed bird on perch noticeably squat, with flat crown and often little sign of ears; posture of alarmed bird noticeably attenuated as in Long-eared Owl *Asio otus*. Gait restricted, with shuffling walk and occasional lope after prey.

Territorial call characteristic sound of south European summer—'tyuu', pitched a little higher by ♀ (difference noticeable in duet). Beware confusion with calls of midwife toad *Alytes obstetricans* and *G. passerinum* (see Voice).

Habitat. Breeding in west Palearctic in warm dry lowlands in middle and lower middle latitudes, mainly continental temperate and Mediterranean, but also steppe and oceanic. Occurs up to 1400–1500 m in Alpine region and to 2000 m in southern USSR; in Himalayas also locally up to *c.* 1500 m; in juniper *Juniperus* forest in Baluchistan to *c.* 2500 m (Dementiev and Gladkov 1951a; Ali and Ripley 1969; Glutz and Bauer 1980). As a nocturnal arboreal species hunting in the open, requires ample cover of trees providing quiet shaded roosting sites and nest-holes adjoining open ground rich in large insect prey; accordingly, avoids both closed forest and extensive open tracts, preferring broad-leaved and mixed open woodland with underbrush and old hollow trees in USSR (Dementiev and Gladkov 1951a). In Camargue (France), favours riverain forest dominated by white poplar *Populus alba* with common elm *Ulmus campestris*, willow *Salix*, alder *Alnus*, and other broad-leaved trees; also groups of elms, poplars, and tamarisk *Tamarix* flanking neglected drainage ditches (Hoffmann 1958). Such natural sites have, however, progressively been largely superseded by managed situations such as olive and almond plantations, orchards, vineyards, farmland spinneys, parks, gardens, roadside trees, avenues, and even boulevards and squares in small towns and some cities. Normally infrequently inhabits conifers, but *pulchellus* in USSR lives in pine-forests with burnt-over slashes and wide glades, as well as junipers and other trees in extralimital mountain forests up to *c.* 1800 m or more (Dementiev and Gladkov 1951a). In Tunisia, pinewoods and mimosa thickets less favoured than palms in south and olive groves in north, with many breeding, as elsewhere, in old nests of Magpie *Pica pica*, often built in acacias (Bannerman 1955). Also uses nest-boxes.

In winter in Africa, inhabits wide range of vegetation types, in which however dense clumps of bushes for daylight roosting are essential, and penetration into deep forest seems to be avoided. In Sénégal, in dense bamboos; in southern Nigeria and Sudan, in tall grass belt; in Eritrea and southern Somalia, up to *c.* 1400–1500 m (Moreau 1972). In East Africa, frequents bush country, acacia belts along dry river beds, savanna woodland, and areas where there are baobabs *Adansonia digitata* (Williams 1963). Descends to ground freely but moves little on it, and has no special attachment to water. Flies in lower airspace, except possibly on migration.

Distribution. Marked contraction in northern part of European range, especially Switzerland, Austria, and Czechoslovakia, due mainly to habitat changes and reduction in numbers of large insects (Glutz and Bauer 1980); some retreat south in France.

FRANCE. Some retreat of range to south; since 1936, no longer breeding in Brittany north of Loire, nor in Normandy, Picardy, and Lorraine (Yeatman 1976). WEST GERMANY. Bred 1902 and 1960–1 (Bezzel *et al.* 1980; Glutz and Bauer 1980). AUSTRIA. Range decrease; bred until *c.* 1905 in Lower Austria and to *c.* 1960 in parts of Upper Austria; also rare breeder in Tyrol in late 19th century (Glutz and Bauer 1980). SWITZERLAND. Extinct or apparently so in many former breeding areas (Glutz and

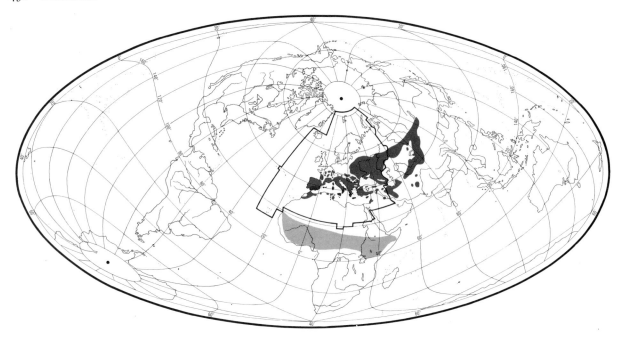

Bauer 1980). SYRIA. No proof of breeding (Kumerloeve 1968b).

Accidental. Iceland, Faeroes, Britain, Ireland, Belgium, Netherlands, West Germany, Denmark, Norway, Sweden, Poland, East Germany, Canary Islands, Madeira.

Population. FRANCE. Probably a little under 10 000 pairs (Yeatman 1976). Decreased Camargue (Blondel and Isenmann 1981). CZECHOSLOVAKIA. About 50 pairs (Mikkola 1983); some increase after 1950, but decreased since 1965 (KH). See also Randík (1959). HUNGARY. Fairly common; no recent change in numbers (LH). AUSTRIA. About 50–150 pairs (Mikkola 1983). SWITZERLAND. Marked decrease last 20 years; now under 20 territorial ♂♂ (Glutz and Bauer 1980). SPAIN. Estimated over 30 000 pairs 1975 (AN). USSR. Common in Ukraine and Caucasus, rare elsewhere (Dementiev and Gladkov 1951a). LEBANON. Abundant Bekaa valley (Tohmé and Neuschwander 1974). ISRAEL. Marked decrease in 1950s with heavy use of pesticides; some recovery since (HM). CYPRUS. Common (PRF, PS).

Oldest ringed bird 6 years (Glutz and Bauer 1980).

Movements. Northern populations migratory, southern ones partially so or resident.

Of 6 west Palearctic races, only one (*cyprius*, endemic to Cyprus) wholly resident. 2 others are partial migrants. *O. s. mallorcae* of Iberia present there all year in south, though winter numbers considerably reduced (Bernis 1967; Bannerman and Bannerman 1983); emigrants presumably cross to Africa, whence single specimens attributed to *mallorcae* collected Haut Atlas (Morocco) on 19 March and Bilma oasis (Niger) on 6 November (Vaurie 1965). Aegean and Near Eastern race *cycladum* largely migratory (some winter in southern Greece), but winter range of emigrants unknown; no African records.

Remaining 3 races considered here are long-distance migrants, most of which reach Afrotropical region in winter: nominate *scops* (breeding central and southern Europe and north-west Africa), *pulchellus* (breeding east European USSR to central Siberia and Mongolia), and *turanicus* (breeding Iraq to Pakistan and southern Soviet Central Asia). Winter distributions poorly known, however. A few European birds winter in Mediterranean basin and North Africa. Asiatic birds possibly winter in part (perhaps only irregularly) in Pakistan and north-west India (Ali and Ripley 1969), and in Arabia where, however, recorded chiefly in passage periods September–October and February to mid-April (e.g. Bundy and Warr 1980). On present knowledge, winters (October–March) mainly in scrub habitats in northern Afrotropics, where not readily separable in the field from African Scops Owl *O. senegalensis*. Probably occurs all across this zone from Sénégal to Ethiopia and Kenya (Moreau 1972; Backhurst *et al.* 1973; Morel and Roux 1973); latitudinally, from Niger inundation zone of Mali (scarce: Lamarche 1980) and central Chad (reasonably common on *Acacia* steppes: Newby 1980) southwards to Gulf of Guinea in Ghana and Togo, and to equator or a little beyond it in Kenya. As to be expected, only European nominate *scops* identified in West and central Africa (apart from 1 *mallorcae* in Niger), and specimens of Asiatic *pulchellus* and *turanicus* extant only from north-east Africa (Vaurie 1965; Friedmann and Keith 1968; Britton 1980). However, African data sparse in all respects in relation to putative numbers of winter

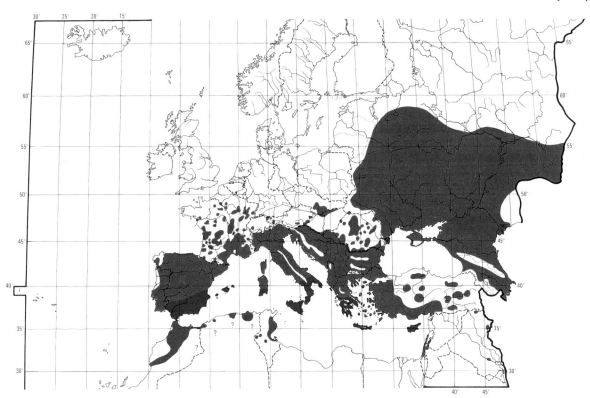

visitors and large breeding range from which they are assumed to originate; Siberian birds must travel 7000–8000 km to wintering areas (Mikkola 1983).

Migrates on broad front, birds regularly overflying Sahara and (especially in autumn) apparently Mediterranean also, in unbroken flight. Seen regularly in spring (but not autumn) in Mauritania and Western Sahara (Rio de Oro), and trans-Saharan passage perhaps has diagonal element, since spring migrants ringed Cap Bon (Tunisia) recovered essentially east of north: in Malta (8), Sardinia, Corsica, Sicily, Italian mainland (7), Yugoslavia (8), and Greece. Also, 2 ringed Morocco in spring found to northeast: in Var, France (June) and Sicily (September). No recoveries south of Sahara, or elsewhere in winter. 2 shot Kenya in March were very fat, suggesting that eastern birds deposit pre-migratory fat at low latitudes prior to trans-Arabian flight (Moreau 1972). Paucity of Arabian peninsula records (Meinertzhagen 1954) suggests that region also largely overflown by migrants.

Food. Largely insects and other invertebrates; a few small birds, reptiles, amphibians, and mammals. Hunts at night; occasionally also by day, perhaps especially when feeding young. One breeding pair active mostly 19.00–24.00 hrs and 02.00–05.00 hrs (Koenig 1973); see also final paragraph. Hunts mainly from perch, flying down to grip prey with feet; may land alongside it first (Koenig 1973). Catches moths (Lepidoptera) in feet in rapid direct aerial-pursuit—not undulating like normal flight; moths taken to perch for eating, wings discarded (Meinertzhagen 1959). Said by Koenig (1973) to force flying insects to land rather than take them in flight. Rarely hovers. Hunts on foot, seizing small items with bill; also pulls earthworms (Lumbricidae) out of ground with bill. Small insects swallowed whole. Larger ones and vertebrates pulled apart; birds plucked. Insects to be fed to young have legs, etc., pulled off. (Kadochnikov 1963; Koenig 1973; Glutz and Bauer 1980.) Small stores of surplus food may accumulate in nest-hole (Glutz and Bauer 1980). For cannibalism, see Social Pattern and Behaviour.

Pellets mainly of insect chitin; decompose quickly (Glutz and Bauer 1980). Measurements: $c.\ 36 \times 17$ mm (Desfayes 1949a); $20-35 \times 10-12$ mm (Uttendörfer 1952).

In west Palearctic, insects comprise mostly crickets, etc. (Orthoptera: Gryllidae, Tettigoniidae, Gryllotalpidae), beetles (Coleoptera: Carabidae, Dytiscidae, Geotrupidae, Scarabaeidae, Elateridae, Bostrichidae, Tenebrionidae, Cerambycidae), and adult and larval moths (Lepidoptera: Noctuidae, Sphingidae, Geometridae); also dragonflies (Odonata: Aeshnidae, Corduliidae, Libellulidae), cicadas (Cicadidae), earwigs *Forficula*, wasps (Vespidae), ants (Formicidae), flies (Diptera). Other invertebrates: spiders (Araneae), harvestmen (Opiliones), crustaceans (Amphipoda, Isopoda), millipedes (Diplopoda), earthworms (Lumbricidae), nematode worms (Nematoda). Vertebrates comprise lizards, frogs (including tree-frog *Hyla meridionalis*), and birds up to size of Redwing *Turdus iliacus* (but mostly smaller than this); mice (Murinae) taken rather rarely, and bats (Chiroptera) and shrews (Soricidae) exceptionally. (Moltoni 1932, 1937; Grote 1933a; Priklonski 1958; Meinertzhagen 1959; Mebs 1960; Kadochnikov 1963; Herrera and Hiraldo 1976; Glutz and Bauer 1980; Mikkola 1983.) Will take lettuce leaves regularly in captivity (see below); plant material recorded in pellets in the wild (Uttendörfer 1952).

159 items, southern Spain, comprised 94·3% (by number) invertebrates, 2·5% reptiles, 1·2% amphibians, 1·2% Murinae, and 0·6% birds (Herrera and Hiraldo 1976). Other data scanty—see below; also Desfayes (1949a) for France, Mebs (1960) for West Germany, Priklonski (1958) and Kadochnikov (1963) for USSR, Valverde (1967) for Spain, Moltoni (1932, 1937) for Italy, and Blanchet (1951) for Tunisia.

Captive adults fed *ad lib* ate average 20 g of mice, 7 g of heart, 8 g of mealworms, and 20 cm^2 of lettuce leaves per bird per day (Koenig 1973). For consumption rate in young, see below.

Data very limited, but no indication that food of young differs substantially from that of adults (e.g. Ternovski and Ternovskaya 1959). At Volkach (West Germany) identified items brought to nest were 11 bush-crickets *Tettigonia* and 1 large caterpillar (Geometridae) (Mebs 1960). In Voronezh region (USSR), neck-collar samples comprised 7 large moths (*Agrotis, Acronycta*), 3 beetles (*Geotrupes stercorarius, Monochamus galloprovincialis, Prionus corarius*), and 5 large spiders (Kadochnikov 1963). Most other sources with data on food (see above) include at least some items brought to nest (not necessarily all for young). Outside west Palearctic, in Novosibirsk (USSR), 94 items comprised 56·4% Orthoptera, 33·0% Lepidoptera, 5·3% Coleoptera, 3·2% spiders, 2·1% Odonata. Over 12 days, when being fed by both parents, fed for average 6 hrs per night, starting 20.23–20.40 hrs and finishing 04.15–04.50 hrs; peak frequency early in night with average 3 min (0.5–18) between visits; average 91 visits (76–105) per night (Ternovski and Ternovskaya 1959). At Volkach, in week up to fledging, young fed from *c.* 20.30 hrs onwards; every 11 min (5–18) for first *c.* 1½ hrs, later about every ½ hr (Mebs 1960). In Hungary, at nest with large young, fed about every 5 min 19.43–20.50 hrs; for rest of night, fed very seldom if at all (Ferguson-Lees 1958). In Novosibirsk, nestlings received average 30 'food portions' (*c.* 20 g total) per bird per day (Ternovski and Ternovskaya 1959).

DJB

Social pattern and behaviour. Most comprehensive study by Koenig (1973) on captive population of 60 birds in large aviary, making results unusually applicable to the wild.

1. No information on degree of gregariousness outside breeding season, but for communal roosting in winter quarters, see below. According to Hekstra (1973), family may stay together on migration. BONDS. Monogamous mating system. Though no direct evidence, pair-bond probably renewed annually; not known if bond maintained outside breeding season (Koenig 1973), but unlikely. In captive study, if bird died, partner did not usually pair again the same year; some ♂♂, however, tended 2 ♀♀ at once, sometimes starting a clutch with 2nd ♀ while 1st incubating. One ♂ Allopreened (see part 2), fed, and copulated in turn with 3 ♀♀. ♂♂ believed to be occasionally polygamous in the wild (Koenig 1973). Age of first breeding in the wild not known. As in other Strigidae, ♀ takes dominant share of nest-duties: performs almost all incubation and brooding up until young well-developed (see Roosting, and Relations within Family Group, below). Captive young still partially dependent on parents for food up to *c.* 60 days, independent thereafter (Koenig 1973; see also Relations within Family Group). BREEDING DISPERSION. Solitary and territorial, though locally concentrated in some areas: e.g. in Reghin (Rumania), 5 nests in field copse of 0·6 ha, shortest distance between 2 simultaneously occupied nests 8 m (Kalabér 1971). In Krk (Yugoslavia), 10 calling birds in *c.* 2 ha of park and avenue (Böck and Walter 1976). In southern Tyrol (Austria), up to 5 pairs per ha, shortest distance between 2 nests 10 m (see Glutz and Bauer 1980). In southern France, up to 5 pairs per km^2; on island of Port-Cros, 1·8–2·3 pairs per km^2 or 3·6–4·6 pairs per km^2 of suitable habitat (see Glutz and Bauer 1980). In central Europe, generally more sparse: e.g. in 60–65 km^2 between the Hron and Ipel' (Czechoslovakia), 44 birds responded to territorial calls (Randík 1959; see also Glutz and Bauer 1980). Nest-territory (no information on size) serves for courtship, copulation, and hunting. ♂ establishes territory (König 1965; Koenig 1973). Same nest-site may be used in successive years, though not known if by same pair (Kadochnikov 1963). In 2 successive years, pair used sites 350 m apart; in 2nd year, 1st nest destroyed during breeding, and pair re-laid *c.* 800 m from previous 2 sites (Barthos 1958). ROOSTING. Outside breeding season, birds typically roost together in dense foliage (Morel and Roux 1966). In April, Egypt, up to 14 in a tree (Meinertzhagen 1930). In Chad, 5 birds (probably nominate *scops*) roosted side by side in a low *Acacia* (Newby 1980). In breeding season, paired and unpaired birds roost alone at all times. ♀ signals selection of nest-site by roosting in it (Koenig 1973; see also part 2). In one case, ♂ mate roosted by day nearby in territory, on a regular perch in scrub *c.* 50 m from nest (Desfayes 1949a). Typically roosts close up against trunk or branch for camouflage (Ferguson-Lees 1958; see also part 2). In resting posture, typically draws one foot up into plumage, bill pointing forward and not tucked into plumage (Koenig 1973). When young in nest well-developed, ♀ stops passing them food received from ♂, and leaves nest to roost separately from him. Older young lean against sides of nest-cavity when sleeping. Up to *c.* 3 weeks after leaving nest, young try to roost in bodily contact with siblings or ♀. Thereafter (at least when awake), keep 10–15 cm apart (Koenig 1973). When breeding, pair sometimes slept for 10–20 min between 24.00 and 02.00 hrs; less so when feeding young (Koenig 1973).

2. If disturbed in day-roost, adopts remarkable Sleeked-upright posture (Fig A); camouflage then facilitated by perhaps most cryptic plumage of any Strigidae in west Palearctic. When suddenly aware of being watched, adults make the following adjustments, all within *c.* 1 s: foot drawn up in resting posture usually put down; bird tenses, sleeking body plumage and raising head plumage, notably ear-tufts; eyes almost shut. Bird thus faces source of danger while turning slightly away from it, and draws folded wing across breast so that head looks over carpal joint. Bird then remains motionless and so typically resembles bark or stump of tree on which perched (Koenig 1973). If observer does not move too quickly, bird may maintain posture for several minutes. As observer approaches, bird flicks upper eyelids alternately or together, and will allow approach to 1 m before flushing, whereupon all plumage relaxed and rounded

A

appearance of head resumed. If flushed and followed, continues to change perch rather than adopting Sleeked-upright posture again. Bird disturbed by day often adopts only low-intensity Sleeked-upright posture, with ear-tufts only slightly raised. Full posture never performed in dark (Koenig 1973). For development of Sleeked-upright posture in young, see Anti-predator Responses of Young (below). For apparent Sleeked-upright and Forward-threat postures (see below) of captive bird confronted with Nightjar *Caprimulgus europaeus*, and Sleeked-upright posture when faced with dog, see Nagy (1931–4). For other alarm reactions, see Parental Anti-predator Strategies (below). ANTAGONISTIC BEHAVIOUR. Rival ♂♂ mainly exchange long bouts of Advertising-calls (see 1a in Voice). ♂'s song-perch usually exposed and/or elevated; when calling intensely, ear-tufts lowered, eyes wide open (Koenig 1973). According to Dementiev (1952), calls also in flight, though this emphatically denied by Stadler (1932); see also Koenig (1973). Vocal exchanges between rival ♂♂ distinguished from ♂–♀ duetting (see below) by lack of any synchrony in former (see Voice). Rival ♂♂ seldom appear to encroach on resident's territorial boundary (Koenig 1973). From start of laying, ♂ calls less frequently, especially during incubation and chick-rearing when mostly silent unless provoked; may then call even during day, but usually 19.00–05.00 hrs during territory-establishment and pair-formation; in some places, no calling until well after dusk (Koenig 1973). Heard regularly dawn and dusk (Kadochnikov 1963). For calling during day, especially in dull warm weather, at start of breeding season, see Pukinski (1977). After young independent, ♂ sometimes begins to call again, and may do so up until October (Koenig 1973, which see for diurnal and seasonal variations). On Crete, where resident, occasionally calls on warmer evenings from November or January, and vigorously from late March (Stresemann 1943). Elsewhere, typically begins April (e.g. Dementiev 1952, Moore and Boswell 1956b). No physical contact between rivals known in the wild, but, in captivity, dominant bird may make diving attacks, striking with talons, on subordinates engaged in courtship activities (Koenig 1973). Paired ♀ may also react aggressively to close approach of strange ♂; ♀ may give Anger-call (Koenig 1973: see 4 in Voice, also subsection 6 of Heterosexual Behaviour). HETEROSEXUAL BEHAVIOUR. (1) Pair-bonding behaviour. ♂ ready to pair gives Advertising-call (see above). At start of breeding season, unpaired ♀♀ also give Advertising-call (see 1b in Voice) and may thus attract ♂♂ from quite a distance. Paired ♀♀ may also sometimes respond to Advertising-call of distant ♂. Once ♀ attracted to ♂, she responds to him with variant of her Advertising-call, both birds performing duet (see 1c in Voice) which may continue for several minutes (Koenig 1973). ♂ said to call first in duet (Mebs 1960). (2) Nest-site selection. ♂ performs Nest-showing as in (e.g.) Pygmy Owl *Glaucidium passerinum* (König 1965). Following account by Koenig (1973) typical of numerous observations at nest-box sites. Often after bout of duetting, ♂ flies to potential nest-site, sits in entrance with hind-body sticking out. Enters after 1–2 min and starts nibbling at inner walls, and begins to give Advertising-call, typically then poking his head out of hole and calling loudly and persistently, occasionally pausing if ♀ calls in response. ♂ then enters again and resumes scratching and calling. When ♀ approaches, ♂ drops down into cavity (see also König 1965). ♀ looks in and, after 30 s–1 min, enters and begins to scrape vigorously. After a few minutes, ♂ leaves, and ♀ continues to scrape for 5–10 min before also leaving. Procedure repeated several times during the night. Over several days, ♂ may show ♀ several holes. ♀ signals choice by sitting in cavity with head out of hole during day (see Roosting, above). In captive pair, 1st egg laid 3 days after nest-box selected. (3) Mating. Following account from Koenig (1973). ♀ sometimes invites copulation with Soliciting-call (see 2 in Voice). ♂ sometimes feeds ♀ (see below) immediately before copulation. ♂ ready to mate stares at ♀ for a few seconds, then gives 1 or more Advertising-calls leading to duetting; after 15–30 s of duetting, ♂ utters especially loud calls to which ♀ replies with Soliciting-call. During last unit of Advertising-call, ♂ flies on to ♀'s back. She adopts horizontal posture and ruffles plumage. ♂ grips with feet, pushes bill into side of her face, spreads wings for balance, and, on cloacal contact, gives Copulation-call (see 3 in Voice). About 0·5 s later, ♂ dismounts; duration of mounting not more than c. 2–3 s. Copulation occurs near nest (but apparently not in nest-hole), mostly 18.00–19.00(–22.00) hrs, occasionally after 01.00 hrs, 7–10 times per night. Last copulation observed 2 days before laying final egg of clutch. (4) Courtship-feeding. ♂ feeds ♀ initially off nest (e.g. Mebs 1960), but especially on nest once laying begins (Koenig 1973). ♂ ready to feed ♀ looks across at her and gives Advertising-call. Flies and perches near to her, and gives Soliciting-call, and feeds her bill-to-bill. Often feeds her several times in succession, in captivity 6–15 times per night up to hatching (Koenig 1973). Captive ♀♀ occasionally fed themselves during laying, but never after clutch completed. Sitting ♀ typically sticks head out of nest-hole and gives quiet Advertising-call or Soliciting-call and, at least in captivity, ♂ usually feeds her fairly promptly (Koenig 1973). ♂ carries food in bill and passes it quickly through entrance, not entering however (Kadochnikov 1963). ♀ begins to forage for herself when oldest young 18 days old; fed by ♂, therefore, for c. 6 weeks. When replacement clutch laid, ♂ feeds ♀ much less than for 1st clutch (Koenig 1973). (5) Allopreening. ♂ may Allopreen ♀ for up to 1 min or more, usually on head or face (Fig B). ♀ remains quiet or gives brief Anger-call.

B

Allopreening may follow directly after Courtship-feeding, or occur separately; may serve to strengthen pair-bond. Allopreening rare after clutch completed, and ceases after hatching; often none associated with replacement clutch. ♀ rarely Allopreens ♂ (at any time) (Koenig 1973). (6) Behaviour at nest. Periodically during day, ♂ gives quiet Advertising-calls and ♀ responds in brief duet until ♂ falls silent and dozes off again. Occasionally after calling, ♂ flies to nest-hole, looks inside, and returns to roost-site. Sometimes (e.g. in wet weather), ♂ settles down beside incubating ♀ who then tries to drive him away with Anger-call. ♀ tends to leave nest only once per night, ♂ incubating during her absence (Koenig 1973). In study, France, following sequence repeated night after night during presumed incubation. ♂ approached nest, calling ever more urgently, and sitting ♀ responded by half-emerging from hole, looking around, and calling to ♂. After 5 min, on 'another' call (possibly Soliciting-call) of ♂, ♀ flew towards mate and the 2 birds flew off. Returned c. ½ hr later and ♀ entered nest-hole (Mouillard 1939). RELATIONS WITHIN FAMILY GROUP. Young call from egg c. 12 hrs before hatching. After hatching, rather immobile (Kadochnikov 1963); sleep for c. 6–7 hrs before giving food-calls (see Voice). ♀ broods from outset, usually assiduously (Mouillard 1939), but when weather very warm, young 2–3 days old not brooded by day, ♀ staying outside nest (Roget 1971). Brooding ♀ summons ♂ to bring food with Advertising- or Soliciting-call. ♂ tends to bring small items which newly hatched young receive bill-to-bill via ♀, and can swallow whole (Kadochnikov 1963; Koenig 1973). ♀ transfers such items to young without delay, but divides larger prey into portions first. Young less than 3 days old are fed for c. 10 hrs per day, and take rest for 14 hrs. Eyes begin to open at 3 days (allowing retrieval of dropped food), fully open at c. 5 days (Koenig 1973). By 3 days, also able to stretch neck upwards (Kadochnikov 1963) and begin to beg by wagging head at rate of c. 2–4 sideways movements per s, also pecking at gape flanges of parents. When very hungry, rapidly snap bill and tremble outspread wings. At 3–5 days, especially when ♀ gives Soliciting-call, young rush towards her with outspread wings; call serves to encourage young reluctant to feed. Snapping when hungry ceases at 4 days, and from 6th day young generally more passive. Begin to eject pellets at 8 days (Koenig 1973). At 11 days, able to stand properly on tarsi, and by 15 days come to nest-entrance to snatch food (Kadochnikov 1963). Exercise wings frequently from 14 days (Koenig 1973). From 12 days, food items transferred to and held rather clumsily in one foot; from 16 days, held firmly and bits torn off, sometimes with aid of wings as supports. From 19th day, young begin to perform Body-swaying (Fig C, right) an exaggerated, slow (1–2 movements per s), rhythmic sideways motion of head and upper body; serves to supplement food-calls, helping approaching adult to locate young. Body-swaying at first

D

of low intensity, but well-developed by 21 days when young leave nest. Young also cock head by 90° (Fig C, left), apparently to facilitate hearing. Both Head-cocking and Body-swaying decline by age of 50 days, and rare in adults. Young also perform bouts of staring for up to 1 min (Koenig 1973; see also Heinroth 1926–7). After oldest young 18 days, ♂ gives progressively less food for transfer to young, and usually gives none when oldest young leaves nest. ♀ also leaves then, and stops brooding, though will roost in nest by day until young fledge if only 1–2 in brood (Koenig 1973). At 18 days, young jostle to look out of nest-hole; at 21 days push body half out (Kadochnikov 1963). In disputes between them, young give juvenile Anger-call (see Voice); also call thus when cold (Koenig 1973). For fledged young fighting over perch, see Roget (1971). Before leaving nest, young sit for a long time at nest entrance and call. Parents fly to entrance from time to time, evidently trying to entice young to leave (Kadochnikov 1963). Of brood of 5, 3 young left one day, the other 2 4 days after 3rd (Roget 1971). Young quit nest at dusk, initially flapping clumsily only 4–5 m to perch-site in tree. By next night, had moved 15–20 m from nest (Kadochnikov 1963). Do not return to nest after departure (Mebs 1960; Roget 1971). Outside nest, young continue to give food-calls and are fed about equally by both parents. Captive young initially moved around, parrot-fashion, with aid of bill; started flying properly at c. 30 days (see also Meade-Waldo 1921), and then started making sorties to the ground as for hunting. Though self-feeding at 35 days, fed unremittingly by both parents up to 40 days, then increasingly less up to 50 days, by which time self-sufficient. However, parents may feed them up to 60 days, while young continue to beg occasionally up to c. 80 days (Koenig 1973); different conditions may prevail in the wild, where migratory demands likely to hasten onset of independence. Young markedly social up to independence, e.g. preening one another; activity wanes at c. 45 days, birds then keeping at least 10–15 cm apart. ANTI-PREDATOR RESPONSES OF YOUNG. Widely studied, both in captivity, and in the wild. Young actively defend themselves from c. 6 days. Face source of danger and start Bill-snapping, alternating from 8th day with slight hissing sound, Body-swaying, and ruffling of plumage (Koenig 1973; see also Mebs 1960). Young perched on finger reacted with feather-ruffling and Body-swaying in rather crouched posture (Mebs 1960). When frightened, 12-day-old young crouched and sleeked feathers; at 14 days, Bill-snapped and tried to hide in corner of nest-cavity if observer tried to touch them (Kadochnikov 1963). From 13 days, begin to react to any disturbance with Sleeked-upright posture, and posture fully developed by fledging; same as in adult (see above), except that, in young, wing nearest source of alarm is drawn further across breast so that bird looks broadside at observer. From c. 16th day, threatened young adopt Forward-threat posture (Fig D), enhanced by extreme Body-swaying, staring with wide open eyes, contracting pupils, flicking upper eyelids, ruffling plumage, and tilting forward open wings. Newly fledged birds react thus when closely threatened by man and dog (Koenig 1973). Newly fledged young also scratched with claws when hand extended towards it (Mouillard 1939); may also bite (Roget 1971). From

C

18th day, young may give Suspicion-call (see 6 in Voice, also Calls of Young) during high intensity threat when predator nearby; also give juvenile Anger-call (Koenig 1973). PARENTAL ANTI-PREDATOR STRATEGIES. (1) Passive measures. Even on closest approach, ♀ typically sits tightly while incubating, and until young well-grown (Koenig 1973; see also Fellay 1949, Kadochnikov 1963). For threat by sitting bird, see below. When taken out of nest, ♀ sat in observer's hand (Grote 1933a). ♀ flushed initially, but sat tightly when intrusion continued over several days (Mebs 1960). Flushed ♀ sat quietly on nearby branch until intruder left (Kadochnikov 1963). More easily flushed at night, but returns just as quickly as by day. If ♂ in nest is disturbed, he typically flees (Koenig 1973). (2) Active measures: against man. When disturbed at nest, ♀ may adopt Forward-threat posture: ruffles plumage, raises or spreads wings, makes irregular, slow rocking movements c. 10 cm from side to side. Opens eyes wide and closes them, together or alternately, every 1–1·5 s with emphatic slowness. Periodically makes Hissing sound (see 7 in Voice) and Bill-snaps 6–10 times per s. Sometimes give Suspicion-call (Koenig 1973). From start of laying, ♂ near nest no longer adopts Sleeked-upright posture but instead adopts round-headed appearance and stares. If intruder goes up to nest, ♂ typically mounts flying attack, trying to rake head, less commonly hands, with claws (Koenig 1973). Makes dive-attacks, always from behind (Kadochnikov 1963). Attack may be repeated several times, at intervals of several seconds or minutes. ♂ may give single Advertising-calls or short bouts of them; also Alarm-calls (see 5 in Voice), usually from perch, sometimes in flight, presumably warning mate and young. When ♀ ceases brooding and roosting in nest, she becomes much more aggressive, perhaps more so than ♂: gives Alarm-calls and mounts flying attacks (Koenig 1973). For attacks by both parents simultaneously, see Roget (1971); in one such attack, pair brushed intruder's head with wings (Mebs 1960). One or both members of pair often move through vegetation to perch as little as 1 m from intruder, calling (see 1 and 5 in Voice) and/or Bill-snapping (Mouillard 1939; Fellay 1949).

(Figs A–D from photographs in Koenig 1973.)   EKD

I P A D Hollom Israel April 1980

II C König Spain May 1977

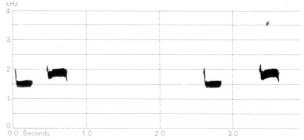
III C Chappuis Rumania May 1967

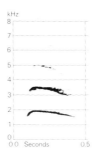
IV C König Spain June 1974

Voice. Diverse repertoire in breeding season widely studied. Largely, if not entirely, silent in winter quarters (Elgood et al. 1966; Morel and Roux 1966). For calling, allegedly by nominate scops, in Nigeria in winter, see Sharland (1966); also Meinertzhagen (1954) for Arabia. For Bill-snapping, see Social Pattern and Behaviour. Apart from Bill-snapping, 7 adult calls distinguished by Koenig (1973) which see for sonagrams of both adults and young.
  CALLS OF ADULTS. (1) Advertising-call. (a) ♂'s call. Very like rather sonorous human whistle—'tyuu' (E K Dunn: Fig I). A clear, plaintive bell-like whistle: 'kyüü' given with seemingly endless, clockwork regularity (Weijden and Ginn 1973). Given usually at rate of 22–26(–28) units per min, with pauses of 1·8–2·3 s between units (Koenig 1973). For longer pauses, usually 2·5 s or more, see Weijden (1973). Pitch usually 1·18–1·35 kHz, but call sometimes a disyllabic 'kuyuu', and 2nd syllable then higher, up to 1·45 kHz (Weijden 1973, 1975). In recording by C Chappuis, France, 't Tyuu' (J Hall-Craggs). Different ♂♂ have distinctive pitch and delivery rate (Koenig 1973). ♂ may call up to 30 min or more without interruption; longest bout 900 calls in 40 min (Koenig 1973). In European part of range, little or no geographical variation (Weijden 1973, which see for sonagrams). Remarkably similar to calls of Pygmy Owl Glaucidium passerinum and midwife toad Alytes obstetricans; call of latter purer and shorter (Thönen 1968; Burton and Johnson 1984). Call serves to demarcate territory, attract ♀, and to contact ♀ mate—e.g. when Nest-showing, just before copulation, or when intruder at nest (Koenig 1973). Adult called softly to young just out of nest, apparently to encourage them to move further (Roget 1971). For further details, notably of diurnal and seasonal variation, see Social Pattern and Behaviour. Call develops in young at c. 6 months (Koenig 1973). (b) ♀'s call. Similar to ♂'s call, but higher pitched and sharper (Desfayes 1949a; see also call 1c below). Given by unpaired ♀♀ at beginning of breeding season. Though said to be very different from calls given in duet (Koenig 1973: see call 1c, below) differences not very apparent in recording by C König (Fig II). (c) Duetting. ♀ answers Advertising-call of mate with softer, higher-pitched sound than his, producing duet of often several minutes' duration. ♀ answers ♂'s 'tjüt tjüt...' with a 'questioning' 'düi düi...', sometimes a descending 'düa düa...' or harsher 'krüa krüa...' (Koenig 1973). Although sometimes said to be 'adapted' to ♂'s rhythm (Koenig 1973)

and antiphonal (Pukinski 1977), ♀'s contribution also described as irregular (Hekstra 1973) and less frequent than ♂'s (Weijden 1973). Pair alternate regularly in Fig III, but no synchrony apparent in recording by C König. Duet serves as contact between pair-members, notably during Nest-showing and just prior to copulation, and also when ♂ feeding ♀ on nest (Koenig 1973). (2) Soliciting-call. Soft, rolling 'drrr-drrr...', usually repeated several times, somewhat lower pitched and less clear than call 1. Units each 0·1–0·4 s, with intervals of 0·2–0·8 s. Given from perch with bill closed, by both sexes inviting close approach of partner or young: e.g. by ♀ inviting copulation or begging food from ♂, by ♂ announcing readiness to feed ♀, or by either sex about to feed young, especially if latter reluctant to feed. Develops in young at c. 4 months (Koenig 1973). (3) Copulation call of ♂. A loud, shrill, twittering 'züzüzüzüzüzü', comprising 6–8(–10) units, total duration 0·6–1·0 s. Volume increases then decreases, pitch rises then falls. Given by ♂ when cloacal contact achieved (Koenig 1973). (4) Anger-call. A shrill, trilling 'tschrr-tschrr...', comprising series of 3–5 units, each 0·4–0·7 s, with pauses between units of about the same duration. Strident, prolonged 'tsirirriririri' of captured ♂ (Desfayes 1949a) probably the same (see also call 5). Expresses annoyance in adults of both sexes, e.g. when ♀ resisting sexual advances of ♂ (Koenig 1973). See also Calls of Young. (5) Alarm-call. A clear, loud, shrill, and penetrating 'piiääää', c. 0·4 s long, variously modulated, with rising or slightly falling pitch, given at irregular intervals of up to 1 min or, in great excitement, much more rapidly with only c. 1 s between calls (Koenig 1973). In recording (Fig IV), sounds like mew or yelp of sparrowhawk *Accipiter* (E K Dunn). Probably same or variant described as a soft, plaintive, weak mewing 'mia-ô-ô-ô', or 'ui-ôô', like Little Owl *Athene noctua* (Mouillard 1939; P J Sellar); also mewing 'ououou' of captured ♂ (Desfayes 1949a). Short, sharp, lengthily repeated 'hi-hi-i-i-i', probably same or variant of high-intensity call; sound sometimes appears to originate from great distance (Mouillard 1939). Given by both sexes, usually from perch, sometimes in flight (Koenig 1973). Expresses anxiety, notably when nest disturbed, and especially by ♀ when intruder close to newly fledged young (Mouillard 1939; Koenig 1973). Also when predator, e.g. Tawny Owl *Strix aluco*, flew overhead. Otherwise, not often heard, though common during migration (Koenig 1973). (6) Suspicion-call. A quiet, clear 'psieeee', 0·3–0·5 s long, given once or repeated at intervals of 10 s to several minutes. Given rarely as threat by ♀ with young up to 8 weeks old, especially when intruder lingered at nest; sometimes given by brooding ♀ (Koenig 1973; see also Calls of Young, below). (7) Hissing-call. A sharp, hissing, expiratory 'ffffff' sound, c. 0·2 s long, given at intervals of 0·5 to several seconds, with bill open and tongue against palate. Given by both sexes in threat (Koenig 1973; see also below).

CALLS OF YOUNG. Main study by Koenig (1973, see also for development of adult calls). From hatching, food-call a hoarse, whispering 'tschp tschp' (Koenig 1973) or 'tischup tschup' (Nagy 1931–4); units each 0·15 s long at intervals of 0·5–1·0 s, often repeated tirelessly. Food-call given in response to Soliciting-call of parent; more fulsome at 30 days, and from 2½ months begins to give way to Advertising- and Soliciting-calls. From shortly before hatching, when cold, ailing, or discontented, young give trill similar to but slightly higher in pitch than Anger-call of adult (Koenig 1973). In small young, rendered 'wiwiwiwi' (Mebs 1960) or 'hihihihihi' (Desfayes 1949a). Given shrilly during disputes between brood-members, also when predator nearby or when handled (Koenig 1973). From 8th day, threatened young give Hissing-call, as in adult but initially weaker and at somewhat faster rate; by 18th day, volume as in adult. In threat, Hissing alternates with Bill-snapping (Koenig 1973). From 18th day, young express general unease with Suspicion-call, as in adult, either while staring motionless at something (mild threat) or in high-intensity threat. Adult Alarm-call develops in young at 5–7 months (Koenig 1973); sharp rattling whistle 'phtrrrrüü', similar to Kestrel *Falco tinnunculus*, but quieter and higher pitched, given by juvenile bird when handled (Nagy 1931–4), probably Alarm-call.

EKD

Breeding. SEASON. Southern Europe: see diagram. North Africa: eggs laid May and early June (Heim de Balsac and Mayaud 1962). SITE. In hole in tree, building, or wall; sometimes in old nest of other species, especially crows (Corvidae); will use nest-box. Nest: nothing done to existing cavity; diameter of one hole 8 cm, depth 38 cm (Mebs 1960). EGGS. See Plate 97. Short elliptical, smooth and slightly glossy; white. 31×27 mm ($28–33 \times 22–29$), $n = 100$ (Witherby *et al.* 1938). Weight 13 g (12·6–13·3), $n = 4$ (Koenig 1973). Clutch: 4–5 (3–7). One brood. Replacements laid after egg loss. Laying interval 1–3 days (Glutz and Bauer 1980). INCUBATION. 24–25 days. By ♀ only (Witherby *et al.* 1938). Starts on laying of last egg according to Witherby *et al.* (1938), but with 2nd or 3rd

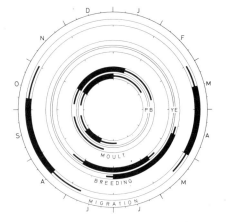

egg and hatching asynchronous, 4–5 young over 3 days, according to Koenig (1973). YOUNG. Altricial and nidicolous. Cared for and fed by both parents, though ♀ may feed small young with food brought by ♂ (Witherby *et al.* 1938; Ferguson-Lees 1958). FLEDGING TO MATURITY. Fledging period 21–29 days. Become independent at 30–40 days (Koenig 1973). No further information.

Plumages (nominate *scops*). ADULT. Entire upperparts including ear-tufts very finely vermiculated and freckled dark grey and rufous-cinnamon or white; grey and rufous usually dominates on central crown, ear-tufts, hindneck, inner scapulars, and back to upper tail-coverts (showing as deep rufous-cinnamon patches), grey and white on sides of crown, sides of head, upper mantle, outer scapulars, and tertials (showing as pale silvery-grey patches), but strong individual variation in extent of each colour, some birds appearing largely rufous-cinnamon with limited silvery-grey on outer scapulars, others mainly silver-grey with some rufous visible near feather-bases on central crown, mantle, or inner scapulars. All feathers of upperparts (except back and rump) with sharp and narrow black shaft-streaks *c*. 1 mm wide (up to 2 mm on central crown); feathers of hindcrown and mantle with broad white subterminal bar narrowly bordered black at both sides and narrowly divided in middle by dark shaft, forming white-spotted band from base of ear-tufts across upper nape and broader spotted band from sides of neck across upper mantle; outer scapulars with large rounded cream-buff to cream-white spot on outer web, broadly bordered by black at tip, forming pale-spotted row along scapulars. Facial disc poorly defined, not bordered by ruff of specialized feathers; narrow line from sides of bill to just above eye white with some black-tinged bristles projecting along sides of bill, remainder of face faintly speckled pale buff, grey, and white, appearing dull grey, bordered at rear by coarsely cinnamon-and-black marked bar just behind ear-opening. Chin pale cream, remainder of underparts finely freckled and vermiculated on cinnamon-rufous, pink-buff, or white ground (some birds with much rufous on chest, sides of breast, and legs, and with pink-buff breast to vent; others pale grey with limited rufous near feather-bases on chest), each feather with sharp black shaft-streak $\frac{1}{2}$–1 mm wide (2–2$\frac{1}{2}$ on sides of breast and belly) and with 1 or several uniform white or pale cream subterminal bars (accentuated at both sides by dark grey speckling); bars rather indistinct and narrow on chest, wider and distinct on sides of belly, lower flanks, and under tail-coverts (but less distinct in birds with white ground-colour). Vent almost uniform white or cream, legs coarsely spotted brown or black. Tail cream-buff to cream-white with fine grey speckling and vermiculation, marked with 5–8 virtually unspeckled buff or cream bars, each basally bordered by irregular narrow black rim. Flight-feathers marked with cream and buff bars (*c*. 8 of each on longer primaries, *c*. 4 on inner primaries and secondaries), pale bars partly broken along shafts and all bars finely speckled dusky grey, in particular on inner webs, obscuring barred pattern; only distinct large spots along outer edges of outer 5–7 primaries uniform pale cream-buff or (when bleached) cream-white, accentuated above and below by freckled dusky borders; tips of flight-feathers pale buff to cream with narrow black-brown shaft-streak and close dusky speckling. Greater upper primary coverts like primaries, remainder of upper wing-coverts like mantle and inner scapulars, finely speckled and with narrow dark shaft-streaks and narrow uniform cream bars; outer webs of outer median and greater upper wing-coverts with large rounded cream or white subterminal spots, bordered by black towards shafts and bases and by some dark speckling towards tips. Lesser upper wing-coverts buff-brown to deep rufous with fine black shaft-streaks and fine dark speckling; marginal coverts cream-buff or cream-white. Under wing-coverts cream-buff or white, partly marked with rufous bars and black arrows at shafts, mainly near leading edge of wing and on under primary coverts. Abrasion and bleaching make both rufous and grey tinges distinctly paler, upperparts appearing pale grey, grey-buff, or buff-brown by July–August, not as contrastingly grey and rufous as in spring. Sexes similar (but see length of bill to cere in Measurements). DOWNY YOUNG. 1st down (neoptile) short, soft, and dense; white, slightly mottled grey once juvenile starts to grow below it from 7th day. JUVENILE (mesoptile). Structure closely similar to adult, only base of neck, belly, lower flanks, vent, thighs, and under tail-coverts fluffy; no soft and woolly feathers as in (e.g.) Little Owl *Athene noctua* or Pygmy Owl *Glaucidium passerinum*, nor almost down-like as in *Strix* and *Asio* owls. Juvenile feathers visible from 8th day, on upperparts and chest first; neoptile gradually lost from day 10–11, but some still clinging to tips of mesoptile until 3rd week; juvenile plumage completed in *c*. 4th week, but flight-feathers and tail not full-grown until 45th day (Desfayes 1949*a*; Koenig 1973). Colour of plumage closely similar to adult, but ground-colour paler, cream-buff or white, less rufous; upperparts and upper wing-coverts less closely freckled dusky, marked with rather irregular fine vermiculation; dark shaft-streaks shorter, narrower, and paler; upperparts thus appear uniform pale grey or pale grey-buff without obvious white spots on hindcrown and mantle and with much less distinct streaking. Rounded white spots on outer scapulars and upper wing-coverts small or virtually absent, not as boldly bordered by black as adult, centres of longer upper wing-coverts more extensively marbled dusky grey; ear-tufts shorter. Underparts with paler ground-colour and paler grey well-spaced vermiculations; dark shaft-streaks narrower and shorter than adult, often forming small triangular spots on chest instead of distinct streaks; dark streaks on belly and flanks narrow, almost similar in width to dark and irregular subterminal bars; fluffy feathers of vent and under tail-coverts almost uniform white or pale cream with single narrow dark subterminal bar and faint dusky shaft-streak. No obvious differences in tail- or flight-feathers, but tail of juvenile tends to show irregular and partly freckled barring up to tip (tips in adult more uniformly freckled); dark bars of flight-feathers gradually narrower and more irregular towards tips (more equal in width in adult, except on finely freckled tip), tips of flight-feathers often slightly pointed, less broadly rounded, tertials often with bold dull black subterminal dot or cross-mark. FIRST ADULT. Similar to adult, but juvenile flight-feathers, tertials, all tail or outer tail-feathers, greater primary coverts, often greater upper wing-coverts, and sometimes part of median coverts retained; for characters, see juvenile. Colour, pattern of marks, and wear of retained juvenile wing-coverts and tertials often distinctly different from fresh neighbouring scapulars and shorter upper wing-coverts (all uniform in adult); primaries all of similar age in autumn (adult moulting or with arrested moult then), often distinctly worn in spring (still fresh in adult); all secondaries similar in age during 1st year (though inner secondaries and in particular tertials more abraded than others in late spring and summer), while adults show mixture of old and new secondaries due to serial moult from 3 centres (age differences occasionally difficult to see).

Bare parts. ADULT AND FIRST ADULT. Iris lemon-yellow, sulphur-yellow, or bright yellow, sometimes almost golden-yellow, often with orange tinge; rarely pale lemon-yellow, yellow-green, or grey. Eyelids greenish-brown. Bill bluish-black, brown with black tip, or dark horn-olive; base paler dark green, olive-grey, or yellow-brown; cere dark brown or dull olive-grey. Toes grey, light green-grey, olive-grey, brown-grey, or yellow-brown; claws light horn-grey, dusky green, or yellow-horn, tips black. DOWNY YOUNG AND JUVENILE. Bare skin pink at hatching, shining through down; bill whitish or lilac-pink with grey tinge and pale cutting edges, cere and eyelids flesh-red, toes and claws flesh-pink or whitish. Eyes open from 5–7 days, pupils tinged opaque blue until 3rd week; bill and claws shade to black from c. 6 days; iris grey-yellow 9th day, green-yellow c. 18th day. Similar to adult from fledging. (Hartert 1912–21; Chapin 1939; Desfayes 1949a; Koenig 1973; Glutz and Bauer 1980.)

Moults. (1) Nominate *scops* and *pulchellus*. Based on Stresemann and Stresemann (1966), Piechocki (1969), Koenig (1973), Glutz and Bauer (1980), and specimens in RMNH, ZMA, and ZMM. ADULT POST-BREEDING. Complete, except for part of secondaries; primaries complete, descendant. Mainly starts late June to late July, sometimes not until late August; p1 first, soon followed by p2–p3; moult suspended from mid-August or mid-September when pre-migration fattening starts, birds migrating with 4·6(10)2–7 inner primaries new. In Iberian adults, primary moult suspended during autumn migration with p1 (1), p1–p2 (2), p1–p5 (2), or p1–p6 (1) new (Mead and Watmough 1976). ♀ may start earlier than ♂: of 8 birds starting before 15 July, 5 ♀; of 11 starting after 15 July, 10 ♂. Primary moult resumed in winter quarters, completed late December–February. One captive ♀ did not arrest primary moult; p10 shed mid-August, 52–55 days after shedding of p1. Secondaries moult from 3 centres: ascendantly and descendantly from s11 or s12, ascendantly from s5 and s1; starts c. 1 week after shedding of p1. Arrested in autumn as in primaries; in contrast to primaries, moult not resumed in winter quarters and variable number of secondaries retained until next moulting season (or even 3rd), when also new series starts with s11–s12; in captive bird, centres at s1 and s5 start 26 days after centre at s11–s12 (33 days after p1). Tail lost simultaneously or almost so, between shedding of p3 and p5; all feathers lost within 2–11 days with centrifugal tendency; regrown after 40 days; no tail-moult during migration, when tail either all old or all new, but occasionally central feathers new and outer old. Body at same time as tail; moult arrested during migration (if started), completed in winter quarters. POST-JUVENILE. Partial: head and body (not always all vent or tail-coverts), lesser and median upper wing-coverts (sometimes inner median only), and t1–t2 or all tail. Starts on c. 40th day; completed in 3rd month, September–October (August–November). In captive birds, partial 1st pre-breeding moult follows in December–February, involving head, body, and lesser and median upper wing-coverts—all or only parts of feathering. FIRST POST-BREEDING. Partial. Timing and sequence as in adult pre-breeding; part of juvenile secondaries retained until 2 years old.

(2) *O. s. cyprius* and *mallorcae* and non-migratory birds of other races. Few details. Single ♂, Mallorca, had p3 already growing on 9 June (Stresemann and Stresemann 1966). January–February adults from Cyprus and Middle East had all primaries new. Extent of post-juvenile as nominate *scops*, but frequently a greater number of juvenile median and greater upper wing-coverts retained and at least occasionally all tail. No evidence for partial 1st pre-breeding moult.

Measurements. Nominate *scops*. France, Italy, Switzerland, and Balkans, March–August; skins (BMNH, RMNH, ZMA). Bill (F) to forehead, bill (C) to cere.

WING AD	♂ 160	(1·42; 11)	158–162	♀ 161	(2·90; 10)	158–165	
JUV	156	(1·43; 5)	154–158	158	(2·44; 5)	154–161	
TAIL AD	66·1	(2·94; 11)	62–70	68·1	(3·57; 10)	63–72	
BILL (F)	18·3	(0·57; 11)	17·8–19·5	18·1	(0·81; 10)	17·1–18·9	
BILL (C)	10·9	(0·51; 10)	10·0–11·4	11·7	(0·76; 9)	11·0–12·7	
TARSUS	26·8	(1·04; 11)	25·7 × 28·1	27·6	(1·08; 10)	26·3–29·3	

Sex differences significant for bill (C). Juvenile wing significantly shorter than adult, other measurements similar.

Caucasus and Transcaucasia, ages combined, April–August; skins (ZMM).

WING	♂ 153	(5·99; 15)	146–165	♀ 155	(3·17; 6)	150–158

O. s. mallorcae. Spain, April–July; skins (BMNH, RMNH, ZMA).

WING AD	♂ 153	(3·80; 9)	146–157	♀ 156	(2·41; 5)	153–158
TAIL AD	65·2	(3·32; 8)	61–70	64·7	(1·92; 5)	62–66
BILL (F)	17·3	(1·30; 9)	15·7–18·5	18·5	(0·68; 5)	17·8–19·3
BILL (C)	10·7	(0·68; 9)	9·8–11·4	11·7	(0·47; 5)	11·3–12·3
TARSUS	25·6	(1·21; 9)	23·8–27·1	25·9	(0·94; 5)	25·1–26·9

Sex differences significant for bill (C). Wing and tarsus significantly shorter than nominate *scops* from France to Balkans. Wing of juvenile ♂ 149 (5) 142–154.

O. s. cycladum. Southern Greece, Crete, southern Turkey, and Levant, February–August; skins (RMNH, ZMA).

WING AD	♂ 160	(4·61; 5)	156–167	♀ 161	(3·22; 7)	158–166
BILL (F)	18·0	(0·34; 4)	17·6–18·4	19·0	(0·44; 4)	18·4–19·4
BILL (C)	10·9	(0·92; 4)	9·8–11·6	12·2	(0·26; 4)	12·0–12·6
TARSUS	27·6	(0·32; 4)	27·4–28·1	28·0	(1·07; 4)	27·1–28·7

Sex differences significant for bill.

O. s. cyprius. Cyprus, January–May; skins (RMNH, ZMA).

WING AD	♂ 159	(2·25; 4)	157–162	♀ 162	(2·68; 8)	159–167
TAIL AD	67·8	(—; 3)	62–71	70·6	(3·50; 8)	66–76
BILL (F)	18·0	(—; 3)	17·5–18·5	18·7	(0·59; 8)	18·0–19·6
BILL (C)	10·9	(—; 3)	10·6–11·2	11·6	(0·72; 8)	10·9–12·5
TARSUS	27·6	(—; 3)	27·2–28·2	27·7	(1·19; 8)	26·2–29·0

Sex differences not significant.

O. s. turanicus. Turkmeniya (ZMM), Iran, and Afghanistan (Vaurie 1960d), age unknown.

WING	♂ 155	(4·83; 10)	149–162	♀ 158	(3·53; 7)	154–165
TAIL	63·6	(2·14; 10)	60–67	65·9	(2·85; 7)	63–70

O. s. pulchellus. Wing: ♂ 151–164, ♀ 154–169 (Hartert 1912–21; RMNH).

Weights. ADULT AND FIRST ADULT. Nominate *scops*. Average in captive pair, ♂ 83, ♀ 98; range of monthly average usually 77–81 in ♂, 90–95 in ♀; slight peak February (♂ c. 86, ♀ c. 103), distinct peak October (♂ 105, ♀ 119) when very fat (Piechocki 1969).

On migration: (1) southern France (Glutz and Bauer 1980); autumn (a) August–September, (b) October; (2) Malta (J Sultana and C Gauci); (3) south-east Morocco (Ash 1969b); (4) Libya (Erard and Larigauderie 1972); (5) central Nigeria (Smith 1966).

	Autumn			Spring		
(1) (a)	92·0	(14·3; 169)	64–135	78·1	(8·00; 136)	60–118
(b)	99·6	(14·8; 86)				
(2)	101	(8·3; 16)	77–112	85·5	(11·3; 34)	66–127
(3)				73·4	(—; 45)	54–91
(4)				81·4	(13·8; 22)	58–109
(5)	86	(—; 2)	78–94	109	(8·36; 6)	98–122

Isle of Man (England), April: 85 (BTO). On ship off Norway, June: ♀ c. 80 (Haftorn 1971). Breeding, south-west Switzerland, July: ♂ 76·6, ♀ 102 (Desfayes 1949a).

O. s. pulchellus. USSR and Mongolia: ♂♂ 78, 80, 80·5; ♀♀ 80, 82·5 (Dementiev and Gladkov 1951a; Gavrin et al. 1962; Piechocki 1968c).

Race unknown. Turkey, August and early September: 72, 74, 76 (BTO). Kuwait: January–February 77·4 (11·2; 5) 62–92; March–May 75·0 (9·79; 8) 59–90 (V A D Sales, BTO).

Downy Young and Juvenile. Nominate *scops*. At hatching, 9·3 (0·68; 5) 8·6–10·4; on 3rd day 21·7 (3) 21–22 (Desfayes 1949a). Captive birds on 5th day 34 (7), 10th 55 (7), 15th 71 (7), 20th 81 (7), 35th 90 (4) (Koenig 1973; Glutz and Bauer 1980).

Structure. Wing rather long and broad, tip slightly rounded. 10 primaries: p8 longest, p9 0–3 shorter, p10 12–19, p7 2–5, p6 9–15, p5 18–25, p4 26–33, p1 43–45, independent of race or age. In all races, tip of p10 between tips of p5 and p6 in closed wing or equal to p6; rarely (5 of 50 examined) between p6 and p7. In nominate *scops* from central Europe and in *mallorcae*, tip of p10 on average 4 mm shorter than tip of p6; in *cyprius*, *cycladum*, *turanicus*, and nominate *scops* from Caucasus, p10 1·5 mm shorter than p6. Inner web of (p8–)p9–p10 emarginated, outer web of p8–p9 slightly so. Outer web of p10 with short serrations. Tail square or slightly rounded; 12 feathers, t6 3–8 shorter than t1. Bill strong, short; base relatively heavy, gape wide, tip laterally compressed; sides of bill with relatively longer and stronger bristles than other west Palearctic owls. Facial disc rather indistinct, not surrounded by ruff of specialized feathers; distinct ridge of feathers above eye, ending in ear-tufts above ear-opening, each tuft consisting of c. 10 feathers up to 20–25 mm long. Tarsus and toes rather short, slender. Tarsus covered with rather short but close feathering down to base of toes. Middle toe without claw 17·3 (0·73; 9) 16–18 in ♂ of nominate *scops*, 17·0 (0·42; 7) 16–17 in ♀; 16·4 (0·58; 10) 16–17 in ♂ of *mallorcae*, *cyprius*, and *cycladum*, 16·6 (0·55; 16) 16–17 in ♀; outer toe without claw c. 77% of middle, inner c. 89%, hind c. 58%. Claws slender, rather short; middle claw 8·6 (0·49; 15) 8–9 in ♂ of nominate *scops*, *cyprius*, and *cycladum*, 8·6 (0·45; 18) 8–10 in ♀; 7·9 (0·83; 5) 7–9 in ♂ of *mallorcae*, 8·3 (0·65; 5) 8–9 in ♀; outer claw c. 84% of middle, inner c. 97%, hind c. 86%.

Geographical variation. Slight in size, more marked in tinge of ground-colour and in width and intensity of dark vermiculation and shaft-streaks; tinge of ground-colour also subject to individual variation, general colour varying between rufous and grey in some areas (though without real morphs, as all types of intermediates occur), but remarkably constant in others. Nominate *scops* markedly variable over much of geographical range, though some local populations rather uniform; *mallorcae* from Balearic Islands and Iberia rather constant in colour, vermiculation of both upperparts and underparts slightly darker and heavier, black streaks on average slightly broader; limited amount of buff or cinnamon tinge above, upperparts and upper wing-coverts appearing slightly darker and purer grey with more distinctly streaked crown, mantle, and scapulars; underparts slightly greyer and more broadly streaked. Some grey variants of nominate *scops* closely similar, however, and *mallorcae* mainly maintained here because of its smaller size—in particular, wing (if known-age specimens compared) and tarsus shorter. Only a few specimens from north-west Africa examined; part of these were migrant nominate *scops*, but at least some breeders from Morocco close to *mallorcae* in size, though plumage variable as in nominate *scops*. Nominate *scops* gradually merges into *pulchellus* east from Volga; *pulchellus* poorly differentiated from nominate *scops*, but black shaft-streaks average slightly narrower, grey vermiculation less distinct, and larger white spots show on hindneck and upper mantle, spotting often extending to crown and scapulars; grey birds average paler and more uniform grey-buff on upperparts than grey nominate *scops*, but more rufous birds closely similar apart from white spotting; pale rounded spots on scapulars and wing-coverts purer white, less cream or buff; ground-colour of underparts slightly paler and with narrower and more contrasting black shaft-streaks. O. s. cycladum inhabits Peloponnisos, Cyclades, Crete, Rhodos, southern Turkey south from southern Anatolia and east to at least Adana area, and from western Syria to Israel; grey vermiculations on upperparts coarser and blacker than in *mallorcae* and nominate *scops*, streaks broader and tending to form cross-marks (not as straight as in *mallorcae*), white spots on hindneck, mantle, outer scapulars, and longer upper wing-coverts pure white and distinct; upperparts appear darker grey than *mallorcae*, underparts more coarsely vermiculated or narrowly barred with darker grey; rufous birds more similar to rufous nominate *scops*, but streaks on crown broader and less straight and white spots on hindneck and mantle larger. No birds from northern half of Asia Minor examined; birds from Caucasus and Transcaucasia (ZMM) had marks on upperparts intermediate between nominate *scops* and *cycladum*; colour of upperparts mainly rufous, not grey as in most *cycladum*; size apparently small (see Measurements). O. s. cyprius even darker grey than grey *cycladum*; black streaks on upperparts heavier, extending laterally in narrow black bars, white spots on hindneck and mantle larger and more contrasting, spotting often extending to crown and scapulars; underparts with much wider black streaks and coarser black vermiculation and narrow bars, often isolating distinct white spots on breast, belly, and flanks; no rufous morphs, but most birds have ground-colour of crown, inner scapulars, back, and rump slightly buff, appearing dark brownish-grey; buff bars on inner webs of primaries usually almost absent (faint in other races), but white spots along inner borders more extensive. O. s. turanicus from Iraq and perhaps south-east Turkey eastward rather similar to *pulchellus*, but upperparts paler and more silvery-grey than any other race, black shaft-streaks narrower and sharper, and white spots on hindneck, mantle, and outer scapulars more conspicuous.

For relationships with Oriental Scops Owl O. sunia and African Scops Owl O. senegalensis, see Striated Scops Owl O. brucei.

CSR

Bubo bubo Eagle Owl

PLATES 45 and 55
[between pages 494 and 495, and facing page 567]

Du. Oehoe   Fr. Hibou grand-duc   Ge. Uhu
Ru. Филин   Sp. Buho real   Sw. Berguv

Strix Bubo Linnaeus, 1758

Polytypic. Nominate *bubo* (Linnaeus, 1758), Europe south to France, Sicily, western Greece, Yugoslavia, Rumania (except south and east) and northern Ukraine, east in European USSR to about Arkhangel'sk, Gorkiy, and Voronezh; *ruthenus* Zhitkov and Buturlin, 1906, central European USSR, east of nominate *bubo*, east to foothills of Ural mountains, south to *c.* 50°N and to lower Volga basin; *sibiricus* (Gloger, 1833), western foothills of Urals east to about Ob river, south to *c.* 50°N; *hispanus* Rothschild and Hartert, 1910, Iberian peninsula, and perhaps formerly northern Morocco and northern Algeria; *interpositus* Rothschild and Hartert, 1910, Turkey, Levant, and north-west Iran, north through Caucasus to Moldavia, southern Ukraine, lower Don basin, and Volga delta, west to eastern and southern Rumania and probably Bulgaria; *turcomanus* (Eversmann, 1835), steppes between lower Volga and lower Ural rivers, east through Transcaspia to lowlands of Soviet Central Asia; *nikolskii* Zarudny, 1905, eastern Iraq, Iran (except north-west), Afghanistan (except north-east), and western Pakistan; *ascalaphus* Savigny, 1809, North Africa and Middle East, east to western Iraq, south through Sahara and to northern Sudan and Arabia. Extralimital: *bengalensis* (Franklin, 1831), central Pakistan and India, north to foothills of Himalayas, and *c.* 11 other races in Transcaspia, central Asia, eastern Siberia, and China.

Field characters. 60–75 cm; wing-span 160–188 cm. Largest owl of west Palearctic (though smaller ♂♂ overlap with ♀ Snowy Owl *Nyctea scandiaca* and ♀ Great Grey Owl *Strix nebulosa*); much larger and bulkier than Buzzard *Buteo buteo*. Large to huge, almost barrel-shaped, eared owl, with heavily marked brown plumage in taiga forms and more uniform buff plumage in desert forms. Facial disc quite dark, dominated by large, glowing orange eyes in taiga forms; pale, with yellower eyes, in desert forms. Chiefly nocturnal. Flight recalls that of *Buteo*. Sexes similar; no seasonal variation. Juvenile separable. 8 races in west Palearctic, but only smaller southern forms easily distinguished from others.

ADULT. (1) Races of Europe, Turkey, and Levant, nominate *bubo*, *ruthenus*, *hispanus*, and *interpositus*. All essentially dark but warm brown, with ochre ground-colour; thickly marked above with broad black streaks and spots, bars (on flight-feathers), and vermiculations; marked below with long broad black droplets on chest, and narrower black streaks and close wavy bars on flanks and underbody. Uniformity of pattern relieved on head by black-brown ear-tufts, paler buff edges to crown (meeting over bill), and by buff-white narrow ruff and broad white patch on throat and upper breast; cheeks brown-grey containing intensely orange eyes, narrowly rimmed black. Eyes and broad pale throat dominate face, creating gentle expression in repose. In flight, wing shows dusky leading edge and primary coverts, contrasting with yellow-buff ground-colour to primaries and outer secondaries (outer secondaries more strongly barred than inner primaries); underwing ochre, with black-brown barring on coverts and obvious black edge to primary coverts. Feathers of tarsus buff-white barred brown; of toes buff-white (and often sparse). Bill black. Birds of north-west Europe (nominate *bubo*) darkest in this group, those of Iberia (*hispanus*) less tawny but still brown and more sharply marked, and those of southern Russia and northern Middle East (*interpositus*) increasingly yellower towards south of range and again more sharply marked. (2) North African and southern Middle East race, *ascalaphus*. About 20% smaller than those races in west Palearctic; shorter eared and shorter tailed. Plumage less heavily streaked but more mottled. Tarsal feathers uniform buff-white. Lighter birds have plumage ground-colour yellow-buff, streaking more distinct, and dark-edged facial disc and upper breast pale buff-white (creating distinctly pale face); these birds sometimes treated as distinct race, but complete intergradation with darker birds occurs. Eyes paler, yellower than in more northerly races. JUVENILE. Lacks more than rudimentary ear-tufts. Head fluffier and less marked, greyer than adult but with more obvious black eyebrows over relatively larger orange eyes. Breast lacks obvious spread of black droplets.

Unmistakable if seen well, but confusion otherwise possible with Great Grey Owl *Strix nebulosa* (in northern Europe) and Brown Fish Owl *Ketupa zeylonensis* (in Middle East). Distinction from former not difficult, since *S. nebulosa* lacks ears and has essentially grey plumage, round lined face with white 'nose' and black 'chin', long, more square-cut, and strongly barred wings with striking pale patch on bases of outer primaries above, and dark band on end of tail. Separation from *K. zeylonensis* not easy; see that species. Flight of *B. bubo* powerful and fast, with rather shallow, regular wing-beats; wings raised more above body than depressed below it. Action often recalls diurnal raptor such as *Buteo buteo* and lacks wavering quality of most owls, with bursts of wing-beats interspersed with long, straight, fast glides on slightly arched wings. Attacks prey with low fast glide of considerable momentum. Usually stands upright on ground or perch; body barrel-shaped (narrowing towards head) at most times, though neck surprisingly long (but still thick) in alert posture. Ear-tufts always visible, even at distance, except in flight. Gait a shuffling walk and occasional lope, but more agile on ground than most owls.

Territorial call a booming 'oo-hu', repeated every c. 8–10 s; in alarm, a harsh 'ka ka-kau'. For possibility of confusion with Long-eared Owl *Asio otus* and Ural Owl *Strix uralensis*, see Voice.

Habitat. Surviving vestiges of wilderness, immune from human exploitation and from significant human disturbance, in boreal, temperate, steppe, and Mediterranean zones, where such a powerful predator can find within a compact hunting area sufficient and reliable supply of large prey, together with sites for secure nesting and roosting. Such sites exist only in regions of sparse human settlement or in topographically inaccessible or forbidding terrain, tending to be steep, and rocky, with detritus or broken ground and frequent caves or crevices, and mossy marshlands or streams liable to flood. Groups or rows of trees (often conifers) provide varied cover, even on rock-faces, against adverse weather and offer daytime roosts combining concealment with view of intruders. Hunting takes place sometimes within forests, but open or sparsely wooded terrain preferred, as well as floodlands, heaths, and farmed valley bottoms or levels with grassland, small arable fields, and even refuse dumps. Remote inner mountain valleys less suitable than main valleys under farming, or upland fringes, or sites of rockfalls just beneath treeline. In Switzerland, regular breeding occurs up to c. 1800–1900 m, and isolated examples above 2000 m occur in central Europe, with hunting up to c. 2800 m. Breeding in hollow trees and old buildings exceptional; in plains, often uses deserted tree-nests of other species and nests on ground, e.g. among fallen trees or in alder *Alnus* carr. Vegetation growing by nest welcomed, and sites below overhang favoured (Glutz and Bauer 1980). In central Asia and Himalayas, nests at up to 4200–4500 m; *bengalensis* in bush-covered rocky hills and wooded country with outscoured ravines, old mango *Mangifera* orchards, and groves of ancient trees in neighbourhood of cultivation and villages; partial to steep earth banks and clay cliffs of dry nullahs and rivers. Not uncommon in eroded semi-desert thorn jungle but avoids both pure desert tracts and humid evergreen forest. Spends daytime in seclusion of foliage and rock fissure but often sits on rock pinnacle or other lookout in daylight (Ali and Ripley 1969).

Flies mostly in lower airspace except for longer distance and when soaring on rising air-currents.

Distribution. Marked range reductions in Europe, with extinction in Luxembourg and Denmark.

FRANCE. Some decrease in range, e.g. now extinct Juras, Vosges, and Burgundy (Yeatman 1976). BELGIUM. Last proved breeding 1905 and 1907 (Lippens and Wille 1972). Bred 1982 (Doucet *et al.* 1982) and 1983 (PD). LUXEMBOURG. Last bred 1930s (Glutz and Bauer 1980). WEST GERMANY. For extinctions in west and north, and recent breeding following re-introduction, see Glutz and Bauer (1980) and Bauer and Thielcke (1982). DENMARK. Last bred 1884 (Dybbro 1976). SWEDEN. Widespread until late 19th century, but became extinct in many areas up to c. 1920; further reductions in range since, especially in north (Curry-Lindahl 1950; Olsson 1976). Sparse (10–20 pairs) in northern region and only general area mapped (LR). POLAND. Marked range decrease; bred throughout up to mid-19th century (Tomiałojć 1976a). EAST GERMANY. Became extinct in many areas from mid 19th century; some recolonized since mid-1950s (for details, see Glutz and Bauer 1980). ITALY. Recently rediscovered breeding in Sicily (PB, BM). GREECE. Formerly bred more widely, including in Peloponnisos (Bauer *et al.* 1969). TURKEY. Under-recorded; probable breeding area shown, although breeding rarely proved (MB, RFP). SYRIA. Present distribution imperfectly known, but rare and scattered (Kumerloeve 1968b; HK). LEBANON. Few recent records (HK). CHAD. May breed Tibesti, but no proof (Simon 1965).

Accidental. Britain, Belgium, Luxembourg, Netherlands, Denmark.

Population. Decreased many areas in 19th century, due mainly to human persecution; some increases recently, aided by protection and reintroductions.

FRANCE. Under 100 pairs (Yeatman 1976); considered too low by Cheylan (1979)—also by Bergier and Badan (1979) who estimated 100 pairs in Bouches-du-Rhône. Total now probably 500–700 pairs (R Cruon). May represent real increase in some areas, e.g. Massif Central (Choussy 1971; Faurc 1978). PORTUGAL. Decreasing (RR). SPAIN. At least 400 pairs 1965; marked decline since c. 1950 (AN). WEST GERMANY. Around 1900, after marked decline, 130–135 pairs, then decreased further to c. 35 pairs by c. 1935. Considerable increase in main stronghold (Bayern) beginning c. 1965, mainly due to nest protection; elsewhere (e.g. Baden-Württemberg), aided by re-introductions, reached 150–185 pairs c. 1975–7 and c. 170–210 pairs 1981 (see details in Glutz and Bauer 1980 and Bauer and Thielcke 1982; also Herrlinger 1973 and Rockenbauch 1978a). NORWAY. 500–600 pairs 1960 (Hagen 1964); over 500 pairs 1982, following some re-introductions in east since 1980 (GL, VR). SWEDEN. Marked decline since late 19th century due mainly to human persecution and from 1950s due mainly to poisonous seed-dressings. In 1943–8, 291 breeding pairs, 78 probably breeding and 86 sites where single birds seen (Curry-Lindahl 1950); in 1964–5, comparable figures were 56, 28, and 91, and in 1974–5, 46, 68, and 57 (Olsson 1976). Since 1965, some recovery south of c. 60°N, aided by re-introductions and ban on seed-dressings (Odsjö and Olsson 1975; Broo 1982), but still decreasing markedly

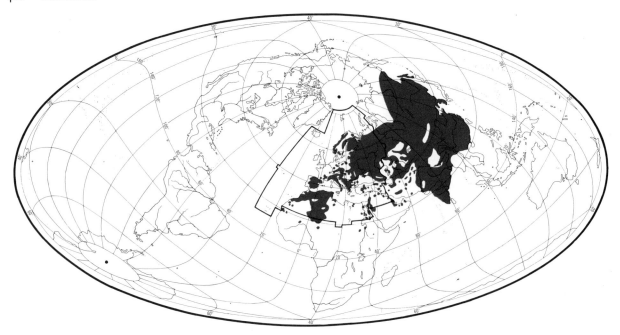

further north (LR). FINLAND. Estimated 200 pairs; marked decrease due to human persecution (Merikallio 1958); now increasing, at least locally, with over 1000 pairs (OH). EAST GERMANY. 35–42 pairs (Mikkola 1983). POLAND. Marked decline in second half of 19th century. Estimated 60–70 pairs 1952 (Ferens 1953); apparently little recent change (Tomiałojć 1976a; Glutz and Bauer 1980). CZECHOSLOVAKIA. Increased after protection in 1929, still continuing some areas, but also local declines (KH). Estimated 270 pairs Bohemia and Moravia 1970, and 80–250 pairs Slovakia (Glutz and Bauer 1980). HUNGARY. Estimated 25–30 pairs (Glutz and Bauer 1980); little change in recent years (LH). AUSTRIA. Probably c. 200 pairs (Glutz and Bauer 1980). SWITZERLAND. About 60 pairs (Glutz and Bauer 1980; see also Haller 1978). ITALY. Decreased some areas (PB, BM). GREECE. Marked decrease since c. 1960, due to human persecution (WB, HJB, GM). Estimated 200 pairs (B Hallmann). BULGARIA. About 50 pairs (Mikkola 1983); decreased in last 30 years (JLR). RUMANIA. Decreased (Vasiliu 1968). USSR. Common north of c. 55°N and in Caucasus; rare in central areas, especially Ukraine (Dementiev and Gladkov 1951a). Estonia: 50–60 pairs (Curry-Lindahl 1950); 100–120 pairs (Randla 1976). Latvia: 400–420 pairs (Curry-Lindahl 1950); decreased, now rare (Baumanis and Blums 1972); c. 20–30 pairs (Viksne 1983). Lithuania: decreased markedly, almost extinct (B Šablevičius). KUWAIT. Only 1 known breeding pair (PRH). ALGERIA. Scarce to rare in north; more plentiful in desert (EDHJ).

Survival. Data inadequate for precise calculations (see, e.g., Olsson 1979); for details of recoveries of wild and re-introduced birds, see Glutz and Bauer (1980). Oldest ringed bird at least 21 years (Olsson 1979).

Movements. Resident in Europe, though considerable nomadism reported in USSR (especially Asiatic part) where winter conditions harsh.

12 Norwegian recoveries were 8–220 km (average 95) from ringing site, and suggested tendency for inland birds to disperse towards coasts (Haftorn 1971). Swedish study (Olsson 1979) confirmed that, as in other Strigiformes, fledged young spend considerable period in parental care, and do not start dispersing until August–September. Between then and following summer, c. 50% of young birds recovered more than 75 km from birthplace, directions random (maximum recovery distance c. 180 km NNE); thereafter, however, 75% found within 50 km of birthplace (maximum 86 km). Hence immatures dispersing furthest either have higher 1st winter mortality, or they return towards natal area (as found in one study on Little Owl *Athene noctua*—see that species); possibly both apply (Olsson 1979). Recoveries for oldest birds all quite near ringing site: 2 nest siblings found 16 years later 16 km and 21 km from birthplace; one 15-year-old found 15 km from birthplace; not known whether these had returned from distant dispersal or never left vicinity. Once adult, fidelity to territory is high—for 20 years in one case (Lettesjo 1974; see also Social Pattern and Behaviour).

Fewer data available for west-central European population, but these consistent with Scandinavian findings. Recoveries showed movement of 11–205 km; from May of 3rd calendar year ($n=22$) median distance $47 \cdot 5 \pm 31 \cdot 6$ km (Glutz and Bauer 1980). Longest: bird ringed Baden-Württemberg as chick found Lorraine (France) in March of following year. 205 km north-west (Rockenbauch 1978a). In initial stage of dispersal, one juvenile moved 18 km SSW by 24 September, 3 months after fledg-

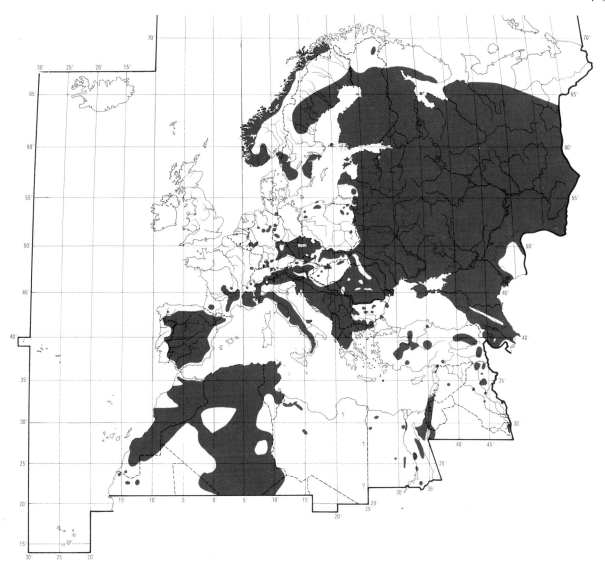

ing (Mebs 1972). 2 nest siblings ringed Oberbayern (West Germany) found respectively 100 km south-east in (Tirol) Austria in August of 2nd year, and 7 km from birthplace in neighbouring territory 10 years later (Glutz and Bauer 1980). 52 released captive-bred birds dispersed 0–110 km, plus one released Switzerland August 1973 found Salzburg (Austria) April 1976, 400 km east (Glutz and Bauer 1980).

In all populations, individuals wander outside their normal range to some extent, rarely in some though more regularly in others (Dementiev and Gladkov 1951a; Vaurie 1965). Nominate *bubo* (central and northern Europe) has straggled to Scotland, Finnmark, Murmansk, Kandalaksha (White Sea), and Timanskaya tundra. Pale race *sibiricus* (Urals and western Siberia) shows considerable winter nomadism, and has straggled west to Mezen basin (Arkhangel'sk) and Karelia, south to Caspian and Syr-Dar'ya (Kazakhstan). Transcaspian and Kazakhstan race *turcomanus* makes more or less regular southward movements, notably towards Caspian coasts including Astrakhan, occurring there mid-October to February–March. Also, *ruthenus* (central European USSR) has occurred exceptionally in winter on Volga delta (Astrakhan) and Orenburg steppes. Various southern races from Caucasus eastwards show some altitudinal movement in mountainous regions, with higher birds descending in autumn to winter in valleys. However, only for nominate *bubo* in Europe has there been ringing study to quantify the scale of dispersal.

Food. Largely mammals from size of water vole *Arvicola terrestris* to adult hares *Lepus*, and birds from size of Jay *Garrulus glandarius* to adult Mallard *Anas platyrhynchos*. Hunting essentially nocturnal, starting immediately after sunset (Meinertzhagen 1959); active also in daylight in summer at northern edge of range (Dementiev and Gladkov 1951a). In Sörmland (Sweden), June–August, food brought to nest between c. 1 hr after sunset and c. 1 hr before dawn (Kranz 1971); at 59°N, Norway, summer, nocturnal activity peaks c. 22.30 hrs (mostly) and 02.00 hrs (Hagen 1950); at 62°N, Norway, mid-June, food brought to nest 22.0–04.00 hrs (Mysterud and Dunker 1983); active till dawn, Poland (Bocheński 1960); see also Blondel and Badan (1976). Few observations on hunting in the wild, but apparently involves much time waiting on open perch; closed woodland usually avoided (Willgohs 1974). When hunting voles (Microtinae), waits on average only 5 min (range 10 s to 32 min) before moving to next perch; searching flights hardly longer than 50–100 m (Glutz and Bauer 1980). Not clear whether some information in Glutz and Bauer (1980) refers to behaviour seen in the wild, observations in captivity, or inferences from nature of prey taken: said to surprise prey while flying close to ground or tree-tops, to kill them in their sleep, and to make systematic searches of rock-crevices for roosting birds. Will kill Eiders *Somateria mollissima* on nest; recorded 'jumping around' in nesting area looking for one, but usually watches from nearby, waiting for bird to reveal itself. Puffins *Fratercula arctica* captured regularly at colony, apparently near nest-burrows. Other auks (Alcidae) apparently taken largely from sea (Willgohs 1974). Remains at nests indicate that all occupants of a nest of (e.g.) crows *Corvus* or buzzards *Buteo*, often including adults, are frequently killed over 1–2 nights (Olsson 1979). Capable of seizing shot birds before they reach ground (Baumgart *et al.* 1975), and said by Thiollay (1968) to take birds in full flight. Fish said to be taken by plunging into water like Osprey *Pandion haliaetus*, by hovering over water, or from nets and fishing lines (Glutz and Bauer 1980). Some items, e.g. crabs (Brachyura), presumably taken by hunting on ground. Wild bird recorded climbing into hen house to take hens (Willgohs 1974). For proportions of disabled individuals among prey taken, see Frey (1973); also Olsson (1979). For actions of (apparently captive) bird in seizing prey, see Glutz and Bauer (1980). Can carry adult *S. mollissima* in flight for at least 500 m (Willgohs 1974), and prey weighing 3 kg found at nest. Once flushed carrying full-grown fox *Vulpes vulpes* in feet (Fossheim 1955). Prey usually taken to exposed rock or tree for eating (Glutz and Bauer 1980); probably killed largely with feet, but bill also used; frequently decapitated (head sometimes left). Prey up to size of thrushes *Turdus* and rats *Rattus* may be swallowed whole. Birds sometimes plucked. Hedgehogs *Erinaceus* sometimes have skin with spines ripped off first, but spines often eaten. (Willgohs 1974; Glutz and Bauer 1980.) Caches food in hollows and under snow (Dementiev and Gladkov 1951a); see also Social Pattern and Behaviour. Recorded returning the next night to finish half-eaten prey (Willgohs 1974). Cannibalism occurs: usually weakest young being eaten by parents or siblings, but full-grown bird also recorded being eaten.

Pellets usually have one end square, other more pointed, but variation enormous. Average 72×34 mm ($30-178 \times 20-60$), $n = 210$ for length, 102 for width (Willgohs 1974); average $77 \times 30 \times 27$ mm, 3·2 prey per pellet (Mikkola 1973). For proportions of different bones of various prey appearing in pellets, see Bezzel *et al.* (1976).

The following prey recorded in west Palearctic. Mammals: hedgehogs (*Erinaceus europaeus*, *E. algirus*, *Hemiechinus auritus*), shrews (*Sorex araneus*, *Neomys fodiens*, *N. anomalus*, *Crocidura leucodon*, *C. suaveolens*, *C. russula*), Russian desman *Desmana moschata*, mole *Talpa europaea*, bats (*Rhinolophus ferrumequinum*, *R. hipposideros*, *Myotis myotis*, *M. blythii*, *M. oxygnathus*, *Vespertilio serotinus*, *Nyctalus noctula*), rabbit *Oryctolagus cuniculus*, hares (*Lepus europaeus*, *L. timidus*), pika *Ochotona*, flying squirrel *Pteromys volans*, red squirrel *Sciurus vulgaris*, Siberian chipmunk *Tamias sibiricus*, European suslik *Citellus citellus*, alpine marmot *Marmota marmota*, Lataste's gundi *Massoutiera mzabi*, garden dormouse *Eliomys quercinus*, forest dormouse *Dryomys nitedula*, edible dormouse *Glis glis*, dormouse *Muscardinus avellanarius*, mouse-like hamster *Calomyscus bailwardi*, common hamster *Cricetus cricetus*, golden hamster *Mesocricetus auratus*, migratory hamster *Cricetulus migratorius*, gerbils (*Gerbillus campestris*, *G. nanus*, *G. gerbillus*, *G. pyramidum*, *Pachyuromys duprasi*, *Meriones persicus*, *M. vinogradovi*, *M. shawi*, *M. libycus*, *Psammomys obesus*), wood lemming *Myopus schisticolor*, Norway lemming *Lemmus lemmus*, mole-voles (*Ellobius talpinus*, *E. lutescens*), bank vole *Clethrionomys glareolus*, grey-sided vole *C. rufocanus*, water vole *Arvicola terrestris*, pine voles (*Pitymys subterraneus*, *P. duodecimcostatus*), common vole *Microtus arvalis*, short-tailed vole *M. agrestis*, root vole *M. oeconomus*, snow vole *M. nivalis*, social vole *M. socialis*, muskrat *Ondatra zibethicus*, harvest mouse *Micromys minutus*, striped field mouse *Apodemus agrarius*, yellow-necked mouse *A. flavicollis*, wood mouse *A. sylvaticus*, black rat *Rattus rattus*, brown rat *R. norvegicus*, house mouse *Mus musculus*, lesser mole-rat *Spalax leucodon*, northern birch mouse *Sicista betulina* jerboas (*Allactaga elater*, *A. williamsi*, *A. major*, *Jaculus jaculus*, *J. orientalis*), red fox *Vulpes vulpes*, raccoon dog *Nyctereutes procyonoides*, young badger *Meles meles*, stoat *Mustela erminea*, weasel *M. nivalis*, mink *Lutreola lutreola*, polecat *Putorius putorius*, pine marten *Martes martes*, beech marten *M. foina*, sable *M. zibellina*, genet *Genetta genetta*, wild cat *Felis silvestris*, wild boar *Sus scrofa*, young roe deer *Capreolus capreolus*; also domestic cat, young dog,

PLATE 37 (*facing*).
Clamator jacobinus pica Jacobin Cuckoo (p. 388): **1–2** ad intermediate morph, **3** juv intermediate morph.
Clamator glandarius Great Spotted Cuckoo (p. 391): **4–5** ad (fresh plumage), **6** juv.
Chrysococcyx caprius Didric Cuckoo (p. 400): **7–8** ad ♂, **9** ad ♀.
Centropus senegalensis aegyptius Senegal Coucal (p. 427): **10–11** ad, **12** juv. (DIMW)

PLATE 38. *Cuculus canorus canorus* Cuckoo (p. 402): **1** ad ♂, **2** ad ♀ grey morph, **3** ad ♀ rufous morph, **4** rufous juv, **5** grey juv. (DIMW)

PLATE 39. *Cuculus saturatus horsfieldi* Oriental Cuckoo (p. 417): **1** ad ♂, **2** ad ♀ grey morph, **3** ad rufous morph, **4** rufous juv, **5** grey juv. (DIMW)

PLATE 40. *Coccyzus erythrophthalmus* Black-billed Cuckoo (p. 423): **1** ad, **2–4** juv in autumn. *Coccyzus americanus* Yellow-billed Cuckoo (p. 425): **5** ad, **6–8** 1st winter. (DIMW)

PLATE 41. *Chrysococcyx caprius* Didric Cuckoo (p. 400): **1** ad ♂, **2** ad ♀. *Centropus senegalensis aegyptius* Senegal Coucal (p. 427): **3** ad, **4** 1st ad, **5** juv. (DIMW)

young sheep, and cow (see, especially, Janossy and Schmidt 1970; also Höglund 1966, Thiollay 1969, Wagner and Springer 1970, Willgohs 1974, Baumgart 1975, Hiraldo *et al.* 1975, Bezzel *et al.* 1976, Randla 1976, Vein and Thévenot 1978, Perez Mellado 1980). Most larger mammals (e.g. hares *Lepus*, fox *Vulpes*) often taken as young (e.g. Olsson 1979); no firm evidence that roe deer *Capreolus* ever taken alive when full- or even half-grown (Willgohs 1974; Glutz and Bauer 1980). Enormous range of bird species recorded, up to size of full-grown Grey Heron *Ardea cinerea*, adult ♂ Capercaillie *Tetrao urogallus*, and full-grown bird of own species (see, especially, Janossy and Schmidt 1970, Willgohs 1974, Hiraldo *et al.* 1975, Olsson 1979). Small passerines taken rather infrequently: e.g. except for crows (Corvidae), only 11·3% of 2282 birds taken in Norway were passerines (Willgohs 1974); 2·6% of 2913, southern Sweden (Olsson 1979). Important groups include ducks (Anatinae, especially *Anas platyrhynchos* and *Somateria mollissima*), grouse (Tetraonidae), rails (Rallidae, especially Coot *Fulica atra*), waders (Charadrii), gulls (Laridae), auks (Alcidae), and crows (Corvidae). Frequently takes raptors (Accipitriformes and Falconiformes) and other owl species, apparently more often than their abundance warrants; for review, see Mikkola (1976a). By weight, birds taken are mostly full-grown. Will take birds' eggs (Olsson 1979). Reptiles: turtle *Clemys caspica*, lizards (*Lacerta lepida, L. hispanica, Psammodromus algirus*), slow-worm *Anguis fragilis*, grass snake *Natrix natrix* (Hiraldo *et al.* 1975; Bezzel *et al.* 1976; Olsson 1979). Amphibians: frogs (*Rana temporaria, R. ridibunda, R. esculenta*), toads (*Bufo bufo, Pelobates fuscus, P. cultripes*) (Bocheński 1960; Marián and Marián 1973; Hiraldo *et al.* 1975). Fish: trout *Salmo trutta*, pike *Esox lucius*, carp, etc. (*Cyprinus carpio, Barbus barbus, Chondrostoma polilepis, Rutilus rutilus, Leuciscus cephalus, L. idbarus*), eels (*Anguilla anguilla, Conger conger*), cod, etc. (*Pollachius pollachius, P. virens, Gadus morhua, Brosme brosme, Lota vulgaris, L. lofa*), roach *Gymnocephalus cernua*, perch *Perca fluviatilis*, wrasse *Labrus bergylta*, bullhead *Cottus*, lumpsucker *Cyclopterus lumpus*, angler fish *Lophius piscatorius*; in Norway, most fish 'small or medium', but 2 weighed *c*. 1·5 kg (see, especially, Willgohs 1974; also Curry-Lindahl 1950, Bocheński 1960, Höglund 1966, Emmett *et al.* 1972, Hiraldo *et al.* 1975, Bezzel *et al.* 1976, Olsson 1979). Invertebrates usually occur infrequently: include beetles (Carabidae, Dytiscidae, Lucanidae, Scarabaeidae, Tenebrionidae, Cerambycidae, Curculionidae), crickets, etc. (Tettigoniidae, Gryllotalpidae), mantis *Mantis religiosa*, centipede *Scolopendra*, scorpions *Buthus*, crayfish *Astacus fluviatilis* (Moltoni 1937; Bocheński 1960; Höglund 1966;

PLATE 42 (*facing*).
Cuculus canorus canorus Cuckoo (p. 402): 1–9 ad ♂, 10 ad ♀ grey morph, 11 ad ♀ rufous morph, 12–13 grey juv, 14 rufous juv. *Cuculus saturatus horsfieldi* Oriental Cuckoo (p. 417): 15 ad ♂, 16 ad ♀ rufous morph, 17 grey juv, 18 rufous juv. (DIMW)

Thiollay 1969; Hiraldo *et al.* 1975; Vein and Thévenot 1978); marine crabs (*Cancer pagurus, Carcinus maenas*) may be taken regularly where available; molluscs (*Buccinum, Littorina*) in pellets probably from guts of bird prey (Willgohs 1974). Grass, grain, etc., recorded in pellets (Willgohs 1974; Olsson 1979). In captivity, recorded sometimes swallowing green leaves, apparently not for lack of material for pellet formation (Räber 1950).

For major review of food analyses throughout Palearctic range, see Janossy and Schmidt (1970); for 5 west Palearctic studies, see Table A. See also the following significant work. Norway: Hagen (1950), Willgohs (1974), Mysterud and Dunker (1983). Sweden: Curry-Lindahl (1950), Höglund (1966), Schaefer (1970, 1971), Emmett *et al.* (1972); see also sources in Olsson (1979). Finland: Sulkava (1966); see also sources in Mikkola (1983). France: Thiollay (1968, 1969), Choussy (1971), Mebs (1972). East and West Germany: Uttendörfer (1939b, 1952), März (1954), Bezzel *et al.* (1976); see also sources in Glutz and Bauer (1980). Switzerland: Wagner and Springer (1970). Austria: Frey (1973), Frey and Walter (1977). Poland: Schnurre (1954), Bocheński (1960, 1966), Banz and Degen (1975). Czechoslovakia: Schaefer (1972), Vondráček (1978), Vondráček and Honců (1978). Bulgaria: Baumgart (1973, 1975). USSR: Neufeldt (1958b), Randla (1976); see also sources in Janossy and Schmidt (1970). Spain: Ruiz Bustos and Camacho Muñoz (1973), Hiraldo *et al.* (1975), Ferreira (1981). North Africa: Janossy and Schmidt (1970), Saint Girons *et al.* (1974). For discussion of trophic niche breadth in Spain compared with central and northern Europe, see Hiraldo *et al.* (1976).

For year-to-year study over 13 years, Sweden, see Olsson (1979): proportion of rodents in diet followed their population levels — by number, 23–65% in different years; shortfall in rodents largely made up for by increase in birds, other prey groups staying fairly constant. In Finland, one pair took largely *Arvicola terrestris* when these abundant, switching to *Rattus norvegicus* from nearby rubbish dump at other times (Mikkola 1983). Seasonal changes difficult to study due to inadvisability of frequent visits to nests and difficulties of pellet collection outside breeding season. Olsson (1979) found adult hares *Lepus* increased in importance in late winter; other prey (especially newly fledged birds) became important as they became available. In central Spain, some suggestion that proportion of birds increased in late spring and early summer (Perez Mellado 1980), though Hiraldo *et al.* (1975) found no clear pattern. See also Willgohs (1974), Förstel (1977), and Wickl (1979). In autumn at Col de Bretolet (France), seems to specialize on concentrated stream of migrant birds which pass through (Thiollay 1968). Major differences occur in diets recorded from different (nearby) habitats (see, especially, Olsson 1979; also Willgohs 1974, Bezzel *et al.* 1976).

Daily food consumption of captive birds 300–400 g in winter, 200–300 g in summer, much more for young birds

472 Buboninae

Table A Food of Eagle Owl *Bubo bubo* as found from nest-remains and pellet analyses, largely from breeding season. Figures are percentages of total live weight of vertebrates only.

	SE Sweden	Nordbayern (W Germany)	Provence (S France)	Central Spain	Moyen Atlas (Morocco)
Hedgehogs *Erinaceus*	0	23.7	8.3	1.0	15.2
Rabbits, hares (Lagomorpha)	8.8	26.2	68.7	83.7	1.2
Squirrels (Sciuridae)	1.8	1.7	0.7	0	0
Dormice (Muscardinidae)	0	0.1	0.5	0.2	0.1
Hamsters (Cricetinae)	0	+	0	0	0
Water vole *Arvicola terrestris*	11.9	1.4	0.5	0.6	0
Rats *Rattus*	9.6	6.1	9.1	4.3	0.1
Other mice, voles, gerbils (Muridae)	2.8	1.2	0.1	+	2.5
Jerboas (Dipodinae)	0	0	0	0	73.2
Red fox *Vulpes vulpes*	1.1	0.5	3.8	0.9	0
Martens (*Mustela*, *Martes*)	0.6	0.3	0.2	0	0.1
Other mammals	0.8	0.9	0	0.7	0.3
Waterbirds	46.7 } 59.5	7.9 } 37.9	0.6 } 8.1	0 } 8.4	0 } 6.1
Other birds	12.8	30.0	7.5	8.4	6.1
Reptiles	+	0	0.1	0.1	0.2
Amphibians	1.4	+	0.1	+	0.8
Fish	1.4	0.1	+	+	0
Total no. of vertebrates	6450	14185	389	1436	917
Source	Olsson (1979), data reanalysed	Wickl (1979) and Glutz and Bauer (1980), data reanalysed	Blondel and Badan (1976).	Perez Mellado (1980).	Vein and Thévenot (1978).

(Averin and Ganya 1966); usually 450 g (Heinroth and Heinroth 1926–7); c. 500 g (nominate *bubo*), apparently c. 800 g (*ruthenus*) (Dementiev and Gladkov 1951a); average 230 g (Herrlinger 1971); average 234 g, peaking at 310 g in September–October (see Glutz and Bauer 1980).

Studies discussed above contain large proportion of food brought to nests, but no detailed work done on differences between diet of adults and young. In Nordbayern (West Germany), apparently fewer hedgehogs *Erinaceus* brought to nest (10.8% by weight) than eaten by adults (23.8%); more birds brought (42.9% rather than 34.1%) (Bezzel *et al.* 1976).   DJB

Social pattern and behaviour. Well studied. Major accounts reviewed by März and Piechocki (1980) and Mikkola (1983). Account includes notes supplied by V Olsson.
  1. Solitary. Pairs sedentary and strongly territorial throughout the year, resorting to areas of better food supply only in hard winters (Blondel and Badan 1976; Mikkola 1983; V Olsson). BONDS. Pair-bond monogamous and life-long (Blondel and Badan 1976; Mikkola 1983). Divorce not recorded. After loss of mate, 5 birds (3 ♂♂, 2 ♀♀) continued to hold territory and took on average c. 10 months (1–23) to replace mate; one widowed ♀ moved 1.5 km within 3–4 weeks and bred successfully the same year (Haller 1978). Age of first breeding 2–3 years, ♂♂ probably at 2 (Ferens 1960; König and Haensel 1968). ♀ performs all incubation and brooding until young well-grown; until ♀ stops brooding, ♂ alone provisions her and young (e.g. Frey 1973, Blondel and Badan 1976; see also Relations within Family Group, below). Young dependent on parents for up to 5 months; leave parental territory thereafter (Glutz and Bauer 1980; see also Relations within Family Group). Though young sometimes eaten by siblings, aided by ♀ parent, during food shortage (Brdicka 1969; Mikkola 1983), no evidence that young actually killed (März and Piechocki 1980). BREEDING DISPERSION. Solitary. Birds occupy home-range with size dependent on food supply, etc., and difficult to determine. Birds hunt outside strictly defended area, and home-feeding ranges of neighbouring pairs thus overlap (König and Haensel 1968; Thiollay 1969). No precise measurements of defended area, and the following refer to home-range. Probably 12–20 km^2 typical, but activity during breeding season often concentrated in 1–1.5 km^2 (Glutz and Bauer 1980), or in radius of c. 2 km around nest (Baumgart *et al.* 1973; Frey 1973; Haller 1978). In Provence (France), average c. 14 km^2 (Blondel and Badan 1976). In Franken (West Germany), c. 15 km^2 (Mebs 1972). In Bulgaria, where prey less abundant, 12–40 km^2 (Baumgart *et al.* 1973). Average distance between 29 nests 8.5 km, radius of home-range c. 4–5 km (Olsson 1979). Other estimates of radius 2.3–7 km (Thiollay 1969; Mebs 1972; Förstel 1977; März and Piechocki 1980). In Elbsandsteingebirge (East Germany), radius of home-range apparently 3–4 km in summer, 5–7 km in winter (März 1940), exceptionally up to 13 km (Desfayes 1951). Distance between nests varies widely: in southern France, minimum 100–600 m apart (Thiollay 1969; Choussy 1971; Blondel and Badan 1976; Bergier and Badan 1979), average 1.4 km ($n=32$) (Bergier and Badan 1979), usually 2–8 km (Thiollay 1969). In Lower Austria, average 2.4 km (0.4–3.5) (Frey 1973); in Fränkischer Jura (Bayern, West Germany) average 4.4 km, minimum 2.9 (Mebs 1972). Where prey abundant, average typically 5.5–8.5 km (Förstel 1977; Haller 1978; Olsson 1979). In optimum conditions, density locally high, e.g. 3 pairs per 10 km^2 (Choussy 1971); in Provence, up to 28 pairs in 140 km^2; in Bulgaria, 12 pairs in 35-km stretch (see Glutz and Bauer 1980). In Poland, 0.03–0.06 pairs per 100 km^2 ($n=79$) (Lagerström 1978); in southern France, 1 pair per 20–80 km^2 (Thiollay 1969); in Fränkischer Jura, 8 pairs in c. 100 km^2 (Mebs 1972). For other densities, see especially Haller (1978) and Glutz and Bauer (1980). Fidelity to breeding area shown over several years (e.g. Mebs 1972); see also Movements.

Several nest-sites within territory may be used (see Breeding) but birds usually strongly attached to 1 or 2; one scrape used for 15 out of 16 years (Olsson 1979). Birds attended potential nest-sites 1·2–2·3 km apart within year, up to 3·3 km between years (Haller 1978). Shares hunting range amicably with Golden Eagle *Aquila chrysaetos* (Richard 1923) and White-tailed Eagle *Haliaeetus albicilla*; in 3 cases, nested 600 m and 900 m (twice) from nests of *H. albicilla* (Olsson 1979). Same pair said to have bred in middle of Grey Heron *Ardea cinerea* colony for 30 years, feeding offspring on young of that species (see Schnurre 1936a). ROOSTING. Diurnal and solitary. Usually on rock ledge with natural roof offering good all round view, also in upper half of evergreen tree, as near as possible to trunk, or on ground at foot of thickly foliaged tree or under bush (Guichard 1956b; Ferens 1960; Blondel and Badan 1976, which see for tree species; Haller 1978). Pair-members roost separately, usually a few hundred metres apart (Guichard 1956b); use 4–5 regular sites which may also serve as alternative nest-sites (Blondel and Badan 1976). From January, ♂ typically makes patrolling flights by day, especially in afternoon, returning to different roost-site afterwards (Guichard 1956b). ♀ roosts at nest-site from 1–2 weeks before laying (März and Piechocki 1980). While ♀ incubates and broods, ♂ roosts nearby, often as close as 10 m to nest (König and Haensel 1968; Blondel and Badan 1976) or up to c. 100–200 m away (Schnurre 1936a). Usually roosts all day, but may leave well before dark (Blondel and Badan 1976). For periods of hunting activity, see Food. Roosting posture upright, bird motionless, eyes half closed. Likes to sunbathe at rock sites (Guichard 1956b). Fond of bathing in water, often getting too soaked to fly; sand-bathing also reported (Frey 1973). For rain-bathing in captivity, see Scherzinger (1974d). Fledged young adopt daytime roost-sites on cliff (etc.); fed there until independence (Frey 1973).

2. Not especially shy, though wary and quickly alert to disturbance (Ferens 1960); allows close approach by human at roost-site, and indifferent to taunts (Gugg 1934; Guichard 1956b). When flushed, may soar upwards, like buzzard *Buteo*, high into sky (Guichard 1956b). One roosting ♂ always flushed at 50 m, another allowed closer approach then crouched flat on branch (Schnurre 1936a), this probably incipient Forward-threat posture. Full posture (Fig A) adopted when confronted by predator: bird leans forward, stares at source of threat, markedly ruffles plumage (especially on breast, back, and face), half raises ear-tufts, fans tail, lifts and cants wings forward such that they form shield-like arc over back. Accompanies posture with Excitement- and Hissing-calls (see 2 and 8 in Voice), and Bill-snapping (März and Piechocki 1980). Full Sleeked-upright posture, as in many other Strigiformes, of alarmed bird not reported, though mildly disturbed bird may draw itself up, erect ear-tufts, and half-close eyes (e.g. März and Piechocki 1980, in which see Plates 40 and 43). See also Parental Anti-predator Strategies. ANTAGONISTIC BEHAVIOUR. Main territorial defence performed by ♂ who is much more vocally demonstrative than ♀. ♂ starts demarcating territory with Advertising-call (see 1a in Voice) from (August–) October, peaking December–January (e.g. Desfayes 1951, Blondel and Badan 1976). Young, unpaired birds probably start calling earlier in autumn than established birds (Pukinski 1977). Said never to call from roost in Provence (Blondel and Badan 1976), though elsewhere ♂ may start calling from roost-site (König and Haensel 1968; Frey 1973), and then move between favourite song-posts, often on territorial boundaries. Song posts usually high trees and ridgetops (Frey 1973; Blondel and Badan 1976); in Valais (Switzerland) also in flat open country, perhaps sometimes from roof of isolated farmhouse (Desfayes 1951). Rarely stays more than 5–10 min at each song-post; may also call in flight between song-posts, especially December–January (Blondel and Badan 1976). In Auvergne, starts calling at dusk (17.30–18.00 hrs) in January, progressively later through February–March (Choussy 1971). In Valais, ♂♂ vary individually in nightly calling patterns, but typically start c. 18.00–18.30 hrs, then, after hunting, have another bout between 22.00 and 23.00 hrs, often with another brief bout about dawn (Desfayes 1949b); usually calls in bouts of 5–20 min; exceptionally, one bird called 600 times in 3½ hrs, with maximum 3 min silence (Desfayes 1951; see also Voice). Once nest occupied, ♂ typically follows bout of calling with patrolling flight near nest, then hunting trip (Gugg 1934; König and Haensel 1968). Little vocal advertisement after hatching (Burnier and Hainard 1948; Desfayes 1951). ♂ and ♀ typically give single Advertising-calls in direct confrontations with rivals (Glutz and Bauer 1980), but no reports of fighting (e.g. Fischer 1959). ♀ intruding on nest-area pursued by resident ♀ for 1·5 km (Haller 1978). ♀ also gives Alarm-calls (see 6 in Voice) in response to calls of rival ♀♀ (Glutz and Bauer 1980). HETEROSEXUAL BEHAVIOUR. (1) Pair-bonding behaviour. Advertising-call of unpaired ♂ (for seasonal pattern of calling, see above) attracts ♀♀. In Provence, heterosexual activity intense December to late January, especially in calm weather (Blondel and Badan 1976); in south-east Thüringen (East Germany), February–March, especially on moonlit nights (Fischer 1959). Courtship ends when laying begins (Frey 1973). During peak period, ♀ may initiate vocal exchanges (König and Haensel 1968: see 1b and 4 in Voice). In Meeting-ceremony, ♂ and ♀ Duet (see 1 and 4 in Voice) from respective vantage points and fly towards each other, sometimes calling, before separating again and renewing Duet (König and Haensel 1968; Blondel and Badan 1976). Calling ♂ typically perches horizontally, head, body, and tail in line, and rocks back and forth in rhythm with units of disyllabic call, exposing white throat-patch as he does so; thought to be recognition signal between pair-members perched near and facing each other (Choussy 1971; Blondel and Badan 1976). Duetting usually followed by mating (see below) or visit to nest-site. Once nest-site chosen, activity centres there, usually starting before sunset and ending c. 30 min after (König and Haensel 1968). In apparent display (Driving-flight), pair flew out from nest-area, ♂ following ♀, and circled before returning; this repeated 8 times (Fischer 1959); see also subsection 3 (below). Wing-clapping reported once during evening courtship at beginning of March (König and Haensel 1968). (2) Nest-site selection. In Switzerland, pairs visit potential nest-sites in winter (Haller 1978). At start of January, birds make several scrapes in favourable sites in territory (Blondel and Badan 1976; see also Breeding Dispersion, above, and Breeding). Scraping activity mostly by ♂ (Olsson 1979). In captive study, ♂ made several scrapes in January; from March, ♂ started Nest-showing, running to site and scraping alternately with feet and bill, while giving rapid Excitement-calls—also, rarely, gave variant of Advertising-call (Steinbacher 1956: see 1c in Voice). For luring behaviour of another captive ♂, see Baumgart *et al.* (1975). Once site chosen, ♀ scraped in it, like ♂, and spent a lot of time sitting

A

in it and settling as if on eggs, also sometimes giving Excitement-calls (Steinbacher 1956). Both sexes also give Soliciting-call while scraping (Gugg 1934; Schnurre 1936a; Glutz and Bauer 1980). (3) Mating. Occurs at dusk shortly before laying (März and Piechocki 1980). Often preceded by ♂ following ♀ in Driving-flight, accompanied by Excitement-calls (Fischer 1959). Mating usually preceded also by lengthy Duetting and other calling by ♂ and/or ♀. ♂ may fly straight on to ♀ or display from perch: stands horizontally, giving Excitement-call, with tail half-cocked, wings slightly drooped and flapping (Steinbacher 1956; Frey 1973). In one case, ♀ flew to join ♂ displaying thus, Duetted with him, then flew c. 100 m, ♂ following and copulating; during copulation, 'screeching' and one soft 'who' (♂) heard; after dismounting, ♂ flew off with slow wing-beats, ♀ staying to preen (Frey 1973). In 2 cases, after having called lengthily from ridge, ♂ flew and landed by crouching ♀, mounted, flapping wings for balance, and immediately flew off. ♀ then straightened up slowly, preened, and waited for ♂ to return with prey (Blondel and Badan 1976). For similar sequence in which ♂ landed on ♀'s back in gliding flight, see Choussy (1971). See Fig B (adapted to suggest heterosexual copulation) for markedly 'puffed', somewhat aggressive appearance of captive ♂, just after mounting and trying to copulate with forearm of human on whom bird imprinted (Baumgart et al. 1975). Apart from those above, variety of calls associated with mating: ♂ invites copulation with Soliciting-call (März and Piechocki 1980). In recording of copulating birds by S Palmér, ♀ gives apparent variant of Advertising-call (see 1c in Voice) every 5 s, while ♂ gives grunting sounds, Squeaking-calls (see 3 in Voice), and, once or twice, variant of Advertising-call (see 1c in Voice). One sequence accompanied by loud Bill-snapping (Schnurre 1936a). Copulation brief (lasting 3–6 s) and probably repeated several nights before laying: captive birds copulated twice in 25–30 min, and again shortly before laying (Leibundgut 1973). (4) Courtship-feeding and food-caching. ♂ feeds ♀ from shortly before laying until young c. 1 month old (Blondel and Badan 1976). Initially, food transferred at regular site, often tree perch, well away from nest (Schnurre 1936a; Frey 1973). ♂ carries prey to perch in one foot (Frey 1973). ♀ begs with Contact-call (Mebs 1972; Scherzinger 1974d: see 4 in Voice), apparently also, during incubation, with Soliciting-call (Fischer 1959). Arriving ♂ gives Excitement-calls (Steinbacher 1956). On arrival of ♂, ♀ gives Squeaking-call (Brandt 1941; Fischer 1959; Mebs 1972); see also 1c in Voice. During breeding, food-caching known only in chick-rearing phase. At that time, surplus prey at nest cached by ♀ under bush or in crevice near nest, to be withdrawn later for young. Large, freshly killed prey commonly cached more than once until completely eaten (Blondel and Badan 1976). (5) Allopreening. None reported. (6) Behaviour at nest. ♀ stays at or near nest by day from 1–2 weeks before laying (Frey 1973; März and Piechocki 1980). Sitting ♀ regularly adds own pellets to nest-lining (Blondel and Badan 1976); reaches out to draw material in towards nest (Haller 1978). RELATIONS WITHIN FAMILY GROUP. Eyes of young open at 6 days; young stand fully upright at 16 days (Scherzinger 1974d, which see for details of development). Young brooded almost constantly for 2 weeks, less so thereafter. At 1 month, left alone by day and for much of night; ♀ however stays on guard not far from nest. At (2–)4 weeks, young make walking excursions from nest (Gugg 1934; Blondel and Badan 1976), up to 80 m at 4 weeks, 200 m at 7 weeks (König and Haensel 1968). Almost certain that parent (presumably ♀) will carry young from nest if badly disturbed: in 1st case, at regularly checked nest, unfledged young found c. 900 m away in neighbouring valley; in the other, 3 young thought to have been removed after ♀ became distressed by intrusions (Choussy 1971). Young start giving food-calls (see Voice) near sunset. ♂ transfers usually decapitated prey to ♀ at or away from nest (Brdicka 1969; Blondel and Badan 1976). Brings 1–3 prey per night; after passing prey to ♀, leaves immediately, never feeding young himself. ♀ dismembers and transfers food to young until 1 month or older. Invites young to feed with Bowing-display (Fig C): faces young and bows up and down, giving Soliciting-call. Mealtimes very noisy and animated, accompanied by vigorous food-calls, cracking of bones (of prey), wing noises, Bill-snapping, and much jostling; ♀ always closes eyes when dismembering and distributing prey. When young sated, ♀ seizes left-overs and caches them (see above) or more often departs to eat them elsewhere. Young able to dismember prey at 6 weeks (Blondel and Badan 1976). Between feeds, young intersperse period of sleep with exploration and 'play'. Sleep leaning against one another (Fig D); sometimes bill one another gently. During active spells, exercise wings, scrape, bite at feet, feathers, etc., and make incipient hunting movements: e.g. catch grass in claws, focus on nearby objects with staring and wide head-circling movements (Burnier and Hainard 1948); practise hunting from 3 weeks, pouncing on feathers or stones (Desfayes 1949b). Young start flying at 50–60 days (Hagen 1950; Mebs 1972). Begin to catch prey for themselves at 70 days (Mikkola 1983). Stay in parental territory, largely dependent on parents, often until October, exceptionally November, when 20–24(–26) weeks old (Glutz and Bauer 1980; Mikkola 1983). When in territory, young typically restless, flying hither and thither in broad daylight, and giving persistent food-calls (Guichard 1956b). ANTI-PREDATOR RESPONSES OF YOUNG. From 15 days, young Bill-snap and hiss in response to any disturbance (Scherzinger 1974d), and until 4 weeks also typically crouch (Mikkola 1983). When approached, young in one nest leaned against one another, slowly turned their heads from left to right, blinked their wide eyes, and Bill-snapped (Guichard 1956b). May raise wings in incipient Forward-threat posture (Desfayes 1949b). When threatened, half-grown young may lie on back and raise claws (Schnurre 1936a). Out of nest, often hide in bushes by day and, if closely threatened, adopt Forward-threat posture, as in adult, and rock from one foot to the other, while Bill-snapping and hissing (Desfayes 1949b; König and Haensel 1968; Mikkola 1983). Fledged young out of nest crouched and closed eyes when approached, flushing on closer approach (Choussy 1971). PARENTAL ANTI-PREDATOR STRATEGIES. (1) Passive measures. When spotted from quite far away,

B

D

sitting ♀ crouches and lowers ear-tufts (Schnurre 1936a); closes eyes and compresses facial disc, thus making herself less visible (Choussy 1971). Sitting ♀ may also freeze when mobbed by raptors (Haller 1978). ♀ sits tightly, but this varies: one allowed approach to 10 m, then flew off (Gugg 1934). When young 6–7 weeks old, ♀ left nest when intruder c. 200 m away (Brandt 1941). Flushing ♀ often gives a cooing call (see 9 in Voice) when flying off (März and Piechocki 1980). Also gives Warning- and Alarm-calls (see 6–7 in Voice), silencing young (Choussy 1971; H Delin). (2) Active measures: general. ♂ markedly passive, even if young in danger, while ♀ shows marked individual variation in response, especially towards human intruders (Mikkola 1983). (3) Active measures: against birds. Does not tolerate Peregrine *Falco peregrinus* in territory (e.g. Blondel and Badan 1976). When incubating ♀ mobbed by Mistle Thrushes *Turdus viscivorus* (Burnier and Hainard 1948), or brooding ♀ dive-bombed by Ravens *Corvus corax*, Kestrels *Falco tinnunculus*, etc., typically responds with low-intensity Forward-threat posture (Haller 1978), but may adopt full posture (Brdicka 1969). Another ♀ with 14-day-old young launched succession of direct attacks on pair of *C. corax*, flying out 40 m (Haller 1978, which see for calls). Also hissed when mobbed by crows Corvidae (Gugg 1934). Bird nesting in middle of Capercaillie *Tetrao urogallus* lek regularly Bill-snapped at them (Brandt 1941). (4) Active measures: against man. Most ♀♀ relatively passive; minority highly aggressive, more so at night, and when young about to leave nest (Choussy 1971; Blondel and Badan 1976). Outright attack rare (V Olsson). Some (probably most) ♀♀ flee from danger, giving Alarm-calls; often alight nearby and scold (Choussy 1971). May adopt Forward-threat posture (V Olsson). In 133 daytime visits to nests, only 5 elicited direct attack (Lagerström 1978), though exact nature not clear. Mock-attacks to within 3 m of intruder's head, accompanied by Hissing-calls (Choussy 1971) and Alarm-calls (Gugg 1934). One ♀ studied for 7 years always scolded harmlessly by day but attacked at night, once striking intruder violently on neck, probably with her breast, stunning him (Blondel and Badan 1976). Some ♀♀ perform mild distraction-lure display of disablement type: when observer c. 40 m from nest, one ♀ flushed, perched clumsily on tree-top, stared, drooped wings, turned, appeared to lose balance, and flew off in laboured fashion for c. 20 m to another tree-top; repeated display 3 times before flying off (Frey 1973). (5) Active measures: against other animals. Sitting ♀ Bill-snapped at hare *Lepus* which came close to nest. Bird attacked dog near nest, drawing blood (Gugg 1934). Successful defence of young against ibex *Capra ibex* reported (Haller 1978).

(Fig A from photograph in März and Piechocki 1980; Fig B adapted from photograph and drawing in Baumgart *et al.* 1975; Figs C–D from photographs in Blondel and Badan 1976.) EKD

Voice. Freely used in breeding season, less so at other times. Repertoire highly complex, with much individual variation, also gradation between calls of given individual, so that literature isolates several calls likely to be variants. Much overlap in renderings and ascribed functions of (especially) calls 2 and 3, also 4 and 6, and divisions somewhat arbitrary. Following scheme therefore tentative and others possible. For Wing-clapping (probably rare) and Bill-snapping, see Social Pattern and Behaviour. Account includes notes supplied by H Delin.

CALLS OF ADULTS. (1) Advertising-call. (a) ♂'s call. A deep (average 384 Hz), sonorous, booming 'buho' or 'oohu', with emphasis on 1st syllable, and pitch descending to 2nd (Glutz and Bauer 1980). In recording (Fig I), initial frequency mainly c. 350 Hz, dropping to below 260 Hz at end (J Hall-Craggs). Although rather soft and muffled, carries up to 1·5 km, exceptionally 4 km. Repeated monotonously, typically at intervals of 8–10 s (Glutz and Bauer 1980), not uncommonly 13–30 s (Desfayes 1951), making it readily distinguishable from Long-eared Owl *Asio otus* (interval c. 3 s) which it otherwise resembles at a distance when only 1st syllable heard; confusion possible, however, with start of Advertising-call of Ural Owl *Strix uralensis* (H Delin). May be heard at any time of year but mainly autumn to spring; in peak period (see Social Pattern and Behaviour), given up to (60–80)–200 times in succession, in excitement at much shorter intervals (1–3 s) (März and Piechocki 1980); rapid series of quieter calls given in flight. Very variable in pitch, volume, timbre, and rhythm, allowing individual recognition. Rhythm maintained in antiphonal Duets with ♀ (see calls 1b and 4, below). Call attracts ♀♀ and rebuffs ♂♂; in confrontations, often only single (or well spaced) calls given, also thus directed at predators and young approaching independence (Glutz and Bauer 1980). For further details, see Social Pattern and Behaviour. (b) ♀'s call. An 'u-hu' or 'uh-ju' (Glutz and Bauer 1980). Slightly higher pitched than ♂; 2nd syllable about same pitch as 1st of ♂ (Desfayes 1949b); also more distinctly disyllabic than ♂, 1st unit more compressed (Guichard 1956b; Glutz and Bauer 1980); more plangent (Gugg 1934), harsher, and more variable than ♂, sometimes ending in hissing sound (Fischer 1959). No so far-carrying as ♂'s call but used in same contexts, notably in Duets with ♂ and to repel rival ♀♀; given much less often (Gugg 1934; Glutz and Bauer 1980). Often given as drawn-out (c. 1 s) 'uuuuuuh' (März and Piechocki 1980). (c) Nasal variant. Given by both sexes; highly variable and difficult to render, but usually described as short sequence of soft, muffled, nasal, or wheezy 'uh' or 'uhju' sounds, sometimes at intervals of c. 1 s. Given during Nest-showing, food transfer from ♂ to ♀ or ♀ to young, and often grading into call 2 (Glutz and Bauer 1980). Single calls given by ♂ sometimes during, and often after mating (Schnurre 1936a; H Delin): e.g. in recording by S Palmér, 1 or 2 deep swallowed 'chu-OO' sounds (E K Dunn) during copulation; in same recording, powerful harsh barking 'CHUo' sounds given by ♀ at intervals of 5 s. (d) Other variants. In recording of *ascalaphus* (Fig II), bird gives 'hooWAha' and 'WAha' calls (P J Sellar), apparently associated with hunting. (2) Excitement-call. Rapidly repeated and descending 'hohohoho...' (♂) and higher pitched 'huhuhuhu...' (♀) sounds, often referred to as giggling or laughing (Glutz and Bauer 1980; März and Piechocki 1980). Used in various heterosexual contexts, mostly by ♂: e.g. in Meeting-ceremony (Gugg 1934), often when ♂ flying towards ♀ (Desfayes 1951); by ♂ before copulation (Schnurre 1936a; H Delin), often during Driving-flight

(Fischer 1959); by Nest-showing ♂ (Steinbacher 1956); by ♂ during food transfer (Brandt 1941; Steinbacher 1956). Also given during confrontations (März and Piechocki 1980). (3) Squeaking-call. Rapid series of wheezy, squeaking 'wi-wi-wi...' sounds (Glutz and Bauer 1980), given in rapid rhythmic fashion by ♂ during copulation. Evidently transitional from call 2, and in recording of mating sequence, ♂ changes from low grunting 'cho' sounds, through 'chwi' sounds, to higher pitched 'wi' sounds; Fig III shows 'cho-cho-chwi-chwi-chwi' (E K Dunn, J Hall-Craggs). Squeaking-call thus high-intensity Excitement-call. ♀ at nest gives presumably homologous 'pipipipipipie' or 'kjikjikjikji...' on greeting food-bearing ♂ (Fischer 1959); also rendered 'ikkie ikkie...' (März and Piechocki 1980). (4) Contact-call of ♀. A harsh sound, difficult to render, but strongly reminiscent of Jay *Garrulus glandarius*, or sudden tearing of coarse cloth (Desfayes 1949b, 1951; Choussy 1971). Carries 400–500 m (Desfayes 1951). Variously rendered 'kvèck' (Desfayes 1951; Choussy 1971), 'gwäch' or 'gwäng' (Mebs 1972; März and Piechocki 1980), or 'chriä' (König and Haensel 1968). For other renderings, see Glutz and Bauer (1980). Often given in rather monotonous series near time of laying, not uncommonly in Duet with call 1a of ♂. More rapid series, perhaps variant, associated with food transfer and copulation (Glutz and Bauer 1980). ♀ arriving to feed young gives 'grä grä grää' sounds (März and Piechocki 1980). Sound evidently similar to call 6, but harsher. (5) Soliciting-call. Continuous series of clucking 'tucka' sounds (Fischer 1959), 'glugg-glugg-glugg' (Leibundgut 1973), 'couloucoulouc...' (Desfayes 1949b), 'godegode...' (Gugg 1934). Given by both sexes (mainly ♀) in various situations to invite close approach: e.g. by Nest-showing ♂, by ♀ soliciting food or copulation, by both adults (especially ♀) to encourage young to take food (Gugg 1934; Schnurre 1936a; Fischer 1959; Glutz and Bauer 1980; März and Piechocki 1980). Adult proffering food to young accompanies call with Bowing-display (Blondel and Badan 1976). (6) Alarm-call. A sharp croaking 'gräck', like Grey Heron *Ardea cinerea* (Glutz and Bauer 1980). Startlingly loud, ♂'s call deeper than ♀'s (H Delin). In recording of ♀ (Fig IVa), a loud, barking 'chwa' or 'kwa', like fox *Vulpes*; presumably directed at recordist near fledged young. May be given perched or in flight, often in series of (3–)4–5 units: 'ka-ká-ka ka-ka' (H Delin); 'ke-kē-ke', or more raucous 'grä-grä-gra'; given especially to warn young of approaching danger, and silences them (Choussy 1971). Also given during attacks on intruders (see Social Pattern and Behaviour). Similar to call 4, but typically has fierce, nasal quality, rather than harsh, hoarse quality of that call (H Delin). (7) Warning-call. A succession of high-pitched, drawn-out piping sounds with wavering, finely vibrant quality, with merest suggestion of pause between units (thus more continuous than, e.g., call 3); reminiscent of distress cry of small mammal. Typically given when intruder walking straight towards hidden young (H Delin), and presumably stimulates them to crouch and fall silent. Described as high-pitched whistle, seldom heard (V Olsson). (8) Hissing-call. Hissing sound given in anger and threat: e.g. when mobbed (Gugg 1934), in dive-attacks on humans (Choussy 1971), and in Forward-threat posture (März and Piechocki 1980). (9) Other calls. (a) ♀ flying away from nest during incubation often gives soft cooing sound (März and Piechocki 1980). Curious sounding 'rrro rrro rrro rrro rrro', alternate units lower pitched, like noise of a small saw on wood, heard from ♀ at nest above food-calls of young (Burnier and Hainard 1948), perhaps same or related call. (b) Numerous reports of growling, grunting, jarring, rattling, moaning, screeching, etc., likely to be mostly same or variants of those listed above.

CALLS OF YOUNG. For additional sonagrams, see Scherzinger (1974d). When eggs turned 1 day before hatching, young inside give soft 'bibbering' sounds. On

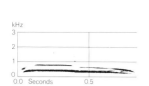

I S Palmér/Sveriges Radio (1972) Sweden March 1957

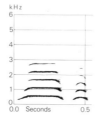

II T C White Tunisia May 1977

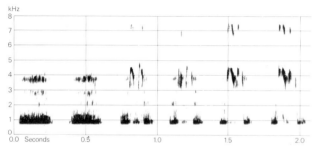

III S Palmére/Sveriges Radio (1972) Sweden April 1957

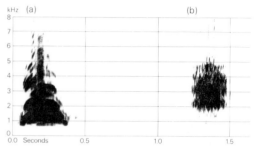

IV S Palmér and S Johansson/Sveriges Radio (1972) Sweden September 1969

hatching, food-call a thin high-pitched 'sri', also 'djchü', given mostly when adults away from nest, usually at intervals of 5–10 s (Scherzinger 1974d; Glutz and Bauer 1980). During food transfer, a rapid 'hihihihi-ia' (Desfayes 1949b), 'kli-kli-kli...', at rate of c. 10 per s (Hagen 1950). At shrillest, resembles Squeaking-call of adult. Distress-call (e.g. in disputes between siblings) a variety of chirping, gackering, and screeching sounds (Scherzinger 1974d; Glutz and Bauer 1980). With age, food-call develops through series of subtle changes, first harsh and compressed, then hissing and more drawn out; by 24 days, a dry rasping 'chwätch' (Scherzinger 1974d, which see for descriptions of changes, also März and Piechocki 1980). From 4th week, audible over 300 m (Glutz and Bauer 1980). By 7th week, food-calls a higher-intensity 'chüjü' and 'chjüjöo' (Scherzinger 1974d). Other renderings: 'chü-eet' (Hagen 1950), 'chui' (Choussy 1971), 'chii' (Burnier and Hainard 1948). In recording (Fig IVb) of well-fledged young, a hissing 'chee' or 'chach' (E K Dunn). Now carries up to 1 km (Desfayes 1949b). Calls become steadily more like adult Advertising-call, and continue up to independence (Scherzinger 1974d). At c. 4 months, young started to give hoarse 'houououo-hou', and one gave a deep resonating 'houom' (Desfayes 1949b). Full Advertising-call first heard at 5 months (März and Piechocki 1980). Young out of nest confronted intruder with barking 'hia hia hia', perhaps incipient Alarm-call of adult (Desfayes 1959b). Threatened young also Bill-snap and make cat-like hissing noises with gaping bill and tongue raised, as in adult (Scherzinger 1974d). EKD

Breeding. SEASON. Scandinavia: see diagram. Southern France: laying begins in last few days of December and continues to mid-March (Blondel and Badan 1976). Austria and southern West Germany: laying begins end of February and early March (Frey 1973; Förstel 1977; Rockenbauch 1978a). North Africa: laying late February to early May (Heim de Balsac and Mayaud 1962). USSR: laying begins first half March in Moscow region, end of March further south (Dementiev and Gladkov 1951a).

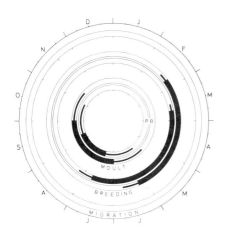

SITE. On cliff-ledge or in crevice; on ground on steep slope; in cave; in hole in tree; sometimes in old tree nest of other species. In southern Sweden, chooses cliffs and nests at average height 30 m (10–60), $n=41$; ledge aspect important with 60% of 181 sites facing south-west or SSW; prefers slight overhang or vertical face behind ledge, but also chooses site beside vegetation, tree-stump, etc. (see Olsson 1979). Each pair has average 3·3 sites (1–5) in territory, and these used over average of 9·3 years, so average consecutive use of site 3 years, but range 1–15 (Olsson 1979). Nest: shallow scrape or none. Building: scraping mostly by ♂ (Olsson 1979). EGGS. See Plate 97. Short elliptical, smooth though finely pitted, and slightly glossy; white. 60 × 50 mm (53–66 × 46–53), $n=56$ (Witherby et al. 1938). Weight 73 g (Makatsch 1976). Clutch: 2–4 (1–6). Of 119 clutches, Sweden: 1 egg, 6%; 2, 47%; 3, 34%; 4, 10%; 6, 3%; average 2·59 (Curry-Lindahl 1950). Of 27 clutches, south-east Sweden: 1 egg, 1; 2, 19; 3, 6; 4, 1; average 2.26 (Olsson 1979). Of 36 clutches, southern West Germany: 2 eggs, 16; 3, 17; 4, 3; average 2·64 (Mebs 1972). Of 23 clutches, Bayern (West Germany): 1 egg, 2; 2, 6; 3, 8; 4, 5; 5, 2; average 2·96 (Förstel 1977). Of 15 clutches, southern France: 2 eggs, 1; 3, 11; 4, 3; average 3·13 (Blondel and Badan 1976). Of 42 clutches in same area: 2 eggs, 22; 3, 17; 4, 3; average 2·55 (Bergier and Badan 1979). In Algeria and Tunisia, clutch size 2–4, but only 2 in southern Algeria (Heim de Balsac and Mayaud 1962). One brood. Replacement clutch laid 7–11 days after loss of eggs (Blondel and Badan 1976). Of 4 replacement clutches, Bayern (West Germany): 2 eggs, 3; 4, 1 (Förstel 1977). Laying interval 2·4 days. INCUBATION. 34–36 days. Begins with 1st or 2nd egg; hatching asynchronous. By ♀ only. YOUNG. Altricial and nidicolous. Cared for and fed by both parents; brooded more or less continuously by ♀ while small. FLEDGING TO MATURITY. Fledging period 50–60 days or more; may leave nest before fully fledged (Mebs 1972). Independent at 20–24(–26) weeks (Glutz and Bauer 1980). First breeding at 2–3 years (see Social Pattern and Behaviour). BREEDING SUCCESS. Extremely sensitive to disturbance, often abandoning eggs and even small young (Mikkola 1983). Of 219 occupied territories, south-east Sweden, 1961–77, 66% had active nests of which 60% produced fledged young; 136 young produced—1·6 (1·0–2·5) per successful nest, 0·9 (0·4–1·5) per active nest, 0·6 (0·3–1·5) per occupied territory; production varied according to food supply; main losses due to starvation, predation, and human disturbance including egg collectors (Olsson 1979). In southern France, 35 young produced from 13 nests, average 2·69 (Blondel and Badan 1976); in same area, 72 young produced from 50 nests, average 1·44, including 13 nests with no young produced (Bergier and Badan 1979). In southern West Germany, 59 young produced from 37 nests, average 1·85, or 1·17 per nest started; losses due to starvation, addling, and predation (Mebs 1972). Of 98 nests, Austria, 62 successful, producing average 1·85 young, or 1·10 young per nest started (Frey 1973). Of 46 nests, Bayern, 29 successful, producing average 1·79 young, or 1·13 young per nest started (Förstel 1977). Of 26 nests, Baden-Württemberg (West Germany), 23 successful, producing average 1·91 young, or 1·69 per nest started (Rockenbauch 1978a). Of 50 nests, Switzerland, 46 successful, producing average 1·80 young, or 1·32 per nest started (Haller 1978).

Plumages (nominate *bubo*). ADULT. Forehead, ear-tufts, and crown black, sides and tips of feathers on forehead finely mottled buff or cream-white, sides of feathers of crown tawny-buff or cream-buff with narrow black bars, crown appearing black with buff streaks, lateral bases of ear-tufts similarly patterned buff; occasionally, forehead, tufts, and crown almost completely black. Hindneck and upper mantle broadly streaked black and tawny-buff, black lateral barring of feathers reduced or absent. Feathers of lower mantle and scapulars with black central streak widening to bold black spot on tip, sides of feathers rich tawny-buff to cream-white (tinge depending on bleaching) with rather irregular dark grey or black vermiculations and bars, appearing buff with bold black blotches, not as regularly streaked as hindneck and upper mantle. Back, rump, and upper tail-coverts tawny-buff with rather indistinct grey bars and vermiculations, tips of upper tail-coverts more heavily mottled and vermiculated dark grey or black. Rather poorly defined facial disc dull greyish-buff with indistinct and fine concentric dark grey bars and with blackish bristle-like feather-tips; patch between eye and base of bill white with projecting black bristles; patch above eye black merging into largely black ear-tufts. Sides of head behind disc finely barred cream-buff and dark grey, some rich tawny-buff of feather-bases sometimes visible. Chin white; chest, breast, and sides of breast tawny-buff with broad black shaft-streaks (8–20 mm wide) and some dark grey irregular bars or spots on sides of feathers; some uniform white of feather-bases sometimes visible on central chest and breast. Belly, flanks, and vent like breast, but black central streaks narrower (2–6 mm), and dark grey bars on feather-sides regular, narrow, and rather close, appearing rather narrowly streaked and finely barred on tawny or cream ground; some uniform tawny-buff of feather-bases sometimes partially showing. Feathering of leg and foot tawny-buff with fine grey mottling, least so on sides and tips of toes. Under tail-coverts pink-cinnamon or tawny-buff, rather regularly marked with narrow black bars, longest coverts sometimes with bars up to 5 mm wide, tips of longest coverts mottled black. Central pair of tail-feathers dark olive-grey with *c.* 6 narrow and rather irregular buff bands or buff with *c.* 5 dark olive-grey bands; olive-grey finely mottled or vermiculated buff, buff with some coarse dark vermiculation or mottling; broad olive-grey bands gradually narrower and more solidly black towards t6, intervening buff gradually more extensively marked with finer mottling towards t6, especially on tip and more distal bars; t6 tawny-buff with 7–9 rather narrow greyish-black bars (2–7 mm wide), only *c.* 2 cm of tip and 2–3 distal buff bands extensively clouded by grey mottling. Flight-feathers tawny-buff, outer webs with 4–6 regularly spaced and broad olive-black or greyish-black bands; similar bars on inner webs, but these narrower and less regular, reduced towards inner edge and base; base uniform tawny-buff; dark bands on p9–p10 broad and irregular, outer webs especially mainly black with limited buff. Buff bands of outer webs of flight-feathers finely peppered grey, more coarsely spotted and marbled grey on tertials; tips and distal 2(–3) bars heavily clouded grey, buff hardly visible, tip and distal bars only slightly paler than intervening black bars. Greater upper primary coverts as primaries; median upper primary coverts and bastard wing like primaries, but feather-tips more extensively black. Greater, median, and longer lesser upper wing-coverts like scapulars, but feather-tips dark olive-grey rather than black and more speckled with buff; ground-colour of large subterminal spot on outer web soon bleached to cream or white, forming rows of pale spots across wing. Shorter lesser upper wing-coverts black with some buff mottling, forming distinct black bar across forewing, extending from upper scapulars to base of bastard wing. Under wing-coverts and axillaries cinnamon-buff (partly white on longer coverts), narrowly barred dark olive-brown or black; tips of greater under primary coverts broadly black, contrasting with mainly uniform tawny-buff primary-bases. Sexes similar. Tawny-buff ground-colour rather strongly influenced by bleaching and wear, grading to off-white, especially on scapulars, upper wing-coverts, tertials, t1, and feather-tips of breast and belly. DOWNY YOUNG. 1st down (neoptile) short, soft, and dense, extending to tips of toes, but base of bill, area round cloaca, undersurface of tibio-tarsal joint, and soles naked. Down uniform white or dirty cream-white, slightly buff on forehead, under eyes, on wing, and on rump; naked skin shines through after a few days. Gradually replaced by brown 2nd down, shining through from age of *c.* 8 days. (Scherzinger 1974d; Harrison 1975; RMNH.) JUVENILE. 2nd down (mesoptile) long and soft, covering all head, body, and wing, but longer scapulars and greater upper wing-coverts rather feather-like (though less so than in Snowy Owl *Nyctea scandiaca*), stiff along shafts, loose at sides. Flight-feathers, tail-feathers, greater primary coverts, and bastard wing with structure as adult, but often narrower and tips more pointed, in particular innermost secondaries (tertials) rather gradually attenuated and with loose sides. Down of head, body, and wing rusty-buff to pale ochre with narrow black-brown bars, but barring obscured until age of *c.* 5 weeks by uniform cream-buff neoptile clinging to tips of mesoptile; tarsus and toes uniform buff. Mesoptile fully developed from *c.* 25th day, when short and downy ear-tufts also start to appear. White triangular patch round base of bill develops from *c.* 15th day, reaching maximum extent between eyes, on cheeks, and on throat in 5th week; contrasting black patch round eye and black eye-brow develop 4th–7th week. Juvenile flight-feathers start to grow from 12th day, tail from 15th; both full grown at 8–10 weeks. Mesoptile gradually covered by 1st adult 38th–82nd day (mantle, scapulars, upper wing-coverts, and chest first); feathers of ear-tufts develop during 11th to 16th week. (Heinroth and Heinroth 1926–7; Scherzinger 1974d; RMNH.) FIRST ADULT. Like adult, but juvenile flight-feathers, greater primary coverts, bastard wing, and tail retained. Flight-feathers as in adult, but dark bands relatively narrower and less attenuated towards edge of inner webs, intervening tawny bands broader; both dark and tawny bands reach closer to feather-tip than in adult, tip not as extensively dark. Tail as in adult, but central feathers show almost uniform white fringe along tip, 8–10 mm wide; this less pronounced on outer feathers. Juvenile characters of flight-feathers and tail often difficult to ascertain. Up to October–November of 1st autumn usually easily recognized by partly retained mesoptilous greater upper wing-coverts, tertial coverts, or tail-coverts; these distinctly more bleached, narrower, and with fluffier sides than neighbouring fresh 1st-adult median upper wing-coverts, scapulars, and rump. In 1st winter and spring, best distinguishing characters are: (1) juvenile tertials (s11–s16), especially innermost, which are rather short and narrow and taper to narrowly rounded tip, patterned with rather diffuse and irregular fine bars and vermiculations (in adult, with broadly rounded tip and more distinctly patterned with broad dark bands); (2) juvenile primaries, which are all of similar age, all equally fresh in midwinter, inner fresh and outermost gradually more worn by spring; adults show scattered mixture of old (greyish-tipped and frayed) and new (black-tipped and smooth) primaries. SECOND ADULT AND SUBSEQUENT PLUMAGES. Like adult, but at 1 year old only juvenile tertials replaced; juvenile primaries and most secondaries retained—all rather worn (especially outer primaries), not a scattered mixture of newer and older feathers as in adult; hence, similar to 1st adult except for new tertials. In summer of 3rd calendar year, primary moult starts with p7 and secondary moult continued: in 3rd winter, p6–p8 new, occasionally also p1, p5, and p9, remainder of primaries and often at least s1 still juvenile, thus showing a group of new primaries, p6–p8 (p5–p9), contrasting with an abraded and brown-bleached old group, p2–p4 (p1–p5); in adult, never more than 2 neighbouring feathers of similar age. In 4th autumn, winter, and spring, one or a few juvenile primaries (often p10 or p3) usually still present, more heavily abraded than is usual among old adult feathers, but age-pattern of primaries otherwise indistinguishable from adult. Outline above based on specimens of northern races with help of data in Glutz and Bauer (1980); Mediterranean and Middle East populations may start with p7 when 1 year old rather than 2, and *ascalaphus* specimens examined with p6–p8 (p5–p9) new and remaining primaries juvenile more probably in 2nd winter than in 3rd.

Bare parts. ADULT AND FIRST ADULT. Iris bright golden-yellow, orange-yellow, or orange; in *ascalaphus*, tawny-yellow to orange. Bill greyish-black or black. Cere dark olive-grey or slate-grey. Claws black. DOWNY YOUNG AND JUVENILE. Skin flesh-pink. Eyes open from 6th day; iris pale yellowish-grey or pale yellow at first, attaining adult colour at 4–5 weeks. Cere, bill, and claws greyish-pink at hatching; cere grey and bill and claws bluish-grey to slate-black in 2nd–5th week. Eyelid conspicuously pink at end of 3rd month. (Lavauden 1920; Heinroth and Heinroth 1926–7; Scherzinger 1974d.)

Moults. Mainly based on data of captive birds from T Mebs (in Glutz and Bauer 1980), with some additional data from specimens examined and from Dementiev and Gladkov (1951a). ADULT POST-BREEDING. Partial; flight-feathers serially, from centres on p7, p1, s16, s5, and s2. Sequence and timing of primaries:

```
        8—9——10
     7<
        6—5—4
        1—2—>3
     ———————————>
```

Sequence and timing of secondaries:

```
  16–15–14–13–12–11–10–9–8
                          >—7
                    5<—6
                          >—4
                    2<—3
                          >—1
  ———————————————————————————>
```

Only few feathers replaced each moulting season, and moult continued in nest; while moult of previous series is continuing, new series may start; primaries start every 2–3 years with p7, secondaries every 1–2 years with s16. Due to slow moult and short moulting season, replacement of single series of primaries takes 6–8 years in breeding adult, secondaries 8–12 years or perhaps longer; and 2–4 series of primaries and 3–6 series of secondaries are active on each wing in single moulting season, but in each series only 1–2 feathers replaced, and in all series together not more than 2–5 primaries and 5–9 secondaries (including tertials) per wing replaced annually. Moult of left and right wing often not symmetrical. First flight-feathers in breeding adults shed early June to mid-July, when young 7–8 weeks old; time needed for growth of single feather 56–84 days; no flight-feathers lost after late August and all feathers full-grown by late October. Non-breeding adults start flight-feather moult from early May, halting at same time as breeders; 3–6 primaries and 6–11 secondaries replaced annually. Tail moults at same time as wing; sequence centrifugal, but t6 sometimes earlier; moult often serial, t1 (–t2) moulted in first year, (t2–)t3–t6 in next, but t1 sometimes every year. Head, body, and wing-coverts at same time as wing, but longer scapulars, some longer wing-coverts, and perhaps some other feathers not every year. Duration of moult probably subject to geographical and local conditions and more rapid in southern populations. POST-JUVENILE. Partial: involves head, body, and wing-coverts, but not flight-feathers, tail, greater primary coverts, or part of bastard wing; occasionally, not all tertial coverts or greater upper wing-coverts. Mesoptile down gradually lost from age of *c.* 40 days, first on scapulars and upper wing-coverts; at $2\frac{1}{2}$ months, body and wing-coverts mainly 1st adult, but not yet all head; head and body in complete 1st adult at end of 3rd month, ear-tufts at end of 4th. Moult completed at $5\frac{1}{2}$ months, October–November. (Heinroth and Heinroth 1926–7; Scherzinger 1974d.) FIRST IMMATURE POST-BREEDING. Partial, May–October. Involves head, body, wing-coverts, tertials, and often all tail. Sequence as in adult; in 2nd winter, s1–s9(–s11) and all primaries still juvenile. SUBSEQUENT MOULTS. In May or June of 3rd calendar year, moult of secondaries continued from (s11–)s9, and 2nd series starts with s1; primaries start with p7; at end of moulting season, moult suspended with many primaries and 1 or a few secondaries still juvenile. Last juvenile flight-feathers (mainly p3 or p10) lost in summer of 5th or 6th calendar year. As more series started than completed in younger bird, later series gradually need more time for completion: e.g. 1st series of primaries and secondaries each completed in 3 years, but 4th series not completed until 7–8 years later. Also, in a bird of (e.g.) 16 years old s16 already replaced 11 times, s1 only 5 times; p7 6 times, p3 and p10 4 times.

Measurements. Nominate *bubo*. Central and northern Europe, all year; skins, for tarsus also some skeletons (RMNH, ZMA). Bill (F) to forehead, bill (C) to cere.

WING	♂ 444	(8·02; 9)	430–453	♀ 482	(14·3; 12)	463–513
TAIL	240	(7·17; 8)	231–252	266	(12·6; 12)	248–288
BILL (F)	51·1	(1·75; 9)	49–53	55·1	(2·17; 11)	52–58
BILL (C)	32·2	(1·25; 9)	30–34	36·0	(1·84; 13)	33–40
TARSUS	79·3	(3·23; 9)	74–82	82·2	(4·37; 14)	76–88

Sex differences significant, except tarsus. Juvenile tail similar to adult, but juvenile wing on average 9 mm shorter; too few examined to test significance, and all ages combined in tables.

Wing of large samples from Vaurie (1963): (1) nominate *bubo*, (2) *ruthenus*, (3) *sibiricus*, (4) *hispanus*, (5) *interpositus*, (6) *turcomanus*, (7) *nikolskii*, (8) *ascalaphus*, North Africa only.

(1)	♂ 448	(23)	435–480	♀ 473	(29)	455–500
(2)	453	(7)	440–468	482	(6)	476–490
(3)	456	(9)	435–480	491	(8)	472–515
(4)	430	(7)	420–450	453	(8)	445–470
(5)	451	(16)	425–475	466	(15)	440–485
(6)	450	(12)	440–470	482	(21)	445–512
(7)	419	(9)	405–430	438	(9)	410–465
(8)	346	(20)	324–368	367	(20)	340–390

Wing of *ascalaphus*: north-west Africa, ♂ 338 (2) 330–345, ♀ 383 (5) 374–390; Egypt, ♂ 355 (4) 344–366, ♀ 390 (2) 390–391; Middle East, ♂ 359 (8) 335–380, ♀ 395 (8) 350–430. Wing of *interpositus* from Middle East: ♂ 425 (3) 420–430, ♀ 448 (4) 440–460. Tail of *ascalaphus*: North Africa, ♂ 175 (12) 160–190, ♀ 203 (3) 201–206; Middle East, ♂ 186 (3) 160–203, ♀ 207 (5) 190–220. Tail of *interpositus*: ♂ 261 (10) 240–290 (all geographical range), 247 (3) 242–250 (Middle East); ♀ 254 (4) 240–265 (Middle East). (Hartert 1912–21; Vaurie 1960*b*; RMNH, ZMA.) Tarsus of *ascalaphus* 62–75 mm, middle toe without claw 38–50, middle claw 23–32, bill to forehead 37–43, bill to cere 23–28.

Weights. ADULT AND FIRST ADULT. Nominate *bubo*. Netherlands: ♂ 1880 (ZMA). Switzerland: ♂♂ 1570, 2010; ♀ 2332 (14), maximum 3000. East Germany: ♂ 1890 (3), ♀ 2554 (13). Minimum of living exhausted birds: ♂ 1166, ♀ 1380. (Glutz and Bauer 1980.) Norway, mainly from winter: ♂ 2380 (14) 1835–2810, ♀ 2992 (12) 2280–4200; weight increasing from November, very fat January–February(–March), especially ♀, decreasing April–May; one exhausted ♀, 1490 (Hagen 1942). Italy, mainly October–January: ♂♂ 1500, 1950; ♀ 2276 (246; 9) 1820–2650 (Moltoni 1949). Greece: ♂ 1550 (Makatsch 1950). USSR: ♂ 2458 (238; 6) 2100–2700, ♀ 3164 (76; 6) 3075–3260 (Dementiev and Gladkov 1951*a*).

DOWNY YOUNG AND JUVENILE. At hatching, on average 52 (Scherzinger 1974*d*) or 45·6 (6·5; 8) 37–55; weight increases slowly in 1st week, strongly from 2nd to end of 4th (Glutz and Bauer 1980). On leaving nest at (28–)35 days, on average *c*. 1560 (Scherzinger 1974*d*). Increase slow later on (sometimes slightly decreasing temporarily at end of 2nd month); adult weight reached after 4 months (Rosnoblet and Menatory 1975; Glutz and Bauer 1980).

Structure. (1) Nominate *bubo* and other European races. Wing long and broad, tip rounded. 10 primaries: p8 longest, p9 3–12 shorter, p10 32–50, p7 0–10, p6 20–30, p5 50–63, p4 70–92, p1 122–148. Outer web of (p6–)p7–p9 and inner web of (p6–) p7–p10 deeply emarginated. Outer web of longest feather of bastard wing, outer web of reduced outermost greater upper primary covert, outer web of p10, and emarginated parts of outer webs of p8–p9 serrated. Tail rather short, slightly rounded; 12 feathers, t6 20–30 shorter than t1. Bill heavy, deep and wide at base, laterally compressed at tip; culmen strongly arched; except for tip, largely hidden under long bristle-like feathers growing from sides of base. Facial disc distinct, but only upper half bordered by ruff of stiff feathers; a pronounced feather ridge extends from base of culmen to above eye, ending in long ear-tuft; ear-tuft consists of *c*. 10 elongated feathers up to 70–90 mm long. Tarsus and toe strong, but rather short; densely covered with hair-like feathers down to base of claws, soles naked. Middle toe without claw 53·9 (3·60; 9) 49–58 in ♂, 57·9 (3·32; 13) 52–62 in ♀; outer toe *c*. 71% of middle, inner *c*. 86%, hind *c*. 60%. Claws very strong, though not as heavy as in Brown Fish Owl *Ketupa zeylonensis* and not as strongly curved as in *Nyctea scandiaca*; middle claw 32·0 (1·73; 9) 29–34 in ♂, 35·7 (3·37; 12) 31–40 in ♀; outer claw *c*. 86% of middle, inner *c*. 103%, hind *c*. 85%.

(2) *B. b. ascalaphus*. 10 primaries: p8 longest, p9 0–2 mm shorter, p10 24–41, p7 2–9, p6 20–32, p5 42–55, p4 62–72, p1 96–112. Rather indistinct emarginations to outer web of p8–p9, inner web of p7–p10 more deeply emarginated. Serrations of outer web of p10 and of emarginated part of outer web of p9 longer than in nominate *bubo*. Tail relatively shorter than in nominate *bubo*, *c*. 50% of wing length; toes, claws, and bill relatively shorter and more slender, but tarsus relatively long; feathering of tarsus and toes shorter. Ear-tufts proportionally shorter, 35–55 mm.

Geographical variation. Pronounced, predominantly clinal, chiefly correlated with climatic factors. Birds in far north and from high altitudes are larger; those of humid areas darker and browner, those of open or arid regions paler and more yellowish. In Europe, gradually smaller and paler southward, larger and paler from central and northern Europe east to Yenisey; also gradually paler from Caucasus to northern Iran. Position of *ascalaphus* difficult to establish; see below. Nominate *bubo* is darkest race; ground-colour of upperparts golden-brown to tawny-buff, on underparts tawny-buff; crown, mantle, scapulars, and upper wing-coverts with large black feather-tips, hindneck and underparts (especially chest and breast) broadly streaked black. Birds from Finland through west European USSR to Carpathians, western Greece, and southern Italy average paler than those of Scandinavia and central Europe, streaks slightly narrower, less black on upperparts, and under tail-coverts less heavily barred; birds of south-west Norway darkest. *B. b. ruthenus* from central European USSR has slightly paler cream-buff or buff ground-colour, especially on hindneck, outer scapulars, chest, and breast; borders of ear-tufts buff, not largely black as in nominate *bubo*; black marks of head and body slightly narrower, lower mantle and scapulars showing some more vermiculation. In western part of range, *ruthenus* grades clinally into nominate *bubo*, in western foothills of Urals it grades into *sibiricus*. Latter race very pale, ground-colour mixed cream and white or almost fully white; crown, hindneck, and underparts only narrowly streaked black; limited black spots on centres of feather-tips of scapulars and upper wing-coverts, these mainly indistinctly vermiculated grey and cream or white; belly and flanks hardly streaked, marked only by fine grey bars on cream-white or white ground. In Pyrénées, nominate *bubo* grades into *hispanus* of Iberia; smaller than nominate *bubo* and with slightly paler ground-colour and narrower and sharper dark marks, especially on underparts; in past, occurred on northern slopes and summits of Atlas mountains in northern Algeria (Heim de Balsac and Mayaud 1962). Nominate *bubo* replaced by *interpositus* in eastern and southern Rumania (Paşcovschi and Manolache 1970), south European USSR, Asia Minor, and north-west Iran, intergrading with nominate *bubo* in eastern and southern foothills of Carpathians, in Moldavia, and in Ukraine at *c*. 50°N, with *ruthenus* along middle course of Don and west of lower Volga, with *turcomanus* in Volga delta, and with *nikolskii* in Elburz mountains and northern Iraq; also occurs in Levant south to at least Haifa and perhaps Gaza, but not east from Jordan river and Syrian desert, where replaced by *ascalaphus*. *B. b. interpositus* rather similar to *ruthenus* in depth of ground-colour, but only centres of feather-tips of lower mantle, scapulars, and upper wing-coverts black (not almost whole tips), and bases and sides of feathers coarsely vermiculated cream-white, buff, and greyish-black, upperparts appearing grey-vermiculated rather than black-blotched; dark bars on tertials and tail less solid, partly vermiculated grey; broad black streaks on underparts restricted to narrow chest-band; remainder of underparts narrowly streaked only, with barring finer than in *ruthenus*. *B. b. turcomanus* from steppes north of Caspian Sea and across much of Transcaspia east to plains of eastern Kazakhstan rather like *interpositus*, especially on underparts, but ground-colour of upperparts yellow-buff, less vermiculated grey, and dark marks of mantle, scapulars, and upper wing-coverts slightly larger; upperparts appear mainly yellow, mixed with some dusky and white, underparts white with limited amount of dark marks. *B. b. nikolskii* from southern and eastern Iraq through central and southern Iran to Afghanistan and western Pakistan rather like *turcomanus*, but with slightly more rusty tinge to upperparts and with more limited dark marks on both upperparts and underparts; much smaller. Not known which race inhabits central Iraq; *ascalaphus* occurs western Iraq. *B. b. ascalaphus* from North Africa and Middle East rather variable in colour: some birds (typical *ascalaphus*) have ground-colour rather like nominate *bubo* but slightly more pinkish, black marks rather extensive (though different in pattern from nominate *bubo*); others are pale sandy-pink with black marks restricted to triangular spots on feather-tips of upperparts, upper wing-coverts, and chest, and dark bands on tail and flight-feathers sharp and rather narrow; paler birds generally occur in deserts while darker typical *ascalaphus* predominate in well-vegetated areas, and hence paler birds are sometimes separated as *desertorum* Erlanger, 1897. However, both colour-types occur together in several places, intermediates can be encountered anywhere, and some pale birds occur in relatively wet places like Algerian Kabylie mountains and some dark birds in deserts; recognition of *desertorum* thus not warranted (Vaurie 1960*b*). Main character of *ascalaphus* is small size (see Measurements and Structure); though wing and tail increase in length towards Middle East, still much smaller here than neighbouring *interpositus* and *nikolskii*. Unlike in other races, black marks not a straight streak along entire centre of feather, but restricted to streak on tip which widens into black subterminal bar of varying width (absent on paler birds) and into black triangular mark on tip, bar and triangle isolating paired buff or pink spots; in darker birds, crown black with buff spots, in paler ones crown pink with black triangles, and remainder of upperparts and chest equally variable; breast, flanks, and belly similar to *hispanus* or *interpositus* in darker birds, unstreaked in paler birds, latter showing fine white and pink-buff barring only. *B. b. ascalaphus* formerly overlapped with *hispanus* without interbreeding in northern Algeria; it differs from other races of *B. bubo* in size, colour pattern, voice, and structure (short ear-tufts, long legs, short tail; slender bill, toes, and claws—closer in these respects to African Spotted Eagle Owl *B. africanus*), and hence seems to warrant recognition as a separate species (Pharaoh Owl *B. ascalaphus*). However, some intergradation occurs with *interpositus* in Middle East, where some intermediate specimens are known—mainly from western shore of Dead Sea, just west of Jordan river, and in Anti-Lebanon mountains (Vaurie 1960*b*). Considering uncertain relationship with *interpositus* in northern Syria or south-east Turkey or with *nikolskii* in Iraq, trend of most recent authors to incorporate *ascalaphus* in *B. bubo* is followed here. *B. b. bengalensis* from India is about as small as *ascalaphus* and sometimes considered a separate species also; however, plumage characters and structure similar to *B. bubo* and intergradation with some large and pale races of latter occurs in Kashmir. CSR

Ketupa zeylonensis Brown Fish Owl

PLATES 45 and 55
[between pages 494 and 495, and facing page 567]

Du. Bruine Visuil Fr. Ketupá brun Ge. Wellenbrust-Fischuhu
Ru. Рыбная сова Sp. Búho pescador Sw. Brun fiskuv

Strix zeylonensis Gmelin, 1788

Polytypic. *K. z. semenowi* Sarudny, 1905, Middle East through Iran to western and northern Pakistan. Extralimital: nominate *zeylonensis* (Gmelin, 1788), Ceylon; *leschenault* (Temminck, 1820), India, Burma, and southern part of south-east Asia; *orientalis* Delacour, 1926, southern China and northern part of south-east Asia.

Field characters. 54–57 cm; wing-span 145–150 cm. Size approaches that of smallest Eagle Owls *Bubo bubo*; 30–35% larger than Long-eared Owl *Asio otus*. Large, flat-headed, heavy-billed, eared owl, with rather loose and uniform pink-buff plumage. Ears less close-formed than in other eared owls, appearing as ragged sprays. Facial disc less developed than in most owls, and expression thus less pronounced, mostly morose, with bright yellow eyes set against brown upper cheeks. In flight, wings make slight singing noise. Commonest call a trisyllabic moan. Sexes similar; no seasonal variation. Juvenile difficult to separate.

ADULT. Plumage basically rather pale pink-buff, with upper cheeks browner, ears streaked black-brown, and crown, back, and wing-coverts streaked and spotted black-brown. Flight- and tail-feathers broadly barred black-brown and pale buff. Underwing bright buff, with paler, almost white-based greater coverts and broad black tips to carpal patch. Scapulars and some outer wing-coverts have buff-white margins, forming obvious marks. Underparts less well marked, with throat distinctly paler than rest of body; lower breast and flanks evenly and narrowly streaked black-brown (lacking more than fine bars). When worn, ground-colour of plumage becomes yellower, with pale bands on wings and tail approaching white. Bill pale horn. Legs usually blue-grey. JUVENILE. Little studied, but apparently less heavily streaked, with darker, spotted upper wing-coverts.

Size and general colouration invite confusion with *B. bubo* but, in good view, lesser bulk, duller plumage, weaker (and not noiseless) flight, and flat-browed facial expression all combined into unique generic character; top of bill level with centre of eyes (in *B. bubo*, level with bottom edge of eyes). Sprayed ears and loose body plumage also obvious at close range. Flight slow, with loose, flapping wing-beats producing slight but audible singing noise on down-stroke; when hunting over water, dangles relatively long and unfeathered legs (emphasizing relative shortness of tail).

Commonest call a dismal 'oomp-ooo-oo' with emphasis on middle syllable.

Habitat. In Mediterranean, subtropical, and tropical

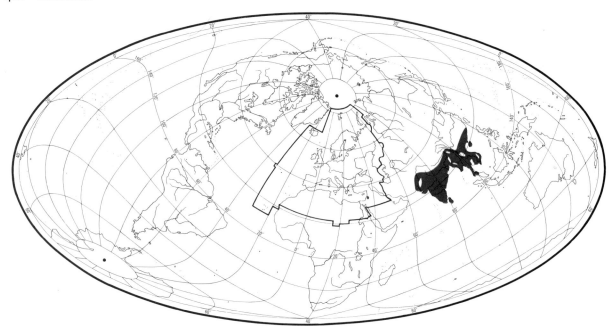

lower middle and lower latitudes, chiefly in plains but in India in hills and mountain foothills up to c. 1400–1500 m and c. 1800 m in Ceylon. In Israel, in upper Jordan valley and in Golan Heights, in wadis with shrubs but few large trees. In India, in well-wooded well-watered country, especially overgrown eroded ravines and steep river-banks and densely foliaged trees along forest streams and ponds; also in old mango groves, bamboo clumps, roadside and canal avenues, at rock pools, and commonly near human habitations. Skims low over water and waddles in shallows for bathing and feeding. Breeds in fork of tree-trunk, rocky cleft, or ruined building (Ali and Ripley 1969). Content with narrow strip of forest by water provided there is access to larger strips for hiding by day (Henry 1955).

Distribution. TURKEY. In late 19th century, 3 collected near Mersin and 1 near Aydin (Kumerloeve 1961). Present status uncertain; may still breed along rivers by Mediterranean coast (Vittery et al. 1971), but no recent records (MB, RFP). SYRIA. Perhaps bred formerly outside Golan Heights. 2 collected in north (Kebir river) 1879 (HK). ISRAEL. Formerly bred Acre (Vaurie 1965), and in Galilee; disappeared from there in 1950s when thallium sulphate widely used as rodenticide (HM). JORDAN. Said to breed but no details (Benson 1970); no recent observations (PADH, DIMW). IRAQ. Only record a pair collected in south, October (Allouse 1953).
Accidental. Lebanon.

Population. ISRAEL. Including Golan Heights, now probably under 10 pairs (SC).

Movements. Resident throughout range (Vaurie 1965), though no studies of local dispersals.

Food. In Israel, freshwater crabs *Telphusa fluviatila* and fish *Capoeta damascina* important in diet (A Boldo). In river gorges of Zagros mountains (Iran), main prey apparently freshwater crab *Potamon* (Paludan 1938). In India, largely freshwater crabs, fish, and frogs; also rodents, reptiles, birds, and large beetles (Wait 1931; Dharmakumarsinhji 1955). Remains of whistling ducks *Dendrocygna* and pond heron *Ardeola* found in nest (Eates 1938). Recorded feeding on carcass of crocodile (Ali and Ripley 1969). Locates aquatic prey from low perch or by flying back and forth; takes fish from near surface while skimming over water (Ali 1953; Ali and Ripley 1969). Feet and claws show unusual adaptations for fishing (see Structure); flight-feathers not adapted for silent flight.   DJB

Social pattern and behaviour. Very little known, especially for west Palearctic. Following account based chiefly on extralimital nominate *zeylonensis* and *leschenault*.
1. Said to stay in pairs (Ali and Ripley 1969), though not known if this applies throughout the year. BONDS. No information on pair-bond. ♂ thought to assist in incubation (Baker 1934). BREEDING DISPERSION. Birds return to same nest-site for many successive years (Baker 1934); not clear if this means marked site fidelity of pair, which is probable, or continuous use by different birds of traditional sites. ROOSTING. Rests and sleeps by day in large, solitary, densely leaved trees (Baker 1927; Whistler 1941), or perched on cliff-face (Whistler 1941). One roost-site in cave frequented for long period, and ground beneath littered with feathers, pellets, and prey remains (Herklots 1967). Pair-members typically roost separately (Henry 1955), but off-duty ♂ often stays close to nest when ♀ incubating (Baker 1934). Activity semi-diurnal, birds frequently flying, even hunting, by day, especially in cloudy weather (Ali and Ripley 1969; see also Baker 1927). At or before dusk, typically flies to water, giving Mewing-call (Whistler 1941; Ali and Ripley 1969: see 2 in Voice). Fond of bathing: wades into shallows, shuffling plumage in water, and preening afterwards (Ali and Ripley 1969).

2. Birds hunting in daylight frequently mobbed by small birds (Baker 1927). Startled bird raised crest and ruffled plumage (see Herklots 1967). Captive bird, when alarmed, Bill-snapped (see Voice) and gave Hissing-call (see 3 in Voice: Vidal 1880; see also Layard 1851, Dharmakumarsinhji 1955). In India, bird disturbed on roosting bough by Malabar giant squirrel *Ratufa indica* awoke and adopted a crouched posture, body horizontal and head lowered towards intruder. When squirrel called and rushed forwards, bird shuffled backwards for *c*. 1 m toward main trunk, and adopted Forward-threat posture: lowered head, drooped wings, raised mantle feathers, and gave Hissing-call. When squirrel halted, bird maintained Forward-threat posture for several seconds. When squirrel then continued to advance, bird turned its back on it and resumed more normal upright stance. From time to time, turned head towards squirrel and once uttered a growling call (see 4 in Voice). Finally bird shook itself and glided away (T Cleeves). ANTAGONISTIC BEHAVIOUR. Nothing known about responses to conspecific birds. Perhaps include behaviour described in confrontation with *R. indica*, above. HETEROSEXUAL BEHAVIOUR. On emerging from roost, pair-members call to one another (see 1 in Voice) for some time. When calling, throat swells markedly (Henry 1955). RELATIONS WITHIN FAMILY GROUP. No information. ANTI-PREDATOR RESPONSES OF YOUNG. No information. PARENTAL ANTI-PREDATOR STRATEGIES. On close approach of human intruder, bird sat tightly but did not mount any other defence of eggs or young (Baker 1934). EKD

Voice. Often heard in breeding season, though in Israel, upper Jordan, and Golan Heights, where pairs widely dispersed, said to call rarely (S Cramp). Following account based on extralimital nominate *zeylonensis* and *leschenault*. Bill-snapping of caught or wounded bird described as a snapping 'tuck-tuck' (Dharmakumarsinhji 1955; see also Social Pattern and Behaviour).

CALLS OF ADULTS. (1) Advertising-call. No good description for *semenowi*. In *leschenault*, a deep hollow 'boom-boom' or 'boom-o-boom' with a reverberating ventriloquial quality, repeated at intervals, and suddenly 'exploding' in quiet surroundings (Ali and Ripley 1969); uttered continuously during breeding season (Baker 1927). Also rendered a loud dismal 'haw-haw-haw' or a deep 'hu-who-hu' (Jerdon 1877; Whistler 1941). In nominate *zeylonensis* call likewise trisyllabic (Baker 1927); middle unit slightly higher than other two (Lushington undated). At dusk, pair-members call to each other for some time before going off to hunt; one calls 'oomp-OOO-oo', the other, 'oo' (Henry 1955). Several accounts agree in describing call as doleful or moaning, with human-like intonation (see, e.g., Vidal 1880, Henry 1955). In recording by J T Marshall, disyllabic unit dominant in call at nest—a deep diabolical laugh (P J Sellar), 'oof uh-oof uh-oof uh-oof uh-oof uh-oof u-uh-h-HA-oo-oo-oof', lasting 4.8 s (J Hall-Craggs), and repeated after short interval. (2) Mewing-call. A strange screaming sound, like an eagle *Aquila* or Stone Curlew *Burhinus oedicnemus*, given when flying to water at dusk (Whistler 1941); very like mewing of cat (Baker 1927). This possibly the harsh descending cry, 'ooooahrrrr' which starts with a tone-crescendo and ends with a noise-diminuendo (J Hall-Craggs), in recording by J T Marshall of bird away from nest. (3) Hissing-call. Sound like 'engine blowing off steam', given by captive bird when alarmed (Vidal 1880 and by bird in Forward-threat posture (T Cleeves; see Social Pattern and Behaviour, also call 4, below). (4) Growling-calls. Low growls followdd loud Hissing-call of bird alarmed during day (Layard 1851). Muted growls given by bird threatened by squirrel *Ratufa indica* (T Cleeves: see Social Pattern and Behaviour). 'Groans of displeasure' and low chucklings when feeding young (Baker 1927) probably the same or related calls.

CALLS OF YOUNG. No information. EKD

Breeding. SEASON. No information from west Palearctic. SITE. In cleft or on ledge on rocky or mud cliff, in tree-hole or cradled in fork, or in deserted nest of bird of prey; in India, sometimes in abandoned building (Whistler 1941; Ali and Ripley 1969). Nest: open sites, especially those appropriated from other birds, sometimes lined with a few sticks, leaves, and feathers (Baker 1927; Ali and Ripley 1969). Building: no information on role of sexes. EGGS. See Plate 97. Broad oval, pitted and slightly glossy; white with faint creamy tinge. Average 58×49 mm, $n = 10$ (Ali and Ripley 1981). Clutch: 1–2, rarely 3. INCUBATION. About 5 weeks. ♂ thought to assist in incubation (Baker 1934). No further information.

Plumages (*K. z. semenowi*). ADULT. Forehead, ear-tufts, crown, hindneck, sides of neck, mantle, and upper scapulars bright pink-cinnamon with sharply defined black shaft-streaks, *c*. 2–3 mm wide on ear-tufts, crown, and hindneck, *c*. 5–8 mm on upper scapulars; some faint pink-buff marbling on feather-sides of lower mantle and shorter outer scapulars. Outer and lower scapulars paler than shorter scapulars and with narrower black streaks, feathers indistinctly marbled pale pink-buff and cream laterally, soon wearing to white on outer webs. Back, rump, and upper tail-coverts pink-cinnamon with narrow and sharp black streaks. Face cinnamon-buff, feathers with black bristle-like shafts projecting; slightly paler cream on lores and just above eye; facial disc not distinctly defined. Chin and throat cream-white or white; feather-tips of lower throat and all chest closely and indistinctly barred cinnamon and pale pink-buff and with narrow and sharp black shaft-streaks 1–2 mm wide. Breast, belly, vent, and flanks similar to chest, but slightly paler, pale cream-pink or cream-white with close and narrow pink-cinnamon bars. Under tail-coverts cream-white with narrow black-brown shaft-streaks. Tail dark brown; *c*. 2 cm of tip and 3–4 well-separated bands *c*. 1 cm wide pink-buff, latter in part slightly mottled dusky. Outer web of p10 and tips and middle portions of other flight-feathers dark brown with pink-buff bands *c*. 1 cm wide; latter in part faintly vermiculated dusky, least so on outer webs of primaries; pink-buff grades to white on inner webs of secondaries; *c*. 2 cm of tip of all flight-feathers almost uniform pink-buff with faint dusky markings, bases of flight-feathers uniform pale buff. Inner tertials marbled cream and olive-grey and with traces of irregular dark brown bands on middle portions and bases. Lesser and median upper wing-coverts bright pink-cinnamon, like shorter scapulars; median with rather bold black centre (tapering towards base and tip), lesser with limited black in centre, appearing almost uniform rufous-cinnamon. Greater upper wing-coverts paler than lesser and median, more like inner webs of outer and longer scapulars; marbled pink-buff, cream, and olive-grey and with black shaft-streak, black extending irregularly towards sides of feather-tip, isolating pale pink-buff spots. Upper primary coverts and bastard wing dark brown with pink-buff bands, like primaries. Under wing-coverts and axillaries bright cinnamon with narrow black shaft-streaks; ground-colours almost white on greater and marginal coverts, greater under primary coverts broadly tipped black. In worn plumage, bright cinnamon of upperparts fades to yellow-buff; pink-buff bands on tail and flight-feathers almost white. DOWNY YOUNG. No information. JUVENILE. No *semenowi* in full juvenile plumage examined. Retained juvenile feathers in moulting 1st adult rather similar to those of adult, but feathers softer and looser, especially on underparts; feathers of upperparts pale buff or buff with narrow and sharp shaft-streaks (narrower than in adult), not as marbled laterally as in adult; juvenile upper wing-coverts dark olive-brown with broken, well-separated pink-buff bars (forming spots) and cream or off-white tips; narrow shaft-streaks black. Primaries and tail-feathers slightly narrower than in adult and with slightly pointed tip (not broadly rounded). FIRST ADULT. Like adult, but juvenile flight-feathers and tail retained; slightly pointed tips of primaries diagnostic, as well as all primaries showing similar wear, not a mixture of new and old feathers with broadly rounded tips as in adult. SUBSEQUENT PLUMAGES. Start of replacement of juvenile primaries with p7 at *c.* 1 year old; as in Eagle Owl *Bubo bubo*, some juvenile innermost and outermost primaries probably retained for several years.

Bare parts. ADULT AND FIRST ADULT. Iris yellow, bright yellow, sulphur-yellow, or bright golden-yellow. Eyelids pale flesh-brown. Bill dirty pale yellow, pale horn-colour, pale greenish-grey, or pale blue-grey; cere, base of upper mandible, and culmen dusky horn-grey, dark horn-brown, or black. Leg and foot blue-grey; occasionally dirty grey-yellow, dusky yellow, dull orange-brown, or dusky grey with yellow front of tarsus and yellow soles; claws light or dark horn-brown or black. DOWNY YOUNG AND JUVENILE. No information. (Hartert 1912–21; Ali and Ripley 1969; T Cleeves, BMNH, RMNH, ZMA.)

Moults. As in *B. bubo*, primary moult serially ascendant and descendant from centre on p7, probably with 2nd centre at innermost primaries. Each series very slow, only 1–2 feathers replaced each moulting season. Moult from p7 probably starts every other year, resulting in moult patterns such as: year 1, p7 shed; year 2, p6 and p8–p9 shed; year 3, p5–p4 and p10 shed, as well as following series starting with p7; occasionally, p6 perhaps moulted before p7. Active flight-feather moult encountered in birds from July, September, and October; some still moulting part of body December. A Syrian bird from 14 November was in 1st adult except for all flight-feathers, greater primary coverts, and tail and for scattered juvenile feathers in scapulars, upper tail-coverts, belly to under tail-coverts, outer lesser and median upper wing-coverts, and many greater upper wing-coverts.

Measurements. *K. z. semenowi*, Lebanon, Syria, and Iran, all year; skins (BMNH, RMNH). Bill (F) to forehead, bill (C) to cere.

	♂			♀		
WING	412	(15·1; 5)	396–429	402	(2)	399–404
TAIL	202	(4·40; 4)	197–207	206	(2)	199–214
BILL (F)	51·4	(2·45; 4)	49–54	49·8	(2)	48–51
BILL (C)	33·1	(0·50; 4)	32–34	34·2	(2)	33–35
TARSUS	77·0	(2·99; 4)	74–80	74·6	(2)	74–76

Samples too small to test whether sex differences are significant. Levant birds average slightly larger than those of Iran: wing Levant 417 (3) 410–429, Iran 396 (17·7; 7) 369–425; tarsus Levant 79·6 (2) 78–80, Iran 74·6 (0·69; 6) 74–76 (Hartert 1912–21; BMNH, RMNH).

Weights. No information for *semenowi*. *K. z. leschenault*. Nepal: ♀, March, 1105 (Diesselhorst 1968). Fat captive ♀, 1308 (ZMA).

Structure. Wing rather long, broad, tip rounded. 10 primaries: p6 and p7 longest, about equal; p8 2–11 shorter, p9 22–40, p10 71–82, p5 4–10, p4 21–31, p3 51–60, p2 76–83, p1 96–110. Outer web of p4–p9 and inner web of p5–p10 emarginated. Outer webs of primaries not serrated at edges, unlike eagle owls *Bubo*. Tail rather short, slightly decurved, 12 feathers; t6 *c.* 10 shorter than t1. Bill heavy; deeper, more strongly hooked, and more protruding than in *Bubo*; covered at sides by stiff bristles up to 3–5 cm long, projecting forwards from lores, base of both mandibles, and upper chin. No facial disc; eye protected above by ridge of crown-feathers and by short bristle-like feathers on eyelid; feathers projecting back and downwards from behind and below eye stiff and bristle-like with reduced barbs. Feather-ridge at sides of forecrown extends into dense, soft, ragged tuft of 10–15 longer (up to 50–60 mm) and many shorter feathers, projecting laterally, not above level of crown. Tibia feathered down to tibio-tarsal joint, tarsus bare (upper front of tarsus feathered in some extralimital races). Tarsus and uppersurface of toes covered with small knob-like scutes, undersurface of toes with slightly spiny scutes, especially on pronounced wart-like cushions. Middle toe without claw 42.8 (7) 39–46 mm; outer toe *c.* 79% of middle, inner *c.* 90%, hind *c.* 62%. Claws strong, deep, laterally flattened, with sharp cutting edge on undersurface; middle claw 27·8 (4) 26–29 mm in ♂, 31·3 (3) 30–33 in ♀; outer claw *c.* 90% of middle claw, inner *c.* 117%, hind *c.* 97%.

Geographical variation. Marked, but mainly clinal. Involves size and colour saturation. *K. z. leschenault* from India on average slightly smaller than *semenowi* from Levant and Iran, ground-colour of upperparts and upper wing-coverts dark brown-cinnamon, less pinkish-cinnamon; feathers of crown and hind-neck finely barred dusky laterally, black shaft-streaks on mantle, scapulars, and upper wing-coverts irregularly extending to sides of feathers, isolating small pale spots, white tips of outer webs of outer scapulars and median and greater upper wing-coverts more contrasting; underparts more contrastingly marked with close pink-cream and rufous-cinnamon bars; flight-feathers and tail deeper brownish-black with slightly narrower pale bands. Birds from north-west India slightly paler than those of north-east India, but darker and greyer than *semenowi*, those of Pakistan nearly as pale as *semenowi* but size as in north Indian birds; birds of peninsular India slightly smaller. Nominate *zeylonensis* from Ceylon smaller than *leschenault*, upperparts slightly darker, underparts with slightly broader black shaft-streaks. *K. z. orientalis* from northern Indo-China and southern China also described as darker (Delacour 1926); distinctions between races of India, Ceylon, and south-east Asia not marked and individual variation in colour large; perhaps only 2 races (*semenowi* and nominate *zeylonensis*) should be recognized.

Forms superspecies with Blakiston's Fish Owl *K. blakistoni* from Manchuria, eastern Siberia, Sakhalin, Hokkaido, and Kuril Islands.   CSR

Nyctea scandiaca Snowy Owl

PLATES 46 and 55
[between pages 494 and 495, and facing page 567]

Du. Sneeuwuil   Fr. Harfang des neiges   Ge. Schneeeule
Ru. Белая сова   Sp. Buho nival   Sw. Fjälluggla

Strix scandiaca Linnaeus, 1758

Monotypic

Field characters. 53–66 cm; wing-span 142–166 cm. 2nd bulkiest owl of west Palearctic; over 50% larger than Barn Owl *Tyto alba*. ♀ up to 20% larger than ♂. Large, earless, predominantly white owl, with relatively small but round head, barrel-shaped body, and less square-cut wings than most other owls. Legs and toes fully feathered. Facial expression cat-like. Does not shun daylight. Flight recalls buzzard *Buteo*. Sexes dissimilar; no seasonal variation. Juvenile separable.

ADULT MALE. Cold or pale creamy-white overall; all show small black-brown spots on tips of primaries and many have a few black-brown spots, chevrons, or bars on crown, scapulars, and larger wing-coverts, but all these marks invisible at distance. Facial disc incomplete but ruff visible below level of eyes; golden eyes obvious in shadowy pits. Dense feathering over black bill also casts shadow across throat. Appears wholly white in flight. ADULT FEMALE. Ground-colour of plumage as ♂ but with crown and nape spotted and rest of upperparts and most of underbody heavily chevroned and barred dark brown. Head-on, pure white face and centre of breast contrast obviously with barred body. In flight, these areas and little-marked vent, tail, and underwing all appear white but most of body and all upperparts appear obviously barred, with markings strongest on primary coverts, flight-feathers, and tail. JUVENILE. Up to October, may retain loose downy feathers on head and underbody. Crown, nape, back, wing-coverts, and body noticeably dark grey, with indistinct grey-white speckles of downy plumage still protruding. Against uniformity of this, white face obvious and white (barred brown) flight- and tail-feathers contrast boldly in flight. FIRST WINTER. Both sexes resemble adult ♀ but young ♀ much more heavily barred and less white-faced, while young ♂ less heavily barred and smaller (differing from young ♀ in size and lacking heavy bars on crown and neck). Adult plumage assumed at 1 year old.

In prolonged observation, adult of either sex unmistakable but potential confusions haunt this species. Wishful observers in temperate Europe frequently mistake white-breasted *T. alba* for ♂ but that species always shows buff and grey upperparts and is much smaller. In subarctic and temperate Europe, beware also mistakes arising from brief view of 'large white bird', e.g. pale Rough-legged Buzzard *Buteo lagopus*, extensively white individuals of Buzzard *B. buteo*, and even pale Short-eared Owl *Asio flammeus*. Flight accomplished, often more active than that of most other owls; wing-beats usually deep and regular but accelerated during aerial chases, which may culminate in falcon-like strike at flying or grounded prey or pass at intruder, e.g. Raven *Corvus corax*; glides and side-slips frequent. Flight actions generally recall those of (e.g.) *Buteo*, though remarkable buoyancy typical of family still evident. Rests mostly on ground but will perch on stumps or even higher. Carriage on ground sometimes upright, sometimes half so; often sits down, resting on breast (strongly recalling cat). Has rolling walk and lurching lope. Usually suspicious and difficult to approach.

Adaptation to tundra clearly reflected in habitat preference of winter vagrants which occur on short vegetation of islands, coasts, and inland moors and ridges, and shun forests. Solitary except when breeding. Breeding ♂ calls with single, loud, hollow hoot; both sexes mob intruders with grating bark. Silent outside breeding season.

Habitat. Breeding beyond treeline at highest terrestrial latitudes, mainly on arctic tundra in open, exposed, and very cold situations, from sea-level to uplands. Prefers low tundra, often near coast, though ranging between 1100 m and 1500 m in Norway (Bannerman 1955). Avoids damp hollows, requiring dry hummocks, tussocks, or rocks, where snow melts early, as look-out posts and nest-sites, commanding wide all-round views. Otherwise differences in type of tundra less significant than its capability for sustaining ample supply of prey animals. On Baffin Island (Canada), 1953, predominant prey species lemmings *Lemmus* most abundant on wet grass-heath, becoming more uncommon on dry heath and hillsides, and breeding birds concentrated accordingly. In the same year, Bylot Island (Canada), breeding birds favoured dry heathy tundra, the preferred summer habitat of the locally predominant lemming *Dicrostonyx* (Watson 1957). Access to water also relevant. In Alaska, some nests are on coastal tideflats, others on boulders. In Scandinavian low Arctic, breeds in upland birch belt, typically in 'moraine landscapes' (Bannerman 1955). Breeding on Fetlar (Scotland), 1967–75, was on bare moorland of rough grass and heather with rocky outcrops and boulder-strewn slopes 150 m above sea-level, grazed by ponies and sheep and disturbed by man (Sharrock 1976).

In Europe, in open landscapes, favouring coast and coastal islands, sand-banks, lakes, river levels, harvest fields, mown meadows, and steppes; also on extensive moors and marshes, airfields, golf courses, and by human settlements. Avoids forests, but may occur on fringes; similarly with mountains (Glutz and Bauer 1980). Winters in open country such as fields, prairies, marshes, coasts,

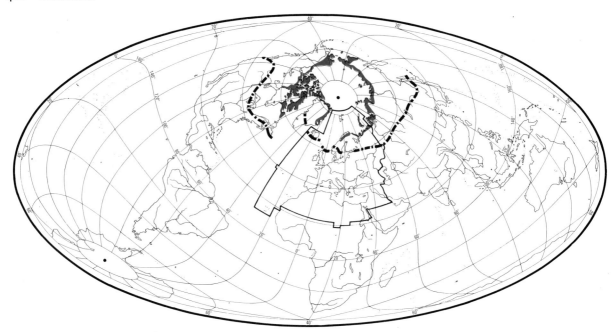

shores of lakes, and large rivers in Canada, perching on ground and also on fence-posts, straw stacks, radio towers, and infrequently on buildings and trees (Godfrey 1979). Wanderers off western Greenland often roost on icebergs, or fly far out to sea through drift-ice (Salomonsen 1950–1). Often persecuted where in contact with man, and liable to heavy losses when surplus populations are displaced after bountiful lemming year. Although flying strongly, does not ascend freely to upper airspace and spends much time on ground, usually stationary.

Distribution. Marked fluctuations in numbers, with nesting often sporadic; breeding area mapped shows where nesting may occur in good years but in poor years probably little breeding in most of west Palearctic. Winter distribution still less known—black line shows approximate southern limit of more or less regular winter occurrences, but birds move further south in eruption years (see Movements).

ICELAND. Seen annually, but doubtful if it has bred in last 25 years (AP). BRITAIN. Bred Fetlar (Shetland) 1967–75 (Sharrock 1976); only ♀♀ present since (Sharrock et al. 1982). NORWAY. Bred Hardangervidda 1948, 1952, 1956, 1959–60, 1963, 1966–7, and 1974 (maximum nests found 12 in 1959, but only 1 in 1966–7 and 1974, and now apparently extinct in south); at least 10 pairs in north (Nordland and Troms) in 1978 (GL, VR). SWEDEN. May breed annually, but only 10 nests found 1961–75, then c. 75 nests 1978, none 1979–80, a few 1981 (Risberg 1982; LR). FINLAND. Probably does not breed regularly when lemmings *Lemmus* scarce; good years 1880–1, 1903–4, 1907, 1910, 1930–1, and 1934 (Merikallio 1958). No breeding proved from 1930s until 1974, when several pairs nested (OH).

USSR. Breeding everywhere more or less sporadic (VG).

Accidental. Spitsbergen, Bear Island, Jan Mayen, Faeroes (perhaps annual), Ireland, France, Belgium, Netherlands, West Germany, Denmark, East Germany, Poland, Czechoslovakia, Austria, Hungary, Yugoslavia, Albania, Azores.

Population. Numbers fluctuate markedly with population levels of lemmings *Lemmus*; some suggestion of decline in Palearctic in 20th century (Portenko 1972).

NORWAY. Tendency to decrease in last 50 years (Hagen 1964). USSR. Marked fluctuations, probably due to variation in food supply; in some areas, no nesting in years when lemmings scarce or absent (Dementiev and Gladkov 1951a).

Oldest ringed bird 9 years 5 months (Kennard 1975).

Movements. Partially migratory and nomadic; also eruptive at intervals. Though morphologically adapted to withstand sub-zero temperatures, prey availability and (possibly) winter darkness limit scope for wintering in high-arctic latitudes.

Certainly in USSR majority withdraw from northernmost areas in autumn (though has been found in winter even on Novaya Zemlya), and move to open country further south—to scrub tundra, and even to Siberian steppes as in northern Kazakhstan (Dementiev and Gladkov 1951a; Portenko 1972). Otherwise nomadic in Eurasia, dispersing westwards as well as southwards through tundra zone, presumably in response to food availability. Normal European winter range lies north of 60°N, but irrupting birds occur irregularly south to 53–55°N in Scotland, Denmark, northern plains of West and East Ger-

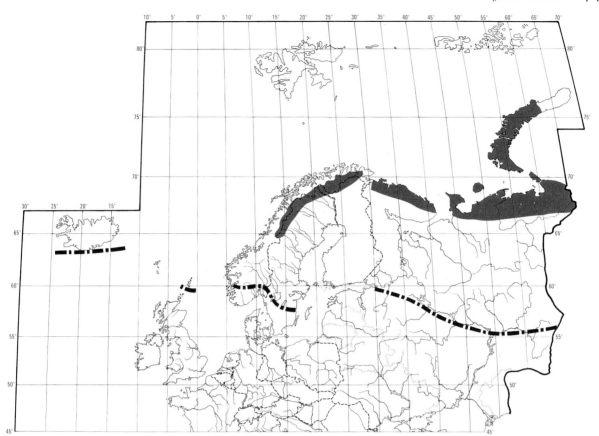

many, northern Poland, and European USSR, or to c. 48°N in Siberia. Nomadic life-style also apparent from impermanent breeding range limits, and from changes in densities between summers as birds move to regions where rodent numbers higher (see below). In Canada and Alaska, where breeding numbers also fluctuate regionally, majority probably withdraw from high-arctic zone of 24-hr winter darkness (Snyder 1957; Godfrey 1966); in non-eruption years, apparently few individuals go further south than c. 55°N. More or less regular migratory movements noted also in northern Greenland where, however, some birds are resident at least as far north as Germania Land; movements there more marked in eruption years, when birds occur south to Cape Farewell (Salomonsen 1950–1, 1967a). Greenland emigrants probably responsible for impermanent presence and irregular breeding in Iceland.

In North America and Greenland, lemmings (*Lemmus*, *Dicrostonyx*) are the most important prey, and these show abundance cycles of c. 4 years. However, lemming cycle not synchronous all over these vast areas of tundra, so birds move into areas with temporary abundance of prey (Chitty 1950; Lack 1954). It is when lemming numbers crash over very large areas that eruptions occur (Watson 1957). Gross (1947) demonstrated regular cycles of 3–5 (mainly 4) years for autumn irruptions into southern Canada and northern USA, these correlating well with lemming crashes. Irruptions in western Greenland, at least, also correlated with those in eastern Canada, indicating involvement of same population (Salomonsen 1950–51). Mortality among irrupting birds is high (Gross 1947). See also Shelford (1945), Snyder (1947, 1957), and Meade (1948).

In Europe, eruptions less regular, which Lack (1954) linked to smaller size of tundra area involved. However, ultimate reason is probably that European birds are not all as dependent on lemmings as is usually implied in the literature (see Food), voles (*Microtus*, *Clethrionomys*) often also being important. Possibly infrequent eruptions in Europe triggered as much by cold weather as by rodent population levels. For 1960–3 irruptions into southern Fenno-Scandia, see Nagell and Frycklund (1965), where numbers involved and passage through Finnish islands considered indicative of eastern (Russian) origin of many. Majority of birds were juveniles, and adult ♂♂ especially scarce. Began arriving September–October, though not in strength until November–December; some individuals stayed for lengthy period at a site, disappearing about March. One ringed 1963 Hordaland (Norway) as nestling found July 1964 in Finnmark, 1380 km NNE, and full-grown bird ringed Pori (Finland) in December 1960 found February 1962 in Nordland (Norway), 810 km NNW (Holgersen 1965).

In contrast, geographically isolated birds which bred

Shetland (Scotland), 1967–75, were resident (adults—some ♀♀ still so, 8 years later) to dispersive (juveniles). One ringed Fair Isle, June 1972, found in Outer Hebrides in January 1975, 310 km WSW.

Food. On tundra, almost wholly lemmings (*Lemmus*, etc.) or voles (*Microtus*, *Clethrionomys*); elsewhere, mammals (e.g. rabbit *Oryctolagus cuniculus*) and medium-sized birds as available. Most hunting done in twilight of morning and evening, though also during day; in summer, southern Norway, mostly 21.00–06.00 hrs (Hagen 1960), and on Shetland (Scotland) most 16.00–04.00, especially 22.00–03.00 (M Robinson). In winter, at least occasionally active in darkness (Nagell and Frycklund 1965; Tulloch 1969), though at one site in Canada birds made fewer hunting attempts in dusk than in daylight (Höhn 1973). Captive birds active to some extent at all hours (Scherzinger 1974c). In daylight at least, prey evidently located largely by sight. Most hunting apparently done from open perch on ground or (where available, usually in wintering areas) often c. 4·5–6 m high. Watches in one direction for a few seconds, then moves head round c. 20°, thus watching all round; in winter, Alberta (Canada), scanned thus for 1 hr or more, but on average located prey 22 min after starting to scan (see also below). Makes low flight to prey which is often distant: in 12 cases, Alberta, c. 27–159 m, average c. 89 m (Höhn 1973); c. 700 m recorded in Shetland, where 14 flights averaged 4 min long (M Robinson). May fly directly to prey (Tulloch 1969) or fly close to it and wait before jumping on it (H Delin). Searching flight used only occasionally in Shetland (M Robinson), but apparently described as main method on Baffin Island (Canada) where hovering (at c. 15 m) also used extensively. Drops on to prey vertically or by gliding down; sometimes carries prey in feet straight to perch without landing (Watson 1957). When attacking large hare *Lepus*, seen to grasp it with one foot and use other as brake in snow or herbage; will also use wings as brakes (Meinertzhagen 1959). Sometimes dives under snow for small mammals (Nagell and Frycklund 1965). Will pursue and capture small and large birds in flight (Nero 1964; Tulloch 1969), and take birds from surface of water (King *et al.* 1966; Campbell and MacColl 1978). Recorded walking about in field, intermittently jumping into reeds after prey (King *et al.* 1966). To catch fish, said to lie flat along edge of water with head laid down and turned towards water; when fish appears, thrusts foot out to catch it (Meinertzhagen 1959); also taken by hovering above water (Glutz and Bauer 1980). For actual and apparent food-piracy on Glaucous Gull *Larus hyperboreus*, Hen Harrier *Circus cyaneus*, and Short-eared Owl *Asio flammeus*, see Pitelka *et al.* (1955), Duffy *et al.* (1976), and Lein and Boxall (1979). Prey usually eaten where killed (Meinertzhagen 1959). With rabbit, usually just stands on it, occasionally biting it, until it stops kicking (Tulloch 1969). Lemmings usually swallowed whole, head-first (Watson 1957); ♀ once ate rabbit (size not stated) whole (Tulloch 1968). Prey for ♀ and brood generally carried by ♂ in feet and transferred to bill before passing to ♀ (Tulloch 1969). For food-caching, see Social Pattern and Behaviour. Feathers of own species found in pellets not uncommonly; one ♀ several times seen to deliberately eat her own feathers during preening (Tulloch 1968). Once, ♀ flew from nest with damaged or infertile egg and ate contents. ♀ will feed dead young to siblings. (M Robinson.) Of 12 flights to prey from perch in winter, Alberta, 4 successful, 4 probably successful, 4 unsuccessful (Höhn 1973). In 10 observations on one ♂, Baffin Island, summer, always caught lemming within 5 min of starting to hunt (see also above); over 24 hrs (before eggs hatched), caught 10 lemmings, 8 within 90 min; dense fog had no apparent effect on hunting, and rate of weight increase of young did not appear linked to weather (Watson 1957).

Pellets average 92 × 33 mm (56–153 × 26–40); 4·3 prey per pellet, $n = 19$ pellets (Hagen 1960). 66 × 26 mm (49–84 × 18–35), $n = 108$; vegetation 2·2% of pellet dry weight (Kennedy 1981). Sand found in all of 7 pellets (Marquiss and Cunningham 1980). Captive bird produced 1·3 pellets per day ($n = 16$ pellets), and all large bones of prey reappeared in pellets (Mikkola 1983).

The following prey recorded in west Palearctic. Mammals: common shrew *Sorex araneus*, rabbit *Oryctolagus cuniculus*, hares (*Lepus europaeus*, *L. timidus*), Norway lemming *Lemmus lemmus*, bank vole *Clethrionomys glareolus*, grey-sided vole *C. rufocanus*, water vole *Arvicola terrestris*, short-tailed vole *Microtus agrestis*, root vole *M. oeconomus*, wood mouse *Apodemus sylvaticus*, weasel *Mustela nivalis*. Birds range in size from Redpoll *Carduelis flammea* to full-grown Black Grouse *Tetrao tetrix*; waders (Charadrii), waterbirds, and grouse *Lagopus* recorded most frequently. Frogs *Rana*, fish, beetles (Coleoptera), and offal also recorded. (Fisher 1893; Hagen 1952, 1960; Nagell and Frycklund 1965; Tulloch 1968; Andersson and Persson 1971; Mikkola 1983; M Robinson.) For vegetation, etc., in pellets, see 2nd paragraph.

Diet not well studied in west Palearctic, especially on tundra; see Table A for Scandinavia. In Shetland, all of 25 items brought to nest before hatching were rabbits; after hatching, 102 items comprised 58% (by number) rabbit, 23% fledgling Oystercatcher *Haematopus ostralegus*, 3% fledgling curlews *Numenius*, 1% fledgling Lapwing *Vanellus vanellus*, 1% fledgling Arctic Skua *Stercorarius parasiticus*, 4% unidentified birds, 11% other unidentified items (Tulloch 1969). 7 pellets from Outer Hebrides (Scotland), July, contained only rabbits (Marquiss and Cunningham 1980). ♀ on Isles of Scilly (England) in winter ate mostly rabbits, including dead ones (King *et al.* 1966). 4 stomachs from Rybinsk (USSR) contained hares *Lepus* and grouse *Lagopus* (Spangenberg 1972). For North American breeding-season studies, see Pitelka *et al.* (1955), Watson (1957), Taylor (1974), and Kennedy (1981); in summer on Agattu Island (Aleutian Islands) where no small mammals present (and birds studied not

Table A   Food of Snowy Owl *Nyctea scandiaca*. Figures are percentages of numbers of vertebrates only.

	Breeding season				Autumn and winter	
	Hardangervidda (S Norway)		N Sweden	N Finland	N Finland	S Finland
	1934	1959	1969–70	1974–5	1975	1961–74
Norway lemming *Lemmus lemmus*	17·9 ⎱ 98·6	84·6 ⎱ 96·9	90·3 ⎱ 98·5	30·7 ⎱ 95·6	34·5 ⎱ 98·7	0 ⎱ 84·8
Other voles (Microtinae)	80·7[1] ⎰	12·3 ⎰	8·2 ⎰	64·9 ⎰	64·2 ⎰	84·8 ⎰
Shrews (Soricidae)	0·4	0	0·5	0·4	0·4	5·1
Weasel *Mustela nivalis*	0·1	0	0	0·1	0	0
Squirrel *Sciurus vulgaris*	0	0	0	0·2	0	0
Hares *Lepus*	0	0	0	0·4	0·5	2·0
Birds	0·8	3·1	1·0	1·8	0·4	8·1
Frogs *Rana*	0	0	0	1·1	0	0
Fish	0·1	0	0	0·5	0	0
Total no. of vertebrates	1395	287[2]	206	834	226	94[3]
Source	Løvenskiold (1947)	Hagen (1960)	Andersson and Persson (1971)	Mikkola (1983)	Mikkola (1983)	Mikkola (1983)

1. Includes 79·0% root vole *Microtus oeconomus*.   2. Also 3 beetles (Coleoptera).   3. Also 1 insect.

breeding), took exclusively birds, largely auks (Alcidae) (Williams and Frank 1979). For Siberia, see Bolshakov (1968). Wintering birds in coastal Vancouver (Canada), took exclusively birds, largely ducks (Anatinae), grebes (Podicipedidae), and gulls (Laridae) (Campbell and Mac-Coll 1978). In breeding season, proportions of different age and sex classes of lemmings taken are roughly in line with their abundance and activity above ground, i.e. more ♂♂, and more young animals later in season (Thompson 1955; Watson 1957). Note that pellet studies may underestimate amount of large prey taken as these often have only soft parts eaten.

Observations suggested hunting ♂ in summer ate average 280 g per day; less active ♀, 220 g (Watson 1957). Captive 4-week-old given average 450 g per day; at 5½ weeks, when weight constant, 326 g (Pitelka *et al.* 1955); another bird ate 338 g per day (Mikkola 1983). For further studies in captivity, see Glutz and Bauer (1980). On Baffin Island, young a few days old often ate half their own weight (sometimes more than their own weight) in food per day. Just before leaving nest (i.e. well before fledging) when weighing *c.* 1 kg, young at one nest each ate *c.* 195 g per day. During period of total dependence on parents (*c.* 60 days), each young ate *c.* 11 kg or 140 full-grown *Lemmus*. In area of high *Lemmus* density, numbers taken over whole breeding season (May to early September) amounted to 20–31% of those present in late June or early July, or 8–20% of those present mid-August (Watson 1957). In Shetland, in year when unfledged waders 21% of food by number, ¼–⅓ of all available young *Haematopus ostralgeus* and Whimbrels *Numenius phaeopus* were taken (M Robinson).

No details on difference between food of adults and young (but see 4th paragraph). ♀ once fed pellet to chick (Tulloch 1968). At one nest, first prey brought 23.00 hrs, then every 10–15 min until 01.00 hrs (Hagen 1960). For food consumption of young, see penultimate paragraph.

DJB

Social pattern and behaviour. Studied mostly in Nearctic, especially by Watson (1957) and Taylor (1973). Contribution from study of birds breeding Shetland (Scotland), 1967–75, also important: Tulloch (1968, 1969), and data collected by Royal Society for the Protection of Birds and supplied by M Robinson.

1. Essentially solitary outside breeding season, and, in some situations at least, defends fixed individual feeding territory in winter quarters. In Wisconsin (USA), *c.* 0·5–2·6 km², *n* = 5 (Keith 1964; see also for literature review). At Uppsala (Sweden), one 'territory' 1 × 2 km (Nagell and Frycklund 1965); in Vancouver (Canada), 5 birds in 'distinct territories' over one winter, e.g. 2 birds occupying different ends of island of 1 ha (Campbell and MacColl 1978); not clear, however, if these areas defended. Summer non-breeders at Barrow (Alaska) present at up to 2 or more per km² (Pitelka *et al.* 1955). BONDS. Essentially monogamous mating system, though polygyny (♂ and 2 ♀♀) not rare, and 2 ♂♂ once found defending one nest simultaneously (Hagen 1960). Of 11 ♂♂, Hardangervidda (Norway), 1 bigamous (Hagen 1960); of 5 ♂♂, Sweden, all monogamous (C G Wiklund); of 5 ♂♂, Baffin Island (Canada), 1 bigamous (Watson 1957). One ♀ of bigamous ♂ may lay 2–3 weeks later than other (Watson 1957; Hagen 1960); sometimes, at least, ♂ does not feed 2nd ♀ and nest fails; relationship between the 2 ♀♀ hostile (M Robinson; see part 2). Pair-bond apparently for duration of breeding season only, but ♂ and ♀ seen keeping together over 1 week in November–December, Wisconsin (Keith 1964). Not known if same birds normally renew bond each year, though one pair in Shetland kept together for 8 years (Tulloch 1975). Age of first breeding not known; probably not until at least 2 years old (Portenko 1972); 1 year in captive-bred pair (Callegari 1970a). ♀ broods and feeds young at nest, provisioned by ♂; later, both hunt and feed young. Young dependent for 2½–3 months after hatching (see Relations within Family Group, below). BREEDING DISPERSION. Solitary and territorial; territories may be clumped (Miller and Russell 1973), even in apparently uniform habitat (Darling 1956). Territory size and thus density of pairs linked to abundance of lemmings (*Lemmus*, etc.). At high lemming density, Baffin Island, territories *c.* 1·3–3·9 km² (including bigamous ♂ with *c.* 2·6 km²); with fewer lemmings, one territory at least *c.* 8–10 km²; area with fewer still was occupied by non-breeding pair ranging over *c.* 23 km² (Watson 1957, which see for other sources; also Miller and Russell 1973). At Khromskaya Guba (USSR), *c.* 1 pair per km² (Uspenski *et al.*

1962b). Breeding ♂ in Shetland hunted over c. 4 km² (M Robinson). On Hardangervidda, 12 nests separated by average 2·1 km (1·2–3·7); ♀♀ of bigamous ♂ 1·3 km apart (Hagen 1960). On Southampton Island (Canada), in lemming year, 11 nests separated by average 4·5 km, 1 nest per 22 km²; in following year, no birds seen (Parker 1974). Territory defended by ♂ only; used for all hunting and raising of young; ♀♀ of bigamous ♂ keep to separate parts of territory (Watson 1957). ROOSTING. Since activity essentially crepuscular (see Food), birds presumably roost at night (during winter), but no information. Especially in winter, often loafs and/or sleeps during middle of day, e.g. 10.00–15.00 hrs at Uppsala in winter (Nagell and Frycklund 1965). Even when feeding young, ♂ usually inactive for hours at a time (e.g. Watson 1957). In breeding season, ♂ uses favourite perches scattered over territory; in winter, uses open ground, ice, or exposed perch (e.g. Keith 1964, Lein and Boxall 1979).

2. Generally rather shy in winter, but may become habituated to presence of man within c. 6 m (Keith 1964). ANTAGONISTIC BEHAVIOUR. In territorial Hooting-display, ♂ bows forward at each call (see 1 in Voice), with tail lifted (Taylor 1973: Fig A). Used to advertise presence to other ♂♂ and in direct boundary conflicts. In latter, ♂♂ face each other across border—usually not close, e.g. c. 200 m (C G Wiklund), c. 800 m (Watson 1957); also use Mantling-display and Display-flight (see Heterosexual Behaviour, below) (Taylor 1973). Use of Hooting-display decreases after early June, birds staying within own territories, but ♂ may fight fiercely with transient intruders (Watson 1957; Taylor 1973). ♀♀ of bigamous ♂, Shetland, not visible to each other while incubating, but when 1st-established ♀ caught sight of 2nd ♀, 1st flew towards 2nd, calling (see 5 in Voice); often chased her, sometimes fighting briefly in mid-air. If one ♀ saw other copulating with ♂, 1st would chase off other ♀, or ♂; 1st ♀ sometimes then flew to ♂ and adopted Soliciting-posture (see below). Young ♀♀ present during breeding elicited no reaction from nesting ♀♀. (M Robinson; see also Hume 1975.) In conflicts over winter feeding territories (see part 1), one bird called (see 5 in Voice) while attacking; one encounter lasted c. 5 min, birds

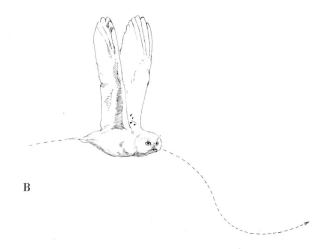

B

continually clashing in mid-air; tethered bird placed in territory was attacked viciously with feet and wings (Keith 1964). HETEROSEXUAL BEHAVIOUR. (1) Pair-bonding behaviour. No firm information on early stages. ♂ and ♀, after maintaining separate territories in winter quarters, Ontario (Canada), seen together one evening in March; rose in air together, apparently fighting, then flew off northward, side by side (Mitchell 1947). Display-flight of ♂ most frequent and complete early in season. On Bathurst Island (Canada), performed 2 May–19 July, and only when ♀ present—even c. 1 km away. ♂ flies with wings held in deep V at top of stroke (Fig B), causing bird to drop c. $\frac{1}{2}$ m; height regained with subsequent down-stroke, producing undulating flight; eventually climbs c. $1\frac{1}{2}$–3 m then drops vertically with wings in V, flapping or not. Flight may cover up to c. $1\frac{1}{2}$ km. In May (less often later), ♂ usually carried lemming or other prey in bill, perhaps indicating to ♀ sufficiency of food supply. (Taylor 1973.) Mantling-display of ♂ may or may not follow Display-flight. On landing, ♂ often drops prey. Stands with wings partly spread, carpal joints raised (Fig C); sometimes walks or turns around, facing away from ♀; leans forward with tail partly fanned (Fig D) as ♀ approaches, frequently peering round wings at her; wing nearer ♀ usually higher. Fate of lemming not known. Display normally lasts 1–2 min. Display-flight and Mantling-display may occur in sequence for over 30 min, ♀ following ♂ around. (Taylor 1973.) (2) Nest-site selection. No information on selection process. On Shetland, more than one scrape made; 11–19 days between finding of 1st scrape and laying of 1st egg (M Robinson). One territory on Baffin Island apparently contained only 1 fresh scrape (Watson 1957). (3) Mating. Of 102 complete copulations, Shetland, 81% followed ♂ flying to ♀ after ♀ left nest, 19% followed ♀ flying to ♂ (M Robinson). At Khromskaya Guba, most copulations occurred when ♀ left

A

C

D

E

nest briefly or raised herself up off it (Uspenski and Priklonski 1961). Display-flight or Mantling-display often triggered in ♂ by ♀ leaving nest. On landing, ♀ may raise tail, hold wings loosely at sides, and lean forward with body feathers usually slightly ruffled (Soliciting-posture: Fig E); ♂ may then mount, ♀ maintaining posture during and after. Once, after attempted copulation, ♀ followed ♂ around—♂ displaying (presumably Mantling-display) and ♀ giving Mewing-call (see 3 in Voice). (Taylor 1973; see also Watson 1957, Tulloch 1968, Portenko 1972, Scherzinger 1974c.) For calls given during mating, see 1, 3, 6 and 8 in Voice. At Khromskaya Guba, 3–4 copulations per hr up to c. 5–6 June (Uspenski and Priklonski 1961). Copulation infrequent after clutch complete (Tulloch 1968; Taylor 1973), but may occur up to hatching (M Robinson). (4) Courtship-feeding and food-caching. During and after incubation, ♂ provides all food until young wander from nest. ♂ delivers it to ♀ at nest or nearby perch; either bird caches surplus at one of several open sites within c. 180 m of nest. ♀ begs for food by swaying lowered head from side to side, speed increasing as ♂ approaches; finally straightens up and often beats wings; ♂ may also beat wings; both give Cackling-call (see 6 in Voice). ♂ often rubs ♀'s face and breast with food item, especially if not taken at once; ♀ then often closes eyes and moves head. (Watson 1957; Tulloch 1968; M Robinson.) Food transfer in mid-air recorded (Sutton and Parmelee 1956). (5) Behaviour at nest. ♀ leaves nest infrequently and usually briefly, but sometimes for 5–25 min. On nearby perch, may preen, defecate, or eject pellet. May give Cackling-call on leaving and returning to nest. (Watson 1957; M Robinson.) During incubation, Shetland, ♀ often reached forward with bill to pull lumps of turf from rim of nest and fling them aside; most frequent 12.00–21.00 hrs, and peaking around hatching; no cause apparent (M Robinson); see also Parental Anti-predator Strategies (below). One ♀ (with young in nest) seen frequently deepening nest with bill, pushing material toward nest-rim (Barth 1949). RELATIONS WITHIN FAMILY GROUP. Conflated largely from Watson (1957) and Tulloch (1968). For development of young in captivity, see Scherzinger (1974c). Young call from egg for up to 1 day before hatching. After 4 days, can lift heads. Eyes begin to open about day 5(–7); fully open by 8–10 days; later if food supply poor. Newly hatched young said to eat half-digested food from adult (Collett and Olsen 1921). For at least first few days, fed by ♀ on soft parts of prey; from 5 days or earlier, parts with bones also given; at 11 days, fed on whole or part-rodents of c. 20 g. ♂ recorded passing food to chicks when ♀ away from nest. Most demonstrative young tend to be fed first; thus, large young fed early in day, smaller later. With very small chicks present, ♀ may feed them while ignoring others. No hostility between young. When ♀ not brooding, small young crawl under larger siblings. Allopreen from 18–20 days and, mutual Allopreening occurs. ♀ allopreens young; young which do not call during this are treated as food (Glutz and Bauer 1980). Young beg by nibbling at ♀'s bill, calling, or in same manner as ♀ begs from ♂; feed themselves off prey in nest from c. 23 days. Leave nest-site and scatter at 14–28 days (before able to fly); ♀ continues to brood smaller young at nest, but stops c. 1 month after hatching. When nest abandoned, ♂ starts passing food to young and ♀ starts hunting; in one year, Shetland, ♀ brought 21% of items from then up to fledging (M Robinson). Just before fledging, adult with prey may encourage young to chase it. When able to, young fly to parent with food, and thus more often found together than initially after leaving nest. In one case, first flight at c. 35 days. Young fly strongly at 45 days (43–50) (M Robinson). After fledging, stay loosely together within territory. Start to attempt hunting at c. 60 days, but apparently dependent on adult for all food until at least 9–11 weeks old. In one case, brood apparently dispersed by c. 14 weeks. ANTI-PREDATOR RESPONSES OF YOUNG. May move out of nest when danger threatens (Hagen 1960). From 8–10 days, will Bill-snap when handled and at c. 2 weeks also start to give Hissing-call (see 7 in Voice), but lie stiff and motionless when turned on to back. Forward-threat posture first shown at 20–25 days: as in adult (see below), but may also lie on back, striking with claws and bill. (Watson 1957.) Young gathered together after leaving nest scatter at warning call (not described) from adult and squat with head on ground; at close approach, threaten as above then run away (Tulloch 1968). PARENTAL ANTI-PREDATOR STRATEGIES. (1) Passive measures. ♂ appears to perform sentinel role, but not clear whether he actually warns ♀ of danger. When man up to 800 m or more away, ♀ flies low and silently from nest; sometimes goes 800 m (Watson 1957; Wiklund and Stigh 1983). Adult may warn young not in nest by calling (see above). Birds disturbed at nest often perform mating, especially before clutch complete. Treated as displacement behaviour by Watson (1957) and others, but apparently normal for mating to be triggered by ♀ leaving nest, even when birds undisturbed (Taylor 1973). (2) Active measures: against man. Mainly uses perched or aerial demonstration (calling), distraction-lure display, distraction-threat (Forward-threat posture, Hooting-display, Grass-pulling), and expelling attack (dive-attack); see below. However, some birds keep at distance even when young handled (Watson 1957). ♂ usually more aggressive than ♀ (in 45 of 46 pairs: Wiklund and Stigh 1983); in Shetland, ♀ took over dominant role from ♂ c. 1 month after hatching (M Robinson). of 46 pairs with large unfledged young, northern Sweden, dive-attack recorded from 20% of ♂♂ and 2% of ♀♀, distraction-display (distraction-lure and/or distraction-threat) 22% ♂ and 76% ♀, Alarm-call (see 2 in Voice) 98% ♂ and 61% ♀, Mewing-call 0% ♂ and 98% ♀, Advertising-call 12% ♂ and 0% ♀ (Wiklund and Stigh 1983); see also Hagen (1960). Intensity of reaction increases as intruder nears nest, young, or adult's favourite perches; decreases with greater number of intruders (Watson 1957). (a) Demonstration. Both sexes give Alarm-call, both when perched and in flight; will also hover over intruder (e.g. Tulloch 1968). (b) Distraction-lure display (disablement type). In ♂, flank and back feathers ruffled, wings trailed (sometimes raised almost in swimming motion), head lowered and often swayed from side to side, and tail spread; at times, turns round, exposing back or side. ♀ generally more prostrate, beating wings and threshing about; may call (see 8 in Voice). Performed at between 50 m or less (♂, late in season) and c. 800 m (♀, early

F

in season). (Watson 1957; Tulloch 1968.) Apparently less common than Forward-threat posture. (c) Forward-threat posture. Bird lowers front of body and stretches head forward, with feathers of head, back, and flanks ruffled; wings only partly extended, or arched so that tips drag on ground (Fig F); calls (see 2–4 and 7 in Voice) or Bill-snaps; may advance thus, to 5–20 m. Performed by ♀ only occasionally. (Watson 1957; Tulloch 1968; Wiklund and Stigh 1983.) (d) Hooting-display. As in territorial circumstances (see Antagonistic Behaviour, above), but body more upright and tail not lifted; also hoots in flight (Tulloch 1968; Taylor 1973). (e) Grass-pulling. By ♂ (largely) and ♀, mostly when unfledged young present. Sometimes or always associated with Forward-threat posture. Bird tears at ground with feet and (sometimes) bill, facing intruder; occasionally shakes turf held in bill; wings usually slightly extended and front of body lowered; may sway from side to side. (Watson 1957; Tulloch 1968.) See also subsection 5 of Heterosexual Behaviour (above). (f) Dive-attack. By ♂ more than ♀. Usually a long shallow glide at intruder's head from behind, bird silent or Bill-snapping; often no contact made, but may strike with wings, back of feet, or viciously with claws. Attacks usually (not always) wane after 5–10 min, stopping after 15–20 min (Watson 1957; Wiklund and Stigh 1983). (2) Active measures: against other animals. Dogs and foxes *Alopex* dive-attacked and driven off (Wiklund and Stigh 1983). Sheep (etc.), dive-attacked and otherwise threatened; threats often ignored (Tulloch 1968). Large falcons *Falco* may be chased and struck. Large gulls *Larus* chased infrequently and usually without effect (Watson 1957; M Robinson).

(Figs A–E from drawings in Taylor 1973; Fig F from photograph in Andersson and Persson 1971.)   DJB

Voice. Noisy when disturbed at nest, but otherwise usually silent, especially outside breeding season. ♂ more vocal than ♀, probably because of greater share in defence of territory and nest (Watson 1957). Both sexes perform loud Bill-snapping in threat and attack. For idiosyncratic treatment of repertoire, not clearly agreeing with that developed here (especially for call 1), see Glutz and Bauer (1980).

CALLS OF ADULTS. (1) Advertising-call. Loud, hollow, booming hoot—'hoo'; may have element of roughness ('hoorh'), be drawn out ('hooooo'), or suggest 'aaow' of Great Black-backed Gull *Larus marinus*. Usually 2 units given together (sometimes 1 or up to 6 or more), 1–2 s between each. Frequently audible 3 km away, sometimes from 10 km; loudest in territorial defence. Given in Hooting-display, in territorial and anti-predator circumstances, and before and after copulation. Call given mostly by ♂, rarely by ♀. (Watson 1957; Hagen 1960; Tulloch 1968; Taylor 1973; H Delin.) (2) Alarm-call. From ♂, a loud, grating bark—'kre-', 'crac-', or 'ergh-' (Fig I) given in bouts of 5 (3–6) units and similar to call of upset ♀ Mallard *Anas platyrhynchos*; ♀'s call (given less often) similar but higher pitched (Fig II). Given when disturbed at nest; commonest call heard by human intruders. (Watson 1957; Tulloch 1968; H Delin.) (3) Mewing-call of ♀. Variable, shrill whistling, mewing, squealing, or yelping (Fig III), sometimes more rasping and screeching—'biw', repeated less rapidly than call 2. Call normally given in alarm when intruder at nest (once, also by ♂); also used before and after being fed by ♂ and during copulation. (Watson 1957; Tulloch 1968; Taylor 1973; Wiklund and Stigh 1983; H Delin.) (4) ♀ frequently intersperses bouts of Mewing-call with a drawn-out, high-pitched piping—'see-üü', 'süü-üü', or 'see-üü-ee' (H Delin: Fig IV). In recording by M Sinclair, given also during Forward-threat display of ♀; on nearing intruder, bird also gives 3 sharp loud squeals. (5) Attack-call. 3 descriptions perhaps relate to essentially same call (or may be linked to call 3): hoarse 'keeeeea' from ♀ in anger against ♂ and people (Barth 1949); high piercing cry given by one bird attacking another in winter territorial dispute (Keith 1964); long squealing wheezing call from one ♀ attacking another—both paired to same ♂ (M Robinson). (6) Cackling-call. Apparently associated with routine activity at or near nest; rather common and variable. Low, rapid cackling 'ka-ka-ka-ka-ka-ka'; ♀'s call often a higher-pitched 'ke-ke-ke-ke-ke-', or 'kuk-kuk-kuk-kuk' like broody hen (Watson 1957); continuous guttural clucking and croaking noises (Tulloch 1968). Given by both sexes: e.g. when ♀ comes off nest, before and during copulation, or during feeding (of ♀ or young). (7) Hissing-call. Both sexes hiss in threat (Watson 1957). (8) Other

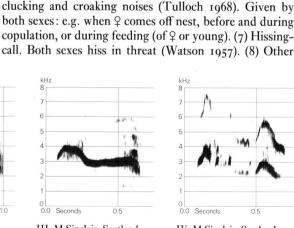

I M Sinclair Scotland
August 1967

II M Sinclair Scotland
August 1967

III M Sinclair Scotland
August 1967

IV M Sinclair Scotland
August 1967

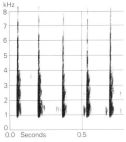
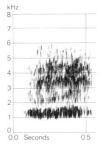

V M Sinclair Scotland August 1967

VI M Sinclair Scotland August 1967

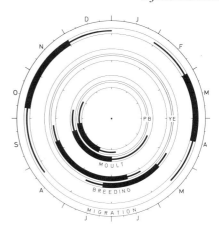

calls, some probably relating to those listed above: (a) raucous 'ca-ca-oh', like crowing rooster, from ♀ attacking (Sutton and Parmelee 1956); (b) at nest, 'tjag-tjag-tjag-gooh' (Collett and Olsen 1921); (c) loud shriek 'krow-ow' in flight when excited (Witherby et al. 1938); (d) 'yah yah yah' from angry ♀; (e) 'ou ou ou' or 'gou gou' from ♂—perhaps calling to ♀; (f) hoarse 'kra kra' from ♂ in warning; (g) rapid 'gau gau gau gau' from ♂—'call of mating' (Barth 1949); (h) wader-like trills from ♀ during copulation (Uspenski and Priklonski 1961); (i) low 'kaaarrr' from ♀ during copulation; (j) loud 'ko-ah' from ♂ approaching perch after distraction-lure display (Watson 1957); (k) in recording by M Sinclair of ♀ performing disablement-type distraction-lure display, a high-pitched squeaking similar to homologous call of Arctic Skua *Stercorarius parasiticus* (P J Sellar).

CALLS OF YOUNG. When very small (and from egg), a faint cheeping whistle. From just over 1 week old, food-call a penetrating, high-pitched, whistling squeal—rather humming in tone and (from older birds) falling in pitch towards end; lasts 1·5–2 s. Still given 1 month after starting to fly, and then audible at over $c.$ $1\frac{1}{2}$ km. In threat, Bill-snapping (Fig V) and Hissing-call (Fig VI) given (see above). (Watson 1957; Tulloch 1968; Glutz and Bauer 1980; H Delin.) DJB

Breeding. SEASON. West Palearctic range: see diagram. SITE. On ground, normally on raised hummock, rock outcrop, or ridge, or any protuberance free of snow at time of laying. Solitary. Sites not usually re-used in successive years (Watson 1957). Nest: slight scrape; unlined or with small pieces of vegetation, probably blown in. Building: probably by ♀; for method in captivity, see Scherzinger (1974c). EGGS. See Plate 97. Short elliptical, smooth and slightly glossy; white. 57×45 mm ($55–63 \times 42–48$), $n = 100$ (Witherby et al. 1938). Weight 60·3 g (47·5–68·0), $n = 66$ (Uspenski and Priklonski 1961); 53 g (50–59), $n = 15$ (Hagen 1960). Clutch: varies with food availability, usually 3–9 (2–14). Of 9 clutches, Shetland: 4 eggs, 1; 5, 4; 6, 3; 7, 1; average 5·4. In years of food abundance, average 6·3, $n = 3$ clutches; in years with poor food supply average 4·8, $n = 4$ (M Robinson). Of 24 clutches, Alaska, average 7·5 (Murie 1929); see Watson (1957) for examples of smaller samples. One brood. No information on replacement clutches. Laying interval $c.$ 50 hrs (M Robinson). INCUBATION. Average 31·6 days (30–33), $n = 8$ (M Robinson). Begins with 1st egg; eggs hatching at similar intervals to those of laying (Watson 1957). By ♀ only; fed on nest by ♂; ♀ usually leaves nest only briefly (Watson 1957). YOUNG. Altricial and nidicolous. Cared for and fed by both parents. Brooded more or less continuously by ♀ for $c.$ 21 days after hatching, intermittently thereafter, and fed by ♀ with food brought by ♂. At $c.$ 16 days, leave nest and hide in surrounding vegetation, where fed by ♂, while ♀ continues to look after young left in nest (Taylor 1973). FLEDGING TO MATURITY. Fledging period average 45·4 days (43–50), $n = 7$ (M Robinson). Not independent until at least 9–11 weeks old. Age of first breeding not known (see Social Pattern and Behaviour). BREEDING SUCCESS. Of 49 eggs laid, Shetland, 44 (89·8%) hatched, and 23 young (46·9% of eggs laid) fledged. Overall success 2·5 young per pair; in years of food abundance, 5·3 young per pair hatched and 3·1 young fledged ($n = 3$ pairs); in moderate food years, 4·5 young per pair hatched and 3·0 young fledged ($n = 2$); in poor food year, 4·8 young per pair hatched and 1·5 young fledged ($n = 4$). Of 21 young dying, 12 died in first 10 days after hatching, 5 before leaving nest, and 4 afterwards; main cause of death starvation when bad weather interrupted hunting—also disease and damp (M Robinson). In good lemming (*Lemmus*, *Dicrostonyx*) year, Canada, 17 of 19 pairs probably bred, producing 32 eggs, from which 31 young fledged, but such high productivity unusual (Watson 1957, which see for further records of small samples); breeding success governed not only by lemming abundance in July, but by timing of spring thaw in May, permitting feeding and nesting by adults.

Plumages. ADULT MALE. Completely white or cream-white, usually with some small brown spots or small broken bars on sides of crown, longer or outer scapulars, flanks, sides of belly, or upper wing-coverts. Tail fully white or with single dark broken bar subterminally on t1(–t3). Flight-feathers white, tertials and outer primaries with 1(–3) dark broken bars on tips, inner primaries and secondaries fully white or with small dark subterminal spot. Some birds have tips of upper wing-coverts, outer primaries, and tertials well marked, but body fully white; others fully white except for a few small and rounded brown spots on some primaries or t1. Variation in marking apparently independent of age, birds described above being all over 3 years old. Brown marks subject to bleaching, fading to buff in worn summer plumage and then very indistinct. ADULT FEMALE. Crown and sides of head white with rounded dark brown dots. Hindneck, outer webs of scapulars, and rump usually unmarked white. Feathers of mantle, scapulars, back, and upper tail-coverts white, each with dark brown regular half-moon-shaped subterminal crescent (9–11 mm wide in middle), some feathers with straight 2nd dark brown bar nearer base, but this usually hidden under other feathers, upperparts appearing regularly spotted rather than barred; no dark mottling on tips of longer scapulars. Face down to chest and central breast white, sides of breast, belly, and flanks white with regular and straight dark brown bars c. 4–5 mm wide and at least 1 cm apart; vent and under tail-coverts white. Tail white with 3 broken dark brown bars on distal half of t1, reduced outwards to some spots on tip of t4(–t5), t6 fully white. Upper wing-coverts like mantle and scapulars, but 2nd bar may show on some longer coverts; outer webs of greater and median coverts fully or largely white in some birds. Flight-feathers white with 2–4 dark brown bars on distal parts, bars usually broken or reduced to rounded spots, especially those on middle portions of feathers; inner primaries and secondaries sometimes with traces of single subterminal bar only. Under wing-coverts white, axillaries white or with some dark brown subterminal marks. As in ♂, brown marks subject to bleaching—also, white of plumage becomes rather dirty, especially in breeding season. DOWNY YOUNG. 1st down (neoptile) white or slightly cream, rather short, plentiful. From day 5–10, dark grey 2nd down (mesoptile; see below) starts to grow, with neoptile still clinging to tips; during 3rd week, white neoptile still visible, reduced afterwards, but bird appears grey from 8–14 days. JUVENILE (mesoptile). Feathers loose and downy on body and most wing-coverts, especially on head and underparts; structure of flight-feathers, greater primary coverts, and tail normal. All body and all upper wing-coverts (except greater primary coverts) dark grey-brown, feathers tipped and speckled greyish-white; feathers of mantle and scapulars not as loose as remainder of body feathers, indistinctly barred white. On 10th day, face dark grey with white X-mark between eyes and down sides of bill; from 18–20 days, large black surrounds to eyes, with white X-mark more conspicuous; from 30 days, white 1st adult starts to grow round eyes, and black round eyes limited to sides of face from 38–40 days, fully white from c. 50 days. Grey down on body gradually replaced by 1st adult plumage from 28th day, on wing-coverts from 30th day; some grey mesoptile retained for several months on head, neck, sides of chest, and vent (Tulloch 1968; Scherzinger 1974c). Flight-feathers, greater primary coverts, and tail as in adult, but with more numerous and broader dark brown bars, and tips of tertials, greater upper primary coverts, and outer primaries extensively mottled, freckled, or vermiculated white. In ♂, tail with 2–3 rather narrow bars on distal half, broken or reduced to spots, gradually reduced outwards, (t3–)t5–t6 fully white; distal halves of tertials and outer primaries with 3–5 broken bars, some more spots occasionally on outer webs nearer feather-bases; inner primaries and all secondaries white with some spots near tips only, mainly on outer webs; basal halves of tail and flight-feathers white or almost fully so. In ♀, tail and flight-feathers barred to base, though bars usually reduced to spots near coverts; on tail, c. 5 bars on t1 (distal 3 complete), reduced outwards to 2–3 broken bars on distal half of t6; flight-feathers with c. 6–8 bars, these complete in distal part of feather, broken on middle portion, reduced to spots (mainly on outer webs) on basal part. FIRST ADULT MALE. White ground-colour slightly greyish, not as pale cream or ivory as in adult. Extent of unmarked white parts on wing, tail, and body about as in adult ♀, but dark bars markedly narrower. Front and sides of crown spotted dark brown; central crown, hindneck, sides of neck, and chin to central upper breast usually uniform white, but occasionally partly spotted. Small patch of short and soft mesoptile feathers retained on central hindneck, dark grey-brown with narrow white bases, sometimes hidden under longer 1st adult feathers and difficult to detect. Mantle, scapulars, back, upper tail-coverts, and upper wing-coverts white with rather narrow grey-brown crescent-like bars c. 3–5 mm wide and 15–25 mm apart, subterminal bar on each feather not as half-moon-shaped as in adult ♀, margins of crescents rather irregular; lesser upper wing-coverts sometimes with rather large spots; outer webs of outer scapulars and of outer median and greater coverts sometimes uniform white. Breast, flanks, and belly with narrow dark bars 1·5–3 mm wide and c. 10–15 mm apart. Under tail-coverts, under wing-coverts, and axillaries white, under tail-coverts occasionally with a faint and narrow brown subterminal bar. Flight-feathers, tertials, tail, and primary-coverts juvenile (see above for barring); tertials, tertial coverts, greater primary coverts, bastard wing, and longer primaries freckled and mottled dark on tips. Primaries all equally fresh or slightly worn, tips rather pointed (inner web near tip not as broadly rounded as in adult, flight-feathers not a mixture of differently aged feathers). For ageing and sexing of adult and 1st adult, see Josephson (1980). Brown marks strongly subject to bleaching, especially head and body appearing fully white by May–June of 2nd calendar year. FIRST ADULT FEMALE. Rather like adult ♀, but body more extensively and closely barred brown; juvenile flight-feathers and tail retained, barred almost to bases; freckled and mottled tips of tertials, tertial coverts, primary coverts, bastard wing, and longer primaries and retained mesoptile patch on central hindneck as in 1st adult ♂. Crown, hindneck, sides of neck, and chest usually completely barred, only face, throat, cheeks, foot, under wing-coverts, and often patch on nape uniform white. Feathers of mantle and scapulars with subterminal dark bar less regular and smooth-sided than half-moon-shaped mark of adult ♀, other dark bars 6–9 mm wide, closer together than in adult ♀; dark bars on underparts 4–6 mm wide, almost as wide as intervening white. Under tail-coverts with at least 6 bars (adult ♀ unbarred or with less than 6 bars); axillaries partly marked grey-brown. SECOND AND THIRD ADULT (in 2nd and 3rd winter respectively). Within each sex, plumage in 2nd adult rather like juvenile but without dark freckling and mottling on tips of tertials, and dark barring slightly sharper, less irregular; in 3rd adult, pattern of bars on body, tail, and upper wing-coverts intermediate between 1st adult and adult. No patch of mesoptile feathers on central hindneck. Best distinguished by flight-feathers. In 1st summer, tertials and frequently p7 replaced; remaining feathers still juvenile, contrasting in wear, shape, and pattern with new feathers in following autumn, winter, and spring. In 2nd summer, some more juvenile secondaries replaced, as well as juvenile p8, p9, p6, and possibly p5 and p1, again retaining some heavily worn juvenile flight-feathers throughout 3rd autumn, winter, and spring, showing pointed tips and heavier barring than neighbouring feathers. Last juvenile flight-feathers replaced in 3rd summer (4th calendar year) or perhaps occasionally in 4th. Full adults also show mixture of old and new flight-feathers, and though these also often show slight mutual differences in pattern and wear, differences not as marked as in 2nd and 3rd adult.

PLATE 43. *Tyto alba* Barn Owl (p. 432). Nominate *alba*: **1** ad ♂, **2** ad ♀, **3** nestling juv (mesoptile) moulting into 1st ad, **4** nestling juv (mesoptile). *T. a. gracilirostris*: **5** ad ♂. *T. a. erlangeri*: **6** ad. *T. a. guttata*: **7** ad. *T. a. detorta*: **8** ad. (HD)

PLATE 44. *Otus brucei* Striated Scops Owl (p. 450). *O. b. exiguus*: **1** ad. *O. b. obsoletus*: **2** ad.   *Otus scops* Scops Owl (p. 454). Nominate *scops*: **3** brown ad, **4** grey ad, **5** juv, **6** downy young moulting into juv. *O. s. mallorcae*: **7** ad. (HD)

PLATE 45. *Bubo bubo* Eagle Owl (p. 466). Nominate *bubo*: **1** ad, **2** nestling juv (mesoptile) moulting into 1st ad, **3** nestling juvs (mesoptile) at *c.* 35–45 days. *B. b. sibiricus*: **4** ad. *B. b. hispanus*: **5** ad. *B. b. ascalaphus*: **6** ad. *Ketupa zeylonensis semenowi* Brown Fish Owl (p. 481): **7–8** ad. (HD)

PLATE 46. *Nyctea scandiaca* Snowy Owl (p. 485): **1** ad ♂, **2** ad ♀, **3** 1st ad ♂ (1st year), **4** 1st ad ♀ (1st year), **5** nestling juv (mesoptile) moulting into 1st ad ♂, **6** nestling juv (mesoptile), **7** nestling juv (mesoptile) with some 1st down (neoptile) remaining. (HD)

PLATE 47. *Surnia ulula ulula* Hawk Owl (p. 496): **1** ad, **2** juv, **3** unfledged juv at *c*. 4 weeks. *Glaucidium passerinum* Pygmy Owl (p. 505): **4–5** ad, **6** juv, **7** unfledged juv. (HD)

PLATE 48. *Athene noctua* Little Owl (p. 514): *A. n. vidalii*: **1** ad (England), **2** ad (Switzerland, intergrade with nominate *noctua*), **3** juv, **4** unfledged juv. *A. n. lilith*: **5** ad. *A. n. glaux*: **6** ad. *A. n. indigena*: **7** ad. *Aegolius funereus funereus* Tengmalm's Owl (p. 606): **8** ad, **9** juv moulting into 1st ad (face and wing-coverts), **10** unfledged juv, **11** nestling juv (mesoptile). (HD)

PLATE 49. *Strix aluco* Tawny Owl (p. 526). Nominate *aluco/sylvatica*: **1** pale grey ad (Fenno-Scandia and eastern Europe only), **2** intermediate ad, **3** deep rufous ad, **4** fledged juv at *c.* 6 weeks, **5** grey nestling juv (mesoptile) with 1st down (neoptile) remaining, **6** rufous nestling juv. *S. a. mauritanica*: **7** ad. *Strix butleri* Hume's Tawny Owl (p. 547): **8** ad. (HD)

PLATE 50. *Strix uralensis liturata* Ural Owl (p. 550): **1–2** ad, **3** fledged juv at *c.* 7 weeks, **4** unfledged juv at *c.* 3½ weeks. *Strix nebulosa lapponica* Great Grey Owl (p. 561): **5–6** ad, **7** unfledged juv, **8** nestling juv (mesoptile). (HD)

Bare parts. ADULT AND FIRST ADULT. Iris lemon-yellow, golden-yellow, or orange-yellow. Eyelids black. Bill and claws blackish-horn, brown-black, or black; tips of claws paler horn. DOWNY YOUNG AND JUVENILE. Iris bluish-white, cold pale grey, light grey, pale yellow-grey, or greenish-yellow when eye starts to open at 5–7 days, golden-yellow from 4–6 weeks. Eyelids bluish-black. Bare skin of face blackish. Bill bluish-black, paler blue-grey on cere and culmen. Gape and mouth flesh-pink. Skin (including that of leg and foot) pink at hatching, greyish-black later on. Claws white at hatching, soon darkening to blue-grey and black. (Witherby *et al.* 1938; Dementiev and Gladkov 1951a; Tulloch 1968; Portenko 1972; Scherzinger 1974c; Wotton 1976; RMNH, ZMA.)

Moults. ADULT POST-BREEDING. Flight-feathers moulted serially; primaries descendantly and ascendantly from centre at p7, later on followed descendantly from centre on p1, as in Eagle Owl *Bubo bubo*. One set formed by p7 to p10, one by p6 to p4, one by p1 to p3; in each set, only 1–2 feathers replaced each year, and moult suspended with 2–6 primaries new in each wing at end of moulting season; in next season, moult continued but usually not completed; in 3rd season continued but usually also starting again with p7. Moult occurs late June to early September (occasionally early October) in breeders, from early June in non-breeders; in this period, body, wing-coverts, and tail completely replaced or almost so (some scapulars, wing-coverts, and tail-feathers frequently retained to next season); head moulted last, August–September (Watson 1957; Portenko 1972). Moult of secondaries not fully ascertained; starts descendantly with centre on innermost, but not known whether other moult centres become active later on as in *B. bubo* and not known how many years needed for complete replacement. In captive ♀, starts with some feathers of breast at laying of 1st egg, soon followed by some primaries; moult suspended when eggs hatch, resumed when young independent; in this 2nd period, also feathers of foot and tail replaced (latter centrifugally); head moulted last; in captive ♂, moult of body feathers starts when feeding young, flight-feathers replaced when young able to fly; moult mainly mid- and late August, extending to mid-September (Scherzinger 1974c). POST-JUVENILE. Complete, except for patch on central hindneck, tail, flight-feathers, and greater primary coverts; often also some juvenile longer scapulars, tertial coverts, or other wing- or tail-coverts retained during 1st winter. Starts on face when 1 month old, followed by head, body, and wing-coverts, and completed by October(–November). FIRST IMMATURE POST-BREEDING. Partial, in summer of 2nd calendar year, starting at about same time as non-nesting adult, completing up to October–November. Involves head, neck, and all or nearly all body, tail, wing-coverts, and tertials (about s14–s18), usually p7 and perhaps occasionally p6–p8 and some other secondaries. Of 5 2nd-winter birds examined, all had replaced (s12–)s14–s18 by 2nd adult, but no other secondaries; 3 had p7 new, 2 retained all juvenile primaries. Captive ♂ replaced p6–p9 (in sequence 6–7–8–9) in 2nd year, remainder (in sequence 10–5–1–4–3–2) in 3rd; tail-feathers shed within 13–26 days, sequence t1 to t6, growing largely simultaneously (Haarhaus 1983). No further information for subsequent moults (but see Plumages); presumably as in *B. bubo*.

Measurements. Northern Europe and Greenland, all year; skins (also some skeletons for legs) (RMNH, ZMA). Bill (F) to forehead (to feathers, on average 6 mm less), bill (C) from tip to cere.

WING	♂ 409	(6·73; 13)	400–421	♀ 448	(8·85; 12)	434–460
TAIL	212	(4·85; 12)	206–222	231	(8·59; 11)	217–241
BILL (F)	42·5	(1·87; 12)	40–45	46·1	(1·30; 9)	44–48
BILL (C)	26·0	(1·69; 12)	24–28	27·5	(1·06; 10)	26–29
TARSUS	52·3	(2·24; 11)	48–55	56·6	(3·27; 10)	53–62

Sex differences significant.

Wing of larger samples from Portenko (1972, 1973): all geographical range, (1) adult, (2) juvenile; ages combined, (3) Europe, (4) western and central Siberia, (5) eastern Siberia, North America, and Greenland.

(1)	♂ 412	(— ; 31)	395–429	♀ 445	(— ; 32)	430–470
(2)	414	(— ; 37)	395–430	449	(— ; 29)	432–471
(3)	415	(8·93; 28)	395–430	445	(9·83; 19)	0–460
(4)	414	(7·66; 32)	395–430	448	(9·35; 30)	430–471
(5)	405	(6·27; 8)	396–415	446	(8·65; 13)	435–462

(5) significantly shorter than (3) and (4) for ♂♂, but not for ♀♀.

Weights. ADULT AND FIRST ADULT. Shetland, June: ♂ 1752 (BTO). Norway, December–April: ♂ 1462 (3) 1430–1515, ♀ (very fat) 2409 (4) 2290–2580 (Hagen 1942). Several sources of data combined: ♂ 1730 (36) 710–2500, ♀ 2120 (23) 780–2950; birds with lowest weights starving, weight of ♂ normally not below *c.* 1200, ♀ not below *c.* 1500 (Watson 1957; Portenko 1972). Some (presumably North American) sources combined: ♂ 1642 (27) 1320–2013, ♀ 1963 (30) 1550–2690 (Earhart and Johnson 1970). Some USSR sources, combined: ♂ 1491 (11) 873–2000, ♀ 1961 (17) 1262–2500 (Dementiev and Gladkov 1951a; Uspenski and Priklonski 1961; Portenko 1972). ♀ in heavy moult, late August, Canada: 2025 (Parmelee *et al.* 1967). DOWNY YOUNG AND JUVENILE. On 1st day 44·9 (6·08; 12) 30–55; occasionally some decrease to 20 in first few days; on 4th–5th day, average 118 (5); on 10th day, *c.* 360; on 20th, *c.* 930, with some over 1000; on 28th day, 1300–1600 (Watson 1957; Uspenski and Priklonski 1961; Parmelee *et al* 1967; Portenko 1972). For growth curves, see Watson (1957).

Structure. Wing long and broad, tip rounded. 10 primaries: p8 longest, p9 1–10 shorter, p10 34–48, p7 4–10, p6 35–52, p5 64–82, p4 90–102, p1 142–162. Outer web of (p6–)p7–p9 and inner web of (p6–)p7–p10 deeply emarginated. Outer web of p10 and emarginated portions of other outer primaries slightly serrated. Tail rather short, tip rounded, 12 feathers; t6 25 (9) 18–32 shorter than t1 in ♂, 30 (9) 23–37 in ♀. Facial disc poorly developed, only feather-ridge from base of upper mandible to above eye distinct, ending in rudimentary ear-tufts consisting of *c.* 10 feathers 20–25 mm long. Bill short, strong, laterally compressed; culmen strongly curved. Basal sides of bill with stiff bristle-like feathers, projecting forward at rest and almost concealing bill except for extreme tip. Tibia long, tarsus very short; tibia, tarsus, and toes thickly covered with wool-like feathers, even claws sometimes largely hidden; length and density of feathers dependent on abrasion. Middle toe without claw 39·0 (2·12; 12) 36–43 in ♂, 42·1 (2·42; 10) 39–47 in ♀; outer toe *c.* 67% of middle, inner *c.* 81%, hind *c.* 60%. Middle claw in ♂ 27·3 (1·55; 11) 25–29, in ♀ 31·6 (1·58; 10) 30–34; outer claw *c.* 89% of middle, inner *c.* 105%, hind *c.* 87%.

CSR

Surnia ulula Hawk Owl

PLATES 47 and 53
[between pages 494 and 495, and facing page 543]

Du. Sperweruil Fr. Chouette épervière Ge. Sperbereule
Ru. Ястребиная сова Sp. Lechuza gavilana Sw. Hökuggla N. Am. Northern Hawk-Owl

Strix ulula Linnaeus, 1758

Polytypic. Nominate *ulula* (Linnaeus, 1758), northern Eurasia, in central Siberia south to Tarbagatay; *caparoch* (P L Statius Müller, 1776), North America, straggling to Britain and Canary Islands. Extralimital: *tianschanica* Smallbones, 1906, Tien Shan and Dzhungarskiy Alatau mountains in Soviet central Asia and bordering Sinkiang (China).

Field characters. 36–39 cm; wing-span 74–81 cm. Head and body smaller and wings shorter than those of *Asio* owls but tail proportionately much longer (making *c*. 40% of overall length); overall size and structure comparable to ♀ Sparrowhawk *Accipiter nisus*. Distinctive, medium-sized owl, with hawk-like silhouette, and partly diurnal behaviour. Plumage strongly patterned: dark brown above, with pale hindneck and scapulars; almost white, with black barring below. Almost white face strongly outlined black, giving fierce expression. Habitually perches conspicuously and flies low over trees. Flight action swifter and more agile than any other west Palearctic owl. Sexes similar; no seasonal variation. Juvenile separable. 2 races occur in west Palearctic; distinguishable at close range.

Adult. (1) North Eurasian race, nominate *ulula*. Crown black-brown, thickly spotted white; nape black-brown, with a few white spots; broad hindneck grey-white, with lateral dusky patches opposing lower nape. Back mostly dark brown, with white spots particularly predominant on outer scapulars (forming obvious pale panels over folded wing). Rump and tail black-brown but with dull white bars (most obvious on spread tail). Face and sides of head strongly patterned, with white margin to crown, and almost black line running from bill through and behind eye, and then dividing to create black band round side of grey-white facial disc, and long almost black smudge down side of nape. Black band round facial disc may join with dull black-brown bars across upper breast and almost black 'chin' to complete bold dark outline to face and to contrast with only faintly barred white breast. Lower underparts and undertail barred black-brown on white. Wing-coverts dark brown, spotted white (in transverse lines) on outer feathers. Flight-feathers dark brown, spotted and barred white (most noticeably on primaries). Underwing white, narrowly barred brown on coverts and more broadly barred dusky on flight-feathers. Bill yellow-horn. Eyes yellow; legs and toes fully feathered, white. (2) Nearctic race, *caparoch*. Usually, blacker above than nominate *ulula*, with spotting on crown less obvious but rest of white marks accentuated. Best distinguished by broader, tawnier bars on underparts, particularly over flanks and lower belly.

Length of tail, and basically black, dark brown, and white plumage prevent confusion with other owls. More likely to be mistaken for small hawk, both in flight and when prominently perched; at closer range, however, broad head, rather thickset body, and plumage pattern of perched bird rule out confusion. Note particularly that while wing-tip more pointed than in any other owl, whole wing still relatively short; also, long tail tapers, often narrowing to wedge shape with rounded end. Flight recalls that of ♀ *A. nisus*, being straight, direct, and unhurried; action normally combines bursts of wing-beats with glides but is varied to include markedly propulsive, bounding progress (like Little Owl *Athene noctua*) and marked agility when mobbing larger raptors or hunting small birds; can hover. Uses rather low approach to perch, sweeping abruptly up to pitch on topmost branch of tree. Stance on landing rather flat, with characteristic sudden upward jerk and then noticeably slower lowering of tail. Normal posture (e.g. while watching for prey) quite upright, with hawk-like silhouette. Gait restricted, with waddling walk and bouncing lope. Fearless, attacking intruders near nest.

Vagrants usually appear in wooded farmland or coastal localities. Song of ♂ a prolonged, rapid trilling 'prullul-lullu...' with bubbling quality; also utters angry scream and harsh chattering 'ke(ki)-ke-ke-ki...' of alarm, recalling Merlin *Falco columbarius*.

Habitat. Arboreal, northern limits closely paralleling southern limits of Snowy Owl *Nyctea scandiaca*, occupying fringes of forest tundra and boreal taiga as far as treeline, and ranging south to edge of forest steppe and cultivated lands. Also in some mountain forests further south, as high as timberline to 1800–2000 m in Altay (Dementiev and Gladkov 1951a). Seeks ready access to clearings, burnt areas, open peatlands (in North America, muskegs), dry eminences or ridges and sparse woodland, including birch, aspen, and mixed woods, with some preference for pine *Pinus*, larch *Larix* and stunted krummholz trees, especially those terminating in broken-topped stumps or bare branches suitable as look-outs for hunting (Bannerman 1955). Avoids dense coniferous forest. Sometimes breeds by farmed smallholdings within forest, but in Europe infrequently ranges over more extensive cultivated lands. In North America, however, in winter inhabits prairies, even perching on haystacks, and also open heathland (Bent

Surnia ulula 497

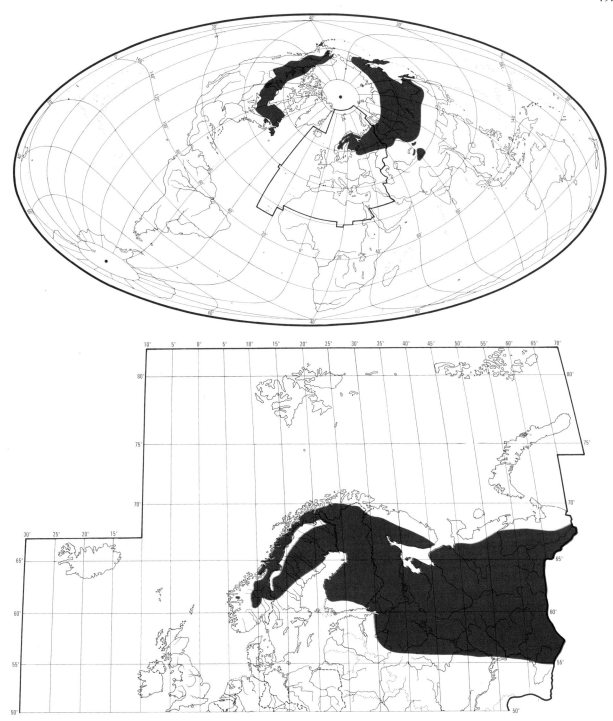

1938). Flies strongly but usually in lower airspace. Habitat occupancy much influenced by state of prey populations, irrespective of all other factors.

Distribution. Some periodical variations in breeding range with population fluctuations. For winter eruptions, see Movements.

SWEDEN. Bred Småland 1972, well south of main range (LR). FINLAND. Breeds further south when numbers high (AJL). NORWAY. Sporadic breeding further south in good rodent years (Haftorn 1971). USSR. Estonia: bred on 4 occasions 1936–51 (Randla 1976).

Accidental. Britain, France, Belgium, Luxembourg, Netherlands, West Germany, Denmark, East Germany,

Poland, Czechoslovakia, Hungary, Austria, Switzerland, Yugoslavia, Rumania, Canary Islands.

Population. Numbers fluctuate markedly with abundance of small rodents.

SWEDEN. Up to 10000 pairs in good rodent years (Ulfstrand and Högstedt 1976). FINLAND. Declined since 19th century, due to disappearance of hollow trees and also human persecution. Numbers fluctuate enormously with rodent numbers; tentative estimate *c.* 3600 pairs (Merikallio 1958). Marked fluctuations in numbers, from almost absent to fairly common (OH). USSR. Common in northern taiga, numbers probably fluctuating with rodent populations (Dementiev and Gladkov 1951a).

Movements. Dispersive and eruptive. Like Great Grey Owl *Strix nebulosa*, leads an essentially nomadic life, dispersing extensively within coniferous (taiga) zone in response to regional variations in food availability. Population fluctuations follow closely cycles of rodent prey. When voles (*Microtus, Clethrionomys*) are at normal levels the birds breed and winter well north, while vole abundances can lead to large though temporary southward extensions to breeding range (see Distribution); but vole populations crash at intervals of 3–5 (usually 4) years, and eruptions of *S. ulula* may then occur (e.g. Mikkola and Sulkava 1969b, Glutz and Bauer 1980, Mikkola 1983). However, vole cycles vary geographically (Hagen 1956); hence owl eruptions less predictable than vole crashes, and only extensive when prey density falls over very large areas. Only large irruptions involve all 3 Fenno-Scandian countries simultaneously.

Most recent European eruptions (in 1956–8, 1961–2, 1964–5, 1971–2) have been on reduced scale compared with some earlier ones (this century: notably 1914–15, 1928–9, 1930–1, 1931–2, 1942–3, 1950–1), possibly because of reduced population size since early 1950s (Mikkola 1983), but further large eruption occurred 1976–7 and 1983–4. Erupting birds only infrequently cross Baltic, apparently a fairly effective water barrier, though during larger invasions vagrants have reached Britain, Denmark, and continental countries south to France, Yugoslavia, and Rumania. Also, an irregularity occurs in southern winter limits in central USSR, e.g. not annual (though sometimes numerous) in Leningrad, Pskov, and Kuybyshev (Dementiev and Gladkov 1951a).

1950–1 irruption in Scandinavia studied by Edberg (1955) and Hagen (1956). South of breeding range, arrivals began early September and numbers peaked during October–November, then fell progressively December–March. Hagen (1956) suggested birds originated northern Fenno-Scandia, though Edberg considered origin in USSR more likely. Latter applied in 1957–8. 1957 was a good vole year in northern USSR and *S. ulula* nested in large numbers (Bianki and Koshkina 1960), but only 3 nests found in Finland that year; hence big autumn 1957 influx into Finland, and stragglers south-west to Germany, almost certainly had eastern origin. Numbers remained to breed in Finland in 1958, when spring very cold and voles were numerous (Mikkola 1983). Large irruption in autumn 1976 was most marked in Finland, and in clear association with other eastern species (Hildén 1977).

Scandinavian population likewise dispersive, and certainly involved in eruptions. 1961 noted as eruption year in Sweden, and 6 nestlings ringed there that year recovered in Finland (Oulu, 10 October) and USSR (Murmansk, Arkhangel'sk, Leningrad, Yaroslavl, and Perm, October 1961–April 1962); hence movements north-east to south-east and distances up to 1860 km (Österlof 1969). Other Swedish nestlings have shown comparable movements: Norrbotten 1946, found Murmansk in November (no other details); Västerbotten 1948, found Nordland, Norway (September 1948); 4 ringed Dalarna 1963, found Swedish Lapland (December 1963), Norway (September 1964, November 1964), and Murmansk (January 1965); Norrbotten 1967 (another eruption year), found Vologda, USSR (March 1968); Småland 1972, found Murmansk (October 1974). One Finnish nestling (Turku ja Pori, 1962) found Oppland, Norway (autumn 1962).

Of 80 irrupting birds handled in Sweden, 1950, 85% juveniles, 15% adults; neither age-class showed any differences in geographical or temporal distribution, and all seemed in good nutritional condition (Edberg 1955). 52 live captures in Finland in 1976 invasion were all juveniles, and 150 Finnish museum skins (various dates) from south of breeding range comprised 88% juveniles, 12% adults; most 'adults' considered 2nd-year birds (Forsman 1980).

Comparable eruptive movements occur in Alaska and Canada, in which birds reach northern USA (e.g. Godfrey 1966, Green and Janssen 1975). 4 birds attributed to Nearctic race *caparoch* collected in Europe: western Britain (August, December, March) and Canary Islands (autumn).

Food. In breeding season, almost wholly small voles with a few small birds and larger mammals; perhaps more birds in winter. Diurnal, and will hunt in bright sunlight; at least when young in nest, will also hunt at night (see final paragraph)—though not then necessarily dark. Little information on hunting methods. Mainly hunts from exposed perch, changing frequently between several habitually-used ones; pounces on prey or skims down over ground. Hovers frequently and sometimes quite persistently; occasionally seizes prey in flight. (Dementiev and Gladkov 1951a; Meinertzhagen 1959; Glutz and Bauer 1980.) Recorded plunging into soft snow, North America, apparently hunting by sound, but habit evidently rare (Nero 1980). Recorded hunting using searching flight at height of 2–4 m (Sellin 1965). Cannibalism (adults eating own young) occurs (Mikkola 1972c).

Pellets average $41 \times 22 \times 19$ mm ($30–76 \times 17–35 \times 13–23$), with up to 4 prey per pellet, average 1·7, $n = 40$ (Mikkola 1983).

Table A Food of Hawk Owl *Surnia ulula* as found from pellets, nest-remains, and stomach analyses. Figures are percentages of live weight of vertebrates only.

	Breeding season			Outside breeding season
	Central Norway 1949	All areas of of Finland 1964–70	Murmansk (USSR) 1957	Finland and Murmansk 1912–71
Mice (small Murinae)	0	0.3	0	0
Small voles (Microtinae)[1]	96.3	88.8	75.6	5.1
Larger mammals[2]	1.4	3.5	13.8	3.2
Shrews (Soricidae)	0.1	0.7	0	1.6
Birds	2.3	6.0	10.1	90.2[3]
Frogs, fish	0	0.7	0.5	0
Total no. of vertebrates	400	563	174	21
Source	Hagen and Barth (1950), data reanalysed	Mikkola (1970–1)	Bianki and Koshkina (1960), data reanalysed	Mikkola (1972c), data reanalysed

1. Largely *Clethrionomys* and *Microtus*; lemmings (*Lemmus*, *Myopus*) occurred only in breeding season in Norway (14.6% by weight of total vertebrates) and Finland (2.8%).
2. Largely water vole *Arvicola terrestris* and Mustelidae.
3. Composed of 5 Willow Grouse *Lagopus lagopus*, 1 Hazel Grouse *Bonasa bonasia*, and 1 Redpoll *Carduelis flammea*.

The following prey recorded in west Palearctic. Mammals: shrews (*Sorex minutus*, *S. araneus*, *Neomys fodiens*), wood lemming *Myopus schisticolor*, Norway lemming *Lemmus lemmus*, red-backed vole *Clethrionomys rutilus*, bank vole *C. glareolus*, grey-sided vole *C. rufocanus*, water vole *Arvicola terrestris*, common vole *Microtus arvalis*, short-tailed vole *M. agrestis*, root vole *M. oeconomus*, harvest mouse *Micromys minutus*, yellow-necked mouse *Apodemus flavicollis*, weasel *Mustela nivalis*. Birds mostly small passerines up to size of thrushes *Turdus*, but recorded up to size of (apparently full-grown) Willow Grouse *Lagopus lagopus*. Other prey apparently unusual: frogs *Rana*, small fish, and beetles *Carabus*. (Mikkola 1972c.)

For summary of major west Palearctic studies, see Table A and review by Mikkola (1972c). See also Hagen (1952) for Norway, Mikkola (1971a), Hublin and Mikkola (1977), and Pulliainen (1978) for Finland, Vladimirskaya (1948) and Parovshchikov and Sevastyanov (1960) for USSR, and Uttendörfer (1952). Breeding-season diet essentially similar in all areas studied (e.g. Mikkola 1970–1) unless prey availability restricted: thus on island in Baltic Sea (Finland), where *Arvicola* the only suitable species present, it comprised 99.4% by weight of 41 items (Pulliainen 1978). Lemming numbers low in diet, Lappland, despite high populations during study periods, so evidently not important food source (Mikkola 1972c). In Murmansk (Table A), shrews known (through trapping) to be present in study area but not present in pellets (Bianki and Koshkina 1960).

No information on year-to-year variations in diet and little on seasonal changes: small sample size of non-breeding-season data (Table A) makes conclusions on seasonal changes impossible, but increase in bird prey possibly real as data collected in periods when voles less available through low numbers and heavy snow cover (Mikkola 1972c). Vladimirskaya (1948) and Uttendörfer (1952) also reported more birds eaten in winter.

No specific data on food of young, though some studies detailed above include items brought to nest. In Sweden, pair fed young once per hr during bright day; after 17.00 hrs, feeding less frequent but visits recorded in daylight 02.00–03.00 hrs (Mikkola 1972c). In Finland, fed usually 04.00–19.00 hrs; 0.8–2.5 feeds per hr, with no obvious peaks (Leinonen 1978); brood of 1 fed every 30–45 min at night, every 1 hr 35 min to 3 hrs by day (Simon and Simon 1980).   DJB

Social pattern and behaviour. Includes some information from North American race, *caparoch*.
1. Little studied outside breeding season. After young leave nest, family may remain in nest-territory for 6–8 weeks (Mikkola 1972c; see also Pukinski 1977 and Bonds, below). BONDS. Probably monogamous mating system (Glutz and Bauer 1980). In captivity, ♂ mated with 2 ♀♀, and both laid fertile eggs (Böhm 1982a). No hard evidence on duration of pair-bond or age of first breeding, though probably breeds at 1 year old (Hagen 1956). As in other Strigidae, ♀ takes greater share of nest-duties, performing all incubation, and all brooding until young well-grown (see Heterosexual Behaviour and Relations within Family Group, below). ♂ sometimes said to incubate a little (Dementiev and Gladkov 1951a; Parslow 1969), but elsewhere said rarely to enter nest after start of incubation by ♀, and never to incubate (Leinonen 1978; see also Glutz and Bauer 1980). While ♀ incubating and brooding, ♂ brings food to nest and helps to feed young. On leaving nest, both parents care for young (Leinonen 1978). Young independent at c. 2½ months (Hagen 1956; Mikkola 1972c). BREEDING DISPERSION. Little studied. Solitary and territorial. In Norway, 4 pairs in 200 km² (Hagen 1956). In good habitat, Sweden, 0.2 pairs per 100 km² (Enemar et al. 1965; Ulfstrand 1965). Nest-territory thought to be quite large (Mikkola 1972c), though at nest with single young, parents hunted entirely within 100 m of nest (Simon and Simon 1980); serves for courtship, copulation, and provides food for self-maintenance and raising young (see above). ♂ establishes territory and attracts

A

♀ to nest-site (see Heterosexual Behaviour, below). No information on site-fidelity, though nomadic habits render fidelity unlikely. ROOSTING. Mostly at night (activity markedly diurnal and crepuscular). Typically on branches of trees, not in cavities (Dementiev and Gladkov 1951a; Pukinski 1977). While ♀ incubating, ♂ stayed in immediate vicinity of nest (Leinonen 1978). By day, typically rests perched on exposed tree-top, in inclined posture like Sparrowhawk *Accipiter nisus* (Holgersen 1951), sometimes preening (Köpke 1969). Captive birds fond of bathing (Böhm 1982a). For snow-bathing, see Cade (1952). See also Food.

2. When alert or excited, perched bird typically raises and slowly lowers tail, and looks around actively, turning head jerkily through 180° (less commonly c. 90°), calling (Smith 1970; Glutz and Bauer 1980: see 4 in Voice). Also said to bob head (Fridzén 1959). When aware of aerial predator or distant ground predator, adopts concealing Sleeked-upright posture (Fig A): flattens plumage, stands erect, and stares at source of alarm (Hagen and Barth 1950; Yunick 1965). When mobbed by passerines, adopts Sleeked-upright posture but with eyes reduced to slits, and may extend neck forwards, while giving Screeching-call (see 3a in Voice); on 2 occasions, Sleeked-upright posture seemed to appease mobbing bird which flew on (Köpke 1969). When fiercely mobbed, ♂ (which had just killed young of mobbing parents) was more aggressive, giving Screeching-calls with quivering red tongue conspicuous (Smith 1922). Bill-snapping, as defensive-threat reaction, apparently rare (see Glutz and Bauer 1980). For bird fleeing into shelter of a thick spruce *Picea* on sighting Goshawk *Accipiter gentilis* in territory, see Mikkola (1976a). For Head-cocking (Fig B), presumably to aid location of prey, see Smith (1922), and especially drawings in Butsch (1963). ANTAGONISTIC BEHAVIOUR. ♂ establishes territory a few weeks before start of nesting, proclaiming ownership with Advertising-call (see 1a in Voice), given mostly during Display-flight over tree-tops in territory (Parslow 1969; see also Pukinski 1977). Wing-clapping, as bird flies among trees, also reported (Parslow 1969). Advertising-call also given while perched, with head raised slightly, drawing attention to white, black-bordered throat (Glutz and Bauer 1980). ♂'s body and (especially) tail vibrate, apparently with effort of calling (Leinonen 1978). In 1 year, central Finland, ♂♂ called 17 February–13 April (Mikkola 1972c); in USSR, from mid-March (Dementiev and Gladkov 1951a). No information on direct confrontations in the wild. In captivity, both ♂ and ♀ reacted to calls or sight of rivals with Screeching-calls; rivals confronted each other, sleeking face plumage and staring. Advertising-call of ♂ usually first elicited Screeching-calls from ♀, sometimes more aggressive Forward-threat posture: belly feathers markedly ruffled, tail fanned, and wings held far back and extended such that undersides turned towards ♂. Once, ♀ attacked ♂, pushing against him with breast (Glutz and Bauer 1980). HETEROSEXUAL BEHAVIOUR. Based largely on captive studies. (1) Pair-bonding behaviour. ♂ ready to pair gives Advertising-call (see above). ♀ responds with her Advertising-call (Pukinski 1977; Glutz and Bauer 1980: see 1b in Voice). Once paired, birds perform antiphonal duet of Trilling-calls (see 2 in Voice) at almost every meeting. Sometimes bow to each other so that foreheads touch, while giving Trilling-calls or variant; Billing also reported (Glutz and Bauer 1980). (2) Nest-site selection. ♂ performs Nest-showing, as in other Strigidae. ♂ sits below prospective site or on nearby branch, and attracts ♀ with Advertising-call. Pair inspect site together, alternating Trilling-calls with gnawing and scraping movements. While ♀ inspecting, ♂ may stare fixedly at site and give Advertising-call. Before entering site, ♀ may also direct Lure-call (see 5 in Voice) at ♂. ♂ and ♀ thus share initiative in Nest-showing, and both perform Scraping, etc. Final choice, however, probably by ♀. Nest-site inspection observed around midday, also at dusk (Glutz and Bauer 1980). (3) Mating. Usually preceded by loud duet. Mating may be initiated by ♂ or ♀. ♀ gives vigorous bout of Soliciting-calls (see 6 in Voice) in a horizontal posture, with tail cocked and wings drooped. Soliciting ♂ nudges ♀'s flank while fluttering his wings and giving Trilling-call. During copulation, Trilling- and other calls exchanged between partners. After mating, ♂ remains in rather stiff posture before flying off silently. ♀ also remains stiffly for up to ½ min and makes slow pumping movements with tail. Copulation observed around midday, also at dusk (Glutz and Bauer 1980). (4) Courtship-feeding and food-caching. ♂ begins to feed ♀ towards end of courtship, around same time as copulation begins. Provisions ♀ both at nest and away from it. ♀ may summon ♂ with Soliciting-call, given with wide circling movements of head. Food-bearing ♂ announced his arrival with Trilling-calls (Glutz and Bauer 1980). When ♀ fed off nest, she typically reciprocated Trilling-call as she landed on branch near ♂, ruffling plumage and drooping wings (Smith 1922)—presumably, as before copulation, a soliciting posture. While mouse (Murinae) pinned beneath ♂'s talons, pair tore bits off with their bills. Pair then flew off, and ♀ continued to call and hustle ♂ as he eviscerated prey and cached it in fork of tree (Smith 1922). At Fairbanks (Alaska), food-caching frequently related to courtship-feeding. ♂ brought prey to perch, fed on and/or eviscerated prey, and summoned ♀ with Trilling-call. If ♀ responded with Trilling-call, ♂ transferred prey to her, either on nest or on perch to which ♀ had flown. ♀ not uncommonly ate only part of prey and cached the rest. If ♀ did not answer ♂'s Trilling-call, ♂ usually cached prey. Once, ♀ gave Trilling-call when ♂ active near nest; ♂ did not respond, and ♀ then flew to tree and retrieved prey (Ritchie 1980, which see for discussion of function of food-caching). (5) Behaviour at nest. ♀ stays in nest, where fed by ♂, several days before egg-

B

laying. ♂ feeds her from edge of nest-cup or at hole entrance; may enter for a few hours before laying, but only rarely after start of incubation (Glutz and Bauer 1980). RELATIONS WITHIN FAMILY GROUP. Young start calling from within egg (Böhm 1982b). In 8 nests, ♀ stayed with and brooded young until oldest 13–18 days old (Leinonen 1978). When feeding young at nest, parent spread and flapped wings for several minutes (Smith 1970). At nest, parent disembowelled mammalian prey before offering it. Begging behaviour of young seemed to regulate rate at which they were fed. Often, young gave food-call (see Voice) to which, each time, parent appeared to respond with delivery of food after c. ½ hr. Young only fed if they responded thus to call (probably Trilling-call) of incoming parent. Food-cache apparently important in regulating chick-feeding rate (Simon and Simon 1980). ♂ once cached food after young failed to beg (Ritchie 1980). For call of sated young (Böhm 1982b), see Voice. When food in short supply, adults may eat own small young (Hagen 1969; Mikkola 1972c). At nest, northern USSR, youngest chick left 1 week after oldest (Bianki and Koshkina 1960). On leaving nest, young initially unable to fly, and spend much time clambering around branches of nest-tree (Mikkola 1972c; Leinonen 1978). Sometimes fall to ground, but can jump and flutter up to low branches again (Leinonen 1978). Stayed by nest for 10 days, until able to fly short distances, one achieving this at 32 days, shortly after leaving nest (Leinonen 1978). At 40 days, able to fly 20–30 m (Pukinski 1977). ANTI-PREDATOR RESPONSES OF YOUNG. On hearing Yelping-call (see 4 in Voice) of parents, young at nest initially replied, but fell silent when nest-tree climbed (Smith 1970). Thus warned by parents, fledged young 'froze' (Pukinski 1977). When approached by human, young adopted an upright posture, and remained motionless (Leinonen 1978). On close approach, fledglings ceased giving food-call, became quiet, and stood erect, with feathers loose, and folded wings slightly out from body ('fear-posture'); able to remain thus for up to ½ hr (Hagen and Barth 1950). Young perched on ground made hissing sound (see Voice) when approached, but did not move away; when picked up, began to 'screech' (Smith 1970). PARENTAL ANTI-PREDATOR STRATEGIES. (1) Passive measures. No details. (2) Active measures: against man. Notoriously fearless and aggressive when human intruder near nest. Perched birds allow themselves to be closely approached, and often mount fierce flying attacks, especially after hatching (Mikkola 1972c), and—by both members of pair—after first young leaves nest; aggression increases when last young fledged (Leinonen 1978). Clear division of roles said to exist between sexes; one bird of pair, perhaps ♂, always adopted sentinel role near nest, likewise after young had left, immediately starting to give Yelping-call on approach of intruder; bird constantly shifted position, gliding steeply from tree-top and swooping up to new perch. Behaved thus for as long as disturbance lasted. Rarely attacked, but, when danger imminent, glided to ground to perform distraction-display (see below). Meanwhile, other bird, perhaps ♀, typically attacks, trying to fly unseen at intruder (Hagen and Barth 1950). When observer came near perched ♂, bird thrust head forward, gaped wide (displaying quivering red tongue), and gave Screeching-calls (Smith 1922). When intruder climbed nest-tree, bird flew around, periodically perching, within 30 m of nest for 70 min, giving Yelping- and Screeching-calls, but did not land at nest itself (Smith 1970). Aerial attack especially likely if nest-tree closely approached or climbed: pair fly repeatedly at head of intruder, raking with talons (often drawing blood), and giving Yelping-calls (e.g. Fridzén 1959, Simon and Simon 1980, Böhm 1982a). After all young had left nest, but rarely before, ♀ sometimes performed distraction-lure display of disablement type when intruder

C

approached: bird flew off in apparent difficulty giving 'sree' sound (see 7b in Voice), landed on ground or stump, drooped wings (Fig C), and continued along ground, fluttering and calling. Display continues for several minutes, and may be repeated (Leinonen 1978). When young newly fledged from nest, central Sweden, agitated parents behaved much as above; one jumped away along ground, the other uttered hoarse, moaning sounds (presumably 7b in Voice) and feigned injury, dragging its wings (Fridzén 1959). Distraction-display resorted to by sentinel bird (see above) when danger to young imminent, or by attacking bird after prolonged (usually 30–45 min) disturbance: glided to ground, 'waved' wings, and uttered 'piping' sounds (Hagen and Barth 1950).

(Fig A from photograph in Hagen and Barth 1950; Fig B from drawing in Butsch 1963; Fig C from photograph in Mikkola 1983.)   EKD

Voice. Freely used during breeding season, when repertoire diverse; much less vocal at other times, though see call 3b. For Wing-clapping and Bill-snapping, see Social Pattern and Behaviour. Adults, unlike many Strigidae, apparently make no hissing sounds in threat. For additional sonagrams, see Bergmann and Helb (1982).

CALLS OF ADULTS. (1) Advertising-call. (a) ♂'s call. A sonorous, trilling, vibrant and rolling 'hu hu hu uuuuuuu' or a clearer 'hu hu hu üüüüüüü', said to last 2–3 s (Glutz and Bauer 1980); 10–14 s (Leinonen 1978); comprises c. 60 syllables (H Delin). Also rendered 'ulyu lyu lyu lyu lyu lyu' (Pukinski 1977). In recording (Fig I), a bubbling rising ripple, ending in short, sharp sound of higher pitch; comprises 95 units, c. 12 per s (J Hall-Craggs, P J Sellar); each call c. 8–9 s, with similar gap between them. Also given at intervals of c. 2 s, notably during Display-flight (Parslow 1969; Pukinski 1977); also when perched, e.g. when Nest-showing (Glutz and Bauer 1980). Call also used to summon incubating ♀ (Leinonen 1978). (b) ♀'s call. Compared with call 1a, less constant in rhythm and pitch, and rather hoarse (Glutz and Bauer 1980); less sonorous, shriller and shorter than call 1a (Bergmann and Helb 1982); a melancholy, wheezing sound (Pukinski 1977). (2) Trilling-call. A short, sharp trilling 'kiiiiiiiirrl' (c. 1 s), resembling sound of ungreased machine shaft, given when human intruder present at nest (Hagen and Barth 1950; Hagen 1952). Tonal quality varies from a soft, purring, rolling sound (♂) to a harsh bleating or shrill, metallic sound (♀) (Glutz and Bauer 1980). Former, but attributed to both sexes, presumably the more muted kitten-like 'pllllllrr'itt', typically heard when pair both at nest during

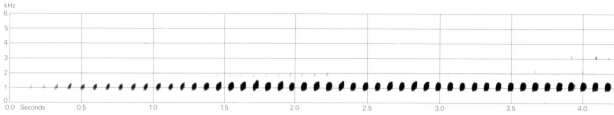

I B Kristiansson/Sveriges Radio (1981) Sweden February 1976

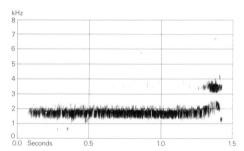

II I Hills Finland June 1975

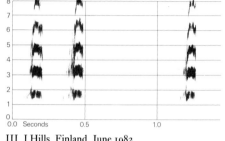

III I Hills Finland June 1982

disturbance (Hagen and Barth 1950; Hagen 1952). Serves as a contact call during pair-formation, when given at almost every meeting, e.g. during nest-site inspection, during Billing, or when summoning mate. Also as duet when prey being passed, and during copulation (Glutz and Bauer 1980). Presumably same or similar call, lasting c. 3 s, and starting with prolonged twitter, becoming lower and softer, finally rising to strong chirp, given by ♀, apparently in distress, perched near human intruder (Leinonen 1978). (3) Screeching-call. (a) A repeated, drawn-out (1·2–1·3 s) wheezy 'aaaaa-ik', with emphasis on final unit (P J Sellar: Fig II), expressing alarm when human intruder near nest with well-grown young; ♂ meanwhile gave call 4; a furious 'tschi-itsch' given by ♀ between attacks, whilst bobbing head and tail (Fridzén 1959). In caparoch, a 'screeeeee-yip' or 'schreeeeee-yup' (Smith 1970). Given with gaping bill when fiercely mobbed (Smith 1922). Also given by ♂ and ♀ at sight of rivals (Glutz and Bauer 1980). Same call (described as 'pss-ih', c. 1·5 s long, with emphasis on final unit) occasionally used as contact-call, e.g. by ♂ bringing prey to nest, or by ♀ calling to young, especially when latter out of nest (Leinonen 1978; Mikkola 1983). (b) Clear whinnying sound, gradually descending in pitch, and similar to call 3a, given by bird flying excitedly around in autumn and winter, when probably serving as long-distance contact (Glutz and Bauer 1980). May also function thus in breeding season, when ♀ may give a hoarse, drawn-out 'kshée-lip', apparently when soliciting food, in answer to call 1 of ♂ (H Delin). In higher-intensity variant, a penetrating shrill screech, like Barn Owl *Tyto alba*, frequently ending in tremolo. Function not clear, but often heard in combination with calls 1a and 2 (Glutz and Bauer 1980). (4) Yelping-call. A sharp, strident yelping 'kwitt kwitt' (Fig III), 'kwi kwi', 'ki ki ki', or 'ki ki kikikikiki' (Glutz and Bauer 1980), like Merlin *Falco columbarius* (H Delin), given as warning by alarmed bird near threatened nest or fledged young (Glutz and Bauer 1980); 'kvi kvi kvi', by either sex, ♂'s call higher pitched (Leinonen 1978), or 'kvitt kvitt' at rate of 2 per s (Hagen and Barth 1950; Hagen 1952). Typically given in flight, notably during aerial attack (Fridzén 1959; Simon and Simon 1980), also more querulously when perched, and then accompanied by raising and lowering of tail (Köpke 1969; Smith 1970). Usually gives way to call 3a (Smith 1970). (5) Lure-call. A shrill, hard sound, like nest-lure call of a falcon *Falco*, mostly heard during nest-site inspection. Given with gaping bill, especially by ♀ or ♂ before she enters prospective nest-site (Glutz and Bauer 1980). (6) Soliciting-call. A hoarse, rasping 'chät', 'chüt' or 'dchüji', usually ending in a loud squeaking, given by ♀ with wide circling movements of head when inviting copulation (Glutz and Bauer 1980). The 'monkey-chattering' sound of ♀ caparoch soliciting food (Smith 1922), probably the same. (7) Other calls, probably of contact-alarm type. (a) 'Kuk kuk', given by ♀ bringing food to nest, once by ♂ as intruder approached nest (Leinonen 1978). Single, soft muted 'gjuhk', heard rarely and only when close (Glutz and Bauer 1980), presumably same. (b) ♀ answered ♂'s call 1a with a rather muted 'mui' (Glutz and Bauer 1980). Possibly same call rendered a soft 'cuee' (J Hall-Craggs, P J Sellar: 1st call in Fig IV), or strained delicate 'sreee sreee' (Leinonen 1978), given by ♀ at nest while young giving food-calls; perhaps invites close approach. Apparently same call also given by ♀ performing distraction-lure display (Leinonen 1978: see Social Pattern and Behaviour).

CALLS OF YOUNG. Food-call of young a wheezy hiss (E K Dunn) or hiss-like peep; see Fig IV, in which 1st low sound is 'cuee' call of ♀, and calls of young—which have a tonal ('peep') sound halfway up their spectra—partly overlap (J Hall-Craggs); a long-drawn (c. 1·5 s)

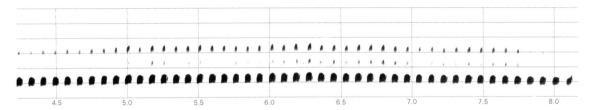

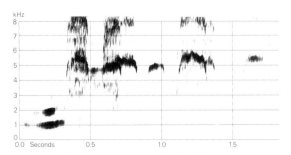

IV S Palmér/Sveriges Radio (1972) Sweden June 1963

'ksssssss'itt' given continuously at all times of day and audible at 200–300 m (Hagen and Barth 1950); a hoarse crescendo 'pssssssssssitt', final 'itt' higher pitched and stressed, and 'i' sound often drawn-out; similar calls also given by older young in flight when mobbed by passerines (Köpke 1969), and by newly fledged bird at approaching intruder; resembled weaker variant of call 3a of adult (Smith 1970; Leinonen 1978). Similar to juvenile Short-eared Owl *Asio flammeus* (H Källander). Doleful descending thin cries of fledged young, in recording by T von Essen/Sveriges Radio, probably the same (E K Dunn). When more excited, young gave a 'kssäkssäkssä' or 'tchäpptchäpp', very like call 4 of adult (Köpke 1969). Nestling which was sated, and refusing food from ♀, made 'hüp hüp' sound (Böhm 1982b). EKD

Breeding. SEASON. Finland: eggs laid end of March to late June (Mikkola 1972c). USSR: laying begins end of March (Dementiev and Gladkov 1951a). SITE. In hole in tree, top of tree-stump, or nest-box; also uses old nests of raptors and crows (Corvidae). One nest, Finland, 13 m above ground (Pulliainen 1978). Nest: available hole or platform of twigs; no material added (Mikkola 1972c). Building: none. EGGS. See Plate 97. Blunt elliptical, smooth and glossy; white. 40×32 mm ($36-44 \times 29-34$), $n=100$ (Witherby *et al.* 1938). Calculated weight 21 g (Makatsch 1976). Clutch: 6–10 (3–13); larger in good vole years. Of 134 clutches, Finland: 3 eggs, 4%; 4, 12%; 5, 23%; 6, 20%; 7, 17%; 8, 12%; 9, 6%; 10, 2%; 11, 3%; 13, 1%; average 6·3. Average clutch size 6·56 ($n=101$) in Lappland, 5·94 ($n=18$) in central Finland, and 5·13 ($n=13$) in southern Finland (Mikkola 1972c). One brood. Replacements laid after egg loss. Laying interval 48 hrs (Makatsch 1976). INCUBATION. 25–30 days. By ♀ only (Mikkola 1972c). Begins with 1st egg; hatching asynchronous. Youngest chick left nest 1 week after oldest (Bianki and Koshkina 1960). YOUNG. Altricial and nidicolous. Cared for and fed by both parents. FLEDGING TO MATURITY. Fledging period 25–35 days, chicks typically scrambling out on to nearby branches before true fledging. Remain in vicinity of nest for some weeks after. Age of independence at least $2\frac{1}{2}$ months (Mikkola 1972c). Age of first breeding probably 1 year (Hagen 1956). BREEDING SUCCESS. Little information, though clutch size larger in good vole years so likely to be influenced by food supply. At one nest, Kajnani (Finland), 8 young fledged from 9 eggs (Leinonen 1978).

Plumages (nominate *ulula*). ADULT. Crown and hindneck sooty-black, each feather with white terminal spot *c.* 3 mm across and with white subterminal bars, broken by black along shaft; spot and bars often partly joined and width of bars rather variable, appearance of crown either mottled with equal amounts of black and white, or black with rounded white spots, or white with close and short black bars; sides of crown at border with facial disc almost fully white. Central hindneck usually with solid black patch (in birds with whitish crown, sometimes spotted white), sides of hindneck whiter than crown, only slightly barred or grizzled black. Mantle dark brown-grey (tinged plumbeous when plumage fresh), upper mantle with large white V-marks, forming mottled half-collar, lower mantle uniform dark. Scapulars dark brown-grey with large white paired squarish spots; inner webs of shorter inner scapulars mainly brown-grey on tips, appearing uniform, tips of outer webs of outer scapulars white, together forming white band. Back and rump dark grey-brown with some white spots laterally; upper tail-coverts closely and evenly barred white and brown. Poorly defined facial disc white with some faint black shaft-streaks; bordered above by narrow black streak from base of upper mandible just over eye, ending in very short and mainly black ear-tufts (rarely visible in life), bordered behind by broad black crescent extending from base of ear-tufts to sides of chin. Some black-tipped bristles projecting forward round base of bill. Central chin dusky buffish-grey. Broad black crescents at sides of facial disc bordered behind by mainly white streak down sides of hindneck, this streak in turn bordered by conspicuous black spot at upper sides of neck, surrounded by mainly white feathers at sides of hindneck and on lower sides of neck. Throat immediately below dusky chin and lower border of facial ruff white, bordered below by narrow black band across upper chest, which is broken in centre by cream or white feathers with black bars. Sides of chest and lower chest white with some short grey-brown bars or spots; remainder of underparts closely barred white and dark brown-grey; dark bars *c.* 7 mm apart, each *c.* 1·5–2·5 mm broad (dark bars sometimes slightly closer and broader on sides of breast, usually broader on longer under tail-coverts, sometimes faint or almost absent on vent, shorter under tail-coverts, and feathering of leg and foot). Tail dark grey-brown, t1 with *c.* 8 narrow and well-spaced white bars visible beyond upper tail-coverts (usually broken in middle and slightly tinged or mottled pale buff-grey) and with white tip 3–6 mm wide; pale bars of other tail-feathers indistinct, pale buff-grey, less contrasting with intervening dark grey-brown, buff-grey replaced by wider white spots along border of inner webs, slightly also on basal borders of outer webs. Flight-feathers dark grey-brown with rather narrow white fringes along tip (forming smoothly rounded white arcs) and with faint paler grey-brown bars, latter ending in large and conspicuously white spots along borders of inner webs; smaller white spots along outer webs (spots virtually absent on tips of outer 2 primaries and on outer webs of innermost primaries); white spots on inner 1–2 tertials join to form complete bars (sometimes of slightly irregular shape). Greater upper primary coverts like primaries, other upper wing-coverts dark brown-grey or plumbeous-brown with white subterminal spots on sides of greater, median, and longer lesser coverts, largest on outer webs of median coverts and outer larger lesser coverts, forming rows of white spots. Under wing-coverts and axillaries white with short and broken dark grey-brown bars, contrasting rather with heavily barred undersurface of flight-feathers. Abrasion and bleaching have some influence: plumbeous-brown and black colour of upperparts and upper wing-coverts changes to duller brown; crown, hindneck, and chest whiter; brown of scapulars, tertials, and bases of outer webs of tail-feathers often strongly bleached to pale greyish by spring and early summer, white spots and bars hardly contrasting and partly lost by abrasion. DOWNY YOUNG. 1st down (neoptile) short, rather thick, white or white with a yellow-buff tinge. JUVENILE (mesoptile). Plumage rather feather-like, not as soft and loose as mesoptile of (e.g.) *Strix* and *Asio* owls, more like that of *Athene* and *Glaucidium*. Rather like adult, but upperparts grey-brown (not as black or plumbeous as in adult), less extensively marked white: feathers of crown tipped grey-white, nape to upper mantle mouse-brown with some indistinct off-white marks on feather-centres; scapulars and upper wing-coverts brown with white tips. Facial disc largely black at age of 2–3 weeks, but contrastingly marked with white on leaving nest at *c*. 4 weeks—disc black-brown with large white V-mark between and above eyes and with large white patch on lower cheek. Chest to belly and flanks dirty white with brown bars; vent and under tail-coverts downy, off-white. (Witherby *et al.* 1938; Mikkola 1970–1.) FIRST ADULT. Like adult, but juvenile flight-feathers, greater primary coverts, and tail retained. Tail differs from adult in more pointed feather-tips (not as broadly rounded), tip of t1 forming white triangle 8–12 mm long (in adult, more even white fringe *c*. 5 mm wide); t5–t6 with subterminal pale bar (next to black terminal bar below white tip) fully grey (in adult, partially white towards inner edge); 2–3 innermost tertials gradually attenuated towards rather soft and pointed tip, sides with loose rami (worn by autumn, heavily worn and bleached by spring), dark brown with white spots at sides (in adult, broad with rounded tips, fresh in autumn, slightly worn in spring, and with distinct white bars on 1–2 innermost); secondaries all equally new or worn (in adult, often a mixture of old and new ones); outer primaries with rather irregular grey-and-white V-shaped fringes along tips (not forming even white arcs as in adult) (Forsman 1980; C S Roselaar).

Bare parts. ADULT AND FIRST ADULT. Iris bright yellow. Bill pale yellow, light horn-yellow, or greenish-yellow. Claws blackish-brown or black. DOWNY YOUNG AND JUVENILE. No information on newly hatched chick. Similar to adult on leaving nest.

Moults. ADULT POST-BREEDING. Complete or almost complete; primaries descendant. Start earlier than most other Holarctic owls; p1 shed late April to early June when eggs just hatched or bird still breeding, completed with regrowth of p10 mid-August to late September (Dementiev and Gladkov 1951a; Stresemann and Stresemann 1966; BMNH, RMNH). In 1-year-olds, secondaries start about July from 4 centres, at s1, s5, about s10, and s15; at end of moulting season, September or early October, s3–s4 and s6–s8 usually still juvenile, and frequently also s2 and s9 (in 17 birds, 3–10 secondaries still juvenile) (Forsman 1980); moult continued in next season with old feathers and s10–s15, but centre on s5 not always activated then. Pattern of old and new secondaries more complicated in older birds and difference in age more difficult to see than in 2nd winter, when old juvenile feathers are heavily worn and differently shaped. According to Forsman (1980), some birds replace all secondaries in single moulting season; none examined, however, no specimens in BMNH, RMNH, or ZMA to support this, not even in resident race *tanschanica*. Tail-feathers lost in quick succession shortly after loss of p7, growing almost simultaneously early July to early September; body completely moulted at same time. POST-JUVENILE. Partial: all head, body, and wing-coverts, but no flight-feathers, tertials, greater primary coverts, and tail. June to August or September (Dementiev and Gladkov 1951a); all in full 1st adult by mid-September, except for one still replacing mesoptile greater upper wing-coverts in late October (Forsman 1980).

Measurements. Nominate *ulula*. Northern Eurasia, mainly autumn (RMNH, ZMA). Bill (F) to forehead, bill (C) to cere.

WING	♂	234	(4·57; 15)	224–239	♀ 238	(5·70; 15)	230–249
TAIL		177	(6·05; 12)	164–187	178	(7·43; 12)	166–191
BILL (F)		24·1	(0·74; 10)	22·8–25·1	25·2	(1·23; 10)	23·6–26·8
BILL (C)		17·5	(0·92; 12)	16·0–18·4	18·5	(0·60; 12)	17·5–19·3
TARSUS		25·8	(0·95; 11)	23·9–27·0	25·5	(1·11; 10)	24·4–27·2

Sex differences significant for bill. Wing of Siberian birds on average slightly longer than European: in west, ♂ 234 (5·18; 10), ♀ 237 (4·98; 12); in east, ♂ 235 (4·28; 7), ♀ 240 (5·91; 10) (Hartert 1912–21; Eck 1971; RMNH, ZMA).

Wing of (1) *caparoch* from North America and single bird off Canary Islands (BMNH, RMNH, ZMA), and (2) *tianschanica* from Tien Shan (Dementiev and Gladkov 1951a).

(1) ♂ 231 (2·89; 8) 226–235 ♀ 239 (3·84; 6) 235–246
(2)   243 (— ; 11) 238–251   248 (— ; 7) 243–252

Weights. ADULT. Nominate *ulula*. USSR: ♂ 314 (6) 247–375, ♀ 348 (3) 323–371 (Dementiev and Gladkov 1951a). Norway, August–March: ♂ 296 (242–325), ♀ 300 (285–325); another sample 280 (230–330) (Haftorn 1971). Finland: ♂ 270 (16) 215–310, ♀ 320 (17) 270–380 (Haartman *et al.* 1963–72).

S. u. caparoch. North America: ♂ 299 (16) 273–326, ♀ 345 (14) 306–392 (Earhart and Johnson 1970). Alaska and Yukon territory (Canada), February–May: ♂♂ 322, 350; ♀♀ 310, 335, 350, 384 (Irving 1960).

DOWNY YOUNG AND JUVENILE. Finland, 6 young in single nest: on 28 May, 50, 72, 90, 114, 130, 135; on 6 June, 180, 186, 206, 214, 220, 228; all left nest with weight over 200 between 7 and 13 June (Pulliainen 1978).

Structure. Wing rather long, broad, tip rather pointed. 10 primaries: p8 longest, p9 7–15 shorter, p10 35–52, p7 0–5, p6 20–24, p5 37–43, p4 49–55, p1 75–84. Outer webs of (p7–)p8–p9 and inner web of p8–p10 emarginated. Outer web of p10 with very short serrations. Tail long, strongly rounded, 12 feathers; t6 32–42 mm shorter than t1. Bill heavy, rather protruding. Tarsus and toes rather short and slender, closely covered by hair-like feathers; middle toe without claw 22·0 (1·29; 12) 20–24 in ♂, 23·6 (0·86; 9) 22–25 in ♀; outer toe without claw c. 73% of middle, inner c. 91%, hind c. 60%. Claws long, rather slender; middle claw 16·9 (0·89; 11) 15·4–17·6 in ♂, 17·5 (0·45; 11) 16·7–18·0 in ♀; outer claw c. 82% of middle, inner c. 98%, hind c. 89%.

Geographical variation. Rather slight, in both colour and size, rather obscured by pronounced individual variation. In northern Eurasia, populations in Siberian part of range of nominate *ulula* average slightly paler and larger and these sometimes separated as *pallasi* Buturlin, 1907, but difference in size very slight (see Measurements), while some Scandinavian birds equally pale and some from eastern Siberia and Lake Baykal dark. North American *caparoch* distinctly darker than nominate *ulula*, though a few nominate *ulula* perhaps not separable from palest *caparoch*; dark colour of upperparts and upper wing-coverts blacker and white marks smaller; feathers of crown with single terminal white spot and single white subterminal bar (both sometimes joined), base of feathers largely black (in nominate *ulula*, several white subterminal bars and feather-base mainly white); large solid black patch on central hindneck, often connected with large black patches at upper sides of neck; only limited amount of white on hindneck and upper mantle; black-barred band across upper chest distinctly broader, sides of breast also barred black, only centre of lower chest uniform white; dark bars on breast, belly, and flanks broader, especially those on sides of breast, which show as dark patches (unlike nominate *ulula*), width of bars on belly and flanks 2·5–3 mm (mainly 1·5–2·5 in nominate *ulula*), dark bars on lower flanks and belly tinged rufous; undersurface of tail-feathers more contrastingly barred. *S. u. tianschanica* from central Asia rather similar to *caparoch* on upperparts or with even blacker ground-colour and smaller white marks; underparts rather like nominate *ulula* or intermediate between nominate *ulula* and *caparoch*; larger than these races (see Measurements).

CSR

Glaucidium passerinum Pygmy Owl

PLATES 47 and 54
[between pages 494 and 495, and facing page 566]

Du. Dwerguil   Fr. Chouette chevêchette   Ge. Sperlingskauz
Ru. Воробьиный сыч   Sp. Mochuelo chico   Sw. Sparvuggla

Strix passerina Linnaeus, 1758

Polytypic. Nominate *passerinum* (Linnaeus, 1758), Europe. Extralimital: *orientale* Taczanowski, 1891, central and eastern Siberia, intergrading with nominate *passerinum* in western Siberia and Altay mountains.

Field characters. 16–17 cm; wing-span 34–36 cm. Between $\frac{2}{3}$ and $\frac{3}{4}$ size of Tengmalm's Owl *Aegolius funereus* and Little Owl *Athene noctua*, with proportionately small head and narrow tail; hardly more bulky than Hawfinch *Coccothraustes coccothraustes*. Tiny, rather small-headed owl, with plumage pattern recalling that of dark races of *A. noctua* but differing in barred chest-sides (in adult) and narrower and relatively longer, white-barred tail. Stern facial expression; bill noticeably long (50% more so than in *A. noctua*). Fearless killer of passerines and small rodents, hunting particularly at twilight. Flight bounding and dashing, like woodpecker *Dendrocopos*. Sexes similar; no seasonal variation. Juvenile separable.

ADULT. Upperparts dark dull brown, spotted white-buff overall but less boldly than in *A. noctua*; lacks bold white spots on scapulars but shows pale set of 'reversed brackets' behind head. Face lacks full disc of most small owls, being dusky with short, almost white eyebrows; dull white bill feathering and chin, and indistinct grey-white half-circles on outer cheeks. Small yellow eyes appear closer set than in other small owls. Pale horn bill often protrudes from feathering. Flight-feathers dark brown, narrowly barred white-buff. Under wing-coverts noticeably white, contrasting with dark grey, faintly barred flight-feathers. Tail dark brown, narrowly but distinctly barred buff-white. Throat almost white but inconspicuous; upper breast and sides of chest noticeably barred dark brown, but most of underbody white, irregularly and sparsely spotted and streaked dark brown. Legs and toes feathered white. JUVENILE. Closely resembles adult, but darker, with duskier brown, unspotted upperparts and wing-coverts, unbarred patches on chest-sides, and thinner streaks on pale underparts, but similar flight- and tail-feathers. Facial disc more uniform.

When true size evident, unmistakable. When not so, can be confused with *A. funereus* but small head, stern (not astonished) expression, absence of obvious pale spotting on scapulars, and presence of sharp white bars on rather long and square-cut tail rule out confusion. Flight active and dashing, with fast wing-beats, frequent closure of wings, and sudden turns in pursuit of small passerines conveying impression of great energy and purpose. Markedly bounding over longer distances, recalling woodpecker *Dendrocopos*. Carriage variable, sometimes upright but usually only half so; fidgets, often waving tail and flicking it upwards; will sit with tail cocked (like flycatcher *Ficedula*).

Song a flute-like 'dü' repeated monotonously 5–10 times per 10 s.

Habitat. In taiga and montane forest, mainly coniferous, especially of silver fir *Picea abies*, sometimes to treeline. In lowlands in higher and upper middle latitudes and in temperate zone ranging from 250–300 m in narrow cool and moist ravines to above 1000 m in Alps—highest breeding station, in Valais (Switzerland), at 2150 m. In

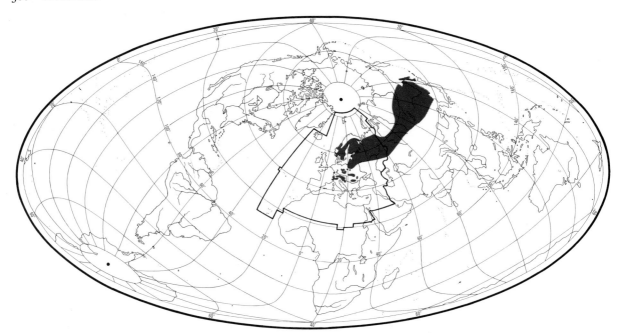

forests of Swiss Jura at between 1100–1400 m. Mainly in tall forest interior, dominated by conifers but often interspersed with beech *Fagus*, aspen *Populus tremula*, birch *Betula*, and other broad-leaved trees also used for nesting. For hunting, favours ready access to clearings, moors, meadows, or avalanche pathways, and to water. Requires ample choice of suitable holes for breeding and other purposes, although roosting largely in cover of foliage or by tree-trunks. As miniature predator of largely crepuscular habits, depending much on surprise in taking prey and itself hunted by larger raptors, is concerned as much with structure of forest as with its composition of tree species. In Alps occupies habitat from highest breeding territories of Tawny Owl *Strix aluco* upwards, and must tolerate lower temperatures and higher precipitation.

In winter may shift to broad-leaved mixed woodland with some conifers, or to neighbourhood of human settlements. In breeding season infrequently in contact with man, but sometimes, especially in lowlands, will occupy managed plantations and make use of nest-boxes. Flight normally confined to lower airspace and to short distances. An arboreal bird, little seen on ground. (Based mainly on Glutz and Bauer 1980 and Glutz von Blotzheim 1962.)

Distribution. Precise distribution often not yet determined.

WEST GERMANY. Has disappeared from large parts of its range (Bauer and Thielcke 1982). NORWAY. Has occasionally bred further north in 1970s (GL, AR). POLAND. Distribution imperfectly known; probably more widespread in 19th century (Tomiałojć 1976a). USSR. No longer breeds in Carpathians (VG).

Accidental. Belgium, Denmark.

Population. Fluctuating in some areas; marked variations reported in Scandinavia, related to changes in populations of voles (Microtinae) (Hagen 1969). Has probably decreased in West Germany and parts of USSR (see also Distribution).

SWEDEN. Tentative estimate 20 000 pairs (Ulfstrand and Högstedt 1976). FINLAND. Estimate of 200–300 pairs (Merikallio 1958) probably too low; now some thousands of pairs with no evidence of decline, despite large-scale destruction of old forests (OH). EAST GERMANY. Fluctuating, probably 120–150 pairs (Glutz and Bauer 1980); *c.* 70 pairs 1978 (Makatsch 1981). WEST GERMANY. No total estimate; for local estimates and fluctuations, see Glutz and Bauer (1980) and Bauer and Thielcke (1982). POLAND. Scarce or very scarce breeder (Tomiałojć 1976a). CZECHOSLOVAKIA. Perhaps at least 25–30 pairs (Mikkola 1983); some recent increase in southern Bohemia (KH). SWITZERLAND. Highest density in spruce *Picea abies* forest of northern Alps (Glutz and Bauer 1980). FRANCE. 100–1000 pairs (Yeatman 1976). USSR. Scarce, numbers fluctuate in some areas (Dementiev and Gladkov 1951a). Estonia: 100–200 (Mikkola 1983). Latvia: rare (HV). Lithuania: very rare, apparently decreased (Ivanauskas 1964).

Movements. Mainly resident. Degree of dispersal higher in northern than in central Europe, and north European birds eruptive at irregular intervals.

NORTHERN POPULATIONS (Fenno-Scandia and USSR). Marked tendency to leave forest territories in autumn, to winter nearer human habitations where food more readily found in cold weather (Dementiev and Gladkov 1951a; Mikkola 1983). Not known how far birds move in normal

Glaucidium passerinum 507

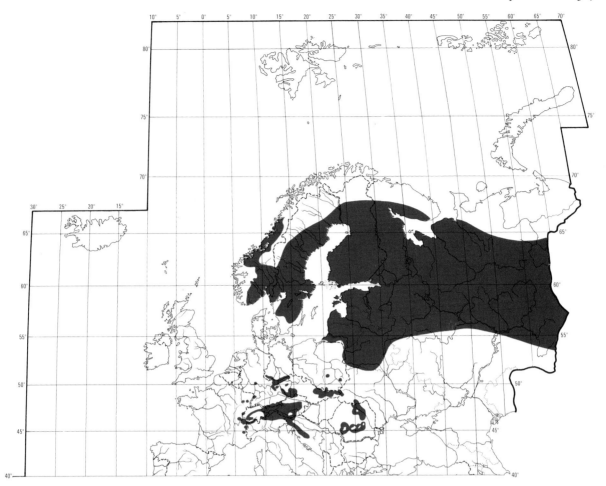

winters (they remain within breeding range), and no data on post-fledging dispersal of 1st-year birds; the surprisingly few ringing recoveries relate to eruption years.

Eruptions probably triggered by combination of cold weather and rodent scarcity (Mikkola 1983); invasions not synchronous over all of Fenno-Scandia, though larger ones manifested in all 3 countries at once. Irruptions recorded in winters 1903–4, 1907–8, 1914–15, 1918–19, 1921–2, 1926–7, 1931–2, 1932–3, 1934–5, 1935–6, 1938–9, 1942–3, 1946–7, 1947–8, 1949–50, 1950–1, 1954–5, 1955–6, 1960–1, 1963–4, 1965–6, 1968–9, 1969–70, 1970–1, 1971–2, and 1974–5 (Mikkola 1983); hence irregular intervals of 1–6 years. Most of these quite small, but large in 1955–6, 1963–4, 1968–9, 1971–2, and 1974–5, when stragglers reached Denmark. Reluctant to cross open water, so erupting birds seldom penetrate south of Baltic (Lindberg 1966), but pass eastern end into Soviet Baltic territories (e.g. Meshkov and Uryadova 1972). Eruptive movements begin September–October, and birds that winter in southern Sweden disappear in March (Lindberg 1966). Recoveries of birds ringed as nestlings: June 1966–September 1966 (230 km south in Finland); June 1963–March 1964 (124 km south in Sweden); June 1974–November 1974 (72 km east in Sweden). Ringed as autumn migrants: October 1974–February 1975 (106 km east in Sweden); September 1974–February 1975 (290 km south-east in Sweden); October 1959 (south Finland)–December 1959 (Leningrad, USSR), 300 km ESE. Following the 1963–4 invasion into southern Sweden, distribution of spring sightings suggested withdrawal towards north-east, probably into Finland where same irruption had been experienced (Lindberg 1966).

CENTRAL EUROPEAN POPULATIONS. Adults (♂♂ at least) stay on territory all year, though young birds disperse to limited extent while seeking territories of their own. Austrian nestling found following summer 9 km SSW, and Baden-Württemberg juvenile (August 1968) found Switzerland (October 1969) 90 km south. Unringed individuals recorded exceptionally, September–February, up to 250 km from nearest breeding place. See Schönn (1976, 1978) and Glutz and Bauer (1980).

Food. Small mammals, especially voles (Microtinae), and small birds. Hunting takes place mostly at dusk and dawn, but also during day. Apparently little or no hunting at night; prey located by sight (Mikkola 1983), and night vision poor (Lindblad 1967). Breeding-season activity as follows: in Austria, mostly 03.00–05.00 and 19.00–21.00 hrs, reducing to zero during 22.00–03.00 and 12.30–14.30 hrs (Bergman and Ganso 1965; Scherzinger 1970); in southern Norway, active 01.00–23.00 hrs, mostly before and after resting period and at 09.00–11.00 hrs (Seierstad *et al.* 1960); in central Finland, active at all times, mostly 22.00–01.00 and 09.00–10.00 hrs (Mikkola 1970c); see also Jansson (1964) and Mikkola (1972d). Summer activity pattern in Finland broadly follows that of bank vole *Clethrionomys glareolus* (Mikkola 1970c). In winter, Fenno-Scandia, hunts throughout day (Mikkola 1983; H Delin). Hunts mammals from open perch, dropping or gliding on to them; flicks tail and rattles wings during take-off. Hunts birds from concealed perch, snatching them from branches or striking them from below in the air. (Schönn 1978; Glutz and Bauer 1980; Mikkola 1983.) Will hunt at bird-tables (H Delin). Apparently relies on surprise and does not pursue prey if first attack unsuccessful (Kellomäki 1966). Strikes small mammals with feet and immediately bites them on snout; while prey still alive, spreads wings for support. Carries prey in foot to concealed place for eating (Glutz and Bauer 1980). Items for delivery to nest or for caching often have head end eaten first. Birds apparently killed with feet and/or bites to head or neck; may be plucked or (at least in captivity) not (Glutz and Bauer 1980; Zschoke 1982). Prey eaten piecemeal and bones thus broken up; intestines usually rejected (Glutz and Bauer 1980). Food-caching behaviour better developed than in any other west Palearctic member of Strigiformes. In summer, caches are small (frequently 1 item) and mainly in open sites—e.g. on branches, exceptionally on ground (Glutz and Bauer 1980). In northern Europe, caching peaks over short period in November–January, caches being then larger (200 items recorded) and usually in holes (including nest-boxes). Prey numbers cached vary greatly between years, and large caches made only in years with dry and cold November/early December; apparently no link with availability of prey species. During winter (when frost preserves prey), caches not added to, and items gradually eaten. In south-east Norway, average 26·3 items per cache in January, declining to 3·4 in April. (Likhachev 1971; R Solheim; see also Shilov and Smirin 1959, Scherzinger 1970, Ahlbom and Carlsson 1972, Solheim 1973.)

Pellets of adults, in winter, average 28×12 mm, $n = 40$; those of young smaller (Mikkola 1983). Bones mostly fragmented.

The following prey recorded in west Palearctic. Mammals: shrews (*Sorex minutus*, *S. araneus*, *S. caecutiens*, *S. minutissimus*, *S. alpinus*, *Neomys fodiens*), Daubenton's bat *Myotis daubentonii*, garden dormouse *Eliomys quercinus*, dormouse *Muscardinus avellanarius*, wood lemming *Myopus schisticolor*, bank vole *Clethrionomys glareolus*, grey-sided vole *C. rufocanus*, water vole *Arvicola terrestris*, pine vole *Pitymys subterraneus*, common vole *Microtus arvalis*, short-tailed vole *M. agrestis*, root vole *M. oeconomus*, snow vole *M. nivalis*, harvest mouse *Micromys minutus*, striped field mouse *Apodemus agrarius*, yellow-necked mouse *A. flavicollis*, wood mouse *A. sylvaticus*, brown rat *Rattus norvegicus*, house mouse *Mus musculus*, weasel *Mustela nivalis*. Birds almost wholly up to size of finches (Fringillidae), but recorded up to size of full-grown Great Spotted Woodpecker *Dendrocopos major* and Mistle Thrush *Turdus viscivorus*. Other prey uncommon: common lizard *Lacerta vivipara*, slow-worm *Anguis fragilis*, perch *Perca fluviatilis*, grasshoppers (Acrididae), and beetles (Coleoptera) recorded. (See, especially, Schönn 1978; also Moltoni 1937, Mikkola 1970c, Likhachev 1971, Kellomäki 1977, Glutz and Bauer 1980.)

For 2 north European studies, see Table A; for summary of some other studies (showing similar results) and for details of further sources not listed below, see Mikkola (1983). Significant studies as follows. Norway: Lund (1951), Sonerud *et al.* (1972). Sweden: Ulvblad (1962), Jansson (1964), Ahlbom (1970, 1973b). Finland: Mikkola (1970c,d), Mikkola and Jussila (1974). East and West Germany: Uttendörfer (1939b, 1952), März (1964), Scherzinger (1974a), Schönn (1976, 1978), Bezzel *et al.* (1977). Switzerland: Mattes (1981). Austria: Bergman and Ganso (1965), Scherzinger (1970). USSR: Karpovich and Sapetin (1958), Shilov and Smirin (1959).

Feeding ecology studied most thoroughly by Kellömaki (1977), in southern Finland; see Table A. Towards end of breeding season, diversity of bird prey and its proportion in diet increased (similarly in work by Jansson 1964 and Sonerud *et al.* 1972), apparently because of reduced availability of small mammals (due to growth of vegetation) and increased availability of birds, especially fledglings. In first half of nestling period, young birds 18% (by number) of total birds; in second half, 31% (see also Lund 1951, Bergman and Ganso 1965). Compared with trapping

Table A Food of Pygmy Owl *Glaucidium passerinum* as found from nest-remains, caches, and pellets. Figures are percentages of live weight of vertebrates only.

	S Finland 1962–72 Breeding season	Belorussiya and Moscow region (USSR), 1950–69	
		Summer	Winter
Mice (small Murinae)	0·3 ⎫ 59·0	1·8 ⎫ 57·2	1·1 ⎫ 43·1
Small voles (Microtinae)	58·7 ⎭	55·4 ⎭	42·0 ⎭
Larger mammals	0	0	0·5
Shrews (Soricidae)	1·2	15·1	9·5
Birds	39·5	27·7	47·0
Other	0·3 (reptiles, bats)	0	0
Total no. of vertebrates	2234	273	839
Source	Kellömaki (1977)	Likhachev (1971), data reanalysed	

results, shrews taken less often than predicted by their abundance, *C. glareolus* taken more often, and *M. agrestis* in line with abundance. From year to year, proportion of voles in diet varied with their abundance, proportion of birds varying inversely with this. Feeding ecology thus resembles that of Tengmalm's Owl *Aegolius funereus*, but if voles scarce *A. funereus* switches to shrews, *G. passerinum* to birds.

Food consumption in captivity 25–40 g per day, depending on temperature (Scherzinger 1970; Eck and Busse 1973). See also Glutz and Bauer (1980).

No detailed studies of differences between diets of adult and young, but young evidently given more birds than eaten by adults before eggs hatch (see above). At one nest with 5 young, central Sweden, in main (morning and evening) feeding periods ♂ typically brought 6–12 items, number increasing through nestling period (Jansson 1964; see also Mikkola 1970c, Schönn 1976). DJB

Social pattern and behaviour. Major studies by Scherzinger (1970) and Schönn (1976, 1978).

1. Solitary or in pairs at all times. In central Europe, ♂ (at least) defends territory all year (see Breeding Dispersion, below); such behaviour presumably less prevalent in (dispersive) northern population. BONDS. Monogamous mating system, pair-bond lasting for 1 season. However, after temporary severance in late autumn, birds may renew bond (Glutz and Bauer 1980)—once, for 4 years, using same nest throughout (Schönn 1978). First breeding at 1 year old (e.g. König 1968, Scherzinger 1970). Only ♀ incubates, broods, and feeds young in nest; ♂ normally does all hunting during incubation and most of nestling phase. Young dependent for c. 4 weeks after fledging (see Relations within Family Group, below). BREEDING DISPERSION. Solitary. ♂ defends fixed territory all year, sometimes assisted by ♀; closeness of association between ♂ and ♀ outside breeding season not clear. All hunting and breeding activity takes place within territory, though nest often in extension from main area. Boundaries usually follow landmarks (woodland edges, streams, etc.). (Glutz and Bauer 1980.) Territory sizes: Alps and Böhmer Wald (West Germany) 0·45–1·9 km², average c. 1·25 km² (Scherzinger 1970); Bayerischer Wald average 1·4 km², $n=50$ (Scherzinger 1974a); Belovezhskaya forest (USSR) 2·5–4 km² (Golodushko and Samusenko 1961); Stazer Wald (south-east Switzerland) one territory 0·45 km² (Mattes 1981). In uniform habitat, territories evenly distributed (e.g. Scherzinger 1970). No correlation between population density and territory size (Glutz and Bauer 1980). Population densities: eastern Alps (Austria) 1·4 pairs per 10 km² ($n=8$), 6 nests 0·6–1·6 km apart (Scherzinger 1970); Bayerische Alpen (West Germany) 1·2–1·7 pairs per 10 km² over c. 864 km² (Glutz and Bauer 1980); Bayerischer Wald (when population high) 4·2 pairs per 10 km² over 120 km², shortest distance between nests 1 km (Scherzinger 1974a); Norway 0·15–0·23 pairs per 10 km²; southern Finland 0·017 territories per 10 km²; northern Finland 0·4 territories per 10 km² (Glutz and Bauer 1980). At least when rodent supply good, may breed alongside Tengmalm's Owl *Aegolius funereus*: in north-central Sweden, 3 pairs on 1·5 km² with 6 pairs of *A. funereus*; nests of the 2 species as close as 14 m (Lindberg 1966). ROOSTING. Presumably largely nocturnal; for periods of hunting activity, see Food. Mostly uses dense spruces *Picea* (C König); sometimes uses holes in winter (Glutz and Bauer 1980; see also for roosting postures, and bathing behaviour—in water, rain, snow, and sun).

A

2. When agitated, sometimes waves tail from side to side (H Delin); movement may also be vertical and in loop or figure-of-eight (Glutz and Bauer 1980). If approached slowly by man, may close eyes and ruffle feathers; usually flies at 3–4 m (Scherzinger 1970). In concealing Sleeked-upright posture, stretches body upwards with feathers sleeked, carpal joint of one wing drawn in front of body; at higher intensity, head feathers sleeked so that stiff ear-tufts project sideways (Fig A); may close eyes to slit, or repeatedly jerk body upwards and bend legs (Scherzinger 1970). If threatened and unable to escape, adopts Threat-posture: head and body plumage ruffled and tail spread (Schüz 1957; Scherzinger 1970, 1972a); see Fig B. When excited, ♂ may land on perch with wings extended and slightly trembling (Glutz and Bauer 1980). ANTAGONISTIC BEHAVIOUR. Territory advertised and defended with Advertising-call and Scale-song (see 1–2 in Voice), though difference in function of these calls not clear. Scale-song used most in autumn, when 1st-year birds are attempting to establish themselves; territorial activity peaks then—also in spring. Apparently, ♂ performs major role in spring but ♀ also involved in autumn (see Voice). In interactions between ♂♂, bird on territory may adopt Threat-posture (Klaus et al. 1976). May attack from behind, often striking at nape with claws; attacked bird often adopts posture similar to Sleeked-upright (Glutz and Bauer 1980). Pursuit-flights of over 500–1000 m occur (Schönn 1976; Glutz and Bauer 1980), and birds will also make uncoordinated take-offs and fly up in loops (Klaus et al. 1965; Scherzinger 1970). HETEROSEXUAL BEHAVIOUR. (1) General. ♂ and ♀ tend to be shy of contact with each other, even during breeding season (Glutz and Bauer 1980). (2) Pair-bonding behaviour. Involves alternating fear of and aggression towards mate—especially at start, even between birds renewing bond in autumn; pursuit and attacks occur. Mostly

B

starts in spring, even from mid-January. Unpaired ♂♂ give Advertising-call persistently from open perch; paired ♂♂ call only briefly. (Glutz and Bauer 1980.) Nest-showing begins 6–8 weeks before laying and continues up to laying or ends 3 weeks earlier (Jansson 1964; Scherzinger 1970). ♂ flies to hole, enters, and looks out; sometimes takes food in; gives Trill-call (see 4 in Voice) and single units of Advertising-call. ♀ approaches hole, giving Begging-call (see 3 in Voice). ♂ leaves hole, flying to ♀, and sometimes attempts to copulate. Shortly, ♀ enters hole and often spends long time looking out; calls (see 12 in Voice) each time she leaves. (König 1965; Schönn 1976.) Not clear whether ♂ offers several holes to ♀. Once hole selected, ♀ cleans it out and stays nearby during day (C König). (3) Mating. Takes place on relatively open branch near nest (Bille 1972). ♀ ready to copulate holds body horizontal with head well forward and tail raised; may solicit with Kiu-call (see 9 in Voice). ♂ bows forward, and takes off giving Trill-call; hovers briefly over ♀ and lands on her back to copulate, keeping balance by fluttering, and sometimes gripping ♀'s nape feathers. ♀ calls; perhaps ♂ also (see 6 in Voice). ♂ flies off giving 'düo' call (see 5 in Voice), usually to self-preen and shake. (Jansson 1964; Klaus et al. 1965; König 1968; Scherzinger 1970.) One pair copulated c. 60 times over 25 days, starting 10 days before 1st egg laid (Jansson 1964); once, 4 times in 30 min c. 4 days before 1st egg laid (Bille 1972). Ends after laying of last eggs or declines gradually up to hatching (Glutz and Bauer 1980). (4) Courtship-feeding. Starts after Nest-showing. ♂ gives Advertising-calls and 'düo' calls, lifts prey high in bill, and 'trills'. ♀ flies in, giving Begging-calls, and takes prey from ♂'s bill; ♀'s calls develop into shrill screeching, sometimes accompanied by wing-flapping (Jansson 1964). See also sub-section 2 (above). (5) Allopreening. Mutual Allopreening occurs (Glutz and Bauer 1980). (6) Behaviour at nest. Only ♀ incubates, remaining in or near nest also before and during laying; appears clumsy and sluggish. Once incubation starts, leaves nest only briefly, e.g. to take food from ♂ (König 1965; Schönn 1976; Glutz and Bauer 1980). Such transfers always away from nest (usually 30–50 m), or ♀ may take item from ♂'s food-cache (C König). Relations within Family Group. Young give discomfort calls from 1–2 days before hatching. 3–4 days after hatching sit up on heels; can stand upright at 15–16 days. Open eyes at 9–10 days. (Glutz and Bauer 1980.) From 3–5 days after hatching, ♀ starts to keep nest clear of pellets, food remains, and faeces (Scherzinger 1970; Schönn 1976). ♀ broods young in first 2 weeks, but will leave them alone after 10 days; in 4th week, spends night in hole only in cool weather. Normally, only ♀ feeds young in nest; ♂ passes prey to ♀ outside hole (Glutz and Bauer 1980), and ♀ sometimes stimulates ♂ to hunt by apparent attacks (Schönn 1976). Young beg with food-calls from hatching onwards, especially in response to Begging-call of ♀, Advertising-call of ♂, or Trill-call. ♀ usually feeds one nestling until sated, showing no favour towards more demonstrative young. Young able to deal with prey independently from c. 1–2 weeks after fledging. (Glutz and Bauer 1980.) At 3 weeks, start to climb up to nest-hole, and leave at (28–)30–34 days, usually in morning—within a few hours of each other or over 3–4 days (Bergmann and Ganso 1965; Scherzinger 1970; Schönn 1976). ♀ immediately lures young into cover. Fed by both parents until driven out of territory c. 4 weeks after fledging. (Glutz and Bauer 1980.) Mutual Allopreening between fledged young occurs (Klaus et al. 1982). Anti-predator Responses of Young. Fall silent on hearing Kiu-call of adult or scratching on nest-tree (C König). Parental Anti-predator Strategies. (1) Passive measures. Unlike *A. funereus*, ♀ does not put head out of hole when nest-tree scratched (C König). (2) Active measures. In nest, ♀ stands over young with head and body feathers ruffled and wings raised with underside towards intruder; sometimes Bill-snaps. If hand put into hole, strikes with claws or tries to hide under young. Some ♀♀ allow themselves to be lifted out. Will make dive-attacks on man, dog, Jay *Garrulus glandarius*, and Nutcracker *Nucifraga caryocatactes* near nest: leaves perch 'with wing-rattling and tail-whipping', approaches in silent glide, and brushes intruder; may draw blood. No distraction-lure displays. (Bergmann and Ganso 1965; Scherzinger 1970; C König.) Will use Threat-posture in defence of fledged young (Klaus et al. 1982).

(Figs A–B adapted from photographs in Scherzinger 1970 and Klaus et al. 1982.) DJB

Voice. Advertising-call of ♂ (call 1a), Scale-song (call 2), and Begging-call of ♀ (call 3) heard most frequently. Calls intermediate between those described here also occur (W Thönen). Calls of ♂ and ♀ largely the same, though voice of ♀ higher pitched. Hissing-call (given in threat by many Strigidae) does not occur. Both sexes and young Bill-snap in threat (Glutz and Bauer 1980); for wing-rattling, see Social Pattern and Behaviour and Food. Account based on outline supplied by W Thönen. Major studies by König (1968) and Schönn (1976, 1978). For additional sonagrams, see König (1968) and Glutz and Bauer (1980).

Calls of Adults. (1) Advertising-call. (a) ♂'s call. A flute-like 'dü' (Fig Ia) repeated monotonously at 5–10 units per 10 s (mostly 7–8 per 10 s), rate depending on level of excitement; audible at up to 1 km (C König). Bouts of calling may last several minutes, usually with short gap (omission of 1–2 units) every 5–15 units, though excited bird may give 50 or more units without pause. Suggests Advertising-call of Scops Owl *Otus scops*, but units of that species' call are longer, repeated more slowly, and have characteristic modulation at beginning. Calls of midwife toad *Alytes obstetricans* even more similar but shorter, like stroke with hammer. In common stuttering variant of Advertising-call (given in greater excitement), groups of 1–3(–5) similar but shorter, weaker, and slightly lower-pitched units—'ü-ü-ü' (Fig Ib)—are interspersed with normal ones (H Delin, W Thönen); such insertions vary from persistent trill to voiceless 'tj' (Glutz and Bauer 1980). Softer version of Advertising-call given on bringing food to ♀ when eggs or dependent young present. Calling occurs all year except during moult and bad winter weather, but peaks during pair-formation (from mid-March) and in autumn. Most regular in $\frac{1}{2}$ hr before sunrise and after sunset, but also occurs during day; rarely at night. (W Thönen; see also König 1968, Thönen 1968, Burton and Johnson 1984.) (b) ♀'s call. Similar to stuttering variant of ♂'s call but usually higher pitched (by up to a third) and rather cackling. Given usually in response to Advertising-call of ♂ (W Thönen); also in defending territory against another ♂ or ♀, or when man climbs nest-tree (Scherzinger 1970). (2) Scale-song. Series of 5–11 whistles recalling bicycle pump with finger over hole: 'cheek-cheek- ...', notes ascending in pitch and increasing in delivery rate; 1st note about same pitch as Advertising-call (sometimes much higher), last note(s) often squeaking. In recording (Fig II), crescendo up to c. 6th unit, then slight

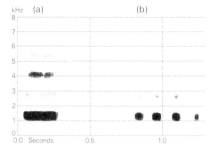
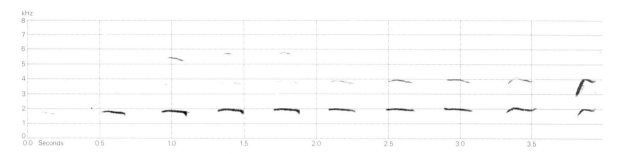

I C König West Germany April 1981

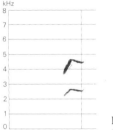

II C Chappuis/Alauda (1979) France October 1968

diminuendo, and last 2 units suddenly louder (J Hall-Craggs). Usually given at intervals of 10 min or more, though can be 1–3 min in disputes. Given by ♂ and ♀—all year, but especially in autumn when connected with territorial advertisement and driving out of offspring. (H Delin, W Thönen.) (3) Begging-call of ♀. High-pitched, not-quite-pure 'siiiih', 'siiht', or 'tseeh' lasting almost 1 s and recalling Blackbird *Turdus merula*, Robin *Erithacus rubecula*, or distant Swift *Apus apus*, but with demanding character; pitch rises at start and has short drop at end. Given all year but especially in pair-formation (e.g. in answer to ♂'s Advertising-call) and in begging food from ♂—and may continue while distributing food to young. (König 1968; H Delin, W Thönen; see also Bergmann and Wiesner 1982.) Infrequently-heard ♂ version, lower pitched 'tsüüüüüh' (H Delin); weaker and shorter than ♀'s call (C König). (4) Trill-call. A rapidly repeated (*c*. 8 per s), short, soft 'tü', usually higher pitched than Advertising-call (W Thönen); hard bübübübüb-bübübüb...' (König 1965). Given by ♂ and ♀, usually when carrying food (W Thönen); also by ♂ before copulation (Scherzinger 1970) and during Nest-showing (König 1965). (5) Soft, attenuated 'düö', or 'züüü', falling slightly in pitch; given by ♂ in association with food transfer and copulation (König 1965; Scherzinger 1970). (6) Shrill, twittering 'kjikjikji...' given by ♀ during copulation (König 1968); perhaps also by ♂, developing into screeching 'squie' (Glutz and Bauer 1980). (7) High-pitched squeaking or creaking, probably by both sexes. Given usually when food transferred; rarely, by ♂ in response to imitations of its call. (W Thönen.) (8) Contact-call. Soft whistling 'uit' or 'üh', rising in pitch; like soft human whistle. Given by ♂ and ♀ (Glutz and Bauer 1980; W Thönen). (9) Kiu-call. Accelerating series of (usually) 5–10 'kiu' or 'kjäw' units with falling pitch and increasing volume; rather mewing. In recording by C König, 'kiu' initially, but final 3–4 units more like 'kyup' (D J Brooks). Given in conflicts between birds (including mates), and (1 or more units) to young as warning of danger; also by ♀ to solicit copulation. (König 1968; W Thönen.) (10) Alarm-call. Single note, similar to call 8 but shorter, rather clear, and not mewing. Given by ♂ and ♀ when potential predator near nest or fledged young (W Thönen). (11) Sharp 'zr' given by birds fighting; also by ♀ attacking Nutcracker *Nucifraga caryocatactes* attempting to steal prey (Thönen 1965; W Thönen). (12) Quiet 'zirl' or 'zrl' from ♀ leaving hole during Nest-showing (König 1965).

CALLS OF YOUNG. Food-call similar to Begging-call of adult ♀ (call 3), with same short drop in pitch at end, but shorter (rarely as much as $\frac{1}{2}$ s), more pure, and not rising in pitch at start. Given during period in nest and after leaving it; probably also until independence according to W Thönen, or changing to Contact-call (adult call 8) at *c*. 7 weeks according to Glutz and Bauer (1980). High-pitched twittering given in discomfort (Glutz and Bauer 1980). Call similar to adult call 11, but weaker, given in nest. (W Thönen.)

DJB

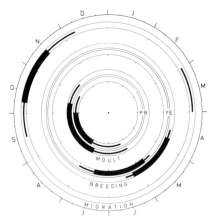

Breeding. SEASON. Baltic region: see diagram. Austria and southern West Germany: laying begins early April. Western USSR: laying begins late April. (Dementiev and Gladkov 1951a; Scherzinger 1974; Makatsch 1976.) SITE. In hole in tree, natural or excavated by woodpecker (Picidae); will also use nest-box. Of 13 nests, West Germany, average height above ground 5·5 m (3·5–8), facing in any direction (Scherzinger 1974a). Nest: available hole; no material added but debris may be removed. Building: cleaning of nest by ♀ (Schönn 1976). EGGS. See Plate 97. Blunt elliptical, smooth and slightly glossy; white. 29 × 23 mm (27–32 × 22–25), $n=80$; calculated weight 8·3 g (Schönwetter 1967); fresh weight 8 g, $n=4$ (Glutz and Bauer 1980). Clutch: 4–7 (3–10). Of 24 clutches (various countries): 3 eggs, 3; 4, 6; 5, 5; 6, 5; 7, 3; 8, 2; average 5·2 (Schönn 1978). Of 13 clutches, West Germany: 3 eggs, 3; 4, 6; 5, 2; 6, 1; 7, 1; mean 4·3 (Scherzinger 1974a). One brood. Replacements probably laid after egg loss (Makatsch 1976). Laying interval 2 days (Schönn 1976); at one nest, first 4 eggs laid in 5–6 days, then 3 in 6 days (Sonerud et al. 1972). INCUBATION. 28–30 days. Begins with last egg (Schönn 1976); hatching synchronous or nearly so. By ♀ only (Schönn 1976). YOUNG. Altricial and nidicolous. Cared for by ♀ and brooded while small. ♂ brings food exclusively to start with, though ♀ joins in towards end of fledging period. FLEDGING TO MATURITY. Fledging period 27–34 days. Become independent c. 4 weeks later (Glutz and Bauer 1980); leave vicinity of nest from 1–11 weeks after fledging (Scherzinger 1974a). Age of first breeding 1 year (Bergmann 1939). BREEDING SUCCESS. In West Germany, 57 young fledged from 17 nests, average 3·3 per nest (Scherzinger 1974a). 17 young fledged from 44 eggs in unspecified number of nests (Schönn 1978). At one nest, Norway, 7 young flew from 7 eggs laid (Sonerud et al. 1972). In different years and habitats, south-east Norway, average number of fledged young per egg 0·52–0·94, due to variations in clutch size (5·8–7·3) and food supply (R Solheim).

Plumages (nominate *passerinum*). ADULT. Crown, hindneck, and sides of neck dark olive-grey, densely marked by fine white or cream spots or short bars, each spot or bar narrowly bordered black; spots or bars on sides of crown and sides of neck sometimes less distinctly defined and cream-buff, central crown and hindneck sometimes almost uniform, only indistinctly spotted. A broad white streak obliquely running down from behind each ear and another across side of neck join at upper mantle, forming > <-mark when seen from behind; colour inside each V rather dark black-brown, somewhat resembling pair of dark eyes and, together with white streaks, forming rather indistinct face-like pattern; white streaks often partly mottled dusky and tinged cream or buff, especially on upper mantle. Lower mantle, scapulars, tertials, back, rump, and upper tail-coverts dark olive-grey with short white or cream-buff bars, each bar narrowly accentuated by black-brown border. Poorly defined facial disc closely spotted and barred white and black-brown; narrow white line over eye, widening into wider white streak along sides of bill and over lower cheeks. Upper chin white; lower chin, central throat, and bar across throat to sides of neck mottled grey, olive-brown, and white, bordered below by mainly white bar. Sides of breast and upper flanks dark olive-brown with rather contrasting white or cream-buff bars; pale-and-brown barred patches on sides of breast almost meet on central chest. Feathering of legs white with close brown and buff mottling. Remainder of underparts largely white, feathers of breast, belly, and lower flanks and longest under tail-coverts with rather narrow and sharp black shaft-streaks, slightly widening on feather-tips; lower belly, vent, and shorter under tail-coverts only indistinctly streaked grey. Tail dark cinnamon-brown grading to dark olive-grey on tip, marked with 3–4 narrow and equally-spaced bars, each narrowly bordered by black (1–2 more bars hidden under tail-coverts); white of bars tinged cream-buff, especially towards t6 and when plumage fresh; t1–t4 with narrow paired white spots on tips, reduced to white spot on tip of inner web only on about t5–t6. Flight-feathers brown-black with 4–6 rather narrow, well-spaced and rather indistinct pale grey-buff bars; pale bars end in distinct white or cream spots at border of both inner and outer webs, on outer web often partly accentuated by black; brown-black intervening bars partly tinged cinnamon-brown towards base of outer web. Upper wing-coverts cinnamon-brown grading to olive-grey on tips, closely barred and spotted white or cream-buff like mantle and scapulars; lesser upper wing-coverts almost uniform dark olive-grey, contrasting with white leading-edge of wing. Under wing-coverts and axillaries white, longer feathers with some black marks; undersurface of flight-feathers contrastingly dark grey, marked with narrow white bars. Some individual variation in colour of upperparts and upper wing-coverts and in extent and colour of marks; general colour between olive-brown and olive-grey (usually not as cinnamon-brown as 1st adult), marks either fine specks or distinct though narrow bars, colour of marks between cream-buff and white (in part depending on wear of plumage). Conspicuousness of white streaks on face, across throat, and of face-like pattern on hindneck strongly dependent on carriage of bird and angle of light. DOWNY YOUNG. 1st down (neoptile) short and dense, white; down tinged greyish from 6th–7th day, caused by breaking through of dark juvenile (mesoptile) feather-pins hidden under neoptile; on 8th day, pins visible; from 12th day, mesoptile pins open and produce feathering, but appearance largely spiny on 10th–18th day; still much white neoptile clinging to mesoptile feather-tips during 3rd week, nestling appearing mottled grey-white (Scherzinger 1969b; Schönn 1978). JUVENILE. 2nd down (mesoptile) not as downy as in (e.g.) *Strix* and *Asio* owls, more feather-like (as in *Athene*), but still distinctly softer and more woolly than adult plumage. Mesoptile fully developed by 18–20th day, but belly still largely naked until 4th–5th week and flight-feathers and tail not full-grown until 6th–7th. Mesoptile rather similar to adult plumage, but upperparts, facial disc, chest, and upper wing-coverts dull chocolate-brown or cinnamon-brown (not as olive-grey or olive-brown as adult), only indistinctly spotted buff on forecrown, scapulars, and longer upper wing-coverts (not extensively and distinctly spotted white or cream-buff); no white streaks on hindneck and sides of neck; uniform dark chest indistinctly defined from off-white throat and breast (not barred and well-demarcated as in adult), dark grey-brown streaks on off-white breast, belly, and flanks rather short and broad, poorly defined; vent and under tail-coverts uniform off-white or indistinctly tipped grey, fluffy; on leaving nest, face dark brown (not as distinctly spotted and barred white as adult) with contrastingly white X-mark above and between eyes reaching down to sides of throat; as in adult, extent of white also dependent on carriage and angle of light. Flight-feathers, greater primary coverts, and tail mainly like adult, but see description of 1st adult. (Heinroth and Heinroth 1926–7; König 1978; Schönn 1978; Baumann 1982; RMNH.) First Adult. Like adult, but upperparts often more rufous-brown or cinnamon-brown, less tinged with olive-grey; spots on crown often smaller, forming rounded buff specks (not cream-white short bars); hindneck almost uniform brown, white marks (forming face-like pattern in adult) hardly developed; no contrasting barring on sides of neck; mantle and scapulars with ill-defined and rather rounded pale buff-brown subterminal spots (not with distinct short bars); sides of breast and chest with rather poorly defined cream-buff and rufous-brown bars, meeting on centre of breast (in adult, bars contrastingly cream-white and black-brown, scarcely meeting in centre of chest or breast); streaks on belly and flanks browner and less sharp than in adult; flight-feathers (including tertials), greater primary coverts, and tail still juvenile, tips of inner tertials more rufous-brown and with smaller and buffier spots at sides, t6 with narrow ill-defined rufous-buff fringe on tip of outer web. Some birds have upperparts and chest rather olive-grey (rather similar to adult) and these may show more contrasting and paler cream marks; in these, rufous-brown tinge of tips of juvenile tertials and tail-feathers show some contrast in colour with upperparts, but indistinguishable in spring when tail and tertials worn.

Bare parts. Adult and First Adult. Iris pale yellow, sulphur-yellow, or bright yellow. Eyelids black. Bill wax-yellow, base tinged dusky olive-grey or greenish; cere blue-grey. Soles bright yellow, yellow-grey, or pale blue-grey; claws black-brown with paler brown bases. Downy Young and Juvenile. Eyes open at 9th–10th day; iris pale grey when just open, soon changing to yellow. Bare skin, including that of cere and legs, pink, shading to grey in 1st week. At fledging, iris yellow, eyelids black, bill olive-yellow, cere and foot pale yellowish-flesh or bluish-grey, claws black. (Heinroth and Heinroth 1926–7; Scherzinger 1969b; König 1978; Schönn 1978.)

Moults. Adult Post-breeding. Complete, primaries descendant. Starts July, soon after fledging of young; one from European USSR, 12 August, not yet moulting (ZMM). In one captive bird, complete moult early June to early September. Tail-feathers lost almost simultaneously, with centripetal tendency, at about same time as shedding of p3. Secondaries moulted from 3 centres (see Haarhaus 1983). Moult completed September–October. Post-juvenile. Partial: all head, body, and wing, but not flight-feathers, tertials, greater primary coverts, or tail. Starts with face late July to mid-August, a few weeks after fledging; moult more general mid-August to mid-September; completed at age of c. 10 weeks, or October(–November). (Stresemann and Stresemann 1966; Scherzinger 1969b; Glutz and Bauer 1980; Haarhaus 1983; BMNH, ZMM.)

Measurements. Nominate *passerinum*. Central and northern Europe, all year; skins (RMNH, ZFMK, ZMA, ZMM). Bill (F) to forehead, bill (C) to cere.

WING	♂ 97·5	(2·14; 11)	93–100 ♀ 105·3 (2·57; 23)	101–109
TAIL	56·1	(2·37; 10)	53–60 61·9 (1·99; 17)	58–65
BILL (F)	14·4	(0·60; 6)	13·5–15·0 15·7 (0·62; 8)	14·8–16·3
BILL (C)	10·5	(0·35; 7)	9·9–11·0 11·5 (0·43; 8)	10·8–12·2
TARSUS	16·5	(0·37; 6)	16·0–17·0 17·4 (0·78; 7)	16·8–18·5

Sex differences significant.

Weights. Adult and First Adult. Norway: ♂ 58 (50–65), ♀ 73 (67–77) (Haftorn 1971); ♂ 61, ♀♀ 67, 77 (Hagen 1942). Finland: ♂ 47–62 (5), ♀ 55–70 (12) (Haartman *et al.* 1963–72). East Germany: ♀, June, 80 (Schönn 1978); wounded ♀, April, 53 (Eck 1971). USSR: ♂♂ 51, 56, 62; ♀♀ 58, 64, 70, 79 (Dementiev and Gladkov 1951a; Gavrin *et al.* 1962; Fedyushin and Dolbik 1967). Alps: ♂ 65–72, ♀ 74–83 (Schönn 1978).

Downy Young and Juvenile. At hatching, average 5·6; approximate further growth (read from graphs): average 10·5 on 5th day, 27 on 10th, 45 on 14th, 60 on 20th, 65 on 25th–27th; ♀ 4–10 heavier than ♂ from 6th day (Schönn 1978).

Structure. Wing rather short, broad, tip rounded. 10 primaries: p7 longest, p8 0·5–2 shorter, p9 7–14, p10 27–35, p6 1–4, p5 9–14, p4 14–19, p1 21–26. Outer web of p6–p9 and inner web of p7–p10 emarginated. Outer web of p10 and emarginated portions of outer webs of p8–p9 faintly serrated. Tail rather short, square, 12 feathers; t6 0–3 shorter than t1. Bill relatively heavy, deep and wide at base. Facial disc poorly defined, not surrounded by stiffened ruff-feathers. Leg and toes feathered to base of claws, only soles bare. Middle toe without claw 13·7 (0·70; 7) 12·7–14·4 in ♂, 14·5 (0·54; 7) 13·8–15·2 in ♀; outer toe without claw c. 65% of middle, inner c. 82%, hind c. 55%. Claws rather long, slender; middle claw 9·1 (0·72; 7) 8·0–9·8 in ♂, 9·9 (0·54; 8) 9·5–10·6 in ♀; outer claw c. 79% of middle, inner c. 95%, hind c. 80%.

Geographical variation. Slight, in colour only. Siberian race *orientale* differs from European nominate *passerinum* in paler, greyer, and less brown upperparts which show purer white and more sharply defined spots; breast and flanks more strongly marked brown (Hartert 1912–21; Vaurie 1965); difference mainly caused by different proportion of colour morphs, grey morph predominating in central and eastern Siberia, brown in western Siberia and Europe, though proportion varies rather with locality (Dementiev and Gladkov 1951a). As grey or brown colour and sharpness of marks also dependent on age (see Plumages), validity of *orientale* in need of rechecking with known-age specimens.

Forms species-group with Pearl-spotted Owlet *G. perlatum* from Afrotropical savannas, Red-chested Owlet *G. tephronotum* from Afrotropical rain forests, American Pygmy Owl *G. gnoma* from western North and Middle America, Least Pygmy Owl *G. minutissimum* from Mexico south to northern and eastern South America, Ferruginous Pygmy Owl *G. brasilianum* from southern USA south to Tierra del Fuego, and Cuban Pygmy Owl *G. siju* from Cuba (Snow 1978; C S Roselaar). CSR

Athene noctua Little Owl

PLATES 48 and 54
[between pages 494 and 495, and facing page 566]

Du. Steenuil   Fr. Chouette chevêche   Ge. Steinkauz
Ru. Домовый сыч   Sp. Mochuelo común   Sw. Minervas uggla

Strix noctua Scopoli, 1769

Polytypic. Nominate *noctua* (Scopoli, 1769), Sardinia, Corsica, mainland Italy, south-east Austria, north-west Yugoslavia, southern Czechoslovakia, Hungary, and Rumania north and west of Carpathian mountains, intergrading over wide area with *vidalii* in southern France, Switzerland, southern Germany, Austria, and Czechoslovakia and with *indigena* in Yugoslavia; *vidalii* Brehm, 1858, Iberia north to Netherlands, east through Denmark and West and East Germany to Poland and Baltic States, introduced Britain; intergrades with *indigena* in European USSR north of Kiev, Voronezh, and Orenburg; *indigena* Brehm, 1855, Albania, south-east Yugoslavia, southern and eastern Rumania, and south European USSR, south to Crete, Rhodes, Turkey (except south-east), Levant south to about Haifa, Transcaucasia, and south-west Siberia; *glaux* (Savigny, 1809), North Africa and coastal Israel north to Haifa; *saharae* (Kleinschmidt, 1909a), northern and central Sahara, south of *glaux*; *lilith* Hartert, 1913, Cyprus and inland Levant from Sinai to south-east Turkey, intergrading with *bactriana* in Iraq and with *saharae* in Saudi Arabia; *bactriana* Blyth, 1847, south-east Azerbaydzhan and Iran east in Transcaspia to Afghanistan and Lake Balkhash. Extralimital: 3 races in central Asia, 2 along Red Sea coast from Sudan to Somalia.

Field characters. 21–23 cm; wing-span 54–58 cm. 15% larger than Scops Owl *Otus scops* and much bulkier in most attitudes; 10% smaller and markedly less large-headed than Tengmalm's Owl *Aegolius funereus*. Rather small, non-eared, compact, and tubby owl, with frowning, baleful expression, spotted upperparts, splashed underparts, and proportionately long legs. Plumage colour variable, from grey- or rufous-brown to ochre-buff. Facial disc obvious on dark birds, creating stern expression. Flight consists of alternating bouts of flapping and closing wings, producing conspicuously undulating progress. Partly diurnal, frequently perching on artefact or rock. Commonest call sharp and complaining in tone. Sexes similar; no seasonal variation. Juvenile separable. 7 races in west Palearctic, subject to variation and intergradation; 2 races, easily distinguished at end of clines, treated here (see also Geographical Variation).

(1) West European race, *vidalii*. ADULT. Darkest race, with ground-colour of upperparts dark umber-brown and that of flight-feathers sepia. Crown closely flecked white; mantle, scapulars, and larger wing-coverts spotted white, boldly on hindneck and edges of scapulars. Flight-feathers show 3–4 pale buff bars overall (these most conspicuous of all small owls); tail shows at least 3 buff bars (but these not conspicuous). Facial disc obvious: basically white on edges of crown, around outer cheeks, and under bill and eyes, but dusky from bill through eye and in half-circle in centre of cheeks; general effect is to add to fierce, baleful stare of yellow eyes. Ground-colour of underparts white, but band across throat black-brown and most of lower breast and body broadly splashed and streaked dark umber-brown (in pattern as obvious as spots of upperparts). Underwing also white, with dark spots on coverts and dusky bars on flight-feathers showing through. Legs feathered white, toes sparsely. Bill pale green-yellow. JUVENILE. Distinctly paler (greyer) and much more uniformly patterned than adult; spots buff (not white) and streaking narrower and paler brown. Crown virtually uniform. (2) North-west Arabian race, *lilith*. Palest race, with plumage pattern as *vidalii* and other races but ground-colour of upperparts pale buff, white markings of upperparts more extensive, and markings of underparts rufous (not brown) in tone and less extensive. Facial disc even less distinct but eyes have noticeably dark rim. Other races intermediate between these two, with variations in ground-colour, spotting, and streaking almost random.

South of Fenno-Scandia, commonest small owl of west Palearctic and likely to be seen before any other. Shows some resemblance in character and appearance to much smaller Pygmy Owl *Glaucidium passerinum*, smaller *Otus* owls, and larger *Aegolius funereus* and thus chances of confusion exist. Close observation essential to separation of these species (see *G. passerinum*, *O. scops*, Striated Scops Owl *O. brucei*, and *A. funereus*). Important to remember that *Athene noctua* is the most boldly spotted small owl, lacks ears or sharp corners to head, and has distinctive calls and flight. In south-east of west Palearctic, care should be taken not to overlook possible occurrence of Spotted Little Owl *A. brama*: similar to darker races of *A. noctua* but with boldly barred underparts with pale central division, and darker crown and face; race of *A. noctua* in area of potential overlap in Iraq, *bactriana*, almost as pale as *lilith*. Flight usually low and with commonest action (see above) strongly recalling large woodpecker *Picus*, but when hunting or in escape, action less bounding, wing-beats less fitful, and flight more direct. Long legs allow nimble walk and fast loping run; can hop. Carriage usually erect but, particularly in excitement or alarm, also adopts half-upright, twisted, or hunched attitudes, from which bird bobs up and down in demented manner.

Vagrants occur in coastal scrub and ground cover. Territorial call a mellow hoot, 'goooek'; more commonly, however, gives a sharp, clear '(k)weew'.

Habitat. In upper middle, lower middle, and regionally down to lower latitudes, both continental and marginally oceanic, mainly temperate, steppe, and Mediterranean, but extending to boreal and tropical also. While adapting to windy and rainy climates favours warm, even semi-arid conditions and is vulnerable to severe frosts and snow cover. Not a forest species, and tends even to avoid margins or enclaves between forests.

Less arboreal and more terrestrial than most Strigiformes, and particularly addicted to use of conspicuous perches as look-outs, as well as to frequent brief low flights in daylight in the open. Avoids tall and dense stands of trees and any dense vegetation, as in wetlands and croplands. Shows little attraction to water. In northern and middle parts of range a lowland species, ascending even in mountainous regions of central Europe only rarely above c. 600 m (Glutz and Bauer 1980). Further south however found up to 1900–2000 m in Georgia and Armeniya, 2000–2800 m in Altay, and 4200 m in Pamir, occupying ravines, gorges, gullies, walls of river terraces, precipitous cliffs, and dry unwooded mountains, as well as dry hilly steppes, semi-deserts, sandy or clay deserts, and wastelands in southern USSR (Dementiev and Gladkov 1951a). Contrastingly, recently colonized habitats in Britain are agricultural countryside well endowed with hedgerow trees and farm buildings, old orchards with parkland, drained fenland with lines of pollard willows *Salix*, and marginal types such as industrial waste ground, sand-dunes, moorland edges, old quarries, seacliffs and inshore islands, treeless rising ground, and settlements, although only infrequently and impermanently within major cities (Sharrock 1976). A special liking for old orchards is widely recognized, and even under intensive management occupancy may be continued provided a nucleus of old trees remains standing. Also in central Europe, pastures and meadows flanked by pollard trees affording ample nest-sites and hunting look-out posts, and year-round short herbage with plenty of invertebrate prey provide optimal habitat, characteristic of conservative small farming economies. Carrying capacity can however be prejudiced by intensification of agricultural methods, ground clearance, reduced availability of nest-holes through tree-felling and demolition or repair of old buildings, excessive use of toxic chemicals, traffic deaths, etc. Such injurious factors can be partially offset by conservation and, where necessary, by regular pollarding of old nesting trees, provision of nest-boxes in substitution for lost natural holes, and other remedial measures (Glutz and Bauer 1980). Among many alternative nesting-places adopted are crevices in ruins and down wells, adobe buildings in Sahara, holes in quarries, walls, sand-pits, and disused rabbit burrows (Bannerman 1955), while in USSR breeding recorded in windmills, mud tombs, granaries, haylofts, haystacks, below overhanging rocks, in burrows of large jerboas, and in nests of rock nuthatches *Sitta*; some are tunnelled by the birds themselves (Dementiev and Gladkov 1951a).

Essential elements of habitat, including open hunting ground rich in small prey, hunting perches, day-roosts, and nest-holes, and with benign climate and land management regimes which give reasonable long-term continuity without too-drastic changes, can be met within wide diversity of landscapes and ecosystems. Recent declines in numbers and range in much of Europe (see Distribution and Population) show that tolerable limits are tending to be exceeded.

Distribution. Range decreased locally in many parts of Europe (see also Population).

BRITAIN. Introduced late 19th century mainly in Kent and Northampton (England), then spread rapidly over much of England and Wales, aided by further introductions, until 1930, with slower spread north into Scotland to 1950 (Parslow 1967; Sharrock 1976). SWEDEN. Bred 1939 (LR). AUSTRIA. Extinct in some areas; see Glutz and Bauer (1980). SWITZERLAND. Marked and continuing decrease in range (Juillard 1980). KUWAIT. Present all year, but no confirmed breeding (PRH).

Accidental. Ireland, Norway, Sweden, Finland, Malta.

Population. Fluctuating, especially in north of range where marked decreases after severe winters. Recent decreases, often marked, over much of Europe; ascribed mainly to habitat changes, including loss of suitable nest-sites (offset in some areas by provision of artificial sites—see, e.g., Ullrich 1980, Juillard 1980). BRITAIN. Rapid increase after introduction in late 19th century to c. 1930, then more slowly, with local decreases in 1940s, perhaps due to hard winters, and further local decreases, especially in south-east, perhaps due to pesticides, from 1955 to early 1960s. Peak populations perhaps in early 1930s; estimated 1000–10000 pairs 1967 and 7000–14000 pairs 1976. (Parslow 1967; Sharrock 1976.) FRANCE. 30000–80000 pairs, decreased (Yeatman 1976; Mikkola 1983). SPAIN. At least 50000 pairs (AN). BELGIUM. Decreased markedly, due to habitat destruction and pesticides; c. 12000 pairs 1950, c. 4000 pairs 1972 (Lippens and Wille 1972). NETHERLANDS. Declined since c. 1935, probably mainly due to habitat changes; estimated 6000–8000 pairs 1974–5 and likely to have decreased since (Teixeira 1979; CSR). See Glutz and Bauer (1980) for local counts. WEST GERMANY. Marked decline, probably due to habitat changes. Numbers severely affected by hard winters; see Glutz and Bauer (1980) and Bauer and Thielcke (1982), where many area estimates. DENMARK. Decreased (Dybbro 1976). EAST GERMANY. Marked recent decline (Makatsch 1981); rare or very rare (Glutz and Bauer 1980). POLAND. Scarce in most areas, with marked declines after severe winters

516 Buboninae

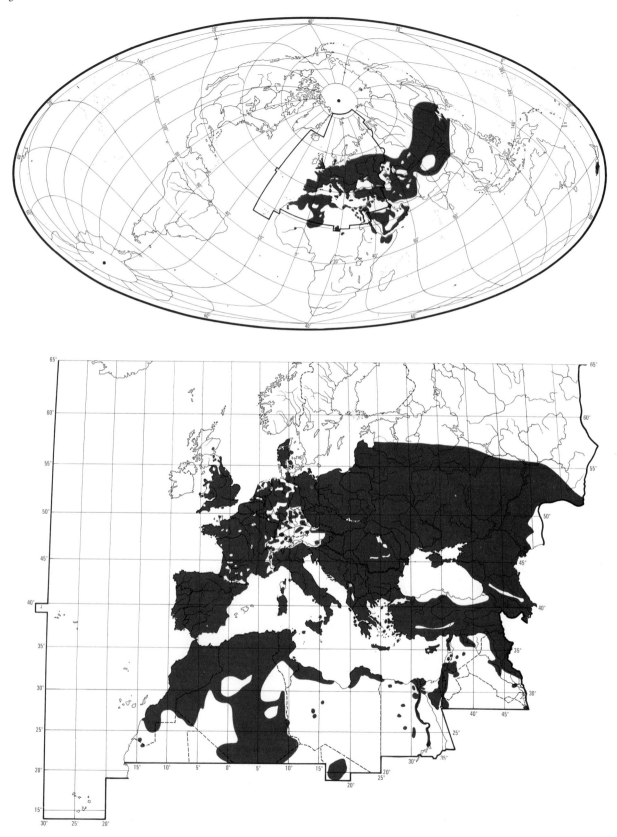

(Tomiałojć 1976a). CZECHOSLOVAKIA. Marked decrease since c. 1960 (KH). AUSTRIA. Decreasing (Glutz and Bauer 1980). SWITZERLAND. Marked decrease, due especially to habitat changes; estimated 185 pairs 1980 (Juillard 1980). ITALY. Marked decrease in some areas, though signs of recent recovery (SF). ALBANIA. Rare (EN). GREECE. Marked decline since c. 1970 (HJB, WB, GM). BULGARIA. Common (Patev 1950); no apparent recent change (JLR). USSR. Latvia: rare (Baumanis and Blūms 1969). Lithuania: common, especially in south (Ivanauskas 1964; Valius et al. 1977). CYPRUS. Common (PRF, PFS). LIBYA. Common in north (Bundy 1976). TUNISIA. Common (MS). MOROCCO. Common (JDRV).

Survival. West Germany and Netherlands: mortality in 1st year of life 70·1%, annual adult mortality 35·2% (Exo and Hennes 1980). Switzerland: 1st-year mortality 74%, 2nd-year 64% (Glutz and Bauer 1980). For seasonal mortality, see Glue (1973). Oldest ringed bird 15 years 7 months (Rydzewski 1978), but this considered dubious by Glutz and Bauer (1980), who gave at least $9\frac{1}{2}$ years.

Movements. Resident. 1st-year birds disperse (randomly) somewhat, but mostly settle within 20 km of birthplace (see below); only small proportion of ringing recoveries show movement above 50 km. Some adults also make (normally) short dispersals in autumn and winter.

In introduced British population, birds disperse on average further than sympatric Tawny Owl *Strix aluco* (see that species). Those ringed as nestlings and May–July juveniles, and recovered August onwards ($n = 174$), moved as follows: 0–10 km 64%, 11–20 km 17%, 21–50 km 12%, 51–100 km 4%, over 100 km 3% (maximum 150 km). A September juvenile recovered 2 years later 175 km NNE of ringing site. Unlike *S. aluco*, dispersal does not cease in late autumn; several full-grown birds ringed in December–January recovered later some distance away, including 1 December bird found 16 months later (April) 110 km north-east. Most of those ringed as adults recovered within 10 km (usually locally), but several up to 45 km, and one exceptional adult ♀ moved 182 km NNW (May, Dorset, found 13 months later in Hereford) (BTO). Stragglers (none ringed) have crossed Irish Sea to Isle of Man and eastern Ireland.

Continental birds, also, generally disperse over short distances only. Review of West German and Dutch ringing data (Exo and Hennes 1978, 1980) found juvenile dispersal begins in August. From November–March (inclusive) of 1st winter, median distances were 15 km for West Germany ($n = 30$) and 10–19 km for Netherlands ($n = 18$); after reaching breeding age (April onwards), equivalent figures were 7·5 km ($n = 53$) and 0–9 km ($n = 25$) respectively. Overall, c. 55% had settled within 10 km of birthplace, and 9% at distances above 100 km; c. 74% of adults were recovered within 10 km of ringing site.

In population study in Nordwürttemberg, West Germany (Ullrich 1980), 1st-year birds moved 0·5–220 km, mainly under 40 km; included 3 cases of juveniles dispersing in autumn but later returning to near natal area, one from 36 km away. Recoveries for 5 pairs of siblings showed different degree of dispersal within brood: e.g. 3 km west, 36 km north-east; 3 km WSW, 38 km north; 3 km north-east, 190 km south-west. In 21 cases where breeding site known for birds ringed as nestlings, 18 were 0–16 km (average 6·6) from birthplace, and 1 each at 22, 55, and 190 km; the 20 cases at 0–55 km involved 16 ♂♂ and 4 ♀♀, suggesting ♀♀ more likely to move away (though sex-ratio at ringing unknown). In adults, territorial fidelity normal, though 6 cases found of birds changing territory: 5 moved 1·5–19 km (average 10), and 1 (♀) 150 km south-west. Many adults stay in territory through autumn and winter, but some found residing outside it (up to 20 km distant) for considerable periods, such birds possibly searching for mate (Ullrich 1980).

Longest recoveries: Mittel-Franken, West Germany (June nestling) to Bas-Rhin, France (January), 230 km WSW; Hessen, West Germany (October) to Halle, East Germany (April), 270 km north-east; Dresden, East Germany (July, full-grown) to Austria (February), 297 km south; 2 Nordwürttemberg nestlings (June) found October in Switzerland (220 km south-west) and Zielona Góra, Poland (600 km north-east) (Glutz and Bauer 1980). Such erratic movements as these consistent with vagrancy to Helgoland and southern Fenno-Scandia.

No data on dispersal for resident southern populations.

Food. Largely small mammals and birds, reptiles, amphibians, beetles (Coleoptera), crickets, etc. (Orthoptera), earwigs (Dermaptera), and earthworms (Lumbricidae). Hunting essentially nocturnal: mainly dusk to midnight, then break of 2 hrs before resumption to dawn; little or no hunting done during day, even when young in nest (Glue and Scott 1980; see also Hibbert-Ware 1937–8, Haverschmidt 1946, Glutz and Bauer 1980, Laursen 1981). Most hunting done from perch, bird dropping on to prey below or nearby; occasionally hovers at 1–2 m, rather clumsily compared with Barn Owl *Tyto alba* (Haverschmidt 1946; Glue 1979). Also hunts on ground, largely for beetles and earthworms. One bird took beetles landing on lawn by approaching in short low flight, rapid run, or 2–3 hops, depending on distance (Beven 1979). Hunts earthworms by hopping over ground; stops, bends forward, and pulls worm with bill, often flapping wings and sometimes falling on to back as worm comes loose; on meadow in evening, took at least 23 in 45 min, 5 in 15 min, 19 in 20 min (Haverschmidt 1946). One bird regularly stole worms from Blackbird *Turdus merula* (Tricot 1968). Will pursue flying insects, twisting and turning (Haverschmidt 1946; Beven 1979). Birds taken from roost-sites (e.g. Gilbert 1979), from water (Barber 1925), and from nests and nest-holes (Hibbert-Ware 1937–8; Burton 1983). In east Palearctic, said to pursue desert rodents in burrows underground (Stegmann 1960). Large prey taken with feet, smaller prey (beetles, etc.) usually with bill. During attacks on live prey in captivity, ruffles feathers and mantles prey with wings; mantling may continue after initial attack (Ille 1983). In captivity, unless food supply excessive, mice and sparrows *Passer* eaten completely (Hibbert-Ware 1936). Beetles swallowed whole or (sometimes) held up in foot to be bitten 2–3 times (Beven 1979); earthworms 'swallowed in a single snap' (Haverschmidt 1946). For development of prey-catching and prey-handling skills in young captive birds, see Ille (1983). Cannibalism of young by siblings in nest occurs (e.g. Haverschmidt 1946); also, adult recorded being eaten by (apparently) own mate and young (Mills 1981). According to Hibbert-Ware (1937–8) food-caching behaviour poorly developed, most so-called caches probably being sites used for hand-ling prey; however, accumulations (presumed to be true caches) of 30 or more items recorded in Glutz and Bauer (1980), and cache of 167 headless Storm Petrels *Hydrobates pelagicus* recorded (Lockley 1934a). For caching behaviour in captivity, see Angyal and Konopka (1975, 1977).

Pellets pale grey when damp, rather fragile. Typically 20–40 × 10–20 mm; both ends usually rounded, though occasionally narrowed almost to a thread at one end and then resembling pellets of Kestrel *Falco tinnunculus*; often contain plant material, etc. (Hibbert-Ware 1937–8; Thiollay 1963b; Géroudet 1965). For descriptions of various types, see Hibbert-Ware (1937–8). In one study of captive birds (Hibbert-Ware 1936), all bones of mouse and bird prey recovered in pellets, but this not always so (Rose 1982).

The following prey recorded in west Palearctic. Mammals: hedgehog *Erinaceus europaeus*, shrews (*Sorex minutus, S. araneus, Neomys fodiens, N. anomalus, Suncus etruscus, Crocidura leucodon, C. suaveolens, C. russula*), moles (*Talpa europaea, T. caeca*), bats (*Rhinolophus, Myotis myotis, Plecotus austriacus, Pipistrellus, Nyctalus noctula*), rabbit *Oryctolagus cuniculus* (up to ¾-grown: Tinbergen and Tinbergen 1932), spotted suslik *Citellus suslicus*, garden dormouse *Eliomys quercinus*, common hamster *Cricetus cricetus*, migratory hamster *Cricetulus migratorius*, bank vole *Clethrionomys glareolus*, water vole *Arvicola terrestris*, pine voles (*Pitymys subterraneus, P. savii, P. duodecimcostatus*), common vole *Microtus arvalis*, short-tailed vole *M. agrestis*, snow vole *M. nivalis*, social vole *M. socialis*, harvest mouse *Micromys minutus*, striped field mouse *Apodemus agrarius*, yellow-necked mouse *A. flavicollis*, wood mouse *A. sylvaticus*, black rat *Rattus rattus*, brown rat *R. norvegicus*, house mouse *Mus musculus*, jerboa *Alactagulus pumilio*, weasel *Mustela nivalis*. Birds mostly up to size of thrushes *Turdus*; larger species (apparently full-grown, but not stated) include Magpie *Pica pica*, Moorhen *Gallinula chloropus*, and 'juvenile' Lapwing *Vanellus vanellus*; numbers of young gamebirds (Phasianidae) taken insignificant (e.g. Hibbert-Ware 1937–8); recorded taking *Hydrobates palagicus* in large numbers (e.g. Lockley 1938). Reptiles: tortoise *Testudo graeca*, lacertid lizards (*Acanthodactylus erythrurus, Psammodromus algirus, Lacerta muralis, L. vivipara, L. agilis*), slow-worm *Anguis fragilis*, skink *Chalcides bedriagai*, snakes (*Natrix natrix, N. maura*). Amphibians: salamander *Pleurodeles waltl*, newt *Triturus boscai*, frogs and toads (*Discoglossus pictus, Pelobates fuscus, Hyla arborea, Rana temporaria, R. esculenta, R. ridibunda*). Fish: carp *Cyprinus carpio*, minnow *Phoxinus phoxinus*. Insects: crickets, etc. (Orthoptera: Gryllidae, Tettigoniidae, Gryllotalpidae, Acrididae), earwigs (Dermaptera: Labiduridae, Forficulidae especially *Forficula auricularia*), cockroaches and mantises (Dictyoptera: Blattidae, Mantidae), bugs, etc. (Hemiptera: Pentatomidae, Reduviidae, Pyrrhocoridae), lacewing *Chrysopa*, adult and larval moths (Lepidoptera: Hepialidae, Tortricidae, Notodontidae, Noctuidae, Lasiocampidae, Sphingidae), crane-flies (Tipulidae), Hymenoptera—sawflies (Tenthredinidae), ichneumons (Ichneumonidae), wasps (Vespidae), bees (Apoidea), ants (Formicidae) including pupae, beetles down to 2 mm long (Coleoptera: Cicindelidae, Carabidae especially *Pterostichus madidus*, Dytiscidae, Hydrophilidae, Silphidae, Staphylinidae, Histeridae, Buprestidae, Byrrhidae, Lucanidae, Geotrupidae especially *Geotrupes stercorarius* and *Typhaeus*, Scarabaeidae especially *Rhizotrogus* and *Melolontha melolontha*, Cantharidae, Elateridae, Dermestidae, Anthicidae, Tenebrionidae, Meloidae, Coccinellidae, Cerambycidae, Chrysomelidae, Curculionidae, Scolytidae). Other invertebrates include millipedes (*Julus, Polydesmus, Glomeris*), centipedes *Scolopendra*, woodlice (*Oniscus, Porcellio*), scorpions *Buthus*, spiders (Araneae), mites (Acari), snails *Helix*, and earthworms *Lumbricus*. (Information largely from Hibbert-Ware 1937–8 and Máñez 1983a, b; also Collinge 1922, Witherby et al. 1938, Haverschmidt 1946, Dementiev and Gladkov 1951a, Kulczycki 1964, Haensel and Walther 1966, Valverde 1967, Mebs 1966, Krzanowski 1973, Lovari 1975a, Libois 1977, Melendro and Gisbert 1978, Kochan 1979, Mienis 1979, Glutz and Bauer 1980, Laursen 1981, Gerdol et al. 1982.) Plant material perhaps taken deliberately at times; chiefly grass and other leaves, but includes small fruits, berries, and maize *Zea*. In England, comprised 6·5% of bulk of food from 212 stomachs collected throughout the year (Collinge 1922; Thiollay 1968).

For summary of major west Palearctic studies, see Table A. Other contributions as follows. Britain: Collinge (1922), Hibbert-Ware (1937–8). France: Madon (1933), Festetics (1959). Poland: Kulczycki (1964), Kochan (1979). Czechoslovakia: see Glutz and Bauer (1980). Hungary: Marián and Schmidt (1967). Rumania: Barbu and Sorescu (1970). USSR: Dementiev and Gladkov (1951a), Averin and Ganya (1966), Bashenina (1968). Spain: Valverde (1967), Herrera and Hiraldo (1976). Italy: Moltoni (1937), Gerdol et al. (1982). Tunisia: Vernon et al. (1973) and *Alauda* 1974, **42**, 236. Egypt: Koenig

Table A Food of Little Owl *Athene noctua* as found from pellets and gut contents. Figures are percentages of total number of items; soft-bodied invertebrates are under-represented, and earthworms (Lumbricidae) largely excluded.

	Denmark all year	East Germany all year	Belgium all year	N France Sep–Jun	Spain all year	Central Italy Sep–Mar	Moldavia (USSR) all year
Mice (small Murinae)	1·2	4·8	0·6	2·6	2·1	2·1	51·1
Small voles (Microtinae)	5·3	21·0	9·4	17·4	1·2		
Larger mammals	+	0·3	0·3	0·1	0·1		
Shrews (Soricidae)	0·2	0·4	0·2	1·6	0·8		
Birds	0·3	0·8	0·6	2·5	0·5	0·1	1·7
Reptiles, amphibians	+	0·5	0·1	0·3	0·7	0·3	1·0
Beetles (Coleoptera)	16·5	58·5	66·8	58·1	39·8	22·1	46·3
Crickets, etc. (Orthoptera)	0	2·6	—	0·9	27·8	3·9	
Earwigs (Dermaptera)	76·4	—	9·7	15·2	14·2	49·2	
Millipedes (Diplopoda)	0	—	—	0	1·8	0·6	
Other invertebrates	0	11·1	12·4	1·3	11·1	21·7	
Total no. of items	8635	2993	1780	5501	16598	723	1261
Source	Laursen (1981)	Haensel and Walther (1966)	Libois (1977)	Thiollay (1968)	Máñez (1983a, b)	Lovari (1975a)	Mikkola (1983)

(1917). See also Uttendörfer (1939b, 1952). Important extralimital study (Turkmeniya, USSR) by Sukhinin *et al.* (1972).

Although Table A suggests preponderance of invertebrates (especially beetles) in diet, proportion by weight much less. Thus, in data for Spain, invertebrates 94·7% by number but 33·6% by weight (Máñez 1983a, b); see also Glutz and Bauer (1980) for analysis by weight. Earthworms not precisely quantified in diet, but evidently important—at least in northern areas, at times. Comprise considerable proportion of diet during summer in Denmark (Laursen 1981).

Diet varies seasonally, though changes sometimes slight; in any case, no universal trends, pattern varying markedly between studies. In most, however, invertebrates important all through the year. In Britain, birds at peak of importance May–July, and few taken October–January; earwigs most important autumn and winter; adult craneflies *Tipula* taken in abundance August–November (Hibbert-Ware 1937–8; see also for discussion of seasonal and other influences on occurrence of individual invertebrate groups in diet). In Belgium, insects and earthworms most important March–August (Libois 1977); similarly (April–October) in Denmark (Laursen 1981). In Moldavia (USSR), invertebrates 56·5% ($n=855$) of total prey numbers in spring and summer, 24·9% ($n=406$) in autumn and winter (Mikkola 1983). See also Barbu and Sorescu (1970), and Lovari (1975a).

Average daily food requirement 50–80 g (Glutz and Bauer 1980).

Food of young apparently contains more invertebrates than that of adults (e.g. stomach analyses by Collinge 1922). In north-west Switzerland, of 207 items brought to nest, 54·6% earthworms (at least 34% by weight), 18·4% Lepidoptera, 11·1% Orthoptera, 6·8% beetles, 5·8% various invertebrates, 1·9% small mammals (33·4% by weight), and 1·0% birds (Glutz and Bauer 1980). At one nest, 22.00–23.00 hrs, parents brought food every few minutes (for details, see Haverschmidt 1946). DJB

Social pattern and behaviour. No thorough study.
1. Solitary or in pairs outside breeding season. Some birds move away from breeding territory in winter, but many do not, staying in pairs (Glue and Scott 1980; Ullrich 1980); in latter case, territory presumably defended through winter (see Breeding Dispersion, below). BONDS. Monogamous mating system, pair-bond often persisting all year and for up to 4 years (Glue and Scott 1980), perhaps usually until one partner dies. 1 record of divorce in study by Knötzsch (1978), but none in study by Ullrich (1980), who considered (despite not-uncommon occurrence of territory changes between years) that bond persists through attachment to territory rather than to mate. First breeding at 1 year old (Glue and Scott 1980; Glutz and Bauer 1980), though not all birds breed every year (see Breeding Dispersion, below), and these presumably often include 1-year-olds. Only ♀ incubates; ♂ usually does all hunting during incubation and first part of nestling phase, then ♀ hunts also. Young dependent for up to 1 month after fledging (see Relations within Family Group, below). BREEDING DISPERSION. Solitary and territorial. In 'optimal habitats' of central Europe, 0·3–0·5 pairs per km², rarely up to 1·5 or more pairs per km²; on area of 1·2 km² near Geneva, 20·8 pairs per km² in one good year, but more typically *c.* 4–5 pairs per km² (see Glutz and Bauer 1980, also Exo 1983). Closest nests, Britain, 240 m and 320 m (Glue and Scott 1980); in Switzerland, 50 m recorded (Glutz and Bauer 1980). In southern England, territory averages 35 ha ($n=8$) on water meadows, 38 ha ($n=11$) on mixed farmland (Glue and Scott 1980); 'hunting area' *c.* 0·5 km² (Mebs 1966). Where birds remain in same area all year, territory said to be maintained from end of moult (about September) until young independent in following year (Glutz and Bauer 1980); see also Antagonistic Behaviour (below). Not uncommonly, non-breeding or unpaired birds hold territory through breeding season (Glue and Scott 1980). ♀ may or may not use same nest-hole from year to year (Ullrich 1980; Exo 1981); successive use of 5 years recorded (Glutz and Bauer 1980). Recorded breeding 2 m away from nest of Barn Owl *Tyto alba* (Glue and Scott 1980). ROOSTING. Diurnal; for periods of hunting activity, see Food. Usually roosts on perch offering cover but also clear view; thus, in tree, bush,

A

loft, or rock crevice or other cavity (Averin and Ganya 1966; Glutz and Bauer 1980). In breeding season at least, will also perch habitually by day on more exposed site—perhaps on look-out when young in nest (Haverschmidt 1946; Glutz and Bauer 1980). For roosting posture, and bathing behaviour (in sun, dust, and rain—said never to bathe in standing water), see Glutz and Bauer (1980).

2. When agitated, bobs body up and down, moving between horizontal and vertical positions, and makes sideways movements with tail; sometimes, briefly jerks wing-tips. Has concealing Sleeked-upright posture (see, e.g., Pygmy Owl *Glaucidium passerinum*), with one wing drawn up in front of body. (Witherby *et al.* 1938; Scherzinger 1971b; Glutz and Bauer 1980.) ANTAGONISTIC BEHAVIOUR. Territorial advertisement occurs mostly October–November and (especially) late January to mid-April; ♂ gives Advertising-call (see 1 in Voice) from exposed perch (Petzold and Raus 1973; Exo and Hennes 1978; Glue and Scott 1980), and white throat-patch sometimes conspicuous then (Haverschmidt 1946); see Fig A. Rival ♂♂ perch nearby, giving Advertising-call (Haverschmidt 1946; Petzold and Raus 1973); ♀♀ said also to 'participate' (Glutz and Bauer 1980). ♂♂ finally pursue each other giving shrill keckering calls (see 9 in Voice) but not fighting. In captivity, rivals threaten by ruffling head and body feathers (white throat-patch conspicuous), and both ♂ and ♀ will attack vigorously. (Glutz and Bauer 1980.) Subordinate to *Tyto alba* and Tawny Owl *Strix aluco* in competition for nest-sites (Petzold and Raus 1973). HETEROSEXUAL BEHAVIOUR. (1) Pair-bonding behaviour. No information, either on behaviour or timing; not clear whether Nest-showing (see below) forms part of process. Advertising-call presumably serves to attract ♀ if ♂ unpaired. (2) Nest-site selection. In Nest-showing, ♂ calls (see 7 in Voice) from inside hole; ♀ flies into hole. Frequently follows copulation (Glutz and Bauer 1980), or ♂ may fly to hole after copulation as if to show it to ♀ (Haverschmidt 1946). (3) Mating. ♂ gives Advertising-call to attract ♀ for copulation, but procedure following this evidently variable. ♂ may bob up and down, 'weaving and dancing' while facing ♀ (Glue and Scott 1980), ♀ may solicit by crouching low and shivering wings (Hosking and Newberry 1945; Glue and Scott 1980), or ♂ may mount without preliminaries (Haverschmidt 1946).

During copulation, ♀ holds body horizontal and ♂ may grasp ♀'s head feathers with bill (Glutz and Bauer 1980). Both may call—little (Haverschmidt 1946) or almost continuously (Glutz and Bauer 1980: see 8 in Voice). Occasionally, ♂ and ♀ chase each other in flight and ♂ sometimes hovers over perched ♀ (Glue and Scott 1980); see also above. Copulation can occur several times (up to 15) in quick succession. Said to be performed sporadically July–September, but mostly from late January to start of incubation—almost daily March–April (Glutz and Bauer 1980). (4) Courtship-feeding and food-caching. Outside hole, ♀ begs like fledged young bobbing body up and down, flapping wings, and calling (presumably 4 in Voice) continuously (Haverschmidt 1946). If food delivered to ♀ in hole, ♂ approaches silently or gives Kiew-call (Glutz and Bauer 1980; P A D Hollom: see 2 in Voice). ♀ sometimes fetches food from site where ♂ deposits it (Glutz and Bauer 1980). (5) Allopreening. Mutual Allopreening of head, hindneck, breast, and feet occurs, sometimes before and after copulation; bouts last up to 20–30 min (Haverschmidt 1978b; Scott 1980). Allopreening also common outside breeding season (Glutz and Bauer 1980). (6) Behaviour at nest. Only ♀ incubates. Sometimes, at least, ♀ does some hunting during incubation (Haverschmidt 1946), but otherwise breaks last only a few minutes at start of incubation, *c.* 30 min towards end (Glutz and Bauer 1980). RELATIONS WITHIN FAMILY GROUP. Young give food-call from egg. Can stand on bent legs from 5 days old. Open eyes at (8–)10 days. ♀ does not keep nest clean. ♀ broods intensively only during 1st week, and after *c.* 10 days stays outside hole during day (Glutz and Bauer 1980); leaves nest only briefly until day 14–16 according to Glue and Scott (1980). In one case, however, ♀ hunted regularly from hatching (Haverschmidt 1946). ♀ feeds young in hole (rarely assisted by ♂), giving Feeding-trill (see 5 in Voice) and Kiew-call when dividing up food. At 14 days, young start to tear off pieces of food and at 25 days can divide up whole rodents. At 3 weeks, climb up to nest-entrance, and may venture outside hole at night before fledging. Fledge at 30–35 days, individual young 0–5 days apart. (Glutz and Bauer 1980.) Able to fly well within 1 week of leaving, but still fed by parents for up to 1 month (Haverschmidt 1946; Glue and Scott 1980). ANTI-PREDATOR RESPONSES OF YOUNG. Fledged young seek cover on hearing adult's warning calls (Glutz and Bauer 1980: see 9 in Voice). PARENTAL ANTI-PREDATOR STRATEGIES. (1) Passive measures. Some ♀♀ allow themselves to be seized on nest (Glutz and Bauer 1980). (2) Active measures. ♀ more aggressive than ♂; sometimes bites and claws hand put into nest; threatens by ruffling head and belly feathers, also by spreading wings so that undersides turned towards intruder (Fig B), Bill-snapping, snoring, and hissing. Sometimes (not always: Haverschmidt 1946) attacks other Strigiformes, raptors (Accipitriformes, etc.), cats, and humans near nest. Usually flies at head of man, clawing and biting. (Scherzinger 1971b; Glutz and Bauer 1980.) Demonstrates at presence of intruder near nest or fledged young by calling (see 9 in Voice).

(Fig A from drawing in Hainard 1955; Fig B from drawing in Scherzinger 1971b.) DJB

B

Voice. Information sparse and confusing. No thorough study, and some calls perhaps duplicated in treatment given here. Bill-snapping performed in threat. For additional sonagrams, see Glutz and Bauer (1980).

CALLS OF ADULTS. (1) Advertising-call. (a) ♂'s call. A full, mellow hoot, 'goooek', suggesting Advertising-call of Scops Owl *Otus scops* but more drawn out and rising

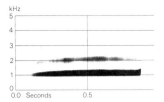

I C König West Germany May 1964

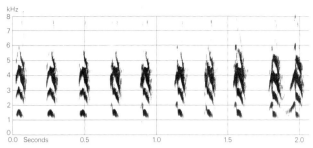

III P J Sellar England June 1973

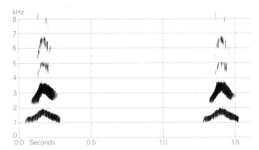

II R Margoschis England March 1980

IV V C Lewis England June 1974

slightly but sharply in pitch at end; similar to Low-whistle call of Curlew *Numenius arquata* (D J Brooks, H Delin: Fig I); loud, questioning 'hoo' (Haverschmidt 1946); 'huui' or 'ghu(k)', sometimes given in crescendo series of 4–10(–60) units, with 2–11 s between units or 12–20 units per min; towards end of series, units change to an excited 'hooi', 'guhah', 'guiau', or 'kwiau', and series ends abruptly with shrill 'hoo-ee', 'miju', or 'míau'—probably call 2 (Stadler 1945–6; Haverschmidt 1946; Glutz and Bauer 1980; P J Sellar). (b) ♀'s call. Higher pitched than ♂'s call, nasal, and less full (H Delin). No such call from ♀ described by Haverschmidt (1946), who also considered descriptions of apparently duetting pairs to refer to rival ♂♂. (2) Kiew-call. Commonest call. Recording (Fig II) suggests a clear '(k)weew' (D J Brooks); sharp, complaining, falsetto 'kee-cw' (H Delin); call repeated irregularly. Said to be contact-call to mate and young, and given during copulation and Nest-showing. (3) Tremblingly quiet, whispering 'schrie' also used as contact-call. Given especially by ♀ during heterosexual behaviour (Glutz and Bauer 1980). (4) Begging-calls of ♀. Extend from juvenile-like 'tsiech' and thin 'siej' sounds to a screeching (Glutz and Bauer 1980). In recording by P A D Hollom, presumed ♀ gives snoring screech; after ♂ has departed, ♀ gives soft, wheezy, rasping sounds (D J Brooks). (5) Feeding-trill of ♀. When feeding young, a rapid, nasal 'gek-gök-gök-gök', or hoarse cackling; sometimes interrupted by 'uuhg' call (see call 2) (Glutz and Bauer 1980). (6) Excitement-call. Penetrating and variable. Mentioned, but not fully described, by Glutz and Bauer (1980), though 'jau', 'míau', 'iwidd', and 'kuwitt' apparently variants. Given in rapid sequence (28–36 calls per min), mainly in association with copulation and other aspects of heterosexual behaviour—also in fights between rivals. (7) Nest-showing call of ♂. Call similar to 'zick zick' of Kestrel *Falco tinnunculus* or 'tjuck tjuck' of domestic cockerel given during Nest-showing; occasionally also during mating (Glutz and Bauer 1980). (8) Calls associated with copulation. Soft 'oo oo' from birds sitting close together, before or after copulation. Shrill 'shricking' from ♀ during copulation. Advertising-call of ♂ often precedes mating and sometimes occurs during mounting. (Haverschmidt 1946.) See also calls 2, 6, and 7. (9) Calls associated with alarm, anxiety, warning, etc. Loud, chattering 'kek kek' (Fig III) when disturbed at nest (Haverschmidt 1946); probably the same as the short, explosive, high-pitched 'chi chi chi-chi. . .', rather like tern (Sternidae), given apparently in alarm (H Delin). Recording (Fig IV) of bird disturbed approaching nest with food suggests a hollow 'whoorp' (D J Brooks). Rather grating 'shar' from ♀ associated with alarm at nest (Glue and Scott 1980). Calls based on 'queb' and 'keck' sounds said to express unease or fear; shrill, short 'kja' or 'kju' as warning in contact with rival or predator; 'quip' in fright; 'quijep' as warning to young at fledging. Snoring or screeching given in displeasure or fear. (Glutz and Bauer 1980.) (10) Hissing-call. Hissing given in threat (Scherzinger 1971b). (11) Other calls. (a) Squeaky 'uik' during allopreening; (b) compressed, sometimes hissing or rasping sounds given by ♂ and ♀; said to function probably as begging- and contact-calls (Glutz and Bauer 1980); (c) curious faint snoring given persistently by day in spring; (d) sound like exhalation of person in deep sleep (Witherby *et al.* 1938; see also for further descriptions); (e) various shrieking, yelping, grumbling, and rasping calls given during heterosexual behaviour (Glue and Scott 1980; Glutz and Bauer 1980); (f) in recording by P A D Hollom, adult in nest-hole with small young gives continuous clucking barks (D J Brooks).

CALLS OF YOUNG. Food-call from egg and small young a monosyllabic 'psiep', 'szip', or 'srie' and a disyllabic

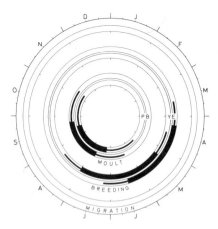

'uiiet'. In 2nd week, become harsh, 'chsij', 'chriie', or hissing 'schwö'; in 4th week, rasping snoring like Barn Owl *Tyto alba*. Several adult calls—e.g. 'gjuu' (call 2), 'kek', snoring (call 9), hissing (call 10)—develop in nest or shortly afterwards; Advertising-call given during 1st autumn. (Ullrich 1973; Glutz and Bauer 1980.) DJB

Breeding. SEASON. North-west Europe: see diagram. Rather little variation across range with similar dates in Spain, Greece, and North Africa (Heim de Balsac and Mayaud 1962; Makatsch 1976). SITE. Hole in tree, also in building, wall, or ground (e.g. burrow of rabbit *Oryctolagus cuniculus*). Of 482 nests, Britain, 24% in oak *Quercus*, 23% in ash *Fraxinus*, 18% in fruit trees *Prunus*, and 15% in willows *Salix*; of 267 nests, 37% in main trunk, 23% in lateral branch, and 14% in pollarded tree (Glue and Scott 1980). 0·3–12·2 m above ground, average 3·0, $n=482$ (Glue and Scott 1980). For details of tree-sites on lower Rhine (West Germany), see Exo (1981). Nest-box readily used; for details of different types and usage, see Juillard (1980). Nest: available hole; no material added but chamber may be cleared and scrape formed in bottom (Ullrich 1973). 10 chambers, Britain, averaged 20 cm wide (10–50); reached by passage averaging 80 cm long (50–130), $n=74$ (Glue and Scott 1980). Different types of hole distinguished, including horizontal and vertical (Exo 1981). Building: ♀ cleans nest and may form shallow scrape. EGGS. See Plate 97. Short sub-elliptical, smooth, not glossy; white. *A. n. vidalii*: Britain 36 × 30 mm (33–40 × 27–31), $n=100$ (Witherby *et al.* 1938); central Europe 35 × 29 mm (32–37 × 28–31), $n=22$ (Makatsch 1976). *A. n. indigena*: Greece 34 × 28 mm (32–38 × 27–30), $n=38$ (Makatsch 1976). Fresh weight 15·6 g (12·5–19·0), $n=95$ (Glutz and Bauer 1980). Calculated weight, all races, 14–15 g (Makatsch 1976). Clutch: 2–5 (1–7). Of 268 clutches, Britain: 1 egg, 1%; 2, 10%; 3, 35%; 4, 39%; 5, 13%; 6, 1%; 7, 1%; average 3·6 (Glue and Scott 1980). Of 80 clutches, northern France: 3 eggs, 23%; 4, 62%; 5, 15%; average 4·0 (Labitte 1951). Of 14 clutches, West Germany; 3 eggs, 3; 4, 7; 5, 3; 7, 1; average 4·2 (Ullrich 1973). Normally one brood; single record of 2 broods, Britain (Glue and Scott 1980), and also occur in central Europe (Glutz and Bauer 1980). Replacements occasionally laid after egg loss. Laying interval 1 day (Ullrich 1973; Glue and Scott 1980). INCUBATION. 27–28 days (23–35), exceptionally less; 19·5 days for one egg (König 1969a). By ♀ only, beginning with 1st or 2nd egg, or sometimes when clutch complete (Ullrich 1973; Glue and Scott 1980). Hatching asynchronous or nearly synchronous. YOUNG. Altricial and nidicolous. Cared for by ♀ and brooded while small, ♂ bringing food, though ♀ also brings food late in rearing period. FLEDGING TO MATURITY. Fledging period 30–35 days, exceptionally longer; up to 43 days (Glue and Scott 1980). Young may leave nest before fledging and hide in surrounding branches or vegetation. Young fed for up to 1 month after fledging (Haverschmidt 1946). Age of first breeding 1 year (see also Social Pattern and Behaviour). BREEDING SUCCESS. Of 477 eggs laid (Britain), 56·4% hatched and 49% fledged. Of 156 clutches, 28% failed before completion of laying or during incubation, 26% during fledging period, and 26% produced flying young; of 52 failures, 54% due to man taking eggs or young, destroying nest, or shooting adults, 16% due to predation, and remainder miscellaneous. Of 241 broods at fledging, Britain: 1 young, 21%; 2, 33%; 3, 36%; 4, 8%; 5, 3%; mean 2·4 (Glue and Scott 1980). Of 129 eggs laid in 30 clutches, West Germany, 57 young fledged (König 1969b). Overall success from 13 nests, West Germany, 58%, with mean 2·4 young fledged per nest (Ullrich 1973).

Plumages (*A. n. vidalii*). ADULT. Ground-colour of upperparts and upper wing-coverts dark chocolate-brown or dark fuscous-brown when plumage fresh, usually with slight olive-brown tinge, sometimes nearer dark olive-brown; slightly paler dark greyish-brown or olive-brown when plumage worn. Forehead and crown closely marked with cream-buff (when fresh) or white (when worn) shaft-streaks, each streak often widening slightly towards feather-tip, and crown then showing rounded spots when plumage fresh, but more usually crown shows elongate spots. Distinct white V-mark on hindneck from behind ear down to centre of neck, only slightly tinged buff when plumage fresh, slightly mottled by narrow dark feather-tips. Mantle and scapulars with large white or pale cream rounded subterminal spots, often joined to form single broad subterminal bar, spots separated by dark shaft only; white spots largely hidden when plumage fresh, except on outer webs of outer scapulars, where only narrowly bordered by dark brown terminally, often conspicuously spotted white when plumage worn. Back, rump, and upper tail-coverts with large rounded or spade-shaped subterminal spots tinged cream-buff. Facial disc poorly developed, not surrounded by distinct ruff; ring round eye white, widest between eye and base of bill where mixed with black bristles; outer and lower border of disc buff (when fresh) or white (when worn), mottled with grey or black specks, latter joining to dark crescents along rear and lower border of white round eye. Chin white, patch extending narrowly towards lower border of ear-opening; bordered below by buff-and-grey mottled band across throat to sides of neck. Upper chest white; visibility of this and white chin patch strongly dependent on carriage of bird. Lower chest, breast, and upper flanks dark fuscous-brown to olive-brown, each feather with paired subterminal pink-buff or white spot or short bar; pale spots gradually more elongated further down, only feather-centres brown, belly, upper vent, and lower flanks appearing boldly and evenly streaked brown and cream-buff (when fresh) or white (when worn). Lower vent, under tail-coverts, and feathering of leg and foot white, tinged pink-buff when fresh, in particular on thighs. Ground-colour of tail as upperparts or slightly paler olive; rather poorly defined fringes along tips white, rather narrow and usually broken bars buff (when fresh) to cream-buff or white (when worn), terminal bars or dots in particular subject to bleaching; bars reduced to spots or short bars on outermost feathers. Pattern of pale bars on t1 highly variable: one straight bar just visible at tips of longest upper tail-coverts, only narrowly interrupted by brown along shaft, a similar 2nd and 3rd bar across middle portion of t1, and traces of a 4th terminal bar often present in form of spot on tip of 1 or both webs, sometimes contiguous with white fringe along feather-tip, but occasionally developed into complete 4th bar and then sometimes traces of 5th bar visible. 4th bar frequently absent and 3rd bar occasionally not reaching webs at sides or more widely interrupted at shaft, forming rounded spot on 1 or both webs; more rarely, no bars reach sides of 1 or both webs and then reduced to rounded spots, 3rd bar and (when present) 4th reduced to single spot only. Flight-feathers dark olive-brown or fuscous-brown, darkest along shafts and on outer primaries; primaries with 3–4 well-spaced shallow pale spots along outer webs, much larger spots or short bars on inner webs; spots on outer webs pink-buff or white (depending on wear) and each slightly accentuated by partial blackish rim, spots on inner webs grading from cream near shaft to white along inner edge, spots of inner and outer webs on outer primaries faintly connected by buff-brown bar; tips of primaries with partial pale buff or white fringe along 1 or both webs. Secondaries similar to primaries, but only 2–3 rows of spots or bars; pink-buff spots on outer webs more bar-like, faintly connected with large rounded or triangular spots on inner webs by buff-brown bar; tips of secondaries with large or small white paired dots or these completely absent. Upper wing-coverts like mantle and scapulars, but white spots of greater, median, and most primary coverts smaller and often restricted to one web only, lesser coverts faintly spotted only; greater primary coverts with paired pink-buff subterminal spots; only white spots on outer webs of median coverts and of outer greater coverts distinct. Under wing-coverts and axillaries cream-white, faintly spotted and mottled grey; shorter under primary coverts more heavily spotted black, greater under primary coverts boldly tipped and marked black. Sexes similar, but ♂ tends to have whiter face. DOWNY YOUNG. 1st down (neoptile) short and dense, white; mottled grey from $c.$ 1 week when mesoptile starts to grow under 1st down. JUVENILE. 2nd down (mesoptile) rather feather-like, not as soft and downy as in (e.g.) *Strix* and *Asio* owls; feathers still distinctly softer and shorter than in 1st adult and adult; structure of flight-feathers, tail, and greater primary coverts as in adult. Neoptile clings to tips of mesoptile up to 3–4 weeks (retained longest on crown, flanks, and thighs). Mesoptile rather like adult, but ground-colour paler, dull dark grey-brown, white or cream spots on upperparts and upper wing-coverts less distinctly defined, less contrasting; crown, back, rump, and upper tail-coverts almost uniform grey-brown or with faint buff spots. Face as adult, but greyer and less contrastingly marked; no distinct white patch on upper chest; indistinctly marked grey and white instead; breast largely grey-brown with ill-defined white spots, grey streaks on remainder of white underparts diluted (not sharp and contrasting as in adult). For method of ageing juvenile with growing primaries, see Juillard (1979). FIRST ADULT. Closely similar to adult; retained juvenile flight-feathers and tail as in adult, pattern variable at all ages. Main differences: (1) shape of primary tips, especially p10—tips rather pointed in 1st adult, inner web not strongly curved (in adult, tip almost square, tip of inner web almost perpendicular to shaft); (2) relative wear of flight-feathers, tail, and (in particular) tertials—slightly worn in autumn, distinctly in spring, heavily in summer (in adult, fresh until onset of breeding season); (3) shape of tertials—rather soft, narrow, and tapering to top in 1st adult, broader and squarer-tipped in full adult (shape affected by wear in both, however).

Bare parts. ADULT AND FIRST ADULT. Iris lemon-yellow. Eyelids dark slate-blue. Bill lemon-yellow, sulphur-yellow, or greenish-yellow, base tinged olive-green or grey. Cere slate-grey or olive-black. Tarsus yellow or greyish-yellow, toes grey-brown, greyish-black, or olive-black. Claws dark brown to brownish-black. DOWNY YOUNG AND JUVENILE. At hatching, skin under down, cere, bare patch below tibio-tarsal joint, and toes pink; bill and claws whitish- or greyish-pink, corner of mouth pink-yellow. Eye opens at $c.$ 10th day, iris pale yellow or amber-yellow at first. Skin partly darkens to grey after a few days, first on toes and cere. At fledging, iris and eyelids as in adult; bill olive-yellow with pale yellow tip of culmen and some dusky grey towards base; cere slate-blue, dark violet-grey, or blackish violet-green; corner of gape pale yellow or pale orange; tarsus yellow or yellow-grey, toes flesh-grey or dusky grey, claws black with grey-blue base. (Heinroth and Heinroth 1926–7; Glutz and Bauer 1980; ZMA.)

Moults. ADULT POST-BREEDING. Complete, primaries descendant. Usually starts with p1 when young fledge or independent (Stresemann and Stresemann 1966), but late breeders (e.g. with replacement clutches) may start at about hatching (Ullrich 1970). P1 shed mid-May to mid-July, moult completed with p10 early September to early November; not much variation with locality, though most Mediterranean birds moult about mid-June to mid-September and bulk of British, Netherlands, and German birds start mid-June or early July and complete mid-October; northern birds starting from mid-May perhaps non-breeders. In 2 captives, p10 shed 98–99 days after p1 (Piechocki 1968b). Secondaries moult from 3 centres: s12 (starts with shedding of p3–p4), s5 (starts with p6–p7), and s1 (starts with p5–p7); s4, s7, and s8 moulted last, shed at about same time as p10 (Piechocki 1968b); most secondaries replaced during last stages of primary moult (at about moult scores 30–50, early August to mid-October). Tail-feathers shed almost simultaneously, sequence irregular; shed 20–30 (15–35) days after p1 (Piechocki 1968b) or when p4–p5 growing, completed during growth of p7 (Stresemann and Stresemann 1966); in Netherlands sample, tail shed at primary moult score 6–22, completed at 20–35 (RMNH, ZMA). Body moult starts at primary moult score 10–25, lesser and median upper wing-coverts first, soon followed by mantle and scapulars; heavy moult all over at score 25–35, completed at $c.$ 45, toes last. POST-JUVENILE. Partial: head, body, and wing-coverts; not flight-feathers, tertials, tail, or greater primary coverts. Starts soon after fledging; face and lesser and median upper wing-coverts first (early July to mid-August), largely completed 6–7 weeks later at age of $3\frac{1}{2}$–$4\frac{1}{2}$ months (September or early October), crown, neck, and feathering of leg and toes last (toes up to late November).

Measurements. *A. n. vidalii*. Netherlands, all year; skins (ZMA). Bill (F) to forehead, bill (C) to cere. Adult and 1st adult combined, except for retained juvenile wing and tail of 1st adult.

WING	AD	♂ 163	(3·66; 13)	158–169	♀ 166	(3·90; 13)	161–173
	JUV	160	(2·98; 35)	155–166	163	(4·12; 39)	157–171
TAIL	AD	75·9	(2·00; 11)	74–79	79·6	(2·06; 9)	77–83
	JUV	73·1	(2·86; 29)	69–78	74·1	(2·70; 27)	71–80
BILL (F)		20·6	(0·99; 20)	19·0–22·3	20·4	(0·92; 18)	18·9–21·9
BILL (C)		14·0	(0·49; 20)	13·3–14·7	13·9	(0·78; 21)	13·0–15·2
TARSUS		34·8	(1·19; 21)	33·4–36·3	35·6	(1·09; 20)	34·0–36·8

Sex differences significant for tarsus and juvenile wing. Juvenile wing and tail significantly shorter than adult.

In following samples, ages combined and hence range relatively larger; in all samples, average of 1st adult wing and tail *c*. 1 mm below average given, adult wing and tail *c*. 2 mm over. All data from skins (BMNH, RMNH, ZMA), all year.

A. n. vidalii. Spain and Portugal.

WING	♂ 158	(2·43; 9)	154–161	♀ 161	(3·09; 13)	157–166
TAIL	71·8	(3·24; 9)	68–76	72·9	(3·38; 9)	68–77
BILL (C)	13·4	(0·49; 8)	12·8–14·1	13·5	(0·64; 12)	12·8–14·4
TARSUS	34·6	(1·04; 8)	33·0–35·9	34·8	(1·06; 12)	32·8–36·3

West and East Germany and north-west Austria.

WING	♂ 161	(2·20; 9)	158–164	♀ 164	(4·12; 7)	160–172
TAIL	74·4	(2·39; 9)	70–77	74·4	(1·62; 7)	73–77
BILL (C)	13·5	(0·53; 9)	13·0–14·3	13·9	(0·86; 7)	12·8–14·6
TARSUS	34·5	(0·85; 9)	33·5–35·6	35·0	(1·29; 7)	33·4–36·2

Nominate *noctua*. Northern Italy and Sardinia.

WING	♂ 158	(2·22; 8)	156–162	♀ 161	(3·44; 10)	156–166
TAIL	75·5	(2·27; 8)	73–79	77·0	(1·80; 8)	75–80
BILL (C)	13·8	(0·75; 8)	12·8–14·6	14·7	(0·47; 8)	14·1–15·4
TARSUS	31·9	(1·08; 8)	30·6–33·6	31·8	(1·46; 8)	29·4–33·3

A. n. indigena. Greece (including Cyclades), Albania, and southern and eastern Rumania.

WING	♂ 164	(4·14; 12)	158–171	♀ 167	(3·85; 13)	162–174
TAIL	80·0	(4·46; 6)	75–87	81·6	(4·03; 7)	76–89
BILL (C)	14·4	(0·38; 6)	13·9–15·0	14·9	(0·46; 7)	14·5–15·4
TARSUS	33·2	(0·67; 6)	32·1–34·1	32·7	(1·04; 7)	31·8–34·4

Crimea and northern shore of Caspian Sea, USSR.

WING	♂ 166	(3·88; 6)	160–170	♀ 172	(— ; 2)	171–173

Western Asia Minor.

WING	♂ 166	(4·37; 9)	160–173	♀ 169	(2·70; 6)	166–173

A. n. lilith. Cyprus.

WING	♂ 157	(3·06; 15)	152–162	♀ 156	(2·11; 11)	153–160
TAIL	74·6	(2·95; 4)	71–78	74·9	(2·22; 6)	72–78
BILL (C)	14·3	(0·50; 4)	13·8–15·0	14·2	(0·46; 6)	13·8–15·0
TARSUS	30·0	(0·80; 4)	29·3–31·1	30·2	(0·83; 6)	29·2–31·6

Syria, Lebanon, Israel, Jordan, and Sinai peninsula.

WING	♂ 160	(2·99; 17)	154–164	♀ 162	(4·10; 14)	157–172

A. n. glaux. Nile valley, Nile delta, and Faiyum (Egypt).

WING	♂ 159	(3·56; 8)	157–165	♀ 159	(2·79; 7)	156–163

North-west Morocco.

WING	♂ 161	(2·87; 4)	157–164	♀ 161	(2·73; 6)	157–164

A. n. glaux and *saharae*, combined; Algeria (including Ahaggar), Tunisia, Libya, Tibesti (Chad), north-west Egypt, and Dakhla oasis (Egypt).

WING	♂ 154	(3·63; 34)	146–161	♀ 157	(4·11; 23)	151–165
TAIL	72·8	(2·59; 14)	69–77	74·2	(3·63; 12)	70–79
BILL (C)	13·7	(0·62; 11)	12·7–14·6	14·1	(0·56; 14)	13·2–15·2
TARSUS	32·0	(0·79; 10)	31·1–33·4	32·3	(1·49; 14)	29·6–34·6

A. n. bactriana. Iran to western Pakistan, Afghanistan, and Kazakhstan (Vaurie 1960*c*, 1965).

WING	♂ 166	(— ; 40)	159–174	♀ 169	(— ; 21)	159–177

Weights. ADULT AND FIRST ADULT. *A. n. vidalii* Netherlands: (1) November–March, (2) April–June, (3) July–October, (4) died from starvation, all year (ZMA).

(1)	♂ 177	(8·73; 8)	164–187	♀ 206	(14·6; 6)	189–226
(2)	160	(18·6; 5)	139–178	176	(19·3; 5)	153–197
(3)	162	(14·7; 9)	146–193	166	(12·3; 12)	148–188
(4)	127	(15·2; 8)	99–141	129	(13·1; 9)	112–155

Nominate *noctua*. Italy: (1) December–March, (2) April–November (Moltoni 1949). Czechoslovakia and Hungary: (3) December–March, (4) April–June, (5) July–November (Keve *et al.* 1962). North-west Rumania: (6) all year (Keve *et al.* 1962).

(1)	♂ 161	(— ; 3)	118–190	♀ 168	(9·70; 6)	155–180
(2)	140	(24·7; 4)	105–160	159	(— ; 3)	152–170
(3)	182	(11·0; 12)	166–196	179	(20·6; 8)	138–207
(4)	136	(23·2; 5)	108–168	140	(— ; 2)	120–161
(5)	175	(17·8; 6)	160–210	180	(14·3; 6)	163–200
(6)	158	(26·8; 9)	115–198	191	(20·9; 8)	150–215

A. n. indigena. Greece and Turkey, March–June: ♂♂ 162, 175; ♀♀ 130, 146, 260 (Niethammer 1943*a*; Makatsch 1950; Kumerloeve 1961).

A. n. bactriana. Iran, Afghanistan, and Kazakhstan, March–July(–September): ♂ 151 (14·7; 11) 118–172, ♀ 199 (36·6; 5) 165–260 (Paludan 1938, 1959; Schüz 1959; Gavrin *et al.* 1962).

DOWNY YOUNG AND JUVENILE. At hatching, 10–13 (Heinroth and Heinroth 1926–7; Juillard 1979). In Switzerland, on 1st day 15·5, on 5th 31 (28–38), 10th 81 (61–98), 15th 112 (99–127), 20th 131 (114–150), 25th 134 (109–159), 30th 141 (129–162), 35th (on leaving nest) 147 (129–183) (Juillard 1979; data approximate, read from graph). Development in Westfalen (West Germany) similar at first, but average on 20th day 133, 25th 146, 30th 155, 35th 158, in 1st week after fledging 154 (36) 112–175 (Glutz and Bauer 1980). Downy young from southern West Germany 104 (9·82; 9) 92–125 on 11th–13th day (Ullrich 1973); captive birds *c*. 160 at 3 weeks and 170–180 at 4 (Heinroth and Heinroth 1926–7).

Structure. Wing rather short and broad, tip rounded. 10 primaries: p7 or p8 longest or either one 0–2 shorter than other; p9 2–6 shorter than longest, p10 18–26, p6 3–8, p5 16–21, p4 25–30, p1 38–45. Outer web of (p6–)p7–p9 and inner web of (p6–)p7–p10 emarginated. Outer edge of p10 with short serrations. Tail short, nearly square; 12 feathers, t6 about similar in length to t1. Facial disc distinctly bordered by feathered eyebrow above, but indistinctly at sides and below; no ruff of specialized feathers—indicated by colour only. Bill rather heavy, wide and deep at base, laterally compressed at tip; partly covered by stiff bristles at sides, extending from lateral base. Tarsus rather long, toes rather short; tarsus covered by short woolly feathers, but lower part bare (except for bristles) in some southern races; toes thinly covered with bristles, but basal part often feathered when plumage fresh (least so in ♀); in some eastern races, uppersurface of toes fully feathered. Middle toe without claw 18·9 (0·80; 32) 17–20 in ♂ nominate *noctua*, *saharae*, *glaux*, *lilith*, and Iberian *vidalii*, 19·4 (0·86; 41) 18–21 in ♀; 20·1 (0·91; 32) 19–22 in ♂ *vidalii* (from Britain, Netherlands, and central Europe), *indigena*, and *bactriana*, 20·4 (0·81; 38) 19–22 in ♀; outer toe without claw *c*. 70% of middle, inner *c*. 82%, hind *c*. 55%. Claws rather short and slender, strongly decurved; middle claw 11·3 (0·68; 33) 10–12 in ♂ of smaller-sized group (see middle toe), 11·4 (0·60; 40)

10–12 in ♀; 11·8 (0·69; 37) 11–13 in ♂ of larger group, 12·3 (0·83; 45) 11–14 in ♀; outer toe *c.* 77% of middle, inner *c.* 99%, hind *c.* 77%.

Geographical variation. Marked and complex. Involves size of wing, tail, and tarsus, colour of upperparts and of streaks on underparts, width of streaks of underparts, pattern of pale tail-bars, and size of white spots on crown, mantle, and scapulars. Variation clinal, boundaries between races in general very indistinct, intergrading over wide areas. Colour not obviously related to environment; though in general paler towards east and south, some races from arid areas quite dark. Individual variation in colour marked in some areas (birds often tending to be dimorphic, some more rufous, others more olive-grey), but virtually absent in others. As colour also strongly influenced by bleaching and wear (birds palest and with white spots on upperparts and dark streaks on upperparts most distinct in breeding season) as well as by discoloration with age in museum skins, basic division as given here mainly based on size and not on colour as usually done elsewhere (e.g. Vaurie 1960c). Long tarsus (average 34–36) and intermediate wing is shown by *vidalii* from Iberia north to Britain and then east to Poland and Baltic States, south to northern Switzerland and north-west Austria. Short tarsus (average 30–31) is shown by *lilith*; in this race, Cyprus birds have wing on average shorter than those of Levant and south-east Turkey (marked in ♀♀). All remaining races have tarsus intermediate (average 31–34)— wing short (average 154–157) in *glaux* and *saharae* from Algeria and Tunisia, as well as Sicilian birds; intermediate (average 158–161) in nominate *noctua* from Sardinia, Corsica, mainland Italy, northern Yugoslavia, south-east Austria, southern Czechoslovakia, Hungary, and north-west Rumania, and in *glaux* from Morocco and Egypt; large (average 164–172) in *indigena* from Greece and Albania through southern and eastern Rumania to south European USSR and Asia Minor (except south-east), and *bactriana* from south-east Azerbaydzhan and Iran eastward. *A. n. vidalii* from Iberia north to Netherlands darkest, rather cold dark fuscous-brown with some variable olive tinge, but no russet-brown; pale tail-bars on t1 complete, usually reaching sides of feathers and at most narrowly interrupted at shaft, usually 3 bars (occasionally 4) showing beyond tip of longest upper tail-coverts with trace of partial bar at tip. Typical nominate *noctua* from type locality Italy dark russet-brown on upperparts; dark streaks of underparts slightly narrower, russet-brown; tail-bars in about half of birds more reduced than in any from Netherlands or Iberia, nominate *noctua* showing 2 pale bars on t1 with traces (1–2 dots) of a 3rd, or even 1 bar and traces of a 2nd; colour rather constant. Populations similar to Italian nominate *noctua* extend eastward to Hungary and north-west Rumania, latter sometimes separated as *daciae* Keve and Kohl, 1961 (Keve *et al.* 1962); individuals inseparable from nominate *noctua* occur south-east France (once Vendée) and Yugoslavia. Sicily birds different from mainland Italy and Sardinia; pale olive-grey and small, similar to Tunisia. Populations from West Germany through Poland and north-west Austria often slightly browner than *vidalii*, but generally not as russet as nominate *noctua*; as tail-pattern and size similar to *vidalii* and colour closer to *vidalii*, included in latter. Colour of introduced British birds more similar to German than to Dutch birds, but size similar to Dutch ones from which thought to originate. Population in northern part of range in European USSR usually included in nominate *noctua* (e.g. Vaurie 1960c), but colour of upperparts in general more olive-brown (not russet), size larger, and tail-bars usually complete; probably best considered intermediate between *vidalii* or nominate *noctua* and *indigena*. Individual variation in south-east Europe and Middle East (*indigena*, *bactriana*, and *lilith*) marked: colour of upperparts between pale russet-brown (paler than nominate *noctua*) and greyish olive-brown (in general, more russet in *indigena* and more olive-grey in *lilith* and *bactriana*); pale bars on t1 reduced in *indigena*, showing 1–2 bars broken into spots and sometimes a few more spots terminally, but occasionally virtually unspotted; bars of *bactriana* complete as in *vidalii*, in *lilith* slightly reduced, bars often not reaching sides of feathers; toes of *bactriana* thickly feathered (least so in breeding season); white spots on upperparts of *lilith* on average larger than in other races; see also Measurements. Birds from Dalmatian coast, Albania, and western Greece approach nominate *noctua* in colour of upperparts and tail-bars. For individual variation and races in Balkans, see Keve *et al.* (1962); for Levant, Harrison and Hovel (1964). *A. n. glaux* from Nile valley and delta of Egypt rather dark rufous-brown on upperparts (paler and less saturated russet-olive than *indigena* or nominate *noctua*), underparts rather heavily streaked with same colour, pale bars on t1 reduced to (1–)2 broken bars and some spots; similar birds occur Faiyum, coastal Cyrenaica, northern Tunisia, and Algeria north of Atlas, while those of north-west Morocco are slightly darker russet-brown (warmer russet than in nominate *noctua*, without olive-grey tinge) and some from Tanger approach *vidalii*. Remainder of range in North Africa inhabited by *saharae*, which is generally paler than *glaux* and shows more narrowly streaked underparts; colour of *saharae* highly variable, some more rufous or sandy-pink, others pale olive-grey, and 2 morphs perhaps occur, more rufous morph predominating in some localities, olive-grey in others (situation in south-east Europe and Middle East perhaps similar). *A. n. saharae* and *glaux* occur together in a number of localities, as on Hauts Plateaux of Algeria and Figuig in south-east Morocco, and *saharae* perhaps better considered a colour morph of *glaux* only, especially as some populations from central Sahara are unexpectedly dark and near *glaux* (e.g. those of Ahaggar and Dakhla oasis). Iraq birds near *bactriana* in colour, but near *lilith* in size and probably best considered intermediate; birds from Arabia rather variable, some resembling *saharae*, some *lilith*, others intermediate. For survey of all races (including extralimital ones), see Hartert (1912–21) and Vaurie (1960c, 1965).

CSR

Subfamily STRIGINAE wood owls and allies

6 genera: (1) *Strix* (11–12 species, Eurasia and the Americas); (2) *Rhinoptynx* (1 species, Central and South America); (3) *Asio* (5 species, world-wide except Oriental and Australasian regions); (4) *Pseudoscops* (1 species, Jamaica); (5) *Nesasio* (1 species, south-west Pacific); (6) *Aegolius* (4 species, northern Eurasia and the Americas). 4 species of *Strix*, 3 *Asio*, and 1 *Aegolius* in west Palearctic, all breeding.

For general features, moults, etc., see Strigidae. Ears relatively larger and more complex than in Buboninae, with asymmetrical ear-flaps; sometimes (e.g. *Aegolius*), even whole skull asymmetrical. Facial disc well developed.

Strix aluco Tawny Owl

PLATES 49 and 53
[facing pages 495 and 543]

Du. Bosuil Fr. Chouette hulotte Ge. Waldkauz
Ru. Обикновенная неясыть Sp. Cárabo Sw. Kattuggla

Strix Aluco Linnaeus, 1758

Polytypic. Nominate *aluco* Linnaeus, 1758, Europe from Belgium, Netherlands, German Rheinland, Vosges, Jura, and Alps east to *c.* 35°E in central European USSR, Ukraine, and Crimea, south to northern Italy and Balkans; *siberiae* Dementiev, 1933, Ural mountains and western Siberia, intergrading with nominate *aluco* between 35° and 55°E in central European USSR; *sylvatica* Shaw, 1809, Britain, France (except extreme east), Iberia, western and central Turkey, Levant, and (probably this race) southern Italy and Greece; *mauritanica* (Witherby, 1905), north-west Africa; *willkonskii* (Menzbier, 1896), Caucasus, Transcaucasia, north-east Turkey, and north-west Iran east along southern Caspian to western Kopet Dag (Turkmeniya); *sanctinicolai* Zarudny, 1905, Zagros mountains in western Iran, intergrading with *sylvatica* in northern Iraq. Extralimital: 5 races in central and eastern Asia from eastern Afghanistan and Tien Shan through Himalayas to China, Korea, and Taiwan.

Field characters. 37–39 cm; wing-span 94–104 cm. Almost twice size of Little Owl *Athene noctua*; over 10% larger than Barn Owl *Tyto alba*; bulkier and proportionately shorter winged than Long-eared Owl *Asio otus* and Short-eared Owl *Asio flammeus*. Dark-eyed, kind-faced, medium-sized, relatively broad-winged owl, with rather uniform mottled brown plumage, varying from rufous-brown to grey in ground-colour. Large round head and noticeably tubby body obvious, even in flight. Chiefly nocturnal. Flight medium-paced, with noticeably level glides. Gives long quavering hoot. Sexes similar; no seasonal variation. Juvenile separable. 6 races in west Palearctic, difficult to distinguish; 3 described here—see also Geographical Variation.

ADULT. (1) North and east European race, nominate *aluco*. Up to 10% larger than British and west European race *sylvatica*. Facial disc complete, noticeably round and, together with rotund feather cloak to head, as broad as body; usually grey-brown, even grey-white, more rarely rufous-brown, divided by conspicuous pale cream-white forecrown stripes, and grey-white eyebrows, lores, and moustaches, all these forming conspicuous pale area setting off large jet-black eyes; disc rim well marked black on lower half. Ground-colour of underparts varies as face but always marked by broad barred streaks, heavy on chest but becoming light on belly and rear flanks. Body plumage loose, so that grey- or brown-white legs and feet rarely visible. Underwing grey- to brown-white, appearing dull and relieved most by dark-brown tips to primary coverts (forming bold but less obvious patch than in *Asio* owls) and dull grey or brown barring over primaries. Relatively darker head and body make bird appear front-heavy in flight. Ground-colour of upperparts varies as face and underparts, softly streaked on back and wing-coverts and relieved by broad white fringes and tips to scapulars and outer median and greater coverts (all forming conspicuous spots or stripes, paler than face-divide) and soft but conspicuous barring of primaries, tertials, and tail. (2) British and south-west European race *sylvatica*. Smallest race. Plumage pattern as nominate *aluco* but rufous-brown or tawny-chestnut ground-colours commoner (predominant in Britain). (3) North-west African race, *mauritanica*. Larger, with wing-span up to 20% greater than European birds. Plumage pattern as nominate *aluco* but ground-colour always grey-brown; feather-marks heavier, with stronger bars and vermiculations above and below. JUVENILE. Up to September (in Europe), noticeably loose, even shaggy, body plumage obvious in close view when greater incidence of finely-barred feathers should also show.

Although, with *A. noctua*, one of the two most frequently encountered owls in temperate regions of west Palearctic, by no means unmistakable. In sudden, starkly illuminated view, e.g. alongside road, important to ignore illusory plumage contrasts and concentrate on tubby, large-headed, and relatively broad-winged form (shared by much larger Eagle Owl *Bubo bubo* and Great Grey Owl *S. nebulosa*) and characteristic glides of flight action. In normally illuminated and unobstructed view, black eyes set in divided, dark face give kind expression and bold pale spots on upperparts catch eye. Hunched, barrel or ball shape of perched bird always pronounced; erect, slimmer postures adopted under threat, or in thin cover. Flight usually undertaken at rather greater height than other owls, with classically silent, measured, and direct progress only occasionally broken by wheels and tilts; action much less flapping than in *T. alba* or *Asio*, with wing-beats more contained, less uneven, and stronger, recalling *A. noctua* at times. Gait restricted, with shuffling walk and occasional lope or hop.

Song 'hoo', given first in long, drawn-out monosyllable, then a long pause, then a faint monosyllable, a short pause, and finally an extended, soft quaver, which falls in pitch. Other common call a sharp 'ke-wick', heard most often in spring and early summer.

Habitat. In breeding range mainly complementary to Ural Owl *S. uralensis* but in oceanic as well as continental upper and lower middle latitudes, from fringe of boreal through temperate and steppe to Mediterranean and

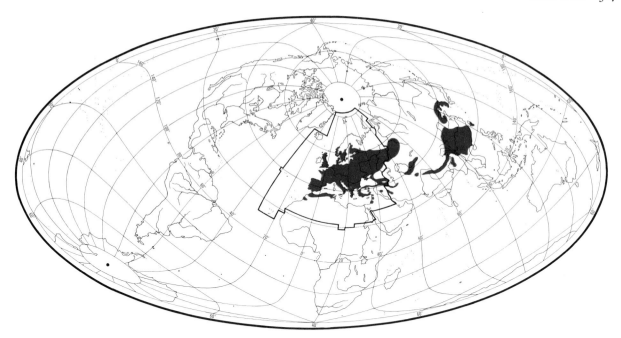

related montane zones, avoiding unduly windswept, rainy, frosty, or arid climates and large unbroken forests, wetlands, or naked rocky terrain, or treeless plains. Mainly in lowlands, but in Scotland nests up to c. 550 m (Baxter and Rintoul 1953), and in Alps to c. 1600 m, exceptionally even higher (Glutz and Bauer 1980). Ranges up to 2350 m in Turkey. In Turkestan (USSR), in broadleaf mountain forests up to similar altitude, nesting in rock clefts (Dementiev and Gladkov 1951a). In Britain, mostly in deciduous or mixed woodlands, tree-dotted farmland, parks, large gardens, and churchyards; often in built-up areas, even inner London. Not uncommon in mature coniferous plantations, and exceptionally in open country. In USSR, rare in taiga zone, where occurring in coniferous forests; abundant in broadleaf forests and parks, orchards, and cultivated lands in central parts of range, nesting chiefly in hollow trees, especially oak *Quercus*, lime *Tilia*, and birch *Betula*, up to c. 10 m above ground (Dementiev and Gladkov 1951a). Requires richly structured habitat with plenty of look-out posts for hunting in sparse woodland, clearings, avenues, cemeteries, hedges, and gardens, especially among mature trees, and with some preference for access to water. In southern Asia ranges up to 4250 m in Himalayas and affects oak, pine, and fir forest, sometimes living above treeline and nesting on ground under overhanging rock by streams, or in holes in cliffs in ravine (Bates and Lowther 1952; Ali and Ripley 1969). Adaptability to close proximity of man shown in various parts of range.

Distribution. Colonized Finland in late 19th century; some range increases Netherlands and northern USSR.

BRITAIN. Bred Isle of Man 1961 (Sharrock 1976). NETHERLANDS. In 19th century, restricted to east, south, and centre; from c. 1925 colonized much of remainder of country and still spreading in north and coastal dunes (CSR). NORWAY. Bred near 66°N, 1969 (Haftorn 1971). FINLAND. First proved breeding 1878; range expanded in 1920s and especially 1930s (Merikallio 1958). GREECE. May breed Samos, Rhodes, and Kos but no proof (HJB, WB, GM). USSR. Spreading north in southern Karelia (Neufeldt 1958a). CYPRUS. Record of breeding in 1966 (Stewart and Christensen 1971) now considered doubtful (PRF, PFS).

Population. Data limited. Some increases in Finland, Britain, and Netherlands; fairly stable in central Europe (Glutz and Bauer 1980).

BRITAIN. Declined 19th century due to human persecution (Alexander and Lack 1944). Increased over much of England, Wales, and southern Scotland from c. 1900 to 1930 (until c. 1950 some areas); apparently stable since (Parslow 1973). 10000–100000 pairs (Parslow 1973); probably 50000–100000 pairs (Sharrock 1976). FRANCE. 10000–100000 pairs (Yeatman 1976). SPAIN. Over 50000 pairs 1975 (AN). BELGIUM. About 3000 pairs; some recent increase (Lippens and Wille 1972); little recent change (PD). LUXEMBOURG. About 800 pairs (Lippens and Wille 1972). NETHERLANDS. Estimated 2500–3000 pairs 1977 (Teixeira 1979); increasing, mainly because of range extension but formerly aided by afforestation (CSR). WEST GERMANY. Estimated over 20000 pairs (Rheinwald 1982). SWEDEN. About 10000 pairs (Ulfstrand and Högstedt 1976); decreased in 1960s, probably due to pesticides, but has recovered since (LR). FINLAND. Estimated 2000 pairs; increased (Merikallio 1958). POLAND. Scarce to fairly

528 Striginae

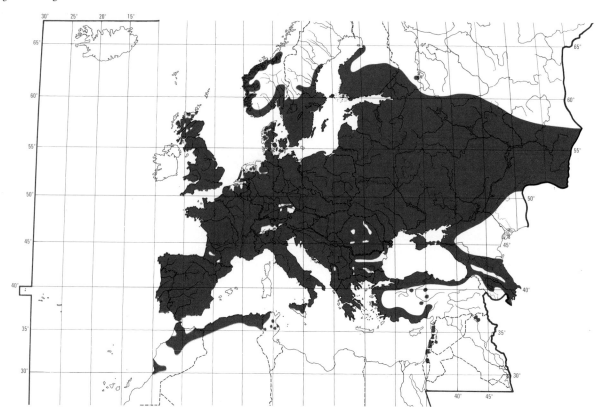

numerous; increasing in centres of large towns in recent years (Tomiałojć 1976a). EAST GERMANY. About 25 000 pairs (Mikkola 1983). CZECHOSLOVAKIA. Numbers apparently stable (KH). ALBANIA. Rare (EN). GREECE. Not common (HJB, WB, GM). USSR. Common in central areas; rare in taiga, Crimea, north-west Caucasus, and Talysh plain (Dementiev and Gladkov 1951a). Estonia: *c.* 500 pairs (Randla 1976).

Survival. Sweden: mortality in 1st year of life 71·2%, in 2nd year 44·3%, in 3rd year 47·6% (Olsson 1958). Switzerland: mortality in 1st year of life 49·4%, in 2nd and 3rd years 24·5% (Glutz and Bauer 1980). England: mortality in 1st year of life 52·6%, in 2nd year 22·2%, in 3rd year 31·8% (small sample, $n = 19$) (Southern 1970). West Germany: mortality in 1st year of life 48%. Belgium: mortality in 1st year of life 58%. (Delmée *et al.* 1978.) For life-table see Southern (1970), for seasonal mortality see Glue (1973), and for deaths of juveniles from starvation, see Southern (1970) and Hirons *et al.* (1979). Oldest ringed bird 18 years 10 months (Delmée *et al.* 1980).

Movements. Mainly resident. Breeding adults sedentary, remaining in territory all year. Juveniles have protracted period of 2½–3 months under parental care, and disperse in period August–November; mortality highest then, for those not replacing dead adults usually starve rapidly (Hansen 1952; Southern *et al.* 1954; Southern 1970; Hirons *et al.* 1979). See also Social Pattern and Behaviour.

Dispersal complete by late autumn, juveniles by then either dead or established in territory; very little subsequent movement, and then usually under stress of cold winter weather in northern parts of range (Olsson 1958).

In temperate Europe, dispersals normally short, within radius of a few km from birthplace; but proportion of longer movements is higher in Fenno-Scandia. Sample of British birds ringed as nestlings and recovered after dependence period ($n = 458$) moved: 0–10 km 83%, 11–20 km 11%, 21–50 km 4·5%, 51–100 km 1%, over 100 km (maximum 133 km) 0·5% (BTO). In contrast, Swedish and Finnish birds ($n = 370$) moved: 0–19 km 68%, 20–50 km 16%, 51–100 km 12%, longer 4%; maximum 745 km—a northward straggler into Lappland (Olsson 1958). Birds disperse in all directions. In Sweden north of 58°N, more northwards (ENE–WNW) movement (55% of recoveries) than in southern Sweden (45%); interpreted as adaptation for range expansion (as occurred, e.g., in Finland), normally constrained by external factors (Olsson 1958).

In central and west-central Europe, birds behave more like those in Britain, with recoveries seldom showing movement above 50 km. Furthest are: Swiss nestling found Loire (France), 280 km SW; Belgian nestling found Cher (France), 450 km SSW; Czechoslovakian nestling found in Hungary, 460 km SE; January-ringed bird recovered August, 350 km ENE within Poland (Glutz and Bauer 1980). Irregular nomadic wanderings by immature birds (e.g. under stress of cold weather) reported from northern parts of USSR range; October migrant in Kaliningrad recovered Latvia in May, 210 km N (Dementiev and Gladkov 1951a). *Strix aluco* involved in small-scale owl irruptions into Pskov region, USSR, in autumns 1960 and 1963 (Meshkov and Uryadova 1972).

Food. Wider range of prey than other medium-sized Strigiformes of west Palearctic. In woodland, mainly small rodents; also larger mammals (up to size of young rabbit *Oryctolagus cuniculus*), birds, amphibians, shrews (Soricidae), earthworms (Lumbricidae), and beetles (Coleoptera). In towns, largely birds; also small rodents and other prey as available. Normally hunts almost wholly between dusk and dawn; occasionally in full daylight (even in the open), and this not uncommon when young in nest (e.g. Brown 1936a, Hardy 1977, Hirons 1977; see final paragraph). Most prey evidently located by ear. For head movements in this and other Strigidae and their possible use in localization of prey, see Lindblad (1962). Much hunting done by waiting on perch and gliding or dropping on to prey; on impact, will extend wings to cover victim or strike it (Meinertzhagen 1959). While hunting from perch, bird turns body occasionally, and makes frequent short flights (at least one every 5 min), returning to same perch or another. Where habitat includes open ground, often also hunts in flight, 2–3 m high between bushes in marsh or grassland; flight then slow and frequently interrupted with glides; after flying in zigzag pattern over area $c.\ 30 \times 50$ m, flies 100–200 m before resuming zigzag search (Nilsson 1978); may hover (Ruthke 1935). Of 2 birds radio-tracked for 54 and 33 hrs, late February to early May, southern Sweden, one spent 27% of time sitting still, 40% hunting from perch, and 34% hunting in flight; other bird, 49%, 9%, and 42% (Nilsson 1978). Birds apparently taken mostly from roosting sites: may dive into bushes, etc., containing roost (Green 1905; Fitter 1949; Meinertzhagen 1959); said also to drive birds out by beating wings against twigs (Schnurre 1934), or to make them call by wing-clapping nearby (Uttendörfer 1939a); sometimes investigates such sites by hovering near them for a few seconds (Witherby *et al.* 1938). Recorded catching Sand Martin *Riparia riparia* hovering briefly at $c.$ 10 m (after being released following ringing) by flying out of trees and taking it in feet (Mather 1979). Will take bats (Chiroptera) in flight (Uttendörfer 1939a). Takes young and adult birds from open nests (e.g. Snow 1958, Harrison 1960) and from nest-holes and nest-boxes; may check holes regularly, especially when rain or heavy snow cover make other forms of hunting difficult (Schnurre 1934; Stülcken 1961). Fish said to be taken from water surface while bird in flight, or by wading in shallows (Glutz and Bauer 1980); once seen hovering over coastal surf, East Germany (Ruthke 1936). Said to be able to carry prey of up to $c.$ 320 g in flight (Bernhoft-Osa 1973). Usually eats on elevated perch and will search for pieces of food dropped accidentally (Glutz and Bauer 1980). Empty nest-box sometimes used for plucking prey (e.g. Berndt 1943). In captivity, generally removes head of mice (Murinae) and swallows this first; small mice sometimes swallowed whole. Often tears out thighs of frogs (Anura) before eating them (Räber 1950). Birds are plucked, though feathers may occur in pellets (Thiollay 1963b); captive bird dealt with sparrow *Passer* by holding tail down with foot, pulling at head to remove tail, and then removing wing-feathers; swallowed head-first (Bütikofer 1909–10). Remains of squirrel *Sciurus* indicated that it had been dealt with by leaving head intact and pulling skin back in one piece from neck towards rear; flesh had then been removed without breaking bones or severing tendons (Brichambaut 1978). Bird hunting for earthworms typically flies down to grass from nearby perch, sits still until it appears to hear a sound, then rotates head and stretches neck; may hop forward, repeating head movements, finally taking 2–3 long leaps with wings partially spread to peck up worm. If worm on surface, immediately eaten; if not, bird pulls up and back with wings partially open and worm commonly breaks. May sit still for 10 min before moving after another worm. In 6 observations totalling 1 hr, average feeding rate 0·39 worms per min (Macdonald 1976). Bird on lawn (presumably hunting for worms, though none seen to be taken) gave 4–5 patters with feet then moved $c.\ 1\cdot5$ m and repeated; similar pattering (with both feet) seen from bird on freshly dug earth (Meinertzhagen 1959). Recorded waiting in hedge while field being ploughed, flying out each time plough passed to search fresh furrow, mixing with gulls *Larus* (Beven 1965). Sometimes runs after small prey (Bösiger and Faucher 1962). In captivity, most large terrestrial insects caught with feet, but some with bill (Bütikofer 1909–10; Räber 1950). Presence of pellets containing soil, plant roots, and remains of *Typhaeus typhoeus* near excavated burrows of this beetle suggests that bird will dig for such food (Burton 1950). Insects may be taken in flight: e.g. captive bird took large moths (Lepidoptera) in feet, flying to perch to eat them (Bütikofer 1909–10). Will take cockchafers *Melolontha melolontha* from trees by flying up, turning over in flight, and hanging by one or both feet from ends of twigs, taking beetles in foot or bill (Schnurre 1934; Oldenburg 1954b; Schneider 1979); caterpillars (Lepidoptera) may also be taken in trees (Uttendörfer 1936). For behavioural study of reactions to various prey types, living and dead, see Räber (1950). Recorded feeding on carrion in the wild (e.g. Roberts 1944, Gruzdev and Likhachev 1960)—sometimes, at least, when unable to find other food (Heurn 1954), though habit apparently regular in some individuals (Vásárhelyi 1966–7); for items taken, see below. Cannibalism between full-grown birds recorded, involving juveniles as prey (Wendland 1972a); in one case, at least (Gloger 1861), apparently killed and eaten by same bird. For cannibalism among nestlings, see final paragraph and Social Pattern and Behaviour. Will cache surplus food in or on tree (Walpole-

Bond 1938; Philipson 1948), throughout the year according to Glutz and Bauer (1980); see also Social Pattern and Behaviour. Little data on hunting success and none by direct observation. Bad weather (wind, rain) likely to reduce success through noise interference; may also reduce activity, and thus availability, of prey species. At Wytham (Oxford, England), twice as many rodents brought to nest on dry nights as on wet; worms brought mainly in rainy periods (Hirons 1976). In Ythan valley (Aberdeen, Scotland), some evidence suggested hunting finished earlier (having presumably been more successful) on fine nights. Poor hunting success when young in nest (measured in terms of weight changes of young) often, but not always, coincided with low temperature or rain (Hardy 1977), and, at Wytham, more clutches deserted in rainy weather than at other times (Southern 1970). Wind, rain, and amount of moonlight found not to lead to any marked change in areas used for hunting in Ythan valley (Hardy 1977).

Most pellets consist of mammal fur and bones, and nearly all of these are medium grey when dry, with slightly elastic texture like felt; more matted if held in stomach for short time, dusty and friable if held for long time (Guérin 1932; Southern 1954). Texture similar to those of Long-eared Owl *Asio otus*, but looser and more fragile than Barn Owl *Tyto alba* (Thiollay 1963b; Glue 1970). Large prey cause large and loose pellets with characteristic colour—pale grey (rabbit *Oryctolagus cuniculus*) or black (mole *Talpa europaea*). Shrews *Sorex* cause small black pellets. Fur normally little changed, but bones more or less disarticulated and broken (Thiollay 1963b). Skulls usually broken (less so for shrews), upper jaw with part of cheek often becoming detached. After feeding on earthworms, bird produces brownish fibrous pellets consisting of decaying leaves and sand and containing numerous earthworm chaetae (Southern 1954). Chitinous parts of beetles (Coleoptera) usually little damaged, but pellets formed almost wholly of these in highly fragmented state occur occasionally (Guérin 1932). Pellets often contain soil, plant remains, etc. (e.g. Gruzdev and Likhachev 1960, Thiollay 1963b); may even be composed entirely of soft wood or sawdust (Southern 1954). See Guérin (1932) for further descriptions. Pellets normally measure: 30–65 × 18–24 mm (Guérin 1932); 33–67 × 23–30 mm, rarely less than 40 mm long, $n=22$ (Witherby *et al.* 1938); 30–60 × 20–25 mm (Thiollay 1963b); 30–70 × 18–26 mm (Glue 1970); average 55 × 24 × 20 mm (34–84 × 17–30 × 11–28), $n=45$ (Mikkola 1983). Record of pellet containing complete skull (with long bill) of Snipe *Gallinago gallinago*, total length c. 100 mm (Gurney 1929), and pellets containing long feathers recorded up to 163 mm (Krauss 1977). Dry weight of pellets containing only small rodents averaged 2·59 g ($n=384$) from captive ♂ and 2·57 g ($n=377$) from wild birds; average 0·80 prey per g of pellet (captive) and 0·72 (wild), or 2·1 prey per pellet (captive) and 1·9 (wild) (Southern and Lowe 1982): 2–5(–9) items per pellet (Thiollay 1963b); average 2·2 items per pellet, from 1·8 in winter to 2·6 in spring (Contoli and Sammuri 1978). In captive ♂, number of pellets produced per day (averaged over 1-month periods) 0·9–1·5; average 1·03 April–September, 1·27 October–March (*contra* Guérin 1932). Significant proportion of items eaten were too well digested to reappear in pellets (as identifiable skulls): average 22·0% of mice *Apodemus*, 20·1% of bank voles *Clethrionomys glareolus*, 13·7% of short-tailed voles *Microtus agrestis*; losses greatest (up to twice these averages) in summer, when digestion most efficient (see penultimate paragraph) (Lowe 1980; see also Raczyński and Ruprecht 1974, Hardy 1977). Pellets cast from scattered and frequently changing sites in rather dense vegetation; through the winter, sites become fewer, more consistently used, and higher—to c. 9–12 m (Southern 1954). Empty nest-box occasionally used (Berndt 1943), but apparently not nest-site at which breeding in progress (Witherby *et al.* 1938).

Morphological adaptations include short broad rounded wings (see Structure) with high wing-loading which, used in rapid shallow strokes, give high manoeuvrability necessary in woodland (Smeenk 1973; Hardy 1977, which see for structural comparison with *Asio otus* and Short-eared Owl *A. flammeus*). For experimental study of low-volume low-frequency sound produced by wings in flight, and for adaptations of wing-feathers, see Neuhaus *et al.* (1973).

The following prey recorded in west Palearctic. Mammals: young hedgehog *Erinaceus europaeus*, shrews (*Sorex minutus, S. araneus, Neomys fodiens, N. anomalus, Suncus etruscus, Crocidura leucodon, C. suaveolens, C. russula*), moles (*Talpa europaea, T. caeca*), bats (*Rhinolophus ferrumequinum, R. hipposideros, Myotis daubentonii, M. mystacinus, M. nattereri, M. bechsteinii, M. myotis, M. blythii, Plecotus auritus, Pipistrellus pipistrellus, P. nathusii, Vespertilio serotinus, V. nilssoni, V. murinus, Nyctalus noctula, N. leisleri*), young rabbit *Oryctolagus cuniculus*, young brown hare *Lepus*, red squirrel *Sciurus vulgaris*, grey squirrel *S. carolinensis*, garden dormouse *Eliomys quercinus*, forest dormouse *Dryomys nitedula*, edible dormouse *Glis glis*, dormouse *Muscardinus avellanarius*, common hamster *Cricetus cricetus*, wood lemming *Myopus schisticolor*, bank vole *Clethrionomys glareolus*, water vole *Arvicola terrestris*, pine voles (*Pitymys subterraneus, P. savii, P. duodecimcostatus*), common vole *Microtus arvalis*, short-tailed vole *M. agrestis*, root vole *M. oeconomus*, snow vole *M. nivalis*, muskrat *Ondatra zibethicus*, harvest mouse *Micromys minutus*, striped field mouse *Apodemus agrarius*, yellow-necked mouse *A. flavicollis*, wood mouse *A. sylvaticus*, black rat *Rattus rattus*, brown rat *R. norvegicus*, house mouse *Mus musculus*, northern birch mouse *Sicista betulina*, stoat *Mustela erminea*, weasel *M. nivalis* (Skuratowicz 1950; Simeonov 1963; Kulczycki 1964; Beven 1965, 1982; Smeenk 1972; Wendland 1972b, 1980; Krzanowski 1973; Holmberg 1976; Schmidt 1976; Corbet and Southern 1977; López Gordo *et al.* 1977; Contoli and

Sammuri 1978; Gerdol *et al.* 1982). Wide range of birds: from size of Goldcrest *Regulus regulus* to those as large and unlikely as adult Mallard *Anas platyrhynchos* (Schnurre 1934) and full-grown Kittiwake *Rissa tridactyla* (Saunders 1962); mostly, however, species that roost and/or feed in trees or are associated with man—thrushes *Turdus*, tits *Parus*, Starling *Sturnus vulgaris*, sparrows *Passer*, and finches (Fringillidae). See, especially, Smeenk (1972), Wendland (1980), and, for review of birds of prey taken (up to size of *Asio otus*), Mikkola (1976). Reptiles: lizards (*Lacerta viridis*, *L. muralis*, *L. agilis*, geckos), slow-worms *Anguis*, and grass snakes *Natrix* (Simeonov 1963; Wendland 1963; Glutz and Bauer 1980). Amphibians: frogs and toads (*Rana temporaria*, *R. ridibunda*, *R. esculenta*, *R. arvalis*, *Alytes*, *Hyla*, *Bombina*, *Bufo bufo*, *Pelobates fuscus*, *P. cultripes*), and great crested newt *Triton cristatus* (Smallcombe 1934; Kulczycki 1964; Smeenk 1972; Wendland 1972a, b; López Gordo *et al.* 1977; Glutz and Bauer 1980). Fish: trout *Salmo trutta*, roach *Leuciscus rutilus*, goldfish *Carassius auratus*, perch *Perca fluviatilis*, and miller's thumb *Cottus gobio* (Williams 1964; Beven 1965, 1982; Glue 1969; Delmée *et al.* 1979). Insects: mostly beetles, including Carabidae (especially *Carabus nemoralis*), Dytiscidae, Hydrophilidae, Histeridae, Silphidae, Staphylinidae, Lucanidae, Geotrupidae (especially *Geotrupes* and *Typhaeus typhoeus*), Scarabaeidae (especially cockchafer *Melolontha melolontha*), Elateridae, Meloidae, Tenebrionidae, Coccinellidae, Cerambycidae, Chrysomelidae, Curculionidae, and Scolytidae; also dragonflies (Aeshnidae, Libellulidae), stoneflies (Plecoptera), crickets *Gryllus*, mole-cricket *Gryllotalpa gryllotalpa*, grasshoppers (Acrididae), earwig *Forficula auricularia*, shieldbug (Pentatomoidea), Neuroptera, Lepidoptera (hawkmoths Sphingidae, adult and larval Noctuidae, Lymantriidae), horse-fly *Chrysops*, ants (Formicidae), wasps (Vespidae), and bees *Bombus* (see especially Uttendörfer 1939b, Southern 1954, Haensel and Walther 1966, Smeenk 1972; also Witherby *et al.* 1938, Moltoni 1940, Barruel 1958, Simeonov 1963, Beven 1965, 1982, Hagn-Meincke 1967, Shchegolev 1974, López Gordo *et al.* 1977, Glutz and Bauer 1980). Other invertebrates include earthworms (*Lumbricus terrestris*, *Allolobophora*), cockle *Cardium edule*, snails *Lymnaea*, slugs (*Agriolimax laevis*, *Limax agrestis*), crayfish *Austropotamobius*, wolf-spider (Lycosidae), and Myriapoda (Witherby *et al.* 1938; Uttendörfer 1939a; Bergman 1961; Zanola 1966; Macdonald and McDougall 1970; López Gordo 1974; Fryer 1976; López Gordo *et al.* 1977; Delmée *et al.* 1979). Items recorded as carrion include *Rattus rattus*, *Lepus*, *Mustela nivalis*, polecat *Putorius putorius*, young sheep, and trout (Salmonidae) (Roberts 1944; Gruzdev and Likhachev 1960; Vásárhelyi 1966–7; see also Kochan 1979). In captivity, sometimes swallows green leaves, apparently not for lack of material for pellet formation (Räber 1950); for other inanimate material in pellets, see 2nd paragraph.

Extensive literature of food analyses, mostly based on pellet examination. Diet and most related aspects of ecology more thoroughly studied than in any other species of west Palearctic Strigiformes. For summary of major studies from various rural areas, see Table A. See also the following significant studies. Britain: Beven (1965). Denmark: Andersen (1961). East and West Germany: Uttendörfer (1939a, b, 1952), Wendland (1975); see also sources in Glutz and Bauer (1980). Poland: Skuratowicz (1950), Kulczycki (1964). USSR: Gruzdev and Likhachev (1960), Averin and Ganya (1966). Bozsko (1967). France: Guérin (1932), Thiollay (1968). Spain: Elosegui (1974), López Gordo (1974), Herrera and Hiraldo (1976). Italy: Moltoni (1940). Bulgaria: Simeonov (1963). Important to remember in all these studies that (1) proportion of food provided by invertebrates is not accounted for (see further, below), and (2) no allowance made for possible differential digestion (i.e. non-appearance in pellets) of the various vertebrate prey types (see 2nd paragraph).

Diet strongly influenced by the species' sedentary territorial habit which favours a diverse diet (yet restricts number of prey species available) and necessitates dietary shifts of emphasis between various prey types as changing conditions affect their relative profitabilities. Apparent specialization of individual birds on certain prey types thus often due to their local abundance, though Southern (1954) found year-to-year comparisons between food taken in different (nearby) territories to indicate that real individual preferences do occur. Local habitat differences will also affect diet: at Wytham (Table A), birds in pure woodland took fewer large mammals and more small rodents than those hunting also over open ground—*Talpa* 21·6 and 15·6% (by weight) respectively, *Oryctolagus* 17·4 and 11·0%, *Rattus* 6·3 and 1·6%, *Apodemus* 21·4 and 23·2%, *Clethrionomys* 16·6 and 27·5%, and *Microtus agrestis* 7·8 and 11·6% (Southern 1954). In rural areas (see below for diet in towns), small rodents usually form predominant prey type by weight (Table A). Commonly held to be preferred prey, with other food of more constant availability being utilized instead when small rodents less readily caught; no real evidence for this, however, and it seems more likely that there are conditions (e.g. increased availability of other prey, notably young *Oryctolagus*, usually in early spring) under which small rodents become suboptimal prey, even though their absolute availability is unchanged. Little concrete data on prey availability which would permit assessment of preferences, even for small rodents (but see review by Hardy 1977). However, at Wytham, relative to their abundances, *Apodemus sylvaticus* found to be taken more than twice as often as *Clethrionomys glareolus*, *Apodemus* making greater use of open habitat (Southern and Lowe 1982); *Clethrionomys* taken from various densities of ground cover in proportion to its abundance in such cover types (greatest density in thicker cover), while *Apodemus* taken more from open areas despite being equally abundant in all cover types;

Table A Food of Tawny Owl *Strix aluco* in rural areas as found from pellets. Figures are percentages of total live weight of vertebrates only.

Area	Wytham (S England)	Ythan valley (N Scotland)	Kvismaren (S Sweden)	Twente (E Netherlands)	Oignies (Belgium)
Period[1]	1945–52, all year	1972–5, mostly Dec–June	1961–70	1965, winter–July	1960–74, all year
Mice (small Murinae)[2]	22.5 ⎫ 56.4	11.0 ⎫ 60.7	29.2 ⎫ 65.5	8.6 ⎫ 27.4	13.7 ⎫ 53.6
Small voles (Microtinae)	33.9 ⎭	49.7 ⎭	36.3 ⎭	18.8 ⎭	39.9 ⎭
Larger mammals[3]	34.7	19.5	9.6	20.6	23.9
Shrews (Soricidae)	5.2	9.4	5.3	8.5	16.0
Birds	3.7	10.4	5.3	34.2	1.8
Amphibians	0	0	14.3	9.2	4.1
Other	+ (bats)	0	+ (bats)	0	0.6 (fish)
Total no. of vertebrates	9494	922	3385	2624	9656
Source	Southern (1954)	Hardy (1977)	Källander (1977b), data reanalysed	Smeenk (1972)	Delmée et al. (1979), data reanalysed

1. Pellets more difficult to find in summer and autumn, and these seasons thus usually less well represented.
2. Largely *Apodemus*; also includes a very few dormice *Muscardinus avellanarius* (Muscardinidae).
3. Mostly moles *Talpa*, rabbit *Oryctolagus cuniculus*, water vole *Arvicola terrestris*, and brown rat *Rattus norvegicus* (but largely edible dormouse *Glis glis* at Trieste).
4. Comprises 4 bats (Chiroptera) and 34 unnamed items.

no evidence for selection between sexes or size classes of either species (Southern and Lowe 1968). Shrews usually average less than c. 8% of food by weight, but may greatly exceed this in the short term, at Oignies (Table A) reaching c. 29% by weight in one year (recalculated from Delmée et al. 1979; see also Thiollay 1968). Possible also that relatively high levels (14–15% by weight) may be normal in some areas of France (Guérin 1932; Southern 1954). Birds often taken in quantities similar to shrews, but can also be much more important, and while lack of alternative prey readily explains predominance in towns (see below), empirical variations between apparently well-studied woodland areas (Table A) more problematical. Perhaps due to seasonal bias in pellet collections, in some areas to longer snow cover which might cause birds to be favoured while small mammals thus less available, or perhaps simply to local differences in abundance of various prey types (e.g. Delmée et al. 1979). Presence of suitable habitat crucial in occurrence of frogs and toads in diet. In spring and summer, Grunewald (Table A), food of one individual comprised 45% (presumably by number of items) *Pelobates fuscus*; *Bufo bufo* much commoner in the area but only 5 occurred in 31 986 prey items (Wendland 1972a). No amphibians recorded in diet at Wytham, where hardly any habitat suitable for frogs (Southern 1954).

For year-to-year studies of proportions of major items in diet, see Southern (1954: 7 years), Wendland (1975, 1981: 21 years), Delmée et al. (1979: 15 years), and Goszczyński (1981: 6 years). Such proportions subject to pronounced fluctuations in line with prey population levels (in case of preferred prey at least): e.g. in Grunewald, 1959–79, normally a 3-year cycle in proportions of *Apodemus flavicollis* (main prey)—from 36.2–48.3% (by number) at peak, to low of 13.8–23.8% 2 years later (Wendland 1981); strong negative correlation between proportions of birds and of *Apodemus*, presumably indicating replacement of *Apodemus* by birds in year when *Apodemus* population low (Wendland 1972a). At Oignies, similar negative correlation presumably indicating replacement of small rodents by shrews; small rodents varied from c. 18% to c. 68% by weight, shrews c. 2–29% (recalculated from Delmée et al. 1979). At Wytham, however, proportions of birds and shrews stayed more or less constant between years while large mammals appeared to make up for deficiencies in small rodents; small rodents varied from 46.8 to 73.7% by weight, large mammals 16.4–39.1% (Southern 1954).

Seasonal analyses provided by Skuratowicz (1950), Gruzdev and Likhachev (1960), Beven (1965, 1982), Smeenk (1972), Hardy (1977), Contoli and Sammuri (1978), and Goszczyński (1981); most complete year-round analysis still the classic study by Southern (1954). Pattern of seasonal change varies widely (Fig II), depending presumably on local prey availabilities. However, constant feature in all well-studied areas is decline in proportion of small rodents in diet in spring and summer (at Wytham, at least, lasting through until autumn). These may be replaced largely by birds (e.g. Gruzdev and Likhachev 1960, Wehner 1962, Wendland 1963, Goszczyński 1981), large mammals (e.g. Southern 1954,

Table A continued.

Area	Grunewald (West Berlin)	Turew (Poland)	Madrid (cent. Spain)	Trieste (NE Italy)	Farma valley (cent. Italy)
Period[1]	1974–8	1970–5, all year	1973–6, autumn, winter, spring	1978, all year	1975–7
Mice (small Murinae)[2]	28·9 } 37·2	16·7 } 72·6	18·9 } 20·4	52·0 } 67·1	55·4 } 96·7
Small voles (Microtinae)	8·3	55·8	1·5	15·2	41·3
Larger mammals[3]	12·9	6·4	67·8	28·4	0
Shrews (Soricidae)	0·3	0·6	1·2	1·8	3·3
Birds	46·6	19·2	10·2	2·7	0
Amphibians	0·5	1·2	0·2	0	0
Other	2·5[4]	0·2 (bats, lizards)	0	0	0
Total no. of vertebrates	2739	1600	303	283	381
Source	Wendland (1980), corrected	Goszczyński (1981)	López Gordo et al. (1977)	Gerdol et al. (1982), data reanalysed	Contoli and Sammuri (1978)

Hardy 1977), or both (Smeenk 1972); at Amsterdamse-Waterleidingduinen (Fig II), both main winter prey groups (birds and small rodents) decline while *Oryctolagus* increases (F J Koning). At Wytham, change occurs fairly rapidly around c. 7–8 May when young half-grown; at this time, ground cover suddenly increases (making small rodents more difficult to catch) and young of large mammals readily available. Rapidity of change perhaps linked to appearance of young *Talpa* at this time which, at Wytham, quickly become a major food for young in nest (Southern 1954, 1969; see also final paragraph); increased numbers of young *Oryctolagus* and *Rattus* available from earlier on, however, and this may account for earlier and more gradual change in diet in areas where *Talpa* less important (Fig II). Insects, especially beetles, also become plentiful at this time and contribute further to spring change in diet. At Wytham and Twente (Table A), proportions of small rodent species stay fairly constant relative to each other, whereas at Amsterdamse-Waterleidingduinen most of decrease in small rodents accounted for

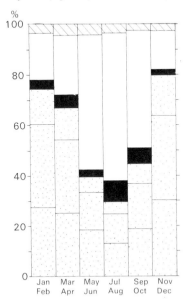
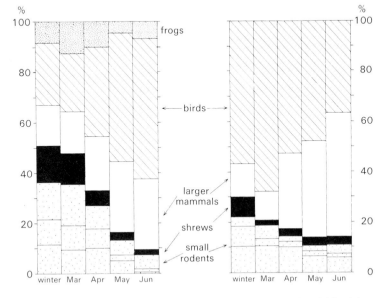

(1) Wytham (southern England), 1949–52; total 9357 vertebrates (Southern 1954).

(2) Twente (eastern Netherlands), 1965; total 2624 vertebrates (Smeenk 1972).

(3) Amsterdamse-Waterleidingduinen (western Netherlands), 1965–79; total 2214 vertebrates (F J Koning).

Fig II Seasonal variation in food of Tawny Owl *Strix aluco* at 3 sites in Europe. Data refer to proportion by weight of vertebrate prey only. Subdivisions within small rodents are (from bottom to top): mice (largely wood mouse *Apodemus sylvaticus*); bank vole *Clethrionomys glareolus*; *Microtus* voles. At Amsterdamse-Waterleidingduinen, April–June, larger mammals comprise mostly rabbit *Oryctolagus cuniculus*.

Table B Food of Tawny Owl *Strix aluco* in urban and suburban areas as found from pellets. Figures are percentages of total live weight of vertebrates only.

Area	Poznań (Poland)	Holland Park (London, England)	West Berlin	Morden (London)
Habitat	Central city area	Urban park	Parks, gardens	Suburbs
Period	Spring	1957–64, all year	Not stated	1955–6, all year
Birds	90·1	93	66·4	45
Mice (small Murinae)	5·8 ⎱ 6·3	3 ⎱ 3	10·4 ⎱ 15·4	19 ⎱ 28
Small voles (Microtinae)	0·5 ⎰	0 ⎰	5·0 ⎰	9 ⎰
Larger mammals[1]	2·1	4	16·5	27
Shrews (Soricidae)	0	0	0·1	0
Amphibians	1·5	+	1·3	0
Other	0	+ (fish[2])	0·3[3]	0
Total no. of vertebrates	191	335	2647	262
Source	Bogucki (1967)	Beven (1965)	Wendland (1980), corrected	Beven (1965)

1. In more built-up habitat, almost wholly brown rat *Rattus norvegicus*.
2. Goldfish *Carassius auratus*, evidently from park and garden ponds.
3. Comprises 2 bats (Chiroptera) and 11 unnamed items.

by *Apodemus*. Also at Amsterdamse-Waterleidingduinen, proportion by weight of medium-sized birds (*Sturnus vulgaris* to Jay *Garrulus glandarius*) whose remains were found in nests stayed at 54·0–62·7% through March–May, while smaller birds decreased from 21·1 to 3·1% and larger birds (pigeons *Columba*, Woodcock *Scolopax rusticola*) increased from 16·2 to 37·7% (*n* = 303). In Farma valley (Table A), where large mammals said to be unavailable, Murinae comprised *c*. 85% of total food by weight in summer, *c*. 35% in winter; Microtinae *c*. 8% summer, *c*. 60% winter (sample size for whole study only 500 items, however). In deep snow, Bulgaria, Simeonov (1963) found proportion of *Microtus arvalis* in diet decreased while *Apodemus* increased.

Importance of invertebrates difficult to quantify due to paucity of stomach analyses. Individuals sometimes feed on one such prey type continually if abundant. Production of fibrous pellets after ingestion of earthworms (see 2nd paragraph) can indicate their frequency: in Ythan valley, 26·8% of 257 pellets were fibre type December–March, 17·0% of 165 April–June (Hardy 1977; see also Andersen 1961, Southern and Lowe 1982).

Compared with *Asio otus* (an often-migratory vole specialist), diet of *S. aluco* thus much more diverse—both in comparisons between individual birds of each species and between the 2 species as a whole. For major work of comparison, and review, see Smeenk (1972). For further comparisons with diet of other Strigiformes in same or nearby area, see Hardy (1977: *Tyto alba, Asio otus, A. flammeus*, with seasonal data), Källander (1977b: *A. otus*), López Gordo et al. (1977: *T. alba, A. otus*), Contoli and Sammuri (1978: *T. alba*), Delmée et al. (1979: *T. alba, A. otus*), Glutz and Bauer (1980: *A. otus*), Lundberg (1980b: Ural Owl *S. uralensis*), and Gerdol et al. (1982: Little Owl *Athene noctua, Asio otus*). For the only year-to-year and seasonal analysis involving comparisons between species (*T. alba, S. aluco, A. otus*), see Goszczyński (1981): at Turew (Poland), in years with superabundance of *Microtus arvalis*, diets of *T. alba* and *A. otus* became very similar; same tendency in *S. aluco*, but diet remained rather more general; diets of the 3 species most similar autumn and winter, due largely to *S. aluco* taking more *M. arvalis* and fewer birds.

In towns, diet characterized by predominance of birds, though this less so as habitat becomes more rural or open ground more extensive (Table B; see also Guichard 1967, Harrison 1960, Bozsko 1967, Busse 1966). Birds comprise largely House Sparrow *Passer domesticus*; also, commonly, feral Rock Dove *Columba livia*, thrushes *Turdus*, Starling *Sturnus vulgaris*, and Greenfinch *Carduelis chloris*. Invertebrates also occur, but little data on quantification. 47% of 381 pellets at Morden (Table B), including 10 consecutive ones, were fibre type (resulting from consumption of earthworms); 14 ground beetles *Carabus* also recorded. At Esher (Table B), also 39 fibre pellets and 29 Coleoptera; at Hampstead Heath, also 41 fibre pellets and 21 Coleoptera; at Richmond Park, also 8 fibre pellets and 810 Coleoptera. At Tierpark (East Berlin, East Germany), 4·4% of 136 items in pellets were insects (Busse 1966). No remains of invertebrates (including earthworms) found

Table B continued.

Area	Esher (Surrey, England)	Manchester (England)	Hampstead Heath (London)	Richmond Park (London)
Habitat	Suburban gardens	Suburbs, some woodland	Suburban woodland, open ground	Suburban woodland, open ground
Period	1971–7, June–Dec	Jan–July	1967–80, all year	1966–9, all year
Birds	48.2	88.7	35.8	33.6
Mice (small Murinae)	5.1 ⎫ 18.6	4.1 ⎫ 6.2	9.0 ⎫ 31.6	5.9 ⎫ 18.1
Small voles (Microtinae)	13.5 ⎭	2.1 ⎭	22.6 ⎭	12.2 ⎭
Larger mammals[1]	15.1	2.6	16.8	45.3
Shrews (Soricidae)	0.3	0	0.1	0.2
Amphibians	15.2	2.6	15.5	2.9
Other	2.7 (fish[2])	0	0	0
Total no. of vertebrates	367	122	519	109
Source	Beven (1982)	Yalden and Jones (1970)	Beven (1982)	Beven (1982)

among remains of 102 vertebrates from one urban site in London (Harrison 1960). Seasonal analyses (though with small sample sizes) given by Beven (1965, 1982) and Yalden and Jones (1970). At Holland Park (Table B), proportions of small and large mammals, and birds, stayed rather constant through the year, though large birds (pigeons *Columba*, *Garrulus glandarius*) increased in second half of year while smaller birds decreased. At Hampstead Heath, frogs *Rana temporaria* taken mostly March–April (when moving to breeding ponds), but also frequent at Esher August–November.

Food consumption of captive ♂ greatest autumn and early winter: averaged 57.6 g live weight per day August–January, 44.9 g per day February–July; this reflected in bird's weight (Lowe 1980); see also Weights. 6 captive birds each ate average 55.5 g per day in winter, 49.8 g per day in summer (recalculated from Ryszkowski *et al.* 1971). Other captive birds averaged 64.2 g per day in winter (Hardy 1977). In the wild, over 14 24-hr periods, ♂♂ seen to bring average 61 g of prey per day to incubating ♀♀ (Hirons 1976); over 3 nights, average 3.3 items per night (Southern and Lowe 1982). Calculated average requirement in the wild in winter, 73.5 g per day (Simeonov 1963). Bird feeding only on small rodents thus normally takes 2 or (perhaps more usually) 3 per day; however, compare estimates for (smaller) *Tyto alba* and *Asio otus* (see those species). This may have major impact on prey populations. In each 2-month period at Wytham, 1954–6, $c. \frac{1}{5}$–$\frac{1}{3}$ of standing crop of *Clethrionomys glareolus* removed by *S. aluco*; *A. sylvaticus* less abundant but $c. \frac{1}{3}$–$\frac{3}{4}$ removed (Southern and Lowe 1982). From 148 ha of woodland and fields, Wielkopolska (Poland), several *S. aluco* in 1 year removed estimated 2213 rodents, i.e. 15.0 rodents per ha per year (rodents 55.3% of food by weight) (Ryszkowski *et al.* 1971). Digestion efficiency of captive ♂ (see above) inversely proportional to weight of food

Table C Food of Tawny Owls *Strix aluco* at Wytham (Oxford, England) assessed by 3 different methods. Figures are percentages of total numbers of items (Southern 1969).

	1. Adult pellets		2. Night-time nest-watching		3. Surplus food in nests	
	Up to 7 May	8 May onwards	Up to 7 May	8 May onwards	Up to 7 May	8 May onwards
Mole *Talpa*	2	9	0	0	2	40
Small rodents	69	48	56	12	52	15
Other vertebrates	18	20	20	2	46	35
Beetles	11	23	0	57	0	10
Earthworms	+	+	25	29	0	0
Total no. of items	996	420	39	239	279	124

eaten: thus in November–December, 1 g of pellet represented 15–17 g of food; in July, up to 26 g (Lowe 1980). For rate of food intake of young, see below.

Little specific data on food of young, and even in studies of food brought to nest, some presumably eaten by incubating or brooding ♀. According to Schnurre (1935), does not differ significantly from that of adults, but observations at Wytham (Southern 1969) indicated *Talpa* and *Oryctolagus* (caught largely in daytime) eaten mostly by chicks, while small rodents and birds tended to be eaten more by adults. Up to 7 May (when on average chicks about half-grown) small rodents predominated in diet at Wytham; then fairly rapid swing to *Talpa* and beetles (mainly *M. melolontha*), both then being readily available (Table C; see also Hirons 1976). Similarly in Ythan valley (Table A), prey brought to nest included fewer small mammals and more large birds than found in pellets of adults. 247 items recorded from nest-watching and (mostly) surplus prey comprised 39% by weight large mammals (mostly *Orcytolagus*), 15·2% small rodents, 1·8% Soricidae, and 44·2% birds; only 2 earthworms and no insects recorded; 2 *S. aluco* chicks found in nests half-eaten and, as 38% of 21 hatched chicks disappeared, more may have been eaten (Hardy 1977). 1245 items brought to chicks, Tambov (USSR), included 41·0% (by number) small rodents, 1·3% Soricidae, 14·7% birds, and 41·0% *M. melolontha* (recalculated from Shchegolev 1974). Prey brought on 429 visits to nests (perhaps not all with young), Wytham, included 35% (by number) *C. glareolus*, 13·5% *M. agrestis*, 11% *A. sylvaticus*, 10·5% *Talpa*, 14% Soricidae, 4% earthworms (Hirons 1977). Adults said to feed separate feathers to young (Beven 1969). Surplus prey first appears in nest a few days before eggs hatch; little present after chicks $c.$ 2 weeks old (Southern 1969). See Holmberg (1976) and Lundberg (1980b), however, for analyses of prey remains left in nests after breeding, central Sweden. Young produce first pellets at $c.$ 8 days old (Guérin 1932) or perhaps later; loose in texture, disintegrating readily, and parents rapidly remove them from nest (Southern 1969). In Tambov, $c.$ 2 feeding visits to chicks per hr around sunset, rising to $c.$ 15 per hr at 24.00 hrs, then falling to $c.$ 2 per hr by 03.00 hrs (Shchegolev 1974). At Wytham, 21% of feeding visits to nests were in daylight (sunrise to sunset), but frequency increased sharply after sunset, continuing at high level until sunrise (Hirons 1977). Feeding rate greatest $c.$ 11–25 days after hatching — $c.$ 45–55 visits per night, though this depressed (e.g.) to 15 per night by heavy rain. Items usually brought singly: in 35 nights, only twice brought 2–3 *M. melolontha* at once. Large prey (rodents, birds, etc.) given to chicks in pieces (Shchegolev 1974). At nest with 3 young (oldest 15–25 days old), each received average 88 g food per day; immediately before fledging, 124–141 g each per day (Ritter 1972). For most detailed study of feeding behaviour during nesting, see Hirons (1976). After fledging, young at Wytham apparently make little or no attempt at hunting until quite independent of parents, i.e. after 2½–3 months (Southern *et al.* 1954). However, at Strødam (Denmark), young attempt to hunt June–July, apparently earlier than at Wytham (Andersen 1961), and young said by Shchegolev (1974) to be fed for only $c.$ 1½ months. At Wytham, fledged young fed mostly 24.00–02.00 hrs; based on pattern of calling, 3 broods of 1, 2, and 3 were apparently fed, respectively, 2·5 times per night ($n=4$ nights), 0·6 times ($n=5$), and 2·8 times ($n=5$) (Muir 1954); figures for broods of 2 and 3 seem unrealistically low unless juveniles being fed consistently on large items. Very few young die during period of dependence in parents' territory, but often starve quickly after being driven out (Southern 1970; Hirons *et al.* 1979). DJB

Social pattern and behaviour. Major studies by Stülcken (1961), Wendland (1963), Southern (1970), Hirons (1976), and Delmée *et al.* (1978). Account includes material by V Wendland.
1. Typically solitary or in pairs throughout the year. Established pairs defend fixed territory all year, and from year to year (Hansen 1952; Wendland 1963; Dambiermont *et al.* 1967; Southern 1970; Hirons 1976). Family parties remain on territory for considerable period after fledging (see Bonds, below). BONDS. Mating system usually monogamous (but see below). Pair-bond life-long, and maintained all year (Räber 1954; Wendland 1963; Delmée *et al.* 1978). In Wytham Wood (Oxford, England), 2 of 34 ♂♂ apparently bigamous throughout 3-year study; caused by resident ♂ annexing territory and mate of neighbouring ♂ who had died. One bigamous ♂ divided his attention very unevenly between the 2 ♀♀ in his territory, at least as far as Courtship-feeding (see below) concerned. Successful breeding never achieved by both ♀♀ in one season (Hirons 1976). In another case of bigamy, both ♀♀ bred in same season (Scherzinger 1968; see also Breeding Dispersion, below). Age of first breeding occasionally 1 year, more often 2–3; of 10 ♀♀, 1 bred first at 1 year, 3 at 2 years, 5 at 3 years, and 1 at 4 years (Delmée *et al.* 1978). 8-month-old ♀ paired with father and, at 1 year, reared 1 young (Wendland 1963). In captive study, 1-year-old ♂ bred with 5-year-old ♀ (Smith 1976). ♂ and ♀ divide breeding duties: ♀ incubates, broods, and performs most defence of young in nest; also helps to provide food for young after 2–3 weeks old; ♂ provisions ♀ and young in nest; both sexes feed young after fledging, ♂ probably more than ♀ (Southern 1969, 1970; Hirons 1976; H N Southern). In nest, fratricide sometimes occurs when food short: e.g. after much squabbling, weakest 4th chick tipped over, attacked, and eaten by 3rd sibling (Haller 1939). Young dependent on parents for up to 2 months after fledging (Wendland 1963); for 2½–3 months after fledging (Southern *et al.* 1954; Delmée *et al.* 1978). BREEDING DISPERSION. Solitary and territorial. Territorial boundaries remarkably stable between years, especially at high densities where they may remain fixed over long period despite changes of ownership (Andersen 1961; Southern 1970; Hirons 1976; see Antagonistic Behaviour, below); e.g. at Wytham, boundaries changed relatively little over 20 years (Southern 1970; Hirons 1976). Shortest distance between nests 100–150 m (Glutz and Bauer 1980), though nests of 2 ♀♀ paired with same ♂ only 50 m apart (Scherzinger 1968). Size of territory varies with habitat, and generally smaller in denser woodland. In central and western Europe, territories in optimum habitat 25–30(–50) ha, and defended boundaries $c.$ 2–3 km long (Southern 1954; Andersen 1961; Mebs 1966; Wendland 1972b). At Wytham, average 7·3 ha ($n=16$) in fairly sparse ground cover on limestone, 13·8 ha ($n=24$) in dense ground

cover on clay, where most territories contiguous (Southern 1970). In mainly deciduous wood, 18·2±2·06 ha (*n*=31), in mixed farmland 37·4±14·04 ha (*n*=10), in mature spruce *Picea* 46·1±7·12 ha (*n*=17) (Hirons 1976). In open woodland, Denmark, 3 territories 27, 35, and 50 ha (Andersen 1961). In mixed oak *Quercus*, Belgium, 65–75 ha (*n*=10) (Delmée *et al.* 1978); in Huy region (Belgium), 25 territories in *c.* 1800 ha of forest and park, i.e. *c.* 1 pair per 72 ha (Dambiermont *et al.* 1967). In Scandinavia, territories usually larger, e.g. in central Sweden, 2 ♀♀ in 89 and 146 ha (Nilsson 1977). Considering only wooded areas, density in central Europe typically 0·5–1·0(–1·6) pairs per km² (Glutz and Bauer 1980; see also for densities in urban areas); e.g. in lower Reuss Valley (Switzerland), 0·5 pairs per km² or 11 pairs per km² of woodland (Fuchs and Schifferli 1981); in Bodanrück/Bodensee (West Germany), 0·5–1·8 pairs per km² of woodland (Schuster 1971). In 2 years, West Germany, 42 territories in 50 km² and 21 territories in 25 km² (Ziesemer 1979). In Grunewald forest (West Berlin), 18–20 pairs (for 20 years) in *c.* 31 km², i.e. *c.* 0·6 pairs per km²; actual density lower since many areas unsuitable for hunting; birds hunt up to 2 km beyond territorial boundaries, though to far lesser extent than within territory (Wendland 1963, 1980; V Wendland). Birds occasionally trespass on territories relatively distant from their own (Southern and Lowe 1968). Territory established by ♂♂ and, to lesser extent, by ♀♀ (Southern 1970). Apart from feeding, territory serves for nesting and rearing young; fledged young may range widely within territory but seldom outside (Southern and Lowe 1968). Fidelity of both sexes to territory very strong. Of 34 ♂♂, only 1 moved territory during 3 years, and then only due to interference by observer; ♀♀ similarly sedentary (Hirons 1976). On death of mate, whether ♂ or ♀, other holds territory and attracts new mate, often by the following year (e.g. Delmée *et al.* 1978). Nest-site fidelity also marked (e.g. Wendland 1963), though changes occasionally occur if site deteriorates or disturbed (Delmée *et al.* 1978). Young settle to breed near natal site, but only if unoccupied territory (or territory with 1 bird) available—otherwise starve rapidly (Southern 1970; Delmée *et al.* 1978). In competition over nest-sites, Finland, Ural Owl *S. uralensis* expels *S. aluco*; however, Great Grey Owl *S. nebulosa* does not compete and may nest peaceably nearby (Lahti and Mikkola 1974). ROOSTING. Diurnal. Especially in warm weather, uses branch of tree, typically in cover near trunk; also in hole (usually in tree), in chimney (Stülcken 1961), or, once, a sand-pit (Wendland 1963). Perches with ruffled plumage, neck retracted, and eyes closed. From July–October, typically roosts alone, and changes site frequently; from late autumn up until breeding season, sites become fewer and more consistently used. Over same period, pair members increasingly roost together, in *c.* 60% of cases by January–February (Southern 1954, 1970; see also Guérin 1932). Outside breeding season, pairs often roosted together in former nest site (Wendland 1963). After February, proportion roosting in pairs steadily decreases (Southern 1970), coinciding with daily occupation of nest-sites by ♀♀. Some ♂♂ who lack sufficiently good cover for roosting share site with ♀ (Southern 1954, 1970; see also Spencer 1953). While ♀ incubating, ♂ usually seeks roost-site nearby, sometimes alongside nest (Delmée *et al.* 1978). Activity mostly nocturnal; for periods of hunting activity, see Food. From *c.* 2 weeks after fledging, young typically roost close together in dense cover high in tree; shortly before leaving territory, roost separately (Stülcken 1961; Wendland 1963), this coinciding with moult (Räber 1954). Young sometimes use same site for several nights, or may change frequently from day to day; probably make quite long flight from roost at dusk, and return at dawn (Southern *et al.* 1954). Adults sunbathe with one or both wings outstretched, once lying prone

A

on branch with foot dangling. Readily bathe in light rain (Wendland 1963). See also Räber (1954) for dust-bathing in captive birds.

2. When perched bird alarmed and aware of being watched, adopts concealing Sleeked-upright posture and freezes until danger passed (Wendland 1963); posture similar to that of *S. uralensis*—see that species. Bird may use 'emergency' refuge hole in territory when mobbed (Stülcken 1961). Bird prevented from fleeing roost hole often hisses (Andersen 1961: see 11 in Voice). Captive bird confronted with polecat *Putorius* adopted threat posture (Fig A): ruffled plumage (especially on back), lowered and fanned nearer wing, and Bill-snapped (Scherzinger 1969a). In January, bird attacked fox *Vulpes* in daylight, raking its back with claws; when fox crouched, bird continued diving and calling (Paterson 1964: see 9 in Voice). ANTAGONISTIC BEHAVIOUR. ♂ markedly territorial. Whether established or new claimant, demarcates territory mainly by giving Advertising-call and, to lesser extent, Contact-call (see 1 and 3a in Voice), also by threat and supplanting attacks (e.g. Andersen 1961). Calls throughout the year, but especially October–November. Calls from roost-sites for 15–20 min after dusk before beginning to hunt, etc. (Southern and Lowe 1968; Southern 1970). Also uses various song-posts 250–300 m from roost (Wendland 1963). Calling highly contagious among neighbours; ♀♀ also give Advertising-call (see 1b in Voice), but much less than ♂♂. From December to February, as territories consolidated, calling diminishes, until seldom heard except from failed breeders. Equally silent during moult, mid-May to early September (Southern 1970; Hirons 1976; see also Wendland 1963). In Sweden, said to be most vocal in early spring, often also in summer, occasionally in autumn (H Delin). Birds may call at any time until dawn, occasionally also by day; less in cold windy weather (Hansen 1952), and when food supply poor (Delmée *et al.* 1978). In autumn, boundary disputes sometimes occur, typified by loud but usually brief exchanges of excited and disjointed Advertising-calls, mixed with wailing and screaming sounds ('caterwauling' Southern 1970: see 7b in Voice). Boundary disputes between pairs, or single birds, usually of same sex, though ♀ may support mate against rival. Paired ♀♀ also occasionally evict intruding ♂♂; ♀ asserts ownership more by direct intervention than vocally. Where ♂'s territory contained 2 ♀♀, each ♀ defended part of territory against the other; such ♀♀ also cooperated with ♂ in boundary disputes. In deciduous woodland, September–December, 0·42 boundary disputes per hr, in open farmland 0·14 per hr; in open land, the 4 recorded cases of caterwauling occurred when rivals less than 3 m apart. Resident evicts trespasser with supplanting attack (Hirons 1976). Rivals approach each other, exchanging short Advertising-calls (Wendland 1963: see 1 in Voice). In encounters between neighbouring ♂♂, resident chased intruder into intruder's territory, whereupon roles reversed, resulting in chase in opposite direction. Such chases violent and noisy, aggressor caterwauling loudly and rivals frequently colliding audibly with

branches (Hirons 1976), occasionally fighting on ground (Wendland 1963). Long-eared Owl *Asio otus* similarly evicted (Andersen 1961). In rivalry over ♀, resident ♂ called, Bill-snapped, and made furious attacks on younger ♂ (Bäsecke 1942). As boundaries become hardened, neighbouring ♂♂ fly parallel to each other, sometimes close together, along boundary, both calling loudly and continuously. Occasionally, after boundary dispute, ♂ supplants own mate from perch. More often, boundary disputes—especially if prolonged and fought by both members of pair—followed by pair duetting, interpreted as territorial proclamation (Hirons 1976; see Voice). For duetting of pair after neighbouring broods met near boundary, see Southern *et al.* (1954). After death of owner, territory usually occupied very quickly, in one case 2 days or less, in 5 of 8 cases 2 weeks or less. Replacement ♂♂ begin calling in centre of vacant territory, and gradually extend to neighbouring boundaries. Long-established boundaries tend to be respected by neighbours even if territory suddenly vacated; recently established neighbours are thus more likely to absorb suddenly vacated adjacent territory (Hirons 1976; see also Heterosexual Behaviour, below). HETEROSEXUAL BEHAVIOUR. (1) General. Comprises aerial display, Courtship-feeding, and mating, also nest-site selection and Allopreening. Though perhaps common, aerial display, and displays when perched, not widely described, and typical sequence remains tentative. ♀♀ whose mates disappeared did not resist encroachment by neighbouring ♂♂, and not uncommonly flew into adjacent territory to solicit resident ♂ until evicted by his mate (Hirons 1976). (2) Pair-bonding behaviour. Main period of pair-formation in autumn, when ♂ attracts ♀ with Advertising-call. At this time, much intimate behaviour associated with nest-site selection (see below). Display also conspicuous during February–March, when laying imminent (Wendland 1963; Southern 1970). Apparent display-flight recorded rarely. In one case (late March in Grunewald), ♂ left roost 39 min after sunset and flew silently in broad spirals to *c.* 200–250 m; ♀ sitting not far away in nest-hole also silent (Wendland 1963). In 4 other cases, ♂ descended towards ♀ on stiff, quivering wings, while ♀ gave Trilling-call (Wahlstedt 1959; Andersen 1961: see 4 in Voice). Wing-clapping reported by Witherby *et al.* (1938), but not by other observers, and must occur rarely, if at all, in heterosexual context (see also Food). In Meeting-ceremony, ♂ delivers Advertising-call; ♀, sometimes answering with repeated Contact-calls (see 3a, b in Voice), flies around ♂ but not above canopy height. After a while, pair repeat the ceremony elsewhere (Uttendörfer 1937; Wendland 1963; Southern 1970). Failed or non-breeders may repeat sequence without subsequent Courtship-feeding and copulation (Wendland 1963). In apparent courtship-display, ♂ alighted in tree and 2nd bird, presumed ♀, landed *c.* 1 m below. ♂ began swaying from side to side, then up and down, alternately raising each wing, finally both wings together. ♂ also alternated between extreme ruffling and sleeking of plumage, meanwhile walking leisurely a short way along branch and back again, giving grunting sounds (see 6 in Voice). ♀ watched silently without moving. Display lasted *c.* 1½ min after which ♀ flew off, ♂ following (Churchill 1939). (3) Nest-site selection. ♂ and ♀ sometimes inspect sites independently (Spencer 1953), but final choice made by ♀ (Wendland 1963; Delmée *et al.* 1978). Established pairs may begin to inspect potential site daily in October (Stülcken 1961). In Surrey (England), pair prospected nest-box site frequently from October until late March when squirrel *Sciurus* visited box; ♂ or ♀ entered box and attracted mate with Bubbling-call (P A D Hollom: see 2 in Voice for details). In typical sequence, ♂ gave Bubbling-call from hole in tree, and ♀ gave Contact-call as she flew to join him, calling more excitedly (see 12b in Voice) as she neared. Both then gave Bubbling-call near entrance (Stülcken 1961; see also Harrison 1947*b*). During early stages of Courtship-feeding, ♂ feeds ♀ wherever she roosts, and eventual roost-site may become nest-site (Wendland 1963). In chosen site, ♀ begins to make shallow scrape in floor of cavity. ♂ never entered hole when ♀ scraping (Stülcken 1961). (4) Mating. Typically begins with vocal communication between pair from respective roost-sites. ♂, perched or in flight, contacts ♀ with Advertising-call, ♀ answering with Contact-call. ♂ then flies to ♀ or vice versa (e.g. Gebhardt 1948–9). ♀ may also take initiative by summoning ♂ with Contact-call. Once, after inspecting nest, ♀ summoned ♂ and copulation followed (Spencer 1953). Copulation occurs near, but never in, nest-site (Stülcken 1961). No other preliminaries, and copulation often follows in silence (Wendland 1963, but see below). ♀'s Contact-call (see 3c in Voice) may become more excited as ♂ approaches. ♀ adopts a horizontal posture and droops wings. Once alongside her, ♂ sometimes ruffles plumage and droops wings. ♂ mounts, trampling and beating wings to gain position; once gave Advertising-calls as he mounted (Krauss 1974). For other calls associated with copulation, see 2, 4, and 12b in Voice. Copulation brief (once *c.* 7 s), but may be repeated several times in succession. After copulation, ♀ typically gives Soft Contact-call (Wendland 1963: see 3b in Voice); may ruffle plumage and both birds may self-preen before flying off (Krauss 1974). Once, mutual Allopreening followed copulation (Wendland 1963). (5) Courtship-feeding and food-caching. ♂ begins to feed ♀ in January (Hirons 1976: Fig B).

B

Feeding frequent before laying, starting before dark; diminishes markedly after incubation begins. Early on, ♂ typically feeds ♀ near her roost-site, but sometimes ♂ attracts ♀ to his (Wendland 1963). During incubation, and until young *c.* 2 weeks old, ♀ seldom strays far or long from nest, and catches no prey for herself. Incubating ♀ summons ♂ with Contact-call or, if very hungry, sometimes with Bubbling-call. ♂ announces arrival when *c.* 150–200 m from nest, and ♀ flies out to receive prey away from immediate vicinity of nest (Uttendörfer 1937; Wendland 1963; Hirons 1976). However, ♂ brings prey to nest at hatching time when ♀ sitting tight (Andersen 1961: Stülcken 1961). As prey transferred, ♀ adopts a horizontal posture, quivers wings, and calls excitedly (Andersen 1961; see 3c in Voice). ♂ gives short Advertising-call. ♂ typically brings incubating ♀ 2–4 prey per night (Géroudet 1965; Ritter 1972; Hirons 1976). Feeding rate evidently influences success of nesting attempt: e.g. nest at which rate highest (*c.* 3 prey per 24 hrs) was only one to fledge young (Hirons 1976; see also Behaviour at nest, below). Shortly before, and also after, hatching, ♀ may cache prey (received from ♂) in nest-site, as apparent insurance against future prey shortage (Stülcken 1961; Wendland 1963; Delmée *et al.* 1973). (6) Allopreening. Mostly associated with Courtship-feeding, but also occurs at other times; once, mutual Allopreening followed copulation. Allopreening pair perch side by side, and ♀ periodically gives Soft Contact-call (Wendland 1963). In September, ♂ flew to join ♀ in tree, ran along branch with tripping steps, head bowed, and mutual Allopreening, including around bill, followed; one bird, thought to be ♀, gave apparent Soft Contact-calls (Krauss 1974). In captive study, ♂ often nibbled at ♀'s breast and neck, after feeding her (Smith 1976). For another report of Allopreening in captive birds, see Scott (1980). (7) Behaviour at nest. Before laying, ♀ occupies nest-site by day, while ♂ guards nearby (see Roosting). When laying begins, ♂ notably silent (Wendland 1963). At dusk, ♂ sometimes visits ♀ briefly at nest (Spencer 1953; Wendland 1963). Incubating ♀ leaves nest only for 1–2 brief spells of a few minutes each night, and not thought to catch any prey at this time, at least where nest successful. Compared with other pairs, ♀ fed least consistently by ♂ left nest most often and for longest periods (Hirons 1976, which see for details of activity rhythm of ♀♀). In 2 cases, incubating ♀ left for ¾ hr, duration apparently depending partly on time taken to regurgitate pellet (Uttendörfer 1937). Occasionally, ♀ left nest carrying prey recently brought by ♂ (Hirons 1976). RELATIONS WITHIN FAMILY GROUP. For details of development of young, see Scherzinger (1980). Young begin to call from egg a few hours before hatching (Delmée et al. 1978). Brooded assiduously by ♀ day and night, not emerging from beneath her until 2–3 days old (Stülcken 1961); closely brooded until 10–15 days (Uttendörfer 1937; Stülcken 1961; Delmée et al. 1978), while ♂ brings food which ♀ transfers to young. Young said by Ritter (1972) to be fed only at night, but according to Stülcken (1961) may receive cached prey from ♀ by day. ♀ tears off tiny portions of prey for small young, and feels, with her eyes closed, for head of nestling beneath her (Fig C), this procedure continuing until young c. 12

C

days old (Stülcken 1961), by which time eyes of young fully open (Räber 1954), enabling them to take more initiative in begging. Older young beg in prone posture with quivering wings and intense food-calls (Andersen 1961). ♀ continues to dismember prey up to 4th week, after which young handle their own (Herberigs 1954; Géroudet 1965). ♀ eats faeces of young until 11 days old (Stülcken 1961). Young begin casting pellets, initially of rather liquid consistency, from 10–12 days (Guérin 1932; Delmée et al. 1978). ♂ brings prey throughout nestling period, ♀ only making significant contribution after brooding ceases (Hirons 1976). ♂ may announce arrival with short Advertising-call, and alights only long enough to drop prey beside or inside nest (Uttendörfer 1937; Stülcken 1961; Hirons 1976). When young well-grown, ♂ flies silently and directly to nest, and without perching, throws prey into cavity (Andersen 1961b). ♂ never enters nest to feed young (Wendland 1963), but see Stülcken (1961) for apparent exception. Especially when brooding less, ♀ may receive prey from ♂ away from nest, and announce return to nest with Soft Contact-call (Uttendörfer 1937; Stülcken 1961; Wendland 1963). In one case, ♀ first brought prey, probably received from ♂, when chick 11 days old, but only regularly brought more than 1 item, presumably caught by herself, when chick 19 days old (Hirons 1976, which see for detailed activity rhythms). ♀ initially hunts near nest (Stülcken 1961). From 21–25 days, young spend much time at nest-entrance, and begin to emerge 3–4 days later (Stülcken 1961; Wendland 1963). Oldest young evidently first to leave nest (Weller 1944). ♀ flies up to branch in tree and gives Soft Contact-calls, apparently encouraging young to climb (Stülcken 1961). Young climb awkwardly with aid of claws, bill, and flapping wings, as in other *Strix*, and often fall down (Stülcken 1961; Delmée et al. 1978). On leaving nest, stay near it for several days. By day, typically sit close together (Stülcken 1961). During moult, captive young recorded Allopreening (Räber 1954). Begin to fly at 32–37 days, c. 1 week after leaving nest; fly strongly c. 2 weeks later (Southern et al. 1954; Stülcken 1961; Southern and Lowe 1968). Young spend time exploring territory; usually spend most time in one part, but also range widely, retreating from boundaries on hearing neighbouring brood (Southern et al. 1954). Until evicted from territory at 2½–3 months, young entirely dependent on parents for food, and make little or no effort to feed themselves (Southern et al. 1954). Occasionally hunt if food short (Stülcken 1961). ANTI-PREDATOR RESPONSES OF YOUNG. On hearing repeated Alarm-calls (see 8 in Voice) or Distraction-call (see 9 in Voice, also below) of ♀, young fall silent (Southern et al. 1954). In nest, frightened young crouch; older young Bill-snap (Andersen 1961). Out of nest, young aware of being watched adopt concealing Sleeked-upright posture, as in adult (Wendland 1963). In presence of intruder, fledged young gave food-call and Bill-snapped (Uttendörfer 1937). When closely approached, fledged young on ground adopted Forward-threat posture: ruffled plumage until bird looked about twice normal size, spread wings, and Bill-snapped (Wenner 1911; Räber 1954). Captive young changed response from crouching to Forward-threat posture when c. 20–24 days old (Räber 1954). When garment thrown over bird, then withdrawn, bird apparently feigned injury: lay motionless, eyes closed, plumage sleeked; suddenly opened eyes wide, struck out with claws, and beat wings vigorously (Wenner 1911). PARENTAL ANTI-PREDATOR STRATEGIES. (1) Passive measures. When man, or mammalian predator, enters territory, bird gives repeated Alarm-calls (Andersen 1961; V Wendland). Incubating and brooding ♀ often a tight sitter (e.g. Cassidy 1946); one ♀ allowed herself to be pushed off brood and turned over on her back, where she lay motionless as long as she was being watched (*Br. Birds* 1911, **5**, 195). ♀ sat tightly, especially on intruder's first visit to nest, but, after several visits, flushed when intruder approached (Labitte 1952). Much individual variation reported, however, in wariness and aggression (e.g. Labitte 1952, Delmée et al. 1978). (2) Active measures: against man. ♀ prevented from leaving nest-hole may hiss (see 11 in Voice) and Bill-snap. Most nest-defence after hatching by ♀ who stays on guard near nest when not brooding. When young 2–3 weeks old, ♂ also in close attendance (Andersen 1961; Stülcken 1961; Wendland 1963; Delmée et al. 1978). At approach of intruder, ♀ usually flies silently into cover, where she often gives one or more brief muffled hoots (Delmée et al. 1978) or Alarm-call (Stülcken 1961). ♀♀ reacted thus when alerted by noise of fledged brood in vicinity. Less often, ♀ approaches intruder, alighting in trees nearby, Bill-snaps and caterwauls, but does not attack (Delmée et al. 1978). Outright attacks often mistakenly thought to be common, but typically occur only in built-up areas where birds regularly disturbed (Wendland 1963, 1972a; Southern 1970). Spectacular, dangerous attacks of particularly aggressive ♀♀ (and occasionally ♂♂) widely reported (e.g. Read 1918, Steffen 1953a, Dobbs 1966, Dambiermont et al. 1967). One pair mobbed intruder regularly for 3 weeks, when 50 m from nest containing young, usually at night, once by day; swooped from behind at head (Boswall 1965). ♀ approached silently from behind, but only attacked when intruder between her and unfledged young on ground (Stülcken 1961). Attacking bird also buffets shoulders and back, usually when back turned (Cadman 1934; Delmée *et al.* 1978), and may draw blood with claws (e.g. Uttendörfer 1937, Stülcken 1961). ♀ often veers off short of physical contact, giving Anger-calls (Andersen 1961: see 7a in Voice). When newly fledged young closely threatened, ♀ occasionally performs distraction-lure display of disablement type: flutters on ground and tries to entice intruder from young; display accompanied by Distraction-call (see 9 in Voice) which immediately silences young (V Wendland). (3) Active measures: against other predators. One bird attacked marten *Martes* 3 times, once nearly knocking it off tree, and finally flying after it as it left (Szomjas 1952–5). ♀ drove off fox (Andersen 1961) and cat (Stülcken 1961) with diving attacks. For accompanying calls, see 7a and 8 in Voice.

(Fig A from photograph in Scherzinger 1969; Figs B–C from photographs in Stülcken 1961.) EKD

Voice. Freely used, especially at start of breeding season and at other times for territorial demarcation. ♂ and ♀ share remarkably diverse repertoire, as in other *Strix* (see, e.g., Scherzinger 1980). For Bill-snapping, see Social Pattern and Behaviour. Important studies by Andersen (1961), Wendland (1963), and, for captive birds, Arvola (1959). Following account includes extensive recordings and notes by P A D Hollom.

CALLS OF ADULTS. (1) Advertising-call. (a) ♂'s call. Familiar, distinctive hooting with pure ocarina-like tone (Niethammer 1938; Stadler 1945–6), regularly repeated 2–4 times per min (Andersen 1961). Rendered 'hoōo', then pause of 2–3 s before abrupt, subdued 'hŭ', followed almost at once by prolonged, resonant, wavering 'hŭhŭhŭ hoōo' (Fig I: Southern 1970). Also rendered 'oohoohoooo oo oohoohoohoohoooooo', the 2 long phrases having distinct vibrato and steeply falling pitch in the terminal, long-drawn 'hoooo...' (Andersen 1961). In our recordings, gap before brief 'hŭ' more often 4–5 s. In recording by V C Lewis, 3 calls delivered *c*. 14 s and 10 s apart. Calls of individuals vary, and readily distinguishable to human ear (Wendland 1963; Southern 1970; Hirons 1976). Given mostly from perch, but also in flight, to demarcate territory and contact ♀, sometimes in duet with, or overlapping, call 3 of ♀. Variants include notably short, incomplete calls; hoarser, more compressed, and less consistent in structure than long phrases of complete call, e.g. often rising in pitch; 1–2 calls given together (Andersen 1961); in recording by P A D Hollom, 'hu-huuuuu' (E K Dunn). Function apparently similar to call 2; given (e.g.) by ♂ announcing arrival when visiting ♀, often to feed her, when quality muffled and fluting (Southern 1970), wavering or crooning (Hirons 1976); sometimes a hissing 'chruuuuuu' (Wendland 1963). Numerous gradations between complete and incomplete calls occur typically when bird excited or agitated, e.g. during confrontations (see also call 7b), copulation, or courtship-feeding (Arvola 1959; Andersen 1961; Stülcken 1961; Wendland 1963; P A D Hollom). (b) ♀'s call. Less consistent than ♂'s, with wailing quality in last phrase ('wow-wow-hoōo': Southern 1970).

Said to be similar to incomplete call of ♂, but slightly higher pitched (Andersen 1961). In recording, hoarser than ♂: 'cher oooOOooo' (Fig II), followed by 'chr chro cherEEooooooooooo coooEEooooo' (J Hall-Craggs, P A D Hollom). Used especially to assert territorial ownership in autumn (Southern 1970), but also, as in Fig I, in communication with mate (P A D Hollom). (2) Bubbling-call. A quiet tremulous trill (P A D Hollom). A bubbling 'loo-loo-loo...' (Fig III), likened to human making 'oo' sound while moving tongue rapidly from side to side (Harrison 1947*b*; E K Dunn). Soft rolling sounds, constant or fluctuating in pitch and strength, often compared to Drumming of Snipe *Gallinago gallinago* (e.g. Spencer 1953, Andersen 1961). Given by both sexes: ♂'s call (Fig III) softer and lower than ♀'s, which has hoarser quality (P A D Hollom). Given at intervals of 4–12 s by both birds together, for 5–10 min, along with call 3 (Harrison 1947*b*). Audible over relatively short distance, perhaps 30–40 m and therefore often thought to be uncommon (e.g. Witherby *et al.* 1938). On the contrary, frequently used, especially in autumn and winter, notably during nest-site inspection. Either sex may give call, presumably as high-intensity mate-summons, from potential nest-site, and be joined by mate who often takes up call on arrival, the pair then calling together for 1–2 min, followed by several minutes of chirping sounds (P A D Hollom). According to some reports (see Southern 1970), given throughout the year, mostly by ♂ in breeding season when mate nearby, but other evidence supports belief that main use is in autumn and winter (Stadler 1945–6; Harrison 1947*b*) by nest-prospecting birds (Spencer 1953; Stülcken 1961). Once given by ♂ during copulation (Wendland 1963), once heard when eggs hatched (Andersen 1961). (3) Contact-calls. (a) Familiar sharp disyllabic 'ke-wick' (P A D Hollom). For similar renderings, and variants, see Boase (1950), Arvola (1959), and Andersen (1961). In recording (Fig IV), 'ke-wick' followed by 'oooooo', probably of answering ♂, this combination widely known and popularly ascribed to single bird (J Hall-Craggs). Used as Contact-call by both sexes, and especially by ♀ in roost or nest to solicit food from ♂. Also by both sexes to announce arrival with food for fledged young (Andersen 1961). Sometimes directed by ♂ at other ♂♂ (Wendland 1963). (b) Soft Contact-call. As breeding season approaches, ♀ increasingly gives soft variant, rendered 'oo-wip' (Southern 1970) or 'kwik' (P A D Hollom: Fig V); 'kwitt kwitt' by sitting ♀ when eggs newly hatched (Uttendörfer 1937). A soft 'ui' by both sexes, but mainly by ♀ when pair perched close together, often when Allopreening; ♀ may call thus, sometimes in nest, up to $\frac{1}{2}$ hr. Also sometimes given by ♀ to young on arrival at nest with food ('iüi'), and to undisturbed fledged young (Wendland 1963). For closely related variant, see call 8. (c) When ♂ approaches with prey for ♀, her disyllabic call changes to excited high-pitched 'kiv-kiv-kiv-kiv-kiv...', culminating in a long peeping, whistling 'sii-sii-sii-sii-siiiii' as she receives prey (Andersen 1961). Excited 'witt

Strix aluco 541

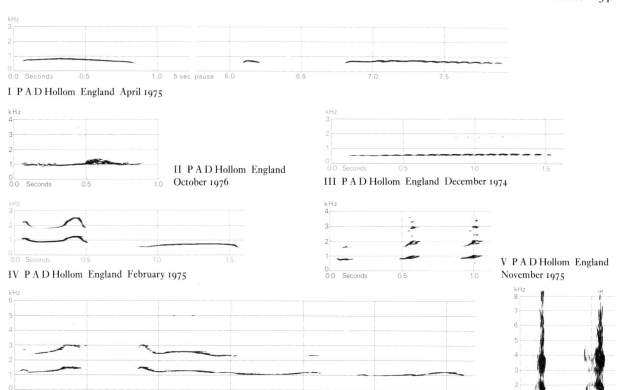

I P A D Hollom England April 1975
II P A D Hollom England October 1976
III P A D Hollom England December 1974
IV P A D Hollom England February 1975
V P A D Hollom England November 1975
VI P A D Hollom England November 1975
VII V C Lewis England May 1972

witt' of ♀ before copulation (Krauss 1974), and rapid 'kikikiiiii' of copulating ♀ (Wendland 1963) perhaps the same. See also call 4. (4) Trilling-call. Sexually excited ♀, sometimes during copulation, gives trilling sound with 'i' component dominant, likened to green toad *Bufo viridis* (Wendland 1963). Trilling 'lee-lee-lee...', given by ♀ when ♂ descended from apparent Display-flight (Wahlstedt 1959; Andersen 1961) probably this call. (5) Feeding-call. A clicking 'hüng hüng' given by ♀ to young when reluctant to accept food (Scherzinger 1980). Also described as 'üng-üng-üng-haüg-haüg', and used also by brooding ♀ to comfort young (Stülcken 1961). Also given when young fell out of tree (Stülcken 1961). (6) Grunting-call. Soft pig-like grunting sounds given by ♂ displaying to ♀ (Churchill 1939). Same or similar sound also reported by Scherzinger (1980). (7) Anger-calls. (a) When young imminently threatened, bird (usually ♀) gives monotonous, though not especially loud, rapid 'goov-...' sounds, or alternatively composed of 'koo-...' or 'boo-...' sounds. Given when young threatened by man or mammalian predator, and often accompanies mock-attacks (Andersen 1961). (b) In boundary disputes, rivals give brief exchanges described as caterwauling—a mixture of short hoots, screams, and wails (Southern 1970); these undoubtedly include incomplete Advertising-calls with, at peak intensity, a cat-like mewing quality (Wendland 1963). (8) Alarm-call. Rapidly repeated 'wick wick wick' (Southern 1970), 'kwik kwik kwik' (P J Sellar), 'kvik', shorter and harder than call 3a (E K Dunn). ♂'s call markedly deeper than ♀'s (H Delin). Softer variants, similar to call 3b, e.g. 'goo-i goo-i', in accelerating series, also reported (Andersen 1961), also 'owui' when ♀ mobbed (Stülcken 1961). Main call and variants given as warning by both sexes when enemy approaches nest, or young outside nest. When given by ♀, ♂ approaches in silence, and young fall silent (Andersen 1961). Also serves as threat to intruder of incipient attack (P A D Hollom). (9) Distraction-call. Wavering 'iiiiii', resembling twittering of small passerine, given as warning by ♀ when intruder threatening young, also by ♀ before performing distraction-display (Wendland 1963; V Wendland). Also described as a long-drawn, unsteady flute-like piping of despair (H Delin). High-pitched, vibrant 'kree-ee' as bird attacked fox *Vulpes*, outside breeding season (Peterson 1964), perhaps the same or a related call. (10) Long-call. Long-drawn, moaning call given apparently when mate absent, possibly to announce arrival at nest, but function obscure (P A D Hollom). In recording (Fig VI), 'keeeeee keeuuuh keeuhkuhkuh', 1st unit ascending slightly in pitch, 2nd dropping in pitch in middle and reminiscent of Mew-call of Herring Gull *Larus argentatus* (E K Dunn). Seldom heard, and typically given only once, with no calls of any sort immediately before or after (P A D Hollom). (11) Hissing-call. Angry snake-like sound, given to threaten intruder, as in other Strigidae (Andersen 1961); also when mobbed by Corvidae (Wendland 1963). (12) Other calls. (a) Hissing 'chochochocho' or 'chrochrochrochro' of ♂. Function obscure; sometimes seems to express gentleness, but when given after incomplete Advertising-call, apparently anger; once heard during fierce fight (Wendland 1963). (b) Great variety of calls given when pair in intimate contact, including copulation; hard to describe but may well be low-intensity variants of those above. In Meeting-ceremony, ♀ gives variety of crowing, screeching, and mewing sounds, usually preceded by 'hissing' sounds (Wendland 1963). Between autumn and spring, when pair together in potential nest-site, typically after excited bout of Bubbling-calls, ♂ gave soft fluting toots, and chittering calls, typically 2 at a time. ♀ gave soft, plaintive squeaks. Other quiet conversational sounds included chirps and partial hoots (P A D Hollom). Similar sounds associated with mating include chirruping of ♀ before copulation, and chittering squeaks from bird of unknown sex during copulation (Spencer 1953). High-pitched screams and cackles also almost certainly associated with copulation (Southern 1970).

CALLS OF YOUNG. Small young give a delicate piping 'bi-bi-bi-bi-...', especially when ♀ interrupts brooding, and therefore probably indicating discomfort (Andersen 1961). In recording by P A D Hollom, high-pitched twittering sound possibly the same. Older young give call in response to jostling with siblings, often combined with Bill-snapping (Andersen 1961); also given when handled (Räber 1954). Food-call of small young a faint, squeaking, long-drawn 'sjiii', gradually increasing in strength and becoming a more-or-less distinct 'sziii-ii' or 'psji-ii' (Andersen 1961), or 'kzik' (Wendland 1963) or 'tsjuk' (P J Sellar: Fig VII); hoarse, complex, and richly modulated, and often distinguishable between young. With age, call of some young becomes shriller (Andersen 1961). In 18-day-old young, 'zoui-wi zoui-wi' (Uttendörfer 1937). 2 variants distinguished by Muir (1954): 'ke-serp' and higher pitched 'ke-suip'; individual young in brood seem to use one or other, but not both. From late nestling stage onwards, young give call rhythmically and monotonously throughout much of night (Andersen 1961), usually from 21.00 to 03.00 hrs (Muir 1954). Once able to swallow whole prey, food-call grades into more urgent, excited squeaky 'siii siii siii' at moment of food transfer (Andersen 1961); also rendered 'ksew' (Muir 1954) or 'psirr' (Wendland 1963). When agitated, young give rapid 'kwik kwik kwik'; when handled, hiss and Bill-snap (Wendland 1963). EKD

Breeding. SEASON. North-west Europe and Britain: see diagram. Few eggs laid before mid-March in Norway and Sweden, and season broadly similar through much of

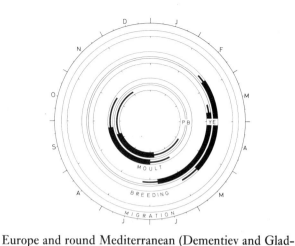

Europe and round Mediterranean (Dementiev and Gladkov 1951a; Heim de Balsac and Mayaud 1962; Makatsch 1976). Breeds much earlier in West Berlin suburbs than in wooded hinterland (Wendland 1972b). SITE. In hole in tree, also in cliff or building; often in old nest of Magpie *Pica pica*, or occasionally drey of squirrel *Sciurus*. Up to 25 m above ground, but more usually up to 12 m, especially in tree. Sometimes in hole in ground, or among tree roots. Nest-box taken readily. Nest: available hole, which may be cleaned and slight scrape made; no material added; once, pellets broken up to form nest-lining (Löhrl 1962a). Building: ♀ cleans hole and makes scrape. EGGS. See Plate 97. Blunt elliptical, smooth and slightly glossy; white. 47×39 mm ($42–50 \times 36–41$), $n = 100$ (Witherby et al. 1938). Fresh weight 40 g, $n = 16$ (Glutz and Bauer 1980). Clutch: 2–5 (1–6); varies with food supply (e.g. voles *Clethrionomys*, *Microtus*), and possibly with age (Delmée et al. 1978). In Britain, 1949–72, average $2 \cdot 70 \pm 0 \cdot 81$ (Hirons 1976). Of 113 clutches, Belgium: 1 egg, 4%; 2, 30%; 3, 36%; 4, 24%, 5, 4%; 6, 1%; average 2·95 (Dambiermont et al. 1967). Of 256 nests, also Belgium: 1 egg, 9%; 2, 32%; 3, 29%; 4, 25%; 5, 4%; 6, 1%; average 2·88 (Delmée et al. 1978). In Finland, in poor vole years, average clutch size $3 \cdot 0 \pm$ SD $0 \cdot 9$, $n = 28$, and in good years $4 \cdot 2 \pm$ SD $1 \cdot 2$, $n = 70$ (Linkola and Myllymäki 1969). One brood. Replacements occasionally laid after egg loss; on 6 occasions in 16 years (average 16 pairs breeding each year), Belgium, average clutch 2·2 (1–3) (Delmée et al. 1978). Laying interval 3–4 days. INCUBATION. 28–30 days. By ♀ only, beginning with 1st egg; hatching asynchronous. YOUNG. Altricial and nidicolous. Cared for by ♀ and brooded for c. 15 days after hatching, ♂ bringing food (Delmée et al. 1978). FLEDGING TO MATURITY. Fledging period 32–37 days, but usually leave nest at 25–30 days and hide on nearby branches. Become independent 3 months after fledging (Southern 1970). Age of first breeding 1–2 years. BREEDING SUCCESS. Of 256 eggs laid, Belgium, 24% did not hatch, of which 10% infertile, 51% addled, and 39% deserted or destroyed by ♀ when food supply low; of 195 young hatching, 94% fledged, 6% dying in nest, probably from starvation; average brood size

PLATE 51. *Asio otus* Long-eared Owl (p. 572). Nominate *otus*: **1–3** typical (pale) ad ♂, **4** typical (dark) ad ♀, **5** fledged juv moulting into 1st ad at *c.* 6 weeks, **6** nestling juvs (mesoptile) at *c.* 3 weeks. *A. o. canariensis*: **7** ad. (HD)

PLATE 52. *Asio flammeus* Short-eared Owl (p. 588): **1** ad ♂ (fresh plumage), **2–3** ad ♀ (fresh plumage), **4** fledged juv moulting into 1st ad at *c.* 6 weeks, **5** unfledged juv (mesoptile) at 3½ weeks, **6** nestling juv (mesoptile) with some 1st down (neoptile) remaining at 2 weeks. *Asio capensis tingitanus* Marsh Owl (p. 601): **7–8** ad. (HD)

at fledging 2·06 varying through 16-year period from 0 to 3·25 depending on vole numbers (Delmée et al. 1978). Also in Belgium, 146 nests produced 2·52 young likely to fledge per nest (Dambiermont et al. 1967). In Finland, brood size when young at least 1 week old $2·4 \pm SD$ 0·8, $n = 44$ in poor vole years, and $3·4 \pm SD$ 1·2, $n = 145$ in good years (Linkola and Myllymäki 1969). Of 562 eggs laid, Britain, 44% lost before hatching, and of 314 young hatching, 2% lost before fledging (Southern 1970). Of 357 pairs holding territory in woodland, West Berlin, 1958–78, 160 pairs bred producing 333 young, average 2·08 per successful pair; 13 pairs in city parkland reared 47 young, average 3·3 young per pair; number of pairs breeding each year, and success of those pairs, wholly dependent on cyclical abundance of main prey species, yellow-necked mouse *Apodemus flavicollis* (Wendland 1980). In Britain, over 10 years, average 44·6% (15–100%) of pairs did not breed in any one year (Southern 1970). Success inversely dependent on density of breeding pairs (Southern 1970; F J Koning).

Plumages. ADULT. Highly variable. In southern and eastern part of range, 2 morphs (rufous and grey) rather constant in character; in Scandinavia and western and central Europe, all kinds of intermediates occur between rufous and grey, without distinctly defined morphs; many birds even more extensively rufous than rufous morph of southern Europe. 2 types of grey morph occur in Europe, pale-grey morph with limited black marking predominating in eastern Europe and dark-grey morph with heavy and distinct black barring (as in North African race *mauritanica*) from Iberia east to Asia Minor. Colour independent of age or sex. In following account, emphasis laid on western and central Europe; as no morphs separable here, only generalized colour types described (in sequence of decreasing amount of rufous colour saturation), but note that intermediates occur and some birds of (e.g.) more saturated rufous colour type may show underparts of a paler rufous type and vice versa. DEEP RUFOUS BIRDS. Upperparts and upper wing-coverts rufous-chestnut to rufous-cinnamon (paler tawny-cinnamon when plumage worn); crown, hindneck, and sides of neck with broad and sharply black shaft-streaks; mantle, scapulars, and back with narrower black shaft-streaks; upper wing-coverts with less distinct dark grey shaft-streaks widening subterminally on median and greater. Outer webs of outer scapulars and of outer greater and median upper wing-coverts with large white or cream spot on tip, bordered terminally by narrow black or rufous rim or speckles. Facial disc deep rufous-cinnamon with 2 contrasting white arcs from above each eye and from lower cheek below eye joining at base of bill. Disc bordered above and on sides by chestnut-and-black mottled ruff of narrow and rather stiff feathers; lower border of ruff less distinct, cinnamon or rufous with some black marks, bordered below by mainly white patch on throat. Chest, breast, belly, and flanks deep rufous-cinnamon with rather narrow and sharp black shaft-streaks, some white or pale tawny of feather-bases visible on chest; distinct white spots on sides of feathers of lower breast, belly, and flanks, contrasting with black streaks and cinnamon bars. Vent, shorter under tail-coverts, and feathering of legs white with cinnamon or partly black bars and mottles, longer under tail-coverts white with black or dark rufous streaks or arrowheads. Central pair of tail-feathers (t1) almost uniform deep rufous-cinnamon, only faintly speckled dusky on sides and tip and with faint traces of dusky bands along shaft near tip; t2 similar but dark bands on inner webs more pronounced; ground-colour gradually paler cinnamon on outer webs and paler buff on inner webs towards t6, and dark grey bands purer and more contrasting; tips and inner borders of t2–t6 mottled and speckled dusky grey. Primaries with broad black or brown-black bands (tinged partly buff towards outer edges), usually narrowly connected mutually by black along shafts; outer webs with rows of large and contrasting rufous-cinnamon spots (soon bleaching to buff or cream along outer edge), inner webs with indistinct dark greyish-buff bars shading to paler buff on basal inner edges; c. 1·5–2 cm of tips of primaries cinnamon with dusky grey mottling, bleaching to buff and sometimes cream or off-white when abraded. Secondaries similar to primaries, but black bands narrower and partly freckled buff near tips of outer webs; tertials mainly rufous-cinnamon with narrow dusky shaft-streak and incomplete dusky bars. Greater upper primary coverts and bastard wing banded black and cinnamon like primaries. Smaller outer upper primary coverts white with partial buff tinge and some black streaks, similar to under wing-coverts; axillaries cinnamon or buff, greater under primary coverts white with broad black tips. AVERAGE RUFOUS BIRDS. Like deep rufous birds, but ground-colour of upperparts slightly less deep rufous-cinnamon, sometimes pale cinnamon-brown or tawny-brown, on crown and hindneck usually paler tawny-buff, pale cinnamon, or cream-buff; differs in particular by less sharply black shaft-streaks, these laterally extending into dark bars; mantle, scapulars, and upper wing-coverts faintly and narrowly barred grey; facial disc less deep cinnamon, white crescents between bill and eye less contrasting, sometimes with white wash or dusky mottling on outer sides of disc; underparts more extensively white, but still with much rufous along borders of black shaft-streaks, in particular breast, flanks, and feet with rather close black or rufous barring, underparts appearing white with many rufous bars and spots rather than rufous with white spots as in deep rufous birds. Tail similar to deep rufous birds, but t1 slightly paler rufous with more extensive dusky specks on sides and tip and with slightly more distinct dusky bars at centres; dark bars on other tail-feathers slightly broader though rather diluted; flight-feathers similar but rufous ground-colour slightly paler and tips more heavily speckled dusky (not with diluted dusky wash); tertials with broader and longer though often more diluted bars. INTERMEDIATE BIRDS (usually called 'grey morph' in western Europe, but not as grey as birds from eastern and southern part of range). Ground-colour of crown and hindneck buff to pale cream-yellow, of mantle, scapulars, upper wing-coverts, and back to tail buff-brown, pale brown, or grey-brown; broad black shaft-streaks on feathers of crown and hindneck extend laterally into short black bars, remainder of upperparts with less distinct narrow and close dusky grey or black bars and vermiculations. Facial disc off-white, rufous (if any) restricted to chin, patch above eye, or outer border of disc; upper and outer border of disc often marked with fine dusky grey concentric bars. Underparts white with some tawny-yellow of feather-bases often

PLATE 53 (*facing*).
Tyto alba Barn Owl (p. 432). Nominate *alba*: **1** ad. *T. a. guttata*: **2–3** ad.
Surnia ulula ulula Hawk Owl (p. 496): **4–5** ad.
Strix aluco aluco/sylvatica Tawny Owl (p. 526): **6** rufous ad, **7** grey ad.
Strix butleri Hume's Tawny Owl (p. 547): **8–9** ad.
Asio otus otus Long-eared Owl (p. 572): **10–11** ad.
Asio flammeus Short-eared Owl (p. 588): **12** ad ♂, **13** ad ♀.
Asio capensis tingitanus Marsh Owl (p. 601): **14–15** ad. (HD)

visible on chest, flanks, and thighs; rufous (if any) restricted to slight wash along borders of black shaft-streaks of chest and flanks; often with short and rather close black bars on sides of feathers of chest, breast, and flanks (occasionally almost absent), feathering of leg and foot white with coarse dusky specks. Tail as in average rufous birds but ground-colour buff or pale brown and t1 more distinctly speckled dusky and more distinctly banded with diluted grey on terminal half; flight-feathers similar, but pale tips and bands paler cinnamon-buff or grey-buff (in part soon fading to off-white); under wing-coverts as in deep rufous birds, but ground-colour white, buff wash limited. PALE GREY BIRDS (occurring from Scandinavia and central Europe eastward). Ground-colour of upperparts, upper wing-coverts, tail, and flight-feathers as in intermediate birds or slightly paler buff or grey-buff, on crown and hindneck sometimes mixed with much white, differing from intermediate birds by lacking distinct dark barring on sides of each feather, only indistinctly vermiculated grey instead, appearing much paler; facial disc white with some buff and grey mottling along outer border; underparts white with pronounced black shaft-streaks, only tips of feathers of lower chest, breast, and flanks sometimes showing a few narrow dark bars or slight cinnamon wash. DARK GREY BIRDS (occurring mainly southern Europe and Asia Minor, but a few north to Britain and West Germany). Ground-colour of upperparts cream-buff to off-white, but this and pattern of black shaft-streaks largely suppressed by close and narrow black bars on crown and hindneck and dense dark grey vermiculation on remainder of upperparts and upper wing-coverts; even white spots on tips of outer webs of outer scapulars and outer upper wing-coverts sometimes partly speckled and vermiculated dusky and these spots not as distinct as in intermediate and rufous birds; lesser upper wing-coverts mainly dark brown. Facial disc pale grey with much dark grey barring and mottling, surrounded by black ruff mottled with white or rufous. Underparts white with limited cinnamon or pale tawny-yellow on breast, flanks, and thighs, heavily barred with narrow black bars and black shaft-streaks (least so on lower flanks, vent, and under tail-coverts). T1 and tertials grey-buff to off-white with intricate pattern of dark grey bands, speckles, and marbling; other tail-feathers and flight-feathers with more discrete pink-buff to off-white and brown-black bands, pale bands on secondaries, on tips of primaries, and on tail-feathers often speckled dusky, uniform only along borders of brown-black bands. Under wing-coverts and axillaries white with some black spots and streaks, tips of greater under primary coverts black. DOWNY YOUNG. 1st down (neoptile) thick, rather short and soft; white, or buff; appears mottled grey once 2nd down starts to grow under it from 8th–12th day. JUVENILE. 2nd down (mesoptile) long, slightly feather-like (in particular on longer scapulars and greater upper wing-coverts) but distinctly softer and looser than normal feathers. Pins of mesoptile appear from 8th day, open 10th–12th, pattern visible from 11th; fully covered in mesoptile from 18th day, but some neoptile clinging to tips of mesoptile up to c. 3–4 weeks (Scherzinger 1980; ZMA). Colour as variable as in adult. Crown dark grey with white or buff feather-tips; remainder of upperparts and upper wing-coverts rather evenly barred dark grey and white, yellow-buff, or pale rufous-cinnamon, feathers with broad white tips (in particular on scapulars and upper wing-coverts), conspicuous in buff- and cinnamon-coloured birds. Facial disc poorly defined; greyish-white to pale cinnamon. Underparts white, yellow-buff, or pale rufous-cinnamon with rather indistinct grey bars; feather-tips white, soon abrading. For juvenile tail and flight-feathers, see 1st adult. First adult plumage on head, body, and wing-coverts starts to grow from 5th–6th week, first on face, ruff, thighs, and lesser upper wing-coverts.

FIRST ADULT. Like adult, but juvenile flight-feathers, tertials, primary coverts, and tail retained. Most easily distinguished by (1) shape and colour of tail-tip, less so by (2) shape of inner tertials and (3) lack of mutual difference in abrasion of primaries and secondaries. (1) Tips of tail-feathers more pointed than in adult, less broadly rounded; tip of t1 uniform white or pale rufous (depending on colour morph), distinctly paler than remainder of tail, usually bordered subterminally by straight and narrow dark subterminal bar, tip forming uniform triangle c. 1 cm long and c. 2 cm wide at base, marked at most with faint darker shaft-streak or a few dark specks along shafts; outer feathers usually also with mainly uniform tips in rufous and pale grey birds, but speckled in intermediate and dark grey birds (in adult, tips broader and heavily speckled dusky; in a few rufous birds, tip of t1 almost uniform, but tip then similar in colour to remainder of tail, not paler, and usually not with a distinct dark subterminal bar). (2) Innermost 1–2 tertials narrower and with looser sides than adult, rather mesoptilous in character, soon affected by wear and then appearing to taper to tip; in rufous birds, tip uniform white when not too worn (in adult, broader and with rounder tip, less strongly subject to wear). (3) All flight-feathers equally fresh in first few months after fledging, inner secondaries and tertials and p4–p8 gradually more worn later on (adults between 1 and c. 5 years old show variable mixture of old and new flight-feathers as not all feathers are moulted every year; in older adult, mutual difference in age between feathers difficult to see without experience, except when an occasional feather retained for more than 2 years). SUBSEQUENT PLUMAGES. In 2nd year, p1 to p3 (p2–p4), often p9–p10, and usually some outer secondaries still juvenile, worn, often narrower with more pointed tip, and with dark pattern different from neighbouring new feathers.

Bare parts. ADULT AND FIRST ADULT. Iris deep brown. Eyelids blue-grey, thickened borders flesh-red, bright red, or dark red. Bill yellow with greenish or greyish base, tip of culmen pale yellow or almost white; cere blue-grey. Soles of feet bluish-grey or dull yellow-grey; claws dark slate-grey or brown-black with pale olive or grey base. DOWNY YOUNG AND JUVENILE. At hatching, bare skin, including leg, foot, and cere flesh-pink, pale blue on some future apteria; bill pale blue-grey or bluish-pink; soles and claws cream-white or pale grey. Eye opens from 8th–11th day, turbid blue until 16–20 days. At 14th day, rim of eyelid pink-red; cere greyish purple-flesh; bill pale grey-blue with darker base and whitish tip; leg pink, gradually shading to pale blue on toes and soles, tips of claws dark horn-brown. At fledging, like adult. (Heinroth and Heinroth 1926–7; Räber 1954; Scherzinger 1980; ZMA.)

Moults. ADULT POST-BREEDING. Complete, but not all flight-feathers moulted every year; primary moult serially descendant, apparently starting from a centre on p2 but p1–p3 often lost almost simultaneously. Flight-feather moult starts early May to late June, suspended early September to late November; groups of neighbouring feathers sometimes almost simultaneously shed and not always easy to tell which feather shed first; in 16 birds examined with active primary moult, only 2 had a single growing feather, 6 had 2 neighbouring feathers growing, 3 had 3, 4 had 4, and 1 had 5 adjacent feathers growing at same time. Though, when moult suspended, several groups of new primaries often separated by an old one, moult apparently does not occur at 2 different places in one wing—one group completed before next started. Average 6·0 (4–9) primaries replaced every year ($n = 44$). In birds with suspended primary moult, 21 had single group of 5·5 (4–8) primaries new, 24 had 2 groups of new primaries separated by 2·3 (1–5) old ones, 10 had 3 groups of new primaries each separated by 1·7 (1–3) old feathers. Tail moult almost simultaneous or disjunct (see Ural Owl *S. uralensis*), mainly (June–)July–August. Secondaries moult at same time as primaries, but sometimes completed slightly later; sequence approximately 10–11–12–9–5–6–2–3–8–7–4, hence with moult-centres at about s10, s5, and s2; timing of s1 irregular (Piechocki 1961), or sequence 12–13–11–14–10–1–9–5–2–8–6–3–7–4, with centres at s12 (moulting towards s14 and to s8), s1 (moulting towards s4), and s5 (moulting towards s7); not all secondaries replaced every year, but tertials (about s10–s14) often are (ZMA). POST-JUVENILE. Partial, mainly 45th–140th day, May–October. All head, body, and wing-coverts, but not flight-feathers, tertials, greater primary coverts, and tail; lesser and median upper wing-coverts first, crown and upperparts last (Heinroth and Heinroth 1926–7; Räber 1954; ZMA). FIRST POST-BREEDING. Complete, usually except for some inner and outer primaries and some secondaries; starts April–June, completed September–October. Primaries start with p1, p2, p3, or p4, halting with p8, p9, or p10: 8·1 (5–10) primaries replaced when 1 year old (n = 14) and juvenile innermost and outermost usually retained during 2nd winter (Glutz and Bauer 1980); in another sample, moult descendant with centre mainly at p4 or p5, several feathers often shed almost simultaneously shed and 4·7 (3–6) primaries replaced (n = 8), mainly p4–p8 (p2–p10), but birds with more complete moult perhaps overlooked (RMNH, ZMA). Secondaries and tail start at same time as primaries; tail lost almost simultaneously June–July; secondaries moult from centres on s11, s2, and s5, but only tertials replaced completely, usually not all secondaries, and 1 or both outer centres (s2 and s5) may not become active (Glutz and Bauer 1980; ZMA). In following moult season, regular moult series starts with p2, retained juvenile innermost primaries shed first, followed by retained outermost; see also *S. uralensis* for data on Spotted Owl *S. occidentalis*.

Measurements. Nominate *aluco*. Netherlands, all year; skins (RMNH, ZMA). Bill (F) to forehead, bill (C) to cere.

WING	AD	♂ 267	(5·12; 35) 259–275	♀ 278	(5·48; 30) 269–287
	JUV	262	(6·23; 14) 253–270	271	(5·63; 14) 263–282
TAIL	AD	155	(6·36; 24) 148–166	162	(5·06; 19) 154–171
	JUV	151	(4·19; 13) 146–158	160	(5·88; 12) 150–167
BILL (F)		30·5	(1·04; 38) 28·4–32·3	32·0	(1·30; 26) 29·8–34·3
BILL (C)		19·6	(0·88; 34) 18·4–21·1	20·6	(0·92; 26) 19·4–22·4
TARSUS		46·5	(1·14; 36) 44·8–48·3	48·3	(2·14; 23) 46·2–52·8

Sex differences significant. Juvenile wing significantly shorter than adult. Sexes best separated by length of middle claw (from base above to tip): ♀ 17·8 and over in Netherlands, ♂ below (11% mis-sexed by this), or by a combination of several measurements, but as geographical variation in size pronounced, no character valid for all populations can be given; for sex discriminant in British *sylvatica*, see Hardy *et al.* (1981).

WING. Combined data from skins (BMNH, RMNH, ZMA) and from various authors (Hartert 1912–21; Dementiev 1933a; Mayaud 1939a; Dementiev and Gladkov 1951a; Vaurie 1965; Eck 1971; Haftorn 1971; ages combined, as these usually not given). (A) nominate *aluco*: (1) Netherlands, (2) West Germany (mainly Rheinland), (3) East Germany, (4) Poland south through Czechoslovakia and eastern Austria to Yugoslavia and Rumania, (5) Italy (probably mainly northern part), (6) Sweden, (7) Norway, (8) west European USSR. (B) *siberiae*: (9) Urals and western Siberia. (C) *sylvatica*: (10) Britain, (11) France (except north and east), (12) Iberia, (13) western and central Asia Minor and Levant. (D) *willkonskii*: (14) Caucasus, Transcaucasia, and north-east Turkey, (15) Lenkoran area (Azerbaydzhan) and coast of southern Caspian. (E) *sanctinicolai*: (16) eastern Iraq and Iran. (F) *mauritanica*: (17) north-west Africa.

(A)	(1) ♂ 265	(5·86; 49) 253–275	♀ 276	(6·48; 44) 263–287
	(2) 268	(4·66; 9) 260–275	278	(7·57; 11) 266–288
	(3) 271	(5·89; 25) 261–284	280	(8·59; 18) 266–299
	(4) 280	(6·58; 12) 270–291	289	(8·05; 14) 275–302
	(5) 268	(5·76; 5) 261–273	280	(5·43; 6) 275–287
	(6) 274	(— ; 25) 265–283	284	(— ; 25) 272–298
	(7) 280	(— ; —) 272–294	292	(— ; —) 285–303
	(8) 283	(— ; 53) 268–295	296	(— ; 66) 277–311
(B)	(9) 291	(— ; 4) 280–300	303	(— ; 3) 301–307
(C)	(10) 257	(4·81; 47) 248–268	267	(6·01; 32) 255–272
	(11) 260	(5·02; 17) 252–269	268	(6·44; 17) 253–278
	(12) 261	(7·65; 8) 251–273	266	(3·38; 6) 260–270
	(13) 261	(6·88; 5) 254–271	269	(3·00; 9) 264–274
(D)	(14) 274	(— ; 14) 266–296	295	(— ; 25) 282–304
	(15) 288	(9·33; 6) 279–301	299	(— ; 12) 290–307
(E)	(16) 266	(— ; 12) 255–273	279	(— ; 7) 270–285
(F)	(17) 282	(— ; 1) —	296	(— ; 4) 292–303

Other measurements show similar trends. Range of averages from smallest (Britain) to largest (west European USSR) population: tail 155–175, bill (F) 30–33, bill (C) 18–22, tarsus 45–50. Wing of live or freshly dead British *sylvatica*: ♂ 262 (4·8; 90), ♀ 274 (7·03; 81) (Hirons *et al.* 1979); of birds from southern Belgium ♂ 265 (6) 250–270, ♀ 276 (17) 260–290 (Delmée *et al.* 1978). Wing of *mauritanica*, sexes combined and unsexed birds, skins: 289 (7·21; 13) 276–303 (BMNH, RMNH, ZMA).

Weights. ADULT AND FIRST ADULT. Nominate *aluco* (A) Netherlands: (1) November–February, (2) March–June, (3) July–October, (4) exhausted or freshly dead birds, all year (RMNH, ZMA). (B) Italy: (5) November–February, (6) March–June, (7) July–October (Moltoni 1949).

(A)	(1) ♂ 452	(47·3; 9) 380–540	♀ 530	(56·1; 8) 467–650
	(2) 397	(36·4; 11) 342–493	486	(57·3; 4) 418–537
	(3) 407	(35·3; 5) 359–456	456	(27·7; 7) 420–503
	(4) 287	(16·1; 12) 264–309	354	(32·4; 14) 301–397
(B)	(5) 426	(34·6; 11) 385–500	524	(76·7; 7) 415–620
	(6) 410	(25·5; 5) 370–440	514	(51·4; 9) 420–595
	(7) 391	(72·0; 7) 310–540	494	(62·8; 4) 432–560

Belgium: 12 live weights of 6 different ♂♂, 440 (385–475); 58 weights of 17 different ♀♀, 553 (480–660) (Delmée *et al.* 1978). East Germany, all year: ♂ 412 (63·6; 6) 301–460, ♀ 459 (148·2; 5) 314–662 (Eck 1971). Norway: ♂ 390 (295–480), ♀ 553 (336–695) (Haftorn 1971). Belorussiya (USSR): ♂ 327–727 (13), ♀ 335–730 (21) (Fedyushin and Dolbik 1967). USSR: ♂ 472 (3) 450–490, ♀ 642 (4) 590–682 (Dementiev and Gladkov 1951a).

S. a. sylvatica. Britain, all year: (1) live, (2) found dead (Hardy *et al.* 1981).

(1) ♂ 409	(33·3; 20) 352–465	♀ 533	(74·8; 22) 435–716
(2) 384	(37·0; 63) 304–464	484	(64·2; 79) 385–700

Normal weight of British 1st adults mainly 325–470 in ♂, 390–575 in ♀; starvation weight 215–345 in ♂, 310–395 in ♀ (Hirons *et al.* 1979). France, all year: ♂ *c.* 370 (11) 331–450, ♀ *c.* 430 (8) 395–475 (Mayaud 1939a).

S. a. willkonskii. South Caspian, April: ♂ 510, ♀ 582 (Schüz 1959).

DOWNY YOUNG AND JUVENILE. Nominate *aluco*. At hatching, 26–30; captive birds at age of 3 weeks 308 (78·7; 8) 160–395, at 5 weeks 383 (60·9; 11) 282–500; at 45 days with start of post-juvenile moult, 425 (49·3; 11) 355–510; at about end of moult at 90 days, 459 (45·7; 15) 375–520 (Räber 1954). At hatching, *c.* 35; at fledging 300–350 (Delmée *et al.* 1978).

Structure. Wing rather short, broad; tip rounded. 10 primaries: p6 and p7 longest, about subequal; p8 5–13 shorter, p9 25–39, p10 65–84, p5 10–17, p4 30–43, p3 49–59, p1 65–78. Outer web of (p6–)p7–p9 and inner web of p7–p10 emarginated. Outer web of longest feather of bastard wing, outer web of p10, and emarginated part of outer web of (p8–)p9 with rather long and curved serrations. Tail rather short, slightly rounded; 12 feathers, t6 18–32 mm shorter than t1. Bill heavy, wide and deep at base, laterally compressed at tip; basal half rather straight (not as bulbous and tapering as in *Asio* owls), tip of upper mandible strongly decurved (distinctly heavier than in *Asio*). Facial disc distinct, but ruff of narrow and rather stiff feathers surrounding it not very marked. Tarsus and toes rather short; tarsus and upper surface of toes (except distal digit) densely covered with feathers. Middle toe without claw 29·6 (1·24; 33) 28–32 in ♂ of nominate *aluco* from Netherlands, 31·6 (1·06; 22) 30–33 in ♀; outer toe without claw *c.* 74% of middle, inner *c.* 88%, hind *c.* 54%. Claws rather long and slender, curved; middle claw 16·8 (0·86; 38) 15·4–18·4 in ♂ from Netherlands, 18·2 (0·59; 26) 17·2–19·0 in ♀; outer claw *c.* 84% of middle, inner *c.* 107%, hind *c.* 90%. Relative to birds from Netherlands, middle toe and claw each on average 1·0 mm shorter in Britain, 0·8 longer in Iberia and eastern and northern Europe, 1·3 longer in north-west Africa.

Geographical variation. Marked, in both colour and size. As colour variation also subject to large individual variation, recognition of races in account below mainly based on size differences (mainly following Dementiev 1933a and Vaurie 1965), though treatment should perhaps be slightly different when plumage also taken into account. *S. a. sylvatica* smallest race (see Measurements); occurs Britain, France (except east), Iberia, Asia Minor (in north, east to Trabzon, in south at least to Taurus mountains), and Levant. Scottish birds similar to English ones—average wing Scotland 256 (♂), 268 (♀); England 258 (♂), 265 (♀); samples 8–18 birds. Birds from Sardinia are probably *sylvatica*—wing of single ♀ 272 (Hartert 1912–21); probably this race also in Sicily, southern Italy, and Greece. Birds intermediate in size between *sylvatica* and nominate *aluco* inhabit Netherlands, Belgium, West German Rheinland, and Vosges and Jura in France. Nominate *aluco* from central and northern Europe, northern Italy, and Balkans clinally larger towards east; smallest in Denmark, West Germany, and northern Italy, largest in west European USSR; Scandinavian birds rather large also (but Swedish birds perhaps smaller than Norwegian ones—see Measurements and Hartert 1912–21); in western USSR, clinally smaller towards south, range of ♂ and ♀ 278–295 and 282–305 respectively from Leningrad area to Belorussiya, 267–281 and 288–291 in Ukraine, and 264–275 and 273 (1) in Crimea (Dementiev 1933a). *S. a. siberiae* from Ural mountains eastward larger and paler than nominate *aluco*, birds from central European USSR between *c.* 35°E and 50°E considered intermediate (Dementiev 1933a). *S. a. sanctinicolai* from eastern Iran and Zagros mountains in Iran about as small as *sylvatica*, and birds of similarly small size inhabit northern Iraq and probably southeast Turkey; these birds rather abruptly replaced by large *willkonskii* from Caucasus, Transcaucasia, north-west Iran, south Caspian east to western Kopet Dag mountains in Turkmeniya, and north-east Turkey, in latter area occurring west to Rize (Jordans and Steinbacher 1948) and (approximately) Erzurum; within *willkonskii*, Caucasus birds average smaller than south Caspian ones and latter (east from Lenkoran area) sometimes separated as *obscurata* Stegmann, 1926. North-west African *mauritanica* about as large as nominate *aluco* from USSR and *willkonskii*, but separated from these by smaller populations. 2 main groups discernible in colour. (1) Northern group from Britain, France, Norway, and northern Italy east to western Siberia: no distinct colour morphs; all colour variants from deep rufous to dark grey (see Plumages) occur in west of range, and pale grey predominates in east, with all rufous and intermediate variants absent east from Volga and Kama basins. (2) Southern group from Iberia and north-west Africa east through Asia Minor to Caucasus, Levant, and Iran: 2 rather constant colour morphs occur, rufous and grey (rufous absent in north-west Africa, Iraq, and Zagros mountains), with an additional dark brown morph in Caucasus, Transcaucasia, and southern Caspian, this dark morph exceptionally also west to Turkey and Italy. In Britain ($n=57$) and Netherlands ($n=104$), 65% deep rufous or average rufous variants, 30% intermediate, 5% dark grey, 8% pale grey. East from Scandinavia, West Germany, Austria, and Balkans, none dark grey; rufous, intermediate, and pale grey variants occur in rather equal numbers Scandinavia to Balkans. About 34% rufous, 16% intermediate, and 50% pale grey in west European USSR; 20%, 5%, and 75% respectively in central European USSR (Dementiev 1933a); all pale grey further east. *S. a. siberiae* from Urals eastward even paler than pale grey variants of nominate *aluco* from central and northern Europe; dark transverse marks strongly reduced; much white on hindneck, scapulars, and upper wing-coverts; underparts predominantly white with rather narrow dark streaks, hardly any bars. Southern group characterized by close dark barring on crown, hindneck, and underparts and by heavily vermiculated and speckled mantle, scapulars, upper wing-coverts, and t1. Grey morph similar to dark grey variant from Britain and Netherlands, but brown-buff ground-colour of upperparts paler, cream-buff to off-white, ground-colour of underparts white with narrower black streaks but more distinct black bars. Rufous morph of southern group similar to average rufous variant from Britain and Netherlands, but rufous of upperparts often slightly browner when plumage fresh and dark bars on sides of feathers of upperparts slightly more distinct, but often hard to distinguish, in particular when plumage worn. In Iberia, Asia Minor, and Levant, dark grey and rufous morphs occur in about equal proportions; though grey morph distinguishable from (uncommon) dark-grey variant of British *sylvatica*, rufous morph hardly separable from British rufous variants and hence populations of Iberia, Asia Minor, and Levant provisionally included in *sylvatica* rather than being treated as separate race *clanceyi* (Jordans, 1950). *S. a. mauritanica* from north-west Africa has dark-grey morph only; similar to dark-grey morph of Iberia or with slightly finer black marks; much larger in size. *S. a. willkonskii* from Caucasus and north-east Turkey eastward more heavily barred, streaked, and vermiculated than *sylvatica* from Asia Minor and Levant or than *mauritanica*; *c.* 17% of 35 in 3rd dark brown morph, where pale ground-colour replaced by rufous-brown to dark chocolate-brown, facial disc uniform dark brown, and dark flight-feathers and tail hardly show paler bands. *S. a. sanctinicolai* from eastern Iraq and Zagros mountains in Iran nearly always grey; black shaft-streaks on crown and body narrow, black bars narrow and well-spaced, appearing paler grey and with more white of ground-colour showing (in particular on hindneck and underparts) than in Asia Minor. Grey birds examined from Dihok (northern Iraq) rather intermediate between *sanctinicolai* and dark-grey morph of *sylvatica* from Asia Minor; perhaps *sylvatica* and *sanctinicolai* intergrade in south-east Turkey and neighbouring part of Iraq. Extralimital races from mountains of central Asia larger than any west Palearctic race, in general rather closely barred and streaked like *mauritanica* or *willkonskii*; races from eastern China, Korea, and Taiwan smaller and heavily barred or scalloped.

CSR

Strix butleri Hume's Tawny Owl

PLATES 49 and 53
[facing pages 495 and 543]

Du. Palestijnse Bosuil Fr. Chouette de Butler Ge. Fahlkauz
Ru. Африканская неясыть Sp. Carabo oriental Sw. Klippkattugla

Asio butleri Hume, 1878

Monotypic

Field characters. 37–38 cm; wing-span 95–98 cm. Close in size to smaller Tawny Owls *S. aluco*; 10% larger than Barn Owl *Tyto alba*. Pale-faced, medium-sized owl with character close to *S. aluco*, being less bulky but with similar broad, round head. Upperparts buff, mottled brown and grey and showing pale collar; face and underparts buff-white, only faintly marked. Eyes have orange-yellow iris (dark in *S. aluco*). Strictly nocturnal. Sexes similar; no seasonal variation. Distinction of juvenile not known.

ADULT. Facial disc pale buff, with almost white half-circles abutting over bill and setting off eyes; recalls that of *T. alba* but shows dark divide above bill, formed by narrowing extension of crown. Crown and nape golden-buff, with dark brown-black feather-tips giving strongly mottled appearance. Collar round lower hindneck pale golden-buff and distinct. Back cream-buff, with indistinct bars and spots on mantle and clear brown and pale cream spots obvious on scapulars. Upperwing patterned as *S. aluco* but with paler buff ground-colour and obvious brown spots on coverts. Underwing buff-white, patterned as *S. aluco* with obvious dark barring on tips of primary coverts and all flight-feathers. Tail buff, barred dark brown. Underparts buff-white, with faint grey-buff bars and streaks over chest and flanks and unmarked vent appearing paler. JUVENILE. Not studied in the field, but apparently resembles adult.

Little known, but increasingly observed in Middle East in recent years. In north of range, liable to confusion with pale birds of *S. aluco* (though lack of broad chest-streaks a clear difference), and throughout its range mistakable for *T. alba* when generic structure and shape not apparent (though short legs, goggle-like divide to facial disc, and more barred wings and tail provide clear differences in good view). Confusion with pale Short-eared Owls *Asio flammeus* also possible (though dark eye-pits and long wings of that species should prevent mistake). Flight buoyant and silent. Stands on rocks with body leaning forwards; also sits on posts.

Hoot a diagnostic 'hoooo huhu-huhu'.

Habitat. In warm arid lower middle and subtropical latitudes; thought to be desert form replacing Tawny Owl *S. aluco* in that habitat. Habitat data limited. In Dead Sea region (Israel) occurs towards breeding season in gorges with springs or long-lasting rain pools, at some distance from known localities for *S. aluco* (Mendelssohn *et al.* 1975); in Sinai (Egypt) in palm oasis (Meinertzhagen 1930). Heard at *c.* 2800 m on steep slope of mountain in Saudi Arabia with trees of juniper *Juniperus* and other

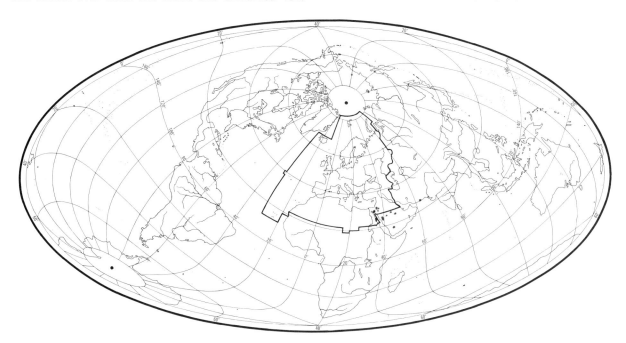

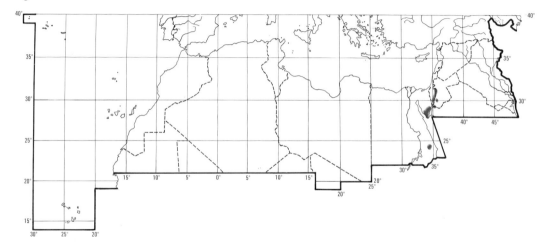

vegetation, but more frequently on rocky, more or less treeless, terrain (King 1978; M C Jennings).

Distribution. Still inadequately known.
SYRIA. Despite Vaurie (1965), no definite records (HK). ISRAEL. Occurs from near Jericho southwards in almost every wadi with water; 1 nest found, and elsewhere fledgling seen in hole in rock (Mendelssohn et al. 1975; Leshem 1981; Subah 1983; HM). JORDAN. Heard near Petra 1936 (HM) and feathers found near Aqaba 1978 (Steinbacher 1979). EGYPT. Almost certainly breeds Sinai; 2 calling Wadi Feiran, and specimens from there and 3 other localities (Meinertzhagen 1930; Leshem 1981; HM). One bird collected in Red Sea mountains, 1983 (SMG).

Population. Little information.
ISRAEL. 47 sites known (Leshem 1981; J Leshem); 15 killed by cars 1973–8 (Aronson 1979).

Movements. Resident in Israel (HM) and Saudi Arabia (Jennings 1980). No further information.

Food. In Israel and Sinai (Egypt), includes: gerbils (*Gerbillus dasyurus*, *G. henleyi*, *Meriones crassus*), jerboas *Jaculus*, jird *Sekeetamys calurus*, spiny mice (*Acomys russatus*, *A. cahirinus*), white-toothed shrew *Crocidura suaveolens*; gecko *Ptyodactylus hasselquistii*, agama *Agama stellio*; Desert Lark *Ammomanes deserti*, House Sparrow *Passer domesticus*; grasshopper *Sphodromerus pilipes*, beetles (including Tenebrionidae), scorpion *Nebo hierochonticus*, scolopenders (Chilopoda) (Mendelssohn et al. 1975; Aronson 1979; Leshem 1979, 1981; H Mendelssohn). 183 items in pellets and food remains from nest in Israel comprised (by number) 63·4% *Gerbillus*, 18·0% *Meriones*, 13·1% *Jaculus*, 0·5% birds, and 4·9% unidentified (Subah 1983). 52 pellets, March, contained 48 mammals (mostly *Gerbillus* and *Meriones*), 3 reptiles, 3 birds, and 47 arthropods (including 29 *Sphodromerus*) (Leshem 1981). Hunts at night from posts, etc., and will take insects in flight under artificial light (Mikkola 1983). DJB

Social pattern and behaviour. Very little known. Following account includes notes supplied by M C Jennings for Saudi Arabia.
1. Apparently essentially solitary. BONDS, BREEDING DISPERSION. No firm information, but see part 2. ROOSTING. One bird found roosting in cave (M C Jennings). Activity highly nocturnal (Leshem 1981).
2. Roosting bird allowed approach to within 2 m (M C Jennings), and captive juvenile remained passive in the hand (Jennings 1977). However, another juvenile made defensive Bill-snapping sounds when observer approached to *c.* 2 m below narrow crevice in which it was sitting (Leshem 1979, 1981). In Forward-threat posture (Fig A), captive adult leaned forward towards observer, ruffled plumage (notably on face, breast, and

A

mantle), and Bill-snapped (Leshem 1981). Calling (see 1 in Voice) suggests ♂♂ take up breeding territories January–April in Israel (Leshem 1981; see also Aronson 1979); same period in Saudi Arabia (mostly March–April), but also heard September (Jennings 1980; M C Jennings). Usually calls from a perch, occasionally in flight, and frequently changes perch between bouts of calling. Calling by 1 bird may cause another to call also, or (if already calling) to call more vigorously, move closer, and follow the other (M C Jennings); see also Voice. Bird may call, with breaks, for several hours; in Saudi Arabia, earliest 20 min after sunset, latest 2 hrs after sunrise (Jennings 1977; M C Jennings).

(Fig A from photograph in Leshem 1981.) EKD

Voice. Used freely in presumed breeding season. Following account based largely on notes supplied by M C Jennings for Saudi Arabia. For Bill-snapping, see Social Pattern and Behaviour.

I B King Saudi Arabia January 1977

CALLS OF ADULTS. (1) Advertising-call. See Fig I. Diagnostic 'hoooo huhu-huhu', i.e. a long unit, followed by 2 shorter disyllabic units (Jennings 1981b; M C Jennings); also rendered 'whooo woohoo-woohoo', the 'woo' being higher pitched than the 'hoo' (King 1978). For same or similar renderings, see Aronson (1979) and Gallagher and Rogers (1980). These sources in agreement, but at variance with the often-quoted (and probably faulty) rendering of Meinertzhagen (1930, 1954). Calls typically repeated at intervals of 20–60 s, though single calls also heard. One series of 25 calls given in 15 min, shortest interval 17 s, longest 85 s. Long gap sometimes indicates change of perch (M C Jennings). 2 or more birds may call simultaneously (Meinertzhagen 1930; Gallagher and Rogers 1980; Jennings 1981b). 2 birds may perform duet, calling alternately, or one completing the call of the other, i.e. one calls 'hoooo', the other 'huhu-huhu' (M C Jennings). (2) A rapid, agitated 'hu-hu-hu-hu-hu-hu-hu-hu-hu', lasting 2–3 s, given when 2 birds giving Advertising-call drew closer together. May be stimulated by imitation or playback of Advertising-call (Jennings 1977, 1981b). According to Meinertzhagen (1930, 1954), Advertising-call sometimes varied by a tremulous and more throaty hoot, as in Tawny Owl *S. aluco*. Throaty coughs given commonly in Israel, January–April (J Leshem); possibly the same call.

CALLS OF YOUNG. Chicks quiet, not making sound even when handled by man (Subah 1983). EKD

Breeding. SEASON. Israel: main song period January–April; well-grown young bird found March (Leshem 1975, 1981). In central Saudi Arabia, bird out of nest, still with some down showing, in mid-July (Jennings 1980). Fresh juveniles recorded Saudi Arabia in August and Sinai in September (C S Roselaar). SITE. Only one nest recorded: in ancient cistern (Subah 1983). In central Saudi Arabia, probably in rock crevice or cave (Jennings 1980). EGGS. Not described. Clutch: in one case, 5 (Subah 1983). INCUBATION. In one case, 35 days. By both sexes according to Subah (1983), though this would be unusual for west Palearctic owls generally. YOUNG. Altricial and nidicolous. FLEDGING TO MATURITY. Fledging period in one case 37 days (Subah 1983). No further information.

Plumages. ADULT. Only 2 examined. (1) Bird from near Jerusalem (Israel) mainly pale sandy-rufous in appearance. Crown, mantle, scapulars, and back to upper tail-coverts pale sandy-rufous, each feather with faint grey-brown or rufous-brown central streak and ill-defined brown patch at tip, latter most pronounced on sides of crown to nape (forming streak) and on lower mantle and outer scapulars, where accentuated by sharply dark brown shafts. Tertials and central tail-feathers (t1) coarsely barred pale sandy-rufous and dark grey-brown or sepia; 4–5 dark bars visible on tertials, 5–6 on t1; remainder of tail-feathers with similar dark bars, but ground-colour of outer edges and tips almost white. Facial disc pale cream or isabelline, nearly white at sides of bill, contrastingly black-shafted bristles projecting from sides of bill; edge of facial disc rufous-brown. Underparts pale pinkish-buff with narrow and faint brown shaft-streaks (most pronounced on belly) and faint pale grey-brown bars on chest (subterminal) and on belly (terminal). Vent, legs, and under tail-coverts cream-white. Flight-feathers broadly and regularly banded sepia (often slightly freckled cream on outer web); ground-colour pale grey on inner web (more rufous towards base), pale sandy-rufous on outer web (grading to cream-white on edge). Lesser upper wing-coverts almost uniform dull grey; median and greater coverts dull grey with broken dot-like pale sandy-rufous subterminal bars, some outer median and most greater coverts with large rounded cream-white dots, partly bordered at both sides by narrow dark sepia line. Under wing-coverts yellowish-cream, axillaries pale sandy-rufous. (2) Bird from Wadi Kelt near Dead Sea markedly greyer above than Jerusalem bird. Crown and upper mantle darker cinnamon-buff, dark grey to dull black of feather-tips more extensive. Lower mantle, scapulars, and upper wing-coverts grey with limited sandy-buff of feather-bases showing. Outer median and lesser coverts with large rounded subterminal white or pink-buff spots; facial disc and underparts as in Jerusalem bird. Tail marked with *c.* 5 dark grey-brown bars 1·5–2 cm wide, separated by buff *c.* 1 cm wide; width and colour of bars on flight-feathers and tertials as in tail. DOWNY YOUNG AND JUVENILE. No information. FIRST ADULT. Only 2 examined, and information on a 3rd from Jennings (1977). Not known whether characters shown are due to age or just individual. (1) Bird from Mahd adh Dhabab (Saudi Arabia) mainly similar to Wadi Kelt adult described above, but with large black spots on crown (forming streak on centre), broader and blacker feather-tips on upperparts, and broader and blacker bars (*c.* 2 cm wide) on tail, tertials, and flight-feathers; facial disc completely and finely barred white and dark grey, edge of disc blackish at border with crown, cinnamon at border with underparts; underparts cinnamon-pink with broad but faint grey-brown feather-tips and dark shafts on chest and flanks; leg white with narrow buff and dark grey-brown bars. Some down still adhering to primary-tips. (2) Bird from Wadi Feiran (Sinai) almost completely cinnamon-rufous above, with markedly black feather-tips and with uniform pinkish-rufous patches on sides of neck; facial disc uniform cream; underparts rufous-sandy; wing and tertials mainly uniform dull brownish-grey, only flight-feathers, greater primary coverts, outermost median and greater coverts, and tertials coarsely barred rufous. (3) Bird from central Saudi Arabia. Upperparts generally light warm brown, mixed yellow, especially on sides of head and hindneck; dark brown streak on central crown; facial disc buffish-grey with fine brown bars below bill; underparts buffish-white with some faint dark bars; some dark shafts on flanks. Feathering of leg uniform whitish. Tail with 5–6 dark bars. (Jennings 1977.)

Bare parts. ADULT. Iris yellow. Bill horn-yellow. Claws black. OTHER PLUMAGES. No information on small chick. In large chick and just-fledged bird, iris orange or orange-yellow, bill light horn or slate horn-coloured, claws bluish-grey. (Mendelssohn *et al.* 1975; Jennings 1977; J Swaab, BMNH.)

Moults. None in active moult examined. Plumage fresh in 1st adults from August–September, rather fresh in January adult, slightly worn in March adult (BMNH, ZFMK).

Measurements. (1) Adult ♂, Jerusalem (Israel), January (ZFMK); (2) 1st adult ♂ Mahd adh Dhabab (Saudi Arabia), August (BMNH); (3) 1st adult, possibly ♂, Wadi Feiran (Sinai), September (ZFMK); (4) adult ♀, Wadi Kelt near Dead Sea (BMNH). Bill (F) to forehead, bill (C) to cere.

	(1)	(2)	(3)	(4)
WING	251	252	238	255
TAIL	137	134	137	140
BILL (F)	26·7	29·0	27·5	26·3
BILL (C)	15·0	16·8	15·8	15·1
TARSUS	51·4	55·4	—	53·5

Makran (western Pakistan) and Sinai: wing 252, 255; tail 150, bill 30, tarsus 50 (Hartert 1912–21). Wadi Suenit (Jordan Valley), November: wing ♂ 245, ♀ 240 (Hüe and Etchécopar 1970).

Live 1st adult, central Saudi Arabia, July: wing 247 (Jennings 1977).

Weights. Israel: ♀♀ 214, 220; emaciated ♀ 162 (Mendelssohn *et al.* 1975). 1st adult, Saudi Arabia, July: 225 (Jennings 1977).

Structure. Wing rather long, broad, tip rounded. 10 primaries: p7 longest, p8 0–7 shorter, p9 12–22, p10 47–58, p6 3–8, p5 23–28, p4 42–48, p1 73–77. Tail short and broad, slightly rounded, 12 feathers; t6 8–12 mm shorter than t1–t2. Tarsus and toes rather short and slender; tarsus feathered to toes, upperside of toes either covered with spines or feathered; middle toe without claw 19·5–23·0 (38–43% of tarsus length), middle claw 12·2–14·0; tarsus relatively longer than in Tawny Owl *S. aluco*, but middle toe and middle claw shorter and more slender. Remaining structure as in *S. aluco*.

CSR

Strix uralensis Ural Owl

PLATES 50 and 55
[facing pages 495 and 567]

Du. Oeraluil   Fr. Chouette de l'Oural   Ge. Habichtskauz
Ru. Длиннохвостая неясыть   Sp. Cárabo uralense   Sw. Slaguggla

Stryx uralensis Pallas, 1771

Polytypic. Nominate *uralensis* Pallas, 1771, European USSR east from *c.* 45°E and western Siberia; *liturata* Lindroth, 1788, northern Europe and European USSR east to *c.* 40°E, south to northern Poland and Belorussiya, intergrading with nominate *uralensis* between 40° and 45°E; *macroura* Wolf, 1810, central Europe from Carpathian mountains and formerly Erzgebirge and Böhmerwald south to Yugoslavia and Bulgaria. Extralimital: 6–7 further races east from Yenisey basin.

Field characters. 60–62 cm; wing-span 124–134 cm. About 15% smaller than Great Grey Owl *S. nebulosa*, with 30% shorter wings; *c.* 60% larger than Tawny Owl *S. aluco*, with relatively longer tail. Large, rather long, round-headed, dark-eyed owl, suggesting overgrown *S. aluco*; with long tail, recalls juvenile Goshawk *Accipiter gentilis* in flight. Plumage rather pale, ochre-grey copiously streaked on back, back of head, and underparts. Facial disc pale ochre-grey, with relatively small dark eyes and indistinct paler divide giving rather vacant, kind expression. Dark bars across flight-feathers obvious. Nocturnal. In flight, wing-beats slower than *S. aluco* and length and wedge shape of tail obvious. Sexes similar; no seasonal variation. Juvenile separable. 3 races in west Palearctic, north-eastern birds probably separable in the field (see Geographical Variation).

ADULT. Ground-colour of head, back, and underbody ochraceous grey-white, with noticeable uniform pattern of dark brown streaks on crown, broad nape, back, and all underparts except vent; ochre tone most obvious around head and on chest. Circular facial disc, narrowly outlined black-brown and divided above bill with dark brown extension of crown; face grey-white, with ochre tone strongest around eyes and on outer parts of disc, and white tone strongest on short eyebrows and feathers around bill (creating much less obvious divide in centre than in congeners). Eyes black-brown and relatively small, not dominating face as much as in most other owls. Edges of long back feathers and scapulars distinctly fringed white, forming obvious pale braces. Wings basically dark ochre-grey, with smaller coverts showing rows of small dark spots, larger ones conspicuously edged white and flight-feathers conspicuously barred almost black. In flight, extended wing more heavily marked than in any other west Palearctic owl, with heavy black-brown barring over flight-feathers (particularly over base of outer primaries where white ground-colour creates chequered patch). On underwing, buff to cream-white coverts contrast with heavily barred flight-feathers, and broad black tips to primary coverts form carpal band. Tail ochre-grey, with 5–6 bold dark bars of even width; not only long but noticeably wedge-shaped, with outer feathers up to 4 cm shorter than central ones. Legs and feet densely feathered. Bill deep straw-yellow. If pair seen, ♂ distinguishable by smaller bulk. JUVENILE. Facial disc noticeably pale; head feathers heavily tipped white, bird thus appearing pale-headed. Underbody grey, barred brown.

Unmistakable if seen well, but in brief glimpse or when hunting in the open, may suggest Short-eared Owl *Asio flammeus* as much as paler morphs of *S. aluco*. Basic plumage pattern of *A. flammeus* not dissimilar but its smaller size, proportionately longer and narrower wings, plumage, and heavily marked face soon rule out confusion. Can also be mistaken for *S. nebulosa* (larger and essentially grey; see that species). Flight with sustained glides recalls that of *S. aluco* but overall action much slower, suggesting buzzard *Buteo* though with slower beats of more rigidly held wings. Head–body–tail line level, but dangling feet add to tail-down appearance noted by some observers. At nest, very aggressive to intruders.

Territorial call a soft, deep, staccato hoot—'wo-ho wo-ho oh-who-ho', with pause of 2–3 s between first 2 disyllables and slight terminal acceleration. Alarm-call a loud sharp bark—'kark' or 'kark-kark'.

Habitat. Similar to that of Great Grey Owl *S. nebulosa* but penetrating less far above Arctic Circle and extending into somewhat lower temperate latitudes, with relict outliers in more southerly mountain ranges of west Palearctic (Voous 1960). Breeds in Slovakia at 450–850 m, in beechwoods of Soviet Carpathians to treelimit at 1100–1200 m, and in Rumania to 1600 m (Glutz and Bauer 1980) but in north mainly in lowlands, avoiding extensive dense forests, especially of pure conifers. In former East Prussia (Poland/USSR), favoured oaks *Quercus*, hornbeam *Carpinus*, lime *Tilia*, aspen *Populus tremula*, birch *Betula*, alder *Alnus*, maple *Acer*, fir *Abies*, and pine *Pinus*, especially near glades or fire-breaks and on margin of meadows and fields (Niethammer 1938). In summer, hunts in forest glades and fringes of woods; in winter in parks, open areas around human communities as well, including vicinity of villages where rodents and birds collect around threshing floors, and even cities. Open water often present in territory. Indifferent to specific composition of forest, provided it offers open hunting grounds with tall stands of trees offering suitable nest-sites. Prefers more open woodlands than *S. nebulosa* and more commonly found in moist than in dry woods, with preference for conifers north of Baltic and deciduous woods further south. Often near human dwellings, in use or abandoned, especially favouring extensive cultivation, and also pastures (Glutz and Bauer 1980). For comparison with habitat of Tawny Owl *S. aluco*, see Lundberg (1980b).

Distribution. NORWAY. Has occasionally nested outside main range (Haftorn 1971). POLAND. Bred Lower Silesia, 1909 (Tomiałojć 1976a). CZECHOSLOVAKIA. Nested in southern Bohemia until end of 19th century (KH). WEST GERMANY. Known as breeding in late 19th century in Böhmerwald, Bayern (Glutz and Bauer 1980). HUNGARY. Record of nesting in 1906 considered doubtful; bred 1973, and perhaps since, in Zempliner mountains (LH, A Keve), but see also Glutz and Bauer (1980). AUSTRIA. Sporadic breeder in second half 19th century; more recent reports of nesting unconfirmed (Glutz and Bauer 1980). BULGARIA. Very probably bred 1979–80, and 5 breeding season records earlier in 20th century (TM). USSR. Latvia: bred 1974 and 1980 (Viksne 1983). Lithuania: bred 19th century (B Šablevičius).

Accidental. West Germany, East Germany, Hungary, Italy.

Population. Data limited. Fluctuating breeding numbers linked with rodent fluctuations; some recent increase in Finland and perhaps Sweden.

SWEDEN. Perhaps *c*. 3000 pairs (Ulfstrand and Högstedt 1976); probably increased in recent decades (LR). In central Sweden, almost constant at *c*. 100 pairs 1969–78; territories maintained even when no breeding (Lundberg 1981). FINLAND. Marked fluctuations; highly tentative estimate *c*. 700 pairs (Merikallio 1958). Increased, partly due to provision of nest-boxes (Lahti 1972; OH). Fluctuations in study area southern Tavastia, with 60 nests in peak year 1973, compared with 8 in low year 1971 (P Saurola).

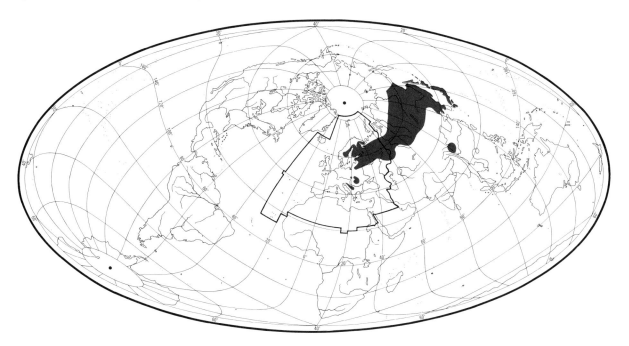

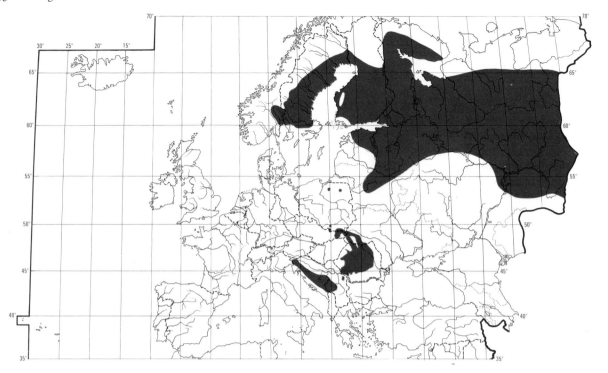

POLAND. Very scarce breeder (Tomiałojć 1976a). CZECHOSLOVAKIA. Fluctuating; no clear long-term trends (KH); at least 100 pairs in good years (Glutz and Bauer 1980). RUMANIA. Decreased, now very scarce (WK). USSR. Fluctuating population, numbers not high, no nesting in some areas in poor rodent years (Dementiev and Gladkov 1951a). Estonia: estimated 200–300 pairs in 1970s, marked fluctuations (Randla 1976). Belorussiya: much decreased (Fedyushin and Dolbik 1967). Bashkiriya: declined (Ilyichev and Fomin 1979).

Sweden. Annual adult survival c. 90% (Lundberg 1981). Oldest ringed bird at least 14 years (A Lundberg).

Movements. Mainly resident, even sedentary. Most data for Fenno-Scandian population (*liturata*). Like Tawny Owl *S. aluco*, survival linked to possession of exclusive year-round feeding territory, which juveniles acquire in autumn or die. Young birds disperse September–November; thereafter, all age-classes strongly territorial and hence sedentary, even in poor rodent years, though may not breed then (Mikkola 1983). Ringing in Finland showed that about ⅔ of 1st-year birds dispersed within 30 km of birthplace, though up to 200 km recorded.

Likewise resident or sedentary in central Europe (*macroura*). Longest ringing recovery: 145 km north-west, from Czechoslovakia into Poland. Vagrants to Italy (none ringed) showed concentration in north-eastern provinces of Udine and Trieste (Glutz and Bauer 1980).

Autumn and winter wandering apparently more extensive in western Siberia (nominate *uralensis*) and central Siberia (*yenisseensis*), though no data on ages of such birds.

Nominate *uralensis* reaches lower reaches of Volga and Ural rivers, and specimens collected south and south-east to Altay and Krasnoyarsk (Dementiev and Gladkov 1951a). Small-scale irruptions (race unknown), among other Strigiformes, recorded in Pskov (European USSR) in autumns 1960 and 1963 (Meshkov and Uryadova 1972).

Food. Includes some information from summary provided by A Lundberg.

Largely mammals from size of small rodents to water vole *Arvicola terrestris*, and birds from size of finches (Fringillidae) to Woodpigeon *Columba palumbus*. Hunting essentially nocturnal; during breeding season, in some cases at least, often occurs also during day (Teiro 1959; Lundberg 1980b; Mikkola 1983). Most hunting done from perch, but probably also uses searching flight (A Lundberg). Can take rodents under 20–30 cm of snow (Lundin 1961). Will take carrion (Collett and Olsen 1921). Larger prey decapitated before being given to nestlings: 71% ($n=756$) of *A. terrestris* brought decapitated, 8% ($n=1037$) of voles *Microtus* and *Clethrionomys* (A Lundberg). For rate of prey delivery to nest, see final paragraph.

Pellets no different from larger ones of Tawny Owl *S. aluco*; difficult to find as roost-site changes constantly (Glutz and Bauer 1980). Average size $62 \times 25 \times 22$ mm, $n=100$ (Mikkola 1981). 4·0 prey per pellet, $n=46$ (Mikkola 1973); average 4·2 per pellet, representing 115 g of prey (Lundberg 1981).

The following prey recorded in west Palearctic. Mammals: hedgehog *Erinaceus europaeus*, shrews (*Sorex minutus, S. araneus, Neomys fodiens, Crocidura leucodon, C.*

suaveolens), mole *Talpa europaea*, young hares (*Lepus timidus*, *L. europaeus*), red squirrel *Sciurus vulgaris*, flying squirrel *Pteromys volans*, dormouse *Muscardinus avellanarius*, wood lemming *Myopus schisticolor*, bank vole *Clethrionomys glareolus*, grey-sided vole *C. rufocanus*, water vole *Arvicola terrestris*, pine vole *Pitymys subterraneus*, common vole *Microtus arvalis*, short-tailed vole *M. agrestis*, striped field mouse *Apodemus agrarius*, yellow-necked mouse *A. flavicollis*, brown rat *Rattus norvegicus*, weasel *Mustela nivalis*, cat *Felis*. Birds range in size from tits *Parus* to Black Grouse *Tetrao tetrix*, but commonly only up to size of Woodpigeon *Columba palumbus* (Lundberg 1981); Capercaillie *T. urogallus* recorded, but injured or dead (Mysterud and Hagen 1969). Amphibians: frog *Rana esculenta*, toad *Bufo*, newt *Triturus*. Lizards and fish recorded by Mikkola (1972a). Insects rather infrequent (see Mikkola 1983); include beetles (Carabidae, Geotrupidae, Cerambycidae) and crickets, etc. (Tettigoniidae, Gryllotalpidae). (See, especially, Kohl and Hamar 1978; also Mysterud and Hagen 1967, Mikkola and Mikkola 1974, Lundberg 1976, 1981.) Leaf of alder *Alnus* recorded in stomach (Sládek 1961); captive young readily took lettuce (Heinroth and Heinroth 1931-3).

For summary of major west Palearctic studies, see Table A. Other contributions as follows. Norway: Hagen (1968), Mysterud and Hagen (1969). Sweden: Ahlbom (1976), Lundberg (1976, 1977, 1979b, 1980b). Finland: Teiro (1959), Mikkola (1970b, 1971a); see also sources in Mikkola (1983). Germany: Schäfer and Finckenstein (1935). Poland: Kochan (1979). Czechoslovakia: Sládek (1961). USSR: Parovshchikov and Sevastyanov (1960). See also Uttendörfer (1939b, 1952).

Arvicola terrestris much more important in Fenno-Scandia than in studied areas of central Europe, though importance decreases again towards north (Mikkola 1972a; Ahlbom 1976; Lundberg 1981). In poor vole years, Uppland (Sweden), took significantly more *A. terrestris* (37·7% by number, rather than 31·3% in good vole years), birds (12·0%, 6·4%), and amphibians (4·0%, 1·9%) (Lundberg 1981). However, Lundberg (1979b) claimed that *A. terrestris* preferred as prey, and other voles preyed on at rate dependent on abundance of *A. terrestris*. In eastern USSR, when small rodents scarce in summer, will switch to birds, amphibians, reptiles, and insects (Vorobiev 1954; Kislenko 1967). No detailed information on prey availability for west Palearctic, but for disparity in Japan between proportions of small mammal species eaten and those found (by trapping) to be available, see Imaizumi (1968).

Little information on seasonal variation in diet, and no clear pattern apparent. In Uppland, stomachs of 49 birds showed significantly more birds taken November-April than May-October (26% and 2% by number) and fewer shrews *Sorex* (17%, 52%) (Lundberg 1981); small voles more important before eggs hatch than after (Lundberg 1979b). Due to subterranean habits, *A. terrestris* not available until late spring in central Sweden (Lundberg 1981) and this must be reflected in diet outside breeding season. See also Table A and Mikkola (1983).

General ecology, and strategy of maintaining territory all year (Lundberg 1979a), evidently rather similar to that of *S. aluco*. In Uppland, diet similar to that of *S. aluco* breeding nearby, differing only in detail: *S. uralensis* took more squirrels *Sciurus* (*S. aluco* took none) and rather

Table A Food of Ural Owl *Strix uralensis* as found from pellets, nest-remains, and stomach analyses. Figures are percentages of total live weight of vertebrates only.

	Uppland (central Sweden)	Central Finland	Central Europe		Rumania
			Breeding season	Outside breeding season	
	1969-78, mostly breeding season	1966-70, breeding season			All year, mostly winter
Mice (small Murinae)	0·2 ⎱ 19·4	0·2 ⎱ 18·1	7·9 ⎱ 40·2	5·4 ⎱ 23·3	10·7[3] ⎱ 64·7
Small voles (Microtinae)	19·2 ⎰	17·9 ⎰	32·3 ⎰	17·9 ⎰	54·0 ⎰
Larger mammals[1]	65·1[2]	51·0	12·2	32·2	12·9
Shrews (Soricidae)	0·7	1·6	2·1	1·3	3·4
Birds	13·7	26·5	41·9	42·9	16·7
Other	1·1	2·9 (amphibians)	3·6 (frogs, lizard)	0·4 (frogs)	2·3 (frogs)
Total no. of vertebrates	2309	1014	271	192	262
Source	Lundberg (1981)	Mikkola and Mikkola (1974), data reanalysed	Mikkola (1972a), data reanalysed		Kohl and Hamar (1978)

1. In Uppland, largely water vole *Arvicola terrestris*; elsewhere, also mole *Talpa europaea*, young hares *Lepus*, squirrels (Sciuridae), brown rat *Rattus norvegicus*, and Mustelidae.
2. Includes 2·1% unidentified mammals.
3. Includes 0·5% dormouse *Muscardinus avellanarius* (Muscardinidae).

fewer mice and birds; ecologies separated by habitat (Lundberg 1980b).

In captivity, each young ate c. 3 kg during whole period in nest; c. 30 g per day at 1–5 days, 80 g at 10 days, 105 g at 20 days, 90 g at 30 days (Scherzinger 1974). In the wild, based on prey remains from 30 nests, each young averages c. 100 g per day between 10 and 28 days old (A Lundberg). In Uppland, during breeding season (April–June) of 1976 (poor vole year) and 1977 (peak vole year), pair (with young) estimated to consume respectively 321 and 510 items, including 62 and 289 Microtus agrestis, 70 and 77 Arvicola terrestris, and 72 and 32 birds—i.e. 3·6 and 5·7 items per day (Lundberg 1981).

Studies discussed above contain large proportion of food brought to nests, but no detailed work done on differences between diet of adults and young. Young possibly eat more insects (Kohl and Hamar 1978). Incubating ♀ fed by ♂ once every c. 5 hrs at night (72 hrs observation) (A Lundberg and B Westman). Nestling or newly fledged bird fed on average once every 5 hrs (129 hrs observation) (Teiro 1959). Rate of prey delivery to nestlings reduced by bad weather—wind, rain, snow (Lundberg 1980b). For rate of food consumption of young, see above. DJB

Social pattern and behaviour. Closely similar to Tawny Owl Strix aluco. Account based largely on outline supplied by A Lundberg, also on notes by T Holmberg and important study of captive birds by Scherzinger (1980).

1. Typically solitary or in pairs throughout the year. Always hunts alone. Established pair markedly sedentary, remaining in breeding territory all year, due perhaps to scarcity of nest-sites and consequent need for continuous defence (Lundberg 1979a, 1981). Individuals, however, may be forced to hunt more widely outside breeding season. In central Sweden, family parties persist in territories until August–September, when juveniles evicted by parents (A Lundberg; see also Scherzinger 1980). BONDS. Monogamous mating system. Pair-bond probably life-long, unless mate dies, and maintained throughout the year (A Lundberg). Divorce recorded once, when ♀ left ♂ to breed elsewhere; ♂ then paired with 11-year-old ♀ who joined him from territory 1 km away where she had, perhaps, been unpaired, having hitherto laid only infertile eggs (Saurola 1980). Age of first breeding usually 3–4 years (Lundberg 1979a); birds occasionally pair and breed in 1st year (Lagerström 1969), most likely if born the year before a peak in local prey population (A Lundberg); breeding in 1st year also reported in captivity (Scherzinger 1975, 1980). ♂ and ♀ divide breeding duties: ♀ incubates, broods, and performs most defence of young in nest; also helps to feed young out of nest. ♂ feeds ♀ and young in nest, and probably provides most food for young out of nest (Teiro 1959; Holmberg 1974a; Lundberg 1980b). After leaving nest, young dependent on parents for c. 10–12 weeks (Teiro 1959). BREEDING DISPERSION. Solitary and territorial. Minimum distance between nests, central Sweden, 1–3 km, more often c. 5 km (A Lundberg). In southern Finland, usually c. 3·5 km (Lammin-Soila and Uusivuori 1975). In one year, Czechoslovakia, 3 pairs 0·3 km and 0·5 km apart, but more often 2–5 km (Glutz and Bauer 1980). In areas of high density, spacing of pairs quite regular (Nilsson et al. 1981). In Scandinavia, density probably declines northward (A Lundberg). In central Sweden, c. 100 pairs averaged c. 0·05 pairs per km², maximum c. 0·25 pairs per km² (Lundberg 1974, 1979b, 1981). In southern Finland, average 55 km² per pair (Lammin-Soila and Uusivuori 1975); in central Finland, 250 km² per pair (Heinonen and Kellomäki 1971). In Leningrad region (USSR), up to 6 pairs per 25 km² of woodland (Pukinski 1977). All breeding and hunting activities carried on within territory. Birds probably attempt to establish territories in 1st autumn (Lundberg 1980b; Mikkola 1981a; see Antagonistic Behaviour, below). Territories and nest-sites mostly traditional, and, given marked site fidelity, occupied by same pairs and successors over many years: e.g. pairs occupied one territory, central Sweden, for at least 40 years, during which 5 nest-sites used; one nest-site used for 34 years (Lundberg 1979b, 1981; A Lundberg). Pairs may occupy for several years, but not breed in, territories lacking suitable nest-sites (Lundberg 1981). If territory changed, usually close to former one: in 3 cases 700 m, 800 m, and 1300 m away (Lundberg 1979b). Sometimes breeds in unusually close proximity to Tawny Owl S. aluco (Lammin-Soila and Uusivuori 1975; A Lundberg); once, 200 m from Great Grey Owl S. nebulosa (Wahlstedt 1969a). ROOSTING. Diurnal. Outside breeding season, probably solitary, typically in conifers. In breeding season, ♀ roosts on or near nest, ♂ elsewhere in territory (A Lundberg). Both sexes constantly change roost-site, and consequently no accumulation of pellets (Glutz and Bauer 1980). In captive study, once young hatched, ♂ switched from site in cover to sentinel post nearer nest (Scherzinger 1980, which see for roosting behaviour of pair and family). See also Relations within Family Group. For periods of hunting activity, see Food. In April–May, short but intense bout of vocal communication between ♂ and ♀ (see Heterosexual Behaviour, below), often occurred just before sunrise, probably signalling occupation of daytime roost (Lundberg 1980a).

2. Following account of alarm reactions from Scherzinger (1980, see for drawings). When perched bird alarmed and aware of being watched, may adopt concealing Sleeked-upright posture, as in other Strigidae: effects slim appearance by stretching upwards on extended legs, and sleeking plumage; 'folds' facial disc and almost shuts eyes; turns thus half sideways to source of threat. For posture in young, see Anti-predator Responses of Young (below). In presence of dog, perched adult adopted Attack-posture: ruffled plumage (especially on back), and drooped and spread wing nearest dog. When Buzzard Buteo flew overhead, bird cocked head backwards to gaze up at it, and drooped a wing, as above. For Attack-posture (Fig B) when about to swoop at man, and other alarm reactions, see Parental Anti-predator Strategies (below). ANTAGONISTIC BEHAVIOUR. ♂♂ markedly territorial (Pukinski 1977). Territory-owners challenge intruding ♂♂ with Advertising-call (see 1a in Voice), less often Contact-alarm call (see 2 in Voice); these frequently heard October–November, central Sweden. During breeding season, February–April, calling between neighbouring resident ♂♂ occurs only rarely. ♀♀ not heard to participate in territorial dis-

A

B

putes. Fighting, by either sex, not reported in the wild (Lundberg 1980a; A Lundberg). In captive study, threat display and pursuit-flights also occurred. In late winter, ♂ directed aggression mainly at mate, using Forward-threat posture (Fig A): bowed forwards, ruffled back and belly plumage, spread facial disc, lowered wings slightly, fanned tail, gaped, Bill-snapped, and sometimes gave Alarm-calls (see 8 in Voice); at higher intensity sometimes launched attack-flight. Attacked bird either fled with a twittering call (see 4 in Voice), or retaliated with Forward-threat posture. Aggressive behaviour similar towards territorial rivals (Scherzinger 1980); rarely, threatened bird gives Hissing-call (Wells 1912; Scherzinger 1980: see 10 in Voice). HETEROSEXUAL BEHAVIOUR. (1) Pair-bonding behaviour. Vocal advertisement (see 1 in Voice) frequent in autumn. Unpaired ♂♂, and ♀♀ which have lost mate, call more than paired ♂♂, bouts often lasting more than ½ hr (Lundberg 1980a). (2) Nest-site selection. Following account from Scherzinger (1980). Birds started prospecting sites in autumn, did so only sporadically in winter, and resumed in spring. During pair-formation, usually only one bird at a time entered nest-hole. From late February, ♂ (either perched near site or in flight) gave Contact-alarm call, thought to attract ♀'s attention to site (Nest-showing). ♀ gave similar call from March onwards, sometimes from inside hole, at entrance, or on nearby branch. From end of January, either ♂ or ♀ may initiate scraping in nest, both sexes participating; continues up to laying, and recurs if eggs or young lost. Either ♂ or ♀ flies silently to hole, and while one enters to scrape, other stares fixedly at entrance. (3) Mating. Occurs mostly in vicinity of nest-site (A Lundberg). Vocal exchange between mates typical. ♂ begins giving Advertising-call and ♀ answers with Soliciting-call (see 3 in Voice), but may switch to Advertising-call (see 1b in Voice). Duet may last for less than ½ min or up to 20 min; in one sequence involving 2 copulations, ♂ called 56 times and ♀ gave 256 Soliciting-calls (A Lundberg). In captive study, copulation usually preceded by duet (see 1, 2, 3 in Voice); after his last call, ♂ flies directly toward ♀ and lands on her back, flapping until cloacal contact achieved. ♀ adopts horizontal posture with head and tail lowered, sometimes raising head towards ♂; may give Twittering-call (see 4 in Voice) on cloacal contact. After copulation, ♀ may hold horizontal posture for 1–60 s, occasionally much longer, then start self-preening. Copulation quite brief, c. 3–5 s from approach of ♂ to dismounting. Usually 1–2 (–3) copulations in evening phase of activity, on average 41 min between copulations ($n = 17$ intervals) (Scherzinger 1980). Copulation occurs mid-March to mid-April (Lundberg 1976). (4) Courtship-feeding. Mostly confined to close vicinity of nest (A Lundberg), to which ♀ begins to restrict her activities from a few weeks before egg-laying (Holmberg 1974a). Shortly before laying, ♀ stops self-feeding and sits by nest where fed entirely by ♂ (Lundberg 1979b, 1980a). Vocal duets, similar to those at mating, associated with Courtship-feeding (Holmberg 1974a; Lundberg 1976, 1980a; see above). Prior to laying period, food-bearing ♂ announces arrival at nest with Advertising-call. ♀ answers with Soliciting-call, the two birds duetting. ♂ continues to feed ♀ during incubation and chick-rearing, latterly more often announcing his arrival with Soliciting-call. For distribution of prey between ♀ and young during brooding period, see below. When prey abundant, Courtship-feeding more frequent (Lundberg 1980a; A Lundberg). Captive, nest-showing ♂♂ cached prey at nest-site, apparently as indirect form of Courtship-feeding; ♂ typically fed ♀ after Nest-showing, after copulation, or after Soliciting-calls of ♀ or food-calls of young. During incubation, ♂ fed ♀ inside or outside nest-hole (Scherzinger 1980). In the wild, ♂ does not feed ♀ after copulation (A Lundberg). (5) Allopreening. Description here from captive birds, though recorded also in the wild (Mikkola 1983). Pair stand in close contact, side by side or face to face. Either ♂ or ♀ may initiate preening, concentrated on edge of facial disc, and around ears, eyes, and bill. Recipient sometimes ruffles back and belly plumage; either accepts Allopreening passively or may reciprocate; if unreceptive, gives Grumbling-call (see 6 in Voice) and Twittering-calls and flies off. Mutual Allopreening may last up to 10 min (Scherzinger 1980). (6) Behaviour at nest. ♀ enters nest shortly before laying, and thought to stay there throughout laying; incubates assiduously. During incubation, ♂ sometimes enters hole briefly when ♀ absent, but does not incubate (Scherzinger 1980). RELATIONS WITHIN FAMILY GROUP. From hatching, young beg with food-calls and by snapping bill, craning upwards and swaying head from side to side, initially in response to both acoustic and tactile stimuli. Before feeding young, ♀ gives Soliciting-call, less commonly Advertising-call. ♀ typically dismembers prey for small young; may give Feeding-call (see 5 in Voice) if young reluctant to feed. Young first cast pellets from 8–9 days (Scherzinger 1980). ♀ broods closely, leaving nest only once or twice per day, until young c. 7–10 days old (Wels 1912; Holmberg 1974a), at which stage (at least in captivity) ♂ begins visiting nest more frequently, staying up to 1 hr in hole; from 2nd week, ♀ broods more sporadically (Scherzinger 1980). During first 2 weeks, ♀ interrupts brooding mostly to be fed by ♂ (see above). Of 64 prey delivered to 3 nests, 30% eaten by ♀, rest by young. If ♀ off nest, she sometimes flies to nest to receive prey from ♂ or to dismember prey left there by ♂ (T Holmberg). In 3rd week after hatching, ♀ spends only a few hours per day with young, spending most time on guard, day and night, near nest. ♀ may now do some hunting, but still partly dependent on food from ♂ (Teiro 1959; Holmberg 1974a), From 25 days, captive ♀ came to nest only to feed young (Scherzinger 1980). In 4th week, in the wild, ♀ visited nest only at night (see Scherzinger 1980). At 5–6 weeks, young nuzzled up to parent when fed. From 7 days, young use bill parrot-fashion, from 10

C

days also wing-tips, to facilitate support and movement. From 20 days, start making flutter-jumps, and at 27 days can jump on to and land confidently on branches (Scherzinger 1980). Leave nest at average 25 days (A Lundberg), well before fledging, and, by alternate use of wing-flapping, claws, and bill, scale nearby trees (Scherzinger 1980, which see for details; A Lundberg). After young out of nest, ♀ stops transferring prey from ♂ to young and ♂ typically delivers food directly to them (Fig C), while ♀ (at least in first few days) remains on guard nearby. Almost all vocal communication by adults now ceases. Later, ♀ helps to feed young, but contribution not known (Teiro 1959; Holmberg 1974a; T Holmberg). In captivity, ♀ preened young out of nest (Scherzinger 1980). After leaving nest, young stay in territory at least 10–12 weeks before being evicted by parents (Teiro 1959; Scherzinger 1980). ANTI-PREDATOR RESPONSES OF YOUNG. Known mainly from captive birds. From 14 days, Bill-snap at anything unfamiliar; from 15 days, give Hissing-call (see Voice) if observer makes any sudden movement. Until 3 weeks old, however, mostly passive in nest, crouching if threatened. From 3 weeks, show elements of adult Forward-threat posture: at 20 days, alarmed hand-reared bird spread primaries and raised secondaries, hissed, and Bill-snapped. At 31 days, young threatened by dog adopted low hunched posture and raised wings (Scherzinger 1980). In last week in nest, young may apparently defend more actively, turning on backs and raising claws (Wels 1912). On hearing Warning-call, young out of nest typically fall silent (Holmberg 1974a). If discovered, may adopt Sleeked-upright posture similar to adult (see above), but turn side-on to intruder and look over bend of wing; full posture adopted by 73-day-old young in presence of dog. May also threaten by drooping and fanning one wing, as in adult (Scherzinger 1980; see above). For calls associated with fear, defence, etc., see Voice. PARENTAL ANTI-PREDATOR STRATEGIES. (1) Passive measures. No details. (2) Active measures: against man. ♀ rarely aggressive when incubating, but markedly so once young hatch, and after young leave nest (A Lundberg). However, considerable individual variation reported (e.g. Danko and Švehlík 1971, Kislenko and Naumov 1972); some incubating ♀♀ very aggressive (A Lundberg). In central Sweden, when intruder approaches nest containing well-grown young, or young out of nest, ♀ often performs distraction-flight: flies away from nest area and perches in clear view, such that intruder between her and nest or young. Often watches in silence, but sometimes gives Alarm-call or Distraction-call (see 8 and 9 in Voice). If approached, ♀ flies conspicuously away, perches again, and so on, finally disappearing from view when intruder well away from nest or young. If nest closely approached, ♀ Bill-snaps and often gives Alarm-calls. When attack imminent, adopts Attack-posture (Fig B): ruffles plumage, notably on breast, droops one wing, and fans tail; gives Alarm-calls and Bill-snaps. Attack usually withheld until intruder climbs nest-tree or closely threatens young. ♀ directs fierce dive-attack at head and shoulders, usually striking with claws and capable of inflicting serious injury (Holmberg 1974a; T Holmberg, A Lundberg). Attacking bird may give a twittering call (Scherzinger 1980: see 4 in Voice). Rarely, ♂ joins ♀ in attack (Holmberg 1974a). In Finland, ♀ made several attacks, then perched c. 20 m away, slowly turned head away, and closed eyes (Otto-Sprunck 1967). In Czechoslovakia, bird flew from nest-area towards intruder 80–100 m away; bird stayed perched 30–40 m away as long as intruder looked at bird but swooped as soon as back turned (see Scherzinger 1980). In captive study, ♂♂ usually shyer, often adopting Sleeked-upright posture when nest inspected, but attacking from behind when intruder's back turned; ♂ cornered in nest-hole pecked, clawed, and Bill-snapped (Scherzinger 1980).

(Fig A from photograph in Teiro 1959; Fig B from photograph in Pukinski 1977; Fig C from photograph in Mikkola 1972a.)   EKD

Voice. Freely used in breeding season, when perched or in flight, especially for pair-formation and food transfer; also outside breeding season during territorial establishment. ♂ and ♀ share diverse repertoire, ♀'s calls harsher and lower pitched (A Lundberg). Repertoire similar to that of Tawny Owl *S. aluco* (see Scherzinger 1980 for differences). Bill-snapping sound produced as in Great Grey Owl *S. nebulosa* (Scherzinger 1980); for context, see Social Pattern and Behaviour. Important studies by Holmberg (1974a), Lundberg (1980a), and, for captive birds, Scherzinger (1980). Following account includes translations supplied by T Holmberg and A Lundberg. In all renderings from Swedish, 'v' pronounced 'w'.

CALLS OF ADULTS. (1) Advertising-call. (a) ♂'s call. A deep hooting 'vóhu voho-hovóho' (Holmberg 1974a); in tone and pitch, resembles Stock Dove *Columba oenas* (A Lundberg). Also rendered 'buhūū būhulo-būhu', the first 2 units typically separated by c. 2–3 s (Scherzinger 1980). Recording (Fig I) suggests 'HU-OOO HOOO hu-HOO-hoo' (J Hall-Craggs, P J Sellar). Typically c. 15–50 s between calls (Holmberg 1974a). Calling rate increases near nest, and prior to copulation; maximum reported 75 calls in single bout, but usually less than 15 (A Lundberg). Given in territorial defence, but most often in communication with ♀, e.g. during Courtship-feeding and copulation (Holmberg 1974a; Lundberg 1980a; see call 3). Hollow and echoing towards rivals, often soft and sonorous towards mate (Scherzinger 1980; T Holmberg). Audibility varies from a few metres to c. 2 km (Lundberg 1980a). (b) ♀'s call. Similar to ♂'s but a much harsher 'vehä vehä hävähä' (Holmberg 1974a). Harsh, croaking sound; sometimes comprises only 1st, or else only 2nd and 3rd units (Scherzinger 1980). Given less often than ♂'s call; recorded sporadically in communication with ♂, especially during copulation; sometimes also in extreme anger, e.g. if disturbed at nest (Holmberg 1974a). (2) Contact-alarm call. ♂'s call 'wowowowowowo' (Holmberg 1974a); series of 3–5(1–8) 'buh' sounds, longest bout 5 min (Scherzinger 1980). Recording (Fig II) suggests 'hoo hoo hoo hoo hoo hoo hoo', with marked crescendo up to 4th unit, then falling at end, repeated at intervals of c. 13 s (E K Dunn, J Hall-Craggs). Audible up to 1·5 km (Lundberg 1980a), or soft and sonorous (Scherzinger 1980). ♀'s call a hoarser, more subdued 'chächächächächächä', given less often than ♂'s call (Holmberg 1974a); a rather harsh 'chrochro...' (Scherzinger 1980). Function obscure; given in variety of contexts, not uncommonly alternating with Advertising-call, especially when bird agitated, e.g. if human imitates Advertising-call of ♂; also used in communication between ♂ and ♀, sometimes during Nest-showing, and by ♀ during distraction-flight (Lindblad 1967; Hedvall and Wahlstedt 1969; Holmberg 1974a; Lundberg 1980a;

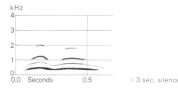
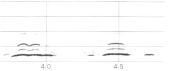

I P J Sellar Sweden May 1970

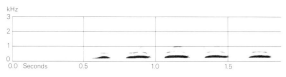

II P J Sellar Sweden May 1970

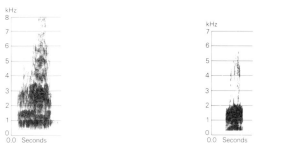

III S Palmér/Sveriges Radio (1972) Sweden June 1964

IV P J Sellar Sweden May 1973

T Holmberg). In captive study, given mainly by ♂, less often ♀, during nest-site prospecting (Scherzinger 1980). (3) Soliciting-call. Harsh, barking, typically disyllabic 'kuwäck' (Fig III), repeated at intervals of 1–5 s (Scherzinger 1980); also rendered 'veäv' or 'chäu' (Holmberg 1974a). Frequently heard, mostly from ♀, often monotonously for up to ½ hr, usually in response to—and sometimes duetting with—Advertising-call of ♂ (Holmberg 1974a), sometimes also Contact-alarm call of ♂ (Scherzinger 1980). Given when begging (and often after receiving) food from ♂, and during copulation. ♂ sometimes gives similar call to ♀, especially during chick-rearing, less often before egg-laying (Holmberg 1974a). (4) Twittering-call. High-pitched, twittering 'vivivivivi' (Holmberg 1974a). Given by both sexes, probably usually ♀, in various contexts, notably by copulating ♀, when quality shrill; also when Allopreening. Probably expresses discomfort or discontent, as in young. Soft, whining, twittering sound when attacking or fleeing from intruder perhaps the same (Scherzinger 1980). (5) Feeding-call. Loud, but muffled rattling 'tschrrt', up to 1 s long, not uncommonly given by ♀ to young, typically when young are reluctant to accept food (Scherzinger 1980). (6) Grumbling-call. Quiet grumbling sound up to 1 s long given, mainly by ♀, when inspecting nest-site, Allopreening, or after copulation. Thought to contain aggressive elements, and may change to calls 1 or 7 (Scherzinger 1980). (7) Anger-calls. In captive study (Scherzinger 1980),

where closer contact perhaps promoted more intense confrontations, a variety of shrill, penetrating, screeching sounds, mainly by ♂ during territorial disputes, mixed together with Advertising-call. (a) 'korah', stress always on 2nd syllable, associated with Forward-threat posture; (b) 'chroooh', up to 2 s long, usually delivered in crescendo series; (c) clear, echoing 'guō', sometimes given with increasing rapidity and ending in cackling 'gogogok'. (8) Alarm-call. ♂'s call a brief, forceful barking 'vå' or 'våvā'; ♀'s call 'kwå, or 'kwå-vå' (Holmberg 1974a). Recording (Fig IV) suggests a rather nasal 'wuk' (J Hall-Craggs, P J Sellar). When disturbed at nest, and especially by sentinel bird when young out of nest threatened, call typically disyllabic, with 2–4 s between units, and 3–8 s between calls (Scherzinger 1980); on hearing call, young fall silent (Holmberg 1974a; Scherzinger 1980). When threat imminent, accompanied by Bill-snapping and threat postures, and may precede attack (Scherzinger 1980; A Lundberg). Several closely related variants reported (Holmberg 1974a; Scherzinger 1980). (9) Distraction-call. Long-drawn, rather low whistling 'schiiiiii' (Holmberg 1975a) or 'eeeeeh' (T Holmberg), sometimes given, at least by ♀, near nest during distraction-flight. (10) Hissing-call. Hissing sound, given with bill slightly open. Used rarely—when threatened by conspecific birds, etc. (Wels 1912; Scherzinger 1980).

CALLS OF YOUNG. From day of hatching, contact and food-calls a mixture of shrill, cheeping 'ptzipp', similar to *S. aluco*, and thicker, wheezy 'ptjääp' (Holmberg 1974a). Among several variants reported by Scherzinger (1980), equivalent sounds probably 'SZip' or 'kiSZip', and 'psjiet'. For other renderings, see Schmidt (1966). Food-calls of young, c. 40 days old, audible at 300–400 m (Steinfatt 1944). By 12th week, 'tschwök', similar to food-call of fledged Eagle Owl *Bubo bubo* (Scherzinger 1980). Small young express discomfort (e.g. when cold), with twittering 'sisisisi'; if alarmed or squabbling over food, when in or out of nest, a powerful chirping 'tsrtsrtsr tsrtsrtsr...'. In Sleeked-upright posture, one bird gave a high-pitched, warning 'chiep'. From 14 days, alarmed young Bill-snapped loudly, and from 15 days, gave Hissing-call (Scherzinger 1980, which see especially for development of adult calls in young). EKD

Breeding. SEASON. Dependent on population cycles of voles *Microtus*. For average period in Scandinavia and western USSR, see diagram; little variation over rest of range. SITE. In hole in tree or stump, nest-box, old nest of other species, especially raptor (*Accipiter, Buteo, Pernis*)

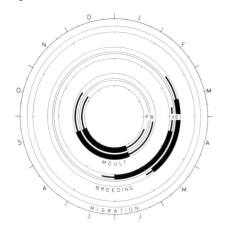

or crow *Corvus*, also squirrel *Sciurus*; also recorded nesting in building, on ground, and on rock face (Lahti 1972). Of 250 nests, Finland, 1870–1969: nest-box 33%, old nest of other species 28%, stump 28%, hole in tree 9%, building 1%, on ground 0·5%, on rock face 0·5%; use of nest-boxes increased over this period, replacing natural holes; use of old nests of other species also increased, with Goshawk *Accipiter gentilis* and Buzzard *Buteo buteo* providing over 50% of such sites; nest-boxes more commonly used in southern Finland, stumps in central and northern Finland; of 54 trees in which old nests of other species were used, 82% conifers and 18% broad-leaved; nests averaged 9·5 m (3–16) above ground (Lahti 1972). Nest: available hole or old nest of other species; no material added, though hole may be cleaned. Building: see above. EGGS. See Plate 97. Blunt sub-elliptical, smooth and not glossy; white. *S. u. macroura*: 50 × 42 mm (48–53 × 40–43), $n = 36$; *liturata* not significantly different. Weight 46 g (39·8–51·1), $n = 37$ (Glutz and Bauer 1980). Clutch: 2–4 (1–6), varying with food supply, especially numbers of voles. In poor vole year, Finland, clutch size $2·5 \pm SD\ 0·6$, $n = 4$; in good year $3·3 \pm SD\ 1·2$, $n = 24$ (Linkola and Myllymäki 1969). Of 87 clutches, central Sweden: 1 egg, 5%; 2, 42%; 3, 30%; 4, 18%; 5, 5%; average 2·8 (A Lundberg). Clutch size declines through season. One brood. Replacements laid after egg loss in *macroura* but not recorded in *liturata* (A Lundberg). Laying interval 1–3 (–5) days (Glutz and Bauer 1980; A Lundberg). INCUBATION 2·7–29 days (*macroura*), 31–34 days (*liturata*) (Glutz and Bauer 1980; A Lundberg). By ♀ only. Begins with 1st or 2nd egg; hatching asynchronous. YOUNG. Altricial and nidicolous. Cared for mainly by ♀, and brooded continuously for up to 2 weeks and intermittently thereafter (A Lundberg). FLEDGING TO MATURITY. Fledging period *c*. 40 days, but young leave nest before that, at *c*. 28 days (24–32) (*liturata*) and 34–35 days (*macroura*), or earlier if disturbed (Glutz and Bauer 1980; A Lundberg). Age of independence not certainly known, but young remain in parental territory for 2–3 months (A Lundberg). May breed at 1 year, but most not until 3–4 (Lagerström 1969; Scherzinger 1975). BREEDING SUCCESS. Greatly affected by food supply. In poor vole years, average brood size of young older than 1 week $1·5 \pm SD\ 0·6$, $n = 4$; while in good vole years $2·7 \pm SD\ 1·0$, $n = 40$ (Linkola and Myllymäki 1969). Of 203 eggs, central Sweden, 80% hatched, with most losses of complete clutches (15% of 62) probably due to food shortage causing ♀ to cease incubation for prolonged periods; of 162 young hatched, 98% fledged, giving overall productivity of 0·78 young per egg laid. In period 1969–82, 0–88% of pairs attempted to breed and 0–75% of pairs successful. Average number of young produced per successful pair 2·4 (1·0–3·4), $n = 86$, but of all pairs 1·1, $n = 212$ (Lundberg 1981; A Lundberg).

Plumages (*S. u. liturata*). ADULT. Crown, hindneck, and sides of neck broadly streaked dark olive-brown or black-brown and cream-buff or white, brown streak on centres of feathers sometimes widening on tips, and pale feather-sides then reduced to large spots. Mantle and scapulars with broad and ill-defined dark brown or black-brown shaft-streaks, streaks wider and paler olive-brown or grey-brown on feather-bases and sometimes also widening on feather-tips; sides of feathers cream-buff or white, often partially with indistinct brown or pale grey speckling or marbling; central streak on outer scapulars dark and sharp, contrasting with large uniform white spot on tip of outer web. Back, rump, and upper tail-coverts uniform dark brown, olive-brown, or tawny-brown, longer upper tail-coverts indistinctly and irregularly banded off-white and grey. Facial disc silvery-white, cream-buff, or buff-brown, fully white between eye and base of bill and on chin, feathers with brown-black shafts ending in bristle-like tips; fully black bristles at sides of bill. Disc surrounded by distinct ruff, consisting of narrow black-brown feathers with rounded tips, latter spotted or fringed white or pale cream-buff. Chest, breast, belly, and flanks white or pale cream-buff, feathers with broad dark olive-brown shaft-streaks on chest and breast, slightly narrower ones on belly and flanks; feather-tips sometimes faintly barred grey, uniform tawny-buff or pale cinnamon of feather-bases often partly visible. Vent and shorter under tail-coverts uniform cream-white or tawny-buff, longer under tail-coverts white with pale brown or grey bands and terminal arrow-mark. Feathering of leg and foot pale brown, tawny-buff, or cream-white, closely and narrowly barred with darker brown or rufous-brown, but sometimes scarcely spotted only. Tail black-brown, dark sepia-brown, or dark olive-brown with 4–5 broad pale grey-brown or grey bands, latter sometimes reduced to large spots or tending to form streaks on terminal half of feathers; grey of bands often grading to white on inner webs and often partially surrounded by white line at border of dark brown; tips of feathers pale grey-brown or grey for 2–3 cm, with narrow white terminal fringe, indistinct darker brown bar, and paler grey or off-white speckling or marbling. Flight-feathers rather evenly banded black-brown or dark sepia-brown and pale grey-brown or pale grey, only basal parts of inner webs uniform pale grey or white without dark bands; grey bands often narrowly broken by dark brown along shafts, grey of distal halves usually slightly darker than that of bases, sometimes slightly tinged rufous or with off-white line at border of dark brown; feather-tips rather uniform grey for 1·5–3 cm, except for ill-defined white fringe along tip and often for paler grey or off-white bar at border of dark brown subterminal band. Lesser upper wing-coverts mainly uniform dark grey-brown, dark olive-brown, or dull rufous-brown, some indistinct grey or off-white spots showing occasionally; median coverts streaked like scapulars, greater banded like secondaries, both median and greater with large and contrasting white spot near tip of outer web; upper primary coverts and bastard wing banded like primaries. Under wing-coverts and axillaries tawny-buff to cream-white, rather narrowly streaked black-brown, rather contrasting with broadly white and black-brown banded undersurface of flight-feathers; tips of greater under primary coverts broadly black. Some individual variation in tinge of ground-colour and in extent of dark shaft-streaks; some birds are largely dark olive-brown above, with rather narrow tawny-buff streaks or spots on crown, hindneck, mantle, and scapulars, larger white spots restricted to outer scapulars and median and greater upper wing-coverts, facial disc pale buff-brown, and underparts mainly tawny-buff with rather broad dark streaks and white spots on feather-sides; others have dark streaks on upperparts rather narrow with much white or pale cream on sides of feathers, disc mainly silvery-white, and underparts white with rather narrow dark streaks. ♂ reported to be slightly darker than ♀ (Mikkola 1971b), but no evidence for this in small series examined; tawny-buff ground-colour subject to bleaching, changing to cream-white or white in spring and summer. DOWNY YOUNG. 1st down (neoptile) rather short, scanty, buff-white or dirty grey; not as white, dense, and short as in Tawny Owl *S. aluco* (Scherzinger 1980). JUVENILE. 2nd down (mesoptile) starts to grow from 5th–8th day, completed 12th–16th day, first on scapulars and upper wing-coverts; neoptile clings to down-tips of mesoptile, but gradually lost in 3rd week and only traces remain up to 4–5 weeks on cheeks, brows, belly, lower flanks, vent, and under tail-coverts. Mesoptile rather variable in colour, whitish-grey, hazel, dirty yellow, or dull brown; tipped with white and with broad but ill-defined dark barring showing from 11th–16th day; barring gradually fades and hardly visible from age of *c.* 2 months, last on crown and hindneck. Flight-feathers start to develop from 5th day, but tail not until leaving nest on day 29–35; both full-grown at *c.* 80th day. Differs from *S. aluco* in darker, broader, and less distinctly defined barring, distinctly defined ruff (from 26th day) with more contrastingly patterned facial disc, and more protruding bill (Scherzinger 1980). FIRST ADULT. Like adult, but juvenile tail, flight-feathers, greater primary coverts, and frequently some greater upper wing-coverts, tertial coverts, or some under tail-coverts or feathers of vent or thigh retained. Juvenile flight-feathers as in adult, but all primaries of similar age, or else longer ones slightly more abraded than inner 3–4 and outer 2–3 (in adult, mixture of old and new feathers). Retained juvenile innermost tertial (s13) narrower, more tapering, and with more pointed tip than adult, sides fluffier; usually 3–4 cm shorter than longest tertial (s11) (mainly 2 cm shorter in adult), tip often with partial traces of narrow mesoptile barring, not with regular broad dark bands. Juvenile tail-feathers narrower and more pointed at tip than in adult, t1 with almost uniform white triangular tip 2–3 cm long and 3–4 cm wide at base (in adult, tip of t1 broader and smoothly rounded, terminal 2–3 cm usually grey with traces of a mottled dark bar and with narrow white fringe). Retained mesoptile wing-coverts, tail-coverts, or other scattered feathers quite different from neighbouring fresh 1st adult feathers: narrower, shorter, loose-sided, and with close and narrow dark bars; some of these retained until autumn of 2nd calendar year. SECOND ADULT. Like adult, but some primaries, secondaries, occasionally a few feathers elsewhere, and perhaps sometimes tail still juvenile; retained juvenile flight-feathers more heavily abraded than usual among adults and often slightly narrower and with more pointed tips; many birds indistinguishable.

Bare parts. ADULT AND FIRST ADULT. Iris deep dark brown. Rim of eyelids dull pink-red to deep vinous-red. Bill horn-yellow, wax-yellow, or orange-yellow. Soles grey with yellow tinge, yellow, or orange-yellow; claws black or horn-brown with black tip. Intensity of colour of bare parts depends on state of excitement and food. DOWNY YOUNG AND JUVENILE. At hatching, skin flesh-pink, including foot, claws, and cere; eyes closed, bill grey-blue. During 1st week, all flesh-pink shades to grey; claws black in 2nd week, bill-tip turns to pale horn-yellow after 3 weeks. Eyes start to open from 3rd day, fully open from 7th; eye turbid milky-blue until 5th week, dark brown later on; rim of eyelids pink. Adult colours obtained during post-juvenile moult, August–September. (Glutz and Bauer 1980; Scherzinger 1980.)

Moults. Mainly based on Kohl (1975) and Scherzinger (1980), with additional data from specimens in RMNH and ZMA. ADULT POST-BREEDING. In breeding ♀, starts late April to late May, some primaries and upper wing-coverts first, followed by upperparts and part of chest and breast; some more primaries shed when juveniles fledge and then all tail-feathers lost (usually almost simultaneously—see below) late May to August; head and underparts moulted last and all head, body, and wing-coverts new August–October(–November). Adult ♂ starts 6–7 weeks later than ♀, non-breeders start May–June. Not all flight-feathers replaced every year; sequence of primaries not fully established, but moult patterns in skins examined not unlike those of Spotted Owl *S. occidentalis* from western North America (see below). Secondaries moult serially from centres at s1, s5, and s10; only tertials replaced every year, other secondaries on average once every other year. Tail-feathers not always lost simultaneously: some birds show several growing feathers with old or new ones in between, others suspend moult with scattered feathers still old or show single growing feathers as late as October (–December); method of tail-moult apparently similar to *S. occidentalis*, where tail (1) usually almost simultaneously lost (within 3–15 days) and all feathers grow at about same time, but moult occasionally either (2) disjunct, with a few scattered feathers lost May or early June and remainder following simultaneously late June or July, or (3) partial, with scattered feathers retained until next year (Forsman 1981). POST-JUVENILE. Partial: all head, body, and wing-coverts, but no flight-feathers, tertials, tail, greater primary coverts, and occasionally not all greater upper wing-coverts, tail-coverts, or feathers of vent and leg. Starts on 45th day with lesser upper wing-coverts and scapulars; in heavy moult (wing-coverts, upperparts, and face) on 60th day; underparts moulted from day 72–76; moult largely completed at day 100, about September–October (August–November); crown, neck, belly, vent, tail-coverts, and feathering of leg moulted last. FIRST POST-BREEDING. Few details for *S. uralensis*, and these not in contradiction with data of related *S. occidentalis* closely studied by Forsman (1981). Captive birds of that species replaced tail every other year, starting at age of 2 years. Primaries moulted serially from a centre on p3; sequence descendant outwards, but either 3–2–1 or 3–1–2 inwards. In 1-year-olds, also a 2nd centre sometimes active, starting somewhere on longer outer primaries (starting point perhaps dependent on state of wear, as in some immature waders Charadrii which also may start from a variable point somewhere on primaries); moult from additional 2nd centre descendant; main centre at p3 sometimes not active in 1-year-old. For instance (theoretically), in 1-year-old bird p3–p2 (1st main series) and p6 (additional series) moulted, in next year p1 and p4–p5 (1st main series) as well as p7–10 (additional series); when 3 years old, p3 and p1 (2nd main series) and p6–p9 (1st main series); in next year, p2 and p4 (2nd main series), p3 (3rd main series), and p10 (1st main series), etc. Completion of single series p3–p10 may take 4 moulting seasons particularly in older birds, but sometimes (1–)2 years in younger birds. Moult frequently asymmetrical between wings.

Measurements. *S. u. liturata.* Sweden, Finland, and north-west USSR, all year; skins (BMNH, RMNH, ZFMK, ZMA). Bill (F) to forehead, bill (C) to cere.

WING	♂ 358	(8·27; 14)	340–370	♀ 363	(7·54; 15)	349–376
TAIL	272	(9·67; 8)	253–282	271	(5·26; 13)	264–280
BILL (F)	39·6	(1·12; 7)	38–41	40·4	(2·58; 9)	38–45
BILL (C)	23·8	(1·71; 8)	20–25	25·6	(0·93; 11)	24–27
TARSUS	51·4	(1·52; 8)	50–54	53·9	(1·67; 11)	51–56

Sex differences significant for bill (C) and tarsus; see also Structure.

Sample of (1) Swedish *liturata*, and (2) nominate *uralensis* (Vaurie 1965):

WING (1)	♂ 355	(15)	343–360	♀ 362	(16)	345–373
(2)	353	(7)	340–375	370	(10)	360–380

S. u. macroura. (1) Soviet Carpathians, (2) eastern Czechoslovakia (Mošanský 1958); (3) Czechoslovakia, (4) Rumania, (5) Yugoslavia (Kohl 1977); skins, but also some freshly dead birds for Rumania.

WING (1)	♂ 380	(8·23; 6)	370–390	♀ 394	(9·76; 13)	377–415
(2)	378	(7·24; 12)	370–392	390	(8·49; 19)	376–410
(3)	370	(8·62; 14)	354–395	377	(9·18; 10)	360–393
(4)	373	(12·5; 63)	358–393	381	(8·18; 85)	360–400
(5)	369	(4·47; 9)	362–377	374	(8·45; 28)	355–391
TAIL (1)	293	(4·13; 5)	290–300	303	(7·93; 13)	290–315
(2)	304	(12·4; 10)	279–320	311	(11·4; 18)	280–326
(3)	292	(8·81; 11)	270–300	300	(6·83; 11)	285–314
(4)	297	(7·61; 55)	283–312	299	(7·02; 86)	283–315
(5)	293	(4·47; 10)	285–297	291	(8·43; 26)	277–313

One ♂, Rumania, had wing 403 (Eck 1971). Wings of (3)–(5) probably not fully stretched, in contrast to (1)–(2); small sample from RMNH, ZFMK, and ZMA similar to (1)–(2).

Weights. *S. u. liturata.* Finland: ♂ 590 (12) 451–825, ♀ 870 (12) 520–1020 (Mikkola 1971b). USSR: ♂ 720 (4) 680–750, ♀ 888 (5) 820–970 (Dementiev and Gladkov 1951a).
Nominate *uralensis*. USSR: ♂ 657 (3) 560–712, ♀ 950 (1) (Dementiev and Gladkov 1951a).
S. u. macroura. (1) Uzhgorod (Soviet Carpathians), September–April (Hrabár 1926); (2) Czechoslovakia and (3) Rumania, all year (mainly winter) (Kohl 1977).

(1)	♂ 663	(50·4; 22)	540–730	♀ 950	(109; 22)	720–1200
(2)	725	(102; 10)	538–900	797	(71·1; 9)	673–930
(3)	706	(112; 40)	503–950	863	(137; 57)	569–1307

At hatching, 28–32 (Heinroth and Heinroth 1931–3). Approximate weights of 2 captive juveniles on 1st day 38–40, on 5th 58, on 10th 150, on 15th 270, on 20th 410, on 24th ♂ 480, ♀ 520; adult weight reached at *c.* 6 months (Scherzinger 1980).

Structure. Wing rather long, broad, tip rounded. 10 primaries: p6 longest, p7 0–6 shorter, p8 13–23, p9 40–58, p10 96–122, p5 9–13, p4 39–45, p3 59–68, p2 75–88, p1 90–106. Inner web of (p5–)p6–p10 and outer web of (p5–)p6–p9 emarginated. Outer web of longest feather of bastard wing, of reduced outer greater upper primary covert, and of p10, and emarginated parts of outer webs of (p5–)p7–p9 with rather long and curved serrations. Tail long, tip rounded; 12 feathers, t6 45–70 mm shorter than t1. Large facial disc surrounded by distinct ruff. Eyes smaller than in Tawny Owl *S. aluco*. Bill heavier and more protruding than in *S. aluco*. Tibia rather short, tarsus and toes short; tarsus and toes feathered except soles and beyond outermost joint; middle toe without claw 36·6 (0·80; 8) 36–38 in ♂, 40·1 (2·08; 9) 38–43 in ♀; outer toe without claw *c.* 75% of middle, inner *c.* 89%, hind *c.* 65%. Claws long, but rather slender, not as strongly curved as in (e.g.) *S. aluco* and Eagle Owl *Bubo bubo*; middle claw 22·6 (1·37; 6) 21–24 in ♂, 24·0 (1·30; 10) 22–26 in ♀; outer claw *c.* 78% of middle, inner *c.* 104%, hind *c.* 84%; for structure of claws, see Höglund and Lansgren (1968).

Geographical variation. Rather slight. Nominate *uralensis* from east European USSR and western Siberia paler than Scandinavian *liturata*; ground-colour white, dark central streaks of feathers narrower, and limited grey mottling or clouding on bases and tips of feathers; scapulars, most upper wing-coverts, and underparts appearing white with narrow and contrasting streaks, face white; size similar to *liturata*; intergrades with *liturata* in Arkhangel'sk and area of Vologda, Moscow, Gorkiy, and Ryazan (Dementiev and Gladkov 1951a; Vaurie 1965). Isolated *macroura* from central and south-east Europe distinctly larger than both nominate *uralensis* and *liturata*—wing usually over 370, tail over 285; colour rather variable, on average with ground-colour similar to *liturata*, but dark marks of feathers often more extensive; however, some as pale as pale *liturata*, especially in Yugoslovia. Most birds (except some pale ones) show concentric dark marks on lower part of facial disc, especially near base of bill (prominent in 39% of 108 birds, absent in 10%; this rarely in *liturata*). A melanistic morph occurs frequently in *macroura*, e.g. in 10 out of 25 in northern Carpathians, 5 out of 194 Rumania, 20 out of 95 Yugoslavia; colour entirely deep brown or chestnut-brown, including face, often with traces of paler spotting on upperparts and faintly banded tail and flight-feathers; mesoptile dark also. (Kohl 1977.) Extralimital races gradually smaller towards east, those of Japan with average wing length 311–319; colour slightly to distinctly darker than nominate *uralensis*—in particular, populations from southern Japan, and isolated race in south-west China as dark as *macroura* or darker (Vaurie 1965).

Probably forms superspecies with Barred Owl *S. varia* from eastern North America (Eck 1971) and with latter's western counterpart Spotted Owl *S. occidentalis*. CSR

Strix nebulosa Great Grey Owl

PLATES 50 and 55
[facing pages 495 and 567]

Du. Laplanduil   Fr. Chouette lapone   Ge. Bartkauz
Ru. Бородатая неясыть   Sp. Cárabo lapón   Sw. Lappuggla

Strix nebulosa Forster, 1772

Polytypic. *S. n. lapponica* Thunberg, 1798, Eurasia. Extralimital: nominate *nebulosa* Forster, 1772, North America.

Field characters. 65–70 cm; wing-span 134–158 cm. Almost as long as Eagle Owl *Bubo bubo* but less barrel-shaped; at least 10% larger than Ural Owl *S. uralensis* with 25% longer wings. ♀ 20–25% larger than ♂. Huge, with large round head and relatively long wings and tail. Plumage almost completely brown-black and grey-white, merging into dusky grey at distance. Facial disc uniquely marked, with concentric barring around eyes, grey-white eyebrows and moustaches, and conspicuous black 'beard'. Often active in daylight. In flight, shows large pale patch at base of primaries above, and broad, dark terminal band to tail. Flight majestic, with slower action than other owls which emphasizes huge span of wings. Sexes similar; no seasonal variation. Juvenile separable.

ADULT. Ground-colour of head, back, wings, and tail grey-white, profusely marked with black-brown streaks and bars on head and hindneck; quite obvious brown blotches and fine mottles on back, dense and darker brown mottling on wing-coverts, and almost black barring over folded primaries. Pattern of marks fairly uniform, relieved most by black-streaked, white-edged scapulars, white tips to longer inner wing-coverts, and pale buff ground-colour to primaries. Facial disc large and most decorated of west Palearctic owls, with dark outline and almost concentric circular brown-black barring (forming remarkable pattern at close range), obvious, almost white, eyebrows and moustaches (creating striking central divide), black patches above and on inner side of small yellow eyes, and almost black patch below bill (with pure white patches on each side of it). Facial pattern confers both surprised and frowning expression. Ground-colour of underparts dusky-white, with heavy black-brown streaks on chest and belly, and lighter but still sharp bars obvious on chest-sides and vent. In flight, upperwing shows striking pale patch across base of outer primaries, where predominantly yellow-buff ground-colour contrasts with dark carpal patches and dominates over broad dusky bands. Underwing pale grey, finely barred on coverts but broadly barred on flight-feathers, particularly on primaries; darker carpal patch not obvious. Grey tail profusely mottled; broad dark terminal band above contrasts markedly. Legs and feet densely feathered and faintly barred. Bill deep yellow or pinkish. Eyes yellow, relatively small. If pair seen together, ♂ appears less bulky. JUVENILE. Much darker than adult, with upperparts sooty-brown (mottled with grey-white), facial disc at first confined to solid black patches around eyes and bill, and underparts dull brown (closely barred).

Unmistakable when seen well, but in brief glimpse (particularly of retreating bird not showing dark tail-band) confusable with *B. bubo* and *S. uralensis*. *S. nebulosa* greyer than either, longer tailed and less tubby than *B. bubo*, and longer winged and less barred on primaries than *S. uralensis*. Large head appears vertically cut off in profile. Flight action recalls Grey Heron *Ardea cinerea*, being slow and lacking frequent wavers of smaller species. Fearless near nest or young, often attacking intruder.

Territorial song extremely deep and booming, recalling Bittern *Botaurus stellaris*, but not far-carrying—'buoo buoo...' (given up to 12 times), with some acceleration and fall in pitch.

Habitat. In west Palearctic, in continental higher latitudes, boreal and arboreal, matching Snowy Owl *Nyctea scandiaca* as a large owl fully adapted to diurnal hunting, and Hawk Owl *Surnia ulula* in inhabiting dense mature tall forest of pine *Pinus* and fir *Abies*, sometimes interspersed with birch *Betula*. Predominantly in lowlands but also mountain forests. Hunts over adjoining moors, bogs, glades, and clearings, nesting often in old eyries of larger diurnal raptors, especially Goshawk *Accipiter gentilis*, mainly in pines in Europe but in poplars *Populus* in Siberia. (Niethammer 1938; Dementiev and Gladkov 1951a; Glutz and Bauer 1980.) In Finland, however, great majority of nests examined were in forest dominated by spruce *Picea*. In North America, studied near southern fringe of breeding area by Nero (1980) on which the following based. Ranges from stunted transition forest to subalpine and montane forests. Nests in stands of mature poplar adjacent to muskeg, and also in islands of black poplar *Populus deltoides*, aspen *P. tremuloides*, or tamarack *Larix laricina* amid a sea of spruce *Picea* and pine. Near Winnipeg (Canada), inhabits a band of black spruce *Picea mariana* and tamarack, forming a boreal forest lowland bordered with aspen parkland, partly subjected to marginal cropping, abandoned farm fields, grassy meadows, cleared forest, and old burns, with surviving pine and spruce woods, bogs, and streams affording relatively sparse forest cover; breeding occurs in stunted spruce tamarack stands less than 10 m high. Transitional aspect of such forest, altered by fire and cutting, thought to make such sites especially attractive. Experiments with supply of man-made nests in Finland and Canada have successfully increased breeding in suitable areas where scarcity of disused nests of large raptors was limiting factor (Nero 1980; Mikkola 1983).

In Finland, although nests often found in severely

cleared forests, occasionally even in solitary trees left in middle of clear-felled areas, large-scale destruction of virgin stands by modern forestry poses serious threat; on the other hand, clear-felled areas and abandoned fields are favoured hunting grounds so rapid increase of such land may at least partly counterbalance continued contraction of mature forest (Hildén and Helo 1981).

Distribution. Fluctuating with population changes; in Sweden and Finland some shift southwards in recent years.

NORWAY. Scarce and irregular breeder, mainly in north; last nested 1978. Bred Nordland 1879 (Haftorn 1971; Mikkola 1981a; CL, VR). SWEDEN. Breeding range shifted south recently, with occasional nesting even further south than mapped (Mikkola 1981a). FINLAND. In 19th century and until 1930s mainly confined to Lappland; since mid-1960s, range much more to south (Hildén and Helo 1981) with occasional nesting further south than mapped (OH). POLAND. Bred 1929–30 in east of Białowieża forest (Tomiałojć 1976a); see also Mikkola (1983). USSR. Latvia: breeds irregularly (Baumanis and Blums 1969; HV). Lithuania: bred 1825 (Ivanauskas 1964). Estonia: bred 19th century; in 20th century summering birds but no nests found (HV).

Population. Markedly fluctuating with changes in vole (Microtinae) numbers (Mikkola 1983). Apparently decreased in Sweden and Finland from late 1930s to early 1960s, then recent increase, but this may partly reflect changes in observer activity.

SWEDEN. About 100 pairs (Ulfstrand and Högstedt 1976). Marked fluctuations with numbers of pairs breeding varying from 0 to 100 in any one year, highest number of nests recorded 76 in 1981; trends unknown—apparent increase since 1960s perhaps due to increased observation (Risberg 1982; Stefansson 1983; LR). In eastern Norrbotten, rodent populations tend to peak every fourth year; when rodent numbers low, territories often maintained but few breed (Stefansson 1979, 1983). FINLAND. Marked fluctuations, with recent increase; after late 1930s, apparently almost extinct for over 20 years; since mid-1960s, population in peak years (following 3–4 year rodent cycle) perhaps several hundreds of pairs (Hildén and Helo 1981; Mikkola 1981a). USSR. Marked fluctuations; scarce (Dementiev and Gladkov 1951a).

Oldest ringed bird 6 years 11 months (Clapp et al. 1983).

Movements. Resident and nomadic. Less dependent on an exclusive year-round feeding territory than other European *Strix* owls; degree of normal winter movement uncertain, but markedly eruptive in some years.

Mikkola (1983) suggested that birds lead basically nomadic life outside breeding season, erupting to greater or lesser extent westwards and southwards through coniferous zone in response to food availability; Finnish food study did not indicate adaptability to declining prey that would facilitate sedentary existence. More conservative view (Hildén and Helo 1981) is that birds normally remain close to breeding area, moving short distances only, except when rodent populations crash from previous peak and birds then erupt to avoid starvation. Some birds, at least, remain on breeding territory in winter (Wahlstedt 1976). In USSR, has occurred in winter far to the north as well as south of taiga breeding range, and in Russian Lappland is fairly common in some years but not nesting at all in others (Dementiev and Gladkov 1951a). Winter conditions generally less severe in western extremity of range (Fenno-Scandia), where studied most thoroughly, and results perhaps not applicable to USSR taiga population.

17 Finnish juveniles moved 0–226 km (Mikkola 1983), 16 Swedish juveniles moved 0–490 km (9 up to 40 km, 7 at 100 km or more) (O Stefansson). These included nest-siblings that moved (a) 20 km ENE and 220 km north-east; (b) 40 m, 12 km south-east, and 100 km north-east. Also, 2 nest-siblings (1969) found as a breeding pair in 1981, 4 km from birthplace (O Stefansson). Longest movement (490 km north-east) from Norrbotten, Sweden (June 1970) to Murmansk, USSR (November 1970). Swedish results showed no correlation between extent of movement and rodent population levels (O Stefansson); post-fledging dispersal thus random, as in other European Strigiformes.

While a proportion of Fenno-Scandian adults certainly faithful to territory between years (e.g. Wahlstedt 1976)—for 5 years at one nest—others found up to 36 km distant and 2 Norrbotten ♀♀ further: (a) ringed June 1961, found June 1965 in Finland, 110 km north (Höglund and Lansgren 1968), (b) ringed June 1977, found May 1979 in Finland, 430 km south-east (O Stefansson). See also Social Pattern and Behaviour. ♀♀ seem more prone to death from starvation, with peak mortality in April (O Stefansson). At present too few recoveries for birds ringed as adults to indicate how abnormal are such adult displacements between breeding season. Certainly, some adults remain on territory without breeding in poor rodent summers (Stefansson 1983).

In Finland, 10 invasion winters during 1895–1943, and subsequent ones in 1955–6, 1962–3, 1963–4, 1964–5, 1968–9, 1970–1, 1974–5, 1976–7, and 1980–1. Latter the largest recorded, with hundreds reaching central and southern provinces; such numbers indicate birds coming out of USSR (Hildén and Helo 1981; Mikkola 1981b, 1983). Eruptions also reach Sweden and (fewer) Norway, but do not extend south of Baltic.

For invasions in North America, see Godfrey (1967), Green (1969), and Nero (1980).

Food. Largely small voles; also water vole *Arvicola terrestris*, shrews *Sorex*, and birds. Hunting crepuscular or diurnal (see, especially, Mikkola 1983). In breeding season, northern Sweden, hunts throughout 24-hr period (Wahlstedt 1969a). Said to be primarily nocturnal in arctic Norway (Blair 1962). At nest in northern Finland, most food brought 18.00–06.00 hrs during incubation, but all day after hatching (Pulliainen and Loisa 1977). In winter, Finland, hunts in full daylight (Hildén and Helo 1981); in south-east Canada, hunting peaks in early morning and from late afternoon to dusk; normally inactive around midday, but hunting continues throughout daylight after bad weather (Brunton and Pittaway 1971; also Nero 1980). Takes prey from beneath snow, evidently locating it by sound alone. Moves between open perches, scanning area until prey located. If near, glides straight down; if further away, gives 2–4 high wing-beats then glides directly to spot located. Often hovers at *c*. 0·5 m with legs dangling sometimes for over 10 s) before plunging vertically or at an angle into snow with wings half-closed and feet outstretched; often strikes snow head-first instead, and may completely disappear into it (see photographs in Nero 1980 and Hildén and Helo 1981). Typically then looks around, in bright sunshine holding wings over surface of snow like umbrella. Eating takes only 2–3 s: reaches down, takes prey from feet, and swallows it whole and head-first. May stay on surface of snow 5 min or more after eating. Recorded plunging through snow crust hard enough to support 80-kg man. (Brunton and Pittaway 1971; Nero 1980; Hildén and Helo 1981.) Also recorded apparently using similar means to break into burrow of gopher *Thomomys* and take it (Mikkola 1983). Hunts from perch in breeding season also (Mikkola 1976b). Once, flew directly from c. 180 m away (in daylight) to snatch small mammal from surface of snow then land c. 8 m further on (Brunton and Pittaway 1971); bill may be used to take prey in such an attack (McNicholl and Scott 1973). Flight usually below tree-top level, involving bouts of 2–3 wing-beats and much gliding; in gliding, wings recorded as level, with outer primaries uptilted (Brunton and Pittaway 1971), or as somewhat raised (Hildén and Helo 1981). Will hunt from surface of snow when no suitable perch (Brunton and Pittaway 1971). In Finland, also uses continuous searching flight like harrier *Circus* (Hildén and Helo 1981), though this never seen in south-east Canada (Brunton and Pittaway 1971). In June–July, northern Sweden, hunting success apparently least in strong sunshine and in middle of night (Wahlstedt 1969a). Cannibalism of nestlings by parents and/or siblings occurs (Höglund and Lansgren 1968).

Pellets average 63 × 29 × 25 mm, n = 100 (Mikkola 1981a); 67 × 26 mm (38–101 × 20–35), with 6·3 (2–10) vertebrates per pellet, n = 50 (calculated from Höglund and Lansgren 1968).

The following prey recorded in west Palearctic (see, especially, Mikkola 1981a; also Pulliainen and Loisa 1977). Mammals: shrews (*Sorex minutus*, *S. araneus*, *S. caecutiens*, *S. minutissimus*, *Neomys fodiens*), mole *Talpa europaea*, young hare *Lepus timidus*, red squirrel *Sciurus vulgaris*, wood lemming *Myopus schisticolor*, bank vole *Clethrionomys glareolus*, grey-sided vole *C. rufocanus*, water vole *Arvicola terrestris*, common vole *Microtus arvalis*, short-tailed vole *M. agrestis*, root vole *M. oeconomus*, muskrat *Ondatra zibethicus*, harvest mouse *Micromys minutus*, yellow-necked mouse *Apodemus flavicollis*, brown rat *Rattus norvegicus*, house mouse *Mus musculus*, weasel *Mustela nivalis*. Birds range in size from tits *Parus* to adult Hazel Grouse *Bonasa bonasia* (Mikkola and Sulkava 1970; Mikkola 1973). Also frogs (*Rana arvalis*, *R. temporaria*) and, apparently rarely, beetles (Coleoptera) and snails (Gastropoda). Recorded eating own addled eggs (Mikkola 1973) and, in irruption winter, fat put out for tits (Hildén and Helo 1981).

For summary of diet in west Palearctic, see Table A. Studies up to 1974 summarized by Mikkola (1981a); these

Table A Food of Great Grey Owl *Strix nebulosa* in Fenno-Scandia[1] as found from analysis of pellets, nest-remains, and stomachs. Figures are percentages of live weight of vertebrates only. (Mikkola 1981a, b, data reanalysed.)

	Breeding season	Outside breeding season
Mice (small Murinae)	0·1	0·9
Small voles (Microtinae)[2]	86·5	69·1
Larger mammals[3]	9·2	10·0
Shrews (Soricidae)	1·2	12·4
Birds	2·2	7·4[4]
Frogs *Rana*	0·8	0·2
Total no. of vertebrates	5177	389

1. Breeding-season data include also 130 items from Murmansk (USSR).
2. Largely short-tailed vole *Microtus agrestis* (except in Murmansk, where largely grey-sided vole *Clethrionomys rufocanus*).
3. Largely water vole *Arvicola terrestris*.
4. Composed of 1 Willow Grouse *Lagopus lagopus*.

and other studies as follows. Sweden: Höglund and Lansgren (1968). Finland: Mikkola (1970b, 1973, 1974, 1976b, 1981b), Mikkola and Sulkava (1970), Alaja and Lyytikäinen (1972), Eskelinen and Mikkola (1972), Pulliainen and Loisa (1977). USSR: Parovshchikov and Sevastyanov (1960), Mikkola (1972b).

Diet varies rather little—either between individuals or within the year. Even changes between years consist largely of alterations in proportions of small vole species, rather than in overall proportion of this group; such changes tend to follow their population levels (e.g. Höglund and Lansgren 1968, Mikkola 1973). Table A indicates shrews more important outside breeding season, though much of these data apparently from years with low vole numbers (e.g. Mikkola and Sulkava 1970). Numbers of shrews eaten can reach high levels (though weight contribution, as used in Table A, much lower): e.g. 48·7% of 154 items outside breeding season in Finland (Mikkola 1981a). Small-mammal trapping in northern Finland indicated *Microtus*, *Clethrionomys*, and *Sorex* taken approximately in line with their abundance, but *M. agrestis* taken more often than abundance indicated, *M. oeconomus* less often (Pulliainen and Loisa 1977); elsewhere, *M. agrestis* more often, *Clethrionomys* less often (Mikkola 1976b).

Not clear why *S. nebulosa* should be so large when its prey is so small (Eagle Owl *Bubo bubo* about same size but typically takes much larger prey). Even when small voles scarce or other (larger) prey abundant, diet does not substantially alter. Thus, birds found starving in winter in irruption years, Sweden, had eaten nothing except small rodents and shrews; in breeding season, water voles *Arvicola* and young hares *Lepus* almost completely ignored despite abundance in one summer (Höglund and Lansgren 1968). Possible that large size purely an adaptation to permit hunting of voles beneath heavy snow cover in winter (for techniques required, see 1st paragraph).

One incubating ♀ ate average 3·6 (3–4) voles per day (brought by ♂) over 9 days during incubation (calculated from Pulliainen and Loisa 1977), i.e. *c*. 60–80 g live weight per day. Feeding experiments indicated adult ♀ eats 162 g per day (Craighead and Craighead 1956). Estimated consumption 150 g per day (Mikkola 1983).

No information on differences between diets of adult and young. In Sweden, ♂♂ (♀♀ not hunting) delivered prey to nests with young as follows: 20.00–07.30 hrs, 3 visits (4 young in nest); 13.10–18.30, 2 (5); 11.00–18.30, 3 (?); 23.00–21.30, 5 (3); 21.00–08.00, 10 (6) (Höglund and Lansgren 1968). At nest in Finland with 4 young at least 14 days old, ♂ (mostly) and ♀ brought average 10·3 voles per day over 9 days (Pulliainen and Loisa 1977). DJB

Social pattern and behaviour. Includes information supplied by O Stefansson; also data from studies of Nearctic race, nominate *nebulosa*, especially by Nero (1980); see also Oeming (1955). For major review, see Mikkola (1981*a*).

1. Essentially solitary in winter, and many birds probably nomadic (see Movements), though less so than Snowy Owl *Nyctea scandiaca* (Höglund and Lansgren 1968; see below). If food abundant, several birds may settle in same area, and relatively high density may persist into breeding season (Wahlstedt 1974; Nero 1980). Occupy breeding territories a few weeks before laying (Wahlstedt 1969*a*). In winter, perhaps defends feeding territory, sometimes same as, or including, breeding territory (e.g. Wahlstedt 1976); in one case, winter 'home range' *c*. 45 ha (Brunton and Pittaway 1971). Family parties may remain near breeding territory long after young leave nest; exodus from nest-area faster if food short (Nero 1980; Mikkola 1981*a*; O Stefansson; see also Bonds, below). BONDS. Mating system probably monogamous; pair-bond of unknown duration, but not maintained outside breeding season (Glutz and Bauer 1980). ♂♂ perhaps sometimes bigamous (Mikkola 1981*a*, which see for evidence). In one case, siblings paired and bred (O Stefansson). Average age of first breeding not known, but one ♀ bred at 1 year (O Stefansson). Captive ♀ bred at less than 1 year (Enehjelm 1969). ♂ and ♀ divide duties: ♀ incubates, broods, and largely defends young; ♂ feeds ♀ and young in nest, and probably provides most food for young out of nest (see below). In nest, fratricide sometimes occurs (Ingram 1959), perhaps not uncommonly (Mikkola 1981*a*), though youngest chick often dies without intervention of siblings (Höglund and Lansgren 1968). Young dependent on parents until *c*. 130–150 days old (O Stefansson). BREEDING DISPERSION. Usually solitary and well spaced, but in favourable areas of mature forest, well supplied with old raptor nests suitable for occupation (see Breeding) and plentiful food, several pairs may breed in relatively small area: e.g. in Norrbotten (Sweden), 7 pairs in 20 km², 8 pairs in 50 km² (in 2 years), 9 pairs in 100 km² (O Stefansson), 7 pairs in area *c*. 3 km in diameter (Wahlstedt 1974; see also Mikkola and Sulkava 1969*a*). Distance between nearest nests sometimes remarkably little; for several cases of nests 100–400 m apart, see Höglund and Lansgren (1968), Wahlstedt (1969*b*), Mikkola (1973, 1976*b*, 1981*a*). According to Wahlstedt (1974), such concentrations 'socially induced', but more likely reflect locally abundant food (Hildén and Helo 1981). Territory size dependent on food supply (Mikkola 1981*a*); may be relatively small, e.g. territory of unpaired ♂ *c*. 800 m in diameter (Berggren and Wahlstedt 1977). In Canada, one 'territory', which presumably included feeding area, *c*. 2·6 km² (Craighead and Craighead 1956). Territory comprises essentially nest-site and immediate vicinity; used for pair-formation, mating, and nesting; often feeds outside territory, and ♂ does not exclude conspecific or other birds from such feeding areas (Mikkola 1981*a*). Relatively tolerant of neighbours during breeding season and territorial defence appears minimal (Nero 1980). Territorial boundaries strongly respected and ♂, ♀, or young do not trespass on neighbouring territory (O Stefansson). If food locally abundant in successive years, birds often show marked site-fidelity. 3 adult ♀♀ returned to same nests in 2 successive years (Wahlstedt 1976; see also Kislenko and Naumov 1972, Nero 1980, Mikkola 1981*a*). ♀ ringed breeding 1977 bred at same site 1978, 1980, and 1981 (O Stefansson); others, however, move away (see Movements). Young not uncommonly return close to natal area to breed. 1-year-old ♀ (see above) bred at natal site; 11-year-old ♀ bred 2 km from natal site, 4 km from it the following year; 7-year-old ♂ bred 40 km from natal site. Others bred up to 100 km from natal area (O Stefansson). Not uncommonly breeds close to other birds of prey (Hildén and Helo 1981, Mikkola 1981*a*, which see for examples and discussion). ROOSTING. Mostly diurnal, though often active by day (see Food). Perches close to tree-trunk, facilitating camouflage. Lowers head, closes eyes, and 'folds' facial disc, thus dividing face with vertical dark line that often matches adjacent bark pattern. During breeding season, ♂ loafs on branch near nest, but alert to any sounds, e.g. from ♀. Perched birds regularly preen; sequence often head-scratching, then body-preening, comfort-shaking, ruffling head plumage. Bird perched in open sometimes sunbathes, facing sun with eyes closed. Droops wings when too hot (Nero 1980).

2. Allows quite close approach by man in winter, but not in breeding season (e.g. Nero 1980). When aware of being watched, adopts concealing Sleeked-upright posture (Fig A): sits bolt upright, sleeks plumage, and remains motionless, as in other Strigidae (Höglund and Lansgren 1968; Wahlstedt 1969*a*). For photograph of same in young out of nest, see Mikkola (1981*a*). Sleeked-upright posture also elicited by raptor overhead (Mikkola 1981*a*). Alarmed bird typically Bill-snaps; sound made by lower mandible being thrust forward beyond bill-tip then slipped back off tip to snap against upper mandible. Hissing sound (see 7 in Voice) also made in same context (Nero 1980). For threat posture adopted by captive bird alarmed by human or cat, see Parental Anti-predator Strategies, below, also 1c in Voice. ANTAGONISTIC BEHAVIOUR. Advertising-call of ♂ (see 1 in Voice) serves partly to repel rivals. Sometimes, tail wags downwards with each unit of call (Wahlstedt 1969*a*). In Sweden, may begin calling from first mild weather in January, more usually late March, and most vocal April to mid-May (Wahlstedt 1969*a*;

A

Berggren and Wahlstedt 1977). At one site, ♂ called regularly autumn and winter, then bred there (Wahlstedt 1976). ♂ starts calling at sunset, stops around sunrise, and does not call by day (Berggren and Wahlstedt 1977). Threat-postures well-developed but not reported to be directed at conspecific birds (see Parental Anti-predator Strategies, below). In Canada, confrontations observed late February and March were possibly disputes over breeding territories. Once, ♀ perched not far from nest the latter occupied, flew at approaching bird when c. 100 m away, and birds grappled with feet in mid-air. In another case, 2 birds circled each other before closing; one made a pass at the other, then flew above it, and the two clashed in mid-air, tumbled to ground, and flew off separately (Nero 1980). Defence of winter territory uncommon, and rarely vocal. Territory-owner gave short rasping call, then Advertising-call, at bird approaching in flight, apparently causing it to change course (Brunton and Pittaway 1971). Similar incident, and aerial clash, reported by Godfrey (1967). HETEROSEXUAL BEHAVIOUR. (1) Pair-bonding behaviour. Comprises aerial display, Courtship-feeding, and Allopreening. In years when food supply reduced, pairs may occupy territories and perform some display, but make only tentative breeding attempt (e.g. Hildén and Helo 1981). Aerial display associated with Courtship-feeding which, in Canada, occurs January–April. ♂, with food in bill, approaches ♀ in slow undulating flight, alternately flapping and then gliding with wings in V, though not as high as in Short-eared Owl *Asio flammeus* (Nero 1980; O Stefansson). According to Nero (1980), aerial display also perhaps includes ritualized hunting behaviour (see Food), indicated by frenzied, repeated stoops to snow-covered ground, and by presence of 2nd bird. Aerial display also included pursuit of ♀ by ♂; once, pair—with flapping flight—spiralled upwards 10–15 m in circle c. 6–10 m across, during which each touched other's wings, then both descended steeply to trees where, after perching for c. 5 min, performed undulating circular flight, beating wings audibly against branches. Once, ♂ following ♀ braked, and fell momentarily upon ♀ in mid-air, as if trying to mount. (Nero 1980.) (2) Nest-site selection. In Canada, birds visit nests from mid-February, at least in mild winters (Nero 1980). ♂ at nest gives Advertising-call (Nest-showing) while ♀ calls (probably 4 in Voice: Nero 1980). If ♀ joins ♂ at nest, often sits and makes scraping movements (Höglund and Lansgren 1968; O Stefansson). ♂ may then fly off, followed by ♀ (Nero 1980). ♂ may thus show ♀ several sites, but final choice probably by ♀ (Höglund and Lansgren 1968; Nero 1980). (3) Mating. Little studied. Said often to occur on ground (Karalus and Eckert 1974) but, in the only detailed account available, occurred in tree. ♀ perched c. 5 m above ground on tip of dead branch. ♂ mounted, flapping vigorously for balance. At same time, one or both birds gave peculiar rasping screech, audible at c. 300 m. Shortly after, ♂ flew off and ♀ resumed hunting (Nero 1980). For calls during mating, see 2a, 2b, 4, and 5 in Voice. (4) Courtship-feeding and food-caching. As breeding season approaches, ♂ begins feeding ♀; provides all her food from c. 1 week before laying (O Stefansson). ♂ may approach ♀ with food, or ♀ may beg for food with Soliciting-call (Nero *et al.* 1974). Before presenting food, ♂ transfers it from feet to bill. Food presented, and accepted, with brief shutting of eyes. In one pair, ♂ often released food at nest only when ♀ gripped and pulled hard with head turned sideways. When prey abundant, ♂ caches appreciable amounts, once 26 lemmings *Lemmus* together, beneath favourite perch (Nero 1980). Throughout incubation, and until oldest young c. 14 days old, ♀'s entire food intake supplied by ♂ (Pulliainen and Loisa 1977). (5) Allopreening. Frequently performed by adults and immatures of both sexes. Allopreening bird alters facial disc, drawing back plumage, raising rictal bristles, stretching neck to full length, then tilting head to one side with partly closed eyes. In late February, 2 birds perched very close, and bent towards each other until markedly horizontal, whereupon each, especially ♂, Allopreened other on face and head for 15–20 min (Nero 1980). In captive study, ♂ flew to ♀'s perch and, standing breast to breast, began rubbing bill over hers, while calling softly (see 8 in Voice). Often ♂ Allopreened ♀'s face, his bill describing circular path. Allopreening occurred regularly for 8 days, at any time of day or night (Oeming 1955). Captive ♂ also lifted one of ♀'s feet in his, then groomed it with his bill. Also groomed ♀'s breast with claws of one foot. When ♀ Allopreened ♂, he adopted somewhat hunched posture, plumage half ruffled, eyes closed. Allopreening may be interspersed with self-preening and wiping of bill (Nero 1980). (6) Behaviour at nest. ♀ sits very tightly, especially when eggs hatching, also when young small; moves off very occasionally, and then only briefly, for short distance from nest, usually around midnight (Pulliainen and Loisa 1977, which see for details of comfort movements, etc., at nest). Sitting ♀ sometimes picks up and draws in twigs around nest, presumably giving rise to some erroneous reports of nest-building (O Stefansson). For calls during incubation and brooding, see especially 2, 3, and 4 in Voice. RELATIONS WITHIN FAMILY GROUP. Cheeping sounds audible from egg at least 1 day before hatching (Mikkola 1976b). When eggs hatch, ♀ not uncommonly eats shells (Grote 1933a; Pulliainen and Loisa 1977; Mikkola 1981a). Young hatch blind; eyes open on 7th day (Dementiev and Gladkov 1951a). At one nest, ♀ brooded and protected young until 14 days old, spending 99% of time at nest. When oldest young 15–24 days, ♀ off nest 13·1% of time, usually sitting nearby. In rain, ♀ protects young by spreading her wings. ♀ also preens young (and vice versa), and eats their pellets (Pulliainen and Loisa 1977). Small young, eating only flesh torn from prey by ♀, cast no pellets, but do so when older and fed whole prey; ♀ will also eat faeces of young (Nero 1980). ♂ attended only momentarily when feeding ♀, but present for longer when eggs hatched (Nero *et al.* 1974). Young beg by stretching upwards and giving food-calls. ♀ plunges flesh food into gape of chick which clamps its mandibles on each side of ♀'s bill and pulls back, stimulating ♀ to release food (Nero 1980). Up until oldest young c. 14 days, only ♂ brought food, transferring it from bill to ♀'s bill for distribution. If ♀ refused prey, ♂ usually left it at nest or flew off with it (Pulliainen and Loisa 1977). After c. 14 days, ♂ increasingly gives prey directly to young. If young reluctant to feed, ♂ sometimes offers it to ♀ who typically sits near nest, guarding young (O Stefansson). At c. 14 days, young start exercising wings at edge of nest, and fixing stare on moving objects (Mikkola 1981a). One young first clambered from nest at 20·5 days, but returned, and finally left nest the next day. Young flightless on leaving nest, but make flutter-jump to ground. Out of nest, young beg noisily and persistently (Pukinski 1977). ♀ in constant attendance (Mikkola 1981a; see below for role of sexes). ♀ lured young away from nest area with a repeated contact-call (Nero 1980: see 2a in Voice). Young out of nest

PLATE 54 (*facing*).
Otus brucei Striated Scops Owl. (p. 450). *O. b. exiguus*: 1 ad. *O. b. obsoletus*: 2 ad.
Otus scops scops Scops Owl (p. 454): 3 grey ad, 4 brown ad.
Glaucidium passerinum Pygmy Owl (p. 505): 5–6 ad.
Aegolius funereus funereus Tengmalm's Owl (p. 606): 7–8 ad.
Athene noctua Little Owl (p. 514). *A. n. vidalii*: 9–10 ad. *A. n. lilith*: 11 ad. (HD)

quickly and readily climb sloping tree-trunks, stumps, etc., for safety (Pulliainen and Loisa 1977; see also Höglund and Lansgren 1968, Mikkola 1976b, 1981a, Nero 1980). Young visited in perch-sites by parents for feeding. When young satiated, they fell into apparently deep sleep, lying crosswise on branch, with head and tail hanging limply down (Wahlstedt 1969a). Following account from O Stefansson. In dense forest, young leave nest at earlier age than in open forest; also leave earlier if food scarce. Oldest young may leave first, so that brood of (e.g.) 4–6 may depart over 1 week. Disperse only at night. Older young disperse fastest and climb highest. During first few days, stay low (2–5 m) in trees, but after 1 week or more, not uncommonly reach 10–15 m. Sometimes, each takes refuge in separate tree, but often 2 young on same branch. In c. 2 weeks, young may be spread c. 100 m around nest. In open forest, young more mobile, and may disperse 300–800 m before reaching safety. Often move towards ♂'s feeding area, and may switch refuge site appropriately if ♂ moves to new area. Until young c. 45–60 days old, only ♂ hunts for them, but thereafter ♀ may also help to feed them. Young start flying to meet parent approaching with food. Young able to fly properly at c. 60–65 days. Not independent until c. 130–150 days old. ANTI-PREDATOR RESPONSES OF YOUNG. Warned by calls (see 6a, b in Voice) and Bill-snapping of parent, young fall silent (Höglund and Lansgren 1968). Young said to show no aggression, even when handled, up to 10 days old (Mikkola 1981a), though small young reported to rear up in nest and Bill-snap and give Hissing-call, as in adult (Nero et al. 1974; Nero 1980). Increasingly aggressive from 21 days, pecking and clawing when handled (Mikkola 1981a). On approach of human intruder, 3 fledged young, initially 30–60 m apart, gathered until only 5–10 m apart; finally flew when intruder 30 m away (Jensen et al. 1982). PARENTAL ANTI-PREDATOR STRATEGIES. (1) Passive measures. ♀ sits tightly (see above), sometimes even when touched (Nero et al. 1974; Nero 1980). (2) Active measures: against birds. Raptors such as Goshawk *Accipiter gentilis*, Peregrine *Falco peregrinus*, and crows (Corvidae) flying near nest

B

caused ♀ to fly immediately to perch on branch of nest-tree. Her call (probably 6c in Voice) sometimes summoned ♂ to nest (Pulliainen and Loisa 1977). ♂ mounted flying attack on Rough-legged Buzzard *Buteo lagopus* from below, and chased it off (Wahlstedt 1969a). ♂ defended nest against attacking *A. gentilis* by adopting Forward-threat posture (Pulliainen and Loisa 1977: Fig B): leans forward and directs lowered head at source of threat, partly spreads and holds wings forward, fans tail, ruffles back plumage (especially scapulars), Bill-snaps, and gives Hissing-call (Oeming 1955; Nero 1980). ♀ Bill-snapped at encroaching Ural

PLATE 55 *(facing)*.
Bubo bubo Eagle Owl (p. 466). Nominate *bubo*: **1–2** ad. *B. b. ascalaphus*: **3** pale ad.
Ketupa zeylonensis semenowi Brown Fish Owl (p. 481): **4–5** ad.
Nyctea scandiaca Snowy Owl (p. 485): **6** ad ♂, **7–8** ad ♀, **9** 1st ad ♀ (1st year).
Strix uralensis liturata Ural Owl (p. 550): **10–11** ad.
Strix nebulosa lapponica Great Grey Owl (p. 561): **12–13** ad. (HD)

Owl *S. uralensis* (Wahlstedt 1969a). (2) Active measures: against man. Extremely hostile to human intruders at nest, but response differs markedly between pairs, even at same stage of breeding (e.g. Blair 1962, Höglund and Lansgren 1968), varying from Bill-snapping or distraction behaviour (see below), to flying attacks, capable of inflicting serious injury. ♀ typically more aggressive than ♂, often defending nest and young so vigorously as to prevent close approach (Höglund and Lansgren 1968). Outright attack, however, often withheld unless young badly harassed (Nero 1980). ♀ less aggressive during incubation and when young in nest able to handle prey for themselves; very aggressive at hatching and after young leave nest. On first awareness of human intruder, ♀ gives Growling- and Warning-calls (see 6a, b in Voice). May Bill-snap and give Warning-call while flying from branch to branch (Wahlstedt 1969a). When young closely approached, ♀ often attacks quite quickly (Höglund and Lansgren 1968). May attack from 50–100 m (Wahlstedt 1969a). ♀ may indicate readiness to attack by adopting Attack-posture (Fig C):

C

sits upright, droops wings, and fully exposes bill (Nero 1980); may also pant and tramp on perch (Mikkola 1981a); for associated calls, see 6c, d in Voice. Bird swoops swiftly and silently at intruder's head (Fig D); may swoop close but not strike, or

D

strike with claws, and buffet with wings or breast (Blair 1962; Pukinski 1977; Nero 1980). Sometimes ♂ and ♀ attack simultaneously, from different directions (Höglund and Lansgren 1968; O Stefansson). Distraction behaviour varied, but perhaps performed only by shyer birds (Nero 1980). Simplest form, though seldom reported, of vocally demonstrative kind. When intruders approached nest which young had recently vacated, ♀ initially circled overhead, calling (not described) vigorously (Nero 1980). (For distraction-lure display by same ♀ 2 weeks earlier, when young in nest, see below.) On close approach to nest (stage not described), ♀ uttered strong 'distress' calls, and ♂ then performed apparent distraction-flight of impeded type: when approached, ♂ turned and flew off low and heavily, just above shrub layer, periodically and suddenly dropping down, almost to ground, as if alighting. When ♂ eventually perched, he looked backwards at pursuer, and flew off in same manner when gap narrowed again (Nero 1980). Alleged distraction-display of disablement type, perhaps combining elements of both Forward-threat display and lure, reported from ♂ and/or ♀ at 3 out of 5 nests, northern Sweden, when well-grown young closely approached and handled. ♂, arriving with prey for young, flew to low stump, drooped wings, and stared fixedly at intruder while calling (see 6d in Voice); on closer approach, bird moved *c.* 10 m away to repeat display (Wahlstedt 1969*a*). In northern Sweden, when young handled at nest, ♀ sometimes performs mobile distraction-lure display of disablement type: flies to ground and begins calling (see 6d in Voice) while flapping wings rapidly, sometimes shuffling along ground, movements of body and wings feigning injury (O Stefansson; see also below for same, or similar, display). (4) Active measures: against other animals. Most observations include close human presence. When dog or cat near young, adult often adopted Forward-threat posture (Nero 1980), also Bill-snapped (Mikkola 1981*a*). When young taken out of nest, and dog barked and jumped towards them, ♀ performed intense mobile distraction-lure display of disablement type: calling incessantly (see 6d in Voice), flew down near man and dog, flinging herself wildly, sometimes headlong, against branches, wings outspread or flapping (Nero 1980).

(Figs A–B from photographs in Wahlstedt 1969*a*; Figs C–D from photographs in Mikkola 1973.) EKD

Voice. Diverse repertoire in breeding season; freely used by both sexes which share several calls, though timbre of ♀ somewhat different (Höglund and Lansgren 1968). Mostly silent outside breeding season. For Bill-snapping, see Social Pattern and Behaviour. For additional sonagrams, see Berggren and Wahlstedt (1977), Mikkola (1981*a*), and Bergmann and Helb (1982). Repertoire of Nearctic race, nominate *nebulosa*, apparently the same.

CALLS OF ADULTS. (1) Advertising-call. (a) ♂'s call. Regularly repeated series of 8–12 very deep, muffled, pumping sounds—'ho-ho-ho-', *c.* 1·5 units per s (Wahlstedt 1969*a*: Fig I, which shows 1st 7 of 12 units); 'whoo whoo whoo' (O Stefansson); deep booming 'buoo buoo buoo' (H Delin). Each series on average 6·5 s long and given at intervals of, on average 23 s (Mikkola 1976*b*), but often much shorter; according to Höglund and Lansgren (1968), at intervals of *c.* 8 s. Unpaired ♂ gave, in course of night, 310 series in 3 hrs 40 min (Mikkola 1976*b*). Towards end of series, units descend in pitch, become quieter, and are given at faster rate (Höglund and Lansgren 1968). For individual variation, see Berggren and Wahlstedt (1977). Volume relatively weak, but audible at up to 400–800 m (Höglund and Lansgren 1968; Wahlstedt 1969*a*). Typically given during territorial establishment, pair-formation, and Nest-showing; in recording (V Berggren), also after ♂ fed ♀. Sometimes outside breeding season, especially if ♂ retains territory (Wahlstedt 1976). (b) ♀'s call. Similar to ♂'s, but harsher—'chro chro chro' (Glutz and Bauer 1980); somewhat screechier and in shorter series than ♂ (Oeming 1955). Occurs rarely, and then before egg-laying (Höglund and Lansgren 1968). (c) Variants. In anger, ♂ and ♀ give weaker, often disyllabic 'ho ho' and 'chro chro' respectively, up to 100 'ho' or 'chro' units in sequence, at rate of *c.* 3 per s, and audible at up to *c.* 50 m (Wahlstedt 1969*a*). This perhaps the rapidly repeated faint 'who who who who', given by captive ♀ confronted by human or cat (Oeming 1955). (2) Contact-calls. (a) Muffled 'woo woo' by ♀ or stronger 'woo-oo woo-oo' or 'woo-oo oo-ooo-woo' by ♂ (Höglund and Lansgren 1968); low, weak 'whoo', given 1–3 times, by food-bearing ♂ to ♀ mate or young, often eliciting call 4 from ♀ or food-calls from young; also given by ♂ before copulating (O Stefansson). Same or similar call (muted 'wowk' or 'whoop') used by ♀ to lure young on ground away from

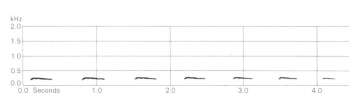

I S Palmér/Sveriges Radio (1972) Sweden May 1961

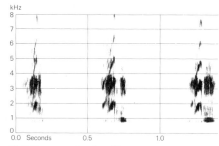

II V Berggren Sweden May 1973

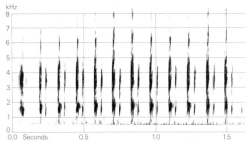

III V Berggren Sweden May 1973

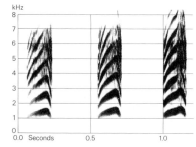

IV V Berggren Sweden September 1974

nest-site (Nero 1980). This perhaps call of ♀ attributed soft cooing quality (Oeming 1955). (b) Mewing 'njau' or 'niow' (Höglund and Lansgren 1968), given 1–2 times by ♀ at nest, presumably to contact ♂; also given before copulation (Wahlstedt 1969a). See also call 4. (c) A 'cheeh' or 'keeah' (Höglund and Lansgren 1968), given by ♀ with head bent somewhat up and back, audible at up to c. 100 m, and always answered by young (Höglund and Lansgren 1968). For variants, see Wahlstedt (1969a). ♂ gives similar call, though higher pitched and very weak (Höglund and Lansgren 1968). (d) Weak, melodious 'go-go-go-go' (Höglund and Lansgren 1968) or soft low 'kockorrock' (Wahlstedt 1969a) by ♂ on arrival at nest with food for ♀ or young. Equivalent call of ♀ a clucking 'tötötötötötö' (Wahlstedt 1969a), heard during incubation (Höglund and Lansgren 1968); in recording (Fig II), ♀ being fed by ♂ gives rapid 'tuk-tuk- . . .' (P J Sellar) following Soliciting-call (call 4). (3) Brooding-call. Purring 'korr-korr-krrr-krrr-krrr', sometimes given by ♀ brooding young after ♂ has left prey at nest (Höglund and Lansgren 1968). (4) Soliciting-call. Fairly high-pitched 'nje nje nje' or 'kje kje kje', typically comprising 3 units (Wahlstedt 1969a); also rendered 'njeet' or 'nnjet' (pronounced like Russian 'nyet'), the 'nn' sound drawn out c. 1–2 s, audible at c. 200–300 m (O Stefansson). In Fig III, transients audible between units, thus a yapping 'ap a(p)oo a(p)oo' (J Hall-Craggs). Call resembles food-call of older young (Berggren and Wahlstedt 1977), also of Tawny Owl *S. aluco* (H Delin). Given regularly by ♀, with units at intervals of c. 0·35 s, to beg food from ♂, or invite copulation: when fed by ♂, calling rate much faster, with intervals of less than 0·1 s between units (Berggren and Wahlstedt 1977). (5) Copulation-call of ♀. Twittering or whinnying sound given by ♀ during copulation (O Stefansson); probably closely related to rapid variant of call 4. Series of squealing sounds (Höglund and Lansgren 1968) also reported. (6) Alarm- and Warning-calls. Often interspersed with Bill-snapping when threat imminent. (a) Growling-call. Very deep, long-drawn, porcine grunting 'grrrrrrrrrok' like sound of grand piano without wheels being pushed over parquet floor; given by ♀ when human encroaches on nest-area; serves as warning to young which immediately fall silent (H Delin). Also described as deep, growling, vibrating 'hrrmm hrrmm', given usually by ♀, sometimes also by ♂, when young handled (O Stefansson). This probably the steady rumbling call said to be given with pumping motion of throat and apparently closed bill (Oeming 1955). (b) Warning-call. Harsh, low 'hoch-hoch-hoch' (Wahlstedt 1969a), 'grrook-grrook-grrook', thought to express greater alarm than call 6a (H Delin). Given sometimes by ♂, but mostly by ♀, e.g. flying from branch to branch when intruder threatening nest, or young out of nest (Wahlstedt 1969a). (c) Shriek-call. Loud, extended shriek said to resemble intense version of call 4, given by ♀, especially when young threatened (Nero 1980). Very rapid, high-pitched, sharp but harsh 'éééééééé', directed by ♂ at intruder (H Delin), is perhaps the same or a related call. (d) During distraction-display, a wide variety of calls, presumably same or variants of above, including vigorous mournful hooting, mewing, high-pitched wails and squeals, often ending in heron-like squawk or bark; the greater the distress, the more intense the calls (Nero 1980). ♀ treading on branch prior to attack, and ♀ injury-feigning, made snoring, whimpering, and twittering sounds (O Stefansson). In mild injury-feigning display, ♂ made wailing, moaning sounds (Wahlstedt 1969a). (7) Hissing-call. Harsh hissing sound, associated with Bill-snapping and threat (Nero 1980), as in other Strigidae. (8) Other calls. (a) Continuous, high-pitched 'oooooooooo', lasting up to c. 1 min, given by ♀ when fed by ♂ (see Wahlstedt 1969a). (b) Powerful, trumpeting 'hoyyyy', with vibrating ending, occasionally given by ♂ (Blomgren 1958). (c) ♂ Allopreening ♀ made faint humming or droning sound (Oeming 1955).

CALLS OF YOUNG. Food-call of small young, given typically in response to call 2c of ♀ (Wahlstedt 1969a), initially a feeble cheeping sound, growing stronger and more raucous with age—'che-a', most often given 3 times by older chicks, twice when younger, and more loudly when ♂ arrives with prey (Höglund and Lansgren 1968); also rendered a hoarse, hissing 'psiit' or 'pijeep' (Mikkola 1981a), 'ziepp' or sharp 'zäv' (Wahlstedt 1969a; Berggren and Wahlstedt 1977). Recording (Fig IV) suggests squealing 'ark ark ark' (J Hall-Craggs) in periodic bouts of 3–4, given in mid-September by juvenile still fed by parents. Twittering sounds may be given by young squabbling in nest (Höglund and Lansgren 1968). After leaving nest, young develop contact-call, a shrill 'ooo-ih' (Wahlstedt 1969a). For development of Bill-snapping and Hissing-call, see Social Pattern and Behaviour. EKD

Breeding. SEASON. Finland and northern Sweden: see diagram. No apparent geographical variation in USSR (Dementiev and Gladkov 1951a). Timing of start of laying varies with food supply and weather (O Stefansson). SITE. In old tree-nest of other species, especially raptors *Accipiter* and *Buteo*; also on stump and occasionally on ground; in recent years has successfully taken to artificial twig nests built in trees (O Stefansson). Of 93 nests, Finland: old tree-nest 90%, stump 5%, rock face 5%; of 84 nests of other species, 64% were Goshawk *Accipiter gentilis* and 20% buzzards *Buteo* (Pulliainen and Loisa 1977). Of 117 nests, Finland: 72% in old nest of other species (principally raptors), 6% in artificial nest, 18% on stump (1–4 m high), 2% on ground, 2% in open nest-box (Hildén and Helo 1981). Nest: available twig-nest or hole; no material added. Building: normally none, but scrape may be made on ground (Mikkola 1981a). EGGS. See Plate 97. Sub-elliptical, smooth and slightly glossy; white. 53 × 43 mm (49–58 × 40–46), $n = 153$ (Wasenius 1929). Clutch: 3–6 (1–9). Of 241 clutches, Finland: 1 egg, 1%; 2, 3%; 3, 17%; 4, 29%; 5, 35%; 6, 13%; 7, 1%; 9, 1%;

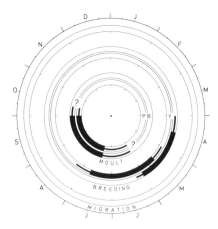

average 4·4 (Mikkola 1981a). Much variation depending on food supply, especially voles (Microtinae). One brood. Replacements laid if eggs lost soon after laying (O Stefansson). Laying interval 2–4 days (O Stefansson); one day for small clutches, but longer, even 6–12 days, between later eggs of large clutches (Blair 1962). INCUBATION. 28–30 days (O Stefansson), but 36 days recorded (Pulliainen and Loisa 1977). By ♀ only, beginning with 1st egg; hatching asynchronous. During 9 24-hr watches, ♀ off nest for only 0·6% of time, and usually between 21.00 and 01.00 hrs; normally 1 long incubation period per 24 hrs and 1–2 shorter ones; average of 4 long periods 997·5 min (651–1329) and of 21 shorter periods 181·2 min (43–487); longest absence from nest only 5 min, and average of 31 absences 2·8 min (Pulliainen and Loisa 1977). YOUNG. Altricial and nidicolous. Cared for by ♀ and brooded almost continuously for first 14 days (Hildén and Helo 1981). FLEDGING TO MATURITY. Fledging period 60–65 days, but young leave nest at 20–28 days; become independent at 130–150 days (O Stefansson). Age of first breeding probably 2 years, but one ♀ bred at 1 year (O Stefansson). BREEDING SUCCESS. Of 147 eggs laid in 42 nests, Finland, 95% hatched and 69% fledged (Mikkola 1981a). Average number fledging, Finland, in different years, 2·7–3·9 young per nest (no sample sizes given); all figures for years with adequate food supply; in poor vole years, no young fledged (Hildén and Helo 1981). Of 11 nests, 10 successful, fledging 22, perhaps 27, young (Wahlstedt 1974). In Sweden, average 3·6 young fledged from each of 21 nests (Wahlstedt 1976).

Plumages (*S. n. lapponica*). ADULT. Crown, hindneck, and sides of neck yellow-buff, cream, or white, closely barred and streaked dark grey or brownish-grey. Mantle, scapulars, tertials, back, and rump with narrow dark grey shaft-streak widening to large blob on feather-tip; sides of feathers yellow-buff or cream at base, grading to cream or white on tip, rather coarsely mottled with ill-defined grey vermiculation; extent of vermiculation rather variable, some birds appearing largely plumbeous-grey above, others white or cream with grey blobs and streaks; tips of outer webs of outer scapulars mainly uniform white, forming conspicuous row of spots except in palest birds. Upper tail-coverts closely vermiculated grey and white or cream. Facial disc well-developed, feathers greyish-white with narrow black bars, forming huge pale grey circle marked with 8–11 concentric black lines. Black semicircle round each eye broken by grey laterally, bordered by broad white band from above eye along lateral base of bill to lower cheek; chin black. A distinct circular ruff formed by rather stiff and narrow brown-black feathers, spotted and barred tawny-yellow to cinnamon-buff; portion of ruff below chin mainly black, portion below lower cheeks cream-white with narrow black streaks. Chest, breast, belly, and upper flanks white or cream with broad and rather ill-defined dark grey shaft-streaks and variable amount of indistinct barring, latter often restricted to chest and sides of breast. Lower flanks, vent, feathering of leg and foot, and shorter under tail-coverts cream-yellow or off-white with poorly defined dark barring, longer under tail-coverts white with narrow dark grey bars and vermiculation and dark grey shaft-streaks on tips. Central pair of tail-feathers (t1) grey-brown with *c.* 5 cream bands and black-brown tip; grey-brown broken into grey and cream-white vermiculations and mottling, however, and cream marbled and vermiculated grey, hence bands poorly defined; bands more distinct on t2 (especially on outer web) and tip more extensively dusky; bands gradually less vermiculated towards t6, dark bands on basal half of t5–t6 uniform grey-brown, intervening greyish-white or cream, only slightly speckled on outer web, more heavily on inner web; distal bands of t5–t6 black-brown or black, intervening paler bands and broad tip dark grey-brown, separated from black-brown by irregular cream or grey-white surround. Flight-feathers with 5–8 broad and well-spaced brown-black bars, intervening areas cinnamon-buff on basal halves of feathers, where partly speckled and clouded grey, but dark grey on distal parts of feathers and on tips, hardly contrasting with brown-black; cinnamon-buff of primary-bases slightly speckled grey only, cinnamon of outer webs soon bleaching to cream or white; pale intervening bands on secondaries more heavily mottled and clouded grey, often with pale surround at border of brown-black. Upper wing-coverts like scapulars, but sides of feathers more heavily mottled and vermiculated grey, appearing less pale; tips of outer webs of median and greater coverts often with large and contrasting uniform white spots; bastard wing and greater upper primary coverts banded brown-black and dusky buffish-grey. Under wing-coverts and axillaries white or cream with rather narrow black bars and spots, contrasting somewhat with broadly dusky-grey and cream-banded undersurface of flight-feathers. Subject to marked individual variation, and bleaching has some influence: ground-colour of both upperparts and underparts tawny-yellow, cream, or white; tawny-yellow or cream of feather-tips soon bleached to cream or white respectively, crown and hindneck appearing chequered white with only limited amount of cream feather-bases visible; width of grey bars on crown and hindneck and width of shaft-streaks and terminal blobs on all body rather variable; density of grey vermiculation and mottling on feather-sides of mantle, scapulars, upper wing-coverts, chest, and breast strongly variable. ♀♀ on average have ground-colour more yellow, ♂♂ whiter, but sexing by this character generally considered untrustworthy in view of large individual variation and influence of bleaching. DOWNY YOUNG. 1st down (neoptile) pale grey or grey-white above, white below; rather short, scanty on back and underparts. JUVENILE. 2nd down (mesoptile) develops from age of *c.* 1 week, but at first hidden under uniform greyish neoptile clinging to tips of mesoptile. Mesoptile of upperparts rather feather-like, but softer and shorter than adult feathers; on underparts loose and downy. Mesoptile of upperparts dark olive-brown to black-brown with white tips and rather narrow off-white sub-terminal bars; underparts evenly barred white and brown. During c. 2nd–5th week, facial disc forms contrasting black mask (unlike Ural Owl *S. uralensis*); disc pale grey with concentric dark bars (like adult) later on, though remainder of head and body still largely in mesoptile in 5th–6th week. Flight-feathers appear from c. 10th day, full-grown by c. 6th week. 1st-adult feathers start to grow from 4th week (at first hidden under mesoptile); 1st-adult plumage fully developed during 7th week, but head, vent, lower flanks, thighs, and tertial coverts may show some mesoptile until age of 2–3 months and occasionally up to November. FIRST ADULT. Like adult, but juvenile flight-feathers, tertials, greater primary coverts, and tail retained. Small differences from adult in colour pattern of primaries and tail-feathers bridged by strong individual variation and strong effect of bleaching and wear. Best distinguishing characters are: (1) innermost tertials rather regularly banded and barred black-brown (not with broad black-brown shaft-streak and irregular lateral vermiculation as in adult); (2) tips of tail-feathers narrowly fringed white (soon abraded on t1 but usually readily visible on t2–t5 until spring, especially from below; tail-tips not uniform dusky grey as in adult); (3) tail-feathers rather narrow and tapering to pointed tip, width (at 7–8 cm from tip) 45–55(–60) mm (in adult, broad and straight, tip broadly rounded, width 55–70); (4) all flight-feathers similar in colour and all equally abraded (in spring, longest primaries slightly more worn and bleached than outer and inner primaries; in adult, flight-feathers a mixture of old and new feathers, though difference in abrasion difficult to detect in some birds). Some heavily abraded and narrow juvenile flight-feathers retained until 2 years old.

Bare parts. ADULT AND FIRST ADULT. Iris bright yellow. Eyelids black. Bill wax-yellow or pale pink-horn, probably subject to seasonal change (Nero 1980). Cere and base dusky bluish-grey. Claws black or dark brown. DOWNY YOUNG AND JUVENILE. At 10–14 days, iris brown-grey or yellow-grey, bill brown with pink base, cere greenish-yellow with distinctly orange central part. From 3rd to 6th week, iris dull pale yellow, bare patch below tibio-tarsal joint orange; from 6th week, iris bright lemon-yellow, bill green with yellow tip, cere greenish with slight orange tinge. (Hartert 1912–21; Dementiev and Gladkov 1951a; Mikkola 1981a.)

Moults. ADULT POST-BREEDING. Flight-feathers moulted serially, primaries apparently descendantly from centre on p5 or p6. Only limited number of primaries replaced each year. For instance, may replace p5–p9 in one year, p1–p4 and p10 in next, apparently as in Tawny Owl *S. aluco* or Spotted Owl *S. occidentalis* (see *S. uralensis*). Tail either lost almost simultaneously, or only part of feathers replaced in irregular sequence (as in *S. uralensis*). No information on time of moult; no active moult in birds examined October–May, hence moult probably June–September. POST-JUVENILE. Partial: all head, body, and wing-coverts; not flight-feathers, tertials, greater primary coverts, or tail. Mainly mid-July to September, some feathers retained until about November. FIRST POST-BREEDING. As in adult post-breeding; not all juvenile flight-feathers replaced. Single bird with p5–p6 new and remainder juvenile probably in 2nd winter, another retaining heavily worn juvenile p1–p2 perhaps in 3rd.

Measurements. *S. n. lapponica.* Northern Europe, all year; skins (BMNH, RMNH, ZFMK, ZMA). Bill (F) to forehead, bill (C) to cere.

	♂			♀		
WING	446	(11·5; 8)	430–466	452	(8·30; 13)	441–467
TAIL	295	(6·28; 8)	285–303	305	(9·84; 13)	287–323
BILL (F)	41·1	(2·45; 7)	38–44	43·2	(2·19; 10)	40–45
BILL (C)	23·0	(1·40; 8)	22–25	24·3	(1·88; 11)	22–27
TARSUS	53·9	(1·73; 7)	52–56	55·1	(1·93; 7)	53–58

Sex differences not significant. Birds from Finland only: wing, ♂ 449 (17) 432–477, ♀ 463 (24) 443–483; tail, ♂ 298 (17) 276–319, ♀ 308 (26) 279–322 (Mikkola 1981a). North American nominate *nebulosa* similar to *lapponica* (Mikkola 1981a).

Weights. *S. n. lapponica.* ADULT AND FIRST ADULT. Finland: ♂ 884 (30) 660–1100, ♀ 1186 (44) 977–1900; exhausted birds during March–May, ♂♂ 585, 590, 603, 646, ♀♀ 680, 710, 870, 940 (Mikkola 1981a). Sweden and Finland (Höglund and Lansgren 1968):

	♂			♀		
Aug–Nov	846	(138; 13)	500–1050	1125	(183; 16)	700–1450
Dec–Mar	789	(175; 17)	568–1100	1159	(306; 21)	680–1900
Apr–Jun	778	(183; 16)	490–1095	1005 (217; 11)		700–1250

♀ only exceptionally over 1550; weights below 650 in ♂ and 870 in ♀ probably all from exhausted birds.

DOWNY YOUNG AND JUVENILE. All data read from graphs and approximate only. On 1st day, 37·5 (3·04; 9) 35–43; on 5th, 115 (34·6; 8) 75–167; on 10th 253 (58·4; 12) 145–330; on 15th, 418 (66·1; 10) 310–525; on 20th, 527 (83·8; 10) 375–640; often decreases slightly towards c. 25th day, increasing slightly thereafter; on leaving nest (18th–38th day, mainly 20th–25th) 553 (58·4; 9) 475–645 (Höglund and Lansgren 1968; Mikkola 1981a). Increase in 2 captive birds less rapid in first 3 weeks, c. 150 at 10 days, c. 335 at 20, c. 580 at 30, c. 690 at 40, c. 750 at 50, c. 840 at 70; sexes similar to at least age of 60 days (Mikkola 1981a).

Nominate *nebulosa.* North America: ♂ 935 (7) 790–1030, ♀ 1298 (6) 1144–1454 (Earhart and Johnson 1970).

Structure. Wing long and broad, tip rounded. 10 primaries: p6 longest, p7 0–7 shorter, p8 12–22, p9 40–55, p10 105–125, p5 8–18, p4 45–58, p3 75–90, p2 98–115, p1 112–130. Outer web of (p5–)p7–p9 and inner web of p5–p10 emarginated. Outer edge of longest feather of bastard wing, outer edge of p10, and emarginated portion of outer web of p9 deeply serrated. Tail very long, tip rounded; 12 feathers, t6 32–54 mm shorter than t1. Facial disc large, pronounced, surrounded by distinct almost circular ruff of rather stiff and narrow feathers. Eyes small. Feathers of head and body longer and softer than in any other owl, e.g. up to 17 cm on breast (Mikkola 1981a), together with relatively long tail and flight-feathers greatly exaggerating size. Bill rather strong; largely hidden by rather soft bristles growing from basal sides. Tarsus short, compared with length of tibia shorter than any other Palearctic owl except Snowy Owl *Nyctea scandiaca*; toes rather short, both tarsus and toes densely covered by hair-like feathering, except soles. Middle toe without claw 35·8 (1·92; 7) 33–38 in ♂, 38·7 (2·13; 9) 35–41 in ♀; outer toe without claw c. 66% of middle, inner c. 80%, hind c. 55%. Claws rather long and slender (see Höglund and Lansgren 1968 for curvature); middle claw 24·2 (1·44; 8) 23–26 in ♂, 26·8 (1·00; 12) 25–28 in ♀, outer claw c. 89% of middle, inner c. 107%, hind c. 98%.

Geographical variation. Rather slight, in colour only. Populations of *lapponica* from eastern Eurasia slightly darker and more contrastingly marked than those of western Eurasia, those of Sakhalin slightly darker and duller and tending to nominate *nebulosa* (Hartert 1912–21), but individual variation too large and overlap in colour too great to warrant recognition of more than 1 race in Eurasia. Nominate *nebulosa* from North America darker than *lapponica*, grey streaks on upperparts less distinctly defined and pale sides of feathers more extensively barred grey; underparts more heavily barred and mottled, less distinctly streaked.

CSR

Asio otus Long-eared Owl

PLATES 51 and 53
[between pages 542 and 543]

Du. Ransuil Fr. Hibou moyen-duc Ge. Waldohreule
Ru. Ушастая сова Sp. Buho chico Sw. Hornuggla

Strix Otus Linnaeus, 1758

Polytypic. Nominate *otus* (Linnaeus, 1758), Azores and Eurasia south to north-west Africa and Levant; *canariensis* Madarász, 1901, Canary Islands. Extralimital: *wilsonianus* (Lesson, 1830), eastern North America; *tuftsi* Godfrey, 1948, western North America east to Alberta, southern Saskatchewan, south-west Manitoba, and *c.* 100°W in USA; *abyssinicus* (Guérin-Méneville, 1843), highlands of Ethiopia; *graueri* Sassi, 1912, mountains of central and East Africa.

Field characters. 35–37 cm; wing-span 90–100 cm. Size confusing, with bulk altering markedly with posture; noticeably long and slender when alarmed but also fluffed out when at ease. Slightly smaller than both Tawny Owl *Strix aluco* and Short-eared Owl *A. flammeus*, with form and flight silhouette between those of these species. Has fairly long but not narrow wings, rather narrow head, and long ear-tufts. Plumage mixed grey and rufous-buff overall, finely marked. Facial disc warm buff, with black patches and obvious, almost white divide; dominated in daylight by glowing orange eyes. When hunched on perch, ruffled plumage, raised ears, narrowed eyes, and oblique eyebrows convey strikingly feline appearance. Flight action similar to *A. flammeus*, but wing-pattern different, lacking obvious black point. Chiefly nocturnal when breeding but often seen by day on migration. Song quiet and moaning. Sexes similar (but see Plumages); no seasonal variation. Juvenile separable.

ADULT. General tone of plumage less brown than *S. aluco*, less yellow than *A. flammeus*. Ground-colour markedly uniform, varying from golden-buff to rufous brown-buff above and only slightly paler below; most feathers broadly fringed paler buff or pale grey, freckled and well streaked over head, nape, and (more diffusely) on back where wavy bars break up pattern. Rump only dully marked and this paler than back. Square tail buff, lined with *c.* 7 straight bars on both surfaces. Facial disc complete, outlined black-brown; pale buff on outer cheeks, white on long eyebrows and bill feathering, and almost black around rather soft, glowing golden-orange eyes (darkened by enlarged pupils at twilight). Wings basically as rest of upperparts but wing-coverts not sharply marked (being blotched and freckled rather than spotted). Outer primaries closely barred over distal halves with 5–6 black-brown bars. In flight, both surfaces of wing show pattern reminiscent of *A. flammeus*, but, closely observed, following differences obvious: (1) wing-point not 'dipped' in black, but instead primaries regularly barred on over half of visible length, with no obvious black tips; (2) less wide pale panel over rufous-buff bases of primaries (covering only *c.* $\frac{1}{3}$ of their visible length, not *c.* $\frac{1}{2}$ as in *A. flammeus*); (3) duller, grey-brown carpal patch above; (4) indistinct barring on other flight-feathers, not forming pale trailing edge to secondaries. Ground-colour of under-wing buff-white, contrasting boldly with darker underbody. Underbody broadly streaked or blotched on chest and sharply streaked and barred (both boldly and finely) elsewhere. Bill dark horn. JUVENILE. Ear-tufts downy. Plumage loose and generally greyer, with barring dominant. Facial disc initially smaller and almost black, becoming brown-buff overall except close to eyes; black smudges beside bill obvious. At all ages, ears often conspicuous due to long black central panel and bright rufous-grey edges.

Unmistakable when seen well; orange eyes and complex plumage pattern eye-catching and memorable. Best separated from *A. flammeus* by: (1) broader, more rounded wings; (2) larger head, usually showing long ears; (3) more uniform, little-streaked plumage; (4) more level wing attitude (see below); (5) evenly barred outer primaries, not forming black wing-point; (6) lack of pale trailing edge to inner wing; (7) densely barred tail; (8) eye colour; (9) longer face. Risk of confusion greatest with juvenile *A. flammeus*, which has less strikingly dark chest than adult but still shows diagnostic dark wing-point. See Davis and Prytherch (1976). Relatively broader wings and shorter tail of *A. otus* particularly obvious in side view but such structural differences from *A. flammeus* introduce risk of confusion with *S. aluco*. Latter, however, lacks ears, carpal patch, striking wing pattern, obviously barred tail, and pale eyes, and has different, more direct flight. Hunting flight close to ground and fairly direct, with a few wing-beats alternating with long glides, slight banks, and occasional wavers; all actions have hint of instability (but less so than in *A. flammeus*) and are accomplished with wings set straight and usually not raised above level of body (though at lowest speed, occasionally held in extremely shallow V). Occasionally hovers. Migrant flight more purposeful, with markedly less wavering than in *A. flammeus*. Among trees near nest, or on open perch, adopts upright stance and noticeably tall and slender posture when alarmed. On partly obstructed perch or in bush, adopts less upright, hunched stance; when plumage expanded in such posture, looks much bulkier (like fat cat) until escape flight collapses shape—and much slimmer bird flies off. Normally perches high in trees but will also perch lower like *A. flammeus*.

Less vocal than *S. aluco* with restricted calling season

at peak in early spring. Song a feeble hoot 'oo', repeated every *c*. 2·5 s. Other calls include sharp, barking, disyllabic 'kwek-kwek'. HD

Habitat. In west Palearctic, ranges somewhat farther into boreal coniferous forest and somewhat deeper into Mediterranean zone than apparently competing Tawny Owl *Strix aluco*. In Britain, substantial decline since *c*. 1900 has coincided with expansion of *S. aluco*, while in Ireland and Isle of Man, where *S. aluco* absent, *A. otus* is widely distributed through arboreal habitats, especially deciduous, which it does not usually occupy in Britain. Perhaps has advantages over *S. aluco* in being satisfied with smaller patches of woodland, especially conifer plantations, and in readily using old stick nests of corvids and others. Pairs have also been found in open farmland, among low bushes on sand-dunes and marshes, and in hawthorn and elder scrub. (Sharrock 1976.) More recent study of 200 British nests showed largest group in small plantations, copses, or scattered trees on moorland, heath, or mosses (33%), closely followed by blocks of coniferous, mixed, or deciduous woods (24·5%), and small plantations, shelter belts, or hedgerows on mixed or arable farmland, rough grazing, and other pasture and meadow (24%). Wooded clumps or scrub on coasts or wetlands accounted for 15%, while open parkland figured only in small residue with waste ground and edges of built-up area. Thus appears that a high proportion of sites are in marginal zones, 10% of nests being from 305 m to 533 m in altitude (Glue 1977*b*). In USSR, extends to wooded steppe, preferring conifers but also nesting in aspen *Populus tremula* and birch *Betula* in Siberia, hunting over open spaces and also parks, orchards, and graveyards, but usually avoiding treeless areas; in southern mountains, ascends to 2750 m in Armeniya and even higher in central Asia (Dementiev and Gladkov 1951*a*). In Canary Islands common on lava flows near sea and among *Euphorbias*, but habitat varied; often in cactus-covered terrain in semi-desert, but also on ground, in evergreens and crowns of palms, in towns, or in woodland, and often nesting in holes and crannies in face of steep cliffs (Bannerman 1963*b*). In Indian subcontinent, mainly in winter from plains to *c*. 1800 m, in semi-desert as well as tall grass on waste land, low jungle of stunted trees, and poplar plantations (Ali and Ripley 1969). Optimal breeding habitat includes open spaces with short herbage and abundant prey neighbouring ample cover, especially conifer woods with wide choice of old nests (Glutz and Bauer 1980). Normally flies in lower airspace.

Distribution. Easily overlooked, so distribution imperfectly known.

BRITAIN AND IRELAND. Certainly more widespread than map suggests. Some range declines; see Population (Parslow 1967; Sharrock 1976). Bred Jersey (Channel Islands) 1979–80 (Long 1981*b*). PORTUGAL. Probably more widespread (L Palma). SPAIN. Distribution imperfectly known (AN). Bred Mallorca 1970 and 1973 (Mikkola 1983). NORWAY. Some northward spread (Haftorn 1958); occasional breeding much further north in 1970s (GL, VR). FINLAND. Breeds further north in good lemming *Lemmus lemmus* years (Merikallio 1958). TURKEY. Breeding distribution imperfectly known (MB, RFP). ISRAEL. Bred 1980, 1981; perhaps regular (HM). CYPRUS. Bred 1968–71 and probably 1963; may be regular (PRF, PFS). LEBANON. Probably breeding, but no proof (HK). SYRIA. May breed, but no proof (HK).

Accidental. Bear Island, Iceland, Faeroes, Iraq.

Population. Data limited. Breeding numbers fluctuate with those of small rodents in some areas, but decrease has occurred in Britain and possibly elsewhere, and increase in Netherlands 1850–1900 and recently Belgium.

BRITAIN AND IRELAND. 1000–10000 pairs (Parslow 1973); 3000–10000 pairs (Sharrock 1976). Steady decrease from *c*. 1900 in England, Wales, and south-west Scotland, often attributed to competition with Tawny Owl *Strix aluco*, but no apparent marked change in Ireland or Isle of Man (Parslow 1967; Sharrock 1976). FRANCE. 1000–10000 pairs; probably declined (Yeatman 1976). BELGIUM. Estimated 7000 pairs, increasing (Lippens and Wille 1972). NETHERLANDS. Estimated 5000–7000 pairs 1977 (Teixeira 1979); locally common 19th century, increased 1850–1900 with afforestation, decrease *c*. 1960 but since recovered (CSR). SWEDEN. Estimated 10000 pairs (Ulfstrand and Högstedt 1976); breeding numbers fluctuate markedly with rodent populations (LR). FINLAND. Tentative estimate 2500 pairs (Merikallio 1958). POLAND. Scarce (Tomiałojć 1976*a*). EAST GERMANY. Scarce (Makatsch 1981). CZECHOSLOVAKIA. Possibly some recent decrease (KH). ITALY. Some signs of recent decrease (SF). ALBANIA. Rare (EN). GREECE. Scarce; formerly bred Pelopónnisos (HJB, WB, GM). RUMANIA. Common (Vasiliu 1968). USSR. Abundant. Fluctuating (Dementiev and Gladkov 1951*a*). Estonia: *c*. 1000 pairs, marked fluctuations (Randla 1976; HV). Latvia: fairly numerous (Baumanis and Blūms 1972). TUNISIA. Scarce (MS). CANARY ISLANDS. Fairly common, possibly increasing (KWE, AM).

Mortality. West Germany, East Germany, and Switzerland: mortality in 1st year of life *c*. 52% and in subsequent years *c*. 31% (Glutz and Bauer 1980). Oldest ringed bird 27 years 9 months (Rydzewski 1978).

Movements. Mainly migratory in Fenno-Scandia and in USSR north of 50°N where, however, some birds winter in favourable years (when small rodents abundant and weather not severe). Further south in Europe, mainly resident (apart from post-fledging dispersal of young), birds either staying on territory all year or congregating in autumn/winter (roosting communally) in places where food abundant (see Social Pattern and Behaviour);

574 Striginae

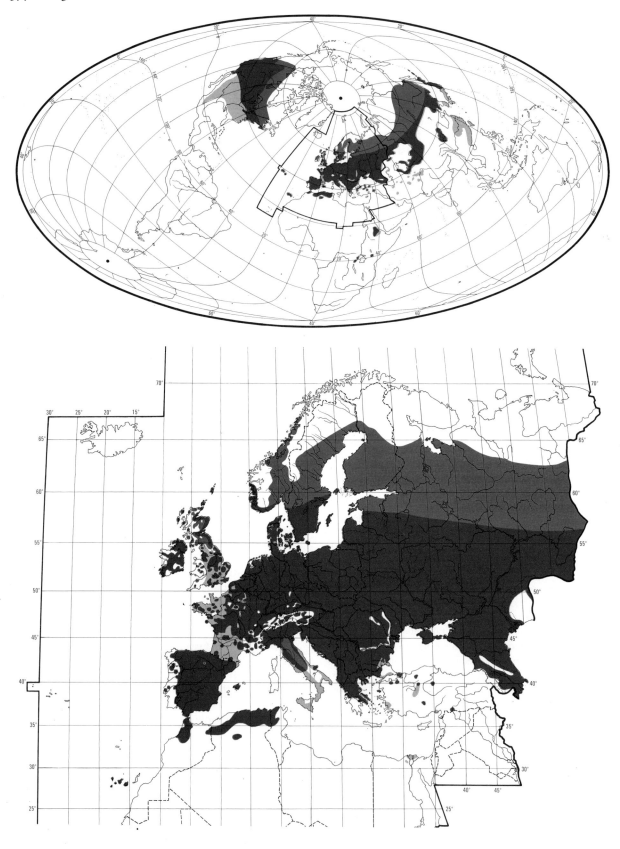

communal strategy perhaps applies especially to 1st-winter birds, though no ringing study. Situation comparable to that in North America, where, in autumn, birds withdraw completely only from northernmost breeding areas (central Canada). Winters mainly within southern two-thirds of breeding range (southern Canada and USA, temperate and southern Europe, southern Palearctic Asia), and south of it only to limited extent, a few reaching southern USA and northern Mexico, Levant (exceptionally Egypt), Persian Gulf states, northern Pakistan and India, and southern China. Separate races in Canary Islands and Afrotropical region are resident.

Mainly migratory populations of northern Europe are also subject to fluctuations linked to rodent numbers (as in other sub-arctic owls), with oscillations in breeding numbers and clutch sizes related to density of *Microtus* vole populations. In Finland, *c.* 80% of birds nomadic, breeding only when and where food sufficiently abundant (Juvonen 1976). As consequence, numbers of autumn migrants larger in some years than in others; numbers of migrants passing through Helgoland (West Germany) peak irregularly at intervals of 2–5 years (Schmidt and Vauk 1981). On Soviet Baltic coast, migrants pass mid-September to early December (peak October and early November) and late March to mid-June (peak in April and early May) (Belopol'ski 1975). On Helgoland, passage recorded early September to mid-December (peaking October) and early March to early June (peaking April) (Schmidt and Vauk 1981). At British coastal observatories also, late birds still transient in early June; these presumably non-breeders.

Swedish ringing recoveries mapped in Fig III, and those for Kaliningrad (USSR) and Finland shown in Glutz and Bauer (1980). Patterns comparable: recoveries widely scattered, but mainly south to south-west of ringing site with furthest in northern Spain; hence these birds migrate regularly to and through all North Sea countries. Migrants ringed on Helgoland found in winter between Denmark and northern France and in Britain, and in subsequent years in Norway, Sweden, Finland, and western part of European USSR (Schmidt and Vauk 1981). Continental immigrants ringed or found in Britain mainly Fenno-Scandian birds, though also minority east to 48°30′E in USSR (Mariyskaya) and south-east to Czechoslovakia (a May recovery).

In west and central Europe, main element to movement is randomly directed 1st autumn/winter dispersal by young, which travel further than those of Tawny Owl *Strix aluco* (e.g. Rockenbauch 1978b). British birds ringed as nestlings or summer (May–July) juveniles (*n* = 76) moved as follows: 0–10 km, 39·5%; 11–30 km, 30%; 31–50 km, 12%; over 50 km, 18·5%; maximum 340 km, but none abroad (BTO). However, some birds ringed as nestlings in Switzerland were found (mainly in winter) moving longer distances—to Spain, southern France, Sardinia and Italy (Glutz and Bauer 1980). Also, 4 East German nest-

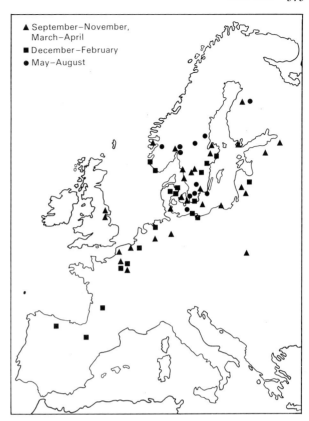

Fig III Recoveries of Long-eared Owls *Asio otus* ringed in Sweden (Österlof 1977).

lings found October–March in France (1200–1250 km south-west), and another in Portugal (November, 2140 km south-west), while West German birds have reached France and Italy, mainly in 1st winter (Hückler 1970). Of nestlings from Netherlands up to 1970 (*n* = 318), 20% recovered abroad: 2·5% West Germany, 2% Britain, 9% Belgium, 6·5% France (Speek 1973).

Rockenbauch (1978b) noted fluctuating densities between years (involving immigration) in upland breeding population of Schwäbische Alb (southern West Germany), hence indicating nomadism comparable to that of Short-eared Owl *A. flammeus*. Moreover, 5 ringed as nestlings or juveniles in Switzerland, West Germany, and East Germany (all in 1961) found in subsequent summers in USSR (Moscow, Kalinin, Tambov, Kaluga, Tula), 1700–2300 km ENE (Hückler 1970; Glutz and Bauer 1980).

Food. In typical circumstances, largely small rodents, especially voles *Microtus arvalis* or *M. agrestis*; also a few birds, larger mammals, and shrews (Soricidae); diet often more diverse in summer. Essentially nocturnal, but may hunt in daylight, especially when young in nest—then recorded, sometimes regularly, up to *c.* 4 hrs before sunset and after dawn (Dennis 1968; Tindal 1968; Hardy 1977;

Bayldon 1978). Able to locate prey in total darkness (Marti 1974; see also for vision in poor light) with accuracy of 1° in both horizontal and vertical planes (Chernyi 1974; see also for head movements involved). Typically, uses slow, steady searching flight over open ground, flying back and forth 0·5–1·5 m up; wing-beats quite fast and interrupted by frequent (usually short) glides; occasionally hovers. Flies lower than Short-eared Owl *A. flammeus*, with faster wing-beats, and without raising wings above level of body (Armitage 1968; Dennis 1968; Glue and Hammond 1974; Scott 1975). Takes prey in shallow glide (Glutz and Bauer 1980), by stalling and dropping feet-first (Glue and Hammond 1974), or by wheeling back on flight-line (Dennis 1968). Searching flight frequently used also in plantations, under tree canopy (Glue and Hammond 1974). Hunting from perch occurs more in windy weather, when hunting concentrated in sheltered areas; perches 2–4(–5) m above ground (Armitage 1968; Voronetski 1974a; Glutz and Bauer 1980). Rain and wind may reduce hunting success by 2–3 times, to 1 successful attack in 6–8(–11)—or hunting may stop altogether (Voronetski 1974a). Seen to glide low over bushes containing roosting birds, clap wings together below body, then drop into bushes (König 1963). For capture of roosting Starling *Sturnus vulgaris*, see Westin (1980). 2 birds reported flying on each side of a hedge, one taking birds disturbed by other (Flegg 1969). Once seen chasing harvest mice *Micromys* on foot in afternoon (Godin and Loison 1975). Recorded gliding over water and pushing feet below surface, possibly to take fish (Reichholf 1973). Said to kill prey in precisely the same way as Barn Owl *Tyto alba*: nips through skull just posterior to hard palate and teeth rows, so that cranium always recovered crushed and separate, whereas palatal portion of skull and jaws entire. Small mammals usually eaten whole, so leg bones and pelvis unbroken; head may be discarded and rest eaten in pieces. Small birds usually decapitated and pulled apart, wings and tail plucked; entrails usually eaten first, then breast (Ticehurst 1939). For behavioural study of reactions of captive bird to various prey types, living and dead, see Räber (1950). Food-caching behaviour poorly developed (see Social Pattern and Behaviour).

Pellets resemble those of Tawny Owl *Strix aluco*, though normally paler grey (whether dry or not) and slightly narrower; friable, with fur much more digested and broken up. As in *T. alba*, covered with mucus when fresh which dries to hard film. Pellets contain leaves, soil, etc., though less often than in *S. aluco*. (Ticehurst 1939; Thiollay 1963b; Armitage 1968; Glue 1970.) Samples of pellet measurements: usually 36–50 × 19–21 mm (22–62 × 17–24) (Ticehurst 1939); 38 × 20 mm (20–60 × 14–27), $n = 220$ (South 1966); 46 × 21 mm (20–70 × 14–27) (Araújo *et al.* 1974); average 33 × 19 mm, $n = 2484$ (Glue and Hammond 1974); see also Mikkola (1983). Each pellet contains remains of 2–3 (0–8) items (Thiollay 1963b; Hagen 1965; Armitage 1968; Araújo *et al.* 1974; Glutz and Bauer 1980); average 2·3, $n = 11\,390$ pellets (Cătuneanu *et al.* 1970). Full-grown captive bird produced, over 1 year, average 1·4 pellets per day (1·0–1·7), each pellet representing 26·7 g live weight of food (Marti 1973). Young captive bird, fed at night, produced 1 pellet per day—average $17\frac{3}{4}$ hrs (*c.* 10–$22\frac{1}{2}$) after feeding (Hagen 1965). Due to use in winter of traditional and communal roost-sites (unlike *S. aluco* in both aspects), large accumulations of pellets occur, but these unlikely ever to be complete, others being ejected during night hunting (Mikkola 1983; Wijnandts 1984). According to Raczyński and Ruprecht (1974), 21% of prey individuals consumed are not represented in pellets, but Hardy (1977) and Mikkola (1983) found no loss of small mammals fed to captive birds.

Compared with *S. aluco*, longer and narrower wings with lower wing-loading, higher aspect-ratio, and deeper and slower wing-beats form adaptation to hunting in the open; see *S. aluco*. For adaptations of auditory system, see Ilyichev (1977).

For review of prey species taken in Europe, see Czarnecki (1956), Cătuneanu *et al.* (1970), Smeenk (1972), and, especially, Schmidt (1973–4). The following recorded in west Palearctic. Mammals: shrews (*Sorex minutus, S. araneus, Neomys fodiens, N. anomalus, Suncus etruscus, Crocidura leucodon, C. suaveolens, C. russula*), moles (*Talpa europaea, T. caeca*), bats (*Rhinolophus ferrumequinum, Myotis daubentonii, M. nattereri, M. myotis, M. blythii, Plecotus auritus, P. austriacus, Pipistrellus pipistrellus, Vespertilio serotinus, Nyctalus noctula*), rabbit *Oryctolagus cuniculus*, young blue hare *Lepus timidus*, brown hare *L. europaeus*, red squirrel *Sciurus vulgaris*, grey squirrel *S. carolinensis*, European suslik *Citellus citellus*, garden dormouse *Eliomys quercinus*, edible dormouse *Glis glis*, dormouse *Muscardinus avellanarius*, common hamster *Cricetus cricetus*, golden hamster *Mesocricetus auratus*, migratory hamster *Cricetulus migratorius*, Norway lemming *Lemmus lemmus*, bank vole *Clethrionomys glareolus*, grey-sided vole *C. rufocanus*, steppe lemming *Lagurus lagurus*, water vole *Arvicola terrestris*, pine voles (*Pitymys subterraneus, P. savii, P. duodecimcostatus*), common vole *Microtus arvalis*, short-tailed vole *M. agrestis*, root vole *M. oeconomus*, snow vole *M. nivalis*, harvest mouse *Micromys minutus*, striped field mouse *Apodemus agrarius*, yellow-necked mouse *A. flavicollis*, wood mouse *A. sylvaticus*, brown rat *Rattus norvegicus*, house mouse *Mus musculus*, lesser mole rat *Spalax leucodon*, birch mice (*Sicista betulina, S. subtilis*), stoat *Mustela erminea*, weasel *M. nivalis* (Dementiev and Gladkov 1951a; Jiráčková 1963; Hagen 1965; Jensen 1968; Krzanowski 1973; Schmidt 1973–4; Araújo *et al.* 1974; Glue and Hammond 1974; Magalhães 1974; López Gordo *et al.* 1977; Veiga 1980; Radetski 1981; Gawlik and Banz 1982; Gerdol *et al.* 1982). Birds range in size from Goldcrest *Regulus regulus* to adult Moorhen *Gallinula chloropus*; however, mostly sparrows *Passer* and finches (Fringillidae), also thrushes *Turdus* and starlings *Sturnus*—unlike in *S. aluco*, larger species taken only exceptionally (e.g. Araújo *et al.* 1974, Glue and Hammond 1974). Young gamebirds (Phasianidae) rare in diet (e.g. Glue and Hammond 1974, Hardy 1977). Unlike in *S. aluco*, reptiles (lizards *Lacerta*, slow-worm *Anguis*), amphibians (frog *Rana temporaria*, toad *Pelobates fuscus*), and fish (carp *Cyprinus carpio*) eaten only rarely (Uttendörfer 1939b, 1952; Huber and Wüst 1956; Smeenk 1972; Marián and Marián 1973; Glutz and Bauer 1980; Veiga 1980). Insects: mostly beetles (including Carabidae, Dytiscidae, Hydrophilidae, Lucanidae, Geotrupidae, Scarabaeidae, Tenebrionidae, Cerambycidae) and crickets, etc. (Gryllidae, Tettigoniidae, Gryllotalpidae, Acrididae); also dragonflies, etc. (Odonata), mantises *Mantis*, Hemiptera, earwigs *Forficula*, adult and larval moths and butterflies (Lepidoptera), larval and pupal flies (Diptera), and ants (Formicidae). Other invertebrates include spiders (Araneae), centipedes, etc. (Myriapoda), crabs (Decapoda), snails (Gastropoda), mussels (Lamellibranchia), and earthworms (Lumbricidae). (Hagen 1965; Araújo *et al.* 1974; Glue and Hammond 1974; Rey 1975; Degn 1976; Jonghe 1979; Glutz and Bauer 1980; Veiga 1980; Amat and Soriguer 1981; Wijnandts 1984.) Fruits of yew *Taxus baccata* (Fairley 1967) and own eggshell (Hagen 1965) recorded in pellets; for other inanimate material, see 2nd paragraph.

Extensive literature of food analyses from west Palearctic, based almost wholly on pellet examination. However, diet and related aspects of ecology not studied with thoroughness devoted to *S. aluco*. For summary of major studies, see Table A and Schmidt (1973–4). See also the following significant studies. Britain: Ticehurst (1939), South (1966), Hardy (1977), Village (1981). Sweden: Lundin (1960), Gerell (1968), Jönsson and Schaar (1970), Källander (1977b), Ahlbom (1979), Nilsson (1981). Finland: Soikkeli (1964), Sulkava (1965). Denmark: Jensen (1968), Degn (1976). Netherlands: Tinbergen (1933), Wijnandts (1984), and review by Smeenk (1972). Belgium: Delmée *et al.* (1979). France: Thiollay (1968), Chaline *et al.* (1974). East and West Germany: Uttendörfer (1939b, 1952), Deppe (1979, 1982), Wendland (1981), Gawlik and Banz (1982); see also sources in Glutz and Bauer (1980). Poland: Harmata (1969), Kochan (1979). Czechoslovakia: Folk (1956), Jiráčková (1963). Hungary: Schmidt (1965a, 1972a, b). Rumania: Homei and Popescu (1969), Schnapp (1971), Barbu and Gál (1972). Bulgaria: Simeonov (1964, 1966). USSR: Averin and Ganya (1966), Bozsko (1967), Abelentsev and Umanskaya (1968), Bashenina (1968). Spain: Araújo (1971), López Gordo *et al.* (1977), Corral *et al.* (1979), Veiga (1980), Amat and Soriguer (1981). Portugal: Magalhães (1974). Italy: Gerdol *et al.* (1982). In southern Iraq, pellets from area of gardens and open ground, February and April, contained 83 vertebrates: by live weight, 46·0% birds, 37·2% *Mus musculus*, 16·4% *Rattus*, and 0·4% *Suncus etruscus* (recalculated from Hartley 1947). On Canary Islands, where no indigenous rodents, takes mainly *Mus musculus*; also small passerines, *Rattus*, and insects (largely crickets *Gryllus*); takes more birds than local *T. alba* (K W Emmerson and A Martin). On Tenerife, August, 81 vertebrates in pellets comprised, by weight, 58·1% *Mus*, 38·5% *Rattus*, and 3·4% birds; also 18 Tettigoniidae, 18 *Gryllus*, 1 *Mantis*, and 3 spiders (2·7% of total by weight) (recalculated from Rey 1975). Insects evidently sometimes predominate here, as stomachs recorded full of them (Bannerman 1963b).

At least in northern and central Europe, where territoriality usually limited to breeding season, with winter often being spent elsewhere, birds can afford to be adapted to rather specialized diet consisting largely of small voles (over much of west Palearctic, *Microtus arvalis* or *M. agrestis*). Of 47 sites on mainland Britain, at only 7 was *M. agrestis* not most important prey species: at 4 sites birds, at 2 *Apodemus*, at 1 *Rattus* (Glue and Hammond 1974). In Poland, proportion by weight of *M. arvalis* recorded up to 97·5% (recalculated from Harmata 1969) and 93·8% (Goszczyński 1981). Small voles widely assumed to be constantly preferred prey type, and this evidently so under most circumstances, at least outside breeding season. Little hard evidence via prey availability however, and, as small voles can become much less important food in summer (e.g. Nilsson 1981; see below), possibility remains that seasonal or other changes in prey availabilities could make other species more profitable at times, even though availability of small voles unchanged (such a situation less likely than in *S. aluco*, however). Thus, in absence of detailed studies, possible that *Microtus* in fact taken only in proportion to its availability (relative to other small rodents) in habitat used for hunting; small-scale work at Loch Mallachie, northern Scotland (Wooller and Triggs 1968) suggested this to be the case. For other studies involving some discussion of prey availability, see Hagen (1965), Armitage (1968), Hardy (1977), Goszczyński (1981), Village (1981), and Wijnandts (1984). Prey groups other than small rodents do seem to be discriminated against (perhaps except *Arvicola* and *Rattus*), and all are often taken much less than by *S. aluco*. At Amsterdamse-Waterleidingduinen (Netherlands) young *Oryctolagus* hardly taken at all despite great abundance in spring (F J Koning). Shrews rarely over 2% of diet by weight (Table A); 9·6% ($n=99$ vertebrates) recorded by Gerdol *et al.* (1982) and 12·5% ($n=191$) by Village (1981, recalculated). In area and at time when *T. alba* took 48·5% shrews by number ($n=916$) and *S. aluco* took 44·4% ($n=225$), *A. otus* took only 1·8% ($n=110$) (Thiollay 1968). Data from small-mammal trapping also suggest shrews selected against as prey (Village 1981). Perhaps captured without being identified first: one bird twice seen to catch one, pass it from foot to foot, hold it in bill, then drop it and fly off (Thiollay 1968); seems to signify active dislike rather than recognition of suboptimal prey. In purely rural areas, proportion of birds rather variable but

Table A Food of Long-eared Owl *Asio otus* as found from pellets. Figures are percentages of total live weight of vertebrates only.

Area	All areas of Britain (mostly) and N Ireland	N E Ireland	Norway	S Sweden	Amsterdamse-Waterleiding-duinen (W Netherlands)	Somme (N France)
Period[1]	1964–73, all year	1963–4, all year	1880–1959, mostly Apr–Jun	1961–74, all year	1965–80, mostly Mar–Jun	1963–71, winter
Mice (small Murinae)	15·5 ⎫ 66·7	75·3 ⎫ 75·3	10·1 ⎫ 63·7	27·2 ⎫ 91·8	23·5 ⎫ 73·2	20·9 ⎫ 93·6
Small voles (Microtinae)	51·2 ⎭	0[3] ⎭	53·6 ⎭	64·6 ⎭	49·7 ⎭	72·7 ⎭
Larger mammals[2]	19·1	20·4	31·7	4·9	8·2	0·4
Shrews (Soricidae)	1·3	0·1	1·5	1·0	1·2	+
Birds	12·8	4·2	3·1	2·2	17·4	5·9
Other	0·1 (frogs, bats)	0	0	+ (bats)	+ (frogs)	0[4]
Total no. of vertebrates	7761	1157	3629	13917	18788	19237
Source	Glue and Hammond (1974)	Fairley (1967), corrected	Hagen (1965), data reanalysed	Källander (1977a), data reanalysed	F J Koning	Saint Girons and Martin (1973), data reanalysed

1. Most pellets collected from winter roosts.
2. Mostly water vole *Arvicola terrestris* and brown rat *Rattus norvegicus*.
3. No voles present over most of Ireland.
4. Only mammals and birds included in this study.

generally less than c. 10% by weight. However, evidently (with larger mammals) major alternative food source when (or where) small rodents less available, and in more built-up habitat proportion can be much higher than normal—in Netherlands up to 80·3% by weight, n=220 (recalculated from Tinbergen 1933) and in Berlin (East Germany) 80·2% by weight, n=7125 (recalculated from Gawlik and Banz 1982); see also Hartley (1947), Stiefel (1970), Hillarp (1971), Elvers *et al.* (1979), Deppe (1982), and Wijnandts (1984). Change in rural habitats may also affect proportion of birds taken: when grazing increased at Amsterdamse-Waterleidingduinen, decreasing attractiveness of area for small mammals, birds increased in diet from c. 20% to c. 35% by weight (F J Koning). Invertebrates (largely insects) sometimes taken in significant numbers; pellets may consist of nothing else (e.g. Garzón Heydt 1968, and see above for Canary Islands). Usually, however, taken over short period only and generally of no importance (e.g. Araújo *et al.* 1974, Källander 1977a). Not stated in most studies whether pellets examined for chaetae of earthworms (Lumbricidae), but evidently eaten infrequently—found in 1 pellet of 164, Eskdalemuir, Scotland (Village 1981), and in pellets from 2 out of 51 sites from all over Britain (Glue and Hammond

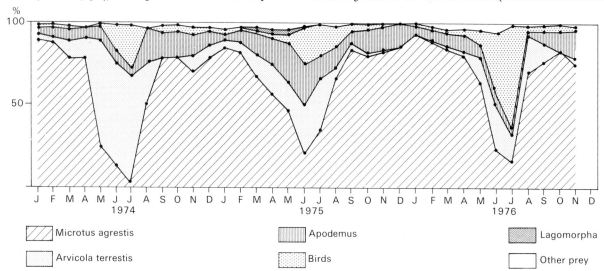

Fig IV Monthly variation in food of Long-eared Owl *Asio otus* in Revinge area (southern Sweden). Data refer to proportion by weight of total 7811 vertebrates. (Nilsson 1981.)

Table A continued.

Area	Berlin (East Germany)	Turew (Poland)	Hungary	All areas of Rumania	N Ukraine (USSR)	Central Spain
Period[1]	1949–62	1970–5, all year	Period not stated			1970–3, Mar–Nov
Mice (small Murinae)	11·3 ⎫ 96·2	8·5 ⎫ 97·8	26·2 ⎫ 91·7	52·8 ⎫ 78·6	3·8 ⎫ 98·3	10·1 ⎫ 90·9
Small voles (Microtinae)	84·9 ⎭	89·3 ⎭	65·5 ⎭	25·8 ⎭	94·5 ⎭	80·8 ⎭
Larger mammals[2]	1·1	0·4	0·3	11·0	1·2	2·2
Shrews (Soricidae)	1·0	0·2	0·2	0·4	0·1	1·9
Birds	1·7	1·7	7·8	10·0	0·4	4·7
Other	+ (bats)	0	+ (bats)	0	0	0·3 (bats)
Total no. of vertebrates	24 308	5400	23 057	26 346	5896	6945
Source	Zimmermann (1963), data reanalysed	Goszczyński (1981)	Schmidt (1973–4), data reanalysed	Cătuneanu et al. (1970), data reanalysed	Pidoplitschka (1937), data reanalysed	Araújo et al. (1974), data reanalysed

1974). Few stomach analyses (which, unlike pellets, would permit quantification of invertebrates): 44 alimentary tracts from southern Poland, 1969–74, all year, contained (by dry weight) 94·8% mammals, 5·0% birds, 0·03% insects, 0·1% plant material, and 0·1% non-organic material (Kochan 1979); see also Newstead (1908).

For year-to-year study over 12 years, see Zimmermann (1963); shorter year-to-year studies by Saint Girons and Martin (1973), Goszczyński (1981), and Gawlik and Banz (1982). Seasonal variation less well studied than in *S. aluco*; no universal pattern, and diet variously found to change rather little during the year (Saint Girons and Martin 1973; Goszczyński 1981; Deppe 1982; Wijnandts 1984), to experience reduction in small rodents in autumn and/or winter (Fairley 1967; Glue and Hammond 1974), to become more diverse in spring and summer (Degn 1976; Hardy 1977; Nilsson 1981), or to become less diverse in breeding season with increase in birds (Gawlik and Banz 1982). See also Gerell (1968). By far the most thorough study that by Nilsson (1981); see Fig IV.

For comparisons with diet of other Strigiformes in same or nearby area, see *S. aluco*; also Thiollay (1968: *T. alba*, *A. flammeus*), Saint Girons and Martin (1973: *T. alba*, *A. flammeus*), and Amat and Soriguer (1981: *T. alba*).

Energy budget of wild birds, Netherlands, studied in detail by Wijnandts (1984). ♂ and ♀ together found to eat equivalent of 3·8–6·1 (monthly averages) small voles per day (most in late winter/spring and in autumn), total 1854 per year; family of ♂, ♀, and 3 young surviving to end of year eat average total equivalent of 3313 small voles per year. Outside breeding season, equivalent to energy intake of 391–415 kJ per bird per day August–November, falling to 292 kJ in January and rising to 400 kJ by March; basal metabolic rate 114 kJ per day. In USA, requirement of wild adults in winter calculated as 665 kJ per day (Graber 1962). Consumption of full-grown captive bird averaged 37·5 g live weight per day (25·4–42·8) over 6 4-week periods during 1 year; highest in winter, least in spring (Marti 1973). For higher figures (63 g, and spring/autumn maximum 73·5 g), see Glutz and Bauer (1980).

No specific information on diet of young. In Netherlands, 2–4 (0–6) prey delivered to nest per night during incubation, 2–8 (1–12) after hatching (Wijnandts 1984). When *c*. 1 week old, parents bring *c*. 20–24 small animals per day (Averin and Ganya 1966). 6 chicks 10–20 days old each ate average 79·5 g over 3 successive nights; birds 4–9 weeks old each ate 65 g per day (Glutz and Bauer 1980).   DJB

Social pattern and behaviour. Major accounts by Richter (1952), Wendland (1957, 1958), and Hawley (1966). For review, see Mikkola (1983). Following account includes some material, notably by Armstrong (1958), on North American *wilsonianus*. Account also includes information supplied by D A Scott, D E Glue, and I N Nilsson.

1. Essentially solitary (except for roosting) outside breeding season, though dispersion varies markedly between regions, also between years in given region, depending notably on food supply. In southern parts of range, pairs often (probably typically) remain on territory all year (D A Scott). Further north, greater tendency to disperse and birds may congregate to roost where food locally abundant (Mikkola 1983; D E Glue). Such birds may migrate in small parties (Witherby et al. 1938), sometimes mixed with Short-eared Owls *A. flammeus* (see Moritz 1979), and often roost communally in appreciable numbers (see Roosting, below). In Netherlands, communally roosting birds dispersed nightly into nearby 'home ranges' to hunt; these on average 20·3 km², though only *c*. 20% of range used each night; these smaller areas possibly used by individuals exclusively (Wijnandts 1984). BONDS. Monogamous mating system the rule. Pair-bond renewed annually, at least among sedentary birds (D A Scott); this also interpreted from marked site-fidelity (see below). One report of bigamy: in Wageningen-Hoog (Netherlands), ♂ paired with 2 ♀♀ in his territory; each occupied separate nest-sites in

territory, but 1 ♀ left when other started laying (Bijlsma 1977; see also Heterosexual Behaviour, below). Both sexes breed first at 1 year (Burckhardt 1952; Mebs 1966). ♀ takes dominant share of nest-duties, incubating and brooding until young well grown (see below). Young independent at c. 60 days (Mikkola 1983; I N Nilsson), but may remain in parental territory much longer (D A Scott; see Relations within Family Group, below). Cannibalism amongst siblings not uncommon (D A Scott), though not clear whether young ever kill each other. Also one case of parent eating 2 of own young, not known if dead or alive (Werner 1958). BREEDING DISPERSION. Solitary and territorial. Size of hunting area varies markedly with food supply, and birds may feed well outside defended area: e.g. birds with young in nest hunted up to 2·5 km from nest (Kramer 1937). Where food abundant, territory 50–100 ha in Finland (Koskimies 1979), c. 50 ha in Dovrefjell, Norway (Haftorn 1971). In Grunewald (West Berlin), average distance between 5 nests 1250 m (600–2000) (Wendland 1957); in Tambov (USSR), for 5 nests, c. 800–950 m (Radetski 1981); minimum 50–150 m (Glutz and Bauer 1980). In Danube delta (USSR), 9 breeding pairs along 1·5 km transect (Anisimov and Tikhonov 1983). Estimates of density complicated by proportion of resident but non-breeding pairs, this varying with food supply: e.g. in 4 years, Eskdalemuir (Scotland), 9–18 pairs present on 10 km², of which average 17% did not breed each year (Village 1981). In optimum conditions, territories contiguous and density high: in Denmark, 3 pairs bred in barely more than 2 ha (Trap-Lind 1965); in 2 cases, Kreis Altenburg (East Germany), 5 pairs on 10 ha; in Drenthe (Netherlands), 39 pairs in 1717 ha of mixed woodland (see Glutz and Bauer 1980). In southern Sweden, average 0·25 pairs per km² (I N Nilsson). In 2 years, lower Reuss Valley (Switzerland), 0·9 and 1·1 pairs per km² (Fuchs and Schifferli 1981). Overall density, central Europe, probably c. 10–12 pairs per 100 km² (Rockenbauch 1968, 1978b; Ziesemer 1973; Glutz and Bauer 1980; see these sources for variations with food supply). According to Mikkola (1983), typical density probably 10–50 pairs per 100 km². Repeated use of same nesting area in consecutive years often reported (Cairns 1915; Wendland 1957, 1958), and, though no studies of ringed birds known, evidence from individually recognizable birds strongly suggests same pairs often involved each year (D A Scott). Same nest-site may be used in successive years (though not certainly known if by same pair). Otherwise, nest may be as little as 1–3 m from previous year's. Replacement clutches sometimes in same nest or nearby (Glue 1977b). 1 nest used for 7 consecutive years, including at least 2 replacement clutches (Paulussen 1955). Birds disturbed during building or incubation relaid up to c. 1 km away (Labitte 1956a). Occasionally, will nest close to Kestrel *Falco tinnunculus*; in 3 cases, nests only 6–10 m apart (Labitte 1956a; Ouweneel 1968; Rockenbauch 1968). ROOSTING. Mostly diurnal. Typically in densest cover available, favouring trees covered in ivy *Hedera* (Sharrock 1976); often close to trunk in conifer or in hawthorn *Crataegus*, gorse *Ulex*, bramble *Rubus*, or even long grass (Witherby *et al.* 1938; Flegg 1969; Kemp 1982); ground roosting, however, much less common than in *A. flammeus*. In central Europe, communal roosting (see below), especially in hard winters, reported in suburban gardens, orchards, etc., once on buildings, not uncommonly open to view in bare trees; such aggregations of 20–60 reported (Bruns 1965; Schelcher 1965), once c. 100 (Bruns 1965). In Britain, communal roosts typically 6–20 birds, often in traditional sites (Glue and Hammond 1974). Similar numbers occur elsewhere, e.g. Denmark (Glass 1971), Spain (Araújo *et al.* 1974), and smaller assemblages reported in Turkey (see Mikkola 1983). Roost occasionally includes a minority of *A. flammeus* (März 1965; Kemp 1982); once, 4 *A. otus* and 2 Barn Owls *Tyto alba* (Oldenburg 1954a). Birds typi-

A

cally perch a short distance (e.g. 1–2 m) from one another. Where pair remain on territory throughout the year, birds roost near each other, sometimes in same tree, outside breeding season (D A Scott). Probably resident birds also reported roosting up to 70 m apart (Armitage 1968). Roosting posture similar to *A. flammeus* (see that species). Sometimes sunbathes, lying on ground with one wing raised, exposing that side to sun (D A Scott). Captive birds often sunbathed with both wings spread and tail fanned (Fig A, in which head turned in response to observer disturbance: Räber 1950, 1954). Communal roosts dwindle typically in February as birds resort to breeding territories (Blanc 1958; Glass 1971; Wijnandts 1984). For onset of pair-formation among roosting birds, see Heterosexual Behaviour (below). From occupation of territory, pair roost near each other (Wendland 1957) and ♀ may share ♂'s roost-site (Richter 1952). Once nest-site chosen, however, ♀ roosts there, ♂ roosting 10–70 m away (Richard 1914; Richter 1952; Glue and Hammond 1974), and may use same site until young fledged (Wendland 1957). For periods of hunting activity, see Food; also, notably for daytime activity, Mikkola (1983).

2. Outside breeding season, communally roosting birds may allow approach to c. 10 m before flushing, but perhaps less tolerant of disturbance than *A. flammeus*, readily abandoning roost-site (Kemp 1982, which see for flushing behaviour). Roosting bird quickly alert to danger: adopts Sleeked-upright posture, effecting remarkable change in face from relaxed (Fig B, left) to alarmed (Fig B, right) expression: contracts facial disc, highlighting dark markings and half closing eyes, raises ear-tufts and stares at intruder, following every move (Steffen 1955); to human observer, change intimidating (Hosking 1941). On close approach, flushes silently (Bryant 1905). Birds disturbed at roost mobbed nearby dog, making low passes with legs extended, but not striking (Holland 1974). For similar behaviour in breeding season, see Parental Anti-predator Strategies (below). During breeding season, ♀ approached observer's hide, glared, and bowed low so that chin almost touched perch, then suddenly stood erect and swayed from side to side; facial disc then drawn forwards (as in Fig B, right) and ear-tufts laid flat (Hosking 1941). When confronted with rat *Rattus*, perched captive bird ruffled plumage, notably on breast and back, raised ear-tufts, etc., and drooped wing nearest rat (Räber 1950, which see for drawing). ANTAGONISTIC BEHAVIOUR. Rather tolerant of conspecific birds (D A Scott). Apart from calling (see below), no overt aggression reported: e.g. aggressive aerial displays and boundary disputes, as in *A. flammeus*, not known. ♂ demarcates territory by giving Advertising-call (see 1 in Voice), typically from high perch, sometimes during flight, exceptionally from ground

B

(Witherby *et al.* 1938; Wendland 1957). In Britain, ♂♂ resident on territory all year begin calling late October or early November; activity low from early December to early January, high from late January to mid-March and after (D A Scott). Other ♂♂ occupy and begin defending territories March–April (Glue 1977*b*; see also Roosting, above). Start calling after sunset but before dark. In Sweden, ♂♂ began calling in April at *c*. 20.00 hrs and continued doing so for 4 hrs, then fell silent, with resurgence 03.00–04.00 hrs (Wahlstedt 1959). In Michigan (USA), ♂ continued calling in lengthy but well separated bouts during night when young in nest (Armstrong 1958). One unpaired ♂ called regularly during day, but not by day once paired (Richter 1952), indicating that temporal pattern of calling likely to have strong heterosexual component. Subordinate in disputes with *S. aluco* (D A Scott). Typically does not retaliate when mobbed (e.g. Armstrong 1958), but when ♀ attempted to commandeer nest occupied by crows (Corvidae), she Bill-snapped during confrontation (Richter 1952). HETEROSEXUAL BEHAVIOUR. (1) Pair-bonding behaviour. In Netherlands, pairs left communal roost from early January to visit breeding territories in hinterland, gradually spending greater proportion of each day in territory (Wijnandts 1984). In Jutland, from 20 January, sometimes 2 birds leaving communal roost flew towards each other and repeatedly passed closely, with seemingly exaggerated wing-beats and occasional swoops; this perhaps rudimentary pair-formation (Glass 1971). For territorial advertisement by unpaired ♂, see above. ♂ occupies territory first, followed 3 (1–8) days later by ♀. Occupation signals marked increase in vocal and aerial display. ♂ regularly performs Display-flight, beginning at dusk and ending abruptly at dawn (Richter 1952). Displaying ♂ flies typically in zigzag, dipping and rising, through trees, sometimes at or above tree-tops. Flies with deep slow wing-beats, interspersed with glides and occasional Wing-claps below body (Hosking 1941; Wendland 1957, 1958; see also Voice): Wing-claps about once every 3rd wing-beat (H Delin), never several at a time as *A. flammeus* does; perhaps 6–20 Wing-claps during Display-flight, more at higher intensity (Hawley 1966). At peak intensity, ♂ displays while circling prospective nest-site in which ♀ sits responding with Nest-call (see 2 in Voice); display wanes markedly after nest chosen. ♀ also occasionally Wing-claps, though much less than ♂ (Wendland 1958; see also below). In territory of bigamous ♂, dominance of 1 ♀ inferred from frequency of her Wing-clapping which greatly exceeded that even of the ♂ (Bijlsma 1977). (2) Nest-site selection. By ♀, as soon as she joins ♂ on territory. ♀ runs and manoeuvres through branches, looking for nests, ♂ circling nearby. Sometimes pair fly briefly together, and ♂ may call persistently as if enticing ♀ towards nest. Once ♀ selects site, she makes circular flights around it, up to 20–30 m radius, for 15–20 min, and sometimes Wing-claps. ♀ advertises choice (Nest-showing) by flying repeatedly to nest and sitting there for long periods; once, ♂ appeared to select site (Richter 1952). ♀ now remains on or near nest day and night (Wendland 1957); then gives Nest-call (see Voice) at night for *c*. 4 weeks until laying (Hawley 1966). Once, ♀ laid in nest which contained 2 eggs of Woodpigeon *Columba palumbus* (Ziesemer 1973). In exceptional case, ♀ laid 1 egg in nest of Sparrowhawk *Accipiter nisus*, and *A. otus* young raised by host parents, at least to fledging, along with 1 of host's young (Mortensen 1965). (3) Mating. Takes place on branch or on ground near nest. Of 13 matings by 1 pair, 11 on branch (Fritz *et al.* 1977); of numerous matings between bigamous ♂ and ♀♀, all on ground (Bijlsma 1977). Mating sequence usually initiated by ♂ (see below), but dominant ♀ paired with bigamous ♂ (see above) initiated copulation on 60% of occasions (Bijlsma 1977). Birds may copulate several nights in succession shortly before

C

and during laying, and several times each night (Mannes 1971; Bijlsma 1977). In 13 copulations at one nest (Fritz *et al.* 1977, which see for drawings) sequence as follows: prior to copulation ♂ called to ♀ and displayed aerially (as above), then perched and made vigorous bucking movements, while raising and lowering himself on his legs, and swivelling closed wings up and down (Fig C); displayed thus for 15–20 s. Meanwhile, ♀ lay flat on perch with wings slightly drooped. Usually ♂ then flew to ♀, twice vice versa. ♂ mounted, bending forward until head just behind ♀'s (see also subsection 5), both birds extending wings and flapping for balance, and copulation took *c*. 3 s. Copulation on ground described by Wendland (1958), Mannes (1971), Lauermann (1975), and Bijlsma (1977). Typically, in these accounts, pair perform pre-copulatory duet (Advertising-call of ♂, Nest-call of ♀), which gradually increases in loudness and rate of calling. ♂ displays aerially, then glides to ground, shortly followed by ♀, or vice versa. ♀ adopts a horizontal posture, ruffling breast feathers and under tail-coverts, and quivering wings. After brief copulation, accompanied by Copulation-calls (see 6 in Voice), ♂ flies off, and ♀, after shaking herself, follows suit, often to nest. Pair resume calling, as before, but more quietly, and ♂ often returns soon to feed ♀ (see below). For Allopreening prior to copulation, see subsection 5. (4) Courtship-feeding and food-caching. ♂ deposits prey at nest before laying starts, and delivers food to ♀ throughout incubation and until young well-grown. Though ♂ may leave prey excess to immediate requirements of young, apparently does not cache food systematically as practised by (e.g.) *A. flammeus* (D E Glue). ♀ in nest solicits ♂ with Nest-call. ♂ may arrive in silence (Wendland 1958; Mannes 1971), but typically calls when ♀ brooding. Following account after Hosking (1941): ♂ announces arrival with Contact-alarm call (see 3a in Voice), ♀ answering with Nest-call. During subsequent exchange of calls (which, in our recordings, frequently includes Contact-alarm call of ♀: see 3b in Voice), ♀ very excited, quivering and vibrating wings. As ♀ reaches up towards prey in ♂'s bill he slowly flaps wings and rocks body from side to side, clutching and unclutching perch. Calling wanes, ♀ takes prey and lays it on nest, and ♂ leaves. ♀ once brought prey apparently received from ♂ outside nest. For other calls associated with food-transfer, see 4 and 5 in Voice. (5) Allopreening. Not widely reported, but perhaps often overlooked. Following account based on 2 observations by Scott (1980). ♂ alighted beside ♀ on branch and both began to make head-circling movements (in vertical plane), uttering Twittering-calls (see 7 in Voice), and nibbling each other's crown and facial feathers (mutual Allopreening). Copulation followed, during which ♂ seemed to nibble back of ♀'s head and nape. Immediately after copulating, birds resumed mutual Allopreening on breast and face before flying off separately. In 2nd case, when young 16–18 days old, ♀ parent perched near nest, began to 'sigh', and make clockwise head-circling movements, whereupon ♂ alighted beside her and began Allopreening her head and nape, while ♀ Allopreened his breast. Pair Allopreened for 4 minutes (Scott 1980). (6) Behaviour at nest. Though only ♀ usually incubates and broods, ♂ occasionally incubates for short periods and broods tiny young (D A Scott). ♀ usually a tight sitter. At one nest, early in incubation, ♀ left nest for short spells during day; later, absent only in evenings when apparently hunting near nest for 8–10 min (Armstrong 1958). Another sitting ♀ did not leave nest until c. 40–50 min after sunset, flying a short distance to preen for c. 5–10 min before returning to nest (Wendland 1957). RELATIONS WITHIN FAMILY GROUP. On hatching, ♂ brings young decapitated prey (Hosking 1941). On his arrival, ♀ shields young with arched wings (Glue 1977b, which see for photograph). ♀ dismembers prey for small young, stimulating them to gape (when blind) by touching bill flanges with strips of food (Hosking 1941). Young fed more directly when able to stand up and open eyes at 8 days (Räber 1954). Ailing young sometimes eaten by older siblings (Mikkola 1983; D A Scott), presumably when food scarce; not known if actually killed. Young brooded assiduously by ♀ for 1st week (Gurney and Turner 1915), less so thereafter, when ♀ often stands alongside. ♀ stops brooding young at c. 2 weeks old, but continues to feed them with prey received from ♂, and to guard them, never straying far from nest (Wendland 1957). At c. 3 weeks, young leave nest by climbing and jumping along branches with aid of bill and wings. As soon as 1st young leaves, ♀ tends to station herself closer to it. ♀ thought to start hunting as soon as young leave nest, and both parents then feed fledglings. When all young out of nest, tend to stay close together near nest for several days, associating amicably (Wendland 1957; Mikkola 1983). Young now intensely curious, exploring surroundings and examining objects with rapid head-circling movements, first either clockwise or anticlockwise, then reversed (Armstrong 1958). Brood of 5 reported 'playing like kittens', rolling around on ground, bowling into one another, etc. (Herschel 1968). Young able to make short flights 5–6 days after leaving nest and fully fledged at 4 weeks (Wendland 1957). Now beg vigorously and persistently for food (see Voice) each night from dusk onwards, food-call persisting until c. 100 days olds (Räber 1954). Fed by parents until they leave territory, though may sometimes hunt unsuccessfully for themselves (Wendland 1957). Period of dependence varies. Probably usually until c. 60–70 days old (Wendland 1957; Mikkola 1983; I N Nilsson), but young reported still in territory 2 months after fledging (Garling 1944), and occasionally staying until following year (D A Scott). ANTI-PREDATOR RESPONSES OF YOUNG. At c. 7 days, young start Bill-snapping if disturbed (Mikkola 1983). From c. 2 weeks old, adopt Forward-threat posture (Fig D) when threatened: bow forward, ruffle plumage, spread and droop wings, and arch secondaries over back, framing head; glare with wide eyes, blink them, and may flap wings, hiss, and Bill-snap (Witherby et al. 1938; Wille 1970; Mikkola 1983). May also peck and strike with claws when handled. Frequency of these responses increases with age. One bird 25–26 days old jumped out of nest on approach of intruder, and, when approached more closely, lay on ground on back with claws raised; young still in nest backed away when confronted (Armstrong 1958). Young out of nest may adopt Sleeked-upright posture, as in adult but with eyes open (Wille 1970). PARENTAL ANTI-PREDATOR STRATEGIES. (1) Passive measures. ♀ mostly sits tightly, but this varies, some flushing more readily than others (Glue 1977b). ♀ may flush less readily as incubation proceeds (Armstrong 1958). When young threatened, parents, and especially sentinel ♀, give Warning-calls (see 9 in Voice), immediately silencing young (Wendland 1957). (2) Active measures: against man. Responses vary from aerial demonstration, through distraction-display, to (rarely) dive-attacks. When confronted, ♀ mostly gazes with half-closed eyes, ear-tufts half-raised. If more disturbed, stares widely and ruffles plumage (Gurney and Turner 1915, which see for photographs; see also introduction to this paragraph). May also use Forward-threat posture (Mikkola 1983; for photograph, see Walker 1974). ♀ typically much more demonstrative than ♂, and birds more aggressive around time young leave nest (e.g. Wille 1970). However, marked individual variation occurs: some birds relatively passive, others (perhaps most) not exceeding aerial and vocal demonstration. Thus, distraction-displays never recorded by Wendland (1957) during close study. In commonest response, bird flies agitatedly around, landing from time to time, staring fixedly and head-bobbing; gives Contact-alarm and alarm-calls (especially Barking-calls: see 8a in Voice), and may also Bill-snap and occasionally Wing-clap (Richard 1914; Armstrong 1958; Hawley 1966; Wille 1970). Such demonstration may precede mobile distraction-lure display of disablement type, in which ♀ plummets headlong to ground (less often into scrub, etc.), ruffles plumage, and droops, spreads, and sometimes beats wings, flopping around, typically accompanied by long Warning-calls (Richard 1914; Whitman 1924; Wille 1970: see 9a in Voice). Once ♀ plunged headlong to ground repeatedly for c. 15 min (Labitte 1956a). Another bird flew directly at intruder and, when only 3–4 m away, tumbled to foot of nest-tree, beat wings and legs feebly, and dragged itself c. 1 m before flying off (Hawley 1966). Another ♀, lying on one side, pushed herself along ground, alternately flapping and dragging one wing (Walker 1974). If intruder approaches, bird may move to repeat display further away (Richard 1914). 2 reports of display by ♂. One landed on ground, ruffled plumage, drooped wings, stamped feet, and Bill-snapped; performed thus every day until young left nest (Weesenbeeck 1941). Another ♂ displayed with wing-spreading, Bill-snapping, and long Warning-calls (Armstrong 1958). Direct attack, usually by ♀, uncommon. When closely provoked, one brooding ♀ lunged, hissed (see 11 in Voice), and grasped with claws and bill; also perched nearby and gave long Warning-calls and alarm calls (Armstrong 1958). Before flying at intruder, perched bird adopts Attack-posture (Fig E): ruffles plumage, holds wings loosely by body, and stares. Dive-attacks occur but physical contact rare (e.g. Whitman 1924, Weesenbeeck 1941); may strike with wings or claws (Armstrong 1958; Walker 1974; Scott 1975). In exceptional case, intruder at nest with eggs mobbed by 5 birds together (Yeates 1941a). (3) Active measures: against other animals. In late February, bird made low passes at fox *Vulpes*, and in May at a trio of fox cubs, scattering them. Also in May, ♀ made apparently more intense attack on fox, diving at it with almost continuous 'whick-whack' calls (Scott 1975). When some of brood had left nest, ♀ attacked squirrel (no details) on nest-tree, knocking it to ground with feet (Armstrong 1958).

(Fig A from drawing in Räber 1950; Fig B from drawings in Mikkola 1983; Fig C from drawings in Fritz et al. 1977 and Mikkola 1983; Fig D from photograph in Matthews 1983; Fig E from photograph in Mikkola 1983.) EKD

D   E

Voice. Both sexes have remarkably complex repertoire in breeding season, but are relatively silent at other times.

Compared with some Strigiformes, voice typically subdued and not far-carrying, with exception of Advertising-call and food-call of young. Published accounts often underestimate repertoire (Hawley 1966); much variation between individuals and within given calls of a given bird, depending on circumstances. Most calls difficult to render, and descriptions from different studies hard to reconcile completely. Following scheme (based largely on recordings and notes supplied by P A D Hollom and P J Sellar) therefore provisional, and further variants and interpretations likely. Wing-clapping a whip-like sound (P J Sellar); also likened to muffled snap of a heavy towel being cracked like a whip (Grant 1959). Unlike Short-eared Owl *A. flammeus*, which Wing-claps in rapid series, *A. otus* Wing-claps only once at a time, and irregularly (Hawley 1966; P J Sellar). For further details of Wing-clapping, and Bill-snapping, see Social Pattern and Behaviour.

CALLS OF ADULTS. (1) Advertising-call of ♂. A slow, evenly-spaced series of long-drawn, low cooing 'ōō' or 'hu' sounds (Fig I: single call), quiet but penetrating, and audible up to 1 km or more (Witherby *et al.* 1938; Wendland 1957; Mikkola 1983; H Delin). Calls delivered at intervals of *c.* 2·5 s, the first few in series usually of lower pitch and volume than the rest (H Delin). Pitch and delivery rate vary significantly between individuals; in recordings of 3 different birds, pitch varies individually—*c.* 329–415 Hz (J Hall-Craggs). Prior to Display-flight, ♂ usually gives *c.* 30 calls, but from less than 10 to more than 200 reported (Hawley 1966). Call not uncommonly given in duet with Nest-call of ♀ (see below). (2) Nest-call of ♀. Commonest call of ♀. A nasal buzzing sound, remarkably rich in harmonics (see Fig II); difficult to render and variable between individuals, but aptly likened to sound produced by blowing through paper-covered comb (Hosking 1941). May also sound like lamb or high-pitched sheep (Hawley 1966); a feeble, slack, nasal, cracked 'péh-ev' (H Delin). Sound *c.* 1 s long, often repeated at intervals of 2–8 s, and barely audible at 60 m. Given almost exclusively when perched—typically from nest, serving to attract ♂; often then alternating with call 1 of distant or approaching ♂, or given in response to his Wing-clapping sound (Wendland 1957, 1958; P J Sellar). Between nest-site selection and laying, one ♀ called nightly at intervals of *c.* 8 s, much more rapidly on detecting approach of ♂ in Display-flight (Hawley 1966). Also given by brooding ♀ until young 8 days old, apparently as contact-call to young (Wendland 1957, 1958; see also call 3b, below). (3) Contact-alarm call. So-called 'sharming' of *A. otus* devotees. A descending, long-drawn harsh wheezing sound (E K Dunn, P J Sellar), difficult to render. (a) ♂'s call. A cat-like hissing 'chwau' (Wendland 1957; Mikkola 1983). In recording of birds alarmed by man at nest, 'chēēēeee' (E K Dunn). However, most often given as announcement of arrival at nest, especially when passing food to ♀ (Wendland 1957). (b) ♀'s call. Somewhat more disyllabic, more querulous, and higher pitched than ♂'s call. Rendered 'quicho' (Wendland 1957), 'chäu' (Wendland 1963), 'shoo-ogh' (Moffat 1905). In recording of alarmed birds, as above, 'CHOOO-oooh' (E K Dunn). In Fig III, this call given apparently by ♀ in response to same call of food-passing ♂, but according to Wendland (1957, 1958), her call directed at young, and this the only call given in this context by ♀ after young are *c.* 8 days old. (4) Mewing-call. In recording by P J Sellar, a repeated broken mewing sound by ♀, described as a pleading squeal (P J Sellar), when ♂ at nest giving Contact-alarm call but apparently reluctant to pass food.

I C König West Germany March 1968

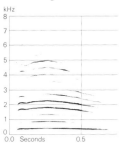

II C Chappuis/Alauda (1979) France February 1966

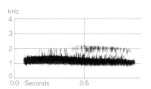

III P A D Hollom England June 1976

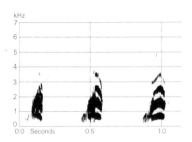

IV P A D Hollom England June 1976

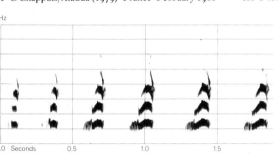

V P A D Hollom England June 1976

VI P J Sellar England June 1976

This perhaps the repeated 'pi-e' sounds given by ♀ on nest after ♂ left (Wendland 1957). (5) Departure-call of ♂. Harsh 'woch woch' sounds given after feeding ♀ (Wendland 1963). In recording by P A D Hollom, ♂ gives a low croaking 'choh choh choh choh' as he flies off after transferring prey (E K Dunn, P J Sellar). (6) Copulation-call. During copulation, a clear 'drdrdrdrdr' sound, like young when handled; not known if given by ♂ or ♀ (Lauermann 1975); a persistently repeated soft high 'srie-srie-srie' always accompanies copulation (Bijlsma 1977). Not definitely known whether one or both sexes give call, but thought possibly to be ♀ only (D A Scott). (7) Twittering-call. Soft twittering sounds presumably expressing contentment; accompany Allopreening (Scott 1980) but also heard in a variety of other situations, sometimes outside breeding season (D A Scott). (8) Alarm-calls. Variable group of calls. See also call 3. (a) Barking-call. Given mainly by sentinel ♀ (also sometimes by ♂) perched or in flight, when nest or young threatened; does not silence young (Wendland 1957): sharp 'ooak ooak ooak' (Harwood 1950; Mikkola 1983: Fig IV). Barking sounds also express agitation in other contexts: e.g. in recording by P J Sellar, ♂ which elicited Mewing-call from ♀ left nest with gruff 'wak wak wak'. In recording by P A D Hollom, ♀, perhaps impatient for ♂ to return with food, gave similar call. (b) Yelping-call. In recordings by P A D Hollom, ♀ gives remarkably puppy-like, yelping 'wup' and squealing 'yaow' sounds, apparently in excitement or agitation. High-pitched 'yip-yip-yip' given in flight (Hawley 1966) perhaps this call. (c) A gruff 'wawo' given by concerned adult, probably ♀, at nest in which well-grown young begging loudly, apparently intended to silence them (P J Sellar); interspersed with mild Barking-calls. (d) Kvik-call. A sharp 'kvik kvik kvik' (Fig V), very similar to 'wick' (or 'koo-ick') Alarm-call of Tawny Owl *Strix aluco* (Witherby *et al.* 1938; P J Sellar), and given in similar circumstances; 'whick whack' sounds given by ♀ diving at fox (Scott 1975) possibly the same or similar. (9) Warning-calls. These instantly silence young. (a) Long Warning-call. A long-drawn, plangent 'psii psii psii . . .', said to be given only by ♀, in extreme fear or anger, e.g. while making dive-attack on intruder (Wendland 1957). Presumably the same or similar calls (variously described as wailing, squealing, or mournful cries) given during distraction-display. Likened to squeal of small mammal in distress (Walpole-Bond 1932; Labitte 1956a). (b) Short Warning-call. Loud, rather compressed 'chwäit', like Grey Heron *Ardea cinerea* or ♀ Peregrine *Falco peregrinus*, given in similar circumstances to Long Warning-call (Wendland 1957). (10) Clucking-call. Quiet, intimate 'cu cu cu cu' sounds, like domestic fowl, given by ♀ feeding young, possibly as encouragement to eat (P J Sellar). Loud 'gock', like domestic cock, given 5 times by ♀ returning to nest after brief excursion (Wendland 1957) perhaps a related call. (11) Hissing-call. Sharp hissing sound, similar to that given by other Strigiformes, in anger and defiance (Walpole-Bond 1932; Armstrong 1958). (12) Other calls. For further descriptions, some (at least) probably same as, or variants of, those listed above, see Walpole-Bond (1932) and Witherby *et al.* (1938).

CALLS OF YOUNG. During most of nestling period, food-call and contact-call a short quiet 'szi' (Glutz and Bauer 1980) or a high-pitched piping 'pzeei', often repeated at intervals of $c.$ 5 s (Haartman *et al.* 1967). Brood of small young calling together likened to sound of jingling small coins. Smallest young's call highest pitched and the most frequently given (Walpole-Bond 1932; Hawley 1966). From near leaving nest until $c.$ 3 months old (see Social Pattern and Behaviour), food-call a shrill, mournful, penetrating pipe, likened to noise of gate with squeaky hinges (Witherby *et al.* 1938); audible up to 1 km or more (Mikkola 1983; H Delin). Duration $c.$ 0·5 s (Voronetski 1974b, which see for details of structure). Described as a melancholy 'pi-e' or 'psie-e', given at intervals of 5–8 s (Wendland 1957), or 'pi', more regular than, and lacking sibilant sound of, young *S. aluco* (Wendland 1963); further carrying, less hoarse, and less disyllabic than *S. aluco* (Kemp 1981). In available recordings, often sounds more trisyllabic than disyllabic, though distinction difficult unless heard close by, and varies with age: in 2½-week-old young, a trisyllabic fluting 'pee-ee-ee' (Fig VI), units delivered in rapid succession; in newly fledged 4-week-old young, a longer, slightly descending 'peeeeeee'. In recording by K Biggadyke, fledged young intersperse food-call with a yelping 'pyep pyep pyep' (E K Dunn), answered by parent's Barking-calls. Weak, high-pitched, slightly hollow sounding 'eewick', audible up to 15 m, and given by fledged young curious about human intruders (Hawley 1966), probably the same or a related call. Distress-call of handled young apparently similar to call 6 of adult (Lauermann 1975).   EKD

Breeding. SEASON. North-west Europe and Scandinavia: see diagram. Remarkably little variation in timing across range, with laying in Britain from late February about the earliest (Dementiev and Gladkov 1951a; Heim de Balsac

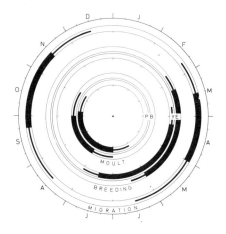

and Mayaud 1962; Makatsch 1976; Glue 1977b). SITE. In old tree nest of other species, principally Magpie *Pica pica* and crows *Corvus*, and sometimes squirrel *Sciurus*; less often on ground, usually in thick vegetation or at base of tree. Of 256 nests, Britain: old nests of *P. pica* 44%, of *Corvus* 34%, other species 15%, on ground 4%, miscellaneous 3% (Glue 1977b). Also uses nest-boxes. In Canary Islands, nests in cliff-holes (Bannerman 1963b). Nest: old nest of another species, occasionally adding material. On ground, shallow scrape formed. Building: scrape formed by ♀. EGGS. See Plate 97. Blunt elliptical, smooth and slightly glossy; white. 40 × 32 mm (36–45 × 30–34), n = 100 (Witherby *et al.* 1938). Calculated weight 22 g (Makatsch 1976). Clutch: 3–5 (1–6). Of 87 clutches, Britain: 1 egg, 1; 2, 4; 3, 22; 4, 30; 5, 26; 6, 4; average 4·15 (Glue 1977b). Average of 16 nests, Sweden, 4·56 ± SD 1·03 (I N Nilsson). Typically one brood, but thought occasionally to have 2 (Reinsch and Warncke 1968; Rinne 1981). Replacements occasionally laid after egg loss. Laying interval 2 days. INCUBATION. 25–30 days. Begins with 1st egg; hatching asynchronous. By ♀ only, though rarely ♂ sits for short periods (Glue 1977b). YOUNG. Altricial and nidicolous. Cared for and brooded more or less continuously while small by ♀; ♂ has been recorded as brooding tiny young (Glue 1977b). FLEDGING TO MATURITY. Fledging period probably over 30 days, but young leave nest at 21–24 days. Become independent at about 2 months (I N Nilsson). Age of first breeding 1 year. BREEDING SUCCESS. Of 287 eggs laid, Britain, 30% hatched and 24% produced fledged young, though these probably underestimates; of 78 nests, 41% produced at least 1 fledged young (Glue 1977). Of total 58 pairs over 4 years, Eskdalemuir (Scotland), 83% laid, 63% hatched young, and 57% fledged young; 23% of all completed clutches deserted. More than half the nests in which young hatched lost at least 1 young, in most cases apparently due to death of youngest. Average 3·2 young per successful nest, 1·7 young per nest started (Village 1981). In Sweden, average 2·85 ± SD 1·02 young left 34 nests; of 87 young leaving nests, 75% survived to independence (I N Nilsson). Ability to breed and breeding success dependent on food supply with most losses after hatching occurring through starvation in first few days (I N Nilsson). In poor vole *Microtus* years, Finland, average 2·4 ± SD 0·9 (n = 9) young in nest at age of 1 week or more; in good vole years, average, 3·6 ± SD 1·5 (n = 22) (Linkola and Myllymäki 1969).

Plumages (nominate *otus*). ADULT MALE. Forehead, crown, hindneck, sides of neck, and upper mantle pale golden-buff to pale cream, feathers often paler cream-buff or white towards tips (especially when worn); broad but poorly defined shaft-streaks dark grey to brown-black (darkest on crown), buff or cream feather-sides slightly speckled or sparsely vermiculated dusky grey. Ear-tuft at either side of forecrown black, feathers broadly bordered golden-buff to cream on outer webs, cream to white on inner. Lower mantle, scapulars, and upper tail-coverts like crown and hindneck, but distal part of feather-sides broadly pale cream-buff to white with rather heavy dusky grey or black speckling and vermiculation; in some birds, much uniform buff of feather-bases visible and lower mantle and scapulars hardly closer marked than hindneck, in others scapulars virtually completely and closely vermiculated black and white, appearing dark grey with black shaft-streaks, contrasting with buff hindneck; outer webs of outer scapulars uniform white, occasionally tinged cream-buff or with black subterminal bar and black freckling at tip. Back and rump golden-buff with indistinct dusky marks and often some grey of feather-bases visible. Broad band over eye down sides of bill to chin greyish-white or pale cream-white, narrowly bordered by black above and below eye and at basal sides of bill; tips of feather-like bristles at sides of bill black; remainder of facial disc buff, pale greyish-buff, or greyish-white, sometimes faintly speckled dusky grey at outer border. Facial disc surrounded by distinct ruff of rather stiff and narrow feathers, white with tips black or finely barred and speckled black and white. Chest, breast, belly, and flanks white with black or dark grey shaft-streaks (3–6 mm wide on chest, 2–4 mm elsewhere), feather-bases pale golden-buff to yellow-cream, feather-tips with grey speckling sometimes joining to form narrow irregular vermiculation or faint broken bars; in general, underparts appear white with dark streaks, some buff or cream visible on chest and flanks only and dusky speckling indistinct, but some birds have chest and flanks more extensively golden-buff or have greater amount of speckling on breast and sides of belly (though never as much as ♀). Central belly, vent, and under tail-coverts cream or white, uniform except for narrow grey-and-buff shaft-streak on longer under tail-coverts and (occasionally) vent; feathering of leg and toes uniform buff, pale cream-buff, or buff partly washed white. Central pair of tail-feathers (t1) buff with *c.* 5 rather broad greyish-black bands visible beyond longest upper tail-coverts; dark bands partly freckled buff, buff intervening bands often with faint dusky shadow or freckling in centres (more distinct towards feather-tip); most distal dark bar rather narrow and irregular, often hardly noticeable on closely speckled and vermiculated terminal 2 cm of tail-tip. Other tail-feathers similar, but dark bands gradually narrower and purer black-brown towards t6 (bands sometimes reduced or virtually absent on basal half of outer feathers), buff intervening bands and tips less speckled dusky and grading to white on inner edge and on tip of inner web; on t6, usually tip of outer web speckled dusky only and buff often tinged grey, but unspeckled in some pale birds; t6 frequently with dark grey shaft-streak. Basal halves of outer webs of outer 6 (5–7) primaries uniform deep golden-buff, inner webs similar but with broad white inner edge (usually except for dark spot on basal outer web of p10 and basal inner web of p3–p5); terminal halves buff with increasing amount of grey wash and speckling towards tip and with 4–6 rather evenly spaced brown-black bands; distal dark bands of p8–p10 often irregular in shape, hardly noticeable on closely speckled dusky grey feather-tip. Inner primaries and secondaries pale greyish-buff (purest buff near base) with increasing amount of grey speckling towards tips; broad border of base and middle portion of inner webs white; outer webs with 4 well-spaced brown-black bands visible beyond greater upper wing-coverts (band nearest coverts broadest and blackest), inner webs with narrower bands of nearly uniform width, gradually reduced in length towards featherbases. Tertials similar to secondaries, but pale grey-buff heavily speckled dark grey and dark bands broken into freckling, sometimes hardly noticeable. Upper primary coverts and bastard wing with broad black bands, buff of tips and intervening bands almost completely freckled and washed greyish-black (least so on outer greater primary coverts and outer edge of bastard wing), appearing very dark. Remainder of upper wing-coverts like lower mantle and scapulars; lesser mainly dark grey with some buff streaks and spots, median and greater with closely speckled black-and-white tips and variable amount of golden-buff of feather-bases visible; tips of outer webs of outer median coverts often with large but indistinct uniform white spots. Under wing-coverts and axillaries white; some narrow black shaft-streaks bordered by some buff often visible on shorter outer under primary coverts; outer greater under primary coverts with boldly black tips 1–2 cm wide; undersurface of basal and middle portion of flight-feathers white, often except for slight buff bars just below 2(–3) narrow black bars across secondary-tips. ADULT FEMALE. Most birds readily sexed by plumage, especially by underparts and underwing. Of c. 300 skins examined, most ♀♀ had darker golden buff ground-colour and heavier black marks; however, c. 5% of ♀♀ as pale as ♂♂ (no ♂ as dark as average ♀), and c. 10% of birds (both sexes) intermediate in plumage characters. Dark streaks on upperparts and upper wing-coverts of ♀ slightly broader and blacker on average, intervening ground-colour deeper golden-buff; tips of feathers of lower mantle and scapulars more heavily vermiculated black and deep buff, not as pale grey as ♂; paler inner edges of ear-tufts often narrowly barred black, not uniform cream or white or with fine specks only. Facial disc usually darker buff or rufous-buff than ♂, contrasting more with white bands over eyes down sides of bill, less so with black feathering at inner border of eye; facial disc frequently buff also in ♂, but ♂ never rufous-buff and ♀ never cream-buff or greyish-white. Chest deep golden-buff with black streaks 4–8 mm wide, virtually without white (unlike ♂). Breast, sides of belly, and flanks golden-buff; feather-tips white, broadly divided by black shaft-streak, and bordered terminally by black freckling and basally by 1–2 black bars, white showing as squarish white spots; ♂ also often has some black freckling at feather-tips, but white more extensive and not bordered basally by bars. Underparts of ♂ thus appear white with dark streaks and usually limited amount of yellow-buff on chest, breast, and flanks; ♀ appears golden-buff with black streaks, bars, and cross-marks, and with paired squarish white spots on belly and flanks. Central belly, vent, and feathering of legs and toes deeper golden-buff than ♂; under tail-coverts golden-buff with white tip and bold black shaft-streak or arrow-mark (mainly white with faint streak in ♂). Tail as in ♂, but dark bands on t1 on average broader, blacker, and more heavily freckled buff; other tail-feathers with narrow bars as in ♂, but these blacker also; inner edge of t6 buff, not white or pale yellow. Flight-feathers as ♂, but broad borders of inner webs pale buff or cream-buff (white in ♂); undersurface shows distinct buff tinge (in ♂, silvery-white or pale cream with traces of buff bars across secondary-tips). Under wing-coverts and axillaries deep buff (occasionally, with much white of longer coverts visible), shorter primary coverts and inner coverts often with black shaft-streaks or spots (in ♂, white, usually except for marginal coverts, and virtually unstreaked); greater number of greater primary under wing-coverts with black tips, these tips usually broader than in ♂ (up to 3–3½ cm rather than up to 1–2). DOWNY YOUNG. 1st down (neoptile) short and soft, rather dense, white; appears slightly mottled grey when feather-pins of 2nd down start to grow from c. 1 week old. JUVENILE. 2nd down (mesoptile) long, soft, and dense; that of greater upper wing-coverts and scapulars rather feather-like, but soft, and rami loose. Cream-white or greyish neoptile clings to tips of mesoptile up to 3 weeks, obscuring pattern of mesoptile; tiny fragments of neoptile still present at fledging, mainly on crown, flanks, and thighs. Mesoptile cream-white or buff, rather closely and narrowly barred dusky grey or grey-brown; much dull grey of bases of down occasionally visible, especially on crown, hindneck, and chest. Structure and colour of juvenile flight-feathers, tail, and greater primary coverts as in adult (but see 1st adult). Mesoptile gradually replaced by 1st adult during 3rd–10th week, last on lower flanks, under tail-coverts, legs, and toes. FIRST ADULT. Like adult, but juvenile flight-feathers, tertials, greater upper primary coverts, longer feathers of bastard wing, and tail retained. Most readily distinguished by extra bar across terminal half of flight-feathers and tail, dark bars slightly narrower and often gradually slightly closer towards tips; as consequence, tips of (e.g.) p8–p9 have 5 (4–6) black bands, adult 3–4 (in ♀, rarely 5), and secondaries show 5–6 bars beyond tips of greater upper wing-coverts in spread wing (occasionally, with another bar half-hidden under tips of coverts; pattern of inner and outer web sometimes different, unlike adult; adult shows 4 bars, a 5th occasionally half-hidden); distance from tip to 4th bar on outer web of s1 (measured along shaft to basal border of 4th bar) 38·5 (4·39; 141) 27–53 mm in juvenile, 54·4 (4·31; 141) 46–69 mm in adult (ZMA); adult only exceptionally below 48, juvenile rarely over 45 (over 45 mainly in rather pale birds with heavily speckled secondary-tips; narrow terminal bar then often showing on inner web of s1 or on other secondaries). T1 with 5–6 bars beyond longest upper tail-coverts (4 in adult); narrow black sub-terminal bar often contrasting with narrow white dusky-speckled tip (in adult, broad subterminal bar dusky with buff speckling, broader tip buff speckled dusky, less contrasting); tips of outer webs of t4–t6 virtually uniform off-white or grey (in adult, speckled dusky, least so on t6); distance of 4th bar on inner web to tip of t6 (measured from basal border of bar 3 mm from shaft) 37·6 (4·15; 99) 29–43 mm in juvenile, 46·1 (3·66; 109) 42–55 mm in adult (4th bar on inner web of t6 occasionally absent; then measured on t5). Additional ageing characters by wear: body distinctly bleached in spring (adult still fresh and golden-buff; most marked in ♀♀); secondaries all slightly abraded in midwinter, more distinctly in spring and following summer (adult still fresh and smoothly edged in spring; besides this, some adults retain some old secondaries, unlike 1st adult). SECOND ADULT. Like adult, but some birds (estimated c. 30% in sample from Netherlands) retain part of juvenile secondaries (mainly s3–s4 and s7–s8), these differing in number of bars from fresh neighbouring adult secondaries (adults retaining old secondaries have bar-pattern of these similar to new ones). Exceptionally, also juvenile s1 retained but in these birds new s5 or s9 has adult bar-pattern, unlike 1st adult.

Bare parts. ADULT AND FIRST ADULT. Iris bright golden-yellow, orange-yellow, or orange-red. Eyelids slate-grey. Bill dark horn-brown or blackish-grey; cere dark flesh-colour. Toes brown-grey or yellow-grey, claws greyish-black or black. DOWNY YOUNG AND JUVENILE. Skin under down and bare patch at rear of tarsus and undersurface of toes pink at hatching; cere pink-yellow, bill bluish-black, claws greyish-pink. Eyes open from 5th day; pupil opaque blue at first, iris chrome-yellow or orange-yellow from 2nd–3rd week. At 4 weeks, cere, culmen ridge, and cutting edges of bill dull olive-grey, remainder of bill olive-black; toes yellow, claws black. (Hartert 1912–21; Heinroth and Heinroth 1926–7; ZMA.)

Moults. ADULT POST-BREEDING. Complete, often excepting some secondaries; primaries descendant. Starts with p1 early June to early July in ♂ and non-breeding ♀, late June to mid-July in ♀ with young. Inner primaries shed rapidly, p2 on average 3 days after p1, p4 6 days after p3, etc., and up to 5 inner primaries may grow at same time, but p7 and p8 each lost c. 17 days after previous feather and only a few of longer primaries grow simultaneously; p10 shed only c. 5 days after p9 and outer 4 feathers may grow at same time (Piechocki 1968a; ZMA). Small sample of ♂♂, Netherlands, had primary moult score 23 (7) 14–32 by about 1 August, ♀ 17 (5) 13–22. Moult completed with p10 (score 50) mid-September to early November, mainly October. Moult of secondaries from 3 centres: s11 or s12 (moulting ascendantly and descendantly), s5, and s1 (both moulting ascendantly); s11–s12 shed with loss of p4–p6, soon followed by s10 and s13; s5 with p7–p8, s1 at same time or slightly later; secondary moult arrested after shedding of p10, regardless of stage reached. Of 107 birds of 2nd winter and older examined, 57 had all secondaries new; of remaining 50 retaining some old secondaries, 12 were in 2nd winter and had average 3·4 (1–6) old (juvenile) secondaries, and 38 were older and retained average 2·5 (1–6) old (adult) secondaries. Frequency of retention of individual secondaries (in ascending order) gives indication of sequence of secondary moult: 12–11–13–10–9–5–6–1–8–2–3–7–4. Tail-feathers shed within a few days, growing simultaneously; shed with p5–p7 or at primary moult score 20–25, completed at score 40. Moult of body mainly (July–)August–September; lesser upper wing-coverts first, crown and underparts last. POST-JUVENILE. Partial: all head, body, and wing-coverts; not flight-feathers, tertials, greater upper primary coverts, or tail. Starts in 3rd week with facial disc, followed by lesser upper wing-coverts in 4th and by first feathers of mantle, scapulars, and median coverts in 5th. At end of 6th week, facial disc and ruff 1st adult, upperparts (except crown, hindneck, and upper tail-coverts), and upper wing-coverts in full moult, underparts starting. At c. 7 weeks, wing and tail full-grown and only crown, neck, sides of breast, flanks, vent, toes, and under tail-coverts partly mesoptile; some down remains here until 10th–11th week. FIRST POST-BREEDING. As adult post-breeding, but moult of secondaries and tail starts 2–3 weeks later relative to primaries, and moult period longer—103 days between shedding of p1 and p10 in captive bird rather than 78–97 days (Piechocki 1968a; Haarhaus 1983). Despite later start of secondary moult, all secondaries replaced about as often as in adult (see adult post-breeding). See Wijnandts (1984).

Measurements. Nominate *otus*. Netherlands, all year; skins (ZMA). Bill (F) to forehead, bill (C) to cere. Juvenile wing and tail refer to retained juvenile of 1st adult.

WING AD	♂ 294	(6·04; 57)	282–310	♀ 299	(5·95; 64)	287–309	
JUV	290	(5·80; 41)	279–302	299	(5·72; 50)	286–316	
TAIL AD	137	(3·71; 35)	130–144	141	(4·49; 52)	132–149	
JUV	139	(4·24; 38)	130–147	143	(3·52; 41)	136–148	
BILL (F)	27·4	(1·10; 21)	25·8–29·1	28·9	(0·98; 15)	27·3–30·0	
BILL (C)	16·1	(0·96; 20)	14·5–17·5	17·6	(1·27; 14)	15·7–19·3	
TARSUS	38·2	(0·95; 20)	36·9–40·0	39·9	(1·19; 16)	38·4–42·3	

Sex differences significant. Juvenile wing of ♂ significantly shorter than adult, juvenile tail significantly longer. For sexual differences, see also Winde (1977).

East Siberian birds slightly larger than European ones: eastern China, winter; skins (ZMA).

WING AD	♂ 308	(— ; 2)	303–312	♀ 309	(7·88; 6)	298–319
JUV	295	(3·29; 8)	290–300	305	(8·32; 15)	293–314
TAIL AD	138	(— ; 2)	138–139	144	(6·35; 6)	136–152
JUV	140	(2·88; 8)	136–145	146	(3·56; 15)	140–154

A. o. canariensis. Tenerife and Gran Canaria, Canary islands; all year (BMNH, RMNH, ZFMK, ZMA).

WING	♂ 266	(6·40; 10)	252–276	♀ 276	(5·41; 5)	269–284

Tail on average 13 shorter than nominate *otus* (sample 5), bill (F) 0·8 shorter, bill (C) 0·7, and tarsus 0·7.

Weights. ADULT AND FIRST ADULT. Nominate *otus*. Netherlands: (1) December–March; (2) April–July; (3) August–November; (4) starved, found dead or dying, all year (ZMA). (5) Switzerland, adult, December–March (Glutz and Bauer 1980). (6) Italy, (October–)November–March (Moltoni 1949). See also Wijnandts (1984).

(1)	♂ 256	(20·8; 21)	221–303	♀ 308	(37·4; 24)	262–435
(2)	233	(21·6; 6)	207–268	278	(35·3; 8)	235–334
(3)	277	(31·9; 5)	246–331	288	(27·6; 12)	243–352
(4)	177	(13·8; 33)	151–198	202	(14·7; 34)	181–225
(5)	247	(21·1; 14)	220–280	304	(37·1; 19)	250–370
(6)	242	(2·03; 19)	210–280	288	(28·1; 69)	230–349

DOWNY YOUNG AND JUVENILE. 3 at hatching 19 each (Heinroth and Heinroth 1926–7). Average at 5th day c. 50, 10th c. 120, 15th c. 195, 20th c. 200, 25th c. 210 (155–260); on leaving nest 219 (29; 72) 160–300 (I N Nilsson). See also Wijnandts (1984).

Structure. Wing long, rather broad, tip rounded. 10 primaries: p9 longest, p10 20–31 shorter, p8 0–7, p7 16–25, p6 33–46, p5 55–66, p1 116–128. Outer web of p9 and inner web of (p9–)p10 emarginated. Outer web of longest feather of bastard wing, outer web of p10, and emarginated part of outer web of p9 with long curved serrations. Tail rather long, square; 12 feathers, t6 0–10 mm shorter than t1. Facial disc distinct, surrounded by circular ruff of rather stiff and narrow feathers; ruff extends into 2 rather long ear-tufts above each eye, each tuft consisting of 6–8 soft feathers up to c. 45 mm long. Bill rather long, heavy and bulbous at base, laterally compressed at tip; protected by bristle-like feathers at sides. Tarsus and toes rather long; covered with short woolly feathers, except terminal digit and undersurface of toes. Middle toe without claw 26·6 (1·05; 22) 25–28 in ♂, 29·2 (1·05; 16) 28–31 in ♀; outer toe without claw c. 75% of middle, inner c. 87%, hind c. 69%. Claws long, slender, and sharp, decurved (but less so than in many other owls); middle claw 16·2 (1·06; 21) 15–18 in ♂, 17·6 (0·69; 16) 17–19 in ♀; outer claw c. 81% of middle, inner c. 99%, hind c. 84%.

Geographical variation. Rather slight within Palearctic, more marked between continents. East Asiatic populations of nominate *otus* differ from European ones in slightly larger size (see Measurements, and Eck 1973) and slightly paler ground-colour with slightly less heavy dark speckling and vermiculation, but difference too small to warrant recognition of separate race. *A. o. canariensis* distinctly smaller than nominate *otus*; upperparts, upper wing-coverts, and breast much more heavily mottled and vermiculated (in particular on crown and hindneck), belly and flanks with slightly broader dark shaft-streaks and more barring; however, depth of ground-colour of head, body, and wing-coverts (including facial disc and under wing-coverts) as in nominate *otus*. Heaviness of marks of ♂ *canariensis* rather similar to ♀ nominate *otus*, but ground-colour paler; ♀ *canariensis* distinctly darker than any nominate *otus*. *A. o. wilsonianus* from eastern North America heavily mottled and vermiculated black on upperparts and upper wing-coverts (similar to *canariensis*, but marks coarser and with larger white spots on crown, hindneck, and scapulars); underparts with narrower dark shaft-streaks than nominate *otus*, more distinctly barred dark instead; sex for sex, depth of ground-colour as in nominate *otus* (but facial disc often slightly deeper golden-rufous and ruff blacker and more distinct), ♂ appearing white below with rather bold dusky streaks on chest and rather sparse and narrow dusky streaks and bars on belly and flanks, ♀ heavily streaked dusky black, golden, and white on chest, with belly and vent closely and heavily barred and streaked dusky, showing rather small paired white spots on golden-yellow ground; in both sexes, tips of flight-feathers and tail with more numerous and narrower bars than nominate *otus* (number dependent on age, as in nominate *otus*); iris yellow. West North American *tuftsi* reported to be paler than *wilsonianus* (Godfrey 1948, 1966). Isolated African races larger than Holarctic ones—wing 330–365 (BMNH, ZFMK); *abyssinicus* from Ethiopia deeper yellow (♂) or golden-brown (♀) than nominate *otus*, hardly speckled or vermiculated black, but instead marked with broad and sharp black streaks and bars on head, body, and upper wing-coverts, narrower and more sharply barred black on flight-feathers and tail; *graueri* from isolated mountains in central and eastern Africa similar, but black streaks wider and black bars narrower, appearing streaked rather than evenly cross-marked; ground-colour more similar to nominate *otus*.

Forms superspecies with Madagascar Long-eared Owl *A. madagascariensis* (Snow 1978). CSR

Asio flammeus Short-eared Owl

PLATES 52 and 53
[between pages 542 and 543]

Du. Velduil   Fr. Hibou des marais   Ge. Sumpfohreule
Ru. Болотная сова   Sp. Lechuza campestre   Sw. Jorduggla

Strix Flammea Pontoppidan, 1763

Polytypic. Nominate *flammeus* (Pontoppidan, 1763), Eurasia and North America. Extralimital: *c*. 8 further races in Neotropics, Hawaii, and Caroline islands (Micronesia).

Field characters. 37–39 cm; wing-span 95–110 cm. Same length as Tawny Owl *Strix aluco* but bulk less striking, with much longer wings and smaller head; wings longer and narrower than Long-eared Owl *A. otus*. Long-winged owl, with heavily streaked, pale yellow-brown and ochre-white plumage. Short ear-tufts usually invisible but dark wing-tips and dark, broadly streaked chest of adult obvious. Bulging facial disc buff-white, with striking black patches round glaring yellow eyes; expression baleful. Flight typically 'rowing' in action, with long wings rigidly set forward and held up in shallow V at low speed. Most diurnal of owls in west Palearctic; often perches on ground. Sexes differ slightly; no seasonal variation, though often strongly bleached in summer. Juvenile separable.

ADULT. Ground-colour of upperparts variable, tawny-buff to ochre-buff; increasingly larger black-brown streaks from crown to back, with scapulars more barred and blotched and often fringed off-white. Rump paler, due to lack of copious marks. Slightly wedge-shaped tail coloured as back, with 4–5 quite obvious black-brown bars on both surfaces. Facial disc completely surrounded by dark-spotted ruff enclosing buff-white cheeks which, with grey-white eyebrows and bill feathering, enclose distinct black areas round yellow eyes. Wings basically as rest of upperparts but smaller coverts spotted and blotched rather than streaked, while greater coverts and flight-feathers strongly barred. In flight, both surfaces of wing show obvious dark carpal patch and wing-point, both contrasting with intervening broad pale panel formed by virtually unmarked, pale buff bases to most primaries. Contrast of carpal panel and wing-point sharper against pale cream to pale buff underwing. Pattern of wing-point important in separation of *A. otus*: from tip inwards, shows long black tips to 5 longest primaries, an irregular pale bar (or set of 'mirrors') on all but outermost of these feathers, an irregular but still obvious black bar, and finally some dark half-bars and blotches. Rest of flight-feathers show indistinct brown bars (coalescing on tertials as darker patch) and distinct pale trailing edge to secondaries. Pale underwing not separated by dark body as in *A. otus*, since at least centre of belly, rear flanks, and vent appear cream or palest buff. Fore-underparts yellow to tawny buff, broadly streaked dark brown under throat and on chest, but more narrowly streaked on fore-belly and flanks. No freckling or vermiculation anywhere. ♀ deeper buff than ♂ (especially on underbody), and more heavily marked. JUVENILE. Plumage loose and generally darker than adult, with less defined pattern of streaks and with cream tips of upperpart feathers forming loose bars. Facial disc indistinct, with dark areas larger and pale central divide little developed. Ground-colour of underbody noticeably more buff than adult, with less obvious streaks on chest. Moult to adult plumage complete by October. At all ages, ear-tufts inconspicuous and usually discernible only in attenuated postures.

Main risk of confusion is with congeneric *A. otus* which shares similar wing pattern and flight action and, during migration, is frequently found in habitat normally associated with *A. flammeus*; see *A. otus* for discussion. May also be initially mistaken for Barn Owl *Tyto alba* (smaller, more compact except for dangling legs, with much less marked plumage pattern and wholly pale face) or even larger species, e.g. Ural Owl *S. uralensis* (such species all broader winged and much larger). Wings of *A. flammeus* proportionately the longest of any owl in west Palearctic. Hunting flight usually higher than in *A. otus* (see Food); markedly wandering, with a few measured, 'rowing' beats on rigid wings alternated with wavering glides, sudden banks, side-slips, and turns; all actions appear rather unstable and accomplished at lowest speed with wings set forward and usually held up in shallow V. In some regions has habit of flying up and hovering above disturbing observer; also uses slow hover while searching for food. On migration, flight more direct but still wavering, with action uneven and appearing uncertain. Unworn tail slightly wedge-shaped. Adopts rather horizontal posture on ground but more upright, head-up stance on perch. Most often seen on low perch but will also use branches or high posts.

Song of ♂ a low-pitched, hollow, oft-repeated hoot, 'boo-boo-...' (at least 6 syllables). Alarm-call 'chef-chef-chef'.

Habitat. Breeding in west Palearctic from high to middle latitudes, continental and oceanic, in arctic tundra, boreal, temperate, steppe, and Mediterranean zones, overlapping into certain mountain ranges, but predominantly in lowlands. Tolerates correspondingly wide range of climates, but shifts away from extreme cold of northern winters. Ascends in Armeniya and in Altay to 2000–2350 m (Dementiev and Gladkov 1951a) but in central Europe not above 650 m (West Germany; Glutz and Bauer 1980). A dominant factor in breeding habitat selection is abundance or super-abundance of small mammal prey, in absence of which many otherwise suitable areas may be unused for a period or even indefinitely. A terrestrial and aerial rather than an arboreal bird, nesting and often roosting on ground and little dependent on commanding look-out perches.

In Britain, favours substantial tracts of open country remote from excessive human disturbance and well populated with rodents, such as moorland heaths, newly afforested hillsides, extensive rough grazings, marshes, bogs, sand-dunes, and inshore islands. Has taken during past half-century to coastal nesting in England, but main factor in recent increase has been spread of young forestry plantations where grass grows long, encouraging multiplication of voles (Microtinae), and grazing and persecution are excluded (Sharrock 1976). In optimal area of southwest Scotland favours grassy type of moorland, and especially ungrazed areas recently planted with trees, or fenced for planting (Bannerman 1955). In Orkney, rarely on hilltops, preferring valleys and slopes clad in rushes and heather (Walpole-Bond 1914). On various Scottish isles breeds in long heather *Calluna*, or on flat moist mosses (Baxter and Rintoul 1953). Association with margins of wetlands, reclaimed marshes, tundra, and waterlogged pastures, as well as in places with croplands, farms, and other open land with rough herbage relates to need for simultaneous satisfaction of requirements for resting and nesting cover and for productive hunting territory.

In winter, shifts to harvested fields, stubble, crops of turnips and potatoes, and coastal salt-marshes, taking shelter from bad weather beneath overhanging banks, in willows *Salix*, turf cuttings, hedgebanks, scrub, and plantations. In USSR in winter, prefers open cultivated land, but also occurs in deserts (Dementiev and Gladkov 1951a).

Distribution. Range varies with population fluctuations; south of main range breeds irregularly, probably in years with local high rodent density or when rodent numbers low in main range (Hölzinger *et al.* 1973). Range of regular breeding has contracted in Netherlands and Poland, and extended slightly in Britain. Map attempts to show main breeding areas when numbers high, though sporadic breeding records are not included.

ICELAND. First bred in 1920s (Gudmundsson 1951). BRITAIN. Some range extensions, e.g. in parts of eastern and southern England, Wales, and southern Scotland; breeds occasionally elsewhere (Parslow 1967). IRELAND.

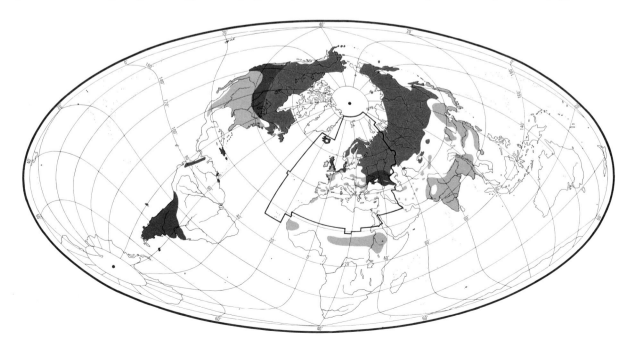

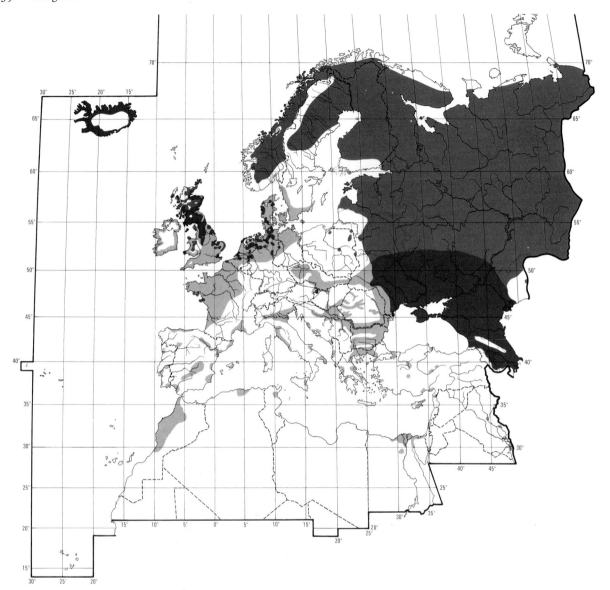

Bred 1959, 1977, and 1979–80 (CDH). FRANCE. Breeding regular, or almost so, in areas mapped (RC); irregular in many other places (see Yeatman 1976). SPAIN. Bred 1962 and 1976 (AN). BELGIUM. Bred on 13 occasions 1945–75 (Glutz and Bauer 1980); probably annual in recent years (PD). NETHERLANDS. Marked contraction of range 1850–1940, due to habitat changes; occasionally breeds elsewhere (CSR). WEST GERMANY. Range varies with population; in pre-peak and poor rodent years mostly in north, especially East Friesian islands (Glutz and Bauer 1980). SWEDEN. In years of high population, breeds south to coast (LJL). EAST GERMANY. Irregular and local breeder (Makatsch 1981); nests relatively often on Baltic coast, irregularly inland (see Glutz and Bauer 1980 for details). POLAND. Now irregular breeder, mainly in north. More widespread in 19th century (Tomiałojć 1976a; Glutz and Bauer 1980). CZECHOSLOVAKIA. Irregular breeder, perhaps annual. Sometimes tens of pairs (KH); see also Glutz and Bauer (1980). HUNGARY. Irregular breeder, a few pairs in some years (Glutz and Bauer 1980; LH). SWITZERLAND. Last bred 1953 (Mikkola 1983). YUGOSLAVIA. Has bred at least 5 times in past (Matvejev and Vasić 1973; VFV). GREECE. Bred 1962 and 1971 (WB, HJB, GM). BULGARIA. Bred 1950 and 1953, possibly bred 1951 and 1960 (Boev 1962). RUMANIA. Has bred (Vasiliu 1968; VC). TURKEY. Bred 1970 (MB, RFP). ISRAEL. Bred 1936, 1963–6, 1975–6, 1980, and probably 1981 (HM). MALTA. Bred 1906, 1909, and 1983 (Sultana and Gauci 1982; JS). MADEIRA. May breed occasionally, but no proof (Bannerman and Bannerman 1965; PAZ).

Accidental. Spitsbergen, Jan Mayen, Bear Island, Canary Islands.

Population. Marked fluctuations in breeding numbers with changes in small rodent populations; decline in Netherlands but probably some increase in Britain. ICELAND. About 150–300 pairs (Mikkola 1983). BRITAIN. Marked fluctuations with changes in vole populations, especially short-tailed vole *Microtus agrestis* (see, e.g., Lockie 1955). Estimated 1000–10 000 pairs; perhaps slight general increase due to spread of young conifer plantations (Parslow 1967); perhaps *c.* 1000 pairs in poor vole years (Sharrock 1976). FRANCE. 10–100 pairs (Yeatman 1976). NETHERLANDS. Fluctuating strongly (e.g. Bakker 1957) but estimated 130–185 pairs 1977, probably more in good rodent years (Teixeira 1979; CSR). WEST GERMANY. Marked fluctuations. Estimated 300–350 pairs (exceptionally higher in good rodent years) but only 30–100 pairs in pre-peak and poor rodent years; for details, see Glutz and Bauer (1980). DENMARK. Marked fluctuations; 5–30 pairs (TD). NORWAY. Fluctuating (Haftorn 1971). SWEDEN. Estimated 8000 to over 10 000 pairs in good vole years (Ulfstrand and Högstedt 1976). FINLAND. Fluctuating. Merikallio (1958) estimated 9000 pairs, and this probably correct in good vole years (Mikkola 1983). AUSTRIA. Very small numbers breeding Hansag area (HS). USSR. Abundant; fluctuating (Dementiev and Gladkov 1951a). Estonia: *c.* 1000 pairs (HV).

Oldest ringed bird 12 years 9 months (Rydzewski 1978).

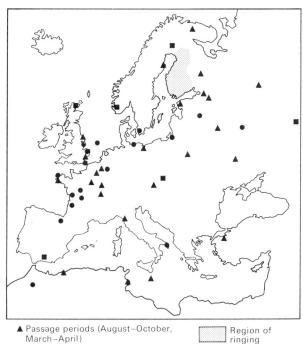

▲ Passage periods (August–October, March–April)
● Winter (November–February)
■ Subsequent summer (May–July)
▨ Region of ringing

Fig V Foreign recoveries of Short-eared Owls *Asio flammeus* ringed as nestlings in Finland (Saurola 1983).

Movements. Migratory (in north) to partially migratory; the only owl, other than Scops Owl *Otus scops*, to include an element of trans-Saharan passage. Also highly nomadic at all seasons; migrants move further south and west, and in larger numbers, in some winters than in others, while breeding densities fluctuate regionally from year to year according to changes in prey abundance. Races of South America and of Caribbean and Pacific islands are resident; not considered further here.

In autumn, normally withdraws from Alaska and most of Canada, Fenno-Scandia (except southernmost Sweden), and most of USSR. Winters within southern portions of breeding range (e.g. in northern USA and temperate Europe) but also south of it as far as Mexico, northern Afrotropics, Middle East, Indian subcontinent, and south-east Asia (Dementiev and Gladkov 1951a; Vaurie 1965; Ali and Ripley 1969). Only small numbers (which probably vary annually) cross desert barrier of Sahara into Afrotropics; occurs there in Sénégal and Mali, and in Sudan and Ethiopia, irregularly to Somalia and Kenya (Moreau 1972; Britton 1980; Lamarche 1980). West African winter visitors presumably from Europe (occurs on passage in Mauritania), but in north-east Africa immigrants include not only Nile valley migrants but also eastern birds that have crossed Red and Arabian Seas; various ship records, southernmost at 3°30′ S off Kenyan coast (*Bull. E. Afr. nat. Hist. Soc.* 1971, 77).

For recoveries of birds ringed in Finland, see Fig V; wide scatter evident, though also a preponderance of recoveries south-west from Finland. The few recoveries from ringing in Scandinavia conform to this pattern, and an April migrant in Tunisia found 21 days later in Arkhangel'sk (USSR), 3345 km NNE. Extent of emigration from Iceland uncertain: one juvenile found Scotland (Solway Firth) in October. Small-scale ringing in Britain has shown that some from Scotland and northern England cross into Ireland for winter, while others move south within Britain or further: 5 recoveries in Spain, and one (a Norfolk nestling) found Malta in October. From southern parts of continental breeding range, emigration mainly south-west, including West German birds found Portugal (November) and Spain (January), Czechoslovakian bird in Sardinia (October), and Ukrainian bird found Hungary (January); however, another West German bird found Voronezh (USSR) in January.

Ringing results also show evidence of nomadism, with subsequent summer records far from ringing site. Fig IV shows such summer records of Finnish-bred birds east to Orenburg (52°17′E in USSR) and west to Scotland (also Spain, where does not breed). Several Dutch and West German birds, ringed as nestlings, found later years north to Finnmark (Norway), Swedish Lappland, Finland, and Arkhangel'sk, and east to Sverdlovsk (61°35′E) in USSR (Kuhk 1961; Hölzinger 1974; Glutz and Bauer 1980; Saurola 1983). June adult from Netherlands found August, 3 years later, in Finland; Swedish nestling found

Netherlands in June (2 years later); French nestling found 6 years later (August) in Astrakhan by Caspian Sea; English nestling found 2 years later (September) in Vologda (USSR).

In northern Europe, breeding numbers can increase dramatically, over large or small areas, during vole (Microtinae) plagues (e.g. Lockie 1955, Picozzi and Hewson 1970, Mikkola 1983). In southern West Germany, also, breeding numbers peak at 3–4 year intervals (e.g. 1964, 1967, 1971), and Hölzinger et al. (1973) concluded that birds leave northern breeding areas when food supply scarce, to settle temporarily (in adequate vole years) further south, but return north subsequently. In general, vole populations peak at 3–4 year intervals in arctic and subarctic Europe, which accounts for periodic southward invasions in autumn/winter, some birds staying to breed (Hörnfeldt 1978; Mikkola 1983). Irregular peaks of 3–5 years apparent in numbers of migrants crossing Helgoland, West Germany (Schmidt and Vauk 1981). In wintering areas, large aggregations may appear where voles abundant, e.g. 1948–9, 1951–2, 1952–3, and 1973–4 on polders of Netherlands; an exceptional c. 2000 on Noordoost polder in 1948–9 (Bakker 1957).

Food. In many areas, almost wholly small voles (Microtinae). Hunting apparently occurs at all times of day and night, throughout the year, but typical pattern, especially in winter, not clear. In Britain, 'at several sites, a peak of crepuscular hunting activity, with birds generally staying out of winter roosts until just before dawn' (Glue 1977a). In winter, North America, hunts from late afternoon to sunrise (Short and Drew 1962; see also Kemp 1982, Mikkola 1983). At nest with young in Scotland, 3 activity peaks: 2–3 hrs after sunrise, early afternoon, and 0–2 hrs after sunset; visits continued throughout night (Hardy 1977). For activity pattern of captive bird, see Erkinaro (1973a). Hunts largely by slow flap-and-glide searching flight low over vegetation; in flapping, raises wings above level of body, unlike Long-eared Owl *A. otus* (Dennis 1968). Said by Dennis (1968) to fly at c. 2–10 m or more, higher than *A. otus*—though this difference not found by Hardy (1977); 0·3–2(–3) m according to Clark (1975). Frequently hovers momentarily before pouncing or continuing. Perhaps when prey less readily available, also hovers more persistently (up to c. 5 min according to Parker 1977) at 2–30 m or more (Clark 1975); c. 90 m recorded (Meinertzhagen 1959). Descends, sometimes in stages, by raising wings (Clark 1975). Hunts from perch infrequently (Clark 1975; Hardy 1977). Will also watch for prey from ground (Averin and Ganya 1966) and chase voles (Glue 1977a); said to capture prey on ground after lying in ambush (Dementiev and Gladkov 1951a). On Scolt Head Island (Norfolk, England), regularly takes many unfledged young terns (Sternidae); will hold one in foot while capturing a 2nd (Chestney 1970; R Chestney). Seen to take Redshank *Tringa totanus* in flight, striking it with wing, perhaps accidentally (Meinertzhagen 1959), and recorded persistently chasing Meadow Pipit *Anthus pratensis* c. 9–12 m up (Dickson 1971). Recorded stealing food from Kestrel *Falco tinnunculus* in flight (Reese 1973; see also Social Pattern and Behaviour), and from stoat *Mustela erminea* (Wood 1976). Seen hovering over water and possibly fishing (Campbell 1969). After catching prey, usually flies elsewhere to eat it. In flight, in winter, carries prey whole distance in feet, but during breeding season may or may not transfer it to bill before landing (Chislett 1941; Clark 1975). When carrying prey in flight, sometimes reaches down to feet with bill to kill it (Clark 1975); may even swallow it in flight (Henry 1982b). Small voles and mice swallowed whole, head-first; larger mammals torn into pieces. In captivity, sparrows *Passer* eaten 'in one or two meals'; thrushes *Turdus* and starlings *Sturnus* 'in two meals with considerable wastage of large wing and tail feathers, hind quarters and intestines' (Glue 1977a). Caecum of prey often discarded. Young tear up food more than adults (Clark 1975). For food-handling, see also Short and Drew (1962). For food-caching, see Social Pattern and Behaviour. Cannibalism of young by parents and/or siblings occurs (Mikkola and Sulkava 1969; Pulliainen 1978). For drinking behaviour, see Clark (1975). Apart from Snowy Owl *Nyctea scandiaca*, this is the only owl occurring in west Palearctic for which significant data on hunting success available. In Scotland, 2 birds of a pair successful in 13% and 18% of pounces ($n = 45$ and 28); each took 2–135 min to catch prey, averages 45 min and 17 min respectively (Lockie 1955). Of 628 pounces, North America, 20·7% successful, 8·1% outcome unknown (Clark 1975); see also Table A. For effect of weather on success, see final paragraph.

Table A Hunting success of one ♂ Short-eared Owl *Asio flammeus*, Manitoba (Canada). Figures are percentages of total pounces (Clark 1975).

	Flap-and-glide	Hovering	Perch hunting
Successful	19	26	33
Unsuccessful	73	62	67
Outcome unknown	8	12	0
Sample size	101	82	9

Pellets dark grey (Glue 1977a), confusable with those of Barn Owl *Tyto alba* but lacking varnished appearance and often segmented (Glue 1970); pale grey, rather similar to those of *A. otus* and Tawny Owl *Strix aluco* but bigger (Thiollay 1963b). More compact, containing more bones than pellets of Hen Harrier *Circus cyaneus* (Clark 1972). Average size 45×22 mm, mostly $35–70 \times 18–26$ mm, $n = 740$ (Glue 1977a); average $48 \times 22 \times 18$ mm, $n = 200$ (Mikkola 1981a); for seasonal variation, see Erkinaro (1973b). In winter, averages of 1·7–2·3 prey per pellet (Bakker 1957; Glutz and Bauer 1980). Not clear how many pellets ejected per day in the wild; for data and discussion,

Table B Food of Short-eared Owl *Asio flammeus* as found from pellets, nest-remains, and observed kills. Figures are percentages of live weight of vertebrates only.

	Britain and Ireland, 1964–73, all year	E Ireland several years, winter	S Finland 1964–7, breeding season	Noordoost polder (Netherlands) 1947–54, winter	Vendée (W France) 1965–6, winter
Mice (small Murinae)	6·9 ⎫ 54·6	15·6 ⎫ 15·6	1·3 ⎫ 91·3	0·1 ⎫ 99·9	0·3 ⎫ 98·3
Small voles (Microtinae)	47·7 ⎭	0² ⎭	90·0 ⎭	99·8 ⎭	98·0 ⎭
Larger mammals¹	37·3	77·8	4·0	0	0
Shrews (Soricidae)	1·2	0·1	1·7	0	0·1
Birds	7·0	6·5	3·0	0·1	1·4
Other	+ (frogs, bat)	0	0	0	0
Total no. of vertebrates	4120	105	1209	4102	14012
Source	Glue (1977a)	Glue (1977a), data reanalysed	Mikkola and Sulkava (1969), data reanalysed	Bakker (1957), data reanalysed	Thiollay (1968), data reanalysed

1. Largely brown rat *Rattus norvegicus* in Britain and Ireland, water vole *Arvicola terrestris* in S Finland.
2. No voles present over most of Ireland.

see Clark (1975). For experimental study of pellet formation, see Chitty (1938). Of 83 items fed to captive bird, 2·4% did not reappear in pellets (Clark 1975); see also Short and Drew (1962) and sources in Mikkola (1983).

For asymmetry in skull and external ear structure, see Kuroda (1967). For data on wing-loading (low), aspect-ratio (high), and power-to-weight ratio (high), see Clark (1975).

The following prey recorded in west Palearctic. Mammals: hedgehogs (Erinaceinae), shrews (*Sorex minutus, S. araneus, S. caecutiens, Neomys fodiens, Crocidura russula*), mole *Talpa europaea*, pipistrelle *Pipistrellus pipistrellus*, rabbit *Oryctolagus cuniculus*, hamsters (Cricetinae), bank vole *Clethrionomys glareolus*, water vole *Arvicola terrestris*, pine vole *Pitymys subterraneus*, common vole *Microtus arvalis*, short-tailed vole *M. agrestis*, root vole *M. oeconomus*, harvest mouse *Micromys minutus*, striped field mouse *Apodemus agrarius*, wood mouse *A. sylvaticus*, brown rat *Rattus norvegicus*, house mouse *Mus musculus*, stoat *Mustela erminea*, weasel *M. nivalis*. Birds range in size from Goldcrest *Regulus regulus* to adult Moorhen *Gallinula chloropus* (Heubeck and Okill 1981), but mostly open-country species from size of finches (Fringillidae) to thrushes *Turdus* (e.g. Glue 1977a); once attempted to take adult ♂ Pheasant *Phasianus colchicus* (Evans 1961b). Insects infrequent: mostly beetles (Carabidae, Dytiscidae, Gyrinidae, Hydrophilidae, Geotrupidae, Scarabaeidae, Chrysomelidae); also earwigs (Dermaptera), craneflies (Tipulidae), and wasps (Vespidae). Other prey rare: lizards (*Lacerta vivipara, L. muralis*), frog *Rana temporaria*, earthworms (Lumbricidae), snail (Gastropoda). (Moltoni 1937; Dementiev and Gladkov 1951a; Kulczycki 1966; Mikkola and Sulkava 1969; Glue 1977a; Palmer 1982; Mikkola 1983.)

For summary of major west Palearctic studies, see Table B. Other contributions, largely pellet analyses, as follows. Britain and Ireland: Lockie (1955), Buckley (1973), Jeal (1976), Hardy (1977); see also sources in Glue (1977a). Finland: Aho (1964), Mikkola (1971a, 1983), Pulliainen (1978). Sweden: Goransson et al. (1975). Norway: Hagen (1952), Klemetsen (1967). Belgium: Van Gompel (1979). France: Martin and Saint Girons (1973), Saint Girons and Martin (1973). East and West Germany: Deppe (1982); see also sources in Glutz and Bauer (1980). Poland: Kochan (1979). Hungary: Szlivka (1959a), Kulczycki (1966). Yugoslavia: Schmidt and Szlivka (1968). USSR: Dementiev and Gladkov (1951a), Parovshchikov and Sevastyanov (1960), Bashenina (1968), Tarasov (1979). See also Uttendörfer (1939b, 1952). Not clear how much bias introduced by type of site from which pellets collected: pellets found at roosts will relate to different hunting period (and thus perhaps different prey) from those found elsewhere; for possible bias in North America, see Clark (1975).

Although small voles typically predominate in diet (Table B), *Oryctolagus, Arvicola, A. sylvaticus, R. norvegicus*, or birds may each be most important if local conditions make them more profitable or restrict supply of voles (Glue 1977a; Pulliainen 1978). Insects comprised 14·0% (by number) of 74 items in 36 stomachs from Finland in autumn 1961–77 (Mikkola 1983).

Proportion of small voles in diet often assumed to follow their population cycles (e.g. Mikkola and Sulkava 1969, Glue 1977a), but no good data yet produced to demonstrate this. In southern Finland, even in year in which population of *M. agrestis* at minimum, it comprised 90·1% (by number) of items eaten ($n = 765$); 88·8% ($n = 196$) in previous year (Mikkola and Sulkava 1969). For most detailed (but short-term) study of relationship with prey populations, see Lockie (1955) for Scotland: from 15 April to 1 June, up to 6 pairs ate 3–13% of number of voles present on 15 April, accounting for 8–51% of reduction in vole numbers over the period; on same area, 2 June–1 August, 2 pairs removed 2–6% of number of voles present on 2 June, accounting for 4–15% of reduction in vole numbers. In July (but not in April–May) young voles taken out of proportion to abundance in population. No year-round study at one site to demonstrate seasonal variation in diet, and studies at different seasons in different places do not indicate any consistent differences. For studies involving comparisons with diet of other Strigiformes in same or nearby area, see *S. aluco* and *A. otus*.

Consumption of one wild pair averaged 76 g (2·8 voles) per bird per day; another pair, 85 g (3·1 voles). Nestlings over 12 days old ate average 54–101 g (2·1–3·9 voles) per bird per day. (Lockie 1955.) Energy requirement of wild adults in winter, USA, 786 kJ per day (Graber 1962). See also Chitty (1938), Clark (1975), and Glutz and Bauer (1980).

No details on differences between food of adults and young. In one year in Manitoba (Canada), c. 7% of 59 pellets of unfledged young which had left nest contained insects, presumably caught by young themselves (Clark 1975). Heavy rain may lead to death of young in nest, presumably via effect on hunting of adults (Hardy 1977). Adults brought average 3·9 voles per day to nest on 13 days with light winds, 2·1 per day on 7 days with strong winds, but difference not statistically significant (Lockie 1955). DJB

Social pattern and behaviour. Most comprehensive study by Clark (1975) in Manitoba (Canada). Important studies in Europe include Christoleit (1931), Lockie (1955), Gerber (1960), and Hölzinger *et al.* (1973); useful early study by Hesse (1912*b*).

1. Mostly solitary outside breeding season, but migrants may travel in small parties, sometimes mixed with Long-eared Owls *A. otus* (Moritz 1979); may occur together in large numbers (see also Roosting, below) where food locally abundant. Near Leipzig (East Germany), c. 30 together, and reports elsewhere of c. 100, January (Schmidt 1958; Gerber 1960), and c. 200 (März 1965). In western Baltic (West Germany), c. 50 birds 'along 2 km stretch' (Babbe 1953). Once, in Dumfriesshire/Galloway (Scotland), 'scores' together, and 36 at one site (see Watson 1972). Defends winter feeding territory where food supply abundant, otherwise ranges freely over area around communal roost (D E Glue). In Manitoba, some birds defend individual winter feeding territories; birds seldom trespassed, but occasionally hunted outside territory (Clark 1975); average area of 5 territories 6·3 ha. Some birds subsequently enlarged their winter territories for breeding purposes (Clark 1975). BONDS. Typically monogamous mating system. Pair-bond of seasonal duration (Glutz and Bauer 1980); bond between individual birds probably not renewed in successive years (D E Glue). ♂♂ thought occasionally to be bigamous, partly on evidence of some exceptionally large clutches probably laid by 2 ♀♀ (Mikkola 1983, which see for other evidence); once 2 ♀♀ incubated clutch of 8 eggs (Haartman *et al.* 1967). Age of first breeding 1 year (Glutz and Bauer 1980). As in other Strigiformes, ♀ takes dominant share of nest-duties (see Roosting, and Relations within Family Group). Young dependent on parents for a few weeks after fledging. Cannibalism among siblings probably not uncommon during food shortage (see Relations within Family Group, also Breeding). BREEDING DISPERSION. Solitary and territorial but birds not uncommonly feed outside territory (see below); territory size and density depend largely on food supply (Lockie 1955; Pitelka *et al.* 1955; Clark 1975). In area of conifer plantation, Stirlingshire (Scotland), 7 territories averaged 17·8 ha; later in same year, and in response to apparently dwindling food supply, whole area occupied by 2 pairs, with average territory size of 137·2 ha (see Lockie 1955). Other average sizes: in Bayern (West Germany), 15 ha (9–22), $n=7$ (Hölzinger *et al.* 1973); in Finland, c. 50 ha, $n=40$ (Grönlund and Mikkola 1969), 25·5 ha, $n=9$ (Rikkonen *et al.* 1976), c. 200 ha, $n=33$ (Korpimäki *et al.* 1977). For territory size in Alaska and Manitoba, see Pitelka *et al.* (1955) and Clark (1975). For local densities in Europe, see Glutz and Bauer (1980). Minimum distance between nests 145 m (Hölzinger and Schilhansl 1968), 300 m (Mikkola and Sulkava 1969), c. 790 cm (Goddard 1935). Hunting apparently confined to territory when food adequate (Lockie 1955) but, if short, and especially during chick-rearing, birds regularly hunt outside territory (e.g. Miera 1976); in one year, Oulu (Finland), hunted up to 1·5 km from nests (Mikkola and Sulkava 1969; Mikkola 1983); 2 km or more (Christoleit 1931). One pair thus bred and hunted in c. 4·5 km² (Becker 1978*b*). ♂ establishes territory and attracts ♀. Once, territory occupied in summer by non-breeding pair (Clark 1975). No information on site-fidelity, but given that territorial dispersion not rigid from year to year, fidelity not likely to be strong. ROOSTING. Widely studied. Mostly diurnal, though activity varies with season and food supply (see Food, and below). Outside breeding season, birds typically roost communally, often 6–12 birds, sometimes 30–40 (Glue 1977*a*); in Flevoland (Netherlands) over 100 in good vole years (C S Roselaar). Roost in Picardie (France) occupied about end of November, and numbers fluctuated from 8 to 18 during winter; roost persisted until end of March when birds took up breeding territories (Martin and Saint Girons 1973). In Manitoba, numbers increased throughout winter (Clark 1975). Roost close together, on ground or up to 2 m above it (Hendrickson and Swan 1938; Clark 1975; Kemp 1982). For photographs, see Gerber (1960). Roost-site typically provides good cover: e.g. evergreen thicket, pile of lumber, hawthorn scrub; in boggy area, raised tussocks, tree stumps, etc., used (Buckley 1973; Clark 1975; Davis and Prytherch 1976; Kemp 1982); in south-west Scotland commonly in woods (Mikkola 1983). Some winter sites used year after year. Roost may be used throughout winter or, if weather or food supply deteriorate (birds typically hunt near roost), birds occupying roost not uncommonly move *en masse*—in colder weather to denser cover. One bird roosted 1½ m from nearest of 3 *A. otus* (Kemp 1982), and 2 birds recorded roosting with 50 *A. otus* (März 1965). In cold weather, bird adopts a hunched posture with ruffled plumage, and stands on one leg, withdrawing other into plumage; thought to sleep much of day. In hot weather, stands with neck and legs more extended, wings drooped, and may perform gular fluttering; brooding ♀ may adopt this posture. Sunbathing bird closes eyes, stretches neck forwards, and spreads wings with undersides tilted forwards, usually for at most a few minutes. Roosting birds commonly preen, especially before hunting in evening, wing-stretch (Fig A), etc. (Clark 1975, which see for other comfort behaviour; also Gerber 1960). During winter, and early in breeding season, often leave roost to hunt at c. 15.00 hrs (Christoleit 1931; Hendrickson and Swan 1938; Gerber 1960); also often hunt by day when rearing young, but at other times usually not until near dusk (Clark 1975; but see Schaub 1937). Birds defending winter territories, Manitoba, usually roosted

A

within them, but one territorial bird roosted communally. At break up of communal roost, February–March, birds started roosting in breeding territories (Clark 1975). Territories nearest communal roost-site first to be occupied (Hölzinger et al. 1973). During breeding season, ♀ roosts at nest-site, at least until young leave nest, ♂ typically 40–100 m from nest (Clark 1975).

2. Though wary, not markedly shy. When agitated or curious, raises ear-tufts (Chislett 1941; Boyle 1974; see also Parental Anti-predator Strategies). Bird disturbed from roost rarely flies far before alighting. One bird disturbed from roost performed Wing-clapping (Chislett 1941; see Voice, and below). Compared with A. otus, flushed sooner from winter roost (at c. 50 m compared with c. 10 m), less agile in escape, and more readily sought open ground, frequently flying high (Kemp 1982). In August, bird mobbed fox Vulpes, launching dive-attacks at its head but apparently not striking (Evans 1961a). As in other Strigiformes, bird may adopt Forward-threat posture when strongly provoked, perhaps when cornered, though never used against conspecific birds. With increasing intensity, wings progressively extended, and tipped forward to expose uppersides; accompanied by Bill-snapping and Hissing-call (Clark 1975: see 4 in Voice). ANTAGONISTIC BEHAVIOUR. Vigorously defends boundaries of breeding or winter feeding territory. In breeding season, defence chiefly by ♂. Various displays, well described by Lockie (1955) and Clark (1975), on which the following based. In Underwing-display (Fig B), bird flies with slow, deep wing-beats, bringing wings high over back to expose pale undersides to rivals. 2 ♂♂ performed Underwing-display while flying towards shared boundary, turning away from each other when they neared it. ♂ often flies directly towards and pursues trespasser with rapid shallow wing-beats, often expelling it; roles sometimes reversed when trespasser re-enters its own territory. Boundary disputes often result in skirmish in which rivals rear up in flight, each attempting to get above other; in intense confrontation, raise feet and sometimes briefly lock claws with opponent. Occasionally, grappling rivals spin downwards (Clark 1975, which see for references). Skirmishing often accompanied by short Barking-calls (Clark 1975: see 2 in Voice). After skirmishing, returning

B

bird may Wing-clap, striking wings repeatedly and in rapid succession below body before making upwards recovery stroke (Lockie 1955; Clark 1975). Territorial behaviour wanes as breeding season progresses (Clark 1975). No antagonism between hunting ♂♂ during chick-rearing period (see Miera 1976). Confrontations over food with Kestrels Falco tinnunculus often reported, and regularly mobs this species when it enters territory (Goddard 1935); possibly competes for food in winter (Van Gompel 1979). F. tinnunculus may rob A. flammeus of food (Mascher 1963; Clegg and Henderson 1974), or vice versa. Aerial contests may be fierce, birds grappling with claws (Balfour 1973; Reese 1973; Boyle 1974). Fight with Hen Harrier Circus cyaneus also recorded (Dickson 1971). Roosting assemblage of 10–15 birds collectively mobbed tethered Eagle Owl Bubo bubo, accompanied by Barking-calls (Gerber 1960). HETEROSEXUAL BEHAVIOUR. (1) Pair-bonding behaviour. Begins in late winter. In communal roost, early February, 18 birds flushed in twos, suggesting already paired (Martin and Saint Girons 1973). In perhaps related case, bird arriving at communal roost greeted by another with lively head movements and Trilling-calls (Gerber 1960: see 5 in Voice). Early in breeding season, ♂ Wing-claps in low flight in rapid pursuit of presumed ♀. As season progresses, main form of ♀-attraction by territorial ♂ is elaborate Display-flight (Clark 1975, on which the following based). Typically, ♂ climbs quite rapidly with characteristically rhythmic wing-beats—at peak of upstroke, wings appear to pause momentarily, and at end of downstroke, when approximately horizontal, appear to 'bounce' back into upstroke; climbs in fairly tight circles, Wing-clapping quite often, and typically losing height as it does so (unlike A. otus—see that species); Wing-clapping bird described as 'dropping like a stone' (Mikkola 1983). With increasing height, Wing-claps less, sometimes soars, and may hover, delivering Advertising-call (see 1 in Voice). Intersperses hovers with series of quite shallow descending glides which end with Wing-clap, then starts climbing again. Display culminates in spectacular near-vertical stoop in Rocking-flight: rolls from side to side with wings held in deep V; descent may be continuous or interrupted by 1–2 rapid level glides. 3 Display-flights lasted 32, 60, and 64 min, in the last the bird climbing to c. 350 m (Clark 1975). After Display-flight, one ♂ alighted on top of tall tree and gave Advertising-call for almost 1 hr (Spencer 1945). Perched bird calls in a horizontal posture, head bobbing slightly, throat moving in rhythm with call (Clark 1975). When ♂ displaying, ♀ may hunt, or perch and watch ♂, occasionally giving Barking-call. ♂ displays at any time of day or night, more often in early morning (starting before dawn) and in late afternoon or evening (Clark 1975). In Lapua (Finland), usually from 17.00 or 18.00 hrs to 02.00 or 03.00 hrs, but no display during cold weather (Grönlund and Mikkola 1969). Display focused on relatively small area (mostly nest-site). Wing-clapping during Display-flight wanes with nesting. Display-flights also reported during autumn and spring migration (Mead 1969). On 12 December, Borkum (West Germany), performance of aerial display by 3 pairs apparently associated with sudden rise in temperature (Schoennagel 1979). (2) Nest-site selection. No information. (3) Mating. Takes place on the ground. Described only once, Courtship-feeding (see below) forming preliminary. In evening, late May, ♂ performed Underwing-display, then caught prey, landed, and dropped it. ♀, c. 40 m away on ground, gave Barking-call. ♂, with plumage markedly ruffled, stood with body horizontal and began swivelling it c. 30° from side to side. ♀ then made flying approach, and ♂ picked up prey in bill, partially opened wings, and passed it to ♀. ♀ turned and ♂ mounted, spreading wings for balance. ♂ dismounted after c. 4 s and pair flew off separately, ♂ to settle on post, ♀ to nest-scrape where she settled as if eggs present. (Clark 1975.) (4) Courtship-feeding and food-caching. Sequence of Courtship-feeding much as above. ♂ lands in territory with prey, calling repeatedly. ♀ replies with Barking-call and always flies to ♂ who performs Begging-display, as in young, with wings spread and fluttering (see Relations within Family Group, below), then resumes hunting, etc. In 6 cases of Courtship-feeding, 1 followed by copulation (Clark 1975; see above). Until young leave nest, ♂ feeds ♀ at nest, or she flies to collect it from him nearby (Clark 1975; Mikkola 1983). After exchanging calls, as above, ♂ lands at nest with quivering wings and ♀ takes prey from his foot or bill (Chislett 1941). When food abundant, ♂ leaves surplus food by nest for young. Adult, presumably ♂, cached prey side by side c. 7–10 cm from nest in neatly made tunnels (c. 12–15 cm deep) in grass (Ingram 1959). (5) Allopreening. None reported. (6) Behaviour at nest. Nest-scrape made first, then lined with material (Clark 1975), initially brought mainly from outside immediate nest-area (Hölzinger et al. 1973, which see for amounts of material used). Sitting ♀ adds to nest by reaching out and tearing up grass, etc., from around nest (Schuster 1930). Only ♀ incubates; evidently exceptional incubation by ♂ claimed by Forster (1955). RELATIONS WITHIN FAMILY GROUP. For development of young, see Scherzinger (1974d). ♀ broods attentively, especially if food supply good, until almost all young have left nest. Until then ♂ does all hunting for ♀'s brood, transferring prey to ♀ to dismember for small young (Clark 1975). ♀ also dismembers cached prey (Chislett 1941). From 3–4 days, when able to maintain upright posture, young beg by flapping wings, gaping, and giving food-call (Clark 1975). ♀ feeds small (blind) young by touching gape-flanges with food (Gerber 1960); for vocal communication with young, see 6 in Voice. Eyes open at 8–9 days, at which age hand-reared young also started casting pellets. Nestlings defecate by backing up to 1 m away from nest (Clark 1975). If parents cannot supply enough food, strongest chick often eats weakest (Mikkola 1983). Not known whether victims usually dead already or killed before being eaten; in one case, however, restless older chick grabbed smaller sibling by head and swallowed it alive (Ingram 1962). Young leave nest at 12–17 days (before able to fly) and scatter widely, typically 50–200 m from nest (Hölzinger et al. 1973); one young 175 m from nest 4 days after leaving it (Clark 1975). Such rapid dispersal possibly explains earlier belief that parents remove young from nest. At night, young wander considerably (Clark 1975), typically making 'runs' in vegetation (Watson 1972). Usually spend most of day hiding in one place. While in nest, and initially after leaving it, sleep stretched prone; once fledged, switch to upright sleeping posture. Both parents now feed young until latter capable of self-feeding; one partly-self-feeding young successfully begged for prey 25 days after fledging. Young advertise positions to food-bearing adults with Begging-display (Fig C): give food-call, ruffle plumage, and quiver wings close to body; at full intensity, rotate wings so that undersides face forwards, and flutter, almost flap them. Wing action attracts parents' attention, but also helps balance when begging from raised perch such as post. Transfer of prey brief, parent passing it bill-to-bill, then leaving (Clark 1975). ♀ may deliver prey received from ♂ some distance from young (Mikkola 1983). Parents also drop prey for young to retrieve (Watson 1972). Roving young intensely curious about any animals (etc.) they encounter; pick up and mandibulate objects, and presumably eat some insects (see Food). No detailed information on age at which family bonds broken, though in one case (see above) young not fully independent at c. 7 weeks (Clark 1975). ANTI-PREDATOR RESPONSES OF YOUNG. On close approach by man, unfledged young out of nest feigned death: lay still on ground, with staring eyes, and allowed itself to be turned over on back; extended foot upwards but did not strike, even when handled (Densmore 1924). When danger imminent, fledged young ruffle plumage, Bill-snap, and hiss (Gerber 1960). Young c. 30 days old adopted Forward-threat posture when teased by observer (Clark 1975). PARENTAL ANTI-PREDATOR STRATEGIES. (1) Passive measures. Sentinel ♂ uses prominent look-out perches (Becker 1978b). ♀ sits tightly, and often flushes only when almost stepped on (Christoleit 1931; Clark 1975). Sitting ♀ half closes eyes, hiding conspicuous yellow irises (Mikkola 1983). On close approach, ♀ effects startling transformation in facial pattern (Fig D: compare with Fig A): contracts facial disc, greatly dilates pupils, and raises ear-tufts, darkening face and giving appearance of A. otus; transformation alarming to man, and perhaps other would-be predators (Watson 1972). Parents warn unfledged young out of nest with short Barking-call (Densmore 1924: see 2 in Voice). (2) Active measures: against man. Diverse response of attack and distraction-display, especially by ♂ (see, e.g., Christoleit 1931, Zukowsky 1964, Clark 1975); ♀♀ usually less demonstrative, but some as aggressive as ♂♂. Reaction distance of ♂ varies between individuals and stage of breeding. Especially aggressive when young in nest. When ♀ incubating, one ♂ attacked intruder at c. 200 m from nest, but once young hatched, at 300–400 m (Christoleit 1931). Typical sequence as follows. On first detecting distant intruder, ♂ flushes and soars or flies towards him. First challenge relatively mild: may fly directly towards intruder with high wing-beats (Underwing-flight) before veering off to perch or soar. If several pairs breeding near each other, sometimes 5 or more circle over intruder (as in Marsh Owl A. capensis), but too close approach by neighbours to a given nest may elicit aggression from resident ♂ (Mikkola 1983). On closer approach of human intruder, ♂ may bark agitatedly (see 2 in Voice) and mount repeated dive-attacks with wings in V, accompanied by Wing-clapping and Bill-snapping; may lower feet, raking or thumping intruder's head, sometimes drawing blood (Urner 1925; Gerber 1960; Miera 1976; Mikkola 1983). Bird stoops from a height and pulls out of dive just over ground, gliding low with wings horizontal or just below, sometimes quivering, and feet extended forwards. At higher intensity, lands on ground, neck erect, wings outstretched, and gives Distraction-call (see 3 in Voice). One bird climbed on anthill to spread and droop wings (Clark 1975), another perched on tree (Fontaine 1969). Especially if nest or young imminently threatened, such behaviour by ♂, and sometimes by ♀, may develop into

C

D

distraction-lure display of disablement type (e.g. Armstrong and Phillips 1925, Christoleit 1931, Miera 1976). Diving bird may thus crash-land away from nest: with seemingly self-damaging force, rolls over, looks back at intruder, with wings outstretched, and gives appearance of serious injury (Mikkola 1983). At nest in which both members of pair displayed, ♂ fanned tail, shook and trembled outspread wings, and moved about intermittently with tripping steps, giving Distraction-call; ♀ feigned injury with small fluttering jumps, whilst Bill-snapping (Christoleit 1931). If approached, displaying bird typically flushes, circles, and repeats performance further away, often finally following retreating intruder out of territory. ♂ continues to perform distraction-display until young fledged (Clark 1975). (3) Active measures: against birds. Sentinel ♂ vigorously drives off crows (Corvidae) and other raptors (Chislett 1941; Gerber 1960; Hölzinger et al. 1973), pursuing them up to 1 km from nest (Becker 1978b). ♂ aerially attacked Hooded Crows *C. corone*, striking with claws and Bill-snapping; Wing-clapped during stoop after successful expelling attack (Christoleit 1931). (4) Active measures: against other animals. When closely confronted by sheep, ♀, which had 1 young sitting on her back, brooded it below, and squared up to sheep, crouching as if ready to spring, and staring fixedly (Chislett 1941).

(Figs A and D from photographs in Gerber 1960; Fig B from photograph in Fontaine 1969; Fig C from photograph in Heinroth and Heinroth 1926–7.)   EKD

Voice. Relatively restricted repertoire, used throughout the year, but especially in breeding season; not especially powerful (Clark 1975). Wing-clapping (made by striking wing-tips repeatedly in rapid succession below body before making upward recovery stroke) makes rattling sound like subdued handclaps (Mead 1969; Clark 1975). Given by ♂ in low-intensity threat and notably in his heterosexual Display-flight. During latter, bird Wing-claps *c.* 15–20 times, in regularly repeated bursts of 2–6 claps, average length of burst 0·8 s (0·5–1·0), $n = 13$ (Clark

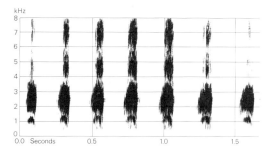

III C Chappuis/Alauda (1979) France July 1971

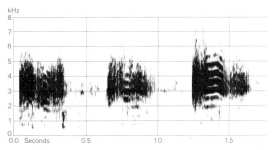

IV P A D Hollom Scotland May 1979

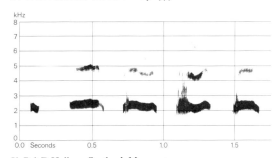

V P A D Hollom Scotland May 1979

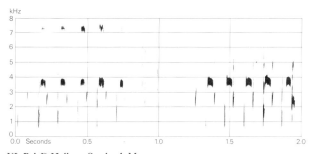

VI P A D Hollom Scotland May 1979

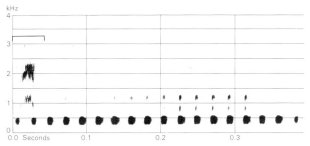

I C Chappuis/Alauda (1979) France March 1968

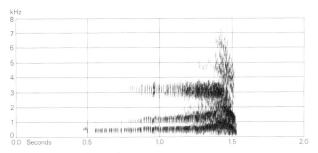

II P A D Hollom Scotland May 1979

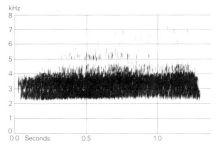

VII C Chappuis/Alauda (1979)

1975). In recording by R Margoschis (England), rate of clapping c. 10 per s (J Hall-Craggs). For further details of Wing-clapping, and for Bill-snapping, see Social Pattern and Behaviour.

CALLS OF ADULTS. (1) Advertising-call of ♂. A low-pitched, hollow 'boo-boo-boo-boo...' (Fig I, in which 1st upper unit is call 2 of ♀), resembling distant puffing of steam engine. Series by one perched ♂ comprised 16–20 units, given at rate of 2 per s, and series repeated 5–6 times per hr (Mikkola 1983). Series may be as short as 6 units (Witherby et al. 1938). Most often given during heterosexual Display-flight, often when hovering into wind, also to announce arrival with food for ♀ or young (Clark 1975; Mikkola 1983). ♀ answers with call 2. (2) Barking-call. Commonest but most variable call of both sexes, used in variety of contexts and serving equally variable functions. In ♀, serves as contact- and soliciting-call, inviting Courtship-feeding or copulation by ♂; rendered 'schieh' (Becker 1978b), or 'ché-ef', sometimes more drawn out when begging intensely (H Delin). In recording (Fig II, also 1st upper unit of Fig I), this call a shrill, hoarse rasping 'chooo AH', with hissing quality (E K Dunn), given by ♀ in response to call 1 of approaching ♂; also given by ♀ when ♂ giving Advertising-call in Display-flight (Clark 1975). More clipped variants given in alarm, also as warning to young, when adults aware of intruders; often a trisyllabic 'chef-chef-chef' (H Delin). In one recording (Fig III), a rapidly repeated, harsh, yelping 'che che che ...' (E K Dunn) directed by flying bird of unknown sex at human intruder; in another recording (Fig IV), an urgent repeated nasal 'nyeh' (E K Dunn) by ♂ in flight when intruder at nest. Similar call also heard from ♂ in territorial disputes throughout the year (Clark 1975). Sounds described as 'wuw', 'hek', and 'kew' (Schaub 1937) presumably the same or variants. For other renderings and variants, see Gerber (1960). (3) Distraction-call. A monotonous rasping squeal, given from the ground or in flight, by either sex engaged in distraction-display (Clark 1975). A repeated petulant squeaking or whimpering sound (Fig V) given by ♂ after dropping to ground after giving alarm variant (Fig III) of call 2 (E K Dunn, P A D Hollom). (4) Hissing-call. Sound similar to that of other Strigiformes, given by closely threatened bird (Clark 1975). In captive bird, sound sometimes preceded by a soft grunting 'whu' (see Mikkola 1983). (5) Trilling-call. A pleasant sounding 'trill' of apparent contentment, given by bird greeting arrival of another in communal winter roost (Gerber 1960). (6) Crooning-call. In recording by P A D Hollom, relaxed soft crooning sounds, like those of domestic fowl (P J Sellar), given by ♀ brooding young, and interspersed with squeaking food-calls (see below) of young. Guttural croaks given by food-bearing ♀ to young just outside nest (see Mikkola 1983) perhaps the same or related. (7) Other calls. Sounds given by captive bird (see Mikkola 1983) included: (a) an explosive 'kook-ook-a-rook', reminiscent of red squirrel *Sciurus vulgaris*;

(b) long-drawn 'sku-r-r-t' or 'whu-r-r-t', beginning with whistling sound and grading into a hoarse sound, like that made by water in tap half turned on; (c) rapidly repeated whistling piping sound.

CALLS OF YOUNG. While still in egg, young give single high-pitched squeaks (Clark 1975). On hatching, food-call a similar cheeping or chirping sound; also give trembling whimpering of apparent contentment (Gerber 1960). In young 4–10 days old, food-call a repeated high-pitched sucking, squeaking sound (P J Sellar: Fig VI). In recording by P A D Hollom, interspersed with call 6 of ♀. As young grow, food-call develops into a long-drawn wheezing 'pssssss-sip' (Clark 1975; Mikkola 1983: Fig VII), resembling snoring food-call of young Barn Owl *Tyto alba* but thinner and higher pitched (Gerber 1960). Wheezing food-calls given in series of c. 10–12, with variable pauses between, but series usually lasting more than 1 min. Pitch gets progressively lower with age, stabilizing by time young leave nest. Given more often after leaving nest, when young roam widely; call has apparent ventriloquial quality (Clark 1975). During begging from parent with food, call described as a loud screeching 'chie' (Gerber 1960). Young also have a high-pitched chittering 'psseee', expressing discomfort. Also Bill-snap and hiss, much as in adult (Clark 1975). EKD

Breeding. SEASON. Britain and north-west Europe: see diagram. Iceland: laying begins May (Hantzsch 1905). Northern Scandinavia and arctic Russia: laying begins mid-May to early June depending on weather (Dementiev and Gladkov 1951a; Makatsch 1976). Southern USSR: laying begins mid-April (Dementiev and Gladkov 1951a). Exceptionally, breeds in autumn and winter months (Mikkola 1983). SITE. On ground, usually in thick cover of grass, reeds, heather, etc., occasionally more in the open. Nest: shallow scrape, roughly lined with pieces of available vegetation. Average length of lining material 3–6 cm; average weight of 6 nests 74 g (31–132) (Hölzinger et al. 1973). Building: scrape in lining by ♀ (see Social Pattern and Behaviour). EGGS. See Plate 97. Elliptical, smooth and not or slightly glossy; white. 40 × 31 mm (37–45 × 30–34),

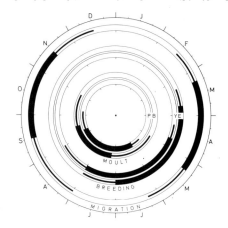

$n = 100$ (Witherby *et al.* 1938). Calculated weight 21 g (Makatsch 1976). Clutch: 4–8(–16); very variable depending on food supply, especially voles *Microtus*, but large clutches perhaps due to 2 ♀♀ (Mikkola 1983). Normally one brood but 2 occasionally reared in exceptional vole years. Replacements laid after early egg loss. Laying interval 1–2 days (Witherby *et al.* 1938); 21–32 hrs (Grönlund and Mikkola 1969). INCUBATION. 24–29 days, average 25·7 ($n = 4$ nests) (Grönlund and Mikkola 1969). By ♀, beginning with 1st egg; hatching asynchronous. YOUNG. Altricial and nidicolous. Cared for by ♀ and brooded more or less continuously while small. FLEDGING TO MATURITY. Fledging period 24–27 days but young leave nest at 12–17 days (Witherby *et al.* 1938). According to Heinroth and Heinroth (1926–7), able to fly well at *c.* 35 days. Become independent some weeks later. Age of first breeding 1 year (Glutz and Bauer 1980). BREEDING SUCCESS. Of 121 eggs laid in 17 nests, West Germany, 44 young hatched and 33 fledged; those that disappeared thought to have been killed and eaten by siblings (Hölzinger *et al.* 1973). In period of decreasing vole numbers, Britain, only 5 broods hatched from 24 nests, and only 2 fledged. Heavy predation by crows (Corvidae) and foxes *Vulpes* recorded (Lockie 1955); once, 8 adults and 68 young in fox earth (Watson 1972).

Plumages (nominate *flammeus*). ADULT MALE. Forehead, ear-tufts, crown, hindneck, sides of neck, and upper mantle tawny-buff or pale cinnamon-buff, marked with straight black-brown shaft-streaks 1–2 mm wide on forehead and forecrown, 4–7 mm wide on hindneck and sides of neck. Mantle and scapulars black-brown with buff fringes and irregular subterminal spots or broken bars, rather variable in size, appearing mottled about equally buff and brown or largely black with limited buff streaks (except for outer webs of outer scapulars which are largely uniform pale buff or cream). Back, rump, and upper tail-coverts buff or cream-buff with rather indistinct dark grey spots or arcs. Facial disc mainly white or pale cream; ring round eye black, widening towards sides of head into broad buff arc marked with narrow black shaft-streaks, extending from ear-tuft down to lower cheek; tips of rather soft bristles between eye and bill black. Distinct ruff of rather stiff and narrow feathers round facial disc deep rufous-buff or cinnamon with fine black streaks and specks. Chest and upper flanks tawny-buff or yellowish-buff with black-brown shaft-streaks *c.* 5 mm wide; remainder of underparts pale pink-buff or cream, grading to cream-white or white on under tail-coverts; breast, lower flanks, and sides of belly with narrow black-brown shaft-streaks $\frac{1}{2}$–$1\frac{1}{2}$ mm wide, remainder uniform. Central pair of tail-feathers (t1) rather evenly banded buff and black-brown (*c.* 4–5 bands of each showing beyond upper tail-coverts); black-brown bands narrowly connected by same colour along shaft, buff bands with black-brown mottling in middle, in particular on distal ones, buff feather-tip more heavily mottled or marked black-brown. Other tail-feathers similar, but black-brown bands gradually narrower towards t6 and reduced or absent on feather-bases, buff gradually paler towards t6; t4–t5 cream-buff with whitish base and outer edge, *c.* 3 dark bars on distal feather-half, and faint dark shaft-streak; t6 with 2–3 narrow dark bars on distal part of inner web only, remainder cream-white or white. Ground-colour of primaries cinnamon-buff on outer webs (soon fading to pale buff), pale tawny-buff to white on inner webs; outer 5–6 primaries with dusky grey tip 3–6 cm long and with 2–3 broad black bands on middle and distal portions of outer webs, only a single black subterminal band on distal portion of inner web, outer webs showing 2–3 buff mirrors subterminally, inner webs usually a single one only; buff of tips partly speckled dusky. Secondaries and inner primaries with white tip and almost uniformly white inner webs; outer webs with 3–4 black-brown bands separated by buff bands of about equal width, but buff often heavily speckled and clouded dusky; black-brown bands and dusky speckling often extend slightly across shafts to inner webs, in particular near feather-tips and on inner primaries and inner secondaries. Tertials intermediate in pattern between inner secondaries and longer scapulars. Upper primary coverts and bastard wing black with small and partly dusky-speckled buff marks, contrasting markedly with pale bases of primaries and with white leading edge of wing; upper wing-coverts mainly black-brown, narrowly streaked buff on lesser coverts and irregularly spotted and barred buff on median and great coverts; median and longer lesser upper wing-coverts and part of greater show rather large rounded white or pink-buff spots near tips of outer webs. Underwing and axillaries almost uniform white or cream, except for boldly black tips of outer 4–6 greater under primary coverts and black tips of outer 4–5 primaries (latter in part with pale and rounded subterminal spots or mirrors); 1–3 rather narrow and indistinct dusky bands extend along tips of inner primaries and inner secondaries, sometimes along other secondaries also. Bleaching and wear have marked influence: buff, tawny, and cinnamon change to white, in particular on exposed parts of body and wings; ground-colour of crown, hindneck, face, and underparts almost fully white, mantle, scapulars, and upper wing-coverts black-brown with strongly contrasting white mottling and spotting. ADULT FEMALE. Similar to adult ♂, but ground-colour distinctly deeper buff. Difference on upperparts and upper wing-coverts not marked, ground-colour on average only slightly deeper and dark streaks of crown and hindneck in most ♀♀ wider and blacker than in most ♂♂; dark arcs on rump and upper tail-coverts often slightly broader and more distinct, less diluted. Facial disc more extensively and deeper buff, usually without white border along sides in front of ear-opening (unlike ♂). Underparts distinctly deeper buff, chest to vent and foot markedly uniform deep tawny-buff when plumage fresh, pale buff when worn (in ♂, belly paler than chest in fresh plumage, but some birds then close in colour to ♀ in worn plumage, however; ♂ in worn plumage has ground-colour almost uniform white; ♂ only rarely as uniform and deep buff as ♀); dark streaks on flanks and sides of belly often broader than in ♂; under tail-coverts usually narrowly streaked dusky, not uniform whitish as in ♂. Under wing-coverts and axillaries pale buff, usually with slight and sometimes with heavy dark spots, in particular on under primary coverts and greater under wing-coverts (not almost uniform cream-white). T6 more heavily marked with 4 (3–5) bars on both webs, sometimes rather irregularly shaped (♂ has 2–3 narrower bars, mainly restricted to inner web). Inner webs of secondaries more distinctly marked with 3–4 dark bars. DOWNY YOUNG. 1st down (neoptile) short and dense, covering whole body except cere, bill, patch below tibiotarsal joint, and soles; upperparts pink-buff to buff (darkest on part of wing and along sides of mantle); general colour perhaps in part dependent on sex); underparts white with slight buff tinge, particularly on chest. Feather-pins of mesoptile appear from *c.* 5 days, first on face, legs, wing-coverts, and scapulars, showing as dark mottling under neoptile. JUVENILE. 2nd down (mesoptile) soft and downy, longer and denser than neoptile; rather like loose feathers on mantle, scapulars, and upper wing-coverts, fluffy elsewhere. Upperparts and upper wing-coverts dark brown or dark grey-brown, feathers tipped buff; feathers of mantle, scapulars, and upper wing-coverts with incomplete buff subterminal bars on distal halves, about equal in width to intervening grey-brown. Facial disc brownish-black, feathers narrowly tipped and mottled buff; conspicuous pale cream or white patch projects laterally from side of bill below eye from late in 3rd week to 6th week. Feathers of throat grey-brown with buff tips; chest, breast, flanks, and sides of belly warm buff, rather finely barred brown; remainder of underparts and feathering of legs uniform buff. Uniform pale buff neoptile clings to tips of mesoptile during 2nd–3rd week, retained longest on crown and hindneck. Growing feathers markedly tinged rose-pink towards base, specially on under wing-coverts. First Adult. Like adult, but juvenile flight-feathers, tertials, greater primary coverts, and tail retained. Probably best distinguished by t1 (but sample of known-age adults examined rather small): 10–15 mm of tip of t1 uniform buff or white, except for black-brown shaft-streak which tapers to point at tip, frequently with some fine speckling in middle of pale tip or at border of shaft-streak (in adult, broad pale tip heavily marked dusky or tip almost completely dark, except for narrow pale fringe; dark shaft-streak, if present, not tapering and not quite reaching tip); juvenile tertials bleached from October–November, and remainder of plumage from December–January (adult plumage still fresh December–January, including tertials).

Bare parts. Adult and First Adult. Iris pale sulphur-yellow to bright chrome-yellow or deep golden-yellow, sometimes tinged brown. Eyelids greyish. Bill black or dark horn-black. Cere slate-black. Toes slate-grey on uppersurface, yellowish on soles. Claws black. Downy Young and Juvenile. Eyes open from 5th–7th day, clouded blue during 2nd week, pale yellow later on. Bare skin under down (including leg and foot) pink, partly grey on head and body; cere greyish-pink or yellow, bill slate-grey or bluish-black, gape yellowish. Claws grey, soon darkening to black. Similar to adult at fledging. (Hartert 1912–21; Heinroth and Heinroth 1926–7; RMNH, ZMA.)

Moults. Adult Post-breeding. Usually complete, only rarely some secondaries retained until following moult; primaries descendant. In captive birds, Finland, moult started with shedding of p1 mid-May or late May; moult of body started about mid-June; flight-feather moult completed mid- or late August (shedding of primaries taking 84–89 days), moult of body feathers halted at same time (Erkinaro 1975). In captive bird followed over 7 years, p1 shed 25 May to 9 June, p10 24 July to 3 August, shedding of all primaries taking 53–61 days; secondaries moulted from 3 centres, descendantly from s12 (starting at about shedding of p3), ascendantly from s1 and s5 (start later but variable); tail completely replaced between shedding of p6 and p8, lost almost simultaneously with centripetal tendency (Haarhaus 1983). In Netherlands and USSR, primaries start June (sometimes second half of May or early July); primary moult score of 25 (5) 17–33 reached 1 August, 37 (5) 32–40 on 1 September (in one, already completed then, score 50); moult completed September or early October (Streseman and Stresemann 1966; RMNH, ZMA). Tail-feathers shed approximately simultaneously when p6 about full-grown; completed during regrowth of p9. Occurrence of partial pre-breeding moult, January–March (e.g. Witherby *et al.* 1938), not certain; none of those examined were in body moult January–February, but some moulted March–May, extent not established. Post-juvenile. Partial: head, body, and wing-coverts; not flight-feathers, greater primary coverts, tertials, or tail. Starts late in 4th week, largely completed in 7th (Heinroth and Heinroth 1926–7); mantle, scapulars, and upper wing-coverts first, crown, hindneck, sides of chest, lower flanks, and legs last. Most birds attain adult appearance in August; some growing feathers present until October.

Measurements. Nominate *flammeus*. Europe, all year; skins (RMNH, ZMA). Bill (F) to forehead, bill (C) to cere.

WING	♂ 315	(6·29; 39)	304–326	♀ 319	(6·38; 28)	309–331
TAIL	142	(5·65; 39)	134–152	144	(4·27; 32)	137–154
BILL (F)	28·5	(0·66; 16)	27·5–29·4	29·0	(1·21; 12)	27·7–29·8
BILL (C)	16·7	(0·72; 16)	16·0–17·5	16·9	(0·60; 12)	16·2–18·0
TARSUS	44·5	(1·01; 16)	42·8–46·0	46·2	(1·31; 14)	44·5–48·1

Sex differences significant for wing and tarsus. 1st adult wing and tail each on average *c.* 1 mm shorter than full adult, but difference not significant and ages combined in tables.

North American birds (1) similar to European ones (Eck 1971; RMNH, ZMA), Siberian birds (2) larger (migrants eastern China, ZMA).

WING (1)	♂ 314	(7·37; 5)	303–323	♀ 320	(7·19; 5)	312–330
WING (2)	321	(5·97; 16)	312–332	322	(6·57; 21)	321–338

Weights. Nominate *flammeus*. Adult and First Adult. Europe, mainly Netherlands and Italy (Moltoni 1949; Fedyushin and Dolbik 1967; Eck 1971; RMNH, ZMA).

Dec–Feb	♂ 362	(24·5; 4)	344–396	♀ 346	(56·9; 4)	290–420
Mar–May	278	(23·1; 3)	260–304	312	(27·0; 5)	280–350
Jul–Sep	290	(— ; 2)	281–300	298	(27·6; 6)	259–331
Oct–Nov	324	(27·7; 7)	286–350	350	(44·4; 10)	280–425

Exhausted birds, Netherlands: ♂♂ 198, 204, 299; ♀♀ 229, 245 (ZMA). Undated samples: Norway, ♂ 352 (8) 303–427, ♀ 419 (350–505) (heaviest ♀ from September: Hagen 1942; Haftorn 1971); USSR, ♂ 350 (10) 320–385, ♀ 410 (4) 400–430 (Dementiev and Gladkov 1951*a*); North America, ♂ 315 (20) 206–368, ♀ 378 (27) 284–475 (Earhart and Johnson 1970).

Downy Young and Juvenile. 2 captive siblings: at hatching, 16, 18; at 10–11 days, 120, 182; at 20–21 days, 290, 352; at 25–26 days, final weight of 340 (♂) and 405 (♀) reached (Heinroth and Heinroth 1926–7). In a brood from Finland, 14·2 (139; 9) 13–17 on 1st day, 48·0 (5·02; 6) 40–54 on 5th, 113 (21·2; 5) 93–146 on 10th; 190 (23·1; 4) 166–221 on 14th–15th day, when about to leave nest (Pulliainen 1978). In West Germany, 4 on hatching 15 each, on 5th day 44·0 (4·00; 8) 40–50, on 10th 132 (12·8; 9) 115–152, on 12th–14th (last day in nest) 190 (26·6; 9) 155–235 (Ziegler 1974). In North American samples, range 60–70 on 5th day, 60–175 10th, 175–280 15th, 260–345 20th; maximum of 390–415 reached 27th day, decreasing slightly afterwards once capable of flight; 380–400 on 30th day (Clark 1975).

Structure. Wing long, rather narrow, tip slightly rounded. 10 primaries: p9 longest, p10 6–20 shorter, p8 5–11, p7 26–33, p6 42–53, p5 62–77, p1 129–148. Outer web of (p8–)p9 and inner web of p10 emarginated. Outer edge of longest feather of bastard wing, outer edge of p10, and emarginated portion of outer web of p9 with short, fine serrations. Tail rather long, tip slightly rounded; 12 feathers, t6 8–14(–26) shorter than t1. Bill rather heavy, bulbous at base, laterally compressed at tip. Facial disc rather distinct, surrounded by distinct ruff of rather stiff and narrow feathers; short ear-tuft at border of ruff above eye, consisting of 2–4 feathers up to *c.* 25 mm long (extending up to 10 mm above remainder of feathering). Leg and foot densely feathered, except for distal digit of toe and soles. Tarsus and toes rather short; middle toe without claw 25·8 (1·41; 16) 24–28 in ♂, 27·4 (1·12; 14) 26–29 in ♀; outer toe without claw *c.* 71% of middle, inner *c.* 84%, hind *c.* 61%. Claws long and slender, decurved;

middle claw 16·7 (0·69; 16) 16–18 in ♂, 17·7 (0·89; 14) 17–19 in ♀; outer claw *c*. 84% of middle, inner *c*. 98%, hind *c*. 86%.

Geographical variation. Rather slight. Variation within nominate *flammeus* from Eurasia and North America limited, involving size only (see Measurements). Most extralimital races darker than nominate *flammeus*, more heavily streaked black-brown on crown, hindneck, and often on underparts; mantle, scapulars, and upper wing-coverts with limited number of buff spots and marks; flight-feathers with broader dark bands; size slightly larger in races from southern South America and Falkland Islands, slightly to distinctly smaller in other insular races, though some with relatively heavy bill and legs. CSR

Asio capensis Marsh Owl

PLATES 52 and 53
[between pages 542 and 543]

Du. Afrikaanse Velduil Fr. Hibou du Cap Ge. Kap-Ohreule
Ru. Капская сова Sp. Lechuza mora Sw. Kapuggla

Otus capensis A Smith, 1834. Synonym: *Asio helvola*.

Polytypic. *A. c. tingitanus* (Loche, 1867), north-west Africa. Extralimital: nominate *capensis* (A Smith, 1834), savanna of mainland Afrotropics; *hova* Stresemann, 1922, Madagascar.

Field characters. 29–31 cm; wing-span 82–99 cm. Averages 20% smaller than Short-eared Owl *A. flammeus* but largest ♀♀ overlap with ♂♂ of that species. Medium-sized, fairly long-winged, dark-eyed owl, with strikingly dark brown upperparts, dark brown and white face, and dark brown upper chest contrasting with pale buff-white underwing and rear underbody. Flight action supple but hesitant. Frequently diurnal and notably inquisitive when disturbed. Sexes similar; no seasonal variation. Juvenile separable at close range.

ADULT. Top and back of head, back, wing-coverts, and chest warm umber-brown, with paler rufous-buff tone on short ears, below head, and along scapulars; fine pale freckling (on all feathers) obvious only on wing-coverts at close range. Facial disc pale buff-white, with white eyebrows and bill feathering and black rings around black-brown eyes; produces distinctly sad expression. Flight-feathers dark brown, softly barred with rufous-buff. Tail dark brown in centre, outer feathers buff with at least 3 almost black bars. In flight, upperwing shows pale patch at base of primaries and pale tips to greater coverts; underwing strongly marked, with pale buff-white ground relieved by dark carpal patch, almost black primary-tips, and dark bars on distal parts of all flight-feathers. Underbody below chest cream-white with faint buff spots and bars. Contrast between upper- and underparts strongest of all owls in west Palearctic. Legs and feet well feathered, white. Eyes dark brown. Bill black. JUVENILE. Closely resembles adult but brown plumage deeper and warmer, with obvious barring above and on chest. Face pattern less distinct.

Likely to be seen in habitats associated with *A. flammeus*, but smaller size and dark unstreaked chest and upperparts allow immediate separation. Flight typical of genus; action rather lighter and even more erratic than *A. flammeus*, with bird wandering at low level in manner of harrier *Circus*. Has habit of investigating disturbance, frequently flying towards and closely inspecting human observer with intense stare; hence reputation of tameness.

Distinctive croak, uttered freely at dusk, recalls Raven *Corvus corax*.

Habitat. In west Palearctic, mainly in Mediterranean zone; also on fringe of Afrotropical region, mainly in lowlands, in long grass, and on marshy ground or in swamps and damp places in open country, nesting in hollows amongst grass. Like Short-eared Owl *A. flammeus*, of which it is the African counterpart, is among least arboreal of owls, spending time mostly on ground. Apparently differs from *A. flammeus* in being more restricted to wet or moist habitats, and to coarse grasslands. (Williams 1963; Prozesky 1970; Serle *et al.* 1977.)

Distribution. MOROCCO. Local; previously bred as far south as Essaouira, but no recent records from that area (JDRV). ALGERIA. 2 collected near Alger in mid-19th century, no proof of breeding (Heim de Balsac and Mayaud 1962); no records since (EDHJ).

Accidental. Portugal, Spain, Canary Islands.

Population. MOROCCO. Total not high (JDRV). Decreased due to habitat changes (Pineau and Giraud-Audine 1979).

Movements. Dispersive and partially migratory. Wanders to uncertain extent outside breeding season (which varies geographically), and thus reported from time to time well away from known breeding areas.

In Mali, either becomes secretive or is absent in wet season following December–April breeding (Lamarche 1980). Small influxes noted coastal Gambia July–December in some recent years (Gore 1981); now confirmed as non-breeding wet-season visitor there, with Mali as nearest known potential source (Smalley 1983*a*). Breeds

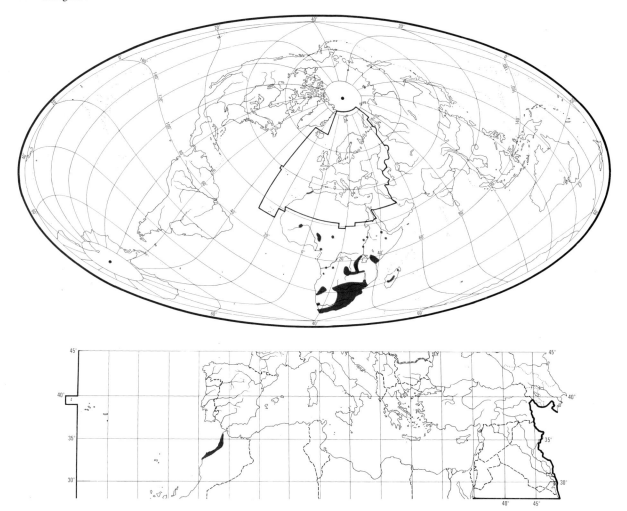

Nigeria north of great rivers October–January, and rare southern records occurred April–May (Elgood 1982). Comparable movement also occurs in localized population breeding north-west Morocco, including rare November–January records in Meknes area below Moyen Atlas (Heim de Balsac and Mayaud 1962) and a series of autumn–winter records from Spanish marismas (Bernis 1967).

Food. Beetles (Coleoptera), shrews *Crocidura*, and mice *Mus musculus* found in pellets in Morocco (Heim de Balsac and Mayaud 1962); no further information on diet in west Palearctic.

South of Sahara, diet consists of small mammals, birds, lizards, frogs, and invertebrates. At Barberspan (South Africa), seen hunting 16.00–09.00 hrs; daylight hunting more frequent autumn and winter (Dean 1978). Emerges *c.* 1 hr before dark (Brown 1970). Breeding birds in northern Nigeria started hunting after dusk (Smith and Killick-Kendrick 1964). Hunting methods evidently similar to Short-eared Owl *A. flammeus*. Flies slowly over grassland, frequently changing direction, and diving on prey with a turn and swoop; will also hover for short periods (Masterson 1973; Dean 1978; Hustler 1978; Steyn 1982). Seen on exposed perches at night, so hunting perhaps done from them (Smith and Killick-Kendrick 1964). Vagrant on Canary Islands recorded taking insects by aerial-pursuit in light of street lamp (Deppe 1984). Recorded searching clump of long grass from ground (Meinertzhagen 1959). Hunts at grass fires (W R J Dean).

Pellets thinner and more elongated than in Barn Owl *Tyto alba*; prey remains highly fragmented (Vernon 1972).

The following prey recorded in Afrotropical region. Mammals: shrew *Crocidura bicolor*, bat *Eptesicus capensis*, gerbils (*Desmodillus auricularis*, *Gerbilla paeba*), mice and rats (*Rhabdomys pumilio*, *Mus minutoides*, *Praomys natalensis*, *Otomys angoniensis*, *Malacothrix typica*, *Mystromys albicaudatus*), mole-rat *Cryptomys hottentotus*, zorilla *Ictonyx striatus*. Birds range in size from Fan-tailed Warbler *Cisticola juncidis* to full-grown Hottentot Teal *Anas hottentota*, but mostly small passerines. Also recorded: frog, lizards, scarab beetle, locust *Nomadacris septemfasciata*, termites (Isoptera), and scorpion (Scorpiones) (Jackson 1938; Vesey-Fitzgerald 1955; Meinertzhagen 1959; McLachlan and Liversidge 1970; Vernon 1971a; Masterson 1973; Craib 1974; Dean 1978.)

At Barberspan, took largely *Rhabdomys*, *Praomys*, *Otomys*, and birds, with arthropods present in over 80% of pellets; average prey size over 3 years 41·7–50·4 g (Dean 1978). In Namibia, pellets contained 17 small birds, 2 *Desmodillus*, 1 *Gerbilla*, and 1 *Malacothrix*; *Tyto alba* nearby fed almost exclusively on gerbils (Vernon 1971a).

Young said to be frequently fed on birds (Craib 1974). In northern Nigeria, items brought to nest were almost exclusively rodents of body length *c.* 2·5 to 13 or 15 cm; once, a locust (Smith and Killick-Kendrick 1964). DJB

Social pattern and behaviour. Little known, especially for west Palearctic. Following account based mainly on studies of nominate *capensis* in South Africa and Nigeria, where density, at least in dry season, evidently much higher than in west Palearctic.

1. Outside breeding season, Nigeria, occurs in ones and twos wherever habitat is suitable, but may concentrate (see Roosting, below) as preferred habitat dries out and contracts, dispersing again on resumption of rains (Smith 1964; Smith and Killick-Kendrick 1964). In South Africa, almost always occurs in small parties, rarely singly (Mackworth-Praed and Grant 1952). In Rhodesia, usually in parties of 5–6 (see Priest 1934). In dry season, South Africa, when cover restricted, *c.* 30–40 birds may occur in *c.* 2 ha (Masterson 1973). In Nigeria, 30 in *c.* 12 ha (Smith 1964; Smith and Killick-Kendrick 1964). At Barberspan (South Africa), 24 flushed together (Milstein 1975). BONDS. No information. BREEDING DISPERSION. Solitary and territorial. In Nigeria, large swamps may support several pairs: territories well-defined, though each pair has access to perimeter of swamp, and feeding area beyond (Smith 1964, 1971). At least 4 territories in *c.* 12 ha (Smith and Killick-Kendrick 1964). In parts of Morocco, 'pairs' only 100–200 m apart (Naurois 1961). ROOSTING. Singly or communally by day in hollowed bower or depression in the grass (Smith 1964; Masterson 1973), with tunnels in surrounding vegetation (Smalley 1983a); distinguished from nest by thinner ground layer (Farkas 1962). Site usually temporary (Masterson 1973), birds moving from day to day (Smalley 1983a). At one site in Gambia, up to 28 birds at average density 1 per 16 m^2, maximum 1 per 2 m^2 (Smalley 1983a). For roosting of young, see Relations within Family Group, below. Often hunts by day (Masterson 1973; Milstein 1975), especially if overcast and dull (Craib 1974), typically from 16.00–16.30 hrs onwards (Craib 1974; Dean 1978; Hustler 1978). Daylight hunting more frequent in autumn and winter, South Africa (Dean 1978).

2. Though wary, not especially shy, and even inquisitive (e.g. Butler 1908). When flushed by day, does not fly far, and may circle inquisitively, watching intruder; after resettling, difficult to flush again (Jackson 1938b; Smith 1964; Masterson 1973). ANTAGONISTIC BEHAVIOUR. Breeding birds usually chase away intruding conspecifics (Smith and Killick-Kendrick 1964; for one exception, see Parental Anti-predator Strategies, below). HETEROSEXUAL BEHAVIOUR. Display-flight evidently similar to other *Asio*. At dusk, or on moonlit nights, members of pair, or perhaps prospective pair, fly in wide circles, with deliberate wing-beats and periodic Wing-clapping. Participants chase each other, occasionally flying up to meet each other feet to feet (Smith 1964, 1971), though not seen to pass food (Smith and Killick-Kendrick 1964). During aerial chase, Croaking-call given (Smith 1964; Smith and Killick-Kendrick 1964: see 1 in Voice). Display-flights apparently continue at least up to chick-rearing, performed in evening prior to hunting (Smith and Killick-Kendrick 1964). No further information. RELATIONS WITHIN FAMILY GROUP. Eyes of young open after 6–8 days (Smith 1964, 1971; Smith and Killick-Kendrick 1964). Once all young hatched, they are left on their own, parent spending day nearby in grass. When feeding brood of 3 in nest, parent stood with youngest between its feet, and others at each side, one under each wing (Smith and Killick-Kendrick 1964). After leaving nest, young thought to be fed for 2–3 weeks; one young *c.* 5–6 weeks old, barely able to fly, found outside nest with abundant food laid beside it (Smith 1964, 1971; Smith and Killick-Kendrick 1964; see also below). Young apparently continue to use nest as roost-site for a few months after fledging, thereafter choosing separate sites (Farkas 1962). ANTI-PREDATOR RESPONSES OF YOUNG. Departure from nest at *c.* 2–3 weeks to crawl into surrounding vegetation may be response to human disturbance (Joubert 1943; Masterson 1973; Dean 1978). PARENTAL ANTI-PREDATOR STRATEGIES. (1) Passive measures. Incubating bird sits tightly and, if flushed, returns rapidly to nest (Smith and Killick-Kendrick 1964). In Zimbabwe, incubating bird put off nest flew *c.* 50 m and dropped into grass; when approached, flushed 2–3 times more and eventually returned directly to nest (Vincent 1946b). (2) Active measures: against man. On close approach of intruder, birds with well-incubated eggs or newly-hatched young perform aerial demonstration and vigorous distraction-lure display. In aerial demonstration birds fly around in tight circles. Once, when intruder at nest with newly-hatched young, aerial display of both parents attracted 3 adults from neighbouring territories, the 5 birds flying closely around nest with no signs of antagonism between them (Smith and Killick-Kendrick 1964). In typical distraction-lure display, bird flies rapidly around in tight circles, then flings itself to the ground, flaps in the grass, and gives Mewing-call (Smith 1964, 1971; Smith and Killick-Kendrick 1964: see 3 in Voice). While nest being examined, parent flew around calling, and finally landed *c.* 10 m away where it started flapping as if wing broken, and giving Mewing-call. When approached, bird flew further on, and repeated performance (Dean 1969). In similar performance, aerial demonstration accompanied by much Bill-snapping and calling (Joubert 1943: see 4 in Voice). Both members of pair may perform distraction-display at same time (Masterson 1973). All accounts above refer to intrusion by day; for report of distraction-display at night only, see Dean (1969). EKD

Voice. No information for west Palearctic; following account based on nominate *capensis*. For Bill-snapping when alarmed, see Social Pattern and Behaviour.

CALLS OF ADULTS. (1) Croaking-call. Deep frog-like 'kaaa' given in flight or on ground (Smith 1964, 1971); 'quark', like crow (Corvidae) or frog (Masterson 1973). In Fig I, a harsh unmusical rasping bark, 0·9 s long, similar to Short-eared Owl *A. flammeus* (P J Sellar). During Display-flight, may be given several times in quick succession (Smith 1964): 'kaa-kaa-kaaa-aa-aa' or variant (Smith and Killick-Kendrick 1964). In recording by C Chappuis of bird in flight, a wheezy double bark (P J Sellar). Given softly in flight, e.g. when hunting in evening (Smith and Killick-Kendrick 1964; Masterson 1973). Described as 'normal' call, i.e. that most often heard during breeding season (Smith 1964, 1971). Probably same or closely related call during aerial demonstration (see

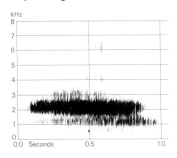

I C Chappuis Mali February 1969

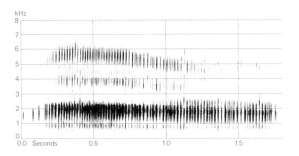

II T Harris South Africa (captive) May 1980

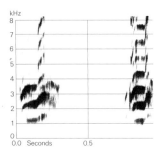

III J-C Roché (1967) Morocco March 1966

call 4). (2) Threat-call. A prolonged (1–2 s) rasping frog-like 'krrrrrrrrrr krrrrrrrrrr' (E K Dunn: Fig II), likened to slowly winding up a clockwork toy (J Hall-Craggs), given in threat by captive bird. Presumably extended variant of call 1. (3) Mewing-call. A mewing (Dean 1969) or more often described as a squealing sound, given during injury-feigning display and at no other time (Smith 1964, 1971; Smith and Killick-Kendrick 1964; see also Masterson 1973). A high yelping 'ooee kuik' (J Hall-Craggs: Fig III), perhaps by 2 birds, as observer approached nest. (4) Other calls. A feeble whistle, perhaps same or variant of call 3, given by birds demonstrating aerially when human intruder at nest; accompanied by croaks and Bill-snapping (Joubert 1943).

CALLS OF YOUNG. Small young in nest give a husky 'queeeep'; by the time they leave nest, call has changed to a soft, more musical 'too-eeee', with rising inflection, and audible at considerable distance (Smith 1964; Smith and Killick-Kendrick 1964). 2 hand-reared young gave 'whee-o' sound when they saw cat or dog, perhaps expressing fear or alarm (Smith and Killick-Kendrick 1964).

EKD

Breeding. Information from South Africa except as shown. SEASON. North Africa: eggs found from end of March to late May; possibly some breeding from late February (Heim de Balsac and Mayaud 1962). SITE. Usually on ground, in grass tussocks or other long vegetation, but also found in old corvid nest, 4 m above ground (Naurois 1961). Nest: shallow depression, unlined or with slight pad of vegetable matter placed in centre of tussock; often under overhanging long grass helping to conceal nest from above, and sometimes with well-formed entrance tunnel up to 1 m in length leading to nest-chamber (Smith 1964; Masterson 1973). Building: no information on role of sexes. EGGS. See Plate 97. Blunt elliptical, smooth and not or slightly glossy; white. No measurements from North Africa. 40×34 mm (38–43×31–37), $n = 50$ (McLachlan and Liversidge 1970). Clutch: in North Africa, 3–4(–5) (Heim de Balsac and Mayaud 1962; Etchécopar and Hüe 1967); in Morocco, 4 cases of 2 eggs or young, 6 of 3, 2–3 of 4, 1 of 5 (Naurois 1961). In South Africa, 2–3(–5) (McLachlan and Liversidge 1970). In Nigeria, 2 clutches of 3 eggs, 2 of 4, 1 of 6 (definitely only 1 pair involved); laid at 2-day intervals (Smith 1964; Smith and Killick-Kendrick 1964). One brood, but possibly 2 in years of prey abundance. In Nigeria, replacement clutch laid if 1st lost (Smith and Killick-Kendrick 1964). INCUBATION. About 28 days, Nigeria (Smith 1964). ♂ said to assist in incubation (Ayres 1871). Begins with 1st egg; hatching asynchronous (2-day intervals) (Smith and Killick-Kendrick 1964). YOUNG. Altricial and nidicolous. Cared for by both parents (McLachlan and Liversidge 1970). FLEDGING TO MATURITY. Young spend 2–3 weeks in nest, and probably fed nearby for further 2–3 weeks (Smith 1964; Smith and Killick-Kendrick 1964). BREEDING SUCCESS. In South Africa, 19 young reared from 45 eggs laid; success markedly higher in 3 dry years (17 young from 32 eggs) than in 3 wet years (2 young from 13 eggs) (Dean 1978).

Plumages (*A. c. tingitanus*). ADULT. Crown, mantle, scapulars, tertials, back, and rump almost uniform dark sepia-brown, deep earth-brown, or dull chocolate-brown (paler greyish-brown when plumage worn), feathers partly and indistinctly vermiculated or speckled dull pink-brown or rufous-brown; only short ear-tufts, feathers just behind facial disc, and hindneck more coarsely vermiculated brown and cream-buff or pale grey, appearing lighter. Upper tail-coverts coarsely vermiculated and spotted olive-brown and cream-buff, longest feathers with paired pale cream spots on tips. Distinct ring round eye black; remainder of facial disc cream-buff to greyish-white with some indistinct dull grey spots or shaft-streaks, feathers of ruff round disc similar but with more distinct dark brown tips; disc at sides of chin and at ear almost white. Throat coarsely barred brownish-black. Chest dark buff-brown, indistinctly vermiculated pink-buff or cream-buff, breast and flanks similar but with traces of pale partly vermiculated bars on feathers, belly similar but feathers with white spots subterminally, appearing spotted white (sometimes on lower belly only). Vent, under tail-coverts, and legs buff, usually barred red-brown on under tail-coverts, occasionally on vent and legs. Central pair of tail-feathers (t1) with 3–4 broad brownish-black bars; narrow bars in between and broader tip cream-buff to cinnamon-buff with much brown-black vermiculation. Other tail-feathers with slightly narrower black-brown bars, pale buff or cream-white bars in between about equal in width and without dark vermiculation (except t2–t3), sharply contrasting; broad tips of outer feathers uniform off-white. Tips of outer primaries black, subterminally bordered by broad and contrasting black and cinnamon-rufous bars; black bars reduced towards bases, in particular middle and basal portions of inner webs uniform cinnamon-rufous, contrasting sharply with dark upper wing-coverts; inner primaries broadly banded black and cinnamon-rufous (black bands reduced on basal inner webs), cinnamon-rufous bands partly vermiculated or speckled dusky; tips of inner primaries cream-buff or off-white, partly speckled dusky; width of pale tips reduced in extent outward, absent from about p7. Secondaries dark olive-brown to sepia-black, outer webs with traces of dusky-speckled cinnamon bars; broad bands on inner webs and broad tips cinnamon-buff or yellowish-white. Upper primary coverts black, contrasting with rufous-cinnamon bases of primaries, coverts along leading edge of wing and bastard wing dark sepia-brown with some buff spots. Remaining upper wing-coverts dark like upperparts, but pale vermiculation slightly coarser and paler cream-buff, slightly more contrasting, especially on median coverts, tending to form pale bands. Under wing-coverts and axillaries buff or cream with some coarse black spots or bars; tips of greater under primary coverts black, contrasting markedly with uniform pinkish-white or yellowish-white undersurface of basal halves of primaries. Sexes almost similar, but ♀♀ on average have broader rufous-brown bars on belly and flanks, white spots averaging smaller, and breast about as dark as chest; ♂♂ on average have narrower barring on belly and flanks, appearing white with brown bars, not buff-brown with white spots, breast not as dark as chest, showing more white barring. DOWNY YOUNG. 1st down (neoptile) short, soft, and dense; buff (Clancey 1964). JUVENILE (mesoptile). Feathering of head, body, and wing-coverts down-like, especially on head and underparts; only longer scapulars and greater wing-coverts feather-like, though with downy sides. Rather like adult, but feathers of crown, mantle, scapulars, and tertials black-brown with narrow and broken buff to off-white bars, these far more contrasting than irregular and vermiculated bars of adult, upperparts appearing barred rather than almost uniform sepia-brown; scapulars and tertials with distinct white tips when plumage fresh, but these soon wear off. Upper wing-coverts partially barred with coarse brownish-black and buff bands; pale tips of coverts soon bleaching to white and quite conspicuous, especially on greater coverts (pale tips of smaller coverts, if any, rapidly abrading); pale bands and tips virtually unmarked with dusky, unlike adult. Chest and breast finely barred dusky and off-white rather than almost uniform sepia-brown with indistinct pale vermiculation. Tail as in adult, but pale tips less broad and pale bands and tip of t1 hardly speckled or vermiculated dusky; width of white tip in middle of inner web of t4–t6 10–16 mm (21–28 in adult). Flight-feathers as in adult, but pale tips of secondaries and inner primaries narrower, fully white, not or hardly speckled, more sharply defined. FIRST ADULT. Like adult; rather easy to identify when some pale-barred and white-tipped longer scapulars, tertials, or upper wing-coverts still present, more difficult when only juvenile flight-feathers and tail-feathers remain, most of these showing rather narrow and virtually unmarked white tips, unlike adult (see Juvenile).

Bare parts. ADULT AND FIRST ADULT. Iris dark brown or umber-brown, often with red outer ring. Bill black. Leg and foot dark greenish-grey, grey-brown, or olive-black, claws dark horn-brown or brown-black. (Hartert 1912–21; BMNH, RMNH, ZMA.) DOWNY YOUNG AND JUVENILE. No information.

Moults. ADULT POST-BREEDING. Complete; primaries descendant. No information for North African birds. In West Africa, where breeding October–December, primary moult starts late January to mid-April and primary moult score 30 reached mid-March to late May; moult probably completed May–July. In South and East Africa, where breeding mainly March–July, birds in early stages of primary moult recorded late September and October, in final stages November–December. Moult of head, body, and tail during primary moult; sequence as in Short-eared Owl *A. flammeus*. POST-JUVENILE. Partial: involves head, body, and wing-coverts, but no greater primary coverts and not always all greater upper wing-coverts. Starts at age of *c*. 1 month, completed at 2–3 months. A single Spanish bird, early November, still showed scattered juvenile on body and upper wing-coverts. See also Smith and Killick-Kendrick (1964).

Measurements. Nominate *capensis* from South and East Africa and *tingitanus* from north-west Africa, combined, all year; skins (BMNH, RMNH, ZMA). Bill (F) to forehead, bill (C) to cere.

WING	♂ 292	(5·55; 11) 284–300	♀ 298	(5·66; 10) 290–306
TAIL	138	(3·75; 9) 132–142	145	(3·27; 9) 142–153
BILL (F)	30·7	(2·01; 9) 28·2–32·9	30·5	(1·25; 9) 29·1–33·1
BILL (C)	17·4	(0·79; 6) 16·5–18·5	17·9	(0·56; 7) 16·8–18·5

Sex differences significant for wing and tail.

Sexes combined, including unsexed birds (1) *A. c. tingitanus*: Spain, Morocco, Algeria. (2) Nominate *capensis*: South and East Africa.

WING	(1) 294	(6·02; 14) 284–303	(2) 296	(6·48; 13) 286–306
TAIL	142	(4·97; 13) 133–153	141	(4·06; 13) 132–145
BILL (F)	30·8	(1·35; 13) 29·0–33·0	30·4	(1·57; 13) 27·8–33·1
TARSUS	56·7	(2·66; 12) 52·7–60·0	54·1	(2·04; 11) 50·8–56·5

Differences significant for tarsus.

Wing of birds from West Africa east to Lake Chad area rather large: 306 (11·2; 9) 290–318. Wing of juvenile on average *c*. 6 below adult, but samples too small to test whether difference significant and all ages combined in tables.

Weights. Nominate *capensis*. ADULT AND FIRST ADULT. 310 (38·7; 16) 227–355 (Anon 1968b; Biggs *et al.* 1979; W R J Dean). DOWNY YOUNG. On 1st day, 19·5 (W R J Dean). One adult ♀ of large Madagascar race *hova*, 485 (ZMA).

Structure. Wing long, rather broad, tip rounded. 10 primaries: p8 longest, p9 1–8 shorter, p10 21–34, p7 4–10, p6 21–34, p5 42–55, p4 58–74, p1 96–115. Outer web of p7–p9 and inner of (p8–)p9–p10 emarginated. Outer web of p10 and emarginated parts of outer webs of p8–p9 serrated; serrations slightly longer and more curved than in *A. flammeus*. Tail rather long, tip slightly rounded; 12 feathers, t6 10–20 shorter than t1. Facial disc marked, bordered by distinct ruff; small tufts at border of ruff above eye, consisting of 3–5 feathers up to 20–35 mm long. Tarsus completely covered by feathers, but feathering on toes variable; usually, at least central uppersurface of basal 2–3 digits of toes feathered, but feathering sometimes lost by abrasion; occasionally, all uppersurface of toes feathered, except for distal digits. Middle toe without claw 27·9 (1·80; 6) 26–30 in ♂, 29·7 (1·47; 8) 28–32 in ♀; outer toe *c*. 69% of middle, inner *c*. 82%, hind c. 58%. Claws slender and sharp, base slightly wider than in *A. flammeus*; middle claw 16·3 (1·48; 6) 15–18 in ♂, 16·3 (0·37; 8) 15–18 in ♀; outer claw c. 83% of middle, inner c. 96%, hind c. 88%. Remainder of structure as in *A. flammeus*.

Geographical variation. Rather slight on African mainland. Measurements of all populations rather similar, only tarsus of north-west African birds and wing of birds in belt south of Sahara slightly longer. *A. c. tingitanus* of north-west Africa rather dark; upperparts, chest, and breast dark sepia-brown with slight rufous tinge; belly and flanks heavily marked red-brown with distinct rounded white spots; vent and under tail-coverts buff, well-marked with rufous-brown. Nominate *capensis* from southern and eastern Africa generally slightly colder dark olive-grey on upperparts and chest; breast and belly less densely marked, appearing barred white rather than spotted; vent and under tail-coverts often paler cream or white and less marked with brown. In both races, ♂ paler than ♀, and north-west African ♂ rather similar to South African ♀ in density of marks of underparts. Birds from West Africa east to Lake Chad somewhat intermediate in colour and pattern between Maghreb and South African birds, but nearest latter and hence best included in nominate *capensis*. Single bird from south-west Sudan similar to *tingitanus*, one from south-east Sudan like nominate *capensis*. Madagascar race *hova* spotted buff on crown and hindneck and more distinctly and heavily barred on scapulars, chest, and breast; large, with wing 320–340, tail 155–170, bill to forehead 34–36, tarsus 60–70.

CSR

Aegolius funereus Tengmalm's Owl

Du. Ruigpootuil   Fr. Chouette de Tengmalm   Ge. Rauhfusskauz
Ru. Мохноногий сыч   Sp. Lechuza de Tengmalm   Sw. Pärluggla   N. Am. Boreal Owl

Strix funerea Linnaeus, 1758

Polytypic. Nominate *funereus* (Linnaeus, 1758), Europe, except Caucasus; *caucasicus* (Buturlin, 1907), Caucasus. Extralimital: *pallens* (Schalow, 1908), western Siberia (intergrading with nominate *funereus*), Tien Shan, and southern Siberia east to Sakhalin; *magnus* (Buturlin, 1907), north-east Siberia from Kolyma to Anadyr and Kamchatka, intergrading with *pallens* between northern Transbaykalia and Verkhoyanskiy range; *beickianus* Stresemann, 1928, Lahul (north-west India) and Tsinghai (south-west China); *richardsoni* (Bonaparte, 1838), North America.

Field characters. 24–26 cm; wing-span 54–62 cm. Total bulk close to Little Owl *Athene noctua* but large head closer in size to that of Hawk Owl *Surnia ulula*; tail longer than in *A. noctua*. Quite small but noticeably deep- and broad-headed owl, with well-spotted dark brown upperparts and blotched white underparts. Facial disc dull white, with almost black upper corners and raised white eyebrows giving characteristic astonished expression. Flight recalls that of flushed Woodcock *Scolopax rusticola*. Chiefly nocturnal. Song distinctive. Sexes similar; no seasonal variation. Juvenile separable. 2 races in west Palearctic; difficult to separate.

ADULT. (1) European and west Siberian race, nominate *funereus*. Ground-colour of upperparts umber-brown, relieved by copious white spots on crown, fewer and larger white spots and blotches on broad nape and centre of mantle, bold white edges to outer mantle feathers and scapulars, and discrete but less numerous white spots on wing-coverts and flight-feathers; no more than 5 rows of spots on primaries (7 in *A. noctua*, 9 in *S. ulula*). Facial disc pale buff-white, outlined with brown-black (particularly above and at side of eyebrows, forming dark corners to head and lower cheeks) and with whiter raised eyebrows and bill feathering emphasized by dull brown-black patches between bill and yellow eyes; area under yellow bill shadowed but not forming black throat-patch. Tail umber-brown, with 3 rows of white spots remaining isolated even when tail closely folded (unlike all other owls). In flight, upperwing less obviously barred than in *A. noctua*, with white spots forming dotted lines; underwing white, spotted brown on primary and lesser coverts and indistinctly barred on flight-feathers which also lack prominent dark tips. Underparts basically white, blotched and spotted warm brown, with markings coalescing as broad mottling on chest and soft streaks on body. Legs and feet thickly feathered white. (2) Caucasian race, *caucasicus*. Darker and less white-spotted than nominate *funereus*. JUVENILE. Much darker, more chocolate-brown than adult. Virtually unmarked on crown and back; heavily blotched and suffused on chest. Facial disc complete, but brown plumage of head extends over cheeks with white only on eyebrows and moustaches. Wing and tail patterns like adult but wing-coverts less spotted.

Unmistakable when seen well: large rectangular head and perplexed expression eye-catching. In brief glimpse or incomplete view, liable to be confused with *A. noctua* or even Pygmy Owl *Glaucidium passerinum*. Important to recognize that *A. funereus* has not dissimilar flight action to Tawny Owl *Strix aluco* and lacks, in particular, rather stupid, frowning look and bounding flight of *A. noctua* and stern expression and rapid, dashing flight of *G. passerinum*. Flight includes rapid wing-beats typical of small owls, and also glides on straight wings; this combination more reminiscent of *Scolopax rusticola* or small *Strix aluco* than other small owls. Markedly agile in flight, being capable of tight turns in dense cover.

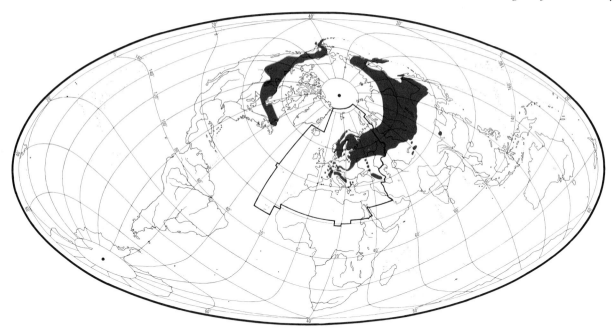

Song usually 5–7 soft, rapid 'po-' notes in repeated phrases lasting 1–2 s. In summer and autumn, often gives a short, smacking 'yiop'.

Habitat. Like Pygmy Owl *Glaucidium passerinum* and Ural Owl *Strix uralensis*, inhabits within west Palearctic a wide belt of northern taiga lowlands and detached series of montane coniferous forests in temperate zone. Shows preference for spruce *Picea*, but often occurs in mixed forests of pine *Pinus*, birch *Betula*, and poplar *Populus* (Mikkola 1983). On Scandinavian fells, ascends to lower fringes of birch belt and occasionally beyond (Bannerman 1955). In USSR, up to 2000 m in Altay; in north-east Siberia, in broad-leaf (e.g. poplar) forests of river valleys (Dementiev and Gladkov 1951a). In central European mountains, typically from *c.* 1100–1800 m. In uplands down to *c.* 400 m or lower and sparsely also in lowlands where winter frosts are prolonged and summer temperatures low (Glutz and Bauer 1980). Highest density in Jura found in mixed woodland of silver fir *Abies* and beech *Fagus*. Appears at least in parts of range so dependent on using holes made by Black Woodpecker *Dryocopus martius* as to be almost inseparably linked to it (Glutz von Blotzheim 1962). Apparently disadvantaged in competition with Tawny Owl *Strix aluco*, which prefers steep slopes in montane regions, leaving flat plateaus and ridges to *A. funereus*. Hunts indiscriminately in dense forest or along rides, in clearings and young plantations, on moors and meadows, along forest-edges and in dwarf growth, and even near forest dwellings or small settlements. Wandering juveniles may occur in parks and gardens. Avoids scrub and extensive bare ground (Niethammer 1938; Glutz and Bauer 1980). Geographically and ecologically similar to *G. passerinum*, and available data do not enable clear distinction to be made between their respective habitat needs.

Distribution. Some recent extension of range in France, Belgium, and Netherlands. For winter eruptions, see Movements.

FRANCE. Some recent extension in Var, Bourgogne, and Lorraine; breeding first recorded eastern Pyrénées 1962 and Mont Ventoux 1965 but probably overlooked previously (Yeatman 1976). BELGIUM. First bred 1963 and 1965; now annual (Tricot 1977). NETHERLANDS. First proved breeding 1971, but song heard some years earlier (CSR). DENMARK. First confirmed breeding 1979–80: 1 pair Bornholm (TD). FINLAND. Range extends slightly less far north than mapped except in years of high population (LJL). BULGARIA. Bred Rila mountains 1980 (Simeonov 1980); first nesting record since 1914, 1922, and 1927, but may be regular breeder (TM). GREECE. Local; probably breeds elsewhere in mountainous areas of north and centre (WB, HJB, GM). RUMANIA. Decreased; breeds occasionally, last 1981 (Vasiliu 1968; VC). USSR. Formerly bred Crimea (Stegmann 1938); bred 1966 near Rostov (Minoranski 1976).

Accidental. Britain, Hungary.

Population. Fluctuating; some evidence of recent increase in north-west Europe.

FRANCE. 100–1000 pairs (Yeatman 1976). BELGIUM. 12 pairs 1969; at least 10 pairs 1972 (Tricot 1977); 1–6 pairs since (PD). NETHERLANDS. 5–10 ♂♂ in north 1975–7, declined 1978–9; 1 heard in south 1977 (CSR). WEST GERMANY. At least 800 territorial ♂♂ in good food years, significantly above some recent estimates (Glutz and Bauer

608 Striginae

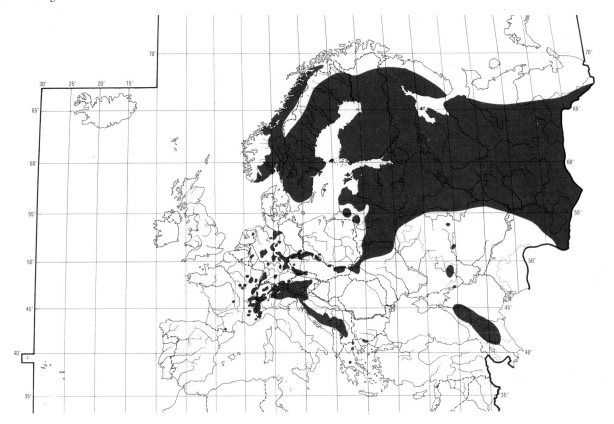

1980); *c*. 900–2400 pairs (Bauer and Thielcke 1982). SWEDEN. Estimated 35 000 pairs (Ulfstrand and Högstedt 1976). No changes in population known (LR). FINLAND. Fluctuating; estimated 1500 pairs (Merikallio 1958); probably too low (Mikkola 1983). Fluctuations significantly positively correlated with those of *Microtus* voles (Korpimäki 1981). EAST GERMANY. Over 135 pairs (Mikkola 1983). POLAND. Scarce breeder in mountain areas (Tomiałojć 1976a). CZECHOSLOVAKIA. About 100 pairs (Mikkola 1983). USSR. Common in northern taiga, very rare in Caucasus (Dementiev and Gladkov 1951a). Fairly numerous in Estonia, with up to 500 pairs in good years (Randla 1976), but rare Lithuania and Latvia (HV).

Oldest ringed bird 8 years 2 months (Glutz and Bauer 1980).

Movements. In north, largely resident (adult ♂♂) or nomadic (♀♀ and immatures), becoming eruptive at times; resident but with limited dispersal in central Europe.

Formerly regarded as wholly resident in Europe but now known that, in northern boreal zone, shows cyclic movements at 3–4 year intervals that are linked to mammalian prey densities (Mysterud 1970). Finnish study (Saurola 1979) indicated 4-year cycle there as follows. Year 1: peak production of young in southern Finland (e.g. 1973, 1977), and in autumn part of population moves northwards into central and northern Finland; low numbers of transient birds at southern bird observatories (Sweden and Finland) then, while ringing recoveries tend to be to north of ringing site. Year 2 (e.g. 1974, 1978): peak breeding success in northern Finland, with some southerly dispersal in autumn. Year 3 (e.g. 1975): few young reared as rodent population crashes, and birds irrupt (to greater or lesser extent) into southern Fenno-Scandia. Year 4 (e.g. 1976): rodent numbers begin to recover, as does breeding success of *A. funereus*, in southern Finland. Hence breeding densities can change regionally between years, according to variations in prey density, as in (e.g.) Hawk Owl *Surnia ulula*; some birds certainly resident though non-breeding in poor rodent summers (Linkola and Myllymäki 1969); probably ♂♂ at least (see below). Ringing recoveries within Fenno-Scandia of up to 900 km displacements, including westwards (into Norway) as well as north and south. Erupting Finnish birds also recovered (October–November) in Orel, USSR (1310 km south-east), north Ukraine, USSR (1220 km south-east), and Mecklenburg, East Germany (1350 km south-west); one Norwegian bird found in Durham, England (1015 km south-west). Despite normal reluctance to cross wide waterbodies, more British records (over 50) than for other northern irruptive owls except Snowy Owl *Nyctea scandiaca*; also reaches countries bordering southern Baltic. One October migrant in Kaliningrad found $1\frac{1}{2}$ years later (February) in Smolensk,

USSR (750 km east). For comparable eruptive behaviour in North America, see Catling (1972b).

Probably younger birds predominate in eruptive movements; longer Norwegian and Finnish ringing recoveries involved that age-group, though no clear age distinction in Swedish results (Wallin and Andersson 1981). Among adults, marked sexual difference apparent in movement pattern (Lundberg 1979a), with adult ♂♂ resident in restricted home range—visiting prospective nest-holes from December onwards (T Holmberg)—while adult ♀♀ nomadic, generally without nest-site attachment. This strategy of partial migration regarded as a compromise between conflicting pressures of need to defend nest-holes (a scarce resource) and periodic food shortage (Lundberg 1979a), but not yet proven that ♂♂ invariably defend home-range during years of low vole abundance as well as years of plenty (Mikkola 1983); see also Social Pattern and Behaviour. In Sweden a small fraction of adult ♀♀ also show home-range attachment, but majority are nomadic with proven breeding sites up to 200 km apart in different years (T Holmberg); one case of 510 km NNE displacement between 1968 and 1970 nests (Wallin and Andersson 1981). In Finland, one 1967 nestling found breeding 1968, 36 km south-east of natal site; in 1969, this ♀ found breeding in May (2·7 km from 1968 nest) and in July (10·6 km from previous nest); hence demonstrated ♀'s unfaithfulness to nesting area as well as probable polyandry (see Social Pattern and Behaviour) and age of maturity (Mikkola 1983). No evidence of sexual difference in juvenile dispersal (T Holmberg).

CENTRAL EUROPEAN POPULATION. Under conditions of less extreme fluctuations in food supply (compared with northern Europe), birds show higher degree of year-round adherence to home range. Ringing data mainly for ♀♀, due to nest-trapping method. Recaptures showed that first-time breeders usually settled within c. 20 km of natal area, and thereafter remained relatively site-faithful; 57 recaptures in later years (adult ♀♀) were 0–18·7 km distant, including 13 recaptured at same nest hole. Longest movements (all by birds ringed as nestlings): 125 km east (September juvenile) and 194 km north-east (breeding in subsequent year), both within southern West Germany; 235 km WNW (October juvenile) from West Germany to Netherlands. See Glutz and Bauer (1980). Dispersing young thus probably involved in modern range extensions into western Europe.

Food. Largely small voles (Microtinae); also mice (small Murinae), shrews (Soricidae), and small birds. Hunting essentially nocturnal, though birds often active also in early morning in breeding season; birds with large young active at low level throughout day (Klaus et al. 1975; Korpimäki 1981). 2 peaks of activity at nest in central Europe, merging into 1 further north: at Jena (East Germany), 20.00–22.00 hrs and 02.00–05.00 hrs; at Konnevesi (southern Finland), 23.00–24.00 hrs and 01.00–02.00 hrs; at Oulu (central Finland), 23.00–03.00 hrs (Klaus et al. 1975)—similar single peak in northern Sweden (Norberg 1970). For activity pattern of captive bird, see Erkinaro (1973a). Hunts from perch (average height 1·7 m, $n = 154$), usually in woodland or other cover. Twice seen hovering for 2–3 s at 1–1·5 m. In 244 min of observation, average c. $1\frac{4}{5}$ min on each perch; average length of flight between perches 17 m ($n = 153$). On perch, turns head rapidly in various directions and, once prey located, stares straight at it, sometimes making small lateral or upward head movements or occasionally rotating head c. 30° about longitudinal axis; moves feet close together or treads on the spot, and lowers head almost to level of feet. Makes shallow wing-beats in first part of flight to prey (in short, steep descent may use tail as brake at first). When c. 1 m from prey, glides head-first, directly towards it. About 50 cm from prey, moves feet forward and close to body, legs bent. At c. 25 cm, raises body, holds wings in shallow V, and spreads tail. Just before impact, draws head back, closes eyes, extends legs forward, and spreads claws (outer, as well as hind, pointing backwards). On striking, immediately lowers spread wings and tail to ground for support (struggling mouse may carry bird 10–20 cm); looks around, and within a few seconds kills prey by biting head or back of neck; sometimes stays on ground for 20–30 s. (Norberg 1970, which see for photographs of attack-flights; see also Lindblad 1967.) Always begins to eat prey from head (Korpimäki 1981); rarely swallowed whole; intestines discarded. Said to take birds by stalking them on branch or by hovering over them; occasionally robs nests (Glutz and Bauer 1980). In flight, carries prey in one foot, transferring it to bill at intermediate perch just before nest (Norberg 1964, 1970). Can carry Fieldfare Turdus pilaris in flight (Korpimäki 1981). In captivity, frozen small mammals 'brooded' to thaw them out before eating (Bondrup-Nielsen 1977), and fresh ones similarly prevented from freezing during day, before being eaten at night (Scherzinger 1979). Weak or dead young eaten by siblings or fed to them by ♀ (Glutz and Bauer 1980; Korpimäki 1981). Food-caching (in crevices, forks of branches, etc.) occurs in winter as well as in breeding season (Schelper 1972a; see also Social Pattern and Behaviour). Eats snow (Glutz and Bauer 1980).

Pellets generally smaller than those of Little Owl *Athene noctua*, but darker and more solid (Thiollay 1963b). Average measurements: 22 × 12 mm, $n = 1142$ (Erkinaro 1973b; see also for seasonal variation); average 30 × 13 × 12 mm, $n = 42$ (Mikkola 1981a). Captive newly fledged bird produced average 1·2 (0–2) per day (Mebs 1966). Roost-sites change frequently and pellets thus difficult to find.

For structure of ears (set with bilateral vertical asymmetry to aid sound localization), see Norberg (1973, 1978).

The following prey recorded in west Palearctic. Mammals: shrews (*Sorex minutus*, *S. araneus*, *S. caecutiens*, *S. minutissimus*, *Neomys fodiens*), bats (*Plecotus*, *Vespertilio nilssoni*), red squirrel *Sciurus vulgaris*, flying squirrel *Pteromys volans*, forest dormouse *Dryomys nitedula*, dormouse *Muscardinus avellanarius*, wood lemming *Myopus schisticolor*, Norway lemming *Lemmus lemmus*, ruddy vole *Clethrionomys rutilus*, bank vole *C. glareolus*, grey-sided vole *C. rufocanus*, water vole *Arvicola terrestris*, pine vole *Pitymys*, common vole *Microtus arvalis*, short-tailed vole *M. agrestis*, harvest mouse *Micromys minutus*, striped field mouse *Apodemus agrarius*, yellow-necked mouse *A. flavicollis*, wood mouse *A. sylvaticus*, brown rat *Rattus norvegicus*, house mouse *Mus musculus*, northern birch mouse *Sicista betulina*. Birds range in size from Goldcrest *Regulus regulus* to (commonly taken) large thrushes *Turdus*. Invertebrates apparently taken only rarely: beetles (*Carabus*, *Geotrupes*) and Ichneumonidae recorded. Frogs *Rana* also recorded. (Moltoni 1937; Kadochnikov 1962; Fredga 1964; Lindhe 1966; König 1969b; Sulkava and Sulkava 1971; Klaus *et al.* 1975; Glutz and Bauer 1980; Korpimäki 1981; Mikkola 1983.)

Feeding ecology most thoroughly examined by Korpimäki (1981). For summary of representative European studies, see Table A. Other contributions as follows. Norway: Hagen (1959). Finland: Linkola (1963), Sulkava and Sulkava (1971), Klaus *et al.* (1975). Sweden: Fredga (1964), Ahlbom (1973, 1975), Kampp (1982). East and West Germany: Gasow (1968), Haase (1969), König (1969b), Klaas (1971), Schelper (1972a, b). USSR: Kadochnikov (1962). See also Uttendörfer (1939b, 1952) and sources in Korpimäki (1981).

In south-west Finland, 1973–9, *Sorex araneus* taken in proportion to its abundance (as found from trapping); *Microtus* taken in higher proportions than predicted by abundance, and *Clethrionomys* taken in lower proportions. Among birds, *Turdus pilaris*, Song Thrush *T. philomelos*, and Bullfinch *Pyrrhula pyrrhula* taken in higher propor-

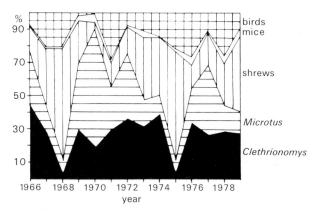

Fig VI Year-to-year variation in food of Tengmalm's Owl *Aegolius funereus* at end of breeding season, Kauhava (south-west Finland). Data refer to proportion by number of total 6689 vertebrates. (Korpimäki 1981.)

tions than predicted by abundance. During breeding season, proportion of *Microtus* increases in March–April when melting snow in fields forces them above ground where cover lacking, *Clethrionomys* and shrews increase in May after snow melts in woods, and birds increase in June as their young become available and cover for small mammals grows up (Table A). In spring, both mammal and bird prey contain significantly more ♂♂ than ♀♀, due to ♂♂'s greater activity at this time. (Korpimäki 1981; see also for proportions of different prey caught at different times of night.) Year-to-year variation considerable (Fig VI—but note that shrews would be much less important in an analysis by weight, as in Table A); see also Sulkava and Sulkava (1971) and Klaus *et al.* (1975); for analysis of data in Fig V with respect to phase of population cycle of *A. funereus*, see Korpimäki (1981). During February–June 1973–9, south-west Finland, ate 9·1%

Table A Food of Tengmalm's Owl *Aegolius funereus* as found from food remains in winter roost-sites, stored prey, pellets, and nest-remains. Figures are percentages of total live weight of vertebrates only.

	SW Finland			N Sweden	E Germany
	1970–9, winter	1968–79, start of breeding	1966–79, end of breeding	1964, summer	1966–74, summer
Mice (small Murinae)	2·3 } 67·8	1·4 } 82·6	1·8 } 62·5	0 } 89·1	11·0[1] } 86·0
Small voles (Microtinae)	65·5	81·2	60·7	89·1	75·0
Larger mammals	0	0·8	1·8	0	0
Shrews (Soricidae)	2·9	6·1	11·5	9·8	5·9
Birds	29·2	10·6	24·2	1·1	8·1
Other	0	0	0	0	+ (bats)
Total no. of vertebrates	217	2050	6689	489	1430
Source		Korpimäki (1981)		Lindhe (1966), data reanalysed	Klaus *et al.* (1975), data reanalysed

1. Includes a very few dormice *Muscardinus avellanarius* (Muscardinidae).

of small mammals present in habitat in spring, including 17·2% of *Microtus* present (Korpimäki 1981). For comparison of diet with that of 7 other Strigidae in southwest Finland, see Korpimäki (1981).

Food consumption of captive newly-fledged bird 55·6 g per day over 44 days, gaining 26 g; adult 65 g per day (Mebs 1966). In captivity, average 28·0 g of meat per day, September–March; in breeding season *c*. 21·3 g of meat per day. Over period of rearing to independence, young ate average 30·5 g of meat per day. (Glutz and Bauer 1980.) Over whole period in nest, wild young ate average 21·0 g per bird per day (Korpimäki 1981).

No details of differences between diets of adults and young, but evidently substantially similar. Studies are often based on analysis of nest-remains and, as ♀ does not usually eject pellets inside nest-hole, these primarily indicate food of young. Feeds last 5–10 min in 1st week and take place at ½-hr intervals through day and night (Glutz and Bauer 1980). At 5 nests, average numbers of visits to nest per day when young in nest: 13·1, 13·0, 8·6, 5·3, 4·7; number of visits correlated positively with number of young, negatively with rainfall (Klaus *et al.* 1975). Average 6·0 feeding visits to nest per night ($n = 17$ nights) during first 3 weeks when ♀ in nest continuously, 4·3 ($n = 12$) after ♀ leaves (Korpimäki 1981). On one night, young fed 7 times 22.35–01.00 hrs (Pynnönen 1949). DJB

Social pattern and behaviour. Major studies by Schelper (1972a) and Korpimäki (1981). Account includes information supplied by T Holmberg.

1. Solitary outside breeding season (Glutz and Bauer 1980). BONDS. Mating system perhaps normally monogamous, but polygyny and polyandry occur, perhaps commonly. 6 published records of polygyny (♂ and 2 ♀♀), 2nd ♀ laying 18–30 days after 1st; of the 12 nests, only 2 (both of 2nd ♀♀) produced no fledged young (Solheim 1983). In 21 nesting attempts in Sweden, estimated 14 ♂♂ involved, i.e. 50% of ♂♂ bigamous; cases of ♂ paired to 3 ♀♀ probable but not proved; ♂ sometimes appears to desert one ♀ and nest thus fails (T Holmberg). 4 records of polyandry, involving successive pairings between ♀ and 2 ♂♂, ♀ abandoning 1st brood when *c*. 3 weeks old (as often happens also in monogamous pairings) and then breeding with 2nd ♂; 2nd clutch started 50–64 days after 1st; of the 8 nests, all produced fledged young (Solheim 1983). In one case, ♂ and ♀ of a pair both bred subsequently in same season with different ♀ and ♂ (Kondratzki and Altmüller 1976). Long breeding season and large distance between nests of polygamous birds (see Breeding Dispersion, below) could mean polygamy more frequent than literature suggests (Solheim 1983). Pair-bond apparently does not persist outside breeding season—certainly not usually, as most ♀♀ nomadic and ♂♂ resident. One record of ♂ and ♀ breeding together in 2 consecutive years (Korpimäki 1981), but as ♀♀ less faithful to particular breeding area than ♂♂, this probably not typical. ♀ does all incubation and brooding; ♂ does all hunting, at least until young *c*. 3 weeks old. Once, however, ♀ which lost mate reared brood of 5 alone (Holmberg 1974). Fledged young tended for 5–6 weeks (see Relations within Family Group, below). First breeding (for both sexes) commonly at 1 year old in central Europe (Glutz and Bauer 1980); perhaps later further north if food supply unfavourable. BREEDING DISPERSION. Solitary and territorial. Pairs often in loose groups of 2–4 (Korpimäki

1981). Breeding density highly variable, both with time and space, and even if nest-boxes evenly spaced. In good year, Sweden, 1·5 nests per km² over 14 km², locally (in areas of woodland and fields) 5 nests in 0·25 km² or 250–300 m between nests; in low-density areas (pure woodland with bogs), nests averaged 3·5 km apart (T Holmberg). In western Finland, 1973–9, 0·08–0·33 pairs per km² over 24 km² (Korpimäki 1981). In peak years: Sweden 0·48 pairs per km² (Källander 1964), East Germany 0·83 pairs per km² (Feuerstein 1960), West Germany 1·3 pairs per km² (Schelper 1972a). See also sources in Glutz and Bauer (1980) and Korpimäki (1981). Territory apparently limited to immediate vicinity of nest-hole; defended by ♂ only, and then only until hole occupied by ♀ (Solheim 1983). Closest recorded nests 25 m apart (T Holmberg). Such cases perhaps involve polygamy, but close nesting not typical here. Nests known as close as 100 m without link (via polygamy) between them (Solheim 1983). In polygyny: of 5 cases, nests of the 2 ♀♀ in each case averaged 0·5 km apart (T Holmberg); of 6 cases, nests 0·5–2·1 km apart. In polyandry: of 4 cases, successive nests 0·5–10 km apart (Solheim 1983). Ringing in Sweden suggests ♂♂ mostly nest within home-range of *c*. 2 km diameter which they occupy all year. Small fraction of ♀♀ show similar tenacity, but majority seem highly nomadic (T Holmberg; see Movements). ROOSTING. Diurnal. In dense vegetation, usually close to trunk in crown of tree; bird changes site frequently. Outside breeding season, nest-hole used only exceptionally in central Europe (Glutz and Bauer 1980)—this perhaps more common further north (Sulkava and Sulkava 1971), though not found by Norberg (1964) in south-central Sweden. During breeding, ♂ usually roosts within 100 m of nest (Glutz and Bauer 1980). For bathing activity (perhaps relating to captive birds), see Glutz and Bauer (1980). For periods of hunting activity, see Food.

2. When hunting, flushes from approaching man at 3–5 m (Norberg 1970). When excited, will jerk tail and wing-tips, though rarely. In alarm, adopts concealing Sleeked-upright posture (Fig A): body elongated with feathers sleeked; carpal joint of wing nearer danger drawn up in front of and to level of bill; feathers at edge of facial disc above wide-open eyes raised; feathers between eyes fanned out (Catling 1972a). If more directly threatened may threaten intruder (see Parental Antipredator Strategies, below) or perform mock-sleeping (Glutz and Bauer 1980). ANTAGONISTIC BEHAVIOUR. ♂ advertises territory with Advertising-call (see 1 in Voice), though only up to pair-formation. Approaches intruding ♂♂, calling (König 1969b: see 7 and 12 in Voice). Observer imitating Advertising-call often similarly approached by resident ♂ who intensifies calling; exceptionally, swoops low over observer or even Bill-snaps and

A

wing-claps (Knobloch 1958; Schelper 1972a; Zapf 1973). In central Europe (if food supply good), resurgence in territorial behaviour occurs September–October; involves Advertising-call and, exceptionally, also pair-formation calls from ♂ and ♀ (Glutz and Bauer 1980). In years of poor food supply, ♂♂ largely silent, but will still respond to imitations of Advertising-call (Lundberg 1979a). Heterosexual Behaviour. (1) General. ♂ advertises territory (see Antagonistic Behaviour, above), and after attracting ♀, offers her one or several nest-holes. Final choice made by ♀. (Glutz and Bauer 1980.) (2) Pair-bonding behaviour. When unpaired ♀ approaches ♂ giving Advertising-call, she identifies herself by calling (see 5a in Voice) from 100–300 m away or more (if not, ♂ may drive her off). ♂ stops calling and flies into nest-hole where he gives Trill-call (see 3 in Voice). If ♀ remains silent or continues with same call, ♂ breaks off further courtship, but if ♀ responds (sometimes only after days of contact) with Gickering-call (see 5b in Voice) ♂ continues Trill-call until ♀ moves away from nest-tree. ♂ changes to Stutter-call (see 2 in Voice) and then, having flown to more open area, to Advertising-call. In fresh approach, if ♀ gives Contact-calls (see 6 in Voice) as well as Gickering-calls, she then normally enters hole. ♂ may give several phrases of quiet version of Advertising-call then fetch food for ♀ from cache; ♀ gives Begging-call (see 9 in Voice). Time between start of pair-formation and 1st mating may be 1–2 days or over 1 month. (Glutz and Bauer 1980.) (3) Mating. Occurs when ♀ leaves nest in evening—usually on regular branch near hole. ♂ flies in giving short phrases of Trill-call; ♀ bows forward, cocks tail steeply, and ruffles belly feathers. Copulation lasts 4–6 s, during which ♂ gives quiet short Trill-calls and ♀ gives Copulation-call (see 10 in Voice). ♂ flies off giving short Trill-calls. If ♀ not initially cooperative, ♂ chases her in flight, giving soft short Trill-calls. (Kuhk 1949; Glutz and Bauer 1980.) (4) Courtship-feeding and food-caching. After ♀ occupies nest-hole, ♂ provides all her food until she ceases brooding young and leaves nest. ♂ may (Korpimäki 1980) or may not (T Holmberg) cache prey in nest-holes outside breeding season. Some food usually stored in nest before hatching. Each item kept for 1–9 days (average 1·6 days, $n = 148$ items), and items eaten in order of storage. Up to 35 items per nest: number greatest at laying (average 5·4 items over 58 visits) and at hatching (average 5·1 items over 91 visits); normally none during last 2 weeks of nestling period after ♀ stops brooding. (Korpimäki 1981.) Number of prey stored at nest can change greatly over a few days, and cache may thus serve as buffer against changes in prey availability (T Holmberg). (5) Allopreening. None (Glutz and Bauer 1980). (6) Behaviour at nest. ♂ and ♀ perform scraping in nest-hole (Schelper 1972a). Some ♂♂ visit nest-hole only when ♀ absent, others will spend some time in hole with ♀ (Glutz and Bauer 1980). ♀ stays in hole for up to 6 days before laying (Kuhk 1949; Plucinski 1966). Only ♀ incubates, provisioned by ♂; prey transferred inside nest or at entrance. ♀ leaves nest for 3–9 min only once or twice per night to defecate, eject pellet, and preen (Glutz and Bauer 1980). Relations within Family Group. Young give food-calls before hatching; first fed soon after (Glutz and Bauer 1980). Only ♀ feeds young; may be typical for ♀ either to call (Glutz and Bauer 1980: see 11 in Voice) or to remain silent (T Holmberg). Weak or dead young eaten by siblings or fed to them by ♀ (Glutz and Bauer 1980; Korpimäki 1981). Young Allopreen each other (Glutz and Bauer 1980). Until oldest young 15–23 days old, ♀ leaves nest only briefly (Korpimäki 1981); ♂ may also do some brooding (Glutz and Bauer 1980), though this apparently unusual. ♀ then leaves nest and sometimes plays no further part (may start new clutch with 2nd ♂—see Bonds, above). ♀ sometimes brings food, however (also before brooding phase over), but apparently only if food supply poor

or ♂ bigamous: during brooding phase, 4 of 60 deliveries to nest made by ♀; after ♀ leaves nest, 38 of 54 deliveries (at 38 nests) made by ♀ (T Holmberg). ♀ keeps nest clean during incubation and brooding, but faeces and food remains accumulate after she leaves (Kuhk 1969; Korpimäki 1981). Young open eyes at c. 1 week (T Holmberg) or 8–11 days (Kuhk 1969). Can raise head at 3 days, stand upright at 9 days, and clamber about at c. 20 days; climb up to nest-entrance at 26–30 days (Glutz and Bauer 1980; Korpimäki 1981). Can fly well on leaving nest; not influenced to leave by parents (T Holmberg). ♂ gives quiet version of Advertising-call or short Trill-calls to move young to different areas: one brood moved over 600 m in first 10 days (Knobloch 1958); 250 m in first 2 weeks (Korpimäki 1981). About 14 days before independence, young of one brood were wandering up to 1 km from regular site of prey transfer (Glutz and Bauer 1980). Young fed by parents for 5–6 weeks after fledging (Haase and Schelper 1972). Anti-predator Responses of Young. If threatened, react by Bill-snapping from 9 days, hissing from 14 days or earlier. At 10–20 days, play dead when handled. At 3 weeks, ruffle feathers, lean back (even right over on to back), and strike with claws and bill. (Glutz and Bauer 1980; T Holmberg, C König.) Parental Anti-predator Strategies. (1) Passive measures. ♀ normally sits tight, usually putting head out of hole when nest-tree tapped. Often stays on nest even if nest-box opened, apparently as adaptation to predation by (especially) martens Martes. Both sexes will give Warning-call (see 13 in Voice) to silence nestlings or fledged young. (T Holmberg.) (2) Active measures: against man. ♀ on nest Bill-snaps, gives Hissing-call (see 14 in Voice), bows forward, ruffles feathers (especially on head and nape), and may fan wing nearer intruder; a few ♀♀ press themselves on nest, not moving. Will strike with claws and bill. If ♀ flies out, usually waits 2–20 m away. Sometimes Bill-snaps or gives Warning-call. ♀ seldom makes flying attacks and even then rarely strikes, touching intruder only lightly. When brooding phase over, ♀ usually does not appear in vicinity of nest when intruder present, but may give Warning-call from some distance. (Glutz and Bauer 1980; T Holmberg, C König.) (3) Active measures: against other animals. Attacks recorded against squirrel Sciurus on neighbouring tree (Glutz and Bauer 1980) and deer (Cervidae) near fledged young (Schelper 1971).

(Fig A adapted from photographs in Hagen 1935 and Scherzinger 1969a.) DJB

Voice. Advertising-call (call 1) and Smacking-call (call 7) heard most frequently. Calls 1–4 have common basis and evidently intergrade. Both sexes perform Bill-snapping in threat (T Holmberg). Wing-clapping in display apparently exceptional (König 1964, 1968). Account includes information provided by T Holmberg. For additional sonagrams, see Glutz and Bauer (1980).

Calls of Adults. (1) Advertising-call. Soft, regularly repeated 'po-po-po-' (T Holmberg: Fig I), often audible 2 km away (Mikkola 1983). Bout of calling often begins with long phrase (usually more than 12 units); following phrases usually of 5–7 units, each phrase lasting 2–3 s (Mikkola 1983) or 1–2 s, and separated by 2–3 s (Glutz and Bauer 1980). Phrases often of 25 or more units. Overall pitch and tempo of calling varies—low-pitched calls slower, high-pitched faster (H Delin); c. 5 units per s fairly typical. Unpaired ♂ may call continuously through the night (Glutz and Bauer 1980). Call given mostly by ♂;

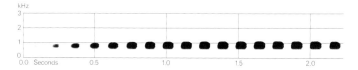

I P J Sellar Sweden May 1970

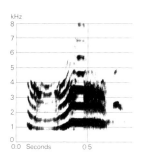

II C Chappuis/Alauda (1979) France October 1967

rarely by ♀—in connection with pair-formation or as reaction to ♂'s arrival at nest; harsher than ♂'s call (T Holmberg), or like quiet short phrase of call 3 (Glutz and Bauer 1980). Advertising-call given mostly January–April, though unpaired ♂♂ may continue until July (Glutz and Bauer 1980; Korpimäki 1981). Quiet version of call, consisting of a few softer, more muffled phrases, given (e.g.) to summon fledged young (Glutz and Bauer 1980). (2) Stutter-call of ♂. Similar to Advertising-call but quieter, with abbreviated or slurred pauses and irregular intervals. Given when potential mate has been attracted to territory but then left again; phrases longest immediately after ♀'s departure (Glutz and Bauer 1980). (3) Trill-call of ♂. Like Advertising-call but continuous, with units shorter and delivered more rapidly (8 per s in one recording, though this apparently rather fast); a soft 'dududu...'—almost uninterrupted for up to 10 min or more; audible only over a few hundred metres. Used to attract ♀ into nest-hole. Also given in short phrases in early stages of ♀'s nest-occupancy, during mating, and before and after delivering food to ♀ in nest early in incubation. (Kuhk 1949; Glutz and Bauer 1980; T Holmberg.) (4) Roll-call of ♂. Like Advertising-call, but lower pitched and more sonorous or full-sounding. Given when chasing ♀ before mating. (Kuhk 1949; Glutz and Bauer 1980.) (5) Pair-formation calls of ♀. (a) Loud, short, hoarse, sometimes slightly shrill 'kjäck'; normally first call given by ♀ in pair-formation. (b) Gickering-calls: single or series of 'guikguikguik...' calls, softer 'guiguigui...' (Glutz and Bauer 1980), or weak 'pee pee...' (T Holmberg). (c) Soft, attenuated, disyllabic 'muid' (Kuhk 1949). (6) Contact-calls. Use decreases as breeding progresses: 'uhd', 'muid', 'iuätt' (Glutz and Bauer 1980). (7) Smacking-call. Commonest call in summer and autumn, but given all year. Short, forceful, smacking 'yiop' (T Holmberg) or 'chiák' (H Delin) recalling squirrel *Sciurus*; given 1 or a few units at a time. Occurs in response to Advertising-call of another ♂ or in presence of man near nest, but cause not always apparent. Given by both sexes (T Holmberg). (8) Soft 'keuw' or 'mew-eh' from ♂ arriving at nest with food for ♀ or young. Perhaps given more often when ♂ hesitating to go to nest because man present (T Holmberg). Given by ♀ flushed from nest containing young (Kuhk 1953). ♀'s call harsher than ♂'s (T Holmberg). (9) Begging-call of ♀. High-pitched whistling 'speeh', like Redwing *Turdus iliacus* or Reed Bunting *Emberiza schoeniclus* but sharper (Kuhk 1953; T Holmberg). (10) Copulation-call of ♀. A shrill twittering during copulation (Glutz and Bauer 1980; C König). (11) ♀ sometimes encourages nestlings to feed with harsh trills (Glutz and Bauer 1980). (12) Forceful 'kuwäck', 'cuwáke', or 'ko-weck' (Fig II), similar to 'kewick' of Tawny Owl *Strix aluco*. Given by ♂, but rarely; given once as reaction to calling of another bird in spring, and once in association with call 15a (König 1969*b*; H Delin, T Holmberg). (13) Warning-call. Sharp 'weck-weck-'; used by both sexes to silence nestlings or fledged young (T Holmberg). (14) Hissing-call. Weak hissing given by ♀ on nest in threat (T Holmberg). (15) Other calls. (a) Drawn-out hoarse whistling from ♂ after being released near nest following ringing; once, in association with call 12 (T Holmberg). (b) Twittering from ♂ during ringing (T Holmberg). (c) A 'psitt psitt' from presumed ♀ near nest (T Holmberg). (d) Various barking, crowing, wailing, and yelping calls described (Glutz and Bauer 1980).

CALLS OF YOUNG. Food-call a short hissing 'psee' (T Holmberg) or 'ksí' (H Delin). Twittering given in discomfort; softer 'djü-djü-' when content. Hissing given in threat. Fledged young give a croaking shrill 'zrik' or chirping 'rick' in alarm (Glutz and Bauer 1980).   DJB

Breeding. SEASON. Scandinavia: see diagram; first laying date very variable, late February to early June, depending on food availability (T Holmberg). Similar periods and variation in rest of range. SITE. Hole in tree; either natural or, especially, that of woodpecker (Picidae). Readily uses nest-box. Rarely in building. Of 200 nests, Finland, 183 in boxes, 17 in holes of Black Woodpecker *Dryocopus martius* (Korpimäki 1981). Of 19 nests, Sweden, 9 in hole in deciduous tree, 5 in hole of *D. martius* in conifer, 5 in nest-box (Norberg 1964). Wide range of nest-box sizes used, with opening 5·7–18 cm diameter, with smallest possible hole 5·0–5·4 cm (Korpimäki 1981). Height above ground average 5·8 m (2·5–8), $n=7$ (Baudvin 1974). Site may be re-used in successive years. Nest: available hole; no material added. Building: none. EGGS. See Plate 97. Blunt elliptical, smooth and fairly glossy; white. 33 × 26 mm (32–37 × 24–29), $n=107$ (Witherby *et al.* 1938). Fresh weight 12·5 g (12–13), $n=15$ (Glutz and

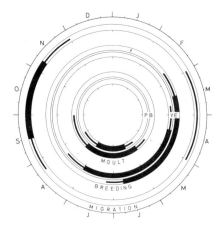

Bauer 1980). Clutch: 3–7 (1–10), varying with food abundance. Of 402 clutches, Finland: 1 egg, 1%; 3, 4%; 4, 19%; 5, 30%; 6, 28%; 7, 13%; 8, 3%; 9, 1%; 10, 1%; average 5·35 ± SD 1·26; clutch size increased from south to north; average clutch size correlated significantly with size of available vole *Microtus* populations; average clutch size also varied through season (Korpimäki 1981). Average of 138 clutches, Sweden, 6·1 (T Holmberg). One brood. Replacements very occasionally laid after egg loss. Laying interval 2 days; average 1·7 days for 4-egg clutches, 2·2 days for 3-egg clutches, $n = 29$ (Korpimäki 1981). INCUBATION. 28·5 days ± SD 2·0 (25–32), $n = 46$; 29·2 days ± SD 1·7 (26–32) for 1st egg, $n = 34$; 26·6 days ± SD 1·3 (25–29) for last egg, $n = 12$ (Korpimäki 1981). By ♀ only, beginning normally with 2nd egg; hatching asynchronous (Korpimäki 1981). YOUNG. Altricial and nidicolous. Cared for by ♀ and brooded more or less continuously for average 21·1 days (15–23), $n = 9$ (Korpimäki 1981). FLEDGING TO MATURITY. Fledging period 31·7 days ± SD 2·3 (28–36), $n = 19$; last-hatched young in 30·3 days ± SD 1·4 (28–33), $n = 12$ (Korpimäki 1981). Become independent after 5–6 weeks (Korpimäki 1981). Age of first breeding 1 year in central Europe, perhaps sometimes later further north. BREEDING SUCCESS. Of 701 eggs laid, Finland, 85·4% hatched, and 68·2% of young fledged, giving overall 53·6% success; hatching and fledging success closely correlated with clutch size, and linked with food supply; main losses of eggs and young from desertion and predation (Korpimäki 1981). Of 211 broods, Sweden, 60% produced at least 1 flying young (52–82% over 4 years); 4·6 young reared per successful pair, 2·8 per breeding attempt ($n = 120$); main losses to predation, particularly during incubation, and to desertion during food shortage (T Holmberg).

Plumages (nominate *funereus*). ADULT. Crown russet-brown or olive-brown, feathers with small white terminal droplets on forecrown and side of crown, larger white spots on central crown. Hindneck, sides of neck, and upper mantle white with faint brown bars or mottling, but some feathers with much russet-brown or olive-brown on tips, joining to form almost uniform brown patch on nape and broad brown streak from behind ear to centre of mantle. Lower mantle and shorter inner scapulars almost uniform russet-brown or olive-brown, paired white spots or broad short bars on feather-centres mainly hidden under brown feather-tips; large white spots on outer and longer scapulars more conspicuous, those on outer scapulars joining to form white oblique streak with only slight brown bars or mottling. Back and rump brown with indistinct small white spots, upper tail-coverts more conspicuously barred white. Facial disc white or pale cream-buff, indistinctly tinged and mottled grey towards sides, pure white only from above eye to base of upper mandible and down to below eye; narrow ring round eye black, broadest and with some black bristles projecting forward along side of bill at inner corner. Distinct ruff of narrow, rather stiff, and slightly curved feathers round disc, forming blunt feather-ridge round face when raised; black-brown with contrasting white bars and spots, darkest behind middle of ear, palest at border of facial disc. Underparts white or cream-white; feathers of chest, breast, and flanks with olive-brown or russet-brown shaft-streaks widening towards tip on sides of breast and flanks or with brown terminal bar on chest; central belly, vent, and shorter under tail-coverts uniform silky-white or with faint narrow grey or brown shaft-streaks; longer under tail-coverts with heavier brown streak or arrow-mark. Feathering of leg and foot white with variable amount of brown or russet mottling. Tail russet-brown or olive-brown, each feather with 4–5 rather narrow bars at sides, broadly interrupted by brown in middle, 1–2 more bars hidden underneath tail-coverts; distal bar narrowest, restricted to small spot on edge of outer webs of outer tail-feathers. Flight-feathers dark grey-brown or dark olive-brown, outer web with row of contrastingly white spots along edge (absent on terminal half of p9–p10; faint or absent on innermost primaries and outer secondaries), inner web with large white bars (partly tinged grey towards shaft, especially on tips of primaries). Upper primary coverts and lesser coverts uniform russet-brown or olive-brown; bastard wing and remainder of upper wing-coverts with large and rounded subterminal white spots on outer webs; small coverts at base of bastard wing with small white spots. Under wing-coverts and axillaries white with some fine brown or russet marks; tips of greater under primary coverts more heavily marked. Brown of outer scapulars, tertials, and tail-tips bleached and abraded to pale grey-brown or whitish in spring and early summer. According to Frutiger (1973), ♂ has facial disc pale tawny-brown with grey tinge towards outer border, depth of grey increasing with age; ♀ uniform tawny-brown or (when older) tawny-yellow without grey; this difference not perceptible in small series of skins examined. Limited number of known-age skins seem to show also that olive-brown or dark greyish-olive birds are fully adult, more russet-brown or reddish-chocolate-brown ones are 1st adult; white spots on scapulars and upper wing-coverts on average larger and squarer in olive birds, smaller and rounder in russet ones; olive ones less heavily marked with olive-brown on underparts, russet ones more heavily with cinnamon-brown or russet-brown. DOWNY YOUNG. 1st down (neoptile) sparse, short, and soft; white. From c. 7th day, feather-pins of mesoptile start to grow, hidden under neoptile, down appearing white with grey mottling; at 2 weeks, covered in growing mesoptile, but still much white neoptile clinging to feather-tips, this partly retained until age of c. 3 weeks. (Kuhk 1969.) JUVENILE. 2nd down (mesoptile) rather feather-like, though looser and softer than adult plumage. At 3 weeks, growing mesoptile on upperparts completely dull chocolate-brown; facial disc dark brown with broad black fringe along outer border, some white round base of bill; upper wing-coverts and shorter and median primary coverts short and fluffy, uniform dull chocolate-brown; underparts mainly brown, only limited white of bases of growing mesoptile feathers visible; vent, under tail-coverts, and feathering of leg and foot pale buff-brown, mixed with some cream-white. At fledging (4–5 weeks), mesoptile full-grown, including some longer white-marked feathers; flight-feathers and tail still growing; mesoptile as at 3 weeks, but some faint and poorly defined white marks on feather-bases on hindneck, some clearer and larger white subterminal spots on scapulars (partly hidden under feathers); facial disc dark greyish-brown with conspicuous and broad white X from above eyes to base of upper mandible and along gape; narrower white streaks along outsides of facial disc and from lower rim of ear-flaps to base of lower mandible; chest and breast dark brown with some faint pale marks of feather-bases visible, gradually merging into cream belly and flanks, which are marked with poorly defined dark grey-brown streaks; vent, under tail-coverts, and rather short feathering of leg and foot cream with indistinct brown mottling. Mesoptile wing-coverts moulted soon after fledging, largely replaced by 1st adult when flight-feathers full-grown, but mesoptile of head and body retained until age of $c.$ 3 months. FIRST ADULT. Closely similar to adult and sometimes hard to distinguish. Upperparts apparently more russet-brown than adult, white spots smaller and rounder, underparts more heavily marked (see adult). Juvenile flight-feathers, tertials, greater primary coverts, and tail retained: juvenile primaries sometimes with indistinct whitish fringes along tips (soon abraded, adult uniform dark brown); all flight-feathers of similar age and appearing equally worn (but in spring, least protected p5–p8 often more heavily worn); adult shows mixture of old and new feathers—in particular, single old feathers with frayed off-white tips conspicuous between fresh ones; however, difference in age between feathers often very difficult to see when birds over $c.$ 4–5 years old); retained juvenile tertials narrower and softer, more heavily worn in spring, more russet in colour than primary-tips (in adult, broad, with smoothly rounded tip, and same colour as primary-tips); short white bars on sides of each tail-feather taper to point towards shafts (in adult, bars have squarer ends, except for narrow subterminal one).

Bare parts. ADULT AND FIRST ADULT. Iris bright sulphur-yellow. Eyelids dark grey or blue-grey. Bill pale yellow, horn-yellow, or bluish-yellow, base and cere olive-grey or greyish-yellow. Soles greyish-yellow, claws black. DOWNY YOUNG AND JUVENILE. Eyes open at 8–11 days; iris greyish-yellow at first, soon deepening to bright yellow or sulphur-yellow. Bare skin under scanty down, eyelid, cere, leg, and foot pale red until 6th day, pink-red at 10th, pink-red to dark violet-red at about fledging. Cere and base of bill grey or olive-slate, ridge of culmen and tip pale greyish-horn or pale yellow, corner of mouth pink. Similar to adult at fledging. (Heinroth and Heinroth 1926–7; Kuhk 1950, 1969; Glutz and Bauer 1980.)

Moults. ADULT AND FIRST ADULT POST-BREEDING. Partial: all head, body, wing-coverts, and tail, but only part of flight-feathers; sequence of primary moult not fully understood, but apparently serial, starting from a centre somewhere on middle primaries (as in Palearctic *Strix* owls). Moult starts with body late May (♀ slightly earlier than ♂), last feathers shed $c.$ $2\frac{1}{2}$ months later, late July or early August; flight-feathers and tail start $c.$ 2 weeks after first body feathers, late May or early June, last feathers shed late June to late July (Erkinaro 1975), tail-feathers growing simultaneously or nearly so. Moult of primaries complicated: in 2nd calendar year, 3–6 outermost replaced (4·3 on average, $n=20$), mainly (p6–)p7–p10; in 3rd calendar year, neighbouring middle ones follow, but juvenile innermost, p1 or p1 to p3(p2–p4), retained; in next year, last juvenile feathers lost, as well as variable number of adult outer feathers, and in following year some middle primaries (Glutz and Bauer 1980); pattern increasingly complicated and variable with age. Secondaries moulted serially, apparently starting from centres on s9–s10 and s1; one bird in 3rd winter retained juvenile s5–s7; tertials usually replaced every year, but not always completely. Captive adult ♀ replaced p7–p10 in one year, p3–p6 in next, p1–p2 and p7–p10 in 3rd; p1–p2 shed early June, p3–p6 mid-June to mid-July, p7–p10 early July to mid-August; tail almost simultaneously shed, sequence t6 to t1, but moult sometimes incomplete (Haarhaus 1983). Single adult *caucasicus*, 18 August, had p6–p7 growing, remainder of primaries old; tail simultaneously growing; remainder of plumage old except new median upper wing-coverts (ZMM). POST-JUVENILE. Partial: all head, body, and wing-coverts, but not flight-feathers, tertials, greater primary coverts or tail. Starts when $c.$ 2 months old, face and wing-coverts first; bird may leave nesting area before mesoptile head and body replaced. Completed late July to early December, mainly August–September.

Measurements. Nominate *funereus*. Scandinavia, all year; skins (BMNH, RMNH, ZMA). Bill (F) to forehead, bill (C) to cere; to feathers, on average 2·0 less than to forehead.

WING	♂ 172	(3·22; 13)	167–178	♀ 176	(3·57; 12) 170–182
TAIL	96·0	(3·86; 12)	89–101	101	(3·02; 12) 97–105
BILL (F)	20·2	(0·77; 11)	19·4–21·4	21·4	(0·82; 11) 20·4–22·8
BILL (C)	14·0	(0·48; 10)	12·9–14·5	14·6	(1·16; 12) 13·7–15·9
TARSUS	22·2	(0·79; 13)	21·1–23·4	22·6	(1·18; 12) 20·7–24·0

Sex differences significant for wing, tail, and bill (F).

Wing of Swedish birds according to Vaurie (1960, 1965): ♂ 167 (16) 162–173, ♀ 176 (25) 167–182; tail ♂ 94 (9) 91–102, ♀ 95 (12) 90–104. Wing of birds from Alps: ♂ 167 (6) 162–175, ♀ 175 (5) 173–177 (Glutz and Bauer 1980), ♀ from Ardennes 182 (RMNH); Rumanian ♂ 175, ♀ 180 (Dombrowski 1912). Live breeding adults, Sweden: ♂ 171 (50) 166–178, ♀ 178 (127) 168–188 (T Holmberg).

A. f. caucasicus. Sexes combined, Caucasus, all year; skins (ZMM).

WING	159	(4·06; 7)	154–166	TAIL 85·0	(5·93; 6) 75–92

Of these, largest sexed bird was ♂ and smallest ♀ (remainder unsexed), but both were newly-fledged juveniles and sexing probably not correct; more probably, ♀ larger than ♂ as in other races of *A. funereus*. Isolated *beickianus* from north-west India and south-west China sometimes included in *caucasicus* (Vaurie 1960d), but size sex for sex probably distinctly larger: wing ♂ 166, ♀ 182, unsexed 165 (Vaurie 1960d), hence probably as in nominate *funereus*.

Weights. ADULT AND FIRST ADULT. During breeding, Sweden (T Holmberg) and Lüneburger Heide, West Germany (Glutz and Bauer 1980): (1) ♂, (2) ♀ during incubation and hatching, (3) ♀ early in nestling period when brooding young continuously, (4) ♀ late in nestling period when out of nest.

	Sweden			West Germany	
(1)	107	(51)	98–121	101 (4·37; 99)	90–113
(2)	179	(102)	134–215	167 (15·7; 96)	126–194
(3)	156	(47)	120–180	161 (16·4; 26)	—
(4)	131	(8)	122–145		

Migrants: Col de Bretolet (south-west Switzerland), August–October, 113 (13·5; 70) 87–160; Falsterbo (southern Sweden), late October and November, 122 (11·7; 130) 99–153 (Glutz and

Bauer 1980). Northern Italy, January–February: ♂ 105, ♀ 115 (Moltoni 1949).

DOWNY YOUNG AND JUVENILE. At hatching, 7·5–8 (Kuhk 1969). On 1st day 8·8 (0·65; 9) 7·5–9·5, on 4th 21·8 (4·12; 8) 18–29, on 10th 61·6 (9·02; 5) 46–68, on 15th 93·6 (13·8; 8) 70–111; maximum of 129 (8·00; 16) 113–142 reached on day (17–)21–27, weight decreasing slightly during period of maximum growth of mesoptile plumage; fledge on about 29th–34th day at 119 (15·6; 10) 94–140 (Kuhk 1970); for growth curves, see also Klaus et al. (1975).

Structure. Wing rather short, broad; tip rounded. 10 primaries: p8 longest, p9 6–10 shorter, p10 26–38, p7 1–3, p6 7–11, p5 16–21, p4 24–29, p1 45–50. Inner web of p9–p10 and outer web of (p8–)p9 emarginated. Outer web of longest feather of bastard wing, outer web of p10, and outer web of emarginated portions of p8–p9 serrated. Tail rather short, slightly rounded; 12 feathers, t6 3–11 mm shorter than t1. Bill rather heavy, largely hidden by bristles laterally projecting forward from base. Facial disc surrounded by distinct ruff of rather stiff and narrow feathers. Tarsus and toes short and slender, covered by dense hair-like feathering, except on soles. Middle toe without claw 17·9 (1·07; 13) 16–19 in ♂, 18·4 (1·05; 12) 17–20 in ♀; outer toe without claw c. 69% of middle, inner c. 84%, hind c. 61%. Claws rather short, slender; middle claw 11·1 (0·54; 13) 10·5–12·1 in ♂, 12·5 (0·63; 12) 11·6–13·2 in ♀; outer claw c. 72% of middle, inner c. 98%, hind c. 75%.

Geographical variation. Distinct in southern and eastern part of range, otherwise slight. Involves size, depth of brown ground-colour, and size of white marks on upperparts and brown marks on underparts. *A. f. caucasicus* from Caucasus distinctly smaller than nominate *funereus* from central and northern Europe, *beickianus* apparently similar in size to European birds (see Measurements); wing of populations from western and central Siberia east to Yakutsk area on average 2 mm longer than European nominate *funereus*, that of isolated Tien Shan population 4 mm shorter; *magnus* of north-east Siberia and Kamchatka much larger, wing ♂ 181 (8) 172–188, ♀ 187 (3) 180–192 (Dementiev and Gladkov 1951a; Vaurie 1960d; ZMA); North American *richardsoni* slightly larger, wing ♂ 174 (6·57; 10) 168–189, ♀ 179 (5·68; 8) 168–185 (Vaurie 1960d), both sexes combined 167–188 (RMNH). *A. f. pallens* from western and central Siberia and Tien Shan reported to be greyer and with slightly larger white marks above than more russet-brown European nominate *funereus*, but in western Siberia and Tien Shan both greyish and russet-brown birds occur (Dementiev 1933). Same difference also present between greyish-brown adult and russet-brown 1st adult of nominate *funereus*, however, and difference between Europe and Siberia perhaps mainly caused by different age composition of samples, as in Hawk Owl *Surnia ulula* (Forsman 1980) and Pygmy Owl *Glaucidium passerinum* (C S Roselaar): in Europe, most birds are collected during irruptions in autumn and these are mainly juveniles; in Siberia, most birds are collected by expeditions in spring and summer on breeding grounds, where higher proportions of adults present. Recognition of *pallens* thus requires confirmation. *A. f. magnus* of north-east Siberia and Kamchatka pale grey-brown above with markedly large white spots, white predominating over grey-brown on hindneck, scapulars, median upper wing-coverts, and chest, with remainder of underparts virtually all white (Dementiev 1933b; ZMA); populations from Yakutsk area (northern Transbaykalia to Verkhoyanskiy range) intermediate in colour between *magnus* and *pallens*, but size as in *pallens*, and these perhaps separable as *jakutorum* (Buturlin, 1908). North American *richardsoni* similar to nominate *funereus* but ground-colour of upperparts and chest colder fuscous-brown, less olive- or russet-brown (RMNH). *A. f. caucasicus* also darker than nominate *funereus*, but ground-colour warmer chocolate-brown; white spots on upperparts, upper wing-coverts, tail, and flight-feathers smaller; facial disc more heavily mottled brown, dark throat-patch larger; chest, breast, and sometimes flanks and belly chocolate-brown, feathers with rather small paired white subterminal spots (of nominate *funereus*, white with brown streaks and some indistinct brown bars); newly-fledged juvenile darker than juvenile of nominate *funereus*, head and body completely dark chocolate-brown with faint white spots on outer scapulars, sides of breast, belly, and vent only (ZMM). Colour of isolated populations of *beickianus* apparently similar to *caucasicus* (Vaurie 1960d). For subspeciation in *A. funereus*, see Mysterud (1970). CSR

Order CAPRIMULGIFORMES

Rather small to rather large birds, mainly insect-feeders. 2 suborders, one consisting of family Steatornithidae (single species, Oilbird *Steatornis caripensis* of South America) and the other of 4 families (of which 1 represented in west Palearctic): (1) Podargidae (frogmouths, 12 species; north-east Asia to Indonesia and Australia); (2) Nyctibiidae (potoos, 5 species; Neotropical region); (3) Aegothelidae (owlet-frogmouths, 7–8 species; Indonesia, New Guinea, Australia, and New Caledonia); (4) Caprimulgidae.

Family CAPRIMULGIDAE nightjars, nighthawks

Rather small to medium-sized aerial insect-feeders, mostly nocturnal or crepuscular. 2 subfamilies, Caprimulginae and Chordeilinae, both represented in west Palearctic.

Bodies mainly round. Necks short. Skulls flattened; very wide gape. Eyes and ears large. Wings usually long, rather narrow, tips pointed; 10 primaries; wing diastataxic. Feathers with aftershaft. Tails mostly of medium length, tip square or forked; 10 feathers. Legs short and slender. Feet with 3 toes pointing forward and 1 back; claw of middle toe with comb-like ridges. Oil-gland small and naked. No crop. Caeca large, but intestines rather short. Skin very thin. Plumages soft and loose, mainly buff, brown, rufous, grey, or blackish, often with white patches on wings or tail of ♂♂. Young semi-altricial and nidicolous, down cryptic. Juveniles rather similar to adults. Post-breeding moult complete, interrupted during migration in some species; pre-breeding moult partial. Adult plumage acquired at 5–10 months.

Subfamily CAPRIMULGINAE nightjars

About 69 species in *c.* 15 genera, widely distributed in warmer regions of the world. Over half the species are of genus *Caprimulgus*, to which belong the 4 species (all breeding) in west Palearctic.

For general features, moults, etc., see Caprimulgidae. Rictal bristles large.

Caprimulgus nubicus Nubian Nightjar

PLATES 56 and 57
[between pages 662 and 663]

Du. Nubische Nachtzwaluw Fr. Engoulevent de Nubie Ge. Bajudanachtschwalbe
Ru. Нубийский козодой Sp. Chotacabras núbico Sw. Nubisk nattskärra

Caprimulgus nubicus Lichtenstein, 1823

Polytypic. *C. n. tamaricis* Tristram, 1864, Dead Sea depression and western Arabia to Oman. Extralimital: nominate *nubicus* Lichtenstein, 1823, Nile valley of northern Sudan; *torridus* Phillips, 1898, northern Somalia; *jonesi* Grant and Forbes, 1899, Socotra; *taruensis* van Someren, 1919, southern Ethiopia, northern and eastern Kenya, and north-east Uganda.

Field characters. 21–22 cm; wing-span 46–53 cm. About 20% smaller than Nightjar *C. europaeus*. Smallest and slightest nightjar in west Palearctic, mainly grey above and ochre below, with conspicuous white band across outer wing, bold white spots on ends of outer tail-feathers, and rufous patch on carpal area. At close range, pale buff collar on hindneck and white cheek-bar and throat-patches may show. Flight action lighter than *C. europaeus*. Sexes similar; no seasonal variation. Juvenile separable at close range.

ADULT MALE. Plumage colours and pattern less variegated than *C. europaeus*, with ground-colour of upperparts pale, rather sandy-grey, and that of underparts grey-buff. Markings complex but not bold, except for pale buff-yellow band on hindneck (not forming as contrasting a half-collar as on Red-necked Nightjar *C. ruficollis*) and narrow white patches on lower cheeks and on sides of lower throat. Underpart-barring diffuse. In flight, broad white primary-spots placed across middle of outer wing create more obvious panel than on any other congener and white tips to 2 outermost pairs of tail-feathers larger than on any west Palearctic congener; in daylight, rufous bases to inner primaries and outer secondaries and rufous tone to primary coverts create warm patch over carpal area. Under wing-coverts pale rusty-buff, combining with pale underbody to make bird appear much lighter below than all congeners except Egyptian Nightjar *C. aegyptius* and Golden Nightjar *C. eximius*. ADULT FEMALE. White primary-spots and tips to outer tail-feathers less distinct. JUVENILE. Paler than adult, lacking collar and having even more indistinct primary-spots and tips to outer tail-feathers.

In fortunate circumstances, combination of small size, pale grey and buff plumage, bold white wing-panel and tail-tips, and rufous bases to secondaries and inner primaries allow certain identification. Flight lighter, more flitting than *C. europaeus*. Behaviour apparently much as *C. europaeus* but fond of sitting on open ground.

Song a hollow, resonant 'koww koww', repeated every 1–4 s.

Habitat. Very imperfectly known, but confined to hot dry low latitudes and appears to fit in between Red-necked Nightjar *C. ruficollis*, which ranges into closed woodland as well as dwarf semi-arid open scrub, and Egyptian Nightjar *C. aegyptius*, which demands, at most, sparse shrub cover and is content even with bare desert. *C. nubicus* tolerates sparse desert scrub of tamarisk *Tamarix*, thorn, etc., and will rest by day in shade of rocks or bushes and on passage under low thorns on hard sand among dunes along the coast. In East Africa frequents desert thorn scrub in close proximity to water, inhabiting open *Acacia* bush surrounding wells affording a permanent overflow of water, or hawking at dusk above pools or brief stretches of running water in sandy river beds. When rains come, shifts to open bush where rainwater collects in hollows, ranging up to *c.* 1400 m (Archer and Godman 1961). In other parts of range, must learn to live with lesser and more irregular water resources.

618 Caprimulginae

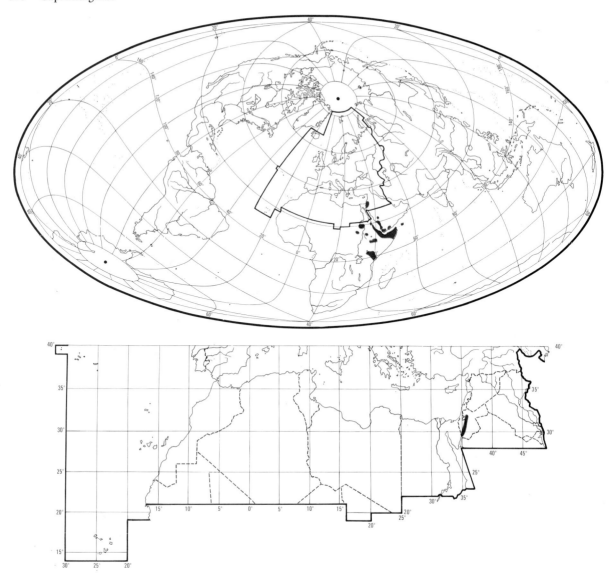

Distribution. JORDAN. Few records; breeding not proved recently, but almost certain in west north to Jericho (Vaurie 1965; HM, DIMW).

Population. ISRAEL. Common in rift valley in south (A Zahavi).

Movements. Poorly known. Perhaps mainly resident, since recorded in winter even in Israel, at northern end of breeding range. However, several seen in large fall of migrants (after sandstorm) at Port Sudan harbour, 11 September (Elliott and Monk 1952), 3 seen on Aden coast, 18 August (Ennion 1962), and 5 specimens of Arabian race *tamaricis* obtained northern Somalia in October and February (Archer and Godman 1961). Evidently some movement in southern Red Sea/Gulf of Aden region, even involving small parties, but scale uncertain.

Food. Probably only insects. These are taken in flight at dusk, often near water (Archer and Godman 1961). In Saudi Arabia, sometimes also forages on ground around animal droppings. Moths (Lepidoptera) apparently main prey and wings said to be rejected before swallowing; 3 stomachs contained no Coleoptera (Meinertzhagen 1954).
MGW

Social pattern and behaviour. Very poorly known.
 1. Degree of association outside breeding season not known. In Somalia, 4 birds (probably migrants) quite close together late February (Archer and Godman 1961). BONDS, BREEDING DISPERSION. No information. ROOSTING. Diurnal. On ground, in shade among rocks or bushes; also fond of sitting on sandy roads and other open places (Meinertzhagen 1954). 4 apparent migrants (see above) rested on ground under thorn bushes (Archer and Godman 1961). Foraging activity begins after sunset (Meinertzhagen 1954).

2. Of 2 birds flushed from ground at *c.* 10 m, Hazeva (Israel), mid-August, 1 performed Wing-clapping like Nightjar *C. europaeus* (A G Gosler and R Vroman). 2 of 3 birds near Dead Sea (Israel), late March, landed on track *c.* 15 m from car in headlights; 1 stood briefly then lowered body to ground; flew at *c.* 7–8 m when approached on foot. No calls heard (P A D Hollom). Persistent Advertising-calls given from ground in 3 areas of Saudi Arabia, 2nd week of April. Throat conspicuously puffed out at each call (King 1978: see 1 in Voice). No further information. MGW

Voice. For Wing-clapping, see Social Pattern and Behaviour.

CALLS OF ADULTS. (1) Advertising-call (Song). A hollow, resonant 'koww koww' or 'poww poww', usually 2(–3) units, given at intervals of 1–4 s (King 1978). Recording (Fig I) suggests a rather staccato 'kiu-kiu', 2nd unit (*c.* 90 ms) of each pair being *c.* 10 ms shorter than 1st (J Hall-Craggs); like soft, slightly muffled yelping or yapping of (somewhat agitated) small dog (M G Wilson). A quiet barking, 2nd unit being slightly shorter than 1st; carries for *c.* 100–200 m but invariably gives impression of distant low-volume sound (A Zahavi). Irregular temporal pattern, with rate sometimes accelerated (J Hall-Craggs, P J Sellar). Also rendered 'kroo-kroo' (Gallagher and Woodcock 1980). Remarkably similar, both to ear and in sonagram, to extralimital Freckled Nightjar *C. tristigma*, though illustration reveals differences in timing and in structure of units (see Chappuis 1981). (2) Call-note. Said to resemble Nightjar *C. europaeus* (Tristram 1866). No further information. MGW

Breeding. No information from west Palearctic. SEASON. Somalia: eggs found early June (Archer and Godman 1961). SITE. On ground in the open. Nest: shallow scrape. Building: no information. EGGS. See Plate 94. Elliptical; ivory-white with faint lilac marbled markings mainly at broad end. 25 × 19 mm (*torridus*) (Archer and Godman 1961). Clutch: 2. No further information.

Plumages (*C. n. tamaricis*). ADULT MALE. Central forehead and central crown pink-buff with broad streaks on feather-centres and faint dark grey vermiculation; sides of forehead, sides of crown, nape, and lower ear-coverts finely vermiculated grey and white, appearing pale grey. Lores, cheeks, and upper ear-coverts mottled buff and grey. Broad band round hindneck extending to sides of neck cinnamon-buff with some dark grey or black marks, forming contrasting half-collar. Mantle, back, rump, and upper tail-coverts pale grey, like nape and sides of crown, feathers with slightly darker shaft-streak and sometimes short black bars; occasionally, nape, sides of crown, and mantle to upper tail-coverts pale buff-brown rather than pale grey. Scapulars and tertials finely vermiculated buff and grey and with distinct black shaft-streak, like centre of crown, but with uniform pink-buff or pink-cinnamon spot subterminally, bordered by black bar towards feather-base. Chin finely barred cream and grey, throat and chest finely barred buff and grey; single patches on each side of throat uniform white, lower chest with some indistinct pink-cinnamon spots. Breast, flanks, belly, and vent pale cinnamon-buff with dark grey bars; latter gradually further apart towards lower flanks and vent; under tail-coverts uniform cinnamon-buff. Central pair of tail-feathers (t1) pale grey or pale buff, finely vermiculated grey and with 5–7 narrow, well-spaced broken black bars; t2 similar but with heavier black bars; t3 and basal and middle portions of t4–t5 evenly barred with heavy black and pale grey or pink-buff bars; tips of t4–t5 white for 31–41 mm (measured on inner web along shaft). Outer primaries, (p4–)p6–p10, black with faint grey tinge and fine specks on distal outer web and tip and with broad white band *c.* 1·5 cm wide across middle portion, forming conspicuous bar; outer web of p10 black with elongated white patch along middle portion of outer edge; inner primaries and secondaries broadly barred rufous-cinnamon and black, tips uniform grey with some specks; rufous bars on p5–p6 rather irregular; rufous marks extend sometimes to base of p10. Greater upper wing-coverts and primary coverts coarsely barred rufous and dull black like secondaries, but tips rufous-buff with coarse black speckling. Median upper wing-coverts finely vermiculated white or pale buff and grey and with dark shaft-streaks, like scapulars, rounded spot on tip of outer web contrastingly uniform pink-cinnamon, subterminally bordered by rather coarse rufous and dull black bars. Lesser upper wing-coverts finely vermiculated white and grey like mantle, coverts along leading edge of wing cinnamon-buff with some black or grey specks. Under wing-coverts and axillaries cinnamon-buff with black specks or streaks. In worn plumage, grey ground-colour of upperparts and buff of underparts paler; rufous half-collar and cinnamon-pink spots on scapulars and upper wing-coverts bleached to buff or cream-white, less contrasting. ADULT FEMALE. Like adult ♂, but t4–t5 have white on only 16–26 mm of tip, often with slight buff tinge or grey freckling on outer and terminal edges; white spots on outer primaries often slightly less sharply defined, borders tinged cinnamon, patch on middle portion of outer web of p10 less distinct, patch on inner web of p10 slightly more rounded in shape, reaching shaft for short distance only. DOWNY YOUNG. No information. JUVENILE. Paler and more finely vermiculated than adult, no rufous neck-collar. All upperparts and upper wing-coverts finely vermiculated pink-buff and grey, without black streaks on crown, scapulars, tertials, and wing-coverts; no rounded cinnamon spots on scapulars and outer webs of wing-coverts; only small dark grey subterminal arrow-marks on crown, scapulars, and wing-coverts; rather small

I B King Saudi Arabia April 1976

cream-white or off-white patches on wing-coverts. Chest finely vermiculated grey and white; remainder of underparts cream-white, very faintly barred. T1–t2 without darker bars, completely vermiculated sandy-buff and grey; t3 with 8–9 indistinct dark grey bars; t4–t5 pale cinnamon with *c.* 10 narrow, well-spaced black bars, except for base of inner web and for unbarred whitish cinnamon tip 2–4 cm wide (width probably depending on sex). Flight-feathers as in adult, but borders of white primary-patches tinged cinnamon; outer edge of p10 cinnamon without white to middle portion; some cinnamon marks on primary-tips; pale spot on inner web of p10 rather small and rounded, not reaching shaft. In worn plumage, ground-colour dirty off-white on upperparts and chest, whitish on belly. FIRST ADULT. Like adult, but rounded white patch on inner web of p10 not quite reaching shaft; white spot on outer web of p10 not sharply demarcated from buff edge, in ♀ hardly any white on outer web.

Bare parts. ADULT. Iris dark brown. Bill dark horn-coloured, brown-black, or pink with dark horn tip. Leg and foot greyish-pink, brown-olive, fleshy-brown, or light brown. DOWNY YOUNG. No information. JUVENILE. Iris brown. Bill flesh-coloured, ridge of upper mandible and tip dark horn-coloured; mouth flesh. Leg and foot grey, soles yellow. (BMNH.)

Moults. Few in moult examined. Sequence of moult probably as in other *Caprimulgus*. Of late November to mid-June *tamaricis* examined, none in active moult; a single ♀, 1 July, had primary moult score 26, tail all old and body mainly new, but part of belly, under tail-coverts, and many upper wing-coverts old; another adult, 21 September, completely new; moulting season probably June–September. Single 1st-calendar-year bird, 22 October, mainly in juvenile plumage, but crown, some lesser and median upper wing-coverts, and scattered feathers of mantle, shoulders, and chin to breast new. Probable 1st adults with plumage entirely new examined January–March.

Measurements. *C. n. tamaricis.* ADULT. Western and southern Arabia, mainly February–May; skins (BMNH). Bill is exposed culmen; to forehead, 9–12 mm more.

WING	♂ 151	(4·54; 11)	145–157	♀ 149	(2·63; 7)	146–154
TAIL	107	(4·04; 11)	102–114	102	(4·92; 7)	96–107
BILL	10·0	(1·05; 11)	8·9–11·8	10·4	(1·00; 6)	9·0–11·2
TARSUS	19·4	(0·93; 11)	18·5–21·3	19·6	(0·81; 7)	18·5–21·2
TOE	20·4	(0·92; 11)	18·7–22·0	20·6	(0·62; 7)	19·8–21·5

Sex differences not significant.

JUVENILE. Only a few examined. Apparently similar to adult.

Weights. Juvenile ♂, Oman: October, 46 (BMNH).

Structure. 10 primaries: p8 longest, p9 1–3 shorter, p10 4–10, p7 9–15, p6 30–36, p5 41–46, p1 64–70. Bill relatively slightly longer than in Nightjar *C. europaeus*, feathering of forehead not extending as far as nostrils. Tarsus bare, except for upper front. Outer toe *c.* 53% of middle; inner *c.* 58%, hind *c.* 35%. Remainder of structure as in *C. europaeus*.

Geographical variation. Pronounced, involving colour only. *C. n. tamaricis* is pale and grey, with golden-yellow collar on hindneck. Nominate *nubicus* from northern Sudan has upperparts and chest rufous-sandy, golden neck-collar less conspicuous, and spots on scapulars and upper wing-coverts larger, pink-cinnamon. *C. n. torridus* of Somalia grey on upperparts and chest (slightly darker and more heavily marked with black than *tamaricis*) with contrasting large chestnut or rufous-cinnamon spots on crown, scapulars, chest, and upper wing-coverts. Kenyan race *taruensis* like *torridus*, but rufous spots and grey ground-colour slightly paler; ground-colour less rufous-buff and with rufous spots more extensive than in nominate *nubicus*. *C. n. jonesi* of Socotra greyer than *torridus* and collar distinct (Mackworth-Praed and Grant 1952).

Recognition. Rather similar in colour to Egyptian Nightjar *C. aegyptius*, but adults of latter differ by complete absence of rufous half-collar on hindneck, no pure white on outer tail-feathers, no large white patches on outer primaries, no cinnamon-rufous on bases of flight-feathers, and coarser and more widely spaced vermiculation on upperparts. Juveniles more closely similar, but *C. nubicus* always much smaller than *C. aegyptius*: wing of *C. aegyptius* over 180, *C. nubicus* below 160; tail lengths overlap. Pale races *unwini* and *plumipes* of *C. europaeus* also rather like *C. nubicus*, ♂♂ showing similar tail- and wing-spots, but lacking rufous half-collar of *C. nubicus* and with wing over 174, tail over 125.

CSR

Caprimulgus europaeus Nightjar

PLATES 56 and 58
[between pages 662 and 663]

Du. Nachtzwaluw   Fr. Engoulevent d'Europe   Ge. Ziegenmelker
Ru. Обыкновенный козодой   Sp. Chotacabras gris   Sw. Nattskärra

Caprimulgus europaeus Linnaeus, 1758

Polytypic. Nominate *europaeus* Linnaeus, 1758, central and northern Europe and northern Asia to Lake Baykal area; *meridionalis* Hartert, 1896, southern Europe, North Africa, and Asia Minor, north to Pyrénées, central Italy, Yugoslavia, Hungary, southern and eastern Rumania, Crimea, and Caucasus, east to north-west Iran; *unwini* Hume, 1871, eastern Iraq and Iran (except north-west) east to western Tien Shan and Kashgaria, north to southern Turkmeniya and Uzbekistan; *sarudnyi* Hartert, 1912, Kazakhstan from east shore of Caspian east to Kirgiziya, Tarbagatay, and Altay. Extralimital: *plumipes* Przhevalski, 1876, northern China east from Sinkiang and eastern Tien Shan, and southern and north-west Mongolia; *dementievi* Stegmann, 1949, north-east Mongolia and southern Transbaykalia.

Field characters. 26–28 cm, of which tail 10–11 cm; wing-span 57–64 cm. Size close to that of ♂ Merlin *Falco columbarius*, exceeding that of all other nightjars except Red-necked *C. ruficollis*. A nocturnal insectivore, with structure suggesting both small falcon *Falco* and Cuckoo *Cuculus canorus* and flight recalling larger storm-petrel *Oceanodroma*. Large-headed, with bill hardly visible. At distance, plumage appears mottled dusky brown, lacking obvious marks except white wing- and tail-spots on ♂. At close range, plumage pattern extremely intricate, with myriad small bars below and blotches and freckles above; white throat band and grey-white spots along shoulders form most obvious marks. Except during aerial feeding at dusk and dawn, most often seen in car headlights, with glimpse of red-brown retinas followed by ghostly flit, as bird wheels up and away. Advertises presence in breeding habitat by churring song, usually from perch. Sexes dissimilar; no seasonal variation. Juvenile not easily separable from ♀. 3 races in west Palearctic, not separable in the field.

ADULT MALE. Colours of plumage vary from silver-grey (most evident as ground-colour to head, scapulars, and central tail-feathers) through brown to dull rufous (most evident as ground-colour to central crown, cheeks, throat, scapular-bands, rear underbody, and under wing-coverts). Markings extremely complex; those visible at close range are white patch on lower throat, broken line of black and white streaks on sides of neck, black lines on nape and mantle, wavy brown-black bars on scapulars, primaries, and tail, and line of almost white or buff spots at shoulder. All other feathers densely vermiculated, freckled, and, on underbody, finely barred. In flight, large white spots on (1) distal halves of 3 outermost primaries and (2) tips of 2 outermost pairs of tail-feathers form distinct, eye-catching marks, even in poor light. ADULT FEMALE. White spots on wing and tail replaced by buff, or absent. JUVENILE. Resembles adult ♀ but paler, with less well marked scapulars and underparts; chest lacks vermiculation of adult; no buff primary-spots or tail-tips.

Nightjar identification as much a matter of fortune as of effort or knowledge; all species essentially crepuscular and observation difficult even in occasional diurnal discoveries, as birds use astonishingly cryptic plumage to disappear against many backgrounds, even at point-blank range. Before fussing over detailed description of complex feather markings, important to judge size (medium-large in *C. europaeus*), general plumage tone (dusky brown), wing- and tail-marks (striking on ♂), and throat-patch (small). When seen well, *C. europaeus* can be confused only with Common Nighthawk *Chordeiles minor*. For fuller distinctions between *C. europaeus* and other nightjars, see accounts of latter. Important to recognize that in west Palearctic north of Iberia, North Africa, and Iraq, *C. europaeus* is the only breeding nightjar. Flight totally silent and accomplished, as befits persistent aerial insectivore; actions of hunting bird very different from those of escaping one, with deep deliberate wing-beats, easy wheels, and floating glides (with wings extended), all combined with sudden twists at moment of prey capture. Able to hover; lands with characteristic sudden disappearance of long wings and tail. Display flight includes wing-clapping and glide, with wings raised and tail spread (see Social Pattern and Behaviour). Gait restricted, but able to walk with short steps (and with body in horizontal attitude).

Migrants seek low, tangled cover and, in hot regions, shade. Song a long-sustained churring trill; flight-call 'koo-ik' (mostly from ♂); alarm-call 'chink-chink-chink' (♂) or 'chuck' (♀).

Habitat. Ranges beyond continental into oceanic zone, and northward through and beyond upper middle latitudes, from temperate into boreal, as well as southward to steppe, Mediterranean, and semi-desert zones. During recent years, however, even after allowing for losses of habitat, has substantially declined in more oceanic temperate and boreal parts of range, for reasons not yet clear. As a food-gatherer, wholly aerial but, so far as observation goes, mainly concentrated in lower airspace and otherwise mostly a ground bird, not only for nesting but for day-time resting, using tree branches and other perches less as bases for foraging than as song-posts and look-outs. Generally avoids mountains, forests, and most dense tall stands of vegetation from mature plantations to field crops, reed-beds, and tall grassland. Sensitivity to day-time disturbance also implies avoidance of human settlements and their associated activities. Prefers dry, open, well-drained habitats, such as well-spaced conifer woods, birch *Betula* and poplar *Populus* spinneys, scrub oak *Quercus*, heathery glades and clearings in coniferous or mixed woodland, heather moors with self-sown pines *Pinus*, burnt patches and sunny woodland margins, steppe woodlands, shrubby steppes, and semi-desert (Voous 1960). In Britain, mainly on lowland heath with scattered trees, but also on recently felled woodland, new forestry plantations, moorland (including even wet blanket bog in west), commons, sand-dunes, shingle, chalk downland, industrial waste tips, and open woodland, and occasionally even dense coppice (Sharrock 1976). Intensive study in Yorkshire (England) showed clear requirements for open ground to minimum extent of 2 ha, and commonly for presence of tall marginal trees, but forest sites preferred to heath (Leslie 1981). In eastern England, favours patches of heather *Calluna*/*Erica* with low birch (Berry 1979). In USSR, avoids true steppes and unwooded plains, preferring forest fringes, glades, clearings, tree-clad slopes of steppe ravines or steppe birch copses, and, in temperate zone, dry pinewoods, mixed forests, and small broad-leaf coppices, foraging also over meadows, farmlands, and stagnant ponds. In Caucasus and Armeniya, often inhabits mountains in glades, clearings, and scrub to timberline, up to c. 2200–2500 m, foraging also over fields and meadows (Dementiev and Gladkov 1951a). In central Europe, usually below c. 700–800 m, but locally up to c. 1800 m; ability of the soil to absorb and release solar warmth seen as a critical factor in distribution and density (Glutz and Bauer 1980). Habitat in African winter quarters similarly varied, embracing clearings in wet evergreen forest, dry *Acacia* steppe near the coast, and high ground from 1500 m to 5000 m (Moreau 1972).

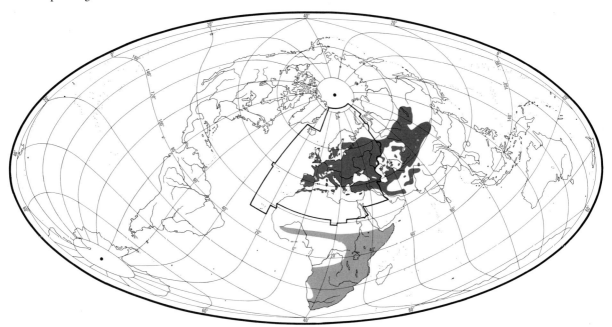

Distribution. Decreased in north-west Europe and almost certainly elsewhere (see Population).

BRITAIN AND IRELAND. General decrease in range, still continuing (Stafford 1962; Parslow 1967; Sharrock 1976; F C Gribble). FRANCE. Apparent decrease in range, especially in north (Yeatman 1976). NETHERLANDS. Widespread in 19th century in heaths and dunes, now mainly in higher areas (CSR). WEST GERMANY. Extinct in many areas (GR)—compare with 1975 distribution shown in Rheinwald (1977). IRAQ. Probably breeds (Allouse 1953). SYRIA. Probably bred formerly (Kumerloeve 1968b); no recent evidence (Macfarlane 1978; HK). LEBANON. Bred formerly (HK); no recent evidence (Tohmé and Neuschwander 1974; Macfarlane 1978).

Accidental. Iceland, Faeroes, Azores, Madeira, Canary Islands.

Population. Marked decreases, first noted in Britain and Ireland in late 19th or early 20th century, but widespread in recent years in rest of north-west and northern Europe, parts of central Europe, and even in Italy and Bulgaria. Possible causes include habitat changes, disturbance, and decrease of large insects due to pesticides, but no proof of link with climatic changes.

BRITAIN AND IRELAND. Decreased, probably since late 19th century, markedly since 1930s and still continuing; reasons unknown—probably habitat loss and disturbance, perhaps pesticides, but no clear link with climatic changes; in 1972, 3000–6000 pairs (Parslow 1967; Sharrock 1976). Britain: estimated 2100 ♂♂ 1981 (F C Gribble). FRANCE. Probably 1000–10000 pairs; apparently decreased (Yeatman 1976). BELGIUM. About 750 pairs, decreased (Lippens and Wille 1972); c. 200 pairs 1982 (P D), and see also Matthé (1982). LUXEMBOURG. About 35 pairs (Lippens and Wille 1972). NETHERLANDS. Common in 19th century, decreased with habitat changes; since c. 1960 further marked decrease of 80% or more; estimated 500–600 pairs 1977 (Teixeira 1979; CSR). WEST GERMANY. About 5000 pairs (Glutz and Bauer 1980); steady decline in most areas probably due to changes in forest management and reduced numbers of large insects due to pesticides (GR). DENMARK. Marked decrease (Dybbro 1976). SWEDEN. Some decrease since 1940s (Stolt 1972; LR); c. 5000 pairs (Ulfstrand and Högstedt 1976). FINLAND. Decreased in recent years (OH); estimated 4300 pairs (Merikallio 1958). EAST GERMANY. About 10000 pairs (Glutz and Bauer 1980). POLAND. Generally scarce, although numerous in some localities (Tomiałojć 1976a). CZECHOSLOVAKIA. Marked decrease since c. 1960 (KH). AUSTRIA. Over 500 pairs (Glutz and Bauer 1980). SWITZERLAND. Decreased (RW). ITALY. Marked decline since c. 1960 (SF). BULGARIA. Very common (Patev 1950); some recent decline (JLR). CYPRUS. Common (PRF, PJS).

Oldest ringed bird 8 years (Dejonghe and Czajkowski 1983).

Movements. Migratory. Details poorly known, due to nocturnal habits and paucity of ringing data. Rare winter specimens from Pakistan; otherwise all populations have winter quarters in Africa, so that easternmost (Mongolian) birds cross nearly 100° of longitude (Moreau 1972).

WEST PALEARCTIC POPULATIONS (nominate *europaeus*, *meridionalis*). Identified in winter all across northern Afrotropics (away from equatorial forest), and in eastern half of continent south to South Africa and Namibia. Relatively few certain records from West Africa, where known from Gambia (but not Sénégal), Guinea-Bissau, and Sierra Leone to Nigeria and Cameroun; numbers so

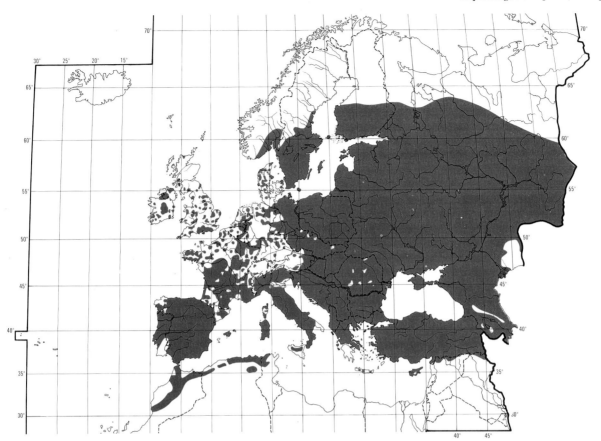

far found inconsistent with strength of passage through western Mediterranean and north-west Africa. Migration broad-front across Europe, Mediterranean, and Sahara (various oasis records from Sahara), and the few ringing recoveries are consistent with this. British-ringed birds found on passage in France (6) and Spain (1), and 1 Dutch bird in France; from Sweden, migrant (May) found in Czechoslovakia 2 years later (also in May), and bird ringed as nestling found Switzerland in October; Czechoslovakian adult (June) recovered southern Spain (date uncertain); from USSR, a Kaliningrad migrant (October) found Yugoslavia (November), and Vitebsk juvenile (June) recovered Hungary (next May). 4 spring migrants ringed Tunisia found in Italy, and Cyprus migrant (September) killed 7°45′S in Tanzania (March). No information on routes used by Siberian nominate *europaeus*.

CENTRAL ASIAN POPULATIONS (*unwini*, *sarudnyi*, *plumipes*, *dementievi*). Reach Africa via broad-front passage across Middle East; *sarudnyi* and *unwini* (from Soviet Central Asia) both identified in Iraq, and *unwini* to Turkey and Sinai. Common autumn migrant through southern Iran, Pakistan, and north-west India, but these regions evidently bypassed to the north in spring (Ali and Ripley 1970). Various autumn ship records off Arabia indicate some direct overwater crossing to East Africa then. All Asiatic populations winter in eastern half of Africa, from Sudan and Ethiopia to Cape Province. Birds recorded in northern parts of this range perhaps mainly on passage since commonest East Africa in October–November and March–April (Britton 1980); *plumipes* considered to winter Zimbabwe–Mozambique–eastern South Africa, occurring only on passage further north (Borrett and Jackson 1970). All races overlap in winter. 73 skins from south-east Africa attributed to *plumipes* (11), *unwini* (5), *sarudnyi* (31), *meridionalis* (7), and *europaeus* (19) (Clancey and Mendelssohn 1979)—*dementievi* not separated in that study. Pale specimens of *unwini*-type outnumber *europaeus* in Zimbabwe (Benson *et al.* 1971); *unwini*-type predominate in eastern Tanzania also, but accounted for *c*. 30% of birds handled by ringers at Ngulia (Kenya) in November–December (Britton 1980).

Autumn movement begins second half July, probably 1st-brood juveniles; main exodus from breeding areas late August to early October, with only stragglers left in Europe in November. Arrivals south of equator October–November, and return passage there begins March. Exceptional European records in late March, but main immigration April–May, northern breeding areas occupied last. Migrates nocturnally, singly or in small parties. Spring flocks in East Africa can comprise one sex only (Moreau 1972), and in European breeding areas ♂♂

arrive 4–5 days earlier than ♀♀ (Stülcken 1962). 122 counted in 10 min (only part of total movement) in big northward departure from north-east Somalia at dusk, 3 May 1980 (Ash and Miskell 1983).

Food. Includes material supplied by I H Alexander on behalf of Stour Ringing Group (England) cited as SRG.

Insects, mainly moths (Lepidoptera) and beetles (Coleoptera). Particularly when food plentiful takes prey in sustained lower-level aerial-pursuit (rarely above trees) like swallow (Hirundinidae), normally approaching prey from below or, less often, from side; exceptionally, hovers then swoops down (Schlegel 1967). Low over (e.g.) bracken or heather, flapping alternates with descending glides in which wings held in V like harrier *Circus*. In rides or along forest edge, bird flies close to top of canopy and hovers for c. 3s every 5–20 m, apparently to take food items from amongst (but not from surface of) vegetation (SRG). Bill opened only immediately prior to making prey capture; bird does not 'trawl' for food with open bill (Heinroth 1909; Bühler 1970, which see for detailed morphological study; M H Brinkworth); for further discussion, see Madon (1934). When fewer insects flying (and probably only after dark: SRG, C J Cadbury), also makes short flights from low perch or ground like flycatcher (Muscicapidae): holds head at angle, occasionally turning it before suddenly taking off; flight usually up to c. 10–15 m from perch. Prey may be approached as in sustained hunting flight (Schlegel 1967; SRG) or as often from above as from below (M H Brinkworth). Capture (sometimes of 2–3 items) followed by return to same perch (or 2–3 may be used) where prey swallowed live and whole (Schlegel 1967; Howes 1978). Insects presumably spotted against light of sky (Schlegel 1967). Use of echo-location suggested for foraging Common Nighthawk *Chordeiles minor* (Griffin 1958); neither this nor auditory location of sounds produced by insects shown experimentally in *C. europaeus* but further study recommended (M H Brinkworth; also Schlegel 1967). Many older and some more recent reports of ground feeding should be viewed with caution. Observations on captive birds suggest this method is at least rare in the wild (Schlegel 1967; also Heinroth 1909, Heinroth and Heinroth 1924–6). See also Stanford (1954) and comments on grit-taking (below). Occasionally makes short forward dart on ground to seize Orthoptera and Coleoptera (Glutz von Blotzheim 1962); once seen to pick up flightless ♀ glow-worm *Lampyris noctiluca* (Owen 1954); at Ockham Common (Surrey, England), pellets contained mainly flightless worker ants (Formicidae) (M H Brinkworth). According to Neufeldt (1958b), occasionally takes food from vegetation. Food carried to young in gullet; chick seizes adult's bill which is then opened to prize chick's further apart. Saliva-enveloped food-ball then transferred. Prey items neither predigested nor dismembered (Heinroth 1909; Schlegel 1969). Undigested chitin fragments ejected as pellets measuring c. $10 \times 14 \times 17$ mm (Schlegel 1969). Few data on prey capture rates. One ♀ hunting from perch caught 15 insects in 15 min; another took 31 items in 28 such flights during a similar period. ♂ hunting by aerial-pursuit caught 12 insects in c. 1 min. Experiments on captive birds indicate ability to conserve energy in bad feeding conditions by falling into lethargic state with reduced body temperature, pulse, and breathing rate. In autumn, birds normally have high fat levels and are able to live off these for some time; a captive bird fasted for 8 days without apparent ill effect (Schlegel 1967). May forage up to several km from nest territory prior to laying, often exploiting abundant resources in open areas, by herds of grazing animals or by water (Glutz von Blotzheim 1962; Schlegel 1967); in Budapest (Hungary), birds took insects attracted by street lighting (Vasvári 1931–4). Particular sites used repeatedly. Generally forages in nesting territory after laying (Schlegel 1967; see also Koenig 1952). Adults forage for c. $1-3\frac{1}{2}$ hrs per night to meet own needs, incubating ♀ (also ♂ in case of overlapping broods) for only c. 1 hr. Generally rests briefly between bouts of foraging and none normally done c. 23.00–01.00 hrs; otherwise hunts at any time dusk to dawn (Schlegel 1967). According to Neufeldt (1958b), peak foraging activity at dusk, also before dawn if evening bout not very productive; feeding on dull afternoons also occurs (Glutz von Blotzheim 1962). Drinking recorded only exceptionally (Glutz von Blotzheim 1962), once in flight (Pettet 1982, which see for references to other Caprimulgidae).

Reduced number of oil droplets in eyes probably important in providing contrast for foraging bird (Peiponen 1964; Lythgoe 1979; M K Brinkworth). Further special adaptations for hunting are exceptionally wide gape and prominent rictal bristles which unfold to form natural trap (Heinroth 1909), and probably cleaned with pectinated middle claw (Schlegel 1969). Effectiveness probably further increased by membranous and highly elastic gape. Conspicuously vascular nature of gape probably also enhances sensitivity to contact with prey (Howes 1978; see also Red-necked Nightjar *C. ruficollis*).

Composition of diet mainly dependent on supply of insects available; varies according to year, season, weather, and temperature (e.g. exploitation of high Lepidoptera populations, main flying time of particular Diptera and Coleoptera). Adults foraging for own needs probably do not select prey (but see below). Prey size ranges from mosquitoes (Culicidae) and Microlepidoptera up to large Lepidoptera (e.g. *Cossus cossus*) and Coleoptera e.g. Neufeldt 1958b, Schlegel 1967). Lepidoptera include occasional butterflies but overwhelmingly moths: Hepialidae, Cossidae, Pyralidae, Arctiidae, Lymantriidae, Geometridae, Noctuidae, Cymatophoridae, Drepanidae, Lasiocampidae, Notodontidae, Sphingidae, Tortricidae, and Microlepidoptera. Wide range of Coleoptera: Carabidae, Dytiscidae, Scarabaeidae, Geotrupidae, Elateridae, Coccinellidae, Cerambycidae, Curculionidae,

Chrysomelidae, Silphidae; occasionally, Lucanidae and Anthicidae. Other insects recorded include mayflies (Ephemeroptera: particularly *Siphlonurus lacustris*), dragonflies (Odonata: *Lestes*, *Aeshna*), crickets, etc., (Orthoptera: Tettigoniidae, Gryllidae, Acrididae, Gryllotalpidae), cockroach *Ectobius lapponicus*, bugs (Hemiptera: Corixidae, Miridae, Cercopidae, Jassidae, Ledridae, winged Aphidoidea); a few Hymenoptera (Ichneumonidae and various ants Formicidae). Also takes Neuroptera (antlions Myrmeleontidae, lacewings Hemerobiidae and Chrysopidae), caddis flies (Trichoptera: Hydropsychidae, Phryganeidae, Leptoceridae, Limnophilidae), and flies (Diptera: Tipulidae, Culicidae, Anisopodidae, Limnobiidae, Cecidomyiidae, Scatophagidae, Tachinidae, Empididae). Only other invertebrates recorded are spiders (Araneae), also mites (Acari) probably ingested with Formicidae (Naumann 1833; Csiki 1905; Rey 1908; Collinge 1924–7; Vietinghoff-Riesch 1928; Marples 1939; Escherich 1942; Mal'chevski and Neufeldt 1954; Neufeldt 1958b; Schlegel 1967; Korenberg et al. 1972; Kolesnikov 1976; M H Brinkworth.) Few references to vegetable matter (e.g. Marples 1939, M H Brinkworth); all items probably taken accidentally (see Schlegel 1967). For further details, including discussion of some older references, see Madon (1934). Small stones, etc., taken regularly (e.g. Rörig 1900, Marples 1939, Remmert 1953, Neufeldt 1958b). Also sometimes fed to young (mortar fragments up to 10 × 10 × 15 mm recorded in neck-collar sample). Grit not found in all stomachs though present in pellets (Schlegel 1967, 1969).

62 stomachs (May–October) from various localities in Britain contained (by number) 49·0% Lepidoptera (mainly Hepialidae and Noctuidae), 38·0% Coleoptera (Scarabaeidae, also Geotrupidae), and 13·0% Diptera. 1 July stomach held 15 beetles *Rhizotrogus solstitialis* and moths: 67 *Hepialus lupulina*, 40 *Agrotis segetum*, and 8 *Noctua pronuba*; 1, August, contained 163 craneflies (Tipulidae) as well as many moths and beetles (Collinge 1924–7). 10 stomachs obtained over 5 years (1 from April, 5 from August, 3 from September, 1 date unknown) from Oberlausitz (East Germany) contained 534 Lepidoptera, 137 Coleoptera (including 94 *Aphodius* and 19 *Criocephalus rusticus*), 4 lacewings *Chrysopa*, 2 Tipulidae, 2 ichneumons Ophinioninae, and 1 unidentified insect. Faecal samples contained other Coleoptera and, in 1, at least 500 winged Formicidae (Schlegel 1967). Stomachs from Hungary (14 from May–June, 2 from September) contained 44 Scarabaeidae, 15 Geotrupidae, 5 Silphidae, 9 Noctuidae, and 1 *Gryllotalpa* (Csiki 1905), 24 stomachs from southern Kareliya (USSR) contained 368 Lepidoptera (Geometridae, Sphingidae, Lymantriidae) in 91·6%, 76 Coleoptera (many *Criocephalus rusticus*) in 58·3%, 88 Tipulidae in 25·0%, 6 dragonflies *Aeshna grandis* in 12·3%, 53 winged ♂ ants *Lasius niger* in 4·1%, and 14 Trichoptera in 4·1% (Neufeldt 1958b). For analysis of 10 adult stomachs from south-east Ukraine (USSR), see Kolesnikov (1976). 2 stomachs (January and February) from Avakubi (Zaïre) contained 1 grasshopper (Acrididae), 2 cockroaches (Blattidae), 1 small praying mantis (Mantidae), bugs (Hemiptera), 4 moths, and small beetles (Chapin 1939).

In Oberlausitz, 74 neck-collar samples from 11 broods (young 2–17 days old) produced 3935 prey items: primarily Lepidoptera (62·4%; mainly Noctuidae, followed by Microlepidoptera and Geometridae), Diptera (12·2%), Coleoptera (7·7%), Trichoptera (7·2%), and Neuroptera (5·7%). At 2–4 days, prey size 7·9–10·2 mm ($n = 584$); at 5–17 days, 11·2–13·9 mm ($n = 3351$). Diptera decreased from 31·1% (by number) at 2 days to 9·2% at 5–17 days, Microlepidoptera from 25·2 to 14·7%, while other Lepidoptera increased from 24·3% to 45·5%, Cerambycidae from 1·0 to 5·8%, and other insects from 18·4 to 21·2% (though at 3 days 17·5 and at 4 days 8·7%). Younger chicks thus given larger proportion of delicate, soft-skinned Microlepidoptera and Diptera which adults obtain in lower-than-usual hunting flight. Otherwise no major differences between diet of young and adults (Schlegel 1967). For further detailed study in Voronezh region (USSR), see Mal'chevski and Neufeldt (1954); see also Kolesnikov (1976). Chicks recorded eating own faeces at c. 5–6 days (Stülcken 1962). Given whole items from 1st day (Schlegel 1967). Only 1 chick fed per visit (Stülcken 1962); both, according to Selous (1899) and Tutt (1955). Each fed on average c. 10 times per night receiving c. 10–11 g (c. 150 insects) per day between 1 and 17 days (Schlegel 1967; see also Kolesnikov 1976). MGW

Social pattern and behaviour. For detailed studies, see especially Lack (1930b, 1932, 1957), Stülcken and Brüll (1938), Stülcken (1962), and monograph by Schlegel (1969). For study of captive birds, see Heinroth (1909). Following account based on detailed outline supplied by C J Bibby and R Berry; includes data supplied by I H Alexander on behalf of Stour Ringing Group (England) cited as SRG.

1. Not markedly gregarious, though loose aggregations may occur in favourable feeding areas, e.g. up to 15 at rubbish-tip (Reinsch 1970a); 14 over bog of c. 3 ha, mid-June, though most gatherings smaller (Berry 1979); groups (mainly juveniles) of up to 10 not uncommon, August–September (SRG; also Seifert 1961). Not particularly gregarious on migration even where large numbers may concentrate (e.g. Meinertzhagen 1954 for Saudi Arabia). Spring migrant flocks noted, however, in Kenya and these apparently single-sex (Moreau 1972). Probably mostly solitary also in winter but in Zimbabwe several recorded roosting in close proximity (Jackson 1978). BONDS. Pair-bond presumably monogamous and lasting for 1 season, though possibility of partner change for 2nd brood and the extent and nature of reported helping by extra ♂♂ require more detailed investigation (Schlegel 1969; Glutz and Bauer 1980; see also below). Suggested by (e.g.) Stülcken (1962) that repeated use of same nest- or roost-sites perhaps an indication of pair-bond lasting beyond single season (see also Bauer 1976), though may merely reflect more exacting habitat requirements than presently understood (C J Bibby and R Berry). Age of first breeding in both sexes apparently 1 year: single cases of marked birds of both sexes caught on nest at 1 year old (SRG). Young cared for by both

parents until independence; in case of 2nd (overlapping) brood, ♂ tends young of 1st brood from c. 14–16 to 35 days old, i.e. for period corresponding almost exactly with complete incubation of 2nd clutch (Schlegel 1969; see also Lack 1930a, Stülcken and Brüll 1938, Reinsch 1961, Stülcken 1962, and Löhrl 1980). Bond between siblings maintained at least to some extent after fledging (P A D Hollom), at which time young occasionally fed by extra ♂ (Schlegel 1969). Young normally disperse when independent (at 35 days) and adults generally depart at same time as those of 2nd brood (Stülcken 1962; Schlegel 1969). BREEDING DISPERSION. Solitary and territorial. In Oberlausitz (East Germany) nests c. 200–400 m apart in open heathland with scattered mature pines *Pinus*, much natural regeneration of smaller ones, and bare patches of varying size (Schlegel 1969). At 2 sites in East Anglia (England), mean distances to nearest neighbour 164 m and 172 m (Berry 1979); see also Lack (1932) for similar examples. Separation between adjoining pairs often greater because ideal habitat fragmented (C J Bibby and R Berry). Pairs often dispersed rather linearly along woodland edges which have scattered old trees and invading younger ones in otherwise open heathland (Berry 1979; Cadbury 1981). Emphasized by Glutz and Bauer (1980) that singing migrants may give false impression of occupied territories up to late May or early June. Trapping of marked birds in one year suggested fluid pattern of dispersion until c. 5 June (SRG). Density varies markedly with vegetation and, as optimal habitat mostly linearly dispersed, rarely reaches expected close-packed maximum of c. 20 pairs per km² except locally (Schlegel 1969). Densities at 3 East Anglian sites: average 7·8 pairs per km², range 5·1–17·0 (or more likely 11·4) pairs per km² over 30 years on 176 ha; average 11·2 pairs per km², range 8·3–13·6 over 8 years on 169 ha; mean 5·7 pairs per km², range 7·0–14·0 over 10 years on 126 ha (Berry 1979). 7 pairs within and 7 on the edge of 86 ha; 12·0–14·0 pairs per km² (Lack 1932). In Oberlausitz pinewood, 34–36 pairs over 4 years on 350 ha (i.e. 9·7–10·3 pairs per km²); 19·4 pairs per km² in best areas (Schlegel 1969). In pine forest near Steckby (East Germany), 33 pairs on 900 ha (Steinke 1981). On more extensive areas such as Lüneburger Heide (West Germany), 1–1·5 pairs per km² (Glutz and Bauer 1980 which see for additional central European data). On northern edge of range, in Finland, 0·1–0·2 pairs per km² (Merikallio 1946, 1958), but 2–5 pairs per km² in optimal habitat (Glutz and Bauer 1980). Territory apparently selected by ♂, and patrolled and defended by him, at least early in season. Sites often traditional for short run of years: probably due to limited number of suitable nest- or roost-sites or, especially, song perches (C J Bibby and R Berry). Several records of fidelity to nesting area or actual nest-site (e.g. Junge 1938, Glutz and Bauer 1980). Territory used for courtship, copulation, nesting, and rearing of young, but adults frequently forage in loose aggregations away from territories at good sites: e.g. round an old oak *Quercus*, over bogs and open heathland (Lack 1932; Berry 1979); a marked ♂ ranged up to c. 500 m from nest, i.e. well outside territory (SRG). Suggested, however, by Schlegel (1969) that birds forage mainly within territory once incubation started. Breeding ♂♂ may roost outside their territories (see below). Territoriality much less pronounced after laying when foraging strangers may be tolerated and extra ♂♂ may occasionally feed young (Schlegel 1969; see below). Territory size can be assessed by plotting perches of ♂ but likely to give slight underestimate as flight range often extends beyond outermost song perches. In Oberlausitz, 4 such measurements 3·8 ha and 6·7 ha in sparsely populated area and 5·0 ha and 4·3 ha where density higher (Schlegel 1969). At Steckby, c. 3·5 ha and c. 3·1 ha, unaltered for 1st brood but later extended to c. 4·8 ha and c. 4·7 ha respectively (Steinke 1981). In East Anglia, song-posts encompassed ranges of 5–6 ha (Berry 1979). 2nd nests (presumably same pair involved, though not proved) sited within same territory, often within c. 100 m of 1st and rarely more than c. 200 m (Lack 1932; Stülcken and Brüll 1938; Stülcken 1962), though up to 500 m recorded (Glutz and Bauer 1980). Woodland glades of less than 1–1·5 ha unlikely to support a pair, while from 3·2 ha upwards, 2 ♂♂ may hold territories (Brünner 1978); in Karlsruhe area (West Germany), clearings occupied (each by 1 ♂) 5–32 ha (Leibig 1972). ROOSTING. Diurnal; activity nocturnal. Most information relates to breeding season but essential features probably similar at other times (see, e.g., Jackson 1978). Before laying, roosting ♀♀ rarely found and apparently roost preferentially in trees, perhaps out of territories (R Berry). After c. 3 weeks, ♀ invariably roosts at nest with eggs or small young; ♂ may take on this role shortly before fledging if ♀ incubating 2nd clutch. Few records of both adults roosting at nest with young near fledging (Schlegel 1969; Reinsch 1970a; Berry 1979). ♂ generally roosts low down in territory, often on ground, sometimes on branch up to c. 1 m high (Lack 1930b); site c. 50–150 m from nest, on 4 occasions within a few metres (Lack 1932), or usually c. 50 m, exceptionally up to 100 m away (Berry 1979); up to 200 m recorded (Burkitt 1916). A radio-tagged breeding ♂ roosted well outside his territory and in that of another bird; thus closer to 1, possibly 2 nests than his own (SRG). Typical roost-sites include bare patches on ground (6 cases), tree stumps (14), dead branches on ground (11), roots (2), or piles of twigs on ground (3) (Schlegel 1969); also on thick horizontal branches of tall trees or old bits of cardboard or sacking (Stülcken 1962). Alternatively, ♂ may roost mainly in cover of pine forests, or birch *Betula* or *Rhododendron* scrub (SRG). Dependent juveniles usually roost on ground (R Berry). Roost-sites not concealed; bird relies on crypsis. Sites often used for long periods during breeding season but birds may change if disturbed. May also be used from year to year but unclear whether this due to adult site-fidelity or scarcity of ideal places in particular area; record of 3 birds being shot on 3 consecutive days from identical spot on branch in garden (Naumann 1833) apparently suggests latter. Roosting bird usually has body pressed tight to ground or perch in elongated posture, rarely projecting over such features as branch or stump. Often lands crossways on branch then moves round to parallel (Lack 1930b). Head not tucked beneath scapulars; eye closed to slit, thus enhancing cryptic appearance. Roosting birds apparently not fully asleep; move (e.g.) occasionally from shadow to sun-bathe (Lack 1932). Captive birds unwilling to roost on a white surface; further reported to have sat facing the light with body tapering away and thus casting no shadow (Heinroth 1909), apparently confirmed for wild birds by Wälti and Locher (1952) but not by Lack (1932) or Schlegel (1969). First call (usually 1 or more Quaw-eek calls: see 2 in Voice) good indicator of start of evening activity but converse not so in morning. Waking times vary with time of sunset (i.e. season and latitude), weather, and phase of moon. Wakes c. 11 min before to c. 26 min after sunset (Schlegel 1969); Advertising-call first given c. 15–30 min after sunset (Wynne-Edwards 1930: see 1 in Voice; see also Ashmore 1935); nest-relief averaged 13 min after sunset (Lack 1957; see below). In Britain (Owen 1949) and central Europe at least, usually a break in calling around midnight (Schlegel 1969; Leibig 1972). In Finland, in high summer, continues throughout night until c. 03.00 hrs (Ehrström 1955); in autumn (mid-September), restricted to c. ½ hr at dusk and flight activity begins from when sun 5°30′–8°10′ below horizon and ends with it 8°10′–12°15′ below (Lehtonen 1951), i.e. with more light at start of activity than at end (see also Schlegel 1969). Activity thus starts later nearer summer solstice and at higher latitude and relatively earlier on dark and wet nights. First

Advertising-call up to 20 min later in full moon than new; suggests possible breeding synchronization with lunar cycle, since full moon does not rise until sunset and thus has no effect on light intensity to create the effect (Wynne-Edwards 1930). Few (not fully substantiated) records of wild birds apparently in torpid state (see Jaeger 1950, Novrup 1956, Anon 1957). Experimental work indicated that such lethargy (bird stiff, cold, and motionless, wings slightly spread, plumage ruffled, eyes closed, pulse and breathing rate slowed) occurs only when fat reserves exhausted and bird loses weight (Schlegel 1969), and not induced just by low ambient temperature (Fog and Petersen 1957). Bird apparently able to awake spontaneously from slight hypothermia (body temperature not below 15°C) but not from deep hypothermia unless ambient temperature rises above 15°C. Body temperature shows 2 daily minima (midday and midnight) and 2 maxima (coinciding with morning and evening peak activity). For further details, including experimentally induced torpidity through fasting combined with lowering of temperature, see Peiponen and Bosley (1964), Peiponen (1965, 1966, 1970), and Schlegel (1969); see also Schlegel (1967) for observation on 10-day-old bird.

2. When threatened at roost, bird relies on crypsis, adopting Cigar-posture (Fig A): head moved forward and down with eyes

A

closed to slit. Remains motionless thus, usually allowing approach to within a few metres (c. 5 m: Wadewitz 1956). Finally flies up suddenly and may give alarm-call or Wing-clap at departure; usually lands c. 20–40 m away (Heinroth 1909; Schlegel 1969: see Voice); may also perform rather erratic Butterfly-flight in which wings held high over body and beaten jerkily, then often returns to land nearby before renewed (final) departure made (Wadewitz 1956). Most cases of Advertising-call being given in daytime probably due to disturbance at roost (Reinsch 1961; N Cleere: see 1 in Voice). Even early in season (Stülcken 1962), potential predators and other intrusions in territory investigated in close approach and hover with body held nearly vertical (Fig B); often within a few metres of man. Also

B

elicited by deer (Cervidae), flashlight, stick thrown into the air, etc. (Schlegel 1969). On leaving roost, *C. europaeus* may be mobbed by other birds (Lawson 1951) or bats (Chiroptera) (Bauer 1976); perhaps because of superficial resemblance to cuckoo (Cuculidae) or small raptor. One waking ♂ was motionless for first hour, then twice jerked head forward and back and swayed from side to side; preened under wings, stretched each wing once, then was still again for a few minutes before it flew off and gave Quaw-eek call (c. 13 min after first signs of awakening) (Schlegel 1969: see 2 in Voice). In captive birds, actions rarely made suddenly, being normally preceded by similar swaying and jerking motions; interpreted as protective adaptation (Heinroth 1909; see also Wadewitz 1956); considered unlikely by Schlegel (1969) who suggested that it probably only expressed agitation (see also below). FLOCK BEHAVIOUR. No information. ANTAGONISTIC BEHAVIOUR. (1) General. ♂ takes up territory shortly after arrival. Cases of Advertising-call being given in late April followed by interval before start of regular activity (Berry 1979) possibly refer to passage migrants or to relative temporary quiet of early-arriving ♂♂. For record of ♂ giving Advertising-call whilst actually migrating, see Glutz and Bauer (1980). Possible tendency for territories to be occupied in similar order in successive years (Berry and Bibby 1981). Territory delineated by ♂ who patrols and gives Advertising-call (which also attracts ♀: see below); typically delivered from top of bushy tree or from bare horizontal branch high up larger tree, less often lower down (e.g. stump) or even on ground (Schlegel 1969); cottage roofs also recorded (Christian 1949). See also photograph in Ferguson-Lees (1961) which shows that ♂ may stand rather than squat when giving Advertising-call. Patrolling flights also performed throughout season by unpaired ♂; in one case, presumed patrolling flight unusually high at c. 10–15 m above trees (N Cleere). Boundaries of patrolled range, and of neighbours, usually clearly defined and obvious. After disturbance, human intruder possibly followed to boundary but not beyond; attempts to chase incubating ♀ beyond this range always unsuccessful (Schlegel 1969). (2) Threat and fighting. ♂♂ on adjoining territories often give Advertising-call simultaneously from nearest perches. One bird invariably faced nest, even when rival singing in opposite direction (Christian 1949). Intruding ♂♂ often driven off early in season: territory owner approaches intruder with vigorous Wing-clapping and Quaw-eek calls. Normal flight often interspersed with stiff-winged glide (presumably similar to or part of Butterfly-flight) which shows off white wing- and tail-markings. Territory owner may alight (on the way) and give Advertising-call or sway from side to side with tail fanned. Pursuit-flights in which territory owner apparently attempts to overtake intruder also occur; intruding ♂ quick to leave territory in most cases (Schlegel 1969; Leibig 1972; C J Bibby and R Berry). However, neighbouring ♂♂ often fly together with slow wing-beats, glides, and Wing-clapping, much as in courtship (Lack 1932; see below). Earlier reports of vigorous fights between ♂♂ (e.g. Brehm 1876) not confirmed more recently apart from one observation open to various interpretations (see Stülcken and Brüll 1938) and brief scrimmage (birds tugging at each other's feathers) followed by departure of one bird (Schlegel 1969); at Steckby, regular and at times violent boundary disputes involved only ♂♂ and were more pronounced for 1st brood (Steinke 1981). Fights (not damaging) said to occur in the air and on ground (Reboussin 1924). 2 ♂♂ may face each other on bare ground with tails slightly raised and wings drooped. One or both may make short leaps into the air or towards the other; Wing-clapping sometimes performed. Interaction ends with one bird leaping at the other which flies off and is pursued (SRG); see also Stülcken and Brüll (1938). After expulsion of intruder, territory owner may give prolonged Advertising-call from part of boundary where other first appeared or departed (Schlegel 1969). Reacts to mirrors (see Derim 1962 and Parental Anti-predator Strategies). In response to playback of recorded Advertising-calls, paired ♂ (from late May or June) approached sound source and gave Advertising-calls from different perches on territory; often joined by ♀. Unpaired (but territory-holding) ♂ showed little response (N Cleere). Particularly later in season, may be indifferent to intruders. Conspecific neighbours responding to alarm-calls of another territory-owner tolerated as long as danger present (Schlegel 1969); see also Leibig (1972). 3 ♂♂ silently circled territory for $c.$ 10 min in Butterfly-flight just after one had copulated; both intruders then landed near ♀ and remained in silence for $c.$ 5 min (R Berry). From early July (i.e. when young of 1st brood about half-grown) increasing number of observations of 2 (once 3) ♂♂ together at nest-site and showing no intolerance (Schlegel 1969). 2 ♂♂ reported together at roost on 14 June (Reinsch 1961); in one case, 2 ♂♂ regularly roosted $c.$ 2·5 m apart (Lack 1932). In at least 3 cases, 2nd ♂ helped to feed young, though vigorous chase seen as late as 17 July (Schlegel 1969); renewed antagonism perhaps linked with fledging of young (N Cleere). ♀ joined by neighbouring ♀ on ground for $c.$ 10 min showed no reaction; ♂ hovered and Wing-clapped nearby (R Berry). For responses of captive pair to various small birds near nest, see Heinroth (1909). HETEROSEXUAL BEHAVIOUR. (1) General. Early in season, ♂ gives Advertising-call in brief bursts with long pauses; frequency and duration later increased. Advertising-call given not uncommonly in brief bursts during day; probably mainly when disturbed (see above, also Parental Anti-predator Strategies, below). Courtship may begin immediately after arrival of ♀♀ or may be delayed for a period during which ♀♀ may be most unobtrusive (Wadewitz 1956; Stülcken 1962; Schlegel 1969; C J Bibby and R Berry). In Suffolk (England), ♀♀ arrived 1–20 days (mean 10·9) after ♂♂ on 12 territories (Berry and Bibby 1981). Most activity takes place at dawn (more so) and dusk and markedly weather-dependent, being depressed in cool or wet conditions. In Oberlausitz, peak courtship activity for $c.$ 1 week at end of May. Preliminary to 2nd brood essentially similar but of lower intensity. Advertising-call may be given throughout season up to departure but only in sporadic short bouts after leaving roost from mid-July (Schlegel 1969). After laying, ♂ gives Advertising-call less often (may cease completely with hatching: M H Brinkworth) and courtship Butterfly-flights with Wing-clapping cease but resume shortly after hatching (C J Bibby and R Berry). (2) Pair-bonding behaviour. ♂ may give Quaw-eek calls when flying from roost to song perch and also frequently Wing-claps when moving to another perch (Schlegel 1969). In absence of intruding ♂♂ or predators, generally indicates presence of ♀ (R Berry). In slow Butterfly-flight (interspersed with Wing-claps), ♂ glides with wings raised and tail depressed and fanned (Fig C); common feature of ♂'s display, especially early in season. In early stages of courtship, ♂ follows ♀. In early-morning courtship chases, ♂ followed ♀ in circles, giving frequent Quaw-eek calls; ♀ generally silent though often climbed to $c.$ 25 m and there circled rapidly, ♂ not following; ♀ finally descended in fast glide (Owen 1949). Once pair-formation completed, ♀ tends to follow ♂ who apparently leads her to ground for copulation and (possibly) nest-site selection. If ♀ flies directly towards ♂, his Advertising-call has characteristic clearer-sounding ending (see 1 in Voice); this given before take-off (or in Butterfly-flight: Vollbrecht 1938). ♂ then descends to ground, Wing-clapping and giving Quaw-eek calls; Butterfly-flight, with white wing- and tail-markings prominently displayed, performed in last stage of descent. ♀ follows $c.$ 1–4 m behind ♂ and may also Wing-clap and give Quaw-eek calls (Schlegel 1969; see also Horst 1938); ♀ also performs Butterfly-flight in this context. Once, $c.$ 10 s after birds had landed on ground, ♂ flew up and circled ♀; wing-tips quivered rapidly and tail held at angle so that white markings pointed at ♀; ♂ then landed by ♀ again (Leibig 1972). (3) Mating. Believed to take place on ground after Butterfly-flight sequence as described. One interaction initiated by ♂'s arrival at (2nd-brood) nest: ♀ flew up with Wing-clapping and landed at roost-site with Quaw-eek call, ♂ landing opposite her. ♀ swayed body, ♂ following suit. ♀ then lay still and ♂ changed to vertical bobbing movement: rear body and tail jerked up and down, wings partially opened; Bobbing accelerated then ceased simultaneously with sudden erection of fanned tail on which white spots strikingly obvious. ♂ rose with single flick of wings and glided on to ♀'s back. During copulation, ♂'s wings raised and quivered, head lowered and lower mandible vibrated. ♀ then moved forward and forced ♂ off by raising wings slightly. ♂ twice returned to Bobbing with terminal tail-flash but ♀ thwarted further mounting attempts by raising wings and running forward; ♂ glided off with remarkably fast quivering wing-beats (Lack 1932). In captive birds, willing ♀ cocked tail after ♂ had flown on to her back; during copulation, ♂ gave peculiar murmuring sound (Heinroth 1909: see 6 in Voice). On another occasion, ♀ on ground had wings open and tail spread; circled by ♂ who had fanned tail and drooped wings and gave quiet bubbling phase of Advertising-call. Copulation ensued after several minutes of circling (R Berry: see 1 and 6 in Voice). See also description by Reboussin (1924) who additionally mentioned Wing-clapping by ♂ on ground. Only ♀ called (see 6 in Voice) in possibly unusual (but see photograph in Ferguson-Lees 1961) case of copulation (3 times in succession) on ♂'s elevated song-perch to which ♀ had flown after ♂ had landed on the ground and given bubbling call (R Savage). Display and copulation sequence may be performed several times per evening; less frequent and obvious for 2nd brood (Schlegel 1969). (4) Nest-site selection. Several days before laying, probably by ♂. Cases of ♂ landing on ground and there giving quiet bubbling call (muffled churring: Leibig 1972; see also 1 in Voice) probably attempt to attract ♀ to potential nest-site (R Berry). In captive pair, ♂ chose site: supported himself on carpal joints and made vigorous scraping movements with both feet; called almost incessantly (see 8 in Voice). ♀ constantly followed ♂ at this stage and often urged ♂ away and settled where he had been scraping (Heinroth 1909). Belief that ♂ chooses site supported by several

C

D

cases of eggs laid at spot occupied by ♂ a few days earlier (e.g. Lack 1932, Stülcken and Brüll 1938, Stülcken 1962); this invariably so (R Berry). However, in another study, pair reported near eventual nest-site 1–2 days before laying but in every case exact site was that used by ♀ (Schlegel 1969); ♀ also found on nest 3 days before laying (Reinsch 1961). (5) Behaviour at nest. Daytime incubation by ♀. At start of every evening activity, ♂ gives Quaw-eek calls and Advertising-call for a while. Then flies to nest Wing-clapping and gives bubbling call (Schlegel 1969: see 1 in Voice) or Quaw-eek calls (Lack 1932). ♂ may give Advertising-call (perhaps bubbling variant) in flight when approaching nest, and ♀ said to respond similarly from ground; ♂ lands with fanned tail, and ♀'s rear body and wing-tips tremble (Stülcken 1962). Relieved bird normally departs abruptly before incoming bird has landed. On landing, wings typically raised high momentarily (Fig D), then carefully folded before bird proceeds to nest (Hosking 1950, which see for photograph of departing bird; see also Ferguson-Lees 1961 for photograph of changeover). Probable bubbling calls and waggling movements of tail and rear body before ♀'s departure described by Selous (1899), which see for detailed observations on nest-relief and other interactions at nest; see also Burkitt (1916). When ♀ leaves, ♂ moves to nest with tripping gait and usually incubates for c. 15 min, while ♀ absent (and presumably foraging) for c. 30–60 min (Schlegel 1969); thus ♂ usually leaves before ♀ returns or she may call him off with Wing-claps (Stülcken 1962). In one case, ♂ did not always relieve ♀ who left on her own at dusk (Lack 1957). RELATIONS WITHIN FAMILY GROUP. Young call up to 2 days before hatching (Heinroth 1909; Lack 1930b). Incubating bird becomes more restless and gives Grunting-call (Stülcken 1962: see 7 in Voice). Chicks active and eyes open on 1st day. Brooded by ♀ all night until a few days old; short nest-reliefs by ♂ at dusk and dawn much as during incubation. With chicks 2–10 days old, adults bring food in alternation and one usually broods until other returns (Schlegel 1969). Brooded by day until c. 16–20 days (Lack 1932), though ♂ not always very attentive in later stages (Lack 1957). Young respond to swaying of adult with similar movements before being brooded (Wadewitz 1956). Beg with food-calls (see Voice) and tugging at rictal bristles or bill of adult. Transfer of food may take 10 s or more. Heads of both parent and young move up and down; chick opens wings and may hold them forward over parent's head while its body trembles markedly (Stülcken 1962; Schlegel 1969). During feed, parent gives Grunting-call and chick gives food-call (P A D Hollom; see also Selous 1899). One chick fed in evening by regurgitation from ♀ who had been on nest all day. Chick may wave or flap wings while being fed (Selous 1899). After feeding, chicks often run backwards to defecate initially c. 20 cm and later up to c. 1 m from nest so a circle of faeces accumulates. Chick may move short distances (e.g. to shelter from sun or rain, or if disturbed: (Schlegel 1969). Some movements may be more permanent, as indicated by accumulated faeces, but difficult to ascertain to what extent influenced by disturbance. Most young stay within a few metres of nest; in 1 case moved c. 25 m (Berry 1979). Siblings may perform rocking movements and call (see Voice) while together at nest (P A D Hollom); see also Tutt (1955). One captive brood performed greeting ceremony to one another (rarely to adults) up to independence: ran towards one another and called; neck extended vertically, wings often widely spread and raised (Heinroth 1909: see Voice). Not clear in what circumstances such behaviour occurs in the wild, but for wing-raising by young, possibly as greeting to ♂ arriving with food, see Hainard (1955). Young can rise from ground by wing-flapping by c. 2 weeks and take short flights, with successful prey captures, at c. 3 weeks. Such activities rare by day. If 2nd clutch laid, ♂ cares for young of 1st brood from c. 14–16 days until independence, feeding them and roosting with them by day for variable period. Also continues to relieve ♀ briefly on other nest at dusk and dawn (Tutt 1955; Lack 1957; Stülcken 1962; Schlegel 1969). Young of 2nd brood tended mainly by ♀ (Stülcken 1962). When only 1 brood reared, ♀ brooding young by day (until fledging) sometimes joined at nest by ♂ (e.g. Reinsch 1970a). From c. 19 days, young may follow foraging ♂ who may Wingclap and give food-calls; fed on nest or elsewhere on ground (Lack 1957; Stülcken 1962). A captive ♂ started threatening and chasing his ♂ offspring at c. 5 weeks (Heinroth 1909) which coincides with time of independence and dispersal in wild young (Schlegel 1969). In one case, when 1 young bird had left nest area, other stayed put but the 2 birds maintained vocal contact. Young later reunited—returning bird called and snuggled up against sibling (P A D Hollom: see Voice). For suggestion that young stay together with one or both adults for (part of) autumn migration, see Lack (1930b) and Lehtonen (1951). ANTIPREDATOR RESPONSES OF YOUNG. When only a few days old, usually crouch and remain motionless, eyes closed to slit; posture thus resembling Cigar-posture of adult (Wadewitz 1956); crouch and stare up when Grunting-call of adult gives warning of raptor (Stülcken 1962: see 7 in Voice). At only 1 day old (Weber 1957), or at c. 6 days old and not earlier despite provocation (Schlegel 1969), or not until 10 days old (Wadewitz 1956), give Gapingthreat display. Bird crouches still lower, then slowly raises head and wings from Cigar-posture; neck may be extended in spiralling movement and, at peak, hissing sound (or cluck: Lack 1930b) given for c. 5 s, revealing wide red mouth (Fig E); sometimes leaps up a little, creating effect very startling to man, much like young Cuckoo *Cuculus canorus*. May subside and repeat performance several times. Pecks at hand if near enough (Lack 1930b;

E

Wadewitz 1956; Schlegel 1969; N Cleere). One bird of brood often more active (e.g. Wadewitz 1956); in one case, young alternated: one performed Gaping-threat display, other crouched, and so on (Schlegel 1969). When less than 10 days old, bird may remain crouched and hiss with bill closed (Krambrich 1954). Disturbed young may scrabble a short way from nest and crouch in cover or lie flat with extended wings, thus increasing impression of size; hiss when approached or handled (Lack 1930b; N Cleere); see also Burkitt (1916). Newly fledged young rely more on flattened Cigar-posture and at less than 3 weeks old do not attempt to escape until caught (Schlegel 1969). PARENTAL ANTI-PREDATOR STRATEGIES. (1) Passive measures. At nest usually relies considerably on crypsis adopting Cigar-posture (Fig A). For eye movements, see Derim (1962), and for freezing in presence of Jays *Garrulus glandarius*, see Tutt (1955). Usually flushes only when approached to within a few metres (more if frequently disturbed and if approached directly rather than tangentially); usually at greater ranges by night than by day (Schlegel 1969). Old reports of eggs or young being deliberately transported (e.g. Liebe 1887) not believed by Schlegel (1969) whose studies involved much disturbance but did not elicit such behaviour; bird may however run backwards using bill to roll eggs to another site following disturbance (Lehtonen 1972). (2) Active measures: against birds. One flew at Tawny Owl *Strix aluco* in a tree, giving an alarm-call and followed when it left (Schlegel 1969: see 3 in Voice). Both *S. aluco* and Little Owl *Athene noctua* chased (once by pair) with much Wing-clapping and Quaw-eek calls; aggression increased after laying (Owen 1949, 1951). One *C. europaeus* caught with claw marks in wing, and 5 cases of owls (Strigidae) being mist-netted low over tape-recorder on ground suggest these may be serious potential predators (SRG). Woodcocks *Scolopax rusticola* apparently ignored (Owen 1951). ♀ performed distraction-lure display of disablement type in presence of *G. glandarius* (R Berry). For reaction to stuffed birds (Strigidae and Magpie *Pica pica*), see Derim (1962) and below. (3) Active measures: against man. Various distraction-threat and distraction-lure displays performed especially by day but also at night (Lack 1932; Schlegel 1969). In impeded flight from nest, inner wing held forward and primaries thrust back, tail spread, usually depressed, and often twisted, and dangling legs appear weak. Leaves nest with violent flaps but makes slow progress and may fall to ground as if exhausted with wings and tail spread. May briefly stay still thus on ground, or wings may be shaken and tail waved from side to side. One bird rolled from side to side, shaking one wing at a time (Lack 1932). One ♀ lay on her side 'frozen' for *c*. 5 min (Tutt 1955). In response to an imitated Quaw-eek call, bird landed *c*. 2 m from observer and leaped several times into the air, each leap accompanied by a single jerking wing-beat; also waved wings (Smith 1955). Bill often wide open and bird hisses (Fig F); ♀ may also utter faint alarm-call (Lack 1932; Schlegel 1969: see 3a and 4 in Voice); Distress-call also noted (Steinke 1981: see 5 in Voice). Sometimes flies to a branch and may perch crossways with drooping wings and tail or wings shaking (Lack 1932). One ♂ gave impression of difficulty in landing at all (Tutt 1955). One ♀ occasionally called and Wing-clapped (N Cleere: see 3c in Voice); see also photograph of extralimital Jungle Nightjar *C. indicus* in Neufeldt (1982). Flushed bird sometimes flies off in normal flight or with series of glides in which wings held stiffly up as in Butterfly-flight (more often from ♂). May come back calling (not described) or land on ground. When ♀ flushed from nest and ♂ close by, only ♀ performed distraction-lure display; ♂ generally departed quickly (Lack 1932). Much individual variation in extent of display and may also diminish at frequently visited nests. Given with greatest vigour at hatching, only some start earlier (Lack 1932). When distraction-lure display apparently unsuccessful, bird may perform distraction-threat display much as Gaping-threat display of young (see above). Hissing sound given, also when handled (Schlegel 1969) and apparently in flight; or bird may swoop so that wings brush intruder's head, and give shrill call (Turner 1914). Once, bird leaped up from nest and struck at intruder with wing (Booth 1914). Division between distraction-lure and distraction-threat displays perhaps not always clear-cut. One bird (habitually) allowed approach to *c*. 30 cm, then slowly half-raised body, plumage being ruffled simultaneously; called almost incessantly (see 3c in Voice). Wing nearer danger raised, spread, and underside turned to face intruder while bird moved feet without shifting position; then gradually retreated with head held forward and wings raised alternately as described; finally flew off (Derim 1962). Distraction-lure display of disablement type generally briefer (if given at) all after dark. Bird more likely to fly around very close; repeated Quaw-eek and alarm-calls given (see 2, 3a, 3b in Voice). Sometimes lands or hovers (Fig B; see also above) and bird's mate often appears and exhibits similar behaviour (Lack 1932); ♂ may also Wing-clap (Selous 1899). (4) Active measures: against other animals. ♂ flew close to intruding fox *Vulpes* and gave Quaw-eek and higher-intensity alarm calls (Stülcken and Brüll 1938: see 2 and 3c in Voice). ♂ approached by viper *Vipera* rocked from side to side and slowly raised and lowered wings repeatedly while looking at snake, which did not attack (Stülcken 1962: Fig G). Bat (Chiroptera) once chased by pair (Owen 1949).

(Fig A after photograph in Ferguson-Lees 1961; Figs B and D after photographs in Hosking 1950; Fig C based on photograph in Ferguson-Lees 1961; Fig E from photograph in Krambrich 1954; Fig F after photograph in Schlegel 1969; Fig G from photograph in Stülcken 1962.) MGW

Voice. Rich and varied repertoire freely used during breeding season—mainly between dusk and dawn. For renderings and relative frequency of calls given by autumn migrants in Finland, see Lehtonen (1951). Calls 1 and (probably) 2 also heard in winter (Walker 1969); see also Heinroth (1909) for captive birds. For details of sexual variation, see below. Wing-clapping given in variety of contexts (see Social Pattern and Behaviour), mainly by ♂ (Lack 1932; Schlegel 1969). A captive ♂ first attempted to Wing-clap at c. 5 weeks old (Heinroth 1909). Each sound of Wing-clapping like quick snapping of thin dry branch or wooden stake; at times more like hand-clap (J Hall-Craggs: Fig I). Produced by striking wings together over back or by sudden simultaneous upward or downward movement of both wings (Coward 1928; Guggisberg 1941; Burnier 1979). Wing-clapping given up to 25 times in succession by ♂ in courtship (Selous 1899).

CALLS OF ADULTS. (1) Advertising-call (Song) of ♂. Well-known churring sound difficult or impossible to render satisfactorily. Vibrant, churring trill like very rapid but slightly 'wooden' or slightly muffled tapping (Witherby et al. 1938); formerly often compared to sound of spinning wheel ('wheel bird': Bannerman 1955), more recently to small motorbike passing at distance (Glutz and Bauer 1980). Easily confused with stridulation of mole-cricket Gryllotalpa gryllotalpa, but call of C. europaeus slightly lower pitched and slower, and fluctuates more (see below) (Burton and Johnson 1984; see also for similarity to wood-cricket Nemobius sylvestris and natterjack toad Bufo calamita). Recording (Figs II–IV) shows tremolo lasting c. 6 s with c. 29–30 units per s and notable overall crescendo; changes abruptly to faster, lower-pitched tremolo (42 units per s) lasting c. 1 s (0·5 s, $n=11$: Hunter 1980). This merges (via 4 slower units) into a quiet bubbling trill at 2 pitches with c. 36 units per s and lasting c. 3 s, followed by pause of less than 1 s and ends with just audible continuation of trill (c. 1 s); final section a short pianissimo continuation of (or coda to) the bubbling (J Hall-Craggs). For further details of essential components and experimental study, see Abs (1963); see also Jany (1951). For individual variation shown by voice-printing, see Collyer et al. (1982). Bubbling trill which may be given as an independent call in a variety of situations (see Social Pattern and Behaviour) is normal ending to Advertising-call and further described as a quiet 'quorre-quorrequorre' (Schlegel 1969); muffled churr (as if given under water) during heterosexual interaction on ground (Schlegel 1969; see also Stülcken 1962) unlikely to be distinct. In high-intensity sexual excitement, tremolo changes to clearer-sounding 'djill...djilll...djillll' (Schlegel 1969), 'tjull tjull tjull tjull...turrrrr' (Vollbrecht 1938) or 'djiüürrrr-dürrrdürrrdürrrr-rrrrrrrrrrr' (Horst 1938); a long-drawn 'dii-düü' followed by bubbling trill interspersed with Wing-clapping (Bergmann and Helb 1982, which see for sonagrams). In recording by D Sutton, June, more rasping brief churrs given after Advertising-call with ending as described. Advertising-call given for up to 5 (Witherby et al. 1938) or even 9 min 'without pause' (Schlegel 1969); 19½ min with a few short interruptions (Leibig 1972). First of a series of bouts often shorter; longer after change of perch and return (see Goddard 1938). One bird regularly gave 10–11 bouts in first hour after sunset (Christian 1949). Bill appears closed but careful observation reveals very rapid vibrating movement of mandibles (mainly lower against upper) (Heinroth 1909; Schlegel 1969; see also photograph in Ferguson-Lees 1961). Claim that change to brief, lower-pitched, and fast tremolo perhaps due to head movement (Witherby et al. 1938) considered unlikely (P A D Hollom); possibly due to breathing pattern with inhalation at faster section (Wadewitz 1956; Hunter 1980; J Hall-Craggs). According to Stülcken (1962), full Advertising-call given also by ♀; other reports suggest ♀ more likely to give brief (up to c. 20 s) bubbling trill in various contexts (e.g. Owen 1949, Schlegel 1969). Advertising-call normally audible at c. 200 m, in favourable conditions up to c. 600 m (Brünner 1978). (2) Quaw-eek call. A 'quaw-eek', 'kooik', or 'quoik', at times more like 'keweep' or virtually monosyllabic 'kweep'; with nasal or twangy timbre, sometimes frog-like (Selous 1899; P A D Hollom, M G Wilson: Fig V). Given mainly in flight but also just before take-off (Lack 1932), primarily by ♂ but probably more frequently by ♀ than sometimes suggested (in ♂–♀ interactions, including at nest) (P A D Hollom; see also Schlegel 1969). ♀'s call a more muffled, hissing 'schräit', compared with 'rhuit' of ♂ (Leibig 1972). Contact function apparently confirmed in captive birds (Heinroth 1909); alarm element also present (Lack 1932). (3) Alarm- and warning-calls. (a) Recording (Fig VI) suggests for ♀ irregular (in speed, pitch, and volume) series of 'chuk' or 'chek' sounds, with some units doubled (i.e. 'chek-ek', etc.). Not unlike 'chink' or hard (pre-roost) 'chik' calls of Blackbird Turdus merula but sound energy concentrated in lower part of frequency spectrum, end transient more abrupt, and resulting sound harder (J Hall-Craggs, M G Wilson); 'chuck' sounds louder and sharper when flying close to intruder than in distraction-lure display (Lack 1932). A usually intermittent and quiet rasping 'chack', dying away almost to inaudibility without bird moving (P A D Hollom). Blackbird-like 'dack' sounds said to be given by both sexes (e.g. Wadewitz 1956, Stülcken 1962, Schlegel 1969). (b) ♂ gives higher-pitched sounds (Fig VII) though with similarly irregular temporal pattern: 'chink', closer than ♀'s call to T. merula with rapid ascent and descent of pitch in each unit but more mellow owing to relatively low energy at high frequency (J Hall-Craggs). A 'krittkritt' (Wadewitz 1956) or 'gritt-gritt-gritt' (Schlegel 1969). A quiet clucking in flight or on perch (P A D Hollom). (c) A muffled 'oak-oak' or 'uok-uok' given by both sexes when disturbed at nest and apparently expressing higher-intensity alarm (Stülcken 1962). A deep, guttural sound given by ♀ in distraction-lure display on branch (N Cleere)

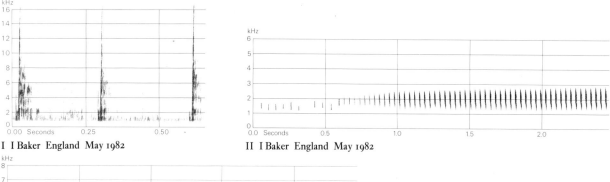

I I Baker England May 1982

II I Baker England May 1982

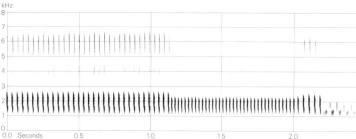

III I Baker England May 1982

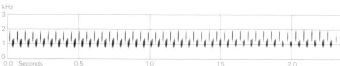

IV I Baker England May 1982

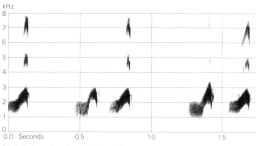

V I Baker England May 1982

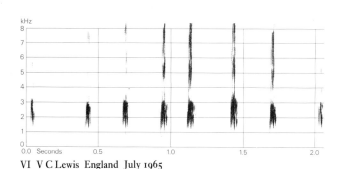

VI V C Lewis England July 1965

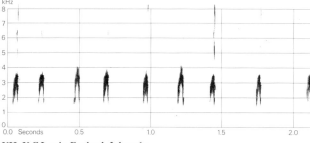

VII V C Lewis England July 1965

presumably the same. See also Dcrim (1962). (4) In various forms of distraction-display, a hissing, spitting, or feline snarling sound given with bill wide open. Sometimes accompanied by slapping sound possibly produced by wings being beaten against ground (J Hall-Craggs, M G Wilson). A 'hch' and 'schd' given by both sexes when disturbed at nest and threatening intruder, in flight and on ground (Stülcken 1962); like domestic goose (Schlegel 1969). (5) Distress-call. A long-drawn, plaintive, not loud 'ko-iiiek' given by ♂ (probably also ♀) seeking lost young (Stülcken 1962). Also rendered 'schuiet'; given as higher-intensity brood-summons when young trapped and unable to respond normally (Steinke 1981). (6) Copulation-calls. In recording by R Savage, ♀ gives a 'kwik wik wik wik', each unit with a sharp end transient. Rather like a soft, liquid variant of call 2 and some units have same frequency range and structure as 2nd sub-unit of that vocalization (J Hall-Craggs). Captive ♂ gave peculiar murmuring sound during copulation (Heinroth 1909); apparently distinct from bubbling trill phase of call 1 which was once given by wild ♂ prior to copulation (R Berry). (7) Grunting-call. A low-pitched, grunting, or puffing 'wuff-(w)uff-(w)uff' (Stülcken 1962; Schlegel 1969) or 'schut schut schut' (Steinke 1981) given by both sexes on hatching eggs and later when brooding, tending, feeding, or summoning young. Sometimes has alerting function (Stülcken 1962). A gruff 'wuk wuk wuk', at times louder and more quacking by ♀; ♂ gives lower-pitched 'uok uok' sounds, perhaps different from call 3c, as brood-summons and feeding call. ♂'s call a quiet 'puik', resembling 2nd syllable of call 2 but much lower pitched and more gentle. May be speeded up when brooding, developing remarkable 'chuffer-train' rhythm with softer quality (J Hall-Craggs, P A D Hollom, P J Sellar). A low croodling sound (Selous 1899); see also Tutt (1955). (8) Other calls. (a) A rather soft, grumbling 'rrrrrrrrrr...' given by captive ♂ while scraping, apparently as nest-lure call; at other times possibly with aggressive character (Heinroth 1909).

CALLS OF YOUNG. A high-pitched piping or cheeping sound given from egg (Heinroth 1909); a soft, whispered 'brüh-brüh' given as food-call, apparently unchanged, up to fledging (Stülcken 1962). According to Schlegel (1969), older young give a 'brrrüh brrrüh' in various parent-young interactions; probably also with locatory function. A harsh 'see-arr' at c. 20–21 days (Lack 1957). Distinction perhaps to be drawn between food-call—wheezy, squeaky or sucking sounds (P A D Hollom), rasping squeal (J Hall-Craggs)—and call given when unattended or separated from sibling: a 'treub' or 'treep' becoming lower pitched with age; fledged young (c. 30 days old) alternated between the two (P A D Hollom). Murmuring sounds given in greeting ceremony by captive birds (Heinroth 1909). A hissing, wheezing or gasping sound in Gaping-threat display (Krambrich 1954; Wadewitz 1956; Schlegel 1969); may also give 'cluck' sound (Lack 1930b). A dry bubbling sound given by fledged young on return to sibling at nest (P A D Hollom). Captive ♂ first gave bubbling trill variant of adult call 1, also adult call 2, at c. 35 days (Heinroth 1909). Adult alarm-call (see call 3) given at 17 days (Wadewitz 1956).   MGW

Breeding. SEASON. Britain, Ireland, and western Europe: see diagram; little variation across range, with similar dates in central Europe, western USSR, and southern Scandinavia (Dementiev and Gladkov 1951a; Makatsch 1976). North Africa: laying begins second half of June (Heim de Balsac and Mayaud 1962). SITE. On ground, in the open or in woodland clearings, or among low scrub or tall vegetation. Nest: shallow scrape; unlined. Building: none. EGGS. See Plate 94. Long elliptical, smooth and

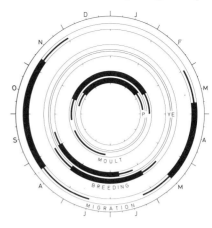

fairly glossy; grey-white to cream, irregularly marked with spots and blotches of yellow-brown, dark brown, and grey; rarely unmarked. 32 × 22 mm (29–37 × 20–25), n = 100 (Witherby et al. 1938). Weight of fresh eggs: 8·2 g (7·5–9·0), n = 8 (Lack 1957); 7·9 g (7·0–9·9), n = 6 (K Hudec). Clutch: 2, exceptionally 3, though as with rare clutches of 4, 2nd ♀ probably involved. 1–2 broods, 2nd usually laid when young of 1st brood c. 2 weeks old (Lack 1957). Laying interval 36–48 hrs (Witherby et al. 1938; Lack 1957). INCUBATION. 17–18 days; up to 21 days if nest disturbed (Lack 1932). Mainly by ♀, relieved by ♂ at dawn and dusk. Begins with 1st egg; hatching asynchronous (Lack 1932). YOUNG. Semi-altricial and nidicolous. Cared for and fed by both parents. Brooded by ♀ for first 13 days, then by ♂ if ♀ begins 2nd brood, otherwise ♀ continues (Lack 1932). Fed bill-to-bill. FLEDGING TO MATURITY. Fledging period 16–17 days. Become independent c. 16 days after fledging (Lack 1932). Age of first breeding 1 year. BREEDING SUCCESS. Of 28 broods, England, 16 fledged at least 1 young; average brood size of successful nests 1·69, and of all nests 0·96, with average 1·35 young reared per pair per season; of 56 eggs laid, 66% hatched and 48% fledged with losses possibly due to adder *Vipera berus* (Berry and Bibby 1981).

Plumages (nominate *europaeus*). ADULT MALE. Forehead, crown, and nape finely vermiculated pale cream or greyish-white and grey, feathers on central forehead, crown, and nape with black pointed shaft-streaks up to 3–4 mm wide, shaft-streaks narrower towards side of crown and faint or absent on lores and above eye; shaft-streaks on central crown bordered by cinnamon-buff and grey vermiculation; top of head appears buff-brown with bold black streaks on centre, almost uniform pale grey-buff or pale grey on line from lores over eye. Patch in front of eye, sides of head, lower cheeks, and chin narrowly barred or speckled cinnamon-buff and greyish-black; a narrow streak below gape to below eye uniform cream-yellow or white; black bristles bordering bill; indistinct half-collar from below and behind ear-coverts to sides of hindneck cream-buff or cream-yellow, boldly spotted black, sometimes extending across hindneck as an indistinct black-and-cinnamon-barred band. Mantle, back, rump, and upper tail-coverts more coarsely vermiculated and barred greyish-white or grey-buff and blackish-grey than crown, appearing slightly darker; greyish-black or black shaft-streaks less sharply defined and usually narrower than on crown; back often partly barred cinnamon and dull black. Scapulars with bold and pointed black shaft-streaks, sides vermiculated cinnamon-buff or cream-white and dull black; outer webs of inner scapulars with broad uniform cream-buff lateral streak (forming pale streak over scapulars); outer webs of outer scapulars with coarse oblique cinnamon and black bars and broad black border along tips. Throat with large white lateral patches, usually divided by grey-and-cinnamon barred feathers in between; patches often partly tinged buff or cream or partly spotted black; occasionally, almost absent. Upper chest closely barred and vermiculated cinnamon-buff or greyish-white and dark grey, appearing dusky grey; lower chest similar, but feather-tips on centre with large rounded-triangular uniform cream-buff to off-white spot. Rest of underparts closely barred cinnamon-buff and dull black; black bars furthest apart and cinnamon brightest on under tail-coverts; tips of longest coverts sometimes washed white. Central pair of tail-feathers with 6–8 narrow and partly broken black bars (shaped like shallow V), coarsely vermiculated pale buff-grey or cream-white and black in between; other tail-feathers with broad and rather irregular black bands, coarsely speckled or vermiculated cinnamon-buff or grey-buff and black in between (least so towards inner margins); tip of t4–t5 uniform white for 23–36 mm (measured along shaft), often with buff tinge or some faint grey specks along outer edges or tips. Primaries black or brownish-black, 1–2 cm of tips tinged grey and with coarse black vermiculation and traces of dark bars, p1–p7 and bases of p8–p10 with cinnamon broken bars (broadly interrupted by black along shaft, cinnamon usually partly speckled dusky); large rounded-triangular patch of white on inner web of p8–p10 (c. 45–55 mm from tip), sometimes partly suffused cinnamon at borders. Shape and extent of white patches on p8–p10 rather variable: on p10 often rather rounded and not or just reaching shaft, maximum length (measured parallel to shaft) 14–20 mm, but occasionally (in 7 of 35 birds examined) reached shaft over c. 10 mm and patch then more squarish; on p8–p9, patch always touches shaft over fair length and patch rather square, occasionally continued into a cinnamon band on outer web; rarely (1 of 35 west European birds examined), white of p8–p9 forms broad band across both webs (this more common in eastern races). Secondaries brownish-black with broken and dusky-freckled cinnamon bars. Upper primary coverts and bastard wing brownish-black with narrow and broken cinnamon bars. Lesser upper wing-coverts black, tips freckled cinnamon; tips of longest with large and distinct cream-buff to off-white patch. Median and greater upper wing-coverts and tertials closely vermiculated (as in mantle and rump), appearing distinctly paler and greyer than neighbouring blackish lesser coverts, primary coverts, and secondaries; outer webs each with small cinnamon-buff or cream-buff spot on tip, subterminally bordered black. Under wing-coverts, axillaries, and small coverts along leading edge of wing closely barred black and cinnamon-buff. In worn plumage, upperparts duller grey and contrasts less marked; cinnamon-buff ground-colour fades to off-white, especially below; cream-buff spots on scapulars, upper wing-coverts, and chest bleach to off-white, often partly abraded. ADULT FEMALE. Similar to adult ♂, but throat-patches often smaller and usually more extensively tinged buff. No white patches on p8–p10; usually a rather broad and partly rounded cinnamon bar or spot with some dusky freckling on place where ♂ shows white, well-separated from more regular and narrower cinnamon bars on bases of inner webs; frequently, all bases and middle portions of inner webs of p8–p10 barred cinnamon, most distal spot in regular row only slightly enlarged. T4–t5 without white; tips usually without bold dark bars, terminal 10–22 mm appearing uniform (though actually cinnamon-buff with extensive grey freckling, only border often uniform cream; in ♂, 23–36 mm of tip uniform); occasionally (3 of 15 examined), dark bars extend almost to tip, though still some contrast in density of vermiculation between distal 20 mm and remainder of feather. DOWNY YOUNG. Sparsely covered with dense patches of long cream-buff down; longest tufts on upperparts and thighs red-brown. JUVENILE. Like adult, but generally paler and no white or dusky-freckled cinnamon-buff patch on outer primaries and outer tail-feathers. Ground-colour of crown, bases of outer scapulars, all longer scapulars, chest, tertials, and median and greater upper wing-coverts grey-white or buff-white (paler than in adult) with finer and less extensive vermiculation; appears paler grey than adult. Black marks on scapulars less extensive; cream-buff streaks on outer webs of inner scapulars virtually absent. Upper tail-coverts tipped cream with subterminal black border, unlike adult. Pale throat-patches small; chest narrowly barred buff and dull black (in adult, closely vermiculated grey and contrasting with remainder of underparts), feather-tips on lower chest narrowly off-white (in adult, boldly spotted). Dark bars on flanks and from breast to under tail-coverts poorly defined, brown (in adult, black and sharply contrasting), virtually absent on vent and shorter under tail-coverts; feathers of vent and under tail-coverts softer and looser than in adult. Tail as adult, but ground-colour of t1 usually paler and bars on other feathers often narrower and further apart; tips of feathers rather narrowly fringed cream-buff, widest (3–8 mm) on t5, no almost uniform white or buff patches to tip of t4–t5 as in adult. Flight-feathers as adult, but tips narrowly fringed cream-buff or white (soon abraded), tips of primaries slightly paler grey; usually no larger white or cinnamon patches on p8–p10, though c. 30% of ♂♂ show small white spot on outer primaries (up to 15 mm long, often partly suffused cinnamon), unlike any ♀ (Piechocki 1966; RMNH, ZMA); basal and middle portions of inner webs serrated with regular cinnamon bars, only most distal bar on p9–p10 sometimes slightly enlarged (similar in this respect to some adult ♀♀). Median and greater upper wing-coverts with poorly defined cream tips (not restricted to outer web and boldly bordered black subterminally as in adult). FIRST ADULT. Like adult, but some probably retain a few juvenile flight-feathers or wing-coverts when 1 year old (see Moults). By autumn, some juveniles largely in 1st adult already, retaining juvenile tail, flight-feathers, greater wing-coverts, and scattered other wing-coverts and upper tail-coverts only. Birds on breeding grounds with plumage new late June and July are perhaps newly-arrived 1-year-olds. Amount of white on t5 in ♂ probably dependent on age: less than 27 mm in 1-year-olds, 28–35 mm in 2- and 3-year-olds, over 37 when older (Glutz and Bauer 1980).

Bare parts. ADULT AND JUVENILE. Iris deep brown. Bill black. Mouth deep red; greyish-white when bird in torpid state. Leg and foot brown or flesh-brown, soles and edges of scutes whitish-horn. DOWNY YOUNG. Iris deep brown. Eyelids grey. Bill pinkish-horn or flesh-brown. Leg and foot greyish-pink or flesh-grey. (Hartert 1912–21; Heinroth and Heinroth 1924–6; Witherby et al. 1938; Schlegel 1969; BMNH, RMNH, ZMA.)

Moults. ADULT POST-BREEDING. Complete; primaries descendant. Starts on breeding grounds July–September, but extent rather variable; some birds replace all head, body, and tertials, others parts only, and some depart with all plumage unmoulted and worn. Exceptionally, p1 replaced on breeding grounds in August (Stein 1982). In Kareliya (USSR), starts 9–18 July with feathers round eye, neck, and breast; in full moult of head and body 18–21 July, and moult largely completed with all head, body, and rictal bristles new 21 July–12 August (Neufeldt 1958a). Moult arrested during migration, resumed in winter quarters late September to October(–November) with remainder of old body feathers and p1. Apparently, a partial pre-breeding moult follows in winter quarters, as birds with primary moult score c. 30 (late October to late December) showed heavy moult of underparts and median upper wing-coverts, and head and body new at score 45–50. Tail replaced between primary moult scores 30 and 45–50, sequence 1–2–3–5–4. Secondaries start at score c. 35, moulting ascendantly from s1 and descendantly from s12–s13 (innermost); s7–s8 last, completed with p10 at score 50 (Stresemann and Stresemann 1966). All moult completed early January to late March. POST-JUVENILE. Complete. Starts from late July or August in early-hatched birds, part of crown, outer scapulars, and part of chest and flanks first, some showing much of head, neck, mantle, scapulars, chin to breast, and lesser and median upper wing-coverts new on departure for winter quarters in September–October; late-hatched birds still fully juvenile as late as October–November. Moult continued or started after arrival in winter quarters; primaries start with p1 from November–December, later than in adult post-breeding. Moult completed March–April or perhaps occasionally later; one bird examined showed arrested primary moult, retaining 2 old juvenile primaries in summer of 2nd calendar year; some birds which have retained a few old secondaries or outer upper wing-coverts perhaps also 1 year old.

Measurements. Nominate *europaeus*. Netherlands and West Germany, April–October; skins (RMNH, ZMA). Bill is exposed culmen; to forehead, 12·3 (16) 11–14 more.

WING AD	♂ 192	(5·50; 33)	184–201	♀ 195	(6·38; 19)	184–202
JUV	190	(2·90; 8)	185–194	190	(4·90; 9)	183–197
TAIL AD	137	(5·03; 34)	129–146	136	(4·63; 23)	129–144
JUV	131	(3·12; 7)	127–135	129	(6·34; 8)	123–136
BILL	8·8	(0·56; 12)	8·0–9·5	8·9	(0·69; 16)	7·5–9·7
TARSUS	16·8	(0·67; 10)	16·1–17·8	17·2	(0·72; 12)	16·3–18·2
TOE	20·5	(0·92; 13)	19·3–21·7	21·0	(0·80; 14)	19·7–22·1

Sex differences not significant. Wings of 208 (♂) and 211 (♀) excluded from ranges. Juvenile tail significantly shorter than adult, other differences not significant. In sample from central and eastern Europe, juvenile wing averaged c. 6 shorter than adult, tail c. 11 (Piechocki 1966).

Adult wing of nominate *europaeus* from (1) Britain, (2) north-west European USSR, Scandinavia, and Germany, (3) eastern Romania and southern European USSR; of *meridionalis* from (4) Iberia and Balearic Islands, (5) Morocco and Algeria, (6) Greece (Vaurie 1960a; BMNH, RMNH).

(1)	♂ 191	(4·03; 10)	185–195	♀ 189	(3·34; 9)	184–194
(2)	196	(3·70; 12)	190–200	195	(4·15; 11)	187–201
(3)	201	(3·99; 5)	198–208	194	(6·47; 8)	185–202
(4)	186	(3·24; 7)	183–192	187	(1·71; 4)	185–189
(5)	181	(3·91; 12)	175–186	183	(4·16; 5)	176–186
(6)	180	(4·40; 7)	175–186	180	(— ; 3)	179–181

Wing of *meridionalis* from USSR c. 181 (168–190), of *sarudnyi* c. 185 (171–196) (Dementiev and Gladkov 1951a; Stepanyan 1975). Wing of adult ♂ of *unwini* from Iran, Afghanistan, and Pakistan 184 (4·90; 10) 178–192 (Vaurie 1960a).

Weights. Nominate *europaeus*. Central Europe: adult, ♂ 87 (4) 72–101, ♀ 83 (3) 81–87; juvenile ♂ 70 (6) 61–86, ♀ 67 (5) 61–86; exhausted and lean birds 35–50 (Piechocki 1966). Kareliya (USSR), June–August: ♂ 73·3 (5·67; 23) 62–86, ♀ 83·2 (2) 78–88 (Neufeldt 1958a). Belorussiya (USSR), summer: ♂ 71 (10) 64–78, ♀ 74 (6) 66–79 (Fedyushin and Dolbik 1967). Generally, below 80 during breeding season, but often over 100 just prior to autumn migration; July, adult ♀♀, 66, 75, 78 (Schlegel 1969). Netherlands, adults May–July: ♂♂ 62, 64, 66, 71; ♀ 76 (RMNH, ZMA). Migrants, Camargue (France): April–May, ♂ 65 (9·2; 25) 41–84, ♀ 65 (9·0; 8); September 67 (8·1; 36) 56–85 (Glutz and Bauer 1980). Nestlings at hatching 6·0 (8) 5·1–7·1 (Lack 1957; Schlegel 1969). Averages of 7–10 birds on 5th day 21·2, on 11th 45·4 (38·9–50·3), on 15th 54·3; 3 on 18th day 57·9 (Lack 1957). Afrotropics, October–February: ♀♀ 54, 66, 70, 75 (Moreau 1944; Verheyen 1953; Jackson 1969).

C. e. meridionalis. Greece, May: ♂ 71, ♀ 55 (Makatsch 1950). Turkey, late July: ♂ 52 (Kumerloeve 1961).

C. e. unwini. Iran and Afghanistan, May–June: ♂♂ 51, 54, 57, 61; ♀ 74 (Paludan 1938, 1940, 1959; Desfayes and Praz 1978). Turkey, October: ♀ 70 (Kumerloeve 1967). Tanzania, March: 54 (Moreau 1944).

C. e. plumipes. Mongolia, late June to mid-August: ♂♂ 54, 64, 66; ♀♀ 58, 70, 72 (Piechocki 1968c).

Race unknown. Afrotropics, April: ♀ 60; sex unknown, 98 (Fry 1970; Colston 1971). Kazakhstan, probably mainly *sarudnyi*: May, ♂ 49–84 (7), ♀ 62–82 (5); June–July, ♂ 55–74 (9), ♀♀ 64, 65; August–September: ♂♂ 75, 83; ♀♀ 66, 68, 75 (Dolgushin *et al.* 1970).

Structure. Wing long, rather narrow, tip pointed. 10 primaries: p9 longest, p10 4–10 shorter, p8 3–8, p7 22–38, p6 42–54, p5 52–72, p1 84–106; outer web of p8–p9 and inner web of p9–p10 slightly emarginated. Edges of inner webs of primaries rather loose, slightly serrated. Tail rather long, tip slightly rounded; 10 feathers, t5 3–15 shorter than t1. Bill weak, flattened; mouth very wide, with gape almost reaching to below hind corner of eye, only small but strong and curved nail and cutting edges unfeathered; long and stiff bristles along gape of upper mandible, directed forwards, smaller ones round nostril, small tuft of hair-like feathers on chin. Nostrils rather small, rounded, surrounded by raised operculum. Eyes very large; ear opening large. Skin of body thin, plumage loose. Tarsus and toes short and slender; upper and middle front of tarsus feathered; base of front toes connected by small web; inner edge of middle claw serrated (scarcely in juvenile). Outer toe c. 59% of middle, inner c. 56%, hind c. 35%.

Geographical variation. Marked, mainly clinal; involves size, general colour, and extent of white on outer primaries in adult ♂. Size greatest in northern part of range of nominate *europaeus*, from Scandinavia through Siberian taiga to north-west of Lake Baykal, gradually decreasing towards south; smallest birds are *meridionalis* from North Africa, Greece, and Turkey, and *unwini* from eastern Iraq and Iran to Pakistan (see Measurements). Colour darkest in birds from humid areas, palest in deserts. Wide intergradation zones between various races, and boundaries as defined in species heading mainly arbitrary. In Europe, variation in size and colour clinal, but not quite parallel; *meridionalis* from southern Europe and North Africa in general smaller and paler than nominate *europaeus*, ground-colour of sides of crown, mantle, scapulars, tertials, median upper wing-coverts, and chest paler and grey vermiculation narrower and sharper, these parts giving more silvery-grey appearance than dull pale brownish-grey of nominate *europaeus*; however, in both *meridionalis* and nominate *europaeus*, western birds duller than eastern birds, eastern nominate *europaeus* (e.g. from south European USSR and eastern Rumania) about as pale as Iberian *meridionalis* (but much larger), and nominate *europaeus* from Britain, Netherlands, and probably north-west France markedly dull brownish above (these also smaller than typical Scandinavian nominate *europaeus*). Birds from southern France (birds from Garonne and Rhône valleys examined) dark as nominate *europaeus*, but nearly as small as Iberian *meridionalis*, Hungarian birds pale but rather large. Note that juveniles distinctly paler and smaller than adults, and for instance juvenile nominate *europaeus* from north-west Europe rather similar to adult *meridionalis* from Tunisia, southern Italy, or Greece. Situation in Asiatic part of distribution rather complex. Following account based mainly on Hartert (1912–21), Dementiev and Gladkov (1951a), Johansen (1955), Vaurie (1960, 1965), Stepanyan (1975), Clancey and Mendelsohn (1979), and skins in ZMM. All adult ♂♂ of eastern races (except most birds of nominate *europaeus* from Siberian taiga) show larger white primary-patches than European birds, white on inner webs of p8–p9 extending broadly across whole outer web. Siberian nominate *europaeus* on average slightly paler grey on upperparts and chest than those of western Europe, similar in this respect to birds from eastern Rumania and south European USSR. *C. e. sarudnyi* from forested steppes of Kazakhstan rather variable in colour and size, more or less intermediate between nominate *europaeus* and *unwini*; primary-patches large as in *unwini*, upperparts more buff-brown than in nominate *europaeus*. *C. e. unwini* from eastern Iraq east to western Tien Shan and Pakistan distinctly paler and more greyish than previous races, less heavily streaked black, less extensively vermiculated grey; general colour rather sandy-grey, pale patches on throat whiter and larger, under tail-coverts pale cream and virtually unbarred. *C. e. plumipes* from Sinkiang (China) and parts of Mongolia pale as *unwini*, but ground-colour cinnamon-buff rather than grey; uniform spots on tips of scapulars, tertials, median upper wing-coverts, and chest larger than in other races, pale pink-buff or cream; light bars on inner webs of primaries and secondaries broader; dark bars on tail narrower, virtually absent on t1. *C. e. dementievi* of north-east Mongolia and southern Transbaykalia rather pale, but more greyish and more heavily vermiculated than *plumipes* and *unwini*; ground-colour of underparts more yellowish-loam than in *unwini*. CSR

Caprimulgus ruficollis Red-necked Nightjar

PLATES 56 and 58
[between pages 662 and 663]

Du. Moorse Nachtzwaluw Fr. Engoulevent à collier roux Ge. Rothalsziegenmelker
Ru. Красношейный козодой Sp. Chotacabras pardo Sw. Rödhalsad nattskärra

Caprimulgus ruficollis Temminck, 1820

Polytypic. Nominate *ruficollis* Temminck, 1820, Portugal, Spain, and Morocco (except north-east); *desertorum* Erlanger, 1899, north-east Morocco, Algeria, and Tunisia.

Field characters. 30–32 cm, of which tail 11–13 cm; wing-span 65–68 cm. 15% larger than Nightjar *C. europaeus*, averaging slightly longer winged and 15% longer tailed, with noticeably larger head. Largest nightjar in west Palearctic with more robust structure than *C. europaeus* and browner, more variegated plumage. In close view, large white throat, rufous half-collar on nape, bold white spots on outer tail-feathers, and smaller white spots on outer primaries may all show. Flight more powerful than in *C. europaeus*. Song distinctive. Sexes almost similar; no seasonal variation. Juvenile separable at close range. 2 races in west Palearctic, separable in the field.

(1) Iberian and Moroccan (except north-east) race, nominate *ruficollis*. ADULT MALE. Plumage colours and pattern close to those of *C. europaeus*, but rufous ground-colour more obvious especially on sides of head, hindneck (with broad, dark-lined, rufous-buff, pale-looking half-collar), and along scapulars. Tends to look paler above than *C. europaeus*. Markings extremely complex, with narrow white moustache, large white throat, rufous-buff chest, and broader barring across wing-coverts all more obvious than on *C. europaeus*. In flight, white spots on 3 outer primaries resemble those of *C. europaeus* but longer white tips to 2 outermost pairs of tail-feathers are more evident. ADULT FEMALE. As ♂, but primary-spots and tail-tips only slightly duller (and thus more prominent than on ♀ *C. europaeus*). JUVENILE. Resembles ♀ but duller, with pale buff, not rufous, chest and half-collar on hindneck, and distinct primary- and tail-spots. (2) Algerian, Tunisian, and north-east Moroccan race, *desertorum*. Distinctly paler, greyer, and sandier above, and less clearly barred below. Black markings on head, back, and scapulars more obvious than on nominate *ruficollis*.

In fortunate circumstances, separable from *C. europaeus* on larger size, longer tail, paler and more rufous plumage, and large white throat. Flight like *C. europaeus*, but all actions more powerful and swoops apparently more pronounced; larger head and longer tail often noticeable to experienced observer. In strong light, rufous hindneck may show. Behaviour and gait similar to *C. europaeus*.

Voice differs distinctly from *C. europaeus*. Song 'cut-ock cut-ock . . .'; loud and echoing, delivered more slowly than *C. europaeus* so that each disyllable distinct, reminiscent of sound of wood-chopping.

Habitat. Breeds in lower middle latitudes, in dry warm Mediterranean and oceanic climates, mainly in lowlands below 800–1000 m. Basic requirements include patches of bare or sandy soil, scattered ground cover, and some bushes or trees for song-posts. In parts of range, resorts to closed stands of trees, including woods of stone pines *Pinus pinea* on moist ground (Coto Doñana, Spain), *Eucalyptus* groves, olive gardens, cork oak *Quercus suber* scrub, and dense thicket of *Halimium* and gorse *Ulex* or of bramble *Rubus* and tree heath *Erica arborea* up to 3 m high. Also locally favours more open areas of low scrub or prickly pear *Opuntia* with scattered trees, and even semi-desert or arid hillsides with dwarf vegetation, but not

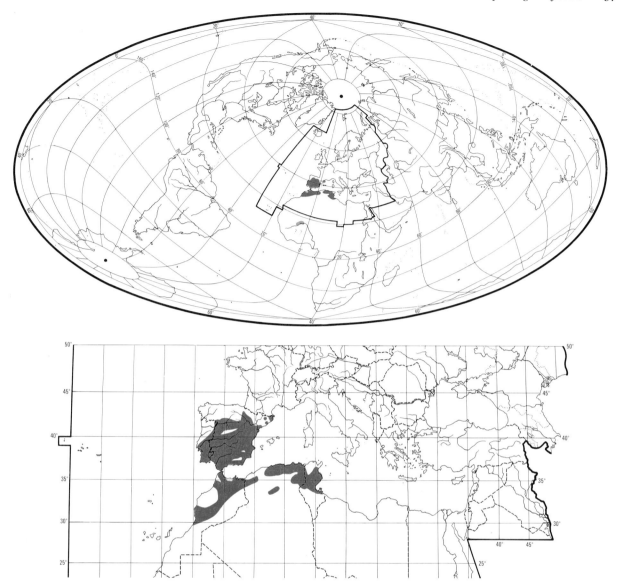

treeless dunes or steppe country. Habitat separation from or overlap with Nightjar *C. europaeus* not clear in view of degree of opportunism in adapting to superficially distinct types of terrain which in different ways fulfil basic requirements. On migration occurs in atypical situations as with other migrants. (Valverde 1958; Beven 1973.) Winter habitats almost unknown.

Distribution. Little information on current distribution in Algeria and Tunisia (EDHJ, MS).

Accidental. Britain, France, Italy, Libya, Malta, Madeira.

Population. SPAIN. Rather numerous (AN).

Movements. Migratory. Winter quarters lie in West Africa, though not yet delineated, and few passage observations south of breeding range.

Movements evidently broad-front. Reported mid-March to early May in Western Sahara, across Morocco (sometimes common around Tangier), and in western Algeria, arriving Tunisia and Spain in late April (e.g. Heim de Balsac and Mayaud 1962, Smith 1968, Beven 1973). Most have left Iberia by late October, though stragglers occur in November; regularly on passage on Gibraltar (Bernis 1970; Lathbury 1970). In Morocco, Smith (1965) saw 12 obvious migrants on coast between Temara and Cap Rhir, 18 October–13 November; rare records from Mauritania also coastal (Banc d'Arguin, Nouakchott), in October, November (long dead), and early May (Knight 1975; Browne 1981). Both races collected Sénégal in November, but not thought to winter there

(Morel and Roux 1966). Scarce visitor to Niger inundation zone (Mali), with various records spanning October–March; 2 skins (October) of nominate *ruficollis* (western race) and suspected that *desertorum* occurs also (Malzy 1962; Lamarche 1980). No records Chad or Nigeria, but singles obtained northern Ghana in March (Bannerman 1953) and northern Ivory Coast in January (Traylor and Parelius 1967). Hence winter range probably extends further south than that of Egyptian Nightjar *C. aegyptius* (Moreau 1972).

Food. Insects. Prey taken mainly in silent floating and wheeling flight with at times bouts of fluttering, also sudden erratic twists and turns (Beven 1973). Such foraging activity predominantly crepuscular and nocturnal (Witherby *et al.* 1938; Cowles 1967). Prey apparently stored in extended pharyngeal cavity during foraging bout, and digested when bird at rest (Madon 1934). Less commonly forages on ground, taking (e.g.) young locusts (Orthoptera) (Witherby *et al.* 1938), and beetles (Coleoptera) by animal droppings (Koenig 1895); larvae of Lepidoptera also recorded (Etchécopar and Hüe 1967). Said to show preference for larger insects (Ceballos and Purroy 1977).

Apart from prominent rictal bristles (e.g. Schuhmacher 1971), bird endowed with highly vascular and membranous palate lining; extremely sensitive and thus easily stimulated by contact with flying insect (see Cowles 1967). For further details of morphology, see Bühler (1970).

Prey recorded includes moths (Lepidoptera), beetles (Coleoptera), locusts and grasshoppers (Orthoptera), and flies (Diptera) (Koenig 1895; Witherby *et al.* 1938; Blanchet 1957; Etchécopar and Hüe 1967; Ceballos and Purroy 1977).

In Spain, takes mainly moths, including many species harmful to man. August stomachs contained processionary moth *Thaumetopoea pityocampa* and other Noctuidae (Ceballos and Purroy 1977). Stomach of adult ♂ from Tunisia, May, contained 1 large orthopteran, small beetles, and earth; ♀ obtained at same time contained beetles, grasshoppers, and earth. Of 2 ♂♂, June, one contained unidentifiable debris and sand grains, the other contained small beetles, mosquitoes (Culicidae), and sand grains. Stomach of adult ♂ from early October was almost empty apart from unidentifiable debris and *c.* 10 small seeds (presumably taken accidentally). A presumed juvenile of 13 August contained beetles, some quite large (Blanchet 1957).

No information on food of young. MGW

Social pattern and behaviour. No major studies; for summary see Beven (1973).
1. Probably mainly solitary, though some evidence of birds assembling to feed (Erlanger 1899), when may also associate with Nightjar *C. europaeus* (Koenig 1895). BONDS. No detailed information on mating system or pair-bond. Young cared for by both parents, then by ♂ alone if ♀ begins 2nd clutch (Ferguson-Lees 1969b). BREEDING DISPERSION. In Tunisia, *desertorum* may form small neighbourhood groups of 4–12 pairs; nests *c.* 2–15 m apart in (e.g.) cactus plantations. In southern Spain, nominate *ruficollis* apparently always solitary (Marinkelle 1959). Beven (1973) recommended caution in any attempt to deduce density from number of ♂♂ heard simultaneously as sound far-carrying (see also 1 in Voice). ROOSTING. Diurnal. Lengthwise on branch; also on ground in woods or scrub. In Portugal, bird roosted regularly on stone in scrubby cover. Also rests on roads at night (Beven 1973). All vocal activity (and most other activity) occurs from dusk to dawn (Ferguson-Lees 1969b). Tends to give Advertising-call (see 1 in Voice) and to feed later in evening than *C. europaeus*. In southern Spain, on clear evenings in May, call began *c.* 30 min after sunset or slightly earlier (Beven 1973).

2. In Tunisia, foraging birds showed no fear of man (Erlanger 1899). An injured bird assumed an aggressive threat posture (see *C. europaeus*): bill opened wide to display bright red mouth, and loud hissing sound given (Cowles 1967; see also Koenig 1895). FLOCK BEHAVIOUR. No information. ANTAGONISTIC BEHAVIOUR. Wing-clapping given in flight, sometimes apparently as threat towards conspecific birds (Bergmann and Helb 1982; see also Heterosexual Behaviour and *C. europaeus*). Advertising-call—as in *C. europaeus* given mainly at dusk and dawn but also at night (Chappuis 1979)—probably also has territorial function and may be combined with Wing-clapping. In response to playback of Advertising-call, one ♂ flew towards source calling at about twice normal speed and also performed *c.* 8 Wing-claps as it passed near (Beven 1973). HETEROSEXUAL BEHAVIOUR. Courtship-chase resembles that of *C. europaeus*, ♂ following ♀ and Wing-clapping (Schuhmacher 1971). Wing-claps tend to be more rapid than usual. Once, ♂ gave 12 very rapid Wing-claps in succession when pursuing ♀ (Mountfort 1958). At one nest in Spain, only ♀ seen to incubate; ♂ often hovered over incubating ♀ in late afternoon though no change-over took place (Beven 1973). RELATIONS WITHIN FAMILY GROUP. See also Bonds. Young brooded initially in nest. At one Spanish site, a recently hatched young bird leapt 'like a clumsy frog' *c.* 1 m from nest into shade. summoned back (call not described) by parent. At *c.* 4 days, adult and young *c.* 3·5 m from nest (Beven 1973). At another site, adult and chick *c.* 6 days old were *c.* 10 m from nest (Hidalgo 1974). See also Schuhmacher (1971) where indicated that birds move well away from nest when young *c.* 5–6 days old. Young make short flights at *c.* 2½ weeks; independent at *c.* 4–5 weeks (Beven 1973). No detailed information on break-up of family but family parties seen in flight at Ledesma (Spain) in July (Bernis Madrazo 1945). ANTI-PREDATOR RESPONSES OF YOUNG. Chick *c.* 3 days old moved towards observers (Schuhmacher 1971). PARENTAL ANTI-PREDATOR STRATEGIES. (1) Passive measures. Birds relatively confiding at nest. At one site, hovered close to humans or hide at night; at another, bird was photographed without difficulty during day (Beven 1973). When incubating or brooding, bird usually remains motionless, with eyes closed to narrow slit, and allows close approach (to *c.* 1 m) (Schuhmacher 1971). Bird flushed from eggs returned after *c.* 20 min (Beven 1973). At another site, bird returning to chick when observer had entered hide, landed close by, apparently Wing-clapping, then ran in over last section. When flushed, flew only short distance and soon returned (Schuhmacher 1971). (2) Active measures. In Tunisia, birds with eggs said always to have attacked human intruders furiously (Marinkelle 1959), but no details. At Ledesma, parent bird accompanying young in flight once dropped to ground and there performed distraction-lure display of disablement type (Bernis Madrazo 1945). MGW

Voice. Wing-clapping probably produced by striking carpal joints together over back (Ferguson-Lees 1969b). Less frequently performed than by Nightjar *C. europaeus* (J-C Roché). For situations in which Wing-clapping performed, see Social Pattern and Behaviour.

CALLS OF ADULTS. (1) Advertising-call (Song) of ♂. Sometimes preceded by brief and muffled chuckling or gurgling sound, but otherwise usually a long series of loud, resonant, and rather low-pitched double knocking (or drumming) sounds: 'cut-ock cut-ock cut-ock...', 'toktok toktok...' or 'kotok kotok...' (Mountfort 1958; Mountfort and Ferguson-Lees 1961; Beven 1973; Bergmann and Helb 1982, which see for additional sonagram). Recording (Fig I) suggests an irregular and hollow wooden-sounding 'chettir', with accent on 1st syllable (P J Sellar) or 'tut-tut' at c. 140 units per min (J Hall-Craggs). Often compared to hard rapping with knuckle on hollow wood; 1st syllable appears to echo into 2nd which then apparently cut short as knuckle pauses briefly. 2 syllables still distinct at rate of c. 100 per min (Ferguson-Lees 1969b), though in recording by P A D Hollom, Spain, May, some units sound monosyllabic; see also (e.g.) Witherby (1928) and Etchécopar and Hüe (1967). Speed of delivery said by Mountfort (1958) to vary considerably, normally c. 90–100 times per min including pauses, sometimes accelerated to c. 200 per min to produce almost continuous sound, though at other times slow and widely spaced 'cut-ock' sound uttered; a faster and higher-pitched variant, also with changing rhythm in middle of bout, normally given at dawn (C Chappuis). Audible up to c. 400 m (Bergmann and Helb 1982). At very close range more like 'kuituk kuituk kuituk' and both loud and penetrating (Beven 1973); unpleasant in such circumstances (Witherby 1928). Other renderings—'whitchoo whitchoo whitchoo' given many times (Tuke 1953); 'tiogo tiogo tiogo' (Bernis Madrazo 1945); a persistent and monotonous 'dugdug

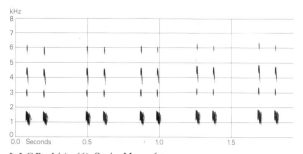

I J-C Roché (1966) Spain May 1965

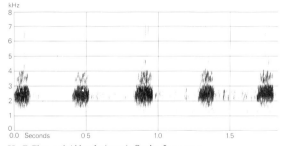

II C Chappuis/Alauda (1979) Spain June 1970

dugdug...' (Schuhmacher 1971). Given from perch or in flight (Bergmann and Helb 1982). According to Chappuis (1979), difference in tonality of rhythm between 2 sequences (one from one ♂, other from 2 ♂♂ in vocal contact) may possibly indicate existence of antagonistic variant. Recording by G Jarry, Mali, January, probably *C. ruficollis*, thus indicating that Advertising-call may be given also in winter quarters: lower pitched and slower than that normally given on breeding grounds but this type of variation noted in *C. europaeus* (Chappuis 1981; see also above). Advertising-call of ♂ differs markedly in pitch and especially in rhythm from *C. europaeus* (Abs 1963, which see for additional sonagram). (2) Main call of ♀. In response to playback of ♂ Advertising-call at dawn, probable ♀ gave a harsh and regular sound (Fig II), analogous in rhythm to call 1, of which it was presumably the equivalent (Chappuis 1979); a regular scraping or sawing sound not unlike the chuffing of a steam-engine (P J Sellar, M G Wilson). Rendered 'tuk-tuk-...'; lower pitched than call 1 and delivered with a metronome-like regularity of rhythm c. 150–400 times per min (Bergmann and Helb 1982, which see for additional sonagram). Whilst incubating, one bird uttered a long series of single knocking sounds; accelerated to rate of c. 400 per min (Mountfort 1958). Probably associated with nest-relief (Beven 1973; Bergmann and Helb 1982). (3) Flight-calls. Mono- or disyllabic sounds, probably similar to *C. europaeus* (Ferguson-Lees 1969b; Bergmann and Helb 1982), though perhaps less freely given (J-C Roché). (4) Alarm-calls. High-pitched, sharp sounds (Beven 1973; Bergmann and Helb 1982). (5) For hissing sound given as threat, see Social Pattern and Behaviour.

CALLS OF YOUNG. No information. MGW

Breeding. SEASON. Spain: breeds early May to July (Witherby *et al.* 1938); eggs recorded late August (Hidalgo 1974). Algeria and Tunisia: eggs found mid-May to mid-June, though 2nd broods may prolong season into August. Central Morocco: eggs found July (Heim de Balsac and Mayaud 1962). SITE. On ground in the open or among scattered low bushes. Nest: shallow scrape. Building: no information. EGGS. See Plate 94. Long elliptical, smooth and fairly glossy; grey-white, marbled and blotched yellow-brown, 32 × 23 mm (29–34 × 21–25), $n = 50$; calculated weight 8 g (Witherby *et al.* 1938; Makatsch 1976). Clutch: 2. Probably 2 broods. No further information.

Plumages (nominate *ruficollis*). ADULT MALE. Superficially rather similar to Nightjar *C. europaeus* (especially Mediterranean race *C. e. meridionalis*), but more boldly marked and with golden-rufous collar round neck. Forehead, crown, and nape paler and more silvery-grey than in *C. europaeus*; black streaks on central forehead and central crown slightly broader, more distinctly bordered pale cinnamon at sides. Lores, sides of head, and chin rufous-cinnamon or cinnamon-buff with narrow black bars or streaks, not as dusky as in *C. europaeus*; pale patches below eye and on sides of throat more distinct, larger, and white, throat-patches more distinctly bordered black below. Obvious golden-rufous collar round hindneck and down sides of neck c. 1 cm wide, narrowly barred black at border with mantle. Remainder of upperparts and all upper wing-coverts as in *C. europaeus*, but pale pink-buff or cream spots on outer scapulars and longer lesser and median upper wing-coverts larger, more boldly bordered black on scapulars; grey-vermiculated parts of mantle, inner scapulars, tertials, and upper wing-coverts slightly paler, though not as silvery as forehead and crown. Upper chest below black borders of white throat-patches cinnamon with narrow black bars or spots instead of vermiculated dull grey and black as in *C. europaeus*, lower chest vermiculated pale grey, usually without buff or cream spots which are often prominent in *C. europaeus*. Remainder of underparts including under tail-coverts warm buff with narrow dull black bars, as in *C. europaeus*. Tail as in *C. europaeus*, but t1 often with narrower and sharper black bars, and intervening parts sometimes slightly paler grey with less black marbling; tip of outer feather (t5) white for 35–50 mm, occasionally with some partial grey freckling; slightly smaller amount of white on tip of t4. Flight-feathers and upper greater coverts as in *C. europaeus*, but rufous barring often deeper, chestnut rather than cinnamon, often slightly wider and less regular; white spots on inner webs of p8–p10 relatively larger, 18–30 mm long, extending as pink-buff spots on outer web of p8–p9, joining shaft for 7–15 mm on p10. Under wing-coverts and axillaries as in *C. europaeus*. ADULT FEMALE. Similar to adult ♂, including pale spots on p8–p10 and t4–t5 (and thus differing from adult ♀ *C. europaeus*). White tips on t4–t5 usually with slight buff tinge and often with partial grey freckling on outer web, unlike ♂; extent of white or pale buff on t5 (14–)17–23(–30) mm, less than in adult ♂. White patches on p8–p10 smaller than in adult ♂, often with partial buff fringe, not extending into pink-buff spots on outer web of p8–p9, usually not reaching shaft on p10; greatest length of white spot on inner web of p10 15–19 mm, rounded-triangular in shape rather than square. DOWNY YOUNG. Down long and sparse, pale yellow-brown (Hartert 1912–21). JUVENILE. Rather like adult, but paler and greyer on upperparts, sometimes resembling adult and juvenile of North African race *desertorum* superficially. Black streaks on crown and scapulars narrower and duller black than in adult; rufous half-collar round neck barred and vermiculated dusky grey, sometimes hard to see (very similar then to juvenile *C. europaeus*, distinguishable only by larger and whiter spots on outer primaries); scapulars and tertials finely vermiculated dull grey and white, without bold pink-buff or cream and black marks of adult; white patches on throat speckled and barred black; chest evenly and narrowly barred rufous and dusky grey; breast to under tail-coverts as in adult, but dark bars narrower, less deep black, less distinctly defined. Longer lesser and median upper wing-coverts tipped white across whole tip (in adult, cinnamon-buff or cream spot on outer web only), basally vermiculated dusky grey and white, without solid black bars of adult. Tail as in adult, but feathers narrower (see Structure), tips rounded rather than almost square; tips of t4–t5 white or with slight buff tinge and partial grey freckling in juvenile ♂ (t5 pale for 14–29 mm), fully buff and grey freckled in juvenile ♀, border to tip of t5 only 5–10 mm wide and usually not uniformly pale. Flight-feathers as in adult, but white spots on p8–p10 smaller. In juvenile ♂, pattern similar to adult ♀ — spot on p10 rounded-triangular, not reaching shaft, greatest length (parallel to shaft) 12–20 mm; juvenile ♀ similar but spots even smaller and white more extensively invaded by buff, greatest length of spot of p10 10–16 mm. Unlike adults, tips of primaries narrowly bordered white. FIRST ADULT. Like adult, but juvenile outer primaries, sometimes a few or all tail-feathers, and occasionally some feathers of body or upper wing-coverts retained (see Moults). Outer primaries with smaller white spots than adult (see juvenile), usually still with traces of white fringes on tip; 3–8 inner primaries contrastingly new. Some probably indistinguishable from adult.

Bare parts. ADULT AND JUVENILE. Iris dark brown. Bill dark horn-brown or black-brown with flesh-coloured base; mouth bright red. Leg and foot grey-brown. DOWNY YOUNG. As *C. europaeus*. (Beven 1966; Cowles 1967; BMNH.)

Moults. ADULT POST-BREEDING. Complete; primaries descendant. Starts with innermost primary late June to late August, followed by head and neck from primary moult score of *c.* 15. At score 20–30, head, neck, chest, and median upper wing-coverts mainly new, body and tail in full moult. Sequence of tail-feather replacement about 1–2–(or 2–1–)3–5–4. Tail new and secondaries moulting at moult score of 40, all moult completed with regrowth of p10. Duration of primary moult not known; birds with completely new plumage February–March probably adult, those with p10 still growing late March or April either adult or advanced immature. In Iberia, moult suspended at start of autumn migration; 5 birds had 2, 3, 4, 5, and 6 primaries new (Mead and Watmough 1976). POST-JUVENILE. Partial or perhaps complete. Starts July–October with head, neck, and some scapulars; most birds in 1st adult by September–October, only juvenile flight-feathers, tail, tertials, and part of median and all greater upper wing-coverts retained. Moult continued in winter, including tail and inner primaries, but no birds in active moult examined. During May–August of 2nd calendar year, primary moult suspended with score 15 (3 inner primaries new, 1 bird examined), 20 (4 new, 2 birds), 30 (2 birds), 35 (4), and 40 (1); 3 had not started (score 0; 2 of these retained juvenile tail also), some perhaps had completed moult (see adult post-breeding). Usually part of juvenile secondaries and occasionally a few tail-feathers, part of belly, or some wing-coverts retained. One, late September, completed moult with p10 growing after summer suspension with score 30; at same time, p1–p3 new and p4 growing, thus showing serially descendant moult; not known whether this is usual practice in birds which suspend moult during summer.

Measurements. (1) Nominate *ruficollis*, adult, Spain, Portugal, and north-west Morocco, March–September; (2) *desertorum*, adult, Algeria and Tunisia, February–September; other measurements of both races combined; skins (BMNH, RMNH, ZFMK, ZMA, ZMM). Bill is exposed culmen; to forehead, 14·5 (7) 12–15 mm more.

WING	(1)	♂ 206	(5·18; 15)	198–214	♀ 204	(3·23; 11)	199–210
	(2)	207	(5·02; 11)	200–214	204	(4·43; 7)	198–210
	JUV	202	(3·36; 7)	196–207	199	(4·77; 5)	194–206
TAIL	(1)	160	(4·78; 15)	154–168	155	(5·42; 10)	148–162
	(2)	163	(6·10; 11)	153–171	159	(6·77; 7)	148–166
	JUV	154	(5·61; 6)	145–162	148	(9·96; 5)	135–157
BILL		11·3	(0·50; 14)	10·5–12·2	10·8	(0·78; 12)	9·7–11·9
TARSUS		21·9	(1·40; 14)	20·8–24·3	21·0	(0·78; 12)	20·1–22·7
TOE		24·5	(1·45; 14)	22·6–26·5	23·7	(1·00; 12)	22·3–24·8

Sex differences not significant. Juvenile tail and wing (♀ only) significantly shorter than those of adult. For wing of adults, Portugal, see Soares (1973).

Weights. West-central Algeria, May: 75 (Dupuy 1970a). Migrant, south-east Morocco, April: 62 (BTO).

Structure. 10 primaries: p9 longest, p10 3–4 shorter, p8 1–4, p7 23–34, p6 44–56, p5 57–71, p1 91–104. Outer web of p8–p9 and inner web of p9–p10 emarginated; edge of emarginated portion slightly serrated. Tail rather long, tip slightly rounded, 10 feathers; t5 11–15 shorter than t1. Maximum width of t5 25–29 in adult, 21–25 in juvenile. Outer toe c. 56% of middle, inner c. 59%, hind c. 46%. Tarsus bare, except for upper front. Rest of structure as in *C. europaeus*.

Geographical variation. Considerable, in colour only. *C. r. desertorum* of north-east Morocco, Algeria, and Tunisia differs from nominate *ruficollis* of Iberia and rest of Morocco in distinctly paler ground-colour and narrower black marks; pattern of primaries and tail-tips as in nominate *ruficollis*, however. In adult, sides of forehead and crown, mantle, inner webs of scapulars and of upper wing-coverts, and back to upper tail-coverts rather more finely vermiculated or speckled grey than nominate *ruficollis*, appearing paler silvery-grey; black streaks of central forehead, central crown, and especially scapulars and upper wing-coverts narrower and less prominent; half-collar on hindneck wider (c. 2 cm), paler golden-cinnamon, without partial black barring of nominate *ruficollis*; grey-vermiculated parts of scapulars, tertials, and upper wing-coverts distinctly suffused pink-cinnamon or buff; chin and sides of head almost uniform deep rufous-cinnamon, less marked with black; chest below prominent white throat-patches rufous-cinnamon with finer black bars and vermiculations than in nominate *ruficollis*, some feathers tipped white; dark bars on breast and belly duller, narrower, and further apart; vent and under tail-coverts virtually without black marks, tips of longer under tail-coverts washed white. Black bars of tail-feathers narrower; grey of t1 suffused pale cinnamon, each black bar on t1 often inclined to be bordered by uniform pale cinnamon bar distally. Black bars of flight-feathers slightly narrower than in *C. ruficollis*, sometimes strongly reduced towards feather-bases; grey of primary-tips paler. Upperparts of juvenile appear vermiculated cinnamon-buff and pale grey rather than grey and black as in juvenile nominate *ruficollis*; general tinge rather similar to Egyptian Nightjar *C. aegyptius*, but pattern of primaries different. CSR

Caprimulgus eximius (Temminck, 1826) Golden Nightjar

Fr. Engoulevent doré   Ge. Prachtnachtschwalbe

A little-known non-migratory species, which inhabits dry savanna along southern edge of Sahara: in northern and central Sudan west to Darfur (nominate *eximius*), and from c. 10°E in Niger to c. 10°W in Mauritania (*simplicior*) (Heim de Balsac and Mayaud 1962; Vaurie 1965). Western race may extend into Western Sahara (Rio de Oro), whence a possible sight record from Guelta du Zemmour in June 1955 (Valverde 1957). For eggs, see Plate 94.

Caprimulgus aegyptius Egyptian Nightjar

PLATES 56 and 57
[between pages 662 and 663]

Du. Egyptische Nachtzwaluw   Fr. Engoulevent d'Égypte   Ge. Ägyptischer Ziegenmelker
Ru. Буланый козодой   Sp. Chotacabras sahariano   Sw. Ökennattskärra

Caprimulgus aegyptius Lichtenstein, 1823

Polytypic. *C. a. saharae* Erlanger, 1899, North Africa east to Nile delta; nominate *aegyptius* Lichtenstein, 1823, Suez Canal area, Sinai, Israel, and Arabia east through Iraq to Iran, Afghanistan, Turkmeniya, Uzbekistan, and upper Syr Darya river in USSR.

Field characters. 24–26 cm; wing-span 58–68 cm. Slightly smaller than Nightjar *C. europaeus*; c. 10% larger than Nubian Nightjar *C. nubicus*. Medium-sized nightjar, with pale pink-grey-buff plumage usually relieved only by white throat-patches and, when visible, pale buff vent. In flight, lacks bold isolated marks on tail and primaries, but undersurface of both noticeably pale-barred. Flight appears slower than in congeners. Sexes similar; no seasonal variation. Juvenile separable at close range. 2 races in west Palearctic, distinguishable in the field.
  ADULT. (1) Middle East and Russian race, nominate *aegyptius*. Plumage colours paler and pattern even less variegated than in *C. nubicus* with ground-colour of upperparts pale pink-grey-buff and that of underparts pink-buff. Markings complex but indistinct; most obvious at rest are white patch on each side of lower throat, black freckles on scapulars, tertials, and wing-coverts, and wavy black bars across flight-feathers and tail. Barring of underparts diffuse, absent on belly and vent. Under wing-coverts pale pink-grey-buff, merging with white-spotted bases of flight-feathers to create, with underbody, strikingly pale bird from below in flight; upperwing shows small white spots on primary-bases. Tail lacks white tips to outer feathers; appears grey, barred pink-buff. (2) North African race *saharae*. Differs distinctly in generally more sandy (less greyish) plumage, with more rufous tone obvious above. JUVENILE. Both races. Plumage somewhat paler than adult, mainly lacking contrasting marks. Best distinguished by lack of black freckles on scapulars and by complete pale tips (not just on outer web) to median upper wing-coverts and tertials.
  In fortunate circumstances, rather small size, pale buff colour, and (particularly) white-spotted bases of flight-feathers and pale-barred tail allow certain identification. In some lights, *C. nubicus* may appear as pale but shows white band across middle of outer primaries and white

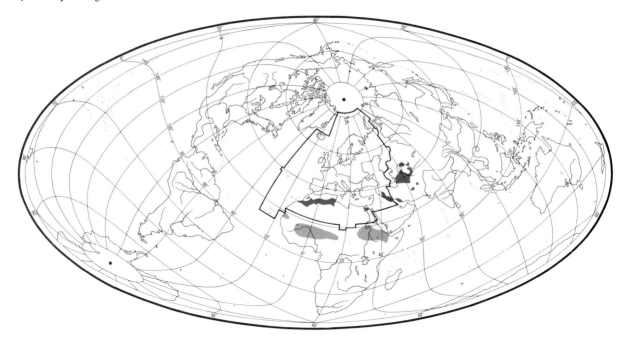

tail-spots (fully distinct on adult ♂ only). Flight as *C. europaeus*, but actions appear slower and most ghostly (perhaps caused by pale plumage). Gait and behaviour as other *Caprimulgus*.

Song of ♂ a series of rapid 'kowrr' sounds.

Habitat. Breeds in lower middle and low latitudes, in open mainly flat and very warm terrain. In southern central Asia in low-lying hot deserts and semi-deserts with soft soil, but locally adjacent to rivers and other waters; avoids mountains, river floodplains, and wooded areas. Prefers low sand-dunes with sparse trees and herbage, or hot low-lying sandy and clay flats with scattered tamarisks *Tamarix* and other shrubs. Favours wastes near fresh water with reeds and meadows, or dry patches of soil surrounded by water. Occasionally nests in clay deserts with patches of windblown sand, but rarely in deep desert far from water (not beyond 100 mm isohyet) and moves towards rivers in hot weather. Catches insects near wells and bushes, and hunts over still pools, or among nomad tents, encampments, and grazing livestock. In daytime, rests on ground in shade of bush or clay bank; after sundown rests readily on dusty roads and near irrigation ditches. Normally flies in lower airspace, but on migration at high altitudes (Dementiev and Gladkov 1951a). In southern Morocco, where sand rare, frequents vehicle tracks and nests on steppe with cover of *Artemisia* and *Haloxylon* or on desert-like alluvial plain at average altitude of c. 500 m between 200 and 300 mm isohyets. Leaves area when autumn rains change habitat (Robin 1969). In Afghanistan, found in desolate lowland area between cultivated fields, and in area with dunes and scattered scrub c. 1 m high (Paludan 1959). At Azraq (Jordan), bred on open salt-encrusted ground in sparse *Tamarix* at edge of wetland (Nelson 1973). Accordingly, appears the least dependent on trees and the most tolerant of semi-desert or desert of Caprimulgidae occurring in west Palearctic, but like congeners is fairly opportunist in choice of habitat according to available options, and is attracted to water whenever possible. Wintering birds in West Africa have been found in *Salsola* and *Tamarix*, and over rice-fields, but also occur in very dry country (Moreau 1972).

Distribution. ISRAEL. Formerly, not rare as a breeder in coastal plain; no records for over 30 years, apparently due to agricultural and urban development (HM). JORDAN. Nested Azraq, 1969 (Nelson 1973). EGYPT. Distribution imperfectly known (SMG, PLM, WCM). TUNISIA. Current distribution uncertain (MS). ALGERIA. May breed Ahaggar (Heim de Balsac and Mayaud 1962).

Accidental. Britain, West Germany, Sweden, Italy, Malta, Syria, Jordan, Kuwait, Libya.

Population. TUNISIA. Considered fairly common by Blanchet (1955), but apparently much rarer now (Thomsen and Jacobsen 1979; MS). IRAQ. Common (Allouse 1953).

Movements. Mainly migratory, with African and Asiatic populations partly overlapping in winter.

NORTH AFRICAN POPULATIONS (*saharae*). Winter in Sahel zone below Sahara, alternating in time from Mali eastwards with Plain Nightjar *C. inornatus*, an intra-African migrant (Moreau 1972). Only partially migratory in Egypt (Meinertzhagen 1930), though certainly summer visitor to north-west Africa, arriving March and leaving

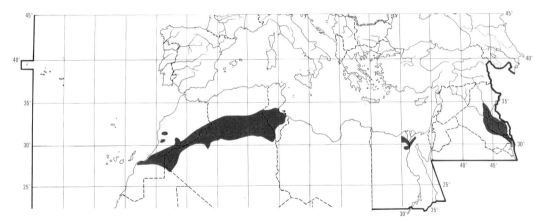

September–October (Robin 1969). Numbers present December–February in Sénégal, in *Salsola* and *Tamarix* areas and hawking over ricefields (Morel and Roux 1966). Also not scarce in Sahel and Niger inundation zones of Mali, November–February (Duhart and Descamps 1963; Lamarche 1980); mainly western race *saharae*, though also (in eastern Mali) some long-winged birds (>210 mm) within size range of Asiatic *aegyptius*. Probably not uncommon winter visitor to northern Nigeria, where frequent at Malamfatori by Lake Chad (Elgood 1982); however, few records from Chad (Salvan 1968b). During October–March, present over larger part of Sudan, south to 9°30′N (Lynes 1925a; Cave and Macdonald 1955; Macleay 1960), but these include Asiatic birds (see below) as well as North African emigrants. See also Geographical Variation.

ASIATIC POPULATIONS (nominate *aegyptius*). Summer visitor to Soviet Central Asia, arriving early April to mid-May and leaving in September; large flocks seen in Seistan 23 September–9 October (Dementiev and Gladkov 1951a). Arrives Iraq from mid-March and plentiful in April; numbers there increased by passage migrants in August–September, with departures in September (Ticehurst *et al.* 1922; Allouse 1953). These winter in northeast Africa, and migrate on broad front across Arabia from September to early November and March to mid-May (Bundy and Warr 1980; Jennings 1981a); found occasionally in autumn on ships in Persian Gulf, northern Arabian Sea, and Gulf of Aden (e.g. Tuck 1965, Casement 1974). December–January records exist from Oman and Aden (Meinertzhagen 1954; Walker 1981b), but not known to winter regularly in Arabia. African winter range not delineated. Includes Sudan (Vaurie 1965), and birds may penetrate further west within Sahel zone (see status in Mali, above); sparse Gulf of Aden passage suggests occurrence as far south in Africa as Somalia, but only one (December) record from there (Baird 1979). Rare Ethiopian winter records (Smith 1957; Ashford and Bray 1977) also probably of Asiatic origin.

Food. Entirely insects taken mainly in low flight, though at times ascends rapidly and flies higher than normal in pursuit of prey. Frequently forages near or over water (Dementiev and Gladkov 1951a), often dipping down like swallow (Hirundinidae) to drink (Dolgushin *et al.* 1970). Insects also hunted over agricultural crops (Koenig 1919), including at night in winter quarters (Morel and Roux 1966), and over towns (Geyr von Schweppenburg 1918). Less commonly forages on ground (Dementiev and Gladkov 1951a).

Insects recorded chiefly beetles (Coleoptera) large and small (Scarabaeidae, Elateridae, Carabidae); also bugs (Hemiptera), moths (Lepidoptera), Orthoptera, ants (Formicidae), termites (Isoptera), and mosquitoes (Culicidae) (Geyr von Schweppenburg 1918; Lynes 1925a; Bates 1937a; Witherby *et al.* 1938; Dementiev and Gladkov 1951a; Dementiev 1952; Rustamov 1954; Sopyev 1967; Robin 1969; Dolgushin *et al.* 1970).

2 ♂♂ from Saudi Arabia, March–April, both contained mainly Scarabaeidae, 1 also other Coleoptera (Bates 1937a). The few other more detailed studies refer to extralimital parts of the range in USSR. In Transcaspia, takes various beetles, including at times the quite large *Mouodon nentodou*, also Orthoptera and Lepidoptera (primarily Noctuidae); bugs *Pentatoma* apparently a favoured prey (Dementiev and Gladkov 1951a). In Kazakhstan, where wide range of insects indicated, prey list includes flightless beetles *Discoptera komarowi* and *D. eilandi* (Dolgushin *et al.* 1970); see also Dementiev (1952) for Turkestan. Stomach from Kara-Kum contained beetles—Elateridae, *Ditomus*, and *Grugonocnemis* (Rustamov 1954).

No information on food of young. MGW

Social pattern and behaviour. Poorly known, though some aspects said to resemble Nightjar *C. europaeus* (e.g. Meinertzhagen 1930).

1. More gregarious than other west Palearctic Caprimulgidae (e.g. Heuglin 1869, Meinertzhagen 1930), particularly prior to and, to some extent, during migration: e.g. in Seistan region (USSR), 'huge' aggregations in advance of autumn emigration, though singly, in twos, or up to 10 on spring passage (Dementiev and Gladkov 1951a; see also, e.g., Koenig 1919 for Egypt, Moore and Boswell 1956b for Iraq). Members of such small spring 'flocks' not very closely associated (Dolgushin *et al.* 1970). At Akkarkuf (Iraq), 'hundreds' at roost, August–September

(Ticehurst *et al.* 1922), and *c.* 50 at diurnal roost, Sénégal, mid-December (Morel and Roux 1966). May also congregate to feed (e.g. Rustamov 1954, Robin 1969), with up to 10 together (Dolgushin *et al.* 1970). Some evidence that migrant 'flocks' single sex: 6 obtained from flock of *c.* 50 all ♀♀ (Heuglin 1869); party of 4 ♂♂, Faiyum (Egypt), March (Shelley 1872); 5 ♂♂ together at Biskra (Algeria), late April (Koenig 1895); see also Koenig (1919). BONDS. No information on mating system or pair-bond. Young tended by both parents (Dolgushin *et al.* 1970). BREEDING DISPERSION. Solitary; no groups recorded (Marinkelle 1959; Robin 1969). ROOSTING. Diurnal and sometimes communal (see above). On ground, generally in shade of bush or tree (e.g. Heuglin 1869, Ticehurst *et al.* 1922, Rustamov 1954), but not in tree (Heuglin 1869). Commonly rests on roads and tracks like some other Caprimulgidae (Robin 1969). Said to make depression in sand for roosting (Adams 1864). Sites may be close to water where birds feed at other times (Dementiev and Gladkov 1951*a*). At Karradah Sharqiyah (Iraq), site on river bank used traditionally by up to 15 birds over several years, July–October; birds roosted 'close together' (Marchant 1961*b*, 1962*b*). Like other Caprimulgidae, activity generally crepuscular and nocturnal, though birds may be relatively active throughout day in dull, misty conditions (Rustamov 1954). At Ali Gharbi (Iraq), early August, birds flying but not feeding at *c.* 16.30 hrs: with shade temperature of *c.* 120°F, ground probably too hot for resting (Ticehurst *et al.* 1922; see also Meinertzhagen 1954).

2. Not timid in presence of man, and will fly close when hawking insects; also forages around villages, encampments and amongst grazing animals. ♂ giving Advertising-call (see 1 in Voice) ceases only when man very close (Dementiev and Gladkov 1951*a*; Rustamov 1954). In Algeria and Egypt, birds flew up silently, often from under observer's feet, and quickly landed again (Koenig 1895, 1919). At Faiyum, 4 ♂♂ resting on bare sand (see above) said to have called when flushed and to have made for cover (Shelley 1872: see 2b in Voice). Said by Heuglin (1869) to be loath to fly when disturbed, tending rather to run from bush to bush, and to call with throat inflated (see 2c in Voice); such behaviour not confirmed elsewhere (Witherby *et al.* 1938). Wing-clapping occurs (Hüe and Etchécopar 1970), though circumstances not clear; apparently much less loud than in *C. europaeus* (Robin 1969). FLOCK BEHAVIOUR, ANTAGONISTIC BEHAVIOUR. No information. HETEROSEXUAL BEHAVIOUR. In Kara-Kum (USSR), ♂ gives Advertising-call from first few days after arrival (Rustamov 1954); continues up to laying (Dementiev and Gladkov 1951*a*). In Morocco, given at dawn and dusk for less than 1 hr during laying (Robin 1969; also Dupuy 1970*b*); given from ground (Chappuis 1981). In Egypt, pair-formation said to take place only shortly before breeding (Shelley 1872); in Kazakhstan (USSR), soon after arrival (Dolgushin *et al.* 1970). Incubation mainly by ♀, and ♂ reported to sit nearby when ♀ on nest (Dementiev and Gladkov 1951*a*; Dolgushin *et al.* 1970). RELATIONS WITHIN FAMILY GROUP. Young brooded for 2–3 days. Family said by Robin (1969) sometimes to move a few metres away from nest soon after hatching, parents transporting young between feet and pressed against belly; possibly refers only to shading of young (Glutz and Bauer 1980). Parents and young roost separately during day, remaining motionless by clumps of *Artemisia*. Young start to fly at *c.* 1 month (Robin 1969)—but compare *C. europaeus*. Siblings reported to stay together after family otherwise breaks up (Dresser 1871–81). ANTI-PREDATOR RESPONSES OF YOUNG. No information. PARENTAL ANTI-PREDATOR STRATEGIES. Generally a tight sitter, remaining motionless with eyes almost completely closed (see *C. europaeus*) and allowing approach to *c.* 2 m (Robin 1969). Said often to run only a few steps when flushed (Heuglin 1869). MGW

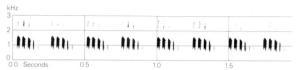

I J-C Roché (1967) Morocco March 1966

Voice. Mostly silent apart from Advertising-call (Dementiev and Gladkov 1951*a*). For Wing-clapping, see Social Pattern and Behaviour.

CALLS OF ADULTS. (1) Advertising-call (Song) of ♂. Recording (Fig I) suggests a regular and rapidly repeated series (*c.* 3–4 per s) of 'kowrr' or 'powrr' sounds, with slowing ('winding down') and slight irregularity towards end of a bout; chugging quality of distant motor-boat (M G Wilson); a regular glottal 'purr purr purr' (P J Sellar). Each unit comprises 5 brief (*c.* 5–20 ms) sub-units, descending in pitch and separated by pauses of *c.* 20 ms (J Hall-Craggs). Other renderings: a series of rapid 'kre-kre-kre' sounds, at times changing to 'kru' (a clear 'u' sound) or 'kro', and with metallic timbre (Dementiev and Gladkov 1951*a*); 'kre-kre-kre-u-o' (Dementiev 1952); see also Rustamov (1954). In Iraq, at night, a 'toc-toc-toc-toc', like a rapid hammering on wood (Ticehurst *et al.* 1922). (2) Other calls. (a) An occasional disyllabic and guttural 'tuk-l tuk-l' uttered in flight (Lynes 1925); subdued croaking sound given at intervals in flight (Allen 1864), probably the same. Very quiet calls (and Wing-clapping) given in pauses between bouts of Advertising-calls (J-C Roché); in recording by J-C Roché, Morocco, March, very frog-like guttural and rolling croaking sounds (M G Wilson). (b) 'Little snapping sound' given when flushed (Shelley 1872). (c) Grumbling and quacking sounds given by birds running about and seeking shelter from heat (Heuglin 1869); possibly the same as call 2a.

CALLS OF YOUNG. No information. MGW

Breeding. SEASON. Algeria and Tunisia: eggs found mid-April to mid-June (Heim de Balsac and Mayaud 1962). Morocco: laying begins between 15 March and 15 April with earliest dates in dry years; in wet years breeding may last from April to August with peak in May and June (Robin 1969). Turkmenistan (USSR): laying begins mid-May (Dementiev and Gladkov 1951*a*). SITE. On ground in the open, or under low bush. Nest: shallow scrape, or just bare space cleared of stones (Robin 1969); footprint of camel (Sopyev 1967). Building: no information. EGGS. See Plate 94. Elliptical, smooth and glossy; cream or white, marbled and blotched pale olive and grey. 32×22 mm ($30-35 \times 21-23$), $n = 22$ (Witherby *et al.* 1938); average of 53 eggs, 31×22 mm (Robin 1969). Weight of 2 clutches, 8·7 and 9·2 g, 7·2 and 7·6 g (Rustamov 1954; Sopyev 1967). Clutch: (1–)2 (Dementiev and Gladkov 1951*a*). 2 broods probable (Dementiev and Gladkov 1951*a*). Begins with 1st egg; hatching asynchronous. YOUNG. Semi-altricial and nidicolous. Cared for and fed by both parents. FLEDGING TO MATURITY. Fledging period *c.* 1 month (Robin 1969). No further information.

Plumages (nominate *aegyptius*). ADULT. Forehead, crown, and nape pale pink-buff, appearing slightly grey due to very fine grey vermiculation; feathers of central crown with small triangular black terminal spots, thin black shaft-streak, and larger uniform pink-buff subterminal spots. Sides of head, chin, upper throat, and centre of lower throat pale pink-buff with some fine grey specks or short bars, each side of lower throat with rather small white patch. Hindneck, sides of neck, mantle, back, and rump almost uniform pale pink-buff, fine grey vermiculation and thin dark shaft-streaks or arrow-marks very faint. Scapulars pale pink-buff with slightly more distinct grey vermiculation (darkest and coarsest on tips), appearing about as greyish-buff as crown; each scapular with narrow wavy black bar subterminally and with black shaft-streak on tip, both sides of bar bordered by uniform pink-buff patch, widest on outer web. Upper tail-coverts like scapulars, but longer ones with 2 well-separated narrow black bars. Chest pink-buff with fine grey vermiculation, each feather subterminally with uniform pink-buff spot. Breast and flanks pink-buff with regular narrow grey bars; ground-colour paler and bars more faint towards vent; under tail-coverts uniform cream-buff. Central pair of tail-feathers (t1) pale pink-buff with *c*. 6 well-separated wavy black bars (each 1–2 mm wide), each bar bordered on basal side by extensive greyish-black vermiculation, on distal side uniform pink-buff; other tail-feathers similar, but black bars slightly wider, more distinct, and more numerous, 7–9 bars on t5; outer web of t5 hardly vermiculated. Hardly any white on tips of tail-feathers; only *c*. 8–12 mm of tips of t1–t4 almost uniform pink-buff with faint grey specks; tip of t5 almost uniform for 10–20 mm (see below). Outer web and 40–50 mm of tip of primaries boldly barred brownish-black and pink-buff (terminal point of each more uniform greyish-buff), pink-buff bars with brownish-black specks (least so on outer web of p10); basal and middle portion of inner web of primaries white with coarse black transverse wedges (smallest near base), appearing serrated. Secondaries and greater upper wing-coverts boldly barred pink-buff and brown-black, pink-buff speckled brown-black as on primaries; ground-colour of barred base of inner web almost white; *c*. 5 mm of tip of all secondaries almost uniform pinkish-cream. Tertials and median and lesser upper wing-coverts pale pink-buff with fine grey vermiculation (like scapulars), tips with uniform pink-buff spots except for shorter lesser coverts, spots largest and accentuated by black subterminal border on outer webs. Axillaries and under wing-coverts pale pink-buff with narrow grey bars, like belly. Sexes similar, but in ♂ 12–20 mm of tip of t5 uniformly white; in ♀, 10–15 mm of tip pale buff with slight grey freckling. Abrasion and bleaching strongly influence colour. In fresh plumage, general tinge pale pink-buff, brightest on underparts and flight-feathers, uniform spots on crown, scapulars, upper wing-coverts, and chest not marked; in worn plumage, ground-colour fades to pale cream-yellow on upperparts, upperwing, and chest and to dirty white on belly to vent, fine grey vermiculations more distinct and whole upperparts and chest more sandy-grey with more marked uniform pale spots on feather-tips of crown, scapulars, tertials, and upperwing (pale spots almost absent when heavily abraded, however) and less prominent black marks; warmer buff only retained on inner webs of secondaries, inner primaries, and greater upper wing-coverts. White of throat-patches often almost completely lost by abrasion. DOWNY YOUNG. Down thick; pale buff or sandy-yellow, bleaching to pale cream-yellow (Harrison 1975; ZMM). JUVENILE. Rather similar to adult, but feathers of head, upperparts, and in particular underparts softer and looser, almost downy on vent and under tail-coverts. Crown, upperparts, and upperwing pale pink-buff (soon bleaching to pale cream-yellow), faintly barred pale grey and without narrow dark grey subterminal bars of adult, no black spots or shaft-streaks on crown and scapulars (unlike adult); upper wing-coverts and tertials with tip uniform cream-white, poorly defined from pale pink-buff base, no pink-buff spot accentuated by black on tip of outer web (unlike adult); underparts pale pink-buff (bleaching to pale cream), grading to cream-white on vent and under tail-coverts; white throat-patches (if any) poorly defined; grey bars on underparts more faint and less regular than in adult, no bars from lower belly and lower flanks to under tail-coverts. Flight-feathers as adult, but inner primaries indistinctly fringed white on tips when fresh; tail as adult, but dark tail-bars perhaps closer together, tip of t5 partly freckled in ♂, fully in ♀. FIRST ADULT. As adult, but some retain worn juvenile outer primaries or all primaries in spring (see Moults).

Bare parts. ADULT AND JUVENILE. Iris dark brown. Bill dark horn-brown, dark slate-grey, or black; base flesh-coloured or pale horn-brown. Leg and foot pale lilac, pale reddish-grey, pale red-brown, or black. DOWNY YOUNG. Iris dark brown. Bill, leg, and foot flesh-pink; tip of bill grey. (Hartert 1912–21; BMNH, ZMM.)

Moults. ADULT POST-BREEDING. Complete; primaries descendant. Starts with p1 late April to mid-June, irrespective of breeding area; mainly mid-May. By mid-July, average primary moult score 21, by mid-August 36; moult completed with p10 (score 50) either on breeding grounds early September to late October, or (probably after suspension during migration) in winter quarters October to early December. Body starts at score 25–35 (about July–August), scattered feathers of throat and chest and some scapulars and median upper wing-coverts first, soon followed by remainder of head, body, and wing-coverts; all largely new by score 45–50. Moult of tail regularly centrifugal (t1 to t5), starting at score *c*. 25, completed at *c*. 40; up to 8 feathers sometimes grow simultaneously. Secondaries replaced at score 30–50. POST-JUVENILE. Partial or perhaps sometimes complete. Starts from July with scattered feathers of chest to belly, head, and some scapulars. From late September to late October, head, body, and most wing-coverts in 1st adult, but flight-feathers, greater wing-coverts, tail, and some lesser or median upper wing-coverts or tail-coverts still juvenile. Further moult not fully elucidated; only a few 1st adults from winter quarters examined, none in active moult. On return to breeding grounds in spring, all 1st adults had new tail and some also new inner primaries, suspending with up to p7 or p8 new in 3 examined; at least 4 birds retained all juvenile flight-feathers, these heavily worn; some birds perhaps moult all flight-feathers in winter quarters, but this cannot be ascertained in spring, as they would then be indistinguishable from adult.

Measurements. ADULT. Nominate *aegyptius*. Turkmeniya and Uzbekistan (USSR), Iraq, and neighbouring south-west Iran, March–September; skins (BMNH, ZMM).

WING   ♂ 206 (4·07; 17) 201–216   ♀ 207 (4·72; 17) 200–216
TAIL   126 (4·56; 17) 118–133   127 (3·12; 14) 122–132

Israel, Jordan, western Arabia, and Sinai west to Suez Canal area, March–August; skins (BMNH).

WING   ♂ 200 (4·90; 11) 193–208   ♀ 198 (6·86; 4) 193–204
TAIL   121 (5·78; 11) 114–129   122 (5·45; 4) 114–127

C. a. saharae. Nile delta, Wadi el Natrun, and Faiyum (Egypt), March–September; skins (BMNH, ZFMK, Giza Zoological Museum per S Goodman and P L Meininger).

WING   ♂ 191 (3·17; 10) 186–196   ♀ 189 (5·29; 8) 185–199
TAIL   112 (4·04; 8)   106–119   112 (7·89; 6) 103–122

Algeria to Sénégal, all year; skins (BMNH, RMNH, ZFMK, ZMA).

WING	♂	205	(1·10; 6)	203–207	♀	201	(—; 3)	197–203
TAIL		125	(5·58; 5)	119–133		124	(—; 2)	122–126

Nominate *aegyptius* and *saharae*, combined. Iraq, south-west Iran, western Arabia, and Algeria, summer; skins (BMNH, ZFMK, ZMA). Bill is exposed culmen; to forehead, 12–13 mm more.

BILL	♂	9·3	(0·94; 16)	8·1–11·2	♀	8·8	(0·32; 6)	8·4–9·3
TARSUS		21·0	(1·12; 16)	19·4–23·1		21·5	(0·54; 6)	20·5–22·0
TOE		22·6	(0·74; 16)	21·2–23·8		22·8	(1·38; 6)	21·2–24·3

Sex differences significant.

Sexes combined, sources as above. Nominate *aegyptius*: (1) Kazakhstan; (2) Turkmeniya and Uzbekistan; (3) Iraq and south-west Iran; (4) western Arabia, Jordan, Israel, and Sinai; (5) Egypt and Sudan, winter. *C. a. saharae*: (6) Nile delta, Wadi el Natrun, and Faiyum; (7) Lake Chad, February; (8) Algeria to Sénégal.

(1)	WING	212	(4·34; 5)	208–218	TAIL 127	(4·85; 4)	124–134
(2)		208	(4·32; 22)	201–216	127	(3·74; 22)	119–133
(3)		205	(4·25; 12)	200–212	124	(4·19; 9)	118–130
(4)		200	(5·28; 15)	193–208	121	(5·51; 15)	114–129
(5)		203	(4·61; 11)	201–209	120	(3·60; 11)	114–125
(6)		190	(4·17; 18)	185–199	112	(5·73; 14)	103–122
(7)		190	(6·66; 3)	183–196			
(8)		204	(2·78; 9)	197–207	125	(4·83; 8)	119–133

JUVENILE. Similar to adult, but wing on average 4 shorter.

Weights. Nominate *aegyptius*. Iran and Afghanistan, March–April: ♂♂ 68, 85, 93; ♀ 70 (Paludan 1959; Diesselhorst 1962).

C. a. saharae. Italy, April: ♀, 81 (Moltoni 1968). Morocco, April: 72 (BTO).

Structure. Wing rather long and broad, tip pointed. 10 primaries: p9 longest, p10 2–14 shorter, p8 1–8, p7 22–34, p6 42–54, p5 58–70, p1 91–107; outer web of p8–p9 and inner of p9–p10 slightly emarginated. Tail rather shorter than in other Palearctic *Caprimulgus*, tail/wing ratio, *c.* 0·6 in this species, against *c.* 0·7 in Nightjar *C. europaeus* and Nubian Nightjar *C. nubicus*, and *c.* 0·8 in Red-necked Nightjar *C. ruficollis* (in Common Nighthawk *Chordeiles minor*, *c.* 0·5 only); 10 feathers, t5 3–7 shorter than t1. Tarsus and toes rather short, slender; tarsus bare, except for upper front. Outer toe *c.* 57% of middle, inner *c.* 53%, hind *c.* 32%. Remainder of structure as in *C. europaeus*.

Geographical variation. Rather slight in colour, pronounced in size. Nominate *aegyptius* from Middle East shows more extensive grey vermiculation on pink-buff feathers of upperparts, upperwing, and chest, appearing greyish sandy-pink in fresh plumage, more isabelline-grey when plumage worn and pink-buff ground-colour bleached. North African *saharae* has ground-colour slightly brighter pink-yellow or buff-yellow, and grey vermiculation fainter and less extensive, appearing more rufous-sandy in fresh plumage, yellowish- or pinkish-isabelline when worn. Boundary between grey and pinkish races in Egypt not fully established; according to Vaurie (1960), greyish nominate *aegyptius* occurs west to Nile delta and Cairo region, pinkish birds only west of Nile, but all Egyptian birds examined from breeding season (April–July) were pinkish, only those of Suez Canal area and north-east fringe of Nile delta greyish like typical nominate *aegyptius*. From September to March, greyish birds common in Nile delta. Variation in size mainly clinal in nominate *aegyptius*, less so in North Africa. Nominate *aegyptius* smallest in Suez Canal area, Sinai, Levant, and Arabia; though few breeding records here, birds examined March–August too small to belong to any other known breeding population; Iraq and western Iran on average slightly larger, Turkmeniya larger still, Kazakhstan largest (see Measurements); larger populations sometimes separated as *arenicolor* Severtzov, 1875 (e.g. Vaurie 1960*a*), but as variation clinal with no sharp boundary, included here in nominate *aegyptius*. *C. a. saharae* from Algeria and Tunisia fairly large, from Nile delta distinctly smaller. From September to March, greyish nominate *aegyptius* occurs commonly all over Egypt (average wing 201) and Sudan (average wing 208); larger *saharae*, similar to north-west African population, examined from Sénégal and Sudan; no smaller *saharae* seen away from breeding grounds, except for 3 birds from Lake Chad in February which were more saturated yellow-pink than typical *saharae* and, if proved breeding there, may be a distinct race. CSR

Subfamily CHORDEILINAE nighthawks

7–8 species in 4 genera found in the Americas. About half the species are of genus *Chordeiles*, to which belongs the single species occurring (accidentally) in west Palearctic.

For general features, moults, etc., see Caprimulgidae. Rictal bristles absent. Bills project less than in Caprimulginae and plumages less soft.

Chordeiles minor Common Nighthawk

PLATES 56 and 58
[between pages 662 and 663]

Du. Amerikaanse Nachtzwaluw Fr. Engoulevent d'Amérique Ge. Nachtfalke
Ru. Малый козодой Sp. Chotacabras yanqui Sw. Nattfalk

Caprimulgus minor J R Forster, 1771. Synonym: *Chordeiles virginianus*.

Polytypic. Nominate *minor* (J R Forster, 1771), south-east Alaska, Canada, and north-east USA east of Great Plains and south to northern Georgia and southern Virginia, accidental in Europe. Extralimital: *chapmani* Coues, 1888, south-east USA, south of nominate *minor*, south to Florida and Louisiana and east of Great Plains; 7 further races in south-central Canada south to Central America and Florida Keys.

Field characters. 23–25 cm, of which tail 8–9 cm; wing-span 59–68 cm. About 15% shorter than Nightjar *Caprimulgus europaeus* (with most difference accounted for by 25–30% shorter tail), but looks larger in flight, due to

10% longer wings. Plumage colours and pattern similar to *C. europaeus*, differing most in large white or buff patch under chin, more grey-and-black upperparts, more heavily barred underparts, and (in ♂) almost complete white bar near end of shorter, slightly forked tail. Flight differs in both action (often strong and hawk-like) and use of airspace (often at considerable height, chasing faster-flying insects). Sexes dissimilar; no seasonal variation. Juvenile separable at close range.

ADULT MALE. Ground-colour of upperparts almost black, conspicuously spotted and freckled grey and buff; ground-colour of underparts cream-white, closely barred black-brown. Markings complex but less so than in *C. europaeus*; most obvious are grey-spotted supercilium, large, pure white throat, black-lined rear cheeks, buff-spotted black breast, and bold white spots on outer 5 primaries (nearer carpal joint than in *C. europaeus*). In flight, white spots on primaries form bold band on both surfaces of outer wing, and white bands on all but central pair of tail-feathers create distinct, shallow V-shaped subterminal bar under end of tail. In contrast to basically streaked pattern of upperparts in *Caprimulgus europaeus*, *Chordeiles minor* more marbled and uniformly marked. ADULT FEMALE. Throat-patch smaller, buff. White primary-band narrower, appearing broken at close range; tail lacks subterminal white bar. JUVENILE. Resembles ♀ but distinguished by conspicuously greyer and paler upperparts (in fresh plumage) and whitish fringes to flight-feathers.

Appearance of bird on ground initially recalls both *C. europaeus* (in general appearance) and Red-necked Nightjar *C. ruficollis* (in striking throat-patch), but clear observation of smaller body and tail, different structure (with wing-points extending to or beyond end of tail), black upper breast, and evenly marked upperparts allows certain identification. In flight, proportionately longer wings and shorter, slightly forked tail create noticeably more rakish silhouette, strongly recalling long-winged falcon *Falco*; if light permits, bolder, more proximal primary-spots and (in ♂) fuller tail-bar show clearly. Flight generally more active than *Caprimulgus*, with quicker, even more erratic bursts of wing-beats, faster wheels and dives (recalling swift *Apus*), and fewer slow glides than *C. europaeus*. Gait as *C. europaeus*.

Flight-call a distinctive harsh 'paint', 'peent', or 'peeik', but not known from vagrants.

Habitat. Breeds in Nearctic from upper middle latitudes to subtropics, mainly in lowlands but up to *c.* 2400 m in Rocky Mountains, there inhabiting barren hillsides and ridges and breeding in rocky exposed situations (Niedrach and Rockwell 1939). Further west in Washington prefers open logged or burnt-over land, rocky brush areas, open coniferous forests such as yellow pine *Pinus ponderosa* or lodgepole pine *P. contorta*, and, for foraging, open areas, lakes, rivers, valleys, and meadows, cruising high in twilight (Larrison and Sonnenberg 1968). On central Great Plains, seems highly adaptable to using varied habitats, but prefers open grasslands, sparse woods, burnt areas, and cities. Recent large-scale and widespread colonization of flat, preferably gravelled, roofs of city buildings has transformed pattern of habitat and distribution. On passage through Venezuela frequents open country near water up to 1000 m and is often found roosting on tile roofs. An outstandingly aerial species, often foraging in middle airspace in daylight over inhabited areas.

Distribution. Breeds in North and Central America, from south-east Alaska and Canada (British Columbia, southern Yukon, Alberta, southern Mackenzie, and Manitoba east to southern Labrador) south to Panama. Winters in South America from Colombia to Argentina.

Accidental. Iceland, Faeroes, Britain, Azores.

Movements. Migratory in North America, and this may also apply to little-studied population of Central America (Eisenmann 1962); winters in South America, over most of subcontinent south to central Argentina. Often gregarious on passage, forming loose flocks (typically of 20–40 birds), which pass in waves: e.g. up to 1000 birds per hr through Duluth (Minnesota) at peak of autumn passage (Green and Janssen 1975), and 2500 in 25 min over one Tennessee site on 31 August (Nicholson 1975). Movements diurnal and nocturnal.

Fairly regular autumn migrant on Bermuda, where sometimes very common (Bent 1940), and regular then on Lesser Antilles, including Barbados (ffrench 1973). This indicates that in autumn part of eastern population uses transoceanic route between south-east Canada/New England and South America; for transatlantic vagrancy conditions, see Yellow-billed Cuckoo *Coccyzus americanus* (p. 426).

Autumn exodus from USA begins late July and lasts to early October, with main movements in second half of August and early September; returns late April to late May, with main influx in second half of May (Bent 1940; Stewart and Robbins 1958; Green and Janssen 1975). All 9 British records in period 12 September–25 October, rather later than main passage period in North America.

Voice. See Field Characters.

Plumages (nominate *minor*). ADULT MALE. Crown, hindneck, and sides of neck down to lower cheeks and ear-coverts mainly black with fine pale cinnamon or pink-buff spots on forehead, crown, nape, and cheeks, coarser cinnamon streaks on ear-coverts, and many white spots on sides of crown in front of, over, and extending backwards from eye, latter forming mottled pale supercilium; cinnamon or buff spots on hindneck form broken collar. Mantle, scapulars, tertials, and back to upper tail-coverts mainly black also; sides of feathers with indistinct and broken pale grey or off-white bars or spots, vermiculated or marbled black; scapulars and innermost tertials with more sharply defined cinnamon or pink-buff spots or bars subterminally. Conspicuous white triangular patch on throat with apex on chin, extending up to sides of neck. Foreneck deep black with rather large rounded or triangular cinnamon spots; chest dull black with rather narrow and irregular white bars and vermiculations. Remainder of underparts down from breast and flanks evenly and coarsely barred pale buff and dull black, pale buff gradually paler towards vent; bars white and dull black on under tail-coverts. Central pair of tail-feathers (t1) black with rather broad pale grey or pale grey-buff bars, latter vermiculated or marbled black; bases and middle portions of t2–t5 and all outer web of t5 with similar but paler and narrower bars; distal part of t2–t4 and distal inner web of t5 black with broad conspicuous white band *c.* 15 mm from feather-tip, latter forming broad straight white bar subterminally across tail when tail spread, except for t1; extent of bar on t5 variable. Flight-feathers black, p6–p10 with broad white band across middle portion; on p10, band extends to shaft but hardly to outer web; trace of white sometimes along shaft of p5. Tips of p4–p8 slightly tinged grey; secondaries and p1–p3 with paler grey tip, indistinct and widely spaced pink-buff bars on inner web and traces of such bars subterminally on outer web. Greater upper primary coverts deep black; lesser primary coverts and lesser upper wing-coverts deep black with fine cinnamon to pale buff spots; outer median and all greater upper wing-coverts dull black with pale buff or pink-buff spots or short bars; inner median coverts and tertial coverts similar but with many pale grey or off-white vermiculations on tips. Under wing-coverts and axillaries black with narrow buff bars; leading edge of wing at carpal joint white. In worn plumage, cinnamon marks of crown, nape, ear-coverts, scapulars, upper wing-coverts, and underparts paler, pale buff to off-white. ADULT FEMALE. Like adult ♂, but throat cinnamon to pale buff, often slightly spotted black; ground-colour of breast to vent deeper buff, under tail-coverts barred black and pale buff instead of black and white. Tail completely black with rather narrow buff to pale grey-buff bars, latter more finely mottled dull black than paler and purer bars of ♂; no broad white subterminal band on t2–t5. Wing as in adult ♂, but white band across middle portions of p5–p10 slightly narrower, often finely peppered black on outer web, hardly reaching shaft of p10 (not just across shaft as in ♂). JUVENILE. Much greyer and less black than adult ♂ and ♀. Whole upperparts including upper wing-coverts finely vermiculated dull black and off-white or pale buff, solid black restricted to larger patches on crown, scapulars, upper primary coverts, and lesser upper wing-coverts. Buff or off-white and dull black marks form broken collar round hindneck; scapulars with small cream-buff or off-white spots instead of broader cinnamon bars. Longer lesser and median upper wing-coverts with white spot on tip, black of centre not reaching tip (unlike adult). Cheeks and ear-coverts dull black, barred and streaked pale buff or off-white (in adult, black with bold cinnamon marks). Triangular patch on throat and foreneck poorly defined; white with pale buff feather-tips or completely pale buff, marked with fine dull black spots or broader dull black bars. Chest and sides of breast dull black with cream-buff or off-white bars and vermiculation, not deep black with distinct cinnamon spots as in adult. Remainder of underparts barred as in adult, but dark bars duller black and narrower on vent and under tail-coverts, on latter sometimes irregular, broken, or T-shaped. Tail as in adult ♀ (lacking white subterminal band of adult ♂), but pale bars narrower and more numerous, sometimes inclined to form marbling rather than regular speckled bars; tips of feathers narrowly fringed white; maximum width of t5 15–18 mm (17–21 mm in adult). Flight-feathers as in adult ♀, but tips with distinct cream or white fringes. Sexes mainly similar: throat of ♀ tends to be less white and more heavily marked than ♂; white band on inner web of p10 square and extending along shaft for 10–16 mm in ♂, narrower, more rounded and hardly touching shaft in ♀. FIRST ADULT. Like adult; indistinguishable when last juvenile upper wing-coverts (white-tipped), tail-feathers (narrowly barred), and primaries (with traces of white fringes) lost.

Bare parts. ADULT AND JUVENILE. Iris dark brown. Bill black, pale towards base. Leg and foot dark horn-coloured with greyish tinge. (Ridgway 1914; BMNH, ZMA.)

Moults. Similar to *Caprimulgus* nightjars (Stresemann and Stresemann 1966). ADULT POST-BREEDING. Complete, primaries descendant. Probably occurs wholly in winter quarters: the single autumn migrant examined had no new feathers; a few birds from spring had plumage completely new. One *chapmani*, Kentucky (USA), started with inner primaries mid-July (Mengel 1965). Sequence of tail-feather replacement: 1–2–3–5–4 (Stresemann and Stresemann 1966). POST-JUVENILE. Complete or almost complete. Probably mainly in winter quarters, but one mid-September migrant from Curaçao had forehead, forecrown, scattered feathers of nape, sides of breast, and flanks, and some lesser upper wing-coverts new. Some summer birds retain a few heavily worn upper wing-coverts and these perhaps 1 year old; occasionally, some juvenile outer primaries retained.

Measurements. Nominate *minor*. ADULT. Canada and northeast USA, April–August; skins (BMNH, RMNH, ZMA). Tail is to longest feathers, fork is depth of tail-fork. Bill is exposed part of culmen; to forehead, 7–9 mm more.

WING	♂ 208	(3·32; 9)	202–213	♀ 198	(4·68; 9)	194–204
TAIL	108	(3·64; 9)	103–113	108	(4·18; 9)	102–115
FORK	16·1	(3·62; 9)	11–23	15·1	(3·23; 8)	11–20
BILL	7·6	(0·60; 9)	6·8–8·5	7·5	(0·74; 9)	6·5–8·6
TARSUS	14·9	(0·72; 9)	14·0–15·9	15·3	(0·88; 8)	14·0–16·4
TOE	19·3	(0·97; 9)	18·0–21·0	19·2	(1·12; 8)	18·2–21·0

Sex differences significant for wing.

JUVENILE. Only a few examined; ranges fall within those of adults, averages similar.

Weights. Nominate *minor* and races of similar size (see Geographical Variation). USA: ♂♂, April–May, 77, 82, 88; ♀♀, June–July, 64, 69; unsexed, September, 81 (Grinnell *et al.* 1930; Murray and Jehl 1964; Mengel 1965). Belize, July: ♂♂ 69, 75; ♀♀ 63, 67, 81 (Russell 1964). Exhausted juveniles after crossing Caribbean, early October: ♂ 44, ♀ 40 (ZMA).
C. m. chapmani. USA, mid-June to early August: ♂ 55; ♀♀ 61, 67, 77 (Norris and Johnston 1958; Mengel 1965).

Structure. Wing long and narrow, tip pointed. 10 primaries: p10 longest, p9 0–4 shorter, p8 12–19, p7 25–35, p6 43–55, p5 59–73, p1 105–120. Primaries stiff; not emarginated, but gradually tapering to narrowly rounded tip. Tail rather long, slightly forked (see Measurements); 10 feathers, tips rounded (less broad and square-tipped than in *Caprimulgus*). Bill relatively much smaller than in *Caprimulgus*; nostrils as in *Caprimulgus*; gape wide as in *Caprimulgus*; rictal bristles small and weak, indistinct (unlike *Caprimulgus*). Feathering of body more compact and less soft than in *Caprimulgus*. Tarsus short and slender, upper front of tarsus feathered. Toes rather long and slender; outer toe *c.* 55% of middle, inner *c.* 58%, hind *c.* 35%. Claws as in *Caprimulgus*, middle claw serrated.

Geographical variation. Considerable, in both colour and size. Races of eastern USA (including nominate *minor* of Canada and north-east USA and *chapmani* of south-east USA), of USA west of Rocky Mountains, and of Panama darker; in particular, nominate *minor* predominantly black above, others with finer or coarser greyish, buff, or rufous marks, depending on race. 5 races inhabiting Great Plains south to southern Mexico paler, upperparts mainly grey or buff, in some with coarse marks. All races from central and western USA south to western Mexico large, as in nominate *minor*; *chapmani* from south-east USA and birds from southern Texas through Central America to Panama smaller, average wing of ♂ 190 or less, of ♀ 186 or less. For variation in western USA and Middle America and for relationship with Antillean Nighthawk *C. gundlachii*, see Selander (1954), Selander and Alvarez del Toro (1955), and Eisenmann (1962).

CSR

Order APODIFORMES

Minute to medium-sized, highly aerial birds, feeding on insects and/or nectar. 3 families: (1) Apodidae (swifts); (2) Hemiprocnidae (crested swifts); (3) Trochilidae (hummingbirds); doubted, however, by Lack (1964) whether Trochilidae should be included. Apodidae and Hemiprocnidae united in suborder Apodi. Only Apodidae represented in west Palearctic.

Family APODIDAE swifts

Rather small to medium-sized aerial insect-feeders. 3 subfamilies (2 represented in west Palearctic): (1) Cypseloidinae (12 species, Americas); (2) Chaeturinae; (3) Apodinae. Position of *Collocalia* (swiftlets, c. 16 species, south-east Asia and Oceania) and of aberrant *Schoutedenapus* (2 species, Africa) uncertain. See (e.g.) Brooke (1970).

Bodies compact. Necks short; 13–14 cervical vertebrae. Wings long and pointed; 10 long primaries; 8–11 short secondaries, diastataxic in Chaeturinae and Apodinae, eutaxic in Cypseloidinae. Feathers with long aftershaft. Tails vary from short and square, to long and forked; 10 feathers. Bills short and broad, with wide gape. Legs extremely short and feathered; feet small and strong, with sharp, pointed toes sometimes feathered; in Cypseloidinae and Chaeturinae, hind toe directed backwards; in Apodinae, all 4 toes directed forwards. Oil-gland naked. No crop. No caeca. Salivary glands large, and enlarged further in nesting season when (except in Cypseloidinae) saliva used in nest construction. Food collected for young is formed into balls in mouth. Plumages strong and compact, mostly black, blue, or blackish-brown, sometimes with white patches, most often on rump and underparts. Post-breeding moult complete, starting soon after nesting or not until arrival in winter quarters in some species, interrupted during migration. Juveniles either similar to adult or duller with slightly wider white fringes to feathers (fringes rufous in Palm Swift *Cypsiurus parvus*). Young altricial and nidicolous, usually partly covered with down in Chaeturinae and Apodinae, naked in Cypseloidinae. Adult plumage acquired at 4–9 months.

Subfamily CHAETURINAE spine-tailed swifts

23 species in 7 genera from the Americas, Asia, and Africa. Represented in west Palearctic by 1 (accidental) species of *Hirundapus*.

For general features, see Apodidae. Spiny tips on tail-feathers. Hind toe directed backwards.

Hirundapus caudacutus Needle-tailed Swift

PLATES 59 and 63
[between pages 662 and 663, and facing page 686]

Du. Stekelstaartgierzwaluw Fr. Martinet épineux Ge. Stachelschwanzsegler
Ru. Иглохвостый стриж Sp. Rabitojo mongol Sw. Taggstjärtseglare

Hirundo caudacuta Latham, 1801. Synonym: *Chaetura caudacuta*.

Polytypic. Nominate *caudacutus* (Latham, 1801), central and eastern Siberia, Sakhalin, northern Mongolia, northern China, and North Korea, straggling to Europe. Extralimital: *nudipes* (Hodgson, 1837), Himalayas from northern Pakistan to south-west China, and hills of Assam.

Field characters. 19–20 cm; wing-span 50–53 cm. Body bulkier than other swifts of west Palearctic but wings and square tail shorter than in Alpine Swift *Apus melba*. Powerful swift, with distinctive broad-beamed silhouette. Dark plumage interrupted by white U-shaped mark around vent, large dun patch on back, and white chin.

Flight includes slow, level planing. Sexes similar; no seasonal variation except by wear. Juvenile separable at close range.

ADULT. Upperparts dark brown glossed green on crown, nape, and upper tail-coverts, relieved by contrasting cream forehead and large pale dun-brown patch on mantle. Flight- and tail-feathers and wing-coverts all metallic bottle-green, but usually appear almost black and show little relief except for white inner webs to innermost tertials (latter forming pale patch at rear of wing-base). Chin and whole of throat white, contrasting with dark brown breast and belly; latter area surrounded by bold white U-shaped mark (formed by lower rear flanks, vent, and under tail-coverts), contrasting again with black undertail. Underwing dark brown with green gloss (but gloss virtually invisible, serving only to give blacker appearance). When seen head or half-side on, usually appears uniformly dark; from above, pale mantle amazingly distinct; from below, side, or rear, white U similarly so. Gloss becomes purple with wear and may disappear on very worn birds, which show pale nape-patch and even paler mantle contrasting with apparently black wings. JUVENILE. Differs from adult in having pale forehead reduced to lateral brown-white patches, less green gloss, and dark brown tips to wider tail-coverts. From behind, white U narrower.

At a distance, no swift unmistakable, but *H. caudacutus* most distinctive of all in silhouette, while U-shaped surround to rear belly instantly diagnostic. As much if not more commanding of airspace than other swifts, with unusually slow but powerful flowing flight (with few wing-beats, and level glides on slightly bowed wings) on breeding grounds, fast level flight (with alternate bursts of fast wing-beats and glides) when on migration, and extremely rapid, high, and circling flight (with most frequent wing-beats) in display, etc.

Wanders at all seasons but shows no particular habitat preference outside breeding range. Rather more solitary than other swifts but forms large flocks on migration and in winter. Vagrants unlikely to call but scream less piercing than that of Swift *Apus apus*.

Habitat. Breeds in continental upper middle and lower middle latitudes, in dry warm climates. Apart from roosting and nesting in holes or crevices of rocks or trees, is entirely aerial, ranging from lower to upper airspace and covering great distances at exceptional speed, thus rendering inapplicable conventional habitat descriptions. Changing air conditions are more relevant, both directly in influencing avoidance of rain, thunderstorms, and thick cloud by causing descent to near ground or water, and indirectly by effects upon concentrations of insects, which are often hawked in fine weather high over river valleys and upland pastures, in India normally between *c.* 1250 and 4000 m (Ali and Ripley 1970). In USSR, in wooded areas of hills and plains with open stretches, but, while feeding, visits equally mountain meadows almost to snowline and marshy lowlands; also forages over gorges, river valleys, and water bodies (Dementiev and Gladkov 1951a). In Australian winter quarters, prefers undulating and mountainous country, more often near coast than in interior (Slater 1970).

Distribution. Breeds in central Siberia (Tomsk region), then east through southern Siberia to Sakhalin, Kuril Islands, Japan, northern China, and Taiwan; also outer Himalayas, Assam, and probably northern Burma and western China. Resident and migratory, northern populations wintering in Australia.

Accidental. Britain, Ireland, Norway, Finland, Malta.

Movements. Largely resident (southern race *nudipes*) or wholly migratory (northern race, nominate *caudacutus*). Latter, to which rest of account refers, breeds central Siberia to Japan and winters in Australia. Not recorded from Indian subcontinent; hence central Siberian birds must move east to south-east in autumn (reverse in spring) within USSR, passing to north of major mountain chains of central Asia. Southward passage then occurs through China, eastern Thailand, and Indochina (King and Dickinson 1975; Cheng 1976); western fringe of broad-front movement touches Malay peninsula in autumn and spring (Medway and Wells 1976), and (further east) migrants cross New Guinea, where some may winter in south since mid-December skins exist (Vaurie 1965). In Australia, present mainly eastern and south-eastern regions (including Tasmania), and rare to absent in arid centre and Western Australia. Irregular visitor to New Zealand; vagrant to Fiji and subantarctic Macquarie Island.

Begins to leave USSR late August and departure virtually complete by 20 September; passes through south-east Asia September–November, and present Australia October to March or April. Northward migration April–May, with first arrivals south-east Siberia in early May but elsewhere in USSR not until later in May. See Dementiev and Gladkov (1951a), Neufeldt and Ivanov (1960), Condon (1975), and Medway and Wells (1976). European records May–July, though November in Malta.

Voice. See Field Characters.

Plumages (nominate *caudacutus*). ADULT. Forehead and lores white, forming band *c.* 4 mm wide over upper mandible, at side reaching black crescent in front of eye; occasionally some fine brown mottling on centre of forehead. Crown, nape, and hind-neck down to just below eye, ear-coverts, and sides of neck dark olive-brown, feather-tips of crown, nape, and sides of neck strongly glossed greenish-blue, these parts appearing fully glossy when plumage fresh, but much brown of feather-bases visible when worn; basal halves of feathers of nape contrastingly white, but these usually not revealed by wear. Mantle and back pale yellow-brown when fresh, silvery-white when worn (sometimes faintly streaked or with limited grey-brown of feather-bases visible), forming conspicuous pale saddle, contrasting with dark olive-brown scapulars and central rump and with sides of rump and upper tail-coverts which are strongly glossed dark greenish-blue. Chin and throat white, contrasting markedly with dark olive-brown ear-coverts, sides of neck, and chest. Upper flanks and chest to vent dark olive-brown, sometimes with slight bronze gloss on central belly; lower flanks and under tail-coverts contrastingly white. Tail black with strong green gloss. Flight-feathers and upper primary coverts black, glossed with dark blue, green-blue, or green when fresh, dull black when worn; basal and middle portions of inner webs of primaries pale brown-grey. Inner webs of inner 3 tertials white (except for broad dark streak along shaft of longest), outer webs glossy green; tips of inner webs either fully white or with narrow dusky border, tip of innermost nearly always white. Lesser, median, and greater upper wing-coverts strongly glossed dark blue on outer webs and tips, green on inner webs; deep black without gloss when worn. Under wing-coverts and axillaries brownish-black. JUVENILE. Mainly like adult, but central forehead dark grey-brown, lores paler grey-brown, hardly contrasting with forecrown; crown, hindneck, sides of neck, sides of rump, and upper tail-coverts black, only faintly glossed bronze-green; often some white on bases of feathers of rump, usually hidden; mantle and back pale olive-brown, less silvery and less contrasting than in adult; upper chin and lower cheeks smoke-grey, white throat-patch slightly less contrasting; lower flanks streaked and spotted black, less obviously white; white under tail-coverts with black rims to tips, widest (c. 2–4 mm) on longest. Tail, flight-feathers, and all upper wing-coverts black, only faintly glossed green. White inner webs of longer 2 tertials more broadly bordered dull black than in adult, that of shortest inner tertial with narrow black rim. Some variation in colour of forehead, lores, and throat, in width of dark rims on tertials, and in width and amount of dark tips on under tail-coverts; in some, chin, throat, and lores fully white; occasionally, no black tips on under tail-coverts except for longest. FIRST ADULT. In early winter, plumage worn juvenile; white feather-bases of nape and rump and grey ones on rest of body sometimes show, body appearing mottled; no gloss on wing and tail. From about April to August, head and body as in adult, but faint black rims sometimes on shorter tertials and on under tail-coverts, forehead often still grey-brown (occasionally, these features also in full adult), and flight-feathers, greater coverts, and primary coverts still juvenile, heavily abraded (adult flight-feathers fresh or rather fresh then). After completion of 1st post-breeding moult (about July of 2nd calendar year to February of 3rd), indistinguishable from adult.

Bare parts. ADULT AND JUVENILE. Iris dark brown. Bill black. Leg and foot flesh-colour with purple tinge or dark bluish-flesh; claws black. (Hartert 1912–21; Witherby *et al.* 1938; BMNH, ZMM.)

Moults. ADULT POST-BREEDING. Complete; primaries descendant. Only a few specimens in moult examined. Some start moult in or near breeding area (perhaps non- and failed breeders only). A few had some wing-coverts new in mid-July; some showed slight moult of body, tail, or tertials in August; others still in complete slightly worn adult plumage in late August and thus would probably not have started moult until reaching winter quarters. Adult from eastern Siberia and Japan with body and tail worn and in active primary moult had moult scores 2 (5 August), 10 (13 August), 13 (5 September), and 23 (2 October). Only 1 dated Australian wintering bird examined (January): moult score 39, body and tail mainly new. Moult probably suspended during migration, but no migrating nominate *caudacutus* examined; single migrant *nudipes* (26 November, Java) had suspended with inner 8 primaries and all body and tail new. POST-JUVENILE. Complete. Starts in winter quarters from December–February of 1st winter; some upper wing-coverts, tail-coverts, tail-feathers, and tertials first. By March–May, body, tail, and wing-coverts new, flight-feathers heavily worn. In those examined summering near breeding area, flight-feathers replaced from July–August onwards; at same time, wing-coverts, tail-coverts, or tail moulted, as in adult post-breeding. Specimens had primary moult scores 2 (14 August), 4 (July), 9 (5 August), 23 (30 August), and 28 (September), perhaps thus moulting slightly earlier than adult. Moult probably suspended during autumn migration; completed in winter quarters. (BMNH, RMNH, ZFMK, ZMM.)

Measurements. Adult wing of (1) central Siberia (Altay to Lake Baykal), (2) Amur area, Sakhalin, Anadyr, and Ussuriland, (3) north-east China and Japan; other measurements combined for these areas; May–September, skins (BMNH, RMNH, ZFMK, ZMA, ZMM). Tail is t1, including terminal spine; bill is exposed culmen; to forehead, 6–8 mm more.

		♂			♀		
WING	(1)	208	(4·35; 7)	200–213	206	(4·01; 10)	198–211
	(2)	209	(3·62; 23)	202–216	210	(4·55; 9)	204–217
	(3)	211	(3·19; 8)	207–216	203	(5·45; 4)	196–209
	JUV	209	(3·96; 11)	203–215	206	(6·21; 11)	195–215
TAIL	AD	49·7	(2·15; 11)	47–53	48·8	(1·32; 7)	47–51
	JUV	50·3	(2·76; 9)	47–55	50·6	(1·77; 10)	48–53
BILL		8·6	(0·69; 7)	8·0–9·7	8·5	(0·50; 4)	8·0–9·0
TARSUS		17·1	(0·48; 7)	16·4–17·5	17·1	(0·50; 4)	16·6–17·6

Sex differences not significant. Juvenile not significantly different from adult. Wing of unsexed adult, Britain: 200 (BMNH). In large sample of unsexed Japanese trade-skins, wing of adult 201 (6·77; 206) 184–218, juvenile 202 (6·79; 3814) 184–218; in both, range mainly 190–210 (Kleinschmidt 1970).

Weights. Nominate *caudacutus*. Siberia: ♂ 122 (11) 109–140, ♀ 114 (5) 101–125 (Collins and Brooke 1976). Central Siberia, ♂ 126 (June); Ussuriland, ♀ 173 (June); Sakhalin, ♂ 138 (May), ♀ 102 (July) (ZMM). Manchuria: ♂ 110 (July) (Piechocki 1958).

Structure. Wing long and narrow (though relatively shorter and broader than in Alpine Swift *Apus melba*), tip sharply pointed. 10 primaries: p10 longest, p9 1–4 shorter, p8 10–15, p7 18–32, p6 34–52, p5 51–70, p1 126–137, s1 138–150; outer primaries slightly curved, all gradually tapering towards narrowly rounded tip. Secondaries very short, especially middle ones; s1 longer and shaped like inner primaries. Tail very short, almost square; 10 stiff feathers with rounded tips, but with stiff spine-like shaft projecting 3–5 mm beyond webs. Under tail-coverts long, reaching to within 0–3 mm of webbed tail-tips in adult, 5–8 in juvenile. Tarsus and toes short, but relatively longer than in *Apus* swifts; tarsus bare; foot with 1st toe directed backwards, other 3 forwards (unlike *Apus* swifts). Length of middle toe 16·9 (10) 16–18 including claw; outer toe c. 89% of middle, inner c. 90%, hind c. 64%. Claws heavy, sharp; relatively longer and more strongly curved than in *Apus*.

Geographical variation. Mainly in colour of upperparts. Himalayan race *nudipes* differs from nominate *caudacutus* in uniform black forehead, crown, nape, and sides of head, glossed with dark blue or dark greenish-blue (no white on forehead or lores). Mantle and back slightly browner, less silvery-grey; more extensive dark blue gloss on side of rump and upper tail-coverts; breast, belly, and under wing-coverts slightly darker olive-brown, central breast and belly slightly glossed green; wing on average slightly shorter, 206 (6) 194–214 mm.

Forms superspecies with White-vented Needletail *H. cochinchinensis* from Nepal, Assam, southern China, Taiwan, and south-east Asia. This species treated either as monotypic (Mees 1973, 1977a) or split into 3 poorly defined races (Collins and Brooke 1976); in overlap area with *H. caudacutus*, *H. cochinchinensis* inhabits lower elevations. Plumage closely similar to *nudipes*, but throat smoke-grey or grey-brown, not contrastingly white, inner webs of tertials often dull grey instead of white, and size smaller—wing 185·5 (5·16; 23) 176–194 (Biswas 1951; Mees 1977a; RMNH). CSR

Subfamily APODINAE typical swifts

About 25 species in 5–6 genera worldwide, mainly in tropics. Represented in west Palearctic by 7 breeding and 1 accidental species of *Apus* and by 1 accidental species of *Cypsiurus*.

For general features, moults, etc., see Apodidae. Hind toe directed forwards.

Apus alexandri Cape Verde Swift

PLATE 60
[between pages 662 and 663]

Du. Kaap Verden Gierzwaluw   Fr. Martinet d'Alexander   Ge. Alexandersegler
Ru. Стриж Зеленого мыса   Sp. Vencejo de Cabo Verde   Sw. Kap Verde-seglare

Apus alexandri Hartert, 1901

Monotypic

Field characters. 13 cm; wing-span 34–35 cm. About 20–25% smaller than Swift *A. apus*; *c.* 10% smaller and shorter winged than Plain Swift *A. unicolor*. Small swift with shallow tail-fork. Most uniformly coloured of west Palearctic swifts, with plumage tone more grey-brown than black. Known only from Cape Verde Islands. Sexes similar; no seasonal variation. Juvenile not studied in the field.

ADULT. Head and body grey- or mouse-brown, palest below and relieved only by dull white throat. Wings and tail darker, but apparently not contrasting noticeably with rest of plumage.

Little studied in the field but structure gives bird lighter, rather more fluttering flight action than *A. apus* (vagrant to Cape Verde Islands) and *A. unicolor* (potential straggler there); both these species larger and darker bodied, with marked tail-fork. When breeding, most obvious in dusk gatherings near colonies, spending much of day high over cliffs.

Voice thinner than *A. apus*, with rather reeling quality.

Habitat. Breeds in small group of tropical oceanic offshore islands, ranging from sea-level to crater summit at 2800 m. Prefers foraging over *ribeiras*—deep warm gullies sheltered from wind but with strong updraughts, extending from sea-level to high ground. Also seen in numbers dashing over pines *Pinus*, cypresses *Cupressus*, and eucalyptus *Eucalyptus* in pass at *c.* 1600 m. Flies regularly along the shore and cliffs. Nests in crevices in rock faces, in caves, and in house roofs. Ecologically a counterpart of Plain Swift *Apus unicolor* in Canary Islands. (Bannerman and Bannerman 1968.)

Distribution and population. Breeding confined to Cape Verde Islands, where occurs on most islands. No information on numbers or population trends. (Bannerman and Bannerman 1968; Naurois 1969b.)

Movements. Recorded Cape Verde Islands in most months and assumed resident there (Bourne 1966; Bannerman and Bannerman 1968), though White (1960b) failed to find any in his early July (dry-season) visit and suspected absence then. Certainly makes inter-island movements at least, e.g. visiting arid São Vicente where not known to breed (Bannerman and Bannerman 1968; Lambert 1980), but not known away from Cape Verde Islands.

Food. Insects taken in flight (Bannerman and Bannerman 1968; Naurois 1969b). No detailed information. BDSS

Social pattern and behaviour. Virtually nothing known. Performs communal Screaming-displays round nesting colonies in evening, as in Swift *A. apus* (Bourne 1955; see also Bannerman and Bannerman 1968). EKD

Voice. No information outside breeding season.
CALLS OF ADULTS. Screaming-call resembles that of Swift *A. apus* but is shorter, more abrupt, harsher, and weaker, with distinctive metallic reeling quality (Bourne 1955). Earlier reports of a trilling call (Bourne 1955, 1957) subsequently repudiated (Bourne 1966).
CALLS OF YOUNG. No information. EKD

Breeding. SEASON. Cape Verde Islands: birds visit nest-sites August–September; egg reported found March (Bannerman and Bannerman 1968). SITE. Natural crevice or cave in cliff from sea-level to 1600 m; also in house roofs (Bannerman and Bannerman 1968). Nest: not described. EGGS. Long sub-elliptical; white, finely freckled red-brown, forming faint zone at broad end. Clutch: 2 nests, each with 2 eggs, reported by Naurois (1969b). No further information.

Plumages. ADULT. Lores, forehead, crown, and nape dark grey-brown, feather-tips of lores and forehead fringed pale grey-brown, appearing slightly paler than crown. When plumage fresh, crown-feathers narrowly fringed pale, appearing faintly scaled. Black spot in front of eye. Sides of head down to gape and ear-coverts medium grey-brown. Mantle, scapulars, and back dark grey-brown, similar to crown and nape or slightly darker, almost brownish-black, faintly scaled with pale grey-brown when plumage fresh, some grey-brown of feather-bases visible when plumage worn. Rump and upper tail-coverts slightly paler, medium grey-brown, feather-tips narrowly and faintly fringed pale grey. Chin and central throat pale grey or greyish-white, with faint dusky shaft-streaks, gradually darker towards medium grey-brown sides of throat and chest; pale throat-patch hence small and poorly defined. Whole underparts down from lower throat medium grey-brown, feather-tips indistinctly fringed pale grey-brown on lower throat and chest, fringes slightly paler and more distinct from belly to under tail-coverts. Feather-bases on underparts paler grey-brown than feather-tips, often partly visible, especially on chest when plumage worn, giving faintly barred appearance. Tail dark olive-brown with slight green gloss, hardly darker than upper tail-coverts. Outer webs and tips of most primaries black, innermost primaries and inner webs of other primaries dark olive-brown. Secondaries dark olive-brown, tips almost black with faint white fringes. Greater and medium upper wing-coverts dark grey-brown (slightly paler than mantle and back, slightly darker than rump), lesser coverts and primary coverts gradually darker and almost black near leading edge of wing. Under wing-coverts and axillaries medium grey-brown, greater and median faintly tipped white, those along leading edge of wing with narrow white fringes, leading edge appearing mottled grey-brown and white. NESTLING. No information. JUVENILE. Closely similar to adult and indistinguishable when plumage worn. In fresh plumage, inner primaries and secondaries faintly and narrowly edged white at tip (tips of secondaries similar in some fresh adults); t5 slightly less pointed (differences about the same as that between adult and juvenile Swift *A. apus*), but distinction reliable only when adult available for comparison.

Bare parts. ADULT AND FEMALE. Iris deep brown. Bill black. Leg and foot dark flesh colour. (Bannerman and Bannerman 1968.) NESTLING. No information.

Moults. ADULT POST-BREEDING. As in Plain Swift *A. unicolor*, primary moult sometimes suspended: only 1 examined showed complete moult, 3 had 8 inner primaries new (score 40) and 2 outer old; 1 had all old except p9. No active primary moult in 2 birds from September and 2 from March; 2 from November had primary moult score of 1 (p1 just shed) and 24 (p5 growing, p6 shed). Body fresh in single September bird; worn in another bird from September, 1 from November (score 1), and 2 from March; another November bird (score 24) had body and tail moulting, hindneck, upper tail-coverts, and chin to chest still old. POST-JUVENILE. Single bird examined, November, had wing still juvenile, worn; body and tail in fresh 1st adult except forehead, chest, and a few tail-feathers.

Measurements. Cape Verde Islands, September–April; skins (BMNH). Tail is to tip of t5; fork is tip of t1 to tip of t5; bill is exposed culmen.

WING	♂ 140	(0·67; 5)	139–141	♀ 140	(0·82; 4)	139–141
TAIL	57·5	(1·58; 4)	56–59	58·2	(1·44; 4)	57–60
FORK	17·1	(0·85; 4)	16–18	17·2	(1·26; 4)	16–19
BILL	5·5	(0·73; 4)	4·5–6·0	5·0	(0·09; 4)	4·9–5·2
TARSUS	8·6	(0·55; 3)	8·2–9·2	8·6	(0·43; 4)	8·1–9·0

Sex differences not significant.

Weights. No information.

Structure. 10 primaries: p9 longest, p10 3–5 shorter, p8 6–7, p7 21–23, p6 38–40, p5 52–54, p1 88–92. Tail shallowly forked, fork 30% of tail-length, similar to Alpine Swift *A. melba* (36% in *A. apus* and Pallid Swift *A. pallidus*, 41% in *A. unicolor* and Pacific Swift *A. pacificus*, 45% in White-rumped Swift *A. caffer*); outer middle toe with claw *c.* 7·7 mm. Remainder of structure as in *A. apus*.

Geographical variation. None.

Often thought to be conspecific with *A. unicolor* (e.g. Hartert 1901, Lack 1956b, Vaurie 1965), differing from it in smaller size, slightly shallower tail-fork, paler colour of body and wing-coverts, and less marked pale scaling on underparts. However, considered to be only distantly related to other *Apus* by Brooke (1971a), mainly on grounds of aberrant egg colour (freckled red-brown instead of more or less uniform white as in all other *Apus*). Treated separately here pending further study of biology.

Recognition. Closely similar to *A. unicolor*, but upperparts dark grey-brown instead of mainly black (black, if any, restricted to mantle; in *A. unicolor*, extends from nape to back), throat grey-white rather than grey-brown (similar to tinge of chest) as in *A. unicolor* (a few *A. unicolor* have throat as pale and extensive as *A. alexandri*, however), and chest to vent medium rather than dark grey-brown or almost black, with pale scaling less distinct. *A. pallidus* rather similar also, but much larger with forehead and crown paler grey-brown, white throat patch larger and more distinct, and breast and belly slightly darker and with more distinct white feather-fringes. CSR

Apus unicolor Plain Swift

PLATE 60
[between pages 662 and 663]

DU. Madeira Gierzwaluw   FR. Martinet unicolore   GE. Einfarbsegler
RU. Тусклый стриж   SP. Vencejo unicolor   SW. Enfärgad Seglare

Cypselus unicolor Jardine, 1830

Monotypic

Field characters. 14–15 cm; wing-span 38–39 cm. About 20% smaller than Swift *A. apus* and Pallid Swift *A. pallidus*, but with similar structure to *A. apus* except for relatively deeper fork to tail (*c.* 3 cm). Rather small,

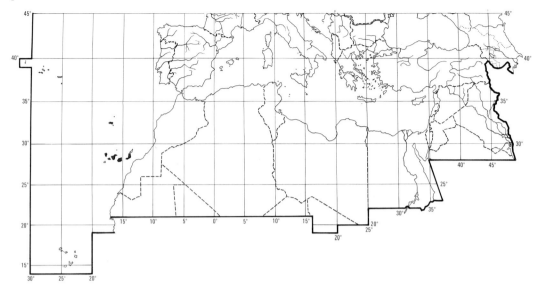

dark grey-brown swift, with grey chin and throat difficult to see. Sexes similar; no seasonal variation. Juvenile perhaps separable at close range.

ADULT. Dark grey-brown above and below, unrelieved except for small grey or dirty white patch on chin and throat and (except when worn) faint pale feather-margins on under-body. Primary-bases slightly paler than rest of wing when fully extended. At distance, appears uniformly black in any light. JUVENILE. Not studied in the field, but apparently paler, more ashy, than adult.

Except in late autumn and early winter, *A. unicolor* restricted to Madeira and Canary Islands, which it shares with migrant *A. apus* and breeding and migrant *A. pallidus*. Overlap of 3 similar *Apus* swifts creates severe test in field identification, though this passable given lengthy comparison in good light and at close range. Distinction from *A. pallidus* not difficult, since *A. unicolor* distinctly smaller, slighter in head and body, and has relatively narrower wings and shorter and more forked tail, as well as being almost black (*A. pallidus* dark brown, with grey forehead and large white throat). Separation from *A. apus* much more difficult but *A. unicolor* differs as follows: (1) smaller in all measurements except depth of tail-fork; (2) less sickle-shaped wings; (3) indistinct or obscure throat-patch never paler than pale grey (in *A. apus*, except when worn, distinct and off-white); (4) plumage less black (except when worn); (5) flight action more erratic, with rapid twists and turns, and even faster, more 'twinkling' wing-beats. Also, *A. unicolor* usually confined to higher altitudes, using cliffs and caves for nesting (see Habitat).

Highly vocal: scream indistinguishable from *A. apus*; also a rapid trill and a liquid disyllable.

Habitat. Breeds in 2 small groups of subtropical mountainous oceanic islands, in which it forms counterpart to Cape Verde Swift *A. alexandri*. In Canary Islands breeds at higher altitude than congeners, in caves and crevices in walls of higher *barrancos* or ravines, foraging above pinewoods and skimming over *Retama* bushes in full bloom which attract insects. Usually at above 300 m, and at midday up to *c*. 600 m; observed flying round volcano crater disregarding hot sulphurous vapours, and over rim of giant caldera at nearly 2500 m; a high flier, mounting to become a mere speck in the sky or even beyond visual range. In Madeira, also frequents higher peaks and gullies or *ribeiras* in summer, nesting in deep clefts of rocks, but commonly nesting and roosting in some sea-cliffs, and even breeding on solitary rock in sea. Observed sitting on tree, and taking refuge in hole in wall. (Bannerman 1963b; Bannerman and Bannerman 1965.)

Distribution. MADEIRA. Occurs on all islands, but breeding not proved on Porto Santo group and Desertas (Bannerman and Bannerman 1965). CANARY ISLANDS. Found on all major islands, but not proved breeding on Lanzarote and Hierro (KWE).

Accidental. Morocco.

Population. MADEIRA. Common (Bannerman and Bannerman 1965). CANARY ISLANDS. Abundant Tenerife and Gran Canaria, very common La Palma, fairly common Gomera, and local Fuerteventura (KWE).

Movements. Largely migratory. Present all year on Madeira, numbers considerably reduced in winter though still quite common (Buxton 1960; Bannerman and Bannerman 1965). Further south, on Canary Islands, largely migratory though winter status not clear: odd birds present at any time, but, at least occasionally, flocks occur in late December (Meade-Waldo 1889a,b; Wallace 1964); migratory birds return January–March (main arrivals in February) and depart September to mid-October (Bannerman 1963b).

Winter quarters of emigrants unknown, but presumably

in Africa. Recent records of up to 60 *Apus* in Agadir–Oued Massa area (south-west Morocco), December–January, attributed to *A. unicolor* (Thévenot *et al.* 1980); Smith (1965) had previously seen small numbers of dark *Apus* in that region in winter (late November to late January). Former assertion (e.g. Murphy 1924, Etchécopar and Hüe 1967) that *A. unicolor* occurs on Cape Verde Islands in winter is wrong (Bourne 1966; Bannerman and Bannerman 1968).

Food. Insects taken in flight (Bannerman 1963*b*). No detailed information. BDSS

Social pattern and behaviour. Little known. Except where indicated, the following based on information from Canary Islands supplied by K W Emmerson.
  1. Gregarious at all times, occasionally occurring singly. Report of being more gregarious than Pallid Swift *A. pallidus* (Bannerman 1963*b*) misleading, and arises from relative scarcity of latter. On Canary Islands, feeding flocks mostly 5–30 birds, up to 100–300 or more where food concentrated. Forms large flocks increasingly from March onwards, sometimes associating with *A. pallidus* (Koenig 1890; Cullen *et. al.* 1952), also with migrating Swifts *A. apus* and House Martins *Delichon urbica*. BONDS. No information. BREEDING DISPERSION. Colonial. Size of colony depends on availability of nest-sites: 3–5 pairs to well over 30. Nests may be closely grouped (as little as 1 m apart) or widely spaced. ROOSTING. Resident birds roost at breeding sites throughout the year; for first suggestion that birds may use old nests outside breeding season, see Morphy (1963). Outside breeding season, birds enter roost singly, or in small groups of up to 8 birds. Begin roosting *c.* 1–2 hrs before dusk, and not uncommonly enter roost in almost complete darkness. As in other Apodidae, exit from roost relatively late: outside breeding season, leave colony 10 min to 1 hr before sunrise, usually in groups of 2–12.
  2. FLOCK BEHAVIOUR. Forms flocks for feeding, courtship, pre-migratory behaviour, and migration. Communal Screaming-display, apparently similar to *A. apus*, performed throughout the year in vicinity of breeding colonies. Small groups of birds begin to ascend, initially in manner not different from usual flight. Soon, however, begin to spiral on outstretched wings, and intermingle, calling frequently (see 1 in Voice), which attracts others. Become increasingly vociferous and tightly bunched, some birds 'lunging' at others. Groups of 2 may break from flock to perform rapid chase, accompanied by Screaming-calls. Others follow in tight formation, constantly changing position. Further twos or threes break away to start chases. Screaming-display most common during late afternoon or evening, less often early morning. Most frequent December–February, when associated with pair-formation and courtship, and late August–October, prior to migration (as in *A. apus*). In variant of display, ascending birds much more active from outset, darting about rather than gliding; break-away chases also more animated, birds zigzagging, twisting, and turning in close pursuit before suddenly separating. In another variant, birds ascend loosely intermingled, as above, and those higher in spiral suddenly rush down, screaming, upon lower birds, almost colliding with them, whereupon latter start rapid pursuit of former. For further aerial display, associated with flock formation, see Heterosexual Behaviour (below). ANTAGONISTIC BEHAVIOUR. Aggressive aerial swoops sometimes directed at Alpine Swift *A. melba* and *D. urbica*. Once, several birds swooped repeatedly at *A. melba* and drove it from vicinity of their colony. HETEROSEXUAL BEHAVIOUR. (1) General. So-called 'slip-streaming' occurs December–February; aerial display perhaps related to pair-formation and/or mating. Begins with group of birds gliding gently down in echelon formation. Leading bird of apparent pair gives half, flicking wing-beat prior to beating wings in rapid, quivering 'bat-like' fashion. Partner pursues closely, the two almost touching, before they swoop up, gliding and then separating. Leader and pursuer may change position. (2) Mating. Occurs in the air, ♀ soliciting ♂ with Wings-high display, exactly as in *A. apus*; not known whether mating also occurs at nest. RELATIONS WITHIN FAMILY GROUP. ANTI-PREDATOR RESPONSES OF YOUNG. No information. PARENTAL ANTI-PREDATOR STRATEGIES. Birds on eggs or small young do not flush until observer puts hands near nest (Schmitz 1905; K W Emmerson). Then usually flutter around, flying at great speed when threatened, but making less noise and fuss than *A. apus*. ♂ gripped with claws when handled, drawing blood (Schmitz 1905). EKD

Voice. Used freely throughout the year in Canary Islands, but especially December–February and late August to October (K W Emmerson).
  CALLS OF ADULTS. (1) Screaming-call. Usually a somewhat hoarse 'sriii' or 'srriä'; in pursuit flights, a shorter and rapidly repeated 'srii' or 'sri', indistinguishable from Swift *A. apus* but not as rasping or deep as Pallid Swift *A. pallidus* (Bergmann and Helb 1982, which see for sonagrams). (2) Other calls. (a) A 'zkick zkick zkick' given by shot bird when picked up (Koenig 1890), possibly a distress-call. (b) Rapid trill, different from call 1 (Cullen *et al.* 1952; K W Emmerson). (c) Liquid disyllabic call (K W Emmerson).
  CALLS OF YOUNG. Food-call evidently strong, as audible to human observer outside nest-site (Cullen *et al.* 1952). EKD

Breeding. SEASON. Canary Islands and Madeira: March–August (Bannerman 1963*b*; Bannerman and Bannerman 1965). SITE. In cave, crack, or fissure of cliff or gully, coastal or inland; also under roof tiles, in cleft in building, or under bridge. Nest: irregularly-shaped shallow saucer, *c.* 1–2 cm deep, 10–12 cm across; main material downy seedheads of Compositae (also sometimes dry grass, leaves, moss, algae, ferns, and shreds of man-made material), compacted with saliva and stuck to rock; lined with a few small feathers (Schmitz 1905). Building: no information on role of sexes. EGGS. See Plate 98. Long sub-elliptical, smooth but not glossy; dull-white and translucent, often with longitudinal or spiral milk-white lines at pointed end. 22×15 mm (22–24×14–15), $n = 14$ (Schmitz 1905). Clutch: 2. Frequently 2 broods. INCUBATION. Probably by both sexes, as both have brood-patch (Schmitz 1905). No further information.

Plumages. ADULT. Forehead, lores, and forecrown dark grey-brown, extending in narrow line to just above eye; feathers indistinctly fringed pale grey-brown when plumage fresh. Hindcrown, nape, mantle, scapulars, and back dark grey-brown, feathers with poorly defined black tips and narrow grey-brown fringes. Rump dark grey-brown with narrow pale grey-brown feather-fringes; upper tail-coverts dark olive-brown with faint green gloss, narrowly fringed grey-brown. Sides of head dark grey-brown; black spot in front of eye. Cheeks, sides of neck, and lower throat dark grey-brown, gradually paler grey-brown towards central throat and chin, feather-tips with poorly defined pale grey-brown (on lower throat and cheeks) or pale grey (on upper throat and chin) fringes, giving faintly barred appearance; chin and upper throat sometimes uniform pale grey (except for some faint dusky shaft-streaks), forming indistinct and poorly defined small pale throat-patch. Underparts below throat (including under tail-coverts) dark grey-brown, feathers with poorly defined black tips; tips narrowly but distinctly fringed white, giving scaled appearance. Tail dark olive-brown with slight green gloss. Primaries blackish olive-brown with slight green gloss, outermost feathers almost black; basal borders of inner webs paler grey-brown. Secondaries dark olive-brown with indistinct white edges along tips. Greater upper wing-coverts dark grey-brown, coverts gradually darker towards leading edge of wing; primary coverts and shorter lesser coverts brownish-black with slight green gloss. Under wing-coverts and axillaries medium grey-brown, tips of coverts narrowly fringed white when fresh; small coverts along edge of wing at carpal joint pale grey-brown. In worn plumage, pale fringes of feather-tips almost completely lost, except sometimes for traces on rump, belly, flanks, and under wing-coverts; nape to back uniform glossy brownish-black, chin and throat pale grey-brown with some darker brown mottling, chest to vent faintly barred dark grey-brown and brownish-black. NESTLING. No information. JUVENILE. Closely similar to adult, hardly distinguishable. In fresh plumage, pale fringes to feather-tips slightly more marked, especially below; outer web of t5, and tips of secondaries, inner primaries, tertials, greater upper primary coverts, and (especially) median and greater upper wing-coverts with narrow white or pale grey edges more prominent and more uniform in appearance than in fresh adult; pale fringes and edges strongly subject to wear and appearance soon similar to adult. Distal border of inner web of t5 straight or slightly convex, gradually tapering towards tip; in adult, border slightly concave, forming faint emargination (useful only when tail fresh and direct comparison possible).

Bare parts. ADULT AND JUVENILE. Iris deep brown. Bill black. Leg and foot pinkish-brown to brownish-black. (BMNH.) NESTLING. No information.

Moults. ADULT POST-BREEDING. Of many examined February–September, none in active flight-feather moult; thus, probably moults in winter, as in Swift *A. apus*. Winter moulting season apparently shorter than in *A. apus*, as in that species $c.$ 70% of birds complete moult of all primaries between August–September and March–April, only $c.$ 30% retaining old p10 in one or both wings (exceptionally p9–p10); in contrast to this, 26 specimens of *A. unicolor* examined had all primaries new, 3 suspended with 4 outer primaries old, 9 with 3, 7 with 2, and 6 with 1. In following winter, these old outer feathers probably replaced descendantly, while inner moult at same time (serially descendant moult), but one or both of these moulting series sometimes not completed, and thus various old primaries may be retained into following summer. Most common patterns shown in those examined were: (1) after probable suspension with 8 inner primaries new in summer, 2 outer feathers and 6–7 inner replaced in following winter, retaining old p8 or p7–p8; (2) similarly, only p10 old in summer, this feather replaced in winter together with 8 inner primaries, showing old p9 in following summer; (3) after suspension with 8 inner primaries new in winter, p9 and 6 (or 7) inner primaries replaced next winter, retaining old p7–p8 (or p8 only) and p10 in following summer. Differences in wear sometimes difficult to detect, and perhaps more birds may show suspended primary moult than the $c.$ 50% suggested here. Body and tail new February–April and timing of moult probably as in *A. apus*. POST-JUVENILE. Moult not started in September–October specimens. Body and tail probably moulted in winter, as in *A. apus*, but no information on timing of primary moult. Flight-feathers perhaps moulted in 2nd autumn and winter, with worn juvenile primaries thus retained during summer of 2nd calendar year: some summer birds showed heavily worn primaries combined with rather fresh body and tail, similar to 1-year-old *A. apus*, but age not certain in this case as no distinctly juvenile characters retained and no details of gonads available.

Measurements. Gomera, La Palma, Gran Canaria, and Tenerife, February–September; skins (BMNH, RMNH, ZFMK, ZMA). Tail is to tip of t5, fork is tip of t1 to tip of t5; bill is exposed culmen.

WING AD	♂ 153	(2·22; 26)	150–158	♀ 155	(2·54; 12)	152–159
TAIL AD	70·3	(1·88; 22)	67–74	69·6	(1·63; 9)	68–72
FORK AD	29·4	(1·93; 22)	27–34	28·7	(0·90; 9)	28–30
BILL	5·4	(0·20; 7)	5·1–5·7	5·4	(0·30; 5)	5·1–5·8
TARSUS	9·9	(0·34; 7)	9·4–10·4	9·6	(0·26; 5)	9·2–9·9

Sex differences not significant. Juvenile wing, tail, and fork $c.$ 2–3 less than adult. Only single bird from Fuerteventura examined: ♂, wing 160 (ZFMK).

Madeira, April–September; skins (RMNH, ZFMK, ZMA).

WING AD	♂ 154	(1·33; 6)	152–156	♀ 155	(1·40; 5)	153–156
TAIL AD	69·7	(— ; 3)	68–72	69·8	(— ; 2)	66–73

Weights. No information.

Structure. 10 primaries: p9 longest, p10 1–6 shorter, p8 7–10, p7 22–28, p6 38–45, p5 54–62, p1 99–106. Tail relatively more deeply forked than in *A. apus*, fork 41% of tail length. Outer middle toe with claw 8·1 (7) 7·3–9·1; outermost toe $c.$ 83% of outer middle, inner middle $c.$ 93%, innermost $c.$ 64%. Remainder of structure as in *A. apus*.

Geographical variation. None.

Sometimes considered to be conspecific with Cape Verde Swift *A. alexandri* (see that species). Sometimes included in *A. apus*, mainly on zoogeographical grounds (Lack 1956b; Brooke 1970, 1971a), but (without denying probable close relationship) treated separately here because of distinctly smaller size, relatively more deeply forked tail, and paler body without contrasting white throat.

Recognition. Rather similar to Swift *A. apus*, which differs in having: slightly deeper and more glossy black upperparts in fresh plumage (closely similar when plumage worn); distinctly blacker underparts, without grey-brown of feather-bases showing, even in worn plumage; small but distinct throat-patch markedly paler, off-white; pale scaling on underparts faint or absent, even in fresh plumage. Underparts of *A. unicolor* closely similar to those of Pallid Swift *A. pallidus*, showing same width and extent of scaling when plumage fresh and similar amount of grey-brown feather-bases, but throat-patch of *A. pallidus* usually much paler and larger (darkest *A. pallidus* similar to palest *A. unicolor*, however), and upperparts distinctly paler, in particular median and greater upper wing-coverts, tertials, secondaries, and innermost primaries.
CSR

Apus apus Swift

PLATES 60, 61, and 63
[between pages 662 and 663, and facing page 686]

Du. Gierzwaluw   Fr. Martinet noir   Ge. Mauersegler
Ru. Чёрный стриж   Sp. Vencejo común   Sw. Tornseglare

Hirundo apus Linnaeus, 1758. Synonym: *Cypselus apus*.

Polytypic. Nominate *apus* (Linnaeus, 1758), Europe and North Africa, east and south to Turkey, Kazakhstan, Russian Altay, and Lake Baykal; *pekinensis* (Swinhoe, 1870), Iran (and perhaps eastern Iraq) east to Mongolia, northern China, and western Himalayas, migrating through Middle East.

Field characters. 16–17 cm; wing-span 42–48 cm. Measurements overlap with Pallid Swift *A. pallidus* but slightly slimmer, rounder headed, narrower winged, and more fork-tailed; up to 20% larger than Plain Swift *A. unicolor* but proportionately less fork-tailed. Medium-sized, black-brown swift, with almost white chin and throat, paler (browner) uppersurfaces to flight-feathers. Flight powerful and rapid but often noticeably light with wings appearing to 'twinkle'. Groups scream on breeding grounds. Sexes similar; no seasonal variation. Juvenile separable at close range. 2 races occur in west Palearctic, hardly distinguishable in the field.

(1) West Palearctic and Siberian race, nominate *apus*. ADULT. Ground-colour black-brown, with (in fresh plumage) green gloss most obvious on mantle and relieved only by faintly grey forehead, white or brown-white chin and throat (only obvious head-on), faint dull white or grey tips or markings on rump, underbody, and under wing-coverts, and paler (browner) inner webs on flight-feathers. Except in full sunlight and at close range, only pale throat and uppersurface of flight-feathers constitute field characters; at a distance, often not visible and, against pale background, bird appears wholly black. JUVENILE. Closely resembles adult but plumage generally darker, with throat-patch cleaner and larger, pale tips to body feathers and coverts cleaner white (particularly on underwing), and all flight- and tail-feathers paler brown and narrowly margined almost white. At close range, white fringes to upper wing-coverts form most certain character. At middle distance, juvenile still shows distinctly rougher-looking plumage. (2) South-central Asian and Chinese race, *pekinensis*. Browner in all plumages, especially on secondaries. Forehead paler and throat-patch larger. Easily confused with *A. pallidus*.

Commonest and most widespread swift of west Palearctic and the only one breeding in northern Europe. Overlaps with *A. pallidus* in Mediterranean region, and most confusion arises with that species; see *A. pallidus*. Occasional albinistic birds (when white-bellied) suggest larger Alpine Swift *A. melba* and (when white-rumped) White-rumped Swift *A. caffer* and Pacific Swift *A. pacificus*. *A. pacificus* of similar size to *A. apus* and, apart from broad white rump, shows slight but distinct plumage pattern of grey-brown foreparts and wings contrasting with apparently black tail; distinct white feather-margins on underbody and (particularly) under wing-coverts often obvious. *A. caffer* much smaller, with narrow white rump-band, silvery flight-feathers (pale-tipped on secondaries), and relatively longer and more deeply forked tail. Flight of *A. apus* dramatic, showing complete mastery of open airspace and marked ability in gliding, wheeling, diving, accelerating, or stalling (when feeding), and climbing; wing-beats rapid and made usually with wings in distinctive backward curve, and at times seeming to be beaten alternately or unevenly (most obvious head-on or from behind). Scything action and long shearing track typical of swifts, and quite unlike swallows and martins (Hirundinidae) or any other aerial-feeding birds. Flight in close surroundings much less masterful; bird entering nest-site may require several circuits before accurate pitch. Shuns landing on flat surfaces but fit bird can take off from one. Gait restricted to shuffling in nest cavity. Clings vertically, with tiny feet brought up to chest.

Commonest call in full flight a long, harsh scream, 'sree', but breeding birds also chirrup when visiting nests.

Habitat. In west Palearctic, breeds over much broader geographic band than congeners, from lower and upper middle latitudes to above Arctic Circle, in desert, steppe, Mediterranean, temperate, and boreal climatic zones, both continental and oceanic. This expanded range, unique among Apodidae, is maintained only through a very short breeding season and through adaptations enabling it to be interrupted during spells of bad weather by sudden long-distance displacements, during which young survive for a period unfed. Over most of range, former nest-sites on crags, sea-cliffs, and in caves, and in east of region in hollow trees and nest-holes made by other species, have been largely replaced by use of buildings, spread of which has facilitated expansion of the species. Apart from momentary contact with water in flight, all activity is aerial, shifting from lower airspace in windy, chilly, or wet weather to upper airspace in warmer and more settled summer conditions. Also uses airspace for roosting in flight at night. Nature of underlying terrain is of secondary relevance except in so far as it dictates local abundance or scarcity of insect foods and their readiness to rise to preferred feeding heights above *c.* 50 m. Unlike Alpine Swift *A. melba*, generally prefers to remain close to nesting locality during breeding season unless displaced by adverse weather conditions. In USSR, habitat differs markedly from that in west and south of range, embracing lowland and mountain

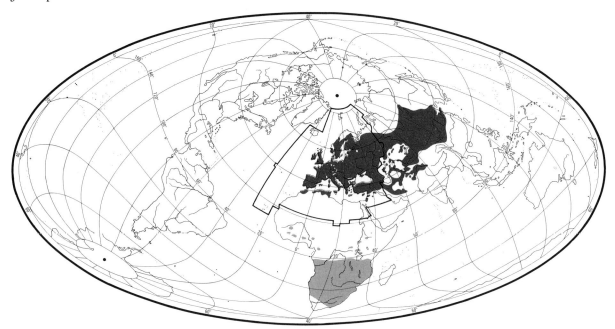

forests, old high or sparse forests with hollow trees, clearings, and burnt areas, bare mountains, and open plains with vertical outcrops, ridges, and steep river banks, as well as mountain crags, sea-cliffs, and buildings. In many areas requires trees suitable for nesting and open spaces, but is indifferent to size of woods and their species composition, or to altitude, ascending to or above 2000 m in Caucasus (Dementiev and Gladkov 1951a). For review of observed altitudes and durations of flight and responses to extremes of weather, see Glutz and Bauer (1980). Recent replacement of suitable tenanted tall buildings by more modern structures has severely reduced availability of man-made breeding sites in some areas, especially in city-centres which, where air pollution levels permit, are often preferred to suburbs and villages. In Amsterdam (Netherlands), an ordinance forbids re-roofing of buildings concerned unless access for *A. apus* retained (Sharrock 1976). Migration usually in lower airspace. In African winter quarters, found hunting in higher airspace than resident Apodidae and does not share aversion to airspace above forests (Moreau 1972).

Distribution. Few recorded changes, although adoption of buildings for nesting must have led to considerable local extensions, especially in 19th century, continuing in USSR in 20th century.

BRITAIN AND IRELAND. Few significant changes in range in Britain; perhaps some spread in western Ireland (Parslow 1967). MALTA. Formerly bred (Sultana and Gauci 1982).

Accidental. Spitsbergen, Iceland, Faeroes, Azores.

Population. No precise details of numbers or trends, though increases in last 100 years reported from many parts of central Europe (Glutz and Bauer 1980).

BRITAIN AND IRELAND. Thought to be over 100000 pairs (Parslow 1967) or *c.* 100000 pairs (Sharrock 1976). Perhaps some increase in Ireland since 1932; little known change in Britain (Parslow 1967). FRANCE. Up to 1 million pairs, perhaps increasing (Yeatman 1976). BELGIUM. About 30000 pairs (Lippens and Wille 1972); *c.* 29000 pairs 1982 (PD). LUXEMBOURG. About 4000 pairs (Lippens and Wille 1972). NETHERLANDS. Estimated 50000–85000 pairs (Teixeira 1979); no known changes (CSR). WEST GERMANY. Perhaps up to 1 million pairs (GR). SWEDEN. Estimated 400000 pairs (Ulfstrand and Högstedt 1976). FINLAND. Estimated 25000 pairs (Merikallio 1958). CZECHOSLOVAKIA. No known changes (KH).

Mortality. England: in Oxford breeding population, annual adult mortality 15–16%, apparently highest in first 2 years after ringing; 21–23% from remainder of birds ringed under national ringing scheme (Perrins 1971). Sweden: annual adult mortality said to be 19% (Magnusson and Svärdson 1948), but chances of recapture not properly allowed for (Perrins 1971). Switzerland: annual adult mortality (from sample of 18) *c.* 17% (Weitnauer 1947). Czechoslovakia: mortality in 1st year of life 29%, in 4th year 12% (Beklová 1976). Oldest ringed bird 21 years (Rydzewski 1978).

Movements. All populations migratory. Winters in small numbers or irregularly in northern India (Ali and Ripley 1970) and Arabia (Bundy and Warr 1980; Walker 1981a), but otherwise in Africa.

Nominate *apus* (breeding west Palearctic and Siberia) winters mainly Zaïre and Tanzania south to Zimbabwe and Moçambique, and *pekinensis* (breeding Iran to north-

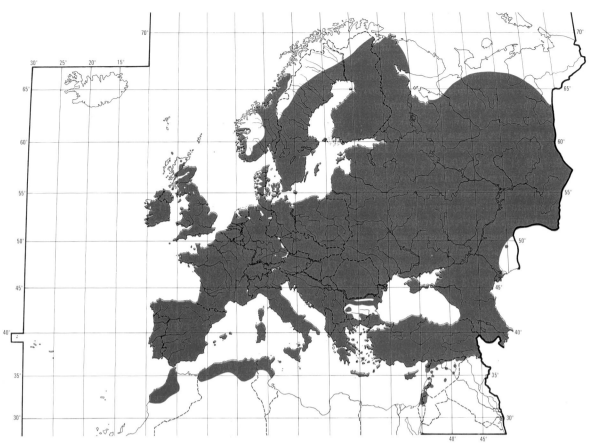

ern China) winters especially in arid south-west Africa—southern Angola, Namibia, and Botswana (Brooke 1975; Clancey 1981b). Segregation far from complete, however, since *pekinensis* also found sparingly in winter north to Zaïre, Uganda, and Sudan (Moreau 1972; Britton 1980). Birds wintering south-east Africa attributed in part to an intergrade population (presumed Asiatic), sometimes upheld as race *marwitzi* (see Geographical Variation). Majority of birds (all populations) winter south of equator, but not exclusively as formerly supposed (e.g. Lack 1956a). Now scattered winter records across West Africa from Gambia and Liberia to Nigeria and Cameroun, and northwards to Mali (Moreau 1972; Lamarche 1980; Gore 1981; Elgood 1982); also in comparable latitudes east to Ethiopia. Winter records from Morocco probably refer to Plain Swift *A. unicolor* (see that species), which may reach West Africa also (though none collected).

Foreign recoveries from British ringing (to 1982) summarized in Table A. Note that those in Iberia are mainly in autumn, while those in France mainly in spring (and 2 Tunisian recoveries both in spring); this suggests a more eastern trend to return movements through Europe. Absence of recoveries from West Africa suggests that region largely bypassed by diagonal movements across Sahara; very large passages through Mali inundation zone in autumn (Curry and Sayer 1979) and Nigeria in spring (Elgood 1982) have no known equivalents further west. Recoveries in Zaïre in most months (only August–September excepted) must reflect summering there by some pre-breeders; this is an established, regular feature in Africa (Moreau 1972), though many in fact return to breeding range. The south to south-west autumn movement within Europe shown by British ringing applies also to other European ringing data (Glutz and Bauer 1980):

Table A Foreign recoveries of British-ringed Swifts *Apus apus* to December 1982.

Recovery area	Apr–Jun	Jul–Sep	Oct–Dec	Jan–Mar	Total
Scandinavia	1	1	—	—	2
North-west Europe*	1	6	—	—	7
France	11	4	—	—	15
Spain	2	12	2	—	16
Portugal	—	1	—	—	1
Morocco	7	9	—	—	16
Tunisia	2	—	—	—	2
Congo Republic	—	1	1	—	2
Zaïre	5	1	3	4	13
Tanzania	—	—	1	—	1
South-east Africa†	—	—	6	6	12

*Includes Denmark (4), West Germany (1), Netherlands (1), eastern North Sea (1).
†Includes Zambia (1), Malawi (9), Zimbabwe (1), Moçambique (1).

e.g. Swiss analysis showed movement across southern France and south-east Spain towards Morocco (Weitnauer 1975). Exceptional recoveries of north European birds in Yugoslavia, Bulgaria, and USSR (Stavropol). No ringing data for eastern population, but *pekinensis* found on passage in Kenya and Uganda (Britton 1980) and once (October) as far west as Chad (Salvan 1968b).

Autumn migration begins late July or early August, soon after young have fledged; juveniles infrequently seen in feeding flocks in north-west Europe. 1 English juvenile which fledged 31 July was found 4 days later in Madrid, Spain, minimum 1275 km distant (Spencer 1959). Only minority (of any age) left in Europe in September, though stragglers remain until November and (very exceptionally) later. Main arrivals in northern Afrotropics from 2nd week August, reaching southern wintering areas from September. Large-scale weather-related movements occur within Africa during non-breeding season—many reports of large numbers appearing during disturbed conditions in areas where few seen earlier (Moreau 1972; Elgood 1982). Over much of eastern Africa, big southward movements can continue into December, and northwards from mid-January, indicating major redeployment during winter months (Tree 1961; Moreau 1972; Britton 1980). Spring passage of presumed *pekinensis* noted Persian Gulf from mid-February (Bundy and Warr 1980; Walker 1981a,b), when also arrivals begin in Near East breeding areas (Safriel 1968). Spring passage through northern Afrotropics most conspicuous in April, though vanguard reaches southern Europe in second half of March, with northward movement continuing to early June.

Comparable to African situation (see above), big weather movements can occur in Europe during breeding season, involving thousands or even tens of thousands of birds. Movements around depressions (to avoid rainfall) reported in many parts of Europe, but especially in Baltic, Scandinavia, and North Sea region. Such movement pronounced in south-east sector of approaching depression; birds considered to fly into wind to reach warm sector with least delay. See Koskimies (1947, 1950), Svärdson (1951), Lack (1955, 1956a), and Glutz and Bauer (1980). Birds ringed on Öland (Sweden) during weather movements recovered in presumed breeding areas up to 600 km away, and a complete movement around a depression will be much longer, perhaps 1000–2000 km (Svärdson 1951). For Swiss radar study, see Bruderer and Weitnauer (1972). Summer recoveries well north of ringing area, such as those in Table A, will include instances of weather movement. Such movements probably involve mainly non-breeders (Koskimies 1950), but certainly include some breeding adults (Svärdson 1951).

Of 148 ringed as nestlings in Switzerland and recovered there in subsequent years, 19 were found breeding 5–122 km from natal area; only 4 ringed as adults changed breeding site, by 4–36 km (Weitnauer 1975). Bird ringed as nestling at Gelderland (Netherlands) found breeding Buckinghamshire (England) 455 km away. Site-fidelity weak in young birds, strong in established breeders (see Social Pattern and Behaviour).

Food. Almost exclusively flying insects and airborne spiders of small to moderate size. Prey taken in flight. Avoids stinging insects; these appear not to be recognized by colour, as insects with warning coloration or wasp (Vespidae) mimics, e.g. hoverflies (Syrphidae) may be taken (Lack and Owen 1955). Record of 8 adults feeding around beehive in Zaïre, where stingless drones selectively taken (Lacey 1910). Recorded taking warningly coloured wood tiger moths *Arctia plantaginis* in large numbers in Cumbria, England (Hale Carpenter 1938). However, large numbers of ladybirds (Coccinellidae) avoided during plague season (Bromhall 1980). Considerable vertical range for feeding flights. Often as low as *c.* 2 m or less (e.g. Harding 1979). Insect numbers decrease up to *c.* 100 m and amount of predation decreases proportionally (Freeman 1945). In turbulent conditions, insects may be swept to high altitudes. Birds rise to *c.* 1000 m but no evidence of high-altitude feeding (Gustafson *et al.* 1977). Readily attracted by swarms of insects including Lepidoptera (Hale Carpenter 1938), Hymenoptera (Chapin 1939), Coleoptera (Meylan 1974), and Diptera (Heyder 1927; Schacht 1979; Davenport 1981). Not seen to pass through large swarms and appears to take individuals on periphery (Davenport 1981). Gatherings of several thousand birds may occur, feeding on clouds of midges (Chironomidae) (Lack and Owen 1955). Tends to feed over water in rough weather when aquatic insects more likely to emerge than terrestrial ones. During breeding season, usually feeds close to colony, so limited to areas of reasonably abundant prey (Lack and Owen 1955). Depending on weather, foraging range may extend to 7–8 km (Koskimies 1961), exceptionally 40–50 km (Gladwin and Nau 1964; Jacoby *et al.* 1970). Insects selected by size. Largest that can be comfortably ingested are preferred; size in good weather normally 5–8 mm (Britain) and up to 12 mm (Gibraltar). In poorer weather (Britain), prey of 2–5 mm mainly caught (Lack and Owen 1955; Finlayson 1979). Insects accumulate in back of throat and bind together with salivary product to produce ball. Average weight 2·53 g, Switzerland (see Weitnauer 1947); 1·16 g, England (Lack and Owen 1955). In Czechoslovakia, average in 1972 0·73 g, $n = 196$; in 1973, 0·64 g, $n = 161$ (Pellantová 1981). In England, food-balls contain 300–1000 insects (Lack and Owen 1955; Bromhall 1980); at Gibraltar, most over 200 with one containing 427 (Finlayson 1979); in Switzerland, average 450 (Poncy 1928). Rarely lands on walls to pick off insects (Meiklejohn 1928; Gilbert 1944). Glow-worm *Lampyris noctiluca* picked off woodwork by incubating bird (Moreau and Moreau 1939).

Takes bugs (Hemiptera: Aphididae, Cicadellidae, Cercopidae, Hydrometridae, Cicadidae, Delphacidae), beetles (Coleoptera: Hydrophilidae, Staphylinidae,

Nitidulidae, Mycetophagidae, Coccinellidae, Chrysomelidae, Curculionidae, Lampyridae, Elateridae, Scarabaeidae), flies (Diptera: Bibionidae, Mycetophilidae, Stratiomyidae, Tabanidae, Empididae, Dolichopodidae, Lonchopteridae, Phoridae, Syrphidae, Sepsidae, Chamaemyiidae, Ephydridae, Sphaeroceridae, Drosophilidae, Agromyzidae, Chloropodidae, Tipulidae, Cordilurinae, Muscidae), Hymenoptera (Formicidae, drone Apidae, Ichneumonidae, Chalcididae), moths and butterflies (e.g. Arctiidae, Tortricidae), thrips (Thysanoptera), lacewings (Neuroptera), termites (Isoptera), grasshoppers (Acrididae), earwigs (Dermaptera), mayflies (Ephemeroptera), dragonflies (Odonata), and spiders (e.g. Linyphiidae) (Lacey 1910; Hale Carpenter 1938; Chapin 1939; Moreau and Moreau 1939; Weitnauer 1947; Moltoni 1950; Lack and Owen 1955; Schacht 1979; Bromhall 1980; Glutz and Bauer 1980).

In Europe, over 500 prey species recorded, mainly aphids, Hymenoptera, Coleoptera, and Diptera (Glutz and Bauer 1980). In May, Hungary, *Tortrix viridana* caterpillars caught as they descended on silk threads from tree canopy (Turček 1956-8). In Gibraltar, 78·2% of items taken by breeding birds Hymenoptera (mainly 5–8 mm), 11·5% Diptera (mainly under 4 mm), 6·5% Hemiptera (mainly under 4 mm), and 3·5% Coleoptera (mainly under 4 mm) (Finlayson 1979). Diet more specialized than found in England—more Hymenoptera taken and no arachnids (compare with Owen and Le Gros 1954, Lack and Owen 1955). This probably reflects prey availability. Estimated that with average 300 insects per food-ball, each nest receives 3000 per day; assuming 3000 pairs in Gibraltar plus food taken by adult breeders, then 18 000 000 insects taken in the area per day (Finlayson 1979). In Oxford (England), food-ball in July at time of aphid (Aphididae) plague contained 726 aphids from a total of 898 items: also 48 leafhoppers (Cicadellidae), 22 spittle bugs (Cercopidae), 23 craneflies (Tipulidae), 18 dung flies (Cordilurinae), 13 ladybirds (Coccinellidae), 10 ants (Formicidae) and other flies, beetles, bugs, Hymenoptera, and thrips (Thysanoptera). Samples from 2 nests close together, July, suggested different feeding areas. In one, bugs (Delphacidae) made up 279 of 543 items, in the other, hoverflies (Syrphidae) made up most of sample. In August, abundance of beetles noted but aphids and plant bugs main prey in June (Bromhall 1980). In extensive surveys in and around Oxford, wide variety of Hemiptera, Coleoptera, and Diptera taken and 7 species of spiders (Linyphiidae). Results probably reflect changes in aerial fauna over season, e.g. in July at both locations 4 aphids (*Acyrthosiphon pisum, Metopolophium dirhodum, Aphis fabae, Macrosiphum avenae*) represented in all meals. Fly *Platypalpus* absent from meals on 2 July at Oxford, yet by 12 July 1131 recorded in meals (Lack and Owen 1955). 3 food-balls, Oltingen (Switzerland), July–August, contained 188 flies (Diptera), c. 160 mayflies, 88 cicadas, 74 aphids, 31 beetles, 20 flying ants, 18 ichneumons, 3 spiders, and 1 bug (see Weitnauer 1947). For review of prey in Italy, see Moltoni (1950). In Thüringen (East Germany), 11 403 items found to include 6080 bugs, 2963 beetles, 1125 flies, 207 butterflies and moths, 130 lacewings, 17 spiders, and 3 dragonflies (Glutz and Bauer 1980, which see also for diet in Basses-Pyrénées, France).

Wintering flocks in Zaïre, August–February, took flying termites (Isoptera), flies, and beetles (Herroelen 1953). Of 17 stomachs, Zaïre (no dates), 12 contained winged ants, 5 contained termites (Chapin 1939).

Most feeding studies already cited represent analyses of food-balls presented to young. In Geneva (Switzerland), stomachs of 2 young in early July contained 10 beetles, 1 bug, several Hymenoptera, and flies. In late July stomach of one young contained 80 ants *Formica fusca*, 24 beetles, and 2 Hymenoptera. In excreta around a nest, 72 beetles (mainly *Phyllotreta*) found with ants (Formicidae) (Poncy 1928). Young fed 13·5 times (6–26) over 24 hrs, Czechoslovakia (Pellantová 1981). At Oltingen, broods of 3 fed 7–33 times per day, depending on weather (Weitnauer 1947). In Gibraltar, on average, young fed hourly for 14 hrs per day (Finlayson 1979). In England, for brood of 2, average 14 feeds in 10 hrs in good weather, 7 in poor weather. Up to 42 feeds per brood (size unspecified) per day. In poor weather, feeding rate peaks 16.00–17.00 hrs; in fine weather, when feeding more frequent, 11.00–12.00 hrs (Lack and Owen 1955). BDSS

Social pattern and behaviour. Major studies in Oxford (England) by Lack and Lack (1952) and Lack (1956a), and in Oltingen (Switzerland) by Weitnauer (1947, 1980). Following account includes notes supplied by T Fagerström.

1. Gregarious in breeding season, and mostly throughout the year. During breeding season, travels and feeds in small groups (Lack and Lack 1952), but large numbers, probably mostly non-breeders, may travel together or assemble where food locally abundant, e.g. in June, Hanningfield reservoir (England), at least 80000 together (Smith 1980). Summer movements of birds on English east coast may produce continuous stream of 1000 birds per hr or more (e.g. Harrison 1947a). At end of breeding season, large flocks comprising birds from several colonies form prior to migration (Lack 1956a; see also Flock Behaviour, below). One such flock Syria, contained thousands (Hutson 1946). Typically migrates in small groups: of 255 flocks passing Belen (south-east Turkey) in autumn, average size 17·7 (Sutherland and Brooks 1981). At beginning of August, Zaïre, migrating flocks 30–50 birds, later in autumn more than several hundred (Herroelen 1953). At the beginning of breeding season, arriving flocks comprise both 1-year-old and older birds (Weitnauer 1947). At end of breeding season, Oxford, non-breeders leave at same time as breeding birds (Lack and Lack 1952). BONDS. Monogamous mating system, and pair-bond maintained from year to year (Weitnauer 1947; Lack 1956a). Mate fidelity marked between years (Weitnauer 1947): at Oltingen, ringed pairs persisted for up to 12 years of study (Weitnauer 1975). For pair-bonds maintained for 6 years, see Weitnauer (1947), Zimmermann (1956), and Hladik (1958). In colony of c. 12 pairs, Örskär (Sweden), only 23 birds in 10 years changed mate when partner of previous year alive and breeding (T Fagerström). Pair-members probably do not associate outside breeding season; tend not to leave and

return to colony on same day (Lack and Lack 1952; Lack 1956a). Age of first breeding 2 years at some colonies at least; few colonies studied, and average age of first breeding at some (see below) may be older. At Oltingen, 1-year-olds formed pairs and built nests, but did not breed until following year (Weitnauer 1947, 1949b, 1960; see also Magnusson and Svärdson 1948). A 1-year-old ♂ paired temporarily with widowed older ♀ that had young, and while ♂ fed young once when small, thereafter attended nest erratically, did not feed them again, and did not renew pair-bond the following year (Weitnauer 1947, 1980). On average, 25·4% of nest-sites, Oltingen, occupied by non-breeding 1-year-olds (5–27 pairs, 41 years: see Glutz and Bauer 1980). In study at Oxford, only 8 out of 621 ringed young returned to breed; from this small sample, 1-year-olds thought not to secure sites, and attempted breeding at 2–3 years unsuccessful; successful breeding not likely until 4 years (Perrins 1970). Average age of first breeding at given colony perhaps varies with availability of nest-sites there and in neighbourhood (Perrins 1970; C M Perrins). Both sexes share nest-duties equally (Weitnauer 1947; Lack and Lack 1951b); for suggestion that ♀ does more incubation, see Cutcliffe (1951). For departure of one parent before fledging, see Relations within Family Group, below. Young independent at fledging (Bertram 1908; Weitnauer 1947; Lack and Lack 1951b). BREEDING DISPERSION. Sometimes solitary, but mostly gregarious and colonial. Colonies usually not more than 30–40 pairs, rarely up to 100 pairs (Glutz and Bauer 1980). In woodland, northern Europe, mainly solitary or in small groups, distributed over several trees, though exceptionally up to 80 pairs, Finland (Koskimies 1956; see also Glutz and Bauer 1980). At colony in building, Örskär, minimum distance between nests 1 m (T Fagerström). Nest-territory confined to immediate surroundings of nest; serves for part of courtship, some copulation, and raising of young (Lack 1956a). ♂ usually selects site (Weitnauer 1947), which is then defended by pair (Lack and Lack 1952). 2-year-olds return to site they established the previous year (see above). Breeding birds show strong fidelity to colony and to nest-site (Weitnauer 1947; Lack and Lack 1952; Dachy 1954). Of 389 site-owners ringed and recovered, Switzerland, only 4 changed colony, these moving 4–36 km. 213 breeding birds were controlled at their nests for 1–13 years (Weitnauer 1975). For fidelity of ♀ to same nest for 11 years, see Hladik (1958). At Örskär, where c. 12 pairs studied for 10 years, 22 cases of change of nest where both members of pair still alive. In 12 cases, pair moved together to another nest; in rest—due to change of mate—only one bird moved (T Fagerström). Within colony, change of nest-site almost invariably to immediately adjacent site (Dachy 1954). Vacant nests readily occupied by new pairs (T Fagerström). Fidelity of young birds to colony relatively weak (Perrins 1970): at Örskär, none of 152 young returned to breed in colony, and only breeding record c. 70 km away (T Fagerström). ROOSTING. Widely studied. Behaviour complex, varying with age, breeding status, and time of year. Both terrestrial and aerial roosting occur. On arrival at colony, roosting in nests by owners initially erratic, especially if mate yet to arrive (Lack and Lack 1952), but after a few days each pair roost regularly in nest-site throughout breeding season (Weitnauer 1947), often until a few days after young fledged (Lack 1956a). Pair lie side by side, usually one in nest, other just outside, facing away from entrance. In cold weather, bodies hunched, plumage ruffled (Fig A), and 1 bird often on top of other (Lack and Lack 1952; see also Bacmeister 1919, Tucker 1936). Occasionally, in spring or late summer, 3rd bird, thought to be a passing migrant, may enter nest-box and roost with resident pair (Lack and Lack 1952; Lack 1958), once lying between them (Lack 1956a). Once, in cold weather, 6 huddled in nest (Nash 1924); at Örskär, during migration, up to 4 roosted

A

together, always in unoccupied box (T Fagerström). Immatures and other non-breeders with nest-sites may occupy them throughout the season (Cutcliffe 1951) but also roost aerially (see below). Migrating birds adopt variety of other roost-sites. May roost singly while hanging from pendulous branch of tree (e.g. Cox 1953, Church 1956, Schmidt 1959, Holmgren 1978) or even hang upside down from wire (Matthes 1966). May roost on window-ledge (Lack 1956a); one bird roosted thus for c. 1 week, in each of 2 successive years (S Cramp). May also cling vertically to wall (Lowe 1962), sometimes in dense clusters in cold weather (Lack 1956a; Törnlund 1979); masses, like 'bunches of grapes' (one of 200 birds), on wall at Lake Constance, West Germany (Kuhk 1948); 6 clusters, each of 100–200 birds, Basle, Switzerland (Burckhardt 1948; Weitnauer 1949a). Each bird tries to avoid outside of cluster by crawling inwards (Lack 1956a). In breeding season at least, aerial roosting occurs regularly, especially among 1-year-olds. Early evidence came from ground observations (e.g. Poncy 1928) and chance encounters at night with aircraft; substantiated by airborne and radar studies (Weitnauer 1952, 1954, 1960; Bruderer and Weitnauer 1972; see also Graaf 1947, 1950, Brander 1950). Evening ascent typically begins with birds circling and giving Screaming-calls (see 1 in Voice), then bunching more tightly and climbing with rapid, almost quivering wing-beats. Breeding birds initially join ascent but most break away to return to nests (Poncy 1928; Lack 1956a; Rothgänger and Rothgänger 1973; Semago 1974). Birds ascend to often warm layer, typically at 1000–2000 m (Weitnauer 1954, 1955, 1960). For large flock, facing into wind at 3000 m, see Guérin (1923). 27 min after last of breeding birds went to roost in holes, flocks of 3–50 found at up to 1550 m (Weitnauer 1952). On clear nights, flocks fly in all directions, periodically joining and scattering; in moderate air current, birds drift with it (Weitnauer 1960). Relatively slow wing-beats alternate with a few seconds of gliding (Bruderer and Weitnauer 1972, which see for flight speeds). Flocks assemble over colonies before descending at dawn (Weitnauer 1960); come down at fairly gentle angle, rapidly and silently (Semago 1974). Once, descended from at least 2000 m, 20 min before first birds emerged from nests (Weitnauer 1952). On breeding grounds, aerial roosting occurs from onset of fine weather in mid to late May, though may roost in prospective nest-sites in bad weather. Late in chick-rearing period, breeding birds also participate—initially ♂♂, but also ♀♀ just prior to fledging of offspring. Earlier in season, breeding birds rarely roost aerially unless accidentally benighted (Weitnauer 1960; see also Graaf 1950, Koskimies 1950). Fledged

PLATE 56 (facing).
Caprimulgus nubicus tamaricis Nubian Nightjar (p. 617): **1–2** ad ♂, **3** ad ♀.
Caprimulgus europaeus europaeus Nightjar (p. 620): **4–5** ad ♂, **6** ad ♀.
Caprimulgus ruficollis ruficollis Red-necked Nightjar (p. 636): **7–8** ad ♂, **9** ad ♀.
Caprimulgus aegyptius saharae Egyptian Nightjar (p. 641): **10–11** ad ♂.
Chordeiles minor minor Common Nighthawk (p. 646): **12–13** ad ♂, **14** ad ♀. (CETK)

PLATE 57. *Caprimulgus nubicus tamaricis* Nubian Nightjar (p. 617): **1** ad ♂, **2** ad ♀, **3** juv. *Caprimulgus aegyptius* Egyptian Nightjar (p. 641). Nominate *aegyptius*: **4** ad ♂, **5** juv. *C. a. saharae*: **6** ad ♂. *Chordeiles minor minor* Common Nighthawk (p. 646): **7** ad ♂, **8** juv ♂ (CETK)

PLATE 58. *Caprimulgus europaeus* Nightjar (p. 620). Nominate *europaeus*: **1** ad ♂, **2** ad ♀, **3** juv. *C. e. meridionalis*: **4** ad ♂. *Caprimulgus ruficollis* Red-necked Nightjar (p. 636). Nominate *ruficollis*: **5** ad ♂, **6** ad ♀, **7** juv. *C. r. desertorum*: **8** ad ♂. (CETK)

PLATE 59. *Hirundapus caudacutus caudacutus* Needle-tailed Swift (p. 649): **1** ad breeding (plumage slightly worn), **2** juv.   *Apus melba* Alpine Swift (p. 678). Nominate *melba*: **3** ad, **4** juv, **5** juv before fledging. *A. m. tuneti*: **6** ad. (DIMW)

PLATE 60. *Apus alexandri* Cape Verde Swift (p. 652): **1–3** ad, **4** juv, **5** juv before fledging.   *Apus unicolor* Plain Swift (p. 653): **6–8** ad, **9** juv, **10** juv before fledging.   *Apus apus apus* Swift (p. 657): **11–12** ad.   *Apus pallidus* Pallid Swift (p. 670). **13** ad. (DIMW)

PLATE 61. *Apus apus* Swift (p. 657). Nominate *apus*: **1** ad, **2** juv, **3** juv before fledging. *A. a. pekinensis*: **4** ad. *Apus pallidus* Pallid Swift (p. 670): *A. p. brehmorum*: **5** ad, **6** juv, **7** juv before fledging. *A. p. illyricus*: **8** ad. (DIMW)

PLATE 62. *Apus pacificus* Pacific Swift (p. 676): **1** ad. *Apus caffer* White-rumped Swift (p. 687): **2** ad, **3** juv, **4** juv before fledging. *Apus affinis galilejensis* Little Swift (p. 692): **5** ad, **6** juv, **7** juv before fledging. *Cypsiurus parvus* Palm Swift (p. 698): **8** ad, **9** juv. (DIMW)

young also thought to roost aerially (Graaf 1950). Aerial roosting also likely in winter quarters (Weitnauer 1955): e.g. in Namibia, late December, after drinking in evening, c. 80 birds giving Screaming-calls ascended spirally in groups of 3–4 to c. 1000 m (Sauer and Sauer 1960). Breeding birds are relatively late risers, and roost up to c. 1 hr after sunset. Exit from roost later at lower latitudes. At 50°N, West Germany, emerged c. 10–30 min before sunrise (Scheer 1949); at 60–63°N, Finland, 1 hr before (Haartman 1949). See also Koskimies (1950) for variations in rising and roosting times with latitude, season, and weather. At arctic latitudes, usually roost for c. 3 hrs at night (Haartman 1949, 1950). For time and energy budgets, USSR, see Dol'nik and Kinzhevskaya (1980). Leaves roost earlier, and enters later, on fine than on dull days (Lack 1956a), but may enter relatively late in cold weather (Chislett 1947). Shortly before migrating, August, stayed in nests up to 1 hr after sunrise (Chislett 1947). At Oxford, those with young roosted up to 20 min later than those without (Lack 1956a). May enter and leave nest 2–3 times before finally staying (Cutcliffe 1955; Lack 1956a). Never leaves or enters nest after dark (Weitnauer 1947), though frequently shuffles around in nest at night, giving Nest-call (see 2 in Voice: Lack and Lack 1952; see also Haartman 1940). Regularly stays in nest by day if wet and cold, especially if not breeding (Lack and Lack 1952). In nest, frequently scratches, shakes, and self-preens to get rid of parasites (Lack and Lack 1952). Preening in flight, and other flight manoeuvres, thought to dislodge parasites (Oehme 1968; Rothgänger and Rothgänger 1973). Bathes and drinks by descending in shallow glide to skim water (Slijper 1948; Lack 1956a), commonly fluttering wings as it touches surface (Rothgänger and Rothgänger 1973). For 'smoke-bathing', see Alder (1951). For roosting of young shortly before fledging, see Relations within Family Group, below. Despite some earlier reports, no proof that young ever return to nest to roost after fledging (Weitnauer 1947; Lack 1956a).

2. Flying birds often perform Wing-clapping: wings meet over back, producing clapping sound, then quickly lowered until they meet below body. May be repeated several times (once, c. 6) in rapid succession (Bundy 1975; see also Rothgänger and Rothgänger 1973). Function not known. Almost immediately after separating from aerial copulation (see below), ♂'s raised wings produced brief, vibrating 'trrr . . . t' sound (Mester 1959). FLOCK BEHAVIOUR. Forms flocks for feeding, social display at colony, pre-migratory behaviour, and migration. In breeding season, colony members frequently perform communal Screaming-display. Usually initiated by 1 bird dashing towards colony giving variant of Screaming-call (Rothgänger and Rothgänger 1973: see 1 in Voice). Others follow, and flock races around colony area, giving Screaming-calls. Birds in nest often respond with same call (Jourdain 1901) and leave nest to join display, especially in evening (Lack 1958). All colony members may thus participate. After several (typically 5–6) circuits, birds disperse and re-ascend to resume feeding, etc. At Örskär, each performance lasted only a few seconds (T Fagerström). After a time, Screaming-display performed again, and may recur several times in fairly quick succession (Jourdain 1901). May occur at any time of day but, at Oxford, most commonly for 1–2 hrs around 07.30 hrs and more especially around 18.00 hrs; restricted to fine weather (Lack 1956a). Occurs throughout the season, but rare at start and very end when few birds present (Lack and Lack 1952). Prior to migration, birds from several colonies often unite in Screaming-display, especially 07.00–09.00 hrs, then circle up to great height before departing (Lack 1956a; Rothgänger and Rothgänger 1973). Function of Screaming-display not well understood, but thought to influence social cohesion (Lack 1956a). For reactions of flock to stooping Hobby *Falco subbuteo*, see Slijper (1948); see also Parental Anti-predator Strategies, below. ANTAGONISTIC BEHAVIOUR. Following account of nest defence from Lack and Lack (1952) and Lack (1956a). If strange bird prospecting sites, residents often return to defend them; sit at nest-entrance and advertise ownership with intensive Screaming-call, in duet if mate also present (see 1 in Voice). If stranger enters occupied site, owner threatens by giving Screaming-call, rising up on its feet, lowering head, and advancing with wings held partly out and raised. In heightened threat, raises wing nearest intruder, tipping body sideways to expose legs and feet (potential weapons), this often inducing intruder to retreat. Intruder may retaliate, especially early in season: flicks wings and walks on raised legs, then both birds prance around each other giving Screaming-calls, periodically pausing with wing nearest rival upraised. After evicting intruder, pair may briefly threaten each other. If intruder holds ground, fight ensues. Rivals rush each other, tipping body sideways to expose feet; grip each other's legs and body with claws, cuff with wings, and peck violently, calling throughout. Bouts of furious struggling alternate with spells of rest and silence. Victor is often the bird which ends up lying on its back underneath opponent, thus gaining extra purchase. Beaten bird commonly utters Piping-call (see 3 in Voice). Fight may end with combatants tumbling out of nest-entrance, sometimes landing on ground still interlocked, motionless and inseparable for up to 25 min (see also Roper 1960, Reuter *et al.* 1980). Fights often prolonged, 16 encounters lasting 20–343 minutes. Mostly develop when only one member of resident pair present; if both present, one bird, presumably ♂, fought much more than the other which sometimes retired to rear of nest. Birds probably fight over sites rather than mates (Lack and Lack 1952; Lack 1956a). Once, 3 birds entered temporarily empty box and fought, suggesting fights occasionally start in the air, and then possibly over mates (Weitnauer 1947). At Oxford, fighting occurred invariably before egg-laying, in May and early June (Lack and Lack 1952). Unpaired birds landing at potential nest-sites may be pursued and attacked by other, presumably competing, birds which simultaneously strike out with claws at alighted bird until it flies away (Farina 1980). For aerial fight after one bird pulled another off wall, see Boley (1961). Usually wins contests for nesting space with House Sparrow *Passer domesticus*, often evicting its nest, eggs, and young (Dachy 1954; Ehrlich 1982); less commonly successful with Starling *Sturnus vulgaris* (e.g. Lack 1956a). Also steals nest-material from other birds, including conspecifics (Weitnauer 1947; Cutcliffe 1951). HETEROSEXUAL BEHAVIOUR. (1) General. Attraction of ♀ mate probably dependent on prior acquisition of nest-site by ♂. If ♂ or ♀ dies outside breeding season, mate usually returns to same nest-site; in case where ♂ disappeared, ♀ retained site and attracted new ♂ to it (Weitnauer 1947). Birds whose mates left well before them at end of breeding season sometimes formed temporary association with bird in neighbouring box; thought to facilitate pair-formation in event of death of mate outside breeding season (Lack 1958). Sometimes, ♂ arriving with known pair at start of breeding season, and persistently following them around, cuckolded established ♂. In one case, ♀ then bred with the new ♂ at new site, presumably established by him (Cutcliffe 1955). At Oltingen, 1-year-olds started pairing from first half of June onwards (Weitnauer 1947, 1949b). Birds may pair at any stage during breeding season. Single birds rarely remain unpaired for more than a few days (Lack and Lack 1952; Lack 1956a, on which subsections 2–4, below, largely based; see also Lack 1958). Other than mating (see below), role of aerial display, if any, in pairing not well understood. For aerial display (Screaming-display, Wings-high display posture: see below) in Natal (South Africa), late January, see Brooke (1975b). (2) Site-

B

prospecting. Performed by ♂♂ to establish sites, perhaps also by ♀♀ to seek sites occupied by bachelor ♂♂. Prospecting bird typically circles colony silently and unhurriedly, passing close to one or more nest-holes or brushing or 'banging' against them with its wing, before flying on. Occasionally alights and looks in; rarely enters, but if so, acts nervously and is often attacked by occupier (see above). Sometimes several prospecting birds fly in single file, each brushing the same site as they pass. Sometimes follows a resident bird to its nest. Prospecting occurs throughout breeding season in fine calm weather, mostly around 08.00 hrs. (3) Pair-bonding behaviour. Courtship at nest most pronounced before egg-laying; also serves, in less intense form, as Meeting-ceremony once breeding under way (see below). Behaviour initially similar between birds establishing new pair-bond, and those renewing old one. In either case, but especially when unpaired, first reaction of site-occupier to arrival of potential mate usually hostile: birds advance towards each other, give Screaming-calls, partially raise wings, and may scuffle briefly, but do not grapple with claws. If unpaired, bird enters occupied site tentatively; may retreat and quickly return, or stay put; makes submissive approach, cocking head back, and inviting occupier to touch throat with its bill and, with greater familiarity, to Allopreen; exposure of pale throat thought to signal appeasement, and Allopreening thus reduces hostility between pair (J Desselberger). Thereafter, mutual Allopreening the most common interaction between birds in nest, and, in established pairs, may occur frequently from first night together. Pair approach each other, sometimes almost bowing, and start to preen each other vigorously on head (Fig B) and especially throat. In intense bout, pair sit close together in a rather humped posture, raise plumage, and may quiver wings. Allopreening also accompanied by Nest-call. Resurgence of Allopreening and Nest-calling occurs if eggs or young lost (Magnusson and Svärdson 1948; Lack 1956a), also after young fledged (Lack 1958). (4) Mating. Occurs both on nest and in the air. Relative frequency of these alternatives not known, though mating in nest probably more common. (a) At nest. Following account from Lack and Lack (1952) and Lack (1956a). ♀ sits in resting posture, then raises tail and twists body round to present rear to ♂ who often gives Screaming-call before mounting. ♂ grips ♀'s back with his claws and her nape with his bill. Usually copulates 3–4 times in succession, sometimes only once. Before and afterwards, pair often perform mutual Allopreening. At Oxford, copulation in nest usually at 06.30–07.30 hrs and 14.30–18.30 hrs, confined to period just before or during laying. For several cases of apparent copulation while ♀ clinging in the open, see Cutcliffe (1951) and King (1972a). Copulation at nest may follow aerial preliminaries (quivering flight: see below) (Weitnauer 1947). (b) In the air. Though no proof of insemination, successful copulation thought likely. Widely reported (e.g. Christoleit 1934, Dupond 1943, Daanje 1944, Mester and Prünte 1957; for earlier references, see Wächtler 1932). Following account after Lack (1956a), based on his, and other, sources. Often preceded by aerial chase, though this need not lead to copulation. One member of pair, presumably ♀, flies 20–30 m (10–60) above ground, in front of the other, and if soliciting ♂, suddenly performs Wings-high display: holds wings stiffly in V above back, sometimes almost vertically. Sometimes, but not usually, gives Pre-copulatory call (see 4 in Voice). Then glides downwards, rapidly pursued by ♂. ♀ then often beats wings in quivering fashion, and shortly both resume normal but slow flight, one closely behind the other. ♂ then alights gently on ♀'s back, sometimes after 1–2 unsuccessful attempts. Pair then descend in shallow glide, ♀ holding wings horizontally, ♂ angling his upwards (Fig C), while both birds spread and twist tails. One (perhaps both) sometimes gives Screaming-call. During glide, both may hold wings still, but sometimes one beats them rapidly in a small arc, and, if they lose too much height, both may do so. Separate after a few seconds (Lack 1956a). For alleged 2nd copulation after brief separation, see Hesse (1916) and Laubmann (1916). Aerial copulation most frequent at c. 07.00 and 18.00 hrs, as at nest (Lack 1956a). (5) Behaviour at nest. Mates meeting at nest greet each other with Screaming-call, perhaps with partly raised wings (Meeting-ceremony). Intensity varies, presumably with strength of pair-bond. May be absent or negligible from outset, and tends to diminish as season progresses (Lack and Lack 1952; Lack 1956a). At nest-relief during incubation, arriving bird often brings nest-material in bill, working it into nest when it sits. Arriving bird (presumably ♂) said occasionally to feed ♀ (Weitnauer 1947), but this not observed by other authors, and possibly mistaken for Billing. If sitting bird reluctant to move off eggs, returning bird may prod mate or gradually insinuate itself under mate's body. Relieved bird usually sits in box for 2–3 min before flying out. In unusually cold or hot weather, sitting bird sometimes departs to feed before mate returns, leaving eggs uncovered; once, for at least 6½ hrs (Lack and Lack 1951b, 1952). In hot weather, clutch sometimes left uncovered all day (Ehrlich 1982). Incubating bird spends much time pecking around nest, sticking down loose bits, and drawing material towards rim. 1-year-olds continue nest-building throughout summer, but less constructively than breeders. Fertile and infertile eggs not uncommonly ejected from nest—at any time (Lack and Lack 1952), mainly in bad weather (Weitnauer 1947; O'Connor 1979). Many viable eggs thus discarded; if one egg of clutch ejected, other usually follows 1–2 days later (O'Connor 1979). RELATIONS WITHIN FAMILY GROUP. Young can grip strongly with claws from day of hatching. In 1st week, brooded almost all day, in 2nd week for about half the day, thereafter rarely by day, though covered at night until much older. If bad weather persists, young become cold, clammy, and semi-torpid, thus reducing energy demands (Weitnauer 1947; Lack and Lack

C

1952; Lyuleeva 1981). One just-hatched chick survived without brooding for 48 hrs (Lack and Lack 1951*b*). At all ages, young almost always fed in nest. At return of parent, dash to nest if outside it, and beg with food-call. Small young wave gaping bill, older young flap wings and chase parent around, trying to grasp its bill. Beg continuously between feeds if parent remains in nest (Lack and Lack 1952; Lack 1956*a*). Up to 2 weeks old, young of brood share food-ball, thereafter each capable of swallowing entire ball (Bromhall 1980). Parent may preen throat of young which fails to beg (Lack 1956*a*). Young may preen each other (Bertram 1908). Shuffle around from 2–3 weeks, jump up and down, and exercise wings. Also perform 'press-ups' on wing-tips, eventually raising themselves for *c*. 10 s at a time (Lack and Lack 1952). May crawl up to 6 m from one nest to another (Gustavsson 1973). Near fledging, young roost side by side with parents (Nash 1924). Shortly before fledging, stay at nest entrance, and usually fledge before 08.00 hrs when parent absent. Other brood members usually leave on later days (Lack and Lack 1952; Lack 1956*a*). In 2 out of 21 cases, both young left on same day; in others, 2nd young left 1–6 days after 1st (Lack and Lack 1951*b*). Despite some earlier reports (e.g. Cutcliffe 1951), young rarely return to nest after fledging (Weitnauer 1947; Lack and Lack 1952; Koskimies 1950). Never seen to be fed by parent outside nest (T Fagerström). At Oltingen, parents usually left colony same day as last young, one member of pair leaving earlier, especially if weather fine and only one young remaining or if only one from outset (Weitnauer 1947). At Oxford, parents usually left colony a few (often 5) days after young, up to 26 days later in wet summer. In 4 cases, one left 1–2 days before last remaining young. One parent left 5 days before single young (Lack and Lack 1952; Lack 1956*a*). Bird leaving prematurely said to be ♂ (Dachy 1954; Cutcliffe 1955). In 4 out of 17 pairs, both parents left colony on same day; in rest, 1 left 1–5 days after mate (Lack and Lack 1952). ANTI-PREDATOR RESPONSES OF YOUNG. Some older young perform defensive-threat display when about to be handled in nest: suddenly lunge forward, flicking wings partly open; performed usually by those young whose parents behaved similarly (Lack and Lack 1952; see below). Perhaps same reaction described as beating with wings if touched or if any strange movement nearby (Bertram 1908), and striking out with wings when handled (Cutcliffe 1951). Older young may crawl away from nest if disturbed (T Fagerström). PARENTAL ANTI-PREDATOR STRATEGIES. (1) Passive measures. Most adults described as tame and loath to leave nest, but some flush readily (Lack and Lack 1952). May crawl away from nest to escape from another exit (T Fagerström). (2) Active measures: against birds. Often follows or mobs raptors such as *Falco subbuteo*, Sparrowhawk *Accipiter nisus* (Daanje 1944; Slijper 1948), once, Buzzard *Buteo buteo* (Lack 1956*a*). After one bird taken by *F. subbuteo*, others may chase and swoop at latter (Weitnauer 1947; A M Rackham). Often spiral upwards with—but staying safely above—*F. subbuteo* or *A. nisus* (Daanje 1944; Slijper 1948), sometimes accompanied by circling Swallows *Hirundo rustica* and House Martins *Delichon urbica* (Daanje 1944; Weitnauer 1947). Circling birds criss-cross one another's path, and remain silent (Daanje 1944). Dense flock of *c*. 30 birds, giving Mobbing-call (see 5 in Voice), seen to fly slowly behind ♀ *A. nisus* (N J Collar). (3) Active measures: against man. Some birds will fiercely attack hand inserted into nest, or perform defensive-threat display, as in young (Lack and Lack 1952; see above).

(Figs A–B from photographs in Lack 1956*a*; Fig C from drawing in Mester and Prünte 1957.) EKD

Voice. Freely used in breeding season, though much less so away from breeding areas. Little known outside breeding season, but see calls 1 and 6a; calls infrequently on migration (D J Brooks). For musical interpretation of calls, see Hoffmann (1917) and Stadler and Schmitt (1917). For Wing-clapping, etc., see Social Pattern and Behaviour. No complete description, especially of call 1, available, and following scheme therefore tentative. For additional sonagrams, see Bergmann and Helb (1982).

CALLS OF ADULTS. (1) Screaming-call. Shrill 'sree' of variable volume and pitch (Fig I); also rendered 'sriih' (Bergmann and Helb 1982) or 'srie' (Weitnauer 1980). Typically heard in communal Screaming-display, when delivered in rapid series, e.g. 'sisisi...' (Bergmann and Helb 1982); 'ssie-ssie-ssie', or more mute 'si-si-si' by bird initiating display (Rothgänger and Rothgänger 1973). According to Stadler and Schmitt (1917), 4 main variants, alone or in combination: (a) rolling, drawn-out 'sirrr', (b) quite short, sibilant 'i', (c) 'sri', (d) longer, and often repeated, 'sIi', also rendered 'swee-ree' (Lack and Lack 1952) or 'swii-srii' (Haartman 1940), representing duet by ♂ and ♀; one half of duet higher pitched than other. In recording by P J Sellar (Fig II), presumed pair call antiphonally, one *c*. 5–5·5 kHz higher than the other, at combined rate of 9 pairs of units in 6·5 s (J Hall-Craggs). Given in aerial pursuit and, according to Stadler and Schmitt (1917), pursuer gives the lower-pitched sound. Also given by pair at nest threatening strange, usually site-prospecting, birds flying past (Lack and Lack 1952). Solo calls also given in variety of contexts: during fights, often at great intensity; by ♂ just prior to mating; sometimes during aerial copulation (not known by which sex); briefly as greeting in Meeting-ceremony (Lack and Lack 1952; Lack 1956*a*); after birds separate in aerial copulation (Laubmann 1916). For quiet Screaming-calls given by 2 birds performing Screaming-display in winter quarters, January, see Brooke (1975*b*), also call 6, below. For further variants, see Bergmann and Helb (1982). (2) Nest-call. Soft, low-pitched 'cluck' (Lack and Lack 1952); 'chick chick chick...' (Tucker 1936), given by breeding birds during mutual Allopreening, and at other times when pair together on nest; occasionally by lone bird in nest. Given only by one sex, probably ♀. Also given by adult non-breeders and 1-year-olds (Lack 1956*a*). (3) Piping-call. Plaintive piping, uttered periodically by defeated bird in latter stages of fight; not heard at other times (Lack and Lack 1952; Lack 1956*a*). (4) Pre-copulatory call. Typically subdued sound, mid-way between calls 1 and 2, usually but not always given by leading bird (presumed ♀) in aerial chase just prior to copulation (Lack and Lack 1952). (5) Mobbing-call. Dry, hirundine-like 'chiroo' or 'schroo', given by birds in tight flock following Sparrowhawk *Accipiter nisus* (N J Collar). (6) Other calls. (a) Single 'chk' sounds given by 2 birds performing Screaming-display in winter quarters (Brooke 1975*b*; see also call 1, above). This perhaps the low-pitched 'zrip' when highly excited (Bergmann and Helb 1982) and the hard, low-pitched sound which sometimes follows 'sirrr' variant of call 1 (Stadler

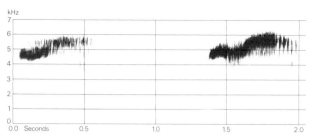

I S Palmér/Sveriges Radio (1972) Sweden July 1959

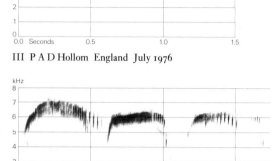

III P A D Hollom England July 1976

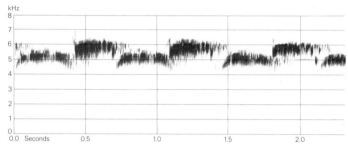

II P J Sellar Scotland June 1978

IV V C Lewis England July 1976

and Schmitt 1917). (b) Hissing 'fffsiiii' given when adult knocked to ground and stunned by hail (Stadler and Schmitt 1917); probably variant of call 1.

CALLS OF YOUNG. A range of whistling sounds, variable in intensity and duration. From within egg shortly before hatching, and in first few days after hatching, young give a plaintive but penetrating, high-pitched whistle, much feebler than call 1 of adult (Lack and Lack 1952; Lack 1956a): 'pee' (E K Dunn) or 'i-i i-i. . .' (Weitnauer 1947); for other renderings, see Glutz and Bauer (1980), who described it as a contact-call. After c. 1 week, similar sounds, described as a quiet high-pitched murmuring, given by young throughout time parent stays with them between feeds (Lack 1956a); in recording by R Prytherch, given while self-preening. In recording (Fig III), a soft, tremulous, liquid 'pee-ee-ee' (E K Dunn, P J Sellar); for similar sonagram (call rendered 'tsriih'), see Bergmann and Helb (1982). Such calls typically given by young struggling out from beneath brooding parent (R Overall). One fledged bird gave similar call, probably signalling distress, when handled (Lack and Lack 1952). Food-call of young a repeated, shriller variant of contact-call, rendered a rolling or chirping 'zjirr', at higher intensity 'zjierr zjierr' (Glutz and Bauer 1980); especially loud and high-pitched when parent enters nest, or when any other sound disturbs hungry young (Lack and Lack 1942; Dachy 1954). When older, young may give a distinctly explosive variant ending in a faint hiss (H Myers), perhaps as greeting or threat. Young near fledging gave regularly-repeated call (Fig IV) similar to call 1 of adult. Newly-fledged young also said to give a descending 'bit bit bit' (Stadler and Schmitt 1917).

EKD

Breeding. Main references Lack and Lack (1951b, 1952). SEASON. Britain and Ireland: see diagram. Switzerland: c. 1 week earlier. Finland and Sweden: c. 1 week later (Weitnauer 1947; T Fagerström). Laying delayed in cold weather. SITE. On top of flat surface under eaves of building or in hole in wall; occasionally in crevice in cliff or quarry, or in tree. Will use nest-box. 63% of nests, Gibraltar, under eaves, 27% in hole in wall, 8% under gutter, and 2% in hole in building (Finlayson 1979). Sometimes evicts House Sparrow *Passer domesticus*, Starling *Sturnus vulgaris*, etc., and uses their nest-sites (Weitnauer 1947; Lack 1956a). Nest: shallow cup of straw, grass, leaves, and feathers, cemented together with saliva; average external measurements 12·5 × 11 cm, with cup 4·5 cm across (Weitnauer 1947). Building: by both sexes collecting material in the air; cup shaped by pressing with body and pushing with feet (Lack 1956a). EGGS. See Plate 98. Long elliptical, smooth and not glossy; white. 25 × 16 mm (22–28 × 14–18), $n=100$ (Witherby *et al.* 1938). Weight 3.6 g (3.0–4.3), $n=113$, England (Lack and Lack 1951b). Average 3·5 g, $n=175$, Czechoslovakia (Pellantová 1975). Clutch: 2–3 (1–4), varying with location and season. Of 168 clutches, England: 1 egg, 2%; 2, 72%; 3, 26%; average 2·4. Of 79 clutches, Switzerland: 1 egg, 3%; 2, 28%; 3, 66%; 4, 3%; average 2·7. Of 57 clutches, Finland: 1 egg, 7%; 2, 73%; 3, 18%; 4, 2%; average 2·1. Of 72 clutches, Sweden: 2 eggs, 79%; 3, 21%; average 2·2. Average of 99 clutches laid up to 8 June, England, 2·29; of 26 clutches laid from 9 June 1·92. Some apparent variation in clutch size with temperature at time of laying (Weitnauer 1947; Magnusson and Svärdsson 1948; Lack and Lack 1951b; see also Dachy 1954, Pellantová 1975).

One brood. Replacements sometimes laid after early egg loss, after interval of 2–3 weeks (Lack 1956a). Laying interval 2–3(–6) days, with longer interval for larger clutches (Weitnauer 1947). Most eggs laid in morning, 08.00–11.00 hrs. INCUBATION. 19·6 days (18·5–24·5), $n=38$, England (Lack and Lack 1951b). 21·7 days (19–27), $n=29$, Czechoslovakia (Pellantová 1975). Prolonged during spells of poor weather. By both sexes in equal shares beginning with last egg, though occasionally begins with 2nd in 3-egg clutch; 1st egg covered at night. Hatching at 24-hr intervals. Incubation spells $c.$ 2 hrs long (2 min to $5\frac{3}{4}$ hrs), with longer spells before 11.00 hrs when off-duty bird having more difficulty in finding food; at these times, eggs may also be uncovered for periods of a few min to $6\frac{1}{2}$ hrs. YOUNG. Altricial and nidicolous. Cared for and fed by both parents. Brooded continuously for 1st week, and for about half the time for 2nd week, but left in periods of poor weather when adults having difficulty finding food. FLEDGING TO MATURITY. Fledging period variable, mean 42·5 days (37–56), $n=61$, with longer periods in poor weather. Independent from fledging. Age of first breeding probably 4 years. BREEDING SUCCESS. Affected by weather, with larger broods (as well as better success overall) in dry warm summers than in cold wet ones. Of 258 eggs, England, 78% hatched; little variation with brood size. In poor summers, England: of 7 young in broods of 1, 86% fledged, 0·9 young per nest; of 48 young in broods of 2, 50% fledged, 1·0 young per nest; of 36 young in broods of 3, 31% fledged, 0·9 young per nest. In good summers, England: of 24 young in broods of 1, 83% fledged; of 118 young in broods of 2, 95% fledged, 1·9 young per nest; of 15 young in broods of 3, 90% fledged, 2·4 young per nest. Overall, 58% of eggs produced fledged young (Lack and Lack 1951b). Of 213 eggs in 79 nests, Switzerland, 76% hatched and 65% fledged (Weitnauer 1947). Average number of young raised by $c.$ 12 pairs over 10 years, Sweden, $1·13 \pm$ SD 0·50 (0·64–1·93) (T Fagerström). Of 99 eggs laid, Czechoslovakia, 20·2% addled, 9·1% destroyed or disappeared; of 70 young hatched, 11·4% died, mostly of cold or starvation after crawling out of nest; overall 62 young fledged from 99 eggs laid (Pellantová 1975).

Plumages (nominate *apus*). ADULT. Forehead, lores, and thin line to above eye dark greyish olive-brown, gradually darkening towards crown, not or hardly contrasting with black spot in front of and just above eye. Hindcrown, nape, sides of head and neck, mantle, scapulars, and back black, glossed greenish-blue or oily-bronze depending on light; some sooty olive-brown of feather-bases often visible, especially when plumage worn. Rump and upper tail-coverts sooty or dark olive-brown, feathers slightly darker towards tips, tips narrowly and faintly fringed pale olive-grey or off-white. Lower cheeks and lower throat dark grey-brown (feather-tips fringed greyish-white), merging gradually into pale grey or dirty white patch on chin and upper throat. Pale throat-patch $c.$ 2 cm long, $c.$ 1 cm wide (poorly defined and difficult to measure), narrowly streaked by dark brown shafts and mottled by grey feather-centres. In worn plumage, grey and off-white feather-fringes of cheeks and throat abraded, and throat-patch heavily mottled and hardly contrasting with underparts, only chin and central throat showing traces of off-white. Chest, breast, belly, flanks, and vent black, slightly glossed oily-bronze; fresh feathers with white fringes up to 0·5 mm wide at tip, soon lost by abrasion, lasting longest on lower flanks and vent; some sooty olive-brown of feather-bases often visible, especially when plumage worn, all chest to vent then appearing glossy blackish olive-brown. Under tail-coverts dark greyish olive-brown, slightly darker subterminally, tips usually narrowly fringed white. Tail dark olive-brown with slight oily-bronze sheen. Outer webs and tips of outer 4–5 primaries and greater upper primary coverts black with blue-green or bronze-green gloss (depending on light); inner webs of outer primaries and their coverts and both webs of inner primaries and their coverts dark olive-brown with slight bronze-green gloss, broad paler grey-brown border along inner webs. Secondaries and tertials blackish olive-brown with slight bluish-green or bronze-green sheen, slightly darker and more glossy than neighbouring dark olive-brown greater upper wing-coverts, inner primaries, and inner greater primary coverts. Lesser and median upper wing-coverts, smaller upper primary coverts, and bastard wing black with bluish or greenish gloss. In fresh plumage, whole upperwing appears uniform blackish olive-brown or black, though greater wing-coverts, inner primaries, and inner greater primary coverts actually slightly browner; in worn plumage, upperwing more greyish-brown and less glossy, only lesser coverts, outer primaries, and outer primary coverts black. Longer and median under wing-coverts dark grey with white fringe along tips up to $c.$ 1 mm wide; shorter coverts olive-black or sooty-black, those along leading edge of wing narrowly and indistinctly bordered pale grey-brown or off-white. Axillaries black, glossed bronze-green. Heavily worn flight-feathers show row of tiny white spots along shafts. P10 occasionally old—browner and more abraded than fresh neighbouring p9. For rare partial albinistic birds, see Catley (1978) and Sharrock (1978b). NESTLING. Naked at hatching. On 10th day, small spines appear on upperparts and flanks; on 13th, spines open to produce short greyish-white down from mantle to rump and on lower flanks; on 17th, down longer, denser, and browner, and head, all wing, tail, and underparts still spiny; on 20th day, feather-tips break out of spines; from 25th day, down on upperparts and flanks pushed out by developing feathers, nestling appearing fully feathered (except for bare nape) from $c.$ 4 weeks (Heinroth and Heinroth 1924–6). JUVENILE. Body and wing deeper black than adult, less glossed with bluish or bronze-green; many feathers narrowly but conspicuously

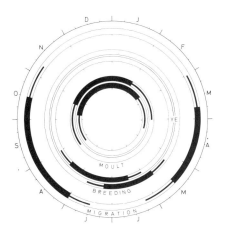

fringed white. In fresh plumage, forehead and lores white, sometimes partly mottled brown; feathers of crown, cheeks, and lower throat narrowly fringed white; feathers of upperbody narrowly and indistinctly fringed pale grey-brown, most marked on rump and upper tail-coverts; chin and upper throat white, throat-patch larger, more uniform, and more clearly defined than in adult, extent sometimes approaching that of Pallid Swift *A. pallidus* (though narrower and more rounded on lower throat, less broad and triangular there than *A. pallidus*); remainder of underparts as in fresh adult (but visible feather-bases duller grey-brown, not shiny dark olive-brown), white fringes along more narrowly rounded feather-tips slightly wider. Tips of tail-feathers and outer web of t5 narrowly margined white; inner web of t5 wider and more curving towards rounded tip than in adult, inner web not as straight and tip less acute than in adult (see Fig VI). All flight-feathers, greater upper wing-coverts, and

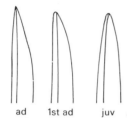

Fig VII Shape of outermost tail-feather (t5) in Swift *Apus apus apus*.

small coverts along leading edge of wing narrowly but distinctly fringed white; tertials and other coverts fringed off-white or pale grey-brown (often less conspicuous); white fringes along outer primaries narrow, soon abraded. Under wing-coverts more broadly fringed white than in adult, fringes along tips of greater and median coverts 2–4 mm wide. In fresh plumage, easily distinguished from adult by numerous and extensive white feather-fringes; when plumage slightly worn (in some, from August) rather similar to slightly worn non-breeding adult (breeding adult heavily worn by August), differing mainly in whiter forehead and lores, traces of white fringes on forecrown, larger and purer white throat-patch, narrow white margin along outer edge of t5, shape of t5, white fringes to secondaries, inner primaries, and greater upper wing-coverts, and broader white fringes on under wing-coverts. By midwinter, wing distinctly worn, and no flight-feather moult (unlike adult). FIRST ADULT. Head, body, and tail as adult (shape of t5 intermediate between adult and juvenile, see Fig VI). Wing still juvenile, except sometimes for part of smaller coverts; slightly or heavily abraded, especially outer primaries; brown with dull black leading edge; no traces of white fringes left, except for greater under wing-coverts which can often still be seen to be broader than those of adult. SECOND ADULT. Like adult, and often indistinguishable. Difference between worn inner and fresh outer primaries sometimes more distinct than in adult; shape of t5 about as in 1st adult; occasionally, juvenile p10 retained, extremely abraded (more so than in adult with retained p10), sometimes only bare shaft remaining. THIRD ADULT. Some probably have primary abrasion and tail-shape similar to 2nd adult.

Bare parts. ADULT AND JUVENILE. Iris deep brown. Bill black. Leg and foot blackish flesh-colour in juvenile, black with occasionally slight pink or grey tinge in adult; claws black. NESTLING. Skin, including that of base of bill and of leg and foot, flesh-colour or bluish-flesh. Tip of bill black, lateral edges of bill basally yellowish-white. Mouth flesh-colour, tip of tongue with brown spot. Upperparts and flanks downy from $c.$ 13 days. (Hartert 1912–21; Heinroth and Heinroth 1924–6; Harrison 1975; ZMA.)

Moults. ADULT POST-BREEDING. Complete (occasionally partial); primaries descendant. Starts immediately on arrival in winter quarters. Rarely, some moult occurs on breeding grounds: of 90 summer birds, Netherlands, only 3 (July–August) showed some moult on neck and chest, others none even as late as mid-September. Exceptionally, moult of inner primaries starts in summer: one with growing p1 in July, Gibraltar (Finlayson 1979). In winter quarters, starts with body and p1 between mid-August and late September. Primary moult score increases rapidly at first (as inner primaries are short): score of 25 reached late September to early November, 35 early November to early January. Increase in score slower when long outer primaries moulted: 45 reached early January to early April, moult completed (score 50) mid-February to late April (De Roo 1966; Stresemann and Stresemann 1966; BMNH). Moult of both wings symmetrical in earlier stages, but sometimes one wing more advanced than other during growth of outer primaries. Frequently, late moulters do not complete primary moult and suspend with p10 old (score 45) from late March or April, either in both wings or in one only; this old p10 retained during summer. Age has no influence on retention: may occur in adult every 3–5 years (Weitnauer 1977). In sample of 174 summer adults, Netherlands, 22% showed old p10 in one or both wings (in one, both p9 and p10 old) (RMNH, ZMA), in Belgium and East Germany 30% ($n=60$ (De Roo 1966), in Switzerland 25%, $n=142$ (Glutz and Bauer 1980), in Zaïre 33%, $n=30$ (De Roo 1966). Occasionally, growth of p10 in winter quarters stopped before full length attained; exceptionally, p10 replaced on breeding grounds, perhaps after accidental loss only. Old p10 usually replaced in following winter during moult of inner primaries, thus showing serially descendant moult; lost at score of inner series 0–10, fully grown at 25–35 (De Roo 1966; RMNH). Moult of head and body completed January–March; tail and secondaries (both from outermost feather inwards) moulted November–March; occasionally, single tail-feather or secondary retained during summer. POST-JUVENILE. Partial; in winter quarters, about December–March. Involves head, body, tail, and lesser and shorter upper wing-coverts; no other coverts or flight-feathers. FIRST IMMATURE POST-BREEDING. Complete; in 2nd winter, timing as in adult post-breeding. Juvenile flight-feathers, tertials, and remaining juvenile wing-coverts replaced in this moult. Rarely, starts with p1 in late summer when still close to breeding area (ZMA). As in adult, moult occasionally suspended before p10 replaced and juvenile p10 then not replaced until 3rd winter.

Measurements. Mainly Netherlands, a few Britain and West Germany, April–August; skins (RMNH, ZMA). Adult wing includes wing of birds with old p10, as these not significantly different from those with wing fully moulted; juvenile measurements only from full-grown birds mid-August to October, wing including measurements of 1-year-olds. Tail is to tip of t5; fork is tip of t1 to tip of t5; bill is exposed culmen.

WING AD	♂ 173	(3·22; 85)	167–179	♀ 173	(3·64; 81)	164–180
JUV	166	(5·88; 14)	157–174	168	(3·15; 13)	162–173
TAIL AD	75·8	(2·72; 50)	71–82	74·3	(3·06; 35)	69–79
JUV	69·1	(2·57; 7)	65–73	67·2	(2·15; 11)	64–70
FORK AD	31·6	(2·15; 48)	28–36	29·7	(3·10; 34)	25–34
JUV	27·6	(3·17; 7)	24–32	24·5	(1·62; 11)	22–27
BILL	6·6	(0·42; 10)	6·0–7·2	6·6	(0·31; 10)	6·2–7·2
TARSUS	10·9	(0·60; 10)	10·2–11·7	11·3	(0·52; 10)	10·6–12·1

Sex differences significant for fork. Tail and fork of 1-year-old similar to adult. Juvenile wing, tail, and fork significantly shorter than adult. Wing length of juvenile fledging at normal time rather variable, in part depending on amount of food received during feather-growth; averages of some British samples 160–169, range of all birds examined (146–)155–172; wing may grow 1–3 mm after fledging (Lack and Lack 1951b); for birds fledging prematurely, see Weights.

Weights. Nominate *apus*. England, captured in nesting holes, 42·7 (2·5; 102) 36–52 (Lack and Lack 1951b). England, captured when flying low over feeding area: (A) on days with maximum temperature 16°C or over, (B) below 16°C (Gladwin and Nau 1964).

	(A)			(B)		
1ST HALF MAY	44·2	(98)	36–50	43·5	(521)	36–55
2ND HALF MAY	44·0	(280)	35–56	40·0	(414)	31–50
1ST HALF JUNE	44·7	(47)	38–51	41·5	(233)	36–49
2ND HALF JUNE	37·6	(218)	31–43	36·9	(533)	31–46
1ST HALF JULY	39·0	(80)	33–44	37·0	(107)	33–43

Rather heavy upon arrival at breeding grounds, independent of weather; in England, late April, 47·4 (5) 45–51 (Gladwin and Nau 1964). Weight low 2nd half June and July when feeding young; non-breeders maintain heavier May weight then (Lack and Lack 1951b). Migrants usually rather heavy, e.g. spring and autumn migrants, Brittany and England, 53·9 (6) 50–58; lower if bad weather inhibits feeding during migration, e.g. August, Sweden, 32·5 (19) 28–39 (Lack and Lack 1951b). Weight increases during day, except perhaps in unfavourable weather: e.g. 40·7(122) 05.00–09.00 hrs to 42·1(57) 15.00–19.00 hrs (Gladwin and Nau 1964). During bad weather, average of adult may fall to 27·6 (Lack and Lack 1951b).

Sexed birds, Netherlands, hit by traffic when flying low during unfavourable weather: (1) May, (2) June, (3) July and early August; (4) exhausted birds found dying, summer (RMNH, ZMA).

	♂			♀		
(1)	39·0	(3·33; 12)	34–47	40·3	(6·42; 9)	33–52
(2)	40·8	(4·19; 28)	34–51	43·8	(7·07; 7)	37–56
(3)	40·2	(3·87; 14)	33–45	37·2	(4·76; 10)	32–43
(4)	27·6	(2·59; 5)	26–32	30·6	(3·10; 6)	27–35

Gibraltar: summer, 44·9 (4·90; 24); average and standard deviation (sample size not given), June 45·1 (1·91), July 44·8 (2·43), August 40·9 (4·38) (Finlayson 1979). East Africa, November–April: 39·0 (13) 35–44 (Moreau 1933; Lack and Lack 1951b; Friedmann and Williams 1969).

NESTLING AND JUVENILE. At hatching, Britain, 2·7–3·1; in fine weather, rapidly increases to peak of c. 50 at end of 4th week, but increase slow and irregular in poor weather, losing up to 6·6 per day and feather-growth halted when not fed; increase up to 10 per day when weather favourable (Lack and Lack 1951b). In England, 73 birds reached peak of 52·1 (41–58) on 28th (19th–50th) day; in Switzerland, 30 had peak of 58·5 (52–66) on 24th (20th–32nd) day; in England, 73 birds fledged with weight of 41·4 (34–52) on 42nd (37th–56th) day; in Switzerland, 30 fledged at 53·5 (48–56) on 42nd (38th–46th) day (Weitnauer 1947; Lack and Lack 1951b). Newly-hatched young may survive 2–3 days without food; large young (at peak weight) can survive up to 21 days (one decreased from 57 to 21 g in this time); older young fledge prematurely when not fed (smallest one had wing of 104), probably succumbing if wing still below c. 145; weights of those succumbing 20–28 (Hugues 1907; Lack and Lack 1951b; RMNH, ZMA). Minimum of exhausted fledged full-grown juveniles: 21 (RMNH).

A. a. pekinensis. ADULT. Mongolia, late May to August: ♂♂ 38, 39, 44; ♀♀ 42, 42, 45 (Piechocki 1968c). Afghanistan, late March and early April: ♂ 41; ♀♀ 39, 46, 49 (Paludan 1959).

Structure. Wing long and narrow, tapering gradually to sharp tip. 10 primaries. P9 usually longest, but of 72 birds with all primaries fresh, 8 had p10 0·5–2 longer, 5 had p10 subequal to p9, 41 had p10 1–2 shorter than p9, and 18 had p10 3–5 shorter. In birds with retained old p10, this feather 0–8 shorter in adult, 5–15 in 2-year-old. P8 usually 7–10 shorter than tip, but in 6 of 86 examined 4–6 shorter, in another 6 11–12 shorter; p7 23–29 shorter than tip, p6 41–48, p5 56–66, p1 104–120. Wing formula of juvenile similar to adult (but p10 never old); tips of outer 3 primaries about subequal in 1-year-olds (due to abrasion). Primaries very narrow, curved backwards and with pointed tips, no emarginated webs. 9 secondaries, very short; s7–s9 form tertials. Tail forked, 10 feathers; for depth of fork, see Measurements; for shape of t5 in relation to age, see Fig VI. Patch of erect plush-like feathers in front of and just above eye; upperside of eye protected by ridge on skull. Bill very flat and broad at base, only small tip projecting beyond feathers; tip of upper mandible strong and curved; corner of gape almost below posterior edge of eye. Nostrils large, oblong. No bristles on base of bill. Tarsus short, strong, feathered at front. Toes short, strong; 'hind' toe directed forwards to form innermost; able to grip with 2 inner toes opposite 2 outer, but all toes directed forward when clinging to vertical surface. Claws strong, curved, and sharp; outer middle claw almost equal length of its toe: toe 6·9 (20) 6·5–8·0, claw 6·1 (19) 5·4–6·8. Innermost toe with claw c. 68% of outer middle toe with claw, inner middle c. 90%, outermost c. 83%.

Geographical variation. In colour only; clinal. Nominate *apus* from Europe and North Africa east to Lake Baykal dark; *pekinensis* from Iran east to Mongolia and northern China distinctly paler. Races grade into each other over wide area in southern Kazakhstan, Kirgiziya, and north-west Sinkiang and Mongolia (Vaurie 1965; ZMM). Adult *pekinensis* differs from nominate *apus* in paler grey-brown forehead and narrow line over eye, often bleaching to off-white; slightly browner upperparts; purer white and larger throat-patch; distinctly paler and less deep black chest to vent; greyer under tail-coverts; broader white fringes on under wing-coverts; distinctly paler inner primaries, secondaries, and greater upper wing-coverts. Juvenile *pekinensis* often shows more white on forehead, lores, and line over eye than juvenile nominate *apus*, white of chin extending laterally to gape; crown and body slightly browner and with slightly wider white fringes; inner flight-feathers and greater coverts paler, as in adult; white patch on lores diagnostic (Brooke 1975a). Based on migrants and winter specimens from Tanzania and southern Africa, race *marwitzi* Reichenow, 1906, sometimes recognized (Clancey 1981a, b): deeper black head and body (including face) than nominate *apus*, but whiter chin and throat; blacker wing than nominate *apus*, with green rather than bronze sheen, but greater upper wing-coverts and inner flight-feathers paler than rest of wing (like *pekinensis*), and white fringes on under wing-coverts wider. Supposed to breed Turkey, Cyprus, Transcaucasia, and perhaps elsewhere in Middle East (Hartert 1912–21; Clancey 1981a), but birds examined from these areas (BMNH, ZMM) are within range of colour variation of typical nominate *apus*, and *marwitzi* not recognized while breeding area unknown.

Forms species-group with Plain Swift *A. unicolor*, Pacific Swift *A. pacificus*, and Dark-backed Swift *A. acuticaudatus* of south-east Himalayas, latter 2 comprising superspecies (Brooke 1970).

Recognition. Pale central Asiatic race *pekinensis* only slightly darker than Yugoslavian race of Pallid Swift *A. pallidus illyricus*, latter differing in paler olive-brown crown, rump, and greater upper wing-coverts, with more contrasting dark saddle on mantle, slightly wider white fringes on chest to vent, paler under tail-coverts, less deeply forked tail, broader head, and wider gape.

CSR

Apus pallidus Pallid Swift

PLATES 60, 61, and 63
[between pages 662 and 663, and facing page 686]

Du. Vale Gierzwaluw	Fr. Martinet pâle	Ge. Fahlsegler
Ru. Бледный стриж	Sp. Vencejo pálido	Sw. Blek tornseglare

Cypselus pallidus Shelley, 1870

Polytypic. *A. p. illyricus* Tschusi, 1907, east coast of Adriatic Sea (and perhaps this race on western coast also); *brehmorum* Hartert, 1901, Canary Islands, Madeira, coastal North Africa east to north-west Egypt, Iberia, southern France, western and southern Italy, southern Greece, Cyprus, and (perhaps this race) Turkey; nominate *pallidus* (Shelley, 1870), Banc d'Arguin, hills of Sahara, Egypt, and from Levant to Pakistan.

Field characters. 16–17 cm, wing-span 42–46 cm. Close in size to Swift *A. apus*, but under careful observation shows distinctly broader and flatter head, slightly greater girth to body, broader rump, slightly broader wings (particularly outer primaries), and slightly shorter and more rounded tail-forks; noticeably larger than Plain Swift *A. unicolor*, with differences in outline just described all more pronounced. Medium-sized brown swift, with pale grey-white forehead, conspicuous white throat, slightly darker saddle on back, paler (greyer or sandier) upper-surface to flight-feathers, and pale margins visible on body, sides of rump, and upper wing-coverts on both adult and juvenile (pale margins giving plumage characteristic rough appearance). Flight much as *A. apus* but action often slower, lacking 'twinkling' appearance. Sexes similar; no seasonal variation. Juvenile inseparable. 3 races in west Palearctic, hardly distinguishable in the field.

(1) Mediterranean (except Levant) races, *brehmorum* and *illyricus*. ADULT. Up to 5% larger and darker than nominate *pallidus* (see below) with brown ground of plumage tinged sandy rather than grey. Observation of differences in plumage pattern compared to *A. apus* requires prolonged views in constant light but the following form most certain basis for separation: (a) broader, flatter head, with more obvious grey forehead and sides of forecrown (emphasizing shadowy eye-pit), and larger, usually white throat (extending below most of head and not just to eye as in *A. apus*); (b) less uniform upperparts, with brown saddle between wings contrasting slightly with paler head, rump, and flight-feathers, and wing-coverts occasionally showing pale fringes and tips (unlike adult *A. apus*); (c) mottled underparts, with pale margins obvious on most of body and particularly so on sides of rump (as in juvenile *A. apus*); (d) slightly broader wings (lacking narrow point of *A. apus*), usually showing distinct contrast between relatively dark coverts and relatively pale flight-feathers (latter particularly pale over uppersurface of secondaries, adding to contrast of saddle); (e) slightly shorter and blunter tail-forks, most obvious when spread in turning.

JUVENILE. Little studied in the field, but pale margins to feathers more distinct than those of adult and juvenile *A. apus*. (2) Saharan and Middle East race, nominate *pallidus*. At all ages, closely resembles *brehmorum* but shows noticeably grey or dun cast to brown plumage, appearing paler in good light. Thus tonal differences from *A. apus* in south-east of west Palearctic should be greater though this not critically established.

Diagnosis made difficult not only by close similarity to *A. apus* but also by effect of light on appearance of all medium-sized, uniformly coloured swifts. Thus separation of these species where they occur together commonly possible with practice but still risky with all migrant or vagrant individuals outside normal range. Most obvious differences from *A. apus apus* covered above, but see also Plumages and Geographical Variation; obvious danger of confusion with *A. apus pekinensis* not studied in the field. Differences in structure discernible not only in silhouette but also in flight action, which includes more planing than *A. apus*, slower turns, and heavier wing-beats (not producing characteristic 'twinkling' action of *A. apus*). Nevertheless, capable of similarly sudden and dramatic aerial manoeuvres, so that difference in flight not constant. Behaviour as *A. apus*.

Voice similar to *A. apus* but deeper and less shrill; sometimes at least, disyllabic.

Habitat. Difficult to distinguish from that of Swift *A. apus* except in being restricted to southern fringe of latter's breeding range, in lower middle and lower latitudes, and in being largely linked with coastlines and coastal or riparian lowlands of Mediterranean or subtropical climates. Overlapping extends not only to common range for aerial foraging but locally to mixed nesting colonies, although elsewhere mingling of neighbouring groups seems to be avoided (Voous 1960). Breeding distribution pattern suggests some preference for maritime surroundings. In Madeira, flies all afternoon over an arm of the sea, almost touching the water, apparently roosting on

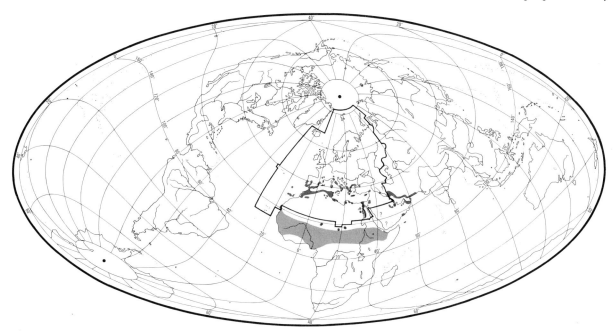

cliffs of inshore islands. In Canary Islands, also nests in crevices of rocks and caves on sea-cliffs as well as in steep walls of ravines, preferring plains but sometimes ascending to c. 1200 m; also nests in high buildings such as tower of Las Palmas cathedral. Forages low over fields of maritime plains, but also over towns and villages, occasionally ascending to great heights in air (Bannerman 1963b; Bannerman and Bannerman 1965). In Cyprus, breeds at all elevations to c. 1750 m or more, choosing sites apart from *A. apus* (Bannerman and Bannerman 1971). Apparently transfer from cliff-nesting to use of buildings has developed less far than in *A. apus*, although not less than in Alpine Swift *A. melba*.

Distribution. Inadequately known owing to possible confusion with Swift *A. apus*.

FRANCE. First bred Corsica 1932, and on mainland (Banyuls) in 1950, with first inland colony in 1966 (Toulouse); may have been overlooked earlier in some cases (Yeatman 1976). ITALY. Inland, breeds only in Turin area (Boano 1979). SYRIA. Status uncertain; may breed on coast (Kumerloeve 1968b; HK). LEBANON. Present status uncertain; bred 1956 (Kumerloeve 1968b; HK). KUWAIT. Occasional reports of breeding of swifts probably refer to *A. pallidus* (PRH). ALGERIA. Probably more widespread in north (EDHJ). CANARY ISLANDS. Precise breeding distribution uncertain; recent proof only on Gran Canaria and Tenerife (KWE).

Accidental. Britain.

Population. FRANCE. 100–1000 pairs; probably increased—see Distribution (Yeatman 1976). CYPRUS. Locally fairly common (PRF, PJS) LIBYA. Common, locally abundant (Bundy 1976). CANARY ISLANDS. Not common (KWE). MAURITANIA. Banc d'Arguin: 90–110 pairs (Naurois 1969a).

Survival. Gibraltar: mortality 67·3% in 1st year of life; annual adult mortality 26% (Finlayson 1979).

Movements. Largely migratory, but few data available; no ringing recoveries, and regular use of higher airspace in African winter quarters has influenced number of reliable identifications there. Differs from other European *Apus* in being double-brooded, so remains later into autumn in breeding areas; consequently, higher proportion (than in *A. apus*) commence moult before onset of autumn migration (J C Finlayson).

Winter records from breeding range not rare: e.g. Laferrère (1974) for France, Thouy (1978) for Morocco, Bundy (1976) for Libya, Goodman and Watson (1983) for Egypt, Meinertzhagen (1954) and Walker (1981a) for Arabia; however, majority migrate. In Canary Islands, bulk arrive in February (some in January) and most depart in September (Bannerman 1963b); present Morocco and Gibraltar late February or March to October (Heim de Balsac and Mayaud 1962; Smith 1965; Thouy 1978; J C Finlayson); arrives southern France in early April, leaving mid-November (Affre and Affre 1967; Yeatman 1976); present mainly March–October in Libya (Bundy 1976) and Iraq (Allouse 1953). Spring passage through Elat (Israel) peaks in first half of March; daily maximum (in 1977) of 1000 on 3 March (Pihl 1978). Hence present in breeding areas for 7–8 months (c. 4 months in *A. apus*).

On present evidence (certainly incomplete) principal wintering areas lie in northern Afrotropics, where numbers of non-breeders also summer. Known from Gambia

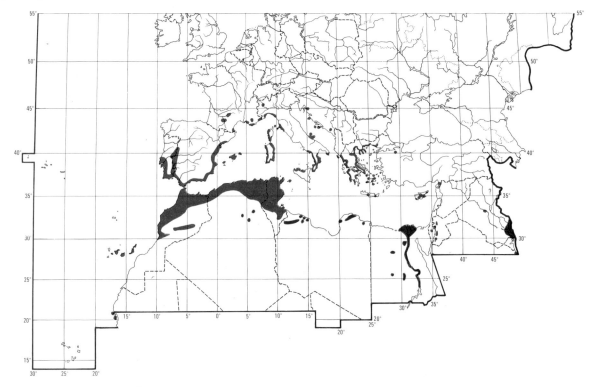

and Sierra Leone east to Sudan, Somalia, and northern Uganda (Lack 1956b; Moreau 1972; Gore 1981), being quite common in Niger inundation zone of Mali, Nigeria, and Sudan (north to Khartoum) at least (Macleay 1960; Lamarche 1980; Elgood 1982). Passage through Sahel zone of Mali and central Chad, with first arrivals July–August (Lamarche 1980; Newby 1980), indicates broad-front Saharan crossings. Specimens (both west Mediterranean race, *brehmorum*) also obtained Zambia (December) and north-west Cape Province (February); these indicate sporadic or thinly-spread winter occurrences over large part of southern Africa, where it would be easily overlooked among big flocks of *A. apus* (Benson *et al.* 1971; Moreau 1972). Like other European swifts, probably makes extensive intra-seasonal movements within Afrotropics (Moreau 1972).

Food. Mainly flying insects of moderate size. Feeding flocks apparently denser than those of Swift *A. apus* (see Finlayson 1979). Said to fly lower than *A. apus*—down to *c*. 1·5 m (König and König 1973; Boano 1979; see also Affre and Affre 1967). For apparent defence of feeding area, see Social Pattern and Behaviour. Will concentrate on locally abundant prey, e.g. flying ants (Formicidae) in Algeria, June (Geyr von Schweppenburg 1918). 4 food-balls, Gibraltar, weighed 1·1–1·2 g (Finlayson 1979).

Takes dipteran flies including mosquitoes (Culicidae), Hymenoptera including flying ants (Formicidae), Hemiptera, Coleoptera, Lepidoptera, dragonflies (Odonata), and spiders (Araneae).

In Tunisia, of 8 stomachs, May–June, 6 contained mosquitoes, 1 small flies, and 1 empty; 5 also contained small beetles (Blanchet 1957). On La Palma (Canary Islands), seen hunting with Plain Swift *A. unicolor* over arid fields with an abundance of grasshoppers (Cullen *et al.* 1952). At Gibraltar (in study of food brought to nestlings) took wider range of food than Swift *A. apus* in same area, but dietary overlap still evident. Prey taken was larger than for *A. apus*; maximum size 27 mm. Of 1293 items, 56·2% Hymenoptera, 25·2% Hemiptera, 12% Diptera, 4·8% Coleoptera, 1·2% Lepidoptera, and 0·8% others, including spiders (Araneae) and dragonflies (Odonata) (Finlayson 1979).

At Gibraltar, on average, birds returned to nest with food once per hr or more frequently (Finlayson 1979). In Toulouse (France), of 21 nests, mean of 4 feeds per hr recorded in October, but only 0·86–0·9 per hr when colder (Affre and Affre 1967).   BDSS

Social pattern and behaviour.
1. Relatively gregarious throughout the year. In breeding season, Canary Islands, usually encountered singly or in small parties, but feeding flocks sometimes of thousands (Bannerman 1912). From colonies of *c*. 30 pairs, Gibraltar, mean size of feeding flocks 7·3 ± 3·86 birds (Finlayson 1979). In Canary Islands, forms flocks with Plain Swift *A. unicolor* for feeding (Koenig 1890; Cullen *et al.* 1952); in Marrakech (Morocco), with Little Swift *A. affinis* (Meise 1959). Mixed flocks, Mali, October–May, 40% *A. pallidus*, 55% Alpine Swift *A. melba*, 5% Swift *A. apus* and *A. affinis* (Thiollay 1974). BONDS. No information. Monogamy likely, as in *A. apus*. BREEDING DISPERSION. Colonial and gregarious. Of 92 colonies, Gibraltar, 47·8% of less than 10 pairs;

some colonies only 3–4 m apart (Finlayson 1979). In Sardinia, 2 colonies, each of c. 10 pairs, c. 300 m or more apart (Mathieu 1965). For diagram of dispersion (no measurements given) of nests in colony of 59 pairs, Toulouse (France), see Affre and Affre (1967). For apparent communal defence of feeding territory, see Antagonistic Behaviour (below). Breeding birds show marked fidelity to colony and to nest-site (Boano 1979; F Hiraldo). 1 incubating adult ringed on nest, Carmagnola (Italy), found incubating in same nest 2 years later (Boano 1979). Mixed colonies with *A. apus* and *A. melba* recorded (Laferrère 1972; Finlayson 1979). ROOSTING. From occupation of nest-site, pair roost on nest, sitting side by side (Affre and Affre 1967; see also Boano 1979). Other birds reported roosting by clinging to wall near nests (König 1962), and in crevices possibly containing nest (Géroudet 1961a). In some places, probably roosts in flight, as in *A. apus* (Lack 1969). Time of roosting varies with weather, but most birds re-entered nests, Carmagnola, from c. 15 min before to 30–40 min after sunset (Boano 1979). At Toulouse, most returned to roost ½ hr before to ½ hr after sunset, peak from 10 min after to ½ hr after sunset (Affre and Affre 1967; Boano 1979). Birds in nest, March, called occasionally during night, and highly vocal from about dawn onwards (P A D Hollom). On 18 May, Gibraltar, little flight activity at colony for first 2 hrs after sunrise, then sudden increase up to peak exodus at 10.00 hrs. In evening, rate of return greatest a few hrs before sunset, and all apparently returned by sunset (Finlayson 1979; see also Géroudet 1961a). Birds returned to colony, Toulouse, alone or in small groups, and went directly into nests. No social or vocal display accompanied roosting (Affre and Affre 1967). In Corsica, mid-July, evening roosting preceded by communal Screaming-display (see part 2).

2. FLOCK BEHAVIOUR. Forms flocks for migration, feeding, and courtship (e.g. Lack 1956a, Finlayson 1979). At Gibraltar, feeding flocks apparently more dense than those of *A. apus* (Finlayson 1979, which see for details). Communal Screaming-displays ('screaming-parties': Lack 1956a), in which birds gather to chase one another while uttering Screaming-calls (Géroudet 1961a; see 1 in Voice), occur almost exclusively on calm mornings and evenings, and are rare early and late in season when few birds present (Finlayson 1979; see also Antagonistic Behaviour, below). ANTAGONISTIC BEHAVIOUR. The following information from Finlayson (1979). At Gibraltar, birds from a given colony (see Breeding Dispersion, above), appeared to defend feeding area above colony against intrusion by conspecific birds from other colonies, and against *A. apus*. Birds from neighbouring colonies usually avoided each other. Strange *A. pallidus* released by observer into flock were chased off by 2–10 flock members. Once, entire flock (of 30 birds) chased *A. melba* approaching colony. Since mixed colonies of *A. pallidus* and *A. apus* at Gibraltar performed communal Screaming-displays, antagonism thought to be more between colonies than between species. Suggested that Screaming-display may advertise feeding areas. HETEROSEXUAL BEHAVIOUR. Little known. (1) General. On arrival from winter quarters, birds immediately take possession of colonies, and perform Screaming-display in evening. Pair occupy nest-site regularly 4–7 days before they lay eggs: e.g. at 17.00–18.00 hrs, when most birds still airborne, one or both members of pair typically in nest when due to lay eggs in 1–2 days (Boano 1979). (2) Mating. Aerial copulation occurs (Heymer 1977). At Sousse (Tunisia) in May, birds in flocks seen to perform Wings-high display, as in *A. apus*: raised wings in V in flight (M G Wilson); in *A. apus*, thought to be prelude to copulation, and possibly solicits it (Lack 1956a). In Mali, October–May, where birds thought to be breeding, at least 3 cases of aerial copulation occurred after lengthy pursuit-flights (Thiollay 1974). On arriving back from feeding trip, breeding birds usually chased in circles for several seconds or minutes before entering their nest-holes (Affre and Affre 1967). RELATIONS WITHIN FAMILY GROUP. Young in nest raise heads when parent arrives with food (Affre and Affre 1967). Parent dispenses food from gular pouch, as in other Apodidae (Lack and Lack 1951a; Affre and Affre 1967). Small young brooded regularly at night; older young still brooded but not completely covered (Affre and Affre 1967). ANTI-PREDATOR RESPONSES OF YOUNG. No information. PARENTAL ANTI-PREDATOR STRATEGIES. On close approach of human intruder, birds circled overhead (König 1962). EKD

Voice. Heard mostly in breeding season; at Carmagnola (Italy), less so from end of July onwards (Boano 1979). Said to be less freely used than that of Swift *A. apus* (Affre and Affre 1967).

CALLS OF ADULTS. (1) Screaming-call. See Fig I. Rising and falling 'hüiii' (Westernhagen 1957); 'srieh' (Meise 1959); rasping, nasal 'tschriih' (Bergmann and Helb 1982); loud, thin 'seeyrr' (Gallagher and Woodcock 1980). From nest, thin 'seer' and loud insistent 'see-yer' (P A D Hollom). Deeper, and not as shrill, as same call of *A. apus* (Westernhagen 1957; Meise 1959; König 1960, 1962; Bergmann and Helb 1982). Some of these descriptions, and that by Burges (1983), indicate call tends to be disyllabic, unlike *A. apus*. For additional sonagrams, see König and König (1973) and Bergmann and Helb (1982). Call given repeatedly by birds in communal Screaming-display (see Social Pattern and Behaviour—also *A. apus*) and (singly) on approaching nest when human intruder nearby (König 1962). In recording (Fig II), apparently of 2 birds at nest, marked drop in pitch in middle of some calls (J Hall-Craggs, P J Sellar). In recording by P A D Hollom

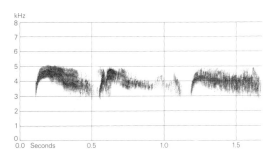

I E D H Johnson/Sveriges Radio (1972) Algeria February 1968

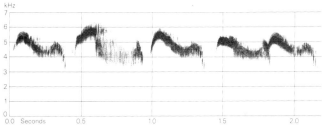

II P A D Hollom Morocco March 1978

of birds in flight, some very short screams, sounding like squeals, are perhaps a variant. (2) Other calls. A 'pseek-pseek', shorter than Screaming-call; not further described (Gallagher and Woodcock 1980). In recording by P A D Hollom of flying birds, single 'cheeoik' (J Hall-Craggs) after series of Screaming-calls. In March, birds roosting in nest at night gave brief, quiet calls (P A D Hollom).

CALLS OF YOUNG. In recording by C Chappuis of young in nest, a harsh 'szree szree' (E K Dunn). EKD

Breeding. SEASON. Gibraltar: first eggs laid second half of April, with peak 3rd week May; 2nd clutches from end of July, with last young fledging 2nd week October (Finlayson 1979). North Africa: first eggs end of March, main laying season April–May, 2nd clutches in July (Castan 1955; Heim de Balsac and Mayaud 1962). SITE. On building, especially under eaves, also in hole in wall, and in cave or cliff crevice. Reported nesting in hole in palm tree, Algeria and Portugal (Laferrère 1972; M Calvert). In Gibraltar, 70% of nests under eaves, 12% in hole in building, 7% under gutter, 7% on cliff, 3% in hole in wall; sample size not given (Finlayson 1979). Nest: shallow cup of straw, grass, and feathers, cemented together with saliva. Nests in Tunisia averaged 10–12 cm long, 8–10 cm wide, 2–5 cm deep inside, 4–5 cm deep externally (Castan 1955). Building: no information on role of sexes. EGGS. Long sub-elliptical, smooth and not glossy; white. 25×16 mm ($22-27 \times 15-17$), $n=11$; average weight of 4 eggs, 3·3 g (Castan 1955). Clutch: 2–3 (1–4). Of 32 clutches, Tunisia: 2 eggs, 14; 3, 18 (Castan 1955). Average clutch size, Gibraltar: 1st clutch $2·89 \pm SD\ 0·46$, $n=19$; 2nd clutch $1·95 \pm SD\ 0·58$, $n=22$ (Finlayson 1979). 2 broods. INCUBATION. $21·40 \pm SD\ 1·4$ days, $n=8$ (Finlayson 1979). By both sexes. Begins with 1st egg; hatching asynchronous. YOUNG. Altricial and nidicolous. Cared for and fed by both parents, brooded while small. FLEDGING TO MATURITY. Fledging period $46·4 \pm SD\ 2·18$ days, $n=10$ (Finlayson 1979). Age of independence and first breeding not recorded. BREEDING SUCCESS. Of 55 eggs in 1st clutches, Gibraltar, 87% hatched and 53% fledged; of 43 eggs in 2nd clutches, 65% hatched and 30% fledged (Finlayson 1979). In one year, Piedmont (Italy), 18 eggs from 8 clutches produced 9 young in the nest (average 1·1 young per nest); in another year, 25 eggs from 9 clutches produced 16 young (average 1·8 young per nest) (Boano 1979).

Plumages (*A. p. brehmorum*). ADULT. Forehead and lores pale drab-grey, gradually darkening to grey-brown on crown and ear-coverts. Large black spot in front of eye, extending to just above eye; together with dark eye, forms contrastingly black patch on side of head. Nape, side of neck, mantle, scapulars, and upper back dark olive-brown or olive-black, forming dark saddle which contrasts with paler crown, rump, tertials, and median upper wing-coverts; some paler olive-brown of feather-bases often visible, especially when plumage worn; fresh feathers with faint off-white fringes along tips. Lower back and rump greyish olive-brown; feathers with slightly darker olive-brown arcs sub-terminally, narrowly fringed off-white along tips when plumage fresh. Upper tail-coverts dark olive-brown. White or greyish-white patch on chin and throat; rather variable in size and usually poorly defined from grey-brown feathers of lower cheeks, which show white fringes. Throat-patch either uniform or with faint dark shaft-streaks; occasionally, some pale grey-brown of feather-bases shows; patch $c.$ 2 cm long, greatest width (1–)1·5–2 cm; in birds with wide patch, white or greyish white usually extends to cutting edge of lower mandible below nostrils, sometimes contiguous with small off-white patch on forehead. Chest, belly, flanks, and vent olive-brown, feathers with olive-black tips showing as indistinct dark crescents and with narrow white fringes along tips showing as pale scaling; under tail-coverts pale grey-brown with olive-brown subterminal crescents and narrow white fringes along tips. Tail dark olive-brown with slight green lustre. Outer 2–4 primaries black with dark olive-brown inner webs; inner primaries gradually paler, innermost, secondaries, and tertials greyish olive-brown with narrow white edges along tips. Lesser upper wing-coverts dark olive-brown like mantle and scapulars; median and greater coverts contrastingly paler, greyish olive-brown, narrowly fringed pale greyish. Greater upper primary coverts like primaries, median primary coverts and bastard wing black, distinctly darker than inner and middle primaries, greater primary coverts, and median and greater secondary coverts. Small coverts along leading edge of wing, under wing-coverts, and axillaries pale olive-brown with rather broad but poorly defined pale grey fringes. In worn plumage, white or off-white fringes of body and wing largely disappear, but traces usually still visible on rump, belly, and under tail-coverts; more olive-brown of feather-bases visible on hind-neck, mantle, scapulars, and chest to belly, appearing more mottled, less uniform olive-black. NESTLING. Naked at hatching. Dense grey down develops on upperparts and flanks from age of $c.$ 1 week (Kainady 1976). JUVENILE. Closely similar to adult, but forehead off-white, feathers of crown narrowly fringed white; throat-patch white without shaft-streaks, gradually merging into pale grey-brown of cheeks; chest to vent and flanks paler, dark subterminal crescents on feathers narrower and duller olive-brown, pale fringes wider but less white, and less distinctly defined; upper wing-coverts and flight-feathers with more distinct white fringes; in particular, fringes of greater primary coverts and primaries more marked, faintly visible up to p10; tail-feathers with narrow white edges along tips and along outer web of t5; tip of t5 wider, both webs curved towards narrowly rounded tip (in adult, webs have straight taper towards rather acutely pointed tip). Hardly distinguishable from adult once plumage worn, except by difference in moult (see Moults) and in shape of t5. Some non-juvenile birds have shape of t5 about intermediate between juvenile and adult; these perhaps in 2nd or 3rd calendar year.

Bare parts. ADULT AND JUVENILE. Iris deep brown. Bill black. Leg and foot flesh-brown, dull purplish-brown, or black (Ali and Ripley 1970; BMNH, ZMA). NESTLING. No information.

Moults. ADULT POST-BREEDING. Complete; primaries descendant. In Gibraltar, starts during raising of 2nd brood, suspended during migration October–November; of 38 adults, July, 4 had moulted inner 2 primaries; of 31 adults, August, 8 had moulted inner 2; single September bird had also moulted inner 2 primaries, a single October one had suspended with inner 4; other July–August birds had not started (Finlayson 1979). Of large sample of *brehmorum* and *illyricus* examined (BMNH, RMNH, ZFMK, ZMA), none from March–June in active moult, nor single from late July or 2 from August (Italy and

Cyprus); also, no January–April nominate *pallidus* were moulting. Only 2 examined from other times of year, both nominate *pallidus* from Sudan, late June (BMNH): one had wing just completed (primary moult score 50), other had score 19, mantle, back, and part of belly new, remainder of body, wing, and tail old. As in *A. apus*, arrested moult of primaries during breeding season frequently encountered: 11 out of 22 adults from Europe retained old primaries (mainly p10 only, one each p9, p9–p10, and p8–p10), but only 3 out of 42 North African birds (p10 only), probably indicating longer moult season of latter (perhaps more often starting during nesting). More information on timing and duration of moult needed. POST-JUVENILE. Of those juveniles examined, July–November, none in moult (presumably unlike adults). Not known whether juvenile flight-feathers replaced in summer of 2nd calendar year (as in Alpine Swift *A. melba*) or in 2nd winter (as in *A. apus*).

Measurements. *A. p. brehmorum*. Canary Islands, Madeira, northern Algeria, Tunisia, Spain, southern France, and Italy, March–July; skins (RMNH, ZFMK, ZMA). Tail to tip of t5; fork is tip of t1 to tip of t5; bill is exposed culmen.

WING AD	♂ 171	(3·10; 18)	168–178	♀ 171	(2·87; 20)	167–176
TAIL AD		71·0	(2·51; 19)	68–75	70·0	(2·25; 20)	67–74
FORK AD		26·3	(2·39; 19)	23–30	25·1	(1·35; 20)	23–27
BILL		6·6	(0·27; 16)	6·4–7·0	6·7	(0·46; 10)	6·2–7·3
TARSUS		11·1	(0·37; 11)	10·6–11·6	11·4	(0·43; 6)	10·6–11·8

Sex differences not significant. Juvenile wing on average 2 below adult, tail and fork 4·5.

Samples from various localities, sexes combined, February–July; skins (Vaurie 1959e; BMNH, RMNH, ZFMK, ZMA). (1) *A. p. illyricus*, Yugoslavia. *A. p. brehmorum*: (2) Canary Islands and Madeira; (3) northern Algeria and Tunisia; (4) Spain, including Balearic Islands; (5) southern France and Capraia (Italy); (6) Aegean Greece (Goodman and Watson 1983); (7) Cyprus; (8) Bahig (north-west Egypt) (Goodman and Watson 1983). Nominate *pallidus*: (9) Nile Valley (Cairo to Luxor), Faiyum, and Kharga oasis (Egypt); (10) Iraq.

	WING			TAIL		
(1)	174	(2·26; 4)	170–176	71·2	(1·94; 4)	68–73
(2)	173	(3·76; 11)	168–180	71·1	(2·27; 6)	68–74
(3)	172	(3·59; 22)	166–178	69·9	(2·48; 17)	67–75
(4)	171	(1·86; 11)	169–174	71·5	(2·32; 11)	68–74
(5)	170	(1·12; 5)	168–171	69·4	(2·04; 5)	67–73
(6)	168	(— ; 12)	162–174	—		
(7)	165	(— ; 2)	165–166	69·0	(— ; 2)	68–70
(8)	166	(— ; 28)	161–173	—		
(9)	166	(2·77; 25)	162–172	66·2	(2·76; 8)	63–70
(10)	165	(1·93; 4)	162–166			

Wing of nominate *pallidus* from Banc d'Arguin (Mauritania), 167 (8) 161–172 (Naurois and Roux 1974); southern Algeria, Niger, and Chad 165 (15) 157–174 (Vaurie 1965; Naurois and Roux 1974); Oman, Bahrain, southern Iran, and Pakistan 170 (3·79; 11) 164–176 (BMNH).

Weights. Gibraltar, breeding season, 41·3 (3·63; 100); average weight and standard deviation (sample sizes not given), April 43·9 (3·69), May 41·5 (3·48), June 41·3 (3·06), July 40·9 (3·59), August 41·5 (4·13); one September bird 36·5, one October 40·0 (Finlayson 1979). Southern Spain, early May; ♂♂ 32, 34, 37, 38; ♀♀ 30, 32 (ZMA). Morocco, April, 40, 46 (BTO). Pakistan, February: ♂♂ 32, 34 (BMNH).

Structure. 10 primaries: in birds with all primaries new, p9 and p10 about equally likely to be longest or either one 0–3 shorter than other; in birds with retained old p10, p9 longest and p10 0–3(–5) shorter; p8 7–12 shorter than longest, p7 22–30 shorter, p6 42–49, p5 60–67, p1 112–119 ($n = 30$); thus, no constant difference from *A. apus*. Tail has slightly shallower fork than *A. apus*, but some overlap: fork (tip of t1 to tip of t5) 22–28(–30) in 60 adult *A. pallidus* of all races examined, (25–)27–36 in 90 adult *A. apus*. Difference between lengths of t4 and t5 5(1·51; 45)3–8 in adult *A. pallidus*, 9·5(1·20; 25)7·5–11·5 in adult *A. apus* (Casement 1963; BMNH). Head of *A. pallidus* broader than *A. apus*: measured across crown between ridges above eye, *A. pallidus* 17·5(2·05; 47)15–20·5, *A. apus* 14·5(2·18; 37)11–17 (Casement 1963); broader head results in wider gape, angle between rami of lower mandible as seen from below almost 90° in *A. pallidus*, *c.* 75° in *A. apus*; width of bill at gape 11·7 (1·08; 13) in *A. pallidus*, 9·6 (0·31; 10) in *A. apus* (Finlayson 1979). Outer middle toe (with claw) longest, 11·1 (17) 10·5–11·9; innermost toe *c.* 58% of outer middle, inner middle *c.* 90%, outermost *c.* 85%. Remainder of structure as in *A. apus*.

Geographical variation. Clinal—rather slight in wing length, more pronounced in depth of colour; clines do not run parallel and boundaries between races hard to define. *A. p. illyricus* of eastern Adriatic (and perhaps western also) slightly larger and distinctly darker than typical *brehmorum* from Canary Islands and Madeira; crown, hindneck, and back deeper olive-brown, olive-black of mantle slightly more glossy and extending up to crown and back; feathers of rump with slightly wider dark olive-brown subterminal crescents, rump appearing less pale olive-brown; white throat-patch large, as in *brehmorum*, but more sharply defined; chest, belly, vent, and flanks distinctly darker, with only limited amount of grey-brown feather-bases visible; under tail-coverts slightly darker; wing and tail as in *brehmorum*, but lesser upper wing-coverts slightly darker and outer primaries slightly deeper black. Compared with *A. apus*, sides of head, upperparts, and upperwing of *illyricus* distinctly paler—various shades of olive-brown instead of glossy black; white throat-patch larger; remainder of underparts almost equally dark, however, but with distinct white fringes on feather-tips. In *brehmorum*, birds from coastal North Africa similar to or slightly paler than topotypical *brehmorum* from Canary Islands and Madeira; birds from other Mediterranean countries (except Levant, inland Egypt, and Adriatic) slightly darker, more or less intermediate between typical *brehmorum* and *illyricus*; some of latter group sometimes included in *illyricus*—e.g. eastern Libya and Cyprus (Vaurie 1965), Greek Aegean and coastal north-west Egypt (Goodman and Watson 1983)—but here all arbitrarily included in *brehmorum*. Size in this broad-concept *brehmorum* gradually decreases towards east, average wing 173 on Canary Islands and Madeira, 171 in western Mediterranean, 165–166 in Cyprus and north-west Egypt (see Measurements). Nominate *pallidus* from inland Egypt smaller than most populations of *brehmorum* (except easternmost) and distinctly paler; both upperparts and underparts grey-brown rather than olive-brown; forehead and crown pale grey-brown; dark saddle on mantle less marked than in *brehmorum*; white of throat slightly more extensive; chest, belly, vent, and under tail-coverts with much pale grey-brown of feather-bases showing; wing paler—in particular, secondaries, inner primaries, and greater and median upper wing-coverts paler grey-brown. Sahara populations of nominate *pallidus* (including those of Banc d'Arguin) similar in size to typical nominate *pallidus* from inland Egypt, but slightly darker; birds from southern slopes of Algerian Atlas (Biskra) tend slightly in colour towards nominate *pallidus*, but large in size and thus included in *brehmorum* (Vaurie 1959e; Naurois and Roux 1974). Populations of nominate *pallidus* from Middle East similar in colour to African birds; those from Iraq and Saudi Arabia similar in size to Egyptian birds, those from southern Iran, Pakistan, Bahrain, and Oman slightly larger (see Measurements).

Forms superspecies with Nyanza Swift *A. niansae* from highlands of north-east and East Africa (Lack 1956*b*; Brooke 1969*a*; Snow 1978).

Recognition. Adult differs from Swift *A. apus* in distinctly paler and less uniform upperparts; in particular, forehead, crown, rump, tertials, secondaries, inner primaries, and median and greater upper wing-coverts contrastingly pale, greyish olive-brown rather than black; pale throat-patch usually wider, extending further down (but *A. pallidus* with narrowest patch similar to *A. apus* with widest); chest to vent and flanks less uniform black, appearing mottled olive-brown and with distinct white feather-fringes, under tail-coverts usually contrastingly paler. Pacific Swift *A. pacificus* also has large white throat-patch and white feather-fringes on belly, but upperparts deep black with contrasting white rump. Plain Swift *A. unicolor* closely similar to *A. pallidus* in general tinge, but upperparts slightly darker with blacker saddle on mantle; median and greater upper wing-coverts and inner flight-feathers hardly paler than lesser coverts and outer primaries; underparts with mottling and fringing similar to *A. pallidus*, but chin and throat hardly paler than remainder of underparts, not contrastingly white or pale grey.

CSR

Apus pacificus Pacific Swift (Fork-tailed Swift)

PLATES 62 and 63
[facing pages 663 and 686]

Du. Siberische Gierzwaluw   Fr. Martinet du Pacifique   Ge. Pazifiksegler
Ru. Белопоясный стриж   Sp. Vencejo asiático   Sw. Orientseglare

Hirundo pacifica Latham, 1801

Polytypic. Nominate *pacificus* (Latham, 1801), USSR from Altay to Kamchatka, northern Mongolia, northern China, and Japan. Extralimital: *kanoi* (Yamashina, 1942), south-east Tibet, southern China, and Taiwan; *leuconyx* (Blyth, 1845), outer Himalayas and hills of Assam; *cooki* (Harington, 1913), south-east Asia, east from eastern Burma, south of *kanoi*.

Field characters. 17–18 cm; wing-span 48–54 cm. Size close to Swift *A. apus*, but overall length and wing-span greater; tail also more distinctly forked; up to 40% larger than White-rumped Swift *A. caffer* and Little Swift *A. affinis*. Elegant, sooty swift, with large white rump and noticeably black tail. Flight silhouette slightly more rakish than *A. apus*. Sexes similar; no seasonal variation. Juvenile separable at close range.

ADULT. Foreparts and wings dark sooty-grey, at close range rarely appearing as black as in *A. apus*; divided above from black tail by obvious broad and deep white rump and (in good light) contrasting below with blacker vent and undertail. At close range, underbody feathers (when fresh) and larger under wing-coverts show white margins, while almost white chin and upper throat more obvious than on any other white-rumped *Apus*. JUVENILE. Resembles adult, but at close range white margins to all body feathers and wing-coverts may show.

With *A. apus* and Pallid Swift *A. pallidus*, forms trio of similarly-built, medium-sized *Apus* swifts but instantly distinguished from them by white rump. Beware, however, partially albinistic *A. apus* and also *A. caffer*; though much smaller, latter has somewhat similar structure, including long tail-fork. Appearance in flight close to *A. apus* but silhouette and action more rakish due probably to slightly greater attenuation of wings and tail. Feeding behaviour as *A. apus*.

Solitary vagrant unlikely to call, but notes of breeding birds, though of similar structure to *A. apus*, slightly softer and less noisy.

Habitat. Breeds from temperate through boreal and subarctic to low Arctic in both continental and oceanic zones, including small offshore islands. Occurs over low farmlands and settlements, plains and steppes, sometimes taiga, and up to *c.* 4000 m in mountains. Breeding colonies, whether in crags or rock cliffs or in crevices or under eaves of tall buildings, require ready access to stretch of water. In good weather, hunts over mountain peaks, but during fog, rain, or snow shifts to low valleys and foothills. In northern latitudes will hunt at very late hour, even beyond midnight. On migration crosses mountain ranges and seas. (Dementiev and Gladkov 1951*a*.) Winters largely over lowlands, in aerial habitat as always; in Australia over open country generally, semi-deserts to coast and islands, sometimes over cities (Pizzey 1980).

Distribution. Breeds from Siberia eastwards to Kamchatka and Japan, and south through China to Thailand, Burma, outer Himalayas, and Khasai hills. Northern races highly migratory, wintering in Malaysia, Greater and Lesser Sunda Islands, New Guinea, and Australia (see Movements).

Accidental. North Sea: one (nominate *pacificus*) captured 19 June 1981 on Leman Bank gas platform (53°06′N 2°12′E); taken ashore and released in Norfolk, England (Rogers *et al.* 1983).

Movements. Largely migratory, but few details known due to absence of ringing and frequent high-flying outside breeding season. Broad outline knowledge based mainly on museum specimens.

Himalayan race, *leuconyx*, is at best a short-distance migrant; movements capricious and unpredictable outside breeding season, though certainly reaches peninsular India. South-east Asian race, *cooki*, present all year in breeding range (Burma, Thailand, Indochina), but specimens exist from Malay peninsula, and birds with narrow rump-band (probably this race) locally common there in some winters. Central and south China race, *kanoi*, also a short-distance migrant, wintering south to Philippines, Malaysia, and Indonesia. Nominate *pacificus*, breeding Siberia to northern China and Japan, long-distance migrant; winters Indonesia, Melanesia (notably New Guinea), and Australia (south to Tasmania), straggling to New Zealand and subantarctic Macquarie Island. A few specimens of nominate *pacificus* exist from northern India, though status there unclear; likely that, as in Needle-tailed Swift *Hirundapus caudacutus*, main movements between breeding and wintering ranges occur further east, through China and East Indies. See Vaurie (1965), Ali and Ripley (1970), Condon (1975), and Medway and Wells (1976). Birds attributed to nominate *pacificus* recently found occurring sparingly but regularly on Seychelles (Indian Ocean), October–November and May; dates suggest passage to/from unknown winter quarters, possibly in Madagascar or Africa (Feare 1979c). North Sea vagrant (see Distribution) also attributed to nominate *pacificus* (C S Roselaar).

Bulk of birds arrive Mongolia and Siberia during May, departing August and early September (Dementiev and Gladkov 1951a). Passage through Malaya, which includes nominate *pacificus*, mid-September to mid-November and late February to late May (Medway and Wells 1976). Widespread and common in Australia between October–April, though with local numerical fluctuations linked to thunderstorms and tropical cyclones (Condon 1975).

Voice. See Field Characters.

Plumages (nominate *pacificus*). ADULT. Forehead, lores, crown, nape, and upper mantle, down to lower cheeks and sides of neck blackish-brown, darkest on crown, with some dark grey-brown of feather-bases showing on sides of head and neck and on nape to upper mantle; feather-tips narrowly fringed pale grey or white, especially when plumage fresh, most conspicuously so on feathers just above eye, on lower cheeks, and on sides of neck. Black spot in front of eye. Mantle, scapulars, and back deep black, contrasting with black-brown upper mantle; feather-tips narrowly and faintly fringed grey on mantle and scapulars, white on back. Rump contrastingly white, extending into large white patches on rear flanks to form conspicuous U-shaped band *c.* 1·5 cm wide; white of rump and rear flank with small and narrow dusky shaft-streaks, but these occasionally almost absent. Upper tail-coverts deep black. Throat-patch white with some fine black shaft-streaks; patch rather narrow on chin, widening downwards; contrasts markedly with dark lower cheeks and chest; shape of patch as in Pallid Swift *A. pallidus* and sometimes almost as large as in that species, but more sharply defined than in both *A. pallidus* and Swift *A. apus* and wider on throat than in latter. Sides of chest, upper flanks, and chest down to under tail-coverts black, feather-tips with sharply defined white fringe 1–1·5 mm wide; fringes larger and wider than on any other Palearctic *Apus* (including *A. pallidus* and Plain Swift *A. unicolor*), producing marked scaling on underparts. Some dark grey-brown of feather-bases often visible on underparts, sometimes almost greyish-white on under tail-coverts. Tail black. Outer primaries, greater primary coverts, and lesser and median upper wing-coverts deep black with slight blue gloss; inner primaries, secondaries, tertials, and other coverts slightly paler, dull black; inner webs of flight-feathers broadly bordered dark grey-brown. Under wing-coverts and axillaries dull black with contrasting white fringes on tips, forming conspicuous scaly pattern along leading edge of wing. In worn plumage, pale feather-fringes on upperparts and wings abraded and lost, but usually still readily visible on underparts and longer under wing-coverts; white throat and rump often more distinctly streaked and white of feathers partly lost through abrasion; throat and rump sometimes less contrasting than in fresh plumage. JUVENILE. In fresh plumage, closely similar to fresh-plumaged adult; differs from adult mainly in showing fresh plumage when adult slightly or heavily worn, difference most marked on coverts of leading edge of wing. Often more extensively white on throat and rump than adult (rump up to *c.* 2 cm wide), virtually without dusky shaft-streaks. Distal border of inner web of t5 near tip straight or slightly convex (in adult, concave or with slight notch). FIRST ADULT. Similar to adult between late 1st winter and autumn of 2nd calendar year, but juvenile flight-feathers (except some tertials) and greater coverts retained, heavily worn and faded to brown. Shape of t5 as in juvenile or intermediate between juvenile and adult. See also Moults.

Bare parts. ADULT AND JUVENILE. Iris deep brown. Bill black or dark blackish-horn. Mouth flesh-coloured. Leg and foot purplish-black or black, soles purplish-flesh (Dementiev and Gladkov 1951a; C S Waller; RMNH.)

Moults. Mainly based on birds from Indonesia (RMNH); few birds available after November. ADULT POST-BREEDING. Complete; primaries descendant. Starts with p1 between late September and late October; completed with p10 late January to early March. Primary moult score 4 (11) 0–10 in first half October, 7 (10) 0–20 in second half October, 16 (6) 6–23 in first half November; 3 birds from first half October showed suspended moult or had just started moult after suspension (scores 15, 20, 20), with inner primaries probably replaced in breeding area. September–October birds which had not yet started primary moult (score 0) on average had 30% of feathers of upperparts new (mainly nape, lower mantle, back, outer scapulars, and some upper tail-coverts) and 65% of underparts (mainly breast and flanks, part of vent and chest), with tail old; these feathers perhaps already replaced in breeding area. At score 25, *c.* 60% of feathers of upperparts and 80% of underparts new, as well as some lesser upper wing-coverts or tertial coverts; tail still old. By late January, body, tail, and wing-coverts nearly all new; by about February, moult completed. No indication of occasional retention of old p10 as in *A. apus*, but perhaps overlooked. POST-JUVENILE. Some still fully juvenile in November–December, others start body moult from late November, scattered feathers of nape, mantle, back, and underparts first. In May–August of 2nd calendar year, head and body fresh or slightly worn 1st adult, tail, lesser upper wing-coverts, tertial coverts, and a few tertials fresh 1st adult; remainder of wing-coverts and flight-feathers still juvenile, worn. Juvenile flight-feathers probably replaced in 2nd winter, as in *A. apus*. Some October birds showing adult moult pattern had heavily worn outer primaries; these probably in 2nd calendar year.

Measurements. Nominate *pacificus*. Ussuriland and Kamchatka (USSR), May–July, and Indonesia, September–

February; skins (RMNH, ZFMK). Tail is to tip of t5; fork is tip of t1 to tip of t5; bill is exposed culmen.

WING AD	♂ 180	(3·74; 23)	176–186	♀ 177	(2·46; 13)	173–182
TAIL AD	79·4	(2·36; 26)	75–83	79·1	(4·06; 16)	76–88
FORK AD	33·3	(2·17; 26)	30–38	32·7	(3·18; 16)	29–40
BILL	7·1	(0·35; 11)	6·8–7·6	7·0	(0·27; 11)	6·7–7·4
TARSUS	11·4	(0·34; 10)	10·8–11·9	11·2	(0·57; 11)	10·6–12·2

Sex differences significant for wing. Juvenile wing on average 4 mm shorter than adult, tail 9, fork 6.

Weights. Nominate *pacificus*. Summer, USSR: ♂ 42·4, ♀ 44·7 (Dementiev and Gladkov 1951a). Mongolia: early June, ♂ 48·1 (5·87; 7) 38–54, ♀ 42·5 (2) 38–47; early July, ♂ 37 (Piechocki 1968). New Guinea, January: ♂ 33 (RMNH).

A. p. kanoi. Taiwan, April: ♂ 56, ♀ 50 (RMNH).

Structure. 10 primaries: p9 longest, p10 3–6 shorter, p8 8–14, p7 28–38, p6 48–59, p5 66–82, p1 117–131. Tail rather long, rather more deeply forked than in *A. apus* (fork 41% of tail length, against 36% in *A. apus*). Outer middle toe without claw 6·4 (7) 5·6–6·8, claw 6·5 (7) 5·9–7·0; innermost toe with claw c. 69% of outer middle toe with claw, inner middle c. 83%, outermost c. 80%. Remainder of structure as in *A. apus*.

Geographical variation. Within nominate *pacificus*, variation in colour slight; among large sample from USSR, China, and Japan (ZMM), USSR birds on average slightly less deep black, but still overlapping completely with others. Birds of China and Japan apparently slightly larger than those of USSR, but only a few measured: adult wing 187 (5) 182–192. *A. p. kanoi* from south-east Tibet, southern China, and Taiwan deeper black than nominate *pacificus*, especially on crown, mantle, and belly; crown not paler than mantle as in nominate *pacificus*; pale throat-patch smaller, grey rather than white, sometimes faintly streaked dark grey; white rump slightly narrower, occasionally with faint dusky streaking; pale fringes of feather-tips sometimes less conspicuous; length of wing intermediate between USSR and Japan populations of nominate *pacificus*, tail less deeply forked. *A. p. leuconyx* from outer Himalayas (Kashmir to Arunachal Pradesh) and hills of Assam dark as in *kanoi* but much smaller, wing 147–171 (Lack 1956b; Vaurie 1965; Ali and Ripley 1970). *A. p. cooki* from eastern Burma throughout south-east Asia (south of *kanoi*) darker than both *kanoi* and *leuconyx*, upperparts glossy bluish-black, white feather-fringes faint or absent, white rump c. 1·0 cm wide rather than 1·5 cm, numerous and distinct dusky shaft-streaks on throat and rump; wing 162–172, p10 about equal to p9 (Delacour and Jabouille 1931; Lack 1956b; Medway and Wells 1976).

Forms superspecies with Dark-backed Swift *A. acuticaudus* from hills of Assam (India), where range overlaps with *A. p. leuconyx*; wing close to nominate *pacificus*, but tail less deeply forked; plumage similar to *kanoi* and *leuconyx*, but rump dark like back, not white (Lack 1956b; Brooke 1969c). *A. acuticaudus* and *A. pacificus* form species-group with *A. apus* and Plain Swift *A. unicolor* (Brooke 1970). CSR

Apus melba Alpine Swift

PLATES 59 and 63
[between pages 662 and 663, and facing page 686]

Du. Alpengierzwaluw Fr. Martinet alpin Ge. Alpensegler
Ru. Белобрюхий стриж Sp. Vencejo real Sw. Alpseglare

Hirundo melba Linnaeus, 1758. Synonym: *Tachymarptis melba*.

Polytypic. Nominate *melba* (Linnaeus, 1758), northern Morocco, southern Europe, Asia Minor, and north-west Iran; *tuneti* Tschusi, 1904, central and eastern Morocco, Algeria, Tunisia, Libya, coastal Levant (except, perhaps, Dead Sea Depression), and from Iran (except north-west) east to western Pakistan; *archeri* Hartert, 1928, northern Somalia, south-west Arabia, and (probably this race) western Arabia north to Dead Sea Depression in Jordan. Extralimital: 4 races in Afrotropics and Madagascar, 3 races in Indian subcontinent.

Field characters. 20–22 cm; wing-span 34–60 cm. About 20% larger and longer winged than Swift *A. apus*, with greater bulk as evident in powerful flight as in structure. Large, robust swift differing from all fork-tailed west Palearctic congeners in underpart pattern. Silhouette thick-set, and tail relatively shorter than in *A. apus* Plumage mainly umber-brown, distinctly paler above than all *Apus* except Pallid Swift *A. pallidus*, sharply relieved below by brown breast-band and white central underbody. Flight impressive, much more vigorous and powerful than smaller species. Sexes similar; no seasonal variation. Juvenile separable from worn adult. 2(–3) races in west Palearctic, doubtfully separable in the field.

(1) European race, nominate *melba*. ADULT. Head, upperbody, upper and under wing-coverts, broad breast-band, upper flanks, vent, and under tail-coverts umber-brown, with variable grey to sandy tone; all upperparts thinly tipped grey-white but becoming uniform when worn, all dark underparts tipped white, most noticeably on flanks and under tail, and not becoming uniform with wear, always retaining noticeably rough appearance. Patch in front of eye black, obvious at close range. Chin and central throat white, rough-edged under face. Lower breast, belly, and lower flanks white, rough-edged under wing but not across breast or vent. Flight and tail-feathers black-brown, with green gloss occasionally shining in sunlight. JUVENILE. Resembles adult in fresh plumage, but white tips to wing-coverts and flight-feathers more prominent. Leaves nest from late July and, in close view, distinguishable from worn adults in August and early September; also, adults (unlike juveniles) show missing primaries July–November(–December). (2) North African and south-west Asian race, *tuneti*. At all ages, brown paler and greyer than in nominate *melba*. (3) For possible

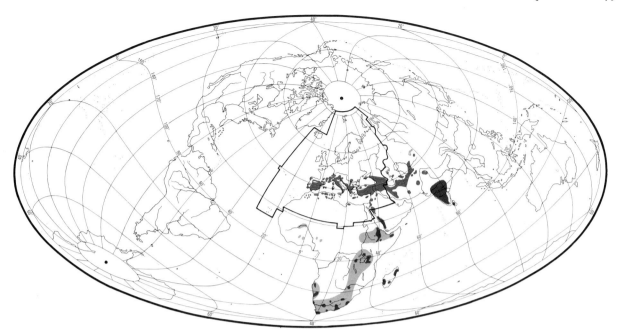

presence of (even paler) *archeri* in Jordan and Sinai, see Geographical Variation.

Normally unmistakable in west Palearctic but northern races confusable in African winter quarters with local races and, in brief view, Mottled Swift *A. aequatorialis* (darker, with barred, not pure white, underbody). More serious danger of confusion with occasional aberrant *A. apus* showing underpart pattern recalling *A. melba* only prevented by patient observation of size and structure. Important to note that a few *A. melba* show only small or virtually no white on throat in the field (at least in autumn); not known if due to moult, abrasion, or immaturity. Flight action slower than *A. apus* and *A. pallidus* (lacking characteristic 'twinkle' of *A. apus*) and much more powerful, with individual wing-beats more easily discernible (and creating audible swooshing at extreme close range) and all aerial manoeuvres less sudden but not less confident. In brief glimpse or distant view, silhouette has sufficient bulk to suggest small falcon *Falco*, unlike other swifts occurring in west Palearctic except Needle-tailed *Hirundapus caudacutus*. Speed of flight deceptive, sometimes appearing slow but actually rapid; outpaces *A. apus* and *A. pallidus*. Behaviour much as *A. apus*, but appears to feed in higher airspace.

Main call a high-pitched trilling 'trihihihihi', quite different from main call of *A. apus*; frequent at breeding areas, much less so elsewhere.

Habitat. In west Palearctic, breeds in lower middle latitudes, from Mediterranean and steppe to temperate zones, including boreal montane regions. Basic habitat aerial and of immense extent and volume. Birds cover an estimated 600–1000 km daily and ascend at least to limit of range of powerful binoculars, apparently influenced by temperature, time of day, air currents, and density of insect food, which may be too low to account for presence at highest altitudes frequented (Lack 1956a). Nature of terrain underlying this aerial habitat apparently only significant in so far as it affects pattern of air currents and abundance of flying insects; at lower levels, however, usually avoids areas with frequent obstructions such as trees and buildings (unlike Swift *A. apus*), and seems less often to skim low over water, although occasional access to it is necessary. Mobility often involves much smaller proportion of breeding population being present in vicinity of colony than with *A. apus*, and tendency to maintain longer individual distance or to operate individually away from colony. Partial shift, which must have occurred within recent historic times, from exclusive use of mountain rock ledges, crevices, caves, coastal cliffs, and occasional holes in trees to colonial use of suitable (largely free-standing) tall buildings has enabled extension of range over lowlands and northwards—also in some areas to increases in colony size. Unless interfered with, interiors of suitable buildings tend to be used over decades or even centuries, being resorted to not only for breeding and roosting but as refuges in case of intolerably wet or chilly weather. Such breeding habitats are dry, cool, and ill-lit. (Dementiev and Gladkov 1951a; Arn 1959, 1960; Glutz von Blotzheim 1962; Lüps *et al.* 1978.) Migrates in upper airspace; noted in tropical Africa feeding low above a grass fire on small grasshoppers, and skimming a river among swarms of bugs (Hemiptera), dipping occasionally in the water (Bannerman 1955). In southern Africa, also noted as coming down low during overcast weather; in East Africa, as moving in response to thunderstorms (Williams 1963; Prozesky 1970).

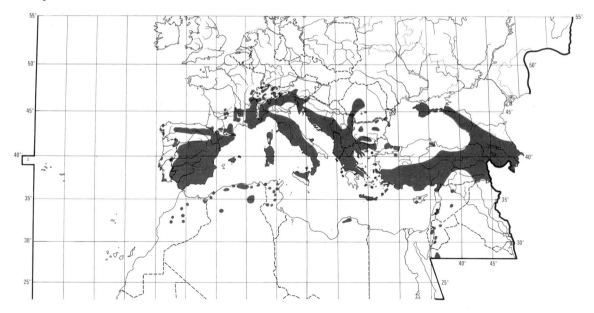

Distribution. Some recent spread in France; West Germany colonized in 1955.

FRANCE. Some recent range increase in Provence and Hautes-Maritimes (Yeatman 1976); spreading north in Massif Central (Cochet 1978, 1982). WEST GERMANY. First proved breeding 1955 (Glutz and Bauer 1980). SWITZERLAND. First nested in human settlements mid-18th century; gradual spread since (see Glutz and Bauer 1980). RUMANIA. First bred 1966; spreading (Talpeanu and Paspaleva 1979; Cătuneanu and Fuhn 1980). ALGERIA. Probably more widespread in north than mapped (EDHJ).

Accidental. Britain, Ireland, Belgium, Netherlands, Denmark, Norway, Sweden, Finland, East Germany, Czechoslovakia, Hungary, Iraq, Madeira.

Population. Limited information. Marked fluctuations in Switzerland due to bad weather (see Dizerens 1982).

FRANCE. 1000–10000 pairs (Yeatman 1976). SPAIN. No complete counts; 328 pairs in Pyrénées (Purroy 1973). WEST GERMANY. Sole colony increased from 3–4 pairs 1955 to over 30 pairs 1979 (Glutz and Bauer 1980). SWITZERLAND. Not less than 1250 pairs 1970–4, of which at least 760 pairs nesting in buildings; after mass mortality caused by severe weather in late autumn 1974, average only of 320 pairs in buildings 1975–8 (Glutz and Bauer 1980); now increasing again (RW). LEBANON. Decreasing Beirut (Tohmé and Neuschwander 1974). CYPRUS. Common, with colonies up to c. 150 pairs (PRF, PJS).

Survival. Switzerland: annual adult mortality in 3rd–7th years 20·9%; exact calculations for 1st and 2nd year not possible, but probably 21–24% (Glutz and Bauer 1980). Oldest ringed bird 26 years (Arn 1967).

Movements. Migratory in west Palearctic, partially so in Indian subcontinent and Afrotropical region.

India and Ceylon races make local movements at least, especially in monsoon season, and assumed to mingle in southern India, but details not worked out (Ali and Ripley 1970). 5 races (including *archeri*) in Afrotropics, with southernmost populations partially migratory: majority absent from Cape Province May–August, with passage through Zimbabwe May–June and August–October (Brooke 1971d; Curry-Lindahl 1981).

West Palearctic populations (*melba, tuneti*) migratory but not separable in sparse data from African wintering areas; indeed, no Afrotropical specimens of *tuneti* known (Brooke 1971d), though North African birds at least must occur. Winter distribution poorly known due to use of higher airspace then, and also (in East Africa) to occurrence of endemic races. Passage noted Mali October–November and April–May (Laferrère 1968); also January–February records from Niger inundation zone, but biggest concentrations in June and breeding suspected (Lamarche 1980). Further south, recorded Sierra Leone, once in December (G D Field); Liberia, around Mt Nimba in February (Moreau 1972); Ghana and Togo, various records January–May, including 1 recovery of Swiss-ringed bird (Glutz and Bauer 1980; Walsh and Grimes 1981; Cheke 1982); Nigeria, a few November–May records, largest flock 200 (Elgood 1982); Sudan, common migrant, some wintering in south (Cave and Macdonald 1955); Ethiopia, common migrant and winter visitor (Urban and Brown 1971); northern Zaïre, regular winter visitor to Uelle region and nominate *melba* collected (Chapin 1939); Uganda, where regular September to early April, Palearctic birds collected January and March, and northward passage (up to 1000 per day) over Chobe each spring (Britton 1980). Hence probably occurs all across northern tropics, but doubtful whether Palearctic birds reach Kenya or Tanzania (Britton 1980), while reexamination of skins found that (contrary to previous assertions) there was no museum evidence of them south of equator (Brooke 1971d; Moreau 1972).

Passage movement on broad front through Mediterranean basin, North Africa, and Middle East. Winter specimens extant from north-west India (Ali and Ripley 1970), and from Dead Sea and Sinai (possibly *archeri*—see Geographical Variation); some birds winter regularly in Arabia (Jennings 1981a), otherwise movement directed towards Africa. Ringed only in Switzerland, whence birds recovered on passage in France (29), Mediterranean Spain (8), Italy (6), Algeria (3), and Morocco (3). Autumn passage 6 weeks later than in Swift *A. apus*, hence, unlike that species, adults begin moulting in breeding area. Migration through south Palearctic largely concentrated September to mid-October, though some southward movement in August by juveniles at least (Glutz and Bauer 1980). Spring vanguard reaches North Africa in second half of February, Switzerland late March to early April; some northward movement continues to mid-May. For age/sex variation in return to colony, see Social Pattern and Behaviour.

In Europe, midsummer movements around depressions (to avoid rainfall) occur as in *A. apus*, though on smaller scale; possibly involve mainly pre-breeders. Recoveries of Swiss-ringed birds in Czechoslovakia (2), Austria, southern West Germany (6), Denmark, and England will include cases of such movement (Glutz and Bauer 1980). This phenomenon likely to constitute major cause of vagrancy north of breeding range, including unprecedented flock of *c.* 100 in Kent (England) on 15 July 1917 (Lack 1956a).

In Swiss study, 83 certain cases of birds settling to breed away from natal area involved displacements of 7–112 km, average 25 km (Glutz and Bauer 1980; see also Social Pattern and Behaviour).

Food. Flying insects and spiders of moderate size. Prey caught entirely in flight. Foraging behaviour compares closely with Swift *A. apus*. Reported feeding at night on moths attracted to light (Freeman 1981; see also Blatti 1947). Readily takes advantage of locally abundant food sources, e.g. bugs *Agnoscelis nubila* found crammed in guts of birds in India (Ali and Ripley 1970). Chitinous remains ejected as pellets (Dementiev and Gladkov 1951a).

Takes mostly flying insects including stoneflies (Plecoptera), caddis flies (Trichoptera), dipteran flies (Muscidae, Tipulidae, Culicidae, Tabanidae, Syrphidae), beetles (Scolytidae, Staphylinidae, Carabidae, Coccinellidae, Curculionidae), Hymenoptera (Siricidae, Apoidea, Vespidae, Chalcidoidea, Formicidae, Ichneumonidae, Braconidae, Proctotrupoidea), butterflies and moths (Lepidoptera), bugs (Aphididae, Cicadidae), lacewings (Neuroptera), mayflies (Ephemeroptera), grasshoppers (Acrididae), and dragonflies (Odonata); also spiders (Araneae) (Hess 1924, 1927; Hartert 1927; Bartels 1931; Arn 1960; Glutz and Bauer 1980; D W Yalden).

Most studies of diet based on food-balls from adults at colony; most of this food presumably intended for young. In USSR, takes midges and mosquitoes (Culicidae) (Dementiev and Gladkov 1951a), also winged termites (Isoptera) (Ivanov 1969). Composition of food-balls (fed to young) varies markedly (Glutz and Bauer 1980). In Switzerland, June–August, food-balls contained 156–220 insects each; included dragonflies, aphids, ichneumons, Diptera (especially *Tabanus bovinus*), beetles, and butterflies (Zehntner 1890). At Solothurn (Switzerland), food-ball contained 25 insects: 1 stonefly (Perlidae), 8 flies (Diptera, including 2 Tipulidae), 3 bugs, 1 lacewing, 2 beetles (Staphylinidae, Coccinellidae), 2 ants (Formicidae), 6 ichneumons, 1 moth, and 1 butterfly (Hess 1924). One food-ball, regurgitated by nestling, Solothurn, included wood-wasp *Sirex gigas* and 7 drone honeybees *Apis mellifera* (Hess 1927). At 2 colonies, Solothurn, 21–23 July, 6 food-balls contained mostly Hemiptera, Hymenoptera, Coleoptera, and Araneae; mean 219 items (11–626) per food-ball, mean weight 2·53 g (2–3) (Arn 1960, which see for details). 17 food-balls from birds near Berne (Switzerland) contained 308 flies, 125 aphids, 100 beetles (including Scolytidae and Curculionidae), 94 wasps (Vespidae), 2 ants, 21 spiders, 16 cicadas (Cicadidae) and 1 other bug, 15 lacewings, and 2 butterflies (Bartels 1931). In Akrotiri (Cyprus), one bird found to have taken 1 dragonfly (Libellulidae), 28 drone *Apis mellifera* (presumed to be a mating swarm), beetles (2 Curculionidae, 1 Carabidae), and 2 bugs (D W Yalden). Stomachs of 2 birds feeding in huge flock over grass fire, Zaïre, contained grasshoppers, bugs, beetles, and a leaf-hopper (Cicadellidae); 11 stomachs contained almost exclusively bugs (Chapin 1939).

Food of young also includes crickets (Orthoptera), ants, and beetles, often presented alive (Glutz and Bauer 1980). In good weather, young received, on average, 10 feeds per day (Arn 1935). BDSS, EKD

Social pattern and behaviour. Major studies by Bartels (1931) and Arn (1945, 1960) in Solothurn (Switzerland). In many respects, behaviour similar to Swift *A. apus*.

1. Gregarious in breeding season, and mostly throughout the year. During migration and when feeding, in or out of breeding season, typically travels in flocks of variable size. On autumn migration, Belen (south-east Turkey), 223 flocks averaged 7·4 birds, smaller than in *A. apus* (Sutherland and Brooks 1981). Migrating flocks in spring also smaller than those of *A. apus*, related to more gradual return of *A. melba* to colonies (Arn 1959). At end of March, Iraq, migrating flocks of 20–30 (Johnson 1958). At start of breeding season, Solothurn, first arrived in small groups, exceptionally up to 50 together (Arn 1945). Adult ♂♂ tend to return before ♀♀, 1-year-olds later still. At end of breeding season, juveniles tend to depart before adults (Arn 1960); departure (of all ages) gradual unless weather bad, when mass exodus occurs (Arn 1945). BONDS. Monogamous mating system, and pair-bond maintained from year to year (Arn 1960). Mate fidelity marked between years, facilitated by strong site fidelity (see Breeding Dispersion, below). Same mate retained for 2–11 years (Arn 1960). 3 pairs bred respectively for 4, 5, and

6 (perhaps 7) years, each pair using same site every year (Arn 1945). Age of first breeding usually 2–3 years, sometimes 1. Of 56 birds first controlled at colony in 1st year, 7·1% bred; of 87 in 2nd year, 65·5%; of 116 in 3rd year, 94·8%. 1-year-olds typically form pairs and build nests, but most do not breed (Arn 1960). Both sexes share nest-duties equally (Bartels 1931), ♀ perhaps incubating slightly more (Arn 1960). For accidental fostering of young, see Relations within Family Group (below). Unlike *A. apus*, parents at colony, Solothurn, never departed before young fledged. Young independent at fledging (Arn 1960). BREEDING DISPERSION. Typically gregarious and colonial. At Solothurn, 1967–82, average size of 2 colonies 93·7 (31–170) and 39·4 (17–62) pairs (Dizerens 1982). In Pyrénées (Spain), 322 pairs bred in 22 colonies, average colony size 14·9 (2–88) pairs (Purroy 1973). In Israel, usually in large colonies, though 1 solitary pair occurred (Tristram 1866). At Solothurn, minimum distance between nests 1 m (Bartels 1931). Nest-territory confined to immediate vicinity of nest, and said to be smaller than in *A. apus*; probably ♂ selects site, and first to return to former one (Arn 1960). 2-year-olds return to site established in previous year (Arn 1960). Breeding birds show strong fidelity to colony and nest-site (Arn 1942, 1945, 1960): in 8-year study of 62 birds at Solothurn, 33 used same nest throughout, 13 changed nest, and status of rest not clear (Arn 1942). Records of birds breeding 11 and 12 years in same nest (Arn 1945); one for 17 years (Arn 1959, 1960). Young birds may settle at or away from natal colony; of 4561 nestlings ringed, 1932–56, 520 later controlled at natal colony, one bird returning to breed in nest-site where reared (Arn 1960). ROOSTING. Typically in colony; unlike *A. apus*, aerial roosting, at least in breeding season, thought to be exceptional. From arrival at colony until young fledged, breeding pairs roost at nest; pair may lie side by side in nest, or 1 just outside nest, sometimes clinging vertically, head upwards, on wall nearby. Even when young near fledging one parent often outside nest, but family may roost, criss-cross fashion together in nest, young sometimes on parents' backs (see also Relations within Family Group). After young fledged, not all breeding birds roost in nest, some staying outside on walls (Arn 1945, 1959, 1960, 1968). Pairs which have lost young may continue to roost on nest (Arn 1960), but after nests robbed, many birds roosted on wall outside colony, some inside (Bartels 1931). Non-breeders roost close together, plumage ruffled, on outside walls of colony usually near entrance holes or under eaves, less commonly inside (Arn 1945, 1959, 1960, 1968). In cold weather, non-breeders typically cling to wall in dense clusters (Brüllhardt 1969; Dizerens 1982: Fig A); at Wildegg (Switzerland), 25 birds thus bunched like moving ball on floor of loft, birds on outside trying to crawl into middle (Brüllhardt 1969). Unlike *A. apus*, no records of birds ascending in evening and descending in morning to suggest that aerial roosting is usual among non-breeders (Arn 1960, 1968; Sudhaus 1972). At Solothurn, active for fewer daylight hours than average for *A. apus*. From first arrival at colony, mid-March, until end of April, emerged from roost 09.00–10.00 hrs, singly or in small groups, dispersed from colony, and returned at 18.00 hrs. Once breeding started, most left nests at 06.00–07.00 hrs to feed away from colony, and most returned before 19.00 hrs (Arn 1960); colony may appear completely deserted during day (D J Brooks). By comparison, *A. apus* out before sunrise and often not back until after sunset: *A. melba* therefore active, on average, 2 hrs less in both morning and evening (Arn 1960). In August, when most young fledged, adults and juveniles remained in colony until 10.00 hrs, returned *c.* 17.00 hrs (Arn 1960; see also Bartels 1931, Arn 1959). Immatures, however, often did not roost outside building until 30–45 min after sunset (Arn 1960). For birds still active, even feeding, after sunset, see Sudhaus (1972) and Freeman (1981)—also Food. In bad weather, adults leave nest infrequently, and restless at night (Häari 1931; Arn 1945, 1959); once, roosted from 16.00 hrs (Zehntner 1890). Once, in heavy rain, 8 birds of which 2 were 1-year-olds, flew into nest-entrance, and clung there, calling loudly (Arn 1945). At time of roosting, birds highly vocal (Bartels 1931; see 1 and 4 in Voice). At night, mostly silent, unless bird suddenly returns, when fighting may occur (Arn 1945; see Antagonistic Behaviour, below). Between waking and leaving roost, birds self-preen assiduously for *c.* 1 hr (Arn 1959, 1960).

2. FLOCK BEHAVIOUR. Forms flocks for feeding, social display, pre-migratory behaviour, and migration. In breeding season, colony members frequently perform communal Screaming-display. No detailed study, but said to be same as *A. apus*, except for timing (see above) and calls (see 1 in Voice); display most common in fine weather for *c.* 1–2 hrs before roosting (Arn 1959). At very start of breeding season, when few birds present, no display occured. Once, however, when 50 birds arrived together, displayed from outset (Arn 1945). ANTAGONISTIC BEHAVIOUR. Compared with *A. apus*, more tolerant of conspecific birds, ritualized threat postures poorly developed, and fighting rare (Arn 1960). Early in season, may defend nest with Threat-call (see 3 in Voice), but this becomes less common, perhaps either through greater familiarity with neighbours, or because of being occupied feeding young. If bird entering colony passes 20–30 cm from another's nest, owner gives Threat-call but offers no other challenge; at 10–30 cm, owner stretches neck and gives Threat-call; at 10–15 cm, may briefly attack and pursue intruder a short way. Confrontations most frequent, though by no means common, on return to roost, e.g. when pair flew in to wrong nest, resident leaped with a squeal on to back of nearer of pair, and the two tumbled down *c.* 1 m before separating (Arn 1960). Associates amicably with feral Rock Doves *Columba livia* nesting immediately adjacent (Arn 1960, which see for photograph; for fierce aerial attack by pair, however, see Ferguson-Lees 1969*c*). HETEROSEXUAL BEHAVIOUR. (1) General. As in *A. apus*, ♂ assumed to take initiative in attraction of mate (Arn 1960). (2) Site-prospecting. ♂ presumed to select site, or to re-claim former one (Arn 1960). Behaviour apparently as in *A. apus*; single birds or small groups fly up under eaves and cling briefly to wall, apparently looking for sites. If in group, birds fly in single file. Site-prospecting occurs mainly in first few weeks of breeding season, typically on fine evenings. Birds prospecting at end of June presumably immatures (Vleugel 1952; see also Arn 1960).

A

(3) Pair-bonding behaviour. At Solothurn, courtship at nest most evident late May and early June, before egg-laying; sequence also serves as Meeting-ceremony. Behaviour apparently similar whether establishing new pair-bond or renewing former one. Before nest-building, pair regularly perform mutual Allopreening (especially in neck region) and Billing (perhaps Allopreening around base of bill, as in *A. apus*), accompanied by Whispering-calls (see 2 in Voice). In days before laying, established pairs also repeatedly alternate occupancy of nest over short period. Immatures spend much of season nest-building in preparation for following season; may begin or complete nest (Arn 1960). (4) Mating. Occurs both on or beside nest, and in the air; former probably more common, though no detailed study (Arn 1945, 1960). Also reported in birds clinging to wall in colony (Arn 1960). (a) At nest. No details, but preliminaries said to resemble Meeting-ceremony: pair lie pressed close together, Allopreening, Billing, and giving Whispering-calls (Arn 1960). Copulation usually in morning, or evening after 18.00 hrs (Zehntner 1890). (b) In the air. Frequently reported (Koenig 1886; Ali 1943; Arn 1960; Thiollay 1974; Heymer 1977). In one account, pair flew, one after the other, c. 30 m up, giving Whispering-call. ♀ swooped down and, when she made horizontal glide, ♂ mounted for c. 10 m, during which wings of both birds beat in shallow arc (Arn 1960, which see for diagram). Copulating pairs also described as whirling round and round, falling slowly downwards for c. 30 m with motionless, outstretched wings; then separated and each flew off alone (Ali 1943). In pre-copulatory sequence, presumed ♀ crouched on ledge by nest-hole with partially spread wings while ♂ crouched beside her and stroked her back with one wing. Both trilled, and eventually ♀ flew off with half-open trembling wings. ♂ instantly followed, and mounted ♀ in the air; ♀'s wings then extended and motionless, ♂'s raised and beating, both birds slowly losing height (Ferguson-Lees 1969c). (5) Behaviour at nest. In Meeting-ceremony (see above) at nest-relief, bird at nest typically gives apparent mild Threat-call to incoming mate, then both exchange Whispering-call (Arn 1960). Meeting-ceremony presumably most vigorous early in incubation. If weather fine, incubating bird may leave nest for several hours in afternoon. In late incubation, off-duty bird often sits alongside nest (Arn 1960). Birds begin building nest, or repairing former one, soon after arrival at colony (Arn 1935). Nest-building typically continues after laying, exceptionally up to hatching; sitting bird regularly tends nest-rim, and returning mate may bring nest-material (Bartels 1931; Bloesch 1931). RELATIONS WITHIN FAMILY GROUP. Following account mostly from Arn (1945, 1960). From 1st day, young lift heads, wave them around, and if disturbed, e.g. by strange sound, give quiet food-calls (see Voice, also for changes with age). Later, young able to distinguish such sounds from parents arriving (Arn 1945). Arriving bird divides food-ball between young when small, giving whole ball to one young when older. Parent pushes bill far into gape of young to transfer food-ball. As in *A. apus*, small young able to survive perhaps 2–3 days without food; when older, able to fast perhaps 8–10 days. By 28 days, young silent in nest when parents present, even if not fed for several hours. Until c. 10 days old, young brooded almost continuously, thereafter less so, and not at all once fully feathered (Arn 1945, 1960). However, parents may stay with young by day if weather bad (Häari 1931). Once brooding period over, one parent will raise young alone if mate dies. Adults may preen young from 8–12 days onwards; young also self-preen from 12 days. Eyes open at 12–15 days. From c. 30 days, able to crawl around in immediate vicinity of nest, though strongly attached to it. Once, parent arriving and not finding young in nest regurgitated food-ball into nest. From 40–45 days, young exercise wings vigorously, sometimes for 10 min at a time. From c. 45 days, young are fed either near or in entrance to nest-cavity, and some adults, perhaps not always their parents, feed them from outside. Parents apparently do not recognize own offspring at any age. Well-grown young regularly visit neighbouring nests, and, if same age as neighbours' young, are readily accepted and reared. Such young may expel or smother smaller young in neighbouring nest, and are then fostered. For accidental fostering of young *A. apus* with 3 *A. melba* young, see Arn (1960). Both members of pair—rightful or foster—rear young until fledging, unlike *A. apus* in which one parent may leave before other (Arn 1960). Independent young may beg from adults, occasionally in roost (Glutz and Bauer 1980), exceptionally in the air (Arn 1970), but are not fed. Young regularly roost in colony after fledging, staying several weeks before migrating (Arn 1945). ANTI-PREDATOR RESPONSES OF YOUNG. If uneasy in nest, young cease calling and crouch down (Arn 1945, 1960). Older young crawl out of nest at approach of human intruder (Arn 1945), then crouch on ground or up against wall (Arn 1935). When handled, some 3-week-old young responded more aggressively: made brief but vigorous upward fluttering movements, accompanied by Bill-snapping to one side (Bartels 1931). 1 young responded thus at 20 days, another at 40 (Arn 1960). PARENTAL ANTI-PREDATOR STRATEGIES. (1) Passive measures. Compared with *A. apus*, relatively tolerant of man except early in season (April) when markedly shy (Arn 1960), most flushing readily on approach of intruder (Bartels 1931). Disturbed birds fly around colony giving Alarm-call (see 4 in Voice). Birds sit more tightly late in incubation, and some allowed themselves to be picked up and replaced on nest, without flushing (Arn 1945, 1960; see also Hartert 1927). (2) Active measures: against man. In a few cases when brooding, sitting bird responded to close approach with upward fluttering and Bill-snapping (as above); procedure repeated several times until bird apparently exhausted (Bartels 1931). One bird beat wings rapidly and noisily against walls (Arn 1935). When intruder 1 m from nest, bird leaned forward, gaped widely, beat folded wings, and leaped at outstretched hand. Another leaped 1 m at intruder's chest, clinging there, beating wings and Bill-snapping. Generally quiet when handled, but may give Distress-call (Arn 1945, 1960; see 5 in Voice).

(Fig A from photograph in Dizerens 1982.) EKD

Voice. Freely used in breeding season, much less so away from breeding areas. No detailed study at colony (Glutz and Bauer 1980), and little known outside breeding season. Isolated birds, while feeding or on migration, stay silent (Géroudet 1961c). In breeding season, Solothurn (Switzerland), mostly silent 10.00–16.00 hrs, when mainly feeding (Arn 1960). Though behaviour closely similar to *A. apus*, vocalizations very different. For additional sonagrams, see Bergmann and Helb (1982).

CALLS OF ADULTS. (1) Trilling-call. Loud chittering sound, quite different from homologous Screaming-call of *A. apus* (Arn 1945). High-pitched trilling 'tritritri ririri' and 'trihihihihi' (Arn 1945, 1960); also rendered 'trrr-tritititi' (Moltoni 1966; see also Géroudet 1961c). Such renderings, however, greatly simplified. Thus, in recording (Fig I), call consists of c. 124 units, of which 117 shown; trill (1 bird) a crescendo comprising 2 tremolos—1st peaks at 4·6 kHz and begins to fade after c. 1·3 s, whereupon 2nd begins, entirely displacing 1st c. 1·5 s later; 2nd tremolo initially lower than 1st, reaches c. 3·5–

684 Apodinae

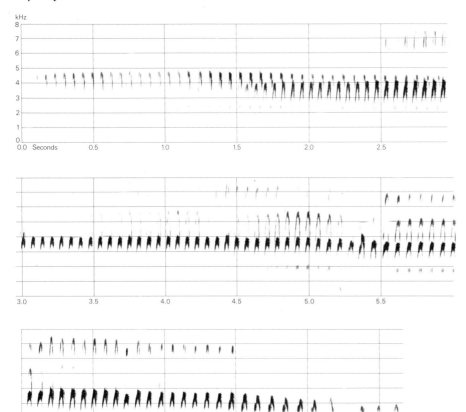

I J-C Roché (1966) Greece April 1963

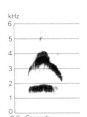

II J-C Roché (1966) Greece April 1963

4·0 kHz, dropping to less than 3 kHz at end (J Hall-Craggs). Trill falls in pitch at end, and becomes slower 'tititjitjü' (Bergmann and Helb 1982). Typically heard during Screaming-display, also when clinging to walls outside nests (Arn 1945, 1960). Call variously combined with: (a) single 'ziiu' (Fig II) or 'zri' sounds, commonly by bird flying alone; (b) lower-pitched 'ziäh ziäh ziäh zrrrr'; (c) short phrases, starting slowly and steadily accelerating, rendered 'sgirsgigirsgirsgirr ... gigigi' (Arn 1960) or 'sghisghisghisghishihihi' (Arn 1945; see also Bergmann and Helb 1982). Main vowel component of basic call and variants pronounced 'ee'. Variants mixed, depending on circumstances (Arn 1945, 1960). (2) Whispering-call. A quiet, whispering 'swilswilswil' (pronounced 'sveelsveelsveel') given, typically after call 3, in Meeting-ceremony at nest, probably also before copulation; also in aerial display between pair when they come together to fly for a short distance, notably before aerial copulation (Arn 1960). Presumably homologous to Nest-call of *A. apus*. (3) Threat-call. Loud trilling 'gigigigigi', usually ending in 'wiuh' sound, given by resident birds to intruding conspecific, typically when latter walks too close to nest (Arn 1960). In Meeting-ceremony, bird on nest often directs moderately loud 'gigigigi' at incoming mate, probably expressing initial aggression, before reverting to more intimate call 2 (Arn 1960). (4) Alarm-call ('Warning call'). Sharp 'zig zig zig' (Arn 1945) or 'ziu ziu ziu' (pronounced 'tseeu tseeu tseeu') (Arn 1960), given by disturbed birds flying around colony, and taken up by nearby conspecifics. Also heard when many birds land at close-packed roost (Arn 1945, 1960; Glutz and Bauer 1980). Series of 5-20 rapid 'kek' sounds, given by *c*. 50 birds flying around together at colony (Meise 1959), probably this call. (5) Distress-call. Loud, sharp trilling 'zigezigezige' (Arn 1945), or 'ziguziguzigu' (Arn 1960). Given sometimes when handled or stuck; also when 2 birds fly into colony together and almost collide at nest-entrance, or during thunderstorms (Arn 1945, 1960; Glutz and Bauer 1980).

CALLS OF YOUNG. Up to *c*. 10 days, food-call a quiet

cheeping or piping, with hissing, whimpering quality, given whether parents present or not. In older young, a loud, hoarser, screeching or hissing sound (Bartels 1931; Arn 1945, 1960), rendered 'chfff chfff chfff(ü)' or 'fchfchfch' (Glutz and Bauer 1980), given only when parents arrive. Young just fed, and settling down in presence of parent, call 'chwäd chwäd', apparently expressing contentment (Glutz and Bauer 1980).   EKD

Breeding. SEASON. North Africa: first eggs laid mid-April (Heim de Balsac and Mayaud 1962). Switzerland: first eggs laid in mid- or last 10 days of May, exceptionally in early June, depending on weather (Arn 1959). SITE. Ledge or hole in cliff or tall building. 80% of sites, Switzerland, on flat surface (Arn 1959). Nest: shallow saucer of straw and feathers, cemented together with saliva. Average diameter of 12 nests 12.5×13.0 cm ($10.5-15.3 \times 11.5-15.3$); depth of cup 2·8 cm ($1.5-3.75$); of 8 nests on flat surface, average overall height 3·9 cm and of 4 on slope 4·9 cm, overall range 1·5–6·5 cm (Bartels 1931). Building: material gathered in flight, by both sexes. EGGS. See Plate 98. Long elliptical, smooth and not glossy; white. 30×19 mm ($27-34 \times 18-21$), $n = 115$ (Arn 1945). Weight 6·2 g (5·7–7·0), $n = 15$ (Arn 1960). Clutch: 3 (1–4). Of 2661 clutches, Switzerland: 1 egg, 6%; 2, 29%; 3, 64%; 4, 1%; average 2·6 (Arn 1959). Experienced breeders lay 3–4 eggs, first-time breeders usually 2. Of 970 clutches during main laying period: 1 egg, 2%; 2, 26%; 3, 71%; 4, 1%; average 2·7. Of 100 clutches late in season: 1 egg, 4%; 2, 75%; 3, 21%; average 2·2 (Lack and Arn 1947). Replacement clutches laid after egg loss. Of 33 repeat layings: 1 egg, 3; 2, 24, 3, 6; average 2·1 (Lack and Arn 1947). Laying interval 1 day for 3-egg clutches, but 2 days for 2-egg clutches (Arn 1959). $\frac{2}{3}$ of eggs laid before 09.00 hrs, remainder 09.00–18.00 hrs (Arn 1959). INCUBATION. 20 (17–23) days. By both sexes. Begins with last egg according to Arn (1959), though earlier reported as with 1st egg (Bartels 1931). Hatching synchronous, almost always within 24 hrs (Arn 1959). YOUNG. Altricial and nidicolous. Cared for and fed by both parents. Brooded for first 3 weeks after hatching, more or less continuously for first 10 days. At c. 3 weeks may climb around area of nest (Arn 1959). FLEDGING TO MATURITY. Fledging period 45–55 days. Become independent at fledging or soon after. Age of first breeding normally 2–3 years, rarely 1 (Lack and Arn 1947; Arn 1960). BREEDING SUCCESS. In long-term study in Switzerland, hatching success found to be independent of clutch size but survival to fledging decreased with size of brood. Of 323 clutches of 2 eggs, 94% of eggs hatched; of 658 clutches of 3 eggs, 94% of eggs hatched; of 5 clutches of 4 eggs, all eggs hatched; of 81 broods of 1, 98% of young fledged or 0·98 young per brood; of 351 broods of 2, 86% of young fledged, 1·7 young per brood; of 557 broods of 3, 78% of young fledged, 2·3 young per brood; of 5 broods of 4, 60% of young fledged, 2·4 young per brood. Clutches laid in main laying season more successful than those laid late, with average 2·1 young per brood from 885 broods in main season, and 1·5 young per brood from 99 late broods (Lack and Arn 1947). Success can vary between colonies in same town, with 1992 nests at one site in Switzerland during 1932–56, producing 2·56 eggs per nest, 2·10 hatched per nest, and 1·53 young fledged; at another colony, 668 nests produced 2·72 eggs per nest, 2·41 hatched per nest, and 1·79 young fledged (Arn 1959).

Plumages (nominate *melba*). ADULT. Forehead and lores pale grey-brown, gradually darker towards crown. Black spot in front of eye; side of head and neck down to lower cheek, hindcrown, hindneck, mantle, scapulars, and back dark greyish olive-brown, each feather narrowly fringed white at tip when plumage fresh and with dark olive-brown tinge subterminally. Rump and upper tail-coverts slightly paler grey-brown than mantle and back, with dark olive-brown subterminal border and terminal white fringe of rump feathers more distinct. Chin and throat white, often with faint dark shaft-streaks, contrasting sharply with dark greyish olive-brown cheeks and dark olive-brown band across chest (chest-band 0·5–1 cm wide in middle, widening slightly towards side of neck). Tips of feathers of chest-band fringed white when plumage fresh. Flanks, vent, under tail-coverts, and feathering on legs dark olive-brown, feather-tips fringed white; vent and under tail-coverts with much pale olive-brown of feather-bases showing. Breast and belly white, contrasting sharply with dark chest-band, flanks, and vent; some faint and narrow dark shaft-streaks often visible, especially on lower belly and at border with flanks. Tail olive-black with slight bronze lustre; base of outer web of t5 tinged grey-brown and narrowly fringed off-white. Flight-feathers and upper primary coverts olive-black on outer webs and tips, deep black on outermost primaries; inner webs paler dull grey-brown; fresh secondaries and inner primaries narrowly fringed white along tip and inner web, outer web of p10 narrowly edged off-white. Lesser upper wing-coverts dark greyish olive-brown, similar to mantle, scapulars, and back. Tertials and median and greater upper wing-coverts grey-brown, slightly paler than bordering secondaries, lesser coverts, and primary coverts; tertials narrowly fringed white at tip when fresh; tertial coverts with dark subterminal border. Small coverts along leading edge of wing, under wing-coverts, and axillaries olive-black, tips rather broadly fringed white. NESTLING. Naked at hatching. From c. 2 weeks, pale grey down appears on upperparts and flanks, and feather-pins on head, wing, and tail; down on body gradually pushed out by feathers from 4th to 5th week (Heinroth and Heinroth 1924–6; Arn 1960; Harrison 1975). JUVENILE. Rather similar to adult, differing mainly by showing extensive white fringing on feather-tips at a time when adults are mainly worn and show only limited fringing (in particular, adult wing worn); sides and tips of greater and median upper wing-coverts narrowly fringed white (in fresh adult, tips only); ground-colour of upperparts slightly darker than in fresh adult, narrow white fringes of crown and hindneck contrasting more. Both webs of distal half of t5 curve gradually towards narrowly rounded tip (in adult, webs slightly emarginated to form rather narrow and more sharply pointed tip). Median and greater upper wing-coverts narrower and with more narrowly rounded tip than in adult, not as broadly rounded or almost square. In 1st winter, when worn or body moulting, differs from adult in rather worn flight-feathers (not moulting or new) and in shape of upper wing-coverts and t5. FIRST IMMATURE. In summer of 2nd calendar year, like adult, but feathers of head, body, and most upper wing-coverts less worn, flight-feathers more worn; primary moult often more advanced than adult; part of greater upper wing-coverts still juvenile, narrower than adult (especially at tip); shape of t_5 intermediate between adult and juvenile (edges of both webs near tip of t_5 straight, not slightly convex as in juvenile or slightly emarginated or concave as in adult).

Bare parts. ADULT AND JUVENILE. Iris deep brown. Bill black or dark horn-brown with black nail. Leg and foot flesh-pink, pink-yellow, or pinkish horn-colour, tips of toes dark horn or horn-black, claws black. NESTLING. At hatching, bare skin (including bill and foot) pale pink; colour gradually darkens to bluish-grey over feather-tracts and to horn-brown on bill, tips of toes, and claws; bill horn-black at c. 3 weeks, claws black at fledging. (Hartert 1912–21; Harrison 1975; Glutz and Bauer 1980; RMNH, ZMA.)

Moults. ADULT POST-BREEDING. Complete; primaries descendant. In a small sample of nominate *melba* from Yugoslavia, Crimea, and Caucasus (BMNH, RMNH, ZMM), moult started with loss of p1 mid-June to early July; primary moult score 15 reached mid-July to mid-August, 30 mid-August to mid-September. Small sample of *tuneti* started from early June, reached 15 from early July, and 30 early to late August. Moult completed in winter quarters, but hardly any details known. 2 birds, August and September, showed suspended primary moult (both score 30), indicating that at least some may suspend moult during migration; 2 *tuneti* from November were nearing completion (score 50) with scores 44 and 47. One possible *archeri* from Sinai had score 12, October (BMNH). In Switzerland, p1 lost mid-June to early July at hatching of eggs; p6 shed c. 60 days after loss of p1 (10 August–10 September), p7 at c. 75 days (first half September); leaves for winter quarters without suspending, with up to p6 or p7 new; in large sample of unusually late Swiss birds (early November), brought down by adverse weather (p6–) p7–p8(–p9) growing (Stresemann and Stresemann 1966; Glutz and Bauer 1980). Primary moult probably completed December–February. Non-breeding adults and sub-adults start primary moult from May, reaching primary moult score 30 mid-July to mid-August (Glutz and Bauer 1980; BMNH, ZMM). Timing of body moult rather variable: some start with scattered feathers of upperparts from May before p1 lost, others still in old plumage at score 25 in August; at score 30, all birds had head and upperparts mainly new, chest and wing-coverts partly new; upper tail-coverts, belly, and vent replaced last, usually in winter quarters. Tail moults from t_5 to t_1; t_5 lost at score 30–40, t_1 completed at same time as p10. Moult of secondaries convergent: descendantly from s9 (lost with p4) and ascendantly from s1 (lost with p6); s5 and s6 replaced last (Stresemann and Stresemann 1966; Glutz and Bauer 1980; BMNH). POST-JUVENILE. Data on timing limited. Head, body, and part of wing-coverts replaced in winter quarters, starting from September–October; tail new on arrival in breeding area in spring and thus apparently moulted late winter. Flight-feathers and remaining juvenile wing-coverts replaced in summer of 2nd calendar year; p1 shed from May as in non-breeding adults. Feathers of upperparts, head, and chest replaced for 2nd time during late stages of post-juvenile primary moult.

Measurements. ADULT. Nominate *melba*. (1) Switzerland, Italy, and Balkans, April–September; skins (BMNH, RMNH, ZFMK, ZMA); (2) Caucasus and Crimea, May–September; skins (ZMM). Tail is to tip of t_5; fork is tip of t_1 to tip of t_5; bill is exposed culmen.

WING (1) ♂ 227 (5·65; 11) 220–240 ♀ 221 (5·34; 16) 214–230
      (2)   226 (3·92; 11) 221–234   225 (4·69; 16) 218–233
TAIL (1)   84·4 (4·51; 10) 79–94   80·7 (4·23; 17) 75–89
      (2)   85·8 (3·06; 6) 83–91   83·1 (4·67; 5) 78–87
FORK (1)   25·8 (1·78; 10) 24–28   22·9 (2·65; 17) 18–27
      (2)   25·8 (1·79; 5) 24–28   24·0 (1·58; 5) 22–26
BILL (1)   8·9 (0·62; 7) 8·0–9·5   8·8 (0·52; 13) 7·8–9·5
TARSUS (1)   13·9 (1·07; 7) 12·8–15·1   14·1 (0·70; 13) 13·4–15·3

Sex differences significant for wing and fork of (1).

A. m. tuneti. Algeria and Tunisia, March–August; skins (BMNH, RMNH, ZFMK, ZMA).

WING ♂ 227 (2·80; 12) 222–231 ♀ 222 (4·00; 12) 216–227
TAIL   82·5 (2·41; 10) 79–86   79·7 (3·00; 9) 75–83
FORK   24·8 (2·28; 10) 22–28   22·6 (1·32; 9) 21–25

Sex differences significant.

A. m. archeri. Dead Sea Depression and Sinai, October–February, and south-west Arabia, April: wing 207 (5·97; 6) 199–215 (BMNH).

JUVENILE. Few examined. Wing similar to adult, tail and fork average c. 3 shorter.

Weights. ADULT. Nominate *melba*. Switzerland: adult, April–August, 104 (45) 76–120; heavier in evening than in early morning; evening of 22 September 108 (24) 95–125 (Arn 1960); exhausted birds, November, 56 (96) 45–72 (Glutz and Bauer 1980). Crimea (USSR): May, ♂ 94, ♀ 98; July, ♂ 95; September, ♂ 98, ♀ 98 (ZMM). *A. m. tuneti*. Mangyshlak (Kazakhstan, USSR): ♂ 71–100 (4), ♀ 82–115 (5) (Dolgushin *et al.* 1970). Iran and Afghanistan, May–July: ♂♂ 68, 77, 80; ♀♀ 79, 84 (Paludan 1940, 1959).

JUVENILE. Switzerland: at hatching c. 6, on 24th day c. 90, on 35th c. 110, at fledging c. 93 (Arn 1960).

Structure. 10 primaries: p10 longest, p9 0–5 shorter (occasionally, p10 1 shorter than p9), p8 10–20, p7 29–38, p6 52–61, p5 73–83, p1 147–158. Tail has rather shallow fork (see Measurements), 10 feathers; in adult, t_5 slightly emarginated on both webs to form rather acutely pointed tip; in juvenile, tip more rounded (see Plumages). Tarsus and toes very short but strong; tarsus feathered except for lower part and rear; all toes directed forward, even in small nestling (unlike most other *Apus*—see Brooke 1972c). Outer middle toe with claw 14·0 (20) 13–16, of which claw comprises about half; innermost toe with claw c. 67% of outer middle, inner middle c. 92%, outermost c. 88%. Claws sharp and very strong. Remainder of structure as in Swift *A. apus*.

Geographical variation. Involves depth of colour of upperparts, chest-band, flanks, vent, and under tail-coverts, width of

PLATE 63 (*facing*).
Hirundapus caudacutus caudacutus Needle-tailed Swift (p. 649): 1 ad (fresh plumage), 2–4 ad, 5 juv.
Apus apus Swift (p. 657). Nominate *apus*: 6–8 ad. *A. a. pekinensis*: 9 ad.
Apus pallidus Pallid Swift (p. 670). *A. p. brehmorum*: 10–12 ad. *A. p. illyricus*: 13 ad.
Apus pacificus Pacific Swift (p. 676): 14–16 ad, 17 juv (autumn, worn plumage).
Apus melba Alpine Swift (p. 678). Nominate *melba*: 18–20 ad. *A. m. tuneti*: 21 ad.
Apus caffer White-rumped Swift (p. 687): 22–24 ad, 25 juv.
Apus affinis galilejensis Little Swift (p. 692): 26–29 ad, 30 juv.
Cypsiurus parvus Palm Swift (p. 698): 31–32 ad. (DIMW)

PLATE 64. *Halcyon smyrnensis smyrnensis* White-breasted Kingfisher (p. 701): **1** ad, **2** juv, **3** nestling. *Halcyon leucocephala acteon* Grey-headed Kingfisher (p. 705): **4** ad, **5** juv, **6** nestling. (NA)

PLATE 65. *Alcedo atthis* Kingfisher (p. 711): *A. a. ispida*: **1** ad ♂, **2** ad ♀, **3** juv. Nominate *atthis*: **4** ad ♂. (NA)

PLATE 66. *Ceryle rudis* Pied Kingfisher (p. 723): 1 ad ♂, 2 ad ♀, 3–4 juv, 5 nestling. (NA)

PLATE 67. *Ceryle alcyon* Belted Kingfisher (p. 731): 1 ad ♂, 2 ad ♀, 3 juv ♂, 4 juv ♀. (NA)

chest-band, and size (as expressed in wing-length and weight). In general, colour darkest and chest-band widest in areas of high rainfall (e.g. Madagascar, Ceylon), palest in semi-desert areas (e.g. breeding grounds of *tuneti* and *archeri*) (Brooke 1971b). Birds of Ruwenzori mountains race in Africa large and heavy; nominate *melba*, *tuneti*, and race inhabiting Himalayas fairly large, Somalian *archeri* and race of Madagascar small; other races of Afrotropics and Indian subcontinent intermediate; for details, see Abdulali (1965) and Brooke (1971b). *A. m. tuneti* from central and eastern Morocco east to Libya, coastal Levant, and from Iran eastwards distinctly paler than nominate *melba*: forehead, crown, mantle, scapulars, and back dull drab-grey rather than dark greyish olive-brown; rump and upper tail-coverts pale grey-brown; sides of head and neck, cheeks, chest-band, flanks, and under wing-coverts slightly paler than in nominate *melba*; vent and under tail-coverts distinctly paler grey-brown with pronounced olive-black subterminal crescents; tertials and median and greater upper wing-coverts drab-grey like mantle; lesser upper wing-coverts, secondaries, and inner primaries dark olive-brown, slightly paler than nominate *melba*. However, some variation in depth of colour—caused partly by bleaching and wear, but also occurring in fresh-plumaged birds—and boundary between *tuneti* and nominate *melba* not sharp. Birds from Sardinia, Cyclades, Crete, Rhodes, and Cyprus intermediate in colour between *tuneti* and nominate *melba*; some nearer former, others nearer latter, while a few dark birds similar to nominate *melba* breed in colonies of *tuneti* (e.g. Algeria, Afghanistan), and some birds as pale as *tuneti* occur in summer in breeding area of nominate *melba* (e.g. Morocco, southern Spain, France, Turkey). Of 11 breeders from Crimea, USSR (ZMM), 4 almost as pale as *tuneti*. Situation in Middle East and Arabia not fully elucidated. Winter birds examined from Dead Sea Depression and Sinai were smaller than both nominate *melba* and *tuneti* and some even paler than *tuneti*; 2 April birds from south-west Arabia were similar, and perhaps all birds supposedly breeding from Dead Sea and Sinai south along western Arabia (see Jennings 1981a) to Aden are small and pale. These birds provisionally included in *archeri* of northern Somalia which is also pale though even shorter winged—197 (9) 193–208 (Brooke 1971c); however, perhaps form separate race, for which name *petrensis* Bangs, 1911, perhaps available.

A. melba forms species-group with Mottled Swift *A. aequatorialis* of Afrotropics; sometimes separated from other *Apus* swifts in genus or subgenus *Tachymarptis*, mainly because of large size, different structure of foot of nestling, and different species of feather-lice (Mallophaga) (Brooke 1972c). CSR

Apus caffer White-rumped Swift

PLATES 62 and 63
[facing pages 663 and 686]

Du. Kaffergierzwaluw   Fr. Martinet cafre   Ge. Kaffernsegler
Ru. Кафрский стриж   Sp. Vencejo cafre   Sw. Kafferseglare

Cypselus caffer Lichtenstein, 1823

Polytypic. Nominate *caffer* (Lichtenstein, 1823), southern Africa north to about Namibia and Zambezi river; *streubelii* (Hartlaub, 1861), Afrotropics north of nominate *caffer*. Not known which race breeds Morocco and Spain.

Field characters. 14 cm; wing-span 34–36 cm. About 15% smaller than Swift *A. apus*, but with relatively longer and more forked tail; head and body size close to Little Swift *A. affinis*, but wings average longer and tail *c.* 40% longer and deeply forked. Small, rather slim-bodied swift, with noticeably attenuated rear body and tail. Black-blue plumage relieved by grey-brown face, pale underside to flight-feathers, and narrow white band over upper rump. Flight free and graceful, with high ascents characteristic. Sexes similar; no seasonal variation. Juvenile separable at close range.

ADULT. Plumage mainly blue-black, with blue tone most obvious on back, forward wing-coverts, and (when lit) underbody. Most obvious mark is narrow white band over upper rump, less than half as wide as square patch of *A. affinis*. At close range in good light, the following characters also visible: (1) white-brown forehead and line over eye and round black eye-patch linking with small white chin (and creating at distance grey or pale brown face); (2) silvery underside to primaries and secondaries, contrasting with black under wing-coverts; (3) browner-toned uppersurface to wing, with faint green gloss to flight-feathers occasionally shining (and, at least in races occurring in West Africa, faint bar on outer secondaries caused by their paler brown tips). JUVENILE. Not fully studied in the field. Flight-feathers show white tips at close range; blue tones less intense.

Differs markedly from *A. affinis* in tail shape and rump-band. In Africa, subject to confusion with Horus Swift *A. horus*, of closely similar appearance but *c.* 10% larger though also fork-tailed. Important to realize that outer tail-feathers of *A. caffer* are not only relatively longer but also distinctly narrower than those of any other *Apus* swift occurring in west Palearctic. Flight action recalls *A. apus* and Plain Swift *A. unicolor* but action even more 'twinkling', with marked fluttering and manoeuvrability in tight spaces (clearly contributing to exploitation of low nest-sites); appears lighter in flight than any congener. Behaviour typical of genus, but ascends higher than *A. affinis*; associates closely with hirundines.

Voice rather low in pitch, particularly when compared

PLATE 68 (*facing*).
Halcyon smyrnensis smyrnensis White-breasted Kingfisher (p. 701): 1–2 ad.
Halcyon leucocephala acteon Grey-headed Kingfisher (p. 705): 3–4 ad.
Alcedo atthis Kingfisher (p. 711): 5–6 ad ♂.
Ceryle rudis Pied Kingfisher (p. 723): 7–8 ad ♂, 9 ad ♀.
Ceryle alcyon Belted Kingfisher (p. 731): 10–11 ad ♂, 12 ad ♀.
(NA)

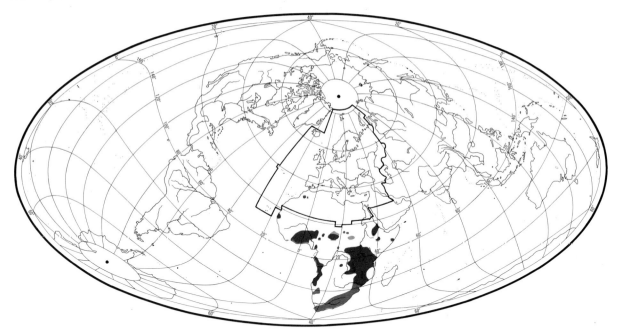

to common notes of *A. apus*; commonest call a rather guttural twitter or chatter, recalling a bat.

Habitat. In Afrotropical low latitudes and very recently (since 1966) breeding locally in Spain and Morocco. To a remarkable degree, takes over nests of hirundines (Hirundinidae) and Little Swift *A. affinis*. Bred originally in caves and on rock faces in hills, but now largely switched to artefacts such as concrete or iron road or rail culverts or bridges and corrugated iron verandas or roofs and under house eaves (Serle *et al.* 1977). In East Africa, often occurs over inland lakes and swamps (Williams 1963). In southern Africa, also found in towns, feeding usually at considerable height (Prozesky 1970). At Asni (Morocco), mixed flocks with Red-rumped Swallow *H. daurica* ascend to high crags in heat of day, descending again in late afternoon (Chapman 1969). While congeners such as Little Swift *A. affinis*, Swift *A. apus*, and Alpine Swift *A. melba* have to varying degrees adapted from natural nesting sites, especially on rocks, to human artefacts, *A. caffer* is remarkable for having done so at one remove, through nest-parasitism on swallows which have been making a similar switch. With this change of nest-site more or less extensive habitat changes must have followed. Fuller study of these would be desirable, especially in view of new colonization of south-west Europe, where breeding occurs in nests of *H. daurica* under rocky outcrops, and birds forage over adjacent farmland (Allen and Brudenell-Bruce 1967).

Distribution and population. Recent major expansion of range to Spain and Morocco.

SPAIN. First recorded 1964–5, breeding proved 1966 Cádiz province (Junco and Gonzalez 1966; Allen and Brudenell-Bruce 1967): at first incorrectly identified as Little Swift *A. affinis* (see Ferguson-Lees 1967, Benson *et al.* 1968, and *Br. Birds* 1968, **61**, 36–41). Has since spread to provinces of Córdoba and Almería (AN). MOROCCO. First recorded Imlil valley (High Atlas) July 1968 when *c.* 30 seen and thought to be nesting (Chapman 1969); smaller numbers seen since in summer in this area and Ijoukak (JDRV). Breeding first proved near Ouarzazate in 1979 (Blankert 1980).

Accidental. Finland: young ♂ found dead, Kestilä, November 1968 (Ojanen 1983).

Oldest ringed bird at least 10 years (R K Schmidt).

Movements. Migratory in southern Africa south of River Zambezi; absent May–August, moving northwards into equatorial belt (Clancey 1964). Elsewhere in Afrotropical region, mainly resident but with local movements, though partially migratory or dispersive (extent unclarified) in Sahel and northern savanna zones. Thus present all year in savanna zone of Chad, while small part of population penetrates north into Sahel in wet season, July–September (Salvan 1968a), and this may apply elsewhere on southern edge of Sahara, e.g. in Darfur (Sudan) where found only in midsummer and does not breed (Lynes 1925a). Resident in Nigerian savanna (Elgood 1982), though reported absent from Zaria November–January and from Borgu between mid-November and mid-March (Elgood *et al.* 1973).

Movements of birds breeding Morocco and Spain little known, since numbers very low; no winter data from Morocco. Believed summer visitor to southern Spain, arriving May and leaving by October (Ferguson-Lees 1967); various records of apparent migrants on Spanish side of Straits of Gibraltar, 11 August–13 October (*Ardeola* 1974, **20**, 368). Yet rare December–January

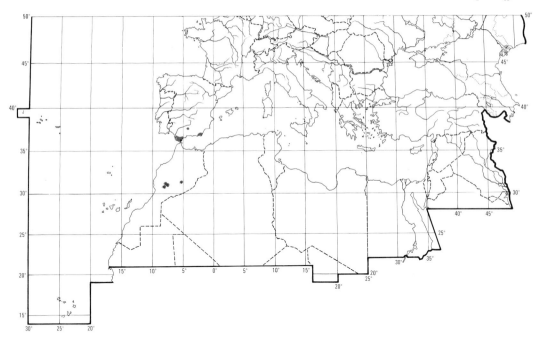

records known from Spain (*Ardeola* 1975, 22, 147; 1978, 24, 262). Winter quarters of emigrants unknown (Bernis 1970).

Food. In Zimbabwe, takes beetles (Coleoptera), small bugs (Hemiptera), small wasps (Vespoidea), spiders (Araneae), and weevils *Bataninus* (Borrett 1973). In Zäire, 3 out of 4 stomachs filled with winged ants (Formicidae); also a bug (Chapin 1939). Feeds at considerable height (Prozesky 1970). BDSS

Social pattern and behaviour. Little information for west Palearctic, and following account includes material on nominate *caffer*, partly from notes supplied by R K Schmidt.
 1. Rather gregarious in breeding season and probably throughout the year, though less so than Little Swift *A. affinis* (Rudebeck 1980). In September, Cadiz (Spain), flock of 10 along with 80–100 Pallid Swifts *A. pallidus* (Hiraldo and Alvarez 1973). For ones and twos migrating, often with other *Apus*, Cadiz, August–October, see GEMRA (1973). 3 migrant flocks, Windhoek (Namibia), 35, 40, and 100 birds (Tree 1973). BONDS. Monogamous mating system and maintenance of pair-bond from year to year probably typical. In South Africa, pair recorded breeding together (in same nest) for 3 years, another pair for 2 (R K Schmidt). Age of first breeding not known, but 1 ringed bird first found breeding at 2 years (R K Schmidt); thought not to breed until at least 2 years old (R K Brooke). Parents take nearly equal share in all nest duties (Brooke 1957; Schmidt 1965b). BREEDING DISPERSION. Loosely colonial in west Palearctic, also solitary. In southern Africa, not markedly colonial (Brooke 1969e). Closeness of spacing determined by availability of nests of swallow and martins (Hirundinidae) for breeding in. In Spain (typically breeding in nests of Red-rumped Swallow *Hirundo daurica*) known colonies of less than 5 pairs (Junco 1969; Ferguson-Lees 1969d; García Rúa 1974); 3 nests *c*. 25 m and *c*. 100 m apart (Ferguson-Lees 1969d). In South Africa (using nests of Larger Striped Swallow *Hirundo cucullata*), closest pairs 2·5 m apart (R K Schmidt). Breeding birds, South Africa, showed marked fidelity to nest-site though this facilitated by provision of artificially reinforced nests. Of 48 birds caught at same nest in 2 or more seasons, 13 in 2 seasons, 14 in 3, 9 in 4, 5 in 5, 2 in 6, 2 in 7. Of 3 birds controlled for 8 seasons, 2 changed nests (R K Schmidt); moved 100 m and 1 km (R K Schmidt; see also Schmidt 1965b). Change of nest-site may be caused by successful challenge of another pair (R K Schmidt: see Antagonistic Behaviour, below). No ringed young recovered in later years at colony, but 2 found breeding 1 km and 10 km away (R K Schmidt). ROOSTING. At night, members of pair roosted together in own nest, Tanzania, before eggs laid (Moreau 1942). At Gatooma (Zimbabwe), within 1 week of arrival at colony, 1 bird roosted on nest, joined shortly by mate. Thereafter, pair usually roosted together. Sometimes one, sometimes both roosted on nest or just beside it. Roosting thus started well before laying, and continued during incubation and chick-rearing, and in gaps between successive broods (Brooke 1957). In study, South Africa, always roosted in their nests, from arrival in spring until departure in autumn (R K Schmidt). After 2nd brood fledged, parents probably used nest for 1–2 nights before departing (Schmidt 1965b). At Gatooma, birds started roosting 5–20 min after sunset. Evening roosting preceded by Screaming-display (see part 2). If disturbed, flew off and did not return to roost at nest that night. Usually left nest 5–20 min before sunrise (Brooke 1957). Birds remained on nest by day when winds strong (Schmidt 1965b). Never seen to self-preen on nest by Brooke (1957), but this often seen by Schmidt (1965b). Solitary birds occasionally roost in site without nest (Brooke 1957). For roosting of young, see Relations within Family Group, below.
 2. FLOCK BEHAVIOUR. At colony, Gatooma, communal Screaming-display typically performed by *c*. 6 birds, of which 1 (2 at most) gave Screaming-call (see 1 in Voice). Flock flew around and then said to have entered a shallow dive to come up under eaves of building, whether nest there or not (Brooke 1957). For response of nest-occupants, see Antagonistic

Behaviour (below). Screaming-displays occurred at any time of day, but most often in evening (Brooke 1957). Towards end of breeding season, Transvaal (South Africa), Screaming-displays occurred in early morning; birds flew fast and low near colony, calling continuously (Cumming 1952). ANTAGONISTIC BEHAVIOUR. When birds performing Screaming-display passed near sitting bird, it gave Screaming-call and sometimes rose from nest to sit in entrance, always with raised wings (Brooke 1957). May take over nests of other species for breeding, ejecting occupiers and their eggs (e.g. Someren 1944, Schmidt 1965b, Ferguson-Lees 1969d. Intraspecific fighting for nests more frequent when relatively few nests of *H. cucullata* available (R K Schmidt); occupied nests may be assailed simultaneously by 4–6 birds which cling to entrance while owner inside calls (see 2d in Voice), this sometimes causing assailants to pause momentarily; attacks sometimes last 3–5 min, repeated several times per day; no physical contact seen. In one case, closest neighbours (2·5 m apart) fought regularly, eventually swapping nests. (R K Schmidt.) Record of 2 birds falling interlocked from nest, apparently copulating (Moreau 1942b), probably refers to fighting as in Swift *A. apus*. HETEROSEXUAL BEHAVIOUR. Not well known. (1) Pair-bonding behaviour. At Kasai (Zaïre), birds often flew slowly in pairs, and close together. No description of response to Wings-high display (raising wings in V in flight) of another bird, but *A. affinis* seen to react to and follow *A. caffer* thus displaying (De Roo and Deheegher 1966). Following account of meeting at colony perhaps describes pairing behaviour, as birds had no nest. Bird said to have repeatedly fed another sitting on stonework (potential nest-site) in colony. Next day, 2 birds seen at same spot sitting face to face. One uttered soft calls (see 2 in Voice) and quivered wings. Other responded with louder call which almost became a scream, and started to mock-feed and Allopreen ('caress') head of partner. Meanwhile, both shuffling about (Brooke 1957). Billing and Allopreening also reported by Schmidt (1965b). (2) Mating. Occurs possibly both in flight and in nest. Apparent attempted copulation seen in flight, and twice at nest-entrance (Moreau 1942b). (3) Behaviour at nest. Bird arrived silently at nest if mate present, otherwise gave 1 or more Screaming-calls before entering. Relieving bird Billed head and bill of mate for variable time. After feeding or tending young, incoming adult left immediately or settled down to brood, its mate then leaving within 5 min (Brooke 1957). Departure of relieved bird sometimes almost instantaneous, sometimes prolonged, pair remaining at nest for up to 137 min. Sometimes both arrived together and one stayed only briefly while other sat; at other times, one arrived and both then left (Moreau 1942a). RELATIONS WITHIN FAMILY GROUP. From day of hatching (if site permits), young highly active, crawling and climbing readily (Moreau 1942a; Schmidt 1965b), but fed only in nest (Brooke 1957). Initially, any stimulus elicits food-call (see Voice) but, later, young respond only to parents (Brooke 1957). Give food-call intermittently when fed, otherwise said to be silent (Moreau 1942a). However, young 1–2 days old heard calling when parents absent, and young 32–33 days old continued calling for some time after parent left, attempting to beg from each other (Schmidt 1965b). Young brooded for at least 1 week, sometimes more than 2 (Moreau 1942a; Brooke 1957). Brooded for most of day. Defecate out of nest from 10–12 days (Moreau 1942a; Schmidt 1965b). Near fledging, young roosted sitting on top of parents (Brooke 1957). Leave nest with no coaxing, and immediately fly strongly. At one nest, the 2 young fledged 4 min apart, 16 min after parent last visited them. At another, 1 young flapped wings vigorously at nest-entrance before launching off. Occasionally, fledging young follow parent out (Moreau 1942a). At one nest, 1st young of brood flew at 47 days, 2nd 22 days later (Schmidt 1965b). Do not return to nest after fledging (R K Schmidt). ANTI-PREDATOR RESPONSES OF YOUNG. Young grip nest-lining hard with claws when handled (Moreau 1942a; Brooke 1957). PARENTAL ANTI-PREDATOR STRATEGIES. For mobbing of Lammergeier *Gypaetus barbatus* by mixed flock of *A. caffer* and *H. daurica*, see Chapman (1969). EKD

Voice. Said to be little used compared with Little Swift *A. affinis* (Rudebeck 1980). Little information, and this for nominate *caffer*.

CALLS OF ADULTS. (1) Screaming-call. No detailed description, but has trilling quality (Brooke 1957). In recording by J-C Roché, Morocco, initial series of 5–6 rapidly repeated 'sip' sounds breaks into a trill (J Hall-Craggs). Given during Screaming-display (Cumming 1952; Brooke 1957) and Wings-high display (J-C Roché). Same or related call also given: (a) by sitting bird when closely approached by birds flying outside giving Screaming-calls; (b) by bird arriving to relieve mate; (c) by one bird of pair during courtship inside building (see Social Pattern and Behaviour). (2) Other calls. (a) Soft calls given by bird during courtship (context as in call 1c). (b) Nasal, resonant 'pah-pee-ti-ti pah-pee-ti-ti' (J Hall-Craggs, P J Sellar: Fig I) given by bird trying to evict Lesser Striped Swallow *Hirundo abyssinica* from its nest, probably expressing threat or alarm; recording also includes single 'pa' sounds, with buzzing quality (E K Dunn). Guttural and buzzing sounds (De Roo and Deheegher 1966). (c) A chattering crescendo 'pi-pi-pi-pi-pi-pi-pee-pee' (J Hall-Craggs: Fig II); has a bat-like quality, and is lower pitched than Little Swift *A. affinis* (C Chappuis). Function not known. (d) Shriek, given by resident when its nest attacked by conspecific birds (R K Schmidt); probably variant of call 1.

CALLS OF YOUNG. Food-call of small young described as a twitter (Moreau 1942a); soft chirping (Schmidt 1965b). Pitch becomes lower with age. Older young in nest occasionally gave guttural hiss (Brooke 1957). EKD

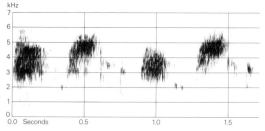

I M E W North Kenya November 1969

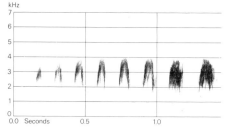

II C Chappuis/Sveriges Radio (1972) Spain January 1970

Breeding. SEASON. Spain: from end of May (Brudenell-Bruce 1967, 1969). SITE. In Spain, only found in old nests of Red-rumped Swallow *Hirundo daurica*; in Africa, sometimes breeds on ledge or in crevice of rock or building, as well as in nests of Hirundinidae and Little Swift *A. affinis* (Brooke 1957; Junco and Gonzalez 1969; Hiraldo and Alvarez 1973). Nest: lines old nests of Hirundinidae with feathers, which usually protrude from entrance (Brudenell-Bruce 1967). On ledges, constructs own nest of straw and feathers cemented together with saliva to form shallow cup (Brooke 1957). Building: material collected in the air. EGGS. Long elliptical, smooth and not glossy; white. 23×15 mm ($21–26 \times 14–16$), $n = 40$ (McLachlan and Liversidge 1970). Clutch: 2(1–3). Of 93 clutches, Tanzania (*streubelii*): 1 egg, 1; 2, 92; average 1·99 (Moreau 1942a). 3 broods in Tanzania and Zimbabwe (Moreau 1942a; Brooke 1957). Interval between broods $c.$ 1 week. Laying interval 38–48 hrs. INCUBATION. About 21 days ($20 \pm 1 – 26 \pm 1$), $n = 18$, Tanzania (Moreau 1942a). 22·6 days (21–25), $n = 20$, South Africa (Schmidt 1965b). By both sexes in approximately equal spells, varying from a few min to 3 hrs or more; eggs often left uncovered for long periods (Moreau 1942a). Probably starts with 1st egg, which hatches up to 30 hrs before 2nd (Moreau 1942a; Schmidt 1965b). YOUNG. Nidicolous and altricial. Cared for and fed (by regurgitation) by both parents. Brooded for whole of 1st week and sometimes through 2nd (Brooke 1957). FLEDGING TO MATURITY. Fledging period 42 days (35–47), $n = 19$ (Moreau 1942a). 46 days (41–53), $n = 8$ (Schmidt 1965b). Age of independence and of first breeding not recorded (but see Social Pattern and Behaviour). BREEDING SUCCESS. Of 97 eggs laid, Tanzania, 11 did not hatch, disappearing or being lost from nest; of 86 young hatched, 74 fledged (Moreau 1942a). Of 11 eggs laid in 2 seasons, Zimbabwe, 10 hatched and 6 young fledged; 1 egg damaged accidentally by observer, 3 chicks wandered from nest inside buildings and starved to death, 1 died of unknown causes (Brooke 1965). 100 eggs, South Africa, produced 81 chicks of which 57 survived; average 2·5 chicks reared per pair per year (Schmidt 1965b).

Plumages (*A. c. streubelii*). ADULT. Forehead and lores pale grey-brown, extending in narrow line to just above eye. Crown, nape, and sides of head down to lower cheeks dark grey-brown, feathers indistinctly fringed paler grey-brown on forecrown and lower cheeks, feather-tips gradually blacker towards hindcrown and sides of neck, slightly glossed blue-black on nape and lower sides of neck. Black spot in front of eye. Mantle, scapulars, and back deep glossy bluish-black, sharply contrasting with white band $c.$ 1 cm from front to back across rump, scarcely extending on to rear flanks. White of rump with faint sharp shaft-streaks. Lower rump feathers tinged dark grey-brown, upper tail-coverts glossy bluish-black. Chin and throat white, forming sharply contrasting patch $c.$ 1·5 cm long; occasionally, part of chin and throat shows some faint and thin dusky shaft-streaks. Remainder of underparts deep glossy bluish-black, some brown-black of feather-bases visible on vent and under tail-coverts. Tail dark olive-brown with slight green gloss, undersurface glossy dark grey-brown. Flight-feathers dark olive-brown, slightly paler grey-brown along borders of inner webs, faintly glossed green on outer webs and tips of primaries; secondaries and sometimes innermost primaries narrowly fringed white on tips; undersurface glossy dark grey-brown. Tertials, primary coverts, and greater upper wing-coverts slightly darker olive-brown than flight-feathers; median and lesser upper wing-coverts gradually darker towards leading edge of wing—median coverts brownish-black with slight bluish-green gloss, lesser coverts black with tips glossed deep blue, those near leading edge deep bluish-black as mantle. Leading edge of wing pale grey at carpal joint. Under wing-coverts and axillaries dark grey-brown, slightly paler and with indistinct off-white or pale fringes on greater coverts and on smaller coverts along leading edge of wing, slightly darker blackish-brown on innermost coverts. In worn plumage, forehead, lores, and supercilium paler grey; crown mottled brown, contrasting more with black mantle; some dark grey-brown of feather-bases visible on chest to belly; rump and throat sometimes with more pronounced dusky streaks; white fringes on tips of secondaries and inner primaries abraded. In fresh plumage, black feathers of underparts sometimes have faint narrow white fringes on tips. NESTLING. Naked at hatching; from $c.$ 14th day, covered with thick grey down clinging to closed feather-sheaths of upperparts and thighs; from $c.$ 20 days, down gradually disappears as feathers emerge (Moreau 1942a). JUVENILE. Closely similar to adult. At fledging, primaries and small coverts on leading edge of wing narrowly fringed white at tips, and secondary-tips more broadly fringed white than in adult, but fringes on primaries and coverts soon disappear and fringes of secondaries narrower due to abrasion, and these then similar to adult. Rump and throat-patch with no indication of shaft-streaks (adults usually show slight streaking). Inner web of t5 tapers gradually to narrowly rounded tip (in adult, inner web strongly emarginated and tip more sharply pointed). FIRST ADULT. As adult, but emargination of t5 intermediate between adult and juvenile (Brooke 1969b).

Bare parts. ADULT AND JUVENILE. Iris deep brown. Bill black. Leg and foot dark sepia, pinkish-black, or black with flesh-coloured soles; claws black. NESTLING. At hatching, generally pale pink; head, wing, foot, and claws almost white. Mouth pale pink, tip of bill and edges of gape pink-white or white (tip occasionally black). From 7–8 days, eye opens slightly, bill blackish, and skin darkened, appearing blue on feather-tracts; from $c.$ 14th day, eye fully open (iris dark brown), head, body, and wing fully covered by down and feather-sheaths, except on hindneck. On fledging, iris dark brown, bill black, leg and foot dark flesh-colour. (Moreau 1942a; BMNH, ZFMK.)

Moults. No details known for west Palearctic and hardly any for Afrotropics. In Sudan, where birds breed December–July (Brooke 1971e), 2 moulting September birds had primary moult scores of 12 and 23, and body and tail new except for part of head, underparts, and t5. No moult, and plumage fresh or slightly worn in April–July adults from Nigeria, Chad, Sudan, and Ethiopia; thus, unlike Little Swift *A. affinis*, apparently no moult during breeding season, though some South African breeding birds start primary moult towards end of breeding season, e.g. while still feeding 2nd brood (R K Schmidt). Juveniles examined from Sudan were fresh in March, worn in June. (BMNH, ZFMK.) For details of moult in Zaïre, see De Roo and Deheegher (1966).

Measurements. *A. c. streubelii*. Sudan and Ethiopia south to East Africa, all year; skins (BMNH, ZFMK). Tail is to tip of t5; fork is tip of t1 to tip of t5; bill is exposed culmen.

WING AD	♂ 141	(2·98; 13)	137–145	♀ 139 (2·90; 10)	135–143
TAIL AD		70·3 (3·33; 11)	66–75	69·9 (2·50; 10)	68–74
FORK AD		32·6 (2·59; 10)	30–36	30·0 (1·84; 10)	27–32
BILL		5·4 (0·47; 9)	4·8–6·0	5·5 (0·49; 5)	4·9–5·9
TARSUS		9·2 (1·07; 4)	8·5–10·2	9·2 (0·21; 4)	8·9–9·4

Sex differences significant for fork. Juvenile wing on average *c*. 3·5 shorter than adult, tail and fork both *c*. 6 shorter.

Weights. *A. c. streubelii*, Afrotropics, all year, sexes combined: 22·1 (54) 18–28 (Brooke 1971b). Nominate *caffer*, South Africa, 25·3 (3) 23–28 (Anon 1968b; Brooke 1971b).

Structure. 10 primaries: p9 longest, p10 2–5 shorter, p8 5–8, p7 18–22, p6 30–37, p5 44–51, p1 81–94. Tail long, deeply forked, 10 feathers; t5 longest, t4 9–12 shorter, t3 17–22, t2 22–29, t1 27–36 (depth of fork); in juvenile, t1 21–32 shorter than t5. Outer middle toe with claw 8·5 (5) 7·8–9·3; innermost *c*. 66% of outer middle, inner middle *c*. 79%, outermost *c*. 88%. Claws relatively longer and less strong than in Swift *A. apus*. Remainder of structure as in *A. apus*.

Geographical variation. Slight and clinal, mainly in size; differences in colour negligible. In Afrotropics, populations north of 5°S smaller, average wing 138; smallest in lowlands bordering Gulf of Guinea; populations from southern Zaïre and southern Tanzania south to southern Angola, Botswana, and *c*. 26°S in South Africa and Moçambique slightly larger, average wing of various populations 138–143, all combined 140·7 (98) 133–148 (Brooke 1971b). Birds south from Namibia and *c*. 28°S distinctly larger, adult wing 149·6 (61) 143–157 (Brooke 1971b), wing up to 170 according to Chapin (1939). Variation in size slight over most of Afrotropics and only southern population much larger, separated by apparently narrow intergradation zone. Recognition of 2 races thus seems warranted—*streubelii* as a slightly variable smaller-sized population inhabiting most of Afrotropics, nominate *caffer* as the large southern race. Not known which race breeds Morocco and Spain. Single bird from Spain said by Del Junco and Gonzales (1969) to have wing 149, but Zimbabwe bird also measured was said to be an improbable 151 and thus measuring technique perhaps not comparable with that commonly used (Brooke 1971b).

CSR

Apus affinis Little Swift

PLATES 62 and 63
[facing pages 663 and 686]

Du. Huisgierzwaluw   Fr. Martinet des maisons   Ge. Weissbürzelsegler
Ru. Малый стриж   Sp. Vencejo culiblanco   Sw. Stubbstjärtad seglare

Cypselus affinis Gray, 1832

Polytypic. *A. a. galilejensis* (Antinori, 1855), north-west Africa south to Niger, Chad, central Sudan, and northern Somalia, and Middle East east to Uzbekistan (USSR) and western Pakistan; *aerobates* Brooke, 1969, Banc d'Arguin (Mauritania), and West and central Africa south of *galilejensis*. Extralimital: nominate *affinis* (Gray, 1832), peninsular India south of Himalayas, except Assam and Kerala; also, coastal East Africa from southern Somalia to northern Moçambique; 3 other races in Africa and 5 others in southern Asia.

Field characters. 12 cm; wing-span 34–35 cm. Slightly larger bodied than White-rumped Swift *A. caffer*, but marginally shorter due to short square tail; thus more compact in silhouette than any other swift in west Palearctic. Small but chunky swift, lacking forked tail; square white rump and pale face contrast with otherwise sooty plumage. Flight not as graceful as other *Apus* species, with fluttering and steady sailing actions well developed. Sexes similar; no seasonal variation. Juvenile separable at close range.

ADULT. Plumage mainly sooty, with brown and green tones occasionally showing at close range (green most obvious on back and underbody). Square pure white patch over whole of rump and lateral coverts striking (particularly from behind); gives 'broad-beamed' appearance to rear body. Head or side on, white or grey-white forehead (occasionally looking speckled), and grey line over eye and round black eye-patch link with large white chin and throat to create obvious pale-faced appearance; in some lights, this is extended by brown crown into pale-headed look. Undersurface of flight-feathers paler than under wing-coverts but contrast less marked than in *A. caffer*; upperwing almost uniform with back. JUVENILE. Inner flight-feathers show white tips; rest of plumage duller, less dark, with paler tips on underbody and wing-coverts occasionally visible.

Short tail, appearing square in the field, unique in swifts of west Palearctic, but in brief view *A. affinis* confusable with *A. caffer*; separable on size of white rump, shape of tail, and flight actions. (South of Sahara, also confusable with spine-tailed swifts (Chaeturinae), particularly Mottled-throated Spinetail *Telacanthura ussheri*.) Flight less free and graceful than *A. caffer*, and tends to fly more constantly at one level, not indulging so frequently in marked accelerations, and alternates fluttering and gliding; gliding most distinctive, with wings often set straight out from body (allowing notably steady glide) and also suddenly raised above body, with tail simultaneously spread (creating slightly accelerated and downward glide). Speed of normal flight less than usual in *Apus*, but excited parties dash about in rapid chases. Behaviour typical of *Apus* but, in most of range, flies at lower levels.

In flight near nest, calls with high-pitched silvery rippling sound, much more feeble than scream of Swift *A. apus*.

Habitat. Breeds in lower middle and subtropical as well as widely in tropical latitudes beyond the equator. In

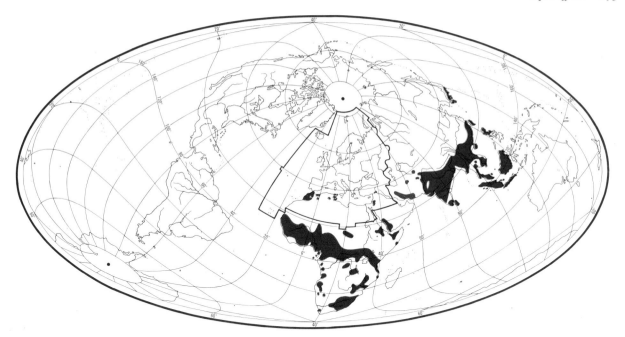

North Africa, nests in towns, sometimes villages or even isolated houses (Etchécopar and Hüe 1967), as well as cliffs. In USSR, at north of range, inhabits mountains, breeding even in cool damp and shady gorges, occupying nests of House Martin *Delichon urbica*; observed obtaining flying insects from favourable grassland area 25 km from breeding place (Dementiev and Gladkov 1951a). In India, in plains and up to *c.* 2000 m mountains and hills but now largely adapted away from cliff nesting to sites under eaves of houses, angles between wall and ceiling, arched gateways, bridges, old mosques, and ruins; even in noisy bazaars crowded with traffic and people. In Ceylon, keeps chiefly in neighbourhood of big rock-masses in foothills and low country (Ali and Ripley 1970). Hawks insects in flocks, often at an immense height; wet weather or a cold snap may send birds half torpid to nests (Whistler 1941). At Delhi, favourite nest-sites are ceilings and roofs of chambers in ancient tombs, as well as occupied buildings (Hutson 1954). In West Africa, much as in India, breeds in towns and villages and rarely seen far from them, but found in old forest clearings in Ghana. Nests inside buildings, including colony of over 1000 birds at Sekondi power station, Ghana (Bannerman 1951; Serle *et al.* 1977). In Somalia, however, has remained a cliff dweller, not nesting in towns; differs from other Apodidae there in being often on the wing through the heat of the day (Archer and Godman 1961). In southern Africa, mainly town-dwelling, building colonial nests against high buildings or water towers, but also under overhanging ledges (Prozesky 1970). Widespread shift from dependence on natural sites to variety of traditional and modern buildings has made possible substantial expansion of range and population, and exploitation of greater diversity of habitats.

Distribution. Marked expansion in Morocco, slight expansion in Algeria, and perhaps retreat after expansion in Tunisia; recently found nesting Turkey.

MOROCCO. Marked expansion. At end of 19th century known only at Essaouira and Marrakech; spread since to (e.g.) Le Sous and Rabat 1925, Tanger 1952, Larache 1953, and elsewhere on coast and inland, though colonies may be unstable (Heim de Balsac and Mayaud 1962; JDRV). ALGERIA. First discovered 1924, spread to Messad, Aïn Sefra, and Oran (Heim de Balsac and Mayaud 1962); nesting Massif du Chelia 1970 (EDHJ). TUNISIA. At end of 19th century bred only at some localities in south; by 1925 many colonies in hilly areas and breeding Gabès and Tunis (Blanchet 1955; Heim de Balsac and Mayaud 1962), though apparently now more restricted (Thomsen and Jacobsen 1979; MS). TURKEY. First found breeding 1971 (Beaman *et al.* 1975). LEBANON. Has bred occasionally in past (HK); may still nest (Macfarlane 1978). SYRIA. May breed, at least occasionally (Macfarlane 1978; HK).

Accidental. Britain, Ireland, Sweden, Italy, Egypt, Malta.

Population. Information limited, but numbers probably increased in parts of North Africa and perhaps Turkey (see Distribution).

TUNISIA. Estimated 140–210 pairs (Thomsen and Jacobsen 1979). MAURITANIA. Banc d'Arguin: estimated 90–110 pairs (Naurois 1969a); in 1980, bred on 5 or more islands; 110–120 nests on one, perhaps over 500–1000 pairs in total (J Trotignon, R A Williams).

Movements. Mainly resident, but migratory or partially migratory in northern (Palearctic) parts of range. Data sparse, however.

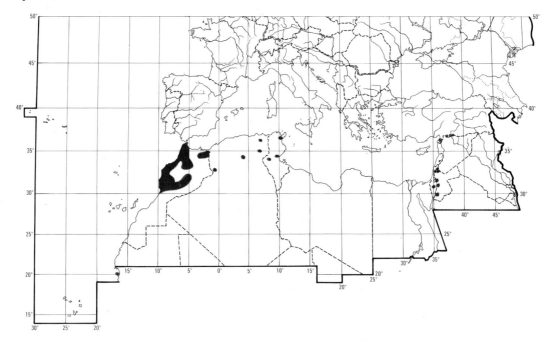

AFRICAN POPULATIONS. Resident in Afrotropical region, though local movements reported from some places: e.g. almost deserts Freetown (Sierra Leone) in dry season, January–March (G D Field). Though said by Heim de Balsac and Mayaud (1962) to be largely absent from northwest Africa in December–January, established since that many overwinter in Morocco and Tunisia at least (Smith 1965; Thouy 1978; M Smart). Race concerned (*galilejensis*) collected several times, October–June, in Niger, northern Nigeria, Sahel zone of Chad, and Darfur (Sudan); also in Eritrea and northern Somalia, where *galilejensis* also breeds (Brooke 1971c; Moreau 1972). Considered of regular occurrence in Chad, with important October passages at Ouaddai attributed to northern birds (Salvan 1968b). Migrant *galilejensis* known only from southern edge of Sahara, and numbers wintering there probably very small (Moreau 1972), but not separable in the field from local (resident) populations which occur alongside it there. Libyan status obscure: known only as non-breeding visitor to Tripolitania in February–October, mainly February–April (Bundy and Morgan 1969; Bundy 1976).

ASIATIC POPULATIONS. Apparently summer visitor, March–September, to recently discovered breeding areas in southern Turkey (Beaman *et al.* 1975). Probably at least partially migratory elsewhere in Near East and northern Middle East, though data sparse (Hüe and Etchécopar 1970): e.g. absent in winter from Israeli breeding areas (though seen in Syria), but wintering areas unknown and not found migrating through Gulf of Aqaba (Safriel 1968). Summer visitor to USSR (Tadzhikistan, Uzbekistan, Turkmeniya), though not known whether these winter Africa, Arabia, or India. Resident in Indian subcontinent, though in Pakistan and northern India numbers much reduced (or birds absent) during coldest months; presumed then to enter peninsular India, though no certain identifications there of Palearctic race *galilejensis*, to which Pakistan birds belong (Ali and Ripley 1970). Arabian status obscure, with records more widespread (including Persian Gulf states as vagrant) than known breeding sites; some of these may be migrants from further north and east.

Food. In India, takes midges (Chironomidae), small bugs (Hemiptera), beetles (Coleoptera), winged ants, and airborne spiders (Araneae) (Ali and Ripley 1970). In Yunnan (China), insects constitute 97·8% of food brought to young, chiefly flies, ants, and Homoptera (together *c.* 57·3% of total); also termites, hoverflies (Syrphidae), mosquitoes, mayflies (Ephemeridae), and aphids (together *c.* 28·8%). Adults averaged 25·8 (8–55) visits to nest per day; average 1·7 (0–16) visits per hr, with maximum feeding rate 19.00–20.00 hrs (3·3–3·7 visits per hr). At each visit, bring *c.* 168 (3–627) insects, or 1·47 (0·03–3·8) cm^3 (Chih-Tung 1973).

BDSS

Social pattern and behaviour. Little information for west Palearctic. Based largely on studies of nominate *affinis* in India and of *aerobates* south of Sahara.

1. Gregarious in breeding season, and probably throughout the year. In breeding season, readily associates with other *Apus*: e.g. Pallid Swift *A. pallidus* in Morocco (Meise 1959). BONDS. Little known. Mating system presumably monogamous, with pair-bond maintained from year to year, as in other *Apus*. Evidence from Morocco and Baroda (India) indicates pairs remain together, in association with nest, throughout the year (Brosset 1957, 1961; Razack and Naik 1965; Naik and Razack 1967; see also Roosting, below). Age of first breeding not known.

Parents share nest-duties (e.g. Moreau 1942b). BREEDING DISPERSION. Usually colonial, nests often densely packed. In North Africa, often c. 50 pairs or less (e.g. Heim de Balsac 1925, Brosset 1961, Pineau and Giraud-Audine 1979). In Morocco, one colony of c. 60 nests (P A D Hollom), though also often nests singly (Heim de Balsac 1954); colonies may be dense or dispersed (Brosset 1961). In colony, Algeria, compact masses of 10–12 nests beneath balcony of minaret (Heim de Balsac 1926). At Baroda, nests in clusters of 2–12: highest density 12 nests on 0·2 m^2 (Naik and Razack 1963). Colony of c. 95 nests, Kenya, in one dense overlapping mass (D J Brooks). Breeding territory confined to entrance and interior of nest. Probably highly site-faithful, even where nest not occupied all year round. At Tanger (Morocco), nests of previous year used, though not known if by same birds (see Ferguson-Lees 1967). In India, some nest clusters known to have been used continuously for 80 years (Ali and Ripley 1970). ROOSTING. Widely studied. A relatively inactive bird, spending much of its life in the nest, which may be occupied for most of the day (Heim de Balsac 1925, 1926, 1954; Brosset 1957, 1961), this pattern persisting all year in some places (see above). After breeding season, Morocco, some pairs built nest for roosting (Brosset 1961). During breeding season, Kasama (Zimbabwe), roosted either in own nest, whether or not complete, less commonly in hole or other crevice nearby; roost-site varied (Brooke and Vernon 1961), though this presumably more typical of non-breeders. During breeding season, Baroda, some birds roosted regularly at prospective nest-site. In summer, some clung to outside of a nest or vacant site, but in winter such birds thought to roost inside building (Naik and Razack 1967). In Morocco, pairs started roosting in nest from beginning of its construction at start of June, and continued until December (Brosset 1957). At start of breeding season, Baroda, birds usually returned to their nests only at night, to roost; nests never contained more than 2 birds, and seldom less (Razack and Naik 1965). Activity markedly crepuscular, especially in breeding season, birds feeding in morning and evening and remaining in nests (in pairs) in heat of the day (Heim de Balsac 1925, 1926; Brosset 1957, 1961); in Morocco, however, feed in the middle of the day in winter (Brosset 1957, 1961). Morning exodus from roost in Morocco on 12 April started 06.15 hrs, most birds not active until $\frac{1}{2}$–$\frac{3}{4}$ hr later (Meise 1959); on 20 March, first birds left 06.30 hrs, c. 15 min before sunrise (P A D Hollom). In evening, Morocco, started roosting in daylight, and all in roost by sunset (Meinertzhagen 1940). At Kasama, left roost long after sunrise and went to roost just after sunset, earlier if overcast; all went to roost within 5–10 min (Brooke and Vernon 1961). At Baroda, some did not leave roost until 10.00 hrs, later on cold mornings. Pair-members often left together. Exit from roost preceded by much vocal activity in nests. Before initiator finished 'chuck-' of Nest-call (see 2 in Voice), others joined in, and continuous waves of calling thus passed through colony. After lull of one to several minutes, another series followed. Larger colonies called more often, and for longer, than smaller ones, while very small ones were silent (Razack and Naik 1965). For behaviour of birds after leaving roost, see Flock Behaviour, below. Roosting time apparently determined by light intensity around nests (Naik and Razack 1963). In poorly-lit colonies, or parts of colonies, birds roosted 10–13 min later than in well-lit ones (Razack and Naik 1965). For daily and seasonal variations in roosting behaviour, see Razack and Naik (1965) and Naik and Razack (1967). Roosting typically preceded by Pre-roosting flight display (Young 1946; Razack and Naik 1965). At Baroda, birds returned to vicinity of colony c. 20–30 min before roosting time, and quickly coalesced into tight flock which circled ever higher, only to scatter again within a few minutes. Within 10 min this sequence usually repeated twice. About 7 min before roosting, birds descended almost to colony and suddenly began faster, circling flight, from which a few birds at a time periodically flew close past colony. Sometimes part of flock bunched and ascended again for c. 1 min before scattering and descending (Razack and Naik 1965). In Cameroun, ascending birds fluttered slowly in circles, calling (see 1 in Voice: Young 1946). At Baroda, there was a further increase in flight speed just before roosting. Birds entered roost-sites with sudden downward swoop, first by a few, then the rest (Razack and Naik 1965; see also Moreau 1942b). For a few seconds after entering nests, birds silent, then started calling, this increasing in intensity for a time before fading out. Birds from different colonies do not combine in pre-roosting flights (Razack and Naik 1965). At Kasama, some birds left roost-sites within a few minutes of entering, and did not reappear that night (Brooke and Vernon 1961).

2. FLOCK BEHAVIOUR. Upon leaving roost, Baroda, birds of several neighbouring colonies joined into one or several flocks, birds sometimes moving from one to another. Occasionally a few birds separated from flock and performed Screaming-display: following circular path, flew past colony once or more giving Screaming-call (see 1 in Voice), to which birds still in colony sometimes responded with Nest-call. Soon after, birds dispersed from immediate colony area (Razack and Naik 1965; Naik and Razack 1967). Where nesting dispersed, birds flock for feeding (Brosset 1961). For Screaming-displays by mixed parties of *A. affinis*, White-rumped Swift *A. caffer*, and Palm Swift *Cypsiurus parvus*, see Brooke and Vernon (1961). According to latter, display appears to combine, depending on time and circumstances, search for nest-sites, roost-sites, 'pure play', migration preliminaries, maintenance of social cohesion, and possibly other functions. Screaming-display observed in November at colony in Tunisia (P A D Hollom). ANTAGONISTIC BEHAVIOUR. 2 birds found interlocked and motionless on ground under nest-site (remained thus for c. 15 min) had presumably been fighting as in *A. apus* (Reuter et al. 1980). At Kasama, other species of *Apus* sometimes landed on back of nest-building bird, whereupon latter called (see 3b in Voice). HETEROSEXUAL BEHAVIOUR. Little information. (1) General. In the evenings, nest-owners flew in close formation in pairs or trios, often close to colony, giving Screaming-calls. Flight then butterfly-like, fluttering with wings raised in V above back (Wings-high display), only tips vibrating rapidly (Ali and Ripley 1970); described as sailing flight with wings raised at 45° (Moreau 1942b). For *A. affinis* attracted to Wings-high display of *A. caffer*, see De Roo and Deheegher (1966). At Baroda, aerial chases of 1 bird by 1–2 others occurred both in morning and evening; rare in winter, and during the 4 evenings in June when chasing was relatively intense, tight pre-roosting flocks did not form (Naik and Razack 1967). (2) Mating. For aerial copulation, see Lack (1956a). (3) Behaviour at nest. When intending to land (not clear whether at own nest) may call on approach (Brooke and Vernon 1961; P A D Hollom: see 1 in Voice). At Tanga, nest-relief during incubation variable in pattern and duration. Often, when bird entered nest another left immediately, suggesting rapid change-over (Moreau 1942b, which see for other patterns of arrival at, and departure from, nest). RELATIONS WITHIN FAMILY GROUP. Eyes of young start to open 7 days after hatching (Chih-Tung 1973); fully open at c. 12 days (Moreau 1942b). Brooding assiduous for first few days, ceasing some time during first 2 weeks (Moreau 1942b). Young appear to achieve thermoregulation by 20–24 days (Collins 1975). Fledge on own initiative, or else follow parent. Siblings follow soon after 1st young to fledge. Of 6 young, all fledged 2–4 hrs after sunrise. Fledged young apparently do not return to nest for roosting (Moreau 1942b). ANTI-PREDATOR RESPONSES OF

YOUNG. No information. PARENTAL ANTI-PREDATOR STRATEGIES. (1) Passive measures. Sitting birds loath to leave nest, even when handled (Lynes 1925b; Heim de Balsac 1926); easily caught, even on uncompleted nest (Heim de Balsac 1954). (2) Active measures: against birds. Members of colony, Cameroun, united to follow encroaching Black Kite *Milvus migrans* or Harrier-Hawk *Polyboroides typus* (Young 1946). EKD

Voice. Used chiefly in flight (Ali and Ripley 1970). The following based largely on studies of nominate *affinis* in India.

CALLS OF ADULTS. (1) Screaming-call. A rapid, high-pitched rippling sound, with a light, silvery quality (J Hall-Craggs, P J Sellar). Quite different from scream of some other *Apus* (Heim de Balsac 1925, 1926). In Fig I, frequency up to 8 kHz, but in recording by P A D Hollom in Morocco, reaches 10 kHz (J Hall-Craggs); pitch typically drops at end (P A D Hollom). Rendered 'chee-ch-ch-ch-ch' (Razack and Naik 1965); shrill, spirited musical scream—rapid 'siksiksiksik sik-sik siksiksiksik' or variant (Ali and Ripley 1970); thin shrieks (De Roo and Deheegher 1966). Given by birds performing Screaming-display; shrill trills of birds ascending in pre-roosting flight (Young 1946) probably the same. This, or similar call, described as shrill trill, also given by bird intending to land at nest (P A D Hollom; see also Brooke and Vernon 1961). Birds in nest sometimes answered with same call, and also called thus at other times, apparently independently of any birds outside nest (P A D Hollom). (2) Nest-call. A 'chuck-chrrr' is usual call of birds before leaving nest in morning, also when recently settled in nest at night (see Roosting in Social Pattern and Behaviour), and contagious among roosting birds, culminating in chorus of prolonged 'chrrrr' sounds. Apparently incomplete variants of call also heard when roosting activity subdued. Nest-call sometimes given in response to call 1 of birds outside (Razack and Naik 1965). In Morocco, first part of call, rendered 'tick' or 'sik', predominant, producing ticking sound; 'chrrr' sound only twice heard, and then as trisyllabic 'churrerer' and 'churrer-chur'. Up to 12 'ticks' given in 6 s; delivery sometimes slower ('ticks' up to 1 s apart) and more erratic, e.g. only 1 or 2 at a time; occasionally developed into the pre-landing shrill trill (see call 1). First heard before sunrise, prior to exit from roost. (P A D Hollom.) (3) Other calls. (a) Bird engaged in nest-building gives 'scream of rage', whenever another species of *Apus* lands on its back (Brooke and Vernon 1961). (b) In recording, flying birds give staccato 'chip-chip-chip...' (Fig II) in addition to call 1.

CALLS OF YOUNG. No definite information. Whispered 'swee-swee' sounds, also described as quiet sucking squeaks, heard rather persistently from nests, especially after arrival of parents (P A D Hollom), possibly food-calls of young. EKD

Breeding. SEASON. North-west Africa; first eggs laid mid-April, continuing to at least June, presumably with 2nd broods (Heim de Balsac and Mayaud 1962). SITE. On or in building, on cliff, or under bridge. Also makes use of old nest of Red-rumped Swallow *Hirundo daurica* or House Martin *Delichon urbica*. Nest: near-globular structure of feathers, grass, and straw, cemented with saliva, with entrance as narrow slit or short tunnel; own nests and those built by other species lined with feathers. For more details of construction, India, see Naik and Razack (1963). Building: by both sexes, gathering material in the air; see also Naik and Razack (1963). EGGS. See Plate 98. Long elliptical, smooth and not glossy; white. 23×15 mm ($21-25 \times 14-16$), $n=18$ (McLachlan and Liversidge 1968). Clutch: 2–3(–4). No clutch size data from west Palearctic. Of 514 clutches, India: 2 eggs, 30%; 3, 70%; average 2·7 (Razack and Naik 1968). Probably 2 broods. Average laying interval, India, in 2-egg clutches 2·5 days ($n=358$), and in 3-egg clutches 2·2 days between 1st and 2nd, 2·4 days between 2nd and 3rd ($n=156$) (Razack and Naik 1968). INCUBATION. In India, average 22·1 days (18–26), $n=160$ (Razack and Naik 1968). Begins with 1st egg; hatching asynchronous, with intervals of 1–2 days (Razack and Naik 1968). By both sexes. For details of incubation shifts, Tanzania, see Moreau (1942b). YOUNG. Altricial and nidicolous. Cared for and fed by both parents. FLEDGING TO MATURITY. Average fledging period, India, 40·5 days (33–49), $n=195$; 12% of nestlings left at 33–36 days, 71% at 37–43 days, and 17% at 44–49 days (Razack and Naik 1968). Age of independence not recorded. Age of first breeding not known. BREEDING SUCCESS. No information from west Palearctic. Of 1193 eggs laid, India, 28·2% lost, most ejected by birds themselves, some accidentally broken by observer; of 846 eggs remaining after these losses, 84·7% hatched; average nestling losses in different seasons 61·1–79·7%, with majority of losses due to young falling out of nest (Razack and Naik 1968).

Plumages (*A. a. galilejensis*). ADULT. Forehead and lores pale grey or pale grey-brown, nearly white in narrow line extending over black patch in front of eye. Forecrown pale grey-brown; crown gradually darker towards nape and upper mantle, where dark grey-brown with black feather-tips. Lower cheeks and sides of head medium grey-brown, darkening towards brownish-black and black on sides of neck. Feather-tips of forehead, crown, nape, and cheeks narrowly and faintly fringed off-white or pale grey-brown when plumage fresh. Mantle, scapulars, and back deep

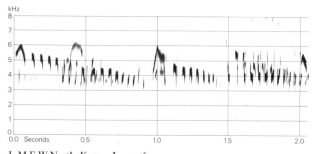

I M E W North Kenya June 1963

glossy bluish-black, some dark grey-brown of feather-bases often visible, especially when worn. Broad white band across rump c. 1·5 cm from front to back in middle, 2 cm at sides, extending on to white rear flanks. Some lower rump feathers white with pale grey-brown suffusion on tip. Upper tail-coverts dark or medium grey-brown, feather-tips narrowly but distinctly fringed white when fresh. Large patch on chin and throat white, some feathers with faint and narrow shaft-streak; underparts from breast to vent contrastingly black, slightly glossy and with some narrow but distinct off-white feather fringes when plumage fresh, duller and with some dark grey-brown of feather-bases visible when worn. Shorter under tail-coverts medium grey-brown, longer pale grey-brown or pale grey, lateral ones sometimes greyish-white; all coverts narrowly but distinctly fringed white on tips. Tail pale grey-brown above (paler than upper tail-coverts), pale silvery-grey below (about similar in tinge to longer under tail-coverts). Primaries and greater upper primary coverts dark grey-brown, tips and outer webs almost black; secondaries and other upper wing-coverts darker, dull black, darker and more glossy bluish-black on lesser coverts; fresh flight-feathers narrowly edged white along inner webs and tips when fresh, except for 5–6 outer primaries. Under wing-coverts and axillaries dull black or dark grey-brown, feather-tips narrowly fringed white; white fringes of smallest coverts form almost white leading edge to wing. NESTLING. Naked at hatching. At c. 12 days, back covered with grey down, feathers on head, wing, and tail developing (Moreau 1942b). JUVENILE. Like adult, but crown, mantle, and scapulars duller black, not glossed bluish-green, faint olive-green lustre only. Faint and narrow grey-brown feather-fringes from chest to under tail-coverts, no gloss from chest to vent. Narrow white edges to tips of secondaries and inner primaries (faintly up to p8); faint pale grey-brown fringes to tips of median and greater upper wing-coverts. Similar to adult when plumage worn. FIRST ADULT. Only separable from adult by early primary moult in 2nd calendar year, or, when primaries moult at same time as adult, by more heavily abraded outer primaries (see Moults).

Bare parts. ADULT AND JUVENILE. Iris dark brown. Bill horny-black or black. Leg and foot pink-brown or pink-horn. NESTLING. At hatching, bill blackish and claws white; on 12th day, bill and claws black. (Moreau 1942b; Ali and Ripley 1970; BMNH.)

Moults. Based mainly on c. 40 actively moulting birds from Palearctic (BMNH, RMNH, ZFMK, ZMM). ADULT POST-BREEDING. Complete; primaries descendant. Starts with p1 from late April to early June or early July; primary moult score 10 (16) 0–23 mid-June, 22 (2) 19–26 mid-July, 35 (3) 31–39 late August, 43 (2) 42–44 late October, probably completed (score 50) December–January; moult occasionally suspended (e.g. with score 10 in June). Body starts with some feathers of belly and flanks from score c. 15, but moult mainly at score 30–45; tail and secondaries mainly during last stages of primary moult, starting at score 30–40, completed with p10. For details of moult in Indian nominate affinis, see Naik and Naik (1965). POST-JUVENILE. Complete; timing probably variable, but not fully established because some immatures are difficult to age. Head and body first, probably starting soon after fledging; some largely in 1st adult on head, mantle, scapulars, and underparts by late July, others still in worn juvenile in October. Moult of head, mantle, scapulars, and underparts followed by back, rump, tail-coverts, and tail; when tail moulting or completed, flight-feathers follow with p1, often when body still fresh (in adult, body feathers usually worn when p1 lost). Primaries start mainly December–May, completing July–November of 2nd calendar year, but one specimen had started September; unlike adult post-breeding, body apparently not replaced during primary moult, at least not in 7 birds examined which were in last stages of primary moult (score 40–49) June–August: these had head and body old, tail and wing-coverts still in fairly new 1st adult, and only a few had some feathers of head new.

Measurements. *A. a. galilejensis*. ADULT. North-west Africa, February–November; skins (BMNH, RMNH, ZFMK, ZMA). Tail is to tip of longest feathers; bill is exposed culmen.

WING	♂ 137	(3·26; 11)	133–140	♀ 136	(2·26; 14)	134–141
TAIL	41·4	(1·07; 11)	40–43	40·9	(1·38; 14)	39–43
BILL	5·7	(0·31; 9)	5·3–6·1	5·9	(0·37; 8)	5·4–6·3
TARSUS	9·7	(1·00; 9)	8·7–11·0	9·5	(0·47; 5)	9·1–10·1

Israel, Jordan, and Lebanon, March–October; skins (BMNH, RMNH, ZFMK).

WING	♂ 137	(1·57; 7)	137–138	♀ 135	(2·09; 7)	132–137
TAIL	42·0	(1·25; 10)	40–44	41·8	(1·91; 8)	39–44

Uzbekistan, Turkmeniya, Afghanistan, and north-west Pakistan, May–October; skins (RMNH, ZMM).

WING	♂ 136	(2·68; 6)	132–140	♀ 133	(— ; 1)	—

Southern Arabia, south-east Iran, and Baluchistan (Pakistan), February–October; skins (BMNH).

WING	♂ 131	(2·36; 5)	128–134	♀ 128	(1·26; 4)	126–129

Sex differences not significant. Juvenile wing on average 2·5 shorter than adult, tail similar to adult.

A. a. aerobates. Banc d'Arguin (Mauritania), January–March: wing 131 (5) 130–132, tail 37·6 (5) 35–39 (Naurois 1972). Sénégal, Mali, Guinea-Bissau, and Cameroun: wing 129 (14) 124–135, tail 38·0 (14) 36–40 (Naurois 1972; RMNH).

Weights. *A. a. galilejensis*, Afghanistan, May: ♂ 21 (Paludan 1959). Morocco, April: 25·3, 27·5 (BTO). Combination of several other African races (shorter winged than *galilejensis*): 25·0 (64) 18–30 (Britton 1970; Brooke 1971c). Taiwan race *kuntzi* (wing length similar to *galilejensis* from Levant), April–July: ♂ 22·6 (13) 18–26, ♀ 26·2 (9) 22–35 (RMNH). For weight study of small nominate *affinis* from India, see Naik and Naik (1966).

Structure. 10 primaries: p9 longest, p10 1–2 shorter, p8 4–6, p7 15–19, p6 28–35, p5 41–47, p1 80–87. Tail short, square; 10 feathers, t3–t4 longest, t1 2–5 shorter, t2 and t5 0–1 shorter. Toes short, strong; outer middle toe without claw, and claw, each 5·3 (5·0–5·8); innermost toe with claw c. 68% of outer middle toe with claw, inner middle c. 80%, outermost c. 79%. Remainder of structure as in Swift *A. apus*.

Geographical variation. Complex; involves depth of brown of body, wing length, and shape of tail. Perhaps too many races recognized in central and southern Africa and too few in North Africa and Arabia; however, Brooke (1971c) and Clancey (1980) followed here. In general, colour darkest and rump- and throat-patches narrower in high-rainfall areas (islands in Gulf of Guinea, Kerala, Ceylon, south-east Asia) and palest and with larger rump- and throat-patches in the most arid climates (*galilejensis*), though there are some exceptions (Himalayan and south-west African races both intermediate, instead of dark and pale respectively). Nominate *affinis* from plains of peninsular India and from south-west Somalia to north-east Moçambique closely similar to *galilejensis*, but slightly darker on forehead and sides of crown, white rump slightly narrower, white throat-patch slightly smaller, and under tail-coverts and undersurface of tail slightly darker. *A. a. aerobates* of West and central Africa slightly blacker on mantle, back, belly, and wing-coverts than both nominate *affinis* and *galilejensis*. Population of Banc d'Arguin (Mauritania) slightly paler than typical *aerobates*, but still nearer to *aerobates* than to *galilejensis* (Naurois 1972). Wing of *galilejensis* from north-west Africa and Middle East long (average 136); of similar size in birds from Taiwan (136) and south-east Asia (134), and larger still in Java (142). Birds from Banc d'Arguin, Afrotropics, and peninsular India short-winged (average *c.* 130); wing slightly longer in southern Africa (134) and in Himalayas (133). Populations from Himalayas and from Assam to Taiwan and Java have shallow tail-fork: t5 5–14 longer than t1, depending on population. Palest race, *galilejensis*, not split by Brooke (1971c) or Clancey (1980); though colour equally pale throughout North Africa, Arabia, and Middle East, size far from uniform. Birds from north-west Africa, Israel, Jordan, Lebanon, western Soviet Central Asia, Afghanistan, and north-west Pakistan large, wing 136 (47) 132–141; those from Niger to northern Somalia, southern Arabia, south-east Iran, and Baluchistan (Pakistan) distinctly smaller, wing 129 (13) 124–134 (RMNH, BMNH), latter perhaps warranting recognition as *abessynicus* (Streubel, 1848) (in the past, this name has been used erroneously for *aerobates*: Brooke 1969d). CSR

Cypsiurus parvus Palm Swift

PLATES 62 and 63
[facing pages 663 and 686]

Du. Afrikaanse Palmgierzwaluw   Fr. Martinet des palmes   Ge. Palmsegler
Ru. Пальмовый стриж   Sp. Vencejo palmero   Sw. Palmseglare

Cypselus parvus Lichtenstein, 1823

Polytypic. Nominate *parvus* (Lichtenstein, 1823), arid northern Afrotropics from Sénégal to Sudan and western Ethiopia. Extralimital: 6 races in Africa south of nominate *parvus*, on Comoro Islands, and on Madagascar.

Field characters. 16 cm, of which tail up to 7 cm; wing-span 33–35 cm. Least bulky of family in west Palearctic, with head, body, and wings as long as those of Little Swift *Apus affinis* but all much slimmer; *c.* 10% slighter than Plain Swift *A. unicolor* and Cape Verde Swift *A. alexandri*. Marked attenuation of form culminates in streamers of outermost tail-feathers. Small, umber-brown swift, with long tail-streamers, slim build, and rapid wing-beats. Sexes similar; no seasonal variation. Juvenile separable at close range.

ADULT. Plumage umber-brown, with grey or faintly green tones occasionally visible (depending on light intensity); shows little relief, except for darker flight-feathers and tail and paler, almost white, throat. JUVENILE. Feathers of upperparts tipped buff, creating mottled appearance; throat grey. Tail shorter than in adult, but still elongated.

Unmistakable in west Palearctic; confusion with even smallest uniformly coloured *Apus* instantly ruled out by exceptional length of outermost tail-feathers (less so in juvenile) and remarkably slim build. Flight astonishingly rapid: bird hurtles along, with noticeably flickering wing-beats allowing sudden turns, fast wheels, and accelerated dashes and half-loops and giving mastery of manoeuvrability in tight spaces. Behaviour recalls hirundine as much as swift; most active in last hours of daylight.

Less gregarious than *Apus* but regularly forming flocks. Highly vocal, with birds around nest-sites constantly uttering high-pitched chittering calls (though not loud in chorus).

Habitat. In tropical West Africa coincides with thorn-scrub and northern half of grass-woodland belt, birds being common wherever there are palms. Mostly shows strong preference for breeding in borassus palm *Borassus flabellifer*, growing by streams, on edges of marshes, and in depressions, but also uses drooping fronds of dom palm *Hyphaene thebaica* and, in forest, oil palm *Elaeus*; West African race *brachypterus* nests on beaches in coconut palms, and in Liberia prefers coastal towns. Elsewhere hunts over neighbouring grasslands or drinks at pools, but also rises fairly high. (Bannerman 1933, 1951; Serle *et al.* 1977.)

Distribution. Breeds in Africa south of Sahara, from Sénégal east to Ethiopia and southern Somalia, south to Angola, northern Namibia, north-east Transvaal, and Natal. Mainly resident (see Movements).

Accidental. Egypt: formerly occasional visitor in extreme south (Abu Simbel), where a few seen February 1928 (Meinertzhagen 1930); habitat now disappeared due to filling of Lake Nasser (PLM, WCM).

Movements. Non-migratory, and with only local dispersals. Even Abu Simbel occurrence (see Distribution) probably represented just short displacement from Wadi Halfa (northern Sudan) breeding area.

Voice. See Field Characters.

Plumages (nominate *parvus*). ADULT. Forehead, crown, upperparts, tertials, and upper wing-coverts brownish mouse-grey, darkest on crown and lesser upper wing-coverts, almost black on upper greater primary coverts. Feather-tips of crown and lesser coverts faintly bordered grey when fresh, those of mantle to upper tail-coverts, median and greater upper wing-coverts, and tertials narrowly fringed greyish-white. Lores, sides of head down from ear-coverts, and all underparts pale mouse-grey; gradually paler toward greyish-white chin and central throat, which show varying amount of darker grey shaft-streaks.

Feathers of underparts narrowly fringed off-white on tips when fresh. Tail and flight-feathers olive-brown with slight green lustre, darkest on outer webs of outer primaries; broad borders along basal inner webs of primaries pale grey; tips of secondaries and inner borders of primaries narrowly edged white. Under wing-coverts and axillaries pale mouse-grey. Sexes mainly similar, but chin and throat of ♀ often pale grey (white in ♂), dark grey streaks thus less contrasting. JUVENILE. Like adult, but all feathers of head, body, tail, and wing (except secondaries) narrowly but conspicuously fringed rufous on tips. Tail shorter and less deeply forked than in adult, t5 with broadly rounded tip rather than strongly elongated with thread-like tip (see Structure). In worn plumage, rufous fringes largely lost through abrasion, but traces of bleached buff fringes usually still readily visible on back and rump, under tail-coverts, and some wing-coverts; t5 still short and with rounded tip. FIRST ADULT. Like adult, but tail slightly shorter and less deeply forked; length and shape of t5 intermediate between adult and juvenile, inner web slightly emarginated (as in adult), but tip narrowly rounded and not as sharply pointed as in adult.

Bare parts. ADULT AND JUVENILE. Iris deep brown. Bill black. Leg and foot reddish-grey. (Reichenow 1902–3; BMNH.)

Moults. Data limited. In Sudan, plumage of adults fresh December–January, worn March–May (May bird breeding); January–March juveniles slightly worn, some March birds attaining scattered 1st adult feathers on crown and back (BMNH); undated birds had body and tail worn at primary moult score 12 (inner 2 primaries new, p3 growing), and body worn but tail new (t1–t2 still growing) at score 45 (RMNH).

Measurements. Nominate *parvus*. Mainly Sudan, some Eritrea and Chad, December–May; skins (BMNH, ZFMK). Tail to tip of t5; fork is tip of t1 to tip of t5.

WING AD	♂ 133	(3·41; 9)	128–138	♀ 132	(2·40; 7)	129–136	
JUV	128	(4·57; 4)	123–133	124	(3·82; 4)	119–128	
TAIL AD	92·4	(2·60; 9)	89–96	90·8	(2·97; 7)	87–94	
FORK AD	59·2	(2·86; 9)	55–63	58·7	(3·15; 7)	54–62	

Sex differences not significant.

Weights. Average weight of nominate *parvus* c. 13·0. Data from several Afrotropical races combined (including some of larger size than nominate *parvus*) 13·6 (61) 10–18 (Brooke 1972b).

Structure. Wing long and narrow, pointed. 10 primaries: p9 longest, p10 3–6 shorter, p8 5–10, p7 19–25, p6 32–39, p5 46–52, p1 83–91. Tail long, deeply forked, 10 feathers; in adult, t5 60·4 (23) 54–67 longer than t1, sharply pointed, tip thread-like, inner web strongly emarginated; in juvenile, t5 37·0 (6) 35–40 longer than t1, tapering to rounded tip; in 1st adult, shape and length intermediate (Brooke 1969b, 1972b). Bill very small, exposed culmen c. 4·5 mm. Tarsus and toes very short, relatively rather strong; tarsus c. 8·8 mm, middle toe without claw c. 4·5. Front of tarsus feathered. Claws strong and sharply curved, middle claw about equal to length of middle toe. Remainder of structure as in Swift *Apus apus*.

Geographical variation. Slight; only Madagascar race well differentiated. Size (expressed in wing length) largest in southern Africa (south from 15°S) and on eastern plateau of central Africa from Uganda and eastern Zaïre south to Zambia (average wing 132–134); smaller in remainder of African mainland (including northern Afrotropics inhabited by nominate *parvus*) and on Comoros (average wing 126–129); smallest on Madagascar (average wing 123) (Brooke 1972b). General colour darker and throat more heavily streaked in races of high-rainfall areas, paler and throat faintly streaked in races of arid areas: races of forest block of western and west-central Africa, and those of Comoros and Madagascar darkest, nominate *parvus* of Sahel zone and race of Botswana, southern Angola, and Namibia palest. Tail in juvenile of Malagasy race more deeply forked than in juveniles of African mainland races; tail of 1st adult similar in shape to adult, not intermediate between juvenile and adult (Brooke 1969b, 1972b).

Forms superspecies with Asian Palm Swift *L. balasiensis* from Pakistan east to Philippines and western Indonesia; differs from *C. parvus* in small size, shallow tail-fork, unstreaked throat in adult, and white instead of rufous feather-edges in juvenile (Brooke 1972b, 1974). CSR

Order CORACIIFORMES

Very small to very large perching or ground-living birds showing great diversity in structure and way of life. Greatest numbers and diversity in tropics. Composition most generally followed (e.g. Wetmore 1960) recognizes 10 families (4 occurring in west Palearctic): (1) Alcedinidae; (2) Todidae (todies, 5 species; Greater Antilles); (3) Momotidae (motmots, 9 species; tropical America); (4) Meropidae; (5) Coraciidae; (6) Brachypteraciidae (ground-rollers, 5 species; Madagascar); (7) Leptosomatidae (single species—Cuckoo-roller *Leptosomus discolor*; Madagascar); (8) Upupidae; (9) Phoeniculidae (wood hoopoes, 8 species; Afrotropics); (10) Bucerotidae (hornbills, 44 species; Afrotropics and southern Asia). One of the most heterogeneous of all avian orders, and long subject to controversy. Very probably polyphyletic, and has been split into as many as 6 orders (Stresemann 1927–34, 1959). Wetmore's (1960) classification essentially derived from work of Fürbringer (1888) and Gadow (1892–3), and recognizes 4 suborders: (1) Alcedines (families 1–3); (2) Meropes (family 4); (3) Coracii (families 5–9); (4) Bucerotes (family 10). Recent work on biochemistry, anatomy, and phylogeny of this order has examined egg-white proteins (Sibley and Ahlquist 1972), bony stapes (Feduccia 1977), appendicular myology

(Maurer and Raikow 1981), and feeding apparatus (P J K Burton). Current views fairly unanimous that families 1–4 constitute a natural group, and families 5–7 another (Cracraft 1971). Relationships of families 8–10 are less well agreed, though several workers have regarded these, too, as constituting a natural group.

Relationships to other orders are equally disputed, though some general alliance with Piciformes usually accepted (see that order). Recent studies (e.g. Feduccia 1977) have also tended to support inclusion of Trogonidae (trogons), treated by Wetmore (1960) and others as a distinct order Trogoniformes.

Relatively few characters shared by all members of order. Feet often show more or less syndactyly and have 3 toes pointed forwards (outer reversible in *Leptosomus*; inner lacking in some Alcedinae) and hind toe present. Basipterygoid processes absent or rudimentary, hypotarsus complex, and syrinx tracheobronchial (bronchial in *Leptosomus*).

Family ALCEDINIDAE kingfishers

Small to large perching birds, with large heads, long dagger-shaped bills, and small legs and feet. 3 subfamilies, all represented in west Palearctic: Daceloninae, Alcedininae, and Cerylinae. Fish or other aquatic prey form principal food of last 2 subfamilies, but many Daceloninae are terrestrial feeders.

Sexes of same size. Necks short. 14–15 cervical vertebrae. Wings rather short, rounded; flight generally rapid and direct, with fast wing-beats especially in smaller species. 10 primaries; p11 much reduced, and absent in Alcedininae. 11–14 secondaries. Mostly eutaxic, some Daceloninae diastataxic. 12 tail-feathers, except in *Tanysiptera* (Daceloninae) with 10. Oil-gland tufted. Aftershafts lacking. Bills usually long, laterally compressed in Alcedininae and Cerylinae. A basisphenoid notch (unique to this family) present in all Cerylinae, many Alcedininae, and some Daceloninae (Burton 1978). Tongue and hyoid much reduced. Tarsus short; feet with 3 toes directed forward and 1 back, strongly syndactyle, with middle and outer toes united for more than half their length, and inner joined to middle for its basal third (inner toe sometimes lacking); soles much flattened. No caeca.

Plumages of Daceloninae and Alcedininae often bright, with brilliant blues in many species; Cerylinae generally duller, or pied. All Cerylinae, some Daceloninae, and a few Alcedininae show sexual dimorphism. Irises dark. Bills dark, bright red, or combinations thereof, less often yellowish or horn. 1 complete moult and 1 partial per cycle. Primaries moulted descendantly in Daceloninae, descendantly in 2 synchronized groups in Alcedininae and (less synchronized) in Cerylinae. Secondaries moulted ascendantly from outside to s4 and descendantly from inner to s5. Nestlings altricial, naked at first, acquiring spiny appearance later as feathers remain sheathed until well developed.

Subfamily DACELONINAE forest kingfishers and allies

Small to large kingfishers, many occupying terrestrial, even arid, habitats, though others frequenting water margins and coasts. 58 species in 8 genera: (1) *Pelargopsis* (stork-billed kingfishers, 3 species; southern Asia); (2) *Lacedo* (single species—Banded Kingfisher *L. pulchella*; south-east Asia); (3) *Dacelo* (kookaburras, 4 species; Australasia); (4) *Clytoceyx* (single species—Shovel-billed Kingfisher *C. rex*; New Guinea); (5) *Melidora* (single species—Hook-billed Kingfisher *M. macrorrhina*; New Guinea); (6) *Cittura* (single species—Celebes Blue-eared Kingfisher *C. cyanotis*); (7) *Halcyon* (forest kingfishers, 39 species; Africa, Asia, Australia); (8) *Tanysiptera* (paradise kingfishers, 8 species; Australasia). Represented in west Palearctic by 2 species of *Halcyon*, both breeding. Fry (1980) recognized only 55 species, merging *Pelargopsis* with *Halcyon*, but separating 6 south-east Asian species from *Halcyon* in *Actenoides*.

For general features, moults, etc., see Alcedinidae. P10 usually much shorter than p5, longer only in a few species of *Halcyon*. Wing diastataxic in some species of *Halcyon*. Tail fairly long, always more than half wing length. Bills generally less laterally compressed than in other 2 subfamilies (*Pelargopsis* an exception), culmen rather rounded and strongly depressed at base; bill very short and deep in *Clytoceyx*, hooked at tip in *Dacelo* and *Melidora*. Skull relatively broad. Tibiotarsus feathered to distal end, tarsometatarsus rather long (longer than inner toe without claw); toes long, 2nd shorter than 3rd and, excluding claw, just shorter than or equal to 4th.

Sexes usually differ in colour of upperparts or tail, rarely (*Lacedo* and some *Halcyon*) in colour of underparts. Plumage always with blue or turquoise, primaries never spotted or barred with white.

Halcyon smyrnensis White-breasted Kingfisher

PLATES 64 and 68
[between pages 686 and 687]

Du. Smyrna Ijsvogel   Fr. Martin-pêcheur de Smyrne   Ge. Braunliest
Ru. Красноносый зимородок   Sp. Martin pescador turco   Sw. Smyrnakungsfiskare

Alcedo smyrnensis Linnaeus, 1758

Polytypic. Nominate *smyrnensis* (Linnaeus, 1758), Asia Minor to north-west India. Extralimital: *fusca* (Boddaert, 1783), India (except north-west) and south-east Asia, south to Ceylon and Java; *saturatior* Hume, 1874, Andaman Islands; *gularis* (Kuhl, 1820), Philippines.

Field characters. 26–28 cm; wing-span 40–43 cm; bill 6–7 cm. Twice the size of Kingfisher *Alcedo atthis*. Largest resident kingfisher in west Palearctic, with massive red bill, and predominantly dark wine-chestnut and blue plumage relieved by white throat and central breast. Usually hunts away from water. Laughs noisily. Sexes similar; no seasonal variation. Juvenile separable.

ADULT. Massive coral-red bill. Top and sides of head, back and sides of neck, upper mantle, lesser wing-coverts, sides of breast, and all lower underparts deep vinaceous-chestnut, looking dark brown in shade and relieved only by white fleck behind eye and bold white bib from chin and throat over foreneck to centre of breast. Back, rump, tail, and most of wing light blue, strongly iridescent (with green tone obvious on back and tail, and turquoise tone noticeable on upper tail-coverts and wing), and relieved by black median coverts and black outer webs to primaries. In flight, upperwing shows pale blue band across base of primaries, fading out over secondaries, and underwing shows bold white band in same position. Legs orange-red. JUVENILE. Plumage pattern as adult but paler and duller; breast often shows crescentic marks. Bill at first orange-yellow with dusky tip and base.

Unmistakable; Grey-headed Kingfisher *H. leucocephala* has pale head and partly black back (and does not enter range of *H. smyrnensis*). Flight powerful, with bursts of wing-beats and looping of glides and turns ending in somewhat untidy pounce or dive on prey; action often more reminiscent of large woodpecker *Dryocopus* than of *A. atthis*. Behaviour typical of family, with bird alternating periods of secretive perching and energetic hunting; most conspicuous in breeding season, often calling from open perch.

Very vocal, with shrill call and noisy laugh, loud and far-carrying.

Habitat. In warm dry tropical, subtropical, and neighbouring low latitudes, mainly in lowlands but locally in India up to *c*. 1800 m. Ranges from seashore (and mangrove swamps in Andaman Islands), fish-curing yards, along canals, drains, streams and ditches, by ponds, pools, flooded borrow-pits, and wet paddyfields to dry terrain such as deciduous forest, gardens in towns, forest clearings, and, in Burma, even the depths of teak forests. Away from water forages more like shrike *Lanius* from such elevated perches as telegraph wires, branches, or

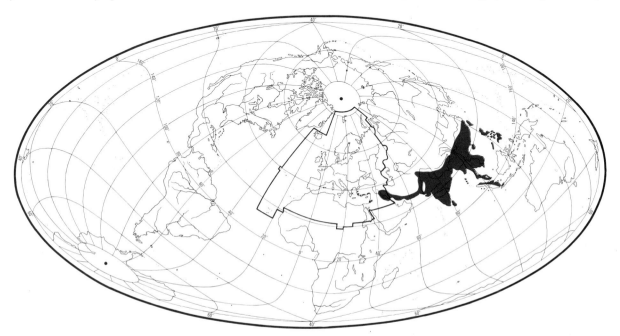

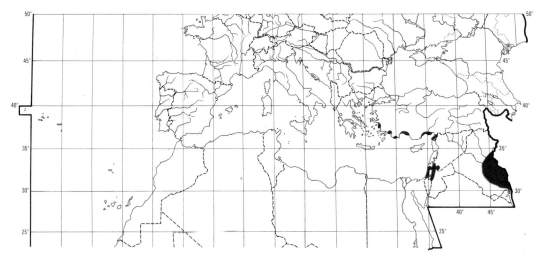

fence-posts, stooping to ground for prey. Will sing from exposed tree-top, and nest in tunnel bored in steep bank of dry nullah, roadside cutting, or similar vertical face (Smythies 1953; Hutson 1954; Ali and Ripley 1970). Although evidently aquatic in origin, has adapted widely to terrestrial dry habitats, short of deserts and interior of dense forests.

Distribution. LEBANON. Probably bred formerly but no recent proof (HK). SYRIA. Believed to breed in the north, north-east, and east though no recent proof (Kumerloeve 1968b; HK); 10 sightings 1975–7, August–March, with none in valley of Euphrates, despite frequent visits (Macfarlane 1978).
Accidental. Greece, USSR, Lebanon, Kuwait, Egypt, Cyprus.

Population. TURKEY. Not uncommon, though rather local (Beaman *et al.* 1975); at least 100 pairs (RFP). IRAQ. Very common, especially in centre and south (Allouse 1953). ISRAEL. Has increased with growth of agriculture, which produces one of its main foods, mole-cricket *Gryllotalpa gryllotalpa*; apparently resistant to heavy pesticide use in these areas (HM).

Movements. Basically resident throughout range, but individuals wander to uncertain extent outside breeding season.
May still breed Syria (see Distribution) but latest records (Kattinger 1970; Macfarlane 1978) all August–March, which is equally consistent with their having been visitors from elsewhere; modern sightings in Lebanon have similar temporal distribution (Benson 1970; Tohmé and Neuschwander 1978). Some evidence for southward movement in Iraq in autumn; chiefly seen around Al Kut November–April (Moore and Boswell 1956b), and individuals seen in southern desert at Shaibah October–April (Chapman and McGeoch 1956). Also scarce visitor (usually singly) to Kuwait, most mid-July to early November, with single records in January, March, and April (Bundy and Warr 1980); stragglers have reached Riyadh area (east-central Saudi Arabia) May and September (Jennings 1981a). Pakistan and Indian population typically resident, though individuals may move more than supposed: often taken nocturnally at lights during monsoon season in areas where otherwise absent (Ali and Ripley 1970).

Food. Insects, fish, amphibians, and reptiles; occasionally, mammals, birds, crustaceans, and worms. Mainly takes prey on ground, stooping from perch (e.g. post, tree, wire); prey battered on perch if lively. Sometimes dives into water for prey, and takes crabs, etc., on seashore. Will take insects from tree-trunks and in flight, and keep close to Cattle Egret *Bubulcus ibis* and grazing cattle, picking up insects disturbed by them. (Ali and Ripley 1970; Mukherjee 1976.) In June–July, Iraq, often active until well over 1 hr after dusk, pursuing and catching beetles *Scarabaeus sacer* in flight (Sage 1960b).
In USSR, takes aquatic insects, crabs, other invertebrates, frogs, lizards, and small quantities of fish (Dementiev and Gladkov 1951a). In Iraq, often takes Orthoptera, including grasshopper *Eremopeza gibbera* and mole-cricket *Gryllotalpa gryllotalpa* (Sage 1960b). In India, takes substantial numbers of large insects, including grasshoppers (Acrididae), crickets (Gryllidae), mantises (Mantidae), beetles (including Scarabaeidae and Dytiscidae), ants (Formicidae), and emerging winged termites (Isoptera); on sandy seashore takes crabs *Ocypode*, and in flooded paddyfields inland takes crustacean *Paratelphusa*; also scorpions, centipedes, frogs, lizards (e.g. *Mabuya*, *Calotes*), mice, and small birds including munias *Lonchura*, white-eyes (Zosteropidae), sparrows *Passer*, and a chick of Red-wattled Plover *Vanellus indicus*; occasionally fish (Ali and Ripley 1970). In West Bengal, 192 stomachs from all seasons contained (by weight) 31·2% fish (43·0% in wet season, 19·4% in dry season), 19·1% amphibians, 13·9% reptiles, 6·9% mam-

mals, 11·2% crustaceans, 16·2% insects, and 1·5% annelid worms (Mukherjee 1976).

Little information on food of young. At one nest, Ceylon, young—from soon after hatching—fed mostly on freshwater crabs, less often frogs and fish; crabs brought to young had their stout shells already smashed (Zylva 1973). BDSS

Social pattern and behaviour. Little information, and that mostly extralimital. Based partly on material supplied by C H Fry for Carey Island (Malaysia).
1. Usually solitary or in twos; seldom more gregarious (Sharpe 1868–71; Ganguli 1975). Outside breeding season on Carey Island, pairs, rarely groups of 3 birds, defend linear feeding territories c. 200–500 m long (C H Fry; see also Dharmakumarsinhji 1955, Ali and Ripley 1970). BONDS. Monogamous mating system; evidence that pair-bond maintained outside breeding season, and probably all year. Existence of groups of 3 birds outside breeding season (see above) suggests probability of helpers at nest (C H Fry), as known for Pied Kingfisher *Ceryle rudis*. Both sexes care for young, continuing to feed them for c. 1 month after fledging (Inglis 1937; Phillips 1946; Ali and Ripley 1970). BREEDING DISPERSION. Solitary and territorial. Territory typically includes habitual sentinel perch near nest (e.g. Hutson 1954). 5 pairs on c. 1 km of ditch, Turkey; nearest nests c. 100 m apart (Warncke 1964). 5 pairs on c. 3·2 km of canal, India (Hutson 1954). 8 nests within less than c. 2·6 km², Ceylon (Zylva 1973). 17 pairs on 2·1 km of dirt track with adjacent drainage ditch, Carey Island; probably c. 3 times greater than average density along 200 km of similar habitat. Site fidelity marked, birds returning to same nesting area, sometimes same hole, year after year (Murton 1969; C H Fry; see also Dharmakumarsinhji 1955). ROOSTING. Parents and recently fledged young roosted overnight in nest-tunnel (Inglis 1937). At Dal Lakes (Kashmir) most active dawn and dusk, with distinct midday lull (Pring-Mill 1974). For activity at dusk, Iraq, see Food. In Iraq, shelters in palms in heat of day (Moore and Boswell 1956b). Bathing bird jumped or glided down to dip in shallow pool from perches 0·5–1·25 m high; returned to perch almost immediately to shake itself (but not preen) vigorously and to wipe bill on branch; procedure repeated 4–5 times (Ganguli 1975; C H Fry; see also Dharmakumarsinhji 1955).
2. Described as sluggish and wary; in alarm, retires to protection of foliage rather than taking direct flight (Tristram 1866). While surveying surroundings (e.g. for prey) from perch, swings lowered tail up and down, and Head-bobs (Ali and Ripley 1970). FLOCK BEHAVIOUR. No information. ANTAGONISTIC BEHAVIOUR. No details. HETEROSEXUAL BEHAVIOUR. Sequence of pair-bonding behaviour little studied. On approach of breeding season, ♂ gives Advertising-call (see 2 in Voice), bill c. 30° above horizontal (C H Fry; also Phillips 1946); given for long periods, mainly in early morning, usually from exposed perch (Phillips 1946; Ali and Ripley 1970), sometimes from ground (C H Fry). Sits with tail turned under perch, from time to time flicking wings stiffly open (Wings-spread posture) for c. 1–2 s, exposing conspicuous white wing-patches. Typically spreads wings in vertical plane, but at low intensity nearer horizontal. Though usually directed at conspecific birds, also displays at other species (C H Fry), sometimes when no other bird apparently nearby (Ali and Ripley 1970). In courtship, also said to droop wings (exposing wing-patches) and cock tail (Inglis 1937). In early morning, May and early June, Basrah (Iraq), displaying ♂ ascended with 'lolloping' flight to c. 50–60 m, screaming loudly, before descending in steep fairly tight spiral to below tree-top level, then repeating procedure (see Moore and Boswell 1956b). Report by Finn (1902) of bird occasionally flying around slowly and aimlessly high in the air, calling (see 4a in Voice) probably also aerial advertisement, as in Grey-headed Kingfisher *H. leucocephala*; apparently, both members of pair may participate, ascending initially until almost out of sight, then flying around, calling, for up to 1 hr (Herklots 1967). ♀ likely to be courtship-fed, as in other *Halcyon*, though autumn-recrudescence copulations, Carey Island, not preceded by courtship-feeding (C H Fry). ♀ inviting copulation uttered distinctive call (see 3 in Voice) while shivering partly open wings, whereupon ♂ perched a few cm away (Ali and Ripley 1970). RELATIONS WITHIN FAMILY GROUP. Young born blind; eyes open on 4th–5th day. Parents bring fish lengthwise in bill with head at bill-tip, and transfer it (likewise frogs) to young head-first. Adults enter nest-chamber to feed small young; older young venture along nest-tunnel to meet arriving parent, and near fledging come right to entrance. Young defecate towards tunnel-entrance, ejecting faeces to well over 30 cm, thus keeping nest-chamber (but not tunnel) clean (Zylva 1973). Family party (2 parents, 3 young) remained in vicinity of nest-site for 1 week after fledging (see Ganguli 1975; also Bonds, above). ANTI-PREDATOR RESPONSES OF YOUNG, PARENTAL ANTI-PREDATOR STRATEGIES. No information. EKD

Voice. CALLS OF ADULTS. (1) Contact-alarm call. Usual call a loud defiant rattling or cackling sound (Ali and Ripley 1970); described as a raucous stropping or pumping 'chuk-chuk-chukkeruk ...' (Moore and Boswell 1956b); 'chake ake-ake-ake-ake' (Hollom 1959: Fig I), likened to alarm call of Blackbird *Turdus merula* (P J Sellar); 'kenk-kenk-kenk' (Phillips 1946). Given mainly from high vantage point (C H Fry; see also Hollom 1959), commonly on taking off (Finn 1902; Phillips 1946; Ali and Ripley 1970; Ganguli 1975). Given repeatedly by bird excavating nest-tunnel (Phillips 1946). (2) Advertising-call of ♂.

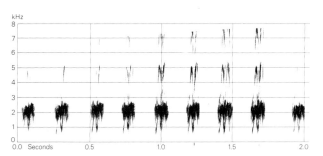

I T C White Malaysia March 1967

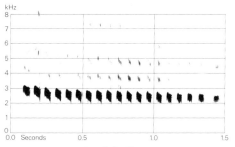

II P A D Hollom Israel April 1979

A penetrating, drawn-out, tremulous musical whistle 'kililili', persistently repeated for long periods from tree-top perch; each phrase ends in a detached harsh undertone like the 'pench' of a snipe *Gallinago*, audible only at close range (Ali and Ripley 1970). A tittering, descending 'ti-ti-ti-ti-ti-ti-ti-ti-ti-tieu' (Fig II) of which the first ⅔ sounded at a distance rather like a Whimbrel *Numenius phaeopus* (Hollom 1959). Described as a high-pitched rapid descending call, given from dense foliage (Moore and Boswell 1956b; see also Phillips 1946). (3) Pre-copulatory call of ♀. A prolonged 'kit-kit-kit-kit . . .' by ♀ inviting copulation (Ali and Ripley 1970). (4) Other calls. (a) A peculiar wailing cry, very different from call 1, given by bird flying slowly high in the air (Finn 1902; see Social Pattern and Behaviour). (b) A quite harsh call of anger (Dharmakumarsinhji 1955). EKD

Breeding. SEASON. Iraq: eggs found late April to mid-May (Tomlinson 1916; Ticehurst 1922; Marchant 1963b). SITE. In vertical bank of river or pit, not always over water; also in rock crevice or tree roots overhanging water (Baker 1934). Nest: burrow 60–150 cm long, ending in chamber 15–20 cm across (Dementiev and Gladkov 1951a). Burrow 7 cm wide, with chamber 60 cm in from entrance, offset (Marchant 1963b). In Assam (India), Baker (1934) described oval structure of moss wedged into hollow, but this not reported by any other observer and needs substantiating. Building: tunnel excavated by adults. According to Baker (1934), layers of moss pressed into hollow with bill, but without intertwining of material. EGGS. See Plate 98. Almost spherical, smooth and glossy; white. 29 × 26 mm (28–32 × 25–26), $n = 30$ (Baker 1927). Clutch: 5–6 (4–7). INCUBATION. By both parents. YOUNG. Altricial and nidicolous. Tended by both parents. FLEDGING TO MATURITY. Fledging period 18–20 days (Zylva 1973). No further information.

Plumages (nominate *smyrnensis*). ADULT. Forehead, crown, hindneck, upper mantle, and sides of head and neck to sides of chest deep chestnut; feather-tips often faintly bordered white when fresh; occasionally, a faint paler chestnut, cinnamon, or off-white line on lores. Lower mantle and scapulars between greenish cerulean-blue and turquoise-green, depending on light; back, rump, and upper tail-coverts bright cerulean-blue. Chin, throat, and central chest to central upper belly white with cream-buff tinge; remainder of underparts grading from deep chestnut on sides of chest and breast to slightly paler cinnamon-chestnut on under tail-coverts. Tail turquoise-blue, less glistening than glossy green-blue and blue of lower mantle and scapulars to upper tail-coverts; faintly and narrowly barred green and blue; shafts black; undersurface of tail greyish-black. Primary-tips black—c. 35 mm on p10, c. 50 mm on p9–p8, and then gradually less towards p1 (10–15 mm); remainder of outer web of p1–p9 sky-blue, slightly deeper cerulean at border with black tip, outer web of p10 black; remainder of inner web of all primaries white. Pale primary-bases form conspicuous pale blue patch on upperwing, white on underwing. Secondaries, tertials, greater and median upper primary coverts, and bastard wing between greenish cerulean-blue and turquoise-green, depending on light; inner webs of secondaries broadly bordered black (almost completely black on outer secondaries), tips narrowly so. Greater upper wing-coverts dull turquoise-green with black base; median coverts deep black, lesser coverts deep chestnut. Shorter upper primary coverts and leading edge of wing along carpal joint off-white, sometimes partly mixed with chestnut. Axillaries and under wing-coverts rufous cinnamon-chestnut, about similar in tinge to lower flanks and vent, not as deep as head and sides of chest. In worn plumage, chestnut of head, neck, underparts, and lesser upper wing-coverts slightly paler and more rufous, less uniform; lower mantle and scapulars with some duller blue-green of feather-bases visible, less uniform, occasionally with blackish shafts showing; chin to central upper breast purer white; blue of tail and flight-feathers less uniform. Sexes mainly similar, but ♀ has blue of lower mantle, scapulars, tail, and flight-feathers on average slightly paler and more greenish than in ♂; blue of back to upper tail-coverts slightly paler cerulean; much overlap, however. NESTLING. Naked at hatching; later on, spiny as in Kingfisher *Alcedo atthis*, but no information on timing. JUVENILE. Like adult, but chestnut of head, neck, and upperparts paler rufous-cinnamon, some dark grey of feather-bases often visible; lower mantle and scapulars duller greenish-blue; back, rump, and upper tail-coverts paler turquoise-blue, less deep cerulean-blue; chestnut of underparts less deep, grading to pale rufous-cinnamon on vent and under tail-coverts, some white of feather-bases often visible, especially when worn; in fresh plumage, often some black crescents to feather-tips on breast and belly. Primaries as in adult, but black of tips duller, border with blue and white less sharp and straight; often some slight buff suffusion at border of black tip and white inner web. Blue of tail, flight-feathers, and tertials distinctly greenish; black borders to inner webs and tips of secondaries and tertials slightly wider and less sharply defined. Lesser upper wing-coverts paler rufous-cinnamon than adult, often with some blue spots or freckling and with dark grey of feather-bases visible; median coverts greenish-blue with dull black base (all deep black in adult). FIRST ADULT. Like adult, but flight-feathers and apparently tail still juvenile; juvenile characters of retained feathers often hard to see, and then separable only by more heavily worn tail and outer primaries than adult at same time of year.

Bare parts. ADULT. Iris brown. Bill bright coral-red. Leg and foot orange-red, paler pink-red on rear of tarsus and sole. Claws black-brown or black. NESTLING. Bare skin pink-flesh, shading to blue-grey on future feather-tracts within a few days of hatching. Basal ⅔ of upper mandible and basal ⅓ of lower bluish-black, middle of lower mandible greyish-flesh, tip of bill pink; flanges at gape pale yellowish-ochre. Leg pink-flesh, rear of tarsus and soles pink-white, uppersurface of toes blue-grey. Claws bluish-black with pink tip. (ZMA.) JUVENILE. At fledging, iris brown, bill pale orange-yellow with dusky grey base and tip, leg and foot greyish-orange. Later on, bill orange-red with some dusky wash at tip, leg and foot dusky orange-red; adult colours obtained during post-juvenile moult. (Ali and Ripley 1970; BMNH.)

Moults. ADULT POST-BREEDING. Complete; primaries descendant. In Asia Minor and Middle East, apparently August–October; none of many examined November to mid-July in moult—November birds had plumage new, late May to mid-July birds had plumage heavily worn. One bird from early September, Turkey, had primary moult score 33; forehead, crown, and cheeks new, much active moult on nape, mantle, chest, and lesser upper wing-coverts, tail with all feathers growing simultaneously except t5; remainder old, underparts in particular heavily worn. One, mid-September, Israel, had score 40, about half of head, body, and wing-coverts new, and tail new except for growing t6. One, early October, Jordan, had score 46 (primary moult almost completed), and wing and tail new except for a few scattered feathers growing on body and growing t6. In *Halcyon* generally, secondaries moulted ascendantly and descendantly from s11–s12 (starting at same time as p1) and ascendantly from s1 (starting with p6–p7); s4 moulted last, growing with p10; tail moulted centrifugally, starting with t1 at loss of p2–p3 (Stresemann and Stresemann 1966). POST-JUVENILE. Partial; all head, body, and wing-coverts, no flight-feathers and apparently no tail-feathers. Starts soon after fledging, head and underparts first; completed October–November.

Measurements. Nominate *smyrnensis*. ADULT. Turkey, Lebanon, Israel, Jordan, and Iraq, all year; skins (BMNH, RMNH, ZFMK, ZMA, ZMM). Bill (F) to forehead, bill (N) to distal edge of nostril.

WING	♂	128	(3·00; 15)	124–134	♀ 127	(1·95; 11)	124–131
TAIL		86·7	(3·87; 11)	82–93	86·6	(3·95; 7)	84–92
BILL (F)		64·1	(3·99; 11)	58·7–70·7	64·4	(3·04; 7)	60·3–69·4
BILL (N)		51·2	(3·41; 14)	47·3–56·9	52·1	(2·66; 9)	48·3–56·7
TARSUS		16·6	(0·48; 13)	15·9–17·2	17·0	(0·93; 8)	15·8–18·0

Sex differences not significant.

Sexes combined. Nominate *smyrnensis*: (1) Southern Turkey (Izmir to Amik Gölü); (2) Lebanon, Israel, and Jordan; (3) Iraq. *H. s. fusca*: (4) Nepal to central India; (5) Ceylon; (6) Malaya and Sumatra; (7) Fukien (south-east China).

		WING			BILL (N)		
(1)	130	(2·05; 9)	128–134	51·4	(1·76; 9)	48·9–54·6
(2)	128	(2·70; 16)	125–133	50·5	(3·25; 14)	47·3–56·9
(3)	126	(1·87; 24)	123–129	52·1	(2·01; 20)	49·3–56·9
(4)	122	(3·50; 6)	118–126	50·6	(3·94; 4)	46·8–55·1
(5)	114	(2·92; 21)	109–118	47·1	(2·41; 22)	44·7–51·9
(6)	120	(2·52; 9)	116–123	47·3	(2·57; 8)	44·2–51·5
(7)	127	(3·27; 5)	123–132	48·9	(3·63; 4)	44·9–53·4

JUVENILE. Wing and tail in Middle East birds both on average *c.* 3 shorter than adult; bill not full-grown until after post-juvenile moult.

Weights. Nominate *smyrnensis*. Iraq, February: ♀ 110; sex unknown, 92, 98, 104 (BMNH). Iran, ♂♂: February, 85; March, 88 (Paludan 1938; Diesselhorst 1962).
 H. s. fusca. India: 3 ♂♂, 78–83; sex unknown, 79 (Ali and Ripley 1970). Malaya: ♂♂ 76, 78, 87 (ZMO). Nepal, ♂♂: 82, 83 (Diesselhorst 1968).

Structure. Wing short and rather broad, tip rounded. 10 primaries: p8 longest, p9 4–7 shorter, p10 20–29, p7 0–1, p6 1–3, p5 8–11, p4 12–15, p1 20–25. Inner web of (p6–)p7–p10 and outer web of p6–p9 slightly emarginated. Tail rather short and narrow, tip rounded; 12 feathers, t6 16–22 shorter than t1. Bill large and massive; base broad and wide, pentagonal in cross-section, tip laterally compressed; culmen and gonys with blunt ridge, both curving slightly to sharply pointed tip. Length of exposed culmen 8 mm less than length from tip to forehead. Nostrils small, rounded-triangular, bordered at rear by loral feathering. Leg short and rather weak; tibia feathered, tarsus bare. Toes rather long, slender; outer toe *c.* 88% of middle, inner *c.* 58%, hind *c.* 56%. Claws short, sharp, strongly curved.

Geographical variation. Rather complex and mainly clinal; involves size, depth of chestnut and blue colouring, and extent of white on underparts. Nominate *smyrnensis* from Asia Minor to north-west India large and relatively pale; head, neck, upper mantle, and much of underparts deep chestnut; blue of lower mantle, scapulars, wing, and tail turquoise- or greenish-blue; blue of back to upper tail-coverts bright cerulean-blue. Isolated *saturatior* from Andaman Islands also large; chestnut darker, almost blackish; blue of upperparts, wing, and tail deep violet-blue. *H. s. gularis* from Philippines differs from all others in completely chestnut underparts (except for cream-buff or pale cinnamon chin) and in restricted area of chestnut on lesser upper wing-coverts, these mainly black like median. Other populations all attributed to *fusca*, following Vaurie (1965), though far from uniform in size and colour, and 1–2 more races perhaps recognizable. Typical *fusca* from southern India and Ceylon small (see Measurements), chestnut very dark, almost blackish (except vent and under tail-coverts), and blue of body, tail, and wing turquoise or pale cerulean as in nominate *smyrnensis*; grades clinally into nominate *smyrnensis*, birds from Nepal to central peninsular India being intermediate in depth of chestnut and in size. Further east, from eastern India and Bangladesh through south-east Asia, depth of chestnut and size more or less intermediate between nominate *smyrnensis* and typical south Indian *fusca*, but blue of upperparts, tail, and wing bright dark blue or violet-blue, only slightly greenish in some lights on tertials and t1, not as pale turquoise- or greenish-blue as in both those races. Within south-east Asia, birds from south-east China largest and often with deep bill-base and markedly angled gonys; those from Malaya southward smallest and brightest blue.
 Forms superspecies with Java Kingfisher *H. cyanoventris* from Java and Bali; sometimes considered conspecific (e.g. Dammerman 1929–30), but breeding ranges show recent overlap in western Java (Somadikarta 1973). CSR

Halcyon leucocephala Grey-headed Kingfisher

PLATES 64 and 68
[between pages 686 and 687]

Du. Grijskopijsvogel Fr. Martin-chasseur à tête grise Ge. Graukopfliest
Ru. Сероголовый зимородок Sp. Martin pescador de cabeza gris Sw. Gråhuvad Kungsfiskare

Alcedo leucocephala P L Statius Müller, 1776

Polytypic. *H. l. acteon* (Lesson, 1831), Cape Verde Islands. Extralimital: nominate *leucocephala* (P L Statius Müller, 1776), northern Afrotropics from Sénégal to Ethiopia and northern Somalia, south to northern Zaïre and Lake Victoria; *semicaerulea* (Gmelin, 1788), south-west Arabia; *hyacinthina* Reichenow, 1900, coastal Kenya and north-east Tanzania, Pemba, and Zanzibar; *pallidiventris* Cabainis, 1880, southern Afrotropics north to southern Zaïre, Tanzania, and inland Kenya.

Field characters. 21–22 cm; wing-span 32–34 cm; bill 4 cm. About 20% smaller than White-breasted Kingfisher *H. smyrnensis*, with slighter build and structure. Medium-sized kingfisher, with large red bill, grey head and breast,

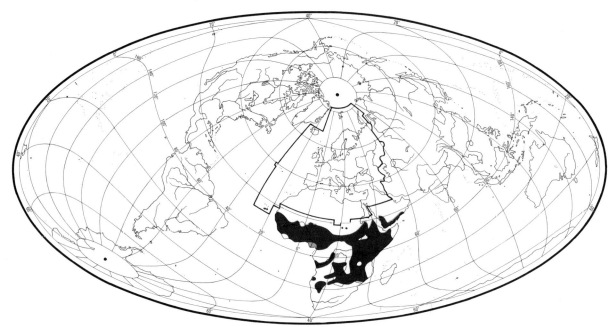

blue and black back and wings, blue tail, and chestnut underbody. A bird of dry habitats. Sexes similar; no seasonal variation. Juvenile separable.

ADULT MALE. Large vermilion bill. Head and breast pale grey (on some ♂♂, head stone-white), becoming white on throat and foreneck, and (unless plumage very worn) sharply divided from chestnut belly. Back mainly black. Rump and tail and most of wing deep blue, with black primaries and larger wing-coverts. In flight, blue centre to back exposed but wing does not show pattern as bold as *H. smyrnensis*. Legs coral-red. ADULT FEMALE. As ♂, but head often tinged brown, lacking clear grey-white tone. JUVENILE. Chestnut underbody paler (especially in centre) and less extensive than on adult; breast shows crescentic marks and is not sharply divided from belly. Bill all or partly black; legs pink-grey.

Unmistakable. Flight direct and fast, but action less rapid than that of Kingfisher *Alcedo atthis*, with spread of bigger wings more obvious. Hunts like shrike (Laniidae) by pouncing from perch.

Commonest calls a rapid chatter and (when breeding) a shrill twitter.

Habitat. In tropical Africa, including oceanic islands (Cape Verde Islands), in lowlands but also mountain plateaux up to *c.* 1600 m. Avoids mangroves and closed forest, although quickly colonizes clearings. Has widest habitat tolerance of its genus in West Africa, occurring on driest savanna and anywhere in open country with ground cover. Commonest garden kingfisher in Sierra Leone (G D Field). In rains, ranges furthest towards desert. Not particularly attracted to water, but found in swamps and in Nigeria often seen by tree-fringed margin of stream. Breeds in tunnels in sandy banks of river or dried up streambed, or sides of irrigation ditch in cultivated fields, or in borrow-pit (Bannerman 1951). In Cape Verde Islands, numerous in dry ravines and in vineyards; generally in driest valley, sitting on branches of castor oil plant *Ricinus communis* or coral tree *Erythrina*. Common also in neighbourhood of dwellings, sitting on concrete wall of water tank in public garden, and diving breast first to bathe in a shady pool in heat of day (Bannerman and Bannerman 1968). In mainland Africa, shifts after dry season from forest clearings to savanna zone. In East Africa, frequents wooded areas, *Acacia* savanna, and dry semi-desert bush (Williams 1963). In Somalia, often lives far from water in dry open *Acacia* scrub, or perches in dead tree in middle of patch of dry cultivation, or along dry water-courses or marsh-edges, watching for insects. Near coast, favours gardens with groves of date palms and little irrigation channels, or edges of mangrove swamp (Archer and Godman 1961).

Distribution and population. In west Palearctic, breeds only in Cape Verde Islands on São Tiago, Brava, and Fogo. No information on numbers, but human persecution, formerly rare, said to have increased in recent years (Bannerman and Bannerman 1968; Naurois 1983b).

Movements. *H. l. acteon* of Cape Verde Islands resident. In Africa, populations on both sides of equator move seasonally into higher latitudes either before or after breeding.

In northern tropics, movements studied most closely in Nigeria, where birds make unusual 2-stage northward progression during first half of year (Skinner 1968; Elgood *et al.* 1973; Elgood 1982). Present southern Nigeria in dry season (November–February), then moving northwards into central Nigeria (Guinea savanna zone) to breed March–May. Following this, moves further north to become non-breeding wet-season visitor (May–October) to Sudan savanna and Sahel zone and even further into Niger. Some presumed sub-adults stay there for ensuing dry season, but most birds return south in October–November; many (perhaps not all) overfly breeding range to reach southern Nigeria again by mid-November.

Elsewhere in Sahel zone (e.g. Mali, central Chad, Darfur), arrives and breeds during scanty summer rains, May–October (Lynes 1925a; Salvan 1968b; Lamarche 1980; Newby 1980). See also Britton (1980) for East Africa, including 2 Kenyan recoveries of birds ringed in Ethiopia. Population breeding southern Arabia absent December–March, and presumably migrates to Africa (Jennings 1981a) as there are a few records of birds at sea in Gulf of Aden and southern Red Sea in November and April–May (Moreau 1938; Bailey 1966b).

Food. On Cape Verde Islands: lizards (including geckos), one recorded c. 12 cm; mice; insects, including grasshoppers and locusts (Acrididae, e.g. *Catenops*) and beetles (Coleoptera). Small mouse struck against branch until dead, then held for 15 min before being swallowed head-first (Bannerman and Bannerman 1968). In Saudi Arabia, seen to catch fish by flying low over water and dipping bill in (King 1978). In Africa, mole-crickets (Gryllotalpidae), ants (Formicidae), and frogs also recorded (Mackworth-Praed and Grant 1970; McLachlan and Liversidge 1970). DJB, BDSS

Social pattern and behaviour. Little information for west Palearctic (see Sharpe 1868–71, Bannerman and Bannerman 1968). Based largely on information for nominate *leucocephala*, East Africa, provided by H-U Reyer.

1. Dispersion outside breeding season appears to vary regionally. After breeding, family parties of up to 6 seen, East Africa (Someren 1956). On Cape Verde Islands, formerly occurred in flocks of 8–10 'outside insect season' (see Sharpe 1868–71). Said to be strictly solitary, Somalia, outside breeding season (Archer and Godman 1961). At Lake Nakuru (Kenya), solitary, but at Lake Victoria (Kenya) in pairs (see Bonds, below). On migration, Nigeria, several birds per ha may occur (C H Fry). Bonds. Little information. Mating system probably basically monogamous. At Lake Victoria, pairs typically remain together throughout the year (H-U Reyer). Both ♂ and ♀ incubate and care for young (Sharpe 1868–71; Someren 1956; H-U Reyer). Although young remain with parents for a long time after fledging, more than a few weeks' dependence for food unlikely (H-U Reyer; see Relations within Family Group, below). While prolonged family bonds offer the potential for nest-helpers to occur, as in Pied Kingfisher *Ceryle rudis*, none yet reported for *H. leucocephala* (H-U Reyer). Breeding Dispersion. Defends breeding territory which serves for nesting and care of young at least up until fledging. During breeding season, Sierra Leone, pairs widely separated (Sharpe 1868–71). In Kenya, territories strung out, sometimes only 100 m apart along both sides of river, and extending 200–300 m back from it; territory size therefore c. 2–3 ha (H-U Reyer). In many places, mainland Africa, said to breed 'almost in colonies' (Mackworth-Praed and Grant 1970). In Ghana, birds arriving from migration took over breeding territories vacated by departing Senegal Kingfishers *H. senegalensis* (Greig-Smith 1978a). In some areas (not in, e.g., Kenya), pairs typically continue to defend territories outside breeding season (H-U Reyer). Roosting. At night, said to assemble (numbers not given) at roost in dense woodland cover (see Sharpe 1868–71). By day, regular hunting perch also used for loafing (e.g. Sharpe 1868–71). During heat of day, often bathes (Alexander 1898; see also Ruwet 1964). Performs series of dives from perch; between dives returns to perch to shake and preen (Alexander 1898). After bathing and preening, sometimes rests to sunbathe on perch or on sandy river-bank, crouching on belly with bill half-open; tail fanned and wings outspread (H-U Reyer; see also Bannerman and Bannerman 1968). Wings spread horizontally and for much longer periods than in the different Wings-spread posture (P W Greig-Smith; see below).

2. On Cape Verde Islands, said to be very tame, flushing only on close approach with Contact-alarm call (see 1 in Voice); in alarm, ♂ erects crown feathers (Sharpe 1868–71). As in other *Halcyon*, commonly Head-bobs on perch, possibly expressing alarm or excitement (Chapin 1939; see also Greig-Smith 1978a, b). For timid behaviour at nest, however, see Parental Anti-predator Strategies, below. Flock Behaviour. No information. Antagonistic Behaviour. Markedly aggressive at start of breeding season when establishing and defending territories. Defence and advertisement of territory mainly (perhaps exclusively) by ♂. Advertising ♂ may sit upright on prominent perch, wings slightly lowered and tail held up at c. 45°, and give Advertising-call (see 2 in Voice). May also fly around territory, circling high in the air, giving Advertising-call; aerial advertisement (see Heterosexual Behaviour, below) often provokes confrontation with neighbours, leading to a chase. Rivals may perch close to each other and threaten by facing each other in an upright posture with open bills. Bird thus perched may spread its wings in vertical plane (Wings-spread posture) and splay tips to highlight bold black and white underwing markings (H-U Reyer, C H Fry). In Ghana, birds of various other species driven off by chasing, swooping attacks (P W Greig-Smith). Heterosexual Behaviour. Sequence of courtship and pair-formation not well known. Comprises aerial and ground behaviour, apparently same displays and vocalizations serving to attract mates and repel rivals (H-U Reyer). In Sierra Leone, aerial display involved a lot of calling (see 2 in Voice) in high semi-hovering flight, and circling round, followed by plunging dives to tree (G D Field). In Oman, birds performing display-flight circled close to ♀ giving Advertising-call; apparent Duetting-calls (see below) also heard, mostly from perched birds (Walker 1981b). On ground, courting birds of both sexes commonly adopt Wings-spread posture, more readily so than in antagonistic encounters (H-U Reyer). May display thus alone or facing mate 2–10 m away. Mate may reciprocate, calling, with whole body vibrating slightly, and sometimes pivoting body and wings through c. 90° without shifting stance on perch. At end of bout of such Duetting-calls (see 3, also 2, in Voice), folds wings abruptly. May extend them again equally rapidly, opening and closing them once or twice more without calling (C H Fry). Display between pair probably serves as Meeting-ceremony. When Duetting-calls delivered in antiphonal duet, may represent territorial display of established pair (H-U Reyer), as proposed for Striped Kingfisher *H. chelicuti* (Wickler 1976). When nest-chamber almost complete, ♂ starts courtship-feeding ♀, and may continue to do so well after egg-laying (H-U Reyer). ♂ seen to catch insects in flight to feed to ♀ on nest (Sharpe 1868–71). Relations within Family Group. Young in nest start begging on hearing either wing-beats of approaching parent, or Trilling-call (see 5 in Voice) of parent

perched with food near tunnel entrance. Chick in nest-chamber sitting directly at end of nest-tunnel often pecks and bites siblings until they fall silent. Chick thus positioned is usual recipient of food (see also Kingfisher *Alcedo atthis*). Parent enters nest-chamber to feed small young, then turns and leaves tunnel head-first. From *c*. 2 weeks, 1 chick meets parent half-way along tunnel to snatch food, parent then retreating tail-first; chick thus fed returns to nest-chamber where some jostling occurs for most favourable position—that directly facing nest-tunnel (H-U Reyer). From *c*. 1–2 days before fledging, young make progressively longer sorties to tunnel entrance; initially look around briefly before retreating, but gradually remain longer, making wing-stretching movements. Fledging partly forced by siblings pushing from behind. Birds hand-reared in captivity caught live grasshoppers (Acrididae) at 2 weeks, and continued begging until *c*. 4 weeks old. Young chased away by parents at *c*. 3½–4 months. One 2½-month-old juvenile seen with parents who had 2nd clutch (H-U Reyer). ANTI-PREDATOR RESPONSES OF YOUNG. No information. PARENTAL ANTI-PREDATOR STRATEGIES. Timid and retiring at nest, sitting bird readily flushing from hard-set eggs well before close approach of man (Lynes 1925*a*; Broughton-Leigh 1932; Jourdain and Shuel 1935; Vincent 1946*a*). H-UR, EKD

Voice. Freely used during breeding season. Most information for nominate *leucocephala*; nothing known about possible differences between races. Motivation and function of different calls poorly understood. Following account compiled from recordings, sonagrams, classification, and notes provided by H-U Reyer.

CALLS OF ADULTS. (1) Contact-alarm call. Very rapid, reeling 't-t-t-t-t-t-....' with harsh, strident quality (E K Dunn: Fig I). May contain up to 150 't-' units. May give single call or more often a series with *c*. 4–20 s between calls. Birds may call alone or alternate with mate. Typically given when disturbed or excited, e.g. by observer suddenly appearing from hide, by juvenile leaving nest, or when chased by birds of other species. (2) Advertising-call. Rapid (6 per s), staccato 'chi-chi-chi-chi-chi-...' (E K Dunn: Fig II). Call contains *c*. 10–40 'chi-' sounds, each separated by *c*. 120 ms. Probably serves both to attract ♀♀ and rebuff ♂♂. Mostly heard at start of breeding season, when it occurs in long bouts with *c*. 1 s between calls. Given mainly, perhaps exclusively, by ♂, both in flight and from perch (see Social Pattern and Behaviour). Neighbouring ♂♂ respond with same call, either overlapping or alternating with 1st ♂. Advertising-call elicits call 3 (see below) in ♀♀, whereupon advertising ♂ may also switch to call 3. (3) Duetting-call. Similar to call 2, serving closely related function. Given by both sexes, either unaccompanied or in antiphonal duet. Recording (Fig III) of unaccompanied bird begins with rapid warble, followed by 3–15 'chee'-like units, delivered in rapid series. Often given in Meeting-ceremony; also directed (in antiphonal duet) by pair at neighbouring pairs, then probably serving as joint territorial display of established pair. (4) Hunting-call. A brief, high-pitched, rapid trill, with fluting quality—'tititit' or 'trrrrt' (E K Dunn). Given often, but not always, as bird pounces on prey, and also while returning to perch. (5) Trilling call. Rather faint call (not des-

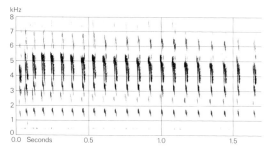

I H-U Reyer Kenya May 1976

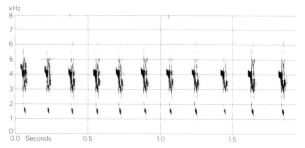

II H-U Reyer Kenya May 1976

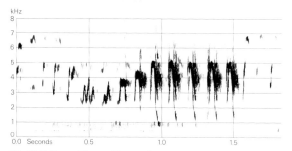

III H-U Reyer Kenya May 1976

cribed) sometimes given by food-carrying bird, e.g. just before entering nest. Function not clear, but possibly a contact-call as ♂ Striped Kingfisher *H. chelicuti* gives similar call just before courtship-feeding (H-U Reyer).

CALLS OF YOUNG. Food-call changes in structure and volume from hatching up until *c*. 4 weeks after fledging, when begging ceases. At *c*. 2 weeks, a hoarse, throaty 'schri-schri-schri-schri-...' given in long bouts (E K Dunn); increases in volume when parent approaches nest, or when young jostling for position, and then can be clearly heard outside nest. At *c*. 3 weeks (shortly before fledging), young acquire call similar to call 1 of adult, and which later develops into it. Call 3 of adult first heard in juvenile at *c*. 4 months, when family breaks up (H-U Reyer).

H-UR, EKD

Breeding. SEASON. Cape Verde Islands: mid-July to mid-December and February–April (Naurois 1983*b*). In East Africa, clearly associated with rainy season (H-U Reyer); see also Movements. SITE. Hole in cliff, or bank, of river, road cutting, or pit; 15 cm to *c*. 20 m above ground, but usually in upper part of cliff or bank (H-U Reyer). Nest: tunnel with chamber at end. Usually 40–100 cm long, 5–

6 cm in diameter, with chamber 15 cm across (Serle 1939; H-U Reyer). Building: tunnel excavated by both adults. EGGS. See Plate 98. Almost spherical, smooth and glossy; white. 24 × 22 mm (22–25 × 20–22), $n = 14$ (Nigeria) (Serle 1939). Clutch: 3–4(–5). 2 broods recorded (Bannerman 1953). INCUBATION. About 18 days (H-U Reyer). By both parents. YOUNG. Altricial and nidicolous. Fed by both parents. FLEDGING TO MATURITY. Fledging period $c.$ 25 days (H-U Reyer). Probably become independent at $3\frac{1}{2}$–4 months (H-U Reyer). Age of first breeding not known. BREEDING SUCCESS. No information.

Plumages (*H. l. acteon*). ADULT MALE. Forehead, crown, and ear-coverts pale buff-grey, remainder of head, hindneck, sides of neck, and upper mantle pale cream-grey or greyish-white. Lower mantle, scapulars, and upper wing-coverts deep black; some contrastingly white bases of scapulars sometimes visible; occasionally, rufous-cinnamon border where whitish upper mantle meets black. Back, rump, and upper tail-coverts glossy bright dark blue; tail slightly paler, more cerulean-blue, fresh feathers narrowly edged dusky grey (broader on inner webs of t2–t5), t1 laterally slightly tinged greenish-blue; tail dull black below. Chin to chest and centre of breast white or with slight buff or cream wash, remainder of underparts bright rufous-chestnut. Bases of primaries pale cerulean-blue on outer web, white on inner; 50–55 mm of tip black on p9, gradually less on inner primaries, until 10–15 mm on p1; tip subterminally bordered by dark blue or violet-blue. Secondaries and tertials dark glossy blue, contrasting strongly with deep black upper wing-coverts; narrow tip and broader border along inner web of secondaries dull black, blue deepest along shaft, often slightly violet-blue. Greater and median upper primary coverts turquoise-blue with tips darkening to dark blue; lesser primary coverts cinnamon. Bastard wing dark blue, feather-tips and shafts dusky. Small coverts along leading edge of wing chestnut, grading to pale cinnamon and white at carpal joint. Under wing-coverts and axillaries rufous-chestnut; greater under primary coverts white with black tips. White of primary-bases forms triangular patch on undersurface of wing, white extending 10–15 mm beyond black tips of greater under wing-coverts on p10, $c.$ 40 mm on p1. In fresh plumage, feathers of crown, hindneck, and cheeks show faint dusky terminal fringes (narrower than those of juvenile); in worn plumage, head and neck mainly white in some birds, but much grey on forehead and crown in others, or with white forehead grading to grey hindcrown; often shows dusky shaft-streaks when crown worn. Chestnut of underparts bleached to orange-cinnamon when worn. ADULT FEMALE. Like ♂, but forehead and crown slightly darker pale grey-buff, less whitish, giving brown-grey appearance when plumage worn rather than almost white as in some ♂♂; back to tail slightly paler blue, less bright dark blue; secondaries cerulean-blue or turquoise-blue rather than dark blue; chestnut of underparts sometimes less deep; white of chest sometimes extends slightly further down. NESTLING. Naked at hatching, becoming spiny later on as in Kingfisher *Alcedo atthis*; no information on timing. JUVENILE. Like adult ♀, but tips of feathers of crown with fine buff and dull black bars, appearing mottled; cream-buff hindneck with grey subterminal bars; black of mantle, scapulars, and upper wing-coverts duller and more greyish, less velvety black; much dark grey of feather-bases showing on back and rump, dark blue feather-tips relatively narrow. White of chin, throat, and chest tinged buff; cheeks, chest, and upper breast with distinct but narrow black fringes on feather-tips, appearing scaled, often some fine grey specks, streaks, or spots subterminally; chestnut of remainder of underparts paler, more yellow-cinnamon, partly washed white, especially on mid-belly. Wing as in adult ♀, but black of primary-tips duller and more extensive; blue of primaries paler, more greenish; white of bases of inner webs of primaries extends to just across shaft on outer webs; inner webs and tips of secondaries with broader dark grey borders; small coverts along leading edge of wing yellow-buff. Sexes similar, but back to tail of ♂ more often dark blue, of ♀ greenish-blue. See also Bare Parts. In worn plumage, forehead and crown heavily streaked dusky; much off-white or dull grey visible among greyish-black of mantle, scapulars, and upper wing-coverts; cinnamon of underparts strongly bleached, more cream or white of feather-bases visible. FIRST ADULT. Like adult, but flight-feathers still juvenile; juvenile character of these often difficult to see, especially when worn. Fresh cheeks and chest occasionally with dark crescents, as in juvenile.

Bare parts. ADULT. Iris warm sepia or dark hazel-brown. Bill brilliant scarlet or vermilion. Leg and foot coral-red or crimson-red. NESTLING. Skin flesh-colour. Iris brownish-grey. Bill black, tip of bill and corner of mouth dull orange-yellow. Leg and foot dusky brown, rear of tarsus and sole yellow. (Chapin 1939.) JUVENILE. Iris sepia. Bill black at fledging; gradually becomes more red, at base first; red with dusky tip during last stages of post-juvenile body moult, fully red when body moult completed. Leg and foot pinkish-grey with dusky edges to scutes; soles and rear of tarsus pale flesh-white; gradually brighter pinkish-red and orange-red during post-juvenile body moult. (BMNH, RMNH.)

Moults. (1) *H. l. acteon* (Cape Verde Islands). ADULT POST-BREEDING. Complete, primaries descendant. Starts with loss of p1, followed by scattered feathers of forehead and crown and some feathers of belly and flanks from primary moult score 15–25; much of head and underparts and part of mantle, scapulars, back, rump, and upper wing-coverts new at score 30–35, all new at $c.$ 45. Tail started at score 20–30, completed with primaries (score 50); sequence approximately 1–2–5–3–4. Secondaries replaced during last stages of primary moult. Timing in relation to breeding season difficult to establish due to prolonged breeding. Probably starts with p1 late October to early January, completing April–June. In late November and early December, scores 0, 13, 14, and 22 recorded, but also 31 and 35; in mid-January, 14, 17, and 20; in mid-February, 16, in mid-April, 49; in May and June, moult completed (score 50). 3 birds, December–February, suspended primary moult with scores 25, 25, and 30; these perhaps started moult after fledging of 1st brood, suspending when 2nd raised. POST-JUVENILE. Partial: head, body, and most wing-coverts, not flight-feathers and apparently not tail. Crown and nape first, followed by underparts and mantle; vent, tail-coverts, and part of wing-coverts last. 2 groups of juveniles discernible, probably corresponding with young of 1st and 2nd broods. In 1st group, juvenile plumage fresh October–November; moult started November–December; head, neck, and underparts, and scattered feathers of upperparts and wing-coverts new January–February; all new except tail, flight-feathers, and a few scattered feathers on vent and wing March–April. In 2nd group, juvenile plumage fresh January–February and moult $c.$ 3 months later than in 1st group.

(2) Nominate *leucocephala*. In Nigeria, where breeding February–May (late in dry season), adults and juveniles have complete moult May–November (in wet season). Onset of primary moult variable, adult starting mainly late May to late June,

juvenile early June to early August, but early-starting birds moult slower than late ones and moult in all birds completed with p10 within a short period late October and early November. In birds starting May, complete replacement takes up to 176 days and on average 1·7 primaries grow simultaneously; in those starting July–August, as little as 92 days needed and average of 2·3 primaries grow simultaneously. Only occasionally is moult suspended and a few outer primaries then retained until next moulting season. Secondaries moult ascendantly from s1 to s4 (s1 lost when p5–p6 growing) and descendantly from s12 to s5 (starting when p2 full-grown); moult of both series completed at about same time as p10 in adult, but s2–s4 completed later than p10 in juvenile (juvenile otherwise similar to adult from October). (Jones 1980.)

Measurements. *H. l. acteon*. Cape Verde Islands, November–June; skins (BMNH, RMNH, ZMA). Bill (F) is to forehead, bill (N) to nostril, both in adult only.

WING AD	♂ 105	(1·84; 14)	102–109	♀ 103	(2·16; 13)	99–106
JUV	103	(2·99; 4)	100–106	103	(4·16; 5)	97–106
TAIL AD	65·2	(2·58; 14)	62–70	63·4	(2·60; 10)	60–67
JUV	62·1	(1·65; 4)	60–64	62·0	(2·27; 4)	59–64
BILL (F)	47·2	(1·85; 14)	45–51	46·9	(2·71; 13)	44–52
BILL (N)	37·7	(1·57; 14)	36–41	37·7	(2·46; 13)	35–42
TARSUS	14·9	(0·40; 15)	14·3–15·5	15·2	(0·63; 10)	14·3–15·8

Sex differences significant for adult wing. Juvenile wing and tail not significantly shorter than adult.

Weights. No information for *acteon*. Afrotropical races (similar in size to *acteon* except for slightly shorter bill). Nigeria: adult 40·4 (2·7; 8) 36–44, juvenile 38·5 (2·0; 12) 35–42 (Fry 1970). Ghana: 44 (Greig-Smith and Davidson 1977). Zaïre: ♂♂ 41, 42, 44; ♀ 41 (Verheyen 1953). Kenya: ♂♂ 34, 42, 43, 45; ♀♀ 37, 40 (Britton 1970; Colston 1971). Tanzania: 2 ♂♂ and 2 ♀♀, 41–52 (Meise 1937). Zambia: 42, 45 (Britton and Dowsett 1969). Botswana: 47 (Jackson 1969).

Structure. Wing rather short and broad, tip rounded. 10 primaries: p8 longest, p9 1–3 shorter, p10 12–18, p7 0–1, p6 1–3, p5 7–9, p4 9–14, p1 17–22. Inner web of p8–p10 and outer web of p7–p9 slightly emarginated. Tail rather short, slightly rounded; 10 feathers, t5 5–9 shorter than t1. Bill as in White-breasted Kingfisher *H. smyrnensis*, but much smaller and less massive. Tarsus short and slender; toes rather short and slender. Middle toe with claw 21·6 (20–23); outer toe with claw *c.* 87% of middle, inner *c.* 56%, hind *c.* 53%. Remainder of structure as in *H. smyrnensis*.

Geographical variation. Mainly in colour, slight in size. 2 widely distributed races in Afrotropics, rather different in colour—nominate *leucocephala* in belt south of Sahara and *pallidiventris* in southern third of Africa; 3–4 more local races in Arabia and East Africa combine characters of these 2 in varying degrees. Size of all races similar to *acteon*, but bill shorter: e.g. in nominate *leucocephala*, wing 104 (14) 100–108, tail 65 (14) 60–69, bill to forehead 43·2 (13) 40–46, bill to nostril 34·0 (13) 31–37. Colour of West African nominate *leucocephala* as in *acteon*, but head, neck, upper mantle, and breast slightly darker buffish-grey, whitish only on indistinct supercilium and on throat, often with rufous band at border of black lower mantle, occasionally also across back of head; blue of back to tail and on wing paler greenish-blue or turquoise-green. *H. l. semicaerulea* from south-west Arabia similar to nominate *leucocephala*, but blue darker cobalt-blue; Ethiopian birds intermediate between *semicaerulea* and nominate *leucocephala* or nearer latter. *H. l. pallidiventris* from southern Africa differs from previous races by uniform medium grey head, neck, upper mantle, and breast with distinct white supercilium; belly to tail-coverts, flanks, and under wing-coverts tawny-cinnamon instead of rufous-chestnut; blue of upperparts and wing more violet or purple-blue than in other races. *H. l. hyacinthina* of coastal East Africa and perhaps south into Moçambique combines colour of head and underparts of nominate *leucocephala* with purple-blue flight-feathers and back to tail of *pallidiventris*. On other hand, birds breeding inland East Africa (west to eastern Zaïre) combine grey head and neck of *pallidiventris*, paler blue of nominate *leucocephala*, and variable depth of chestnut on underparts; these either included in *pallidiventris* (Mackworth-Praed and Grant 1952), included in nominate *leucocephala* (Britton 1980), or separated as *centralis* Neumann, 1905.

Considered by Fry (1980) to form superspecies with Black-capped Kingfisher *H. pileata* of eastern and southern Asia. CSR

Subfamily ALCEDININAE small kingfishers

Very small to medium-sized kingfishers, mainly frequenting watersides. 23 species in 4 genera: (1) *Alcedo* (9 species; Eurasia, Africa); (2) *Myioceyx* (single species—African Dwarf Kingfisher *M. lecontei*); (3) *Ispidina* (pygmy kingfishers, 2 species; Afrotropics); (4) *Ceyx* (far-eastern dwarf kingfishers, 11 species; southern Asia, Australasia). Represented in west Palearctic by 1 species of *Alcedo*, breeding. Fry (1980) recognized 22 species in 3 genera (*Ceyx*, *Corythornis*, *Alcedo*).

For general features, moults, etc., see Alcedinidae. P10 never shorter than p5. Tails very short (less than half wing length). Bills strongly compressed laterally, culmens narrow, not depressed at base. Skulls narrow. Lower end of tibio-tarsus bare for a short distance, tarso-metatarsus short, but always clearly longer than inner toe without claw. Middle and outer toes long; inner short, vestigial, or absent.

Sexes alike in colour of upperparts; in 2 species, some differences in underparts; bill colour differences in *Alcedo*. Primaries and tail without white markings. Plumage always with strong blues or violaceous, crown spotted or barred with blue. Bills black and/or reddish.

Alcedo atthis **Kingfisher**

PLATES 65 and 68
[between pages 686 and 687]

Du. IJsvogel Fr. Martin-pêcheur d'Europe Ge. Eisvogel
Ru. Обыкновенный зимородок Sp. Martín pescador Sw. Kungsfiskare

Gracula atthis Linnaeus, 1758

Polytypic. Nominate *atthis* (Linnaeus, 1758), north-west Africa, southern and eastern Spain, Corsica, central and southern Italy, south-east Europe south from coastal and southern Yugoslavia, southern, eastern, and central Rumania, and European USSR (except Baltic States, Leningrad area, and extreme western Ukraine), Turkey, and Levant east to Pakistan, western Himalayas, Soviet Central Asia, and north-west Sinkiang (China); *ispida* Linnaeus, 1758, Europe north and west of nominate *atthis*. Extralimital: *bengalensis* Gmelin, 1788, northern India east to south-east Asia, north to Lake Baykal, Sakhalin, and Japan; *taprobana* Kleinschmidt, 1894, southern India and Ceylon; 3–4 other races from Celebes and Lesser Sunda Islands eastward.

Field characters. 16–17 cm; wing-span 24–26 cm but tail only 2–2.5 cm; bill 4 cm. Smallest kingfisher in west Palearctic, with noticeably shorter tail than those of other species. Small, compact kingfisher, with proportionately long bill and large head, bright blue upperparts, and orange-chestnut underparts. At close range, head shows complex pattern of blue, orange-chestnut, and white marks. Flight rapid and shooting with wing-beats noticeably whirring. Restricted to open-water habitats. Sexes similar; no seasonal variation. Juvenile separable at close range. 2 races in west Palearctic, not easily separable in the field.

(1) West and central European race, *ispida*. ADULT. Long dagger-shaped bill; in ♂, all-black, sometimes with some reddish at base of lower mandible; in ♀, basal $\frac{1}{3}$–$\frac{5}{6}$ of lower mandible usually reddish. Long crown and nape, long and broad moustache, and all upperparts blue, with predominant tone varying with light intensity and angle; pale blue spotting obvious at close range on crown and wing-coverts; astonishing pale sheen flashes from centre of back and rump both at rest and in flight. Scapulars, tip of tail, and primaries usually look darker—bluer or blacker. Upper lores, ear-coverts, foreneck, underbody, and under wing-coverts orange-chestnut, paler by throat and in centre of belly. Small mark in front of eye, chin, throat, and patch between nape and end of moustache white. Legs coral-red. JUVENILE. Lacks brilliant iridescence and pure blue tones of adult, appearing greener in most lights. Breast usually washed blue-grey. Legs dull flesh to orange. (2) Mediterranean and south-west Asian race, nominate *atthis*. Slightly smaller than *ispida*, with more slender bill. Plumage slightly paler.

Unmistakable in west Palearctic (but range overlaps with even smaller, much darker blue Malachite Kingfisher *Corythornis cristata* on coasts of southern Red Sea). Plumage, especially broken head pattern, markedly cryptic in shade. Flight typically rapid and low over water, with whirring wing-beats allowing sudden jinks and dashes, remarkable deceleration leading to accomplished hover (or rapid perch), and forceful plunges after fish. When moving between waters, flight somewhat slower and often higher, with shooting rises and falls over obstacles. Behaviour includes quiet patient spying for prey from both open and enclosed perches, sudden fishing (with plunge often creating audible splash), and energetic courtship chases (over land and through woods as well as by water). Excited or wary bird may bob head and flick tail upwards.

Commonest call a shrill 'ch(r)ee' not particularly loud but extremely penetrating, carrying far; also a disyllabic 'chi-kee' (this or first call often extended in short series approaching trill) and other short notes. Song a sweet whistling.

Habitat. In west Palearctic, in upper and lower middle latitudes, from boreal through temperate and steppe to Mediterranean climate, oceanic or continental, wherever there is clear ice-free water, preferably still or gently flowing, and in breeding season fresh rather than brackish or salt. Ample supplies of small fish and availability of look-out perches of some kind for locating prey are other essentials. Streams, small rivers, canals, drains, and ditches with shady patches and shallow clear water are preferred to open, exposed waterbodies such as lakes, estuaries, and reservoirs, but prime consideration is ease of obtaining aquatic prey, and, in breeding season, ready access to suitable banks for nest-tunnel. If necessary, however, will accept nest-site 250 m or more from foraging water. Mainly confined to lowlands below *c*. 650 m, and only exceptionally above 900 m in continental Europe, but in USSR to *c*. 2000 m. At other seasons more ready to widen habitat choice to larger or smaller smooth waters, natural or artificial, and to coastlines, but cannot readily adapt to fast-running broken mountain streams or turbid conditions, and is sensitive to various types of pollution. Flies almost invariably in lowest airspace, and usually above or alongside water, but makes short cuts overland, and will also travel overland on long journeys. Although shy and wary can adapt to human settlements or to rivers much used for recreation when other conditions are sufficiently favourable. Vulnerability to severe winters partly offset by high reproductive potential and wide-ranging habits which facilitate recolonization of sites where local stocks have been eliminated. (Dementiev and Gladkov 1951a; Voous 1960; Sharrock 1976; Glutz and Bauer 1980.)

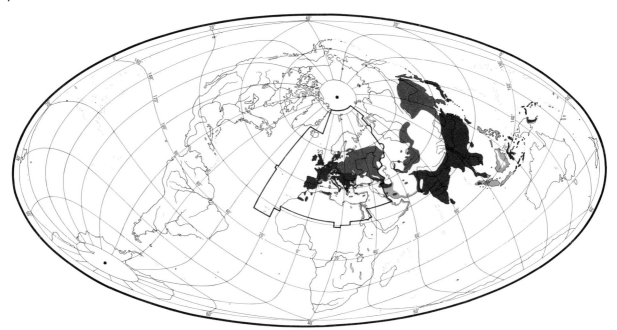

Distribution. Some range spread in north, with Sweden and Estonia colonized around 1900, and Finland in 1939.

BRITAIN. Range decreased in Scotland, especially since 1947, due to hard winters and possibly river pollution (Parslow 1967); some recent recovery (Sharrock 1976). Channel Islands: bred 1959 and 1968 (Long 1981b). NORWAY. Bred 1962, 1969, 1974, and 1975 (Haftorn 1971; SH). SWEDEN. Breeding annually since c. 1900 and spread since (LR) FINLAND. First proved breeding 1939 (Merikallio 1958). ITALY. Sicily: no proof of breeding (BM). USSR. Estonia: colonized in late 19th or early 20th century (HV). GREECE. May breed Pelopónnisos and some islands but no proof (WB, HJB, GM). TURKEY. May be more widespread than mapped (MB, RFP). LEBANON. Rare breeder (Tohmé and Neuschwander 1974); perhaps sporadic breeder but no proof (HK). SYRIA. May breed, but no proof (Kumerloeve 1968b; HK). IRAQ. Has bred (Allouse 1953). EGYPT. Bred Sinai 1982 (PGM). LIBYA. Bred 1974-9 and probably 1978-81 (GB). ALGERIA. Probably more widespread in north than mapped (EDHJ).

Accidental. Madeira.

Population. Marked fluctuations in northern and central Europe due to hard winters. Recent overall decrease in several countries, ascribed mainly to water pollution, river management, and persecution by man.

BRITAIN AND IRELAND. Marked fluctuations due to hard winters in Britain; less noticeable in Ireland. Decreased 19th century due to human persecution but probably no general decline since; 1000–10000 pairs (Parslow 1967), 5000–9000 pairs (Sharrock 1976). FRANCE. Marked fluctuations after hard winters, with general decline due to river management, pollution, and human persecution; 1000–10000 pairs (Yeatman 1976). BELGIUM. About 450 pairs, with fluctuations after hard winters but no general decline (Lippens and Wille 1972); declined in south since 1976 (when 100–150 pairs) due to weather, habitat loss, and persecution (Hallet and Doucet 1982). LUXEMBOURG. About 140 pairs (Lippens and Wille 1972); c. 65–90 pairs 1965, declined further since, due mainly to river pollution (Glutz and Bauer 1980). NETHERLANDS. Marked fluctuations due to severe winters, but some overall decrease due mainly to pollution and river management; perhaps 250–400 pairs 1955–6, 9–14 pairs 1963 after hard winter, recovered slowly to 275–325 pairs 1975, then after winter frost fell to 90–140 pairs 1976 (Timmerman 1970; Meininger et al. 1978). WEST GERMANY. No total population figures, but the many local surveys detailed in Glutz and Bauer (1980) suggest in excess of 1000 pairs; c. 1200 pairs (Rheinwald 1982). DENMARK. Increased 20th century, now 100–200 pairs (Dybbro 1976; TD). SWEDEN. Rare in 19th century (only 3 breeding records, but increased since c. 1920s, and c. 200 pairs 1976 (Ulfstrand and Högstedt 1976; LR). Marked fluctuations due to hard winters (Svensson 1978). FINLAND. Estimated 10 pairs (Merikallio 1958); very rare (LJL). EAST GERMANY. Declined due to water pollution and river management (Makatsch 1981); for some local counts, see Glutz and Bauer (1980). POLAND. Scarce or very scarce; recent marked decrease (Tomiałojć 1976a). CZECHOSLOVAKIA. Marked fluctuations but decreasing trend (KH). SWITZERLAND. At least 200 pairs (Glutz and Bauer 1980); c. 180 pairs (RW). BULGARIA. Not numerous (Patev 1950). Probably decreasing due to water pollution (JLR). USSR. Estonia: marked fluctuations but now fairly common (HV).

Survival. Britain: mortality in 1st year of life $77.8 \pm SE8.0\%$, $n=21$; average annual adult mortality $76.2 \pm SE6.6\%$, $n=32$ (Morgan and Glue 1977). Central

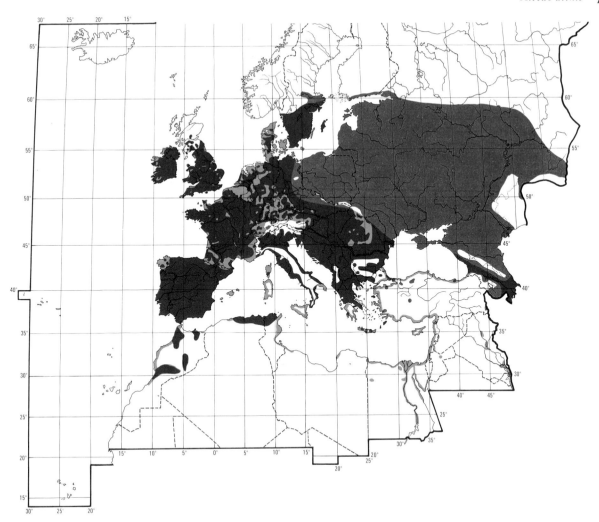

Europe: 80% died in 1st year to May, 15% in 2nd year, and 4% in 3rd year (Glutz and Bauer 1980). Oldest ringed bird 15 years 5 months (Dejonghe and Czajkowski 1983).

Movements. West Palearctic races show progression from being mainly migratory in northern and central USSR and west to Finland and Poland (where water closed by ice in winter), through partially migratory across central Europe (so that population levels affected by severe winters), to dispersive or even mainly resident in western maritime countries and in Mediterranean basin. Immatures move more often, and on average further, than adults; among adults, ♀♀ move rather more than ♂♂. Winters from west Baltic (including southern Sweden) southwards, often by coasts in northern wintering areas. Migrants move into western and southern parts of breeding range, and small numbers beyond: to Mediterranean islands, North Africa (including Libya and Egypt), Gulf of Aqaba, Red Sea and Persian Gulf coasts of Arabia, and coastal mangroves of Pakistan and north-west India. In Afrotropics, known only as rare visitor, October–March, to extreme northern Sudan (Cave and Macdonald 1955). Migratory eastern Siberia and northern China birds (*benghalensis*) penetrate south in winter through southern Asia to Philippines and Indonesia, overlapping various resident races there.

No ringing analysis for USSR, though presumed this the source of birds which winter in Middle East. Czechoslovakian study (Hladík and Kadlec 1964) found birds dispersing in all directions, at all seasons, but with longer movements strongly oriented south to south-west, with furthest birds in Yugoslavia, Italy (including Sicily), Malta, France (mainly south), and south-east Spain, the latter at 1820 km from ringing site; listed recoveries of over 200 km included only 2 ringed as adults—movements of 610 km and 790 km into Italy. This the broad pattern from ringing elsewhere in continental Europe; see data in Glutz and Bauer (1980). East German birds found November–January west to Denmark and Belgium, south-west to France, and south to Italy. 85 recoveries of West German birds ringed as nestlings, juveniles, and summer adults included 28 which moved over 100 km; included 1 January recovery in East Germany, and furthest birds (to maximum 1450 km) in Spain (including Balearic Islands), France, and Italy (including Sicily). In 83 Dutch recoveries of birds ringed as nestlings, 14 moved over 100 km, longest being a November recovery from France (890 km); only 1 of 15 summer-ringed adults moved over 100 km. Similarly, Belgian adults move only short distances but young birds disperse further and in all directions, with furthest in Netherlands (300 km), England (350 km), France (over 900 km), and northern Spain (1100 km) (Glutz and Bauer 1980; Hallet and Doucet 1982). In British recoveries up to 1981 ($n=406$), 60% moved 0–9 km, 34·5% 10–99 km, 5·5% further, including 4 July–August birds found in Channel Islands, Belgium, and France (2) (BTO). Dispersal pattern shown by British birds is without preferred direction, mainly within 50 km, and most movement occurs in autumn, indicating general dispersal of young birds away from parental territory; longer movements in winter tend to be towards coasts in hard weather, extent of such movements varying greatly between years according to severity of weather (Morgan and Glue 1977). In France, also, some birds make long autumn movements between south and south-west, though some of these may have been birds (already transient) that originated from further north or east; recoveries between western Italy and Spain, and one Finistère bird (July) found in Wales (March).

Dispersal begins in summer, direction random, with juveniles leaving parental territory within days of achieving independence (e.g. Hallet and Doucet 1982). One Dutch nestling (May) found next month 170 km NNE. In Ryazan (USSR), independent young moved 7–30 km per day (Numerov and Kotyukov 1979). Dispersal most pronounced July–October, with gradual transition into autumn migration; Czechoslovakian juveniles had reached Sicily, Malta, and Spain (1340–1820 km) by September (Hladík and Kadlec 1964). In partially migratory/dispersive populations, adult ♂♂ show more attachment to nesting territory than do adult ♀♀; hence latter more involved in dispersal (Glutz and Bauer 1980). Return movement begins February, and in western Europe territories reoccupied then or in March; migratory USSR populations return late April or in May. In Ryazan, 1st-time breeders settled within 20 km of natal area (Numerov and Kotyukov 1979); elsewhere, maximum recorded displacements 115 km and 128 km (Glutz and Bauer 1980).

Food. Principally freshwater fish, also aquatic insects and marine fish; more rarely, crustaceans, molluscs, terrestrial insects, and amphibians. Feeds either by diving from perch or by hovering prior to dive (Eastman 1969; Hollick 1973; Pring-Mill 1974; Frost 1978). Dives typically from low perch (e.g. 1–3 m), but up to 11 m reported (Frost 1978). Plunge from perch to water accelerated by a few wing-beats, and wings extended back alongside body whilst entering water. If diving from low perch into deep water, first flies upwards from perch before plunging (Boag 1982). Maximum depth reached c. 1 m (Glutz and Bauer 1980). Bird flaps wings to return to surface, though own buoyancy also helps; leaves water on downstroke of wings. Nictitating membranes cover eyes during immersion (Eastman 1969; Kniprath 1969c). For detailed drawings of diving procedure, see Glutz and Bauer (1980). Prey swimming near surface preferred but capture success independent of depth. Prey recognized by the way it moves as well as by shape. Small slender fish preferred, and weaker fish also selected; living prey preferred to dead (Kniprath 1969c). Reported diving for fish from horizontal flight (Rowan 1918; Radetzky 1976), and also diving through ice (Shooter 1978). Sometimes takes prey from water surface without submersion (Kniprath 1969c). Will batter larger prey, degree of battering dependent on length. May grasp fish by tail and swing head against hard object (Eastman 1969; Kniprath 1969c). Hovering observed where there are no ready perches—especially on open estuaries and sea-shores (Bennett 1965; Eastman 1969). Reported hovering and taking insects aerially like flycatcher (Muscicapidae). Hovering may be normal procedure for capture of dragonflies (Odonata) and spiders (Araneae) (Hodson 1961; Sharrock 1962; Ruthke 1968). In Nepal, c. 54% of dives for fish from hovering flight successful, 38% from perches (Pring-Mill 1974). In hard winters, may rob water shrews *Neomys fodiens* of fish (Weise 1968). Also reported harassing Dippers *Cinclus cinclus* for fish (Grist 1934; Eastman 1969).

Fish include, most commonly, bullhead *Cottus gobio*, loaches *Cobitis* and *Noemacheilus*, minnows *Phoxinus*, roach *Rutilus rutilus*, barbel *Barbus barbus*, grayling *Thymallus thymallus*, sticklebacks *Gasterosteus*, trout *Salmo*, perch *Perca fluviatilis*, pike *Esox lucius*, carp *Carassius*, ruffe *Acerina cernua*, bleak *Alburnus*, nase *Chondrostoma*; also *Gambusia affinis*, *Dicentrarchus*, *Gobio*, *Pungitius*, *Lota lota*, *Scardinius erythrophthalmus*, *Leuciscus*, *Blennius*, *Lepadogaster*, *Boops boops*, and *Atherina hepsetus*; occasionally, eels *Anguilla* and lampreys *Lampetra*. Insects include adults and nymphs of dragonflies (Odonata), mayflies (Ephemeroptera), stoneflies (Plecoptera), caddis fly (Trichoptera) adults and larvae, adult bugs (Hemiptera), and beetles (Coleoptera). Other invertebrates include spiders (Araneae) and crustaceans (e.g. crayfish *Astacus*, prawns *Palaemon*, shrimps *Gammarus*, Isopoda). Also young frogs *Rana*. Sometimes dives for pieces of vegetation. (Kumari 1939b; Kinzelbach 1963; Doucet 1969b; Kniprath 1969c; Hallet 1978; Glutz and Bauer 1980; Iribarren and Nevado 1982.) Stomachs sometimes contain molluscs (Lymnaeidae, Ancyclidae, Physidae, Planorbidae, Unionidae, *Pisidium*), mostly thought to be derived from stomachs of fish ingested (Hallet 1977).

On River Lesse (Belgium), 90% of 14 475 elements of fish in pellets belonged to bullhead *Cottus gobio* and Cyprinidae, 6% loaches (Cobitidae), and remainder stickleback *Gasterosteus aculeatus*, trout *Salmo*, grayling *Thymallus thymallus*, and perch *Perca fluviatilis* (Hallet 1977). In various other Belgian localities 36·4% (by number) of fish taken were Cyprinidae, 28·7% *G. aculeatus*, 27·7% *C. gobio*, 4·08% trout *Salmo*, 2·32% *P. fluviatilis*, 0·5% ruffe *Acerina cernua*, and 0·2% loach *Cobitis barbatula* (Doucet 1969b). In Estonia (USSR), catches mainly minnows *Phoxinus*, bleak *Alburnus*, and gudgeon *Gobio gobio*; usual size for all species 3–5 cm, maximum 7 cm (Kumari 1978); when feeding young, 4–7 cm, maximum 10 cm (Hallet 1982). In Spain, 96 birds from 43 sites contained wide variety of prey types. Fish 77·5% of total weight; of fish, 51·3% (by weight) Cyprinidae (including 22·2% roach *Rutilus arcasii*, 11·7% *Chondrostoma polylepis*), 11·6% loaches (Cobitidae), and 6·9% trout *Salmo trutta*. Also 3·22% Anura (those identified were frogs *Rana*), 10·4% crustacean *Austropotamobius pallipes*, and 8·49% insects (4·67% Odonata and 1·61% Coleoptera). (Iribarren and Nevado 1982.) Marine prey included bogue *Boops boops* and sand smelt *Atherina hepsetus* off Crete, and 3 prawns *Palaemon serratus*, 2 tompot blennies *Blennius gattorugine*, 1 goby *Gobius*, and 1 Cornish sucker *Lepadogaster lepadogaster* off south-west France (Kinzelbach 1963). In Brittany (France), bleak *Alburnus alburnus* most frequently taken prey; also takes other surface-feeding Cyprinidae, gobies, barbels *Barbus*, and loaches; less commonly, bullheads and larvae of Odonata (Fjerdingstad 1937). In River Test (southern England), mainly bullheads, minnows *Phoxinus phoxinus*, and sticklebacks; once tried unsuccessfully to feed on eel *Anguilla anguilla* (Eastman 1969).

During first 10 days of rearing young, adults catch fish up to 5 cm long; larger prey caught once young can ingest them (Swanberg 1952; Hallet 1982). Fish presented to young head-first (Eastman 1969). In one brood, nestlings fed 88 times in 846 min; on average, each individual every 50 min (Svensson 1978). In other observations, nestlings fed regularly every 45 min at 6 days old, every 20–25 min at 12 days, and 15–20 min at 18 days (Ruthke 1968; see also Reinsch 1962, Schulz-Waldmann and Dominiak 1971, Zöller 1975). At another nest, estimated 80 minnows delivered to 17-day-old brood during 1 day (Swanberg 1952). For development of feeding skills in young, see Social Pattern and Behaviour. BDSS

Social pattern and behaviour.
1. Seldom gregarious. Mostly solitary outside breeding season when many birds defend feeding territories (Kumari 1939a; Papadopol 1965; Eastman 1969; Leibrich and Uhlenhaut 1976; see also Fjerdingstad 1937). Adult ♂ usually defends breeding territory of previous summer, while winter territory of ♀ (mate) may be adjacent or nearby (Eastman 1969), occasionally partly shared (Boag 1982, which see for details). Family parties persist only briefly after young fledge; in study of one family group, Hampshire (England), birds solitary by September (Eastman 1969). Small parties seen heading out to sea, eastern Scotland, August and September (see Eastman 1969). Especially in hard winters, density may increase locally at favourable feeding sites, e.g. 30 birds along 4 km of river, Hungary (Nagy 1921). BONDS. Mating system essentially monogamous. Though pair-bond sometimes lasts from one breeding season to next, change of mate (and territory) during breeding season not uncommon (e.g. Gentz 1940, Numerov and Kotyukov 1979, Zöller 1980). Pair-members behave as individuals during winter (see above). Polygamy—possibly associated with surplus of ♀♀ in population (see Podolski 1982)—occurs in some cases, leading to variation in parental roles. In monogamous pair, both excavate nest-tunnel (e.g. Ingram 1933, Heinroth 1939), ♂ often doing most (Clancey 1935a; Zöller 1975). Pair usually share incubation and care of young where nesting attempts not overlapping (see below). ♂ may, however, do less feeding of young, sometimes none (Kumari 1939a; see also Brown 1934). For case of ♂ hatching eggs and rearing young alone after death of mate (4 days before hatching), see Ruthke (1968). If monogamous pair has 3 (rarely 4) broods, however, 2nd and 3rd (and subsequent) broods typically overlap. ♂ may then do most (or all) brooding and feeding of 2nd brood while ♀ lays and incubates 3rd (Persson 1934; Brown 1935; Kumari 1939a, 1978; Swanberg 1952; Reinsch 1962; Heyn 1963; Svensson 1978; Zöller 1980). For ♂ helping to incubate 3rd clutch while also feeding 2nd brood, see Swanberg (1952). In 4 pairs, ♂ incubated 2nd and 3rd clutches while ♀ raised young of preceding broods (Podolski 1982). For division of labour in 2 pairs that raised 4 broods, see Zöller (1980). Usually both members of pair help to incubate final clutch of season and raise brood (Kumari 1978; Zöller 1980; Podolski 1982). Polygamy usually involves ♂ paired to 2 ♀♀. At Saratov (USSR), 53 nesting associations included 42 with 1 ♀, 10 with 2 ♀♀, 1 with 3 ♀♀ (Podolski 1982). In southern Sweden, 116 nesting-associations comprised 109 with 1 ♀, 7 with 2 ♀♀ (S Svensson). At Oka reserve (USSR), on average 35% of breeding ♂♂ polygamous each year (3 years, 26 cases of polygamy); once, ♂ bred with 3 ♀♀, each of whom had 2 clutches, so that ♂ shared incubation and rearing in 6 nests. Presence of surplus ♂♂ indicates polygamy not attributable to unbalanced sex-ratio (Numerov and Kotyukov 1979). Bigamy may be simultaneous, both ♀♀ incubating (separate) clutches at same time while ♂ helps to incubate 1 or both (Heyn 1968; Podolski 1982). Perhaps more often overlapping, ♂ taking 2nd ♀ when 1st incubating or brooding (e.g. Svensson 1960, 1978, Heyn 1963, 1968, Podolski 1982). ♂ may leave 1st ♀ to hatch and/or raise brood alone. In case of synchronous mating with 2 ♀♀, ♂ assisted in feeding both broods (Heyn 1968). In overlapping polygamy, ♂ may be associated with 4–6 clutches in season (Podolski 1982, which see for details). In one case, same 3 birds known to be involved in bigamous association over 2 successive years (Svensson 1978). Most birds breed first at 1 year (S Svensson; see also Clancey 1935b). For report of juvenile ♀ of 1st clutch in season starting to nest in same season, see Gentz (1940). BREEDING DISPERSION. Solitary and territorial. Assessment of density complicated by differences in habitat and marked annual fluctuations in populations. Pairs usually well spaced. Up to 15 pairs on 18 km, Ahja river (Estonia); neighbouring pairs 0·3–1 km apart (Kumari 1939a, 1978). In 2 years, Pra river (Ryazan, USSR), pairs usually 1–2 km apart (Kartashev 1962). Closest, Klippan (Sweden); 125 m, in 3 cases 200 m, but usually much further (Svensson 1978). For pairs nesting near each other (in 1 case *c.* 150 m), Scotland, see Brown (1935). For other densities in Europe, see Glutz and Bauer (1980). Breeding territory serves for courtship, copulation, nesting, and care of young until a few days after fledging (see Relations within Family Group, also Roosting, below). Territory also contains central area to which ♂ especially attached (Kumari 1939a). Most or all food for young sought in separate territory, close to or, sometimes, quite far from breeding territory, though

usually not more than 1 km from nest; breeding and feeding territories together occupy up to 0·8–1·5 km of watercourse (Kumari 1939a). Breeding territory established in spring, probably by ♂; in 2 years when sex determined, ♂ arrived at breeding site 9 and 14 days before ♀ (Svensson 1978). Both members of pair help to defend territory. Marked fidelity to breeding territory (Kumari 1939a; Hunziker-Lüthy 1971; Zöller 1975). Same nest sometimes used for consecutive clutches, but many pairs use 2–3 or more for successive (especially overlapping) attempts within a season and between years: e.g. in a given year, pair commonly nest in 1 hole, move to another for 2nd clutch, and back to original one for 3rd (e.g. Reinsch 1970b, Schmidt 1981). Nest-sites of pair typically up to 20 m apart (Kumari 1939a); 2nd hole not uncommonly excavated c. 1 m or less from 1st, especially if 1st nesting attempt fails, or if broods overlap (Brown 1935; Poulsen 1937; Kumari 1939a; Hunziker-Lüthy 1971; see also Reinsch 1962, Eastman 1969, Numerov and Kotyukov 1979). For 2nd (overlapping) clutch started 5·5 km from 1st, see Swanberg (1952). Winter territory usually smaller than summer breeding/feeding territory (Kumari 1939a). ROOSTING. Usually alone at all times of year, in dense cover such as bushes or trees near water; once in maize fields (Fjerdingstad 1937; Kumari 1939a, 1978; Ruthke 1944; Weise 1968; Eastman 1969; Görner 1974; see also Glutz and Bauer 1980). Roost-site in breeding season some distance from nest-site (e.g. Horst 1938). In winter, sometimes roosts in holes in bank (Kumari 1939a, 1978); 1 (occasionally 2) birds roosted overnight in nest-tunnel, February–March; bird arrived after dark, and left before dawn (Kumari 1978). In winter, one bird roosted 6–8 m high in spruce *Picea* trees 18 m from lake shore; in another case, in spruce 2–3 m high (Ruthke 1944). In August–September, 1st-year bird flew in silence to roost-site in belt of alder *Alnus* at 18.40 hrs (Görner 1974). Roosts with bill tucked backwards under wing (Boag 1982, which see for photograph). At Dal Lakes (Srinagar, Kashmir), most active (based on sightings) at 09.00–12.00 hrs and 17.00–18.00 hrs, with midday lull (Pring-Mill 1974). Between nest-duties, birds loaf on habitual perch near nest-site; perch near nest during laying but further away after onset of incubation (Rivière 1933a). Loafing and dozing birds said to bob body and flick tail up and down (Rivière 1933a). Bathing frequent during day, especially after nest-building, copulation, nest-relief, feeding young, prey capture, or several unsuccessful attempts at prey capture (Kniprath 1969b). One bird bathed by diving vertically, 4 times, from perch, preening between each dive and after bathing bout over (Rowan 1918; see also Ris 1937). Bird preens probably for c. 2 hrs per day, often in bouts of 15–20 min. After each bout, typically yawns and stretches before flying off (Boag 1982). A cleaning movement 'unknown in other species of birds' occurs in which underside of the wing used to massage upper part of head (Kniprath 1969a). Numerous accounts of bathing after feeding young; especially frequent when young older and nest-hole gets sullied (e.g. Kniprath 1969a, Plucinski 1969, Zöller 1975). On leaving nest-tunnel, belly-flops nearby to submerge lower abdomen and tail (Brown 1923); seen to enter water with wings outspread, move around on surface, and duck underwater (Ris 1937; Feiler 1957). One adult dipped to bathe thus usually 1–5 times after each visit to young, but during longer pauses between feeding up to 14–15 times (Svensson 1978); up to 20 times (Kumari 1978). In middle of bout of feeding, only 2–3 quick dips, but longer spell, with preening, after young apparently satiated (Eastman 1969). For apparent bathing by young on leaving nest, see Kniprath (1969b); for bathing by juveniles, see Swanberg (1952) and Kumari (1978).

2. Alarmed or excited bird bobs and flicks tail up and down; may raise crown feathers and give Contact-alarm call (Kerr 1918;

Rivière 1933a; Clancey 1935a; Kumari 1939a, b; Eastman 1969; see 1 in Voice). Same behaviour occurs in newly fledged young (Clancey 1935a, Kumari 1939a). When raptor flies overhead, sitting bird tracks movement with bill skywards (see drawing in Glutz and Bauer 1980). Often allows close approach by man when feeding (e.g. Hollick 1973). Bird disturbed by human intruder lay flat along perch (Eastman 1969; see also Antagonistic Behaviour, below). When handled, bird invariably holds head at c. 30° to vertical and slowly rotates head almost 180° to one side, back to mid-point, then through similar arc to other side; movement continues until released (Campbell 1966; see also Kuhk 1954, and Antagonistic Behaviour, below). ANTAGONISTIC BEHAVIOUR. Vigorous in defence of breeding territory; also of feeding territory, in or out of breeding season (Sharpe 1868–71; Horst 1938; Kumari 1939a). Rivalry between pairs may be intense early in season, especially when trying to nest close together; birds then fight frequently and sometimes enter neighbours' nest-hole to puncture eggs (Brown 1935, 1936b; Clancey 1935a; Heyn 1963). ♂ especially pugnacious in second half of chick-rearing when required to do less brooding; strongly defends central area ('play area': Kumari 1939a) of breeding territory. Especially in middle of breeding season, offspring forcibly expelled from territory by one or both members of pair (e.g. Svensson 1978; see Relations within Family Group, below). May harry and drive off some passerines (Creutz 1956; Reinsch 1962; Eastman 1969), especially if trespassing on favourite perches (Boag 1982). For tolerance of passerines nesting nearby, see Creutz (1956), Hübner (1965), Reinsch (1968), and Kumari (1978). For harassment of Little Grebe *Tachybaptus ruficollis*, see Greenwell (1952). Bird confronts conspecific rival—often perched c. 1 m apart (Boag 1982)—with Upright-threat display (Fig A): stands upright, sleeks plumage, hunches shoulders, often raises carpal joints or droops wings; points bill forward or slightly up (never directly at rival), often half-open, and slowly and repeatedly sweeps it in arc from side to side while watching rival intently (Eastman 1969; Leibrich and Uhlenhaut 1976; Boag 1982). Exaggerated movement draws attention to bill as potential weapon (Eastman 1969). One bird stretched neck so that bill vertical, and delivered Threat-call (Boyle 1952: see 2 in Voice). When rivals meet on common perch, one or both may express higher-intensity threat by mounting Forward-threat display (Fig B): typically turn sideways-on to each other (Boag 1982) and slowly bow forward until lying, body hunched but neck extended, almost along perch (Boyle 1952; Eastman 1969). Bird sometimes 'sways' from side to side (Boag 1982). Bird in Forward-threat posture made forward jerking motions of whole body while uttering intermittent (undescribed) calls (Boyle 1952). In apparently aggressive encounter between 2 birds sitting not more than 1 m apart, they frequently and simultaneously exchanged perch-sites, and slowly alternated between Upright-threat and Forward-threat postures. At times, one bird, facing somewhat away from other, bowed and fanned wings as if drawing attention to bright blue plumage on its back; display also included occasional tail-bobbing and spreading of wings—

A

B

C   D

usually with flicking movement—outwards and backwards (Wings-spread posture: Fig C) (Goodfellow and Dare 1955); latter thought to indicate strongest threat (Glutz and Bauer 1980). Threat postures often form prelude to flying attack and aerial chase (Eastman 1969) accompanied by Threat-calls; both members of pair sometimes evict intruder together (Kumari 1939a). In flying attack, rivals may try to dislodge each other from perch; may clash physically, grappling with bills (Brown 1934). For attacks in which bird stabbed and grasped at stuffed bird's breast, neck, and head, see Eastman (1969), Leibrich and Uhlenhaut (1976), and Boag (1982). Combatants may topple into water, there to continue fight (Brown 1934), birds often trying to 'duck' each other (Forster 1962; Eastman 1969). In one dispute, apparently over winter territory, bird tried to escape by diving into water; aggressor then repeatedly landed on its back and the birds splashed about, beating wings, and continued to rise and dive. Eventually aggressor seized rival by bill and tried to force its head under water; rapid zigzag pursuit over water followed 2nd ducking (Steffen 1953b). Disputes usually short, but 1 fight lasted at least 8 hrs before intruder left (Svensson 1978). In breeding season, share of nest-duties partly determines role of sexes in rebuffing intruders. In 1 day at Dal Lakes, of 37 defences of territory made by pair, ♀ made 23 (62%) while ♂ more engaged in feeding young (Pring-Mill 1974). At Dal Lakes, aggressive chases followed by apparent site-ownership display: owner hovered at territorial boundaries, in turn followed by display-dives which seemed designed to attract attention rather than catch prey. After confrontations, territory-owner flew out in straight line from some landmark on lake shore, hovered at edge of territory for c. 10 s, then turned 90°, swooped down, and rose to hover again c. 15 m further on; this repeated until bird reached other lakewards corner of territory, from which it flew straight back to bank (Pring-Mill 1974). Antagonistic behaviour develops from early age in nest; fledged young defend favoured fishing perches against intrusion by siblings (Heinroth 1939; Eastman 1969) and begin to squabble over territories at c. 6 weeks old (Boag 1982). For drawings of aggressive postures of juveniles, see Glutz and Bauer (1980). When fledged young assembled on one perch, tried to rob each other of fish (Svensson 1978). HETEROSEXUAL BEHAVIOUR. (1) Pair-bonding behaviour. Pair-formation begins with vociferous aerial-chasing, always near prospective nest-site (Clancey 1936a). 2 birds interspersed chasing with calling from various perches (Zöller 1980). In one study, sequence began with one bird (presumably ♂) sitting on branch giving Advertising-call (see 3 in Voice) every 2–3 min. After 1 hr, 2nd bird approached, and the nearer it came, the more frequently the other called. Eventually a chase ensued, with Contact-alarm calls (Heyn 1963). Height of flight-path typically fluctuates, at one moment low over water, at the next high in tree-tops (Heyn 1963; Boag 1982), where participants sometimes perch (Clancey 1936a). Report of 2 birds flying rather hesitantly in circles of 15–20 m diameter, and singing (see 4 in Voice) continuously, occasionally alighting in bushes (Marsden 1927). For similar calls of ♂ circling perched ♀, see Beneden (1930). ♂ may occasionally fly in semicircular path away from ♀ and glide back towards her with stiff wings. ♂ and ♀ also sometimes perform weak, fluttering flight just above water (Boag 1982: see Courtship-feeding, below). Chasing may continue for hours (Horst 1938) and at various times of day (Heyn 1963). Usually only 2 birds involved; 5–7 (Clancey 1936a) exceptional. Chasing observed in autumn and winter (Gentz 1940; see also Eastman 1969) but mostly February–June (Clancey 1936a; Gentz 1940; Heyn 1963). Courtship-flights recur between 1st and 2nd broods though less intensely than before 1st (Reinsch 1962; Heyn 1963). Bigamous ♂ seen chasing 2nd ♀ over lake (Heyn 1963). Early pairing behaviour frequently involves elements of aggression including threat postures. ♂ and ♀ perched close to each other also adopt Head-up posture (Fig D) similar to that used in Upright-threat display but less hunched and more relaxed: stand side-on to each other, back almost vertical, wings drooped so that primaries level with feet; head tilted stiffly and slightly up, bill closed except when giving Contact-call (see 1 in Voice); pair perched thus may alternate calls. Head-up posture sometimes adopted later in season, apparently as Meeting-ceremony (Boag 1982). (2) Nest-building. Chasing often leads to nest-inspection or excavation which may have significant display value; early on, ♀ usually sits and watches initiative taken by ♂, while both give Contact-alarm calls (Heyn 1963; Boag 1982). If nest-tunnel already present, ♂ flies in and out; otherwise ♂ starts digging one (e.g. Eastman 1969). When young of 1st brood 11 days old, ♂ flew 2–3 m from perched ♀ to hover and strike bank near nest-site, returning to sit by ♀ between blows. As ♀ shortly departed from site to start overlapping clutch some distance away, ♂'s building activities interpreted as 'ceremonial' (Swanberg 1952). Role of sexes in excavation varies with nesting regime. Almost equal shares often taken for 1st clutch (Ingram 1933; Heinroth 1939; Plucinski 1969; Zöller 1980). In one case, ♂, almost unaided by ♀, dug nest-tunnel for 2nd clutch in 3 days (Heyn 1963), though usually takes at least 4–7 days (Gentz 1940; Heyn 1963); c. 3 days if pair make late start to breeding, up to 14 days if early (Svensson 1978). Excavation mostly in morning (Heyn 1963; Plucinski 1969; Zöller 1975). (3) Courtship-feeding. Begins shortly before completion of nest-tunnel (Heyn 1963). Often follows bout of courtship-chasing. For apparent aggression before Courtship-feeding, see Arthur and Cutter (1949). When ♀ perched, ♂ often flies off and returns with fish. When fish-bearing ♂ approaches ♀, both excitedly exchange contact-calls. On his arrival, ♀ may adopt Head-up posture (Boag 1982), or make tripping steps towards him (Zöller 1975; see also Gentz 1940). ♂ crouches low with drooped wings and stretches forward to present food (Clancey 1936a; Gentz 1940; Heyn 1963). After passing food, ♂ may adopt Head-up posture. For variant, apparently of greater intensity, with bill vertically up and tail fanned, see Glutz and Bauer (1980); after presentation, ♂ may also preen own breast (see Glutz and Bauer 1980). One ♀ seized and swallowed fish with little or no ceremony while both birds called quietly and continuously (Clancey 1936a); ♀ may also shiver wings when receiving food (Arthur and Cutter 1949; Boag 1982). If ♀ has not been fed for some time, she begs by drooping wings, extending neck, and giving Begging-call (Gentz 1940: see 5 in Voice). ♀ may also follow ♂ to his perch-sites at feeding grounds (Heyn 1963; Boag 1982). After accepting fish, ♀ seen to fly off low over water with weak fluttering flight, ♂ following suit (Boag 1982); one ♂, flying thus ahead of ♀, periodically dipped lightly into water (Burns 1944). For other accounts of Courtship-feeding, see Kerr (1918), Pettit (1952), and Eastman (1969). ♀ fed several times per day (Eastman 1969; Zöller 1975). Feeding continues until start of incubation, then declines rapidly (Heyn 1963). For ♂ feeding ♀ when young newly hatched, see Brown (1934). ♂, calling as he arrived, brought sitting ♀ up to 5 fish in morning, at intervals of 30–75 min, during which ♀ never left nest. ♀ received fish at tunnel entrance or inside (Zöller 1975, which see for size of fish; see also Zöller 1980). ♀ digging new nest-tunnel for 2nd clutch spent a lot of time sitting on perch, and was periodically fed by ♂ engaged in feeding 1st brood (Zöller 1980). (4) Mating. Nearly always preceded by Courtship-feeding, and thus regular towards end of nest-excavation (Heyn 1963; Zöller 1975, 1980); according to Boag (1982), preceded by Courtship-feeding in $c.$ 50% of cases. Initially, ♀ sometimes unreceptive, and rebuffs ♂. If ♂ and ♀ land on perch simultaneously, ♀ often threatens soliciting ♂ with open bill and hissing call (Zöller 1975, 1980: see 7 in Voice). ♀ sits in Head-up posture, shivering drooped wings and calling (probably Begging-call). ♂ hovers briefly over ♀ who leans forward almost horizontally, then lands on her back, grasping nape with bill, and beating wings for balance. ♀ then moves tail sideways and pair make brief cloacal contact (Eastman 1969; Boag 1982; see also Gentz 1940). Twice when pair-members side by side, ♂ flew up and hovered over ♀'s back before mounting. On 2 other occasions, ♂ descended from branch above ♀. ♂ flew off after copulating (Rivière 1933a). ♂ (but not ♀) may also bathe afterwards (Kniprath 1969b). ♂ usually flew away after copulation, but after a vigorous bout sat alongside ♀; pair usually copulated on different perch each time (Zöller 1980). Before 2nd clutch laid, ♂ mounted with no preliminary Courtship-feeding or invitation by ♀ (Eastman 1969). Copulation much more frequent prior to 1st clutch than to subsequent ones in same season; high frequency maintained for $c.$ 8 days (Zöller 1980). In 1 morning, 4 copulations recorded within 3 hrs, each preceded by Courtship-feeding (Zöller 1975). Copulation declines after laying, but recurs during chick-rearing if ♀ about to start overlapping clutch (Zöller 1980; see also Eastman 1969). (5) Behaviour at nest. No special ceremony at nest-relief. Arriving bird often calls (up to 2–3 times: Plucinski 1969) on approach, and perches near nest; sitting bird emerges, sometimes also calls as it flies off, and awaiting bird quickly enters nest-tunnel (Reinsch 1962, 1968; Zöller 1980). At start of incubation, ♀ left nest voluntarily without signal from ♂ who arrived, within 10 min, to relieve her; once rhythm established, however, relieving bird called on approach and change-over rapid (Eastman 1969). Clutch thus not closely incubated until 3rd day (Zöller 1980). Rapid nest-relief continues into chick-rearing, so both parents seldom in nest-chamber simultaneously (Heyn 1963). From laying onwards, pellets (formerly ejected outside nest-tunnel) increasingly deposited in nest-chamber which becomes littered with fish-bone debris (Kumari 1939a; Heyn 1963; Zöller 1980). ♀ cleans out old fish-bones before laying 2nd clutch in same hole (Reinsch 1968; Eastman 1969). Thus no evidence that fish-bones deliberately laid as nest-lining. Adult emerges from nest head- or tail-first, more often tail-first when young older (Kumari 1939a; Swanberg 1952; Heyn 1964; Schulz-Waldmann and Dominiak 1971). RELATIONS WITHIN FAMILY GROUP. Young hatch at intervals of a few hours (Heyn 1963); first fed within a few hours of hatching. For 1st week, brooding parent sits on top of young; thereafter broods more under wing (Heyn 1964). Brooding ceases at 6–10 days (Kumari 1939a; Swanberg 1952; Heyn 1963, 1964; Schulz-Waldmann and Dominiak 1971). For allocation of time to brooding and feeding, see Heyn (1963) and Zöller (1980). For calls of young seeking brooding or feeding, see Voice. Parent bringing fish often gives Contact-call on approaching nest (Zöller 1975: see 1 in Voice). Arriving ♂ once gave fish to ♀ outside nest-tunnel for transfer to young (Ris 1937). Equal distribution of food to young ensured by highly disciplined behaviour (Heinroth and Heinroth 1924–6; Kumari 1939a, 1978). After hatching, young sit in nest-chamber in radial arrangement with bills pointing outwards (Creutz 1956). Arrangement maintained by 2-day-old brood removed from nest (Kumari 1939a, 1978). When adult enters tunnel, food proffered (fish head-first) rather aimlessly, bird pushing prey towards nest-chamber and releasing it when grasped by chick (Heyn 1963; Schulz-Waldmann and Dominiak 1971). Prey passed to chick facing nest-tunnel; only this chick gapes (Heyn 1964) and gives food-call, calling continuously until parent arrives to feed it. Other young usually passive until due to be fed. Once fed, chick retires backwards (Eastman 1969) and next chick in rota moves into line with tunnel entrance, to be fed on next visit, and so on (Heinroth and Heinroth 1924–6). Broods differ in direction of rotation. Any chick trying to get food out of turn vigorously pecked by siblings; sometimes even tossed backwards (Schulz-Waldmann and Dominiak 1974). Rota said to break down in curved tunnel where no light reaches chamber from entrance to orientate young (Boag 1982). As young grow, configuration changes. After 2–3 weeks, young increasingly adopt 'roof-tile' pattern in which they bunch in forward-facing, overlapping rows (Creutz 1956, which see for other variants; also Heyn 1964). From $c.$ 14 days, young beg like adult ♀ with shivering wings (Kumari 1939a, b, 1978; Heyn 1963). Towards fledging, as inter-sibling rivalry increases, sequential feeding tends to break down and young fed more in tunnel than in chamber (Schulz-Waldmann and Dominiak 1977; Kumari 1978). Young defecate after being fed, towards light from $c.$ 1 week; after 2 weeks, sometimes out of nest-entrance (Heinroth and Heinroth 1924–6; Heyn 1963; Puschmann 1976). Towards fledging, young self-preen (Kumari 1939a), wing-stretch (Heyn 1963, 1964), and, in last 3–4 days before fledging, fall silent, except when about to be fed (Schulz-Waldmann and Dominiak 1971; Zöller 1975). By this time, young have developed a contact-call (Heinroth and Heinroth 1924–6; Kumari 1939b, 1978; Heyn 1963: see Voice). Near fledging, parents increasingly enter nest without food and in silence (Heyn 1964), though may continue to bring fish until nest empty (Eastman 1969). First of young fledged when ♀ arrived with fish and gave Contact-call (Zöller 1975). Parents said to entice young to fledge with Excitement-call (Heyn 1963; see also Reinsch 1962: see 6 in Voice); according to Boag (1982), however, parents give young no encouragement to fledge. Brood usually fledges on same day (at intervals of 10–20 min), sometimes over 2 days (Zöller 1975; Svensson 1978). On fledging, young fly short distance to convenient perch where they wait to be fed (Kumari 1939a: Zöller 1975), though also follow adults to fishing sites (Kortner 1979). Fledged young never return to nest (Heyn 1964). In one study, ♂ attended to fledged young while ♀ fed those still in nest (Zöller 1975). Though fed by parents for 2–4 days after fledging (Boag 1982), young begin diving on first day out of nest (Eastman 1969). Only 1 hr after fledging, young caught pond-skaters *Gerris* (Svensson 1978). Fishing success improves rapidly and young independent within a few days (Heinroth 1939; Eastman 1969; Svensson 1978), at which time they abandon chick contact-call (Eastman 1969). Captive young began to bend over and knock bill against substrate (action which kills captured prey) at 29 days (Tardent 1951). For other observations on captive young, see Heinroth (1939). 50-day-old young dived repeatedly to retrieve small twig from water surface (Swanberg 1952). Young disperse rapidly up to 300 m from nest the day after fledging (Guggisberg 1937; Zöller 1975), up to 4 km in a few days (Ptushenko and Inozemtsev 1968)—and have left parental territory, often by forcible eviction, in 1–4 days (Clancey 1935a; Kumari 1939a; Heyn 1963). ♂ may apparently lure young away from territory by offering food (S Svensson). Dispersing young infiltrate other birds' territories and may fish successfully by keeping silent (Boag 1982). ANTI-PREDATOR RESPONSES OF

YOUNG. Fall silent in nest if danger threatens (Boag 1982). For reactions of young 22–23 days old to handling, see Kumari (1939a). PARENTAL ANTI-PREDATOR STRATEGIES. (1) Passive measures. Accounts vary as to degree of attentiveness to nest when danger threatens (e.g. Kumari 1939a, 1978). Probably usually a tight sitter, leaving only if seriously provoked; often continues to nest in burrows manipulated by man (Kumari 1978). While clutch being laid, ♀ sits over—but does not incubate—eggs, perhaps as defence against damage by conspecific intruders (Heyn 1963). Apart from brief contact-calls, little calling after clutch complete (see Glutz and Bauer 1980), and especially quiet near fledging (see above). (2) Active measures. Little information. Bird flushed by man flies back and forth in agitated state, some distance from nest (Kumari 1939a).

(Fig A from photograph in Leibrich and Uhlenhaut 1976; Figs B and D from photographs in Boag 1982; Fig C from drawing in Goodfellow and Dare 1955.) EKD

Voice. Freely used in spring and summer, relatively silent in autumn and winter (Kumari 1978). Little information outside breeding season. Most calls are variants of Contact-alarm call, and following scheme therefore tentative.

CALLS OF ADULTS. (1) Contact-alarm calls. Variously described as combinations of high-pitched 'ti', 'tit', and 'che'. (a) Contact-calls. Rendered 'ti titi' (Ris 1937), 'tji tjii', and 'tjii tit tit tit tit' (♂ in flight) (Kumari 1939a, b); see also Heyn (1963), Zöller (1975), and Kumari (1978). In recording (Fig I), 'teee ti-tee ti-tee ti-tee' (J Hall-Craggs) given by bird flying towards nest with fish; once perched, call changed to more rapid 'tititi' (J Hall-Craggs: Fig II), probably corresponding to 'greeting call' of Zöller (1975). Contact-call given when perched or more often when flying to and from nest, with or without fish (e.g. Zöller 1975), or while excavating nest (Heyn 1963). Pair often give it alternately, e.g. when in Head-up posture (Boag 1982). Calls of bird flying back and forth while mate excavating had soft whispering quality; when perched, sometimes a protracted plaintive 'tiiiit', interspersed as soft contact-call (Zöller 1975). Excited ♂ to ♀ (mate) emerging from nest called 'tit-tüht tit-tüht' (Horst 1938). ♂ and ♀ often have distinguishable calls (Kumari 1939a; see Kumari 1939b for other variants). (b) Alarm-call. Apparently a sharper, more emphatic variant of contact-call. Rendered 'kee kee' (Kerr 1918) or protracted 'tjii', heard almost invariably in flight (Kumari 1939b). (2) Threat-call. Rasping or grating 'kritritrit', given by both sexes during territorial defence (whether threat, pursuit flight, or fight) against conspecific birds including ♀ mate and offspring (Kumari 1939a, b, 1978); also by parent harassed by begging young (Kumari 1939a). Also rendered as a guttural 'shrit-it-it' (Eastman 1969; see also Clancey 1935a); a furious trilling 'trrr trrr trrrtrrrtrrrtrrrtrrrtrrr trrr trrrt' (Horst 1938). Probably the loud trilling call given by bird in Upright-threat posture (Boyle 1952). (3) Advertising-call. Perhaps only by ♂ (Kumari 1939b, 1978). Renderings usually comprise repeated units, often disyllabic. Thus 'tee titi titi titi' (Rivière 1933a); 'tji-tii-ih tji-tii-ih' (Kumari 1939b); 'che-tee chee-tee chee-tee' (Kumari 1978); ringing 'chi-kee chi-kee' (Clancey 1935a, 1936). In recording by P A D Hollom, similar calls given by ♂ passing food to ♀. ♂ on perch apparently advertising for ♀ gave protracted 'tieht' at intervals (Heyn 1963). Frequently given in flight, less often from perch (Kumari 1939b); in courtship-flight (by most birds in group of 5–7) (Clancey 1936a); by ♂ passing ♀ repeatedly on perch (Rivière 1933a). Given throughout the year (Boag 1982). (4) Song. A rich mixture of pipings, whistles, and warbles (Carter 1947); an explosive, bubbling, gurgling song likened to that of Starling *Sturnus vulgaris* (Greenwell 1935). Also likened to song of Dipper *Cinclus cinclus* or Greenfinch *Carduelis chloris* (Marsden 1927; Beneden 1930). According to Boag (1982), such accounts excessively elaborate, and so-called song more a sequence of whistles of varying emphasis. Given by ♂ circling perched ♀ (Beneden 1930); perhaps by both ♂ and ♀ circling each other (Marsden 1927). See also Zöller (1975). (5) Begging-call of ♀. Plaintive sound said to resemble food-call of young (Gentz 1940; Heyn 1963). Rendered 'chee-chee' by ♀ soliciting food; a staccato bleating by ♀ inviting copulation (Eastman 1969). (6)

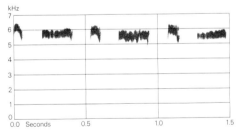

I P A D Hollom England July 1981

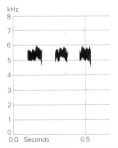

II P A D Hollom England July 1981

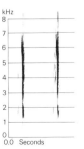

III P A D Hollom England June 1974

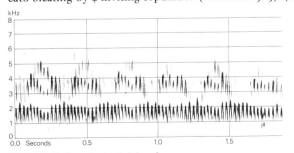

IV P A D Hollom England July 1981

Excitement-call. In recording (Fig III), 2 short sharp 'tsuk' sounds by bird at nest with well-grown young; call, accompanied by bobbing movements, accelerated when another bird approached (P A D Hollom; see also Calls of Young). Function not clear, but this probably the 'zütt zütt' ascribed a warning function by Horst (1938), and 'tüt-tüt' of parents apparently enticing young to fledge (Heyn 1963); however, no firm evidence of lure function. (7) Other calls. (a) ♀ resisting courting ♂ gave call described as a soft hissing (Zöller 1975; see also Zöller 1980). (b) Hoarse 'kreah' said to be given by parent feeding young in nest (Heyn 1963). In recording by P A D Hollom, similar call highly distinctive: brief, harsh and likened to winding-on of motor-driven camera (P J Sellar); occasionally given by bird before it flies away from its mate (P A D Hollom). (c) In January, ♀ defending feeding territory against rivals (including stuffed bird) 'whistled' before mounting flying attack (Leibrich and Uhlenhaut 1976).

CALLS OF YOUNG. Varied repertoire, changing with age. Small young (e.g. 2 days) solicit brooding with squeaking 'vrhüii vrhüii' (Kumari 1939a) or 'rrüü rrüü' (Kumari 1978). Plaintive 'peep' or 'seep' (Clancey 1935a) probably the same. From 10 days, food-call a loud purring 'uirr uirr' (Fig IV), audible up to 30 m from nest (Kumari 1978; see also Clancey 1935a). Rendered 'rüerüerüe' (Heyn 1963). According to Schulz-Waldmann and Dominiak (1971), given only by chick in line for feeding, and almost continuously until parent arrives with food (see Social Pattern and Behaviour). Young also give loud, clear 'bjüll' (Kumari 1939a), or quiet 'biu' (Kumari 1978); function unknown. Young near fledging or fledged, give harsh growling 'grräed grräed' (Kumari 1939a) or 'gred gred' (Kumari 1978) while being fed, accompanied by fluttering of wings. At 17–18 days, young develop sharp 'zück' (Kumari 1939a) or 'chick' (Kumari 1978); rendered as a sneezing cough—'zipp' (Heinroth and Heinroth 1924–6); 'tschick' (Heyn 1963). Serves as contact-call out of nest, and persists until 1 week after fledging (Kumari 1978).

EKD

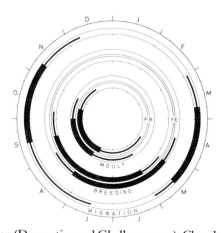

Breeding. SEASON. For eastern Europe, see diagram. Starts up to 1 month earlier in Britain and north-west Europe (Dementiev and Gladkov 1951a; Makatsch 1976). Prolonged by multiple broods. SITE. Tunnel in steep or vertical bank of stream, river, or gravel-pit, normally over water, occasionally not; has used hole in wall, among tree-roots, or even burrow of rabbit *Oryctolagus cuniculus* (Witherby *et al.* 1938). Most nests 90–180 cm above water; of 21 nests, extremes 60 cm and 36 m (Brown 1934). Nests re-used in successive years. Nest: excavated tunnel with enlarged chamber at end, horizontal or slightly rising from entrance. Average length of 21 burrows, 54·5 cm (33–92) (Kumari 1939a); mostly 45–90 cm, with extremes 23–135 cm (Clancey 1935). Diameter of tunnel normally 5–5·5 cm (Zöller 1975). Chamber averages 16 × 17 cm across, 11 cm high (12–23 × 11·5–20 × 9–14), sample size not given (Dementiev and Gladkov 1951a). Chamber often offset to one side or upwards. Unlined, though fishbones accumulate in chamber. Building: by both sexes, mainly ♂ (Clancey 1935), loosening soil with bill and ejecting with feet; time taken varies with soil but usually 7–12 days. EGGS. See Plate 98. Almost spherical, smooth and glossy; white. 23 × 19 mm (21–24 × 17–20), n = 100 (Witherby *et al.* 1938). Fresh weight 4·2 g (3·6–4·7), n = 34 (Glutz and Bauer 1980). Clutch: 6–7(4–8). Of 12 clutches, Britain: 4 eggs, 1; 6, 1; 7, 10; average 6·7 (Clancey 1935). Normally 1–2 broods, occasionally 3 (e.g. Clancey 1935, Plucinski 1970, Lambie 1974, Schmidt 1981) or 4 (Zöller 1980); up to 6 if ♂ bigamous (Podolski 1982). Broods commonly overlap with ♂ beginning excavation of new burrow while first brood still unfledged; recorded starting when previous brood 6 days old (Swanberg 1952), usually 10–16 days (e.g. Swanberg 1952, Reinsch 1968, Ruthke 1968, Zöller 1980), and up to fledging at *c.* 27 days (Ruthke 1968). For 3rd brood, may revert to original hole (Schmidt 1981). Laying interval 1 day. INCUBATION. 19–21 days. By both sexes about equally (Clancey 1935), with ♀ incubating in morning and ♂ in afternoon (Reinsch 1968); in shifts of *c.* 2·5–4·5 hrs (Zöller 1980). Begins with last egg, though penultimate recorded (Puschmann 1976). Hatching synchronous. YOUNG. Altricial and nidicolous. Cared for and fed by both parents, except when ♂ bigamous (see Social Pattern and Behaviour) or ♀ starts 2nd brood. At 4 days old, brooded more or less continuously in equal spells of 450–475 min (Swanberg 1952). When ♀ starts 2nd brood, ♂ usually carries on feeding alone but occasionally helped by ♀ (e.g. Zöller 1980). FLEDGING TO MATURITY. Fledging period 23–27 days, becoming independent within a few days. Age of first breeding 1 year. BREEDING SUCCESS. Of 69 eggs hatched, 55 reared to fledging, average 5·5 young per nest; overall success 80% (Clancey 1935). Of 619 eggs laid by 39 pairs, Switzerland, 333 young (53·8%) fledged (Glutz and Bauer 1980). In Sweden, main nest predators fox *Vulpes vulpes* and mink *Mustela vison* (Svensson 1978).

Plumages (*A. a. ispida*). ADULT. Forehead, crown, and nape dark bluish-olive or olive-black, closely marked with narrow and contrasting bright cerulean-blue bars. Patch on lores rufous-chestnut or rufous-cinnamon, bordered by black line above gape. Patch from just below and behind eye over ear-coverts rufous-chestnut, bordered behind by contrasting pale cream or silky-white stripe along sides of neck to hindneck; cream or white often slightly washed orange-chestnut, in particular towards rear. Broad malar stripe from base of lower mandible to upper sides of chest cerulean-blue or cobalt-blue, faintly barred dusky olive on lower cheeks. Central mantle, back, rump, and upper tail-coverts glistening cerulean-blue, often slightly darker cobalt-blue towards longer upper tail-coverts; sides of mantle and scapulars duller dark bluish-green or dark green, sometimes with faint paler blue streak along shafts. Chin and throat cream or silky-white with orange-buff to cinnamon tinge (similar in colour to stripe on side of neck); sides of chest with dull cobalt-blue or greenish-blue patch extending from sides of mantle; remainder of underparts deep rufous-cinnamon or orange-chestnut, often slightly paler yellow-buff on lower belly and flanks, deepest on chest and upper flanks. Tail dark blue with black shafts on upper-surface, dull black on undersurface. Outer webs of flight-feathers rather dull blue or greenish-blue, but tips of primaries and outer web of p10 dull black; inner webs of flight-feathers and narrow line along shafts dull black, basal inner borders with variable amount of buff or cream-pink tinge (mainly visible from below). Upper wing-coverts and tertials dark greenish-blue or bluish-green (tertials and longer coverts more bluish, shorter coverts more greenish), similar in general tinge to scapulars and outer webs of flight-feathers; tertial coverts, median upper wing-coverts, and longer lesser coverts with small glistening cerulean-blue spot on tip; centres of primary coverts and of feathers of bastard wing black. Under wing-coverts and axillaries rufous-cinnamon. Some variation in depth of blue of crown-bars, malar stripe, mantle to upper tail-coverts, and spots on upper wing-coverts, in part depending on angle of light; generally, colour cerulean-blue tending to cobalt-blue on nape, malar stripe, and upper tail-coverts, but sometimes turquoise-blue or greenish-blue, in particular on bars of forehead and on central mantle to rump; rarely, all blue entirely cobalt-blue. Plumage looks worn from about April: dark bars of crown and scapulars duller and blacker; pale stripe on sides of neck and chin and throat purer white; rufous of underparts and patches in front and behind eye paler, in particular central underparts more yellowish-buff. NESTLING. Naked at hatching. Spiny feather-sheaths gradually appear, but do not open until day 20–22, bird appearing fully covered with spines late in 2nd and during whole of 3rd week. Fully feathered by day 21–24, except for some spines round eye, on forehead and chin, and over ear; flight-feathers and tail still growing at fledging. (Heinroth and Heinroth 1924–6; Glutz and Bauer 1980.) JUVENILE. Like adult, but forehead, crown, and nape more olive-green, less deep olive-black; bars narrower and sharper, paler greenish-blue rather than cerulean-blue; malar stripe duller greenish-blue or bluish-green; central mantle, back, rump, and upper tail-coverts paler turquoise-green or turquoise-blue; outer mantle and scapulars sometimes more olive-green, less blue. Underparts similar to adult, but feathers of chest and breast with blue-grey fringes, sometimes hard to detect but occasionally giving marked blue-grey wash to breast. Blue of tail and flight-feathers sometimes more greenish than in adult. Upper wing-coverts often greener than in adult, paler blue spots more greenish and smaller, less inclined to extend into blue fringe along tips of coverts. Fresh juvenile rather similar to fresh adult (except in particular for blue fringes on breast), but adult distinctly worn when juvenile fresh; juvenile plumage usually worn by October–November, when adult fresh; in worn plumage, glistening blue parts of body distinctly tinged sea-green, central belly and vent occasionally even paler than in worn adult, buffish-white. Unlike adult, all flight-feathers equal in age, no active or suspended moult. FIRST ADULT. Similar to adult, but variable number of juvenile body feathers, tail-feathers, or wing-coverts, and all flight-feathers retained (see Moults). Indistinguishable by plumage when all body feathers replaced (but all flight-feathers equally new, while many adults retain some old ones, mainly p5, p6, p10 and s3–s5); often some or many body feathers retained, however, those on breast with faint pale grey tips, those on crown with narrow sea-green bars.

Bare parts. ADULT MALE. Iris brown. Bill jet-black, sometimes with horn-brown tinge or with some indistinct brownish-orange or dull red spots near base of lower mandible. Leg and foot orange-red or coral-red, soles pink-red or pale orange-red, claws horn-brown; tarsus often with brown spot or entirely brown during moult in autumn. ADULT FEMALE. Similar to adult ♂, but lower mandible brownish-red, orange, or pinkish-red, with black tip; black tip covers $\frac{1}{6}-\frac{1}{4}(-\frac{2}{3})$ of bill length, occasionally no black on tip at all; sometimes a slight amount of dull brown at base of lower mandible. NESTLING. Pink at hatching, including bill, leg, and foot. From $c.$ 7th day, future feather tracts darken to grey-blue. Eye gradually opens at day 8–12. At day 14, bill black with pink basal cutting edges, bare eye-ring flesh-grey, leg and foot greyish-flesh with blue-grey front of tarsus and uppersurface of toes, claws horn-black. JUVENILE. At fledging, iris brown, bill black with whitish-horn tip, gape violet-pink, leg and foot dark brown or dull flesh-brown with orange-red rear of tarsus and soles. Some dull pink spots develop on base of lower mandible from $c.$ 1 month after fledging: in ♂, spots never occupy more than $\frac{1}{3}$ of bill length; in ♀, spots become gradually larger—$c. \frac{1}{3}$ of bill-base orange-pink from $c.$ 2 months after fledging, $c. \frac{1}{2}$ of lower mandible pale at 4 months, similar to adult ♀ from 8–12 months. Whitish-horn bill-tip often retained until spring. Leg and foot dull brown-red with grey cast in 1st autumn and winter, grey gradually lost in 1st spring and summer. (Hartert 1912–21; Heinroth and Heinroth 1924–6; Clancey 1936b; Doucet 1971; Glutz and Bauer 1980; RMNH, ZMA.)

Moults. Based mainly on Doucet (1971), with some additional information from specimens (BMNH, RMNH, ZMA). ADULT POST BREEDING. Complete or almost complete; primaries descendant from 2 centres. Starts with scattered feathers of body from June–August, first on underparts. Primaries start simultaneously with p1 and p7 between end of June and mid-August, moult slow, each feather not shed until previous one $\frac{2}{3}$- or full-grown. Both centres moult descendantly, finishing with p6 and p10, full-grown mid-October to late November. However, often stops moult late October or November, retaining old p6; more rarely, also p10 and p5(–p4). These feathers replaced from May–June onwards in following year, either before new series starts with p1 or p7 or at same time. Secondaries moult from 2 centres: both ascendantly and descendantly from s10, ascendantly only from s1; s10 starts at same time as p1 and p7, but s1 not until shedding of about p4, p9, and p8, at about same time as completion of innermost tertial (s12); s3–s5 replaced last, after p6 and p10. Moult often arrested before all secondaries new, s3–s5 often retained throughout winter, as well as sometimes s2 and s6–s7 and rarely s1, s8, or s12; these feathers replaced first in next moulting season. Tail starts at same time as primaries, completed about half-way through primary moult at about same time as centre on s1 becomes active; sequence irregular, several alternating or neighbouring feathers often growing at same time. Moult between each wing and between tail-halves often asymmetrical and apparently increasingly irregular with age. POST-

JUVENILE. Partial: all or part of head and body and variable number of tail-feathers. Timing highly variable due to protracted fledging period (May to early October): some start from July–August and head and body of these largely in 1st adult by October, others still fully juvenile in late December. Moult arrested during winter, when many immatures retain at least some scattered juvenile feathers on chest, breast, or vent; moult continued in spring, when sometimes some tertials also replaced, but a few juvenile feathers on underparts, part of tail, and all wing retained until 1st post-breeding moult starts (June–)July–August. In 1st autumn, some tail-feathers often replaced, mainly t1 only or inner 2–3(–4) pairs; some more feathers replaced in spring, some occasionally for 2nd time. As in adult, some moult of body may occur throughout year, except during adverse weather; not known whether this is a protracted post-juvenile or post-breeding moult or if some feathering in part replaced twice a year (i.e. whether or not a partial 1st pre-breeding is involved).

Measurements. *A. a. ispida*. Britain, Netherlands, West Germany, and France, all year; skins (RMNH, ZMA). Bill (F) to forehead, bill (N) to distal corner of nostril, both of adult and 2nd calendar year only.

WING	♂ 78·3	(1·35; 51)	76–81	♀ 78·2	(1·38; 63)	76–81
TAIL	36·5	(1·46; 13)	35–39	37·1	(1·66; 32)	34–40
BILL (F)	44·6	(2·53; 24)	40–47	42·9	(1·82; 33)	40–46
BILL (N)	33·5	(1·75; 26)	30–36	32·7	(1·50; 34)	30–35
TARSUS	10·2	(0·58; 12)	9·4–10·9	10·2	(0·40; 23)	9·6–11·2
TOE	17·3	(0·66; 11)	16–18	17·5	(0·95; 15)	16–19

Sex differences significant for bill (F). Juvenile similar to adult, combined above, but full bill length not attained until midwinter. Average length of bill (F) in fledged juvenile ($n = 4$–10): June–July, ♂ 34·0, ♀ 35·3; August, ♂ 39·5, ♀ 40·4; September, ♂ 41·4, ♀ 41·5; October, ♂ 42·5, ♀ 41·2; November, ♂ 42·5, ♀ 42·6; December, ♂ 45·3, ♀ 42·8 (RMNH, ZMA). Bill shorter in summer than in winter, caused by abrasion when digging nest-hole: Switzerland, May–August, ♂ 40·6 (68) 35–45, ♀ 39·4 (67) 35–43; February–April, ♂ 42·1 (31) 38–47, ♀ 41·0 (21) 38–46 (Glutz and Bauer 1980).

In Rumania, wing of *ispida* from westernmost part of country 76·8 (1·07; 8) 75–78; nominate *atthis* from centre (west of Carpathians) 74·7 (1·33; 6) 73–76, from south-east and east 74·1 (1·53; 16) 72–76 (Papadopol 1965).

Nominate *atthis*. Southern Spain, north-west Africa, Sardinia, and southern Italy; all year; skins (BMNH, RMNH, ZMA).

WING	♂ 76·8	(1·72; 12)	74–79	♀ 77·5	(1·72; 13)	75–80
BILL (N)	36·2	(1·99; 10)	35–39	34·9	(1·60; 10)	33–37

South European USSR, summer, and Cyprus, Levant, Iraq, and Egypt, all year; skins (BMNH, RMNH, ZMA).

WING	♂ 74·2	(1·49; 30)	72–77	♀ 74·2	(1·67; 20)	71–76
BILL (N)	34·3	(2·19; 22)	30–38	34·0	(2·00; 15)	30–37

Wing of west Mediterranean birds close to those of north-west Europe, but bill significantly longer (mainly over 35, north-west Europe mainly below 35); bill of eastern birds close in length to bill of *ispida* (though more slender, as in west Mediterranean birds), but wing distinctly shorter.

A. a. bengalensis. China (summer) and western Indonesia (winter); skins (BMNH, RMNH).

WING	♂ 72·1	(2·12; 19)	69–76	♀ 72·4	(1·95; 41)	69–76
BILL (N)	33·4	(1·23; 18)	31–35	32·8	(1·08; 32)	31–35

Wing and bill (N) of Japanese birds slightly longer, of Taiwan birds intermediate: sexes combined, Japan, 74·3 (2·16; 16) 72–77 and 34·5 (1·34; 14) 33–36; Taiwan, 73·8 (1·61; 24) 72–77 and 33·6 (1·72; 24) 31–36 (RMNH).

Weights. *A. a. ispida*. Hertfordshire (England) (Reynolds 1975); data approximate, read from graph.

JUN–AUG	35·8	(2·11; 162)	NOV	40·8 (2·75; 41)
SEP	36·4	(1·97; 46)	DEC–FEB	43·7 (3·22; 44)
OCT	38·3	(2·88; 32)	MAR–APR	40·2 (3·78; 22)

Netherlands, all year (ZMA).

♂ 39·3 (4·43; 6) 34–43   ♀ 39·2 (3·84; 10) 34–46

Switzerland, 1st adult and adult (Glutz and Bauer 1980).

JAN–MAR	♂ 42·8 (13)	♀ 41·4 (7)
APR–MAY	41·5 (40)	46·2 (41)
JUN–JUL	40·9 (41)	43·8 (38)
AUG–SEP	39·1 (9)	38·8 (12)

Lean full-grown birds Netherlands, mainly frost-killed: 30·1 (3·01; 12) 25–35 (ZMA). Switzerland: juvenile, July–September, 36·0 (234); juvenile and adult, October–November, 39·9 (28); range (all year) 35–55 (Glutz and Bauer 1980). Camargue: adult, July–September, 35·5 (44); juvenile, July–October (a few from June and November), 35·1 (806) (Glutz and Bauer 1980). Portugal: July 32·3 (1·73; 16) 30–36; August 30·9 (1·22; 8) 29–33; September and early October 31·5 (1·50; 12) 28–34 (C J Mead). Malta (perhaps including some nominate *atthis*): autumn, 33·7 (3·8; 43) 26–40 (J Sultana and C Gauci). Belorussiya (USSR), where intermediate with nominate *atthis*: ♂ 39·6 (5) 36–45, ♀ 43 (Fedyushin and Dolbik 1967).

Nominate *atthis*. Turkey, August: 36·0 (4·46; 9) 32–45 (D I Sales; BTO). Northern Iran, second half of August: 27·0 (2·35; 15) 23–33 (P J K Burton, BTO). Southern Iran, late August and early September: 29·0 (3·33; 16) 26–36 (D J Garbutt, BTO). Sexed birds, Turkey and Iran: adults, February–May, ♂ 32·6 (2·77; 5) 28–35, ♀♀ 30, 34·5; juvenile, July, 25·6 (Paludan 1938, 1940; Schüz 1959; Rokitansky and Schifter 1971). Afghanistan and Kazakhstan, April–October: 29·9 (2·71; 9) 26–34 (Paludan 1959; Dolgushin *et al.* 1970).

A. n. bengalensis. Taiwan, mainly June–September: ♂ 25·4 (4·12; 52) 19–40, ♀ 25·7 (3·31; 46) 20–30 (RMNH). Manchuria, July–September: 27·2 (5) 25–30 (Piechocki 1958).

Structure. Wing short, broad at base, tip rounded. 10 primaries: p8 and p9 longest, p10 and p7 1–4 shorter, p6 4–6, p5 7–9, p4 10–12, p1 17–22. Inner web of p10 slightly emarginated. Tail short, slightly rounded; 12 feathers (exceptionally up to 15: Doucet 1971), t1 0–2 shorter than t2–t3, t6 2–5 shorter; tips of tail-feathers rounded. Under tail-coverts nearly reach tail-tip. Bill long, straight, rather gradually tapering to sharply pointed tip; rather wide and deep at base, laterally compressed at middle and end. Nostrils narrow, slit-like, almost covered by membrane. Feathers of nape, mantle, and rump, and upper tail-coverts rather narrow, elongated, rather loose at tips. Tarsus very short, bare. Toes rather short, bases partly joined to form flattened sole; outer toe with claw *c.* 90% of middle toe with claw, inner *c.* 53%, hind *c.* 50%. Claws short, slender, and sharp.

Geographical variation. In Palearctic, slight in size, very slight in colour; some extralimital races more strongly divergent in colour. Both nominate *atthis* (from Mediterranean and central European USSR eastward) and *bengalensis* (from northern India to eastern Asia) have chin in adult slightly whiter, contrasting more sharply with rufous-cinnamon underparts than in *ispida*; colour often stated to be darker blue on upperparts, but variation large in all races and mainly individual or due to age differences; juveniles average paler below than *ispida*, in particular *bengalensis* cream or white from chin to vent and often with extensive and contrasting blue-grey tinge on chest and breast, forming more or less complete band. Bill of *ispida* slightly heavier than in nominate *atthis* and *bengalensis*, both upper and lower mandible slightly more bulging towards tip, culmen especially; base and middle of bill appear deeper, but difference sometimes hard to see. Nominate *atthis* from European USSR, Rumania, Yugoslavia, Bulgaria, and Greece have wing shorter than in *ispida* (almost always 76 or less; in *ispida*, 76·5 and over) and bill on average longer (but much variation). Populations from southern Spain, north-west Africa, Corsica, Sardinia, southern Italy, and probably Dalmatia agree with typical nominate *atthis* from Middle East in slender bill; however, both wing and bill decidedly longer (in particular, little overlap in wing), but differentiation not marked enough to warrant recognition of separate west Mediterranean race. Position of populations inhabiting southern Soviet Central Asia, Afghanistan, and Kashmir problematical; here included in nominate *atthis* following Vaurie (1965), though wing slightly and bill apparently distinctly shorter (bill to nostril as low as 28 in some adults examined), underparts on average paler, and chest and breast of juvenile more extensively blue; sometimes separated as *pallasii* Reichenbach, 1851 (see also Tschusi zu Schmidhoffen 1904, Johansen 1955, Ali and Ripley 1970). Birds inseparable from '*pallasii*' or *bengalensis* occur in Iraq and Arabia in winter. *A. a. bengalensis* poorly differentiated from nominate *atthis* from Middle East or from '*pallasii*'; upperparts on average slightly brighter blue (but extensive overlap in colour), adult ♀ apparently often has completely pale lower mandible, juveniles on average whiter below with extensively blue breast; slightly smaller in size, in particular in populations from northern India, south-east Asia, and China, but larger again towards north, with populations of Japan and perhaps south-east Siberia similar in size to those of Middle East. Other extralimital races differ in small size and generally much darker blue or violet-blue upperparts; Australasian races have ear-coverts partly or fully blue instead of rufous.

Forms superspecies with Half-collared Kingfisher *A. semitorquata* of Afrotropics, which differs only in slightly smaller size, blue ear-coverts, and (in both sexes) black bill. Blyth's Kingfisher *A. hercules* from northern India and south-east Asia perhaps also rather closely related, but breeding range overlaps widely with *A. atthis bengalensis* (Snow 1978; Fry 1980). CSR

Subfamily CERYLINAE pied kingfishers and allies

Very small to large kingfishers, frequenting waterside habitats. 9 species in 2 genera: (1) *Ceryle* (pied kingfishers, 5 species; Asia, Africa, the Americas); (2) *Chloroceryle* (green kingfishers, 4 species; tropical and subtropical America). Represented in west Palearctic by 2 species of *Ceryle*, 1 breeding, 1 accidental. Fry (1980) recognized 3 genera, including only Pied Kingfisher *C. rudis* in *Ceryle* and assigning remaining 4 species to *Megaceryle*, while retaining *Chloroceryle* as constituted here.

For general features, moults, etc., see Alcedinidae. P10 always longer than p4 except in some individuals of Pygmy Kingfisher *Chloroceryle aenea*. Tails rather long, always more than half wing length. Bills strongly compressed laterally, culmen narrow, not depressed basally. Skull narrow. Lower end of tibio-tarsus unfeathered for some distance, tarso-metatarsus short or very short, no longer than inner toe without claw. Toes short or very short, but inner relatively longer than in other subfamilies, equal to or longer than outer.

Sexes differ in colour of underparts only. Primaries usually barred or spotted with white. Bills dark.

Ceryle rudis Pied Kingfisher

PLATES 66 and 68
[between pages 686 and 687]

Du. Bonte IJsvogel   Fr. Alcyon pie   Ge. Graufischer
Ru. Малый пегий зимородок   Sp. Martín pescador pío   Sw. Gråfiskare

Alcedo rudis Linnaeus, 1758

Polytypic. Nominate *rudis* (Linnaeus, 1758), Middle East, Egypt, and Afrotropics. Extralimital: *leucomelanura* Reichenbach, 1851, eastern Afghanistan and western Pakistan east to southern China, south to southern Burma and Ceylon, except south-west India; *travancoreensis* Whistler and Kinnear, 1935, Kerala to Cape Comorin (south-west India).

Field characters. 24–26 cm; wing-span 45–47 cm; bill *c*. 6 cm. Medium-sized to large kingfisher, with long black bill, pied black and white plumage, noticeable crest, and rather long tail. Flight agile but action more flapping than most kingfishers; hovers persistently. Restricted to open-water habitats. Sexes dissimilar; no seasonal variation. Juvenile separable at close range.

ADULT MALE. Long, black, dagger-shaped bill. Plumage predominantly black above and white below, with most obvious marks white supercilium contrasting with white-streaked black crown and black cheeks, 1 broad and 1 narrow black band across chest, bold black and white barring across back and wings, and white-barred black tail. In flight, white-bases to flight-feathers create striking panel.

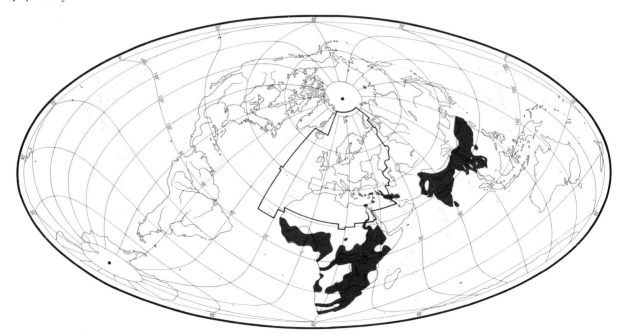

Legs black. ADULT FEMALE. As ♂, but lower chest-band absent, and upper restricted to broad black side-patches, usually totally separated. JUVENILE. Resembles ♀ but chest-marks grey, not black, and usually joined across chest; throat feathers fringed grey.

Unmistakable. Flight freer and looser than Kingfisher *Alcedo atthis*, with wing-beats noticeably more flapping; hunts fish mostly by hovering followed by powerful plunge. Often conspicuous, perching in the open and forming small noisy parties. Flicks tail frequently.

Vocabulary includes a sharp, penetrating disyllabic 'quick-ick' and a vibrating whistle.

Habitat. From lower middle latitudes of west Palearctic, in Mediterranean and desert zones through subtropical to tropical regions. In Iraq, breeds near rivers, canals, and marshes (Allouse 1953) and in Israel widespread where suitable water (S Cramp). In Egypt, fishes along all watercourses, and even on coast (Etchécopar and Hüe 1967). In West Africa, frequent on coasts, even preferring tidal waters and sometimes fishing in surf, hawking over waves, or inhabiting mangrove belts (Bannerman 1933). In Sierra Leone, favours coast, where muddy and mangrove-lined, and tidal reaches of rivers; also lakes (G D Field). Numerous in lagoons and creeks in southern Nigeria, sitting on posts, and in Mali, at ponds in interior, fishing in wet season on tiny pools and in roadside ditches (Bannerman 1951). Also other habitats offering fish, such as marshes or ricefields (Serle *et al.* 1977). In southern Africa, on dams, lagoons, and seashore, fishing just beyond breakers (Prozesky 1970). In India, from sea-level to *c.* 1800 m, on every kind of standing fresh water, such as canals, pools, irrigation reservoirs, village tanks, and flooded ditches, and also sluggish rivers and streams; apparently only occasionally on tidal creeks and on intertidal rock pools on sea-shore (Ali and Ripley 1970). As a breeding species, largely confined to banks of rivers, although found in plains wherever there is water, except in midst of forest (Whistler 1941). In Cyprus, watched in winter frequenting harbour and neighbouring coast, but sometimes shifting inland to fresh waters, even roadside ditch, or sitting on telegraph wire (Bannerman and Bannerman 1971). Forages in lower airspace, especially by hovering over water.

Distribution. LEBANON. Still breeding at Anjar (Tohmé and Neuschwander 1974). EGYPT. Faiyum: common in 1957 (Horváth 1959) but no recent observations there (PLM, WCM).

Accidental. Poland, Greece, USSR, Cyprus.

Population. TURKEY. Rather local but not uncommon (Beaman *et al.* 1975); numbers unknown, possibly 200 pairs (Parslow and Everett 1981). IRAQ. Very common Fao to Mosul (Ticehurst *et al.* 1922). LEBANON. Almost disappeared (Tohmé and Neuschwander 1974). SYRIA. Not rare in north (Kumerloeve 1968b); apparent decrease since 1940s with only 2 records in breeding season 1975–7 (Macfarlane 1978). ISRAEL. Decreased considerably (HM). EGYPT. Common, but major decrease some areas, e.g. environs of Cairo and Giza (PLM, WCM).

Survival. East Africa: in ♂, 1st-year mortality 51%; average annual adult mortality 45% in ♂, 54% in ♀; oldest ringed bird at least 5 years (H-U Reyer).

Movements. Resident and to some extent dispersive.
Somewhat more widespread in Turkey in winter (Vit-

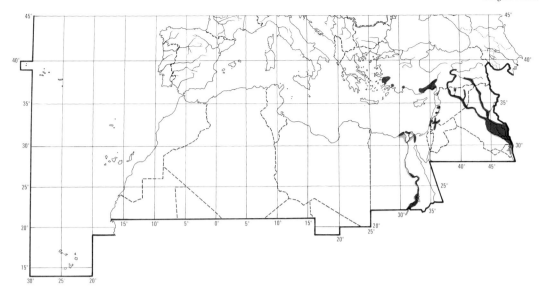

tery et al. 1972; Beaman et al. 1975), and Turkish birds (presumably) occur as rare and irregular winter visitors to Cyprus, October–April (Stewart and Christensen 1971). In Lebanon, known mainly as autumn and winter visitor to coasts, estuaries, and rivers (Benson 1970). In Iraq and south-west Iran, birds make local movements with season and river height (Moore and Boswell 1956); some disperse southwards for longer distances to become scarce visitors, mainly October to early April, to Kuwait and (more rarely) Persian Gulf coast as far as United Arab Emirates (Bundy and Warr 1980; Jennings 1981a). No well-defined movements in Africa either, though seasonal fluctuations of numbers in some places show that dispersals occur (e.g. Douthwaite 1973, Elgood 1982). One ringing movement of 760 km from Ethiopia to Uganda (Britton 1980).

Food. Principally fish, occasionally crustaceans, frogs, and aquatic insects. Dives for prey—either after hovering or from perch, e.g. tree branch or boulder. Flies parallel to shore, swooping up intermittently to hover at 2–10 m. On detecting prey, may drop in stages before diving (Whitfield and Blaber 1978). Fish held at right angles to bill and gripped behind gill covers. Small fish may be swallowed in flight; larger prey carried to perch, battered, shaken, and crushed dorso-ventrally before being swallowed headfirst. Degree of battering appears to be linked to stoutness of fish, e.g. more hits given to *Tilapia zillii* weight for weight than to slimmer *Hemihaplochromis multicolor*. At Lake Victoria, only fish longer than 55 mm battered before consumption (Douthwaite 1971b). In South Africa, seen to beat small crab against rock (Cooper 1981). Recorded catching 2 *Pranesus pinguis* in one dive from above (Whitfield 1978). Usually feeds within 50 m of shore, needing to return to perch to ingest large prey. Exceptionally feeds up to c. 3·2 km from shore (Junor 1972). 4–5 peaks in fishing activity per day, including peak up to dusk (Tjomlid 1973). Never fishes in heavy rain (Douthwaite 1976). Recorded perching on and diving from hippopotamus *Hippopotamus amphibius* (Pitman 1961; Pooley 1967). Recorded hovering above clawless otter *Aonyx capensis*, possibly to observe disturbed prey (Boshoff 1978). Insects, especially winged termites (Isoptera), taken in aerial pursuit; other termites taken from ground (Douthwaite 1976; Every 1976).

In Africa, fish include Cichlidae (*Tilapia zillii*, *Hemihaplochromis multicolor*, *Sarotherodon mossambicus*), silversides *Pranesus pinguis*, glassies *Ambassis*, mullets (Mugilidae), thornfish *Terapon jarbua*, sardine *Limnothrissa miodon*, round herrings *Gilchristella*, and halfbeaks *Hyporhamphus*; also *Barbus*, *Alestes*, *Nannocharax*, and *Aplocheilichthys*. In India, fish include Cyprinidae (*Puntius*, *Chela*) and Bagridae (*Mystus*); also Mugilidae (*Rhinomugil*, *Mugil*), Sciaenidae (*Pseudosciaenia*, *Johnius*, *Pama*), *Ambassis*, *Polynemus*, *Oryzias*, and *Harpodon*. Crustaceans (all areas) include *Macrobrachium*, *Palaemon*, *Cardina*, and *Metapenaeus*. Insects include larvae of dragonflies (Odonata), water-beetles (Dytiscidae, Gyrinidae), water-scorpions (Nepidae), water-bugs (*Belostoma*, *Notonecta*, *Corixa*), and termites (Isoptera). Also small frogs (Anura). Gastropod molluscs *Bellamya* and bivalve *Corbicula* recorded in stomachs but perhaps secondarily ingested. (Dementiev and Gladkov 1951a; Douthwaite 1976; Mukherjee 1976; Whitfield 1978; Whitfield and Blaber 1978).

In rare Palearctic observations in USSR, small fish, large insects, shrimps, tadpoles, and small frogs recorded (Dementiev and Gladkov 1951a). Stomachs of 299 adults from West Bengal (India) contained (by weight) 57% fish, 17% crustaceans, and 26% insects. Fish included (by number) 26% *Mystus*, 17·5% *Ambassis*, 16·9% *Puntius*, 16·3% *Mugil parsia*, and 7·1% *Oryzias melastigma*; of crustaceans, *Metapenaeus brevicornis* comprised 31·9%, *Macrobrachium lamerrei* 23·7%, *Cardina gracilipes* 18·1%, *Metapenaeus monoceros* 17·8%; of insects, *Eretes stictus* 21·9%, *Belostoma* 20·1%, *Corixa* 18·4% (Mukherjee 1976). In Uganda, pellet analysis showed diet almost totally fish; *Haplochromis* and *Engraulicypris argenteus* most important. Parasites and prey of fish taken also found in pellets, including larval midge *Chaoborus*, gastropod *Bellamya unicolor*, and bivalve *Corbicula africana*. Winged termites *Macrotermes* eaten throughout year, soldiers and workers taken occasionally (Douthwaite 1976). On Kafue flats (Zambia), takes mostly fish, 44% by number being Cichlidae, despite others being more abundant (Denton and Nicole 1962). Cichlidae easier to spot and, within preferred length group, are plumper, giving more food than (e.g. *Barbus* and *Alestes* (Tjomlid 1973). At Lake St Lucia (South Africa), fish comprised (mostly up to 15 g) 13 species but 80% by weight Mozambique tilapia *Sarotherodon mossambicus*; crustaceans also taken (Whitfield and Blaber 1978).

In Uganda, during first 9 days, preferred food for young is *Engraulicypris argenteus*; after this, the larger *Haplochromis* preferred; both adults and young discriminate in prey size selection. Chicks digest most bone consumed (Douthwaite 1976). BDSS

Social pattern and behaviour. Based on outlines supplied by R J Douthwaite and H-U Reyer, also on studies by Douthwaite (1978) in Uganda and Reyer (1980, 1982) in Kenya.

1. Mostly rather gregarious throughout the year, especially when roosting, breeding, and sometimes feeding (Robinson 1974; R J Douthwaite). In Bahrain, flock of over 80 reported in December (Meinertzhagen 1954). Outside breeding season in East Africa, occurs singly, in twos, or in small groups along lake shores, most groups representing family parties. (Reyer 1980.) BONDS. Monogamous mating system. Pair-bond lasts as long as mates survive, unless ♂ cuckolded by secondary helper (see below). Due to high adult mortality, only 23% of 56 pairs remained intact for more than 1 season, and none lasted more than 3. Little information about bonds outside breeding season, but 1 marked pair stayed together throughout the year (H-U Reyer). Both parents incubate (though mainly ♀), and care for young. Fledglings usually feed independently at 2–3 weeks though contact with parents may continue much longer (H-U Reyer; see also Relations within Family Group). In Kenya, probably most ♀♀ breed at 1 year, but only c. 11% of ♂♂ (n=18). Shortage of ♀♀ prevents most ♂♂ from breeding until 2–3 (H-U Reyer). In Uganda, Kenya, and Zambia, local variation in sex-ratio 1·5–2·5 ♂♂ : 1 ♀, overall 1·8 : 1 (n=1684); ratio apparently equal in young (Reyer 1980) and bias in older birds possibly caused partly by higher mortality of dispersing juvenile ♀♀ (H-U Reyer), partly by higher predation on incubating ♀♀ (Douthwaite 1973, 1978; Reyer 1980). Surplus of ♂♂ evidently linked to flexible system of 'nest-help'. Breeding birds can have 'primary' helpers only, primary and 'secondary' helpers, or secondary helpers only (Reyer 1980, 1982; see also Douthwaite 1978). Primary helpers are mostly 1-year-old, (sometimes 2-year-old) sons of at least 1 bird of resident breeding pair, which they accompany from start of breeding season; primary helpers feed resident ♂, assist him to feed resident ♀ during courtship and incubation, assist in defending nest-site against rivals and potential predators, and in feeding young. At 2 colonies, Kenya, c. 1 in 3 pairs had 1 primary helper; 2 primary helpers recorded only twice (n=82 helpers: Reyer 1980, 1982; H-U Reyer). Secondary helpers are ♂ non-breeders and failed breeders, apparently unrelated to resident pair, and not firmly associated until 3–4 days after the latter's young have hatched; secondary helpers are apparently attracted mainly by adult ♀♀ which, in the case of surplus non-breeders, they try to feed before egg-laying. Would-be helpers are persistently driven off by resident ♂ until 3–4 days after his young have hatched. Thereafter they are accepted in 2–4 days during which they first bring fish to resident ♀ (H-U Reyer) but then assist in feeding young and defending nest. Tolerance of secondary helpers by resident ♂♂ varies between pairs and colonies depending on number of offspring in nest, and their ability to raise them; number of helpers greater (up to 3–4) when local conditions produce low feeding rates and demand high energy expenditure of parents (Douthwaite 1978; Reyer 1980, 1982, which see for details). Secondary helpers frequently breed in place they helped at the previous year, sometimes with the ♀ they helped (Reyer 1980, 1982). BREEDING DISPERSION. Varies with availability of food and nest-sites. Along permanent rivers where both plentiful, pairs solitary, defending linear territories for both breeding and feeding. For defence of area around feeding-perch, see Antagonistic Behaviour, below. Where feeding and breeding sites separated, and latter scarce (e.g. in large freshwater lakes of East Africa), usually colonial, pairs defending only small territory around nest-entrance (Reyer 1980; H-U Reyer). In Kashmir, *leucomelanura* forms numerous colonies of not more than c. 12 pairs each (Phillips 1946). Regularly colonial on River Tigris (Iraq), one colony 'consisting of hundreds' (see Ticehurst et al. 1922). In Sierra Leone, largest colonies in sea-cliffs—of at least 30 pairs; single pairs occur where suitable sites widely separated (G D Field). In East Africa, most colonies less than 20 pairs, but some, Uganda, may exceed 100 pairs (Reyer 1980; R J Douthwaite). In colonies, mean distance between adjacent nest-holes 5·2 m at Lake Victoria, 1·6 m at Lake Naivasha (both Kenya); minimum 0·5 m (Reyer 1980; see also Jourdain 1935). Nest usually abandoned after fledging; rarely, used for re-nesting same season, either by same or different pair (Reyer 1980; H-U Reyer). In some areas, same holes used year after year, though not known if by same pairs; in other areas, new holes excavated every year, even if previous ones intact. In cases of divorce between seasons, ♂ returns to former nest-site while ♀ moves (H-U Reyer). ROOSTING. Adults and immatures share common nocturnal roost throughout the year. In Kenya, roost of c. 100 birds on fallen tree (*Scopus* 1980, 3, 107–20); in Uganda, one regularly (November–December) over 200 (Douthwaite 1973; see also Meinertzhagen 1954, Douthwaite 1982). Roost usually in date palms, or in banks (Meinertzhagen 1954), also papyrus swamps (H-U Reyer). At Lomé (Togo), apparent roost area, October–December, comprised up to 50 holes excavated high on sandstone cliff; although no direct evidence, birds thought to breed in same holes at other times of the year (Robinson 1974). In Kashmir, nest-holes likewise used for roosting in winter (Phillips 1946). In Uganda (see above), birds approached roost-site by stages, often attempting to feed at each staging post (R J Douthwaite; see also Sugg 1974 for evidence of pre-roost feeding); almost all birds arrived at roost within 20 min of sunset. Birds loaf by day in much smaller groups, usually in shady tree on lake shore (R J Douthwaite). May bathe by dipping in and out of water, resorting to loafing site for preening (e.g. Dharmakumarsinhji 1955). Off-duty bird, most often ♂ during laying and incubation, guards and loafs on ground outside nest-entrance (Douthwaite 1978) or on sentinel perch nearby (Priest 1934; Greaves 1937). At Dal Lakes (Kash-

A

mir), most active 08.00–09.00 hrs and 17.00–18.00 hrs, but no clear peaks; most feeding 08.00–09.00 and 16.00–18.00 hrs (Pring-Mill 1974).

2. Alarmed or excited bird typically flicks tail up and down (Vincent 1946a; McLachlan and Liversidge 1970; Robinson 1974). Birds disturbed at communal roost scatter as when at colony (see Parental Anti-Predator Strategies). On 2 occasions, group consisting of pair and 4-month-old ♂ offspring pursued and mobbed Marsh Harrier *Circus aeruginosus* (H-U Reyer). FLOCK BEHAVIOUR. No details but see Heterosexual Behaviour (below). ANTAGONISTIC BEHAVIOUR. Birds defend individual distance (e.g. in display groups: see below), nest-holes, prey items, and sometimes territories, by threatening, fighting, and chasing. Disputes may arise over favoured perches and area surrounding them, especially when perches essential for hunting. Where fish caught predominantly by hovering, no feeding territories held (Dharmakumarsinhji 1955; Reyer 1980; H-U Reyer). Bird has 2 threat postures: in defensive Wings-spread posture (Fig A), stands upright and half-extends wings in vertical plane; in aggressive Forward-threat posture (Fig B), leans forwards, directing bill towards opponent, with wings half raised and tail fanned (Douthwaite 1978; H-U Reyer). Both postures usually accompanied by Advertising-calls (see 5 in Voice), Forward-threat posture also by Aggressive-calls (see 6 in Voice). Rival may submit by turning away, often with bill pointing downwards, and giving Appeasement-call (see 7 in Voice). Alternatively, threatened bird may adopt a threat posture: if Wings-spread posture adopted, the 2 birds may then jump at each other; if Forward-threat posture, the 2 peck and snap, leading to a fight (H-U Reyer). Outright fighting rare except during grounddisplay (see Heterosexual Behaviour, below) at start and end of breeding season. Fighting birds, usually ♂♂, may briefly grasp each other's wing, or grab and twist at bill (R J Douthwaite). Birds seeking secondary helper status at nest (see Bonds, above) often threatened and chased off by resident ♂♂; chasing prolonged if pair have few nestlings and no need of help (Reyer 1980). In period of food shortage, Botswana, flying birds often attacked and disrupted feeding activity of others (Douthwaite 1982). Dispute over food may lead to brief tussle over fish in mid-air (H-U Reyer). HETEROSEXUAL BEHAVIOUR. (1) Pairbonding behaviour. Return to colonies marked by groups of 3–8 birds chasing high over colony area (Douthwaite 1978; R J Douthwaite). At roost-site, Sierra Leone, 12 birds performed aerial display (not described), accompanied by calling (G D

B

Field). When burrows being excavated, flying flocks land and display on open ground; ♂♂ outnumber ♀♀, but both sexes display in similar fashion. 1 or more birds adopt Wings-spread posture and turn about within group, giving Advertising-calls. Display lasts a few seconds and is usually repeated several times. If display directed at particular bird, latter responds with same display or else Appeasement-call (Douthwaite 1978; R J Douthwaite). Ground-display here thought to establish and maintain bonds with mates and primary helpers. When Advertising-call given (usually without any threat posture), serves as Meetingceremony—during and after pair-formation—between members of pair and with accepted helpers (H-U Reyer). Behaviour of excited bird, apparently in courtship, said to include raising of crown feathers, wing-quivering, tail-flicking, and calling (Phillips 1946). In alleged courtship display, bird flew to and fro above perched bird (presumed mate) with gradually lengthening swoops, just above bird's head (Hutson 1954). (2) Courtshipfeeding. Occurs most often outside nest-hole (H-U Reyer). Begins during nest-excavation and ends when young hatch. Immediately before and during laying, ♀♀ become very passive, seldom flying except to receive food from mate (Douthwaite 1978) who may supply all her food during laying (R J Douthwaite). Soliciting ♀ gives Begging-call (see 8 in Voice) and often also adopts Begging-posture: body upright, with crest raised and bill pointing almost vertically down (H-U Reyer). ♂ holds fish such that head nearest bill-tip, and, usually after brief tussle, transfers it to ♀ (R J Douthwaite), often to accompaniment of calls (H-U Reyer: see 9 in Voice). ♀ then pecks at ♂ who leaves (H-U Reyer). ♂ may hover to pass fish to perched ♀ (Robinson 1974). Primary helpers regularly feed ♀♀ (Reyer 1980, 1982). One 4-month-old ♂ offered fish to ♀ parent outside breeding season (H-U Reyer). (3) Inter-♂ feeding. Resident ♂ sitting outside nest-entrance during laying and incubation sometimes approached by primary helper or intending secondary helper offering fish. Resident ♂ usually seized fish and most often ate it after a tussle, whereupon other departed. Occasionally ♂ refused fish, responding with Appeasement-call (Douthwaite 1978). If accepted, fish sometimes passed to mate (Reyer 1980). ♂–♂ feeding (and ♂–♀ feeding: H-U Reyer) occurs throughout the year, sometimes well away from colony, but often outside nest in which ♀ incubating (Douthwaite 1978). (4) Mating. Usually occurs near nest-hole, often after courtship-feeding (H-U Reyer); most often in first 2 hrs of daylight (R J Douthwaite). Soliciting ♀ adopts Begging-posture but with tail slightly raised and body less upright; also gives Begging-call. ♂ approaches ♀ in Forward-threat posture, uttering Aggressive-call, and ♀ then turns away and crouches. ♂ mounts and, during copulation, grasps ♀'s forehead with bill, and beats wings. During copulation, calls given (see 10 in Voice). Copulation lasts *c*. 7–10 s, and afterwards, ♀ often drives ♂ away by pecking (H-U Reyer). (5) Behaviour at nest. No information. RELATIONS WITHIN FAMILY GROUP. Eyes of young open over 5–9 days. Food-call of young changes after a few days (Douthwaite 1978; R J Douthwaite: see Voice). Young fed from 1st day by ♂, and by helpers if present. Later, when brooding declines, ♀ usually takes increasing share in feeding (Reyer 1980). However, parental role varies with degree of helper support; sometimes, where up to 4–5 ♂♂ involved (including ♂ of pair), ♀ plays no part in feeding young (Douthwaite 1969); at other nests, where brood small, pair may raise young unaided by secondary helpers (Douthwaite 1969; Reyer 1980). When inadequate food given to captive brood of 5, strong inter-sibling rivalry arose, heavier nestlings pushing others aside, and thereby receiving most food. Young accept food as readily from helpers as from parents (Reyer 1980). For evidence of helpers feeding recently fledged young, see Douthwaite

(1978). Parents coax young to fledge with Contact-call (see 4 in Voice). Newly fledged young remain near nest-hole for 1–2 days during which they usually perch to receive food. Thereafter they fly towards parents approaching with fish; often after long chase, parent transfers fish in mid-air but offspring returns to perch to swallow it. Within 3 days of fledging, young capable of shaking and battering fish; in colonial dispersion, young then leave colony-area and stay near fishing grounds, and within 2 weeks can dive and successfully catch fish (Douthwaite 1978; Reyer 1980; H-U Reyer). Juveniles fed by adults usually for less than 1 month after fledging and can be independent at 2–3 weeks (H-U Reyer). When food became short after young self-feeding at 2 months, they begged with 'pitchek' call (see Voice), sometimes successfully, from presumed parents (R J Douthwaite). Juvenile ♀♀ appear to leave parents and natal area at 3–4 months, when they start moulting. In one case, juvenile ♂ remained with parents throughout the year; others arrive in colony with parents at start of breeding season, suggesting family bonds maintained during the year after fledging; such offspring typically serve as primary helpers if they fail to find mates (Reyer 1980; H-U Reyer). ANTI-PREDATOR RESPONSES OF YOUNG. No information. PARENTAL ANTI-PREDATOR STRATEGIES. (1) Passive measures. No information. (2) Active measures: against birds. If predator approaches colony, alerted bird gives Alarm-call (see 2 in Voice), whereupon birds in the open perform silent, low-level rush out over water, often dipping in then flying off in a different direction (R J Douthwaite). (3) Active measures: against man. On approach of intruder, alarmed birds gave Alert-call (see 1 in Voice) while circling slowly with bursts of 3–4 rapid and exaggerated wing-beats alternating with glides (Douthwaite 1978; R J Douthwaite). Parents with young also give Alarm-calls (H-U Reyer). (4) Active measures: against other animals. Domestic dog treated as man. In Uganda and Kenya, snakes, ground squirrel *Xerus*, and mongoose *Herpestes* elicit Alert-call and are swooped on by 2 or more birds (only those with eggs or young: H-U Reyer), and occasionally struck on tail; monitor lizard *Varanus niloticus* also struck on head (Douthwaite 1978; R J Douthwaite, H-U Reyer).

(Figs A–B from drawings in Douthwaite 1978.)   EKD

Voice. Freely used, especially in breeding season. Following scheme compiled from outline and recordings supplied by H-U Reyer from studies in Kenya, supplemented by outline supplied by R J Douthwaite from studies in Uganda.

CALLS OF ADULTS. (1) Alert-call. High-pitched 'quick' or 'quick-ick', repeated irregularly (Douthwaite 1978); rendered 'kwik-kwik' (Mackworth-Praed and Grant 1962) or 'kik-kik' (Serle *et al.* 1977). In recording by P A D Hollom of 2 birds feeding, call has liquid quality (P J Sellar). In recording (Fig I), a staccato 'trit' at sporadic, fairly short intervals (E K Dunn). Frequently given by single birds shortly before take-off, while flying, and on alighting; with increasing disturbance and threat, progressively merges with call 2 (H-U Reyer). (2) Alarm-call. Low-pitched 'jerp' (Douthwaite 1978). In recording by H-U Reyer, sound resembles 'trrr trrr trrr' (E K Dunn). At low intensity, given irregularly but rate of repetition and volume increase when danger greater (H-U Reyer). Often given by parents with young of any age in presence of intruder (Douthwaite 1978; H-U Reyer). (3) Distress-call. Shrill, rapidly repeated 'preepreepreepree' (E K Dunn), given by trapped bird (H-U Reyer). (4) Contact-call. 'TREEtiti TREEtiti' (E K Dunn: Fig II) or 'kittle te ker' (Douthwaite 1978), repeated every 0·5–2 s by bird arriving at or departing from nest; also given by parents coaxing young to fledge, and by birds flying towards roost. Probably signals that bird is arriving or leaving (or intends to do so), and may also express invitation to follow (H-U Reyer). (5) Advertising-call. High-pitched, staccato 'CHICKkerker' (Fig III), repeated irregularly and given in defence of nest-site or perch, often in a threat posture; also often in Meeting-ceremony (R J Douthwaite, H-U Reyer; see Social Pattern and Behaviour). (6) Aggressive-call. Shrill, repeated 'shreeur', usually given by ♂ in Forward-threat posture when confronting rival, or approaching mate with intention of copulating. Similar call (in recording, an extended 'shreeee': E K Dunn) given by bird in front of nest-hole as mate approaches it to excavate. May express conflict between approaching and staying/retreating (H-U Reyer). (7) Appeasement-call. Loud, crescendo then diminuendo 'werk ... werk ... werk werkwerkerkerk erk' or 'sooip ... sooip ... sooip sooipsooipipipipip' (Douthwaite 1978). In recording (Fig IV), begins with a few 'werk-' sounds, then breaks into series of 'sooip' sounds which increase in rate of delivery, ending in volley of 'ip-' sounds (E K Dunn). Given in response to call 5, or by perched birds after call 4, apparently mainly by subordinate individuals, thus more often by ♀♀ and helpers than by resident ♂♂ (H-U Reyer; see also Calls of Young, below). (8) Begging-call. In recording by H-U Reyer a sharp, brief 'pi-chee' (E K Dunn), repeated a few to several times. Regularly given by ♀ to solicit food from mate, primary helpers, and other birds. Also given by ♀ inviting copulation, when probably serves partly to appease ♂ (see call 6), as it may also when given by subordinate ♂ on approach of more dominant ♂ (H-U Reyer). (9) Courtship-feeding calls. In recording, a complex sequence of soft warbled sounds and chirps, given probably by both birds during transfer of fish. One sound, slightly similar to call 6, increases in occurrence when pair tussle over fish (H-U Reyer). (10) Copulation-call. Very soft 'pre-' or 'pirree-' sounds given sporadically during copulation; not known whether by one or both participants (H-U Reyer).

CALLS OF YOUNG. Food-call of young 1–2 days old a grating, repeated 'scare'. After *c.* 2 days, and up to a few

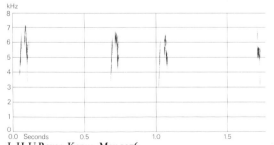

I H-U Reyer Kenya May 1976
For Figs II–IV, see Volume V, pp. 1054–5.

days old, young give repeated, high-pitched 'choop' when hungry. Thereafter, and up to a few days after fledging, call changes to a loud, penetrating, continuously repeated 'cherr erh'; 1st 'cherr' produced on exhalation, 'erh' on inhalation; given just before being fed. From c. 10 days after fledging, young repeatedly call 'pip weep' or 'pitchek' at intervals of c. 5 s when flying, or when perched and another bird flies past; probably a food-begging call (R J Douthwaite). For ages at which adult calls develop in fledged young, see Douthwaite (1978). H-UR, EKD

Breeding. SEASON. Egypt: eggs laid March–May (Dementiev and Gladkov 1951a; Etchécopar and Hüe 1967). SITE. Tunnel in bank, over water or dry ground. Colonial or solitary. Nest: excavated tunnel 80–250 cm long, shorter in hard ground, with chamber at end c. 45 × 24 × 15 cm high (Douthwaite 1978). Unlined, but with increasing litter of fish-bones. Building: by both sexes. EGGS. See Plate 98. Short elliptical, smooth and fairly glossy; white. 29 × 24 mm (26–32 × 22–25), $n = 133$; weight 8·2 g (6·7–10·4), $n = 34$, Uganda (Douthwaite 1978). Clutch: 4–5 (1–7); average of 22 clutches, Uganda, 4·9 (Reyer 1980). Small and large clutches may be replacements, latter containing eggs from previous clutch (R J Douthwaite). One brood, possibly 2. Replacements laid after egg loss. Laying interval 1 day. INCUBATION. About 15 days. Begins with 1st egg; hatching asynchronous, over c. 3 days (Douthwaite 1978). By both sexes, but more by ♀, who sits through night and during part of day (Douthwaite 1978). YOUNG. Altricial and nidicolous. Cared for and fed by both parents. FLEDGING TO MATURITY. Fledging period 23–26 days, Uganda (Douthwaite 1978). Become independent within 2 months of fledging. Age of first breeding 1 year, though some ♂♂ not until 2 (Douthwaite 1978). BREEDING SUCCESS. No data from west Palearctic. Of 58 clutches, Uganda, 52% hatched, and 50% fledged, with average 4·5 young per brood. Fledging success significantly increased by helpers (see Social Pattern and Behaviour): 39% of hatched birds fledged with no helpers involved, 78% with 1 helper, and 100% with 2 (Douthwaite 1978; Reyer 1980).

Plumages (nominate *rudis*). ADULT MALE. Forehead and crown black; sides of feathers narrowly edged white when fresh, slightly wider towards hindcrown, appearing streaked white when plumage fresh, black with limited traces of white streaks on forehead and hindcrown when plumage worn. Large triangular white patch on lores, extending into white streak over eye towards nape (narrow just above eye, wider above ear-coverts); nape white with limited black streaking, but white usually hidden below mainly black elongated crown feathers. Black patch in front of eye and below eye from gape over ear-coverts, narrowly streaked white below eye and on ear-coverts; remainder of head and neck white with narrow black streak from ear-coverts down sides of neck and dusky streaks on central hindneck. Feathers of mantle, back, and rump, scapulars, and upper tail-coverts with basal half white and terminal half black, latter with white fringe 1–4 mm wide at tip; black of tail-coverts and shorter scapulars with 1 white blob on each side, longer scapulars with 1–2 blobs. In fresh plumage, upperparts appear black with marked white scaling; when worn, white feather-fringes partly lost by abrasion and upperparts appear more uniform black, but much white of feather-bases sometimes then exposed. Chest with distinct black band, wide at sides, narrowing slightly towards centre (feathers narrowly tipped white when fresh; some grey or white of feather-bases visible when worn), a narrower 2nd black band below it across breast; remainder of underparts white, but some black blobs and streaks on lower flanks and thighs, sometimes giving hint of 3rd partly developed band on sides of lower belly. Basal half of central tail-feathers (t1) white with some black blobs or broken bars, distal half black with terminal 5 mm white; black on other tail-feathers gradually reduced towards outer, forming black band of 15–20 mm wide on tip of outermost, again with terminal 5 mm white and with white blotches at side, neighbouring 1–2 feathers often also blotched white in middle of black of inner web (giving indication of white band in middle of black on undertail); white middle portion of variable number of outer feathers with 1–2 black blotches (giving indication of black bars proximal to broad black band on uppertail). Basal $\frac{2}{3}$ of primaries white, terminal $\frac{1}{3}$ black with narrow white tip (widest on innermost); black of tip extends in tapering point to shaft; outer web of (p8–)p9–p10 black except for narrow white outer edge on p10 and similar edge to bases of others; basal $\frac{2}{3}$ of outer web of p6–p8 either all white or partly spotted black. Secondaries white, terminal $\frac{1}{3}$ of outer web of middle ones with 2–3 black blotches narrowly connected by black along shaft; terminal half of outer web of s3–s4 black with some white blotches, black extending partly to inner web; outer web of s1–s2 largely black, and black on terminal half of inner web extensive; tertials with terminal halves largely black; all secondaries with white tip, c. 0·5 cm wide on middle ones, narrower on outer and inner. In closed wing, secondaries white with 2–3 black subterminal bands, bordered by mainly black outermost secondaries and tertials; in spread wing, middle secondaries largely white with bands broken into 2–3 rows of black blotches. Greater upper primary coverts and bastard wing largely black, lesser mainly white. Greater upper wing-coverts white with black subterminal blotches; outermost and tertial coverts more extensively tipped black like corresponding secondaries and tertials; longer median upper wing-coverts white (forming white panel across wing), outermost with large black tips; remaining smaller upper wing-coverts black with broad white fringes. Marginal coverts, under wing-coverts, and axillaries white. ADULT FEMALE. Like adult ♂, but lower band across breast absent and upper band on chest restricted to large black patches at sides, either interrupted by white in middle or (occasionally) connected by narrow black line. NESTLING. Naked at hatching. Feather-pins appear at 7th day (at 3 weeks on belly and vent); flight- and tail-feathers breaking out of sheaths at 11–13 days; scapulars, wing-coverts, and tail-coverts fully grown at 15 days; eye well open by 9 days (Douthwaite 1978). JUVENILE. Like adult ♀, but white patch on lores and white supercilium partly speckled black; loral patch appearing larger but less sharply defined and supercilium less distinct, latter occasionally almost absent; mantle, scapulars, back, rump, and upper tail-coverts more extensively black, rump and upper tail-coverts especially with less white of feather-bases visible; white cheeks, lower throat, and sometimes upper throat and upper breast with black fringes or spots on feather-tips (occasionally, almost absent); black band across chest with much grey of feather-bases visible, not solid black and usually not interrupted in middle as in many adult ♀♀; no trace of adult ♂'s 2nd band across breast; flanks more profusely streaked, without adult's limited number of bold black spots; flight-feathers as adult, but less white on tips of primaries (but abraded soon at any age), more extensively black on outer web of p6–p10; secondaries largely black, only limited amount of white on base and on inner border of inner web of middle and inner secondaries, outer webs with 2–4 rather small white blobs along outer edge only, usually absent on s1–s3(–s4). Upper wing-coverts as adult, but black often more extensive; median coverts usually blotched black, black on tips of innermost reaching tips, unlike adult. Much variation in amount of black on upperparts and upperwing; some juveniles as white as adult, while a few adults show relatively limited white, approaching dark juveniles; ageing on amount of white alone not reliable, except for extremes. FIRST ADULT. As adult, separable only when some juvenile upper wing-coverts (darker than fresh neighbouring ones) or all juvenile primaries (all equally fresh or worn, not mixture of old and new as in adult) retained. Usually inseparable once primary moult started at c. 6 months old.

Bare parts. ADULT AND JUVENILE. Iris dark brown. Bill black, paler at base of lower mandible; mouth blue-grey in adult, pink in juvenile at fledging. Leg and foot black, soles pink, yellow, or black. NESTLING. Pink at hatching; future feather-tracts of crown, nape, scapulars, tail, and wing darken to grey from 4th day, remainder (except belly and vent) from 6th; bill, leg, and foot (except soles) blackening from c. 1 week. (Sugg 1974; Douthwaite 1978; BMNH, RMNH.)

Moults. ADULT POST-BREEDING. Partial; primaries ascendant and descendant, starting from p5. Sequence of primaries 5-6-7-4-8-3-9-2-10-1; starts with p5 at end of nesting period; as moult slow (within each ascendant and descendant series, each feather usually not shed before neighbouring one full-grown), usually not completed when following nesting period starts and moult then suspended, e.g. with p9–p10 and p1–p2 still old. In following moult period, resumed from point of suspension and fresh series also starts again with p5, e.g. moult may start after nesting with p9, p2, and p5. (Douthwaite 1971a.) In sample of c. 100 west Palearctic birds (BMNH, ZFMK), active wing moult mainly May–January in Egypt, July–March in Middle East; 1–2 primaries growing simultaneously in one wing, rarely 3–4 (average 1·6, $n=37$); upon suspension, 4–10 primaries new (average 6·1, $n=48$); old primaries retained after suspension usually replaced in next moulting season, and only 2 birds showed moult pattern explainable only by retention of some feathers until 3rd moulting season; after suspension during nesting, new series did not start automatically, in contrast to findings of Douthwaite (1971a)—p5 of new series usually lost when p8 or p9 (or p10) of previous series growing, and when moult suspended with (e.g.) up to p6 (new series) and up to p9 (old series) new, moult continued in next moult season with p7 and p10 only, without starting a fresh (3rd) series again with p5. Moult sequence thus probably

$$5 \begin{cases} 6\text{--}7\text{--}8\text{--}9\text{--}10 \\ 4\text{--}3\text{--}2\text{--}1 \end{cases}$$

$$5 \begin{cases} 6\text{--}7\text{--}8\text{--}9\text{--}10 \\ 4\text{--}3\text{--}2\text{--}1 \end{cases}$$

→

and this can be interrupted by suspension at any point. Replacement of single set of primaries takes 180–193 days when not interrupted (Douthwaite 1971a). Secondaries replaced ascendantly and descendantly from s13, and ascendantly from s1, sequence 13-12-11-10-9-14-8-7-1-6-2-5-3-4; when not suspended (though it usually is), whole replacement takes c. 260 days; a new series usually starts before previous one completed (Douthwaite 1971a). In west Palearctic birds, body and tail apparently completely renewed June–December. POST-JUVENILE. Complete, but moult suspended once nesting started and completion may require 2 moulting seasons. In west Palearctic, head and body first, starting with scattered feathers of neck, mantle, and outer scapulars from April–August; head, body, and tail new October–December, but some wing-coverts usually still juvenile then. Sequence and duration of flight-feathers as in adult. In captive birds, secondaries started c. 160 days after hatching, primaries c. 180 days (Douthwaite 1971a). In west Palearctic, p1 shed July–February; many 1-year-olds retain old outer and inner primaries and all retain part of outer secondaries when moult suspended during breeding season.

Measurements. Nominate *rudis*. Turkey, Cyprus, Levant, and Iraq, all year; skins (BMNH, RMNH, ZFMK, ZMM). Bill (F) from tip to forehead (exposed culmen on average 3·5 less), bill (N) from tip to distal corner of nostril, both in adult only.

WING AD	♂	146	(1·50; 13)	145–149	♀ 147	(1·61; 19) 144–151
BILL (F)		62·2	(2·64; 15)	58–67	62·1	(1·94; 14) 60–66
BILL (N)		49·3	(2·37; 15)	47–53	50·5	(1·63; 13) 48–54

Sex differences not significant. Juvenile wing on average c. 3 shorter than adult; juvenile bill full-grown once 1st adult plumage on body attained. In East Africa, bill full-grown at 3–4 months (Douthwaite 1978).

Egypt from Nile delta and Faiyum to 25°N, all year; skins (BMNH, RMNH, ZFMK; Giza Zoological Museum *per* S Goodman and P L Meininger).

WING AD	♂	140	(2·03; 25)	137–144	♀ 141	(2·85; 24) 138–145
JUV		140	(3·69; 8)	134–144	139	(3·18; 8) 135–143
TAIL AD		73·2	(2·33; 17)	69–77	74·4	(2·82; 16) 70–79
JUV		74·2	(3·12; 6)	70–78	70·9	(2·02; 4) 68–73
BILL (F)		61·8	(2·83; 22)	58–68	60·7	(2·46; 22) 57–65
BILL (N)		49·3	(2·77; 22)	46–55	48·6	(2·36; 22) 45–54
TARSUS		11·3	(0·38; 13)	10·9–12·2	11·6	(0·71; 12) 11·0–12·6
TOE		20·3	(0·62; 11)	19·7–21·4	20·0	(0·97; 11) 18·5–21·2

Sex differences not significant. Wing of Egyptian birds significantly shorter than those of Middle East ones, 5 adults from Middle East below 146 and 2 below 145 (mainly from Iraq), and only 3 Egyptian birds over 144.

Sexes combined, all year, sources as before. Nominate *rudis*: (1) Middle East; (2) northern and Middle Egypt; (3) Upper Egypt (Aswan area) and northern Afrotropics (Sénégal to Ethiopia); (4) Liberia to Zaïre and northern Angola; (5) southern Kenya and Tanzania; (6) South Africa (McLachlan and Liversidge 1970; RMNH). *C. r. leucomelanura*: (7) India, Ceylon, and southern China.

	WING AD			BILL (F) AD		
(1)	147	(1·56; 32)	144–151	62·2	(2·29; 29)	58–67
(2)	141	(2·45; 49)	137–145	61·2	(2·68; 44)	57–68
(3)	139	(1·10; 14)	137–141	58·4	(2·76; 13)	56–60
(4)	134	(2·70; 32)	128–138	62·7	(4·06; 15)	60–69
(5)	132	(4·10; 11)	128–137	61·7	(3·18; 11)	58–66
(6)	140	(— ; 38)	132–146	61·0	(1·32; 4)	59–62
(7)	139	(4·11; 7)	135–145	66·4	(3·34; 7)	63–72

Weights. Nominate *rudis*. Iran, February: ♂ 95 (Diesselhorst 1962). Adult, Kenya, March–November: ♂ 82·4 (6·03; 189) 68–100, ♀ 86·4 (7·38; 96) 71–110 (Sugg 1974). Adult, Zaïre, August–January: ♂ 65 (4) 56–72, ♀♀ 61, 69 (Verheyen 1953). At hatching, average c. 8; at 10th day, 55·1 (15·2; 9) 33–80; at 15th, 101·0 (7·44; 4) 93–111; peak reached on c. 19th day, 122 (3) 110–134, fledging on 24th–25th at 94·6 (3·78; 5) 89–99 (Douthwaite 1978).

Structure. Wing rather long and broad, tip fairly rounded. 10 primaries: p8 longest, p9 0·5–3 shorter, p10 and p6 9–13, p7 1–5, p5 19–24, p4 26–30, p1 38–46. Outer web of (p7–)p8–p9 and inner web of p8–p10 slightly emarginated. Tail rather long, tip square; 10–12 feathers. Bill long, straight, sharply pointed; gradually tapering towards tip or with slightly convex gonys; wide and deep at base, but not as bulbous as in *Halcyon* kingfishers, with middle and tip more strongly compressed laterally. Nostrils rather small, narrow, partly covered by thin flap above. Feathers of hindcrown and nape narrow and elongated, forming ragged crest. Leg and foot short and slender, lower tibia and tarsus bare. Soles flattened, front toes partly joined at base; outer toe *c.* 88% of middle, inner *c.* 68%, hind *c.* 54%. Middle claw rather long, others short, strongly curved.

Geographical variation. Within Middle East and Africa, no variation in colour (except between individuals), but marked in size (see Measurements). Middle East birds distinctly larger than those of Africa (wing 144–151, tail 78–82), bill slightly heavier at base and distinctly deeper in middle, gonys more markedly curved. Birds of northern Afrotropics (including those of southern Egypt) tend to have short bill. Birds of West, central, and East Africa have distinctly shorter wing than elsewhere in Africa, and should perhaps be separated as *bicincta* (Swainson, 1837). *C. r. leucomelanura* from southern Asia differs by completely white tail-base, without traces of bars proximal to broad black band across tip, except sometimes on 1–2 outer feathers; no white spots in middle of black band on outer tail-feathers (thus no broken white bar in black tip of undertail). White feather-tips on upperparts slightly wider, crown more heavily streaked white; fewer black blotches on white of outer web of p5–p8; black marks on upper wing-coverts and underparts often larger. Some geographical variation within *leucomelanura*: birds from north-east China (sometimes separated as *insignis* Hartert, 1910) average larger than Indian birds, with bill deeper at base and middle (Vaurie 1959e). *C. r. travancoreensis* from Kerala south to Cape Comorin (south-west India) similar to *leucomelanura*, but upperparts appear darker, white being less extensive and tinged grey; flanks more profusely spotted black (Ali and Ripley 1970). CSR

Ceryle alcyon Belted Kingfisher

PLATES 67 and 68
[between pages 686 and 687]

Du. Bandijsvogel   Fr. Alcyon ceinturé   Ge. Gürtelfischer
Ru. Ошейниковый зимородок   Sp. Alción   Sw. Bälteskungfiskare

Alcedo alcyon Linnaeus, 1758. Synonym: *Megaceryle alcyon*.

Monotypic

Field characters. 28–35 cm, wing-span 47–52 cm; bill 5 cm. Largest kingfisher to occur in west Palearctic; size of Jackdaw *Corvus monedula*, and up to 30% larger than Pied Kingfisher *C. rudis*. Huge, powerful, broad-winged kingfisher, with large crested head, dark grey upperparts contrasting with white throat and almost complete neckcollar, dark chest, and white underbody. Flight powerful, with action like that of *C. rudis*. A noisy bird. Sexes dissimilar; no seasonal variation. Juvenile separable.

ADULT MALE. Large, black, dagger-shaped bill. Head, broad chest-band, and upperparts dusky blue; chin, throat, almost complete collar round neck, and most of underbody white. At close range, small white marks visible above and below eye. Scapulars and all wing-feathers except outer primaries show white tips and notches on inner webs, notches large enough on middle primaries to form white patch at base of feathers. When visible, dusky blue flanks show barred lower edge. ADULT FEMALE. Differs from ♂ in having flanks and a 2nd lower and narrower chest-band markedly rufous. JUVENILE. Bill distinctly shorter and chest-band(s) mixed dusky blue and red-brown. ♂ may show vestiges of 2nd chest-band.

Unmistakable, resembling only extralimital Giant Kingfisher *Ceryle maxima* (confined to Africa south of Sahara), which is larger (*c.* 40 cm) and has extensive chestnut below. Flight of *C. alcyon* powerful, with actions closest to *C. rudis* but wing-beats even more flapping; although heavy in build, mastery of sudden hover, bounding dashes, and headlong plunges (producing loud splash) as marked as in smaller kingfishers. Behaviour typical of Alcedinidae, but will feed on invertebrates away from water. Not shy, perching conspicuously on both branches and artefacts like telephone wires.

Advertises presence with loud, harsh rattle, sounding almost mechanical and carrying far.

Habitat. In Nearctic, from arctic Alaska through boreal, temperate, subtropical, and tropical zones, ranging in Rocky Mountains above 2500 m. Breeds almost anywhere near water supporting aquatic animal populations, where bluffs, road cuts, gravel-pits, sand-banks, or other similar nearly vertical earth exposures provide suitable nest locations. Forages up to *c.* 8 km from nest-site (Johnsgard 1979). Prefers clear water, either fresh or salt: lakes, ponds, rivers, streams, and water near shore or islands, not too far from elevated perches such as trees, posts, or telephone wires. Breeding populations in rocky areas limited by availability of nesting sites (Godfrey 1966). Avoids open, arid, treeless country. Some winter as far north as there are still open streams or tidal creeks (Forbush and May

1939). In Venezuela, on sea coasts, in mangroves, rivers, lakes, and swamps to 450 m (Schauensee and Phelps 1978). Flies in lowest airspace except on long journeys.

Distribution. Breeds in North America, except in arid areas, south from treelimit (Alaska to Labrador) to California and across to Florida. Winters within breeding range where waters remain open, south to Caribbean and Panama.

Accidental. Iceland, Ireland, Britain, Netherlands, Azores.

Movements. Partially migratory. Winters as far north as water remains ice-free; hence normally withdraws from most of Canada and from northernmost inland states in USA, but some winter further north on coasts (where open water available)—regularly to south-east Alaska in west and New England in east. Migrants winter in southern half of USA, West Indies, and Central America, with small numbers reaching Panama and Caribbean coast of South America. (Godfrey 1966; ffrench 1973; Ridgely 1976; Schauensee and Phelps 1978.) Birds ringed as nestlings in Indiana and Wisconsin found November in (respectively) Texas and South Carolina, both in 1st winter (Bent 1940); a New York migrant, September, found Dominica (West Indies) in March (Bull 1974). Common as a migrant on Atlantic coasts; occasional ship records in late autumn offshore from eastern USA (Scholander 1955); occurs regularly (with overwintering) on Bermuda (Bradlee et al. 1931).

Post-breeding dispersal begins late July; gradual autumn passage at peak in USA late September to October, but continues through November; vanguard reaches southernmost wintering areas in first half October (e.g. Ridgely 1976, Schauensee and Phelps 1978), though as far north as Maryland many still transient in early November (Stewart and Robbins 1958). Return movement March to early May. 6 European (including Icelandic) records occurred September–October (2) and November–December (4), latter including 2 which wintered; Azores record in March.

Voice. See Field Characters.

Plumages. ADULT MALE. Forehead, crown, hindneck, and sides of head down to lower cheek medium blue-grey or plumbeous-blue, feathers with narrow and faint black shaft-streaks on forehead, sides of head, and hindneck; more boldly streaked black on elongated feathers of crown and nape; feathers of hindneck partially white. Small but distinct white spot on lores in front of eyes, another white spot just below eye. Chin to foreneck and sides of neck white, slightly streaked blue-grey on sides of chin and at border of blue-grey lower cheeks, tending to form dark malar streak. Remainder of upperparts including upper wing-coverts blue-grey like top of head, but virtually unstreaked, except for faintly blackish shafts; longer and outer scapulars, back to upper tail-coverts, and median and greater upper wing-coverts with narrow white tip when fresh and with narrow, well-spaced broken white subterminal bars, appearing speckled white; outer greater coverts with black wash on base of inner web. Chest with broad uniform blue-grey band (feathers narrowly edged white when fresh), no traces of rufous or dusky brown. Flanks blue-grey with variable amount of white patches and bars, sometimes largely white; grey-blue sometimes extends slightly towards sides of breast and often to thighs; remainder of underparts white. Central pair of tail-feathers (t1) blue-grey with broad black shaft-streak, latter bordered by short, narrow, and incomplete white bars, tip narrowly bordered white; black more extensive on other feathers and narrow white bars more prominent; on t6, blue-grey reduced to narrow border along outer web, white bars interrupted by black at shaft and by plumbeous-black along feather-edges. Outer web and distal $\frac{1}{3}$ of primaries black, basal $\frac{2}{3}$ of inner web white; primaries narrowly tipped white (narrowest and soon abrading on outermost), basal and middle portions of outer webs (except on outer 1–2 primaries) with rows of rather small and irregular white spots, forming bands on closed wing; basal outer edges of inner primaries tinged blue-grey; black of outer web of primaries extends in shallow sawtooth pattern on to white middle and basal portion of inner web, often forming 1–2 complete dark bars subterminally on innermost primaries. Secondaries black with narrow white tips and broad blue-grey border along outer web; inner border of innermost plumbeous-grey; tertials blue-grey with black central streak; blue of outer web with traces of narrow white bars (especially on outermost secondaries), black of inner web with distinct white bars. Greater upper primary coverts black with narrow white tip, other primary coverts blue-grey; bastard wing blue-grey with much black on feather-centres. Marginal coverts, under wing-coverts, and axillaries white; some axillaries partly marked blue-grey. In worn plumage, white feather-tips of upperparts, upper wing-coverts, tail, and chest-band largely disappear through abrasion, plumage appearing more uniform blue-grey; faint white barring restricted to rump, upper tail-coverts, t1, outer median and greater upper wing-coverts, and outer secondaries, more conspicuous barring of other tail-feathers, inner webs of secondaries, and outer webs of primaries showing in flight only. ADULT FEMALE. Like adult ♂, but additionally has rufous-cinnamon band across breast, separated from uniform blue-grey chest-band by white. Rufous band narrower in middle, sometimes partly mixed with white on central breast; extends broadly towards rufous-cinnamon flanks, blue-grey on flanks usually restricted to thighs; axillaries often all or partly suffused rufous-cinnamon. JUVENILE MALE. Like adult ♂, but blue-grey feathers of chest-band with variable amount of rufous-cinnamon on tips and centres (least so towards sides of chest); flanks usually with some rufous suffusion, especially near sides of breast and thighs. In worn plumage, hard to separate from adult, as rufous-cinnamon bleached or lost by abrasion: abraded feather-tips of chest-band tinged buff-brown or dull grey-brown (uniform blue-grey in adult), and feather-centres in centre of chest-band often still have pale buff suffusion. JUVENILE FEMALE. Feathers of blue-grey chest-band tipped and partly suffused cinnamon-rufous as in juvenile ♂ (uniform blue-grey in adult ♀), but remainder of underparts similar to adult ♀, showing cinnamon-rufous band across breast and much rufous tinge on flanks and often axillaries. Rufous breast-band often partly interrupted by white in middle and white of flanks sometimes intermixed white, ♀♀ with least amount of rufous hardly separable from those few juvenile ♂♂ which show much rufous, but feather-tips of chest-band of ♀ usually more extensively rufous than in ♂. In worn plumage, rather difficult to separate from adult ♀, as rufous of chest-band disappears through wear (see juvenile ♂). FIRST ADULT. Rather variable; either fully juvenile and heavily worn (especially tail and outer primaries, unlike adult), still with traces of buff-brown in chest-band, and feathers of head and body occasionally partly mixed with contrastingly new ones, or plumage fresh as in adult, but with some old and abraded primaries (mainly p10 or some or all of p1–p6) and secondaries (mainly s4 or s3–s5) retained (see Moults).

Bare parts. ADULT AND JUVENILE. Iris dark brown. Bill black or greyish-black, base occasionally grey or lead colour, especially on lower mandible. Leg and foot livid-slate or slate-black; in autumn juvenile, dusky bluish-grey, occasionally partly tinged bluish-flesh, rear of tarsus and soles ochre, yellow-grey, or dull flesh colour. (Ridgway 1914; RMNH, ZMA.)

Moults. Based on Stresemann and Stresemann (1966) and on *c*. 50 skins in BMNH, RMNH, ZMA. ADULT POST-BREEDING. Complete, primaries starting from 2 centres: both descendantly and ascendantly from p7, and descendantly from p1. Moult starts with loss of p7 late June to mid-July, probably shortly before young fledge, soon followed by p8, p9, and p10 (sequence occasionally 8–7–9–10, however). During growth of p9 or at about loss of p10, p6 shed, followed by p1, p2, and p5 (sequence rather irregular); p3–p4 last. By late August and early September, primary moult score 20–35 reached; in birds wintering near breeding area, moult continued and all primaries new by October or early November; in long-distance migrants, moult suspended from mid-September at score 20–35, resuming from late October on reaching winter quarters, and completing November–December. Autumn migrants suspend with (e.g.) p1–p5 and p10 (score 20) or p2–p4 old (score 35). Tail, secondaries, and body feathers start at about loss of p9 (score 11). Sequence of tail approximately 2–1–3–5–4–6, many feathers often growing simultaneously; tail new at score 20–30. Secondaries moult from 2 centres: ascendantly and descendantly from s13 or s14 (starting at about loss of p9), and ascendantly from s1 (starting at about loss of p10); s4 replaced last, at about same time as last primaries; birds suspending in autumn may have (e.g.) s12–s16 (tertials), s1 and s11–s16, or s1 and s10–s16 new. Head and body in heavy moult August and early September at score 12–30, largely completed at score 30; thus, in some suspending autumn migrants, head and body largely already new, like tail. POST-JUVENILE. Partial, extent varying. Fully juvenile until October–November. In birds wintering near breeding area, either all juvenile retained until 1st post-breeding moult, or all head and up to 60–80% of feathering of body replaced by March–May. Moult in birds wintering Caribbean more extensive; starts December–January with part of head, mantle, scapulars, and upper chest. On return to breeding area, April–May, head and body often new, usually except for scattered feathers of lower chest and breast and some tail-coverts; sometimes a few tail-feathers new; occasionally, some primaries also new (p7–p8, p7–p9, p6–p9, p6–p10, and p2–p10 recorded), and these birds had also replaced part or all tail and tertials and a few (rarely all) secondaries.

Measurements. Eastern USA and eastern Canada, April–September; skins (BMNH, RMNH, ZFMK, ZMA, ZMM). Bill (F) from tip to forehead; bill (C) exposed culmen (includes data from J V Remsen); bill (N) from tip to distal corner of nostril; all for adult only.

	♂			♀		
WING	160	(3.91; 33)	154–170	161	(2.01; 16)	157–164
TAIL	86.1	(2.42; 19)	83–90	87.0	(3.00; 14)	83–92
BILL (F)	58.6	(2.29; 16)	56–63	59.1	(2.76; 10)	56–62
BILL (C)	51.8	(2.33; 36)	48–57	53.0	(2.69; 21)	49–58
BILL (N)	44.7	(1.64; 15)	42–47	45.9	(1.94; 10)	43–48
TARSUS	11.8	(0.42; 11)	11.3–12.4	11.8	(0.45; 12)	11.2–12.5
TOE	21.5	(0.92; 9)	20.5–23.2	21.8	(1.06; 10)	20.4–23.7

Sex differences not significant. Juvenile wing, tail, tarsus, and toe similar to adult, combined above; juvenile bill not full-grown until October(–November).

Weights. Eastern North America: adult ♂♂ 154, 160; adult ♀ 142. Western North America: adult ♂ 158 (15.9; 4) 140–173; adult ♀ 157 (3) 138–178; immature ♂ 141 (4.19; 4) 137–147; immature ♀ 157 (18.3; 8) 130–190. (J V Remsen.) New Jersey, mainly September: 154.4 (14) 127–175 (Murray and Jehl 1964). California, October: adult ♀ 184 (Grinnell *et al.* 1930). Pribilof Islands, August: adult ♂ 166 (Thomson and DeLong 1969). Lean juveniles, mid-October, Curaçao (Antilles): ♂ 102, ♀ 97 (ZMA).

Structure. Wing rather long and broad, tip fairly rounded. 10 primaries: p8 and p9 longest or either one 0–2 shorter; p10 and p7 6–11 shorter, p6 16–23, p5 25–32, p4 30–38, p1 44–54. Outer web of p7–p9 and inner web of p8–p10 slightly emarginated. Tail rather long, tip slightly rounded or almost straight; 12 feathers, t6 3–8 shorter than t1. Bill rather long, straight, heavy; relatively shorter and with deeper base than Pied Kingfisher *C. rudis*, culmen and gonys slightly curved towards pointed tip, sides of upper mandible slightly more bulbous. Feathers of crown and back of head narrow and elongated, forming full crest (not on hindcrown only as in *C. rudis*). Outer toe *c*. 90% of middle, inner *c*. 71%, hind *c*. 46%. Remainder of structure as in *C. rudis*.

Geographical variation. Slight. Birds of western North America often separated as *caurina* Grinnell, 1910, differing by wing averaging *c*. 7 mm longer than in birds from eastern North America, bill *c*. 3 mm longer (Ridgway 1914); size overlap large, however, and separation impracticable. Exposed culmen of adults from western North America: ♂ 52.1 (2.37; 40) 48–58, ♀ 53.7 (3.75; 33) 46–62 (J V Remsen), thus similar to those of eastern North America cited in Measurements, and wing of some Massachusetts and Florida birds as large as largest western ones (BMNH, RMNH).

Only distantly related to Pied Kingfisher *Ceryle rudis*, differing in moults, anatomy, and behaviour. With 3 related species, perhaps better placed in separate genus *Megaceryle*, leaving *Ceryle* with single species, *C. rudis* (Fry 1980). CSR

Family MEROPIDAE bee-eaters

Small to medium-sized perching birds, with long, slightly decurved bills, small legs and feet, and usually long tails. Aerial insect prey (especially Hymenoptera) the principal food source, though *Nyctyornis* forage largely from flowers and foliage of trees. 24 species in 3 genera: (1) *Nyctyornis* (2 species, Oriental region); (2) *Meropogon* (single species—Celebes Bearded Bee-eater); (3) *Merops* (21 species, southern and central Eurasia, Africa, and Australasia). Represented in west Palearctic by 3 species of *Merops*, breeding.

Sexes of same size. Necks short. 14 cervical vertebrae. Wings pointed, flight graceful and agile, wheeling and gliding while feeding, slightly undulating in direct flight, with rapid wing-beats interspersed with brief pauses. 10 primaries, p10 much reduced or absent. Wing eutaxic. 12 tail-feathers, middle pair often elongated. Oil-gland naked. Aftershafts vestigial. Bills long, gently decurved, and laterally compressed for much of length; weak rictal bristles. Tarsus very short, feet strongly syndactyle, middle and outer toes connected to beginning of last joint, inner and middle by basal joint only. Caeca long. A single (left) carotid, except in *Nyctyornis* which has paired carotids.

Plumages generally bright, green frequent as overall colour, bright reds to yellows common in throat region, often delimited by black breast-band. Sexual dimorphism slight. Irises dark or crimson, bills dark. Post-breeding moult complete, starting shortly after breeding season and interrupted during migration in migratory species. A restricted non-breeding plumage present in some migratory species. Nestlings altricial, naked at first, appearing spiny later due to late retention of feather-sheaths.

Merops orientalis Little Green Bee-eater

PLATES 69 and 71
[between pages 734 and 735]

Du. Kleine Groene Bijeneter   Fr. Guêpier d'Orient   Ge. Smaragdspint
Ru. Малая щурка   Sp. Abejaruco pequeño   Sw. Grön dvärgbiätare

Merops orientalis Latham, 1801

Polytypic. *M. o. cleopatra* Nicoll, 1910, Nile delta and Faiyum (Egypt), south along Nile valley to northern Sudan; *cyanophrys* (Cabanis and Heine, 1860), southern Israel and western and southern Arabia. Extralimital: *viridissimus* Swainson, 1837, arid belt south of Sahara from Sénégal to Ethiopia, north to Aïr (Niger), Ennedi (Chad), and Gebel Elba (Sudan); *beludschicus* Neumann, 1910, southern Iran and east to Pakistan and north-west India; nominate *orientalis* Latham, 1801, India (except north-west and Assam); *ferrugeiceps* Anderson, 1878, eastern Assam and south-east Asia; *muscatensis* Sharpe, 1886, central and eastern Arabia.

Field characters. 22–25 cm, of which tail-streamers up to 9 cm; wing-span 29–30 cm. Head and body size only *c.* 60% that of Bee-eater *M. apiaster* and Blue-cheeked Bee-eater *M. superciliosus*; bill noticeably shorter. Small, pale green bee-eater, with slim build and long or short tail-streamers. Plumage noticeably more uniform than other Palearctic *Merops*, with throat not contrasting with body, no supercilium, and bronze tone on head, neck, and across bases of flight-feathers indistinct except at close range. Sexes similar; no seasonal variation. Juvenile separable. 2 races in west Palearctic, separable in the field in fresh plumage (2 others also occur on passage—see Geographical Variation).

ADULT. (1) Egyptian race, *cleopatra*. Basic plumage tone grass-green; head marked with narrow black eye-stripe, pale blue stripe below eye-stripe, and short black bar across centre of upper breast. At close range, bronze- or golden-brown on inner webs of flight-feathers forms distinct sheeny panel across wing, which shows black trailing edge. Under wing-coverts and underside to base of flight-feathers pale golden- or bronze-buff, creating distinctly paler fore-underwing than on west Palearctic congeners. Central tail-feathers form noticeable streamers extending 4–9 cm beyond other feathers. (2) Israeli and Arabian race, *cyanophrys*. Basic plumage tone less golden, bluer. Forehead, supercilium, and chin to upper breast brilliant blue, with breast-bar dark blue to purple-black and more triangular in shape than in *cleopatra*. Central tail-feathers extend by 1–2·5 cm, forming only points. JUVENILE. All races. Central tail-feathers not elongated. Plumage bluer in tone than adult, with head-marks less distinct.

Can be confused with *M. superciliosus* at distance or in glimpse but smaller size, rather pale underwing, and lack of pale supercilium diagnostic. Flight typical of genus but less graceful than larger congeners, usually lacking slow wheels and sudden accelerations; hawks insects in manner of flycatcher (Muscicapidae), with similar habit of sudden sally from and return to perch and frequent descents to prey on ground. Gait restricted to shuffle, due to short legs.

Social but not gregarious, forming only small parties even on migration. Commonest call a soft repeated 'teerp' or 'tree', with chittering tone in chorus; shrill chatter in alarm.

Habitat. In lower subtropical latitudes, avoiding cooler and wetter conditions and mainly inhabiting lowlands, although locally ascending to *c.* 1500–2000 m in India. In open country interspersed with light forest and cultivation, roosting communally in bamboo clumps or leafy trees and making sorties after winged insects from perches on telegraph wires, fence-posts, dead branches, bare ground, or, sometimes, backs of grazing cattle. Perhaps less attached to water than Blue-cheeked Bee-eater *M. superciliosus*, but in Israel found in irrigated agricultural areas in south, and in Egypt breeds commonly along the Nile. On sea-coast prefers sandy zone immediately above high-tide mark. In Pakistan, occurs in open semi-desert

PLATE 69. *Merops orientalis* Little Green Bee-eater (p. 734). *M. o. cleopatra*: **1** ad, **2** juv. *M. o. cyanophrys*: **3** ad.   *Merops superciliosus* Blue-cheeked Bee-eater (p. 740). *M. s. persicus*: **4** ad, **5** juv. *M. s. chrysocercus*: **6** ad. (CETK)

PLATE 70. *Merops apiaster* Bee-eater (p. 748): **1** ad ♂ breeding, **2** ad ♀ breeding, **3** ad ♂ non-breeding, **4** juv. (CETK)

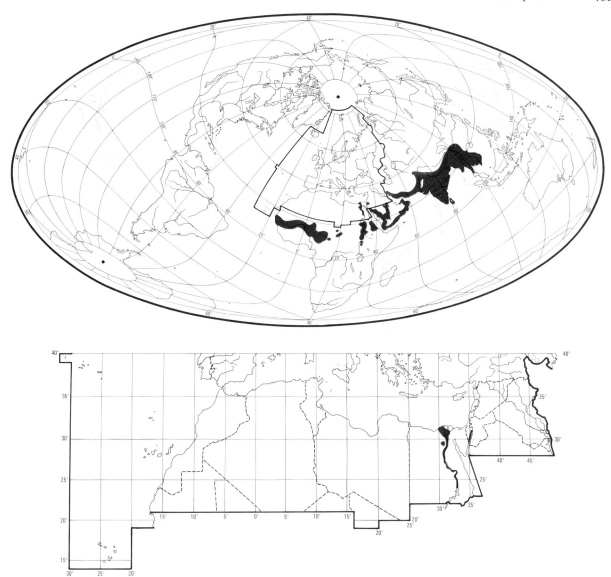

near cultivation and grazing (Ali and Ripley 1970). Avoids heavy forest; frequents both cultivation and desert areas (Whistler 1941). Near Delhi on banks of rivers, canals, and lakes, on railway embankment and tree-bordered roadsides and in a plant nursery. Fond of dust baths but also of bathing, dipping into water during flight (Hutson 1954). More tolerant of arid conditions than other *Merops* except *M. superciliosus* and one extralimital species (Fry 1969b).

PLATE 71 (*facing*).
Merops orientalis cleopatra Little Green Bee-eater (p. 734): **1–2** ad. **3** juv.
Merops superciliosus persicus Blue-cheeked Bee-eater (p. 740): **4–5** ad, **6** juv.
Merops apiaster Bee-eater (p. 748): **7–8** ad ♂ breeding, **9** juv.
Upupa epos Hoopoe (p. 786): **10–11** ad ♂ **12** juv. (CETK)

Distribution. EGYPT. Has expanded north in Nile delta (Flower 1933). JORDAN. No records, though breeds in adjacent area of southern Israel.

Accidental. Iraq.

Population. EGYPT. Still common, but probably some decline in recent decades (PLM, WCM). ISRAEL. Increased; perhaps now 100–200 pairs (HM).

Movements. Basically resident or locally dispersive, but evidently some longer movements in Africa.

AFRICAN POPULATIONS. Sub-Saharan race *viridissimus* breeds in second half of dry season; majority resident, but limited northward and southward dispersal occurs January–May (Thiollay 1971). In central Chad, small resident population seems to be augmented in wet season by birds that have already bred further south (Newby 1980), and moulting birds present Darfur (western Sudan), January–July, disappeared in autumn (Lynes 1925a). Irregular visitor to Sénégambia, December–May (Morel 1968; Gore 1981), and *viridissimus* has been collected Gebel Elba (Sudan) in April (Meinertzhagen 1930). Egyptian race *cleopatra* is partially migratory, some remaining in winter as far north as Nile delta (Flower 1933; P L Meininger and W C Mullié) while others ascend Nile into Sudan, where identified south to Torit, close to Ugandan frontier (Cave and Macdonald 1955). Birds (perhaps *cleopatra*) occur as scarce non-breeding visitors in dry season (August–September) in northern Uganda (Britton 1980).

ASIATIC POPULATIONS. Resident in southern and western Arabia and southern Israel (Meinertzhagen 1954; Inbar 1977; Jennings 1981a), though some movement (perhaps local) takes place. In Saudi Arabian interior, likely that some birds depart for winter (Jennings 1980); small coasting movement detected northern Oman, with parties flying north-west in April and south-east in September–October (Walker 1981a). Only a vagrant to Eastern Province of Saudi Arabia and to southern Iraq, and not recorded Kuwait; no data from Iran. In Pakistan and India, *beludschicus* and nominate *orientalis* both resident though with marked local seasonal movements, withdrawing from wet areas during monsoon and from northern areas in winter (Ali and Ripley 1970).

Food. Account includes material supplied by C H Fry.

Small flying insects. Like other small Meropidae (especially Little Bee-eater *M. pusillus*), generally hunts like flycatcher (Muscicapidae) from low perch, sometimes overhead wire or outer branch of tree (C H Fry); in India, recorded using backs of grazing animals (Ali and Ripley 1970; Bastawde 1976). Rapid pursuit of prey as in Bee-eater *M. apiaster*. 2–3 insects may be taken before return to perch and, in evening, more sustained flights undertaken, often in company of swallows (Hirundinidae) (Jerdon 1877), though systematic aerial-pursuit characteristic of some other Meropidae not recorded by Meinertzhagen (1930). Prey taken after upward swoop (Ali and Ripley 1970). After return to perch, smaller items swallowed immediately; larger ones immobilized, and sometimes dismembered, by being beaten against perch (Jerdon 1877; C H Fry); dragonflies (Odonata) repeatedly beaten but wings not detached (Ganguli 1975). Devenoming as in *M. apiaster* (Fry 1969a). May occasionally (frequently, according to Mackworth-Praed and Grant 1952) take insects from vegetation in flight (Jerdon 1877), e.g. caterpillars of *Sylepta derogata* from cotton crop (Husain and Bhalla 1937). Said sometimes to attempt to flush prey, but no details (Dharmakumarsinhji 1955). Said frequently to swoop down to take insects from water surface (Cunningham 1904; see also Hutson 1954) and may dive into water like kingfisher (Alcedinidae) presumably also to feed (Jerdon 1877), though record of 5 birds, Egypt, mid-September clearly relates to bathing (White 1960a; see also *M. apiaster* and Blue-cheeked Bee-eater *M. superciliosus*). Sometimes occurs in large numbers on freshly ploughed fields where possibly attracted by insects brought to surface (Cunningham 1904); further indications of possible ground-feeding in Meinertzhagen (1930) and Dharmakumarsinhji (1955) but no definite proof that this method employed (C H Fry). Undigested items ejected as pellets (e.g. Ganguli 1975). Only report of prey capture rate refers to bird making sallies from perch and catching 12 insects (9 probably bees *Apis*) in 12 min (Fletcher and Inglis 1924). In breeding season, forages all day from *c*. 30 min after dawn to *c*. 30 min before dusk (less in winter or when cloudy or wet); in hot weather, much less active around midday and early afternoon. Towards evening, rate increases until departure for roost (Bastawde 1976; C H Fry).

Like other Meropidae, probably hunts opportunistically, taking any suitable insects that pass so that diet relatively varied (C H Fry). No detailed information on diet in west Palearctic; following list based entirely on extralimital studies. Insects include dragonflies (Odonata: e.g. *Potamarcha obscura* and *Crocothemis servillea*), few crickets and grasshoppers (Orthoptera), termites (Isoptera), bugs (Hemiptera: Reduviidae, Coreidae, Pentatomidae), butterflies and moths (Lepidoptera: Nymphalidae, Pyralidae, Noctuidae, Microlepidoptera), flies (Diptera: Drosophilidae, Muscidae, Asilidae, Tabanidae, Calliphoridae), Hymenoptera (bees Apidae, Halictidae, wasps Vespidae, Pompilidae, Sphecidae, Scoliidae, Chalcididae, Chrysididae, winged ants Formicidae), beetles (Coleoptera: Bruchidae, Scarabaeidae, Carabidae, Curculionidae, Coccinellidae, Chrysomelidae). Once a spider (Araneae) (Mason and Maxwell-Lefroy 1912; Fletcher and Inglis 1924; Baker 1927; C H Fry).

At Lake Chad (northern Nigeria), pellets contained *c*. 400 insects: 75% (by number) mostly small Hymenoptera of over 50 species (flying ants most numerous, followed by bees Halictidae but only 6 *Apis mellifera*), 17% Coleoptera, mostly *Spermophagus* (Bruchidae); remaining 8% comprised termites, bugs, Microlepidoptera, and Diptera from small *Drosophila* up to Asilidae and Tabanidae (C H Fry). In eastern parts of range, e.g. in Delhi area, August (Ganguli 1975), commonly takes dragonflies (including large *Crocothemis servillea*) and bees in beekeeping districts (C H Fry). 850 stomachs from Pakistan contained up to 31 bees *Apis cerana*, average of 3·45 in each (Latif and Yunus 1950); birds probably killed at apiaries and stomach contents thus not representative of year-round diet (C H Fry). At Pusa (Bihar, India), takes mainly bees and other Hymenoptera; small insects—Microlepidoptera, Diptera, and small Coleoptera (especially *Myllocerus*)—probably taken more frequently than studies indicate. Also in India, recorded catching Noctuidae and Pyralidae but Lepidoptera often difficult to identify in stomachs. 30 stomachs, January–November, contained 284 insects, mainly bees *Apis florea* and *A. cerana*, also other Hymenoptera, Lepidoptera, Coleoptera, Diptera, a few crickets, 1 dragonfly, and 1 spider. 13 stomachs, April–July, obtained near hives contained almost exclusively bees. Considered serious pest, visiting hives in rains and in hot weather preceding them. Hymenopteran prey also includes fierce wasp *Polistes hebraeus*, *Sphex lobatus*, and *Chrotogonus*; large butterfly *Melanitis ismene* (Nymphalidae) also recorded (Mason and Maxwell-Lefroy 1912). In Burma, takes formidable rock honey-bee *A. dorsata* (Smythies 1953).

No information on food of young. MGW

Social pattern and behaviour. No major study but clearly similar to other west Palearctic Meropidae. Account includes material supplied by C H Fry.

1. In pairs or generally small, loose-knit feeding flocks throughout the year. Sometimes solitary outside breeding season (C H Fry). In Oman, usually in pairs or families, though larger groups formed for passage or roosting (Gallagher and Woodcock 1980); see also (e.g.) Heuglin (1869). Rather more gregarious in eastern parts of range: e.g. in Saurashtra (India), flocks of up to 500 (Dharmakumarsinhji 1955) and, especially in winter, large and compact roosting assemblies typical (Bastawde 1976). Adults and young band together after fledging (e.g. Hutson 1954, Koenig 1956). BONDS. Mating system evidently monogamous though stability and duration of pair-bond not known (C H Fry); however, in Nile valley (Egypt), said to occur in 'pairs' throughout winter (Etchécopar and Hüe 1967). Helpers not recorded (see Bee-eater *M. apiaster*), and occurrence in pairs rather than trios suggests these do not occur (Fry 1972; C H Fry). Both adults tend young (e.g. Ali and Ripley 1970), also after fledging (Salvan 1968b), but exact duration not known. BREEDING DISPERSION. In Africa, solitary; where number of suitable sites restricted, semi-colonial concentrations may occur (C H Fry); see also (e.g.) Meinertzhagen (1930) for Egypt, where said never to form such large colonies as other Meropidae. In eastern parts of range, often in isolated pairs (e.g. 1 pair per borrow-pit in Assam), but perhaps less commonly colonial in India than in south-east Asia where colonies sometimes well defined with up to 30 pairs. In scattered colonies of *beludschicus*, holes may be *c*. 10–100 m apart (Baker 1927, 1934). Near Delhi (India), pairs breed solitarily after break-up of flock but possibly within a 'flock territory' (Hutson 1954). Home-range less than 5 ha, probably *c*. 1 ha, weakly defended against conspecific birds during feeding times at all seasons (Fry 1972; C H Fry). Birds said normally to feed within *c*. 2 km of roost (Bastawde 1976). ROOSTING. Nocturnal and usually communal at all times of year. In leafy or evergreen trees including those standing in water, or in bamboo *Bambusa* thicket (Jerdon 1877; Hutson 1954; Dharmakumarsinhji 1955; Koenig 1956). May also roost in nest (presumably when breeding) but not known if both sexes involved (C H Fry). Birds arrive at roost around sunset, earlier in overcast conditions when assembly takes *c*. 30–45 min (*c*. 15–20 min on clear days); go directly to roost if they arrive after sunset. Depart *c*. 30 min before sunrise, later in dull, wet weather (Bastawde 1976; also Ali and Ripley 1970). Little available information on resting at other times but birds make no attempt to seek shade on perches (C H Fry). For water-bathing, flies down from perch, not submerging completely, then returns to perch to shake plumage (Hutson 1954); may splash about on surface (White 1960a); rain-bathing also recorded (Ganguli 1975). See also Flock Behaviour (below).

2. Bold and confiding (e.g. Baker 1927, 1934, Meinertzhagen 1930); at Pusa (Bihar, India), birds quite fearless of man, coming close to take bees *Apis* when hives being examined (Mason and Maxwell-Lefroy 1912). May sit upright on perch with slow up-and-down swinging of tail (Ali 1953). Shortly before landing, wings widely spread and erect, closed abruptly as bird alights (Koenig 1956). FLOCK BEHAVIOUR. Flock may break up during day, birds foraging singly or in small groups within 'territory'. Congregate again noisily around sunset on wires or tree-top then suddenly depart for roost (Hutson 1954). May also fly to roost in pairs or small flocks; assembly at roost noisy and protracted. Early arrivals may leave roost repeatedly; flocks ascend to some height and give (undescribed) calls, then quickly return. *M. orientalis* tends to roost apart from other birds. At approach of (e.g.) Black Drongo *Dicrurus macrocercus*, crows (Corvidae), or birds of prey (Accipitridae), high-pitched warning call given (see 2 in Voice) and all birds may leave temporarily or chase off intruder *en masse*. Some may leave to roost elsewhere if disturbed, though return in morning when, in clear weather, departure normally in flocks of 5–30 birds (Bastawde 1976). Huddle in groups of 3–7 on cold mornings (Smythies 1953). Birds roost in compact rows, keeping physical contact; typically, much bickering and supplanting of occupants from favoured perches in middle of row (Ali and Ripley 1970). Following frequent brief upflights, birds make characteristic sideways-tripping movements to reform compact row (see also *M. apiaster*). In mixed roosting flock of *c*. 20–30 adults and young, Mali, late June, some adult pairs roosted apart (Koenig 1956; see also Antagonistic Behaviour, below). Often in close-packed flocks for dust-bathing on earth roads in evening (Jerdon 1877; Hutson 1954; Smith 1957). ANTAGONISTIC BEHAVIOUR. Birds usually sleep shoulder-to-shoulder in roosts (C H Fry), though pairs sleeping apart from other birds (see above) threatened any coming too close (Koenig 1956). Threatening behaviour apparently frequent in small aggregations but details of social context lacking (C H Fry) and birds apparently tolerant of one another when resting or foraging from perch (e.g. Gillet 1960). In high-intensity threat, bird adopts a hunched posture with mantle feathers ruffled; tail spread and vibrated; wide-open bill directed at threatened object and call given (C H Fry: see 1 in Voice). HETEROSEXUAL BEHAVIOUR. (1) General. Scant data indicate resemblance to extralimital Red-throated Bee-eater *M. bullocki* (C H Fry); for study of *M. bullocki*, see Fry (1973). Members of a pair keep in close proximity for much of day and year; habitually perch touching and, even when actively feeding, often return to perch within a few cm of each other (C H Fry). (2) Pair-bonding behaviour. In Lahore (Pakistan), circular flights, said to be part of courtship, performed soon after arrival (Dewar 1909). Rapid tail-vibrating also recorded and 2 birds (possibly a pair) occasionally fly to ground (Hutson 1954). Greeting-ceremony resembles behaviour shown in threat (see above) but involves contrasting elongated and erect posture: legs straightened and neck extended; neck and crown feathers ruffled (C H Fry). (3) Nest-site selection. No information. (4) Nest-building. By both members of pair. While one bird digs for *c*. 2 min, other gives Contact-calls (see 1 in Voice) from nearby perch; then change over (Dewar 1909). As in other Meropidae, 2–3 tunnels abandoned after varyingly short distance before definitive one excavated nearby. Usually digs new tunnel each season but may use old one (Baker 1934; C H Fry). (5) Courtship-feeding. Few details, but clearly as in other Meropidae. In one case, took place for 5 consecutive evenings in garden; may continue during incubation (Ganguli 1975). (6) Mating. May follow Courtship-feeding. In Delhi area,

first recorded 20 February. May take place on wire. After copulating, birds sit briefly side by side, then ♂ flies up almost vertically, turns over, and descends to ♀ in erratic glide; birds may then copulate again and, following another pause, fly down to ground (Hutson 1954). (7) Behaviour at nest. Incubation by both sexes, more by ♀ who does so at night, when ♂ roosts in tree (e.g. Ganguli 1975), though not clear how far from nest. RELATIONS WITHIN FAMILY GROUP. Very little information and none on nestling period. After fledging, family sticks close together and young, which apparently show no begging behaviour after leaving nest, fed by parents probably for several weeks (C H Fry; also Salvan 1968b). No further information.

MGW

Voice. Account includes material from C H Fry.

CALLS OF ADULTS. (1) Contact-call. Monotonous, subdued, repeated trilling or buzzing: 'tree' (Smythies 1953); 'teerp' (Ali and Ripley 1970), or 'zürr' (Koenig 1956); in recording by P A D Hollom, Iran 'PEEer PEEer' (P J Sellar). Varies in intensity, and probably duration, according to mood: e.g. shriller in threat (C H Fry), excited trilling given by birds disputing perches at roost (Bastawde 1976; also Hutson 1954), and in recording by P A D Hollom, Iran, a very soft, plaintive 'pee pee pee' from ♀ after mating (P J Sellar). Given mainly in flight and frequently throughout the day, especially in gregarious situations (C H Fry). (2) Alarm-call. A staccato 'ti-ic' or 'ti-ti-ti' lacking rolling quality of call 1 (Salvan 1968b; C H Fry); 'tit-tit-tit' interspersed in bouts of Contact-calls (Ali and Ripley 1970). Other renderings: 'trititit' (Gallagher and Woodcock 1980); 'dick dick' (Koenig 1956; see also Bee-eater *M. apiaster*); 'chit chitty chit' (Walker 1981b) presumably the same. (3) Other calls mentioned by Gallagher and Woodcock (1980) but not assigned to any particular situation: 'krrrew' and 'krreeuk'.

CALLS OF YOUNG. Food-call of younger nestlings not described but that of fledged young an insistent low 'tcheeb' (Walker 1981b).

MGW

Breeding. SEASON. Egypt: April (Meinertzhagen 1930); February–April (Mackworth-Praed and Grant 1952). SITE. Tunnel in bank, natural or artificial. Nest: excavated tunnel 0·3–2·0 m long with enlarged chamber at end; unlined, though food remains accumulate. Building: by both sexes. EGGS. See Plate 98. Blunt elliptical, smooth and glossy; white. Average of 30 eggs 20 × 18 mm (Hüe and Etchécopar 1970). Clutch: 3–5. INCUBATION. By both sexes, beginning with 1st egg; hatching asynchronous. No further information.

Plumages (*M. o. cleopatra*). ADULT. All upperparts, tertials, upper wing-coverts, and sides of neck bright and shiny grass-green, sometimes tinged golden-cinnamon or bluish-green, depending on light; some pink-cinnamon of feather-bases often visible, especially on hindneck, mantle, tertials, and upper wing-coverts. Narrow black stripe across lores and just below eye to ear-coverts; black bristles along gape at upper mandible. Black stripe narrowly bordered below by pale sky-blue on cheeks. Chin to breast bright grass-green; slightly greener and less golden than upperparts, interrupted by rather narrow black bar at border of lower throat and upper chest. Remainder of underparts bluish-green, less glossy; pale blue-green on under tail-coverts; flanks with much pink-cinnamon of feather-bases visible or wholly pink-cinnamon. Central pair of tail-feathers (t1) and outer webs of others pale green, duller and less golden than upperparts; tip of t1 black, shafts of all feathers dark horn-brown or black; inner webs of t2–t6 pale pinkish-green; undersurface of tail pale brown, shafts white. Flight-feathers bright pale rufous-cinnamon, outer webs with slight green tinge and narrow pure green outer edge; outer webs of outer primaries pale blue near tips; secondaries and inner primaries with contrastingly deep black tips c. 5–8 mm wide, black gradually duller and less extensive on outer webs towards outer primaries. Under wing-coverts and axillaries pink-cinnamon, similar in tinge to undersurface of basal and middle portions of flight-feathers. Bleaching and abrasion have marked influence, colours of birds in worn plumage differing strongly from fresh birds described above. In slightly worn birds, tips of green feathers of body tinged pale blue, first on tertials, rump, sides of t1, and tips of outer webs of outer primaries. In heavily worn birds, crown, mantle, scapulars, breast to under tail-coverts, and upper wing-coverts rather dull bluish-green with much pink-cinnamon of feather-bases visible; forehead, supercilium, back, rump, upper tail-coverts, sides of head down from black stripe through eye, chin, throat, and chest uniform glossy pale greenish-blue with sharp black bar across upper chest; outer edges of flight- and tail-feathers pale greenish-blue. NESTLING. Naked at hatching. No information on development of feathering. JUVENILE. Rather like adult, differing most conspicuously by absence of black gorget, lack of elongated t1, and larger p10 with rounded instead of pointed tip. Upperparts, tertials, and upper wing-coverts bluish-green, soon changing to pale bluish; much pink-cinnamon or pink-buff of feather-bases visible, even in fresh plumage. Chin and throat cream-white with limited pale blue on feather-tips below black eye-stripe and on lower throat; no black gorget, though patch often indicated by some dusky or bluish feather-tips; chest to under tail-coverts pink-buff to cream with limited blue-green or pale blue on feather-tips of chest, breast, and under tail-coverts. Tail: t1 bluish-green, other feathers greenish-cinnamon; see also Structure. Flight-feathers as in adult, but rufous-cinnamon less bright and black tips duller and less sharply defined, in particular tips of inner secondaries and all primaries washed dull grey rather than marked solid black. FIRST ADULT. Like adult; indistinguishable when last juvenile feathers (e.g. p10) replaced.

Bare parts. ADULT AND JUVENILE. Iris crimson or blood-red. Bill greyish-black, brownish-black, or black; mouth pink. Leg and foot olive-brown, grey-brown, or brownish leaden-grey; sole paler yellowish or olive-grey. NESTLING. Bare skin (including leg and foot) pink. (Hartert 1912–21; Ali and Ripley 1970; Harrison 1975; BMNH, ZMA.)

Moults. ADULT POST-BREEDING. Complete; primaries descendant. Few Palearctic birds in moult examined. Most December–April birds examined were in fresh plumage, May–June ones worn. Moult of body, flight-feathers, and tail probably starts June or July. Birds in advanced primary moult examined August (primary moult scores 35–40; body largely new, tail new, but t5 or t6 still growing); one had moult just completed except for growing p10 in early October. Indian birds (including *beludschicus*) moult completely July–September; sequence of moult 1–6–3–2–4–5 (Marien 1952). POST-JUVENILE. Complete. Only very few birds in moult examined. Birds from late September and early October still in juvenile plumage, but inner primaries moulting (scores 10, 16, and 20); t1 growing in 2 birds. If speed of moult similar to adult, primaries probably start August–September and all moult completed about January. Indian birds have complete moult at same time as adult (Marien 1952).

Measurements. ADULT. *M. o. cleopatra*. Nile valley and Faiyum (Egypt), February–April; skins (ZFMK, ZMA). Tail is fresh t1, tip is from tip of t1 to 2nd (non-elongated) longest. Bill (F) is from tip to forehead, bill (N) from tip to distal corner of nostril; exposed culmen on average 5·5 mm less than bill (F), 3·5 more than bill (N).

WING	♂	94·2 (1·15; 15)	92–96	♀	91·2 (1·68; 11)	89–94
TAIL		145 (6·22; 10)	138–158		130 (7·82; 7)	121–144
TIP		74·9 (6·54; 12)	66–88		58·2 (9·73; 11)	42–74
BILL (F)		31·1 (1·37; 13)	29–33		28·8 (0·94; 11)	28–30
BILL (N)		22·0 (1·28; 13)	21–24		19·8 (0·87; 11)	19–21

Sex differences significant. Sexes combined: tarsus 9·6 (0·73; 11) 8–10; middle toe with claw 14·4 (0·95; 10) 13–16.

M. o. cyanophrys. Western Arabia, mainly spring; skins (BMNH).

WING	♂	92·2 (1·47; 10)	90–95	♀	93·1 (2·42; 10)	90–97
TAIL		92·6 (3·21; 10)	89–97		91·3 (3·57; 9)	86–96
TIP		18·4 (3·10; 10)	15–23		16·8 (2·39; 9)	12–20
BILL (F)		32·0 (1·06; 10)	30–34		31·6 (1·01; 10)	30–33
BILL (N)		23·0 (1·07; 10)	21–24		22·7 (1·05; 10)	22–24

Sex differences not significant.

M. o. viridissimus. Sudan, Chad, and Niger, bill also Sénégal to Ethiopia; sexes combined (mainly ♂♂): wing 88·6 (3·28; 16) 83–94; bill (F) 30·8 (1·58; 46) 27–33; tail as in *cleopatra* (Niethammer 1955; Vaurie 1959e; RMNH, ZFMK).

M. o. beludschicus. Iran and Pakistan. Wing ♂ 96·4 (2·33; 8) 94–100, ♀ 93·7 (3·89; 10) 86–99; tail ♂ 110–134, ♀ 107–114; elongated tip of t1 ♂ 40–70, ♀ 35–45; bill (F) 29–32; bill (N) ♂ 24·2 (1·20; 9) 22–26, ♀ 22·3 (1·87; 9) 19–25 (Marien 1952; Ali and Ripley 1970; BMNH, ZFMK).

JUVENILE. Wing on average 4 mm shorter than adult; length of t2–t6 about similar to adult, but t1 4–10 shorter than others, not elongated. Bill not full-grown until post-juvenile moult started.

Weights. *M. o. viridissimus*. Ennedi (Chad), April: ♂ 13 (ZFMK). Aïr (Niger), February: 12–13·5 (4 ♂♂) (Niethammer 1955).

M. o. beludschicus. 19–20 (3 ♂♂), 18–27 (3 ♀♀) (Ali and Ripley 1970).

Nominate *orientalis*. ♂♂ 17, 18; sex unknown 15 (Ali and Ripley 1970).

Structure. Wing rather short, broad at base, tip rather pointed. 10 primaries: in adult, p8–p9 longest, subequal or either one 0–2 shorter than other, p10 53–59 shorter; in juvenile, p8 longest, p9 1–2 shorter, p10 42–50; at all ages, p7 4–7 shorter than longest, p6 11–15, p5 16–20, p4 20–24, p1 24–34. P10 reduced; narrow and pointed in adult, slightly larger and with rounded tip in juvenile. Secondaries, inner 5–6(–7) primaries, and tail-feathers (except t1) with notch at tip. Tail rather long, 12 feathers; in adult, t1 slightly (*cyanophrys* and *muscatensis*) or strongly (other races) elongated, middle portion gradually tapering, distal portion (extending beyond other feathers) 1–2·5 mm wide, tip slightly wider and narrowly rounded when fresh; tips of t2–t6 form square tail-tip, t6 usually slightly longer than t2; in juvenile, t3–t4 longest, t2 and t6 often slightly shorter, t1 4–10 shorter than t3–t4, tail appearing shallowly forked. Proportion of bill about as in Blue-cheeked Bee-eater *M. superciliosus*, not as decurved and heavy at base as in Bee-eater *M. apiaster*. Outer toe *c.* 83% of middle, inner *c.* 69%, hind *c.* 50%. Remainder of structure as in *M. apiaster*.

Geographical variation. Marked. 3 subspecies-groups separable: (1) *viridissimus* group in Africa with *viridissimus* in northern Afrotropics and *cleopatra* in Egypt, with plumage mainly bright grass-green or golden-green and t1 strongly attenuated; (2) *cyanophrys* group in Arabia with *cyanophrys* in southern and western Arabia north to southern Israel and east to Hadhramawt, and *muscatensis* in central and north-east Arabia, showing mainly blue face and underparts, purplish-black rather than black gorget, and only slightly elongated t1; (3) nominate *orientalis* group with *beludschicus* from southern Iran and possibly Iraq east to north-west India, nominate *orientalis* in India and Ceylon, and *ferrugeiceps* in south-east Asia, with golden-green to rufous crown, blue chin and throat, black gorget, bluish-green underparts, and well-elongated t1. Within each of these groups, races only recognizable in fresh plumage, as all strongly affected by abrasion which causes colour shift towards blue-green and reveals part of cinnamon feather-bases. *M. o. viridissimus* distinctly brighter golden-yellow or bronze-green than *cleopatra*, not as grass-green or bluish-green, but both similar in worn plumage; perhaps grades into *cleopatra*, as some birds from Sudan, January–March, are near *cleopatra* in colour but near *viridissimus* in size. *M. o. cyanophrys* has forehead and stripe over eye bright pale blue; remainder of upperparts and tail as in *cleopatra*, but slightly duller green, less golden on crown, more bluish on lower scapulars, tertials, and rump; black gorget broader in middle, triangular rather than forming a narrow bar, and tinged purple-blue; chin and throat down from black eye-stripe bright pale blue, grading to purple-blue on central throat and where bordering gorget; breast bluish-green, belly to under tail-coverts pale greenish-blue. *M. o. beludschicus* differs from *cleopatra* by slightly purer green upperparts with (in similar stage of wear) slightly more extensive pale blue on tertials and rump and more pink-cinnamon shining through on crown, hindneck, and upper mantle; light blue on cheeks not restricted to narrow strip below black eye-stripe as in *cleopatra*, but extending to chin and sides of throat and sometimes down to gorget; underparts slightly bluer than in *cleopatra*. Arabian *muscatensis* similar to *cyanophrys*, but throat slightly paler blue and less or no purple-blue at border of gorget and on rear of lower cheeks; nominate *orientalis* similar to *beludschicus* but sides of crown and nape more extensively tinged golden-brown, upperparts slightly darker green, chin and throat slightly deeper blue, underparts less bluish; *ferrugeiceps* similar to nominate *orientalis*, but crown to upper mantle still more extensive golden-chestnut or red-brown, and throat greener (see also Marien 1952). CSR

Merops superciliosus Blue-cheeked Bee-eater

PLATES 69 and 71
[between pages 734 and 735]

Du. Groene Bijeneter   Fr. Guêpier de Perse   Ge. Blauwangenspint
Ru. Зеленая щурка   Sp. Abejaruco papirrojo   Sw. Grön biätare

Merops superciliosus Linnaeus, 1766

Polytypic. *M. s. persicus* Pallas, 1773, northern Egypt and Middle East, east to Kazakhstan (USSR) and north-west India; *chrysocercus* Cabanis and Heine, 1860, western Sahara from south-east Morocco and central Algeria south to Mauritania, Mali, and Nigeria. Extralimital: nominate *superciliosus* Linnaeus, 1766, Madagascar, Comoro Islands, Pemba, and eastern Africa from Ethiopia and Somalia south to Zimbabwe and Moçambique; *alternans* Clancey, 1971, Angola and northern Namibia.

Field characters. 27–31 cm, of which tail-streamers up to 10 cm; wing span 46–49 cm. Averages slightly larger than Bee-eater *M. apiaster*, with 15% longer bill, bulkier head, and tail-streamers up to 3 times longer. Medium-sized, intensely green bee-eater, with bold head pattern (particularly yellow-chestnut throat) contrasting with uniform body, as does bright copper-chestnut underwing. Call slightly different from *M. apiaster*. Sexes similar; no seasonal variation. Juvenile separable. 2 races in west Palearctic, separable at close range.

ADULT. (1) North-west African race, *chrysocercus*. Tail-streamers exceptionally long: 7–10 cm in ♂, 5–6 cm in ♀. Basic plumage tone golden-green, but rapidly becoming bluer as plumage wears. Head decorated by white to bright pale blue forehead, bright pale blue supercilium, quite broad black eye-stripe, and pale blue line under eye-stripe. Chin yellow, throat cinnamon to copper-chestnut; both form obvious warm patch below head. Upperparts show blue tone on rump; underparts have distinctly yellow tinge, particularly on flanks. Wing appears almost uniform at rest, but in flight shows blue-sheened bases and dark tips to flight-feathers. Under wing-coverts and axillaries copper-cinnamon, and bases of underside of flight-feathers similarly toned. Tail noticeably dull beneath. (2) Egyptian and south-central Asian race *persicus*. Tail-streamers 3·5–6·5 cm—shorter than in *chrysocercus*, but still at least twice as long as those of *M. apiaster*. Basic plumage tone grass-green, becoming bluer than *chrysocercus* with wear. White patch on forehead larger, blue supercilium and upper cheeks darker, and latter divided from black eye-stripe by white line. Rump and secondaries with obvious blue sheen. Underparts lack yellow tone. JUVENILE. Both races. Central tail-feathers not elongated. Plumage duller and bluer in tone than adult, with head marks less distinct (lacking obvious blue and white outlines to eye-stripe). Throat less chestnut, more saffron.

Confusable with Little Green Bee-eater *M. orientalis* only when size and head pattern obscured; immediately separable from *M. apiaster* in good view but, when colours obscured, identification on silhouette and flight-call tricky. Beware particularly lack of multi-coloured appearance in *M. apiaster* in juvenile and brief non-breeding plumage; see that species. Flight as *M. apiaster*, with most actions identical but wheels noticeably lazier at times; frequently launches itself from low outcrop of ground, using rapid take-off and upward glide to perform characteristic 'shooting' capture of prey.

Commonest call similar to that of *M. apiaster* but distinguishable with experience: 'greep' or 'treet'; higher pitched, less liquid, more husky, thinner, and occasionally almost disyllabic; does not carry as far.

Habitat. Breeds in lower middle dry very warm latitudes, in Mediterranean, steppe, and desert zones, down to subtropics but, unlike Bee-eater *M. apiaster*, with which it infrequently overlaps, not in temperate or oceanic zones. In Jordan, breeds on marsh at Azraq, largely on islands where soil soft beneath hard crust, in fringe zone of tamarisk *Tamarix* with spiky grass, sedges, and halophytes (Nelson 1973). In Iraq, breeds near rivers and irrigation canals (Allouse 1953). In USSR, in deserts and semi-deserts dissected by hills, cliffs, ravines, river valleys, canals, and lakes with shores covered by sparse arboreal and bushy vegetation or reeds; also on seashores, at springs and wells, and in oases or by human settlements; avoids places bare of vegetation, and for nesting prefers sandy to clay soil, occasionally using clefts in rocks (Dementiev and Gladkov 1951a). In India, in neighbourhood of water, such as canals, irrigation reservoirs, pools, and sandy seashores in more arid regions, using telegraph wires as perches and breeding in sand, even on seashores above high-tide level (Ali and Ripley 1970). In Baluchistan (Pakistan), at *c*. 1200 m, also proved closely attached to water and especially to tamarisk trees and shrubs, sitting on tops rather than on sides like *M. apiaster*, and flying less high and less far (E M Nicholson). In Afghanistan, regarded as a bird of the plains, where it replaces *M. apiaster* which prefers hilly country (Paludan 1959). Sharing a common diet with *M. apiaster*, largely avoids interspecific competition by preference for more arid and more low-lying regions, and within these by closer association with water, which in turn enables it nevertheless to find sufficient woody vegetation and often also suitable nest-sites. Will occasionally roost in reeds, and even dive into water. Frequently also along roads and railways, making use of posts and overhead lines as look-outs, as well as using treetop perches, for example on acacias in African savanna, its main winter habitat. In Sierra Leone, passes dry season

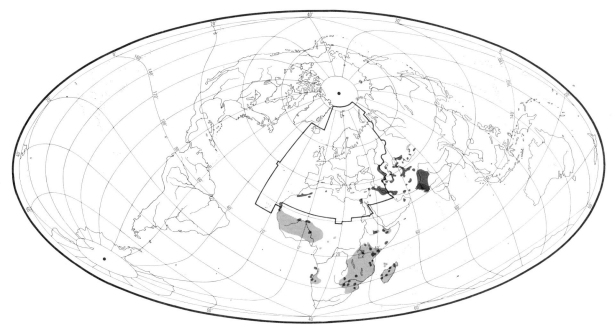

mainly on open low-lying ground, swamps, and river valleys near coast, some spreading over hills to forest edge and to scrub forest, on summits at *c*. 600–900 m; flocks roost in mangroves (G D Field).

Distribution. TUNISIA AND LIBYA. Despite Etchécopar and Huë (1967), no evidence of breeding (MS; see also Bundy 1976). LEBANON. Bred 1945, but no evidence of nesting since (HK). ISRAEL. Bred in Jordan valley until *c*. 1950 (HM). USSR. Has bred occasionally north of main range (Dementiev and Gladkov 1951*a*).

Accidental. Britain, France, Netherlands, Sweden, Italy, Yugoslavia, Greece, Malta.

Population. MOROCCO. Local, perhaps 40–60 pairs (JDRV). ALGERIA. Probably nowhere common (EDHJ). MAURITANIA. Banc d'Arguin. Breeds in small numbers, at least in some years (Dick 1975; JT). TURKEY. Perhaps 40–80 pairs (Parslow and Everett 1981). USSR. Abundant Lenkoran area (Dementiev and Gladkov 1951*a*). EGYPT. Several thousand pairs in Nile delta 1983 (PLM).

Movements. Mainly migratory. 2 Palearctic races both winter in Africa; no ringing data, but discrete winter ranges determined from identifications of specimens. However, Asiatic population overlaps with a 3rd (Afrotropical) race in winter quarters in eastern half of Africa.

WEST SAHARAN POPULATION (*chrysocercus*). In North Africa, part of population makes northward post-breeding movement in August–September (probably from late July), Morocco to Tunisia, though rarely reaches Mediterranean coast (Heim de Balsac and Mayaud 1962); followed by southward migration in October. Winters in West Africa from Sénégambia, Guinea-Bissau, and Sierra Leone east to Nigeria. Some present all year in western Sahel zone, breeding there at least sporadically from Mauritania to northern Nigeria (status needs clarification); numbers augmented in passage and winter periods. In most wintering areas, migrants arrive October–November and depart March–April (e.g. Gore 1981, Elgood 1982, Richards 1982), and returning migrants noted Moroccan and Algerian Sahara in April (Smith 1968); in Sierra Leone, however, winter visitors do not arrive until December (G D Field). Few records of the species in Chad (Salvan 1968*b*) and not recorded in western Sudan; hence gap in winter distribution between *chrysocercus* and *persicus*. However, April migrants through Kufra (Libyan Desert) resembled *chrysocercus* (Cramp and Conder 1970), these being unexpectedly far to the east.

ASIATIC POPULATION (*persicus*). Mainly migratory throughout breeding range (Egypt to Soviet Central Asia and north-west India), though minority are resident in Pakistan and India (Ali and Ripley 1970). Winters in eastern half of Africa from Ethiopia and eastern Sudan to South Africa (Orange Free State, Natal). Migrates on broad front across Middle East to enter Africa through Nile valley and across Red Sea; big spring and autumn passages through Eritrea provide major food source for Sooty Falcons *Falco concolor* nesting on Dahlak archipelago (Clapham 1964). Also, scattering of ship records from Arabian Sea, which indicate some eastern birds make direct crossing in autumn towards East Africa (Moreau 1938; Tuck 1961*b*). Autumn passage across Middle East mid-August to mid-November (chiefly September–October), and spring return mid-March to early June (chiefly April and early May) (e.g. Ali and Ripley 1970,

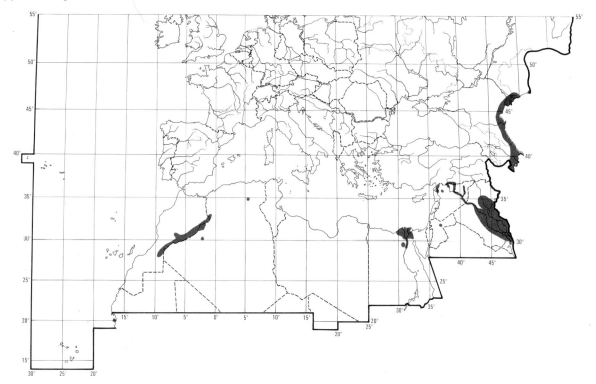

Bundy and Warr 1980, Walker 1981b); present in winter range October to early April (e.g. Britton 1980). Migration both diurnal and nocturnal (Hutson 1947; Ali and Ripley 1970).

EASTERN AFROTROPICS AND MADAGASCAR POPULATION (nominate *superciliosus*). Movements not so clear cut as formerly supposed: now found breeding (perhaps sporadic in some regions) Somalia and Moçambique, and no confirmation for large-scale migration between Madagascar and continental Africa (Moreau 1966; Fry 1981b). However, extensive intra-African movements indicated by temporal changes in abundance. Thus most numerous East Africa May–September, when also common non-breeding visitor to eastern Zaïre; transient birds pass through Zambia, Malawi, and northern Zimbabwe April–May and August–September (e.g. Moreau 1966, Britton 1980).

Food. Account includes material supplied by C H Fry.

Entirely winged insects, with dragonflies and damsel flies (Odonata) important throughout the year. Often gregarious for feeding. Normally hunts from perch—dead branch of tree or, especially, telegraph wire (Koenig 1953; Fry 1981a); not uncommonly from ground (e.g. Hutson 1947, Dolgushin *et al.* 1970). Usually returns to perch after single capture, especially if prey large (Hutson 1947; C H Fry). Vagrant, Netherlands, made flights 1–1·5 m above ground up to 100–200 m from perch (Meeth 1962). At Zaria (Nigeria), early February, 6 birds sometimes flew up to *c.* 100 m from perches *c.* 3 m high; 3–4 insects captured before return to perch and many taken less than 10 m from ground. Prey typically seized from below; taken in bill-tip with head momentarily thrown back, bill vertical (Dyer 1980; see also Fry 1981a for detailed description of aerial hunting in closely-related Blue-tailed Bee-eater *M. philippinus*). At Jumna river (near Delhi, India), mid-October, *c.* 100 birds (possibly *M. philippinus*) taking butterflies (Lepidoptera) flew back and forth over water seizing prey, or hunted from jetsam or sand-bank flying directly at low-flying insects. Upward swoop or glide (in some cases after dive) invariably preceded capture; no second attempt made where first failed but several birds frequently chased same insect and some butterflies escaped after capture (Hutson 1947, 1954). In Sierra Leone, often hunts high in the air (G D Field). Sometimes swoops down on to prey; exceptionally takes insect from ground in flight (Dharmakumarsinhji 1955). Will fly behind vehicles for several hundred metres to feed on flushed insects (Boswall 1970). Said to take food (insects, possibly also fish) from water occasionally, swooping down to snatch insects from surface or plunging straight in (see Sage 1960a, Took 1963, Reynolds 1965). However, bathing behaviour similar (e.g. White 1960a, Ganguli 1975); for review and discussion, see Fry (1981a, 1982). Captive birds spent more time on ground than Bee-eater *M. apiaster* (Koenig 1953) and ground-feeding possibly not unusual in wild: in Algeria, birds took insects on ground particularly when these forced down by rain (Koenig 1895). During longer flights from perch (see above), insect crushed by being passed back and forth in mandible tips for *c.* 3–4 s, sometimes tossed up and crushed again. At least once, prey treated thus was venomous wasp (Sphecidae) (Dyer 1980). Treatment of prey at perch much as in *M. apiaster*, with beating, squeezing, and biting as required (Koenig 1953; also Hutson 1947). At end of feeding bout last insect brought to perch and there beaten and rubbed (Dyer 1980). No detailed description of de-venoming (see above for possible de-venoming in flight) but presumably as in *M. apiaster* with which said to share immunity to stings of venomous Hymenoptera (Koenig 1953). Dragonflies usually swallowed quickly by Dutch vagrant (see above), butterfly tossed up beforehand (Meeth 1962); dragonflies and beetle *Blaps* swallowed head-first (Moore and Boswell 1956a). Wings of ant-lions *Palpares solidus* (Neuroptera) always stripped off and only body eaten (Sage 1960b). Virtually all insect hard parts regurgitated as well-formed, blackish, elongated, cylindrical pellets (Koenig 1895; Fry 1981a).

Probably feeds opportunistically like (e.g.) *M. apiaster*, but dragonflies and damsel flies (Odonata) evidently of primary importance at least in extralimital parts of range; the few data from west Palearctic allow no firm conclusions. Other insects in diet include grasshoppers and locusts (Acrididae), mantises (Mantidae), termites (Isoptera), bugs (Hemiptera: Cicadidae, Reduviidae, Gerridae, Lygaeidae, Mononychidae, Coreidae, Pentatomidae, Cydnidae, Gelastocoridae, Nepidae), ant-lions (Myrmeleontidae), butterflies and moths (Lepidoptera: Pieridae, Nymphalidae), flies (Diptera: Tabanidae, Syrphidae), Hymenoptera (bees Apidae, Halictidae; wasps Vespidae, Sphecidae, Pompilidae, Scoliidae, Chalcididae, Chrysididae, also Ichneumonidae and ants Formicidae), and beetles (Coleoptera: Scarabaeidae, Platypodidae, Bostrychidae, Curculionidae, Tenebrionidae, Carabidae, Buprestidae, Chrysomelidae, Cerambycidae) (Koenig 1895; Chapin 1939; Hutson 1947, 1954; Hartley 1948; Koenig 1953; Pek and Fedyanina 1961; Meeth 1962; Dolgushin *et al.* 1970; Atakishiev 1971; Fry 1981a; C H Fry).

In Iraq, April–May, takes butterflies (Pieridae), although none found in pellets; also grasshoppers and bees (Hartley 1948). Microscopic examination of pellets from 2 localities in Nigeria showed them to contain even insignificantly small parts (e.g. compound-eye fragments) so that pellets considered to give unbiased and accurate picture of diet (*contra* Hartley 1948). In Nigerian study, 96 pellets, March–May, contained 924 insects: 65·5% (by number) Odonata (*c.* 30–80 mm long), and 13·4% Hymenoptera (of which *c.* $\frac{2}{3}$ stinging workers of bee *Apis mellifera*); also 6·9% Hemiptera, 6·3% Diptera, 4·0% Lepidoptera, 3·3% Coleoptera, 0·5% Orthoptera, 0·1% Mantidae. 2 stomachs (December) from Bangui (Central African Republic) contained 13 Odonata, 4 flying ants, 1 bug (Coreidae), 1 butterfly, and 1 fly. Odonata probably even more important by weight (Fry 1981a). Pellets from Biskra (Algeria) contained mainly Hymenoptera, also locust *Schistocerca gregaria*, beetles, and Lepidoptera (Koenig 1895). More recent analysis of pellets and nest contents from same locality revealed mainly wasps (Sphecidae) and 2 beetles (Curculionidae, Buprestidae) (Koenig 1953). 4 stomachs from Zaïre contained dragonflies and termites (Chapin 1939). Honey-bees may be eaten to exclusion of other prey when plentiful. 1 stomach from Zimbabwe contained 40–50 *Apis mellifera* and 1 other small bee (Borrett 1973). Locusts may be avoided even when abundant (Bates 1934). In Azerbaydzhan (USSR), takes Orthoptera, Coleoptera, Lepidoptera, etc., but apparently few hive-bees: maximum of 9 in 49 stomachs compared with 35–139 in 401 stomachs of *M. apiaster* (Atakishiev 1971). In Kazakhstan (USSR), takes locusts (particularly *Schistocerca gregaria*), Odonata, Hymenoptera, and Coleoptera. 5 stomachs, April–May, from near apiaries on Golodnaya steppe contained *S. gregaria*, Scarabaeidae, Sphecidae, and bees; 1 stomach (mid-July), Keles river, contained 5 locusts, 2 cicadas, and a bee. Of 47 stomachs, August–September, from near apiaries, Mirzachul', 61% contained hive-bees, 29·8% wild bees, 25·2% wasps, 4·2% hornets, 21% beetles (Scarabaeidae, Buprestidae, Curculionidae, Carabidae), 21·3% locusts, 6·3% bugs, and 2·1% mantises (Dolgushin *et al.* 1970). For analysis of 2 stomachs from Kirgiziya (USSR), see Pek and Fedyanina (1961).

Little available information on food of young. In Golodnaya steppe, given 26·9% dragonflies and 21·3% wasps. Peak feeding rates around 09.00 hrs (*c.* 15 visits per hr) and midday (*c.* 20 per hr) for brood of 4 nestlings *c.* 16 days old (Sagitov and Fundukchiev 1980). At Biskra, bouts of feeding every few minutes sometimes followed by pauses of *c.* 30 min to 1 hr (Koenig 1953; see also *M. apiaster*). MGW

Social pattern and behaviour. Fullest study by Koenig (1953), of wild birds near Biskra (Algeria) and of captive birds, indicates similarity of most aspects to Bee-eater *M. apiaster*; see also Hartley (1966). Account includes material supplied by C H Fry.

1. Loosely gregarious, occurring in small flocks for foraging throughout the year. Overall more gregarious than *M. apiaster*. On migration, sometimes in large though loose-knit flocks (Chapin 1939; Hutson 1947; Dolgushin *et al.* 1970; C H Fry). In southern USSR, non-breeders possibly remain in flocks (Potapov 1959; Dolgushin *et al.* 1970). In Egypt, large flocks prior to autumn exodus contained adults and young (Meinertzhagen 1930); other reports refer to separate juvenile flocks on migration (e.g. Dharmakumarsinhji 1955, Clapham 1964). For association with *M. apiaster* on passage, see that species, also Karcher (1940), Moore and Boswell (1956a), Morel and Roux (1966), and Reynolds (1974). BONDS. Monogamous mating system. Pair-bond perhaps life-long but no detailed study. Birds of a pair associate more with each other all year than with other flock members (C H Fry). Age of first breeding not known exactly but possibly not at 1 year old (Dolgushin *et al.* 1970). Young tended by both parents (Ali and Ripley 1970), also after fledging (Dementiev and Gladkov 1951a). Helpers at the nest (see *M. apiaster*) not recorded (C H Fry). BREEDING DISPERSION. Usually forms small, loose-knit colonies; sometimes solitary. Near Biskra, 9 nests (7 occupied) on *c.* 0·5 km²; at least 43 m

between nests (Koenig 1953). In Mali, nests in colony of similar size were c. 15–100 m apart (Koenig 1956). Colonies at times more compact. In Iraq, 60 pairs as close-packed nucleus on c. 50 × 50 m of flat ground, with c. 30 pairs more dispersed to c. 500 m (Marchant 1963c); 641 nests in area c. 60 × 27 m and 125 nests along c. 50 m (Moore and Boswell 1956a). In Golodnaya steppe (Kazakhstan/Uzbekistan, USSR), 2–3 nests per metre of canal bank (Sagitov and Fundukchiev 1980). In Iraq, colonies sometimes number thousands of birds (Ticehurst et al. 1922) and in Syria, c. 300 nests recorded though not certain if all occupied (Misonne 1956). Near Biskra, each pair apparently had foraging and resting area c. 10–20 m long as well as other favoured perches near nest (Koenig 1953). Some antagonistic interactions near nest-entrance possibly indicate immediate vicinity of nest defended but no other suggestions of territoriality (C H Fry). Some colony sites apparently traditional over many years (e.g. Koenig 1895, Koenig 1953, Dharmakumarsinhji 1955), but nest-tunnel normally used only once (Dementiev and Gladkov 1951a). Sometimes nests close to, or forms mixed colonies with, *M. apiaster* (Tristram 1884; Heim de Balsac and Mayaud 1962; C H Fry). For nesting associates in USSR, see Dolgushin et al. (1970) and Sagitov and Fundukchiev (1980). ROOSTING. Nocturnal and, at least outside breeding season, communal. In thorny or leafy trees, including those standing in water (Dharmakumarsinhji 1944; C H Fry); on fronds of date palms *Phoenix dactylifera* (Koenig 1895); cliff-sites used in Oman (Gallagher and Woodcock 1980) and reed-beds by migrants (e.g. Dolgushin et al. 1970). Will roost with *M. apiaster* (see that species). ♂ said to roost away from nest-site (Hartley 1966) though not clear how far or at what stage of breeding cycle. Near Biskra, during nestling phase, both adults roosted in nest, and possibly sheltered there during sandstorms (Koenig 1953; see also *M. apiaster* and Parental Anti-predator Strategies, below). Breeding birds active from about sunrise to sunset (Koenig 1953). In Kazakhstan (USSR), late summer and autumn, birds sometimes visit reed-beds to rest around midday, otherwise fly there well before sunset for nocturnal roosting and leave at sunrise (Dolgushin et al. 1970). In India, migrates mostly from dusk to midnight and c. 04.30–09.30 hrs, feeding and resting during day (Ali and Ripley 1970). In Egypt, flock of c. 500 recorded resting on ground in cold, windy weather (Meinertzhagen 1930) and may also roost nocturnally on ground (Glutz and Bauer 1980). Frequently dust-bathes (Ganguli 1975). Wild birds not seen to sun-bathe but captive birds do so as readily as *M. apiaster* (Koenig 1953, which see for further details of comfort behaviour).

2. At approach of observer in white sun-helmet, perched birds gave Alarm-calls (see 5 in Voice) and fled; allowed approach to c. 15 m when no headgear worn, but shyer near nest (Koenig 1953). Feeding flocks of adults and young, late summer, relatively approachable (Eates 1939); see also Karcher (1940) for example of remarkably tame pair at noisy site. Vagrant, Netherlands, remained on perch when Sparrowhawk *Accipiter nisus* appeared; when raptor overhead, momentarily adopted an erect and elongated posture, with bill pointed up and plumage sleeked (Meeth 1962). In captive birds, panic-flight even more pronounced than in *M. apiaster*. Prior to roosting, bird assumes a steeply erect posture, gives Roosting-call, and vibrates tail (Koenig 1953: see 2 in Voice). FLOCK BEHAVIOUR. In small foraging flocks, birds generally maintain visual and acoustic contact (C H Fry). Migrants arriving on breeding grounds or at roosts may perform aerial evolutions likened to those of swifts *Apus*, including close-packed communal circling; also before departure in morning (Moore and Boswell 1956a; Hartley 1966). Often noisy when settling at roost (Heuglin 1869); birds observed by

A

Moore and Boswell (1956a) turned to face sun after landing. May remain at roost for some time in morning and give quiet Contact-calls, though noisy on leaving (Heuglin 1869: see 1 in Voice). Birds roost in compact rows, shoulder to shoulder (Koenig 1895; Koenig 1953; C H Fry: Fig A). Members of small flocks sat short distance apart on wires (Moore and Boswell 1956a). For association at mixed roosts with *M. apiaster*, see that species. Captive *M. superciliosus* and *M. apiaster* show little or no response to each other's Contact-calls (Koenig 1953). ANTAGONISTIC BEHAVIOUR. Generally peaceable and tolerant in flocks (Koenig 1895). Disputes arise on breeding ground when strange ♂ tries to feed ♀ whose mate excavating tunnel below ground. Antagonism usually indicated by vertical raising of wings; bird may also lunge towards opponent with pale facial feathers ruffled (Hartley 1966). Captive birds raised both wings, or sometimes only wing facing away from opponent, and gave Threat-call (see 6 in Voice). As in *M. apiaster* (which is dominant over and displaces *M. superciliosus*), single or repeated leaps (through 180°) during sideways-tripping motion on branch associated with both threat and courtship. Appeasing birds adopt posture identical to that of food-begging *M. apiaster* and give juvenile food-call (Koenig 1953, which see for full details, including illustrations of interactions with *M. apiaster*). HETEROSEXUAL BEHAVIOUR. (1) General. Very similar to *M. apiaster*. In Golodnaya steppe, flocks split up into pairs after arrival, late April (Sagitov and Fundukchiev 1980), and in Iran, courtship and copulation noted early May (Erard and Etchécopar 1970). (2) Pair-bonding behaviour. Birds may wrestle with locked bills. Also perform display in which wings raised at carpal joint and coppery undersurface exhibited; nape feathers sometimes ruffled. In contrast to aggressive lunging (see above) may lean away from partner in strained fashion (Hartley 1966). (3) Nest-site selection. Near Biskra, pair flew around calling excitedly; excavated briefly at various sites then departed. Possibly same pair performed similarly on following 2 days (Koenig 1953). (4) Nest-building. By both sexes, evenly shared (Hartley 1966); mainly in morning (Sagitov and Fundukchiev 1980). While 1 bird digging, other may kick away sand

B

from entrance (Hartley 1966). General features as in *M. apiaster*. Habitually-excavated false-start tunnels usually abandoned after a few cm, but may attain *c.* 1 m (C H Fry). (5) Courtship-feeding. Process as in *M. apiaster*. In Tadzhikistan (USSR), ♂ brought food while ♀ apparently excavating and summoned her to tunnel entrance with gentle Contact-calls. ♀ took food back into hole (Golovanova 1978: Fig B). ♀ frequently fed by ♂ during incubation, receiving food at entrance or flying out to take it (Hartley 1966). (6) Mating. Preceded by Courtship-feeding or initiated by call from ♀ (Hartley 1966: see 4 in Voice). Takes place on perch. Near Biskra, late May, ♂ approached ♀ in sideways-tripping motion and twice made beating or jabbing movements against her belly plumage (see *M. apiaster*), whereupon ♀ solicited (not described). ♂ called (see 4 in Voice), leaped on to ♀'s back, seized feathers on her forehead, and copulated. Vigorous beating movements performed by ♂ afterwards. Process similar to *M. apiaster* but lasted only *c.* 5 s (Koenig 1953, 1956). (7) Behaviour at nest. No information. RELATIONS WITHIN FAMILY GROUP. Eyes of young begin to open at *c.* 5 days, fully open by *c.* 14 days, when birds becoming alert and mobile. Nestling supports itself on heels until fledging (or shortly thereafter) and can shuffle forwards and backwards with surprising speed (C H Fry). Duration of brooding not known. Near Biskra, change-overs (at nest with young) took place every *c.* 1–1½ hrs and off-duty bird foraged or perched nearby and self-preened (Koenig 1953). According to Sagitov and Fundukchiev (1980), 1 adult brings food and passes it to mate in nest for feeding to young (of uncertain age). After fledging, young fed for *c.* 2½–3 weeks until finally proficient in prey capture and handling and thus independent (Dementiev and Gladkov 1951a). ANTI-PREDATOR RESPONSES OF YOUNG. No information. PARENTAL ANTI-PREDATOR STRATEGIES. During disturbance at nest, birds often await developments at regularly used perch nearby. When nests opened to remove some nestlings for captive rearing, one pair circled and gave ceaseless Alarm-calls (see 5 in Voice); another resumed feeding after nest-chamber re-sealed (Koenig 1953). At colonies in flat ground, lower Syr-Darya (USSR), birds said often to block entrance-holes at night or when they leave during day (Spangenberg and Feygin 1936); not confirmed elsewhere (Koenig 1953; C H Fry).

(Fig A after photograph in Koenig 1953; Fig B from photograph in Golovanova 1978.) MGW

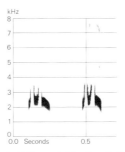

I C Chappuis/Sveriges Radio (1973) Sénégal January 1969

Voice. Used freely during all day-time activities, with almost continuous noise emanating from larger breeding colonies and foraging or migrating flocks; also calls at night. Repertoire similar to and homologous with that of Bee-eater *M. apiaster* (Hutson 1947; Koenig 1953; C H Fry). Fullest study by Koenig (1953), based partly on observations of captive birds, shows similarity of many calls to *M. apiaster*; not easy to distinguish between 2 species in high-flying flocks (Reynolds 1974). Higher pitch of *M. superciliosus* calls said to be a more or less general feature (Koenig 1953); see also below. Account includes material supplied by C H Fry.

CALLS OF ADULTS. (1) Contact-calls. A clear-sounding, more or less distinctly disyllabic 'diripp' or 'dripp'; clearer than the sonorous 'rüpp' of *M. apiaster* (Koenig 1953); trilling 't-r-r-r-p' (Moore and Boswell 1956a); repeated musical and interrogative 'tetew' as contact-call from migrants (Ali and Ripley 1970). Recording (Fig I) of most usual call given by perched bird suggests a reedy or husky sounding 'p-rr-eer', i.e. with division into 3 sub-units (J Hall-Craggs, P J Sellar). Sonagraphic analysis of this call shows pitch to rise *c.* 1·0 kHz higher than Contact-call of *M. apiaster*, but to descend (in 3rd sub-unit of 2nd unit) *c.* 200 Hz lower (J Hall-Craggs). Higher pitch of this call in *M. superciliosus* also noted by (e.g.) Christensen (1962) and Feeny *et al.* (1968); belief of some authors that this call lower pitched than equivalent of *M. apiaster* possibly due to latter's timbre, brevity, and general structure (J Hall-Craggs). In recording by J-C Roché, Morocco, March, initial sounds of bout strongly reminiscent of Dunlin *Calidris alpina* and followed by possible Alarm-calls (see call 5, below): 'wheer wheer wheer ti ti-ti-ti pip pip' (J Hall-Craggs, P J Sellar). Such 'wheer' sounds probably main reason why *M. superciliosus* voice sometimes described (e.g. Chapin 1939, Dementiev and Gladkov 1951a, Anon 1952) as less resonant and attractive, hoarser and harsher, harder in tone than *M. apiaster*. For further descriptions and comparisons, see (e.g.) Whistler (1941), Dharmakumarsinhji (1955), and Dolgushin *et al.* (1970). (2) Roosting-call. A slightly rolling 'dirippdiripp-diripp' (Koenig 1953). (3) Courtship-calls. A short 'tjup' and a rolling 'diripp' (Koenig 1953); 2nd of these presumably same as call 1 (above). (4) Calls associated with mating. A plangent sound given by ♀ prior to copulation (Hartley 1966); perhaps similar to Distress-call uttered by ♀ *M. apiaster* in same situation. A clear 'ping-ping-ping' from ♂ mounting ♀ (Koenig 1953). (5) Alarm-call. A 'dick dick' very like *M. apiaster* but slightly higher pitched; captive *M. apiaster* respond immediately with similar calls (Koenig 1953). (6) Threat-call. A clear-sounding, almost shrill, 'drüüü', falling slightly in pitch (Koenig 1963). (7) Food-call of young (see below) used by adults in appeasement (Koenig 1953).

CALLS OF YOUNG. Food-call of older young a trilling 'drüüüü' or 'dirrüüüü'; unlike *M. apiaster*, no marked changes during development (Koenig 1953). MGW

Breeding. SEASON. Iraq: main laying period first half of May (Marchant 1963c). North-west Africa: eggs found early May to early June (Heim de Balsac and Mayaud 1962). SITE. Tunnel in sloping ground, vertical bank, or even nearly flat ground. Colonial. Nest: excavated tunnel 1–2 m long, ending in enlarged chamber out of sight of entrance; in slightly undulating ground, tunnel inclined down by 10° (Marchant 1963c). Building: excavation by both sexes; no material added. EGGS. See Plate 98. Blunt elliptical, smooth and glossy; white. *M. s. persicus*: 26 × 21 mm (24–27 × 20–23), $n = 100$ (Baker 1934); calculated weight 6·5 g (Makatsch 1976). *M. s. crysocercus*: no information. Clutch: 5–7. INCUBATION. By both birds during day, but in longer spells by ♀ (up to 3 hrs or more); longest spell by ♂ *c*. 30½ min. Only ♀ at night, for up to 12½ hrs (Hartley 1966). No further information.

Plumages (*M. s. persicus*). ADULT BREEDING. Forehead white, bordered behind by bright pale blue band which continues in broad pale blue supercilium, sometimes mixed with some white at border of black stripe from lores to just in front of and below eye. Crown, hindneck, sides of neck, all uppersurface of body, tertials, and upper wing-coverts uniform bright grass-green, similar in tinge to upperparts of Egyptian race of Little Green Bee-eater *M. orientalis cleopatra*. Ear-coverts black with variable amount of green on feather-tips; in some birds, largely dark green, others mainly black. Chin bright yellow, merging into rufous-cinnamon of central throat, bordered at sides by grass-green; a broad bright pale blue area from gape backwards below black or green-black stripe from lores to ear-coverts; pale blue often separated from green-black ear-coverts by narrow white line. Remainder of underparts down from chest grass-green grading to bluish-green on belly and pale greenish-blue on under tail-coverts. Uppersurface of tail green with dark horn shafts (slightly duller than upperparts), undersurface dusky grey with pale horn shafts. Primaries and outer webs of secondaries grass-green, like upperparts and upper wing-coverts; broad borders of inner webs of primaries dusky grey, grading to cinnamon at base and to black on tip, where black narrowly extends to outer webs; inner webs of secondaries mainly deep pinkish-cinnamon, tips narrowly black (much narrower than in Bee-eater *M. apiaster*). Under wing-coverts, axillaries, and bordering feathers of flank rufous-cinnamon, similar in tinge to undersurface of flight-feather bases; small under wing-coverts below leading edge of wing and below wrist slightly greenish-yellow. Abrasion and bleaching have marked influence: all grass-green changes to light blue, first on feather-tips of rump, on greater upper primary coverts, and on tips of primaries (by about April), soon followed by tertials, tips of scapulars, t1, and tips and edges of other tail-feathers (by about May). In worn plumage, white on forehead and just above and below black (rather than greenish-black) ear-coverts limited; pale blue on forehead and sides of head paler, more restricted; chin yellowish-white; crown, hindneck, mantle, scapulars, tertials, and upper wing-coverts pale blue, mixed with variable amount of grass-green on feather-bases; rump, upper tail-coverts, tail, and flight-feathers mainly pale blue, except for some grass-green on bases or centres of feathers hidden under other feathers at rest; green of underparts slightly to markedly bluer, greenish-blue or deep turquoise-blue, sometimes closely similar to underparts of *M. apiaster* (though with different throat pattern, e.g. lacking black band down across throat). Sexes similar. ADULT NON-BREEDING. Mainly like adult breeding. New feathers on forehead and in pale blue on sides of head mainly grass-green; on chin yellowish-green; on throat rufous with greenish feather-tips; on crown, neck, scapulars, chest, and breast duller green than in fresh adult, slightly more olive-green, on crown often brownish towards feather-bases. NESTLING. Similar to *M. apiaster*, but more slender in appearance (Koenig 1953). JUVENILE. On fledging, upperparts, sides of neck, tertials, and upper wing-coverts green, slightly duller than in adult breeding and with dusky feather-bases shining through (tips with slight golden lustre in some lights), forehead and broad supercilium yellow-green or pale blue-green (not white or pale blue), feather-tips of lower scapulars, tertials, rump, and upper tail-coverts already abraded to pale blue. Dark stripe from lores through eye to ear-coverts as in adult. Chin pale yellow, merging more gradually into rufous-cinnamon at throat than in adult; colours less deep than adult, some white of feather-bases often exposed, yellow and cinnamon extend laterally to dark lores and ear-coverts; unlike adult, virtually no white, pale blue, grass-green on cheeks. Remaining underparts pale blue-green, some grey of feather-bases shining through. Tail as in fresh adult, but t1 not elongated (see Structure). Flight-feathers as adult, but green slightly duller, feather-tips dusky grey rather than black and less sharply defined; inner webs of secondaries with cinnamon paler and more restricted to base. Under wing-coverts, axillaries, and bordering part of flanks slightly less deep rufous-cinnamon than adult. As in adult, influence of abrasion and bleaching marked: all green of exposed feather-tips and -sides changes to blue. In extreme cases, all upperparts glossy dark turquoise-blue, rump and upper tail-coverts brighter blue, underparts down from chest and outer webs of outer primaries pale blue, tail and outer webs of secondaries greenish-blue; chin and throat slightly paler than in fresh juvenile. FIRST ADULT NON-BREEDING. Like adult non-breeding, but probably much worn juvenile retained. As in adult, new feathers on forehead and supercilium grass-green with variable amount of pale blue suffusion; those on throat tipped yellow-green, on crown and remainder of body grass-green with brownish or olive base. FIRST ADULT BREEDING. Like adult breeding, and usually indistinguishable. A few birds examined retained a single heavily worn secondary or tail-feather, and these probably 1 year old.

Bare parts. ADULT. Iris crimson or blood-red. Bill black. Leg and foot brownish-flesh or dark brown, edges of scutes whitish. NESTLING. As Bee-eater *M. apiaster*; bare skin pink, but pale yellow in one bird in poor condition (Koenig 1953). JUVENILE. As *M. apiaster* (Koenig 1953); iris dark brown, leg and foot dark flesh (Dementiev and Gladkov 1951a).

Moults. Based on specimens examined in BMNH, RMNH, ZFMK, ZMA, and ZMM. ADULT POST-BREEDING. Complete, primaries descendant. Starts from mid-July in North Africa (including Egypt), from August in Middle East and USSR; p1 and scattered feathers of mantle, scapulars, rump, and chin to chest first. Moult suspended during autumn migration, when inner 3–5 primaries new North Africa and 0–4 in USSR. Moult resumed in winter quarters, October–November; completed December–February. Tail starts with loss of p6–p7, completed at about same time as p10; sequence of replacement is 1–2–6–3–5–4. Secondaries start from 2 centres (inwards from s1, in both directions from s10–s11) at about same time as t1; completed (s4 last) with p10 and t4–t5. ADULT PRE-BREEDING. Partial; in winter quarters. Probably involves only those parts of feathering of head and body which were replaced by non-breeding during early stages of post-breeding. POST-JUVENILE. Complete. Starts at age of *c*. 3 months (Koenig 1953); p1 and scattered feathers of head and body first. As in adult post-breeding, *chrysocercus* and Egyptian *persicus* start when near breeding area, replacing 10–90% of feathering of head and body (by non-breeding) and 3–4 inner primaries August–September; moult suspended during autumn migration and resumed in winter quarters, where remaining old juvenile plumage apparently directly replaced by fresh breeding, as in adult; moult completed January–March (including partial 1st pre-breeding). No juvenile of *persicus* from USSR examined showed moult August, and these birds probably migrate in full or almost full juvenile plumage, starting in winter quarters, completing March–May. As in adult post-breeding, tail and secondaries start with loss of about p6–p7 (primary moult score 17–26); sequence of replacement similar. A few 1-year-olds from USSR retain a single juvenile secondary or tail-feather during summer.

Measurements. *M. s. persicus*. USSR and Middle East, April–August; skins (RMNH, ZMA, ZMM). Tail is to tip of t1; tip is tip of t1 to tip of 2nd longest (usually t3–t4). Bill (F) is to forehead, bill (N) to distal corner of nostril; both in adult only.

WING AD	♂	157	(2.77; 24)	151–161	♀ 151	(3.84; 20)	146–158
JUV		145	(5.18; 13)	139–155	138	(1.70; 7)	136–140
TAIL AD		146	(7.97; 23)	134–162	132	(5.47; 15)	123–141
JUV		90.9	(2.47; 10)	87–97	88.5	(2.24; 6)	86–91
TIP AD		55.4	(6.87; 23)	43–67	41.4	(5.26; 15)	35–52
BILL (F)		48.4	(2.46; 12)	47–50	47.5	(2.53; 10)	44–50
BILL (N)		36.0	(1.59; 22)	33–38	34.0	(2.50; 19)	30–37
TARSUS		13.0	(0.56; 10)	12.5–13.9	12.6	(0.29; 7)	12.2–13.1
TOE		19.9	(0.97; 9)	19.1–21.3	19.7	(1.83; 6)	18.5–21.8

Sex differences significant, except juvenile tail, bill (F), tarsus, and toe.

M. s. chrysocercus. Morocco and Algeria, March–August; skins (BMNH, RMNH, ZFMK, ZMA).

WING AD	♂	152	(2.99; 20)	148–159	♀ 145	(2.42; 15)	142–150
TAIL AD		168	(10.9; 16)	150–192	141	(6.76; 10)	130–151
TIP AD		78.2	(11.1; 16)	65–105	54.6	(6.56; 9)	46–64
BILL (F)		45.2	(1.74; 6)	42–47	43.9	(2.46; 9)	40–47
BILL (N)		32.3	(1.65; 7)	29–34	31.5	(2.61; 9)	28–35

Sex differences significant, except bill. Birds from Sahel zone (involving wintering birds from north-west Africa as well as local breeders) have significantly longer bills: e.g. bill (N), ♂ 34.7 (11) 33–37, ♀ 34.0 (5) 31–37.

Weights. *M. s. persicus*. Afghanistan: April, ♂♂ 49, 53; ♀♀ 45, 49; late July and early August, ♂ 46, ♀ 48 (Paludan 1959). Migrants southern Caspian, late April: ♂♂ 45, 45, 46 (Schüz 1959). Southern Iran, August: 38, 42, 44 (D J Garbutt, BTO). Zaïre and Tanzania, December to early February: ♂♂ 47, 51, 52, 52, 56; ♀♀ 50, 51 (Meise 1937; Verheyen 1953).

M. s. chrysocercus. Algeria, breeding: 38 (probably ♀) (Dupuy 1970a). South-east Morocco, April: 45.5 (BTO).

Nominate *superciliosus*. Zaïre and Tanzania: ♂♂ 44, 48; ♀ 41.2 (4) 40–43 (Verheyen 1953; Britton 1970).

M. s. alternans. Angola: ♂ 45.4 (1.45; 16) 43–49, ♀ 41.2 (3.77; 4) 37–46 (Clancey 1971b).

Structure. 10 primaries: p9 longest, p10 97–106 shorter in adult, 82–96 in juvenile; in both, p8 6–16 shorter, p7 18–26, p6 26–36, p5 35–47, p1 58–80; p10 reduced, narrow and acutely pointed in adult, sometimes slightly broader and with less sharp tip in juvenile. Tail rather long, 12 feathers; in adult, t1 elongated, distal half attenuated towards sharply pointed or narrowly rounded tip; t3–t5 2nd longest, t2 often slightly shorter than these, t6 2–5 shorter; in juvenile, t4–t5 longest, t6 and t2 2–5 shorter; t1 not elongated, though tips not as rounded as in other feathers, either about equal in length to t4–t5 or slightly shorter, frequently up to 12 mm longer. Bill longer and more gradually decurved than in *M. apiaster*, relatively less deep at base and with finer tip. Outer toe c. 90% of middle, inner c. 79%, hind c. 56%. Remainder of structure as in *M. apiaster*.

Geographical variation. Rather slight, mainly in size. *M. s. chrysocercus* from north-west and West Africa differs from *persicus* in particular by combination of shorter wing with relatively longer t1; differences in colour less marked, those described elsewhere attributable mainly to differences in abrasion. In general, adult *chrysocercus* shows less white on forehead, sometimes virtually none; pale blue supercilium and pale blue band on forehead narrower (*persicus* may show similar narrow streak in non-breeding) and often no white below greenish-black ear-coverts (some *persicus* also without this, and white may disappear when plumage worn); upperparts and especially t1 and outer webs of primaries perhaps slightly more yellow-green when plumage fresh and paler blue when worn, but practical use doubtful in view of rapid colour change through wear; no constant difference in bill shape, but note longer bill length of Sahel birds (see Measurements), perhaps in part caused by limited abrasion of bill during non-breeding season. Nominate *superciliosus* from eastern Africa (Ethiopia and Somalia to Zimbabwe and Moçambique—see Fry 1981b), and Madagascar differs from both *persicus* and *chrysocercus* in narrow white band along forehead and over eye (white often slightly green, yellow, or blue, sometimes absent on forehead; not broad and bright pale blue); dark brown crown; duller green upperparts and tail; limited pale yellow on chin, extending into narrow pale yellow or white strip below black stripe from lores to ear-coverts, with remainder of chin and throat rather dull rufous-cinnamon; remainder of underparts duller olive-green, less bluish; size smaller, adult wing 128–140, elongated tip of t1 45–63 (♂) or 32–44 (♀). However, sometimes difficult to tell nominate *superciliosus* from *persicus*/*chrysocercus*: *persicus* and *chrysocercus* in non-breeding plumage show narrower bands on forehead and above eye, browner crown, and duller green body, while some breeding adults of nominate *superciliosus* have crown only slightly browner green than upperparts and in some East African non-breeding birds umber-brown on crown is replaced by green and vice versa; also, *alternans* from northern Namibia and Angola has greener crown than nominate *superciliosus*, more grass-green body (more bluish than nominate *superciliosus* when worn), wider white chin and malar streaks, and slightly larger size (Clancey 1971b), tending towards *persicus* or *chrysocercus*.

Forms superspecies with Blue-tailed Bee-eater *M. philippinus* from India, south-east Asia, Indonesia (local), New Guinea, and Bismarck archipelago; rather similar to nominate *superciliosus*, agreeing in size (wing adult ♂, India and Pakistan, 133 (8) 131–137, against 154.4 (5) 150–158 for *persicus* from same area: Marien 1952) and olive-green general colour, but crown similar in colour to mantle (tinged green-buff like chest); pale streak along forehead and above eye narrow or almost absent; tertials, rump, upper tail-coverts, and tail bright blue (only slightly greenish when fresh; distinctly brighter than dull pale blue tail of *persicus* in abraded plumage); pale line below black streak through eye pale blue and rather narrow; under tail-coverts pale blue, contrasting with bluish-green belly. For races and distribution of *M. philippinus*, see Deignan (1955) and Fry (1969b). *M. philippinus* separated from *superciliosus* on account of slight overlap in breeding range with *persicus* (without interbreeding) (Marien 1952). *M. s. persicus* (with *chrysocercus*) sometimes considered separate species from nominate *superciliosus* (with *alternans*), superspecies then comprising 3 species; occasionally, *philippinus* combined with nominate *superciliosus*, while *persicus* considered separate species. However, in view of intergradation of *chrysocercus* and *persicus* into nominate *superciliosus* and *alternans*, these last 4 better considered races of single polytypic species.

CSR

Merops apiaster Bee-eater

PLATES 70 and 71
[between pages 734 and 735]

Du. Bijeneter   Fr. Guêpier d'Europe   Ge. Bienenfresser
Ru. Золотистая щурка   Sp. Abejaruco común   Sw. Biätere

Merops Apiaster Linnaeus, 1758

Monotypic

Field characters. 27–29 cm, of which tail-streamers up to 2·5 cm; wing-span 44–49 cm. Medium-sized insectivore close in size to Golden Oriole *Oriolus oriolus*, with long decurved bill, long tail, and sharply pointed wings in flight. Commonest and most widespread bee-eater of west Palearctic, with yellow throat and pale green-blue body dominant below and chestnut crown and wing-coverts and yellow lower back contrasting with sheeny blue flight-feathers above. Persistent aerial hunter of insects, with flight including characteristic sailing and sudden accelerations. Call characteristic. Sexes almost similar and not always separable in the field; marked seasonal variation. Juvenile separable.

ADULT MALE BREEDING. Plumage multicoloured, more vivid and contrasting than any other bird in west Palearctic. Crown, nape, and mantle dark chestnut shading into golden-chestnut and buff-yellow on back, scapulars, and rump, with last two distinctly paler than rest and obvious in flight. Upper tail-coverts dark blue-green; tail dark sheeny green above, grey-brown below, with central 2 feathers extended into long, almost black points. Wing shows chestnut patch on most greater and median and innermost lesser coverts, surrounded by blue-green outermost lesser coverts, primary coverts, and innermost greater and median coverts and mainly sheeny blue, almost black-tipped flight-feathers. Face and sides of head strongly patterned, with white forehead, pale green and blue fore-supercilium, bold black eye-stripe, pale blue-green line under eye-stripe, and vivid yellow chin and throat, bordered below with narrow black band reaching chestnut hindneck. Rest of underparts sheeny green-blue. Underwing pale orange-chestnut. ADULT FEMALE BREEDING. At close range, usually shows more green on forecrown, across back, on rump, and within chestnut area of wing. ADULT NON-BREEDING. Plumage of body renewed from August and retained until December with striking increase in green blue-tipped feathers on back and rump serving to isolate chestnut rear head. Black throat-band restricted to black-speckled green line. Sexes not separable in the field. JUVENILE. Resembles winter adult but whole appearance less vivid and variegated. Most striking differences: (1) green tone to crown and nape; (2) blue to green back, with dull grey- or buff-toned scapulars; (3) blue to green rump; (4) dull blue tail, central feathers not elongated; (5) uniform dull green wing-coverts; (6) duller blue-green flight-feathers, lacking chestnut outer webs on inner secondaries.

Breeding adult unmistakable, but adult from August to December and juvenile from fledging in mid-June to February lose vivid contrasting colours and assume largely green backs, thus inviting confusion with Blue-cheeked Bee-eater *M. superciliosus*. At all times, however, yellow throat and brown-toned rear of head diagnostic; *M. superciliosus* shows dark coppery underwing—also however, when worn, blue underbody, but blue then darker than that of *M. apiaster*. Flight variable in action and speed: when direct, bursts of rather stiff wing-beats alternate with momentary wing closures and produce now fast, now shooting, now slow and slightly undulating track; when hunting insects in flight, planing on rigid wings alternates with graceful wheeling, interrupted by bursts of wing-beats to maintain momentum or to capture prey; when on migration, bursts of noticeably regular wing-beats maintain rapid progress often at great height. On migration, distant compact wheeling flocks closely resemble those of Levant Sparrowhawk *Accipiter brevipes*. Perches freely but gait restricted to active shuffle.

Commonest call, constantly given by breeding and migrant birds, a liquid, rather cheerful 'quilp' (rather similar to call of *M. superciliosus*—see that species).

Habitat. Breeds in middle and lower middle dry warm latitudes of west Palearctic, in Mediterranean, steppe, and desert zones, and marginally into temperate continental regions, up to July isotherm of 21°C, but exceptionally to *c.* 17°C. Although mainly in lowlands, ascends in Indian subcontinent to 2000 m or more and to *c.* 2500 m in Armeniya (USSR). On migration and in African winter quarters occasionally even higher, up to *c.* 3000 m. Requires sunny warm open landscapes with scattered trees, sheltered valleys, steppe plains, or banks of rivers, taking abundant large invertebrates in flight, and rising freely to upper airspace at all seasons. Less aerial, however, than swifts (Apodidae), preferring to rest for long periods on commanding perches such as telephone wires (Voous 1960). In Spain, lives mainly in valleys where cultivation mingles with little woods, but also in vast treeless tracts and in clearings in pine and oak forests, nesting sometimes in small cuttings or in burrows under almost level soil; very commonly along rivers (Valverde 1953*a*). In Camargue (France), a principal habitat is the high *sansouire*, a dry steppe dominated by glasswort *Salicornia*, giving way with declining salinity to sea lavender *Statice* and grasses, often colonized by philaria thicket *Phillyrea* and tamarisks *Tamarix gallica*; this provides hunting area, and also breeding place where sandy cliffs occur (Hoffman

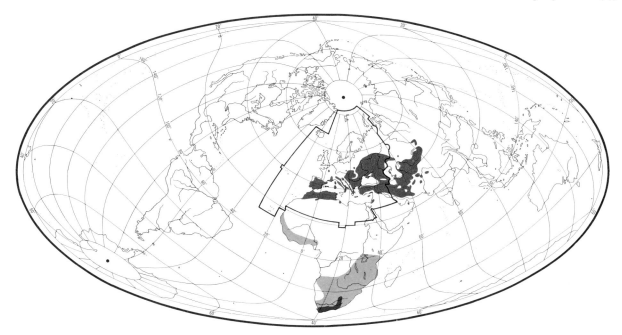

1958). More favourable habitat, however, found in reclaimed and irrigated ricefields, with margins of canals, *étangs*, ponds, and, to lesser extent, vineyards and grasslands where dragonflies (Odonata) abundant. In stony plains of the Crau, hunting extends to bare hills and above all, over olive groves. In Gard, vineyards favoured on hills and farmland, or pasture lower down; in Petite Camargue, dunes covered with pines (Swift 1959). In USSR, on open steppeland or in sandy semi-deserts, along precipitous clay river banks, shrub-clad ravines, gullies, cliffs with woods or scattered trees, and around orchards and villages, especially those with bee-hives. In mountainous areas, ascends valleys to 1500–2500 m. Occupancy of otherwise suitable habitat often limited by lack of nesting sites (Dementiev and Gladkov 1951a). Burrows are dug mainly in cliffs and steep banks of dry loamy clay, firm sand, soft sandstone, laterite, or ground. Humanly modified habitat must afford blend of dry grassland, tree or shrub stands, and uneven terrain with banks or exposed faces comparable to that found in natural wooded steppe or steppe heath, and must include some rich source of invertebrate food, but not necessarily water or wetland. In south of range, cork oaks *Quercus suber* and palms often favoured. In African winter quarters, occupies savannas, plains, grasslands, dry forests, lake shores, banks of large rivers, and cultivation, but avoids dense forest. In Sierra Leone, winter habitat differs from that of Blue-cheeked Bee-eater *M. superciliosus* in being further inland, in central grasslands or savanna woodlands and hillier areas, including main mountain plateau levels up to *c*. 1300 m (G D Field). Migrates normally in upper airspace, but sometimes at 50 feet or so along coasts (S Cramp). When crossing elevated land at hours when it is not warmed by sun (e.g. Baluchistan at *c*. 2000 m and over) frequently experiences stalled conditions due to insufficient lift (E M Nicholson). Otherwise strongly aerial. Hunts generally within *c*. 1 km of nest, but sometimes up to 4 km or more distant. Will sometimes plunge momentarily from flight into still water.

Distribution. Range fluctuating, with occasional breeding outside normal limits at any one period. Some signs of northward expansion in 1840s, followed by recession from *c*. 1875; better-documented increase beginning in 1920s and 1930s, with some expansion in central Europe in later 1940s, followed by more marked expansion further west from mid-1960s (reaching even Denmark and Sweden) still to some extent continuing (Glutz and Bauer 1980). However, some of these gains (e.g. Scandinavia) were temporary, and even further south (e.g. parts of Czechoslovakia and Austria) nesting often irregular.

BRITAIN. Bred 1955 Sussex and attempted breeding 1920 Midlothian (Parslow 1967). Channel Islands: bred Alderney 1956 (Le Sueur 1951). FRANCE. Before *c*. 1930, bred mainly in south-east (Gard) and Corsica with isolated nesting elsewhere: e.g. Somme 1840, 1910, Lorraine 1910, Brenne 1908 and 1911. Marked spread began soon after to other parts of south-east (Hérault, Bouches-du-Rhône, Vaucluse, Var, etc.) and more recently further north to Ardèche and Isère and regularly Paris region since 1968, with isolated breeding further west, e.g. Normandy and Finistère (Swift 1959; Yeatman 1976; Glutz and Bauer 1980; Tostain and Siblet 1981; R Cruon). BELGIUM. Bred 1956, with unsuccessful attempt 1933 (Lippens and Wille 1972). NETHERLANDS. Bred 1964, 1965, and 1983 (CSR). WEST GERMANY. Most breeding records from southern Germany, but occasionally elsewhere. In 19th century

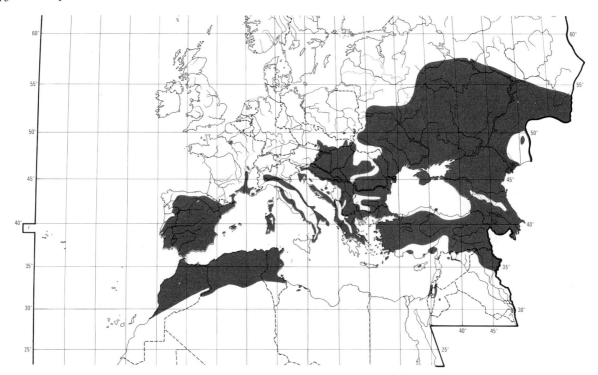

bred 1834, 1843, 1854, 1858, and 1873–8 in south, and 1889 in Hessen. In 20th century, after nesting 1912 and 1916 in south, many more recent breeding records: in south, 1956, 1964, 1966–7, 1975–7 (perhaps continuously at one site from c. 1965); 1964 Schleswig-Holstein, 1971 Hessen, and 1978 Westfalen (Glutz and Bauer 1980); also bred near Bonn 1974–5 and Niedersachsen c. 1976–9, and thought likely to have bred every year in small numbers (usually at different sites) in last 25 years (GR). DENMARK. Bred 1948, 1961–2, 1966, and 1973 (Dybbro 1976). SWEDEN. Bred 1976–8 (LR). FINLAND. Probably bred 1954 (Merikallio 1958). EAST GERMANY. Bred 1973–6 and possibly 1971 and 1977 (Glutz and Bauer 1980; Makatsch 1981). POLAND. Bred 1792, 1887, c. 1950, and c. 1963–5; from 1965 has apparently nested regularly (Tomiałojć 1976a; AD). CZECHOSLOVAKIA. Not breeding annually in southern Moravia; has bred occasionally west and south of main areas (Glutz and Bauer 1980; KH). AUSTRIA. Breeds fairly regularly, especially in northern Burgenland, more irregularly elsewhere (Glutz and Bauer 1980; Gamauf and Haar 1981; HS). ITALY. Sicily: bred 1869 and 1982 (BM, A Ciaccio and M Siracusa). USSR. Sporadic nesting outside main range (Dementiev and Gladkov 1951a). Although increased reporting may lead to some bias, signs of range increase to north in recent years (Dementiev and Gladkov 1951a; Glutz and Bauer 1980; G Buzzard). SYRIA. Has bred south of area mapped (Kumerloeve 1968b), but no evidence 1974–7 (A M Macfarlane). LEBANON. May have bred formerly, but no recent evidence (Tohmé and Neuschwander 1974; HK). CANARY ISLANDS. Bred Gran Canaria 1856 (Bolle 1857).

Accidental. Britain, Ireland, Belgium, Netherlands, Norway, Sweden, Finland, East Germany, Azores, Madeira, Canary Islands, Cape Verde Islands.

Population. No detailed figures for most of range. Numbers have probably fluctuated with changes in range (see Distribution); European population may be at temporary peak, but recent local declines in Israel and Cyprus. SPAIN. Numerous, locally abundant (Valverde 1953b; AN). FRANCE. Marked increase (see Distribution); estimated 100–1000 pairs (Yeatman 1976). Corsica: nowhere abundant; may have decreased (J-C Thibault). CZECHOSLOVAKIA. Increased 1946–50, since then both increases and decreases locally (KH). AUSTRIA. About 20 pairs 1959–60, none in 1965, 30 pairs in 1978 (Glutz and Bauer 1980). HUNGARY. Estimated 1271 pairs 1949, c. 2000 pairs 1955, and 1350 pairs 1977; these may be underestimates, with fluctuations in some areas (Glutz and Bauer 1980). ITALY. Decreasing in some northern areas (PB, BM). GREECE. Locally common in north and centre, very rare in Pelopónnisos, rare on some islands (WB, HJB, GM). ISRAEL. Marked recent decrease (HM). CYPRUS. Locally fairly common (Stewart and Christensen 1971); now scarce (PRF, PS). MOROCCO. Common, except in high mountains (JDRV).

Movements. Migratory; exclusively a summer visitor to breeding range. A few winter records from India (Ali and Ripley 1970) and Arabia (Bundy and Warr 1980), but normally winters entirely within Africa.

Recorded all across northern Afrotropics, though over

large north-central parts (Nigeria, Chad, western Sudan) the species is chiefly a passage migrant and (with the exception of Chad) scarce even then. Hence winter range in 2 distinct segments (Moreau 1972).

(1) West Africa and Sénégal to Ghana, marginally to Nigeria. These likely to be birds from south-west Europe (including France) and north-west Africa, judging from large-scale passage at Straits of Gibraltar and across western Sahara. No ringing recoveries south of Sahara, but south-west/north-east movement in Mediterranean basin indicated by French birds found Balearic Islands (August), south-east Spain (September), and Morocco (April); also Tunisian migrant (May) found next day in Italy, 520 km NE. Not known whether any which enter West Africa continue further south.

(2) Eastern and southern Africa, mainly south of equator and Congo basin forests. Forms the larger and more important part of winter range, presumably taking birds from central and eastern Europe, western Siberia, and south-west Asia. One ringed Ryazan, USSR (41°E) found in Zimbabwe (18°S), February; 3 other Ryazan birds killed on passage (October) through Cyprus. Other east Mediterranean recoveries are Hungarian bird in Greece and 2 Bulgarian birds in Cyprus, all found September. Important passages along Nile and Rift Valleys, and along and across Red Sea. Also, this is one of the few species which regularly overflies the main equatorial forest block (Moreau 1972), and these may include the large numbers of migrants which occur southern Chad in September–October and early May (Salvan 1968b).

Passages broad-front overland, with unbroken overflying of Sahara and Arabian deserts. Migrant flocks favour thermal conditions; hence some tendency to concentrate at narrows for Mediterranean crossing: Straits of Gibraltar, Sicilian Channel, and Cyprus/Levant. Sicilian Channel passage much more pronounced in spring, in line with more conspicuous spring (than autumn) passage through North Africa generally (e.g. Smith 1968 for Morocco, Bundy 1976 for Libya). In Persian Gulf states, also, spring passage is the larger (Bundy and Warr 1980). Not clear whether this variation is due to route changes or (as is more likely) to less long-range overflying in spring.

Family parties begin congregating in second half of July, and main exodus from Europe spans mid-August to early October. In Straits of Gibraltar study (López Gordo 1975), 35 850 birds crossed in autumn 1974, 80% in September; usually in parties of 20–30 birds, passing from $\frac{1}{2}$ hr before sunrise to $\frac{1}{2}$ hr before sunset but with lull around midday; see also Sutherland and Brooks (1981) for south-east Turkey. Some nocturnal movement occurs (Hutson 1947; Telleria 1979). See also Social Pattern and Behaviour for migrant flock sizes and sociability. When not overflying inhospitable terrain, diurnal migrants often break their journey to hawk insects before continuing (Hutson 1947). Present in African winter quarters from mid-September (early October in South Africa); return movement begins March in south, continuing through April. In lower Egypt, where first arrivals in 2nd week of April, passage 3 weeks later than Blue-cheeked Bee-eater *M. superciliosus*; European spring passage mid-April to late May. Spring migrants often overshoot in anticyclonic weather, regularly reaching north-west Europe and exceptionally nesting.

Small numbers which breed in South Africa believed not to be Palearctic birds nesting a second time in winter quarters, but a discrete population which moves northwards within southern Africa after breeding (Clancey 1964). This supported by freshly moulted birds collected Angola in July–August (Moreau 1972).

Food. Account includes material supplied by C H Fry.

Flying insects, with Hymenoptera usually predominant and bees and wasps (Aculeata) preferentially selected. Often gregarious when foraging. Hunts from perch (bare branch, wire, etc.), rarely lower down on (e.g.) animal droppings (where no other perch available) making flights of a few to several hundred metres in length. Apparently able to spot flying insect *c*. 50–100 m away (C H Fry; 80–95 m in Blue-tailed Bee-eater *M. philippinus*: Fry 1981a). Head moves jerkily to fix both eyes on prey (bee *Bombus* at *c*. 60 m: Helbig 1982); follows movements of insect, then takes off abruptly (Koenig 1951) and makes fast, direct approach before curving away to come at insect tangentially from behind (Helbig 1982; also Stichmann 1964). According to Glutz and Bauer (1980) direction of attack apparently depends more on position of hunter and prey than on other factors; butterflies (Lepidoptera) generally seized from below (e.g. Sassi and Zimmer 1941), perhaps in order to avoid wings. Alternates gliding on rigid wings with rapid wing-beats and will dive, twist, and turn in pursuit of prey taking evasive action (C H Fry); rarely, will fly upside down (Fintha 1968). Audible snapping marks prey-capture and insect then carried in bill-tip back to (usually same) perch (Koenig 1951; Helbig 1982; C H Fry). Normally 1 item taken but may swallow prey in flight and catch 2nd insect immediately if opportunity arises while returning to perch (Schumann 1971). Also uses continuous low searching flight, which may involve capture of several items. As young become older, bird more often flies thus directly from and to nest, returning as soon as single item captured (Koenig 1960; Helbig 1982; also Barham *et al.* 1956, Krimmer *et al.* 1974). Also ascends, sometimes using thermals, to *c*. 150 m or more (may be invisible, only calls heard); such flights may last up to *c*. 20 min and several items caught and swallowed while still aloft; sustained high-level flights recorded on breeding grounds and in winter quarters, birds presumably feeding on (e.g.) swarming ants (Formicidae) and termites (Isoptera) (Glutz and Bauer 1980; Helbig 1982). Venomous Hymenoptera (see below) possibly not taken at this time (Thiollay 1971) but Carmine Bee-eater *M. nubicus* apparently capable of de-venoming prey in flight

(Fry 1973). In 19 hrs of observation, Westfalen (West Germany), mainly before hatching, hunting from perch recorded 140 times out of 158, but when young *c.* 23–24 days old, short, low searching flights used 27 times out of 47 (Helbig 1982). In another study, sustained pursuit-flights more common than brief flights from perch (Ursprung 1979). Not recorded using animate perches much favoured by *M. nubicus*, but will follow cattle, man, or vehicles to take insects disturbed from long grass (Cunningham-van Someren 1970; C H Fry). Less commonly employed methods include hovering to take insects from vegetation (e.g. beetles *Cetonia*, *Anisoplia*), particularly when fewer insects flying (Homeyer 1863; Petrov 1954), or to take wasps (Aculeata) at their nests in earth-banks or under roof guttering (9 nests systematically plundered thus over 2 days) (Lokcsánszky 1934). Dragon-flies (Odonata) and others may be snatched from ground in flight (Fintha 1968; also Kowalski and Kowalski 1957), and bees (Apidae) picked off hive (Schumann 1931). Wild birds not uncommonly seen on ground (Koenig 1951) and apparently feed there exceptionally: stomachs from lower Volga (USSR) contained spiders and harvestmen (Arachnidae), also a woodlouse (Oniscoidea) (Osmolovskaya and Formozov 1955). Some spiders may, however, be taken in the air (C H Fry); in Spain, birds seen trailing silk threads (Mountfort 1958) and in Zimbabwe, pellets contained caterpillars (Lepidoptera) which similarly often hang on silk threads and were perhaps also caught by flying birds (C H Fry). 1 stomach from Siebenbürgen (Rumania) contained ant pupae; possibly ingested with predatory insect (Glutz and Bauer 1980). In Denmark, flightless ground beetle (Carabidae) found in pellet (Larsen 1949); possibly eaten incidentally when bird on ground to play with leaves or to take grit or sand (C H Fry). See also Petrov (1954) and Helbig (1982). In Kazakhstan (USSR), larvae of flies (Diptera) probably taken from dung heaps where birds sometimes perched (Korelov 1948). Some records of birds hovering or making short glide before plunging into water and not submerging completely, probably refer to feeding as items seen in bill and fallen item once retrieved (see Barham *et al.* 1956, Sage 1960*a*, Swift 1960*a*, Tree 1960, 1961*a*); however, although other Meropidae known to take fish, many similar reports for *M. apiaster* probably relate to bathing (Fry 1982; C H Fry). If prey (e.g. dragonfly) snapped at and then drops to water injured, bird retrieves it without difficulty (Fintha 1968). Also often skims surface to drink like swallow (Hirundinidae) (Sage 1960*a*; White 1960*a*); hovers briefly (Swift 1959), or exceptionally settles on water for same purpose (Ticehurst *et al.* 1926). Captive birds showed no need of water; possibly snapped at droplets when rain-bathing (Koenig 1951). While returning to perch with prey, insect may be mandibulated and tossed up (Barham *et al.* 1956; Schumann 1971). Non-venomous prey beaten or rubbed against perch; number and vigour of movements increase roughly in proportion to size and strength of insect (Rivoire 1947; Koenig 1951). Hard elytra and thorax of rosechafer *Cetonia* removed by hard knocking (Matoušek 1951); captive birds apparently unable to handle this beetle (Koenig 1951) and great and sudden increase in number of whole prey items beneath perches after fledging suggests juveniles experience considerable initial difficulties with these Coleoptera and large dragon-flies *Aeshna* (Ursprung 1979). If prey dropped, bird normally plunges after it (Ursprung 1979; Helbig 1982), but shows no further interest if it reaches ground (Koenig 1951). Wings of Lepidoptera bitten off in flight or detached by knocking against perch (Baum and Jahn 1965; Schumann 1971). Odonata swallowed whole (Swift 1959). No detailed description of treatment of venomous insects by *M. apiaster* but captive Red-throated Bee-eater *M. bullocki* handled worker bees *Apis* as follows. Bee dextrously mandibulated until held in bill-tip at petiole of abdomen, then head beaten sharply 1–2 times against perch; grip very rapidly changed to abdomen tip which is rapidly rubbed 5–10 times against perch to discharge venom. Eyes closed during each rubbing movement and bill apparently rotated slightly about axis, with prey also being nibbled between bouts. Stings often stick into perch and thus torn out. Head then hammered less violently 1–4 times and bee tossed to back of throat to be swallowed head-first. Whole treatment lasts *c.* 10 s. Drone bees apparently distinguished but not instantly recognized as non-venomous: rubbed and beaten at low intensity. Captive *M. apiaster* handled wasps *Vespa vulgaris* similarly (C S Roselaar). Other insects, including bee-like but non-venomous hover-flies (Syrphidae), recognized and merely immobilized (Fry 1969*a*; C H Fry). See also Chaplin (1937); for further detailed account of de-venoming in Rainbow Bee-eater *M. ornatus*, see Nicholls and Rook (1962). Stings frequently found in food remains of *M. apiaster* and are thus swallowed, although may be damaged during handling (Helbig 1982). Apparently partially immune to venom; birds show pain when stung but (except when stung on eyelid) no other effect evident (Naumann 1826; Koenig 1951; Fry 1969*a*). Drone honey-bees *Apis mellifera* probably selected preferentially (e.g. Matoušek 1951). In eastern Austria, *c.* 60% of bees taken were drones—fed particularly to young. Natural ratio 50–100 workers: 1 drone (Ursprung 1979). In neck-collar samples from Banat (Rumania), drones 11·4% by number (11·7% by weight) and workers 17·5% (8·9%) of total prey (Korodi Gál and Libus 1968). At Zaria (northern Nigeria), adult *M. bullocki* apparently select large, non-venomous insects as food for young (Fry 1973). This not proven for *M. apiaster*; however, although an opportunistic forager, probably selects prey according to size and mode of flight—large Hymenoptera and probably Coleoptera preferred, and far fewer Diptera taken than expected from relative abundance (Helbig 1982 *contra* Herrera and Ramírez 1974). Experimental study also showed that inexperienced young birds apparently select black and yellow insects in preference to plain ones; also prefer larger black and yellow insects (size of hornet *Vespa crabro*) to smaller ones, but smaller black and yellow to larger plain ones (Koenig 1950); findings apparently confirmed by field observations (Ursprung 1979). Captive birds reject some aposematically coloured and noxious insects: tiger-moths Arctiidae, sawflies Tenthredinidae, and beetles (Chrysomelidae, Coccinellidae, Cantharidae, Lampyridae, *Metoecus*), but aposematic burnet-moths (Zygaenidae) and noxious beetle *Lytta vesicatoria* taken in small numbers (Callegari 1970b); *L. vesicatoria* eaten commonly in the wild (Fintha 1968) and malodorous bugs (*Aelia, Eurygaster*) also recorded (Csörgey 1934). Of 153 flights in hunting from perch, 68% successful (Helbig 1982). 2–4 medium-sized insects taken (presumably by same method) in intensive feeding bout of c. 5 min (Korelov 1948). Birds will feed at bee-hives if near nesting area—otherwise, mostly in cold, dull weather or after rain; flocks (up to c. 250) arrive in morning and evening (Korelov 1948; Atakishiev 1971; also Fintha 1968). Bees apparently much less active when *M. apiaster* present (Korelov 1948). In Asia and South Africa, often considered serious pest by apiculturists (C H Fry); for combative measures, see (e.g.) Kraft and Korelov (1938). In Kazakhstan, most damage done after breeding and during autumn migration (Korelov 1948). For diurnal feeding rhythms and factors affecting them, see Social Pattern and Behaviour.

Pellets characteristically elongated (Dolgushin *et al.* 1970) and composed mainly of insect chitin. Ejected every 1½–4 hrs by captive birds (Koenig 1951). Average dry weight 0·15–0·30 g, range 0·10–0·55 g, $n = 28$ (Herrera and Ramírez 1974); see also Lomont (1946), Swift (1959), and Baum and Jahn (1965). Measure c. 18–35 × 8–15 mm: e.g. 25–30 × 10–14 mm, $n = 38$ (Schumann 1971), average 23·6 × 11·7 mm, $n = 24$ (Glutz and Bauer 1980); see also Swift (1959), Baum and Jahn (1965), Dolgushin *et al.* (1970), and Ursprung (1979). In Zimbabwe, winter, average 12·5 items per pellet, birds taking c. 60% Formicidae and Isoptera (C H Fry). In eastern Austria, 1–14 items, though pellet size relatively constant; average 4·9 items, $n = 319$, with prey c. 55% large bees *Bombus* (Ursprung 1979). Pellets regularly contain small stones and occasionally quite large quantities of fine sand (Ursprung 1979; see also Baum and Jahn 1965); young birds recorded eating sand (Barham *et al.* 1956).

Wide range of prey recorded, but analyses invariably show preponderance of bees and wasps. Hymenoptera primarily bees and bumble bees (Apoidea); also wasps (Vespidae, Sphecidae, Pompilidae, Scoliidae, Pteromalidae, Tiphiidae, Chrysididae, Mutillidae), sawflies (Siricidae, Tenthredinidae), Ichneumonidae, and ants (Formicidae). Other insects include mayflies (Ephemeroptera), dragonflies and damsel flies (Odonata: Aeshnidae, Libellulidae, Gomphidae, Coenagriidae, Agriidae, Lestidae), stoneflies (Plecoptera), grasshoppers, etc. (Orthoptera: Acrididae, Tettigoniidae, Tetrigidae, Gryllidae, and Gryllotalpidae), earwigs (Dermaptera), mantises (Mantidae), termites (Isoptera), bugs (Hemiptera: Naucoridae, Pyrrhocoridae, Coreidae, Pentatomidae, Lygaeidae, Reduviidae, Cydnidae, Notonectidae, Nepidae, Cercopidae, Scutelleridae, Miridae, Cicadidae), Neuroptera (Raphidiidae, Chrysopidae), scorpion flies (Panorpidae), butterflies and moths (Lepidoptera: Pieridae, Lycaenidae, Satyridae, Hesperiidae, Nymphalidae, Papilionidae, Sphingidae, Noctuidae, Geometridae; larger species rather uncommon in analyses, perhaps because wings detached before swallowing), caddis flies (Trichoptera), flies (Diptera: Stratiomyidae, Tabanidae, Syrphidae, Asilidae, Scatopsidae, Tachinidae, Calliphoridae, Muscidae, Tipulidae, Nemestrinidae), and beetles (Coleoptera: Carabidae, Haliplidae, Dytiscidae, Hydrophilidae, Histeridae, Geotrupidae, Scarabaeidae, Silphidae, Staphylinidae, Elateridae, Dermestidae, Meloidae, Chrysomelidae, Curculionidae, Cerambycidae, Cicindelidae, Tenebrionidae, Byrrhidae, Buprestidae, Oedemeridae, Cleridae, Lagriidae, Prionidae). Few non-insects recorded: harvestmen and spiders (Arachnida), woodlouse (Oniscoidea), and snails (Gastropoda). (Pitman 1929; Korelov 1948; Matoušek 1951; Petrov 1954; Osmolovskaya and Formozov 1955; Maŕan 1958; Yakubanis and Litvak 1962; Yatsenya 1966; Fintha 1968; Korodi Gál and Libus 1968; Schumann 1971; Ursprung 1979; Sackl 1981; Helbig 1982; Fry 1983; C H Fry, A Gretton.) Most prey over 10 mm long (Helbig 1982); 28·2% 5–10 mm, 50·8% 10–15 mm, 17·5% 15–20 mm (Petrov 1954). Plant food taken only exceptionally, 5 of 25 birds (average weight c. 15% less than normal) found dead, Zimbabwe, after cold spell in November, had berries of *Cedrella tuna* stuck in throats (Steyn and Brooke 1970, where also report of Vitaceae fruit as food). Numerous studies based on analysis of pellets, nest-contents, and stomachs; for list of most significant of these (also covering some extralimital parts of USSR and exceptional breeding outside normal range in north-west Europe), see Glutz and Bauer (1980). See also for USSR: Petrov (1954), Pek and Fedyanina (1961), Yakubanis and Litvak (1962), and Belskaya (1976). For Austria, see Sackl (1981); for West Germany, Helbig (1982). Odonata tend to predominate over Coleoptera in observational studies and stomach analyses, being easier to identify; Coleoptera tend to predominate in studies of pellets, Odonata being represented only by mandibles and wing fragments (C H Fry). For difficulties regarding specific identification of items in pellets, see Hoffrichter and Westermann (1969); for validity of different methods, see Swift (1959) and Helbig (1982).

Variations in diet mainly reflect seasonal, annual, and geographical changes in insect fauna (Rivoire 1947; Swift 1959; Fintha 1968; Ursprung 1979). Temporary and local exploitation of particular species probably typical (Ursprung 1979): e.g. locusts (Acrididae) or swarming termites in Africa (Pitman 1929; C H Fry), or, on breeding grounds, beetles, dragonflies, or wasps (e.g. Lokcsánszky

1934, Swift 1959). A summary of 12 heterogeneous European studies apparently showed Hymenoptera most important overall, followed by Odonata, Coleoptera, Lepidoptera, Diptera, and Orthoptera (Swift 1959; see also Fry 1973, 1983 for approximate percentage values). Lepidoptera (e.g. Matoušek 1951) and Orthoptera (Thiollay 1967 for Corsica, Korelov 1948 for Kazakhstan) more important in some areas. In southern France, Coleoptera and Hymenoptera apparently predominate in May, Coleoptera and Odonata in June, Odonata and Lepidoptera in July (Swift 1959). Hymenoptera rarely below 50% by number at any time of year and can reach over 90%. In temperate zone, bumble-bees *Bombus* single most important prey for adults and young living near clover fields and areas with many flowers (C H Fry; see, e.g., Larsen 1949, Yatsenya 1966). Pellets from 3 colonies, Andalucia (southern Spain) contained 2141 items: 69.4% by number Hymenoptera (51.6% *Apis mellifera*), 21.0% Coleoptera (10.6% Scarabaeidae); remaining 9.5% comprised Dermaptera, Odonata, Orthoptera, Hemiptera, Lepidoptera, and Diptera (Herrera and Ramírez 1974). 319 pellets, June–July, eastern Austria, contained 1560 items: 82.8% by number Hymenoptera (*Bombus* 55.3%, Vespidae 13.6%, *A. mellifera* 12.5%, others 1.4%), 6.6% Odonata, 6.5% Coleoptera, 4.1% Hemiptera, Lepidoptera, and Diptera (Ursprung 1979). In Denmark 45 pellets contained 660 items: 59% Hymenoptera (mainly *A. mellifera*) in June when birds foraged over turnip fields, increasing to 91% (mainly *Bombus*) from clover fields in August (Larsen 1949). In Zimbabwe, winter, pellets yielded 860 insects: 27.8% *A. mellifera*, 1.0% other Apoidea, 0.7% Vespidae, Sphecidae, and Pompilidae, 29.9% Formicidae, 2.7% Ichneumonidae, 27.4% Isoptera, 3.1% Coleoptera, 2.4% Hemiptera, 0.1% Odonata, 4.9% Lepidoptera (Fry 1983; C H Fry). Daily food requirement *c.* 39 g of insects (Priklonski and Lavrovski 1974); *c.* 225 bee-size insects (Korelov 1948). For nestling (19–30 days old), average consumption 13 g (43 items) per day; calculated from neck-collar samples and assumed 12-hr feeding period (Korodi Gál and Libus 1968).

Studies indicate important differences between nestling diet. and that of parents at same time (see, however, above for possibility of bias in analyses). In eastern Austria, nest contents (mainly nestling food) revealed larger proportions of wasps, bees, and other Hymenoptera and *c.* 35% more Coleoptera, but *c.* 20% fewer *Bombus* and no Odonata (3.3% taken by adults) (Ursprung 1979). See also (e.g.) Yatsenya (1966) for virtual absence of Odonata in nestling diet In Camargue (southern France), however, *c.* 50% of 642 items brought were Odonata (Biber 1971). Preponderance of Odonata (in some cases *Bombus*) and larger prey items in general (see, e.g., Matoušek 1951, Hachler 1958, Krimmer *et al.* 1974, Priklonski and Lavrovski 1974) accords with observations on other Meropidae that young given larger items and adults eat smaller ones themselves (C H Fry). In Turkmeniya, nestlings (8–24 days old) given mainly locusts (Belskaya 1976). 40 neck-collar samples from Banat (Rumania), 19 July–6 August, contained 1161 items: Hymenoptera (66.2% by number, 64.0% by weight: *A. mellifera* 28.8%/20.7%, *Bombus* 33.2%/43.5%), Odonata (13.7%/17.8%), Lepidoptera (5.5%/5.7%), Coleoptera (2.9%/4.8%), Diptera (4.8%/3.3%), Hemiptera (5.7%/2.6%), Orthoptera (1.1%/0.8%) (Korodi Gál and Libus 1968). In England, fledged young seen to catch bees and smaller insects, while butterflies (sometimes refused) fed to them by adults (Barham *et al.* 1956). Rate of prey delivery to nest varies markedly with type of prey, distance to feeding grounds, and weather. Rate of 6–8 feeds per min (Tapfer 1957) recorded; also pauses of up to 3 hrs (e.g. Barham *et al.* 1956, Krimmer *et al.* 1974). In England, *c.* 1 week before fledging, young fed 249 times in 8½ hrs (Barham *et al.* 1956). In Hungary, broods fed for 13 out of 14 daylight hours, with rate often exceeding 50 visits per hr; at one nest with helper (see Social Pattern and Behaviour), 143 per hr (Dyer and Fry 1980); provisioning rate increases with brood size and age (Dyer and Demeter 1982). Peak feeding times 06.00–07.00 to 08.00–09.00 hrs, 10.00–12.00 hrs, and 14.00–16.00 hrs (Fintha 1968). In Camargue, pair with large young (18 July) fed them 175 times 16.00–19.00 hrs (Lomont 1946). For rate of food consumption of young, see penultimate paragraph. For further details on feeding rates, see (e.g.) Baum and Jahn (1965) and Belskaya (1976). MGW

Social pattern and behaviour. Includes material supplied by C H Fry. For detailed aviary study of 2 hand-reared sibling groups, see Koenig (1951). For field study in Camargue (southern France), see Swift (1959).

1. Gregarious throughout the year, especially on migration and in winter quarters. Flocks often large outside breeding season, comprising hundreds of birds (Glutz and Bauer 1980; C H Fry); flocks typically loose-knit (e.g. Hutson 1947) but often compact on migration (D J Brooks). In Camargue, after fledging, all birds from colony remain together, mostly within *c.* 5 km of nesting area; eventually depart in flocks of *c.* 50–60 (Swift 1959). On migration, usually in flocks of *c.* 20 birds; those of over 100 less common, except at halting and roosting places (Hutson 1947); at Belen (southern Turkey), autumn, average 19.7, largest 77, $n = 1928$ birds (Sutherland and Brooks 1981); see also Movements. In Nairobi (Kenya), roost contained *c.* 80% juveniles (Pelchen 1978). Associates loosely with Blue-cheeked Bee-eater *M. superciliosus* on migration, but mainly at staging posts and roosts; generally travel separately (Hutson 1947). BONDS. Monogamous mating system the rule, though casual bigamy recorded (J Krebs). Perhaps pairs for life (as in, e.g., Red-throated Bee-eater *M. bullocki*), but not adequately studied (Fry 1972; C H Fry). Persistence of bond outside breeding season not known: captive pair—formed from different sibling-groups (see further below)—disbanded after breeding, each bird returning to own group and showing hostility to each other later (Koenig 1951). For hybridization in captivity with Carmine Bee-eater *M. nubicus*, see Callegari (1970b). As in other open-country colonial Meropidae, non-breeding birds sometimes attach themselves (singly or several together) to breeding pairs as 'helpers'; no detailed study of helper phenomenon in *M. apiaster*. Sex-ratio

skewed towards ♂♂ (as in all colonial Meropidae: Fry 1972) and these likely to form majority of non-breeders reported at colonies (see Glutz and Bauer 1980). Sexed specimens in museums comprised 60% ♂♂ (Fry 1972); in other counts, 59% ♂♂ (Glutz and Bauer 1980). Studies of other (extralimital) Meropidae (e.g. Fry 1972, Reynolds 1972) suggest that supernumerary (probably 1-year-old) ♂♂ form majority of helpers. Observations on more or less isolated pairs apparently indicate such trios relatively common in *M. apiaster* (see Glutz and Bauer 1980), but low incidence of trios considered by C H Fry to be more likely everywhere. Highly probable that helper related to one member of pair (offspring or sibling). Helper recorded at 1 of 8 nests studied in Hungary (Dyer 1979), at 2 of 5 nests in Camargue (A Biber); also reported from Spain (Cano 1960), on Corsica (C H Fry), and in West Germany (Schumann 1971). In *M. bullocki*, bond between pair and helper may last beyond 1 season (Fry 1972). Helper tolerated at nest-hole and to some extent on territory perches (see below), but not as sexual partner (Glutz and Bauer 1980); however, helper will feed ♀ (see part 2) and may attempt to copulate with her, though mating attempts likely to be discouraged (Schumann 1971). Apparently in addition to helper phenomenon, and as in other Meropidae (see, e.g., Emlen and Demong 1980 for White-fronted Bee-eater *M. bullockoides*), some birds visit several nests in colony over course of breeding season (Fry 1972; C H Fry). Both sexes incubate and tend young. Any helper also feeds young, but extent of helper participation varies, some bringing food only when young call loudly or appear at nest entrance (Glutz and Bauer 1980). Possible that helper assists at times in all nesting functions, including incubation (proved for captive birds: Koenig 1959), judging from studies on other Meropidae (C H Fry). Helper sometimes accompanies foraging parent but does not help to feed young (Baum and Jahn 1965). Successfully breeding captive pair comprised 1-year-old ♂ and 2-year-old ♀ (Koenig 1951); see also Relations within Family Group (below). ♂ recorded breeding at 1 year old in the wild (Gehlhaar and Klebb 1979), though existence of non-breeding flocks (e.g. Dolgushin *et al.* 1970) suggests not all do, see also discussion of helpers and sex-ratio (above). Juveniles fed for 2–3 weeks after fledging (Dementiev and Gladkov 1951a); in Camargue, at least until departure, i.e. for 3–5 weeks (Swift 1959); see also Relations within Family Group (below). Fed only by own parents and helper, not by other non-breeders in colony (Swift 1959). Bonds among fledged young evidently strong. Captive birds maintained separate sibling-groups after fledging and through 1st winter (Koenig 1951). Such associations evidently exist in the wild: 2 sibling juveniles from Camargue recovered together 2 days later on Mallorca, Spain (Glutz and Bauer 1980). Captive ♀ did not pair-up with own brother (dominant bird in aviary) but with low-ranking ♂ of other sibling-group (Koenig 1951). BREEDING DISPERSION. Typically colonial, though solitary nests not uncommon. Pair (mainly ♂) defends 1 or more perches (or small area of ground) near nest (Swift 1959, which see for diagram showing location of perches in section of colony). In another study, birds apparently defended section of bank containing perches and nest; first-established territories large, contracting as newly arriving pairs asserted themselves (see Hahn 1981; see also Antagonistic Behaviour, below). Perches probably reduce interference during copulation (Swift 1959). In Camargue, nest-hole and immediate vicinity defended throughout breeding season, though defence of perches wanes rapidly after start of incubation (Swift 1959). Competitors for nest-hole, e.g. Rock Sparrows *Petronia petronia*, may be aggressively driven off (Olioso 1974). Colonies range from a few to many pairs, reaching several hundred in USSR (C H Fry). In Camargue colonies, 75% of 2–6 nests each, 25% of over 6 (average 17); only 8%

of nests solitary (Swift 1959). French colonies seldom exceed 25–30 nests (C H Fry); in Hungary, colonies of 40–50 nests frequent, once 400 (Tapfer 1957; Fintha 1968). Nests in colony typically loosely grouped: in flat ground, seldom have entrances more densely packed than 1 per m², more usually, several metres apart; on cliffs, sometimes closer together, often in a row a given distance below cliff-top, entrances 50 cm apart (Besson 1967; C H Fry). Occupied holes sometimes as close as 50–60 cm, but typically further: (2–)5–10 m (Glutz and Bauer 1980). Colony of 102 pairs spread over 1 km of bank (Pérez Chiscano 1975); 12 pairs over 400–500 m (Lomont 1946). In France, ground colony of at least 19 pairs in area 300 × 150 m (Besson 1964); for dispersion in similar colony, Spain, see Álvarez and Hiraldo (1974). For limited data on overall densities, see Glutz and Bauer (1980). Distribution of colonies determined by availability of nesting sites (sand cliffs preferred to flat ground) and food (C H Fry). In Palearctic west of *c.* 20°E, colonies irregularly dispersed or of sporadic occurrence (C H Fry). Camargue birds generally forage within *c.* 1 km of nests; range extended to *c.* 5 km after fledging (Swift 1959). In Kazakhstan (USSR), birds rarely forage more than *c.* 6–7 km from colony (Korelov 1948). A solitary pair, Westfalen (West Germany), used area of *c.* 14 ha (Helbig 1982). Extension of foraging range recorded during incubation (Bastien 1957). ROOSTING. Typically nocturnal and communal. In groves, orchards, waterside shrubbery, etc. (C H Fry); usually amongst vegetation in the open, normally in tall isolated trees or ones at edge of wood—those with open canopy preferred, or with dead top or lateral branches; roosting in cavities reported (but apparently abnormal) outside breeding season (Glutz and Bauer 1980). In Zimbabwe, wintering flock of *c.* 100–150 moved between 3 roosts, using each for up to 5–6 nights at a time; sited in tall trees, with perches at least 7 m above ground (Finch 1971); in Nairobi (Kenya), each site used for maximum 2 weeks (Pelchen 1978). During halts on spring and autumn migration, traditional sites used by many flocks at a time and by succession of flocks, night after night. Roosts shared with *M. superciliosus*, species keeping apart (Hutson 1947). For clumping during roosting, see Flock Behaviour (below). In breeding season, nest-hole used for roosting by ♀ or both of pair (Glutz and Bauer 1980). Off-duty bird may sleep in the open *c.* 100–150 m (Lomont 1946) or up to *c.* 2–3 km away (Swift 1959; Krimmer *et al.* 1974). Captive pair roosted together on perch apart from rest of flock using same site; ♀ used hole as soon as it was completed but ♂ continued to roost in open at first, joining ♀ later (Koenig 1951). Hole may be used by both members of pair for up to *c.* 14 days after hatching, both also sheltering there in bad weather during day. ♀ may continue to roost in hole until young *c.* 24 days old (König and Wicht 1973; Glutz and Bauer 1980; Helbig 1982). Juveniles continue to sleep in nest at night for several days after fledging (e.g. Barham *et al.* 1956), or may roost with adults (Swift 1959), or separately in trees (Fintha 1968). In Camargue, birds arrive at roost well before dark but may take up to *c.* 1 hr or more to assemble. Leave relatively late in morning and little activity evident at colony until well after sunrise (Swift 1959). Feeds much less in cold, wet weather (e.g. Bastien 1957), also around midday and early afternoon in hot climates (e.g. Pitman 1929, Korelov 1948). Patterns of alternating intense feeding and resting recorded: 2-hr cycle in England (Barham *et al.* 1956), 4-hr cycle in Hungary (Fintha 1968). Sun-, dust- or sand-, and rain-bathing occur regularly and birds self-preen often and for long periods (Swift 1959; Dolgushin *et al.* 1970; C H Fry). Will bathe like swallow *Hirundo*, immersing briefly following swooping glide or hover (see Food). For details of comfort behaviour in captive birds, see Koenig (1951, 1953).

2. Alarm-calls (see 10 in Voice) usually given only if human

intruder appears unexpectedly. Normally flies at c. 30–40 m or more, but this distance may be reduced (in case of pedestrians and moving vehicles) to c. 10–12 m during breeding season (Glutz and Bauer 1980). Can be difficult to scare away from beehives (Atakishiev 1971). Frightened captive birds adopt an upright posture with plumage sleeked and carpal joints often held away from body; give Alarm-calls and may suddenly take off and rapidly spiral upwards in panic-flight (Koenig 1951; see also below). Generally shyer in colonies (Rivoire 1947), especially smaller ones (Swift 1959). Birds of prey (Accipitridae) and Cuckoo *Cuculus canorus* may be chased away from feeding grounds (Fintha 1968). Sometimes mobbed by passerines (e.g. Kowalski and Kowalski 1957). FLOCK BEHAVIOUR. Strong social bond evident in flocking, colonial breeding, and communal roosting. Preening, feeding, bathing, sleeping, and some aspects of courtship behaviour all highly infectious (see also Voice). Bird may alert others to food source with 'joyful' calls and lowest-ranking bird forced to feed alone gave frequent Summoning-calls (Koenig 1951: see 2 in Voice). Migrant flocks comparatively tight-packed when resting (Hutson 1947). May circle roost noisily before settling and again before final departure in morning (Marchant and Macnab 1962). In Kazakhstan (USSR), ♂♂ circle colony as flock before sunset and some may enter holes briefly before departure for roost (Dolgushin *et al.* 1970). Prior to roosting, 1 bird gives Roosting-call (see 4 in Voice) from an upright posture whilst tail-vibrating. Others arriving in response to this make sideways-tripping movement on perch and give quiet variant of Contact-call (see 1 in Voice); sideways-tripping and Roosting-calls continue until all have settled (Koenig 1951). Birds roost in tight-packed rows of 2–8 (Karcher 1940; Finch 1970; Pelchen 1978). Such clumping, also used during day if cold, contrasts with individual distance of c. 5–10 cm normally maintained outside breeding season by captive birds and increased at start of breeding when social hierarchy evident (Koenig 1951). In non-foraging flight, wild birds normally stay c. 1–2 m apart (C H Fry). At roost in Camargue, birds (apart from 2 members of pair far from colony) said always to remain a little apart and unpaired individuals not to tolerate another close by (Swift 1959). Birds usually silent at roost from dusk (Dolgushin *et al.* 1970). In Nairobi, winter, flocks of 5–10 assemble at traditional site c. 1½ hrs before sunset, then move to roost at sunset; silent c. 5–10 min later (Pelchen 1978). In Zimbabwe, after slight activity for c. 20 min, depart from roost noisily at c. 06.00 hrs (Finch 1971). ANTAGONISTIC BEHAVIOUR. (1) General. Defence of particular perches near nest increases during nest-building, reaches peak when nest-hole completed, then wanes rapidly (Koenig 1951; Swift 1959; Hahn 1981). Captive ♂ also defended mate (Koenig 1951). Even at peak, birds may assemble at any of several perches in and around colony without showing aggression (C H Fry). Defence breaks down when necessary to expel common enemy (Swift 1959). In Greece, disputes between established pairs rare; fights frequent only when new arrivals trying to acquire territory. Territory-owners more aggressive than intruders and, at peak (lasting c. 1–2 days), make up to 25 attacks on them per hr (Hahn 1981). In disputes over perches, ♂ attacks ♂, and ♀ other ♀. Intruding pair initially nervous and often flee but become gradually more aggressive if they gain possession of a perch (Swift 1959). Captive birds frequently fight bitterly in summer and fledged young markedly aggressive from c. 50 days old (Koenig 1951). (2) Threat and fighting. Bird landing at some distance from another, then approaching stealthily in sideways-tripping motion, less likely to provoke threat or attack (Koenig 1951). Adjacently perched birds frequently bicker without fighting properly (C H Fry). Bird in aggressive mood may assume an upright posture with feathers

A

ruffled, particulary on nape and mantle; black gorget prominent. In low-intensity threat, gapes silently and may briefly open wings. At higher intensity, tail spread (as often in antagonistic interactions), and bird bows towards opponent with feathers ruffled and gives Threat-call (see 12 in Voice) whilst gaping (Fig A, right); may also make very fast bill-snapping lunges and briefly raise wings. Threatened bird may lean slightly backwards and gape silently, with plumage sleeked (Fig A, left). In appeasement, bird may give Greeting-call and/or Bleating-call (see 3 and 14 in Voice). Bill-wiping and Mock-preening not uncommonly performed in apparent attack–escape conflict. Equally aggressive bird seizes other's bill, leading to bouts of tugging and shaking with bills interlocked; wings raised (Fig B); bill-snapping and fluttering occur. Fights also take place in the air. Fleeing bird often chased and attacked again (Koenig 1951; Swift 1959). For changing pattern of tolerance and aggression in pair with helper, see Schumann (1971). When mobbed by hirundines (Hirundinidae) in flight, may turn head back and call (Barham *et al.* 1956: see 15a in Voice), or lunge with bill (Baum and Jahn 1965; Krimmer *et al.* 1974), or use abrupt jinking manoeuvres to evade various small passerines (Kowalski and Kowalski 1957). For interactions with *Petronia petronia*, see (e.g.) Olioso (1974). HETEROSEXUAL BEHAVIOUR. (1) General. Observations on both wild and captive birds indicate marked similarity in behaviour between sexes and frequent interchange of sexual roles (Koenig 1951; Swift 1959). Though adults can often be sexed by plumage (e.g. Reid 1974), Courtship-feeding (see below) apparently the only behaviour in which roles of sexes can be reliably attributed (Koenig 1951; Baum and Jahn 1965); thus captive fledged young performed mock-copulation, then immediately reversed roles (Koenig 1951). Only simple form of Greeting-ceremony (see below) recorded in Africa in winter (C H Fry), but birds arrive on breeding grounds already paired (Fry 1973), though some partner changes may occur during 1st week (Swift 1959). Camargue birds move into colony c. 10 days after arrival (Biber 1971); see also Bauer (1952) for Austria. Graceful flight manoeuvres with loud calls possibly part of courtship soon after arrival

B

(Swift 1959; Alleijn *et al.* 1966). Heterosexual behaviour evident in captive birds in first half of March; individuals show abrupt changes from aggression to courtship behaviour towards particular members of flock (Koenig 1951). Pronounced synchrony in start of laying noted in one Greek colony (Hahn 1981). (2) Pair-bonding behaviour. In winter quarters, simple Greeting-ceremony often performed by 2 birds together on perch: stand rather upright, tail fanned and vibrated; crown feathers slightly ruffled and Summoning-calls (see 2 in Voice) given (C H Fry). On breeding grounds, more elaborate. When pair together on perch, one bird (probably ♂ in most cases) makes short flight and returns to perch by 2nd bird. Latter greets 1st with tail-vibrating and often opens wing facing away from incoming bird or, if approach from front, both wings (see illustrations in Koenig 1951, reproduced in Glutz and Bauer 1980). Bird lands by partner with emphatically stiff and angular movements; wings spread momentarily before being closed abruptly (Koenig 1951). Landing bird may strike other with downward movement of wing; other submissive (Goodwin 1952b). Then stands erect, with plumage sleeked, tail closed and pointed more or less straight down (at highest intensity, wing-tips beneath tail), and neck extended; crown feathers sleeked, those on nape ruffled giving angular head shape. Rapid, brief upward jerking movements of whole body made at intervals of a few seconds and accompanied each time by Courtship-call (see 5 in Voice); black throat feathers ruffled. Often concludes with one or several slow and emphatic beating or jabbing movements (resembling those used for prey treatment—see Food) against perch by partner's feet (Koenig 1951). Presumed ♀ may adopt a hunched posture with ruffled feathers (particularly on mantle) or remain still; one may lean against the other and beat with offside wing. Birds may fly from branch to branch and briefly display at each as described. Often, both adopt a hunched posture afterwards. Bill-wiping against perch and Mock-preening are common features of heterosexual activity. Birds of a pair spend remarkably long periods perched close together, especially early in season (Swift 1959). (3) Nest-site selection. Birds (pairs and single birds) may glide back and forth above a locality over 3 days; members of a pair often perch and call excitedly (Mountfort 1957). ♂ reported to initiate excavation, thereby choosing site (Glutz and Bauer 1980). Said sometimes to clear out old nest-holes (Swift 1959; Fintha 1968), but old holes not reoccupied ac rding to Glutz and Bauer (1980). (4) Nest-building. For process of excavation, see Breeding; also Koenig (1951) and Mountfort (1957). Reported to have pecked through obstructing root (Glutz and Bauer 1980), and to carry out pebbles and soil clumps which are beaten like prey (Koenig 1951); such behaviour unknown in other Meropidae (C H Fry). In so-called 'Tunnelling-duet', one member of pair gives persistent Summoning-calls from perch near nest-site; other approaches and calls—not described, but sonagram in Hahn (1982a) indicates similarity to Nest-call (see 7 in Voice)—given while excavating. Mate continues with Summoning-call and rate of delivery influenced by Nest-calls from digging bird. Roles in duet change as birds swap over (see Hahn 1982a for details). 2–3 holes usually started at same time within *c.* 1–2 m of definitive one (C H Fry). When breeding over, fledged young sometimes perform rudimentary digging (Barham *et al.* 1956). (5) Courtship-feeding. Usually starts after several days' nest-building (Koenig 1951; Swift 1959). Continues during and (decreasingly) after laying, ♀ receiving food at nest-entrance or leaving to take it (Swift 1959; Biber 1971). Food sometimes passed in flight (Rivoire 1947; Trippmacher 1983). Typically, ♀ gives Distress-call from perch (Koenig 1951: see 6 in Voice) and performs Greeting-ceremony (see above) as ♂ arrives with food (Swift 1959). Large items (e.g. Odonata) frequently brought (Matoušek 1951). ♂ may land some distance from ♀, perform Courtship-jerking several times, give quiet Contact-calls, and approach in sideways-tripping motion. If facing opposite way to ♀ on arrival, jumps around before offering food (Koenig 1951). For full sequence, see Figs C–E, and note that ♂ may also perform Courtship-jerking (Fig E, left) after ♀ has accepted food. Courtship-feeding may take place up to 8 times in rapid succession (Hahn 1981). For participation by helper, see Bonds (above). (6) Mating. Takes place throughout nest-building, more frequently towards end (Swift 1959). In captive birds, rate increased up to 10 times daily after completion of building and ceased with laying of last egg (Koenig 1951). In wild pair, occurred 5 times in 2 hrs; interspersed with Courtship-feeding (Hejl-Mračovský 1958), which normally precedes it. Takes place on perch or ground, sometimes at site defended by pair (see above). ♀ usually swallows food quickly (may not succeed before ♂ mounts: C H Fry), then adopts Soliciting-posture: body more or less horizontal, head and throat feathers ruffled, eyes usually closed (see illustrations in Koenig 1951 and Glutz and Bauer 1980); Soliciting-call given (see 8 in Voice). ♂ moves sideways, often giving quiet Contact-calls, and, when touching ♀, stands

C

D

E

F

fairly erect (similar to posture in Courtship-jerking). ♂ gives Copulation-call (see 11 in Voice) before and during copulation, also afterwards as he takes off from ♀'s back. ♂ usually grips feathers on ♀'s forehead during copulation (Koenig 1951: Fig F). ♀ gives Copulation-call when ♂ mounts (see Jilka and Ursprung 1980, and 15b in Voice). Copulation lasts c. 2–10 s (Schumann 1971; Reid 1974). Only shaking of plumage by both birds noted afterwards (Koenig 1951), though 'aerial display' (no details) reported by Réz (1932). (7) Behaviour at nest. ♀ may spend long periods in nest at start of laying, ♂ then taking over, initially only briefly, until regular alternation established. Incoming bird gives Nest-call from perch or *en route* to nest and both may give guttural calls (presumably Nest-calls) inside hole (Koenig 1951; also Swift 1959). Change-over usually in nest, at least initially (Baum and Jahn 1965). Relieved bird quick to emerge. When change-over outside, bird leaving nest may perform Mock-beating (as in prey treatment) against perch while mate still present or afterwards (Koenig 1951). Bird may also flutter by nest and give Contact-calls prior to change-over (Schumann 1971). Change-overs 15–20 min apart (Lomont 1946; Koenig 1951; Alleijn *et al.* 1966); in one case, ♂ took over 4–8 times daily for periods of (4–)30–60 min, and only ♀ incubated at night (Schumann 1971). For activity rhythms of off-duty birds, see (e.g.) Dolgushin *et al.* (1970). RELATIONS WITHIN FAMILY GROUP. Parent responds with quiet Nest-call to sounds of young in eggs (Koenig 1959). After hatching, eggshells moved aside in chamber (Belskaya 1976) or taken out and broken up (Koenig 1959 for captive birds). Nestling unable to lift head for some hours after hatching but gives food-call on 1st day (see Voice for development of food-calls); eats strenuously at c. 3 days old (Koenig 1951; C H Fry). Adult arriving with food gives Nest-call into tunnel and also persistently whilst entering; young responding with food-calls. Young do not gape or show other begging movements until c. 21 days old when fully alert and mobile and normally wait for adult at entrance; in early stages, adult therefore locates nestling by its call, and releases food when contact with sensitive corner of nestling's mouth elicits snapping response (Koenig 1951). Eyes begin to open at c. 5 days (C H Fry); exceptionally, nestling comes to entrance when still blind (White *et al.* 1978). Nestling fed at entrance usually moves backwards when sated and another takes its place (e.g. Tapfer 1957, Koenig 1959); waiting bird may tug sibling's tail causing it to vacate front position. Nestlings generally described as quarrelsome (Glutz and Bauer 1980), even at c. 1 week old (C H Fry). However, damaging fights between captive nestlings reported by Koenig (1951) clearly resulted from unnatural opening-up of tunnel. Young brooded by both parents in spells of c. 10–30 min (for $c. \frac{1}{3}$ of daytime) until c. 1 week old; adult may spread wings over young (Barham *et al.* 1956; C H Fry). ♂ may brood for longer periods than ♀ (Krimmer *et al.* 1974; Glutz and Bauer 1980). Young defecate in nest, running backwards until raised tail touches wall of chamber; faeces absorbed by dry bed of insect remains (Koenig 1951; Glutz and Bauer 1980). Although ammonia concentrations rise as high as 700 ppm and CO_2 levels as high as 6%, neither appears to affect young adversely. Temperature in nest-chamber c. 3·5°C above average ground temperature at same depth. Nests long enough to minimize temperature variation; temperature of nest-chamber remains near birds' thermal neutral zone (White *et al.* 1978; C H Fry); for physiological aspects, see White *et al.* (1978). See also Alvarez and Hiraldo (1974) for 'chimneys' which perhaps help with ventilation. Claimed by Valverde (1953b) that young able to fall into lethargic state and thus survive cold spells and food shortage; however, chilled and moist young elicit no parental care, pushed aside, and may even be tossed out (Koenig 1951); runts also not infrequently left in nest, so that fuller investigation required (Glutz and Bauer 1980). Whole brood normally fledges within 1–2 days (e.g. Belskaya 1976). Young may leave independently or be called out by parent perched nearby with food (Barham *et al.* 1956). One adult may tend fledged young while mate feeds any still in nest (Tapfer 1957). Catch some prey for themselves 1–2 days after fledging, then fed with decreasing frequency (also by helper if present: Schumann 1971) until largely independent by week before emigration. Adults eventually threaten or drive off food-begging young, and fledged juveniles also show adult-type threat behaviour to one another (Barham *et al.* 1956; Swift 1959). Near Weissenfels (East Germany), birds moved to feeding area c. 2 km from nest c. 4–5 days after fledging; stayed there at least c. 10–11 days (Krimmer *et al.* 1974). In England, young initially accompanied back to nest-hole in evening for roosting; apparently encouraged to enter tunnel by parents' Alarm-calls (see 10 in Voice) and called out each morning by adults arriving from separate roost (Barham *et al.* 1956). In Zambia, late September, adult twice fed juvenile in flock of 11 birds (Robinson and Robinson 1975). Captive juveniles showed vigorous courtship activity and marked aggression at c. 50 days old (Koenig 1951; see also Koenig 1953, 1959, 1960 for further details). ANTI-PREDATOR RESPONSES OF YOUNG. Generally quiet when parents absent, or when adult gives Alarm-call (Krimmer *et al.* 1974); rapidly retreat into hole on hearing this call (Reid 1974; Jilka and Ursprung 1980) and attempt to withdraw still further if tunnel opened up (Rivoire 1947; Krimmer *et al.* 1974). When handled, may regurgitate (Randík 1961) or excrete (Koenig 1951) a yellow-brown fluid. At c. 24–25 days old may also peck at hand (Krimmer *et al.* 1974). PARENTAL ANTI-PREDATOR STRATEGIES. Shyness at nest varies considerably. May flee when man at c. 120 m and stay well away during disturbance (Krimmer *et al.* 1974), or carry on feeding young when observer at c. 15 m, then fly about with no extreme agitation if closer approach made (Bauer 1952). For report of apparent death-feigning (feathers ruffled, head and wings drooped) during serious disturbance, see Krimmer *et al.* (1974). Birds of prey elicit various reactions. Kestrel *Falco tinnunculus* attacked c. 800 m from nest; Red Kite *Milvus milvus* and Buzzard *Buteo buteo* ignored (Wiegank 1977); see also Kowalski and Kowalski (1957). Great Grey Shrike *Lanius excubitor* also chased off when very close to nest (Schumann 1971). Flock (in which birds give Alarm-calls) may strive to climb above (e.g.) Hobby *F. subbuteo* (Barham *et al.* 1956; Besson 1970)—apparently serious predator in some areas (Atakishiev 1971). Similar response elicited by man, other raptors, and snake *Malpolon monspessulanus*, though reaction less intense towards Marsh Harrier *Circus aeruginosus* and Magpie *Pica pica*. Dive-attacks (not against man) made when threat more serious (Swift 1959): e.g. repeatedly against stoat *Mustela erminea* and susliks *Citellus* (Schumann 1971; Reid 1974).

(Figs A–B based on drawings in Koenig 1951; Figs C–F from photographs in Hahn 1982b.) KELS, MGW

PLATE 72. *Coracias garrulus* Roller (p. 764). Nominate *garrulus*: **1** ad breeding, **2** ad non-breeding, **3** juv. *C. g. semenowi*: **4** ad breeding. *Coracias abyssinicus* Abyssinian Roller (p. 776): **5** ad, **6** juv moulting (body largely 1st ad). (CETK)

PLATE 73. *Coracias benghalensis benghalensis* Indian Roller (p. 778): **1** ad, **2** juv moulting into ad, **3** juv. *Eurystomus glaucurus* Broad-billed Roller (p. 783): **4–5** ad, **6** juv. (CETK)

Voice. Liquid and pleasant-sounding; employed constantly at all times and seasons, calling being frequent during most activities (C H Fry). Vocabulary large but consists mainly of variants of the most usual vocalization (Koenig 1951). This rendered 'rüpp' (etc.) in German and variously in other languages, with vowel-sound often modified to 'i'; in English, 'prruip', 'crick-wicka', 'pruuk-pruuk', etc. (C H Fry). Very slight variations in overall frequency pattern, onset (of sound), fundamental frequency, duration, and harmonics probably important for information content—see (e.g.) calls 10–11. Various sounds may be combined in patterned sequences, particularly during pair-formation (Jilka and Ursprung 1980; see also Fig I). No real advertising call. Birds of pair often call reciprocally (e.g. in courtship display, when excavating nest-hole), using same or different call, but this hardly constitutes duetting as sometimes stated (see Glutz and Bauer 1980, Bergmann and Helb 1982, Hahn 1982a), and existence of true antiphonal singing not established (C H Fry), 'Duets' embracing 'di' sequences of varying pitch performed only in flight (Jilka and Ursprung 1980, which see for wider general definition of 'duetting'). No difference between sexes in shared calls, but some others characteristic of one sex only. Some evidence of individual differences, at least in call 7 (see below), perhaps allowing older young to recognize parents by voice (Jilka and Ursprung 1980). In aviary study of tame, hand-reared birds (Koenig 1951), minimum of 14 calls recognized on which this account mainly based. First 9 of these of 'rüpp' type (see above), all except call 9 associated with friendly contact; calls 10–11 of 'dick' type; calls 12–13 of 'dääää' type; call 14 like that of newly fledged young. This scheme further modified and expanded in major study (including sonagraphic analysis) of wild birds in eastern Austria by Jilka and Ursprung (1980). For further details and extra sonagrams, see Nikol'ski (1979), Glutz and Bauer (1980), and Bergmann and Helb (1982).

CALLS OF ADULTS. (1) Contact-call. A sonorous 'rüpp', or sometimes 'güpp'; attractive, liquid, and mellow 'quil(u)p' only just perceptibly disyllabic (Anon 1952). Uttered singly at quite long intervals or often several times in succession—almost incessantly at times, particularly in flight. Used to maintain contact with conspecific birds. Usually shows as 2-peaked unit in sonagrams. More rolling variants (up to 5 peaks) probably express higher-intensity excitement (Jilka and Ursprung 1980); recording (Fig I) which suggests a 'pi-pi-pi purr purr pi-rr purr' (J Hall-Craggs) evidently contains such variants. Quieter 'güpp-

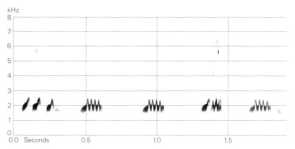

I P J Sellar France May 1977

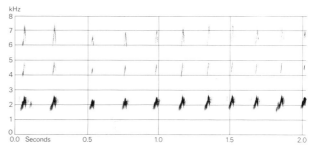

II J-C Roché (1966) France June 1965

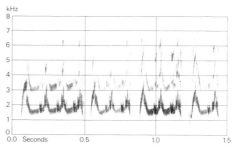

III M Orszag/BBC Hungary August 1977

güpp-güpp-güpp' interspersed during Roosting-calls (see 4, below). In flying flock particularly, birds answer each other with same call and tend to stay close to calling individuals. For other renderings, see above. For differences from equivalent call of Blue-cheeked Bee-eater *M. superciliosus*, see that species. (2) Summoning-call. A clear 'rüpp-didi-rüpp', 'rüpp-di-rüpp', or 'didi-rüpp'; uttered singly at intervals or several times in succession. Used to call up mate or members of flock, especially when separated; also by ♂ carrying food for ♀ (similar but lacking overtones: Jilka and Ursprung 1980). Component of so-called 'tunnelling duet' (see Glutz and Bauer 1980, Hahn 1982a): given by bird perched near hole as mate digs within, giving Nest-call (see call 7). Rendered 'pruik' or 'pruük' by Mountfort (1957). (3) Greeting-call (Appeasement-call). A very clear, ringing 'rüpp'; uttered several times rapidly in succession, sometimes interspersed with Summoning-call (see call 2). Used typically as call by previously lone bird when joined by, or when joining, mate or flock member and as answer by other to it, with reciprocal and simultaneous calling as birds perform

PLATE 74 (*facing*).
Coracias garrulus Roller (p. 764): **1–2** ad breeding, **3** juv.
Coracias abyssinicus Abyssinian Roller (p. 776): **4–5** ad, **6** juv moulting (body largely 1st ad).
Coracias benghalensis Indian Roller (p. 778): **7–8** ad, **9** juv.
Eurystomus glaucurus Broad-billed Roller (p. 783): **10–11** ad.
(CETK)

Greeting-ceremony (see Social Pattern and Behaviour); also as response at times to threat by conspecific bird (see call 12). (4) Roosting-call. A low-pitched, guttural, rolling 'rüpp-rüpp - rüuuuuuuuu - rüpp-grrruuuuuuuuuprüpp-rüüüuuuuuuuuuüp-rüpp', etc.; uttered fairly persistently with only short intervals. Used to call other flock members to particular roosting site (see Koenig 1951 for details; see also call 1). (5) Courtship-calls. Noticeably throaty, rolling 'rüpp', grrrp', or 'gruuup', sounds; uttered singly at short intervals or in rapid sequence. Used during courtship, mainly during breeding season, with answering calls from mate—and sometimes from other members of flock. (6) Distress-call. A clear, rather troubled-sounding 'rüppüpp' or also 'rüpp'; given singly and irregularly at quite long intervals. Used generally when bird in some discomfort, distress, etc. (e.g. when nest-tunnel collapsed); also by ♀ when ready to copulate—thus at times inducing ♂ to bring food and initiate mating sequence. See also Jilka and Ursprung (1980) who noted that, in higher-intensity excitement, ♀'s calls become rapid, throaty, gobbling, and vibrato. (7) Nest-call. A quiet, soft, noticeably guttural 'rüprüprüprüprüp'; uttered repeatedly. Used when digging nest-tunnel, prior to nest-relief, and when bringing food for young—typically only when close to nest-site, at entrance, and within hole. Hence parental feeding call, inducing food-seizing in nestlings. Elicits same call in answer by mate when both in nest. See also call 2. According to Jilka and Ursprung (1980), more variable, may be combined with soft 'di' sounds, and given also when leaving nest; 'drüüü' sounds noted by Glutz and Bauer (1980). (8) Soliciting-call of ♀. A muted, long-drawn 'grrrrüüüp grrrrrüüüüp', given many times in succession, only when mate close by. Used to induce mounting, but also noted in Courtship-feeding not followed by copulation; similarly structured to ♀ Copulation-call (see call 15b) but with different sound (Jilka and Ursprung 1980). (9) Babbling-call of ♀. A persistent, hard, rapid, almost rolling 'rüppuppidüppirüuuüppigrrrp . . .' in middle-frequency range and of moderate volume. Used by ♀, mostly in breeding season, when perched alone (apparently on territorial perch) for some purpose of self-advertisement; also as threat to approaching individual of own species (not mate). Probably of territorial significance. (10) Alarm-call. A clear, piercing 'dick-dick-dick-dick-dick'. Used when seeing something visually unfamiliar or frightening. Birds apparently give same call for aerial and ground predators (Jilka and Ursprung 1980). Also uttered (e.g.) when attacked by bee *Apis* or when molested by swallows (Hirundinidae) (Jilka and Ursprung 1980), and on hearing screech of pain from conspecific bird (see call 13). May lead to panic-flight (see Social Pattern and Behaviour). Induces similar wary mood in members of own species—which sometimes also give same call in response. Recording (Fig II)—presumably of this call—suggests a 'puik' or 'quit' given at c. 5–6 units per s; end of each unit much sharper and briefer than start, 'p' sound longer and softer than final 'k' (J Hall-Craggs, P J Sellar). Also described as a shrill, monosyllabic 'wit wit' or, occasionally, a perceptibly disyllabic 'wit(a) wit(a)' or 'duit duit'; a rapid 'wit-wit-wit-wit-wit' assembled colony into compact flock wheeling above predator (Barham *et al.* 1956; C H Fry); 'wuit wuit' in high-intensity alarm (Swift 1959; see also call 15a). (11) Copulation-call of ♂. A loud, clear, piercing 'dick-dick-dick-dick-dick-dick' (lower pitched than call 10: Jilka and Ursprung 1980) or 'ping-ping-ping-ping-ping-ping'. Used by ♂ only when in presence of ♀ actively sought out by him as pair sit close together, typically in response to soliciting of ♀ (see call 9 and Social Pattern and Behaviour). Rate of calling rapid as ♂ mounts, slower during copulation itself, and accelerates again as ♂ leaves. Sound of call often induces other members of flock to give same call (see also Hahn 1981 for possible role in breeding synchrony of colony). (12) Threat-call. A quiet, hoarse, muted, long-drawn 'däääa'. A noise rather than a tone (Jilka and Ursprung 1980, which see for sonagram). Used during attack-threat against conspecific encroaching on individual distance (see Social Pattern and Behaviour); also during fights in courtship period. Antagonist either reacts silently and defensively, becomes actively aggressive exactly like first bird, or gives Greeting-call (see call 3) or Bleating-call (see call 14)—sometimes both at once—in appeasement. (13) Pain-call. A rapid, high-pitched, piercing 'dääa-dääa-dääa-dääa-. . .', each unit descending slightly in pitch. Used in extreme fear, e.g. when seized by keeper. For sonagram, see Jilka and Ursprung (1980). (14) Bleating-call. Like call given by fledged young (see below). Given in appeasement (see call 12); also when begging food from keeper. (15) Other calls not mentioned by Koenig (1951). (a) A slow, wheezy 'ki-ew', 'pruik ki-ew', 'crruic quir', or 'crruic quiiou' in low-intensity alarm, e.g. when harried by small passerines (Barham *et al.* 1956; Swift 1959; C H Fry). (b) Copulating ♀ gives a quiet, long-drawn gurgle of very low pitch descending to c. 700 Hz as soon as ♂ mounts (Jilka and Ursprung 1980, which see for sonagram). (c) A 'raaaaah' structurally similar to call 8 (above) but with different sound; given during Courtship-feeding but exact significance not known (Jilka and Ursprung 1980, which see for sonagram).

CALLS OF YOUNG. (1) Food-calls (mainly after Koenig 1951; for sonagrams, see Jilka and Ursprung 1980). (a) From 1st day, nestlings rapidly utter a clear, slightly whispering 'djüpdjüpdjüpdjüp', becoming more distinct and louder each day. (b) From 9th day, 1st call gradually superseded by next call over next few days: a sonorous, fluting, trill—'dluiudluiudluiudluiu' or 'dluiuiuiuiuiu'; later 'dülülülülülü'; see also Jilka and Ursprung (1980) who mentioned a curious tapping sound coming from nest at this time. (c) When young begin to gape for food accurately at c. 21–23 days, give new food-call: a clear, long-drawn, bleating 'drüüüüüüü-drüüüüüüü'—each unit ascending slightly in pitch. (d) A 'dickada dickada'

(Jilka and Ursprung 1980) or 'whicka-ca-whicka whicka-ca-whicka whicka-ca-wew-wew' (Barham et al. 1956); given by older young at hole entrance. (e) After fledging, young now replace last nestling food-call with juvenile version (same call as given by adults in appeasement—see adult call 14): a piercing bleating—'eeeeeeeee-eeeeeeeee'—each unit descending in pitch. (2) Other calls. (a) Contact-call: adult-like 'rüpp' sounds (see adult call 1) interspersed among food-calls from c. 10 days. (b) Pain-call: adult-like 'däää . . .' uttered when one nestling attacked by another in nest-hole. (c) Roosting-call: adult-like version (see adult call 4) first heard on 18th day (Koenig 1951). (d) A soft, musical 'wit-wieu' or, less often, 'wit-wit', 'wu-wu', or 'wi-wi-wiew', possibly expressing contentment (Barham et al. 1956). Adult-type calls used by newly fledged young sound rather different from those of adults, being strongly frequency-modulated with 2 quite significant overtones (Jilka and Ursprung 1980); calls of this type possibly illustrated in Fig III.

KELS, MGW

Breeding. SEASON. Iberia and southern France: see diagram. Similar timing in north-west Africa but 2–3 weeks later in Greece, Yugoslavia, Rumania, and southern USSR (Dementiev and Gladkov 1951a; Heim de Balsac and Mayaud 1962; Makatsch 1976). SITE. Tunnel in sloping or flat ground, or vertical bank (natural or artificial). Old nest-holes sometimes re-used (C H Fry). Colonial. Nest: excavated tunnel ending in enlarged chamber, on axis of tunnel or to one side; in sloping or vertical bank, tunnel horizontal or inclined very slightly upwards; in flat ground, vertical hole 10 cm deep, from bottom of which is horizontal tunnel. Average length 118 cm (70–200), $n = 109$, shorter in hard ground (Swift 1959); one measured 50 cm (Barham et al. 1956); up to 275 cm (Witherby et al. 1938). Diameter of tunnel 6·5–7 cm, entrance hole 10–12 cm wide (Rivoire 1947; Mountfort 1957). Chamber $32 \times 25 \times 12$ cm high (Rivoire 1947). Unlined, though layer of food-remains 2·5–3 cm thick builds up during nesting (Barham et al. 1956). Building:

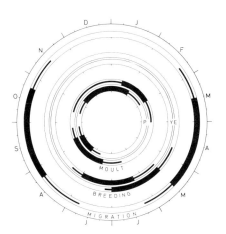

by both sexes; on vertical bank, starts with short forward and backward flights attacking with bill 15–18 times then resting while mate takes over; when foothold established, bird clings to face, pressing spread tail, and excavating with bill; as soon as hole large enough, feet used to kick loose soil out behind; pair work together in 20-min stints before resting and feeding; on flat ground, excavation started with feet and then bill; c. 3 weeks to complete (Mountfort 1957); 10–14 days (Swift 1959). Pair may start 2–3 holes at once and may excavate these to as much as 60 cm before concentrating solely on eventual nest-hole (C H Fry). EGGS. See Plate 98. Blunt elliptical, smooth, and glossy; white, though fresh eggs show rosy tinge. 26×22 mm (24–28×20–24), $n = 100$ (Witherby et al. 1938). Calculated weight 6 g (Makatsch 1976). Clutch: 6–7 (4–9). Of 41 clutches, Hungary: 5 eggs, 9; 6, 24; 7, 7; 8, 1; mean 6·0 (Makatsch 1976). One brood. Replacements laid after egg loss. Laying interval 24–48 hrs. INCUBATION. 20 days (Swift 1959). By both sexes, beginning with 1st egg; hatching asynchronous. Incubation in spells of 10–30 min; ♀ sits overnight (Swift 1959). YOUNG. Altricial and nidicolous. Cared for and fed by both parents; brooded while small. FLEDGING TO MATURITY. Fledging period given as 20–25 days, young fledging 1–2 days apart (Swift 1959), or 31–33 days (Koenig 1951, 1959; Matoušek 1951). Age of independence not recorded but young return to nest-holes for roosting, and families may depart on migration together (Swift 1959). Age of first breeding probably 1–2 years (see Social Pattern and Behaviour). BREEDING SUCCESS. Apparently less in colonies of under 6 pairs than in larger, because birds more wild in small colonies and more susceptible to disturbance (Swift 1959). In Ryazan region (USSR), 51 eggs laid produced 32 fledged young, 3·5 per successful brood (Belski 1958). In Hungary and Rumania, 11 successful broods produced average 4·6 fledged young each (Fintha 1968; Korodi Gál and Libus 1968).

Plumages. ADULT MALE BREEDING. Forehead white, feather-tips yellow or (at border of crown) pale blue; short supercilium from near nostril to just above eye pale blue; blue of forehead and supercilium merges into bright grass-green on forecrown and on narrow line over eye and above ear-coverts; lores, bristles at gape, narrow stripe below eye, and ear-coverts black. Remainder of crown and nape deep chestnut, merging into slightly paler rufous-chestnut on hindneck, sides of neck, and mantle, this in turn merging into bright golden-yellow on scapulars, back, and rump; golden-yellow sometimes with slight green tinge, especially on lower mantle and on lower rump, but usually less so than adult ♀. Upper tail-coverts and tail rather dull bluish-green, not as bright green as forecrown; shafts of tail-feathers black, undersurface dusky grey. Chin and throat rich yellow; a narrow line of pale blue just below black stripe on lores and through eye, whitish below rear ear-coverts; some white of feather-bases sometimes visible in yellow. Yellow throat conspicuously bordered below by black band 2–6 mm wide. Remaining underparts bright turquoise-blue, deepest blue on chest and sides of breast, paler greenish-blue towards vent and under tail-coverts. Primaries cerulean-blue, inner web with broad dull black tip and grey inner border, outer web with narrow dusky fringe at tip and narrow grass-green edge at base; secondaries bright rufous-cinnamon with broad black tip and green tinge on outer webs of outermost and innermost; tertials bluish-green, outer tipped dusky. Greater upper primary coverts bright cerulean-blue; lesser upper primary coverts and lesser upper wing-coverts grass-green; greater and median upper wing-coverts rufous-chestnut, greater and outer often partially fringed or edged green; tertial coverts bluish-green. Under wing-coverts and axillaries pale cinnamon or cream-buff, slightly darker than pink-buff of undersurface of flight-feather bases. Bleaching and abrasion affect colour: from about June, white of forehead partly disappears through abrasion; crown to mantle less bright chestnut, some grey of feather-bases visible; golden-yellow of scapulars, back, and rump bleached to pale yellow or partly yellow-white; green of tertials, upper tail-coverts, and tail changed to greenish-blue or blue (edges and tips first); chin and throat paler yellow with more white of feather-bases visible, pale blue above and below black loral stripe partially bleached to whitish; underparts deeper turquoise-blue or cerulean-blue (some grey of feather-bases often visible); primaries bluer, no traces of green; lesser upper wing-coverts dusky bluish-green. ADULT FEMALE BREEDING. Similar to adult ♂ breeding, and sometimes indistinguishable. Often differs in less deep chestnut on crown; chestnut reaches less far down on mantle, and in particular lower mantle tinged grass-green or golden-green; golden-yellow of scapulars often partly tinged green, especially on inner ones; rump golden-green, grass-green, or blue-green, not as yellow as in ♂; throat sometimes slightly paler yellow; black bar below throat sometimes narrower, less solidly black, and less sharply defined; chest to belly slightly paler, cerulean-green to greenish-turquoise, not as bright bluish-turquoise as in ♂ (but colour in both sexes much influenced by wear). Chestnut on upperwing often more restricted and slightly browner, less rufous; usually only median coverts chestnut and chestnut on outer webs of secondaries limited to middle ones, remainder of median coverts and outer webs of secondaries extensively green; occasionally, chestnut restricted to suffused patches on outer webs of some median coverts and some middle secondaries. ADULT NON-BREEDING. As adult breeding, but scattered new grass-green feathers between heavily worn feathers of breeding plumage on crown, hindneck, mantle, scapulars, back, rump, and some longer lesser upper wing-coverts; some new bluish-green feathers on chest and breast; feathers of black band below throat tipped green, band less sharply defined. New feathers soon bleach and abrade to green-blue and pale blue, tips first. NESTLING. Naked at hatching. From c. 10 days, spiny sheaths appear, first on back, tail, and wing; by 18th day, fully spiny except for bare ring round eye and belly, while tips of sheaths of tail and wing start to open to produce feather-tips; by c. 22 days, bird appears fully feathered, except for largely bare belly; completely feathered on fledging at 31 days, but tail- and flight-feathers still growing (Koenig 1951). See also Belskaya (1976). JUVENILE. Rather like adult, but differing most conspicuously in dull green hindneck, mantle, and upper wing-coverts (virtually no chestnut), mainly green scapulars (no glistening golden-yellow), faint dusky band below yellow throat (not contrastingly black), and absence of elongated t1. Forehead and supercilium a mixture of pale blue and grass-green, some white of feather-bases visible; crown dull chestnut with variable amount of green feather-tips. Remainder of upperparts and upper wing-coverts rather dull green, scapulars often paler bluish- or yellowish-green, inner webs occasionally greenish-yellow or ochre, feather-tips on rump cerulean-blue, part of median and greater upper wing-coverts often suffused rufous. Hardly any pale blue below black eye-streak; yellow of throat paler than in adult and with much white of feather-bases visible. Bar below yellow throat dusky grey, narrow, poorly defined; sometimes partly suffused green or slightly bordered by rufous above. Underparts pale greenish-blue, under tail-coverts sometimes pale sky-blue; sides of breast often suffused grass-green or slightly rufous. Tail as in adult, but tip of t1 rounded, not or hardly exceeding others in length. Flight-feathers as in adult, but black tips duller, less sharply defined, and less extensive; outer webs of secondaries buffish-green, virtually without chestnut tinge, inner webs with less deep and extensive rufous-chestnut tinge. Wear has some influence: chestnut of crown duller, upperparts more bluish-green (in particular, rump, upper tail-coverts, and tail blue); throat whiter; underparts paler and bluer with more greyish-white of feather-bases exposed; outer webs of primaries blue rather than green. Sexes mainly similar, but crown and hindneck deeper and more extensively chestnut in ♂, scapulars mainly golden-buff (not pale greenish-yellow or blue-green as in ♀); underparts of ♂ rather like adult ♀ (though less glossy) with almost black (though poorly defined) bar below throat (belly paler blue-green and hardly any throat-bar in juvenile ♀); in ♂, often more rufous tinge to inner median upper wing-coverts, outer webs of middle greater wing-coverts, and outer webs of middle secondaries (sometimes resembling adult ♀); in ♀, upperwing mainly dull green, except for outer webs of some secondaries. FIRST ADULT NON-BREEDING. Like adult non-breeding, but tail and flight-feathers still juvenile. FIRST ADULT BREEDING. As adult breeding; indistinguishable when last juvenile flight-feathers replaced.

Bare parts. ADULT. Iris crimson or blood-red in summer, grey-red or pale red in winter (but depth of red colouration partly depends on light intensity). Bill black or greyish-black. Leg and foot brownish-grey, purple-brown, brownish-black, or purplish-black. NESTLING. Eye opens from 6th day. Iris black or dark brown. Bare skin of body, bill, leg, foot, and mouth pale pink or flesh-pink at hatching, edges of gape yellowish-pink; bill darkens from tip from c. 7th day, when also skin of back and wings starts to darken to grey. JUVENILE. At fledging, iris dark brown, bill black, and leg and foot pinkish-grey or grey. During post-juvenile moult of body, iris changes through red-brown to red, and leg to dark flesh-brown and purple-brown. (Koenig 1951; RMNH.)

Moults. Based on data in Marien (1952) and Glutz and Bauer (1980) and on specimens in BMNH, RMNH, and ZMA. ADULT POST-BREEDING. Complete; primaries descendant. Starts from mid-June to late July. Scattered feathers of forehead, neck, mantle, and throat first, followed by p1 from August. Moult suspended during autumn migration, when much of head and body may show non-breeding and up to p3 new; some birds examined had not started primary moult, however. In Iberia, 19 of 24 adults showed suspended moult at start of autumn migration, 5 had not yet started; 2 birds suspended with p1 new, 12 with p1–p2, 4 with p1–p3, 1 with p1–p4 (Mead and Watmough 1976). Moult resumed in winter quarters from mid-October or November, remaining old breeding feathers on head and belly and wing-coverts directly replaced by fresh breeding. Tail replaced in winter quarters, starting with loss of about p6; starts from t1 outwards and from t6 inwards, t6 falling shortly after t3 (t2–t4); t4 or t5 replaced last, full-grown at about same time as p10. Secondaries start with s13 towards both sides (lost with p3) and with s1 inwards (lost with p6); s5–s6 last, completed at about same time as p10. All moult completed February. ADULT PRE-BREEDING. Partial; December–March. Involves those parts of head, body, and wing-coverts which were replaced by non-breeding during post-breeding moult. POST-JUVENILE. Complete. Sequence as in adult, but timing slightly later. Some start with scattered feathers from August when near breeding area, many others still fully juvenile during autumn migration and a few up to late October. Birds starting in winter quarters obtain breeding plumage directly, no non-breeding. Primaries replaced from October or early November, completed February–March; timing of tail and secondary moult in relation to primaries as in adult post-breeding. FIRST PRE-BREEDING. More limited than adult pre-breeding and probably no 1st pre-breeding at all in birds starting post-juvenile in winter quarters.

Measurements. All parts of range, April–August; skins (RMNH, ZMA). Tail is to tip of t1; tip is tip of t1 to tip of 2nd longest (t2 in adult). Bill (F) is to forehead, bill (N) to distal corner of nostril, both in adult only; exposed culmen 5 longer than bill (N).

		♂			♀		
WING	AD	152	(2·93; 30)	148–159	146	(2·89; 18)	140–151
	JUV	144	(3·37; 11)	139–149	141	(3·55; 5)	137–145
TAIL	AD	115	(3·31; 29)	111–121	107	(3·14; 17)	101–114
	JUV	90·8	(5·15; 10)	86–101	87·4	(2·38; 5)	84–91
TIP	AD	23·0	(2·37; 29)	19–28	17·2	(1·95; 17)	14–20
BILL (F)		43·1	(2·95; 26)	37–47	41·2	(2·98; 17)	34–45
BILL (N)		30·7	(2·64; 25)	25–35	29·7	(2·88; 15)	24–34
TARSUS		12·8	(0·68; 25)	11·7–13·8	12·6	(0·55; 15)	11·6–13·5
TOE		18·9	(0·96; 19)	17·3–20·3	18·4	(0·71; 14)	17·4–19·3

Sex differences significant for adult wing, adult tail, and tail-tip. Western birds (north-west Africa, Iberia, France, and Italy) slight smaller than eastern ones (Balkans eastward): e.g. for adults,

		♂			♀		
WING	W	152	(3·01; 14)	148–158	143	(2·67; 7)	140–146
	E	153	(2·85; 16)	149–159	147	(2·09; 11)	144–151
BILL (F)	W	42·0	(2·94; 14)	37–45	39·2	(3·16; 7)	34–42
	E	44·3	(2·54; 12)	40–47	42·6	(1·97; 10)	39–45

Differences significant for ♀♀.

Weights. ADULT. Camargue (France): May–June, ♂ 55·8 (4·44; 93), ♀ 54·4 (4·52; 97); July, ♂ 52·7 (0·8; 4); 51·1 (2·5; 19); August, ♂ 58·9 (2·8; 10), ♀ 57·0 (4·8; 14); total ranges, ♂ 48–78, ♀ 44–72 (Glutz and Bauer 1980). Turkey, Iran, Afghanistan, and Kazakhstan (USSR): April, ♂ 50, ♀ 51; May, ♂ 53·1 (4·11; 7) 48–60, ♀♀ 45, 50; June, ♀ 62; July, ♂ 56, ♀ 50·6 (3) 49–52 (Paludan 1938, 1940, 1959; Schüz 1959; Dolgushin et al. 1970; Rokitansky and Schifter 1971). Turkey, August: 54·1 (1·39; 6) 52–56 (D I Sales, BTO). Inland in Algeria and Libya, April: 54·5 (4) 45–63 (Dupuy 1970a; Erard and Larigauderie 1972). South-east Morocco, spring: 48·7 (46) 36–60 (Ash 1969b); 52·4 (4·36; 36) 41–60 (BTO, in part involving same birds as in Ash 1969b).

JUVENILE. On 1st day, 3·5 (1); on 6th 10 (1), rapidly rising to 44·5 (4) on 15th day and 63 (4) on 19th to reach peak of 68 (4) on 23rd day, decreasing to 56 (4) at fledging on day 32 (Koenig 1951). Averages of 2 samples, Turkmeniya (USSR): both 4·8 on 1st day; one sample 21·3 on 6th, 52·7 on 12th, peaking at 61·5 on 16th; other, 18·5 on 7th, 56·5 on 15th, peaking at 60·3 on 21st, decreasing to 54·7 at fledging on 27th (Belski 1958; Belskaya 1976). Camargue, July–August: 54·2 (5·46; 191) (Glutz and Bauer 1980). Turkey and Afghanistan, mid-July to early August: ♀♀ 47, 48, 52; unsexed 40, 45, 49 (Paludan 1959; Rokitansky and Schifter 1971).

Structure. Wing rather long, broad at base, tip pointed. 10 primaries: p9 longest, p10 89–101 shorter in adult, 85–90 in juvenile; in both, p8 1–7 shorter, p7 13–20, p6 24–31, p5 32–41, p1 57–71; primaries not emarginated, but tip of p1–p6(–p7) and of all secondaries with shallow notch in adult, up to p8–p9 in juvenile; p10 reduced, sharply pointed with concave inner web in adult, sometimes more rounded at tip and with convex border to inner web in juvenile. Tail rather long, 12 feathers; in adult, t1 elongated, narrow tip projecting 14–28 mm beyond other feathers (see Measurements); t2 2nd longest, others gradually shorter, t6 2–6 shorter than t2; tips of t2–t6 with shallow notch, tip of t1 narrowly rounded; in juvenile, t4–t5 usually longest, t2, t3, and t6 1–8 shorter, but t1 longer with more rounded tip (sometimes slightly attenuated), about equal to t4–t5 or (occasionally) 0–5 mm longer. Bill long, gradually decurved to sharply pointed tip; deep and wide at base and with pronounced culmen ridge, laterally compressed at tip. Short bristle-like feathers at gape and sides of bill, almost covering nostril. Tarsus and tibia short, lower part of tibia and all tarsus bare. Hind toe short, others long and slender, joined basally to form flat sole. Outer toe c. 87% of middle, inner c. 75%, hind c. 51%.

Geographical variation. Negligible; see Measurements. South African population similar in size and plumage to Palearctic birds, but separable by breeding and moulting in different seasons.

CSR

Family CORACIIDAE rollers

Large perching birds of crow-like build, with robust bills and rather short legs. Food mainly large insects or small vertebrates captured after flight from perch. Prey taken largely from ground by *Coracias*, in the air by *Eurystomus*. 11 species in 2 genera: (1) *Coracias* (8 species, southern and central Eurasia and Africa); (2) *Eurystomus* (3 species, southern and eastern Asia and Afrotropical and Australasian regions). Represented in west Palearctic by 3 species of *Coracias* (2 breeding, 1 accidental) and 1 species of *Eurystomus* (accidental).

Sexes of same size. 13–14 cervical vertebrae. Wings fairly long (especially in *Eurystomus*); flight buoyant with fairly rapid wing action in *Coracias*, and especially agile, with wheeling and gliding, in *Eurystomus*. 10 primaries, p9 longest, p10 only slightly shorter; wing diastataxic. Tails fairly long, 12 feathers. Oil-gland naked. Well developed aftershafts. Bills stout, very broad at base in *Eurystomus*; slightly hooked at tip, with a few rictal bristles. Tarsus quite short, feet syndactyle, with inner and middle and also middle and outer toes united by basal joint. Long caeca. Paired carotids.

Plumages mostly bright, with brilliant blues on wings especially. Sexes similar. Irises pale to dark brown. Bills blackish in *Coracias*, yellow or red in *Eurystomus*. Adult post-breeding moult complete, starting shortly after breeding, interrupted in migratory species; partial pre-breeding moult in some species, involving variable amount of feathers of body and upper wing-coverts. Nestlings naked at first, later acquiring spiny covering of feathers in retained sheaths as in Alcedinidae and Meropidae.

Coracias garrulus Roller

PLATES 72 and 74
[between pages 758 and 759]

Du. Scharrelaar Fr. Rollier d'Europe Ge. Blauracke
Ru. Сизоворонка Sp. Carraca común Sw. Blåkråka

Coracias Garrulus Linnaeus, 1758

Polytypic. Nominate *garrulus* Linnaeus, 1758, North Africa and Europe east to Levant, Asia Minor, north-west Iran, and south-west Siberia; *semenowi* Loudon and Tschusi, 1902, Iraq and Iran (except north-west) east to western Pakistan, Kashmir, and western Sinkiang (China), north to Turkmeniya and southern Kazakhstan (USSR).

Field characters. 30–32 cm; wing-span 66–73 cm. Similar in size to Jay *Garrulus glandarius* but with 5% longer wings and 20% shorter tail. Rather corvine in appearance, with rather pale green-blue head, body, and wing-panel set against dark black-blue flight- and tail-feathers and chestnut-brown saddle. Flight confident, usually direct, recalling large flapping pigeon *Columba* or plover *Vanellus* as much as crow *Corvus*. Sexes similar; some seasonal variation. Juvenile separable.

ADULT BREEDING. General appearance as above, but at close range shows brown-white forehead, very pale blue streaks on throat and breast, purple-blue lower back and rump, pale green-blue outer webs to tail-feathers, and short, black-blue points to outermost tail-feathers. Striking pattern of upperwing in flight created by pale green-blue primary, lesser, and greater coverts and similarly coloured bases of flight-feathers contrasting with black-blue to purple-blue lesser coverts and ends to all flight-feathers; pale central panel shows strong sheen at certain angles, 'flashing' at times. Under wing-coverts pale but vivid green-blue, in stark contrast to dark black-blue flight-feathers. Some show cinnamon tinge on breast. ADULT NON-BREEDING. Loses full vivid colours, with crown and nape dull brown-green, mantle less chestnut, and breast sullied green-brown. JUVENILE. Resembles winter adult but much duller overall—face dull white, cheeks brown, and throat and breast distinctly brown, obviously streaked white.

Confusion with *G. glandarius* unlikely since *C. garrulus* always shows large areas of blue (not just on wing bend) and never bold white rump. Confusion with other *Coracias*, particularly Abyssinian Roller *C. abyssinicus*, much more likely; see those species. Flight actions include fairly deep and regular wing-beats over distance and faster, more plucked wing-beats when hunting. Perches freely but gait restricted. Hunting bird suggests lethargic bee-eater *Merops*, with most prey taken by dive from perch. Migrants form small parties which pass in characteristic long procession.

Commonest call 'ack-er-ack', 'rack-kack', or 'kacker'; also a softer, slow 'kruk' and sharper, mournful 'kraa'.

Habitat. Breeds throughout temperate, steppe, and Mediterranean zones enjoying reliable warm summer climates; accordingly concentrated in continental interior, avoiding oceanic influence and adjusting extensively to changing climatic trends by decline and withdrawal from adversely affected regions (see Distribution). Predominantly a lowlands species, but ascends in Moroccan High Atlas to 2000 m: in Caucasus no higher than 800–1000 m and in central Europe only to that level in isolated cases, 400–600 m being normal limit. Normally avoids deserts, semi-deserts, and treeless grasslands, and shows no attachment to water but will inhabit lines or groups of poplars *Populus* and other trees along banks of rivers traversing open steppe. Over most of breeding range, favours open old oak forest with plenty of hollow trees, and pinewoods, especially of *Pinus sylvestris* with clearings, heaths, and other interruptions. In Morocco, especially in cork oaks, at 700–1200 m. Also inhabits old parks, avenues, tree-lined river banks, orchards, copses, willow stands, and dry plains with scattered thorn trees, but avoids intensive cultivation. Nests in artefacts (e.g. cracks in masonry) only in abandoned structures. In Spain, found concentrating in stone pine *Pinus pinaster* woodland, where dependent for nest-sites on operations of resin tappers creating scars which are eventually opened up by Green Woodpeckers *Picus viridis*, whose holes, however, are barely large enough (England 1963). In northern Europe, holes of Black Woodpecker *Dryocopus martius* freely used, and even, in places, nest-boxes. Will make use of alternative nest-sites in exposed sandy or other banks, walls, or fissured rock-faces where trees lacking. Hunts from suitably commanding lookout posts on trees, overhead wires, etc., above bare or sparsely vegetated ground or short vegetation providing little cover for prey, moving only clumsily for brief distances on ground. Flies freely at varying heights up to 300 m or more. Turkestan race *semenowi*, in extreme east of west Palearctic, inhabits steppe, semi-desert, and desert localities with cliffs and ravines sometimes entirely bare of vegetation, nesting or roosting in burrows and under eaves of roofs. In African winter quarters, particularly favours dry bushy country with

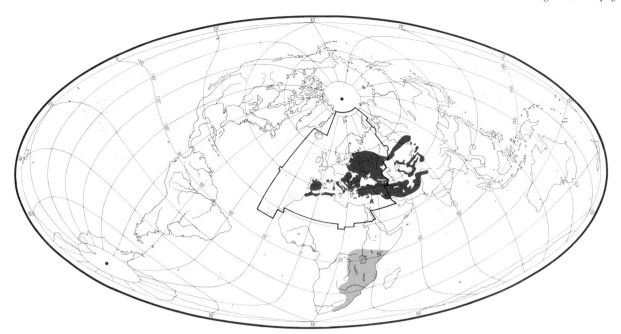

sprinkling of dead trees, especially areas prone to grass fires where it can feast on charred remains of invertebrates and can snap up survivors from bush top or hillock. On coast hunts locusts from mangrove swamps, but avoids forests. (Dementiev and Gladkov 1951a; Bannerman 1955; Voous 1960; Archer and Godman 1961; Glutz and Bauer 1980.)

Distribution. Much more widespread in Europe in late 18th century and early 19th century (Glutz and Bauer 1980); since ceased breeding in Denmark, West Germany, and Sweden, and range decreased in East Germany, Czechoslovakia, and Austria, though slight recent spread in southern France.

FRANCE. Slight increase in range since 1936 (Yeatman 1976). WEST GERMANY. Probably bred almost throughout in 18th century and 19th century, but extinct around 1900; from 1943 to 1949, several pairs bred or probably did so, with last attempted nesting in 1965 (Glutz and Bauer 1980). DENMARK. Formerly bred in eastern Denmark and some islands; last certain nesting 1868 (Salomonsen 1963). SWEDEN. In mid-19th century bred along Baltic coast from Skåne to c. 60°N and parts of south-central area; by end of 19th century much reduced, but still breeding in odd coastal areas. Last nested on mainland 1943; in 1930s started breeding Fårö (north of Gotland), reached 14 pairs in 1951 and 1954, decreased to 3 pairs in 1965, and last bred in 1967. FINLAND. Has bred (or probably bred) sporadically—around 1785, in 1845, 1931, and 1938, and around 1940 (Merikallio 1958; Salomonsen 1963). CZECHOSLOVAKIA. Formerly bred over much of country (KH). AUSTRIA. Last bred Lower Austria 1925 and Carinthia 1974 (Glutz and Bauer 1980). SWITZERLAND. Bred 1896 (Glutz and Bauer 1980). ITALY. Breeds occasionally further north, most frequently Emilia Romagna (Lovari 1975b; PB, BM). GREECE. Probably breeds Crete, Rhodes, and Kos; rare Pelopónnisos, breeding not confirmed (WB, HJB, GM). USSR. Spread north in 19th century in some areas (Dementiev and Gladkov 1951a). LEBANON. No recent evidence of breeding (Tohmé and Neuschwander 1974).

Accidental. Faeroes, Britain, Ireland, Belgium, Netherlands, Norway, Azores, Madeira, Canary Islands.

Population. Declined in north-west and north-central Europe (see Distribution, and below), also central Italy and Israel. Former attributed largely to climatic changes affecting food supply (Durango 1946; Rudloff 1967), but habitat changes and other factors also involved there and elsewhere.

SPAIN. Not scarce (AN). FRANCE. Estimated 10–100 pairs (Yeatman 1976). EAST GERMANY. Decreased steadily in last 30 years: 95–134 pairs 1961, 20–27 pairs 1976 (Creutz 1979); 15–17 pairs 1981, 10–12 pairs 1982 (Robel 1982; D Robel and S Bude). POLAND. Scarce or very scarce; for some local counts, see Tomiałojć (1976a). CZECHOSLOVAKIA. Decline some areas from end of 19th century; major decrease since 1950 (KH). HUNGARY. Not common (Glutz and Bauer 1980). ITALY. Decreased in central area due to hunting and to effect of pesticides on food supply (Lovari 1975b). USSR. Abundant in many southern areas, with numbers decreasing further north (Dementiev and Gladkov 1951a). Estonia: marked decrease since c. 1960; now rare (HV). Latvia: decreased some areas (Transehe 1965). IRAQ. Common (Allouse 1953). ISRAEL. Marked decrease (HM). CYPRUS. Scarce to

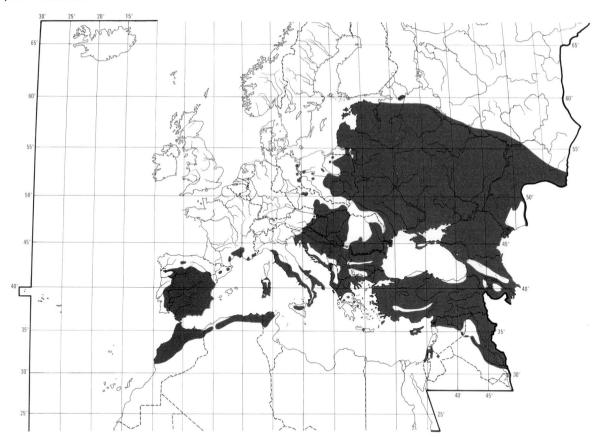

fairly common, possibly decreased (PRF, PS). ALGERIA. Nowhere common (EDHJ).

Oldest ringed bird 9 years 1 month (Rydzewski 1978).

Movements. All populations migratory, wintering in Afrotropical region, especially in eastern half of continent. South-eastern race *semenowi* (breeding Iraq to Soviet Central Asia and Pakistan) does not winter Iraq or in any numbers in Arabia, as previously suggested, and presumably reaches eastern Africa (Marchant 1963b; Moreau 1972), where specimens known from Kenya and Tanzania (Britton 1980).

Said by Dorst (1962) that breeding birds from Iberia and north-west Africa make diagonal crossing of Sahara to winter in East Africa; this now thought unlikely, as birds since found in various West African localities, where not readily separable in the field from young and moulting adult Abyssinian Roller *C. abyssinicus* (Moreau 1972; Smalley 1983b). Numbers pass through Sénégal in August–September (Morel and Roux 1973); parties of up to 50 occur Niger inundation zone (Mali) August–November (Lamarche 1980), and birds found as thinly-distributed winter visitors in Gambia (Smalley 1983b), Ivory Coast (Thiollay 1971), Ghana (Macdonald 1978), Dahomey and southern Haute Volta (Green and Sayer 1979), and Nigeria (Elgood 1982). More numerous from Chad eastwards, and south to Zimbabwe and Moçambique (irregular in South Africa: Broekhuysen 1974), reflecting higher breeding density in eastern half of Europe and western Asia. Estimated 2–3 million birds winter eastern Kenya (Brown and Brown 1973), at northern end of main winter range; even larger numbers must occur further south (Ash and Miskell 1980). Often solitary in winter, but loose flocks form where food, e.g. swarming termites (Isoptera), abundant; concentration of 5000 at Dodoma (Tanzania), 26 December (Meiklejohn 1940). Probably to some extent nomadic within winter range. Numbers fluctuate in Kenya, within and between seasons (Brown and Brown 1973); movements in East Africa continue into December (Britton 1980); only small numbers remain in Zambia during mid-season rains (Benson *et al.* 1971), and numbers fluctuate in southern Africa generally, with birds penetrating further south in some years (Clancey 1965; Broekhuysen 1974).

Migrates diurnally, singly or in small parties, birds following each other in steady stream. Small-scale ringing in east-central Europe (East Germany, Poland, Latvia, Lithuania, Hungary, Bulgaria) produced April–May and August–October recoveries in Greece (9), Turkey (2), Cyprus (1), Israel (1), and lower Egypt (4), emphasizing important east Mediterranean passage; 1 Bulgarian bird found Tanzania in December (Glutz and Bauer 1980).

C. g. semenowi common as autumn migrant through north-west India (Punjab to Kutch), heading WSW, but this region bypassed to north in spring (Ali and Ripley 1970). Passage continues across Arabia, including ship records from Arabian Sea and northern Indian Ocean; birds presumably enter Africa across Red Sea and Gulf of Aden.

Visible movement through Chad, Sudan, and Eritrea much stronger in autumn than in spring (Moreau 1972); while in Tunisia, Libyan coastal plain, Malta, and Cyprus, more common in spring (Heim de Balsac and Mayaud 1962; Bundy 1976; Sultana and Gauci 1982; Flint and Stewart 1983). Pattern suggests that Mediterranean and Sahara crossed in unbroken flight in autumn, while in spring birds fatten at low latitudes then overfly northern Afrotropics and Sahara to pause in Mediterranean basin (Moreau 1961, 1972). Different spring and autumn strategies may also apply in Middle East, though fewer data available; commoner in autumn in Aden (Browne 1950), but in spring in Iraq (Marchant 1963b). See also Indian status (above). Birds will feed opportunistically on passage, as in ENE mass migration of 40 000–50 000 birds seen in Somalia, 13 April (Ash and Miskell 1980). In mass north to north-east migration in eastern Tanzania, 28 March–4 April 1982, tens of thousands passed on peak days, indicating that evacuation of wintering area can be concentrated into short period (Feare 1983).

Food. Mainly insects, particularly medium- to large-sized beetles (Coleoptera) and crickets, etc. (Orthoptera). Normally hunts from open perch *c.* 1–5 (–10) m above ground, especially long, dead branch below crown of tree (Haensel 1966; Glutz and Bauer 1980). Glides to ground folding wings slowly after landing (Schinz 1960), or flapping them constantly whilst seizing prey (Vahi 1963). May make short, rather clumsy, flapping pursuit of prey on ground, but often gives up without chase (Glutz and Bauer 1980). In steppe, USSR, recorded hunting crickets (Gryllidae) on ground (Frank 1950), but no further details. Prey seized in bill and bird returns to same or nearby perch against which larger items normally beaten once or several times before swallowing. May also be tossed up beforehand (Glutz and Bauer 1980); this activity 'practised' by captive young using pieces of wood or lumps of earth (Frisch 1966). At *c.* 35 days old, captive young softened up mice (Murinae) before swallowing by repeatedly shaking and beating them (Heinroth and Heinroth 1924–6). In winter, Africa, *Zonocerus elegans* (Orthoptera) crushed in mandibles, intermittently tossed just clear apparently to change its position, then swallowed; legs not removed (Moreau and Moreau 1941b). Recorded breaking open snail (Gastropoda) by smashing it against stone (Norman 1980). In Iraq, birds waded in marshy ground to take small frogs *Rana* (J A McGeoch); captive birds seized these by back legs and beat them repeatedly against ground before swallowing (Naumann 1901). In India, said to take fish fry trapped in shallow pools (Baker 1927), but see comments in Indian Roller *C. benghalensis* on possible confusion with bathing behaviour. Also flies from perch to seize prey in the air like flycatcher (Muscicapidae); high proportion of prey sometimes caught using this method (D Robel and S Bude). Bird collecting food for young accompanied plough on 2 successive days and took large Orthoptera mainly in 'flycatching' sallies from ground (Bernáth 1959). More sustained hunting flight like swallow (Hirundinidae) used regularly and prey then swallowed in flight—as it may be on other occasions (D Robel and S Bude); may ascend to *c.* 200 m (V Wendland). Gleaning of insects in flight from seed-heads of standing cereal crops (Hammling 1917), birds fluttering but not landing (Barabash-Nikiforov and Semago 1963), possibly exceptional. Occasionally hovers like Kestrel *Falco tinnunculus* (Dolgushin *et al.* 1970), and will plunge suddenly from level flight to take prey on ground (Durango 1951). In Hungary, no food taken from plants, always from ground (Szijj 1958). In winter quarters, prolonged hunting flights when preying on termites (Isoptera) and ants (Formicidae) comparable to those of broad-billed rollers *Eurystomus* (Glutz and Bauer 1980). May also fly back and forth through termite swarm, perching intermittently (Moreau and Moreau 1946). Attracted to brush fires (e.g. Lippens and Wille 1976). Rarely takes young of other birds: e.g. brood of 5 Starlings *Sturnus vulgaris* once extracted from nest-box over 3 hrs (Barabash-Nikiforov and Semago 1963). Powerful bill enables bird to deal with Coleoptera having extremely hard exoskeletons—e.g. *Morimus asper* and *Oryctes nasicornis* (Cassola and Lovari 1979); see also Szijj (1958) for further examples. Also readily consumes several insects which produce pungent secretions: e.g. beetles *Procrustes coriaceus* and *Calosoma sycophanta* (Cassola and Lovari 1979), glow-worms (Lampyridae) (Floericke 1919), larvae of goat moth *Cossus cossus* (Rörig 1900), and, in Africa, aposematically coloured *Zonocerus elegans* (Moreau and Moreau 1941b). Large (up to *c.* 15 cm) and poisonous centipedes (Myriapoda) also much exploited by birds of one colony (Meade-Waldo 1907). Food carried to young in bill (Hesse 1910; D Robel and S Bude); delivered whole (Prokofieva 1965).

Undigested items (mainly chitinous parts of insects) ejected as pellets by adults and young. Dry weight 3·5 g ($n=24$); 2–4 pellets regurgitated 2–3 times per day (Ganya *et al.* 1969). Specimens found under perches used habitually by adults measured *c.* 20–35 × 10–15 mm; 15–24 items per pellet (Haensel 1966). Captive juvenile ejected 1–2 smooth, clay-like pellets daily (Hüe and Rivoire 1947).

Composition of diet reflects above all differences in distribution of prey species, but most abundant terrestrial or slow-flying large arthropods usually predominate (Szijj 1958; Glutz and Bauer 1980). Coleoptera apparently more important in north of range, Orthoptera in south (e.g. Prokofieva 1965). Of 2231 prey, 90·6% measured 10–30 mm, 48·4% 10–15 mm (Haensel 1966); similar results in

Klausnitzer (1960). Insects mostly crickets, etc. (Orthoptera: Gryllidae, Tettigoniidae, Acrididae, Gryllotalpidae), and beetles (Coleoptera: Scarabaeidae, Geotrupidae, Carabidae, Tenebrionidae, Chrysomelidae, Silphidae, Curculionidae, Lucanidae, Cerambycidae, Elateridae, Cicindelidae, Dytiscidae, Staphylinidae, Histeridae, Buprestidae, Hydrophilidae, Byrrhidae, Coccinellidae, Lampyridae, Colydiidae); also larvae of dragonfly *Agrion*, earwigs (Dermaptera), praying mantises (Mantidae), termites, bugs (Hemiptera: Pentatomidae, Scutelleridae, Notonectidae, Cicadidae), Lepidoptera (adult *Vanessa polychloros* and Pieridae, larvae of Noctuidae, Cossidae, and Sphingidae), flies (Diptera: Tipulidae, Anthomyzidae, also other pupae), Hymenoptera (wasps Vespidae, bees and bumble bees Apoidea, ants Formicidae, adult and larval sawflies (Cimbicidae, Tenthredinidae). Non-insect part of diet always much less important: millipedes and centipedes (Myriapoda), spiders (Araneae, including jumping spiders Salticidae), scorpions (Scorpiones), pseudoscorpions (Chelonethi), Solifugae; also snails and slugs (Gastropoda), mussels (Unionidae), earthworms (Lumbricidae), frogs *Rana* (including tadpoles), slow-worms *Anguis fragilis*, lizards (*Lacerta*, *Podarcis*), snakes (e.g. young *Natrix natrix* of *c.* 20 cm); very few small rodents (Muridae), shrews (Soricidae), and young or adult birds (exhausted small migrants on Greek islands in May when no other food available: Reiser 1905). Plant food represented solely by various fruits, particularly grapes *Vitis* in southern USSR and figs *Ficus* in Mediterranean region. (Rörig 1900; Naumann 1901; Reiser 1905; Bartos 1906; Meade-Waldo 1907; Baer 1910; Floericke 1919; Witherby *et al.* 1938; Prekopov 1940; Hüe and Rivoire 1947; Dementiev and Gladkov 1951a; Pokrovskaya 1956; Szijj 1958; Klausnitzer 1960, 1963; Pek and Fedyanina 1961; Barabash-Nikiforov and Semago 1963; England 1963; Vahi 1963; Prokofieva 1965; Haensel 1966; Ganya *et al.* 1969; Ivanov 1969; Khanmamedov and Gasanova 1971; Cassola and Lovari 1979; Glutz and Bauer 1980; Robel and Bude 1981; J G Walmsley.) Small stones found in nest-hole and object once brought by adult was probably also a stone; significance not known (England 1963).

75 stomachs (April–September) from Hungary (see Csiki 1905 for earlier analysis of 23 of these) contained mainly *Gryllus* (19·6% of total number of items; high proportions throughout season, particularly May 32·4% and September 34·4%, dropping to 6% in August), 8·6% small Carabidae (most important in May 15·4% and September 26·6%), 8·4% grasshoppers Acrididae (present in stomachs from June and reaching peak of 15·8% in August), 7·4% bugs *Eurygaster* and *Aelia* (48·1% August), 4·1% adult cockchafers *Melolontha* (30·4% April), 3·8% bush-crickets Tettigoniidae (peaking at 7·7% in July, before Acrididae), 3·8% beetle *Silpha obscura* (20·4% June), 2·7% other (larger) Carabidae (9·5% June), 1·8% rosechafer *Cetonia aurata* (6·4% July), 1·6% mole-crickets *Gryllotalpa* (21·7% in April), 1·2% weevils *Cleonus* and *Bothynoderes*, 0·9% chafer *Tropinota hirta* (Szijj 1958; see also Csörgey (1934) for importance of *Eurygaster* in (July–) August. At Ihárosberény (Hungary), birds reported to take 70% frogs *Rana esculenta* and 30% Coleoptera, with *Melolontha* important in some years (Bartos 1906). 14 stomachs from Danube delta (Rumania) contained at least 135 items: mostly smaller Scarabaeidae and wetland Carabidae (particularly *Amara*), also 20 Orthoptera (Kiss *et al.*1978). 400 stomachs (mainly June–July), Voroshilov (northern Caucasus, USSR), contained mainly beetles (92·7% by number; predominantly Scarabaeidae with many *Pentodon*, *Gymnopleurus*, and *Potasia*), also Orthoptera and a few slugs (Limacidae); only 0·5% vertebrates (Prekopov 1940). 310 stomachs from Azerbaydzhan (USSR) also contained primarily Coleoptera, as well as Orthoptera and Heteroptera; 2–3 stomachs held Muridae, 1 stomach contained a lizard and a snake; molluscs also recorded (Khanmamedov and Gasanova 1971). For detailed list of other studies in USSR, see Prokofieva (1965); see also Pek and Fedyanina (1961), Ganya *et al.* (1969). Analysis of pellets collected under regularly used perches and of nest-hole contents in Frankfurt/Oder district (East Germany) produced 2231 identifiable items (97% insects): 77·6% beetles (Cerambycidae 30·0%, Scarabaeidae 19·3%, Carabidae 10·6%) and 16·4% Orthoptera (Acrididae, Gryllotalpidae). Other items included snails (Helicidae, 0·9% of total prey) and, only from nests near water, freshwater mussels (Unionidae, 0·7% of total); also 1 *Vanessa polychloros*, 7 frogs, 1 slow-worm, 8 Muridae (probably mostly *Microtus arvalis*), and 1 bird (Haensel 1966). See also (e.g.) Rörig (1900) and Naumann (1901) for further German studies.

Few details on winter diet. 2 stomachs (♂ late March, ♀ mid-November) from Zaïre contained large grasshoppers (Orthoptera), 2 dung beetles (Scarabaeidae), and other Coleoptera (Chapin 1939); see also (e.g.) Moreau and Moreau (1946) for sight records involving Isoptera and Orthoptera.

Captive juvenile ate up to 40 fair-sized grasshoppers and 4–5 lizards (6–10 cm long) per day (Hüe and Rivoire 1947). See also Frisch (1966).

Food of young does not differ significantly from that of adults (Szijj 1958); nest-content analyses—see Haensel (1966, above), and Klausnitzer (1960, 1963, below)—likely to refer to food of both adults and young. According to Maglio (1976), young nestlings are given caterpillars (Lepidoptera) and earthworms if vicinity of nest-site moist, otherwise mostly insects. In Tuscany (central Italy), food remains collected from nest when young 10–27 days old; as none found before this, younger nestlings possibly given delicate, partly digested food. 129 items comprised large Coleoptera (60·5% Scarabaeidae, 11·6% Carabidae), 8·5% Cicadidae, and 1·6% Orthoptera (Cassola and Lovari 1979). In Lausitz (East Germany), 10 nests from 1959–61 contained 1461 items: 96% beetles—44·7% Geotrupidae, 19·6% Carabidae, 16·2% *Spondylis buprestoides* (Cerambycidae); of Orthoptera, only *Gryllotalpa* important (3·4%); also amphibians (Anura), lizards, rodents, and shrews (Klausnitzer 1963). For earlier study (511 items from 8 nests) with closely similar results, see Klausnitzer (1960). 45 items from neck-collar samples, Voronezh (USSR), comprised 57·8% Orthoptera, 13·3% molluscs *Anisus*, and 9·0% Coleoptera; also 3 bugs *Eurygaster austriacus*, 1 caterpillar (Noctuidae), 2 spiders, 1 frog *Rana arvalis*, and 2 lizards *Lacerta agilis*. 107 neck-collar samples and 3 stomachs, Leningrad (USSR), contained 131 items: 60·3% beetles (including 42 *Spondylis buprestoides*), 18·3% Orthoptera (mainly *Decticus verrucivorus*), 6·8% Hymenoptera, 1 caterpillar *Mimas tiliae*, 1 earthworm, 15 *Rana temporaria*, 1 *Lacerta vivipara*, and 1 shrew. Observations at 2 nests, mid-July, showed no differences in diet of nestlings 4–7 and c. 21 days old. Food portion consisted usually of 1, rarely 2, items. Young apparently not able to cope with snails *Anisus* more than 1·5 cm across (Prokofieva 1965). For further study from Leningrad region, see Pokrovskaya (1956). In Azerbaydzhan, nestlings fed 15–17 times in 2 hrs (at c. 1 day old), 18–20 times at c. 2 days, 15 times at c. 4 days, 10–12 at c. 18 days, 10 at c. 24 days, and 7 times at c. 26 days (Khanmamedov and Gasanova 1971). At 2 sites on Fårö (Sweden), 17 and 23 feeds by both parents in 2 hrs in fine weather, 4 and 7 in damp conditions, significantly more visits made by $\male$ (Wigsten 1955); see also Durango (1951). Relative frequency of visits by $\male$ and $\female$ may fluctuate during day but overall roughly equal. No marked reduction around midday but quite long pauses occur especially when only 1 nestling present; visits more frequent for larger broods and adults forage more near nest (D Robel and S Bude). In Touloubre valley (southern France), when young apparently well-grown and given almost entirely Orthoptera, visits made by both parents 'without pause', though rate dropped off at midday (Hüe and Rivoire 1947). MGW

Social pattern and behaviour. Many aspects still inadequately known. For recent summary, see Glutz and Bauer (1980) and for study of captive birds, see Frisch (1966). Following account includes material supplied by D Robel and S Bude from study in Niederlausitz (East Germany).

1. Normally solitary in winter quarters (Moreau and Moreau 1946). In eastern Kenya, c. 23–25 per km² (Brown and Brown 1973). Birds usually c. 100–200 m apart over c. 2 km; show no intra- or interspecific (with Lilac-breasted Roller *C. caudata*) territorial behaviour, normally staying c. 1–2 weeks in one area before moving on (Sassi and Zimmer 1941; Sauer and Sauer 1960). May be attracted to favourable food supply—e.g. c. 5000 at brush fire, Tanzania (Lippens and Wille 1976)—but large aggregations otherwise typical only of pre-migration period: e.g. 'hundreds' at Amani (Tanzania), with birds arriving in loose flocks of up to c. 80 (Moreau and Moreau 1946). In later stages of migration, usually singly or in flocks of up to 10 (Moreau and Moreau 1941a; see also Ticehurst *et al.* 1922, Meinertzhagen 1930, Stresemann 1944, Dementiev and Gladkov 1951a). Larger flocks reported for early period following arrival, Iraq, where migrants associated with Bee-eater *Merops apiaster* (Marchant 1963b); communal roosts formed in Israel (Tristram 1866; see below). In southern France, up to c. 20 occur together at good feeding grounds (Hüe and Rivoire 1947) and, in East Germany, 6–8, even in nesting territory where young not yet fledged (Rutschke 1983); see also Wigsten (1955) for Fårö (Sweden) and Vahi (1963) for Estonia. In Niederlausitz, birds feeding together often 5–15 m apart; less in flight (D Robel and S Bude). Flocks of c. 10–15 formed, southern USSR, prior to autumn emigration; birds depart singly or in small flocks (Dementiev and Gladkov 1951a). BONDS. Monogamous mating system (Dementiev and Gladkov 1951a); 3rd bird said sometimes to help with rearing of young (Glutz and Bauer 1980; Rutschke 1983) but such reports considered erroneous by D Robel and S Bude who, however, state that pairs with no offspring often associate with family parties (in which young fledged but still dependent) though indifferent towards juveniles (i.e. do not feed them). Exact duration of pair-bond not known. Repeated use of same nesting territory and nest-site (D Robel and S Bude), also hunting area (and perches) for up to 3 years (Hüe and Rivoire 1947), possible indication of longer-term bond; see also Wigsten (1955). In Niederlausitz, birds normally paired on arrival; single $\male\male$ taking up territories possibly 1-year-olds (D Robel and S Bude). Both parents care for young up to and beyond fledging (Hüe and Rivoire 1947); in Niederlausitz, young fed regularly for c. 14 days after fledging, then no longer after c. 21–24 days, though loose family-bond maintained until departure (D Robel and S Bude). Age of first breeding in captive birds 1 year (Geitner 1979). Regular occurrence of non-breeders—single birds and pairs—suggests most do not breed at 1 year old (Glutz and Bauer 1980; see also Frisch 1973). BREEDING DISPERSION. Solitary and territorial, though small loose-knit neighbourhood groups occur, perhaps due to shortage of suitable nest-sites in some cases (Dementiev and Gladkov 1951a). In Grunewald (West Berlin), formerly 8–10 pairs on 3150 ha, of which 2720 ha woodland, mainly *Pinus sylvestris* (Wendland 1971); in one year, 4 pairs on barely 1 km², normally 1–2 pairs (Hesse 1921). On small (113 km²) island of Fårö, 2–14 pairs over 13 years (Wigsten 1955). Local concentrations typical of dwindling East German population: e.g. 10 pairs along 12 km of forest in Spree valley, i.e. 0·8 pairs per km. Nests quite often 80–500 m apart, once 36 m in house gardens (Rutschke 1983). Even where density high, nests normally c. 70–200 m apart (Naumann 1901). In loose groups (typically found in south of range) of 2–4, rarely up to 10 pairs, distance between nests may be reduced to c. 5–10 m (e.g. Zarudnyi 1896, Reiser 1905). Near Karlovo (Bulgaria), c. 70 pairs with c. 100 pairs of Lesser Kestrel *Falco naumanni* in oak *Quercus* grove (Reiser 1894; see also Reiser 1905); for report of c. 300 pairs in ruins (locality not given), see Meade-Waldo (1907). In Camargue (southern France), 22–27 pairs in study area of c. 17000 ha (J G Walmsley); at Vergière (southern France), c. 6 pairs on 1 km² (Frisch 1966). Highest density (0·9 pairs per 10 ha) in Middle Atlas (Morocco) in cork oak *Quercus suber* woods (many tree cavities) at c. 700–1000 m (Glutz and Bauer 1980). In Israel, mostly dispersed solitarily with nests c. 1·6 km from communal roost; one colony of c. 20–30 pairs (Tristram 1866). In Azerbaydzhan (USSR), average 5 birds along 3-km transects in 2 areas (Khanmamedov and Gasanova 1971); see also Drozdov (1963). Nesting territory of c. 50-m radius defended mainly by $\male$; used for some courtship and copulation (which may also take place up to c. 100 m away; see also Heterosexual Behaviour, below), and rearing of young, including (rarely) post-fledging care if territory offers favourable feeding conditions (D Robel and S Bude). Foraging range up to 1–2 (–3) km from nest. At one site, birds foraged mainly near nest, then switched temporarily to feeding grounds c. 3 km away and

used regular route (Hüe and Rivoire 1947). On Fårö, initially foraged at c. 1 km, later (once nesting had begun) at c. 500 m from nest-site (Wigsten 1955). See also Durango (1951), Vahi (1963), Prokofieva (1965), Haensel (1966), and Rutschke (1983). In Niederlausitz, breeding birds feed in area of c. 100–300 ha but do not use all of it, and other conspecific birds tolerated there; tend rather to visit particular sites repeatedly if food supply favourable (D Robel and S Bude; also J G Walmsley for Camargue); such sites (and same perches) may be used for several days or weeks (Glutz and Bauer 1980). ROOSTING. Nocturnal. In southern Kenya, winter, large communal roost established in dense thorn clump (Mackworth-Praed and Grant 1952). At Jericho (Israel), trees used by 'large flocks' for several evenings from early April, birds assembling shortly before sunset; communal roosting ceased with start of breeding (Tristram 1866; see also Flock Behaviour, below). Similarly, in Iraq, c. 45 presumed non-breeders roosted together in date palms *Phoenix dactylifera*, mid-June to early August, arriving at about sunset in groups of 3–4 (Sage 1960b). In Niederlausitz, breeding birds roost in crowns of tall trees, away from nest-site (D Robel and S Bude); on Fårö, ♂ roosted in hole of Black Woodpecker *Dryocopus martius* c. 50 m away from nest when ♀ incubating (Wigsten 1955). Fledged young roost in thick cover near nest (Vahi 1963). Rests during day on posts, telegraph wires, etc.; *contra* Glutz and Bauer (1980) no indication that relatively well concealed perches used preferentially for this purpose. Often very little activity in rain; in fine weather, generally active throughout day with no pronounced morning or late-afternoon peak, and no midday pause (D Robel and S Bude). For development of comfort behaviour by captive birds, see Heinroth and Heinroth (1924–6) and Frisch (1966).

2. Generally rather shy and secretive, particularly in north of range (Glutz and Bauer 1980). When alarmed, captive birds adopt Concealing-posture: bill pointed at danger, tail extended directly to rear so that head, back, and tail in straight line; feathers sleeked so that bird appears strikingly slim. Wild birds probably act thus in tree canopy but difficult to ascertain as plumage there has cryptic effect. Strange objects (particularly blue or green) elicit Warning-call and Wing-fluttering (see 3 in Voice). Rapid sequence of 'rak' calls given in High-erect posture (Fig A), particularly after change of perch (Frisch 1966: see 1d in Voice); up-and-down tail movements also typically performed at this time (Heinroth and Heinroth 1924–6; Moore and Boswell 1956b). FLOCK BEHAVIOUR. At Amani, early April, most birds highly vocal and excited; some perched quietly high or low and were often displaced (without protest). All flew up, following take-off of one bird which had previously occupied high perch; loose flock formed before birds moved away (Moreau and

A

B

Moreau 1946). At Jericho roost (see above), birds assembled in large, noisy flocks; volley of 'discordant screams' preceded take-off by a few individuals which then performed somersaults over roost; all others soon followed suit and Aerial-display (see below for full description) given 12 or more times (Tristram 1866). In Iraq, birds (perhaps not all) ascended to c. 10–15 m shortly before dark, then glided, twisted, and spiralled down (Sage 1960b). See also Moore and Boswell (1956b) for early-morning activity. ANTAGONISTIC BEHAVIOUR. (1) General. Lack of detailed study on wild birds makes it difficult to ascribe functions to particular displays and postures. (2) Palaver and Bowing-display. Based on observations of captive birds. One of 3 ♂♂ gave 'rak' calls in High-erect posture; others joined in and turned to face one another, with tail fanned, wings opened at carpal joints and slightly drooped (Palaver: Fig B, left; see also subsection 5 in Heterosexual Behaviour). Call given (see 6 in Voice) whilst executing deep bow (at lowest point, bill below perch: Fig B, right) during which birds made knocking or wiping movements with bill against branch. Performance continued for several min, with increasing excitement. Dominant bird finally chased others but social hierarchy constantly changing and sometimes only 2 birds participated. ♀ placed with 3 ♂♂ acted similarly (Frisch 1966). (3) Aerial-display. See also Flock Behaviour (above). Probably serves primarily to demarcate territory and given mainly before start of breeding, almost exclusively at appearance of strange ♂ and both birds then give high-intensity performance. Much less frequently performed during later stages of cycle (D Robel and S Bude), but see also below. According to Wigsten (1955), Aerial-display given also by ♀ (see also, e.g., Hesse 1910); reports of both birds of a pair in Aerial-display together probably refer to neighbouring ♂♂ (Glutz and Bauer 1980). Observations on Rufous-crowned Roller *C. naevia* and Racquet-tailed Roller *C. spatulata* in Transvaal (South Africa), suggested Aerial-display directed by one pair to a neighbouring pair; probably functions in pair-bond maintenance as well as territorial delineation and advertisement (C H Fry). In *C. garrulus*, ♂ makes steep ascent (up to c. 200–250 m according to Wigsten 1955) with deep wing-beats and gives slow series of 'rak' calls; almost stalls, tips over forward to dive steeply down, flapping wings and tilting body from side to side; descent phase accompanied by characteristic series of rapid rattling calls (see 2 in Voice). Descent, almost to ground at times, may be followed immediately by renewed ascent (Wahn 1937; Wigsten 1955; Glutz and Bauer 1980); otherwise lands on high perch (Naumann 1901). (4) Threat and fighting. Nesting territory defended by ♂, including during nestling period; ♀ involved at times, but no details. Failed breeding pairs often attempt to associate with other pairs which still have eggs or young; frequently leads to disputes with territory owner (D Robel and S Bude). According to Glutz and Bauer (1980), intruding bird usually flees without a fight. In Niederlausitz, however, quite long disputes occur and contenders in fights may become interlocked; territory owner normally performs Aerial-display after successful eviction (D Robel and S Bude); see Naumann (1901) for further indications of pugnacity; also 1e and 5 in Voice for calls given when attacking intruders. Other observations refer to disputes over nest-holes: with e.g. Kestrel *Falco tinnunculus* or Starling *Sturnus vulgaris* (though if traditional site occupied by this species, *C. garrulus* may wait until young fledge). Subordinate to Jackdaw *Corvus monedula* and, at least in case of pair with young, to Stock Dove *Columba oenas* (but see Wigsten 1955); even Redstart *Phoenicurus phoenicurus* and Great Tit *Parus major* reported to have successfully defended nest-box against *C. garrulus* (Wahn 1939; Glutz and Bauer 1980). In Kecskemét (Hungary), *C. garrulus* mobbed *D. martius* at their nest; eventually took over and laid on top of abandoned eggs and young (Mészáros 1948–51). For interactions with captive Great Spotted Cuckoo *Clamator glandarius*, see that species and Frisch (1973). HETEROSEXUAL BEHAVIOUR. (1) General. Pair-formation and (some elements of) courtship apparently often, perhaps invariably, take place in winter quarters or *en route* for breeding grounds (Glutz and Bauer 1980; see below). In south of range, *c*. 2 weeks pass between arrival and occupation of territory and nest-site; 2 events virtually simultaneous further north (Glutz and Bauer 1980). In Niederlausitz, birds arrive already paired and initially spend little time near future nest-site, most on feeding grounds (D Robel and S Bude). On Fårö, longer periods spent at nest-site *c*. 8 days before laying (Wigsten 1955). Recrudescence of courtship activity often occurs after fledging (Rutschke 1983). (2) Pair-bonding behaviour. At Amani (Tanzania), possible courtship-chases recorded in late February. Twice, bird landed close by another on ground; first bird apparently took little notice, merely prodded at ground, whereupon new arrival walked in semi-circle around first, threw up head (presumably akin to High-erect posture: Fig A) and gave rapid series of 'cruk' sounds (Moreau and Moreau 1941*a*, 1946: see 1d in Voice). In central Spain, up to 6 birds seen in pursuit-flights through trees ('nuptial chase'; see below). 'Rattling' duets also recorded but not clear whether performed by pair or 2 ♂♂ (England 1963; see subsection 2 of Antagonistic Behaviour, subsections 4 and 6, below, and Indian Roller *C. benghalensis*). Before laying, pair perform (usually brief) Bowing-display as meeting-ceremony, no chases recorded following this (D Robel and S Bude; see also section 2 in Antagonistic Behaviour, above). Mutual Allopreening performed by pair near nest close to laying (Wigsten 1955). In Iraq, mid-April, birds seen Billing at time when others inspecting potential nest-sites, chasing, and fighting (J A McGeoch). For Aerial-display, see Antagonistic Behaviour (above). (3) Courtship-feeding. At Amani, noted on several occasions, late March and early April (Moreau and Moreau 1941*a*, 1946). On Fårö, ♂ once arrived with snake; ♀ eventually came to nest-opening and double 'rak' call given (probably by ♂); ♀ joined ♂ on perch, took food, at which time treble 'kra' sound heard; birds then flew off (Durango 1949, see 1b and 4 in Voice). In Niederlausitz, Courtship-feeding takes place before laying but also after hatching and fledging (D Robel and S Bude). According to Wigsten (1955), ♀ provisioned entirely by ♂ from (at earliest) a few days before until completion of laying; not so in Niederlausitz where ♀ also forages for her own needs at this time. Courtship-feeding most pronounced in pairs which have lost eggs or young (D Robel and S Bude). Sporadic cases up to end of June perhaps refer to non-breeding pairs (possibly 1-year-olds; see Bonds); some reports of fledged but still dependent young being fed very early in season may in fact have been Courtship-feeding (Glutz and Bauer 1980). (4) Mating. Prior to copulation, birds normally perch close together and mutually Allopreen; bills then raised jerkily and various chattering calls given (see 1b and 7g in Voice); at higher intensity, ♂ gives different call (Wigsten 1955: see 1e in Voice).

Few other details. In one case, at Amani, late March, ♂ had previously fed ♀ several times (Moreau and Moreau 1941*a*). Usually takes place on perch near nest (D Robel and S Bude; see also Drechsler and Meyer 1964). In southern USSR, occurs *c*. 1 week before laying (Ivanov 1969). (5) Nest-site selection. In Niederlausitz, pair moving into territory occupied in previous year immediately show interest in traditional site where ♂ repeatedly performs Nest-showing behaviour (also at other holes in territory); ♂ particularly active in first few days and will Nest-show also at hole occupied by (e.g.) *S. vulgaris*. Birds usually reoccupy old site; may change if other suitable holes available nearby (D Robel and S Bude). In Spain, chases (see above) occasionally interrupted by birds apparently Nest-showing at holes of woodpeckers (Picidae). Significance uncertain as none of holes visited was later used for nesting and at least some pairs already had eggs or young at this time (England 1963). In captive group, Nest-showing performed by dominant ♂ of the moment (never by ♀) several times during Palaver (see Antagonistic Behaviour, above). Remained *c*. 10–15 s at hole entrance, poked head in (once entered box), and called (see 6 in Voice); other birds apparently uninterested. All such behaviour (including Palaver) possibly directed at keeper; birds generally silent when man not present. Palaver and Nest-showing declined during May and ceased towards end of month (Frisch 1966). In Grunewald, pair said to have prospected together (Hesse 1911). (6) Behaviour at nest. No material collected; birds sometimes seen with bark in bill — possibly part of a display. ♀ spends long periods in nest prior to laying (Wigsten 1955). D Robel and S Bude found both sexes to incubate in roughly equal proportions during day with regular change-overs; only ♀ incubated at night. Elsewhere stated that most incubation done by ♀ (e.g. Wigsten 1955), who leaves for *c*. 10–30 min to forage, etc. (Khanmamedov and Gasanova 1971). Bird may emerge from nest, then self-preen and give single or double contact calls; 'prrra' sounds follow if no response forthcoming (see 1a, 1b, and 5 in Voice) and bird eventually returns to nest. Once, mid-June, ♂ arrived from roost in early morning, chased off *C. oenas* and remained briefly near nest; returned to roost where 'visited' by ♀ at *c*. 05.35 hrs. Later in morning, ♀ left nest to self-preen; ♂ arrived in response to ♀'s calls but soon left again and ♀ went back to nest after further bout of self-preening. ♂ arriving for nest-relief gives various contact calls (see 1a, 1b, and 7b in Voice) from perch near nest (♀ may respond similarly); may also fly to nest and peer inside, when calls again given by ♂ or both birds; ♀ later emerges and birds sit together for *c*. 2–3 min and call; after ♀'s departure, ♂ moves to nest and may give 'prrra' sounds at entrance (Wigsten 1955). After nest-relief, off-duty bird usually stays briefly near nest and self-preens, then flies off to feed. *Contra* Glutz and Bauer (1980),

C

does not take up sentinel post by nest and often apparently takes no notice of intruders (D Robel and S Bude). RELATIONS WITHIN FAMILY GROUP. See also Bonds (above). After hatching, ♂ passes food for young to ♀, though he may also feed nestlings directly. During first *c.* 14 days, adults enter nest-hole to feed young; subsequently feed them only from outside (D Robel and S Bude: Fig C). On Fårö, favourite perch often used briefly before final approach, though birds also dived down directly to nest from some height (Durango 1951). Eyes of nestlings open at *c.* 7–8 days (Vahi 1963). Young defecate in nest-hole but faeces dry rapidly or absorbed by sand or wood-dust (Naumann 1901; Wigsten 1955). In Spain, presumed ♀ spent long periods in nest presumably brooding small young, other adult not coming near at this time. Captive young initially gave loud food-calls (see Voice); 1 bird immediately assumed dominant position by pecking viciously at others' heads (especially near eyes). Siblings offered no resistance, attempted only to hide heads under each other's bodies. Having fed, dominant bird usually sank back and defecated, and another then took its place; youngest of brood apparently able to assert itself just as well as others (England 1963). 4 captive young (taken from nest at *c.* 10 days old) gave food-calls with simultaneous gaping and orientated by sight. Beating of wings and also fairly vigorous pecking at siblings occurred at *c.* 14 days old (Frisch 1966, which see for development of comfort behaviour). On day before fledging, Touloubre valley (southern France), both parents arrived together in morning and called (not described) in apparent attempt to entice young out. One juvenile made bowing movements and opened wings. 2 young fledged in evening of one day, 3rd in morning of next. Fledged young fed initially on ground (Hüe and Rivoire 1947). Young normally fledge within 2 days of each other; first to fledge tended by one parent and splitting of brood may continue for post-fledging care or whole family stay together. Birds normally soon move to good feeding grounds where juveniles attempt to take prey independently *c.* 5 days after fledging: spend much time on ground, hopping after food. Young continue to beg after *c.* 14 days when parents much less inclined to feed them; fully independent at *c.* 21–24 days. Loose association maintained for further *c.* 3–5 weeks ; departure (juveniles first) usually singly (D Robel and S Bude). For Aerial-display by adult in presence of young to which it afterwards flew, see Durango (1951). ANTI-PREDATOR RESPONSES OF YOUNG. Call when disturbed by man (Vahi 1963). When handled, excrete faeces of tar-like consistency (Wigsten 1955) or regurgitate brown fluid (Drechsler and Meyer 1964). Fledged young attempted to hide in hole at base of nest-tree when approached by man (Hüe and Rivoire 1947). Tend to stay quietly in cover, and adopt more horizontal posture than adults (Wigsten 1942). PARENTAL ANTI-PREDATOR STRATEGIES. (1) Passive measures. A tight sitter (Dementiev and Gladkov 1951*a*); one bird could be photographed through nest entrance; generally silent and unobtrusive when feeding young (Frisch 1966); see also Wigsten (1955) for secretive habits. Perhaps some variation from pair to pair. Initial alarm exhibited when man present, birds perching nearby for some time until feeding resumed, typically first by bolder ♂. Birds normally peered out first before leaving nest; continued to feed young when *F. tinnunculus* nearby (Hüe and Rivoire 1947). In another study, ♂ described as shyer than ♀ (Wigsten 1955, which see for habituation to man). In Grunewald, birds soon resumed normal feeding, sometimes coming within *c.* 10–12 m of man (Hesse 1921). (2) Active measures. Attacks crows (Corvidae) in nesting territory (Wigsten 1955; D Robel and S Bude); dive-attacked Jay *Garrulus glandarius* when it attempted to enter nest-hole (Hüe and Rivoire 1947; see also Phillips 1946). Raptors also normally attacked only in nesting territory, though Goshawk *Accipiter gentilis* (apparently a serious predator) attacked immediately even at *c.* 100 m or more from nest. Birds most aggressive at fledging and for few days thereafter (Robel and Bude 1981; D Robel and S Bude).

(Figs A–B from photographs in Frisch 1966; Fig C after photograph in England 1963.) MGW

Voice. Used most freely in early part of breeding season in association with various displays, also after fledging (Wigsten 1955). Generally silent in winter but very noisy in flocks prior to migration (see Moreau and Moreau 1946). Several calls have corvine quality (e.g. Hüe and Rivoire 1947, Bergmann and Helb 1982).

CALLS OF ADULTS. (1) Contact-alarm calls. (a) Various 'rak', 'cruk', harsh 'chak', 'tack', or 'kack' sounds (e.g. Hesse 1910, Hammling 1917, Moreau and Moreau 1946, Hüe and Rivoire 1947, Frisch 1966). (b) Such sounds often doubled and reference may then be specifically to alarm calls: e.g. dry, hard-sounding 'kack-ack', like 2 wooden blocks being knocked together (Durango 1951), 'rak-rak ...' (Bergmann and Helb 1982); see also Hesse (1910, 1912*a*). Recording (Fig I) suggests a hard 'ka-kak' or 'kak-kak' (J Hall-Craggs) or 'kika' (P J Sellar) (c) 3-unit variants also occur: a 'ka-kack-ack' in alarm (Durango 1951); 'kikoka' (Bergmann and Helb 1982)—situation not given. In recording by J-C Roché, Turkey, a 'tikika' (P J Sellar) preceding call 2. (d) Rapid sequences of 'rak' or 'kack' sounds given in High-erect posture in various antagonistic and heterosexual interactions, including as greeting between birds of a pair when arriving at nest (Frisch 1966), also as part of Advertising-call (see below). (e) A 'rakkiri-rakkiri-rakkiri', in which last 'ri' like soft car horn, given by ♂ at high intensity prior to copulation, also when expelling intruders (Wigsten 1955); see also call 5. (2) Advertising-call. Begins with various single, double, treble, or polysyllabic 'rak' or 'kuk' units: e.g. in recording (Fig II), a 'KUK-uk-uk-uk-UK' (J Hall-Craggs); then develops into a remarkable accelerando and crescendo cranking or rattling sound like a football fan's rattle (J Hall-Craggs, M G Wilson: Fig III). Single 'rack' units leading, in descent phase of Aerial-display, to a rapid, wooden-sounding 'rärrärrärrärrärr ...' (Glutz and Bauer 1980); a rhythmically repeated 'kera-grarah-grarah ...', often preceded by 'rak' excitement calls, alternates with a more rasping or grating 'rra-rra ...' (Bergmann and Helb 1982 where extra sonagram). (3) Warning calls. Loud 'äääääärrrr' (Frisch 1966); screeching, crowing 'krrähähäh' (Hesse 1921). Recording (Fig IV)—presumably of bird disturbed by man—suggests a harsh, nasal, rasping or screeching like Jay *Garrulus glandarius*; also given in flight (E D H Johnson) and may be preceded by contact-alarm calls with pauses of irregular length, and 'eow' or 'ow' sounds, latter probably abbreviated variant of screech (M G Wilson). Sonagram of long-drawn 'rraa' (Bergmann and Helb 1982) indicates that call probably also belongs here. (4) Long-drawn 'kraah' said to be given in fright (Heinroth and Heinroth 1924–6); 'kar' and 'krah' sounds

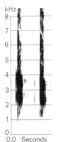
I E D H Johnson Morocco April 1966

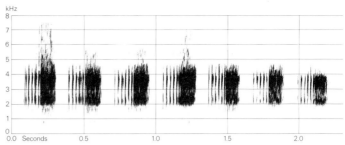
III J-C Roché (1966) Turkey May 1966

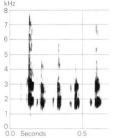
II J-C Roché (1966) Turkey May 1966

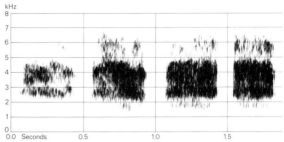
V S Palmér/Sveriges Radio (1972) Sweden July 1965

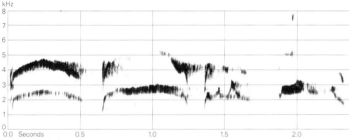
IV E D H Johnson Morocco April 1966

given by one bird (perhaps part of a family group), apparently when disturbed by man (Hammling 1917) presumably the same or related but 'kra' sounds also given in heterosexual interaction (Durango 1949). (5) Rattling 'prrra-prrra-prrraaa', rather like Magpie *Pica pica*, said to express annoyance or discontent. Used in various situations at the nest but also, primarily by ♂, when driving off or attacking intruders (Wigsten 1955). A 'k-k-k-k-k-k' or 'k-k-k-k-k-k-krak-ra' given similarly when expelling intruders (Durango 1951). (6) Persistent and hoarse-sounding 'rääääb-grääääb-gräääää-rääb' given by captive birds in Bowing-display and Nest-showing (Frisch 1966). (7) Other calls. (a) In recording by S Palmér, Sweden, July, 'kr-r-r-r-ak' (Palmér and Boswall 1972); perhaps a variant contact-alarm (see call 1) or related to call 5. (b) Guttural 'tjattri' and 'tjaari' sounds given in pair contact; a shriller variant of the 'tjaari' as a warning call from adult tending fledged young (Wigsten 1955). (c) Explosive 'chaa', one of several contact calls given by birds in winter (Moreau and Moreau 1946). (d) Low-pitched 'grå' when chasing Carrion Crow *Corvus corone* (Hesse 1910); 'kro' of Bergmann and Helb (1982) perhaps the same. (e) A 'kräh-kräh' given by a 3rd bird in flight near pair on territory (Durango 1951). (f) A cawing like Rook *Corvus frugilegus* but higher pitched; given in flight (Moore and Boswell 1956b). (g) A 'krrr' sound given occasionally during a bout of 'tjakk-tjakk' or 'tnakk-tnakk' contact calls prior to copulation (Wigsten 1955).

CALLS OF YOUNG. Food-call of young nestlings 'tiou tiou' like domestic chicks (Hüe and Rivoire 1947). Captive young (c. 10 days old when removed from nest) gave a whispering 'wüe-wüe' (Frisch 1966); 'schui', with whimpering, wailing quality (Heinroth and Heinroth 1924–6); see also (e.g.) Hesse (1910) and Durango (1951). At 18 days, 'wia-wüe' used as food-call and apparently also to express contentment and 'wüe-wüe' changed to 'rü-rü-rü-rü' similar to food-call of Bee-eater *Merops apiaster* (Frisch 1966) or, at c. $3\frac{1}{2}$ weeks old and exercising wings, 'rürr rürr' like Black Woodpecker *Dryocopus martius* (Heinroth and Heinroth 1924–6); a 'krr-krr' when bigger and moving around in nest (Durango 1951). Recording (Fig V), presumably of well-grown nestlings, suggests wild wailing and yelping; initially 'whee-ou', then faster and more excited 'quee-a' sounds (M G Wilson). Distress call like that of hare *Lepus*, but louder (Vahi 1963). A few days after fledging, 'rü-rü-rü-rü' changes to 'rak-rak-rak-rak' of adults (Frisch 1966; see adult call 1d), though higher pitched (Wigsten 1942); juvenile food-call given for long time after birds able to fly well (England 1963; see also Heinroth and Heinroth 1924–6, Durango 1951). A quiet 'buäää' or 'miäää', recalling Buzzard *Buteo buteo*, given at rest and apparently expressing contentment, in post-fledging period (Frisch 1966). In recording of captive (and injured) juvenile by W B Braughton, a 'wheeyup' when nestling in hand after being fed. MGW

Breeding. SEASON. Central and southern Europe: see diagram. Little apparent variation throughout range (Dementiev and Gladkov 1951a; Heim de Balsac and Mayaud 1962: Makatsch 1976). SITE. Hole in tree, often that of Black Woodpecker *Dryocopus martius*, or sometimes in rocks or building, occasionally in bank or nest of other species, e.g. raptor. Of 141 nests, East Germany, 38% in pine *Pinus sylvestris*, 21% in oak *Quercus robur*, 10% in limes *Tilia*, 9% in alders *Alnus*, 22% in other species. Height above ground of 104 nests: under 2·5 m, 5%; 2·5–5 m, 18%; 5–7·5 m, 27%; 7·5–10 m, 29%; 10–12·5 m, 10%; 12·5–15 m, 8%; 15–17·5 m, 2%; over 20 m, 1%; average 9·0 m; no directional tendency (Creutz 1964a). Nest: available hole or old nest. Interior height of hole 7·5 cm, diameter 12 cm, diameter of nest cup 7·5 cm and height of cup 4 cm (Dementiev and Gladkov 1951a). Tunnels dug in banks *c*. 60 cm long and 10 cm diameter with enlarged nest-chamber *c*. 25 cm diameter and 15 cm high (Dementiev and Gladkov 1951a). No material added (Wigsten 1955), though Dementiev and Gladkov (1951a) indicated that material formed into nest-cup. Building: no information on role of sexes in hole excavation. EGGS. See Plate 98. Short elliptical to nearly round, smooth and glossy: white. 35 × 28 mm (32–40 × 26–32), n = 208 (Dementiev and Gladkov 1951a). Weight 12 g (10·2–13·6), n = 120 (Khanmamedov and Gasanova 1971); fresh weight 14·3 g (12·4–16·7), n = 27 (Glutz and Bauer 1980). Clutch: 3–5 (2–7). Of 49 clutches, Sweden: 2 eggs, 3; 3, 17; 4, 18; 5, 9; 6, 2; average 3·8 (Durango 1946; Wigsten 1955). One brood. Replacements laid after egg loss. Laying interval 2 days, occasionally 3 days. INCUBATION. 17–19(–20) days. Begins before completion of clutch; hatching takes place over shorter period than laying (Wigsten 1955). By both sexes, but ♀ takes larger share, sitting at night, and during day in bad weather (Wigsten 1955). YOUNG. Altricial and nidicolous. Cared for by both parents, though ♀ mainly transfers food brought by ♂ (Wigsten 1955). FLEDGING TO MATURITY. Fledging period 26–27 (25–30) days. Continue to be fed by parents for some time after fledging (Dementiev and Gladkov 1951a). Independent at *c*. 21–24 days (D Robel and S Bude). Age of first breeding probably (1–)2 years (see Social Pattern and Behaviour). BREEDING SUCCESS. No information.

Plumages (nominate *garrulus*). ADULT BREEDING. Forehead, crown, hindneck, ear-coverts, and sides of neck pale greenish-blue or pale turquoise green (intensity of colour and gloss depending on light, forehead palest), feathers bordering nostril and (sometimes) those just above lores cream-buff or white. Sparse bristles and feathers on lores and round eye black and dusky grey. Mantle, scapulars, and tertials rufous-cinnamon (occasionally, slightly greenish in some lights), inner webs of longer and inner tertials partly dark olive-grey. Back and rump bright violet-blue (rump-feathers partly green basally), upper tail-coverts bluish-green with some darker blue suffusion. Upper chin cream-buff or white with some black bristles laterally; lower chin, cheeks, and throat bright pale greenish-blue or pale turquoise-blue with narrow glistening pale blue shaft-streaks. Remainder of underparts pale blue or pale greenish-blue. Central pair of tail-feathers (t1) dusky grey with slight olive or blue tinge; t2 with pale blue tip 2–3 cm wide, remainder dusky dark blue with greenish tinged fringe along outer web and broad black fringe along inner web; other tail-feathers like t2, but pale blue of tip gradually more extensive towards outer (3–5 cm wide on t5), and tips often slightly darker greenish-blue; t6 like t5, but tip emarginated and slightly elongated, black for 14–22 mm; shafts of all tail-feathers black; from below, middle portions and bases of inner web of t2 bright violet-blue, gradually more restricted to bases towards outer feathers and changing to less contrasting greenish-blue on t6. Flight-feathers black, outer web of (p7–)p8–p9 slightly tinged greenish-blue, of p10 extensively so; outer webs of secondaries tinged dark violet-blue, tinge sometimes hardly visible; undersurface of flight-feathers bright violet-blue, except for tips. Extreme bases of inner primaries and mainly hidden basal halves of secondaries pale blue, forming pale blue bar across wing together with pale blue greater upper primary coverts; pale blue tinged bright violet at border of black on flight-feathers; greater upper primary coverts of p8–p9 tipped bright violet-blue, others similar or tipped dusky grey. Lesser upper wing-coverts bright violet-blue, forming bar along forewing of 1·5–2 cm wide, remaining upper wing-coverts pale greenish-blue, about similar in tinge to hindneck, slightly duller than pale blue primary coverts. Under wing-coverts and axillaries pale blue, like belly and vent. Bleaching and wear have some influence: crown, neck, and underparts appear more cerulean-blue, less greenish, pale shaft-streaks on throat hardly visible; much grey of feather-bases exposed, especially on face; mantle, scapulars, and tertials paler cinnamon; pale blue of upper wing-coverts slightly duller and more olive. Sexes similar, but back

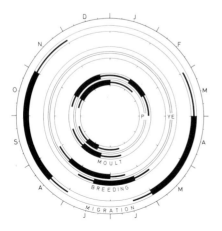

and rump of ♀ more extensively tinged pale green-blue, sometimes hardly showing any bright violet-blue; variable number of worn non-breeding longer lesser and median upper wing-coverts retained, showing olive-brown or buff-brown suffusion; olive or buff tips sometimes showing on retained non-breeding feathers on chest or breast to belly; tips of greater upper primary coverts often less extensively violet-blue. ADULT NON-BREEDING. Like adult breeding, but crown, hindneck, sides of neck and upper mantle duller olive-green, less bright pale blue, feathers tipped olive-brown; lower mantle, scapulars, and tertials darker rufous-brown, less bright cinnamon; feathers of throat, chest, and breast, as well as longer lesser and median upper wing-coverts extensively tinged pale olive-brown or cinnamon-brown on tips. NESTLING. Naked at hatching. On 13th day, covered with closed sheaths, except for skin round eye, rump, and belly, appearing spiny; from 17th day, sheaths open to produce feather-tips, large feathers first, throat, belly, and rump last; appears feathered from 22nd day (Heinroth and Heinroth 1924–6). JUVENILE. Rather like adult non-breeding, but head, neck, and breast browner, hardly bluish-green; upperparts more olive-brown, less rufous; hardly any violet-blue on rump and upper wing-coverts; t6 shorter than other tail-feathers, not emarginated and without black tip. Crown, hindneck, and sides of head and neck brownish olive-green, grading to pale buff-brown on forehead, at gape, and on lower sides of neck; some grey of feather-bases often visible, especially when plumage worn; sometimes an indistinct off-white supercilium. Mantle, scapulars, and tertials cinnamon-buff with grey or olive tinge. Back and rump pale greyish-blue or greenish-grey; upper tail-coverts slightly brighter bluish-green. Chin off-white, throat to breast pale cinnamon-buff; narrow shaft-streaks white with pale blue tinge, feather-tips sometimes slightly tinged pale blue. Remaining underparts pale greenish-blue, often slightly suffused buff, less bright pale blue than adult. Tail as in adult, but tips and outer webs of t2–t5 dull olive-green, hardly contrasting with dull blue and dark grey bases and middle portions of inner webs (tips pale blue and markedly contrasting with violet-blue bases in adult, also on outer webs); t6 similar in colour to t5 (unlike adult, not slightly elongated and without contrasting black or violet-blue emarginated tip); undersurface of tail hardly showing any bright violet-blue towards base when spread, unlike adult. Flight-feathers as in adult, but black less deep and tips tinged grey, bordered by poorly defined white fringes when fresh. Greater upper primary coverts with ill-defined greyish tips, not all or partly tipped bright violet-blue as in most adults; lesser upper wing-coverts and tips of median coverts cinnamon-buff or greenish-grey, violet-blue on shorter lesser coverts more restricted than in adult and often more greenish-blue, less deep violet. FIRST ADULT NON-BREEDING AND BREEDING. Like adult, but juvenile flight-feathers, greater upper primary coverts, and sometimes part of tail-feathers retained; primaries in particular more heavily worn than those of adult at same time of year, showing greyish tips, and inner primaries with traces of white fringes on tips. Sexes often indistinguishable, as 1st breeding ♂ usually retains as much olive-brown or buff feather-tips on breast and upper wing-coverts as 1st adult and adult ♀.

Bare parts. ADULT. Iris hazel, light brown, or brown. Bare skin round eye greyish-black. Bill black or brownish-black. Leg and foot ochre-yellow, dark yellow, yellow-brown, greenish-brown, or olive-brown; claws black. NESTLING. Bare skin, including bill, leg, and foot, pink at hatching; small flanges at gape yellowish-white. At 13th day, bare skin, bill, and foot greyish-flesh. JUVENILE. Iris grey-brown or yellow-brown. Bare skin behind eye grey, below eye bluish with yellow spot. Bill dark horn-grey, paler flesh-pink towards base; basal cutting edges yellowish-white, tip greyish-black. Leg and foot greyish-flesh, pale brown, yellow-brown, or yellow horn. (Heinroth 1924–6; Baker 1927; Glutz and Bauer 1980; BMNH, RMNH, ZMA.)

Moults. Based on *c.* 50 actively moulting birds, mainly from Rumania eastward (ZMA, ZMM) and from East Africa (BMNH, RMNH, ZMA), and on data from Zaïre (Herroelen 1962). ADULT POST-BREEDING. Complete; primaries descendant. Starts with p1 in or near breeding area, mid-June to early August. Moult suspended mid-August to mid-September with score 10–35 (inner 2–7 new) when leaving breeding area; average score of suspended moult on arrival in Zaïre 26 ($n=7$). Primary moult resumed in winter quarters between mid-October and late November, completed with p10 early December to early February. Moult of head and body starts at about same time as primaries; scattered feathers of mantle, throat, and scapulars first, soon followed by head, neck, tertials, and median upper wing-coverts; at score 25, over half of feathers in these parts replaced, and moult on remainder of body and wing started. Tail moulted from primary score *c.* 25; sequence t1–t6–t2–t3–t4–t5; completed with p10 or slightly later. ADULT PRE-BREEDING. Partial; January–March. Extent rather variable, in ♂ apparently more extensive than in ♀. Involves at least throat and scattered feathers of mantle, chest, breast, and scapulars; in ♂ apparently all head, neck, mantle, and scapulars, much of underparts and upper wing-coverts. Breeding plumage also obtained by abrasion of buff-brown non-breeding feather-tips; these latter narrower in ♂ than in ♀, disappearing more rapidly, and this process rather than moult perhaps responsible for sexual difference in colour of underparts and upper wing-coverts. POST-JUVENILE. Partial. Starts from September or, after arrival in winter quarters, from October–December. Involves head, body, lesser and median upper wing-coverts, and part of tail (e.g. t1 and t6 only or t1–t3 and t6) or all tail; juvenile flight-feathers and greater primary coverts retained, but exceptionally inner primaries replaced, probably in late winter: one, Rumania, mid-May, had suspended primary moult, score 15. Tail replaced from late October onwards, sequence as in adult post-breeding. FIRST PRE-BREEDING. Rather limited, as in adult non-breeding ♀, or perhaps occasionally no moult at all. Involves scattered feathers of mantle, scapulars, throat (often complete), chest, breast, and belly, and sometimes a variable number of upper wing-coverts. FIRST POST-BREEDING. As in adult post-breeding, but starts often slightly earlier, juvenile p1 lost between late May and late July.

Measurements. Nominate *garrulus*. North-west Africa, central Europe, Balkans, south European USSR, and Turkey, summer; skins (RMNH, ZFMK, ZMA). Tail is to tip of t1. Bill (F) to forehead, bill (N) to distal corner of nostril, both in adult and 1-year-old only.

WING	AD	♂ 203	(2·80; 10)	200–207	♀ 203	(3·68; 6)	198–208
	JUV	198	(5·49; 7)	192–204	196	(3·95; 7)	191–200
TAIL	AD	122	(3·28; 12)	117–130	124	(4·31; 16)	118–129
	JUV	130	(2·59; 5)	128–133	126	(3·40; 4)	122–129
BILL (F)		42·5	(1·80; 14)	41–44	42·5	(2·04; 11)	41–45
(N)		26·2	(1·14; 13)	25–28	25·8	(1·31; 11)	24–28
TARSUS		24·4	(0·84; 18)	23–26	24·0	(0·60; 18)	23–25
TOE		30·7	(1·78; 13)	29–33	30·1	(1·38; 17)	29–32

Sex differences not significant. Juvenile wing significantly shorter than adult, t1 longer (significant in ♂); for t6, see Structure. Juvenile bill not full-grown until late autumn or winter.

C. g. semenowi. South-central USSR (Turkmeniya to Kazakhstan), summer; skins (ZMA, ZMM).

WING AD ♂ 202 (4·16; 26) 196–210 ♀ 199 (3·78; 9) 196–205
JUV 194 (4·29; 11) 189–200 191 (6·36; 8) 184–200

Sexes combined, adult: bill (F) 44·7 (1·54; 8) 43–46, bill (N) 27·2 (1·35; 8) 26–29. Iraq birds considerably shorter winged, Kashmir slightly longer, but bill of both similar to USSR sample (see Geographical Variation).

Weights. ADULT. Nominate *garrulus*. Belorussiya (USSR), summer: ♂ 146 (13) 127–160, ♀ 148 (4) 141–158 (Fedyushin and Dolbik 1967). Hungary, Turkey, and Iran, combined: April, ♂ 137 (18·1; 5) 110–154, ♀ 129 (3) 125–132; May–June, ♂ 131, ♀♀ 117, 120; July, ♂♂ 121, 130; ♀♀ 150 (3) 123–189 (Paludan 1940; Vasvari 1955; Schüz 1959; Rokitansky and Schifter 1971; Glutz and Bauer 1980). Czechoslovakia, August, 195 (Glutz and Bauer 1980). South-east Morocco, April: 107 (BTO). Azraq (Jordan), April: 139 (BTO).

C. g. semenowi. Iran, Afghanistan, and Kazakhstan, combined: April–May, ♂♂ 132 (3) 127–135, ♀ 140 (34·1; 4) 113–190; June–July: ♂♂ 124, 151, ♀♀ 127, 136; August, ♂ 153, ♀ 152 (Paludan 1938, 1959; Dolgushin *et al.* 1970).

Races combined. Zaïre and Kenya, November–February: ♂ 125, ♀♀ 126 (3)103–146 (Verheyen 1953; Herroelen 1962; ZMA).

JUVENILE. Races combined. One at hatching, 10; another on 13th day 110, on 17th 140, on 23rd 160; lighter again at fledging on 28th day (Heinroth and Heinroth 1924–6). Afghanistan, ♀♀: 105 (July), 100 (September) (Paludan 1959). Netherlands, ♂: 126 (August) (RMNH). Camargue (France): 158 (October) (Glutz and Bauer 1980). Zaïre, winter: 120 (Herroelen 1962).

Structure. Wing rather long, broad at base, tip slightly pointed. 10 primaries: p9 longest, p10 3–8 shorter, p8 1–6, p7 9–13, p6 22–31, p5 36–44, p1 63–77. Inner web of p8–p10 and outer web of (p7–)p8–p9 emarginated. Tail rather long, 12 feathers; tip slightly rounded, t5 3–15 mm shorter than t1; t6 11–25 shorter than t1 in juvenile, shape of tip normal, but t6 longer than t5 in adult and 1st adult, tip slightly elongated and with emarginated inner web, width of feather at constriction 6–11 mm, length of narrow part of tip 11–18 mm. In adult ♂, t6 0–13 longer than t1; in adult ♀ t6 4 longer to 6 shorter than t1; in 1st adult, t6 0–9 shorter than t1. Bill deep and wide at base, laterally compressed at tip; culmen slightly decurved, but with strongly hooked tip. Nostrils narrow, rather large; bordered above by feathers. Short bristles at gape. Leg and toes short, but strong; tibia feathered down to tarsus joint; front of tarsus and upperside of toes covered with large scutes. Outer toe *c*. 87% of middle, inner *c*. 72%, hind *c*. 58%. Claws short, strong, and sharp, decurved.

Geographical variation. Rather slight, both in colour and size. Nominate *garrulus* constant in size over entire range, birds of Maghreb and Iberia similar in wing length to those of western Siberia and Altay, but south-eastern race *semenowi* smaller in western part of range, clinally larger towards east: adult wing (sexes combined) in Iraq, Iran, and Afghanistan 193 (5·59; 13) 184–202, Turkmeniya 200 (4·53; 12) 191–206, western Kazakhstan 200 (3·64; 11) 196–207, Tadzhikistan 201 (3·64; 18) 194–208, eastern Kazakhstan 202 (4·34; 8) 198–211, Kashmir 203 (8·47; 7) 190–211; bill of all *semenowi* slightly larger than nominate *garrulus* (BMNH, RMNH, ZFMK, ZMA, ZMM). One exceptional Kashmir bird had wing 222 (Hartert 1912–21). Colour of *semenowi* similar to nominate *garrulus*, but blue of head, neck, underparts, and upperwing slightly paler, especially on throat, sometimes more greenish (many indistinguishable, however); bright violet-blue band on shorter lesser upper wing-coverts of adult narrower, 12–16 mm wide in middle of arm when wing spread rather than 15–20 mm as in nominate *garrulus*. Boundary between both races apparently not sharp; some birds from Transcaspia (in breeding area of *semenowi*) indistinguishable from nominate *garrulus*; some birds from Turkey and northern and southern shores of Caspian Sea similar to *semenowi*, though most birds here similar to nominate *garrulus*.

Forms superspecies with Abyssinian Roller *C. abyssinicus* and Lilac-breasted Roller *C. caudatus*; see *C. abyssinicus*. CSR

Coracias abyssinicus Abyssinian Roller

PLATES 72 and 74
[between pages 758 and 759]

DU. Sahelscharrelaar FR. Rollier d'Abyssinie GE. Senegalracke
RU. Абиссинская сизоворонка SP. Carraca etíope SW. Savannblåkråka

Coracias abyssinica Hermann, 1783

Monotypic

Field characters. 40–45 cm, of which tail 12–28 cm; wing-span 58–60 cm. Slighter and up to 20% shorter winged than Roller *C. garrulus*, with similar plumage but (in adult) exceptionally long outer tail-feathers. Noticeably more agile in flight, and dark areas on wings show strong purple tone. Sexes similar; slight seasonal variation. Juvenile separable.

ADULT BREEDING. Differs from *C. garrulus* in paler turquoise-blue tone to head, underparts, and wing-coverts, more obvious pale streaks on head and forebody, and deeper purple or violet tone to all dark blue feathers. Tail pattern distinct, with central feathers dark olive-green contrasting with pale blue outer ones, and outermost elongated and black on distal half. ADULT NON-BREEDING. Shows delicate fawn suffusion on crown, breast, and sides of lower neck. JUVENILE. Outer tail-feathers not elongated. Closely resembles *C. garrulus* and best separated by much whiter face (particularly forehead), purple tone on dark wing-feathers, and paler blue, fawn-tinged body.

At distance, separation from *C. garrulus* sometimes difficult. At closer range, swallow-tail of adult unmistakable (but note that in central Africa, 3 other *Coracias* have elongated outer tail-feathers); during moult, tonal differences in main plumage colours allow identification even if tail-streamers absent. At any range, distinction of juvenile from similarly aged *C. garrulus* requires close observation of size and characters given above. Flight fast with wing-beats at times lacking 'plodding' quality of *C. garrulus*. Feeding behaviour more active than *C. garrulus*.

Noisier than *C. garrulus*, with voice less corvine in tone; calls both cackling and screeching in quality.

Habitat. In tropical low latitudes, across thorn-scrub savanna and grass woodland belt, in open country and grasslands and cultivated areas round villages, but usually shuns habitations and does not occur in forests. Also in dry marshland and in forest clearings; commonly perches on telegraph wires. Like Senegal Coucal *Centropus senegalensis*, ranges further north into desert and further south into savanna than most birds of dry West Africa (Bannerman 1933, 1951). Frequents orchard bush, thorn scrub, and open grass provided there are tree perches. Descends on prey on ground from conspicuous perch on wire, tree, or shrub. Nests in hole in tree or building (Serle *et al.* 1977). Often attracted by grass fires to take fleeing insects.

Distribution. Breeds in Africa south of Sahara from Sénégal east to Ethiopia and western Somalia, and occurs as non-breeder south to savannas north and east of equatorial forest; also breeds south-west Arabia. Resident and migratory (see Movements).

Accidental. Egypt, Libya, Mauritania.

Movements. Some resident, others are well-defined rains migrants. Present all year in Sahel and savanna zones, breeding there March–June; however, many then spread south as far as Gulf of Guinea, northern Zaïre, Uganda, and Kenya, being present as non-breeding visitors for duration of dry season there.

Most detailed study in Nigeria, where present all year (and breeds) in far north (Sokoto, Kano, Malamfatori); at lower latitudes, absent in wet season for progressively longer periods until south of great rivers it is essentially a dry-season non-breeding visitor, September–March; as far north as Malamfatori on Lake Chad it is commoner in dry season than in rains, and thought that in rainy season some birds move north into Chad (Elgood *et al.* 1973). Comparable movements described from Togo (Douaud 1957) and Ivory Coast (Thiollay 1971). Further north, resident population in wooded wadis of central Chad supplemented by large numbers of rains migrants, which depart in September–October after breeding (Newby 1980); also basically a summer visitor (during rains) in Khartoum area of Sudan (Macleay 1960). On southern edge of range, a non-breeding dry-season visitor (November–March) to Uelle region of north-east Zaïre (Chapin 1939), and in Uganda and Kenya it is mainly a non-breeding migrant from the north, occurring October–May (Britton 1980).

Voice. See Field Characters.

Plumages. ADULT. Closely similar to Roller *C. garrulus*, differing in broader white or cream band along forehead, extending laterally to above eye; cream-white on chin more extensive; shaft-streaks on throat occasionally white instead of pale blue; white of chin and of streaks on throat sometimes slightly lilac. Back, rump, and upper tail-coverts brighter, more glistening violet-blue; black part of outer pair of tail-feathers (t6) strongly elongated, forming narrow streamers (see Measurements and Structure); (t3–)t4–t5 sometimes narrowly tipped violet-black, tip of t5 sometimes emarginated; pale greenish-blue on outer tail-feathers more extensive, *c.* 4 cm of tip of t2 pale blue, *c.* 7 cm on t5 (*c.* 1·5–3·5 cm in *C. garrulus*), violet-blue of base of tail-feathers more restricted. Distal halves of flight-feathers glossy violet-blue (mainly black in *C. garrulus*, only slightly dull violet on outer webs of secondaries and inner primaries). Some variation in blue of crown, hindneck, underparts, and median and greater upper wing-coverts, in part depending on abrasion: some birds pale cerulean-blue, others pale turquoise-green. Size smaller than in *C. garrulus* (see Measurements). NESTLING. Naked at hatching; further development probably as in *C. garrulus*. JUVENILE. Differs from adult in same way as juvenile *C. garrulus* differs from adult (Bannerman 1933); crown, hindneck, and chin to chest more olive-brown; mantle, scapulars, tertials, and lesser upper wing-coverts pale greyish-brown; outer tail-feathers not elongated; violet-blue to tips of secondaries not sharply defined from pale blue bases. Differs from juvenile *C. garrulus* by smaller size, broader cream-buff band along forehead and above eye, brighter violet-blue rump and flight-feathers (though not as glistening as adult), and more extensive pale turquoise-green on tips of tail-feathers. FIRST ADULT. Differs from adult in same way as 1st adult *C. garrulus* differs from adult.

Bare parts. ADULT AND JUVENILE. Iris brown or greyish-brown. Bill black or dark horn-brown. Leg and foot pinkish-brown, greenish-brown, olive-brown, grey-green, greenish-slate, or dirty olive-yellow. (Bannerman 1933; Moltoni 1939*b*; RMNH, ZFMK.) NESTLING. No information.

Moults. ADULT POST-BREEDING. Complete, primaries descendant. Only a few in moult examined, and no information about relation of breeding season and moult. In birds from Ethiopia, Sudan, as well as West Africa, moult starts with p1 between late April and late July, completing with p10 September–November. Moult of body and tail mainly between primary moult scores 10 and 40; sequence apparently as in *C. garrulus*. All plumage fresh in birds from November–December, slightly worn from February onwards. Apparently no pre-breeding moult, or perhaps limited only. POST-JUVENILE. Partial; perhaps occasionally complete. In November–December Sénégambian birds, head, body, and tail in moult, with fresh 1st breeding prevailing and flight-feathers still juvenile. One from Ethiopia, February, had started primary moult, suspending with score 25; head, body, and tail worn 1st breeding, but t6 new and p6–p10 worn, juvenile. Others perhaps may have complete flight-feather moult in non-breeding season.

Measurements. ADULT. All breeding range, November–February; skins (RMNH, ZFMK, ZMA). Tail is to tip of t1; tip is tip of t1 to tip of t6; bill (F) to forehead, bill (N) to distal corner of nostril; exposed culmen 6 mm longer than bill (N), 9 shorter than bill (F).

WING	♂ 162	(3·51; 20)	156–169	♀ 160	(6·84; 15)	151–170
TAIL	133	(4·54; 21)	126–140	133	(6·76; 14)	124–144
TIP	122	(17·1; 18)	98–155	111	(32·0; 11)	78–181
BILL (F)	42·2	(1·39; 13)	41–45	38·7	(1·89; 10)	36–41
(N)	26·0	(1·08; 13)	25–28	24·5	(1·50; 10)	23–26
TARSUS	22·6	(0·33; 14)	22–24	21·6	(1·08; 10)	20–23
TOE	27·7	(1·72; 10)	26–30	27·3	(1·62; 8)	25–29

Sex differences significant for bill and tarsus.

JUVENILE. Wing on average $c.$ 4 mm shorter than adult; tip of t6 not elongated or slightly only; bill smaller than adult until about post-juvenile moult.

Weights. Nigeria, May: 99·5 (Fry 1970).

Structure. Wing rather long, broad, tip rounded. 10 primaries: p8–p9 longest, about subequal; p10 6–12 shorter, p7 0–6, p6 7–15, p5 20–25, p1 38–48. Outer web of p6–p9 and inner web of (p7–)p8–p10 emarginated. Tail long, tip indented either side of centre, 12 feathers; t3–t4 shortest, t2 and t5 2–4 mm longer, t1 7–12 longer; t6 slightly shorter to up to 30 mm longer in juvenile, strongly elongated in adult (see Measurements), inner web emarginated below tip of other tail-feathers, elongated part 3–5 mm broad, ending in narrowly rounded tip; in some 1st adults, slightly broader (up to 7 mm) and shorter than in older adults. Bill relatively slightly less heavy at base than *C. garrulus*. Outer toe $c.$ 87% of middle, inner $c.$ 71%, hind $c.$ 59%. Remainder of structure as in *C. garrulus*.

Geographical variation. None; see Friedmann (1930).

Forms superspecies with *C. garrulus* and with Lilac-breasted Roller *C. caudatus* from eastern and southern Africa (Snow 1978); breeding ranges contiguous. *C. caudatus lorti*, breeding eastern Ethiopia and Somalia, differs from *C. abyssinicus* in slightly greener crown, vinous-purple cheeks and throat streaked with white, and in showing hardly any violet-blue on bases of tail-feathers (these differences bridged by a few *C. abyssinica*, which show slightly greenish crown, slightly lilac throat with more marked white/streaks, and more limited and greener violet-blue on tail-base); *C. caudatus caudatus* from southern and eastern Africa north to Kenya shows even greener crown, and vinous-purple of throat extends to breast. Racquet-tailed Roller *C. spatulata* from Tanzania, Moçambique, and west to Angola also close to this superspecies, but crown browner, tail forked rather than with double-indented tip, elongated t6 with widened spatulate tips, and median upper wing-coverts rufous like scapulars; kept separate because of structural differences and large overlap in breeding range with *C. caudatus* (Snow 1978).

CSR

Coracias benghalensis Indian Roller

PLATES 73 and 74
[between pages 758 and 759]

Du. Indische Scharrelaar   Fr. Rollier indien   Ge. Hinduracke
Ru. Индийская сизоворонка   Sp. Carraca india   Sw. Indisk blåkråka

Corvus benghalensis Linnaeus, 1758

Polytypic. Nominate *benghalensis* (Linnaeus, 1758), Iraq, Iran, and north-east Arabia, east to northern India. Extralimital: *indicus* Linnaeus, 1766, India south from 20°N, and Ceylon; *affinis* Horsfield, 1839, south-east Asia west to eastern Nepal, Assam, and Bangladesh.

Field characters. 32–34 cm; wing-span 65–74 cm. Slightly bulkier than Roller *C. garrulus* with rather broader and more rounded wings and slightly longer, square-ended tail (at all ages). Roller with mainly brown head and body, relieved by intensely purple-black and dark green-blue wings, blue rump, and blue band across outer tail-feathers; shows much less blue at rest than *C. garrulus* but wing pattern even more striking. Sexes similar; no seasonal variation. Juvenile separable at close range.

ADULT. Plumage predominantly dull vinous-brown below and brown-olive above, with rufous-white face, buff forecrown, almost-white streaks over cheeks, throat, and (less noticeably) forebody, blue-green cap to rear head, indistinct purple rear collar, and pale blue belly and vent. Long rump turquoise-blue. Wing and tail colours much as *C. garrulus* but patterns more contrasting with greater extent of pale blue across primary coverts and outer section of primaries obvious in flight. Pale blue under wing-coverts contrast with dark undersurface of flight-feathers. Pale blue on outer tail-feathers restricted to central band. Worn birds have buff-brown upperparts and paler, pinker chest. JUVENILE. Resembles adult but plumage duller, with green crown, virtually no collar, dark wing-feathers, and green tone to belly.

At a distance and when colours indistinct, inseparable from *C. garrulus*. Adult at close range readily distinguished from *C. garrulus* by darker crown and upper wing-coverts, less rufous mantle and sides of back, and noticeably streaked lower face and breast. Distinction of juvenile not studied in the field (see Plumages). Flight and behaviour as *C. garrulus*, with slight differences in structure noted above not tested in the field.

Commonest call a loud 'chack'.

Habitat. In subtropical and tropical lower latitudes, mainly in lowlands, below $c.$ 1200 m, preferring open, often cultivated, country and light deciduous forest. Sits habitually on elevated open perch (e.g. dead bough, ruined building, telegraph post or wire, or, failing all else, thorn bush or heap of stones) from which it watches for prey on ground. Nests in hole, usually in tree but frequently

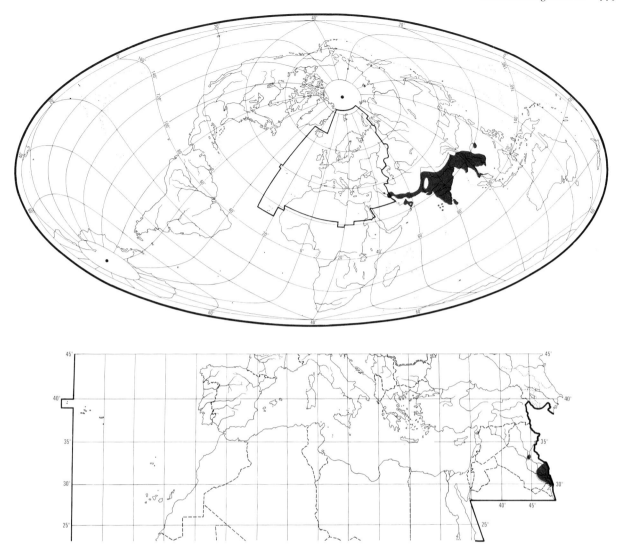

in building or wall. Although markedly aerial does not usually pursue insects in flight, and does not travel far. (Whistler 1941.)

Distribution. IRAQ. Has extended range recently to just north of Baghdad (Marchant 1963a).
Accidental. Syria, Kuwait.

Population. IRAQ. Fairly common in southern date palm areas (Allouse 1953).

Movements. Nominate *benghalensis* (breeding Persian Gulf and Iraq to northern India) resident throughout its range, but also wanders to an unclarified extent.
No firm data from Iraq, though capacity for local movement (at least) indicated by modern range expansion along River Tigris (Marchant 1963a); previously reported by Ticehurst *et al.* (1922) as straggling northwards outside breeding season. Various Kuwait records of single birds in non-breeding period (late July to mid-March), and stragglers reach eastern Saudi Arabia then (Bundy and Warr 1980); former, at least, seem more likely to be from Iraq than elsewhere. Resident in cultivated areas of United Arab Emirates and Oman, but young birds in particular disperse into surrounding plains and lower foothills from late June, and occasional records in October–April on west coast of Emirates and on Masirah Island (Griffiths and Rogers 1975; Bundy and Warr 1980; Walker 1981a). In India, both nominate *benghalensis* and southern race *indicus* are subject to little-understood local movements (Ali and Ripley 1970).

Food. Primarily large insects with fewer amphibians, reptiles, and small rodents. Hunts mainly from elevated perch (e.g. wire, post, tree, or lower down on clod of earth), descending slowly and silently. Prey may be killed and consumed on ground or taken back to same or nearby perch where battered before being swallowed whole (Jerdon 1877; Fletcher and Inglis 1920; Ali and Ripley 1970). Insects less often taken in the air, sometimes in prolonged aerial-pursuit; termites (Isoptera) may be exploited thus almost exclusively when they emerge with onset of rainy season (Jerdon 1877; Mason and Maxwell-Lefroy 1912). Briefer sortie from perch with spiralling or direct ascent also used (Dharmakumarsinhji 1955). In Burma, cicadas (Cicadidae) picked off bark of trees (Baker 1927). Small fish taken skilfully from near surface of water (Ali and Ripley 1970). At Bhamo (Burma), seen plunging 4 times into water like kingfisher (Alcedinidae); bird briefly submerged completely, each time returning to perch where prey (probably frog: Smythies 1953) quickly swallowed and plumage shaken vigorously (Radcliffe 1910); bird observed by Dalgliesh (1911) hovered beforehand, remained in water for a few seconds then flew off—not clear whether actually feeding. Such feeding methods probably confined to certain individuals (Ali and Ripley 1970); see also Tiwary (1911) who considered birds more likely to be bathing. Occasionally plunders nests of other birds (e.g. Mason and Maxwell-Lefroy 1912, Smith 1942). Will feed at any time of day (Baker 1927); brief aerial-pursuit of insects from perch said to occur mainly at midday (Dharmakumarsinhji 1955). Termites often taken until dusk well advanced (Henry 1955), and 1 bird habitually waited for emergence of crickets (Gryllidae) at dusk (Fletcher and Inglis 1924). Frequently forages near jungle fires (Blanford 1895) and 5–6 birds once took insects attracted by powerful airport lights at c. 04.40 hrs (Ganguli 1975). Undigested food ejected as pellets (Fletcher and Inglis 1924). Said never to drink (Dewar 1906).

Birds seen in aerial-pursuit of insects, Iraq (Christensen 1962), but all detailed information on diet extralimital (mainly from India). Insects include crickets, etc. (Gryllotalpidae, Gryllidae, Acrididae), earwigs (Forficulidae), mantises (Mantidae), termites (Isoptera), bugs (Fulgoridae, Cicadidae), adult and larval moths (Noctuidae), Hymenoptera (wasps Eumenidae, ants Formicidae), adult and larval beetles (Trogidae, Scarabaeidae, Carabidae, Tenebrionidae, Chrysomelidae, Curculionidae). Non-insect food generally much less important: exceptionally spider (Araneae); scorpions (Scorpiones), frogs (at times important), toads, lizards, occasionally small snakes, mice (Muridae), shrews (Soricidae), small fish, and rarely young (probably also eggs) of birds. (Jerdon 1877; Macdonald 1906; Mason and Maxwell-Lefroy 1912; Fletcher and Inglis 1920; Baker 1927; Ali and Ripley 1970; Gallagher and Woodcock 1980.)

18 stomachs (January–October) from Pusa (Bihar, India) contained 412 insects, notably Orthoptera (particularly *Gryllotalpa africana* and *Chrotogonus*), Lepidoptera larvae, ants (many *Myrmecocystus setipes*), and beetles (Tenebrionidae, Trogidae, Scarabaeidae); also 1 spider and 2 frogs (Mason and Maxwell-Lefroy 1912). In Burma, frogs important in diet when available (Smythies 1953).

In India, young given mainly crickets and large larvae of Coleoptera or Lepidoptera (Mason and Maxwell-Lefroy 1912). At Yamethin (Burma), April, 2 almost full-grown nestlings given c. 90% large frogs (Smith 1942). In Pakistan, ♀ seen to kill lizard and carry it to nest (Mountfort 1969). MGW

Social pattern and behaviour. Following account based almost entirely on extralimital material from India.

1. Not markedly gregarious; often seen singly or in twos (e.g. Lamba 1963, Ali 1968). In Iraq, winter, usually in twos (Moore and Boswell 1956*b*). Near Delhi (India), October–January, normally encountered singly, possibly 2–3 together at roosts. Roosting parties of 5 and 3, September, perhaps families (Hutson 1954). Several sometimes occur together where food abundant (e.g. Baker 1927). BONDS. Mating system presumably monogamous as in Roller *C. garrulus* but see Stonor (1944) and Antagonistic Behaviour (below). No detailed information on duration of pair-bond but see comment for Iraq (above). No hybridization with *C. garrulus* in area of sympatric breeding at Samawa, Iraq (Christensen 1962). Both sexes care for young (Ali and Ripley 1970) and family apparently stays together for some time after fledging (Hutson 1954). BREEDING DISPERSION. Solitary and territorial. No detailed information on densities. In Lodi Gardens, Delhi, 3 pairs apparently fairly close together (Hutson 1954). In one case, Madras (India), nest c. 150 m from tree (apparently also defended) where pair regularly displayed; territory established though extent not known (Stonor 1944). Other hole-nesting species driven away from immediate vicinity of nest-hole (Lamba 1963), but, in one case, occupied nest only c. 1 m from that of Hoopoe *Upupa epops* (Dewar 1909). Regular perch (possibly larger feeding territory) defended outside breeding season (Dharmakumarsinhji 1955). ROOSTING. Nocturnal and generally solitary (see above for possibility of roosting in small groups). Normally in trees (Hutson 1954). Spends much of day perched rather sluggishly (Fletcher and Inglis 1920). Most active during morning and evening but will feed even at hottest time of day and, at this time, still uses exposed perches (Baker 1927); frequently sun-bathes with wings spread (Oates 1890). During incubation, bird often leaves nest around midday and perches nearby (Lamba 1963). Birds rain-bathing on wires had wings raised over backs (Ganguli 1975).

2. Not shy (Stonor 1944; Ali 1968). When perched for long period, bird usually has feathers on head and neck ruffled (Jerdon 1877); tail occasionally jerked and call given (Dewar 1909; Fletcher and Inglis 1920: see 1 in Voice), though tail movement also described as slow and swinging (Ali and Ripley 1970). Inconspicuous at rest and sudden exhibition of brilliant colours on moving may startle potential predator (Fletcher and Inglis 1924). Often able to escape from raptor in fast, high ascent (Dharmakumarsinhji 1955). Pursued by trained Red-headed Falcon *Falco chicquera*, said to fly off at angle, then to dive perpendicularly and to call whilst heading for cover (Jerdon 1877: see 2 in Voice); see also Dharmakumarsinhji (1955) for pursuit by Peregrine *F. peregrinus*. ANTAGONISTIC BEHAVIOUR. (1) General. Aerial-display treated fully under Heterosexual Behaviour (see below); exact significance not known, but for suggestion that it may be primarily antagonistic, see Stonor (1944). Quartets of 2 pairs sometimes formed, with members described as excited and restless. In a trio comprising a pair and another bird, one individual (of uncertain identity) performed Aerial-display, each time ending in fast dive to land in tree near

other two. Trio thrice performed apparent communal display-flight (equal participation by all) with shallow dives and bouts of gliding; considered distinct from Aerial-display (Stonor 1944), though latter apparently sometimes performed by 3–4 birds together in winter (Hutson 1954). (2) Threat and fighting. 2 ♂♂ often fight over ♀. Fiercely attack each other using bill and wings and call; may continue in the air. Sometimes fall to ground entangled and fight there. ♀ takes no part. Another ♂ may challenge victorious ♂ almost immediately (Sharma 1969: see 3 in Voice). Conspecific birds approaching pair engaged in courtship usually driven off by one or both of pair: first give threatening calls (see 2 in Voice), then pursue intruder; more vigorous repulsion elicited by intruder's apparent interest in a potential nest-site near displaying pair. Once, when one bird of a pair flew near tree regularly used by neighbouring pair for display, both birds of this pair launched attack and chased fleeing intruder in apparent low-intensity Aerial-display (Stonor 1944, which see for full details of interaction between neighbouring pairs). Immediate vicinity of nest defended vigorously against other hole-nesters (Lamba 1963). HETEROSEXUAL BEHAVIOUR. (1) General. At Samawa (Iraq), heterosexual activity well advanced in March (Christensen 1962). Pair-formation in Delhi area from February (Hutson 1954). (2) Pair-bonding behaviour (including Aerial-display). In Delhi area, early pair-formation said to involve chases in which birds sweep through trees and call (Hutson 1954: see 1 in Voice); significance not certain but presumably distinct from full Aerial-display. In Madras (India), Aerial-display performed frequently by one bird (perhaps ♂) of pair up to laying; resumed c. 3 weeks after young of 1st brood had fledged but no indication that it led to 2nd brood (Stonor 1944). Near Delhi, performed throughout the year (Hutson 1954). Considered by Lamba (1963) to play important role in attracting ♀; sometimes apparently directed at perched ♀ (Ali and Ripley 1970), and ♂ may launch himself into Aerial-display from perch by ♀, returning there afterwards (Fletcher and Inglis 1920). Both birds of a pair may participate simultaneously (MacDonald 1960; Ali and Ripley 1970). Aerial-display consists of near-vertical ascent, erratic steep undulations in rather loose, flapping flight, tumbling, somersaulting, nose-dives, looping the loop, hovering, and rolling from side to side; after steep descent with closed wings, bird may pull out close to ground then 'sail away' or climb to repeat performance. Frequent and loud calls given (Stonor 1944; Macdonald 1962; Lamba 1963; Ali and Ripley 1970: see 2 in Voice). After Aerial-display, birds may sit close together side by side and Bill briefly (Lamba 1963); may call when sitting close together early in season (Fletcher and Inglis 1920: see 5 in Voice). Bowing-display performed frequently in early morning, occasionally in late afternoon, after laying, apparently at particular display perch (Stonor 1944): birds face one another on (high) perch, stand erect with bill pointed up and tail partially spread, then bow slightly. Both birds may call and occasionally open wings (Stonor 1944; Dharmakumarsinhji 1955: see 2 in Voice). Often, one bird then flies to another perch where performance repeated or silent pause may be interposed; tail-waving accompanied by calls from both birds also occurs (Dewar 1909: see 2 in Voice). (3) Courtship-feeding. No details. Said to occur frequently during breeding season (Dharmakumarsinhji 1955). (4) Mating. May follow Aerial-display (MacDonald 1962; Lamba 1963), frequently occurs after Bowing-display (Stonor 1944), sometimes after fight between rival ♂♂ (Sharma 1969). Not confined to any particular time of day and takes place on elevated perch or sometimes on the ground (Lamba 1963). In Burma, pair copulated when young still in nest (Smith 1942). ♂ said to hop around ♀ with 'peculiar gestures' and to call. ♀ then crouches and ♂ flies on to her back, flapping wings, apparently only for balance, during copulation lasting c. 15–20 s and accompanied by calls (see 3–4 in Voice) from both birds or ♀ only. Afterwards, ♂ may fly off while ♀ moves to nearby branch to self-preen (Sharma 1969; also Lamba 1963); ♀ may later join ♂ on nearby perch (Lamba 1963). Pursuit may occur, pursuer giving loud calls, or one bird may fly down to feed (Stonor 1944: see 2 in Voice). (5) Nest-site selection. Few details. Mating said to be indication that nest-site already chosen (Lamba 1963). In one case, pair apparently prospected together initially, thereafter only one bird (MacDonald 1960). (6) Behaviour at nest. Both sexes bring material and building may continue after laying. Most incubation done by ♀, ♂ usually perched unobtrusively nearby (Baker 1927; Stonor 1944: Lamba 1963). RELATIONS WITHIN FAMILY GROUP. Few details; see Bonds. Young fed by both parents; fledge at c. 4–5 weeks (Lamba 1963). Give incessant and loud food-calls when just out of nest (Smythies 1953: see Voice). ANTI-PREDATOR RESPONSES OF YOUNG. No information. PARENTAL ANTI-PREDATOR STRATEGIES. (1) Passive measures. Incubating ♀ said to allow close approach by man and even to lay replacement clutch in nest previously robbed (Baker 1927, 1934). However, elsewhere claimed to be not normally a tight sitter. Birds with young cautious in return to nest when man present (Lamba 1963); may wait for hours, giving melancholy call (Oates 1890; Dewar 1909: see 6 in Voice). (2) Active measures. Chases crows (Corvidae) away from nest and shows extreme aggression towards raptors (Lamba 1963), e.g. repeatedly dive-bombed Egyptian Vulture *Neophron percnopterus* (Hutson 1954). In Pakistan, when man near nest, ♀ apparently unconcerned and continued to feed young, while ♂ expressed agitation in wild and noisy Aerial-display (Mountfort 1969). MGW

Voice. Used more freely in breeding season (Hutson 1954). Some confusion arises from multiplicity of individual descriptions and fuller study required to provide firmer basis for differentiation of calls; following scheme therefore tentative.

CALLS OF ADULTS. (1) Contact-calls (sometimes perhaps with element of alarm or antagonism). Recording (Fig I) of bird being mobbed by crows (Corvidae) suggests a 'chack' like Jackdaw *Corvus monedula* (J Hall-Craggs, P J Sellar). Other descriptions: a sharp, harsh 'tjock' given, perhaps only by ♂, while perched for long period (Fletcher and Inglis 1920); 'tschow' said to accompany tail-jerking (Dewar 1909); a staccato 'k'yow ... k'yow ...' with explosive quality, as if clearing throat (Henry 1955); loud 'kak' or 'chack', like knocking 2 stones together; sometimes several given in succession (Ali and Ripley 1970; Gallagher and Woodcock 1980), e.g. when chasing off intruder (Ali and Ripley 1970), or during chases in early pair-formation

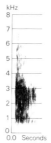

I F M Gauntlett India April 1968

(Hutson 1954). (2) In Aerial-display, excited, discordant screeching and shrieking sounds, perhaps simply attenuated variants of call 1 (Ali and Ripley 1970); clearly subject to variation as may be harsh but short in this context (Fletcher and Inglis 1920). Cackling and screeching sounds also given by both birds during Bowing-display (both give 'unmusical' sounds whilst tail-waving: Dewar 1909), and loud screeching by one member of pair in pursuit following mating (Stonor 1944), or when fleeing from raptor (Jerdon 1877). Cackling sounds probably express antagonism as used in intraspecific disputes (Stonor 1944). Harsh chattering and screeching 'kack kacka kacka kerrr kerrr' given by bird chasing Indian House Crow *C. splendens* in recording by F M Gauntlett, India, April (M G Wilson). (3) A sharp sounding 'kri ... kri' from fighting ♂♂; also given by both birds during copulation (Sharma 1969); single, rather prolonged call said to be given only by ♀ during copulation (Lamba 1963) probably the same. (4) A buzzing sound given by ♂ about to mount ♀ (Sharma 1969). (5) Chuckling sounds from both birds sitting close together early in breeding season (Fletcher and Inglis 1920). (6) A mournful croaking sound given by bird alarmed at presence of man near nest and reluctant to approach (Dewar 1909). (7) Following calls difficult to classify owing to lack of details: 'shrark' (possibly belongs under call 1), 'tseeek', and a plaintive 'kew' (Gallagher and Woodcock 1980); a harsh 'kurrr' (Dharmakumarsinhji 1955); an 'eh-eh' said to be given in October (Hutson 1954).

CALLS OF YOUNG. Food-call reported to be loud with 'distressed' quality (Dewar 1909). Recently fledged young give a loud 'screaming gobble' when fed (Smythies 1953). When foraging, independent young give a cat-like mewing (Ali and Ripley 1970). MGW

Breeding. SEASON. Iraq: breeds end of April (Ticehurst *et al.* 1922). Hole in tree or wall, less often inside chimney, or in hole in mud-bank. Nest: available hole, sometimes enlarged, or may be excavated completely, e.g. in rotten wood; feathers, straw, and grass form mat in bottom of hole, usually rather thin, occasionally up to 5 cm thick, rarely nil (Lamba 1963). Building: by both sexes, for 5–7 days before laying 1st egg, and often also after laying (Lamba 1963). EGGS. See Plate 98. Broad oval, smooth and glossy; white. Average 33×27 mm, $n = 47$ (Lamba 1963). Clutch: 4 (3–5). Of 9 clutches, India: 3 eggs, 2; 4, 5; 5, 2 (Lamba 1963). Possibly 2 broods, Iraq (Ticehurst *et al.* 1922). INCUBATION. 18 days (17–19), $n = 9$ (Lamba 1963). Mainly by ♀, beginning with 1st egg; hatching asynchronous. YOUNG. Altricial and nidicolous. Cared for and fed by both parents. FLEDGING TO MATURITY. Fledging period 30–35 days. No information on age of independence or first breeding. BREEDING SUCCESS. Of 36 eggs laid in 9 nests, India, 32 hatched and 24 young (66·7%) fledged; average clutch size 4·0, at hatching 3·6; average brood size at fledging 2·7 (Lamba 1963).

Plumages (nominate *benghalensis*). ADULT. Forehead, lores, and small feathering round eye pink-buff, slightly tinged magenta or lilac. Crown and nape rather dull dark bluish-green. Hindneck, sides of neck, mantle, scapulars, and tertials dull brownish olive-green; feathers on hindneck and sides of neck partly tinged purple or vinous-red, forming indistinct collar. Back, rump, and tips of tail-coverts bright turquoise-blue, remainder of upper tail-coverts deep violet-blue. Chin pink-buff (some white of feather-bases visible), ear-coverts dark rufous-brown with narrow cream-pink shaft-streaks, both grading into magenta or lilac-purple on throat, where marked with narrow and contrasting pink-white or cream-white shaft-streaks. Chest and breast vinous red-brown, faintly and narrowly streaked cream-white. Remainder of underparts pale blue. Central pair of tail-feathers (t1) rather dull bluish olive-grey, shafts black; t2–t6 deep violet-blue with slight green tinge for *c.* 2 cm of tip, bordered by broad and contrasting pale blue band subterminally (*c.* 3 cm wide on t2, *c.* 5 on t6); bases of t2–t6 deep violet-blue. Flight-feathers mainly brilliantly deep violet-blue; broad borders along inner webs and narrow ones along tips dull black; (p4–)p5–p10 with contrasting pale blue or sky-blue band subterminally on both webs (*c.* 3 cm wide on p9–p10, gradually narrower towards innermost, reduced to rounded spot on p4 or p5); tips of p6–p10 violet-blue and greenish-blue for 1·5–3 cm, less brilliantly coloured than inner primaries and basal portions of outer primaries and contrasting less with pale blue band; tips of p5–p10 dull bluish-grey below, hardly contrasting. Basal portions of outer webs of secondaries rather dull turquoise-blue, mainly hidden under greater coverts. Upper wing-coverts mainly rather dull greenish-blue or olive-blue (intensity of blue partly depending on light); tips of greater coverts pale blue; primary coverts pale blue, tips of greater primary coverts dull green or violet-blue; lesser coverts along leading edge of wing deep violet-blue. Under wing-coverts and axillaries pale blue, like flanks and belly, contrasting with dark violet-blue of undersurface of secondaries and primary-bases. Abrasion and bleaching affect plumage slightly: in fresh plumage, longer lesser and median upper wing-coverts tinged olive-brown or buff-brown on tips; when worn, upperparts duller, scapulars and tertials more buff-brown, chin to breast paler pink-brown with less distinct purple and vinous tinge. NESTLING. Naked at hatching. JUVENILE. Like adult, but crown dull green, hardly bluish, some grey of feather-bases shining through; hardly any purple or vinous on hindneck; mantle, scapulars, and tertials drab-brown (not slightly glossy olive-green as in adult); throat more vinous-buff with narrower white shaft-streaks, not purple; belly dull greenish-blue with slightly buff edges on feather-tips; slightly more extensive violet-blue on tail-tips, but fringes of tips washed olive-brown or buff-brown; violet-blue of shorter lesser upper wing-coverts often partly tinged pale blue; longer lesser and median upper wing-coverts with pale buff-brown suffusion on tips; no violet-blue on tips of greater upper primary coverts as in some adults; greenish-blue of bases of outer webs of secondaries more extensive; tips of p6–p10 duller and with less extensive violet-blue. FIRST ADULT. Like adult; indistinguishable when last juvenile tail- and flight-feathers (narrower and more heavily worn than those of adults) replaced.

Bare parts. ADULT. Iris grey-brown or brown. Rim of bare skin round eye dull orange. Bill black, tinged horn-brown on base. Leg and foot dirty brown-yellow or yellow-brown. NESTLING. Bare skin pink at hatching. JUVENILE. Iris grey-brown. Bill dark horn-brown, yellowish-horn on base. Leg and foot pale brown or yellow-brown. (Ali and Ripley 1970; BMNH.)

Moults. Mainly based on skins of nominate *benghalensis* and *indicus* (BMNH, RMNH). ADULT POST-BREEDING. Complete, primaries descendant. Starts with p1 from mid-June to mid-August, completed November to early March. As in Roller *C. garrulus*, primary moult occasionally suspended, though birds supposedly non-migratory: e.g. one with score 25 mid-October, Iran. Body moulted from primary score *c.* 20, scapulars, tertials, and chin to chest first; largely completed at score 40, crown and hindneck apparently last; birds which suspended moult of primaries do not necessarily stop body moult. Tail started from primary score 20–25, completed with p10 or slightly later; sequence of replacement 1–6–2–3–4–5, as in *C. garrulus*. Not known whether head and body undergo pre-breeding moult; some January adults fairly abraded already, and these will perhaps show pre-breeding moult later on, but others still new then. POST-JUVENILE. Partial or (probably in *indicus* only) complete. Starts with scattered feathers of forehead, crown, cheeks, neck, and chest from August–October, and later followed by remainder of head and body, wing-coverts, and all or part of tail, showing fresh 1st adult with worn juvenile flight-feathers and occasionally t4–t5 from January or February onwards. Some birds from southern India and Ceylon started primary moult from November; one bird suspended primary moult (score 25) at start of breeding season in February; all flight-feathers still worn juvenile in 1-year-old nominate *benghalensis* from Iraq, Iran, and Pakistan.

Measurements. Nominate *benghalensis*. ADULT. Mainly Iraq, Iran, and north-east Arabia, a few northern India and Nepal, all year; skins (BMNH, RMNH, ZFMK, ZMA). Bill (F) to forehead, bill (N) to distal corner of nostril; exposed culmen *c.* 7 longer than bill (N).

WING	♂ 191	(9·16;	6)	184–200	♀ 183	(6·25; 8)	175–190
TAIL	134	(6·59;	6)	129–145	126	(5·87; 8)	118–132
BILL (F)	44·6	(2·45;	6)	44–47	41·7	(2·49; 8)	39–45
(N)	28·0	(1·75;	6)	26–30	26·2	(1·50; 8)	24–28
TARSUS	26·6	(0·96;	6)	26–27	25·4	(1·08; 8)	24–27
TOE	32·4	(1·90;	6)	31–35	31·5	(2·34; 8)	30–35

One ♂ with wing 209 (Nepal) excluded from range. Sex differences significant for tail only.

JUVENILE. Wing on average 4 shorter than adult; tail, tarsus, and toe similar, bill similar to adult once post-juvenile moult in progress.

C. b. indicus. Ceylon (BMNH, RMNH).

WING	♂ 182	(5·16; 8)	175–188	♀ 181	(0·96; 4)	180–182	
TAIL	121	(3·13; 8)	118–125	120	(— ; 3)	116–124	
BILL (F)	44·2	(1·34; 7)	42–46	41·6	(2·30; 4)	39–44	

Weights. Nominate *benghalensis*: ♂ 166; ♀♀ 166, 176 (Ali and Ripley 1970).

Structure. Wing relatively shorter, broader, and with more bluntly rounded tip than in *C. garrulus*. 10 primaries: p8 longest, p7 and p9 0–2 shorter, p10 11–17 shorter, p6 5–8, p5 17–21, p4 24–29, p1 40–50; outer web of p6–p9 and inner web of p8–p10 emarginated. Tail long, tip almost square; 12 feathers, tips rather square and broad; tip of t6 not emarginated and narrow as in *C. garrulus*. Bill as in *C. garrulus*, but culmen rather less strongly decurved, bill slightly more gradually tapering towards hooked tip of upper mandible. Remainder of structure as in *C. garrulus*.

Geographical variation. Clinal in central India, marked in eastern Indian subcontinent. *C. b. indicus* from southern India and Ceylon clinally grades into nominate *benghalensis* in central India, boundary arbitrarily drawn at *c.* 20°N; differs in slightly shorter wing and tail (see Measurements), darker bluish crown and upper wing-coverts, browner (less olive) mantle, scapulars, and tertials, and, in particular, in more marked rufous-chestnut collar on hindneck. *C. b. affinis* from south-east Asia markedly different, but intergrades with nominate *benghalensis* in eastern Nepal, Sikkim, north-east Bihar, western Assam, and northern and eastern Bangladesh (Ali and Ripley 1970), and hence treated as conspecific: forehead blue-green like crown; no collar on hindneck; different pattern on back, rump, and upper tail-coverts; chin and upper throat dark violet-blue with blue streaks, remainder of underparts purple-brown, only vent and under tail-coverts pale blue; tips of tail-feathers broadly pale blue, basal and middle portions violet-blue, upper wing-coverts more extensively violet-blue; size slightly larger.

Forms species-group with Rufous-crowned Roller *C. naevia* of Afrotropics and with Temminck's Roller *C. temminckii* of Celebes, Indonesia (Snow 1978).

Recognition. Differs from adult Roller *C. garrulus* by darker greenish-blue crown and upper wing-coverts, duller olive (less rufous) mantle and scapulars, and rufous-brown cheeks, throat, and breast, streaked with off-white; colour pattern of flight-feathers and tail quite different; note, however, that juvenile of *C. garrulus* is also duller above and shows pale buff-brown cheeks, throat, and breast.   CSR

Eurystomus glaucurus Broad-billed Roller

PLATES 73 and 74
[between pages 758 and 759]

DU. Geelbekscharrelaar   FR. Rolle violet   GE. Zimtroller
RU. Желтоклювый широкорот   SP. Carraca piquigualda   SW. Brednäbbad blåkråka

Coracias glaucurus P L Statius Müller, 1776. Synonym: *Eurystomus afer*.

Polytypic. *E. g. afer* (Latham, 1790) northern Afrotropics from Sénégal to Sudan and Ethiopia, straggling to Cape Verde Islands. Extralimital: nominate *glaucurus* (P L Statius Müller, 1776), Madagascar, migrating to East and central Africa; *suahelicus* Neumann, 1905, eastern Africa from southern Somalia and Uganda to northern Zambia and Tanzania; *pulcherrimus* Neumann, 1905, southern Africa from Natal north to Angola, southern Zambia, and Moçambique.

Field characters. 29–30 cm; wing-span 55–58 cm. Noticeably smaller than *Coracias* rollers, with slimmer rear body and noticeably shorter tail. Rather compact, relatively short-tailed roller, with noticeably dark plumage, hunched silhouette, and hawk-like flight. Body plumage strongly chestnut above and deep lilac below; wings lack obvious contrasting pattern of *Coracias*. Short bill orange-yellow. Feeds only in flight. Noisy. Sexes similar; no seasonal variation. Juvenile separable.

ADULT. Top of head, back, scapulars, and centre of rump

chestnut, with intense, almost chocolate tone; lower face and underbody to rear belly deep purple-lilac, contrasting with pale blue vent, sides of rump, and base of black-tipped tail. Wing chestnut on inner wing-coverts, dark violet-blue elsewhere; leading under wing-coverts purple-lilac, contrasting with pale blue greater coverts and bases of flight-feathers. JUVENILE. Plumage noticeably dirtier brown than adult on upperparts and from chin to chest (latter streaked almost black); breast and belly mainly green-blue.

Unmistakable in west Palearctic (though in central Africa easily confused with Blue-throated Roller *E. gularis*); character, flight, and behaviour all markedly distinct from *Coracias* rollers. Flight recalls both those of small falcon *Falco* or hawk *Accipiter* and bulky bee-eater *Merops*, incorporating characteristic sudden take-off and accelerated climb in persistent mobbing of passing large birds or capture of flying insects; also indulges in more sustained hawking at height, with actions including bursts of fluid wing-beats, wheeling turns, and swooping glides, and ending in plummet to perch.

Commonest calls a loud cackling 'sar-a-roc sar-a-roc' and in excitement a screamed 'crik-crik-crik-crik'.

Habitat. In tropical low latitudes, mainly lowland and sometimes mainly along coasts but elsewhere more wide-ranging and extending up to bamboo zone on highest hills (Mackworth-Praed and Grant 1952). In West Africa, widely distributed in savanna and forest clearings, provided there are tall trees such as baobab *Adansonia*, mahogany *Khays*, and locust bean *Ceratonia*. In all types of country with fair- to large-sized trees, in forest clearings, and in fringing forest, usually in tree-tops at *c*. 9–18 m (Bannerman 1933). Will sit hunched up on branch awaiting prey, or hawk high in the air from bare bough (Mackworth-Praed and Grant 1952). In East Africa, in woodland areas and coastal and riverine forest (Williams 1963). In southern Africa, prefers denser to more open bush country; feeds from prominent perch, hawking insects in the air, sometimes at great height (Prozesky 1970). Over much of range, migrates or shifts habitat seasonally—in Sierra Leone, for example, a dry-season migrant to most of forest zone, but remaining during rains in more open parts, and common in northern savanna woodlands in April, breeding before onset of rains in holes in trees, particularly tops of dead oil palms, reached by diving in vertically. Foraging entirely aerial, sometimes in flocks with bee-eaters *Merops*, or at dusk with nightjars *Caprimulgus*. Will drink by skimming water surface like swallow *Hirundo* (G D Field).

Distribution. Breeds in Africa south of Sahara from Sénégal to Eritrea and south (except forest zones) to northern Transvaal, northern Zululand, northern Zimbabwe, and Angola; also Madagascar. Largely migratory or nomadic (see Movements).

Accidental. Cape Verde Islands (2).

Movements. Migratory or partially so.

Nominate *glaucurus* breeds Madagascar (present October–March); these birds then cross Moçambique Channel, and present February–November in continental Africa, north-west to Zaïre. Probably the latter is the main wintering region, since records in (e.g.) Malawi and Zambia are concentrated in passage periods February–April and October–November (Chapin 1932, 1939; Moreau 1966; Benson *et al.* 1971). Tropical African races *suahelicus* and *pulcherrimus* also strongly migratory over much of their range. Breed Zaïre to Kenya and south to northern South Africa; at least as far north as Katanga (10°S), however, they are present only in breeding season (September–April), withdrawing for rest of year into equatorial belt (Chapin 1939; Clancey 1964). In Tanzania, an intra-African migrant (present October–April) even as far north as Kibondo (3°30′S) (Britton 1980).

Northern race *afer*, breeding northern savannas from Sénégal and Sierra Leone to Ethiopia and western Kenya, is a partial rains migrant. Thus effectively absent from Ivory Coast late June to early September, having withdrawn north for duration of wet season; adults return in September–November, and (after breeding) depart quickly during May, with juveniles withdrawing more slowly (Thiollay 1971). Present all year in Nigeria, though breeds (May–July) only in narrow savanna zone; extends further south, some entering coastal and forest areas, only as dry-season visitor from October to early May, while (in contrast) occurs in arid areas north of breeding zone only during rains from May to early October (Elgood *et al.* 1973; Elgood 1982). Likewise resident though with seasonal shifts in Gambia; breeds in eastern savanna, but part of population moves west as far as forest edge in dry season (Jensen and Kirkeby 1980). Present all year in Sudan and Mali, but numbers and range restricted until rainy season influx (May–November) of birds withdrawing northwards from wetter areas further south (e.g. Lynes 1925, Lamarche 1980). See also Habitat. 2 Cape Verde Islands specimens, April and November, belong to this race (Bannerman and Bannerman 1968).

Voice. See Field Characters.

Plumages (*E. g. afer*). ADULT. Entire upperparts, including tertials and lesser and median upper wing-coverts, rich rufous-chestnut, except for longer and lateral upper tail-coverts which are pale cerulean-blue or greenish-blue, sometimes tinged darker blue on tips of longest; often an indistinct lilac-purple collar round upper hindneck. Broad supercilium bright lilac-purple, poorly defined from rufous-chestnut crown. Entire underparts including ear-coverts and cheeks deep lilac-purple, except for pale sky-blue or pale greenish-blue lower vent and under tail-coverts; some rich rufous of feather-bases shining through, especially on chin, sides of chest, and belly; purple deepest and strongly glossy on throat and central chest, especially on feather-shafts. Central pair of tail-feathers (t1) dusky greyish-black with blue suffusion on base and along shafts; t2 sky-blue or greenish-blue with rich dark violet-blue tip 2½–3 cm wide, latter with dusky black border on inner web and tip (tip mainly dusky grey on under surface of tail); other tail-feathers similar to t2, but violet-blue tip gradually narrower outwards, c. 1 cm wide on t6; shafts of tail-feathers black. Flight-feathers, greater upper wing-coverts, upper primary coverts, and bastard wing rich dark violet-blue, markedly contrasting with rufous-chestnut remainder of upperwing; distal parts of outer webs of outer primaries paler greenish-blue. Narrow bar along leading edge of wing violet-blue. Lesser and median under wing-coverts and axillaries purple like underparts; greater under wing-coverts and undersurface of basal and middle portions of flight-feathers pale blue or pale greyish-blue; undersurface of tips of flight-feathers dusky grey or slightly violet-blue. Bleaching and abrasion cause slight change, upperparts becoming slightly less deep rufous-chestnut and underparts more rufous (less purple) in worn plumage. Sexes mainly similar, though ♂ on average slightly brighter coloured than ♀, especially on underparts. JUVENILE. Much duller than adult, and quite differently coloured: upperparts and chest mainly brown, remainder of underparts greenish-blue. Forehead, crown, hindneck, mantle, scapulars, back, rump, and lesser and median upper wing-coverts dark greyish-brown, feathers tipped bright rufous-chestnut (narrow and hardly visible on forecrown, rather wide on mantle, scapulars, and wing-coverts); slight blue suffusion on lesser and median upper wing-coverts, especially outer. Tertials dark buff-brown or rufous-cinnamon with some grey or blue suffusion on tips. Central upper tail-coverts grey-brown, tipped buff-brown; longer and lateral coverts pale greenish-blue. Sides of head and neck grey-brown, feathers tipped cinnamon-brown on lores, just below eye, and on sides of neck; some pale blue tinge on cheeks and ear-coverts. Entire underparts pale blue or pale greenish-blue with much grey-brown of feather-bases exposed; feathers of chin to chest tipped buff, some purple occasionally on centres of belly feathers. Tail as in adult, but t1 duller and greyer; tips of t2–t6 broadly dull grey (not narrowly black), subterminal violet-blue less sharply defined from pale blue middle portions and bases. Primaries and upper primary coverts mainly dull violet-blue (partly paler greenish-blue on outer webs of outer primaries and on greater upper primary-coverts; not as bright and glossy deep violet-blue as in adult); secondaries duller violet-blue than in adult; tips and broad borders of inner webs of flight-feathers more extensively and duller black than adult. Under-wing and axillaries largely pale blue (less extensively lilac-purple). FIRST ADULT. Indistinguishable from adult once post-juvenile moult completed; easily recognized and piebald when part of differently coloured juvenile plumage still present. Juvenile flight-feathers and tail last to be replaced; relatively more worn than those of adult, tips more extensively grey (c. 5–20 mm, rather than a few mm only).

Bare parts. Iris brown or hazel. Bill bright chrome-yellow or orange-yellow. Leg and foot greenish-brown, olive-horn, dirty yellow, or greenish-yellow. JUVENILE. Iris brown or dark brown. Bill dark horn-brown or horn-black with yellow base of lower mandible and some yellow below nostril on upper mandible; yellow gradually spreading during post-juvenile moult, dark horn-colour retained longest on top of culmen. Leg and foot grey-black, grey, olive-brown or brownish-white. (Meise 1937; BMNH, RMNH.)

Moults. (1) *E. g. afer*, West Africa (BMNH, RMNH). ADULT POST-BREEDING. Complete, primaries descendant. Starts just after fledging of young or slightly earlier; p1 first, lost late July to mid-September. Moult of tail, head, and body apparently starts immediately after loss of p1; birds with primary moult score 10–20, August–September, in heavy moult. All moult completed December–March. POST-JUVENILE. Partial: head, body, most upper wing-coverts, and tail. Starts 1–2 months after fledging, September–November, scattered feathers of head, mantle, chin to chest, and scapulars first. Largely in 1st adult from January–April, except for flight-feathers, greater wing-coverts, and sometimes a few tail-feathers (mainly t4–t5), some outer median and lesser upper wing-coverts, a few tertials, or some upper tail-coverts or belly feathers.

(2) Nominate *glaucurus*, breeding Madagascar, spending non-breeding season in central Africa (Herroelen 1964; BMNH). ADULT POST-BREEDING. Apparently starts soon after hatching of young, November–December, suspending primary moult with scores 20–35 during migration to central Africa about March. Primary moult resumed in non-breeding area, completed late June to late August. All tail and apparently much of body replaced in non-breeding area, hence at much higher primary moult score than *afer*. POST-JUVENILE. Partial; extent as in *afer*. In non-breeding area, July–October.

Measurements. *E. g. afer*. West Africa (Sénégal to Ghana), all year; skins (BMNH, RMNH). Tail is to tip of t6 (to t1, 7–15 less); bill (F) to forehead, bill (N) to distal corner of nostril (exposed culmen 3·3 longer than to nostril), both in adult only.

WING AD	♂	178	(3·16; 13)	175–183	♀ 171	(4·69; 7)	163–177
JUV		170	(3·77; 8)	165–176	167	(3·75; 6)	161–171
TAIL AD		91·7	(2·98; 10)	88–97	88·4	(4·63; 6)	83–93
BILL (F)		28·0	(1·63; 10)	26·2–30·1	27·6	(1·06; 5)	26·6–28·9
(N)		15·4	(0·75; 10)	14·4–16·3	15·5	(0·55; 6)	14·7–16·1
TARSUS		17·5	(0·47; 10)	16·8–18·2	17·9	(0·86; 6)	16·7–18·7
TOE		22·9	(0·90; 9)	22·0–24·2	22·5	(0·84; 5)	21·9–23·0

Sex differences significant for adult wing only. Juvenile wing significantly shorter than adult; juvenile tail on average c. 6 shorter; bill not full-grown until post-juvenile moult in progress. Juvenile ♂, Cape Verde Islands, November: wing 172 (BMNH).

Weights. Afrotropics, several races similar in size to *afer* combined: adult ♂ 111 (6·27; 11) 98–112; adult ♀♀, 101, 110; juvenile ♀♀, 99, 135 (Meise 1937; Verheyen 1953; Jackson 1969).

Much larger-sized nominate *glaucurus*, Zaïre (where non-breeding visitor), adult and 1st adult. On arrival, February: ♀ 140. May–October: ♂♂ 91 (exhausted), 164; ♀ 159 (11·2; 5) 145–174. On departure, September–November: ♂ 205; ♀♀ 212, 217 (Herroelen 1964).

Structure. Wing rather long, broad at base, tip bluntly pointed. 10 primaries: p9 longest, p10 2–10 shorter, p8 0–4, p7 9–18, p6 25–36, p5 40–48, p1 63–75 (*afer* only). Inner web of (p7–)p8–p10 and outer of (p7–)p8–p9 emarginated. Tail rather short, slightly forked; 12 feathers, t1 8–15 shorter than t6 in ♂, 6–13 in ♀, independent of age. Bill short, deep and very wide at base; culmen decurving to strong hook at tip of upper mandible; sides of upper mandible swollen; bill much shorter, wider at base, and less strongly compressed laterally at tip than in *Coracias* rollers, no stiffened bristles at gape. Leg and foot relatively shorter and more slender than *Coracias*. Outer toe c. 86% of middle, inner c. 75%, hind c. 67%.

Geographical variation. Races breeding on African mainland similar in size to *afer* (see Measurements) or slightly larger; colour similar, except for upper tail-coverts of adult: in *afer* (of northern Afrotropics), central upper tail-coverts brown like rump, longer and lateral ones pale greenish-blue; in *suahelicus* of eastern Africa, all upper tail-coverts deep blue; in southern race, *pulcherrimus*, all upper tail-coverts (including central) pale greenish-blue (Mackworth-Praed and Grant 1970). Nominate *glaucurus* of Madagascar similar in colour to *pulcherrimus*, but much larger—e.g. adult wing 190–220, juvenile wing 177–203, adult tail 101–126, juvenile tail 97–111 (Herroelen 1964; BMNH)—and much heavier (see Weights).

Forms superspecies with Blue-throated Roller *E. gularis* from West and central African rain forests (Snow 1978); closely similar to *E. glaucurus*, but upper tail-coverts black, central throat blue, and remainder of underparts rufous-chestnut. CSR

Family UPUPIDAE hoopoes

Medium-sized perching bird, with long decurved bill, prominent crest, and rather short legs. Single species, Hoopoe *Upupa epops*, breeding in west Palearctic.

Sexes of same size. 14 cervical vertebrae. Wings rounded, flight appearing butterfly-like and wavering, but can fly strongly. 10 primaries, p6 and p7 longest; wing eutaxic. Tail rather long, 10 feathers. Oil-gland tufted.

Aftershafts vestigial or absent. Bill long, narrow, and tapering, decurved, strongly compressed laterally. Skull and jaw musculature highly modified for opening bill while probing. Tongue reduced. Tarsus short. Middle and outer toes united by basal joint. No caeca. A single (left) carotid. Nestlings with long down. For other details, see species account.

Upupa epops Hoopoe

PLATES 71 and 75
[facing pages 735 and 806]

Du. Hop   Fr. Huppe fasciée   Ge. Wiedehopf
Ru. Удод   Sp. Abubilla   Sw. Härfågel

Upupa Epops Linnaeus, 1758

Polytypic. Nominate *epops* Linnaeus, 1758, north-west Africa and Europe east to north-west India, Sinkiang (China), and Ob–Yenisey watershed in USSR; *major* Brehm, 1855, Egypt, northern Sudan, and Ennedi (Chad); *senegalensis* Swainson, 1837, dry belt south of Sahara, from Sénégal to Ethiopia and Somalia, north to Ahaggar (southern Algeria). Extralimital: *saturata* Lönnberg, 1909, eastern Asia east of nominate *epops*, south to Tibet and central China; *longirostris* Jerdon, 1862, south-east Asia east and south from Bangladesh, Assam, and southern China; *ceylonensis* Reichenbach, 1853, plains of Pakistan and India south to Ceylon; *waibeli* Reichenow, 1913, Cameroun and northern Zaïre, east to northern Kenya and Uganda; *africana* Bechstein, 1811, eastern and southern Africa south from central Kenya; *marginata* Cabanis and Heine, 1860, Madagascar.

Field characters. 26–28 cm, of which bill 5–6 cm and tail 8–9 cm; wing-span 42–46 cm. Slimmer than Jay *Garrulus glandarius* on ground but of similar size in flight, when large expanse of wings and tail fully visible. Long- and curve-billed, pinkish, ground-feeding bird with bold cream-white bars over black wings, rump, and tail-base, and remarkable crest. When excited, fans crest and then looks large-headed. In flight, recalls large butterfly. Sexes similar; no seasonal variation.

ADULT MALE. Head, crest, neck, upper mantle, and underbody to flanks pale brown-pink, with almost chestnut tone to rear crest, and vinous tinge to throat and breast. Black and white tips to crest feathers obvious but dusky-brown streaks on whiter rear flanks usually hidden. Back shows 3 orange-cream bands on black; crescentic rump-band white; at rest, barring of back contiguous with 5 cream and white bands across black coverts and flight-feathers. Tail black, with white band near base, curving back at edges. Vent white. In flight, white and black bands on wing visibly incomplete, with wholly black primary coverts and primary bases forming broad wedge between white bands across inner wing and extension of rearmost band across tips of primaries. Underwing pink-white, showing obvious barring only against strong light. Bill and legs dark grey. ADULT FEMALE. Some distinguishable from ♂ (in comparison) by white-tipped throat and browner-toned breast. JUVENILE. Duller than adult, usually lacking any vinous tone on forebody. White wing-bands strongly tinged cream, and black plumage less glossy. Bill short and less decurved.

Unmistakable (but beware escapes of imported birds from southern Asia, where races are smaller and much redder). Flight as characteristic as plumage pattern and structure, usually suggesting erratically flitting butterfly (when wings alternately spread and closed, exposing and hiding barring) but also recalling Lapwing *Vanellus vanellus* (when wing-beats less exaggerated and bird intent on long-distance movement or evasion of predator); capable of rapid looping climb and long wavering plummet. Gait well developed, as befits ground feeder, with walk and short runs. Perches freely and can climb up rough surfaces.

Commonest call a trisyllabic 'poo-poo-poo', soft in tone

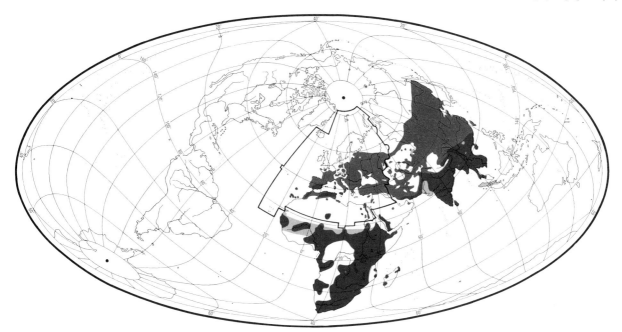

and low in pitch but carrying far; not diagnostic, since ♂ Oriental Cuckoo *Cuculus saturatus* utters closely similar phrase; also chatters and utters cat-like mew.

Habitat. Breeds in upper middle to lower latitudes of west Palearctic through warm boreal, temperate, steppe, Mediterranean, and desert subtropical zones, but much more thinly in more humid or chilly regions and those of uncertain warmth and sunshine. Basically a bird of warm, dry, level or gently undulating terrain with much exposed bare surface (sandy, silty, or rocky), but numerous incidental upstanding features offering perches, shade, and breeding cavities. Avoids woodland, reedbeds, or any tall dense vegetation or wetland; also precipitous terrain, beaches, water margins, cities, and extensive featureless open tracts, including cultivation and pasture. In characteristic Asian haunts, preferred habitats include steppes, plains, river valleys, and foothills, semi-desert, and stony or talus deposits and vicinity of cliffs, together with glades and clearings in sparse open pine or oak woods, or mixed landscapes where woods alternate with open steppe-land (Dementiev and Gladkov 1951*a*). In Afghanistan, present thinly up to 3100 m, but normally below *c*. 2000 m; in barren foothills but also poplar groves, tamarisk scrub, and fertile valleys (Paludan 1959). In southern Europe, fond of vineyards, orchards, olive groves, and fringes of maquis and woodland, as well as farmland with walls and freestanding trees, bare or sparsely vegetated soil being in every case an essential for ground feeding (Bannerman 1955). In temperate Europe, favours open country with enough old trees, such as pollard willows *Salix* lining watercourses alongside meadows, orchards, avenues, spinneys, and woodland edges, or even large clearings and glades in broad-leaved or pine woods (Niethammer 1938; Münch 1952). In Switzerland, avoids heavily wooded and mountainous regions but favours damp meadows and cattle pastures with some old timber; also parkland and neighbourhood of carr woodland, and locally vineyards, sometimes penetrating into city suburbs. Altitudinal limits in Jura under 1000 m, but on slopes above upper Rhone regularly observed up to 1200–1300 m (Glutz von Blotzheim 1962). In Low Countries sometimes in wooded dunes. (Lippens and Wille 1972.) In Canary Islands, most frequent on semi-desert islands of group; on Tenerife from coastal plains to *Retama* zone at 2000 m (K W Emmerson). Winter migrants in Sierra Leone feed on ground in gardens, along roadside verges, and on bare patches, flying into trees when disturbed (G D Field). Flies freely for some distance but normally in lower airspace, although climbing to considerable height when pursued by falcon *Falco* (Bannerman 1955).

Distribution. Marked fluctuations in range in 19th and 20th centuries. In mid 19th century widespread and common over whole of central Europe (Glutz and Bauer 1980) with regular breeding in Denmark and southern Sweden. Then (though earlier in France and Netherlands) range declined in Europe, especially on north-western periphery, until mid 20th century, when some recovery in many areas, but retraction of range again since *c*. 1955–60. Climatic change sometimes suggested as main cause, but connection far from established and earlier data especially often inadequate for precise conclusions; other factors, especially habitat changes and, more recently, increased use of insecticides, probably involved.

BRITAIN. 30 cases of proved breeding in last 140 years,

788 Upupidae

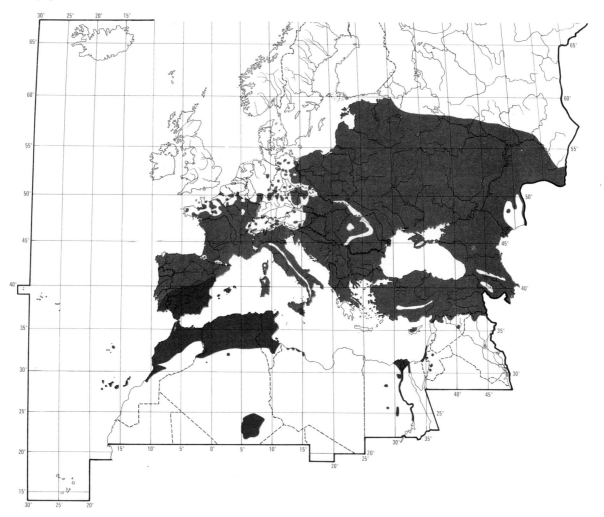

with peaks in 1895–1906 and 1950s (Sharrock 1976); since bred 1977 (4 pairs) and possibly 1978 and 1980 (Sharrock *et al.* 1983). FRANCE. See Yeatman (1976) for details of range changes. LUXEMBOURG. Became extinct *c.* 1930, then bred from *c.* 1952, reaching *c.* 15 pairs 1960; no breeding since attempt in 1971 (Glutz and Bauer 1980). NETHERLANDS. Common over large areas in 18th century and early 19th century. By late 19th century mainly restricted to south; only occasional pairs since 1925 until sudden increase (perhaps 3–10 pairs) from 1943, but only irregular breeding recorded from 1967 (CSR). WEST GERMANY. For details of range declines and occasional breeding, see Glutz and Bauer (1980) and Bauer and Thielcke (1982). DENMARK. Widespread in some areas 1835–50, then nested sporadically until 1876; since bred 1949, 1970, and 1977 (3), and perhaps 1975–6 (Rosendahl 1977; Glutz and Bauer 1980; TD). SWEDEN. In mid 19th century, bred south and centre, locally common. Became rare by 1880; small numbers in south-east until *c.* 1920, then only 10 breeding records 1921–70, but nested almost every year Öland and Gotland in 1970s (Risberg 1979; LR). FINLAND. Bred 1940 (Merikallio 1958). EAST GERMANY. Marked decrease in range continuing (Klafs and Stübs 1977; Glutz and Bauer 1980; Rutschke 1983). SWITZERLAND. Disappeared from most areas in north since 1953 (see map in Glutz and Bauer 1980). USSR. Occasional and sporadic breeding in north of range (Dementiev and Gladkov 1951a); expanded northwards in 1950s in Bashkiriya (Ilyichev 1959). SYRIA. Present distribution not fully known (HK). LEBANON. Formerly probably regular breeder, now perhaps occasional (Tohmé and Neuschwander 1974; HK). ISRAEL. Marked expansion since 1938, with spread to human settlements (HM). MADEIRA. Formerly regular breeder Madeira (last 1937) and Porto Santo before 1890 (Bannerman and Bannerman 1965). Bred Porto Santo again from 1965 (Zino 1969a).

Accidental. Spitsbergen, Iceland, Faeroes, Azores, Cape Verde Islands.

Population. Limited data on populations, but trends

probably reflect range fluctuations (see Distribution), with general decline since mid 19th century, interrupted only by short-lived recovery in mid 20th century.

FRANCE. Marked fluctuations: declined early 19th century, local increases from c. 1950, since decreased. Up to 100 000 pairs (Yeatman 1976). SPAIN. Numerous except in north, where local (AN). BELGIUM. Increased 1942-52 when reached c. 120 pairs; declined to c. 20 pairs c. 1972 (Lippens and Wille 1972; Glutz and Bauer 1980); 1-2 pairs 1982, perhaps now irregular (PD). WEST GERMANY. Fluctuating (Glutz and Bauer 1980); marked decline to c. 60-100 pairs (Bauer and Thielcke 1982). EAST GERMANY. Marked decline (Glutz and Bauer 1980); c. 150 pairs (Makatsch 1981). POLAND. Widespread, but everywhere scarce (Tomiałojć 1976a). CZECHOSLOVAKIA. Marked decline since 1950, now generally scarce (KH). AUSTRIA. Decline in some areas after 1955 (Glutz and Bauer 1980). SWITZERLAND. Major decline since c. 1953; estimated 138 pairs 1977-9 (Glutz and Bauer 1980). USSR. Abundant in south, rare or occasional in north of range (Dementiev and Gladkov 1951a). Estonia: scarce; marked fluctuations (HV). Lithuania: decreasing (Ivanauskas 1964). ISRAEL. Marked increase since 1938 (HM). EGYPT. Common (PLM, WCM). CYPRUS. Fairly common (PRF, PJS). MOROCCO. Not common (JDRV). MADEIRA. Increased Porto Santo, now fairly common (Zino 1969a; PAZ, GM). CANARY ISLANDS. Most numerous on Lanzarote and Fuerteventura (KWE).

Movements. Migratory in northern parts of range (Europe, Siberia), partially migratory to resident elsewhere in Palearctic. Extralimital races make short-distance dispersals in southern Asia, which is also wintering region for migratory *saturata* from central and eastern Siberia (Vaurie 1965; Ali and Ripley 1970), while Afrotropical races all make intra-African movements of unclarified extent (Moreau 1972; Britton 1980). Egypt and north Sudan race *major* is resident; collected once, 13 April, in Ennedi (Chad)—possibly a migrant (Niethammer 1955), but *major* may breed further west than supposed (Vaurie 1965).

Nominate *epops*, to which rest of account refers, winters in small numbers in North Africa and Mediterranean basin, only exceptionally further north. Probably most European migrants winter in Africa south of Sahara, where not distinguishable in the field from local *senegalensis*. Migrants found in savanna and scantily wooded country all across northern tropics from Sénégambia to Somalia, and southwards into Kenya and Uganda; nominate *epops* also identified a few times in Tanzania and once in northern Malawi (Benson *et al*. 1971; Britton 1980). Also present all year across Middle East to India, though (in absence of ringing) not known whether breeding birds are resident or replaced in winter by migrants from further north. One ringed Ethiopia, March 1971, recovered Saudi Arabia in August 1973. In hills and plains of northern India, numbers greatly augmented by immigrants in winter (Ali and Ripley 1970).

Migration (much of which nocturnal) occurs on broad front across Europe and Mediterranean, and probably across Sahara also. Regular at oases in Morocco, Algeria, and Libya, and in Niger inundation zone (Mali), during passage periods (e.g. Smith 1968, Bundy 1976, Lamarche 1980), while in central Chad 300 flew north past Arada in 3 days, 11–13 April (Newby 1980). In Egypt, loose flock of 103 presumed migrants at Sohag (Nile valley) in mid-October (Horváth 1959); migrants more typically occur singly or in small parties.

Available European recoveries detailed by Glutz and Bauer (1980); none south of Sahara, and all in passage periods (March–May, August–October) except 1 Spanish bird found south-east Morocco on 2 February (*Ardeola* 1974, 20, 87). Birds ringed west-central Europe (West and East Germany, Switzerland, Austria) recovered France and Spain (10), Italy, Sicily, and Malta (12), Balkans (4), Algeria (1), and western Ukraine (1). Smaller sample of birds ringed east-central Europe (Poland, Czechoslovakia, Hungary, Yugoslavia, Rumania) recovered France (1), Italy and Malta (7), and Balkans (13); Morocco migrant (April) found 3 months later in Czechoslovakia. Hence west-central European birds migrate mainly south to south-west, east-central European birds mainly south to south-east. No clear migratory divide; European mountains (Alps, Carpathians) probably no barrier, since recorded to 6400 m in Himalayas (Ali and Ripley 1970).

Migration seasons notably protracted. Autumn dispersal spans mid-July to late October or even into November (Hudson 1973; Glutz and Bauer 1980); normal termination period masked by overwintering in southern breeding areas. Begin arriving south of Sahara in second half of August, with main arrivals there September–October. Return movement detected from early February in Morocco and Malta (Smith 1965; Sultana and Gauci 1982); at peak mid-March to April, tailing off during May. In Britain and Ireland more recorded in spring (77% of modern records) than autumn (23%); spring records involve overshooting in anticyclonic weather and arrivals in overcast conditions with fronts moving into English Channel and North Sea (Sharrock 1974). East German bird (June 1961) recovered on ship at 49°50′N 11°00′W the following April (*Vogelwarte* 1964, 22, 178).

Food. Almost entirely animal, primarily larger insects and especially their larvae and pupae. Normally forages alone (Löhrl 1977b), mainly on ground, walking about and jabbing with bill to left and right (Bussmann 1950). Prey usually located by sight; sometimes flushed out and seized after brief, rapid pursuit (Löhrl 1977b); also perhaps by sound, as head sometimes held at slight angle (Bussmann 1950; Münch 1952). Probing in ground frequently employed. Bill inserted, then rapidly opened and closed again (Löhrl 1977b). During intensive foraging may run

1–2 m, then probe 3–5 times at one spot (Glutz and Bauer 1980). Also frequently probes in animal droppings and dung heaps as well as in accumulations of rubbish and even carrion which attract insects (Naumann 1901; Greaves 1936; Münch 1952; Kaczmareck 1976). In case of mole-crickets *Gryllotalpa* and larvae of cockchafers *Melolontha*, recognition of visual cues probably precedes probing; then seizes prey in bill-tip and pushes against ground with legs, often turning in a circle several times; pressure of bill and movement of bird create wider funnel-shaped hole facilitating prey extraction (Löhrl 1957; see also Löhrl 1977b for observation on captive juvenile). Sometimes delivers blows at ground (e.g. Greaves 1936, Panov 1973), excavating large hole (Skead 1950). Stones and leaves never picked up to be tossed aside but by inserting bill beneath them and lifting upper mandible (Löhrl 1977b); cow-pats flicked over with sideways or upward movement of bill (Skead 1950 for *africana*). Relatively slow-flying *Melolontha* and some Diptera sometimes captured in the air (Münch 1952; Hirschfeld and Hirschfeld 1973), also termites (Isoptera) (Skead 1950). According to Smith (1887), frequently hangs from tree branches, foraging for insects on underside of foliage; *africana* once recorded foraging amongst lichen on tree bough and once on stonework of arch (Skead 1950). Larger prey items usually hammered, or seized and beaten against hard surface (traditional site may be used: Bussmann 1950). Elytra, wings, and legs of insects usually removed (Bussmann 1950; Münch 1952; Hirschfeld and Hirschfeld 1973; Löhrl 1977b). Lizard *Lacerta agilis* pushed and beaten against ground; detached tail tossed up and swallowed, rest carried away (Bäsecke 1951); sometimes killed by hammering head and puncturing skull (Hirschfeld and Hirschfeld 1973). Prey (held in bill-tip) normally swallowed by jerking bill up and opening it, while slightly lowering then immediately raising head (Hirschfeld and Hirschfeld 1973; see illustration in Glutz and Bauer 1980). Prey brought for young in bill-tip (Meade-Waldo 1907), almost always singly (Löhrl 1977b); for exceptional cases of 2–3 *Gryllotalpa* at once, see Stirnemann (1940). Centipede (Chilopoda) of c. 6·5 cm folded into c. 4 loops (Meade-Waldo 1907). Usually stuffed well down into throat of young (Bussmann 1950). Near Budapest (Hungary), in severe drought, adults several times attempted to place empty shells of snails *Helix pomatia* into bills of young (see Koffán 1957, 1964). Adult and young eject chitinous material as pellets (Heinroth and Heinroth 1924–26; Jourdain and Harrison 1936). Foraging range markedly dependent on food supply. Sometimes 1·5–2 km from nest (Jacobs 1943; Bussmann 1950; Münch 1952; Hedegaard-Christensen 1970; Hirschfeld and Hirschfeld 1973); in one case, birds regularly flew several hundred metres uphill to exploit large numbers of *Gryllotalpa* (see Aellen 1942).

Mole-crickets *Gryllotalpa gryllotalpa* (e.g. Stirnemann 1940, Aellen 1942, Bussmann 1950, Heldmann 1951) and larvae of cockchafers *Melolontha melolontha* and *M. hippocastani* (e.g. Koffán 1957, Hirschfeld and Hirschfeld 1973) often favoured. Insects comprise mainly larvae and pupae of moths and butterflies (Lepidoptera: Noctuidae, Sphingidae, Geometridae, Lymantriidae, Arctiidae, Pieridae) and adults and larvae of beetles (Coleoptera: Scarabaeidae, Geotrupidae, Carabidae, Elateridae, Chrysomelidae, Silphidae, Staphylinidae, Lucanidae, Tenebrionidae, Cerambycidae, and Curculionidae); also takes dragonflies (Odonata), crickets, etc. (Orthoptera: Gryllidae, Tettigoniidae, *Gryllotalpa*, Acrididae), cockroaches (Blattidae), bugs *Eurygaster*, cicadas (Cicadidae), larvae of ant-lions (Myrmeleontidae), earwigs (Dermaptera), adults and pupae of ants (Formicidae), adults and larvae of flies (Diptera: Calliphoridae, Syrphidae, Tipulidae). Non-insect food includes spiders (Araneae), woodlice (Isopoda), centipedes and millipedes (Myriapoda), earthworms (Lumbricidae), small terrestrial and aquatic snails and slugs (Gastropoda), mussels (Unionidae), lizards (including slow-worm *Anguis fragilis*), frogs and toads (Anura), and exceptionally birds' eggs. Only vegetable food recorded small seeds, berries of elder *Sambucus nigra*, and various parts of *Poa* and *Potentilla* (Floericke 1919; Baillie 1956; Ryabov 1965). Sand and small stones sometimes recorded in stomachs. (Naumann 1901; Csiki 1905; Meade-Waldo 1907; Rey 1908; Bussmann 1934; Stirnemann 1940; Bäsecke 1951; Herberg 1952; Münch 1952; Ryabov 1965; Ganya *et al.* 1969; Peltzer 1969; Hirschfeld and Hirschfeld 1973; Kolbe 1976; Glutz and Bauer 1980.)

26 stomachs from Hungary (April–September) contained 81 Carabidae (mainly *Harpalus*), 26 Scarabaeidae (including 18 adult and 24 larval *Melolontha*), 2 Silphidae, 11 Elateridae, 4 Curculionidae, 8 unidentified Coleoptera larvae, 65 Lepidoptera larvae, Formicidae (with 150 *Formica fusca* in 1 stomach), 5 ant-lion *Myrmeleon* larvae, 3 *Gryllotalpa*, 1 Acrididae, 1 spider *Epeira*, and 1 frog *Rana* (Csiki 1905). 24 stomachs from Moldavian SSR (USSR) contained 247 items: included 17·4% by number Cerambycidae larvae, 11·3% Scarabaeidae larvae, 7·7% Geometridae larvae, 6·1% *Eurygaster*, 5·7% Noctuidae larvae, 5·3% *Myrmeleon formicarius* larvae, 5·3% Carabidae, 4·5% Elateridae, 4·0% *Gryllotalpa*, 3·6% Gryllidae, 0·8% earthworms *Lumbricus*, and 0·4% *Lacerta agilis* (Ganya *et al.* 1969). 25 stomachs from Kustanay region (northern Kazakhstan, USSR), May–July (including 8 fledglings) contained 99% animal food: 51 larvae of Noctuidae and Arctiidae, 25 Coleoptera (mostly Carabidae and Scarabaeidae), 15 Orthoptera (mostly *Calliptamus italicus* and *Locusta*), 2 Cicadidae, and 11 small Araneae. 2 stomachs contained plant matter: leaves, stems, and rhizome of *Poa bulbosa*, also leaves and flower of *Potentilla*. Similar results in Kirgiziya (Pek and Fedyanina 1961); in both regions, mainly took abundant, less mobile insects (Ryabov 1965). For further data from extralimital parts of USSR, see (e.g.) Kekilova (1970) and Ataev (1974).

Food of young well studied in Kyffhäuser district (Nordthüringen, East Germany). 385 items seen to be brought over 2 days in June included 152 larvae and 3 pupae of Lepidoptera, 1 adult butterfly, 63 Scarabaeidae larvae, 2 adult Coleoptera, and 3 Gryllidae. Further visual observations on 4 broods over 2 years showed nestling diet to consist mainly of Lepidoptera larvae and pupae (particularly larvae of *Celerio euphorbiae*); also many *Melolontha* larvae, a few adult Coleoptera, and Gryllidae; rarely, *Gryllotalpa*, Acrididae, and 1 adult brimstone butterfly *Gonepteryx rhamni*. 249 items of nestling food obtained over 2 years by neck-collar method (fitted after c. 14 days old) comprised 85·9% (by number) Lepidoptera larvae and pupae (53·4% Noctuidae and 32·1% Sphingidae with many *Celerio euphorbiae*), 1·6% Coleoptera (larvae and pupae of Scarabaeidae, Elateridae, Cerambycidae), Orthoptera (Acrididae, Tettigoniidae, Gryllidae), 0·4% Tipulidae pupae, 5·6% spiders (mainly Lycosidae), and 2·0% lizard *Lacerta agilis*. (Hirschfeld and Hirschfeld 1973.) In Luzerner Seetal (Switzerland), nestlings given larvae of Diptera, Lepidoptera, and others for first few days, first *Grylloptalpa* at c. 7–8 days, and almost exclusively *Gryllotalpa* (some *Melolontha* larvae) from c. 13–14 days (Bussmann 1950). See also (e.g.) Bussmann (1934), Stirnemann (1940), Bäsecke (1951), Jacobs (1943), and Kolbe (1976). Large items generally brought (Löhrl 1977b); *Lacerta agilis* up to c. 16 cm (Hirschfeld and Hirschfeld 1973). Feeding period light-dependent; begins c. 03.00–05.00 hrs and ends c. 19.00–20.30 hrs. During brooding phase, ♂ brings food c. 5–8 times per hr, more frequently later (Hirschfeld and Hirschfeld 1973); up to 20(–60) times per hr (Kumari 1941). ♂'s feeding rate peaks in early morning, midday, early afternoon, and (less marked) late evening. ♀ usually begins to bring food from 11th day; share then increases but remains lower than ♂'s. On 22 June, 04.05–20.20 hrs, ♂ brought food 161 times, ♀ 95 times (Hirschfeld and Hirschfeld 1973). In Luzerner Seetal, 30–50 feeds per day during early nestling period, 70–80 in middle, and decline towards end: e.g. 40–50 times at 23 days (Bussmann 1950). See also (e.g.) Bussmann (1934), Kumari (1941), Söding (1961), Kolbe (1976), and Reinsch (1979); for *africana*, see Skead (1950) and Winterbottom (1952). MGW

Social pattern and behaviour. Literature large but no definitive study in west Palearctic. For monograph, see Münch (1952); for observations on captive birds, Löhrl (1977b); for recent summary, Glutz and Bauer (1980). Extralimital *africana* studied in some detail by Skead (1950), South Africa.

1. Usually seen singly or in pairs. According to Glutz and Bauer (1980), never gregarious—always feeding solitarily, even in captivity (see Löhrl 1977b). Small flocks occur on migration. In family parties for a while after young fledge (see below). 8–10 seen feeding together in maize fields, Hungary, August and early September (Homonnay 1938), and 12 together, Switzerland, in July (Stirnemann 1941). Noted in small parties at Saharan oases in winter (Tristram 1886). May also associate in small flocks after spring arrival—e.g. 4 feeding amicably, France, in early April (Hüe 1947b). Several birds may assemble to feed, India (Everitt 1964). In South African study, birds in flock of up to 9 (♂♂, ♀♀) for c. 4 months prior to breeding; fed as fairly compact group at times (see Skead 1950 for further details). BONDS. Monogamous mating system the rule. Breeding trios of 2 ♂♂ and 1 ♀ reported (Skead 1950) but evidently exceptional. Pair-bond of seasonal duration, even where resident (see Greaves 1936, Skead 1950). Probably breeds first at 1 year old. Only ♀ incubates (fed by ♂) and does all or most of brooding; both sexes feed young. Juveniles remain together with parents for variable period after leaving nest: for up to 1 month after starting to feed themselves c. 6 days after fledging (see Glutz and Bauer 1980). According to Greaves (1936), however, remain in family party for some months after fledging, Egypt, parental feeding continuing for a long time (no details); young gradually drift away in autumn, 1–2 sometimes remaining longer until chased off. In pair nesting Somerset (England), single juvenile with parents for 4–5 weeks, then parents migrated first; fed for at least c. 14 days after fledging (Slade 1978). If pair double brooded, ♂ may feed young alone for last few days (Glutz and Bauer 1980); hostility by incubating or brooding ♀ to young of 1st brood coming near nest reported by Hirschfeld and Hirschfeld (1973). BREEDING DISPERSION. Solitary and territorial, though shortage of suitable holes can lead to 2 pairs nesting close at times: e.g. 2 nests at opposite ends of same verandah (Scott 1866); 2 nest-boxes occupied in same garden c. 30 m apart (Puhlmann 1912), 2 nests under adjoining roof joists with birds of both pairs apparently using same entrance (Greaves 1936). Same hole often used year after year: 4 years in succession, Egypt, and again later after gap of 1 year (Greaves 1936); 5 years, West Germany (Bäsecke 1943); exceptionally, 17 years (Glutz and Bauer 1980). Little information, available on breeding densities. In one area, East Germany, maximum of 7 pairs on 25 km^2; in two others, 6 and 5–6 pairs on c. 50 km^2; in woodland with nest-boxes, up to 10 pairs on 25 km^2 (Dornbusch 1968a). Size and nature of territories in west Palearctic uncertain; only defence of nest-site and immediate vicinity perhaps involved. Birds will feed up to 2 km from nest (see Food). In South Africa, breeding pairs strongly territorial after leaving winter pre-breeding flock (see above), occupying distinct feeding territories; these not always near occupied nest (see Skead 1950). ROOSTING. Nocturnal and usually solitary, but little information. According to Fjerdingstad (1939), roosts in holes, though this seems exceptional—sites typically in trees, etc., close to trunk (Glutz and Bauer 1980). When tending young ♀ typically roosts in hole with them, ♂ elsewhere (e.g. Hirschfeld and Hirschfeld 1973); both members of one pair, however, roosted apart from young in large disused nest-box (Stirnemann 1948). Solitary roosting in banana trees reported from Egypt, birds clinging to trunk like woodpeckers (Picidae) at junction of down-pointing dead leaves 1–2 m from ground (Crowe and Brownlow 1951). In South African study, birds found to roost individually in same general area, one to a tree or shrub, members of same flock all leaving feeding area loosely together at sunset (Skead 1950). Sites quite exposed; some birds used same perch for weeks on end, even if disturbed, but others less conservative. Breeding ♀♀ took to tree-roosting again once young fledged, juveniles then using same trees and bushes they frequented during day.

2. Tame and confiding in some areas where not persecuted (e.g. Egypt, India), but often shy, wild, and elusive elsewhere (e.g. France)—particularly ♀♀ (Skead 1950). Erectile crest usually kept lowered while feeding, etc. (e.g. Witherby *et al.* 1938, Whistler 1941), but erected when alarmed (e.g. Smith 1887, Skead 1950); bird feeding close to family of Magpies *Pica pica* continually expanded and contracted crest (Slade 1978). Crest

invariably raised briefly on landing (see also, e.g., Greaves 1936, Glutz and Bauer 1980). Large and boldly patterned wings probably also have passive signal function in flight, when bird particularly conspicuous (Kipp 1961). If disturbed on ground, often merely flies up into tree, etc. (Whistler 1941). Will, however, freeze at times when man near. One did so on ground, squatting with neck held a little forward, remaining motionless for c. 12 min until flushed by dog (King 1980a). Another did so in tree, adopting upright posture on branch with back to source of danger but head turned to one side; surprisingly inconspicuous, and remained motionless thus for several minutes except for occasional slight turn of head; flew off when observer looked away briefly then froze in same manner when relocated, again for several minutes (Vinicombe 1975). 2 others alarmed by same observer also froze when approached closely (one on ground, other in tree), but in normal posture more or less front-on. On ground, sometimes adopts elaborate concealing posture closely similar to that assumed when sunning and said to be elicited by passing bird of prey (e.g. Smith 1887, Jourdain 1911) and at times by man (Jourdain 1911): bird lies spread-eagled with tail fanned and wings fully extended, so that tips almost meet in front, and head thrown back with bill pointing vertically up; disruptive pattern of plumage can make it almost invisible, especially on sandy or rocky ground. References to such behaviour go back to last century (see Münch 1952, Glutz and Bauer 1980), but modern confirmation needed. Shows remarkable ability in avoiding aerial pursuits of falcons *Falco* and other birds of prey, making quickly for cover if possible or evading attacks (see Glutz and Bauer 1980). See also below. FLOCK BEHAVIOUR. No information available for west Palearctic. In winter flock, South Africa (see part 1), birds fed together peacefully until pre-breeding period when hostilities broke out at times (see further, below). When flock disturbed and flew up, a few sometimes engaged in fluttering, moth-like flights: would dash erratically up to height of about 3 m, wobble at top of flight, then drop almost to ground, and swoop up again (Skead 1950). ANTAGONISTIC BEHAVIOUR. (1) General. In west Palearctic, most disputes assumed to be linked with territorial demarcation in spring (e.g. Hirschfeld and Hirschfeld 1973). Constant disputes reported, especially where several pairs close together (Münch 1952). In study in suburbs of Cairo (Egypt), where birds resident, Greaves (1936) found that some fighting occurred in autumn (second half October), when nest-sites re-visited and pairs started to form again, such activities ceasing when weather colder; at beginning of breeding season in spring, intense competition for sites resulted in some fighting—this fatal at times (one bird blinded in both eyes). (2) Pooping-display of ♂. Comprises well-known, far-carrying, monotonous Pooping-call (song: see 1 in Voice) typically given persistently from elevated song-post in ritualized manner, though accounts of posture and movement differ as following accounts show. Bird while singing bows head and inflates neck with crest lowered (Jourdain 1911). Each time song-phrase given, ♂ bows head, inflates neck, and rests bill against breast (giving false impression of tapping perch with bill), then raises head again (Hüe 1947b). Bird blows out throat, flattens crest, and lowers head—which throbs as each unit of song given; after slight pause, next phrase of song uttered (Skead 1950). ♂ nods, raises crest, and calls (Bäsecke 1951). ♂ adopts an erect posture then bows when calling with throat swollen, and feathers ruffled, tossing head to right and left and up and down, with whole body moving at times and tail whipped up and down; crest partly raised each time (Münch 1952). In South African study, Pooping-display first given when birds still in flock well in advance of breeding season; bouts of calling initially spasmodic and brief, becoming increasingly intense and prolonged. At first, song given from any elevated perch in vicinity of flock, occasionally on ground (see below); when singing bird returned to flock, sometimes initiated hostilities with second ♂. Later, after break-up of flock during establishment, defence, and advertisement of territory, given persistently from regular song-posts. In immediate pre-breeding period, especially, intensity of calling remarkable, birds singing for long periods at a time and engaging in song-duels (see 1 in Voice); amount of song depended on whether territorial neighbours present—if so, could continue until young fledged (Skead 1950). In Cairo, singing by ♂♂ occurred during brief autumn resurgence of reproductive behaviour (see above), then again in spring (Greaves 1936). In Europe, song starts more or less on spring arrival or soon after (Hüe 1947b; Guichard 1949–50; Hirschfeld and Hirschfeld 1973). Before pair-formation, ♂ often calls from dead top of tall tree, especially early morning; continues to call while ♀ incubating, often near nest, but singing much reduced after hatching of young, ceasing when they fledge (Bäsecke 1951). Peak of ♂ song in period up to completion of nest-building (Glutz and Bauer 1980). (3) Threat and chasing. Encounters between ♂♂ in winter pre-breeding flock, South Africa, mild at first, becoming increasingly intense (Skead 1950). Aggressor ♂—sometimes after giving Pooping-call from ground—would suddenly ruffle feathers, open wings, thrust bill slightly forward, and run towards another; chase then ensued on ground and, if pursued bird took wing, in flight. During such aerial chases, birds might fly back and forth several times or continue pursuit on foot in tree, encounter then petering out with participants returning to ground to feed again. A few weeks later, ♀♀ started similar period of aggression. Little information available from west Palearctic. When one ♂ in post-arrival flock of 4 started singing, others froze and listened for some time, not moving even when calling ceased (Hüe 1947b). According to Glutz and Bauer (1980), Threat-posture (Fig A) of nominate *epops* same as that recorded for *africana* by Steyn (1967): bird stands erect with crest raised, tail fanned, and wings raised and widely spread (see further under Parental Anti-predator Strategies, below). In disputes recorded by Münch (1952; see above), birds uttered harsh screeching (see 2 in Voice) with crest-raising and chased one another. (4) Fighting. In flock, South Africa, fights broke out eventually when 2nd ♂ resisted threat, etc., of aggressor (see above): the two would approach, raising crest and bobbing head, then grab bills and flutter up breast-to-breast 2–3 m in the air, break apart, and drop back to ground; sequence might be repeated 2–3 times then participants continued feeding (Skead 1950). ♀♀ would also fight thus breast-to-breast equally aggressively but without preliminary crest-raising and head-bobbing. Fighting also occurred later when flock disbanded, e.g. between ♂♂ on edges of territory—though usually disputes settled without combat, intruder retreating when chased away. If fight did ensue, birds would run towards each other, head-bobbing with flattened crest, then fight much as described above. In west Palearctic, such fighting between ♂♂ said to be common and sometimes severe, combatants leaping and fluttering in the

A

air like gamecocks *Gallus* (Jourdain 1911). According to Glutz and Bauer (1980), fighting follows from Threat-posture described above, accompanied at times by rasping calls (see 2 in Voice). Both members of pair made pecking attacks on stuffed conspecific bird with open bill, giving scolding, snoring sounds with bill wide open and crest raised; ♂ and ♀ sometimes attacked simultaneously, ♂ the more aggressive (Hirschfeld and Hirschfeld 1973). HETEROSEXUAL BEHAVIOUR. (1) General. In study, South Africa, pair-formation observed while birds still in flock (Skead 1950; see above). After initial period of threatening and chasing each other, ♂♂ started chasing ♀♀ as well—for up to 5 min at a time. ♂–♀ pursuits less aggressive than ♂–♂ ones, with ♂ making a rattling sound (see 5 in Voice) and also pausing to utter a few quiet Pooping-calls with throat puffed out. Later, while birds still flocking for much of day, Courtship-feeding began in immediate pre-breeding period (see further, below); this coincided with outbreak of aggression between ♀♀ (see above). Next, ♂♂ (and to lesser extent ♀♀) started searching for sites, resulting in break-up of flock and establishment of territories. In Egyptian garden study, birds thought to pair-up in autumn during temporary resurgence of breeding activity (Greaves 1936; see above). Where migratory, as in Europe, usually arrive singly in spring, ♂♂ apparently ahead of ♀♀ with subsequent pair-formation (Münch 1952; Hirschfeld and Hirschfeld 1973; Glutz and Bauer 1980). Much calling and chasing during courtship, ♂ raising crest and fanning tail, followed by territorial establishment site-seeking (Hirschfeld and Hirschfeld 1973). Sometimes 5–6 or more birds at a time may be involved in sexual chase early in cycle; nest-site sometimes chosen during period of pairing-up and courtship (Münch 1952) though usually after (see also Bäsecke 1951). When site selected, pair familiarize themselves with area for *c.* 2 days, then start nest-building and mating—this phase lasting 1–14 days (Kubík 1960). In captive pair studied by Löhrl (1977b), ♂ started singing in 1st week of April and first attempted Courtship-feeding at end of April, retreat by ♀ causing ♂ to chase her on ground and in flight—with increasing hostility until ♀ finally accepted offering (3 May), pursuits then becoming more and more ritualized and ♂ increasingly vocal. Next (4 May), ♂ started showing ♀ nest-box (Nest-showing: see below); Courtship-feeding now frequent, copulation starting on 9 May and continuing until end of egg-laying (see further, below). (2) Courtship-feeding. In flocking birds, South Africa, first occurred when couples temporarily fed apart from rest of flock, ♂ on finding food running over and presenting it to ♀; ♀ might refuse offering at first, running away from ♂ (Skead 1950). Courtship-feeding mainly non-ritualized except when performed as part of Mating-ceremony (see further, below); ♂ gives repeated Feeding-calls (see 3 in Voice) on approaching ♀ with item of food held in bill-tip, ♀ taking it or occasionally allowing ♂ to place it inside her open gape (Löhrl 1977b). At times, ♂ attracts ♀ by uttering Pooping-call repeatedly with food in bill: in incident observed by Hartley (1945), Egypt, mutual flight followed when ♀ appeared, then food passed over with much calling—♂ having tail spread widely; in incident observed by Hüe (1947b), France, ♂ called for *c.* 10 min before ♀ came, copulation following the feeding. Passing of food in flight recorded by Heldmann (1951: see 2b in Voice) but probably atypical. Later in cycle, ♀ fed regularly on nest by ♂ until day-brooding of young over—usually through nest-entrance (e.g. Martin 1974) though ♂ may call ♀ off nest (e.g. Kubík 1960). (3) Nest-showing by ♂. For active searching for sites by ♂ on ground, and associated calls, see Skead (1950); one ♂ attracted attention of ♀ to hole by moving excitedly in and out of it and walking round and round outside (see also Lint 1964). For pair inspecting tree-hole site, see Fig B. Unpaired ♂♂, South Africa,

B

also sought sites, with much use of Pooping-call. In captive pair, ♂ would approach ♀ without food in bill and call persistently (Short-cawing call; see 2b in Voice), then enter nest-box and give same call from inside, attracting ♀'s attention; ♀, now fed frequently by ♂, entered box same day (Löhrl 1977b). (4) Mating-ceremony. Link between Courtship-feeding and mating indicated by several observers (e.g. Hüe 1947b) but details available only from study of captive pair (Löhrl 1977b), as follows (see also Hüe 1947b, Glutz and Bauer 1980). Repeated Short-cawing calls from ♂ form invariable introduction as, with food in bill, he persistently attempts to approach ♀ from behind; at first, ♀ turns away, causing ♂ to circle. Series of Mock-feeds follows, ♂ offering food item repeatedly but withdrawing it each time—up to 50 times in early days, more than 100 later—as ♀ leans to one side with head held to same side and bill open. After series of Mock-feeds, birds pause, then copulation follows in next few minutes when ♀ solicits with front of body bowed low and rear elevated. ♂ mounts and copulates, with or without use of extended wings in balance, cloacal contact lasting up to 5 s. No post-copulatory display. Mating-ceremony usually occurs on ground, sometimes on large branch of tree. (5) Other behaviour. For details of behaviour leading to 2nd brood, see Löhrl (1977b). Tame hand-reared ♂ displayed to keeper, bowing with neck inflated, tail spread, and crest erected while making bill-clattering sound (Heinroth 1944). In incident, Egypt, early April, ♂ gave Pooping-calls from tree while ♀ fed below, then flew down behind ♀, slow running chase following for *c.* 30 m; ♂ more upright than usual, neck elongated and distended, tail fanned and depressed, and almost fully open wings flapping slowly; ♀ periodically opened and waved wings and erected crest; couple fed together peacefully afterwards (Simmons 1950). On many

C

occasions, March–April, often at dusk, same birds seen flying slowly round and round together at height of up to c. 12 m, one behind the other, with periodic opening of crest. When young in nest, ♀ may prevent ♂ from entering, raising crest and, with ruffled wing and body feathers, blocking entrance (Lint 1964). RELATIONS WITHIN FAMILY GROUP. Direct contact with nestlings mainly by ♀: though ♂ provides most of food in early stages, seldom enters hole, takes little or no part in brooding, and usually feeds young at entrance to nest (Fig C) or passes food to ♀ (see, e.g., Bussmann 1950 and Hirschfeld and Hirschfeld 1973 for further details). Young gape wide and grasp bill of parent, adult placing food well down gullet (Bussmann 1934); feeding done quickly, each nestling running forward, seizing parent's bill, and disappearing again (Bussmann 1950). Young of brood of 2 fed alternately (Bussmann 1934) and operation of 'feeding carousel'—as in Kingfisher *Alcedo atthis*—described by Münch (1952): brood adopts radial arrangement with one facing entrance, this one being fed next, another nestling taking its place, and so on as group performs gradual clockwise rotation. Circle formation in nest also noted by Hirschfeld and Hirschfeld (1973) and Löhrl (1977b). White gape-flanges of young provide strong signal effect in poor light of nest-hole (Dorning 1930b; Münch 1952). ANTI-PREDATOR RESPONSES OF YOUNG. Nestlings have several responses in face of danger threatening them in nest-hole. Able to raise crest as soon as it is partly grown, expanding it both laterally and vertically, e.g. when handled (Greaves 1936; Münch 1952)—presumably producing frightening effect against small mammals, etc., entering hole. Also: puff out feathers making themselves look large (Fjerdingstad 1939); hiss (see Voice) and gape (Stirnemann 1943); aim rapid upward dart of head while bill-clattering, and make lightning-fast wing movements against wall of cavity (Löhrl 1977b). Some broods docile when nest examined, others highly aggressive—hissing, and attacking with sharp lunges and pecks (Skead 1950). Main anti-predator response, however, combines: (a) exudation of evil-smelling fluid in droplets from temporarily enlarged and modified oil-gland, producing intolerable stink (like rotten meat or musk); (b) copious and forceful ejection of liquid faeces and contents of gut from cloaca—see (e.g.) Dorning (1930a, 1943), Bussmann (1934), Stirnemann (1943), Heinroth (1944), Sutter (1946), Münch (1952), Hirschfeld and Hirchfeld (1973), and Glutz and Bauer (1980). Secretion from gland first occurs at 4th day; gland reaches maximum size when young 12 days old, then special structure gradually reabsorbed. Ejection of material from cloaca first occurs at 6th day, continuing throughout nestling period; more pronounced in older young, though brood may habituate to repeated handling by observer—and (e.g.) try to hide (see Dorning 1932). Not directed at first, but aimed accurately later—often by several of brood at same time: young hiss with crest raised, lean forward on breast, elevate rear with tail cocked and expanded over back (Fig D), and squirt out faecal material repeatedly while simultaneously releasing fluid from rump-gland. Fanned out tail and erected under-tail coverts and feathers round cloaca form pattern of 3 rosettes (Skead 1950). Faeces may be discharged 25–30 cm (Ivanov 1969) or further—50–60 cm, usually more or less steeply upwards (Münch 1952); evacuation also occurs (as does secretion from gland) when young handled (Löhrl 1977b). PARENTAL ANTI-PREDATOR STRATEGIES. (1) Within nest-hole. When incubating and brooding, ♀ sits tight (e.g. Guichard 1949–50, Kubík 1960) and does not leave hole when disturbed. May remain on eggs or young—having to be removed at times by observer in order to examine nest contents (Ivanov 1969)—or move aside or to back of hole (Bäsecke 1951; Hirschfeld and Hirschfeld 1962; Löhrl 1977b); may hiss and raise crest (see also Puhlmann 1912). Some ♀♀ move neck from side to side, peck at entrance hole, and make sneezing sounds; one incubating bird had whole body plumage ruffled and crop swollen (Kubík 1960). When stoat *Mustela erminea* in mouth of nest-hole, sitting ♀ apparently remained motionless and silent (Yeates 1941b). During incubation and nestling period, foul-smelling secretion from rump-gland develops in ♀ as in young (Sutter 1946); may be emitted (e.g.) when nest examined (Kubík 1960). ♀ said also to discharge faecal material over 24–30 cm (Ivanov 1969). Reputation for deliberately keeping evil-smelling nest-hole by not practising nest-sanitation in order to discourage predators from entering is ill-deserved; ♀ regularly removes droppings if site permits, tossing them out through entrance (e.g. Puhlmann 1912, Bussmann 1934, Heinroth 1944, Sutter 1946); young defecate away from centre of nest and later, if possible, through entrance hole. Nest stinks usually only when young and/or ♀ have cause to discharge defence fluids, some smell remaining (see also Skead 1950). (2) Outside nest-hole. Accounts vary; no detailed studies. Remains nearby when man present, flying off as he moves away—may perhaps have previously buried eggs in nest-debris (Ivanov 1969). When man approaches nest containing young, may demonstrate vocally, screeching a lot, but more usually is extraordinarily shy, not coming anywhere near nest, even if young removed (Stein 1928). Nervous when people present: moves about, raising and lowering crest, flying past or around nest-tree, perching briefly but not entering nest, giving alarm calls (see 2 and 5 in Voice); if disturbance prolonged, finally disappears without feeding young; after finally going into nest, very cautious before emerging again (Jacobs 1943). Birds driving off Tree Sparrows *Passer montanus* near nest adopted posture with body erect, wings spread, and ran at them (Hirschfeld and Hirschfeld 1973). Adult *africana*, slightly alarmed near nest containing young, approached cautiously and several times raised crest fully, raised and spread wings, and spread tail in same distraction-threat display (see Steyn 1967).

(Fig A based on photograph in Hirschfeld and Hirschfeld 1973; Figs B–C from photographs in England 1969; Fig D after photograph in Herberg 1952.) KELS

D

Voice. Apart from well-known song of ♂ (call 1), most calls harsh and unmusical. Largely silent outside breeding season (Kipp 1961). In spite of extensive literature, full vocabulary of west Palearctic birds still imperfectly known, different accounts being difficult to reconcile. Only 4 adult calls described for nominate *epops* studied in captivity by Löhrl (1977b), but 8 by Skead (1950) for wild *africana* in South Africa. Young make noise by striking wings against sides of hole to intimidate intruder (Löhrl 1977b). Bill-clattering reported from tame bird displaying to keeper (Heinroth 1944) and from young in nest as anti-predator response (Witherby *et al.* 1938; Löhrl 1977b).

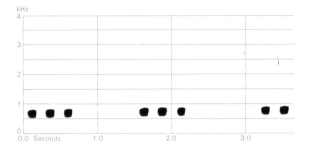

I P A D Hollom France June 1974

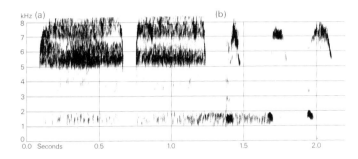

II P A D Hollom England July 1977

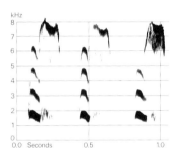 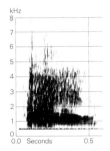

III P A D Hollom England July 1977   IV P A D Hollom England July 1977

CALLS OF ADULT. (1) Pooping-call of ♂. Advertising-call (song). A low but far-carrying, monotonous 'oo oo oo', often uttered frequently and persistently for many minutes on end in Pooping-display. Renderings include: a low, soft 'hoop-hoop-hoop' (Witherby *et al.* 1938, Europe); 'upupup' (Löhrl 1977b, Europe); 'poop poop poop' (Hartley 1945, Egypt); a loud, rather mellow 'hoot' or 'hud' (Whistler 1941, India). Call has a lovely purity of tone and sounds like noise made by blowing into bottle (P J Sellar). Quieter version given at beginning and end of seasonal calling period. Usually 2–4 notes (units) in a phrase but most commonly 3, with 2 next, rarely 1 or 5 (see also, e.g., Dorning 1930b, Bäsecke 1951, Münch 1952, Löhrl 1977b). According to Witherby (1928) and Guichard (1949–50), 2-unit calls predominate; 4- and 5-unit calls given when bird particularly excited (Hirschfeld and Hirschfeld 1973); single-unit call, usually long-drawn, typically uttered when ♂ perched near nest while ♀ incubating (Löhrl 1977b). Recording of bird giving series of 2- and 3-unit phrases shows definite temporal pattern of calling, with total of 15 phrases falling into 5 3-phrase sentences; pauses between phrases *c*. 1 s, between sentences *c*. 2 s (J Hall-Craggs). Sonagram (Fig I, from double-speed playback) illustrates phrases 10–12 (sentence 4) of this series. Rate of calling *c*. 24 calls per min, duration of each unit a fairly constant 0·4–0·5 s (Glutz and Bauer 1980). Our recording, however, indicates rate of *c*. 30 calls per min, with duration of each unit less than 0·2 s. For periodicity, see Social Pattern and Behaviour. Sometimes given while holding food in bill (Hartley 1945; Hüe 1947; Löhrl 1977b). Sometimes interspersed with other calls. (2) Cawing-calls. Given in fights between conspecific birds; also when disturbed by predators (Bussmann 1934; Bergmann and Helb 1982). (a) Long-cawing call: unmistakable, impure sound—long-drawn and harshly rasping 'schräa' (Löhrl 1977b; Bergmann and Helb 1982; see these sources for sonagrams); uttered when excited, e.g. during fights between rival and during courtship chases. For other renderings, see Jacobs (1943) and Münch (1952). (b) Short-cawing call: rasping in timbre, same as last, but call distinctive (no rendering given); uttered by ♂ when showing selected nest-hole to ♀, invariably during preliminaries of copulation, and sometimes (in rapid sequence) during copulation itself (Löhrl 1977b); 'chrä chrä' of bird, presumably ♂, passing food in flight (Heldmann 1951), perhaps the same. (3) Feeding-call. A sonorous, rolling call like that of mole-cricket *Gryllotalpa*; invariably uttered by ♂ when offering food to ♀ or young if food not accepted immediately, also (e.g.) persistently while pursuing ♀ with food in bill early in cycle while ♀ still unresponsive (Löhrl 1977b; see also for sonagram); see also Social Pattern and Behaviour. According to Glutz and Bauer (1980), uttered by both sexes when bringing food to nest. Adult coming to nest with food calls thus: quiet 'gru-gur-grugrugru' changing to high-pitched version when young answer (Bussmann 1934); low-pitched 'charrr' or, at higher intensity, 'charrr ... gurrr' (Stirnemann 1948); 'kirr' or 'kurr' sound by ♂ (Dorning 1930c); long-drawn 'tirr' or 'terre' (Reuver 1959). (4) Food-call of ♀ from nest a low chattering (Martin 1974) or a rather plaintive 'huut-huut' (Everitt 1974, *longirostris*). (5) Rattling-call. Quiet, rattling 'tr' sound, sometimes interspersed with Cawing-calls (Bergmann and Helb 1982), and therefore probably expressing excitement and alarm. Probably the same as sound like whirling rattle, given especially by ♂ chasing ♂♂ or ♀♀, in the air or on the ground, occasionally also in courtship-feeding (Skead 1950). (6) Huff-call. A snake-like 'huffing' sound made by ♀ (and young) when disturbed in nest-hole. A hissing sound reported by some observers (e.g. Smith 1887) probably the same: by ♀ (and young) in nest when disturbed, like other hole-nesting birds—e.g. tits (Paridae) and woodpeckers (Picidae) (Fjerdingstad 1939); by ♀ when taken out of nest—'pf pf' (Hirschfeld and Hirschfeld 1962); excitedly by ♂, with food in bill when following ♀ in flight and after landing (Hartley 1945).

CALLS OF YOUNG. Food-calls of tiny nestlings a faint mouse-like squeaking (Lint 1964). Calls louder later and given mainly only in response to calls of parents approaching with food, in contrast to almost incessant food-calls of certain Picidae (Hüe 1947; Münch 1952; Hirschfeld and Hirschfeld 1973). Variously described: a cheeping, piping sound followed by a soft bark at *c*. 18 days—rendered 'sisisisgö' at 21–22 days (Bussmann 1934); a clear, high-pitched 'sie-si-sie' (Jacobs 1943); a hissing, whistling, wheezy sound—'tssi tssi tssi' (Hüe 1947). Recording of older nestlings includes a peculiar hissing 'ssst-sisst' (Bäsecke 1951: Fig IIa) and a noisy hissing 'psihb' delivered repeatedly at variable intervals (Fig IIb); in recording (Fig III) of same young 4 days later, 'psihb' has developed much more tonal quality (J. Hall-Craggs). Call of fledged but still-dependent young described as a thin, whistling 'czirirri czrirri' (Dorning 1930b). Hissing (Fig IV) with predominantly low frequencies (compare Fig IIa) reported from nestlings by several observers as anti-predator response (see adult call 5), and often associated with ejection of faeces (see Social Pattern and Behaviour); described as cat-like (Bussmann 1934) and likened to hissing of young Wrynecks *Jynx torquilla* (Hirschfeld and Hirschfeld 1962). KELS

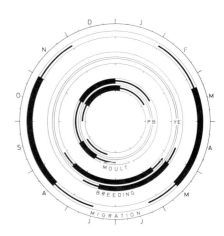

Breeding. SEASON. Central and southern Europe: see diagram. North Africa: laying from mid-February to end of May (Heim de Balsac and Mayaud 1962). Canary Islands: laying in January (Bannerman 1963b). SITE. In hole in tree, building, or ground; will use nest-box. Of 31 nests, Hungary: 25 in trees, 3 in banks, 1 in nest-box. Height above ground of 25 tree-nests: 5 under 1 m, 8 1–2 m, 9 2–3 m, 2 3–5, 1 12 m (Kubík 1960). May be unlined or have grass, leaves, moss or pine needles brought to form lining (Bäsecke 1951; Herberg 1952). Building: ♂ and ♀ clear out cavity and make scrape; may hack away rotten wood (Hirschfeld and Hirschfeld 1973; Löhrl 1977b). EGGS. See Plate 98. Elliptical, smooth and not glossy, marked with conspicuous pores; grey, pale yellow, or olive, sometimes pale green or brown, becoming very stained in nest. 26×18 mm ($23-29 \times 16-19$), $n=124$; average weight 4·45 g, $n=15$ (Kubík 1960). Clutch: 7–8 (4–10, rarely –12). Of 59 clutches, Hungary: 5 eggs, 8; 6, 11; 7, 20; 8, 16; 9, 2; 10, 2; average 7·0 (Kubík 1960; Makatsch 1976). Normally 1 brood, sometimes 2 in south of range. Replacements laid after egg loss; in one case, after 12 days (Labitte 1954a). Laying interval 1 day (Bussmann 1950), but can be 2 days in *africana* (Skead 1950). INCUBATION. 15–16 (14–20) days. By ♀ only, starting with 1st egg; hatching asynchronous, at same interval as laying (Bussmann 1950), though Skead (1950) found near-synchronous hatching in *africana*. YOUNG. Altricial and nidicolous. Cared for and fed by ♀ with food brought by ♂; later both feed. Brooded continuously while small. FLEDGING TO MATURITY. Fledging period 26–29 days. Fed by parents for some while after fledging. Age of first breeding probably 1 year. BREEDING SUCCESS. Of 150 eggs laid in 21 nests, 80% hatched and 52·6% fledged, with average 5·7 young hatching from average clutch of 7·1, and average 3·4 young fledging per nest (Kubík 1960).

Plumages (nominate *epops*). ADULT MALE. Forehead and crest bright rufous-cinnamon; shorter feathers of forepart of crest with small black tips, longer crest-feathers with black tips 6–12 mm long, longest subterminally bordered by white or cream band (usually broken by cinnamon along shaft). Sides of head and neck and hindneck pink-cinnamon, grading into greyish-cinnamon, cinnamon-drab, or pale grey-brown on mantle. Scapulars pink-buff with broad oblique subterminal band and base black. Lower mantle and back with black, cream, and drab bands, rump white; shorter upper tail-coverts white with broad black subterminal band, longer black. Chin to breast deep vinaceous-pink or vinaceous-cinnamon, grading to vinaceous-pink on flanks and sides of belly, to white on central belly; sides of belly with a few rather ill-defined black streaks. Vent and under tail-coverts cream-pink or white. Tail black with broad white band across; on central pair (t1), white band *c.* 12 mm wide and 40–45(–55) mm of tip black; band gradually closer to tip and slightly wider towards outer feathers; *c.* 15 mm wide with 25–35 mm of tip black on inner web of t5 (where black basal and distal borders of tail-band approximately parallel), widening on outer web (where white extends along outer edge, reaching to 10–20 mm from tip). Rarely, a 2nd white band close to tail-base. Primaries black, a broad white band (15–30 mm wide) across (p2–)p4–p9 (*c.* 3–4 cm from wing-tip), reduced to white subterminal spot on inner web of p10 and on p1–p2(–p3). Secondaries black with 5 broad white bands across; bands reduced to broken bars or spots on s2–s3, s1 fully black except for small white subterminal spot on inner web (continued in white band across primaries); white bands often partly tinged cream or with cinnamon edges at border with black, especially on inner secondaries; basal 1–2 bands on secondaries hidden under median and greater upper wing-coverts which show similar broad black and white or cream bands. Rather narrow strip of lesser upper wing-coverts rufous-cinnamon. Upper primary coverts and bastard wing black; lesser primary coverts along leading edge of wing white. Tertials dull black shading to dark brown terminally; broad streak over inner web and basal outer web and broad fringe of outer web and tip pale cinnamon-rufous or cream-buff. Under wing-coverts and axillaries pale pink-cinnamon or pink-white; under primary coverts white, shorter with narrow black tips, greater largely black. Plumage strongly influenced by bleaching and wear. In worn plumage, forehead and crown paler buff-cinnamon or yellow-cinnamon, rather similar in colour to sides of head and upper mantle (not distinctly deeper rufous); longer crest feathers with pale cream-buff subterminal band. Sides of head and neck and hindneck cinnamon-buff, less pinkish; mantle duller and greyer; pale bands on lower mantle, scapulars, back, tertials, inner secondaries, and upper wing-coverts bleached to cream-white, less cinnamon. Chin to chest pale greyish-buff, much less pinkish than in fresh plumage; belly cream-buff, vent, flanks, and under tail-coverts off-white, black streaks on sides of belly more distinct than in fresh plumage. Tips of tail-feathers, flight-feathers, and tertials abraded to greyish-black or greyish-brown. ADULT FEMALE. Similar to adult ♂, but slightly duller; forehead and crest slightly less bright cinnamon-rufous; mantle and lesser upper wing-coverts duller greyish-drab, less cinnamon; chin partially tinged white; sides of head, chest, and breast duller cinnamon-rufous with hardly any pink tinge, except on chin and throat, where less deep vinaceous-pink; belly and vent pale buff to dirty white; more extensive black streaks on sides of belly and breast. In worn plumage, mantle and chest duller grey, remaining underparts still more extensively streaked dusky. Adult ♂ in worn plumage closely similar to adult ♀ in rather fresh plumage, and sexing considered impossible unless known pair seen together and ♂ known to be fully adult, not 1st adult. NESTLING. Covered with long and fluffy white down, close on crown and upper back, rather scanty on lower back and underparts. After *c.* 1 week, closed feather-sheaths appear, first on wing and tail, and eye opens; during 2nd week, closed sheaths cover all head and body (except belly), giving spiny appearance with down clinging to sheath-tips; from *c.* 15 days, feathers break through tips and down gradually lost. Appears largely feathered except for belly from *c.* 18 days, fully feathered from 22–23 days. (Bussmann 1950; Münch 1952; Hirschfeld and Hirschfeld 1973; Löhrl 1977*b*.) JUVENILE. Closely similar to adult ♀ at fledging, when colour still fairly bright and head, neck, and chest rather bright cinnamon-rufous (slightly tinged with vinaceous-pink on throat and upper chest in ♂), but colours rapidly fading through greyish-cinnamon to pale buff-grey and drab, only forehead and crown retaining rufous-cinnamon. Worn plumage often duller still than worn adult ♀, and belly and breast more streaky, but often indistinguishable from worn adult, except for showing new flight-feathers and tail when adult has these worn or in moult. Also, tail-feathers often narrower than those of adult and pattern of white tail-band different, especially on t5: distal and basal border of white band on inner web of t5 less parallel and less sharply defined, sometimes forming rounded patch not quite reaching feather-edge; border between black and white on outer web (or both webs) mottled or freckled black and white, sometimes forming extensive mottled patches; outer web of t5 often partly fringed white, fringe sometimes extending to tip of t5. FIRST ADULT. Similar to adult, but 1st adult ♂ occasionally slightly duller than full adult ♂, more similar to adult ♀. Part or all of juvenile tail frequently retained, feathers narrower and more worn than in adult at same time of year; in about half of birds, juvenile t5 with characteristic partial mottling still present. Flight-feathers juvenile until 1st complete post-breeding moult in 2nd autumn and winter; tips slightly worn in 1st winter (adult moulting or new), distinctly worn in summer (adult usually still with smoothly edged tips of longer primaries).

Bare parts. ADULT AND JUVENILE. Iris brown. Bill black or horn-black with bluish-slate base of upper mandible and middle portion of lower, and flesh or pale greyish-flesh base of lower. Leg and foot pale grey, slate-grey, brown-grey, or greyish-black, occasionally with slight pink tinge; soles slightly paler grey or greyish-flesh, claws, dark horn-grey or black. NESTLING. Iris bluish-black; brown on fledging. Bill blue-grey with large swollen white, yellow-white, or pink-white flanges at gape (largest on lower mandible); flanges half or more of bill length up to 10th day, size decreasing later on, lost at *c.* 3 weeks. Bill darkens to black from *c.* 1 week. Bare skin on lores and round eye blue-grey; mouth bright red. Leg and foot greyish-pink or grey. (Bussmann 1950; Münch 1952; RMNH, ZMA.)

Moults. Based on Stresemann and Stresemann (1966) and specimens from BMNH, RMNH, and ZFMK. See also Ginn and Melville (1983). ADULT POST-BREEDING. Complete; primaries descendant. Starts in breeding area from July or August; face, crest, cheeks, mantle, throat, and chest first, occasionally followed by a few inner primaries or some tail-feathers; moult suspended during migration. Moult of wing and tail usually starts in winter quarters; moult of body continued there (or started if not started in breeding area). Primaries: p1 shed early September to mid-November, all moult completed with p10 late December to late February. Tail starts at same time as primaries with t1; further sequence apparently irregular, t5 often last. Secondaries start ascendantly and descendantly from s7 (longest tertial) at loss of p1; a 2nd centre starts ascendantly from s1 at about loss of p5; s4 moulted last, usually slightly before completion of p9–p10. Not known whether any pre-breeding moult occurs; some birds in spring have head (including crest) and chin to breast markedly newer than remainder of body, though these parts moulted first in post-breeding and thus expected to be older: perhaps a partial pre-breeding moult in January–February, involving head, neck, and much of underparts in ♂, only limited part of head and chest in ♀. POST-JUVENILE. Partial: head and body, occasionally also all or part of tail and tertials; juvenile flight-feathers retained. Usually starts in winter quarters, but advanced birds replace much of head, neck, and chest and some other feathers from August, and these birds perhaps have partial pre-breeding in spring.

Measurements. Nominate *epops*. Europe, north-west Africa, and south-central USSR, February–November; skins (RMNH, ZFMK, ZMA). Juvenile wing includes 1st adult; bill (N) to distal corner of nostril (adult and 1st adult only); exposed culmen 4·2 (38) 3–6 mm more, to forehead (skull) 10·2 (48) 8–13 more.

WING AD	♂	151	(2·20; 23)	147–153	♀ 146	(3·21; 12) 142–151
JUV		146	(3·42; 34)	141–152	142	(3·85; 34) 135–149
TAIL AD		103	(2·42; 14)	99–107	101	(4·61; 10) 95–107
JUV		100	(3·83; 18)	94–105	97	(3·22; 25) 91–102
BILL (N)		50·2	(3·32; 46)	45–57	45·1	(2·66; 39) 41–49
TARSUS		23·3	(1·01; 16)	22–25	21·7	(1·26; 13) 20–23
TOE		22·1	(1·26; 14)	21–24	21·4	(1·08; 13) 20–23

Sex differences significant, except adult tail and toe. Juvenile wing and tail significantly shorter than adult; juvenile tarsus and toe similar to adult from fledging, juvenile bill from about midwinter. No geographical variation in wing length; variation in bill slight, largely attributable to different months of collection in different areas, as bill tends to wear down during nesting. Bill (N) of adult ♂ (including some winter birds from northern Afrotropics): January–February 51·6 (3·27; 13) 46–58; March 50·9 (2·53; 21) 46–54; April 51·8 (3·48; 12) 46–57; May–June 49·2 (2·50; 10) 46–55; July–August 48·9 (2·83; 10) 45–54; September–November 49·7 (3·06; 7) 46–54. Bill (N) of fledged juvenile ♂: July 33·9 (3) 33–37; August 43·9 (4·05; 9) 36–48; September 47·3 (1·79; 5) 45–50; November 47·8 (3) 46–49. Trend in ♀♀ similar, but average and range 5 mm less.

U. e. major. ADULT. February–May. Egypt and Ennedi (Chad): wing, ♂ 151 (4·19; 9) 145–156, ♀ 148 (1); bill (N), ♂ 57·1 (1·80; 7) 54–60, ♀ 52·3 (1). Northern Sudan: wing, ♂ 147 (3) 142–152, ♀♀ 138, 145; bill (N), ♂ 53·4 (3) 48–57, ♀♀ 46·3, 53·2 (BMNH, ZFMK).

U. e. senegalensis. ADULT. Sénégambia, Mali, Aïr (Niger), and Lake Chad area, all year (BMNH).

WING AD	♂	137	(2·50; 10)	134–141	♀ 131	(1·67; 5) 130–134
BILL (N)		44·6	(2·24; 10)	42–48	40·6	(1·70; 5) 39–43

Single adult ♀, Ahaggar (Algeria), May: wing 127, bill (N) 45·6 (ZMA).

Weights. Nominate *epops* (including some *saturata*). Eurasia south to North Africa, Afghanistan, and Nepal, adult, combined: (1) March–April; (2) May; (3) June–July; (4) August–September; (5) October–November (Paludan 1938, 1959; Makatsch 1950; Münch 1952; Niethammer 1955; Piechocki 1958, 1968c; Schüz 1959; Diesselhorst 1968; Dolgushin *et al.* 1970; Kumerloeve 1970; Rokitansky and Schifter 1971; RMNH, ZMA). (6) Czechoslovakia and Belorussiya (USSR), adult, summer (Fedyushin and Dolbik 1967; Glutz and Bauer 1980).

(1)	♂ 70·7	(8·29; 10)	51–80	♀ 63·4	(5·09; 7)	56–72
(2)	66·9	(8·40; 9)	47–74	65·5	(5·75; 6)	57–72
(3)	66·9	(6·94; 8)	57–80	60·3	(—; 3)	57–63
(4)	69·2	(9·32; 9)	57–87	64·2	(7·19; 6)	55–75
(5)	80·0	(—; 1)	—	68·5	(—; 2)	68–69
(6)	69·0	(—; 14)	55–80	64·5	(—; 9)	57–80

Minimum weight of exhausted birds: adult 40 (ZMA), juvenile 38 (Glutz and Bauer 1980). France, April: 68·6 (9·42; 8) 59–83 (R L Swann, BTO). South-east Morocco, (March–)April: 59·8 (7·52; 35) 46–80 (Ash 1969b; BTO). Kuwait: (February–)March 63·0 (7·64; 14) 52–78; April 57·1 (5·25; 5) 50–62; August 55·8 (12·5; 6) 45–80; September(–October) 61·2 (8·44; 15) 49–82 (V A D Sales, BTO).

Camargue (France), migrants and breeders, combined (Glutz and Bauer 1980, which see for details):

MAR	63·8 (10·2; 15) 41–83	JUN	63·3 (5·0; 28) 52–72
APR	61·4 (8·1; 75) 41–83	JUL	61·8 (6·5; 15) 3–72
MAY	64·8 (9·1; 51) 45–97	AUG	67·6 (10·5; 25) 53–100

At hatching 2·6–3·8, increasing to c. 76 at 16–18 days, decreasing to 60–70 at fledging on c. 26th day (Heinroth and Heinroth 1924–6; Bussmann 1950; Münch 1952). Camargue juveniles similar to adult from July; both combined for July and August in table above.

U. e. major. Ennedi (Chad), April: ♂ 77 (Niethammer 1955; ZFMK).

U. e. senegalensis. Aïr (Niger), February: ♂ 64 (Niethammer 1955).

U. e. africana. Kenya and Zaïre: ♂ 55·2 (2·59; 5) 52–59, ♀ 52·0 (3) 46–59 (Verheyen 1953; Britton 1970; Skead 1974).

Structure. Wing short and broad, tip broadly rounded. 10 primaries: p7 longest, p10 60–70 shorter, p9 16–22, p8 1–4, p6 0–2, p5 2–3, p4 9–15, p3 18–25, p2 24–31, p1 28–36. Inner web of p6–p9 and outer of p5–p8 emarginated; p10 reduced, rather narrow with rounded or slightly pointed tip. Tail rather long, straight or slightly forked at tip; 10 feathers, each slightly widening towards broadly rounded tip, t1 2–8 mm shorter than t4 or t5. Bill long, slender, c. 2 times head length; gradually decurved and tapering towards sharp tip; base narrow, laterally compressed; both mandibles triangular in cross-section. Bill of juvenile at fledging short and almost straight; that of nestling very short and broad with large swollen flanges at base of both mandibles. Nostrils small, elliptical; situated just in front of loral feathering. Large erectile crest on crown, consisting of 2 rows of broad feathers up to c. 5 cm long, also in juvenile. Tarsus short and slender; unfeathered, covered with scutes in front and behind. Toes short and slender, soles slightly flattened; outer and middle toe joined at extreme base. Claws of front toes short, that of hind toe about twice as long. Outer toe with claw c. 80% of middle with claw, inner c. 75%, hind c. 95%.

Geographical variation. Marked. Involves size, depth of rufous of forebody and underparts, and pattern of black and white on wing and tail. In general, size smaller and colour deeper in tropical races. Populations of nominate *epops* from north-west Africa and Europe east to south-central USSR and north-west India rather uniform in size and colour, but those of south-central USSR and Sinkiang (China) average slightly larger than those of neighbouring north-west India and Middle East— former have wing of adult ♂ 153 (7) 148–158. East Siberian race *saturata*, grading into nominate *epops* east from Yenisey river and Tibet, similar in size to west Siberian birds, but mantle slightly greyer and throat to chest less pinkish (Vaurie 1959e). *U. e. ceylonensis* from India and Ceylon smaller than nominate *epops* (with which it intergrades in Himalayan foothills and plains of northern India), wing mainly 120–140; head, neck, mantle, and underparts deeper and richer rufous-cinnamon, hardly vinaceous; no white subterminally on longest crest-feathers; wing- and tail-pattern as in nominate *epops*. *U. e. longirostris* from south-east Asia (east from Bangladesh, south from southern China) similar to *ceylonensis*, but chin and chest more pink-vinaceous, less deep rufous, wing mainly 130–150, and white bands across secondaries narrower than black ones (rather than converse), wing appearing blacker. *U. e. major* from Nile delta (Egypt) south to northern part of Sudanese Nile valley and perhaps in Ennedi mountains (Chad) similar to nominate *epops*, but colours slightly duller, belly more streaked, tail-band narrower, and less white in secondaries; size larger, especially bill (length to nostril mainly over 55 in ♂, 50 in ♀; nominate *epops* usually below this), and bill deeper at base and middle, depth at nostril $7\frac{1}{2}$–9 mm ($6\frac{1}{2}$–8 in nominate *epops*), appearing slightly straighter and less gradually attenuated. *U. e. senegalensis* from belt south of Sahara (Sénégal to Somalia) distinctly smaller than nominate *epops* (see Measurements); about similar in colour to nominate *epops*, but head deeper rufous and less subterminal white on crest feathers; wing- and tail-pattern similar, but adult ♂ often with reduced amount of black on basal halves of secondaries, showing large white patches, and white bars on tips of secondaries partly reduced. *U. e. africana* from eastern and southern Africa much more deeply coloured rufous than *senegalensis*, especially adult ♂; basal half or $\frac{2}{3}$ of secondaries fully white in ♂, largely white in ♀ and juvenile; no white band across primaries; white tail-band closer to tail-base; size slightly larger. Sometimes treated as separate species (e.g. Mackworth-Praed and Grant 1952), but populations with colour, size, and secondary pattern intermediate between *africana* and *senegalensis* occur from Cameroun to northern Kenya and are often separated from *senegalensis* as *waibeli* (Vaurie 1959e); see also Lawson (1962), Cunningham-van Someren (1977), and Snow (1978). A paler cinnamon race *minor* in southern and eastern South Africa separable according to Lawson (1962) and Clancey (1965). Malagasy *marginata* similar to nominate *epops*, but white tail-bar narrower and nearer tail-base, and p10 relatively longer (reaching white on p9); size large.   CSR

Order PICIFORMES

Very small to large perching or climbing birds. Distribution worldwide except Australasia, with greatest numbers and diversity in tropics. 6 families recognized by Wetmore (1960), 1 occurring in west Palearctic: (1) Galbulidae (jacamars, 17 species; tropical America); (2) Bucconidae (puffbirds, 34 species; tropical America); (3) Capitonidae (barbets, 81 species; Old and New World tropics); (4) Indicatoridae (honeyguides, 14 species; Africa and southern Asia); (5) Ramphastidae (toucans, 33 species; tropical America); (6) Picidae. Order possibly a polyphyletic one; recent reviews of phylogeny and relationships have examined egg-white proteins (Sibley and Ahlquist 1972), hind-limb anatomy (Swierczewski and Raikow 1981), osteology (Simpson and Cracraft 1981), and feeding apparatus (P J K Burton). Wetmore (1960) recognized 2 suborders: Galbulae (families 1–5) and Pici (family 6). He further divided Galbulae into 3 superfamilies: Galbuloidea (families 1–2), Capitonoidea (families 3–4), and Ramphastoidea (family 5). Close relationship of families 1–2 (Galbuloidea) has been generally accepted since work of Steinbacher (1937), but recent work, noted above, views them as more distantly related to rest of order than implied by Wetmore's (1960) classification. Moreover, families 3–6 currently regarded as a good natural unit, in which Ramphastidae are most closely related to Capitonidae, and Picidae to Indicatoridae.

A relationship between Galbuloidea and Alcedinidae (of Coraciiformes) suggested by Sibley and Ahlquist (1972), but more recent studies do not support this. However, Galbuloidea share many derived features of feeding apparatus anatomy with families 5–7 of Coraciiformes (P J K Burton). Of all non-passerine orders, Piciformes appear closest to passerines, but at present no strong evidence linking them to any particular passerine group.

Principal morphological features uniting the 6 families are zygodactyl foot ('outer' toe reversed, usually permanently) and unique type of hind-limb osteology and myology. In particular, families 3–6 share important features of hind limb anatomy (Simpson and Cracraft 1981; Swierczewski and Raikow 1981) as well as that of feeding apparatus, origins of specialized tongue and hyoid of Picidae being discernible in other families, especially Indicatoridae (P J K Burton).

Family PICIDAE wrynecks, woodpeckers, and allies

Very small to large perching or climbing birds with zygodactyl feet and highly extensible tongues. 3 subfamilies (2 represented in west Palearctic): (1) Jynginae; (2) Picumninae (piculets, *c.* 29 species; South America, central Africa, and south-east Asia); (3) Picinae.

Necks rather long, 14 cervical vertebrae. Wings

rounded, flight usually strongly undulating. 10 primaries, p10 much reduced in adults, larger in juvenile plumage; wing eutaxic. 12 tail-feathers, modified as climbing support in Picinae; t6 reduced. Oil-gland usually tufted, occasionally naked. Feathers with well-developed aftershaft. Bills quite short to rather long, pointed or chisel-tipped; modified for wood excavation. Tongue proper greatly reduced, but carried on enormously elongated hyoid skeleton, allowing protrusion to great distances. Salivary glands very highly developed. Tarsus short. Feet zygodactyl; hallux absent in some genera. No caeca. A single, left, carotid.

Plumages variable, but often pied or greenish. Red crests and other striking head markings frequent in Picumninae and Picinae, and sexual dimorphism in these families commonly involves differences of head pattern. Adult post-breeding moult complete, starting directly after nesting; no pre-breeding moult (except for partial moult in Jynginae). Post-juvenile moult almost complete, usually excepting secondaries, greater coverts, and greater upper primary coverts; starts before fledging or shortly after, completed within a few months (except *Sphyrapicus* of Picinae). Nestlings naked, with well developed rough pad at rear of tibio-tarsal joint.

Subfamily JYNGINAE wrynecks

Medium-small perching birds with short bills, rather long, soft-feathered tails, and cryptic brown plumage. Single genus, *Jynx*, with 2 species: Wryneck *J. torquilla*, widespread in Palearctic and Oriental regions, and Red-breasted Wryneck *J. ruficollis* in Afrotropics.

For general features, moults, etc., see Picidae. Tail-feathers not pointed or stiffened, outermost pair (t6) very small, but positioned normally. Bills rather short, sharply pointed, considerably less hard and robust than those of Picinae. Nostrils dorsally situated, not covered by stiff feathers. Skull very lightly constructed and lacking hammering adaptations of Picinae. Tongue not barbed, but hyoid skeleton greatly elongated, horns extending partway down neck, before curving up and over skull. Inner front and rear toes much shorter and thinner than outer front and rear.

Jynx torquilla Wryneck

PLATES 76 and 78
[facing pages 806 and 830]

Du. Draaihals   Fr. Torcol fourmilier   Ge. Wendehals
Ru. Вертишейка   Sp. Torcecuello   Sw. Göktyta

Jynx Torquilla Linnaeus, 1758

Polytypic. Nominate *torquilla* Linnaeus, 1758, Europe east to Urals, south to Pyrénées, Alps, Yugoslavia (except western Slovenija to northern Dalmatia), Bulgaria, Caucasus area, and (perhaps this race) northern Portugal, Spain, and Balearic Islands; *sarudnyi* Loudon, 1912, western Siberia east to Yenisey, intergrading with nominate *torquilla* in southern Urals, migrating through Middle East; *tschusii* Kleinschmidt, 1907, Italy, Corsica, and coastal Yugoslavia south to northern Dalmatia, and (perhaps this race) Greece; *mauretanica* Rothschild, 1909, north-west Africa. Extralimital: 3–5 further races in Asia.

Field characters. 16–17 cm, of which tail 4·5 to 6 cm; wing-span 25–27 cm. Size and shape close to that of larger chats (e.g. Nightingale *Luscinia megarhynchos*) and even largest warblers (e.g. Barred Warbler *Sylvia nisoria*). Aberrant, mainly migratory relative of woodpeckers, with initially confusing but subsequently unmistakable character, and plumage pattern and colours recalling nightjar *Caprimulgus*. Passerine resemblance strongest in long-tailed silhouette, rather ordinary flight action, and frequent perching and feeding on ground. Sexes similar; no seasonal variation. Juvenile inseparable.

ADULT. At distance appears as mottled grey and buff-brown bird with no striking characters. At close range, exhibits complex plumage pattern strongly reminiscent of nightjar *Caprimulgus* with the following distinctive features: (1) grey crown and back, with dark bars on sides of crown, dark brown centre to nape, neck, and mantle, and dark centres to pale-edged scapulars (this pattern most obvious from behind, with grey panels alongside mantle and up sides of neck appearing as braces); (2) grey lower back, rump, and tail, with short bars on lower back and rump and 3 brown to almost black bands across tail;

(3) buff-brown wings, closely barred rusty, pale ochre, and dark brown on flight-feathers and loosely dark-brown on tertials; (4) pale yellow-white or buff-white underparts, with profuse, short dark bars from throat to chest and dark chevron spots on rest of underparts. Sides of chest often rufous so that bird looks dark-chested. Underwing appears dusky brown. Face marked by dark panel running from eye down into side of neck. Bill and legs pale brown. Eye bright hazel. JUVENILE. Grey ground-colour of upperparts paler and bars on underparts less prominent and less dark.

With experience, unmistakable, but time needed to appreciate that *J. torquilla* frequently gives impression of bulky, lengthy passerine. Confusion stems most from indistinct plumage pattern and colour, and rather tired, hesitant, and undulating flight. Flight may suggest several species (e.g. small thrush *Turdus*) but all doubt removed by clear sight of strange, sinuous character of bird at rest. On flat surface, hops jerkily, often with long tail raised. Perching and climbing behaviour contain both passerine- and woodpecker-like actions, with fully vertical, tail-supported clinging much less common than horizontal or oblique attitudes. Does not bore holes, and often feeds on ground. Combination of sinuous head and neck movements, crest erection, and jerky gait creates distinctly reptilian appearance, further compounded by flicking tongue.

Commonest call, 'quee-quee- . . .', is markedly reminiscent of small falcon *Falco* and Lesser Spotted Woodpecker *Dendrocopos minor* but more fluty than latter and given in longer sequence. Migrants silent.

Habitat. Breeds in west Palearctic from boreal subarctic through temperate to Mediterranean zones, strongly favouring continental rather than oceanic climates but avoids true steppe, desert, mountains, and wetlands. A lowland bird, but in Switzerland a few breed in favourable valleys above 1000 m (Glutz von Blotzheim 1962). Has bred in Kashmir above 3000 m (Bates and Lowther 1952). Does not favour dense or tall forest, preferring fringes, open woodlands, clearings, or, especially, parks, orchards, cemeteries, large gardens (even in towns), avenues, riverside trees, and heaths with colonizing pines. Prefers deciduous to coniferous trees, and is less interested in trunks than in branches, often fairly close to ground. Importance of ants (Formicidae) in diet leads to frequent occurrence on warm dry ground, either bare or with short herbage; presence of such foraging areas as well as of suitable nest-holes (which it is unable to excavate) is critical for choice of breeding habitat. In parts of range freely uses nest-boxes or old woodpecker holes. In USSR, also uses natural hollows and cavities, burrows in clay banks and even hollows in wooden buildings. Readily becomes accustomed to human presence when other conditions suitable.

On migration, occurs in variety of strange habitats with little or no tree cover, even in deserts and low scrub, while wintering birds even found in broad-leaved or thorn scrub, semi-desert, and cultivation. Flight normally in lower airspace. (Niethammer 1938; Dementiev and Gladkov 1951a; Bannerman 1955; Glutz von Blotzheim 1962.)

Distribution. In Britain, major decline from early 19th century in England and Wales leading to near extinction, though somewhat offset by colonization of Scotland in small numbers since 1969. Elsewhere, smaller changes; in 20th century, mostly decreased in range, especially in north-west Europe, but some northward expansion in Norway, and spread in Denmark.

BRITAIN. England: in early 19th century, bred in all counties of England and Wales except Cornwall and Northumberland; from 1830s decline began in north and became marked in most areas in second half of 19th century; by 1956 largely restricted to south-east England, and now almost extinct. Scotland: first bred 1969. (Monk 1963; Parslow 1967; Peal 1968; Sharrock 1976.) See maps of range decline in Monk (1963) and Sharrock (1976); current distribution not mapped. FRANCE. Range decreased in north and north-west since 1936 and perhaps since 1900 (Peal 1968; Yeatman 1976). SPAIN. Bred Mallorca 1977 (AN). BELGIUM. Range decreased (Lippens and Wille 1972; Glutz and Bauer 1980). NETHERLANDS. Bred also in some coastal areas in early 20th century (Peal 1968). WEST GERMANY. See Glutz and Bauer (1980), for local range changes and irregular breeding; also Peal (1968). DENMARK. Spread in Jutland since *c.* 1960 (TD). NORWAY. Some expansion in north in 20th century (Peal 1968); bred Troms 1976 (GL, VR). TURKEY. Probable breeding area mapped, but only once proved to nest (MB, RFP).

Accidental. Iceland, Faeroes, Ireland, Chad, Madeira.

Population. Information limited. Most marked decline in Britain from early 19th century (see also Distribution). Elsewhere in Europe, suggested by Glutz and Bauer (1980) that declines had occurred, on much smaller scale, from late 19th century and early 20th century, with further declines in 1930s, and then recovery to about 1945–52 — but data only scanty. Declines since in Switzerland, Czechoslovakia, and Estonia, and locally elsewhere. Climatic change suggested as major cause (see Glutz and Bauer 1980), but no detailed analyses, and in Britain, where most marked decline, not considered proven (Peal 1968); other possible factors include, more recently, pesticides and habitat changes.

BRITAIN. England: marked decrease, beginning *c.* 1830, now almost extinct (see Distribution); estimated 150–400 pairs 1954, 100–200 pairs 1958, 26–54 pairs 1964, 0–5 sites 1973–80. 2–22 sites 1973–81, maximum in 1978, mainly in Scotland. No confirmed breeding anywhere 1981, though birds present at some sites. (Monk 1963; Peal 1968; Sharrock 1976; Sharrock *et al.* 1982, 1983.) FRANCE. 10000–100000 pairs (Yeatman 1976). PORTUGAL. Decline since *c.* 1928 (Peal 1968). BELGIUM. Declined from *c.* 1920, continuing decrease from *c.* 175 pairs 1951 to *c.* 35 pairs

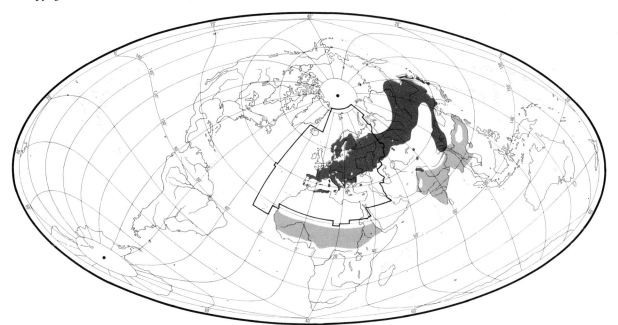

(Peal 1968; Lippens and Wille 1972); now less than 5 pairs (PD). LUXEMBOURG. Scarce (Glutz and Bauer 1980), perhaps 500–800 pairs (Peal 1968). NETHERLANDS. Marked annual fluctuations; probably declined, 125–250 pairs 1977 (Teixeira 1979). WEST GERMANY. Perhaps c. 1200 pairs (Rheinwald 1982); fluctuating, with declines in Niedersachsen, Nordrhein-Westfalen, Rheinland-Pfalz, Saarland, and Baden-Württemberg (Bauer and Thielcke 1982); near Braunschweig, downward trend, with marked fluctuations over 1954–78 (Berndt and Winkel 1979). DENMARK. Marked decrease in 20th century; fluctuating at least locally, 150–300 pairs (Peal 1968; TD). SWEDEN. Fluctuating, but no known marked trends (LR); about 20000 pairs (Ulfstrand and Högstedt 1976). FINLAND. Estimated 19000 pairs (Merikallio 1958); probably some recent decrease (see Linkola 1978), but marked fluctuations during 1926–77 suggested by Järvinen and Väisänen (1978) considered most unlikely (OH). EAST GERMANY. Some local declines (see Glutz and Bauer 1980). Near Halle, downward trend, with marked fluctuations 1929–40 (Berndt and Winkel 1979). POLAND. Fairly numerous but locally scarce (Tomiałojć 1976a). CZECHOSLOVAKIA. Marked decrease since c. 1960 (KH). SWITZERLAND. Marked decrease in north and centre since c. 1950 (Peal 1968; RW). ITALY. Increasing recently in some areas, especially in north, aided by provision of nest-boxes (SF; see also Peal 1968). USSR. Estonia: fluctuating, decrease since c. 1970 (HV).

Oldest ringed bird at least 10 years (Glutz and Bauer 1980).

Movements. Northern and eastern populations migratory, but partially migratory to resident in Mediterranean countries.

Nominate *torquilla* (breeding east to Urals and Caspian) migrates through Europe and Middle East into Africa; reputedly also in small part through Turkestan into Pakistan and western India, but these birds more likely *torquilla–sarudnyi* intergrades or due to authors combining these taxa. *J. t. sarudnyi* (breeding west and central Siberia) winters in Indian subcontinent (overlapping migratory extralimital races *chinensis* and *himalayana*); has been collected in Turkey and Iraq (C S Roselaar), and may reach north-east Africa (Britton 1980). Central Mediterranean race *tschusii* is in part resident, though occurs regularly on Malta (Sultana and Gauci 1982) and known to reach coastal plains of Algeria and Tunisia (October–March skins); probably accounts for most winter records in southern Europe between southern France and Greece (Glutz and Bauer 1980). This race twice collected (March, April) in north-west Egypt (Goodman and Watson 1983) and once (October) at Entebbe, Uganda (Britton 1980); its migratory status in north-east Africa needs clarification. Present all year in Spain (where birds are *torquilla–mauretanica* intergrades), but numbers much reduced in winter; winter quarters of emigrants unknown. Northwest African race *mauretanica* resident, with local altitudinal movement, and some individuals wander, e.g. to central Tunisia in winter; single records from El Oued in north Algerian Sahara (April) and Khenifra below Moyen Atlas (October), but no proof of Saharan crossing (despite Etchécopar and Hüe 1967). See Vaurie (1959d, 1965), also Heim de Balsac and Mayaud (1962), Ali and Ripley (1970), and Bernis (1970).

European *torquilla* to which rest of account refers, winters in very small numbers or irregularly in Mediterranean basin and Middle East; otherwise in Africa south of Sahara. Found there in acacia steppe across northern

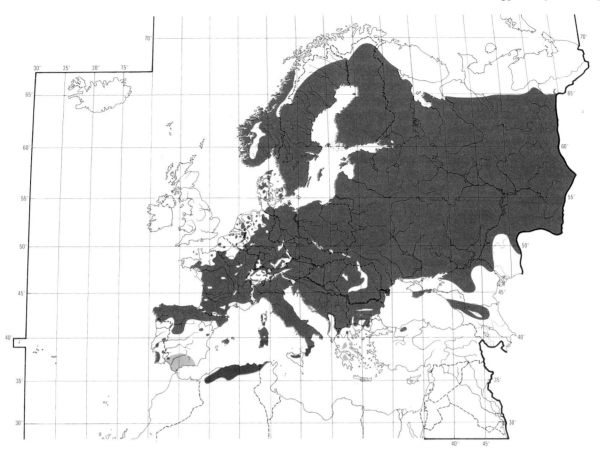

tropics from Sénégambia and Sierra Leone east to Ethiopia, and south to c. 3°N in Cameroun and Zaïre. Quite common in Ethiopia, on passage and even locally in winter (e.g. Smith 1957, Tyler 1979), but much scarcer further south in eastern Africa (reaching Uganda and northern Kenya: Britton 1980); proportion of Red Sea migrants may continue south-west heading into Sudan and Zaïre (Moreau 1972).

Migrates on broad front across Europe and occurs regularly at Alpine cols, though recoveries indicate tendency to take shortest sea-crossings of Mediterranean (see below). Occurs along full length of North African coast, however; quite common on both passages (Moreau 1961) though possibly more so in spring (e.g. Meinertzhagen 1930, Bundy 1976). Migratory departures normally nocturnal; not notably social, though loose parties of transient birds seen resting up during daytime (Meinertzhagen 1930). Immatures start to disperse, with gradual transition into autumn migration, as soon as post-juvenile moult complete or nearly so, so that June-hatched young move significantly earlier than later broods (Peal 1972). Main autumn passage period through west Palearctic is mid-August to early October (stragglers into November or even later); from early September onwards south of Sahara. No recoveries from Afrotropical region, and only 1 elsewhere in winter: Bayern (West Germany) autumn migrant found Drôme (France) on 22 January (Schloss 1975).

Within Europe, directions of movement shown by recoveries are south to south-west in autumn, reverse in spring. In southern Europe, concentrations of recoveries in France–Iberia and Italy–Balkans, but this possibly biased by greater hunting pressure in those countries. Believed long sea-crossings selected against, however (Glutz and Bauer 1980). Recoveries provide suggestion of trend towards migratory divide across Europe, with those from west and north-west Europe generally taking Iberian route, those from further east passing through Italy and Balkans. However, birds from wide north–south band across Europe (Finland and eastern Sweden, East Germany, Austria) found on both south and south-west routes (details in Glutz and Bauer 1980). Scandinavian drift migrants occur regularly in eastern Britain in autumn, with recoveries to and from Norway and Sweden. Birds passing through Near East probably from unringed Russian population; one ringed Azraq (Jordan), 18 April, retrapped 23 April, had increased weight from 33 g to 42 g, and was recovered Ukraine on 9 May, having averaged 130 km per day (Ferguson-Lees 1966). Spring passage begins early March, though minority still in Afrotropics in early May. Vanguard commonly reaches west and central Europe in second half of March (occasionally earlier), but major reoccupation of European breeding range early April to mid-May, averaging later to north and east.

Young birds tend to return in spring to general vicinity of birthplace; 75% of recoveries then within 24 km, 90% within 70 km, and nearly all adults found close to previous breeding place (Peal 1972). Exceptions include Fenno-Scandian birds recovered in spring 110 km ENE, 160 km WNW, 360 km ENE, 415 km north-east, and 480 km WNW of ringing site (Glutz and Bauer 1980), and recent breeding in Scotland believed from Scandinavian stock (Sharrock 1976). Dutch breeding adult found breeding next year in Niedersachsen (West Germany) 425 km ENE. See Rüppell (1937) for homing ability in spring by transported birds.

Food. Principally, and sometimes exclusively, ants (Formicidae), but also other insects. Ants usually taken from nests; may be dug out with bill, or nest broken up. Prey in holes adhere to long, glutinous tongue and are drawn out. Usually pecks at immediately available adults and larvae and uses tongue for more inaccessible prey (King 1974a; Löhrl 1977e). Also uses tongue to extract prey from under stones, from tree trunks, or from small holes made in ground (Claudy 1951; Glutz and Bauer 1980). Sometimes sits for lengthy period by ant trails in bushes, etc., to pick up ants as they pass (e.g. Löhrl 1977e). Also darts out tongue to take flies (Dekhuijzen-Maasland et al. 1962); may also stun flies thus, then pick them up (Thomas 1844). In Leningrad region (USSR), often perches on low branches over water to collect insects from water surface; also seen to dip bill full of ants in water before feeding young (Pokrovskaya 1963). Often brings miscellaneous (usually shiny) objects to nest (e.g. plastic, stones, shells). Some objects ingested; 29% of nestlings examined in Finland had ingested objects (Terhivuo 1977). For other items found in nests, see Menzel (1968) and Heuer and Krägenow (1973). Mollusc shells and small bones perhaps supplement mineral requirements of young (Löhrl 1978). Small stones (perhaps used in grinding food in gizzard) and bone splinters seen to be fed to young and also found in faeces. Young regurgitate larger pieces (Klaver 1964). Adults regurgitate indigestible parts of insects to some extent as pellets (Siewert 1928; Glutz and Bauer 1980).

Prey includes ants (Formicidae), beetles (Coleoptera: including Carabidae, Chrysomelidae, Cantharidae, Scolytidae, Curculionidae, Scarabaeidae, Elateridae), flies (Diptera: including Tipulidae, Muscidae), grasshoppers (Acrididae), sawflies (Tenthredinidae), cicadas *Cicadina* and other bugs (Hemiptera), mayflies (Ephemeroptera), butterflies and moths (Lepidoptera), spiders (Araneae), woodlice (Isopoda), molluscs (including Planorbidae and *Succinea putris*), tadpoles of frog *Rana temporaria*, birds' eggs. Plant food includes berries of bilberry *Vaccinium myrtillus* and elder *Sambucus* (Pokrovskaya 1963; Klaver

1964; Menzel 1968; Löhrl 1978; Glutz and Bauer 1980). Analysis of faeces of single bird at Portland Bill (Dorset, England) revealed 800 ants *Lasius niger* in a sample of 1·5 cm^3; small numbers of *L. flavus* also found (Speight 1974). In Altay (USSR), grasshoppers (Acrididae) found to be major food item, elsewhere in USSR aphids (Aphidae) and other insects found (Dementiev and Gladkov 1951a). In Switzerland, ants (especially adults and pupae) main prey; species included *Formica rufa*, *Lasius flavus*, and *Myrmica*; other prey spiders and Lepidoptera (Bussmann 1941). In East Germany, of 9302 invertebrate items, 4057 were adults and 5245 other stages. 90·4% of all adults made up of ants *Tetramorium caespitum* and *Myrmica lobicornis*; rest made up of *Lasius niger*, *Formica fusca*, and *F. rufa*. Other prey included beetles and spiders (Dornbusch 1968b). No information on diet outside breeding season.

In detailed observations on feeding of young in Leningrad region, young at one nest fed pupae of ants *Camponotus* for first 4 days, on 5th day received adult *Camponotus* and *Formica*. From day 7, other insects offered including beetles (Carabidae, *Rhagium*, Elateridae larvae, *Melolontha*) and craneflies (Tipulidae). At another nest, young fed almost exclusively on aquatic invertebrates taken from surface of river by adults (see above)—mainly mayflies (Ephemeroptera); terrestrial invertebrates, usually brought in evening, included Carabidae and *Formica*. At 3rd nest, young fed on most abundant prey species—sawflies (Tenthredinidae) and beetles *Melasoma alnea*. At 2 other nests, young fed mainly on pupae of *Formica* but regularly fed on eggs taken from nests of Great Tit *Parus major*. In 2 woodland nests, basic food was eggs, larvae, pupae, and adults of beetles *Ips* which were locally abundant. Bilberries *Vaccinium myrtillus* also eaten, possibly for moisture content. 2 pairs often fed young on tadpoles *Rana temporaria*, also Tenthredinidae and pupae and winged adults of *Formica*. When adults feeding young on *Myrmica* had this food source experimentally cut off after day 9, diet became more varied and included caterpillars (Lepidoptera), spiders, Tenthredinidae, beetles (Chrysomelidae), small Acrididae, and leaf hoppers (Cicadellidae). (Pokrovskaya 1963.) In Netherlands, 13 food-balls fed to young were composed of pupae and adults of *Lasius niger* and 1 *Formica fusca*; stones and pieces of bone identified in faeces of young (Klaver 1964).

BDSS

Social pattern and behaviour. Most important studies by Bussmann (1941) in Switzerland, Steinfatt (1941b) in Rominter Heide (Poland/USSR), and Ruge (1971a) mainly in south-west of West Germany. For monograph, see Menzel (1968). For detailed study of closely related Red-breasted Wryneck *J. ruficollis*, including comparisons with *J. torquilla*, see Tarboton (1976).

1. Normally solitary outside breeding season. Up to 20 recorded together in post-fledging period (Pokrovskaya 1963) or on passage (Meinertzhagen 1954), but such gatherings probably always accidental or due to environmental factors (food, cover).

In breeding season, in pairs or family parties (Glutz and Bauer 1980). For association on breeding grounds with Red-backed Shrike *Lanius collurio*, see Ronsil (1949). BONDS. Monogamous mating system; for possible polygamy, see Linkola (1978). Pair-bond normally lasts for duration of one breeding season only (Szöcs 1942; Creutz 1964b). Many cases of fidelity to nesting area (also nest-site) or natal area (including 2 siblings of a brood) (e.g. Steinfatt 1941b, Szöcs 1942, Creutz 1964b, Peal 1972) but pair-fidelity proved only once (Creutz 1976). For pairing with progeny of previous year, see Spengemann (1975). Care of young, by both parents, continues after fledging (e.g. Boutillier 1914, Siewert 1928). Age of first breeding 1 year (Creutz 1964b). BREEDING DISPERSION. Solitary and territorial. Overall density low everywhere and caution necessary in dealing with some data; important to remember that both sexes give Advertising-call (see part 2) and may do so far from nest (e.g. 800 m or more). In study areas with nest-boxes (for survey, see Menzel 1968), pairs using natural cavities often overlooked and recorded pairs not individually marked. In good years, 2nd broods (often overlapping with 1st) may give false impression of high population (see Glutz and Bauer 1980; also Monk 1955, 1963). Pairs may be fairly close together in favourable areas. In Rominter Heide, at least 4 nests on c. 1 ha, as close as c. 40–50 m apart; dispersion generally dependent on food supply and availability of suitable nest-sites (Steinfatt 1941b). In Leningrad region (USSR), up to 13 pairs on 25 ha, mainly in holes of woodpeckers (Picinae); shortest distance between nests c. 20 m; up to 8 pairs in boxes on 1 ha. Normally more dispersed, with nests c. 150–270 m apart (Pokrovskaya 1963). Exceptionally, c. 30 pairs on 235 ha, Laxenburger Park, Niederösterreich, Austria (Glutz and Bauer 1980). For further details, see Strokov (1963), Creutz (1964b), and, especially, Menzel (1968). Early in season, birds occupy large home-range (up to c. 500–1000 m across: Durango 1942; Dekhuijzen-Maasland *et al.* 1962; Smit 1970) which then (generally after pair-formation) contracts to smaller nesting territory—in one case c. 0·42 ha (Stephan 1961). May contain several favoured song-posts (Ruge 1971a; Glutz and Bauer 1980). Birds forage within c. 150–1000 m of nest (Boutillier 1914; Steinfatt 1941b), but generally close to nest once breeding started (Glutz and Bauer 1980). 2nd-brood nest c. 12–400 m from 1st, or same nest may be used (Štusák 1958; Creutz 1964b; see also Relations within Family Group, below). ROOSTING. Little information and that relating only to breeding season. Prior to, or during, laying, one pair-member (sex unknown) roosts nocturnally in nest-hole; sleeps in posture similar to that used for incubating (Bussmann 1941; Steinfatt 1941b) or clings to cavity wall (Menzel 1968). Other bird roosts in cavity close (c. 10 m) to nest (Bussmann 1941; Steinfatt 1941b), though may do so much further away, possibly more than 200 m (Menzel 1968). Fledged young may roost in tree in the open or return initially to nest (Delamain 1931; Steinfatt 1941b; see also Relations within Family Group, below). During day, often remains perched motionless for long periods, tail pointed vertically down, head drooped (Glutz and Bauer 1980). Said to water-bathe readily (Glutz and Bauer 1980). Movements suggesting anting recorded by Stone (1954) and King (1974a).

2. Often approachable (e.g. Ternier 1913, Steinfatt 1941b); especially if not stared at, may allow approach to c. 2 m (Puhlmann 1914a). Otherwise makes for nearest cover in danger or may crouch and rely on crypsis (Kipp 1954; Glutz and Bauer 1980). Migrant confronted by Great Grey Shrike *L. excubitor* at c. 2 m performed apparent Head-swaying: turned head vigorously from side to side as if preening, but without actually touching feathers; *L. excubitor* mock-preened rather than attacked in response (Bates and Brown 1977); see also Antagon-

A

istic Behaviour, Anti-predator Responses of Young, and Parental Anti-predator Strategies (below). *J. torquilla* may be mobbed or even attacked by small birds (Löhrl 1950; Ruge 1971a). ANTAGONISTIC BEHAVIOUR. (1) General. Advertising-call (see 1 in Voice) used in territory-demarcation and pair-formation (Glutz and Bauer 1980). In some cases, birds silent for first few days after arrival on breeding grounds, and react to playback of tape-recording only when ready for courtship (Stephan 1961; Ruge 1971a). Advertising-call given roughly up to hatching (Steinfatt 1941b) or beginning of incubation (Peal 1968), though unpaired bird may call until late June (Peal 1973). Most Advertising-calls given in (early) morning (Bussmann 1941), to some extent in evening (Steil 1957), but see also Voice. Resurgence of courtship activity with renewed Advertising-calls occurs prior to 2nd brood (Ruge 1971a). Calling bird normally raises head on extended neck (Glutz and Bauer 1980: Fig A). Isolated pairs in particular may fly round territory and visit favoured perches (see Breeding Dispersion, above). Where several birds occur together (e.g. 6–7 in courtship display in orchard) frequent calling typical and birds may not breed in such circumstances (Menzel 1968; Ruge 1971a). (2) Threat and fighting. In presence of rival, bird often reacts as follows: head (especially crown) feathers ruffled, bill pointed at opponent with neck extended and tail fanned to varying degree depending on situation: wings usually slightly drooped (Fig B). Posture may be adopted towards other bird species. Often leads to Head-swaying in which head moves horizontally back and forth and neither turns (movements may appear circular: Ruge 1971a) nor withdraws; see also Anti-predator Responses of Young and Parental Anti-predator Strategies (below). Normally accompanied by Contact-calls (see 2 in Voice). Once, ♂ and ♀ c. 8 m apart and 3rd bird (possibly ♂) c. 1 m from resident ♂; when resident ♂ gave Advertising-call, 3rd bird raised head and Head-swayed (Ruge 1971a; Löhrl 1978; Glutz and Bauer 1980). Head-swaying corresponds to that of some Picinae and evidently contains threat component, but see below for its role in heterosexual encounters. Fights evidently rare in the wild and no damaging ones recorded. After mutual threat between one member (possibly ♂) of a territory-owning pair and a 3rd bird, both pair-members chased off the intruder and

B

aggressive pecking also ensued (Ruge 1971a, which see for experiment with stuffed specimen). For antagonistic interactions between captive birds, including older brood siblings, see Thomas (1844) and Ruge (1971a). Even after nest-site selected, often destroys nests of other hole-nesting birds, throwing out material and sometimes breaking eggs and killing young (Löhrl 1940; Ruge 1971a; Olšanik 1975). All accessible holes may be subjected to such treatment over a wide area (Creutz 1964b). Behaviour may be locally common, depending on stage of reproductive cycle, supply of holes, and population level (e.g. Löhrl 1940, 1978, Dekhuijzen-Maasland et al. 1962). Nests of conspecific birds sometimes destroyed (e.g. Löhrl 1978). Behaviour performed by both sexes (Ruge 1971a), but see Löhrl (1940) for indication that ♂ perhaps more active. Bird may drum inside hole between bouts of tossing out material (Puhlmann 1914a). Normally enters warily and only when occupier absent so that fights comparatively rare (Löhrl 1940). Where disputes occur, *J. torquilla* may appear passive but even vigorous defence by owners may ultimately fail, although contest of several days not always successful for *J. torquilla* (Ruge 1971a; Glutz and Bauer 1980). In case of tits *Parus*, fight may end fatally for either party (Oppenoorth 1938). For further details, see Menzel (1968), Ruge (1971a), Olšanik (1975), and Glutz and Bauer (1980). HETEROSEXUAL BEHAVIOUR. (1) General. Courtship probably begins fairly soon after arrival (see above for initial silence of some birds), intensifies up to laying and wanes around hatching. Resurgence preceding 2nd brood (which may not always follow) includes copulation, Nest-showing, and calling (with intermittent feeding of 1st brood young), though less time may be spent at 2nd nest-hole prior to laying. Pair-formation typically proceeds slowly and inconspicuously and 1–4 weeks may pass between arrival and laying (Ruge 1971a; also Steinfatt 1941b, Pokrovskaya 1963, Menzel 1968, Hald-Mortensen 1971). ♂ and ♀ show no marked behavioural differences (Ruge 1971a). (2) Pair-bonding behaviour. Bird spends much time near or in potential nest-hole (Fig A) giving Advertising-call or more compressed variant (Ruge 1971a; see 1 in Voice; also Buxton 1950, Menzel 1968). Such behaviour performed by both sexes even simultaneously at separate sites (Ruge 1971a); serves to attract potential mate and advertise site (see subsection 3, below). Swift response leads to immediate approach and increased excitement evident in accelerated rate of calling. In one case, bird called for several days; when vocal response finally came, turned abruptly in the direction of the call, and the 2 birds then moved gradually closer together (Ruge 1971a). In Meeting-ceremony, birds often Head-sway rhythmically as in antagonistic encounters (see above); bill pointed up rather than directly at other bird may indicate appeasing component dominant. Contact-calls (see 2 in Voice) normally given by both birds (Ruge 1971a; also Löhrl 1978). Short, hopping pursuits recorded (Siewert 1928). For further details, see (e.g.) Boutillier (1914) and Bussmann (1941). (3) Nest-site selection. Search for nest-site begins soon after arrival (Creutz 1964b); typically prolonged (Pokrovskaya 1963), birds inspecting virtually all cavities over large area (Creutz 1964b). Some earlier reports on role of sexes in nest-site selection (e.g. Puhlmann 1914a, Siewert 1928) should be viewed cautiously as authors do not state how sexes distinguished (Ruge 1971a); see also Steinfatt (1941b). As mentioned above, both sexes give Advertising-call or variant and thus Nest-show. May do so first in hole, then outside (Menzel 1968). One pair displayed at 7 nest-boxes before finally choosing woodpecker hole advertised by ♀ while ♂ had been apparently more interested in site occupied by Nuthatch *Sitta europaea*. Nest-showing may last for several days, even at a site eventually abandoned. First inspection of a site always hesitant (Ruge 1964, 1971a). Presence

C

of apparently interested bird frequently causes site-advertiser to Tap or Drum near or inside hole (Schneider 1961; Glutz and Bauer 1980). Birds then enter hole and remain inside for quite long period. Nest-clearing (see above) may begin when 2 birds first meet at potential site (Glutz and Bauer 1980). (4) Courtship-feeding. Few details, including on timing, but reported to occur very rarely during incubation (Menzel 1968). Bills turned at right angles to each other, as when feeding young (Fig C). (5) Mating. Takes place on perch or ground, mainly during courtship and laying with sporadic attempts early in incubation. Both birds give excited Advertising-calls as they approach, then Contact-calls and finally a muted variant of Advertising-call. ♂ may Tongue-dart at ♀'s head before mounting. Birds may flit together through trees or ♂ may fly around near ♀ ready to copulate (Ruge 1971a). ♀ adopts crouched Soliciting-posture, in some cases having first flown to ♂ who mounts by jumping up from side or may fly, tail spread, straight on to ♀'s back. ♀ cocks tail to permit cloacal contact; ♂ grasps ♀'s nape feathers. Lasts c. 4–5 s; afterwards, ♂ may remain with wings spread and drooped for c. 5 s before flying off (Claudy 1951). Birds copulate several times a day, once at interval of only c. 15 min (Ruge 1971a). (6) Behaviour at nest. Before or during laying, one member of pair spends night in nest (see Roosting, above). ♀ normally lays in early morning. ♂ frequently gives Advertising-call from nearby perch, ♀ responding similarly or with muted Contact-call. Bouts of antiphonal calling and copulation occur when ♀ leaves (Ruge 1971a; also Bussmann 1941). Before incubation, either bird makes brief visits to nest during day (Velichko 1963). Still not clear whether ♂ (Ruge 1964, 1971a; Menzel 1968) or ♀ (Bussmann 1941; Steinfatt 1941b) incubates at night; in early stages of incubation bird sometimes crouches by eggs or clings to cavity wall (Ruge 1964; Menzel 1968). Incubation shared more or less equally during day, with shifts of 1–2 hrs (Ruge 1971a); 3–5 change-overs daily (Steinfatt 1941b), more frequent towards end of incubation (Ruge 1971a). Short pauses of c. 1–3 min, used for self-preening, etc., apparently regular (Ruge 1971a). Much longer breaks—eggs uncovered for maximum 351 min and total of 7 hrs during day (Steinfatt 1941b)—perhaps due to disturbance (Ruge 1971a). Nest-relief (also during brooding—see below) normally accompanied by Contact-calls from one or both birds. Incoming bird often flies first to branch of nest-tree, then hops to hole, there awaiting mate's exit, or may enter immediately so that pair together inside for up to c. 5 min. Low-intensity Head-swaying occurs prior to change-over. Especially when nest-relief due, bird in nest may clamber around, occasionally peck at cavity wall and frequently look out, sometimes neck bent like tortoise *Testudo* in pre-departure alertness (Bussmann 1941; Ruge 1971a); see also Velichko (1963) for cautious approach and departure. Bird not uncommonly leaves before other arrives, although calls given beforehand (Bussmann 1941); incoming bird approaching after c. 3 min Tapped by entrance to hole before going inside (Menzel 1968). RELATIONS WITHIN FAMILY GROUP. Young call at least 12 hrs before hatching (Ruge 1971a). Eggshells not removed and are later fed to or eaten by young (Löhrl 1978; Glutz and Bauer

PLATE 75. *Upupa epops* Hoopoe (p. 786). Nominate *epops*: 1 ad ♂, 2 ad ♀, 3 juv. *U. e. major*: 4 ad ♂. (CETK)

PLATE 76. *Jynx torquilla* Wryneck (p. 800). Nominate *torquilla*: 1 ad, 2 juv, 3 nestling. *J. t. tschusii*: 4 ad. (NA)

1980). Young immediately form 'warmth pyramid' typical of Picidae: sit on heels, belly to belly, and rest necks on each other's shoulders (Creutz 1943). Brooded more or less continuously (change-over when mate brings food) for c. 5 days; day-time brooding then reduced (resumed for longer periods if cold) and ceases at c. 12–14(–16) days. Adult broods young at night up to 17–18 days—see Menzel (1968) for description of posture. Enters hole c. 18.00–21.00 hrs, remaining there c. 7–8 hrs (Bussmann 1941; Menzel 1968; Ruge 1971a; Glutz and Bauer 1980). In one case, ♀ brooded young one night, ♂ the next (Hald-Mortensen 1971). Young acquire thermoregulating ability mainly in period 3–12(–13) days old when air temperature c. 22°C (Böni 1942). Small nestlings left alone generally quiet (Steinfatt 1941b) but call when brooded (Sutter 1941). At c. 9–14 days change from 'pyramid' (see above) to tile-formation, all facing same way; radial ('star') formation, young resting necks against wall (as in other Picidae), also occurs (see Creutz 1943, 1955, 1964b; also Sutter 1941). Young first fed only a few hours after hatching; adult initially always enters nest-hole to feed them and may continue to do so until c. 17–18 days old (Ruge 1971a) or start feeding by hanging down from entrance-hole at c. 6–7 days (Bussmann 1941); differences possibly due to variations in nest-chamber dimensions (Menzel 1968). Younger nestlings stretch up in pyramid, gape widely, and give loud food-calls (see Voice); sway heads about and snap at parent's bill when contact made. Sometimes attempt to grasp each other or parent's tail (Sutter 1941). Adult may encourage gaping response by touching nestling's head (Hoogsteyns 1959) or gape flanges (Klaver 1964). Eyes fully open at c. 8–9 days (Sutter 1941; Velichko 1963) but nestlings normally gape even before this when nest-chamber darkened by arrival of adult, and beg also particularly when brooding bird gets up, normally indicating other's arrival with food (Sutter 1941; also Ruge 1971a). Saliva-enveloped food-ball divided up between several small nestlings, but from c. 13–14 days old given to only one. Food stuffed well down into nestling's gape; bill of adult twisted at 90° to that of young. Adult moves head up and down, while nestling opens and closes bill with sucking motion (Sutter 1941). Older young seek food remains on nest-floor, picking up items with sticky tongue (Henze 1962). Often poke heads out and may examine immediate vicinity of entrance by tongue-darting (Bussmann 1941), but often withdraw when parent arrives (Dekhuijzen-Maasland *et al.* 1962). Towards end of nestling period (when fed at entrance), 1 young bird often by entrance and gives persistent food-calls while others silent (Steinfatt 1941b). Apparently no inter-sibling disputes over best place for receiving food (Boutillier 1914). Young defecate between feeds, on edge of nest or against cavity wall. Faecal sacs probably swallowed by adults up to c. 6–7 days (exceptionally at c. 18–19 days: Steinfatt 1941b), later carried out and dropped (even fed to young as conflict behaviour: Dornbusch 1968b). Nest-sanitation may cease during last few days before fledging (Bussmann 1941; Ruge 1971a), perhaps because difficult for adults to reach nest-floor when young well-grown (Menzel 1968), although liquid faeces (produced in pre-fledging excitement or when disturbed—see below) are virtually impossible to remove.

Nest normally cleaned thoroughly before roosting, faeces otherwise removed after feeds (Menzel 1968; Ruge 1971a). Thick layer of faeces may remain after fledging (e.g. Bussmann 1941). Dead nestlings (up to c. 8 days old) normally taken out (Menzel 1968). Clambering around in nest and presumed wing-exercising at latest from c. 17–18 days (Bussmann 1941; Steinfatt 1941b). For further details of nestling development, see Velickho (1963). Young fledge at c. 20–22 (19–25) days (Creutz 1964b; Ruge 1971a); shorter periods (e.g. Szöcs 1942, Creutz 1943) perhaps due to disturbance (Glutz and Bauer 1980). Usually leave nest in morning, over period of 1–3 days, longer in case of larger, 2nd, or late broods, or runts (Ruge 1971a; Glutz and Bauer 1980; see also, e.g., Puhlmann 1914a). Adult may arrive with food but not feed nestlings (possibly encouraging them to leave: Bussmann 1941), or young fledge spontaneously after much moving around, gazing out and tongue-darting (Ruge 1971a). ♀ may lay 2nd clutch before 1st brood fledges (Kalmár 1934; Szöcs 1942); once in same nest 7 days before last nestling fledged (Raab 1961). Young do some self-feeding soon after fledging but constantly give food- and contact-calls so that family generally stays together within c. 100–300 m of nest for 14–21 days, though young more or less independent after c. 10–14 days (Bussmann 1941; Szöcs 1942; Pokrovskaya 1963; Menzel 1968; Ruge 1971a). In one case, 1st-brood young not expelled from territory after 2nd clutch laid (Ruge 1971a). Newly fledged young may hang on vertical structures and even move upwards like woodpeckers (Delamain 1931); see also Ternier (1913) and Boutillier (1914). When whole family returned to nest to roost (see Roosting, above), squabbling took place before all had settled (Bettmann 1972). ANTI-PREDATOR RESPONSES OF YOUNG. For more detailed descriptions of same behaviour patterns alluded to here, see Parental Anti-predator Strategies (below); see also Voice. Some earlier reports inaccurate or vague or indicate that various anti-predator responses may be combined (Glutz and Bauer 1980); see also, especially, Löhrl (1978). Young nestlings show no reaction when adult gives Alarm-call (see 3 in Voice) near nest (Sutter 1941). Perform Head-turning at c. 13–14 days old when also ruffle crown feathers, and by c. 17–18 days have mastered adult Hissing-threat display (Bussmann 1941). Young examined by Steinfatt (1941b) initially gaped and snapped (as if food-begging) when nest-box opened; from c. 17–18 days, first crouched then (after a few seconds) Head-turned and Hissed. Generally retreat into nest-hole if alarmed but may attempt to leap out at c. 15–16 days old (Velichko 1963). Birds may Head-sway with neck extended, legs fully stretched, and body somewhat arched; gape silently (e.g. brood 23 days old) or Hiss (Steil 1957; Henze 1962). Tongue-dart when handled from c. 8 days old (Bussmann 1941). According to Thomas (1844), head may be raised and turned but bird Hisses with bill closed, and may dig claws into hand. Pecking at hand (Schlegel 1925) or excretion of liquid faeces (Ruge 1971a) also recorded. PARENTAL ANTI-PREDATOR STRATEGIES. (1) Passive measures. Birds often become increasingly secretive once nest-site selected (Burton *et al.* 1970). Frequently noted shyness near nest—e.g. Alarm-calls (see 3 in Voice) given when man at c. 28–84 m—sometimes declines with habituation (Stephan 1961; Glutz and Bauer 1980). In panic, bird may seek refuge in nest-hole (Hald-Mortensen 1971). Shyness varies individually (also between pair-members) and with stage of breeding cycle (Ruge 1971a). Bird on nest may stay put (e.g. Schalow 1890, König 1961a) and even allow itself to be touched (e.g. Steinfatt 1941b). Sometimes gives Alarm-call on finally flushing. When disturbance over, may return quickly (Steil 1957) or be slow to do so (Ruge 1971a). In some cases, eggs or small young deserted if adult handled (Creutz 1964b). (2) Active measures: against man. When man near nest with

PLATE 77 (*facing*)
Picus canus Grey-headed Woodpecker (p. 813): 1 ad ♂, 2 ad ♀, 3 juv ♂, 4 nestling ♂.
Picus viridis Green Woodpecker (p. 824). Nominate *viridis*: 5 ad ♂, 6 ad ♀, 7 juv ♂, 8 nestling. *P. v. sharpei*: 9 ad ♂.
Picus vaillantii Levaillant's Green Woodpecker (p. 837): 10 ad ♂, 11 ad ♀, 12 juv ♂, 13 nestling ♂. (NA)

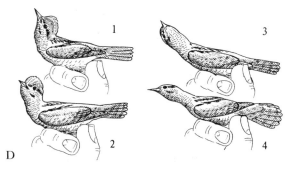

young, birds may fly around giving loud Alarm-calls, repeatedly land on tree, press close to trunk, extend neck, and Hiss, continuing thus until danger passes (Velichko 1963); see also Schlegel (1925) for evidence of variation. When threatened in nest and prevented from escaping, may (repeatedly) perform Hissing-threat display; tail fanned, head feathers ruffled, and neck and body slowly raised towards danger (head bowed or bill pointed up); recoils suddenly from highest point and Hisses (see 5 in Voice); sometimes, strikes with wings (Löhrl 1978; also Müller 1928, Steinfatt 1941b, Janssen 1954, Ruge 1971a). According to Steinfatt (1941b), bill opened wide at maximum stretch when Hissing; however, Löhrl (1978) noted that bill remains (nearly) closed throughout. According to Löhrl (1978), display given only by a few adults when front of nest-box removed so that observer clearly visible. At least during nestling period, may nevertheless be provoked by less severe disturbance (König 1961a; Wendland 1967). Unlike tits *Parus*, bird may break off performance abruptly and escape (Löhrl 1978). Hissing-threat display can have startling effect on man; if disturbance repeated, display may cease to be given (Steinfatt 1941b). Hissing-threat display perhaps mimics snake (Berndt and Winkel 1977), as may Head-turning (Fig D)—probably performed only when freedom of movement inhibited: held in hand, trapped in nest, not yet fully fledged, or with wing injury (Ruge 1971a; see also Warnke 1941). Head feathers ruffled, tail fanned, and head held so that bill pointed up while back horizontal. Head then turned slowly and deliberately through 90° to point sideways while also being pushed far forward and down until bill, head, and back are level. Movement then reverses, head turning from side to side (see Ruge 1971a) in clock-work fashion; bird usually silent. Duration of this behaviour varies individually, as do stimulus thresholds, in both adult and feathered young (Löhrl 1978); see also Kingfisher *Alcedo atthis* for similar behaviour. Head-turning occasionally combined with Hissing-threat display movements (Steinfatt 1941b; Ruge 1971a); see also Thomas (1844) in Anti-predator Responses of Young (above). Injured bird brought in by cat executed movements of Hissing-threat display on ground but was silent (Löhrl 1978). Apparent death-feigning occurs, frequently after Head-swaying: bird hangs limp with eyes closed (Witherby *et al.* 1938; Warnke 1941; Kipp 1954); see also Meinertzhagen (1954) and Antagonistic Behaviour (above). Exceptionally, attacks man near nest (Walpole-Bond 1938; Stephan 1961). (3) Active measures: against other animals. Pair called loudly for *c.* 30 min while cat near nest (Glutz and Bauer 1980). Weasel *Mustela nivalis* which had climbed nest-tree elicited Alarm-calls and was then attacked and driven off (Dekhuijzen-Maasland *et al.* 1962).

(Fig A from photograph in Peal 1973; Fig B based on photograph in Heinroth and Heinroth 1924–6; Fig C from photograph in Reither 1977; Fig D based on photographs in Löhrl 1978.)   MGW

Voice. Normally silent outside breeding season and not very vocal after laying (e.g. Szöcs 1942, Kipp 1954). Fairly restricted repertoire apparently shared by both sexes; for possibility of sexual differences, see below. Several calls typically composed of several units in succession (Ruge 1971a). Tapping and Drumming commonly associated with Nest-showing and nest-reliefs (see Social Pattern and Behaviour). At first (and rarely) an arrhythmic Tapping on a branch ending in an even drumroll (Glutz and Bauer 1980; see Social Pattern and Behaviour). In one case, 4 series of Drumming performed, 6–10 strikes per series (Schneider 1961).

CALLS OF ADULTS. (1) Advertising-call (Song). See Fig I. A 'quee-quee-quee-quee-quee...' recalling small falcon *Falco* (e.g. ♀ Hobby *F. subbuteo*) or Lesser Spotted Woodpecker *Dendrocopos minor* (Witherby *et al.* 1938); more metallic than *D. minor* and, with drop in pitch at end of each unit, has characteristic complaining quality (P J Sellar). According to Steil (1957) and Menzel (1968), 22–28 units given in succession, though Ruge (1971a) noted only 12–14 at peak intensity. When up to 22 units (duration *c.* 5 s) given at higher intensity, sound more like 'kje-kje...' than 'gäh' (Bergmann and Helb 1982). Begins quietly, then becomes louder and ascends slightly in pitch (Bergmann and Helb 1982). In high-intensity excitement, 5–6 bouts per min. ♂'s call sometimes shriller and clearer (Ruge 1971a); not known whether this difference constant—individual variation considerable (Glutz and Bauer 1980). ♂ and ♀ give Advertising-call in isolation or in synchronized duet (Glutz and Bauer 1980; Bergmann and Helb 1982). Modified, muted variant with imploring quality associated with copulation, and a compressed 4–5-unit variant given in higher-intensity excitement or alarm (Ruge 1971a). Advertising-call possibly used infrequently by isolated pair once nest-site chosen (Peal 1973); given around midday and in sultry conditions when many other birds silent, also rarely by migrants in early autumn (Bergmann and Helb 1982). (2) Contact-call. A quiet, guttural cooing or grumbling 'kru' or 'gru' sound; vowel sound more like 'i' in higher-intensity excitement (Steinfatt 1941b; Ruge 1971a); see also (e.g.) Boutiller (1914) and Siewert (1928). Given in direct confrontation between conspecific birds and associated also with copulation, nest-relief, and Head-swaying (Ruge 1971a). A 'graeb' sound also used as brood summons (Steil 1957; also Henze 1937). (3) Alarm-call. Series of loud, rising and falling 'tuck', 'töpp', abrupt and hoarse 'quött', or 'teck' sounds given when predator near nest (Witherby *et al.* 1938; Steinfatt 1941b; König 1961a; Ruge 1971a). Perhaps restricted to period when parents tending young in or out of nest. Heard by Ruge (1971a) towards end of nestling period and after fledging (e.g. in presence of aerial predator). See also call 7b. (4) Fright-call. When handled, a series of shrill screeching sounds ('Schirken': Ruge 1971a): 'tschääp' (Steil 1957), 'tschjääp' or 'jääk' (Glutz and Bauer 1980). (5) Hissing. In Hissing-threat display, a sound like spitting

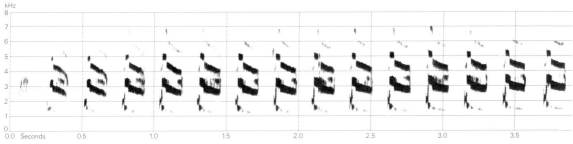

I P J Sellar Sweden May 1971

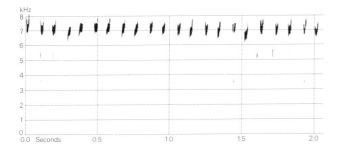

II S Palmér/Sveriges Radio (1972) Sweden July 1967

of frightened cat (Müller 1928) or sizzling of warm mineral water in bottle, including softly explosive sound made when bottle opened (Heinroth and Heinroth 1924–6; König 1961a). Rendered 'tscheck' by Ruge (1971a), though similar onomatopoeia employed by Siewert (1928) and Steil (1957) for an alarm-call which presumably belongs under call 3. (6) Threat-call. Series of discordant 'kschri' or 'wet' sounds given towards rival conspecific or playback (presumably of call 1). See Glutz and Bauer (1980) and Bergmann and Helb (1982)—also for sonagrams. (7) Other calls. (a) Excited 'sgissgissgiss' or 'dübedibdübedib' said to be given in annoyance when mate slow to respond to call 1 or to appear for nest-relief (Bussmann 1941); such calls not heard by Ruge (1971a). (b) Quiet, gentle 'djäk' or 'täck' sounds, sometimes doubled ('djäkdjäk'), allegedly given as greeting to mate by bird approaching nest, or as expression of contentment (Steinfatt 1941b). (c) In Sierra Leone, winter, a harsh 'chchch' as bird flew (G D Field).

CALLS OF YOUNG. Food-call of small young a compressed, rasping noise: 'schäischäischäi' or 'äsch äsch...' (Bussmann 1941; Steinfatt 1941b), later developing into a whirring hissing sound (Bussmann 1941). From c. 18–19 days old (Sutter 1941; Klaver 1964) or from fledging (e.g. Bussmann 1941, Creutz 1964b), young give various clear trilling, tinkling, or bibbering 'tsi', 'zi', metallic-sounding 'zit', or 'tsijek' sounds (Bussmann 1941; Ruge 1971a; for sonagrams, see Glutz and Bauer 1980, Bergmann and Helb 1982). Recording (Fig II) reveals sizzling sounds remarkable in their high pitch and rapid reiteration (J Hall-Craggs, P J Sellar). Short bouts of 10–15 units at even pitch and rate lasting c. 2–3 s typical (Berndt 1941). Given with bill wide open and audible at c. 60 m in still conditions (Menzel 1968). See also Steil (1957) and König (1961a). When being brooded, give quiet and longer-drawn variant of 1st food-call (see above) or, less commonly (although frequently when left alone), a quiet, high-pitched, long-drawn whistling 'pjüht pjüht' (Sutter 1941; Menzel 1968); see also Steinfatt (1941b). Recording by P Szöke contains 2 sounds, both loud and disyllabic: 1st a musical 'pe-too'; 2nd begins similarly but rises rapidly to a higher-pitched tonal sound accompanied by a lower-pitched rasping noise—'te-whee' (J Hall-Craggs). From c. 10 days old, give peculiar quiet yapping sound when handled. Hissing (see adult call 5) given at latest from c. 17–18 days old (Bussmann 1941) and may be combined with sound like scolding alarm of Song Thrush *Turdus philomelos* (Kiersky 1937) or ticking of clock (Henze 1943). Hungry young close to fledging may tap cavity wall with bill (Kipp 1954) or tip of tongue (Klaver 1964). MGW

Breeding. SEASON. Central and northern Europe: see diagram. Similar in western USSR but up to 1 week earlier in southern Europe, and 3 weeks earlier in North Africa (Dementiev and Gladkov 1951a; Labitte 1957; Heim de Balsac and Mayaud 1962). SITE. Natural or artificial hole in tree, wall, or bank, including nest-box; not infrequently takes over hole already in use by other species, e.g. tits *Parus*, sparrows *Passer*, flycatchers *Ficedula* (Peal 1973). Height of 5 nests, Belgium, average 4·5 m (3–6) (Arnhem 1960). Nest: available hole, no lining or material added; old nest of other species may be pulled out of box (Creutz 1976). Building: none. EGGS. See Plate 98. Oval or sub-elliptical, smooth and not glossy; white. 21 × 15 mm (19–23 × 13–17), $n = 100$ (Witherby et al. 1938). Fresh weight 2·6 g (1·9–3·2), $n = 88$ (Glutz and Bauer 1980). Clutch:

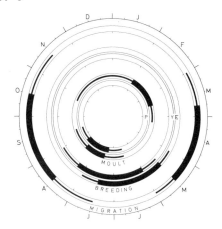

7–10 (5–12). Of 62 clutches, West Germany: 5 eggs, 4; 6, 4; 7, 10; 8, 16; 9, 15; 10, 10; 11, 2; 12, 1; average 8·2 (Creutz 1964b). Of 20 clutches, Luxembourg: 4 eggs, 1; 8, 8; 9, 4; 10, 7; average 8·7 (Menzel 1962). 1–2 broods, rarely 3; 2nd brood in 20% of cases, none at all in some years (Creutz 1976). 2nd and 3rd clutches may overlap previous brood. Of 5 2nd clutches, Hungary, 1 laid 10 days before 1st brood fledged, 1 on day of fledging, others 1, 4, and 6 days after fledging (Szöcs 1942). One pair, Belgium, began 2nd clutch 2 days after hatching of 1st, and laid 3rd clutch 3 days after hatching of 2nd; clutch sizes 7, 6, and 6 (Arnhem 1960). Laying interval 1 day. Eggs laid in morning (Menzel 1968). INCUBATION. 11·5–14 days. By both sexes, though mainly ♀, beginning with last or penultimate egg; hatching almost synchronous. Pair share incubation about equally between 05.00 and 20.00 hrs (Ruge 1971a). YOUNG. Altricial and nidicolous. Cared for and fed by both parents. FLEDGING TO MATURITY. Fledging period 18–22(–25) days. Become independent 7–14 days after fledging (Bussmann 1941; Szöcs 1942). Age of first breeding 1 year. BREEDING SUCCESS. Of 384 eggs laid, West Germany, 75% hatched and 60% fledged; of 345 young, 74% fledged; 78 broods produced 3·3 flying young per pair (Creutz 1964b).

Plumages (nominate *torquilla*). ADULT. Forehead, crown down to eye, and hindneck pale grey with tiny darker grey speckling, each feather with narrow black and rufous-brown subterminal bar and white spot on tip, but size of bar and spot rather variable—often hardly any white and black on forehead and sides of hindneck, appearing grey with rufous-brown bars, white spots and black bars often more distinct on crown; dark and often broken stripe from central crown over central hindneck to mantle black with rufous-brown variegations. A narrow and faint line from nostrils over lores to below eye white or cream with some fine dusky speckling; similar indistinct line from just above eye over ear-coverts. Central ear-coverts mottled black and deep rufous-brown, bordered below by yellow-buff or cinnamon-buff streak from lateral base of upper mandible backwards. Sides of mantle, inner scapulars, back, rump, and upper tail-coverts pale grey with tiny darker speckling, feathers sometimes with fine black or rufous subterminal marks and usually with faint almost uniform terminal off-white bars; back, rump, and upper tail-coverts with narrow black shaft-streaks or narrow long arrowheads. Central mantle with elongate patch of black with rufous-brown mottling and streaking; outer scapulars black with rufous-buff spots on sides near base and large cinnamon-buff or pink-buff spot on tip of outer web. Chin and line over lower cheeks white with narrow black bars; in some, hardly contrasts with buff of upper cheeks and throat, in others forms distinct malar streak. Throat and chest yellow-buff or cinnamon-buff with narrow dark bars $c.\ \frac{1}{2}$ mm wide and $2–2\frac{1}{2}$ mm apart; sides of chest and breast off-white with close and fine brown and dark grey bars and specks. Flanks and under tail-coverts like chest, but ground-colour often paler buff and black bars slightly wider, further apart, and less regular; longer under tail-coverts often partly speckled dusky subterminally and with narrow dusky shaft-streaks. Breast, belly, and vent cream-yellow or white; breast with short and fine black bars, belly and vent with small black arrowheads or specks, sometimes virtually uniform white. Central pair of tail-feathers (t1) closely speckled pale grey or cream and dark grey with $c.\ 2–3$ narrow and irregular bars beyond upper tail-coverts ($c.\ 2$ more bars hidden under coverts); each black bar bordered by rather broad and almost uniform grey band towards base and by yellow-buff band with limited dusky speckling distally, but either grey or buff band sometimes faint or absent. Other tail-feathers similar to t1, but equally spaced black bars slightly wider (in particular towards feather-bases) and more distinct. Flight-feathers greyish- or brownish-black; outer webs with rufous-cinnamon spots about equal in width to intervening black ($c.\ 6–7$ spots beyond upper coverts on longer primaries, $c.\ 4$ on inner primaries and secondaries), spots partly speckled black on secondaries and sometimes slightly on primaries; inner webs with more rounded and paler pink-buff or cream spots along borders. Upper primary coverts and bastard wing like primaries, but greater upper primaries with small cinnamon spot on tip of outer web and with virtually fully black inner webs; marginal coverts mainly white. Tertials and remaining upper wing-coverts closely speckled rufous-buff or buff-brown and grey (like outer scapulars); longer coverts and tertials with narrow black shaft-streaks and with pale cream or buff spot on tip of 1 or both webs, subterminally bordered by irregular black bar. Under wing-coverts and axillaries yellow-buff or cream (paler towards leading edge of wing), finely barred with black, like flanks. Wear and bleaching have some influence: cinnamon spots on tips of outer scapulars, tertials, and upper wing-coverts bleach to off-white; feathers of head, throat, chest, flanks, and under tail-coverts to pale cream-buff or pale yellow; spots on outer webs of flight-feathers to pale cinnamon or buff (especially on outer primaries). In heavily abraded plumage (about June–July), pale spots on feather-tips of upperparts and upper wing-coverts tend to be lost, and upperparts then rather uniform dirty grey with irregular bold black streak from central hindcrown to lower mantle (widest on upper mantle) and broken black streak over outer scapulars; ground-colour of underparts off-white, slightly yellow only on throat and chest, black marks prominent. No constant differences between sexes found (*contra* Dementiev and Gladkov 1951a, Klaver 1964, Menzel 1968). NESTLING. Naked at hatching. Closed feather-sheaths of flight-feathers appear from 6th day, of body on 7th; feather-tips appear from 10th day; appears feathered from 13th–14th day, but all body feathers still growing until $c.\ 20$th day, flight-feathers until $c.\ 31$st day, tail $c.\ 36$th day (Sutter 1941). JUVENILE. Rather like adult, but black of upperparts duller and less sharply defined, forehead and crown indistinctly barred white, buff, and brown-black, dull black patch from central hindneck to mantle poorly defined; outer scapulars and tertials rufous-cinnamon with narrow black bars and fine mottling, less grey than in adult and with pattern of barring rather than of streaking; feathers of back, rump, and upper tail-coverts soft and loose, cream-white with some narrow dusky and buff bars, without intricate speckled pattern of adult; longer upper wing-coverts with poorly defined triangular cream spots on tips, scarcely bordered by black subterminally; sides of head rather evenly mottled dusky grey and buff, only central ear-coverts rufous-brown, no whitish lines from nostrils to below eye, from above eye over ear-coverts, and over lower cheek; underparts like adult but feathering soft and loose, ground-colour of throat and flanks less bright buff, dark marks dull grey rather than black and less sharply defined; tail with more pronounced irregular black bars (see also Structure); flight-feathers similar but pale spots on inner webs small or virtually absent, innermost primaries reduced in length but p10 relatively large (see Structure); greater upper primary coverts more pointed than in adult and tips of both webs usually spotted cinnamon, spots less regular than in adult and often partly speckled black. FIRST ADULT. Like adult, but juvenile secondaries, many or all greater upper primary covers, and sometimes a few central tail-feathers retained. Characteristic large juvenile p10 present up to early August of 1st calendar year. In autumn, retained juvenile secondaries and tail-feathers slightly contrasting in wear with neighbouring fresh feathers and often also slightly shorter and narrower; however, a few adults also may retain some (heavily) worn secondaries or tail-feathers. Juvenile greater upper primary coverts narrower and more sharply pointed than those of adult, tip slightly frayed in autumn (still smooth-edged then in adult); rather greyish-black with rather poorly defined cinnamon spots on both webs and often with narrow inverted cinnamon V on tip, cinnamon often slightly speckled (in adult, deeper black with sharper and usually uniform deep rufous spots on outer web, virtually complete black on inner web, and without V on tip. In spring, some heavily worn juvenile secondaries often still present, mainly s3–s4 (s1–s8); greater upper primary coverts heavily worn and bleached (some outer occasionally no contrastingly new), pale spots on tips largely lost through abrasion, tips appearing sharply pointed (in adult, tips rounded, slightly frayed, cinnamon spots still bright).

Bare parts. Iris light or dark red-brown or reddish pearl-colour. Bill dark greyish horn-brown, often with green tinge. Foot brown-grey or pale brown, often with green tinge, more rarely tinged yellow. NESTLING. Entirely pale flesh-pink at hatching; bill, slightly enlarged flanges at base of lower mandible, and large warty heel-pads whitish-flesh; inside of mouth pink with 2 pale spots. Gape flanges gradually larger until 9–12 days, when $2\frac{1}{2}$–3 mm wide, size then decreasing again until fully reduced at age of 3 months; heel-pads relatively smaller during growth of legs, lost when legs full-grown at c. 14th day. Eyes open from 6th–10th day; iris grey-brown or hazel-brown. At fledging, similar to adult, but bill and leg paler brown with pink tinge in 1st weeks after fledging and iris hazel or yellowish-brown in autumn and occasionally up to spring. (Hartert 1912–21; Heinroth and Heinroth 1924–6; Sutter 1941; Menzel 1968; RMNH, ZMA.)

Moults. Based mainly on Sutter (1941), Stresemann and Stresemann (1966), and Glutz and Bauer (1980). ADULT POST-BREEDING. Complete; primaries descendant. Starts with p1 mid-June to mid-July; heavy moult of body and tail second half of July and first half of August; completed mid-August to mid-September, but part of secondaries occasionally retained and these probably replaced in winter quarters. ADULT PRE-BREEDING. Partial; February–March(–April). Involves head, body, wing-coverts, all or part of tail, and perhaps tertials. POST-JUVENILE. Partial: all head, body, wing, and tail, except for most secondaries, most or all greater upper primary coverts, and some central tail-feathers. Starts at fledging or shortly afterwards with p1 (day 22–24, about mid-June to mid-July, as in adult post-breeding), followed by p2 day 23–26; p3 day 27–33, etc. Replacement of juvenile body feathering starts with shedding of p5 at 40–45 days; p9 shed at c. 60th day, p10 at c. 65th day; all primaries full-grown and body in 1st adult by age of 80–90 days (mainly 20 August–10 September). Migration may start before moult completed, p9 occasionally and p10 often still growing late August or early September. Tail moults from t6 to t1; starts with t6 at loss of p5–p6, t1 at age of 68–70 days; all feathers full-grown at 80th–90th day (often slightly later than p10), but 2–4 central tail-feathers often still old during migration. May start moult of secondaries with s8 from about mid-July, proceeding with s7, s9, and s1, but many birds do not start until arrival in winter quarters and most secondaries still juvenile during autumn migration. Remainder of juvenile tail-feathers and some more secondaries replaced in winter quarters, but 1–6 secondaries retained until spring—mainly s3–s4 (s1–s6), not tertials. FIRST PRE-BREEDING. As adult pre-breeding, but apparently involves part of tail only and no tertials.

Measurements. ADULT AND FIRST ADULT. Nominate *torquilla*. Migrants, Netherlands and West Germany, late August to early October and mid-April to early May; skins (RMNH, ZFMK, ZMA). Bill (F) to forehead, bill (N) to distal corner of nostril; exposed culmen on average 2·6 less than bill (F). Toe is outer front toe.

WING	♂	89·6 (2·06; 45)	86–93	♀	89·0 (2·03; 23)	86–93
TAIL		64·6 (2·09; 37)	60–68		63·8 (2·50; 20)	59–67
BILL (F)		16·1 (0·90; 21)	14·8–17·4		15·8 (0·89; 12)	14·9–17·1
BILL (N)		9·6 (0·57; 22)	8·8–10·5		9·5 (0·39; 12)	8·9–10·0
TARSUS		19·5 (0·49; 18)	18·8–20·5		19·2 (0·49; 10)	18·7–19·7
TOE		20·2 (0·70; 16)	19·2–21·4		19·6 (0·73; 7)	19·2–21·1

Sex differences not significant. Wing rather worn from spring onwards, particularly in breeding season, averaging c. 2 shorter; May–June sample, Netherlands, involving both breeders and migrants, had wing 88·7 (1·79; 20) 85–91. Spain and Portugal, mainly June, sexes combined: 87·3 (1·48; 9) 85–90 (Vaurie 1959; RMNH, ZFMK).

J. t. tschusii. Corsica and Italy, all year; skins (RMNH, ZFMK, ZMA). Sexes combined.

WING	83·9 (2·05; 14)	79–86	BILL (N)	9·6 (0·54; 11)	8·8–10·3
TAIL	61·6 (2·06; 14)	58–65	TARSUS	19·4 (0·54; 12)	18·7–20·1
BILL (F)	16·0 (0·85; 14)	15–17	TOE	19·8 (0·82; 7)	18·8–20·7

J. t. mauretanica. Unsexed, wing 79·2 (15) 76–83 (Vaurie 1959d); wing 81·5 (4) 78–84, tail 63·3 (4) 60–66 (Eck and Geidel 1973).

J. t. sarudnyi. Unsexed, wing 87 (17) 84–90 (Vaurie 1965).

JUVENILE. Wing on average c. 10 shorter than adult and 1st adult, tail c. 5.

Weights. ADULT AND FIRST ADULT. Nominate *torquilla*. On spring migration. Central Nigeria, March: 48·5 (3) 45–52 (Smith 1966). South-east Morocco: 27·7 (41) 23–32 (Ash 1969). Libya, mainly April: 30·6 (3·93; 10) 26–37 (Erard and Larigauderie 1972). Turkey and northern Iran, April–May, ♂♂: 32, 33, 47½ (Schüz 1959; Rokitansky and Schifter 1971). Malta, March–May (including a few *tschusii*) 34 (4·6; 67) 27–46 (J Sultana and C Gauci). Mallorca, ♂♂: 37 (3) 34–40 (ZFMK). Camargue (France), March–May: 33·2 (135) (Glutz and Bauer 1980). Netherlands, late April and early May: ♂ 38·8 (4·83; 6) 31–45, ♀ 35·9 (1) (ZMA).

West Germany: May–June, 40·3 (11) 34–46 (Glutz and Bauer 1980); during nesting 36·5 (2·11; 13) 34–40 (Löhrl 1978). On migration Helgoland (West Germany): 37·7 (26) 34–46, exhausted ♀ 27½ (Hagen 1942).

On autumn migration. Isle of May (Scotland), August and early September: on arrival 33·1 (3·20; 28), on recapture 40·3 (2·01; 5) (Langslow 1977). Netherlands, late August to early October: ♂ 39·0 (2·53; 5) 36–42, ♀ 36·7 (4·32; 5) 32–43; exhausted ♂♂ 22½, 22½; exhausted ♀♀ 22, 23½ (ZMA). Camargue, August–October: 34·6 (258), range (including spring) 22–54 (Glutz and Bauer 1980). Central Nigeria, October: 32½ (Smith 1966).

J. t. sarudnyi. Turkey, April: ♂ 27 (ZFMK). Kazakhstan (USSR): April–June, ♂♂ 32, 40; ♀♀ 34, 39; August–September, ♂♂ 40·3 (3) 35–49 (Dolgushin *et al.* 1970).

J. t. tschusii. Malta: August–November (including some nominate *torquilla*), 33·5 (5·3; 115) 26–50; December–February (including a few nominate *torquilla*) 34 (1·4; 8) 32–36 (J Sultana and C Gauci).

NESTLING AND JUVENILE. At hatching, 1·9–2·1; average on 12th day 25·5; at fledging 24–32 (Sutter 1941). On day 12–13, 25·5 (25) 16–30; on day 15–16, 28·5 (9) 27–30; on day 18–21, 28·4 (19) 20–36 (Löhrl 1978).

Structure. Wing rather short and broad at base, tip rounded. 10 primaries: in adult and 1st adult of nominate *torquilla*, p8 longest, p9 ½–2 shorter, p10 53–60, p7 ½–3, p6 5–7, p5 9–11, p1 22·1 (22) 20–25; in *tschusii*, p8 longest, p9 1½–3 shorter, p10 49–56, p7 0–2, p6 3–6, p5 6–10, p1 19·2 (11) 17–21. Wing more rounded in juvenile, p7–p8 longest; p1–p2(–p3) reduced, shorter than neighbouring s1; juvenile p1 62% of adult length, p2 68%, p3 75%, p4 82%, p5–p6 83%, p7–p8 85%, p9 79%, p10 260% (Glutz and Bauer 1980). P10 strongly reduced in adult, narrow and sharply pointed, almost hidden beneath under primary coverts, tip 12 (16) 9–14 shorter than longest upper primary coverts; juvenile p10 much longer and broader, structure like normal flight-feather, 26–35 shorter than wing-tip, 6 (5) 4–9 longer than upper primary coverts. Outer web of p7–p8 and inner of p7–p9 emarginated. Tail rather long, tip rounded; 12 soft and broad feathers with rounded tips, t1–t3 longest in adult and 1st adult, t4 2–6 shorter, t5 4–10; t6 strongly reduced, a tiny feather hidden at base of t5; tail of juvenile slightly shorter, tip more graduated, feathers slightly narrower; juvenile t1 94% of adult length, t2 95%, t3 88%, t4 68%, t5 50%, t6 38% (Glutz and Bauer 1980). Bill short and straight; rather slender at base, culmen gradually tapering to sharply pointed tip, under mandible with short and angled gonys; tip of bill slightly compressed laterally; no ridges and grooves, unlike woodpeckers (Picinae). Nostrils narrow and slit-like, situated in large depressions, covered below by membrane. Feathering of head and body softer and with different texture than in Picinae. Tarsus rather short and slender, toes rather long and slender; inner front toe with claw *c.* 65% of outer front toe with claw, outer hind *c.* 105%, inner hind *c.* 55%. Claws short and rather slender, sharply pointed.

Geographical variation. Rather slight, involving mainly colour and in part bridged by marked individual variation; clinal in northern part of range, difference more pronounced in Mediterranean basin. Nominate *torquilla* from Scandinavia and central Europe eastward a rather pale and well-marked race: upperparts grey with brown tinge to outer scapulars, tertials, and upper wing-coverts and with broad and contrasting large black marks from central hindcrown to mantle and on outer scapulars; underparts pale with narrow black bars on forebody and flanks and fine spots on breast. In southern Urals and western Siberia, grades into paler *sarudnyi*, which has grey of upperparts purer and paler, throat and chest paler buff with finer and less numerous black bars, and remainder of underparts whiter with less spots. No certain breeders from Asia Minor or Caucasus area examined; these probably inseparable from nominate *torquilla* (see Vaurie 1959d); migrants from this area involved both *sarudnyi* and nominate *torquilla*. *J. t. tschusii* from Corsica, Sardinia, Italy, and coastal Yugoslavia south to northern Dalmatia (Matvejev and Vasić 1973) distinctly darker than other races, especially on underparts; grey of upperparts and tail rather dark, black stripe on central hindcrown to mantle and stripes on outer scapulars heavy and broad, stripe on crown often extending to forecrown; black bars on throat and chest broader, up to 1 mm wide, ground-colour often deeper cinnamon-buff; breast, belly, and vent rather heavily barred or spotted black; flight-feathers blacker with slightly smaller and deeper rufous spots; size similar to nominate *torquilla* but wing shorter and more rounded. A winter bird from Athens (Greece) examined was *tschusii* (RMNH) and this race once found Crete (Eck and Geidel 1973); *tschusii* either breeds in Greece, or winter records may be of birds from Dalmatia, as this race known to be short-distance migrant (see Movements). Situation in western Europe not fully elucidated; birds from southern England, western France, and Iberian peninsula have grey of upperparts slightly browner than in typical nominate *torquilla*, but (unlike *tschusii*) black marks from central hindcrown to mantle and outer scapulars smaller and less distinctly defined, sometimes restricted to upper mantle only; underparts like nominate *torquilla*, but especially in Iberian breeders sometimes as heavily marked as in *tschusii*. As a few similarly marked birds occur in populations of nominate *torquilla* and difference in size slight, all west European birds provisionally included in nominate *torquilla*. North-west African *mauretanica* smaller than other west Palearctic races; colour as in *tschusii*, but upperparts slightly darker and ground-colour of throat and chest paler cream with slightly less heavy marks; in some respects, closer in colour to Iberian populations than to *tschusii*. Races from central Asia and Yenisey eastwards gradually darker and smaller towards Japan; see Hartert and Steinbacher (1932–8), Vaurie (1959d), and Eck and Geidel (1973).

Forms superspecies with Red-breasted Wryneck *J. ruficollis* of Afrotropics (Snow 1978; Short 1982). CSR

Subfamily PICINAE woodpeckers

Small to large perching and climbing birds with hard chisel-like bills and stiff wedge-shaped tails. Most species are climbers, foraging on tree trunks or limbs for timber- or bark-dwelling insects; a few are more or less terrestrial, specializing in ants, and others quite regularly capture airborne insects. 171 species in 23 genera, divided by Morony *et al.* (1975) into 6 tribes. (1) Melanerpini: *Melanerpes* (21 species), *Sphyrapicus* (4 species), *Xiphidiopicus* (1 species). (2) Campetherini: *Campethera* (10 species), *Geocolaptes* (1 species), *Dendropicos* (12 species), *Picoides* (33 species). (3) Colaptini: *Veniliornis* (12 species), *Piculus* (7 species), *Colaptes* (8 species), *Celeus* (11 species). (4) Campephilini: *Dryocopus* (6 species), *Campephilus* (11 species). (5) Picini: *Picus* (14 species), *Dinopium* (4 species), *Chrysocolaptes* (2 species), *Gecinulus* (2 species), *Sapheopipo* (1 species), *Blythipicus* (2 species), *Reinwardtipicus* (1 species). (6) Meiglyptini: *Meiglyptes* (3 species), *Hemicircus* (2 species), *Mulleripicus* (3 species). 10 species of 4 genera breed in west Palearctic. Subfamily of worldwide distribution, excluding Australasia and polar regions. Recent monograph by Short (1982).

For general features, moults, etc., see Picidae. Considerable sex differences in bill and tail length in some species (usually insular or otherwise-isolated forms). Usually 12 tail-feathers; outer pair (t6) short, soft, and sited above t5, absent in some; remaining feathers very stiff, pointed, with broad shafts, functioning as prop when climbing. Bills vary from short to rather long, but always with very hard, strong, rhamphotheca; culmen and gonys sharply ridged, tip pointed or chisel-like. Nostrils covered by stiff, short, forwardly directed feathers. Skulls robustly constructed. For adaptations to hammering and climbing, see Burt (1930), Beecher (1953), Spring (1965), and Jenni (1981). Hyoid skeleton greatly elongated, with horns curving far round over skull, even continuing around orbit or into upper jaw. Tongue barbed, bristle fringed in some genera. Feet zygodactyl, but both hind toes capable of turning sideways or forward in some species (Bock and Miller 1959).

Plumages pied, greenish, brown, or all black, but often with some bright red on head or underparts. Often crested. Sexual dimorphism general, usually involving crest or crown, and malar region.

Colaptes auratus (Linnaeus, 1758) Northern Flicker

FR. Pic flamboyant   GE. Goldspecht

Breeds from treelimit in Alaska and Canada south to Gulf of Mexico and Greater Antilles. Northern populations mainly migratory; in Canada, only a few winter in Ontario, Quebec, and maritime provinces (Godfrey 1966). Common coastal migrant in North America, often passing in loose flocks (e.g. Bent 1939, Stewart and Robbins 1958). Involved relatively frequently in weather-related ship 'falls' in western North Atlantic, and one completed a crossing aboard ship without being handled in October 1962, being seen to fly ashore in Cork, Ireland (Durand 1963, 1972). ♀ in Denmark, May 1972, also presumed to be ship-assisted or an escape (*Dansk orn. Foren. Tidsskr.* 1974, **68**, 138–44).

Picus canus Grey-headed Woodpecker

PLATES 77 and 78
[facing pages 807 and 830]

DU. Grijskopspecht   FR. Pic cendré   GE. Grauspecht
RU. Седой дятел   SP. Pito cano   SW. Gråspett

Picus canus Gmelin, 1788

Polytypic. Nominate *canus* Gmelin, 1788, Europe and western Siberia east to foothills of Altay and northern end of Lake Baykal. Extralimital: *biedermanni* Hesse, 1911, central Asia from Tarbagatay, Altay, and Sayan mountains east to southern Transbaykalia; *jessoensis* Stejneger, 1886, eastern Siberia south to northern China and Hokkaido; 9–10 further races from Korea, central China, and Himalayas south to Sumatra.

Field characters. 25–26 cm; wing-span 38–40 cm. About 20% smaller than adult Green Woodpecker *P. viridis*, with finer bill. Medium-sized woodpecker, with basic plumage colours and pattern recalling *P. viridis* but differing distinctly in lighter flight, greater agility, largely grey, less marked head, and greyer underparts. Sexes dissimilar; no seasonal variation. Juvenile separable.

ADULT MALE. Grey head topped with pure red forecrown and marked only by black line across lores and narrow black moustache above pale throat. Back, scapulars,

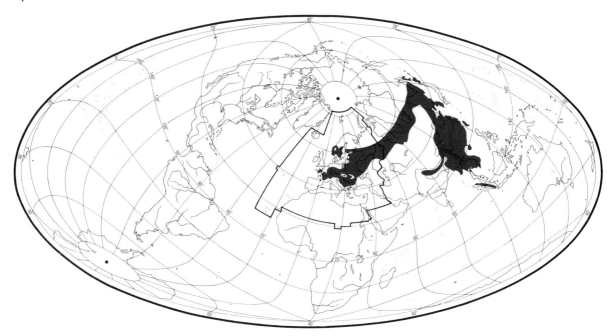

and wing-coverts pale but intense green, lacking yellow tone of *P. viridis*; rump yellower than back but less distinctly so than in *P. viridis*. Breast and underbody pale grey, lacking any barring even on rear flanks (unlike *P. viridis*). Tail green-brown, with indistinct yellower barring. Folded primaries brown-black, barred almost white; inner secondaries green-brown, faintly barred yellow. In flight, primary bars more obvious than in *P. viridis*. ADULT FEMALE. Duller than ♂; crown lacks red but shows black spots and streaks at close range. Black moustache even narrower. JUVENILE. Plumage less clean and browner than adult. Flanks show barring but confusion with *P. viridis* still prevented by lack of striking head pattern. Forecrown of ♂ red but less extensive than adult.

Occurs widely within range of *P. viridis* but absent from that of Levaillant's Woodpecker *P. vaillantii*. At distance, separation from *P. viridis* other than by size not easy but at closer range grey head (and its restricted pattern of marks) allows instant identification. Flight similar to *P. viridis*, but wing-beats lighter and faster. Gait while climbing and on ground as *P. viridis*.

Voice similar to *P. viridis* but song more whistling, less full-throated and laughing: 'kee' sounds, slowing down and becoming lower in pitch towards end. Drums frequently (unlike *P. viridis*): lasts *c*. 1–2 s without any acceleration; thus similar to Lesser Spotted Woodpecker *Dendrocopos minor*, but given less often—only 2–3 times per min.

Habitat. Resident, subject to local seasonal movements, predominantly in temperate and marginally in warm boreal zone of west Palearctic, largely overlapping with Green Woodpecker *P. virdis*, but differing somewhat in more continental and generally more upland and montane distribution, although nature and extent of competition for habitat remain obscure. In lowlands appears content with smaller woodlands than *P. viridis*, and favours moist carrs or trees fringing rivers, even where old timber infrequent: also occurs in open woods of beech *Fagus*, oak *Quercus*, or hornbeam *Carpinus*. Rarely above 600 m in central Europe, but in some regions ascends above 1000 m and locally, especially after breeding season, even slightly above 2000 m, particularly among larches *Larix* where ant (Formicidae) populations are high. This diet leads to characterization of the 2 *Picus* species as 'ground woodpeckers', although dependence on holes in trees for nesting and roosting governs distribution, both species being strongly attached to tree-trunks rather than branches or canopy. *P. canus* occurs unevenly and is often subject to numerical fluctuations, further complicating problem of tracing its relationships with *P. viridis*. Is free of competition with *P. viridis* where conifers predominate and on high ground towards treelimit; also occurs in clearings, rides, and grassy glades within forests. (Niethammer 1938; Bannerman 1955; Voous 1960; Glutz von Blotzheim 1962; Glutz and Bauer 1980.)

Flight almost entirely in lower airspace.

Distribution. LUXEMBOURG. Apparently some spread to west (Glutz and Bauer 1980). FINLAND. Has occasionally nested further north (Merikallio 1958). EAST GERMANY. Bred Mecklenburg 1849 (Glutz and Bauer 1980). SPAIN. 1 pair and 1 juvenile, Belagua, Pyrénées (G Van der Kelen). TURKEY. May breed occasionally (e.g. 4 specimens from Bolu area August–October: Kumerloeve 1961), but no proof (MB, RFP).

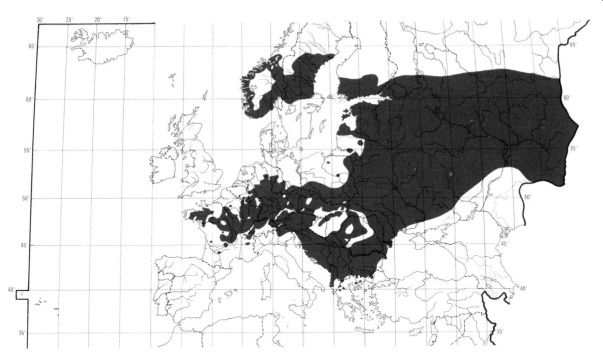

Accidental. Netherlands.

Population. FRANCE. Estimated 1000–10000 pairs (Yeatman 1976). BELGIUM. Rare and fluctuating (Lippens and Wille 1972; Glutz and Bauer 1980). 1–8 pairs (PD). LUXEMBOURG. About 20 pairs (Lippens and Wille 1972). WEST GERMANY. Over 12500 pairs (Rheinwald 1982). SWEDEN. About 500 pairs; formerly considered very rare, probably increasing but may largely reflect increased observer activity (Ahlén and Andersson 1976; LR). FINLAND. Very scarce, fluctuating (Merikallio 1958); no clear evidence of decrease, and in early 1970s some increase due to mild winters (OH). EAST GERMANY. Scarce, fluctuating in lowlands (Glutz and Bauer 1980; WM). POLAND. Scarce; slight recent increase in Silesia (Tomiałojć 1976a). CZECHOSLOVAKIA. No known changes in numbers (KH). HUNGARY. Rare (LH). ITALY. Very scarce (Brichetti 1982). GREECE. Sparse (WB, HJB, GM). USSR. Estonia: locally fairly numberous (HV). Latvia: fairly rare (Baumanis and Blūms 1972). Lithuania: declining, now very rare (Jankevičius *et al.* 1981).

Oldest ringed bird 5 years 4 months (Rydzewski 1978).

Movements. Resident in west Palearctic. Some dispersal occurs within breeding range, but extends beyond it to small (irregular) extent only. In winter, Finland, enters urban areas, visiting garden bird-tables (Vikberg 1982). In Scandinavia, individuals occur irregularly at coastal sites, notably in September–October (Haftorn 1971; Ree 1977; SOF 1978); one exhausted bird on Uppland coast, October 1972, had probably crossed Gulf of Bothnia at Åland narrows (Lundberg and Swanqvist 1973). Almost annual, September–October, on shores of Lake Vänern (Sweden); some show migratory restlessness, though none seen to fly out over lake (Ehrenroth 1973). Extent of eruptive movement unclear, but probably small; in autumn irruption counts, 1962–3, Pskov region (USSR), total of 9 birds, compared with 597 Great Spotted Woodpeckers *Dendrocopos major* (Meshkov and Uryadova 1972). In west-central Europe, many leave deciduous woodlands in late autumn, to winter in riverine cover at lower altitudes (Glutz and Bauer 1980). Longest ringing movement—Finnish bird which moved 65 km SSW, recovered October (Välikangas and Hytönen 1932). Stragglers to eastern Netherlands (4 records: February–April, August) were maximum 150 km from nearest (West German) breeding areas (Blankert and Heer 1982).

In treeless coastal belt of Ussuriland (eastern Siberia), marked southward movement October–November, with less obvious return in March (Panov 1973), but extent of population emigration unclear (Polivanov 1981).

Food. Diet similar to Green Woodpecker *P. viridis* but more varied, with less specialization on ants (Fedyushin and Dolbik 1967; Haftorn 1971). Shorter tongue and differences in leg length, leg musculature, and tail length indicate feeding ecology more like *Dendrocopos* woodpeckers (Rüger 1972). Forages chiefly on ground but also commonly on low trees and buildings and at cracks on walls, rock faces, etc. (Loos 1905; Löhrl 1977; Glutz and Bauer 1980). On ground, probes and digs in soil with bill, licking up ants and other insects with sticky tongue. In trees, searches mainly in broken and rotten areas, tapping only rarely (Short 1973; Löhrl 1977). In Japan, observed following Black Woodpecker *Dryocopus marbius* feeding on insects exposed by its excavations into trees (Matsuoka 1976). Feeds at anthills—but not when snow-covered, and presumed then unable to locate them (unlike *P. viridis*) (Virkunnen 1967). Actively seeks torpid flies from end of August to early November, mainly in early morning when they may be licked out of resting places on walls, etc. (Baier 1973). Commonly visits feeding tables in winter (Löhrl 1962*b*; Ehrenroth 1973; Glutz and Bauer 1980) and occasionally household scrap heaps (Virkunnen 1967; Polivanov 1981). Observed drinking sap of maple *Acer* oozing from natural hole in bark (Panov 1973) and will drink sap from holes drilled in bark by other Picinae, but not known to drill such holes itself (Klima 1959*b*). One pair observed stealing eggs of Coot *Fulica atra* over period of several days: eggs carried by hole in their side to prepared notch in bark; bottom of egg pecked out and contents drunk by holding bill at angle under egg (Niggeler 1968). Apparently eats small quantities of sand from joints in brickwork, presumably to aid digestion (Baier 1973). In winter, Finland, forages 35% of time in mixed forest and 45% in urban areas; 53% of foraging time spent seeking insects on tree trunks, 24% at feeding tables, and 24% at other sites (Vikberg 1982).

Diet given here includes extralimital studies. Food mainly adults, pupae, larvae, and eggs of ants (Formicidae, of various species much as for *P. viridis*, including *Formica rufa* and *Camponotus*), and flies (Diptera). In small numbers, takes crickets (Orthoptera, e.g. *Gryllus*), aphids (Aphidoidea), lacewings (Neuroptera, e.g. *Chrysopa*), adult and larval beetles (Coleoptera), caterpillars (Lepidoptera), and spiders (Araneae). Also single records of amphipods (*Gammarus*) and eggs of Coot *Fulica atra*. Vegetable food includes apples, pears, and cherries, and berries of rowan *Sorbus aucuparia*, elder *Sambucus nigra*, *Philodendron amurense*, and sea buckthorn *Hippophae rhamnoides*; seeds of cedar *Thuja*, *Rhus typhina*, and hazel *Corylus avellana*; seeds, bread, and fat from feeding tables; tree sap. (Naumann 1826; Dementiev and Gladkov 1951*a*; Vorobiev 1954; Klima 1959*b*; Conrads and Herrmann 1963; Fedyushin and Dolbik 1967; Niggeler 1968; Dolgushin *et al*. 1970; Radermacher 1970; Zingel 1970; Haftorn 1971; Baier 1973; Panov 1973; Becker 1978*a*; Glutz and Bauer 1980; Polivanov 1981.)

7 stomachs from Belovezhskaya pushcha (Belorussiya, USSR), October–June, held mainly ants, up to 200–250 in each; 1 also contained 30 flies (Tachinidae). 7 stomachs from elsewhere in Belorussiya, April–November, held only adults and pupae of ants (*Formica rufa*, *Camponotus*). (Fedyushin and Dolbik 1967.) 13 stomachs from Moldavia contained 'black' and 'red' ants and pupae, and larvae of beetles (e.g. Scolytidae, Cerambycidae, Chrysomelidae, and others) (Averin and Ganya 1970). Other stomach analyses extralimital. One stomach from Altay mountains (USSR), contained 57 ladybirds (Coccinellidae) only; others contained other adult insects and larvae and cedar seeds. (Kuchin 1976.) Others from same area contained 'red ants' and their larvae, up to 150 per stomach (Dolgushin *et al*. 1970). One stomach from Primor'e (USSR), March, contained only ants; 2 from same locality, July, also contained ants, 1 other, March, beetle adults (Chrysomelidae) and larvae (Cerambycidae) (Panov 1973). In detailed study of faecal contents found in roost holes during autumn and winter in Japan, items comprised 90·5% ants *Lasius niger*, 0·9% other ant species, 1·6% Ichneumonidae, and 4·7% seeds; beetles, flies, and spiders also present; some seasonal variation, particularly during snow when occurrence of *L. niger* declined (Matsuoka and Kojima 1979).

Young fed by regurgitation, with adult, pupal, and larval ants. Only *Myrmica rubra* and *Lasius flavus* identified (Conrads and Herrmann 1963), but species probably include all those eaten by adults. In study near Lucerne (Switzerland) only pupae recorded (Bussmann 1944). Because of food-carrying technique, feeding visits less frequent than in *Dendrocopos*. In Primor'e, first feed *c*. 06.00 hrs; *c*. 5 feeds per hr in early morning, 1–3 per hr thereafter (Polivanov 1981). Time between visits averaged *c*. 50 min when young half grown, 30 min (22–52) when nearly full grown (Ehrenroth 1973). 9 feeding visits recorded on 2nd day, rising to maximum of 26 on 17th day and 15 visits on day just before fledging, with feeds most frequent morning and evening (Bussmann 1944). At another nest, however, times between visits over 1 day recorded as (in order) 199, 133, 64, 77, 80, 94, and 107 min (Loos 1903*a*). In Teutoburger Wald (West Germany), 6–8 feeding visits on day of hatching, over 10 by day 4; from day 9, average 1·9 per hr. In three consecutive years, another pair averaged 1·7, 1·66, and 2·15 visits per hr; first two rates considered normal. Each visit includes numerous regurgitations to individual chicks; in one pair, ♂ averaged 6·2, ♀ 4·9. In June, first feed of day 04.16–04.51 hrs, last feed *c*. 20.00 hrs. Intervals between feeding visits approximately equal throughout day and not affected by weather (Conrads and Herrmann 1963). ASR

Social pattern and behaviour. Some aspects less well known than Green Woodpecker *P. viridis*, but overall similar to that species. Fullest study by Conrads and Herrmann (1963) in Teutoburger Wald (West Germany); see also Loos (1903*a*, *b*, 1905) and, for other important references, summary by Blume (1981).

1. Mostly solitary. Where local movement occurs, small 'flocks' sometimes seen at passage times: e.g. in Primor'e (eastern USSR), 8 together late November; birds otherwise well dispersed October–November, with 4–6 along 5–6 km of coastal belt (Panov 1973). In Thüringen (East Germany), home-range used throughout the year *c*. 500 ha (Mey 1967); in Hessen (West Germany), *c*. 500–700 ha (Blume 1965*b*). In Primor'e, autumn–winter, birds move only *c*. 200–300 m from roost; 1·5–2 km, February–March (Polivanov 1981). BONDS. Monogamous mating system. Type and duration of pair-bond probably much as

in *P. viridis*. Some contact between paired birds outside breeding season: e.g. apparent pair together at feeding station, late winter–spring (Löhrl 1962b; Schwammberger 1965). For hybridization with *P. viridis*, see that species. Young cared for by both parents up to and to some extent beyond fledging. Feeding of young shared equally or 1 parent does more. Single parent apparently able to cope if mate lost late in nestling period (Loos 1903a; Bussmann 1944; Conrads and Herrmann 1963). Duration of post-fledging care perhaps varies: according to Bussmann (1944), family breaks up c. 16 days after fledging having spent c. 10 days together near nest; Conrads and Herrmann (1963) doubted whether family bond maintained more than a few days after fledging but several records of (part-) family groups (see *P. viridis*) in August (Glutz and Bauer 1980; see also Blume 1965b; 1981). In Primor'e, nomadic groups of 2–5 independent young reported from late June through July (Panov 1973). BREEDING DISPERSION. Solitary and territorial. In Yonne (France), probably c. 1 pair per km² but pronounced flying activity in spring may give false impression of density (Guichard 1954). In open beech *Fagus* and oak *Quercus* woodlands of the Ajoie (north-west Switzerland), 0·6 pairs per 10 ha, in *Quercus* and hornbeam *Carpinus* near Allschwil (Basel, Switzerland), 1·0 pairs per 10 ha (Glutz 1962). Larger-scale studies rarely show more than 0·2 pairs per km². Within territory, apparently defends only critical sites, as in Black Woodpecker *Dryocopus martius*; see that species. Nest-trees of neighbouring pairs at least 1·25 km apart (Glutz and Bauer 1980). In Teutoburger Wald, 1 territory (estimated from territorial and courtship activity) c. 1·3 km²; of 7 territories, only 2 significantly less than 1 km² (Conrads and Herrmann 1963). Near Winterthur (Switzerland), 2 territories 1·5 km² and 2 km² (Glutz and Bauer 1980). In Hessen, also c. 2 km² (Blume 1981). In experimental work (see Antagonistic Behaviour, below), searching flights elicited by playback of calls and orientated towards prospective nest-trees may give early idea of territory's later structure. 1 bird responded up to c. 60 m around Drumming tree, another followed stimuli for c. 900 m (Blume and Jung 1956; Blume 1965b, 1981). Regular sites for Drumming or calling (see *P. viridis*) up to c. 500 m from nest-hole (Koch 1954; Conrads and Herrmann 1963; Erdmann 1973; Blume and Ogasawara 1980). Ranges up to c. 1·2 km from nest when foraging for young (Glutz and Bauer 1980). In one territory, nest-site moved by 50 m and 600 m in successive years. Recorded nesting as close as c. 50 m to *P. viridis* (Conrads and Herrmann 1963). ROOSTING. Essentially as in *P. viridis*. No excavation of special roost-hole reported but readily uses those of other Picidae, especially *D. martius* (Blume and Jung 1959); perhaps less often in building or nest-box than *P. viridis* (e.g. Berndt 1961), but enlarge holes of boxes designed for Starling *Sturnus vulgaris* (Henze 1943). Birds may keep to same general area but change holes more frequently than *P. viridis* (Blume and Jung 1959; Blume 1981). ♀ goes to roost later than ♂ and may have to make do with less favourable site (Blume 1963b). In Primor'e, mid-January, one ♂ in roost c. 18.00–08.00 hrs (Polivanov 1981). ♂ begins to roost in nest-hole several days before laying, entry becoming much earlier shortly before 1st egg laid (Conrads and Herrmann 1963, which see for details of timing). Outside breeding season at least, approach to roost usually direct (i.e. intermediate stations not used) and unobtrusive, unlike *P. viridis* (see, however, Polivanov 1981). In summer, calls (see 2a in Voice) fairly regularly when flying into roosting area (Glutz and Bauer 1980). More restless in roosting area than *P. viridis*, tending to fly about more than to climb (Blume 1965b). Bird normally briefly alert before entering but then does so quickly, without intention movements (Blume and Jung 1959); see also part 2. Like *P. viridis*, typically spends some time apparently dozing below roost-hole after morning exit; may look out, yawn, etc., for c. 10 min beforehand; stretching, etc., especially pronounced (see Blume 1965b, Glutz and Bauer 1980). For further information on beginning and end of activity, see Blume and Jung (1959), Blume (1965b). Comfort behaviour resembles that of Syrian Woodpecker *Dendrocopos syriacus* (Glutz and Bauer 1980); for latter, see Winkler (1972b). Bathing recorded only exceptionally (Conrads and Herrmann 1963).

2. Not particularly shy of man—e.g. allowing approach to c. 5 m (Loos 1905; Zingel 1970)—but has remarkable ability to freeze for up to c. 30 min on opposite side of trunk to danger, or longitudinally on branch, and can then be difficult to detect (Loos 1905; Guichard 1954; Conrads and Herrmann 1963; Ehrenroth 1973). Cryptic colouration most effective on ancient beech *Fagus* (Blume 1965b). Also recorded freezing crossways on branch when Goshawk *Accipiter gentilis* overhead (Hildén 1955). Highly sensitive to disturbance at roost (Blume and Jung 1959), but behaviour depends on (e.g.) degree of familiarity with roost-hole, frequency of disturbance, light intensity, etc. (Glutz and Bauer 1980). Unlike *P. viridis* which may fly off when disturbed before entering roost, *P. canus* sometimes moves only slightly aside then speedily enters hole (Blume 1965b). ♂ threatened squirrel *Sciurus* by Head-swaying, beating wings, and fanning tail; also gave squeaking sounds—presumably threat-calls (Glutz and Bauer 1980: see 2b in Voice). Frequent looking out of roost-hole with rapid Head-swaying (see below) typical. Likely to leave if (only slight) disturbance persists and tends to come to entrance if suspicious, even when dusk well advanced (Blume and Jung 1959; Blume 1965b). ANTAGONISTIC BEHAVIOUR. (1) General. Advertising-call (see 1 in Voice) given with head held up and, unlike *P. viridis*, bill (nearly) closed; bird usually perched vertically but sometimes crossways on branch (Blume and Jung 1959; also Steffen 1956). Call far-carrying and apparently plays some part in advertising and defence of particular sites within territory (see *P. viridis*) though also important in attracting mate and contact thereafter (Conrads and Herrmann 1963). In Yonne, Advertising-call given mainly early March to mid-May, ♂ also performing long flights at this stage (Guichard 1954). In Teutoburger Wald, territorial and courtship activity (including Advertising-call) February–April, mostly when temperature above 10°C (Conrads and Herrmann 1963). Unpaired birds may call later: e.g. in early June, ♂ called persistently near occupied territory but made no intrusion (Blume and Ogasawara 1980). In some areas, display much reduced from about mid-April (*P. viridis* frequently gives Advertising-call at this time), probably indicating completion of nest-site selection or start of excavation (Blume 1981). Territorial defence (see Breeding Dispersion, above) recorded from early April mainly by ♂; ♀ participates but generally silent (Conrads and Herrmann 1963). Brief resurgence of territorial activity with Advertising-calls in autumn (Glutz and Bauer 1980; Bergmann and Helb 1982). In Primor'e, Advertising-call given at roost in winter, morning and evening (Polivanov 1981). Drumming (see Voice), mainly by ♂, serves to advertise territory (Short 1973); typically performed in relaxed posture, plumage ruffled. Birds Drum especially at beginning (when both sexes also give Advertising-calls) and end of activity during pair-formation and nest-site selection. Almost invariably associated with roost- or nest-hole (Conrads and Herrmann 1963; Blume 1965b, 1981). Shorter bursts (maximum 20 strikes; see Voice) of 'threat Drumming' given in autumn and early winter (Radermacher 1970; Glutz and Bauer 1980). (2) Threat and fighting. Less well studied than in *P. viridis*, but most aspects apparently similar. Where several birds displaying, persistent calling and chasing recorded sporadically from March, also in May (Blume 1981). In chases at tree-top height involving up to

3 birds, slower flight performed intermittently, and chirring calls like White-backed Woodpecker *D. leucotos* occasionally given (see 3c in Voice). Birds may land on tree and adopt Bill-up posture with feathers sleeked (Polivanov 1981; see also Nilsson 1942, Panov 1973). After vigorous chase by 3 birds (2 ♂♂), one flew to ground and hopped away (Loos 1905). On hearing Advertising-call of conspecific bird (also imitation or playback) in courtship period, territory-owning ♂ usually responds quickly and vigorously. Performs conspicuous searching flights through tree-tops or above; quickly locates general direction of sound and then makes repeated flights to ascertain distance. Usually then flies in low arc around source, sometimes landing behind it. Moves from tree to tree in deeply undulating flight around sound-source (Conrads and Herrmann 1963). Before landing, sometimes uses Fluttering-flight with rapid and shallower wing-beats (see also Blume 1962c and Heterosexual Behaviour, below) during which excitement and threat-calls given (see 2 in Voice). On tree, ♂ may perform excited, fairly rapid and jerky upward climb, and assume tense, 'bull-necked' posture (see illustration in Glutz and Bauer 1980). Regularly gives Advertising-call (see 1 in Voice for excited variants) with tail spread, and threat-calls, and also Drums. Bird may also throw head back and (Mock-)peck at tree-trunk (possibly same as Excitement-pecking of *P. viridis*; see also Blume and Jung 1958). Wing-twitching and spreading of wings to reveal bright yellow-green rump, Head-swaying (sometimes whole fore-body moved and wings beaten), bobbing movements (moving from crouched to more erect posture), occasional Mock-preening, and Mock-feeding on ground also recorded. ♀ tends to approach sound-source silently, and to spread tail on landing; less commonly, climbs excitedly; 'bull-necked' posture apparently not recorded; see, however, Schmoll (1973) for excited Tapping by ♀ in response to playback. In experiments with playback and imitation, birds call less when nearer holes, but Drumming then more vigorous (Blume and Jung 1958, 1959; Conrads and Herrmann 1963; Conrads 1964; Blume 1965b, 1981; Glutz and Bauer 1980.) In interaction between adult ♀ and juvenile ♀ (Fig A), adult ♀ performed apparent high-intensity threat: Head-swayed from side to side with bill open, bowing simultaneously; gave threat-calls (see 4b in Voice) in rhythmic synchrony with movements. Juvenile ♀ c. 2 m away behind thick branch spread wings sideways, Head-swayed (less so than adult), and gave constant 'äk äk' sounds,

apparently in appeasement. In ensuing fight, birds tumbled out of tree and pursuer had tail well spread when landing crossways on branch (Conrads 1964; see also *P. viridis*). Occasional vocal contact occurs with *P. viridis* (Conrads and Herrmann 1963) and much antagonism, late summer and autumn when roost-holes (re-)occupied. *P. canus* understands Threat-call and 'kjuiuh' call of *P. viridis*, and responds with similar behaviour; generally subordinate to *P. viridis* in disputes over roost- or nest-holes (Blume 1957, 1981; Blume and Jung 1959). Foraging bird ducked under branch when calling *P. viridis* flew over (Short 1973). For further details, see Küchler (1951), Burnier (1961), and Conrads and Herrmann (1963). Generally subordinate also to *D. martius*, but once successfully used louder wing-noise (briefly) to scare off *D. martius* from food-source (Matsuoka 1976: see Voice). HETEROSEXUAL BEHAVIOUR. (1) General. Unusual among west Palearctic Picinae in remarkable contrast between unobtrusive character outside breeding season and conspicuous vocal and other activity from start of that period. See above for further details on timing of various activities. More conspicuous display behaviour wanes at nest-site selection, other features decreasing with start of incubation (Conrads and Herrmann 1963). As in *P. viridis*, ♂ generally dominant over ♀ (Glutz and Bauer 1980). (2) Pair-bonding behaviour. Like (e.g.) *P. viridis*, birds spend more time in morning and evening near roost-hole as breeding season advances. Potential mates make and maintain contact with especially long bouts of Advertising-calls (♂ also Drums), and attempt to locate each other with searching flights over up to c. 900 m (Blume and Jung 1959; Conrads and Herrmann 1963; Blume 1981). For details of closer approach (including Fluttering-flight), see Antagonistic Behaviour (above). When together on tree, birds may climb excitedly and, at short distance, perform rapid, arrhythmic Head-swaying, accompanied by threat-calls. Later, use Contact-calls (Conrads and Herrmann 1963: see 4a in Voice). In Primor'e, ♂ performed Fluttering-flight, and both birds climbed in spirals on tree and adopted Bill-up posture (not leaning as far back as *D. leucotos*; see also *P. viridis*); ♀ then moved down tree holding body close to trunk, half-opened and quivered wings (see also subsection 5, below); ♂ flew in Fluttering-flight to next tree, ♀ following (Polivanov 1981). During pair-formation (probably completed by early April in most cases), birds much livelier and more mobile than *P. viridis*; fly to particular points in territory together, separate after meetings as described, spend time apart, etc. ♂ who lost mate Drummed and called persistently for c. 2 months afterwards (Blume 1965b). (3) Nest-site selection. Essentially as in *P. viridis* (Blume 1981). However, prospective site advertised not by persistent Advertising-calls but by Drumming—mainly by ♂ who thus perhaps chooses site (Conrads and Herrmann 1963). In one case, ♂ and ♀ often together at *D. martius* hole from mid-March, but eventually nested at different site (Blume 1981). New hole normally excavated only if suitable old holes not available or occupied (e.g. by Starling *Sturnus vulgaris*), but apparently more often than in *P. viridis* (Conrads and Herrmann 1963; Glutz and Bauer 1980). Both sexes excavate nest-hole; either may take greater share (Loos 1903a; Conrads and Herrmann 1963). Starts may be made at various sites during pair-formation (Conrads and Herrmann 1963), including 1–2 in tree where birds eventually nest (Bussmann 1944; Niggeler 1968). Excavation takes c. 9 days (Bussmann 1944) or up to 3 weeks (Guichard 1954). Little excavation done in early morning; for further details, see Loos (1903a) and Conrads and Herrmann (1963). Heterosexual activity frequently interrupts excavation. ♂ often gives long (up to c. 10-unit) Advertising-calls to lure ♀ away; threat or Contact-calls given when birds meet near nest-hole (keeping their distance), or when one approaches alone (Conrads and Herrmann

A

1963). Loud Advertising-calls generally cease with completion of nest-hole (Blume and Ogasawara 1980). Nest guarded (see *P. viridis*) most of time during building and (for up to 3 weeks) between completion and laying (Blume and Ogasawara 1980; Blume 1981; see also Guichard 1954). For Head-swaying, etc., directed at man during excavation, see Conrads and Herrmann (1963). (4) Courtship-feeding. In observations on pair seen to feed (unusually) on eggs of Coot *Fulica atra* (see Food), ♂ sometimes called ♀ out of nest with whinnying sound and she took contents of egg wedged in anvil; ♂ also fed ♀ at nest-entrance, allowing liquid to run into her bill, while ♀ made tugging movements (Niggeler 1968). No further information. (5) Mating. Few details, but resembles *P. viridis*—although sexually motivated Head-swaying (see *P. viridis*) apparently rarely recorded in *P. canus* and no indication of its occurrence prior to copulation (Glutz and Bauer 1980). Wing-quivering in ♀ said by Polivanov (1981) to be normal preliminary. Once when both birds feeding on ground, ♀ (possibly making 'dü' sound) approached ♂; ♂ mounted and copulated; both fed or Mock-fed afterwards (Loos 1905). Rapid and loud Contact-calls (see 4a in Voice) given during copulation (Glutz and Bauer 1980). (6) Behaviour at nest. When ♂ enters nest-hole in evening (see Roosting), ♀ tends to stay nearby for up to *c.* 30 min before flying off to her own roost (Conrads and Herrmann 1963). Near laying, ♀ spends quite long periods in nest and may peck or Tap at cavity wall (Blume and Ogasawara 1980). Birds progress gradually from occupation of empty nest-hole to regular nest-reliefs associated with start of incubation proper (Conrads and Herrmann 1963). ♂ takes greater share of incubation (also sitting at night), ♀ doing *c.* 6–7 hrs per day at most (Bussmann 1944); for further details, see Loos (1903*a*) and Conrads and Herrmann (1963). (2–)3–4 nest-reliefs daily. Incoming bird gives Contact-call to announce arrival; ♂ may also Drum or give Advertising-call. While waiting for ♀ to emerge, ♂ adopts Nest-relief posture (Fig B); when ♀ leaves, ♂ spreads one wing and tail slightly. Waiting ♀ more crouched and does not open wing (Conrads and Herrmann 1963; also Loos 1903*a*, Bussmann 1944). ♂ may attempt to drive off ♀ by giving threat-calls (Blume and Jung 1959), but obvious threat noted only once (later in cycle) by Conrads and Herrmann (1963). RELATIONS WITHIN FAMILY GROUP. Parents cease giving loud calls from hatching (Conrads and Herrmann 1963). Young brooded in regular pattern of nest-reliefs (as during incubation) for *c.* 5 days, then in *c.* 15-min periods, adult not waiting for mate to arrive. Daytime brooding may continue up to *c.* 9 days, depending on weather; nocturnal (by ♂) up to *c.* 11–14 days (Loos 1903*a*; Bussmann 1944, which see for physical development of young; Conrads and Herrmann 1963). Up to *c.* 7 days, young huddle together when unattended (Glutz and Bauer 1980). In larger broods, nestlings form 2 layers (Blume and Ogasawara 1980). Adult may give (quiet) Contact-call during approach to nest and when feeding young (Loos 1903*a*; Bussmann 1944; Conrads and Herrmann 1963). Swollen gape-flanges of young sensitive up to *c.* 17–18 days (Bussmann 1944) and adult, entering (partially) to feed young up to *c.* 14 days, seeks contact with these by moving head about (Conrads and Herrmann 1963; see Glutz and Bauer 1980 for illustration and details). Parents rarely in nest together up to *c.* 9 days after hatching. Young open eyes at *c.* 10 days, and from *c.* 9–14 days blocking of light at entrance-hole soon elicits food-calls; scratching noises created by arriving adult later also important stimulus (as in *P. viridis*); eventually beg more intensely on seeing parent on nearby tree, and from *c.* 20 days give food-calls sporadically without stimulus (Bussmann 1944; Conrads and Herrmann 1963; Glutz and Bauer 1980), though less noisy overall than (e.g.) Great Spotted Woodpecker *Dendrocopos major* (Conrads and Herrmann 1963). Feeding process as in *P. viridis* (see also Bussmann 1944, Conrads and Herrmann 1963, and Food). Rarely more than 2 young fed per visit. Fed at entrance from *c.* 15–17 days (occasionally earlier). From *c.* 19 days, ♀ in particular tends to feed young from a distance (Fig C), and may Head-sway, presumably because young sometimes peck aggressively at adult. Fights also occur between older nestlings (Conrads and Herrmann 1963). Adult in nest wards off any aggression from young using free foot and occasional wing-beats (see illustration in Glutz and Bauer 1980). Faeces of young swallowed by adults up to *c.* 7 days, later carried away. ♀ no longer enters to clean nest from *c.* 16–17 days, ♂ from *c.* 22 days, although from *c.* 19 days he Mock-pecks into open bills of young or Head-sways in order to force entry (Conrads and Herrmann 1963; Glutz and Bauer 1980; also Loos 1903*a*, Conrads 1964). Young fledge at *c.* 24–28 days, sometimes over 2 days. Feeds by ♀ may be briefer and from a distance (see above) during last few days or ♂ may cease feeding altogether (see also Bonds). Contact-call from parent (Bussmann 1944) or delay in 1st feed on day of fledging (Conrads and Herrmann 1963) possibly encourages young to leave which they do by climbing or flying quite well. In one case, last to fledge initially accompanied up nest-tree by ♂ (see illustration in Conrads and Herrmann 1963; see also Blume and Ogasawara 1980). No indication of young calling one another together for roosting, unlike *P. viridis* (Blume 1965*b*). Adults soon roost well apart in separate holes; may, however, take up vocal contact with young in morning and briefly accompany them (Blume 1981). Head-swaying possibly involved in break-up of family (see Conrads 1964 and above). For further details, see Bonds (above). ANTI-PREDATOR RESPONSES OF YOUNG. Normally fall silent on hearing alarm-calls of parents (see 2a in Voice). If man near nest may withdraw into hole or remain there and observe (Conrads and Herrmann 1963). If handled, may screech vigorously and peck at hand (Bussmann 1944). Newly-fledged bird adopted apparent defence-posture with wings spread and bill raised in presence of Magpie *Pica pica* (Conrads and Herrmann 1963). PARENTAL ANTI-PREDATOR STRATEGIES. (1) Passive measures. Bird with eggs or small young generally remains in

C

B

D

nest during disturbance but sensitive to strange noises and usually looks out immediately (e.g. Bussmann 1944, Conrads and Herrmann 1963). May flush well before approach by man (Guichard 1954) or sit tight even during excessive disturbance (Polivanov 1981). For pre-departure alertness, see Fig D; bird may freeze thus (or with open bill) for up to c. 30 min (Loos 1905). Particularly towards end of nestling period, alarm-calls may be given (never from nest-tree) in presence of various predators or man. ♀ generally shyer; ♂ may continue to feed young or hide (Conrads and Herrmann 1963; also Loos 1905, Bussmann 1944). (2) Active measures. ♂ directed low-intensity threat at man and dogs before entering nest in evening. One ♀ attacked and chased off squirrel *Sciurus* (Conrads and Herrmann 1963), another ♀ a *D. martius* (Loos 1905).

(Fig A from drawings in Conrads 1964 and Blume 1981; Figs B–C from photographs in Conrads and Herrmann 1963; Fig D from drawing in Conrads and Herrmann 1963.) MGW

Voice. Unlike Green Woodpecker *P. viridis*, rarely used outside breeding season (Loos 1905; Conrads and Herrmann 1963), and much less freely in nestling phase than during incubation (Bussmann 1944). For possible sexual differences, see below.

INSTRUMENTAL SIGNALS OF ADULTS. (a) Drumming. Used more than in *P. viridis* (Ferry 1962). Performed frequently though not regularly (see, e.g., Guichard 1954), mainly by ♂; shorter series given very rarely by ♀ (Glutz and Bauer 1980). Sometimes less associated with holes than Drumming of other Picinae (Blume 1981), and, apart from branches with especially good resonance, often (and persistently) uses sites offering special acoustic effect—metal parts of telegraph poles, sirens, etc. (see, e.g., Peitzmeier 1953, Koch 1954, Steffen 1956). Series comprise 19–40 strikes over c. 1–2 s (Blume 1981; Bergmann and Helb 1982); average 26·4 strikes per series, average duration 1·36 s, $n=16$ series by 5 birds (Zabka 1980); average $26·3 \pm 4·6$ strikes per series (Wallschläger and Zabka 1981). In Austria, 3 series comprised 23–26 strikes, each series 1·14–1·28 s, i.e. 20 strikes per second (Short 1973, which see for comparison with other Asian *Picus*). Recording (Fig I) comprises 33 strikes in c. 1·6 s; interval between onset of successive strikes almost constant at c. 50 ms (see also Zabka 1980, Wallschläger and Zabka 1981). Initial quiet strike followed by c. 12 even quieter, then crescendo to end; very slightly decrescendo c. 4 strikes from end (compare sonagram in Bergmann and Helb 1982 which shows slight decrescendo where Fig I has marked crescendo) (J Hall-Craggs). Bird Drumming on siren gave 17–38 strikes per series; average 42 s between series (Glutz and Bauer 1980). Another (possibly ♀) Drumming on nest-box unusually gave 6 full but not very loud series in c. 1 min (Radermacher 1970). See also Blume (1965b) who recorded 2 series per min, and not infrequently 'broken' series, i.e. apparently 2 close together; normally 2–3 series per min (Glutz and Bauer 1980). (b) Excitement pecking, Tapping. No detailed information, but see Social Pattern and Behaviour for their possible occurrence; see also *P. viridis*. (c) Series of noisy wing-beats in aggressive 'territorial flight' (Bergmann and Helb 1982); see also Social Pattern and Behaviour.

CALLS OF ADULTS. (1) Advertising-call. A series of 5–8 (4–20) slow, fluting, sonorous, musical and far-carrying 'kü' or 'kee' ('dü', 'kli') sounds descending slightly in pitch and with rather melancholy quality. Slower and quieter at end with last (2–3) units separated from others by longer pauses. Easily imitated with human whistle. (Loos 1903a, 1905; Nilsson 1942; Guichard 1948, 1954; Blume and Jung 1956; Conrads and Herrmann 1963; Short 1973; Blume 1981; Bergmann and Helb 1982.) In contrast to ♂'s usually sonorous and loud song (at least during pair-formation), ♀'s Advertising-call quieter, shorter, and hoarser (Loos 1903a, 1905); also lower pitched according to Guichard (1948), but ♂'s call also subject to variation (see, e.g., Panov 1973, Blume and Ogasawara 1980). In recording (Fig II), bout comprises 16 units in just under 4 s, with overall rallentando, and last 3 units have different structure and rougher sound (J Hall-Craggs). In greater excitement associated with territorial defence, etc., pitch may rise then fall, 'ü' change to 'i', or pitch may rise and remain steady (Conrads and Herrmann 1963). Lacks full-throated laughing quality of *P. viridis*, but see that species for possibility of confusion in some variants. Used more widely than equivalent call of *P. viridis*, e.g. as contact-call between family members after fledging (Blume 1981). May be given in short bouts of several series alternating with long pauses (Guichard 1954); up to 3–4 per min (Steffen 1956). Faster, almost clamouring variant said to express alarm (Guichard 1954). (2) Calls expressing excitement, in some cases threat or alarm. (a) Sharp, compressed 'kük', 'kik', 'tück', etc., given in flight or when perched, singly or in loosely connected series (Loos 1903a; Blume and Jung 1959; Short 1973, which see for further sonagraphic analysis; Blume 1981). Further renderings and comment in Guichard (1948, 1954), and Ehrenroth (1973). In recording (Fig III), adult approaching nest with young gives 'piuk' and 'pik' sounds (J Hall-Craggs). For extra sonagrams, see Glutz and Bauer (1980) and Bergmann and Helb (1982). (b) Various 'djäck' or 'jick' (Loos 1903a), 'kjak', 'kjäk', 'kirrk', or 'krrik' sounds, similar to equivalent (Threat-)calls of *P. viridis* and Black Woodpecker *Dryocopus martius*; sounds like Jackdaw *Corvus monedula* (Conrads and Herrmann 1963). May begin quietly and become loud with increasing excitement; given particularly during 'territory demarcation' in autumn, and in searching flights during pair-formation (Glutz and Bauer 1980, which see for sonagram; see also Bergmann and Helb 1982). 4–9

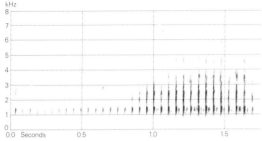
I S Palmér/Sveriges Radio (1972) Sweden May 1964

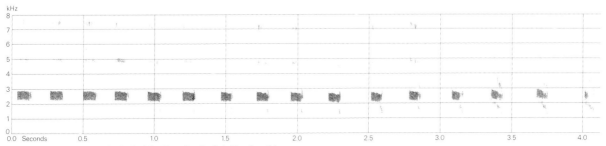
II L-E Olsson/Sveriges Ornitologiska Förening (1982) Sweden May

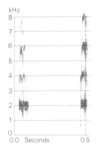
III S Palmér/Sveriges Radio (1972) Sweden May 1964

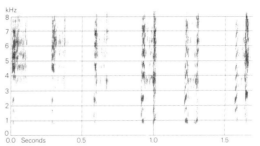
IV S Palmér/Sveriges Radio (1972) Sweden May 1964

units given in 1·10–2·28 s, about equally spaced, slower at end (Short 1973, which see for further analysis). Soft 'kije-kije' or disyllabic 'djüie' sounds given in Fluttering-flight probably also express threat (Blume 1965b, 1981). Sporadically, in autumn, birds may give series of 'kjäck' sounds at take-off (Blume 1981); indistinguishable from *P. viridis* (D Blume). (c) In high-intensity threat (particularly associated with flights in search of, and disputes over, roost-holes, July to early November), 'kück' followed by series of shrill 'keck' sounds (Conrads and Herrmann 1963; Glutz and Bauer 1980). Sharp 'tche-pe-pe-pe tche-pe-pe-pe' in autumn (Nilsson 1942) presumably the same or a related call. See also Guichard (1948), Panov (1973), and Polivanov (1981). (3) Shrill 'quä' or 'gäk' sounds apparently expressing fear (Loos 1903a, 1905), e.g. when handled (Glutz and Bauer 1980). (4) Contact-calls (sometimes apparently containing element of threat). (a) Pair-members maintain contact at close quarters with 'djück' or 'tjük' calls, sometimes 2 or more given together (Conrads and Herrmann 1963); 'sguögg' from ♀ (Bussmann 1944). A softer variant of call 2a (Blume 1981). For

sonagrams and other renderings, see Short (1973), Glutz and Bauer (1980), and Bergmann and Helb (1982). (b) ♀ said sometimes to give hoarse 'gwü' in response to 'djück' from ♂ (Glutz and Bauer 1980), but comparison of sonagrams (see also Bergmann and Helb 1982) indicates this similar to call 4a. Rather slurred 'die die die' or 'wi wi wi' interpreted as short-distance threat-call of ♀ (see Loos 1903a, 1905), and adult ♀ threatening juvenile ♀ gave apparently similar 'wi-(te) wi-(te)' calls (Conrads 1964; see also Social Pattern and Behaviour).

CALLS OF YOUNG. Newly hatched young give quiet cawing or croaking sounds. Food-call from 3 days a peculiar 'ärchärchärch. . .', audible outside nest from c. 6 days (Bussmann 1944), and then rendered 'tschetschetschetschet. . .' (Loos 1903a), 'schätt schätt. . .', or 'zettzett. . .' (Conrads and Herrmann 1963). Recording (Fig IV) reveals abrasive, scraping sounds in 'quick-quick-slow' pattern; such sounds may also show 'paired' pattern (J Hall-Craggs); compare also Fig V in *P. viridis*; apparently a more metallic churring than *P. viridis*, though timbre of nestling Picidae calls likely to be affected by nest-hole

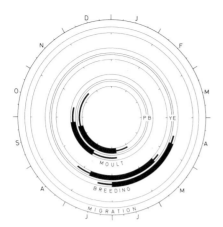

acoustics (P J Sellar). Later, when fed at entrance, young give single rasping 'schatt' or 'tsack' sounds, then 'schatt schatt schätt tschip tschip' (see also Loos 1903a), eventually developing into rougher version of adult call 1. Virtually full adult repertoire mastered by fledging. Querulous 'ki kji' also heard from nestlings when ♂ *P. viridis* at nest-entrance evidently with hostile intent, and fledgling gave a 'kjä kji kji kjä kjä' near Magpie *Pica pica* (Conrads and Herrmann 1963; see also Bussmann 1944, Blume and Ogasawara 1980, Blume 1981). MGW

Breeding. SEASON. Central Europe, southern Scandinavia, and western USSR: see diagram. SITE. Hole in tree, mainly in aspen *Populus*, beech *Fagus*, oak *Quercus*, and lime *Tilia*; also in willow *Salix* and pine *Pinus* (Dementiev and Gladkov 1951a; Makatsch 1976). Average height above ground of 13 nests, France, 5·4 m (1·3–18) (Guichard 1954). Nest: excavated hole, average diameter 5·7 cm (5·4–5·9), $n=7$; depth 28·4 cm (15–36·5), $n=7$; nest-chamber 11 × 12 cm (9–12·5 × 9–12·5), $n=6$; no material added but wood chips left (Guichard 1954). Building: hole excavated by both sexes. EGGS. See Plate 98. Sub-elliptical, smooth and glossy; white. 28 × 22 mm (26–31 × 21–24), $n=52$ (Rosenius 1949). Fresh weight 7–8 g (Guichard 1954). Clutch: 7–9 (4–11). 1 brood. No replacement clutches known (Glutz and Bauer 1980). Laying interval 1 day. INCUBATION. 14–15 days (Blume and Ogasawara 1980); probably not more than 17 days (Conrads and Herrmann 1963). By both sexes. Begins with last egg; hatching synchronous. YOUNG. Altricial and nidicolous. Cared for and fed by both parents. FLEDGING TO MATURITY. Fledging period 24–28 days (Conrads and Herrmann 1963). Become independent soon after (see Social Pattern and Behaviour). BREEDING SUCCESS. In Primor'e (eastern USSR), rarely more than 5 fledged young per brood (Polivanov 1981). No further information.

Plumages (nominate *canus*). ADULT MALE. Feathers at base of upper mandible and bristles covering nostrils pale olive-grey, faintly streaked black, tips of bristles black. Forehead and forecrown scarlet-red, red of feather-tips often yellow subterminally and always grey towards base, but usually not visible unless forecrown heavily worn; some feather-tips at rear of red forecrown often bright yellow or orange. Distinct black streak on lores from sides of frontal feathering to eye, bordered above by light grey supercilium widening behind eye (slightly tinged with green when plumage fresh), bordered below by broad light grey stripe from gape to below ear-coverts. Malar stripe black; narrow but distinct near base of lower mandible, widening slightly and sometimes partly mottled grey towards rear of lower cheeks. Remainder of crown, hindneck, sides of head, and sides of neck light grey or ash-grey, central crown often with narrow black shaft-streaks, crown and hindneck with distinct green tinge when plumage fresh, sides of head and neck slightly so. Mantle, scapulars, tertials, and upper wing-coverts olive-green, distinctly less bright and yellow than grass-green upperparts of Green Woodpecker *P. viridis*, but tertials and upper wing-coverts sometimes with slight yellow or brass tinge as in that species; basal and middle portions of tertials greyish-black with short white bars. Upper back olive-green, grading to yellow on rump and upper tail-coverts; feather-bases olive-green, showing through slightly on rump and readily visible on lower back and upper tail-coverts; rump occasionally greenish-yellow, and generally not as yellow as *P. viridis*. Chin and throat greyish-white with slight olive-grey or buff tinge; remainder of underparts pale olive-green or olive-grey, chest mainly greyish-green, lower flanks and under tail-coverts sometimes slightly brighter olive-green and with indistinct darker grey-green arcs or chevrons subterminally. Central pair of tail-feathers (t1) olive-green or yellow-green at sides, barred dusky black and pale grey-green on centres; shafts and tips dark horn-brown to black. T2–t3 greyish-black or brown-black with slight green tinge; from t4 to t6, gradually paler dark grey-brown on uppersurface, often with green tinge on tips and with faintly paler bars on t5, undersurface uniform pale green-grey on tips and dull black basally, sometimes with some dusky bars on middle portions of feathers. Flight-feathers as *P. viridis*, but outer web of secondaries and rather indistinct fringe along outer web of p1–p3, along greater upper primary coverts, and along basal outer web of p4–p7 duller olive-green (like upper wing-coverts); small pale spots on p6–p9 extend closer to tips. Smaller upper primary coverts below bastard wing often less extensively white than in *P. viridis*. Under wing-coverts and axillaries white with slight grey-green tinge and narrow dusky grey bars. In worn plumage, head duller grey without green or olive tinge; some grey visible between red on forehead and forecrown; black shaft-streaks on central crown and black loral and malar streaks more distinct; upperparts duller and greyer, more olive-green visible amidst yellow of rump; entire underparts dull pale grey, virtually without traces of pale olive-green. ADULT FEMALE. Similar to adult ♂, but forehead and crown light grey with slight olive-green tinge and narrow black shaft-streaks (streaks often faint or absent when plumage fresh, more distinct when worn), without any red on head or (occasionally) with a few red feather-tips on forehead only; black malar stripe often narrower and shorter than in ♂, partly mottled with light grey and often not reaching base of lower mandible. NESTLING. Naked at hatching; feather-pins appear from 12th day and tips open to produce feathers from 15th–16th day, appearing feathered (except for belly) from 19th–21st day (Heinroth and Heinroth 1931–3; Glutz and Bauer 1980). JUVENILE. Unlike *P. viridis*, rather closely similar to adult; in particular, ♀ sometimes difficult to age. In both sexes, olive-green of upperparts slightly duller than adult, some grey of feather-bases often visible on mantle, scapulars, and upper wing-coverts; tertials, outer webs of secondaries, and greater upper wing-coverts with darker and paler green bars (often indistinct); rump more greenish-yellow, back and upper tail-coverts virtually without yellow; black malar stripe narrow and mottled; entire underparts uniform pale olive-grey, soon wearing to dull grey, sometimes with slight buff tinge, indistinctly barred dark grey-brown on lower flanks and under tail-coverts. Tail and primaries like adult, but t1 more narrowly barred dusky and more sharply attenuated, off-white spots reach tips of outer primaries, and juvenile p10 12–17 mm longer than longest greater upper primary coverts (1 mm shorter to 11 mm longer in adult and 1st adult); see also Structure. Forehead and forecrown of ♂ red as in adult ♂, but slightly less extensive, some grey of feather-bases visible, and feather-tips at sides and rear of forecrown extensively green-yellow, yellow, or orange; forehead and crown of ♀ as in adult ♀, but black shaft-streaks sometimes sharper and heavier. FIRST ADULT. Similar to adult, but juvenile tertials, secondaries, and all or part of greater upper primary coverts retained; tertials and secondaries barred with darker and paler green, sometimes indistinct (adults often also show traces of barring, in particular on outer secondaries); inner tertials relatively narrower, less square at tip, and more worn than those of adult at same time of year; greater upper primary coverts with smaller and sharper pale spots on outer webs, hardly bordered by uniform olive-green fringe, feather-tips more pointed and relatively more worn than in adult; occasionally, outer greater upper primary coverts replaced by new adult ones, these contrasting in colour, shape, and abrasion with retained juvenile inner ones (adult sometimes retains one or a few greater primary coverts also, but these similar in shape and colour pattern to new feathers).

Bare parts. ADULT AND FIRST ADULT. Iris white with pink or pale blue tinge, or carmine-red mixed with white. Bill black-brown or greyish-black with slight olive tinge, basal sides of upper mandible and basal $\frac{2}{3}$ of lower mandible pale olive-grey or olive-yellow. Leg and foot olive-grey or green-grey. NESTLING. Pink at hatching, except for blue-grey lids over closed eye, whitish egg-teeth to both mandibles, and white claws. Knob-like flanges at base of lower mandible near gape pink- or yellow-white; small at hatching, largest (c. 6 mm across) on 12th–14th day, reduced afterwards, disappearing by fledging. JUVENILE. At fledging, iris red-brown or red mixed with white; bill like adult, but much yellow on base of lower mandible; leg and foot olive-grey with slate-grey or blackish scutes. (Hartert 1912–21; Heinroth and Heinroth 1931–3; Glutz and Bauer 1980; Short 1982; RMNH.)

Moults. ADULT POST-BREEDING. Timing and sequence as in central European populations of *P. viridis*, but perhaps slightly more rapid, and replacement of secondaries and tail in a few specimens examined slightly earlier relative to primaries (Glutz and Bauer 1980). In USSR, moult on average perhaps slightly earlier than in *P. viridis*, completing about late September or early October (Dementiev and Gladkov 1951a), but in some skins examined (Sweden and Rumania) moult completed later than *P. viridis*, mid- and late November. POST-JUVENILE. Sequence and amount of feathering replaced as in *P. viridis*; as in that species, frequently some outer greater upper primary coverts and exceptionally a few tertials replaced, but remainder of greater upper primary coverts and tertials and all secondaries retained. P1 shed on 22nd day, p2 on 27th shortly after fledging; next feathers follow at intervals of $10\frac{1}{2}$ (8–13) days, completing moult 10–15 days earlier than *P. viridis* (but breeding season slightly later) (Glutz and Bauer 1980). Flight-feather and tail moult halfway through in late July or early August at age of c. $2\frac{1}{2}$ months, body moult just starting; by mid-August (when c. 3 months old) p7–p10, t1, and t5 still juvenile; body in heavy moult early September, completed about mid-September when t1 growing; last juvenile feather (p10) shed 18–19 September (Heinroth and Heinroth 1931–3). Moult rather late in some skins examined (West Germany, Rumania), primary moult only halfway through in mid-September to mid-October.

Measurements. Nominate *canus*. ADULT AND FIRST ADULT. (1) West and East Germany, France, Switzerland, and Austria; (2) Rumania and Bulgaria; (3) Scandinavia and European USSR; all year, skins (RMNH, ZFMK, ZMA). Bill (F) to forehead, bill (N) to distal corner of nostril; toe is outer front toe with claw.

		♂		♀	
WING	(1)	147 (1·81; 23)	145–151	148 (1·74; 7)	146–150
	(2)	146 (2·28; 5)	143–148	145 (2·42; 4)	143–149
	(3)	147 (1·23; 8)	145–149	148 (2·50; 4)	145–150
TAIL		98·9 (2·73; 16)	94–103	98·6 (3·60; 8)	95–104
BILL (F)	(1)	38·1 (1·59; 23)	36–41	36·4 (0·25; 7)	36–37
	(2)	38·1 (1·44; 5)	37–40	38·6 (1·89; 4)	37–41
	(3)	38·3 (1·61; 8)	37–40	37·3 (2·08; 4)	35–40
BILL (N)	(1)	27·2 (1·03; 23)	26–29	25·9 (0·66; 7)	25–27
	(2)	26·7 (0·53; 5)	26–27	27·2 (1·29; 4)	26–29
	(3)	26·4 (0·96; 8)	26–28	25·5 (1·30; 4)	24–27
TARSUS		26·0 (0·60; 20)	25–27	26·1 (0·31; 9)	25–27
TOE		27·8 (1·06; 16)	26–30	27·8 (0·78; 7)	27–29

Sex differences not significant, except bill (1). First adult wing on average 0·6 shorter than adult, other measurements similar; combined above.

JUVENILE. Wing on average 4·7 shorter than adult, tail 5·0 shorter; in Swiss sample, measured freshly dead, juvenile wing and tail on average 7–9 shorter than adult (Glutz and Bauer 1980). Juvenile bill not full-grown until about halfway through post-juvenile moult.

Weights. Nominate *canus*. ADULT AND FIRST ADULT. West Germany: ♂ 122; ♀♀ 127, 138; exhausted ♂ 95 (ZFMK). Switzerland: ♂ 137 (18) 125–165, ♀ 136 (10) 125–160, exhausted birds down to 85 (Glutz and Bauer 1980). Norway: ♂♂ 125, 131; ♀ 123; range, both sexes, 98–122 (Haftorn 1971). Hungary: unsexed 119 (Vasvári 1955).

JUVENILE. On 10th day, 64, 73; on c. 16th, 82, 84, 85; on 19th–20th, 109, 111; at c. 3 months, 125 (Heinroth and Heinroth 1931–3; Bussmann 1944).

Structure. In adult and 1st adult, p6 or p7 longest or either one 0–2 shorter than other; p8 2–6 shorter than longest, p9 24–31, p10 73–82, p5 3–6, p4 14–18, p3 24–28, p2 28–33, p1 32–38; in juvenile, p7 or p8 longest or either one 0–2 shorter, p9 14–17, p10 56–64, p6 4–9, p5 13–19; innermost primaries more strongly reduced than in *P. viridis*, p1–p2 about $\frac{1}{4}$–$\frac{1}{3}$ length of neighbouring s1. P10 strongly reduced in adult and 1st adult, less so in juvenile; p10 1 mm shorter to 11 mm longer than longest greater upper primary coverts in adult and 1st adult (average 5·1 longer, $n=25$), 15·5 (6) 12–17 longer in juvenile; shape as in *P. viridis*. T1 longest; in adult, t2 3–8 shorter, t3 9–16, t4 18–24, t5 29–37, t6 61–71; in juvenile, t2 16–20 shorter, t3 22–26, t4 27–31, t5 34–41, t6 64–71. Bill relatively shorter and with more slender base than *P. viridis*. Inner front toe with claw c. 84% of outer front toe with claw, outer hind toe c. 94%, inner hind c. 42%. Claws shorter and more slender than in *P. viridis*. Remainder of structure as in *P. viridis*.

Geographical variation. Hardly any in Europe; population from south-east Europe (sometimes separated as *perspicuus* Gengler, 1920) sometimes with slightly heavier black streaks on crown than birds from central and northern Europe, wing on average slightly shorter, and bill perhaps slightly longer (see

Measurements), but variation in streaking of crown large and overlap in measurements considerable, hence recognition not warranted. Specimens from Bolu area in Asia Minor are nominate *canus* (Kumerloeve 1961). *P. c. biedermanni* from central Asia similar to nominate *canus*, but fresh plumage slightly greyer olive-green and worn plumage distinctly greyer; *jessoensis* from eastern Asia slightly paler and greyer than nominate *canus* in fresh plumage, distinctly greyer than both *biedermanni* and nominate *canus* in worn plumage (Dementiev and Gladkov 1951*a*); in general, both races closely similar to nominate *canus* and some birds indistinguishable. Remaining races in Himalayas and eastern and south-east Asia have crown and central hindneck heavily streaked black or completely black; upperparts darker greenish-olive, underparts saturated olive-green; size variable (see Danis 1937; Vaurie 1959*a*). Isolated race from mountains of Malay peninsula even deeper green than south-east Asian races, but race of Sumatra has green entirely replaced by maroon-red.

For relationships with *P. viridis* and Levaillant's Green Woodpecker *P. vaillantii*, see those species; several close relatives in south-east Asia, notably Laced Woodpecker *P. vittatus* and Streak-throated Woodpecker *P. xanthopygaeus* (Short 1982). CSR

Picus viridis Green Woodpecker

PLATES 77 and 78
[facing pages 807 and 830]

Du. Groene Specht Fr. Pic-vert Ge. Grünspecht
Ru. Зеленый дятел Sp. Pito Real Sw. Gröngöling

Picus viridis Linnaeus, 1758

Polytypic. Nominate *viridis* Linnaeus, 1758, Europe south to France (except southern Roussillon in eastern Pyrénées), Alps, northern Yugoslavia, and Rumania; *karelini* Brandt, 1841, Italy, south-east Europe south from southern Yugoslavia and Bulgaria, Asia Minor, northern Iran, and south-west Turkmeniya (USSR); *sharpei* (Saunders, 1872), Iberia north to Spanish side of western and central Pyrénées and southern Roussillon. Extralimital: *innominatus* (Zarudny and Loudon, 1905), south-west and southern Iran from Lorestan to Bampur river basin.

Field characters. 31–33 cm; wing-span 40–42 cm. About 25% larger than Grey-headed Woodpecker *P. canus*; *c.* 30% larger and much bulkier than Great Spotted Woodpecker *Dendrocopos major*. Large, strong-billed, green woodpecker, with bright yellow-toned rump, strong head pattern including long red crown, often conspicuous behaviour, and far-carrying laugh. In markedly undulating flight, barred flight-feathers less obvious than glowing rump. Sexes dissimilar; no seasonal variation. Juvenile separable. 3 races in west Palearctic, of which 1 easily separable from other 2 in the field.

(1) Non-Iberian European races, nominate *viridis* and *karelini*. ADULT MALE. Plumage predominantly yellow-green above and pale green-grey below, most relieved on perched bird by grey-mottled red crown and central nape, black eye-patch, and broad black-edged red-centred moustache (set against pale green-buff cheeks), and on flying bird by large glowing yellow rump and cream-barred dark brown primaries. Rear flanks and vent barred green-brown. ADULT FEMALE. Moustache completely black, and eye-patch smaller, hardly extending behind eye. JUVENILE. Plumage short-lived, completely replaced by August–November. Differs distinctly from adult; plumage pattern broken up by almost white spots and bars on upperparts and wing-coverts and obvious broken brown-black bars over underparts and sides of neck. In both sexes, crown and moustache indistinct, being spotted grey and black but ♂ already shows some red tips on moustache. Rump less bright than adult. (2) Iberian race, *sharpei*. ADULT MALE. Differs distinctly in usually having only lores black, rest of face greyer, underparts unbarred. ADULT FEMALE. Lores dusky, bordered by green-grey towards crown, and by greyish-white towards fully black moustache. JUVENILE. Less heavily barred and spotted than in nominate *viridis*.

Confusable only with *P. canus* (since Levaillant's Green Woodpecker *P. vaillantii* isolated in north-west Africa) but noticeably larger and more powerful, with green-yellow wash over almost all plumage and long red crown and strong moustache common to all ages and both sexes. Among other birds, ♀ Golden Oriole *Oriolus oriolus* closest in plumage colour and pattern but much smaller and proportionately narrower winged and longer tailed. Full flight powerful and rapid; undulating, with bursts of 3–4 wingbeats. Appearance in flight striking not only because of vivid colour but also because of markedly oval shape created by strong bill, long body, and pointed tail, most obvious when wings held close to body as bird dips in long undulation. Landing achieved with characteristic last flutter of wings and sudden collapse of bird against trunk or in grass. Hops both on ground and up trees; each bound powerful, ending with brief upright stance. Often on ground, hunting ants.

Voice contains one dominant call—the loud, multi-syllabic laugh given by both ♂ and ♀ in spring and early summer: 'plue plue ...', accelerating in tempo (unlike *P. canus*); amended into several other notes. Tapping of feeding bird loud, but true drumming rather rare (unlike *P. canus*); tempo similar to Great Spotted Woodpecker *Dendrocopos major*, but lasts at least twice as long and astonishingly feeble.

Habitat. Resident in temperate and also marginally in

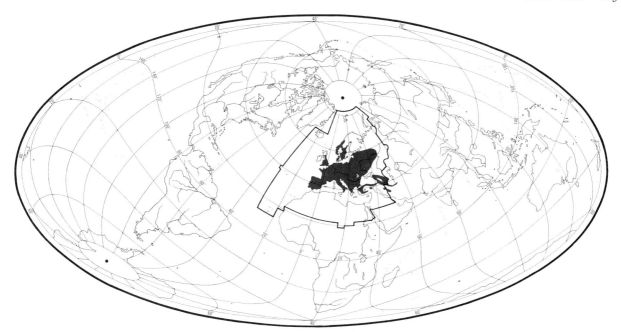

milder boreal and Mediterranean zones of west Palearctic in oceanic as well as continental climates. In most regions, more of a lowland species than Grey-headed Woodpecker *P. canus*. In Switzerland, however, has recently bred up to treeline, 2100 m (Glutz von Blotzheim 1962), and in Caucasus mountains (USSR) as high as 3000 m. Also, apparently, since widespread forest clearance, desertion of unbroken close forests has been taken somewhat further than by *P. canus*, especially in west of range where parkland, orchards, groves, gardens, vineyards, heathland with scattered trees, hedgerow trees, and spinneys, and even treeless dry dwarf scrub and cliff-tops are as much favoured as more traditional open or broken broad-leaved mixed forest with grassy fringes or clearings. Largely a ground feeder, sensitive to changes in close grazing resulting from increases or reductions in stocks of sheep and in rabbit populations, affecting extent of turf short enough to gain greater exposure to sun and consequently support larger and more varied ant populations. New afforestation can have similar effects (Sharrock 1976). Shows consistent avoidance of close stands of conifers, although sometimes tolerates larches and open pinewood where there is leafy understorey and abundance of ants; occurs also in mixed woodlands. Accordingly becoming an arboreal rather than a forest bird, spending much time on ground. Dependent through food requirements on fairly dry conditions, especially in spring when even sunny, rocky situations may be attractive. On trees, strongly prefers trunk to smaller branches, but will use topmost perch when advertising. Flies almost entirely in lower or middle airspace.

Distribution. In 20th century, marked spread in northern Britain and some spread in Denmark and Sweden.

BRITAIN. In most of England and Wales, little change except for colonization of Isle of Wight in 1910; in northern England, after probable decrease in mid 19th century, spread in 20th century (in east from *c.* 1913, in west after 1945), then first bred southern Scotland 1951 and has since spread steadily northwards (Parslow 1967; Sharrock 1976; Dennis 1980). DENMARK. Spread to Jutland 1920–40 (TD); extinct on some islands (Petersen 1945; Løppenthin 1967). SWEDEN. Some recent spread (Curry-Lindahl 1961*a*). ITALY. Formerly bred Sicily (PB, BM).

Accidental. Malta.

Population. Fluctuating in northern areas due to hard winters (see also Glutz and Bauer 1980). Limited data on numbers, but probably increased Britain and declined Netherlands and northern Italy.

BRITAIN. Marked fluctuations due to hard winters. Estimated under 10000 pairs (Parslow 1967); 15000–30000 pairs (Sharrock 1976). FRANCE. Up to 1 million pairs (Yeatman 1976). BELGIUM. About 7500 pairs (Lippens and Wille 1972). LUXEMBOURG. About 2600 pairs (Lippens and Wille 1972). NETHERLANDS. Declined since *c.* 1930, mainly due to habitat changes; estimated 4500–7500 pairs (Teixeira 1979; CSR). WEST GERMANY. Perhaps *c.* 25000–90000 pairs (Rheinwald 1982). DENMARK. Declined 19th century, some recovery in 20th century (Jespersen 1946). SWEDEN. Recent increase (Curry-Lindahl 1961*a*); about 50000 pairs (Ulfstrand and Högstedt 1976). EAST GERMANY. Rare (WM). POLAND. Fairly numerous in most parts (Tomiałojć 1976*a*). CZECHOSLOVAKIA. No known population changes (KH). ITALY. Rapid decrease in north (SF, PB, BM). ALBANIA. Not common (EN). GREECE. Locally not uncommon (WB, HJB, GM). BULGARIA. Very common in wooded areas (JLR).

Oldest ringed bird 7 years 4 months (Dejonghe and Czajkowski 1983).

Movements. Resident. In western and central Europe, dispersal very local in character; individual movements above 20 km unusual. Becomes more widespread within breeding range in autumn–winter, when some birds (especially immatures) leave well-timbered sites to range into (e.g.) farmland, orchards, and along watercourses; adults show high degree of fidelity to home-range (see Social Pattern and Behaviour). No marked altitudinal movement; regularly winters above 1800 m in Alps (Glutz von Blotzheim 1962). Of 92 British recoveries, 75 moved 0–5 km, 11 6–15 km, 3 16–25 km, 3 longer (maximum 69 km). Longest continental ringing movements were 82 km (East Germany) and 170 km (West Germany), these being exceptional (Glutz and Bauer 1980).

Pronounced southward displacements in autumn reported from USSR, where winter climate more severe (Dementiev and Gladkov 1951a), but scale and regularity need clarification.

Food. Largely adult and pupal ants (Formicidae). Feeding ecology thus most specialized of west Palearctic Picinae. Bill consequently weak and usually employed in chiselling on soft wood only. Extremely long tongue (capable of extension to over 10 cm) flat and wide at tip, without barbs of *Dendrocopos* woodpeckers and Black Woodpecker *Dryocopus martius*. Enlarged salivary glands coat tongue with sticky secretion permitting capture of ants, etc., found on surface or in nest galleries. Tip of tongue extremely mobile and capable of independent movement. At all times prefers to feed on ground, though may seek insects on surface of branches (Sielmann 1959). Climbs up or down trees, head always uppermost (Feilden 1908; Witherby *et al.* 1938). Seeks ant nests (particularly of *Formica*) by flying systematically along woodland edges, paths, or embankments, or (for turf-dwelling species, e.g. *Lasius*) by hopping for short distances (Löhrl 1977a). *Formica* nests raided mainly between end of October and May; during summer, *Formica* taken from passages in ground, or from trees or bushes (Glutz and Bauer 1980). At nests, rummages in loose material on top of nest, licking up adults and pupae thus exposed; throat may vibrate rapidly (King and King 1973). Foraging bout may be interrupted if ants attack too vigorously and bird may frequently kick legs to dislodge ants climbing into plumage (Sielmann 1959). Often wipes bill on nearby tree after feeding at anthill, and (for unknown reason) such trees may be vigorously struck before feeding (Niethammer 1938); see also Voice. Anthills readily located under level snow cover (Steinfatt 1944b; Ass 1958; Blume 1958a), and snow up to 30 cm deep may be cleared to allow access (Herschel 1971). Recorded tunnelling 85 cm through snow to ant nest (Glutz and Bauer 1980). In seeking turf-dwelling ants, lifts grass or moss cover and digs funnel-shaped hole up to 8 cm (rarely 12 cm) deep and 2–3 cm wide; licks up ants emerging into hole from exposed passages. May feed thus for over 1 hr on area of 1 m², returning to each of several holes in turn. May visit such a site repeatedly over several weeks (Löhrl 1977a). Often feeds thus on garden lawns (Hendy 1949; Hann 1951; King and King 1973). In winter, other insects may be sought on rock-ledges and in cracks in rocks and walls, weather-boards, shingle roofs, telegraph poles, and house walls. May peck at loose material but bill mainly used for picking up items (Baier 1975). Digs fly larvae out of fungi (Herschel 1971), and seen feeding from animal droppings (Woods 1952). Recorded taking larvae from wasp nest (Schwammberger 1974). In one hard winter, England, numerous attacks on hives of bees *Apis mellifera* recorded: large holes drilled in wood but apparently no bees eaten; birds apparently attracted by humming noise of bees (Ellement 1953). Insects sometimes caught by aerial-pursuit (Braun 1975) and sitting bird observed rapidly flicking out tongue to catch flying insects (Moule 1951). Recorded following the plough with Rooks *Corvus frugilegus* and gulls (Laridae), taking insects from fresh furrows (Mitchell 1948). Sometimes chisels out rotten wood or bark to expose insects (Warham 1950), but attacks on sound wood rare (Collinge 1924–7). However, will remove larvae (probably of hornet moth *Sesia apiformis*) removed from pith of dead and growing osiers *Salix*; located beneath surface (not known how) and rectangular hole chipped with sideways blows of bill to expose larvae (Smalls 1912). Said to drill trees in spring to drink sap though behaviour much less frequent than in Great Spotted Woodpecker *Dendrocopos major* (Osmolovskaya 1946; Turček 1954); confirmation of this behaviour still required according to Glutz and Bauer (1980). Pine cones opened by wedging into enlarged cleft in bark and hammering off scales to reach seeds (Bunyard 1924).

Diet consists largely of ants (Formicidae)—adults, pupae, larvae, and eggs: *Formica rufa, F. polyctene, F. nigricans, F. cordieri, F. fusca, F. exsecta, F. pressibilis, F. picea, Lasius flavus, L. fuliginosus, L. niger, L. alienus, Myrmica rubra, M. scabrinodis, M. laevinodis, M. rubida, Tetramorium caespitum, Serviformica rufibarbis, Camponotus herculeanus, C. pubescens, C. vagus*. Also beetles (Curculionidae, Scolytidae, Cerambycidae, Lucanidae, Coccinellidae), dipteran flies, mostly adults (*Pallenia atramentaria, Scatopse brevicornis, Mediza glabra, Musca domestica, Faunia canicularis*, larvae of *Tipula*), caterpillars (*Sesia apiformis, S. culiciformis, Retinia buoliana, Zeuzera aesculi*), and earwigs *Forficula*. Rarely, woodlice (Malacostraca), spiders (Araneae), millipedes (Myriapoda), larvae of bees, hornets, and wasps (Hymenoptera), mole-crickets (Gryllotalpidae), bugs (Hemiptera), earthworms (Oligochaeta), and snails (Gastropoda); once, newborn adder *Vipera berus*. Plant food occasional: seeds of pines *Pinus*, cedar *Thuja*, and oak *Quercus*, fruits of rowan *Sorbus aucuparia*, yew *Taxus baccata*, and *Pyracanthus*, and cherries, grapes, pears, apples, and buds of various trees. (Csiki 1905; Newstead 1908; Smalls 1912; Bunyard 1924; Collinge 1924–7; Witherby *et al.* 1938; Warham 1950; Dementiev and Gladkov 1951a; Kneitz 1965a; Bährmann 1970; Simms 1971; Skøtt 1971; De Bruyn *et al.* 1972; Speight 1973; Fouage 1975; Glutz and Bauer 1980; Blume 1981; Heer 1982.)

Pellets and faeces from bird feeding on lawn, south-west England, February, contained only ants—*c*. 1000–2000; pellets, *c*. 90% *Myrmica rubra*, with *Lasius* and *M. scabrinodis* present; in faeces, *c*. 90% *L. flavus* (probably), with *L. niger* (probably) and *Myrmica rubra*; differential appearance in pellets and faeces reflects robustness of exoskeletons (Speight 1973, which see also for availability of the various species). Items in faeces, Netherlands, comprised: in summer 1% *Lasius fuliginosus*, 25% *Formica rufa*, 74% smaller species; in winter 10%, 61%, and 29% respectively; in an especially hard winter, 18%, 59%, and 23% ($n=584914$). Apparently seeks smaller species when active on surface in warm weather, switching to larger more conspicuous species in colder weather. *Formica rufa* inhabits large nests which can be found even beneath snow cover. However, in freezing conditions seeks *Lasius fuliginosus* in rotten wood (De Bruyn *et al.* 1972). Stomach from Caucasus (USSR), September, contained over 2000 ants (Dementiev and Gladkov 1951a). 78 stomachs, Britain, contained 75% (by number) larvae of Coleoptera and Lepidoptera, 20% ants, and 5% other invertebrates. One stomach contained over 1300 beetles, another 1100, and 300–800 per stomach was common (Collinge 1924–7).

In Lithuania (USSR), diet of young almost entirely *Formica rufa* (Knystautas and Liutkus 1981). Study using neck-collars at nest in Rumania, throughout nestling period, showed 10 species of ant (adults and pupae) brought to young: *Tapinomma erraticum* (15468 adults, 2160 pupae), 4 *Formica* species (3328 adults, 615 pupae), 2 *Camponotus* species (1345 adults), 2 *Lasius* species (326 adults, 120 pupae), and some *Myrmica ruginodis*. In first 10 days chicks averaged 15 g food each; over days 10–20, 39·5 g; from day 20 onwards, 49·3 g. The 7 young ate an estimated 1·5 million ants and pupae while in the nest (Korodi Gál 1975). Food carried in stomach and regurgitated to young, hence frequency of visits to nest lower than species that carry food in bill. Each visit lasts several minutes and brings sufficient food to provide 6–10 chick feeds. (Tutt 1951; Sielmann 1959; Blume 1961; Knystautas and Liutkus 1981.) ♂ and ♀ take approximately equal share in feeding. Interval between visits 30 min–2 hrs (Tracy 1946; Tutt 1951; Sielmann 1959; Blume 1961; Knystautas and Liutkus 1981); with young 10–11 days old

34 visits per day, at 20–21 days 21 per day (Steinfatt 1944b). 39 visits per day (Loos 1904). Feeding rate falls prior to fledging and one parent may stop altogether (Tutt 1951; Blume 1961).   ASR

Social pattern and behaviour. Most aspects well known. For study in Rominter Heide (Poland/USSR), see Steinfatt (1944b); for major studies in Gladenbach area (Hessen, West Germany), see Blume (1955, 1957, 1961, 1962a). For recent summary, see Blume (1981). Following account based on outline supplied by D Blume. 1. Normally solitary outside breeding season. In Gladenbach area, 8–11 birds may roost in area of c. 16 km^2; 3–4 in wood of c. 1 ha (see also Roosting and Antagonistic Behaviour, below). Home range, roughly estimated from flights to and from roosting area, c. 120–250 ha; home-ranges of individuals overlap (Blume 1961, 1981; D Blume). Birds probably range up to c. 5 km from roost in winter (Horstkotte 1973). Birds normally show pronounced year-to-year fidelity to home-range once this chosen (D Blume). BONDS. Monogamous mating system. Pair-bond of at least seasonal duration, possibly longer-term, even for life (not proved by ringing, but see above for fidelity to home-range), though normally severed during post-fledging care of young in July–August, or at least much weakened outside breeding season (Jourdain 1936a; Steinfatt 1944b; Tutt 1951; Labitte 1953). Loose contact may, however, be maintained between ♂ and ♀ in late summer after breeding: e.g. birds seen feeding together and ♀ fed by ♂, mid-August (Hann 1951). Occasionally, birds (possibly prospective breeding partners) may forage together again from mid-November or December, but 'appropriate individual distance' generally maintained (Blume 1981; D Blume). Occasionally hybridizes with Grey-headed Woodpecker *P. canus* (Salomonsen 1947; Hildén 1955; see also Ruge 1966). Young tended by both parents up to fledging, brood then soon apparently split, 1 parent being often seen with 2–3 young, more often than in other Picinae. Feeding of still-dependent young up to mid-August (Masurat 1966); ♀ and juvenile feeding together and in vocal contact as late as 9 October. Age of first breeding 1 year (C S Roselaar). BREEDING DISPERSION. Solitary and territorial. In Gladenbach area, 2–3 pairs in 16 km^2 of unbroken woodland, 4–5 pairs on 16 km^2 where woods small and scattered. Shortest distance between nests c. 500 m (Blume 1961). In Lithuania (USSR), c. 10 pairs in c. 30-km stretch of riverine woodland; pairs 2–4 km apart (Knystautas and Liutkus 1981). In Rominter Heide, c. 1 pair per 5 km^2 (Steinfatt 1944b). In Grunewald (West Berlin), 1952–62, 1·2 pairs per km^2 over c. 30 km^2; high density probably attributable to decline or absence of main predators and abundance of food and suitable nest-sites (Wendland 1964). In Brandenburg (East Germany), overall density similar to that of Black Woodpecker *Dryocopus martius*: in urban areas and parks c. 0·7–1·6 pairs per km^2, in mixed woodland 3–6 pairs per km^2, in broad-leaved woodland 6–12 pairs per km^2, and in mature beech *Fagus* stands 6 pairs per km^2 (Rutschke 1983). Such values typical of other parts of central Europe. Over larger areas, probably seldom more than 0·25 pairs per km^2 (see Glutz and Bauer 1980). Defends critical sites (see Antagonistic Behaviour, below) within large, all-purpose territory with particular parts preferentially used by ♂ and ♀, though some overlap. Structure of territory as in *D. martius*; see that species. In Gladenbach area 'breeding territories' in meadow, field, and woodland habitat 3·2–5·3 km^2. Location of territories in a particular area determined by position of trees with holes. Some trees used for nesting for 10 years or more, though not always by same birds. Similarly, traditional calling stations (see also further below) usually sited near roosts and used by both sexes early in breeding season (Blume 1961; Glutz and Bauer 1980; D Blume). One ♀ deserting after disturbance re-laid c. 100 m away. Rarely moves more than c. 500 m between seasons (Labitte 1953). ROOSTING. Nocturnal and normally solitary. Exceptionally, up to 10 roosted in church wall, 2 (rarely 3) apparently using same hole (Lugaro 1954); see also below. Usually in old nest-holes (rarely from most recent breeding season); also readily in those of *D. martius*, or nest-boxes (Creutz 1960). Not known to excavate special holes for roosting but occasionally chips away at existing holes in autumn (Blume 1961, 1981; D Blume). Other sites include various cavities in walls, roofs, etc. (Löhrl 1949; Blume 1962b; Menzel 1975), even in burrow of rabbit *Oryctolagus cuniculus* (Catthoor 1954b). Favourable sites used for many years; 18 years recorded (Labitte 1953). Some birds stick to same site for long period, others absent for odd nights or change frequently—sometimes due to disturbance or weather but cause not always clear (Jourdain 1936a; Steinfatt 1944b; Löhrl 1949; Blume 1964a). ♂ and ♀ sometimes roost fairly close together in winter—e.g. c. 60 cm apart in same tree (Labitte 1953). Roosting together occurs in spring—once for several weeks (Löhrl 1949), also with Starling *Sturnus vulgaris* (Creutz 1951; Blume 1963a). In such cases, holes probably allow birds to roost at different levels, or at separate sites within a roomy cavity (Löhrl 1949; D Blume). ♂ begins to roost in nest-hole probably before 1st egg laid but not known whether this the rule; ♀ may retain same site as before or shift nearer nest (Blume 1981); see also subsection 3 in Heterosexual Behaviour, and Relations within Family Group (below). After fledging, juveniles usually roost for first few days in the open on tree—normally cling to trunk under branch (Blume 1962a; Radermacher 1970). From autumn to spring, particularly in evening, birds use regular routes through territory; often means that same activity performed at same spot and roughly same time over several days or (with breaks) weeks (Blume 1981; D Blume). In southern England, ♀ mostly using nest-box active for only c. 7$\frac{1}{4}$ hrs on dark wet days in December and tended to retire early throughout period of observation, November–February (Jourdain 1936a). In Gladenbach area, birds active in early winter for c. 7–8 hrs. In autumn and winter, usually arrives in roosting area after *D. martius* and before Great Spotted Woodpecker *Dendrocopos major*. In spring, often last of these 3 to go to roost following protracted vocal activity in roosting area. Weather (apart from frost) apparently has no influence on time. Like *P. canus* and *D. martius*, but unlike *D. major*, does not feed in transitional phase prior to entering roost. Except in April–May, ♀ generally active for longer period during day than ♂. (Blume 1961, 1962b, 1963b, 1964a, 1981; see also Horstkotte 1973 for influence of light level on roosting times.) In late summer and autumn, new arrivals in a particular area for roosting conspicuous through marked vocal activity and searching flights (may last for days or even weeks). Fly to several holes before entering one, typically after much hesitation and intention movements, and may give Threat-call (see 4 in Voice) to drive off potential competitors. Once established at a particular site, bird sometimes flies direct to roost over up to c. 1 km, but normally approaches by way of intermediate station(s) from which roost or next perch clearly visible. Gives Excitement-call (see 2a in Voice) when landing at intermediate station, and, usually, long series of Flight-calls (see 2b in Voice) when c. 200–250 m from roost. From intermediate station, normally flies direct and low to roost. Sleeps on floor of cavity or clings to wall. Often, c. 20–30 min between waking and departure from roost; normally flies to intermediate station before moving further away (Blume 1955, 1963b, 1964a, 1981; D Blume; see also Löhrl 1949). Like *P. canus* (see

Glutz and Bauer 1980 for illustrations), often apparently dozes on tree, sometimes for c. 15–22 min with bill held up at c. 45° (Tutt 1951); occurs in evening before roosting and in morning after leaving hole, but also at other times, e.g. between bringing feeds to young (Loos 1904; Jourdain 1936a; Tutt 1951; Glutz and Bauer 1980). Occasionally bathes in shallow pools (Glutz and Bauer 1980); once recorded bathing in flight (Stahlbaum 1954), and attempted to bathe in stone-trough (Blume and Jeide 1965). Anting behaviour recorded (Allsop 1949; Stanford 1949).

2. Not particularly confiding (Bannerman 1955), though may allow close approach when feeding in winter (Steinfatt 1944b); see also Baier (1975). Disturbed by man or attacked by sparrowhawk *Accipiter* on tree, attempts to hide like *P. canus* (see that species). If pursued in the open, heads for cover and gives loud Excitement-calls (D Blume); see also (e.g.) Clifford (1947) and Bannerman (1955) for pursuit by Hobby *Falco subbuteo*, and Radermacher (1970) and Baier (1977) for attacks or mobbing by other birds. At roost, some birds do not even come to entrance in response to tapping and scratching on tree, though most rather sensitive to disturbance and leave at any strange sound if not too dark (Blume 1961, 1981). Factors affecting behaviour when disturbed at roost probably as in *P. canus*; not known whether fundamental differences exist (Glutz and Bauer 1980). One bird disturbed at roost apparently directed Head-swaying (see below) at observer, whilst calling (Cowdy 1955); another drove off stoat *Mustela erminea* with blows from bill (Glutz and Bauer 1980).
ANTAGONISTIC BEHAVIOUR. (1) General. Advertising-call (see 1 in Voice) given by both sexes from end of winter to late autumn (Steinfatt 1944b), though mainly late March to early May (Blume and Jung 1956). Use of Advertising-call in central European roosting area, early September and mid-November (D Blume), presumably primarily antagonistic (see further below). Uttered with bill held up and wide open (see *P. canus* for differences); carpal joints held slightly away from body, vibrating with rhythm of call; wings may also be slightly opened and drooped (Fig A).

Usually calls from regular song-posts near prospective nest-hole, rarely further away or in flight; main function to advertise potential nest-site (Blume 1957); see also Heterosexual Behaviour (below). In spring (usually from late February, then indicating that roost possibly becoming future nest-site), 2–4 birds may give Advertising-calls in area where one roosting; may then fly at and chase one another, etc. (Blume 1957, 1962a, 1981; D Blume). Differences in activity rhythms (see also Roosting, above) reduce likelihood of intersexual and interspecific rivalry in roosting area; disputes over roost-holes occur only briefly in autumn when birds settling in new area; disputes occur (less often) on feeding grounds. Excitement-call uttered particularly often during July–November (less so in breeding season, March–June) and associated with territory occupation, selection of roosting area, and disputes over holes. Call 3 given (e.g.) when arriving at an already occupied roost, in pauses between pursuit-flights in roosting area, also during disputes in courtship phase. In central Europe, Drumming (see Voice) performed—if at all—by both sexes late March to mid-May; mainly on branch or section of trunk near completed or prospective nest-hole, e.g. in early morning after leaving roost, March–April. In breeding season, birds Drum also inside hole or just outside, then perhaps near top of another tree (Blume 1962a, b, 1964a, 1981; D Blume). ♂ recorded Drumming while ♀ worked on hole c. 100 m away (Radermacher 1970). (2) Threat and fighting (see, especially, Blume 1955, 1957). In Picidae generally, threat related less to territorial defence than to defence of particular sites in territory (roost- and nest-hole, Drumming post) and of individual distance (Blume 1955; Glutz and Bauer 1980). In *P. viridis*, serious fighting preceded by various degrees of threat which often render fight superfluous. In intra- and interspecific threat at some distance, red head-feathers normally ruffled and Threat-calls given (see 4 in Voice). Bird sometimes freezes in Bill-up posture (Fig B). Rivals then fly or move on tree towards one another. On horizontal branches, characteristically hop in Threat-posture (Blume 1955; Fig C).

On tree-trunks drive one another upwards in spirals and, when close together, begin Head-swaying movements, often accompanied by Threat-calls, around trunk—like game of peek-a-boo (see Labitte 1953 for 'jeu de cache-cache' in heterosexual encounter). In experiments with stuffed specimens, Head-swaying became almost like a dance (Fig D; see Blume 1957 for further details and illustrations). On ground or horizontal

E

branch, birds threaten one another with slow, rhythmic, side-to-side Head-swaying with neck extended (Fig E); bill traces figure-of-eight course or ellipse. Movements recall those employed in pre-regurgitation at nest-hole (see below). More aggressive bird holds head nearly vertical, though pronounced Bill-thrusting typical of threatening *D. martius* does not occur (Blume 1955; D Blume). Birds may be so close as to touch bills (King 1952). In response to playback of Advertising-calls, bird may overfly sound-source repeatedly and give Advertising-call or Threat-call, or occasionally Drum (Blume 1957, 1981). Before proceeding from threat to fight, rivals usually spread wings wide (sometimes almost crouch thus: Selous 1905a) and fan tail; crown feathers ruffled, and Threat-call given with bill open (Blume 1955; D Blume). In experimental study (Blume 1957), birds showed highest-intensity aggression towards stuffed specimen by nest during incubation; performed dive-attacks with tail conspicuously fanned and pecked savagely at model (mainly at head); more ceremonial-type threat around hatching. Wing-twitching, pecking intention-movements, and freezing also recorded. Threat and fighting more intense between ♂♂ than between ♂ and ♀ (D Blume). In a damaging fight, which may end fatally (Catthoor 1954a), birds buffet one another with wings open. Attempt to strike with bill at other's head or neck region, or into bill which is open in defence (D Blume). Tongue darted out and may become entwined with opponent's (Popp 1955). Prolonged bill-tugging and jumping attacks recorded (Selous 1905a). Birds may tumble down from tree together during fight then separate on the ground; brief aerial pursuits with fights also occur (Blume 1955; D Blume). At this stage, no longer try to peck at each other but change (back) to Head-swaying (Fig E). Depending on level of aggression, red head feathers ruffled to variable extent and movements accompanied by Threat-calls. Fights may take place on the ground, tree-trunk, or branch; occur at all times of year. Experimental study indicated that ♀♀ may fight just as aggressively as ♂♂, but under natural conditions, rival ♀♀ probably meet less often than ♂♂ (D Blume); see Ruedi (1983) for fight between 2 ♀♀ when ♂ nearby. Threat and fighting between ♂ and ♀ recorded June and October (Blume 1961; D Blume). *P. viridis* attempts to drive off competitors for nest- or roost-holes (e.g. *P. canus*, *D. major*) by giving Threat-calls, or may pursue such birds up trunk (see above) and give excitement-calls (as recorded in H Sielmann's film *Zimmerleute des Waldes*). Able to evict *P. canus* from hole (see also 4 in Voice), but apparently subordinate to *D. martius* (e.g. Steinfatt 1944b, Radermacher 1970). For other interactions with *P. canus*, see Küchler (1951) and Burnier (1961). Leaps towards *S. vulgaris* in Threat-posture (e.g. Blume 1955); see Fig C. Bird in hole immediately moves to block entrance if competitor appears. In rare cases where hole shared by *P. viridis* and *S. vulgaris* (Blume 1963a), likely to have more than one entrance (D Blume). Battles may be prolonged and *P. viridis* not always ultimately victorious (see Turner 1908 for details). Bird threatened Blackbird *Turdus merula* on lawn with head turning on extended neck apparently like Kingfisher *Alcedo atthis* when handled (Campbell 1966); not clear whether distinct from Head-swaying. HETEROSEXUAL BEHAVIOUR. See, especially, Blume (1955, 1957, 1962a). (1) General. ♂ generally dominant over ♀. Meeting of sexes away from nesting territory and outside breeding season most likely in roosting area, morning and evening, less so during daytime foraging where territories partially overlap (see also part 1); apparently less common than in *P. canus*, but contact-shyness evidently less marked than in *Dryocopus* and *Dendrocopos*, as shown by feeding of ♀ by ♂, as well as bill-touching in various situations (D Blume). Appeasing behaviour involving quiet variants of Advertising-call, also call 5, and Head-swaying with head lowered (see also Cohen 1945, Hainard 1970) presumably the prerequisite for such contacts and possibly for maintenance of pair-bond over several seasons (see Bonds), though either bird may leave area and pair up elsewhere (Blume 1961). (2) Pair-bonding behaviour. Pair-formation takes place from late March to mid-April, though where birds roost in neighbouring trees in territory-overlap, may begin in November. However, mainly from February onwards, prospective breeding partners already give loud and frequent Advertising-calls initially on or near roosting trees, especially in morning at beginning of activity and in evening at its end (Blume 1955; D Blume). 7–14-unit Advertising-calls given at intervals of *c.* 20–40 s; even when response comes, bird keeps to own rhythm; vocal contact may continue for day or weeks over *c.* 1 km (Blume 1955). Occasionally, birds may visit a likely nest-tree together at this stage, but initially advertise their 'own' holes and during any reciprocal visits defend both these and individual distance (Blume 1981; see also Blume 1955, Blume and Jung 1956). Also, first closer approaches may be made only after birds have flown over one another several times (Blume 1955; see also Antagonistic Behaviour, above). Chases may take place up trees, birds may freeze (for up to *c.* 30 min: Creutz 1954) and then perform Head-swaying ceremony with Threat-calls or (variants of) Advertising-call; typical lowering of head (see also Sóvágó 1954, Creutz 1954 for bowing) possibly involves sexual recognition, tolerance, and appeasement (D Blume). Bill-touching also occurs (Sóvágó 1954). Bird may fly intermittently to hole and Drum faintly (Blume 1961). ♂ who lost mate Drummed persistently (Dean 1949; H Löhrl). Pair-bond strengthened through such meetings and breeding behaviour concentrated on future nest-site. Advertising-calls, and Head-swaying at nest-reliefs and in other encounters near nest serve to maintain pair-bond throughout season (D Blume). Excitement-call fades for a time; recurrence probably a sign that pair-formation completed and that birds have become accustomed to nesting area (Blume 1955; Blume and Jung 1956); bird may, however, obtain a mate only when nest-hole near completion (Blume 1981). See also 5 in Voice. (3) Nest-site selection. Advertising-call directed at prospective nest-hole (Nest-showing) and partner. Bird attempts to attract other to chosen site. Persistent vocal contact (see above) has

PLATE 78 (*facing*).
Jynx torquilla Wryneck (p. 800): **1–2** ad.
Picus canus Grey-headed Woodpecker (p. 813): **3–4** ad ♂.
Picus viridis Green Woodpecker (p. 824): **5–6** ad ♂.
Picus vaillantii Levaillant's Green Woodpecker (p. 837): **7–8** ad ♂.
Dryocopus martius Black Woodpecker (p. 840): **9–10** ad ♂. (NA)

PLATE 79. *Dryocopus martius* Black Woodpecker (p. 840): 1 ad ♂, 2 ad ♀, 3 juv ♂, 4 nestling ♂. (NA)

PLATE 80. *Sphyrapicus varius* Yellow-bellied Sapsucker (p. 853): 1 ad ♂, 2 ad ♀, 3 1st-winter ♂, 4 juv. (NA)

important locatory function and may help to synchronize heterosexual behaviour. Once vocal contact established, birds call close by entrance or inside hole (Blume 1981; D Blume); see also Blume and Jung (1956). In Gladenbach area, ♂ apparently attracted more often into ♀'s area and, in more than 50% of cases, ♀'s roost became later nest-site. When selection slow, site chosen was between roosts. ♂'s choice probably decisive when new hole excavated. Birds generally prefer old hole (Blume 1961, 1981; D Blume). In Lithuanian study, birds said always to have excavated new hole (Knystautas and Liutkus 1981). Nest-site may be selected and some rudimentary excavating done in winter (Tutt 1951). Excavation of new hole may take up to 1 month or more (Blume 1961), at least 8 days for replacement (Labitte 1953). More work done by ♂ if hole new (Glutz and Bauer 1980), especially early stage (Steinfatt 1944b); for exception, see Tracy (1933b). Incoming bird first tosses out wood-chips as these not cleared immediately prior to change-over (Tutt 1951, which see for details of nest-hole excavation). Nest-reliefs performed at excavating stage (as later) not highly ritualized: birds may give quiet Advertising-calls and call 5, occasionally brief bouts of quiet Tapping (D Blume: see Voice), though Steinfatt (1944b) noted much calling, especially from ♂. Nest-guarding as follows. Pair often spend long time near hole during day (especially evening) shortly before and after hole completed (or after final nest-site selection in case of ready-made hole) and up to laying. Birds sit quietly and, at quite long intervals, give series of Advertising-calls or quiet, compressed 'kük' sounds (see 3 in Voice). Occasionally, one bird flies to hole and spends some time inside. 1–2 weeks may pass between completion of nest-hole and laying (Blume 1961). For behaviour when potential competitor appears, see Antagonistic Behaviour (above). If ♂ leaves hole for copulation (see below), he returns immediately afterwards (D Blume). (4) Courtship-feeding. Preceded by Head-swaying ceremony with heads lowered, leading to synchronized up-and-down jerking. ♀ adopts crouched posture and pecks at ♂'s bill; whole process may be repeated (Snow and Manning 1954). Said in one case to have occurred repeatedly while ♀ incubating (Feilden 1908); also reported mid-August (Hann 1951). (5) Mating. Usually without preceding ceremony, but may follow Head-swaying or Courtship-feeding (Fish 1943); see also, especially, Cohen (1947) for Head-swaying between 2 juveniles, mid-August. Quiet 'kjuiuh' sounds may be given beforehand (Blume 1962a: see 5 in Voice). Takes place on horizontal branch or on the ground (e.g. Fish 1943), often near nest. ♂ often climbs above, then shuffles down, to ♀ or may fly nearer her. ♀ solicits by crouching. ♂ mounts and Bill-touches with ♀, apparently striving to maintain contact with her bill or head (*contra* Fish 1943, does not grasp ♀'s head feathers: Blume 1962a); sometimes pecks at ♀'s head (Opitz 1953). ♀ has bill open. 5–6 s may pass before ♀ has settled; ♂ makes brief (2–3 s) cloacal contact by beating wings and (possibly not always) tipping over to left so that birds form cross. Actual copulation often silent. ♂ then moves again on to ♀'s back, turns around, jumps or flies off and usually returns rapidly to nest-hole. ♀ may remain crouched with bill closed for 6 s or more when ♂ leaves (for details, including illustrations, see Blume 1962a; also Creutz 1954). Copulation noted by Steinfatt (1944b) only during laying and a few days before. May take place up to 3 times in quick succession (Ritter 1953); up to 6 times per day during nest-excavation (Blume 1962a). (6) Behaviour at nest. Advertising-call (7 or more units) given to invite nest-relief and to confirm that bird ready for change-over (Blume 1981; D Blume). ♀ leaving nest after responding to ♂'s call from inside indulged in chase on tree for c. 2 min, ♂ keeping his distance (Steinfatt 1944b). At nest-reliefs, birds may adopt 'tense' posture and give call 5 (Blume 1955) or give Threat-calls if mate reluctant to leave (Blume 1981). Head-swaying, presumably expressing appeasement or tolerance, also typically performed at change-over (D Blume; see Blume 1961 for full description and interpretation). ♂ generally takes greater share of incubation also sitting at night. Eggs may be left uncovered for c. 20 min (Steinfatt 1944b) and reciprocal calling by pair-members from neighbouring trees may take place (especially later in day) as during pair-formation (Blume 1961). Incubation shifts last 1½–3 hrs (Tracy 1946; Blume 1962a, 1981). RELATIONS WITHIN FAMILY GROUP. During first c. 8 days, young shoot heads up when sensitive and prominent gape flanges touched (Tracy 1946); also seize parent's bill and give food-calls. Scratching on trunk (as parent lands) or brief darkening of nest-chamber may elicit food-calls. Young apparently able to distinguish between scratching sounds made by landing parent and other similar noises (see Blume 1957). Food-calls given more or less constantly up to c. 16 days old (then with longer pauses), though usually only one nestling calling at a time (D Blume). Adult able to deliver up to 15 food portions per visit. Pre-regurgitated with a swinging–slinging movement of the head (see photographs in Blume 1957), then pumped into nestling's gullet (D Blume); tongue may assist (see Tracy 1946). Particularly during middle of nestling period, special secretion used to make thick porridge-like fluid (Blume 1981; D Blume). Adult's bill pushed well inside that of nestling and heads move rapidly back and forth (Tutt 1951). Immediately after feeding, adult encourages defecation by jabbing at nestling's rear end; faecal sac swallowed at once by parent during first few days (Sielmann 1959); according to Tutt (1956), for c. 12–14 days while parents still entering hole to feed young, later often carried out and dropped, though nest-sanitation ceases towards end of nestling phase. Thick layer of faeces may remain (Tracy 1946). Even at 9–10 days old (see illustration in Blume 1962a) young occasionally fed at hole-entrance, then regularly from c. 17 days. Adult moves a little down or aside to pre-regurgitate fresh food portions (D Blume); young sometimes peck at adult (Loos 1904; Tutt 1956). Nestlings brooded by parents alternately until c. 5–7 days old, in shifts of c. 1–2 hrs, longer in cold, wet weather (Blume 1962a; D Blume). May be left for up to c. 1 hr even at 1–2 days old (Steinfatt 1944b) and then huddle together in 'warmth pyramid' (see *D. martius*). Little or no daytime brooding from c. 10–11 days old and ♂ no longer roosts with young from c. 20–21 days old (Steinfatt 1944b). Noisy and restless close to fledging (Tracy 1946) which takes place at c. 23–27 days old (D Blume). Adults perhaps lure young out using call 5 or quieter variant of Advertising-call (Tutt 1956). Initially, both adults feed young together for c. 3–4 days, near nest or up to c. 500 m away. Later, each adult presumably cares for part of brood, such units remaining together for some time (D Blume); 4 birds seen very close together as late as 11 November (Glutz and Bauer 1980; see also part 1). Unlike in *P. canus*, contact between juveniles and with parent generally maintained with Excitement-call, not Advertising-call. At 3–5 days after fledging, young call frequently, with brief bouts up to c. 30 s apart. Apparently strive not to stray more than c. 200 m apart. When sunbathing, often sit together on ground in physical contact. In evening, young call each other together for roosting, at first on neighbouring trees, later in holes. Vocal contact may continue in autumn though some individuals rove about in search of roost-hole from late July (Blume 1961, 1962a, 1981; D Blume). ANTI-PREDATOR RESPONSES OF YOUNG. From c. 7–16 days old tend to crouch and screech (see Voice) rather than give food-calls in response to unspecific scratching sounds or blocking of light at nest-entrance. From c. 16 days, if begging, cease calling if tree tapped or scratched, also on hearing parental alarm (see 2a in Voice); withdraw rapidly into hole if at entrance (Blume 1957, 1961; D

Blume). PARENTAL ANTI-PREDATOR STRATEGIES. (1) Passive measures. Generally a tight sitter; may not leave nest until tree climbed and, after flushing, waits nearby, gives muted Advertising-call, and soon returns (Steinfatt 1944b). When nest threatened by aerial or ground predators, birds normally give Excitement-calls from safe distance (D Blume). On nest-tree, birds with young often alert and make cautious approach, hiding on opposite side of trunk, freezing, etc. (Labitte 1953); see also (e.g.) Tracy (1933b). (2) Active measures. None recorded.

(Figs A–D from drawings by D Blume; Fig E from drawing by D Blume and photograph in Blume 1981.) MGW

Voice. Both sexes equally vocal throughout the year and share same repertoire, including instrumental signals (see below). For seasonal variations in occurrence, see Blume (1961). Vocal utterances most important form of communication in *P. viridis*, while Tapping and Drumming rare (Blume 1981; D Blume; see below and Social Pattern and Behaviour). Calls of Iberian *sharpei* thinner, more tinny, than nominate *viridis* (P A D Hollom).

INSTRUMENTAL SIGNALS OF ADULTS. Ritualized forms associated only with nest-tree. (a) Excitement-pecking. Irregular sequences of strikes given in conflict situations (e.g. when disturbed whilst excavating nest-hole) in all seasons, also on trees near nests of ants (Formicidae) where birds perform mock-pecking before flying down and sometimes also on tree after feeding. Ritualized pecking possibly also occurs in breeding season (Blume and Jung 1958; D Blume; see also Social Pattern and Behaviour). (b) Tapping. Occurs when pair-members meet near nest-hole, in Nest-showing, or during nest-reliefs. Also associated with activities in roosting area, interspecific disputes, and disturbance. Tap sequences usually brief, in rhythmic 4-unit pattern, and not loud (Blume 1955, 1961; Blume and Jung 1958; D Blume). (3) Drumming. Inadequately studied owing to its rarity. In recording (Fig I) of bird apparently foraging between bouts of Drumming, 1 series comprises 18 strikes in 0·74 s, with overall accelerando: 60–40 ms between strikes (approximately even decrease of 2 ms between strikes) and brief diminuendo over last 6 (J Hall-Craggs). Rate of strikes may also slow towards end of series and (rarely) fade away into a Tapping (D Blume; see also above). In study by Zabka (1980), average 25·8 strikes per series, average duration 1·14 s, $n = 5$ series (by one bird); *c.* 40 ms between strikes at start, *c.* 55 ms at end. See also diagram showing similar number of strikes and delivery rate in Glutz and Bauer (1980). Earlier reports based on estimates (see, e.g., Armstrong 1942, Bannerman 1955) indicated fewer strikes and much slower rate; possibly do not refer to true Drumming, though further investigation required. In one case, 4–8 series given per min, with 70 series in 15 min (Hesse 1918). One bird using metal cap of electric pylon Drummed intermittently throughout day with pauses between series of a few seconds to *c.* 5 min (Dean 1949). Burst generally longer than Great Spotted Woodpecker *Dendrocopos major* (Blume 1961), also less clear and sharp-sounding (Hesse 1918). Weaker even than Lesser Spotted Woodpecker *D. minor* and barely audible at *c.* 40 m (Radermacher 1970), though can be loud (Blume and Jung 1958). Drums preferentially on basal section of large tree branch, and Drumming perhaps performed only by certain individuals (Hesse 1918). Advertising-calls (see call 1) may be interspersed between bursts of Drumming (D Blume). In some conflict situations 18–22 relatively slow strikes—apparently a hybrid Tap–Drumming (less common than in other Picinae)—given in 2–2·5 s (Blume 1962a, 1981).

CALLS OF ADULTS. (1) Advertising-call (Figs II–III). Well-known laughing or 'yaffling' sound. See also Grey-headed Woodpecker *P. canus*. Early in breeding season, up to 20 soft 'klü' units given in dynamic sequence (Blume 1981; D Blume); described by Bergmann and Helb (1982) as loud, hard, and full-throated 'kju'. Stress sometimes falls on higher-pitched 2nd unit and sequence becomes slightly faster (Fig III) or slower (Fig II) and quieter towards end, but overall delivery rate varies. Unlike *P. canus*, no appreciable change in pitch (Blume 1961; Bergmann and Helb 1982; D Blume, J Hall-Craggs). Sometimes preceded by brief, quiet series of rasping sounds (possibly variant of call 4), typical Advertising-call following immediately. Later in breeding season, when ♂ and ♀ have approached one another and nest-site selected, normally composed of fewer (2–7), quieter, and less musical units; last units merge one into another and fall slightly in pitch, so that confusion with *P. canus* possible (Blume 1955, 1962b; Blume and Jung 1956; D Blume). According to Bergmann and Helb (1982), ♀'s Advertising-call less hard and slightly hoarse; in sonagram by H-H Bergmann of recording by K Hinrichs, ♀'s call indeed shows strong harmonics, also monosyllabic fundamentals, whereas ♂'s call (Figs II–III, also other published sonograms) apparently has mostly disyllabic fundamentals (J Hall-Craggs). Blume (1981) noted marked individual variation, including in timbre, allowing pair-members and young to recognize one another; see also (e.g.) Steinfatt (1944b). (2) (a) Excitement-call. A series of usually 3–4 (2–7) 'kjäck' sounds, rarely given singly. Expresses general excitement, sometimes alarm, and given when landing on tree (e.g. on roost tree or intermediate station), taking off, or when perched. Indistinguishable from call given by excited *P. canus* when taking off in autumn. Depending on level of excitement, 'kjäck' sounds long drawn and given in slow sequence, or shorter and delivered more rapidly (D Blume). (b) Flight-call. In flight, sharper, slightly higher-pitched 'kjäck' units given in rapid sequence and may sound more like 'kück' or 'kjück'. Usually with definite rhythm, stress falling on 2nd, 4th, etc., unit, or on 4th, 8th, etc. (Blume 1961; D Blume); see also (e.g.) Loos (1904). (3) An abrupt and compressed 'kük', perhaps only a variant of call 2, given when perched or in flight, typically in conflict situations. Very easily confused with equivalent call of *P. canus* (Blume and Jung 1956; Blume 1961; D Blume). (4) Threat-call. A short, sharp, squeak-

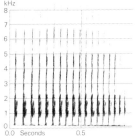
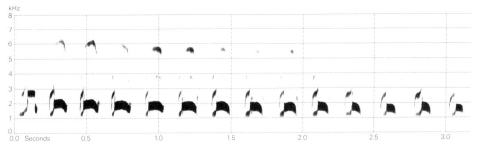

I A J Williams England April 1976   II S Palmér/Sveriges Radio (1972) Sweden April 1957

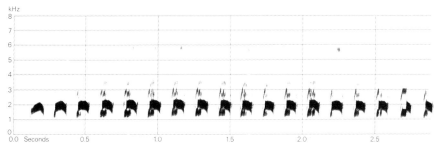

III P A D Hollom England July 1981

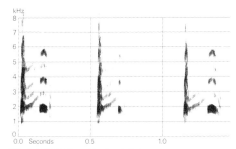
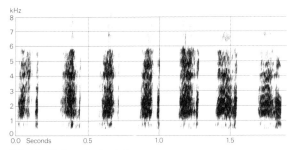

IV P A D Hollom England July 1981   V P A D Hollom England June 1974

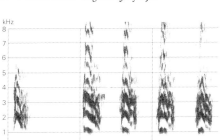

VI P A D Hollom England July 1981

ing 'kjaik' sound, like rubbing window-pane with damp chamois-leather. Sometimes given as a fast, rhythmic series. Given (also in flight) in threat-duels in breeding season, and in intra- and interspecific threat and fights at roosts and on feeding grounds. Understood by *P. canus* whose Threat-call very similar (Cohen 1946; D Blume); for interaction between ♀ *P. viridis* and ♀ *P. canus*, see Küchler (1951). For further descriptions, see Simmons (1947), King (1952), Creutz (1954), and Burnier (1961). (5) A quiet, long-drawn 'kjuiuh' sound given when pair-formation well advanced. Possibly an expression of suppressed threat and therefore perhaps a variant of call 4 (Blume and Jung 1956; Blume 1961; D Blume). Presumably same as 'twie twie twie' which said by Loos (1904) to be given by ♀ only when arriving at nest with young, and quiet, gentle-sounding 'krüth', 'küht', or 'gük' sounds typically associated with nest-relief (Steinfatt 1944b). Similarly, in recording by P A D Hollom, a plaintive 'pwee(p)' or 'puweep(p)' from adult climbing up to nest containing young. In another recording (Fig IV), adult near newly fledged young gives 'ke-eee', 'kay-eee', or 'qu-eee' sounds (M G Wilson), and thin 'KIvik' not unlike Tawny Owl *Strix aluco* (P J Sellar); presumably the same or a related call. For further details, see Tutt (1951), Creutz (1954), and Sóvágó (1954). (6) Trilling-call similar to Nuthatch *Sitta europaea* very rarely heard—only from ♂ flying to nest. Perhaps a more excited variant of call 1 (Blume and Jung 1956; Blume 1961; D Blume). (7) Other calls mentioned by Steinfatt (1944b) possibly only variants, or different renderings, of those given above. (a) A harsh and broad-sounding 'keuk-keuk' as contact-call towards end of breeding season ('autumn call'); presumably variant of call 1. (b) A 'gäp gäp gäp' said to express fear.

CALLS OF YOUNG. From 1–3 days old, give mostly hoarse, rhythmic food-calls: 'räk-äk-äk' or 'räk-äk-äk-äk' (stress on last syllable). Rhythm maintained until c. 7 days old, volume increasing. Presumably same as quiet 'dög dög' or 'diög' of Loos (1904) who also mentioned a long-drawn, hissing 'sch'. Food-calls also described as a continuous, harsh, raucous squawking (Tutt 1951); rasping sounds, quieter than *D. major* (Steinfatt 1944*b*). In recordings by P A D Hollom, young give harsh rasping or scraping sounds, at times like rubbing glasspaper on wood (Fig V); in higher-intensity excitement, some faster, more chattering calls (P J Sellar, M G Wilson). From c. 7–16 days old (though not towards end of nestling period), give long-drawn screeching sounds whilst simultaneously crouching if disturbed. From c. 19 days old, series of food-calls sometimes change to forerunners of single 'kjäck' sounds (see adult call 1) and, shortly before fledging, give 3–4-unit 'kjäck' calls much as adults (D Blume). In recordings by S Palmér (older nestlings) and P A D Hollom (recently fledged young), 'keeyeck' calls reminiscent of Jackdaw *Corvus monedula* (P J Sellar, M G Wilson); sequence from fledgling (Fig VI, which omits 1st unit) suggests 'chik chiuk chi-chi-chi-chi' (J Hall-Craggs). Loud 'ke', 'kjö, or 'kjä' sounds mentioned by Bergmann and Helb (1982), which see for sonagram. Other renderings of calls given around fledging: at 24–25 days old, 'kück-guaga-ga-gack-gack' (Steinfatt 1944*b*); 'kjackkjack-jihk', though birds may revert to earlier food-calls (Blume 1955). A loud 'güep' given by juvenile at probable roost-site (Radermacher 1970). For further details, see Blume and Jung (1956). MGW

Breeding. SEASON. Britain and north-west Europe: see diagram. Up to 2 weeks later in Sweden and western USSR (Dementiev and Gladkov 1951*a*; Makatsch 1976). SITE. Hole in tree, 1–5 m from ground; wide variety of tree species used. Nest: excavated hole with 6 cm entrance diameter; depth 30–50 cm, internal diameter 15–18 cm; unlined (Labitte 1953). May be re-used. Building: by both sexes, though mainly by ♂, taking average 15–30 days, though replacement nest can be finished in 8 (Labitte 1953); for fuller details of techniques, see Tracy (1933*b*). EGGS. See Plate 98. Elliptical, smooth and glossy; white. 32×23 mm (27–35×20–25), $n = 100$ (Witherby *et al.* 1938). Fresh weight 8·5 g (7·9–9·1), $n = 24$ (Glutz and Bauer 1980). Clutch: 5–7 (4–9, rarely 11). Of 30 clutches, France: 5 eggs, 8; 6, 14; 7, 6; 8, 2; average 6·1 (Labitte 1953). One brood. Replacements laid after egg loss, occasionally twice. One pair, France, re-laid 9 days after losing 1st clutch and 6 days after losing 2nd (Labitte 1953). Laying interval 1 day. INCUBATION. 17–19 days, though Tracy (1946) claimed 12 days for one nest, and 15–17 also reported. By both sexes in roughly equal shares, changing over after $1\frac{1}{2}$–$2\frac{1}{2}$ hrs (Blume 1961). Begins with last egg; hatching synchronous. YOUNG. Altricial and nidicolous. Cared for and fed by both parents; brooded for first 3–5 days (Blume 1961). FLEDGING TO MATURITY. Fledging period 23–27 days (D Blume), though Tracy (1946) reported 28 days in one case, while Witherby *et al.* (1938) gave 18–21 days. Become independent 3–7 weeks after fledging (Blume 1961). Age of first breeding 1 year. BREEDING SUCCESS. No information.

Plumages (nominate *viridis*). ADULT MALE. Feathering at base of upper mandible and bristles covering nostrils greyish-black. Forehead, crown, and nape carmine-red, some grey of feather-bases usually visible, especially on centre of crown. Mantle, scapulars, tertials, upper wing-coverts, and sides of neck and chest bright yellowish grass-green, feathers at border of red nape often tipped yellow or orange; tertials and upper wing-coverts sometimes with golden-green tinge; basal inner webs of tertials dull grey with row of short white bars. Upper back like mantle and scapulars, grading to bright yellow on rump and upper tail-coverts; yellow feather-tips sometimes deeper yolk-yellow or slightly orange, feather-bases green (shining through to variable extent), rump sometimes yellow-green (in particular when plumage worn), upper tail-coverts usually with green showing distinctly at bases. Lores and feathers at base of lower mandible black, extending into large black patch round eye and down to broad carmine-red malar stripe, latter bordered black below and behind. Ear-coverts pale green, some greyish-white of feather-bases showing through slightly; green gradually brighter and deeper towards sides of neck. Chin and throat greyish-white with slight green tinge; remainder of underparts pale olive-green, darkest on chest and breast, paler and tinged yellow on belly and vent; lower flanks and under tail-coverts with broad but rather indistinct dark olive-green or olive-grey bars or chevrons, bars sometimes extending up to lower breast and upper flanks as indistinct olive-green spots. Central pair of tail-feathers (t1) black with green sides, black at each side broken by row of pale green or cream-white spots; t2–t4 black with faintly greenish sides and sometimes with traces of pale spots near bases of centres (spots most obvious on undersurface); t5 dull black with indistinct dull grey-green bars on uppersurface, more contrastingly barred pale grey-green and black on undersurface; reduced t6 green with variable black marks on centre and inner web. Outer web of p7–p10 and emarginated parts of outer web of p5–p6

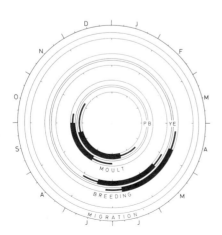

dull black with faint greenish tinge, broken by row of off-white spots except for uniform black terminal part; outer web of p1–p4 and base of outer web of p5–p6 with row of dull black and off-white blobs along shaft and with broad uniform green fringe. Inner webs of primaries greyish-black; basal halves of p7–p10 with broad and incomplete off-white bars, barring gradually more extensive towards inner primaries, almost reaching tip on p1. Outer webs of secondaries grass-green, often with slight brass or golden tinge; narrow streak along shaft dull black, occasionally bordered by row of indistinct pale green or cream-white specks; inner webs similar to inner webs of inner primaries. Greater upper primary coverts and longest feather of bastard wing similar to inner primaries—outer webs with row of black and pale grey-green or off-white spots along shafts, broad outer fringes uniform green; shorter upper primary coverts white with some dark bars and partially with green tips; remainder of upperwing green like mantle and scapulars. Under wing-coverts and axillaries pale green-yellow or green-white, narrowly barred dark grey. In worn plumage, some more grey of feather-bases of crown visible; upperparts and upper wing-coverts duller olive-green, frayed feather-fringes paler yellowish-green; more green visible amidst yellow of rump and upper tail-coverts; chin and throat uniform pale grey or off-white; ear-coverts and underparts down from chest pale greenish-grey, underparts sometimes partly stained dirty grey or brownish by soil particles; dark bars and chevrons on lower flanks and under tail-coverts more contrasting; pale spots on tail, outer webs of primaries, and greater upper primary coverts bleached to white. ADULT FEMALE. Similar to adult ♂, but broad malar stripe competely black; feathers at upper border of black sides of face often slightly tipped green-grey, tips sometimes joining to form narrow and indistinct pale supercilium between red of crown and black above eye; dark olive-green bars of lower flanks and under tail-coverts more often extend as fine specks or small V-marks up to breast and upper flanks. NESTLING. Naked at hatching. Closed sheaths show from age of c. 9 days, opening at tips from c. 14 days; appears feathered from c. 3 weeks, but feathers still growing until 2–3 weeks after fledging. JUVENILE MALE. Crown and nape as in adult ♂, but red of feather-tips slightly less glossy, often slightly paler or more scarlet, more grey of feather-bases visible. Mantle, scapulars, back, and upper wing-coverts duller olive-green, not as bright and yellowish as adult, each feather with white or green-white subterminal spot at shaft, sometimes extending laterally to form pale green bar. Rump and upper tail-coverts barred white and dark olive-green, yellow feather-tips less extensive and less bright than adult. Feathering at base of upper mandible (including bristles covering nostrils), lores, and feathers over eye dark grey or dull black with fine pale green or off-white specks, sometimes joining to form thin lines. Ear-coverts and chin pale grey-green to off-white; indistinctly streaked or mottled grey, feathering along gape, on lateral hindneck, on sides of neck, and on throat more distinctly greyish-black. Malar stripe black, slightly narrower and less distinctly defined than in adult ♂, feather-tips at centre and upper-side of malar stripe tipped red, forming ill-defined patch, but red often restricted to a few feather-tips only and some other feathers tipped green instead of red. Underparts down from chest greenish-white (slightly yellow when plumage fresh, white when worn), distinctly spotted greyish-black on chest and barred greyish-black elsewhere. Tail as adult, but ground-colour paler and dark bars more distinct (see also Structure). Flight-feathers as adult, but off-white bars on inner webs of tertials extend to tips and often slightly to outer web; inner 3 tertials with faint pale terminal bar and more distinct off-white subterminal spot on shaft; pale spots on outer webs of secondaries (along black of shafts) often more distinct; see also Structure. Greater upper primary coverts and longest feather of bastard wing as adult, but outer webs with narrow uniform green fringe only or none at all, pale spots on outer webs average smaller and more distinct; shorter feathers of bastard wing with pale subterminal spots as on upper wing-coverts (all uniform green in adult). In last stages of post-juvenile moult, when characteristic paler-spotted feathers of upperparts and upperwing and heavy barred feathers of underparts replaced by uniform adult feathers, distinguishable by long and broadly rounded juvenile p10 (see Structure) and by absence of secondary moult. JUVENILE FEMALE. As juvenile ♂, but with no trace of red on black malar stripe; malar stripe usually finely speckled pale yellow-green, sometimes hardly differing from similarly speckled sides of head. FIRST ADULT. Like adult, but juvenile secondaries, nearly always tertials, and often all greater upper primary coverts retained; some birds replace outer greater primary coverts only, at least a few replace them all; a few birds retain some juvenile greater upper wing-coverts or some feathers of bastard wing. Best distinguished by juvenile innermost tertials tapering to narrower point than in adult (less square), relatively more heavily abraded, usually still with trace of pale bar across tip and often distinct off-white subterminal mark; juvenile greater upper primary coverts relatively worn and pointed, with rather small and sharp off-white spots on outer webs and virtually without green along outer fringe; difference from adult sometimes not marked, but forms a good character in birds which retain part of coverts only, these contrasting with broadly green-fringed new ones. At least a few birds replace all greater upper primary coverts as well as tertials; these distinguishable by contrast in abrasion between old flight-feathers and new coverts and tertials (in most 1st adults and all adults, tertials and coverts usually wear more rapidly than secondaries and primaries).

Bare parts. ADULT AND FIRST ADULT. Iris white or greyish-white with pink tinge or pinkish outer ring, tinge rarely yellowish. Bill brown-black, greyish-black, or olive-black, base of lower mandible pale green-yellow, yellow-olive, or olive-grey. Leg and foot olive-grey to dirty lead-grey with olive tinge. NESTLING AND JUVENILE. Naked at hatching; completely pink, including gape; eyelids blue-grey, tip of each mandible with white egg-tooth, claws white. Swollen rounded flanges on each side of base of lower mandible pinkish-white; largest in 2nd week. Eye gradually opens from c. 9th day. At fledging, iris grey, pearl-grey, or greyish-white; bill like adult but base of lower mandible yellow; leg and foot like adult. (Hartert 1912–21; Heinroth and Heinroth 1931–3; Short 1982; RMNH, ZMA.)

Moults. Based on Stresemann (1920), Dementiev and Gladkov (1951a), Glutz and Bauer (1980), Ginn and Melville (1983), and c. 60 moulting specimens in RMNH, ZFMK, and ZMA. ADULT POST-BREEDING. Complete, primaries descendant. Starts with p1 between mid-May and late June, occasionally up to late July; mainly in last days of May or first days of June in Mediterranean populations and apparently in USSR, late May to mid-July (mainly late June) in Britain, and mid-June in Netherlands. Primary moult score of 25 reached early August to early September in Britain, mid-July to late August in Netherlands; all moult completed mid-October to late November in Britain, mid-September to early November in Netherlands. Duration of primary moult c. 130 days in Britain. Secondaries moulted from 2 centres: ascendantly and descendantly from s8 (shed with about p3), ascendantly from s1 (shed shortly after p5); s4 replaced last, shed with p9. Tail moults in approximate sequence 2–3–4–5–6–1 (t6 sometimes relatively earlier); starts with t2–t3 at shedding of p3–p5, t1 at about same time as p6; growth of t1 completed at primary score $c.$ 35–40; replacement mainly July–August. Body starts at score $c.$ 20 (July or early August), largely completed at score 40–45 (September or early October); mantle, scapulars, lesser upper wing-coverts, and scattered feathering of underparts first, head (sides in particular) and sides of breast last. Moult of 1st adult similar to adult. POST-JUVENILE. Partial: includes primaries and tail, but no secondaries, usually no tertials, and often not all greater upper primary coverts. Starts with p1 before fledging, 1–5 days later followed by p2 at about fledging; subsequent feathers shed at intervals of 13 (10–16) days, p10 shed at $c.$ 3 months, all primaries full-grown at $3\frac{1}{2}$–4 months. Moult starts with p1 early May to mid-August; mainly mid-May to mid-June in Mediterranean populations, early June to early July elsewhere, mainly completed October or early November (on average, perhaps slightly later than adult). Sequence and timing of tail and body as in adult post-breeding, but regrowth of t1 sometimes slightly later, occasionally still growing after completion of primary moult; moult of body mainly late July to late October. Retained juvenile secondaries, tertials, and greater upper primary coverts replaced in 1st adult post-breeding when $c.$ 1 year old.

Measurements. Nominate *viridis*. Netherlands, all year; skins (ZMA). Bill (F) to forehead, bill (N) to distal corner of nostril. Toe is outer front toe with claw. Adult includes 1st adult, though 1st adult wing on average 3·2 shorter than full adult, 1st adult tail 2·0 shorter, bill (F) and bill (N) 1·2 shorter.

WING	AD	♂ 164 (3·28; 23) 158–170	♀ 164	(2·85; 23)	159–169
	JUV	160 (2·65; 14) 156–164	157	(3·84; 7)	154–161
TAIL	AD	98·9 (2·69; 21) 94–104	100	(2·89; 19)	95–106
	JUV	96·9 (4·55; 6) 94–102	96·5	(— ; 3)	92–102
BILL (F)	AD	45·5 (2·00; 24) 42–48)	45·5	(2·42; 22)	42–49
BILL (N)	AD	33·5 (1·75; 24) 31–36	33·1	(1·91; 21)	31–37
TARSUS		29·7 (0·97; 15) 28–31	29·9	(0·85; 16)	28–31
TOE		32·9 (1·24; 11) 32–35	33·1	(1·46; 16)	31–35

Sex differences not significant. Juvenile wing significantly shorter than adult; juvenile bill reaches adult length when plumage about halfway through post-juvenile moult.

Sexes combined; 1st adult included in adult. Nominate *viridis*: (1) north-west and central European USSR; (2) Sweden; (3) Jutland (Denmark); (4) East and West Germany; (5) Hungary, northern Yugoslavia, Rumania, and Bulgaria; (6) France and north-west Switzerland; (7) Britain. *P. v. karelini*: (8) Italy; (9) Albania, southern Yugoslavia, and Greece; (10) Asia Minor and Transcaucasia; (11) Caucasus; (12) northern Iran. *P. v. innominatus*: (13) south-west Iran. (Stresemann 1920; Niethammer 1936; Dementiev and Gladkov 1951a; Vaurie 1959a; RMNH, ZFMK, ZMA.)

	WING AD	BILL (F) AD
(1)	169 (— ; 28) 161–175	50·7 (— ; 3) 49–53
(2)	167 (2·27; 33) 163–172	49·1 (2·05; 19) 45–52
(3)	166 (— ; 14) 163–169	47·0 (— ; 14) 44–50
(4)	164 (2·50; 48) 157–170	47·1 (2·63; 33) 41–52
(5)	161 (3·10; 41) 156–168	45·7 (2·51; 29) 42–52
(6)	161 (2·98; 14) 156–166	46·5 (1·98; 13) 43–50
(7)	161 (2·57; 47) 157–167	45·5 (2·01; 47) 40–50
(8)	160 (2·63; 49) 154–164	44·2 (1·94; 42) 40–49
(9)	160 (2·10; 26) 156–165	46·6 (1·52; 24) 43–50
(10)	162 (2·17; 14) 159–165	45·4 (1·64; 14) 42–48
(11)	165 (3·00; 8) 160–168	46·9 (1·70; 8) 45–50
(12)	160 (1·98; 14) 157–163	45·8 (1·58; 14) 43–48
(13)	162 (3·24; 8) 158–169	47·4 (1·92; 8) 45–50

Averages of other measurements for some of these localities. Britain ($n=8$): tail 98·1, bill (N) 33·1, tarsus 29·4, toe 31·3. Italy ($n=15$): tail 97·5, bill (N) 32·3, tarsus 28·6, toe 31·7. Rumania and Bulgaria ($n=13$): bill (N) 32·0, tarsus 29·5, toe 32·5. Asia Minor and Caucasus ($n=4$): bill (N) 33·5.

P. v. sharpei. Salamanca, Extremadura, Sevilla, Cadiz, and Galicia (Spain), and Portugal, all year; skins (BMNH, RMNH, ZFMK, ZMA). 1st adult included in adult.

WING	AD ♂ 161 (2·91; 16) 157–166	♀ 159	(2·46; 22)	154–164
BILL (F)	AD 42·6 (1·81; 11) 40–45	41·1	(0·99; 11)	39–42
BILL (N)	AD 30·6 (2·22; 15) 26–34	29·3	(1·37; 20)	27–32

Sex differences significant.

Castellon, Valencia, Teruel, and Madrid, all year.

WING	AD ♂ 169 (— ; 2) 165–173	♀ 166	(2·15; 6)	163–169
BILL (F)	AD 45·3 (— ; 2) 43–47	43·33	(— ; 3)	42–46
BILL (N)	AD 33·4 (— ; 2) 32–35	30·9	(1·34; 6)	29–32

Averages of some other measurements. Salamanca ($n=8$): tail 94·8, tarsus 28·4, toe 30·9. Castellon ($n=5$): tail 97·0, bill (N) 31·7, tarsus 29·8, toe 32·6.

Weights. (1) Britain, adult and 1st adult, mainly June–August (BTO). (2) Netherlands, adult and 1st adult, May–October (RMNH, ZMA). (3) Camargue (France), adult and 1st adult, all year (Glutz and Bauer 1980).

(1)	♂ 189 (9·11; 12) (9·11; 12)	♀ 198	(15·4; 6) 179–217
(2)	196 (8·93; 7) (8·93; 7)	186	(8·81; 4) 178–198
(3)	177 (12·1; 39) (12·1; 39)	174	(12·2; 30) 138–190

Norway: range 185–215 (Haftorn 1971). USSR: range 186–250 (Dementiev and Gladkov 1951a); ♂ 202 (4) 195–210, ♀ 190 (3) 170–205 (Fedyushin and Dolbik 1967). Turkey and northern Iran, March–October: ♂♂ 163, 164, 220; juvenile ♀♀ 142, 176 (Paludan 1940; Schüz 1959; Rokitansky and Schifter 1971; ZFMK). Netherlands, February: ♀ 220 (RMNH). Exhausted adults, Netherlands, ♂♂ 128, 144, 150; ♀♀ 128, 132, 148, 153 (ZMA); a wounded bird, Switzerland, 95 (Glutz and Bauer 1980). Nestling at hatching: 5·7, 5·8 (Heinroth and Heinroth 1931–3). Camargue: juvenile at fledging 152 (9·5; 18) 138–165; near end of post-juvenile moult, September and early October, 173 (8·5; 10) 160–188 (Glutz and Bauer 1980).

Structure. Wing short and broad, tip rounded. 10 primaries: in adult and 1st adult, p6–p7 longest or either one 0–1 shorter than other; p8 3–6 shorter than longest, p9 23–33, p10 82–94, p5 3–6, p4 12–18, p3 24–32, p2 30–38, p1 34–47; in juvenile, p7 longest, p8 $\frac{1}{2}$–3 shorter, p9 16–23, p10 66–76, p6 3–6, p5 11–16, p4 23–29, p3 32–42, and p1–p2 reduced, about half length of neighbouring s1 or less. P10 strongly reduced in adult and 1st adult, less so in juvenile; in adult and 1st adult, narrow, tip rather pointed, inner web deeply emarginated, tip 3 mm shorter to 10 mm longer than tips of longest greater upper primary coverts (on average, 3·5 longer, $n=46$); in juvenile, broader, tip rounded, inner web less deeply emarginated, tip 17·4 (19) 13–23 longer than tip of longest primary coverts. At all ages, outer web of (p4–)p5–p8 and inner web of (p5–)p6–p10 emarginated. Tail rather short, wedge-shaped; 12 feathers, t1–t5 stiff, attenuated towards tips, shafts strong; t6 reduced, tip rather broad and rounded in adult and 1st adult, narrow and pointed in juvenile. T1 longest; in adult, t2 2–5 shorter, t3 6–12, t4 14–19, t5 25–32, t6 62–70; in juvenile, t2 6–9 shorter, t3 10–14, t4 15–22, t5 25–38, t6 64–76. Bill about equal to head length, heavy, base pentagonal in section; upper mandible with strong culmen ridge and short ridge at base below nostril; lower mandible with strong rami and fine gonydeal ridge; culmen ridge slightly decurved towards sharply pointed or finely chisel-shaped tip, lower mandible gradually tapering to tip or with slight gonydeal angle. Nostrils covered by short bristles. Tarsus rather short. Toes rather long, strong, covered with distinct scutes; undersurface of toes covered with rough papillae; inner and outer front toes fused at base; outer front toe longest; inner front toe with claw *c*. 80% of outer front toe with claw, outer hind toe *c*. 86%, inner hind *c*. 41%. Claws very strong, markedly curved, tips sharp.

Geographical variation. Rather slight, mainly clinal; only Iberian race *sharpei* distinct. 4 races separable. (1) Nominate *viridis*. Upperparts bright grass-green with slight yellow tinge, ear-coverts and underparts pale green, hardly tinged grey; when plumage worn, upperparts slightly duller and colder grass-green and ear-coverts and underparts pale green-grey. Occurs from Britain and France east to European USSR, south to west and central Pyrénées, Alps, Hungary, and northern Carpathians. Size clinally larger towards east (see Measurements); overlap too large to warrant recognition of separate races—e.g. of *pluvius* Hartert, 1911, for small British birds, and *frondium* (Brehm, 1831) and *virescens* (Brehm, 1831) for intermediate birds of central Europe. (2) *P. v. karelini*. Upperparts slightly duller green, not as bright and yellowish in fresh plumage as in nominate *viridis*, dull grey-green when worn; cheeks and underparts not as saturated pale green as in nominate *viridis*, only feather-tips pale green or yellowish-green and much grey-white of feather-bases visible when plumage fresh, distinctly grey when plumage worn. Several races have been described: e.g. *pronus* Hartert, 1911, in Italy, *dofleini* Stresemann, 1919, in southern Yugoslavia and northern Greece, and *saundersi* (Taczanowski, 1878), in Caucasus, but differences in colour and size from typical *karelini* of northern Iran too slight to warrant recognition of more than 1 race. All populations small (average wing 160–162) without significant variation in size, only Caucasus birds slightly larger. Populations from northern Yugoslavia, Rumania (named *romaniae* Stresemann, 1919), Bulgaria, and south-west Ukraine about intermediate in colour between nominate *viridis* and *karelini*; size small like *karelini*. See also Matvejev and Vasić (1973). (3) *P. v. innominatus* from Zagros mountains in south-west Iran. Similar to greyish *karelini*, but cheeks, throat, and chest even paler, almost white; bars on tail tend to be sharper, white spots on primaries more conspicuous; juvenile more sharply spotted and barred white and green above and more contrastingly barred grey-black and white below (Vaurie 1959*a*). No birds from south-east Iran (named *bampurensis* Zarudny, 1911) examined; said to show very contrasting bars on tail and spots on flight-feathers, and (in adult) sharp brown bars on underparts from lower breast downwards (Hartert 1912–21); probably best included in *innominatus*, as tail- and wing-markings similar to that race (Vaurie 1959*a*) and similar dark bars on underparts occasionally occur in adults of other races (e.g. distinct in some specimens from Asia Minor examined). (4) *P. v. sharpei* from Spain and Portugal. Similar to *karelini* in colour of upperparts and in showing greyish ear-coverts and chest. Differs markedly from other races in restricted amount of black on sides of face; black present only between base of bill and eye, sometimes restricted to indistinct dark grey line on lores, separated from black or red malar patch by a whitish line (especially in ♀); area above, behind, and below eye dull greenish-grey (not broadly black), sometimes extending into a dull green-grey or grey supercilium, reaching to forehead. Red malar patch of ♂ only narrowly bordered by black below and behind. Red of crown (both sexes) and malar patch (♂) between orange-red and scarlet, not deep carmine-red as in nominate *viridis*; red on feather-tips more restricted, more extensive grey of feather-bases showing on crown, more extensively mixed greyish-black on malar patch. Pale spots on primaries larger and more conspicuous though less numerous; outer tail-feathers more extensively marked black. Dark streaks on head of juvenile and dark bars on underparts less distinct, paler olive-green, contrasting less with pale greenish ground-colour. Size small in Portugal and western Spain, distinctly larger in central and east-central Spain (see Measurements); birds from southern Pyrénées, Cantabrian mountains west to Oviedo, Sierra de Guaddarama, and Sierra Nevada intermediate in size.

Closely related to Levaillant's Green Woodpecker *P. vaillantii*, Wavy-bellied Woodpecker *P. awokera* from Japan, Scaly-bellied Woodpecker *P. squamatus* from Transcaspia east to Himalayas, and Grey-headed Woodpecker *P. canus* (Short 1982); see also Salomonsen (1931). CSR

Picus vaillantii Levaillant's Green Woodpecker

PLATES 77 and 78
[facing pages 807 and 830]

Du. Levaillants Specht Fr. Pic de Levaillant Ge. Vaillants-Grünspecht
Ru. Африканский зеленый дятел Sp. Pito real bereber Sw. Afrikansk gröngöling

Chloropicus Vaillantii Malherbe, 1847

Monotypic

Field characters. 30–32 cm; wing-span 41–43 cm. Woodpecker closely similar to Iberian race *sharpei* of Green Woodpecker *P. viridis*, but all plumages lack red in malar stripe. Drums regularly. Sexes dissimilar; no seasonal variation. Juvenile separable.

ADULT MALE. No red in malar stripe and thus, being greyer below than *P. viridis viridis*, resembles adult ♀ *P. viridis sharpei*. Distinguished from latter only by (1) grey-green (not dusky) area round eye, with indistinct blackish line between bill and eye, and (2) pale line bordering upper edge of malar stripe. ADULT FEMALE. Differs from ♂ in largely blackish crown, with red restricted to rear; rear flanks more barred. JUVENILE. Both sexes resemble adults, but duller; body faintly barred above and below (thus much less distinctly than in juvenile *P. viridis*).

Differs from *P. canus* (in all plumages) in extent and

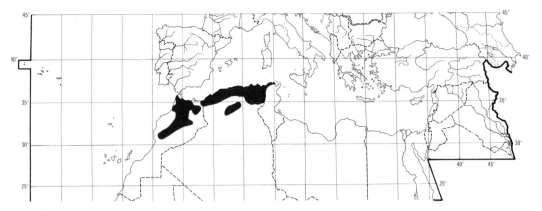

position of red crown-patch: in *P. canus*, absent or on front half of crown only; in *P. vaillantii*, over whole length of crown or on rear only. Flight and other behaviour as *P. viridis*.

Drums frequently, unlike *P. viridis*. Calls closely similar to those of *P. viridis*. DJB

Habitat. Exclusively Mediterranean, in warm dry hilly wooded country, sometimes among thinly scattered trees, and often on ground. Occurs on hilly slopes with open woodland or scattered trees of oak *Quercus*, poplar *Populus*, atlas cedar *Cedrus*, and pine *Pinus*, up to treelimit at altitudes of up to 2100 m, even moving beyond treeline to forage on rocky slopes. Not in coastal lowlands (Harrison 1982).

Distribution. MOROCCO. Very local in woodland above 1000 m (JDRV). ALGERIA. Generally distributed throughout mountain forests (EDHJ).

Population. MOROCCO AND ALGERIA. No information on numbers or trends (JDRV, EDHJ). TUNISIA. Fairly common in forests of cork oak *Quercus suber*, rather rarer in Aleppo pines *Pinus halepensis* further south (Blanchet 1955); no recent records from latter area (MS).

Movements. Resident; no information on local dispersals.

Food. Almost unstudied. Food and feeding behaviour generally regarded as closely similar to Green Woodpecker *P. viridis*. Often feeds on ants (Formicidae) taken on open ground (Dresser 1871–81; Meinertzhagen 1940). May also take adult and larval insects from old, dead trees (Whitaker 1905). One stomach, Tunisia, contained only ants: *c.* 300 *Camponotus nylanderi* and some *Crematogaster scutellaris* (Madon 1930a). ASR

Social pattern and behaviour. Very little known. The following includes notes accompanying recordings by J-C Roché (March, Morocco). Described as very shy (Heim de Balsac 1926; J-C Roché), but perhaps no more so than Green Woodpecker *P. viridis* (P A D Hollom). On close approach usually flies off with Excitement-call (Koenig 1895: see 2 in Voice). Drums frequently, much more so than *P. viridis*, at any time from dawn to dusk (J-C Roché); no information on seasonal variation, though reported March–June (Hartert 1927; E D H Johnson, J-C Roché). In late May, Algeria, bird chased another in flight, both calling; evidently display of pair (Koenig 1895). No further information. EKD

Voice. Differs from Green Woodpecker *P. viridis* mainly in Drumming much more frequently (see Social Pattern and Behaviour). Account based largely on recordings by J-C Roché (Morocco, March), of bird calling in response to playback; recordings of Drumming and Advertising-call include loud vigorous wing-whirring sounds which perhaps have signal function.

DRUMMING OF ADULTS. 2 recordings available, analysed and interpreted by J Hall-Craggs, as follows. In recording (Fig I) of 2 birds, Drumming combined with Advertising-call (see below); Drum series consists of 25 strikes in 1·13 s, interval between strikes *c.* 50 ms almost throughout. Very slight crescendo and decrescendo overall. In recording by J-C Roché, 21 strikes in just less than 1·0 s; other details of structure virtually as in Fig I. Too few recordings of *P. viridis* to allow detailed comparison, although our recording of that species (Fig I in *P. viridis*) differs from both of *P. vaillantii* in showing accelerando throughout.

CALLS OF ADULTS. (1) Advertising-call. A laughing 'whi whi whi whi...', homologous with same call of *P. viridis*, and not audibly different from that species (P A D Hollom). Recording (Fig II) ends with pleasant 'chk chk'; excluding 'chk' sounds, 18 units, delivered at *c.* 7 per s (*c.* 6 per s at end), overall duration 2·54 s; frequency mainly 1·7–1·8 kHz. In all these features, markedly similar to Advertising-call of *P. viridis*, though that species has stronger tendency to disyllabic units (J Hall-Craggs). Another recording by J-C Roché, of perhaps unusually agitated bird, has sharp squeaking quality (P J Sellar). 1st call in recording by E D H Johnson, of 2 birds together, comprises only 5 units (J Hall-Craggs). (2) Excitement-call. Agitated, rapidly repeated, metallic yelping 'chik' sounds, similar to *P. viridis* and presumably given in sim-

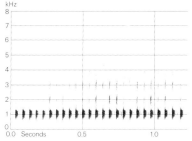

I E D H Johnson Morocco April 1966

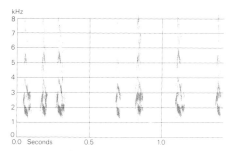

III J-C Roché (1967) Morocco March 1966

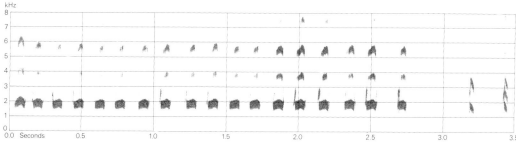

II J-C Roché (1967) Morocco March 1966

ilar circumstances. Recording (Fig III), suggests 'chek-chek-chek chek-chek chek chek' (J Hall-Craggs); rendered 'kück gück' when bird flushed on approach of observer (Koenig 1895), presumably expressing alarm or warning. In another recording by J-C Roché, much more subdued (almost conversational) brief plaintive squeaking sounds, combined with Drumming and wing-whirring sounds (E K Dunn, P J Sellar).

CALLS OF YOUNG. No information. EKD

Breeding. SEASON. Tunisia: eggs found 31 March to 6 June. Morocco: eggs laid early May; many young hatched by 1 June. (Lynes 1920; Heim de Balsac and Mayaud 1962.) SITE. In hole in tree. Nest: excavated hole 40–50 cm deep (Erlanger 1899). Building: no information. EGGS. See Plate 98. Pyriform, smooth and glossy; white. 30 × 22 mm (28–32 × 22–23), $n=11$; calculated weight 7·6 g (Makatsch 1976). Clutch: 6–7 (4–8) (Heim de Balsac and Mayaud 1962). No further information.

Plumages. ADULT MALE. Closely similar to adult ♂ Green Woodpecker *P. viridis*, but forehead and broad supercilium dull grey with slight green tinge (not black); red of crown and nape bright scarlet-red (not deep crimson-red); feathering at gape, below and behind eye, and on ear-coverts pale grey-green, bordered by indistinct dark grey line on lores and just below eye (not black with contrastingly pale green ear-coverts); broad malar stripe fully black without red; chin and throat white with slight buff tinge; dark marks on under tail-coverts and lower flanks slightly broader and deeper olive-green; central tail-feathers less distinctly barred, outer tail-feathers more broadly barred dusky. Iberian race *sharpei* of *P. viridis* rather similar, but forehead and sides of face between red of crown and extensively red malar stripe dark grey or greyish-black without pale line from gape to below eye; lacks broad pale grey supercilium or has faint and narrow one only. ADULT FEMALE. Rather different from adult ♀ *P. viridis*. Forehead and sides of face back to behind eye grey with greenish tinge and faint dark grey shaft-streaks; indistinct dark grey line on lores from nostril to front of eye, bordered below by broad pale grey-green or grey-buff streak from gape to below eye; crown black (feathers narrowly bordered pale grey-green when plumage fresh), bright scarlet-red restricted to patch on nape. Broad malar patch as in adult ♀ *P. viridis*, but greyish-black rather than deep black. Remainder of plumage as in adult ♂. Some ♀♀ of Iberian *P. viridis sharpei* show similar pale line from gape to below eye, but have black streaks on lores wider, forehead and broad streak over eye darker grey, and crown red. Some ♀ *P. vaillantii* have crown grey with rather limited black and a few have forehead and crown grey-green with faint dark shaft-streaks only; these rather similar to adult ♀ Grey-headed Woodpecker *P. canus*, differing in red nape-patch, less distinctly black lores, broader greyish-black malar patch, and more distinctly barred lower flanks, under tail-coverts, and outer tail-feathers. NESTLING. No information. JUVENILE. Rather similar to adult; not heavily streaked black on sides of head and barred black on underparts as juvenile *P. viridis*, but instead differing from adult in about same way as juvenile *P. canus* differs from adult. Head pattern as adult, but crown and nape of ♂ paler orange-scarlet with more grey of feather-bases visible; nape-patch of ♀ as adult ♀ but crown grey or grey-green with limited black along shafts. Green of upperparts, tertials, and upper wing-coverts duller and slightly greyer, tertials and upper wing-coverts with indistinct darker and paler green bars. Underparts down from breast with faint and narrow grey-green bars (in adult, mainly uniform on breast and belly but more boldly barred olive-green on lower flanks and under tail-coverts). Tail and wing differ (see Structure): tail-feathers more sharply pointed, p10 9–15 mm longer than longest greater upper primary coverts (2 mm shorter to 8 mm longer in adult). FIRST ADULT. Similar to adult, and sometimes hardly distinguishable. Retained juvenile tertials have indistinct darker and paler green bars on outer webs and tips, these sometimes showing slightly also on retained secondaries. Retained juvenile greater upper primary coverts relatively more worn and pointed than those of adults, pale spots on outer webs slightly smaller and sharper, outer webs less broadly fringed uniform green; difference from adult often slight and ageing of 1st adults usually possible only for those with retained juvenile inner primary coverts and contrastingly new outer primary coverts (adults sometimes retain 1 or a few coverts, but shape and pattern of these similar to new coverts).

Bare parts. ADULT, FIRST ADULT, AND JUVENILE. Similar to *P. viridis*, but base of lower mandible more extensively pale yellow-green or olive-grey (Hartert 1912–21; Short 1982; BMNH, ZFMK). NESTLING. No information.

Moults. Rather limited number of moulting specimens examined indicate sequence as in *P. viridis* and timing similar to its Mediterranean populations—primary moult of both adults and juveniles starts mid-May to mid-June, completed mid-September to mid-October.

Measurements. ADULT AND FIRST ADULT. North-west Africa, all year; skins (RMNH, ZFMK, ZMA). Bill (F) to forehead, bill (N) to distal corner of nostril. Toe is outer front toe with claw.

WING	♂	164	(2·11; 9)	161–167	♀ 162 (2·92; 7)	159–166
TAIL		101	(3·65; 5)	97–105	101 (3·73; 7)	97–107
BILL (F)		40·6	(2·38; 7)	38–43	38·9 (2·12; 7)	37–42
BILL (N)		28·7	(1·37; 7)	27–30	28·5 (1·93; 6)	27–31
TARSUS		28·7	(1·10; 9)	28–31	28·1 (0·64; 7)	27–29
TOE		31·7	(1·50; 9)	30–33	31·6 (1·19; 7)	30–33

Sex differences not significant.

JUVENILE. Wing on average $c.$ $4\frac{1}{2}$ shorter than adult and 1st adult, tail $c.$ 7; bill similar to adult from about halfway through post-juvenile moult.

Weights. Algeria: ♂, November, 166 (ZFMK).

Structure. In adult, p6 longest, p7 0–1 shorter, p8 3–5, p9 25–30, p10 83–95, p5 2–5, p4 14–18, p3 25–31, p2 32–39, p1 37–42; in juvenile, p7 and p8 longest, p9 17–19 shorter, p10 70–75, p6 3–5, p5 12–15, p4 21–24. Adult and 1st adult p10 8 mm longer to 2 mm shorter than longest greater upper primary coverts, juvenile 9–15 longer. Adult and 1st adult t2 4–7 shorter than t1, t3 11–15 shorter, t4 19–23, t5 30–34, t6 65–70; juvenile t2 10–14 shorter, t3 15–20, t4 20–24, t5 33–36, t6 60–66. Remainder of structure as in *P. viridis*.

Geographical variation. Slight. Birds from Saharan Atlas of Algeria on average duller and more greyish-green on upperparts and upper wing-coverts than birds from northern Algeria and Tunisia (described in Plumages); forehead and crown of ♀ more greenish-grey, less extensively black, tertials slightly barred paler and darker green (even in adult), tail more distinctly barred. Moroccan birds have sides of face slightly darker grey and more distinctly black streak on lores to eye; single ♂ from Tanger rather heavily mottled black on forehead, above eye, and cheeks, approaching *P. viridis*. Only limited number from each region examined and not known whether characters constant or whether recognition of several races warranted.

Often considered a race of *P. viridis* (e.g. Vaurie 1965, Short 1982). Difference from *P. v. viridis* in markings of crown and face and in ventral barring of juvenile more or less bridged by *P. v. sharpei* of Iberia, but head pattern of *P. vaillantii* also rather similar to Grey-headed Woodpecker *P. canus* and barring of upper wing-coverts, secondaries, and tertials of juvenile (and some adult) *P. vaillantii* suggest relationships with both *P. canus* and Scaly-bellied Woodpecker *P. squamatus* from Transcaspia and Himalayas, while some other characters similar to Wavy-bellied Woodpecker *P. awokera* from Japan (Goodwin 1968). *P. vaillantii* perhaps an early offshoot of a common ancestor of *P. viridis*, *P. awokera*, *P. canus*, and *P. squamatus* and hence separated from these, following Voous (1977a). CSR

Dryocopus martius Black Woodpecker

PLATES 78 and 79
[between pages 830 and 831]

DU. Zwarte Specht   FR. Pic noir   GE. Schwarzspecht
RU. Желна   SP. Pito negro   SW. Spillkråka

Picus martius Linnaeus, 1758

Polytypic. Nominate *martius* (Linnaeus, 1758), Eurasia except south-west China. Extralimital: *khamensis* (Buturlin, 1909), mountain valleys of south-west China.

Field characters. 45–57 cm; wing-span 64–68 cm. 50% larger than Green Woodpecker *Picus viridis*. Powerful, crow-sized, almost all-black woodpecker. Flight flopping, mostly lacking undulations. Noisy, with loud and distinctive calls. Drum roll loudest and longest of all woodpeckers. Sexes dissimilar; no seasonal variation. Juvenile separable.

ADULT MALE. Glossy black except for scarlet-red forehead and long crown. ADULT FEMALE. Browner toned and less glossy than ♂, with red on head restricted to small patch above nape. JUVENILE. Both sexes resemble adults but duller with greyish chin; red often less extensive or, occasionally, absent. Bill largely pale horn, legs dark grey.

Unmistakable; crow-like plumage unique among west Palearctic woodpeckers. Flight also differs markedly with 1–2 irregular flaps of wings (not bursts of 4–7) followed by partial closure creating unsteady and unrhythmical

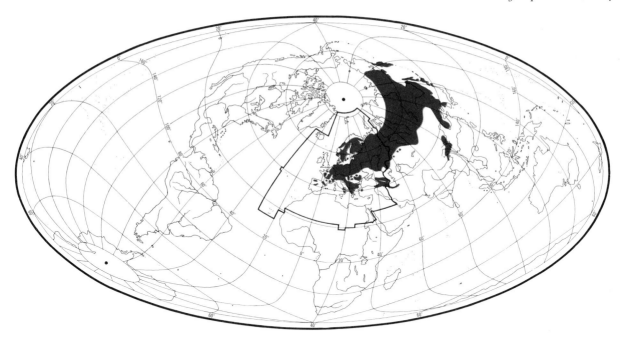

action—recalling Jay *Garrulus glandarius* and (particularly) Nutcracker *Nucifraga caryocatactes*—and culminating in 2 plunging undulations towards perch. Climbs with upward leaps. Holds body away from wood surface, with legs and thighs often visible and cock of head before chiselling or drumming accompanied by distinct angling of neck and head. Not infrequently on ground; hops rather clumsily.

Loud-voiced, with most frequently heard flight-call a grating multisyllable, 'krükrü-...'. When perched, often utters a thin but far-carrying whistle, 'kleea' or 'kee-ee-aaa'. Song similar to *P. viridis*, but higher pitched and even more manic in quality, 'kwick-wick-wick-wick...'.

Habitat. Resident from boreal to warm temperate zones, stopping short of oceanic climates, towards south mainly in mountain forest up to treeline, in Alps to above 2000 m, but in mature northern taiga mainly lowland. Favours tall trunks of climax forest, especially of mixed beech *Fagus* and fir *Abies*, or of pure beech, but also of larch *Larix*, spruce *Picea*, Arolla pine *Pinus cembra*, and other broad-leaved or coniferous trees, preferably well spaced but forming extensive unbroken forest. Also occurs in smaller stands separated from major forest by up to *c*. 4 km, and in winter wanders more widely, even to woodland near towns. Although continuing more attached to forest than *Picus* species, similarly resorts, insofar as ants (Formicidae) figure largely in diet, to more or less open ground for foraging. For nesting, however, holes in large trees remain essential, alternatives such as nest-boxes and pine poles carrying power lines being uncommon.

Outside nesting season, explores burnt and clear-felled areas, stands of trees attacked by bark-beetles, and concentrations of anthills. Flies strongly but usually quite low. Readiness and capability for boring large fresh holes establishes role as major provider of nesting opportunities for other birds (Cuisin 1967–8), where ample stock of adequate-sized trunks left standing. Habitat and wariness do not favour association with man, nor easy tolerance of disturbance. Although recently spreading in some regions, vulnerable to increasing human encroachments, as shown by patchy and islanded pattern of distribution in lowland areas of drastic deforestation. (Niethammer 1938; Dementiev and Gladkov 1951*a*; Voous 1960; Glutz von Blotzheim 1962.)

Distribution. Marked spread in lowland areas of north-west Europe, linked with maturation of deciduous woodlands and rapid spread of Norway spruce *Picea abies* (Glutz and Bauer 1980).

FRANCE. Marked spread into lowlands since 1950s, still continuing; in 1972 breeding proved in 19 départements, in 1979 in 46 (Cuisin 1973, 1980); since then also in Maine-et-Loire and Vienne (RC). PORTUGAL. Seen recently in north-west but no proof of breeding (RR). BELGIUM. First bred 1908, since expanded rapidly (Lippens and Wille 1972; Glutz and Bauer 1980). LUXEMBOURG. First bred 1915 (Yeatman 1976). NETHERLANDS. First bred 1913, then spread slowly, with more rapid spread since 1950, still continuing (CSR). WEST GERMANY. Spread in lowland areas: e.g. Niedersachsen 1890s, Westfalen *c*. 1890–1917, Rheinland and Oldenburger Land around 1900 (Glutz and Bauer 1980). DENMARK. First bred Zeeland 1961 and Bornholm 1971; recently bred Jutland (Dybbro 1976;

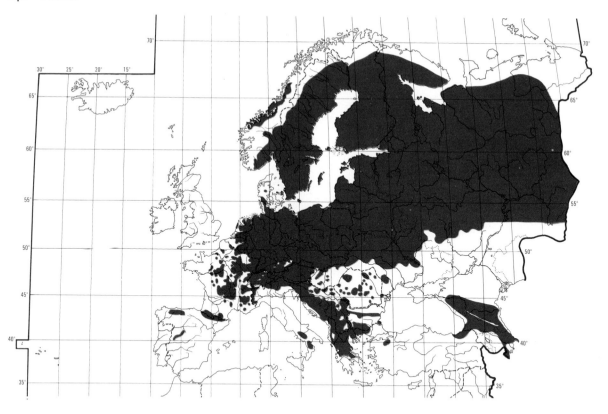

TD). ITALY. Formerly bred Sicily and more widely in Apennines (PB, BM).

Population. Limited data, but increase in lowland areas of north-west Europe with range expansion (see Distribution) and in Hungary; decreased USSR and Finland with destruction of forests.

FRANCE. Recent increase probable in view of range extension; under 1000 pairs (Yeatman 1976). SPAIN. Scarce, except in Cantabria (AN). BELGIUM. About 275 pairs (Lippens and Wille 1972); c. 350 pairs 1982 (PD). LUXEMBOURG. About 60 pairs (Lippens and Wille 1972). NETHERLANDS. Increased since colonization in 1913; perhaps 100–200 pairs c. 1950, 400–600 pairs 1965 (CSR), and 1500–2500 pairs 1977 (Teixeira 1979). WEST GERMANY. No detailed censuses, but presumably increased with spread into lowlands (Glutz and Bauer 1980); perhaps now 6200 pairs (Rheinwald 1982). DENMARK. Increased since first breeding in 1961; over 80 pairs 1974, c. 100 pairs 1980 (Olesen 1974; Dybbro 1976; TD). SWEDEN. About 50 000 pairs (Ulfstrand and Högstedt 1976). FINLAND. Estimated 15 000 pairs (Merikallio 1958); some decrease due to habitat changes (OH). EAST GERMANY. Scarce; disappeared from some areas (Makatsch 1981). POLAND. Scarce; slight local increases in 1920s (Tomiałojć 1976a). CZECHOSLOVAKIA. Scarce; perhaps some recent increase, especially in lowlands (KH). HUNGARY. Slow increase since c. 1935 in lowland woods (Glutz and Bauer 1980). ITALY. Decreased (SF). BULGARIA. Perhaps 1000–1500 pairs (Z Spiridonov). USSR. Decreased with destruction of forests (Dementiev and Gladkov 1951a). Estonia: scarce (HV). Latvia: fairly common (HV); marked decline some areas in 1950s (Viksne 1983).

Oldest ringed bird 7 years (Glutz and Bauer 1980).

Movements. Resident at population level, but some individuals strongly dispersive.

Adults remain all year in neighbourhood of their territory. After loosening of family bond in late summer, young disperse to some extent, though majority settle within (often well within) 40 km of birthplace; see ringing recoveries in Cuisin (1967–8). Some individuals wander to greater extent and have reached (e.g.) Astrakhan (USSR), Helgoland (West Germany), Texel (Netherlands), West Vlaanderen (Belgium), and Brittany (France); such longer dispersals will have facilitated the more notable expansions of range, as in Denmark and France. Longer ringing recoveries include: a movement of 110 km ENE within Finland (found April, 7 years later); 2 Swedish juveniles that moved 117 km south (within Lappland, found October) and 195 km west (into Hedmark, Norway, found September); 2 Czechoslovakian juveniles which moved 110 km NNE (into Poland) and 520 km NW (into Westfalen, West Germany), both December recoveries; Swiss juvenile found September,

160 km WSW into French Jura; 2 Niedersachsen (West Germany) birds found 500 km SSE in Czechoslovakia (December, 1st year) and 1000 km SW in Côtes du Nord, France (December, 2nd year) (Cuisin 1967–8; Glutz and Bauer 1980).

Movement inconspicuous (due to low numbers of transient birds involved) unless birds seen at migration observation stations, which range from coastal sites to mountain cols (Gatter 1977). Pooling such data across years, apparent that movement in Fenno-Scandia occurs mid-September to mid-November; in Europe south of Baltic, dispersal begins late July early August, with small secondary peak in late September and October coincident with Scandinavian movement (Gatter 1977). No obvious link between numbers transient in different autumns and irregular eruptions of *Dendrocopos* woodpeckers, so questionable whether present species is eruptive (Meshkov and Uryadova 1972; Glutz and Bauer 1980); however, more than usual (up to 10 per day) at Falsterbo (southern Sweden) in invasion autumn 1962 (Roos 1965).

Food. Mainly larvae, pupae, and adults of ants (Formicidae) and wood-boring beetles (Coleoptera). Bill bigger and stronger than in west Palearctic *Picus* woodpeckers, but tongue extends only *c.* 50–55 mm beyond bill-tip (compare Green Woodpecker *P. viridis*). Tongue coated with viscous secretion from enlarged salivary glands; horny tip equipped with 4–5 pairs of backward-pointing barbs, enabling bird to impale and extract (e.g.) Cerambycidae larvae (Sielmann 1959; Löhrl 1977d; Papadopol and Mândru 1977). 1 bird, late September, had stomach cuticle sloughed off (see Pynnönen 1943 for discussion). Not infrequently on ground and will spend up to *c.* 2 hrs there, hopping from stump to stump. Rotten tree stumps hacked open with powerful blows of bill throughout the year (Blume 1981); for detailed observations of use of bill and tongue when working on stump, see Sielmann (1959) and Löhrl (1977d); also Callegari (1955). Up to 800 stumps attacked in 32 ha (Olesen and Olesen 1972). In winter, flies purposefully to stumps covered by 20–30 cm of snow; snow—also loose material on ants' nests, soil, and leaf litter—cleared with side-to-side sweeping movements of closed bill (Pynnönen 1943; Löhrl 1977d; Glutz and Bauer 1980). At Werdau (East Germany), winter, bird hacked at 251 stumps on *c.* 28 ha; 100 worked all round, 70 on north, 56 on south, 20 on south-west, 5 on east side. Stumps in undergrowth or long grass avoided. In some cases only roots attacked (Weber 1968); often evident that bird has cleared snow and pecked only at certain spots (Blume 1981). Will peel vertical strips of bark off trees (Pynnönen 1939; Glutz and Bauer 1980), and take bark off dead trees by pecking and levering. First blows struck tangentially to trunk against bark and, after penetration to division between bark and wood, bill inserted repeatedly and bark (sometimes large pieces) levered off. Strikes then made radially into wood, strikes alternating with levering off (Lehmann 1930). In Cottbus (East Germany), ♀ regularly foraged on dead elm *Ulmus*, January–March, initially spending hours there from sunrise. Normally started at 10–18 m and worked down tree to *c.* 4 m where growth of ivy *Hedera*; descent repeated after immediate upward climb. Apart from removing bark, also removed ivy, leaf by leaf (Striegler *et al.* 1982). In Finland, winter, birds forage 26% of time in *Pinus* forest, 5% in *Picea*, 47% in mixed forest, 17% in urban parkland, and 5% in urban woodland; 53% of time spent seeking insects by pecking at trunks, 4% at feeding table, and 42% in other ways (Vikberg 1982). Another Finnish study indicated birds foraged mainly low on trunks of stunted trees, using pecking and tearing as main methods (Alatalo 1978, which see for detailed analysis of habitat use relative to other Picinae). In hard winters, stumps may be less accessible; birds then often favour lower parts of *Picea* trunks where may take larvae of wood wasp *Sirex gigas* (Loos 1910). Also typically work systematically on such trees, starting at bottom. May take insects on surface of timber, and single blow sometimes enough to expose food, but also excavates hollows up to 33 cm deep and up to 1 m from top to bottom (Cuisin 1967–8; Glutz and Bauer 1980). Limits of holes match extent of heart-rot and concomitant presence of ants *Camponotus* and *Lasius fuliginosus* (Lehmann 1930; Bejer-Petersen 1968; Ogasawara and Izumi 1977; Blume 1981). Such holes may be revisited, ants being picked up as they move along or first located with tongue (Löhrl 1977d). Nests of *Formica* ants visited mainly in winter (e.g.) Rendle (1912); after clearing of snow (see above), bird may dig hole up to *c.* 60 cm deep, almost burying itself (Géroudet 1959; Löhrl 1977d). In summer, tends to pick up ants on their regular paths, though will also dig at nests of *Lasius* in clearings (Blume 1981). Less commonly, pecks at fence-posts, detached rotten branches, and blocks of wood; also occasionally on rocks like *Picus* woodpeckers (Glutz and Bauer 1980). Will take bees (Apoidea), including torpid ones, regularly in winter (Glutz and Bauer 1980); said to damage hives in some areas but apparently no recent reports (Madon 1930b; Cuisin 1967–8). Occasionally drills holes in sap-rich trees, mainly May–June; in one case, bird probably took insects adhering to solidified sap (Blume 1981). Other authors (e.g. Klíma 1959a) stated that birds only visit already drilled holes to drink sap, though may peck at them (see also Turček 1954). Seen feeding on growing apples (Liedekerke 1969) and occasionally takes fat, exceptionally seeds at feeding stations (Glutz and Bauer 1980). Removal or puncturing of eggs of other birds (e.g. Goldeneye *Bucephala clangula*: Hasse 1961; Bruchholz 1978) possibly more connected with preparation of holes for roosting or nesting than with feeding. Said, however, to take nestlings of Pied Flycatcher *Ficedula hypoleuca* in Finland (Hortling 1929). Drinks regularly from hollows in trees (Eygenraam 1947).

The following food recorded in west Palearctic. Ants (Formicidae) comprise mainly *Camponotus herculeanus*, *C.*

ligniperda, *C. vagus*, *Formica rufa*, *F. fusca*, *F. exsecta*, and *Lasius niger*; contrary to earlier claims, *Myrmica* not rejected (Blume 1981). Other Hymenoptera include larval wood wasps (Siricidae, especially *Sirex gigas*), sawflies (Tenthredinidae, Xiphydriidae), ichneumons (Ichneumonidae), and larvae, pupal, and adult bees (Apoidea). Particularly larval, but also pupal and adult beetles (Coleoptera) similarly important: mainly Scolytidae and Cerambycidae, also Curculionidae, Anthribidae, Pythidae, Lucanidae, and Elateridae; fewer Carabidae, Silphidae, Staphylinidae, Cleridae, Buprestidae, Pyrochroidae, Scarabaeidae, Nitidulidae, Pterostichidae, and Chrysomelidae. Adults and larvae of flies (Diptera) taken in small numbers, as are moth caterpillars (Lepidoptera: Sphingidae, Cossidae, Lasiocampidae, Limantriidae, Noctuidae) and other invertebrates—spiders (Araneae), centipede *Lithobius forficatus*, and small snails (*Clausilia*, *Helix*, *Patula*). Plant food rarely taken, though sometimes in quantity: fruits of cherry *Prunus*, apple *Malus*, rowan *Sorbus aucuparia*, whitebeam *S. aria*, hawthorn *Crataegus*, bilberry *Vaccinium myrtillus*, and mistletoe *Viscum album*, and seeds of Scots pine *Pinus sylvestris* and Arolla pine *Pinus cembra*. Perhaps very rarely takes nestling passerines (see above). Occasionally eats tree sap. Wood splinters (mainly in stomachs of young birds), bark, leaves (especially pine needles), lichen, feathers, porcelain and bone fragments, eggshells (of own and other species), sand, and small stones also recorded. (Naumann 1826; Csiki 1905; Madon 1930*b*; Meylan 1932; Pynnönen 1943; Bannerman 1955; Pospelov 1956; Neufeldt 1958*b*; Sevastyanov 1959; Cuisin 1967–8, 1975, 1977; Liedekerke 1969; Korodi Gál 1970; Löhrl 1977*d*; Papodopol and Mândru 1977; Glutz and Bauer 1980; Blume 1981.)

47 stomachs (including 13 nestlings) from south-east Finland, May–February, contained 9278 invertebrates: 64·8% (by number) adult and 23·8% pupae, larvae, and eggs of ants, 10·5% larvae and 0·3% adult beetles, 0·5% larval Diptera, and 0·1% Lepidoptera larvae and spiders. In snow-free period of year (May–November), takes 89–97% ants (mainly *Lasius niger*, *Formica rufa*, *F. fusca*, and *Camponotus*). In winter (December–February), ants (only *C. herculeanus*) 54% and beetle larvae 43%; beetles taken in winter probably mainly Scolytidae, with Cerambycidae more important at other times (Pynnönen 1943). In southern Kareliya (USSR), ants important throughout the year. Birds forage mainly on dead trees and rotten stumps by bogs and at fire-affected sites. Beetles (Cerambycidae and Scolytidae) also favoured; one winter stomach contained up to 300 *Scolytus ratzeburgi* (650 reported by Madon 1930*b* elsewhere in USSR), and summer stomachs larvae of the large *Celerio gallii* (Sphingidae) (Neufeldt 1958*b*). 8 stomachs (1 June, 2 August, 2 October, 3 December), Arkhangel'sk (USSR) contained 3249 items: 65·0% (by number) Formicidae (mainly *F. herculeanus*), 34% Coleoptera (almost exclusively Scolytidae), and 0·7% Diptera; one held up to 950 larvae or pupae and 55 adults of beetle *Ips typographus* (Scolytidae) (Sevastyanov 1959). See also Pospelov (1956) who recorded 1712 *Lasius niger* in one June stomach and 112 larval Scolytidae in one from October, Leningrad region (USSR). In Rumania, summer diet mainly wood-boring beetles, including adults, Lepidoptera larvae, and ants. In cold season, larvae and pupae of Cerambycidae and Scolytidae, also conifer seeds (Papadopol and Mândru 1977). 2 faeces (in which readily identifiable parts of ants well preserved) of adults, Aube and Loiret (France), winter, yielded 4419 insects (all ants). Stomachs of 2 ♀♀, Loiret, October and April, contained 651 insects: 96·3% ants, 3·4% Xiphydriidae. Length of ants taken 2 mm (*Lasius*) to 14 mm (*Camponotus*); largest beetle larvae taken *Megopis scabricornis* (Cerambycidae) 37–60 mm (Cuisin 1975, 1977). Particular ant nests may be exploited throughout winter but none destroyed (Rendle 1912; Blume 1961). Bird may eat *c.* 1000 *Formica rufa* at a single meal (Gösswald 1957–8), but colonies of this species hold up to 2 million worker ♀♀ and several thousand queens (Blume 1981).

Food of young studied in Aube and Seine-et-Marne (France). 10 faeces contained 1749 insects (99·2% Formicidae). Neck-collar samples revealed 27205 items: 80·6% ants, 18% Scolytidae; remainder comprised other Hymenoptera, Coleoptera, Diptera, Lepidoptera, small snails (Clausiliidae), and seeds (Cuisin 1975, 1977). Further evidence of importance of ants (particularly *Lasius*) in Papadopol and Mândru (1977) and Cuisin (1981). Wood splinters often found in nestling stomachs averaged 0·5 cm^3, once 3·1 cm^3 (Pynnönen 1943). Up to *c.* 20 days old (when receive food at entrance), young fed total of 16–24 times daily by both parents; feeds average *c.* 90 min apart and last *c.* 3 min. Rate peaks at 20–24 days, then decreases. Adults' 1st foraging flight made *c.* 06.00 hrs (i.e. *c.* 2 hrs after sunrise), last feed at *c.* 20.00 hrs (Sielmann 1959); for further details, see (e.g.) Gebhardt (1940). Adult carries food in glandular stomach and transfers it by regurgitation (e.g. Löhrl 1977*d*); brings 62–1544 items (Cuisin 1969) or up to 26 g of food (Pynnönen 1939) per visit. Up to 12 portions regurgitated per visit; if young sated, adult swallows remainder (Sielmann 1959). Daily food requirements of nestling 2–3 weeks old at least 70 g (Pynnönen 1939), *c.* 60 g or 2370 items (Cuisin 1969). Total number of insects eaten by 3 nestlings over fledging period of *c.* 28 days 150000–180000 (Cuisin 1969) or *c.* 225000 (Korodi Gál 1970). MGW

Social pattern and behaviour. For pioneering work, see especially Loos (1910), also Rendle (1905, 1907, 1914, 1915); for more recent studies, see Eygenraam (1947), Blume (1956, 1959, 1961), Sielmann (1959), and Cuisin (1963, 1975); important summaries by Cuisin (1967–8) and Blume (1981). Account based on detailed outline supplied by D Blume.

1. Normally solitary. Outside breeding season, ♂ and ♀ in separate territories or parts of a territory (D Blume; see also Breeding Dispersion, below, for definition of territory). In favourable roosting area with many tree holes, several territories

may overlap; 7–9 birds may then roost in area of c. 30–35 ha (Striegler et al. 1982). Old birds show pronounced fidelity (confirmed by ringing) to home-range and particular trees with hole (Blume 1981). After break-up of family, juveniles disperse widely and singly or possibly in small groups (D Blume; see also Gatter 1977). BONDS. Monogamous mating system. In most cases, pair-bond probably lasts only for duration of single season, i.e. from winter into summer; attempted copulation recorded from mid-January (Eygenraam 1947; Blume 1961). On Bornholm (Denmark), mate-fidelity for up to 7 years proved by ringing (F Hansen); such cases possibly a result of attachment to territory rather than to mate (D Blume). Sex-ratio skewed towards ♂♂ in both older nestlings (1·29:1, n=94), and breeding adults: in western Erzgebirge (East Germany), 4–8 unpaired birds among population of 21–22 pairs were all ♂♂ (Möckel 1979); see also Cuisin (1967–8, 1981). Under certain conditions (small isolated woodlands, island sites) pair maintain loose contact outside breeding season, presumably encouraged by fact that conspecific birds of different sex more likely to be tolerated in roosting area than those of same sex (D Blume; see Danish record, above). Breeding partners meeting far from roost-site after break-up of family and into autumn may react aggressively (e.g. Bäsecke 1949) or show courtship-type behaviour (see part 2). Birds recorded feeding together at nest of ants (Formicidae) (Glutz and Bauer 1980), but only exceptionally together in same roost- or nest-hole (D Blume) and relationship between sexes strained even in breeding season (e.g. Heinroth and Heinroth 1924–6); ♂ feeds young first if both parents at nest-hole together. Young tended (feeding, brooding, nest-sanitation) by both parents up to fledging; brood then split, and part-family units stay together until independence. Siblings sometimes remain together in late summer and autumn when parental care has ceased (D Blume). Age of first breeding not known. BREEDING DISPERSION. Solitary and territorial. In central Europe, pair normally require wooded area of at least 300–400 ha (Loos 1910; Rendle 1912, 1914), though territories (see below) of less than 100 ha not uncommon in mixed fir *Abies* and beech *Fagus* woodland; in Switzerland 0·75–0·84 pairs in 100 ha (Glutz von Blotzheim 1962); in Provence (southern France), 2 pairs in isolated 68-ha stand of beech *Fagus*, territories extending outside wood (Blondel and Ramadan-Jaradi 1975); see also Peitzmeier and Westerfrölke (1962) and Bannasch (1980). In most woodland types, density less than 0·25 pairs per 100 ha, and nests normally at least 900 m apart (Glutz and Bauer 1980, which see for other higher-than-average densities in mixed woodland). In various states of West Germany, estimated 1 pair per 870 ha of broad-leaved, 1 pair per 650 ha of mixed, and 1 pair per 460 ha of coniferous woodland (see Burow 1970). In Oldenburg (West Germany), 52–60 pairs on 894 km², 1 pair per 1600 ha of total area, 1 pair per 230 ha wooded. Shortest distance between nests 450 m (Taux 1976, which see for other references). In western Erzgebirge (East Germany), 10·8 pairs per 100 km²; study area 171·55 km² of forest (Möckel 1979). Unusually high density of (2–)3–4 pairs in 85 ha, Branitzer Park (Cottbus, East Germany), nests 650–1000 m, in one case only 180 m apart, probably due to isolated concentration of mature *Fagus* (Striegler et al. 1982). On Åland islands (south-east Finland), 0·1 pairs per 100 ha (Haila and Järvinen 1977; see Palmgren 1930 for earlier claim of much higher density). Defends only certain critical sites (see below) in large, all-purpose territory, usually with particular parts preferentially used by ♂ and ♀, though some overlap. In central Europe, territory of breeding pair or (home-range) of single bird outside breeding season c. 150–800 ha (Blume 1981; D Blume); in Finland, where birds apparently need to range much wider for food, c. 800–3000 ha (Pynnönen 1939). Location of territories in a particular area often determined by position of trees with holes, so that even after change of individuals, defended sites and general shape of territory remain unaltered, although shifts do occur. Territory divided into: 1 or more areas of c. 50–200 ha where birds call, Drum, etc.; a breeding area; favoured feeding sites (e.g. forest clearings, coniferous woods with ant nests; for defence of latter, see Striegler et al. 1982); roosting areas; and flight-paths. Territorial defence (♂ presumably more active—see Antagonistic Behaviour, below) concentrated on such key sites and home-range thus contains quite large neutral areas (Blume 1981; D Blume). As in other Picinae, boundaries not clearly delineated (Glutz and Bauer 1980); no boundary disputes recorded by Cuisin (1967–8). Defends area of c. 25 ha around nest-tree against conspecific birds and probably other Picinae. Such a breeding territory (where adults normally forage for young) usually smaller than all-year home-range and may overlap with neighbour's without leading to disputes (Blume 1981). In one case, winter roost-holes of 2 birds 2–5 km apart; these birds subsequently paired, and nest-site then 500 m from ♂'s roost, 2350 m from ♀'s (Tombal 1977). ♂'s main Drumming post (see Voice) less than 100 m or up to 1 km from later nest-site (Eygenraam 1947). Most courtship, also copulation, takes place near (prospective) nest-site (Blume 1962a, 1981). ROOSTING. Solitary and nocturnal. Mostly in old nest-holes (including those from most recent season). Not recorded excavating holes specifically for roosting but, in autumn, occasionally chisels at entrance or in cavity, or removes old nest-material of other bird species (Rendle 1912; Pynnönen 1939; Radermacher 1970; Blume 1981). Probably occasionally roosts on tree trunk, under branch (see Cuisin 1967–8 for discussion); use of hayrick also recorded (Pynnönen 1939). Particular hole(s) used for months, even years; in study by Striegler et al. (1982), ♀♀ changed holes more often than ♂♂. Birds also have emergency holes to which they fly when disturbed (Pynnönen 1939; Blume 1961). In breeding season, ♀ may stick to same roost-site, even up to c. 2 km from nest. Exceptionally, 2 birds may share same hole (only late January to early April) but this will be roomy and/or have at least 2 entrances (Gebhardt 1940, 1950; Eygenraam 1947; Blume and Blume 1981); see also part 2. Fledged juveniles may be led into roosting area where use holes near that of parent (Blume 1981), though often roost in the open clinging vertically to trunk at sheltered spot—head drawn between shoulders, turned to one side, and pressed close to bark (Rendle 1914). 'Ownership' of roost-holes usually settled by October. ♂ generally goes to roost earlier than ♀ and leaves correspondingly later; ♀ may thus roost close to ♂, though often in less good site (Glutz and Bauer 1980). Period of diurnal activity shorter than in small Picinae but similarly synchronized with solar cycle (for details, see Rendle 1907, Pynnönen 1939, Blume 1963b, 1964a). New arrivals in an area (also territory-owners looking for new roost) probably inspect potential sites during day and remember them for roosting (Blume 1961). Such birds more vocal in evening than well-established individuals—tendency to call is much less marked in birds familiar with roost-site. Typically approaches roost giving long series of Flight-calls (see 3 in Voice), then lands on intermediate station (solitary tree, topmost branch—sites traditional over years: Blume 1981); bird alert and gives Excitement-calls (see 2a in Voice), then flies direct to roost-hole giving brief Flight-call. Whole transitional phase c. 30–90 min by new arrival in summer, often less than 20 min in winter if bird familiar with area. Bird makes repeated intention-movements to enter (longer and more intense by new arrivals). After entering, often looks out briefly; gazes out alertly for longer when arrives at roost early on clear, frosty days. Sleeps clinging to cavity wall beneath entrance-hole, or occasionally perhaps on floor. In morning, intention-movements to leave gradually intensify; bird then departs suddenly, gives Flight-call c. 200 m from roost, lands at intermediate station, gives Excitement-call, and self-preens (Blume 1981; D Blume). May go to roost during day in heavy rain or frost (Pynnönen 1939). Few details on comfort behaviour; see Cuisin (1967-8) and Glutz and Bauer (1980) who also recorded sand-bathing.

2. Very wary when feeding on ground (Loos 1910). If disturbed, flies away low, then steeply up to land on tree (Glutz and Bauer 1980). Bird preparing to undertake long flight over open country gives Excitement-call for up to c. 20 min, with fore-body held back, bill wide open, and head drawn in or bent sideways; when giving higher-intensity variant of call, posture more tense (D Blume: see 2 in Voice). May fly to solitary tree before starting flight over c. 500 m of open country and normally hesitates at even slight disturbance or even turns back temporarily from flight once begun giving loud Flight-calls. Unlike Great Spotted Woodpecker *Dendrocopos major*, normally takes off from concealed perch (Gatter 1977). Can evade attacks by raptors and mobbing by Starlings *Sturnus vulgaris* using zigzag manoeuvres, etc. (Loos 1910; Latzel 1972). If attacked by raptor when on tree, climbs in spirals around it. Generally indifferent to other bird species but, if molested, will give Excitement-call or freeze briefly, then fly off silently. Bird disturbed in roosting area may make short flights back and forth and give (variant) Flight-call (see 3 in Voice). On roost-tree, bird holds head well back and performs marked Head-swaying. Behaviour of bird in hole varies individually; shy individuals react swiftly to strange noises, looking out with bill held horizontal. Response less likely in cold and some birds remain in hole throughout disturbance (Blume 1981; D Blume). For behaviour if roost-hole already occupied, see further below. ANTAGONISTIC BEHAVIOUR. (1) General. Various forms of threat and attack used to defend individual distance (reacting to conspecific within c. 100 m) and various critical sites in territory. Aggressive interactions lead to wide dispersion of individuals. In summer and autumn, closer confrontations restricted to vicinity of tree with hole; in breeding season, threat or attack also occur elsewhere (Glutz and Bauer 1980; Blume 1981; D Blume). In Brandenburg (East Germany), vigorous and protracted fights between neighbouring pairs (nest-holes c. 300 m apart) early April, not later in breeding season (Schmidt 1970). Intruding unpaired ♂♂ likely to meet hostile response from territory owners; later, unpaired birds may defend area around roost with much calling and flying about, sometimes until June (Blume 1981). In areas of unusually high density (see part 1), sex- and individual-specific differences in diurnal activity rhythms (for details, see Blume 1963b, 1964a) reduce likelihood of aggressive interactions. Advertising-call and Drumming indicate occupancy of territory and hole (see Voice, and Heterosexual Behaviour, below, for details of functions). In central Europe, Advertising-call given mainly mid-April to May, sporadically to mid-October, not heard November to early January; in Finland, not later than end of August. Adults attracted by imitated calls, February–June, young birds thereafter (Pynnönen 1939; Blume 1961; Glutz and Bauer 1980; D Blume.) Where several territories abut, many birds may call simultaneously shortly before laying; contra Grote (1933b), not a communal display (Glutz and Bauer 1980). When Drumming, bird sits in fairly relaxed posture (see illustrations in Blume (1981). (2) Threat and fighting. Call 4 given when potential rival in view to discourage closer approach; used between rival ♂♂, between ♀♀, and between ♂ and ♀. In antagonistic encounter between birds less than 5 m apart, opponents threaten one another in ritualized fight (Mock-fencing; see also Head-swaying in *Picus* woodpeckers). Occurs at all seasons and most pronounced between rival ♂♂;

A

lower-intensity threat typical of ♂–♀ encounters (see also subsection 2 in Heterosexual Behaviour, below). Rivals sit facing one another on tree-trunk; lower bird apparently has advantage and attempts to drive other upwards. From Bill-up posture (see, e.g., Green Woodpecker *P. viridis*), or tense Mock-sleeping or resting posture, both birds Bill-thrust upwards 4–6 times at intervals of c. 30–70 s, each time giving Threat-call (see 5 in Voice), and simultaneously describe elliptical movement with head, thus presenting red crown-patch to rival (Fig A). Wings slightly spread and twitched (sometimes flapped); tail sometimes fully spread. After Mock-fencing lasting c. 2 s, both birds sink down and remain motionless until next bout. Performance sometimes lasts over 1 hr (Eygenraam 1947; Blume 1956, 1961, 1981; Fridzén 1974; Glutz and Bauer 1980; see also Radermacher 1970). Bird defending ants' nest jabbed at it repeatedly and spread wings (Striegler *et al.* 1982). If 1 bird flees, other pursues it briefly (Blume 1961), and gives call 4. Mock-fencing may lead to (damaging) fight, normally preceded by Excitement-pecking (see Voice) and loud, demonstrative call 4. Dominant bird then flies at opponent and drives it off by buffeting and pecking it. In threat at great distance, birds perform only intention-movements of Mock-fencing (see Eygenraam 1947). Ceremonial threat can be elicited by using stuffed specimens and imitated calls at roost- or nest-site. In summer and autumn, aggressively stimulated bird performs Excitement-pecking; in breeding season, if bird moves away temporarily, gives Drumming typical of conflict situations. Reacts promptly to imitated call 4, making searching flights and becoming more aggressively excited (Blume 1956, 1959). If roost-hole already occupied, usually flies off to spend night elsewhere (see Roosting). Sometimes seizes occupant (e.g. Grey-headed Woodpecker *P. canus* or *D. major*) and pulls it out of hole. In breeding season, may fly over c. 100 m to drive off other Picinae. For close nesting association with *P. canus* and *D. major*, see Schaack (1967), Ferry *et al.* (1957), and Striegler *et al.* (1982). Recorded coexisting peacefully in same hole with *S. vulgaris*, birds using same entrance (but possibly separate nest-chambers: D Blume); see Delmée and Godart (1976). Prior to laying, hole may be usurped by Jackdaw *Corvus monedula* (e.g. Eygenraam 1947, Sielmann 1959), but bird usually able to ward off any later intruders by attacking them at entrance (see, however, Roller *Coracias garrulus*). HETEROSEXUAL BEHAVIOUR. (1) General. Normally associated with roost- or potential nest-holes where pair (or prospective partners) most likely to meet, including outside breeding season. In roosting area, pair-members apparently recognize one another by voice or instrumental signals; bird in hole Taps demonstratively if other flies past, lands nearby and/or gives call 4, but comes alertly to entrance if stranger present. In autumn, calling, Tapping (see Voice), and chiselling at or in holes may occur when conspecific bird attracted to another's roosting area by Advertising-calls; one sexually motivated encounter in late November included copulation attempt (see Jürgens 1978; also Loos 1910). Not known whether firm pair-bond may result from such interactions (Glutz and Bauer 1980). Nest-site selection, nest-hole excavation and successful copulation in western Europe from mid-February (Cuisin 1967–8), in central Europe from March, with peak in first half of April (D Blume), in northern Europe, April to early May (Pynnönen 1939). ♂ more active than ♀ in all aspects of courtship behaviour (Eygenraam 1947; Blume 1981). Intensity may, however, vary from pair to pair; possibly dependent on whether birds pairing up for first time or personally acquainted from previous season or from roosting in same area (D Blume; see also Bonds). Courtship between already acquainted birds brief and of low intensity (D Blume, F Hansen), apparently also when available hole used as nest-site. Individual variation in vocal and Drumming activity perhaps also involved. Period required for sexual synchronization protracted; particularly early in pair-formation, ♀ may solicit copulation (see below) but elicit only Head-swaying threat in ♂. Unpaired birds generally Drum more than paired ones, but one pair-member may still Drum persistently when pair-formation completed. ♀ sometimes also Drums near other potential nest-holes when ♂ has ceased Drumming and occupied definitive site (Blume 1962a, 1981; D Blume). (2) Pair-bonding behaviour. May begin in roosting area (see above). In west and central Europe, call 4 and attempted copulation occur from mid-January; in favourable weather conditions, Drumming and Advertising-calls normally increase in same month (Eygenraam 1947; D Blume). As in (e.g.) *P. canus*, bird gives loud Advertising-calls and Drums (mostly at favoured stations) near roost-hole during increasingly longer stays there, morning and evening. If conspecific bird responds similarly, territory-owner makes high demonstrative searching flight in direction of sound, giving loud Advertising-calls. Both birds may overfly one another, calling loudly. Closer encounters, with Mock-fencing and Threat-calls (see Antagonistic Behaviour, above), apparently allow specific and sexual recognition. Lower-intensity threat typically performed in heterosexual encounter permits closer approach, weaker stimulus of small red patch on ♀'s crown possibly also helping to reduce aggression. Call 4 also typically given in such encounters, either in tense, excited posture by perched bird, or in Fluttering-flight with bill and tail pointed down (Eygenraam 1947: Fig B); tail may also be pointed up. (Glutz and Bauer 1980; Blume 1981; D Blume.) (3) Nest-site selection. ♂ and ♀ may initially call and Drum at separate sites in a territory and make reciprocal visits to holes. ♀ normally attracted into ♂'s area; however, some ♀♀ stay put, advertise site, and cause ♂ to shift, or attract strange ♂ (Blume 1961; D Blume). After several days of display by both birds at separate sites, final choice sometimes made within c. 1 hr (D Blume). Nest-showing performed (mainly by ♂) in high display flight several times to hole and back; (variant) Advertising-call given (see 1 in Voice). Bird may Tap at hole-entrance (see Sielmann 1959, Blume 1981), but typically enters (after intention movements) and Taps inside (see Voice) to advertise site. If other bird approaches, may be allowed to inspect hole or is threatened, even chased off (Blume 1981). Some excavation done throughout pair-formation (for autumn–winter excavation, see Rendle 1912). Mostly uses old holes; some used (not necessarily by same birds) for several years—up to 6 years recorded (Conrads 1967). Where new hole excavated, birds may make several starts until one deepened (14–28 days for completion depending on tree species). Both sexes excavate, but ♂ normally does more, especially towards end (see also Tombal 1977). Division of labour occurs (♂ excavating, ♀ removing wood-chips), but not the rule. Ceremonial nest-reliefs (as later from incubation—see below) gradually develop after initially shorter shifts and low-intensity threat with calls 1, 4, and 5. Nest-guarding as in (e.g.) *P. viridis*; from final excavation-phase until laying, by either bird at night (Rendle 1905; Pynnönen 1939; Eygenraam 1947; Sielmann 1959; Cuisin 1972; Blume 1981; D Blume). (4) Courtship-feeding. No information. (5) Mating. Takes place when new hole completed or (if old hole used) chosen (D Blume); recorded c. 3–4 weeks before laying (Rendle 1905). For copulation attempts outside or very early in breeding season, see subsection 1 (above). Birds usually copulate near nest on horizontal branch free of obstructions; sites may be traditional (Rendle 1905; Eygenraam 1947; Blume 1981). Preceded by flying about of both birds with Advertising-calls; gradually approach, give call 4, and may perform Fluttering-flight (König 1951; D Blume). ♀ driven up tree by ♂ who slinks along in tense and crouched posture (Fig C, left), and may also Head-sway. ♀ gives variant of call 4, Head-sways from side to side in crouched soliciting posture and quivers wings (see illustrations in Eygenraam 1947 and Blume 1961). ♀ ready to copulate keeps head down whilst Head-swaying; if not ready, holds bill up (see Eygenraam 1947, König 1951), but does not Bill-thrust (see Antagonistic Behaviour, above). Immediately prior to copulation, ♀, perched crossways on branch, crouches, cocks tail, and droops wings (Fig C, right). ♂ jumps up and pushes tail under ♀'s; does not grasp ♀'s nape feathers (D Blume); see, however, Glutz and Bauer (1980). At last moment, ♀ may thwart copulation attempt by lowering tail abruptly (Eygenraam 1947). After brief copulation ♂ usually flies far off, giving Flight- or Advertising-call only when some distance away. ♀ remains on branch for a while and self-preens (D Blume). (6) Behaviour at nest. ♂ may begin to roost in (newly-excavated) nest-hole c. 14 days before laying (Rendle 1905). Laying takes place in early morning after ♀ has relieved ♂. Both sexes incubate, at night only ♂ (see Rendle 1914 for exception). Nest-reliefs at intervals of 1–3 hrs after incubation proper has started, ♀ sitting for total of c. 5–6 hrs daily (D Blume; fuller details in Blume 1981). Incoming bird gives Flight-calls (approach sometimes silent in later incubation), then call 4 from perch near nest-tree. Flies to hole, landing beneath it or to one side; gives quieter variant of call 4 (some incomplete), but intensity increases if bird in hole slow to respond. Sitting bird normally Taps (with varying rhythm—see Voice), then

B

C

D

leaves; other bird moves aside and not uncommonly performs (low-intensity) ceremonial threat. Nest-relief ceremony apparently helps to suppress aggression; no longer performed after brooding ceases and call 4 fades for a time (Blume 1959; D Blume; fuller details and illustrations, in Glutz and Bauer 1980, Blume 1981). RELATIONS WITHIN FAMILY GROUP. See, especially, Sielmann (1959); also Pynnönen 1939, Blume 1962a, 1981, Cuisin 1967–8). Eggshells and dead small nestlings usually carried far away, corpses of older young tossed out. Unhatched eggs may remain buried in woodchips for days. For 'warmth pyramid' of young when unattended, see Fig D; for progress of physical development, see Cuisin (1967–8). Over first 2 days, adult wakes young by touching swelling on lower mandible or back several times. For feeding, pushes bill down into nestling's gullet (see illustrations in Blume 1981; see also Lohrl 1977c); further details of feeding, also nest-sanitation, as in *Picus* woodpeckers. From 3rd day, abrupt upward and swaying movement of awakened nestling induces begging in siblings. From 5 days, all nestlings beg following example of sibling touched by adult; adult further stimulated by nestling's pale tip of tongue and white egg-teeth (Sielmann 1959). Later, food-calls of one nestling provide sufficient stimulus; later, only darkening of nest-chamber required, and finally, scratching noises made by landing adult. At *c.* 12 days, eyes open; swelling on lower mandible and white egg-teeth then more or less disappeared (D Blume). Adults exceptionally in nest together: e.g. during rain when young *c.* 6–10 days old (Krambrich 1953). Young brooded more or less uninterruptedly for 6–9 days during day, 10–15 at night. At 18 days, climb up inside wall and may be fed at entrance from *c.* 20 days. At *c.* 27–28 days (last 2 days of nestling period), parents may sit nearby for *c.* 1 hr without feeding, and ♂ no longer roosts with young at this time. Adults tend to feed young from a distance (see *P. canus*) and use call 4 more frequently again. In one case, ♀ fed young alone for last 16 days (Gebhardt 1940). Young leave nest (sometimes over 2–3 days: Papadopol and Mândru 1977) by flying; climb up nearby trees to moderate height and remain there several hours (D Blume). Adult may Drum to encourage young to leave (Vogel 1982). Family usually departs from nesting area by next day. Brood split between parents; adult and offspring use Advertising-call for contact. For roosting habits of young, see Roosting (above). In central Europe, fledged young tended up to mid-August (Loos 1910; Gebhardt 1950) or early September (Blume 1961), though some young begin search for roost-hole in early July, *c.* 5–6 weeks after fledging (Glutz and Bauer 1980). Young do some self-feeding (pecking and inserting tongue) from start, but fed intermittently by adult (Cuisin 1967–8). At end of post-fledging care, adult attempts to 'shake off' young in roosting area in evening by making detours or flying off (Blume 1961). At same time, siblings also threaten each other using ceremonial Mock-fencing and drive one another up trunk (Blume 1956; D Blume). ANTI-PREDATOR RESPONSES OF YOUNG. Few details, but see Cuisin (1981). PARENTAL ANTI-PREDATOR STRATEGIES. (1) Passive measures. Often remarkably insensitive to disturbance, especially ♂♂ in later part of incubation and after hatching (Sielmann 1959). Some can be seized in nest (Blume 1981); Hissing recorded (Bannasch 1980). If young taken by predator, adults completely silent after a few hours, avoid nest-tree, and do not call throughout territory for days (D Blume); see, however, Glutz and Bauer (1980) for report of birds continuing to come to nest in rhythm of previous nest-reliefs and giving call 2 for 3–4 days. (2) Active measures. Give call 2 at appearance of flying raptor. Bird in nest will peck at and drive off *C. monedula*. Squirrel *Sciurus* threatened with wings raised and crown feathers ruffled; bird also leaps forward in attack (Eygenraam 1947; Schaack 1967).

(Fig A from sketch by D Blume; Figs B–C from drawings in Eygenraam 1947 and from sketches by D Blume; Fig D from photograph in Sielmann 1959.)   MGW

Voice. Vocal throughout the year. Both sexes share same repertoire including instrumental signals, but see below for variations in intensity. Individual calls mono- or disyllabic and given several or many times in succession depending on excitement; almost all signals remarkably loud and far-carrying (D Blume). First descriptions and interpretation by Loos (1910) and Rendle (1914). For comparative and experimental studies, see Pynnönen (1939), Eygenraam (1947), Blume (1956, 1959, 1961, 1962a), and Blume and Jung (1956, 1958); see also important summary in Cuisin (1967–8). Following account based on outline supplied by D Blume.

INSTRUMENTAL SIGNALS OF ADULTS. 3 types occur, mainly in spring, occasionally in late summer and autumn, and are of great importance in communication. (a) Drumming. Associated with territory demarcation, attraction of sexual partner, Nest-showing, and sexual stimulation. Given by both sexes, mainly February–August (D Blume). Series given by ♀ shorter and less regular than those of ♂, but apparently no difference in number of series given per minute. In Netherlands, 78% of total Drumming activity by ♂, 13% by ♀, 9% sex unknown (Eygenraam 1947); time spent Drumming also varies individually (Cuisin 1967–8). In study by Tilgner (1976), individual strikes given at slightly slower rate by ♀ (see also Great Spotted Woodpecker *Dendrocopos major*); average 17 strikes per s in ♂, 14–15 per s in ♀. Longest series by ♂ 60 strikes in over 3 s; series by ♀ usually 5–28 strikes (see below). 3 forms of Drumming distinguished (D Blume). (i) Pronounced, loud, far-carrying longer series in quite long bouts; normally 35–43 strikes in *c.* 2·5 s (Blume 1981; D Blume). In southern West Germany, average of 1st bout of series each day by ♂ 29·7 strikes per series in January, 34·5 February, 42·1 early March, 37·3 late March (Tilgner 1976). Average 29·2 strikes per series, average duration 1·60 s, $n=21$ series by 8 birds (Zabka 1980); average $31·1 \pm 9·1$ strikes per series (Wallschläger and Zabka 1981). Rate of delivery of strikes accelerates slightly through series: *c.* 70 ms at start, *c.* 52 ms at end (Wallschläger and Zabka 1981; see also Zabka 1980). Recording (Fig I) comprises 31 strikes in 2·0 s; interval between onset of successive strikes 80 ms at start, 60 ms at end (J Hall-Craggs). Maximum 3 series per min in central Europe, 4 series per min in Netherlands, 7 series per min in Finland (Pynnönen 1939; Eygenraam

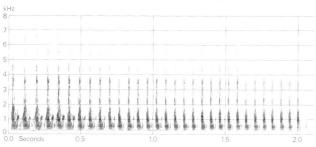

I P Szöke Hungary

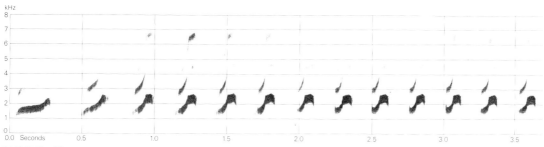

II P Szöke Hungary

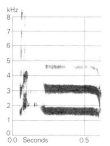

III S Palmér/Sveriges Radio (1972) Sweden April 1963

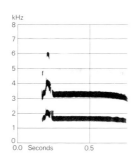

IV H-W Helb West Germany April 1982

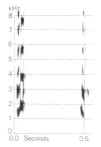

V S Palmér/Sveriges Radio (1972) Sweden April 1963

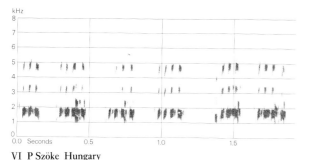

VI P Szöke Hungary

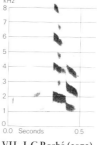

VII J-C Roché (1970) France April 1962

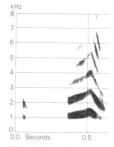

VIII P Szöke Hungary

1947; D Blume). Generally uses particular Drumming trees and special sites on these; resulting timbre possibly important for individual recognition. In Hessen (West Germany), paired ♂ used hollow beech *Fagus* with row of holes; Drumming site at bottom of row throughout season. Signals audible at *c*. 1800 m (3–4 km recorded exceptionally by Tilgner 1976). From 24 March–1 April, ♂ Drummed sporadically during day, during 2–26 April regularly 06.30–08.00 hrs, again towards 15.00 hrs, and 17.00–18.00 hrs; *c*. 400–500 series per day (Blume 1962*a*). Drumming (which may be interspersed with Advertising-calls—see below) normally most intense in morning with renewed activity towards evening (Blume 1981; D Blume); 2nd peak may occur around midday (Pynnönen

1939). Most Drumming during nest-site selection and excavation, briefly later prior to laying, hatching, and fledging (D Blume). Captive ♀ first Drummed (on metal floor of cage) at c. 2 months old (Heinroth and Heinroth 1924–6). (ii) Shorter, single series of up to 20 strikes given in Nest-showing by ♂ at hole-entrance; not recorded in ♀. Audible only up to c. 100 m and directed at nearby mate not on nest-tree (Blume 1962a; D Blume). (iii) Short, irregular series given by both sexes in various conflict situations, also away from nest-hole (Blume 1956). (2) Tapping. Series of quiet or loud taps, 0·75–0·43 s apart; 80–140 strikes per minute (D Blume); sonagram by H-H Bergmann of recording by D Blume shows strikes 0·46–0·22 s apart (J Hall-Craggs). Both sexes Tap to advertise potential nest-holes, to encourage nest-relief (at all stages), and to confirm readiness for change-over (D Blume); sometimes only 2–3 strikes given but may increase up to 7 per s (Tilgner 1976). See Eygenraam (1947), Cuisin (1967–8), and Social Pattern and Behaviour for further details. (3) Excitement-pecking. Irregular series of strikes with fluctuating intensity (see also Eygenraam 1947, Cuisin 1967–8). Unlike in Tapping, pieces of bark or wood, even small branches, may be knocked off (Blume 1956). Occurs in aggressive interactions and various conflict situations (disturbed by man or competitors at tree with hole or on feeding grounds, thwarted in attempt to perform particular behaviour pattern), also in transitional phase from diurnal activity to roosting (D Blume); see also Social Pattern and Behaviour. (4) Distinctly rhythmic wing-noises characteristically given before landing; possibly allow individual recognition, but signal function not proved (Blume 1981; D Blume).

CALLS OF ADULTS. (1) Advertising-call. A series of 10–20 melodious 'kwih' ('quee') sounds; 1st unit often slightly longer drawn and plangent (see below); shorter 'kwih' sounds then following in rapid sequence after short pause (D Blume). Other renderings: 'kui' (Hortling 1929; Pynnönen 1939); 'quae-quiae-quih' (Loos 1910); 'kwoih . . . kwih-kwihkwihkwikwikwikwick' (Glutz and Bauer 1980, which see for extra sonagrams; see also Bergmann and Helb 1982, Striegler et al. 1982). In recording (Fig II), 2 longer, more fluting units followed by series of 11 liquid and whistling 'ouee' or 'ouik' sounds; duration 3·6 s, 1st unit 240 ms, last 120 ms (J Hall-Craggs, M G Wilson). Given in flight or when perched, in and outside breeding season. In breeding season, used to locate, attract, and stimulate sexual partner, and in Nest-showing (in signal flights slightly slower sequences given with more pronounced vowel sounds by ♂: Glutz and Bauer 1980); after fledging, used as contact-call between family members (D Blume). See also Social Pattern and Behaviour. (2) Excitement-call. Loud call with 2 variants given throughout the year and only when perched. (a) Long-drawn, more or less clearly disyllabic 'kijäh' or 'kjäh' (D Blume), 'dliäh' (Loos 1910), or 'klije' (Hortling 1929). In recordings, a 'pikloo' (P J Sellar: Fig III) or apparently slightly differently structured 'klioh' (H-H Bergmann, D Blume: Fig IV). For extra sonagrams, see Glutz and Bauer (1980) and Bergmann and Helb (1982). Functions as contact-call also expressing excitement; used when disturbed or going to roost; given much less during incubation and brooding. At high intensity, c. 4–6 s between calls, c. 8–13 s as excitement wanes (Blume 1981; D Blume). (b) Abrupt, incomplete- and compressed-sounding 'klickje' (D Blume); recording (Fig V) suggests 'pikte' (P J Sellar) or 'kwik-te' (J Hall-Craggs). Characteristic of pronounced conflict situations, particularly in breeding season. Forms intermediate with call 2a occur—'kilche' as fright-call (Rendle 1907); occasionally only 'klick' (D Blume). (3) Flight-call. Long series of loud 'kürr', 'krü', or 'urr' sounds (Blume 1961; D Blume, C S Roselaar); also rendered 'tri' (Loos 1910; Рупнönen 1939) or 'kr' (Hortling 1929). Recording (Fig VI) suggests a musical 'krrri krrri krrri' with rolled 'r' sound (J Hall-Craggs). Given mainly March–May and only in flight; audible up to c. 1000 m depending on terrain (D Blume; also Bergmann and Helb 1982). For extra sonagram, see Glutz and Bauer (1980). Long series given particularly during extended flight over open country, also when flying through open wood to fresh feeding site, or approaching or leaving roost. More like 'kri' when given by agitated bird disturbed in roosting area and up to 20 or more 'kri' units given when flying to more distant emergency roost-hole (D Blume). (4) A 'kijak' (Fig VII) like Jackdaw *Corvus monedula* (Blume 1961; D Blume). Other renderings: 'tsiak-tschiak-tia-gia' (Loos 1910); 'taklleo-djak' (Hortling 1929); 'pijak' (Sielmann 1959). Dynamic and variable—from quiet and soft to hard and almost clicking, long drawn and more markedly disyllabic or short and explosive: a 'kja kuijak' prior to attack (D Blume) and comparison of sonagrams indicates 'te weee . . . o' or 'tikwooto' in recording (Fig VIII) to be similar call (J Hall-Craggs, P J Sellar). Abbreviated variants 'ki' or 'kja' also given by ♀ (perhaps also by ♂: Striegler et al. 1982) prior to copulation (see Glutz and Bauer 1980 for sonagram). Directed at nearby conspecific bird and used for specific and sexual recognition; softer variants given as greeting, harder ones to discourage close approach. Given when perched or in flight and typically in following situations: meetings between future breeding partners near potential nest-site, contact between rivals in nesting or roosting area, in courtship (location and stimulation of partner, approach to tree with hole, nest-site selection, copulation), nest-reliefs during excavation, incubation, and brooding, and for contact between members of family after fledging (D Blume). (5) Threat-call. Series of quiet 'rürr' sounds given in short-range threat (Bill-fencing) between rivals or pair-members (Blume 1956, 1961; D Blume); also rendered 'tuuk-tuuk' (Eygenraam 1947). Given in antagonistic encounters between ♂♂, between ♀♀, and between ♂ and ♀ at start of pair-formation, also during hole inspection and at nest-reliefs. In Netherlands, given late January and

February and in autumn (Eygenraam 1947); in Hessen (West Germany), March–August. Invariably associated with ritualized fight at close quarters, but not well known (D Blume).

CALLS OF YOUNG. Nestlings give slightly rhythmic, chattering, rasping, and smacking food-calls (D Blume); 'tsetse' from newly-hatched young (Glutz and Bauer 1980, which see for sonagrams). Gradually develop into 'äkäkäk' sounds similar to Green Woodpecker *Picus viridis* (see also Loos 1916, Eygenraam 1947). Unlike *D. major*, food-calls given mainly shortly before, during, and after a feed, rather than more or less ceaselessly (D Blume). Older young give peculiar loud screech like Great Spotted Cuckoo *Clamator glandarius* when handled (Cuisin 1981). From *c.* 2–3 days before fledging, nestlings give full repertoire of adult-type calls, but imperfectly—'krrk', 'kwih', 'kijhk', etc. After fledging, fully-developed Advertising-calls (see adult call 1) given as contact-call with siblings and parent (D Blume). MGW

Breeding. SEASON. See diagram for central Europe. Up to 2 weeks later in Scandinavia and northern USSR, but similar in central and southern USSR (Dementiev and Gladkov 1951a; Makatsch 1976). SITE. Hole in tree or, less often, telegraph post. Wide variety of trees chosen, especially beech *Fagus*, pine *Pinus*, spruce *Picea*, poplar *Populus*, birch *Betula*, willow *Salix*, and alder *Alnus* (Cuisin 1967–8). Height above ground 4–25 m. Of 146 holes, East Germany: 2% at 4·5–6m, 21% 7–10m, 57% 11–12 m, 22% 13–18 m; 21% faced north, 27% north-east, 13% east, 14% south-east, 8% south, 12% south-west, 2% west, 3% north-west (Viebig 1935). Nest: excavated hole, with oval entrance 11–12 × 8–11 cm; depth 37–60 cm, internal diameter 19–25 cm wide (Beven 1966). Unlined, or with unremoved chips. Building: by both sexes, though ♂ may complete work in last few days. EGGS. See Plate 98. Blunt elliptical, smooth and glossy; white. 34 × 26 mm (31–37 × 22–27), n = 90 (Witherby *et al.* 1938). Calculated weight 12·4 g (Makatsch 1976). Clutch: 4–6 (1–9). Of 32 clutches, Sweden: 3 eggs, 3; 4, 9; 5, 12; 6, 8; average 4·8 (Makatsch 1976). One brood. Replacements laid after egg loss, occasionally twice. Laying interval 1 day. INCUBATION. 12(–14) days (Cuisin 1967–8). By both sexes. Eygenraam (1947) reported mainly by ♂ but Sielmann (1959) found regular nest-reliefs *c.* 70–90 min apart. Begins with last egg or possibly earlier; hatching near-synchronous. YOUNG. Altricial and nidicolous. Cared for and fed by both parents, but mostly by ♂ (Eygenraam 1947), who sleeps in nest with young. FLEDGING TO MATURITY. Fledging period 24–28 days. Cared for by parents after fledging (see Social Pattern and Behaviour). Age of first breeding not known. BREEDING SUCCESS. 38 broods, western Erzgebirge (East Germany), averaged 2·7 young at fledging (Möckel 1979). 14 broods, Finland, averaged 4·4 young at fledging (Pynnönen 1939).

Plumages (nominate *martius*). ADULT MALE. Forehead and crown bright and glossy carmine-red, some dark grey of feather-bases sometimes visible amidst red. Remaining plumage (including nasal bristles, lores, and supercilium) black; slightly glossed dark blue on sides of head and neck, hindneck, mantle, scapulars, upper wing-coverts, tail, and secondaries, duller with slight grey tinge on breast and flanks down to under tail-coverts and on longer under wing-coverts; primaries tinged sepia-brown. In worn plumage, much grey sometimes visible on top of head, red of feather-tips restricted; black on body duller, underparts and under wing-coverts more distinctly blackish-grey; primaries slightly paler brown, particularly on tips and outer webs of outer ones. ADULT FEMALE. As adult ♂, but forehead and centre of crown black with bluish gloss (like sides of head), red restricted to patch on hindcrown. Wear has same effect as in ♂; red feather-tips on hindcrown occasionally almost completely abraded, patch appearing dark grey with traces of red hardly visible. NESTLING. Naked at hatching. Closed feather-pins appear from 8–10 days; those on crown showing red from 10th day in ♂, bluish-black in ♀ (only nape red); tips of feather-pins open on 16th day, completely covered by feathers on 21st day, but most of these still growing, juvenile plumage completed at *c.* 6 weeks (Heinroth and Heinroth 1924–6; Cuisin 1967–8). JUVENILE. Like adult, but black duller, slightly plumbeous, and not glossy in fresh plumage; chin and throat palest, dark grey; black paler dull brownish-grey in worn plumage, especially on underparts. Red on head as in adult, but slightly duller and not glossy; red feather-tips shorter and narrower than in adult, patch mottled greyish-black. Extent of patch as in adults, but red of ♂ hardly extends to forehead and less extensive on sides of crown, restricted to central crown and hindcrown. Red sometimes partly lost by abrasion. Occasional variants, apparently of either sex, have all crown black without red, even in fresh plumage. FIRST ADULT. Like adult and hardly distinguishable once juvenile p10 replaced. Juvenile p10 rather broad, border of inner web convex or straight, tip extending 16–28 mm beyond longest greater upper primary coverts (in adult, p10 narrower, inner web concave, tip 3–12(–17) longer than primary coverts). Juvenile greater (and occasionally median) upper wing-coverts retained; relatively browner and with more abraded tips than in adult at same time of year; all or part of tertial coverts often replaced, however; these more glossy black and smoother-edged than neighbouring old greater coverts. In spring and early summer, plumage of 1st adult often more distinctly abraded than full adult; in particular, retained juvenile upper wing-coverts bleached to dark grey-brown or show brown fringes, 1st adult primaries browner with

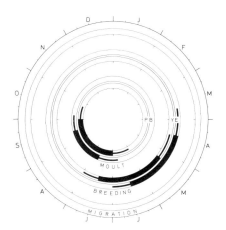

more heavily worn tips. Feathers on sides of neck, upper mantle, back, belly, or lower flanks occasionally heavily abraded, olive-brown, contrasting with less abraded black of remainder of body; not known whether these olive-brown feathers are retained juvenile or extremely abraded 1st adult.

Bare parts. ADULT AND FIRST ADULT. Iris yellow-white, pale yellow, cream-grey, or milky-white, with black spot in front of pupil. Bill greyish-, bluish-, yellowish-, or ivory-white; tip and ridge of culmen dark horn-blue or black. Leg and foot dark grey or blue-grey, claws horn-brown or horn-black. NESTLING. At hatching, bare skin pink, including leg and bill; lids of closed eyes blue-grey, swollen knob-like flanges at base of lower mandible whitish (reaching maximum size in 2nd week), claws white. Eyes open gradually during 7–13th day, iris blue-black at first. At end of 2nd week, iris blue-grey, bill ivory-yellow with white-tipped upper mandible, leg and foot whitish or pale grey. JUVENILE. At fledging, iris pale blue-grey, bill whitish with blue-black tip, leg and foot grey blue; black culmen ridge and yellow-white iris gradually develop in first weeks after fledging. (Hartert 1912–21; Heinroth and Heinroth 1924–6; Cuisin 1967–8; Harrison 1975; Glutz and Bauer 1980; Short 1982; RMNH, ZMA.)

Moults. Based mainly on data from E Sutter in Glutz and Bauer (1980), with additional information from Dementiev and Gladkov (1951a), Piechocki (1968c), and c. 25 moulting specimens examined. ADULT POST-BREEDING. Complete; primaries descendant. Starts with p1 early June to early July; primary moult score of 25 reached in August, 40 late August to mid-September; all moult completed (score 50) late September to late October. Tail and body moulted at same time as in post-juvenile. POST-JUVENILE. Partial: head, body, tail, lesser and median wing-coverts, shorter primary coverts, and primaries; no secondaries, greater upper wing-coverts, or greater upper primary coverts; occasionally, however, some median upper wing-coverts retained, and often some greater tertial coverts or sometimes a few outer greater primary coverts replaced. Starts with p1 at age of c. 3 weeks when still in nest, followed by p2 at about fledging; p7 shed at day 92–100 (late July to early September), p9 at day 125–130, p10 at day 135–140 (September–October). In specimens from Netherlands starts with p1 late May to late June, completed with regrowth of p10 mid-September to early November. Tail starts with shedding of t2 and t6 at about shedding of p4 or at primary moult score 5–18; t3–t4 gradually follow; t1 and t5 shed last, lost with p7–p8, t1 regrown with regrowth of about p7 or at score 35–45. Head, body, and shorter upper wing-coverts start with shedding of p4; in heavy moult during growth of p5–p7; completed with regrowth of p9–p10, parts of belly, vent, rump, and some median upper wing-coverts last. Greater tertial coverts shed with p7–p10.

Measurements. Nominate *martius*. ADULT AND FIRST ADULT. Netherlands, France, West Germany, and Switzerland, all year; skins (RMNH, ZMA). Bill (F) to forehead, bill (N) to distal corner of nostril; toe is outer front toe with claw.

WING	♂ 234	(3·13; 11)	230–240	♀ 234	(3·70; 16)	229–242
TAIL	162	(3·73; 10)	158–170	164	(4·02; 15)	158–173
BILL (F)	62·6	(3·36; 9)	59–69	58·5	(2·57; 15)	55–63
BILL (N)	47·4	(2·71; 11)	44–52	45·2	(1·98; 14)	43–48
TARSUS	36·3	(1·53; 13)	34–38	35·8	(1·52; 18)	34–38
TOE	36·2	(2·02; 14)	34–39	34·0	(1·85; 18)	32–37

Sex differences significant for bill and toe.

Wing of various other populations: (1) Vosges mountains (France), (2) Switzerland and West and East Germany, (3) Norway and Sweden (Voous 1961); (4) northern Iran (Stresemann 1928).

(1)	♂ 237	(4)	232–240	♀ 235	(6) 231–242
(2)	237	(10)	233–243	235	(17) 225–241
(3)	243	(29)	236–250	241	(25) 234–247
(4)	241	(4)	240–242	236	(2) 235–238

Some data of populations from eastern Europe, sexes combined: (1) Poland and Lithuania, (2) Estonia, (3) west European USSR, (4) central European USSR, (5) north European USSR, (6) south-eastern European USSR, (7) Caucasus and Transcaucasia (Dementiev 1939).

(1)	WING	237	(8) 228–243	BILL (N)	46·6	(8) 43–50
(2)		242	(5) 236–248		49·5	(5) 47–53
(3)		240	(10) 233–248		47·9	(10) 43–52
(4)		243	(30) 233–255		48·7	(30) 44–52
(5)		245	(11) 236–257		48·9	(11) 46–52
(6)		247	(26) 236–254		51·3	(26) 47–56
(7)		234	(16) 228–245		46·2	(16) 41–50

JUVENILE. Netherlands: wing ♂ 226 (3) 221–232, ♀ 222 (5·89; 7) 215–232; tail, sexes combined, 132 (4) 126–139 (RMNH, ZMA). Bill full-grown at about completion of post-juvenile moult.

Weights. Netherlands: ♂♂ 272, 287, 315 (June), 294 (December); ♀♀ 255 (April), 226 (frost-killed, February); fledged juveniles, August and early September, ♂♂ 200, 269; ♀ 256 (RMNH, ZMA). Norway, autumn, 318 (4) 296–374; Finland 300–460 (Hagen 1942). Hungary, April: ♀ 300 (Vasvári 1955). Kazakhstan: May, ♂ 300; range, August, 345–361 (Dolgushin et al. 1970). Mongolia: June, ♂ 330; August, adult ♂♂ 342, 342, 350, adult ♀ 340, juvenile ♂ and ♀ each 310 (Piechocki 1968c). Manchuria: July, ♀ 344 (Piechocki 1958). Kuril Islands: ♂♂ 346, 354, 361; ♀♀ 314, 324 (Nechaev 1969).

At hatching, c. 9; on 5th day 83 (3) 80–90, on 7th 118 (6·03; 11) 108–126, on 15th 220 (19·8; 10) 190–240, on 16th–18th 236 (13·4; 32) 204–260, on 19th 254 (9·97; 14) 242–280, on 20th 262 (17·7; 14) 210–276, on 24th 279 (3) 252–302, at 6 weeks 250 (Heinroth and Heinroth 1924–6; Cuisin 1967–8, 1981).

Structure. Wing rather short and broad; tip broadly rounded. 10 primaries: in adult and 1st adult, p6 longest, p7 0–5 shorter, p8 12–18, p9 46–61, p10 121–136, p5 2–5, p4 6–11, p3 20–30, p2 40–50, p1 53–64; in juvenile, p7 longest, p8 5–8 shorter, p9 29–38, p10 94–111, p6 2–10, p5 26–43, p4 44–62, p3 67–89; p1–p2 reduced, c. $\frac{1}{4}$–$\frac{1}{3}$ length of neighbouring s1; p10 reduced (least so in juvenile), see description of 1st adult in Plumages. Outer web of p4–p7 and (indistinctly) inner of p4–p10 emarginated. Tail long, 12 feathers; tip wedge-shaped in adult, t1 longest, t2 10–16 shorter, t3 19–31, t4 34–42, t5 55–65, t6 110–121; shape of juvenile tail peculiar, as early-moulted t2–t3 reduced in length and t1 and t4 longer; t1 longest, t2 37–42 shorter, t3 28–38, t4 21–33, t5 32–49, t6 84–102. Bill slightly longer than head, straight, heavy; wide and rather flattened at base, gradually tapering to chisel-shaped tip; top and sides of culmen, gonys, and rami of lower mandible with marked ridge and grooves, sides of upper mandible thickened. Nostrils oval, hidden under tuft of bristles projecting from base of upper mandible. Feathers of crown and nape rather narrow and slightly elongated in adult and 1st adult. Tarsus short and strong; toes rather short and slender. Outer front toe longest, inner front toe with claw c. 83% of outer front with claw, outer hind toe c. 95%, inner hind toe c. 48%. Claws rather short, heavy, shallowly ridged at sides, sharp, and strongly decurved. For number of feathers, pterylography, and other details of structure, see Cuisin (1967–8, 1976).

Geographical variation. Slight. Variation clinal in nominate *martius*; populations of central and southern Europe east to Caucasus and Transcaucasia on average slightly smaller than populations from Scandinavia and northern USSR eastward (see Measurements), but overlap large and recognition of a smaller race *pinetorum* (Brehm, 1831), as advocated by (e.g.) Dementiev (1939), not warranted. Isolated *khamensis* from south-west China purer and more glossy black than nominate *martius*; bill proportionately shorter and weaker; wing averages slightly longer, 252 (15) 246–260 in ♂ (Vaurie 1959*a*).

Closely related to, and probably forming superspecies with, White-bellied Woodpecker *D. javensis* of eastern Asia (Short 1982; K H Voous).

CSR

Sphyrapicus varius Yellow-bellied Sapsucker

PLATES 80 and 86
[facing pages 831 and 879]

Du. Geelbuiksapspecht   Fr. Pic maculé   Ge. Feuerkopf-Saftlecker
Ru. Желтобрюхий дятел-сосун   Sp. Pico chupador   Sw. Aspsavsugare

Picus varius Linnaeus, 1766

Monotypic

Field characters. 19–20 cm; wing-span 32–34 cm. 10–15% smaller than Great Spotted Woodpecker *Dendrocopos major*; close in size to Middle Spotted Woodpecker *D. medius*. Medium-sized, fine-billed Nearctic woodpecker, with less bulk than similarly sized Palearctic relatives. Plumage pattern of adult recalls Three-toed Woodpecker *Picoides tridactylus* but differs distinctly in red crown and throat, black breast, yellowish-white underbody with dark chevrons on flanks, and white patch on centre of upper wing-coverts. Juvenile mainly mottled brown, lacking striking head pattern but showing white carpal patch and rump. Flight recalls *Dendrocopos* woodpeckers but action lighter. Sexes dissimilar; no seasonal variation. Juvenile separable.

ADULT MALE. Crown and nape black, with fore-centre scarlet. Rear supercilium and most of hindneck white. Bold black eye-stripe (from in front of and below eye), running across cheeks and then down along neck to join black shoulder. Lores and panel under eye-stripe white, dividing latter from black surround to scarlet throat and black bib on upper breast, and joining dark-streaked buff fore-flanks. Mantle black, mottled and loosely barred white; white centre of rump and upper tail-coverts continued down tail by largely white inner webs of innermost tail-feathers. Primaries, secondaries, and tertials black, with white barring. Forewing black at leading edge and base but containing bold white patch across median and central greater coverts; patch appears as long panel on folded wing, as more irregular broad-based T in light. Underbody yellowish-white (brighter yellow in fresh autumn plumage), with dusky chevrons on flanks. ADULT FEMALE. Throat white, not red. Crown sometimes with less red than ♂, especially at rear; occasionally wholly black. JUVENILE. Among true woodpeckers occurring in west Palearctic, uniquely dark or olive-brown, mottled black and white on back, and buff on underbody. Pattern of wing-coverts, rump, and tail as adult but obvious head-marks suppressed to narrow buff-white rear supercilium and stripe running down from lores. Not fully as adult until 1st spring.

Quiet and retiring, so likely to be encountered first in brief glimpse when appearance could suggest *P. tridactylus* or smaller *Dendrocopos* woodpecker. When seen well, juvenile unmistakable and adult quickly identified by head, throat, and breast pattern and carpal panel. Behaviour similar to *Dendrocopos* but both flight and climbing gait less jerky.

Commonest call 'cheerrr', with slurred downward inflexion.

Habitat. Breeds in Nearctic forests and woodlands from subarctic and boreal lowlands to temperate montane region, preferring deciduous trees, especially aspens and poplars *Populus*, alders *Alnus*, willows *Salix*, and birch *Betula*, but also inhabiting mixed stands and even pure conifers, especially pine *Pinus*. Prefers sites near water; also favours orchards (especially of apples) and neighbourhood of small open spaces. Characteristic habit of ringing trees (feeding on sap by making rows of holes) and requirements of nesting cavities, demand easily worked trees, often with decaying cores. Sometimes catches insects in flight from fixed perch like flycatcher. Migrates and winters through lowland woodland areas. (Pough 1949; Gabrielson and Lincoln 1959; Godfrey 1966; Larrison and Sonnenberg 1968; Johnsgard 1979.)

Distribution. Breeds in North America from eastern Canada west to Alberta, north-east British Columbia, southern Yukon, South Dakota, Illinois, Ohio, Pennsylvania and New York, with some southward extension in Allegheny mountains. Winters in central and southern USA, Central America, and Greater Antilles.

Accidental. Iceland: 1 bird collected; no details. Britain: 1st-winter ♂, Scilly, 26 September–6 October 1975 (Hunt 1979).

Movements. Migratory, with virtual separation of breeding and wintering ranges. Winters in central and southern USA, Central America, and Greater Antilles, with small numbers south to Panama and (more rarely) to Netherlands Antilles off Venezuela. Frequent in autumn along Atlantic coasts, but occurs only as a straggler to Bermuda and Lesser Antilles. ♀♀ move on average further south than ♂♂; 210 museum skins collected Central America and West Indies were 77·5% ♀ and 22·5% ♂, while 226 winter specimens from USA (where, however, collecting less random) were 44% ♀ and 56% ♂ (Howell 1953). ♂♂ are first to return to breeding areas in spring (Bent 1939).

Slow autumn passage through USA lasts late August to late October; birds present in southern wintering areas late October to March (e.g. Bond 1960, Ridgely 1976), and return movement occurs late March to mid-May. Peak movements in northern USA late September to mid-October and in last 3 weeks of April (e.g. Stewart and Robbins 1958, Green and Janssen 1975).

Voice. See Field Characters.

Plumages. ADULT MALE. Forehead and crown crimson-red, bordered by narrow black line above tufts covering nostrils and by broad black line from front of eye along sides of crown and across nape. Broad black stripe through eye, extending over ear-coverts and down side of neck; bordered above by white wedge from just above eye widening backwards and extending across hindneck, below by white stripe from tufts over nostrils extending below eye to side of chest. Mantle, inner scapulars, back, upper rump, and innermost tertials white with bold and rather irregular black bars, appearing boldly spotted black and white, bordered on either side by contrastingly uniform black outer scapulars, inner lesser and median upper wing-coverts, tertial coverts, and longer tertials; outer upper wing-coverts contrastingly white, except for black marginal coverts, black bases of outer greater coverts, black median and greater primary coverts, and black bastard wing. Feathers of lower rump and upper tail-coverts white with black outer webs, white forming contrasting line on rear body, joining with white inner webs of t1. Chin and upper throat bright crimson-red (appearing slightly paler than crown, mainly because hidden feather-bases are white instead of black), bordered at sides by black line from base of lower mandible backwards, and below by large and rounded black gorget on lower throat and central chest. Sides of chest, breast, and central belly bright yellow or yellowish-white (usually brightest at border of black gorget); ground-colour of sides of belly, flanks, thighs, and under tail-coverts white, virtually without yellow tinge, closely marked with dusky grey or black streaks or chevrons. Tail black, but inner webs of innermost pair (t1) white with a few black spots or bars and distal parts of outer webs sometimes with a few white spots, (t3–)t4–t5 with some irregular white spots or streaks on distal parts and with white margin along tip and outer web. Primaries and secondaries (including longer tertials, excluding innermost tertials) black with rows of rather small white spots along outer borders (6–8 spots on longer primaries, 3–4 on secondaries), inner web with similar but larger spots, forming bar-pattern. Under wing-coverts and axillaries white with rather narrow dusky streaks or chevrons. White parts in fresh autumn plumage more strongly tinged yellow than in spring plumage described above; pale streaks on sides of head yellow or yellow-buff; mantle, inner scapulars, and back spotted and barred black and yellow; white of upper tail-coverts and tail slightly yellowish (particularly so on fringes of tail-feathers); black of gorget largely hidden under broad yellow feather-fringes; remainder of underparts bright yellow with less distinct dark olive-grey streaks and chevrons; white of wing-coverts and flight-feathers sometimes faintly washed yellow or buff. ADULT FEMALE. Similar to adult ♂, but chin and upper throat white, contrasting with black of lower cheeks and large black gorget on lower throat and chest. Forehead and crown usually similar to ♂, but red often slightly less deep crimson and slightly less extensive, sometimes restricted to forehead and forecrown. A variable number of ♀♀ (usually less than 10%) have forehead and crown completely deep black without red. Black and white stripes on sides of head often slightly mottled and less sharply defined than in ♂; black bars on mantle, back, and rump slightly narrower and closer; black chevrons and streaks on flanks and sides of belly slightly more extensive. Innermost tertials on average more extensively black; white patch on upper wing-coverts often smaller, more black of feather-bases showing. JUVENILE. Rather different from adult; sexes similar. Head and body mainly olive-brown and yellow without solid red or black patches and stripes. Forehead and crown dull olive-brown, feather-tips often with slightly paler olive or off-white spot. Sides of head and neck and hindneck finely mottled with pale and dark olive-brown; indistinct patches above and below eye extending to above and below ear-coverts almost uniform white with slight olive mottling only; patches from behind eye to ear-coverts and from base of lower mandible across lower cheeks dull black with some olive specks. Mantle, scapulars, and back sepia-brown, feathers with paler grey-brown bases and rather narrow off-white fringes or submarginal streaks or spots, appearing browner and more narrowly barred than adult. Rump white, barred or spotted black. Chin off-white or yellow-buff with some fine dusky specks, throat to chest and sides of breast olive-brown, feathers often with darker grey or brown subterminal arcs, chest in particular often appearing closely and indistinctly scaled. Upper tail-coverts and underparts down from flanks and belly similar to fresh-plumaged adult, but yellow of belly and vent often paler. Tail as adult, but inner webs of t1 more closely barred black. Wing as adult, but p9–p10 longer and subterminally much broader than adult. SUBSEQUENT PLUMAGES. Lower mantle, scapulars, tail, and wings similar to adult once autumn migration of northern birds started, but remainder of head and body mainly juvenile (least so in less- or non-migratory southern birds); usually, scattered red feathers show on forehead and crown, and black feathers show on nape, lower throat, or upper chest; sexes indistinguishable, unless some red shows on throat. For further plumage development and ageing, see Moults.

PLATE 81 (*facing*).
Dendrocopos major Great Spotted Woodpecker (p. 856). Nominate *major*: **1** ad ♂, **2** ad ♀, **3** juv, **4** nestling. *D. m. numidus*: **5** ad ♂. *D. m. anglicus*: **6** ad ♂. *D. m. hispanus*: **7** ad ♂. *D. m. canariensis*: **8** ad ♂. *D. m. tenuirostris*: **9** ad ♂.
Dendrocopos syriacus syriacus Syrian Woodpecker (p. 874): **10** ad ♂, **11** ad ♀, **12** juv, **13** nestling. (NA)

PLATE 82. *Dendrocopos medius* Middle Spotted Woodpecker (p. 882). Nominate *medius*: 1 ad ♂, 2 ad ♀, 3 juv, 4 nestling. *D. m. caucasicus*: 5 ad. (NA)

PLATE 83. *Dendrocopos leucotos* White-backed Woodpecker (p. 891). Nominate *leucotos*: 1 ad ♂, 2 ad ♀, 3 juv ♂, 4 nestling. *D. l. lilfordi*: 5 ad ♂, 6 ad ♀. (NA)

Bare parts. ADULT AND FIRST ADULT. Iris brown. Bill slate-grey, blackish-brown, brown-black, or slate-black. Leg and foot greyish olive-green, greenish-grey, or bluish-grey. (Ridgway 1914; Short 1982; ZMA.)

Moults. ADULT POST-BREEDING. Complete; primaries descendant. Starts with p1 late June to late July; all moult completed late August to early October, p10 and s3–s4 moulted last. Body and tail start at primary moult score $c.$ 20, late July or early August, heavy moult August, largely completed early September. POST-JUVENILE. Partial; primaries descendant. Primaries start with p1 shortly after fledging, soon followed by tail; primaries and tail completed before autumn migration starts, but t1 occasionally still growing late September or early October. In contrast to this, moult of body slow and protracted, part of juvenile body feathers often retained until March–April. Lower mantle and inner scapulars moulted first, followed by scattered feathers of forehead, crown, throat, and sides of breast; by October–November, $\frac{1}{4}$–$\frac{3}{4}$ of feathering of forehead and crown new, as well as often all mantle, scapulars, back to upper tail-coverts, all or part of lesser and median upper wing-coverts, and lower breast to flanks and under tail-coverts; part of crown and upper mantle and often much of throat and chest contrastingly juvenile, as well as all greater upper primary coverts, greater coverts, secondaries, and tertials. Moult apparently virtually halted in winter, resumed again in early spring, but this may depend on wintering locality; remaining juvenile of forehead, crown, chin, throat, chest, and upper wing-coverts replaced March–April (scattered feathers occasionally retained). Unlike Palearctic *Picus* and *Dendrocopos* woodpeckers, apparently also tertials and part of secondaries replaced in spring; of 8 late-spring and early-summer 1st adults examined (aged by pointed and heavily worn juvenile greater upper primary coverts), 6 had tertials and secondaries new except for s3–s4 (occasionally, s1–s2 and s5 also), and 2 of these also had 2–3 outer greater upper primary coverts contrastingly new.

Measurements. ADULT AND FIRST ADULT. Eastern North America and migrants in Caribbean, all year; skins (RMNH, ZMA). Bill (F) to forehead, bill (N) to distal corner of nostril. Toe is outer front toe.

WING	♂	126	(2·24; 15)	123–130	♀ 127 (2·54; 13)	122–130
TAIL		72·4	(2·31; 15)	69–77	73·8 (2·90; 12)	70–78
BILL (F)		25·9	(0·86; 15)	25–27	25·4 (1·13; 13)	24–27
BILL (N)		20·5	(1·12; 13)	19–22	20·0 (1·01; 13)	19–22
TARSUS		20·1	(0·65; 13)	19–21	19·7 (0·55; 12)	19–21
TOE		21·3	(0·94; 7)	20–23	21·2 (1·21; 6)	20–23

Sex differences not significant.

JUVENILE. Wing slightly shorter than adult.

Weights. New Jersey (USA), mainly September: 45·8 (7) 40–50 (Murray and Jehl 1964). Kentucky (USA): ♂ 53; ♀♀ 49, 50 (Mengel 1965). Belize, March: ♀♀ 47, 51 (Russell 1964). Mexico, migrant ♂, October: 45 (RMNH). Exhausted migrant ♀, Curaçao (Netherlands Antilles), November: 32 (ZMA).

Structure. Wing rather short, rather broad at base, tip narrowly rounded. 10 primaries: in adult and 1st adult p7 longest, p8 $\frac{1}{2}$–2 shorter, p9 15–22, p10 75–82, p6 2–5, p5 15–22, p4 24–29, p3 30–34, p2 34–38, p1 36–42; in juvenile p8 longest, p9 7–9 shorter, p10 49–53, p7 1–3. Juvenile p1–p2 not reduced, unlike juveniles of other Picinae occurring in west Palearctic. P10 reduced in adult and 1st adult, narrow and pointed, 3–10 shorter than longest greater upper primary coverts; p9–p10 longer and broader in juvenile (Chapin 1921; Short 1982). Outer web of p6–p8 emarginated, inner web slightly. Tail rather long, wedge-shaped, 12 feathers; t1 longest, t2 5–11 shorter, t3 10–17, t4 16–21, t5 20–25, t6 53–59. Bill rather short, straight, and rather slender, gradually tapering to sharp tip; culmen ridge marked, short ridge and groove at basal side of culmen distinct. Nostrils narrow, beneath ridges at basal side of culmen, covered by tuft of short hair-like feathers. Tarsus short, toes rather long and slender; claws short but heavy, strongly curved. Inner front toe $c.$ 81% of outer front toe with claw, outer hind $c.$ 98%, inner hind $c.$ 50%.

Geographical variation. Slight. Birds breeding in southern Appalachian mountains perhaps a trifle darker and smaller than northern ones, but overlap great and recognition of separate race *appalachiensis* Ganier, 1954, not warranted (Short 1982).

Forms superspecies with Red-naped Sapsucker *S. nuchalis* from Rocky Mountains and Red-breasted Sapsucker *S. ruber* from Alaska south to California, the 3 forms often considered single species (e.g. American Ornithologists' Union 1957). Breeding range of *S. ruber* overlaps slightly with *S. varius* and *S. nuchalis*, but relatively few hybrids known; because of this, and because of marked differences in colour pattern of adult head, neck, and back, in juvenile plumage, in amount of sexual dimorphism, in speed of post-juvenile moult, and in migratory habits, all 3 better considered separate species (Ridgway 1914; Short 1982).
CSR

Dendrocopos major Great Spotted Woodpecker

PLATES 81 and 86
[facing pages 854 and 879]

Du. Grote Bonte Specht   Fr. Pic épeiche   Ge. Buntspecht
Ru. Большой пестрый дятел   Sp. Pico picapinos   Sw. Större hackspett

Picus major Linnaeus, 1758

Polytypic. Nominate *major* (Linnaeus, 1758), northern Europe and Siberia from Norway, Sweden, Bornholm (Denmark), and north-east Poland eastward, south in European USSR to *c.* 50°N and in Siberia to Altay mountains, and east to Aldan and Amur river basins; *pinetorum* (Brehm, 1831), central Europe from Netherlands, Denmark (except Bornholm), and East Germany south to France, Alps, northern Yugoslavia, and Carpathians in Rumania; *anglicus* Hartert, 1900, Britain; *italiae* (Stresemann, 1919), mainland Italy, Sicily, and extreme north-west Yugoslavia; *parroti* Hartert, 1911, Corsica; *harterti* Arrigoni, 1902, Sardinia; *hispanus* (Schlüter, 1908), Spain and Portugal; *mauritanus* (Brehm, 1855), Morocco; *thanneri* Le Roi, 1911, Gran Canaria; *canariensis* Koenig, 1889, Tenerife; *numidus* (Malherbe, 1843), northern Algeria and Tunisia; *candidus* (Stresemann, 1919), south-east Europe south from southern Ukraine, eastern and southern Rumania, and southern Yugoslavia; *tenuirostris* Buturlin, 1906, Crimea, Caucasus, and Transcaucasia; *paphlagoniae* Kummerlöwe and Niethammer, 1935, northern Asia Minor; *poelzami* (Bogdanov, 1879), Lenkoran area in south-east Transcaucasia, northern Iran, and south-west Turkmeniya. Extralimital: 7–11 further races in Asia.

Field characters. 22–23 cm; wing-span 34–39 cm. About ⅔ size of Green Woodpecker *Picus viridis*. Quite large, strong-billed woodpecker, with black upperparts boldly pied white (on front and upper part of face and along scapulars) and white underparts ending in large red vent. In markedly bounding flight, shows white-barred black flight-feathers. Sexes dissimilar; no seasonal variation. Juvenile separable. 15 races in west Palearctic, but only 4 described here (see also Geographical Variation).

Adult Male. (1) Continental north European race, nominate *major*. Plumage boldly pied black and white, as in all *Dendrocopos* and *Picoides* woodpeckers. Crimson patch on back of head. Specific characters of *D. major* are: (a) long black moustache which not only joins black shoulder and mantle but also turns up in line behind ear-coverts to join black nape, this line being incomplete or almost absent in all other west Palearctic *Dendrocopos* (though sometimes complete in Middle Spotted Woodpecker *D. medius*); (b) isolated white patch on side of neck behind this black line; (c) unstreaked, almost white underparts which contrast abruptly with strikingly large and wholly crimson red vent (features unique in west Palearctic *Dendrocopos*). (2) British race, *anglicus*. Less robust, with more slender bill. Plumage lacks clean contrast of black and white so evident in nominate *major*, with upperparts browner in tone, cheeks and underparts distinctly buffy, and white of scapulars sullied. (3) North-west African races, *mauritanus* and *numidus*. Show eastward cline of developing red and black band across upper breast. Adult Female. All races. Resembles ♂ but lacks crimson patch on head. Juvenile. All races. Less boldly pied than adult due to duller white plumage areas. Broad red centre to crown, inviting confusion with *D. medius* and ♂ White-backed Woodpecker *D. leucotos*. Vent also pinker and breast-sides may show some fine streaks; wing-barring less obvious than on adult.

No black and white ('spotted') woodpecker instantly identifiable, since all *Dendrocopos* share similar appearance and their ranges overlap widely. *D. major*, however, by far the commonest over most of west Palearctic and hence worthy of close study. For distinctions from other *Dendrocopos*, see those species. In flight, 3–7 flapping wing-beats (in obvious burst) alternate regularly with complete closure of wings, creating distinctive undulating track; landing accomplished with sudden collapse, as in *Picus* woodpeckers. Climbing and perching action and attitudes as *Picus*, but thinner, more retracted neck usually obvious in silhouette. Rarely on ground.

Commonest call a sharp, clear 'tchik', often repeated in slow hammering tempo, and followed by quick harsh chatter, sometimes lengthened to trill in disputes. Drumming usually involves 10–16 strikes lasting *c.* 0·5 s; far-carrying.

Habitat. From arctic taiga through boreal and temperate to Mediterranean and alpine forest zones, wherever there are trees of any sort with sufficient growth to accommodate nest-holes. Isolated and scattered trees in parks, avenues, gardens, orchards, and open or miniature woodlands less favoured, unless adjoined by larger stands of broad-leaved, coniferous, or mixed tree species, latter being commonly preferred. Compared with *Picus*, resorts more to canopy and less to ground feeding, but largely also a trunk climber. For differences in habitat utilization between *Dendrocopos* species, see Food. Adaptability enables successful use to be made even of stands of spruce *Picea*, larch *Larix*, and Arolla pine *Pinus cembra* to above 2000 m in Alps, although at lower altitudes these would come lower in order of preference (Glutz von Blotzheim 1962). Accordingly, distinctions in habitat need to be considered regionally, and not as generalizations valid for entire extensive range, especially where different races or localized populations concerned. Certain habits, such as 'ringing' or 'girdling' trees with series of holes to obtain sap are also common in some parts of range (Gatter 1972) but occasional or rare in others (e.g. Britain). Active in making fresh holes in trees, outside

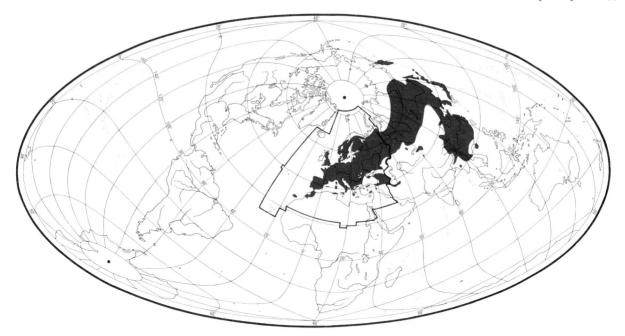

as well as during nesting season, and thus plays major ecological role in developing opportunities for other hole-nesting and hole-roosting species lacking this aptitude. In some settled areas has taken to regularly visiting bird tables to feed. Readiness to experiment leads to numerous cases of strange and unexpected occurrences and behaviour which, when successful, may further diversify habitat and extend range, even when use of artefacts and reconciliation to close human presence involved. Flight normally in lower or lower middle airspace; association with water minimal. (Dementiev and Gladkov 1951*a*; Bannerman 1955; Voous 1960; Sharrock 1976; Harrison 1982.)

Distribution. Marked northward spread in Britain; some in Belgium and Finland.

BRITAIN. Largely disappeared from northern England by early 19th century and from Scotland by *c*. 1860; since *c*. 1890 spread in northern England, and markedly northwards in Scotland, continuing until at least mid-1950s (Parslow 1967; Sharrock 1976). First bred Isle of Wight 1926 (Sharrock 1976) and Jersey (Channel Islands) 1950 (Long 1981*b*). BELGIUM. Some northward spread in 20th century (Lippens and Wille 1972). FINLAND. Apparently some recent northward expansion (Merikallio 1958).

Accidental. Iceland, Faeroes, Ireland.

Population. Limited information. Increased Britain, Netherlands, and Denmark—due to habitat and possibly climatic changes. No indications of large-scale changes in central Europe (Glutz and Bauer 1980).

BRITAIN. Declined in 19th century in northern England and Scotland, then increased there from late 19th century and also in most other parts of England and Wales, with perhaps further increases after 1972, possibly due to spread of Dutch elm disease; estimated 30000–40000 pairs (Parslow 1967; Sharrock 1976). FRANCE. 100000–1000000 pairs (Yeatman 1976). BELGIUM. Estimated 16000 pairs (Lippens and Wille 1972); *c*. 10300 pairs 1982 (PD). LUXEMBOURG. Estimated 4400 pairs (Lippens and Wille 1972). NETHERLANDS. Increased 1850–1940, due to afforestation; 10500–17000 pairs 1977 (Teixeira 1979; CSR). DENMARK. Increased 20th century due to afforestation (Dybbro 1976). SWEDEN. Estimated 200000 pairs (Ulfstrand and Högstedt 1976). FINLAND. Estimated 95000 pairs (Merikallio 1958). POLAND. Fairly numerous (Tomiałojć 1976*a*). CZECHOSLOVAKIA. No known population changes (KH). ALBANIA. Rare (EN). CANARY ISLANDS. Gran Canaria: common in south, scarce in north. Tenerife: rather scarce. (KWE.)

Oldest ringed bird 10 years 9 months (BTO).

Movements. All populations basically resident; southern ones particularly so, but subject to eruptions (occasionally large) in northern coniferous habitats.

NORTHERN POPULATION: nominate *major* (Fenno-Scandia eastward into Siberia). Present all year over breeding range, but has eruptive behaviour more pronounced than in any other west Palearctic Picinae; lesser movements of other species usually associated with it (e.g. Hildén 1969, Meshkov and Uryadova 1972). These eruptions due primarily to poor seed crops of pine *Pinus* and spruce *Picea* in northern boreal zone, but proximate cause still unclear since eruptions begin before these seeds become seasonally important in diet; extent of movement depends largely on degree of seed-crop failure and size of region affected. See Pynnönen (1939), Svärdson (1957),

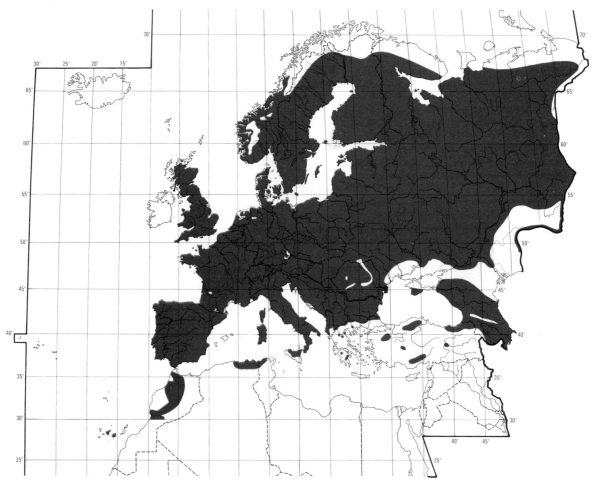

Formozov (1960), Williamson (1963), Hildén (1969, 1971, 1974), and Eriksson (1971). Birds involved are mainly 1st years, and this age-class probably less able to compete for limited food supply (Pulliainen 1963). At Finnish observatories, erupting birds included only 6–12% adults in different autumns, and adult ♂♂ perhaps slightly outnumbered by adult ♀♀ (Hildén 1974). In autumn 1972, Utsira (Norway), only 7 adults in 66 trapped, and these erupting birds had very low fat levels (Ree 1974). Major eruptions less frequent than those of Crossbill *Loxia curvirostra*, but more or less local (geographically restricted) eruptions tend to be more common. In Finland, 12 marked invasion years 1949–1974, though not even the 5 largest (1949, 1958, 1962, 1968, 1972) were always detected in Norway or on Baltic coasts of East Germany and USSR (Meshkov and Uryadova 1972; Hildén 1974; Glutz and Bauer 1980).

Small numbers of northern *major* recorded more or less regularly south to Alps, western France, Britain, and Ireland, and exceptionally to southern France, Faeroes, and Iceland. Significant irruptions into Britain in October 1949 and October 1962 both coincided with anticyclonic conditions over north-west Europe, with light easterly winds over North Sea (Williamson 1963). Lesser numbers reach Britain and Ireland in other years, and this 400-km (minimum) sea-crossing exceeds crossings of Baltic, where birds tend to follow coasts and cross at narrows—Skagerrak, Kattegat, and Gulfs of Bothnia and Finland. On west coast of Finland, movements begin late July, peak in second half of August, and end around 25 October; in big 1962 eruption, numbers on island of Säppi (off Pori) peaked at *c*. 10 000 on 17 August, with peak on 26 August on Åland, 200 km SW. Autumn recoveries during eruptions have been to north as well as west to south-west and south-east; Finnish birds recovered September and early October were from Scandinavia, but from mid-October onwards mainly from USSR, suggesting loop migration eastwards in late autumn (Hildén 1969). Longest recoveries in same autumn–winter: Finland (August) to Orlov, USSR (October), 1280 km south-east; Kaliningrad, USSR (August) to Belgium (November), 1200 km WSW; southern Norway (nestling) to Basses-Pyrénées, France (January), 1830 km south-west. Following irruptions, return movements on much smaller scale detected in spring (late February to mid-April), notable recoveries being: Wangerooge, West Germany (September) to Halsingland, Sweden (February, 5½ years later), 1025 km NNE; Dresden, East Germany (late July) to southern Finland (June, at nest), 1200 km north-east; Bouches-du-Rhône, France (present October 1956–January 1957) to Mordovskaya ASSR (May 1958), 3070 km north-east; Finland (March) to Udmurtskaya ASSR (May, same year), 1940 km ESE; Finland (March) to Gorkiy, USSR (May, 2 years later), 1280 km east. Such May recoveries in USSR suggest that some of the birds involved in European irruptions originate from well to the east (Hildén 1969), as also suspected in Three-toed Woodpecker *Picoides tridactylus* (Hogstad 1983).

True extent of migratory movements in northern USSR remains unclear, but obviously substantial, with post-breeding movement northwards into scrub-tundra (e.g. to Kanin peninsula), and from late autumn penetrating southwards as far as steppes (Dementiev and Gladkov 1951a). Under rigorous winter conditions in northern USSR, regular displacements likely within taiga zone, in response to fluctuations in food supply. Hence irruptions seen in western Europe may be no more than a peripheral phenomenon (Glutz and Bauer 1980).

CENTRAL AND WEST EUROPEAN POPULATIONS: *pinetorum* (breeding Europe south to Pyrénées, Alps, and Carpathians) and *anglicus* (breeding Britain). Resident; also some post-fledging dispersal by 1st-year birds, for the most part within a few km of birthplace. In small samples of recoveries from breeding-season ringing in West and East Germany, Switzerland, and Czechoslovakia, most found within 10 km of ringing site, very few exceeding 50 km, and with adults in particular showing sedentary behaviour (details in Glutz and Bauer 1980). Longest: Czechoslovakian nestling (June) found Vicenza, Italy (September), 570 km SSW; Frankfurt am Main, West Germany (March) recaptured in Czechoslovakia (May, at nest), 485 km east. Movement more pronounced in some autumns than others; at a site in Schwäbische Alb (Baden-Württemberg) more transient birds seen in autumn 1972 (after bumper cone crop the previous year) than in years 1966–1971 combined (Gatter 1973). Stragglers reach Helgoland and West German and Dutch Frisian Islands, and presence of *pinetorum* suspected in southern England in 1962 invasion autumn (Williamson 1963). Such birds may originate from east-central Europe, where recoveries indicate that juvenile dispersal can be more marked (Glutz and Bauer 1980). Breeding populations of western maritime countries show least dispersal. Only 10 of 142 Belgian recoveries were more than 10 km from ringing site (maximum 260 km); while British birds ringed and/or recovered in breeding season ($n=156$) moved 0–9 km 80%, 10–30 km 12%, 31–50 km 5%, over 50 km 3% (maximum 170 km).

SOUTHERN POPULATIONS: 12 races in southern Europe, Mediterranean islands, Canary Islands, North Africa, and Balkans to Caspian Sea. South-western populations sedentary, though south-eastern ones show limited amount of dispersal in response to colder winter conditions there. In Ukraine, some *candidus* occur (September–March) in city gardens and on steppes south to coasts of Black Sea and Sea of Azov. In Caucasus (*tenuirostris*) and Azerbaydzhan (*poelzami*), some birds make altitudinal movements, descending into valleys for winter (Dementiev and Gladkov 1951a). No ringing data, however.

Food. Mainly insects, but tree seeds often staple diet in winter; bird eggs and nestlings may be common in diet during summer. Climbs trees in search of insects using stiff tail-feathers as prop; may hang upside down from branches but never proceeds head downward. In summer, pokes and probes fissures in bark for surface insects and uses bill as forceps to pull away bark (Steinfatt 1937). In winter, seeks insects in decaying trees mainly by hacking and pecking at bark and wood, knocking off loose material with lateral blows of bill and cutting grooves with vertical blows. Chisels holes up to 10 cm deep to expose wood-boring beetles and larvae (Sielmann 1959). In North African *numidus* (found only in deciduous woods), surface probing for insects is main foraging method throughout the year (Heim de Balsac and Mayaud 1962). Morphological adaptations of musculature and skeleton of head and neck facilitating hammering are described in Jenni (1981). Tongue extends up to 40 mm and harpoon-like tip used to impale soft-bodied prey; harder insects adhere to tongue bristles coated with sticky saliva (Sielmann 1959). Apparently decides on promising places to excavate wood by sound of exploratory pecking on bark; also uses visual cues and will investigate areas showing surface evidence of wood-borers (Baer and Uttendörfer 1898; Letswitz 1904; Sielmann 1959). Excavation of wood restricted to dead and rotting trees, hence little or no damage done to forestry interests (Dementiev and Gladkov 1951a). In Switzerland, winter, over 70% of trees chosen are dead; in April–June, when gleaning insects from foliage (etc.) more important, only up to c. 40% dead wood used (Jenni 1983). May strip bark in search of insects (Steinfatt 1937; Dementiev and Gladkov 1951a); up to 15 birds observed feeding thus on single tree (Bilang 1972). In winter, observed stripping bark of trees infected with fungi *Vuilemina comedens* and *Poria xantha*, to feed on fungal mycelium (Glutz and Bauer 1980). Generally no difference between sexes in foraging sites (Hogstad 1978), though ♀ *numidus* forages more often in tree-crowns and on thin twigs than does ♂ (Winkler 1979). Where sympatric with Three-toed Woodpecker *Picoides tridactylus*, avoids competition by foraging 90% of time on trunk and branches in top half of tree; *P. tridactylus* spends only 20% of foraging time in top of tree and then on trunk only (Hogstad 1971a). Forages on thinner twigs than Syrian Woodpecker *D. syriacus* and Middle Spotted Woodpecker *D. medius* (Glutz and Bauer 1980); for comparison with *D. medius*, see that species and Jenni (1983). Rarely, forages for insects on ground (Pulliainen 1963) and at anthills

(Radermacher 1963, 1965) or searches ground for fruit or nuts (Radermacher 1963, 1968, 1970). In many populations, particularly in northern Europe, conifer seeds important in winter. Cones gathered and taken to 'anvil' for extraction of seeds. This behaviour more highly developed than in other *Dendrocopos* and readily learned by young birds (Muckensturm 1971). 3 types of anvil: (1) occasional anvil—any hard surface on which cone or nut may be placed for working; (2) proto-anvil—natural crevice in bark or wood into which cone wedged; (3) true anvil—anvil prepared by opening natural crevice or by cutting bark or wood with blows of bill in near vertical or steeply angled branch or trunk (true anvils not recorded in other west Palearctic Picinae). Proto-anvils used opportunistically when suitably sized cone to hand, but true anvils visited regularly and most used when specializing on one abundant food source (Glutz and Bauer 1980). Territory may contain many anvils—up to 57 on one tree (Venables 1938)—though usually only small proportion used regularly, e.g. of 32 anvils available one bird mainly used 4–5 (Tracy 1933a). During winter, most cones gathered from high in canopy (Skoczylas 1961; Hogstad 1971a). Cones of pines *Pinus* removed from tree by pecking at base, though small cones may be torn or twisted off (Blume 1977; Pflumm 1979); generally takes 1–2 min (Pulliainen 1963). Bird recorded hanging from hazel *Corylus avellana* nut by feet whilst pecking it free to fly off to anvil with nut in feet (Radermacher 1968). Seeds of hornbeam *Carpinus betulus* plucked from tree with bill in flight (Radermacher 1968; Löhrl 1972). Cones carried in bill to nearest suitable anvil within *c*. 20 m (Glutz and Bauer 1980); once recorded carrying cone impaled on bill (Steinfatt 1937). Exceptionally, seeds of birch *Betula* carried several hundred metres to anvil (Haftorn 1971). Adults appear able to choose most suitable anvil available and fly straight to it (Muckensturm 1971). Cones of *Pinus* and spruce *Picea* usually left in anvil after working; bird alighting at anvil holds new cone between breast and trunk while removing and dropping old one; cones dropped accidentally not retrieved. New cone wedged into anvil upright or at *c*. 45° with several blows of bill, and scales hammered off by downward blows enabling removal of seeds with tongue; cone may be removed and turned several times (Tracy 1924a; Steinfatt 1937; Glutz and Bauer 1980). Variations on this general pattern occur: new cone may be lodged in another crevice while old cone removed (Conrads and Mensendiek 1973); if anvil discovered to be wrong shape, cone removed and anvil modified; very occasionally, cone fitted point downwards or worked on one side only (Blume 1964b). In study where seeds of larch *Larix* main food, cones turned up to 7 times during working—more frequently than with *Pinus* or *Picea*; also, in 85% of cases new cone fitted in to anvil before old cone discarded (Pflumm 1979). Large cones of sugar pine *Pinus lambertiana* observed being worked *in situ* on tree (Turček 1961). Time taken to work on cone depends on its species and size; Scots pine *Pinus sylvestris* 3–5 min (Tracy 1924a, 1928; Røskeland 1931; Pynnönen 1939); black pine *Pinus nigra* 10–15 min (Meijering 1967); sitka spruce *Picea sitchensis* 3·5–12 min; Norway spruce *Picea abies* 6–31 min (Røskeland 1931; Steinfatt 1937; Conrads and Mensendiek 1973). In easily worked cones, e.g. *Pinus sylvestris*, 99% of seeds removed (Pulliainen 1963), and in *Picea sitchensis* only 0–2 seeds per cone left (from average 200 seeds per cone); however, in *Picea abies*, 6–91 seeds per cone left (from average 229 per cone) (Madsen 1972). Anvil technique also used for opening hazel nuts (Sielmann 1959), plum stones (Turček 1950), galls to obtain gall-fly larvae (Tracy 1924b; Pfützenreiter 1957), and for holding large insects (Sielmann 1959), and small nestlings for preparation prior to feeding to young (D Blume). Galls may also be opened by holding in bill and beating against branch until split (Howey 1965). Drills rings of holes round trees to drink sap oozing out, or possibly also to eat exposed cambium of tree (Osmolovskaya 1946; Turček 1954; Jenni 1983) or to feed on insects attracted to sap (Kučera 1972). Most active of west Palearctic Picinae in such 'ringing' behaviour (Glutz and Bauer 1980), though habit rare in Britain (Witherby *et al.* 1938; Gibbs 1983). Holes 3–8 mm wide penetrating bark and cambium into outer 1–2 annual rings (Klíma 1959a). In general, holes 3–4·5 cm apart; rings 9–11 cm apart (Keller 1934). Individual trees often used over many years (Leibundgut 1934); thus one elm *Ulmus* had over 400 holes (Gatter 1972). Ringing starts early March (when sugar-rich sap starts to rise), continuing to April and sporadically thereafter. At first, bird drills low down trunk; later on higher up as sap rises (Osmolovskaya 1946; Turček 1954). Drinks sap as it runs into lower mandible. Visits old holes before drilling new ones, working up tree to drill new ring; old scars often re-opened. 2–3 birds may work tree together (Fuchs 1905; Liénhart 1935; Dementiev and Gladkov 1951a; Blume 1977). Wide variety of trees attacked, though *Pinus* generally favoured. Trees usually young, with trunk about as wide as a man's thigh; larger trees also used, though rings may not go right round trunk but be concentrated on sunny side where sap rises fastest (Klíma 1959a). May prey upon eggs and nestlings of wide variety of birds, particularly species using nestboxes; for review, see Perrins (1979). May enter box by enlarging entrance hole, but apparently locates nestlings by sound since new hole often hacked low in side or base of box (Lönnberg 1936; Hickling and Ferguson-Lees 1959; Glue 1975). Nests of Marsh Tit *Parus palustris* and Willow Tit *P. montanus* in rotten timber particularly susceptible to attack, and over half may be destroyed (Ludescher 1973); see also Vilka (1960). Nestlings taken to perch and hammered with bill (Pring 1929; Blume 1977) or dismembered (Löhrl 1972). Commonly visits feeding tables in hard weather. May hang like tit *Parus* to feed on suspended food (Upton 1962) or pull food up to perch (Glutz and Bauer 1980). Other occasional feeding methods include: opening heads of thistles *Cirsium vulgare* to extract larvae of fly *Urophora stylata* (Campbell 1962); opening stalks of mugwort *Artemisia vulgaris* to extract larvae of fly *Oxyna parietina* (Frantzen 1955); removing milk bottle tops to drink milk (Hinde and Fisher 1951; Roberts 1954; Abro 1964); taking insects in the air by flying from a perch, like flycatcher (Muscicapidae) (Sielmann 1959; Zang 1970; Rezanov 1982); plucking insects from a substrate while hovering (Glutz and Bauer 1980). In winter, northern Europe, forages during all daylight hours (Pynnönen 1939; Pulliainen 1963). In Britain, peaks of feeding on cones recorded 06.30–09.30 hrs and 16.30–17.30 hrs (Tracy 1924a).

Such a wide variety of surface-dwelling and wood-boring invertebrates taken that specific diet appears largely dictated by availability; for detailed lists of species, see Csiki (1905), Madon (1930a), Steinfatt (1937), Pynnönen (1943), and Sevastyanov (1959). Invertebrates include Lepidoptera (e.g. larvae of Cossidae, Notodontidae, Lymantriidae, and Sesiidae, and pupae of *Coerura bifida*), Diptera (e.g. *Oxyna parietina*), Coleoptera (e.g. adults and larvae of Cerambycidae, Scolytidae, Lucanidae, Chrysomelidae, Buprestidae, Coccinellidae, Curculionidae, and Carabidae), Hemiptera (e.g. Aphidoidea, Coccoidea), Hymenoptera (e.g. larvae of Siricidae, Tenthredinidae, Cynipidae; also ants Formicidae—some authors list *Myrmica* in diet, others claim these rejected), mayflies (Ephemeroptera), spiders (Araneae), isopods (Isopoda), millipedes (Diplopoda), centipedes (Chilopoda), earthworms (Oligochaeta), and freshwater mussel *Anodonta cygnea* (Rörig 1905; Collett and Olsen 1921; Masefield 1929; Heim de Balsac 1936; Witherby *et al.* 1938; Dementiev and Gladkov 1951a; Frantzen 1955; Pfützenreiter 1957; Kneitz 1960; Gerber 1965; Radermacher 1965; Fellenberg 1969; De Bruyn *et al.* 1972; Blume 1977; Glutz and Bauer 1980; Jenni 1983). Of conifers, takes seed of pines *Pinus*, spruces *Picea*, and larches *Larix*. Seeds and nuts of deciduous trees include hazel *Corylus avellana*, hornbeam *Carpinus betulus*, oak *Quercus*, beech *Fagus*, almond *Prunus amygdalus*, and walnut *Juglans* (Averin and Nazimovich 1938; Hale 1944; James 1945; Dementiev and Gladkov 1951a; Gerber 1965; Löhrl 1972; Glutz and Bauer 1980). Fruit includes gooseberries (skins not eaten) and currants *Ribes*, cherries *Prunus*, plum stones, juniper *Juniperus*, raspberry *Rubus idaeus*, buckthorn, and ash *Fraxinus* (Turček 1950; Dementiev and Gladkov 1951a; Boorman 1972; Creutz 1976). Ringing recorded on 44 species of tree in Europe; in lowland areas, limes *Tilia*, elms *Ulmus*, and oaks *Quercus* favoured; in upland areas, pines *Pinus* used more (Gatter 1972). Other species include Norway maple *Acer platanoides*, larches *Larix*, spruces *Picea*, yew *Taxus baccata*, and birches *Betula* (Paris 1935; Turček 1954; Glutz von Blotzheim 1962; Kucera 1972; Ruge 1973; Glutz and Bauer 1980). Bird eggs and young most commonly of nest-box, nesting species (e.g. Blue Tit *Parus caeruleus*, Great Tit *P. major*, Coal Tit *P. ater*, Redstart *Phoenicurus phoenicurus*, House Sparrow *Passer domesticus*, Spotted Flycatcher *Muscicapa striata*, and House Martin *Delichon urbica*, according to locality), but also wide range of others including ground-nesting precocial species (Belfrage 1893; Pring 1929; Lönnberg 1936; Schnurre 1936b; Barclay 1938; Witherby *et al.* 1938; Watson 1941; Siivonen 1942; Durango 1945b; Tomasz 1944–7; Pfeifer 1952; Bäsecke 1954; Keil 1954; Oldenburg 1954c; Hickling and Ferguson-Lees 1959; Upton 1962; Hague 1967; Arlebo 1975; Asphjell 1975; Glue 1975; Karlsson 1975; Podolski 1981). According to Pring (1929) eggs only eaten if embryonic; fresh ones left. Eggs removed from traps baited for Jays *Garrulus glandarius* (Rivière 1933b). From feeding tables takes nuts, fat, seeds of sunflower *Helianthus*, honey, sugar–water mix, breadcrumbs, currants, and household scraps (May 1943; Upton 1962; Virkkunen 1967; Soper 1969; Wiehe 1970; Creutz 1976). Other foods recorded include *Carpinus betulus* leaves (Naunton 1976), carrion (Glutz and Bauer 1980), a small rodent, apparently taken alive (Cannings 1972), newly-born squirrel (Haftorn 1971), and milk from bottles (Roberts 1954).

Varied diet contains more plant material than other west Palearctic Picinae, only *D. syriacus* being similar. Most populations mainly insectivorous in summer at least, even in north where seeds important in winter, e.g. Finland (Hogstad 1971b). In southern Europe, animal food predominates at most times of year, e.g. southern USSR (Pomerantsev and Shevyrev 1910) and Tuscany, Italy (Giglioli 1891). Ants (Formicidae) vary in importance; form over 80% of animal food in May, northern Karelia, Finland (Pynnönen 1943), but less important in central Europe (Gösswald 1957–8). Plant food, principally conifer seeds, important to some extent for most birds, but more so in more northern populations. In one Finnish study of winter feeding, 42% of foraging time spent opening *Pinus* cones, 11% opening *Picea* cones, 34% apparently seeking insect prey on trunks and branches, 8% at feeding tables, and 5% elsewhere (Vikberg 1982). Finnish birds may forage on green cones July–May. Coniferous seeds form 28–73% of diet August–October, and 100% from November to March or April; few eaten May–July. In Estonia (USSR), conifer seeds form 45–63% of diet by volume October–February, 100% March, and 38% April. (Pynnönen 1943.) In southern Norway in good seed year, 40–50% by volume in October, *c.* 80% November, 80–100% December–February and *c.* 60% March; in poor seed year, 10–25% October, 30–40% November, *c.* 80% December–February, and 50–60% March (figures from stomach analysis and observation: Hogstad 1971b). 21 Swiss stomachs collected throughout the year showed predominance of plant food between September and early April (Glutz von Blotzheim 1962). In France, diet 70% seeds during winter (Madon 1930a). In one locality in West Germany, seeds of *Carpinus* staple food over winter (Löhrl 1972). Elsewhere in East and West Germany, seeds less important and erratically used (Rörig 1900, 1905; Rey 1908, 1910; Baer 1909). Monthly proportions of seeds in diet, Moscow region, USSR (68 stomachs): January–February 100%, March 90%, April c. 10%, May–July c. 5%, August 20%, September 30%, October 25%, November–December 100%; remainder of diet ants, beetles, and other insects (Inozemtsev 1965). For details of stomach contents, May–July, Kirov region (USSR), see Korenberg et al. (1972). Importance of sap little understood, being impossible to sample in stomach; rich in sugars and proteins, however, and must be at least locally important. In one study, March–May, birds spent up to 33% of foraging time taking sap (Osmolovskaya 1946).

Various estimates of quantities of conifer seeds eaten per bird per day during winter when presumed feeding exclusively on cones. *Pinus*: 850–1700 seeds per day, i.e. up to 7–8 g (Pynnönen 1939, 1943; Franz 1943; Turček 1961). *Picea*: 33–51 *P. sitchensis* cones per day, i.e. 7440–10763 seeds (average 9·1, maximum 13·8 g) (Madsen 1972); generally 3 (up to 9) *P. abies* cones per day, i.e. c. 1440 seeds (11·3 g) (Conrads and Mensendiek 1973). *Larix*: observations suggest could eat up to 7000–8000 seeds per day; minimum 5000–6000 estimated necessary for daily food requirement (Pflumm 1979).

Food brought to young in bill; usually up to 4 items at each visit, occasionally 10. In Rominter Heide (Poland/USSR), during plague year of moth *Lymantria monacha*, 2347 prey items recorded in 665 feeding visits: 2209 *L. monacha* larvae, 2 *L. monacha* pupae, 11 other Lepidoptera larvae, 41 Coleoptera larvae, 6 *Phyllobius arborator*, 29 unidentified winged insects, 22 dragonflies *Aeshna*, some larvae of fly *Parasitigena segregata*, 1 *Tipula*, 1 *Haematopota pluvialis*, 4 spiders, 6 small snails, 2 wasps, 4 grasshoppers *Barbistes constrictus*, and 4 eggs of Pied Flycatcher *Ficedula hypoleuca* with large embryos (Steinfatt 1937). At one nest in Norfolk (England), nestling diet mainly small dark gall flies and some Lepidoptera larvae (Tracy 1933b). Detailed study in Korea using nestling collars (Won et al. 1968) showed 78% of diet Lepidoptera and Hymenoptera larvae, 9·4% Lepidoptera adults, and remainder spiders, molluscs, birds' eggs, and some vegetable matter. In Switzerland, young given 85% (by weight) Lepidoptera larvae, 2% beetles, and 9% harvestmen; rest chiefly ants *Lasius* and aphids (Jenni 1983, which see for comparison with *D. medius*). In West Germany, Sielmann (1959) recorded Lepidoptera larvae and cockchafers *Melolontha melolontha*. In southern USSR, moth larvae brought to nest each day, up to 137 *Malacosoma neustra* and 886 *Lymantria dispar* (Korol'kova 1954). Nestlings of other birds (Hodgetts 1943; Löhrl 1972) and cranberries (Pynnönen 1943) also recorded. In Rominter Heide, first feeding visit usually c. 05.10 hrs and last c. 19.30 hrs; feeding rate fairly constant from the outset at c. 8–10 visits per hr (5–17 in first 7 days, 1–20 from 11–23 days); ♂ usually made most visits (Steinfatt 1937). In 3 periods when young 7–8 days old, Norfolk: 11.15–12.00 hrs 16 feeds (7 ♂, 9 ♀); 16.46–17.16 hrs 13 feeds (7 ♂, 6 ♀); 05.02–05.32 14 feeds (10 ♂, 4 ♀) (Tracy 1933a). 43 visits over one 7-hr period, and on another day a visit every c. 7 min; feeding peaked in morning and afternoon with lull at midday (Sielmann 1959). ASR

Social pattern and behaviour. Main studies by Steinfatt (1937) in Rominter Heide (Poland/USSR), Bussmann (1946) in Switzerland, Blume (1958, 1961) in Hessen (West Germany), and Simkin (1976, 1977) in Moscow region (USSR). For summary, see Blume (1977). Following account based on detailed outline from D Blume.

1. Mostly solitary though aggregations occur where feeding conditions favourable: e.g. near Magdeburg (East Germany), mid-January, c. 30–40 in small area and 12–15 simultaneously on one tree (Bilang 1972). Large aggregations sometimes recorded during eruptions of nominate *major*: e.g. numbers peaked at c. 10000 on Säppi island off Pori (Finland), mid-August (see Bergman 1971b and Movements); in such cases, small flocks sometimes formed and maintained for onward passage (Pulliainen 1963; see also part 2). In sedentary populations, outside breeding season, ♂ and ♀ in separate home-ranges or parts of one. Size of home-range of single bird outside breeding season dependent on type of woodland (tree height and girth): varies from 2 ha in mature oak *Quercus* and hornbeam *Carpinus* to 25 ha in 120-year-old *Quercus* and pine *Pinus*. Bird may range up to c. 1 km from roost in winter (Blume 1961, 1977; D Blume); see also Simkin (1976) who reported 'winter territories' of only 0·25–0·8 ha in area of pinewoods and orchards around 2 villages in Moscow region. In roosting area with several holes, 2 or more home-ranges may overlap; up to 12 birds recorded roosting in wood of c. 150 ha (Blume 1961, 1965a; D Blume). In autumn and winter, clusters of several home-ranges of ♀♀ occur without a neighbouring ♂ (see map in Blume 1961). Some studies suggest territorial limits defended in autumn–winter: e.g. in Finnish pinewoods where several territories sometimes contiguous, birds apparently recognize which side of any border tree belongs to their territory (Pynnönen 1939); in Rantasalmi (Finland), October–November, ♂ defended feeding area against 2 others and territory had (at least on one side) clearly delineated boundary (Pulliainen 1963). For further details of territoriality, see Breeding Dispersion and part 2. Feeding association with small birds in autumn and winter varies regionally: in montane and subalpine woods of Alps, large mixed flocks may follow a foraging *D. major* (see Glutz and Bauer 1980). BONDS. Monogamous mating system. Pair-bond lasts for at least one season, i.e. from late January into summer (July). Bond maintained for c. 2½ years in case of isolated pair (Curry-Lindahl 1961b); probably due to attachment to territory rather than to mate (D Blume). In Moscow region, several cases of pair-bond lasting up to 3 years, though one ♀ who showed fidelity to home-range for c. 5–6 years (see above), changed partner 4 times in as many years (Simkin 1977). Former partners sometimes maintain contact outside breeding season, and may feed together (Glutz and Bauer 1980; see also Simkin 1976); in one study of colour-ringed birds, ♂ and ♀ had separate feeding areas and few if any reciprocal visits occurred (D Blume). Aggression also generally likely if birds come too close and ♂ apparently dominant over ♀. Antagonistic encounters occur in evening when roost-holes (re)occupied after breeding season; later, birds make occasional vocal contact, and roosting close together in neighbouring trees not uncommon (Blume 1977; D Blume). Occasionally hybridizes with Syrian Woodpecker *D. syriacus* (see that species), White-backed Woodpecker *D. leucotos* (Curry-Lindahl 1961b), and White-winged

Woodpecker *D. leucopterus* (Vaurie 1959b). Both sexes excavate nest-hole, incubate, and tend young (brooding, feeding, nest-sanitation). Breeding can succeed (though possibly only 1–2 young raised) if ♀ lost, because ♂ incubates and broods at night; ♀ probably cannot cope if ♂ lost. Brood split for post-fledging care which recorded as late as August, though normally lasts only c. 1–2 weeks. Not infrequently 1 parent (usually ♀) stops feeding young for last part of nestling phase and takes no further part in care of young (D Blume; see also part 2). Age of first breeding not known. BREEDING DISPERSION. Solitary and territorial. Pair occupy home-range of 4–60 ha (Kneitz 1961; Ferry and Frochot 1965; Blume 1977). In Russian study, 'breeding territories' c. 0·32–0·8 ha ($n=9$) (Simkin 1976); in Finland, c. 20 ha typical and territories rarely contiguous (Pynnönen 1939). Exceptionally, Brandenburg (East Germany), 1 pair in field copse of c. 1 ha. Shortest distance between 2 occupied nests c. 50–150 m (Rutschke 1983); in Oberlausitz (East Germany), exceptionally c. 25 m (Creutz 1976). Various Swiss and West German studies show highest densities (c. 2 pairs per 10 ha) in mature *Quercus–Carpinus* woodland (Jenni 1977, which see for other references; see also, e.g., Teixeira 1979). For higher density in damp woodland of same composition, Niedersachsen (West Germany), see Schumann (1973a). In Mecklenburg (East Germany), ancient alder *Alnus* favoured, with density of 0·9–5·1 pairs per 10 ha (Klafs and Stübs 1977, which see for densities in other types of woodland). In Brandenburg, density in pure pine plantations and natural pinewoods usually only c. $\frac{1}{5}$–$\frac{1}{10}$ of broad-leaved woods (Rutschke 1983). Above-average densities occur locally in (e.g.) parks with stands of mature timber: e.g. in Heinrich-Laehr-Park (West Berlin), 13–14 pairs on 27·3 ha, i.e. 4·8–5·1 pairs per 10 ha (Matthäs and Schröder 1972); see also Simkin (1976, 1977). Few larger-scale censuses undertaken: in Murnauer Moos (Bayern, West Germany), 14 pairs on 41·7 km^2 (Bezzel and Lechner 1978) and in Danube valley, Regensburg to Straubing (Bayern), 32 pairs on 19 km^2; in south-west Drenthe (Netherlands), 106 pairs on 165·4 km^2, with wooded area of 35·88 km^2 (Glutz and Bauer 1980). In Moscow region, c. 170–200 birds on 6 km^2 (Simkin 1976). Birds show marked fidelity to home-range once chosen; use of same breeding area recorded over 6–8 years (Simkin 1976; Blume 1977; Glutz and Bauer 1980). Home-range usually has parts particularly favoured by ♂ or ♀, though some overlap. Contains certain key sites: areas for calling, Drumming, etc., trees with holes for nesting and roosting, and favoured feeding areas. Outside breeding season, most territorial defence centres on roost and feeding sites—see above for concentrations and evident tolerance where food supply good. In breeding season, (prospective) nest-holes with nearby Drumming-posts (often on prospective nest-tree) more important, and defence more or less confined to immediate vicinity of nest-hole during incubation and nestling period (Glutz and Bauer 1980; see also Antagonistic Behaviour, below). Location and structure of territory often determined by terrain (mountain slopes, strips of woodland, rides), so that approximate limits, areas for calling, Drumming, etc., and breeding, often used traditionally over many years despite change of ownership (D Blume). In breeding season, more especially during nestling period, birds generally move within c. 100–600 m of nest (up to 1100 m recorded: Glutz and Bauer 1980), but pair-members sometimes retain their own separate parts of home-range (Blume 1961; also, e.g., Schuster 1935, Steinfatt 1937). ROOSTING. Solitary and nocturnal; rarely, up to 3 birds in different holes in same tree (Nyholm 1968). Mostly in old nest-holes of own species or other Picinae (D Blume) or, more frequently than other Picinae, in nest-boxes (Simkin 1977; Glutz and Bauer 1980; see also Vauk 1964 for migrants on Helgoland, West Germany). Very rarely in crevices in rocks or buildings, or in the open on tree-trunk (Blume 1977; D Blume). Hole may be excavated (taking up to c. 5 days) specifically for roosting (e.g. Tracy 1938, Bagnall-Oakeley 1972). Hole-excavation coincides with resurgence of territorial activity in late summer and autumn; roost-holes perhaps more rounded than typical nest-hole which may have rather more pointed lower rim (Glutz and Bauer 1980). Pecking at entrance-hole and clearing out of nest-boxes reported by Simkin (1977). As in Black Woodpecker *Dryocopus martius*, birds may have emergency holes to which they fly when disturbed. Sites used for months or years, though bird will sometimes use several within a home-range (Blume 1977). As in other Picinae, ♂ roosts in nest-hole during breeding season; ♀ may also roost in nest-tree during this period if other suitable holes available there. Towards end of nestling phase, ♂ may be forced out of hole at night and then spends some time clinging to trunk nearby; usually moves to another roost c. 2 days before fledging. Fledged young normally roost clinging to tree-trunk in the open for first few days but soon change to holes, usually near that of attendant parent; several may roost in separate holes of same tree (Blume 1965a, 1977; D Blume). Birds perhaps sometimes enter holes to die (Deicke 1957; Blume 1977). Normally active from sunrise to sunset, though period of activity changes less through year than length of daylight; end of activity varies more than start. As in other Picinae, ♀ generally active longer than ♂; in late summer and autumn, ♀ may arrive at roost up to 20 min later than ♂; activity phases of sexes become similar in spring (for full details of activity patterns through the year, see Blume 1961, 1963b; 1964a; see also, e.g., Nyholm 1968 for influence of geographical latitude). Peak activity normally in first hours after leaving roost (e.g. Conrads and Mensendiek 1973). Traditional routes used for flights to and from roost (Blume 1964a). Prior to entering roost in evening, bird typically climbs and flies about, performs Excitement-pecking (see Voice) and gives Contact-calls (see 1 in Voice), alternating with dozing in resting posture in which plumage ruffled and head slightly withdrawn (see illustration in Glutz and Bauer 1980). Transitional phase longer and more conspicuous in new arrivals (especially young birds) than in established older individuals; may last c. 30(–60) min over several weeks until bird familiar with area. Duration of immediate pre-roosting activity becomes shorter from autumn through to winter, though may be longer again in clear, frosty weather. Bird normally flies to particular tree, climbs and pecks, calls from exposed perch, approaches in short flights through canopy, climbs down roost-tree or nearby tree, and makes short flight to hole; briefly alert before entering and also peers out briefly after doing so. In one case, ♂ flew regularly up hill in evening then made steep descent to roost in village garden at foot of hill (Blume 1961, 1963b, 1964a, 1977; D Blume). Bird sleeps on floor of cavity—particularly in nest-box (e.g. Berndt 1960) or clings to cavity wall beneath entrance, head between body and right wing and (at least when cold) feathers of lower back ruffled and covering wings. In morning, first looks out alertly for a while (up to 15–20 min recorded: Simkin 1977), then flies to nearby tree, preens, climbs about, gives Contact-call, and gradually moves away from roosting area. Particularly in winter, may cling to tree by roost-hole for some time (Blume 1964a). For detailed description of comfort behaviour, including bathing and anting, see Gebauer (1982).

2. When threatened, may climb in spirals on tree or freeze on side facing away from source of danger, though will also peer out intermittently from that position (Hammling and Schulz 1911; D Blume). Birds that have escaped from some sort of danger often draw attention through increased vocal activity for some days afterwards. Distress-call (see 6 in Voice) given when pulled out of roost-hole by *D. martius*, also when handled, in

which situation tail characteristically bent forwards and bird may peck and bite fingers (see illustration in Blume 1977). Particularly after dusk, autumn and winter, bird likely to stay in roost-hole even in serious disturbance; Wing-rattling (see Voice) sometimes occurs. Will avoid roost-hole on hearing grumbling sound of edible dormouse *Glis glis*, though also recorded roosting in same tree as this potential predator (Blume 1964a, 1977; D Blume; see also Parental Anti-predator Strategies, below). Often attracted by alarm-calls of various passerines when these mobbing predator; will then join in, using Contact-call (Fleuster 1973; Glutz and Bauer 1980). FLOCK BEHAVIOUR. In small flocks (up to 5), Finland, late September, birds indulged in chases (aerial and on trees) but such behaviour did not disperse flock (Pulliainen 1963). ANTAGONISTIC BEHAVIOUR. (1) General. Aggressive interactions typical of social behaviour throughout the year, more so than in Middle Spotted Woodpecker *D. medius* or Lesser Spotted Woodpecker *D. minor*. In central Europe, most threat and fighting occur mid-March to mid-June and mid-July to mid-October; occasional in January. Aggression more pronounced towards conspecific birds of same sex than those of other sex, and most evident with occupation of breeding territories and defence mainly of certain key sites within them (see Breeding Dispersion, above). In response to acoustic signals, bird makes searching flights with low arcs over sound source; also chases any overflying conspecific birds (Glutz and Bauer 1980; D Blume). For territorial behaviour in population with unusually high density, see Simkin (1976, 1977). Drumming (for contexts, see Heterosexual Behaviour and Voice) given in relaxed posture with plumage often loose or slightly ruffled; bird may self-preen in pauses but alert for predators or Drumming by another conspecific bird. Head raised briefly before 1st strike of series then lowered at right angles to substrate and Dumming given. In dispute over roost-hole, as may occur between ♂ and ♀ of former or prospective pair, bird in hole normally victorious—though this not the rule in interspecific disputes involving larger Picinae (D Blume); species- and sex-specific differences in diurnal activity patterns reduce likelihood of aggression in roosting area (see Roosting, above). (2) Threat and fighting. Following behaviour patterns have threat function, given in order of increasing aggression: calls (5, 4, 3, and 2: see Voice), short bouts of Drumming, driving other bird up tree, pursuit-flights with buffeting, and damaging fights. Fights involve pecking with bill slightly open, biting, and becoming interlocked in flight (D Blume). As in *Picus* woodpeckers, damaging fights rather uncommon, birds using gestures of threat and defence in attempts at intimidation (Glutz and Bauer 1980). Physical contact may be preceded by Bill-up posture (see illustrations in Blume 1958) adopted on vertical or horizontal surface; plumage sleeked, head held forward with bill slightly above horizontal, and tail slightly spread. Opening of bill and ruffling of crown feathers also occur; in ♂, only feathers of red hindcrown ruffled in excitement (Winkler 1972a). Often followed by slightly ritualized, jerky Threat-swaying in which bill thrust forward while forebody moved simultaneously from side to side; typically accompanied by Appeasement-call (see 5 in Voice) with rapid spreading and quivering of wings and spreading of tail (also used interspecifically: see Fig A); for further details, see Eygenraam (1947), Dancker (1958), Smith (1960), Werner (1961), and Glutz and Bauer (1980); see also Voice for other calls and Wing-rattling given during disputes. Antagonistic encounters between rival ♂♂ on territorial boundaries often take place low down on trees or stumps. Each attempts to drive other upwards and thus to chase it off. ♀ less involved in such boundary disputes, but will attack another ♀ vigorously if such a bird approaches her nest-hole or mate; ♀♀ give short Drumming series in pauses between bouts of fighting. ♂ may not participate

A

in these fights (Tracy 1938; Blume and Jung 1956; D Blume). Protracted and noisy fights typically occur between territorial pair and intruding bird (usually strange or neighbouring ♂); such encounters also recorded in other *Dendrocopos* (Lawrence 1967). In pauses between fights, birds perform Excitement-pecking. Scars from such activity obvious on boundary trees or stumps and (like marks created by repeated landings near holes or anvils) possibly serve as visual signals. In conflict situations, in breeding season, birds also perform Tapping and quiet Drumming (see Voice), as well as scratching or Mock-preening mainly on side closer to rival (Blume 1977; D Blume). Antagonistic and conflict responses as described can be elicited experimentally using stuffed specimens (see Blume 1958b for illustrations). Response aggressive when model close to roost- or nest-hole, except when latter contains nestlings more than 12 days old—adults then make strenuous efforts to feed model; rarely, disregard it (see Blume 1977). Experimental playback of 3 short Drumming series and following Contact-calls elicited excited searching flights in both birds of pair, similar Drumming, and Contact-calls changing to Scolding-calls (see 2 in Voice). Unpaired ♂ tends to Drum from favoured site(s) and may also give Kreck-calls (D Blume: see 3 in Voice). For further experimental work with playback of Drumming, etc., see Zabka (1980). *D. major* normally shows aggression towards other *Dendrocopos* in breeding and winter feeding territory (including at bird tables, etc.). At feeding stations will fly at and chase off *D. minor* and Grey-headed Woodpecker *Picus canus*, also tits *Parus* and House Sparrow *Passer domesticus* (see Radermacher 1970, Blume 1977), but some *D. major* wait at a distance until competitors gone (D Blume). Interspecific territoriality exists with *D. syriacus* where the 2 species breed sympatrically (Winkler 1973). In optimal habitat of *D. syriacus*, that species dominant (Keve 1962c); elsewhere, the reverse true (Blume 1977). Main nest-hole competitors are tits, Nuthatch *Sitta europaea*, and, especially, Starling *Sturnus vulgaris*. *D. major* more likely to be successful shortly before laying; removes nest-material and takes possession. In excavation phase, usually successful in driving off any *S. vulgaris* within 50 m of hole, but if *S. vulgaris* manages to place material (often twig with leaves) in completed hole (see Blume 1973), *D. major* moves away to excavate new hole or occupy another—may delay fledging by c. 3 weeks. Once, *S. vulgaris* took over nest of *D. major* containing young; *D. major* less likely to be molested in darker sites (Blume 1977; D Blume, C C Brown). HETEROSEXUAL BEHAVIOUR. (1) General. Aggressive component evident in early stages of relationship between pair-members; possibly connected with defence of roost-hole and own territory conflicting with desire to offer hole as prospective nest-site. Certain antagonism also apparent in later stages, e.g. when both birds arrive simultaneously at nest to feed young: adopt tense posture, and briefly chase one another around trunk; ♂ usually has priority (D Blume). As in other Picinae, sexual contact possible only after long process of synchronization and sexual stimulation. Timing of pair-formation thus difficult to ascertain (Glutz and Bauer 1980). (2) Pair-bonding behaviour. First contact established through Drumming (most important contact signal but initially sporadic and low-intensity: see Voice); given by both sexes, occasionally in December, more often from mid-January. Drumming typical of bird quitting winter territory and moving into another (often nearby) (D Blume). In USSR study, breeding partners found normally to have previously occupied neighbouring home-ranges, pair-bond then sometimes lasting several years (see Bonds). In January, all Drumming (including duets) performed by ♀♀; later, ♂♂ Drum in duet with all birds except future mate (Simkin 1976, 1977). Apart from Drumming, birds also call from tree-tops. In shared roosting area, closer contact more likely at time when sex-specific difference in arrival and departure times (see Roosting, above) least marked: in central Europe, February–March. Initially, aggressive behaviour (see above) includes pursuit-flights, driving one another up trunk, and chases through canopy. For apparent display-flight with wings raised, see Hogg (1975). Closer approach, reduced aggression and sexual stimulation gradually more a feature of day-time activity and concentrated on trees with holes. Birds of a pair may use same Drumming site alternately. Fluttering-flight (Fig B) performed at this stage: tail cocked and widely spread (may appear forked); noticeably shallow wing-beats create impression of fluttering or whirring; normally accompanied by Appeasement-calls (Richter 1950; Berndt 1954: see 5 in Voice), sometimes also louder Kreck-calls (König 1951: see 3 in Voice). Given on sighting (prospective) partner (may then be performed over up to c. 100 m), when both approach or leave (e.g.) branch used for copulation, or Drumming post; also given by bird being chased (Blume 1962c, 1977; Glutz and Bauer 1980; D Blume). (3) Nest-site selection. Both sexes (♂ probably more than ♀) advertise incomplete or completed hole normally as follows: one bird Drums (in long series) and other approaches; advertising bird performs Fluttering-flight (also in response to playback of Drumming) from Drumming post to hole, then Taps demonstratively on hole-rim; other bird attracted nearer and both give Appeasement-calls; hole-owner briefly chases other and drives it up tree-trunk; further Nest-showing with Tapping follows (D Blume; see also drawing in Blume 1973). Repeated intention movements to enter hole may also play a part in Nest-showing— white scapular-patches and ruffled red vent feathers prominent (Blume 1961). If other bird attracted closer, will normally enter hole to inspect it. In two cases, ♂ offering too small a hole (winter roost) visited by different ♀♀ over several days, but none stayed; unpaired ♂ continued Drumming into June. Such birds give long series from several favoured sites in a small area. Both birds may

C

continue to advertise at separate trees in the future breeding territory and work on holes independently until final choice made. Details of selection procedure not known; final choice may not be evident until laying, but ♂ sometimes roosts in nest-hole beforehand. (4) Courtship-feeding. Not recorded. (5) Mating. Recorded before final nest-site selection or beginning of excavation (Pynnönen 1939), but mainly during building, also later during nest-reliefs (e.g. after 6th day of incubation: Steinfatt 1937). Birds usually fly together from nest-tree to suitable, even traditional, branch. ♀ solicits copulation with 3 short, rapid Drumming series (Tracy 1938; Pynnönen 1939; D Blume), Fluttering-flight, or by crouching crossways on branch. After mounting, ♂ holds wings stiffly spread (may flap them for balance: Steinfatt 1937) and then tips over sideways for cloacal contact. Copulation lasts c. 3 s (Steinfatt 1937), accompanied by Appeasement-calls from both birds (see 5 in Voice). ♂ may fly off immediately afterwards while ♀ remains on branch, or both adopt posture shown in Fig C (Steinfatt 1937; Pynnönen 1939; Bussmann 1946; Winkler and Short 1978; D Blume). Birds copulate c. 6 times per day shortly before and during laying, 2–4 times per day early in incubation (Pynnönen 1939). (6) Behaviour at nest. One bird usually near nest before laying and regular change-overs may take place as later (Glutz and Bauer 1980). Incubation proper possibly begins c. 2 days before (Steinfatt 1937) or only when clutch complete (Pynnönen 1939; D Blume). Shifts usually quite short initially (from a few up to c. 30 min), average 11 min (Steinfatt 1937); later, c. 40–50 min. Incubation mostly by ♂ who also sits at night (D Blume). Longest daytime shift by ♂ 121 min, by ♀ 55 min (Durango 1945b). Bird approaching nest for change-over (excavation, incubation, or brooding) gives Contact- or Kreck-call, or Taps; lands on nearby tree, gives Appeasement-call, and flies to nest-tree, landing below and to side of hole, there giving more Appeasement-calls. Other bird leaves and often gives loud series of Kreck-calls some distance from nest. Unlike in *D. martius*, ceremony not strongly ritualized (D Blume; for further details of variations, etc., see Blume 1961, 1977). If mate absent for short period or disappears completely, remaining bird calls and Drums. If ♀ lost during incubation or nestling phase, ♂ will Drum (with or without food in bill) near hole (Blume 1965a, 1977; D Blume). RELATIONS WITHIN FAMILY GROUP. Young call some time before hatching (Ruge 1971b). Eggshells (not necessarily on 1st day) and dead nestlings carried up to c. 40 m away (Steinfatt 1937). Young very sensitive to cold; form 'warmth pyramid' (see *D. martius*) when left alone during first few days (D Blume). Brooded for c. 12 days (Blume 1977) in shifts of c. 10–15 min (Steinfatt 1937); ♂ continues to roost in nest-hole up to c. 2 days before fledging (D Blume). Incubating or brooding adult suffering from heat will poke head out with feathers ruffled and pant with bill wide open, doze in entrance, or leave to self-preen or feed (Haverschmidt 1938; Glutz and Bauer 1980). For physical development of young, see (e.g.) Pynnönen (1939) and Bussmann (1946). Stimuli eliciting increased

B

begging much as in other Picinae (see, e.g., Green Woodpecker *P. viridis*). At least 1 nestling sitting below entrance-hole calls persistently when no adult present (Heinroth and Heinroth 1924–6), and brief increase in volume indicates sibling taking over, though whole brood may also call simultaneously (D Blume). If nest-tree felled or topples over and young survive, they continue to beg and be fed by adults (Weigelin 1960; Schaack 1967). Call less persistently or not at all (except when fed) later in nestling period (Steinfatt 1937). At least at start and in middle of nestling period, ♂ and ♀ take roughly equal share in feeding young (e.g. Stahlbaum 1959); towards end, one bird (usually ♀) may do much less. Unlike *Picus* or *Dryocopus*, food brought to young in bill; nestling takes food with rapid snapping and back-and-forth head movements. Fed at entrance from *c.* 16–18 days (Blume 1961, 1977; D Blume). Nest cleaned after about every 4th feed (Stahlbaum 1959). Faeces (often with covering of wood-shavings) carried away and dropped or deposited on branch *c.* 50 m from nest. Young lean out of nest and give adult-type Contact-calls from *c.* 2 days before fledging (D Blume). Older nestlings can be extremely aggressive, e.g. when jostling for position at nest-entrance; some pecking attacks (commonly directed at head) end fatally. Smallest of brood often tyrannizes siblings (Sielmann 1959). Young fledge at *c.* 20–24 days old (up to 27 days if weather or feeding conditions unfavourable), all on same day, or often over 2–3 days. May be encouraged to leave by parent waiting nearby with food, or be pushed from behind by sibling. Make short, rather clumsy flight to bottom of nearby tree or, particularly weaker birds (which are then often quickly taken by predators), to ground. Peck and tongue-dart frequently and do some self-feeding from 1st day, but give adult-type Contact-calls (also as a food-call) and are still fed by parents; rather squeaky sounds with wing-flapping given during feed. Part-family unit (see Bonds) roves through territory for *c.* 8–10 days (loose bond recorded until August: Blume 1965*a*). Even from 1st day, increasingly towards end of 1st week, parent may threaten young with wing-spreading and driving up tree (Fig D); fights occur in roosting area in evening. Siblings may roam territory together for some time after independence (Blume 1961, 1977; Glutz and Bauer 1980; D Blume), moving further away from general vicinity of nest-hole after *c.* 3 weeks (Skoczylas 1961). ANTI-PREDATOR RESPONSES OF YOUNG. Give louder, different-sounding food-calls at sound of thunder or human voices (Steinfatt 1937). Larger young more or less cease calling on hearing Scolding-call of adult (see 2 in Voice); in serious disturbance, crouch silent in nest. At *c.* 12 days, dig claws into hand and give loud Distress-call if removed from nest; at *c.* 14 days, attempt to hide in darkest corner if nest-hole opened up (Bussmann 1946; Poulsen 1949; D Blume). PARENTAL ANTI-PREDATOR STRATEGIES. (1) Passive measures. Adults generally quick to adapt to disturbance at nest; ♂ bolder than ♀ (Bussmann 1946). When bird prevented from feeding own young (because man by nest), recorded flying to another nest (*c.* 60 m away) and feeding young there (D Blume). Will desert nest-hole in any competition with *Glis glis* which will also take eggs of (D Blume). (2) Active measures. Slight disturbance of bird in breeding territory often elicits Drumming (Glutz and Bauer 1980). Dive-attacks on man recorded (Bussmann 1946). If man near nest, birds (sometimes both pair-members simultaneously) give Scolding-calls in alarm. When brood threatened by marten *Martes martes*, adults give shrill Contact-calls from safe height, but do not attack; will, however, follow predator though remain high in canopy (D Blume). When Jay *Garrulus glandarius* near nest, birds called for *c.* 20 min, ♂ then followed *G. glandarius* for *c.* 60 m as it retreated (Steinfatt 1937). In experimental study, vigorous attacks made on stuffed sparrowhawks *Accipiter* (Kneitz 1965*b*). For vocal and other responses when disturbed in nest, see Voice.

(Fig A from photograph in Smith 1960; Figs B and D from drawings in Blume 1977 and sketches by D Blume; Fig C from drawing in Bussmann 1946 and sketch by D Blume.) MGW

Voice. Vocal throughout the year, particularly in breeding season but also in autumn territorial disputes and fights over roost-holes. In winter, increased vocal activity often associated with feeding on cones and work on anvils (see Food). Both sexes share same repertoire, including instrumental signals which, however, are less intense and less frequently used by ♀. Individual calls short and monosyllabic, hard and clicking; given also in looser or tighter series, depending on excitement. Various multi-unit calls or sequences have rattling or muttering quality. Instrumental signals (see below) given as single series or in short to quite long bouts; also combined with calls. Many descriptions of vocal and instrumental signals: see, especially, Steinfatt (1937), Tracy (1938), Pynnönen (1939), Blume and Jung (1956, 1958), Langelott (1957), Blume (1961, 1965*a*, 1977), Winkler and Short (1978), Zabka (1980); for recordings, see Blume *et al.* (1975), and for extra sonagrams, Glutz and Bauer (1980) and Bergmann and Helb (1982). Account based on outline supplied by D Blume.

INSTRUMENTAL SIGNALS OF ADULTS. 5 types occur. (a) Drumming. Given mainly mid-January to end of June, sporadically in September. Various functions, depending on season: sexual recognition and pair-synchronization, Nest-showing, territorial advertisement and spatial dispersion (both individuals and pairs of conspecific birds as well as other Picinae) (Zabka 1980; D Blume). Suggested by D Blume that number of series given per min perhaps more important for specific identity than number of strikes and duration of series. Apparently able to distinguish Drumming of own species from Black Woodpecker *Dryo-*

D

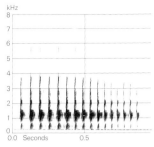
I W Tilgner West Germany April 1974

II P J Sellar Sweden May 1978

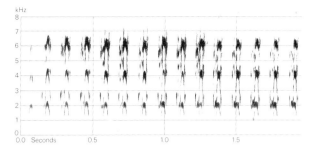
III W Tilgner West Germany October 1980

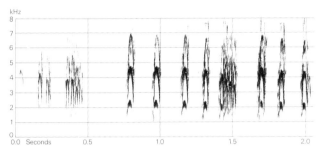
IV W Tilgner West Germany June 1980

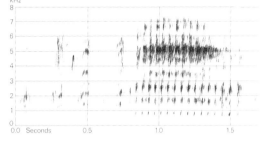
V W Tilgner West Germany October 1980

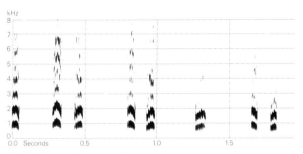
VI W Tilgner Austria April 1982

copus martius (Sielmann 1959), but will Drum in response to that of other *Dendrocopos* (including duet with Lesser Spotted Woodpecker *D. minor*), Three-toed Woodpecker *Picoides tridactylus*, Grey-headed Woodpecker *Picus canus*, and *D. martius* (Winkler and Short 1978; Zabka 1980). ♀ gives shorter series than ♂ (W Tilgner); probably important (possibly the major factor) in sexual recognition (Zabka 1980). Series of both sexes said to become shorter (W Tilgner) or longer (Wallschläger and Zabka 1981) through breeding season (see also below). For further indications of possible geographical variation, see Winkler and Short (1978); for diurnal variation, see Pynnönen (1939) and Langelott (1957). In antiphonal duetting, ♀ starts (Wallschläger and Zabka 1981). Favoured sites for Drumming are dead branches or (mostly top of) trunk; also not uncommonly wooden or metal poles, weather vanes, metal parts of roofs, ceramic insulators, etc. (Glutz and Bauer 1980). Silent Drumming occasionally reported; bill vibrates freely, not touching substrate (Zabka 1980; D Blume). 3 forms of Drumming distinguished (Blume 1977; Glutz and Bauer 1980). (i) Single, usually dull-sounding, also irregular, series. Often shorter than type iii, with pauses of irregular length between strikes. Bird Drums where it happens to be perched, even on thick trunk offering no resonance. Given in various conflict situations (Blume 1958b, 1977), also by either sex on arrival at tree with hole during flight through territory early in season (D Blume). (ii) Bouts of 3(–5) series 1–2(–2·5) s apart, often with long pauses between bouts. Usually energetic and hasty; short bout not uncommonly followed by Contact-call (see below). Bird often Drums thus on short, horizontal, dead branch. Mainly associated with antagonistic and antagonistic–heterosexual interactions between ♀♀ or between ♂♂ (Langelott 1957; Blume and Jung 1958; Blume 1977; Zabka 1980). (iii) Long bouts. Loud, particularly at start, and audible up to c. 800 m. Accelerates markedly through series (from c. 62 to 40 ms between strikes) and fades slightly towards end (Winkler and Short 1978; Glutz and Bauer 1980; Zabka 1980). Usually 10–16 (5–20) strikes per series (Bergmann and Helb 1982). Average 13·11 strikes per series, average duration of series 0·56 s, $n = 104$ series by 33 birds (Zabka 1980). Recording (Fig I) shows 16 strikes in 0·8 s (18 strikes per s); recording by P J Sellar of bird Drumming on metal lamp standard shows 18 strikes in 0·77 s (c. 20 per s), substrate possibly assisting faster rate (J Hall-Craggs) though temporal structure not affected by substrate according to Zabka (1980). Usually given on main Drumming post for quite long period, depending on time of day. At beginning of breeding season (about February in central Europe), c. 3 series per min; 8–10 series per min at peak of courtship before laying (late April); up to 11 reported (Radermacher 1970; see also Langelott 1957). One Drumming with unusual persistency gave 95 series in 20 min (4·75 series per min), ♀ 10 series in 8 min (1·25 series per min). Unpaired ♂ may give up to 600 series per day; 100–200 per day by paired ♂ (Pynnönen 1939). At full intensity, ♂ gives 14·1 strikes per series in 0·598 s, ♀ 10·2 strikes in 0·443 s. Individual variation within same sex also considerable and territorial neighbours probably have series of different length (Zabka 1980). Drumming of *D. major* readily distinguishable from that of other west Palearctic Picinae, except possibly Syrian Woodpecker *D. syriacus* (see that species). (b) Tapping. Short, rhythmic sequences of strikes. Quieter than Drumming, audible within c. 50–100 m. Average 7·2 strikes (3–10 or more), 0·206 s between strikes (Zabka 1980). Usually given (by both sexes) at nest-hole when mate nearby. Used in Nest-showing, perhaps stimulating excavation, but unlike (e.g.) *D. martius* not an integral part of nest-relief (D Blume). Also given away from nest-tree in conflict situations during incubation and nestling period (Blume 1956). In experimentally induced conflict situations, Tapping may accelerate into weak Drumming; conversely, bird provoked by imitation of Drumming may respond with Drumming fading into Tapping—spontaneous occurrence very rarely recorded in April during nest-site selection and excavation (Blume 1958b; D Blume). (c) Excitement-pecking. Bird prevented from flying to anvil or to nest- or roost-hole pecks excitedly, as when feeding but in short series—usually loud and demonstrative. Probably more an irrelevant displacement activity than a signal. At high intensity, pieces of bark (etc.), knocked off; in breeding season, funnel-shaped holes created in a few minutes, as when starting to excavate nest-hole (D Blume). (d) Wing-noise normally has slightly whirring quality, clearly audible at some distance. Such noise much louder from flying birds in various antagonistic interactions in courtship and in disputes over roost-holes (pursuit-flights) (D Blume). (e) Wing-humming and Wing-rattling. Apart from Appeasement-call (see call 5), bird disturbed in nest-hole may give quiet humming sounds, audible only at close quarters: wings raised slightly and vibrated at carpal joint; combined with ruffling of feathers. Similar behaviour described for Hairy Woodpecker *D. villosus* of America (Lawrence 1967). Harder rattling sound given by birds disturbed in roost-hole presumably results from knocking carpal joints against cavity wall (Winkler 1972c; Winkler and Short 1978; D Blume).

CALLS OF ADULTS. (1) Contact-call (Fig II). Commonest call. Sharp, clear 'tchik', given singly or irregularly (D J Brooks); also rendered 'kick' (Steinfatt 1937; Pynnönen 1939). In central European birds, average pitch 2·6 kHz, average duration 34·4 ms (26·4–41·5), $n = 27$. North African *numidus* has slightly softer call than birds of north or central Europe (Winkler and Short 1978); birds from Asia Minor (Schüz 1959) and Japan lower pitched and softer (Winkler and Short 1978). Given when perched or in flight, throughout the year; associated with change of perch or activity, feeding, roosting, and various antagonistic interactions; used to maintain contact between family members after fledging and also functions as warning and alarm-call (e.g. after joining party of passerines mobbing predator). At high intensity, 90–120 (occasionally more) sharper and clearer-sounding calls given per min (this variant considered by Glutz and Bauer 1980 to be Scolding-call; see call 2); at low intensity, less clear, more muffled ('kük': Blume 1977). *D. major* can be attracted experimentally using imitated calls, and small passerines also attracted thereby (Blume 1977; D Blume). (2) Scolding-call (Fig III). Rapid series of slightly higher pitched Contact-calls, up to 600 per min: 'kjettettett'. Similar to equivalent call of Middle Spotted Woodpecker *D. medius* but with different rhythm: in *D. medius*, stress on beginning; in *D. major*, on end. High-intensity calls not easily distinguishable from Kreck-call (but compare Figs III and IV). Series of 'kweek' and 'quuig' sounds considered by D Blume to be variants of Scolding-call (see Winkler and Short 1978 and Glutz and Bauer 1980 for detailed discussion of these and treatment as separate calls). Latter given mainly in breeding season during aggressive flights at rivals and hole-competitors (including *D. medius* and Starling *Sturnus vulgaris*) and when man close to nest containing young; otherwise most likely from bird suddenly frightened or chased. In response to playback of Scolding-call, also Contact-calls and a few Drumming series (see above), territory-owner gives Scolding-call (D Blume). (3) Kreck-calls. Often said to be like Mistle Thrush *Turdus viscivorus*, but this problematical as 2 different calls of *T. viscivorus* resemble 2 different ones of *D. major* (calls 3c and 4). 3a–b correspond to 'Short Rattle Call' and 3c to 'Mistle Thrush Call' of Winkler and Short (1978). Kreck-calls given by both sexes, probably only in flight (compare call 4). (a) 1–2-unit calls. Given particularly by bird leaving vicinity of nest after change-over. (b) 3–5-unit calls. See Fig IV which shows Kreck-calls interspersed with Contact-calls. Given also outside breeding season, e.g. in late-summer or autumn dispute over roost-hole (Blume 1961), or when disturbed. Unpaired (or widowed) ♂ Drumming persistently but in vain often ends with Kreck-call (Stechow 1937); given as it takes off in an apparent display-flight (Blume 1977). (c) Multi-unit calls. Typical expression of aggression used when pursuing conspecific or other rival in breeding season; series then given hastily and with fluctuating volume, although first units normally stressed. Markedly similar to rattle call of *T. viscivorus* (used in aerial attack on predator) and some similarity to chattering of Fieldfare *T. pilaris* also evident (D Blume). A 'tchic-churr' (Richards 1957) given in response to Drumming (also of *D. syriacus*) probably this call (Winkler and Short 1978). (4) Krrr-trill. Like quiet wooden rattle call of disturbed *T. viscivorus* (D Blume);

a hard 'krirr' like Magpie *Pica pica* (Bergmann and Helb 1982). Not loud and rather poorly known. Reminiscent of Drumming (see Fig V which shows Krrr-trill preceded by rather squeaky Contact-calls) and sometimes described as 'vocal Drumming' with tip of tongue rapidly vibrated against bill (Pynnönen 1939; Feindt 1956); more likely vocal signal of Rattle-call type (see Winkler and Short 1978) typical of *Dendrocopos*. Distinct and separate call (*contra* Blume 1977), not derived from series of Appeasement-calls (D Blume: see call 5). Comprises 7–20 units, average 35·5 ms apart ($n=56$), given in regular rhythm (Winkler and Short 1978). Given in antagonistic situations, usually near tree with hole: e.g. when critical distance transgressed between future partners (D Blume), and 30 times in 25 min by ♂ in response to Drumming (Pynnönen 1939). (5) Appeasement-call. Has clicking and rattling components rendered 'rä' or 'wäd' (D Blume), 'träd', 'twäd' (Berndt 1954), 'twitt' or 'twift' (Richter 1950). Recording (Fig VI) suggests 'wad ra-wad ra-wad wad ra-wad' (J Hall-Craggs) with at times squeaky quality of rubber toy (M G Wilson). In study by Winkler and Short (1978), short, clicking and rattling 'wäd' sounds termed 'Wad Call', and longer, more rattling 'rä' sounds separated as 'Mutter Call', but combination and transition occur; probably variants of one call (D Blume). Appeasement-call used intra- and interspecifically and indicates suppressed excitement at varying intensity; may discourage closer approach or be used defensively and express tolerance. Double clicking 'wäd-wäd' given by incubating bird when disturbed; also by paired birds meeting at nest during excavation or incubation, and at nest-relief. In latter, Appeasement-calls (sometimes multi-unit) form integral part of ceremony, while Tapping (see above) not regular. Calls with particularly strong rattling components given by incoming bird to persuade tight-sitting mate to leave nest—also by incubating bird to indicate readiness for change-over. Appeasement-call 'palavers' as in *D. medius* (e.g. at nest-relief) not known in *D. major* (D Blume). During copulation, plaintive variant given by one bird, longer-drawn and noisy variant by other (probably ♀) (Winkler and Short 1978). Early in breeding season, threat element dominant: Appeasement-call given to maintain individual distance of several metres between prospective partners and territorial rivals; also in autumn disputes over roost-holes, between birds of same or different sex (D Blume). (6) Distress-call. Long-drawn, penetrating, at times markedly rhythmic, screech given by bird when handled, or ejected from roost by (e.g.) *D. martius* (Blume 1977; Glutz and Bauer 1980, which see for sonagram). Typically long, loud units show irregular frequency modulation and are rich in harmonics; may culminate in a noise (Winkler and Short 1978, which see for range of sonagrams). In high-intensity aggressive or sexual excitement, Contact-calls (see 1) may develop into short Distress-call. Shorter variants termed 'Squeak Call' and 'Screech Call' by Winkler and Short (1978), whose sonagrams suggest variants of one call or a complex of utterances with similar motivation (D Blume).

CALLS OF YOUNG. In contrast to (e.g.) *D. syriacus*, chirping food-call given virtually without interruption (Winkler and Short 1978; see also Social Pattern and Behaviour). During first few days, nestlings utter a voiceless, harsh (croaking) whirring sound 'krrr-krrr'; from *c.* 6–7 days more like 'kirr-kirr', then changes to 'kirre-kirre' and, from 19 days, to series of 'kiki-kiki' sounds (D Blume). Squeaking elements evident in calls of nestlings from *c.* 12 days (Winkler and Short 1978, which see for sonagrams); in recording by W Tilgner, nestlings *c.* 13 days old give high-pitched and thin 'si si si si si si si si' (J Hall-Craggs). From *c.* 2 days before fledging, young give some adult-type Contact-calls (D Blume). In recording by R Jervis, newly fledged young moving about in foliage give persistent, slightly squeaky 'queek' sounds (M G Wilson); remarkably like roosting calls of Blackbird *Turdus merula* (J Hall-Craggs, P J Sellar). Distress-call given in extreme danger; sonagrams by A Gebauer reveal it to be similarly structured to adult call 6 (D Blume); sonagrams of calls given in discomfort when removed from nest at *c.* 14 days and after last feed at *c.* 21 days (A Gebauer) indicate occurrence of shorter variants (see adult call 6). Soft, musical, and moderately long sounds probably express relief from distress; not well known (Winkler and Short 1978). MGW

Breeding. SEASON. Britain and north-west Europe: see diagram. Rather little apparent variation throughout range, though up to 2 weeks later in Scandinavia (Dementiev and Gladkov 1951*a*; Heim de Balsac and Mayaud 1962; Makatsch 1976). SITE. Hole in tree, of wide variety of species, usually 3–5 m (0·4–20) above ground (Dementiev and Gladkov 1951*a*; Schönfeld 1975). Nest: excavated hole, with entrance diameter 5–6 cm, often slightly elliptical; depth 25–35 cm, internal diameter 11–12 cm. Reused in subsequent years (Blume 1961). Building: excavation takes 14–25 (9–44) days (Glutz and Bauer 1980, which see for other references); equal share taken by sexes, sometimes more done by ♂ (D Blume); for details of excavation

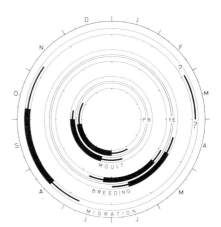

process, see (e.g.) Tracy (1933b) and Sielmann (1959). EGGS. See Plate 98. Elliptical, smooth and glossy; white. Nominate *major*: 27 × 20 mm (24–29 × 19–22), $n = 104$. Fresh weight 4·9 g (4·3–5·4), $n = 17$ (Verheyen 1967). *D. m. anglicus*: 26 × 19 mm (21–29 × 17–22), $n = 100$; calculated weight 5·0 g (Witherby *et al.* 1938; Makatsch 1976). Clutch: 4–7 (3–8). One brood. Replacements laid after egg loss. Laying interval 1 day. INCUBATION. 10–13 days ($8\frac{1}{2}$–16) (Owen 1925; Dementiev and Gladkov 1951*a*; Blume 1961; Glutz and Bauer 1980). Begins with last egg; hatching synchronous. By both sexes, though mainly by ♂. YOUNG. Altricial and nidicolous. Cared for and fed by both parents; brooded while small. FLEDGING TO MATURITY. Fledging period 20–24 days (Blume 1961), though 18–21 quoted by Witherby *et al.* (1938), and Makatsch (1976). Become independent 8–10 days after fledging (D Blume), though Dementiev and Gladkov (1951*a*) stated 25–30 days. Age of first breeding not known. BREEDING SUCCESS. In Czechoslovakia, average clutch 5·6 ($n = 49$), average brood 4·5 ($n = 38$) (Glutz and Bauer 1980). In Moldavia (USSR), of 35 eggs, 26 hatched and 21 young fledged (Averin and Ganya 1970).

Plumages (nominate *major*). ADULT MALE. Tufts covering nostrils black; forehead white, often with slight grey, cream, or buff tinge, sometimes deeper cream-buff or buff, but then usually still whitish along border of forecrown. Crown down to just above eye black with blue gloss, bordered behind by bright crimson-red band across back of head. Lores, ring round eye, area below eye, and ear-coverts white or cream-white, bordered below by black stripe from base of lower mandible across cheek to side of mantle, extending up into black line behind ear-coverts to side of nape and down into large black patch on side of chest. Central hindneck white; white spot on side of neck. Mantle, inner scapulars, back, rump, and upper tail-coverts black, slightly glossed with blue as on crown; some grey of feather-bases often visible on rump. Outer scapulars white, forming large white patches contiguous with mainly white tertial coverts; some black of scapular-bases occasionally visible, mainly when plumage worn. Underparts (apart from black patches on sides of chest and bright scarlet-red vent and under tail-coverts) cream-white or pale grey when fresh, white purest on chin, sometimes pale cream-buff on feather-tips of throat and central chest; dirty pale grey, pale cream, or (occasionally) smoke-grey or pale brown-grey when plumage worn (flanks usually still whitish). Central pair of tail-feathers (t1) black; t2 similar or with some yellow-buff or off-white bars on extreme tip; t3 with larger yellow-buff spot at tip and 1–2 subterminal off-white bars or spots (usually incomplete); t4–t5 white or off-white (except for mainly hidden black basal inner webs), marked with 2(–3) black bars 3–7 mm wide across tip (black often narrower or interrupted at shaft; exceptionally, tip uniform white); reduced t6 (hidden between t5 and upper tail-coverts) black or virtually so. Primaries black, outer web of p9 usually with 2–3 shallow white spots along basal margin, outer web of p6–p8 with 4–6 large and equally spaced white spots, p1–p5 with 3 (p1) to 4–5 (p5) large white spots on outer web and small white spot on tip; inner webs of primaries similar but white spots along borders larger (though not reaching shaft) and distal parts uniform black (extensively so on p6–p9). Exceptionally, white spots on outer webs of outer primaries join partly or completely, after webs then appearing largely white. Secondaries black with (2–)3 large white spots along borders of both outer and inner webs (1–2 more spots hidden under greater coverts); tips of outer secondaries with small white spot or fringe. Tertials more glossy black than secondaries; longer ones with 2(–3) white spots on middle portion and tip fully black, shorter inner 2–3 fully black or with some hidden white near base. Upper wing-coverts glossy black like mantle, except for contrastingly white greater and median tertial coverts; smaller upper primary coverts sometimes partly white, greater with white spots near base and sometimes on middle portion, but this largely hidden underneath black bastard wing. Axillaries and under wing-coverts dirty white or cream-white, longer under primary coverts (and sometimes greater under secondary coverts) spotted dull black subterminally. ADULT FEMALE. Similar to adult ♂, but crown, nape, and hindneck uniform glossy bluish-black, without red band across back of head. NESTLING. Naked at hatching. Closed feather-sheaths appear from 8th day; sheaths open to produce feather-tips from 11th day. Fully feathered on 14th day, but feathers still growing and ring round eye bare; all feather growth completed by 20th day, except flight-feathers and tail. (Heinroth and Heinroth 1924–6.) JUVENILE. Closely similar to adult, differing mainly in red crown (back of head black, unlike adult ♂), restricted pale pink tinge on vent and under tail-coverts, and larger p10. Crown fully red in juvenile ♂ (usually less glossy than red of adult ♂), with hardly any grey or black of feather-bases showing and often only narrowly bordered by black at sides; red of crown of ♀ often slightly less extensive, with more black or grey of feather-bases visible and with broader black line at sides of crown above eye. Overall length of red on crown 27·2 (10) 24–30 mm in ♂, 21·2 (10) 17–25 in ♀. Amount of visible red or grey feather-bases on crown not trustworthy for sexing once plumage worn, especially during eruptions when post-juvenile moult suspended and crown may become heavily abraded. Black of upperparts duller and less glossy blue than adult; white outer scapulars often with some black bars near bases, hidden when plumage fresh but sometimes prominent when worn, white scapular-patches then occasionally appearing heavily barred. Sides of head as in adult, but black stripes often slightly mottled cream or white along borders, black line behind ear-coverts narrower and occasionally not reaching nape. Underparts rather variable: occasionally cream-white or pale grey-buff with black patches on sides of chest, as in adult, but vent and under tail-coverts pink instead of red; more often, underparts dirty grey-white or pale grey-brown with dusky grey streaks on flanks and on sides of breast and belly below black patches on sides of chest; streaks usually indistinct, but sometimes sharp, flanks sometimes broadly barred rather than streaked. Pink of vent and under tail-coverts sometimes rather restricted and with some grey-white of feather-bases visible; occasionally completely absent. Rarely, some feathers on centre of chest with pink or pink-red tips. Tail as adult, but t1 relatively longer and more narrowly pointed; reduced t6 more often spotted white. Wing as adult, but p1–p9 with white or pale buff fringe along tip (in adult, only on p1 to p4 or p5); p10 longer and broader than adult, less reduced, 8 (30) 5–12 mm longer than longest greater upper primary coverts (in adult and 1st adult, 5 shorter to 3 longer, on average 1 shorter, $n = 28$). FIRST ADULT. Like adult and sometimes indistinguishable once juvenile p10 replaced. Juvenile secondaries, tertials, most or all greater upper primary coverts, and often part of greater upper wing-coverts retained; old feathers similar to adult ones, but retained juvenile greater upper primary coverts often slightly more pointed than those of adult and usually relatively more worn at same time of year; some outer greater primary coverts often replaced, however, contrasting with abraded inner ones (adults occasionally retain some old coverts, but then usually 1 or a few scattered among others, not all 6–8 inner ones); greater upper wing-coverts often partly retained (though tertial coverts replaced), duller black and contrasting in colour with newer median coverts; birds with active secondary moult never 1st adult.

Bare parts. ADULT AND FIRST ADULT. Iris red-brown or carmine-red. Bill dark grey, slate-blue, or lead-grey, base of lower mandible pale slate-grey or olive-grey. Leg slate-grey with brown, blue, or olive tinge, foot similar or slightly darker grey. NESTLING AND JUVENILE. At hatching, pink, including foot; closed eyelids blue-grey; slightly swollen and oval flanges at base of lower mandible pink-white or yellow-white; egg-tooth on tip of both shorter upper mandible and longer lower mandible white; claws white. Swollen pink-white heel-pads present at hatching, c. 11 mm long, 3 mm wide, gradually shrinking later on. Bill flanges 6–7 mm long by 3–4 mm wide at hatching, 10 by 4 on 11th day, 4 by 2 on 22nd. Bluish closed feather-sheaths of flight-feathers and tail appear from 6th day, on head and body from 8th–9th day. Eyes gradually open during 8th–14th day; iris brown at first, red-brown at fledging. (Heinroth and Heinroth 1924–6; Bussmann 1946; RMNH, ZMA.)

Moults. Based mainly on *pinetorum* and nominate *major*, with data from Dementiev and Gladkov (1951a), Stresemann and Stresemann (1966), Glutz and Bauer (1980), Ginn and Melville (1983), and c. 80 specimens in RMNH and ZMA. ADULT POST-BREEDING. Complete; primaries descendant. Starts with p1; mid-June to late July in nominate *major*, early June to mid-July in *pinetorum*, *anglicus*, *candidus*, and *tenuirostris*, late May or June in *hispanus* and perhaps other southern races. All moult completed with regrowth of p10; October–November in nominate *major*, mid-September to late October in races from temperate regions, and from early August in *hispanus*. Primary moult takes c. 120 days in *anglicus*. Erupting birds, Norway, had primary moult score 28 (7) 25–33 in second half of September and early October (Ree 1974); moult of these probably slowed down during migration, but apparently no suspended moult recorded, unlike post-juvenile. Tail moults in sequence 2–3–4–5–1; tiny t6 variable, between t2 and t5; starts with t2 at primary moult score 10–20 or with loss of p3–p4, t1 shed shortly after p6 or with p7; all tail new at about regrowth of p7 or score 32–42. Secondaries moult ascendantly and descendantly from s8 and ascendantly from s1, approximate sequence 8–9–7–1–10–6–2–11–5–3–4; s8 shed with p2–p3, s1 with p5–p6, s4 with p8–p10; s4 regrown with p10 or slightly later. Moult of greater upper primary coverts apparently not parallel to corresponding primaries: e.g. one bird with p1–p2 new and p3–p4 growing (score 16), had coverts 1, 2, 4, and 5 new and 6 growing; see also Ginn and Melville (1983). Body at about same time as tail; about half of feathering new at score 25, largely new at score 40; mainly second half of July to early October. POST-JUVENILE. Partial; primaries descendant. Juvenile secondaries, tertials, all or most greater upper primary coverts, bastard wing, and greater upper wing-coverts retained, but no tertial coverts. P1 shed 0–3 days before fledging (on about 20th day), p2 at fledging or shortly afterwards (c. 22nd day); further primaries follow at intervals of 10–15(–20) days, p3 on day 48, p4 on 62, p5 71, p6 89 (Ruge 1969); p10 shed at age of 4 months. Starts late May to early August, completed mid-September to late November, but some variation with latitude and altitude as in adult post-breeding; duration c. 145 days in *anglicus*. During eruptions in nominate *major*, moult suspended or slowed down with only 1 feather growing. Migrants in south-west Norway, second half of September and first half of October, had suspended with score 25 (5 inner primaries new, n=15) or 30 (n=1), or had single feathers growing with score 24·2 (28) 21–29 (Ree 1974). In Netherlands, August–November, moult of some nominate *major* suspended with score 20 (n=2), 25 (n=4), or 30 (n=2); most birds had resumed moult by mid-October, but some later ones still in full moult of primaries, tail, and body as late as December–January (ZMA). Sequence of tail as in adult post-breeding, but timing relatively later; starts at about shedding of p5, completing with t1 at about same time as p10. Head and body at same time as adult or slightly later; starts at moult score 21–26, lower mantle, chest, nape, some scapulars, and back to upper tail-coverts first, from about July (but starting as late as late October in some nominate *major* after eruptions); completed at score 35–45 from early September onwards, part of crown and sides of breast last.

Measurements. ADULT AND FIRST ADULT. Nominate *major*, northern Europe; all year (RMNH, ZFMK, ZMA). Bill (F) to forehead, bill (N) to distal corner of nostril.

	♂			♀		
WING	142	(2·76; 17)	138–147	141	(1·58; 10)	138–144
TAIL	87·7	(2·58; 17)	84–92	85·7	(2·78; 9)	81–90
BILL (F)	28·9	(0·98; 18)	28–31	27·4	(1·04; 10)	26–29
BILL (N)	22·7	(1·00; 17)	20–24	21·6	(0·91; 10)	20–24
TARSUS	24·2	(0·80; 15)	23–25	24·0	(0·53; 9)	23–25

Sex differences significant for bill only.

Sexes combined. Nominate *major*: (1) western Siberia, (2) north European USSR, (3) Scandinavia and Finland, (4) Poland. *D. m. pinetorum*: (5) East and West Germany, (6) Netherlands, (7) France (mainly Vosges mountains). (8) *D. m. anglicus*, Britain. (9) *D. m. hispanus*, central Spain. (10) *D. m. italiae*, Italy. (11) *D. m. parroti*, Corsica. (12) *D. m. harterti*, Sardinia. (13) *D. m. numidus*, Algeria and Tunisia. (14) *D. m. thanneri*, Gran Canaria. (15) *D. m. canariensis*, Tenerife. (16) *D. m. candidus*, Rumania and Bulgaria. (17) *D. m. paphlagoniae*, Turkey. (18) *D. m. tenuirostris*, Caucasus and Transcaucasia (except south-east). (19) *D. m. poelzami*, Lenkoran area (south-east Transcaucasia). (G van Duin, BMNH, RMNH, ZFMK, ZMA, ZMM.)

	WING			BILL (F)		
(1)	142·2	(1·6; 8)	140–145	28·4	(1·82; 9)	27–31
(2)	142·3	(2·6; 12)	138–147	28·4	(1·26; 12)	26–31
(3)	140·7	(2·1; 15)	138–146	28·4	(1·18; 16)	26–31
(4)	137·6	(3·2; 10)	134–145	29·0	(1·31; 10)	27–31
(5)	134·9	(2·7; 23)	129–141	29·1	(1·24; 26)	26–31
(6)	134·0	(2·4; 32)	128–138	28·8	(1·34; 47)	26–32
(7)	134·6	(2·3; 19)	130–139	28·2	(1·37; 19)	26–32
(8)	130·2	(2·6; 28)	126–134	29·9	(1·40; 22)	27–33
(9)	129·5	(3·1; 24)	124–137	30·0	(0·98; 27)	28–32
(10)	128·4	(3·8; 8)	122–134	28·5	(1·46; 8)	26–30
(11)	135·8	(2·7; 5)	133–140	30·2	(0·39; 6)	30–31
(12)	134·4	(2·2; 18)	130–138	28·2	(1·34; 18)	26–31
(13)	128·9	(1·8; 10)	125–131	31·7	(2·03; 12)	28–35
(14)	133·6	(2·0; 15)	130–136	30·1	(1·06; 15)	28–32
(15)	133·8	(1·9; 15)	131–137	30·5	(1·47; 15)	29–33
(16)	133·5	(2·3; 19)	129–138	29·5	(1·40; 17)	26–32
(17)	137·5	(4·5; 8)	131–142	30·8	(2·40; 8)	28–34
(18)	130·2	(5·4; 5)	123–137	28·5	(0·75; 5)	28–30
(19)	124·2	(1·4; 15)	122–127	30·5	(1·54; 15)	29–33

Though bill (F) incorporates ♂ and ♀ combined, differences often significant; bill (F) of ♀ on average 1·2 shorter than ♂ in nominate *major*, *pinetorum*, and *candidus*, 0·9 shorter in *hispanus*, *italiae*, *parroti*, and *harterti*, 1·7 in *anglicus*, *thanneri*, and *canariensis*, 2·6 in *numidus*, 2·3 in *paphlagoniae*, *tenuirostris*, and *poelzami*.

Some data from Vaurie (1959b). Wing of *mauritanus* from coastal lowlands of Morocco (sexes combined) 125·4 (2·04; 25)

121–130, from Atlas mountains 129·6 (3·55; 18) 123–136. Bill (F) of ♂♂: *parroti* 31·0 (1·24; 11) 28–32, *harterti* 29·4 (1·23; 10) 28–32, *numidus* 33·2 (12) 32–36, *mauritanus* 28·5 (12) 27–30, *candidus* 30·4 (10) 29–32, *paphlagoniae* 28·9 (1·36; 9) 28–32, *tenuirostris* 28·8 (1·14; 10) 27–31, *poelzami* 33·8 (1·11; 10) 32–35.

Most measurements directly correlated with wing length: e.g. average tail between 80·5 (*poelzami*) and 87·9 (nominate *major* from European USSR) (but tail of *hispanus* and *italiae* relatively short, 78·9; of Tenerife and Gran Canaria long, 86·6); average tarsus between 22·6 (*poelzami*) and 24·2 (nominate *major* from European USSR); average bill depth (at forehead) 8·1 (*poelzami*) to 9·7 (nominate *major* from USSR); average bill width (at middle of nostril) 9·6 (*poelzami*) to 11·1 (nominate *major* from USSR) (but bill of some races relatively slender—c. 9·9 in *canariensis*, *thanneri*, and *harterti*, 10·1 in *parroti*, and 10·2 in *paphlagoniae*); no correlation between average bill length and bill depth or bill width. Bill grows c. 2·6 mm in length per week (Lüdicke 1933), but this approximately equalled by abrasion at tip and length thus rather constant; length depends on feeding substrate (longer when feeding on soft wood), time of year (shorter during construction of nest-hole, longer in late summer), and age (shorter in less experienced birds which need more pecks to find same amount of food as experienced birds).

JUVENILE. Juvenile wing and tail full-grown on 35th day (Bussmann 1946); in general, wing and tail shorter than 1st adult and adult, tarsus similar, bill similar once post-juvenile moult completed. Thus, in *pinetorum* from Netherlands: adult wing (sexes combined) 133·9 (1·71; 17) 132–137, 1st adult 134·2 (3·02; 15) 128–138, juvenile 129·1 (2·84; 15) 125–133; adult and 1st adult tail 83·1 (2·65; 38) 78–90, juvenile tail 82·5 (3·12; 8) 79–86. However, juvenile wing of nominate *major* from Netherlands, autumn and winter, almost similar to 1st adult: 1st adult (sexes combined) 140·5 (3·22; 15) 136–147, juvenile 139·7 (2·51; 14) 135–144. (ZMA.)

Weights. ADULT, FIRST ADULT, AND JUVENILE. Nominate *major*. (1) USSR (Dementiev and Gladkov 1951a). During eruptions: (2) Netherlands, juveniles September–December. (3) Netherlands, exhausted birds (ZMA); (4) Helgoland (Vauk 1964).

(1)	♂ 89·7	(— ; 7)	70–96	♀ 89·6	(— ;—) 79–97
(2)	89·0	(— ; 3)	86–94	87·8	(3·03; 5) 84–92
(3)	68·5	(5·20; 4)	61–72	65·1	(5·64; 5) 60–73
(4)	89·8	(— ; 5)	80–95	88·9	(— ; 8) 75–100

During eruptions: (5) Utsira (Rogaland, Norway) (Ree 1974); (6) Helgoland (Vauk 1964).

(5)	AD	87·5	(6·97; 7) 75–98	JUV	81·4	(5·68; 59) 70–96
(6)		89·4	(— ; 5) 80–100		83·4	(— ;40) 60–100

Norway: ♂ 91; ♀♀ 80, 84, 93 (Haftorn 1971). Mongolia and Manchuria: 91·5 (6) 80–108 (Piechocki 1958, 1968c).

D. m. pinetorum. Netherlands: (1) March–July, (2) September–January, (3) exhausted birds, all year (ZMA). Switzerland: (4) October–April, (5) May–September. West Germany, all year: (6) Schleswig-Holstein, (7) Hessen. (Glutz and Bauer 1980.)

(1)	♂ 76·0	(87·2; 9) 70–87	♀ 72·7	(3·41; 7) 68–79
(2)	79·8	(10·8; 4) 70–95	75·5	(5·45; 4) 71–82
(3)	59·1	(5·29; 8) 53–65	54·2	(6·44; 4) 45–60
(4)	77·0	(8·1 ; 22) —	74·1	(4·9 ;11) —
(5)	73·1	(3·9 ; 17) —	71·5	(4·7 ; 7) —
(6)	81·7	(— ; 9) 74–93	79·8	(— ;14) 72–87
(7)	83·3	(— ; 7) 76–98	79·7	(— ;12) 73–85

Sachsen (East Germany) summer: 80·6 (18) 75–91 (Niethammer 1938). South-east Europe (in part perhaps involving *candidus*): ♂ 75·8 (4) 70–83, ♀ 80 (Makatsch 1950; Vasvári 1955; Rokitansky and Schifter 1971).

D. m. paphlagoniae. Turkey: ♂ 82; ♀♀ 75, 75, 77 (Kumerloeve 1967; Rokitansky and Schifter 1971).

D. m. poelzami. Northern Iran: ♂♂ 64, 70, 71; ♀ 64; unsexed juveniles 64, 69, 75 (Paludan 1940; Schüz 1959).

NESTLING. *D. m. pinetorum*. At hatching, 3·8 (Heinroth and Heinroth 1924–6) or 4·0 (3) 3·90–4·15; on 3rd day 14·3 (3) 11½–17, on 9th 38½, 40½, on 15th 68, 69; at fledging on 22nd, 73, 75 (Bussmann 1946).

Nominate *major*. On 9th day 56 (3) 48–62, on 15th 77 (3) 69–92; at fledging on 21st 73 (3) 71–75 (Pynnönen 1939).

Structure. Wing rather short, broad at base, tip rounded. 10 primaries: in adult and 1st adult, p6 and p7 longest or either one 0–3 shorter than other; p8 1–5 shorter than longest, p9 20–29, p10 75–88, p5 2–9, p4 14–24, p3 24–33, p2 30–37, p1 33–40; in juvenile, p7 longest, p8 0–3 shorter, p9 12–20, p10 61–74, p6 1–3, p5 8–13, p4 21–29, p3 31–44, p2 46–82. Juvenile p3–p5 slightly reduced, p2 and p1 distinctly so, respectively 47% and 33% of adult length (see Sutter 1974, Glutz and Bauer 1980), but p10 less reduced than in adult, longer and with broader tip (see description of juvenile in Plumages). Outer web of p5–p8 emarginated, inner web of p5–p9 slightly. Tail rather long, graduated; t1–t3 with strong shafts and attenuated tips, t4–t5 with less stiff shafts and more rounded tip; t6 weak, reduced, hidden between t4–t5 and upper tail-coverts; 12 feathers, t1 longest—in adult and 1st adult, t2 3–5 shorter, t3 8–13, t4 14–19, t5 20–28, t6 60–71; in juvenile, t2 6–10 shorter, t3 9–16, t4 13–27, t5 21–28, t6 58–67. Bill strong, straight, heavy at base; culmen with narrow ridge on top, lateral base of upper mandible with distinct groove bordered by ridges above and below; relative length and shape of bill show marked geographical variation (see Measurements). Nostrils situated in lateral groove near base of upper mandible, covered by tuft of rather soft bristles. Tibia feathered down to joint; tarsus bare, short and rather strong. Toes long and rather slender, undersurface with rather rough papillae. Outer front toe with claw 22·9 (10) 21–25 mm (nominate *major* and *pinetorum*, combined); inner front toe with claw c. 85% of outer front, outer hind c. 107%, inner hind c. 51%. Claws rather short, strong, sharply pointed, and strongly decurved.

Geographical variation. Marked; mainly involves size (expressed in wing length or weight), relative length and shape of bill, depth of colour of underparts, and amount of barring on tips of t4–t5. Variation strongly clinal, except for wide areas of secondary intergradation (with strong individual variation) and for some islands; hence boundaries between contiguous continental races often difficult to establish and sometimes arbitrary. In general, northern birds large, bill short and heavy, ground-colour of underparts whitish, and tips of t4–t5 white with rather narrow black bars. Towards south, size smaller, bill long and slender, underparts more greyish or brownish, and bars on t4–t5 equal in width, but some southern races deviate from this general pattern. Following account based on Hartert (1912–21), Hartert and Steinbacher (1932–8), Voous (1947), Vaurie (1959b, 1965), and specimens examined (RMNH, ZFMK, ZMA, ZMM). Nominate *major* from Scandinavia east to north European USSR large (see Measurements); bill short, very broad and deep at base, broad in middle, rather suddenly constricted towards tip as seen from above, both culmen and gonys distinctly curved; underparts greyish- or cream-white (see Plumages), dark bars on tips of t4–t5 rather narrow. Birds from west and central Siberia similar, but average slightly larger with underparts slightly whiter and with slightly denser and fluffier plumage; sometimes separated as *brevirostris* (Reichenbach, 1854), e.g. by Vaurie (1959b), but large overlap in size and colour renders separation unwarranted. Nominate *major* merges into *candidus* in wide zone of secondary intergradation (with much individual variation) in Ukraine at c. 50°N and east to Saratov, and similarly into *pinetorum* in eastern and southern Poland, eastern East Germany, and Bornholm, Denmark (K H Voous). *D. m. pinetorum* from western part of East Germany, Denmark, and Netherlands south to north-west Rumania, northern half of Yugoslavia, northern foothills of Alps, and Pyrénées highly variable, as birds from Denmark and east-central Europe tend towards nominate *major* in broad zone of secondary intergradation, those from Netherlands and western France merge clinally into *anglicus*, and those from southern France into *italiae*. *D. m. pinetorum* smaller than nominate *major*, bill longer and more slender, tip less suddenly constricted, culmen and gonys less curved; white of forehead, sides of head, and underparts pale grey-brown, pale buff-brown, or pale earth-brown, rarely grey-white or deeper buff-brown; width of black tail-bars variable, either narrow as in nominate *major* or equal in width to intervening white. Birds from high altitudes in Alps and Carpathians sometimes separated as *alpestris* (Reichenbach, 1854); these populations problematical, as colour and bill shape similar to nominate *major*, but size intermediate between *pinetorum* and nominate *major*, wing 136·2 (2·79; 16) 132–142 (Voous 1947); following Vaurie (1959b), here included in *pinetorum*. *D. m. anglicus* from Britain forms end of cline running from *italiae* towards north-west; smaller than *pinetorum*, and bill even more slender and more gradually tapering to tip; pale patches on sides of head and underparts darker brown-grey, earth-brown, or buff-brown, especially in worn plumage. *D. m. italiae* from mainland Italy, Sicily, southern slopes of Alps, and extreme north-west Yugoslavia closely similar in size and bill structure to *anglicus*, but pale parts of head and body pale cream-buff, cream-white, or (more rarely) greyish-drab colour. *D. m. parroti* from Corsica rather large; bill slender and gradually tapering as in *anglicus* and *italiae*, but distinctly longer than in *italiae*; ear-coverts and underparts dark brownish-drab or fawn, similar to some Italian and Sicilian birds but colour more saturated; vent and under tail-coverts on average slightly brighter deep red than preceding races; black and white bars on t4–t5 about equal in width. *D. m. harterti* from Sardinia similar to *parroti*, but bill averages shorter and more slender (Goodwin 1968) and white spots on flight-feathers slightly smaller and less numerous, wing appearing blacker. *D. m. hispanus* from Spain and Portugal close to *anglicus* and *italiae* in size and bill shape, but bill averages slightly longer; ear-coverts white or cream; underparts rather pale cream-buff, buff-white, or pale drab, rather similar to *italiae*, paler, less uniform, and less saturated drab than *harterti* and *parroti*; wing with much black as in *harterti*; t4–t5 rather black, with black bars often broader than intervening white; vent and under tail-coverts as in *harterti*, slightly more pinkish and less orange than in other races; rarely, some red-tipped feathers on central chest. *D. m. mauritanus* from Morocco similar to *hispanus*, but ear-coverts and underparts paler cream, isabelline-white, or buff-white; often a scarlet-red crescent on central chest between black patches on sides of chest (unlike *numidus*, no black on central chest); birds from Atlas mountains average larger than those of coastal lowlands, with underparts usually darker sandy-brown, smoke-brown, or drab (Harrison 1944); sometimes separated as *lynesi*. *D. m. numidus* from Algeria and Tunisia differs markedly from all other races in showing broad black chest-band, mixed with glossy scarlet-red feather-tips; wing short but bill longer and more slender than other races of its size; ear-coverts and underparts (apart from chest-band) rather pale, resembling lowland birds of *mauritanus*, *hispanus*, or *italiae*, but red of vent reaches up to mid-belly, t4–t5 white with rather narrow black bars only, and flight-feathers with rather large white spots, unlike these races; for relationships, see Winkler (1979). Canary Islands races *canariensis* and *thanneri* similar to *mauritanus* and *hispanus*, showing small white spots on flight-feathers, narrow white bars on t4–t5, and slender bill, but wing and bill distinctly longer than these races; underparts of *thanneri* (Gran Canaria) pale cream-buff like *hispanus* or *italiae*, and colour and size also rather similar to *candidus*, but bill even narrower and wing and tail blacker; underparts of *canariensis* (Tenerife) saturated buffish-drab similar to *harterti*, but bill longer and ear-coverts whiter. *D. m. candidus* from Ukraine west to eastern and southern Rumania and southern Yugoslavia and south to Greece and European Turkey closely similar to *pinetorum* in measurements, but less variable, and bill on average narrower in middle and more gradually attenuated; ear-coverts and underparts generally white or cream-white, like nominate *major*, but sometimes cream-buff or grey-buff as in *pinetorum*. *D. m. paphlagoniae* from Asia Minor large and with very slender bill; ear-coverts and underparts pale cream or isabelline-white; distinctly larger than *pinetorum* to which this race sometimes assigned (e.g. Vaurie 1959b), bill longer with concave rather than convex sides when seen from above, and underparts paler; close to *candidus* also, but larger and with more elongated bill. Caucasus race, *tenuirostris*, rather similar to *candidus* in size and bill shape, but bill slightly shorter and underparts darker, more ochre-buff or earth-brown; birds from Crimea tend slightly towards *candidus*, birds from south-east Transcaucasia intergrade into *poelzami*, birds from south-west Transcaucasia and perhaps eastern Turkey into *paphlagoniae*. *D. m. poelzami* from Lenkoran area in south-east Transcaucasia smaller but with relatively longer and more slender bill than all previous races; pale patches on head, neck, and underparts dark smoke-brown, darker than other races; birds from northern Iran and south-west Turkmeniya average slightly larger than those of Lenkoran. For extralimital races, see Vaurie (1959b, 1965) and Zheng *et al.* (1975).

Forms superspecies with White-winged Woodpecker *D. leucopterus* from central Asia (with which it occasionally hybridizes and which is perhaps conspecific), Himalayan Woodpecker *D. himalayensis*, Sind Woodpecker *D. assimilis* from Pakistan and south-east Iran, and Syrian Woodpecker *D. syriacus* (Short 1982).

CSR

Dendrocopos syriacus Syrian Woodpecker

PLATES 81 and 86
[facing pages 854 and 879]

Du. Syrische Bonte Specht   Fr. Pic syriaque   Ge. Blutspecht
Ru. Сирийский дятел   Sp. Pico sirio   Sw. Balkanspett

Picus syriacus Hemprich and Ehrenberg, 1833

Polytypic. Nominate *syriacus* (Hemprich and Ehrenberg, 1833), south-east Europe, Turkey, south-west Iran east to Kerman, and Levant south to Israel and Jordan; *transcaucasicus* Buturlin, 1910, Transcaucasia south to Armeniya and Nakhichevan and northern Iran east to Gorgan. Extralimital: *milleri* Sarudny, 1909, south-east Iran.

Field characters. 22–23 cm; wing-span 34–39 cm. Size as Great Spotted Woodpecker *D. major* and appearance closely similar except for more open face pattern and paler vent. Sexes dissimilar; no seasonal variation. Juvenile separable.

ADULT MALE. Closely resembles ♂ *D. major major*, with specific characters restricted to: (1) black moustache merely turning up on rear cheeks and not forming complete line running behind ear-coverts to join black of nape (but note that some juvenile *D. major* show similar lack of complete line); (2) slightly more extensive red hind-crown; (3) flanks sometimes with faint dusky streaks or bars; (4) paler, more pink vent; (5) bolder white barring across flight-feathers; (6) predominantly black outer tail-feathers, not white barred black as in *D. major*. ADULT FEMALE. No red on head; otherwise as ♂. JUVENILE. Central crown red. Downward extension of moustache on sides of lower neck and upper breast marked and breaking up into dark streaks which cover whole of flanks. Centre of upper breast tinged red. Beware confusion, especially, with some juvenile *D. major* which have black line behind ear-coverts absent; also with Middle Spotted Woodpecker *D. medius*, which has wholly red crown, moustache not reaching bill, and streaked flanks, and with White-backed Woodpecker *D. leucotos*, which has white lower back, more extensively barred wings (no bold scapular-patch), and streaked flanks.

Throughout increasing geographical overlap with *D. major*, easily confused with that species in any but close view (see above); absence of black line behind ear-coverts surprisingly difficult to confirm. Flight and behaviour as *D. major*.

Usually drums in longer bursts than *D. major* (*c.* 1 s rather than *c.* ½ s). Contact-call similar to *D. major* but softer.

Habitat. In west Palearctic, in continental warm lower middle latitudes, in Mediterranean, temperate, steppe, and even boreal–montane climates, ranging up to more than 2000 m in south-west Asia but in more recently colonized south-east European areas only to 1000 m in Bulgaria and 350 m in Danube basin, preferring dry regions to humid (Voous 1960; Glutz and Bauer 1980). Habitats broadly complementary to those of Great Spotted Woodpecker *D. major*, which it replaces over most lowlands except treeless plains, especially where its fruit and nut diet can be well met, as in orchards, gardens, along tree-lined roads, in parklands, on farms, and in riverain woodlands. Also in broad-leaved forest, sometimes locally replacing *D. major* even in montane regions, but less attracted to large stands of forest and more at home among scattered trees. In Iraq, typically in scrub oak *Quercus*, and always within 100 m of stream (Chapman and McGeoch 1956). In Azerbaydzhan (USSR), observed leaving trees in early morning to spend autumn days foraging in reedbeds (Dementiev and Gladkov 1951a).

Distribution. Marked spread in Europe since *c.* 1890, beginning in Balkans, then to Rumania, Hungary, USSR, Czechoslovakia, Austria, and recently Poland (frequently at first overlooked, so dates of spread, especially in Balkans, often approximate or largely unknown). Causes unknown.

AUSTRIA. Evidently colonized Neusiedler See area before 1951 (when first recorded), then spread to dry areas of north Burgenland and eastern Niederösterreich; isolated breeding records further south and west (Glutz and Bauer 1980). CZECHOSLOVAKIA. Colonized eastern Slovakia from 1949 and spread west, more slowly after 1960 (Glutz and Bauer 1980; KH). POLAND. First bred 1978–80 in south-east (Ciosek and Tomiałojć 1982). HUNGARY. First recorded 1937, then spread rapidly almost throughout country in next 20 years (Keve 1955, 1960b; Glutz and Bauer 1980). YUGOSLAVIA. Marked spread. First collected Serbia 1899, spreading north, west, and south; reached Vojvodina 1928, then rapidly spread north and west, with westward expansion continuing until 1970, and found Lake Skadar (Montenegro) 1969 (VFV). BULGARIA. First recorded 1890 in north; now widespread (Reiser 1894; TM). RUMANIA. First recorded 1931 in east (Tonani) and 1944 in west (Banat); for details of spread since, see Munteanu (1968). USSR. Marked recent spread into western Ukraine and Moldavia, still continuing: 1948 Chernovtsy region, 1957 Moldavia, 1960 Ternopol' region, 1967 Chernigov region, 1969 L'vov (Marisova 1965; Averin and Ganya 1970; Talposh 1975a; Marisova and Butenko 1976). Recently seen Crimea but breeding not yet proved (VF). LEBANON. Bred formerly, especially in north; no recent records (Tohmé and Neuschwander 1974; Macfarlane 1978; AK). IRAQ. Considered by Allouse (1953) to be possibly a winter visitor, but found breeding

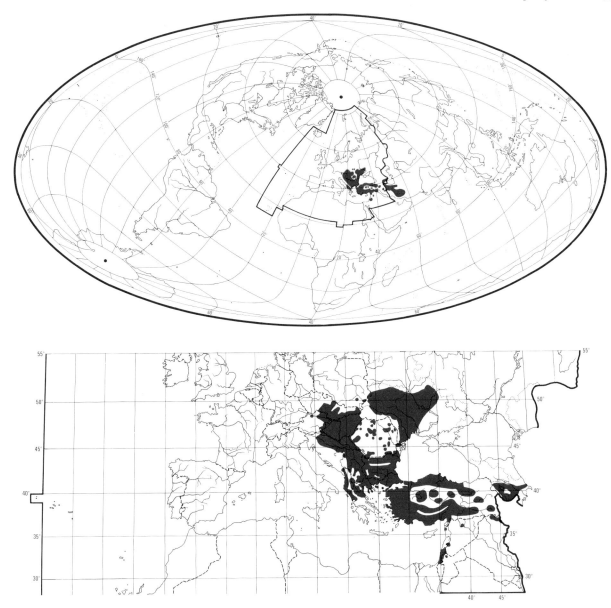

Kurdistan (Chapman and McGeoch 1956; McGeoch 1963).

Accidental. East Germany.

Population. Little precise data, but clearly a marked increase in west Palearctic population with recent spread in Europe (see Distribution); increased Israel.

CZECHOSLOVAKIA. Rather scarce (KH). GREECE. Perhaps several hundred pairs; no known trends (G I Handrinos, GM). USSR. Fairly rare in Transcaucasia (Dementiev and Gladkov 1951a); now common in L'vov and Ternopol' regions (Marisova and Butenko 1976). ISRAEL. Has increased considerably with spread of pecan plantations and orchards (HM).

Movements. Resident. Capacity for dispersal shown in range expansion; Yugoslavian study found c. 80% of colonists were young birds (Szlivka 1957). Usually arrives at new sites in autumn (Keve 1955). Longest ringing movement 13 km (*Larus* 1970, **21–2**, 19), though new (1982) nesting site in Upper Austria was 100 km west of nearest known breeding area (*Br. Birds* 1983, **76**, 275). Rate-of-spread data in Keve (1955) include some apparently comparable figures from east-central Europe.

Food. Chiefly insects, from spiders to large beetles, but also significant amounts of plant food which, unlike in other Picinae, is taken throughout the year; unique in giving fruit to young. Insect prey mainly surface-dwelling on trees, or just under bark, or terrestrial. Unlike Great Spotted Woodpecker *D. major*, rarely seeks deep wood-boring insects. Feeding ecology intermediate between *D. major* and Middle Spotted Woodpecker *D. medius* (Winkler 1967, 1972c, 1973). Searches from ground level to canopy and sometimes airspace beyond. Of 151 sightings, eastern Austria, c. 50% in canopy, c. 30% on trunks, c. 20% on ground; contrasted with c. 80% ($n=66$) by *D. major* in canopy and 0 ($n=52$) by *D. medius* on ground (Winkler 1973). Takes food from ground, tree-trunks, branches, etc., also stonework on buildings. Locates prey by sight and by tongue. 5 feeding methods distinguished by Winkler (1972c, 1973); ground-foraging, excavating, probing, gleaning, and aerial capture of insects. Searches on ground for (e.g.) *Agrotis* (Lepidoptera) pupae, mole-crickets *Gryllotalpa*, and sometimes ants (Formicidae) (Winkler 1973). Excavating comprises hacking, pecking, tapping, and, less often, levering and tearing off bark, accompanied by sideways tossing movements. For details of hacking action, see Winkler (1972c). Probing bird mostly makes rapid darting movements of tongue, often with sideways movements of head; tip of tongue intermediate in structure between *D. major* and *D. medius* (Winkler 1973, which see for details). Probing relatively important and *D. syriacus* thus more mobile than *D. major*; if food dropped, bird flies down after it, unlike *D. major*. Gleaning comprises pecking and snapping from bark and leaves. Also plucks fruits and seizes cockchafers *Melolontha* in flight from perch; may hover briefly to take them off twigs (Winkler 1967, 1972c). May drink at various places on ground, also where water collects in trees. ♂ once seen eating snow (Winkler 1972c). Of 136 sightings, eastern Austria, c. 50% excavating, c. 22% probing, c. 25% gleaning, c. 3% aerial catching; contrasted with c. 70% excavating by *D. major*, c. 60% probing by *D. medius*. When chick-rearing, removes kernels from cherries *Prunus* and gives young only the flesh, but outside breeding season eats only kernels (Molnár 1944–7; Winkler 1973). Extracts kernels of almonds *P. amygdalus in situ* on tree (Schmidt 1973a). Use of anvils well studied (Winkler 1967, which see for comparisons with *D. major* and *D. medius*; see also Ruge 1969b, Winkler 1972c); used throughout the year for breaking up beetles, nuts, kernels, and pieces of bark and wood. Though may improve anvil slightly, does not (unlike *D. major*) prepare and clear out favoured anvil for repeated use; instead often uses several adjacent sites, e.g. along crevice in trunk, fork in branch. May lodge items by hammering with bill, then use variety of movements, including probing with tongue, to extract food. Ripe walnuts *Juglans regia* and almonds found on ground, often split open there and halves treated in anvils (Winkler 1972c). Often takes 3–5 min to open apricot *P. armeniaca* kernels, up to 15–20 min for walnuts (Marisova and Butenko 1976). As in *D. medius*, food-storing possibly a supplementary use of anvil sites (Winkler 1967). One cache, in hollow branch, contained 83 hazelnuts *Corylus avellana*, 7 walnuts, 11 cherry stones, 3 apricot stones (Szlivka 1957, 1959b). Foraging range thought not to exceed 80–100 m from nest (Marisova and Butenko 1976); usually 50–70(–300) m from nest (Talposh 1975a); when raising young, may forage very near nest, at least initially (Szlivka 1957).

Diet in west Palearctic includes the following. Insects mainly adult and larval beetles (Coleoptera: Carabidae, Scarabaeidae, Elateridae, Bostrychidae, Tenebrionidae, Buprestidae, Cerambycidae, Curculionidae, Nitidulidae), and larval and pupal moths, more rarely butterflies (Lepidoptera: Pieridae, Nymphalidae, Nepticulidae, Cossidae, Caradrinidae, Lymantriidae, Pyralidae, Sphingidae, Saturniidae, Phycitidae, Gelechiidae, Noctuidae); less often, mole-crickets (Gryllotalpidae), leaf-hoppers (Cicadellidae), caddisflies (Limnephilidae), flies (Diptera: Bibionidae, Tipulidae), Hymenoptera—ants *Camponotus* and *Formica* (Formicidae), sawflies (Tenthredinidae), and chalcid wasps (Chalcidoidea). Other invertebrates include spiders (Araneae), harvestmen (Opiliones), snails (Mollusca) and worms (Lumbricidae). Plant food includes kernels (and, for young, also flesh of) cherry *Prunus*, plum *Prunus domestica*, apricot *P. armeniaca*, almond *P. amygdalus*, apple *Malus*, pear *Pyrus*, strawberry *Fragaria*, raspberry *Rubus idaeus*, mulberry *Morus*, currant *Ribes*, grape *Vitis vinifera*, seeds of walnut *Juglans regia*, hazel *Corylus avellana*, oak *Quercus*, pistachio *Pistacia vera*, nettle-tree *Celtis occidentalis*, white cedar *Thuja occidentalis*, occasionally also of hibiscus *Hibiscus syriacus*, hemp *Cannabis sativa*, sunflower *Helianthus annuus*, maize *Zea*, and pea *Lathyrus* (Szlivka 1957, 1959b; Stevanović 1960; Szlivka 1962; Marisova 1965; Winkler 1967, 1972c, 1973; Ruge 1969b; Schmidt 1973a; Talposh 1975a; Marisova and Butenko 1976; Glutz and Bauer 1980).

In eastern Europe, importance of walnuts in diet widely reported (Winkler 1967). In Yugoslavia, August–September, stomachs contained up to 75% walnuts (Stevanović 1960). Other important plant foods are almonds, apricots, cherries, plums, and locally mulberries, hazelnuts, and acorns (Szlivka 1959b; Stevanović 1960; Winkler 1967, 1973). In breeding season, Hungary, diet of adults and young mainly mulberries (Szlivka 1962, which see for seasonal variation). In stomachs of 7 birds, collected throughout the year, amounts of plant and animal food about equal; plant food outside breeding season comprised remains of seeds of walnuts, cherries, plums, nettle-tree, and hemp; animal food mainly Lepidoptera larvae and pupae at all times of year: e.g. stomach of ♂, 2 July, contained mulberries, 4–5 larvae of *Homeosoma nebulellum* (Phycitidae), 46 of *Pyrausta nebilalis* (Pyralidae), and remains of *Anarsia lineatella* (Gelechiidae) (Szlivka 1962). In 7 stomachs (no dates), Ukraine (USSR), 29·6% of items weevils (Curculionidae), 52·1% larvae of other Coleoptera, especially Elateridae, 13·2% Lepidoptera larvae, 1% ants,

4·1% spiders; all contained remains of acorns, more so in autumn (Marisova 1965). For similar results for 24 stomachs, Ukraine, see Talposh (1975a).

Ratio of animal to plant food fed to young varies between studies. In eastern Austria, one pair gave young a few almonds, otherwise entirely insects and spiders. Any young still in nest when cherries ripe also given these, minus kernels, and 1 brood given strawberries (Ruge 1969b; see also Molnár 1944–7). In Yugoslavia, young given mainly fruit, e.g. pears, raspberries, mulberries, strawberries, once currants. When young independent, first tended to take hazelnuts and various ripe fruits (Szlivka 1957). Brood observed for c. 1 hr per day for 10 days received mulberries and insects in about equal amounts (Szlivka 1962). In Ukraine, young given mainly insects: over 2 days at 2 nests, 109 adult *Aporia crataegi* (Pieridae), also 56 pupae and 31 caterpillars of other Lepidoptera; also 10 other insects, 9 worms, maize (once), and peas (once) (Talposh 1975a). In another study, Ukraine, young given beetle larvae for first 4–5 days, thereafter also Lepidoptera larvae, etc., and contents of apricot kernels (Marisova and Butenko 1976). In eastern Austria, food of young at one nest from 21 collar samples comprised 15 Lemnophilidae, 49 Bibionidae, more than 10 Diptera larvae, 1 cockchafer, 5 small Carabidae, 2 Noctuidae, and 5 Lepidoptera larvae; also 1 snailshell; plant food consisted of almonds and walnuts. Only small items delivered to small young. When young 12 days old, 2 foodballs weighed 350 mg and 800 mg; at 14 days, 1 weighed 550 mg, another comprised *C. cossus* caterpillar 4·5 cm long and weight 2 g. Peak feeding rate c. 8–9 days. For further information, see Glutz and Bauer (1980). No apparent variation in diurnal feeding frequency, and parental contributions about equal (Ruge 1969b). At nest, Hungary, where young fed mainly on mulberries and animal food, ♂ fed 4 young 54 times and ♀ 40 times in 11 hrs 20 min, i.e. brood visited on average once every 7·2 min; feeding rate higher in morning, starting 03.45 hrs, lower during day, higher again in evening (Szlivka 1962). When 2 weeks old, brood of 3, Ukraine, fed 7 times per hr (Marisova and Butenko 1976). EKD

PLATE 86 (*facing*).
Sphyrapicus varius Yellow-bellied Sapsucker (p. 853): **1–2** ad ♂.
Dendrocopos major Great Spotted Woodpecker (p. 856): **3–4** ad ♂.
Dendrocopos syriacus Syrian Woodpecker (p. 874): **5–6** ad ♂.
Dendrocopos medius Middle Spotted Woodpecker (p. 882): **7–8** ad ♂.
Dendrocopos leucotos White-backed Woodpecker (p. 891): **9–10** ad ♂.
Dendrocopos minor Lesser Spotted Woodpecker (p. 901): **11–12** ad ♂.
Picoides tridactylus Three-toed Woodpecker (p. 913): **13–14** ad ♂. (NA)

Social pattern and behaviour. Based partly on important captive study by Winkler (1972c).

1. Mostly solitary, but non-territorial in winter, and then mildly gregarious where food locally abundant, e.g. 16 together at store of hemp *Cannabis sativa* (Szlivka 1962). BONDS. Probably monogamous mating system, with pair-bond maintained from year to year, as in other *Dendrocopos*. Occasionally hybridizes with Great Spotted Woodpecker *D. major*; for review, see Winkler (1971). In Bandar Abbas region (south-west Iran), hybrids with Sind Woodpecker *D. assimilis* occur (Vaurie 1959b; Haffer 1977). ♂ and ♀ share nest-duties and feeding of young (see Food). At 1 nest, ♂ successfully reared brood after ♀ died (Talposh 1975a). Age of first breeding not known. BREEDING DISPERSION. Solitary and territorial. Said by Szlivka (1957) to be highly territorial, though, as in *D. major*, perhaps often more appropriate to describe territory as home range, since occupied area rarely so surrounded by others as to lead to boundary disputes (Winkler 1972c). No estimates of territory size, but in rural village area, Yugoslavia, 2 pairs nested c. 30 m apart; average density, 1952–60, 8 pairs (5–14) in c. 14 km², i.e. 0·6 (0·4–1·0) pairs per km² (Szlivka 1957, 1962). In fruit-growing region, eastern Austria, 0·5–1·0 pairs per km² (see Glutz and Bauer 1980). These, however, in evidently optimum habitat, and in most areas density appreciably lower (Glutz and Bauer 1980). Territory, especially in varied habitat, typically contains 'core area' in which most self-advertising and confrontations occur. Core area contains one or more tall trees which provide sites for Drumming (see Antagonistic Behaviour, below; also Voice), anvils (see Food), roosting, and nesting. Regular flight-paths link such sites within territory. Outside breeding season, these sites almost the only ones defended, and focus of defence may change if (e.g.) roost-site usurped by another bird (Winkler 1972c). Due to successive nesting failure, pair used 4 nest-holes within season; successive distances 200 m, 180 m, and 80 m to hole where finally bred (Szlivka 1957). Otherwise, site fidelity often marked between seasons; pair use same hole each year, or make new hole in same tree. Up to 7 holes reported in 1 tree (Szlivka 1957; Winkler 1972c). ROOSTING. In pre-existing or freshly excavated holes throughout the year, almost always on underside of angled branch (Szlivka 1957; Winkler 1972c). Once in hole in wall (Szlivka 1962), sometimes in nest-boxes (see Winkler 1972c), rarely in the open (but see below). Before breeding season, ♂ often roosts in prospective nest-hole (e.g. Ruge 1969b). In breeding season, ♂, who typically incubates at night, continues to roost in nest-hole; leaves, apparently due to overcrowding, some days before young fledge (Ruge 1969b; Winkler 1972c). After fledging, ♀, or some other bird, roosts in nest-hole (Winkler 1972c). In breeding season, ♀ roosted in hole 10 m from nest-hole (Ruge 1969b). Pair-members roosted in holes they abandoned after nesting failure (Szlivka 1957). Young and adult ♂♂ seeking roost-holes outside breeding season leads to period of changes in defended areas (Winkler 1972c). Young roosted in nest-hole after fledging, but when too crowded, roosted on underside of branches, pressed close to surface with head tucked into scapulars (Szlivka 1957). In hole, bird sleeps with ruffled plumage, and bill tucked into scapulars. By day, loafs in resting posture with plumage somewhat ruffled and head drawn into body. From noon to c. 15.00 hrs or later, depending on season, generally less active, sitting in resting posture in sheltered place (sometimes roost-hole), and sometimes preening. Pre-roosting behaviour similar to *D. major*. Often bird is near hole up to 1 hr before roosting. Typically sits near top of roosting tree and increases vocal and Drumming activity (see Voice). Pre-roosting activity especially marked early in breeding season when bird often makes repeated,

tentative entries into as yet unfamiliar hole. 2 birds may roost in different holes in same tree. Outside breeding season, birds often fly in from long distance, land directly by hole, and enter immediately. Enter and leave in daylight (Winkler 1972b, which see for seasonal variation in roosting times). Once in hole, rarely looks out. Typically stretches and preens after exit. ♀ reported bathing in pool, wetting belly then splashing water over wings (Dathe 1977). For bathing and comfort behaviour in captive birds, see Winkler (1972c).

2. When alarmed by bird of prey or human intruder, bird in the open typically adopts concealing Erect-posture on trunk, as in other *Dendrocopos* (Szlivka 1957; Winkler 1972c). When closely threatened by predators, captive birds performed erratic escape flight, with Distress-calls (Winkler 1972c: see 9 in Voice). Threatened birds may also bolt into 'emergency' holes in territory and sit tight. In captive study, alarmed bird sat inside hole but mounted a threat display: breast and head visible at entrance, plumage ruffled, bill slightly open; quivered wings and also made quivering movements around body axis, producing humming or strumming sound (Winkler 1972c; Winkler and Short 1978). ANTAGONISTIC BEHAVIOUR. Aggression marked at start of breeding season, again in autumn when birds disperse and contest roost-sites (Winkler 1972c). In breeding season, ♂♂ ward off rivals by Drumming; ♀♀ also Drum. Bird usually Drums close to nest, frequently on regular site, e.g. 40–50 m from eventual nest-site (Ruge 1969b). In Ukraine, Drumming began late February; from mid-March heard almost throughout day (Marisova and Butenko 1976). Continued during incubation and chick-rearing, rarely later. Most Drumming performed 06.00–08.00 hrs and 18.00–19.00 hrs; during incubation, ♂ recorded Drumming 94 times, ♀ 13 (Talposh 1975a). At start of laying, ♂ Drummed through most of day, and a good deal after sunset (Molnár 1944–7). In first half of day, birds Drum 1·7 times more than in second half (Winkler and Short 1978). Drumming also associated with Kweek-call (see 2 in Voice) which perhaps functions partly as self-advertisement (Winkler and Short 1978). Bird also advertises by sitting on conspicuous look-out, and especially by performing High-flights (see 6 in Voice). Any high-flying conspecific bird elicits immediate take-off by territory-owner who chases, with noisy wing-beats, in supplanting attack, accompanied by Zicka-, Scolding-, and Rattle-calls (Winkler 1972c; Winkler and Short 1978: see 4–6 in Voice). Supplanting attacks, often repeated several times, the most frequent form of aggression, also used in heterosexual encounters, disputes with juveniles, and rivalry over roost-sites (Winkler 1972c). During heterosexual encounters, early March, Appeasement-call (see 3 in Voice) also given (Blume 1961). In closer confrontations, perched rivals typically adopt Threat-posture (Fig A): slightly retract neck, ruffle plumage on nape and back, somewhat spread tail, and sometimes open bill. If neither rival flees, higher-intensity threat expressed in one or both birds by Wing-spreading display (Fig B): perched bird spreads wings, primaries open or closed, and may shuffle about; posture may be held briefly while rival approaches in flight. Display may precede outright attack in which birds peck at each other, sometimes short of physical contact; claws may become interlocked. Fighting accompanied by Zicka- and Creaking-calls (Winkler 1972c: see 8 in Voice). When rivals on horizontal branch, both commonly perform Head-swaying display, as in *D. major*: tip of bill, which is held well forward, describes figure-of-eight path; marked synchrony of movements between the 2 birds. For calls, see 7 in Voice. Before fleeing, subordinate may adopt Erect-posture: presses itself close to branch and makes sideways climbing movements, back and forth. Body thus concealed, and often bird sees only one eye of opponent. Rivals may counter each other's movements so that width of branch always between them (Winkler 1972c). 2 ♂♂ sometimes perched only c. 30–50 cm apart in Erect-posture and then made jabbing movements to left and right, as in *D. major* (Ruge 1969b). Confrontations with *D. major* widely reported (e.g. Szlivka 1957, Ruge 1969b); *D. syriacus* said always to be initiator and to chase *D. major* off. Subordinate, however, to Starling *Sturnus vulgaris* in competition for nest-sites, and may even be supplanted by them after egg-laying. At end of breeding season, birds were unable to evict *S. vulgaris* from would-be roost-holes, but successfully evicted, or even killed, smaller passerines (Szlivka 1957, 1962). After several *D. syriacus* had discovered food cache (see Food) of conspecific bird, fierce fighting over cache ensued every day (Szlivka 1959). For Nest-guarding, see subsection 2 of Heterosexual Behaviour. HETEROSEXUAL BEHAVIOUR. (1) Pair-bonding behaviour. Begins at end of winter; partners indulge in much aerial chasing, flying back and forth between trees (Szlivka 1957). In Ukraine, Drumming intense by mid-March, ♀ sometimes duetting with ♂, combined with birds flying around in wide circles, calling (see Voice), and periodically perching. Sit perched close together and said to start regular bowing movements to left and right (Marisova and Butenko 1976); this presumably (antagonistic) Head-swaying display. In early stages, antagonism between partners also reported elsewhere (Ruge 1969b). (2) Nest-site selection. Mainly by ♂, but also sometimes by ♀ (Szlivka 1957; Ruge 1969b). Bird Taps around various holes before choosing and properly excavating one (Szlivka 1957). While Tapping may test suitability of substrate, also thought to have ritual value in advertising prospective nest-site to mate (Winkler and Short 1978). In Ukraine, excavation begins early April, taking 5–7 days in lime *Tilia*, much longer in poplar *Populus*; both members of pair excavate, working quietly and warily (Marisova and Butenko 1976). Site selection may be combined with Drumming nearby, but less so than in *D. major* (Ruge 1969b). ♀ often watches from nearby tree and warns mate with Contact-call (see 1 in Voice) if danger threatens (Szlivka 1957). Chosen site closely guarded (Nest-guarding), often from well in advance of laying, by owners staying inside, and relieving each other periodically; during laying, hole almost continuously occupied (♂ roosting overnight), but no incubation occurs, bird clinging to walls of nest-chamber; sometimes knocks off bits of wood inside, possibly to make layer of chips for eggs (Ruge 1969b; Winkler 1972c). (3) Mating. Performed with little ceremony. ♀ adopts a crouched posture on horizontal branch or fence-post; ♂ flies on to ♀ who spreads wings for balance and

A   B

PLATE 84. *Dendrocopos minor* Lesser Spotted Woodpecker (p. 901). Nominate *minor*: 1 ad ♂, 2 ad ♀, 3 juv ♂, 4 nestling. *D. m. danfordi*: 5 ad ♂. *D. m. comminutus*: 6 ad ♂. *D. m. buturlini*: 7 ad ♂. (NA)

PLATE 85. *Picoides tridactylus* Three-toed Woodpecker (p. 913). Nominate *tridactylus*: 1 ad ♂, 2 ad ♀, 3 juv, 4 nestling. *P. t. alpinus*: 5 ad ♂, 6 ad ♀. (NA)

fans tail during copulation. In one pair, copulation observed 7 times from 9 April, always near nest-hole. Once, ♀ flew to ♂ who was climbing near hole; ♂ flew and hopped towards ♀, and copulation followed. No copulation observed during incubation (Ruge 1969b). In apparently exceptional sequence, ♀ was pecking on tree when ♂ pushed her away with one foot, as if towards place suitable for mating; ♂ then followed ♀ to copulate with her, and did so again 1–2 min later. ♀ laid 4–7 days after copulation (Szlivka 1957). (4) Behaviour at nest. Rhythm of Nest-guarding clearly influences pattern of laying; if ♂ not in hole when ♀ arrives to lay, normal laying interval (1 day) may be disrupted (Ruge 1969b). When ♀ incubating, ♂ said to announce arrival by Drumming 2–3 times. ♀ leaves and the two creep around in tree for a while, gradually nearing nest-hole. ♂ then enters and ♀ departs (Szlivka 1957). According to Ruge (1969b), however, while arriving bird may Drum, or give Contact-calls, these not thought to be part of Meeting-ceremony; rather, while gradually climbing to nest-hole, it gives Appeasement-calls which mate inside reciprocates. Incoming bird leans back, allowing mate to fly out, sometimes with Zicka- or Contact-calls (Ruge 1969b, which see for variants of nest-relief procedure). According to Winkler and Short (1978) departing bird (during incubation) often Drums. RELATIONS WITHIN FAMILY GROUP. Cheeping sounds heard from eggs at least 10 hrs before hatching. Newly hatched young lays neck over another egg, or entwines neck with those of siblings. Young brooded almost continuously for first few days. When food-bearing parent arrives, it usually relieves brooding mate. From 5 days, young brooded less continuously, eventually only at night (by ♂). ♀ performed c. 25% more brooding by day ($n=781$ min). Young markedly vocal until eyes open at 8–9(–13) days (Ruge 1969b). From c. 10 days, soft food-call gives way to contact-call similar to adult (Winkler and Short 1978: see Voice), though never as incessantly as D. major young (Ruge 1969b). From 12–14 days, young often near nest-entrance where fed (Talposh 1975a), retreating only if disturbed (Ruge 1969b). At c. 21 days, young grasped hole entrance (Molnár 1944–7). Parents and young typically twist heads 90° sideways during transfer of food (Ruge 1969b, which see for photograph). On day before fledging, parents noticeably excited, Drumming near nest-site, and flying around giving Kweek-calls. Young equally excited and vocal when adults near nest. When fledging imminent, parents stop feeding young regularly; may enter hole without food, or withhold it once inside. Young sometimes fledge spontaneously, with no vocal encouragement from parents; immediately fly up to 25 m from nest (Ruge 1969b). Brood may fledge within a few hours, or over longer period (Molnár 1944–7; Havlín and Havlínová 1966; Ruge 1969b). After young fledged in afternoon, parents continued to enter hole in evening (Molnár 1944–7). Family stay together for some time after fledging, during which Contact-calls often heard (Winkler and Short 1978). ANTI-PREDATOR RESPONSES OF YOUNG. Young in nest silent when predators nearby (Marisova and Butenko 1976). When disturbed young, which had flown prematurely from nest, was returned to nest, it adopted a 'defensive posture' in nest-entrance and pecked vigorously at fingers or anything else brought close (Havlín and Havlínová 1966). PARENTAL ANTI-PREDATOR STRATEGIES. (1) Passive measures. Incubating bird tends to flush on close approach, and may desert if disturbed repeatedly (Szlivka 1957). Birds tend not to fly directly to hole, at any stage of nesting (Marisova and Butenko 1976). (2) Active measures: against man. When man near nest, ♂ flies around, giving frequent Scolding-calls, while ♀ flushes silently to nearby tree; thought to mock-feed at this time, sometimes for long periods (Marisova and Butenko 1976).

(Figs A–B from drawings in Winkler 1972c.) EKD

Voice. Similar to Great Spotted Woodpecker D. major, but several marked differences (Ruge 1970a). Scheme below based on Winkler and Short (1978), which see for fine structure of calls and seasonal occurrence. For comparison of Drumming with other west Palearctic Picinae, see Zabka (1980). For Tapping and wing sounds, see Social Pattern and Behaviour.

DRUMMING OF ADULTS. Both sexes Drum, ♀ usually less often and in shorter series than ♂ (Ruge 1970a). Length of series and number of strikes per second also vary with acoustic nature of substrate (Winkler and Short 1978). Series commonly comprises 16–31 strikes, total duration 0.8–1.2 s (Winkler and Short 1978); average 20.9 ± 4.3 strikes per series (Wallschläger and Zabka 1981); average 21.6 strikes per series, average duration 0.89 s, $n=11$ series by 6 birds (Zabka 1980). ♂♂ average 27.5 strikes per series ($n=37$ series), ♀♀ 18.0 ($n=6$) (Ruge 1970a). Rate of delivery of strikes accelerates through series: intervals between strikes 60–70 ms at start, 30–40 ms at end (Zabka 1980; Wallschläger and Zabka 1981). In Fig I and other available recordings, 1st strike typically weaker than rest of series; brief diminuendo towards end of series (J Hall-Craggs). Typically 5–6 series per min (Winkler and Short 1978). Drumming, especially of ♂♂, different from D. major in that series of D. syriacus much longer and has relatively bigger gap between 1st and 2nd strike (Ruge 1970a; Zabka 1980), though latter difference scarcely audible to listener. Drumming advertises occupancy of territory and is used in pair-contact; may be elicited by Drumming of another bird, including D. major, also by Contact-calls (see below). For other contexts, see Social Pattern and Behaviour, and D. major, to which similar.

CALLS OF ADULTS. (1) Contact-call. See Fig II. Commonest call. Loud with hard clicking quality (Winkler and Short 1978). Described as a soft 'kjük' or 'gig' or 'dschik', softer than D. major and more like Middle Spotted Woodpecker D. medius (Bergmann and Helb 1982, which see for sonagrams); 'püg' (Winkler 1972c); 'zück' like Water Rail Rallus aquaticus, and more melodious than D. major (Dathe 1977). Average pitch 3.1 kHz ($n=11$). Rate of delivery variable; with increasing excitement, speeds up (also shriller) and grades into call 2. Given throughout the year, in various contexts, e.g. when excited or mildly alarmed, often when changing perch, at nest-relief before entering roost, or when disturbed during incubation (Ruge 1969b, 1970a; Winkler 1972c; Winkler and Short 1978). (2) Kweek-call. Rendered 'kjuig' (Franke 1953), 'quuig' (Winkler 1971, 1972c), 'güg' (Ruge 1970a, which see for sonagram), usually given in series. 2 distinct variants: 1st mainly response, especially of ♀, to Drumming ♂, and presence of other ♂♂, and may be given when following another bird in flight; also given near a fledgling. 2nd variant is loud, sharper sound, reaching 2.3 kHz, usually in rapid series of c. 6 calls. Both variants associated with Drumming and with calls 4 and 6. Serve as main long-distance signal between sexes (Winkler and Short 1978).

The caption for Plate 86 appears on page 877

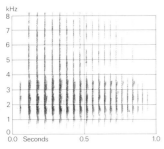

I J-C Roché (1966) Greece April 1965

II J-C Roché (1966) Greece April 1965

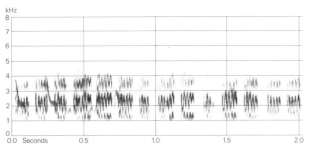
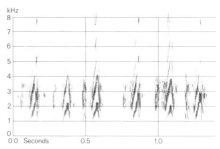

III J-C Roché (1966) Greece April 1965

IV J-C Roché (1966) Greece April 1965

(3) Appeasement-call. Brief, quiet, repeated 'wäd' or 'rä' sound. Of 3 calls, average pitch 0·59 kHz, average duration 111 ms, similar to *D. major* but shorter. Probably serves as appeasement signal, between pair-members near each other, e.g. during copulation, nest-relief, and aggressive encounters (Blume 1961; Ruge 1970a, which see for sonagram; Winkler 1972c; Winkler and Short 1978). (4) Zicka-call. Loud 'zicke' (Winkler 1972c), or 'kreck' (Ruge 1970a, which see for sonagram). Given in short series of usually 3 calls, typically by 2 birds in close combat. Closely related to call 2 and structure may combine elements of both (Winkler and Short 1978, which see for sonagram). (5) Scolding-call. Rapid series of Contact-calls, up to 120–180 per min, given when disturbed by intruders, conspecific or otherwise, at nest (Winkler 1972c). Also given by ♀ whose mate had been caught and held (Winkler and Short 1978). (6) Rattle-call. High-intensity contact call, rendered 'kürr' (Winkler 1972b), 'irrrrr' (Balát and Folk 1956). Recording (Fig III) suggests a harsh 'rr-rr-...' (J Hall-Craggs). In 5 calls, 12–25 units per call, average pitch 1·6 kHz ($n=26$). Variable in duration, volume, and pitch. Used in long-distance contact and territorial demarcation. Typically given during High-flight, after landing in tall tree overlooking territory, and during supplanting attacks (Winkler and Short 1978). In flight, may be preceded by Contact-calls, thus 'püg püg kürr' (Winkler 1972c). For seasonal occurrence, central Europe, see Winkler (1972c). Abbreviated variant, rendered 'kreck', also given, mostly in flight, by highly agitated bird in conflict situations (Winkler 1972c; Winkler and Short 1978). (7) Mutter-call. Like short variant of call 6, but softer and also distinguished by strong frequency modulation at $c.$ 60 Hz.

Bouts given in intense confrontations associated with Head-swaying display (Winkler and Short 1978). (8) Creaking-call. In conflict situations, a compressed creaking or groaning sound, which may develop into call 9 (Winkler 1972c; Winkler and Short 1978). (9) Distress-call. A loud screechy or shrieking sound, given in extreme alarm, e.g. by bird captured at roosting hole, also by ♀ chased by Tree Sparrow *Passer montanus* (Winkler and Short 1978; for sonagram, see Ruge 1970a).

CALLS OF YOUNG. Soft sounds given by 6-day-old young when brought back to normal temperature from overheating or over-cooling, also after squabbles with siblings, thought to express relief from distress. Food-call initially a quiet chirping, barely audible from a distance, given only when parents arrive with food, and thus very different from (e.g.) *D. major*. From $c.$ 10 days until after fledging, young beg with loud squeaking sound similar to call 1 of adult; when parent approaches, call develops excited squeaking quality. After being disturbed or fed, young give melodious 'guit' or 'puit' sounds (Glutz and Bauer 1980); these perhaps the same as calls shown in Fig IV. During fierce fight, 2 nestlings made noisy screeching sounds, similar to call 9 of adult; 2 juveniles fighting over roosting hole made angry twittering sounds (Winkler and Short 1978).   EKD

Breeding. SEASON. Austria: eggs laid from mid-April to early May (Ruge 1969b). Yugoslavia: laying begins mid-May (Szlivka 1957). SITE. Hole in tree, of variety of species. Of 42 nests, Yugoslavia, 24 were in mulberry *Morus*, 5 in walnut *Juglans*, and 13 in other species (Szlivka 1957). Mean height above ground of 22 nests, Yugoslavia, 2·25 m

(0·8–4·0), and of 11 nests, Austria, 3·2 m (1·7–6·0) (Szlivka 1957; Ruge 1969b). Nest: excavated hole 3·5 cm diameter, and average 20 cm deep (15–31), $n=8$ (Ruge 1969a). Building: by both sexes. EGGS. See Plate 98. Elliptical, smooth and glossy white. 25×20 mm (23–28 × 18–21), $n=46$ (Makatsch 1976). Weight of 12 fresh eggs averaged 5·7 g (Ruge 1969b). Clutch: 4–7 (3–8). Of 12 clutches, Hungary: 4 eggs, 1; 5, 10; 7, 1; average 5·1 (Makatsch 1976). Of 7 clutches, Austria: 3 eggs, 1; 5, 4; 6, 1; 7, 1; average 5·1 (Ruge 1969b). In Yugoslavia, young birds typically lay 3–4 eggs, older ones 4–5 (Szlivka 1957). One brood. Replacements laid after egg loss. Laying interval 1 day; sometimes alternate days (Szlivka 1957). INCUBATION. 9–14 days (Ruge 1969b). By both sexes, in roughly equal shares, though with irregular pattern of spells; ♂ at night (Ruge 1969b). By day, longest spell by ♀ 1 hr 35 min, by ♂ 25 min (Talposh 1975a). Begins with last egg; hatching synchronous. YOUNG. Altricial and nidicolous. Cared for and fed by both parents; brooded while small (see Social Pattern and Behaviour). FLEDGING TO MATURITY. Fledging period 24 days (Ruge 1969b); 17–25 days, depending on food supply (Szlivka 1957). Become independent some time after fledging. Age of first breeding not known. BREEDING SUCCESS. Of 31 eggs laid, Austria, 27 hatched (Ruge 1969b).

Plumages (nominate *syriacus*). ADULT. Closely similar to Great Spotted Woodpecker *D. major*, but tufts of bristles covering nostrils white or buff, only tips of longer bristles black (in *D. major*, all bristles fully black); part of forehead just above base of upper mandible buff-brown or rufous-brown, remainder of forehead white or cream-buff (in most races of *D. major*, forehead usually almost uniform buff-white to buff-brown); white of forehead reaches slightly further on to forecrown; red patch on hindcrown of ♂ slightly longer and narrower, reaching further up crown, black of crown slightly less extensive; rarely a few red feathers on crown (Winkler 1972b); central hindneck, mantle, inner scapulars, and back slightly duller black (not glossy bluish-black); no black line behind ear-coverts from nape to lower cheeks—ear-coverts to sides of neck and lateral hindneck cream-white or white, separated from pale buff or off-white chin and throat by black line from lower mandible to side of chest (in adult *D. major*, black line behind ear-coverts prominent); flanks and sometimes sides of belly variably streaked or barred dusky (usually faint; virtually uniform white, cream-grey, or pink-buff in *D. major*); lower belly, vent, and under tail-coverts pink-red (distinctly paler and less saturated than scarlet-red or flame-scarlet of *D. major*); outer tail-feather (t5, excluding reduced t6) black with some white bars or single spot at tip, t4 fully black or with small white bar or spot, t3 virtually black (in *D. major*, t4–t5 white with rather narrow black bars, t3 black with white-marked tip); smaller outer primary coverts (hidden under bastard wing) largely white (in *D. major*, largely black). White spots on flight-feathers slightly larger but fewer in number than in *D. major*—3 white bars 4–7 mm wide on p5–p7 (a 4th bar just underneath tips of greater upper primary covers), only narrowly interrupted by black at shafts (in *D. major*, 5 bars beyond tips of greater upper primary coverts on outer webs of p5–p7, but only 3 on inner webs; basal bars more broadly interrupted by black). Sex differences as *D. major*. NESTLING. As *D. major*. JUVENILE. Rather similar to adult, differing as juvenile *D. major* differs from adult; often a pink-red band across chest, varying from a few red feathers to a rather broad and complete band (juveniles of some races of *D. major* show some red feathers on chest, but rarely as extensive as in *D. syriacus*); dusky streaks on flanks and sometimes sides of belly and breast slightly more extensive than in adult, flanks sometimes partly barred dusky. Differs from juvenile *D. major* in more extensively white bristles on nostril, red feathers on chest, paler pink-red vent and under tail-coverts, mainly black outer tail-feathers, and larger though fewer white spots on flight-feathers; black line behind ear-coverts absent, but occasionally also absent in juvenile *D. major* and lack of this mark thus not a reliable character. FIRST ADULT. Mainly like adult; differs from adult in same characters as 1st adult *D. major* differs from adult.

Bare parts. As *D. major*.

Moults. Largely similar to *D. major*. ADULT POST-BREEDING. Starts with p1 early May to early July; sequence as in *D. major*. Moult completed August–September. POST-JUVENILE. As in *D. major*, but p1 shed at fledging, not when still in nest (Glutz and Bauer 1980). Starts (April–)May–July, completed after $c. 3\frac{1}{2}$–4 months in August–October; moult mainly June–September. Juvenile secondaries, tertials (often), and all or part of greater upper primary coverts and greater upper wing-coverts retained.

Measurements. ADULT AND FIRST ADULT. Nominate *syriacus*. South-east Europe, all year; skins (Stresemann 1920; RMNH, ZFMK, ZMA, ZMM). Bill (F) to forehead; to feathers, on average 1·1 less; to nostril 6·8 less. Toe is outer front toe with claw.

WING	♂ 132	(2·46; 19)	130–137	♀ 130	(2·21; 11)	129–132
TAIL	77·6	(4·23; 7)	73–84	75·6	(0·96; 5)	75–77
BILL (F)	31·7	(0·76; 9)	31–33	29·8	(— ; 2)	29–31
TARSUS	23·3	(0·95; 7)	22–25	22·7	(0·95; 6)	22–24
TOE	22·0	(1·35; 7)	20–24	21·0	(1·07; 6)	19–22

Sex differences not significant, except wing. Wings of 125 (♂), 125, 126 (♀♀) excluded from range.

D. s. transcaucasicus. Kura valley, Azerbaydzhan, USSR, December–January; skins (ZMM).

WING	♂ 128	(1·58; 4)	126–130	♀ 127	(1·41; 12)	126–130
BILL (F)	30·6	(1·35; 4)	29–32	28·2	(0·72; 12)	26–29

Sex differences significant for bill.

About equal numbers of each sex combined. Nominate *syriacus*: (1) Turkey; (2) Syria, Lebanon, and Israel (RMNH, ZFMK, ZMA). *D. s. transcaucasicus*: (3) Armeniya and Nakhichevan, USSR (ZMM).

(1)	WING	130	(2·27; 9)	127–133	BILL (F)	32·0 (1·46; 8) 30–34
(2)		128	(1·12; 8)	126–130		30·9 (1·44; 5) 29–33
(3)		126	(0·96; 5)	125–128		30·5 (1·07; 5) 29–32

JUVENILE. Wing on average $3\frac{1}{2}$ shorter than adult and 1st adult, tail $c.$ 5; bill not full-grown until about completion of post-juvenile moult.

Weights. Nominate *syriacus*. South-east Europe: ♂♂ 75, 80, 80 (January), 83 (April), 77 (December); ♀ 74 (February) (Makatsch 1950; ZMA, ZMM); ♂ 79·5 (2·1; 9) 76–82, ♀ 74·0 (3·2; 11) 70–81 (Glutz and Bauer 1980). Turkey: ♂ 70 (June); unsexed, 62, 64 (July), 74 (August) (BTO, ZFMK). Iran: ♂♂ 60, 74 (May–June), 67 (July); ♀♀ 55, 63 (May–June) (Paludan 1938, 1940; Desfayes and Praz 1978).

D. s. transcaucasicus. Azerbaydzhan (USSR): ♀♀ 61, 62, 63 (ZMM).

Structure. In adult, p6 or p7 longest or either one 0–2 mm shorter than other; p8 0–4 shorter than longest, p9 13–22, p10 68–80, p5 2–4, p4 12–19, p3 22–29, p2 26–34, p1 31–38; in juvenile, p7–p8 longest, p9 12–18 shorter, p10 57–65, p6 1–4, p5 8–12, p1–p2 reduced (though less so than in *D. major*), about half of length of adult feathers. Adult p10 reduced, narrow and pointed; tip 2 shorter to 6 longer than tip of longest greater upper primary coverts (in juvenile, 11–16 longer; broader and with more rounded tip). Bill long and slender, shape about similar to southern races *harterti*, *candidus*, or *tenuirostris* of *D. major*. Outer hind toe with claw *c.* 109% of outer front toe with claw, inner front *c.* 83%, inner hind *c.* 56%. Remainder of structure as in *D. major*.

Geographical variation. Rather slight. European birds have slightly longer wing and tail than those of Transcaucasia, Iran, and Levant; wing of European adults mainly over 130, of others below, but birds from Asia Minor intermediate; bill of birds from Asia Minor on average slightly longer than others, of Azerbaydzhan slightly shorter, but much overlap. Underparts of birds from Europe, Levant, and Iran white or dirty cream, those of Asia Minor and Transcaucasia more pinkish- or brownish-buff; dark marks on flanks of European birds slightly heavier than in other populations, and on this character these birds sometimes separated as *balcanicus* (Gengler and Stresemann, 1919). Small size of *transcaucasicus* shared by some populations of nominate *syriacus*, but *transcaucasicus* unique in showing much white on outer tail-feathers; t5 with 2 broad white bands across tip (frequently interrupted by black at shaft or at sides) or with broad white tip and broad white subterminal bars, nearly always with trace of a 3rd or 4th bar on outer web nearer base; in nominate *syriacus*, t5 mainly black with single white spot or bar on tip (sometimes tiny) and often with trace of 2nd bar formed by subterminal white spot at shaft or on outer web; only 2 out of 25 nominate *syriacus* examined had pattern of t5 similar to darkest birds of typical *transcaucasicus* from Azerbaydzhan. Populations from Armeniya, Nakhichevan, and northern Iran have slightly longer bill and slightly more variable tail pattern than Azerbaydzhan birds, but in general closer to these than to typical nominate *syriacus* and hence included in *transcaucasicus*. Not known whether race *milleri* Sarudny, 1909, from Kuh-e Taftan (south-east Iran) valid, as none examined; described as being larger and with differently shaped red nape-patch in ♂ (Sarudny 1909).

Forms superspecies with Sind Woodpecker *D. assimilis* from south-east Iran and Pakistan, which differs from *D. syriacus* by smaller size (wing mainly below 120), slightly more extensive white on wing and outer tail-feathers, and fully red crown of ♂; the 2 species infrequently hybridize in Bandar Abbas area in southern Iran (Vaurie 1959b; Haffer 1977). *D. syriacus* and *D. assimilis* closely related to Himalayan Woodpecker *D. himalayensis*, White-winged Woodpecker *D. leucopterus* from central Asia, and Great Spotted Woodpecker *D. major* (see also Short 1982). Occasionally hybridizes with *D. major* at boundary of range of *D. syriacus* (Bauer 1957; Kroneisl-Rucner 1957a; Keve 1960b; Winkler 1971). CSR

Dendrocopos medius Middle Spotted Woodpecker

PLATES 82 and 86
[facing pages 855 and 879]

Du. Middelste Bonte Specht   Fr. Pic mar   Ge. Mittelspecht
Ru. Средний пестрый дятел   Sp. Pico mediano   Sw. Mellanspett

Picus medius Linnaeus, 1758

Polytypic. Nominate *medius* (Linnaeus, 1758), Europe south to Spain and Greece, east to European Turkey and west European USSR; *caucasicus* (Bianchi, 1905), northern Turkey to Caucasus, Transcaucasia, and (probably this race) north-west Iran; *anatoliae* (Hartert, 1912) western and southern Turkey, intergrading with *sanctijohannis* in northern Iraq. Extralimital: *sanctijohannis* (Blanford, 1873), Zagros mountains in Iran.

Field characters. 20–22 cm; wing-span 33–34 cm. About 10% smaller than Great Spotted Woodpecker *D. major* but still 50% larger than Lesser Spotted Woodpecker *D. minor*. Rather small woodpecker of similar appearance to *D. major*, distinguishable by pale, mainly red and white head, smaller white patches on scapulars, distinctly streaked flanks, and buff body merging with pink rear belly and vent. Barring on flight-feathers and outer tail-feathers always obvious. Crown appears slightly crested. Commonest call lower pitched than that of *D. major*. Sexes similar; no seasonal variation. Juvenile separable.

ADULT. White plumage less pure in tone than in *D. major* and Syrian Woodpecker *D. syriacus*, with grey, buff, and ochre tinges to fore-face, cheeks, lower chest, and belly. Faded look enhanced by lack of black surround to red crown, by incomplete moustache (not reaching bill), by absence (usually) of black line behind ear-coverts joining cheeks to nape, and by pale, rose-pink vent, which merges with rear belly. Extension of moustache on to chest-side breaks up into streaks which cover flanks. Back and wings resemble *D. major* and *D. syriacus* but, at close range, white patch on scapulars visibly narrower and shorter (restricted to lower feathers), while white barring over extended wing actually more striking, particularly in flight. In direct comparison, ♀ duller than ♂, with red crown-patch pinker and shorter (hindcrown golden-brown in ♀, crimson in ♂). JUVENILE. Resembles adult in diagnostic combination of red crown, incomplete moustache, and streaked flanks. Duller above with more

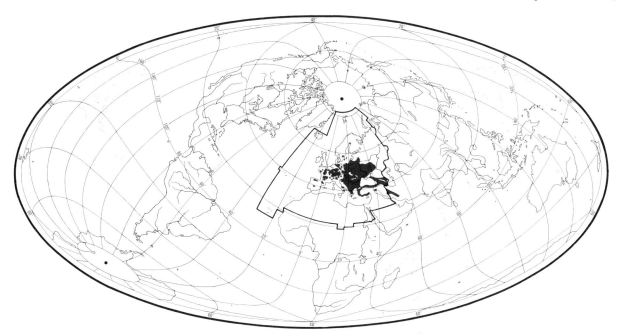

diffuse pattern (and less distinct individual marks). Underparts faintly scaled, with flank-streaking slightly less distinct and buff suffusion reduced. ♀ shows shorter, duller crimson crown in direct comparison with ♂.

Given clear view of head and underbody, unmistakable. In comparison with larger congeners, slighter build and duller, more faded appearance quickly learnt. Face of *D. medius* also noticeably more open and gentle in expression. Separable from *D. major* at all times by incomplete black moustache (not reaching bill); usually also by streaked breast-sides and flanks, though juvenile *D. major* often also has breast-sides streaked, sometimes sharply. Note also that juvenile *D. major* has pink under tail-coverts (like *D. medius*) and occasionally lacks black line behind ear-coverts. When size not apparent, separation from *D. syriacus* needs care, since it can show slight flank-streaking (see that species). Flight and behaviour as *D. major* but markedly restless, frequently changing position in upper branches and so making observation difficult. Uses bill less as axe and more as tweezers. Generally spends time higher in trees than *D. major*.

Contact-call similar to that of *D. major* but lower pitched and often given in quick, trot-like succession. Song (used in place of drumming) a series of nasal 'ähk' sounds.

Habitat. Resident in warm temperate continental climatic zone, not extending to boreal or montane. In Switzerland does not normally breed above 700 m (Glutz von Blotzheim 1962); in canton Zurich prefers the lowest woods at 301–400 m (Müller 1982). Range roughly coincides with that of hornbeam *Carpinus betulus*, and consequently with heartland of European primitive broad-leaved forest. Favours mixed hornbeam–oak *Quercus* woodland, or in parkland elm *Ulmus*; also old orchard and riverain alder *Alnus* woods in floodlands. In canton Zurich prefers pure oakwoods of coppice-with-standards type; but also occurs in oakwoods heavily mixed with spruce *Picea*, beech *Fagus*, or ash *Fraxinus* provided sufficient oaks are interspersed and are not tall enough to create too much shade. Unfitted for such robust excavation as Great Spotted Woodpecker *D. major*, and spends more time in surface gleaning. For similar reasons favours diseased or dead trees and branches. Thus able to coexist with *D. major* in overlapping territories, but fails to spread into many apparently suitable habitats, even within already restricted range of acceptability. For details of habitat utilization and comparison with *D. major*, see Jenni (1983). Prefers oakwoods of 30 ha or more, avoiding those below 5 ha, and not favouring those more than 3 km apart; indifferent to such factors as slope, ground moisture, and presence of water. Flight apparently restricted to lower and lower middle airspace. Relative lack of adaptability and dependence on traditional habitats suggest vulnerability to direct or indirect human impacts, especially through replanting or management unfavourable to requirements. (Niethammer 1938; Voous 1960; Glutz von Blotzheim 1962; Lippens and Wille 1972; Lups 1978; Müller 1982.)

Distribution. No longer breeding Denmark and almost extinct in Sweden.

NETHERLANDS. Bred, or probably bred, in east in 1952, 1955–60, 1962, 1964, and 1973; in central area, nested 3 times in 19th century and probably 1976 (CSR). DENMARK. In 19th century, bred in many areas; declined, last nested 1959 (Dybbro 1978). SWEDEN. In 19th century,

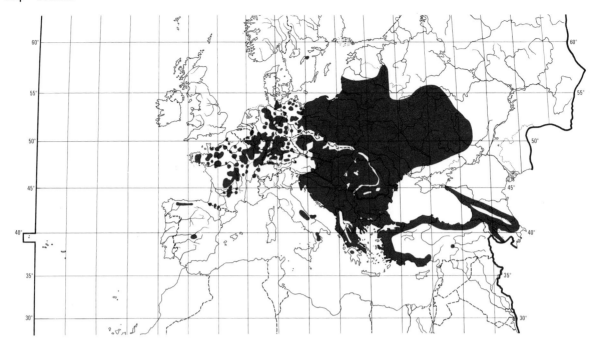

bred mainly in south, with some nesting north to c. 59°N; range decreased latter half 19th century and early 20th century, and from 1940s breeding only Östergötland (LR). AUSTRIA. Range possibly extended into Steiermark in recent years (Brandner and Stani 1982).

Accidental. Portugal, Netherlands, Denmark, Iraq.

Population. Few reliable counts, but almost certainly decreased over much of range, earlier due to forest clearance and more recently because of radical changes in forest management in many areas of central Europe (Glutz and Bauer 1980).

SPAIN. Probably not scarce in Cantabrian mountains (AN). FRANCE. Estimated 100–1000 pairs; density low, but easily overlooked (Yeatman 1976). More abundant than Great Spotted Woodpecker *D. major* in some full-grown forests (Ferry and Frochot 1965; Lovaty 1980). LUXEMBOURG. Estimated c. 100 pairs (Glutz and Bauer 1980). BELGIUM. About 250 pairs (PD). WEST GERMANY. Decreasing some areas; for local estimates of numbers or abundance, see Glutz and Bauer (1980). SWEDEN. Declined from latter half of 19th century to 1940s; rather stable then until late 1960s when sudden decrease to 20 pairs or less 1967–71, 8 pairs 1975, and 2 pairs 1980 (Holmbring 1972; Pettersson 1980; LR). EAST GERMANY. Rare and local breeder (Makatsch 1981); see Glutz and Bauer (1980) for details and local counts. POLAND. Scarce to very scarce breeder (Tomiałojć 1976a; LT). CZECHOSLOVAKIA. Generally scarce and possibly decreased (KH). SWITZERLAND. Estimated 265 pairs 1978; declined in some areas due to forest destruction (RW). HUNGARY. Common in suitable woodlands (Glutz and Bauer 1980). ALBANIA.

Common (EN). GREECE. Quite common in north and centre of mainland, and on island of Lesbos (WB, HJB, GM). RUMANIA. Common, especially in forests of Danube and foothills (Vasiliu 1968). USSR. Mostly quite rare, but more common in some areas, e.g. Voronezh region and parts of Caucasus (Dementiev and Gladkov 1951a).

Oldest ringed bird at least 8 years (Glutz and Bauer 1980).

Movements. Mainly resident. Autumn nomadism reported in European USSR; involves some wandering outside breeding range, e.g. found only in winter in Kaluga and Tula regions. Scale and regularity unclear. Autumn wandering most pronounced in Minsk between late September and early November (Dementiev and Gladkov 1951a); has straggled to Latvia. Considered sedentary in Sweden, with 2 autumn records from Falsterbo on southwest coast (Gatter 1973). Also resident in west-central European breeding areas, though some individuals may disperse over short distances; longest ringing movement 14 km, in West Germany (Glutz and Bauer 1980). For winter occupancy of breeding and feeding territories by adults, see Social Pattern and Behaviour; wandering birds probably immatures.

South-eastern populations (*caucasicus* and *anatoliae* from Turkey to Transcaucasia, *sanctijohannis* in Iran) likewise resident, though winter nomadism reported in Georgian SSR (Dementiev and Gladkov 1951a).

Food. Almost entirely insects throughout the year. During breeding season, feeding behaviour much as *D. major*; searching tree-trunk, branch, twig, and leaf surfaces for insects; acrobatic and hangs upside down like tit *Parus* more than does *D. major* (Rüger 1972; Glutz and Bauer 1980). In winter, Switzerland, over 70% of trees chosen are dead, in April–June only up to c. 40% (Jenni 1983). Excavates wood much less than *D. major* in winter, and surface gleaning remains chief foraging method (Blume 1977); 87% of time gleaning and 6% excavating (10% and 84% in *D. major*) (Jenni 1983). When excavating, confines attention largely to soft and rotten wood (Blume 1977) and is less vigorous than *D. major*, penetrating less deeply. In gleaning, favours furrowed bark, probing with bill and long tongue (Ahlén *et al.* 1978). Apparently uses acoustic as well as visual cues in seeking insects (Ass 1958). Remains less long at a spot before moving on than *D. major*, which often stops to hack at tree; in December–June, 59% of stops last 15–30 s, 23% 30–45 s, 5% 45–60 s, and 13% over 60 s (45%, 16%, 9%, and 30% in *D. major*) (Jenni 1983). In winter, clear preference for foraging on trunk and in lower crown. In winter, Switzerland, 27% of foraging time spent on trunk, 62% in lower crown, and 11% in upper crown (13%, 36%, and 51% in *D. major*); 80% of foraging time spent on branches over 7·5 cm in diameter and 44% over 15 cm (37% and 16% in *D. major*) (Glutz and Bauer 1980; Jenni 1983). In Sweden, only 9% of time on lower ¼ of tree, 26% on crown up to ½ tree-height, 40% on crown up to ¾ tree-height, and 25% on upper ¼; branches generally favoured over trunk (Ahlén *et al.* 1978). In spring, still mainly active on thicker branches (Conrads 1967), but in Sweden forages relatively more on trunk than branches and spends over 70% of time foraging on lower half of tree (Ahlén *et al.* 1978). In Switzerland in winter spends 73% of time on oaks *Quercus*; 60% during April–June (Jenni 1983). Rarely forages on ground and if so mainly in spring; one bird observed rummaging in leaf-litter for c. 7 min (Glutz and Bauer 1980). Ability to use 'anvils' to open nuts and cones only poorly developed. May use existing cracks in bark or flat surfaces in similar way to *D. major*, but never makes own anvils or improves existing ones like that species. Never removes husk, shell, or cone after extracting contents (Bertram 1903). Nuts and fruit stones may be gathered from ground or picked from tree (Marshall 1889; Glutz von Blotzheim 1962; Reichling 1974a). Occasionally takes flesh of fruits still on tree (Conrads 1967; Radermacher 1970), and seen removing seeds from hanging spruce cones (Huber 1953). May visit feeding tables in winter (Radermacher 1970). Though ringing of trees evidently not an important foraging method, observed drinking naturally leaking sap (Löhrl 1972; Jenni 1983) and making new holes of its own (Ruge 1970b). Isolated report of predation on chicks of Starling *Sturnus vulgaris* (Peters 1959). When feeding nestlings, apparently forages exclusively by gleaning (Feindt and Reblin 1959) and not usually more than 100–150 m from nest (Feindt and Reblin 1959; Blume 1977); up to 300–400 m away (Glutz and Bauer 1980). May hawk flying insects at this time (Zollinger 1933).

Animal foods recorded: Coleoptera (adults and larvae of Cerambycidae, Curculionidae, Bostrychidae, Carabidae, Chrysomelidae, Platystomidae, Elateridae), Hymenoptera (Formicidae, Siricidae), Hemiptera (*Eusacoris melanocephalus*, *Idiocercus*), craneflies *Tipula*, larvae of Lepidoptera, earwigs *Forficula*, Orthoptera, harvestmen (Opiliones), spiders (Araneae), and a crab (Csiki 1905; Madon 1930a; Ticehurst and Whistler 1932; Dementiev and Gladkov 1951a; Ahlén *et al.* 1978). Plant food includes barley grains, acorns, seeds of spruce *Picea* and hemp nettle *Galeopsis tetrahit*, hazel nuts *Corylus*, beech mast *Fagus*, cherry, plum, and damson stones, and occasionally grapes, berries, and hemp and sunflower seeds from feeding tables (Bertram 1903; Madon 1930a; Huber 1953; Blume 1977; Glutz and Bauer 1980; Jenni 1983).

Diet 90% by volume animal food (Madon 1930a); largely surface-dwelling insects such as ants (Formicidae) and weevils (Curculionidae), rather than largely wood-boring beetles (e.g. Buprestidae, Cerambycidae) as found in other *Dendrocopos* (Blume 1977). Wood-boring larvae of Coleoptera and Hymenoptera recorded only by Naumann (1826). Vegetable food may be important for brief periods during winter (Glutz and Bauer 1980). For early stomach analyses, see review by Meylan (1932). 9 stomachs from Hungary, July–September and December–March, contained many small beetle and ant adults, mostly the beetle *Anthribus variegatus* (300, 350, and 400 in 3 stomachs from July, September, and December); ants found were *Myrmica laevinodis* (120 in 2 stomachs), *Lasius fuliginosus* (70 in 3 stomachs), *L. alienus* (20 in 1 stomach), *Formica rufibarbis* (10 in 1 stomach), and 1 *F. rufa*; beetles included Carabidae (*Bembidion*, *Amara aulica*), Curculionidae (*Phyllobius oblongus*, *Sitona trivalis*, *Magdalis polydrosus*), and Chrysomelidae (*Lema cyanella*); also Hemiptera (*Eusacoris melanocephalus*, *Idiocercus herrichi*, *I. scurra*) and an unidentified grasshopper (Orthoptera) (Csiki 1905). 11 stomachs from Moldavia, all year, contained acorns, walnuts, and maize (in autumn and winter only), and, in descending order of frequency, ants, small beetles, bugs (Hemiptera), flies (Diptera), and caterpillars (Lepidoptera) (Averin and Ganya 1970). 3 stomachs from Yugoslavia, February, May, and September, contained ants, beetle larvae (Elateridae), adult Curculionidae, other unidentified beetles, pieces of crab, and maize grains (Kovačević and Danon 1950–1, 1957). For diet in Albania, see Ticehurst and Whistler (1932). In Sweden, spiders (Araneae) important in diet (Ahlén *et al.* 1978).

Young generally fed animal food only, gleaned from leaves and twigs (Blume 1977). Many food items carried in bill and passed to young in several feeds (up to 6) in each visit. If food dropped adult flies down to retrieve it (Feindt and Reblin 1959). In one study, flesh of cherries formed considerable proportion of nestlings' diet (Glutz and Bauer 1980; Jenni 1983). Animal foods recorded include adults, pupae, and larvae of Lepidoptera, especially *Tortrix viridana*, ants, aphids, tipulid flies, earwigs, harvestmen, and spiders (Wichmann 1952; Wassenich 1960; Glutz von Blotzheim 1962; Blume 1977; Jenni 1983). A simultaneous comparison with *D. major* showed *D. medius* brought significantly more Lepidoptera larvae and fewer pupae than *D. major*, very few aphids (many brought by *D. major*), but small numbers of soft-bodied cantharid beetles (absent in *D. major*) (Jenni 1983). In several days continuous observation, 238 feeding visits made in 18 hrs; maximum recorded rate 27 visits 08.00–09.00 hrs; lowest rate recorded in early afternoon (Feindt and Reblin 1959). Similar pattern observed by Steinke (1977), who found that young received first feed 04.08 hrs, also by Blume (1977), with last feed *c*. 20.30 hrs. In May, 16 visits observed 10.50–11.50 hrs, 14 visits 16.35–17.45 hrs, and 33 visits 09.45–12.10 hrs (18 by ♂, 15 by ♀) (Wassenich 1960). Feeding visits every 3–5 min 07.10–07.30 hrs, mostly by ♀ (Zollinger 1933); visits every 12 min by ♂, every 14 min by ♀ (Steinke 1977). Several periods of up to 90 min recorded with no feeding visits (Feindt and Reblin 1959). See also Schubert (1978). Striking variation between pairs observed in feeding pattern at end of day; some made only 2–3 visits 18.00–19.00 hrs, others still very active after 20.30 hrs. Adults appear able to judge hunger of young and feed accordingly (Feindt and Reblin 1959). If parents prevented from visiting nest by disturbance will increase feeding rate immediately afterwards (Blume 1977). No tendency for feeding rate to rise with age of young, or to decrease prior to fledging; apparently unaffected by weather (Feindt and Reblin 1959; Wassenich 1960; Blume 1977). ASR

Social pattern and behaviour. Major studies by Steinfatt (1940), Feindt (1956), and Feindt and Reblin (1959). Includes material supplied by B Pettersson for Östergötland (Sweden).
1. Mostly solitary outside breeding season, birds apparently defending fixed feeding territories; in Östergötland, these seemed to be slightly smaller than breeding territories (see below), possibly due to provisions of winter food by man but sample small (2 territories) and further data required. Birds which remained on breeding territories mostly ♂♂ (B Pettersson). Established pairs said to remain faithful to breeding areas outside breeding season, partners maintaining a certain amount of contact (Steinfatt 1940). May form temporary feeding association with other species, occasionally with tits (Paridae), more rarely with Great Spotted Woodpecker *D. major* and Lesser Spotted Woodpecker *D. minor* (Glutz and Bauer 1980). BONDS. Mating system monogamous (B Pettersson). Pair-bond mostly seasonal, presumably renewed from year to year—but see above (Steinfatt 1940). ♂ and ♀ share nest-duties. Family bonds maintained for up to *c*. 2 weeks after fledging (Steinfatt 1940; Feindt and Reblin 1959). Age of first breeding not known. BREEDING DISPERSION. Solitary and territorial. Estimates of territory size vary considerably, from 3·3 ha minimum, Bodensee, West Germany (see Jacoby *et al*. 1970), to *c*. 25 ha, including slightly less than 10 ha of essential oak *Quercus* habitat, Östergötland (B Pettersson). Breeding and feeding territory usually coincide, and may be fairly strictly defended (B Pettersson), or else feeding areas of neighbouring pairs may overlap (see below), in which case area near nest likely to be most closely defended. Territory established by ♂; serves for nesting and rearing of young. Shows marked fidelity to territory (Steinfatt 1940), but apparently always makes new nest-hole each year (Feindt and Reblin 1959). May nest relatively close together, e.g. 2 nests 52 m apart, with nest of *D. major* midway between (Jenni 1977); 2 nests in orchards and meadows *c*. 200 m apart (Schubert 1978); in parkland, 3 nests 350 m, 350 m, and 600 m apart, with large overlap in feeding areas (Feindt and Reblin 1959). In typical woodland habitat, 0·7–1·4 pairs per 10 ha (Jenni 1977, which see for references); e.g. in 11-year study of 25 ha, West Germany, average 0·8 (0–1·2) pairs per 10 ha (Pfeifer and Keil 1961). In 23·4 km^2 of oak woodland, Switzerland, 0·52 pairs per 10 ha (Müller 1982). Up to 2·4 pairs per 10 ha reported (see Glutz and Bauer 1980). In Östergötland, maximum 1 pair per 100 ha, or 0·5 pairs per 10 ha of oak (B Pettersson). For factors influencing density, see Jenni (1977). ROOSTING. Nocturnal. In holes in trees, occasionally also nest-boxes (Zollinger 1933). ♂ roosts in nest-hole before laying (Steinfatt 1940), and continues to roost with young until *c*. 12 days old (Feindt and Reblin 1959); once reported roosting with young at least 17 days old (Steinke 1977). No information from outside breeding season. Like *D. major*, regularly calls (see 4 in Voice) from roost-tree or neighbouring tree before entering roost (Feindt and Reblin 1959; Blume 1977). No detailed studies of activity rhythm, but see Food. Sunbathes with plumage ruffled, eyes closed, and wings drooped (Glutz and Bauer 1980, which see for drawing).
2. Adopts motionless Erect-posture (as in White-backed Woodpecker *D. leucotos*) when disturbed. Mildly disturbed and excited birds also raised crown feathers, more so than in most other Picinae (Glutz and Bauer 1980). For Distress-call of birds held in hand, see 7 in Voice. ANTAGONISTIC BEHAVIOUR. ♂ demarcates territory mainly by Advertising-call (see 1 in Voice), rarely by Drumming (e.g. Voigt 1933, Ferry 1962, Schubert 1978). ♀ also Drums (Haller 1938), but equally rarely. Advertising-call typically delivered while perched on horizontal branch, just like passerine bird (e.g. Holmbring 1972, which see for photograph). Established ♂♂ call from song-posts in territory, younger birds often well away from areas subsequently used for breeding (Feindt 1956; Feindt and Reblin 1959), suggesting self-advertising then serves more to attract mate. Similar dissociation of calling from breeding sites also reported by Schubert (1978) who found (e.g.) that 5 ♂♂ which called in wood later bred in orchards up to 500 m outside it. Advertising-call may also be given in Fluttering-flight: ♂ flies with conspicuous whirring or shivering action, wings moving in light, fairly loose manner, but quite rapidly; at high intensity, tail raised and widely fanned. Not uncommonly, more than 2–4 birds in vicinity of displaying bird, and fighting (see below) may ensue (Feindt 1956; Feindt and Reblin 1959). Birds call mostly early morning and evening (Schubert 1978); from mid-January, mostly March to April or early May, sporadically thereafter (Feindt and Reblin 1959; Ferry 1962; Schubert 1978); resurgence in late May and early June probably unpaired ♂♂ (Schubert 1978). At start of breeding season, association of several birds, especially neighbouring pairs, may lead to fights (Feindt and Reblin 1959; Glutz and Bauer 1980). Confrontations between ♂♂ silent; typically, perched rivals track (visually) each other's every movement. Threaten with Forward-posture: bird crouches low, and extends neck to point bill at rival (Glutz and Bauer 1980, which see for drawing); also uses Wing-spreading display (Fig A, right) similar to other Picinae, though this more often reported in heterosexual context (see below). Head-swaying display of many other Picinae not reported, but since antagonistic behaviour said to be similar to *D. major* (Glutz and Bauer 1980), presumably occurs. In boundary disputes, rival ♂♂ often challenge each other from opposite sides of low branch and try to peck each other;

typically chased for several minutes; in 22 disputes, no ♀♀ recorded (B Pettersson). Sometimes, at high intensity, supplanting attack leads to physical contact in which combatants may fall to ground interlocked (Feindt and Reblin 1959; Conrads 1975). Subordinate to *D. major*, even at own nest-hole (Conrads 1975), but able to assert itself outside breeding season (see Glutz and Bauer 1980). In response to Contact-call of *D. minor*, ♂ gave Advertising-call combined with Wing-spreading display (Steinfatt 1940). Successful in chasing off smaller competitors for nest-sites, including Starling *Sturnus vulgaris*, challenging with threat postures and supplanting attacks (Glutz and Bauer 1980). *D. major* and *S. vulgaris* challenged by soaring in Fluttering-flight, calling or silent, towards intruder (B Pettersson). HETEROSEXUAL BEHAVIOUR. (1) Pair-bonding behaviour. Protracted, beginning in late winter and spring. Advertising-call serves probably more to attract mate than to demarcate territory. Unpaired ♂♂ probably rove widely in breeding areas, calling in various places, and pairs often formed in areas not subsequently used for breeding (Feindt 1956; Feindt and Reblin 1959; Schubert 1978). Duration of calling by individual birds suggests ♂♂ seldom remain unpaired for long. Early contact between ♂ and ♀ indicates conflict between tentative approach and retreat. As in other Picinae, conflict expressed variously by agitated calling, raising of crown feathers, and displacement activities such as pecking and comfort behaviour (Glutz and Bauer 1980). In common sequence, pair appeared to play 'peek-a-boo' on tree trunk: ♂ flew to trunk, landing *c*. 3 m up, and started climbing. ♀ flew to opposite side of tree; ♂ climbed up straight or spiralling (always to left), sometimes jerkily, sometimes with short, slow, and even steps, with wings open and trembling slightly (Wing-spreading display: Fig A, right). ♂ gave presumed Appeasement-calls (see 5 in Voice) during display, and also when he halted periodically with bill forward almost parallel to trunk (perhaps Forward-posture), but not when climbing spirally. ♀ remained silent, wings closed, keeping opposite to ♂ throughout his ascent. Birds usually climbed 1–2 m on tree before ♂ flew to another, ♀ following. 5–6 trees thus climbed in succession (Szemere 1927–8). At outset, also much aerial chasing and display by ♂ to ♀, including especially Fluttering-flights (Feindt and Reblin 1959), sometimes leading to copulation (see below). (2) Nest-site selection. Nest-showing by ♂ involves attracting ♀ to it by Advertising-call, by Tapping around hole, and by performing Wing-spreading display. ♀ responds only near hole. ♂ also occasionally attracts ♀'s attention to nest-hole by performing Fluttering-flight towards it (Glutz and Bauer 1980). Both ♂ (mostly) and ♀ excavate hole (B Pettersson). Time taken for excavation varies with substrate, e.g. 8 days (Weitnauer 1962); ♂ completed hole in *c*. 20 days, apparently without help from ♀ (Steinfatt 1940). More than one hole may be made before one chosen for nesting (Wichmann 1952), though not known which sex makes final choice. Up to 15 days between completion of hole and laying (Feindt and Reblin 1959). (3) Mating. May begin February but said to occur only after hole excavation well advanced, and most frequent during laying (Steinfatt 1940; Conrads 1975). Mostly near nest (B Pettersson) but may also occur quite far from nest-tree, and young birds may copulate outside eventual breeding areas (Feindt and Reblin 1959; see above). Sometimes ♂ attracts ♀ over appreciable distance with Advertising-calls (Steinfatt 1940). Copulation often preceded by many approach attempts, of various kinds, by ♂, e.g. as described in pair-bonding behaviour, above; ♂ may also follow ♀ in Fluttering-flight or chase her more directly in so-called 'Driving' pursuits (Feindt and Reblin 1959; Glutz and Bauer 1980; see also 6 in Voice). Closer confrontations, rich in conflict behaviour (see above, and drawings in Glutz and Bauer 1980), also often precede copulation. When ♀ receptive, she crouches on horizontal branch, quivering wings; sometimes solicits ♂ vocally (see 2 and 5 in Voice) or by Tapping (Steinfatt 1940). ♂ flies to join her. Both birds excited, with crown feathers raised, approach each other until open bills touch. ♀ presses herself, and her fanned tail, flat on branch. ♂ circles ♀ with shuffling, sliding movements, wing slightly open and quivering, or fully spread, then mounts (Feindt and Reblin 1959). Quiet Contact-calls (see 2 in Voice) heard during copulation, which is brief; ♂, crown still raised, finally tips over to one side or hangs with wings outspread before flying off, sometimes with Appeasement-calls; ♀ remains for some time on branch (Steinfatt 1940; Peitzmeier 1956; Feindt and Reblin 1959; Glutz and Bauer 1980). Pair may copulate twice in rapid succession. In one sequence, ♀ flew to nest-hole, looked in, then flew after ♂, and copulation followed; pair copulated again *c*. 3 hrs later, and again 2 hrs after that, both times ♂ flying to ♀; pair copulated up to 6 times per day (Steinfatt 1940). (4) Behaviour at nest. At nest-relief, bird announces arrival with Appeasement-call. ♀ may also give Contact-call on arrival, ♂ answering with Appeasement-calls from inside nest. Incubating bird may leave nest periodically and briefly to preen, etc. ♂ has stronger ties than ♀ with nest, and he incubates at night (Steinfatt 1940, which see for other calls during nest-relief: see also Voice). Pair-members tend to avoid close contact, and meetings at nest-hole during chick-rearing usually arouse mild conflict (e.g. Feindt and Reblin 1959). When ♂ and ♀ arrived at same time with food for young, both raised crown feathers, ♂ looked 'demonstratively' away from ♀, and both flew off almost immediately giving 'zerr' sounds (see 5 in Voice). Once, on such a meeting, ♂ climbed and Drummed or Tapped briefly (3 strikes) before flying off calling, allowing ♀ to feed young. After 1 young fledged, pair apparently more relaxed at nest, giving Contact-calls (Steinke 1977, which see for other interactions and calls between pair-members at nest). RELATIONS WITHIN FAMILY GROUP. Young apparently brooded very little, even during first few days after hatching (Steinfatt 1940). Adults enter hole to feed young for first 14 days, thereafter feed them only at entrance, as in other Picinae. Feeding 'carousel' (as in Kingfisher *Alcedo atthis*) perhaps operates among young to ensure even distribution of food (Feindt and Reblin 1959). On day of fledging, ♀ Tapped weakly near nest-hole, possibly as encouragement to fledge. Shortly before fledging, young not seen at nest-entrance and only quiet 'whispering' sounds heard from inside. Food-bearing ♂ stuck head into hole 24 times and Tapped around hole; as no young came to hole, ♂ swallowed food and flew off with 'zerr' sounds. When 1st young fledged, ♂ fed it near nest while ♀ attended to remaining young in nest. All fledged within 4 hrs 50 min (Steinke 1977). Fledging of whole brood may take up

A

to 1 day (Feindt and Reblin 1959, which see for references). Young fly strongly on leaving nest (Feindt and Reblin 1959; Weitnauer 1962). Give contact-calls between feeds, and exchange contact-calls with parent arriving to feed them (Steinke 1977). Parents continue to feed young for *c*. 14 days after fledging (Steinfatt 1940; Feindt and Reblin 1959). One brood of 6 young perched together in same tree 11 days after fledging (Weitnauer 1962). ANTI-PREDATOR RESPONSES OF YOUNG. For Distress-call, see Voice. PARENTAL ANTI-PREDATOR STRATEGIES. For Scolding-calls when disturbed at nest, see 3 in Voice. No further information.

(Fig A from drawing in Szemere 1927–8.) EKD

Voice. Freely used, especially in breeding season, less often at other times (Ferry 1962). Exceptional among west Palearctic *Dendrocopos* in that Drumming little developed, and vocal advertising (see 1 in Voice) much more important. Tapping described as loud, slow, and rhythmic (Winkler and Short 1978; see Social Pattern and Behaviour for contexts).

DRUMMING OF ADULTS. Only confirmed quite recently that Drumming occurs. Although claimed by Haller (1938) to be performed regularly by both sexes at any time of day up until fledging, this refuted by more recent studies, and evidently rare; in some cases, may have been confused with Tapping (Winkler and Short 1978). Drumming said to be brief and rhythmic, with hollow quality, and not far-carrying (Ferry 1962, which see for review of previous reports, and comparison with other *Dendrocopos*). According to Wallschläger (1980, which see for sonagram), average of 25·6 (18–30) strikes per series ($n=13$); average duration of series 1·29 s (1·03–1·78), *c*. 6–12 series per min. Typically gives 2 series in rapid succession, pausing before giving 2 more, and so on (Wallschläger 1980; S Palmér). In recording (Fig I), 2nd in each pair of series distinctly shorter than 1st (E K Dunn). Strike-rate constant through series, with average 57 ms between strikes, sometimes audibly longer than in Lesser Spotted Woodpecker *D. minor* (49 ms), the only other west Palearctic *Dendrocopos* to Drum with constant strike-rate (Wallschläger 1980; see also Zabka 1980).

CALLS OF ADULTS. (1) Advertising-call. Most characteristic call in spring, said to be given only by ♂ (see Ferry 1962), but according to Géroudet (1948) also rarely by ♀; sometimes by pair together (Blume 1977). Rendered a far-carrying, nasal 'gwääh', like Jay *Garrulus glandarius* (Steinfatt 1940); a throttled 'quäh quäh' (Steinke 1977); 'ähk-ähk-ähk' (B Pettersson); harsh call, whirring and mewing, like axle in need of greasing, and delivered typically in slow series, at rate of 1 per s (Géroudet 1948); see Fig II. May be given singly, but in 92% of 48 cases in series of 2–14, mostly 4–6 (Winkler and Short 1978, which see for details of structure). For series of 31 calls, see Feindt (1956). Rate of calling highly variable, depending on context (Ferry 1962). Given from song-posts, also while feeding, and in low flight between trees, especially during Nest-showing. For details, and seasonal occurrence, see Social Pattern and Behaviour. For additional sonagram, see Bergmann and Helb (1982). (2) Contact-call. A quiet 'teuk' (Ferry 1962); 'djüg' or 'güg' (Steinke 1977). Compared with *D. major*, seldom heard, of lower pitch, and less ringing (Géroudet 1948; Ferry 1962). Average pitch of 3 calls 1·6 kHz, average length 39·0 ms.

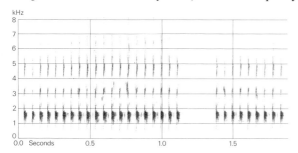

I S Palmér/Sveriges Radio (1972) Sweden June 1962

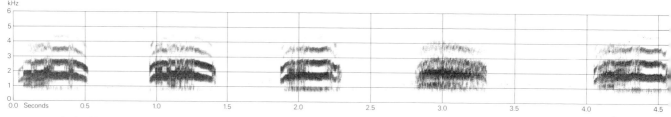

II P Szöke Hungary

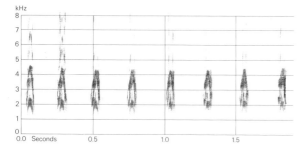

III S Palmér/Sveriges Radio (1972) Sweden June 1962

Given throughout the year as communication signal, and as expression of general excitement, most often while feeding (Steinfatt 1940; Ferry 1962; Glutz and Bauer 1980). (3) Scolding-call. Clear, rapid sequence of units similar to Contact-call, but shorter, louder, and slightly higher pitched. Average pitch of 6 calls 1·8 kHz, average duration of units 54·1 ms (Winkler and Short 1978). Given especially when disturbed at nest (Glutz and Bauer 1980). (4) Rattle-call. Commonest call, rendered 'kik kekekek' (Winkler and Short 1978), 'gíg gegegeg', pitch falling after 1st sub-unit (Glutz and Bauer 1980). Average pitch 1·6 kHz ($n=70$), average duration of units 32·5 ms ($n=55$). Series varies markedly in length (up to 5·5 s) and rhythm, typically speeding up towards middle and slowing towards end (Winkler and Short 1978). In recording (Fig III, which shows end of bout), 'chak chak chak ...', similar to Blackbird *Turdus merula*, pitch mainly *c.* 4 kHz (J Hall-Craggs, P J Sellar). Given throughout the year, notably in disputes and at times of sexual tension (Winkler and Short 1978, which see for variants), also often shortly before entering roost (Blume 1977). Confusion with call 3 not unlikely. (5) Appeasement-call. Similar to 'wäd' or 'rä' of *D. major*. In close contact between pair-members, has quiet, muffled, rasping quality (Szemere 1927–8), at greater distance a 'smacking' quality (Winkler and Short 1978). Given as appeasing signal, and to invite close approach, in variety of contexts, e.g. during pair-formation, copulation, and nest-relief (see Social Pattern and Behaviour). Sounds described as muffled 'krütt' during nest-relief (Steinfatt 1940), also 'ätt' and 'zerr' (Steinke 1977), are perhaps this call or a variant. Soft 'uik-uik' and 'uät-uät' by ♀ soliciting copulation, and 'gäk-gä-ak' during copulation, also considered variants (Glutz and Bauer 1980). (6) Driving-call. Harsh 'grüüüd' or 'grägögö', given by ♂ during 'Driving' pursuits of ♀ and at times of conflict (Glutz and Bauer 1980). (7) Distress-call. Loud screeching sound, *c.* 395 ms long, given by birds held in hand (Winkler and Short 1978). (8) Other calls. For other sounds, probably same or variants of those above, see Steinfatt (1940).

CALLS OF YOUNG. Food-calls before and after fledging similar to *D. major* but much quieter (Glutz and Bauer 1980); comprise a loud chirping, rendered 'tschätschä-tschätschätschä' (Steinfatt 1940), and also squeaking and squealing sounds; a monotonous rhythmic 'zuhzuhzuh ...' (E K Dunn). Near and especially after fledging, young also give a contact-call; in one case, fledgling gave series of 'gück' sounds, in comparison with 'djüg' of ♀ parent arriving to feed it. Perched fledgling also Tapped, and gave quiet 'kück' sounds (Steinke 1977). Young may also give apparent Distress-call, similar to adult (Glutz and Bauer 1980).
EKD

Breeding. SEASON: Central Europe: see diagram. Little apparent variation across range. SITE. Hole in tree (usually decaying, less often healthy: Feindt and Reblin 1959),

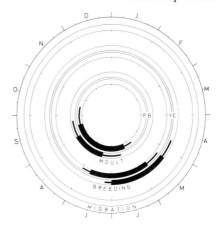

1·25–4·5 m above ground; exceptionally, natural hollow behind bark (Dementiev and Gladkov 1951a; Weitnauer 1962). Nest: excavated hole, *c.* 35 cm deep and 12 cm in diameter, with entrance hole 5 cm in diameter (Weitnauer 1962). Building: by both sexes, perhaps mostly ♂ (Steinfatt 1940; Feindt and Reblin 1959), taking 8–20 days (Weitnauer 1962). EGGS. See Plate 98. Elliptical, smooth and glossy; white 25 × 18 mm (22–28 × 18–21), $n=72$ (Makatsch 1976). Weight 4·0 g (3·75–4·3), $n=6$ (Matoušek 1956). Clutch: 4–7(–8). Of 59 clutches, Hungary: 4 eggs, 8; 5, 17; 6, 24; 7, 8; 8, 2; average 5·64±SD 1·0 (Glutz and Bauer 1980). One brood. No information on replacements. Eggs laid daily (Glutz and Bauer 1980). INCUBATION. 11–14 days (Feindt and Reblin 1959; Weitnauer 1962). By both sexes. YOUNG. Altricial and nidicolous. FLEDGING TO MATURITY. Fledging period 22–23 days, becoming independent 10–14 days later (Feindt and Reblin 1959; Weitnauer 1962). Age of first breeding not known. BREEDING SUCCESS. No information.

Plumages (central European nominate *medius*). ADULT MALE. Forehead and tufts covering nostrils buff-white, pale grey-buff, or buff, tufts sometimes brown-buff; tips of longer bristles of tufts black. Narrow and elongated feathers of crown and back of head bright crimson-red, some grey or buff of feather-bases sometimes visible (crown of juvenile Great Spotted Woodpecker *D. major* sometimes similar, but feather-bases extensively black, back of head black, and crimson extends to forehead). Sides of head and neck silvery-white, usually slightly mottled grey on lores and ear-coverts (in particular when plumage worn); indistinct streak from lower mandible to below ear-coverts buff or pale greyish-buff. Large black patch on rear of lower cheeks, broadly extending downwards along sides of neck to sides of chest, more narrowly up behind ear-coverts towards sides of nape; latter extension variable—in some, just reaches sides of nape, in others extends only half-way across and white of ear-coverts and sides of neck then connected (though this appearance affected also by bird's posture). Central hindneck, mantle, shorter inner scapulars, back, rump, and upper tail-coverts black, feathers with variable amount of fuscous-brown on base, brown often distinctly showing on mantle and back, in particular when plumage worn; never as glossy bluish-black as *D. major*. Outer and longer scapulars white, black-brown of bases sometimes showing through slightly. Chin, throat, and chest white, sometimes with slight yellow or buff tinge, more greyish when plumage worn. Breast and upper flanks (below large black patches on sides of chest) pale yellow or cream-buff, grading into pink on upper belly, sides of belly, and lower flanks; feathers of flanks and sides of breast and belly with distinct black or dark grey shaft-streaks 1–2 mm wide; vent and under tail-coverts uniform deep pink or pink-red (some paler pink or white of feather-bases occasionally visible). Central 3 pairs of tail-feathers (t1–t3) black, t3 usually with white, yellow, or orange border along sides of tip (rarely, also faintly showing on t1–t2); t4–t5 black with conspicuously white distal parts and outer webs, barred with 2–3 black bands (sometimes rather irregular in shape), white of tips on undersurface often tinged yellow-buff or orange; reduced t6 (hidden under tail-coverts) black with white spot. Pattern of white spots and broken bars on primaries as in *D. major*; secondaries as in *D. major*, but with distinct white spots on both webs near tip, forming one extra band of spots (in *D. major*, reduced to small spots on tips of outer webs of outer secondaries only); tertials more profusely spotted, only inner webs of inner 2 virtually without white marks (in *D. major*, inner 2 completely black, others with spots on basal and middle parts of outer webs only). Upper wing-coverts black, but greater tertial coverts white with some hidden black marks, white extending into large white spots on inner greater coverts; inner median coverts fully white (in *D. major*, no white on inner greater coverts, median coverts slightly less extensively white, but lesser upper primary coverts more extensively white). Axillaries and under wing-coverts white, in part indistinctly spotted dusky grey. In worn plumage, much grey of feather-bases shows as grey mottling on crown, sides of head, cheeks, and chin; mantle, back, and shorter upper wing-coverts brown rather than black; chest, breast, belly, and flanks dirty yellow or off-white with more extensive streaking; vent and under tail-coverts paler pink with much grey or off-white of feather-bases visible. ADULT FEMALE. Rather similar to adult ♂ and some indistinguishable. Crown less deep crimson-red, more pinkish-red with slightly more grey of feather-bases visible, bordered by pink-brown, yellow, or golden-brown on back of head (in ♂, back of head similar in colour to crown); forehead, lores, chin, and sometimes sides of head slightly buffier, less white or cream; yellow tinge on lower chest, breast, and flanks on average slightly paler and less extensive. NESTLING. Naked at hatching. Closed sheaths appear from 5th–6th day, feather-tips from 10th–11th (Heinroth and Heinroth 1924–6). JUVENILE. Similar to adult, but feathers of crown less narrow and elongated; red of crown less glossy, less extensive, and with more grey of feather-bases visible; unlike adult ♀, no golden-brown on back of head. Back of head of ♂ black with faint brown or red feather-tips; both hindcrown and back of head black in ♀, only faintly showing brown or red, with brighter pink-red restricted to forecrown. Length of red cap 24·6 (6) 24–25 in ♂, 20·1 (6) 18–22 in ♀; red of ♂ brighter carmine, of ♀ more dirty brown-red (Jenni 1980). White of outer scapulars and inner upper wing-coverts narrowly tipped black. Pale feathers of sides of head, cheeks, chin, sides of breast, and flanks with narrow and diffuse brown tips; streaks on flanks and sides of breast and belly browner and less sharply defined; yellow tinge on breast and flanks and pink of vent and under tail-coverts paler and less extensive (rather similar to worn adult, however). P6–p8 usually with white tips $c.$ 2 mm wide (usually fully black in adult, except for white fringe along tip of outer web); p10 relatively long and broad, tip extending 5–12 mm beyond longest greater upper wing-coverts (reduced in adult, tip 4 mm shorter to 4 mm longer than coverts); see also Structure. Differs from juvenile *D. major* in paler red crown, buff-tinged forehead, sides of head, and chin (partly mottled dusky), browner mantle and upper wing-coverts, spotted inner webs of longer tertials, and more heavily streaked flanks and sides of breast and belly. FIRST ADULT. Similar to adult and many indistinguishable once juvenile p10 replaced. Juvenile tertials, secondaries, and all or part of greater upper primary coverts retained; tertials sometimes more attenuated towards tips (not as broad and square at tip as adult), greater upper primary coverts more pointed, less broad at tip; both tertials and greater primary coverts on average slightly more abraded than those of adult at same time of year, difference most marked in spring, but ageing reliable only for extremes.

Bare parts. ADULT AND FIRST ADULT. Iris bright red-brown or pale bright red. Bill dark lead-grey or grey-black with base of lower mandible lighter grey, greyish-pink, or yellow-horn. Leg and foot lead-grey, sometimes with yellow-brown soles and edges of scutes. NESTLING AND JUVENILE. Similar to *D. major* at hatching: bare skin pink, eyelids bluish-grey; bluish feather-sheaths develop from 5th–6th day. Eye opens narrowly on 7th–8th day, fully 10th–11th. Iris brown once eye open, reddish-brown at fledging. Similar to adult once half-way through post-juvenile moult. (Hartert 1912–21; Heinroth and Heinroth 1924–6; Glutz and Bauer 1980; RMNH.)

Moults. Similar to *D. major* in timing and sequence, but post-juvenile starts with p1 at fledging, not before fledging (Glutz and Bauer 1980). According to $c.$ 20 moulting specimens examined, adult starts primary moult early June to early July, completing about mid-August to mid-September; juvenile starts late May to late July, completing early August to late September; primary moult of juvenile on average $c.$ 12 days later than adult, but sample small and perhaps biased towards retarded juveniles.

Measurements. ADULT AND FIRST ADULT. Nominate *medius*. Wing and bill from (1) Kursk and Voronezh areas (central European USSR), (2) Netherlands, Belgium, Denmark, West Germany, and Austria, (3) Yugoslavia, Rumania, Bulgaria, Albania, Greece, and European Turkey; other measurements central Europe and Rumania only. All year; skins (BMNH, RMNH, ZFMK, ZMA, ZMM). Bill (F) to forehead, bill (N) to distal corner of nostril; toe is outer front toe.

WING (1)	♂ 130	(2·21; 9)	127–133	♀ 129	(3·13; 10)	124–134
(2)	128	(2·24; 8)	126–132	125	(1·49; 10)	122–127
(3)	126	(1·74; 26)	124–130	125	(2·15; 11)	121–129
TAIL	79·0	(3·55; 6)	76–84	78·7	(2·73; 12)	75–83
BILL (F) (1)	26·2	(1·02; 9)	25–28	24·8	(1·09; 9)	23–26
(2)	25·7	(0·85; 7)	25–27	25·0	(1·06; 10)	24–26
(3)	26·4	(1·92; 8)	24–29	25·2	(0·95; 5)	24–27
BILL (N)	19·1	(0·90; 25)	18–21	17·8	(0·75; 22)	17–19
TARSUS	20·5	(0·33; 9)	20–21	20·4	(0·37; 15)	20–21
TOE	20·9	(1·08; 8)	20–22	21·3	(1·28; 12)	20–24

Sex differences significant for wing (2) and (3), bill (F) (1), and bill (N).

D. m. caucasicus. Northern Turkey and northern Caucasus, all year; skins (BMNH, ZFMK, ZMA, ZMM).

WING	♂ 126	(1·76; 13)	124–130	♀ 125	(— ; 2)	122–128
BILL (F)	25·7	(0·80; 13)	24–27	25·2	(— ; 2)	24–26
BILL (N)	18·7	(0·84; 13)	18–20	18·6	(— ; 2)	18–19

Sex differences not significant.

D. m. anatoliae. (1) Western Turkey (Izmir area to Lycia), (2) south-central Turkey (Mersin to Maraş and neighbouring Taurus mountains), all year; skins (BMNH, RMNH, ZFMK).

WING (1)	♂ 122	(1·76; 8)	120–124	♀ 120	(1·60; 7)	117–122
(2)	123	(1·98; 7)	120–125	123	(2·53; 5)	119–125

BILL (N) (1) 18·7 (0·60; 8) 18–20   18·1 (0·77; 6) 17–19
         (2) 18·7 (0·61; 7) 18–20   17·9 (0·92; 5) 16–19

Single ♀, Dihok (northern Iraq), tending towards *sanctijohannis* in appearance: wing 124, bill (N) 18·3 (BMNH).

D. m. sanctijohannis. Zagros mountains (Iran), all year; skins (Paludan 1938; Vaurie 1959*b*; BMNH, ZFMK).
WING   ♂ 122 (2·19; 8) 120–126   ♀ 122 (— ; 2) 122–123

JUVENILE. Wing on average 4–5 shorter than adult and 1st adult, tail 3–4 shorter.

Weights. ADULT AND FIRST ADULT. Nominate *medius*. Central and west European USSR: July, juvenile ♀♀, 54, 55 (ZFMK); late September and October, ♂♂ 53, 57, 58 (ZMM); ♂ 53–61, ♀ 57–59 (Fedyushin and Dolbik 1967). Netherlands, September: ♀, found dead, 50 (RMNH). Central Europe, all year: ♂ 59·2 (2·0; 17) 53–80, ♀ 58·8 (5·0; 14) 50–80 (Glutz and Bauer 1980). West Germany: ♂♂ 63 (January), 56, 58 (March); ♀ 57 (April); juvenile ♀ 70 (July) (ZFMK). European Turkey, May: ♂ 42 (Rokitansky and Schifter 1971).

D. m. anatoliae. Western Turkey, June–July: ♀♀ 47, 48, 55 (Rokitansky and Schifter 1971; ZFMK).

D. m. sanctijohannis. Zagros (Iran), May: ♀ 49 (Paludan 1938).

NESTLING AND JUVENILE. At hatching, *c.* 3·3. 1 day before fledging: ♂ 56·2 (6) 53–60, ♀ 52·7 (6) 47–56 (Jenni 1980).

Structure. In adult and 1st adult, p7 longest, p8 0–3 shorter, p9 14–20, p10 69–80, p6 0–2, p5 3–6, p4 15–19, p3 24–28, p2 29–33, p1 30–38; in juvenile, p8 longest, p9 10–15 shorter, p10 61–67, p7 0–2, p6 2–5, p5 7–11, p1–p2 reduced, about half length of neighbouring s1. P10 reduced in adult and 1st adult, narrow and pointed, 4 shorter to 4 longer than longest greater upper primary coverts (on average, 0·2 shorter, *n* = 16); juvenile p10 longer and broader, 7 (6) 5–12 longer than primary coverts. Outer web of p5–p8 and inner web of (p6–)p7–p9 emarginated. Tail rather long, graduated; t1 longest; in adult, t2 3–7 shorter, t3 8–14, t4 14–23, t5 20–27, t6 61–68; in juvenile, t2 8–11 shorter, t3 10–14, t4 14–19, t5 20–27, t6 56–62; t6 reduced (least so in juvenile), hidden under tail-coverts. Feathers of crown and nape narrow and elongated in adult and 1st adult, and can be raised to form ragged crest. Bill weak, shorter than head, relatively narrow at base, gradually attenuated to sharp pointed tip, without distinct grooves or ridges, except for narrow ridge on top of culmen. Tarsus, toes, and claws relatively longer and more slender than in *D. major*. Inner front toe with claw *c.* 89% of outer front with claw, outer hind *c.* 106%, inner hind *c.* 51%. Remainder of structure as in *D. major*.

Geographical variation. Rather slight, clinal. Nominate *medius* from central European USSR largest and palest, with forehead, sides of head, chin, and chest almost pure white, and yellow on breast, belly, and flanks rather restricted and not as intense as in nominate *medius* from central Europe (described in Plumages); streaks on sides of belly and flanks sometimes rather pale grey and narrow, hardly contrasting, tips of t4–t5 mainly white with limited amount of black marks; individual variation marked, however, and many not separable from those of central Europe. Birds from southern Europe smaller and darker than those from central Europe; readily separable from those of central European USSR, but those of southern Sweden, Poland, West Germany, and France intermediate. Birds from south-east Europe, sometimes separated as *splendidior* (Parrot, 1905), have heavier and blacker streaks on flanks than typical nominate *medius* from Sweden, yellow-buff of breast deeper and extending up to chest, vent and under tail-coverts brighter pink-red; tail-bars as in typical nominate *medius*. Birds from northern Spain, sometimes separated as *lilianae* (Witherby, 1922), similar to those of south-east Europe, but mantle slightly deeper black and flanks slightly barred with dusky grey. Typical *caucasicus* from Caucasus and Transcaucasia rather like nominate *medius* from south-east Europe, but breast and belly brighter golden-yellow (even in worn plumage), pink-red patch on vent and under tail-coverts more restricted (not reaching belly and lower flanks), and tips of t4–t5 distinctly blacker (black and white bars about equal in width). Birds from northern Turkey rather intermediate between typical *caucasicus* and nominate *medius* from south-east Europe; tail sometimes whiter and bars on flanks less heavy than in typical *caucasicus*. *D. m. anatoliae* from western and southern Asia Minor similar to typical *caucasicus*, but distinctly smaller (see Measurements); pink-red of vent and under tail-coverts often slightly paler than typical *caucasicus*, streaks on flanks and sides of breast slightly heavier, in part tending to form bars on flanks. Single bird examined from Dihok (northern Iraq) similar to *anatoliae*, but breast whitish instead of yellow, tending in this respect towards *sanctijohannis* from Zagros mountains in Iran; typical *sanctijohannis* has breast and upper belly white with numerous and sharp but narrow black streaks on flanks and with contrasting yellow band across lower belly at border of deep pink-red vent and under tail-coverts; size of bird from Dihok and of *sanctijohannis* rather small, as in *anatoliae*.   CSR

Dendrocopos leucotos White-backed Woodpecker

PLATES 83 and 86
[facing pages 855 and 879]

Du. Witrugspecht   Fr. Pic à dos blanc   Ge. Weissrückenspecht
Ru. Белоспинный дятел   Sp. Pico dorsiblanco   Sw. Vitryggig hackspett

Picus leucotos Bechstein, 1803

Polytypic. Nominate *leucotos* (Bechstein, 1803), central, northern, and eastern Europe, south to Austria, to northern and eastern Slovenia, northern Croatia, and northern Serbia in Yugoslavia, and to Carpathians in Rumania and Ukraine, east to western foothills of Ural mountains; *uralensis* (Malherbe, 1861), Urals east to central Siberia; *lilfordi* (Sharpe and Dresser, 1871), Europe south of nominate *leucotos*, Turkey, Caucasus, and Transcaucasia. Extralimital: *c.* 10 further races in Asia.

Field characters. 24–26 cm; wing-span 38–40 cm. Biggest *Dendrocopos* in west Palearctic; *c.* 10% larger than Great Spotted Woodpecker *D. major*. Rather large, rather long-billed, and long-necked woodpecker, with black upperparts boldly barred white over wings and usually completely white on lower back and rump, and white underparts copiously streaked. Sexes dissimilar; no seasonal variation. Juvenile separable. 3 races in west Palearctic, 1 separable from other 2 at close range.

ADULT MALE. (1) Northern races, nominate *leucotos* and *uralensis*. Differ distinctly from all other *Dendrocopos* in: (a) wholly white lower back and rump; (b) broad white bands across both wing-coverts and flight-feathers; (c) lack of bold white scapular-patches of *D. major*, Syrian Woodpecker *D. syriacus*, and Middle Spotted Woodpecker *D. medius*. Crown and back of head crimson-red with narrow black edge. Black moustache extends upwards behind ear-coverts (but does not meet nape) and downwards on upper breast, breaking up into long blackish streaks which extend over whole of flanks. Belly tinged buff-pink, merging with pink-red vent. White, narrowly barred outer feathers form conspicuous pale side to black tail. (2) South European and Turkish race, *lilfordi*. Rump invaded by black bars, flank-streaks heavy and more numerous, and outer tail-feathers more heavily barred. ADULT FEMALE. All races. As ♂, but crown black. JUVENILE. All races. In ♂, crown mixed red and black (back of head all-black, unlike adult ♂); less red in ♀, often very little. Red on underparts paler and less extensive than adult.

Unmistakable in flight, when rather large size, white lower back, and broadly barred wings quickly catch eye. ♂ liable to confusion with *D. medius* in brief perched view, though larger size, complete moustache (reaching bill), and other characters evident in prolonged observation. Flight action slower and more powerful than *D. major*; other behaviour similar. Less shy than congeners.

Contact-call 'kiuk', softer than *D. major*; may recall monosyllable of Blackbird *Turdus merula* or Redwing *T. iliacus*. Drumming much longer (c. $1\frac{1}{2}$ s) and slower than in *D. major*, with characteristic terminal acceleration; similar to Three-toed Woodpecker *Picoides tridactylus*.

Habitat. Resident in upper and lower middle continental latitudes, in south of range mainly in upland or montane regions, but preferring warm southerly slopes. In Bayern (West Germany), at 600–1300 m, prefers broad-leaved forest of beech *Fagus*, birch *Betula*, maple *Acer*, ash *Fraxinus*, and elm *Ulmus*, but occurs also in pure conifer stands or in mixed forest with strong coniferous element provided there are enough rotten trees and primitive natural structure (Niethammer 1938); for similar habitat in Norway, see Håland and Toft (1983). In central Sweden, especially in alder *Alnus*, birch, aspen *Populus*, and oak *Quercus* (Anderson and Hamilton 1972). In Finland, favours similar habitat; often on swampy lake shores (Sarkanen 1974). In Norway, most territories on steep or hilly terrain, facing south to west (Håland and Toft 1983). In USSR, also prefers fairly open broad-leaved forests, often in river floodplains, with good sprinkling of fallen birch trunks, but in some areas occupies mixed woods or even conifer stands such as pine *Pinus* and larch *Larix*, and in others accepts open areas with only a few trees; on slopes of Caucasus, ascends to timberline at 1700 m (Dementiev and Gladkov 1951a). In Spanish Pyrénées, prefers beech in pure stands or mixed with silver fir *Abies alba*, from c. 800–1700 m (Purroy 1972). Type of woodland preferred appears to be antithesis of that favoured by forestry managements, including a high proportion of decayed or fallen timber and slow cycle of natural regeneration, with much overmature stock. Increasing spread of economic forestry, particularly with emphasis on conifers, seems unlikely to leave so much room for this species, except in inaccessible or deliberately conserved areas (see Population).

Distribution. Range may have formerly been more extensive in parts of central Europe (Glutz and Bauer 1980), and even now perhaps not fully known as birds may be overlooked because of unobtrusive habits. Marked decrease of range in Sweden and perhaps elsewhere.

FRANCE. Bred 19th century in Pyrénées, then considered extinct, but rediscovered 1936 and 1953; now local (Yeatman 1976; RC). Corsica: specimens collected in 19th century; unconfirmed records in late 1970s (J-C Thibault). WEST GERMANY. May have bred more widely in 19th century (see Glutz and Bauer 1980 for details). SWEDEN. In 19th century bred in some localities to between 65°N and 65°50′N (LR). FINLAND. Range decreased. Breeds occasionally outside area mapped (Sarkanen 1978b). EAST GERMANY. Bred 1872 and 1938 (Makatsch 1981). AUSTRIA. Knowledge of distribution inadequate in some areas (Glutz and Bauer 1980). ITALY. Possibly more widespread in central Appenines than shown on map (PB, BM). TURKEY. Probably more widespread than shown on map (MB, RP).

Accidental. Belgium, Denmark, East Germany.

Population. Information limited. Declined Sweden, Finland, and perhaps elsewhere.

SPAIN. Local but not scarce (AN); see also Senosiain (1978). FRANCE. Estimated 10–100 pairs, threatened by construction of forest roads (Yeatman 1976). WEST GERMANY. Now breeds only Bayern, where 100–300 pairs (Bezzel et al. 1980). NORWAY. About 500–1000 pairs in west (Håland and Toft 1983). Marked decline in east (GL, VR). SWEDEN. Now under 100 pairs (LR). FINLAND. Major decline due to habitat destruction; now estimated 25 pairs or at most 40–60 pairs (Sarkanen 1978b; OH). POLAND. Scarce (Tomiałojć 1976a). CZECHOSLOVAKIA. Scarce, probably decreasing (KH). HUNGARY. Numbers small (Glutz and Bauer 1980). GREECE. Sparse and local breeder (HJB, WB, GM). ALBANIA. Common, less so in coastal areas (EN). USSR. Fairly common; see Dementiev and Gladkov (1951a) for abundance in various regions. Not frequent now in Estonia and Latvia (HV); rare and decreasing in Lithuania (Jankevičius et al. 1981).

Movements. Resident; also to some extent dispersive,

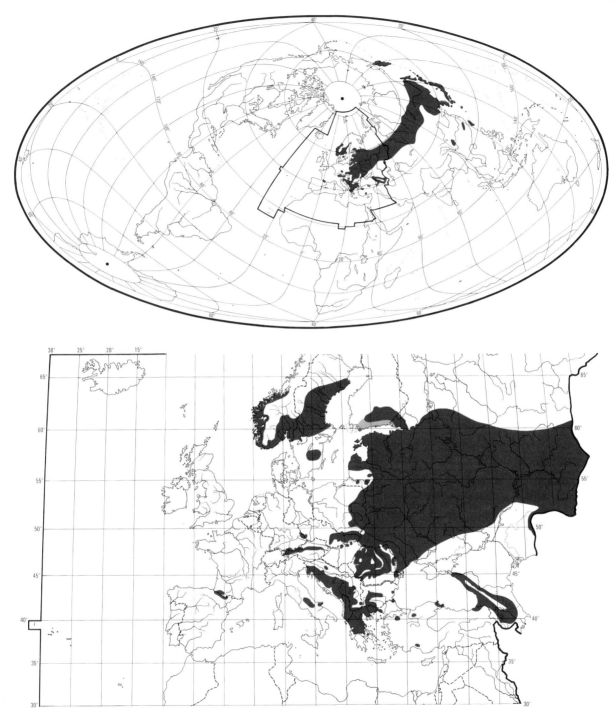

especially in USSR, though unclear how much of this is eruptive.

Reported as nomadic within Russian and Siberian breeding range in autumn–winter, when some individuals also move a little beyond it, apparently more or less regularly. Nominate *leucotos* has then penetrated eastwards across Volga into Bashkirya, south to Yergeni–Sarepta region, and south-west of Dnepropetrovsk; Siberian *uralensis* has occurred in autumn by Taz and Yelaguy rivers, and in winter towards Altay (Dementiev and Gladkov 1951a). In big woodpecker irruptions in Pskov region, autumns 1962–3, very small numbers of *D. leucotos* among many Great Spotted Woodpeckers *D. major* (Meshkov and Uryadova 1972). A few (mainly autumn) records south of breeding range in Norway and Sweden (Haftorn 1971; SOF 1978), and rare stragglers have occurred (autumn–winter) in East Germany, northern West Germany (including Helgoland), and Denmark, (Glutz and Bauer 1980). Probably sedentary in southern breeding areas.

Food. Account includes some material compiled by J R Parrott.

Chiefly insects, especially larvae of wood-boring beetles (Coleoptera), throughout the year; in some places, also nuts and berries outside breeding season. Insect prey obtained mostly by pecking into wood of dead and rotting deciduous trees (e.g. Neufeldt 1958b, Ahlén et al. 1978). In central Sweden, 150 observations of tree use as follows: birch *Betula* 32%, aspen *Populus tremula* 25%, alder *Alnus glutinosa* 15%, oak *Quercus robur* 11%, willow *Salix* 7%, grey alder *Alnus incana* 3%, Scots pine *Pinus sylvestris* 3%, spruce *Picea abies* 3%, old timber 1% (Ahlén et al. 1978). Shows preference for trunks lying on ground, rotten stumps, and lower parts of trunks, e.g. mostly $\frac{1}{2}$–3 m above ground (Virkunnen 1967). In contrast to Great Spotted Woodpecker *D. major* and Middle Spotted *D. medius*, food obtained mainly by pecking deep rather than from surface of wood (Fedyushin and Dolbik 1967; Ahlén et al. 1978). Vigorous pecking action on rotten wood typically gives impression of boring into wood; conical holes thus rapidly created 23 cm or more deep; bird then almost lost to view, peering out alertly from time to time (Franz 1937). May peck at same spot for up to $\frac{1}{2}$ hr (Ruge and Weber 1974b; L L Semago), and will visit heavily infested tree day after day until cleaned out (Neufeldt 1958b). Many trees are completely debarked (Neufeldt 1958b; Ruge and Weber 1974b). Especially favours trees with bark peeling off (Dementiev and Gladkov 1951a); many also peck off strips of green bark, but not thought to 'ring' trees (see *D. major*) or take sap (Franz 1937). Feeding technique varies for different prey. Following methods reported by Ahlén et al. (1978), which see for insects and trees concerned: bark-stripping, pecking fine holes, pecking deep conical holes, pecking off surface of bark or exposed wood. Capable of detecting prey with high degree of accuracy by pecking gently on surface (Ahlén et al. 1978), presumably to detect movements and/or chambers. Some food for young taken from thin twigs in canopy, also by probing in epiphytic mosses and lichens (Franz 1937; Ruge and Weber 1974b). Will take caterpillars from trunk and foliage (Roalkvam 1983), and search leaves by hanging upside down from twigs (Panov 1973). Occasionally also takes insects, especially craneflies (Tipulidae), in flight from perch (Ruge and Weber 1974b). In Spain, autumn, both sexes often perched on branches of hazel *Corylus* to pluck nuts with bill, then flew to insert nut in convenient crevice ('anvil'), typically a vertical crack in hornbeam *Carpinus betulus*. Then cracked open nut by hammering with bill, and ate nut whole, not crushing it as *D. major* would do (Purroy 1972).

In Primor'e (eastern USSR), also uses anvils for handling 'Manchurian nuts' (Panov 1973); as shell impossible to extract once wedged, no repeated use of anvil sites (Polivanov 1981, which see for anvil behaviour). Also picks off and swallows ripening Amur cork-tree *Phellodendron amurense* berries; once tree stripped, may continue to take berries on ground, even digging them vigorously with bill out of snow up to 10 cm deep (Polivanov 1981). Sometimes visits feeding stations to take suet and seeds (see photograph by Ehrenroth 1972; see also Weber 1965, Vikberg 1982). In study of resident pair in autumn, Spain, consistent differences occurred in foraging height of each sex. ♀ preferred to move slowly through the higher branches, searching superficially and periodically lifting small pieces of bark. ♂♂ searched lower and middle parts of trunks, as well as dry stumps; spent long periods pecking holes in wood to obtain larvae. These differences reflected in diet (Purroy 1972; see below). Same habitat separation between sexes appears to hold throughout the year in Spain (Senosiain 1978). No sexual differences in habitat use, Norway (Hogstad 1978). During breeding season, ♂ and ♀ evidently foraged in different parts of territory, as shown by flight-lines (Bringeland and Fjære 1981).

Diet in west Palearctic includes the following. Insects mainly larvae of beetles (Coleoptera: Carabidae, Lymexylidae, Elateridae, Buprestidae, Scarabaeidae, Cerambycidae, Chrysomelidae, Curculionidae, Scolytidae) and larvae and pupae of moths (Lepidoptera: Lymantriidae, Geometridae, Cossidae, Plutellidae); less often, adult and larval Hymenoptera—wood wasps (Siricidae), sawflies (Tenthredinidae), and ants (Formicidae); also some adult and larval flies (Diptera: Tipulidae, Rhagionidae). Plant food includes nuts, berries, and seeds of hazel *Corylus*, oak *Quercus*, buckthorn (Rhamnaceae), blackthorn *Prunus spinosa*, bird cherry *P. padus*, rowan *Sorbus*, pines *Pinus*, and ash *Fraxinus* (Franz 1937; Dementiev and Gladkov 1951a; Dubinin and Toropanova 1956; Neufeldt 1958b; Fedyushin and Dolbik 1967; Dolgushin et al. 1970; Purroy 1972; Ruge and Weber 1974b; Ahlén et al. 1978; Glutz and Bauer 1980; Nuortera et al. 1981). Diet of extralimital birds in USSR also includes grasshoppers, etc. (Acrididae), Cicadidae and Aphrophorinae (Homoptera), centipedes and millipedes (Myriapoda), spiders (Araneae), and molluscs (Mollusca) (Vorobiev 1954; Panov 1973; Polivanov 1981).

In central Sweden, diet mainly wood-boring larvae, comprising up to 50% or more of food delivered to young; prey identified were larvae of beetles *Scolytus ratzeburgi*, *Hylocoetus dermestoides*, *Aromia moschata*, *Necydalis major*, and *Strangalia quadrifasciata*, caterpillars of moth *Argyresthia goedartella*; elsewhere in Sweden, probably also larvae of beetles *Saperda calaris*, *Xylotrechus rusticus*, *Agrilus viridis* and of moth *Cossus terebra* (Ahlén et al. 1978). Of 4 stomachs, Norway, 3 contained *Cossus* larvae (Haftorn 1971). In Pyrénées (Spain), October, stomachs of 1 ♂ and 1 ♀ supported use of deeper probing by ♂: in stomach of ♂, 90% hazelnut fragments, 2 Cerambycidae larvae (over 2 cm long), 7 mandibles, small bark fragments; stomach of ♀ contained 40% hazelnut fragments, remains of a small Carabidae, 1 Buprestidae larva, 75 larvae of fly *Xylophagus* (Rhagionidae), 5 heads of larvae, and 22 mandibles (Purroy 1972). In southern Kareliya (USSR), mainly insects throughout the year; in winter, mostly beetle larvae, especially *S. ratzeburgi* and *S. scalaris*, also *C. cossus* caterpillars, and, rarely, wood-ants (Formicidae). In rest of year chiefly adult and larval beetles *Ips typographus*, *Rhagium inquisitor*, and Elateridae. In summer, also small numbers of larval Chrysomelidae, Curculionidae, Geometridae, and Siricidae. In 1 stomach, 228 larval *S. ratzeburgi*; also larval Siricidae, Cerambycidae, and *C. cossus*. No marked seasonal variation in diet; 10 stomachs contained 475 adult and larval beetles (Elateridae, Cerambycidae, Chrysomelidae, Curculionidae, Scolytidae); 6 contained 23 larval and pupal Lepidoptera (*C. cossus* and Geometridae); 3 contained adult and larval Siricidae (Neufeldt 1958b). Stomachs from European USSR contained up to 80% wood-boring larvae: mainly Cerambycidae and Scolytidae, less often Buprestidae, Siricidae, and ants *Camponotus* (Dolgushin *et al.* 1970). In June, middle reaches of Ural river (USSR), 2 stomachs contained 6 larvae of moth *Lymantria dispar* and 1 each of click beetles *Elater sanguinolentus* and *Lacon murinus* (Dubinin and Toropanova 1956).

Diet of young mostly similar to that of adult. On 2 days at nest in Bavarian Alps (West Germany) where young 6–7 days old, 60% and 68% of feeds ($n = 44$ and 100) consisted of single, or sometimes 2–3, larval Cerambycidae; these comprised 79% and 76% of all food brought by ♂, and 27% and 59% by ♀. Other prey included Tipulidae, caterpillars, and beetles. Stomach contents of 2 10-day-old young included mainly Cerambycidae (adult *Callidium*, 2 adult *Prionus*, and 10 *Rhagium* larvae), adult cockchafers *Melolontha*, other beetle fragments, and a caterpillar (Franz 1937). In Ural valley, young fed on large *C. cossus* caterpillars (Dubinin and Toropanova 1956). In Primor'e, mainly surface-dwelling insects. From 30 neck-collar samples comprising 416 items, and sight observations of 80 items at nest, average weight of items 665 mg (206–1755). On day of hatching, food items so small that scarcely visible in parent's bill (Polivanov 1981). Diurnal feeding rhythm varies with latitude; feeding rate varies partly with age of young. In Finland, adults fed 4 young 51 times from 03.00 to 20.00 hrs, with apparent peak at 11.00–13.00 hrs; adults hunted up to 1·3 km from nest; interval between feeds *c.* 35 min (♂) and *c.* 16 min (♀) (Pynnönen 1939). Elsewhere, feeding rate typically much faster, indicating that prey available nearer nest. At nest, Norway, *c.* 10 min between feeds; first feed at 06.38 hrs (♀), last at 19.55 hrs (♂); peaks at 09.00 hrs and 14.00–16.00 hrs (Bringeland and Fjære 1981). In Bavarian Alps, ♂ and ♀ took about equal shares at feeding young, radiating 50 m or more to hunt; mean interval between feeds 8·7–12·1 min. Brood of 2 7-day-old young received *c.* 100 feeds during 15¼ hr day (Franz 1937). At nest, Bayerischer Wald (West Germany), ♂ and ♀ alternately fed small young at intervals of *c.* 5 min (Verthein 1935). EKD

Social pattern and behaviour. Based mainly on nominate *leucotos*, partly on *uralensis*. For review, see Blume (1977).

1. Mostly solitary throughout the year. Dispersion outside breeding season probably varies with local food supply. Some birds thought to remain on breeding territories throughout the year (Panov 1973; Polivanov 1981), but not clear if these areas defended then; see also Hogstad (1978). BONDS. Monogamous mating system. Pair-bond maintained from year to year, but not known if it persists outside breeding season, though one pair, south-east Norway, often seen in close contact, e.g. feeding near each other, late January to March (Bringeland and Fjære 1981). One pair bred together for 3 years. Another pair together for 2 years; in following year, ♀ thought to have died, and ♂ re-paired; bond with 2nd ♀ lasted 4 years (Bringeland and Fjære 1981). For case of hybrid with Great Spotted Woodpecker *D. major*, see Kolthoff (1920). ♂ and ♀ share nest-duties, though ♂ does more excavation of nest-hole (Weber 1965). Role of sexes in feeding young varies, either sex taking greater share (Verthein 1935; Weber 1965; Sarkanen 1974). In 2 cases, ♀ left care of young to ♂ when 18 and *c.* 14 days old (Pynnönen 1939; Sarkanen 1974). Family bonds maintained for some time after fledging (Verthein 1935; Panov 1973). Age of first breeding not known. BREEDING DISPERSION. Solitary and territorial. In Pyrénées (Spain), one territory apparently *c.* 9·6 ha (Senosiain 1978). In many places, low density makes territory size difficult to determine: e.g. in Finland, where birds feed up to 1·3 km from nest, territories, not including cultivated areas, cover *c.* 2 km² (Pynnönen 1939). In Erzgebirge region (East Germany/Czechoslovakia), *c.* 4 pairs per km². In Steiermark (Austria), 0·7–2·0 pairs per km² over 8 years (Glutz and Bauer 1980). Territory established by ♂ in spring; serves for nesting, and rearing young. Pairs return to same area, perhaps same territory, every year, but excavate new nest-hole each time (Weber 1965; Bringeland and Fjære 1981). ROOSTING. In breeding season, ♂ often roosts in nest-hole with young (Franz 1937; Sarkanen 1978a). One roosting ♂ left nest-site relatively late in morning (Franz 1937). In Finland, pair entered respective roost-sites ½ hr before sunset or earlier, and usually 'awoke' at sunrise (Sarkanen 1978a). Outside breeding season, roosts in holes, perhaps sometimes nest-holes. Sometimes excavates roost-hole (Pynnönen 1939; Panov 1973; Polivanov 1981).

2. When alarmed by bird of prey, typically adopts Erect-posture on trunk, often on side hidden from threat: sits vertically, head upwards, with neck extended and bill in line with trunk, and remains motionless until danger past (Franz 1937; see also Antagonistic Behaviour and Parental Anti-predator Strategies, below). When danger threatens, ♀ warns ♂ excavating nest-hole with Rattle-call (see 6 in Voice); ♂ leaves immediately with Scolding-calls (see 5 in Voice) and may fly up to 1 km away (Weber 1965). When ♂ caught for ringing, ♀ gave harsh alarm calls nearby (Senosiain 1978). Outside breeding season, almost invariably subordinate to *D. major*. In winter, Finland, fiercely attacked and supplanted by *D. major*; one bird, cornered in thicket by 5 *D. major*, reacted very nervously for 20 min, periodically adopting Erect-posture and making no attempt to feed; eventually escaped by climbing to top of tree and flying almost vertically to considerable height before leaving area. Another bird vigorously attacked and chased out of *D. major* territory (Virkunnen 1976). For similar incident, see Purroy (1972).

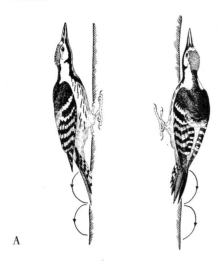

A

ANTAGONISTIC BEHAVIOUR. ♂ demarcates territory by Drumming (see Voice). ♀ also Drums, but to much lesser extent (e.g. Verthein 1935, Schubert 1969), probably more to attract attention of ♂ than to warn off rivals (see Heterosexual Behaviour, below). In Finland, Drumming occurs mainly February–June, peak March–April; not reported in autumn. More frequent in warm than cold springs, and more than 75% before noon (Sarkanen 1978). In Finland, favoured Drumming sites were nest-boxes for Starlings *Sturnus vulgaris* (*c.* 50% of observations), also typically dead birches (*c.* 30%), and electricity poles (*c.* 20%) (Sarkanen 1978a). ♂ twice heard Drumming from inside nest-hole when young in nest (Franz 1937). ♂ continues to defend territory through incubation. Often flies to hole, or beyond it, and sometimes Drums not far away. Drives off any conspecific intruders. Once, incubating ♂ heard another pair on nearby tree, and flew out immediately with Scolding-calls, and drove them off. In similar situation, ♀ gets very agitated but does not attack (Polivanov 1981). Rival ♂♂ may confront each other on opposite sides of tree trunk; adopt Threat-posture (apparently similar to Erect-posture) and reverse down tree trunk with small shuffling jumps (Fig A). Perched confrontations alternate with pursuit-flights: the 2 fly off, twisting and turning in zigzag flight (Fig B), wings making peculiar sound not normally heard (Panov 1973). Hostile encounters accompanied by Wicka-calls (see 4 in Voice). In contrast to winter, ♂♂ successfully expel *D. major* in breeding season, but apparently only from immediate vicinity of nest (Hurme and Sarkanen 1975). ♂ may threaten other large Picinae.

B

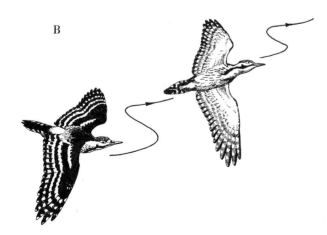

One adopted Threat-posture near Grey-headed Woodpecker *Picus canus*, then began mock-feeding (Panov 1973). For dominance relations between pair-members, see below. HETEROSEXUAL BEHAVIOUR. (1) Pair-bonding behaviour. Few details, but presumably similar to *D. major*. Begins in early March (see Bonds, above), at least in birds which bred together in previous year (Bringeland and Fjære 1981). In Primor'e (eastern USSR), pair-formation frequent in March, comprising aerial chases, display-flights, and display-postures (not described); birds also highly vocal and begin excavating nest-holes (Polivanov 1981). At end of March, Leopoldsteinersee (Austria), ♂ started to make one hole, then another, finally a 3rd which was occupied for nesting *c.* 10 April (Weber 1965). Calls associated with courtship sometimes heard in autumn, Primor'e (Panov 1973). Drumming evidently important in contact between ♂ and ♀: e.g. ♂ may fly to Drumming ♀ (Pynnönen 1939; see also below). Once, ♀ Drummed alternately with ♂ Three-toed Woodpecker *Picoides tridactylus* (Schubert 1969). Pair-members apparently most vocal when one meets other on way to nest. ♂ seems to be subordinate to ♀, and when ♂ meets ♀, flies off or shuffles away with Scolding-calls; this invariably so when ♂ meets ♀ at nest-tree. In encounter between ♂ and 2 ♀♀, repeated 'kirre-' calls heard (see 3 in Voice) and ♂ flew off when ♀ landed on tree where he was perched (Franz 1937). (2) Mating. Performed on tree trunk; few preliminaries. In Primor'e occurred from 13 March onwards (see Bonds, above). In one case, ♀ gave Tcharr-call (see 3 in Voice); ♂ then appeared, mounted, and almost immediately flew off erratically, as in aggressive pursuit-flight (Panov 1973). After copulation on felled tree, calling (see 9 in Voice) heard as birds flew off (Schubert 1969). In mating sequence preceded by Drumming by both sexes, ♀ flew to join ♂ who was *c.* 100 m from nest, and copulation took place straight away (Pynnönen 1939). In attempted copulation, ♂ shuffled up tree stump after ♀ perched on top; ascending ♂ Tapped on trunk, knocking off bits of bark, and sometimes gave soft Contact-call (see 1 in Voice). ♀ neither flew away nor allowed close approach (Panov 1973). (3) Behaviour at nest. Following description by Polivanov (1981). ♀ sits much more tightly than ♂ who frequently looks out of nest-hole and tends to leave rather frequently for breaks of 2–10 min. In incubation spell of 3 hrs 30 min, ♂ looked out 20 times and flew out 10 times. Either sex may incubate at night. Nest-relief may be accompanied by simple Meeting-ceremony. If ♀ takes overnight shift, ♂ flies to nest-hole with Contact-call first thing in morning. ♀ peeps out. Pair then exchange soft Contact-calls, both fly off, and ♀ then returns to nest. ♂ continues to visit her until he takes over incubation. In more elaborate Meeting-ceremony, ♂ flew, calling, to nest; ♀ flew out and pair sat close together on nearby tree, exchanging soft Contact-calls, and touching bills for *c.* 1 min. ♀ then flew off and ♂ went into nest-hole with growling or grumbling sound (see 7 in Voice). Approach to nest always direct during incubation, and incoming bird never stopped first on nearby tree, as sometimes occurred during chick-rearing (see below). RELATIONS WITHIN FAMILY GROUP. On hatching, young brooded closely, but at one nest, 1-day-old young not brooded for almost 3 hrs (Polivanov 1981). ♂ broods while ♀ forages, and vice versa (Pynnönen 1939; Polivanov 1981). When young small, brooding bird usually waited for mate to arrive before departing to feed, though ♂ sometimes left prematurely. When young 8 days old, brooded mainly at night, by either parent. Food-bearing parent sometimes gave Contact-call from nearby tree, but usually silent on arrival at nest, alighting to one side of hole, giving mate room to fly out. Sometimes, ♂ landed a short way above nest-hole and descended with short jerky movements; sometimes landed on opposite side of tree and took spiral path to hole (Polivanov 1981).

At one nest, ♀ which had spent twice as long foraging as ♂, left ♂ to look after 18-day-old young (Pynnönen 1939). At another nest, where ♀ had fed young less than ♂ from outset, she left them when c. 2 weeks old (Sarkanen 1974). Usually, however, both sexes feed young up to and after fledging. Fledged young continue to give food-calls (see Voice). Family parties typically noisy, giving harsh chattering, rarely mixed with high-pitched sounds, and Contact-calls (Schubert 1969). ANTI-PREDATOR RESPONSES OF YOUNG. No information. PARENTAL ANTI-PREDATOR STRATEGIES. (1) Passive measures. After young hatched, parents call less near nest. ♂ reacts less to intruders: e.g. Drumming by strange bird near nest did not cause ♂ to interrupt brooding or even look out of hole (Polivanov 1981). For discrete nature of nest-relief once young hatched, see above. Bird in nest aware of nearby bird of prey made strange head-contorting movements for up to ½ min or more before flying off; if out of nest, often adopted Erect-posture. Once, when crow (Corvidae) settled nearby, ♀ flew out of nest with agitated call, sat on tree near nest, then slipped back into nest, continuing to call from inside (Polivanov 1981). If human intruder present, may call vociferously and adopt Erect-posture; if surprised near nest, may call for a while, and cautious before returning to nest, but carries on feeding (Franz 1937). (2) Active measures. None known.

(Figs A–B from Panov 1973.) EKD

Voice. Used mainly in breeding season; in Primor'e (eastern USSR), sometimes also in autumn (Panov 1973). For comprehensive comparison of Drumming with other west Palearctic Picinae, see Zabka (1980). For fine structure of calls, see Winkler and Short (1978), on whose scheme the following based. For Tapping, see Social Pattern and Behaviour.

DRUMMING OF ADULTS. Much louder and (*contra* Bergmann and Helb 1982) longer than Great Spotted Woodpecker *D. major*. As in Three-toed Woodpecker *Picoides tridactylus*, to which audibly similar, series ('roll') accelerates towards end (Fig I: Pynnönen 1939; Zabka 1980; Bergmann and Helb 1982, which see for further sonagram). Also gets quieter towards end (P J Sellar). As in other *Dendrocopos*, pattern diagnostic, though markedly variable between individuals. Average 29.8 ± 5.0 strikes per series (Franz 1937). Average strikes per series 34·35, average duration 1·64 s (4 birds, 17 series: Zabka 1980); average duration 2·2 s (Sarkanen 1978a). Interval between strikes 80–90 ms at start, 40–50 ms at end (Zabka 1980; Wallschläger and Zabka 1981); frequency increases from 15–19 to 22–25 strikes per s (Winkler and Short 1978). Typically 3–4 series per min (Sarkanen 1978a). In March–April, 3·8 series per min ($n=47$), in June 1·7 ($n=18$). When disturbed from nest, sometimes performs unusually short series—c. 5 strikes in ¼ s (Bergmann and Helb 1982). Used for advertisement, especially early in breeding season, by both sexes, enabling recognition of species and individual (Zabka 1980).

CALLS OF ADULTS. (1) Contact-call. Commonest call (Fig II) a soft 'kjük' (Franz 1937) or 'kiuk' (Thiollay 1963a); average pitch 1·9 kHz, 2 tones lower than *D. major* (Franz 1937; Winkler and Short 1978). Not as hard as 'gick' of *D. major*, or as soft as 'güg' of *P. tridactylus* (Verthein 1935). Given in various contexts: e.g. by ♂ entering nest-hole (Franz 1937), by perched ♂ approaching ♀ prior to attempted copulation (Panov 1973), during Meeting-ceremony (Polivanov 1981), after feeding young ('tet tet tetet': Schubert 1969). Variants include disyllabic 'kjüjuk', with emphasis on 2nd syllable (Franz 1937, which see for other variants). Same or related 'güg-güg-kurr' often combined with Drumming (Bergmann and Helb 1982). (2) Kweek-call. Long-drawn 'gäi', given one or more times;

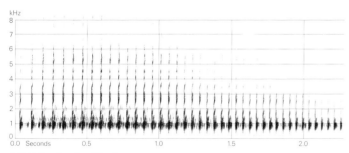

I I Ahlén Sweden March 1979

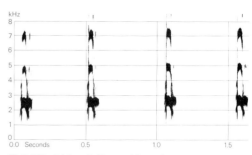

II J-C Roché (1970) France May 1963

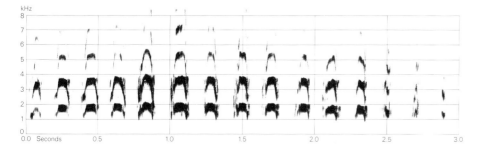

III S Palmér/Sveriges Radio (1972)
Sweden March 1962

associated with Drumming (Verthein 1935). Hoarse, squeaking call, average duration 170 ms (140–200, $n = 11$), thought to serve as long-distance communication between sexes, and so supplement Drumming (Winkler and Short 1978, which see for sonagram; see also Glutz and Bauer 1980). (3) Tcharr-call. Before copulation, ♂ may call 'tcharr tcharr tcharr'; very similar to food-call of fledged young (Panov 1973; see below). Presumably has soliciting and/or appeasing function (E K Dunn), and thus homologous with Appeasement-call of other Picinae (Winkler and Short 1978). Chattering 'kirrekirrekirrekirre' given during aggressive encounters between ♂♂ and ♀♀ (Franz 1937) perhaps the same. (4) Wicka-call. Rapid, hoarse, squeaking 'witsche witsche...' (Bergmann and Helb 1982), or harsh 'viche viche viche', given in confrontations between ♂♂, typically early in breeding season (Panov 1973). (5) Scolding-call. When excited or alarmed (e.g. by intruder, conspecific or otherwise, at nest), variant of call 1 given in unremitting series, with variable spacing between units: series of harsh, sharp 'kek' sounds (Winkler and Short 1978; Bergmann and Helb 1982, which see for sonagrams). Bird disturbed when feeding gave hoarse 'kechkechkech' as it flew off (Schubert 1969). (6) Rattle-call. In higher-intensity alarm, a rapid excited 'kkkkkk' (Franz 1937), or 'ki-ki...' (Bergmann and Helb 1982: Fig III). Given by ♂ in aggressive encounters with ♀, typically as he flees (Franz 1937). Also serves as warning call, e.g. by ♀ to ♂ excavating nest-hole; in response, ♂ fled giving Scolding-call (Weber 1965). Average pitch 2–3 kHz, $n = 8$ (Winkler and Short 1978, which see for other sonagram details, and short variant, comprising 2–3 units). (7) Growling-call. Growling or grumbling sound, given by ♂ entering nest-hole after Meeting-ceremony (Polivanov 1981); function not known, but perhaps low-pitched Tcharr-call. (8) Distress-call. Screeching sound, pitch c. 0.75 kHz, duration c. 280 ms, given repeatedly in extreme danger, e.g. when handled for ringing (Winkler and Short 1978). (9) Bird (sex not known) flying off after copulating called 'dsi dsi dsi' (Schubert 1969). For other sounds, none of which substantially different from repertoire here, see Winkler and Short (1978).

CALLS OF YOUNG. Food-call of 6-day-old young a chirruping 'zerrrrr zerrrrr zerrrrr', with slight difference in pitch between units, and audible at 15 m. Compared with *D. major*, somewhat lower pitched; also, 'e' sound more pronounced and call given less continuously. Begins when incoming adult blocks light at nest-entrance, and tends to persist for c. 10 s after feed. Occasionally given between feeds (Franz 1937). When being fed, call louder and more squeaking (Bergmann and Helb 1982). In fledged young, rendered 'tcharr' (Panov 1973). Calls of independent young apparently same as adult (Verthein 1935). EKD

Breeding. SEASON. Yugoslavia: laying begins late April. Scandinavia: laying begins late April, Norway (Haftorn 1971), to early May, Finland (Sarkanen 1974). USSR: breeding starts late April (Dementiev and Gladkov 1951a; Makatsch 1976). SITE. Hole in tree, usually in rotten wood. Of 37 nests, Norway, 16 in birch *Betula*, 13 in aspen *Populus*, 5 in alder *Alnus*, 1 in rowan *Sorbus*, 2 in posts (Haftorn 1971; Håland and Toft 1983). Height 1.4–6.5 m above ground (Sarkanen 1974); exceptionally higher, even to 28 m (Verthein 1935); average 4.2 m, $n = 19$ (Håland and Toft 1983); average 8.2 m, $n = 16$ (Glutz and Bauer 1980). Nest: excavated hole; entrance hole height and width 56–69 × 47–64 cm (Collett and Olsen 1921; Pynnönen 1939; Krogh 1955; Nævdal and Nævdal 1968); depth of cavity 25–37 cm (Collett and Olsen 1921; Pynnönen 1939; Nævdal and Nævdal 1968). Building: by both sexes, mainly ♂ (Weber 1965). EGGS. See Plate 98. Elliptical, smooth and glossy; white. 28 × 21 mm (26–69 × 20–21), $n = 12$, calculated weight 6.0 g (Makatsch 1976); 26.3–30.2 × 19–27.7 mm (Curry-Lindahl 1961b). Clutch: 3–5. Laying interval c. 24 hrs (Haftorn 1971). One brood. INCUBATION. About $10\frac{1}{2}$ days (average of 1 pair over 3 years) (Bringeland and Fjære 1981). By both sexes. YOUNG. Altricial and nidicolous. Cared for and fed by both sexes until some time after fledging. FLEDGING TO MATURITY. Fledging period 24–28 days (Pynnönen 1960; Bringeland and Fjære 1981). Age of first breeding not known. BREEDING SUCCESS. Of 6 nests, 1 young fledged from 1, 2 from 2, 3 from 2, 4 from 1 (Sarkanen 1974). No further information. MAO, EKD

Plumages (nominate *leucotos*). ADULT MALE. Forehead and tufts covering nostrils white, cream-white, or pale buff (deepest on tufts), longer bristles of tufts black. Crown and back of head bright crimson-red, some grey of feather-bases usually visible (markedly so when plumage worn, especially on crown). Short black streak from sides of forehead to sides of crown above eye. Lores, area round eye, ear-coverts, and sides of neck white, usually with slight yellow-buff tinge (especially on lores and ear-coverts). Broad black stripe from base of lower mandible to below ear-coverts, slightly extending up behind ear-coverts but not reaching back of head, extending down to black patch on side of chest. Central hindneck, mantle, and shorter scapulars deep black (feather-tips slightly glossed blue, bases extensively tinged brown), part of scapulars boldly blotched white on tips; longer scapulars, back, and rump white (feathers narrowly fringed black when plumage fresh; grey feather-bases of back and rump often partly showing). Upper tail-coverts black with some white spots. Chin, throat, and central chest and breast white, upper chin and chest and breast tinged cream-yellow when plumage fresh; flanks and sides of breast and belly below black patches on sides of chest white or cream-white with sharp black shaft-streaks $\frac{1}{2}$–1 mm wide (narrowest on flanks, often slightly widening terminally on sides of breast and belly). Centre of lower belly, vent, under tail-coverts, and often tips of feathers of lower flanks pink-red, deepest on shorter coverts, some white of feather-bases usually visible elsewhere; red distinctly paler pinkish and more extensive than deeper and more saturated red of Great Spotted Woodpecker *D. major*, but similar to that of Syrian Woodpecker *D. syriacus* and Middle Spotted Woodpecker *D. medius* (latter species has entire underparts quite similar to *D. leucotos*, but upperparts strongly different). Central 2 pairs of tail-feathers (t1–t2) black; t3 with white tip (usually tinged orange on underside) and some dark spots or bars; t4–t5 white except for 2–3 rather narrow and often broken black bars and for hidden basal halves of inner webs; reduced t6 (hidden under tail-coverts) white on outer web, black with variable white mark on inner web. Primaries as in *D. major*, but white spots broader, 5–8 mm across rather than 4–5 mm; white spots on outer web of p9 often distinct (virtually absent in *D. major*); white spots on tips larger, usually extending broadly over whole tip of p1–p4(–p5) and broadly on outer web of (p4–)p5–p8. Secondaries as in *D. major*, with 2 rows of white spots across middle portions, a single terminal row, and another row just visible beyond tip of greater coverts or just underneath tips; white spots larger than in *D. major*, however, terminal ones especially prominent (virtually absent on inner secondaries in *D. major*); tertials quite different from *D. major*, broadly banded black and white (not mainly black), innermost sometimes virtually white. Upper wing-coverts black, greater coverts with bold white spots on tips and white bases, median and some longer lesser coverts broadly tipped white (white tips partly fringed black), lesser primary coverts white with some dusky marks (in *D. major* and *D. medius*, upper wing-coverts black except for white tertial coverts and some inner median coverts); strongly white-banded tertials and upper wing-coverts combined with white longer scapulars, back, and rump quite different from mainly black upperparts of *D. major* and *D. medius*, which show large white patch on outer scapulars and tertial coverts only (some juveniles of, in particular, *D. major major* have scapulars and tertial coverts indistinctly barred dusky and these may also show red crown, pale red vent and under tail-coverts, and diffusely barred or streaked flanks, like *D. leucotos*). Under wing-coverts and axillaries white. ADULT FEMALE. Like adult ♂, but crown and rear of head black like central hindneck and mantle, no red. NESTLING. As in *D. major*. JUVENILE. Like adult ♂, but red of crown more restricted and slightly paler orange-red, much black and dull grey of feather-bases showing. In juvenile ♂, red tips on crown 1½–3 mm long (3–5 mm in adult), back of head virtually dull black (uniform crimson-red in adult ♂); in ♀, red tips ½–1½ mm long and often restricted to forecrown or central crown only, often just perceptible, crown and back of head appearing almost completely dull black. Black of upperparts and upper wing-coverts duller and browner than adult; forehead, sides of head, and chin often more extensively tinged buff; streaks on sides of breast and flanks broader but browner and less sharply defined than in adult, more gradually merging into brown-black patch on sides of chest; lower flanks indistinctly barred dusky; red tinge on underparts less extensive and paler, in some confined to vent and under tail-coverts, in others virtually absent or restricted to red-buff wash on under tail-coverts. Tail as in adult, but t1 relatively more sharply pointed and other feathers relatively shorter. Wing as in adult, but ends of longer and outer primaries fringed white on both webs (forming white chevrons), white not restricted to spots on outer webs; greater upper primary coverts often with ill-defined white spots on tips; p10 6–17 mm longer than longest greater upper primary coverts (5 mm longer to 3 mm shorter in adult). FIRST ADULT. Similar to adult, but juvenile secondaries, tertials, and some or all greater upper primary coverts retained, these slightly more worn than in adult at same time of year. Greater upper primary coverts either all juvenile, worn and with ill-defined white spot or fringe on tips (fringe often soon abraded), or inner ones juvenile and outer adult, differing in wear, shape, and colour pattern (adults occasionally retain one or a few scattered primary coverts, but these similar to new coverts in shape and colour pattern).

Bare parts. ADULT AND FIRST ADULT. Iris red-brown or carmine-red. Bill leaden-grey. Leg and foot dark leaden-grey. NESTLING AND JUVENILE. Bare skin pink at hatching. No further information. (Hartert 1912–21; RMNH.)

Moults. According to data in Dementiev and Gladkov (1951a), Purroy (1972), and specimens examined (RMNH, ZFMK, ZMA), moult sequence as *D. major*, but timing earlier. Adult post-breeding starts with p1 from early May to late June, completed with p10 late August to early October; moult of flight-feathers mainly late May to mid-September, tail early June to late August, body late July to late September. Post-juvenile starts early May to mid-June, completed September. Apparently no difference between west Palearctic races in timing, but little data on *lilfordi* available.

Measurements. ADULT AND FIRST ADULT. Nominate *leucotos*. Northern Europe, all year; skins (RMNH, ZFMK, ZMA). Bill (F) to forehead, bill (N) to distal corner of nostril; toe is outer front toe with claw.

WING	♂	148	(2·99; 10)	144–152	♀ 146	(1·90; 10)	144–149
TAIL		88·6	(2·89; 9)	84–92	86·5	(2·93; 6)	83–91
BILL (F)		39·2	(1·16; 7)	38–41	36·7	(1·58; 8)	35–39
BILL (N)		31·0	(1·61; 8)	29–33	29·2	(1·34; 9)	27–31
TARSUS		25·3	(0·83; 10)	24–27	24·9	(0·86; 7)	24–27
TOE		24·3	(1·20; 10)	23–26	24·2	(1·80; 4)	22–26

Sex differences significant for bill.

D. l. lilfordi. Bulgaria, Greece, and Turkey, all year; skins (BMNH, RMNH, ZFMK, ZMA).

WING	♂	149	(3·64; 6)	144–152	♀ 147	(1·75; 6)	144–149

D. l. uralensis. South-west and central Siberia, all year; skins (RMNH, ZFMK, ZMA).

WING	♂	148	(3·23; 5)	143–152	♀ 152	(—; 3)	146–157

Data from Kohl and Stollmann (1968), wing apparently measured with slightly different method: nominate *leucotos*, (1) Scandinavia, Finland, and Estonia, (2) Lithuania, Smolensk area (USSR), and Poland, (3) Carpathians, Dobrogea (Rumania), and northern Yugoslavia, (4) *lilfordi*, mainly from Yugoslavia.

WING (1)	♂	145	(3·55; 16)	137–150	♀ 144	(3·32; 15)	138–150
(2)		144	(4·26; 18)	135–150	143	(2·38; 11)	139–146
(3)		144	(2·65; 34)	139–149	141	(2·41; 30)	134–146
(4)		143	(1·92; 7)	141–147	142	(1·99; 8)	140–147
BILL (N) (1)		30·6	(1·17; 14)	28–32	28·6	(1·32; 15)	27–31
(3)		29·8	(0·87; 35)	28–32	28·0	(1·22; 30)	26–30
(4)		31·9	(1·31; 7)	30–34	29·7	(1·28; 9)	28–32

In another sample of nominate *leucotos* from Scandinavia, wing ♂ 146 (3·48; 10), ♀ 144 (2·26; 8) (Hogstad 1978); in sample from central Europe, wing ♂ 146 (1·98; 9) 143–149, ♀ 144 (2·92; 11) 140–150 (Glutz and Bauer 1980). In small sample of *lilfordi*, bill (F) ♂ 42·2 (3) 41–43, ♀ 37·3 (3) 35–40; bill (N) ♂ 32·5 (3) 31–34, ♀ 29·4 (3) 28–31 (RMNH, ZFMK, ZMA). Wing of *lilfordi* from Pyrénées: ♂♂ 150, 150; ♀ 148 (Purroy 1972). Wing of Corsican *lilfordi*: ♂ 144, ♀ 143 (Voous 1947).

JUVENILE. Juvenile wing on average 5·3 shorter than adult, tail 7; bill full-grown at end of post-juvenile moult.

Weights. ADULT AND FIRST ADULT. Nominate *leucotos*. USSR: ♂ 105–112, ♀♀ 105, 106 (Dementiev and Gladkov 1951a). North-west USSR, ♂ 105 (May); Rumania, ♂ 100 (October); Austria, ♀ 99 (November) (Glutz and Bauer 1980).

D. l. uralensis. Kazakhstan (USSR): ♂♂ 105, 116 (April), 126 (June); ♀♀ 105, 114 (June) (Dolgushin *et al*. 1970). Mongolia and Manchuria, July and early August: ♂ 100, ♀ 100 (Piechocki 1958, 1968c).

D. l. lilfordi. Spanish Pyrénées, October: ♂ 115 (Purroy 1972). European Turkey, May: ♂ 120 (Rokitansky and Schifter 1971).

Extralimital *subcirris* (similar in size to west Palearctic races), Kuril Islands: ♂ 124 (4) 119–128, ♀ 119 (4) 107–128 (Nechaev 1969).

NESTLING AND JUVENILE. No information on nestling. Juvenile ♂♂, Manchuria, late July and early August: 100, 108 (Piechocki 1958). Finland, ♂ 100; Norway, 105 (August), 105 (September) (Hagen 1942).

Structure. In adult and 1st adult, p7 longest, p8 2–4 shorter, p9 21–25, p10 80–87, p6 0–1, p5 3–6, p4 16–24, p3 28–34, p2 35–40, p1 40–45; in juvenile, p7 or p8 longest or either one 0–2 shorter than other; p9 10–16 shorter than longest, p10 62–78, p6 3–7, p5 12–18. P10 reduced in adult and 1st adult, 3 shorter to 5 longer than longest greater upper primary coverts (on average, 1·5 longer, $n=12$); less reduced in juvenile, 13 (6) 6–17 longer. Outer web of p5–p9 and (slightly) inner of p6–p9 emarginated. T2 3–6 shorter than t1 in adult and 1st adult, t3 9–12 mm shorter, t4 18–21, t5 27–30, t6 67–72; t2 8–12 shorter than t1 in juvenile, t3 9–17, t4 12–21, t5 19–29, t6 58–68. Bill relatively longer than in most races of *D. major*; about as heavy at base as in nominate *major*, but more gradually tapering towards long tip; ridge on culmen more pronounced. Inner front toe with claw *c.* 87% of outer front toe, outer hind toe *c.* 115%, inner hind *c.* 57%. Remainder of structure as in *D. major*.

Geographical variation. Slight in size, marked in colour. *D. l. lilfordi* from southern Europe and Asia Minor markedly darker than nominate *leucotos* from northern Alps, northern Yugoslavia, and Carpathian mountains; *lilfordi* sometimes considered separate species in view of apparent overlap with nominate *leucotos* in eastern Serbia (though no actual overlap, as populations breed at different altitudes); see (e.g.) Matvejev and Vasić (1973). However, variation mainly clinal, with typical *lilfordi* from Asia Minor and Caucasus darkest, birds from Greece averaging slightly paler, Bulgaria and southern Yugoslavia paler still, though nearer *lilfordi* than to typical nominate *leucotos* from Carpathians. Birds from northern Croatia, Slovenia, and (formerly) Italian Alps intermediate between both races (Glutz and Bauer 1980). Cline continues to north and north-east: birds from Poland and south-west European USSR slightly paler than nominate *leucotos* from Carpathians, birds from Scandinavia east to north European USSR paler still, and *uralensis* from Urals to central Siberia palest of all. Typical *lilfordi* from Asia Minor differs from typical nominate *leucotos* from Carpathians in more distinct yellow-buff forehead, lores, ear-coverts, and chin; distinct black line behind ear-coverts reaching nape (narrow or absent in nominate *leucotos*); broader black malar stripe from base of lower mandible down sides of neck to sides of chest; longer scapulars and back barred equally black and white (not mainly white); rump black with limited white on feather-tips only (not mainly white); white bars and spots on tertials, flight-feathers, and greater and median upper wing-coverts narrower, 3–5 mm wide rather than 5–8 mm; t4–t5 with rather irregular black and white bars of about equal width (not mainly white with narrow black bars), tip of t3 with only limited amount of white; flanks, sides of breast, and sides of belly broadly streaked black, flanks partly barred black; ground-colour of underparts pale yellow-buff (paler cream in nominate *leucotos*), but pink-red of vent and under tail-coverts similar; upperparts appear closely barred black and white (in nominate *leucotos*, uniform white in centre with some broad white bars extending laterally), underparts heavily streaked (in particular, juvenile almost completely streaked below). Birds from Bulgaria and Greece similar to typical *lilfordi*, but flanks less barred and white bars on outer tail and tertials slightly wider; isolated populations from Pyrénées (Purroy 1972; Senosiain 1978), Corsica (Voous 1947; Moltoni and Brichetti 1977), and Abruzzi, Italy (Moltoni 1959) apparently similar to these. Typical nominate *leucotos* from Carpathians extends east to Dobrogea (Rumania), west to northern Alps (Kohl and Stollmann 1968; Glutz and Bauer 1980). Birds from Poland similar to typical nominate *leucotos* but face and underparts slightly whiter and streaks on underparts slightly narrower; birds from northern Europe have ground-colour virtually pure white (Kohl and Stollmann 1968; RMNH, ZFMK). *D. l. uralensis* similar to nominate *leucotos*, but ground-colour pure white, black line behind ear-coverts absent, tips of longer scapulars broadly white; white bars and spots on tertials, flight-feathers, and upper wing-coverts broader; black streaks on sides of breast and flanks narrower, on flanks occasionally almost absent; t4–t5 on average whiter. Cline of increasing paleness from *lilfordi* to *uralensis* reversed in extralimital races of eastern Asia and offshore islands, where race *sinicus* from Altay to Sakhalin rather similar to nominate *leucotos* but with slightly wider and sharper streaks below, other races in part even darker than *lilfordi*, in particular those from Ryukyu Islands, Taiwan, and south-west and south-east China (see, e.g., Vaurie 1959b).

CSR

Dendrocopos minor Lesser Spotted Woodpecker

PLATES 84 and 86
[between pages 878 and 879]

Du. Kleine Bonte Specht   Fr. Pic épeichette   Ge. Kleinspecht
Ru. Малый пестрый дятел   Sp. Pico menor   Sw. Mindre hackspett

Picus minor Linnaeus, 1758

Polytypic. Nominate *minor* (Linnaeus, 1758), Fenno-Scandia, north-east Poland, and western USSR east to Urals; *kamtschatkensis* (Malherbe, 1861), Siberia east to sea of Okhotsk, south to Altay and northern Mongolia, intergrading with nominate *minor* over wide area in northern Fenno-Scandia and in north, central, and east European USSR; *hortorum* (Brehm, 1831), northern France and Netherlands east to central Poland and Rumania, south to Swiss Alps, Austria, and northern Yugoslavia (except extreme west), intergrading with nominate *minor* in Poland, western Ukraine, and perhaps Moldavia; *comminutus* Hartert, 1907, Britain; *buturlini* (Hartert, 1912), southern France, northern Spain, Italy, western and southern Yugoslavia, Bulgaria, perhaps Albania, and probably northern Greece, intergrading with *hortorum* in southern France and with *danfordi* in southern Yugoslavia and Bulgaria; *ledouci* (Malherbe, 1855), north-west Africa, and intergrading with *buturlini* in Portugal and southern and central Spain; *danfordi* (Hargitt, 1883), Greece and Turkey; *colchicus* (Buturlin, 1909), Caucasus and Transcaucasia (except Lenkoran area); *quadrifasciatus* (Radde, 1884), Lenkoran area in south-east Transcaucasia. Extralimital: *hyrcanus* (Zarudny and Bilkevich, 1913), south Caspian lowlands; *morgani* Zarudny and Loudon, 1904, Zagros mountains in Iran, intergrading with *colchicus* in north-west Iran; 2–3 further races in eastern Asia.

Field characters. 14–15 cm; wing-span 25–27 cm. About $\frac{2}{3}$ size of Great Spotted Woodpecker *D. major*; hardly larger than Nuthatch *Sitta europaea*. Relatively tiny, small-billed, black and white woodpecker, with boldly barred upperparts and well streaked underparts, but lacking coloured vent. Flight silhouette noticeably compact and flight action markedly more fluttering than in any other woodpecker, recalling Woodlark *Lullula arborea*. Sexes dissimilar; no seasonal variation. Juvenile separable. 9 races in west Palearctic; 2 described here (see also Geographical Variation).

ADULT MALE. (1) Continental north European race, nominate *minor*. Least black of genus, with mantle as well as scapulars and wings closely barred white. Crown dull red, not reaching on to back of head. A few most northern birds have white bars across central back so strong and wide as to appear almost white-backed. Narrow edge to crown, nape, and moustache black, moustache turning upwards on rear cheeks (but not reaching nape) and just downwards on side of neck. Sides of breast and flanks streaked black. Under tail-coverts spotted black. (2) British race, *comminutus*. White underparts much sullied with brown and buff tones, with no obvious streaks. ADULT FEMALE. Both races. Forecrown brown-white to pale brown, with red restricted to tips of feathers and usually invisible in the field; rear crown black. JUVENILE. Both races. Crown shows red tips in ♂. Underbody browner, with streaks more numerous and of more spotted form than in adult.

Often elusive but unmistakable when seen well, being so much smaller than any other *Dendrocopos*. Plumage recalls only much larger White-backed Woodpecker *D. leucotos* (but lacks its pink vent). In glimpse, often suggests fluttering passerine before woodpecker, and behaviour includes much searching among topmost branches and twigs as well as trunk-climbing. Flight rather slow and less powerful than other *Dendrocopos*; fluttering in canopy but undulating in open airspace. Climbs less jerkily than congeners, with more creeping progress. At all times, silhouette compact and rather short-tailed for a woodpecker. Usually stays even higher in crowns of trees than Middle Spotted Woodpecker *D. medius*; rarely on ground.

Commonest call (song) a rather weak and slow 'pee-pee...', recalling Kestrel *Falco tinnunculus* or Wryneck *Jynx torquilla*. Contact-call, 'chick', like call of *D. major* but much more feeble. Drums quite frequently: longer (*c*. 1–1$\frac{1}{2}$ s) and more rattling than *D. major*, with lack of power often obvious even at close range; similar to Grey-headed Woodpecker *Picus canus*, but that species drums less often (2–3 times per min, rather than up to *c*. 15 times per min in *D. minor*).

Habitat. Resident almost throughout wooded regions of Europe except for some oceanic fringes in Scandinavia, Britain, and Ireland and some islands in Mediterranean. Also established in Azores, and some woods near coasts of North Africa and Asia Minor and in Caucasus region. Accordingly ranges through greater latitudinal depth than other *Dendrocopos* except Great Spotted Woodpecker *D. major* with correspondingly greater tolerance of high and low temperatures, wind, and rainfall. Also less demanding arboreally, often nesting on underside of branch rather than in trunk, and foraging on branches and foliage of trees or scrub, as well as catching insects in the air. Nest, however, often unusually high for a woodpecker—up to 25 m (Sharrock 1976), although some in stumps or even fence-posts at below 1 m. Availability of easily worked decayed wood may be more essential than tree height or species. Prefers open broad-leaved woodland, edges, spinneys, parkland, riparian and other tree lines or avenues, orchards, and moist woods of oak *Quercus*, hornbeam *Carpinus*, willow *Salix*, alder *Alnus*, or poplar *Populus*. Over

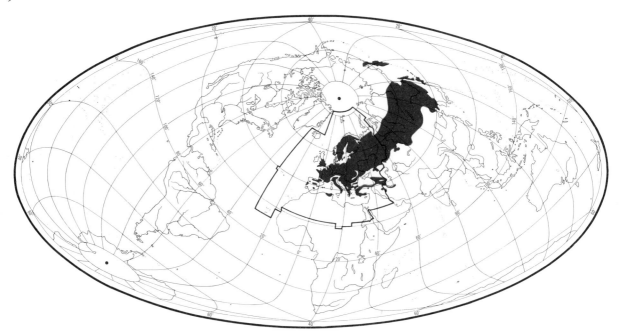

most of range avoids conifers, and any dense, mature stands. Although remarkably unobtrusive, not averse to living at close quarters with man, even in urban parks, cemeteries, and gardens, where presence may be unsuspected until intensive study made, e.g. by mist-netting (Sharrock 1976). Generally a lowland species, in Switzerland mainly below 800 m in winter, and breeding only rarely above 1000 m (Glutz and Bauer 1980). Caucasian race, however, ascends to timber line at 2000 m (Dementiev and Gladkov 1951a). Restless and mobile, showing tendency to appear outside breeding season in unsuitable non-arboreal situations. Not normally observed however making extended flights, especially above lower airspace. Responsive to changes in habitat (e.g. clearance of old orchards, loss of elms through Dutch elm disease: Sharrock 1976), and will occur in marginal terrain such as wooded steppes, tundra birch scrubs, and lower scrub growth with scattered mature trees (Harrison 1982; see also Niethammer 1938 and Bannerman 1955).

Distribution. BRITAIN. No marked changes in range (Sharrock 1976). Bred Jersey 1979 (Long 1981b). FRANCE. Corsica: no proof of breeding (J-C Thibault). DENMARK. First bred 1964 in east and now regular; bred also southern Jutland 1973–4, where now probably regular (TD). ITALY. Formerly bred Sicily (PB, BM). TURKEY. Precise range not fully established (MB, RFP).

Population. Limited information on overall numbers or trends. Local fluctuations due to habitat changes, e.g. declines due to loss of old orchards and temporary increases following death of elms *Ulmus* (Sharrock 1976; Glutz and Bauer 1980). Marked recent decline Finland; possible declines Portugal and Netherlands.

BRITAIN. No known overall variation in numbers, estimated 5000–10000 pairs; some local fluctuations due to habitat changes (Sharrock 1976); now thought to be too high—probably 3000–6000 pairs (BTO). FRANCE. 1000–10000 pairs (Yeatman 1976). PORTUGAL. Local and nowhere common (RR). SPAIN. Scarce except in Asturias (AN). BELGIUM. About 350 pairs (Lippens and Wille 1972); c. 650 pairs 1981 (PD). LUXEMBOURG. About 180 pairs (Lippens and Wille 1972). NETHERLANDS. 1000–2500 pairs 1977 (Teixeira 1979); some local increases 1940–50, but recent tendency to decline due to habitat changes (CSR). DENMARK. Under 10 pairs (TD). SWEDEN. 20000 pairs (Ulfstrand and Högstedt 1976). FINLAND. Over 3000 pairs (Merikallio 1958). Striking recent decline, now very scarce throughout (OH). POLAND. Scarce (Tomiałojć 1976a). CZECHOSLOVAKIA. Scarce; no known trends (KH). GREECE. Rare in Pelopónnisos (HJB, WB, GM). TUNISIA. Rare (Blanchet 1955). ALGERIA. Low density (Heim de Balsac and Mayaud 1962).

Oldest ringed bird 6 years 5 months (BTO).

Movements. All populations basically resident, though considerable nomadism and some eruptive movement in northern and eastern parts of range.

NORTHERN POPULATIONS. Nominate *minor* (breeding Fenno-Scandia and eastern Poland to western Siberia) to some extent nomadic or partially migratory, such birds moving into southern parts of breeding range or short distances beyond. In some autumns, this movement assumes character of small-scale eruption, and then usually coincident with much larger eruptions of Great Spotted Woodpecker *D. major*. Such birds may then reach Black Sea, central Europe south to Baden-Württemberg, and Netherlands, though always in very small numbers

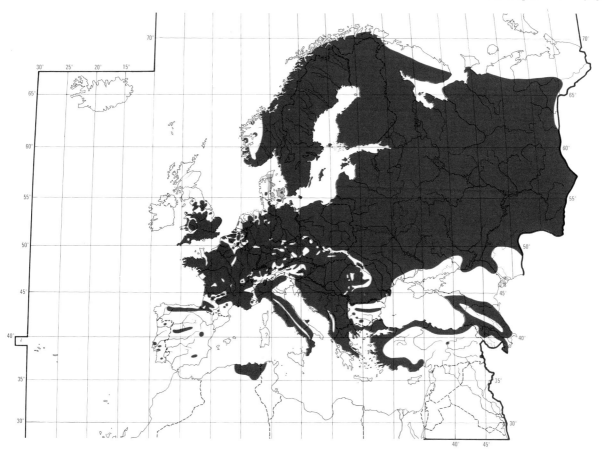

(Dementiev and Gladkov 1951a; Gatter 1973; Glutz and Bauer 1980). In such an eruption in autumn–winter of 1962–3 (for example), at least 20 occurred at Revtangen (south-west Norway) from 20 September to 8 November (Bernhoft-Osa 1963); c. 250, with 5–20 per day, in October–November at Falsterbo, southern Sweden (Roos 1965), and one ringed there 20 October recovered in Denmark in February 1963; total 38, with 354 *D. major*, in migration counts in Pskov, USSR (Meshkov and Uryadova 1972); c. 40 records that winter from Denmark (prior to colonization in 1964); transient birds, believed associated with Scandinavian eruption, occurred south to Dutch Frisian Islands and Württemberg (e.g. Gatter 1973). Subsequent small eruptions in Europe seem to have been most apparent in east Baltic region (Hildén 1969; Meshkov and Uryadova 1972), though in autumn 1972 birds attributed to *minor* (or intergrades) reported from Bremen (West Germany), Switzerland, and Channel Islands (Glutz and Bauer 1980). On Baltic coast of East Germany, migrants occur late August to mid-October (Plath 1976). Longest ringing movements: south Sweden nestling found Ostfold (Norway) on 24 October, 130 km west; migrant ringed Åland (Finland) in August found south-west Sweden in April, 480 km WSW; central Finland migrant (August) found breeding Norwegian Finnmark (June), 600 km NNE. No ringing data for Siberian population (*kamtschatkensis*), which is known to extend, irregularly, further south in some winters, occasionally reaching northern Manchuria and Sakhalin; autumn and winter specimens examined from Orenburg in southern Urals (Vaurie 1965), though possibly from local intergrade population (see Geographical Variation).

CENTRAL AND WESTERN POPULATIONS. Races *hortorum* (breeding south of Baltic to France, Austria, and Hungary) and *comminutus* (Britain) both lack migration element found in northern populations. In autumn and winter a few disperse, but over relatively short distances. In Alps, often seen in autumn higher than breeding altitudes, exceptionally even on treeline (Glutz and Bauer 1980). Has occurred a few times (and bred once) in Channel Islands, and birds of unknown origins have appeared a few times in autumn–winter in Scotland (Sharrock 1976). Longest ringing recoveries: 40 km WSW within England (juvenile, July 1971–October 1973); 60 km N within Switzerland (juvenile, September 1972–June 1978).

SOUTHERN POPULATIONS. 7 races endemic to Mediterranean countries, Balkans, and Asia Minor to Iran are all resident. Some altitudinal movement, descending for winter, reported from Caucasus and Azerbaydzhan, USSR (Dementiev and Gladkov 1951a), but no ringing data.

Food. Almost exclusively insects. Rarely feeds on ground. In summer, chiefly searches for insects on surface of tree-trunks, branches, and leaves; in winter, pecks at rotten wood to find beetle larvae and adults beneath bark, often examining twigs smaller than those used by other west Palearctic Picinae. Agile in gleaning, searching upper and lower surfaces of leaves (Davenport 1897); favours deciduous species such as rowan *Sorbus aucuparia* and aspen *Populus tremula* (Blume 1977), and thin, vertical branches in deciduous tree crowns (Winkler 1972a). Also in summer may take insects by aerial-pursuit (Pynnönen 1939) or fly out at them from a perch, warbler-like (Davenport 1897). When feeding young, forages only by gleaning usually *c.* 100–200 m from nest (Steinfatt 1939). In one pair feeding young close to fledging, ♂ regularly foraged in dead trees near nest, while ♀ foraged in woods *c.* 400–500 m from nest (M G Wilson). In winter, when excavating rotten stumps and branches, rarely digs deeply (Pynnönen 1939; Winkler 1972a) and again often favours slender vertical thickets (Conrads 1967). Reeds may be used all year (Roux 1956). Outside breeding season may extract insect larvae from dry *Artemisia vulgaris* stalks (Meyer 1980; Kummer 1982) and may also work stems of maize *Zea*, teasel *Dipsacus sylvestris*, and burdock *Arctium lappa* (Glutz and Bauer 1980). Insect galls may be opened to obtain larvae (Woolsey *et al.* 1965). Techniques for working cones and seeds poorly developed: items to be worked may be wedged into existing crevices in bark or laid on a flat surface (Wiehe 1976); repeated use of one crevice recorded only in captivity (Winkler 1972a). Winter study in Norway showed no foraging differences between sexes. Spruce *Picea abies* and alder *Alnus incana* used about equally, and birch *Betula odorata* for over 50% of time. 83% of foraging on dead trees; 15% (♂) and 17% (♀) on decaying trees, 2% (♂) on live ones. 68% (♂) and 58% (♀) of trees used were 2–5 m high; 15% and 19% 5·1–10 m high; no taller trees used, remainder less than 2 m. Most foraging low on tree: 75% and 72% up to 2 m above ground; 25% and 28% at 2–5 m. Most foraging on thin branches and twigs: 32% and 28% less than 5 cm diameter; 61% and 64% 5–10 cm diameter. Pecking of small holes into bark and systematic removal of bark used about equally; gleaning very rare (Hogstad 1978). In Finland in winter, spent 12% of time in pine forest, 24% in mixed forest, 18% in deciduous forest, 18% in urban parkland, and 28% elsewhere. 41% of time spent seeking insects on tree trunks, 24% seeking insects on branches, 6% at feeding tables, 6% feeding on pine-cones, and 24% on other feeding activities (Vikberg 1982). For detailed analysis of avoidance of feeding competition with sympatric Picinae, see Alatalo (1978).

Insects include: adults and larvae of Coleoptera (chiefly Cerambycidae, Scolytidae, Curculionidae, and Buprestidae; also Coccinellidae, Chrysomelidae, Elateridae, and Carabidae), larvae of Diptera, Aphididae, Hymenoptera (larvae of Cynipidae, Tenthredinidae, and 'wasps'; adults of Formicidae), and Lepidoptera (larvae of Cossidae, Sesiidae, Tortricidae, and Noctuidae). Occasionally, spiders (Araneae) (Giglioli 1891; Michel 1894; Csiki 1905; Rey 1910; Floericke 1919; Madon 1930a; Witherby *et al.* 1938; Pynnönen 1943; Bannerman 1955; Inozemtsev 1965; Averin and Ganya 1970; Mobbs 1975; Meyer 1980; Kummer 1982). Fruit occasionally taken: pears *Pyrus*, plums *Prunus*, raspberries *Rubus*, and currants *Ribes* recorded (Witherby *et al.* 1938; Averin and Ganya 1970). At feeding tables, prefers seeds of sunflower *Helianthus* to suet and fat (Wiehe 1976).

4 stomachs from south-east Finland, September–October, contained adult aphids and beetles, and larvae of sawflies (Tenthredinidae), Lepidoptera, and wood-boring beetles. 17 collected November–April contained mainly adults and larvae of wood-boring beetles found under tree bark, 50% being larvae of Cerambycidae. One stomach contained 25 larvae of Curculionidae and 26 of Cerambycidae; another, 37 and 36 respectively. 7 stomachs from Estonia (USSR) showed that winter diet of adult and larval wood-boring beetles extends into May (Pynnönen 1943). 14 stomachs from France collected October–April contained 136 wood-boring beetle larvae (Scolytidae, Curculionidae, and Buprestidae being more common than Cerambycidae), 77 unidentified insect larvae, 17 larvae of Lepidoptera, 4 adult beetles, and 1 spider. Of 2 stomachs taken in September, 1 contained only adult beetles and the other contained 145 green oak-leaf aphids, probably *Tuberculatus quercus* (Madon 1930a). In contrast to these studies, ants were important in 8 stomachs collected in Hungary February–April and October–November. In addition to various beetles and larvae, 5 stomachs contained 200 *Lasius alienus*, 50 *L. fuliginosus*, 2 *Dolichoderus*, and 1 *Camponotus sylvestris* (Csiki 1905). Ants also appear in a Russian study: 11 stomachs contained 45 *Camponotus*, 17 *Formica*, and 25 unidentified ants, as well as various beetles (102 larvae of Cerambycidae, 8 adult *Phyllodecta*, 1 adult *Strophosomus rufipes*, 11 adult *Dorytomus*, 2 adult Curculionidae, 15 adult Scolytidae), larvae of Lepidoptera (10 Tortricidae, 1 Noctuidae), and 1 spider (Inozemtsev 1965). Of 9 stomachs from Moldavia (USSR), 1 contained only the flesh of pears, and another that of plums; contents of other 7 included 68·2% Coleoptera (42·3% being wood-boring larvae, 57·5% adults), 22·6% Lepidoptera (age not stated), and 9% Diptera (Averin and Ganya 1970).

Young fed wide variety of adult and larval insects; aphids often particularly important (Pynnönen 1943; Westerfrölke 1955b; Prokofieva 1963). Foods recorded are Homoptera (Aphrophorinae adults and larvae, Psyllidae, Aphidoidea), larvae of sawflies, Diptera (adult Tipulidae and Muscidae), Coleoptera (larvae of Chrysomelidae and Cerambycidae), Formicidae (adults and pupae), Cynipidae (larvae), Lepidoptera (larvae of Nymphalidae, Geometridae, Lymantriidae, and Noctuidae), alder flies (Sialidae), Araneae, and Gastropoda (*Planorbis*, *Discus ruderatus*) (Bäsecke 1932; Steinfatt 1939; Pynnönen 1943; Prokofieva 1963; Polivanov 1981). Stomachs of 6 nestlings, south-east Finland and one independent juvenile (August) contained mainly aphids (Pynnönen 1943). 64 nestling food samples from 2 broods near Leningrad contained mainly aphids plus plant lice (Psyllidae) and alder flies (Prokofieva 1963). At another nest, young fed extensively on larvae of gall wasp (Cynipidae) (Bäsecke 1932). Detailed study of 109 specimens from 34 food portions, Primor'e (eastern USSR), showed diet 34·8% Homoptera, 2·0% Hymenoptera, 4·6% Diptera, 4·9% Coleoptera, 50·4% Lepidoptera, 0·1% other insects, 1·0% Araneae, 2·2% Gastropoda (see Polivanov 1981 for more detailed breakdown). Young in another nest, Rominter Heide (Poland/USSR), fed mainly larvae of Lepidoptera and sawflies, brought 2–5 at a time (Steinfatt 1939). ♂ seen to beat out gut of caterpillar on branch before feeding to young (Garnett 1942). At each of 46 visits, ♂ put head into nest-hole average 7·2 times, ♀ 5·0, suggesting ♂ brings more food per visit (Avery and Cockerill 1982; see also Social Pattern and Behaviour). In Sweden, first feed *c.* 04.00 hrs; last *c.* 18.00 hrs; feeding frequency at peak 04.00–06.00 hrs and *c.* 18.00 hrs (Pynönnen 1939). Last recorded feed in one study, East Germany, 19.15 hrs (Schlegel and Schlegel 1968). Average 15 feeding visits per hr over 4 consecutive hours in morning, chick age not known (Stahlbaum 1960); on day before fledging, visits every 2–16 min in morning and about hourly in afternoon (Steinfatt 1939); 3–11 visits per hr during 1-hr observations over 10-day period, chick age not known (Haverschmidt 1938); average 15·3 per hr over 3 hrs, chick age not known (Avery and Cockerill 1982). ASR

Social pattern and behaviour. No major studies. Some aspects not well known (Blume 1977, which see for summary). No indication of any marked geographical variation and the following account includes material on extralimital *amurensis* from study by Panov (1973) in Primor'e (eastern USSR).
 1. Mainly solitary (e.g. Schumann 1949). As in other Picinae, ♂ and ♀ probably often in neighbouring home-ranges or separate parts of same home-range outside breeding season. Fidelity to home-range typical: e.g. marked ♀ recorded roosting *c.* 70 m from site used by her 2 years previously (Pynnönen 1939; Steinfatt 1939; Labitte 1954*b*). In Rominter Heide (Poland/USSR), *c.* 500 ha available per pair, but such large area probably used only outside breeding season (Steinfatt 1939). In Finland, winter, marked bird ranged up to *c.* 1 km from roost. Said to defend 'feeding territory' possibly also outside breeding season, but no details (Pynnönen 1939). Single *amurensis* ringed at beginning of winter found *c.* 1·5 km away in following May (Panov 1973). Not uncommonly associates with foraging flocks of small passerines, but social bond not strong (Pynnönen 1939). BONDS. Monogamous mating system. ♀ recorded copulating with 2 ♂♂ in succession (Popp 1955; see part 2). Pair-bond possibly longer term through fidelity to home-range (Steinfatt 1939; see above). Contact between pair-members maintained outside breeding season (e.g. Schumann 1949, Labitte 1954*b*; see also Roosting, below); birds sometimes remain closer together for winter foraging than other *Dendrocopos* (D Blume). ♂ often takes greater share of nest-duties, but much variation reported: excavation often mainly (even exclusively) by ♂ (Schuster 1936; Labitte 1945, 1954*b*; Sermet 1973); substantial contribution made by ♀ (Tracy 1933*b*); allegedly by ♀ alone (Groschupp 1885). Most incubation by ♂ (e.g. Schuster 1936, Bussmann 1961, Sermet 1973) or shared about equally (Hachez 1966). ♂ frequently reported to do most (or all: Labitte 1954*b*) feeding of young, particularly after 1st week (Tracy 1942); see, e.g., Westerfrölke (1955*a*) and Stahlbaum (1960) for varying ♀ involvement. Brood possibly split for post-fledging care (Hachez 1966). Recorded helping to feed well-grown young at nest of Great Spotted Woodpecker *D. major* (Driessen 1970); *D. major* (Smith 1972*a*) and Wryneck *Jynx torquilla* (Pynnönen 1939) reported to have fed young *D. minor*; see, however, part 2. Age of first breeding not known. BREEDING DISPERSION. Solitary and territorial. Where habitat suitable, single pairs will make do with smallest field copses, normally inadequate for other Picinae. Density normally lower than that of *D. major*; may approach it at best over only small area in optimal (for *D. minor*) mixed woodland communities—riverine, fen, and damp oak *Quercus*—(Glutz and Bauer 1980, which see for density figures from small-scale censuses, East and West Germany). In south-east England, initially 1–3 pairs, later perhaps *c.* 15 pairs, in wood of 52 ha; increase probably due to increased food supply resulting from Dutch elm disease (Sharrock 1976). Mixed *Quercus* and hornbeam *Carpinus* favoured in Switzerland; up to 0·7 pairs per 10 ha (Schifferli *et al.* 1980). For similar densities, Mecklenburg (East Germany), see Klafs and Stübs (1977). Higher densities in Brandenburg (East Germany): e.g. 1–2 pairs per 10 ha (Rutschke 1983). Town parks and gardens can also hold 0·3–0·5 pairs per 10 ha over a small area. Larger-scale censuses embracing various habitats usually show less than 0·1 pairs per 10 ha (Glutz and Bauer 1980): 35 pairs on 20 km^2 (0·17 pairs per 10 ha) of marshy Unterspreewald, East Germany (Schiermann 1930); in Kreis Gransee (Brandenburg), *c.* 10 pairs on 180 km^2, i.e. 0·005 pairs per 10 ha (Rutschke 1983). In Rominter Heide, *c.* 50 pairs on *c.* 25 km^2 (Steinfatt 1939). Woods along lower Inn (Bayern, West Germany): *c.* 0·1 pairs per 10 ha, compared with 0·37 pairs of *D. major* (Reichholf and Utschick 1972); see also Schumann (1973*b*) for Niedersachsen (West Germany). Extent of territoriality not well known, but probably much as in other Picinae, with roost- and/or future nest-hole being focal point of activity. In Finland, nests at least 1 km apart and no evidence of clearly demarcated boundaries. Drumming (see part 2 and Voice) often performed over much larger area than that eventually used for breeding. In one case, main Drumming post *c.* 100 m from later nest (Pynnönen 1939); up to 300 m recorded by Radermacher (1980). Nesting (not necessarily by same birds) may occur in same tree in consecutive years (e.g. Bussmann 1961), but new hole normally made each season (Blume 1977). Hole for replacement clutch may be excavated in same tree (Labitte 1945) or disturbed birds may shift *c.* 200 m (Schlegel and Schlegel 1968). During nestling period, adults (sometimes using separate feeding areas) forage mainly *c.* 100–200 m from nest-hole (Steinfatt 1939); in Finland, rarely up to *c.* 200 m, *c.* 70% of flights *c.* 40 m from hole (Pynnönen 1939). ROOSTING. Solitary and nocturnal. Usually in hole of own species, including old nest-hole, enlarged partial excavation, or entirely new hole made, in autumn or winter, specifically for roosting (Pynnönen 1939; Glutz von Blotzheim 1962; Schlegel and Schlegel 1968; Radermacher 1970). Occasionally uses hole of *D. major* (Pynnönen 1939), rarely a nest-box (Creutz 1960). Outside breeding season, pair-members may roost fairly close together (Labitte 1954*b*); occasionally use same site, though not simultaneously (Pynnönen 1939). Use of same site recorded for over 2 months (Pynnönen 1939), though generally changes more often than other Picinae

A

(D Blume). Birds may excavate new hole nearby or, if disturbed, c. 100 m away (Pynnönen 1939). ♂ begins to roost in nest-hole before laying and continues to do so until near fledging (Schuster 1936; Steinfatt 1939); ♀ exceptionally roosts with young, ♂ elsewhere (Westerfrölke 1955a). Freshly excavated nest-hole may become ♀'s roost if (for any reason) pair build new one and breed nearby. ♀ then uses site up to about fledging (c. 2 months); also true if nest destroyed (D Blume). In autumn, ♂ may return to roost in nest-hole (Pynnönen 1939). No detailed information on diurnal activity patterns, but see Pynnönen (1939) and Steinfatt (1939). Most Drumming performed in morning. Bird rests (head slightly withdrawn, plumage ruffled) and performs feather-care after morning peak of feeding activity and in late afternoon before roosting (Pynnönen 1939). Captive birds frequently water-bathe, also sun-bathe with plumage ruffled and primaries slightly spread (Winkler 1972a).

2. Generally wary (Groschupp 1885); sometimes allows close approach (to c. 5 m), e.g. when displaying (Hammling and Schulz 1911). Bird disturbed in lower vegetation typically flies up into nearest tree crown (Schuster 1903); ♀ surprised by observer gave Scolding-call (see 3 in Voice) and flew up to join ♂ high in nearby tree (M G Wilson). In presence of raptors, bird may interrupt Drumming for c. 1 min (Hurme and Sarkunen 1975); also recorded flying to lower part of small tree (Schumann 1949); typically hides under or behind branch (Steinfatt 1944b) and there often raises and twitches wings (Winkler 1972a: Fig A). Records of similar hiding (Tracy 1933b) or flying off (Schlegel and Schlegel 1958) may be reactions to Starling *Sturnus vulgaris* (see also further below). Bird disturbed at roost may Tap inside before leaving; in serious disturbance, may avoid roost henceforth (Pynnönen 1939). Captive ♀ occasionally entered hole and trembled vigorously at entrance, creating quiet sound; function not clear (Winkler 1972a; see Voice, also *D. major*). ANTAGONISTIC BEHAVIOUR. (1) General. Drumming performed mainly by ♂ (posture as when resting; see Roosting, above) but also frequently by ♀, though individual variation considerable: more by ♀ (Labitte 1945; Hurme 1973), almost as much as ♂ (Steinfatt 1939), none by ♀ (Schuster 1936). In southern France, Drumming occurs in October (sometimes November), but mainly late December to (at latest) late July (Hüe 1949–50); in Britain, late January to mid-June (Witherby *et al.* 1938); in northern France, March–June (Pynnönen 1939); rare August–September (see Radermacher 1968). In general discussion of *Dendrocopos*, Drumming considered by Winkler and Short (1978) to serve proclamation of territory; in *D. minor*, not known to what extent Drumming sites linked locally to trees with holes (Blume 1977; see also Breeding Dispersion, above). Advertising-call (see 1 in Voice) often combined with Drumming (Radermacher 1980) and similarly given mainly by ♂; far-carrying signal denoting and locating territory-owner (Winkler and Short 1978; also Steinfatt 1939; Hurme 1973). Bird climbing about normally stops and calls with head up at angle (Schuster 1936) or thrown well back (Davenport 1897); also given from inside hole looking out or when hidden in hole (Schuster 1936), exceptionally in flight (Radermacher 1980). Advertising-call given throughout the year, mainly in spring (Bergmann and Helb 1982); not in November–December according to Radermacher (1980). For further details, see Voice and subsection 2 (below). Disputes involving 3–4 birds conspicuous through much display and movement and occur mainly in April (March–May); probably arise, in most cases, from intrusion by ♂ or pair into occupied territory (for details, see below). Threat associated with defence of 'feeding territory' mentioned only by Pynnönen (1939). Damaging fights rare. (2) Threat and fighting. Interactions take place in the air and on trees. Aerial chases, sometimes on circular flight-path, frequent. Birds said sometimes to fight in flight (Labitte 1954b), but not clear how much physical contact involved. ♀ may merely raise crown feathers (Panov 1973), and not otherwise participate when 2 ♂♂ in dispute (Labitte 1954b); once, ♀ performed same movements as 3 disputing ♂♂ but did not allow close approach by any of them (Groschupp 1885). Chases and threat also frequent between ♀♀, however (e.g. Palm 1967, Winkler 1972a). Butterfly-flight (compare Fluttering-flight of other Picinae) probably has aggressive component (Winkler 1972a); in Nearctic *Dendrocopos* (*Picoides*) considered by Short (1971) to be derived from Wing-spreading (threat) display (see below). In Butterfly-flight, bird typically glides on widely spread and horizontal or slightly raised wings. Performed as part of chase among tree branches (e.g. rival ♀♀ keeping c. 2 m apart: Ladhams 1977), also when flying at perched bird in supplanting attack, tail then being fanned (Richardson 1948). Attacker may also fly up and swoop at opponent (Tracy 1933b). In one interaction, bird eventually displaced had bill raised, wings half spread, head withdrawn, and plumage slightly ruffled (Panov 1973); attacked bird may also crouch flat with wings spread in defence. Following also characteristic of antagonistic interactions: simple supplanting attacks; Threat-gaping; frequent ruffling of crown feathers giving markedly different head shape; landing with wings spread and raised, maintaining posture (Wing-spreading display) whilst moving, wings sometimes quivered; hiding behind branch with Wing-twitching (Fig A). In both Butterfly-flight and Wing-spreading display, bird appears bigger and attention drawn to barring on back, wings, and tail (see Fig B where rival below displaying bird). Birds may freeze while perched close together, or peck at one another between bouts of chasing. Quite rapid circling on branch and probable Head-swaying (see Heterosexual Behaviour, below) mentioned by Groschupp (1885). Hacking (possibly Excitement-pecking; see, e.g., *D. major*) and brief bouts

B

of feeding also occur before pursuits, threat, etc., are resumed (may last with pauses for c. 90 min). Antagonistic interactions frequently accompanied by call 4 and call 1 occasionally given by bird perched nearby; dispute may be punctuated by Drumming which may also mark its conclusion. (Groschupp 1885; Tracy 1933b; Pynnönen 1939; Southam 1944; Hurme 1973; Panov 1973; Ladhams 1977.) Playback of Drumming elicits Drumming or loud Advertising-call, only rarely Contact-call (see 2 in Voice). Like *D. major*, bird will make searching flights, sometimes with Butterfly-flight (see above). Will also respond to Drumming of *D. major* and Black Woodpecker *Dryocopus martius* (Winkler 1972a; Zabka 1980). In England, early November, ♂ once gave low-intensity Drumming in response to imitated Advertising-call (Richardson 1948). Stuffed specimens by nest-hole elicit attacks accompanied by call 4b (Blume *et al.* 1975). Once when *D. major* flew in to use regular Drumming site of ♂ *D. minor*, latter moved only c. 1 m (Radermacher 1980); *D. major* represents serious threat to *D. minor* in breeding season (see Parental Anti-predator Strategies, below). Starlings *Sturnus vulgaris* subjected to dive-attacks; though ♀ often more timid (Sermet 1973; M G Wilson); sometimes nests in same tree as *S. vulgaris* (Schlegel and Schlegel 1968). Fights recorded with *J. torquilla* (Pynnönen 1939; see Bonds), and Nuthatch *Sitta europaea* and House Sparrow *Passer domesticus* chased away from nest-hole (Steinfatt 1939; Westerfrölke 1955a). HETEROSEXUAL BEHAVIOUR. (1) General. As in other Picinae, much early contact between prospective partners involves elements of threat and territorial behaviour (probably centred in most cases on trees with holes: see Blume 1977) so that timing of pair-formation difficult to ascertain. For seasonal occurrence of some vocal and instrumental signals, see Antagonistic Behaviour (above). No antagonism apparent between pair-members once bond established (e.g. Schuster 1936, Haverschmidt 1938). In some cases, breeding preceded by little or no courtship (Radermacher 1980). Bouts of low-intensity display outside breeding season (e.g. mid-November) possibly relate more to threat than to courtship but considered by Panov (1973) to be connected with pair-formation; see also Taylor (1977). (2) Pair-bonding behaviour (see also Antagonistic Behaviour, above). First contact normally established through frequent Advertising-calls and Drumming by both sexes (e.g. Steinfatt 1939, Schumann 1949); Drumming in duet occurs, ♂ joining in after ♀ starts (Palm 1967). Closer approach often made (in part) in Butterfly-flight, bird changing to this from (normal) undulating flight (Wilkinson 1977), descending thus at c. 45° (Radermacher 1980), or using it at times in branch-to-branch or tree-to-tree pursuits—latter may be circular (diameter up to c. 50 m) and take birds up to c. 40 m (see Palm 1967). Rolling pursuit flights with wings creating noise (see Voice) reported by Panov (1973) for ♂-♀ encounter in autumn. See also Fig C which shows ♂ about to land near ♀. Further interactions much as described above (see subsection 2 in Antagonistic Behaviour), with birds freezing or moving with wings raised (held aloft for c. 1 s on landing near conspecific) or spread sideways. Call 4 commonly given (Palm 1967; Wilkinson 1977), or soft, chipping Contact-calls during pursuit (M G Wilson). Some driving up tree occurs, and when ♀ moves closer, apparently attracted by slower Drumming or Tapping (see Voice) of ♂, he may chase her away (Steinfatt 1939). ♀ may assume posture with tail fanned, quivering wings half-spread and drooped, and head held forward as ♂ approaches, and both sometimes freeze close together thus (Southam 1944), or face one another with wings close to body and bills directed forward or (possibly as sign of reduced aggression) up for c. 1 min (Beven 1976). Constant restlessness typical so that chasing, etc., soon resumed (Hesse 1909). Behaviour described here does not form normal preliminary to copulation (Steinfatt 1939) and 2 ♂♂ copulating with ♀ in early April (see Bonds) perhaps reacted inappropriately to ♀'s Wing-spreading posture (Glutz and Bauer 1980). (3) Nest-site selection. Details of selection process not known. ♂ and ♀ often move through territory together and Drum or excavate at various sites (D Blume). Excavation by both sexes in bouts of c. 3 min each recorded at site where display (see above) had taken place 3 days previously (Ladhams 1977). Tapping probably used for Nest-showing as in (e.g.) *D. major*; when ♂ Tapped at hole-entrance, ♀ came to inspect site (Winkler and Short 1978). Butterfly-flight performed by bird leaving perch after Drumming and meeting with conspecific bird (Tooby 1943); possibly also involved in Nest-showing (Glutz and Bauer 1980). In one case, birds excavated hole, then moved c. 200 m to start another, but finally nested in 2nd hole in 2nd tree (Schlegel and Schlegel 1968). In one case, ♀ reported to have begun excavation (Schumann 1949). Usually done in morning, then again towards evening (Schuster 1936); such behaviour possibly Nest-guarding (see, e.g., Syrian Woodpecker *D. syriacus*) and undertaken mainly by ♂ (Sermet 1973). ♂ sometimes gives Advertising-call and Drums when ♀ working on hole, flies to hole when ♀ inside, and Taps by hole, pair then giving quiet Contact-calls together for c. 30 s. ♂ working on hole once chased ♀ off, ♀ then starting another hole in same tree (Tracy 1933b, which see for details of excavation process; also Westerfrölke 1955a). (4) Courtship-feeding. Only 1, old, report (Pässler 1856). (5) Mating. Recorded mainly during excavation and laying (e.g. Sermet 1973), ♀ rejecting ♂'s advances after start of incubation (Schuster 1936). Takes place mostly in morning on horizontal branch near nest (Westerfrölke 1955a), though also recorded up to c. 250 m away (Pynnönen 1939). In several cases observed by Schuster (1936), not preceded by any special ceremony but ♀ invariably first entered hole (not seen by Westerfrölke 1955a), then flew up to join ♂ high in nearby tree where copulation took place; see also Tracy (1933b). ♀ may call ♂ out of hole by Tapping (Sermet 1973). Fullest description refers to interaction leading to copulation in early April (thus possibly contains references to more specifically antagonistic behaviour typical of early pair-contact—see above). Drumming ♀ joined by ♂, both then Drummed, suddenly ceased, froze, reared up facing one another with tail spread and mock-pecked at one another 5 times, froze again, changed to a horizontal posture and body-swayed 6–7 times. ♀ spread wings and gave quiet treble Contact-call (see 2 in Voice), encouraging ♂ to fly in semicircle and on to ♀'s back from behind. During copulation lasting c. 6–8 s (c. 2–3 s recorded by Labitte

C

1954b), ♀, not ♂, had wings and tail spread. Afterwards, ♂ gave loud Contact-calls and flew c. 50 m to tree; ♀ held posture and wing-shivered for c. 30 s then also flew off giving Contact-calls. Similar behaviour recorded at same site next day (Schlegel and Schlegel 1971, which see for illustrations). Appeasement-calls may precede or accompany copulation (see 5 in Voice). ♂ often reported to glide (possibly in Butterfly-flight) on to ♀'s back; ♂ keeps wings and tail spread (see Panov 1973) and may grasp ♀'s nape-feathers, ♀ crouching with wings closed (Ladhams 1977), sometimes crossways on branch (Sermet 1973). Pair may perch side by side (and touching) for c. 2 min after copulation (see *D. major*); ♂ normally soon resumes excavation (Tracy 1933b; also Labitte 1954b). (6) Behaviour at nest. ♀ probably lays in early morning, after taking over from ♂ (Pynnönen 1939; Sermet 1973). For incubation shifts, see (e.g.) Schuster (1936) who reported eggs being left uncovered for brief periods, and Bussmann (1961). No special change-over ceremony reported. Sitting bird, perhaps alerted by light impact of mate landing below nest, leaves immediately and other climbs up to enter hole (Hachez 1966). Appeasement-calls (see 5 in Voice) sometimes given however and during excavation and early incubation both birds will Drum near nest if mate close (D Blume). RELATIONS WITHIN FAMILY GROUP. See also Bonds. Young brooded for shifts of up to c. 20 min in early stages (Schuster 1936); not known when daytime brooding ceases, but ♂ normally roosts with young until shortly before fledging (Steinfatt 1939). ♀ rarely roosts with young (e.g. Westerfrölke 1955a). If hot in hole, brooding bird will sit with head out, feathers ruffled, and gape (Haverschmidt 1938). Faeces of young possibly swallowed during early period, later carried away and dropped or deposited on branches, or may be tossed out by ♂ prior to roosting; adults recorded pecking at faeces and swallowing items (Westerfrölke 1955a). Once, dead nestling possibly dismembered in order to remove it (Carlson and Carlson 1978). Adults rarely at nest together (Steinfatt 1939). Normally, incoming bird waits for other to leave; once, when both arrived together, ♂ entered first and ♀ followed but flew out again almost immediately not having fed young and waited for ♂ to leave (Haverschmidt 1938). Loud food-calls given by young when adult landed below and behind nest-hole (T G Easterbrook). Inter-sibling aggression can be marked, even leading to deaths (Blume 1977). Near fledging, young spend much time leaning out and then give Advertising-calls, parents responding similarly (Steinfatt 1939). May be aggressive towards parent when being fed (T G Easterbrook), and ♂ forced to peck at offspring in order to enter nest and clean it (Westerfrölke 1955a). Young fledge after 18 days (Labitte 1954; Westerfrölke 1955a; Blume 1977) or 19–20 days (Weitnauer 1962), on same day (Steinfatt 1939) or sometimes over 2 days (e.g. Martin 1937). ♂ apparently feeding young alone reduced feeding rate near fledging (Garnett 1942), possibly encouraging young to leave. In case of last nestling, ♀ came to nest with food but did not feed young bird which tumbled out in frantic effort to get food, and was then led away (T G Easterbrook). Young do some self-feeding from outset but fed by parents for c. 8–14 days; in most cases, probably move away from nest fairly rapidly (Pynnönen 1939; Steinfatt 1939). One brood of 2 split between parents (Hachez 1966), but not known how typical this is. ANTI-PREDATOR RESPONSES OF YOUNG. No information. PARENTAL ANTI-PREDATOR STRATEGIES. (1) Passive measures. Generally unobtrusive during incubation (Westerfrölke 1955a). ♀ may be reluctant to leave when man nearby and finally flies only short distance, returning quickly after disturbance (Labitte 1945). (2) Active measures. Persistent Scolding-calls given towards human intruder (e.g. Bossiere 1935), e.g. by ♂ perched close to nest (M G Wilson). Loud Scolding-calls given almost ceaselessly in presence of most serious predator *D. major* which also subjected to frequent dive-attacks (see Schumann 1949, Sermet 1973, and, especially, Tracy 1933b whose observations refer to nest-excavation stage). For attacks on *S. vulgaris*, see Antagonistic Behaviour (above). Scolding-call used more by ♂ in post-fledging phase (Weitnauer 1962).

(Fig A from drawing in Winkler 1972a; Fig B from drawing in Ladhams 1977; Fig C from drawing in Panov 1973.) MGW

Voice. Used mainly in breeding season, but calls 1–2 given throughout the year, and Drumming also at other times; for fuller details, see Hüe (1949–50) and Winkler and Short (1978), also Social Pattern and Behaviour; for diurnal variation, see Pynnönen (1939). Repertoire similar to other *Dendrocopos* woodpeckers and shared (including instrumental signals, though only Drumming well studied) by both sexes; for variations in intensity, see below. Fullest study by Winkler and Short (1978) who mentioned no geographical variation (see, however, call 1); for extra sonagrams, see also Glutz and Bauer (1980) and Bergmann and Helb (1982).

INSTRUMENTAL SIGNALS OF ADULTS. (a) Drumming. Usually quieter, also higher pitched and more brittle-sounding than Great Spotted Woodpecker *D. major* (Witherby *et al.* 1938; Sharrock 1976) and differs also in giving significantly longer series: often up to 33 strikes in 1·3 s (Blume 1977); typically, strike rate and volume are constant through the series (Zabka 1980). For exceptionally long series lasting 2–2·5(–3) s (number of strikes not stated), see Radermacher (1980). Average 24·55 strikes per series, average duration 1·2 s, $n = 40$ series by 13 birds; interval between strikes c. 50 ms (Zabka 1980). Average $26·2 \pm 5·9$ strikes per series (Wallschläger and Zabka 1981). Recording (Fig I) shows 26 strikes in 1·56 s, with interval between onset of strikes almost constant at 60 ms (c. 17 strikes per s). In recording by W Pedley, 25 strikes in 1·2 s, with 50 ms between onset of strikes (20 strikes per s) (J Hall-Craggs). Variation most pronounced in length of series, differing (e.g.) between 2 neighbouring ♀♀; individual variation also quite considerable (Zabka 1980, which see for further analysis and experimental study). Of west Palearctic Picinae, confusion most likely with Grey-headed Woodpecker *Picus canus*, but *D. minor* differs significantly in typically short pause between series (Blume 1971; Blume *et al.* 1975); up to 14 (Pynnönen 1939; Hurme 1973), exceptionally 17–20 (Thibaut de Maisières 1940; Radermacher 1980) series per min. In study by Pynnönen (1939), bouts given by ♀ shorter and with longer pauses between series: e.g. ♂ once gave 79 series in 14 min, while maximum from a ♀ was 14 series in a bout. Particularly in case of rivals, ♀♀ may, however, Drum almost as intensely as ♂♂: e.g. 19 series in 4 min (Palm 1967, who also reported alternate and simultaneous Duetting). Paired bird probably gives c. 100–150 series per day (Pynnönen 1939); in mid-March, ♂ (possibly unpaired) gave c. 300 (probably 340–350) series in 25 min, including, however, many of less than 1 s duration

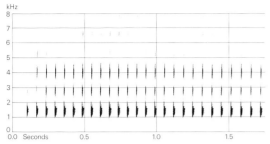
I S Wahlström/Sveriges Ornitologiska Förening (1982) Sweden May

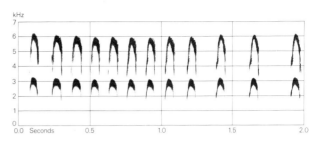
II R W Genever England June 1963

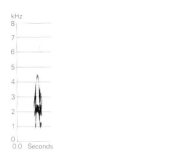
III S Palmér/Sveriges Radio (1972) Sweden April 1962

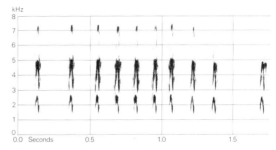
IV P A D Hollom England June 1974

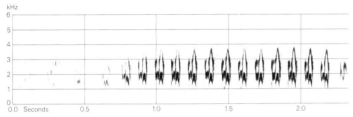
V S Palmér/Sveriges Radio (1972) Sweden April 1962

(Radermacher 1980). Normally favours high tree branches for Drumming, but other sites such as weather vanes, aerials, metal parts of roofs, etc., apparently used more than by *D. major* (Blume 1977); small size allows use of bean-poles (Conrads 1967); on telegraph poles, Drums mainly near large cracks, presumably for better resonance (Bussmann 1946). For contexts, see Social Pattern and Behaviour. (b) Tapping. In fairly loud sound produced by nest-showing ♂ *ledouci*, intervals between strikes 141 (124–155) ms, $n=18$ (Winkler and Short 1978); see also Zabka (1980). Apart from this typical use of Tapping (see, e.g., *D. major*), slow, emphatic Drumming given by ♂ to lure ♀ (Steinfatt 1939) possibly also belongs here. See also Social Pattern and Behaviour. (c) Wing-noise reported for *amurensis* during erratic, rolling pursuit-flights (Panov 1973). (d) As in *D. major* and Syrian Woodpecker *D. syriacus*, wing-quivering combined with feather-ruffling by bird in hole produces quiet humming sound, audible at close quarters (Winkler 1972a for captive ♀).

CALLS OF ADULTS. (1) Advertising-call. Most prominent and frequently heard call; heard also more often than Drumming. A series of 8–20 relatively long units ($c.$ 6 units per s), with soft squeaking quality: 'pee-pee-...'. Often compared to distant Kestrel *Falco tinnunculus*; compare also Advertising-call of Wryneck *Jynx torquilla*. Average pitch 2·7 (2·5–2·8) kHz, duration of single unit 68·3 ms (52·8–86·8), $n=46$, average pause between units 183·3 ms (135·8–256·6), $n=43$ (Winkler and Short 1978, presumably for *hortorum*; also Bergmann and Helb 1982). In recording (Fig II), ♀ *comminutus* gives series of high-pitched 'quee' units with slightly sibilant quality and at faster rate—average pause between units 107·3 (70–220) ms; gradual rallentando at the end (J Hall-Craggs, P J Sellar). Also described as a rather loud shrill 'pee-pee-pee-pee-pee' of uniform tone (Witherby *et al.* 1938). In study by Hurme (1973), calls of ♂♂ comprised average 9·89 units ($n=56$), those of ♀♀ 12·1 units ($n=29$); further investigation required. Frequency of calling most marked in spring: e.g. 21 calls in 40 min (Pynnönen 1939). Normally given when perched, exceptionally in flight (Radermacher 1980). (2) Contact-call. A short, clicking sound of average pitch 2·2 (2·0–2·4) kHz, average length 35·5 (26·4–53·1) ms, $n=10$ (Winkler and Short 1978). Like *D. major*, but much weaker (Schumann 1949) and hence perhaps rarely heard. Recording (Fig III) suggests 'chik' (M G Wilson), or at higher levels 'chuk' and at lower volume 'tuk' (J Hall-Craggs; compare call 3). Quiet, gentle 'gib' when pair meet at nest (Schuster 1936); muted 'kik' from ♂ bringing food to young, and quiet 'kick-kick-kick' from ♀ prior to copulation (Schlegel and Schlegel 1968, 1971). Sometimes given during disturbance at nest (Winkler and Short 1978); also by captive birds when disturbed (Winkler 1972a). (3) Scolding-call. Rapid series of slightly higher pitched Contact-calls. Average pitch 2·3 (2·1–2·4) kHz, average duration 26·1 (22·6–30·2) ms, $n=24$, with units 188·5 (117·0–366·0) ms apart, $n=22$ (Winkler and Short 1978). Sharp and shrill in alarm or aggression (Schuster 1936); see also Garnett (1942). In recording (Fig IV), ♂ disturbed near nest with young gives at times rapid but overall markedly irregular (in rate of calling) series of 'ki' sounds; unit similar in form to call 2 but main frequency c. 500 Hz higher. Rather like agitated Blackbird Turdus merula, though lower pitched (J Hall-Craggs, P J Sellar). (4) Other calls, mainly associated with aggression. (a) Soft 'chwuit' sounds given by ♀ attracted by playback of Drumming; possibly homologous with 'Kweek Call' of other Dendrocopos (Winkler and Short 1978). (b) In attack on stuffed specimens at nest, birds give typically rapid series of noisy 'shwika' calls, c. 172 ms long ($n=2$) and c. 280 ms apart (Winkler and Short 1978). Termed 'Kreck Call' by Blume et al. (1975), but 'kreck' considered to be an inadequate rendering by Winkler and Short (1978), although other descriptions—'terrét terrét' from ♂ with ♀ nearby (Hesse 1909) and loud 'kireck' in pursuits (Pynnönen 1939)—clearly support this rendering. See also discussion of Kreck-call and variants in D. major. (c) In recording (Fig V), quiet 'choo choo choo choo' followed immediately by 14 'ch' units at c. 9 units per s; overall crescendo to 4th unit from end, then slight diminuendo over last 4 (J Hall-Craggs); a chuckling chattering call (P J Sellar) presumably equivalent of long Kreck-call in D. major. Hard 'cht-cht-cht' sounds (Glutz and Bauer 1980; see also for sonagram) presumably a related call. During bout of display involving pursuit of ♀, ♂ gave ascending and descending phrase c. 3 s long and comprising 'keeoo' units delivered staccato and in rapid succession (probably at least closely related to call illustrated in Fig V, if not the same): 3–4 'keeoo' units followed by acceleration and gradual crescendo; at peak, units more like 'kee-er', after which phrase diminished rapidly in volume and pitch and ended abruptly. In occasional variant, a few 'keeoo' units followed by rapid, musical trill and ending with further 'keeoo' units (Wilkinson 1977). (d) Loud churring or screeching sound with slightly hissing quality (Southam 1944), harsh hiss (Ladhams 1977), 'keer-keer-keer' (Beven 1976), or hoarse-sounding 'chshuur-chshuur-chshuur' (Panov 1973); perhaps refer to elements of the longer phrases described above. (5) Appeasement-call. Similar to 'wäd' or 'rä' sounds of D. major but softer; given (usually as a series) at nest-relief, also during copulation when unit c. 50 ms long (Blume et al. 1975; Winkler and Short 1978). Compressed 'vak-vak-vak' from ♀ prior to copulation (Panov 1973) probably the same.

CALLS OF YOUNG. Soft, musical 'puit' given by nestlings (occasionally after fledging) following serious disturbance; probably expresses relief from distress (Winkler 1972a; Winkler and Short 1978). Food-call initially rather harsh chirping and rattling sounds, becoming more musical with age and appearing on sonagrams as short, vertical columns c. 25 ms wide; in older young, c. 60 ms. Calls louder when being fed (Winkler and Short 1978). Gentle trilling and chirping sounds distinguished by Winkler (1972a). See also sonagram of 'gük' series in Bergmann and Helb (1982). In recording by A J Williams, nestlings give very rapid, high-pitched chirruping (M G Wilson). Squeaking quality may become evident close to fledging (Winkler and Short 1978), but particularly prominent when juvenile being fed after leaving nest (Winkler 1972a). In recording by P A D Hollom, juvenile gives higher pitched and hoarser version of adult call 2 (P J Sellar). Close to fledging, young give weaker, thinner and more strained variant of adult call 1, audible at c. 50–100 m (Steinfatt 1939); in recording of fledged juvenile by P A D Hollom, such calls high-pitched and sibilant (P J Sellar). In Oxfordshire (England), nestlings performed Tapping or rudimentary (rapid) Drumming after each feed (T G Easterbrook).

MGW

Breeding. SEASON. North-west Europe: see diagram. Similar in central Europe and southern and central USSR, but up to 3 weeks later in Scandinavia, and 2 weeks later in Tunisia (Dementiev and Gladkov 1951a; Heim de Balsac and Mayaud 1962; Makatsch 1976). SITE. Hole in tree, frequently in side branch, and in rotten wood. 2–8 m (0·5–25) above ground (Dementiev and Gladkov 1951a; Sharrock 1976; Glutz and Bauer 1980). Nest: excavated hole with entrance diameter 3–3·5 cm; depth 10–18 cm, internal diameter 10–12 cm (Pynnönen 1939). Building: by both sexes; normally takes c. 12–16 days (Weitnauer 1962; Sermet 1973), although up to 30 days reported (Labitte 1945) and 5–6 days for late or replacement clutch (Pyn-

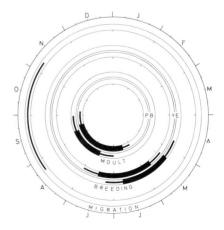

nönen 1939; Labitte 1945). EGGS. See Plate 98. Elliptical, smooth and glossy; white, appearing translucent. 19 × 15 mm (17–21 × 13–16), n = 100 (Witherby et al. 1938). Calculated weight 2·0 g (Makatsch 1976). Clutch: 4–6 (3–8). One brood. Replacements laid after egg loss. INCUBATION. 11–12 days (Weitnauer 1962). By both sexes, with ♂ at night (Witherby et al. 1938). Begins with last egg; hatching synchronous. YOUNG. Altricial and nidicolous. Cared for and fed by both parents (Weitnauer 1962). FLEDGING TO MATURITY. Fledging period 18–20 days (see Social Pattern and Behaviour). Age of independence and first breeding not recorded. BREEDING SUCCESS. No information.

Plumages (*D. m. hortorum*). ADULT MALE. Bristles covering nostrils, lores, and feathering at base of upper mandible cream-buff, buff, or buff-brown, longer bristles black, basal feathering and lores sometimes brown or black-brown. Remainder of forehead cream-buff, pale buff, or off-white. Crown crimson, some white or cream of feather-bases usually visible, some dusky subterminal bars usually showing on hindcrown. Red of crown narrowly bordered by black at sides above eye, more broadly at rear of head. Central hindneck black, stripe from upper rear corner of eye backwards over ear-coverts white or pale cream (sometimes mottled black between ear-coverts and nape), widening towards sides of hindneck and upper mantle. Ear-coverts and area below eye buff, buff-brown, or greyish-buff (rather similar to underparts and forehead, distinctly darker than whitish stripe above ear-coverts). Black stripe from base of lower mandible over lower cheeks (mottled pale grey or buff below gape), widening to large black patch on sides of neck; black sometimes extends upwards to halfway behind ear-coverts, but usually not to nape; black extends downwards to irregular patch on sides of chest, consisting of elongate-triangular feather-centres. Upper mantle white with some black mottling in centre; lower mantle and upper scapulars black, usually with some dark brown of feather-bases visible. Lower scapulars, back, and upper rump white with black bars 2–3 mm wide, intervening white 4–5 mm; lower rump and upper tail-coverts black. Underparts buff or pale buff-brown; chin, vent, and under tail-coverts paler cream-buff or off-white, throat and central chest tinged pink-buff when plumage fresh; black elongate-triangular streaks on sides of chest, gradually narrowing to thin black streaks on flanks and sides of breast and belly, and forming black or dusky arrowheads or short bars on lower flanks and under tail-coverts. Central 2 pairs of tail-feathers (t1–t2) black, t3 black with much white on outer web and some on tip, t4–t5 mainly white or cream with hidden black basal halves of inner webs and 2–3 black bars (2–3 mm wide) across tips; outer web of t4 (–t5) largely white (not barred); reduced t6 (hidden between t4–t5 and upper tail-coverts) black with some white spots. Flight-feathers black or blackish-brown; outer webs of p5–p8(–p9) with 4–6 rather small and equally spaced white spots; towards inner primaries, progressively fewer spots visible beyond upper wing-coverts; 3–4 slightly larger spots on outer webs of secondaries; similar but larger and more rounded white spots on inner webs of flight-feathers (not reaching shafts), but terminal 1–2 spots (though present on outer webs) lacking; tertials as secondaries, but white spots of both webs join across shafts, forming contrasting bars; barred pattern of tertials contiguous with barred lower scapulars and back, but black bars broader. Upper wing-coverts black, but greater and median coverts with broad subterminal white bar or spots contiguous with white bars on tertials and longer scapulars. Under wing-coverts and axillaries white or cream, some shorter coverts spotted black, outer greater under primary coverts largely black. Much variation due to bleaching and wear: white of forehead, of stripe behind eye, and of upperparts purer white, less creamy; ear-coverts and particularly underparts less pure and saturated buff and virtually without pink tinge—ear-coverts, chest, and flanks dirty buff (black streaks more contrasting), throat, belly, and vent dirty grey with slight buff or cream tinge. ADULT FEMALE. Similar to adult ♂, but forehead and forecrown white (tinged buff or cream when fresh, and feathers then sometimes faintly tipped dusky), forming pale patch which is smaller than red patch of ♂, rounded at rear; patch narrowly bordered by deep black at sides, broadly so on hindcrown and nape. Bristles and feathering at base of upper mandible often darker buff-brown or black-brown, as in ♂. Exceptionally, some faintly red feather-tips on forecrown. NESTLING. Naked at hatching. On *c.* 11th day, covered with closed feather-sheaths. Sheaths open to produce feather-tips on *c.* 12th day. Fully feathered on *c.* 17th day. (Heinroth and Heinroth 1924–6; RMNH.) JUVENILE. Rather similar to adult ♀ in showing small pale patch on forehead and forecrown only; forehead and borders of patch buff with dusky grey or brown mottling (in ♀, more uniform off-white or cream), forecrown red in juvenile ♂, but greyish-buff with a few indistinctly red feather-tips or no red at all in ♀; crown patch of ♂ thus considerably smaller than adult ♂, red more pinkish, less bright, with some dusky and grey-buff of feather-bases showing through. Black of upperparts and upper wing-coverts duller and browner than in adult; hindneck, upper mantle, back, and rump more extensively black; sides of breast and flanks with closer and broader but less sharply defined and less deep black streaks, lower flanks indistinctly barred dusky; tail-feathers pointed at tip instead of rounded; tips of p5–p8 often with distinct white fringe along both webs (in adult, some white on outer web only or none at all); p10 longer and more broadly rounded at tip than in adult, 6–9 mm longer than longest greater upper primary coverts (in adult, p10 3 mm longer to 2 mm shorter). FIRST ADULT. Like adult, and hardly distinguishable once juvenile p10 replaced. Retained juvenile tertials with slightly narrower and more rounded tips than adult, tips more worn than in adult at same time of year, and with bars less straight-edged, but character useful only when direct comparison with known-age birds possible. Occasionally, some juvenile greater upper wing-coverts retained, contrasting in wear and colour with bordering new ones. Juvenile greater upper primary coverts retained (except sometimes for outer 1–3), more pointed than those of adult and more worn at same time of year, but birds with heavily worn but rounded coverts or new but slightly pointed ones difficult to age with certainty.

Bare parts. ADULT AND FIRST ADULT. Iris red-brown, brownish-red, or carmine-red. Bill dark grey, greyish-black, or leaden-grey, base of lower mandible paler bluish-grey. Leg and foot greenish-grey, olive-grey, or leaden-grey. NESTLING AND JUVENILE. Completely pink at hatching. At fledging, iris brown-grey or hazel; bill, leg, and foot as in adult, but base of lower mandible more extensively pale grey. (Hartert 1912–21; Witherby et al. 1938; Glutz and Bauer 1980; RMNH, ZMA.)

Moults. ADULT POST-BREEDING. Complete; primaries descendant. Starts with p1 in June, usually completed with p10 in September; duration *c.* 87 days (Ginn and Melville 1983; RMNH, ZFMK, ZMA). Tail starts with t2 shortly after p1, completed with regrowth of t1 at primary moult score 30–40 (at about shedding of p7). Body, tertials, and secondaries start at moult score *c.* 20–25 (July or August), completed with score

c. 40–45 (second half of August or September), but some middle secondaries occasionally up to early October. Sequence of moult as in Great Spotted Woodpecker *D. major*. Rarely, t1 and outer primaries still growing in January (Glutz and Bauer 1980). POST-JUVENILE. Partial; primaries descendant. Juvenile tertials, secondaries, greater upper primary coverts (except sometimes a few outermost), and occasionally greater upper wing-coverts retained. P1 and p2 already shed before fledging; moult of body feathers starts at age of 5–6 weeks, c. 2 weeks later followed by t2; moult virtually completed at age of c. 3 months, about early September (Heinroth and Heinroth 1924–6). In specimens examined, timing of primary moult as in adult, starting June and completing September or early October, but some *danfordi* (and perhaps other southern races) start May; moult of tail and body completed late August or September, but some birds retained juvenile t1 until mid-September, though body and greater upper wing-coverts new (primary score 40–45).

Measurements. ADULT AND FIRST ADULT. *D. m. hortorum*. Netherlands, West Germany, and East Germany, all year; skins (RMNH, ZMA). Bill (F) to forehead, bill (N) to distal corner of nostril; toe is outer front toe with claw.

	♂			♀		
WING	89·9	(1·52; 22)	87–92	90·5	(2·04; 13)	88–94
TAIL	51·6	(1·68; 23)	49–55	52·4	(1·89; 13)	50–55
BILL (F)	17·4	(0·72; 22)	16·4–18·5	17·2	(0·79; 12)	16·3–18·3
BILL (N)	13·0	(0·67; 22)	11·9–14·9	13·0	(0·78; 12)	12·0–14·3
TARSUS	14·3	(0·65; 16)	13·5–15·4	14·4	(0·69; 10)	13·4–15·1
TOE	14·7	(0·70; 15)	13·8–16·1	14·8	(0·77; 9)	14·3–16·3

Sex differences not significant.

Combined data from Domaniewski (1927), Dementiev (1937; for USSR only; wing corrected as measuring method slightly different), RMNH, and ZMA; sexes combined. *D. m. comminutus*: (1) Britain. *D. m. ledouci*: (2) North Africa. *D. m. buturlini*: (3) Spain and Pyrénées; (4) Italy; (5) southern Yugoslavia. *D. m. danfordi*: (6) Greece and Taurus mountains (Turkey). *D. m. colchicus*: (7) Caucasus. *D. m. quadrifasciatus*: (8) Lenkoran area (south-east Transcaucasia). *D. m. hortorum*: (9) Netherlands, (10) West and East Germany, south-west Poland, northern Switzerland, Austria, Czechoslovakia, and Hungary; (11) Rumania. Nominate *minor*: (12) Norway and southern Sweden; (13) northern Sweden and Finland; (14) north-east Poland, Baltic States, and west European USSR; (15) Ukraine; (16) north-west, central, and east European USSR. *D. m. kamtschatkensis*: (17) western Siberia; (18) central and eastern Siberia.

	WING			BILL (N)		
(1)	87·1	(1·20; 34)	84–90	12·7	(0·56; 35)	11·5–13·8
(2)	91·4	(2·29; 4)	88–93	13·1	(0·64; 4)	12·5–14·0
(3)	87·9	(1·48; 8)	86–91	12·6	(0·53; 8)	11·5–13·1
(4)	86·4	(2·24; 16)	84–92	12·7	(0·43; 16)	12·0–13·5
(5)	90·1	(1·86; 8)	88–93	12·6	(0·56; 8)	12·2–13·9
(6)	86·4	(1·61; 9)	84–89	12·7	(0·60; 8)	12·0–13·5
(7)	89·3	(1·87; 44)	85–93	12·6	(0·51; 43)	12·0–14·2
(8)	83·8	(—; 3)	83–85	12·2	(—; 3)	12·0–12·5
(9)	89·9	(1·57; 27)	87–93	13·0	(0·72; 26)	11·9–14·3
(10)	90·5	(1·85; 77)	86–96	12·9	(0·70; 71)	11·5–15·0
(11)	90·2	(1·65; 13)	87–92	12·7	(0·55; 13)	11·9–13·5
(12)	93·7	(1·83; 33)	89–97	13·2	(0·81; 33)	12·5–14·5
(13)	94·4	(2·52; 11)	90–99	13·3	(0·70; 10)	12·3–14·0
(14)	93·9	(1·57; 52)	91–98	13·6	(0·61; 50)	12·3–15·6
(15)	95·8	(1·87; 9)	93–98	13·7	(0·75; 8)	12·6–15·0
(16)	95·6	(1·63; 146)	91–99	13·9	(0·64; 141)	12·6–15·5
(17)	96·9	(1·18; 24)	95–100	14·3	(0·54; 22)	13·2–15·0
(18)	97·3	(1·56; 96)	94–100	14·1	(0·60; 91)	13·0–15·2

Wing of *ledouci* according to Vaurie (1965): ♂ 87·5 (5) 87–88. Average tail of small races *comminutus*, *buturlini*, and *danfordi* 49·4, bill (F) 17·3, tarsus 14·2, toe 14·0; average tail of larger nominate *minor* and *kamtschatkensis* 61, bill (F) 18·1, tarsus 14·5, toe 14·6 (RMNH, ZMA).

JUVENILE. Wing on average 4·2 shorter than adult and 1st adult, tail 2·8 shorter; bill full-grown at completion of post-juvenile moult.

Weights. Nominate *minor*. Norway, July–August: ♂ 25, juvenile 19 (Hagen 1942; Haftorn 1971). USSR: 21–25 (Dementiev and Gladkov 1951a), ♂ 20–22, ♀ 22 (Fedyushin and Dolbik 1967).

D. m. kamtschatkensis. USSR: ♂ 24–32·4, ♀ 25·5 (Dolgushin *et al.* 1970). Mongolia, April–August: ♂ 24 (3) 22–25, ♀ 27; juvenile ♂ 23; juvenile ♀ 22 (Piechocki 1968c).

D. m. hortorum. Netherlands: ♂ 20, exhausted juvenile ♂ (full-grown) 14·8 (ZMA). Central Europe: ♂ 17–25 (11), ♀ 20–25 (10); lean ♂ 14·5, ♀ 16; fledged juveniles second half of June, ♂♂ 15, 21·8, ♀ 13·5 (Niethammer 1938; Glutz and Bauer 1980). Hungary, December: 20 (Vasvári 1955).

D. m. comminutus. Britain, all year (mainly June–August): ♂ 20·1 (1·41; 16) 18–22, ♀ 19·8 (0·74; 6) 19–21 (BTO).

D. m. buturlini. Northern Greece, February: ♂ 20 (Makatsch 1950).

Structure. Wing rather short, broad at base, rounded at tip. 10 primaries: in adult and 1st adult, p7 longest, p8 1–2 shorter, p9 12–19, p10 47–60, p6 0–2, p5 1–3, p4 8–12, p3 15–21, p2 18–24, p1 20–28; in juvenile, p7–p8 longest or either one 0–2 shorter than other, p9 7–10 shorter than longest, p10 36–46, p6 1–3, p5 3–8, p4 10–17, p3 16–28; p1–p2 reduced, about half length of neighbouring s1 or less; p10 less reduced than in adult (for relative length, see description of juvenile in Plumages). Tail rather short, graduated; 12 feathers, t1–t3 more rounded at tip than in other *Dendrocopos*, rather similar in shape to t4–t5, but shafts stiff; in adult and 1st adult, t1 longest, t2 3–7 shorter, t3 6–12, t4 9–15, t5 14–20, t6 38–52; in juvenile, feather-tips pointed, t2 2–5 shorter than t1, t3 4–8, t4 7–12, t5 12–20, t6 33–40; t6 reduced at all ages, hidden between t4–t5 and upper tail-coverts. Bill short, slender, straight; base rather deep but narrow, sides tapering evenly to sharp tip; upper mandible with fine ridge along top, fine longitudinal ridge and shallow groove on side at base. Tarsus and toes short, rather slender; upper front of tarsus feathered. Outer front toe with claw c. 85% of inner front toe with claw, outer hind toe c. 115%, inner hind toe c. 50%. Claws rather long and slender, sharp, strongly curved.

Geographical variation. Marked, involving colour and size. Over 20 races described for west Palearctic alone, but variation probably mostly clinal with hardly any sharp boundaries between races; only 9 races recognized here. Variation in size mainly indicated by length of wing and tail, less so length of bill; 5 groups separable in west Palearctic on size (for colour; see below). (1) *D. m. quadrifasciatus* and (extralimitally) *hyrcanus* along southern Caspian, which are small (wing c. 84) and short-billed. (2) *D. m. buturlini*, *danfordi*, and *morgani* from Italy and Greece through Turkey to Iran, and *comminutus* from Britain, which are rather small (wing c. 86–87) with average or long bill; birds from Iberia and Pyrénées rather intermediate in size between this and next group; size of *ledouci* from North Africa uncertain, as data conflicting (see Measurements). (3) *D. m. hortorum* from France and Netherlands east to south-west Poland, Rumania, and northern Yugoslavia, and *colchicus* from Caucasus and Transcaucasia, which have intermediate wing (c. 90) and bill. (4) Nominate *minor* from north-east Poland, Baltic, west European USSR, and

Fenno-Scandia, with rather long wing (c. 94) and bill. (5) *D. m. kamtschatkensis* from Siberia, with long wing (c. 97) and bill. Birds from Ukraine and north-west, central, and east European USSR intermediate between (4) and (5). Cline in colour runs approximately parallel to size: *kamtschatkensis* (largest) very pale—ground-colour of forehead, ear-coverts, and underparts virtually white, much white on mantle, back, and rump, broad white bars on scapulars, tertials, upper wing-coverts, and flight-feathers, tips of outer tail-feathers virtually white (single black bar on t5), hardly any black streaks on flanks, plumage dense and fluffy; *quadrifasciatus* (smallest) with brown or buff-brown ground-colour on forehead, ear-coverts, and underparts, sometimes with traces of a black band from malar stripe behind ear-coverts up to back of head, upperparts and flight-feathers mainly black with narrow white spots or bars, no white spots on median upper wing-coverts, heavy black streaks and some bars on flanks and sides of breast and belly, and heavy black bars on t4–t5, feathering shorter and less dense. Nominate *minor* from Norway, southern Sweden, north-east Poland, and Baltic similar to *kamtschatkensis*, but forehead, ear-coverts, and underparts slightly tinged buff when plumage fresh (virtually white when worn), white bars and spots on upperparts and flight-feathers slightly narrower, flanks with narrow black streaks, and t5 with c. 2 black bars on tip. Birds from northern Sweden, Finland, Ukraine, and north, central, and east European USSR intermediate in colour between nominate *minor* and *kamtschatkensis* to varying degree; sometimes separated as *transitivus* (Loudon, 1914) or *menzbieri* Domaniewski, 1927. *D. m. hortorum* from central Europe slightly less extensively white on upperparts than nominate *minor*, but difference in colour of underparts more marked: ground-colour buff or pale brown, deepest and sometimes slightly pink on throat and chest, slightly paler on belly (where often off-white when plumage worn); flanks and outer tail-feathers on average with slightly heavier black bars than nominate *minor* (see Plumages); populations which are inseparable from *hortorum* occur in Austria, Hungary, northern Yugoslavia, and Rumania. Caucasus race *colchicus* similar to *hortorum* on upperparts, but underparts on average slightly deeper buff, these rather more heavily streaked black, and always with some traces of black behind ear-coverts; birds from Transcaucasia similar, but upperparts less white, and hence rather similar to *buturlini* from Italy, southern Switzerland, and western Yugoslavia south to northern Dalmatia (which, however, shows virtually no black behind ear-coverts), to *danfordi* from Greece and Asia Minor (which shows full black band behind ear-coverts reaching back of head), and to *quadrifasciatus* from Lenkoran (which is even blacker above and on tail, darker brown on ear-coverts and underparts, and fully black on median upper wing-coverts). Birds from southern Yugoslavia, Bulgaria, perhaps Albania, and probably northern Greece similar to *buturlini*, but often with traces of black behind ear-coverts and perhaps inseparable from (geographically disjunct) *colchicus* (see also Matvejev and Vasić 1973). *D. m. comminutus* from Britain has dark ground-colour as in *buturlini* (hence darker than neighbouring *hortorum*), ear-coverts and underparts buff-brown, pale bars on upperparts slightly narrower than *hortorum*, but only faint dark streaks on flanks and sides of breast, often less than *hortorum* and distinctly less than *buturlini*. Birds from Pyrénées inseparable from Italian *buturlini*, but populations from central and southern Spain and perhaps Portugal even deeper buff-brown on forehead, ear-coverts, and chin to chest (less so on belly), dark streaks on underparts heavier, black bars on t4–t5 broad, black on upperparts more extensive and purer, and often some traces of black behind ear-coverts (not reaching nape), similar in these respects to North African *ledouci*; no direct comparison made between Spanish birds and *ledouci* and thus not certain whether Spanish birds nearer *buturlini* or nearer *ledouci* or whether separable as *hispaniae* Von Jordans, 1938. Extralimital *hyrcanus* from northern Iran similar to *quadrifasciatus* but even darker; *morgani* from north-west Iran and Zagros mountains is buff-brown on chin to chest but contrastingly cream-white with numerous black streaks on breast, belly, and flanks, shows a broad black band behind ear-coverts (like *hyrcanus* and *danfordi*), and has long and narrow bill. Kamchatka and Anadyr birds even paler than *kamtschatkensis*, but populations darker again towards south-east Siberia, northern China, and Hokkaido, where close to nominate *minor*. For details of recognition of races, see Hartert (1912–21), Domaniewski (1927), Hartert and Steinbacher (1932–8), Dementiev (1937), Voous (1947), and Vaurie (1959c, 1965).

Rather closely related to Gray-capped Woodpecker *D. canicapillus* of eastern Asia and Japanese Pygmy Woodpecker *D. kizuki* (Short 1982). CSR

Picoides tridactylus Three-toed Woodpecker

PLATES 85 and 86
[between pages 878 and 879]

Du. Drieteenspecht Fr. Pic tridactyle Ge. Dreizehenspecht
Ru. Трепалый дятел Sp. Pico trídactilo Sw. Tretåig hackspett

Picus tridactylus Linnaeus, 1758

Polytypic. Nominate *tridactylus* (Linnaeus, 1758), northern Europe, south-west Siberia, and south-east Siberia from Altay east to Sakhalin and Ussuriland; *alpinus* Brehm, 1831, central and south-east Europe; *crissoleucus* (Reichenbach, 1854), Siberian taiga from Ural mountains to Sea of Okhotsk, intergrading with nominate *tridactylus* in southern Urals, south-west Siberia, southern Yakutia, southern shore of Sea of Okhotsk, and Sakhalin. Extralimital: 5–8 further races in Asia and North America.

Field characters. 21–22 cm; wing-span 32–35 cm. About 10% smaller than Great Spotted Woodpecker *Dendrocopos major*; c. 50% larger than Lesser Spotted Woodpecker *D. minor*. Medium-sized, sharp-billed, rather large-headed woodpecker, with black and white plumage arranged in different pattern from *Dendrocopos* woodpeckers and always lacking red. Face boldly striped black; wholly black scapulars and wing-coverts isolate long white centre to back and rump. Flanks closely barred, not streaked. Rather tame. Sexes dissimilar; no seasonal

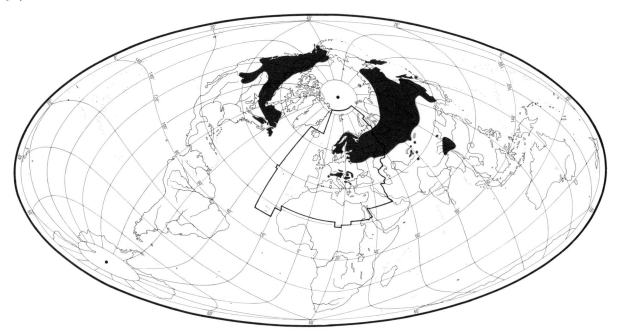

variation. Juvenile separable. 3 races in west Palearctic; the 2 described here are separable in the field.

ADULT MALE. (1) North European race, nominate *tridactylus*. Centre of crown yellow, outlined by and ending in black. Narrow rear supercilium and hindneck below nape white, running into white and black barred upper mantle and white back. Bold black eye-stripe (from in front of and below eye), widening over cheeks and turning down along side of hindneck to join black shoulder. Moustache also black, separated from eye-stripe by white lores, lower cheeks, and centre of sides of neck but also joining black shoulder. Sides of mantle, scapulars, and wing-coverts black, invading white back in form of scattered bars along its edges. Flight-feathers and tertials narrowly barred or tipped white. Outer tail-feathers white, narrowly barred black, contrasting with otherwise black tail. Underparts white, sullied grey, slightly streaked black on sides of breast and prominently but finely barred black on flanks and vent. (2) Alpine and south-east European race, *alpinus*. Noticeably darker than nominate *tridactylus*, with invasive barring on white back (forming ladder pattern) and both streaks and bars on flanks much more pronounced. ADULT FEMALE. Both races. Front and centre of crown spotted white, rear crown and nape forming larger black patch than on ♂. Bill shorter than in ♂. JUVENILE. Both races. Crown yellow in both sexes, as adult ♂. Invasive barring and mottling on back stronger, underparts tinged grey.

Unmistakable when seen well. No other woodpecker in west Palearctic has clearly and deeply barred flanks, black across cheeks, or wings as black. Flight like that of *D minor*. Forages less energetically than other woodpeckers, often content to spend much time on one tree.

Contact-call a soft 'ptuk', lower pitched than *D. major*, but higher than White-backed Woodpecker *D. leucotos*. Drumming noticeably longer than in *D. major* ($c.$ $1-1\frac{1}{2}$ s); also slower (with rattling quality), accelerating towards end; similar to *D. leucotos*.

Habitat. Resident in continental west Palearctic, occupying 2 distinct habitat groups: 1st in high and upper middle latitudes, largely lowland; 2nd in mid-latitudes in mountains between 650 m and 1900 m in central Europe and to 2300 m in Mongolia (Glutz and Bauer 1980). In 1st, boreal/arctic, largely in dense coniferous forests (taiga), with preference for shady, damp, sometimes swampy patches, and areas with much dead wood resulting from fire, lumbering, or windthrow. Favours spruce *Picea*, silver fir *Abies*, larch *Larix*, and sometimes birch *Betula*, willow *Salix*, and others. In 2nd group, mid-latitude subalpine habitat in Switzerland consists of steep inaccessible slopes, often dominated by old spruce, to exclusion of pure pine *Pinus sylvestris*, Arolla pine *Pinus cembra*, or larch stands. Nest-site usually in more open part even of dark closed forest, perhaps created by avalanche, windthrow, or other disturbance, which often goes with higher ratio of decayed or fallen timber. Occurrences also sometimes recorded in separated tree-stands with little fallen timber. For detailed analysis of habitat in Switzerland, see Hess (1983). Owing to unobtrusiveness in winter, little known about seasonal shifts, which appear to involve some movement to higher as well as lower ground. No data concerning use of airspace except in forest interior. Although other west Palearctic woodpeckers penetrate as far into Arctic, and to treeline in mountains, none is so specialized in these directions, and correspondingly in attachment to

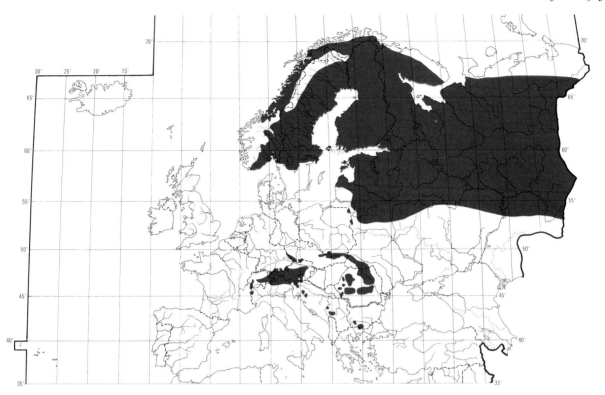

conifers, especially spruce. Habitat of Nearctic population does not differ significantly.

Distribution. WEST GERMANY. Formerly rare breeder in Schwarzwald; last seen 1924 (Glutz and Bauer 1980). POLAND. Bred 1896 near Kudowa (Tomiałojć 1976a). YUGOSLAVIA. Formerly a scattered breeder, except in Makedonija (Matvejev and Vasić 1973); since 1975 recorded from only 4 areas—perhaps overlooked, but certainly disappeared from western Serbia due to habitat degradation (Vasić 1977).

Accidental. Denmark, East Germany, Hungary.

Population. FRANCE. 10–100 pairs (Yeatman 1976). WEST GERMANY. Bayern: 450–700 pairs (Bezzel et al. 1980). SWEDEN. Estimated 80000 pairs (Ulfstrand and Högstedt 1976). FINLAND. Estimated 23000 pairs (Merikallio 1958); scarce in most southern areas (LJL). POLAND. Fairly numerous in Carpathians and Białowieża forest; in small numbers Augustów forest (Tomiałojć 1976a). CZECHOSLOVAKIA. Scarce, and probably decreasing (KH). YUGOSLAVIA. Declined, now rare (VFV). GREECE. Very rare (WB, HJB, GM). USSR. Common, especially in northern parts of European range (Dementiev and Gladkov 1951a). Latvia: rare (HV). Lithuania: very rare, declining (Jankevičius et al. 1981).

Movements. Resident and dispersive. Eruptive to limited extent in northern Europe (nominate *tridactylus*), but apparently partially migratory in Siberia (*crissoleucus*), where it regularly occurs south of breeding range in winter (Johansen 1955).

To some extent nomadic in winter in Russia, where recorded as far south as Kaluga, Tula, and Voronezh (Dementiev and Gladkov 1951a); see also below. In small-scale Scandinavian eruptions, mid-September to mid-November, a few individuals reach south-west Norway (Haftorn 1971) and southern Sweden (SOF 1978), beyond breeding range, and have (exceptionally) even reached Denmark and East Germany. An irregular migrant at Finnish coastal bird observatories, more common in some autumns, e.g. 1956, 1964, 1965, 1968 (Hildén 1969, 1974). Cline of increasing size runs north and east in Europe. 175 Fenno-Scandian museum specimens collected September–March average larger (with bimodal distribution) than 47 collected April–August (distribution normal); this difference not due to moult, and considered that autumn–winter specimens include immigrants from north-east including Russia (Hogstad 1983). For North American invasion status, see West and Speiss (1959).

Upland populations of central Europe (*alpinus*) are sedentary, even in severe winters; only rarely do individuals wander far (Glutz and Bauer 1980).

Food. Chiefly insects, mostly adults and larvae of wood-boring beetles, obtained mainly beneath bark; otherwise prey from surface of bark. 'Ringing' for sap important, but solid plant food rarely taken. Throughout the year, mainly pecks and probes; in breeding season, surface-gleaning more important. Compared with (e.g.) Great Spotted Woodpecker *Dendrocopos major*, does not hack deeply, but works more delicately, even with tapping action. Pecks at trunks and stumps of trees, much more often dead than alive. Levers, knocks, and peels off bark, probes in crevices, mosses, and lichens, also tearing off epiphytes. Typically works away at one spot for long periods, making small holes or removing bark, and may return for several days to same tree until trunk bare of bark (Ryser 1961; Sutter 1961; Ruge 1968; Hogstad 1970, 1977, which see for variation in feeding methods with state of decay of tree). In Switzerland, fed mostly on branches near trunk, sometimes hanging underneath and pecking upwards (Ris 1959); similar feeding method described by Scherzinger (1972b). After snow melts will visit rotten mossy stumps, also branches on ground (Scherzinger 1972b, which see for climbing techniques, postures, etc.). In southern Norway, feeds on dead trunks, never branches (Hogstad 1970), presumably because branches of conifers (see below) not attractive; notably, peels off bark with upward tearing action (Hogstad 1970). In Norway, seasonal and sexual differences occur in choice of height, girth, and condition of tree, and in foraging height. In winter, 60% of observations in lower half of trees, i.e. lower than *D. major*; both sexes almost entirely on dead spruce *Picea abies* (Hogstad 1970, 1977, 1978); of 276 observations, 270 on *Picea*, of which 247 dead, 17 moribund, 6 live (Hogstad 1971a). In summer, ♀♀ visited live trees of other species more often than did ♂♂. In winter, both sexes fed lower than in summer, but ♀♀ usually higher than ♂♂ (Hogstad 1977, which see for other differences). During nestling period, pair-members may divide territory horizontally for feeding, overlapping only near centre (Hogstad 1976a, b; see also Social Pattern and Behaviour). Sexual differences related to size dimorphism, especially of bill (Hogstad 1976a, 1977). In winter, Finland, birds foraged 12% of time on pine *Pinus sylvestris*, 16% on *Picea*, 52% on mixed *Pinus* and *Picea*, 12% in urban parkland, 8% elsewhere; spent 76% of time seeking insects on trunks, 4% on branches, 20% elsewhere (Vikberg 1982). Ground-feeding unusual, but occasionally takes ants (Formicidae) and perhaps their pupae, from anthills (Baumann 1905; Neufeldt 1958b; Meier 1959b). Little information on feeding rates, but once, when feeding fledged young, c. 1 item per minute (Bühler 1959); see also final paragraph. Probably feeds throughout day in winter (Hogstad 1970). From spring onwards, ringing of tree trunks to obtain sap widespread and common (Turček 1954; Schifferli and Ziegeler 1956; Meier 1959a; Ryser 1961; Ruge 1968, 1973), perhaps more so in central Europe than in Norway (see Hogstad 1977). In one study, Switzerland, birds spent as much time ringing as pecking and gleaning; favoured trees recorded being ringed for 100 years. Bird begins with blows of bill from left and right, then pecks more directly to reach cambium for sap and resin. Typically pecks horizontal series of holes, and moves from one to another, returning to inspect ones made earlier (Ryser 1961; Ruge 1968, 1973). Also enlarges old holes (Meier 1959b). Ringing behaviour recorded at all times of day, one bird feeding thus for 48 min, almost without pause. In territory, Switzerland, 28 ringed trees comprised 23 *Picea abies*, 3 Arolla pines *P. cembra*, and 2 larches *Larix*. May ring trees from base to crown. In montane regions, no ringing on branches, but at lower altitudes this reported on lime *Tilia* (Ruge 1968, 1973; see latter for review of tree species used). Holes c. 1 cm apart, made from left to right, spanning 40 cm, i.e. c. $\frac{1}{4}$ circumference of tree (Ryser 1961). On one *P. cembra*, c. 30 m from nest, bands of holes c. 1 cm in diameter, c. 10 cm between bands, over c. 8 m of trunk (Schifferli and Ziegeler 1956).

Diet in west Palearctic includes the following. Insects chiefly larvae and pupae of beetles (Coleoptera: mostly Scolytidae and Cerambycidae; smaller numbers of Chrysomelidae, Buprestidae, Curculionidae, Carabidae, Cleridae, Colydiidae, Elateridae, Lymexylidae, Pythidae, Nitidulidae, and Tenebrionidae), and moths, especially larvae (Lepidoptera: Noctuidae, Geometridae, Tortricidae, Cossidae); in small amounts, stoneflies (Plecoptera: Perlidae), Orthoptera, aphids (Hemiptera: Aphididae), flies (Diptera: Tipulidae, Cecidomyiidae, Rhagionidae), wasps and ants (Hymenoptera: Siricidae, Chalcidoidea, Formicidae). Other invertebrate prey reported: spiders (Araneae) and molluscs. Plant food notably sap (for tree species and references, see above), exceptionally berries of rowan *Sorbus aucuparia* (Baumann 1905; Lanz 1950; Neufeldt 1958b; Meier 1959b; Sevastyanov 1959; Sutter 1961; Ruge 1968; Hogstad 1970; Scherzinger 1972b). For diet in extralimital parts of USSR, see Dementiev and Gladkov (1951a).

Few detailed studies, but indicate a great predominance of wood-boring beetles. Major studies in USSR: in southern Kareliya, 20 stomachs contained mostly larvae of Cerambycidae (75% of stomachs), and Scolytidae (55%), these specially favoured in winter; 1 stomach, late January, contained 268 adults and larvae of *Polygraphus poligraphus* (Scolytidae), also 3 larvae of *Pissodes pini* (Curculionidae). Other groups of lesser importance, not recorded all year round, were Elateridae, Rhagionidae, Siricidae, *Camponotus* (Formicidae), Cossidae, spiders, and molluscs (Neufeldt 1958b, which see for seasonal variation). In Arkhangel'sk, 25 stomachs contained 1166 items, of which 96·2% insects, comprising 97·4% (by number) Coleoptera, 1·9% Hymenoptera, 0·5% Diptera, 0·1% Orthoptera, 0·1% Aphididae; of Coleoptera, 14·9% Cerambycidae (in 80% of stomachs), and 79·4% Scolytidae (in 84% of stomachs), especially *Pityogenes chalcographus* and *P. poligraphus* (Sevastyanov 1959). In southern Norway, 11 stomachs (9 adults, September to February, and 2 juveniles in breeding season) contained entirely insects: in December–February, 90% (by number) Scolytidae, especially larvae and pupae. In autumn, diet includes rarely rowan berries *Sorbus* (Hogstad 1970). Contribution of sap to diet hard to assess, but evidently significant (see above).

Food for young brought in bill, perhaps some also in gullet (Ruge 1971b). ♂ may fly directly from ring holes to nest with sap in bill and let it trickle into half-open bill of young (Thönen 1966; Ruge 1968; see also below). However, main food of young is insect larvae (e.g. beetles, moths); also adult moths and craneflies (Lanz 1950; Schifferli and Ziegeler 1956). Single fledged young, attended by ♂, given large Cerambycidae larvae; young also attempted to ring tree; ♂ then made holes from which young drank, ♂ and young later alternating at one hole (Bürkli *et al.* 1975). 2 stomachs of young each contained 1 large pebble (Dementiev and Gladkov 1951a). In Switzerland, 2 small young fed about twice per hr, faster from 5th day, maximum 6·6 feeds per hr on 11th day. Intervals between feeds up to 1 hr; last feed shortly before 20.00 hrs (Ruge 1971b). In Norway, 8–9 days before fledging, brood fed 05.05–21.45 hrs, on average at 10-min intervals (Hogstad 1969). Well-grown brood fed maximum 11 times in 1 hr (by ♂ only), shortest interval 2 min, longest 3 hrs (Lanz 1950, which see for diurnal variation in feeding rate over several days). Feeding rate often remains high after fledging, e.g. attending ♂ fed single young 8 times in 31 min (Ruge 1971b). Fledged young given about every 4th item of 20–25 collected by ♂ in 20 min (Bühler 1959). Recently fledged young was fed every 3 min; made mild attempts at self-feeding, but still very much dependent on parent (Ryser 1961). EKD, ASR

Social pattern and behaviour. Account includes material supplied by O Hogstad. For detailed information on Nearctic *bacatus*, including displays and postures not reported for nominate *tridactylus*, see Short (1974).
1. Mostly solitary. Outside breeding season, central and southeast Norway, occurs singly or in apparent pairs; some birds remain on breeding territory, though most probably leave it and defend winter feeding territories elsewhere (Hogstad 1970, 1978; O Hogstad). Of 18 winter feeding territories, 13 occupied by 1 ♂ only, 5 others by 1 ♂ and 1 ♀, presumed mates (Hogstad 1978). In one case breeding territory of pair included their shared winter territory (O Hogstad). In south-east Norway, winter territories of ♂♂ usually smaller, and density higher, than breeding territories; minimum size of 4 territories 5·5, 6, 7, and 8 ha, corresponding to 5 ♂♂ per km² (Hogstad 1970). In central Norway, one ♂ occupied 9 ha, another 10 ha. In forest dominated by spruce *Picea* with high proportion of dead trees, territories notably small, in 2 cases *c.* 4 ha and 5·5 ha, suggesting size outside breeding season may be inversely related to density of dead trees, and thus to food supply (O Hogstad). BONDS. Mating system probably always monogamous. Pair-bond maintained seasonally or all year (see above), and from year to year. One pair together for 3 years (Ruge 1974), perhaps 6 years (Bürkli *et al.* 1975). ♂ and ♀ share nest-duties. Family bonds maintained for some time after breeding, in one case for at least 33 days, perhaps up to 2 months (Bürkli *et al.* 1975). Age of first breeding not known. BREEDING DISPERSION. Solitary and territorial. Territory typically separate from, or only partly abuts, those of neighbours (Glutz and Bauer 1980), and size therefore difficult to determine accurately. In Norway, during 3–4 days of nestling period, minimum feeding areas of 3 pairs 11, 17, and 19 ha. Within each territory, ♂ and ♀ foraged in different areas, overlap 5–25% near centre (Hogstad 1976b, 1977). Partitioning of breeding territory between ♂ and ♀ also reported by Ruge (1968) and Sollein *et al.* (1982a). From May to mid-July, pair used *c.* 20 ha (Ruge 1968); probably this same pair stayed in same area for 6 years, but shifted nest-site so that *c.* 1 km² used over whole period (Bürkli *et al.* 1975; see also below). In canton Schwyz (Switzerland), average 'home range' *c.* 0·94 km² (Hess 1983). For other data, see Glutz and Bauer (1980). Little information on breeding density. Over several years, Eisenerzer Alps (Austria), 5 pairs in 11·4 km², i.e. *c.* 0·4 pairs per km² (Ruge and Weber 1974a). In Ammergauer Berge (West Germany), 3 and 4 ♂♂ Drummed in respectively 2 km² and 3 km² (Bezzel and Lechner 1978). In canton Schwyz, 3–4 pairs in 11 km² (Hess 1983). Territory established by ♂; serves for nesting and rearing young. Pairs typically return to same area every year (e.g. Scherzinger 1972b) but excavate new nest-hole each time: e.g. over 6 years, pair annually moved nest-site 300–500 m (Ruge 1974; Bürkli *et al.* 1975). ROOSTING. Nocturnal. In breeding season, ♀ probably roosted in nest-hole from 2 days before laying, and continued to do so during laying. From onset of incubation, ♂ (only) stayed overnight (Sollein *et al.* 1978, 1982b). At 2 nests, ♂ stopped roosting in nest when young 17–21 days old, after which thought to have roosted in holes *c.* 100 m from nest (Ruge 1971a). See also Behaviour at nest, below. No information from outside breeding season.
2. Remarkably fearless of man, and, at any time of year, allows approach to 5 m or less (Lanz 1950; Bühler 1959; Sutter 1961), at most, often raising crown feathers in alarm and giving Contact- or Scolding-calls (Ryser 1961: see 1–2 in Voice). Once aware of intruder, often hides behind trunk rather than flying away (O Hogstad). Captive ♂ confronted observer with aggressive Head-swaying display (Fig A), similar to other Picinae; the more agitated the bird, the steeper the upward angle of bill (Ruge 1968). Birds held in hand gave Distress-calls (Winkler and Short 1978: see 5 in Voice). Mildly alarmed bird may make Tapping sounds with bill on trunk, etc., sometimes developing into weak bout of Drumming (Ruge 1968: see Voice). In response to imitated calls of Pygmy Owl *Glaucidium passerinum*, bird climbed up tree, Tapping and giving Scolding-calls. ♂ pecked at trunk only 30 cm from *G. passerinum* similarly attracted; ♂ once performed vigorous Head-swaying display near the bird, then approached even closer and pecked right in front of it (Scherzinger 1972b). When alarmed by bird of prey, in or out of breeding season, sometimes 'freezes' in Erect-posture (Ruge 1971b; Bürkli *et al.* 1075; O Hogstad). Dominance relations with Great Spotted Woodpecker *Dendrocopos major* vary with resource contested. Outside breeding season, often feeds amicably near *D. major* (Nyholm 1968; Hogstad 1970, which see also for relations with Black Woodpecker *Dryocopus martius*; O Hogstad). However, subordinate to *D. major* in competition for roosting hole (Nyholm 1968) and, once, nest-hole (Sollein *et al.* 1978).

A

For other interactions in breeding season, see below. ANTAGONISTIC BEHAVIOUR. ♂ demarcates territory by Drumming. ♀ also Drums (see Voice) but less often (e.g. Verthein 1935). For Drumming sites, favours dead branches, stumps (Ruge 1968, 1975), and, especially, splintered tops of tall trees which have lost or split crown (Scherzinger 1972b). Also reported Drumming on telegraph poles, etc. (Zimmerman 1956; Meier 1959a). When human intruder imitated Drumming, ♂ approached in stiff, gliding flight, landed in tree above intruder, and made quiet whining sounds (Scherzinger 1972b). Drumming occurs mostly early in morning (P J Sellar) in spring and summer; also sometimes in autumn (Scherzinger 1972b). In Engadin (Switzerland), began mid-April, at lower altitudes sometimes from December (Ruge 1974); most frequent at beginning of pair-formation and during nest-hole excavation; thereafter, sporadic until August (Ruge 1968, 1975). Antagonistic behaviour recorded outside breeding season, in defence of winter territories; where 3 ♂♂ held abutting territories, 1 ♂ twice driven towards centre of its territory by neighbours (Hogstad 1970). In confrontations with rivals, initial response low-intensity Head-swaying display, accompanied by Twitter-calls (see 4 in Voice). In greater excitement, bird Head-sways with bill raised higher. Bird threatening rival more seriously switches to Wing-spreading display (Fig B): opens wings, revealing bold black and white underwing pattern, and fans tail. Head-swaying and Wing-spreading displays also directed at mates (Ruge 1968). Perched bird adopted Wings-spread posture abruptly, 'as if released by a spring', when another (possibly conspecific) bird flew nearby (Barruel 1950). Sometimes, Wing-spreading display proceeds to direct supplanting attack in which bird pecks rival (Ruge 1968). For other calls during confrontations, see Voice. In breeding season, birds successfully drive off *D. major*, but apparently only from immediate vicinity of nest (Hurme and Sarkanen 1975). One pair attacked and expelled *D. major* with much pecking and loud Screeching-calls (Schifferli and Ziegeler 1956: see 5 in Voice). Pursuit-flights, accompanied by Twitter-calls, not unusual; ♀ threatened dummy *D. major* at nest with Wing-spreading display and diving attack (Ruge 1971b). HETEROSEXUAL BEHAVIOUR. (1) Pair-bonding behaviour. ♂ dominant over ♀ and, in course of pair-formation (beginning late in winter or spring) his display to her includes same aggressive elements shown towards ♂ rivals (Schifferli and Ziegeler 1956, Ruge 1968). Once paired (about mid-May in Switzerland), antagonism ceases (Ruge 1974). ♂ attracts ♀ by Drumming; since sexes usually hold separate, albeit often adjacent, territories outside breeding season (see above), carrying power of Drumming is important. When ♂ and ♀ meet, typically engage in aerial chases, Head-swaying displays, and much calling (see especially 1 and 4 in Voice). Reciprocal Drumming heard soon after dawn (Ruge 1974). ♀ Drummed for c. ½ hr with short pauses when ♂ c. 50 m away (Scherzinger 1972b). Bout of excited Drumming may lead to chasing or copulation (Ruge 1974). Typical sequence early in pair-formation, mid-February, as follows: ♀, while feeding, steadily approaches Drumming ♂. When she alights on same tree as ♂, she climbs opposite side from him, and, each time she becomes visible to ♂, he performs Head-

C

swaying display, accompanied by Twitter-calls. ♂ Drums vigorously each time she moves out of sight. In similar sequence, early May, ♂, with crown feathers raised, Head-swayed while calling, and touched ♀'s cloaca or belly with his bill (Scherzinger 1972b). (2) Nest-site selection. Not known which sex chooses site. Both birds often present on tree and take turns at excavating; more than 1 hole may be made, even in same tree, before final choice (Schifferli and Ziegeler 1956; Ruge 1974; Sollein *et al.* 1978). (3) Mating. Only one detailed account (Sollein *et al.* 1978). Pair copulated twice in rapid succession at 05.07 hrs, 27 May, c. 4–6 m from nest-site. First copulation occurred on quite thin branch, close to trunk of tree, c. 2·5 m above ground. During copulation, which lasted maximum 3 s, ♂ beat his wings continuously. ♀ them flew to branch of another tree nearby where she solicited copulation by perching horizontally and swivelling her body from side to side in the arc of a circle. ♂ alighted on trunk c. ½ m below ♀, climbed upwards, then across and on to ♀. ♂ initially beat his wings but latterly stayed quite still, supporting himself on one outspread wing (Fig C). After dismounting, both birds flew off together. Copulation lasted c. 7 s, and interval between 1st and 2nd copulation 15–20 s. Throughout sequence, from 1st copulation until after 2nd, ♂ and ♀ called continuously, apparently with Contact-calls. (4) Behaviour at nest. For roosting relative to onset of laying, see above. ♂ typically incubates at night (Ruge 1971b, 1974; Sollein *et al.* 1978). In study of one pair throughout nesting cycle, ♂ and ♀ only twice entered hole together, once after disturbance (Ruge 1971b). Based on 200 observations, nest-relief typically as follows: incoming bird lands on tree 50–60 m from hole, and pauses at 2–3 others between it and nest-hole. At nest-hole, bird climbs to one side and gives series of Twitter-calls, whereupon mate, reciprocating Twitter-calls, flies out for 50–70 m before perching. On day of hatching, birds notably excited, both Drumming regularly and changing over at nest frequently. Once young hatched, Twitter-call also serves as greeting when birds meet outside nest (Ruge 1971b). RELATIONS WITHIN FAMILY GROUP. During first 4 days, young brooded assiduously, adult leaving nest occasionally for breaks of 4–10 min. Duration of brooding spells without break 5–63 min. From 5th–16th day, when brooding stops, parents brood

B

D

steadily less, with longer foraging breaks. During first few days, nest-relief as in incubation. As brooding wanes, bird often leaves nest without waiting to be relieved. Eyes of young open at 8–10 days (Ruge 1971b). Young begin to beg when they hear adult landing on tree. Adult usually alights above hole and descends to it, calls of young getting louder until fed (Lanz 1950). As young grow, they increasingly cling to walls of nest-cavity and look out of nest-hole. After 16 days, therefore, parents no longer enter hole but feed them at nest-entrance, each with head cocked sideways (Fig D), as in other Picinae. At this stage, parents also stop guarding nest closely. ♂ and ♀ take about equal shares in feeding young until near fledging when ♀ does markedly less (Lanz 1950; Ruge 1971b). Young very excited on day of fledging, leaning out of hole even when not being fed. May fledge over a few hours or up to 24 hrs apart (3 nests) (Ruge 1971b); at 1 nest, 1st and last of brood of 3 fledged 52 hrs apart (Sollein et al. 1982c). Seem to leave with no special encouragement from parents (Ruge 1971b). After fledging, family party often encountered in neighbourhood of nest (Verthein 1935). Parents appear to divide brood between them. Young often sit quietly between feeds, waiting for adult to bring them food (Bürkli et al. 1975), also flying after parent. While being fed, call loudly with gaping bill, and never silent more than 1–2 min (Ryser 1961). Family maintains regular vocal contact, parents with Drumming, Tapping, and Contact- and Twitter-calls, young with food-calls, etc. (Bürkli et al.1975; see Voice). ANTI-PREDATOR RESPONSES OF YOUNG. When danger threatens, young crouch in nest. Young removed from nest tried to escape: at 17–19 days, by climbing tree with jerky hops, twitching wings; at 20–22 days, by moving along ground in search of refuge. When handled, 17-day-old young performed rudimentary Head-swaying display (Ruge 1971b), while another chick jabbed at captor's hand (Lanz 1964). See also Voice. PARENTAL ANTI-PREDATOR STRATEGIES. (1) Passive measures. ♀ flushed when disturbed, giving Scolding-calls; foraged nearby and returned to nest when intruder retreated (Schifferli and Ziegeler 1956). Flushed bird may also Drum, interpreted as expression of mild alarm (Ruge 1975). (2) Active measures: against man. When young handled out of nest, parents gave excited Scolding-calls (Ruge 1971b). Given that birds permit remarkably close approach (see above) when feeding young in nest (Bühler 1959), or after young fledged (Ryser 1961), any further active measures unlikely.

(Figs A–B from drawing in Ruge 1968; Fig C from drawing in Sollein et al. 1978; Fig D from photograph in Schifferli and Ziegeler 1956.)

EKD

Voice. Used mostly in breeding season. Repertoire rather limited by comparison with other Picinae (Ruge 1975, which is major study). Homology with other Picinae not always clear, and following scheme somewhat tentative; see also Winkler and Short (1978). For voice of Nearctic bacatus, see Short (1974). For Tapping sounds, see Social Pattern and Behaviour.

DRUMMING OF ADULTS. Both sexes Drum, ♀ less often. Markedly different from Great Spotted Woodpecker Dendrocopos major in being longer, and having rattling quality, like salvo from machine gun (Ruge 1975). Rendered 'takatakataka...' (Thibaut de Maisières 1943). In Switzerland, 14–26 strikes per series; long series especially frequent early in pair-formation; ♂ gave on average 19·7 strikes (n = 52), ♀ 21·2 (n = 53) (Ruge 1975). Average duration 1·29 ± 0·30 s (0·8–1·8, n = 15). Towards end of series,

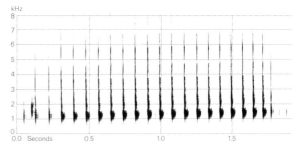

I J-C Roché (1970) Finland May 1963

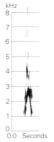

II S Palmér/Sveriges Radio (1972) Sweden June 1962

rate accelerates and pitch descends (Fig I); overall pitch varies significantly with substrate (Ruge 1968, 1975). Volume uniform. Typically gives a few series in succession, usually 3–4 (1–5) series per min (Thibaut de Maisières 1943). Used mainly in territorial demarcation and for pair-contact. For other contexts, seasonal occurrence, etc., see Social Pattern and Behaviour.

CALLS OF ADULTS. (1) Contact-call. Commonest call (Fig II) a soft 'ptuk', sometimes 'ptik', lower pitched than in D. major but higher than White-backed Woodpecker D. leucotos (Thibaut de Maisières 1943). Also rendered 'güg' (Verthein 1935) or 'kjüb' (pronounced 'kyoop'), softer and not as short or shrill as D. major, and with more 'ü' sound (Scherzinger 1972b; Ruge 1975). Given throughout the year, and used in contact between pair (e.g. during nest-relief: W Tilgner) and family (Ruge 1975). See also Calls of Young, below. (2) Scolding-call. Rapid series of Contact-calls, but slower than in (e.g.) D. major and Middle Spotted Woodpecker D. medius. Expresses low-level excitement and mild agitation (Ruge 1975): e.g. given while ♂ drove off, then followed, ♀ (Schifferli and Ziegeler 1956). During copulation, 'tjekk-tjekk-tjekk' by both sexes (Sollein et al. 1978). (3) Threat-call. In confrontations with conspecific birds, 'wätsch wätsch' (Scherzinger 1972b). This call not distinguished by Winkler and Short (1978), though presumably homologous with Wicka-call of other Picinae. (4) Twitter-call. A rhythmic, chattering or twittering sound (Ruge 1975; Winkler and Short 1978): 'hehehehe' (Verthein 1935) or 'quä quä' (Scherzinger 1972b); a muffled 'votvotvotvot' or 'vetvetvetvet', expressing sexual excitement or rivalry (Thibaut de Maisières 1943). Exchanged between pair-members during Head-swaying display and, often in flight, when approaching each other; also in inter- and intraspecific disputes, and in presence of predators (Ruge 1971b, 1975). Sequence of calls sometimes has hissing quality (this perhaps call 3), and gets louder then softer; in short series, typical at nest-relief, change in amplitude not evident (Ruge 1971b). (5) Screeching-call. In warning near nest, 'grügrügrügrü' or 'grägrägrägrä' once given (Lanz 1950). Rapid screeching 'kri-kri-kri', context not given (Scherzinger 1972b), apparently the same. Both sexes give protracted screeching sounds (more than 1 s) from nest-hole (W Tilgner). (6) Other calls. (a) Hoarse, soft 'gyff gyff', heard in spring (Lanz 1950), and thought possibly to be homologous with 'kweek-call' of some other Picinae (Winkler and Short 1978). (b) Quiet whining crying sound given by ♂ landing in tree above intruder who had just imitated Drumming (Scherzinger 1972b). Whining 'kjee kjee kjee' given by incoming bird at nest-relief (Sollein et al. 1982b).

CALLS OF YOUNG. At 6 days, food-call of young 'schrr-schrr-schrr' (Ruge 1971b). In recording by S Palmér, an incessant harsh reeling sound (E K Dunn). Same sound, rendered 'örörörö. . .', heard in response to any extraneous noise, mild disturbance, etc. (Schifferli and Ziegeler 1956). Food-calls of fledged young just after being fed by ♀ also subsided into this sound (Bühler 1959). Food-calls of 2-week-old young a sharper 'gigigigigigigi. . .', louder and faster when being fed, lasting up to 1 min (Lanz 1950; Ruge 1971b). At 17–19 days, young begin to give ceaseless contact-calls (Ruge 1971b), similar to adult (Lanz 1950), but not as fulsome or sonorous (Verthein 1935); continue to be commonly used after fledging. Screeching distress-call given by young c. 18 days old when handled (Ruge 1975) and when observer covered hole of nest containing young 1 week before fledging (Winkler and Short 1978); 'kuiiii' (Sollein et al. 1982b). Twitter-call of adult first given by young 27 days after fledging (Ruge 1974). EKD

Breeding. SEASON. Northern Scandinavia and north-west USSR: see diagram. Southern Europe: laying begins 1–2 weeks earlier (Makatsch 1976). SITE. In hole in tree, often dead or decaying. In Switzerland, mainly spruce *Picea abies*, Arolla pine *P. cembra*, or larch *Larix* (Ruge 1974). In Norway, of 47 nests, 13 in pine *Pinus*, 10 in spruce, 15 in aspen *Populus*, 9 in birch *Betula* (Hogstad 1969; Haftorn 1971; Wabakken 1973; Sollein et al. 1978; O Hogstad). Mainly 1–10 m above ground, but one nest at 21 m (Verthein 1935). In Switzerland: 1–5 m, 22 nests; 6–9 m, 7; 12–15 m, 4; average 5·5 m (Ruge 1974). Average 5·1 m (3·1–7·0), $n=7$, Switzerland (Lanz 1964); average 4 m, $n=32$, Norway (Haftorn 1971). Mostly south-facing (Ruge and Weber 1974a). Nest: excavated hole, new each year; eggs laid on layer of wood chips, etc. Diameter of entrance hole 4·7 cm, depth of cavity 30·5 cm, internal diameter 12 cm (Lanz 1950). Building: by both sexes; in one case, ♀ spent more time excavating than ♂; over 5 days, ♀ performed 1–8 bouts of excavation per day, average 26 (5–50) min per bout; ♂ excavated 3–5 times per day, average 17 min (1–83) each; at one nest, Norway, most excavation 03.20–13.00 hrs, and time to complete hole 12 days (Sollein et al. 1978, which see for detailed time budgets). EGGS. See Plate 98. Sub-elliptical, smooth and glossy; white. 25×19 mm ($22–28 \times 17–20$), $n=86$; calculated weight 5·4 g (Makatsch 1976); average of 3 fresh eggs 5·1 g (Ruge 1974), of 4 eggs 4·51 g (Sollein et al. 1982b). Clutch: 3–5(–7). Of 12 clutches, Switzerland: 3 eggs, 7; 4, 5; average 3·4 (Ruge 1974). Laying interval c. 24 hrs at 1 nest with 4 eggs (Sollein et al. 1978). One brood. INCUBATION. About 11 days (Ruge 1974; Sollein et al. 1982b; O Hogstad). By both sexes; begins with last egg; ♂ at night; shifts by day up to 5 hrs 21 min, averaging $22 \pm SD$ 49 min, partners changing 5–6 times per day 04.30–19.30 hrs (Ruge 1971b). Hatching synchronous. YOUNG. Altricial and nidicolous. Cared for by both sexes. FLEDGING TO MATURITY. Fledging period 22–25 days (Ruge 1971b); $21\frac{1}{2}$ days at 1 nest (Sollein et al. 1982c). Young dependent on parents for more than 1 month (Ruge 1974), probably up to 2 months (Bürkli et al. 1975). Age of first breeding not known. BREEDING SUCCESS. In Switzerland, 1·7 young fledged per brood ($n=16$ broods) (Ruge 1974). Dormice *Eliomys* thought to be significant predator of nests (Ruge 1971b).

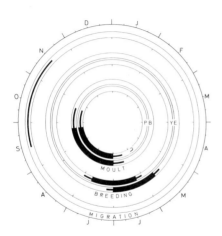

Plumages (nominate *tridactylus*). ADULT MALE. Dense tuft of soft bristles covering nostril grey and white, longest bristles black. Forehead spotted and streaked black and white. Crown yellow, some black and white spots on feather-bases often visible, sides of crown and hindcrown black with some white streaks or spots. Back of head and central hindneck black. Narrow white stripe from upper hind-corner of eye backwards, widening towards sides of hindneck and extending to white central mantle. Area just in front of and below eye black, extending into broad black stripe over ear-coverts and sides of neck, bordered below by narrow white stripe from side of forehead over cheek. A black stripe from base of lower mandible over lower cheek and from there down towards side of chest; usually finely speckled white near lower mandible and rather irregular on side of chest. All black of head, neck, and sides of chest deep and glossed with bluish, contrasting with dull blackish-brown of remainder of upperparts and flanks. Central mantle, inner scapulars, back, and rump white, forming broad white area along central upperbody; some feathers at sides of white area with bold dull black spots or short streaks. Outer scapulars and upper tail-coverts dull blackish-brown or dark sepia-brown. Chin, throat, and centre of chest, breast, and belly white, usually with cream or isabelline tinge; flank and side of breast down from black patch on side of chest white with bold black bars, bars tending to form streaks or spots on sides of belly and towards central breast. Vent and shorter under tail-coverts white with black bars, longer coverts black with broad white tips. Central 3 pairs of tail-feathers (t1–t3) black; t4–t5 black but distal parts white to cream-buff with rather narrow black bars, black often broken into spots near tips; reduced t6 black with white spots on tip and outer web. Primaries and secondaries black-brown, outer webs with rather small square or rounded spots along fringes, these spots largest (3–4 mm across) on p5–p8, smaller elsewhere, and sometimes tiny (less than 1 mm) on secondaries and inner primaries (where sometimes lost by abrasion); c. 7 spots show beyond upper wing-coverts on outer webs of p5–p8, 4–5 on other primaries, 3 on secondaries; inner webs show larger white spots or short bars (not reaching shafts), spaced as on outer webs, but absent from distal parts of outer primaries, and more numerous on basal halves of secondaries. Tertials black-brown with bold white bars or spots on inner webs and tips of inner 3–4, forming barred lateral extension of white rump at rest. Upper wing-coverts and bastard wing uniform black-brown, occasionally with some tiny white specks on some longer outer coverts or bastard wing; marginal shorter primary coverts white. Under wing-coverts and axillaries white, closely barred with black (least so on lesser coverts). Individual variation marked: white stripes on sides of face and white of both central upperbody and central underbody broader and purer in some birds, narrower and more variegated black laterally in others; also, birds tend to become blacker through abrasion, especially on forehead, back of head, mantle, rump, sides of breast, flanks, secondaries, and tertials; yellow of crown subject to both bleaching and wear, yellow paler and less golden in early summer and sometimes virtually lost by late summer. ADULT FEMALE. Similar to adult ♂, but entire forehead, crown, and lores black, marked with rounded white spots 2–3 mm across, contrasting with uniform glossy black of sides of crown, back of head, and central hindneck. NESTLING. Naked at hatching. Closed feather-sheaths break through on day 8–9, bird appearing spiny on day 10; feather-tips appear from sheaths from day 11, feathers growing everywhere on body by day 13–14, fully feathered on day 17–19 (Ruge 1971b). JUVENILE. Similar to adult ♂, but dark areas all dull sepia-brown (in adult, glossy deep black of head and neck contrasts with blackish-brown of remainder of body and wing); white specks on forehead, crown, and lower cheeks smaller; white on central mantle, scapulars, back, and rump more restricted, sometimes partly barred with dusky brown; white of chin, throat, and central underbody often tinged buff, tips of fresh feathers often faintly bordered dusky. Sides of chest and breast, flanks, vent, and under tail-coverts more heavily marked sepia-brown, marks less sharply defined from intervening pale cream or white, flanks not as contrastingly barred as adult. Primaries as in adult but longest primaries with rather broad white or cream fringe along tips (in adult, usually small white speck at tip of outer web only); for primaries and tail, see also Structure and Moults. Forecrown usually yellow, as in adult ♂, but yellow paler and less extensive, feather-bases dusky, less white; some variation in size of patch, perhaps depending on sex; occasionally, crown entirely black-brown without yellow, perhaps in ♀ only (Ruge 1969a). FIRST ADULT. Similar to adult and sometimes indistinguishable. Juvenile greater upper wing-coverts (apparently including tertial coverts) and greater upper primary coverts (sometimes except some outermost) retained, tips slightly softer and more frayed than neighbouring fresh feathers and occasionally slightly paler brown; some birds apparently replace all greater upper wing-coverts and tertials, remaining secondaries and greater upper primary coverts contrastingly worn.

Bare parts. ADULT AND FIRST ADULT. Iris pearl-white or bluish-white. Bill lead-grey with blackish tip and whitish-horn base of lower mandible. Leg and foot dark grey or leaden-grey. NESTLING AND JUVENILE. Pink at hatching. Future feather-tracts tinged grey-blue by 6th day. Iris brown when eye opens. At fledging, iris pale grey-brown, bare eyelids bluish-grey, bill, leg, and foot slate-grey. (Hartert 1912–21; Heinroth and Heinroth 1931–3; Ruge 1971b; RMNH.)

Moults. Mainly based on Stresemann and Stresemann (1966), Piechocki (1968c), Ruge (1969a), and on specimens (RMNH, ZFMK, ZMA). ADULT POST-BREEDING. Complete; primaries descendant. Starts with p1; primary moult about half-way through (score 25) mid-July to late August, completed with p10 mid-September to late October; if linear increase in moult score assumed, p1 should be shed second half of May or June, but in fact none of specimens examined in moult before late June and moult of inner primaries perhaps more rapid, with simultaneous loss of p1–p4 (once recorded in single captive adult and considered aberrant: Ruge 1969a) more regular than usually suggested. Probably also a sexual difference, as moult of ♂ on average 3–4 weeks ahead of ♀; average moult score late August and early September 38(5)36–40 in ♂, and 27(6)22–34 in ♀. Sequence of tail approximately 2–3–6–5–4–1 or 2–6–3–4–5–1; t2 shed at about same time as p5, t1 regrown with p8–p10; moult mainly July–September with t1 sometimes growing up to October. Secondaries moult ascendantly and descendantly from s8 or s9, and ascendantly from s1; s8–s9 shed at about same time as p2, s1 with p5–p6; last secondary (s4 or s11) completed at same time as p8–p9. Head and body start at about same time as shedding of p5, mainly in July; completed September–October. POST-JUVENILE. Partial; primaries descendant. Highly peculiar among Picinae in shedding p1–p6 before fledging, and before juvenile p7–p10 full-grown; reduced p1–p2 shed at 8–10 days old, p3 at 9–12, p4 10–13, p5 11–15, p6 13–20; at fledging (c. 23 days old), 1st adult p1–p2(–p3) full-grown or almost so, (p3–)p4–p6 growing, juvenile p7 and p10 full-grown or almost so, p8–p9 growing. P7 shed at day 34–42, p8 58, p9 69, p10 87. Primary moult generally complete earlier than in adult; moult score 15–35 in mid-July, 25–49 in mid-August, 35–50 in mid-September; primary moult completed (score 50) late August to mid-October. Sequence of tail as in adult post-breeding; starts with shedding of t2 at age of 48 days (between shedding of p7 and p8, hence relatively later than adult); t1 shed last on 90th day, still growing when all primaries and body new (unlike adult), completed September to early November. Head, body, and upper wing-coverts relatively late, starting at about shedding of p7, part of head and lesser coverts first; completed with p10 or shortly afterwards, parts of lower belly and flanks last. Juvenile greater upper primary coverts retained (except sometimes for a few outermost), as well as (at least sometimes) all or part of greater upper wing-coverts. Some non-moulting birds, November–May, have part of secondaries, all tertials, and all greater upper wing-coverts new, but greater upper primary coverts and s2–s5 or s1–s6 old; these birds of uncertain age, but, as single captive juvenile replaced s9 (longer tertial) in early August, other juveniles may perhaps moult more tertials than this.

Measurements. ADULT AND FIRST ADULT. Nominate *tridactylus*. Northern Europe, all year; skins (RMNH, ZFMK, ZMA). Bill (F) to forehead, bill (N) to distal corner of nostril; toe is outer front toe with claw.

WING	♂	125	(2·44; 13)	120–128	♀ 123	(1·36; 8)	120–124
TAIL		80·2	(1·71; 8)	78–82	77·8	(0·93; 6)	77–80
BILL (F)		33·1	(1·35; 12)	31–35	30·3	(0·86; 8)	29–32
BILL (N)		25·5	(0·72; 12)	24–27	23·6	(0·75; 8)	22–25
TARSUS		21·1	(0·62; 8)	20–22	20·4	(0·65; 6)	19–21
TOE		18·7	(0·39; 6)	18–19	17·1	(0·46; 5)	16–18

Sex differences significant. See also Hogstad (1976a, 1978).

P. t. alpinus. Switzerland, Austria, Rumania, and Bulgaria, all year; skins (RMNH, ZFMK, ZMA).

WING	♂	130	(1·73; 6)	127–132	♀ 128	(1·79; 7)	124–129
TAIL		79·8	(0·87; 6)	79–81	81·2	(3·40; 6)	78–85
BILL (F)		33·1	(1·86; 6)	31–36	30·2	(1·75; 7)	28–32
BILL (N)		26·0	(1·13; 6)	25–28	23·6	(1·31; 7)	22–25
TARSUS		21·8	(0·97; 5)	21–23	19·7	(0·50; 6)	19–20
TOE		19·9	(1·05; 4)	19–21	19·1	(0·67; 4)	18–20

Sex differences significant for bill and tarsus. Wing, tarsus, and toe significantly longer than in nominate *tridactylus*.

Some samples of birds measured by J Wattel. (1) *P. t. crissoleucus*: central Siberia to Anadyr. Nominate *tridactylus*: (2) Scandinavia and Finland, (3) Lake Baykal area to Amur basin, (4) Bialystok area in north-east Poland. *P. t. alpinus*: (5) Alps.

WING ♂	WING ♀
(1) 126 (15) 122–130	(1) 124 (13) 120–126
(2) 124 (24) 119–127	(2) 122 (27) 118–126
(3) 127 (3) 126–127	(3) 125 (4) 123–127
(4) 123 (7) 119–127	(4) 119 (2) 118–120
(5) 129 (18) 126–133	(5) 125 (15) 123–130

BILL (F) ♂	BILL (F) ♀
(1) 34·4 (14) 32–36	(1) 31·2 (13) 30–33
(2) 33·0 (12) 32–34	(2) 30·1 (17) 28–32
(3) 34·0 (3) 33–35	(3) 31·4 (4) 30–33
(4) 32·5 (6) 31–33	(4) 29·5 (2) 29–30
(5) 32·8 (14) 31–36	(5) 30·2 (13) 28–32

JUVENILE. Wing and tail on average c. 10 shorter than adult.

Weights. Nominate *tridactylus*. Norway: ♂ 70·1 (3·2; 10) 65–74, ♀ 61·2 (3·8; 5) 57–66 (Hogstad 1976a). Mongolia, August: adult ♂ 55, 62; adult ♀ 62·3 (3) 61–64; juvenile ♀ 35 (Piechocki 1968c). Manchuria, second half of July and early August: ♂ 69·5 (8) 66–73, ♀ 60 (6) 54–65 (Piechocki 1958). Belorussiya (USSR): ♂ 65 (63–69), ♀ 54 (51–59) (Fedyushin and Dolbik 1967).

P. t. alpinus. Central Europe: ♂ 73·0 (4) 71–76, ♀ 60·9 (4) 57–64 (Glutz and Bauer 1980).

P. t. tianschanicus (similar in size to *alpinus*). Tien Shan: ♂ 66·0 (4·68; 16) 52–71, ♀ 64·1 (4·90; 11) 57–75 (Dolgushin et al. 1970).

Structure. Wing rather short, broad at base, tip rounded. 10 primaries: in adult and 1st adult, p7 longest, p8 1½–3 shorter, p9 11–18, p10 71–80, p6 0–2, p5 3–7, p4 14–22, p3 24–29, p2 29–33, p1 32–38. Structure of juvenile wing difficult to establish, as p1–p5 strongly reduced, p6–p7 reduced to variable extent, and p1–p6 shed before p7–p9 full-grown. Length of juvenile p1 1·1–2·1 when shed, p2 2·1, p3 4·0, p4 2·7–3·5, p5 5·2, p6 26·8(6)11–58, p7 84(2)77–91, p8 100, p9 91, p10 34, $n = 1–4(–6)$ (Ruge 1969a), or p1 1·8% of adult length, p2 2·3%, p3 4·3%, p4 3·1%, p5 4·8%, p6 31%, p7 77%, p8 97%, p9 99%, and p10 131% (Glutz and Bauer 1980). Juvenile p10 less reduced than in adult, 6–9 longer than longest greater upper primary coverts (1–8 shorter in adult and 1st adult). Tail rather short (relatively shorter in ♂ than in ♀: Hogstad 1976a), graduated; 12 feathers, t1 longest, t2 6–10 shorter in adult and 1st adult, t3 11–17, t4 17–27, t5 26–33, t6 52–60; length of shed juvenile t1 78 mm, t2 69, t3 68, t4 64, t5 56, t6 27 (Ruge 1969a), or t1 c. 88% of adult length, t2–t5 83–86%, t6 77%. Bill rather short, straight, tapering rather gradually to sharp tip; base wide and bill markedly flattened dorso-ventrally; fine ridge on top of culmen; deep groove at lateral base of upper mandible, bordered at both sides by ridge; nostril situated near base of lateral groove, covered by dense tuft of soft bristles. Tarsus short and rather strong, upper front feathered. Toes rather short and strong; inner front toe with claw c. 94% of outer front toe with claw, hind toe (outer hind toe of other Picinae) c. 112%, inner hind toe absent. Claws very strong, sharp, markedly curved; those of front toes almost longer than toes themselves (unlike *Dendrocopos* woodpeckers).

Geographical variation. Marked, involving proportions of black and white in plumage; also slight variation in size. Variation in Palearctic follows a north–south cline: colour darker and size larger towards south, and southern populations (though in part geographically isolated from each other) closely similar to each other. However, isolated population of mountains of south-west China darker than others, body largely black. *P. t. alpinus* (central and south-east Europe) darkest race in west Palearctic; white spots on forehead (both sexes) and crown (♀ only) smaller, 1–2 mm across; black stripes on sides of head broader; central hindneck uniform white, but central mantle, inner scapulars, and back boldly barred or spotted black, rump black with white spots; flanks, vent, and sides of chest, breast, and belly heavily and closely barred, only a narrow strip from central chest to lower belly uniform white; t4–t5 heavily barred black, black on tips prevailing over white rather than vice versa; under wing-coverts and axillaries more heavily barred black, but not much difference in size of spots on flight-feathers and tertials. Juvenile of nominate *tridactylus* rather similar to adult *alpinus* in general extent of black; juvenile *alpinus* very dark, with white on upperparts restricted to large spots on central mantle, inner scapulars, and back, spots on forehead and crown tiny, white or buff of underparts restricted to chin, throat, and central chest, and remainder of underparts dull black with some white mottling. Birds from north-east Poland usually considered to belong to nominate *tridactylus*, but colour nearer to *alpinus* than to nominate *tridactylus*, differing from *alpinus* only in black-spotted rather than barred upperparts; small in size, however, similar to nominate *tridactylus*. Colour gradually paler towards north-east; birds from southern Urals (Perm, Ufa, Orenburg: Buturlin 1907) and western Siberia east to Ob river are paler than nominate *tridactylus* from north-west and central European USSR to Scandinavia, and these birds sometimes separated as *uralensis* Buturlin, 1907, but individual variation large and Ural birds usually considered inseparable from pale race *crissoleucus* from central Siberia east to Sea of Okhotsk. *P. t. crissoleucus* differs from nominate *tridactylus* in mainly white forehead, broader white streaks on sides of head, more extensive white on central upperparts, less or virtually no barring on flanks, longer and looser body feathering, whiter t4–t5, and larger white spots on flight-feathers and tertials (about equal in size to intervening black on primaries). Kamchatka race even paler than *crissoleucus*, but further south birds gradually darker, those from Altay to Sakhalin and Ussuriland inseparable from nominate *tridactylus*, those from Tien Shan, Korea, and Hokkaido probably inseparable from *alpinus*, though each sometimes given racial status (e.g. Vaurie 1959c). North American races have narrower white streaks on head than Palearctic ones, darker upperparts with generally heavier and closer black barring (except northernmost), and usually fully white distal parts of t3–t5.

Closely related to Black-backed Woodpecker *P. arcticus* of North America; also considered rather close to New World species of *Dendrocopos* (e.g. Hairy Woodpecker *D. villosus*), despite differences in structure of foot and bill, and hence American members now often named *Picoides* (e.g. by Short 1982). *Picoides* now frequently also applied to Old World members of *Dendrocopos* (being an older name) (e.g. Short 1971, Winkler and Short 1978), but Old World *Dendrocopos* differ from both New World '*Dendrocopos*' and *Picoides sensu stricto* in (e.g.) behaviour (Blume 1971), colour pattern (no dark streak behind eye, red under tail-coverts: Goodwin 1968), and absence of marked size dimorphism (Hogstad 1976a, 1978), and hence the 2 genera treated separately here (see also Ouellet 1977). CSR

REFERENCES

ABBOTT, I (1978) *Corella* 2, 40–2. ABDULALI, H (1965) *J. Bombay nat. Hist. Soc.*, 62, 153–60. ABELENTSEV, V I and UMANSKAYA, A S (1968) *Ornitologiya* 9, 331–4. ABRO, A (1964) *Sterna* 6, 81. ABS, M (1963) *Proc. int. orn. Congr.* 13, 202–5; (1975) *Verh. dtsch. zool. Ges. 1974*, 347–50. ADAM, R M (1873) *Stray Feathers* 1, 361–404. ADAMS, A L (1864) *Ibis* (1) 6, 1–36. AELLEN, E (1942) *Vögel der Heimat* 13, 28–9. AFFRE, G and AFFRE, L (1967) *Alauda* 35, 108–17. AHLBOM, B (1970) *Fåglar i Sörmland* 3, 74–7; (1973a) *Fåglar i Sörmland* 6, 26–33; (1973b) *Fåglar i Sörmland* 6, 51–9; (1975) *Fåglar i Sörmland* 8, 58–62; (1976) *Fåglar i X-län* 7, 17–24; (1979) *Calidris* 8, 62–4. AHLBOM, B and CARLSSON, L (1972) *Fåglar i Sörmland* 5, 76–80. AHLÉN, I and ANDERSSON, Å (1976) *Vår Fågelvärld* 35, 21–5. AHLÉN, I, ANDERSSON, Å, AULEN, G, and PETTERSSON, B (1978) *Anser* suppl. 3, 5–11. AHO, J (1964) *Ann. zool. fenn.* 1, 375–6. AKKERMANN, R (1966) *Oldenburger Jahrb.* 64, 43–81. ALAJA, P and LYYTIKÄINEN, A (1972) *Savon Luonto* 4, 38–9. ALATALO, R (1978) *Ornis fenn.* 55, 49–59. ALBERTSEN, W (1934) *Vogelzug* 5, 192. ALDER, L P (1951) *Br. Birds* 44, 281. ALDRICH, H C (1943) *J. Bombay nat. Hist. Soc.* 44, 123–5. ALERSTAM, T (1977) *Vår Fågelvärld* 36, 14–20. ALERSTAM, T and ULFSTRAND, S (1974) *Ibis* 116, 522–42. ALEXANDER, W B (1917) *Emu* 17, 95–100; (1940) *J. Roy. Agric. Soc.* 100 (3), 92–100. ALEXANDER, W B and FITTER, R S R (1955) *Br. Birds* 48, 1–14. ALEXANDER, W B and LACK, D (1944) *Br. Birds* 38, 42–5, 62–9, 82–8. AL-HUSSAINI, A H (1938) *Ibis* (14) 2, 541–4; (1939) *Ibis* (14) 3, 343–7. ALI, S A (1927) *J. Bombay nat. Hist. Soc.* 32, 218–19; (1943) *J. Bombay nat. Hist. Soc.* 44, 9–26; (1953) *The birds of Travancore and Cochin.* London; (1955) *J. Bombay nat. Hist. Soc.* 52, 735–802; (1968) *The book of Indian birds.* Bombay. ALI, S and ABDULALI, H (1938) *J. Bombay nat. Hist. Soc.* 40, 148–73. ALI, S AND RIPLEY, S D (1969) *Handbook of the birds of India and Pakistan* 3; (1970) 4; Bombay. ALLAN, L J (1978) *Can. Wildl. Serv. Wildl. Toxicology Div. Rep.* 38. ALLEIJN, F, BRAAKSMA, S, PEEREBOOM VOLLER, J D G, and SCHENDEL, J A A VAN (1966) *Levende Nat.* 69, 1–13. ALLEN, F G H and BRUDENELL-BRUCE, P G C (1967) *Ibis* 109, 113–15. ALLEN, P (1933) *Br. Birds* 27, 234. ALLEN, S S (1864) *Ibis* (1) 6, 233–43. ALLÉON, A (1886) *Ornis* 2, 397–429. ALLEYN, W F, BERGH, L M J VAN DEN, BRAAKSMA, S, HAAR, M T J F A TER, JONKERS, D A, LEYS, H N, and STRAATEN, J VAN DER (1971) *Avifauna van Midden-Nederland.* Assen. ALLISON, F R (1959) *The Ring* 2, 130–1. ALLOUSE, B E (1953) The avifauna of Iraq. *Iraq nat. Hist. Mus. Publ.* 3. ALLSOP, K (1949) *Br. Birds* 42, 390. ALPERS, R (1973) *Beitr. Naturkde. Niedersachs.* 26 (4), 90–6. ALTENBURG, W, ENGELMOER, M, MES, R, and PIERSMA, T (1982) *Wadden Sea Working Group, Comm.* 6, 1–283. ALTUM, B (1863a) *J. Orn.* 11, 248–60; (1863b) *J. Orn.* 11, 321–6; (1864) *J. Orn.* 12, 97–102. ALVAREZ, F and ARIAS-DE-REYNA, L (1974) *Doñana Acta Vert.* 1, 43–65. ALVAREZ, F and HIRALDO, F (1974) *Doñana Acta Vert.* 1, 61–7. AMIET, L (1957) *Emu* 57, 55–7. AMANN, F (1948) *Orn. Beob.* 45, 227. AMAT, J A and SORIGUER, R C (1981) *Alauda* 49, 112–20. AMERICAN ORNITHOLOGISTS' UNION (1957) *Check-list of North American birds.* Baltimore. ANDERS, K (1975) *SitzBer. Ges. naturf. Freunde Berlin* 15, 21–33. ANDERSEN, J (1945) *Dansk orn. Foren. Tidsskr.* 39, 198–205. ANDERSEN, T (1961) *Dansk orn. Foren. Tidsskr.* 55, 1–55. ANDERSEN-HARILD, P (1969) *Dansk orn. Foren. Tidsskr.* 63, 105–10. ANDERSON, Å and HAMILTON, G (1972) *Vår Fågelvärld* 31, 257–62. ANDERSSON, N A and PERSSON, B (1971) *Vår Fågelvärld* 30, 227–31. ANDREWS, K (1971) B Sc Thesis. Aberdeen Univ. ANGST, R (1975) *Gefiederte Welt* 99, 21–4, 49–52. ANGYAL, L and KONOPKA, H-P (1975) *Gefiederte Welt* 99, 212–17; (1977) *Gefiederte Welt* 101, 66–8. ANISIMOV, V D and TIKHONOV, A V (1983) *Ornitologiya* 18, 186. ANNAEVA, E (1965) *Izv. Akad. Nauk Turkmen. SSR* Ser. biol. 1, 87–90. ANON (1928) *Br. Birds* 22, 23–4; (1936) *Br. Birds* 29, 255; (1952) *Br. Birds* 45, 225–7; (1957) *Orn. Beob.* 54, 203–7; (1961) *Countryside* 19, 203; (1968a) *Ibis* 110, 324; (1968b) *Ostrich* 39, 268; (1976) *Reader's Digest complete book of Australian birds.* Sydney; (1979) *RAF Orn. Soc. Exped. Masirah Island Rep.* ANSINGH, F H, KOELERS, H J, VAN DER WERF, P A, and VOOUS, K H (1960) *Ardea* 48, 51–65. ARAÚJO, J (1971) *Ardeola* 15, 146–8. ARAÚJO, J, REY, J M, LANDIN, A, and MORENO, A (1974) *Ardeola* 19, 397–428. ARCHER, G F and GODMAN, E M (1937) *The birds of British Somaliland and the Gulf of Aden* 1–2. Edinburgh; (1961) 3. London. ARDAMATSKAYA, T B (1977) *Commun. Baltic Commiss. Study Bird Migr.* 10, 87–114. ARIAS-DE-REYNA, L, CORVILLO, M, CRUZ, A, RECUERDA, P, SOLIS, F, and TRUJILLO, J (1981) *Abstr. 17th int. ethol. Conf.* ARIAS-DE-REYNA, L and HIDALGO, S J (1982) *Anim. Behav.* 30, 819–23. ARIAS-DE-REYNA, L, RECUERDA, P, CORVILLO, M, and AGUILAR, I (1982) *Doñana Acta Vert.* 9, 177–93. ARLEBO, S (1975) *Anser* 14, 202. ARMITAGE, J S (1968) *Naturalist* 905, 37–46; (1978) *Br. Birds* 71, 590. ARMSTRONG, E A (1942) *Br. Birds* 36, 37–9; (1947) *Bird display.* Cambridge. ARMSTRONG, E A and PHILLIPS, G W (1925) *Br. Birds* 18, 226–30. ARMSTRONG, W H (1958) *Publ. Mus. Mich. State Univ.* 1 (2). ARN, H (1935) *Orn. Beob.* 32, 73–8; (1942) *Orn. Beob.* 39, 150–62; (1945) *Schweiz. Arch. Orn.* 2, 137–84; (1959) *Br. Birds* 52, 221–5; (1960) *Mitt. naturf. Ges. Solothurn* 19; (1967) *Tierwelt* 77, 1119; (1968) *Orn. Beob.* 65, 23–4; (1970) *Tierwelt* 80, 1508–11. ARNHEM, J (1960) *Gerfaut* 50, 1–10. ARONSON, L (1979) *Dutch Birding* 1, 18–19. ARRIGONI DEGLI ODDI, E (1929) *Ornitologia Italiana.* Milan. ARTHUR, R W and CUTTER, J (1949) *Br. Birds* 42, 247. ARVOLA, A (1959) *Ornis fenn.* 36, 10–20. ASBIRK, S (1978) *Dansk orn. Foren. Tidsskr.* 72, 161–78; (1979a) *Vidensk. Medd. dansk naturh. Foren.* 141, 29–80; (1979b) *Dansk orn. Foren. Tidsskr.* 73, 207–14; (1979c) *Dansk orn. Foren. Tidsskr.* 73, 287–96. ASBIRK, S and FRANZMANN, N-E (1978) In Green, G H and Greenwood, J J D *Joint Biol. Exped. NE Greenland 1974*, 148. Dundee. ASH, J S (1954) *Br. Birds* 47, 84; (1965) *Br. Birds* 58, 1–5; (1969a) *Seabird Group Rep.*, 48; (1969b) *Ibis* 111, 1–10; (1977) *Br. Birds* 70, 504–6. ASH, J S and MISKELL, J E (1980) *Bull. Br. Orn. Club* 100, 216–18; (1983) *Bull. Br. Orn. Club* 103, 107–10. ASH, J S, RIDLEY, M W, and RIDLEY, N (1956) *Br. Birds* 49, 298–305. ASHCROFT, R E (1976) D Phil Thesis. Oxford Univ.; (1979) *Ornis scand.* 10, 100–10. ASHFORD, R W and BRAY, R S (1977) *Bull. Br. Orn. Club* 97, 53–4. ASHMOLE, N P (1962) *Ibis* 103b, 235–73; (1963a) *Postilla* 76, 1–18; (1963b) *Ibis* 103b, 297–364; (1965) *Proc. Nat. Acad. Sci.* 53, 311–18; (1968) *Syst. Zool.* 17, 292–304. ASHMOLE, N P and ASHMOLE, M J (1967) *Bull. Peabody Mus. nat. Hist.* 24, 1–131. ASHMOLE, N P and TOVAR, S H (1968) *Auk* 85, 90–100. ASHMORE, S E (1935) *Br. Birds* 28, 259–60. ASPHJELL, J A (1975) *Sterna* 14, 175–6. ASS, M J (1958) *J. Orn.* 99, 376–7. ASTLEY, H D (1915) *Avic. Mag.* 6, 279–82. ATAEV, K (1974) *Izv. Akad. Nauk Turkmen. SSR Biol.* 6, 48–52. ATAKISHIEV, T A (1971) *Uchen. zap. Kazan. Vet. Inst.* 109, 266–9. AUDUBON, J J (1840) *The birds of America.* London. AUMEES, L (1967) *Orn. kogumik* 4, 32–42. AUSTIN, O L (1940) *Bird-Banding* 11, 155–69; (1945) *Bird-*

Banding **16**, 21–8; (1947) *Bird-Banding* **18**, 1–16; (1949) *Bird-Banding* **20**, 1–39; (1953) *Bird-Banding* **24**, 39–55. AUSTIN, O L, JR (1929) *Bull. N. E. Bird Banding Assn.* **5**, 123–40; (1932) *Bird-Banding* **3**, 129–39. AUSTIN, O L and KURODA, N (1953) *Bull. Mus. comp. Zool., Harvard* **109**, 279–637. AVERIN, Y V and GANYA, I M (1966) *Khishchnye ptitsy Moldavii i ikh rol' v prirode i sel'skom khozyaistve.* Kishinev; (1970) *Ptitsy Moldavii* 1. Kishinev. AVERY, M I and COCKERILL, R A (1982) *Br. Birds* **75**, 378. AXELL, H E (1977) *Minsmere: portrait of a bird reserve.* London. AZZOPARDI, J (1979) *Il-Merill* **20**, 24. AYRES, T (1871) *Ibis* (3) **1**, 147–57.

BABBE, R (1953) *Orn. Mitt.* **5**, 62. BABIN, R (1913) *Rev. fr. Orn.* **3**, 112–14. BACKHURST, G C, BRITTON, P L, and MANN, C F (1973) *J. E. Afr. nat. Hist. Soc.* **140**, 1–38. BACMEISTER, W (1919) *Orn. Monatsber.* **27**, 21–7. BAEGE, L (1967) *Orn. Mitt.* **19**, 19–20. BAEPLER, D H (1962) *Condor* **64**, 140–53. BAER, W (1909) *Orn. Monatsschr.* **34**, 33–44; (1910) *Orn. Monatsschr.* **35**, 401–8. BAER, W and UTTENDÖRFER, O (1898) *Orn. Monatsschr.* **28**, 195–201, 217–24. BAGGERMAN, B, BAERENDS, G P, HEIKENS, H S, and MOOK, M H (1956) *Ardea* **44**, 1–71. BAGLIERI, S and FAGOTTO, F (1977) *Riv. ital. Orn.* **47**, 229–34. BAGNALL-OAKELEY, R P (1972) *Br. Birds* **65**, 399. BÄHRMANN, U (1962) *Falke* **9**, 31; (1970) *Beitr. Vogelkde.* **15**, 203–4. BAIER, E (1973) *Orn. Mitt.* **25**, 97; (1975) *Orn. Mitt.* **27**, 248; (1977) *Orn. Mitt.* **29**, 238–9. BAILEY, A M (1925) *Condor* **27**, 164–71. BAILEY, R S (1966a) *Ibis* **108**, 224–64; (1966b) *Ibis* **108**, 421–2; (1968) *Ibis* **110**, 493–519. BAILLIE, R H (1956) *Br. Birds* **49**, 453. BAILLIE, S R (1982) *Proc. Seabird Group Conf., Uttoxeter, February 1982*, 17–19. BAIRD, D A (1979) *Bull. Br. Orn. Club* **99**, 6–9. BAIRD, P (1978) *Pacific Seabird Group Bull.* **5**, 45–6. BAKAEV, S and SALIMOV, K (1973) *Trudy Samarkand. gos. Univ.* **228**, 26–32. BAKER, E C S (1913a) *J. Bombay nat. Hist. Soc.* **22**, 1–12; (1913b) *J. Bombay nat. Hist. Soc.* **22**, 427–33; (1914a) *J. Bombay nat. Hist. Soc.* **22**, 653–7; (1914b) *J. Bombay nat. Hist. Soc.* **23**, 1–11; (1914c) *J. Bombay nat. Hist. Soc.* **23**, 11–12; (1922) *Bull. Br. Orn. Club* **42**, 93–112; (1923) *Proc. zool. Soc. Lond.*, 277–94; (1927) *The fauna of British India: birds* **4**; (1928) **5**. London; (1934) *Nidification of birds of the Indian Empire* **3**. London; (1942) *Cuckoo problems.* London. BAKER, R H (1951) *Univ. Kansas Publ. Mus. nat. Hist.* **3** (1), 1–359. BAKKER, D (1957) *Levende Nat.* **60**, 104–8. BALÁT, F (1956) *Zool. Listy* **5**, 237–58. BALÁT, F and FOLK, Č (1956) *Zool. Listy* **19**, 281–4. BALDWIN, S P and KENDEIGH, S C (1938) *Auk* **55**, 416–67. BALFOUR, E (1973) *Br. Birds* **66**, 228. BALLANCE, D K and LEE, S L B (1961) *Ibis* **103a**, 195–204. BALTZ, D M, MOREJOHN, G V, and ANTRIM, B S (1979) *Western Birds* **10**, 17–24. BANGS, O (1918) *Bull. Mus. comp. Zool.* **64**, 491. BANKS, P A (1969) *Br. Birds* **62**, 496. BANNASCH, F (1980) *Beitr. Vogelkde.* **26**, 236–7. BANNERMAN, D A (1912) *Ibis*(9) **6**, 557–627; (1931) *The birds of tropical West Africa* **2**; (1933) **3**; (1951) **8**. London; (1953) *The birds of West and Equatorial Africa* **1**. London; (1955) *The birds of the British Isles* **4**; (1959) **8**; (1962) **11**; (1963a) **12**. Edinburgh; (1963b) *Birds of the Atlantic Islands* **1**. Edinburgh. BANNERMAN, D A and BANNERMAN, W M (1958) *Birds of Cyprus.* London; (1965) *Birds of the Atlantic Islands* **2**; (1966) **3**; (1968) **4**. Edinburgh; (1971) *Birds of Cyprus and migrants of the Middle East.* Edinburgh; (1983) *The birds of the Balearics.* London. BANZ, K and DEGEN, G (1975) *Beitr. Vogelkde.* **21**, 258–65. BARABASH-NIKIFOROV, I I and SEMAGO, L L (1963) *Ptitsy yugo-vostoka Chernozemnogo tsentra.* Voronezh. BARBER, W E (1925) *Br. Birds* **19**, 26. BARBU, P and GÁL, I K (1972) *Stud. Cercet. Biol. Ser. Zool.* **24**, 497–503. BARBU, P and SORESCU, C (1970) *Anal. Univ. Bucureşti Biol. Anim.* **19**, 67–72. BARCLAY, M (1935) *Br. Birds* **29**, 217; (1938) *Br. Birds* **31**, 332. BARDARSON, H R (1975) *Náttúrufraedingurinn* **45**, 37–42. BARHAM, K E I, CONDER, P J, and FERGUSON-LEES, I J (1956) *Bird Notes* **27**, 34–43. BARK JONES, R (1946) *Ibis* **88**, 228–32. BARNBY SMITH, C (1910) *Avic. Mag.* **1**, 313–15. BARRIÉTY, L (1957) *Bull. Cent. Ét. Réch. sci. Biarritz* **1**, 567–9; (1963) *Bull. Cent. Ét. Réch. sci. Biarritz* **4**, 363–8; (1964) *Bull. Cent. Ét. Réch. sci. Biarritz* **5**, 197–201; (1971) *Oiseau* **41**, 283. BARRUEL, P (1950) *Oiseau* **20**, 78–82; (1958) *Nos Oiseaux* **24**, 263. BARTELS, M (1931) *J. Orn.* **79**, 1–28. BARTH, E K (1949) *Vår Fågelvärld* **8**, 145–56. BARTHOLOMEW, J (1952) *Scott. Nat.* **64**, 121, 179; (1953) *Scott. Nat.* **65**, 195. BARTHOS, G (1958) *Aquila* **65**, 346–7. BARTLETT, E (1971) *Br. Birds* **64**, 504. BARTOS, J (1906) *Aquila* **13**, 209–10. BÄSECKE, K (1932) *Beitr. Fortpfl. Biol. Vögel* **8**, 232; (1942) *Beitr. Fortpfl. Biol. Vögel* **18**, 143–4; (1943) *Orn. Monatsber.* **51**, 53–4; (1949) *Vogelwelt* **70**, 23; (1951) *Vogelwelt* **72**, 180–3; (1954) *Vogelwelt* **75**, 203; (1955) *Vogelwelt* **76**, 69. BASHENINA, N A (1968) *Ornitologiya* **9**, 49–57. BASTAWDE, D B (1976) *J. Bombay nat. Hist. Soc.* **73**, 215. BASTIEN, P (1957) *Gerfaut* **47**, 45–56. BATES, D J and BROWN, A (1977) *Scott. Birds* **9**, 348–9. BATES, G L (1927) *Ibis* (12) **3**, 1–64; (1934a) *Ibis* (13) **4**, 61–79; (1934b) *Ibis* (13) **4**, 213–39; (1937a) *Ibis* (14) **1**, 47–65; (1937b) *Ibis* (14) **1**, 301–21. BATES, R S P and LOWTHER, E H N (1952) *J. Bombay nat. Hist. Soc.* **50**, 779–84. BATESON, P P G (1961) *Br. Birds* **54**, 272–7. BATTEN, L A (1975) *Bull. Br. Orn. Club* **95**, 127–8. BAU, A (1901) *Orn. Jahrb.* **12**, 20–6. BAUDVIN, H (1974) *Jean le Blanc* **13**, 61–4; (1975) *Jean le Blanc* **14**, 1–51; (1978) *Nos Oiseaux* **34**, 223–31; (1979a) *Alauda* **47**, 13–16; (1979b) *Nos Oiseaux* **35**, 125–34; (1980) *Nos Oiseaux* **35**, 232–8. BAUER, H (1976) *Beiträge zur Biologie der Nachtschwalbe im Erlanger Raum.* Diplomarbeit Univ. Erlangen. BAUER, K (1952) *J. Orn.* **93**, 290–4; (1956) *J. Orn.* **97**, 335–40; (1965) *Egretta* **2**, 35–51. BAUER, K M (1957) *Falke* spec. vol. **3**, 22–5. BAUER, S and THIELCKE, G (1982) *Vogelwarte* **31**, 183–391. BAUER, W, HELVERSEN, O V, HODGE, J, and MARTENS, J (1969) *Catalogus Faunae Graeciae.* Thessaloniki. BAUM, L and JAHN, E (1965) *Corax* **1**, 73–82. BAUMANIS, J and BLŪMS, P (1972) *Latvijas putni.* Riga. BAUMANN, E (1905) *Gefiederte Welt* **34**, 124–5. BAUMANN, P (1982) *Vögel der Heimat* **52**, 108–14. BAUMGART, W (1973) *Zool. Abh. Mus. Tierk. Dresden* **32**, 203–47; (1975) *Zool. Abh. Mus. Tierk. Dresden* **33**, 251–75. BAUMGART, W, SIMEONOV, S D, ZIMMERMANN, M, BÜNSCHE, H, BAUMGART, P, and KÜHNAST, G (1973) *Zool. Abh. Mus. Tierk. Dresden* **32**, 203–47. BAUMGART, W, MAILICK, U, and MAILICK, E (1975) *Falke* **22**, 382–8, 418–21. BAWTREE, R F (1965) *Br. Birds* **58**, 299. BAXTER, A P, BAXTER, R L, and PARRY, P E (1949) *Ibis* **91**, 689. BAXTER, E V and RINTOUL, L J (1953) *The birds of Scotland.* Edinburgh. BAYLDON, J M (1978) *Br. Birds* **71**, 88. BEADON, C (1915) *J. Bombay nat. Hist. Soc.* **24**, 192–3. BEAMAN, M, PORTER, R F, and VITTERY, A (1975) *Orn. Soc. Turkey Bird Rep. 1970–3*. Sandy. BEAVAN, R C (1865) *Ibis* (2) **1**, 400–23. BECKER, K (1963) *Anz. f. Schädlingkde.* **36**, 73–7; (1964) *Bundesgesundheitsblatt* **14**, 213–14. BECKER, L and DANKHOFF, S (1973) *Abh. Ber. Naturk. Mus.-ForschStelle Görlitz* **48** (13), 1–9. BECKER, P (1978a) *Beitr. Naturkde. Niedersachs.* **31**, 21; (1978b) *Beitr. Naturkde. Niedersachs.* **31**, 22–4. BECKING, J H (1975) *Ibis* **117**, 275–84; (1981) *J. Bombay nat. Hist. Soc.* **78**, 201–31. BÉDARD, J (1969) *Can. Wildl. Ser.* **7**. BEE, J W (1958) *Univ. Kansas Publ. Mus. nat. Hist.* **10**, 163–211. BEECHER, W J (1953) *J. Wash. Acad. Sci.* **43**, 293–9. BEGG, G W (1973) *Ostrich* **44**, 149–53. BEJER-PETERSEN, B (1968) *Dansk Skovforen. Tidsskr.* **53**, 266–72. BEKLOVÁ, M (1976) *Zool. Listy* **25**, 147–55. BÉLDI, M (1963) *Aquila* **70**, 209–10. BELFRAGE, J H (1893) *Zoologist* (3) **16**, 310. BELIK, V P (1977) *Ornitologiya* **13**, 186–7. BELL, A P (1965) *Br. Birds* **58**, 150. BELOPOL'SKI, L (1975) *Comm. Baltic Commiss. Study Bird Migr.* **8**, 51–71. BELOPOL'SKI, L O (1957) *Ecology of sea colony birds of the Barents Sea.* Moscow. BELOPOL'SKI, L O, GORAINOVA, G P, MILOVANOVA, N I, PETROVA, I A, and

POLONIK, N I (1977) *Ornitologiya* **13**, 95–9. BELOUSOV,Y A and MAKKOVEEVA, I I (1981) In Flint (1981), 11. BELSKAYA, G S (1976) *Ornitologiya* **12**, 125–31. BELSKI, N V (1958) *Ornitologiya* **1**, 161–4. BENEDEN, A VAN (1930) *Gerfaut*, 24–5. BENGTSON, S-A (1966) *Fauna och Flora*, 24–30; (1971) *Nor. J. Zool.* **19**, 77–82. BENNETT, C G (1965) *Br. Birds* **58**, 196–7. BENNETT, C G and KING, B (1959) *Br. Birds* **52**, 313–14. BENSON, C W (1956) *Occ. Pap. nat. Mus. S. Rhod.* **21**B, 1–51. BENSON, C W, BROOKE, R K, DOWSETT, R J, and IRWIN, M P S (1971) *The birds of Zambia*. London. BENSON, C W, BROOKE, R K, STUART IRWIN, M P, and STEYN, P (1968) *Ibis* **110**, 106. BENSON, S V (1970) *Birds of Lebanon and the Jordan area*. London. BENT, A C (1919) *Bull. US natn. Mus.* **107**; (1921) *Bull. US natn. Mus.* **113**; (1938) *Bull. US natn. Mus.* **170**; (1939) *Bull. US natn. Mus.* **174**; (1940) *Bull. US natn. Mus.* **176**. BENTHAM, H (1957) *Br. Birds* **50**, 75; (1962) *Br. Birds* **55**, 482. BENTZ, P-G (1982) *Anser* **21**, 93–104. BENZON, B (1943) *Dansk orn. Foren. Tidsskr.* **37**, 184. BÉRES, J (1948–51) *Aquila* **55–8**, 274. BERETZK, P (1948–51) *Aquila* **55–8**, 273–4. BERETZK, P and KEVE, A (1971) *Zool. Abh. Mus. Tierk. Dresden* **30**, 227–42; (1973) *Alauda* **41**, 337–44. BERG, A VAN DEN (1979) *Dutch Birding* **1**, 25–6. BERG, A B VAN DEN, and WIJS, W J R DE (1980) *Dutch Birding* **2**, 22–4. BERGGREN, V and WAHLSTEDT, J (1977) *Vår Fågelvärld* **36**, 243–9. BERGIER, P and BADAN, O (1979) *Alauda* **47**, 271–5. BERGMAN, G (1953a) *Acta zool. fenn.* **77**, 1–50; (1953b) *Ornis fenn.* **30**, 83–5; (1955) *Ornis fenn.* **32**, 69–83; (1956) *Vår Fågelvärld* **15**, 223–45; (1961) *Br. Birds* **54**, 307–20; (1971a) *Commentat. Biol.* **42**, 1–26; (1971b) *Ornis fenn.* **48**, 138–9; (1980) *Ornis fenn.* **57**, 141–52. BERGMAN, G, SWAIN, P, and WELLER, M W (1970) *Wilson Bull.* **82**, 435–44. BERGMANN, H-H, and GANSO, M (1965) *J. Orn.* **106**, 255–84. BERGMANN, H-H and HELB, H-W (1982) *Stimmen der Vögel Europas*. Munich. BERGMANN, H-H and WIESNER, J (1982) *J. Orn.* **123**, 315–18. BERGMANN, S (1939) *Beitr. Fortpfl. Biol. Vögel* **15**, 181–9. BERNÁTH, G (1959) *Aquila* **65**, 348–9. BERNDT, R (1941) *Beitr. Fortpfl. Biol. Vögel* **17**, 215; (1943) *Beitr. Fortpfl. Biol. Vögel* **19**, 168–9; (1954) *Orn. Mitt.* **6**, 94; (1960) *J. Orn.* **101**, 364; (1961) *Falke* **8**, 139–40; (1962) *Orn. Mitt.* **14**, 110. BERNDT, R and DANCKER, R (1966) *Vogelwelt* **87**, 48–52. BERNDT, R and WINKEL, W (1977) *Vogelwelt* **98**, 161–92; (1979) *Vogelwelt* **100**, 55–69. BERNHOFT-OSA, A (1963) *Stavanger Mus. Årbok 1962*, 181–8; (1973) *Sterna* **12**, 73–83. BERNIS, F (1967) *Aves Migradoras Ibericas* **1** (5); (1970) **2** (6). Madrid. BERNIS MADRAZO, F (1945) *Bol. R. Soc. Esp. Hist. nat.* **43**, 93–145. BERRY, R (1979) *Br. Birds* **72**, 207–18. BERRY, R and BIBBY, C (1981) *Br. Birds* **74**, 161–9. BERTRAM, G C and LACK, D (1933) *Ibis* (13) **3**, 283–301; (1938) *J. anim. Ecol.* **7**, 27–52. BERTRAM, K (1903) *Orn. Monatsschr.* **28**, 268–7; (1908) *Orn. Monatsschr.* **33**, 415–16. BESSON, J (1964) *Alauda* **32**, 297–303; (1967) *Alauda* **35**, 298–302; (1969a) *Alauda* **37**, 255; (1969b) *Alauda* **37**, 258; (1970) *Nos Oiseaux* **30**, 264. BETTMANN, H (1959) *Orn. Mitt.* **11**, 161–3; (1961) *Orn. Mitt.* **13**, 152; (1965) *Z. Jagdwissensch.* **11**, 136–44; (1966) *Z. Jagdwissensch.* **12**, 99–125; (1970) *Orn. Mitt.* **22**, 39–40; (1972) *Orn. Mitt.* **24**, 214; (1978) *Orn. Mitt.* **30**, 129; (1981) *Orn. Mitt.* **33**, 301. BEVEN, G (1965) *London Bird Rep.* **29**, 56–72; (1966) *Br. Birds* **59**, 233–40; (1969) In Gooders, J (ed) *Birds of the world*, 1340–4. London; (1973) *Br. Birds* **66**, 390–6; (1976) *Br. Birds* **69**, 506–7; (1979) *Br. Birds* **72**, 594–5; (1982) *London Nat.* **61**, 88–94. BEZZEL, E, LANG, M, and LECHNER, F (1977) *Garmischer vogelkdl. Ber.* **2**, 16–19. BEZZEL, E and LECHNER, F (1978) *Die Vögel des Werdenfelser Landes*. Greven. BEZZEL, E, LECHNER, F, and RANFTL, H (1980) *Arbeitsatlas der Brutvögel Bayerns*. Greven. BEZZEL, E, OBST, J, and WICKL, K-H (1976) *J. Orn.* **117**, 210–38. BEZZEL, E and REICHHOLF, J (1965) *Vogelwarte* **23**, 121–8. BIANKI, V V (1967) *Trudy Kandalak. gos. zapoved.* **6**. BIANKI, V V and KOSHKINA, T V (1960) *Trudy Kandalak. gos. zapoved.* **3**, 113–17. BIBER, O (1971) *Alauda* **39**, 209–12. BIČÍK, V and SMĚŠNÁ, H (1971) *Acta Univ. Palackianae Olomoncensis, fac. rer. nat.* **34**, 105–18. BIGGS, H C, KEMP, A C, MENDELSOHN, H P, and MENDELSOHN, J M (1979) *Durban Mus. Novit.* **12**, 73–81. BIJLSMA, R (1977) *Levende Nat.* **80**, 163–70. BILANG, H (1972) *Falke* **19**, 139. BILLE, R-P (1972) *Nos Oiseaux* **31**, 141–9, 173–82. BIRD, C G and BIRD, E G (1935) *Ibis* **13** (5), 837–55. BIRKHEAD, T R (1974a) *Bird Study* **21**, 241–54; (1974b) *Ornis scand.* **5**, 71–81; (1976) D Phil Thesis. Oxford Univ.; (1977a) *J. anim. Ecol.* **46**, 751–64; (1977b) *Ibis* **119**, 544–9; (1978a) *Ibis* **120**, 219–29; (1978b) *Anim. Behav.* **26**, 321–31; (1980) *Ornis scand.* **11**, 142–5. BIRKHEAD, T R and HUDSON, P J (1977) *Ornis scand.* **8**, 145–54. BIRKHEAD, T R and NETTLESHIP, D N (1981) *Auk* **98**, 258–69. BIRKHEAD, T R and TAYLOR, A M (1977) *Ibis* **119**, 80–5. BIRULYA, A (1910) *Ezhegod. Zool. muz. imp. Akad. nauk* **15**, 167–206. BISWAS, B (1951) *Ardea* **39**, 318–21. BJÖRNSSON, H (1976) *Náttúrufraedingurinn* **46**, 56–104. BLACKBURN, H (1895) *Birds from Moidart and elsewhere*. Edinburgh. BLAIR, H M S (1962) *Br. Birds* **55**, 414–18. BLAISE, M (1965) *Oiseau* **35**, 87–116. BLAKER, G B (1934) *The Barn Owl in England and Wales*. London. BLANC, T (1958) *Nos Oiseaux* **24**, 322. BLANCHET, A (1951) *Mém. Soc. Sci. Nat. Tunisie* **1**, 1–92; (1955) *Mém. Soc. Sci. Nat. Tunisie* **3**, 1–84; (1957) *Mém. Soc. Sci. Nat. Tunisie* **1** (2), 93–216. BLANFORD, W T (1870) *Observations on the geology and zoology of Abyssinia*. London; (1895) *The fauna of British India: birds* **3**. London. BLANKERT, H (1980) *Dutch Birding* **1**, 115. BLANKERT, J and HEER, P DE (1982) *Dutch Birding* **4**, 18–19. BLATTI, G (1947) *Orn. Beob.* **44**, 37. BLOESCH, M (1931) *Orn. Beob.* **28**, 183–7. BLOMGREN, A (1958) *Fåglar i Nordanskog*. Stockholm. BLONDEL, J (1962) *Terre et Vie* **3**, 209–51. BLONDEL, J and BADAN, O (1976) *Nos Oiseaux* **33**, 189–219. BLONDEL, J and ISENMANN, P (1981) *Guide des oiseaux de Camargue*. Neuchâtel. BLONDEL, J and RAMADAN-JARADI, G (1975) *Alauda* **43**, 481. BLUME, D (1955) *Vogelwelt* **76**, 193–210; (1956) *Vogelwelt* **77**, 129–51; (1957) *Vogelwelt* **78**, 41–8; (1958a) *J. Orn.* **99**, 377; (1958b) *Vogelwelt* **79**, 65–88; (1959) *Vogelwelt* **80**, 129–42; (1961) *J. Orn.* **102** suppl., 1–115; (1962a) *Vogelwelt* **83**, 33–48; (1962b) *Orn. Mitt.* **14**, 181–7; (1962c) *J. Orn.* **103**, 140–9; (1963a) *Luscinia* **36**, 35–7; (1963b) *Vogelwelt* **84**, 161–84; (1964a) *Vogelwelt* **85**, 11–19; (1964b) *Vogelkosmos* **1**, 89–91; (1965a) *Orn. Mitt.* **17**, 175–80; (1965b) *Vogelwelt* **86**, 33–42; (1971) *Die Spechte fremder Länder*. Wittenberg Lutherstadt; (1973) *Ausdrucksformen unserer Vögel*. Wittenberg Lutherstadt; (1977) *Die Buntspechte*. Wittenberg Lutherstadt; (1981) *Schwarzspecht, Grünspecht, Grauspecht*. Wittenberg Lutherstadt. BLUME, D and BLUME, W (1981) *Vogel und Umwelt* **1**, 234–40. BLUME, D and JEIDE, W (1965) *Orn. Mitt.* **17**, 154–6. BLUME, D and JUNG, G (1956) *Vogelring* **25**, 60–75; (1958) *Vogelring* **27**, 1–13, 65–74; (1959) *Vogelwelt* **80**, 65–74. BLUME, D and OGASAWARA, K (1980) *Orn. Mitt.* **32**, 209–12. BLUME, D, RUGE, K, and TILGNER, W (1975) *Die Sprache unserer Spechte*. Mühlacker. BOAG, D (1982) *The Kingfisher*. Poole. BOANO, G (1979) *Riv. ital. Orn.* **49**, 1–23. BOASE, H (1950) *Br. Birds* **43**, 86. BOCHEŃSKI, Z (1960) *Acta Zool. Cracov.* **5**, 311–33; (1966) *Przegl. Zool.* **10**, 64–5. BÖCK, F and WALTER, W (1976) *Egretta* **19**, 11–22. BOCK, W J and FARRAND, J (1980) *Am. Mus. Novit.* **2703**. BOCK, W J and MILLER, W DE W (1959) *Am. Mus. Novit.* **1931**. BODDY, M (1978a) *Br. Birds* **71**, 309–10; (1978b) *Ring. Migr.* **2**, 50–1; (1981) *Ring. Migr.* **3**, 113–22. BODENSTEIN, G (1949a) *Orn. Mitt.* **1**, 58–9; (1949b) *Orn. Beob.* **46**, 107–16. BOECKELHEIDE, R J (1978) *Pacific Seabird Grp. Bull.* **5**, 52. BOECKER, M (1967) *Bonn. zool. Beitr.* **18**, 15–126. BOEV, H (1962) *Bull. Bulg. Acad. Sc. Zool. Mus.* **11**, 31–46. BOGDANOV, O P (1955) *Priroda* **12**, 114–15. BÖGERSHAUSEN, M (1976) *Beitr. Naturkde. Niedersachs.* **29**, 67–9. BOGUCKI, Z (1967) *Przegl. Zool.* **11**, 71–4. BÖHM, D (1982a)

Gefiederte Welt **106**, 344–6; (1982b) *Gefiederte Welt* **106**, 376–8. BOHNSACK, P (1966) *Corax* **1**, 162–72. BOLAM, G (1889) *Proc. Berwickshire Nat. Club* **12**, 542–51; (1912) *The birds of Northumberland and the eastern Borders*. Alnwick; (1914) *Br. Birds* **7**, 232–3. BOLEY, A (1961) *Natur* **69**, 205–9. BOLLE, C (1857) *J. Orn.* **5**, 305–51; (1863a) *J. Orn.* **11**, 241–8; (1863b) *J. Orn.* **11**, 387–8. BOLSHAKOV, V N (1968) *Ornitologiya* **9**, 336–8. BOLSTER, R C (1922) *J. Bombay nat. Hist. Soc.* **28**, 807–9. BOND, J (1960) *Birds of the West Indies*. London. BONDRUP-NIELSEN, S (1977) *Can. J. Zool.* **55**, 595–601. BÖNI, A (1942) *Arch. suisses Orn.* **2**, 1–56. BOORMAN, K (1972) *Falke* **19**, 247. BOOTH, G A (1914) *Wild Life* **3**, 26–30. BORMAN, F W (1929) *Ibis* (12) **5**, 639–50. BORRETT, R P (1973) *Ostrich* **44**, 145–8. BORRETT, R P and JACKSON, H D (1970) *Bull. Br. Orn. Club* **90**, 25–6, 135. BOSHOFF, A F (1978) *Ostrich* **49**, 89. BÖSIGER, E and FAUCHER, P (1962) *Birds of the night*. Edinburgh. BOSSIERRE, R DE T DE (1935) *Gerfaut* **25**, 196–201. BOSWALL, J (1965) *Br. Birds* **58**, 343–4; (1970) *Bull. Br. Orn. Club* **90**, 92–6. BOTTOMLEY, J B and BOTTOMLEY, S (1975) *Br. Birds* **68**, 514. BOUET, G (1955) *Oiseaux de l'Afrique tropicale* **1**. Paris. BOURNE, W R P (1955) *Ibis* **97**, 508–56; (1957) *Ibis* **99**, 182–90; (1966) *Ibis* **108**, 425–9; (1967) *Sea Swallow* **19**, 51–76; (1968) *Scott. Birds* **5**, 104–7; (1976a) *Br. Birds* **69**, 188–9; (1976b) In Johnston, R (ed) *Marine pollution*, 403–502. London; (1981) *Scott. Birds* **11**, 254–7. BOURNE, W R P and DIXON, T J (1974) *Seabird Rep.* **4**, 1–18. BOURNE, W R P and SMITH, A J M (1974) *Biol. Conserv.* **6**, 222–4. BOUTILLIER, A (1914) *Rev. fr. Orn.* **6**, 346–53, 377–9. BOWEN, W W (1927) *Am. Mus. Novit.* **273**, 1–12; (1928) *Sudan Notes Rec.* **11**, 69–82. BOWER, B P (1970) *Scott. Birds* **6**, 44. BOYD, A W (1933) *Br. Birds* **27**, 166–7. BOYLE, G (1952) *Br. Birds* **45**, 71; (1974) *Br. Birds* **67**, 474. BOZSKO, S and LAJOS, J (1981) *Aquila* **88**, 91–114. BOZSKO, S I (1967) *Aquila* **73-4**, 121–32; (1977) *Aquila* **83**, 173–7; (1979) *Aquila* **85**, 85–92. BRAAKSMA, S (1958) *Vogeljaar* **6**, 3–12. BRAAKSMA, S and BRUIJN, O DE (1976) *Limosa* **49**, 135–87. BRADFIELD, R D (1931) *Ostrich* **2**, 7–9. BRADLEE, T S, MOWBRAY, L L, and EATON, W F (1931) *Proc. Boston Soc. nat. Hist.* **39**, 279–382. BRADSTREET, M S W (1979) *Can. J. Zool.* **57**, 1789–1802; (1980) *Can. J. Zool.* **58**, 2120–40. BRANDER, P W (1950) *Ardea* **38**, 178–82. BRANDNER, J and STANI, W (1982) *Egretta* **25**, 20–2. BRANDOLINI, A (1950) *Riv. ital. Orn.* **20**, 58–61. BRANDT, M (1941) *Orn. Monatsber.* **49**, 104–11. BRANEGAN, J (1959) *Sea Swallow* **12**, 9. BRAUN, M (1975) *Orn. Mitt.* **27**, 219. BRAVERY, J A (1967) *Emu* **66**, 267–71. BRDICKA, I (1969) In Gooders, J (ed) *Birds of the world*, 1314–16. London. BREE, C R (1863) *Zoologist* (1) **21**, 8684. BREHM, A E (1876) *Gefangene Vögel* **2**. Leipzig. BREWER, A D (1969) *Br. Birds* **63**, 282. BRICHAMBAUT, J DE (1973) *Alauda* **41**, 353–61; (1978) *Alauda* **46**, 271. BRICHETTI, P (1976) *Atlante ornitologico Italiano* **1**. Bressica; (1979) *Riv. ital. Orn.* **49**, 197–205; (1982) *Riv. ital. Orn.* **52**, 3–50. BRICHETTI, P and ISENMANN, P (1981) *Riv. ital. Orn.* **51**, 133–61. BRICHETTI, P and MARTIGNONI, C (1981) *Riv. ital. Orn.* **51**, 113–20. BRIEN, Y, BESSEC, A, and LESOUEF, J-Y (1982) *Oiseau* **52**, 87–9. BRIGGS, F S (1932) *J. Bombay nat. Hist. Soc.* **35**, 382–404. BRIGHT, J (1935) *Emu* **34**, 293–302. BRINGELAND, R and FJÆRE, T (1981) *Fauna norv. Ser. C, Cinclus* **4**, 40–6. BRION, A and VACHER, M (1970) *Bull. Acad. Vét.* **43**, 311–17. BRITTON, H A and BRITTON, P L (1976) *Bull. E. Afr. nat. Hist. Soc.*, 52–61. BRITTON, P L (1970) *Bull. Br. Orn. Club* **90**, 142–4, 152–4; (1977a) *Scopus* **1**, 29–34; (1977b) *Scopus* **1**, 48; (ed) (1980) *Birds of East Africa: their habitat, status and distribution*. Nairobi. BRITTON, P L and BROWN, L H (1971) *Ibis* **113**, 354–6; (1974) *Ostrich* **45**, 63–82. BRITTON, P L and DOWSETT, R J (1969) *Ostrich* **40**, 55–60. BRITTON, P L and OSBORNE, T O (1976) *Bull. Br. Orn. Club* **96**, 132–4. BRODKORB, P (1940) *Auk* **57**, 542–9. BROEKHUYSEN, G J (1965) *Ostrich* **36**, 228; (1974) *Ostrich* **45**, 235–50. BROMHALL, D (1980) *Devil birds: the life of the Swift*. London. BROMLEY, F C (1952) *Ostrich* **23**, 131–2. BROO, B (1982) *Vår Fågelvärld* **41**, 24–5. BROOKE, M DE L (1972a) *Bird Study* **19**, 1–6. BROOKE, R K (1957) *Ostrich* **2**, 164–9; (1969a) *Bull. Br. Orn. Club* **89**, 11–16; (1969b) *Bull. Br. Orn. Club* **89**, 78–81; (1969c) *Bull. Br. Orn. Club* **89**, 97–100; (1969d) *Bull. Br. Orn. Club* **89**, 166–7; (1969e) *Ostrich* **44**, 106–10; (1970) *Durban Mus. Novit.* **9**, 13–24; (1971a) *Bull. Br. Orn. Club* **91**, 33–6; (1971b) *Durban Mus. Novit.* **9**, 29–38; (1971c) *Durban Mus. Novit.* **9**, 93–103; (1971d) *Durban Mus. Novit.* **9**, 131–43; (1971e) *Ostrich* **42**, 5–36; (1972b) *Durban Mus. Novit.* **9**, 217–31; (1972c) *Bull. Br. Orn. Club* **92**, 53–7; (1974) *Durban Mus. Novit.* **10**, 127–37; (1975a) *Durban Mus. Novit.* **10**, 239–49; (1975b) *Ostrich* **46**, 119. BROOKE, R K and VERNON, J C (1961) *Ostrich* **32**, 128–33. BROSSET, A (1956a) *Alauda* **24**, 161–205; (1956b) *Alauda* **24**, 303–5; (1957) *Alauda* **25**, 150; (1959) *Alauda* **27**, 36–60; (1961) *Trav. Inst. Sci. Chérifien Sér. Zool.* **22**. BROTHERTON, J F (1965) *Bull. Niger. Orn. Soc.* **2**, 219. BROUGHTON-LEIGH, P W T (1932) *Ibis* (13) **2**, 457–70. BROWN, D J (1981) *Bird Study* **28**, 139–46. BROWN, J C and TWIGG, G I (1972) *J. Zool. Lond.* **165**, 527–30. BROWN, L H (1970) *African birds of prey*. London. BROWN, L H and BROWN, B E (1973) *Bull. Br. Orn. Club* **93**, 126–30. BROWN, R G B (1968) *Ontario Bird Banding* **4**, 144–8; (1969) *Behaviour* **34**, 115–31; (1976a) *Can. Fld.-Nat.* **90**, 166–8. BROWN, R G B, NETTLESHIP, D N, GERMAIN, P, TULL, C E, and DAVIS, T (1975) *Atlas of eastern Canadian seabirds*. Can. Wildl. Serv., Ottawa. BROWN, R H (1923) *Br. Birds* **17**, 155–7; (1936a) *Br. Birds* **30**, 173–4. BROWN, R L (1934) *Br. Birds* **27**, 256–8; (1935) *Br. Birds* **28**, 83–4; (1936b) *Scott. Nat.*, 135–8. BROWN, W Y (1976b) *Auk* **93**, 179–83. BROWNE, P W P (1950) *Ibis* **92**, 52–65; (1981) *Malimbus* **3**, 63–72. BRUCH, A, ELVERS, H, POHL, C, WESTPHAL, D, and WITT, K (1978) *Orn. Ber. Berlin (West)* **3** suppl. BRUCHHOLZ, S (1978) *Beitr. Vogelkde.* **24**, 102. BRUDENELL-BRUCE, P G C (1967) *Ibis* **109**, 114–15; (1969) *Br. Birds* **62**, 122–3. BRUDERER, B, JACQUAT, B, and BRÜCKNER, U (1972) *Orn. Beob.* **69**, 189–206. BRUDERER, B and WEITNAUER, E (1972) *Rev. suisse Zool.* **79**, 1190–1200. BRUEMMER, F (1972) *Beaver* **303**, 40–7. BRUIJN, O DE (1979) *Limosa* **52**, 91–154. BRÜLLHARDT, H (1969) *Orn. Beob.* **66**, 149–50. BRUN, E (1959) *Skokholm Bird Obs. Rep. 1958*, 23–6; (1965) *Sterna* **6**, 229–50; (1966) *Sterna* **7**, 1–17; (1969a) *Sterna* **8**, 209–24; (1969b) *Sterna* **8**, 345–59; (1970a) *Fauna* **23**, 196–7; (1970b) *Astarte* **3**, 45–50; (1971a) *Astarte* **4**, 41–5; (1971b) *Sterna* **10**, 35–56. BRÜNNER, K (1978) *Anz. orn. Ges. Bayern* **17**, 281–91. BRUNS, H (1959) *Biol. Abh.* **17**, 1–36; (1965) *Orn. Mitt.* **17**, 6–9. BRUNTON, D F and PITTAWAY, R, JR (1971) *Can. Fld.-Nat.* **85**, 315–22. BRYANT, C H (1905) *Zoologist* (4) **9**, 265–6. BUB, H (1943) *Orn. Monatsber.* **51**, 138; (1981) *Orn. Mitt.* **33**, 24; (1982) *Orn. Mitt.* **34**, 19. BUCKLEY, F G and BUCKLEY, P A (1972a) *Ibis* **114**, 344–59; (1974) *Ecology* **55**, 1053–63; (1979) In Southern, W E (ed) *Proc. 1979 Conf. Colonial Waterbird Group*, 99–107. Illinois; (1980) In Burger, J, Olla, B, and Winn, H (eds) *Behavior of marine animals* **4**, 69–122. New York. BUCKLEY, J (1973) *Br. Birds* **66**, 143–6. BUCKLEY, J and GOLDSMITH, J G (1972) *Trans. Norfolk Norwich Nat. Soc.* **22**, 320–5. BUCKLEY, P A and BUCKLEY, F G (1969) *Ardea* **57**, 95–6; (1970a) *Auk* **87**, 1–13; (1970b) *Condor* **72**, 483–6; (1972b) *Anim. Behav.* **20**, 457–62; (1976) *Nat. Hist.* **85**, 46–55; (1977) *Auk* **94**, 36–43. BUCKLEY, P A and HAILMAN, J P (1970) *Br. Birds* **63**, 210–12. BUHLER, P (1964) *Vogelwarte* **42**, 153–8; (1970) *Z. Morph. Tiere* **66**, 337–99; (1979) *Vogelwelt* **91**, 121–30. BÜHLER, P and EPPLE, W (1980) *J. Orn.* **121**, 36–70. BÜHLER, R (1959) *Orn. Beob.* **56**, 22–3. BULL, J (1974) *Birds of New York State*. New York. BULLOCK, I D and GOMERSALL, C H (1981) *Bird Study* **28**, 187–200. BULLOUGH, W S (1942) *Proc. zool. Soc. Lond.* (A) **112**, 1–12. BUMP, G and BOHL, W H (1964) *Wildlife* **84**, 27–71. BUNDY, G (1970) *Bull. Br. Orn. Club* **90**, 47–9; (1971) *Br.*

Birds **64**, 32; (1974) *Br. Birds* **67**, 246–7; (1975) *Br. Birds* **68**, 76; (1976) *The birds of Libya*. Br. Orn. Union, London. BUNDY, G and MORGAN, J H (1969) *Bull. Br. Orn. Club* **89**, 139–44, 151–9. BUNDY, G and WARR, E (1980) *Sandgrouse* **1**, 4–49. BUNN, D S (1974) *Br. Birds* **67**, 493–501; (1976) *Br. Birds* **69**, 220–2; (1977) *Br. Birds* **70**, 171. BUNN, D S and WARBURTON, A B (1977) *Br. Birds* **70**, 246–56. BUNN, D S, WARBURTON, A B, and WILSON, R D S (1982) *The Barn Owl*. Calton. BUNYARD, P F (1924) *Br. Birds* **17**, 198–205; (1926) *Bull. Br. Orn. Club* **47**, 45–8. BURCKHARDT, D (1948) *Orn. Beob.* **45**, 205–27; (1952) *Orn. Beob.* **49**, 100. BURGER, J (1972) *Bird-Banding* **43**, 267–75. BURGER, J and LESSER, F (1977) In Southern, W E (ed) *Proc. 1977 Conf. Colonial Waterbird Group*, 118–27. Illinois. BURGES, D J (1983) *Br. Birds* **76**, 350. BURGGRAEVE, G (1977) *Gerfaut* **67**, 75–80. BURKITT, J P (1916) *Irish Nat.* **25**, 157–60; (1933) *Br. Birds* **26**, 308–9. BÜRKLI, W, JUON, M, and RUGE, K (1975) *Orn. Beob.* **72**, 23–8. BURLEIGH, T H and LOWERY, G H (1942) *Occ. Pap. Mus. Zool. Louisiana State Univ.* **10**, 173–7. BURNIER, E (1961) *Nos Oiseaux* **26**, 26–7; (1979) *Nos Oiseaux* **35**, 39. BURNIER, J and HAINARD, R (1948) *Nos Oiseaux* **198**, 217–36. BURNS, P S (1944) *Br. Birds* **37**, 99–100. BUROW, E (1970) *Orn. Mitt.* **22**, 203–4. BURT, W H (1930) *Univ. Calif. Publ. Zool.* **32**, 455–524. BURTON, H, EVANS, T L, and WEIR, D N (1970) *Scott. Birds* **6**, 154–6. BURTON, J A (1973*a*) *Owls of the world*. BURTON, J F (1950) *Br. Birds* **43**, 301–2. BURTON, J F and JOHNSON, E D H (1984) *Br. Birds* **77**, 87–104. BURTON, P J K (1973*b*) *Bull. Br. Orn. Club* **93**, 116–18; (1978) *Bull. Br. Orn. Club* **98**, 68–74; (1983) *Br. Birds* **76**, 314–15. BURTON, P J K, BLURTON-JONES, N G, and PENNYCUICK, C J (1960) *Sterna* **4**, 113–52. BURTON, P J K and THURSTON, M H (1959) *Br. Birds* **52**, 149–61. BURTON, R E (1947) *Br. Birds* **40**, 149–50. BUSSE, H (1966) *Beitr. Vogelkde.* **12**, 197–9. BUSSE, K (1975) Ph D Thesis. Hamburg Univ.; (1983) *Ökol. Vögel* **5**, 73–110. BUSSMANN, J (1925) *Orn. Beob.* **23**, 48–9; (1934) *Orn. Beob.* **32**, 17–24; (1937) *Schweiz. Arch. Orn.* **1**, 377–90; (1941) *Schweiz. Arch. Orn.* **1**, 467–80; (1944) *Schweiz. Arch. Orn.* **2**, 105–23; (1946) *Orn. Beob.* **43**, 137–56; (1947) *Orn. Beob.* **44**, 41–9; (1950) *Orn. Beob.* **47**, 141–51; (1961) *Orn. Beob.* **58**, 199. BÜTIKOFER, E (1909–10) *Orn. Beob.* **7**, 37–40, 50–4. BUTLER, A L (1905) *Ibis* (8) **5**, 301–401; (1908) *Ibis* (9) **2**, 205–63; (1909) *Ibis* (9) **3**, 389–406; (1930) *Ibis* (12) **6**, 383–4. BUTLER, E A (1875) *Stray Feathers* **3**, 437–500. BUTSCH, R S (1963) *Jack-Pine Warbler* **41**, 19. BUTTERFIELD, E P (1915) *Zoologist* (4) **19**, 354–6; (1916) *Zoologist* (4) **20**, 30–5. BUTURLIN, S A (1907) *Orn. Monatsber.* **15**, 9–11. BUXTON, E J M (1960) *Ibis* **102**, 127–9. BUXTON, J (1950) *The Redstart*. London.

CADBURY, C J (1981) *Bird Study* **28**, 1–4. CADE, T J (1952) *Condor* **54**, 360; (1965) *Wilson Bull.* **77**, 340–5. CADE, T J and MACLEAN, G L (1967) *Condor* **69**, 323–43. CADE, T J, WILLOUGHBY, E J, and MACLEAN, G L (1966) *Auk* **83**, 124–6. CADMAN, W A (1934) *Br. Birds* **28**, 130–2. CAIN, A P E and HILLGARTH, N (1974) *Doñana Acta Vert.* **1** (2), 97–102. CAIN, A P E, HILLGARTH, N, and VALVERDE, J A (1982) *Br. Birds* **75**, 61–5. CAIN, W (1933) *S. Austral. Orn.* **12**, 28–30. CAIRNS, D K (1978) M Sc Thesis. Laval Univ.; (1980) *Wilson Bull.* **92**, 352–61. CAIRNS, J (1915) *Br. Birds* **9**, 124–5. CALDERON, J and COLLADO, E (1976) *Doñana Acta Vert.* **3**, 129–35. CALLEGARI, E (1955) *Avic. Mag.* **61**, 168–72; (1970*a*) *Avic. Mag.* **76**, 75; (1970*b*) *Avic. Mag.* **76**, 186–8. CALLION, J (1973) *Scott. Birds* **7**, 260. CAMACHO MUÑOZ, I (1975) *Cuad. Cienc. Biol.* **4–2**, 111–24. CAMPBELL, B (1951) *Bird Notes* **27**, 169–76; (1962) *Br. Birds* **55**, 165–6; (1966) *Br. Birds* **59**, 111–12. CAMPBELL, B and FERGUSON-LEES, I J (1972) *A field guide to birds' nests*. London. CAMPBELL, J M (1969) *Auk* **86**, 565–8. CAMPBELL, R W and MACCOLL, M D (1978) *J. Wildl. Mgmt.* **42**, 190–2. CAMPOS, F (1978) *Ardeola* **24**, 105–19.

CAMPREDON, P (1977) M Sc Thesis. Bordeaux Univ.; (1978) *Oiseau* **48**, 123–50, 263–79. CANNINGS, P A R (1972) *Br. Birds* **65**, 398. CANO, A (1960) *Ardeola* **6**, 324–6. CANTELO, J and GREGORY, P A (1975) *Br. Birds* **68**, 296–7. CARLO, E A DI (1971) *Riv. ital. Orn.* **41**, 86–107. CARLSON, K and CARLSON, C (1978) *Br. Birds* **71**, 360. CARTER, F E (1947) *Br. Birds* **40**, 87. CASADO, M A, LEVASSOR, C, and PARRA, F (1983) *Alauda* **51**, 203–9. CASEMENT, M B (1963) *Ibis* **105**, 266–8; (1974) *Sea Swallow* **23**, 43–69. CASSIDY, J (1946) *Br. Birds* **39**, 155. CASSOLA, F and LOVARI, S (1979) *Boll. Zool.* **46**, 87–90. CASTAN, R (1955) *Oiseau* **25**, 172–82. CASTORO, P L and GUHL, A M (1958) *Wilson Bull.* **70**, 57–69. CATLEY, G P (1978) *Br. Birds* **71**, 222–3. CATLING, P M (1972*a*) *Auk* **89**, 194–6; (1972*b*) *Can. Fld.-Nat.* **86**, 223–32. CATTHOOR, A (1954*a*) *Gerfaut* **44**, 289; (1954*b*) *Gerfaut* **44**, 352. CĂTUNEANU, I I and FUHN, I E (1980) *Verh. orn. Ges. Bayern* **22**, 265–70. CĂTUNEANU, I, HAMAR, M, THEISS, F, KORODI, G, and MANOLACHE, L (1970) *An. Inst. Cercet. Pentru Prot. Plant* **6**, 433–45. CAVE, B (1982) *Br. Birds* **75**, 55–61. CAVE, F O and MACDONALD, J D (1955) *Birds of the Sudan*. Edinburgh. CAYFORD, J (1981) Hons. Thesis. Exeter Univ. CEBALLOS, P and PURROY, F J (1977) *Pajaros de nuestros campos y bosques*. Madrid. CHALINE, J, BAUDVIN, H, JAMMOT, D, and SAINT GIRONS, M-C (1974) *Les proies des rapaces, petits mammifères et leur environnement*. Doin. CHANCE, E P (1920) *Br. Birds* **14**, 218–32; (1922) *The Cuckoo's secret*. London; (1923) *Bull. Br. Orn. Soc.* **43**, 49–56; (1940) *The truth about the Cuckoo*. London. CHANIOT, G E, JR (1970) *Condor* **72**, 460–5. CHANNER, A G (1976) *Br. Birds* **69**, 309. CHAPIN, J P (1921) *Auk* **38**, 531–52; (1932) *Bull. Am. Mus. nat. Hist.* **65**; (1939) *Bull. Am. Mus. nat. Hist.* **75** (2). CHAPLIN, A (1937) *Avic. Mag.* (5) **2**, 103–4. CHAPMAN, E A and MCGEOCH, J A (1956) *Ibis* **98**, 577–94. CHAPMAN, F M (1899) *Bull. Am. Mus. nat. Hist.* **12**, 219–44. CHAPMAN, K A (1969) *Br. Birds* **62**, 337–9. CHAPPUIS, C (1974) *Alauda* **42**, 197–222; (1979) *Alauda* **47**, 277–99; (1981) *Alauda* **49**, 35–58. CHARLEMAGNE, N V (1954) *Zool. Zh.* **33**, 1420–2. CHAVIGNY, J and DÛ, R LE (1938) *Alauda* **10**, 91–115. CHEKE, R A (1982) *Malimbus* **4**, 55–63. CHENG, TSO-HSIN (1963) *China's economic fauna: birds*. Peiping; (1976) *Distributional list of Chinese birds*. Peking. CHERNYI, A G (1974) *Ornitologiya* **11**, 418. CHESTNEY, R (1970) *Trans. Norfolk Norwich Nat. Soc.* **21**, 353–63. CHEYLAN, G (1975) *Alauda* **43**, 23–54; (1979) *Alauda* **47**, 42–3. CHIH-TUNG, K (1973) *Acta Zool. Sin.* **19**, 293–304. CHISLETT, R (1941) *Naturalist*, 205–14; (1947) *Naturalist* **820**, 9–10. CHITTY, D (1938) *Proc. zool. Soc. Lond.* (A) **108**, 267–87. CHITTY, H (1950) *J. anim. Ecol.* **19**, 180–93. CHOUSSY, D (1971) *Nos Oiseaux* **31**, 37–56. CHRISTENSEN, G C and BOHL, W H (1964) *Wildlife* **84**, 1–26. CHRISTENSEN, N H (1962) *Dansk orn. Foren. Tidsskr.* **56**, 56–81. CHRISTENSEN, O and LEAR, W H (1977) *Medd. om Grøn.* **205**, 1–83. CHRISTENSEN, R (1912) *Billeder af dansk fugleliv* **2**. Copenhagen. CHRISTIAN, D (1949) *Country Life* **106**, 749. CHRISTIE, D A (1979) *Br. Birds* **72**, 552. CHRISTOLEIT, E (1931) *Beitr. Fortpfl. Biol. Vögel* **7**, 121–8, 170–4, 216–21; (1934) *Beitr. Fortpfl. Biol. Vögel* **10**, 214–16. CHUNIKHIN, S P (1964) *Zool. Zh.* **43**, 1249–50. CHURCH, H F (1956) *Bird Study* **3**, 217–20. CHURCHILL, W J (1939) *Ool. Rec.* **19**, 35–6. CIOSEK, J and TOMIAŁOJĆ, L (1982) *Przegl. Zool.* **26**, 101–9. CLANCEY, P A (1935*a*) *Br. Birds* **28**, 295–301; (1935*b*) *Scott. Nat.*, 173–4; (1936*a*) *Br. Birds* **29**, 326–7; (1936*b*) *Br. Birds* **30**, 180–2; (1951) *Ann. Natal Mus.* **12**, 139–42; (1960) *Durban Mus. Novit.* **6**, 27–31; (1964) *The birds of Natal and Zululand*. Edinburgh; (1965) *Durban Mus. Novit.* **7**, 305–88; (1966) *Bull. Br. Orn. Club* **36**, 112–15; (1970*a*) *Durban Mus. Novit.* **8**, 375–90; (1970*b*) *Durban Mus. Novit.* **9**, 1–11; (1971*a*) *A handlist of the birds of southern Moçambique*. Lourenco Marques; (1971*b*) *Durban Mus. Novit.* **9**, 39–44; (1973) *Durban Mus. Novit.* **10**, 1–11; (1975) *Durban Mus. Novit.* **10**, 191–206; (1976) *Ostrich* **47**, 228;

(1977a) Bokmakierie 29, 17; (1977b) Ostrich 48, 43–4; (1980) Durban Mus. Novit. 12, 151–6; (1981a) Durban Mus. Novit. 12, 227–9; (1981b) Durban Mus. Novit. 13, 13–20; (1982) Ostrich 53, 102–6. CLANCEY, P A and MENDELSOHN, J M (1979) Durban Mus. Novit. 12, 83–6. CLAPHAM, C S (1964) Ibis 106, 376–88. CLAPP, R B, KLIMKIEWICZ, M K, and FUTCHER, A G (1983) J. Fld. Orn. 54, 123–37. CLAPP, R B, KLIMKIEWICZ, M K, and KENNARD, J H (1982) J. Fld. Orn. 53, 81–124. CLARK, R J (1975) Wildl. Monogr. 47. CLARKE, G V H (1924) Ibis (11) 6, 101–10. CLARKE, W E (1890) Zoologist (3) 14, 1–16, 41–51. CLAUDON, C-J (1951) Oiseau 21, 200–15; (1955) Oiseau 25, 44–9. CLAUDY, M (1951) Nos Oiseaux 21, 107. CLEGG, T M and HENDERSON, D S (1974) Br. Birds 64, 317. CLIFFORD, B (1947) Br. Birds 40, 251. COCHET, G (1975) Grand-Duc 13, 55. COCHET, P (1982) Nos Oiseaux 36, 274–6. CODY, M L (1973) Ecology 54, 31–44. COENEN, L (1954) Gerfaut 44, 357. COHEN, E (1946) Br. Birds 39, 248; (1947) Br. Birds 40, 87–8. COHEN, J (1945) Br. Birds 38, 356. COLLAR, N J (1978) Br. Birds 71, 545–6. COLLETT, R and OLSEN, Ø (1921) Norges fugle. Kristiana. COLLINGE, W E (1921) Scott. Nat. 117–18, 183; (1922) J. Min. Agric. 28, 1022–31, 1133–40; (1924–7) The food of some British wild birds. York. COLLINS, C T (1975) Pavo 11, 1–11. COLLINS, C T and BROOKE, R K (1976) Contr. Sci. Los Angeles County Mus. 282. COLLINS, C T and LeCROY, M (1972) Wilson Bull. 84, 187–92. COLLYER, A A, BEADMAN, J, and HILL, T H (1982) J. Zool. Lond. 198, 177–81. COLQUHOUN, M K (1940) Br. Birds 33, 222–4; (1951) The Woodpigeon in Britain. London. COLSTON, P R (1971) Bull. Br. Orn. Club 91, 110–11. CONDER, P (1949) Ibis 91, 649–55; (1950) Br. Birds 43, 65–9. CONDON, H T (1975) Checklist of birds of Australia 1. Melbourne. CONRAD, R (1979a) Beitr. Naturkde. Niedersachs. 32, 144–8; (1979b) Vogelwelt 100, 155–6. CONRADS, K (1967) Ber. naturw. Ver. Bielefeld 18, 25–115; (1975) Nat. und Heimat 35, 49–57. CONRADS, K and HERRMANN, A (1963) J. Orn. 104, 205–48. CONRADS, K and MENSENDIEK, H (1973) Ber. naturw. Ver. Bielefeld 21, 97–117. CONTOLI, L (1976a) Rich. Biol. Selvaggina Suppl. 7, 237–45; (1976b) In Ragonese, B (ed) Atti I Conv. Reg. Ecol., Noto, 45–60. CONTOLI, L, RAGONESE, B, and TIZI, L (1979) Animalia Catania 5, 79–105. CONTOLI, L and SAMMURI, G (1978) Boll. Zool. 45, 323–35. COOMBS, C F B, ISAACSON, A J, MURTON, R K, THEARLE, R J P, and WESTWOOD, N J (1981) J. appl. Ecol. 18, 41–62. COOPER, A S (1979) Cormorant 9, 135. COOPER, R P (1948) Emu 48, 107–26; (1964) Austral. Bird Watcher 2, 95–8. CORBET, G B and SOUTHERN, H N (eds) (1977) The handbook of British mammals. Oxford. CORKHILL, P (1968) Nat. Wales 11, 85; (1971) Nat. Wales 12, 258–62; (1972) Bird Study 19, 193–201; (1973) Bird Study 20, 207–20. CORNISH, A V (1947) Br. Birds 40, 119. CORNWALLIS, L and PORTER, R F (1982) Sandgrouse 4, 1–36. CORRAL, J F, CORTES, J A, and GIL, J M (1979) Doñana Acta Vert. 6, 179–90. CORTES, J E, FINLAYSON, J C, MOSQUEVA, M A, and GARCIA, E F J (1980) The birds of Gibraltar. Gibraltar. COTT, H B (1954) Ibis 96, 129–49. COTTAM, C (1949) Condor 51, 150. COULSON, J C and HOROBIN, J (1976) J. Zool. Lond. 178, 247–60. COURSE, H A (1943) Br. Birds 36, 162. COVERLEY, W H (1932) Ibis (13) 2, 166–7; (1939) Ibis (14) 3, 149–52. COWAN, P J (1982) Bull. Br. Orn. Club 102, 32–5. COWARD, T A (1919) Br. Birds 13, 139; (1928) Br. Birds 22, 134–6. COWDY, S (1955) Br. Birds 48, 234. COWLES, G S (1967) Ibis 109, 260–5. COX, J R (1980) Br. Birds 73, 413. COX, R A F (1953) Br. Birds 46, 414. CRACKLES, E (1948) Br. Birds 41, 351–2. CRACRAFT, J (1971) Auk 88, 723–52; (1981) Auk 98, 681–714; (1982) Syst. Zool. 31, 35–56. CRAIB, C L (1974) Witwatersrand Bird Club News 86, 13; (1982) Witwatersrand Bird Club News 119, 10–11. CRAIG, A (1974) Ostrich 45, 142. CRAIG, W (1909) J. comp. Neurol. Psychol. 19, 29–80. CRAIGHEAD, J J and CRAIGHEAD, F C (1956) Hawks, owls and wildlife. New York.

CRAMP, S (1958) Bird Study 5, 55–66; (1968) Br. Birds 61, 405–8; (1972) Ibis 114, 163–71. CRAMP, S, BOURNE, W R P, and SAUNDERS, D (1974) The seabirds of Britain and Ireland. London. CRAMP, S and CONDER, P J (1970) Ibis 112, 261–3. CRAMP, S, PARRINDER, E R, and RICHARDS, B A (1964) In Homes, R C (ed) The birds of the London area, 106–17. London. CRAMP, S and TOMLINS, A D (1966) Br. Birds 59, 209–33. CRAWFORD, D N (1977) Emu 77, 146–7. CRAWSHAW, K R (1963) Br. Birds 56, 28–9. CREUTZ, G (1943) Beitr. Fortpfl. Biol. Vögel 19, 115–16; (1951) Vogelwelt 72, 52; (1954) Beitr. Vogelkde. 3, 304; (1955) Beitr. Vogelkde. 5, 6–16; (1956) Vögel am Gebirgsbach. Wittenberg Lutherstadt; (1960) Waldhygiene 5–6, 146–8; (1961) J. Orn. 102, 80–7; (1964a) Falke 11, 39–49; (1964b) Vogelwelt 85, 1–11; (1967) Beitr. Vogelkde. 12, 289; (1970) Falke 17, 416; (1974) Abh. Ber. Nat. Mus. Görlitz 48 (8), 1–23; (1976) Abh. Ber. Nat. Mus. Görlitz 49, 1–24; (1979) Falke 26, 222–30. CROCQ, C (1975) Alauda 43, 337–62. CROWE, R W and BROWNLOW, H G (1961) Br. Birds 44, 23–4. CROZE, H (1970) Z. Tierpsychol. 5, 1–85. CSERNAVÖLGYI, L (1975–7) Aquila 82–4, 206–9. CSIKI, E (1905) Aquila 12, 312–30. CSÖRGEY, T (1934) Aquila 38–41, 253–7. CSORNAI, R (1957) Aquila 64, 361–2. CSORNAI, R, SZLIVKA, L, and ANTAL, L (1958) Aquila 65, 234–9. CUISIN, J and CUISIN, M (1979) Oiseau 49, 81–9. CUISIN, M (1963) Oiseau 33, 36–42; (1967–8) Oiseau 37, 163–192, 285–315, 38, 20–52, 103–26, 209–24; (1969) Nos Oiseaux 30, 66–8; (1972) Oiseau 42, 28–34; (1973) Oiseau 43, 305–13; (1975) Oiseau 45, 197–206; (1976) Oiseau 46, 63–7; (1977) Oiseau 47, 159–65; (1980) Oiseau 50, 23–32; (1981) Oiseau 51, 287–95. CULEMANN, H W (1928) J. Orn. 76, 609–53. CULLEN, J M (1954) Ibis 96, 31–46; (1957) Bird Study 4, 197–207; (1960a) Ardea 48, 1–37; (1960b) Proc. int. orn. Congr. 12, 153–7; (1962) In Bannerman (1962), 80–6. CULLEN, J M, GUITON, P E, HORRIDGE, G A, and PEIRSON, J (1952) Ibis 94, 68–84. CUMMING, R S (1952) Ostrich 23, 116–19. CUNNINGHAM-VAN SOMEREN, G R (1969) Bull. Br. Orn. Club 89, 137–9; (1970) Bull. Br. Orn. Club 90, 120–2; (1977) E. Afr. nat. Hist. Soc. Bull., 132–3. CURRY, P J (1974) Bristol Orn. 7, 67–71. CURRY, P J and SAYER, J A (1979) Ibis 121, 20–40. CURRY-LINDAHL, K (1950) Vår Fågelvärld 9, 113–65; (1961a) Bijdragen Dierkde. 31, 27–44; (1961b) Våra fåglar i Norden 3. Stockholm; (1981) Bird migration in Africa. London. CUTCLIFFE, A S (1951) Br. Birds 44, 47–56; (1955) Br. Birds 48, 193–203. CUTHBERT, N L (1954) Auk 71, 36–63. CVITANIĆ, A and NOVAK, P (1966) Larus 20, 80–100. CZARNECKI, Z (1956) Prace Kom. Biol. Poznań 18, 1–42.

DAAN, S and TINBERGEN, J (1979) Ardea 67, 96–100. DAANJE, A (1944) Ardea 33, 74–84. DACHY, P (1954) Gerfaut 44, 96–173. DALGLIESH, G (1911) J. Bombay nat. Hist. Soc. 20, 853. DAMBIERMONT, J L, FRANCOTTE, J P, and COLLETTE, P (1967) Aves 4, 31–47. DAMMERMAN, K W (1929–30) Treubia 11, 1–88. DANCHIN, E (1983) Biol. Behav. 8, 1–8. DANCKER, P (1958) Vogelring 27, 75–6. DANIS, V (1937) Oiseau 7, 246–71. DANKO, Š and SVEHLÍK, J (1971) Československá ochrana prírody 12, 79–91. DARLING, F F (1938) Bird flocks. Cambridge; (1956) Pelican in the wilderness. London. DATHE H (1967) Beitr. Vogelkde. 13, 216–17; (1977) Aquila 83, 304; (1981) Beitr. Vogelkde. 27, 363–4. DAVANT, P (1967) Actes Soc. Linn. Bordeaux 104, 1–11. DAVENPORT, H S (1897) Zoologist (4) 1, 470–1. DAVENPORT, L J (1981) Br. Birds 74, 537. DAVIES, S (1981) Br. Birds 74, 291–8. DAVIES, S J J F (1970) Behaviour 36, 187–214; (1974) Emu 74, 18–26. DAVIES, S J J F and CARRICK, R (1962) Austral. J. Zool. 10, 171–7. DAVIES, W M (1930) North-west. Nat. 5, 92–5. DAVIS, A H and PRYTHERCH, R (1976) Br. Birds 69, 281–7. DAVIS, J M (1975) Anim. Behav. 23, 597–601. DAY, D H (1975) Ostrich 46, 192–4. DEAN, F (1949) Br. Birds 42, 122–3. DEAN, W R J (1969) Ostrich 40, 23–4; (1977) Ostrich Suppl. 12, 102–7; (1978) Proc. Symp.

African Predatory Birds, 25–45. Pretoria; (1979a) *Ostrich* **50**, 215–19; (1979b) *Ostrich* **50**, 234–9; (1980) *Ostrich* **51**, 80–91. DE BRUYN, G J, GOOSEN-DE ROO, L, HUBREGTSE-VAN DEN BERG, A I M, and FEIJEN, H R (1972) *Ekol. polska* **20**, 83–91. DEGN, H J (1976) *Flora og Fauna* **82**, 59–64. DE GROOT, D S (1931) *Condor* **33**, 188–92. DEICKE, H (1957) *Orn. Mitt.* **9**, 97. DEIGNAN, H G (1955) *Bull. Br. Orn. Club* **75**, 57–9; (1963) *Smiths. nat. Mus. Bull.* **226**. DEJONGHE, J F and CZAJKOWSKI, M A (1983) *Alauda* **51**, 27–47. DEKEYSER, P L (1956) *Mém. Inst. fr. d'Afr. noire* **48**, 79–141. DEKEYSER, P L and DERIVOT, J H (1966) *Les oiseaux de l'ouest africain*. Dakar. DEKHUIJZEN-MAASLAND, J M, STEL, H, and HOOGERS, B J (1962) *Ardea* **50**, 162–70. DELACOUR, J (1926) *Bull. Br. Orn. Club* **47**, 8–22. DELACOUR, J and JABOUILLE, P (1931) *Les oiseaux de l'Indochine française* **2**. Paris. DELAMAIN, J (1931) *Alauda* **3**, 451–2. DELMÉE, E (1954) *Gerfaut* **44**, 193–259. DELMÉE, E, DACHY, P, and SIMON, P (1978) *Gerfaut* **68**, 590–650; (1979) *Gerfaut* **69**, 45–77; (1980) *Gerfaut* **70**, 201–10. DELMÉE, E and GODART, P (1976) *Aves* **13**, 229–34. DEMANDT, C (1954) *Vogelwelt* **75**, 201. DEMENTIEV, G P (1933a) *Alauda* **5**, 331–44; (1933b) *Oiseau* **3**, 501–18; (1937) *Alauda* **9**, 287–99; (1939) *Alauda* **11**, 7–17; (1952) *Ptitsy Turkmenistana*. Ashkhabad. DEMENTIEV, G P and GLADKOV, N A (1951a) *Ptitsy Sovietskogo Soyuza* **1**; (1951b) **2**; (1951c) **3**. Moscow. DEMME, N P (1934) *Trudy Arkt. Inst. Biol. Moscow* **11**, 55–86. DENKER, W (1975) *Vogelwelt* **96**, 68. DENNIS, R H (1967) *Fair Isle Bird Obs. Bull.* **5**, 109–15; (1968) *Scott. Birds* **5**, 108–9; (ed) (1979) *Scott. Bird Rep. 1979*. Edinburgh. DENSMORE, M (1924) *Bird Lore* **26**, 403–4. DENTON, E J and NICOLE, J A C (1962) *J. Physiol.* **165**, 13–15. DENWOOD, J R (1896) *Zoologist* (3) **20**, 383–4. DEPPE, H-J (1979) *Angew. Orn.* **5**, 128–40; (1982) *Orn. Mitt.* **34**, 143–8; (1984) *Orn. Mitt.* **36**, 35–6. DERIM, E N (1962) *Ornitologiya* **5**, 410–13. DE ROO, A (1966) *Gerfaut* **56**, 113–31. DE ROO, A and DEHEEGHER, J (1966) *Rev. Zool. Bot. afr.* **74**, 364–70. DESFAYES, M (1949a) *Nos Oiseaux* **20**, 29–32; (1949b) *Nos Oiseaux* **20**, 49–60; (1951) *Nos Oiseaux* **21**, 121–6; (1974) *J. Bombay nat. Hist. Soc.* **71**, 145–6. DESFAYES, M and PRAZ, J C (1978) *Bonn. zool. Beitr.* **29**, 18–37. DE SMET, W M A (1967) *Gerfaut* **57**, 50–76; (1970) *Gerfaut* **60**, 148–87; (1972) *Gerfaut* **62**, 277–305. DESSELBERGER, H (1929) *Orn. Monatsber.* **37**, 14–18. DEWAR, D (1906) *Bombay ducks*. London; (1909) *Birds of the plains*. London; (1924) *The bird as a diver*. London. DEYL, M (1964) *Plevele poli a zahrad*. Prague. DHARMAKUMARSINHJI, R S (1955) *Birds of Saurashtra, India*. Bombay. DIAMOND, A W (1971a) *Phil. Trans. Roy. Soc. Lond.* (B) **260**, 561–71; (1971b) Ph D Thesis. Aberdeen Univ.; (1976) *Ibis* **118**, 414–19. DICE, L R (1945) *Am. Nat.* **79**, 385–416. DICK, M H and DONALDSON, W (1978) *Condor* **80**, 235–6. DICK, W J A (ed) (1975) *Oxford and Cambridge Mauritanian Exped. 1973 Rep.* Cambridge. DICKENS, R F (1953) *Br. Birds* **46**, 412–13. DICKSON, R C (1971) *Br. Birds* **64**, 543. DICOSTANZO, J (1980) *J. Fld. Orn.* **51**, 229–43. DIEBSCHLAG, E (1941) *Z. Tierpsychol.* **4**, 173–88. DIESSELHORST, G (1962) *Stuttgarter Beitr. Naturkde.* **86**, 1–29; (1968) *Khumbu-Himal* **2**. Innsbrück. DIETRICH, F (1921) *Orn. Monatsschr.* **46**, 33–42. DIJKSEN, A J and DIJKSEN, L J (1977) *Texel Vogeleiland*. Zutphen. DILGER, W C (1954) *Condor* **56**, 102–3. DILKS, P J (1975a) *N. Z. J. agric. Res.* **18**, 87–90; (1975b) *Notornis* **22**, 294–301. DIRCKSEN, R (1932a) *J. Orn.* **80**, 427–521; (1932b) *Orn. Monatsber.* **40**, 133–6. DITTBERNER, H (1966) *Falke* **13**, 282. DIXEY, A E, FERGUSON, A, HEYWOOD, R, and TAYLOR, A R (1981) *Br. Birds* **74**, 411–16. DIZERENS, M (1982) *Tierwelt* **42**, 11–12. DOBBS, A (1966) *Br. Birds* **59**, 108–9. DOBROWOLSKI, K A (1970) *Acta Orn.* **12**, 209–28. DOLGUSHIN, I A, KORELOV, M N, KUZ'MINA, M A, GAVRILOV, E I, GAVRIN, V F, KOVSHAR', A F, BORODIKHIN, I F, and RODIONOV, E F (1970) *Ptitsy Kazakhstana* **3**. Alma-Ata. DOL'NIK, V R and KINZHEVSKAYA, L I (1980) *Zool. Zh.* **59**, 1841–51. DOMANIEWSKI, J (1927) *Ann. Zool. Mus.*

Polonici Hist. Nat. **6**, 60–93. DOMBROWSKI, R (1912) *Ornis Romaniae—Die Vogelwelt Rumäniens*. Bucharest. DOMM, S (1977) *Sunbird* **8**, 1–8. DOMM, S and RECHER, H F (1973) *Sunbird* **4**, 63–86. DONAHUE, J P and GANGULI, V (1965) *J. Bombay nat. Hist. Soc.* **62**, 254–8. DONALDSON, G (1971) *Bird-Banding* **42**, 300. DONALDSON CORMONS, G (1976) *Wilson Bull.* **88**, 377–89. DONNELLY, B G (1966) *Ostrich* **37**, 192. DOR, M (1947) *Mammalia* **11**, 50–4. DORNBUSCH, M (1968a) *Beitr. Vogelkde.* **14**, 122–34; (1968b) *Falke* **15**, 130–1. DORNING, H (1930a) *Aquila* **30**, 355–6; (1930b) *Kócsag* **3** (3–4), 11–16; (1932) *Kócsag* **5**, 102–9; (1943) *Aquila* **50**, 383–5. DORST, J (1962) *The migrations of birds*. London. DORWARD, D F and ASHMOLE, N F (1963) *Ibis* **103b**, 447–57. DOTT, H E M (1974) *Seabird Rep.* **4**, 60–5. DOUAUD, J (1953) *Alauda* **21**, 179–85; (1957) *Alauda* **25**, 241–66. DOUCET, J (1969a) *Aves* **6**, 53–61; (1969b) *Aves* **6**, 90–9; (1971) *Gerfaut* **61**, 14–42. DOUCET, J, FRANCOTTE, J-P, and ROSOUX, R (1982) *Aves* **19**, 212–13. DOUDE VAN TROOSTWIJK, W J (1964a) *Ardea* **52**, 13–29; (1964b) *Trans. 6th Congr. int. Union Game Biol.*, 359–67. DOUGLAS, C E (1950) *Br. Birds* **43**, 258. DOUGLAS-HOME, H (1937) *Field*, 350. DOUTHWAITE, R J (1969) In Gooders, J (ed) *Birds of the world*, 1499–1503. London; (1971a) *Bull. Br. Orn. Club* **91**, 147–9; (1971b) *Ibis* **113**, 526–9; (1973) *Ostrich* **44**, 89–94; (1976) *Ostrich* **47**, 153–60; (1978) *J. E. Afr. nat. Hist. Soc. natn. Mus.* **166**, 1–12; (1982) *J. appl. Ecol.* **19**, 133–41. DOWSETT, R J (1965) *Bull. Br. Orn. Club* **85**, 150–2. DRAGESCO, J (1961a) *Sci. et Nat.* **46**, 1–4; (1961b) *Alauda* **29**, 81–98. DRECHSLER, H and MEYER, F (1964) *Beitr. Vogelkde.* **9**, 433–45. DRENT, R H (1965) *Ardea* **53**, 99–160. DRESSER, H E (1871–81) *A history of the birds of Europe*. London. DRIESSEN, F H (1970) *Limosa* **43**, 163–6. DROST, R (1953) *J. Orn.* **94**, 181–93. DROZDOV, N N (1963) *Ornitologiya* **6**, 216–21; (1968) *Ornitologiya* **9**, 345–7. DRURY, W H (1960) *Bird-Banding* **31**, 63–79. DUBALE, M S and RAWAL, U M (1965) *Pavo* **3**, 1–13. DUBININ, N P and TOROPANOVA, T A (1956) *Trudy Inst. Lesa* **32**, 3–307. DU BOIS, A D (1931) *Ool. Rec.* **48**, 72–3. DUBOIS, P (1979) *Alauda* **47**, 43–5. DUBROVSKI, Y A (1961) *Trudy Inst. Zool. Akad. Nauk Kazakh. SSR* **15**, 188–9. DUFFEY, E and SERGEANT, D E (1950) *Ibis* **92**, 554–63. DUFFY, D (1975) Hons Thesis. Harvard Univ. DUFFY, D C, BEEHLER, B, and HASS, W (1976) *Auk* **93**, 839–40. DUHART, F and DESCAMPS, M (1963) *Oiseau* **33** suppl., 1–107. DUHAUTOIS, L, CHARMOY, M-C, CHARMOY, F, REYJAL, D, and TROTIGNON, J (1974) *Alauda* **42**, 313–22. DUNCAN, C D and HAVARD, R W (1980) *Am. Birds* **34**, 122–32. DUNN, E H (1979a) *Can. Fld.-Nat.* **93**, 276–81. DUNN, E K (1972a) Ph D Thesis. Durham Univ.; (1972b) *Ibis* **114**, 360–6; (1973a) *Nature* **244**, 520–1; (1973b) *Auk* **90**, 641–51; (1975) *J. anim. Ecol.* **44**, 734–54. DUNN, P J (1979b) *Br. Birds* **72**, 337. DUNNET, G M (1956) *W. Austral. Nat.* **5**, 86–8. DUNSIRE, C and DUNSIRE, R (1978) *Scott. Birds* **10**, 56. DUPOND, C (1942) *Gerfaut* **32**, 1–15; (1943) *Gerfaut* **33**, 60–2. DUPUY, A (1966) *Oiseau* **36**, 131–44; (1968) *Alauda* **36**, 27–35; (1969) *Oiseau* **39**, 140–60; (1970a) *Alauda* **38**, 278–85; (1970b) *Oiseau* **40**, 176–7; (1975) *Oiseau* **45**, 313–17; (1976) *Oiseau* **46**, 47–76. DURAND, A L (1963) *Br. Birds* **56**, 157–64; (1972) *Br. Birds* **65**, 428–42. DURANGO, S (1942) *Fauna och Flora* **37**, 228–36; (1945a) *Svensk. Faun. Revy* **4**, 1–13; (1945b) *Vår Fågelvärld* **4**, 4–18; (1946) *Vår Fågelvärld* **5**, 145–90; (1949) *Svensk. Faun. Revy* **11**, 14–21; (1951) *Svensk. Faun. Revy* **13**, 73–8. DYBBRO, T (1976) *De danske ynglefugles udbredelse*. Copenhagen. DYCK, J and MELTOFTE, H (1975) *Dansk orn. Foren. Tidsskr.* **69**, 55–64. DYER, M (1979) Ph D Thesis. Aberdeen Univ.; (1980) *Malimbus* **2**, 76. DYER, M and DEMETER, A (1982) *Aquila* **88**, 87–90. DYER, M and FRY, C H (1980) *Proc. int. orn. Congr.* **17** (2), 862–8. DYRCZ, A (1956) *Zool. Poloniae* **7**, 433–54; (1961) *Przegl. Zool.* **5**, 256–9.

EARHART, C E and JOHNSON, N K (1970) *Condor* **72**, 251–64. EASTMAN, R (1969) *The Kingfisher*. London. EATES, K R (1938)

J. Bombay nat. Hist. Soc. **40**, 750–5; (1939) J. Bombay nat. Hist. Soc. **40**, 756–9. EBER, G (1962) Z. wiss. Zool. **167**, 338–94. ECK, S (1971) Zool. Abh. Staatl. Mus. Tierk. Dresden **30**, 173–218; (1973) Zool. Abh. Staatl. Mus. Tierk. Dresden **32**, 199–202. ECK, S and BUSSE, H (1973) Eulen. Wittenberg Lutherstadt. ECK, S and GEIDEL, B (1973) Zool. Abh. Staatl. Mus. Tierk. Dresden **32**, 257–65. ECKHARDT, R C (1969) Harvard Pap. in theor. Geog., Geog. and Prop. of Surface Ser. **26**. EDBERG, R (1955) Vår Fågelvärld **14**, 10–21. EDELSTAM, C (ed) (1972) Vår Fågelvärld, suppl. **7**. EDDY, J (1961) Flicker **33**, 3–4. EDWARDS, C J and WOODFALL, D L (1979) Northumberland Bird Rep., 106–13. EHRENROTH, B (1972) Vår Fågelvärld **31**, 207; (1973) Vår Fågelvärld **32**, 260–8. EHRSTRÖM, C (1955) Ornis fenn. **32**, 93–9. EHRLICH, G (1982) Falke **29**, 390. EICHNER, D (1954) Vogelwarte **17**, 15–18. EINARSSON, T (1979) Náttúrufraedingurinn **49**, 221–8. EISENMANN, E (1962) Am. Mus. Novit. **2094**, 1–21. ELGOOD, J H (1982) The birds of Nigeria. Br. Orn. Union, London. ELGOOD, J H, FRY, C H, and DOWSETT, R J (1973) Ibis **115**, 1–45, 375–411. ELGOOD, J H, SHARLAND, R E, and WARD, P (1966) Ibis **108**, 84–116. ELKINS, N (1979) Br. Birds **72**, 417–33. ELLEMENT, N (1953) Beds. Nat. **7**, 32. ELLIOTT, C C H (1971) Ostrich suppl. **9**, 71–82. ELLIOTT, H F I (1957) Ibis **99**, 545–86. ELLIOTT, H F I and MONK, J F (1952) Ibis **94**, 528–30. ELOSEGUI, J (1974) Ardeola **19**, 249–56. ELVERS, H, MEICH, P, and POHL, C (1979) Orn. Ber. Berlin **4**, 219–34. EMEIS, W (1932) Vogelzug **3**, 92–3. EMLEN, S T and DEMONG, N J (1980) Proc. int. orn. Congr. **17**, 895–901. EMMERSON, M H (1969) J. Durham Univ. Biol. Soc. **15**, 16–20. EMMETT, R E, MIKKOLA, H, MUMMERY, L, and WESTERHOFF, G (1972) Br. Birds **65**, 482–3. ENCKE, F-W (1963) Beitr. Vogelkde. **8**, 449–56. ENEHJELM, C A F (1969) Vogelkosmos **6**, 11–14. ENEMAR, E, LENNERSTEDT, I, and SVENSSON, S (1965) Fauna och Flora **60**, 46–52. ENGLAND, M D (1963) Br. Birds **56**, 58–62; (1969) In Gooders, J (ed) Birds of the World, 1560–2. London. ENNION, H E (1962) Ibis **104**, 560–2. EPPLE, W (1979) J. Orn. **120**, 226. EPPLE, W and BÜHLER, P (1981) Ökol. Vögel **3**, 203–11. ERARD, C (1969) Oiseau **39**, 268–9; (1970) Bull. Br. Orn. Club **90**, 107–11; (1975) Alauda **43**, 313–15. ERARD, C and ETCHÉCOPAR, R-D (1970) Mém. Mus. natn. Hist. nat. (A) **66**, 1–146. ERARD, C and LARIGAUDERIE, F (1972) Oiseau **42**, 81–169, 253–84. ERARD, C and VIELLIARD, J (1966) Ardeola **11**, 95–100. ERDMANN, G (1965) Beitr. Vogelkde. **11**, 115; (1971) Falke **18**, 165–7; (1973) Beitr. Vogelkde. **19**, 329–41. ERIKSSON, K (1971) Ornis fenn. **48**, 69–76. ERKINARO, E (1973a) Aquilo Ser. Zool. **14**, 59–67; (1973b) Aquilo Ser. Zool. **14**, 84–8; (1975) Beitr. Vogelkde. **21**, 288–90. ERLANGER, C VON (1899) J. Orn. **47**, 449–532. ERN, H (1960) Alauda **28**, 66. ERWIN, R M (1977) Ecology **58**, 389–97; (1978) Condor **80**, 211–15. ESCALANTE, R (1968) Auk **70**, 243–7. ESCHERICH, K (1942) Die Forstinsekten Mitteleuropas. Berlin. ESKELINEN, O and MIKKOLA, H (1972) Pohjois-Karjalan Luonto **2**, 24–5. ESTAVIEV, A A (1981) Sovremennoe sostoyanie, raspredelenie i okhrana avifauny taezhnoy zony basseyna r. Pechory. Syktyvkar. ETCHÉCOPAR, R-D (1957) Oiseau **27**, 309–34; (1969) Oiseau **39**, 178–81. ETCHÉCOPAR, R-D and HÜE, F (1967) The birds of North Africa. Edinburgh; (1978) Les oiseaux de Chine de Mongolie et de Corée: non passereaux. Papeete. EVANS, A H (1922) Bull. Br. Orn. Club **42**, 64–5. EVANS, F C and EMLEN, J T, JR (1947) Condor **49**, 3–9. EVANS, P and WATERSTON, G (1976) Polar Rec. **18**, 283–93. EVANS, P G H (1975) Bird Study **22**, 239–47; (1981) Ibis **123**, 1–18. EVANS, P R (1961a) Br. Birds **54**, 326; (1961b) Br. Birds **54**, 361–2. EVANS, W (1889) Proc. roy. phys. Soc. Edinb. **10**, 106–26. EVERITT, C (1974) Gefiederte Welt **98** (2), 29–31. EVERY, B (1976) Ostrich **47**, 229. EXO, K-M (1981) Vogelwelt **102**, 161–80; (1983) Ökol. Vögel **5**, 1–40. EXO, K-M and HENNES, R (1978) Auspicium **6**, 363–74; (1980) Vogelwarte **30**, 162–79. EYGENRAAM, J A (1947) Ardea **35**, 1–44.

FABER, F (1825–6) Über das Leben der hochnordischen Vögel. Leipzig. FABRICIUS, E and JANSSON, A-M (1963) Anim. Behav. **11**, 534–47. FAIRLEY, J S (1966) Br. Birds **59**, 338–40; (1967) Br. Birds **60**, 130–5. FAIRLEY, J S and CLARK, F L (1972) Irish Nat. J. **17**, 219–22. FAIRON, J (1971) Gerfaut **61**, 146–61. FALLA, R A (1937) Rep. BANZ antarct. Res. Exped. B **2**, 1–304. FALLA, R A, SIBSON, R B, and TURBOTT, E G (1970) A field guide to the birds of New Zealand. London. FARINA, A (1980) Monifore zool. ital. **14**, 106. FARKAS, T (1962) Ostrich Suppl. **4**. FARUQI, S A, BUMP, G, NANDA, P C, and CHRISTENSEN, G C (1960) J. Bombay nat. Hist. Soc. **57**, 354–61. FAURE, R (1978) Nos Oiseaux **34**, 325–6. FAY, F H and CADE, T J (1959) Univ. Calif. Publ. Zool. **53**, 73–150. FEARE, C J (1969) Br. Birds **62**, 237; (1975) Condor **77**, 368–70; (1976) J. Zool. Lond. **179**, 317–60; (1979a) Atoll Res. Bull. **226**, 1–29; (1979b) Atoll Res. Bull. **227**, 1–7; (1979c) Bull. Br. Orn. Club **99**, 75–7; (1983) Bull. Br. Orn. Club **103**, 39–40. FEARE, C J and BOURNE, W R P (1978) Ostrich **49**, 64–6. FEDUCCIA, A (1977) Syst. Zool. **26**, 19–31. FEDYUSHIN, A V and DOLBIK, M S (1967) Ptitsy Belorussii. Minsk. FEENY, P P, ARNOLD, R W, and BAILEY, R S (1968) Ibis **110**, 35–86. FEILDEN, H W (1877) Ibis (4) **1**, 401–12; (1908) Br. Birds **2**, 93. FEILER, B (1957) Vogelwelt **78**, 98. FEINDT, P (1956) Weigold-Festschr. Nat. Jagd Niedersachs. 8 suppl., 99–113. FEINDT, P and REBLIN, K (1959) Beitr. Naturkde. Niedersachs. **12**, 36–48. FELLAY, R (1949) Nos Oiseaux **20**, 32–4. FELLENBERG, W O (1969) Vogelwelt **90**, 108–9. FELLOWES, E C (1967) Br. Birds **60**, 522. FERDINAND, L (1969) Dansk orn. Foren. Tidsskr. **63**, 19–45. FERENS, B (1953) Ochr. Przyr. **21**, 78–114; (1960) Conseil. Nat. Prot. Nature, Pologne **2**. Warsaw. FERGUSON-LEES, I J (1952) Br. Birds **45**, 357–8; (1954) Br. Birds **47**, 393–4; (1957) Br. Birds **50**, 385–9; (1958) Br. Birds **51**, 149–52; (1961) Br. Birds **54**, 69–71; (1964) Br. Birds **57**, 170–5; (1966) Bird Study **13**, 270–2; (1967) Br. Birds **60**, 286–90; (1969a) Br. Birds **62**, 533–41; (1969b) In Gooders, J (ed) Birds of the world, 1393–4. London; (1969c) In Gooders, J (ed) Birds of the world, 1410–12. London; (1969d) In Gooders, J (ed) Birds of the world, 1419–20. London; (1971) Br. Birds **64**, 114–17. FERIANC, O (1947) Sylvia **8**, 53–63. FERIANCOVA, Z (1955) Biologia **10**, 436–49. FERREIRA, A C (1981) Cyanopica **2** (3), 49–54. FERRY, C (1962) Alauda **30**, 204–9. FERRY, C, DESCHAINTRE, A, and VIENNOT, R (1957) Alauda **25**, 296–303. FERRY, C and DUFOUR, M (1959) Alauda **27**, 66–9. FERRY, C and FROCHOT, B (1965) Jean le Blanc **4** (3), 70–6. FERRY, C and MARTINET, M (1974) Jean le Blanc **13**, 11–17. FESTETICS, A (1952–5) Aquila **59–62**, 452–3; (1959) Terre et Vie **106**, 121–7; (1967) Egretta **10**, 32; (1968) Z. Tierpsychol. **25**, 659–65; (1973) Orn. Mitt. **52**, 185–6. FEUERSTEIN, W (1960) Vogelkde. **6**, 408–22. FFRENCH, R (1973) A guide to the birds of Trinidad and Tobago. Wynnewood, Penn. FIALA, V (1974) Zool. Listy **23**, 357–66. FIEBIG, J (1956) Beitr. Vogelkde. **4**, 312–15. FINCH, R S (1971) Honeyguide **65**, 19–20. FINLAYSON, J C (1979) D Phil Thesis. Oxford Univ. FINN, F (1902) Zoologist (4) **6**, 149. FINTHA, I (1968) Aquila **75**, 93–109. FISCHER, W (1959) Beitr. Vogelkde. **6**, 395–407. FISCHER-SIGWART, H (1914) Orn. Jahrb. **25**, 51–3. FISH, W M (1943) Br. Birds **37**, 96–7. FISHER, A K (1893) US Dept. Agric. Bull. **3**, 1–210. FISHER, J (1953) Br. Birds **46**, 153–81. FISHER, J and LOCKLEY, R M (1954) Sea birds. London. FISK, E J (1978) Florida fld. Nat. **6**, 1–8. FITTER, R S R (1949) London's birds. London. FJELDSÅ, J (1976) Vidensk. Medd. dansk nat. Foren. **139**, 179–243; (1977) Guide to the young of European precocial birds. Tisveldeleje. FJERDINGSTAD, C (1937) Alauda **9**, 213–17; (1939) Alauda **11**, 50–4. FLEGG, J J M (1969) In Gooders, J (ed) Birds of the world, 1357–61. London; (1972) Bird Study **19**, 7–17. FLETCHER, M R (1979) Br. Birds **72**, 346. FLETCHER, T B and INGLIS, C M (1920) Agric. J. India **15**, 1–4; (1924) Birds of an Indian garden. Calcutta. FLEUSTER, W (1973) J. Orn. **114**,

417–28. FLINT, P R and STEWART, P F (1983) *The birds of Cyprus*. Br. Orn. Union, London. FLINT, V E (ed) (1975) *Kolonialniye gnezdovya okolovodnykh ptits i ikh okhrana*. Moscow; (1981) (ed) *Razmeshchenie i sostoyanie gnezdoviy okolovodnykh ptits na territorii SSSR*. Moscow. FLOERICKE, K (1919) *Detektivstudien in der Vogelwelt*. Stuttgart. FLOWER, S S (1933) *Ibis* (13) 3, 34–46. FLOYD, C B (1932) *Bird-Banding* 3, 173–4. FOG, J (1970) *Dansk orn. Foren. Tidsskr.* 64, 269–70. FOG, J and PETERSEN, K W (1957) *Dansk orn. Foren. Tidsskr.* 51, 1–6. FOG, M (1979) *Danske Vildund*. 32. FOLK, Č (1956) *Zool. Listy* 5, 271–80. FONTAINE, S (1969) *Aves* 6, 143–7. FORBUSH, E H (1924) *Auk* 41, 468–70; (1925) *Birds of Massachusetts and other New England states* 1. Boston Mass. Dept. Agric. FORBUSH, E H and MAY, J R (1939) *A natural history of American birds*. New York. FORMOZOV, A N (1960) *Proc. int. orn. Congr.* 12, 216–29. FORSHAW, J (1978) *Parrots of the world*. Newton Abbot. FORSMAN, D (1980) *Ornis fenn.* 57, 173–5. FORSMAN, E D (1981) *Auk* 98, 735–42. FORSSGREN, K and SJÖLANDER, S (1978) *Astarte* 11, 55–60. FÖRSTEL, A (1977) *Anz. orn. Ges Bayern* 16, 115–31. FORSTER, B (1955) *Ontario Fld. Biol.* 9, 15–17. FORSTER, G H (1957) *Br. Birds* 50, 239–45; (1962) *Br. Birds* 55, 43. FOSSHEIM, E (1955) *Fauna* 8, 102. FOSTER, R J, BAXTER, R L, and BALL, P A J (1951) *Ibis* 93, 53–9. FOUAGE, J (1975) *Aves* 11, 127. FRANK, F (1950) *Bonn. zool. Beitr.* 1, 144–214. FRANKE, H (1953) *Vogelruf und Vogelsang*. Wien. FRANTZEN, M (1955) *Vogelwelt* 76, 106. FRANZ, J (1937) *Beitr. Fortpfl. Biol. Vögel* 13, 165–74; (1943) *J. Orn.* 91, 154–65. FREDGA, K (1964) *Vår Fågelvärld* 23, 103–18. FREEMAN, C P and BATES, G L (1937) *Br. Birds* 30, 302–4. FREEMAN, H J (1981) *Br. Birds* 74, 149. FREEMAN, J A (1945) *J. anim. Ecol.* 14, 128–54. FREITAG, F and METZ, E (1977) *Luscinia* 43, 139–42. FREY, H (1973) *Egretta* 16, 1–68. FREY, H and WALTER, W (1977) *Egretta* 20, 26–35. FRIDRIKSSON, S (1975) *Surtsey*. London. FRIDZÉN, K-E (1959) *Vår Fågelvärld* 18, 75–6; (1974) *Vår Fågelvärld* 33, 49. FRIEDMANN, H (1930) *Bull. US natn. Mus.* 153; (1948) *Monogr. Washington Acad. Sci.* 1; (1964) *Smiths. Misc. Coll.* 146 (4); (1968) *Bull. US natn. Mus.* 265. FRIEDMANN, H and KEITH, S (1968) *Bull. Br. Orn. Club* 88, 112. FRIEDMANN, H and WILLIAMS, J G (1969) *Contr. Sci. Los Angeles Co. Mus.* 162. FRIELING, H (1932) *Orn. Monatsber.* 40, 175; (1960) *Orn. Mitt.* 12, 120. FRISCH, O VON (1965) *Bonn. zool. Beitr.* 16, 92–126; (1966) *Z. Tierpsychol.* 23, 44–51; (1969a) *Bonn. zool. Beitr.* 20, 130–44; (1969b) *Vogelkosmos* 6, 390–3; (1969c) *Z. Tierpsychol.* 26, 641–50; (1970) *J. Orn.* 111, 189–95; (1973) *J. Orn.* 114, 129–31; (1974) *Vogelwelt* 95, 234–5. FRISCH, O VON and FRISCH, H VON (1967) *Z. Tierpsychol.* 24, 129–36. FRITH, C B (1975) *Ostrich* 46, 251–7. FRITH, H J, McKEAN, J L, and BRAITHWAITE, L W (1976) *Emu* 76, 15–24. FRITZ, R, SANDER, J, SANDER, A, NORGALL, A, and NORGALL, T (1977) *Charadrius* 13, 105–10. FROST, R A (1978) *Br. Birds* 71, 130. FRUGIS, S (1952) *Bird Notes* 25, 186–91. FRUTIGER, P J (1973) *Orn. Beob.* 70, 81–6. FRY, C H (1969a) *Ibis* 111, 23–9; (1969b) *Ibis* 111, 557–92; (1970) *Ostrich Suppl.* 8, 239–63; (1972) *Ibis* 114, 1–14; (1973) *Living Bird* 11, 75–112; (1980) *Living Bird* 18, 113–60;(1981a) *Malimbus* 3, 31–8; (1981b) *Scopus* 5, 41–5; (1982) *Ostrich* 53, 244–5; (1983) *Bee World* 64, 65–78. FRY, C H and ELGOOD, J H (1968) *Br. Birds* 61, 37–40. FRYER, R (1976) *Naturalist* 936, 35. FRYLESTAM, B (1970) *Medd. från Skånes orn. Fören.* 9, 25–8; (1972) *Ornis scand.* 3, 45–54. FUCHS, E (1977a) *Ibis* 119, 183–90; (1977b) *Ornis scand.* 8, 17–32. FUCHS, E and SCHIFFERLI, L (1981) *Orn. Beob.* 78, 87–91. FUCHS, G (1905) *Naturwiss. Z. Land- und Forstwistschaft* 3, 317–41. FUGGLES-COUCHMAN, N R (1962) *Ibis* 104, 563–4. FÜRBRINGER, M (1888) *Untersuchungen zur Morphologie und Systematik der Vögel* 2. Amsterdam.

GABRIELSON, I N and LINCOLN, F C (1959) *The birds of Alaska*. Harrisburg, Pa. GADOW, H (1892) *Proc. zool. Soc. Lond.*, 229–56; (1893) *Klassen und Ordnungen des Thier-Reichs* 6 (4): *Vögel II, Systematisches Theil*. Leipzig. GALL, W (1975) *Regulus* 11, 373–4. GALLAGHER, M D (1960) *Ibis* 102, 489–502; (1977) *J. Oman Stud. spec. Rep.* 1, 27–58. GALLAGHER, M D and ROGERS, T D (1978) *Bonn. zool. Beitr.* 29, 5–17; (1980) *J. Oman Stud. spec. Rep.* 2, 347–85. GALLAGHER, M and WOODCOCK, M W (1980) *The birds of Oman*. London. GAMAUF, A and HAAR, H (1981) *Nat. und Umwelt Burgenland* 4, 3–5. GANDRILLE, G and TROTIGNON, J (1973) *Alauda* 41, 129–59. GANGULI, U (1975) *A guide to the birds of the Delhi area*. New Delhi. GANYA, I M, LITVAK, M D, and KUKURUZYANU, L S (1969) *Voprosy ekol. prakt. znach. ptits mleko.* Moldavii 4, 26–54. GARCÍA RÚA, A E (1974) *Ardeola* 20, 365–8. GARLING, M (1944) *Beitr. Fortpfl. Biol. Vögel* 20, 120–3. GARNETT, R M (1933) *Br. Birds* 27, 167; (1942) *Naturalist*, 159–62. GARRISON, D L (1942) *Bull. Mass. Audubon Soc.* 26, 203–6. GARROD, A H (1875) *Proc. zool. Soc. Lond.* 1875, 339–48. GÄRTNER, K (1981a) Ph D Thesis. Hamburg Univ.; (1981b) *Orn. Mitt.* 33, 115–31; (1982) *Vogelwelt* 103, 201–24. GARZÓN HEYDT, J (1968) *Ardeola* 14, 97–130. GASOW, H (1962) *Ann. Épiphyt.* 13, 225–30; (1968) *Beitr. angew. Vogelkde.* 5, 37–59. GASTON, A J (1976) *J. anim. Ecol.* 45, 331–48; (1980) *Can. Wildl. Serv. Prog. Note* 110. GASTON, A J and NETTLESHIP, D N (1981) *Can. Wildl. Serv. Monogr. Ser.* 6. GÄTKE, H (1900) *Die Vogelwarte Helgoland*. Braunschweig. GATTER, W (1972) *J. Orn.* 113, 207–13; (1973) *Anz. orn. Ges. Bayern* 12, 122–9; (1977) *Anz. orn. Ges. Bayern* 16, 141–52. GATTER, W and PENSKI, K (1978) *Vogelwarte* 29, 191–220. GAUGRIS, Y (1968) *Alauda* 36, 287–8; (1979) *Oiseau* 49, 133–53. GAUTIER, F (1968) *Nos Oiseaux* 29, 235. GAVRIN, V F, DOLGUSHIN, I A, KORELOV, M N, and KUZ'MINA, M A (1962) *Ptitsy Kazakhstana* 2. Alma-Ata. GAWLIK, H M and BANZ, K (1982) *Beitr. Vogelkde.* 28, 275–88. GEBAUER, A (1982) *Zool. Anz.* 208 (3–4), 283–8. GEBHARDT, L (1940) *Beitr. Fortpfl. Biol. Vögel* 16, 52–60; (1948–9) *Luscinia* 22, 41; (1950) *Vogelwelt* 71, 105–10. GEHLHAAR, H and KLEBB, W (1979) *Falke* 26, 88–91. GEHRINGER, F (1979) *Nos Oiseaux* 35, 1–16. GEITNER, H (1979) *Gefiederte Welt* 103, 2–5. GEMRA (1974) *Ardeola* 20, 368. GENTZ, K (1940) *Mitt. Ver. sächs. Orn.* 6, 89–108. GEORGE, U (1969) *J. Orn.* 110, 181–91; (1970) *J. Orn.* 111, 175–88; (1978) *In the deserts of this earth*. London. GERASIMOVA, T D (1962) *Ornitologiya* 4, 11–14. GERBER, R (1960) *Die Sumpfohreule*. Wittenberg Lutherstadt; (1965) *Beitr. Vogelkde.* 10, 325. GERDOL, R, MANTOVANI, E, and PERCO, F (1982) *Riv. ital. Orn.* 52, 55–60. GERELL, R (1968) *Vår Fågelvärld* 27, 193–5. GERMAIN, M (1965) *Oiseau* 35, 117–34. GÉROUDET, P (1948) *Nos Oiseaux* 19, 195–200; (1959) *Nos Oiseaux* 25, 132; (1961a) *Alauda* 29, 147–9; (1961b) *Nos Oiseaux* 26, 133–46; (1961c) *La vie des oiseaux: les passereaux* 1. Neuchâtel; (1965) *Les rapaces diurnes et nocturnes d'Europe*. Neuchâtel; (1976) *Nos Oiseaux* 33, 358–80. GÉROUDET, P and LANDENBERGUE, D (1977) *Nos Oiseaux* 34, 165–7. GEYR VON SCHWEPPENBURG, H (1918) *J. Orn.* 66, 121–76; (1924) *Orn. Monatsber.* 32, 126–8; (1942) *Beitr. Fortpfl. Biol. Vögel* 18, 59–62; (1952) *Vogelwarte* 16, 116–19. GIBAN, J, GATINEAU, M, and GUIBERT, R (1948) *Ann. Épiphyt. Ser. Ent.* 14 (5). GIBB, J (1948) *Br. Birds* 41, 167–73. GIBB, J A and HARTLEY, H J (1957) *Br. Birds* 50, 278–89. GIBB, J-N (1983) *Br. Birds* 76, 109–17. GIBSON, D D (1981) *Condor* 83, 65–77. GIBSON, J A (1950) *Br. Birds* 43, 329–31. GIBSON, J D (1956) *Emu* 56, 131–2. GIBSON, R B (1954) *Notornis* 6, 43–7. GIFFORD, E W (1941) *Auk* 58, 239–45. GIGLIOLI, E H (1891) *Primo resoconto dei resultati della inchiesta ornitologica in Italia* 3. Florence. GILBERT, D C (1979) *Kent Bird Rep.* 26, 84–5. GILBERT, M V (1944) *Br. Birds* 38, 135. GILL, F B (1967) *Proc. US natn. Mus.* 123 (3605), 1–33. GILL, R E (1976) *Calif. Fish and Game* 62, 155. GILLET, H (1960) *Oiseau* 30, 99–134. GILPIN,

A (1962) *Naturalist* 881, 43–4; (1968) *Br. Birds* 61, 529–30. GINN, H B and MELVILLE, D S (1983) *Moult in birds*. Tring. GLADSTONE, H S (1910) *The birds of Dumfriesshire*. London. GLADWIN, T W and NAU, B S (1964) *Br. Birds* 57, 344–56. GLASS, M L (1971) *Dansk orn. Foren. Tidsskr.* 65, 73–9. GLEGG, W E (1925) *Br. Birds* 18, 202–9. GLOE, P (1974) *Orn. Mitt.* 26, 47–51; (1976) *Orn. Mitt.* 28, 117–23; (1977a) *Orn. Mitt.* 29, 107–12; (1977b) *Orn. Mitt.* 29, 231–7; (1978a) *Orn. Mitt.* 30, 91–5; (1978b) *Orn. Mitt.* 30, 107–15; (1979) *Orn. Mitt.* 31, 225; (1980a) *Corax* 8, 13–40; (1980b) *Orn. Mitt.* 32, 24; (1982) *Orn. Mitt.* 34, 29–40. GLOE, P and MØLLER, A P (1978) *Orn. Mitt.* 30, 185–202. GLOGER, C W L (1861) *J. Orn.* 9, 64–6. GLUE, D E (1967) *Bird Study* 14, 169–83; (1969) *Br. Birds* 62, 237; (1970) *Mammal Rev.* 1, 53–62; (1972) *Bird Study* 19, 91–5; (1973) *Ornis scand.* 4, 97–102; (1974) *Bird Study* 21, 200–10; (1975) *Br. Birds* 68, 468–9; (1977a) *Bird Study* 24, 70–8; (1977b) *Br. Birds* 70, 318–31; (1979) *Br. Birds* 72, 595. GLUE, D E and HAMMOND, G J (1974) *Br. Birds* 67, 361–9. GLUE, D E and NUTTALL, J (1971) *Bird Study* 18, 33–4. GLUE, D E and SCOTT, D (1980) *Br. Birds* 73, 167–80. GLUTZ VON BLOTZHEIM, U N (1962) *Die Brutvögel der Schweiz*. Aarau. GLUTZ VON BLOTZHEIM, U N and BAUER, K M (1980) *Handbuch der Vögel Mitteleuropas* 9; (1982) 8 (2). Wiesbaden. GLUTZ VON BLOTZHEIM, U N, BAUER, K M, and BEZZEL, E (1977) *Handbuch der Vögel Mitteleuropas* 7. Wiesbaden. GLUTZ VON BLOTZHEIM, U N and SCHWARZENBACH, F H (1979) *Orn. Beob.* 76, 1–7. GNIELKA, R (1975) *Orn. Mitt.* 27, 71–83; (1978) *Orn. Jber. Mus. Hein.* 3, 31–42. GODDARD, T R (1935) *Br. Birds* 28, 290–1; (1938) *Br. Birds* 32, 46. GODFREY, W E (1948) *Can. Fld.-Nat.* 61, 196; (1966) *The birds of Canada*. Ottawa; (1967) *Can. Fld.-Nat.* 81, 99–101. GODIN, J and LOISON, M (1975) *Aves* 12, 57–71. GODMAN, F DU C (1872) *Ibis* (3) 2, 209–24. GOETHE, F (1932) *Beitr. Fortpfl. Biol. Vögel* 8, 129–34; (1957) *Behaviour* 11, 310–17; (1970) *Br. Birds* 63, 34. GOLODUSHKO, B Z and SAMUSENKO, E G (1961) *Trudy IV Pribalt. orn. Konf. 1960*, 135–40. GOLOVANOVA, E N (1978) *S fotoapparatom za siney ptitsey*. Moscow. GOMPERTZ, T (1957) *Bird Study* 4, 2–13. GOODFELLOW, P F and DARE, P J (1955) *Br. Birds* 48, 35–6. GOODMAN, S M (1982) *Bull. Br. Orn. Club* 102, 16–18. GOODMAN, S M and WATSON, G E (1983) *Bull. Br. Orn. Club* 103, 101–6. GOODWIN, D (1946) *Br. Birds* 39, 146–7; (1947a) *Ibis* 89, 656–8; (1947b) *Br. Birds* 40, 254; (1947c) *Avic. Mag.* 53, 97–103; (1948a) *Br. Birds* 41, 12–13; (1948b) *Br. Birds* 41, 123–4; (1952a) *Avic. Mag.* 58, 205–19; (1952b) *Br. Birds* 45, 32–3; (1954a) *Avic. Mag.* 60, 190–213; (1954b) *Mid Thames Nat.* 7, 14–17; (1955) *Avic. Mag.* 61, 54–85; (1956a) *Bird Study* 3, 25–37; (1956b) *Avic. Mag.* 62, 17–33, 62–70; (1958a) *Bull. Br. Orn. Club* 78, 136–9; (1958b) *Bull. Br. Orn. Club* 78, 139–40; (1959) *Bull. Br. Mus. (nat. Hist.) Zool.* 6, 1–23; (1960a) *Bull. Br. Orn. Club* 80, 45–52; (1960b) *Br. Birds* 53, 201–12; (1963) *Ibis* 105, 263–6; (1966) *Auk* 83, 117–23; (1968) *Bull. Br. Mus. (nat. Hist.) Zool.* 17, 1–44; (1969) *Br. Birds* 62, 77; (1970) *Pigeons and doves of the world*. London; (1978) *Birds of man's world*. London; (1980) *Avic. Mag.* 86, 151–63. GORANSSON, G, KARLSSON, J, and ROOS, G (1975) *Anser* 14, 73–8. GORBUNOV, G P (1932) *Trudy Arkt. Inst.* 4, 1–244. GORDON, S (1948) *Br. Birds* 41, 127; (1951) *Br. Birds* 44, 70. GORE, G (1964) *Birds of the Gambia*. Bathurst. GORE, M E J (1980) *Malimbus* 2, 78; (1981) *Birds of the Gambia*. Br. Orn. Union, London. GÖRNER, M (1974) *Beitr. Vogelkde.* 20, 481–2. GÓRSKI, N and GÓRSKA, E (1979) *Acta. Orn. Warszawa* 16, 513–33. GÖSSWALD, K (1957–8) *Waldhygiene* 2, 234–51. GOSZCZYŃSKI, J (1976) *Pol. Ecol. Stud.* 2 (1), 95–102; (1981) *Ekol. Polska* 29, 431–9. GRAAF, C DE (1947) *Ardea* 35, 157–83; (1950) *Ardea* 38, 165–78. GRABER, R R (1962) *Condor* 64, 473–87. GRANDJEAN, P (1972) *Dansk orn. Foren. Tidsskr.* 66, 51–6. GRANT, J (1959) *Can. Fld.-Nat.* 73, 174–5. GRANT, P J and SCOTT, R E (1969) *Br. Birds* 62, 297–9. GRANT, P R (1971) *Behaviour* 40, 263–81. GRANT, P R and NETTLESHIP, D N (1971) *Ornis scand.* 2, 81–7. GRANVIK, H (1923) *J. Orn.* 71 suppl., 78–80. GRAY, R (1871) *Birds of the west of Scotland*. Glasgow. GREAVES, R H (1936) *Ool. Rec.* 16, 37–40; (1937) *Ool. Rec.* 17, 20–1. GREEN, A A and SAYER, J A (1979) *Malimbus* 1, 14–28. GREEN, C E (1928) *J. Northants. nat. Hist. Soc. Fld. Club* 25, 145–7. GREEN, D (1977) *Br. Birds* 70, 166. GREEN, J C (1969) *Loon* 41, 36–9. GREEN, J C and JANSSEN, R B (1975) *Minnesota birds*. Minneapolis. GREEN, J F (1905) *Zoologist* (4) 9, 253–7. GREENHALGH, M E (1974) *Naturalist* 931, 121–7. GREENHALGH, M E and GREENWOOD, M J (1975) *Naturalist* 935, 145–6. GREENWAY, J C (1958) *Extinct and vanishing birds of the world*. New York. GREENWELL, J (1935) *Vasculum* 21, 121–7; (1952) *Br. Birds* 45, 70. GREENWOOD, J (1964) *Ibis* 106, 469–81. GREENWOOD, J J D (1972) *Proc. int. orn. Congr.* 15, 648. GREIG-SMITH, P W (1978a) *Ostrich* 49, 67–75; (1978b) *Bull. Niger. orn. Soc.* 14, 14–23. GREIG-SMITH, P W and DAVIDSON, N C (1977) *Bull. Br. Orn. Club* 97, 96–9. GRENQUIST, P (1965) *Pap. Finn. Game Res.* 27, 1–114. GRIEVE, S (1885) *The Great Auk or Garefowl: its history, archaeology, and remains*. London. GRIFFIN, D R (1958) *Listening in the dark*. New Haven. GRIFFITHS, C I and ROGERS, T D (1975) *An interim list of the birds of Masirah Island, Oman*. R A F Masirah (cyclostyled). GRIMES, L G (1972) *Bull. Niger. orn. Soc.* 9 (35), 57; (1974) *Ibis* 116, 165–71; (1977) *Ibis* 119, 28–36; (1978) *Bull. Br. Orn. Club* 93, 114. GRINNELL, J, DIXON, J, and LINSDALE, J M (1930) *Univ. Calif. Publ. Zool.* 35. GRIST, W R (1934) *Br. Birds* 27, 304–5. GROCKI, D R J and JOHNSTON, D W (1974) *Auk* 91, 186–8. GROEBBELS, F (1957) *Vogelwelt* 78, 89–97. GRÖNLUND, S and MIKKOLA, H (1969) *Suom. Linnut* 4, 68–76. GROSCHUPP, R (1885) *Orn. Monatsschr.* 10, 198–200. GROSS, A O (1947) *Auk* 64, 584–601. GROSSKOPF, G (1957) *J. Orn.* 98, 65–70. GROTE, H (1933a) *Beitr. Fortpfl. Biol. Vögel* 9, 113–19; (1933b) *Beitr. Fortpfl. Biol. Vögel* 9, 221; (1934) *Beitr. Fortpfl. Biol. Vögel* 10, 66–7; (1936) *Beitr. Fortpfl. Biol. Vögel* 12, 133–9, 195–206; (1944) *Beitr. Fortpfl. Biol. Vögel* 20, 89–91. GRÜLL, F (1979) *Studie ČSAV* 3, 1–224. GRÜNEFELD, A (1952) *J. Orn.* 93, 154–7. GRUZDEV, L V and LIKHACHEV, G N (1960) *Zool. Zh.* 39, 624–7. GUDMUNDSSON, F (1951) *Proc. int. orn. Congr.* 10, 502–14; (1953a) *Náttúrufraedingurinn* 23, 43–6; (1953b) *Náttúrufraedingurinn* 23, 129–32; (1956) *Náttúrufraedingurinn* 26, 206–17. GUÉRIN, C (1924) *Rev. fr. Orn.* 8, 464–5; GUÉRIN, G (1923) *Rev. fr. Orn.* 8, 74–9; (1928) *La vie des chouettes* 1. Paris; (1932) *La vie des chouettes* 2. Fontenay-le-Comte. GUERMEUR, Y and MONNAT, J-Y (1980) *Histoire et géographie des oiseaux nicheurs de Bretagne*. Aurillac. GUGG, C (1934) *J. Orn.* 82, 269–93. GUGGISBERG, C A W (1937) *Orn. Beob.* 35, 7–10; (1941) *Orn. Beob.* 38, 121–2. GUICHARD, G (1948) *Alauda* 16, 200–4; (1954) *Oiseau* 24, 87–95; (1956a) *Alauda* 24, 139–44; (1956b) *Oiseau* 26, 126–34; (1957) *Oiseau* 27, 140–2; (1961) *Oiseau* 31, 1–8. GUICHARD, K M (1947) *Ibis* 89, 450–89; (1949–50) *Alauda* 17–18, 103–7; (1955) *Ibis* 97, 393–424. GUICHARD, K M and GOODWIN, D (1952) *Ibis* 94, 294–305. GURNEY, G H (1929) *Br. Birds* 23, 63–4. GURNEY, J H (1876) *Rambles of a naturalist in Egypt and other countries*. London; (1902) *Zoologist* (4) 6, 81–100; (1905) *Zoologist* (4) 9, 164–9; (1913) *The gannet. A bird with a history*. London. GURNEY, T H and TURNER, E L (1915) *Br. Birds* 8, 58–67. GÜRTLER, W (1973) *J. Orn.* 114, 305–16. GUSH, G H (1979) *Devon Birds* 32, 56–7. GUSTAFSON, T, LINDKVIST, B, GOTBORN, L, and GYLIN, R (1977) *Ornis scand.* 8, 87–95. GUSTAVSSON, L (1973) *Vår Fågelvärld* 31, 191–2. GÜTTINGER, H R (1965) *Orn. Beob.* 62, 14–23. GWYNN, A M (1968) *Austral. Bird Bander* 6, 71–5.

HAARHAUS, D (1983) *Beitr. Vogelkde.* 29, 89–102. HÄÄRI, H

(1931) *Orn. Beob.* **28**, 187–90. HAARTMAN, L VON (1940) *Ornis fenn.* **17**, 7–11; (1949) *Ornis fenn.* **26**, 16–24; (1973) In Farner, D S (ed) *Breeding biology of birds*, 448–81. Washington. HAARTMAN, L VON, HILDÉN, O, LINKOLA, P, SUOMALAINEN, P, and TENOVUO, R (1963–72) *Pohjolan linnut värikuvin* **8**. Helsinki. HAAS, W (1974) *Vogelwarte* **27**, 194–202. HAAS, W and BECK, P (1979) *J. Orn.* **120**, 237–46. HAASE, W (1969) *Beitr. Naturkde. Niedersachs.* **22**, 28–31. HAASE, W and SCHELPER, W (1972) *Vogelkdl. Ber. Niedersachs.* **4**, 65–8. HACHEZ, L (1966) *Aves* **3**, 61. HACHLER, E M (1958) *Sylvia* **15**, 239–46. HAENSEL, J (1966) *Beitr. Vogelkde.* **12**, 129–47. HAENSEL, J and WALTHER, H J (1966) *Beitr. Vogelkde.* **11**, 345–58. HAFFER, J (1977) *Bonn. zool. Monogr.* **10**, 1–64. HAFTORN, S (1958) *Sterna* **3**, 105–37; (1971) *Norges fugler.* Oslo. HAGEN, W (1917) *J. Orn.* **65**, 181–9. HAGEN, Y (1935) *Norsk orn. Tidsskr.* **15**, 71–111; (1942) *Arch. Naturgesch.* **11**, 1–132; (1950) *Stavanger Mus. Årbok*, 93–110; (1952) *Rovfuglene og Viltpleien.* Oslo; (1956) *Sterna* **1** (24), 1–22; (1959) In Blaedel, N (ed) *Nordens fugle i farver* **3**, 160–5. Copenhagen; (1960) *Medd. Stat. Vilt.* (2) **7**; (1964) *Rep. ICBP Conf. Birds of Prey, Caen*, 109–12; (1965) *Medd. Stat. Vilt.* (2) **23**; (1968) *Sterna* **8**, 161–82; (1969) *Fauna* **22**, 73–126. HAGEN, Y and BARTH, E K (1950) *Fauna* **3**, 1–12. HAGERUP, O (1926) *Vidensk. Medd. dansk natur. Foren.* **82**, 127–51. HAGN-MEINCKE, T (1967) *Flora og Fauna* **73**, 11–20. HAGUE, J B (1967) *Naturalist* **903**, 115–16. HAHN, V (1981) *J. Orn.* **122**, 429–34; (1982a) *J. Orn.* **123**, 55–62; (1982b) *Gefiederte Welt* **106**, 327–9. HAILA, Y and JÄRVINEN, O (1977) *Ornis fenn.* **54**, 73–8. HAINARD, R (1955) *Nos Oiseaux* **23**, 33–8; (1970) *Nos Oiseaux* **30**, 266. HAKALA, T and JOKINEN, M (1971) *Ornis fenn.* **48**, 135–7. HÄKKINEN, I, JOKINEN, M, and TAST, J (1973) *Ornis fenn.* **50**, 83–8. HÅLAND, A and TOFT, G O (1983) *Vår Fuglefauna* **6**, 3–14. HALD-MORTENSEN, P (1971) *Flora og Fauna* **77**, 1–12. HALE, J R (1928) *Br. Birds* **22**, 93; (1944) *Br. Birds* **38**, 17. HALE CARPENTER, G D (1938) *J. Soc. Br. Ent.* **1**, 213–14. HALLE, L J (1967) *Atlantic Nat.* **22**, 205–8; (1968) *Nos Oiseaux* **29**, 209–14. HALLER, H (1978) *Orn. Beob.* **75**, 237–65. HALLER, W (1934) *Orn. Beob.* **31**, 111–19; (1938) *Alauda* **10**, 324–6; (1939) *Beitr. Fortpfl. Biol. Vögel* **15**, 29–31. HALLET, C (1977) *Aves* **14**, 128–44; (1982) *Terre et Vie* **36**, 211–12. HALLET, C and DOUCET, J (1982) *Aves* **19**, 1–12. HALLET, J (1978) *Héron*, 84–8. HAMILTON, F M (1957) *Emu* **57**, 147–50. HAMILTON, K L and NEILL, R L (1981) *Am. Midl. Nat.* **106**, 1–9. HAMMLING, J (1917) *Orn. Monatsber.* **25**, 57–8. HAMMLING, J and SCHULZ, K (1911) *J. Orn.* **59**, 384–433. HAMMOND, C E (1958) *Sea Swallow* **11**, 6–13. HANN, C (1951) *Br. Birds* **44**, 134. HANSEN, L (1952) *Dansk orn. Foren. Tidsskr.* **46**, 158–72. HANTSCH, B (1905) *Beitrag zur Kenntnis der Vogelwelt Islands.* Berlin. HARCOURT, E V (1851) *A sketch of Madeira.* London. HARDING, B D (1979) *Br. Birds* **72**, 392. HARDY, A R (1977) Ph D Thesis. Aberdeen Univ. HARDY, A R, HIRONS, G J M, STANLEY, P I, and HUSON, L W (1981) *Ardea* **69**, 181–4. HARDY, J W (1957) *Publ. Mus. Michigan State Univ.* **1**; (1964) *Condor* **66**, 445–7. HARLOW, R A (1971) *Bird-Banding* **42**, 50. HARMAN, I (1974) *Foreign birds* **40** (1), 27–31. HARMATA, W (1969) *Przegl. Zool.* **13**, 98–101. HARMUTH, D (1971) *Falke* **18**, 274–7. HARRINGTON, B A (1974) *Bird-Banding* **45**, 115–44. HARRIS, M P (1970) *Ibis* **112**, 540–1; (1976a) *Ibis* **118**, 115–18; (1976b) *Br. Birds* **69**, 239–64; (1976c) *Trans. nat. Hist. Soc. Northumb.* **42**, 115–18; (1978) *J. anim. Ecol.* **47**, 15–23; (1979) *Bird Study* **26**, 179–86; (1980) *Ibis* **122**, 193–209; (1981) *Br. Birds* **74**, 246–56; (1982a) *Ibis* **124**, 100–3; (1982b) *Scott. Birds* **12**, 11–17; (1983a) *Ibis* **125**, 56–73; (1983b) *Ibis* **125**, 109–14. HARRIS, M P and HISLOP, J R G (1978) *J. Zool. Lond.* **185**, 213–36. HARRIS, M P and MURRAY, S (1977) *Br. Birds* **70**, 50–65; (1981) *Bird Study* **28**, 15–20. HARRIS, M P and OSBORN, D (1981) *J. appl. Ecol.* **18**, 471–9. HARRIS, M P and YULE, R F (1977) *Ibis* **119**, 535–41.

HARRISON, C J O (1969) *Bird Study* **7**, 236–40; (1965) *Behaviour* **24**, 161–208; (1969) *Avic. Mag.* **75**, 97–9; (1971) *Bull. Br. Orn. Club* **91**, 126–31; (1975) *A field guide to the nests, eggs, and nestlings of British and European birds.* London; (1982) *An atlas of the birds of the western Palearctic.* London. HARRISON, J (1947a) *Br. Birds* **40**, 86–7. HARRISON, J G (1955) *Bull. Br. Orn. Club* **75**, 69–70. HARRISON, J M (1944) *Bull. Br. Orn. Club* **64**, 61–3. HARRISON, J M and HOVEL, H (1964) *Bull. Br. Orn. Club* **84**, 91–4. HARRISON, R (1947b) *Br. Birds* **40**, 181; (1974) *Br. Birds* **67**, 514–15. HARTERT, E (1901) *Novit. Zool.* **8**, 328; (1912–21) *Die Vögel der paläarktischen Fauna* **2**; (1921–2) **3**. Berlin; (1924) *Novit. Zool.* **31**, 1–48; (1927) *Mém. Soc. Sci. Nat. Maroc* **16**. HARTERT, E and STEINBACHER, F (1932–8) *Die Vögel der paläarktischen Fauna, Ergänzungsband.* Berlin. HARTHAN, A J (1942) *Br. Birds* **36**, 141. HARTLEY, C H and FISHER, J (1936) *J. anim. Ecol.* **5**, 370–89. HARTLEY, P H T (1945) *Br. Birds* **38**, 334; (1947) *Ibis* **89**, 566–9; (1948) *Ibis* **90**, 361–81; (1966) *XIV Congr. int. orn. Abstr.*, 68. HARTMANN, R (1863) *J. Orn.* **11**, 299–320. HARVEY, W G (1972) *Bull. E. Afr. nat. Hist. Soc.*, 137; (1974) *Bull. E. Afr. nat. Hist. Soc.*, 48–51, 66–9, 80–2. HARWOOD, N (1950) *Br. Birds* **43**, 123. HASSE, H (1961) *J. Orn.* **102**, 368. HAVERSCHMIDT, F (1932) *Beitr. Fortpfl. Biol. Vögel* **8**, 10–11; (1938) *Beitr. Fortpfl. Biol. Vögel* **14**, 9–13; (1945) *Ardea* **33**, 237–40; (1946) *Ardea* **34**, 214–46; (1975) *Orn. Beob.* **72**, 280; (1978a) *Die Trauerseeschwalbe.* Wittenberg Lutherstadt; (1978b) *Br. Birds* **71**, 359–60. HAVLÍN, J (1979) *Folia Zool.* **28**, 125–46. HAVLÍN, J and HAVLÍNOVA, S (1966) *Zool. Listy* **15**, 84–5. HAVRE, R VAN (1951) *Gerfaut* **41**, 288–91. HAWKSLEY, O (1950) Ph D Thesis. Cornell Univ.; (1957) *Bird-Banding* **28**, 57–92. HAWLEY, R G (1966) *Sorby Rec.* **2**, 95–114. HAYMES, G T and BLOKPOEL, H (1978) *Bird-Banding* **49**, 142–51. HAYS, H (1970) *Wilson Bull.* **82**, 99–100; (1975) *Auk* **92**, 219–34; (1978) *Ibis* **120**, 127–8. HAYS, H, DUNN, E, and POOLE, A (1973) *Wilson Bull.* **85**, 233–6. HAYS, H and RISEBROUGH, R W (1972) *Auk* **89**, 19–35. HEADLEY, F W (1919) *Br. Birds* **13**, 57. HEDEGAARD-CHRISTENSEN, J (1970) *Dansk orn. Foren. Tidsskr.* **54**, 270–1. HEDGREN, S (1975) *Vår Fågelvärld* **34**, 43–52; (1976) *Vår Fågelvärld* **35**, 287–90; (1979) *Ibis* **121**, 356–61; (1980a) *Ornis fenn.* **57**, 49–57; (1980b) *Ornis scand.* **12**, 51–4. HEDGREN, S and LINNMAN, A (1979) *Ornis scand.* **10**, 29–36. HEDVALL, O and WAHLSTEDT, J (1969) *Ladrikets fåglar.* Stockholm. HEER, E (1975) *Anz. orn. Ges. Bayern* **14**, 174–80; (1982) *Orn. Mitt.* **34**, 176. HEIDER, E (1953) *Vogelring* **22**, 49–50. HEIM DE BALSAC, H (1924) *Rev. fr. Orn.* **8**, 1–116; (1925) *Rev. fr. Orn.* **9**, 239–48; (1926) *Mém. Soc. Hist. nat. Afr. nord* **1**; (1936) *Alauda* **8**, 263–4; (1954) *Alauda* **22**, 145–205; (1965) *Alauda* **33**, 309–22. HEIM DE BALSAC, H and BEAUFORT, F DE (1966) *Alauda* **34**, 309–24. HEIM DE BALSAC, H and MAYAUD, N (1962) *Les oiseaux du nord-ouest de l'Afrique.* Paris. HEINEN, M and MARGREWITZ, D (1981) *Wir und die Vögel* **13** (4), 14–15. HEINONEN, E and KELLOMÄKI, E (1971) *Suom. Linnut* **6**, 4–10. HEINROTH, O (1909) *J. Orn.* **57**, 56–83; (1911) *J. Orn.* **59**, 168–70; (1939) *Avic. Mag.* **4**, 33–6; (1944) *Orn. Monatsber.* **52**, 45–6. HEINROTH, O and HEINROTH, K (1949) *Z. Tierpsychol.* **6**, 153–201. HEINROTH, O and HEINROTH, M (1924–6) *Die Vögel Mitteleuropas* **1**; (1926–7) **2**; (1927–8) **3**; (1931–3) **4**. Berlin-Lichterfelde. HEINZEL, H, FITTER, R S R, and PARSLOW, J L F (1972) *The birds of Britain and Europe with North Africa and the Middle East.* London. HEJL-MRAČOVSKÝ, F (1958) *Sylvia* **15**, 253. HEKSTRA, G P (1973) In Burton, J A (ed) *Owls of the world*, 94–115. New York. HELBIG, A (1982) *Vogelwelt* **103**, 161–75. HELDMANN, G (1951) *Vogelwelt* **72**, 165–6, HELEŠIC, J (1981) *Acta Sc. Nat. Brno* **15**, 1–39. HEMMINGSEN, A M (1959) *Vidensk. Medd. dansk natur. Foren.* **121**, 1–51. HEMMINGSEN, A M and GUILDAL, J A (1968) *Spolia zool. Mus. Haun.* **28**, 1–326. HENDRICKSON, G O and SWAN, C (1938) *Ecology* **19**, 584–8. HENDY,

E W (1949) *Country Life*, 825. HENNACHE, A (1981) *Oiseau* 51, 127–38. HENRY, C (1982a) *Rev. Ecol.* 36, 421–33. HENRY, G M (1955) *A guide to the birds of Ceylon*. London. HENRY, M (1982b) *Br. Birds* 75, 181. HENS, P (1949) *Limosa* 22, 329. HENSS, M (1963) *Beitr. Naturkde. Niedersachs.* 16, 28–9. HENTY, C J (1961) *Ibis* 103a, 28–36. HENZE, O (1937) *Naturschutz* 18, 40–2; (1943) *Vogelschutz gegen Insektenschaden in der Forstwirtschaft.* Munich; (1962) *Angew. Orn.* 1, 64–72. HERBERG, M (1952) *Beitr. Vogelkde.* 2, 87–93; (1960) *Beitr. Vogelkde.* 6, 430–2. HERBERIGS, H (1954) *Gerfaut* 44, 290–9; (1958) *Gerfaut* 48, 1–4. HERKLOTS, G A C (1967) *Hong Kong birds.* Hong Kong. HERMANSON, W and OTTERHAG, L (1963) *Vår Fågelvärld* 22, 123–30. HERRERA, C M (1974a) *Ardeola* 19, 359–94; (1974b) *Ornis scand.* 5, 181–91. HERRERA, C M and HIRALDO, F (1976) *Ornis scand.* 7, 29–41. HERRERA, C M and RAMÍREZ, A (1974) *Br. Birds* 67, 158–64. HERRLINGER, E (1971) *Rhein. Heimatpfl.* NF 3, 192–200; (1973) *Bonn. zool. Monogr.* 4, 1–151. HERROELEN, P (1953) *Gerfaut* 43, 161–4; (1962) *Gerfaut* 52, 408–15; (1964) *Biol. Jaarb. Dodonaea* 32, 170–6. HERSCHEL, K (1968) *Beitr. Vogelkde.* 13, 462; (1971) *Beitr. Vogelkde.* 17, 165–6. HESS, A (1924) *Orn. Beob.* 21, 175; (1927) *Orn. Beob.* 24, 247. HESS, R (1983) *Orn. Beob.* 80, 153–82. HESSE, E (1909) *J. Orn.* 57, 1–32; (1910) *J. Orn.* 58, 489–519; (1911) *J. Orn.* 59, 361–83; (1912a) *J. Orn.* 60, 298–34; (1912b) *J. Orn.* 60, 481–94; (1916) *Orn. Monatsber.* 24, 89; (1918) *Orn. Monatsber.* 26, 113–15; (1921) *Orn. Monatsber.* 29, 47–9. HEUBECK, M and OKILL, J D (1981) *Br. Birds* 74, 149. HEUBECK, M and RICHARDSON, M (1980) *Scott. Birds* 11, 97–108. HEUER, B and KRÄGENOW, P (1973) *Falke* 20, 103. HEUGLIN, M T VON (1869) *Ornithologie Nordost-Afrika's* 1; (1873) 2. Kassel. HETHKE, H (1968) *Ool. Rec.* 42, 23–32. HEWETT, W (1881) *Zoologist* (3) 5, 65. HEYDER, R (1927) *Orn. Monatsber.* 35, 178–9. HEYDT, J G (1968) *Ardeola* 14, 97–130. HEYMER, A (1977) *Ethologisches Wörterbuch.* Berlin. HEYN, D (1963) *Falke* 10, 153–8; (1964) *Orn. Mitt.* 16, 121–2; (1968) *Falke* 12, 186–7. HIBBERT-WARE, A (1936) *Br. Birds* 29, 302–5; (1937–8) *Br. Birds* 31, 162–87, 205–29, 249–64. HICKLING, R A O and FERGUSON-LEES, I J (1959) *Br. Birds* 52, 126–9. HIDALGO, J (1974) *Ardeola* 20, 363. HILDÉN, O (1955) *Luonnon Tutkija* 95, 26; (1966) *Ann. zool. fenn.* 3, 245–69; (1969) *Ornis fenn.* 46, 179–87 (1971) *Ornis fenn.* 48, 125–9; (1974) *Ornis fenn.* 51, 10–35; (1977) *Ornis fenn.* 54, 170–9; (1978) *Ornis fenn.* 55, 42–3. HILDÉN, O and HELO, P (1981) *Ornis fenn.* 58, 159–66. HILLARP, J-Å (1971) *Medd. Skånes orn. Fören.* 10, 27–31. HINDE, R A and FISHER, J (1951) *Br. Birds* 44, 393–6. HINSCHE, A (1968) *Apus* 1, 198–9. HIRALDO, F and ALVAREZ, F (1973) *Ardeola* 19, 26–7. HIRALDO, F, ANDRADA, J, and PARREÑO, F F (1975) *Doñana Acta Vert.* 2 (2), 161–77. HIRONS, G (1976) D Phil Thesis. Oxford Univ.; (1977) *Game Conserv. ann. Rev.* 8, 58–61. HIRONS, G, HARDY, A, and STANLEY, P (1979) *Bird Study* 26, 59–63. HIRSCHFELD, H and HIRSCHFELD, K (1962) *Falke* 9, 370–6; (1973) *Beitr. Vogelkde.* 19, 81–152. HITCHCOCK, W (1976) In Anon (1976), 221. HITCHCOCK, W B (1965) *Emu* 64, 157–71. HLADÍK, B (1958) *Zool. Listy* 7, 261–71. HLADÍK, B and KADLEC, O (1964) *Zool. Listy* 13, 1–8. HOBBS, J N (1976) *Emu* 76, 219–20. HOCKEY, P A R and HOCKEY, C T (1980) *Cormorant* 8, 7–10. HODGETTS, J W (1943) *Br. Birds* 37, 97. HODGKINSON, J B (1844) *Zoologist* (1) 2, 686. HODSON, N L (1961) *Br. Birds* 54, 430. HOEKSTRA, B (1974) *Levende Nat.* 77, 53–62. HOESCH, W (1934) *Orn. Monatsber.* 42, 68–70. HOESCH, W and NIETHAMMER, G (1940) *Die Vogelwelt Deutsch-Südwestafrikas.* Berlin. HOFFMAN, P W (1927) *Wilson Bull.* 39, 78–80. HOFFMANN, B (1917) *J. Orn.* 65, 459–64; (1927) *Verh. orn. Ges. Bayern* 17, 176–9. HOFFMANN, K (1969) *J. Orn.* 110, 448–64. HOFFMANN, L (1958) *Br. Birds* 51, 321–50. HOFFRICHTER, O and WESTERMANN, K (1969) *Mitt. Bad. Landesver. Naturkde.* 10, 205–7. HOFSTETTER, F-B (1952) *J. Orn.* 93, 295–312; (1954) *J. Orn.* 95, 348–410; (1963) *J. Orn.* 104, 351–6. HOGG, P (1974) *Ibis* 116, 466–76. HOGG, R H (1975) *Br. Birds* 68, 431. HÖGLUND, N H (1966) *Viltrevy* 4, 43–80. HÖGLUND, N H and LANSGREN, E (1968) *Viltrevy* 5, 363–421. HOGSTAD, O (1969) *Sterna* 8, 387–9; (1970) *Nytt. Mag. Zool.* 18, 221–7; (1971a) *Ornis scand.* 2, 143–6; (1971b) *Sterna* 10, 233–41; (1976a) *Ibis* 118, 41–50; (1976b) *Sterna* 15, 5–10; (1977) *Ornis scand.* 8, 101–11; (1978) *Ibis* 120, 198–203; (1983) *Fauna Norv. Ser. C Cinclus* 6, 81–6. HÖHN, E O (1947) *Br. Birds* 40, 187; (1973) *Can. Fld.-Nat.* 87, 468–9. HOLBOURN, J G and GEAR, M (1970) *Scott. Birds* 6, 204–9. HOLGERSEN, H (1951) *Sterna* 1, 1–10; (1952) *Stavanger Mus. Årbok 1950*, 1–12; (1961) *Sterna* 4, 229–40; (1965) *Sterna* 6, 353–94. HOLLAND, T R (1974) *Br. Birds* 67, 212–13. HOLLANDER, W F (1945) *J. comp. Psychol.* 38, 287–9. HOLLICK, K (1973) *Br. Birds* 66, 280–1; (1980) *Br. Birds* 73, 417. HOLLOM, P A D (1940) *Br. Birds* 33, 202–21, 230–44; (1959) *Ibis* 101, 183–200; (1960) *The popular handbook of rarer British birds.* London. HOLMBERG, T (1974a) *Vår Fågelvärld* 33, 140–6; (1974b) *Vår Fågelvärld* 33, 299–300; (1976) *Fauna och Flora* 71, 97–107. HOLMBRING, J-Å (1972) *Vår Fågelvärld* 31, 252–6. HOLMGREN, J (1978) *Anser* 17, 225. HOLMGREN, V (1982) *Anser* 22, 27–42. HOLTZ, L (1863) *J. Orn.* 11, 394–9; (1864) *J. Orn.* 12, 52–60. HOLYOAK, D T (1973) *Bull. Br. Orn. Club* 93, 26–32. HÖLZINGER, J (1974) *Auspicium* 5, 347–50. HÖLZINGER, J, MICKLEY, M, and SCHILHANSL, K (1973) *Anz. orn. Ges. Bayern* 12, 176–97. HÖLZINGER, J and SCHILHANSL, K (1968) *Anz. orn. Ges. Bayern* 8, 277–85. HOLZWARTH, G (1971) *Anz. orn. Ges. Bayern* 10, 180–2. HOMBERGER, D G (1980) *Bonn. zool. Monogr.* 13, 7–192. HOMEI, V and POPESCU, A (1969) *Ocrotirea Nat.* 13, 63–7. HOMEYER, A VON (1863) *J. Orn.* 11, 165–266; (1864) *J. Orn.* 12, 312–14. HOMOKI NAGY, I, JR (1977) *Aquila* 84, 113. HOMONNAY, F VON (1938) *Kócsag* 9–11, 72–9. HONER, M R (1963) *Ardea* 51, 158–95. HOOGERWERF, A and RENGERS HORA SICCAMA, G F H W (1937) *Ardea* 26, 1–51. HOOGSTEYNS, J (1959) *Gerfaut* 49, 269–70. HOPKINS, C D and WILEY, R H (1972) *Auk* 89, 583–94. HOPKINSON, E (1910) *Bird Notes* NS 1, 286–90. HÖRNFELDT, B (1978) *Oecologia* 32, 141–52. HOROBIN, J M (1971) PhD Thesis. Durham Univ. HÖRRING, R (1933) *Dansk orn. Foren. Tidsskr.* 27, 103–5. HORST, F (1938) *Beitr. Fortpfl. Biol. Vögel* 14, 151–2. HORSTKOTTE, E (1973) *Orn. Mitt.* 25, 159–69. HORTLING, I (1929) *Ornitologisk handbok.* Helsingfors. HORVÁTH, L (1959) *Ann. Hist. nat. Mus. nat. Hungarici* 51, 451–81. HOSKING, E J (1941) *Br. Birds* 35, 2–8; (1942) *Br. Birds* 36, 2–4; (1950) *Country Life* 107, 1288–9. HOSKING, E and NEWBERRY, C (1945) *Birds of the night.* London. HOSKING, E J and SMITH, S (1943) *Br. Birds* 37, 55–6. HOVEL, H (1970) *Aquila* 77, 195. HOWARD, R J (1889) *Zoologist* (3) 13, 51–5. HOWELL, T R (1953) *Auk* 70, 118–26. HOWELL, T R and BARTHOLOMEW, G A (1962) *Ibis* 104, 99. HOWES, C A (1978) *Naturalist* 103, 28–9. HOWEY, D H (1965) *Br. Birds* 58, 299–300. HRABÁR, A (1926) *Aquila* 32–3, 166–71. HUBER, J (1953) *Orn. Beob.* 50, 29; (1954) *Larus* 6–7, 183–9. HUBER, J and WÜST, R (1956) *Orn. Beob.* 53, 205. HUBL, H (1952) *Z. Tierpsychol.* 9, 102–19. HUBLIN, P and MIKKOLA, M (1977) *Savon Luonto* 9, 6–8. HÜBNER, G (1965) *Falke* 12, 184–7. HÜCKLER, U (1968) *Auspicium* 2, 338–43; (1970) *Auspicium* 4, 111–137. HUDEC, K (1976) *Acta. Sc. Nat. Brno* 10–11, 1–54; (1977) *Folia Zool.* 26, 355–62. HUDEC, K and ČERNÝ (1977) *Fauna CSSR Ptáci* 2. Prague. HUDSON, P J (1979a) D Phil Thesis. Oxford Univ.; (1979b) *J. anim. Ecol.* 48, 889–98. HUDSON, R (1965) *Br. Birds* 58, 105–9; (1972) *Br. Birds* 65, 139–55; (1973) *BTO Guide* 15. Tring. HÜE, F (1945) *Oiseau* 15, 89–93; (1947a) *Alauda* 15, 177–202; (1947b) *Alauda* 15, 253–6; (1949–50) *Alauda* 17–18, 116; (1952) *Oiseau* 22, 303–16; (1953) *Oiseau* 23, 297–9; (1964) *Oiseau* 34, 272–3; (1970) *Oiseau* 40, 87. HÜE, F and ETCHÉCOPAR, R-D (1957) *Oiseau* 27, 35–58; (1958) *Terre et Vie* 105, 186–219;

(1970) *Les oiseaux du proche et du moyen Orient*. Paris. HÜE, F and RIVOIRE, A (1947) *Oiseau* **17**, 153–66. HUGUES, A (1907) *Bull. Soc. zool. Paris* **32**, 106–8. HULSMAN, K (1974) *Sunbird* **5**, 44–9; (1975) *Sunbird* **6**, 41–3; (1976) *Emu* **76**, 143–9; (1977a) Ph D Thesis. Queensland Univ.; (1977b) *Sunbird* **8**, 9–19; (1977c) *Emu* **77**, 49–60; (1979) *Corella* **3**, 37–40. HUME, A O and MARSHALL, C H T (1878) *The game birds of India, Burmah and Ceylon* **1**. London. HUME, R A (1975) *Bird Study* **22**, 260. HUNT, D B (1979) *Br. Birds* **72**, 410–14. HUNTER, M L (1980) *Condor* **82**, 101–3. HUNTER, N C (1973) *Safring News* **2** (3), 21–3. HUNZIKER-LÜTHY, G (1971) *Orn. Beob.* **68**, 281. HURME, T (1973) *Lintumies* **4**, 1–8. HURME, T and SARKANEN, S (1975) *Lintumies* **10**, 95–9. HURRELL, H G (1980) *Br. Birds* **73**, 413. HUSAIN, M A and BHALLA, H R (1937) *Indian J. Agric. Sci.* **7**, 785–92. HUSTLER, K (1978) *Witwatersrand Bird Club News* **102**, 4–5. HUTCHISON, R E, STEVENSON, J G, and THORPE, W H (1968) *Behaviour* **32**, 150–7. HUTSON, H P W (1946) *Ibis* **88**, 128–9; (1947) *Ibis* **89**, 291–300; (1954) *The birds about Delhi*. Delhi. HUTTON, T (1873) *Stray Feathers* **1**, 331–45. HUXLEY, J S and BROWN, P E (1953) *Br. Birds* **46**, 399–404. HYDE, L B (1937) *Ann. Rep. Bowdoin sci. Stn.* **2**, 30–3.

IJZENDOORN, A L J VAN (1947) *Limosa* **20**, 143–59; (1950) *The breeding birds of the Netherlands*. Leiden. ILLE, R (1983) *J. Orn.* **124**, 133–46. ILYICHEV, V D (1959) *Ornitologiya* **2**, 157–8; (1977) *Zool. Zh.* **56**, 1133–44. ILYICHEV, V D and FOMIN, V E (1979) *Ornitologiya* **14**, 83–96. IMAIZUMI, Y (1968) *Zool. Mag.* **77**, 402–4. INBAR, R (1977) *Guide to the birds of Israel*. Tel Aviv. INGLIS, C M (1937) *J. Darjeeling nat. Hist. Soc.* **12**, 41–50. INGOLD, P (1973) *Behaviour* **45**, 154–90; (1974) *Sterna* **13**, 205–10; (1980) *Z. Tierpsychol.* **53**, 341–88. INGOLD, P and TSCHANZ, B (1970) *Sterna* **9**, 201–6. INGOLD, P and VOGEL, P (1965) *Sterna* **6**, 223–8. INGRAM, C (1959) *Auk* **76**, 218–26; (1962) *Auk* **79**, 715; (1978) *Br. Birds* **71**, 138. INGRAM, G C S (1933) *Br. Birds* **26**, 306–8. INGRAM, G C S and SALMON, H M (1924) *Br. Birds* **18**, 2–10. INOZEMTSEV, A A (1965) *Ornitologiya* **7**, 416–36. INSLEY, H, YOUNG, L, and DUDLEY, B (1980) *Bird Study* **27**, 101–7. IRBY, L H L (1875) *The ornithology of the Straits of Gibraltar*. London. IRIBARREN, I B and NEVADO, L D (1982) *Alauda* **50**, 81–91. IRISOV, E A (1967) *Ornitologiya* **8**, 355–6. IRVING, L (1960) *Bull. US natn. Mus.* **217**. IRVING, N S and BEESLEY, J S S (1976) *Bird pest research project, Botswana. Final rep. 1972–1975*. Centre for Overseas Pest Res., London. ISENMANN, P (1972a) *Nos Oiseaux* **31**, 150–62; (1972b) *Nos Oiseaux* **31**, 297–9; (1972c) *Ardea* **60**, 226–8; (1972d) *Ardeola* **16**, 242–5; (1973) *Alauda* **41**, 365–70; (1975a) *Vogelwarte* **28**, 159–60; (1975b) *Bull. Mus. Hist. nat. Marseille* **35**, 149–51; (1976a) *Alauda* **44**, 92; (1976b) *Alauda* **44**, 319–27; (1976c) *Oiseau* **46**, 135–42; (1980) *Oiseau* **50**, 161–3. ISENMANN, P and CZAJKOWSKI, M A (1978) *Riv. ital. Orn.* **48**, 143–8. IVANAUSKAS, T (1964a) *Lietuvos paukščiai* **2**. Vilnius; (1964b) *J. Orn.* **103**, 488–9. IVANOV, A I (1969) *Ptitsy Pamiro-Alaya*. Leningrad; (1976) *Katalog ptits SSSR*. Leningrad. IVANOV, A I, KOZLOVA, E V, PORTENKO, L A, and TUGARINOV, A Y (1953) *Ptitsy SSSR* **2**. Moscow.

JABLONSKI, B (1976) *Acta Orn.* **16** (1), 61–2. JÄCKEL, A J (1891) *Systematische Übersicht der Vögel Bayerns*. Munich. JACKSON, F J (1926) *Game birds of Kenya and Uganda*. London; (1938a) *Birds of Kenya colony and Uganda Protectorate* **1**; (1938b) **2**. London. JACKSON, H D (1969) *Arnoldia* **4** (24); (1978) *Arnoldia* **8** (28). JACOB, J (1978) In Brush, A H (ed) *Chemical zoology* **10**, 165–211. New York. JACOB, J-P (1979) *Gerfaut* **69**, 425–36; (1983) *Alauda* **51**, 48–63. JACOB, J-P and JACOB, A (1980) *Alauda* **48**, 209–18. JACOBS, J (1943) *Gerfaut* **33**, 37–55. JACOBY, H, KNÖTZSCH, G, and SCHUSTER, S (1970) *Orn. Beob.* **67** suppl., 1–260. JACQUAT,

B (1975) *Orn. Beob.* **72**, 235–79. JAEGER, E C (1950) *Condor* **52**, 90–1. JAHN, H (1942) *J. Orn.* **90**, 1–302. JAHNKE, V (1977) *Corella* **1**, 48–50. JAMES, T O (1945) *Br. Birds* **38**, 274–5. JANKEVIČIUS, K, ZAJANČKAUSKAS, P, BALEVIČIUS, K, KAZLAUSKAS, R, LEKAVIČIUS, A, LOGMINAS, V, MAČIONIS, A, SUKACKAS, V, and TURSA, G (1981) *Red data book of the Lithuanian SSR*. Vilnius. JANOSSY, D and SCHMIDT, E (1970) *Bonn. zool. Beitr.* **21**, 25–51. JANSSEN, P L (1954) *Wielewaal* **20**, 333–43. JANSSON, E (1964) *Vår Fågelvärld* **23**, 209–22. JANY, E (1951) *Vogelwelt* **72**, 157–60. JAPANESE ORNITHOLOGICAL SOCIETY (1958) *A hand-list of the Japanese birds*. Tokyo. JARRY, G (1969) *Oiseau* **39**, 112–20. JÄRVINEN, O and VÄISÄNEN, R A (1978) *J. Orn.* **119**, 441–9. JEAL, P E C (1976) *Bird Study* **23**, 56–7. JENNER, E (1788) *Phil. Trans. Roy. Soc. Lond.* **78**, 219–35. JENNI, L (1977) *Orn. Beob.* **74**, 62–70; (1980) *Orn. Beob.* **77**, 27; (1981) *J. Orn.* **122**, 37–63; (1983) *Orn. Beob.* **80**, 29–57. JENNINGS, M C (1977) *Israel Land Nat.* **2**, 168–9; (1980) *Sandgrouse* **1**, 71–81; (1981a) *The birds of Saudi Arabia: a check-list*. Cambridge; (1981b) *J. Saudi Arab. nat. Hist. Soc.* **2** (1), 8–14; (1981c) *Birds of the Arabian Gulf*. London. JENSEN, A (1968) *Flora og Fauna* **74**, 69–76. JENSEN, A and HAARLØV, N (1963) *Flora og Fauna* **69**, 17–27. JENSEN, J V and KIRKEBY, J (1980) *The birds of the Gambia*. Århus. JENSEN, P V (1946) *Dansk orn. Foren. Tidsskr.* **40**, 80–96. JENSEN, R A C and BERRY, H H (1972) *Madoqua* **1**, 53–6. JENSEN, R A C and JENSEN, M K (1969) *Ostrich* **40**, 163–81. JENSEN, W F, ROBINSON, W L, and HEITMAN, N L (1982) *Jack-Pine Warbler* **60**, 27–8. JERDON, T C (1877) *The birds of India* **1**. Calcutta. JESPERSEN, P (1946) *The breeding birds of Denmark*. Copenhagen. JILKA, A and URSPRUNG, J (1980) *Egretta* **23**, 8–19. JIRÁČKOVÁ, A (1963) *Živa* **11**, 210. JOHANSEN, H (1955) *J. Orn.* **96**, 382–410; (1959) *J. Orn.* **100**, 417–32. JOHNSGARD, P A (1979) *Birds of the Great Plains*. Lincoln, Nebraska. JOHNSON, L R (1958) *Iraq nat. Hist. Mus. Publ.* **16**, 1–32. JOHNSON, R A (1938) *Wilson Bull.* **50**, 161–70; (1941) *Auk* **58**, 153–63; (1944) *Wilson Bull.* **56**, 161–8. JOHNSON, S R and WEST, G C (1975) *Ornis scand.* **6**, 109–15. JOHNSTONE, R E (1978a) *Corella* **2**, 43–5; (1978b) *Corella* **2**, 46–7. JONES, B E (1974) *Ann. appl. Biol.* **76**, 325–66. JONES, L (1903) *Wilson Bull.* **44**, 94–100. JONES, P J (1980) *Ostrich* **51**, 99–106. JONGHE, A DE (1979) *Wielewaal* **45**, 26–7. JÖNSSON, I and SCHAAR, C (1970) *Vår Fågelvärld* **29**, 303–4. JORDANS, A VON (1950) *Syllegomena Biologica*, 165–81. Leipzig. JORDANS, A VON and STEINBACHER, J (1948) *Senckenbergiana* **28**, 159–86. JÓSEFIK, M (1969) *Acta Orn.* **11**, 381–433. JOSEPHSON, B (1980) *J. Fld. Orn.* **51**, 149–60. JOUBERT, H J (1943) *Ostrich* **14**, 42–4. JOURDAIN, F C R (1901) *Zoologist* (4) **5**, 286–9; (1911) In Kirkman, F B *The British bird book* **2**, 436–45; (1925) *Proc. zool. Soc. Lond.*, 639–67; (1935) *Ool. Rec.* **15**, 61–3; (1936a) *Proc. zool. Soc. Lond.*, 251–6; (1936b) *Ool. Rec.* **16**, 62–70. JOURDAIN, F C R and HARRISON, B G (1936) *Ool. Rec.* **16**, 59–62. JOURDAIN, F C R and SHUEL, R (1935) *Ibis* (13) **5**, 623–63. JOY, N H (1943) *Br. Birds* **36**, 176–8. JUCKWER, E-A (1970) *Bonn. zool. Beitr.* **21**, 237–50. JUILLARD, M (1979) *Nos Oiseaux* **35**, 113–24; (1980) *Nos Oiseaux* **35**, 309–37. JUNCO, O DEL (1969) *Ardeola* **14**, 221–3. JUNCO, O DEL and GONZALEZ, B (1966) *Ardeola* **12**, 5–9; (1969) *Ardeola* **13**, 115–27. JUNGE, G C A (1938) *Limosa* **11**, 10–34; (1948) *Zool. Meded.* **29**, 311–26. JUNOR, J F R (1972) *Ostrich* **43**, 185. JÜRGENS, U J (1978) *Corax* **6** (4), 41–2. JUVONEN, A (1976) *Siipirikko* **3**, 13–18. JYRKKANEN, J A (1975) *Can. Fld.-Nat.* **89**, 77–8.

KACZMARECK, L (1976) *Beitr. Naturkde. Niedersachs.* **29**, 70–4. KADOCHNIKOV, N P (1962) *Zool. Zh.* **41**, 465–7; (1963) *Ornitologiya* **6**, 104–10. KAFTANOVSKI, Y M (1938) *Zool. Zh.* **17**, 695–705; (1951) *Mat. pozn. fauny flory SSSR*, N S Otd. zool **28** (13), 1–170. KAHMANN, H and BROTZLER, A (1956) *Biol. Zbl.* **75**, 67–

EGG PLATES

PLATE 87 EGGS

Gelochelidon nilotica Gull-billed Tern, 4 eggs

Sterna sandvicensis Sandwich Tern, 8 eggs

Sterna bengalensis Lesser Crested Tern, 4 eggs

PLATE 88 EGGS

Sterna caspia Caspian Tern, 3 eggs

Sterna bergii Swift Tern 4 eggs

Sterna maxima Royal Tern, 3 eggs

PLATE 89 EGGS

Sterna anaethetus Bridled Tern, 5 eggs

Sterna repressa White-cheeked Tern, 5 eggs

Sterna hirundo Common Tern, 5 eggs

Sterna paradisaea Arctic Tern, 5 eggs

Sterna dougallii Roseate Tern, 5 eggs

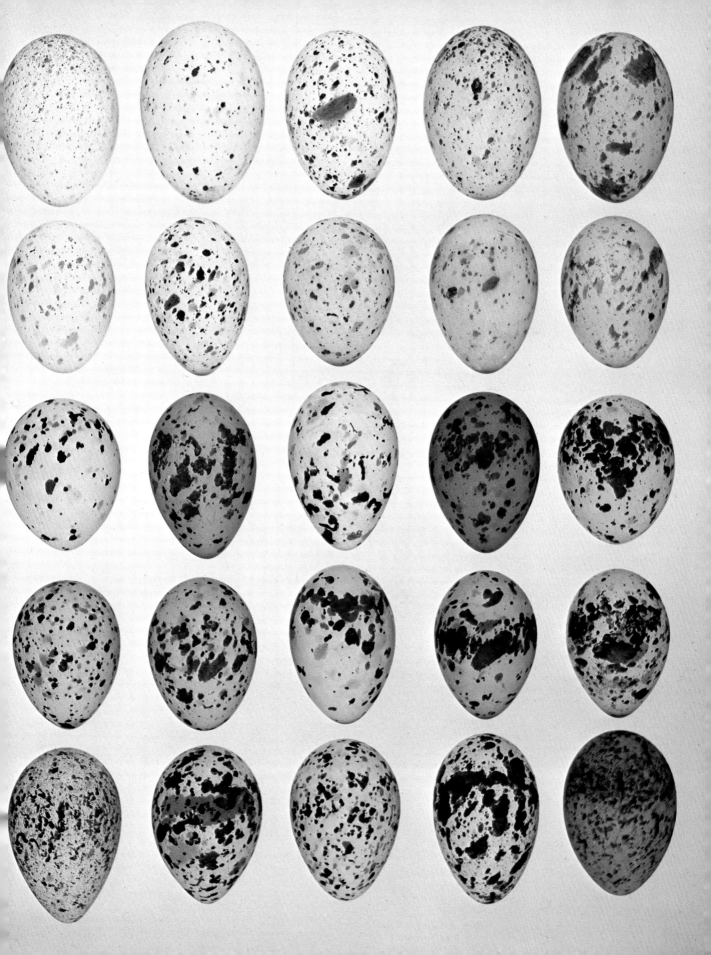

PLATE 90 EGGS

Chlidonias niger Black Tern, 5 eggs

Chlidonias leucopterus White-winged Black Tern, 10 eggs

Chlidonias hybridus Whiskered Tern, 5 eggs

Sterna albifrons Little Tern, 5 eggs

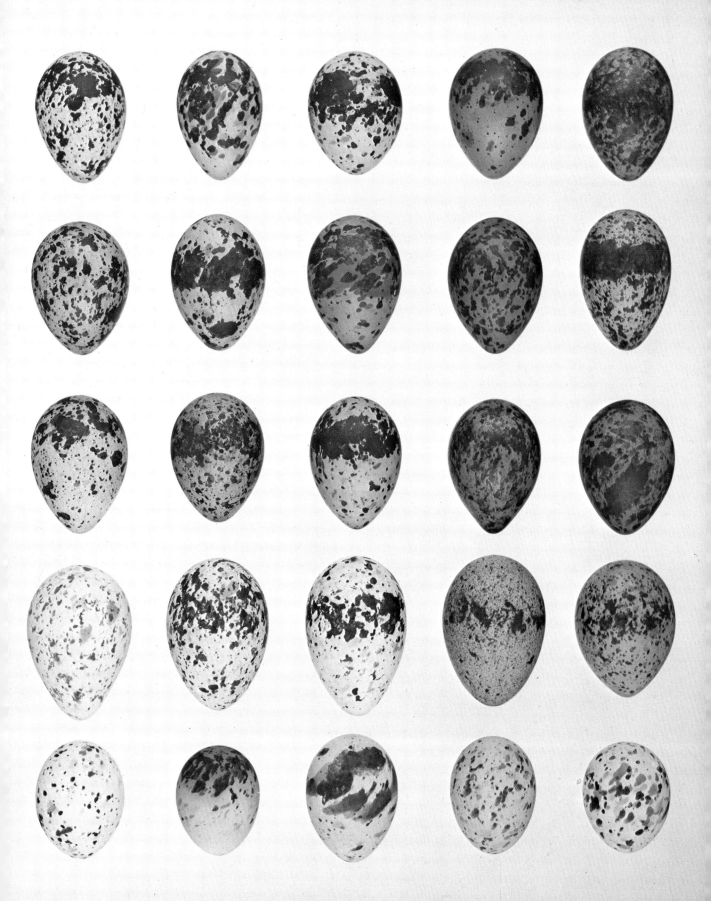

PLATE 91 EGGS

Uria aalge Guillemot, 8 eggs

PLATE 92 EGGS

Uria lomvia Brünnich's Guillemot, 3 eggs

Fratercula arctica Puffin, 1 egg
(centre)

Alca torda Razorbill, 2 eggs
(left and right)

Alle alle Little Auk, 2 eggs
(centre)

Cepphus grylle Black Guillemot, 2 eggs
(left and right)

PLATE 93 EGGS

Pinguinus impennis Great Auk, 1 egg

PLATE 94 EGGS

Pterocles alchata Pin-tailed Sandgrouse, 2 eggs
(left)

Pterocles orientalis Black-bellied Sandgrouse, 2 eggs
(right)

Pterocles senegallus Spotted Sandgrouse, 2 eggs
(left)

Syrrhaptes paradoxus Pallas's Sandgrouse, 2 eggs
(right)

Pterocles coronatus Crowned Sandgrouse, 2 eggs
(left)

Pterocles exustus Chestnut-bellied Sandgrouse, 2 eggs
(right)

Caprimulgus europaeus Nightjar, 2 eggs
(left)

Caprimulgus ruficollis Red-necked Nightjar, 2 eggs
(right)

Caprimulgus nubicus Nubian Nightjar, 2 eggs
(left)

Caprimulgus aegyptius Egyptian Nightjar, 2 eggs
(centre)

Caprimulgus eximius Golden Nightjar, 2 eggs
(right)

PLATE 95 EGGS

(from left to right)
Columba livia Rock Dove, 1 egg
Columba oenas Stock Dove, 1 egg
Columba eversmanni Yellow-eyed Stock Dove, 1 egg
Columba palumbus Woodpigeon, 1 egg

Columba bollii Bolle's Laurel Pigeon, 1 egg
(left)

Columba trocaz Long-toed Pigeon, 1 egg
(centre)

Columba junoniae Laurel Pigeon, 1 egg
(right)

(from left to right)
Streptopelia roseogrisea African Collared Dove, 1 egg
Streptopelia decaocto Collared Dove, 1 egg
Streptopelia turtur Turtle Dove, 1 egg
Streptopelia senegalensis, Laughing Dove, 1 egg
Streptopelia orientalis Rufous Turtle Dove, 1 egg
Oena capensis Namaqua Dove, 1 egg

Psittacula krameri Ring-necked Parakeet, 1 egg
(left)

Centropus senegalensis Senegal Cuckoo, 1 egg
(right)

Clamator jacobinus Jacobin Cuckoo with host Necklaced Laughing Thrush *Garrulax moniliger* on right, 2 eggs
(left)
Clamator jacobinus Jacobin Cuckoo with host Streaked Laughing Thrush *Garrulax lineatus* on right, 2 eggs
(centre)
Clamator jacobinus Jacobin Cuckoo with host Streaked Laughing Thrush *Garrulax lineatus* on right, 2 eggs
(right)

Clamator glandarius Great Spotted Cuckoo with host Magpie *Pica pica* on right, 2 eggs
(left)
Clamator glandarius Great Spotted Cuckoo with host Magpie *Pica pica* on right, 2 eggs
(centre)
Clamator glandarius Great Spotted Cuckoo with host Azure-winged Magpie *Cyanopica cyanus* on right, 2 eggs
(right)

PLATE 96 EGGS

Cuculus canorus Cuckoo (on left) with hosts: (from left to right)

    Meadow Pipit *Anthus pratensis*, 1 egg
    Sedge Warbler *Acrocephalus schoenobaenus*, 1 egg
    Reed Warbler *Acrocephalus scirpaceus*, 1 egg
    Great Reed Warbler *Acrocephalus arundinaceus*, 1 egg

    Icterine Warbler *Hippolais icterina*, 1 egg
    Icterine Warbler *Hippolais icterina*, 1 egg
    White/Pied Wagtail *Motacilla alba*, 1 egg
    Marsh Warbler *Acrocephalus palustris*, 1 egg

    White/Pied Wagtail *Motacilla alba*, 4 eggs

    Robin *Erithacus rubecula*, 4 eggs

    Robin *Erithacus rubecula*, 1 egg
    Wheatear *Oenanthe oenanthe*, 3 eggs

    Bullfinch *Pyrrhula pyrrhula*, 1 egg
    Redpoll *Carduelis flammea*, 1 egg
    Chaffinch *Fringilla coelebs*, 1 egg
    Lesser Whitethroat *Sylvia curruca*, 1 egg

    Wren *Troglodytes troglodytes*, 1 egg
    Wren *Troglodytes troglodytes*, 1 egg
    Dunnock *Prunella modularis*, 1 egg
    Lesser Whitethroat *Sylvia curruca*, 1 egg

Cuculus saturatus Oriental Cuckoo (on left) with hosts: (from left to right)

    Greenish Warbler *Phylloscopus trochiloides*, 1 egg
    Long-tailed Rosefinch *Uragus sibericus*, 1 egg
    Yellow-bellied Wren Babbler *Prinia flaviventris*, 1 egg
    Blyth's Warbler *Phylloscopus reguloides*, 1 egg

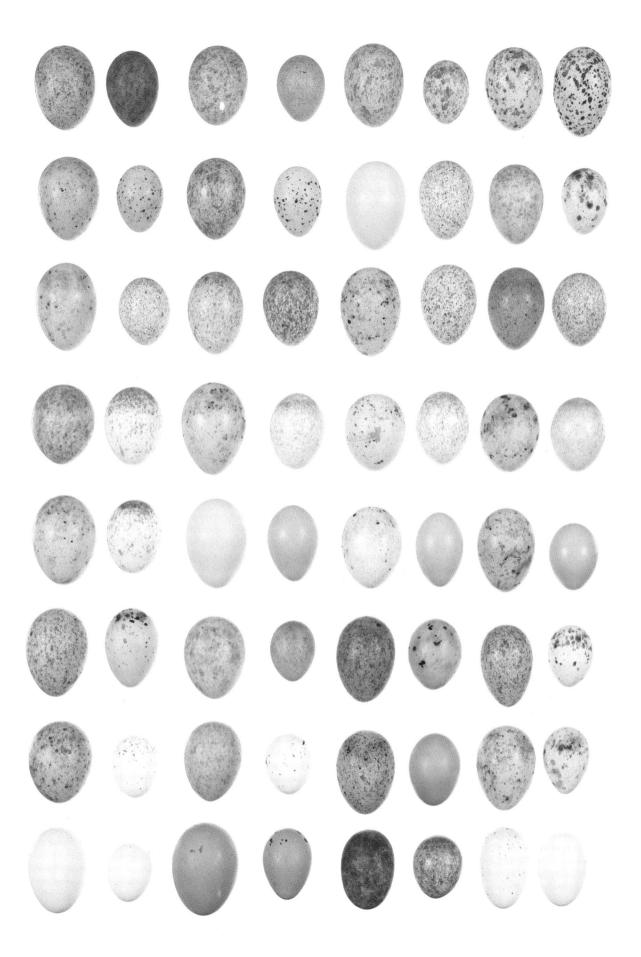

PLATE 97 EGGS

(from left to right)
Otus brucei Striated Scops Owl, 1 egg
Bubo bubo Eagle Owl, 1 egg
Ketupa zeylonensis Brown Fish Owl, 1 egg
Otus scops Scops Owl, 1 egg

Nyctea scandiaca Snowy Owl, 1 egg (left)
Strix nebulosa Great Grey Owl, 1 egg (centre)
Strix uralensis Ural Owl, 1 egg (right)

(from left to right)
Asio otus Long-eared Owl, 1 egg
Asio flammeus Short-eared Owl, 1 egg
Surnia ulula Hawk Owl, 1 egg
Tyto alba Barn Owl, 1 egg

(from left to right)
Athene noctua Little Owl, 1 egg
Glaucidium passerinum Pygmy Owl, 1 egg
Strix aluco Tawny Owl, 1 egg
Aegolius funereus Tengmalm's Owl, 1 egg
Asio capensis Marsh Owl, 1 egg

PLATE 98 EGGS

(from left to right)
Apus unicolor Plain Swift, 1 egg
Apus apus Swift, 1 egg
Apus melba Alpine Swift, 1 egg
Apus affinis Little Swift, 1 egg
Cypsiurus parvus Palm Swift, 1 egg

(from left to right)
Alcedo atthis Kingfisher, 1 egg
Halcyon leucocephala Grey-headed Kingfisher, 1 egg
Halcyon smyrnensis White-breasted Kingfisher, 1 egg
Ceryle rudis Pied Kingfisher, 1 egg
Merops superciliosus Blue-cheeked Bee-eater, 1 egg
Merops orientalis Little Green Bee-eater, 1 egg

(from left to right)
Merops apiaster Bee-eater, 1 egg
Coracias garrulus Roller, 1 egg
Coracias benghalensis Indian Roller, 1 egg
Eurystomus glaucurus Broad-billed Roller, 1 egg
Coracias abyssinicus Abyssinian Roller, 1 egg

Upupa epops Hoopoe, 3 eggs

(from left to right)
Jynx torquilla Wryneck, 1 egg
Picus canus Grey-headed Woodpecker, 1 egg
Picus viridis Green Woodpecker, 1 egg
Picus vaillantii Levaillant's Green Woodpecker, 1 egg
Dryocopus martius Black Woodpecker, 1 egg

(from left to right)
Dendrocopos major Great Spotted Woodpecker, 1 egg
Dendrocopos syriacus Syrian Woodpecker, 1 egg
Dendrocopos medius Middle Spotted Woodpecker, 1 egg
Dendrocopos leucotos White-backed Woodpecker, 1 egg
Dendrocopos minor Lesser Spotted Woodpecker, 1 egg
Picoides tridactylus Three-toed Woodpecker, 1 egg

83. KAINADY, P V G (1976) *Bull. Basrah nat. Hist. Mus.* **3**, 134–41; (1977) *Bull. Basrah nat. Hist. Mus.* **4**, 75–7. KALABÉR, L (1971) *Stud. Comun. Muz. Ştiinţ. Nat. Bacău (Zool.)*, 309–15. KALCHREUTER, H (1979) *Bonn. zool. Beitr.* **30**, 102–16. KALE, H W, II, SCIPLE, W, and TOMKINS, I R (1965) *Bird-Banding* **36**, 21–7. KÄLLANDER, H (1964) *Vår Fågelvärld* **23**, 119–35; (1977a) *Ornis fenn.* **54**, 79–84; (1977b) *Vår Fågelvärld* **36**, 134–42. KALMÁR, Z (1934) *Aquila* **38–41**, 455. KAMPP, K (1982) *Vår Fågelvärld* **41**, 29. KAPITONOV, N I (1962) *Ornitologiya* **5**, 35–48. KAPOCSY, G (1979) *Weissflügel- und Weissbartseeschwalbe.* Wittenberg Lutherstadt. KARALUS, K E and ECKERT, A W (1974) *The owls of North America.* New York. KARCHER, R (1940) *Oiseau* **10**, 361–2. KARLSSON, J (1975) *Anser* **14**, 134. KARPOVICH, V N and SAPETIN, Y V (1958) *Trudy Okskogo zapoved.* **2**, 152–4. KARTASHEV, N N (1960) *Die Alkenvögel des Nordatlantiks.* Wittenberg Lutherstadt; (1962) *Trudy Okskogo zapoved.* **4**, 271–86. KARTASHEV, M N and ILYICHEV, V D (1964) *J. Orn.* **105**, 113–36. KATE, C G B TEN (1966) *Limosa* **39**, 42–81. KATTINGER, E (1970) *Ber. Naturforsch. Ges. Bamberg* **45**, 57–79. KAUS, D (1977) *Anz. orn. Ges. Bayern* **16**, 8–44. KAUS, D, LINK, H, and WERZINGER, J (1971) *Anz. orn. Ges. Bayern* **10**, 69–82. KAY, G T (1947) *Br. Birds* **40**, 156–7. KAZAKOV, B A (1976) *Ornitologiya* **12**, 61–7. KEAST, J A (1942) *Emu* **42**, 133–40. KEAST, A and MORTON, E S (eds) (1980) *Migrant birds in the Neotropics.* Washington DC. KEIGHLEY, J (1950) *Skokholm Bird Obs. Rep. 1949*, 24–6. KEIGHLEY, J and LOCKLEY, R M (1947) *Br. Birds* **40**, 165–71. KEIL, W (1954) *Orn. Mitt.* **6**, 196. KEIL, W and ROSSBACH, R (1979) *Nat. und Mus.* **109**, 344–7. KEITH, L B (1964) *Can. Fld.-Nat.* **78**, 17–24. KEKILOVA, A F (1973) *Izv. Akad. Nauk Turkmen. SSR Ser. Biol.* (4), 63–9. KEKILOVA, A I (1970) *Izv. Akad. Nauk Turkmen. SSR Ser. Biol.* (4), 72–5. KELLER, J (1934) *Orn. Beob.* **31**, 174–6. KELLOMÄKI, E (1966) *Lintumies* **2**, 39–43; (1977) *Ornis fenn.* **54**, 1–29. KEMP, J B (1981) *Norfolk Bird and Mamm. Rep.* **25**, 262–4; (1982) *Br. Birds* **75**, 334–5. KENNARD, J H (1975) *Bird-Banding* **46**, 55–73. KENNEDY, A J (1981) *Nat. can.* **108**, 195–7. KENWARD, R E (1978a) *Ann. appl. Biol.* **89**, 277–86; (1978b) *J. anim. Ecol.* **47**, 449–60. KENWARD, R E and SIBLY, R M (1977) *J. appl. Ecol.* **14**, 815–90; (1978) *Anim. Behav.* **26**, 778–90. KERR, H M R (1918) *Br. Birds* **12**, 36–8. KESSEL, B and GIBSON, D D (1978) *Stud. Avian Biol.* **1**. KEVE, A (1944–7) *Aquila* **51–4**, 116–22; (1948–51) *Aquila* **55–8**, 89–107; (1955) *Aquila* **59–62**, 299–310; (1959) *Aquila* **66**, 310; (1960a) *Nomenclator Avium Hungariae.* Budapest; (1960b) *Vertebr. Hung.* **2**, 243–60; (1962a) *Aquila* **68**, 251; (1962b) *Aquila* **67–8**, 71–8; (1962c) *Falke* **9**, 143–4; (1964) *Aquila* **71–2**, 241–3. KEVE, A, KOHL, I, MATOUŠEK, F, MOŠANSKÝ, A, and RUCNER-KRONEISL, R (1962) *Larus* **14**, 26–67. KEVE, A and KOHL, S (1961) *Bull. Br. Orn. Club* **81**, 41–3. KHANMAMEDOV, A I and GASANOVA, Z R (1971) *Izv. Akad. Nauk Azerb. SSR* **5–6**, 100–6. KHAYUTIN, S N, DMITRIEVA, L P, TARTYGINA, N G, and ALEKSANDROV, L I (1982) *Zool. Zh.* **61**, 1063–77. KHOKHLOVA, T Y (1981) In Flint (1981), 7–9. KIERSKY, W (1937) *Beitr. Fortpfl. Biol. Vögel* **13**, 231. KILHAM, L (1981) *Wilson Bull.* **93**, 29. KING, B (1952) *Br. Birds* **45**, 373–4; (1972a) *Br. Birds* **65**, 32–3; (1972b) *Br. Birds* **65**, 397–8; (1974a) *Br. Birds* **67**, 388; (1974b) *Br. Birds* **67**, 515; (1978) *J. Saudi Arabian nat. Hist. Soc.* **1** (21), 3–24; (1980a) *Bristol Orn.* **13**, 100–1; (1980b) *Bristol Orn.* **13**, 103–4. KING, B and KING, M (1973) *Br. Birds* **66**, 33–4. KING, B, NAYLER, F, and WARDLE, F (1966) *Br. Birds* **59**, 108. KING, B F and DICKINSON, E C (1975) *A field guide to the birds of south-east Asia.* London. KING, H H (1921) *Sudan Notes Rec.* **4**, 39–43. KINZELBACH, R (1963) *Orn. Mitt.* **15**, 181. KIPP, F A (1954) *Vogelwarte* **17**, 183–80; (1961) *J. Orn.* **102**, 273–80; (1976) *J. Orn.* **117**, 457–60. KISLENKO, G S (1967) *Ornitologiya* **8**, 357; (1972) *Ornitologiya* **10**, 337–9. KISLENKO, G S and NAUMOV, R L (1967) *Ornitologiya* **8**, 79–97; (1972) *Ornitologiya* **10**, 339–42. KISS, J B and HÖHN, K (1975) *Vögel der Heimat* **46**, 67–8. KISS, J B and RÉKÁSI, J (1981) *Orn. Beob.* **78**, 13–16. KISS, J B, REKASI, J, and STERBETZ, I (1978) *Avocetta* **2**, 3–18. KJAER, T and ROSENDAHL, S (1975) *Danske Fugle* **27**, 1–21. KLAAS, C (1971) *Nat. und Mus.* **101**, 467–71. KLAFS, G and STÜBS, J (1977) *Die Vogelwelt Mecklenburgs.* Jena. KLAUS, S, KLAUS, M, and BRÄSECKE, R (1982) *Falke* **29**, 330–6. KLAUS, S, KUČERA, L, and WIESNER, J (1976) *Orn. Mitt.* **28**, 95–100. KLAUS, S, MIKKOLA, N, and WIESNER, J (1975) *Zool. Jb. Syst.* **102**, 485–507. KLAUS, S, VOGEL, F, and WIESNER, J (1965) *Zool. Abh. Staatl. Mus. Tierk. Dresden* **28**, 165–204. KLAUSNITZER, B (1960) *Abh. Ber. Naturkde. Mus. Görlitz* **36**, 103–9; (1963) *Abh. Ber. Naturkde. Mus. Görlitz* **38**, 1–4. KLAVER, A (1964) *Limosa* **37**, 221–31. KLEIN, E (1911) *Orn. Monatsber.* **19**, 130–1. KLEINSCHMIDT, O (1970) *Zool. Abh. Staatl. Mus. Tierk. Dresden* **31**, 9–10. KLEMETSEN, A (1967) *Sterna* **7**, 293–4. KLÍMA, M (1959a) *Zool. Listy* **8**, 33–6; (1959b) *Zool. Listy* **8**, 251–66. KLINZ, E (1955) *Die Wildtauben Mitteleuropas.* Wittenberg Lutherstadt. KNEIS, P and GÖRNER, M (1981) *Falke* **28**, 298–308. KNEITZ, G (1960) *Waldhygiene* **3**, 240–5; (1961) *Waldhygiene* **4**, 80–120; (1965a) *Collana Verde* **18**, 187–218; (1965b) *Waldhygiene* **6**, 11–27. KNIGHT, P J (1975) In Dick, W J A (ed) *Oxford and Cambridge Mauritanian Exped. 1973 Rep.*, 52–61. Cambridge. KNIPRATH, E (1969a) *Orn. Mitt.* **21**, 56–8; (1969b) *Bonn. zool. Beitr.* **20**, 200–6; (1969c) *Vogelwelt* **90**, 81–97. KNOBLOCH, H (1958) *Falke* **5**, 39–42, 76–81. KNÖTZSCH, G (1978) *Vogelwelt* **99**, 41–54. KNUDSEN, E I (1981) *Sci. Am.* **245** (6), 82–91. KNUDSEN, E I, BLASDEL, G G, and KONISHI, M (1979) *J. comp. Physiol.* **133**, 1–11. KNYSTAUTAS, A and LIUTKUS, A (1981) *Ornitologiya* **16**, 168–9. KOCH, G (1954) *Orn. Mitt.* **6**, 13. KOCHAN, W (1979) *Acta Zool. Cracov.* **23**, 213–46. KOELZ, W N (1954) *Contr. Inst. reg. Explor.* **1**, 1–32. KOEMAN, J H (1971) *Het Voorkomen en de Toxicologische Betekenis van Enkcle Chloorkoolwaterstoffen aan de Nederlandse Kust in de Periode van 1965 tot 1970.* Utrecht Univ. KOEMAN, J H, OSKAMP, A A G, VEEN, J, BROUWER, J, ROOTH, J, ZWART, P, BROEK, E VAN DEN, and GENDEREN, H VAN (1967) *Meded. Rijksfaculteit Landbouwwetenschappen Univ. Wageningen.* KOENIG, A (1886) *J. Orn.* **34**, 487–524; (1890) *J. Orn.* **38**, 257–498; (1895) *J. Orn.* **43**, 113–238; (1896) *J. Orn.* **44**, 101–216; (1917) *J. Orn.* **65**, 129–60; (1919) *J. Orn.* **67**, 431–85; (1926) *J. Orn.* **74**, 3–152. KOENIG, L (1950) *Zool. Inf. Biol. Stat. Wilhelminenberg* **2**; (1951) *Z. Tierpsychol.* **8**, 169–210; (1953) *Z. Tierpsychol.* **10**, 180–204; (1956) *J. Orn.* **97**, 384–402; (1959) *Mitt. Biol. Stat. Wilhelminenberg* **2**, 50–4; (1960) *Encyclopaedia Cinematographica* E 284/1958; (1973) *Z. Tierpsychol.* suppl. **13**. KOENIG, O (1952) *J. Orn.* **93**, 207–89. KOFFÁN, K (1957) *Aquila* **63–4**, 355–6; (1964) *Vögel vor der Kamera.* Budapest. KOHL, I and STOLLMANN, A (1968) *Aquila* **75**, 193–214. KOHL, S (1975) *Muz. Brukenthal Studii şi Comunicări Şt. nat.* **19**, 319–28; (1977) *Muz. Brukenthal Studii şi Comunicări Şt. nat.* **21**, 309–34. KOHL, S and HAMAR, M (1978) *Orn. Jber. Mus. Hein.* **3**, 67–72. KOLBE, H (1975) *Beitr. Vogelkde.* **23**, 421–3. KOLESNIKOV, A D (1976) *Ornitologiya* **12**, 234–5. KOLI, L and SOIKKELI, M (1974) *Ann. zool. fenn.* **11**, 304–8. KÖLSCH, E (1958) *Vogelwelt* **79**, 53–5. KOLTHOFF, G (1903) *Kung. Sven. Vetensk.-Akad. Handl.* **36**, 1–104. KOLTHOFF, K (1920) *Fauna och Flora* **15**, 1–6. KONDRATZKI, B and ALTMÜLLER, R (1976) *Vogelwelt* **97**, 146–8. KÖNIG, C (1960) *Vogelwelt* **81**, 68–73; (1961a) *Anz. orn. Ges. Bayern* **6**, 81–3; (1962) *Vogelwelt* **83**, 83–4; (1963) *Orn. Mitt.* **15**, 181; (1964) *Alb. Jb. Ver. vaterl. Naturkde. Württemberg* **118–19**, 370–6; (1965) *J. Orn.* **106**, 349–50; (1968) *Vogelwelt Suppl.* **1**, 115–38; (1969a) *Vogelwelt* **90**, 66–7; (1969b) *J. Orn.* **110**, 133–47; (1978) *Wir und die Vögel* **10** (1), 18–20. KÖNIG, C and WICHT, U VON (1973) *Anz. orn. Ges. Bayern* **12**, 52–6. KÖNIG, D (1951) *Orn. Mitt.* **3**, 125–6; (1956) *Orn. Mitt.* **8**, 143–7; (1961b)

Vogelwelt **82**, 1–16. KÖNIG, H and HAENSEL, J (1968) *Beitr. Vogelkde.* **13**, 335–65. KÖPKE, G (1969) *Vogelwelt* **90**, 70–1. KORELOV, M N (1948) *Izv. Akad. Nauk Kazakh. SSR Ser. Zool.* **51** (7), 107–23. KORENBERG, Z I, RUDENSKAYA, L V, and CHERNOV, YU I (1972) *Ornitologiya* **10**, 151–60. KORODI GÁL, I (1970) *Stud. Cercet. Biol. Ser. Zool. București* **22**, 269–76; (1975) *Stud. și Comunicări (Sibiu)* **19**, 329–35. KORODI GÁL, I and LIBUS, A (1968) *Abh. Ber. Mus. Tierk. Dresden* **29**, 95–102. KOROL'KOVA, G E (1954) *Soobshcheniya Inst. Lesa* **2**, 65–106. KORPIMÄKI, E (1980) *Suom. Linnut* **15**, 55–8; (1981) *Acta Univ. Ouluensis* (A) **118**, *Biol.* **13**. KORPIMÄKI, E, IKOLA, S, HAAPOJA, R, and KIRKKOMÄKI, J (1977) *Suom. Linnut* **12**, 100–17. KORTE, J DE (1972) *Beaufortia* **19**, 197–232. KORTNER, W (1979) *Anz. orn. Ges. Bayern* **18**, 82. KOSKIMIES, J (1947) *Ornis fenn.* **14**, 106–11; (1950) *Ann. Acad. Sci. fenn. A IV. Biol.* **15**; (1956) *Ornis fenn.* **33**, 77–96; (1957) *Ornis fenn.* **34**, 1–6; (1961) *Ornis fenn.* **38**, 105–27. KOSKIMIES, P (1979) *Ornis Karelica* **5**, 21–30. KOSTIN, Y V and TARINA, N A (1981) In Flint (1981), 113–15. KOTOV, A A (1974a) *Byull. Mosk. Obshch. Isp. Prir. Otd. Biol.* **79** (6), 36–44; (1974b) *Nauch. dokl. vyssh. shk. Biol. Nauki* **17** (10), 18–25; (1976a) *Byull. Mosk. Obshch. Isp. Prir. Otd. Biol.* **81** (3), 23–30; (1976b) *Byull. Mosk. Obshch. Isp. Prir. Otd. Biol.* **81** (5) 22–9; (1976c) *Ornitologiya* **12**, 132–43; (1978) *Byull. Mosk. Obshch. Isp. Prir. Otd. Biol.* **83**, 71–80. KOTOV, A A and NOSKOV, G A (1978) *Zool. Zh.* **57**, 1202–9. KÖTTER, F (1953) *Orn. Mitt.* **5**, 227. KOVAČEVIĆ, J and DANON, M (1950–1) *Larus* **4–5**, 185–217; (1957) *Larus* **11**, 111–30. KOWALSKI, S and KOWALSKI, H (1957) *Oiseau* **27**, 378–81. KOZLOVA, E V (1930) *Ptitsy yugozapadnogo Zabaykal'ya, severnoy Mongolii i tsentral'noy Gobi.* Leningrad; (1957) *Fauna SSSR. Ptitsy* **2** (3). Moscow. KRAFT, J A and KORELOV, M N (1938) *Bull. Univ. Asiae cent.* **22**, 265–8. KRAMBRICH, A (1953) *Vogelwelt* **74**, 136–9; (1954) *Vogelwelt* **75**, 100–1. KRAMER, H (1937) *Beitr. Fortpfl. Biol. Vögel* **13**, 67–70. KRANZ, P (1971) *Fåglar i Sörmland* **4**, 13–23. KRASOVSKI, S K (1937) *Trudy arkt. Inst. Leningrad* **77**, 33–91. KRAUSS, A (1974) *Beitr. Vogelkde.* **20**, 478–80; (1977) *Beitr. Vogelkde.* **23**, 313–29. KRECHMAR, A V (1966) In Ivanov, A I (ed) *Biologiya ptits*, 185–312. Moscow. KRESS, S W (1981) *Newsl. Fratercula Fund*, 1–4. KRISTOFFERSEN, S (1926) *Norsk orn. Tidsskr.* (2) **7**, 181–95. KRIMMER, M, PIECHOCKI, R, and UHLENHAUT, K (1974) *Falke* **21**, 42–51, 95–101. KRIVITSKI, I A (1965) *Ornitologiya* **7**, 146–52; (1977) In Sludski, A A (ed) *Redkie i ischezayushchie zveri i ptitsy Kazakhstana*, 196–9. Alma-Ata. KROGH, K (1955) *K. norske vidensk. Selsk. Årbok. 1954.* KRONEISL-RUCNER, R (1957a) *Larus* **9–10**, 34–47; (1957b) *Larus* **9–10**, 188–92. KRZANOWSKI, A (1973) *Acta Zool. Cracov.* **18**, 133–40. KUBÍK, V (1960) *Zool. Listy* **9**, 97–110. KUBÍK, V and BALÁT, F (1973) *Zool. Listy* **22**, 59–72. KUČERA, L (1972) *Schweiz. Z. Forstwesen* **123**, 107–16. KUCHIN, A P (1976) *Ptitsy Altaya.* Barnaul. KÜCHLER, W (1951) *Orn. Mitt.* **3**, 276. KUHK, R (1948) *Vogelwarte* **1**, 28–30; (1949) In *Ornithologie als biologische Wissenschaft*, 171–82. Heidelberg; (1950) *Syllegomena biologica*, 220–9. Leipzig; (1953) *J. Orn.* **94**, 83–93; (1954) *Vogelwarte* **17**, 208–9; (1961) *Auspicium* **1**, 212–14; (1969) *Bonn. zool. Beitr.* **20**, 145–50; (1970) *Beitr. Vogelkde.* **16**, 232–8. KULCZYCKI, A (1964) *Acta Zool. Cracov.* **9**, 529–59; (1966) *Przegl. Zool.* **10**, 218–21. KULESHOVA, L V (1968) *Ornitologiya* **9**, 354. KULLENBERG, B (1946) *Ark. Zool.* **38A** (17), 1–80. KUMARI, E (1939a) *Acta Inst. Mus. Zool. Univ. Tartu* **45**, 1–96; (1939b) *Ornis fenn.* **16**, 7–13; (1941) *Ann. Soc. reb. nat. invest. Univ. Tartu* **47**, 7–28; (1961) *Orn. Kogumik* **2**, 9–41; (1967) *Orn. Kogumik* **4**, 61–84; (1978) *Orn. Kogumik* **8**, 99–121. KUMERLOEVE, H (1959) *Alauda* **27**, 26–9; (1961) *Bonn. zool. Beitr.* **12** suppl.; (1962) *Orn. Mitt.* **14**, 105–6; (1967) *Rev. Fac. Sci. Univ. Istanbul* (B) **32**, 74–213; (1968a) *J. Orn.* **109**, 130–1; (1968b) *Alauda* **36**, 190–207; (1969) *Rev. Fac. Sci. Univ. Istanbul* (B) **34**, 245–312; (1970) *Rev. Fac. Sci. Univ. Istanbul* (B) **35**, 85–160; (1980) *Orn. Mitt.* **32**, 16–19. KUMMER, J (1982) *Beitr. Vogelkde.* **28**, 253–4. KURODA, N (1954) *Jap. J. Zool.* **11**, 311–27; (1967) *Yamashina Inst. Orn.* **5**, 106–9. KUSCHERT, H, EKELÖF, O, and FLEET, D M (1981) *Seevögel* **2**, 58–61. KYTZIA, S (1954) *Luscinia* **27**, 19.

LABITTE, A (1940) *Alauda* **12**, 99–118; (1945) *Oiseau* **15**, 118–29; (1951) *Oiseau* **21**, 120–6; (1952) *Oiseau* **22**, 107–12; (1953) *Alauda* **21**, 165–78; (1954a) *Oiseau* **24**, 48–51; (1954b) *Oiseau* **24**, 197–210; (1955) *Alauda* **23**, 67–70; (1956a) *Oiseau* **26**, 139–51; (1956b) *Oiseau* **26**, 194–203; (1957) *Oiseau* **27**, 232–4; (1958) *Oiseau* **18**, 78–93. LACAN, F and MOUGIN, J-L (1974) *Oiseau* **44**, 191–280. LACEY, E (1910) *Br. Birds* **3**, 263. LACHNER, R (1963) *J. Orn.* **104**, 305–56; (1965) *Vogelwelt* **86**, 79–95. LACK, D (1930a) *Br. Birds* **23**, 242–4; (1930b) *London Nat.*, 47–55; (1932) *Ibis* (13) **2**, 266–84; (1954) *The natural regulation of animal numbers.* Oxford; (1955) *Bird Study* **2**, 32–40; (1956a) *Swifts in a tower.* London; (1956b) *Ibis* **98**, 34–62; (1957) *Br. Birds* **50**, 273–7; (1958) *Ibis* **100**, 477–502; (1963) *Bird Study* **10**, 185–203; (1964) *Auk* **73**, 1–92; (1968) *Ecological adaptations for breeding in birds.* London; (1969) In Gooders, J (ed) *Birds of the world* **5**, 1418–19. London. LACK, D and ARN, H (1947) *Orn. Beob.* **44**, 188–210. LACK, D and LACK, E (1951a) *Alauda* **19**, 49; (1951b) *Ibis* **93**, 501–46; (1952) *Br. Birds* **45**, 186–215. LACK, D and OWEN, D F (1955) *J. anim. Ecol.* **24**, 120–36. LACK, D and RIDPATH, M G (1955) *Br. Birds* **48**, 289–92. LADE, B I and THORPE, W H (1964) *Nature* **202**, 366–8. LADHAMS, D E (1976) *Br. Birds* **69**, 410; (1977) *Br. Birds* **70**, 547–8. LAFERRÈRE, M (1956) *Alauda* **24**, 275–86; (1968) *Oiseau* **38**, 175–7; (1972) *Alauda* **40**, 290–2; (1974a) *Alauda* **42**, 343–5; (1974b) *Alauda* **42**, 345–7. LAGERSTRÖM, M (1969) *Ornis fenn.* **46**, 31–2; (1978) *Lintumies* **14**, 34–7. LAHTI, E (1972) *Ornis fenn.* **49**, 91–7. LAHTI, E and MIKKOLA, H (1974) *Savon Luonto* **1**, 1–10. LALAS, C and HEATHER, B D (1980) *Notornis* **27**, 45–68. LAMA, S DE LA (1959) *Ardeola* **5**, 201–4. LAMANI, F and PUZANOV, V (1962) *Bull. U. Sh. Tirane Ser. Sh. Nat.* **16** (3), 87–102, **16** (4), 100–17. LAMARCHE, B (1980) *Malimbus* **2**, 121–58. LAMBA, B S (1963) *Res. Bull. Panjas Univ.* **14**, 21–8; (1966) *Proc. zool. Soc. Calcutta* **19**, 77–85. LAMBERT, K (1980) *Beitr. Vogelkde.* **26**, 1–18. LAMBIE, R M (1974) *Scott. Birds* **8**, 32–3. LAMMIN-SOILA, R and UUSIVUORI, P (1975) *Lintumies* **10**, 109–17. LANCUM, F H (1939) *Br. Birds* **33**, 185–6. LANE, S G (1967) *Austral. Bird Bander* **5**, 57; (1979) *Corella* **3**, 7–10. LANG, E (1981) *Wir und die Vögel* **13** (4), 16. LANGE, H (1948) *Dansk orn. Foren. Tidsskr.* **42**, 50–84. LANGELOTT, N (1957) *Vogelwelt* **78**, 147–53. LANGHAM, N P E (1968) Ph D Thesis. Durham Univ.; (1971) *Bird Study* **18**, 155–75; (1972) *J. anim. Ecol.* **41**, 385–94; (1974) *Auk* **91**, 255–77. LANGSLOW, D R (1977) *Scott. Birds* **9**, 262–7. LANZ, H (1950) *Orn. Beob.* **47**, 137–41; (1964) *Orn. Beob.* **61**, 65–7. LARSEN, A (1949) *Dansk orn. Foren. Tidsskr.* **43**, 129–49. LARSON, S (1960a) *Vår Fågelvärld* **19**, 193–207; (1960b) *Oikos* **11**, 276–305. LARRISON, E J and SONNENBERG, K G (1968) *Washington birds: their location and identification.* Seattle. LATHAM, P C M (1979) *Notornis* **26**, 63–7. LATHBURY, G (1970) *Ibis* **112**, 25–43. LATIF, A and YUNUS, C M (1950) *Bee World* **31**, 91–2. LA TOUCHE, J D D (1931–4) *A handbook of the birds of eastern China* **2**. London. LATZEL, G (1969) *J. Orn.* **110**, 221; (1972) *Orn. Mitt.* **24**, 246. LAUERMANN, H (1975) *Egretta* **18**, 21. LAURSEN, J T (1981) *Dansk orn. Foren. Tidsskr.* **75**, 105–10. LAVAUDEN, L (1920) *Rev. fr. Orn.* **132–3**, 1–8. LAWMAN, J (1975) *Nat. Wales* **14**, 250–4. LAWRENCE, L DE K (1967) *Orn. Monogr.* **5**, 1–156. LAWSON, D F (1951) *Br. Birds* **44**, 281–2. LAWSON, W J (1962) *Durban Mus. Novit.* **6**, 213–30. LAYARD, C C S (1851) *Annal. Nat. Hist.*, 97–107. LAYARD, E L (1854) *Ann. Mag. Nat. Hist.* (2) **13**, 257–64. LEBEDEVA, L A

(1962) *Migr. zhiv.* **3**, 87–91. LEBEURIER, E (1963) *Oiseau* **33**, 212–34. LEBRET, T and OUWENEEL, G L (1976) *Limosa* **49**, 24–7. LEBRETON, P (1969) *Oiseau* **39**, 83–111. LECROY, M (1976) *Am. Mus. Novit.* **2599**, 1–30. LECROY, M and COLLINS, C T (1972) *Auk* **89**, 595–611. LECROY, M and LECROY, S (1974) *Bird-Banding* **45**, 326–40. LEE, SUNG YOON (1969) *An annotated checklist of the birds of Korea.* Seoul. LEEGE, O and WECKMANN-WITTENBURG, P F (1941) *Vögel deutscher Küsten.* Berlin. LEES, J (1946) *Br. Birds* **39**, 136–41. LEHMANN, G (1930) *Naturforscher* **7**, 53–5. LEHMANN, H (1971) *Vogelwelt* **92**, 161–78. LEHMANN, D S (1964) *Sci. Am.* **211** (5), 48–54. LEHTONEN, L (1951) *Ornis fenn.* **28**, 89–109; (1972) In Grzimek, B (ed) *Grzimek's animal life encyclopedia* **8**, 429–36; (1981) *Ornis fenn.* **58**, 29–40. LEIBIG, G (1972) *Beiträge zur Biologie des Ziegenmelkers.* Staatsexamensarbeit. Karlsruhe Univ. LEIBRICH, C and UHLENHAUT, K (1976) *Falke* **23**, 344–51. LEIBUNDGUT, H (1934) *Schweiz. Z. Forstwesen* **85**, 237–40; (1973) *Z. Jagdwiss.* **19**, 122–31. LEIN, M R and BOXALL, P C (1979) *Can. Fld.-Nat.* **93**, 411–14. LEINONEN, A (1978) *Lintumies* **13**, 13–18. LEMMETYINEN, R (1968) *Ornis fenn.* **45**, 114–24; (1971) *Ornis fenn.* **48**, 13–24; (1972a) *Ornis fenn.* **49**, 45–53; (1972b) *Rep. Kevo Subarctic Res. Stat.* **9**, 28–31. (1973a) *Ornis fenn.* **50**, 18–28; (1973b) *Ann. zool. fenn.* **10**, 507–25; (1973c) *Ann. zool. fenn.* **10**, 526–35. LEMMETYINEN, R, PORTIN, P, and VUOLANTO, S (1974) *Ann. zool. fenn.* **11**, 265–70. LESHEM, J (1975) *Wildlife* **17**, 443; (1979) *Nat. und Mus.* **109**, 375–7; (1981) *Sandgrouse* **2**, 100–1. LESLIE, R (1981) *The North York Moors Nightjar (Caprimulgus europaeus) Survey 1980 Rep.* Pickering. LE SUEUR, F (1957) *Br. Birds* **50**, 361–4. LETSWITZ, W (1904) *Verh. orn. Ges. Bayern* **5**, 64–76. LETTESJO, A (1974) *J. Kikaren* **15**, 12–13. LEURQUIN, J (1975) *Aves* **12**, 127–9. LÉVÊQUE, R (1955) *Nos Oiseaux* **23**, 233–46; (1957) *Terre et Vie* **104**, 150–78; (1964) **106**, 52–62; (1968) *Orn. Beob.* **65**, 43–71. LEVER, C (1977) *The naturalized animals of the British Isles.* London. LEVERKÜHN, P (1889) *Monatsschr. Dtsch. Ver. Schutze Vogelwelt* **14**, 398–406. LEVI, W M (1945) *The Pigeon.* Columbia. LEWIS, S (1902) *Zoologist* (4) **6**, 313–15. LEWIS, T (1920) *Br. Birds* **14**, 74–82. LEYS, H M (1964) *Limosa* **37**, 232–63. LIBOIS, R (1977) *Aves* **14**, 165–77. LID, G (1981) *Cinclus* **4**, 30–9. LIEBE, K T (1887) *Monatsschr. Dtsch. Ver. Schutze Vogelwelt* **12**, 236–47. LIEDEKERKE, R DE (1969) *Aves* **6**, 168; (1980) *Aves* **17**, 40–2. LIÉNHART, R (1935) *Alauda* **7**, 498–505. LIKHACHEV, G N (1954) *Byull. Mosk. Obshch. Isp. Prir. Otd. Biol.* **59**, 15–25; (1971) *Trudy Prioksko-Terrasn. gos. zapoved.* **5**, 135–45. LIND, H (1963a) *Vidensk. Medd. dansk natur. Foren.* **125**, 407–48; (1963b) *Dansk orn. Foren. Tidsskr.* **57**, 155–75. LINDBERG, P (1966) *Vår Fågelvärld* **25**, 106–42. LINDBLAD, J (1962) *Zool. Revy* **24**, 17–26; (1967) *I ugglemarker.* Stockholm. LINDHE, U (1966) *Vår Fågelvärld* **25**, 40–8. LINK, H-H (1958) *Orn. Mitt.* **10**, 22–4. LINK, J A (1889) *Monatsschr. Dtsch. Ver. Schutze Vogelwelt* **14**, 439–53. LINKOLA, P (1963) *Ornis fenn.* **40**, 69–72; (1978) *Proc. 1st Nordic Cong. Orn.*, 155–62. LINKOLA, P and MYLLYMÄKI, A (1969) *Ornis fenn.* **46**, 45–78. LINT, K C (1964) *Avic. Mag.* **70**, 119–22. LIPPENS, L (1935) *Gerfaut* **25**, 126–34. LIPPENS, L and WILLE, H (1972) *Atlas des oiseaux de Belgique et d'Europe occidentale.* Tielt; (1976) *Les oiseaux du Zaïre.* Tielt. LIPPERT, W (1975) *Beitr. Vogelkde.* **21**, 372–3. LITTLE, J DE V (1970) *Bokmakierie* **22**, 26. LIVERSIDGE, R (1959) *Ostrich Suppl.* **3**, 47–67; (1968) *Ostrich* **39**, 223–7; (1970) *Ostrich Suppl.* **8**, 117–37. LJUNGGREN, L (1968) *Viltrevy* **5**, 435–504; (1969) *Viltrevy* **6**, 41–126. LLOYD, C S (1972) *Skokholm Bird Obs. Rep. 1971*, 35–9; (1973) *Skokholm Bird Obs. Rep. 1972*, 15–23; (1974) *Bird Study* **21**, 102–16; (1976a) D Phil Thesis. Oxford Univ.; (1976b) *Br. Birds* **69**, 298–304; (1979) *Ibis* **121**, 165–76. LLOYD, C S, BIBBY, C J, and EVERETT, M J (1975) *Br. Birds* **68**, 221–37. LLOYD, C S and PERRINS, C M (1977) *Bird-Banding* **48**, 239–52. LOBB, M G (1983) *Bull. Br. Orn. Club* **103**, 111. LOCKIE, J D (1955) *Bird Study* **2**, 53–69. LOCKLEY, R M (1934a) *Field*, 1365; (1934b) *Br. Birds* **27**, 214–23; (1938) *Br. Birds* **31**, 278–9; (1942) *Shearwaters.* London; (1953) *Puffins.* London. LOFTIN, R W and SUTTON, S (1979) *Wilson Bull.* **91**, 133–5. LOFTS, B and MURTON, R K (1966) *Br. Birds* **59**, 261–80. LOFTS, B, MURTON, R K, and WESTWOOD, N J (1966) *J. Zool. Lond.* **150**, 249–72; (1967a) *Ibis* **109**, 337–51; (1967b) *Ibis* **109**, 352–8. LOHMANN, M (1960) *Orn. Beob.* **57**, 147–9. LÖHRL, H (1940) *Dtsch. Vogelwelt* **65**, 107–8; (1949) *Vogelwelt* **70**, 85–6; (1950a) *Orn. Ber.* **3**, 120–5; (1950b) *Vogelwarte* **15**, 213–19; (1957) *Vogelwelt* **78**, 164–5; (1962a) *J. Orn.* **103**, 487; (1962b) *Orn. Beob.* **59**, 28–9; (1965) *J. Orn.* **106**, 113–14; (1972) *Anz. orn. Ges. Bayern* **11**, 248–53; (1975) *Vogelwelt* **96**, 64–8; (1977a) *Vogelwelt* **98**, 15–22; (1977b) *Vogelwelt* **98**, 41–58; (1977c) *Publ. Wiss. Film., Sekt. Biol.* (10) 29/E2218; (1977d) *Publ. Wiss. Film., Sekt. Biol.* (10) 29/E2218; (1977e) *Publ. Wiss. Film., Sekt. Biol.* (10) 26/E2220; (1978) *Orn Beob.* **75**, 193–201; (1979) *J. Orn.* **120**, 139–73; (1980) *Publ. Wiss. Film., Sekt Biol.* (13) 7/E2441; (1981) *J. Orn.* **122**, 173–80. LOKCZÁNSZKY, A (1934) *Aquila* **38–41**, 179–86. LOMONT, H (1946) *Bull. Mus. Hist. nat. Marseille* **6**, 81–8. LONG, J L (1981a) *Introduced birds of the world.* Newton Abbot. LONG, R (1981b) *Br. Birds* **74**, 327–44. LONGSTAFF, T G (1924) *Ibis* (11) **6**, 480–95. LÖNNBERG, E (1936) *Org. Club Nederland. Vogelk.* **9**, 27–31. LOOS, C (1903a) *Orn. Monatsschr.* **28**, 160–72, 180–98, 231–9; (1903b) *Orn. Monatsschr.* **28**, 457–60; (1904) *Orn. Monatsschr.* **29**, 337–45; (1905) *Orn. Monatsschr.* **30**, 360–7, 412–20. LOOS, K (1910) *Der Schwarzspecht. Sein Leben und seine Beziehungen zum Forsthaushalte.* Vienna; (1916) *Orn. Monatsschr.* **41**, 69–81. LÓPEZ GORDO, J L (1974) *Ardeola* **19**, 429–37; (1975) *Ardeola* **21**, 615–25. LÓPEZ GORDO, J L, LAZARO, E, and FERNANDEZ-JORGE, A (1977) *Ardeola* **23**, 189–221. LØPPENTHIN, B (1951) *Proc. int. orn. Congr.* **10**, 603–10; (1963) *Dansk orn. Foren. Tidsskr.* **57**, 83–95; (1967) *Danske ynglefugle i fortig og notid.* Odense. LORD, J and AINSWORTH, G H (1945) *Br. Birds* **38**, 275. LÖSCHAU, M and LENZ, M (1967) *J. Orn.* **108**, 41–64. LOVARI, S (1975a) *Raptor Res.* **8**, 45–57; (1975b) *Biol. Conserv.* **8**, 19–22; (1978) *Avocetta* **1**, 61–3. LOVARI, S, RENZONI, A, and FONDI, R (1976) *Boll. Zool.* **43**, 173–91. LOVATY, F (1980) *Alauda* **48**, 203. LØVENSKIOLD, H L (1947) *Håndbok over Norges fugler.* Oslo; (1954) *Norsk. Polarinst. Skr.* **103**, 1–131; (1964) *Avifauna Svalbardensis.* Oslo. LOW, G C (1923) *Ibis* (11) **5**, 563–5; (1924) *Ibis* (11) **6**, 590–1. LOWE, C J (1962) *Br. Birds* **55**, 419–20. LOWE, P R (1943) *Ibis* **85**, 499–500. LOWE, V P W (1980) *J. Zool. Lond.* **192**, 283–93. LOWERY, G H (1955) *Louisiana birds.* Baton Rouge. LOWTHER, E H N (1949) *A bird photographer in India.* London. LOYD, L R W (1912) *Br. Birds* **6**, 62–3; (1914) *Br. Birds* **7**, 231–2. LUDESCHER, F-B (1973) *J. Orn.* **114**, 3–56. LÜDICKE, M (1933) *Zool. Jahrb. Abt. Anat. Ontog. Tiere* **57**, 465–534. LUDWIG, J P (1965) *Bird-Banding* **36**, 217–33; (1979) *Mich. Academician* **12**, 69–77. LUGARO, G (1954) *Riv. ital. Orn.* (2) **24**, 56. LÜHMANN, M (1950) *Verh. dtsch. zool. Ges. 1949*, 244–7. LUND, H M (1951) *Nytt Mag. Naturvidensk.* **88**, 247–62. LUNDBERG, A (1974) *Vår Fågelvärld* **33**, 147–54; (1976) *Zoon* **4**, 65–72; (1977) *Fåglar i Uppland* **4**, 11–17; (1979a) *Oecologia* **41**, 273–81; (1979b) *Upps. Diss. Fac. Sci.* **507**, 1–16; (1980a) *Ornis scand.* **11**, 65–70; (1980b) *Ornis scand.* **11**, 116–20; (1981) *Ornis scand.* **12**, 111–19. LUNDBERG, A and SWANQVIST, S-E (1973) *Vår Fågelvärld* **32**, 46. LUNDIN, A (1960) *Vår Fågelvärld* **19**, 43–50; (1961) *Fauna och Flora* **56**, 79. LÜPS, P, HAUVI, R, HERREN, H, MARKI, H, and RYSER, R (1978) *Bern Orn. Beob.* **75** suppl. LYNES, H (1920) *Ibis* (11) **2**, 260–301; (1924) *Novit. Zool.* **31**, 49–103; (1925a) *Ibis* (12) **1**, 344–416, 541–90; (1925b) *Mém. Soc. Sci. Nat. Maroc* **13**. LYTHGOE, J N (1979) *The ecology of vision.* Oxford. LYULEEVA, D (1981) *Commun. Baltic. Commiss. Study Bird Migr.* **13**, 129–44. LUSHINGTON, C (undated) *Bird life in Ceylon.* Colombo. LUTON, W G (1957) *Br. Birds* **50**, 538. LUTTIK, R (1982) *Vogeljaar* **30**, 17–23. LÜTTSCHWAGER, J (1961) *Die Drontevögel.* Wittenberg Lutherstadt. LUZ MADUREIRA, M DA (1979) *Arq. Mus. Bocage* **6**, 343–60.

MCATEE, W L and BEAL, F E L (1912) *US Dept. Agric. Farmers' Bull.* **497**, 1–30. MACDONALD, D (1967) *Scott. Birds* **4**, 374–5. MACDONALD, D W (1976) *Ibis* **118**, 579–80. MACDONALD, J D (1962) *Birds in the sun.* London; (1973) *Birds of Australia.* London. MACDONALD, K C (1906) *J. Bombay nat. Hist. Soc.* **17**, 184–94. MACDONALD, M (1960) *Birds in my Indian garden.* London. MACDONALD, M A (1978) *Bull. Niger. orn. Soc.* **14**, 66–70. MACDONALD, M A and MCDOUGALL, P O (1970) *Scott. Birds* **6**, 175–6. MCDOUGALL, D (1983) *Scott. Birds* **12**, 162–3. MACFARLANE, A M (1978) *Army Bird-Watching Soc. per. Publ.* **3**. MCGEOCH, J A (1963) *Ardea* **51**, 244–50. MACGILLIVRAY, W (1837) *A history of British birds, indigenous and migratory* **1**. London. MACGINLEY, P J (1913) *Irish Nat.* **22**, 211–14. MÄCHLER, G (1955a) *Orn. Beob.* **52**, 96–7; (1955b) *Orn. Beob.* **52**, 202–3. MACIVOR, J and NAVARRO MEDINA, J D (1972) *Ardeola* **16**, 263–4. MACKEITH, T T (1908) *Br. Birds* **2**, 239–40. MCKITTRICK, T H (1929) *Auk* **46**, 529–32. MACKWORTH-PRAED, C W and GRANT, C H B (1952) *Birds of eastern and north eastern Africa* **1**. London; (1962) *Birds of the southern third of Africa* **1**. London; (1970) *Birds of west central and western Africa* **1**. London. MCLACHLAN, G R and LIVERSIDGE, R (1970) *Roberts' birds of South Africa.* Cape Town. MACLAREN, P I R (1954) *Ibis* **96**, 601–5. MACLEAN, G L (1968) *Living Bird* **7**, 209–35; (1969) *J. Orn.* **110**, 104–7; (1976) *Proc. int. orn. Congr.* **16**, 503–16. MACLEAY, K N G (1960) *Bull. Khartoum nat. Hist. Mus.* **1**, 1–33. MCNICHOLL, M K and SCOTT, V H (1973) *Can. Fld.-Nat.* **87**, 184–5. MACPHERSON, A H (1908) *Br. Birds* **1**, 292–3. MACPHERSON, H A (1889a) *The visitation of Pallas's Sand-Grouse to Scotland in 1888 together with an account of its nesting, habits and migrations.* London; (1889b) *Report on Pallas Sand-Grouse (Syrrhaptes paradoxus) in the north-west of England,* 59–75. MADGE, S C and MADGE, P S (1983) *Br. Birds* **76**, 576–8. MADON, P (1930a) *Alauda* **2**, 85–121; (1930b) *Alauda* **2**, 206–40; (1933) *Les rapaces d'Europe.* Toulon; (1934) *Alauda* **6**, 47–86. MADSEN, F J (1957) *Dan. Rev. Game Biol.* **3**, 19–83. MADSEN, G (1972) *Flora og Fauna* **78**, 87–92. MAGALHÃES, C M P DE (1974) *Hábitos alimentares do Bufo-pequeno Asio otus (L.) na Tapada de Mafra.* Dir.-Ger. Serv. Flor. Aquíc., Ser. Zool. Flor. Cin. MAGLIO, M (1976) *Tesi di laurea.* Pavia Univ. MAGNUSSON, M and SVÄRDSON, G (1948) *Vår Fågelvärld* **7**, 129–44. MAGRATH, H A F (1917) *J. Bombay nat. Hist. Soc.* **25**, 149; (1919) *J. Bombay nat. Hist. Soc.* **26**, 672–3. MAINARDI, D (1964) *Riv. ital. Orn.* **34**, 213–16. MAIRLOT, M (1919) *Gerfaut* **9**, 32–5. MAIRY, F (1971) *Ann. Soc. Roy. Zool. Belg.* **101**, 247–54; (1976a) *Biophon* **4** (1), 3–5; (1976b) *Biophon* **4** (1), 6–9; (1976c) *Biophon* **4** (2), 2–5; (1976d) *Biophon* **4** (3), 2–3; (1979a) *Bull. Soc. Roy. Sci. Liège* **48**, 340–54; (1979b) *Bull. Soc. Roy. Sci. Liège* **48**, 378–90. MAKATSCH, W (1950) *Die Vogelwelt Macedoniens.* Leipzig; (1955) *Der Brutparasitismus in der Vogelwelt.* Radebeul; (1957) *Vogelwelt* **78**, 19–31; (1974) *Die Eier der Vögel Europas* **1**; (1976) **2**. Radebeul; (1981) *Verzeichnis der Vögel der Deutschen Demokratischen Republik.* Leipzig. MALBRANT, R (1952) *Faune du Centre Africain français.* Paris; (1957) *Oiseau* **28**, 214–31. MAL'CHEVSKI, A S and NEUFELDT, I A (1954) *Uchen. Zap. Leningrad. Univ. 181 Ser. Biol.* **38**, 61–76. MALUQUER MALUQUER, S (1971) *Ardeola spec. vol.,* 191–334. MALZY, P (1962) *Oiseau* **32** suppl., 1–81. MÄND, E (1982) In Paakspuu, K V (ed) *Eesti N S V Riiklike Looduskaitsealade Teaduslikud Tööd* **3**, 96–108. Tallinn. MÁÑEZ, M (1983a) *Alytes* **1**, 275–90; (1983b) *Actas del XV Congreso Internacional de Fauna Cinegética y Silvestre, Trujillo 1981,* 617–34. MANN, C F (1971) *Bull. Br. Orn. Club* **91**, 41–6. MANNES, P (1971) *J. Orn.* **112**, 231–2. MANNICHE, A L V (1910) *Medd. om Grøn.* **45**, 1–200. MANNING, T H and MACPHERSON, A H (1961) *Trans. Roy. Can. Inst.* **33** (2), 116–239. MANVILLE, R H (1949) *Jackpine Warbler* **27**, 158–61. MAŘAN, J (1958) *Sylvia* **15**, 254. MARBOT, T (1959) *Orn. Beob.* **56**, 8–18. MARCHANT, J (1975) *BTO News* **71**, 1. MARCHANT, S (1941) *Ibis* (14) **5**, 265–95, 378–96; (1961a) *Bull. Br. Orn. Club* **81**, 134–41; (1961b) *Bull. Iraq nat. Hist. Mus.* **1** (4), 1–37; (1962a) *Bull. Br. Orn. Club* **82**, 123–4; (1962b) *Bull. Iraq nat. Hist. Mus.* **2** (1), 1–40; (1963a) *Bull. Br. Orn. Club* **83**, 52–6; (1963b) *Ibis* **105**, 369–98; (1963c) *Ibis* **105**, 516–57. MARCHANT, S and MACNAB, J W (1962) *Bull. Iraq. nat. Hist. Inst.* **2** (3), 1–48. MARDAL, W (1974) *Feldorn.* **16**, 4–7. MARIÁN, M and MARIÁN, O (1973) *Vertebr. Hung.* **14**, 9–18. MARIÁN, M and SCHMIDT, E (1967) *Móra Ferenc Múz. Évk.,* 271–5. MARIEN, D (1952) *J. Bombay nat. Hist. Soc.* **49**, 151–64. MARINKELLE, C J (1959) *Ool. Rec.* **33**, 55–7. MARISOVA, I V (1965) *Zool. Zh.* **44**, 1735–7. MARISOVA, I V and BUTENKO, A G (1976) *Vestnik Zool.* **28**, 29–34. MARPLES, G (1939) *Br. Birds* **33**, 81–2. MARPLES, G and MARPLES, A (1934) *Sea terns or sea swallows.* London. MARQUISS, M and CUNNINGHAM, W A J (1980) *Scott. Birds* **11**, 56–7. MARRIAGE, A W (1914) *Br. Birds* **7**, 268. MARSDEN, W M (1927) *Br. Birds* **21**, 7. MARSHALL, W (1889) *Zool. Vorträge* **2**, 1–76. MARSHALL, W, HEMPELMANN, F, and STRASSEN, O ZUR (1911) *Brehms Tierleben* **7**: *Die Vögel* **2**. Leipzig. MARTEN, D (1952) *J. Bombay nat. Hist. Soc.* **49**, 151–64. MARTI, C D (1969) *Wilson Bull.* **81**, 467–8; (1973) *Wilson Bull.* **85**, 178–81; (1974) *Condor* **76**, 45–61. MARTIN, C and SAINT GIRONS, M-C (1973) *Oiseau* **43**, 51–4. MARTIN, C E (1937) *Br. Birds* **31**, 88. MARTIN, R M (1974) *Int. Zoo Yearbook* **14**, 99–102. MARTINEZ, A and MUNTANER, J (1979) *Alauda* **47**, 29–33. MARTINI, E (1964) *Bonn. zool. Beitr.* **15**, 59–71. MÄRZ, R (1940) *Beitr. Fortpfl. Biol. Vögel* **16**, 125–35; (1954) *Vogelwelt* **75**, 181–8; (1964) *Vogelwelt* **85**, 33–8; (1965) *Beitr. Vogelkde.* **10**, 338–48. MÄRZ, R and PIECHOCKI, R (1980) *Der Uhu.* Wittenberg Lutherstadt. MASATOMI, H (1963) *J. Fac. Sci. Hokkaido Univ. Ser. 6 Zool.* **15**, 284–92. MASCHER, J W (1963) *Vår Fågelvärld* **22**, 293–4. MASEFIELD, J R B (1929) *Br. Birds* **23**, 21–2. MASER, C (1975) *Wilson Bull.* **87**, 552–3. MASON, C R and ROBERTSON, W B (1965) *Auk* **82**, 109. MASON, C W and MAXWELL-LEFROY, H (1912) *Mem. Dept. Agric. India* **3**. MASURAT, H (1966) *Orn. Mitt.* **18**, 77. MASSEY, B W (1974) *Proc. Linn. Soc. NY* **72**, 1–24; (1976) *Auk* **93**, 760–73. MASSEY, B W and ATWOOD, J L (1978) *Bird-Banding* **49**, 360–71. MASTERSON, A (1973) *Honeyguide* **73**, 17–19. MÁTÉ, L (1962) *Aquila* **68**, 251–3. MATHER, J R (1979) *Br. Birds* **72**, 552. MATHIASSON, S (1965) *Göteborgs Naturhist. Mus. Årstryck,* 16–23; (1967) *Vår Fågelvärld* **26**, 297–347. MATHIEU, J (1965) *Nos Oiseaux* **28**, 114. MATOUŠEK, B (1951) *Sylvia* **13**, 122–5; (1956) *Biol. Prace Slov. Akad.* **2** (7), 3–89. MATSUOKA, S (1976) *Tori* **25**, 107–8. MATSUOKA, S and KOJIMA, K (1979) *Tori* **28**, 107–16. MATTES, H (1981) *Orn. Beob.* **78**, 103–8. MATTHÄS, U and SCHRÖDER, H (1972) *Vogelwelt* **93**, 72–4. MATTHÉ, L (1982) *Wielewaal* **48**, 243–5. MATTHES, N (1983) In Hall, D (ed) *The living countryside* **109**, 2161–3. London. MATTHES, W (1966) *Orn. Mitt.* **18**, 123. MATTHEWS, G V T (1947) *Br. Birds* **30**, 31–3. MATVEJEV, S D and VASIĆ, V F (1973) *Catalogus Faunae Jugoslaviae: Aves.* Lubljiana. MAURER, D R and RAIKOW, R J (1981) *Ann. Carneg. Mus.* **50** (18), 417–34. MAY, E S (1943) *Trans. Herts. nat. Hist. Soc.* **21**, 338–9. MAY, R and FISHER, J (1953) *Br. Birds* **46**, 51–5. MAYAUD, N (1939a) *Alauda* **11**, 211–25; (1939b) *Alauda* **11**, 236–55; (1953) *Alauda* **21**, 1–63; (1956) *Alauda* **24**, 206–18; (1958) *Alauda* **26**, 151. MAYR, E (1951) *Proc. int. orn. Congr.* **10**, 91–131; (1963) *Animal species and evolution.* Harvard. MAYR, E and AMADON, D (1951) *Am. Mus. Novit.* **1496**. MEAD, C J (1969) In Gooders, J (ed) *Birds of the world,* 1362–5. London; (1974) *Bird*

Study 21, 45–86; (1978) Ibis 120, 110. MEAD, C J and WATMOUGH, B R (1976) Bird Study 23, 187–96. MEADE, G M (1948) Bird-Banding 19, 51–9. MEADE-WALDO, E G (1889a) Ibis (6) 1, 1–13; (1889b) Ibis (6) 1, 503–20; (1893) Ibis (6) 5, 185–207; (1896) Zoologist (3) 20, 298–9; (1897) Avic. Mag. 3, 177–80; (1906) Avic. Mag. 4, 219–22; (1907) Avic. Mag. 5, 281–2; (1921) Ibis (11) 3, 348–9; (1922) Bull. Br. Orn. Club 42, 69–70. MEBS, T (1960) Anz. orn. Ges. Bayern 5, 584–90; (1966) Eulen and Käuze. Stuttgart; (1972) Anz. orn. Ges. Bayern 11, 7–25. MEDWAY, LORD and WELLS, D R (1976) The birds of the Malay peninsula 5. London. MEES, G F (1973) Zool. Meded. 46, 197–207; (1977a) Zool. Meded. 51, 243–64; (1977b) Zool. Verh. 157, 1–64; (1979a) Mitt. Zool. Mus. Berlin 55 suppl. Ann. Orn. 3, 127–34; (1979b) Zool. Bijdr. 26, 1–63. MEETH, P (1962) Limosa 35, 219–23. MEIER, H (1865) J. Orn. 13, 293–5; (1959a) Orn. Beob. 56, 22–3; (1959b) Orn. Beob. 56, 24–5. MEIJERING, M P D (1967) Ardea 55, 91–111. MEIKLEJOHN, A H (1928) Br. Birds 22, 89. MEIKLEJOHN, M F M (1940) Ostrich 11, 33–40; (1951) Ibis 93, 142. MEINERTZHAGEN, R (1920) Ibis (11) 2, 132–95; (1925) Ibis 12 (1), 600–21; (1930) Nicoll's birds of Egypt. London; (1934) Ibis (13) 4, 528–71; (1938) Ibis (14) 2, 754–9; (1940) Ibis (14) 4, 187–234; (1954) The birds of Arabia. Edinburgh; (1959) Pirates and predators. Edinburgh; (1964) In Thompson, A L (ed) A new dictionary of birds, 711–12. London. MEININGER, P G, KWAK, R, and HEIJNEN, T (1978) Vogeljaar 26, 10–12. MEISE, W (1937) Mitt. Zool. Mus. Berlin 22, 86–186; (1959) Abh. Verh. naturw. Ver. Hamburg 3, 86–104. MELDE, M (1982) Falke 29, 157–63. MELENDRO, J and GISBERT, J (1978) Ardeola 24, 261. MEL'NIKOV, Y I (1977) In Skryabin, N G (ed) Ekologiya ptits vostochnoy Sibiri, 59–92. Irkutsk. MENDALL, H L (1935) The relationship of certain sea birds to the fishing of the state of Maine. Bull. (Maine) Dept. Sea Shore Fish. MENDELSSOHN, H, YOM-TOV, Y, and SAFRIEL, U (1975) Ibis 117, 110–11. MENGEL, R M (1965) Orn. Monogr. AOU 3. MENZEL, H (1962) Regulus 42, 270–5; (1968) Der Wendehals. Wittenberg Lutherstadt; (1975) Beitr. Vogelkde. 21, 344–5. MERCIER, A (1919) Gerfaut 9, 74–7. MERIKALLIO, E (1946) Ann. Zool. Soc. Bot. Fenn. Vanamo 12; (1958) Fauna Fenn. 5, 1–181. MERKEL, M (1957) Orn. Mitt. 9, 215. MES, R and SCHUCKARD, R (1976) Versl. en Techn. Gegev. zool. Mus. Univ. Amsterdam 11, 1–36. MESHKOV, M and URYADOVA, L (1972) Commun. Baltic Commiss. Study Bird Migr. 7, 18–28. MESTER, H (1959) Vogelwelt 80, 119–20. MESTER, H and PRÜNTE, W (1957) Orn. Mitt. 9, 226. MESTRE RAVENTÓS, P (1969) Ardeola 14, 137–42. MÉSZÁROS, G (1948–51) Aquila 55–8, 287–8. METZMACHER, M (1979) Aves 16, 89–123. MEY, E (1967) Thür. Orn. Rdbr. 11, 18. MEYER, H (1965) Beitr. Vogelkde. 10, 451–4; (1980) Beitr. Vogelkde. 26, 302. MEYER-DEEPEN, J (1975) Vogelkdl. Ber. Niedersachs. 7, 60; (1977) Vogelkdl. Ber. Niedersachs. 9, 86. MEYLAN, A (1974) Nos Oiseaux 32, 278–9. MEYLAN, O (1932) Orn. Beob. 29, 120–4. MICHAELIS, H J (1971) Beitr. Vogelkde. 17, 261. MICHEL, J (1894) Orn. Monatsschr. 19, 95–6. MIENIS, H K (1979) Salamandra 15, 107–8. MIERA, C (1976) Falke 23, 403–5. MIKKOLA, H (1970a) Kainuun Linnut 1, 52–5; (1970b) Suom. Riista 22, 97–104; (1970c) Ornis fenn. 47, 10–14; (1970d) Orn. Mitt. 22, 72–5; (1970–1) Angew. Orn. 3, 133–41; (1971a) Päijät-Hämeen Linnut 2, 8–11, 32; (1971b) Orn. Mitt. 23, 40–1; (1972a) Orn. Mitt. 24, 158–63; (1972b) Br. Birds 65, 31–2; (1972c) Br. Birds 65, 453–60; (1972d) Beitr. Vogelkde. 18, 297–309; (1973) Falke 20, 196–204; (1974) Siipirikko 1 (2), 16–18; (1976a) Br. Birds 69, 144–54; (1976b) Savon Luonto 8, 13–22; (1981a) Der Bartkauz. Wittenberg Lutherstadt; (1981b) Savon Luonto 13, 30–3; (1983) Owls of Europe. Calton. MIKKOLA, H and JUSSILA, E (1974) Päijät-Hämeen Linnut 5, 19–21. MIKKOLA, H and MIKKOLA, K (1974) Suom. Linnut 9, 103–7. MIKKOLA, H and SULKAVA, S (1969a) Ornis fenn. 46, 126–31; (1969b) Ornis fenn. 46, 188–93; (1970) Br. Birds 63, 23–7. MILBURN, C E (1915) Br. Birds 9, 95–6. MILENZ, K (1962) Auspicium 1, 444–50. MILLER, A H (1943) Condor 45, 220–5. MILLER, F L and RUSSELL, R H (1973) Can. Fld.-Nat. 87, 180–1. MILLER, W DE W (1912) Bull. Am. Mus. nat. Hist. 31, 239–311. MILLER, W J and MILLER, L S (1958) Anim. Behav. 6, 3–8. MILLS, D H (1957) J. Fish. Res. Bd. Canada 14, 729–30. MILLS, D G H (1981) Br. Birds 74, 354. MILON, P (1950) Ibis 92, 545–53. MILON, P E (1947) Ostrich 18, 183–7. MILSTEIN, P LE S (1968) Br. Birds 61, 36–7; (1975) Ostrich Suppl. 10. MINNEMAN, D and BUSSE, H (1978) Zool. Garten 48, 433–9. MINORANSKI, V A (1976) Ornitologiya 12, 238–9. MIRZOBOKHODUROV, R A (1974) Ornitologiya 11, 233–7. MISONNE, X (1956) Gerfaut 46, 191–7. MITCHEL, J and CAMERON, J M (1979) Glasgow Nat. 19, 510–11. MITCHELL, B L (1964) Puku 2, 129. MITCHELL, K D G (1948) Br. Birds 41, 120. MITCHELL, M H (1947) Can. Fld.-Nat. 61, 68–9. MITROPOL'SKI, O V (1977) In Sludski, A A (ed) Redkie i ischezayushchie zveri i ptitsy Kazakhstana, 201–6. Alma-Ata. MOBBS, F G (1975) J. N. Gloucs. Nat. Soc. 24, 272. MOCCI DEMARTIS, A (1976) Riv. ital. Orn. 46, 133–42. MÖCKEL, R (1979) Jber. Mus. Hein. 4, 77–86; (1981) Veröff. Mus. Naturk. Karl-Marx-Stadt 11, 60–76. MÖCKEL, R and KUNZ, M (1981) Beitr. Vogelkde. 27, 129–49. MOEED, A (1975) Notornis 22, 135–42. MOFFAT, C B (1905) Zoologist (4) 9, 233–4; (1941) Irish Nat. J. 7, 314–15. MØLLER, A P (1975a) Dansk orn. Foren. Tidsskr. 69, 1–8; (1975b) Dansk orn. Foren. Tidsskr. 69, 9–18; (1975c) Danske Fugle 27, 61–77; (1977) Dansk orn. Foren. Tidsskr. 71, 103–11; (1978a) Dansk orn. Foren. Tidsskr. 72, 60; (1978b) Dansk orn. Foren. Tidsskr. 72, 119–26; (1978c) Dansk orn. Foren. Tidsskr. 72, 145–7; (1981a) Vogelwarte 31, 74–94; (1981b) Vogelwarte 31, 149–68; (1981c) Dansk orn. Foren. Tidsskr. 75, 127–30; (1982) J. Orn. 123, 41–53. MOLNÁR, B (1944–7) Aquila 51–4, 189–90; (1950) Aquila 51–4, 100–12. MOLTONI, E (1932) Riv. ital. Orn. 2, 39–46; (1937) Riv. ital. Orn. 7, 13–33, 61–119; (1938) Riv. ital. Orn. 8, 1–16; (1939a) Riv. ital. Orn. 9, 20–31; (1939b) Riv. ital. Orn. 9, 57–70; (1940) Riv. ital. Orn. 10, 176–82; (1945) Riv. ital. Orn. 15, 1–18; (1948) Riv. ital. Orn. 18, 74–86; (1949) Riv. ital. Orn. 19, 95–122; (1950) Riv. ital. Orn. 20, 140–4; (1954a) Natura 45, 137–9; (1954b) Riv. ital. Orn. 24, 147–58; (1959) Natura 50, 77–9; (1965) Riv. ital. Orn. 35, 237–41; (1966) Riv. ital. Orn. 36, 368–71; (1968) Riv. ital. Orn. 38, 355. MOLTONI, E and BRICHETTI, P (1977) Riv. ital. Orn. 47, 149–204. MONARD, A (1940) Arq. Mus. Bocage 11. MONK, J F (1955) Bird Study 2, 87–9; (1963) Bird Study 10, 112–32. MONTAGUE, F A (1926) Ibis (12) 2, 136–51. MONTFORT, F VAN (1947) Gerfaut 37, 98–9. MOON, S J (1983) Br. Birds 76, 335–9. MOORE, H J and BOSWELL, C (1956a) Iraq nat. Hist. Mus. Publ. 9; (1956b) Iraq nat. Hist. Mus. Publ. 10. MOORE, R J and BALZAROTTI, M A (1983) Bull. Br. Orn. Club 103, 65–71. MOORE, T J (1860) Ibis 2, 105–10. MORALES, J A G (1974) Ardeola 20, 361. MOREAU, R E (1933) Ibis 13 (3), 1–33, 187–229, 399–440; (1938) Proc. zool. Soc. Lond. (A) 108, 1–26; (1942a) Ibis 14 (6), 27–49; (1942b) Ostrich 13, 137–47; (1944) Ibis 86, 16–29; (1961) Ibis 103a, 373–427, 580–623; (1966) The bird faunas of Africa and its islands. London; (1967) Ibis 109, 232–59; (1972) The Palaearctic-African bird migration systems. London. MOREAU, R E and MOREAU, W M (1939) Ibis (14) 3, 760; (1941a) Ibis (14) 5, 614; (1941b) Ibis (14) 5, 615; (1946) Ibis 88, 522–4. MOREL, G (1968) Mém. ORSTOM Paris 29, 1–179; (1972) Liste commentée des oiseaux du Sénégal et de la Gambie. Dakar. MOREL, G and ROUX, F (1966) Terre et Vie 20, 19–72, 143–76; (1973) Terre et Vie 27, 523–50. MOREL, G J and MOREL, M-Y (1979) Malimbus 1, 66–7. MOREL, M-Y (1973) Bull. Inst. franç. Afr. noire 35, 180–5; (1975) Oiseau 45, 97–125; (1983) Alauda 51, 179–202. MORGAN, R and GLUE, D (1977) Bird Study 24, 15–24. MORITZ, D (1979) Orn. Mitt. 31, 137–9. MORONY, J J, JR, BOCK, W J, and

FARRAND, J, JR (1975) *Reference list of the birds of the world*. New York. MORPHY, M J (1963) *La Palma (Canary Islands) Exped. Rep.* Univ. Newcastle-on-Tyne; (1964) *Newcastle-upon-Tyne Univ. Explor. Soc. 1964 Orn. Rep.*, 52–9. MORRIS, P (1979) *J. Zool. Lond.* **189**, 540–5. MORSIER, J DE (1947) *Nos Oiseaux* **19**, 141–5. MORTENSEN, P H (1965) *Dansk orn. Foren. Tidsskr.* **58**, 137–8. MÖRZER BRUYNS, W F J and VOOUS, K H (1964) *Ardea* **52**, 117–18. MOŠANSKÝ, A (1958) *Sylvia* **15**, 55–60. MOSELEY, L J (1979) *Auk* **96**, 31–9. MOUILLARD, B (1939) *Alauda* **11**, 55–60. MOULE, C W H (1951) *Br. Birds* **44**, 133–4. MOUNTFORT, G (1957) *Br. Birds* **50**, 263–7; (1958) *Portrait of a wilderness*. London; (1969) *The vanishing jungle*. London; (1981) *Br. Birds* **74**, 265–6. MOUNTFORT, G and FERGUSON-LEES, I J (1961) *Ibis* **103a**, 86–109. MOYNIHAN, M (1955) *Behaviour Suppl.* **4**; (1959) *Am. mus. Novit.* **1928**; (1978) *Terre et Vie* **32**, 557–76. MUCKENSTURM, B (1971) *Rev. Comport. Animal* **5**, 227–48. MUELLER, H C (1976) *Wilson Bull.* **88**, 675–6. MUELLER, H C and BERGER, D D (1959) *Bird-Banding* **30**, 182. MUIR, R C (1954) *Bird Study* **1**, 111–17. MUIRHEAD, G (1895) *The birds of Berwickshire*. Edinburgh. MUKHERJEE, A K (1976) *J. Bombay nat. Hist. Soc.* **72**, 422–47. MULDER, T and TANGER, D (1980) *Watervogels* **5**, 122–6. MÜLLER, A K (1967) *Vogelwarte* **24**, 63–4. MÜLLER, B (1928) *Beitr. Fortpfl. Biol. Vögel* **4**, 224. MÜLLER, H (1959) *Vogelwarte* **20**, 91–115. MÜLLER, W (1982) *Orn. Beob.* **79**, 105–19. MULSOW, R (1979) *Hamb. avifaun. Beitr.* **16**, 25–42. MÜNCH, H (1952) *Der Wiedehopf*. Leipzig. MUNDY, P J (1973) *Ibis* **115**, 602–4. MUNDY, P J and COOK, A W (1974) *Bull. Niger. orn. Soc.* **10**, 1–28; (1977) *Ostrich* **48**, 72–84. MUNTEANU, D (1968) *Lucrările* **1**, 359–65; (1970) *Lucrările* **3**, 347–50. MURIE, A J (1963) *Birds of Mount McKinley Nat. Park, Alaska*. MURIE, O J (1929) *Condor* **31**, 3–12; (1959) *N. Am. Fauna* **61**. US Fish Wildl. Serv., Washington DC. MURPHY, R C (1924) *Bull. Am. Mus. nat. Hist.* **50**, 211–78; (1936) *The oceanic birds of South America* **2**. New York; (1938) *Nat. Hist.* **41**, 164–78. MURPHY, R C and VOGT, W (1933) *Auk* **50**, 325–49. MURRAY, B G and JEHL, J R (1964) *Bird-Banding* **35**, 253–63. MURRAY, J M (1952) *A check-list of the birds of Virginia*. Virginia Orn. Soc. MURTON, P G (1968) *Br. Birds* **61**, 193–212. MURTON, R K (1958) *Bird Study* **5**, 157–83; (1960a) *Ann. appl. Biol.* **48**, 95–106; (1960b) *Br. Birds* **53**, 321–4; (1961) *Bird Study* **8**, 165–73; (1965a) *The Wood Pigeon*. London; (1965b) *Ann. appl. Biol.* **55**, 177–9; (1966) *Bird Study* **13**, 311–27; (1968) *Br. Birds* **61**, 193–212; (1969) In Gooders, J (ed) *Birds of the world*, 1518–19; (1970) *Br. Birds* **63**, 390–3; (1971a) *Man and birds*. London; (1971b) *Behaviour* **40**, 10–42. MURTON, R K and CLARKE, S P (1968) *Br. Birds* **61**, 429–48. MURTON, R K, COOMBS, C F B, and THEARLE, R J P (1972a) *J. appl. Ecol.* **9**, 875–89. MURTON, R K and ISAACSON, A J (1962) *Ibis* **104**, 503–21. MURTON, R K, ISAACSON, A J, and WESTWOOD, N J (1963a) *Br. Birds* **56**, 345–75; (1963b) *Proc. zool. Soc. Lond.* **141**, 747–81; (1966) *J. appl. Ecol.* **3**, 55–96. MURTON, R K and RIDPATH, M G (1962) *Bird Study* **9**, 7–41. MURTON, R K, THEARLE, R J P, and COOMBS, C F B (1974a) *J. appl. Ecol.* **11**, 841–54. MURTON, R K, THEARLE, R J P, and LOFTS, B (1969) *Anim. Behav.* **17**, 286–306. MURTON, R K, THEARLE, R J P, and THOMPSON, J (1972b) *J. appl. Ecol.* **9**, 835–74. MURTON, R K and WESTWOOD, N J (1964) *Trans. 6th Congr. int. Union Game Biol.*, 369–77; (1966) *Bird Study* **13**, 130–46. MURTON, R K, WESTWOOD, N J, and ISAACSON, A J (1964a) *Ibis* **106**, 174–88; (1964b) *Ibis* **106**, 482–507; (1965) *Ibis* **170**, 254–6; (1974b) *Ibis* **116**, 52–73; (1974c) *J. appl. Ecol.* **11**, 61–81. MURTON, R K, WESTWOOD, N J, and THEARLE, R J P (1973a) *Ibis* **115**, 132–4; (1973b) *J. Reprod. Cart.* **19** suppl., 563–77. MUSELET, D (1982) *Oiseau* **52**, 219–30. MUSIL, A (1963) *Gefiederte Welt* **87**, 7–10; (1965) *Gefiederte Welt* **89**, 133–4; (1970) *Falke* **17**, 100–1. MUSSELWHITE, D W and WARE, R (1923) *Bull. Br. Orn. Club* **44**, 30–2. MYRBERGET, S (1959a) *Fauna* **12**, 143–56; (1959b) *Sterna* **3**, 239–46; (1961) *Fauna* **14**, 2–8; (1962a) *Medd. Stat. Vilt.* (2) **11**, 1–51; (1962b) *Fauna* **15**, 1–10; (1962c) *Fauna* **15**, 159–64; (1963) *Nytt Mag. Zool.* **11**, 74–84; (1973a) *Sterna* **12**, 33–40; (1973b) *Sterna* **12**, 307–15; (1981) *Cinclus* **4**, 27–9. MYSTERUD, I (1970) *Nytt mag. Zool.* **18**, 49–74. MYSTERUD, I and DUNKER, H (1983) *Viltrevy* **12**, 71–113. MYSTERUD, I and HAGEN, Y (1969) *Nytt Mag. Zool.* **17**, 165–7.

NADER, I A (1969) *Bull. Iraq nat. Hist. Mus.* **4** (1), 1–7. NADLER, T (1967) *Falke* **14**, 292–4; (1976) *Die Zwergseeschwalbe*. Wittenberg Lutherstadt. NÆVDAL, G and NÆVDAL, G (1968) *Sterna* **8**, 95–6. NAGELL, B and FRYCKLUND, I (1965) *Vår Fågelvärld* **24**, 26–55. NAGY, E (1931–4) *Aquila* **38–41**, 456–7. NAGY, I (1959–61) *Aquila* **66–8**, 253. NAGY, L (1921) *Aquila* **28**, 200. NAIK, R M and NAIK, S (1965) *Pavo* **3**, 96–120; (1966) *Pavo* **4**, 84–91. NAIK, R M and RAZACK, A (1963) *Pavo* **1**, 90–8; (1967) *Pavo* **5**, 57–74. NASH, A (1975) *Bird Study* **22**, 238. NASH, J K (1924) *Scott Nat.* **147**, 77–83. NAUMANN, J A (1826) *Naturgeschichte der Vögel Deutschlands* **5**; (1833) **6**; (1844) **12**. Leipzig. (1901) *Naturgeschichte der Vögel Mitteleuropas* **4**; (1903) (ed Hennicke, C R) **9**; (1905) (ed Hennicke, C R) **11**. Gera. NAUNTON, C R (1976) *Br. Birds* **69**, 311–12. NAUROIS, R DE (1959) *Alauda* **27**, 241–308; (1961) *Alauda* **29**, 241–59; (1969a) *Mém. Mus. nat. Hist. Nat.* (A) **56**, 1–312; (1969b) *Bull. Fond. Afrique Noire* **31**, 143–218; (1972) *Oiseau* **42**, 195–7; (1974) *Oiseau* **44**, 72–84; (1983a) *Riv. ital. Orn.* **52**, 154–66; (1983b) *Cyanopica* **3**, 17–28. NAUROIS, R DE and ROUX, F (1974) *Oiseau* **44**, 72–84. NAVASAITIS, A (1968) *Ornitologiya* **9**, 362–5. NECHAEV, V A (1969) *Ptitsy Yuzhnykh Kuril'skikh Ostrovov*. Leningrad; (1975) *Ornitologicheskie issledovaniye na Dal'nem Vostoke*. Vladivostok. NEEDHAM, J (1942) *Biochemistry and morphogenesis*. Cambridge. NEHLS, H W (1969) *Vogelwarte* **25**, 52–7. NELSON, G E (1962) *Q. J. Fla. Acad. Sci.* **25**, 303–6. NELSON, J B (1973) *Azraq, desert oasis*. London. NELSON, T H (1907) *The birds of Yorkshire*. Hull. NERO, R W (1964) *Blue Jay* **22**, 54–5; (1980) *The Great Gray Owl*. Washington DC. NERO, R W, SEALY, S G and COPELAND, H W R (1974) *Loon* **46**, 161–5. NETTLESHIP, D N (1972) *Ecol. Monogr.* **42**, 239–68. NETTLESHIP, D N and GASTON, A J (1978) *Can. Wildl. Serv. occ. Pap.* **39**. NEUFELDT, I A (1958a) *Trudy Zool. Inst. Akad. Nauk SSR* **25**, 183–254; (1958b) *Zool. Zh.* **37**, 257–70; (1982) *Falke* **29**, 257–74. NEUFELDT, I and IVANOV, A I (1960) *Br. Birds* **53**, 431–5. NEUHAUS, W, BRETTING, H and SCHWEIZER, B (1973) *Biol. Zentralbl.* **92**, 495–512. NEUMANN, E (1942) *Beitr. Fortpfl. Biol. Vögel* **18**, 144. NEUMANN, O (1905) *J. Orn.* **53**, 189; (1909) *Orn. Monatsber.* **17**, 152–5; (1915) *Orn. Monatsber.* **23**, 73–5; (1924) *Verh. orn. Ges. Bayern* **20**, 470–2. NEWBY, J E (1979) *Malimbus* **1**, 90–109; (1980) *Malimbus* **2**, 29–50. NEWTON, A (1861) *Ibis* (1) **3**, 374–99; (1864) *Ibis* (1) **6**, 185–222; (1890) *Ibis* (2) **6**, 207–14. NEWSTEAD, R (1908) *J. Board Agric. Lond. Suppl.* **15** (9). NICHOLLS, C A and ROOK, D A (1962) *West Austral. Nat.* **8**, 84–6. NICHOLLS, D H (1977) *Bokmakierie* **29**, 20–3. NICHOLS, J B (1907) *Br. Birds* **1**, 185–6. NICHOLS, J T (1913) *Auk* **30**, 505–11. NICHOLSON, C P (1975) *Migrant* **46**, 62–3. NICHOLSON, E M (1930) *Ibis* (12) **6**, 280–313, 395–428. NICOLAU-GUILLAUMET, P and SPITZ, F (1958) *Oiseaux de France* **8**, 41–2. NIEDRACH, J and ROCKWELL, R B (1939) *Birds of Denver and mountain parks*. Denver. NIELSEN, B P (1975) *Bull. Br. Orn. Club* **95**, 80–1. NIETHAMMER, G (1936) *Orn. Monatsber.* **44**, 45–52; (1938) *Handbuch der deutschen Vogelkunde* **2**; (1942) **3**. Leipzig; (1943a) *J. Orn.* **91**, 167–238; (1943b) *J. Orn.* **91**, 296–304; (1955) *Bonn. zool. Beitr.* **6**, 29–80; (1962) *Biol. Zentralbl.* **81**, 67–73; (1970) *J. Orn.* **111**, 367–77. NIETHAMMER, G and PRZYGODDA, W (1954) *Vogelwelt* **75**, 41–55. NIETHAMMER, G, KRAMER, H, and WOLTERS, H E (1964) *Die Vögel Deutschlands—Artenliste*. Frankfurt am Main. NIGGELER, E (1968) *Orn. Beob.*

65, 131–2. NIKLUS, M Y (1957) *Proc. 3rd Baltic orn. Conf.*, 205–8. NIKOL'SKI, I D (1979) *Zool. Zh.* **58**, 1511–17. NILSSON, I N (1977) *Fauna och Flora* **72**, 156–63; (1978) *Ibis* **120**, 528–31; (1981) *Ornis scand.* **12**, 216–23. NILSSON, I N, NILSSON, S G, and SYLVÉN, M (1981) *Biol. J. Linn. Soc.* **18**, 1–9. NILSSON, N (1942) *Vår Fågelvärld* **1**, 7–11. NISBET, I C T (1973a) *Nature* **241**, 141–2; (1973b) *Bird-Banding* **44**, 27–55; (1977) In Stonehouse, B and Perrins, C M (eds) *Evolutionary ecology*, 101–9. London; (1978b) *Ibis* **120**, 207–15; (1975) *Condor* **77**, 221–6; (1976) *Bird-Banding* **47**, 163–4; (1978a) *Bird-Banding* **49**, 50–8; (1980) *Rep. US Fish Wildl. Serv., Mass.* NISBET, I C T and COHEN, M E (1975) *Ibis* **117**, 374–9. NISBET, I C T and DRURY, W H (1972) *Bird-Banding* **43**, 97–106. NISBET, I C T, EVANS, P R, and FEENY, P P (1961) *Ibis* **103**a, 349–72. NISBET, I C T, WILSON, K J, and BROAD, W A (1978) *Condor* **80**, 106–9. NÖHRING, R (1965) *J. Orn.* **106**, 390–414. NORBERG, R Å (1964) *Vår Fågelvärld* **23**, 228–44; (1970) *Ornis scand.* **1**, 51–64; (1973) *Yearbook Swedish nat. Sci. Res. Council* **26**, 89–101; (1977) *Phil. Trans. Roy. Soc. Lond.* (B) **280**, 375–408; (1978) *Phil. Trans. Roy. Soc. Lond.* (B) **282**, 325–410. NORDBERG, S (1950) *Acta zool. fenn.* **63**, 1–62. NORDERHAUG, M (1964) *Fauna* **17**, 137–54; (1967) *Fauna* **20**, 236–44; (1968) *Norsk Polarinst. Medd.* **96**, 236–44; (1970) In Holdgate, M W (ed) *Antarctic ecology* **1**, 558–60. London; (1974) *Norsk Polarinst. Årbok 1972*, 99–106; (1980) *Norsk Polarinst. Skrift.* **173**. NORDERHAUG, M, BRUN, E, and MØLLEN, G U (1977) *Barentshavets sjøfuglressurser*. Oslo. NORDSTRÖM, G (1961) *Mem. Soc. Flora Fauna fenn.* **36**, 32–106; (1963) *Ornis fenn.* **40**, 80–124. NORMAN, D M (1980) *Br. Birds* **73**, 264–5. NORMAN, R K and SAUNDERS, D R (1969) *Br. Birds* **62**, 4–13. NØRREVANG, A (1958) *Dansk orn. Foren. Tidsskr.* **52**, 48–74; (1960) *Dansk orn. Foren. Tidsskr.* **54**, 9–35; (1977) *Fuglefångsten på Faeroerne*. Copenhagen; (1978) *Ibis* **120**, 109–10. NORRIS, R A and JOHNSTON, D W (1958) *Wilson Bull.* **70**, 114–29. NORTH, P M (1980) *Bird Study* **27**, 11–20. NOVRUP, L (1953) *Flora og Fauna* **59**, 33–41; (1956) *Dansk orn. Foren. Tidsskr.* **50**, 76–9. NOWAK, E (1965) *Die Türkentaube*. Wittenberg Lutherstadt; (1975a) *Bonn. zool. Beitr.* **26**, 135–54; (1975b) *Orn. Mitt.* **27**, 153–9. NUMEROV, A D and KOTYUKOV, Y V (1979) *Priroda* **6**, 69–73. NUORTERA, M, PATOMÄKI, J, and SAARI, L (1981) *Silva fenn.* **15**, 208–21. NYHOLM, E S (1968) *Ornis fenn.* **45**, 7–9.

OATES, E W (ed) (1890) *Hume's nests and eggs of Indian birds* **3**. London. OBERHOLZER, A and TSCHANZ, B (1968) *Rev. suisse Zool.* **75**, 43–51. O'CONNOR, R J (1962) *Br. Birds* **55**, 481; (1979) *Condor* **81**, 133–45. ODSJÖ, T and OLSSÓN, V (1975) *Vår Fågelvärld* **34**, 117–24. OEHME, H (1968) *Beitr. Vogelkde.* **13**, 393–6. OEMING, A F (1955) M S Thesis. Alberta Univ. OESER, R (1971) *Beitr. Vogelkde.* **17**, 78. OGASAWARA, K and IZUMI, Y (1977) *Misc. Rep. Yamashina Inst. Orn.* **9**, 231–43. OGI, H and TSUJITA, T (1973) *Jap. J. Ecol.* **23**, 201–9; (1977) *Res. Inst. N. Pac. Fish Hokkaido Univ.* spec. vol., 459–517. OJANEN, M (1983) *Lintumies* **18**, 48–9. OLDENBURG, H (1954a) *Vogelwelt* **75**, 152; (1954b) *Vogelwelt* **75**, 203; (1954c) *Vogelwelt* **75**, 204. OLESEN, E M (1974) *Danske Fugle* **26**, 130–8. OLESEN, L L and OLESEN, E M (1972) *Flora og Fauna* **78**, 33–9. OLIOSO, G (1974) *Alauda* **42**, 502. OLŠANÍK, V (1975) *Falke* **22**, 136; (1977) *Gefiederte Welt* **101**, 132–3. OLSON, S L, SWIFT, C C, and MOKHIBER, C (1979) *Auk* **96**, 790–2. OLSSON, V (1958) *Acta Vert.* **1**, 86–189; (1974) *Vår Fågelvärld* **33**, 3–14; (1976) *Vår Fågelvärld* **35**, 291–7; (1979) *Viltrevy* **11** (1). ONNO, S (1966) In Kumari, E (ed) *Ornithological researches in Estonia*, 10–20. Tallinn. OPITZ, M (1953) *Beitr. Vogelkde.* **3**, 198. OPPENOORTH, F (1938) *Levende Nat.* **43**, 33–40. ORTUNO, F and CABALLOS, A (1977) *Spanish woodlands*. Madrid. OSBORNE, P (1982) *Bird Study* **29**, 2–16. OSMOLOVSKAYA, V I (1946) *Zool. Zh.* **25**, 281–8. OSMOLOVKSAYA, V I and FORMOSOV, A N (1955) *Trudy Inst. Geogr. Akad. Nauk SSSR* **66**, 274–86. ÖSTERLOF, S (1969) *Vår Fågelvärld Suppl.* **5**, 1–159; (1977) *Stockholm Ringmärkningscentralen Rep. 1968*, 1–139. OTTOSPRUNCK, A (1967) *Ornis fenn.* **44**, 78. OUELLET, H (1977) *Ardea* **65**, 165–83. OUWENEEL, G L (1968) *Limosa* **41**, 64; (1975) *Limosa* **48**, 197–201; (1979) *Vogeljaar* **27**, 38–9. OWEN, D F (1949) *Bird Notes* **23**, 201–4; (1951) *Br. Birds* **44**, 324; (1954) *Ibis* **96**, 492. OWEN, D F and LE GROS, A E (1954) *Ent. Gaz.* **5**, 117–20. OWEN, J H (1913) *Br. Birds* **6**, 330–3; (1925) *Br. Birds* **31**, 125–8; (1933) *Rep. Felsted School sci. Soc.* **33**, 25–39.

PAIGE, J P (1948) *Br. Birds* **41**, 127; (1960) *Ibis* **102**, 520–5. PALM, B (1967) *Falke* **14**, 424–5. PALMER, M J (1982) *Br. Birds* **75**, 131. PALMER, R S (1941a) *Proc. Boston Soc. nat. Hist.* **42**, 1–119; (1941b) *Auk* **58**, 164–78. PALMÉR, S and BOSWALL, J (1972) *A field guide to the bird songs of Britain and Europe* **1–12**. Swedish Radio, Stockholm. PALMGREN, P (1930) *Acta zool. fenn.* **7**, 1–118. PALUDAN, K (1938) *J. Orn.* **86**, 562–638; (1940) *Danish Sci. Invest. Iran* **2**; (1947) Alken. Copenhagen; (1959) *Vidensk. Medd. dansk natur. Foren.* **122**; (1960) *Alkefugle. Nordens fugle i farver* **3**, 207–51. PANOV, E N (1973) *Ptitsy yuzhnogo Primor'ya*. Novosibirsk. PANOV, E N, ZYKOVA, L Y, KOSTINA, G N, and ANDRUSENKO, N N (1980) *Zool. Zh.* **59**, 1694–1705. PAPADOPOL, A (1963) *Trav. Mus. Hist. nat. 'Grigore Antipa'* **4**, 431–71; (1965) *Trav. Mus. Hist. nat. 'Grigore Antipa'* **5**, 335–46. PAPADOPOL, A and MÂNDRU, C (1977) *Trav. Mus. Hist. nat. 'Grigore Antipa'* **18**, 309–26. PARIS, P (1935) *Alauda* **7**, 502–5. PARKER, A (1977) *Br. Birds* **70**, 546. PARKER, G R (1974) *Can. Fld.-Nat.* **88**, 151–6. PARKIN, D T (1970) *Br. Birds* **63**, 389–90. PARMELEE, D F (1977) In Lano, G A (ed) *Adaptations within antarctic ecosystems*, 687–702. Washington. PARMELEE, D F AND MACDONALD, S D (1960) *Natn. Mus. Canada Bull.* **169**. PARMELEE, D F, STEPHENS, H A, and SCHMIDT, R H (1967) *Natn. Mus. Canada Bull.* **222**. PAROVSHCHIKOV, V Y and SEVASTYANOV, G N (1960) *Ornitologiya* **3**, 122–30. PARSLOW, J L F (1967) *Br. Birds* **60**, 177–202, 261–85; (1969) In Gooders, J (ed) *Birds of the world*, 1327–8. London. PARSLOW, J L F and EVERETT, M J (1981) *Birds in need of special protection in Europe*. Strasbourg. PAŞCOVSCHI, S (1974) *Stud. Comun. Muz. Ştiinţ. Nat. Bacău* **7**, 73–84. PAŞCOVSCHI, S and MANOLACHE, L (1970) *Stud. Comun. Muz. Ştiinţ. Nat. Bacău*, 245–50. PASSBURG, R E (1959) *Ibis* **101**, 153–69. PÄSSLER, W (1856) *J. Orn.* **4**, 34–68. PATERSON, A (1964) *Br. Birds* **57**, 203. PATEV, P (1950) *Ptitzite v Bulgarija*. Sofia. PÁTKAI, I (1966) *Aquila* **73**, 81–107. PATTEN, C J (1906) *The aquatic birds of Great Britain and Ireland*. London. PAULUSSEN, W (1953) *Gerfaut* **43**, 128–31; (1955) *Gerfaut* **45**, 1–5; (1957) *Gerfaut* **47**, 241–58. PAYN, W H (1978) *The birds of Suffolk*. Ipswich. PAYNE, R B (1977) *Bull. Br. Orn. Club* **97**, 48–53. PAYNE, R S (1971) *J. exp. Biol.* **54**, 535–73. PEAKALL, D B (1953) *Br. Birds* **46**, 304. PEAL, R E F (1968) *Bird Study* **15**, 111–26; (1972) *Proc. int. orn. Congr.* **15**, 675–6; (1973) *Br. Birds* **66**, 66–72. PEARSON, T H (1968) *J. anim. Ecol.* **37**, 521–52. PEIPONEN, V A (1964) *Ann. zool. fenn.* **1**, 281–302; (1965) *Ann. Acad. Sci. fenn.* **87**, 1–15; (1966) *Ann. Acad. Sci. fenn.* **101**, 1–35; (1970) *Ann. zool. fenn.* **7**, 239–50. PEIPONEN, V A and BOSLEY, A (1964) *Ornis fenn.* **41**, 40–2. PEITZMEIER, J (1953) *Orn. Mitt.* **5**, 6; (1956) *Orn. Mitt.* **8**, 155; (1974) *Orn. Mitt.* **26**, 22. PEITZMEIER, J and WESTERFRÖLKE, P (1962) *Orn. Mitt.* **14**, 67. PEK, L V and FEDYANINA, T F (1961) In Yanushevich, A I (ed) *Ptitsy Kirgizii* **3**, 59–118. Frunze. PELCHEN, H (1978) *Bull. E. Afr. nat. Hist. Soc.*, 58–9 PELLANTOVÁ, J (1975) *Zool. Listy* **24**, 249–62; (1981) *Folia Zool.* **30**, 59–73. PELTRE, R (1931) *Oiseau* **1**, 66–7. PELTZER, J (1969) *Regulus* **9**, 345–7. PELZELN, A VON (1869) *Novara-Exped., Zool. Theil.* **1** (2), 1–176. PEMBERTON, J R (1927) *Condor* **29**, 253–8. PÉNICAUD, P (1978) *Alauda* **46**, 43–51; (1979) *Terre et Vie* **33**, 591–609. PEN-

NYCUICK, C J (1956) *Ibis* **98**, 80–9. PEREIRA, H C (1979) *Cyanopica* **2** (1), 51–6. PERÉZ CHISCANO, J L (1971) *Ardeola* **15**, 145; (1975) *Ardeola* **21**, spec. vol., 753–94. PEREZ MELLADO, V (1980) *Ardeola* **25**, 93–112. PERRINS, C M (1967) *Br. Birds* **60**, 50–1; (1970) *Bird Study* **18**, 61–70; (1979) *British tits*. London. PERRY, R (1940) *Lundy. Isle of Puffins*. London; (1944) *Geog. Mag. Lond.* **17**, 84–95; (1948) *Shetland sanctuary*. London. PERSSON, F (1934) *Fauna och Flora*, 234. PETERS, H (1959) *Egretta* **2**, 49–50. PETERS, J L (1937) *Check-list of birds of the world* **3**. Cambridge, Mass. PETERS, N (1933) *Orn. Monatsber.* **41**, 5–13. PETERSEN, A (1945) *Dansk orn. Foren. Tidsskr.* **39**, 133–86; (1976*a*) *Astarte* **9**, 43–50; (1976*b*) *Ornis scand.* **7**, 185–92; (1977) *Náttúrufraedingurinn* **47**, 149–53; (1979) *Náttúrufraedingurinn* **49**, 229–56; (1981) D Phil Thesis. Oxford Univ. PETERSEN, N F and WILLIAMSON, K (1949) *Ibis* **91**, 17–23. PETERSON, R T (1960) *A field guide to the birds of Texas*. Boston, Mass.; (1961) *A field guide to western birds*. Boston, Mass. PÉTÉTIN, M and TROTIGNON, J (1972) *Alauda* **40**, 195–213. PETHON, P (1967) *Nytt Mag. Zool.* **14**, 84–95. PETRETTI, F (1977) *Gerfaut* **67**, 225–33. PETROV, V S (1954) *Trudy Nauch.-issled. Inst. Biol. i Biol. Fak. Kharkov. gos. Univ.* **20**, 171–80. PETTERSSON, B (1958) *Vår Fågelvärld* **39**, 316. PETTET, A (1975) *Bull. Niger. orn. Soc.* **11**, 34–40; (1982) *Br. Birds* **75**, 377. PETTINGILL, O S (1958) *Jack-Pine Warbler* **36**, 183–4. PETTIT, R G (1952) *Br. Birds* **45**, 70–1. PETZOLD, H and RAUS, T (1973) *Anthus* **10**, 25–38. PEUS, F (1949) *Vogelwelt* **70**, 52–3. PFEIFER, S (1952) *Vogelwelt* **73**, 141. PFEIFER, S and KEIL, W (1961) *Orn. Mitt.* **13**, 7–12. PFLUMM, W (1979) *J. Orn.* **120**, 64–72. PFÜTZENREITER, F (1957) *Vogelwelt* **78**, 120–3. PHILIPSON, W R (1948) *Birds of a valley*. London. PHILLIPS, B T (1946) *J. Bombay nat. Hist. Soc.* **46**, 89–103. PHILLIPS, W W A (1923) *Ibis* (11) **5**, 604–6. PICOZZI, N and HEWSON, R (1970) *Scott. Birds* **6**, 185–90. PIDOPLITSCHKA, J G (1937) *Veröff. Inst. Zool. Biol. Acad. Ukraine* **19**, 101–70. PIECHOCKI, R (1956) *Falke* **3**, 80–3; (1958) *Abh. Ber. Mus. Tierkde. Dresden* **24**, 105–203; (1961) *J. Orn.* **102**, 220–5; (1966) *Vogelwelt* **46**, 106–12; (1968*a*) *Beitr. Vogelkde.* **13**, 455–60; (1968*b*) *J. Orn* **109**, 30–6; (1968*c*) *Mitt. zool. Mus. Berlin* **44**, 149–292; (1969) *Bonn. zool. Beitr.* **20**, 42–7; (1974) *J. Orn.* **115**, 436–44. PIECHOCKI, R, STUBBE, M, UHLENHAUT, K, and SUMJAA, D (1981) *Mitt. zool. Mus. Berlin* **57** suppl., 71–128. PIENKOWSKI, M W (ed) (1975) *Joint Rep. Univ. E. Anglia Exped. Tarfaya Prov., Morocco 1972, and Cambridge Sidi Moussa Exped. 1972*. Norwich. PIEPER, H (1976) *Z. Säugetierk.* **41**, 274–7. PIERSON, T A, COBB, R G and SCANLON, P F (1976) *Wilson Bull.* **88**, 489–90. PIHL, S (1978) *Birds in Eilat, spring 1977*. Soc. Prot. Nat. Israel, Eilat. PIKULA, J and KUBÍK, V (1948) *Acta Sci. Nat. Brno* **12**, 1–40. PILASKI, J (1967) *J. Orn.* **108**, 84–5. PINEAU, J and GIRAUD-AUDINE, M (1974) *Alauda* **42**, 159–88; (1975) *Alauda* **43**, 135–41; (1979) *Trav. Inst. Sci. Rabat Sér. Zool.* **38**. PITELKA, F A, TOMICH, P Q, and TREICHEL, G W (1955) *Ecol. Monogr.* **25**, 85–117. PITMAN, C R S (1929) *Bull. Soc. Royal ent. d'Egypt* **10**, 93–103; (1961) *Bull. Br. Orn. Club* **81**, 148–9; (1967) *Bull. Br. Orn. Club* **87**, 41–5. PITTMAN, H H (1927) *Condor* **29**, 140–3. PIZZEY, G (1980) *A field guide to the birds of Australia*. Sydney. PLATH, L (1976) *Die Vögel der Stadt Rostock (Nonpasseres)*. Rostock. PLUCINSKI, A (1966) *Orn. Mitt.* **18**, 49–54; (1969) *Orn. Mitt.* **21**, 9–12; (1970) *Vogelwelt* **91**, 199–200. PLUMB, W J (1965) *Br. Birds* **58**, 449–56. PODKOVYRKIN, B A (1977) *Commun. Baltic Commiss. Study Bird Migr.* **10**, 40–52. PODOLSKI, A L (1981) *Ornitologiya* **16**, 181; (1982) In Ilyichev, V D and Gavrilov, V M (eds) *Abstr. int. orn. Congr.* **18**, 265. POKROVSKAYA, I V (1956) *Zool. Zh.* **35**, 96–110; (1963) *Uchen. zap. Leningrad. ped. Inst.* **230**, 19–32. POLATZEK, J (1909) *Orn. Jahrb.* **20**, 1–24. POLIVANOV, V M (1981) *Ekologiya ptits—duplognezdnikov Primor'ya*. Moscow. POLIVANOVA, N N (1971) *Trudy zapoved. 'Kedrovaya Pad'* **3**, 1–235. POMARNACKI, L (1960) *Przegl. Zool.* **4**, 312–13. POMERANTSEV, D V and SHEVYREV, I J (1910) *Trudy lesn. opyt. del. Rossii* **24**, 1–99. PONCY, R (1928) *Orn. Beob.* **26**, 21–3. POOLEY, A C (1967) *Ostrich* **38**, 11–12. POOS, J (1972) *Regulus* **10**, 412–16. POPOV, V A (ed) (1977) *Ptitsy Volzhsko—Kamskogo kraya*. Moscow. POPP, J (1955) *Vogelwelt* **76**, 110–11. PORTENKO, L A (1959) *Aquila* **66**, 129–34; (1972) *Die Schnee-Eule*. Wittenberg Lutherstadt; (1973) *Ptitsy Chukotskogo poluostrova i ostrova Vrangelya* **2**. Leningrad. POSLAVSKI, A N and KRIVONOSOV, G A (1976) *Ekologiya* **3**, 51–6. POSPELOV, S M (1956) *Zool. Zh.* **35**, 600–5. POTAPOV, R L (1959) *Trudy Akad. Nauk Tadzhik. SSR* **115**, 179–201. POTOROCHA, V I (1968) *Ornitologiya* **9**, 354–5. POTTS, G R (1981) *Game Conserv. Ann. Rev.* **12**, 83–7. POTVIN, N, BERGERON, J-M, and FERNET, C (1976) *Can. J. Zool.* **54**, 1992–2000. POUGH, R H (1949) *Audubon bird guide: small land birds*. New York; (1951) *Audubon water bird guide*. New York; (1957) *Audubon western bird guide*. New York. POULSEN, C M (1937) *Flora og Fauna* **43**, 71–85. POULSEN, H (1949) *Ornis fenn.* **26**, 65–7. POWELL, T G (1933) *Br. Birds* **27**, 211. PRAGER, E M and WILSON, A C (1980) *Proc. int. orn. Congr.* **17**, 1209–14. PRATO, S R D DA, DICKSON, J M, and SYMONDS, F L (1981) *Scott. Birds* **11**, 226–7. PREKOPOV, A N (1940) *Trudy Voroshilov. ped. inst.* **2**, 240–1. PRESTON, W C (1968) Ph D Thesis. Michigan Univ. PRESTT, I (1965) *Bird Study* **12**, 196–221. PRICAM, R and ZELENKA, G (1964) *Alauda* **32**, 176–95. PRIEST, C D (1934) *The birds of Southern Rhodesia*. London. PRIKLONSKI, S G (1958) *Trudy Okskogo zapoved.* **2**, 155–7; (1971) *Mat. Pribalt. orn. Konf. 1970* **1**, 83–5. PRIKLONSKI, S G and LAVROVSKI, V V (1974) In Boehme, R L and Flint, V E (eds) *Mat. VI Vsesoyuz. orn. Konf.* **2**, 106–8. PRING, C J (1929) *Br. Birds* **23**, 129–31. PRING-MILL, F (1974) *Bull. Oxford Univ. Explor. Soc.* **23**, 1–49. PROKOFIEVA, I V (1963) *Uchen. zap. Leningrad. ped. Inst.* **230**, 87–91; (1965) *Nauch. doklady vysshey shkoly Biol. nauki* **1**, 37–40. PRÖLSS, D (1956) *Beitr. Naturkde. Niedersachs.* **9**, 53–9. PROVOST, M W (1947) *Am. Midl. Nat.* **38**, 485–503. PROZESKY, O P M (1970) *A field guide to the birds of southern Africa*. London. PSENNER, H (1959) *Orn. Mitt.* **11**, 167. PTUSHENKO, E S and INOZEMTSEV, A A (1968) *Biologiya i khozyaistvennoe znachenie ptits Moskovskoy oblasti i sopredel'nykh territoriy*. Moscow. PUHLMANN, E (1912) *Orn. Monatsschr.* **37**, 430–3; (1914*a*) *Orn. Monatsschr.* **39**, 205–7; (1914*b*) *Orn. Monatsschr.* **39**, 232–4. PUKINSKI, Y B (1977) *Zhizn' sov.* Leningrad. PULCHER, C (1979) *Riv. ital. Orn.* **49**, 234–5. PULLIAINEN, E (1963) *Ornis fenn.* **40**, 132–9; (1978) *Aquilo Ser. Zool.* **18**, 17–22. PULLIAINEN, E and LOISA, K (1977) *Aquilo Ser. Zool.* **17**, 23–33. PURCHASE, D (1973) *CSIRO Wildl. Res. Tech. Pap.* **27**. PURROY, F J (1972) *Ardeola* **16**, 145–58; (1973) *Ardeola* **19**, 89–95. PUSCHMANN, W (1976) *Beitr. Vogelkde.* **22**, 115–21. PYMAN, G A and WAINWRIGHT, C B (1952) *Br. Birds* **45**, 337–9. PYNNÖNEN, A (1939) *Ann. Zool. Soc. Vanamo* **7** (2), 1–171; (1943) *Ann. Zool. Soc. Vanamo* **9** (4), 1–60; (1949) *Ornis fenn.* **26**, 87–8; (1960) *Nordens Fugle i farver* **3**. Munksgaard.

QUINE, D A and CULLEN, J M (1964) *Ibis* **106**, 145–73.

RAAB, K (1961) *Luscinia* **34**, 27. RÄBER, H (1950) *Behaviour* **2**, 1–95; (1954) *Orn. Beob.* **51**, 149–61. RACZYŃSKI, J and RUPRECHT, A L (1974) *Acta Orn.* **14**, 25–38. RADCLIFFE, H D (1910) *J. Bombay nat. Hist. Soc.* **20**, 225–6. RADDE, G (1863) *Reisen im Süden von Ost-Sibirien in den Jahren 1855–1859* **2** *Die Festlands-Ornis des südöstlichen Sibirien*. St Petersburg. RADEMACHER, W (1962) *Orn. Mitt.* **14**, 34; (1963) *Orn. Mitt.* **15**, 205; (1965) *Orn. Mitt.* **17**, 87; (1968*a*) *Orn. Mitt.* **20**, 27–8; (1968*b*) *Orn. Mitt.* **20**, 42; (1970) *Orn. Mitt.* **22**, 179–83; (1980) *Orn. Mitt.* **32**, 69–72. RADETSKI, V R (1981) *Ornitologiya* **16**, 181–2. RADETZKY, J (1962) *Aquila* **68**, 251–3; (1976) *Falke* **23**, 61–5. RADFORD, M C (1961)

Bird Study 8, 174–84. RADU, D (1973) *Ocrotirea nat.* 17, 183–95. RAETHEL, H S, WISSEL, C VON, and STEFANI, M (1976) *Fasanen und andere Hühnervögel*. Melsungen. RAMSAY, A K D (1976) *Seabird Group Rep.* 5, 34–8. RAMSBOTHAM, R H (1906) *Zoologist* (4) 10, 235. RAMZAN, M and TOOR, H S (1973) *J. Bombay nat. Hist. Soc.* 70, 201–4. RANA, B D (1973) *Z. angew. Zool.* 60, 399–403; (1975) *Auk* 92, 322–32; (1976) *Z. angew. Zool.* 63, 25–30. RAND, A L (1951) *Fieldiana Zool.* 32, 561–653. RANDALL, R M and RANDALL, B M (1978) *Cormorant* 5, 4–10. RANDIK, A (1959) *Aquila* 66, 99–106; (1961) *Zool. Listy* 10 (24), 59–66. RANDLA, T (1976) *Eesti Röövlinnud*. Tallinn; (1978) *Metsloomi ja linde Eestis*. Tallinn. RANKIN, M N and DUFFEY, E A G (1948) *Br. Birds* 41 suppl. RANKIN, N, RANKIN, D, and WILLIAMSON, K (1942) *Irish Nat. J.* 8, 19–20. RAO, G S and SHIVANARAYAN, N (1981) *Pavo* 19, 97–9. RAPPE, A (1965) *Aves* 2, 27–31. RASHKEVICH, N A (1965) *Ornitologiya* 7, 142–5. RASMUSSEN, L U (1979) *Dansk orn. Foren. Tidsskr.* 73, 271–9. RAYNOR, G S (1970) *Bird-Banding* 41, 310–11. RAZACK, A and NAIK, R M (1965) *Pavo* 3, 55–71; (1968) *Pavo* 6, 31–58. READ, R H (1918) *Bull. Br. Orn. Club* 38, 82–3. REBOUSSIN, R (1924) *Rev. fr. Orn.* 8, 486–8. REE, V (1974) *Fauna* 27, 39–49; (1977) *Sterna* 16, 113–202. REES, E I S (1983) *Br. Birds* 76, 454. REESE, R A (1973) *Br. Birds* 66, 227–8. REICHENOW, A (1889) *J. Orn.* 37, 1–33; (1900–1) *Die Vögel Afrikas* 1; (1902–3) 2. Neudamm. REICHHOLF, J (1973) *Anz. orn. Ges. Bayern* 12, 81–2; (1976) *Anz. orn. Ges. Bayern* 15, 69–77. REICHHOLF, J and UTSCHICK, H (1972) *Anz. orn. Ges. Bayern* 11, 254–62. REICHLING, L (1974a) *Regulus* 11, 238; (1974b) *Regulus* 11, 239. REID, J C (1974) *Egretta* 17, 15–22. REID, S G (1887) *Ibis* (5) 5, 424–35. REINART, A (1972) *Feltornithologen* 14, 118. REINHARDT, J (1864) *J. Orn.* 12, 339–52. REINKE, E-M (1959) *Z. angew. Zool.* 46, 285–301. REINSCH, A (1961) *Vogelwelt* 82, 115–16; (1962) *Vogelwelt* 83, 74–7; (1968) *Vogelwelt* 89, 137–42; (1970a) *Vogelwelt* 91, 198–9; (1970b) *Vogelwelt* 91, 199–200; (1979) *Anz. orn. Ges. Bayern* 18, 190. REINSCH, A and WARNCKE, K (1968) *Anz. orn. Ges. Bayern* 8, 400–1. REISER, O (1894) *Materialen zu einer Ornis Balcanica* 2, Bulgarien; (1905) 3, Griechenland. Vienna. REITHER, H (1977) *Vogelwelt* 98, 157. RÉKÁSI, J (1975) *Aquila* 80–1, 287–8, 305; (1979) *Aquila* 85, 160. REMMERT, H (1953) *Orn. Mitt.* 5, 231–2. RENAUD, W E, MCLAREN, P L, and JOHNSON, S R (1982) *Arctic* 35, 118–25. RENDAHL, H (1965) *Ark. Zool.* (2) 18, 221–66. RENDLE, M (1905) *Gefiederte Welt* 34, 329–30, 337–8, 345–7, 353–5, 361–2; (1907) *Gefiederte Welt* 36, 213–14, 221–2, 228–9, 236–8, 244–6, 251–3, 258–60, 266–9; (1912) *Gefiederte Welt* 41, 122–4, 132–4, 140–1, 244–6, 251–3; (1914) *Gefiederte Welt* 43, 106–7, 114–15, 122–4, 130–2, 138–40, 146–8, 154–6, 162–4, 170–1, 179–80, 186–8, 194–5, 202–3, 210–11; (1915) *Gefiederte Welt* 44, 131–2, 139–41, 147–8, 155–6, 164–5, 170–1. RENKHOFF, M (1972) *Orn. Mitt.* 24, 63–73. RESSL, F (1963) *Egretta* 6, 9–11. REUTER, T, NEDSTAM, B, and RUOTANEN, T-O (1980) *Ornis fenn.* 57, 132–3. REUTERWALL, O F (1956) *Vår Fågelvärld* 15, 262–8. REUVER, H J A DE (1959) *Limosa* 32, 148–50. REY, E (1892) *Altes und Neues aus dem Haushalte des Kuckucks*. Leipzig; (1908) *Orn. Monatsschr.* 33, 189–97; (1910) *Orn. Monatsschr.* 35, 225–34. REY, J M (1975) *Ardeola* 21, 415–20. REYER, H-U (1980) *Behav. Ecol. Sociobiol.* 6, 219–27; (1982) *Naturwissensch. Rundschau* 35, 31–2. REYMERS, N F (1966) *Ptitsy i mlekopitayushchie yuzhnoy taygi Sredney Sibiri*. Moscow. REYNOLDS, A (1975) *Ring. Migr.* 1, 48–51. REYNOLDS, J F (1965) *E. Afr. Wildl. J.* 3, 129–30; (1972) *Bull. E. Afr. nat. Hist. Soc.*, 116–20; (1974) *Br. Birds* 67, 70–6. RÉZ, E VON (1932) *Kócsag* 5, 112–15. REZANOV, A G (1982) *Ornitologiya* 17, 188. RHEINWALD, G (1977) *Atlas der Brutverbreitung westdeutscher Vogelarten*. Bonn; (1982) *Brutvogelatlas der Bundesrepublik Deutschland*. Bonn. RICHARD, A (1914) *Nos Oiseaux* 1, 79–85; (1923) *Nos Oiseaux* 6, 65–74. RICHARDS, D K (1982) *Malimbus* 4, 93–103. RICHARDS, T J (1957) *Bird Notes* 27, 187–9. RICHARDSON, R A (1948) *Br. Birds* 41, 311. RICHARDSON, R A, SEAGO, M J, and CHURCH, A C (1957) *Br. Birds* 50, 239–45. RICHARDSON, W J (1974) *Ibis* 116, 172–93; (1976) *Ibis* 118, 309–32. RICHTER, H (1950) *Vogelwelt* 71, 58–9; (1952) *Beitr. Vogelkde.* 2, 164–90. RICKLEFS, R E and WHITE-SCHULER, S C (1978) *Bird-Banding* 49, 301–12. RIDDLE, G (1971) *Scott. Birds* 6, 321–9. RIDGELY, R S (1976) *A guide to the birds of Panama*. Princeton. RIDGWAY, R (1901) *Bull. US natn. Mus.* 50 (1); (1914) *Bull. US natn. Mus.* 50 (6); (1916) *Bull. US natn. Mus.* 50 (7); (1919) *Bull. US natn. Mus.* 50 (8). RIDLEY, M W (1954) *Ibis* 96, 311. RIESE, K (1954) *Orn. Mitt.* 6, 95–6; (1968) *Oldenburger Jahrb.* 66, 151–60. RIGGENBACH, H E (1961–2) *Nos Oiseaux* 26, 191–2. RIKKONEN, P, HELSTEN, H, and MOILANEN, P (1976) *Päijät-Hämeen Linnut* 7, 88–93. RINGLEBEN, H (1958) *J. Orn.* 99, 375; (1959) *J. Orn.* 100, 442–3. RINNE, U (1981) *Falke* 33, 62–5. RIPLEY, S D and BOND, G M (1966) *Smiths. Misc. Coll.* 151, 1–37. RIS, H (1937) *Orn. Beob.* 35, 74–7; (1959) *Orn. Beob.* 56, 21–2. RISBERG, L (1978) *Vår Fågelvärld* 37, 193–208; (1979) *Vår Fågelvärld* 38, 221–30; (1982) *Vår Fågelvärld* 41, 361–92. RITCHIE, R J (1980) *Raptor Res.* 14, 59–60. RITTER, F (1972) *Beitr. Vogelkde.* 18, 156–61. RITTER, M (1953) *Beitr. Vogelkde.* 3, 198. RIVIÈRE, B B (1933a) *Br. Birds* 26, 262–70; (1933b) *Br. Birds* 26, 318–29. RIVOIRE, A (1947) *Oiseau* 17, 23–34. ROALKVAM, R (1983) *Vår Fuglefauna* 6, 52. ROBBINS, C S (1974) *Br. Birds* 67, 168–70. ROBBINS, C S, BRUUN, B, and ZIM, H S (1983) *Birds of North America*. New York. ROBEL, D (1982) *Falke* 29, 406–10. ROBEL, D and BUDE, S (1981) *Falke* 28, 386. ROBERSON, D (1980) *Rare birds of the west coast*. Pacific Grove, Calif. ROBERTS, B (1934) *Ibis* (13) 4, 239–64. ROBERTS, C, BROADHURST, K, and GARRETT, P (1963) *Skokholm Bird Obs. Rep. 1963*, 17–24. ROBERTS, E L (1944) *Br. Birds* 38, 56. ROBERTS, P (1954) *Br. Birds* 47, 62. ROBERTSON, W B, JR (1969) *Nature* 223, 632–4. ROBIN, A P (1966) *Alauda* 34, 81–101. ROBIN, P (1968) *Alauda* 36, 237–53. ROBIN, R (1969) *Oiseau* 39, 1–7. ROBINSON, C (1956) *Ostrich* 27, 70–5. ROBINSON, G B and ROBINSON, J M (1975) *Bull. Zamb. orn. Soc.* 7, 107. ROBINSON, N (1974) *Bull. Niger. orn. Soc.* 10, 56–61. ROBINSON, R R (1973) *Camb. Bird Rep.* 47, 59–60. ROBY, D D, BRINK, D L, and NETTLESHIP, D N (1981) *Arctic* 34, 241–8. ROCKENBAUCH, D (1968) *Vogelwelt* 89, 168–76; (1978a) *Anz. orn. Ges. Bayern* 17, 293–328; (1978b) *J. Orn.* 119, 429–40. RODING, G M (1973) *Natuur en Museum* 17, 4–9. ROELKE, M and HUNT, G (1978) *Pacific Seabird Group Bull.* 5, 81. ROGERS, A E F (1969) *Austral. Bird Bander* 7, 36–7. ROGERS, M J and THE RARITIES COMMITTEE (1979) *Br. Birds* 72, 503–49; (1983) *Br. Birds* 76, 476–529. ROGET, S (1971) *Nos Oiseaux* 31, 117–20. ROHWEDER, J (1889) *Monatsschr. Dtsch. Ver. Schutze Vogelwelt* 14, 16–41. ROKITANSKY, G and SCHIFTER, H (1971) *Ann. Naturhist. Mus. Wien* 75, 495–538. ROLNIK, V V (1948) *Zool. Zh.* 27, 535–46. RÖMER, F and SCHAUDIN, F (1900) *Fauna Arctica* 1. Jena. RONSIL, R (1949) *Oiseau* 19, 215–16. ROOS, G (1965) *Vår Fågelvärld* 24, 257–71. ROOTH, J (1958) *Bull. int. Comm. Bird Pres.* 7, 117–19. ROOTH, J and JONKERS, D A (1972) *TNO-nieuws*, 551–5. ROOTH, J and MÖRZER BRUIJNS, M F (1959) *Limosa* 32, 13–23. ROPER, E N (1960) *Br. Birds* 53, 447. RÖRIG, G (1900) *Arb. Biol. Abt. Land- und Forstwirtschaft Kaiserl. Gesundheitsamt Berlin* 1, 1–85; (1905) *Arb. Biol. Abt. Land- und Forstwirtschaft Kaiserl. Gesundheitsamt Berlin* 4, 51–120. ROSENDAHL, S (1977) *Danske Fugle* 29, 59–64. ROSENDAHL, S and SKOVGAARD, P (1968) *Danske Fugle* 20, 4–12; (1971) *Danske Fugle* 23, 97–106. ROSENIUS, P (1949) *Sveriges Fåglar och Fågelbon*. Lund. ROSE, L N (1982) *Naturalist* 107, 15–17. RØSKELAND, A (1931) *Naturen* 55, 51–7. ROSLYAKOV, G E (1979) *Ornitologiya* 14, 196. ROSNOBLET, R and MENATORY, G (1975) *Alauda* 43, 194. ROST, K (1953) *Beitr. Vogelkde.* 3, 211–22; (1957) *J. Orn.* 98,

204–9. ROTHGÄNGER, G and ROTHGÄNGER, H (1973) *Falke* **20**, 124–30. ROTHKOPF, D (1970) *Bonn. zool. Beitr.* **21**, 63–82. ROUTH, M (1949) *Ibis* **91**, 577–605. ROUX, G (1956) *Nos Oiseaux* **23**, 327. ROWAN, M K (1962) *Br. Birds* **55**, 103–14. ROWAN, W (1918) *Br. Birds* **11**, 218–25. RUBINSHTEIN, N A (1981) *Referat. Zh.* **2**, 118. RUCNER, D (1952) *Larus* **4–5**, 56–74. RUDEBECK, G (1974) *Anser* **13**, 149–53; (1980) *Vår Fågelvärld* **39**, 265–74. RUDLOFF, H VON (1967) *Die Schwankungen und Pendelungen des Klimas in Europa seit dem Beginn der regelmässigen Instrumenten-Beobachtungen (1670)*. Braunschweig. RUEDI, M (1983) *Nos Oiseaux* **37**, 80–1. RUGE, K (1964) *Orn. Beob.* **61**, 56–60; (1966) *J. Orn.* **107**, 357; (1968) *Orn. Beob.* **65**, 109–24; (1969a) *Orn. Beob.* **66**, 42–54; (1969b) *Vogelwelt* **90**, 201–23; (1970a) *J. Orn.* **111**, 412–19; (1970b) *J. Orn.* **111**, 496; (1971a) *Orn. Beob.* **68**, 9–33; (1971b) *Orn. Beob.* **68**, 256–71; (1973) *Orn. Beob.* **70**, 173–9; (1974) *Orn. Beob.* **71**, 303–11; (1975) *Orn. Beob.* **72**, 75–82. RUGE, K and WEBER, W (1974a) *Anz. orn. Ges. Bayern* **13**, 300–4; (1974b) *Vogelwelt* **95**, 138–47. RÜGER, A (1972) *Z. Wiss. Zool.* **184**, 63–163. RUIZ BUSTOS, A and CAMACHO MUÑOZ, A (1973) *Cuad. Cienc. Biol.* **2**, 57–61. RUNTE, P (1959) *Orn. Mitt.* **11**, 163. RÜPPELL, G (1969) *J. Orn.* **110**, 161–9. RÜPPELL, W (1937) *J. Orn.* **85**, 120–35. RUPRECHT, A L (1979a) *Acta Orn.* **16**, 493–511; (1979b) *Ibis* **121**, 489–94. RUSSELL, S M (1964) *Orn. Monogr. AOU* **1**. RUSTAMOV, A K (1954) *Ptitsy pustyni Kara-kum*. Ashkhabad. RUTHKE, P (1936) *Beitr. Fortpfl. Biol. Vögel* **12**, 79; (1944) *Orn. Monatsber.* **52**, 112; (1949) *Vogelwelt* **70**, 120–1; (1951) *Orn. Abh.* **11**, 1–40; (1968) *Vogelwelt* **89**, 129–37. RUTSCHKE, E (ed) (1983) *Die Vogelwelt Brandenburgs*. Jena. RUTTLEDGE, R F (1966) *Ireland's birds*. London; (1974) *Br. Birds* **67**, 440. RUWET, J-C (1964) *Rev. zool. bot. Afr.* **69**, 1–63. RYABOV, V F (1965) *Vest. Mosk. gos. Univ.* **20** (*Biol. Poch. Ser.* 6), 13–15. RYDZEWSKI, W (1978) *Ring* **96–7**, 218–62; (1979) *Ring* **98–9**, 8. RYSER, R (1961) *Orn. Beob.* **58**, 199–201. RYSZKOWSKI, L, WAGNER, C K, GOSZCZYŃSKI, J and TRUSZKOWSKI, J (1971) *Ann. zool. fenn.* **8**, 160–8. RYVES, B H (1931) *Br. Birds* **24**, 211–12.

SAARI, L (1979a) *Finnish Game Res.* **38**, 3–16; (1979b) *Finnish Game Res.* **38**, 17–30; (1980) *On the breeding biology of the Wood Pigeon (C. palumbus) in Finland*. Helsinki. SACKL, P (1981) *Natur und Umwelt Burgenland* **4**, 5–12. SAEMANN, D (1968) *Beitr. Vogelkde.* **14**, 176–7; (1975) *Hercynia* **12**, 361–88. SAFRIEL, U (1968) *Ibis* **110**, 283–320. SAGE, B (1960a) *Br. Birds* **53**, 222; (1960b) *Ardea* **48**, 160–78; (1962) *Br. Birds* **55**, 237–8. SAGITOV, A K and FUNDUKCHIEV, S E (1980) *Vestnik zool.* **5**, 88–9. SAINT GIRONS, M-C (1963) *Věstn. Česk. Spol. Zool.* **32**, 185–98; (1964) *Oiseau* **34**, 204–9; (1965) *Mammalia* **29**, 42–53; (1973) *Bull. Soc. Sci. nat. phys. Maroc* **53**, 193–8. SAINT GIRONS, M-C and MARTIN, C (1973) *Bull. Écol.* **4**, 95–120. SAINT GIRONS, M-C, THÉVENOT, M, and THOUY, P (1974) *CNRS Trav. RCP* **249** (2), 257–65. SAINT GIRONS, M-C and THOUY, P (1978) *Bull. Écol.* **9**, 211–18. ST JOHN, O B (1889) *Ibis* (6) **1**, 145–80. ST QUINTIN, W H (1905) *Avic. Mag.* **3**, 64–6. SALES, V A D (1959) *Sea Swallow* **12**, 20; (1965) *Sea Swallow* **17**, 81–2. SALOMONSEN, F (1931) *Proc. int. orn. Congr.* **7**, 413–38; (1935) *Zoology of the Faeroes* **3**. Copenhagen; (1941) *Medd. om Grøn.* **131** (6), 1–21; (1944) *Göteborgs Kungl. Vetenskaps-och Vitterhets-Samhälles Handl., Sjätte Följden*, (B) **3** (5); (1947) *Vår Fågelvärld* **6**, 141–4; (1950–1) *Grønlands Fugle*. Copenhagen; (1953) *Dansk orn. Foren. Tidsskr.* **47**, 126–39; (1955) *Dan. Biol. Medd.* **22**, 1–62; (1956) *Arctic* **9**, 258–64; (1963) *Oversigt over Danmarks Fugle*. Copenhagen; (1967a) *Fuglene på Grønland*. Copenhagen; (1967b) *K. Danske Vidensk. Selsk. Biol. Medd.* **24** (1), 1–42; (1971) *Medd. om Grøn.* **191** (2), 1–52; (1979) *Medd. om Grøn.* **204** (6), 1–214. SALTER, R L (1948) *Murrelet* **29**, 49. SALVAN, J (1968a) *Oiseau* **38**, 53–85; (1968b) *Oiseau* **38**, 127–50. SALVIN, O (1859) *Ibis* (1) **1**, 352–65. SAMORODOV, A V and SAMORODOV, Y A (1972) *Ornitologiya* **10**, 387. SANFORD, R C and HARRIS, S W (1967) *Condor* **69**, 298–302. SANFT, K (1970) *Beitr. Vogelkde.* **16**, 344–54. SANS-COMA, V and KAHMANN, H (1976) *Säugetierk. Mitt.* **24**, 5–11. SANTINI, L and FARINA, A (1978) *Avocetta* **1**, 49–60. SANTOS JÚNIOR, J R DOS (1978–9a) *Cyanopica* **2** (1), 5–27; (1978–9b) *Cyanopica* **2** (1), 57–68; (1979–80) *Cyanopica* **2** (2), 107–11. SARKANEN, S (1974) *Lintumies* **9**, 77–84; (1978a) *Ornis fenn.* **55**, 158–63; (1978b) *Suomen Luonto* **37**, 274–5. SARMENTO, A A (1948) *Vertebrados da Madeira* **1**. Funchal. SARUDNY, N (1909) *Orn. Monatsber.* **17**, 81–2. SARZHINSKI, V A (1977) In Sludski, A A (ed) *Redkie i ischezayushchie zveri i ptitsy Kazakhstana*, 223–5. Alma-Ata. SASSI, M (1906) *Ann. naturh. Mus. Wien* **21**, 45–59. SASSI, M and ZIMMER, F (1941) *Ann. naturh. Mus. Wien* **51**, 236–346. SAUER, F and SAUER, E (1960) *Bonn. zool. Beitr.* **11**, 41–86. SAUNDERS, A A (1926) *Roosevelt Wild Life Bull.* **3** (3), 335–476. SAUNDERS, D R (1962) *Br. Birds* **55**, 591. SAUNDERS, H (1896) *Catalogue of birds of the British Museum* **25**, 1–339. SAUROLA, P (1979) *Lintumies* **14**, 104–10; (1980) *Lintumies* **15**, 121–8; (1983) *Lintumies* **18**, 67–71. SAUTER, U (1956) *Vogelwarte* **18**, 109–51. SCHAACK, K H (1967) *Luscinia* **40**, 52–4. SCHACHT, W (1979) *Anz. orn. Ges. Bayern* **18**, 82. SCHAEFER, H (1970) *Bonn. zool. Beitr.* **21**, 52–62; (1971) *Bonn. zool. Beitr.* **22**, 153–60; (1972) *Ochrana Fauny* **6**, 159–64. SCHÄFER, H and FINCKENSTEIN, G (1935) *Orn. Monatsber.* **43**, 171–6. SCHALOW, H (1890) *J. Orn.* **38**, 1–74. SCHARRINGA, J (1979) *Dutch Birding* **1**, 60. SCHAUB, S (1937) *Orn. Beob.* **34**, 89–93. SCHAUENSEE, R M DE (1966) *The species of birds of South America and their distribution*. Philadelphia; (1970) *A guide to the birds of South America*. Wynnewood, Penn. SCHAUENSEE, R M DE and PHELPS, W H, JR (1978) *A guide to the birds of Venezuela*. Princeton. SCHEER, D (1949) *Vogelwarte* **15**, 104–9. SCHEIFLER, H (1972) *Gefiederte Welt* **96** (3), 43–4; (1979) *Gefiederte Welt* **103**, 176–7. SCHEIN, M W (1954) *Auk* **71**, 318–20. SCHELCHER, R (1965) *Beitr. Vogelkde.* **11**, 102–3. SCHELPER, W (1971) *Vogelkdl. Ber. Niedersachs.* **3**, 11–20; (1972a) Dissertation. Göttingen Univ.; (1972b) *Beitr. Naturkde. Niedersachs.* **25**, 77–83. SCHEMBRI, S P and ZAMMIT, R C (1979) *Il-Merill* **20**, 20–1. SCHENK, H (1976) In Pedrotti, F (ed) *S.O.S. Fauna*, 465–556. Camerino. SCHERZINGER, W (1968) *Egretta* **11**, 56; (1969a) *Vogelkosmos* **6**, 226–9; (1969b) *Vogelkosmos* **6**, 421–3; (1970) *Zoologica* **41** (118); (1971a) *Z. Tierpsychol.* **28**, 494–504; (1971b) *Z. Tierpsychol.* **29**, 165–74; (1972a) *Gefiederte Welt* **96**, 129–33; (1972b) *Orn. Mitt.* **24**, 207–10. (1974a) *Anz. orn. Ges. Bayern* **13**, 121–56; (1974b) *Zool. Gart.* **44**, 59–61; (1974c) *J. Orn.* **115**, 8–49; (1974d) *Bonn. zool. Beitr.* **25**, 123–47; (1975) *Avic. Mag.* **81**, 70–3; (1979) *Anz. orn. Ges. Bayern* **18**, 184–5; (1980) *Bonn. zool. Monogr.* **15**. SCHIERMANN, G (1926a) *Orn. Monatsber.* **34**, 54–5; (1926b) *Beitr. Fortpfl. Biol. Vögel* **2**, 28–30; (1930) *J. Orn.* **78**, 137–80. SCHIFFERLI, A (1955) *Orn. Beob.* **52**, 25–38. SCHIFFERLI, A, GÉROUDET, P, and WINKLER, R (1980) *Verbreitungsatlas der Brutvögel der Schweiz*. Sempach. SCHIFFERLI, I and IMBODEN, C (1972) *Orn. Beob.* **69**, 70–109. SCHIFFERLI, A and ZIEGELER, R (1956) *Orn. Beob.* **53**, 1–5. SCHINZ, J (1960) *Vierteljahrsschr. Naturforsch. Ges. Zürich* **105**, 306–9. SCHLEGEL, J and SCHLEGEL, S (1968) *Beitr. Vogelkde.* **14**, 84–6; (1971) *Beitr. Vogelkde.* **17**, 251–3. SCHLEGEL, R (1915) *Orn. Monatsber.* **23**, 97–111; (1925) *Orn. Monatsschr.* **50**, 184–91; (1967) *Beitr. Vogelkde.* **13**, 145–90; (1969) *Der Ziegenmelker*. Wittenberg Lutherstadt. SCHLENKER, R (1966) *Corax* **1**, 209–16. SCHLOSS, W (1962) *Auspicium* **1**, 395–443; (1966) *Auspicium* **2**, 195–217; (1969) *Auspicium* **3**, 139–52; (1975) *Auspicium* **6**, 91–7. SCHMIDT, A (1977) *Beitr. Vogelkde.* **23**, 233–44. SCHMIDT, E (1965a) *Zool. Abh. Staat. Mus. Tierk. Dresden* **27**, 307–17; (1966–7) *Aquila* **73–4**, 109–19; (1973–4) *Aquila* **80–1**, 221–38; (1972a) *Ornis fenn.* **49**, 98–102; (1972b) *Lounais-Hämeen Luonto* **45**, 1–

10; (1973a) Beitr. Vogelkde. **19**, 175–8; (1973b) Z. angew. Zool. **60**, 43–70; (1976) Aquila **82**, 119–43; (1978) Aquila **84**, 91–100. SCHMIDT, E and SZLIVKA, L (1968) Aquila **75**, 227–9. SCHMIDT, G (1953) Vogelwelt **74**, 63; (1959) Vogelwelt **80**, 117–19; (1966) Vogelwelt **87**, 139–42. SCHMIDT, G A J and BREHM, K (1974) Vogelleben zwischen Nord- und Ostsee. Neumünster. SCHMIDT, H (1972c) Thür. Orn. Rdbr. **19–20**, 56–7. SCHMIDT, H W (1981) Falke **28**, 6–9. SCHMIDT, J (1958) Falke **5**, 141. SCHMIDT, K and SCHÜTZE, H-U (1974) Falke **21**, 85–6. SCHMIDT, R (1970) Veröff. Bez. Heimatmus. Potsdam **21**, 143–53. SCHMIDT, R and SIEFKE, A (1981) Abstr. 17th int. Ethol. Conf. Oxford, 279. SCHMIDT, R C and VAUK, G (1981) Vogelwelt **102**, 180–9. SCHMIDT, R K (1965b) J. Orn. **106**, 295–306. SCHMITT, B (1963) Alauda **21**, 218–21. SCHMITT, C and STADLER, H (1918) J. Orn. **66**, 220–34. SCHMITT, M B, MILSTEIN, P LE S, HUNTER, H C, and HOPCROFT, C J (1973) Bokmakierie **25**, 91–2. SCHMITZ, E (1893) Orn. Jahrb. **4**, 30–2; (1905) Orn. Monatsber. **13**, 197–201; (1910) Z. Ool. Orn. **20**, 68–70. SCHMOLL, H-J (1973) Orn. Mitt. **25**, 56. SCHNAPP, B (1971) Trav. Mus. Hist. nat. 'Grigore Antipa' **11**, 495–510. SCHNEIDER, A (1953) Orn. Mitt. **5**, 96–7. SCHNEIDER, W (1930) Beitr. Fortpfl. Biol. Vögel **6**, 174; (1937) Vogelwarte **8**, 159–70; (1961) Nos Oiseaux **26**, 50; (1964) Schleiereulen. Wittenberg Lutherstadt; (1979) Beitr. Vogelkde. **25**, 364. SCHNOCK, G (1981) Gerfaut **71**, 235–48. SCHNOCK, G and SEUTIN, E (1973) Aves **10**, 182–92. SCHNURRE, O (1934) Beitr. Fortpfl. Biol. Vögel **10**, 206–13; (1935) Beitr. Fortpfl. Biol. Vögel **11**, 58–60; (1936a) Beitr. Fortpfl. Biol. Vögel **12**, 1–12, 54–69; (1936b) Beitr. Fortpfl. Biol. Vögel **12**, 232; (1944) Beitr. Fortpfl. Biol. Vögel **20**, 19–24; (1954) Vogelwelt **75**, 229–33; (1975) Milu **3**, 748–55. SCHNURRE, O and BETHGE, E (1973) Milu **3**, 476–84. SCHOENNAGEL, E (1979) Orn. Mitt. **31**, 198. SCHOLANDER, S I (1955) Auk **72**, 225–39. SCHOLEY, G J (1924) Ool. Rec. **4** (1), 9–11. SCHOMMER, M and TSCHANZ, B (1975) Vogelwarte **28**, 17–44. SCHÖNERT, C (1961) In Schildmacher, H (ed) Beiträge zur Kenntnis deutscher Vögel, 131–87. Jena. SCHÖNFELD, M (1974) Jber. Vogelwarte Hiddensee **4**, 90–122; (1975) Beitr. Vogelkde. **21**, 494. SCHÖNFELD, M and GIRBIG, G (1975) Hercynia **12**, 257–319. SCHÖNFELD, M, GIRBIG, G, and STURM, H (1977) Hercynia **14**, 303–51. SCHÖNFELD, M and PIECHOCKI, R (1974) J. Orn. **115**, 418–35. SCHÖNN, S (1976) Beitr. Vogelkde. **22**, 261–300; (1978) Der Sperlingskauz. Wittenberg Lutherstadt. SCHÖNWETTER, M (1967) Handbuch der Oologie **1**. Berlin. SCHORGER, A W (1955) The Passenger Pigeon: its natural history and extinction. Madison. SCHREIBER, R W and ASHMOLE, N P (1970) Ibis **112**, 363–94. SCHREIBER, R W and DINSMORE, J J (1972) Florida Nat. **45**, 161. SCHUBERT, W (1969) Anz. orn. Ges. Bayern **8**, 515–17; (1978) Anz. orn. Ges. Bayern **17**, 125–31. SCHUHMACHER, A (1971) Gefiederte Welt **95**, 196–7. SCHULZ, H (1978) Falke **25**, 412–17. SCHULZ, H-P (1977) Orn. Mitt. **29**, 14–15. SCHULZ-WALDMANN, K and DOMINIAK, W (1971) Falke **18**, 83–8. SCHULZE, H (1951) Braunschw. Naturk. Schr. **1**, 261–98. SCHUMANN, A (1931) Bull. Inst. Roy. Hist. nat. Sophia **4**, 108–14. SCHUMANN, G (1971) Luscinia **41**, 153–9. SCHUMANN, H (1949) Beitr. Naturkde. Niedersachs. **2** (4), 8–13; (1963) Orn. Mitt. **15**, 31; (1973a) In Ringleben, H and Schumann, H (eds) Aus der Avifauna von Niedersachsen, 67–72; (1973b) 73–8. Wilhelmshaven. SCHÜZ, E (1940) Naturw. Monatsber. **53**, 126–30; (1948) Vogelwarte **15**, 41–2; (1957) Vogelwarte **19**, 138–40; (1959) Die Vogelwelt des Südkaspischen Tieflandes. Stuttgart. SCHUSTER, L (1903) Orn. Monatsschr. **28**, 239–40; (1930) Beitr. Fortpfl. Biol. Vögel **6**, 53–8; (1935) Beitr. Fortpfl. Biol. Vögel **11**, 146; (1936) Beitr. Fortpfl. Biol. Vögel **12**, 221–5. SCHUSTER, S (1971) Anz. orn. Ges. Bayern **10**, 156–61. SCHWAITZER, F (1865) J. Orn. **13**, 291–2. SCHWAMMBERGER, K (1965) Vogelwelt **86**, 185; (1974) Anz. orn. Ges. Bayern **13**, 247. SCHWARZ, M (1938) Orn. Beob. **35**, 145–50. SCHWARZ, R and KRÄGENOW, P M (1968) Falke **15**, 264–7. SCOTT, D (1866) Ibis (2) **2**, 222–3; (1975) Br. Birds **68**, 208–10; (1979) Br. Birds **72**, 436; (1980) Br. Birds **73**, 436–9. SCOTT, J M (1973) Ph D Thesis. Oregon Univ. SCOTT, R E (1971) Br. Birds **64**, 279. SCONE, LORD (1925) Br. Birds **18**, 318. SEALY, S G (1973) Ornis scand. **4**, 113–21. SEALY, S G, CARTER, H R, and ALISON, D (1982) Auk **99**, 778–81. SEARS, H F (1976) Ph D Thesis. North Carolina Univ.; (1978) Bird-Banding **49**, 1–16; (1979) Auk **96**, 202–3; (1981) J. Fld. Orn. **52**, 191–209. SEARS, H F, MOSELEY, L J, and MUELLER, H C (1976) Auk **93**, 170–5. SEEBOHM, H (1884) A history of British birds **2**; (1885) **3**. London; (1895) Classification of birds: an attempt to diagnose the subclasses, orders, suborders, and families of existing birds, supplement. London. SEEL, D C (1973) Br. Birds **66**, 528–35; (1977a) Bird Study **24**, 114–18; (1977b) Ibis **119**, 309–22; (1980) Proc. int. orn. Congr. **17**, 1399. SEEL, D C, WALTON, K C, and WYLLIE, I (1981) Bird Study **28**, 211–14. SEHLBACH, F (1908) Orn. Monatsschr. **33**, 79–80; (1920) Orn. Monatsschr. **28**, 85–7. SEIERSTAD, A, SEIERSTAD, S, and MYSTERUD, I (1960) Sterna **4**, 153–68. SEIFERT, S (1961) Beitr. Vogelkde. **7**, 370–1. SELANDER, R K (1954) Condor **56**, 57–82. SELANDER, R K and ALVAREZ DEL TORO, M (1955) Condor **57**, 144–7. SELIGMAN, O R and WILLCOX, J M (1940) Ibis (14) **4**, 264–79. SELLIN, D (1965) Vogelwelt **86**, 61–2; (1969) Falke **16**, 238–41. SELOUS, E (1899) Zoologist (4) **3**, 388–402, 486–505; (1901) Bird watching. London; (1905a) Bird life glimpses. London; (1905b) The bird watcher in the Shetlands. London. SEMAGO, L L (1974) In Böhme, R L and Flint, V E (eds) Mat. VI Vsesoyuz. orn. Konf. **2**, 128–30. SENGUPTA, S (1976) Pavo **12**, 1–12. SENOSIAIN, N (1978) Ardeola **24**, 236–42. SEPPÄ, J (1969) Ornis fenn. **46**, 78–9. SERGEANT, D E (1952) Br. Birds **45**, 122–33. SERLE, W (1939) Ibis **81**, 654–99; (1943) Ibis **85**, 264–300; (1957) Ibis **99**, 371–418; (1965) Ibis **107**, 60–94. SERLE, W, MOREL, G J, and HARTWIG, W (1977) A field guide to the birds of West Africa. London. SERMET, E (1973) Nos Oiseaux **32**, 3–9. SERVENTY, D L, SERVENTY, V N, and WARHAM, J (1971) The handbook of Australian sea-birds. Sydney. SERVENTY, D L and WHITTELL, H M (1948) Birds of Western Australia. Perth. SERVENTY, V and WHITE, S (1951) Emu **50**, 145–51. SEVASTYANOV, G N (1959) Zool. Zh. **38**, 589–95. SHARLAND, R E (1966) Bull. Niger. orn. Soc. **3**, 46–7. SHARMA, I (1969) J. Bengal nat. Hist. Soc. **35**, 115–16. SHARPE, R B (1868–71) A monograph of the Alcedinidae. London. SHARROCK, J T R (1962) Br. Birds **55**, 134; (1974) Scarce migrant birds in Britain and Ireland. Berkhamsted; (1976) The atlas of breeding birds in Britain and Ireland. Tring; (1978a) Br. Birds **71**, 221; (1978b) Br. Birds **71**, 222–3. SHARROCK, J T R and RARE BREEDING BIRDS PANEL (1980) Br. Birds **73**, 5–26; (1982) Br. Birds **75**, 154–78; (1983) Br. Birds **76**, 1–25. SHARROCK, J T R and SHARROCK, E M (1976) Rare birds in Britain and Ireland. Berkhamsted; SHAW, TSEN-HWANG (1936) The birds of Hopei province **1**. Peking. SHCHEGOLEV, V I (1974) Geogr. Ekol. Nazem. Pozvon. **2**, 111–71. SHELFORD, V E (1945) Auk **62**, 592–6. SHELLEY, G E (1872) A handbook to the birds of Egypt. London. SHEVAREVA, T P (1962) Migr. zhiv. **3**, 92–105. SHIBAEV, Y V (1968) Ornitolgiya **9**, 355. SHILOV, I A and SMIRIN, Y M (1959) Vtoraya Vsesoyuz Orn. Konf. tezisy dokladov **2**, 93–4. SHIVANARAYAN, N, BABU, K S, and ALI, M H (1981) Pavo **19**, 92–6. SHNITNIKOV, V N (1949) Ptitsy Semirech'ya. Moscow. SHOOTER, P (1978) Br. Birds **71**, 130. SHORT, H L and DREW, L C (1962) Am. Midl. Nat. **67**, 424–33. SHORT, L L (1971) Bull. Am. Mus. nat. Hist. **145**, 1–118; (1972) Auk **89**, 895; (1973) Bull. Am. Mus. nat. Hist. **152**, 255–364; (1974) Am. Mus. Novit. **2547**; (1982) Woodpeckers of the world. Greenville, Delaware. SHUGART, G W (1977) In Southern, W E (ed) Proc. 1977 Conf. colonial Waterbird Group, 110–17. Illinois. SHUGART, G W, SCHARF, W C, and CUTHBERT, F J (1978) In Southern, W E (ed) Proc. 1978 Conf. colonial

Waterbird Group, 146–56. Illinois. SIBLEY, C G and AHLQUIST, J E (1972) *Bull. Peabody Mus. nat. Hist.* **39**. SIBSON, R B (1954) *Notornis* **6**, 43–7. SIEGFRIED, W R (1968) *Ostrich* **39**, 39; (1971*a*) *Ostrich* **42**, 155–7; (1971*b*) *Ostrich* **42**, 161–5. SIEGFRIED, W R and UNDERHILL, L G (1975) *Anim. Behav.* **23**, 504–8. SIELMANN, H (1959) *My year with the woodpeckers.* London. SIEWERT, H (1928) *Beitr. Fortpfl. Biol. Vögel* **4**, 47–9. SIIVONEN, L (1942) *Fauna och Flora* **37**, 32–8. SIM, G (1903) *The vertebrate fauna of Dee.* Aberdeen. SIMEONOV, S D (1963) *Acta Mus. Macedon. Sci. nat.* **9**, 35–50; (1964) *God. Sofii. Univ. Biol. Geol. Geogr.* **57** (Zool.), 107–16; (1966) *Fragm. Bak. Mus. Maced. Sci. Nat.* **5**, 169–78; (1980) *Ekologia* **6**, 70–3. SIMKIN, G N (1976) *Ornitologiya* **12**, 149–59; (1977) *Ornitologiya* **13**, 134–45. SIMMONS, K E L (1947) *Br. Birds* **40**, 88; (1950) *Ibis* **92**, 648; (1952) *Behaviour* **4**, 161–71; (1972) *Br. Birds* **65**, 465–79, 510–21; (1974) *Br. Birds* **67**, 442–3. SIMMS, C (1971) *Br. Birds* **64**, 543–4. SIMMS, E (1962) *Br. Birds* **55**, 1–36; (1971) *Woodland birds.* London; (1975) *Birds of town and suburb.* London; (1979) *The public life of the street pigeon.* London. SIMON, P (1965) *Gerfaut* **55**, 26–71. SIMON, S and SIMON, D (1980) *Alauda* **48**, 152–3. SIMPSON, K (1972) *Birds of the Bass Strait.* Sydney. SIMPSON, S F and CRACRAFT, J (1981) *Auk* **98**, 481–94. SINCLAIR, J C (1974) *Bull. Br. Orn. Club* **94**, 57–8; (1976) *Bokmakierie* **28**, 28. SKEAD, C K (1950) *Ibis* **92**, 434–63. SKEAD, D M (1974) *Ostrich* **45**, 189–92; (1977) *Ostrich Suppl.* **12**, 117–31. SKINNER, N J (1968) *Bull. Niger. orn. Soc.* **5** (20), 88–91. SKOCZYLAS, R (1961) *Ekol. Polska* (A) **9**, 229–43. SKOKOVA, N N (1962) *Ornitologiya* **5**, 7–12; (1967) *Trudy Kandalak. gos. zapoved.* **5**, 155–77. SKØTT, C (1971) *Flora og Fauna* **77**, 60–4. SKURATOWICZ, W (1950) *Pr. Biol. Poznansk. Tow. Przyj. Nauk* **12**, 1–10. SLADE, B E (1978) *Bristol. Orn.* **11**, 21–2. SLÁDEK, J (1961) *Biológia Bratislava* **16**, 697–700. SLADEN, A G L (1919) *Ibis* (11) **1**, 222–50. SLATER, P (1970) *A field guide to Australian birds: non-passerines.* Edinburgh. SLATER, P J B (1976) *Nature* **264**, 636–8; (1980) *Ornis scand.* **11**, 155–63. SLATER, P J B and SLATER, E P (1972*a*) *Bird Study* **19**, 105–14; (1972*b*) *Fair Isle Bird Obs. Rep.* **25**, 71–4. SLIJPER, H J (1948) *Ardea* **36**, 42–52. SMALLCOMBE, W A (1934) *Br. Birds* **28**, 205. SMALLEY, M E (1983*a*) *Malimbus* **5**, 31–3; (1983*b*) *Malimbus* **5**, 34–6. SMALLS, J W (1912) *Br. Birds* **5**, 329–32. SMEENK, C (1972) *Ardea* **60**, 1–71. SMET, K DE and GOMPEL, J VAN (1980) *Malimbus* **2**, 56–70. SMIT, A (1970) *Vogeljaar* **18**, 264–6. SMIT, C J and WOLFF, W J (eds) (1981) *Birds of the Wadden Sea.* Rotterdam. SMITH, A and O'CONNOR, L (1955) *Emu* **55**, 255–6. SMITH, A C (1887) *The birds of Wiltshire.* London. SMITH, A J M (1975*a*) *Br. Birds* **68**, 142–56. SMITH, D A (1970) *Can. Fld.-Nat.* **84**, 377–83. SMITH, D G and MARTI, C D (1976) *Raptor Res.* **10**, 33–44. SMITH, D G, WILSON, C R, and FROST, H H (1974) *Condor* **76**, 131–6. SMITH, F N (1922) *Can. Fld.-Nat.* **36**, 68–71. SMITH, F R (1972*a*) *Devon Bird Rep.* **45**, 28. SMITH, G A (1972*b*) *Avic. Mag.* **78**, 120–37; (1975*b*) *Ibis* **117**, 118–68; (1979) *Lovebirds and related parrots.* London. SMITH, H C (1942) *Notes on birds of Burma.* Simla. SMITH, J (1976) *Avic. Mag.* **82**, 194–6. SMITH, J N D (1921) *Br. Birds* **15**, 50–6. SMITH, J T (1980) In Wilson, M (ed) *Essex Bird Rep. 1978*, 14–59. SMITH, K D (1951*a*) *Br. Birds* **44**, 325–6; (1951*b*) *Ibis* **93**, 201–33; (1953) *Ibis* **95**, 696–8; (1957) *Ibis* **99**, 1–26, 307–37; (1965) *Ibis* **107**, 493–526; (1968) *Ibis* **110**, 452–92. SMITH, S (1960) *Br. Birds* **53**, 301–3. SMITH, V W (1964) *Bull. Niger. orn. Soc.* **1** (2), 2–3; (1966) *Ibis* **108**, 492–512; (1971) *Niger. Field* **36**, 41–4. SMITH, V W and KILLICK-KENDRICK, R (1964) *Ibis* **106**, 119–23. SMITHERS, R H N (1964) *A checklist of the birds of the Bechuanaland Protectorate.* Trust Nat. Mus. S. Rhodesia. SMYTHIES, B E (1953) *The birds of Burma.* Edinburgh; (1968) *The birds of Borneo.* Edinburgh. SNOW, D W (1958) *A study of Blackbirds.* London; (ed) (1978) *An atlas of speciation in African non-passerine birds.* London. SNOW, D W and MANNING, A W G (1954) *Br. Birds* **47**, 355–6. SNYDER, D E (1947) *Wilson Bull.* **59**, 74–8; (1953) *Auk* **70**, 87–9; (1957) *Arctic birds of Canada.* Toronto. SOARES, A A (1973) *Arq. Mus. Bocage* (2) **4** (9). SÖDING, K (1961) *Natur und Heimat* **21**, 65–9. SOF (1978) *Sveriges Fåglar.* Stockholm. SOIKKELI, M (1962) *Ornis fenn.* **39**, 60–97; (1964) *Ornis fenn.* **41**, 37–40; (1970) *Ornis fenn.* **47**, 177–9; (1973*a*) *Ornis fenn.* **50**, 47–8; (1973*b*) *Laxforskningsinst. Meddelande* **3**, 1–5; (1973*c*) *Bird-Banding* **44**, 196–204. SOLHEIM, R (1973) *Fjellvåken* **3** (4), 7–8; (1983) *Ornis scand.* **14**, 51–7. SOLLEIN, A, NESHOLEN, B, and FOSSEIDENGEN, J E (1978) *Cinclus* **1**, 58–64; (1982*a*) *Fauna norv. Ser. C Cinclus* **5**, 93–4; (1982*b*) *Fauna* **35**, 121–4; (1982*c*) *Vår Fuglefauna* **5**, 169–74. SOMADIKARTA, S (1973) *Ardea* **61**, 186–7. SOMEREN, V D VAN (1958) *A bird watcher in Kenya.* Edinburgh. SOMEREN, V G L VAN (1944) *Ibis* **86**, 98; (1956) *Fieldiana Zool.* **38**, 1–320. SØNDERGAARD, K (1983) *Dansk orn. Foren. Tidsskr.* **77**, 35–42. SONERUD, G, MJELDE, A, and PRESTRUD, K (1972) *Sterna* **11**, 1–12. SOPER, E A (1969) *Br. Birds* **62**, 200–1. SOPYEV, O (1967) *Ornitologiya* **8**, 221–35. SOUTH, G R (1966) *Br. Birds* **59**, 493–7. SOUTHAM, E V (1944) *Br. Birds* **38**, 55. SOUTHERN, H N (1938) *Proc. zool. Soc. Lond.* (A) **108**, 423–31; (1954) *Ibis* **96**, 384–410; (1962) *Proc. zool. Soc. Lond.* **138**, 455–72; (1969) *Ibis* **111**, 293–9; (1970) *J. Zool. Lond.* **162**, 197–285. SOUTHERN, H N, CARRICK, R, and POTTER, G (1965) *J. anim. Ecol.* **34**, 649–65. SOUTHERN, H N and LOWE, V P W (1968) *J. anim. Ecol.* **37**, 75–97; (1982) *J. Zool. Lond.* **198**, 83–102. SOUTHERN, H N, VAUGHAN, R, and MUIR, R C (1954) *Bird Study* **1**, 101–10. SOUTHWELL, T (1888) *Zoologist* (3) **12**, 442–56. SÓVÁGÓ, M (1954) *Aquila* **55**–**8**, 288–9; (1968) *Aquila* **75**, 220–5. SOVIŠ, B and VALLO, F (1966) *Myslivost*, 28–9. SPANGENBERG, E P (1972) *Ornitologiya* **10**, 139–50. SPANGENBERG, E P and FEYGIN, G A (1936) *Trudy zool. Mus. Mosk. Univ.* **3**, 41–184. SPEEK, B J (1969) *Limosa* **42**, 82–109; (1973) *Limosa* **46**, 109–35. SPEIGHT, M C D (1973) *Br. Birds* **66**, 34–5; (1974) *Br. Birds* **67**, 388–9. SPENCER, K G (1945) *Bird Notes News* **21**, 131–2; (1953) *Bird Notes News* **25**, 141–4. SPENCER, R (1959) *Br. Birds* **52**, 441–92. SPENCER, R and HUDSON, R (1978) *Ring. Migr.* **2**, 57–104; (1980) *Ring. Migr.* **3**, 65–108. SPENGEMANN, H O (1975) *Beitr. Vogelkde.* **21**, 145–6. SPILLNER, W (1975) *Beitr. Vogelkde.* **21**, 172–215. SPINA, F (1982) *Avocetta* **6**, 23–33. SPITERI, N J (1975) M Sc Thesis. Durham Univ. SPRING, L (1971) *Condor* **73**, 1–27; (1965) *Condor* **67**, 457–88. SREBRODOLSKAYA, N I (1975) In Flint (1975), 101. STAAV, R (1977) *Alauda* **45**, 265–70; (1979) *Ornis fenn.* **56**, 13–17; (1980) *Vår Fågelvärld* **39**, 139–48. STAAV, R, AHLMKVIST, B, and HEDGREN, S (1972) *Vår Fågelvärld* **31**, 241–6. STADLER, H (1932) *Alauda* **4**, 407–15; (1945–6) *Vögel der Heimat* **16**, 1–13, 44–7, 53–63, 90–5; (1951) *Alauda* **19**, 178–80. STADLER, H and SCHMITT, C (1917) *Verh. orn. Ges. Bayern* **13**, 152–7. STAFFORD, J (1962) *Bird Study* **9**, 104–15. STAHLBAM, G (1954) *Beitr. Vogelkde.* **3**, 303; (1959) *Falke* **6**, 127–8; (1960) *Vogelwelt* **81**, 95–6. STAINTON, J M (1978) *Br. Birds* **71**, 138. STANFORD, J K (1949) *Br. Birds* **42**, 59; (1954) *Ibis* **96**, 316. STANLEY, E (1838) *A familiar history of birds* **1**. London. STASTNYI, K (1973) *Lynx*, 54–69. STECHOW, J (1937) *Beitr. Fortpfl. Biol. Vögel* **13**, 189–91. STEFANSSON, O (1979) *Vår Fågelvärld* **38**, 11–22; (1983) *Vår Fågelvärld* **42**, 245–50. STEFFEN, J (1953*a*) *Nos Oiseaux* **22**, 73–4; (1953*b*) *Nos Oiseaux* **22**, 97–8; (1955) *Nos Oiseaux* **23**, 82–4; (1956) *Nos Oiseaux* **23**, 179. STEGMANN, B (1968) *J. Orn.* **109**, 441–5; (1969) *Zool. Jb. Syst.* **96**, 1–51. STEGMANN, B K (1978) *Osnovy ornitogeograficheskogo deleniya Palearktiki.* Moscow; (1960) *Ornitologiya* **3**, 315–18. STEIL, W N (1957) *Falke* **4**, 162–5. STEIN, G (1928) *Beitr. Fortpfl. Biol. Vögel* **4**, 197–200. STEIN, H (1982) *Beitr. Vogelkde.* **28**, 191–2. STEINBACHER, G (1931) *J. Orn.* **79**, 349–53; (1956) *Vogelwelt* **77**, 120; (1959*a*) *J. Orn.* **100**, 103; (1959*b*) *Vogelwelt* **80**, 43–7. STEINBACHER, J (1937) *Arch.*

Naturgesch. N S **6**, 417–515; (1965) *Exotische Vögel in Farben.* Ravensburg; (1979) *Natur u. Museum* **109**, 375. STEINFATT, O (1937) *Beitr. Fortpfl. Biol. Vögel* **13**, 45–54, 101–13, 144–7; (1939) *Beitr. Fortpfl. Biol. Vögel* **15**, 9–14; (1940) *Beitr. Fortpfl. Biol. Vögel* **16**, 43–50, 93–9; (1941a) *Beitr. Fortpfl. Biol. Vögel* **17**, 58–63, 90–6; (1941b) *Beitr. Fortpfl. Biol. Vögel* **17**, 185–200; (1944a) *Beitr. Fortpfl. Biol. Vögel* **20**, 31–2; (1944b) *Beitr. Fortpfl. Biol. Vögel* **20**, 48–59, 93–7. STEINKE, G (1977) *Beitr. Vogelkde.* **23**, 72–8; (1981) *Orn. Jber. Mus. Hein.* **5–6**, 37–48. STEIN-SPIESS, S (1956) *Aquila* **63**, 343. STEMPNIEWICZ, L (1981) *Acta Orn.* **18**, 141–65. STENHOUSE, J H (1930) *Scott. Nat.*, 47–9. STEPANYAN, L S (1975) *Sostav i raspredelenie ptits fauny SSSR, non-passeriformes.* Moscow. STEPHAN, B (1961) *Wiss. Z. Humboldt Univ. Math.-Nat. R R* **10**, 147–75. STETTENHEIM, P R (1959) Ph D Thesis. Michigan Univ. STEVANOVIČ, A (1960) *Larus* **12–13**, 55–65. STEVENSON, H (1866) *The birds of Norfolk* **1**. London; (1875) *Zoologist* (2) **10**, 4289–94. STEVENSON, J G, HUTCHISON, R E, HUTCHISON, J B, BERTRAM, B C R, and THORPE, W H (1970) *Nature* **226**, 562–3. STEVENTON, D J (1979) *Ring. Migr.* **2**, 105–12; (1982) *Seabird Rep.* **6**, 105–9. STEWART, P A (1952) *Auk* **69**, 227–45. STEWART, P F and CHRISTENSEN, S J (1971) *A check list of the birds of Cyprus.* STEWART, R E and ROBBINS, C S (1947) *Auk* **64**, 286–74; (1958) *North Am. Fauna* **62**. STEYN, P (1960) *Bokmakierie* **12**, 35–6; (1966) *Bokmakierie* **18**, 83–5; (1967) *Ostrich* **38**, 284–5; (1972) *Ostrich* **43**, 56–9; (1982) *Birds of prey of southern Africa.* Cape Town. STEYN, P and BROOKE, R K (1970) *Ostrich Suppl.* **8**, 271–82. STICHMANN, W (1964) *J. Orn.* **105**, 491–2. STIEFEL, A (1966) *Orn. Mitt.* **18**, 31–3. STIEFEL, A and STIEFEL, R (1970) *Apus* **2**, 148–52. STIRLING, I, STIRLING, S M, and SHAUGHNESSY, G (1970) *Emu* **70**, 189–92. STIRNEMANN, F (1940) *Vögel der Heimat* **11**, 2–6; (1941) *Vögel der Heimat* **12**, 2–3; (1943) *Vögel der Heimat* **13**, 194–9; (1948) *Vögel der Heimat* **19**, 191–3. STÖCKLIN, W (1952) *Vögel der Heimat* **22**, 84–9. STOLT, B-O (1972) *Vår Fågelvärld* **31**, 111–16. STOLT, B-O and RISBERG, E L (1971) *Vår Fågelvärld* **30**, 194–200. STONE, R C (1954) *Br. Birds* **47**, 312. STONEHOUSE, B (1963) *Ibis* **103b**, 474–9. STONOR, C R (1944) *Ibis* **86**, 94–7. STORER, R W (1945) *Ibis* **87**, 433–56; (1952) *Univ. Calif. Publ. Zool.* **52**, 121–222. STORR, G M (1958) *Emu* **58**, 59–62. STOWE, T J (1982) *Ibis* **124**, 502–10. STRANGER, R H (1968) *W. Austral. Nat.* **11**, 4–14. STRAUBINGER, J (1966) *Anz. orn. Ges. Bayern* **7**, 861–3. STRAUCH, J G, JR (1977) *Bull. Br. Orn. Club* **97**, 61–5; (1978) *Trans. zool. Soc. Lond.* **34**, 263–345; (1979) *Bird-Banding* **50**, 283–4. STRESEMANN, E (1920) *Avifauna Macedonica.* Munich; (1926) *Orn. Monatsber.* **34**, 55–6; (1927–34) In Kukenthal and Krombach (eds) *Handbuch der Zoologie* **7** (2) *Sauropsida, Aves.* Berlin; (1928) *J. Orn.* **76**, 313; (1943) *J. Orn.* **91**, 448–514; (1944) *Orn. Monatsber.* **52**, 132–46; (1959) *Auk* **76**, 269–80. STRESEMANN, E and NOWAK, E (1958) *J. Orn.* **99**, 243–96. STRESEMANN, E and STRESEMANN, V (1966) *J. Orn.* **107** suppl; (1969) *J. Orn.* **110**, 192–204. STRESEMANN, V and STRESEMANN, E (1961) *J. Orn.* **102**, 317–52. STRICKLAND, M J and GALLAGHER, M D (1969) *A guide to the birds of Bahrain.* Muharraq. STRIEGLER, R, STRIEGLER, U, and JOST, K-D (1982) *Falke* **29**, 164–70. STROKOV, V V (1963) *Ornitologiya* **6**, 483; (1974) *Ornitologiya* **11**, 274–8. STROMBERG, G (1964) *Vår Fågelvärld* **23**, 256–65. STÜLCKEN, K (1961) *Falke* **8**, 39–45, 79–84, 111–14; (1962) *Falke* **9**, 219–23, 265–71. STÜLCKEN, K and BRÜLL, H (1938) *J. Orn.* **86**, 59–73. ŠTUSÁK, J (1958) *Sylvia* **15**, 259. SUBAH, A (1983) *Torgos* **3**, 21–32. SUCHANTKE, A (1958) *Alauda* **26**, 306; (1960) *Alauda* **28**, 38–44. SUDHAUS, W (1972) *Orn. Beob.* **69**, 301–2. SUDILOVSKAYA, A M (1935) *Oiseau* **5**, 219–35. SUGG, M St J (1974) *Ostrich* **45**, 227–34. SUKHININ, A N, BEL'SKAYA, G S, and ZHERNOV, I V (1972) *Ornitologiya* **10**, 216–27. SULKAVA, P (1965) *Aquilo Ser. Zool.* **2**, 41–7. SULKAVA, P, and SULKAVA, S (1971) *Ornis fenn.* **48**, 117–24.

SULKAVA, S (1966) *Suom. Riista* **18**, 145–56. SULTANA, J and GAUCI, C (1982) *A new guide to the birds of Malta.* Valletta. SULTANA, J, GAUCI, C, and BEAMAN, M (1975) *A guide to the birds of Malta.* Valletta. SUMMERHAYES, V S and ELTON, C S (1928) *J. Ecol.* **16**, 193–268. SUOMALAINEN, H (1939) *Ann. Zool. Soc. 'Vanamo'* **6**, 16–21. SUSHKIN, P P (1908) *Mat. Pozn. Fauny Flory Ross. Imp. Zool.* **8**, 1–803. SUTHERLAND, W J and BROOKS, D J (1981) *Sandgrouse* **3**, 87–90. SUTTER, E (1941) *Arch. suisses Orn.* **1**, 481–508; (1946) *Orn. Beob.* **43**, 72–81; (1961) *Orn. Beob.* **58**, 201–3; (1974) *Rev. suisse Zool.* **81**, 684–9; (1975) *Orn. Beob.* **72**, 199–202. SUTTON, G M (1932) *Mem. Carnegie Mus.* **12** (2) sect. 2, 3–267; (1982) *Living Bird Q.* summer, 17. SUTTON, G M and PARMELEE, D F (1956) *Condor* **58**, 273–82. SVÄRDSON, G (1951) *Proc. int. orn. Congr.* **10**, 335–8; (1957) *Br. Birds* **50**, 314–43. SVENSSON, S (1960) *Vår Fågelvärld* **19**, 333–6; (1978) *Vår Fågelvärld* **37**, 97–112. SVOBODA, S (1961) *Sborn. Kl. přír. Brně* **33**, 93–108; (1968) *Práce zool. Kl. přír. Brně*, 14–23. SWAINE, C M (1945) *Br. Birds* **38**, 329–32. SWALES, M K (1965) *Ibis* **107**, 215–29. SWANBERG, P O (1952) *Vår Fågelvärld* **11**, 49–66. SWANN, R-L and BAILLIE, S R (1979) *Bird Study* **26**, 55–8. SWARTZ, L G (1966) In Wilimovsky, N J and Wolfe, J N (eds) *Environment of the Cape Thompson region, Alaska.* US Atomic Energy Comm., Oak Ridge, Tenn. SWENNEN, C (1977) *Laboratory research on sea-birds.* Texel. SWENNEN, C and DUIVEN, P (1977) *Neth. J. Sea Res.* **11**, 92–8. SWIERCZEWSKI, E V and RAIKOW, R J (1981) *Auk* **98**, 466–80. SWIFT, J J (1959) *Alauda* **27**, 97–143; (1960a) *Br. Birds* **53**, 131; (1960b) *Br. Birds* **53**, 559–72. SZABÓ, L V (1965) *Allatt. Kozlem* **52**, 111–34. SZEMERE, Z (1927–8) *Aquila* **34–5**, 309–12. SZIJJ, J (1958) *Bonn. zool. Beitr.* **9**, 25–48; (1977) *Bull. int. Waterfowl Res. Bur.* **43–4**, 23–4. SZLIVKA, L (1957) *Larus* **9–10**, 48–70; (1959a) *Aquila* **65**, 348; (1959b) *Aquila* **65**, 349; (1962) *Larus* **14**, 121–34; (1965a) *Aquila* **72**, 246–7; (1965b) *Larus* **19**, 107–32. SZÖCS, J (1942) *Aquila* **46–9**, 393–6. SZOMJAS, L (1952–5) *Aquila* **59–62**, 450. SZULC-OLECHOWA, B (1964) *Acta Orn. Warszawa* **8**, 415–43.

TAAPKEN, J (1981) *Vogeljaar* **29**, 323–4. TAIT, C S (1970) *Scott. Birds* **6**, 45; TAIT, G M (1960) *Ardeola* **6**, 259–81; (1961) *Ardeola* **7**, 175–95. TAIT, W C (1924) *The birds of Portugal.* London. TALPEANU, M and PASPALEVA, M (1979) *Trav. Mus. Hist. nat. 'Grigore Antipa'* **20**, 441–9. TALPOSH, V S (1975a) *Nauch. doklady vysshey shkoly Biol. Nauki* **18**, 16–22; (1975b) In Flint (1975), 101–2. TAMANTSEVA, L S (1955) *Trudy byuro kol'ts.* **8**, 91–100. TÅNING, A V (1944) *Dansk orn. Foren. Tidsskr.* **38**, 163–216. TAPFER, D (1957) *Falke* **4**, 3–5. TARASOV, M P (1979) *Byull. Mosk. Obshch. Ispyt. Prir. Otd. Biol.* **84** (4), 79–84. TARBOTON, W (1976) *Ostrich* **47**, 99–112. TARBOTON, W R, CLINNING, C F, and GROND, M (1975) *Ostrich* **46**, 188. TARDENT, P (1951) *Orn. Beob.* **48**, 157–61. TARJÁN, T (1942) *Aquila* **49**, 489–90. TATARINKOVA, I P (1982) *Proc. int. orn. Congr.* **18**, 297. TATARINOV, K A (1965) *Ornitologiya* **7**, 492. TAUX, K (1976) *Vogelkdl. Ber. Niedersachs.* **8**, 65–75. TAVISTOCK, MARQUIS OF (1929) *Parrots and parrot-like birds.* London. TAYLOR, D (1977) *Br. Birds* **70**, 548. TAYLOR, I R (1975) Ph D Thesis. Aberdeen Univ.; (1983) *Ornis scand.* **14**, 90–6. TAYLOR, K (1976) B Sc Thesis. St Andrews Univ.; (1978) *Br. Birds* **71**, 598–9; (1982) Ph D Thesis. St Andrews Univ. TAYLOR, K and REID, J B (1981) *Scott. Birds* **11**, 173–80. TAYLOR, P S (1973) *Living Bird* **12**, 137–54; (1974) M S Thesis. Alberta Univ. TEBBUTT, C F (1942) *Br. Birds* **36**, 115–16; (1977) *Br. Birds* **70**, 546. TEGETMEIER, W B (1888) *Pallas's Sand-Grouse: its history, habits, food and migrations, with hints as to its utility, and a plea for its preservation.* London. TEIRO, H J (1959) *Suom. Riista* **13**, 93–105. TEIXEIRA, R M (ed) (1979) *Atlas van de Nederlandse broedvogels.* Deventer. TELLERIA, J L (1979) *Alauda* **47**, 139–50. TENOVUO, R (1963)

Ann. Zool. Soc. 'Vanamo' **25**, 1–147. TERHIVUO, J (1976) *Ornis fenn.* **53**, 47. TERNIER, L (1913) *Rev. fr. Orn.* **5**, 164–5. TERNOVSKI, D V and TERNOVSKAYA, Y G (1959) *Izv. Sib. Otd. Akad. Nauk SSSR*, 81–9. TERRY, H (1952) *Vie et Milieu* **3**, 451–7. THANNER, R VON (1908) *Orn. Jahrb.* **19**, 198–215. THÉVENOT, M, BERGIER, P, and BEAUBRUN, P (1980) *Docum. Inst. Sci. Rabat* **5**. THIBAUT DE MAISIÈRES, C (1940) *Alauda* **12**, 17–65; (1943) *Aquila* **50**, 372–8. THIOLLAY, J-M (1963a) *Alauda* **31**, 32–5; (1963b) *Nos Oiseaux* **27**, 124–31; (1966) *Oiseau* **36**, 282–3; (1967) *Oiseau* **37**, 104–13; (1968) *Nos Oiseaux* **29**, 249–69; (1969) *Alauda* **37**, 15–28; (1971) *Oiseau* **41**, 148–62; (1974) *Alauda* **42**, 223–5. THOMAS, D H and ROBIN, A P (1977) *J. Zool. Lond.* **183**, 229–49; (1983) *Bull. Br. Orn. Club* **103**, 40–3. THOMAS, G (1982) *Seabird Rep.* **5**, 59–69. THOMAS, G and RICHARDS, P (1977) *Birds* **6** (6), 61. THOMAS, M J (1975) *Bird Study* **22**, 52. THOMAS, W H (1844) *Zoologist* **2**, 433–6. THOMAZ DE BOSSIERRE, R DE (1947) *Gerfaut* **37**, 99–102. THOMPSON, D Q (1955) *Act. Inst. N. Am. Final Rep. Proj. ONR*. THOMPSON, M C and DELONG, R L (1969) *Auk* **86**, 747–9. THOMSEN, P and JACOBSEN, P (1979) *The birds of Tunisia*. Copenhagen. THOMSON, A L (1943) *Br. Birds* **37**, 62–9. THÖNEN, W (1965) *Orn. Beob.* **62**, 196–7; (1966) *Orn. Beob.* **63**, 21–3; (1968) *Orn. Beob.* **65**, 17–22. THORESON, A C and BOOTH, E S (1958) *Walla Walla Coll. Publ. Dept. Biol. Sci.* **23**, 1–37. THORLEY, H M (1963) *J. Durham Coll. nat. Hist. Soc.* **9**, 19–21. THOUY, P (1978) *Alauda* **46**, 87–93. TICEHURST, C B (1923a) *Ibis* (11) **5**, 1–43; (1923b) *Ibis* (11) **5**, 235–75, 438–74; (1923c) *Ibis* (11) **5**, 645–66; (1924a) *Ibis* (11) **6**, 110–46; (1924b) *Ibis* (11) **6**, 643–4; (1926) *J. Bombay nat. Hist. Soc.* **31**, 91–119; (1932) *A history of the birds of Suffolk*. London; (1935) *Ibis* (13) **5**, 329–35; (1939) *Ibis* (14) **3**, 512–20. TICEHURST, C B, BUXTON, P A, and CHEESMAN, R E (1922) *J. Bombay nat. Hist. Soc.* **28**, 210–50, 381–427, 650–74, 937–56. TICEHURST, C B, COX, P, and CHEESMAN, R E (1926) *J. Bombay nat. Hist. Soc.* **31**, 91–119. TICEHURST, C B and WHISTLER, H (1932) *Ibis* (13) **2**, 40–93. TIMA, C B (1968) *Birds of the Baltic region*. Springfield. TIMMERMAN, A (1970) *Limosa* **43**, 31–8. TIMMERMANN, G (1938–49) *Die Vögel Islands*. Reykjavik; (1957) *Parasitol. Schriftenreihe* **8**. TINBERGEN, L and TINBERGEN, N (1932) *Beitr. Fortpfl. Biol. Vögel* **8**, 11–14. TINBERGEN, N (1931) *Ardea* **20**, 1–17; (1933) *Ecol. Monogr.* **3**, 443–9; (1938) *Ardea* **27**, 247–9. TINBERGEN, N and BROEKHUYSEN, G J (1954) *Ostrich* **25**, 50–61. TINDAL, D A (1968) *Scott. Birds* **5**, 108. TISCHLER, F (1934) *Beitr. Fortpfl. Biol. Vögel* **10**, 75. TITAVNIN, A P (1981) In Flint (1981), 13–15. TIWARY, N K (1931) *J. Bombay nat. Hist. Soc.* **34**, 578. TJITTES, A A and KOERSVELD, E VAN (1952) *Ardea* **40**, 119–22. TJOMLID, S A (1973) *Ornis scand.* **4**, 145–51. TOFT, G O (1983) *Fauna norv. Ser. C Cinclus* **6**, 8–13. TOHMÉ, G and NEUSCHWANDER, J (1974) *Alauda* **42**, 243–58; (1978) *Oiseau* **48**, 319–27. TOMASZ, E (1944–7) *Aquila* **51–4**, 199. TOMASZ, J (1955) *Aquila* **59–62**, 101–43. TOMBAL, J-C (1977) *Héron*, 27–48. TOMIAŁOJĆ, L (1976a) *Birds of Poland*. Warsaw; (1976b) *Acta zool. Cracov.* **21**, 585–631; (1979) *Pol. Ecol. Stud.* **5** (4), 141–220. TOMKINS, I R (1942) *Auk* **59**, 308; (1959) *Wilson Bull.* **71**, 313–22; (1963) *Auk* **80**, 549. TOMLINSON, A G (1916) *J. Bombay nat. Hist. Soc.* **24**, 825–9. TOMLINSON, J N (1969) *Br. Birds* **62**, 76–7. TOOBY, H J (1943) *Br. Birds* **37**, 77–8. TOOBY, J (1946) *Br. Birds* **39**, 29–30. TOOK, J M E (1963) *Ostrich* **34**, 176. TOOR, H S and RAMZAN, M (1974) *Pakistan J. agric. Sci.* **11**, 191–6. TORDOFF, H B (1962) *Kansas orn. Soc. Bull.* **13**, 7–8. TORDOFF, H B and SOUTHERN, W E (1959) *Wilson Bull.* **71**, 385–6. TÖRNLUND, N (1979) *Anser* **18**, 139–40. TORTZEN, N J (1962) *Dansk orn. Foren. Tidsskr.* **56**, 135–42. TOSTAIN, O and SIBLET, J-P (1981) *Passer* **18**, 111–24. TOWNSEND, C W (1915) *Auk* **32**, 306–16. TRACY, N (1924a) *Br. Birds* **17**, 276–9; (1924b) *Br. Birds* **18**, 111; (1928) *Br. Birds* **21**, 157; (1933a) *Br. Birds* **26**, 257–8; (1933b) *Br. Birds* **27**, 117–32; (1938) *Beitr. Fortpfl. Biol. Vögel* **14**, 41–8; (1942) *Naturalist Lond.*, 160–1; (1946) *Br. Birds* **39**, 19–22. TRANSEHE, N VON (1965) *Die Vogelwelt Lettlands*. Hannover-Döhren. TRAP-LIND, I (1965) *De Danske Ugler*. Copenhagen. TRAUTMAN, M B (1939) *Wilson Bull.* **5**, 44–5. TRAYLOR, M A and PARELIUS, D (1967) *Fieldiana: Zool.* **51**, 91–117. TREE, A J (1960) *Br. Birds* **53**, 130–1; (1961a) *Br. Birds* **54**, 286; (1961b) *Ostrich* **32**, 86–9; (1969) *Puku* **5**, 181–205; (1973) *Ostrich* **44**, 266; (1974) *Ostrich* **45**, 136. TRELFA, G (1953) *Br. Birds* **46**, 413–14. TRICOT, J (1968) *Aves* **5**, 111; (1977) *Aves* **14**, 1–82. TRINTHAMMER, W F (1859) *J. Orn.* **7**, 387–92. TRIPPMACHER, K-H (1983) *Falke* **30**, 222–4. TRISTRAM, H B (1866) *Ibis* (2) **2**, 59–88; (1884) *The survey of western Palestine*. London; (1889) *Ibis* (6) **1**, 13–32. TROLLOPE, J (1971) *Avic. Mag.* **77**, 117–25. TROTIGNON, E, TROTIGNON, J, BAILLON, M, DEJONGHE, J-F, DUHAUTOIS, L, and LECOMTE, M (1980) *Oiseau* **50**, 323–43. TROTIGNON, J (1976) *Alauda* **44**, 119–33; (1979) *Parc National du Banc d'Arguin, Mauritanie, Rep. 1977–9*. Nouadhibou. TRÖTSCHEL, P (1973) *Vogelwelt* **94**, 65. TROTT, A C (1947) *Ibis* **89**, 77–98. TSCHANZ, B (1959) *Behaviour* **14**, 1–100; (1968) *Z. Tierpsychol.* suppl. **4**; (1979) *Fauna norv. Ser. C Cinclus* **2**, 70–94. TSCHANZ, B and HIRSBRUNNER-SCHARF, M (1975) In Baerends, G, Beer, C, and Manning, A (eds) *Function and evolution in behaviour*, 358–80. Oxford. TSCHANZ, B and SCHOMMER, M (1975) *Vogelwarte* **28**, 17–44. TSCHANZ, B and WEHRLIN, J (1968) *Fauna* **21**, 53–5. TSCHUSI ZU SCHMIDHOFFEN, V (1904) *Orn. Jahrb.* **15**, 93–108; (1909) *Verh. Mitt. siebenbürg. Ver. Naturw.* **58**, 1–41. TUCK, G S (1961b) *Sea Swallow* **14**, 31–40; (1965) *Sea Swallow* **17**, 40–50; (1974) *Sea Swallow* **23**, 7–21. TUCK, G S and HEINZEL, H (1978) *A field guide to the seabirds of Britain and the world*. London. TUCK, L M (1961a) *Can. Wildl. Serv. Monogr.* **1**; (1971) *Bird-Banding* **42**, 184–209. TUCK, L M and SQUIRES, H J (1955) *J. Fish. Res. Bd. Can.* **12**, 781–92. TUCKER, B W (1936) *Br. Birds* **30**, 206–8. TUKE, A J S (1953) *An introduction to the birds of southern Spain and Gibraltar*. Gibraltar. TULL, C E, GERMAIN, P, and MAY, A W (1972) *Nature* **237**, 42–4. TULLOCH, R J (1968) *Br. Birds* **61**, 119–32; (1969) *Br. Birds* **62**, 33–6; (1975) *Birds* **5** (8), 24–7. TURČEK, F J (1950) *Vår Fågelvärld* **9**, 210–12; (1954) *Ornis fenn.* **31**, 33–41; (1956–8) *Aquila* **63–5**, 345; (1961) *Ökologische Beziehungen der Vögel und Gehölze*. Bratislava. TURNER, E L (1908) *Br. Birds* **2**, 141–5; (1914) *Wild Life* **3**, 16–24; (1920) *Br. Birds* **14**, 122–6. TUTMAN, I (1960) *Larus* **14**, 169–79. TUTT, H R (1951) *Proc. int. orn. Congr.* **10**, 555–62; (1955) *Br. Birds* **48**, 261–6; (1956) *Br. Birds* **49**, 32–6. TYLER, J G (1913) *Pacific Coast Avifauna* **9**. Cooper Orn. Soc. TYLER, S (1979) *Scopus* **3**, 1–7.

UDVARDY, M D F (1963) In Gressitt, J L (ed) *Pacific basin biogeography. Symp. 10th Pacif. sci. Congr. Honolulu 1961*, 85–111. ULFSTRAND, S (1965) *Fauna och Flora* **60**, 129–47. ULFSTRAND, S and HÖGSTEDT, G (1976) *Anser* **15**, 1–32. ULLRICH, B (1970) *Vogelwelt* **91**, 28–9; (1973) *Anz. orn. Ges. Bayern* **12**, 163–75; (1980) *Vogelwarte* **30**, 179–98. ULVBLAD, B (1962) *Vår Fågelvärld* **21**, 44–5. UNHOLA, A (1973) MS. Turku Univ., Finland. UPTON, R (1962) *Great Spotted Woodpecker enquiry report 1959/60*. Br. Trust Orn., Tring. URBAN, E K and BOSWALL, J (1969) *Bull. Br. Orn. Club* **89**, 121–9. URBAN, E K and BROWN, L H (1971) *A checklist of the birds of Ethiopia*. Addis Ababa. URNER, C A (1925) *Auk* **42**, 31–41. URRY, D and URRY, K (1970) *Flying birds*. London. URSPRUNG, J (1979) *Egretta* **22**, 4–17. USPENSKI, S M (1956) *Ptich'i bazary Novoy Zemli*. Moscow; (1972) *Ornitologiya* **10**, 123–9. USPENSKI, S M, BOEHME, R L, PRIKLONSKI, S G, and VEKHOV, V N (1962a) *Ornitologiya* **4**, 64–86; (1962b) *Ornitologiya* **5**, 49–67. USPENSKI, S M and PRIKLONSKI, S G (1961) *Falke* **8**, 403–7. USSHER, R J (1913) *Irish Nat.* **22**, 178–9. UTTENDÖRFER, K (1936) *Mitt. Ver. sächs. Orn.* **5**, 67–82; (1937) *Nos Oiseaux*

14, 57–66; (1939a) Ber. Ver. schles. Orn. **24**, 25–36. UTTENDÖRFER, O (1939b) Die Ernährung der deutschen Raubvögel und Eulen, und ihre Bedeutung in der heimischen Natur. Berlin; (1952) Neue Ergebnisse über die Ernährung der Greifvögel und Eulen. Stuttgart. UYS, C J (1978) Ostrich **25**, 50–61.

VAHI, J (1963) Loodusuur. seltsi aastaraamat **55**, 240–54. VÄISÄNEN, R A (1973) Ornis scand. **4**, 47–53. VÄLIKANGAS, I and HYTÖNEN, O (1932) Mem. Soc. Fauna Flora Fenn. **8**, 100–36. VALIUS, M, LOGMINOS, V, PETRAITIS, A, and SKUADIS, V (1977) Ecology of the birds of Lithuania. Vilnius. VALVERDE, J A (1953a) Oiseau **23**, 288–99; (1953b) Nos Oiseaux **22**, 7–10; (1957) Aves del Sahara Español. Madrid; (1958) Br. Birds **51**, 1–23; (1960) Arch. inst. Aclim. Almeria **9**, 5–168; (1967) Monogr. Estac. Biol. Doñana 1; (1971) Ardeola spec. vol., 591–647. VAN DEN ASSEM, J (1954) Levende Nat. **57**, 1–9. VAN DER MEER, G (1930) Org. Club Nederl. Vogelk. **3**, 70–82; (1931) Org. Club Nederl. Vogelk. **3**, 136–45. VAN DER STRAETEN, E and ASSELBERG, R (1973) Gerfaut **63**, 149–59. VAN GOMPEL, J (1979) Gerfaut **69**, 83–110. VAN IERSEL, J J A and BOL, A C A (1958) Behaviour **13**, 1–87. VAN IMPE, J (1977) Alauda **45**, 17–52. VAN ROSSEM, A (1923) Condor **15**, 208–13. VARGA, F (1977a) Aquila **84**, 104–5; (1977b) Aquila **84**, 110–11. VARGAS, J M, ANTUNEZ, A, and BLASCO, M (1978) Ardeola **24**, 227–31. VAROUJEAN, D (1979) Pacific Seabird Group Bull. **6**, 28. VARSHAVSKI, S N (1981) Byull. Mosk. Obshch. Ispyt. Prir. Otd. Biol. **86**, 27–30. VÁSÁRHELYI, I (1966–7) Aquila **73–4**, 196. VASILIU, G D (1968) Systema Avium Romaniae. Paris. VASIĆ, V F (1977) Arch. biol. Sciences **29**, 69–81. VASVÁRI, M (1955) Aquila **59–62**, 167–84. VASVÁRI, N (1931–4) Aquila **38–41**, 432–3. VAUK, G (1963) Vogelwarte **22**, 35–8; (1964) Vogelwarte **85**, 113–20; (1972) Vogelwarte **26**, 285–9. VAUK-HENZELT, E (1982) Gefiederte Welt **106**, 288. VAURIE, C (1959a) Am. Mus. Novit. **1945**; (1959b) Am. Mus. Novit. **1946**; (1959c) Am. Mus. Novit. **1951**; (1959d) Am. Mus. Novit. **1963**; (1959e) Am. Mus. Novit. **1971**; (1960a) Am. Mus. Novit. **1997**; (1960b) Am. Mus. Novit. **2000**; (1960c) Am. Mus. Novit. **2015**; (1960d) Am. Mus. Novit. **2021**; (1961a) Am. Mus. Novit. **2043**; (1961b) Am. Mus. Novit. **2058**; (1961c) Am. Mus. Novit. **2071**; (1963) Am. Mus. Novit. **2132**; (1965) The birds of the Palearctic fauna: non-passeriformes. London; (1972) Tibet and its birds. London. VEEN, J (1977) Behaviour Suppl. **20**. VEIGA, J P (1980) Ardeola **25**, 113–41. VEIN, D and THÉVENOT, M (1978) Nos Oiseaux **34**, 347–51. VELICHKO, M A (1963) Uchen. zap. Leningrad. ped. Inst. **230**, 3–18. VENABLES, L S V (1938) Br. Birds **32**, 148–9. VENABLES, L S V and VENABLES, U M (1962) Br. Birds **55**, 444–5. VERE BENSON, S (1970) Birds of Lebanon and the Jordan area. London. VERHEYEN, R (1950) Gerfaut **40**, 212–31; (1953) Explor. Parc. natn. Upemba Miss. G F de Witte **19**; (1961) Bull. Inst. roy. Sci. nat. Belg. **37** (27), 1–36; (1967) Oologica Belgica. Brussels; (1970) Gerfaut **60**, 327–403. VERHEYEN, R F and DAMME, B VAN (1967) Gerfaut **57**, 365–465. VERICAD, J R, ESCARRE, A, and RODRIGUEZ, E (1976) Mediterranea, 47–59. VERNON, C J (1971a) Ostrich **42**, 153–4; (1971b) Ostrich **42**, 242–58; (1972) Ostrich **43**, 109–24. VERNON, J D R (1971c) Br. Birds **64**, 32–3. VERNON, J D R, CHADWICK, P J, and GRIFFIN, D (1973) Alauda **41**, 345–52. VERTHEIN, J (1935) Orn. Monatsber. **43**, 131–3. VERWEY, J (1922) Ardea **11**, 99–116; (1924) Tijdschr. Ned. Dierk. Veren. **19** (7–8); (1927) Zool. Anz. **71**, 1–4. VESEY-FITZGERALD, D F (1955) Ostrich **24**, 128–33. VIČEK, M and VONDRÁČEK, J (1974) Biologica (CSSR) **29**, 649–56. VIDAL, G W (1880) Stray Feathers **9**, 1–96. VIEBIG, A (1935) Beitr. Fortpfl. Biol. Vögel **11**, 165–9. VIELLIARD, J (1972) Alauda **40**, 63–92. VIETINGHOFF-RIESCH, A VON (1928) Mitt. Ver. sächs. Orn. **2**, 81–93. VIKBERG, P (1982) Lintumies **17**, 60–8. VIKSNE, J (1978) In Kumari, E (ed) Commun. Baltic Commiss. Study Bird Migr. **11**, 76–89; (1983) Ptitsy Latvii. Riga.

VILAGRASA, F X, CARRERA, E, and PARDO, R (1982) Alauda **50**, 108–13. VILKA, J (1960) Orn. Petijumi **2**, 213–14. VILLAGE, A (1981) Bird Study **28**, 215–24. VINCENT, A J (1934) Bull. Br. Orn. Club **54**, 141–3. VINCENT, A W (1946a) Ibis **88**, 48–67; (1946b) Ibis **88**, 306–26. VINICOMBE, K E (1975) Br. Birds **68**, 208. VIRKKUNEN, I (1967) Ornis fenn. **44**, 73–7. VITTERY, A, PORTER, R F, and SQUIRE, J E (eds) (1971) Check list of the birds of Turkey. Orn. Soc. Turkey, Sandy. VITTERY, A, SQUIRE, J E, and PORTER, R F (1972) Orn. Soc. Turkey Bird Rep. 1968–9. Sandy. VLADIMIRSKAYA, M (1948) Trudy Lapland. Gos. Zapoved. **3**. VLAŠÍN, M (1978) Fol. Zool. **27**, 47–56. VLEUGEL, D A (1952) Vogelwelt **73**, 13–16. VOGEL, F (1982) Mitt. Orn. Ver. Hildesheim **6**, 85–9. VOIGT, A (1933) Exkursionsbuch zum Studium der Vogelstimmen. Leipzig. VOLKMANN, G (1959) Orn. Mitt. **11**, 79. VOLLBRECHT, K (1938) Beitr. Fortpfl. Biol. Vögel **14**, 105; (1955) Orn. Mitt. **7**, 9. VONDRÁČEK, J (1978) Beitr. Vogelkde. **24**, 91–3. VONDRÁČEK, J and HONCŮ, M (1978) Sborník severočesk. Mus. Přír. Vedy **10**, 67–71. VOOUS, K H (1947) Limosa **20**, 1–142; (1950) Syllegomena biol., 429–43; (1955) De vogels van de Nederlandse Antillen. 's-Gravenhage; (1960) Atlas of European birds. London; (1961) Bull. Br. Orn. Club **81**, 62–6; (1963a) Limosa **36**, 138–40; (1963b) Proc. int. orn. Congr. **13**, 1214–16; (1964) Nytt Mag. Zool. **12**, 38–47; (1965) Limosa **28**, 68–71; (1977a) List of recent Holarctic bird species. London; (1977b) Bull. Br. Orn. Club **97**, 42–3. VOROBIEV, K A (1954) Ptitsy Ussuriyskogo kraya. Moscow. VORONETSKI, V S (1974a) Mat. VI Vsesoyuz. orn. Konf. **1**, 133–4; (1974b) Ornitologiya **11**, 366–7.

WAARD, S (1952) Br. Birds **45**, 339–41. WABAKKEN, P (1973) Fauna **26**, 1–6. WACHSMUTH, G (1938) Beitr. Fortpfl. Biol. Vögel **14**, 151–2. WÄCHTLER, W (1932) Beitr. Fortpfl. Biol. Vögel **8**, 219–22. WADE, P (ed) (1975) Every Australian bird illustrated. London. WADEWITZ, O (1956) Falke **3**, 39–43. WADLEY, N J P (1951) Ibis **93**, 68–89. WAGNER, G and SPRINGER, M (1970) Orn. Beob. **67**, 77–94. WAGNER, G, TSCHANZ, B, and KÜNG, K (1957) Mitt. Naturforsch. Ges. Bern. **15**, 59–92. WAHLSTEDT, J (1959) Fauna och Flora **54**, 81–112; (1969a) Vår Fågelvärld **28**, 89–101; (1969b) Vår Fågelvärld **28**, 256–7; (1974) Vår Fågelvärld **33**, 132–9; (1976) Vår Fågelvärld **35**, 122–5. WAHN, R (1937) Beitr. Avif. Mitteldeutschlands **1**, 43–7; (1939) Beitr. Avif. Mitteldeutschlands **3**, 27–9. WAIT, W E (1931) Manual of the birds of Ceylon. Colombo. WALDENSTRÖM, A (1976) Calidris **5**, 99–103. WALKER, A B (1949) Br. Birds **42**, 152. WALKER, F J (1981a) Sandgrouse **2**, 33–55; (1981b) Sandgrouse **2**, 56–85. WALKER, L W (1974) The book of owls. New York. WALKER, T (1969) Honeyguide **60**, 27. WALLACE, D I M (1964) Ibis **106**, 389–90; (1973) Ibis **115**, 559–71. WALLER, R H (1955) Ibis **97**, 145. WALLIN, K and ANDERSSON, M (1981) Ornis scand. **12**, 125–6. WALLSCHLÄGER, D (1980) Falke **27**, 310–12. WALLSCHLÄGER, D and ZABKA, H (1981) Abh. Akad. Wiss. DDR 1N, 301–7. WALPOLE-BOND, J (1914) Field studies of some rarer British birds. London; (1932) Br. Birds **25**, 361–2; (1938) A history of Sussex birds. London. WALSH, J F (1980) Ostrich **51**, 191. WALSH, J F and GRIMES, L G (1981) Bull. Br. Orn. Club **101**, 327–34. WÄLTI, E and LOCHER, A (1952) Orn. Beob. **49**, 170–3. WARHAM, J (1950) Field **196**, 635; (1956) Emu **56**, 83–93; (1958) Br. Birds **51**, 303–8. WARMAN, S, WARMAN, C, and TODD, D (1983) Br. Birds **76**, 349–50. WARMAN, S R (1979) Bull. Br. Orn. Club **99**, 124–8. WARNCKE, K (1962) Vogelwelt **83**, 79–80; (1964) Vogelwelt **85**, 161–74. WARNCKE, K and WITTENBERG, J (1958) Vogelwelt **79**, 20–2. WARNKE, G (1941) Orn. Monatsber. **49**, 69–74. WASENIUS, E (1929) Ornis fenn. **6**, 114–15. WASSENICH, V (1960) Regulus **40**, 79–90. WATERS, E (1966) Br. Birds **59**, 341. WATSON, A (1957) Ibis **99**, 419–62. WATSON, D (1972) Birds of moor and mountain. Edinburgh. WATSON, G E (1966) Seabirds of the tropical Atlantic

Ocean. Washington. WATSON, J B (1941) *Br. Birds* **35**, 86. WATSON, P (1969) *Brathay Expl. Group Rep.*, 25–6. WATSON, P S (1981) *Br. Birds* **74**, 82–90. WATT, G (1951) *The Farne Islands.* London. WAVRIN, H DE (1971) *Aves* **8**, 23. WEBER, B (1957) *Falke* **4**, 104–5. WEBER, E (1968) *Falke* **15**, 138. WEBER, W (1965) *Egretta* **8**, 10–11. WEBSTER, J A (1973) *Bird Study* **20**, 185–96. WEESENBEECK, J VAN (1941) *Gerfaut* **31**, 126–9. WEHNER, R (1962) *Orn. Mitt.* **14**, 90–4; (1966) *Vogelwarte* **23**, 173–80; (1967) *Vogelwarte* **24**, 64. WEHRLIN, J (1977) *Z. Tierpsychol.* **44**, 45–79. WEIGELIN, W (1960) *Falke* **7**, 212. WEIJDEN, W J VAN DER (1973) *Bull. IFAN* **35**(A), 716–21; (1975) *Ardea* **63**, 65–77. WEIJDEN, W J VAN DER and GINN, H (1973) In Burton, J A (ed) *Owls of the world*, 197–209. New York. WEISE, H J (1968) *Orn. Mitt.* **20**, 203–10 WEITNAUER, E (1947) *Orn. Beob.* **44**, 133–82; (1949a) *Orn. Beob.* **46**, 80–5; (1949b) *Orn. Beob.* **46**, 86–9; (1952) *Orn. Beob.* **49**, 37–44; (1954) *Orn. Beob.* **51**, 66–71; (1955) *Orn. Beob.* **52**, 38–9; (1960) *Orn. Beob.* **57**, 133–41; (1962) *Orn. Beob.* **59**, 29–30; (1975) *Orn. Beob.* **72**, 87–100; (1977) *Orn. Beob.* **74**, 89–94; (1980) *Mein Vogel.* Oltingen. WELLER, L G (1944) *Br. Birds* **38**, 80. WELLS, D R (1982) *Bull. Br. Orn. Club* **102**, 62–3. WELLS, D R and BECKING, J H (1975) *Ibis* **117**, 366–71. WELS, H (1912) *Z. Ool. Orn.* **22**, 77–84. WENDLAND, V (1957) *J. Orn.* **98**, 241–61; (1958) *J. Orn.* **99**, 23–31; (1963) *J. Orn.* **104**, 23–57; (1964) *Berliner Naturschutzbl.* **8**, 505–7; (1967) *Beitr. Vogelkde.* **12**, 402–11; (1971) *Die Wirbeltiere Westberlins.* Berlin; (1972a) *Vogelwelt* **93**, 81–91; (1972b) *J. Orn.* **113**, 276–86; (1975) *Oecologia* **20**, 301–10; (1980) *Beitr. Vogelkde.* **25**, 157–71; (1981) *Oecologia* **48**, 7–12. WENNER, M V (1911) *Br. Birds* **5**, 194–5. WENZEL, K (1914) *Orn. Monatsschr.* **39**, 457–64. WERNER, E (1889) *Monatsschr. Dtsch. Ver. Schutze Vogelwelt* **14**, 122–6. WERNER, J (1958) *Vogelwelt* **79**, 187–8; (1961) *Vogelwelt* **82**, 121. WERTH, I (1947) *Br. Birds* **40**, 328–34. WEST, J D and SPEIRS, J M (1959) *Wilson Bull.* **71**, 348–63. WESTIN, P (1980) *Vår Fågelvärld* **39**, 44. WESTERFRÖLKE, P (1955a) *Vogelwelt* **76**, 185; (1955b) *Vogelwelt* **76**, 187; (1956) *Vogelwelt* **77**, 189–90. WESTERNHAGEN, W VON (1957) *Bonn. zool. Beitr.* **8**, 178–92; (1970) *Vogelwarte* **25**, 185–93. WETMORE, A (1916) *US Dept. Agric. Bull.* **326**; (1919) *Proc. Biol. Soc. Washington* **32**, 195–202; (1926) *Bull. US natn. Mus.* **133**; (1960) *Smiths. misc. Coll.* **139** (11). WHILDE, A (1979) *Irish Birds* **1**, 370–6. WHISTLER, H (1928) *J. Bombay nat. Hist. Soc.* **33**, 136–45; (1931) *J. Bombay nat. Hist. Soc.* **35**, 189–95; (1941) *Popular handbook of Indian birds.* London. WHITAKER, J I S (1905) *Birds of Tunisia* **2**. London. WHITE, C A (1947) *Br. Birds* **40**, 254; (1960a) *Br. Birds* **53**, 404–5. WHITE, C M N (1960b) *Ibis* **102**, 138–9; (1965) *A revised check list of African non-passerine birds.* Lusaka. WHITE, F N, BARTHOLOMEW, G A, and KINNEY, J L (1978) *Physiol. Zool.* **51**, 140–54. WHITEHEAD, C H T (1911) *J. Bombay nat. Hist. Soc.* **20**, 966–7. WHITFIELD, A K (1977) M Sc Thesis. Natal Univ.; (1978) *Ostrich* **49**, 45. WHITFIELD, A K and BLABER, S J (1978) *Ostrich* **49**, 185–98. WHITFIELD, A K and CYRUS, D P (1978) *Ostrich* **49**, 8–15. WHITMAN, C O (1919) *Carnegie Inst. Washington Publ.* **257** (3). WHITMAN, F N (1924) *Auk* **41**, 479–80. WICKL, K-H (1979) *Garmischer Vogelkdl. Ber.* **6**, 1–47. WICKLER, W (1976) *J. theor. Biol* **60**, 493–7. WICHMANN, H (1952) *Ardea* **40**, 115–19. WIEGANK, F (1977) *Beitr. Vogelkde.* **23**, 229–32. WIEHE, H (1970) *Vogelkdl. Ber. Niedersachs.* **2**, 81–2; (1976) *Vogelkdl. Ber. Niedersachs.* **8**, 50–1. WIGSTEN, H (1942) *Vår Fågelvärld* **1**, 49–55; (1955) *Vår Fågelvärld* **14**, 21–45. WIJNANDTS, H (1984) *Ardea* **72**, 1–92. WIKLUND, C G and STIGH, J (1983) *Ornis scand.* **14**, 58–62. WILDE, N A J (1974) *Br. Birds* **67**, 26–7. WILKINSON, A D (1950) *Br. Birds* **43**, 233–8. WILKINSON, J P (1977) *Br. Birds* **70**, 546–7. WILLE, H-G (1970) *Orn. Mitt.* **22**, 150–2. WILLGOHS, J F (1974) *Sterna* **13**, 129–77. WILLI, P (1972) *Orn. Beob.* **69**, 48–9. WILLIAMS, A J (1971) *Astarte* **4**, 61–7; (1972) M Sc Thesis. Sheffield Univ.; (1974) *Ornis scand.*

5, 113–21; (1975) *Ornis scand.* **6**, 117–24. WILLIAMS, J G (1963) *A field guide to the birds of East and Central Africa.* London. WILLIAMS, J G and ARLOTT, N (1980) *A field guide to the birds of East Africa.* London. WILLIAMS, K (1964) *Br. Birds* **57**, 202–3. WILLIAMS, P L and FRANK, L G (1979) *Condor* **81**, 213–14. WILLIAMSON, K (1948) *Br. Birds* **41**, 31; (1951) *Br. Birds* **44**, 108–9; (1960) *Br. Birds* **53**, 243–52; (1963) *Bird Migration* **2**, 224–51; (1965) *Fair Isle and its birds.* Edinburgh. WILSON, C E (1949) *Sudan Notes Rec.* **29**, 161–73. WILSON, M (1976) *Bristol Orn.* **9**, 127–52. WIMPFHEIMER, D, BRUUN, B, BAHA EL DIN, S M, and JENNINGS, M C (1983) *The migration of birds of prey in the northern Red Sea area.* Holyland Conservation Fund, New York. WINDE, H (1977) *Zool. Abh. Staatl. Mus. Tierk. Dresden* **34**, 143–6. WINGE, H (1892) *Zoologist* (3) **16**, 341–5. WINKLER, H (1967) *Egretta* **10** (2), 1–8; (1971) *Egretta* **14**, 1–20; (1972a) *Egretta* **14**, 21–4; (1972b) *Egretta* **15**, 66–7; (1972c) *Z. Tierpsychol.* **31**, 300–25; (1973) *Oecologia* **12**, 193–208; (1979) *J. Orn.* **120**, 290–8. WINKLER, H and SHORT, L (1978) *Bull. Am. Mus. nat. Hist.* **160** (1), 3–109. WINN, H E (1950) *Auk* **67**, 477–85. WINTERBOTTOM, J M (1952) *Ostrich* **23**, 82–4; (1959) *Ostrich* **30**, 44; (1967) *The farmer's birds.* Cape Town. WISE, F (1876) *Stray Feathers* **4**, 230. WITHERBY H F (1910) *Br. Birds* **4**, 94; (1928) *Ibis* (12) **4**, 587–663. WITHERBY, H F, JOURDAIN, F C R, TICEHURST, N F, and TUCKER, B W (1938) *The handbook of British birds* **2**; (1940) **4**; (1941) **5**. London. WITHERS, G D (1973) *Rep. Univ. Newcastle upon Tyne Iceland Exped. 1973.* WITT, K (1970) *Vogelwelt* **91**, 24–8. WITTENBERG, J (1958) *Vogelwelt* **79**, 107–8; (1969) *J. Orn.* **110**, 30–8; (1980) *J. Orn.* **121**, 96–101. WOLFF, G (1949) *Vogelwelt* **70**, 121. WOLFF, G and GEHREN, R VON (1951) *Vogelwelt* **72**, 14–16. WOLK, R G (1974) *Proc. Linn. Soc. NY* **72**, 44–62. WON, P-O, WOO, H-C, HAM, K-W, and CHUN, M-Z (1968) *Misc. Rep. Yamashina Inst. Orn.* **5**, 363–9. WOLTERS, H E (1975) *Die Vogelarten der Erde* **1**. Hamburg. WOOD, C R (1976) *Br. Birds* **69**, 272. WOODELL, R (1976) *Ibis* **118**, 263–8. WOODS, H E (1952) *Br. Birds* **45**, 374. WOOLFENDEN, G E and MEYERRIECKS, A J (1963) *Auk* **80**, 365–6. WOOLLER, R D and TRIGGS, G S (1968) *Bird Study* **15**, 164–6. WOOLSEY, A K, EVANS, G E, and FROST, B (1965) *Br. Birds* **58**, 150. WOOTTON, M J (1950) *Br. Birds* **43**, 258–9. WOTTON, M (1976) *Audubon* **78** (4), 32–41. WRIGHT, W C (1909) *Br. Birds* **3**, 91. WYLLIE, I (1971) *Br. Birds* **68**, 369–78; (1981) *The Cuckoo.* London. WYNNE-EDWARDS, V C (1930) *J. exp. Biol.* **7**, 241–7; (1935) *Proc. Boston Soc. nat. Hist.* **40**, 233–346.

YAKHONTOV, V D (1968) *Ornitologiya* **9**, 355. YAKUBANIS, V N and LITVAK, M D (1962) *Voprosy ekol. prakt. znach. ptits mleko. Moldavii*, 49–55. YALDEN, D W and JONES, R (1970) *Naturalist*, 87–90. YANUSHEVICH, A I, TYURIN, P S, YAKOVLEVA, I D, KYDYRALIEV, A, and SEMENOVA, N I (1959) *Ptitsy Kirgizii* **1**; (1961) **3**. Frunze. YAPP, W B (1962) *Birds and woods.* London. YARRELL, W and SAUNDERS, H (1882–4) *A history of British birds* **3**. London. YATSENYA, O Z In Voinstvenski, M A (ed) *Ekologiya ta istoriya khrebetnikh fauni Ukraini*, 140–6. Kiev. YEATES, G K (1941a) *Br. Birds* **35**, 82; (1941b) *Br. Birds* **35**, 108–9; (1948) *Ibis* **90**, 425–53. YEATMAN, L J (1976) *Atlas des oiseaux nicheurs de France.* Paris. YERBURY, J W (1886) *Ibis* (5) **4**, 11–24. YOUNG, C G (1946) *Ibis* **88**, 348–82. YUNICK, R P (1965) *EBBA News* **28**, 81–5.

ZABKA, H (1980) *Mitt. zool. Mus. Berlin* **56** suppl *Ann. Orn.* **4**, 51–76. ZAHAVI, A (1971) *Ibis* **113**, 106–9. ZALETAEV, V S (1965) *Ornitologiya* **7**, 469–70. ZANG, H (1970) *J. Orn.* **111**, 107. ZANOLA, R (1966) *Orn. Beob.* **63**, 52. ZAPF, G (1973) *Falke* **20**, 316–17. ZARUDNYI, N A (1896) *Mat. pozn. fauny flory Ross. imp., otd. zool.* **2**. ZEDLITZ, O VON (1909) *J. Orn.* **57**, 241–322. ZEHNTER,

L (1890) *Arch. naturg.* **56**, 189–220. ZELENKA, G and PRICAM, R (1964) *Terre et Vie* **2**, 178–84. ZHELNIN, V A (1959) *Ornitologiya* **2**, 135–7. ZHENG ZUOXIN, XIAN YAOHUA, ZHANG YINSUN, and JIANG ZHIHUA (1975) *Acta Zool. Sinica* **21**, 385–8. ZIEGELER, R and POLL, J (1950) *Nos Oiseaux* **20**, 113–16. ZIEGLER, G (1974) *Alcedo* **1**, 15–26. ZIESEMER, F (1973) *Corax* **4**, 79–92; (1979) *Corax* **7**, 106–8; (1980) *Corax* **8**, 107–30. ZILLMAN, E E (1965) *Austral. Bird Watcher* **2**, 148–51. ZIMMERMAN, D (1956) *Orn. Beob.* **53**, 18–19. ZIMMERMAN, K (1963) *Beitr. Vogelkde.* **9**, 59–68. ZIMMERMAN, R (1931) *Sber. Abh. Naturw. Ges. Isis Dresden 1930*, 25–45. ZIMMERMANN, H (1956) *Orn. Mitt.* **8**, 89–90. ZINGEL, D (1970) *Orn. Mitt.* **22**, 254. ZINO, A (1969a) *Bocagiona* **11**, 1–5; (1969b) *Oiseau* **39**, 261–4. ZÖLLER, W (1975) *Anz. orn. Ges. Bayern* **14**, 196–205; (1980) *Orn. Mitt.* **32**, 171–8. ZOLLINGER, H (1933) *Orn. Beob.* **31**, 33–6. ZSCHOKE, B (1982) *Falke* **29**, 414–20. ZUBAKIN, V A (1975) *Vestnik mosk. gos. Univ. (Biol. Pochvov.)* **3**, 32–6. ZUBAKIN, V A and KOSTIN, Y V (1977) *Ornitologiya* **13**, 49–55. ZUKOWSKY, L (1964) *Beitr. Vogelkde.* **9**, 456–7. ZUSI, R L (1962) *Publ. Nuttall Orn. Club* **3**. ZWEERS, G (1982) *Adv. Anat. Embryol. Biol.* **73**, 1–108. ZYKOVA, L Y and IVANOV, F V (1967) *Ornitologiya* **8**, 355. ZYLVA, T S U DE (1973) *Loris* **13**, 100–1.

CORRECTIONS

CORRECTIONS TO VOLUME I

Page 129. *Pterodroma hasitata* Black-capped Petrel. Column 2, penultimate line. Amend '1852' to read '1850'.

Page 473. *Anas penelope* Wigeon. Heading. Add Russian name 'Ru. Свиязь'.

Page 509. *Anas platyrhynchos* Mallard. **Food**. Line 14. Amend 'Milne' to read 'Mylne'.

References:
Page 702. Amend to read 'BOURNE, W R P and DIXON, T J (1973) *Sea Swallow* **22**, 29–60.'

Page 709. Amend 'MILNE, C K' to read 'MYLNE, C K'.

CORRECTIONS TO VOLUME II

Page 95. *Aegypius monachus* Black Vulture. **Measurements**. Column 1, table, line 1, Amend '735–720' to read '735–820'.

Page 303. *Falco vespertinus* Red-footed Falcon. **Movements**. Column 2, penultimate line. Amend 'through' to read 'though'.

Page 504. *Phasianus colchicus* Pheasant. **Field characters**. Paragraph 1, line 4 from below. Amend to read 'while both melanistic and flavistic forms occur, but ♂ has'.

Page 611. *Fulica americana* American Coot. **Movements**. Column 2, line 10. Amend 'Schauensee' to read 'de Schauensee'.

Page 657. *Ardeotis arabs* Arabian Bustard. **Social pattern and behaviour**. Line 9. Amend 'Nigeria' to read 'Niger'.

References:
Page 677. Insert 'HOESCH, W and NIETHAMMER, G (1940) *J. Orn.* **88** suppl.'.

Page 684. Amend 'SULTAN, J' to read 'SULTANA, J'.

Page 686. Insert 'WILLIAMS, D (1976) BSc Thesis. Edinburgh Univ.'.

CORRECTIONS TO VOLUME III

Page 2. Introduction: Food. Column 2, line 9. Amend to read '(3) Aerial-skimming. (4) Hovering (including hover-feeding). (5) Aerial-pattering . . .'.

Page 7. Acknowledgements. Column 1, paragraph 3, line 18. Amend 'Dr J J D Greenwood' to read 'Dr J G G Greenwood'.

Page 8. Citation. Column 2, line 3. Amend '1982' to read '1983'.

Page 18. *Haematopus ostralegus* Oystercatcher. **Distribution**. Column 2, line 2. Delete 'Egypt'.

Page 90. *Pluvianus aegyptius* Egyptian Plover. **Breeding**. Column 1, line 15. Amend to read 'Calculated weight 8.4–10.1 g (T R Howell). Clutch: 1–3,'.

Page 99. *Glareola pratincola* Collared Pratincole. Heading. Amend to read '[facing pages 113 and 136]'.

Page 115. *Charadrius dubius* Little Ringed Plover. **Distribution**. Lines 6–7. Delete 'Channel Islands . . . (R Long).'.

Page 353. *Calidris maritima* Purple Sandpiper. **Voice**. Column 1, line 19. Delete '(Fig I)'.

Page 379. *Micropalama himantopus* Stilt Sandpiper. **Habitat**. Paragraph 1, lines 2–3 from below. Amend 'Baird's Sandpiper *Calidris bairdii*' to read 'Least Sandpiper *Calidris minutilla*'.

Page 438. *Limnodromus griseus* Short-billed Dowitcher. Heading. Amend French name to read 'Bécasseau à bec court'.

Page 449. *Scolopax rusticola* Woodcock. **Food**. Column 1, paragraph 2, line 6. Amend '(USSR)' to read '(Poland/USSR)'.

Page 478. *Limosa lapponica* Bar-tailed Godwit. **Voice**. Column 2, line 6. Amend to read ' . . . (Fig I; . . .'.

Page 555. *Tringa nebularia* Greenshank. **Voice**. Column 2, line 2. Amend 'IV' to read 'VI'.

Page 588. *Xenus cinereus* Terek Sandpiper. **Habitat**. Column 2, line 8. Amend '200 m' to read '2000 m'.

Page 630. *Phalaropus lobatus* Red-necked Phalarope. **Field characters**. Column 1, line 21 from below. Amend to read 'length of wing-bar (not much extending on to outer wing) and increased vividness of pale markings close to . . .'.

Page 744. *Larus sabini* Sabine's Gull. **Voice**. Column 2, line 5 from below. Delete 'single'.

Page 751. *Larus ridibundus* Black-headed Gull. **Population**. Column 1, paragraph 1, line 13 from below. Amend to read 'in Comachio, 120 pairs . . .'.

Page 912. Index: Noms français. Amend 'Bécasseau roux' to read 'Bécasseau à bec court'.

Plate 81. *Larus fuscus* Lesser Black-backed Gull. Amend caption to read '2 variant ad breeding with white wing-patches, 3 . . .'.

INDEXES

Figures in **bold type** refer to plates

SCIENTIFIC NAMES

aalge (Uria), 170, **16, 22, 23**; eggs, **91**
abyssinicus (Asio), 572
abyssinicus (Coracias), 776, **72, 74**; egg, **98**
acteon (Halcyon), 705, **64, 68**
acuflavida (Sterna), 48
Aegolius funereus, 606, **48, 54**; egg, **97**
aegyptiaca (Streptopelia), 366
aegyptius (Caprimulgus), 641, **56, 57**; eggs, **94**
aegyptius (Centropus), 427, **37, 41**; eggs, **95**
aerobates (Apus), 692
Aethia cristatella, 229, **20, 22, 23**
afer (Eurystomus), 783, **73, 74**; egg, **98**
affinis (Apus), 692, **62, 63**; egg, **98**
affinis (Coracias), 778
affinis (Gelochelidon), 6
affinis (Tyto), 432
africana (Upupa), 786
agricola (Streptopelia), 363
alba (Tyto), 432, **43, 53**; egg, **97**
albidorsalis (Sterna), 27
albifrons (Sterna), 120, **11**; eggs, **90**
albigena (Sterna), 105
albionis (Uria), 170, **16**
Alca torda, 195, **18, 22, 23**; eggs, **92**
Alcedinidae, 700
Alcedininae, 710
Alcedo atthis, 711, **65, 68**; egg, **98**
alchata (Pterocles), 269, **25, 26**; eggs, **94**
Alcidae, 168
alcyon (Ceryle), 731, **67, 68**
alcyon (Megaceryle), 731, **67, 68**
aleutica (Sterna), 100, **15**
alexandri (Apus), 652, **60**
aliena (Oena), 374
Alle alle, 219, **20, 22, 23**; eggs, **92**
alle (Alle), 219, **20, 22, 23**; eggs, **92**
alpinus (Picoides), 913, **85**
alternans (Merops), 740
aluco (Strix), 526, **49, 53**; egg, **97**
americanus (Coccyzus), 425, **40**
anaethetus (Sterna), 109, **9, 10**; eggs, **89**
anatoliae (Dendrocopos), 882
anglicus (Dendrocopos), 856, **81**
Anous stolidus, 163, **9, 10**
antarctica (Sterna), 109
antillarum (Sterna), 120
apiaster (Merops), 748, **70, 71**; egg, **98**
Apodidae, 649
Apodiformes, 649
Apodinae, 652
Apus affinis, 692, **62, 63**; egg, **98**
    alexandri, 652, **60**
apus, 657, **60, 61, 63**; egg, **98**
caffer, 687, **62, 63**
melba, 678, **59, 63**; egg, **98**
pacificus, 676, **62, 63**
pallidus, 670, **60, 61, 63**
unicolor, 653, **60**; egg, **98**
apus (Apus), 657, **60, 61, 63**; egg, **98**
apus (Cypselus), 657, **60, 61, 63**; egg, **98**
arabica (Streptopelia), 336
aranea (Gelochelidon), 6
archeri (Apus), 678
arctica (Fratercula), 231, **21, 22, 23**; egg, **92**
arcticus (Cepphus), 208, **19, 22, 23**
arenarius (Pterocles), 263
arenicola (Streptopelia), 353
arra (Uria), 184
ascalaphus (Bubo), 466, **45, 55**
Asio capensis, 601, **52, 53**; egg, **97**
    flammeus, 588, **52, 53**; egg, **97**
    helvola, 601, **52, 53**; egg, **97**
    otus, 572, **51, 53**; egg, **97**
athalassos (Sterna), 120
Athene noctua, 514, **48, 54**; egg, **97**
atthis (Alcedo), 711, **65, 68**; egg, **98**
atratus (Pterocles), 249
auratus (Colaptes), 813
azorica (Columba), 311, **30**

bactriana (Athene), 514
bakeri (Cuculus), 402
bangsi (Cuculus), 402
bangsi (Sterna), 62
beickianus (Aegolius), 606
beludschicus (Merops), 734
bengalensis (Alcedo), 711
bengalensis (Bubo), 466
bengalensis (Sterna), 42, **1, 4**; eggs, **87**
bengalensis (Thalasseus), 42, **1, 4**; eggs, **87**
benghalensis (Coracias), 778, **73, 74**; egg, **98**
bergii (Sterna), 36, **1, 4**; eggs, **88**
bergii (Thalasseus), 36, **1, 4**; eggs, **88**
biedermanni (Picus), 813
bollii (Columba), 331, **29, 30**; egg, **95**
borealis (Psittacula), 379, **35**
brehmorum (Apus), 670, **61, 63**
brucei (Otus), 450, **44, 54**; egg, **97**
Bubo bubo, 466, **45, 55**; egg, **97**
bubo (Bubo), 466, **45, 55**; egg, **97**
Buboninae, 449
butleri (Strix), 547, **49, 53**
buturlini (Dendrocopos), 901, **84**

caffer (Apus), 687, **62, 63**
californica (Uria), 170
cambayensis (Streptopelia), 366
canariensis (Dendrocopos), 856, **81**

canariensis (Otus), 572, **51**
candidus (Dendrocopos), 856
canorus (Cuculus), 402, **38, 42**; eggs, **96**
canus (Picus), 813, **77, 78**; egg, **98**
caparoch (Surnia), 496
capensis (Asio), 601, **52, 53**; egg, **97**
capensis (Oena), 374, **32, 34**; egg, **95**
Caprimulgidae, 616
Caprimulgiformes, 616
Caprimulginae, 617
Caprimulgus aegyptius, 641, **56, 57**; eggs, **94**
    europaeus, 620, **56, 58**; eggs, **94**
    eximius, 641; eggs, **94**
    nubicus, 617, **56, 57**; eggs, **94**
    ruficollis, 636, **56, 58**; eggs, **94**
caprius (Chrysococcyx), 400, **37, 41**
casiotis (Columba), 311
caspia (Sterna), 17, **1, 3**; eggs, **88**
caucasicus (Aegolius), 606
caucasicus (Dendrocopos), 882, **82**
caudacuta (Chaetura), 649, **59, 63**
caudacutus (Hirundapus), 649, **59, 63**
caudacutus (Pterocles), 269, **25, 26**
Centropodinae, 427
Centropus senegalensis, 427, **37, 41**; egg, **95**
Cepphus grylle, 208, **19, 22, 23**; eggs, **92**
Ceryle alcyon, 731, **67, 68**
    rudis, 723, **66, 68**; egg, **98**
Cerylinae, 723
ceylonensis (Upupa), 786
Chaetura caudacuta, 649, **59, 63**
Chaeturinae, 649
chapmani (Chordeiles), 646
Charadriiformes, 5
Chlidonias hybridus, 133, **12, 13, 14**; eggs, **90**
    leucopterus, 155, **12, 13, 14**; eggs, **90**
    niger, 143, **12, 13, 14**; eggs, **90**
choragium (Clamator), 391
Chordeiles minor, 646, **56, 57**
    virginianus, 646, **56, 57**
Chordeilinae, 646
chrysocercus (Merops), 740, **69**
Chrysococcyx caprius, 400, **37, 41**
Clamator glandarius, 391, **36, 37**; eggs, **95**
    jacobinus, 388, **36, 37**; eggs, **95**
cleopatra (Merops), 734, **69, 71**
Coccyzus americanus, 425, **40**
    erythrophthalmus, 423, **40**
Colaptes auratus, 813
colchicus (Dendrocopos), 901
Columba bollii, 331, **29, 30**; egg, **95**
    eversmanni, 309, **28, 29**; egg, **95**
    junoniae, 334, **29, 30**; egg, **95**
    laurivora, 334, **29, 30**; egg, **95**
    livia, 285, **27, 29**; egg, **95**

oenas, 298, **28, 29**; egg, **95**
palumbus, 311, **29, 30**; eggs, **95**
trocaz (Columba), 329, **29, 30**; egg, **95**
Columbidae, 283
Columbiformes, 283
comminutus (Dendrocopos), 901, **84**
cooki (Apus), 676
Coracias abyssinicus, 776, **72, 74**; egg, **98**
　benghalensis, 778, **73, 74**; egg, **98**
　garrulus, 764, **72, 74**; egg, **98**
Coraciidae, 763
Coraciiformes, 699
coronatus (Pterocles), 249, **24, 26**; eggs, **94**
crissalis (Sterna), 116
crissoleucus (Picoides), 913
cristata (Sterna), 36
cristatella (Aethia), 229, **20, 22, 23**
Cuculidae, 388
Cuculiformes, 388
Cuculinae, 388
Cuculus canorus, 402, **38, 42**; eggs, **96**
　optatus, 417, **39, 42**; eggs, **96**
　saturatus, 417, **39, 42**; eggs, **96**
cyanophrys (Merops), 734, **69**
cycladum (Otus), 454
Cyclorrhynchus psittacula, 230
cyprius (Otus), 454
Cypselus apus, 657, **60, 61, 63**; egg, **98**
Cypsiurus parvus, 698, **62, 63**; egg, **98**

Daceloninae, 700
dakhlae (Columba), 285, **27**
danfordi (Dendrocopos), 901, **84**
decaocto (Streptopelia), 340, **31, 34**; egg, **95**
delalandii (Chlidonias), 133
dementievi (Caprimulgus), 620
Dendrocopos leucotos, 891, **83, 86**; egg, **98**
　major, 856, **81, 86**; egg, **98**
　medius, 882, **82, 86**; egg, **98**
　minor, 901, **84, 86**; egg, **98**
　syriacus, 874, **81, 86**; egg, **98**
desertorum (Caprimulgus), 636, **58**
detorta (Tyto), 432, **43**
divergens (Streptopelia), 366
dougallii (Sterna), 62, **5, 6, 7**; eggs, **89**
Dryocopus martius, 840, **78, 79**; egg, **98**

Ectopistes migratorius, 378
eleonorae (Uria), 184
ellioti (Pterocles), 259
epops (Upupa), 786, **71, 75**; eggs, **98**
erlangeri (Pterocles), 259
erlangeri (Tyto), 432, **43**
ermanni (Streptopelia), 366
ernesti (Tyto), 432
erythrocephala (Streptopelia), 363
erythrophthalmus (Coccyzus), 423, **40**
europaeus (Caprimulgus), 620, **56, 58**; eggs, **94**
eurygnatha (Sterna), 48
Eurystomus afer, 783, **73, 74**; egg, **98**
　glaucurus, 783, **73, 74**; egg, **98**

eversmanni (Columba), 309, **28, 29**; egg, **95**
exiguus (Otus), 450, **44, 54**
eximius (Caprimulgus), 641; eggs, **94**
exustus (Pterocles), 259, **24, 26**; eggs, **94**

faeroeensis (Cepphus), 208
ferrugeiceps (Merops), 734
flammeus (Asio), 588, **52, 53**; egg, **97**
flavirostris (Rynchops), 166, **15**
flecki (Centropus), 427
floweri (Pterocles), 259, **24, 26**
fluviatilis (Chlidonias), 133
forsteri (Sterna), 102, **5, 7, 8**
Fratercula arctica, 231, **21, 22, 23**; egg, **92**
funereus (Aegolius), 606, **48, 54**; egg, **97**
fusca (Halcyon), 701
fuscata (Sterna), 116, **9, 10**

gaddi (Columba), 285
galapagensis (Anous), 163
galilejensis (Apus), 692, **62, 63**
garrulus (Coracias), 764, **72, 74**; egg, **98**
Gelochelidon nilotica, 6, **1, 2**; eggs, **87**
glandarius (Clamator), 391, **36, 37**; eggs, **95**
Glaucidium passerinum, 505, **47, 54**; egg, **97**
glaucurus (Eurystomus), 783, **73, 74**; egg, **98**
glaux (Athene), 514, **48**
grabae (Fratercula), 231, **21**
gracilirostris (Tyto), 432, **43**
gracilis (Sterna), 62
graueri (Asio), 572
grylle (Cepphus), 208, **19, 22, 23**; eggs, **92**
guineae (Sterna), 120
gularis (Halcyon), 701
guttata (Tyto), 432, **43, 53**
gymnocyclus (Columba), 285

Halcyon leucocephala, 705, **64, 68**; egg, **98**
　smyrnensis, 701, **64, 68**; egg, **98**
harterti (Dendrocopos), 856
heckeri (Uria), 184
helvola (Asio), 601, **52, 53**; egg, **97**
hindustan (Pterocles), 259
Hirundapus caudacutus, 649, **59, 63**
hirundo (Sterna), 71, **5, 6, 7, 8**; eggs, **89**
hispanus (Bubo), 466, **45**
hispanus (Dendrocopos), 856, **81**
hoggara (Streptopelia), 353
horsfieldi (Cuculus), 417, **39, 42**
hortorum (Dendrocopos), 901
hova (Asio), 601
hyacinthina (Halcyon), 705
hybridus (Chlidonias), 133, **12, 13, 14**; eggs, **90**
Hydrochelidon leucopareia, 133, **12, 13, 14**; eggs, **90**
Hydroprogne tschegrava, 17, **1, 3**; eggs, **88**

hyperborea (Uria), 170, **16**
hyrcanus (Dendrocopos), 901

illyricus (Apus), 670, **61, 63**
impennis (Pinguinus), 207; egg, **93**
indicus (Coracias), 778
indigena (Athene), 514, **48**
ingramsi (Pterocles), 245
innominatus (Picus), 824
inornata (Uria), 170
insulindae (Cuculus), 417
intercedens (Streptopelia), 340
intermedia (Columba), 285
interpositus (Bubo), 466
iranica (Columba), 311
islandica (Alca), 195
islandicus (Cepphus), 208, **19**
ispida (Alcedo), 711, **65**
italiae (Dendrocopos), 856

jacobinus (Clamator), 388, **36, 37**; eggs, **95**
javanicus (Chlidonias), 133
jessoensis (Picus), 813
jonesi (Caprimulgus), 617
junoniae (Columba), 334, **29, 30**; egg, **95**
Jynginae, 800
Jynx torquilla, 800, **76, 78**; egg, **98**

kamtschatkensis (Dendrocopos), 901
kanoi (Apus), 676
karelini (Picus), 824
kermadeci (Sterna), 116
Ketupa zeylonensis, 481, **45, 55**; egg, **97**
khamensis (Dryocopus), 840
korustes (Sterna), 62
krameri (Psittacula), 379, **35**; egg, **95**

ladas (Pterocles), 249
lapponica (Strix), 561, **50, 55**
laurivora (Columba), 334, **29, 30**; eggs, **95**
ledouci (Dendrocopos), 901
lepidus (Cuculus), 417
leschenault (Ketupa), 481
leucocephala (Halcyon), 705, **64, 68**; egg, **98**
leucomelanura (Ceryle), 723
leuconyx (Apus), 676
leucopareia (Hydrochelidon), 133, **12, 13, 14**; eggs, **90**
leucopterus (Chlidonias), 155, **12, 13, 14**; eggs, **90**
leucotos (Dendrocopos), 891, **83, 86**; egg, **98**
lichtensteinii (Pterocles), 245, **24, 26**
lilfordi (Dendrocopos), 891, **83**
lilith (Athene), 514, **48, 54**
liturata (Strix), 550, **50, 55**
livia (Columba), 285, **27, 29**; egg, **95**
lomvia (Uria), 184, **17, 22, 23**; eggs, **92**
longipennis (Sterna), 71
longirostris (Upupa), 786
luctuosa (Sterna), 116

macroura (Strix), 550
maderensis (Columba), 311
magnus (Aegolius), 606
major (Dendrocopos), 856, **81**, **86**; egg, **98**
major (Upupa), 786, **75**
mallorcae (Otus), 454, **44**
mandtii (Cepphus), 208
manillensis (Psittacula), 379
marginata (Upupa), 786
martius (Dryocopus), 840, **78**, **79**; egg, **98**
mauretanica (Jynx), 800
mauritanica (Strix), 526, **49**
mauritanus (Dendrocopos), 856
maxima (Sterna), 27, **1**, **3**; eggs, **88**
maximus (Thalasseus), 27, **1**, **3**; eggs, **88**
medius (Dendrocopos), 882, **82**, **86**; egg, **98**
meena (Streptopelia), 363, **33**
Megaceryle alcyon, 731, **67**, **68**
melanoptera (Sterna), 109
melba (Apus), 678, **59**, **63**; egg, **98**
melba (Tachymarptis), 678, **59**, **63**; egg, **98**
meridionalis (Caprimulgus), 620, **58**
Meropidae, 733
Merops apiaster, 748, **70**, **71**; egg, **98**
   orientalis, 734, **69**, **71**; egg, **98**
   superciliosus, 740, **69**, **71**; egg, **98**
mexicana (Sterna), 120
migratorius (Ectopistes), 378
milleri (Dendrocopos), 874
minor (Chordeiles), 646, **56**, **57**
minor (Dendrocopos), 901, **84**, **86**; egg, **98**
morgani (Dendrocopos), 901
muscatensis (Merops), 734

naumanni (Fratercula), 231
nebulosa (Strix), 561, **50**, **55**; egg, **97**
neglecta (Columba), 285
nelsoni (Sterna), 109
niger (Chlidonias), 143, **12**, **13**, **14**; eggs, **90**
nikolskii (Bubo), 466
nilotica (Gelochelidon), 6, **1**, **2**; eggs, **87**
noctua (Athene), 514, **48**, **54**; egg, **97**
nubicus (Caprimulgus), 617, **56**, **57**; eggs, **94**
nubilosa (Sterna), 116
nudipes (Hirundapus), 649
numidus (Dendrocopos), 856, **81**
Nyctea scandiaca, 485, **46**, **55**; egg, **97**

oahuensis (Sterna), 116
obsoletus (Otus), 450, **44**, **54**
occidentalis (Coccyzus), 425
Oena capensis, 374, **32**, **34**; egg, **95**
oenas (Columba), 298, **28**, **29**; egg, **95**
olivascens (Pterocles), 259
optatus (Cuculus), 417, **39**, **42**; eggs, **96**
orientale (Glaucidium), 505
orientalis (Ketupa), 481
orientalis (Merops), 734, **69**, **71**; egg, **98**

orientalis (Pterocles), 263, **25**, **26**; eggs, **94**
orientalis (Streptopelia), 363, **33**, **34**; egg, **95**
orii (Streptopelia), 363
otus (Asio), 572, **51**, **53**; egg, **97**
Otus brucei, 450, **44**, **54**; egg, **97**
   scops, 454, **44**, **54**; egg, **97**

pacificus (Apus), 676, **62**, **63**
Palaeornis torquatus, 379, **35**; egg, **95**
palaestinae (Columba), 285
pallens (Aegolius), 606
pallidiventris (Halcyon), 705
pallidus (Apus), 670, **60**, **61**, **63**
palumbus (Columba), 311, **29**, **30**; egg, **95**
pamelae (Otus), 450
paphlagoniae (Dendrocopos), 856
paradisaea (Sterna), 87, **5**, **6**, **7**; eggs, **89**
paradoxus (Syrrhaptes), 277, **25**, **26**; eggs, **94**
parroti (Dendrocopos), 856
parvirostris (Psittacula), 379
parvus (Cypsiurus), 698, **62**, **63**; egg, **98**
passerinum (Glaucidium), 505, **47**, **54**; egg, **97**
pekinensis (Apus), 657, **61**, **63**
persicus (Merops), 740, **69**, **71**
Phaenicophaeinae, 423
phoenicophila (Streptopelia), 366
pica (Clamator), 388, **36**, **37**
Picidae, 799
Piciformes, 799
Picinae, 812
Picoides tridactylus, 913, **85**, **86**; egg, **98**
Picus canus, 813, **77**, **78**; egg, **98**
   vaillantii, 837, **77**, **78**; egg, **98**
   viridis, 824, **77**, **78**; egg, **98**
pileatus (Anous), 163
pinetorum (Dendrocopos), 856
Pinguinus impennis, 207; egg, **93**
plumipes (Caprimulgus), 620
poelzami (Dendrocopos), 856
polaris (Alle), 219
Psittacidae, 378
Psittaciformes, 378
psittacula (Cyclorrhynchus), 230
Psittacula krameri, 379, **35**; egg, **95**
Pterocles alchata, 269, **25**, **26**; eggs, **94**
   coronatus, 249, **24**, **26**; eggs, **94**
   exustus, 259, **24**, **26**; eggs, **94**
   lichtensteinii, 245, **24**, **26**
   orientalis, 263, **25**, **26**; eggs, **94**
   senegallus, 253, **24**, **26**; eggs, **94**
Pteroclididae, 244
Pteroclidiformes, 244
pulchellus (Otus), 454
pulcherrimus (Eurystomus), 783

quadrifasciatus (Dendrocopos), 901

repressa (Sterna), 105, **5**, **7**, **8**; eggs, **89**
richardsoni (Aegolius), 606
ridgwayi (Anous), 163

'risoria' (Streptopelia), 336, **31**, **34**
roseogrisea (Streptopelia), 336, **31**, **34**; egg, **95**
rudis (Ceryle), 723, **66**, **68**; egg, **98**
rufescens (Streptopelia), 353, **33**
ruficollis (Caprimulgus), 636, **56**, **58**; eggs, **94**
ruthenus (Bubo), 466
Rynchopidae, 166
Rynchops flavirostris, 166, **15**

saharae (Athene), 514
saharae (Caprimulgus), 641, **56**, **57**
sanctijohannis (Dendrocopos), 882
sanctinicolai (Strix), 526
sandvicensis (Sterna), 48, **1**, **2**; eggs, **87**
sandvicensis (Thalasseus), 48, **1**, **2**; eggs, **87**
sarudnyi (Caprimulgus), 620
sarudnyi (Jynx), 800
saturata (Upupa), 786
saturatior (Halcyon), 701
saturatus (Cuculus), 417, **39**, **42**; eggs, **96**
saturatus (Pterocles), 249
saundersi (Sterna), 133
scandiaca (Nyctea), 485, **46**, **55**; egg, **97**
schimperi (Columba), 285, **27**
schmitzi (Tyto), 432
sclateri (Chlidonias), 133
scops (Otus), 454, **44**, **54**; egg, **97**
semenowi (Coracias), 764, **72**
semenowi (Ketupa), 481, **45**, **55**
semenowi (Otus), 450
semicaerulea (Halcyon), 705
senegalensis (Centropus), 427, **37**, **41**; egg, **95**
senegalensis (Streptopelia), 366, **33**, **34**; egg, **95**
senegalensis (Upupa), 786
senegallus (Pterocles), 253, **24**, **26**; eggs, **94**
serrata (Sterna), 116
serratus (Clamator), 388
sharpei (Picus), 824, **77**
siberiae (Strix), 526
sibiricus (Bubo), 466, **45**
simplicior (Caprimulgus), 641
sinensis (Sterna), 120
smyrnensis (Halcyon), 701, **64**, **68**; egg, **98**
socotrae (Streptopelia), 366
socotranus (Otus), 450
Sphyrapicus varius, 853, **80**, **86**
staebleri (Sterna), 120
Sterna albifrons, 120, **11**; eggs, **90**
   albigena, 105, **5**, **7**, **8**; eggs, **89**
   aleutica, 100, **15**
   anaethetus, 109, **9**, **10**; eggs, **89**
   bengalensis, 42, **1**, **4**; eggs, **87**
   bergii, 36, **1**, **4**; eggs, **88**
   caspia, 17, **1**, **3**; eggs, **88**
   dougallii, 62, **5**, **6**, **7**; eggs, **89**
   forsteri, 102, **5**, **7**, **8**
   fuscata, 116, **9**, **10**

Scientific names 957

hirundo, 71, **5, 6, 7, 8**; eggs, **89**
maxima, 27, **1, 3**; eggs, **88**
paradisaea, 87, **5, 6, 7**; eggs, **89**
repressa, 105, **5, 7, 8**; eggs, **89**
sandvicensis, 48, **1, 2**; eggs, **87**
saundersi, 133
Sternidae, 5
stimpsoni (Streptopelia), 363
stolickzae (Streptopelia), 340
stolidus (Anous), 163, **9, 10**
Streptopelia decaocto, 340, **31, 34**; egg, **95**
 orientalis, 363, **33, 34**; egg, **95**
 roseogrisea, 366, **31, 34**; egg, **95**
 senegalensis, 366, **33, 34**; egg, **95**
 turtur, 353, **33, 34**; egg, **95**
streubelii (Apus), 687
Strigidae, 449
Strigiformes, 432
Striginae, 525
Strix aluco, 526, **49, 53**; egg, **97**
 butleri, 547, **49, 53**
 nebulosa, 561, **50, 55**; egg, **97**
 uralensis, 550, **50, 55**; egg, **97**
suahelicus (Eurystomus), 783
subtelephonus (Cuculus), 402
sukensis (Pterocles), 245
superciliosus (Merops), 740, **69, 71**; egg, **98**
surinamensis (Chlidonias), 143
Surnia (ulula), 496, **47, 53**; egg, **97**
sylvatica (Strix), 526, **49, 53**
syriacus (Dendrocopos), 874, **81, 86**; egg, **98**
Syrrhaptes paradoxus, 277, **25, 26**; eggs, **94**

Tachymarptis melba, 678, **59, 63**; egg, **98**
tamaricis (Caprimulgus), 617, **56, 57**; eggs, **94**
taprobana (Alcedo), 711
targia (Columba), 285
targius (Pterocles), 245
taruensis (Caprimulgus), 617
tenuirostris (Dendrocopos), 856, **81**
Thalasseus bengalensis, 42, **1, 4**; eggs, **87**
 bergii, 36, **1, 4**; eggs, **88**
 maximus, 27, **1, 3**; eggs, **88**
 sandvicensis, 48, **1, 2**; eggs, **87**
thalassina (Sterna), 36
thanneri (Dendrocopos), 856
tianschanica (Surnia), 496
tibetana (Sterna), 71
tingitanus (Asio), 601, **52, 53**
torda (Alca), 195, **18, 22, 23**; eggs, **92**
torquatus (Palaeornis), 379, **35**; egg, **95**
torquilla (Jynx), 800, **76, 78**; egg, **98**
torresii (Sterna), 42
torridus (Caprimulgus), 617
transcaucasicus (Dendrocopos), 874
travancoreensis (Ceryle), 723
tridactylus (Picoides), 913, **85, 86**; egg, **98**
trocaz (Columba), 329, **29, 30**; egg, **95**
tschegrava (Hydroprogne), 17, **1, 3**; eggs, 88
tschusii (Jynx), 800, **76**
tuftsi (Asio), 572
tuneti (Apus), 678, **59, 63**
turanicus (Otus), 454
turcomanus (Bubo), 466
turtur (Streptopelia), 353, **33, 34**; egg, **95**
Tyto alba, 432, **43, 53**; egg, **97**
Tytonidae, 432

ulula (Surnia), 496, **47, 53**; egg, **97**
unicolor (Apus), 653, **60**; egg, **98**
unwini (Caprimulgus), 620
Upupa epops, 786, **71, 75**; eggs, **98**
Upupidae, 786
uralensis (Dendrocopos), 891
uralensis (Strix), 550, **50, 55**; egg, **97**
Uria aalge, 170, **16, 22, 23**; eggs, **91**
 lomvia, 184, **17, 22, 23**; eggs, **92**

vaillantii (Picus), 837, **77, 78**; egg, **98**
varius (Sphyrapicus), 853, **80, 86**
vastitas (Pterocles), 249
velox (Sterna), 36, **1, 4**
vidalii (Athene), 514, **48, 54**
virginianus (Chordeiles), 646, **56, 57**
viridis (Picus), 824, **77, 78**; egg, **98**
viridissimus (Merops), 734

waibeli (Upupa), 786
willkonskii (Strix), 526
wilsonianus (Asio), 572

xanthocyclus (Streptopelia), 340

yarkandensis (Columba), 298

zeylonensis (Ketupa), 481, **45, 55**; egg, **97**

ENGLISH NAMES

Auk, Great, 207; egg, **93**
 Little, 219, **20, 22, 23**; eggs, **92**
Auklet, Crested, 229, **20, 22, 23**
 Parakeet, 230

Bee-eater, 748, **70, 71**; egg, **98**
 Blue-cheeked, 740, **69, 71**; egg, **98**
 Little Green, 734, **69, 71**; egg, **98**

Coucal, Senegal, 427, **37, 41**; egg, **95**
Cuckoo, 402, **38, 42**; eggs, **96**
 Black-billed, 423, **40**
 Didric, 400, **37, 41**
 Great Spotted, 391, **36, 37**; eggs, **95**
 Jacobin, 388, **36, 37**; eggs, **95**
 Oriental, 417, **39, 42**; eggs, **96**
 Yellow-billed, 425, **40**

Dove, African Collared, 336, **31, 34**; egg, **95**
 Barbary, 336, **31, 34**
 Collared, 340, **31, 34**; egg, **95**
 Laughing, 366, **33, 34**; egg, **95**
 Namaqua, 374, **32, 34**; egg, **95**
 Pink-headed Turtle, 336, **31, 34**; egg, **95**
 Rock, 285, **27, 29**; egg, **95**
 Rufous Turtle, 363, **33, 34**; egg, **95**
 Stock, 298, **28, 29**; egg, **95**
 Turtle, 353, **33, 34**; egg, **95**
 Yellow-eyed Stock, 309, **28, 29**; egg, **95**
Dovekie, 219, **20, 22, 23**; eggs, **92**

Flicker, Northern, 813

Guillemot, 170, **16, 22, 23**; eggs, **91**
 Black, 208, **19, 22, 23**; eggs, **92**
 Brünnich's Guillemot, 184, **17, 22, 23**; eggs, **92**

Hoopoe, 786, **71, 75**; eggs, **98**

Kingfisher, 711, **65, 68**; egg, **98**
 Belted, 731, **67, 68**
 Grey-headed, 705, **64, 68**; egg, **98**
 Pied, 723, **66, 68**; egg, **98**
 White-breasted, 701, **64, 68**; egg, **98**

Murre, Common, 170, **16, 22, 23**; eggs, **91**
 Thick-billed, 184, **17, 22, 23**; eggs, **92**

Nighthawk, Common, 646, **56, 57**
Nightjar, 620, **56, 58**; eggs, **94**
 Egyptian, 641, **56, 57**; eggs, **94**
 Golden, 641; eggs, **94**
 Nubian, 617, **56, 57**; eggs, **94**
 Red-necked, 636, **56, 58**; eggs, **94**
Noddy, Brown, 163, **9, 10**

Owl, Barn, 432, **43, 53**; egg, **97**
 Boreal, 606, **48, 54**; egg, **97**
 Brown Fish, 481, **45, 55**; egg, **97**
 Eagle, 466, **45, 55**; egg, **97**
 Great Grey, 561, **50, 55**; egg, **97**
 Hawk, 496, **47, 53**; egg, **97**
 Hume's Tawny, 547, **49, 53**
 Little, 514, **48, 54**; egg, **97**
 Long-eared, 572, **51, 53**; egg, **97**
 Marsh, 601, **52, 53**; egg, **97**
 Pygmy, 505, **47, 54**; egg, **97**
 Scops, 454, **44, 54**; egg, **97**
 Short-eared, 588, **52, 53**; egg, **97**
 Snowy, 485, **46, 55**; egg, **97**
 Striated Scops, 450, **44, 54**; egg, **97**
 Tawny, 526, **49, 53**; egg, **97**
 Tengmalm's, 606, **48, 54**; egg, **97**
 Ural, 550, **50, 55**; egg, **97**

Parakeet, Ring-necked, 379, **35**; egg, **95**
 Rose-ringed, 379, **35**; egg, **95**
Pigeon, Bolle's Laurel, 331, **29, 30**; egg, **95**
 Laurel, 334, **29, 30**; egg, **95**

Long-toed, 329, **29, 30**; egg, **95**
Passenger, 378
see also Woodpigeon
Puffin, 231, **21, 22, 23**; egg, **92**
  Atlantic, 231, **21, 22, 23**; egg, **92**

Razorbill, 195, **18, 22, 23**; eggs, **92**
Roller, 764, **72, 74**; egg, **98**
  Abyssinian, 776, **72, 74**; egg, **98**
  Broad-billed, 783, **73, 74**; egg, **98**
  Indian, 778, **73, 74**; egg, **98**

Sandgrouse,
  Black-bellied, 263, **25, 26**; eggs, **94**
  Chestnut-bellied, 259, **24, 26**; eggs, **94**
  Crowned, 249, **24, 26**; eggs, **94**
  Lichtenstein's, 245, **24, 26**
  Pallas's, 277, **25, 26**; eggs, **94**
  Pin-tailed, 269, **25, 26**; eggs, **94**
  Spotted, 253, **24, 26**; eggs, **94**
Sapsucker, Yellow-bellied, 853, **80, 86**
Skimmer, African, 166, **15**
Swift, 657, **60, 61, 63**; egg, **98**
  Alpine, 678, **59, 63**; egg, **98**
  Cape Verde, 652, **60**
  Fork-tailed, 676, **62, 63**
  Little, 692, **62, 63**; egg, **98**
  Needle-tailed, 649, **59, 63**
  Pacific, 676, **62, 63**
  Pallid, 670, **60, 61, 63**
  Palm, 698, **62, 63**; egg, **98**
  Plain, 653, **60**; egg, **98**
  White-rumped, 687, **62, 63**

Tern, Aleutian, 100, **15**
  Arctic, 87, **5, 6, 7**; eggs, **89**
  Black, 143, **12, 13, 14**; eggs, **90**
  Bridled, 109, **9, 10**; eggs, **89**
  Caspian, 17, **1, 3**; eggs, **88**
  Common, 71, **5, 6, 7, 8**; eggs, **89**
  Forster's, 102, **5, 7, 8**
  Gull-billed, 6, **1, 2**; eggs, **87**
  Least, 120, **11**; eggs, **90**
  Lesser Crested, 42, **1, 4**; eggs, **87**
  Little, 120, **11**; eggs, **90**
  Roseate, 62, **5, 6, 7**; eggs, **89**
  Royal, 27, **1, 3**; eggs, **88**
  Sandwich, 48, **1, 2**; eggs, **87**
  Saunders' Little, 133
  Sooty, 116, **9, 10**
  Swift, 36, **1, 4**; eggs, **88**
  Whiskered, 133, **12, 13, 14**; eggs, **90**
  White-cheeked, 105, **5, 7, 8**; eggs, **89**
  White-winged Black, 155, **12, 13, 14**; eggs, **90**

Woodpecker, Black, 840, **78, 79**; egg, **98**
  Great Spotted, 856, **81, 86**; egg, **98**
  Green, 824, **77, 78**; egg, **98**
  Grey-headed, 813, **77, 78**; egg, **98**
  Lesser Spotted, 901, **84, 86**; egg, **98**
  Levaillant's Green, 837, **77, 78**; egg, **98**
  Middle Spotted, 882, **82, 86**; egg, **98**
  Syrian, 874, **81, 86**; egg, **98**
  Three-toed, 913, **85, 86**; egg, **98**
  White-backed, 891, **83, 86**; egg, **98**
Woodpigeon, 311, **29, 30**; egg, **95**
Wryneck, 800, **76, 78**; egg, **98**

NOMS FRANÇAIS

Alcyon ceinturé, 731, **67, 68**
  pie, 723, **66, 68**; oeuf, **98**

Bec-en-ciseaux d'Afrique, 166, **15**

Chouette de Butler, 547, **49, 53**
  chevêche, 514, **48, 54**; oeuf, **97**
  chevêchette, 505, **47, 54**; oeuf, **97**
  effraie, 432, **43, 53**; oeuf, **97**
  épervière, 496, **47, 53**; oeuf, **97**
  hulotte, 526, **49, 53**; oeuf, **97**
  lapone, 561, **50, 55**; oeuf, **97**
  de l'Oural, 550, **50, 55**; oeuf, **97**
  de Tengmalm, 606, **48, 54**; oeuf, **97**
Colombe voyageuse, 378
Coucal de Sénégal, 427, **37, 41**; oeuf, **95**
Coucou didric, 400, **37, 41**
  gris, 402, **38, 42**; oeufs, **96**
  jacobin, 388, **36, 37**; oeufs, **95**
  oriental, 417, **39, 42**; oeufs, **96**
Coucou-geai, 391, **36, 37**; oeufs, **95**
Coulicou à bec jaune, 425, **40**
  à bec noir, 423, **40**

Engoulevent d'Amérique, 646, **56, 57**
  cata, 269, **25, 26**; oeufs, **94**
  à collier roux, 636, **56, 58**; oeufs, **94**
  couronné, 249, **24, 26**; oeufs, **94**
  doré, 641; oeufs, **94**
  d'Égypte, 641, **56, 57**; oeufs, **94**
  d'Europe, 620, **56, 58**; oeufs, **94**
  de Nubie, 617, **56, 57**; oeufs, **94**
  à ventre brun, 259, **24, 26**; oeufs, **94**

Ganga de Lichtenstein, 245, **24, 26**
  tacheté, 253, **24, 26**; oeufs, **94**
  unibande, 263, **25, 26**; oeufs, **94**
Guêpier d'Europe, 748, **70, 71**; oeuf, **98**
  d'Orient, 734, **69, 71**; oeuf, **98**
  de Perse, 740, **69, 71**; oeuf, **98**
Guifette leucoptère, 155, **12, 13, 14**; oeufs, **90**
  moustac, 133, **12, 13, 14**; oeufs, **90**
  noire, 143, **12, 13, 14**; oeufs, **90**
Guillemot de Brünnich, 184, **17, 22, 23**; oeufs, **92**
  à miroir blanc, 208, **16, 22, 23**; oeufs, **92**
  de Troïl, 170, **16, 22, 23**; oeufs, **91**

Harfang des neiges, 485, **46, 55**; oeuf, **97**
Hibou du Cap, 601, **52, 53**; oeuf, **97**
  grand-duc, 466, **45, 55**; oeuf, **97**
  des marais, 588, **52, 53**; oeuf, **97**
  moyen-duc, 572, **51, 53**; oeuf, **97**
  petit-duc, 454, **44, 54**; oeuf, **97**
  Huppe fasciée, 786, **71, 75**; oeufs, **98**

Ketupá brun, 481, **45, 55**; oeuf, **97**

Macareux moine, 231, **21, 22, 23**; oeuf, **92**
Martin-chasseur à tête grise, 705, **64, 68**; oeuf, **98**
Martinet d'Alexander, 652, **60**
  alpin, 678, **59, 63**; oeuf, **98**
  cafre, 687, **62, 63**
  épineux, 649, **59, 63**
  des maisons, 692, **62, 63**; oeuf, **98**
  noir, 657, **60, 61, 63**; oeuf, **98**
  du Pacifique, 676, **62, 63**
  pâle, 670, **60, 61, 63**
  des palmes, 698, **62, 63**; oeuf, **98**
  unicolore, 653, **60**; oeuf, **98**
Martin-pêcheur d'Europe, 711, **65, 68**; oeuf, **98**
  de Smyrne, 701, **64, 68**; oeuf, **98**
Mergule nain, 219, **20, 22, 23**; oeufs, **92**

Noddi niais, 163, **9, 10**

Perruche à collier, 379, **35**; oeuf, **95**
Petit-duc de Bruce, 450, **44, 54**; oeuf, **97**
Pic cendré, 813, **77, 78**; oeuf, **98**
  à dos blanc, 891, **83, 86**; oeuf, **98**
  épeiche, 856, **81, 86**; oeuf, **98**
  épeichette, 901, **84, 86**; oeuf, **98**
  flamboyant, 813
  de Levaillant, 837, **77, 78**; oeuf, **98**
  maculé, 853, **80, 86**
  mar, 882, **82, 86**; oeuf, **98**
  noir, 840, **78, 79**; oeuf, **98**
  syriaque, 874, **81, 86**; oeuf, **98**
  tridactyle, 913, **85, 86**; oeuf, **98**
Pic-vert, 824, **77, 78**; oeuf, **98**
Pigeon biset, 285, **27, 29**; oeuf, **95**
  de Bolle, 331, **29, 30**; oeuf, **95**
  columbin, 298, **28, 29**; oeuf, **95**
  d'Eversmann, 309, **28, 29**; oeuf, **95**
  des lauriers, 334, **29, 30**; oeuf, **95**
  ramier, 311, **29, 30**; oeuf, **95**
  trocaz, 329, **29, 30**; oeuf, **95**
Pingouin, Petit, 195, **18, 22, 23**; oeufs, **92**
  Grand, 207; oeuf, **93**

Rolle violet, 783, **73, 74**; oeuf, **98**
Rollier d'Abyssinie, 776, **72, 74**; oeuf, **98**
  d'Europe, 764, **72, 74**; oeuf, **98**
  indien, 778, **73, 74**; oeuf, **98**

Starique à crête, 229, **20, 22, 23**
  perroquet, 230
Sterne aléoute, 100, **15**
  arctique, 87, **5, 6, 7**; oeufs, **89**
  bridée, 109, **9, 10**; oeufs, **89**
  caspienne, 17, **1, 3**; oeufs, **88**
  caugek, 48, **1, 2**; oeufs, **87**
  de Dougall, 62, **5, 6, 7**; oeufs, **89**
  de Forster, 102, **5, 7, 8**

fuligineuse, 116, 9, 10
hansel, 6, 1, 2; oeufs, 87
huppée, 36, 1, 4; oeufs, 88
à joues blanches, 105, 5, 7, 8; oeufs, 89
naine, 120, 11; oeufs, 90
pierregarin, 71, 5, 6, 7, 8; oeufs, 89
royale, 27, 1, 3; oeufs, 88
de Saunders, 133
voyageuse, 42, 1, 4; oeufs, 87
Syrrhapte paradoxal, 277, 25, 26; oeufs, 94

Torcol fourmilier, 800, 76, 78; oeuf, 98
Tourterelle des bois, 353, 33, 34; oeuf, 95
maillée, 366, 33, 34; oeuf, 95
à masque de fer, 374, 32, 34; oeuf, 95
orientale, 363, 33, 34; oeuf, 95
rose-et-grise, 336, 31, 34; oeuf, 95
turque, 340, 31, 34; oeuf, 95

DEUTSCHE NAMEN

Aleuten-Seeschwalbe, 100, 15
Alexandersegler, 652, 60
Alpensegler, 678, 59, 63; ei, 98

Bajudanachtschwalbe, 617, 56, 57; eier, 94
Bartkauz, 561, 50, 55; ei, 97
Bienenfresser, 748, 70, 71; ei, 98
Blauracke, 764, 72, 74; ei, 98
Blauwangenspint, 740, 69, 71; ei, 98
Blutspecht, 874, 81, 86; ei, 98
Brandseeschwalbe, 48, 1, 2; eier, 87
Braunbauchflughuhn, 259, 24, 26; eier, 94
Braunliest, 701, 64, 68; ei, 98
Braunmantelscherenschnabel, 166, 15
Buntspecht, 856, 81, 86; ei, 98

Dickschnabellumme, 184, 17, 22, 23; eier, 92
Dreizehenspecht, 913, 85, 86; ei, 98

Eilseeschwalbe, 36, 1, 4; eier, 88
Einfarbsegler, 653, 60; ei, 98
Eisvogel, 711, 65, 68; ei, 98

Fahlkauz, 547, 49, 53
Fahlsegler, 670, 60, 61, 63
Felsentaube, 285, 27, 29; ei, 95
Feuerkopf-Saftlecker, 853, 80, 86
Flussseeschwalbe, 71, 5, 6, 7, 8; eier, 89

Gelbschnabelkuckuck, 425, 40

Gelbschnabelseeschwalbe, 36, 1, 4; eier, 88
Goldkuckuck, 400, 37, 41
Goldspecht, 813
Graufischer, 723, 66, 68; ei, 98
Graukopfliest, 705, 64, 68; ei, 98
Grauspecht, 813, 77, 78; ei, 98
Grünspecht, 824, 77, 78; ei, 98
Gryllteiste, 208, 19, 22, 23; eier, 92
Gürtelfischer, 731, 67, 68

Habichtskauz, 550, 50, 55; ei, 97
Häherkuckuck, 391, 36, 37; eier, 95
Halsbandsittich, 379, 35; ei, 95
Hinduracke, 778, 73, 74; ei, 98
Hohltaube, 298, 28, 29; ei, 95
Kleine, 309, 28, 29; ei, 95
Hopfkuckuck, 417, 39, 42; eier, 96

Kaffernsegler, 687, 62, 63
Kanarentaube, 331, 29, 30; ei, 95
Kap-Ohreule, 601, 52, 53; ei, 97
Kaptäubchen, 374, 32, 34; ei, 95
Kleinspecht, 901, 84, 86; ei, 98
Krabbentaucher, 219, 20, 22, 23; eier, 92
Kronenflughuhn, 249, 24, 26; eier, 94
Kuckuck, 402, 38, 42; eier, 96
Jakobiner, 388, 36, 37; eier, 95
Küstenseeschwalbe, 87, 5, 6, 7; eier, 89

Lachseeschwalbe, 6, 1, 2; eier, 87
Lachtaube, 336, 31, 34; ei, 95
Lorbeertaube, 334, 29, 30; ei, 95
Bolle's, 331, 29, 30; ei, 95

Mauersegler, 657, 60, 61, 63; ei, 98
Mittelspecht, 882, 82, 86; ei, 98

Nachtfalke, 646, 56, 57
Noddiseeschwalbe, 163, 9, 10

Orientseeschwalbe, 133
Orient-Turteltaube, 363, 33, 34; ei, 95

Palmsegler, 698, 62, 63; ei, 98
Palmtaube, 366, 33, 34; ei, 95
Papageitaucher, 231, 21, 22, 23; ei, 92
Pazifiksegler, 676, 62, 63
Prachtnachtschwalbe, 641; eier, 94

Raubseeschwalbe, 17, 1, 3; eier, 88
Rauhfusskauz, 606, 48, 54; ei, 97
Riesenalk, 207; eier, 93
Ringeltaube, 311, 29, 30; ei, 95
Rosenseeschwalbe, 62, 5, 6, 7; eier, 89
Rothalsziegenmelker, 636, 56, 58; eier, 94
Rotschnabelalk, 230

Rotschnabelseeschwalbe, 27, 1, 3; eier, 88
Rüppellseeschwalbe, 42, 1, 4; eier, 87
Russseeschwalbe, 116, 9, 10

Sandflughuhn, 263, 25, 26; eier, 94
Schleiereule, 432, 43, 53; ei, 97
Schneeeule, 485, 46, 55; ei, 97
Schopfalk, 229, 20, 22, 23
Schwarzschnabelkuckuck, 423, 40
Schwarzspecht, 840, 78, 79; ei, 98
Senegalracke, 776, 72, 74; ei, 98
Senegal-Spornkuckuck, 427, 37, 41; ei, 95
Silberhalstaube, 329, 29, 30; ei, 95
Smaragdspint, 734, 69, 71; ei, 98
Speissflughuhn, 269, 25, 26; eier, 94
Sperbereule, 496, 47, 53; ei, 97
Sperlingskauz, 505, 47, 54; ei, 97
Stachelschwanzsegler, 649, 59, 63
Steinkauz, 514, 48, 54; ei, 97
Steppenhuhn, 277, 25, 26; eier, 94
Streifenohreule, 450, 44, 54; ei, 97
Sumpfohreule, 588, 52, 53; ei, 97
Sumpfseeschwalbe, 102, 5, 7, 8

Tordalk, 195, 18, 22, 23; eier, 92
Trauerseeschwalbe, 143, 12, 13, 14; eier, 90
Trottellumme, 170, 16, 22, 23; eier, 91
Türkentaube, 340, 31, 34; ei, 95
Turteltaube, 353, 33, 34; eier, 95

Uhu, 466, 45, 55; ei, 97

Vaillants-Grünspecht, 837, 77, 78; ei, 98

Waldkauz, 526, 49, 53; ei, 97
Waldohreule, 572, 51, 53; ei, 97
Wandertaube, 378
Weissbartseeschwalbe, 133, 12, 13, 14; eier, 90
Weissbürzelsegler, 692, 62, 63; ei, 98
Weissflügelseeschwalbe, 155, 12, 13, 14; eier, 90
Weissrückenspecht, 891, 83, 86; ei, 98
Weisswangen-Seeschwalbe, 105, 5, 7, 8; eier, 89
Wellenbrust-Fischuhu, 481, 45, 55; ei, 97
Wellenflughuhn, 245, 24, 26
Wendehals, 800, 76, 78; ei, 98
Wiedehopf, 786, 71, 75; eier, 98
Wüstenflughuhn, 253, 24, 26; eier, 94

Ziegenmelker, 620, 56, 58; eier, 94
Ägyptischer, 641, 56, 57; eier, 94
Zimtroller, 783, 73, 74; ei, 98
Zügelseeschwalbe, 109, 9, 10; eier, 89
Zwergohreule, 454, 44, 54; ei, 97
Zwergseeschwalbe, 120, 11; eier, 90

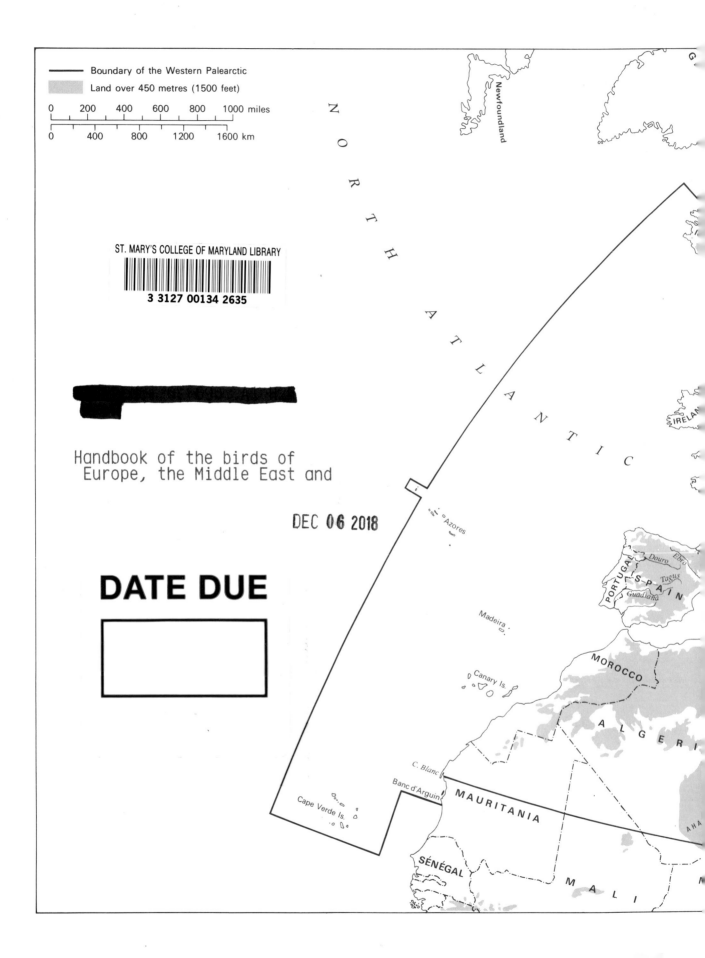

- Boundary of the Western Palearctic
- Land over 450 metres (1500 feet)

Handbook of the birds of Europe, the Middle East and